4th International Building Control Conference (IBCC 2016)

MATEC Web of Conferences Volume 66 (2016)

Kuala Lumpur, Malaysia
7 - 8 March 2016

Editors:

Sr Syahrul Nizam Bin Kamaruzzaman
Azlan Shah Bin Ali
Nur Farhana Binti Azmi
Shirley Chua Jin Lin

ISBN: 978-1-5108-2678-6

Printed from e-media with permission by:

Curran Associates, Inc.
57 Morehouse Lane
Red Hook, NY 12571

Some format issues inherent in the e-media version may also appear in this print version.

Printed by Curran Associates, Inc. (2016)

For additional information, please contact EDP Sciences – Web of Conferences
at the address below.

EDP Sciences – Web of Conferences
17, Avenue du Hoggar
Parc d'Activité de Courtabœuf
BP 112
F-91944 Les Ulis Cedex A
France

Phone: +33 (0) 1 69 18 75 75
Fax: +33 (0) 1 69 28 84 91

contact-edps@webofconferences.org

Additional copies of this publication are available from:

Curran Associates, Inc.
57 Morehouse Lane
Red Hook, NY 12571 USA
Phone: 845-758-0400
Fax: 845-758-2633
Email: curran@proceedings.com
Web: www.proceedings.com

TABLE OF CONTENTS

ISSN (electronic edition): 2261-236X

Preface

The 4th International Building Control Conference (IBCC) is a conference series dedicated to promote the importance, significance and advancement of building control and compliance for construction industry. This conference is deliberately planned for various backgrounds in the Built Environment who are frequently involved in building construction activities. It offers an opportunity for awareness and enhancement of knowledge and to keeping-up with the latest know-how in the field. Several invited guest speaker will be organized for exchanging of idea. Also, a distinguished overseas expert invited to present a keynote lecture on the latest findings on the subject matter.

The organizing committee invites all government officers, practitioners, researchers, academics, professional, students and those interested to participate in the IBCC. This conference is organized by Department of Building Surveying, Faculty of Built Environment, Kuala Lumpur City Hall (DBKL), Royal Institutions of Surveyors Malaysia (RISM), in collaboration with University Kebangsaan Malaysia and University Technology MARA along with its supported organizations.

Opening Address

Associate Professor Dr Sr Syahrul Nizam Kamaruzzaman
4th IBCC2016 Conference Chairman

Dear friends and colleagues,

The International Building Control Conference (IBCC) organized by Department of Building Surveying, Faculty of Built Environment and, Royal Institutions of Surveyors Malaysia (RISM), in collaboration University Kebangsaan Malaysia, Universiti Sains Malaysia and University Technology MARA along with its supported organizations is a biennial conference dedicated to promote the importance, significance and advancement of building control and compliance for construction industry. This conference is deliberately planned for various backgrounds in the Built Environment who are frequently involved in building construction activities. It offers an opportunity for awareness and enhancement of knowledge and to keeping-up with the latest know-how in the field.

This conference which was the third biennial conducted aim to enhance and leverage the effective design and control through knowledge and practice among academia, professionals and participants. The objectives of the programme are to provide a platform for the local practitioners to present the current situation on building control and performance as well as to educate and create an awareness on the latest development of act, regulations, guidelines and good practices.

In addition, it sought to provide needed information and skills in building control and performance practice and implementation. Empowering both professionals and citizens by encouraging engagement, cooperation and team work. Provide a platform to exchange information and opportunity to meet fellow expertises and professionals from other parts of the world to hear their issues first hand and to reflect on solutions and best practiced by others. Last but not least strengthening research collaboration between professional bodies, government agencies, international partners, industry players, other universities and the University of Malaya.

The Conference received 133 papers covering four main themes which are Building Control & Surveying, Building Design, Planning & Real Estate, Building Construction Management and Building Information Modelling (BIM). As an added value, the conference proceeding is indexed with Scopus through the Matec Web of Conference. Furthermore, the conference provide opportunities for all papers to be published in several ISI index journals, Scopus index journal as well as national refereed journals.

It is therefore with great pleasure that I, on behalf of the Organizing Committee, extend our warmest welcome as well as grateful thanks to all our guests, professionals, academicians, speakers, chairpersons, undergraduate, postgraduate and all participants attending this event.

This conference would not have been possible without the dedicated support, sacrifice and generous contributions from various parties, particularly our valued sponsors concerned who have, without fail, given us their regular dedicated support. As such, we would like to thank Kuala Lumpur City Hall (DBKL), Malaysian Association of Registered Building Surveyors, Henry Butcher Malaysia (Mont Kiara) Sdn. Bhd. (HBM), Silcom Marketing & Services, Aras Saintifik Sdn. Bhd., Azamansaidin.blogspot.com, Thye Hong Technology (M) Sdn. Bhd., and Zaisains Instruments & Information Technologies for their generous sponsorship without which we would not have been able to facilitate this Conference at heavily subsidized and affordable cost to participate.

Lastly, it is hope that this Conference will act as a platform to establish contact academia and professional practice in this field either within or outside the country.

To all our eminent and distinguished speakers, presenters and session chairpersons who have come from near and far, please accept our appreciation and thanks for sharing your precious time, knowledge, experiences and wisdom at this Conference.

To those who are from outstation and overseas, we on behalf of the University Malaya, wish you an enjoyable stay in Kuala Lumpur.

Lastly, I would like to take this opportunity to thank the Organizing Committee, the secretariats, the host University, University of Malaya as well as conference assistants for their support, dedication and logistics to make this event a great success.

Wishing all of you an enjoyable and fruitful event.

Thank you.

edp sciences

MATEC Web of Conferences

Statement of Peer review

In submitting conference proceedings to MATEC Web of Conferences, I certify to the Publisher that I adhere to the **Policy on Publishing Integrity** of the journal in order to safeguard good scientific practice in publishing.

1. All articles have been subjected to peer review administered by the proceedings editors.
2. Reviews have been conducted by expert referees, who have been requested to provide unbiased and constructive comments aimed, whenever possible, at improving the work.
3. Proceedings editors have taken all reasonable steps to ensure the quality of the materials they publish and their decision to accept or reject a paper for publication has been based only on the merits of the work and the relevance to the journal.

Title, date and place of the conference

THE 4TH INTERNATIONAL BUILDING CONTROL CONFERENCE 2016 (IBCC 2016)
7th - 8th March 2016
Hotel Pullman Kuala Lumpur Bangsar, Wilayah Persekutuan Kuala Lumpur, Malaysia

Proceedings editor(s):

Associate Prof. Dr. Sr Syahrul Nizam Bin Kamaruzzaman, University of Malaya, syahrulnizam@um.edu.my
Prof. Azlan Shah Bin Ali, University of Malaya, asafab@um.edu.my
Dr. Nur Farhana Binti Azmi, University of Malaya, farhanazmi@um.edu.my
Dr. Shirley Chua Jin Lin, University of Malaya, shirleychua88@um.edu.my

Date and editor's signature

Review on Application of Building Information Modelling in Interior Design Industry

Abu Bakar Abd Hamid[1,a] and Mohamed Rashid Embi[2]

[1]*PhD Candidate, Department of Architecture, Faculty of Built Environment, Univeriti Teknologi Malaysia, Johor Malaysia*
[2] *Associate Professor, Department of Architecture, Faculty of Built Environment, Univeriti Teknologi Malaysia, Johor Malaysia*

Abstract. Building Information Modeling (BIM) is a flourishing device and approach for Architecture, Engineering and Construction (AEC) which assist; the planning, design, constructions and operation with particular to commence the designing in several components of the building, a comprehensive understanding of the activities and needed in all aspects. An aspect of activities and needs is the first step to be analysed in the design and it is the responsibility of professionals' interior designers. To solve the problems, is required to attach with the certain factors such as space planning, specifications fixture, equipment and material, furniture and finishing, lighting fixing and service management. BIM application can provide interior designers such as simulation, analysis, design process and solution with team members. Furthermore, some of the interior designer also using BIM as to organize, documenting data and optimise detail information in solving the problem in the develop process. BIM technology and practices not only to improve the construction and design process development that only the project faster, cost effectively, and prolong but it also helps to increase the capacity of BIM professional by improvise the careers and fields. In the Malaysian context, BIM is still fresh in interior desogn industry.

1 Introduction

Nowadays, every evolution in technology has been achieved with advances in software technology and tools. Building Information Modeling (BIM) is a new a paradigm phrase in the planning and design profession. BIM is the one of tools is changing with a new period and time circulation. Building Information Modeling is the best way to discover about construction and one of the most promising developments in the architecture, engineering, and construction (AEC) industries and is driving an unprecedented revolution in the construction industry [6, 11].

The theory of BIM is to construct a building with the virtually elements, prior to building it physically and theoretical, to work out problems, and to simulate and analyze potential impacts [14]. BIM is the concept of building with emphasizing on the virtual environment to physically building was built with the potential problems can be expected during the construction stage and work processes can be carried out and the analysis can be done early [16].

Malaysian construction industry is very dynamic and complex. The implementation of BIM as a new technology and construction method in major construction for contractors, developer and stakeholders is more accurate. However, using BIM in industry is very limited and adoption was not successful as desired. Presently, Malaysia's construction industries have improved the readiness and awareness for industry players to more concerned using BIM technology.

Defining the BIM in perspective of interior design aspect. Terminology on interior design will be establish to find out the factors and criteria when BIM using in interior design process. Furthermore, there is no description on the interior design in BIM aspect. Identification of the benefits and constraints of BIM aspect in interior design industry and understanding of the relationship between the sizes of interior design firm. The significant on BIM will be verified when the barriers and variables were find out in the literature and survey was achieve.

2 Literature Review

2.1 Definition of Building Information Modeling (BIM)

Building Information Modeling (BIM) is the latest software technology being introduced throughout the Architect, Engineering & Construction (AEC) profession. [4] defines BIM as a process supported by technology of computer generated model used in collaboration to populate information and simulate the planning, design, construction and operation of a facility. But for interior design profession is still lacking and most unpleasant using BIM approach. Skill sets and knowledge necessary for entry-level placement increasingly expands for young

[a] Corresponding author: abah7984@gmail.com

 DOI: 10.1051/matecconf/20166600003

designers seeking employment with firms executing large-scale projects.

According to the [12] a building information model (BIM) is defined as "a digital representation of physical and functional characteristics of a facility. As such it serves as a shared knowledge resource for information about a facility forming a reliable basis for decisions during its lifecycle from inception onward". Building Information Modeling (BIM) is a new and innovative technology, which has emerged in recent years and makes possible the efficient achievement of more sustainable designs. According to the [17] defines BIM in three dimensions:

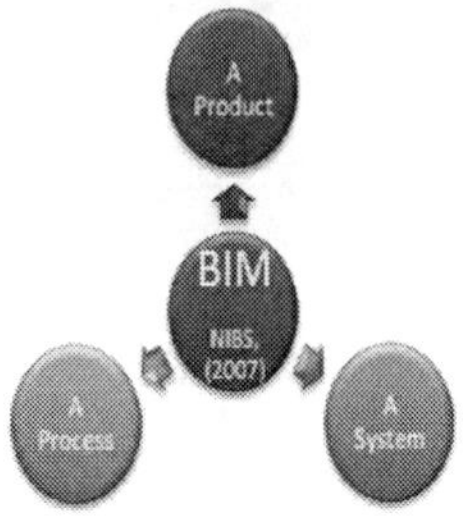

Figure 1. Definition of BIM

Based on Figure 1 is shown a three terminology of BIM which is a product is set up the structure on dataset of building; a process defined the act of creating a BIM and a system is to comprises the business work and communication interm of quality and efficiency.

Table 1. BIM Terminology and Definitions

No.	Terminology	Definitions
1.	Building Information Model	A digital representation of the physical and the functional characteristics of a facility to serves as a shared knowledge resource for information about a facility, forming a reliable basis for decisions during its life cycle from inception onward. [13]
2.	Building Information Modeling	The human activity of using BIM software and other related software, hardware and technologies to create and use in a building information model [8]
3.	3-D BIM	A model that includes three dimensional (3-D) shape information and does not include the 4-D and 5-D characteristics described below [13]
4.	4-D BIM	A 3-D BIM that has objects and assemblies that have a cost dimension added to them. The cost information can be contained in the BIM or can be linked or otherwise associated to the building objects [13]
5.	5-D BIM	A 4-D BIM that has objects and assemblies that have schedule and time constraint data added to them can be contained in the BIM or can be linked or otherwise associated with project design and construction activity scheduling and time sensitivity estimating and analysis systems [13]
6.	2-D/3-D CAD	Two dimensional or three dimensional, Computer Aided Drafting is equivalent to conventional drafting, only performed on a computer.
7.	Integrated Practice	Uses early contributions of knowledge through utilization of new technologies, allowing Architects (Engineers, Owners, Contractors, Manufacturers, Firms, Individuals, Communities) to realize their highest potential as designers and collaborators while expanding the value they provide throughout the project life cycle [8]

Table 2. Understanding of BIM

No.	Technology	No.	Process
1.	Project simulation consist using 3D models integrated with project planning, design, construction and operation [9]	1.	Refining and adjusting according to project specifications and design changes to ensure the model are accurate [3]
2.	Introduced the object-oriented parametric modeling [1]		
3.	The process which is the element modified and adjacent or assembly automatically [15]	2.	To produce a design with optimized for quality, aesthetics, constructability, affordability, timeliness and seamless flow into lifecycle management [10]
4.	Define BIM is supplier, operation and maintanence producers, flow rates and clearence [2, 5]		

According to the Table 2, is devided in two categories of BIM, in technology is define that with 3D project can be simulated all operation and could be more organise. This statement is support by [1]. In process, i scan refine and produce the best way to more construct and the flow of process seemly fine.

2.2 Definition of Interior Design

According to [18], interior design is a multi-faceted profession in which creative and technical solutions are applied within a structure to achieve a built interior environment. These solutions are functional, enhance the quality of life and culture of the occupants, and are aesthetically attractive. Designs are created in response to and coordinared with the building shell, and acknowledge the physical location and social context of the project.

Design must adhere to code and regultory requirements, and encourage the principles of environmental sustainability. The interior design process follows a systematic and coordinated methodology, including research, analysis and integration of knowledge into the creative process, whereby the needs and source of the client are satisfied to produce an interior space that fulfills the project goal.

2.3 The Contextual Understanding BIM in Interior Design

As interior designers, is to be a valued professional, knowledgeable, idealistic and capable to deliver as project team members to incorporate within interior design industry and technology [9]. Interior design industry should take the consideration in the consequence of combination of skills and knowledge as a critical to adopt in design process. Currently, there are no interior design specific set of BIM tools and they are not generating exterior construction or structural systems using by BIM [9]. Building Information Modeling can be beneficial during the design phase due its ability to organize information and data. Is also provides a stronger visualization of the building and interior. BIM is not utilizing during the core and shell phase or the interior design phase.

BIM has already been proven that is valuable resource for construction projects and documentation. With BIM is provided 3D and 4D visualization, detect clashes, manages changes, and database. It also pertains focusing on structural, building enclosure and design phase. BIM can do anything and beneficial to interior design such as 1) easy and speedy to develop an interior space; 2) ability to visualize the design phase; 3) ability to create multiple interior design on one building; 4) organized material information and selections, cost data, schedules, furniture locations efficiency and material quantities.

Interior designers using BIM to facilitate the data document and ultimately disseminate all of the detailed information about the solution interior designers develop [7]. As interior designers, creativity and technical solutions is part of the skill in built interior environment. This is requires for interior designers make a decision to be within a qualitative and quantitative aspect. During the design process, interior designers are responsibility to analyze all of the owner's and occupant's need. At this stage, interior designers will be established with the detail project requirements and project plan with collaborative team member to analysis the consequence. Interior designers must be developed in direct coordination because the interior projects involving within a new building and old building being renovation progress.

Table 3. The Applications BIM for Interior Designers

Stage	Activities
Programming and Schematic Design Development	- To determine space needed to achieve all of the required activities of the occupants. - Virtual models and databases that are interrelated and interconnected
	- Space programs are utilized to initially all of the space elements. - the bubble and diagramming phases are utilized to study the locations and relationships between the spaces. - building design development - Coordination between members of the design team to space planning progress.
Interior Building Components	- assign the information to specify the type, size, construction, materials, finishes, performance, manufacturer and cost. - coordinate the vertical circulation components such as stairs and elevators - to simulate and define the specifying type, size, finish properties and fire rating.
Furniture, Millwork, Fixtures, and Equipment	- provide online BIM catalogues of their products to be used by interior designers. - specific specifications such as size, finish, model number, quality, and cost. - Provided generic models to create new models
Ceilings and Lighting Design	- Integration within interior spaces layout and ceiling system. - To ensure the light fixtures provided appropriate lighting quantities and types. - Fixture types and lamp types are both defined within BIM lighting plans.
Interior Finishes	- allows interior designers to detail each surface of an interior environment and specify how they will be installed, how they will perform, how much they will cost and how to maintain them. - provides interior designers to define all the interior component types and material finishes.

In application BIM for interior design is mentioned that the several processes will establish in the interior design scope of work. With BIM, interior design industry is more manageable and workable to operate the system by coordinate the process.

3 Research Methodology

The research methodology includes a literature review, design tool analysis, case studies and interviews. The literature review studied of the significance of framework as well as the definition and application of BIM in the design practice for interior design discipline. Structured face-to-face interviews were carried out with BIM professionals including architects, clients, engineers, consultants and the government to collect opinions on the benefits of BIM for interior design disciplines and obstacles to its use.

Besides that, analysis structure using software device will be cover to study the implication for interior design in academic and profession context.

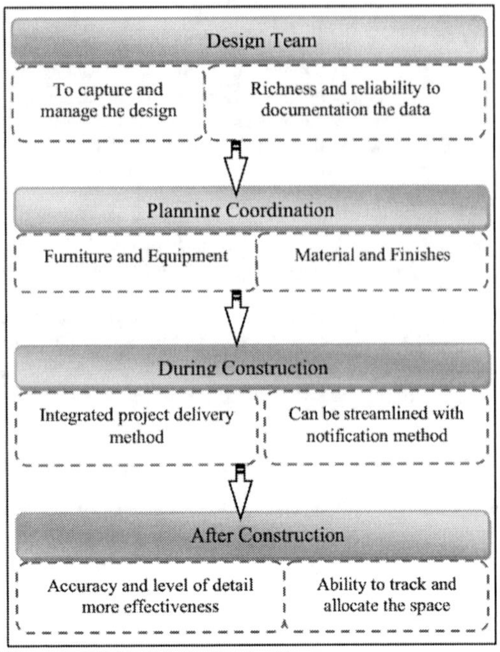

Figure 2. BIM in Interior Design Process

Basically, in interior design process a several stages were identified with the design team where is to capture and manage design. In the planning stage, furniture and equipment were located at the availability of finishes and material. During the construction, integrated is the best method to get the good result in design process. The after the construction phase, accuracy and level of detail were be achieve in the construction. This is also ability to track the defect and problem.

4 Discussion

Team buildings are very important for building owner which is architect, engineers, interior designers and contractors implement the accurate BIM from the beginning project till the hand over. The advantage using BIM in development progress were make the team building easier and speedy in work progress.

From the interior designer's perspective, it can conclude that BIM is more powerful and helping to improve contract documentation, preconstruction estimating, procurement, scheduling, coordination, and cost-efficiency and close-out documents in term of finishes, furniture and equipment that are essential to the environment aspect.

Through the pilot study, Jefree Ismail (2015) has mentioned that BIM in Malaysian context is still small users and this statement was supported by Jamal (2015). They identified that costing is the core barrier to improve the quality of work process. The ability of contractors to deliver the new technology also make it obstacle to achieve the process.

When is the BIM implement for interior design, there is a lot the advantage which is the design process is more speedy and organise. Time and cost can be save without using more manpower. To support this study, several variables was identified as following: costing, time constrain, knowledge, technology, readiness, awareness, conflict in design, and this supportive with some barriers such financial barriers, time barriers, conflict in design barriers, skill workforce barriers and demand from industry are the limitation of the study.

There are many studies about BIM more focus to the architecture, engineering and construction (AEC), facilities management, 3D information model and digital innovations. Although this study about BIM in which it is more to the AEC, but researches on the implementation of BIM for Interior Designers industry is still untapped and lack the information. The limited literature on the implementation and effectiveness of BIM in the Interior Design industry is still far to describe. Constraints and problems in the implementation of BIM is identified the cause why the lack of use in this industry.

Via searching and evaluating which caused constraints and effectiveness is very significance. All results and analysis depends on the success of this study. Thus, the finding and conclusion of this study hopes to help in the research and development in the relevant literature.

References

1. S. Azhar, A. Nadeem, J.Y.N. Mok, and B.H.Y Leung, *'Building Information Modeling (BIM): A New Paradigm for Visual Interactive Modeling and Simulation for Construction Projects'*, Proceedings of the First International Conference on Construction in Developing Countries (ICCIDC-I), Karachi, Pakistan, August 4-5, (2008*)*
2. S. Azhar, and S. Richter, *'Building Information Modeling (BIM): Case Studies and Return-on-Investment Analysis'*, Proceedings of the Fifth International Conference on Construction in the 21st Century (CITC-V), Istanbul, Turkey, 1378-1386, (2009)
3. J. Carmona, and K. Irwin, *'BIM: Who, What, How and Why'*, Building Operating Management, *(*October 2007)
4. CIDB, *Workshop of BIM Portal and Collaboration Platform for Affordable BIM, 9th-11th January 2013*, Holiday Inn Glenmarie, Shah Alam, Selangor. Kuala Lumpur: CIDB Internal Reports, (2013).
5. CRC Construction Innovation, ' *Adopting BIM for Facilities Management: Solutions for Managing the Sydney Opera House*, Cooperative Research Centre for Construction Innovation, Brisbane, Australia, (2007)
6. C. Eastman, P. Teicholz, R. Sacks and K. Liston, "*BIM Handbook: A Guide To Building Information Modeling For Owners, Managers, Architects, Engineers, Contractors, And Fabricators*," 2nd Ed., Wiley, Hoboken, NJ. (2011).
7. P. Elizabeth, *"The Application of BIM for Interior Design and the Implication for Facility Management"*, BIM Academic Symposium,

Building Innovation 2014 the National Institute of Building Sciences Conference & Expo, (2014)
8. F. Jernigan, *ASHRAE Vision 2020.* Atlanta: ASHRAE Vision 2020 Ad Hoc Committee, (2008)
9. W. Kymmell, '*Building Information Modeling: Planning and Managing Projects with 4D CAD and Simulations*', USA: McGraw Hill Construction, (2008)
10. McGraw-Hill Construction,*"Building Information Modeling: Transforming Design and Construction to Achieve Greater Industry Productivity,* McGraw-Hill Construction, New York, (2008)
11. McGraw Hill Construction. *"Building Information Modeling (BIM) Transforming Design and Construction to Achieve Greater Industry Productivity",* (2009)
12. National Building Information Modeling Standard (NBIMS), *Overview, Principles and Methodologies,* Version 1.0 — Part 1, http://www.wbdg.org/pdfs/NBIMSv1_p1.pdf8 (last accessed on August 27, 2010).
13. National Institute of Building Sciences (NIBS), 'United States National Building Information Model Standard,' Version 1. Retrieved May 29, 2013 fromhttp://www.wbdg.org/pdfs/NBIMSv1_p1.pdf. pp 1-2, (2007).
14. D. Smith, *"An Introduction to Building Information Modeling (BIM)."* J. Build. Inf. Modeling, 1(1), 12–14, (2007).
15. J.D. Stine, '*Design Integration Using Autodesk Revit® 2012*,' Mission, KS: SDC Publications, (2011)
16. S. Azhar, *"Building information modeling (BIM): Trends, benefits, risks, and challenges for the AEC industry,"* Leadership and Management in Engineering, vol. **11**, pp. 241-252, (2011)
17. US National Institute of Building Sciences (NIBS) United States National Building Information Modeling Standard, Version 1 – Part 1: Overview, Principles and Methodologies, Facilities Information Council, USA, (2007)
18. National Council for Interior Design Qualification. *NCIDQ Examination Study Guide.* Washington, DC: NCIDQ, pp. 22–23.(2000)

A literature review on maintenance priority - conceptual framework and directions

Alan Chong Kim Wing[1,a], Abdul Hakim bin Mohammed[2] and Mat Naim bin Abdullah[3]

[1,2,3] *Department of Real Estate, Faculty of Geoinformation and Real Estate, University Teknologi Malaysia, Johor Bahru, Malaysia*

Abstract. Proper maintenance prioritization enables organization to ensure their maintenance performance remains in the acceptable level. This paper aims to examine the relevant literature related to maintenance priority. Based on this paper, a conceptual framework and future research directions are presented. A database of 184 publications published during 1990 to 2015 was reviewed for the purpose of this paper. The literature review presented the factors used in maintenance priority, methods for priority setting, and the trend of maintenance priority. A discussion was carried out on the maintenance priority based on organisation goal and balancing maintenance priority. Based on this literature review, a conceptual framework, which presents the facets of factors and methods considered in maintenance priority setting is offered. The understanding of factors and methods considered in maintenance priority setting can aid maintenance managers in maintenance planning and ensure maintenance performance of the assets remains compliant with the needs of the organization.

1 Introduction

Maintenance is crucial in keeping building, infrastructure, and equipment in the best form for normal use [1]. Maintenance is not limited to a certain type of building or location. Regardless of the construction cost of the buildings, they still carries maintenance consequences [2]. In the past, maintenance was viewed as necessary evil because the cost it incurs [3]. Most organisations treat maintenance as a cost burden and demonstrates reluctance to spend in order to preserve the condition of their assets [4]. This reluctance causes a drop in performance and prevents optimum operating capacity.

Maintenance is a costly element in a building's life cycle. Approximately 75% of the total expenditure on life cycle cost of a building is due to maintenance [5]. After construction is completed, more than 90% of the life cycle of building projects requires maintenance [6]. This further proves that maintenance is a critical element in any organisation and it carries dire consequences if ignored. Inadequate maintenance of facilities is a poor economic decision because it generates maintenance backlog which must be addressed in the future [7].

2 Background

[8] stated that "priority management is the allocation of resources, or the expression of preference, to specific order or order groupings (whether supplies, production, or customer orders), in response to current pressures on operational productivity and/or customer service, with the aim of relieving those pressures while at the same time promoting, or minimizing the deleterious impact upon, the wider economic and strategic goals of the company". Instability requires rebalancing of resources and trade-offs. In [8], it is stated that instability causes prioritizing actions. Instability refers to changes in volume, variation and variety.

Demand for maintenance actions is always greater than the resources available [9]. More than often, the budget approved is less than needed [10]. This often leads to budgetary cuts. One of the most common area to suffer budgetary cuts is maintenance. Deferments of maintenance works are carried out in order to reduce cost or give way to areas that the organisation deems to be more important. This solution is not cost-effective and maybe not be the best practice to achieve value for money [11]. However, the reality is that the allocation for maintenance will never match the required amount and thus prioritization is practiced. Prioritization of maintenance is a way to tackle the problem of lack in maintenance allocation [12]. Changes imposed on the management of infrastructure in Scotland caused managers to struggle in prioritizing management and maintenance of bridges as there is limited funding [13]. It is common for property managers to face difficulties in setting maintenance priorities. [14]. Facility managers experience increasing pressure to prioritize the direction of limited resources to address maintenance and capital renewal needs. The results of bad prioritization of the direction of resources and capital appropriately include unscheduled facility maintenance, system unavailability, and additional expense to repair or replace failed components or systems due to short/emergency notice

[a] Corresponding author: alanchong19882003@gmail.com

procurements [15]. In order to decide which maintenance work to prioritize and the consequences of each prioritization, it requires the consideration of multiple factors and thorough understanding of the business [16].

The study attempts to shed some light on important factors and methods used in maintenance priority. For this purpose, the relevant literature is examined, classified and analyzed.

3 Methodology

An exhaustive search of literatures related to maintenance priority and maintenance performance was conducted. The database reviewed were: Emerald, ScienceDirect, Scopus Taylor and Francis, JSTOR, and SpringerLink. Related books and search engines were reviewed as well

In total, 184 literature were reviewed. The reviewed literature consisted of models, framework, techniques and factors. The period of the review is from 1990 to 2015. Literature with redundancy and unsuitable for this paper were excluded. Figure 1 shows the publication related to maintenance priority distributed over the period of 1990-2015.

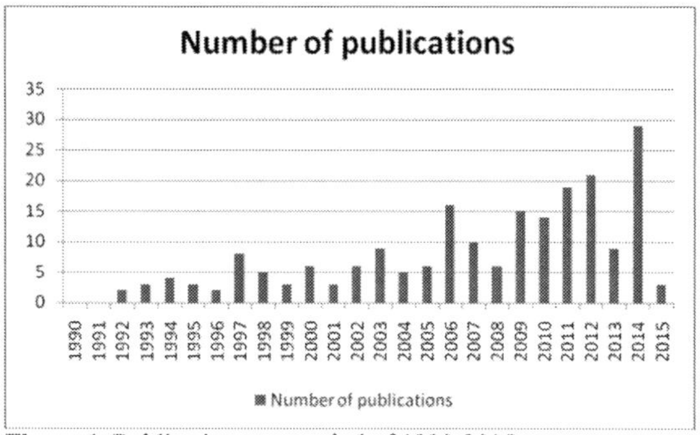

Figure 1. Publication over period of 1990-2015

A total of 70 journals were reviewed. 26 journals which have more than 2 publication related to maintenance priority is shown in Figure 2. The top six journals are Journal of Quality in Maintenance Engineering, Facilities, Journal of Facilities Management, Structural Survey, Construction Management and Economics, and Reliability Engineering and System Safety. Journal of Quality in Maintenance Engineering is where most articles related to maintenance priority are published. 13.5 percent of the total search comes from this journal.

Figure 2. Publication in journals

4 Results

Through detailed and systematic content analysis of the publications, a total of 62 presented framework/models and measurements were reviewed. Analytic Hierarchy Process (AHP) based measurement recorded the highest followed by priority criterion, matrix based measurement, framework and Failure Mode and Effect Analysis (FMEA) based measurement. Some of the least used methods are; simple comparison, criticality index, additive ration assessment (ARAS), fuzzy group Analytic Network Process (ANP), cost-based criticality, Roue's formula, benchmarking, Artificial Neural Network (ANN), probabilistic risk assessment, comparative risk analysis, critical failure analysis, and priority cost FMECA. The AHP group has a few variations which include Goal Programming (GP), Bayesian Tools and decision rules. FMEA based measurement includes Risk Probability Number and Fuzzy sets. The variation of matrix is based on importance and performance matrix, priority matrix, priority category matrix, hybrid structural interaction matrix and risk priority matrix.

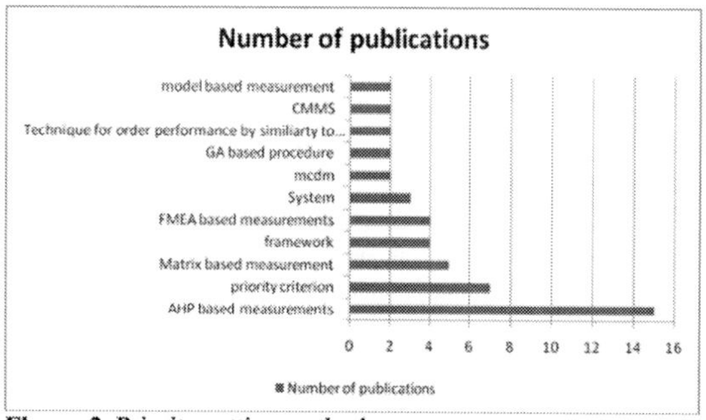

Figure 3. Priority-setting methods

66 literature contain the factor used in maintenance priority setting as shown in Figure 4. The most common indicator used is condition assessment and maintenance cost (direct and indirect) followed by occurrence, severity, importance of usage and safety. The least use factors are urgency, labour, availability, satisfaction, complaints, remaining life cycle, aesthetics, economic motivation, scheduling, comfortable use, facilities, economic performance, managerial problems, damage value, type of material, historical data, pressure from maintenance, ability to response, reliability, long term effect, objectives, applicability, stakeholders, public relation, information system, work measurement and incentives, repairability, social and cultural problems, bottlenecks and indoor environmental quality.

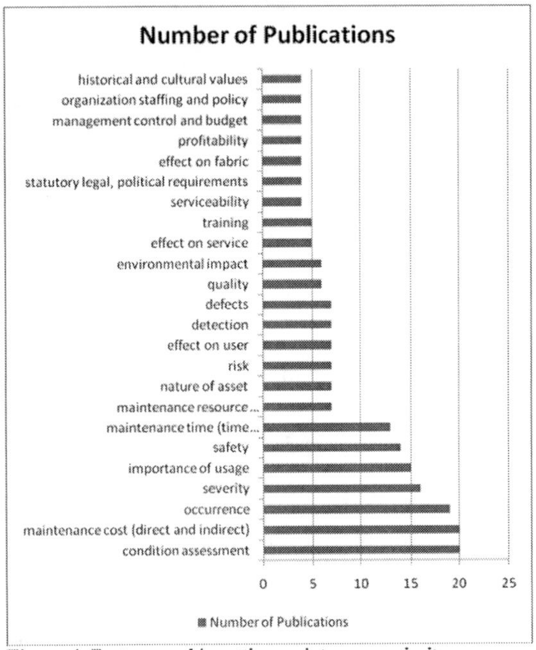

Figure 4. Factors used in setting maintenance priority

An attempt to categorize these factors has been carried out as shown in Table 1. The factors is then categorized into 4 different major categories: Technical, Financial, Social, Political.

Technical factor refers to the performance measurement of an asset [17]. Indicators in form of figures are used to monitor the performance of certain process, systems, equipment and components [18]. [19] included condition, age, life span, energy rating and complexity under the technical factor for planned maintenance prioritization. Physical assessment such as building age and building condition is one of the dominant variables in maintenance planning [20]. In a study to prioritize maintenance, the physical condition of the defective element was examined together with its possibility of breaking down [21]. Elements in critical condition would occupy higher priorities than those in fair condition. Historical maintenance data can also be used as inputs in measuring performance as well [22, 23].

Financial usually refers to budget and cost. There is a need to have a balance between the both for a maintenance plan to succeed. Budget is a strong determinant in any maintenance planning. [20, 24, 25]. A lack of budget can cause shift in priorities and changes in maintenance planning [20]. A sufficient budget is required to procure the needed resources. Resources can refer to equipment, labour, time and cost [26]. The lack of resources causes inability to execute maintenance work. These resources usually translate into monetary needs and can only be satisfied by sufficient budget. In other words, it is the cost required to carry out maintenance. In a study by [24] stated cost of maintenance can be divided into cost of maintenance, cost of recurrent problems, cost of non-functioning, cost of temporary access, and cost of non-maintenance. According to [27], cost can be categorized into direct cost and indirect cost. Direct costs are related to labour costs, spare parts, and other direct cost involved with maintenance works while indirect costs are related to cost from recovering the lost of production and quality insufficiency. While direct cost such as cost of maintenance (manpower and spare parts) are more commonly known, indirect cost such as cost of non-functioning for essential services to the organization is difficult to be measured.

According to [19], the social factors in prioritizing maintenance covers health and safety, building usage, security and building status. In most maintenance management, safety and health requirements are considered as command factors in decision making for managing asset [20]. Safety are closely related to risk. Risk affects the degree of risk of life, physical integrity, health and comfort [28]. The common way to measure risk is through likelihood and consequences [29]. For local authorities in Britain, maintenance priorities related to legal requirements, contracting requirements and safety and health of building occupants and users are considered as the top priority [9]. Legal requirements include legislations, statutory and contracting requirements [19]. Other than that, the other effects on users such as customer satisfaction should not be neglected as they are the main users of the building. [30] concluded that the maintenance works of university buildings need to be prioritized based on perspective of users.

In the political category, it refers to the stakeholders of the organisation. The categories of stakeholders include end-users, managers and others involved in the organizational process [31]. The requests of client or owner of the building must be fulfilled. The requests revolves on the organization strategies, corporate image and identity [20]. Poor communication with the client would influence maintenance performance. As infrastructure assets have multiple functions, there are multiple actors (users, political bodies, private sector) involved or affected by the decisions made on asset management [32]. When defining objectives and planning for asset management, their expectations need to be included and managed even

Table 1. Matrix table for maintenance priority factors

References\Factors	condition	risk	cost	labor	spares	safety	statutory	time	Effect (users)	Effect (fabric)	importance of usage	quality	objectives and policy	budget	information system	aesthetics	Cultural	Complaints	Client request	Sustainability
Taylor (1996)			•	•																
Hassan (1997)			•	•	•	•	•		•											
Pitt (1997)	•								•	•										
Shen et al. (1998)	•							•	•		•									
Labib et al. (1998)		•			•			•				•								
Layzell and Ledbetter (1998)		•																		
Amir et al. (2001)			•									•								
Shohet et al. (2002)	•	•																		
Hassanain et al. (2003)													•							
Li and Brown (2004)		•	•					•												
Oke and Ayomoh (2005)	•			•	•															
Bertolini and Bevilacqua (2006)		•	•					•					•	•	•					
Carnero (2006)		•	•		•	•			•											
Dann et al. (2006)	•					•								•			•	•		
Frangopol and Liu (2007)	•					•														
Jim and Peter (2007)	•	•				•										•				
Chandrashekaran and Gopalakrishnan (2008)			•			•		•												
Gionata (2009)		•														•				
Ali (2009)	•					•								•				•	•	
Flores-Colen et al. (2010)	•		•						•	•	•		•							
Vijay Kumar and Chaturvedi (2011)		•																		
Sadeghi and Manesh (2012)						•						•			•					
Siew-Hong et al. (2014)		•																		
Ali and Hegazy (2014)	•	•																		•

though at times there might be conflicting interest among the stakeholder [32].

The case studies recorded a total of 57. There are 16 categories recorded as shown in Figure 5. Plant has the most recorded case studies. Type of plants range from aluminium, electric, oil refinery, chemical and steel plant. It is followed by manufacturing, building, and education which recorded 6 case studies respectively. Buildings include a combination of intelligent, green and normal buildings. Manufacturing covers food industry, semi-conductor, and steel. Education includes both school and university.

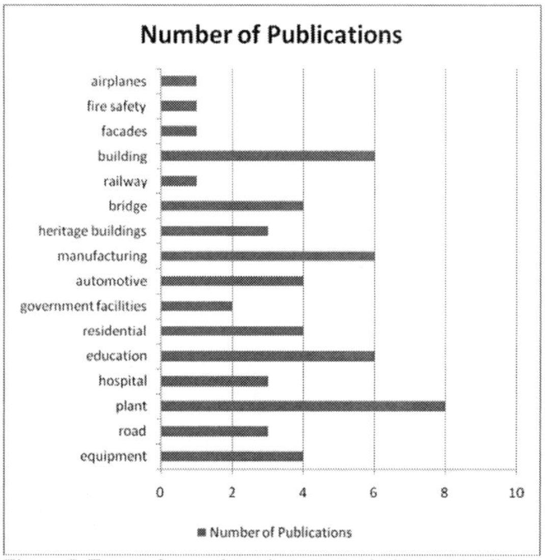

Figure 5. Types of assets in maintenance priority case studies

5 Discussion

Based on further content analysis of these articles, there are two relevant themes to maintenance priority. These themes are possible future research areas.

5.1 Goal-orientated maintenance priority

Maintenance priority starts from the top level of management which is formed by organization vision, mission, goal and objectives as shown in Figure 6. Organizational goal is the focus of different stakeholders and constituents toward achieving a set of goals that are believe to advance the organization [15]. It is common for departments within the same organization to have conflicting interest in regards to maintenance [17]. The purpose of an organization's objective is to ensure the alignment of departmental efforts with the needs of the organization [17]. More than often problems occur when management fails to set goals at the highest level and making sure it is passed down correctly to the lower levels. [33] found that that goals are sometimes negotiated rather than based on stakeholder requirements and strategies. In a study by [32], public agencies faced difficulties in setting infrastructure objectives that aligns with the strategic policy goals. This is due to the infrastructure assets serve multiple interests of different stakeholders. [34] found that objectives and goals in asset management are lacking in Norwegian municipalities. They also lack the direction and knowledge to catch up with maintenance backlog.

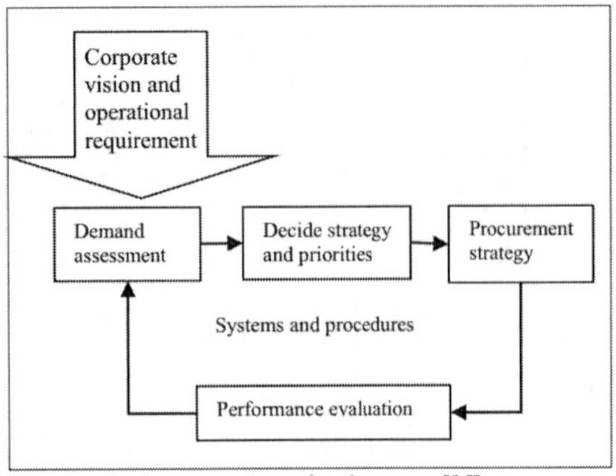

Figure 6. Asset management and maintenance [35]

Conflict is in the nature of business and can be seen in the organization structure. Senior management strives to meet both performance goal and at the same time wants the workforce to feel rewards for its efforts. As such, priority management is used to balance the needs of all party. Maintenance work order priority would have some impact on overall maintenance performance[36]. Maintenance performance could be affected as the consequences of lacking clarity in goals and objective for asset management. Performance measures that are aligned with clear policy goals and objectives help to direct resources to where they are needed to accomplish the agency's mission and can increase outcomes achieved with given levels of resources. Without clear objectives, employees relied on their personal experiences and knowledge in deciding maintenance actions [32]. This would lead to the employees to make decisions based on their personal beliefs and causes the prioritization to be more implicitly assumed than explicitly determined. Defining infrastructure objectives is important because it provides the rationale for evaluation of infrastructure condition and performance, and the prioritization of maintenance repair and renovation activities [32]. In [37], the commonly used objectives in maintenance planning are:

1. Minimizing maintenance cost
2. Maximizing asset performance
3. Minimizing risk of failure

In this context, the following research questions are relevant.

1. How do organization goal and objectives assist in maintenance priority setting?
2. How to effectively prioritize maintenance among conflicting needs?
3. What is the relationship between organization goal, maintenance priority and performance?

5.2 Balancing maintenance priority

From the results, different case studies would prioritize maintenance to suit their own needs. Production based maintenance would focus more on the effectiveness and efficiency of production while building based maintenance focuses on building condition, safety, aesthetics, and user satisfaction. While safety is considered as the most important criteria by some studies [38-40], there are some exception. More than often cost and budget consideration is the first consideration of maintenance managers[41, 42]. In some cases, aesthetic is prioritized so that the building perceived value is high and offers a good perception to the public [39, 43, 44]. Maintenance is often perceived differently depending on the position of the individual [35]. Surveyors would view maintenance needs from the asset protection dimension while building users would emphasize on the working environment. Owners and building managers would be more interested in priorities related to resource allocation and long term development plans. [15] argued that individuals should determine the best solution to their specific circumstances because there is no definite answer to the best way in prioritizing available resources.

In prioritizing, several authors have state the importance of taking into consideration the needs of stakeholders. As such, user satisfaction is used in priority setting. Traditionally, maintenance prioritization is carried out without taking into account the views of tenants. This is supported by [45] which stated that maintenance manager did not discuss priorities set by the client or end-users. [46] stated that tenant participant is crucial in the priority setting process as to enhance tenant's satisfaction with maintenance system. While the satisfaction of user needs to be taken into consideration, caution has to be applied as they do not possess the required maintenance knowledge to make the proper prioritization choices [14]. In a case to prioritize equipments in a hospital, maintenance managers rated medical equipment as the top priority while users rated for the washing machine [47]. A comparison must be carried out between the priorities of different stakeholders as to produce a more effective maintenance plan. In this context, the following research questions are relevant.

1. What are the critical maintenance priorities that should be considered?
2. How do the needs of stakeholders affect maintenance priority?
3. Should maintenance priority be set according to stakeholders' needs?
4. How to apply maintenance priority based on the needs of the circumstances?

6 Conclusion

The literature review has examined the issues relevant to the different aspects of maintenance priority factors and methods in various organizations. Articles published from 1990 to 2015 were classified and analyzed. Based on the findings of this study, the area of maintenance priority requires more systematic research efforts especially in the implementing approaches that includes organizational goals, stakeholder needs and inclusion of critical factors. Priority setting process that aligns with the need of the organization can assist in avoiding losses and improve the maintenance performance. Figure 7 presents a conceptual framework on maintenance priority which encompasses the findings of this study. The findings in this study contributes to the understanding of maintenance priority setting especially in the factors and methods used.

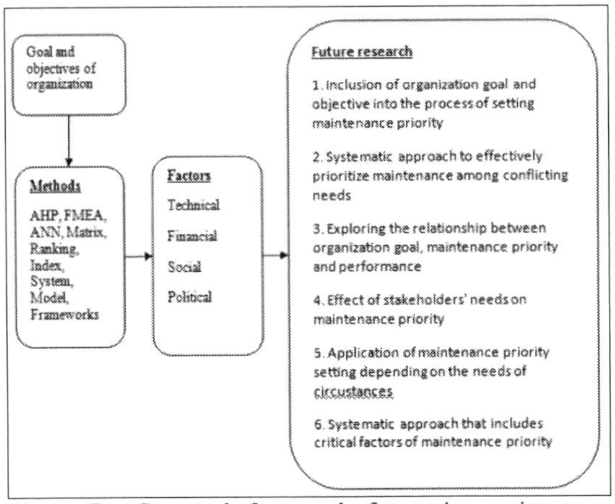

Figure 7. Conceptual framework for setting maintenance priority

References

1. Z. A. Akasah, Shamsuddin, S. H., Rahman, I. A., Alias M., Malaysian Technical Universities Conference on Engineering and Technology (MUCEET 2009) Pahang, 1-5 (2009)
2. M. M. Idris, Constr. Manage. Econ. **16**, 673-679 (1998)
3. B. Moua and J. S. Russell, J. Perform. Constr. Facil. **15**, 46-53 (2001)
4. M. Y. L. T. Chew, S.S. Kang, K.H., J. Archit. Eng. **10**, 80-87 (2004)
5. F. Booty, *Facility management handbook*. 2006, Oxford: Elsevier.
6. V. Kutut, E. K. Zavadskas, and M. Lazauskas, Arch. Civ. Mech. Eng. **14**, 287-294 (2014)
7. A. O. Hopland, F. **33**, 195-205 (2015)
8. W. Roy, Int. J. Oper. Prod. Man. **14**, 4-24 (1994)
9. Q. Shen, F. **15**, 331-339 (1997)
10. T. J. Pitt, F. **15**, 97-104 (1997)
11. O. A. Lateef, J. Build. Apprais. **5**, 201-212 (2009)
12. Q. Shen and A. Spedding, Build. Res. Inf. **26**, 169-180 (1998)
13. B. M. Kamya, *Research update of a typical bridge management system applied to a UK local authority*, in *Bridge Maintenance, Safety, Management and Life Extension*. 2014, CRC Press. p. 1757-1763.
14. N. A. Yusof, et al., Procedia - Social and Behavioral Sciences **62**, 508-513 (2012)
15. S. Lavy, et al., Constr. Manage. Econ. **32**, 1183-1204 (2014)
16. W. J. Moore and A. G. Starr, Comput. Ind. **57**, 595-606 (2006)
17. U. Kumar, et al., J. Q. M. E. **19**, 233-277 (2013)
18. J. S. Mitchell, et al., *Physical Asset Management Handbook*. Clarion Technical Publishers(2002).
19. A. Spedding, R. Holmes, and Q. Shen. *Prioritising planned maintenance in county authorities*. in *International Conference on Planned Maintenance, Reliability and Quality Assurance, Cambridge, UK*.(1995)
20. A. Shah Ali, J. F. M. **7**, 298-306 (2009)
21. Q. Shen, K.-K. Lo, and Q. Wang, Constr. Manage. Econ. **16**, 693-702 (1998)
22. J. H. K. Lai, Constr. Manage. Econ. **28**, 415-427 (2010)
23. A. Ali and T. Hegazy, J. Perform. Constr. Facil. **28**, 528-538 (2014)
24. I. Flores-Colen and J. de Brito, Constr. Build. Mater. **24**, 1718-1729 (2010)
25. J. W. Rogers, F. **31**, 56-67 (2013)
26. F. Taillandier, G. Sauce, and R. Bonetto, E. C. A. M. **18**, 343-362 (2011)
27. S. Antti and D. Mats, J. Q. M. E. **17**, 63-73 (2011)
28. I. Flores-Colen, J. de Brito, and V. Freitas, J. Perform. Constr. Facil. **24**, 337-344 (2010)
29. M. B. Hastak, E., J. Infrastruct. Syst. **7**, 67-76 (2001)
30. A. A. L. Olanrewaju, B. E. P. A. M. **2**, 167-181 (2012)
31. B. Hakim, C. Pierre-Jean, and Z. Nacer Eddine, J. o. S. I. T. **14**, 264-280 (2012)
32. D. Schraven, A. Hartmann, and G. Dewulf, B. E. P. A. M. **1**, 61-74 (2011)
33. M. Bourne, et al., I. J. B. P. M. **5**, 1-24 (2003)
34. M. S. Valen and N. O. E. Olsson, J. F. M. **10**, 287-300 (2012)
35. T. Danny Shiem-Shin, F. **14**, 12-15 (1996)
36. Q. Chang, et al., J. Manuf. Sci. E. **129**, 661-668 (2007)
37. M. Hassanain, T. Froese, and D. Vanier, J. Perform. Constr. Facil. **17**, 51-64 (2003)
38. N. Leveson, Reliab. Eng. Syst. Saf. **136**, 17-34 (2015)
39. J. Smith and P. Stewart, S. S. **25**, 24-38 (2007)
40. A. Sadeghi and R. A. Manesh, Procedia - Social and Behavioral Sciences **62**, 1378-1383 (2012)
41. M. A. Amir, K. P. Andrew, and G. C. Keith, F. **19**, 215-222 (2001)
42. M. Bertolini and M. Bevilacqua, Reliab. Eng. Syst. Saf. **91**, 839-848 (2006)
43. N. Dann, S. Hills, and D. Worthing, Constr. Manage. Econ. **24**, 97-104 (2006)
44. G. Carmignani, Reliab. Eng. Syst. Saf. **94**, 861-871 (2009)
45. M. Nik Elyna, K. Syahrul Nizam, and P. Michael, J. F. M. **9**, 181-199 (2011)
46. N. Ming Yip, P. M. **19**, 10-18 (2001)
47. H. Al-Zubaidi, P. M. **15**, 173-183 (1997)
48. R. W. Taylor,Omega **24**, 217-227 (1996)
49. A. W. Labib, G. B. Williams, and R. F. O'Connor, J. Oper. Res. Soc. **49**, 745-757 (1998)
50. J. Layzell and S. Ledbetter, Build. Res. Inf. **26**, 351-357 (1998)
51. I. M. Shohet, M. Puterman, and E. Gilboa,Constr. Manage. Econ. **20**, 305-314 (2002)

52. L. Fangxing and R. E. Brown, IEEE Trans. Power Del. on **19**, 439-441 (2004)
53. S. A. Oke and M. K. O. Ayomoh, Int. J. Qual. Reliab. Manag. **22**, 607-625 (2005)
54. M. Carnero, Reliab. Eng. Syst. Saf. **91**, 945-963 (2006)
55. D. M. Frangopol and M. Liu, Struct. Infrastruct. E. **3**, 29-41 (2007)
56. A. Chandrashekaran and B. Gopalakrishnan, J. F. M. **6**, 52-68 (2008)
57. G. Carmignani, Reliab. Eng. Syst. Saf. **94**, 861-871 (2009)
58. E. Vijay Kumar and S. K. Chaturvedi, Int. J. Qual. Reliab. Manag. **28**, 109-126 (2011)
59. D. Siew-Hong, K. Shahrul, and A. Ishak Abdul, J. M. T. M. **25**, 415-435 (2014)

The Issues and Considerations Associated with BIM Integration

Anas Bataw[1] Richard Kirkham[1] Eric Lou[1]

[1]School of Mechanical, Aerospace and Civil Engineering. The University of Manchester, UK.

Abstract. The management of data, information and knowledge through the project life cycle of buildings and civil infrastructure projects is becoming increasingly complex. As an attempt to drive efficiencies and address this complexity, the United Kingdom Government has mandated that Building Information Modelling (BIM) methods must be adopted in all public sector construction projects in 2016. Emerging from the US Department of Defence, BIM is an approach to the co-ordination of design and production data using object-oriented principles as described in ISO 29481-1:2010. The underlying philosophy of BIM is to ensure the *"provision of a single environment to store shared asset data and information, accessible to all individuals who are required to produce, use and maintain it"* (PAS 1192-2:2013). A key aspect of BIM lies in the notion of 'interoperability' between various software applications used in the design and construction process and a common data format for the efficient exchange of design information and knowledge. Protagonists of BIM argue that this interoperability provides an effective environment for collaboration between actors in the construction process and creates accurate, reliable, repeatable and high-quality information exchange. This UK government mandate presents numerous challenges to the architecture, engineering and construction (AEC) professions; in particular, the characteristics of BIM Level 2 remain explicitly undefined and this created a degree of uncertainty amongst the promoters and those professionals charged with delivering projects. This research casts a critical lens on the current literature in the domains of object-oriented modelling of infrastructure and the associated implications for procurement and project management. A mixed-methods approach using questionnaire analysis and secondary case study analysis was used to enact an inductive research approach that captures a range of data on the practical issues and considerations associated with the integration of BIM in the industry.

1 Introduction

BIM technology has been in existence, in some form or other, for almost half a century. Within the AEC industry, the first introduction to the theory of BIM was in the late 1970s by scholars at Georgia Institute of Technology. Their research made significant contributions to technology development within the industry, and is perhaps best illustrated by Autodesk under the CAD concept (AutoCAD) in early 1980s. Graphisoft who introduced their initial "Virtual Building Solution" in 1986, now known as ArchiCAD, followed this. This was the start of the software 'revolution' that allowed architects to create virtual, 3D designs of their project instead of the standard 2D drawings.

However, this technology was only limited to designing within the architectural practices; mainly used to illustrate building materials in order to show clients and stakeholders the design of the building in the early stages. This technology allowed the design team to reduce time and increase the quality of design but was also known to be expensive, fragile and complex (Puckett, 2011); designers using 3D software still had to go through the process of producing countless specification sheets in order to express all the required information to the rest of the team.

Since the early 1990s, many other software tools have been developed to assist the roles of other professionals and stakeholders within the industry, such as project management software, programming software, planning software, and developed excel software for pricing etc. These software tools has given the industry an enhanced control of time, cost and quality but it was still a fragmented way of working as each member of the team was working on a separate software. Therefore the creation of a collaborative constructed virtual building model such as BIM was required to complete the missing part of the jigsaw of the virtual design concept as demonstrated in Figure 1.

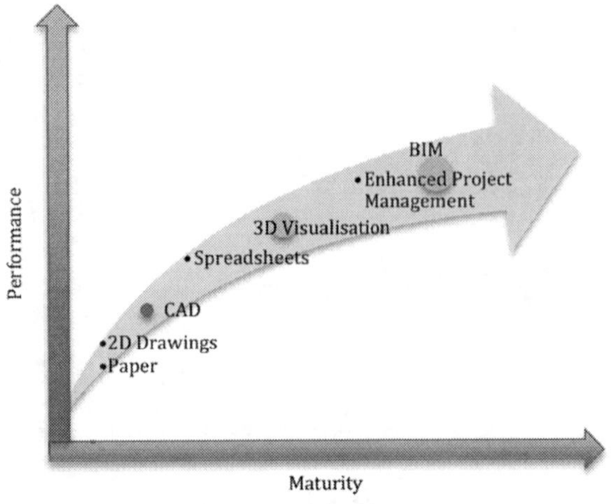

Figure 1. Evolution of technology within AEC industry

The term 'Building Information Modelling' was first used in 2002 to achieve the virtual design concept. BIM has progressed quickly since then, with many construction teams having observed values in using BIM within the construction industry; it has become more attractive with the continuous development of BIM tools. Today BIM is a moving target with a wide progress, becoming increasingly popular within the AEC industry throughout conferences, government strategies, companies strategies, awareness campaigns, and a wide range of journals and researches.

2. The benefits of BIM

Communication problems (whether intended and unintended) remain as much a problem in construction projects as elsewhere in business and commerce. Proponents of BIM argue that it provides a 'bridgehead' enabling reliable and regular knowledge exchange across the project organisation, with the potential to improve relationships between the architects, engineers, construction professionals, facility managers and building owners; this feature could also enable multi-disciplinary teams to remain synchronised – this should improve accuracy and enable a more informed and knowledgeable approach to decision making, which could in return reduce waste and help to achieve a successful project.

BIM is also considered to be a positive transition for designers, where they can be supported by new means of technological tools throughout the design processes, this can make their work easier, smoother and faster while enhancing the quality of designs.

Similarly, contractors can create visual data for the costs, materials and construction sequences within a shared collaborative model. Once the information data is placed within the BIM model, it can automatically present itself in floor plans, elevations, specifications, work sequences, and quantity takeoff, etc. Where all accessible users can view the information and operate on it, instead of working from detached drawings and schedules in the form of many separate paper documents. Information within BIM would also automatically adapt when changes occur to a set of data, together with a clash detection tool if changes are unsuitable. This is different to the traditional fragmented practice of numerous individual sets of drawings, where design and construction teams have to go back to manually change and re-print each set of drawings, elevation, specification, work sequence and quantity takeoff.

Case studies carried out by Kaner *et al.*, (2008) and Eastman *et al.*, (2008) indicated that the use of BIM in projects reduced the number of information requests and order changes; it also improved productivity and efficiency, especially in the early stages of design. A report commissioned by The Department for Business, Innovation and Skills (BIS) and HM Cabinet Office in 2008 recommended that a BIM approach could account for up to 2.5 billion per annum savings in the construction phase alone.

3 Challenges of BIM implementation

The co-ordination of activities to promote and embed BIM in the UK is achieved in part by the 'BIM Task Groups' – and whilst the overall message that has emerged has been one promoting the virtues of BIM, there is an alternative viewpoint that suggests that BIM could introduce problems that, hitherto would not otherwise have been evident using more traditional methods of design and construction co-ordination, these are summarised in Migilinskas et. al. (2013), based on four international case studies;

- Predisposition to software tools and methods of working that are familiar to the project participants;
- Focus on 'difficult' aspects of the project rather than broader application to the whole;
- Some intelligent approaches such as Virtual Project Development utilised for the first time and some concerns regarding the quality of the data used to construct the model.
- Benefits of BIM are directly correlated to the ability to maximize collaboration in project. Piecemeal adoption across the project team leads to problems;
- High costs with software purchases and staff up-skilling
- Lack of standardisation and expectations of contract obligations in certain countries or regions (such as European Union, Americas, Asia and other).

These issues might seem as great barriers to implementing BIM. However, reports and studies illustrate that BIM has been widely used around the world.

4 International perspectives on BIM

Despite the above-mentioned implementation barriers, many countries have showed great interest toward implementing BIM across the Architecture, Engineering and Construction industries. However, each country has its own arrangements and progressed differently:

4.1 United States (US)

The USA was the earliest initiator of BIM, particularly within the public sector. In late 2006 the US General Services Administration (GSA) issued a BIM-guideline outlining an implementation plan to accompany the integration of BIM use within the US AEC sector in general and the Public Building Service (PBS) in particular. Following this, in 2007 the US GSA issued a mandate to obligate all planners to use BIM while applying for GSA funding schemes (GSA, 2007).

In addition to the widely recognised benefits of BIM that the public sector appears to be gaining benefits from, the US AEC sector has established BIM policies by addressing the allocation of risks associated with the implementation of BIM, outlining the roles and responsibilities of each participating party while avoiding any conflicts with the existing construction contracts and policies. This in return has encouraged many stakeholders to use BIM as indicated by the American Institute of Architects' report on the Business of Architecture (2010), confirming that 60% of US architects were using BIM throughout their projects and still increasing yearly.

4.2 Finland

The implementation of BIM within Europe was initiated later than the USA but it has spread more quickly and exposed a wider improvement within the industry, especially in Finland (as shown in Figure 2). According to the Finnish ICT Barometer for architects in Finland (2007), 93% of architects were using BIM in projects with 33% of that usage at BIM level 3. In the same survey it was indicating that nearly 60% of Finland's engineers are using BIM in both the public and private sectors (Kiviniemi, 2007).

This spread of BIM use within Finland is due to increased interest by the AEC and Facilities Management (FM) companies in profiting from the benefits of BIM. Starting from the 1st October 2007, the Finnish FM companies focused on using BIM's modelling technology within common project works. In 2009, they established detailed modelling guidelines to assist with the use of BIM during the design stages. Later in 2009, the governing body of public properties used these guidelines to run several pilot projects; where it had a great impact in making decisions for the Senate Properties' investment processes and enhanced developments within the public sector (VTT, 2007).

BIM has also reached the private sector in Finland, where major companies such as 'Skanska Oy' and 'Tekes' took the lead of adopting BIM within private projects (Kiviniemi, 2009) Giant private companies have also funded a number of BIM related researches with local universities such as Tampere University of Technology; these researches investigated the benefits and outcomes of BIM practice within the industry to promote the integration process of BIM, developing technical tools and investigating the potential of BIM in providing sustainable solutions within the industry (Leicht *et al.*, 2007; Huovila, 2008).

4.3 Singapore

The National Ministry of Singapore has first introduced BIM in Singapore in early 1995. This gave organisations such as 'Development Construction and Real Estate Network' (CORENET) an early involvement to develop and implement BIM within governmental public projects.

Singapore's government has been successful in pushing for BIM implementation and BIM standards on various kinds of projects within the public and private sectors with the help of CORENET's BIM Guideline "Integrated Plan Checking" (Khemlani, 2005). This has noticeably enhanced the number of public-private initiatives to encourage the use of BIM in a large number of pilot projects.

4.4. India

India's fast growth of population and economy have provided a boost to the building environment and provided the perfect platform to implement BIM. India has a strong workforce of qualified, trained and experienced BIM specialists who are not only implementing BIM technology in India's Construction Projects but also assisting on the implementation of BIM in Canada, USA, UK, Singapore and the Middle East regions.

4.5 Canada

The Institution of BIM in Canada (IBC) has taken the responsibility of leading and facilitating the full implementation of BIM into the Canadian built environment where they maintain a keen interest in focusing on the primary stakeholders allowing them the right method and pace to understand their roles and responsibilities and to assess their capacity to contribute in this process.

4.6 France

On April 2014, the French minister of Housing and Development announced the new "Building 2.0" which primarily contains using Building Information Model as the main tool for public projects. Outlining that BIM will become obligatory in all state owned projects by 2017. However, no plan has yet been introduced for these requirements to take place.

4.7 Continental Europe

UK, Netherlands, Denmark, Finland and Norway Governments are already demanding the use of BIM for public projects. Consequently, in November 2013 the European Parliament voted to support the utilisation of electronic tools such as BIM for public works contracts. They have described this convention as an approach to possibly enabling more efficient construction and building projects in Europe and help advance European competitiveness.

Following this decision, leaders from Europe's AEC industries have expressed their support to the European Parliamentary vote on what they described as modernising the European public sector.

This parliamentary vote is perhaps viewed as paving the way for adopting BIM and permitting all 28 EU member states to encourage, specify or mandate the use of BIM for public funded construction and building projects by 2016.

4.8 United Kingdom

The UK Government has already shown their awareness of BIM's benefits in perhaps controlling costs, time and quality and the advantages it could offer to everyone involved in the construction projects, including clients, designers, contractors, suppliers and facilities managers. On 31st May 2011 the UK Government showed its interest in BIM by publishing a construction strategy report that announced that it is aiming to adopt BIM's technologies, processes and collaborative behaviors into all stages of the life cycle of all public projects by 2016. This is expected to advance the use of BIM in the UK, as shown in figure 2. However, the UK industry is not yet ready for the implementation of BIM. The following will establish the barriers that stakeholders within the UK industry will face in seeking to implement change. Most of these issues may only appear during the implementation of BIM level 3. Implementing BIM level 2 should not create significant additional risks, but some amendments might be required to smooth the implementation process.

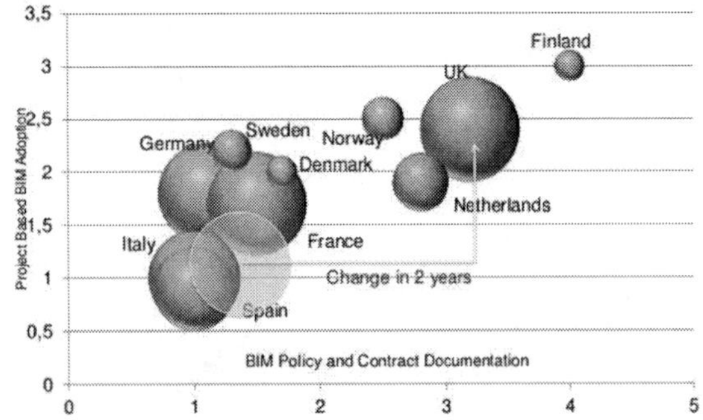

Figure 2. BIM Adoption in Europe

5 Questionnaire survey

In order to achieve BIM expectations and successfully implement BIM around the world, all stakeholders in the industry must reach the required BIM awareness level. To understand the readiness of stakeholders in the UK to the BIM implementation requirements, a questionnaire survey was distributed to a large number of professionals in the UK. 84 participants flagged the concern of misunderstanding BIM and its concepts. The response showing in the graphs below expresses the tardiness of many practitioners and organisations towards BIM awareness and adoption.

- *Are you or is your organization aware of the challenges of implementing BIM?*

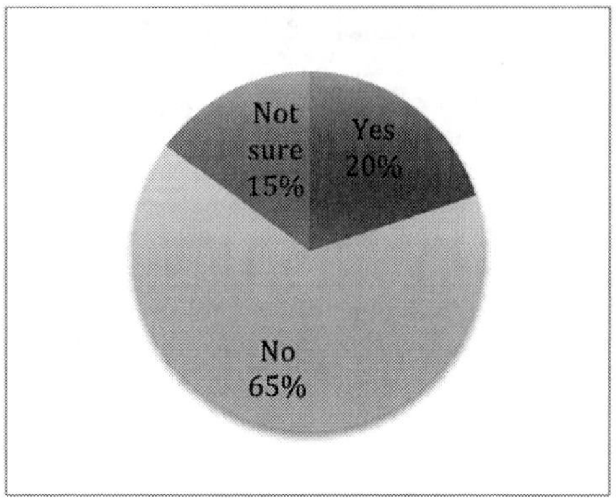

- *Are you concerned about BIM adoption in your organization?*

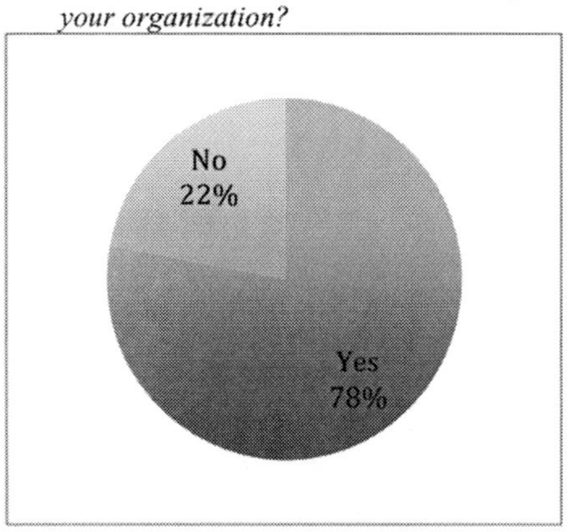

Those participants who were concerned with BIM adoption were asked an additional question to rate their concerns of BIM adoption challenges. From the results showing below and the comments that were obtained from this research and focus groups, an outstanding distress was discovered on the concern of BIM adoption and a large number of professionals seemed to acknowledge BIM but are still unsure of what their organisation is planning to overtake the challenges of adopting BIM level 2 in the very near future. Also, despite their awareness that adopting BIM is a challenging task to many organisations, they still don't seem to know what these challenges are. Therefore, it is highlighted that a comprehensive manuscript is required to outline all the challenges of adopting BIM in the UK.

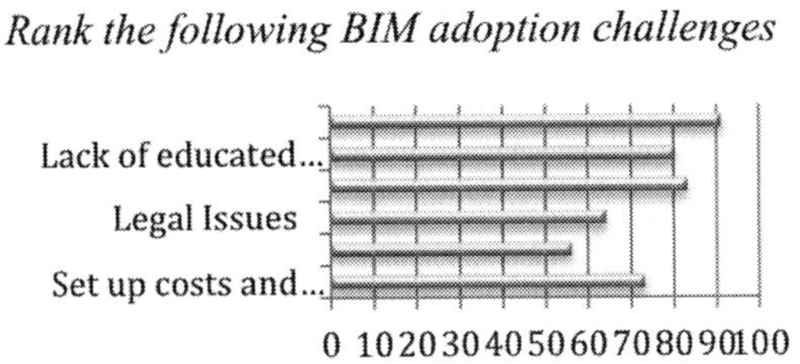

6 Discussion

No doubt, BIM can be of extreme benefit to the industry and potentially improve the way the professionals deal with projects; however, the use of BIM could raise a vast number of issues that deserves serious consideration. In essence, it is only as good as the people using it (Kirkham, 2015). Many clients are still hesitant towards the implementation of BIM as they are still uncertain and puzzled on what BIM really is. This is due to the nature of all participants within the industry and the high costs of BIM implementation owing to the required extensive training of the different professionals, cost of technical expertise, costs of organising protocols and managing a network server to store and access the model.

Other issues preventing the Implementation of BIM are the Legal barriers surrounding liability, uncertainties to the Intellectual Property Rights, digital information exchange and ownership of the program, which could all be resolved in time.

However, most of these issues would only occur while using BIM level 3. Implementing BIM level 2 should not create significant additional risks; nevertheless some amendments might be required to smooth the implementation of BIM.

BIM level 1 only contains the use of the design software features within the design stage; this level is currently used and widespread within the industry without any major implementation issues. On the other hand, BIM level 2 is an increased method in using software technologies within separate disciplines. Therefore, the following topics should be considered before BIM level 2 could be implemented within projects:

- The necessity for intense awareness campaigns and training courses throughout the industry to cover the doubts and debates surrounding BIM and enhance awareness towards the responsibilities and roles of individuals and organisations throughout the use of BIM level 2.
- Implementation of Level 2 BIM may require amendments to the intellectual property legislation.
- Contractual amendments and software measures might require rearrangements to protect users from data corruption and software tool failures especially when different users operate on the same model.
- Enterprises operating on level 2 BIM might become limited during tenders when level 3 BIM is fully implemented by others.
- A BIM protocol must be outlined and agreed during the procurement stage to address risk sharing, detailed responsibilities of all users, technology level of each model, level of definition, and an exclusion of liability. These protocols must be clearly outlined within the agreements between the client and those responsible for the BIM model (Beale and Company Solicitors LLP, 2013)

Implementation of Level 3 BIM is not just a simple step up from level 2 in terms of using software tools; it is an elevation to a very different style working. BIM level 3 will require using advanced tools within one collaborative platform; this will require a number of considerations as detailed below:

6.1 Issues Associates With Integrating BIM On Existing Buildings and Infrastructure

Attempts have been made to use BIM for old and pre-existing facilities. However, this was only possible when the existing facility was rebuilt visually in BIM or converted into the form of BIM. Although, converting an existing building into a BIM model would require numerous assumptions such as the standards and codes of the existing building design, the construction methods used, and the materials used at the time of construction.

6.2 Issues Associates With Integrating BIM On New Buildings and Infrastructure

- Cost – BIM level 3 will require significant investment from across the industry. There would be a need to take into account the costs of BIM's software and hardware as well as other costs, such as the extensive training of the different professionals, cost of technical expertise, costs of organising protocols and organising a network server to store and access the model. These costs raise the concerns of many small/medium enterprises within the industry. Failure of these enterprises in fulfilling the cost requirements will generate a large gap between them and other BIM-using enterprises in terms of recognition, work quality, winning tenders, saving time and money etc.
- Industry 'mind-set' – (Need for teaching and training) The current traditional way of working will not easily adjust to the high-tech collaborative way of working that BIM is introducing to the industry. BIM level 3 will completely change the way that professionals would approach their day-to-day duties, from

the fragmented paper method to having to work within an informational collaborative model that requires regular communication between different participants from early stages. Therefore, All existing and new coming professionals must be trained and educated to fully understand their responsibilities and duties. Also, these responsibilities and duties must be considered and drafted within the contractual documents to ensure services are carried out according to the collaborative nature of BIM.

- Information control - BIM level 3 relies considerably on IT and software systems. This reliance raises many concerns as to the need of various control procedures in order to limit and control access and inputs such as data protection with firewall systems, data backup features in case a corruption of data appears, technical support facilities and professionals etc. The BIM model is the core data platform of the project; one error within the model can be both costly and time wasting.
- Ownership – The issue of model ownership as been widely debated, where many stakeholders within the industry are concerned with who should obtain the final version of the model and the involved data. These debates are mostly due to the misunderstanding of the concept of BIM; if the model generated by BIM was correctly categorise as a product then by law it should only be retained by the buyer, i.e. the client or the building owner. However, the data contained within the BIM model is a separate issue because it is generated from contributions of various team members; they should be authorised to obtain a copy of their contribution for future records. These issues must be considered and discussed by the government to outline and verify the legal regulations towards ownership of the BIM model and the involved data during and after construction.
- Liability exposure - Different professionals from various enterprises contribute towards the BIM model throughout different stages of the product's life cycle through collaborative software systems; this new way of working might create irregular liability issues. BIM software system is protected by "blanket limitation of liability" clauses that generate the question of who is liable for any errors caused by the software tools. Another concern is who is liable if works were carried out incorrectly due to inaccurate information given by a different professional in the early stages. These risks must be dealt with and clarified in contractual protocols and carried out accordingly to distribute risk and liability evenly.
- Insurance - limited insurance companies can currently offer to insure BIM. Due to the limited use of BIM and doubtful impressions that was surrounding BIM's benefits and the risks it may incur, it was incredibly expensive to insure works done with BIM. However, now that BIM has been successfully implemented within different projects, insurance costs have decreased. For the time being, it is important for parties to consider taking out the appropriate insurance to cover their engagement in the BIM process to obtain their usual coverage and protect themselves against liabilities and risks.
- Contractual documents – BIM level 3 offers new roles and responsibilities for existing and new professions such as BIM managers and architects and draftsmen. Project contracts should include a detailed brief of these roles and outline the duties of each professional role to suit the use of BIM within projects. The same set of BIM privileges and requirements should flow through the different contracts to avoid clashes between the clauses of the principal contract and the legal terms of the BIM protocol.

7 Conclusion and recommendations

It is arguably that the construction industry can benefit from the integration of BIM in order to improve the current fragmented way of working, overtake the overpowering issues and possibly provide potential solutions and advantages to the industry. From the literature review and secondary case studies, we can conclude that BIM implementation can possibly offer enhanced products throughout the industry by:

- Reducing errors and omissions, this will make works smoother, reduce RFIs, reduce professional liabilities and insurance costs.
- Provide opportunities to discover errors in early stages, earlier error discovery reduces repair costs in comparison to discovering them once project design progresses.
- Reduce time. Where involved managers, designers and drafters can spend less time developing designs and more time providing creative solutions for clients.
- Have a positive impact on firm's reputations with an increased number, scale and variety of opportunities
- Enhance the reputation of the industry towards sustainability and efficiency
- Increased client satisfaction through visual verification of design intent
- Enhanced way of working with knowledge sharing and virtual Design before construction.

Although these benefits might appear astonishing, they are currently only presented on paper because in reality BIM could just be another idea that could not proceed due to the lack of assessment and the misunderstanding of BIM implementation process. Therefore, detailed implementation plans and arrangements are required to assist with the integration and adoption of BIM within organisations, which is currently realistically unachievable due to the numerous obstacles surrounding BIM and the difficulties of BIM execution. The demand for BIM execution plan was exposed from the questionnaire survey findings, which raised necessity of outlining and applying the following processes:

- Communicate and enhance the understanding of BIM, this could be done by providing a wide range of seminars, conferences, workshops and training courses to existing professionals in all sectors. As well as promoting the publication of articles and carrying researches on BIM.
- Organise and provide many educational and training sessions to allow the new professionals to have the correct knowledge and skills to blend with BIM applications to ensure the new and old professionals within the industry are ready for the 2016 digital BIM switchover.
- Set up clear definitions of roles and responsibilities of each different participant within the new way of working.
- Locate who is responsible for setting up the level of BIM and model standards applied within a project, and when.
- Outline the required outcomes from the use of BIM within projects.
- Examine the contractual and legal issues to find solutions to ownership, sharing, copyright, IP allocation and Insurance and issue a framework to outline the legal process and procedures of BIM.
- Establish BIM guidelines for the UK that can also be integrated with international BIM guidelines.

These themes must be evaluated and investigated by organisations to outline their unique execution plan to make the implementation process of BIM clearer and closer to reality and to overtake the issues and considerations associated with the integration of BIM in the industry

References

1. AGC (2006) "Contractors' Guide to BIM" The Associated General Contractors of America. Agopyan, V. et al. (1998) "Alternatives for Reducing Material Waste on Building Sites", *Research report.*
2. AIA (2007) "Integrated project delivery: a guide", AIA California Council. Baldwin, A. N. et al. (1998) "Planning building design by simulating information flow", *Automation in Construction.* 8. 2. 149-163.
3. AIA (2008), Document E202 - Building Information Modelling Protocol Exhibit, American Institute of Architects, Washington, DC.
4. Alin, P., J. Iorio and J. E. Taylor (2012). Objects as Arbitrators: Spanning Boundaries in Virtual Engineering Networks. Industry Studies Association Conference.
5. Andoh, A. R., X. Su and H. Cai (2012). *A boundary condition-based algorithm for locating construction site objects using RFID and GPS.* Construction Research Congress 2012
6. Arayici, Y and Aouad, G 2010, 'Building information modelling (BIM) for construction lifecycle management', in: *Construction and Building: Design, Materials, and Techniques*, Nova Science Publishers, NY, USA.
7. BIS/Industry Working Group (2010). Building Information Modelling and Management BIM (M): Interim Report from the BIS/Industry Working Group. London
8. Boeykens, S., C. Himpe and B. Martens (2012). "A Case Study of Using BIM in Historical Reconstruction. The Vinohrady Synagogue in Prague." *Digital Physicality| Physical Digitality*: 1-10.
9. Boshyk, Y. and Dilworth, R.L. (Eds.) (2009), "Action Learning: History and Evolution", Basingstoke, U.K.: Palgrave Macmillan
10. Braxton, J. M., & Hargens, L. L. (1996). Variations among academic disciplines: Analytical frameworks and research. In J. C. Smart (Ed 1), Higher education: Handbook of theory and research (Vol. 11, pp. 1-46). New York: Agathon Press.
11. CIRIA (1999), "Standardisation and Pre-assembly Adding Value to Construction Projects", Standardization; Pre-ASSEMBLY; Drivers; Benefits; Implications, London, Report 176
12. Coates, P. Arayici, Y. Koskela, L. Kagioglou, M. Usher, C. O"Reilly, K. (2010), "The limitations of BIM in the architectural process", First International Conference on Sustainable Urbanism, December 2010, Hong Kong
13. Eastman, C. Teicholz, P. Sacks, R. and Liston K. (2008) "BIM Handbook: A Guide to Building Information Modeling for Owners, Managers, Designers, Engineers and Contractors", John Wiley & Sons, Inc., Canada.
14. Edgar, A. (2007) "W15: Introduction to BIM: People, Processes and Tools", NIBS.
15. Environment Agency (1997) "Waste minimisation and waste management: an environmental good practice guide for industry ". Solihull. E.A.
16. Greenwood, D. Lewis, S. and Lockley, S. (2010) Contractual issues in the total use of building information modelling. In: 18th CIB World Building Congress, Salford.
17. GSA (2007) BIM requirements and adaptation. website: www.gsa.gov/bim. [Accessed 22/11/15]
18. Huovila, P. (2008). "Building Information Models and Innovative Sustainable housing", *In proceedings of SB08 Conference*, Melbourne, Australia, 21–25 September 2008.

19. ISO 29481-1:2010, Building information modelling - Information delivery manual - Part 1: Methodology and format, BSI Publications.
20. Jordani, D., (2008), "BIM: A Healthy Disruption to a Fragmented and Broken Process", Journal of Building Information Modelling: Spring 2008, Matrix Group Publishing, Houston.
21. Kaner, I., Sacks, R., Kassian, W. and Quitt, T. (2008). "Case studies of BIM adoption for precast concrete design by mid-sized structural engineering firms." ITcon **13**(Special Issue Case studies of BIM use): 303-323.
22. Kirkham, R. (2015) Ferry and Brandon's Cost Planning Of Buildings (9th Edition 9), Chichester: Wiley-Blackwell, 2015. ISBN: 978-1-119-96862-7.
23. Kiviniemi A. 2007: *Support for Building Elements in the IFC 2x3 Implementations based on 3rd Certification Workshop Results*. VTT, Finland.
24. Kiviniemi, A. (2010), "Challenges of Interoperable BIM in a between organization" http://aarch.dk/fileadmin/filer/Sune/Arto_Kiviniemi.pdf
25. Kiviniemi, M. (2009) "Building Information Model (BIM) promoting safety in the construction site process", SafetyBIM – research project 10/2007 – 2/2009 (TurvaBIM in Finnish).
26. Leicht, R., Fox, S., Mäkelainen, T. & Messner, J. (2007) "Building information models, display media and team performance: An exploratory study", VTT Working Paper No. 88.
27. Migilinskas et. al. (2013), The Benefits, Obstacles and Problems of Practical Bim Implementation. Vilnius Gediminas Technical University, Civil engineering faculty, Saulėtekio al. 11, LT-10223 Vilnius, Lithuania DOI: 10.1016/j.proeng.2013.04.097
28. NIBS (2008), The National Institute of Building Sciences, website: www.nibs.org [Accessed 22/11/15]
29. NIST (National Institute of Standards and Technology) (2004) : Cost Analysis of Inadequate Interoperability in the US Capital Facilities Industry. NIST GCR 04-867 (2004)
30. PAS 1192-2:2013 Specification for information management for the capital/delivery phase of construction projects using building information modeling
31. Puckett, K, 2011, How members of the project team interact through BIM, CPD 2011 Module 4: Building Information Modelling, bdonline.co.uk, Available from: http://www.bdonline.co.uk/business/cpd/cpd-2011-module-4-building-information-modelling/5016713.article [Accessed: 18 Dec 2015].
32. VTT (2007) *Virtual Building Environments II Final Project Report*, http://cic.vtt.fi/projects/vbe-net/data/VBE2_Final_Report.pdf. [Accessed at 10/11/2015]
33. Yan, H., Damian, P. (2011) Benefits and Barriers of Building Information Modelling.

 DOI: 10.1051/matecconf/20166600006

The Influence of Place Making's Attributes on the Resident's Usage and Satisfaction in High-Rise Residential Community: A Case Study

Ardalan Aflaki[1a], Norhayati Mahyuddin[2], Mozhgan Samzadeh[3], Mahsan Mirnezhad[4]

[1,2,3] *Center for Urban Conservation and Tropical Architecture (UCTA), Faculty of Built Environment, University of Malaya, Malaysia*
[4] *Department of Architecture, Faculty of Built Environment, University of Malaya, 50603, Kuala Lumpur,Malaysia*

Abstract. The vitality and viability of open spaces in urban neighborhoods, particularly in high-density residential areas with limited buildable land, have become extremely complex phenomenon. Researchers and designers are seeking the standard of designed environment to successfully meet the human needs and desires. In order to reach this goal, limited studies have been implemented to bridge the gap between research and design and consequently improve the physical and social qualities of open spaces. This paper examines the two specific features of open space - accessibility, social interactions - which contribute significantly to both the vitality and viability with a view to provide an understanding of the relative importance of these elements from users' perceptions and satisfaction. The methods involve quantitative research with questionnaire-based survey, and case study to evaluate residents' perceptions. The research scope is a high-density residential condominium in Kuala Lumpur where the rapid transformation of urban elements in city center has negative impacts on basic qualities of open spaces. The result is a framework for understanding the relative priority of the different elements (variables) for users which vary depending on the location of the residential block they live in; in particular its proximity to the open space. The study concludes with recommendations for how the most valuable elements can be better harnessed in order to promote the satisfaction with open spaces.

Keywords: Open spaces, Accessibility, Social interaction, User satisfaction, high density residential areas

1 Introduction

Open spaces are a significant element contributing to sustainability. They can help create a healthy environment, and ensure the viability and vitality of urban neighbourhoods. Bedmio-Rung et, al.[1] meanwhile noted that, while cities have many elements, open spaces can connect residential districts to all the other parts of a city. Open spaces in neighbourhoods can also bring a sense of community and help forge closer relationships among neighbours because they can act as a node for social gatherings or ceremonies [2]. In addition, the greenery within open spaces can moderate the local micro-climate, create a sense of relaxation, provide welcome shade and increase the positive aesthetic feelings of residents. This study examines the value of selected features in a high density residential area in order to understand which elements are most important for residents in terms of usage and satisfaction levels.

[a] Corresponding author: ar.aflaki@siswa.um.edu.my

2 Factors and their influence on the physical and social quality of open space

There are a lot of potential factors that can influence the quality of open spaces. Accessibility, climate comfort, facilities and amenities, and aesthetic values are among the physical factors, while safety, privacy, social interaction and surveillance are important social factors which can impact on the quality of open space. The current study focuses on accessibility and social interaction in open spaces in a high density residential area which is typical of the recent rapid transformation of residential developments in Kuala Lumpur.

2.1 Accessibility

Levinson [3] measured accessibility with two variables: travel time between two destinations, and the distribution of land use (close proximity). Close proximity and visual accessibility to open spaces are two main indicators of a well-designed urban space [4]. Physical and locational features such as proximity to the park and adequate amount of parks nearby could impact perceived accessibility to urban parks [5]. Location, design and linkages with other types of area are other significant factors which can attract people to open space [6].

Linkages and connections enhance the quality of open spaces and the degree of user satisfaction, by connecting activities which happen within and near open spaces [7]. Distance to and from an open space is one of the indicators of how well that space is connected. There are a lot of methods to measure distance. One such is the shortest network path, which is defined by the shortest (short-cut) distance to an open space[8]. Talen [9] also devised methods to calculate the minimum travel distance from residential units to open spaces, as well as the accessibility of open spaces. There is as yet no consensus over the best method to evaluate accessibility to open spaces. Table 1 shows the variables used in this research to evaluate accessibility from units to the open space.

Table 1. Variables for measurement of accessibility

Variable	Reference
Close proximity Ability to achieve basic needs Good visibility Access to other competing local facilities	Giles-Crti et al.2005[4]
Location	Calthorpe 1993[6]
Travel time	[10]
Context linkage system: location of the space Physical and visual permeability	Heng and Chan 2000[11]
Shortest network paths	Truelove 1993[8]
Diversity	Talen 2000[9]

2.2 Social interaction

Social contacts are another significant factor for attracting people to urban open spaces [12]. Social interaction creates liveliness, vitality and viability in open spaces [13]. There are many variables which can measure social interaction within open spaces. According to Chiesura [14], variables which can help to evaluate the impact of social interaction on the quality of an open space include the volume of interactions and contacts with other people, the use of the land beside the open space, and the reasons users give for using an open space (such as meeting neighbours and social activities).

Social activities such as social interactions are more likely to happen within open spaces when the level of physical and social amenities is good [15]. Urban parks play a vital role to improve human well-being under environmental disorders [16]. Conversely, good quality open spaces can provide better environments for residents' activities and encourage them to use open spaces more [13]. In sum, the amount, type and presence of users and their activities as well as the quality of the open space can impact on social interactions within it. Table 2 shows the variables used in this study to measure social interaction within open spaces.

Table 2. variables for measurement of social interaction

Variable	Reference
Social interaction in spite of activities	Chiesura 2003[14]
Social interaction & being with others Participation in activities	Gedikli 2004[12]
Presence of users in open space	Moirongo 2002[15]

2.3 High density and its effects on open space

High density developments have a number of advantages. They can provide better social interaction, high levels of security and safety, and also a comfortable environment for residents who want to walk, relax or otherwise take their leisure in a given neighbourhood [13]. By providing a balance between residential, commercial and official activities, such developments can increase the sense of security and safety among residents [13, 17]. According to Liao et al. [18] around 40% of the respondents choosing neighbourhoods with friendly environment for walking or biking. On the other hand, some research mentions various negative points related to high density, such as overcrowding, lack of privacy and restricted living space for residents [19].

With regard to open spaces, there are, again, both positive and negative arguments about high density. For instance, high density city life is generally much better than low density suburban living in relation to social community as well as the social quality of open spaces. Rowley [20] argued that, when high density living is accompanied by a good distribution of mixed use in a city; this can offer diversity of use with vitality. Also, high density with its closer social contacts and interaction among residents offers more opportunity for neighbours to do recreational activities in their spare time [13, 17]. The same authors also suggested that high density dwellings can save land, leaving more free land in which to design open spaces and greenery in residential neighbourhoods, although he also noted some negative aspects of high density, such as limited scope to create and enjoy greenery. Overall, negative environmental impacts tend to be greater in urban sprawl, and there is less scope for sustainable design in such environments. Newman [21] mentioned that low density sprawling cities suffer from poor environmental quality for residents, with adverse effects on health. In contrast, in high density cities dependency on cars is much lower, which can have positive impacts on air quality and the natural environment around residential buildings, making these more conducive for residents to use open spaces, breath fresh air, and sit and gather for social interactions.

In conclusion, although high density living can support sustainable, quality living in part, more research is needed into the negative aspects and how to solve these [19, 20] in order to establish if urban high density developments can be a widely viable option for sustainable developments in the future.

2.3.1 The impacts of high density living on accessibility and social interaction in open spaces

There is an expanding base of literature on the implications of high density living for the social quality of open spaces within neighbourhoods. Many of these studies concern the arrangement and distribution of primary and secondary activities. Hillier and Hanson [22] claimed that these arrangements could directly influence the intensity of social interactions. Social activities and contacts could also be persuading when high density of primary and secondary activities happen in suitable and relevant condition [15, 23]. While mixed land use in high density areas creates good opportunities for social interaction, it can on the other hand cause some difficulties such as lack of safety and privacy. Another challenge related to mixed land use developments is encroachment on public areas. For example, where privately owned restaurants or shops are adjacent to public open spaces, it becomes difficult to separate the public space and from the private space and controls may be necessary [13, 21].

3 Research methodology and design

This study adopts a quantitative methodology, using the questionnaire technique to evaluate selected features in open spaces based on user preferences. The data from survey questionnaires is then analysed by comparing cross-tabulations. This focus on users' opinions, and their perceptions of the recreational space, are key sources of information and criteria on open space design [24].
In the current study, the questionnaire consists of three parts, as detailed below:

- General questions on participants, such as their gender and age.
- Information on the residents' (participants') opinions on the physical aspects of the place in question.
- Questions examining the social qualities related to the place and its attributes.

A five-point Likert scale is used for answers, ie: strongly disagree, disagree, neutral, agree, and strongly agree. A content analysis is first used to present established theories relating to the attributes of the place and their influence on the quality of the open space. Consequently, a cross-tabulation from the survey questionnaires is applied to further analyse the findings.
In this study, the residents sampled live in the East Lake condominium in the district of Serdang. The primary field survey in the current study assured that the range of resident's age who were using the open space regularly was between 15 to 65 years old. Thus, the study designed a survey questioner in order to collect the data for the above mentioned range. The survey questionnaires were distributed within open spaces in their respective units to identify and study differences and similarities in ideas among residents. A total of 84 responses were collected out of 140 questionnaires given out.

3.1 Study area

The study area is a high density residential condominium located in the district of Serdang in Kuala Lumpur. It contains five blocks of 19-storey buildings. There is a total of 618 residential units in East Lake condominium. A bird's eye view, showing the layout of all the blocks in the study area, is presented in Figure 1. Figure 2 shows the site plan of the condominiums, with the blocks marked by circles. The arrows indicate the possible routes from the blocks to the open space area, which is marked by a black circle. The pictures in Figures 3 and 4 show the open space and its connections with some of the surrounding facilities, such as a reading room, gymnasium, restaurant, mini-markets and laundry; they also show the residents engaging in activities. A narrow pathway which links Blocks D and E is shown in Figure 4. Residents living in Blocks D and E have to walk alongside two further blocks to reach the open space.

3.2 Scope and limitation of study

The current study was limited to the open spaces in the high-income and high-rise residential building in Kuala Lumpur. There were few studies on the quality of open space with regards to social and physical qualities [16]. As a result, the current study focused on these attributes. Therefore, the main scope of this study was the evaluation of open space quality in the high-rise building where the quite numbers of residents were living in the dense area.
This caused the quality of open space in these residential communities be questionable.
The study faced to certain limitation on selection of study area. In the primary survey, the study selected the total numbers of 5 residential communities with the equipped open spaces. However, due to security and privacy reasons, the management offices did not permit the researcher to collect the data. Among the selected condominium for the study, the researcher got permission from the management office of East lake condominium in the Serdang.

Figure 1. Bird's eye view of case study (Source: Authors, 2013)

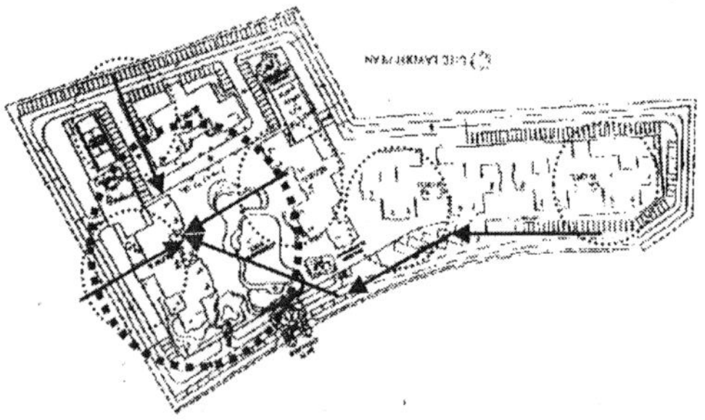

Figure 2 Site plan and arrangement of blocks (Source: Authors, 2013)

Figure 3. Swimming pool and it's alongside activities in open space (Source: Authors, 2013)

Figure 4. Connected Pathway from blocks D and E to open space (Source: Authors, 2013)

4 Results and discussion

As the current research focuses on two most important factors which influence open space use and satisfaction: accessibility and social interaction in the open space, the analysis of results is presented consequently. Accessibility consists of two variables: travel time/distance between two points, and the distribution of land use within reasonable proximity [3, 10]. Location, close proximity, short travel distances, land use surrounding open spaces and the design of open spaces themselves are all significant factors in bringing residents together [6]. Table 3 explains all the above mentioned variables used in this study to examine the degree of accessibility in the Case study. The findings show that the majority of respondents agreed or strongly agreed that accessibility to basic needs (57 respondents), open space location (53 respondents) and close proximity (51 respondents) are important factors for accessibility. In contrast, a majority of respondents disagreed or strongly disagreed that having a variety of accessible ways (54 respondents) and having network paths among blocks (48 respondents) are important factors. Table 4 contains a cross-tabulation of close proximity and four other variables, namely gender, age, the block where respondents live, and travel time. It shows that respondents who stay and live in Blocks A, B or C represent the majority of respondents who agreed or strongly agreed that close proximity and travel time to reach the open space are important. The last columns show the total numbers agreeing about close proximity and also present the percentages of residents who participated in survey in the different blocks. Most of the respondents in Block A (11 of 13 or 84.6%), 94.1% of respondents in Block B (16 out of 17) and all of the dwellers in Block C (100%) in different categories of ages and genders agreed or strongly agreed that distance to the open space is important.

DOI: 10.1051/matecconf/20166600006

Table 3. Evaluation of accessibility degree based on respondent's idea

Measurements of accessibility base on variable	Strongly disagree	Disagree	Neutral	Agree	Strongly agree
Close proximity	6	19	8	27	24
Sidewalk diversity	4	18	23	28	11
Network paths among blocks	13	33	22	15	1
Open space location	5	14	12	42	11
Travel time	4	23	10	30	17
Surrounding's linkages	1	15	24	32	12
Visibility	19	21	4	15	25
Accessibility to basic needs	0	2	20	42	15
Variety of accessible ways	20	34	17	11	2
Permeability	1	20	18	36	9

The last column in Table 4 show the number of residents from the more distant blocks who disagreed or strongly disagreed that close proximity to the open space is important – note that these are people who have to pass other blocks to reach the open space. Fully 65% of the dwellers in Block D (13 of 20 persons) and 84.2% of residents in Block E (16 of 19 persons) fell into these categories. In other words, the results demonstrate that easy access and close proximity to the open space are important aspects for respondents in Blocks A, B and C. However, this raises the question: "Why do those living in Blocks D and E use the open space if they do not generally agree that close proximity is important?" An analysis of the results for greenery and landscape answers this critical question.

In conclusion, the analyses show that respondents in Blocks A, B and C prefer to go to the open space mainly because of its close proximity and short travelling distance. However, respondents in Blocks D and E perceive the landscape features in the open space as important elements which encourage them to go there to spend their spare time. Aside from these reasons, social interaction is another reason for residents going to and staying in the open space.

Getting together, relaxing, appreciating nature, being with children and doing different kinds of activities were further reasons given for going to the open space. Table 5 analyses how far respondents see the open space as an opportunity to be with other neighbours, and the relative degree of agreement in different categories of age, gender and residential block. The figures show that there is broad agreement among respondents that the open space serves as a place for gathering and being with others. This holds for all of the blocks, although the proportions agreeing are slightly higher in the more distant blocks. Thus 61.5% of residents in Block A, 70.5% in Block B, 73.3% in Block C, 80% in Block D and 73.6% in Block E either agreed or strongly agreed that social contacts are one of the main reasons they go to and use the open space.

5 Conclusions and recommendations

The results of this survey show that residents in East Lake condominium come to the open space for four main reasons: closeness/short distance to the open space (physical accessibility); visual accessibility from their residential units to the open space and the opportunity for social interaction, eg with other residents. Although all of these are significant reasons, the degree of agreement on their relative importance varies amongst respondents from different blocks. Residents in Blocks A, B and C, which are located next to the open space, mostly have direct visual access to it from their own windows, and tend to consider visual and physical accessibility (proximity) as the main reasons for going to the open space and social interaction as only secondary reasons. This contrasts with residents in Blocks D and E, who live physically further away from the open space and have no direct visual access to it: they mainly go there because of opportunities for social interaction, with accessibility (eg pathways) and visual permeability as only secondary reasons.

This suggests that the number of dwellers in Blocks D and E using the open space would be likely to drop if for any reason either the aesthetic value or the opportunities for social activities in the open space fell in the future.

Appropriate design could also help to encourage and facilitate the opportunities for holding social events and experiences in the open space. Creating visual accessibility to open spaces, for example by the use of well-designed and located pathways, is another means of enhancing their use. Visual accessibility could also be enhanced by creating a hierarchy of spaces offering glimpses from the private to public areas, thereby encouraging residents to walk from the private space to the public space with different perspectives from different locations.

Table 4. Proximity and travel time evaluation based on survey questionnaire in separated blocks

proximity& travel time		Male	Male	Male	Male	Male	Male	Female	Female	Female	Female	Female	Female	Total
		15-25	25-35	35-45	45-55	55-65	65or above	15-25	25-35	35-45	45-55	55-65	65or above	
BLOCK A	St disagree													**0**
	Disagree					1								**1**
	Neutral						1							**1**
	Agree			2	1				2					**5**
	St agree		1		1			1	1		1	1		**6**
ROW PERCENT		**0.00%**	**7.69%**	**15.38%**	**15.38%**	**7.69%**	**7.69%**	**7.69%**	**23.07%**	**0.00%**	**7.69%**	**7.69%**	**0.00%**	**13/15,51%**
BLOCK B	St disagree													**0**
	Disagree		1											**1**
	Neutral													**0**
	Agree	2			1			1		2				**6**
	St agree		1	1				2	2	2	1	1		**10**
ROW PERCENT		**11.76%**	**11.76%**	**5.88%**	**5.88%**	**0.00%**	**0%**	**17.64%**	**12%**	**23.52%**	**5.88%**	**5.88%**	**0.00%**	**17/21.23%**
BLOCK C	St disagree													**0**
	Disagree													**0**
	Neutral													**0**
	Agree		1	1	1	2		1	1		1			**8**
	St agree	1		3					1	1		1		**7**
ROW PERCENT		**666%**	**666%**	**2666%**	**666%**	**1333%**	**0%**	**666%**	**1333%**	**666%**	**666%**	**666%**	**0%**	**15/17.85%**
BLOCK D	St disagree				1		1					2		**4**
	Disagree	1						3	2	1	1	1	1	**9**
	Neutral			1					1	1				**3**
	Agree	1	1							1				**3**
	St agree			1										**1**
ROW PERCENT		**10%**	**5%**	**10%**	**5%**	**0%**	**5%**	**15%**	**15%**	**15%**	**5%**	**15%**	**5%**	**20/23.80%**
BLOCK E	St disagree				2			1		1	1			**5**
	Disagree	3	1		1	1		1	2		1		1	**11**
	Neutral		1											**1**
	Agree			1					1					**2**
	St agree													**0**
ROW PERCENT		**15.78%**	**10.52%**	**5.26%**	**15.78%**	**5.26%**	**0%**	**10.52%**	**15.78%**	**5.26%**	**10.52%**	**0%**	**5.26%**	**19/22.61%**

Table 5. Social interaction evaluation based on survey questionnaire in separated blocks

Social interaction		Male	Male	Male	Male	Male	Male	Female	Female	Female	Female	Female	Female	Total
		15-25	25-35	35-45	45-55	55-65	65or above	15-25	25-35	35-45	45-55	55-65	65or above	
BLOCK A	St disagree													**0**
	Disagree								1		1			**2**
	Neutral			2				1						**3**
	Agree				2	1	1		2					**6**
	St agree		1									1		**2**
ROW PERCENT		**0%**	**7.69%**	**15.38%**	**15.38%**	**7.69%**	**7.69%**	**7.69%**	**23.10%**	**0.00%**	**7.69%**	**7.69%**	**0.00%**	**13/15,51%**
BLOCK B	St disagree													**0**
	Disagree	1			1					1				**3**
	Neutral	1							1					**2**
	Agree		2	1				2	1	1		1		**8**
	St agree							1		2	1			**4**
ROW PERCENT		**11.76%**	**11.76%**	**5.88%**	**5.88%**	**0.00%**	**0%**	**17.68%**	**12%**	**23.52%**	**5.88%**	**5.88%**	**0.00%**	**17/21.23%**
BLOCK C	St disagree													**0**
	Disagree					2								**2**
	Neutral				1					1				**2**
	Agree			2				2	3					**7**
	St agree		1	2							1			**4**
ROW PERCENT		**0%**	**666%**	**2666%**	**666%**	**1333%**	**0%**	**666%**	**1333%**	**666%**	**672%**	**0%**	**0%**	**15/17.85%**
BLOCK D	St disagree													**0**
	Disagree													**0**
	Neutral	1						1	1		1			**4**
	Agree		1	2	1		1	1		2		2		**10**
	St agree	1						1	2	1			1	**6**
ROW PERCENT		**10%**	**5%**	**10%**	**5%**	**0%**	**5%**	**15%**	**15%**	**15%**	**5%**	**10%**	**5%**	**20/23.80%**
BLOCK E	St disagree													**0**
	Disagree				1						1			**2**
	Neutral		2										1	**3**
	Agree	3			1	1		2	1	1	1			**10**
	St agree			1	1				2					**4**
ROW PERCENT		**15.78%**	**10.52%**	**5.26%**	**15.78%**	**5.26%**	**0%**	**10.52%**	**15.78%**	**5.26%**	**10.58%**	**0%**	**5.26%**	**19/22.61%**

References

1. Bedimo-Rung, A.L., A.J. Mowen, and D.A. Cohen, *The significance of parks to physical activity and public health: a conceptual model.* American journal of preventive medicine, 2005. **28**(2): p. 159-168.
2. Pakzad, j., *The rule of urban design*. 2004, Tehran, Iran: University of Tehran.
3. Levinson, D.M., *Accessibility and the journey to work.* Journal of Transport Geography, 1998. **6**(1): p. 11-21.
4. Giles-Corti, B., et al., *Increasing walking: how important is distance to, attractiveness, and size of public open space?* American journal of preventive medicine, 2005. **28**(2): p. 169-176.
5. Wang, D., G. Brown, and Y. Liu, *The physical and non-physical factors that influence perceived access to urban parks.* Landscape and Urban Planning, 2015. **133**: p. 53-66.
6. Calthorpe, P., *The next American metropolis: Ecology, community, and the American dream.* 1993: Princeton Architectural Press.
7. Bertolini, L. and M. Dijst, *Mobility environments and network cities.* Journal of urban design, 2003. **8**(1): p. 27-43.
8. Truelove, M., *Measurement of spatial equity.* Environment and Planning C: Government and Policy, 1993. **11**(1): p. 19-34.
9. Talen, E., *Measuring the public realm: A preliminary assessment of the link between public space and sense of community.* Journal of Architectural and Planning Research, 2000: p. 344-360.
10. Erkip, F.B., *The distribution of urban public services: the case of parks and recreational services in Ankara.* Cities, 1997. **14**(6): p. 353-361.
11. Heng, C. and V. Chan, *The making of successful public space: A case study of people's park square.* Urban Design International, 2000. **5**(1): p. 47-55.
12. GEDİKLİ, B., *STRATEGIC SPATIAL PLANNING AND ITS IMPLEMENTATION IN TURKEY: ŞANLIURFA PROVINCIAL DEVELOPMENT PLANNING CASE.* 2004, MIDDLE EAST TECHNICAL UNIVERSITY.
13. Jacobs, J., *The Death and Life of Great American Cities.* Readings in Planning Theory, Fourth Edition, 2015: p. 94-109.
14. Chiesura, A., *The role of urban parks for the sustainable city.* Landscape and urban planning, 2004. **68**(1): p. 129-138.
15. Moirongo, B.O., *Urban public space patterns: human distribution and the design of sustainable city centres with reference to Nairobi CBD.* Urban Design International, 2002. **7**(3): p. 205-216.
16. Scopelliti, M., et al., *Staying in touch with nature and well-being in different income groups: The experience of urban parks in Bogotá.* Landscape and Urban Planning, 2016. **148**: p. 139-148.
17. Masnavi, M.-R., *The new millennium and the new urban paradigm: the compact city in practice.* Achieving sustainable urban form, 2000: p. 64-73.
18. Liao, F.H., S. Farber, and R. Ewing, *Compact development and preference heterogeneity in residential location choice behaviour: A latent class analysis.* Urban Studies, 2014: p. 0042098014527138.
19. Williams, K., M. Jenks, and E. Burton, *Achieving sustainable urban form.* 2000: Taylor & Francis.
20. Rowley, A., *Planning mixed use development: issues and practice.* Royal Institution of Chartered Surveyors, London, 1998.
21. Newman, O., N.I.o.L. Enforcement, and C. Justice, *Architectural design for crime prevention.* 1973: National Institute of Law Enforcement and Criminal Justice Washington, DC.
22. Hillier, B., et al., *Natural movement: or, configuration and attraction in urban pedestrian movement.* Environment and Planning B: planning and design, 1993. **20**(1): p. 29-66.
23. Gehl, J., *Life between Buildings (Copenhagen: Arkitektens Forlag).* 1996.
24. Manning, R.E. and W.A. Freimund, *Use of visual research methods to measure standards of quality for parks and outdoor recreation.* Journal of leisure research, 2004. **36**(4): p. 557.

Interval of Routine Maintenance and Maintenance Performance: A Literature Review

Cheong Peng Au-Yong [1,a], Azlan Shah Ali [1] and Shirley Jin Lin Chua [1]

[1]Department of Building Surveying, Faculty of Built Environment, University of Malaya, 50603 Kuala Lumpur, Malaysia.

Abstract. In high-rise residential buildings, the quality of facilities management services is significant to the normal operation of the facilities. Unfortunately, lack of concern towards building maintenance, especially preventive maintenance, happens in domestic housing industry in Malaysia. Majority of the maintenance operations of condominiums suffer from lack of planning, lack of proactive maintenance plan, and lack of proper implementation. Thus, this paper reviews the implementation of preventive maintenance strategy, routine maintenance in specific. An extensive review of literature published in 1987 to 2014 is performed for the purpose of this research. The publications are sourced from journal articles, conference proceedings and books. The literature analysis confirms that the routine maintenance of facilities and building services is vital and it can be influential towards the maintenance performance. Subsequently, a theoretical framework is developed, which shows the relationship between routine maintenance of building facilities & services and maintenance performance. The building facilities & services are divided into two categories. They are essential facilities & services that ensure the safety, health, habitability, and operability of buildings; while value-added facilities & services deal with property value, return on investment, and quality living of buildings. Based on the findings, a future research is proposed, which aims to identify the appropriate routine of maintenance for the facilities and services in high-rise residential buildings to improve the maintenance performance.

1 Introduction

In high-rise residential buildings, the quality of facilities management services is significant to the normal operation of the facilities such as lifts, ventilation system, mechanical and electrical installations, recreational and sport facilities, etc. [1]. Unfortunately, lack of concern towards building maintenance, especially preventive maintenance, happens in domestic housing industry in Malaysia [2, 3]. Hsieh [4] stated that majority of the maintenance operations of condominiums suffer from lack of planning, lack of proactive maintenance plan, and lack of proper implementation. As a result, many complaints of dissatisfaction with the dilapidated condition of the residential buildings and the poor conditions of facilities and services are recorded [5].

High maintenance cost is a common issue in construction industry [6]. Public neglects the importance of building maintenance as a result of the massive costs required for maintenance operations [7]. Meanwhile, ineffective planning of maintenance strategies and low service quality are another issues of maintenance management in Malaysia [8, 9]. Indeed, lack of knowledge about the maintenance strategies, inadequate performance standard, lack of building performance monitoring data, as well as failure to provide appropriate advice on design and planning based on overall performance are the factors that lead to the issues.

Specifically, lack of preventive measure is currently the problem that implicates poor maintenance outcome. Whereby, numerous researches introduce and recommend the scheduled maintenance as a strategy to tackle the issues [10-12]. However, the effectiveness of the maintenance strategy can be jeopardised by the inappropriate or inaccurate routine of inspection and maintenance [13-16]. Long interval between the maintenance actions to perform rehabilitation works as a result of insufficient funds, leads to the rise of deterioration rate and hence dissatisfaction of the residents [17].

The statements demonstrates the effect of poor maintenance management, especially the maintenance interval or maintenance frequency, towards the building performance. Therefore, this paper aims to propose a study about the relationship between frequencies of routine maintenance for the facilities and maintenance performance in high-rise residential buildings. The next section provides a brief background about the maintenance management in building property sector and explores the building maintenance trend in international and local contexts.

[a] Corresponding author: auyongcp@um.edu.my

2 Background

In property sector, the maintenance operation of condominiums in Taiwan suffers from lack of planning. Hsieh [4] analysed that 52 percent of the management committees do not have proactive maintenance plan or any annual operation programmes. However, 16 percent of them do have the plan or programmes but without proper implementation. As a result majority of the condominiums operate their own condominium on an ad hoc basis, which is corrective maintenance.

In Malaysian context, buildings have not been effectively and well maintained throughout the years, though there is commitment from the government towards building maintenance. The building residents neglect the importance of maintenance. Lateef [7] noted that building maintenance in Malaysia is conditionally driven and is usually concerned when there is cost allocation. There is no proper planning on building maintenance. Furthermore, maintenance operation is executed in reactive basis, whereby the maintenance task is performed after the system defect is determined [18].

Currently, the facilities management and maintenance management practices for most of the buildings rely heavily on corrective maintenance, where the item is repaired or replaced as it has damaged or broken [19]. In Malaysia, many buildings are even having lack of adequate maintenance [7]. Corrective maintenance is often carried out instead of other proper maintenance strategies. The statement is proven by Nik-Mat, Kamaruzzaman and Pitt [20], who mentioned that the conventional maintenance strategy is dominantly applied compared to integrated facilities management systems although the new approach has been proven to be advantageous in maintenance operations.

However, higher maintenance cost is required to restore a building to its original function and purpose [21]. Thus, to overcome the issue of high building maintenance cost, the criteria such as cost performance, safety, security and others should be taken into consideration in the review, evaluation and assessment of the maintenance policies and strategies [22]. For example, Malaysian Government has taken the initiative by instructing their agencies to carry out maintenance operations from the early stage to avoid risk of higher maintenance cost due to continuous deterioration through time [23].

Moghaddam and Usher [24] revealed that preventive maintenance involves maintenance tasks such as inspection, monitoring, cleaning, lubrication, adjustment, alignment, repair, replacement and maintenance of building's and systems' components before failures or system breakdowns occur. It is based on component reliability characteristics and aimed to reduce the probability of component failure, as well as to minimise the system downtime [25]. Thus, preventive maintenance is more effective than corrective maintenance in terms of cost and reliability [26].

Scheduled maintenance, as one of the preventive maintenance strategies, is performed at time interval basis to overcome the problem of wear and tear, as well as to minimise the probability of system failure [27, 28]. The maintenance personnel always plan the time interval of inspection or maintenance ahead to ensure the availability or normal operation of the building systems, such as lift, water pump and others. They usually follow the manufacturers' data or standard guides of the service life estimates to determine the maintenance interval and requirement [29].

Basically, scheduled maintenance requires an interruption of the system or temporary disabling the system for part replacement and system testing [30, 31]. It only back into operation when the maintenance task is completed. In some cases, often interruption of the components may affect the efficiency of the system. On the other hand, unavailable or delayed maintenance action at the right time may cause unwanted damages to the system [32].

Therefore, it is necessary to apply appropriate maintenance inspections or treatments at the right time to extend service life of the components [33]. In fact, implementation of scheduled maintenance with minimal maintenance cost might not guarantee the reliability and availability of the systems [34]. In order to enhance the effectiveness of the maintenance strategy, it is compulsory to determine the optimal time interval between the maintenance tasks.

Whereby, residents' satisfaction encompasses the services and facilities provided, as well as the management practices such as the maintenance and repair service [35]. The routine of inspection and maintenance is likely to affect the maintenance performance and also the residents' satisfaction. It ensures the building sustainability by maximising the availability of building services and facilities, extending the lifespan of a building, and reducing the cost of failure [36]. Next, the following section discusses the needs of maintenance in residential building, specifically the high-rise building that equips with complex and comprehensive facilities and services.

2.1 Residential building maintenance

According to Hui [37], residential buildings are subjected to heavy usage and stress under different circumstances, such as wear and tear, climate change and ageing process. As a result of high occupancy rate, proper operation and maintenance are compulsory to reduce the deteriorating rate or restore the buildings and systems to an acceptable operating condition. It helps to sustain the property value and maintain high return on investment [2]. Meanwhile, Yusof, Abdullah, Zubedy and Mohd Najib [3] demonstrated that to keep property habitable and building operable are the important reasons of performing maintenance works.

From demanding for living space, the public now is demanding more towards quality living. Instead of buying a house with complete amenities, the house buyers prefer to buy a home equipped with value added facilities like security, gymnasium, swimming pool, landscaping and others [2]. The available of building services include electricity supply, water supply, sewerage and drainage, telecommunication, lifts and fire services system, as well as the building facilities promotes the residential satisfaction [38]. Inevitably, all the facilities require maintenance to retain their conditions in satisfactory standard. The maintenance works on residential buildings create opportunities to gain economic and social benefits while reducing carbon footprints [39].

Furthermore, building maintenance and building safety are closely related to each other. Maintenance of safety systems like fire services system and security system ensures the safety of occupants. However, unsatisfactory fire safety management and lack of maintenance of fire services system is one of the issues accounted in residential buildings [40]. This would increase the risk of safety issue towards the buildings and occupants. Lai and Yik [1] claimed that the quality of maintenance management is critical to the operation of the building services and facilities. The performance of the services and facilities may affect the health, safety and enjoyment of the occupants. Yusof, Abdullah, Zubedy and Mohd Najib [3] confirmed that the maintenance works are vital to maintain the safety and health of residents.

Therefore, implementation of preventive maintenance is vital to retain the condition of buildings and systems, property values, as well as safety and health of the residents. In order to ensure the effectiveness of preventive maintenance strategy, the frequency of inspection and interval of maintenance must be reviewed. The next section explains the methods utilised for this research.

3 Method

Critical review was conducted to identify the problems in (discussed in Section 1) and needs of (discussed in Section 2) effective maintenance in high-rise residential building that guided to the formulation of research aim. In order to establish the relationship between frequency of routine maintenance and maintenance performance, an extensive search and analytical review of the literature related to the topics were performed. This method was adopted from the researches done by Simoes, Gomes and Yasin [41] and Roda, Macchi, Fumagalli and Viveros [42]. The literature search and review covered publications of journal articles, conference proceedings, and books.

The reviewed publications were published as 36 journal articles, 2 conference proceedings and 2 books from year 1987 to 2014. Table 1 lists the publication sources and number of publications taken into account from each of them. The publications covered either the scopes of building and construction or maintenance management, which were relevant to this research.

Table 1: List of sources for literature review and number of publications considered

Source Title	No. of Publications
Applied Energy	1
Building and Environment	1
Building Engineer Journal	1
Building Research and Information	1
Construction and Building Materials	1
Engineering, Construction and Architectural Management	1
Facilities	1
Gradevinar	1
Habitat International	2
Housing Studies	1
International Journal of Production Economics	1
International Journal of Production Research	1
International Journal of Strategic Property Management	1
Journal of Building Appraisal	1
Journal of Building Property Review	1
Journal of Cleaner Production	1
Journal of Facilities Management	2
Journal of Performance of Constructed Facilities	1
Journal of Quality in Maintenance Engineering	5
Procedia - Social and Behavioral Sciences	2
Procedia Engineering	2
Reliability Engineering and System Safety	1
Renewable Energy	1
Social Indicators Research	1
Structural Survey	2
Structure and Infrastructure Engineering	2
Others (conference proceedings and books)	4
Total	**40**

Then, analysis of literature contents was conducted and it led to the development of theory of routine maintenance and performance. The analysis results are presented and discussed in next section.

4 Result

The researchers throughout the time discuss the building maintenance issues extensively. Based on systematic content analysis of the researches, three main maintenance issues are identified. Table 2 shows the maintenance issues that the researchers discuss and their number of citations. The citation frequency of different maintenance issues has been taken as an indicator of the concern given by the scientific literature to each issue.

Table 2: Citation of the maintenance issues in publications

Citation	Neglect towards importance of maintenance	Lack of preventive maintenance plan	High maintenance cost
Tiun [2]	X	X	
Yusof, Abdullah, Zubedy and Mohd Najib [3]	X	X	
Hsieh [4]		X	
Abdul Karim [5]		X	
El-Haram and Horner [6]			X
Lateef [7]	X	X	X
Kamaruzzaman		X	

and Zawawi [8]			
Ruslan [9]		X	
Olanrewaju, Idrus and Khamidi [18]		X	
Lavy [19]		X	
Nik-Mat, Kamaruzzaman and Pitt [20]		X	
Seeley [21]			X
Mohd Nizar [22]	X		
Mohd-Noor, Hamid, Abdul-Ghani and Haron [23]	X	X	
Moghaddam and Usher [24]		X	
Fouladgar, Yazdani-Chamzini, Lashgari, Zavadskas and Turskis [25]		X	
Batun and Azizoğlu [26]		X	
Total number of citation	**5**	**14**	**3**

The most common building maintenance issue, revealed in 14 publications, relates to the lack of preventive maintenance plan. For instance, Lateef [7] argued that there is a huge backlog of building maintenance in Malaysia due to the ignorance towards proactive and preventive maintenance. Other maintenance issues include neglect towards importance of maintenance and high maintenance cost, discussed in 5 and 3 publications respectively. Thus, the study focuses on the implementation of preventive maintenance strategy.

Based on detailed content analysis of the reviewed publications, the routine maintenance of facilities and building services is vital and it can be influential towards the maintenance performance. In other words, the interval between inspections or maintenance works may affect the maintenance performance significantly. Paulo, Branco and de Brito [17] inadequate interval between maintenance actions leads to the poor consequences like system defect. Table 3 lists the publications that have cited the importance of routine maintenance.

Table 3: Citation of the routine maintenance and its influence towards maintenance performance in publications

Citation	
Importance of routine maintenance	Au-Yong, Ali and Ahmad [10]; Flores-Colen and De Brito [11]; Forster and Kayan [12]; Edward, Holt and Harris [13]; Eti, Ogaji and Probert [14]; Irigaray and Gilabert [15]; Wu, Clements-Croome, Fairey, Albany, Sidhu, Desmond and Neale [16]; Grant and Ries [29]; Paulo, Branco and de Brito [17]; Bevilacqua and Braglia [27]; Hameed, Ahn and Cho [28]; Swanson [30]; Tsang [31]; Narayan [32]; Chen, Lin and Luo [33]; Ahn and Kim [34]; Varady and Carrozza [35]; Ahmad, Nur Azfahani and Nor Haniza [36]
Its influence towards maintenance performance	Edward, Holt and Harris [13]; Eti, Ogaji and Probert [14]; Irigaray and Gilabert [15]; Wu, Clements-Croome, Fairey, Albany, Sidhu, Desmond and Neale [16]; Paulo, Branco and de Brito [17]; Bevilacqua and Braglia [27]; Hameed, Ahn and Cho [28]; Swanson [30]; Tsang [31]; Narayan [32]; Chen, Lin and Luo [33]; Ahn and Kim [34]; Varady and Carrozza [35]; Ahmad, Nur Azfahani and Nor Haniza [36]

Whereby unavailable or delay to carry out maintenance action at the right time may cause damages to the building elements or system components [32]; while tight routine maintenance may require frequent intrusion of the system that causes maintenance downtime that affect the residents' activities [30]. The relationship of the routine maintenance and maintenance performance remains uncertain. Therefore, there is a need to study on the relationship between routine maintenance of facilities and services and maintenance performance. In this context, the following research questions are relevant:

i) What are the building facilities & services that require routine maintenance?
ii) How does the routine maintenance influence the maintenance performance?
iii) What is the appropriate routine of maintenance in order to optimise the maintenance performance?

Based on the research questions, a theoretical framework as shown in Figure 1 is developed for future research. The research aims to identify the appropriate routine of maintenance for the facilities and services in high-rise residential buildings to improve the maintenance performance.

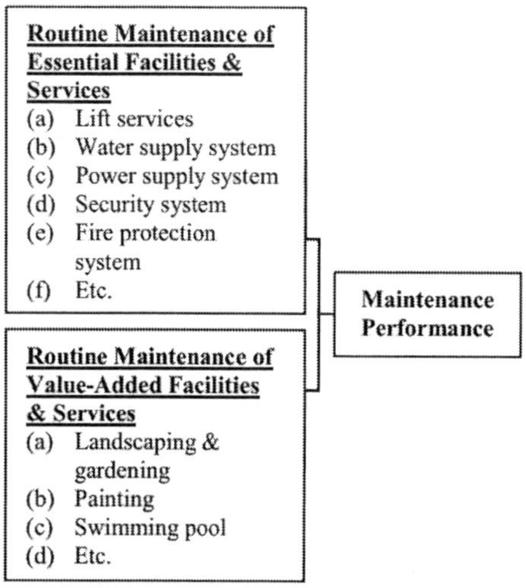

Figure 1: Theoretical framework for future research

5 Conclusion

This research examines the maintenance issues in construction industry, specifically the residential buildings. Literature demonstrates that lack of preventive maintenance plan is the main issue in building maintenance sector. Introduction and implementation of preventive maintenance is necessary to enhance maintenance performance. Scheduled maintenance as one of the preventive maintenance strategies requires routine inspection and maintenance at specific interval. Many researchers argue that the interval of inspection and maintenance brings impact to the maintenance performance. Therefore, there is a need to study on the relationship between routine maintenance of facilities and services and maintenance performance. The study will focus on two aspects, which are routine maintenance of essential facilities & services and value-added facilities & services.

Acknowledgement

The authors gratefully acknowledge the financial support of the University Malaya Research Fund Assistance (BKP), grant no. BK021-2015 established at the Institute of Research Management & Monitoring (IPPP), University of Malaya.

References

1. J. H. K. Lai and F. W. H. Yik, Building and Environment **46** (1), 165-175 (2011).
2. L. T. Tiun, presented at the 2nd NAPREC Conference, Bangi, Selangor, 2009 (unpublished).
3. N. A. Yusof, S. Abdullah, S. Zubedy and N. U. Mohd Najib, Procedia - Social and Behavioral Sciences **62**, 508-513 (2012).
4. H. R. Hsieh, Habitat International **33** (1), 73-80 (2009).
5. H. Abdul Karim, Procedia - Social and Behavioral Sciences **35**, 44-53 (2012).
6. M. A. El-Haram and M. W. Horner, Journal of Quality in Maintenance Engineering **8** (2), 115-123 (2002).
7. O. A. Lateef, Journal of Building Appraisal **4** (3), 207-214 (2008).
8. S. N. Kamaruzzaman and E. M. A. Zawawi, Journal of Facilities Management **8** (1), 75-81 (2010).
9. N. Ruslan, in *National Asset and Facilities Management (NAFAM) Convention, National Asset and Facility Management Development, 13 August* (Kuala Lumpur, Malaysia, 2007).
10. C. P. Au-Yong, A. S. Ali and F. Ahmad, Gradevinar **65** (9), 803-809 (2013).
11. I. Flores-Colen and J. De Brito, Construction and Building Materials **24** (9), 1718-1729 (2010).
12. A. M. Forster and B. Kayan, Structural Survey **27** (3), 210-229 (2009).
13. D. J. Edward, G. D. Holt and F. C. Harris, Journal of Quality in Maintenance Engineering **4** (1), 25-37 (1998).
14. M. C. Eti, S. O. T. Ogaji and S. D. Probert, Applied Energy **83** (10), 1163-1179 (2006).
15. A. A. Irigaray and E. Gilabert, Journal of Quality in Maintenance Engineering **15** (2), 151-166 (2009).
16. S. Wu, D. Clements-Croome, V. Fairey, B. Albany, J. Sidhu, D. Desmond and K. Neale, Engineering, Construction and Architectural Management **13** (2), 136-153 (2006).
17. P. V. Paulo, F. Branco and J. de Brito, Struct. Infrastruct. Eng. **10** (3), 388-397 (2014).
18. A. L. Olanrewaju, A. Idrus and M. F. Khamidi, Structural Survey **29** (5), 397-410 (2011).
19. S. Lavy, Journal of Facilities Management **6** (4), 303-315 (2008).
20. N. E. M. Nik-Mat, S. N. Kamaruzzaman and M. Pitt, Procedia Engineering **20**, 329-338 (2011).
21. I. H. Seeley, *Building Maintenance*, 2nd ed. (Palgrave, New York, 1987).
22. B. J. Mohd Nizar, Journal of Building Property Review **14** (7), 8-9 (1998).
23. N. Mohd-Noor, M. Y. Hamid, A. A. Abdul-Ghani and S. N. Haron, Procedia Engineering **20**, 435-444 (2011).
24. K. S. Moghaddam and J. S. Usher, Journal of Quality in Maintenance Engineering **16** (3), 271-287 (2010).
25. M. M. Fouladgar, A. Yazdani-Chamzini, A. Lashgari, E. K. Zavadskas and Z. Turskis, International Journal of Strategic Property Management **16** (1), 85-104 (2012).
26. S. Batun and M. Azizoğlu, International Journal of Production Research **47** (7), 1753-1771 (2009).
27. M. Bevilacqua and M. Braglia, Reliability Engineering and System Safety **70** (1), 71-83 (2000).
28. Z. Hameed, S. H. Ahn and Y. M. Cho, Renewable Energy **35** (5), 879-894 (2010).
29. A. Grant and R. Ries, Building Research and Information **41** (2), 168-186 (2013).
30. L. Swanson, International Journal of Production Economics **70**, 237-244 (2001).
31. A. H. C. Tsang, Journal of Quality in Maintenance Engineering **1** (3), 3-17 (1995).
32. V. Narayan, *Effective Maintenance Management: Risk and Reliability Strategies for Optimizing Performance*. (Industrial Press Inc, New York, 2003).
33. D.-H. Chen, D.-F. Lin and H.-L. Luo, Journal of Performance of Constructed Facilities **17** (3), 136-143 (2003).
34. S. Ahn and W. Kim, Struct. Infrastruct. Eng. **7** (4), 307-314 (2011).
35. D. P. Varady and M. A. Carrozza, Housing Studies **15** (6), 797-825 (2000).
36. R. Ahmad, A. Nur Azfahani and I. Nor Haniza, Building Engineer Journal **81** (4), 30-33 (2006).
37. E. Y. Y. Hui, Facilities **23** (1/2), 47-62 (2005).
38. M. A. Mohit, M. Ibrahim and Y. R. Rashid, Habitat International **34**, 18-27 (2010).
39. Y. H. Chiang, J. Li, L. Zhou, F. K. W. Wong and P. T. I. Lam, Journal of Cleaner Production, doi: 10.1016/j.jclepro.2014.1007.1069 (2014).
40. Y. Yau, C. D. W. Ho and K. W. Chau, Social Indicators Research **89**, 501-521 (2008).

41. J. M. Simoes, C. F. Gomes and M. M. Yasin, Journal of Quality in Maintenance Engineering **17** (2), 116-137 (2011).
42. I. Roda, M. Macchi, L. Fumagalli and P. Viveros, Journal of Manufacturing Technology Management **25** (4), 528-549 (2014).

Risk management in the construction industry: a new literature review

Berenger Y. Renault[1,a] and Justus N. Agumba[2]

[1]*University of Johannesburg, Johannesburg, South Africa*
[2]*University of Johannesburg, Johannesburg, South Africa*

Abstract. The construction industry is well known as a highly risk prone industry owing to the complexity of its activities and dynamic project environments generating an atmosphere of risks that must be taken into consideration in the decision process. For this reason, we have developed this treatise to elucidate the fundamentals of risk management through a concise new proposal of literature review for risk management in construction. Our explanation of this is venture is that over the years, this area has been acknowledged as a crucial process in the business institutions and the most discussed subject among experts and researchers in the construction sector. We have tried in this report to present the most recent studies considering the impossibility to present all definitions of the concept of risk. This study is mainly a literature review; it looks at the literature relating to the concept of risk, risk management in construction as well as methods used in construction industry.

1 Introduction

All human ventures bring in uncertainty and risk. It is agreed that risk is extreme in the business sector than other sectors. Every entrepreneurial act and a business decision are associated with risk. The risk is an occurrence that has a degree of obscurity and can either be positive or negative. A positive risk is a convenient opportunity, while a negative risk is a threat and hence inconvenient [1]. The more convenient and less convenient risks imply progressive and negative outcomes respectively. However, the CI faces fewer random risks, but these may have adverse consequences for a time, for example, increase cost, time overruns and low-quality work. The factors leading to such an outcome include planning, design and construction intricacy as well as the presence of countless interest groups and material resources [2]. Concerning risk negative impacts, [3] clarified that individuals dislike risk and that each party in the CI is risk averse; consequently, RM becomes significant in eliminating or reducing risk through the various processes of RM. [4] stated that RM has various complicated dimensions relating to the CI, with results that go beyond its direct physical detriment to the financial and cultural processes, and even the way society functions. According to [5], all construction projects are hazardous by nature due to their configuration, financial and organizational arrangements, and technology and resource demands; hence, RM in construction projects is dynamic rather than stagnant. [6] argue that it is crucial for the industry to minimise these risks and uncertainties so as to unearth the impact thereof to determine which part of the project is more expose to risk and less feasible. RM has the task of identifying risks, measuring the probability and the likely impact of events, and treating risks, eradicating or minimising their impact with the minimum investment of resources. RM is being developed and employed in a lot of fields, such as environment, public safety, within enterprise management, and health care. This paper focuses on RM in construction. [7], argued that RM when implemented, protects innovative projects, which are essential to gain competitive advantage and succeed in the market, but inevitably involve risky decisions and activities. Risk management could improve the ability to manage all stages of the innovative projects successfully. It has only been a few years since the management literature started to show an interest in applying RM in construction projects; for this reason, many areas are still understudied. This paper was developed to clarify the basics of risk management through a short new suggestion of literature review for risk management. The research objectives are therefore to analyze the existing literature concerning the concept of risk, RM and methods used in managing risks. The next section is devoted to the theoretical framework, including the definition of risk, risk management and risk management process.

2 Literature review

2.1 Risks in construction

The risk is attached to human life and is present in all human attempts. Nevertheless, the complexity of the work involved in construction activities makes the latter more predisposed to risk [8]. The part of the risk involved

[1] Corresponding author: renault08@yahoo.fr

in construction is also more different and varies in degree owing to this complexity [9]. Several researchers have tried to define risk in various ways; generally these definitions were tailored to the objectives of the projects they were dealing with at a particular time. Nonetheless, the definition of risk is highly dependent on the importance placed on applied management in an establishment [10]. Risk describes a situation where former documentation and experience exist upon which measures are taken by obtaining a possible result [11]. Every time a decision is taken as a continuum of possible outcomes in addition to certain probabilities attached to the outcome, a risk exists [12]. Furthermore, [13] argue that risk is occasionally expansively defined, beyond financial, as a probability of loss, injury, setback, disadvantage or destruction. Similarly, [14] indicates that risk relates to a dearth of information or previous experience in a given situation being managed by a decision maker. Further, the risk is considered as the possibility of an investor's claim in construction work failing to achieve the predictable measures of feasibility [15]. All these authors proposed that risk was disagreeable occurrence; a consequence of imaginable but unexpected circumstances. Imaginable circumstances are events that announce their occurrence, whereas unexpected circumstances are events that occur without any notice.

Various definitions have divulged that risk has an impact or effect on any construction project. The effects thereof may be computed using many terms: financial loss; property damages; injury to people and even a combination of all of these [13]. [16] Indicates that project performance, quality, potency, capacity and financial cost could be altered if an element of risk is involved. According to [2], construction works such as planning, design and development are vulnerable to some factors in an uncertain environment. Construction projects are thus executed in an environment defined by a varying measure of risk and contingencies that could be from known or unknown conditions [12]. Such contingencies are chance occurrences of events where the probability distribution is unknown.

Risks, based on individual views can be categorised in different ways; some are classified based on their probability of occurring while others are classified as a consequence of the impact they may have on construction activities, their types and sources. Despite these various classifications, they are meant to attain a common objective, that is, they are an important aid in risk management and assist in forming risk lists that are used when detecting a risk [17] [18]. [18] furher indicated that that construction risks can be classified based not only on the impact of the risk on the project but also the source of the risk, while [19] stated that risks are mainly grouped into two categories according to their source, namely, internal and external. Risks comprise three groups: known risks, known unknown risks, and unknown risks [12]. The first category includes minimal variations and often occurs and is inevitable features of construction works. The second category is foreseeable risk events with a known probability of occurrence and impact. The last category is those which, the probability of occurrence cannot be predicted. According to [20], the risk could also be grouped based on sources concerned with stakeholders. For example, the time associated risks, environmentally associated risks, cost associated risks and safety associated risks. The sub-categories of human risks associated with construction projects include technical, political, social, economic, legal, financial, health, managerial and cultural risks. [21], grouped risk into ten categories, namely:

- **Design:** Defective design, inaccurate quantities, not coordinated design, rush design, awarding the design to unqualified designers, lack of consistency between bill of quantities, drawings and specifications
- **Physical:** Occurrence of accidents, because of poor safety procedures, supplies of defective materials, security of material and equipment, public security, varied labor and equipment productivity
- **Logistics:** Improper site investigation, inaccurate project program, unavailable labor, materials and equipment, high competition in bids, undefined scope of working, poor communications between the home and field offices (contractor side)
- **Legal:** Ambiguity of work legislations, difficulty to get permits delayed disputes resolutions, legal disputes during the construction phase among the parties to the contract, no specialized arbitrators to help settle fast
- **Environmental:** Adverse weather conditions, difficulty to access the site (very far), environmental factors (floods, earthquakes, etc.)
- **Construction:** Gaps between the implementation and the specifications due to a misunderstanding of drawings and specifications, actual quantities differ from the contract quantities, design changes, lower work quality in the presence of time constraints, rush bidding, undocumented change orders.
- **Management:** Poor communication between involved parties, ambiguous planning due to project complexity, changes in management ways information unavailability (include uncertainty), resource management
- **Cultural:** Religion, cultural custom
- **Financial:** Delayed payments on contract, unmanaged cash flow, inflation, financial failure of the contractor, exchange rate fluctuation, monopolizing of materials due to closure and other unexpected political conditions
- **Political:** New governmental acts or legislations, inflation, unstable security circumstances (invasions).

2.2 Risk management in construction

[8] Stipulates that managing risk makes a significant role in reducing the risks encountered in the CI as it provides a systematic way of assigning risk to construction

 DOI: 10.1051/matecconf/20166600008

projects, thus allowing projects to be managed with a higher degree of expectation and anticipation. According to [10], RM in the CI is the assessment and response to the risk that will be inevitably attached to a project. In parallel, RM is the approach and activities that are put in place to lessen the disturbances that may occur during the action of a project [22]. In essence, RM is required for the effective detection and proper control of risk. Effective risk management means any likely future occurrences thereof in a proactive manner [23]. Consequently, RM that attaches value to proactive measures puts in place an emergency plan that increases the probability of accomplishing the aims and objectives of the project [24]. [18] emphasises that RM involves increasing the results of positive occurrences and curtailing the effects of adverse events. In other words, RM is controlling occurrences that may result in a risk, instead of being passive before such occurrences and reacting later. All the measures that are put in place to control risk are geared towards achieving the project objectives. RM represents a major part of construction works. [11] Asserted that the requirement to manage risk in construction projects is incessantly increasing as a result of the complexity, size, competition, client-consumer requirements, politico-economic challenges and major difficult physical conditions involved in such projects. Comprehensive RM will reduce the probability of an occurrence of an event just as it will reduce the extent of its impact. In accordance with [25], RM refers to a coordinated set of activities and methods that is used to direct an organization and to control the many risks that can affect its ability to achieve objectives. RM ensures that the decision-maker knows and understands the risks and prepares the necessary plan that can prevent disasters or diminish their impact. It covers the process of Identification, Assessment, Allocation, and Management of all project risks [26]. Risk management process (RMP) is the basic principle of understanding and managing risks in a project. All steps in the RMP should be included when dealing with risks, in order to efficiently implement the process in the project. It consists of the main phases [12] [28] [29] [30]:

2.2.1 Risk identification (ID)

This is the first stage in the RMP, and it involves capturing all potentials risk that might occur within the project [31]. Risk Identification (RI) forms the base for the next steps of risk analysis and control and enables organisations to learn about the areas that are exposed to risk. If appropriately performed, RI ensures successful risk management as unknown sources of losses escalate into unmanageable occurrences with unforeseen outcomes [32]. The emphasis is not only aimed at the incapability to identify loss causing risks but also includes the incapacity to determine opportunistic events. The effect of the non-identification of positive risks equates to the effect of non-identification of negative risks [32]. RI involves the identification of all possible risks and circumstances that may affect the organisation, as well as the conditions giving rise to these risks and opportunities. Risk identification, therefore, facilitates the efficient studying of areas and activities where organisational resources are at risk, affecting their ability to achieve their business goals [32]. To conduct RI efficiently, primary project documentation must be in place. The project charter, scope statement, and project management plan (Including the Work Breakdown Structure) need to be available to build an exhaustive list of risks. Without these elements as a frame of reference, it is difficult to assess the risks on a project effectively. The RM plan and the organizational atmosphere also must be undoubtedly understood to carry out RI. These form the environment in which the risks will be evaluated. The RM plan may also identify specific RI practices that are either favoured or banned by the organization as part of their risk culture. According to [33], all this information can encourage thinking about different risk matters and concerns when evaluated using the tools and techniques of RI. The tools and technology that are employed in RI are as varied as the projects they serve. Nevertheless, some groups of tool and technique types are most frequently used. The best known are Brainstorming, Interviews, Questionnaires, Delphi technique, Expert systems, etc.

2.2.2 Risk assessment/analysis

Risk identification follows on risk assessment. [34] [35] identified risk assessment as the next step of the risk management process. According to [8], risk assessment is a method of using available information to determine the frequency of occurrence and the level of consequences in risk management. According to [36], after all, the risks in a project have been identified; the process of qualitative risk assessment must follow which requires further analysis through assessing and estimating the probability of risk occurrence as well as its impact on each risk. Here, different factors need to be considered such as the degree of risk impact on project objectives and its manageability, the timing of an occurrence, and the likelihood of an occurrence and its relationship with other risks. All these factors provide a better understanding of each risk and allow room for a proper and appropriate way of responding to each risk. Two methods are developed for analyzing risk: qualitative and quantitative methods [31] [38]. The qualitative methods are most applicable when risks can be placed somewhere on a descriptive scale from low to high level. The quantitative methods on the other hand are employed to determine the probability and impacts of the risks identified and are based on statistics, arithmetic's estimations [39]. The table below summarises the various techniques used for risk analysis; all these technique are used in construction.

Table 1. Various risk analysis techniques (Ward & Chapman, 1997)

Risk Analysis	
Qualitative	**Quantitative**
Direct judgment	Probability analysis
Ranking options	Sensitivity analysis

Comparing options	Scenario analysis
Descriptive analysis	Simulation analysis

2.2.3 Risk responses (RRs)

Risk response is a central component in the RMP that determines what action (if any) will be taken to address risks evaluated in the identification, qualification, and quantification stages [31]. According to [8], RRs are actions taken to eliminate, reduce or transfer a risk or its consequence. The risk response process encompasses planning for an obligatory action to be considered in case a risk event occurs. It also entails taking planned action if required and following up with the consequences of these actions to ensure that the risk plan results in the required outcome. A risk response is determined by proposing several alternatives to eliminate or mitigate an anticipated risk and assign an optimum alternative as a response [41]. According to [42], RR is the "process of identifying/developing risk response options and determining actions for treating the risk, targeting enhancing opportunities and reducing any threats to projects objectives". RR is, thus, the selection of an appropriate policy to decrease the negative impact of a risk. Given the fact that construction projects are distinctive, the risks involved are massive and dynamic in nature. Hence, it is indispensable to invent appropriate responses to these risks after the initial stages of risk identification, assessment and allocation. [43] defined RR as the planned documents procedure for RM that is used to manage risk throughout the project. It is the stage that involves taking proper measures to accomplish project objectives. Some studies recommend varied response techniques concerning risks in projects. [44] [45], argued that there four distinct ways of responding to risks in a construction project namely: risk avoidance, reduction, transfer and risk retention, whereas [12] divided the response techniques into three by combining avoidance and reduction. [43] on the other hand, indicated that risk responses can be grouped into mitigation or avoidance and acceptance.

2.2.4 Risk control

After risks have been identified, assessed, and appropriate responses have been developed, those findings must be put into action. Risk monitoring and control include implementing the risk plan, which should be an integral part of the project plan. Two key challenges are usually encountered during monitoring and controlling; the first is putting the risk plans into action and ensuring that they are still effective. The second is producing significant documentation to support the process. The step in the description of the RM flow adopted in this study includes ensuring the use of the previous steps. This is to ensure that the identified risks, which are regarded as important, are also controlled exactly the way it was organised in the response step.
In the control step, it is also possible to identify new risks that emerge and the continuous process proceeds According to [46] risk control aims at controlling deviations, minimise risks and increase the project value. This stage of the RMP handles risks in a way that project objectives are achieved effectively. Risk control is based on a proactive approach other than a reactive approach to having the right measures in place and refining on them continually. No ready-made solutions are available to minimise risks. However the following corrective measures can assist in handling the risks associated with construction projects [47] [36]:

- Adjust plans the scope of work and estimates to counter risk implications.
- Monitor risks regularly, evolve alternate plans to manage predictable risks, when needed.
- Make appropriate decisions.
- Keep all concerned informed about possible risks.

3 Methodology

The work methodology included a literature search. The research was conducted with reference to existing theoretical literature, published and unpublished literature. This research is mainly a literature review and looks at the literature relating to risk identification in the construction industry. This is because the concepts of risk and risk management have been on agenda for many years. To ascertain the aim of the study literature on risk management was embarked which considered subtitles including; definition of risk, concept of risk management and methods employed in identifying and analysing risks in constructions.

4 Lessons learned from literature review

This study has revealed that risk management implicates identifying, assessing and prioritizing risks by monitoring, controlling, and applying managerial resources with a coordinated and economic effort so as to reduce the likelihood of adverse events and so as to maximize the achievement of project objectives. Management risks may lead to several benefits such as identification of favorable alternative course of action, increased confidence in realizing project objective, improved chances of success, reduced surprises, more precise estimates, reduced duplication of effort. In additional, the reviewed literature also stressed that the entire risk management process is not only limited to with solving problem in advance but also for the occurrence of any unexpected future problems. The study further shown that issues with possible threats envisaged in a project are not only a means to reduce losses within the project, but also, a means to transfer risks into opportunities, which can lead to economic profitability, environmental and other advantages as postulated by [48]. Undoubtedly, this study has as well shown that compiling of a list of risks necessary for a particular project is a prudent and proactive activity to be embarked by a project team. This implies that if risk attached to a project is not firstly identified it will be almost impossible to respond to it thereby affecting the entire project. RM, as revealed by the study, must not be a

DOI: 10.1051/matecconf/20166600008

standalone process rather a coordinated and integrated task that should involve all the project team to enable a successful risk management. Also, not all risks on projects are entirely of negative influence; some are an opportunity to explore into other segments of that would create profitability to the firm.

5 Conclusion

While most studies have focused on some aspects of project management, this paper has examined literature relating to risk management with the accomplishment of all project objectives regarding cost, time, quality, environment and safety. To achieve the objective of this study, the concepts of risk, RM and methods used in analyzing/assessing risk were reviewed. The literature showed that risk identification, risk analysis and assessment, risk response and control are crucial phases in risk management process. Risks must first be identified before they can be controlled or mitigated. Accordingly, this study concludes that risk management should be considered as the most significant activity for successful completion of a project. Risk identification techniques as concluded from this study includes the following: brainstorming, interviews/expert opinion, questionnaires, Delphi technique, expert systems, checklists and documentation review. In the analysis of risk, two categories of methods are developed: qualitative and quantitative methods. Literature revealed risk avoidance, risk reduction, risk transfer and risk retention as techniques mostly used in responding to risk. The study further concludes that issues with possible threats envisaged in a project are not only a means to reduce losses within the project, but also, a means to transfer risks into opportunities, which can lead to economic profitability.

References

1. Johnson, T. (2008). Project Management Professional (PMP) exam success series: certification exam mannual. Texas: Crosswind Project Management Inc
2. Creedy, G.D. (2005). Risk factors leading to cost overrun in the delivery of highway construction projects. PhD thesis, Queensland University of Technology, Australia
3. Laryea, S. (2008). Risk pricing practices in finances, insurance and construction. In: COBRA Research Conference, September 4th – 5th, Dublin Institute of Technology
4. Loosemore, M., Raftery, J., Reily, C. & Higgion, D. (2006). Risk management in projects. London: Routledge
5. Edwards, P.J. & Bowen, P.A. (2005). Risk and risk management in construction: a review and future direction for research, engineering, *Construction and architectural management*, 5(4): 339-349.
6. Zayed, T. Amer, M. & Pan, J. (2007). Assessing risk and uncertainty inherent in Chinese highway project using AHP, *International journal of project management*, 26: 408-419
7. Vargas-Hernández, J.G. (2011). Modeling Risk and Innovation Management. *Advances in Competitiveness Research*, 19 (3-4), 45-57.
8. Olamiwale, I.O. (2014). Evaluation of Risk Management Practices in the Construction Industry in Swaziland. Master of Quantity Surveying Thesis, Tshwane University of Technology, Pretoria, South Africa.
9. Dey, P.K. & Ogunlana, S.O. (2004). Selection and application of risk management tools and technique for build-operate-transfer-projects, *Industrial management and data system*, 4: 334-336.
10. Simu, K. (2006). Risk management in small construction projects. Licentiate dissertation. Department of Civil and Environmental Engineering, Lulea. LTU
11. Oztas, A. & Okmen, O. (2004). Risk analysis in fixed price design-build construction projects, built and environment, 39, 229-237.
12. Smith, N.J., Merna, T., & Jobbling P. (2006). Managing Risk in Construction Projects. 2nd edition Oxford: Blackwell Publishing
13. Abassi, G. Y., Abdel-Jaber, M. S. & Abu-Khajedart, A. (2005). Risk analysis for the major factors affecting the construction industry in Jordan, Emirate, *Journal of engineering research*, 10: 41-47.
14. Al-Salman, A. (2004). Assessment of risk management perception and practices of construction contractors in Saudi Arabia. Master of Science, Fahd University of Petroleum & Minerals.
15. Warszawski, A., & Sacks, R. (2004). Practical Multifactor Approach to Evaluating Risk of Investment in Engineering Projects, *Journal of Construction Engineering and Management*, 130(3),357-367
16. Mills, A. (2001). A systematic approach to risk management for construction. *Structural survey*, 19(5): 245-252.
17. Wong, J.T.Y. & Huie, E.C.M. (2006). Construction project risks: further considerations for contractors' pricing in Hong Kong. *Construction management and economics*, 24: 425 -438.
18. Klemetti, A. (2006). Risk management in construction project networks. Laboratory of Industrial Management, Helsinki University of Technology.
19. El-KarimiAzari, A., Mousavi, N., Mousavi, F.S., & Hosseini, S. (2011). Risk Assessment model selection in Construction Industry. Iran University of Science and Technology, Islamic Azad University and Tarbiyat Modares University, Tehran, Iran.
20. Zou, P.X.W., Zhang, G. & Wang, J. (2007). Understanding the key risks in construction projects in China. *International journal of projets management*, 25: 601- 614.
21. Kishan, P., Bhavsar, J.J., & Bhatt, R. (2014). A Study of Risk Factors Affecting Building Construction Projects, *International Journal of Engineering Research & Technology*, 3(12), 831-835

22. Skorupka, D. (2008). Identification and initial risk assessment of construction projects in Poland. *Journal of management in engineering*, 24(3): 120-127.
23. Kerzner, H. (1998). Project management: a system approach to planning, scheduling and control. New York, Chi Chester: Johnwill.
24. Adnan, H. (2008a). An Assessment of Risk Management in joint venture projects (JV) in Malaysia, *Asian Social Science*, 4(6),99-106.
25. AS/NZS ISO 31000: 2009 Risk Management-Principles and Guidance. Retrieved from http://www.finance.gov.au/sites/default/files/COV_216905_Risk_Management_Fact_Sheet_FA3_23082010_0.pdf.Accessed:16/01/2016.
26. Association for Project Management. (2000). Project Management Body of Knowledge, 4th edition, Project Risk Analysis and Management, a guide by APM.
27. Tuncel, G., & Alpan, G. (2010). Risk assessment and management for supply chain networks: A case study, Computers in Industry, (61), 250–259
28. Giannakis, M., & Louis, M.(2011). A Multi-agent based frame work for supply chain risk management, *Journal of Purchasing and Supply Management*, (17), 23–31
29. Mazouni, M.H. (2008). Pour une Meilleure Approche du Management des Risques : De la Modélisation Ontologique du Processus Accidentel au Système Interactif d'Aide à la Décision, Thèse de Doctorat de l'Institut National Polytechnique de Lorraine, Université de Nancy, France.
30. Ubani, Chinenye, E., Amade, Benedict, Okorocha, Aku, K., Agwu, A., & Okogbuo,F.(2015). Project Risk Management Issues in the Nigerian Construction Industry, *International Journal of Engineering and Technical Research,* 3(1): 217-232.
31. Tchankova, L. (2002). *Risk identification: basic stage in risk management. Environmental Management and Health,* 13(3):290-297. Towers Perrin. 2008. *Towers Perrin Risk and Opportunity WhitePaper*.Availablefrom: http://www.towersperrin.com[Accessed 05/01/2016].
32. Kuang, Z. (2011). Risk Management in Construction Projects: Application of Risk Management in Construction Period. Bachelor of Architectural Technology and Construction Management. Via University College, Horsens Campus, Denmark.
33. DEAT (2006) Risk Management, Integrated Environmental Management Information Series 23, Department of Environmental Affairs and Tourism (DEAT),Pretoria-SouthAfrica. http://www.deat.gov.za. Accessed [03/01/2016]
34. Cooper, D., Grey, S., Raymond, G. & Walker, P. (2005). Managing risks in large projects and complex procurement. Chichester: John Wiley.
35. Nguyen, N. P. & Li, Y. (2012). "Critical analysis of risk management and significant impacts of its application on Sichuan Post-earthquakes Reconstruction project". Master of Science Thesis. Umea University, Sweden
36. Schatteman, D., Herroelean, W., Vandvonder, S. & Boone, A. (2008). Methodology for integrated risk management and proactive scheduling of construction projects, *Journal of construction engineering and management*, 134(1): 885-893.
37. Winch, G., (2002). "Managing construction projects, an information processing approach". Oxford: Blackwell Publishing.
38. Chapman C. & Ward S. (1997). "Project Risk Management: Processes, Techniques and Insights"
39. El-Dash, K. (2008). Construction risk management application and education selected. Proceeding from the 12th International Congress on Project Engineering, 390-403.
40. Opolot, P.K., Buys, N.S. & Slabber, J.M. 2009. Risk management in the South African construction industry.Availablefrom: http://www.yumpu.com/document/view/11842795/risk-management-in-the-south-africaconstruction. industry.html. Accessed: 10/01/2016
41. Lisa Turnbough, P.E. (2005). Risk management on large capital projects. *Journal of professional issues in engineering and practices*, 365-374.
42. Wang, M. & Chou, H. (2003). Risk allocation and risk handling of highway projects in Taiwan. *Journal of management in engineering*, 19(22): 60-68.
43. PMI. (2008). A Guide to the Project Management Body of Knowledge. Project Management Institute, Newtown Square, PA.
44. Zenghua, K. (2011). Risk Management in Construction Projects. VIA University College, Horsens Campus, Denmark.
45. Dada, J.O. (2010) Strategies for militating risk in construction projects. In proceedings of the 40th Annual general meeting/conference of the Nigerian Institute of Building, 7th- 11th July 2010, Asaba, Delta, Nigeria, 28-35.
46. Lester, A., (2007). Project management, planning and control, 5th edition. Oxford: Elsevier Ltd Light Press.
47.

Motivation in Construction Innovation: Commercial Opportunities, Problem-Solving and Passion for Improvement

Chang Yean Fang[1], Rajah Rasiah[2], Jane E. Klobas[3]

[1]*Department of Business Strategy and Policy, Faculty of Business and Accountancy, University of Malaya, Kuala Lumpur, Malaysia*
[2]*Department of Development Studies, Faculty of Economics and Administration, University of Malaya, Kuala Lumpur, Malaysia*
[3]*Department of Business Strategy and Policy, Faculty of Business and Accountancy, University of Malaya, Kuala Lumpur, Malaysia*

Abstract This article analyses motivations for innovation in construction using the service sector adaptation of the Sectoral System of Innovation (SSI) framework. Interviews and site visits were conducted with four Malaysian firms. Innovation in construction is similar to the service sector. There is evidence of technology-push, capability-push and demand-pull; capability push is the most important. Construction firms innovate to gain commercial opportunities, to solve project-related problems and to improve processes. By simplifying construction work, process innovation saves time and costs, increasing efficiency and productivity, and providing increased competitiveness. Innovation is also motivated by committed and passionate actors within construction firm.

1 Introduction

Construction faces intense competition for international job opportunities (Abdul-Aziz & Wong, 2010) with cross-national trade agreements and globalization providing developing countries with greater freedom of access to markets. Increasing sophistication in societal demands and environmental pressures bring increased demand for safer, higher quality and sustainable construction, creating a further challenge for policy makers. Against this backdrop of challenges, the construction industry worldwide faces problems in safety, quality and delays (Oakland & Marosszeky, 2006), and a poor public image (Samuelsson, 2003).

Historically, innovation in the construction industry has been regarded as very conventional: innovation is incremental, and radical and revolutionary innovations are rare (Slaughter, 1998). In the mid-2000s, lack of investment in research and development (R&D) was identified as the main cause of low innovation in construction in several countries, including the UK, Australia, Hong Kong and Singapore (Lim, 2006). Since then, increased attention has been paid to innovation in construction. Nonetheless, despite substantial investment and research, innovation remains rare in the construction sector. One explanation is that research on construction innovation has relied on theories and methods drawn from manufacturing, but because the construction industry is project-based and requires specific processes (Reichstein et al., 2005), this has hindered rather than aided efforts to understand innovation in construction (Koskela & Vrijhoef, 2001).

To address this shortcoming, we adopt a new approach to study innovation in the construction industry. The Service Sectoral Systems of Innovation framework developed by Tether and Metcalfe (2004) provides a systemic model of innovation for the services industry. We adapt this model to the construction industry to analyse the motivation for innovation of four Malaysian construction firms. The paper is organized as follows. The next section presents the theoretical arguments used to construct the analytic framework. The subsequent sections present the methodology followed by the findings. The paper concludes with a discussion and the final section presents the conclusions of the paper.

2 Literature Review

The need for new construction innovation theory

Shapiro (1999) provides a useful discussion of the differences between construction and manufacturing systems. She describes *complex product systems* (COPS) as systems which are produced on a project basis for specific customers and markets, and notes that their innovation and production processes differ from those assumed to exist in traditional research models based on manufacturing and mass production. COPS production processes stress software development, systems integration and project management over repeated tasks (Shapiro, 1999).

The COPS characteristics of construction mean that innovation in construction needs to be examined differently, using models other than the conventional models based on mass production in the manufacturing sector. Furthermore, the construction sector is more similar to the service sector than manufacturing with regards to innovative behaviour (Reichstein et al., 2005).

Manufacturing models, if applied without modification, are unsuitable, conceptually and in terms of management strategy and practices (Acha et al., 2005; Reichstein et al., 2005; Widén, 2006; Widén, 2010). Existing construction research that is based on manufacturing models cannot provide a comprehensive account of the sector and, as a consequence, risks generating misleading findings about the industry.
Innovation in construction firms is influenced by a diversity of interrelated actors in a complex systemic network. To understand the factors that influence contractors' decisions to innovate one has to understand this complex system of innovation.

2.1 Sectoral systems of innovation (SSI) theory

The sectoral systems of innovation (SSI) approach of Malerba (2002, 2004, 2006) permits the study of multi-faceted non-linear interactions between actors involved in innovation across an industrial sector. The systemic view of the processes of innovation and diffusion of technology originated with Schumpeter (1961) and was later embraced by others. The SSI approach is rooted in the evolutionary economics school with antecedents to Veblen (1898), the founder of institutional economics. Evolutionary theory explains the origin and evolution of different industries and technologies as being shaped by time and locality (Malerba & Nelson, 2011; Nelson, 2008). This approach is a good fit for the study of the construction industry in Malaysia, which differs by sector in different localities with different institutional regimes and by the timing of the evolution of the sector.

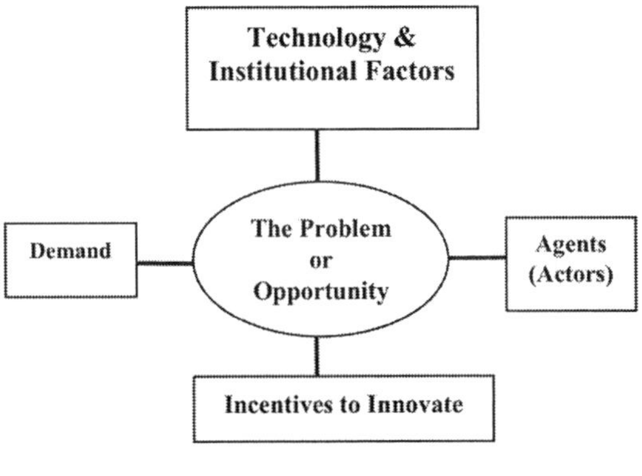

Figure 1: Problem-/Opportunity-Centric Innovation Tether & Metcalfe, 2004)

2.2 Innovation systems in the service sector (ISS) framework

Tether and Metcalfe's (2004) application of the SSI to innovation systems in the service sector (ISS) incorporates propositions about motivations for innovation. Tether and Metcalfe suggest the use of problem and opportunity to examine the interactions and interdependencies, patterns of resistance to innovation, patterns of sources of knowledge, balances of dependency and power, and the relations of these characteristics with patterns of activities and innovation.

This paper adopts this approach. Thus, following Reichstein et al. (2005), we propose that innovations in services are developed around problems (or opportunities) that are framed by a number of contingencies – including the regulatory, cultural and technological context – demand, agents (actors) and incentives (Figure 1).

2.3 Motivations for innovation in construction

The construction industry literature points to three main motivators for innovation: client- or demand- pull; construction firm or capability push; and improvement in project performance. Arditi et al. (1997) Bossink (2004) and Tatum (1989) show that both the availability of technology (technology-push) and client demand (demand-pull) are relevant in the construction industry. However, Gann and Salter (2000) suggest that demand pull is the stronger motivator (Ling, 2003). Toole (1998) suggests that construction firms are motivated to innovate to improve project performance in terms of cost and the project's final structure or system. Empirical research indicates that innovation is pursued in construction generally to reduce costs, improve functionality and sustain market share (El-Mashaleh et al., 2006; Seaden et al., 2003; Thomas et al., 2004; Toole et al., 2013).

3 Methodology

Since the purpose of this paper is interpretative rather than predictive, we use case studies to examine the motivation for innovation in the construction industry (Yin, 2014). We selected four case studies using criteria for differentiation by sector and innovation-type defined by us and criteria for innovativeness suggested by the literature and experts from Malaysia's Construction Industry Development Board (CIDB). Because the firms requested confidentiality, we use pseudonyms to refer to them, namely, the infrastructure developer as InnoInfra, the residential buildings developer as InnoWEBS, and the constructors of commercial buildings as InnoInfo and InnoIBS.

We carried out a total of 19 interviews in Malaysia between October 2012 and January 2013 with personnel engaged in innovation in the selected firms. For analysis, we used the steps suggested by Hsieh and Shannon (2005) for data coding and thematic analysis.

3.1 Case context

Between them, the four firms were responsible for a total of seven innovations: a two-in-one infrastructure, which alleviates flooding with a tunnel which is also used for vehicular traffic; a connecting highway by building a bridge over an existing highway; large-scale public transport infrastructure; a building modelling and project management information system; two industrialized building systems; and a modular house that could be constructed by the consumer (the do-it-yourself or DIY house). Two innovations were more than nine years old, two were completed in 2010, and the other three were still on-going during the interviews.

Table 1: The Case Study Firms and Their Innovations

Case Study Firm	Subsector	No	Innovation	Inno-vation Age
InnoInfra	Civil engineering	1	Joint highway / flood mitigation infrastructure	9 years
		1	Connecting highway bridge	17 years
		1	Public transport infrastructure	6 years
InnoInfo	Building-Commer-cial	1	Building process information system	6 years
InnoIBS	Building-Commer-cial	2	Industrialized building system	1 year
InnoWEBS	Residential	1	DIY house consisting of frame, wall, roof systems	On-going

Two sources of innovations were observed: in two cases, the innovation idea originated from the construction firm itself; in the other two cases, the construction firm purchased the technology or solution for the innovation and customized it. When the idea originated in the construction firm, the owners and senior management initiated the idea then developed the innovation with external consultants. When the idea originated externally, the construction firms customized the technology, either according to company processes or to company or market needs. The extent of initial investment in technology and subsequent customization of the technology varied.

Both firm-led and externally provided innovations used existing solutions with incremental changes in technology. Thus, the innovations were imitations or adoptions of existing innovations with customization to suit the processes of the company and market needs. Contractor leadership was a key feature of innovation in two case study firms, thus, in construction, not all innovations can be deemed supplier-dominated. Further, in the two case studies involving a technology or solution provider, InnoInfo undertook extensive customization with a full-time team of R&D staff and without the support of external solution providers. The observed innovations consisted of innovations in business practices, in the form of business strategy and business contracts (InnoInfra); organizational processes (InnoInfo); and both construction process and product characteristics (InnoIBS and InnoWEBS).

Table 2: Source and Type of Innovation in Case Study Firms

Firm and Innovation	Source of Innovation	Type of Innovation
InnoInfra: Highway (2), Transportation	Construction firm-led	Business practice
InnoIBS: IBS related innovation	Construction firm-led	Construction process & product
InnoWEBS: DIY house consisting frame, wall, roof systems	Technology provider, some customization	Construction process & product
InnoInfo: Information on building process	Solution provider, heavy customization by own full-time R&D staff	Organizational & business process

R&D in the construction firms existed in a different form to the formal conventional R&D of firms with full-time staff and access to testing laboratories. Except for InnoInfo, development of the innovations occurred as part of the firms' technical work on feasibility studies rather than as a permanent, full-time function.

The length of the innovation cycle time depended on three factors. Where the innovation was more complex (infrastructure innovation at InnoInfra), the innovation cycle time was longer, taking up to two years. Where the contractor developed the innovation in-house, with little external support, the innovation cycle time was longer still, e.g. three years in the case of InnoInfo. Where the technology was purchased from a technology provider (InnoWEBS), the innovation cycle was short as the main development work had been completed by other parties. Where the innovation was required urgently (InnoIBS), the innovation cycle time took only a few weeks.

4 Findings

We observed the expected forms of motivation arising from problems and opportunities. In addition, we observed motivations due to the actions of top management, and the personal motivation of innovative owners.

4.1 Problem- and Opportunity-Centric Motivation to Innovate

We found evidence that construction firms, like firms in the service industry, innovate to solve problems or to seize a business opportunities in the market. Technical and business problems are seen as commercial opportunities or as a stimulus for improvement for internal practices and processes.

4.1.1 Commercial problems and opportunities

Firstly, the firm innovates to address a commercial problem or issue in the industry to find better ways of solving business and technical issues (InnoInfra, Corporate Director). Such issues include flood and environmental problems, transportation issues and water supply. Again, here was a problem [referring to the flood problem in Malaysia] so what could we do? This issue triggered a proposal from us. (InnoInfra, Corporate Director). Secondly, innovation is driven by a need for differentiation innovation in the company is driven by the need to differentiate ourselves in the industry. (InnoInfra, Corporate Director) as commercial strategy is driven by the need to show continuous growth in shareholder value necessity is the mother of all invention. We innovate because we have to. We need to show continual improvement in our business and grow shareholder value. To grow, we need to secure new projects, and to do so in a competitive and sometimes uncertain market, we need to think of new opportunities using out-of-the-box approaches. (InnoInfra, Corporate Director) in a fragmented supply chain. Why do we need to innovate software [A]? Because we don't have obvious competitive advantage and we are not differentiated. We are so fragmented; our supply chain is different. (InnoInfo, Innovation Manager)

In the past, members of InnoInfo's workforce, in common with the workforce of the construction industry as a whole, did not require professional qualifications. Its low level of human capital meant that InnoInfo's staff did not always take a professional approach to problem solving and was not sufficiently competitive. The firm was not able to attract new talent. The older generation of construction workers does not need qualifications and bangs tables [is less professional] in their work – this needs to change. Using IT is not to impress talents but overall in an industry we must be more competitive. (InnoInfo, Innovation Manager)

For InnoInfo, innovation enables the industry to remunerate talent. This will then attract talent to join the company, an important factor in ensuring its competitive advantage. Overall, according to the Innovation Manager of InnoInfo, the construction industry has not been able to attract talent due to its negative image of being "dirty, dangerous and difficult" as well as the less than professional behaviour of the older generation construction workers. He expects innovation to provide a more positive branding of the industry which will, in turn, enable InnoInfo to attract higher calibre human capital to the company.

The differentiation strategy of InnoWEBS acts as a competitive advantage strategy to the firm to recoup the expenditure in innovation and ultimately bring revenue to the firm. It is the excitement of knowing, but also we know we can make money. It is the combination of these two factors. At the end of it, it is about monies. We know that if we can capture that knowledge: first is what knowledge can do; two is that we need to recoup for monies spent to commercialize. (InnoWEBS, Managing Director). We believe that we can enter the market with this innovation and get back our monies. We believe in the returns. (InnoWEBS, Managing Director) providing a niche, enter the market and be one of the competitors... and make it as our niche market [as well as providing a new business opportunity]. (InnoWEBS, Managing Director).

The new business opportunity is the venture of InnoWEBS into manufacturing in the future which is expected to provide more business opportunities than construction. We want to be the manufacturer one day. We hope we will be the manufacturer and let someone else be the contractor. Now, the manufacturing part is still under the same entity although we have a different group of people working there (InnoWEBS, Head of Production, Design and Technical).

4.2 Process improvement

Construction firms also innovate because innovation simplifies construction work, which saves time, which in turn translates to cost savings, increased efficiency and productivity, which ultimately provides increased competitiveness. This is the message from our General Manager, as the head of the company, he must always make sure that there is efficiency and productivity – these two things must always be there, or else we can't compete outside. (InnoIBS, Manager Technology and Design). On site, innovative solutions to problems are necessary just to deliver the end product to the customer. For us, innovation is to overcome problems. It's all problem-solving and fire-fighting or so called innovation because we have to deliver the end product. (InnoIBS, Site Manager)

One driver for innovation is the project-based nature of the industry, which often imposes time constraints. InnoInfo is motivated to innovate to improve productivity in order to address four problems. The first is the lack of timely information, which inhibits informed decisions and thus leads to additional costs and delays decisions are often compromised and lead to some 10-20% time and cost overruns. (InnoInfo, Director)

The second problem is a lack of collaboration and coordination amongst stakeholders at the early stage of construction when the ability to control costs is higher and the costs of design changes are lower. The lack of alignment amongst stakeholders further compounds the problem (InnoInfo, Director). Thirdly, inefficient information exchanges between parties in construction projects that are in 2D or hardcopy formats cause rework and risk loss of accuracy (InnoInfo, Director; InnoInfo, Innovation Head). Fourthly, existing work tools and processes are unable to meet the needs of increasingly complex work and compressed time schedules as well as pressures to lower costs and increase productivity (InnoInfo, Director).

The main cause of these problems is the fragmented nature of the industry. Fragmentation is seen in the involvement of many players in any project easily, there can be 100 people [with] different roles and responsibilities without alignment [and] conflicts of interests. (InnoInfo, Innovation Manager). Clients can be one time off to own certain facility [i.e., once only or regular clients]. Consultants that advise developers generally want to make money, want to finish the job within their budget, specifications and time frame. We [construction firms] want things fast. Consultants charge based on the time in designing [and have] no motivation to minimize design time. Subcontractor scope is smaller, and they are specialists. All players' motivations are different. Their motivation is still time and cost. The different role and responsibilities tend to make them draw a line, e.g., this is my liability then this is my problem. (InnoInfo, Innovation Manager). The four problems are seen to cause inefficiency and ineffectiveness in the construction industry. We are not effective [and] have the lowest productivity compared to manufacturing. (InnoInfo, Innovation Manager).

InnoInfo was thus motivated to develop a '5D' innovation based on product life cycle management (PLM) software to achieve high productivity. Manufacturing high productivity as model PLM is the way to go (best possible solution) that we think manufacturing has high productivity, and construction has the lowest productivity, so we want to relook and ask why it is that way for us. (InnoInfo, Innovation Manager)

InnoInfo's motivation to improve productivity is linked to its motivation to obtain competitive advantage. This is because low productivity further affects competitive advantage and branding. Maybe, at the end of the day, your gains are in terms of your branding and in terms of your differentiating between your competitors. When we were in Abu Dhabi, we were asked can you do [X]. We went home two years without any answers. But today we can answer; we are not the experts but we are able to deliver something. (InnoInfo, Innovation Head)

Another driver for innovation is client demand. InnoIBS' innovation resulted from a foreign client's demand for high safety standards (InnoIBS, Site Manager). The motivation was to overcome challenges for safety in construction (InnoIBS, Site Manager). The interaction between the problems caused by the nature of construction industry and firms' motivation to innovate is summarized in Figure 2.

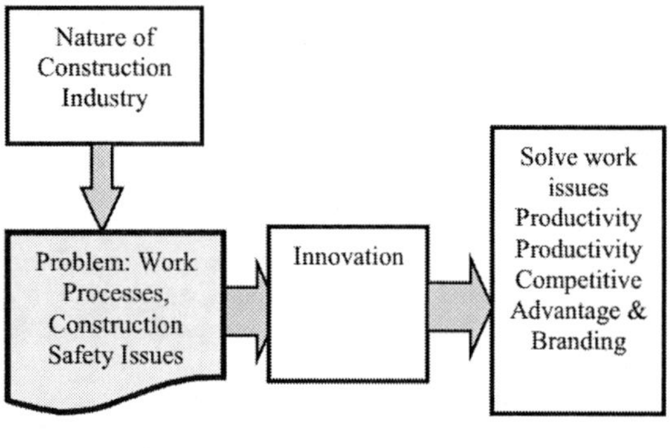

Figure 2: Opportunity- and Problem-Centric Motivation to Innovate, Construction Firms

4.2.1 Positive effects of actors and networks

In additional to the problem- and opportunity-centric motivations for innovation in construction, we found that actors motivated innovation. Top management provided leadership and direction, and showed commitment to innovation. The Board of Director's support for innovation is evident through the allocation of resources and investments in technology and R&D, human capital development and collaboration initiatives with its external users.

In InnoInfo, the existence of a visionary and committed top management drives innovation. Innovation is driven from the top by the Director, through his vision for a virtual construction system, with the support of the company's Board of Directors. He had always had a dream of having something like virtual construction to be implemented. (InnoInfo, Innovation Manager).

The Director is a believer in information technology who has experience in customization. He strategized to keep R&D alive in the company by creating an R&D team. He also sits in the Board of Directors and garners their support. The support of the Board of Directors is evident from their allocation of resources to the innovation. This support was appreciated by subcontractors. More importantly, the developer is committed because the innovation costs lots of money. (InnoInfo, Mechanical Subcontractor) and the site team. The resources are not cheap, so the management itself will have to be committed, willing to spend much money on R&D. (InnoInfo, Site Project Team Member). Board support is important, not only because R&D involves a lot of money, but also because there is negative return in the first few years if at company level you are making RM[XXX] million and RM[X-X] million is in R&D, the question mark is that are we spending the right way. But in technology advancement, the first few years are always negative not positive. (InnoInfo, Innovation Head). The top down vision serves as an important motivator for InnoInfo. You really have got to have top down vision, and enforcement mandatory all the way from the top. (InnoInfo, Innovation Head)

InnoInfra is motivated to innovate because of its Managing Director, who provides leadership in identifying a commercial opportunity and driving the innovation. It is the top management who will actually say there's a problem. Top management will involve a few key people from the functional team. (InnoInfra, Head Design & Technical). Leadership and direction is very important. I greatly respect our MD because the D&T Department only thinks of the technical part, sometimes not on commercial aspects whether the project is viable or not. (InnoInfra, Head D&T)

InnoIBS is motivated to innovate by parent Group policy and practices. Firstly, the parent Group's top management has initiated continuous improvement activities which are implemented by the Group's Quality Unit. Secondly, the Group holds a half yearly quality forum, to which top management is strongly and visibly committed. Specific topics may be initiated by top management (Group Quality Manager), and the Chairman and 40 senior management personnel attend the event. The presence of top management indicates to staff that their ideas are important. This acts as a powerful form of recognition if at the [Group quality forum], staff need to come up with new ideas, but the ideas are not presented to top management, it does not work because the staff may think I have some new ideas and I just present to middle management and my ideas don't matter. This is a strong, good push factor. (InnoIBS, Group Quality Manager). In addition, top management approve staff members' ideas at the forum, and this is supplemented, again, with rewards and recognition of the staff concerned if they come out with new ideas, it ties to their salary bonuses and increment. (InnoIBS, Group Quality Manager). Top management are also involved in tracking the progress of implementation of the new ideas generated at forum other than presenting the status of the previous ideas bosses will also ask the Business Unit's Head at various meetings. The tracking of the progress of these ideas is done by top management, [the forum] and the Group Quality Management Unit. (InnoIBS, Group Quality Manager)

InnoInfra also benefits from internal actors' personal motivations to innovate. The Head of the Design and Technical (D&T) Department innovates for two reasons. Firstly, innovation gives him professional satisfaction. We always look for things to perfect, to do things better, how to make it … for innovation to happen. One thing is professional satisfaction. (InnoInfra, Head D&T). Secondly, innovation challenges him to do things differently, and better. We always think, "The conventional way of doing things is like this, can it be done better?" We always challenge ourselves. (InnoInfra, Head D&T)

InnoWEBS is motivated to innovate due to the personal motivation and training of the innovator. The Managing Director appreciates knowledge, enjoys the excitement of knowing and understands that the skills obtained from his PhD education enabled him to innovate. Likewise, the Head of Production, Design and Technical (HPDT) innovates because he enjoys innovating, it is one of his interests and he sees the result for end users. [I like] things that are different, simple yet serving the purposes, easy for end users to use and they can enjoy it. (InnoWEBS, HPDT)

5 Discussion

The motivations for innovation in the construction industry observed in this study are summarized in Table 3. Construction firms innovate for problem-centric or opportunity-centric reasons, like firms in the service sector. In addition, actors within construction firms motivate innovation through top management commitment, the leadership behaviours and attributes of innovation champions and the personal motivations of the innovators. Thus, the capability push of leaders has a strong effect on construction innovation. We also observed evidence of technology-push (InnoInfo and InnoWEBS). On the other hand, there was little demand pull in the construction subsectors we studied in Malaysia (only InnoIBS), contrary to the propositions made by Gann and Salter (2000).

6 Conclusion

This paper used the Tether and Metcalfe (2004) formulation of the SSI in services to examine the motivation for construction innovation in four Malaysian firms. The evidence showed that, not only is motivation for construction innovation problem and opportunity-centric, like that of the service sector, it is also led by actors within the construction firms themselves. Our results make two important contributions. Firstly, we confirm proposed similarities between innovation in the construction industry and innovation in the service industries. Further research on construction innovation is therefore likely to benefit from adoption or adaptation of models of service industry innovation. Secondly, we observed a personal, almost entrepreneurial commitment

Table 3: Motivation for Case Firms to Innovate, Compared With Extant Literature

Aspect of Motivation	Details	Sources
Problem- and opportunity-centric	Firm seeks competitive advantage–reputation, work processes & ability to attract new employees	Slaughter (1998) This paper
	Firm seeks to improve project and process performance	Project: Toole (1998) Process: New finding in this paper
Actor effects	Leaders' capabilities and formal and informal roles, and personal motivations are strong determinants	New finding in this paper
	Technology-push and demand-pull are both relevant	Arditi et al. (1997), Bossink (2004), Tatum (1989) This paper
	Firm capability capability-push stronger than demand-pull	Gann (2000) This paper, contrary to Gann and Salter (2000)

to innovation in our case study firms. This observation is new and, although it might be sample specific, seems sufficiently interesting to warrant further research. Regardless of firm size, structure and ownership, does successful innovation in the construction industry require entrepreneurial vision and personal commitment of a different kind to that required in other industries?

References

1. Abdul-Aziz, A. R., & Wong, S. S. (2010). Exploring the internationalization of Malaysian contractors: The international entrepreneurship dimension. *Construction Management and Economics, 28*(1), 51-61. doi: 10.1080/01446190903460680
2. Acha, V., Gann, D. M., & Salter, A. J. (2005). Episodic innovation: R&D strategies for

project-based environments. *Industry and Innovation, 12*(2), 255-281. doi: 10.1080/13662710500087990
3. Arditi, D., Kale, S., & Tangkar, M. (1997). Innovation in construction equipment and its flow into the construction industry. *Journal of Construction Engineering and Management, 123*(4), 371-378. doi: 10.1061/(ASCE)0733-9364(1997)123:4(371)
4. Bossink, B. A. G. (2004). Managing drivers of innovation in construction networks. *Journal of Construction Engineering and Management, 130*(3), 337. doi: 10.1061/(ASCE)0733-9364(2004)130:3(337)
5. El-Mashaleh, M., O'Brien, W. J., & Minchin Jr, R. E. (2006). Firm performance and information technology utilization in the construction industry. *Journal of Construction Engineering and Management, 132*(5), 499-507. doi: 10.1061/(ASCE)0733-9364(2006)132:5(499)
6. Gann, D. M., & Salter, A. J. (2000). Innovation in project-based, service-enhanced firms: The construction of complex products and systems. *Research Policy, 29*(7-8), 955-972. doi: 10.1016/S0048-7333(00)00114-1
7. Hsieh, H.-F., & Shannon, S. E. (2005). Three approaches to qualitative content analysis. *Qualitative Health Research, 15*(9), 1277-1288. doi: 10.1177/1049732305276687
8. Koskela, L. J., & Vrijhoef, R. (2001). Is the current theory of construction a hindrance to innovation? *Building Research & Information, 29*(3), 197-207. doi: 10.1080/09613210110039266
9. Lim, J. Y. (2006). *Stimulating construction innovation in Singapore by developing the national system of innovation.* (PhD Dissertation), National University of Singapore, Singapore.
10. Ling, F. Y. Y. (2003). Managing the implementation of construction innovations. *Construction Management and Economics, 21*(6), 635-649. doi: 10.1080/0144619032000123725
11. Malerba, F. (2002). Sectoral systems of innovation and production. *Research Policy, 31*(2), 247-264. doi: 10.1016/S0048-7333(01)00139-1
12. Malerba, F. (2004). *Sectoral systems of innovation: Concepts, issues and analyses of six major sectors in Europe.* Cambridge, UK: Cambridge University Press.
13. Malerba, F. (2006). Innovation and the evolution of industries. *Journal of Evolutionary Economics, 16*(1), 3-23. doi: 10.1017/CBO9780511493270
14. Malerba, F. (2007, September). *Catching up in different sectoral systems.* Paper presented at the GLOBELICS International Conference: Regional and National Innovation Systems for Development, Competitiveness and Welfare: the Government-Academia-Industry Partnership, Saratov State Technical University, Saratov. http://csh.xoc.uam.mx/EII/globelicswp/wpg0801.pdf.
15. Malerba, F., & Nelson, R. R. (2011). Learning and catching up in different sectoral systems: Evidence from six industries. *Industrial and Corporate Change, 20*(6), 1645-1675. doi: 10.1093/icc/dtr062
16. Nelson, R. R. (2008). What enables rapid economic progress: What are the needed institutions? *Research Policy, 37*(1), 1-11. doi: 10.1016/j.respol.2007.10.008
17. Oakland, J. S., & Marosszeky, M. (2006). *Total quality in the construction supply chain.* Burlington, MA: Butterworth-Heinemann.
18. Reichstein, T., Salter, A. J., & Gann, D. M. (2005). Last among equals: A comparison of innovation in construction, services and manufacturing in the UK. *Construction Management and Economics, 23*(6), 631-644. doi: 10.1080/01446190500126940
19. Samuelsson, P. (2003). Improvement processes in construction companies. In B. Atkin, J. Borgbrant & P. E. Josephson (Eds.), *Construction process improvement* (pp. 225-239). Oxford, UK: Blackwell Science
20. Schumpeter, J. A. (1961). *The theory of economic development: An inquiry into profits, capital, credit, interest, and the business cycle.* (R. Opie, Trans.). New York: Oxford University Press. (Original work published 1934).
21. Seaden, G., Guolla, M., Doutriaux, J., & Nash, J. (2003). Strategic decisions and innovation in construction firms. *Construction Management and Economics, 21*(6), 603-612. doi: 10.1080/0144619032000134138
22. Shapiro, G. (1999). Inter-project knowledge capture and transfer: An overview of definitions, tools and practices *CoPS Innovation Centre Working Paper No. 62* (pp. 42). Brighton: CoPS. http://www.researchgate.net/publication/253518640_InterProject_Knowledge_Capture_and_Transfer_An_Overview_of_Definitions_Tools_and_Practices
23. Slaughter, E. S. (1998). Models of construction innovation. *Journal of Construction Engineering and Management, 124*(3), 226-231. doi: 10.1061/(ASCE)0733-9364(1998)124:3(226)
24. Tether, B. S., & Metcalfe, J. S. (2004). Services and systems of innovation. In F. Malerba (Ed.), *Sectoral systems of innovation: Concepts, issues and analyses of six major sectors in Europe.* Cambridge, UK: Cambridge University Press.
25. Thomas, S. R., Lee, S.-H., Spencer, J. D., Tucker, R. L., & Chapman, R. E. (2004). Impacts of design/information technology on project outcomes. *Journal of Construction Engineering and Management, 130*(4), 586-597. doi: 10.1061/(ASCE)0733-9364(2004)130:4(586)
26. Toole, T. M. (1998). Uncertainty and home builders' adoption of technological innovations. *Journal of Construction Engineering and Management, 124*(4), 323-332. doi: 10.1061/(ASCE)0733-9364(1998)124:4(323)
27. Toole, T. M., Hallowell, M., & Chinowsky, P. (2013). A tool for enhancing innovation in construction organizations. *Engineering Project Organization Journal, 3*(1), 32-50. doi: 10.1080/21573727.2012.717531

28. Veblen, T. (1898). Why is economics not an evolutionary science? *The Quarterly Journal of Economics, 12*(4), 373-397. doi: 10.2307/1882952
29. Widén, K. (2006). *Innovation diffusion in the construction sector.* (PhD Dissertation), Lund University, Lund.
30. Widén, K. (2010). Construction Innovation: Statsbygg, the Driving Client. Lund, Sweden: Division of Construction Management, Lund University.
31. Yin, R. K. (2014). *Case study research: Design and methods* (5th ed.). Los Angeles: Sage.

Malaysian Affordability Housing Policies Revisited

Diwa Samad[1], Nurshuhada Zainon[1], Faizul Azli Mohd Rahim[1], Eric Lou[2], Saipol Bari Abd Karim[1]

[1] *Department of Quantity Surveying, Faculty of Built Environment, University of Malaya, Malaysia*
[2] *School of Mechanical, Aerospace, and Civil Engineering, University of Manchester, United Kingdom*

Abstract: Housing has always been a significant aspiration of family expression and distinctly priciest investment by household. It plays a momentous role in the country's economy and so central to the societal well-being that is emplaced in the United Nation Universal declaration of Human rights. Yet in developed and developing world alike, cities struggle to provide decent housing for lower and middle income population. The provision of affordable housing is a major policy concern around the world with Malaysia being no exception; rising income hardly keep pace with price hike of housing unit and housing interventions has majorly concentrated on demand side leading to a non- responsive supply sector. Therefore, this paper highlights affordable housing issues pertaining Malaysia. It formulates Malaysian Map of affordability and conducts an evaluation of global housing schemes to better identify policy priorities for Malaysia. It's significant to harmonize supply and demand side factors in the housing market to ensure that housing supply fits the needs of citizens based on the location, price and target group. In case of Malaysia supply oriented initiative are of urgency in short and medium run. This must be supported by long term demand side schemes in parallel. Convergence of these two factors is essential for a balanced equilibrium and obtaining affordability.

1 Introduction

Housing is decisively rooted in the economic social, and political sphere of any country that it impossible to be explored in isolation from the broader scope of governance and policy. Besides being a remarkably valuable asset, it carries multidimensional significance; It plays an eminent role in accelerating economic growth and it carries social prominence as a spatial locus of personal and familial life (Abd Aziz, Hanif, & Singaravello, 2011; Keivani & Werna, 2001)

Despite its multidimensional implications, access to affordable and adequate housing has been an everlasting challenge globally (Beer, 2007). From slum dwellers in the third world cities to middle-income households in affluent global capitals, millions of people are challenged to find affordable housing without financial pressure. The economic and human cost of the housing affordability issue is massive as 330 million households are affected worldwide. Holding the existing trend constant, total households with unsafe and substandard housing or are financially strained by housing expenditures is estimated to reach 440 million - or 1.6 billion people by 2025 (Woetzel, 2014).

Rapid urbanization combined with population growth has incited a surge in housing prices in many urban areas, mainly in developing countries including Malaysia (Buhaug & Urdal, 2013). This has been supported by the evolution of nuclear families as against extended families brought about by economic development; in 1970 there were 182 households for every 1,000 people, by 2020 there will be 250 households for every 1,000 people (Suraya, 2015). Furthermore, increase in foreigners (expatriate, students and tourists), migration, changing pattern of economic status of the population, change in expectation, and dilapidation of the existing stock has resulted in severe shortage of affordable housing.

Since the Third Malaysia Plan, low cost housing projects has not met its target (Bajunid & Ghazali, 2012; Ramli, Akasah, & Masirin, 2014; Shuid, 2009). For instance, throughout the Eighth Malaysia Plan, only 197,649 low cost housing units were built compared to 230,000 units required (Ramli et al., 2014). Although many private developers were involved to offset the housing need, nonetheless, these developers constructed the low-cost houses merely due to quota requirements as they are non-lucrative projects. Residents of low-cost housing continually encounter many challenges such as maintenance, sub-standard quality, comfort levels, health, safety and security amenities (Zaid, 2011).

Currently there exist 40 per cent gap between the demand for affordable housing and its supply in the country (Khairie, 2013; Lim, 2015). According to the Department of Statistics Housing Income Survey (2014) Malaysian median income stood at RM 4,500 signalling households are unable to afford houses priced higher than RM300, 000 (DOS, 2014). While based on the National Property Information Centre report (2014) only 31.7% of the housing units built in the year 2014, had a price tag lower than RM 250,000 (NAPIC, 2014b). Thus, evidently the income pattern of the middle income household finds it challenging to keep pace with the rising cost of housing unit and thereby the need for affordable housing has become more vital than ever before.

In response to this issue, the Malaysian government undertook numerous initiatives to assuage the rising cost through several projects such as Projek Perumahan Rakyat 1 Malaysia (PR1Ma), Projek Rumah Rakyat Miskin Tegar (PPRT), and Skim Myhome Rumah Pertama outlined in 11th Malaysian plan (EPU, 2015). Despite these attempts, Malaysian people still experience a mismatch between acquiring affordable housing and their household income. This is especially so for the middle-income households who are overqualified for the aforementioned low-cost housing programmes and at the same time unable to afford housing by private housing developers.

Though government initiatives has facilitated housing development considerably, housing interventions have focused primarily on demand which subsidizes a non-responsive supply sector. As a result, the housing price is still higher despite the various government subsidies and loans. Relatively less progress is evident in realizing the elements that restrain supply, which probably thwart developments on the demand side.

1.1 Housing affordability: A definitional controversy

Housing affordability is considered as a household selection decision function between housing and non-housing product expenditure (Suhaida & Tawil, 2010). Many literatures have acknowledged that housing affordability is a complex phenomenon that is tough to pin down in practice, particularly in terms of defining the suitable geographic scope for housing markets, proper definitions of typical households and individuals, and their changing circumstances. (Gan & Hill, 2009; Stone, 2008)

Although there exist no standard definition for this term, in its simplest term, housing affordability is denoted as the rent-to-income ratio or house-price-to-income ratio known as income affordability; more sophisticated measures are purchase affordability, repayment affordability (Gan &Hill, 2009) and life time income affordability (Abeysinghe & Gu, 2011). Purchase affordability studies the ability of household to borrow adequate funds for owning a house. Repayment affordability reflects the burden imposed on a household to repay house mortgage. These concepts encompass additional parameters that include down payment ratio, the interest rate, mortgage to- income ratio, and the length of the mortgage (Gan &Hill , 2009). Studies on income approach includes (Hulchanski, 1995), (Bogdon & Can, 1997), (Chaplin, 1999), (Bunting, 2004) (Chen, 2010) , (Haffner & Heylen, 2011). more complicated measures combining residual and income approach are studied by (Skaburskis, 2004) (Gan & Hill, 2009) (Abeysinghe & Gu, 2011) (Bramley, 2012), (Heylen & Haffner, 2013).

Life time income affordability

The sub-prime mortgage crisis in year 2008 underlined the significance of developing alternative measures addressing long-term housing affordability. Sub-prime lending focused mainly on short-run affordability backed by easy mortgages. Life time income affordability is well established concept under permanent income and life cycle hypothesis which consider the current income added to the discounted present value of expected future income; income includes both labour and non-labour pay (Guest, 2005). In a housing study on Singapore Abeysinghe & gu (2011) defined life time affordability index as the ratio of lifetime income to house price which indicates informative trends and cycles in housing affordability (Abeysinghe & Gu, 2011). Despite recent emphasis in the literature on the significance of long-term affordability, this paper could not benchmark it due to data constraints.

The Median Multiple

In many part of the world definition of affordable housing is centred on the idea of income affordability. According to Demographic international housing affordability survey (2015) the median multiple is a common metric, suggested by the World Bank and the United Nations which rates affordability of housing by dividing the median house price by aggregate annual median household income (Shlomo, 2014). Affordability is rated on a scale of 0 to 5. The 3.0x median multiple signs that the market delivers a distribution of housing that are subjected to minimal distortions – housing supply is responsive and able to match effective demand (Suraya, 2015).

This study benchmarks median multiplier approach as it has found many literatures and cited by the Housing Buyers' Association as well as in the 11th Malaysia Plan as an appropriate threshold for the affordability of the nation's housing market (EPU, 2015).

1.2 Malaysian affordability housing: A geographical analysis

House price has followed an upward pattern in Malaysia even though the Asian financial crisis of 1997 caused majority of high priced housing being left unsold. However during the subsequent plan period, the targets were increased again (Savills, 2008). The resultant excess in supply assisted to retain the prices of relatively moderate houses low thus expedited access to home ownership. Based on UN Habitat's figures 63.4 per cent of households had access to homeownership in 1980s and it reached to a peak of 88 per cent by 1998 (Doling & Omar, 2012). While from 2000 onwards the Malaysian all-house price has raised steadily and accelerated between 2009 and 2014 (Napic, 2000-2014); While the all-house price grew at a CAGR (compound annual growth rate) of 5.6% between 2000 and 2014, between 2009 and 2014, it grew at a CAGR of 10.1% (2000-2009: 3.1%) (Suraya, 2015).

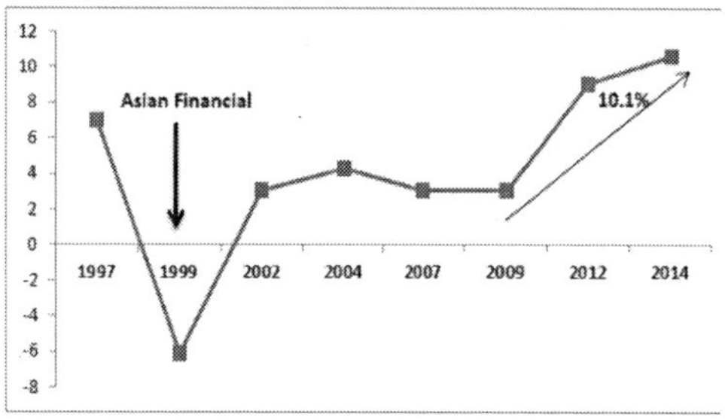

Figure 1: House prices in Malaysia 1997-2014

(CAGR[1])

Given the Malaysian housing price's trend, the question remains as to whether or not Malaysian housing market is affordable. Further answers lie in the formulated map (Figure2) which indicates the geographical distribution of household incomes versus houses offered in the market, otherwise known as the median multiple. Given the heterogeneity of housing markets with location being a key driver of housing prices, it is best analysed according to different submarkets, segmented into different types and localities (Hashim, 2010). For instance, a terrace house in Sabah was three times more expensive in 2014 compared to 2000, while in Selangor, it has grown twice as expensive (NAPIC, 2014b). In general the median price for the Malaysian housing market exceeds the three times median annual household income threshold for affordability. In 2014, it stood at 4.4 times, and has consistently exceeded 4.0 times from 2002 to 2014 (NAPIC, 2014a).

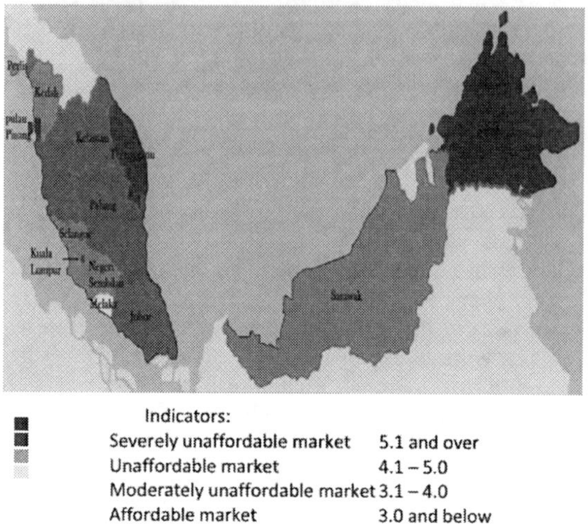

Figure 2: Malaysian Map of Affordability

As highlighted in the figure 2 , Penang and Sabah is above the national statistic whereas Malacca has a median multiple affordability of 3.0. Within the list of state housing markets being assessed, Kuala Lumpur, Pulau Pinang and Sabah stand out as 'severely unaffordable' markets, with median multiples of 5.4 and 5.2 respectively.

One of the reasons behind this extent of unaffordability would be the unresponsiveness of housing supply to effective demand. The lack of houses launched below the three times median multiple prices combined with a high number of high-ended launches contributes towards the severely unaffordable state of housing market (Liew & Haron, 2013; Teck-Hong, 2012). In 2014, of the new properties launched in Kuala Lumpur there were no properties priced under the RM250,000- RM1 million, with the majority of newly launched properties sited in the RM500,000-RM1 million bracket . Given that the three times median multiple price in Kuala Lumpur in 2014 would have been RM274,320, the absence of houses launched below RM250,000 would have skewed the distribution of house prices in the city to the right significantly.

Meanwhile, as highlighted in the map the new housing units launched in Selangor, Negeri Sembilan and Melaka markets are within a more dispersed range of price brackets with a significant

[1] compound annual growth rate based

number of new launches located at or below three times median multiple in Melaka.

2 Malaysian Housing Schemes

Following Malaysian Independence in 1957, housing policy primarily concentrated on public housing, mainly directed as a privilege for public sector officials. During just a few years, it mostly focused on the expansion of a home-owning democracy incorporating all segments of the nation envisioning the future development of the country.

Table 1: existing housing schemes in Malaysia

Schemes/Policies	Objectives/Description	Targeted Income
Low Cost and Affordable Public Housing Program	establish residential amenities to low income household as a way to eradicate poverty and enhance quality of life	<RM1,500
Projek Perumahan Rakyat 1 Malaysia (PR1MA)	Created to develop and contain affordable housing for middle income household in main urban hubs.	2,500-RM7,500
My First Home Scheme	Purposed to help young adults who have newly entered to workforce to buy their first residential property. - Up to 100% financing from financial institutions.	<RM5000
MyHome Scheme	Target low income group- Encourage private developer to build lower cost houses	2500-6000 (Depend on location)
Rumah Mesra Rakyat 1 Malaysia (RMR1M)	Target low income group- Government subsidised between RM15,000 and RM20,000 for the low- income group to create houses valued between RM45,000 to RM65,000	<RM1,500

Housing strategies were established through a consecutive 5 years agenda. Following the overall objective of "adequate, quality and affordable houses to all Malaysians" the policies have realized that the main problem is not in providing housing for higher income households as their purchasing power attracts private developers to safeguard sufficient supply (Agus, 2002). Rather, the policies have focused on attaining sufficient supply of low cost housing that suit the purchasing power of lower income people. Central to this has been the classification of house price involving a targeted range of dwellings identified (Ubale & Martin, 2012)

Within this context, the Malaysian government has undertaken numerous measures for expanding home ownership. establishment of CAGAMAS (National Housing Corporation) in 1986 has been a prominent act which is jointly held by the National Bank and private financial bodies and purposed to provide security to those initiating housing loans (CAGAMAS, 2013). Commercial banks stood for about 80% of housing loans in the 2000s, their businesses is majorly backed by CAGAMAS despite presence of an dynamic private financial sector. Besides, other initiatives ranging from 'Projek Perumahan Rakyat' (PPR) (People's Housing Projects) to My First Home Scheme, Malaysian housing policy has concentrated on either direct provision of low-cost housing, or subsidising the cost of housing for home-buyers.

As portrayed in the table 1, the housing policies have overemphasized the provision of low cost housing that there exist less medium cost house available in the market. Gaps have begun to appear in the system, embodied by the rising concern of middle-income group who are neither entitled for social housing nor capable of affording private sector-supplied houses (Abdullahi & Aziz, 2011). The challenge is highly prevalent in urban areas: while Malaysian home ownership as a whole stood at 72.5% in 2010, urban home ownership was 69.1% and in Kuala Lumpur, it was 53.5%.

3 Methodology

This study was conducted during September 2015-January 2016 to gather literature related to affordable housing in Malaysia and worldwide. Plenty of research has been done in different area of this topic; study has emplaced major concentration on policy dimension of the topic. Articles were retrieved from diverse platform which include book chapters, journals, technical reports, institution's database, news and etc. (N=70). A group of reviewers independently read articles and recorded the main findings on affordable housing policies globally and in Malaysia. The findings are based on the formulated form attached (Appendix).

DOI: 10.1051/matecconf/20166600010

4 Conceptual Framework

Provision of affordable housing cover property rights, regulation, housing finance, and government subsidies. Based on global policy reviews majority of government's initiative can be categorized into supply or demand side measures. Key words searched are within this framework

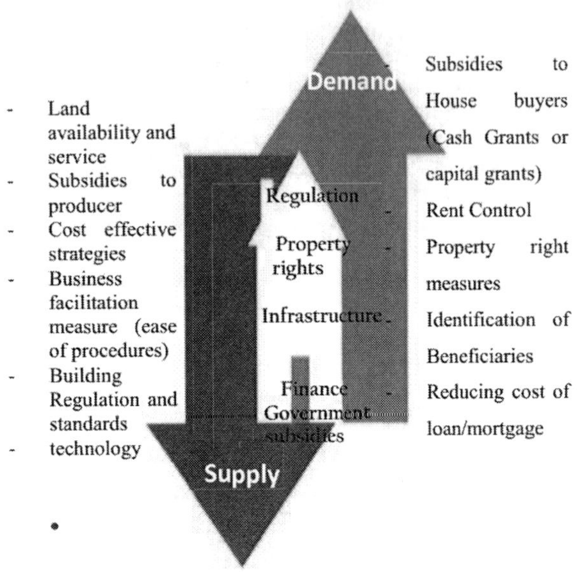

Figure 3: Element of demand and supply side schemes

5 Discussion and Analysis

Based on selected literatures key discussion and analysis can be presented in term supply and demand oriented housing schemes.

5.1 Supply side schemes

Supply side policy practices mainly encompass land and regulation related issues, minimizing housing development cost and operation and maintenance cost. Countries have adopted various approaches, some are discussed below[2]

A number of countries have successfully developed land base instrument to enhance affordable housing for all income segment; among which transit orient development in Hong Kong, releasing government owned land in Turkey, land regulation to unlock private land in Philippine and China, addressing informal land and upgrading land registration methods adopted by Singapore, Japan, South Korea and India are prominent.

1. Land Related Measures		
Key strategies	**Countries case studies**	**Related studies**
Transit-oriented development	Hong Kong , New York	(Loo & Chen, 2010) (Peterson, 2006) (Türel, 2012) (Deng, McMillen, & Sing, 2012) (Usavagovitwong, Pruksuriya, & Mcgranahan, 2013) (Lerman, 2006)
Release of publicly-owned land	Turkey, China	
Land Regulation and taxation	Philippine , China	
Registration of land titles	Singapore	
Land re adjustment	Japan, south Korea, India	
Inclusionary zoning	Spain, U.S	

For instance Honk Kong has built thousands of homes in the new territories across transportation infrastructure to enhance connectivity and address mismatch between housing and jobs. In places where transit facilities are established, land values have raised accordingly (In Honk Kong land values in the immediate area upsurge by 40-60 per cent) therefore more subsidies will be needed to offset its unaffordability.

Meanwhile in Turkey publicly own lands with strategic locations are released by selling or leasing them for affordable housing development. TOKI housing agency has gathered 4 per cent of urban land through government entities which is developed in partnership with private agencies based on revenue sharing schemes. However despite its bright side, stakeholder involvement is challenging, there might be risk of oversupply and market distortion in price setting. Likewise land registration in Singapore, land regulation and taxation practiced in China and Philippine enforce charges on idle land to minimize speculative land hoarding.

With the exception of property taxes, these instruments do not provide a steady stream of income, but they nonetheless have allowed many cities to fund large investments in infrastructure in the absence of other sources of finance.

[2] our result does not include the practices in all countries- it's based on selective articles (there are more than aforementioned countries which might practice it.

2. Cost Effective Housing development (Long term Supply measures)		
Key strategies	**Countries case studies**	**Related studies**
Design and standardization guidelines	Singapore	(Woetzel, 2014)
Labour skill building programs	Malaysia, India	(Barber, 2004)
Productivity improvement programs	United State	(Hilber, 2015)
encourage small and medium enterprises and Engagement of International firms	Saudi Arabia, Egypt , India	(Rust & Koen , 2011)
Innovation in Technology	South Africa	

Cost effective housing development measures requires design to value techniques standardization, capital and labour productivity measures. It generates potential savings for an ever lucrative economics of affordable housing that incites developers to consider housing beyond mid- high income customers.

For instance in Singapore using design and standardization guidelines developer can operate at optimum scale through repetitive procedures. However this approach will require customization of regional nuance and customer needs to minimize monotonousness.

Meanwhile in Malaysia labour skill building programs assist in offsetting shortage of skilled labour. Malaysia caught in an interesting dilemma; on one hand there is huge brain drain of young manpower while on the other hand there is severe shortage of skilled workforce especially in construction industry. Therefore the quick and convenient solutions have been importation of foreign labour and adopting labour skill building programs.

Furthermore United State establishes mechanisms that support creativity, innovation as a knowledge sharing platform. This is carried out through training projects, innovation hubs, demonstration programs and etc. However In adopting this policy ability of industries to adopt new approaches should be assessed against cost of setting up the program.

Nonetheless, China, Saudi Arabia, India , Egypt enhanced support for small and medium industry and engage international players in construction industry which serves as a capacity building platform that improve competitive landscape.

South Africa housing policy reviews indicate innovation in material technology that supports expansion of cost-effective and viable construction materials supplying to the affordable housing sector. For effectiveness of this measure it is significant to ensure that developed products are deployed at optimum rather than being an experimental demonstration.

3. Maintenance and operation cost (Short Term Supply measures)		
Key strategies	**Countries case studies**	**Related studies**
Technical assistance	Brazil, United Kingdom	(Chavez, Khemici, Khater, & Keshishian, 2012) (Clinch & Healy, 2000) (Woetzel, 2014) (Iwaro & Mwasha, 2010)
Energy efficiency measures	United Kingdom , Ireland	
Maintenance quality standards	United Kingdom	
Community management	United Kingdom, China	

Despite incorporating cost saving mechanism in developing houses, additional cost can be saved in operation and maintenance that accounts for a significant percentage of annual housing expenses. It is relatively fragmented industry with relatively subscale and inefficient operator. By pooling demand for this service, business might be encouraged to scale up.

Minimizing such cost by improving energy efficiency and asset management measures can reduce dilapidation and assist in preserving housing stocks.

For instance in United Kingdom and United State government initiatives have subsidized low income groups to furnish homes with energy saving tools to cut energy cost with a two to one return on investment. Besides, The UK social housing buying consortia, UK decent home criteria are prominent programs.

5.2 Demand side schemes

Demand side initiative that is widely adopted by many countries encompasses financial elements, targeted Subsidies, and rental control. Some are discussed below

1. Reduce loan/mortgage cost		
Holistic income assessment	India	(Woetzel, 2014) (Michael & Graham, 2011) (Leong, 2014) (Zenou, 2010)
Digitized mortgage processes	China	
Mortgage liquidity facility	Malaysia	

Mortgage backed securities	United State	

2. Leverage collective Saving		
Housing provident funds	Singapore	(Woetzel, 2014)
Contractual savings schemes	Europe	(Phang S, 2007)

1. Enhance Rental Market		
Rental Subsidies	United State, Netherland	(Monro, 1997) (Woetzel, 2014)
Renter protection measures	Germany	
Rent to own initiatives	United State	

2. Identification of beneficiaries		
Eligibility Database	United State	(André, 2010) (Y. Chen & Sönmez, 2004)
Prioritization scheme	United State	
Allocation Mechanism	Turkey	

3. diversify source of funding		
Tax Increment Financing	United State	(Ezeanya, 2004) (Abdul-Aziz & Kassim, 2011) (Woetzel, 2014)
Public private partnership	Malaysia	
Linkage fee and taxation	United State	

For instance in India holistic income assessment is conducted to target significant proportion of unbanked population or informal sector. Meanwhile in China digital and online networks can initiate a widespread access to banking facilities for low-income households and minimize serving cost. For successful implementation of this scheme massive investment in technology, infrastructure, and education is required. In United States costs Mortgage-guarantee schemes are created to minimize the risk to lenders (offer loans with lower interest rate) by safeguarding them in case of any default.

Malaysia creates wholesale loans to banks given their loan portfolios, or secures the portfolios and converts them into government-backed bonds. Although it might cause market distortion, it assists primary market investors in need of long-term funding.

In Singapore housing provident funds initiated mandatory saving scheme in form of social security. It can offer housing loans at below-market interest rates.

United States has established rental control, diversifying source of funding and appropriate system for identification of beneficiary. Meanwhile Malaysia has proposed Public private partnership in diversifying source of fund which required distributing risks along the housing value chain and . It requires public sector competence and skills in management for its effective implementation.

All of the aforementioned schemes require vigilant design for managing systemic risk. And for a longer term and sustainable schemes, it is vital to have a stable macroeconomic setting that can contain inflation. So where is Malaysia in global affordable housing practices? What is the strength and lessons to be learnt?

Although an interplay of supply-side and demand-side schemes are instrumental in the effective provision of affordable housing, in developing countries experiencing rapid growth including Malaysia; when markets are tight, demand-side subsidies might, indeed, intensify the affordability issue for nonsubsidized low-income households as rents or prices increase. Their benefits will then accrue to landlords or developers, with little impact on the overall housing supply. They perform best at markets where they can incite a supply response.

Evidently analysing, based on the tables above, one can realize that Malaysia has undertaken relatively stronger demand side schemes which is highlighted in the global housing policy reviews. Among which access to home financing via mortgage liquidity facility and diversification of fund sources via public private partnership is prominent. Though Malaysia has undertaken tremendous steps in effective provision of housing, less progress has is seen in addressing the factors that constrain supply, which often thwart improvements on the demand side. Therefore cost effective supply side initiatives mainly driving productivity via increasing technology adoption, modernisation of construction methods and reducing reliance on low-skilled workforce, adaptation of IBS (international building standards) and facilitating construction related business procedures are highly recommended.

6 Conclusion

This paper reviewed affordable housing issues in Malaysia and formulated map of affordability to enable policy makers to consider strategies targeting issues pertaining particular state or geographical area. This enlightens policy makers to carefully design policies and account for urban-rural variance in affordability.

Drawing on literatures from both developed and developing world, this study attempted to provides an overview of the policy schemes that are influential in determining housing market performance. This is done through a supply and demand side policy schemes with the view of safeguarding that policies should reinforce supply and demand side factor toward equilibrium. In case of Malaysia supply oriented initiative are of urgency in short and medium run. This must be supported by long term demand side schemes in parallel. Convergence of this two factors are essential for a balanced equilibrium and obtaining affordability.

Key housing challenge in Malaysia is majorly related to mismatch in demand and supply for affordable housing. Supply side initiatives need more attention in Malaysia mainly in areas related to cost effective housing development measures, maintenance and operation cost and regulation related matters. This paper therefore argues that it is significant to harmonize supply and demand side factors in the housing market to ensure housing supply fits the needs of citizens based on the location, price and target group. Supply and demand side surveys should be conducted in every state to account for regional variations.

References

1. Abd Aziz, D. W., Hanif, N. R., & Singaravello, K. (2011). A Study of Affordable Housing within the Middle Income Households in the Major Cities and Towns of Malaysia. *Australian Journal of Basic and Applied Sciences, 5*(8), 258-267.
2. Abdul-Aziz, A. R., & Kassim, P. J. (2011). Objectives, success and failure factors of housing public–private partnerships in Malaysia. *Habitat International, 35*(1), 150-157
3. Abdullahi, B. C., & Aziz, W. N. A. W. A. (2011). The role of private sector participation in achieving anticipated outcomes for low-income group: A comparative analysis of housing sector between Malaysia and Nigeria. *African Journal of Business Management 5*(16), 6859-6890.
4. Abeysinghe, T., & Gu, J. (2011). Lifetime income and housing affordability in Singapore. *Urban Studies, 48*(9), 1875-1891.
5. Agus, M. R. (2002). The role of the state and market in the Malaysian Housing Sector. *Journal of Housing and nthe Built Environment,* (**17**), 49-67.
6. André, C. (2010). A bird's eye view of OECD housing markets. from http://www.iut.nu/Literature/2010/HousingMarket_inOECDregion_2010.pdfApgar,
7. G., William, C. . (1990). Which Housing Policy Is Best? *Housing Policy Debate, 1*(1), 1-32.
8. Bajunid, A. F. I., & Ghazali, M. (2012). Affordable Mosaic Housing: Rethinking Low-Cost Housing. *Procedia-Social and Behavioral Sciences, 49*, 245-256.
9. Barber, J. (2004). Skill upgrading within informal training: lessons from the Indian auto mechanic. *International Journal of Training and Development, 8*(2), 128-139.
10. Beer, A., Kearins, B., & Pieters, H. (2007). Housing affordability and planning in Australia: the challenge of policy under neo-liberalism. *Housing Studies, 22(1), 11-24., 22*(1), 11-24..
11. Bogdon, A. S., & Can, A. (1997). Indicators of local housing affordability: Comparative and spatial approaches. *Real Estate Economics, 25*(1), 43-80.
12. Bramley. (2012). Affordability, poverty and housing need triangulating measures and standards. *Journal of Housing and the Built Environment, 27*, 133-151.
13. Buhaug, H., & Urdal, H. (2013). An urbanization bomb? Population growth and social disorder in cities. *Global Environmental Change, 23*(1), 1-10.
14. Bunting. (2004). The uneven geography of housing affordability stress in Canadian metropolitan areas. *Housing Studies 19*(3), 361-393.
15. CAGAMAS. (2013). HOUSING THE NATION Policies, Issues and Prospects. from http://www.cagamas.com.my/sites/default/files/publications/Housing%20the%20Nation%20-%20Policies,%20Issues%20and%20Prospects.pdf
16. Chaplin, R., and Freeman,A. . (1999). Towards an accurate description of affordability. *Urban Studies, 36*(11), 1949-1957.
17. Chavez, J., Khemici, O., Khater, M., & Keshishian, P. (2012). *Building Codes and Relative Seismic Vulnerability in Latin American Countries*. Paper presented at the 15WCEE Lisboa.
18. Chen. (2010). Assessing housing affordability in post-reform China A case study of Shanghai. *Housing Studies, 25*(6), 877-901.
19. Chen, Y., & Sönmez, T. (2004). An experimental study of house allocation mechanisms. *Economics Letters, 83*(1), 137-140.
20. Clinch, J. P., & Healy, J. D. (2000). Domestic energy efficiency in Ireland: correcting market failure. *Energy Policy, 28*(1), 1-8.
21. Deng, Y., McMillen, D. P., & Sing, T. F. (2012). Private residential price indices in Singapore: a matching approach. . *Regional Science and Urban Economics, 42*(3), 485-494.
22. Doling, J., & Omar, R. (2012). Home ownership and pensions in East Asia: The case

of Malaysia. *Journal of Population Ageing, 5*(1), 67-85.
23. DOS. (2014). Household Income Survey. from https://www.statistics.gov.my/index.php?r=column/ctwoByCat&parent_id=119&menu_id=amVoWU54UTl0a21NWmdhMjFMMWcyZz09
24. EPU. (2015). Providing Adequate and Quality Affordable House- Strategy Paper 6 *11th Malaysian Plan*.
25. Ezeanya, A. C. (2004). *Malaysian housing policy: Prospects and obstacles of national vision 2020*. Paper presented at the International conference of Adequate and Affordable Housing for All., Kuala Lumpur - Malaysia
26. Gan, Q., & Hill, R. J. (2009). Measuring housing affordability: Looking beyond the median. *Journal of Housing Economics, 18*(2), 115-125.
27. Guest, R. S. (2005). A life cycle analysis of housing affordability options for first home owner-occupiers in Australia. *Economic Record, 81*(254), 237-248.
28. Gyourko, J. (2009). Rethinking Federal Housing Policy How to Make Housing Plentiful and Affordable. Habitat-International. (2014). Global housing indicators: Evidence for action. from http://www.habitat.org/sites/default/files/gov_global_housing_indicators.pdf
29. Haffner, & Heylen. (2011). User costs and housing expenses. towards a more comprehensive approach to affordability. *Housing Studies, 26*(4), 593-614.
30. Hashim, Z. A. (2010). House price and affordability in housing in Malaysia. *Akademika*(**78**), 37-46.
31. Heylen, & Haffner. (2013). A ratio or budget benchmark for comparing affordability across countries. *Journal of Housing and the Built Environment, 28*(3), 547-565
32. Hilber, C. A. (2015). UK Housing and Planning Policies: the evidence from economic research. from http://cep.lse.ac.uk/pubs/download/ea033.pdf
33. Hulchanski, J. D. (1995). The concept of housing affordability: six contemporary uses of the housing expenditure-to-income ratio. *Housing Studies, 10*(4), 471-491.
34. Iwaro, J., & Mwasha, A. (2010). A review of building energy regulation and policy for energy conservation in developing countries. *Energy Policy, 38*(12), 7744-7755.
35. Keivani, R., & Werna, E. (2001). Modes of housing provision in developing countries. *Progress in planning, 55*(2), 65-118.
36. Khairie, H. (2013). 40% gap between affordable housing supply and demand. Retrieved 15 December 2015, from http://www.kinibiz.com/story/quotes/45938/%E2%80%9840-gap-between-affordable-housing-supply-and-demand%E2%80%99.html
37. Leong, c., c. . (2014). *Building Sustainable Housing Finance Systems: Malaysian Experience*. Paper presented at the 6th Global Housing Finance Conference.
38. Lerman, B. R. (2006). Mandatory inclusionary zoning-the answer to the affordable housing problem. *BC Envtl. Aff. L. Rev., 33*, 383-394.
39. Liew, C., & Haron, N. A. (2013). Factors Influencing the Rise of House Price in Klang Valley. International Journal of Research in Engineering and Technology (IJRET), . *International Journal of Research in Engineering and Technology (IJRET), 2*(10), 261-272.
40. Lim, X. Y., Olanrewaju A.L., Tan S.Y.,. (2015). Strategies for Affordable Housing Delivery. *Aust. J. Basic & Appl. Sci., 25*(25), 118-124.
41. Loo, B. P., & Chen, C., & Chan, E. T. . (2010). Rail-based transit-oriented development: lessons from New York City and Hong Kong. *Landscape and Urban Planning, 97*(3), 202-212.
42. Michael, M., & Graham, T. (2011). Affordable land and housing in Asia *United Nation human settlement program* (UN-HABITAT), 2011,.
43. Monro, D. (1997). Public rental housing policy: Learning the lessons from overseas. from http://www.aph.gov.au/About_Parliament/Parliamentary_Departments/Parliamentary_Library/pubs/rp/RP9798/98rp06
44. Naoyuki, Y., & Matthias, H. T., A. . (2015). Housing Policies for Asia: A Theoretical Analysis by Use of a Demand and Supply Model. Tokyo-Japan: Asian Development Bank.
45. Napic. (2000-2014). Property Market Status Report, 2014.fromhttp://napic.jpph.gov.my/portal/web/guest/publication
46. NAPIC. (2014a). Malaysian House Price Index. fromhttp://napic.jpph.gov.my/portal/web/guest/publication
47. NAPIC. (2014b). Summary of Existing Stock for Residential, Commercial, Leisure and Industrial property Sub-sector as at Q1-Q4 2013. from http://napic.jpph.gov.my/portal
48. Peterson, G. E. (2006). Land leasing and land sale as an infrastructure-financing option *Policy Research Working Paper 4043*: World Bank.
49. Phang S, Y. (2007). The Singapore Model of Housing and the WelfareState. from

DOI: 10.1051/matecconf/20166600010

http://ink.library.smu.edu.sg/cgi/viewcontent.cgi?article=1595&context=soe_research

50. Ramli, A., Akasah, Z. A., & Masirin, M. I. M. (2014). Factors Contributing to Safety and Health Performance of Malaysian Low-cost Housing: Partial Least Squares Approach. *esearch Journal of Applied Sciences, Engineering and Technology, 7*(21), 4612.
51. Savills. (2008). Asian Housing Review 2008: Malaysia. Hong Kong. Retrieved 05.01.16, 2016, from http://s3.amazonaws.com/zanran_storage/www.rics.org/ContentPages/40807417.pdf
52. Shlomo, A. (2014). 11th Annual Demographia International Housing Affordability Survey *Demographia* Stern School of Business, New York University.
53. Shuid, S. (2009). Changing structure of low income housing provision in Malaysia: housing allocation under the computerised Open Registration System (ORS) for low cost house buyer. In H. Conference (Ed.): Cardiff.
54. Skaburskis. (2004). Decomposing Canada's growing housing affordability problem Do city differences matter. *Urban Studies, **41***(1), 117-149.
55. Stone, M. E. (2008). What is Housing Affordability? The Case for the Residual Income Approach. *Housing Policy Debate, **17***(1), 151-184.
56. Suhaida, M. S., & Tawil, N. M. H., N. Che-Ani, A. I. Tahir, M. M. . (2010). A Conceptual Overview of housing affordability in Selangor, Malaysia. *International Journal of Social, Behavioral, Educational, Economic, Business and Industrial Engineering, **4***(12), 2289-2291.
57. Suraya, I. (2015). making housing Affordable: khazana Research Institute
58. Teck-Hong, T. (2012). Housing satisfaction in medium-and high-cost housing: The case of Greater Kuala Lumpur, Malaysia. *Habitat International, **36***(1), 108-116
59. Türel, A. (2012). *High housing production under less regulated market conditions in Turkey*. Paper presented at the ERES 19 thAnnual Conference, Turky.
60. Ubale, M. Y., & Martin, D. J., & Seow, T. . (2012). *The current practices of the Malaysian formal low cost housing provision system*
61. Paper presented at the International Conference of Technology Management, Business and Entrepreneurship Kuala Lumpur Malaysia.
62. Usavagovitwong, N., Pruksuriya, A. A., & Mcgranahan, G. O. R. D. O. N. (2013). Housing density and housing preference in Bangkok's low-income settlements. In W. Paper (Ed.). Bangkok: International Institute for Enviroment and Development
63. Woetzel, J. R., S. Mischke, J. Garemo, N. Sankhe, S. . (2014). A blueprint for addressing the global affordable housing challenge. Shanghi: McKinsey Global Institute.
64. Zaid, N., Graham, P. (2011). Low-Cost Housing In Malaysia :A Contribution to Sustainable Development? *Proc., Energy, Environment and Sustainability*, 82-87.
65. Zenou, Y. (2010). Housing policies in China: Issues and options. . Retrieved 01.01.16, from http://www.ifn.se/wfiles/wp/wp824.pdf

Achieving Sustainable Value Planning For Malaysian Public Projects

Faudzi Muhammad[1,a], Rohana Mahbub[1], Abdul Hadi Nawawi[1] and Nazirah Zainul Abidin[2]

[1]*Faculty of Architecture, Planning and Surveying, Universiti Teknologi MARA (UiTM), 40450 Shah Alam, Selangor, Malaysia*
[2]*School of Housing, Building and Planning, Universiti Sains Malaysia (USM), 11600, Penang, Malaysia*

Abstract. Sustainability is the central development issue in the modern economy. Through sustainable development, quality of life can be improved or maintained over time. Since Malaysia is targeting to become a high-income nation by the year 2020, financial investment in public projects should be planned comprehensively so that it will generate immediate and long-term benefits to the country and the people. Within the currently tight financial environment, achieving value for money in public spending is seen as one of the enablers to maintain the right momentum of economic growth. Previous studies have established the importance of integrating sustainability consideration into Value Planning protocol in order to achieve value for money, underpinned by the sustainable development agenda. Despite the establishment of the framework for the integration, the opportunity of such integration within the Malaysian Value Planning protocol for public projects remains unclear. The present state of sustainability consideration within the Value Planning practice should be first evaluated, so that potential interventions to enhance the integration can be introduced. Responding to the gap, this exploratory study was conducted. The data was collected by means of document analysis, interviews and observations; subsequently analysed using the Template Analysis technique. Based on the current practice of Value Planning in Malaysia, ten interventions are proposed to transform the present practice into Sustainable Value Planning. Sustainable Value Planning is seen as a comprehensive concept in achieving value for money in public spending underpinned by the overarching concept of sustainability

1 Introduction

In modern economy, sustainability has become the central issue in improving the quality of life. Sustainability is the result of Sustainable Development; and defined as the capability of continuity [1]. It is formed by the Triple Bottom Line themes of economy, society and the environment [2]. In the global arena, the World Economic Forum introduced the Sustainability Adjusted Global Competitiveness Index in 2012 as complementary information to the traditional Global Competiveness Index (GCI) in comparing the economic performance between countries around the globe [3]. In this evaluation system, the economic performance of a particular country is not measured solely based on its competitiveness score; rather, the social sustainability coefficient and environmental sustainability coefficient are integrated into the competitiveness score in order to get the sustainability adjusted GCI score. Within the Malaysian context, the New Economic Model (NEM) adopted Sustainability as one of its goals beside high-income target and inclusiveness in order to consistently improve the people's quality of life [4].

Malaysia has officially recognised sustainability as its national development agenda since 1981 through the Forth Malaysia Plan [5]. Targeting the right momentum to achieve high-income nation status by the year 2020, the cur Malaysia Plan highlights a few initiatives to encapsulate all efforts towards sustainable development as underpinned by the NEM. Amongst others, Value Management (VM) has been chosen as the strategic management tool to achieve value for money in government development programmes and projects [6]. Within the context of construction projects, the EPU Circular No 3/2009 [7] made it mandatory for all government projects worth RM50 million and above to undergo the VM Study starting from the year 2010. In general, VM is a structured multi-disciplinary value methodology aiming for resources optimisation in achieving the predetermined project functional specification at the acceptable level of quality [8,9]. It focuses on selecting the most economical solution from various alternatives resulting in the same quality of functions. By analysing the functional needs of the project system and its sub-systems, wastages can be identified and eliminated. The main focus of VM is to get the lowest possible investment on a particular project without compromising its functional performance and quality [10].

According to the EPU VM Guide, VM in the context of Malaysian public projects is an umbrella term that covers all types of Value Studies conducted in accordance to the internationally recognised value methodology. These value studies can be categorised as

[a] Corresponding author: faudzimd@gmail.org

Value Planning, Value Engineering and Value Review based on its focus and the moment at which the study is conducted [11]. Based on the project context, Value Planning is the front end VM and conducted during the early project planning stage. Due to various advantages of considering sustainability needs during the early project planning stage [12], Value Planning provides a perfect opportunity to integrate sustainability considerations into project proposals. Despite its strategic opportunity, there is no direct indication found in any major government policies or circulars demanding sustainability integration within Value Planning. However, it might not necessarily mean that Value Planning in Malaysia does not consider sustainability issues at all. Hence, the actual state of sustainability integration within the Malaysian Value Planning remains unclear. Without such knowledge, initiatives to nurture the integration of sustainability and Value Planning could not be effectively planned.

In this case, a dedicated exploratory study that explores the scenario within its original setting is required [13,14]. Responding to the need, this study is conducted; where it aims to answer the questions of (1) "What is the state of sustainability integration within the Malaysian Value Planning?" and (2) "What are the potential interventions that might nurture the integration of sustainability considerations into the Value Planning practice in Malaysia?" In order to achieve the aim of this study, two objectives have been identified to steer the flow of the study; the two objectives are:

- To explore the current state of sustainability integration within Malaysian Value Planning
- To propose potential initiatives towards Sustainable Value Planning in Malaysia

2 Sustainable Value Planning

Sustainable Value Planning introduced by this paper is a conceptual model resulting from the expansion of the original works by Zainul Abidin in integrating sustainability considerations into Value Study practice [1,15,16]. It is learnt that integrating sustainability considerations into Value Study would include input of sustainability, processing sustainability and output of sustainability. Input of sustainability refers to the scenario where sustainability awareness is being imbedded into the scope of the Value Study. Processing Sustainability is where the sustainability themes that reflect the economic benefit, social wellbeing and environmental protection are integrated into the Value Study process. Output of sustainability highlights the quality of the projected outcome of the Value Study's product that consists of tangible and intangible values of the proposed project. In other words, output of sustainability defines the perceived project sustainability performance based on the development proposal conceived through the Value Study. Based on the findings, a conceptual framework for integrating sustainability issues into Value Planning has been introduced by Zainul Abidin and reproduced in Figure 1.

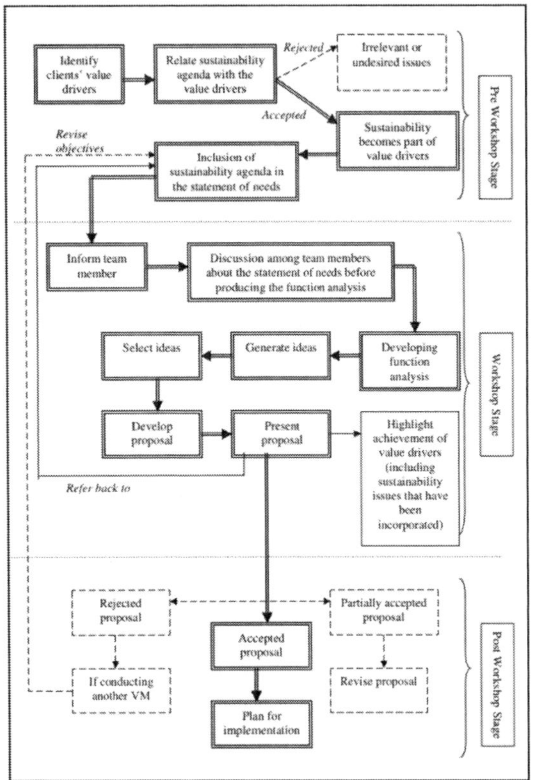

Figure 1. Conceptual Framework For Integrating Sustainability Issues Into Value Planning
Source: Zainul Abidin (2005), Page 179

Through this conceptual framework, the importance of recognising sustainability as part of the value drivers, working with sustainability themes during the course of the study and highlighting the sustainability features for project proposal have been established in order to efficiently integrate sustainability considerations into the Value Study. Although the framework uses the VM terminology to represent the value study, it actually refers to Value Planning since the study was conducted on the VM studies carried out during the early project planning stage. As sustainability deals with meeting present needs without jeopardising the ability of future generations to meet theirs [17], Sustainable Value Planning can be defined as the Value Planning that recognises sustainability themes as its value drivers with the aim to help improve quality of life. Based on this definition, Sustainable Value Planning is concerned about the achievement of the project outcomes on top of traditional Value Planning that focuses on achieving value for money through fulfilling the project objectives. Since Value Planning is conducted during early project planning stage with the structured involvements of key stakeholders as well as experts in various field of knowledge, it is the most suitable moment to integrate sustainability considerations into the project proposal.

The Sustainable Value Planning concept leans towards achieving positive economic, social and environmental impacts of project variables such as design concept, material selection and delivery systems of the construction and facility management to respond towards present and future needs. Due to this nature, problems associated with sustainability issues are ambiguous and unstructured. The whole process of integrating sustainability considerations into the Value Planning process requires dynamic learning processes within the Value Team that involves the real-world view and systems thinking. Due to its complexity, cost savings resulting from the Value Study could not adequately explain the performance of Sustainable Value Planning. In this regard, Simple Multi-Attributes Rating Technique (SMART) VM [18] provides an appropriate framework for Sustainable Value Planning. In SMART VM, multi-attributes are used as the decision criteria in choosing the best project proposal option. Green advocated SMART VM as an innovation to the traditional single attribute VM, where cost is the only determining criterion in selecting options [19].

Just like SMART VM, Sustainable Value Planning should be conducted based on the Soft System Methodology (SSM) paradigm. SSM combines cognitive system thinking and real world perspective in predicting the appropriate solutions for unstructured problems [20]. Figure 2 shows the graphical representative of SSM where the unstructured problem perceived from the real world is analysed and solved through appropriate conceptual models embedded within the individual participant of the Value Study. In this case, Sustainable Value Planning requires the involvement of the right people with the right knowledge and experience. Besides that, the success of Sustainable Value Planning is determined by the ability of the Value Team to nurture dynamic learning processes through effective knowledge sharing amongst the multi-disciplinary Value Team.

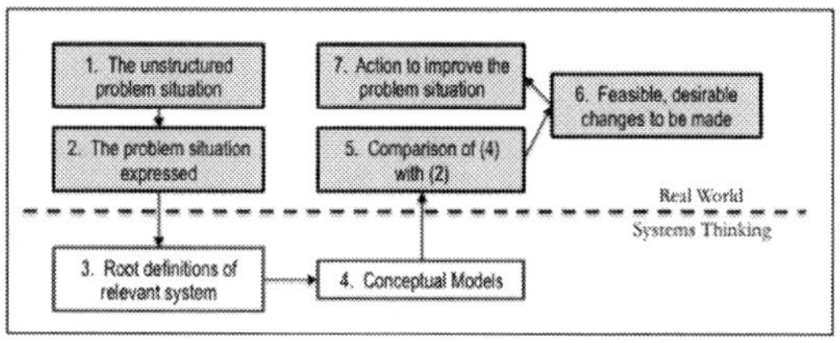

Figure 2. The Soft System Methodology
Source: Adapted from Checkland (1981)

3 Methodology

This study is considered as an exploratory study; an exploratory study is a study "*that aims to seek new insights into phenomena, to ask questions, and to assess the phenomena in a new light*" [13]. To conduct this exploratory study, three data collection methods were adopted namely documentary analysis, observation and interview. Documentary analysis involves desktop analysis of relevant documents and department records relevant to the Value Planning practice for public projects in Malaysia. Five previous Value Assessment Laboratory Reports have been analysed thoroughly besides other department records archived in the Value Management Section of the Economic Planning Unit Malaysia (VMS-EPU). The Economic Planning Unit (EPU) is the central government agency under the Malaysian Prime Minister Department responsible to plan, monitor and evaluate the economic development agenda of the country. Subsequently, controlling the public development expenditure is also one of the agency's responsibilities. A series of semi-structured interviews have been conducted involving the VM practitioners as well as the VMS-EPU officials. In Malaysia, VMS-EPU officials facilitate the Value Assessment Laboratories for public projects. While VM practitioners provide general understanding about the VM concept, VMS-EPU officials provide in-depth information regarding the Value Planning practice for public projects in Malaysia. There were eight participants involved with the series of interview sessions. At the end of the sessions, observations involving five Value Assessment Laboratories for public projects at its practical setting were conducted. These observations covered Value Assessment Laboratories for various natures of projects, from purely building to purely infrastructure construction that were selected for the purpose of this study. As data collected in this study are qualitative in nature, they are analysed using the qualitative data analysis technique. For this reason, the Template Analysis [21] is used. Data collection and analysis in this study have been conducted concurrently to ensure templates are induced dynamically based on the themes emerging throughout the knowledge development process, while available theories provide theoretical underpinning during the analysis process. As validity is one of the five important elements in qualitative research [22], consensus expert validation has been used to ensure the validity of the research's findings (the developed knowledge).

4 The State Of Sustainability Integration Within Malaysian Value Planning

From documentary analysis, it is found that sustainability integration within public project planning in Malaysia is well captured by the Logical Framework Analysis (LFA) Approach introduced by EPU Malaysia through its Circular No 1/2009 [23]. Based on the circular, it is mandatory for every project proponent to conduct the LFA before submitting the proposal for budget approval. Figure 3 shows the summary of LFA for public projects in Malaysia. Based on the circular, the project proponent should develop a project brief for the proposed project to be endorsed at the ministerial level before it is submitted for the budget approval process. The endorsed project brief should be submitted to the EPU through the SPP-II system. A comprehensive project brief should include all items as required by the said circular; the content of the project brief is shown in Figure 3 (Phase 3). The LFA consists of three sequential phases and the project brief is the final product of the LFA. Within Phase 2 of the analysis, the project proponent is required to analyse sustainability considerations of the proposed project that includes financial, economic, social and environmental issues. Before finalising the project brief, a project decision matrix should be established. With in-depth understanding of the sustainability and value concepts, the project decision matrix will induce the project value drivers.

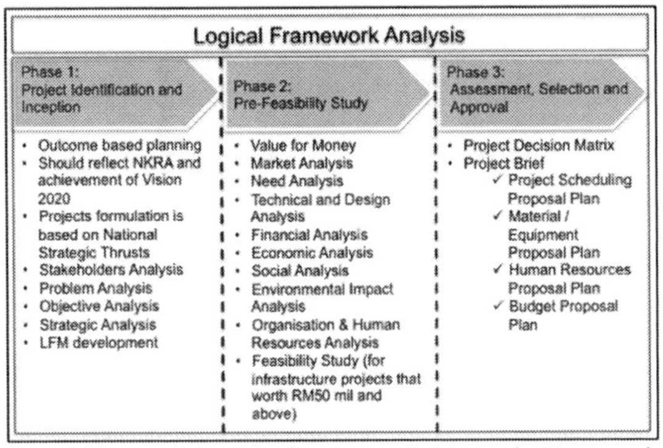

Notes: NKRA = National Key Result Area; LFM = Logical Framework Matrix

Figure 3. Logical Framework Analysis for Public Project Planning in Malaysia
Source: EPU Circular 1/2009

LFA provides good foundation in linking the project objective, expected outcome and the national development agenda. However, through interviews and observations, it is found that the LFA is only used extensively during the Budget Examination Process; which is the first process in evaluating the potential benefits of the proposed project. Although the Budget Examination process involves key stakeholders of the proposed project, it is conducted within a normal meeting setting and the six phases of the value job plan are not involved. Besides that, it is also found that the Budget Examination Process for a particular project is conducted during a meeting lasting just a few hours which is significantly less than the suggested Value Study duration of 40 effective hours. The meeting involves representatives from key stakeholder departments at senior management level. Due to limitation of time, the Budget Examination Meeting does not comprehensively challenge the assumptions made by the proponent while developing the project brief. The main focus of the Budget Examination Meeting is to evaluate the significance of the proposed project at macro level and to identify its potential contributions to the national development agenda. Normally, the meeting will focus on evaluating the expected outcomes and objectives of the project. Technical items such as cost modelling and proposed scope of the project are not comprehensively discussed during the meeting. Projects with high economic potential (normally those directly related to the National Key Economic Areas) and those directly related to the National Key Result Areas of the Government Transformation Programme will be given high priority for budget allocation. Subsequently, projects that worth RM50 million and above or any other projects deemed to be necessary by the committee will be subjected to Value Planning exercise.

Based on the provisional budget approval through the Budget Examination Process, Value Assessment Laboratory will be conducted with the collaboration of the VMS-EPU. The Value Assessment Laboratories are normally conducted at any moment between the stages of 'Preparation & Brief' to 'Developed Design' of the 2013 RIBA Plan of work. However, there are also isolated cases where Value Planning is conducted during early construction stage. In most cases, Value Planning is conducted when VMS-EPU believes that most of the required data, information and knowledge have been accumulated for the proposed project which enables the Value Team to define the perceived problems faced by the project.

During the Value Assessment Laboratory, project parameters are thoroughly discussed and registered in the laboratory templates of 'Project Must', 'Project Given', 'Project Constraint', 'Project Objective' and 'Project Outcome'. Since the 'Project Objective' and 'Project Outcomes' have been established through the LFA, project's sustainability themes can be identified especially within the 'Project Outcome'. Through observations of the five Value Assessment Laboratories, none of them had shown comprehensive analysis of the project outcomes. Normally, the proposed project outcomes will be revisited superficially and endorsed by the laboratory and subsequently registered in the final report. The laboratory emphasises more on the proponent's proposal in achieving the project objectives; this would include evaluating the proposed development cost model as well as the suggested project scope.

Through observations, it is found that the Function Analysis System Technique (FAST) Diagram is not used as a tool during the Function Analysis Phase. FAST Diagramming is a process of logical thinking about the entity as a system and its elements as sub-systems reacting to one another in order to serve the purpose of its existence [24]. With the FAST Diagram, relationships between lower and higher order functions can be explicitly presented. Since Value Assessment Laboratory skips the FAST Diagraming, the Information and Function Analysis Phases of the laboratory seem to be integrated as a single phase where the value team focuses on identifying the project components without considering the importance of the components in achieving the value drivers. Besides that, value drivers are not explicitly discussed and registered during the laboratory. However, since most of the Value Team members are experienced and knowledgeable enough in the nature of the proposed project in all the observed laboratories, the teams have no problem inducing the required project components.

Through observations, it is learnt that the Value Teams induce the project value system based on individual tacit knowledge or solely based on the proponent's original proposal. The danger of using this approach is that public projects in Malaysia might become trapped in the traditional way of doing things, thus limiting the possibility of innovation in the project delivery system. With this approach, the possibility that public projects inherit the recurring problems is high since they are actually adopting the same conceptual

DOI: 10.1051/matecconf/20166600011

models in the project delivery system. This can be seen where most of the ideas accumulated during the Creativity Phase are limited to the project sub-systems level rather than at the project system level. Based on this finding, the Malaysian Value Planning currently leans toward hard system methodology where the ontological aspect of the project system remains untouched.

Since value drivers are not well articulated during the earlier laboratory phases, the Evaluation Phase could only rely on the single-criterion basis in evaluating the accumulated ideas generated during the Creativity Phase; the criterion is the cost implication of the ideas. Although there are efforts identified to evaluate the ideas based on advantages-disadvantages analysis, they are not structured and very much dependent on the subjective interpretation of the team members. Besides that, there are also people with strong characters trying to manipulate the team to accept ideas beneficial to one particular party using the cost sentiment. In the end, cost related comments are found within the advantages and disadvantages columns of the evaluation report such as 'it reduces the project cost' or 'it increases the project cost'. Although there are good ideas accumulated through the Creativity Phase and then evaluated during the Evaluation Phase, it is observed that ideas that challenge the whole project delivery system are limited. From this, it can be concluded that the Value Planning in Malaysia is very much cost oriented.

During the Development Phase, it seems that sustainability features are not well captured within the action plan. There is no clear indication on how the newly conceived project proposal would enhance the achievement of the project's expected outcomes. As most of the sustainability considerations are embedded within the expected project outcomes, thorough analysis on the project proposal towards the achievement of its expected outcome is crucial to highlight the sustainability performance of the new proposal. Within the existing standard template for Value Assessment Laboratory, there are only two items highlighted in the summary sheet; they are the variance of project cost as well as the variance of the project scope by comparing the original proposal and the new proposal conceived by the laboratory. With this information, it is difficult to conclude whether the newly conceived proposal really enhances the project value performance. More importantly, the existing practice of Value Planning fails to highlight contributions of the new proposal towards the national development agenda. In general, it is found that sustainability issues are integrated in some parts of the Malaysian Value Planning. However, the integration in most of the cases is unintentional and unstructured. Hence, Value Planning is Malaysia could not yet to be classified as Sustainable Value Planning.

5 Intervention Needed To Transform Current Practice

Although sustainability integration within the current practice of Value Planning in Malaysia is limited, the potential to achieve Sustainable Value Planning is evident. In order to transform the current practice of Value Planning for public projects into Sustainable Value Planning, ten interventions are identified and will be proposed through this study. The following sub-topics will discuss each intervention separately.

5.1 Establish A Permanent Panel Of Sustainability Experts

In order for Value Planning in Malaysia to address sustainability issues, experts in the field of sustainability should be involved throughout the laboratory sessions. The experts should be identified and a permanent panel of sustainability experts should be formed to provide consultancy regarding sustainability issues for Malaysian Value Planning. These experts should be able to provide accurate knowledge to the Value Team regarding the sustainability concept and are able to forecast the sustainability impact of the proposal conceived by the Value Team. As sustainability is an overarching concept that integrates the three inter-related pillars of economic, social and environmental issues, it involves metacognitive knowledge. Metacognitive knowledge is described as strategic knowledge that involves one's ability to manipulate the cognitive tasks of remembering, understanding, applying, analysing, evaluating and creating [25]. While the fragmented individual knowledge of economics, humanistic and environmental ecology might provide factual, conceptual and procedural knowledge to the Value Team, metacognitive knowledge helps the team to answer the 'how' and 'why' questions to the perceived problems. Current Value Planning practice might involve quantity surveyors as cost estimation experts, officials from the local authority as the local social experts and officials from the Department of Environment as the environmental experts; they might come with pieces of information and knowledge that need to be organised and integrated. It is the task of the sustainability panel to analyse and synthesise the fragmented body of knowledge, thus creating the conceptual models. Using these conceptual models as input, the overall learning process within the Value Team can be enhanced and new knowledge can be created.

5.2 Introduce Sustainability Concept To The Value Team

In order for the Value Team to appreciate the need to integrate sustainability considerations during the laboratory process, awareness regarding the need should be first established. It is found that lack of right knowledge regarding sustainability is one of the main barriers to integrate sustainability into Value Studies [15,26]. Some examples of misconceptions about the sustainability concept are sustainability is all about green buildings, or sustainability is to ensure that the project could be financially sustained in the long-run, or sustainability is achieved when the project uses expensive green technology products, or the overall design is blended with the natural landscape. These are actually the examples of sustainability solutions to the project and not the sustainability themes to be considered during the planning process. It should be emphasised that sustainability solutions are the result of sustainability considerations depending on availability of resources and the actual needs of the project, which are explained by the value themes as part of the project's value drivers.

5.3 Highlight Value Drivers To The Value Team

Value drivers are the key factors that drive the direction of value creation within the value system [27]. They define the client value system and highlight value goals to be achieved by the value study. As Malaysia is pursuing the Sustainable Development agenda [6], the value drivers for public projects in Malaysia should inherit sustainability themes. Value drivers help to explain the relationships between function specifications of the particular project and the definition of project quality; or in simpler terms it helps to articulate the relationships between project objectives and the expected outcomes. Project outcomes for public projects in Malaysia should incorporate sustainability issues [23] as the performance of any Value Planning for public projects in Malaysia is much related to the ability of the laboratory to identify and propose relevant solutions to sustainability problems. While sustainability problems are complex and unstructured in nature, the availability of value drivers might help to define the roots of the problems. With the adoption of SSM and availability of sustainability experts, the Value Assessment Laboratory will be able to identify sustainability solutions based on the interaction of the world-view and systems thinking.

5.4 Establish Value Network That Explains The Project Value System

In order for the Value Team to conceive an efficient project proposal, they should be clearly informed about the project objectives, expected outcomes, constraints, function specifications and boundaries. The present practice successfully captures these pieces of information using the relevant templates. However, the relationship between these variables with the project value system is still vague. It is very helpful for the Value Team if the project value system can be explicitly shown in a single framework. With this framework, the value network can be identified, thus helping the laboratory participants develop new knowledge during the dialogue session. The FAST Diagram and the Value Tree Diagram are examples of Value Network Diagrams that explain the project value system. Despite the success of the current practice in producing laboratory reports at the end of the laboratory sessions without the assistance of the FAST Diagram, the project value system could not be easily understood by people who are not physically involved during the laboratory process. Since Value Planning is conducted at the early planning stage, it is crucial to ensure that the developed knowledge is disseminated throughout the project lifecycle. Besides that, the value network should be shared between laboratories to nurture the continuous learning process in improving the value of public projects.

5.5 Include Project Outcome Analysis As Part Of The Laboratory Scope

In all of the laboratories observed, project outcomes are not comprehensively analysed during the laboratory sessions. The reason is that many facilitators belief that analysing the project outcome is not within the scope of Value Assessment Laboratory and it has been previously endorsed during the Budget Examination Process. Although this perception is to some extent acceptable, it is not entirely accurate. Besides function and cost, value is also related very much to the quality of the project [11]. In the context of public projects, the quality of the project is determined by its outcomes [23]. Thus, it is important for the Value Assessment Laboratory to analyse the expected outcomes of the project through a structured and rigorous process. Towards that, the scope of Value Assessment Laboratory should be expanded to include structure analysis of the project outcomes. This inclusion will help the Creativity Phase of the laboratory to generate radical yet practical ideas in achieving the project outcomes and challenge the suggested objectives. Thus, the new proposal might explore innovations at project system level.

5.6 Determine Evaluation Criteria Based On Value Drivers

Evaluation criteria are important for Sustainable Value Planning since the broadness of sustainability concept might lead to multiple interpretations. Since sustainability involves multiple themes, it is important to determine the weightage of each theme within the project context. The Decision Matrix [19] is one of the evaluation models that can be used during the Evaluation Phase. To ensure coherence within the Value Study, evaluation criteria should be developed based on the project's value drivers and directly related to the Project Decision Matrix developed during the LFA.

5.7 Evaluate The Pereived Sustainability Peformance Of The Project

The quality of the laboratory product should be evaluated to highlight its contributions towards the national development agenda. The Perceived Project Sustainability Performance Indicators (PPSPI) can be used as the measurement indicator for this purpose since it has considered major government policies based on the Triple Bottom Line themes and was specifically developed for Malaysian Value Planning [28]. The PPSPI is shown in Figure 4. The 25 indicators seem to

DOI: 10.1051/matecconf/20166600011

be flexible and comprehensive enough to visualise the sustainability performance for almost all kinds of public construction projects.

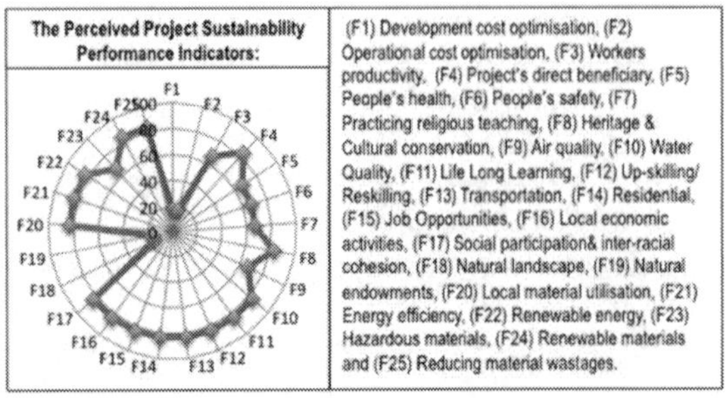

Figure 4. Perceived Project Sustainability Performance Indicator For Value Planning In Malaysia
Source: Wawawi et al. (2015)

5.8 Highlight Sustainability Performance Of The Project To Key Stakeholders

Sustainability performance of the proposed project based on its newly conceived proposal should be highlighted to the key stakeholders during the Presentation Phase. The reason for this is to promote sustainable development awareness for public projects in Malaysia as well as to rationalise wrong perceptions about Value Management. Besides that, it helps to rationalise the variance in project costs as well as the project scope.

5.9 The Perceived Project Sustainability Performance Indicator (PPSPI) To Be Used As Standard Performance Evaluation System For Value Planning In Malaysia

The PPSPI [28] should be used as the standard performance evaluation system for Value Planning in Malaysia. It is suggested that the Sustainability Expert Panel and key stakeholders of the project involved during the laboratory session become the assessors of the newly conceived proposal. The marked radar chart can be used as the graphical representation of the PPSPI result. Figure 4 shows the example of the suggested Perceived Sustainability Performance Report.

5.10 Embed Value Audit Within The Project Action Plan

To ensure that the development proposal conceived through Value Planning transpires in the design, construction as well as operational stages, audits should be conducted at reasonable milestones. Although Value Review (VR) is considered as the value audit for the project, it can only be conducted upon the maturity period of the facility operation [11]. Thus, VR could not be used as a mechanism to ensure that the proposal is accurately interpreted during the design and construction stage. As sustainability initiatives are a continuous process throughout the project life cycle [12], scheduled value audits might be the best platform to ensure sustainability features introduced through Value Planning is well captured throughout the project life cycle.

6. CONCLUSION

It is found that sustainability integration within the current practice of Value Planning for public projects in Malaysia is still limited. In order to nurture the integration, the current practice needs to be transformed. It is believed that Sustainable Value Planning can be achieved in Malaysia with minimum modification of the current practice. Towards that, ten interventions should be strategically integrated into the current Value Assessment Laboratory protocol. Figure 5 shows the overall framework to transform the current practice of Value Planning for public projects in Malaysia into Sustainable Value Planning. Besides the ten interventions, the laboratory facilitator should consistently promote effective knowledge sharing within the multi-disciplinary Value Team to nurture team synergetic relationship as well as enhancing the dynamic learning process. The ten interventions can be strategically embedded into the input, process and product of the Value Planning practice.

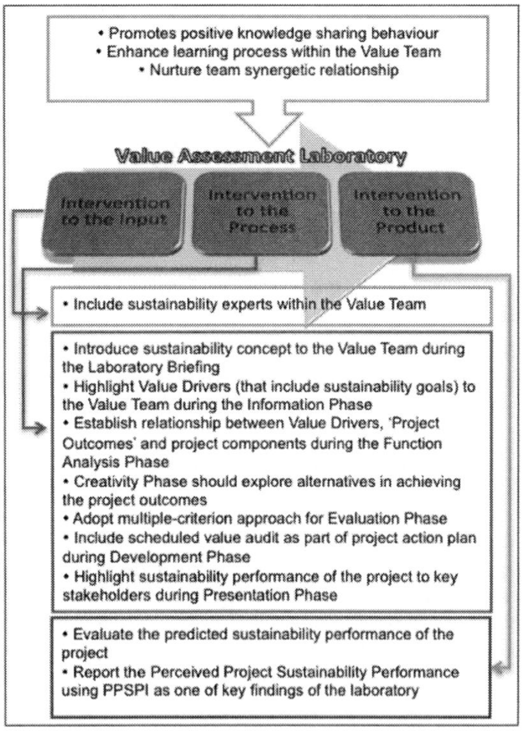

Figure 5. Strategic Intervention To Achieve Sustainable Value Planning For Malaysian Public Projects

References

1 N. Zainul Abidin. *Using Value Management To Improve The Consideration Of Sustainability Within Construction*, PhD thesis, United Kingdom: Loughborough University. (2005)

2 M. Kucukvar, G. Egilmez, O. Tatari. Sustainability Assessment of U.S. Final Consumption And Investments: Triple-Bottom-Line Input-Output Analysis. *Journal of Cleaner Production*, **81**. 234-243. (2014)

3 B. Bilbao-Osorio, J. Blanke, R. Crotti, M. D. Hanouz, B. Fidanza, T. Geiger, C. Ko and C. Serin. Assessing The Sustainable Competitiveness of Nations. *The Global Competitiveness Report 2012-2013*, 49-63. (2012)

4 NEAC Malaysia: National Economic Advisory Council of Malaysia (2010). *New Economic Model for Malaysia – Part 1*. Percetakan Nasional Malaysia Berhad.

5 EPU Malaysia: Economic Planning Unit of Malaysia. *Forth Malaysia Plan: 1981-1985*. (1981)

6 EPU Malaysia: Economic Planning Unit of Malaysia. *Tenth Malaysia Plan: 2011-2015*. (2010)

7 EPU Malaysia: Economic Planning Unit of Malaysia. *Pekeliling 3/2009: Garis Panduan Pelaksanaan Pengurusan Nilai (Value Management)* (2009)

8 SAVE International, *Value Methodology Standard and Body of Knowledge*. (2007)

9 Standards Australia, *AS 4183—2007: Value Management*. (2007)

10. J. Kelly, S. Male and D. Graham,. *Value Management of Construction Projects; Second Edition*, John Wiley & Sons. (2014)

11. EPU Malaysia: Economic Planning Unit of Malaysia. *Panduan Pelaksanaan Pengurusan Nilai dalam Program/Projek Kerajaan*. (2011)

12. RIBA. *RIBA Plan of Work 2013: Overview*. (2013)

13. M. Saunders, P. Lewis, A. Thornhill. *Research Methods for Business Students: Sixth Edition*. Pearson. (2012)

14. J. W. Creswell and V. L. Plano Clark. *Designing and Conducting Mixed Methods Research: Second Edition*. Sage Publications. (2011)

15. N. Zainul Abidin and C. L. Pasquire. Delivering Sustainability Through Value Management: Concept And Performance Overview. *Engineering, Construction and Architectural Management*, **12(2)**, 168 – 180. (2005)

16. N. Zainul Abidin and C. L. Pasquire. Revolutionize Value Management: A Mode Towards Sustainability. *International Journal of Project Management*, **25**, 275-282. (2007)

17. The World Commission on Environment and Development - United Nations. *Our Common Future*. (1987)

18. S. D. Green. A SMART Methodology For Value Management. *Occasional Paper, Chartered Institute of Building* **53**, UK. (1992)

19 S. D. Green. Beyond Value Engineering: SMART Value Management For Building Projects, *International Journal of Project Management*, **12 (1)**, 49-56. (1994)

20. P. Checkland. *Systems Thinking, Systems Practice*. USA: John Wiley. (1981)

21. N. King. Using Templates in the Thematic Analysis of Text. In C. Cassell, and G. Symon, (Eds). *Essential Guides to Qualitative Methods in Organizational Research*. Sage Publications. (2004)

22. J. A. Maxwell. *Qualitative Research Design – An Interactive Approach: Third Edition*. Sage Publications. (2013)

23. EPU Malaysia: Economic Planning Unit of Malaysia. *Pekeliling 1/2009: Garis Panduan Perancangan dan Penyediaan Program dan Projek Pembangunan*. (2009)

24. F. Wojciechowski, *The Various Types and Uses Of The FAST Diagram. Proceedings of the SAVE Conference*. 153-159. (1978).

25. L.W. Anderson, D.R. Krathwohl, P.W. Airasian, Cruikshank, K.A., Mayer, R.E., Pintrich P.R., Raths, J., Wittrock, M.C. *A Taxonomy For Learning, Teaching, And Assessing: A Revision Of Bloom's Taxonomy Of Educational Objectives*. New York: Longman. (2001)

26. Y.M. Al-Saleh, and H.M. Taleb. The Integration of Sustainability Within Value Management Practices: A Study of Experienced Value Managers in the GCC Countries. *Project Management Journal*, **41(2)**, 50–59, (2010)

27. S. Bose, and K.B. Oh. Measuring Strategic Value-Drivers For Managing Intellectual Capital, *The Learning Organization*, **11(4/5)**, 347-356. (2004)

28. Nawawi, A. H., Muhammad, F., Mahbub, R. and Zainul Abidin, N., Perceived Project Sustainability Performance Indicator For Value Planning, *Procedia - Social and Behavioural Sciences*, **202**, 89-97. (2015)

General Adjustment Influence Factor of Malaysian Construction Expatriates Executives Abroad

Halmi Zainol[1,a], Haryati Mohd Isa[2], Kartina Alauddin[2] and Mohmad Mohd Derus[2]

[1]*Town and Regional Planning Department, Faculty of Architecture, Planning and Surveying, Universiti Teknologi MARA, Seri Iskandar Campus, 32610 Perak, Malaysia*
[2]*Quantity Surveying Department, Faculty of Architecture, Planning and Surveying, Universiti Teknologi MARA, Seri Iskandar Campus, 32610 Perak, Malaysia*

Abstract. The success of Malaysian construction companies creates an opportunity to explore abroad. Past studies have shown that the difficulty of expatriates in adjusting to a new environment is the main aspect that leads to failure of assignments. The success in implementing an overseas assignment does not solely depend on an expatriate's technical expertise. The adjustment issues such as the interaction with the host nationals, and adaptability to the host country's culture also exert influence on the assignment. The research was conducted to identify the influence of executive expatriate general adjustment on assignment in host countries. The objective of the study was to identify adjustment influence factors relating to general adjustment abroad. Questionnaires were sent to Malaysian expatriate executives. Sixty four Malaysian expatriate executives from Malaysian construction companies overseas were involved in this study. The findings show interaction, social and living environment influences their adjustment during expatriation. Pre-departure training preparation aspects for expatriates is a good step before their departure to host countries.

1 Introduction

The present internationalization of the construction industry shows that Malaysian firms are getting more involved in construction projects abroad. The involvement of companies to secure projects overseas can strengthen the position of the company. However, to be successful, a company must ensure that individuals selected to work overseas must have the perseverance and must be ever ready to face challenges.

To a great extent, success of the companies depends on the employees assigned to the projects overseas. The efficiency of staff assigned abroad depends not only depends on their technical capabilities, knowledge, and creativity, but also on other non-technical matters that influence the staff in foreign countries. The control over these factors would enable the optimum use of employment for ensuring success of a company [1, 2].

The ability of the expatriates to adjust to general aspects, jobs, and family influences their performance in their assignments [3, 4, 5].The different environmental conditions in the host countries suggest the importance of expatriates of making adjustment so that they can enjoy their normal life. Not all staff sent overseas can adjust themselves to the situation overseas. Therefore, to face the new environmental condition in a foreign country, preparation and adjustment of an expatriate is imperative.

Among the reasons for Malaysian construction companies getting involved in a foreign country is to secure market access in the host country, to increase their market share, to strengthen their networking and to gain experience [6]. The success of a Malaysian construction company to secure a highway project overseas worth RM17.5 billion has opened up opportunities to local Malaysian companies to invest in foreign construction projects [7]. This shows foreign recognition towards the capabilities of Malaysian construction companies. Such opportunities have become the foundation for Malaysian companies to implement foreign projects successfully, apart from increasing their contribution to the national economy. In this connection, the need for Malaysian expatriate to lead this kind of projects has become a priority.

2 Problem Statement

Malaysian multinational construction companies are getting projects in foreign countries such as in the Southeast Asian region, the Middle East, Asia, Africa and other countries [8, 9]. Table 1 shows, up till December 2010, altogether 96 projects with an estimated value of RM40,404.54 million were in operation in foreign countries. The projects were distributed in South Asia (32.3%), Middle East (27.1%), Asian (21%), and the rest

[a] Corresponding author: halmi461@perak.uitm.edu.my

in other countries. According to [6], the fact that 79% of these companies intend to expand their operations overseas within five years is a positive development regarding the involvement of Malaysian multinational construction companies in foreign countries.

Table 1. Involvement of Malaysian Multinational Construction Companies in Foreign Countries

Region	No. of projects	%	Project value (RM mil.)
Asian	21	21.9	6,087.28
Middle East	26	27.1	16,378.80
Africa	6	6.2	7,033.40
South Asia	31	32.3	8,257.52
Others	12	12.5	2,647.54
Total	96	100	40,404.54

Note. Adapted from Zakaria, 2011

Among the problems faced by expatriates when in foreign countries are that they are unable to cope with the situation of being in a new environment and that they are unable to adjust themselves in performing their heavy responsibilities overseas [2], [10]. Therefore, the adjustment aspects of facing the situation in the host country should be taken into consideration by companies so that expatriates can have some expectations about the problems they will be facing overseas.

Initiatives have been taken by the Construction Industry Development Board (CIDB) to encourage local contractors to participate in construction projects abroad. The government through DICB has produced a directory of construction companies involved in construction projects abroad. In addition, the government has published the *Malaysian Builders Go Global* to establish a communication network for the construction industry. With these facilities, Malaysian construction companies can have direct communications with international markets and expand the construction industry in Malaysia, at the same time contributing towards the Malaysian economy. The involvement of the Malaysian multinational construction companies abroad was with their own initiatives [11]. Therefore, when operating abroad, companies have to give priority to the requirements of their staff to facilitate them in performing their tasks.

The general adjustment experienced by executive expatriates affects their work satisfaction and motivates them in performing their assignments [12]. Expatriates not only have to perform their assigned responsibilities well, but also have to face environment changes that require psychological strength to adjust to. Therefore, it is important to understand the changes they have to face when abroad.

This study focuses on ways to solve the problems faced by expatriates in their host countries. The question that needs to be answered in this study is that whether there are any non-technical aspects of general adjustment faced by expatriates of Malaysian multinational construction companies in their host countries, and to what extent these non-technical factors contribute to the adjustment factors.

3 Literature Review

General adjustment involved changes in lifestyle, social status, and comfort faced in the new situation [13, 14]. Life, for every individual is an adaptation process to the existing situation and creating good communication with the environment. Adjustment to new environment condition affects the psychology of the expatriates in the host country [1], [15]. If people are affected psychologically such as if they are feeling discomfort, sadness, and worrisome, when performing their duties, it is a sign that they are unable to adjust to the new situation.

Based on the U-curve adjustment theory, expatriates experience psychological changes after being in a host country. [16], maintained that adjustment is obtained when they have accepted and understood the changes that exist in the host country. At the early stage, the expatriates would not feel the difficulties because they are in a stage of "fantasy". However, after a while being in the host country, differences start to surface as in location of their settlement and culture shock. Nevertheless, the expatriates have to understand and accept these differences that occur. Previous study found that Japanese expatriates require at least four years making adjustment [17]. Expatriates from monoculture countries such as Japan find it difficult to adjust to situation in other countries that have different cultures [18].

General adjustments with regard to food, shopping and banking have also been studied. According to [18], expatriates would face new things, especially food when they are in their host country. Prior exposures to food in the country where the expatriates would be staying will help them get prepared for the changes that they will be facing. Some expatriates often go to work in several countries, and yet they still need this kind of adjustment for their daily comfort [19, 20]. This view is supported by [21] when they agreed that adjustment to food is more crucial compared with adjustment to health facilities, standard of living, and entertainment. With regard to shopping, it was found that shopping is a required daily spending activity for people [20]. Moreover, there are also differences in the way people do shopping in the home country as compared with the host country. [22] observed that expatriates need to adjust to shopping because in some host countries bargaining is practiced while in their home country bargaining may not be practised.

According to [23], the difficulties faced by expatriates to adjust to the living standard in the host country would lead to other difficulties. Some expatriates who live in a host country tend to compare the situation there with that in their home country where it is more comfortable. [20] agreed with this view. They said that one of the reasons staff refuse the offer to work abroad is that the living standard in the host country is different from that of the home country. If the living standard in the host country is higher than that of their home country, it would be difficult for the expatriates to adapt to the situation in the host country [24]. Findings by [25, 26] showed that adjustment to living standards is positively correlated with adjustment of expatriates in their host countries. This shows that adjustment to living standards influences adjustment of expatriates in their host countries. [26, 27], are among researchers, who has taken into consideration the living standards adjustment of expatriates.

The interaction with host country citizens is related closely with the activity of socializing with the local citizens of the host country. This refers to how comfortable the expatriates interact with local citizens when they are performing their tasks and when outside their tasks [23], [26], [28-30]. Previous expatriate research has shown that one main problem of adjustment of expatriates is their difficulty to interact with local citizens during their appointment in the host country [23]. [31] also found same problem faced by expatriates. According to [32] , because expatriates face difficulties to interact, their adjustment would take a longer time, which in turn affects their job performance. In addition, expatriates need to interact with the environment to enable them to take positive actions to facilitate their adaptations with local citizens [30], [35,36].

Adaptation to the culture of the host country is an adjustment to the cultural practice and the customs of the local citizens in the host country [33]. Individuals who can accept the differences in the host country will face less adjustment problems. According to [34], it is difficult for individuals to adjust to a situation with a different socio-economy and society. In addition, [35], [37] believed that language is closely related to culture.The language is the foundation of culture for a country [2]. By having good communication with the local citizens, one shows that he or she is trying to understand and adapt to the local culture. This will facilitate cultural adjustment. Expatriates who have less interactions and making less cultural adaptation to the culture of the host country will require a longer time to make adjustment [26].

4 Methodology

This study uses the questionnaire to collect data. The population is based on the list of Malaysian construction companies operating in foreign countries, which is obtained from the Construction Industries Development Board (CIDB). Questionnaires were distributed to Malaysian expatriates assigned to their various host countries. However, the questionnaires were sent, attached with a covering letter, through e-mails to the human resource managers to be forwarded to their executives working abroad. Three follow-up actions were taken to obtain their responses [36]. Finally, 64 questionnaires were returned by the respondents. Data obtained were analysed using Statistical Package for the Social Science software (SPSS). Data analysis was performed using the quantitative approach as the data obtained were measured using nominal data.

Reliability test were performed using the Cronbach's alpha for determining the reliability of the items tested [38]. Factor analysis procedures were performed based on the interpretation of the research question and findings from statistical analysis [27], [39]. The measurement scale made use of 5-point Likert scale with 1 = *No adjustment at all* to 5 = *Full adjustment* [40, 41]. A mean score less than 2.5 represents no adjustment [6], [42].

5 Results and Discussion

The general dimension consists of 13 items, which include living condition, housing condition, religious practices, and types of local food, shopping, banking, and living standards. Reliability tests performed on the general dimension obtained the value of α = .89. The items with value of Cronbach's α greater than .7 were selected because they have acceptable reliability values.

Using the Principal Component Analysis (PCA), factor analyses were performed on the 13 items. One item, that is, food suitability was dropped for having communalities value less than .5. The factor analysis procedure was repeated and returned a *Kaiser-Meyer-Olkin* (KMO) value of .79 (> .6). The *Barlett's test of Sphericity* was significant at p = .00 (p < .01), and communalities values for all items were greater than .5, as shown in Table 2. The largest eigen value was 5.49 for the first factor, followed by 1.52 for the second factor, and 1.12 for the third factor. The first factor obtained the largest eigenvalue because it gave the largest contribution to the changes in the total variance of the changes in the variables [43]. The results from the factor analyses have grouped three factors which together explained 67.74% of the total variance where the first factor explained 45.76%, followed the second factor, which explained 12.65%, and the third factor, which explained 9.33%.

Table 2, Factor Analysis for General Adjustment

Item	FL	CA	EV
Factor 1 Interaction Adjustment			
Socializing with local	.76		
Interaction with host nationals	.88	.78	5.49
Communication	.87		
Culture adaptation	.75		
Factor 2 Social adjustment			
Shopping	.67		
Banking	.72		
Cost of living	.67	.78	1.52
Entertainment	.60		
Healthcare	.64		
Factor 2 Living environment			
Living condition	.76		
Housing condition	.84	.89	1.12
Religious practice	.75		
KMO Measure of Sampling Adequacy	.79		

Barlett's test of Sphericity	398.39*
Anti-Image Correlation Matrix (MSA)	.67 - .93
Total Variance	67.74%

Note: FL= Factor Loading, CA=Cronbach's Alpha, EV= Eigen Value.

$p < .01$*. Cronbach's alpha coefficients with values less than 0.7 were deleted.

The three factors were called interaction adjustment, general adjustment, and living environment. The interaction adjustment factor comprised items such as socializing with the local society, interaction with local citizens, and adapting to the culture of the host country. The second factor was the social adjustment factor, which include banking, shopping, living standards, recreational and entertainment facilities, and health facilities. The third factor was the living environment, which included living condition, housing condition, and religious practices. After the factors of the items have been identified, reliability tests were performed to determine the reliability of each factor. The Cronbach's alpha reliability test for the general dimension obtained an alpha value for social interaction adjustment factor, α= .78 and environment atmosphere factor, α = .89 Therefore, the reliability of both the factors were acceptable.

The results showed that out the 13 items tested, 12 of them were chosen, showing that those items have an influence on the expatriates' general adjustment in facing the new environment in their host countries. The item on food adjustment was dropped. This result contradicted the results of previous research done by [23], [21]. These authors have included this item on food adjustment. Nevertheless, in the present research, the item was dropped because it did not contribute to factor variance proportion [44, 45]. From the 12 items of general adjustment, three factors were extracted and named interaction adjustment, social adjustment and living adjustment in accordance with the group of items in the factor.

Four items were grouped under interaction adjustment factor. The items were socializing with local society, interacting with local citizens, communication, and adaptation with culture of the host country. These three items were grouped in accordance with [26], [46]. However, [23] did not include the communication item in his research. This shows that the item socializing with local society, interacting with local citizens, communicating and adapting to the culture of the host country is an important aspect which need to be taken into account in expatriates' general adjustment in their host countries.

With regard to general adjustment, five items were grouped under this factor. This was based on the factor analysis procedure performed. The items extracted were shopping, banking, living standards, recreational and entertainment facilities, and health facilities. [23] has outlined three items that contribute to general adjustment, namely shopping, banking and living standards. However, studies by [46, 47] have included recreational and entertainment facilities and health facilities under general adjustment. It seems that there is a difference of opinion between the present research and previous research by [23], [46], who did not include the aspect of health facilities. The combination of all these items grouped under one factor demonstrates the importance of these items to general adjustment. This is in line with previous analysis and literatures which stressed on the comfort of expatriates in adjusting to the living environment in their host countries [36].

Three items were clustered under the living environment factor for general dimension, namely living condition, housing condition, and religious practices. The item living condition and housing condition used by [21] [46, 47-48] are adjustment items that affect adjustment to environment condition. The item religious practices have been considered by [49, 50] as having an influence on expatriates' adjustment. However, the present study contradicted this notion.

6 Conclusion

The study found that three factors, namely interaction adjustment, social adjustment, and living environment contribute to non-technical adjustment in the general dimension. The interaction adjustment aspect, namely communication, cultural adaptation, and socializing with local society need to be considered because expatriates are unable to adjust with those aspects in their host countries.

With regard to general adjustment, those elements associated with health facilities, banking, and recreational and entertainment need to be given serious attention. However, expatriates do not face any problems and they can adjust to the conditions and the living condition. Housing provided by their companies also affects this outcome. Therefore, three factors have been identified, namely interaction adjustment, general adjustment, and living environment. These factors have provided the answer to the fulfilment of the first objective of the study. The first objective is about the non-technical adjustment challenges faced by Malaysian expatriates in the construction industry, in their host countries.

The research findings have identified several factors that affect the general adjustment of Malaysian expatriates in the Malaysian construction industry. Companies and individuals sent abroad need to realise that there are also non-technical factors that would cause the successful implementation of projects abroad. The interaction adjustment and general adjustment provide the largest contribution to expatriates' adjustment in their host countries. Other factors such as social relationship with local society, adaptation to the culture of the host country, and communication are among aspects that should be given more attention by companies so that expatriate executives can make adjustments to a new environment abroad. This study suggests that future research need to focus on family adjustment because this factor also influences expatriate adjustment. In addition, future studies should also pay more attention to

 DOI: 10.1051/matecconf/20166600012

qualitative approach in their studies so that the real issues faced by expatriates can be brought to light.

References

1. A. W. Andreason, K. D. Kinneer, Repatriation Adjustment Problems and the Successful Reintegration of Expatriates and Their Families. J. of Behav. and Appl. Manage., **6**, 109-126 (2005).
2. N. Yang, A Cross-cultural Contextual Model of Work-Family Interfaces in Managing International Assignments. J. of Int. Bus. Res., **6**, 1-10 (2007).
3. B. A. Anderson, Expatriate Management : An Australia Tri-Sector Comparative Study. Thunderbird Int. Bus. Rev., **4**, 33 - 52 (2001).
4. J. P. Katz, D. M. Seiffer, It's a Different World out There : Planning of Expatriate Success through Selection, Pre Departure.... Hum. Resour. Planning, **19**, 32-49 (1996.)
5. Y. Paik, B. Segaud, C. Malinowski, How to Improve Repatriation Management – Are motivations and expectations congruent between the company and expatriates. Int. J. of Manpower, **23**, 635-648 (2002).
6. A.-R. Abdul-Aziz, S. S. Wong, Competitive Assets of Malaysian Int. Contractors. J. Financ. Manage. Prop. Constr., **15**, 176-189 (2010)
7. S. A. Mansor, *Perhimpunan bulanan bersama warga KKR*. Jabatan Kerja Raya, unpublished, Kuala Lumpur (2011).
8. CIDB Malaysian Contractors' *Achievements in the Global Market*, Construction Industry Development Board, Kuala Lumpur (2009).
9. M. Z. Zakaria, *Entry Strategies for Malaysian Construction Related Companies Going Abroad*. Paper presented on National Seminar in Built Environment (NSiBE) Kuala Lumpur (2011).
10. A. A. M. Osman-Gani, S. H. Akmal, Repatriation Readjustment of International Managers; An Empirical Analysis of HRD Interventions. Career Develop. Int., **13**, 456-475 (2008).
11. A.-R. Abdul-Aziz, S. S. Wong, Exploring the Internationalization of Malaysian Contractors: The International Entrepreneurship Dimension. Constr. Manage. and Econ., **28**, 51-61 (2010)
12. R. Takeuchi, P. L. David, V. M. Sophia, Y. Seokhwa, Nonlinear influences of stressors on general adjustment: the case of Japanese expatriates and their spouses. J. of Int. Bus. Stud., **38**, 928-943 (2007).
13. J. S. Black, M. Mendenhall, G. Oddou, Towards a Comprehensive Model of International Adjustment : An Integration of Multiple Theoretical Perspectives. Acad. of Manage. J., **16**, 2, 291-317 (1991)
14. H.-K. Chi, C.-Y. Chiou, The Work Adjustment of Taiwanese Expatriates. The Bus. Rev., **8**, 267-272 (2007).
15. J. Selmer, A. S. M. Leung, Symptom and Problem Focused Coping Strategies of Business Women Expatriates and Their Socio-cultural Adjustment in Hong Kong. Women In Manage. Rev., **22**, 588-605 (2007).
16. G. Pires, J. Stanton, S. Ostenfeld, Improving Expatriate Adjustments and Effectiveness in Ethnically Diverse Countries: Marketing Insights. Cross Cult. Manage. : An Int. J., **13**, 156-170 (2006).
17. Y. Yamazaki, A fit between skills and demands among Japanese expatriates in USA. Manage. Int. Rev., **50**, 81-108 (2010).
18. A. Webb, P. C. Wright, The Expatriate Experience: Implications for Career Success. Career Develop. Int., **1**, 38-44 (1996).
19. H. B. Gregersen, J. S. Black, Antecedents to Commitment to a Parent Company and a Foreign Operation. Acad. of Manage. J., **35**, 65-90 (1992).
20. K. Hutchings, S. W. Ratnasari, Cross-cultural Non-work Transition Stresses: Domestic Transferees in Indonesia. Cross Cult. Manage. : An Int. J., **13**, 114-131 (2006).
21. J. McGingley, Expatriate Adjustment Within A Social Context : Examination of Sample in Russia. J. of Soc., Evolutionary and Cult. Psychology, **2**, 56-68 (2008).
22. P. Caligiuri, M. Lazarova, S. Zehetbauer, Top Managers' National Diversity and Boundary Spanning : Attitudinal Indicator of a Firm's Internationalization. J. of Manage. Psychology, **23**, 848-859 (2004).
23. J. S. Black, Work Role Transitions : A Study of American Expatriate Managers in Japan. J. of Int. Bus. Stud., **19**, 277 – 294 (1988),
24. D. G. Collings, H. Scullion, M. J. Morley, Changing Patterns of Global Staffing in the Multinational Enterprise: Challenges to the Conventional Expatriate Assignment and Emerging Alternatives. J. of World Bus., **42**, 198-213 (2007).
25. J. S. Black, H. B. Gregersen, The Right Way to Manage Expats. Harvard Bus. Rev., **77**, 52-63 (1999).
26. T.-J. Huang, Shu-Cheng & J. J. Lawler, The Relationship Between Expatriates' Personality Traits and Their Adjustment to International Assignments. The Int. J. of Hum. Resour. Manage., **16**, 1656-1670 (2005).
27. A. Fish, Assisting Cross-border Manager Adjustment: Psycho-cultural and Socio-cultural Interventions. Personnel Rev., **24**, 225-245 (2005).
28. P. Bhaskar-Shrinivas, D. Harisson, M. A. Shaffer, D. M. Luk, Input-Based and Time-Based Models of International Adjustment: Meta-Analytic Evidence and Theoretical Extensions. Acad. of Manage. J., **48**, 257-281 (2005).
29. H.-F. Chen, The Relationships of Organizational Justice, Social Exchange, Psychological Contract, and Expatriate Adjustment : An Example of Taiwanese Business Expatriates. The Int. J. of Hum. Resour. Manage., **21**, 1090-1107 (2010).
30. M. A. Shaffer, D. A. Harrison, H. Gregersen, J. S. Black, L. A. Ferzandi, You Can Take It With You: Individual Differences and Expatriate Effectiveness. J. Of Appl. Psychology, **91**, 109-125 (2006).

31. V. Suutari, K. Riusala, *Operating in 'Economics in Transition' – Adjustment and Management Issues Faced by Finnish Expatriate Managers in CEE*. 87-107 (2000).
32. J. Selmer, which Is Easier, Adjusting to a Similar or to a Dissimilar Culture. Cross Cult. Manage., **7**, 185-201. (2007)
33. S. J. Shin, F. P. Morgeson, M. A. Campion, What Do You Depends On Where You Are : Understanding How Domestic and Expatriate Work Requirements Depend Upon The Cultural Context. J. of Int. Bus. Stud., **38**, 1, 64-84 (2007).
34. K. G. Lewis, Breakdown - A Psychology Contract for Expatriates. European Bus. Rev., **97**, 6 (1997).
35. J. S. Black, M. E. Mendenhall, The U-Curve Adjustment Hypothesis Revisited : A Review and Theoretical Framework. J. of Int. Bus. Stud., **22**, 225-247 (1991).
36. J. P. Shay, S. Baack, An Empirical Investigation of the Relationships Between Modes and Degree of Expatriate Adjustment and Multiple Measures of Performance. Int. J. of Cross Cult. Manage., **6**, 275-294 (2006).
37. S. L. O'Sullivan, International performance appraisals: A review of the literature & agenda for future research. Journal of American Academy of Business, Cambridge, **14**, 171-180 (2009).
38. S. J. Coakes, L. Steed, P. Dzidic, *SPSS version 13.0 for Windows Analysis Without Anguish*. Sydney: John Wiley & Sons Australia (2006).
39. T. Roshana, E.M. Ahnuar, *Analysis of Efficiency and Effectiveness Measures of Construction Project Success in Malaysia*. International Conference in Built Environment – iCiBE 2006. Kuala Lumpur 13-15 Jun (2006).
40. E. Babbie, *The Practise of Social Research (12th Edition ed.)*. Wadsworth: Nelson Education Ltd. (2010).
41. W. L. Neuman, *Social Research Methods - Qualitative and Quantitative Approaches (6th ed.)*. Boston: Pearson Education Inc. (2006).
42. F. E. Hurme, *Competencies for rural nursing practise*. Louisiana State University, Louisiana (2007).
43. C. Y. Piaw, *Statistik Penyelidikan Lanjutan : Ujian Regresi, Analisis Faktor dan Analisis Sem- Buku 5*. Kuala Lumpur: Mc. Graw Hill (Malaysia) Sdn. Bhd. (2009)
44. A. Field, *Discovering Statistics Using SPSS (Third ed.)*. London: SAGE Publications Ltd. (2009)
45. C. K. D. Lorio, *Measurement in Health Behaviour - Methods For Research and Evaluation. San Francisco*: Jossey-Bass (2005).
46. L.-Y. Lee, D. V. Vorst, The Influence of Social Capital and Social Support on Expatriates' Cultural Adjustment: An Empirical Validation in Taiwan. Int. J. of Manage., **27**, 628-649 (2010).
46. M. d. G. H. O. d. *Pamich, Cross-Cultural Adjustment and The Expatriate Spouse : A Case Study*. Capella University, USA (2007).
47. S. M. Toh, *Host Country Nationals to the Rescue: A Social Categorization Approach to Expatriate Adjustment*. Unpublished thesis. Texas A&M University, Texas (2003).
48. M. Bastida, M.T. Cancelo, Expatriate Performance and Permanence: Analysis By Gender of 124 Spanish Managers. Appl. Econ. and Int. Devt., **14**, 51-64 (2014).
49. M. F. Tucker, R. Bonial, K. Lahti, The Definition, Measurement and Prediction of Intercultural Adjustment and Job Performance Among Corporate Expatriates. Int. J. of InterCult. Relat., **28**, 221-251 (2004).
50. A. H. M. Tahir, M. Ismail, Cross-Cultural Challenges and Adjustments of Expatriates: A Case Study in Malaysia. Turkish J. of Int. Relat., **6**, 72-99 (2007).

Use Of The Thermo-Hygro Buttons in Tracking The Influence Of Humidity And Temperature On The Deterioration Of The Building Stones Of The Mausoleum Of Imedghassen Batna In Algeria

M. Hamiane[1,a], C.Assafsaf [2]

[1] *University M'Hamed Bougara, (UR-MPE) .Independence Avenue, 35000 Boumerdes, Algeria*

Abstract. The tomb of Imedghassen is a pre-Roman royal mausoleum located at 420 Km from Algiers. It is dated back to the 3rd century B.C. It is built with building stones. The semi-arid continental climate, of the chosen location, is characterized by strong temperature fluctuations and high evaporation. These two factors have a great influence on the building stones' state of degradation. The infiltration of water in this very rainy region, related to time ,had a serious impact on the monument. To get accurate results about the influence of temperature and humidity, we had controlled the microclimate of the mausoleum using PC Thermo track; thermo-hygro buttons. It allowed us to accurately track the values of the average and higher temperature and humidity, and also the dew point. The results clearly show the relationship between these factors and the origin of the building stone deterioration.

Keywords: Degradation, Mausoleum Imedghassen, Temperature, Humidity, Thermo Hygro-buttons

1 Introduction

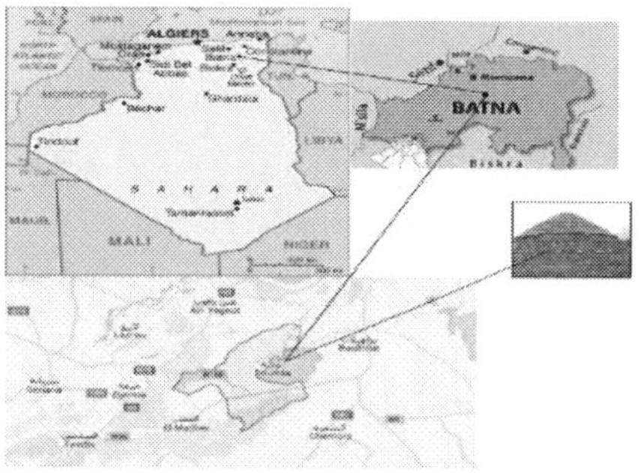

Figure 1. Geographic location.

Figure 2. Stone blocks of the Mausoleum

Imedghassen Mausoleum is a monument listed as a national heritage. It is located in the wilaya of Batna, at the altitude of 914 meters (figure 1). The building measures 18.35 m in hight and 58.86 m in base diameter. The cylindrical base is embellished with decorative columns. The mausoleum is dated back to the 3rd or 4th century BC. It is entirely built from huge blocks of stones together without mortar, joined by lead studs, and small stones (Figure 2).

The monument is in an alarming state of degradation. Several factors are responsible for this deterioration. However we are interested in our article for damages caused by the effect of temperature and humidity.

The semi-arid continental climate of the chosen location, characterized by strong temperature fluctuations and high evaporation, has a great influence on the conservation state of the mausoleum. The infiltration of water in this very rainy region very related to time ,had a severe impact on the monument .To get accurate results about the influence of temperature and humidity, we had controlled the mausoleum microclimate using a PC Thermotrack thermo-hygro buttons logger . It allowed us to accurately track the values of the average and higher temperature, humidity and also the dew point. The results clearly show the relationship between these factors and the origin of the stone deterioration.

[1] hamiane@univ-boumerdes.dz

2 Mausoleum Damage Factors

Among the most significant degradation factors that appear clearly on the mausoleum, are: Plants, lichens, algae, and cracks, color change of stones, Crystallization of salts, corrosion of stones, rising damp (capillary), water infiltration, loss of stone mass and rubble. However the rain, change in humidity and temperature factors have a marked influence on the degradation of the monument. Figures (3 to 6) show these different types of degradation.

Plants

We note the growth of plants of all kinds, in particular between the stones. These plants are abundant in winter because of the availability of water. Over time, these plants have made space between stones, and consequently, the demolition of the walls of the mausoleum as a result.

Microscopic plants

They are represented by algae and lichens, deployed on many stones of the building. They appear as spots or crust, green, yellow or white, depending on lichens. These plants develop on the outer surface of the stones; this alters the stones and influences the aesthetic side of mausoleum.The density of these algae is particularly enhanced in the spaces between the stones, where there are shade, moisture and water.

Figure 3. Plant growth

Figure 4. Lichens growth

Cracks in the rocks

Some stones show visible cracks .We see wide and narrow cracks, vertical and horizontal. The wide variations in temperature and moisture content may be the cause of these cracks. The mechanical pressure of new stones substitued during the restoration, and the presence of plants can also be the cause of these fractures.

Figure 5. Horizontal cracks

Figure 6. Corrosion stones

Corrosion of stones

The mechanical action combined with rain, wind, sand and time has changed the shape of the stones, decreased their size and a mass loss.The surface corroded stone comes off easily.

Efflorescence of salts

The bottom capillary moisture of the wall of the soil, the dissolution of the salts by water from rain and the evaporation of water at high temperature of the surfaces of the stones, are the main cause of salt crystallization. Mausoleum Imedghassen shows very well this phenomenon. White layers are observed on the first sitting of the monument.

 DOI: 10.1051/matecconf/20166600013

Figure 7. Efflorescence salts

3 EXPERIMENTATION

3.1 Thermal shock

The samples are subjected to high temperature (from 60°C to 250°C).After 15 minutes in an oven, the samples were placed in cold water at 5°C. Observations show no change for all samples subjected to the temperatures (60°C to 100°C). For the temperature range (120°C to 180°C), all samples show a slight mass loss. For the temperature 220, slight mass loss for some samples and loss of small pieces for others. The samples subjected to the temperature of 250°C, have become friable, with loss of mass and color change.

3.2 Effect of freeze / thaw

After drying the samples in a drying oven until a constant mass, the samples were immerged for 48 h in water with a temperature (15 ± 5°C) .The samples were then placed in a refrigerating machine with a temperature (-15 ± 5°C) .After a period of 4 hours, the samples were put back into water at (15 ± 5°C) for 4h. The experience is repeated five times (5X) .The samples are then left in the open air for 24 H. They are weighed after drying in an oven at a temperature of (105 ± 5°C) .The results showed that the mass loss (%), for the three samples are : (0.24; 0.26 and 0.30).

3.3 Influence of moisture on the samples

The solutions are prepared in the laboratory to create relative moisture of 45% and 95% .The solutions were placed in desiccators, with perforated partitions, so that the samples are not in contact with the solution. After drying in an oven until a constant weight, they are put into a 45% environment and 95% relative moisture. The masses are weighed after different time (t1, t2, t3..................tn) with tn equal to 30 days. The results are shown by the graphs of Figure 8.

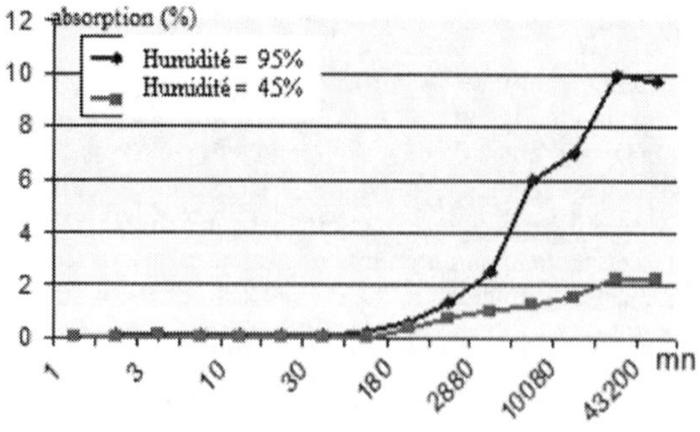

Figure 8. Influence of the H (45% and 95%) on the samples for 30 days

3.4 Monitoring of temperature and relative humidity in the site

This experience is to program the thermo-hygro buttons and set them in representative areas of imedghassen Mausoleum. The experiment was performed during the winter and summer seasons .The choice fell on facade walls whose conservation status is different.

The thermo hygro button is the smallest temperature and humidity logger in the world.

Ideal for monitoring the temperature and humidity, but also for stability testing for the conservation of artifact...Small, rugged, it fits everywhere and supports shock, dust. All records are archived and stored in its protected memory.

Battery life span: up to 10 years.

- Records date and time
- Programmable low and high alarm threshold with temporization parameters
- Free memory for traceability data storage

The Thermo-Hygro Button designation and precision are:

Temperature measurement range -20 / + 85 ° C
Temperature accuracy ± 0.5 ° C
Measuring range humidity 0 / + 100%
RH Accuracy ± 5%
Resolution 0.1 or 0.5 ° C
Number of measurements 4096 or 8192
Measuring rate 1 sec to 273 h
Starting if possible alarm Yes

3.4.1 Placement of the first thermo-hygro-button

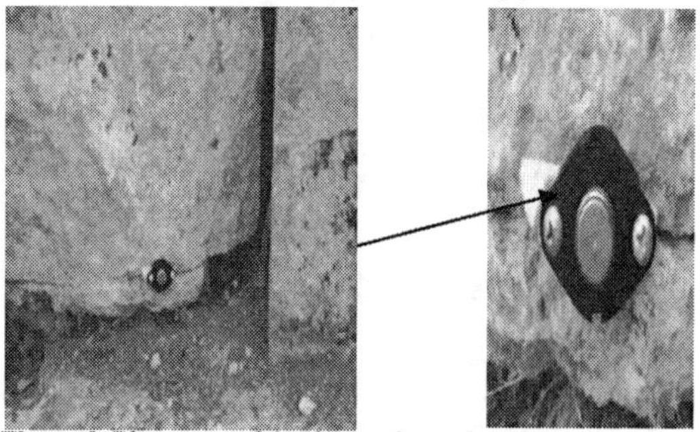

Figure 9. Placement of the thermo-hygro-button

The Thermo-hygro button was placed in the north facade wall, from January 13 to February 12. Figure 9, showing the location of thermo_hygro button in the wall and in Figure. 10, showing the registration of the temperature variation, humidity and dew point during this period. The summary of the recorded mean values are:
H(%) max = 100;H(%) min = 41.8;H(%) Average =85.51
T(°C) max = 27.9 ;T(°C) min = 0.4 ;T(°C) Average =7.88

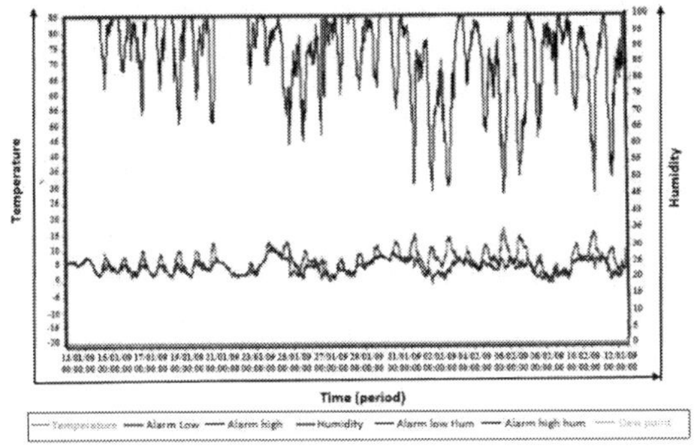

Figure 10. Recording result of H (%) and T°C, the winter period, North façade

3.4.2 Placement of the second thermo-hygro button

The Thermo-hygro button was placed in the East facade of the wall, from January 13 to February 12. The summary of the recorded mean values are:
H(%) max = 100;H(%) min =17.16 ;H(%)Average=76.49
T(°C)max = 37.7 ;T(°C) min = 0.0 ;T(°C) Average=11.17

3.4.3 Placement of the third-thermo-hygro button

The thermo-hygro button was placed in the west facade wall, of 27 June to 9 August. The summary of the recorded mean values are:
H(%) max = 100; H(%) min = 53 ; H(%)Average =28.6 ;
T (°C)max = 61,4; T (°C) min = 15.8 ; T(°C) Average = 31.9

4 Interpretation

4.1 Interpretation of physical results (thermal shock and freeze / thaw)

Samples of stone buildings that are limestone, with an open porosity 35.15% type, have shown the influence of temperature in the analysis of thermal shock, with slight loss of mass from 120 ° C until color change and loss of small pieces to 250 ° C. Samples submitted to the effect of freeze / thaw, show after five cycles, a mass loss of 0.30% .This explains well the effect of changes in temperature and humidity on cracking building stones of mausoleum.

4.2 Interpretation of the results of the influence of humidity (45% and 95%)

The samples, subjected to a 45% environment and 95% relative humidity for a period of 30 days, showed that the limestone-type samples ,with significant open porosity, absorb moisture from the surrounding .This absorption increases with increase of the amount of moisture, increased porosity and the exposure time of the samples in that environment. This clearly shows the effect of changes in humidity, rain and rising damp on the alteration of building stones. The different types of alterations observed in the site (micro organisms, lichens, efflorescence, weight loss, etc.), are accelerated by the presence of moisture.

4.3 Interpretation of humidity and temperature monitoring results in the site

Three thermo-hygro buttons after programming were placed respectively:

- North facade (less exposed to wind and temperate) and best preserved
- Eastern Front (more exposed to wind and temperate), less conserved
  West facade (average state of preservation)

Winter period:

After 30 recording days (from January 13 to February 12), we remove the thermo-hygro buttons from the mausoleum walls and reading them, via the PC software Thermo track. The results are shown in the Figure 10 (thermo-hygro1).It shows the variations in temperature and humidity during attempts (30) days .The values do not show large variations between day and night, against the (thermo_hygro2), that shows that the gap is very important between the values of day and night, with large amplitudes during the 30 days of recording The results clearly show the difference in temperature and humidity between the two facades (North and East) .This clearly explains the influence of climatic parameters on the state of degradation of the east façade against the north wall.

Summer Period

After 42 days of recording (from 27 June to 9 August) .The thermo-hygro buttons after reading the results, clearly show the difference of temperature and humidity values between the winter and the summer period with a temperature difference min (9.8 ± 0.5 ° C) and max (15.8 ± 0.5 ° C) and a standard deviation for the Humidity of max (11.86 ± 0.1%). These large variations in Humidity and temperature may be the main cause of some cracks observed on the mausoleum stones.

5 Conclusion

Building stones of the mausoleum Imadghassen are in advanced state of deterioration. The parameters responsible for that status are many, and they affect each other. We denote particular factors of humidity and temperature. The large magnitude of the variation of humidity and temperature between day and night and between winter and summer, registered in the site, confirm the relationship between these parameters and the degradation state of the monument. The samples subjected to the influence of humidity (45% and 95%) showed that the amount of the absorption increases with the amount of humidity, open porosity and the time of exposure of the samples to this environment. The mass loss recorded in the test of thermal shock and freezing / thawing, confirm the relationship between these factors and the conservation status of mausoleum.

References

1. Adam .J.P, Boussoutrot .A. " restauration architecturale et préservation des sites archéologiques ", dans : conservation restauration en archéologie, France ,1990.
2. Adam .J.P, La construction romaine : matériaux et techniques, troisième édition, édition A et J.Picard, Paris, 1995.
3. Blas J.M, Sintes .C, sites et monuments antiques d'Algérie, France, 2003.
4. C.Assafsaf, M.Hamiane. Approche d'étude technique de l'état de conservation du mausolée d'Imadghasen. Mémoire de Magistère, Algérie. (2010).
5. Camps. G, " nouvelles observation sur l'architecture et l'âge du Medracen mausolée royal de Numidie", dans : compte rendus de l'académie des inscriptions, 1973.
6. Gsell. S, histoire ancien de l'Afrique du nord, tome VI, 1972.
7. ICOMOS," Glossaire illustré sur les formes d'altération de la pierre ", France, 2008, dans : *www.intrnational.icomos.org*
8. M.Hamiane, C.Assafsaf, M.Bouteflika, W.Zekagh, Etat de conservation et lutte contre les graffitis du Mausolée d'Imedghassen de Batna en Algérie. Conférence internationale. Conservation de la pierre. Paris (2011).
9. M.Hamiane, C.Assafsaf, M.Bouteflika, W.Zekagh, M.Saidi, L'état de conservation du Mausolée d'Imedghassen de Batna en Algérie.1er Colloque sur le Patrimoine Maure (Amazigh) du Maroque Antique. (2013).
10. Shaw .T, Voyage dans la régence d'Alger au XVIIIe siècle, traduit par : Carthy (E.M), grande Alger livres éditions, Alger, 2007.
11. Torraca .G, matériaux de construction poreux, iccrom, Italie, 1986.

Public Private Partnership Benefits in Delivering Public Facilities in Malaysia

Sapri M.[1,a], Hariati A.H.[2], Sheau Ting L.[3] and Sipan I.[4]

[1,3,4] *Centre for Real Estate Studies, Universiti Teknologi Malaysia, 81310 UTM Johor Bahru, Johor, Malaysia*
[2] *Faculty of Geoinformation and Real Estate, Universiti Teknologi Malaysia, 81310 UTM Johor Bahru, Johor, Malaysia*

Abstract. The development of infrastructure in developing country such as Malaysia was increasingly founded by the Public –Private Partnership (PPP) scheme. Collaboration with private sector has become popular as a means to improve the delivery of public facilities. Yet, empirical evidence on how PPP initiative has benefits the delivery of public facilities within Malaysia context is lagging. The purpose of this paper is to identify and assess the perception of stakeholders on the benefits of adopting PPP in delivering public facilities in Malaysia. Literature review was carried out to identify PPP benefits, which were then incorporated into the questionnaire. The mean score and mean score ranking was conducted to assess the agreement level of stakeholders towards the PPP benefits. The overall findings show that the implementation of PPP has benefitted the delivery of public facilities in both financial and non-financial aspects. From the analysis, improvement in service quality is perceived as the top advantage followed by innovation in design and transfer of risk. The findings provide more informed basis on the rationale of PPP implementation and its potential in improving the delivery of public facilities within Malaysia context.

1 Introduction

Public facilities for education, recreation, health, safety, and communication are some of the essential community needs. The provision of these facilities is part of the major government responsibilities. Besides making the facilities available to everyone, the government also has to ensure the delivery process is conducted in an efficient way. However, government nowadays has a limited ability to respond to the increasing demand of public facilities. Factors such as scarce sources, macro – economic instability, administrative inefficiencies and large skill gaps have restrained the government to deliver the services at a required level [1].

The pressure felt by the government has led to the implementation of public private partnership (PPP) scheme, which allowed the participation of private sectors in the provision of public facilities. Through this scheme, the government has entrust the private sector with the responsibility of designing, financing, building and operating of public facilities [2]. According to [3] private sector has greater expertise in managing complex projects, capable in delivering them on time and budged, as well as maintaining services thereafter. Therefore, PPP is considered as an effective approach to enhance project productivity through management efficiency and creative skills from business practice [4].

The reasons for adopting PPP may vary in different countries. [5] discovered that United Kingdom (UK) has implement PPP for the financial elements while Hong Kong and Australia were more related to the overall performance of improving public projects. In Malaysia, the principal objectives of PPP implementation are to encourage private sectors participation, reduce government's expenditure and to improve the delivery of public facilities [6, 7]. PPP is adopted to increase efficiency, greater innovation and value-added services [8]. While the rationale for adopting PPP in Malaysia is well understood, it was discovered that assessment on the impact of PPP implementation is remain limited. Since the implementation of PPP is growing in Malaysia, initial investigation on how PPP so far has benefited the delivery of public facilities and services is thus desirable. Hence, this study aims to fill the gap. In order to offer some evidence on the present issues, this study has embarked on quantitative methods.

This paper is structured as follows; Section 2 discuss the development of PPP in Malaysia and Section 3 discuss advantages of PPP in delivering public facilities respectively. Section 4 offers the information regarding research method adopted in this study. Section 5 presents the results and discussion, while Section 6 concludes the findings of this study.

2 Development of public private partnership implementation in Malaysia

The concept of partnership between the public and private sectors in Malaysia has existed since 1980s. Malaysia government has taken a lot of initiatives to

[a] Corresponding author: maimunahsapri@utm.my

 DOI: 10.1051/matecconf/20166600014

foster the involvement of private sectors in delivering public facilities. These include the introduction of economic policies such as The Malaysian Incorporated Policy (1981), Privatisation Policy (1983), Guidelines on Privatisation (1985) and Privatisation Master Plan in 1991 [9]. As in United Kingdom (UK), privatisation programme has triggered the execution of PPP in Malaysia. Operation and Management Licence; Build, Operate and Transfer (BOT); Build, Operate, Own, Transfer (BOOT); and Lease-Sale are among the partnership arrangements that have started to be used [8]. The implementation of PPP until 2010 is mostly in construction, manufacturing, transport, storage and communication. **Figure 1** shows the division of PPP projects according to sectors in Malaysia.

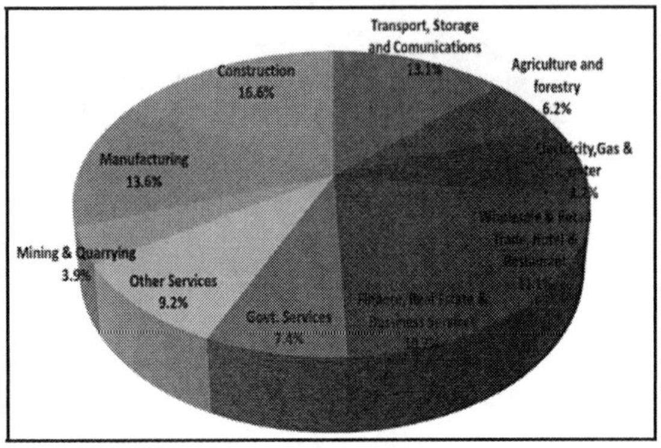

Figure 1. PPP project according to sectors from year 1983 – 2010. (*Source: PPP Unit, Malaysia)*

Malaysian government has officially announced The Private Finance Initiatives (PFI) Programme under PPP in the Ninth Malaysia Plan (2006). The introduction of PFI has become the spotlight for the overall implementation of PPP in Malaysia. It was reported in the Eleventh Malaysia Plan (2016-2020) that the private sectors participation especially via private investment is the lead of economic growth in Tenth Malaysia Plan (10MP). In Malaysia context, PFI is referred to;

'The transfer to the private sector the responsibility to finance and manage a package of capital investment and services including the construction, management, maintenance, refurbishment and replacement of the public sector assets which creates a standalone business. The private sector will create the asset and deliver a service to the public sector client. In return, the private sector will receive payment commensurate with the levels, quality and timeliness of the service provision throughout the concession period' [10]

In assessing the rational of PPP implementation, [11] discovered both public and private stakeholders perceived "to enhance private sector involvement in economic development" as the most important rational in adopting PPP in Malaysia. Besides, [12] found the attractive factors that encourage the adoption of PPP are include "facilitate creative and innovative approaches", "solve the problem of public sector budget restraint", "provide an integrated solution", "benefit local economic development" and "accelerate project development". Another in-depth investigation on PPP implementation by [13] indicated that the implementation of PPP/PFI in Malaysia have several distinctive aspect, which benefits the Malaysian citizen and enhance the sustainability of Malaysian infrastructure.

Parallel with the development, studies in regards to local implementation of PPP is actively being conducted in Malaysia. [14] conducted a research on the possible constrains that hinder the adoption of PPP in Malaysia. The results show that the obstacles that may make PPP less attractive are "lengthy delays in negotiation", "lack of government guidelines and procedures on PPP", "higher charge to direct users", "lengthy delays because of political debate" and "confusion over government objectives and evaluation criteria". A few studies also have been done on the most critical area of PPP which is value for money (VFM). [15] examined the factor that enhancing VFM achievement for PPP projects. The study concludes that five factors that enhance VFM include "competitive tender", "private sector technical innovation", "optimal use of asset/facility and project efficiency", "efficiency risk allocation and performance-based payment mechanism". By stressing the needs for VFM assessment in each phase of PPP project life cycle, study by [16] revealed that both public and private sectors perceived financial and non-financial aspects are vital elements to be captured in that process.

3 Benefits of PPP in delivering public services

PPP scheme has introduced a whole life approach where the involved parties are responsible to design, build, finance and operate public facilities for about 20-30 years. There is a long term consideration for whole life costing that includes capital cost, operating cost, maintenance cost, replacement cost and disposal cost [17]. The ultimate aim is to achieve value for money (VfM), which is 'the optimum combination of whole-life costs, benefit, risks, and quality (fitness for purpose) to meet the user's requirement and getting the best possible outcome at the lowest possible price'[18]. VfM also potrays how much cost savings are achieved when comparing PPP with traditional approach in delivering public facilities [19].

Private parties tend to seek accurate estimation for capital and operational costs as they bear a huge responsibility in the development and operational phase of PPP projects. Private sectors have the advantages in delivering public facilities at lower capital expenditure (Capex) and operational expenditure (Opex). According to [20], private sector was considered better at coordinating the activities for design, construction, finance, operations and maintenance at a lower cost compared to the government. Technical efficiency [21], better risk management, innovation and advance in technology [22] are among the factors that favour the private sectors to deliver an efficient Capex and Opex of projects.

The other reason of adopting PPP is to achieve time savings in project development. The expert owned by the private sectors allows them to expedite the development process. According to the [23] in so called 'fast-tract' design and construction of PPP, certain elements of construction work can be concurrently conducted with ongoing design phase of PPP projects. Construction firms have also been involved since design reviews process to spot any possibility that might delay the complete time of construction.

Besides faster delivery, better project governance is also to be expected in PPP implementation. Project governance is a framework that sets out the structure, resources, communication, reporting and monitoring systems so that the management project will be consistent with the organisation vision [24]. Since the companies involved in PPP usually have a strong track record in development projects, it is believed that the best practice of project governance can be absorbed in PPP projects. In addition, private companies have no option but to scrutinise this aspect as project governance will influence the cost, speed of completion, quality and financial viability of project delivered [25].

The desire for design innovation of public facilities is well recognised. It is clearly shown when the injection of design innovation can be too complex for a certain types of building. In hospital for instance, there are changing patterns in medical care [26] as well as changing requirement for technologies, policies and services, which has made design innovation a challenging task [27]. Therefore, partnership with private sectors could help the government to handle these uncertainties. According to [28], innovation in design and management of public facilities is a factor driven by PPP implementation. Appropriate design innovation could generate savings in future maintenance and operational costs [22]. Thus, superior design innovation is expected from PPP scheme [29].

As PPP has introduced the long term approach, much anticipation has been given towards the services delivered in operational phase of PPP projects. Output specification, payment mechanism and monitoring, deduction of service failures performance are interrelated mechanisms in ensuring a high quality of service delivered under PPP projects [22]. The involvement of facilities management has also raised the quality expectation on service delivered. It is especially when facilities management has to handle lots of crucial aspects related to the operations and maintenance, project planning and management, communication, finance, quality assessment, facility function, human and environmental factors, space and workplace planning, budgeting, renovation or architectural planning and design [30].

In traditional delivery, government is responsible for all risks associated with the delivery of public facilities. PPP allows the government to transfer the risks to the private sectors. Risk transfer is one of the VfM key elements. According to [31], PPP is a medium transferring most of the risks to private parties yet deriving maximum VfM. Risk is transferred to private sectors that seem able to manage the risk better than the government [32]. This includes the risks arose from the building, operation, finance, and management of the PPP project. Therefore, PPP is considered a scheme that could lift the risk of burden bears by the government in delivering public facilities.

Earlier discussion has proved a strong rationale behind the implementation of PPP in delivering public facilities. PPP could improve both financial and non-financial aspect of public facilities. Studies from other countries have shown that there are benefits that successfully delivered and while some are not.

In French, [33] found that PPP projects performs well on issues such as respect of timetable and contracted prices but does not establish that PPP promotes innovation, quality of service delivery and life cycle costing. In accessing the experience of PFI in the UK's National Health Service, [34] has concludes that PFI provides no additional funding, do not deliver better 'value for money', create problems in affordability and increased cost of private finance to the public purse.

[35] have reviewed the elements of cost, quality, flexibility and complexity of PPP hospitals in Australia, Spain and UK. They found that the new facilities have been more expensive. The facilities are more likely to be built on time and within the budget, but at the expense of compromises on quality. The study concludes that PPP seem to further complicates the already difficult task of developing and managing hospital.

By adopting case study approach, [36] discovered that the first PPP hospital in Victoria shows that the hospital facility was built "within budget" and the government have indicate this cases evidence of successful implementation of the PPP policy. In Hong Kong, [37] found that non-privately funded PPP approach is more cost-effective and efficient in the delivery maintenance services. Both the public agent and service provider involve in the study perceive the non-privately funded PPP approach can allow higher flexibility, encourage innovation, enhance cost-effectiveness and improve the efficiency of work when compared with the traditional term contract in the delivery of maintenance services.

It was discovered that the PPP benefits are yet to be tested in Malaysian context. This paper is hence attempted to empirically verify these benefits. Next section will discuss how this study is conducted.

4.0 Research method

4.1 Research instrument

This study is embarked on quantitative methods. Similar to study conducted by [38] and [39], structured questionnaires were used to capture the perception of stakeholders on the topic of the present study. The questionnaire consists of four parts and this paper intends to report the first two parts. Part A consist of the demographic information of the respondents while Part B uses a Likert scale to measure the benefits of implementing PPP in delivering public facilities on a scale of 1-5, where "1" is very disagree to "5" is very agree. **Table 1** provides seven PPP benefits listed in the questionnaire.

Table 1. PPP benefits

No	PPP benefits
1.	PPP leads to the optimum whole life costs of public facilities.
2.	PPP yields lower Capex and Opex compared to traditional procurement.
3.	PPP allows faster delivery of projects
4.	PPP provides better project governance.
5.	PPP projects demonstrate innovative design aspect.
6.	PPP improves the quality of services delivered.
7.	PPP allows transfer of risk to the parties that could manage it better.

4.2 Questionnaire distribution and data collection

The questionnaire survey was distributed to the respondents in a PPP healthcare conference conducted in Malaysia. This approach was adopted by Ismail and Haris [14] to investigate the factors that hinder the successful adoption of PPP in Malaysia. The chosen conference comprises of those who have knowledge, experience or at least interest on PPP scheme in Malaysia. Through the questionnaire, respondents are required to rate their agreements level on the benefits of PPP implementation in delivering public facilities in Malaysia. A total of 110 questionnaires were distributed to the participants. At the end of conference, 45 completed questionnaires were successfully collected representing a response rate of 40.9 per cent.

4.3 Data analysis

All 45 returned questionnaires were properly completed and analysed accordingly by using SPSS statistics software version 22. The demographic information of respondents were analysed using descriptive analysis and the mean score was used to rank the seven identified PPP benefits.

5.0 Findings and discussions

5.1 Response rate and demographic information of the respondents

Demographic information of the respondents is showed in **Table 2**. 47.7 per cent of the respondents are from the public sector whilst the remaining 53.3 per cent are from the private sector. In relation to working experience, majority of the respondents (75.5 per cent) have less than 5 years' experience in PPP projects. In term of number of PPP projects, 62.2 per cent of respondents have been involved in less than five projects and 37.8 per cent have experience in more than five projects. The demographic information of respondents has reflects their credibility in providing reliable data in the survey.

Table 2. Descriptive analysis of respondents

Respondents profile	Proportion of the sample response (%)
Working sectors	
Public	47.7
Private	53.3
Years of experience in PPP projects	
Less than five years	75.5
5-10 years	8.9
11-15 years	8.9
More than 15 years	6.7
Number of projects involved	
Less than 5	62.2
Five and more	37.8

5.2 Result on PPP benefits

Table 3 reveals the frequency of response for each benefits of PPP implementation in delivering public facilities in Malaysia. The first two PPP benefits which are "PPP leads to the optimum whole life costs of public facilities" and "PPP yields lower Capex and Opex compared to traditional procurement" showed that the highest percentage of response is in 'Neutral' category. The answer of respondents was scattered among all five level of agreements. Meanwhile, the responses for the remaining five PPP benefits seem to be centralised at 'Agree' category. The answer of respondent for these non-financial PPP benefits were found to be scattered between 'Disagree' and 'Strongly Agree' categories, while no response fell into 'Strongly Disagree' categories.

Mean analysis was conducted to compare the listed benefits. As shown in **Table 4**, the mean score for the seven benefits range from 3.00 – 3.98, which indicate there is different rank of each benefits. Table 4 show that in rank order, the PPP benefits that agreed by overall respondents were "improve quality of services", "innovative design", "transfer of risk", "faster delivery of projects" and "better project governance". All these benefits represent the non-financial PPP benefits. The result of analysis also showed that financial PPP benefits which are "optimum whole life cost" and "lower Capex and Opex" were neither agreed nor not agreed as the benefits of implementing PPP in Malaysia. Overall, the result indicates that all non-financial PPP benefits are recognised and agreed favourably by the respondent than non-financial PPP benefits.

 DOI: 10.1051/matecconf/20166600014

Table 3. PPP Benefits in Malaysia

PPP benefits	Strongly Disagree		Disagree		Neutral		Agree		Strongly agree	
	n	%	*n*	%	*n*	%	*n*	%	*n*	%
1. PPP leads to the optimum whole life costs of public facilities.	1	2	3	7	19	42	18	40	4	9
2. PPP yields lower Capex and Opex compared to traditional procurement.	2	4	12	27	16	36	14	31	1	2
3. PPP allows faster delivery of projects	-	-	3	7	12	27	22	49	8	18
4. PPP provides better project governance.	-	-	2	4	13	29	26	58	4	9
5. PPP projects demonstrate innovative design aspect.	-	-	2	4	8	18	26	58	9	20
6. PPP improves the quality of services delivered.	-	-	1	2	6	13	31	69	7	16
7. PPP allows transfer of risk to the parties that could manage it better.	-	-	2	4	10	22	25	56	8	18

Table 4. Mean and rank of PPP benefits in Malaysia

PPP benefits	Mean	Ranking
1. PPP improves the quality of services delivered.	3.98	1
2. PPP projects demonstrate innovative design aspect.	3.93	2
3. PPP allows transfer of risk to the parties that could manage it better.	3.87	3
4. PPP allows faster delivery of projects	3.78	4
5. PPP provides better project governance.	3.71	5
6. PPP leads to the optimum whole life costs of public facilities.	3.47	6
7. PPP yields lower Capex and Opex compared to traditional procurement.	3.00	7

The benefits "improves the quality of services" (mean=3.98) has secured the first place thus perceived as the top benefits of PPP implementation in Malaysia. It was expected that PPP will give a positive impact on the quality of services the most. The reformation of facilities management involvement, which is since in the development phase of PPP project, has led to a significant improvement on the quality of services. In contrast, study by [33] found that this benefits cannot be enjoyed in French.

"Design innovation" (mean = 3.93) has ranked as the second benefits in implementing PPP in Malaysia. Nevertheless, this advantage is facing lots of criticism in other countries. In contrast, [27] claimed that PPP was less effective in stimulating design innovation desired by the UK government. Design innovation has also showed a slower progress than process innovation in PPP projects [22]. Considering Malaysia as a developing country and existing design flaws in traditional procurement of public facilities, it is possible for design aspect to improve a lot by the implementation of PPP.

The third benefit of PPP implementation in Malaysia is "transfer of risk to the parties that could manage it better" (mean=3.87). The transfer of some risks, operational and maintenance for instance allow the government to focus on their other core business. Besides, [13] claimed that the encouragement to use Islamic financing in implementing PPP/PFI projects in Malaysia may lead to new perceptive towards the concept of risk transfer and improved risk sharing.

The benefits in terms of cost are less prominent compared to other benefits. In other study, [40] have put cost as the main disadvantage of PPP projects. According to [21], the ability to deliver project at a lower cost depends on private sector partner having the appropriate incentive. This is due to the interest of private, which usually aims for profit maximisation and not cost minimisation. More effort should be taken by the PPP stakeholder to boost financial benefits of PPP implementation. One of the solutions is by running a competitive process that usually offers a better solution at a lower cost [41].

6.0 Conclusion

The study used a questionnaire survey to investigate PPP benefits in delivering public services in Malaysia. Most of the benefits were obtained through private sectors efficiency. The overall results show that "PPP improves the quality of services delivered", "PPP projects demonstrate innovative design aspect" and "PPP allows transfer of risk to the parties that could manage it better" are the top three PPP benefits in Malaysia.

However, there are a few limitations in this study that should be highlighted in order to ensure a fair interpretation of the findings. First, the finding of this is solely based on the perception of local PPP stakeholders. Future studies may consider using other dimension or indicators such as monetary, customer satisfaction etc. to measure the PPP benefits. Second, the used of general questionnaire may not be the best option to grasp the real impact of PPP for specific type of projects. Therefore, case study approach is highly recommended as different projects may show different results. Third, the selection of conference as a medium to approach the sample of study has limited the number of response. In this case, [14] claimed that it would be better to increase the sample size and distribute the survey instrument to larger geographical area or more seminars.

In conclusion, despite the limitations, this study has filling the gap of existing literature in regards to PPP implementation in Malaysia. In particular, this study expose the perception of public and private sectors in regards to PPP benefits in delivering public facilities in Malaysia. The findings has portrayed the impact of PPP and justified the use of this scheme in the future.

Acknowledgement

This research is supported by the Ministry of Education, Malaysia under Fundamental Research Grant Projects (FRGS) (VOT: 4F633).

References

1. Kinder, C. and D. Wright, *Better Public service Delivery Through Public-Private Partnerships: A New Approach to Infrastructure Development*. (2009), Commonwealth Business School and International Financial Services, London (IFSL).
2. Cruz, C.O. and R.C. Marques, *Infrastructure Public-Private Partnerships: Decision, Management and Development*. (2013): Springer Berlin Heidelberg.
3. Yescombe, E.R., *Public-Private Partnerships: Principles of Policy and Finance*. (2011), Butterworth-Heinemann: Elsevier Science.
4. Shen, L.-Y., A. Platten, and X.P. Deng, *Role of public private partnerships to manage risks in public sector projects in Hong Kong*. INT J PROJ MANAGE, (2006). **24**(7): p. 587-594.
5. Cheung, E., A.P.C. Chan, and S. Kajewski, *Reasons for implementing public private partnership projects: Perspectives from Hong Kong, Australian and British practitioners*. JPIF, (2009). **27**(1): p. 81-95.
6. Takim, R., et al., *The Malaysian Private Finance Initiative and Value for Money*. Asian social science, (2009). **5**(3): p. P103.
7. Public Private Partnership Unit, *Public Private Partnership Guideline*, P.M.D. Public Private Partnership Unit, Editor. (2009): Malaysia.
8. Beh, L.S., *Development and Distortion of Malaysian Public-Private Partnerships–Patronage, Privatised Profits and Pitfalls*. AJPA, (2010). **69**(s1): p. S74-S84.
9. Ismail, S., *Critical success factors of public private partnership (PPP) implementation in Malaysia*. APJBA (2013). **5**(1): p. 6-19.
10. Ninth Malaysia Plan, *Ninth Malaysia Plan (2006-2010)*. 2006, Economic Planning Unit: Malaysia.
11. Ismail, S. and F.A. Haris, *Rationales for public private partnership (PPP) implementation in Malaysianull*. JFMPC, 2014. **19**(3): p. 188-201.
12. Ismail, S., *Factors attracting the use of public private partnership in Malaysia*. JCDC, (2013). **18**(1): p. 95-108.
13. Abdullah, N., et al. *PPP/PFI in Malaysian Development Plans: Purpose, Structure, Implementation, Financing and Risk Transfer*. in *Proceedings of 5th Asia-Pacific Business Research Conference*. (2014). Hotel Istana, Kuala Lumpur, Malaysia.
14. Ismail, S. and F.A. Haris, *Constraints in implementing Public Private Partnership (PPP) in Malaysia*. Built Environment Project and Asset Management, (2014). **4**(3): p. 238-250.
15. Ismail, S. *What Drives Value for Money of Public Private Partnership (PPP) Projects Implementation in Malaysia?* in *Persidangan Kebangsaan Ekonomi Malaysia (PERKEM VII)*. (2012). Perak, Malaysia.
16. Takim, R., K. Ismail, and A.H. Nawawi. *A value for money assessment method for Public Private Partnership: A lesson from Malaysian approach*. in *International Conference on Economic and Finance research*. 2011. Singapore.
17. El-Haram, M.A., S. Marenjak, and M.W. Horner, *Development of a generic framework for collecting whole life cost data for the building industry*. JQME, (2002). **8**(2): p. 144-151.
18. Akintoye, A., M. Beck, and C. Hardcastle, *Public-Private Partnerships: Managing Risks and Opportunities*. (2003): Wiley.
19. Siemiatycki, M. and N. Farooqi, *Value for Money and Risk in Public–Private Partnerships*. JAPA, (2012). **78**(3): p. 286-299.
20. Grimsey, D. and M. Lewis, *Public Private Partnerships: The Worldwide Revolution in Infrastructure Provision and Project Finance*. (2007): Edward Elgar.
21. Boardman, A., F. Poschmann, and A. Vining, *North American infrastructure P3s: examples and lessons learned*. The Challenge of Public–Private Partnerships: Learning from International Experience, (2005): p. 162-189.
22. Robinson, H., et al., *Governance and Knowledge Management for Public-Private Partnerships*. (2009): Wiley.
23. U.S. Department of Transportation, *Report to Congress on the Costs, Benefits, and Efficiencies of Public-Private Partnerships for Fixed Guideway Capital Projects*. (2007), U.S. Department of Transportation: Washington, DC.
24. Kumaraswamy, M., et al., *A review of Australian PPP governance structures*. JFMPC, (2010). **15**(3): p. 198-215.
25. Patel, M. and H. Robinson, *Impact of governance on project delivery of complex NHS PFI/PPP schemes*. JFMPC, (2010). **15**(3): p. 216-234.
26. Cruz, C.O. and R.C. Marques, *Integrating Infrastructure and Clinical Management in PPPs for Health Care*. J. Manage. Eng., (2013). **29**(4): p. 471-481.
27. Barlow, J. and M. Köberle-Gaiser, *The private finance initiative, project form and design innovation: The UK's hospitals programme*. Research Policy, (2008). **37**(8): p. 1392-1402.
28. Alshawi, M., *Concept and background to Public Private Partnership / Private Finance Initiative - UK*

Experience. 2009, Iraq Institute for Economic Reforms (IIER).

29. Raisbeck, P., *Considering design and PPP innovation: A review of design factors in PPPresearch*, in *Procs 25th Annual ARCOM Conference*, A. Dainty, Editor. (2009), he Association of Researchers in Construction Management p. 239-47.
30. El-Haram, M.A. and A. Agapiou, *The role of the facility manager in new procurement routes*. JQME, (2002). **8**(2): p. 124.
31. Doloi, H., *Understanding impacts of time and cost related construction risks on operational performance of PPP projects*. INT J STRATEG PROP M, (2012). **16**(3): p. 316-337.
32. Ling, F.Y.Y., et al., *Public Private Partnerships: Overcoming Pre-Contract Problems Faced by Public Clients and Private Providers in Infrastructure Projects in Singapore*. IJCM, (2011). **11**(1): p. 63-77.
33. Bougrain, F., *Energy performance and public private partnership*. BEPAM, (2012). **2**(1): p. 41-55.
34. Liebe, M. and A.M. Pollock, *The experience of the private finance initiative in the UK's National Health Service*. (2009), The Centre for International Health Policy, The University of Edinburgh: Edinburgh.
35. McKee, M., N. Edwards, and R. Atun, *Public-private partnerships for hospitals*. Bulletin of the World Health Organization, (2006). **84**(11): p. 890-896.
36. Sciulli, N., *Public private partnerships: an exploratory study in health care*. Asian Review of Accounting, (2008). **16**(1): p. 21-38.
37. Ng, S.T. and Y.M.W. Wong, *Adopting non-privately funded public-private partnerships in maintenance projects: A case study in Hong Kong*. ECAM, (2006). **13**(2): p. 186-200.
38. Li, B., et al., *Perceptions of positive and negative factors influencing the attractiveness of PPP/PFI procurement for construction projects in the UK: Findings from a questionnaire survey*. ECAM, (2005). **12**(2): p. 125-148.
39. Babatunde, S.O., et al., *Stakeholder perceptions on critical success factors for public-private partnership projects in Nigeria*. BEPAM, (2016). **6**(1): p. 74-91.
40. Eadie, R., P. Millar, and R. Grant, *PFI/PPP, private sector perspectives of UK transport and healthcare*. BEPAM, (2013). **3**(1): p. 89-104.
41. ICA Secretariat, *Attracting Investors to African Public-private Partnerships: A Project Preparation Guide*. World Bank e-Library. (2009), Washington DC: World Bank.

Perceptions Towards Non-Value-Adding Activities During The Construction Process

Haryati Ismail[1,a] and Zakaria Mohd Yusof [2]

[1]*Department of Quantity Surveying, Faculty of Built Environement,Universiti Teknologi Malaysia*
[2]*Department of Quantity Surveying, Faculty of Built Environement,Universiti Teknologi Malaysia*

Abstract. Non-value-adding activities are pure waste during the construction process. However, most of the construction practitioner does not realise that most of the activities performed during the construction process add no value to their project. A total of 375 numbers of questionnaires distributed to the Developer, Jabatan Kerja Raya, Consultants and Contractors. The study found that awareness by construction participants in Malaysia to take actions against non-value-adding activities during the construction process is relatively low. Through analysed by using the Pareto Chart, it has been found that defects and waiting time are two categories of non-value-adding activities that need to be prioritised by the industry. It is also found that non-value-adding activities most frequently occurred during structural and architectural work. This paper also reviewed on the causes of non-value-adding activities and discussed its effect towards time, cost, quality and productivity of the construction project. This paper is also important to give clearness and broader understandings on this form of waste other than material waste.

Keywords: - defect, non-value-adding, Pareto, waiting time, waste,

1. Introduction

The construction term is used to define the activity of creation physical infrastructure which comprises work such as residential projects, non-residential project, civil engineering and specialist project. The construction industry plays a vital role in economic growth, social and national growth of the country [1]. The constructional field provides the socioeconomic infrastructure for economic growth and defines as an economic engine for developed economies. It produces wealth and quality of life and generate a huge employment opportunities. The construction industry is very complicated industry in nature due to various levels of management, complex design, lot of process and labour intensive.

Nevertheless, despite its importance, the construction industry has many challenges, which requires immediate attention. Common problems in this industry, including low productivity, poor safety, poor documentation, hostile work environments, inefficient costs and low quality of production [2]–[6]. Another major problem involved is the great level of waste, which is undesirable, no added value to the product, and also time and cost consume to remove during the construction process.

Waste in the construction industry has gained attention all over the world. According to Rahman et al., (2012)[4], waste elimination is one of the main objectives in lean construction. Skoyles & Skoyles (1987) [7] argued that waste as one of the weaknesses in the construction industry. Alwi, Hampson, & Mohamed (2002) [8] suggested that a proper understanding is needed about waste. Most of the construction practitioners viewed waste as material waste, or debris. After a systematic review of previous researches on waste, Viana, Formoso, & Kalsaas (2012) [9] divided into three (3) categories, namely, construction material waste (physical waste); non-value-adding activities (process waste); and specific waste (such as accidents and rework). This paper will only focus on the second type of waste which is non-value-adding activities.

This paper will discuss the perception by the construction practitioners towards non-value-adding activities during the construction process. Other than that, it also discuss on the causes of non-value-adding activities and its effect towards time, cost, quality and productivity of the construction project. By using Pareto Chart, categories of non-value-adding activities that need to be prioritised and need urgency solution from the industry will be shown at the end of the study.

2. Non-value-adding activties (NVAAs) as pure waste in the construction industry

A study by Felipe et al. (2012) [10] found that construction practitioners, including executive level refer waste as scrap at the construction site. However, according to new philosophy developed by Koskela (1992) [11], waste should be referred as any ineffective usage of equipment, materials, labour, or capital during production. Macomber & Howell (2004) [12] emphasized that anything that's not produces value is waste. Furthermore Formoso & Hirota (1999) [13] refer waste to any activity that from the point of view of the client causes direct or indirect costs, but do not add any value to the product.

The term "non-value-adding activities" (NVVAs) is used by Koskela & Sharpe (1994) [14] to explain waste in the lean production philosophy. According to these researchers, only activities that convert materials and/or information towards what is required by the costumer are value added. Han & Lee (2007) [15] claimed that NVAAs consume time and resources, and need to be avoided. Recently, Ralph & Iyagba (2012) [16] used the term non-value-adding activities to differentiate between physical waste and other waste that occurs during the construction process.

Horman & Kenley (2005) [17] in their research found out that on average, 49.6% of time usage in construction are identified as wasteful activities. Unfortunately, most of the construction practitioner does not realise that most of the activities performed during the construction process add no value to their project. Activities such as waiting, rework, unnecessary movement, overproduction, delay, defects, lack of quality, and inventories are wasted because it's given no value-added to the process. These non-value-adding activities occur throughout the entire construction process and contribute through every participant in the project [6]. Josephson & Saukkoriipi (2007) [18] also supported this phenomenon and therefore further suggest that there is probably a lack of knowledge and size of non-value-adding activities in the construction industry.

Non-value-adding activities will give direct effect to the construction process and project but can be avoided by executing work correctly, close monitoring, controlling and planning. Everyone involved in the construction process has potential to contribute to NVAAs, and therefore, affect the process. Therefore, NVAAs can be referred as activities that consume direct and indirect cost, time, resources, labour and space, but gives no value added to anyone involved during the process.

2.1 Categories of non-value-adding activities

Term waste is used to refer non-value adding activities in lean thinking [18]. The "seven wastes", known as "muda", have been proposed by Ohno (1988) [19]. Waste of defects, waste of transportation, waste of processing and waste of inventory refers to the flow of materials, while waste of waiting and waste of motion refers to workers [11]. As highlighted by Motavallian & Settyvari (2013) [20], many researches have added "unused human talent" as the eighth principle of waste [22] - [23].This research categorized non-value-adding activities into eight categories also known as DOWNTIME as shown in table 1.

Table 1: Defination of non-value-adding activities

Type Of Waste	Define
Defect	Efforts caused by rework, scrap, and incorrect information.
Overproduction	Production that is more than needed or before it is needed.
Waiting	Wasted time waiting for the next step in a process.
Non-utilised talent	Underutilizing people's talents, skills, and knowledge.
Transportation	Unnecessary movements of products and material.
Inventory	Excess products and materials not being processed..
Motion	Unnecessary movements of people.
Extra processing	More work or higher quality than is required for the customer.

Other than that, there are others NVAAs (also referred as waste) by previous researches. For example, poor quality [6], [23], [24], excessive inspection on site [4], [14], [24], waiting for equipment repair [16], [25], activity start delay [16], [24]–[27], not listening and not speaking [12], poor vehicle and truck movement [28], unnecessary over-order of materials [4], [11], [21], excess materials on site [28], too much material handling [8] [16], long walking distances [24], [26], [29], making do [30], excess information [12] and excessive supervision [24], [27], [31].

However, most of these types of waste above can be included in one of the eight categories of non-value-adding-activities in this research. For example, poor quality can be concluded under the defect category, waiting for equipment under the Waiting time category, and excess material on site, under the inventory category. There are few types of NVAAs that maybe the causes of NVAAs rather than NVAAs itself. For example, unreliable equipment [16], [25] is the cause of defect, waiting time and motion.

2.2 Causes of non-value-adding activities

Koskela & Sharpe (1994) [14] indicated that there are three causes of NVAAs, which are designed in hierarchical organisations, ignorance, and nature of production. These researchers suggested that traditional management, improper process during design, errors, and machine breakdown also contributed to NVAAs. Other than that, Koskela & Leikas (1997) [32] mentioned that NVAAs occurred due to failure to recognize or measure waste, missing information and complicated material flows.

Previous researchers found that lack of skills by subcontractors and traders were among the causes of NVAAs. Furthermore a great deal of NVAAs also contributed by changes in design, poor coordination, weather, poor planning and scheduling, poor supervision, design changes, slow decision making, lack of trades and subcontractor skills, incorrect construction method, delay of materials, communication breakdown, lack of coordination and lack of trust among parties [10], [16], [21].

Researches by Alwi et al.,(2002) [6] and Alarcon (1997) [24] found that poor quality of site documentation, weather, unclear site drawing supplies, poor design, design changes, slow drawing revision and distribution, unclear specifications, management, information and resources are among vital factors of NVAAs. All the causes above can be categorised into 8M's; management, measurement, method, man, mother nature, material, machinery and money. For example, poor coordination, poor planning, poor scheduling, and poor supervision fall under the management categories, whereas lack of trades and subcontractor skills fall under the man categories.

3. Research Methodology

This study aims at to identify the perceptions towards non-value-adding activities by the construction practitioners in the Malaysian construction industry. The respondents involved in this study are client (Jabatan Kerja Raya and private developer), consultants (Architect, Engineer, Quantity Surveyor and Project Manager) and contractors. Only Contractors registered G7with CIDB selected for this study due their expertise and experience in handling mega and complex project and have no limit in tendering project.

Before questionnaire distributed to the respondents, it firstly was validated by the academician, field expert and statistical expert. Pilot test also conducted with 30 numbers of respondents. However, only 23 numbers of questionnaires were returned to the researcher. The internal consistency and inter-item correlation (relationship between items) of the questionnaire was assessed by the Cronbach alpha method. The Cronbach Alpha value calculated for pilot test was 0.912. The questionnaire consists of three (3) sections as below:-

Section (A) is structured to obtain general information and background about the respondents such as type of organization, position and years of experience.

Section (B) is structured to identify the perceptions of construction participants towards non-value adding activities during the construction process.

Section (C) is structured to identify the effect of non-value-adding activities.

4. Result and Discussion

A total numbers of 375 questionnaires distributed randomly through postal mail around Malaysia (125 sets to clients, 125 sets to consultants and 125 sets to the contractors). After a few follow up through email, online application and telephone, a total 106 sets of questionnaire returned to the researcher within 4 weeks, giving a 28% response rate. According to Akintoye (2000) [33], in the construction industry, it is normal to get a response rate between 20%–30% for postal questionnaire. This is supported by Hoe (2005) [34] which indicated that the response rate of 28% is acceptable. Teo & Loosemore (2010) [35] on their study were using survey to collect data, got a 29.1% response rate. They believed that data collected from those respondents were representative of the population as a whole. In addition, Love & Smith (2003) [36] suggested that 30% response rate is considered satisfactory. Therefore, 28% response rate for this study is considered acceptable and represents the population.

The calculated Cronbach alpha value is 0.870 which exceeded the minimum Cronbach alpha value of 0.70 [37]. Findings from the data gathered through the questionnaires are as follows:

4.1 Respondent information

Of 106 questionnaires returned, 34 numbers were from the client, 38 numbers from the consultants and 34 numbers from the contractor. Respondents from Client consists of the private sector and the Government, meanwhile, respondent involved represent consultants are from architect, quantity surveyor and engineer. All contractors responds were registered G7 with CIDB. The each percentage of respondents for these three (3) types of respondents is not too different and almost equal which is between 32%-35%.

As shown in the table 2, 67.9% of the respondents have more than 10 years of experience in the construction industry. It can be seen that those with 26 years and above experience contributes 22.6% of the total numbers of respondent. According to the data collected, more than half of the respondents mostly participated in government project (53.8%), whereas 34% and 12.3% of the respondents mainly involved in private project and both project ownership respectively.

Table 2: Respondent information

Type of respondent's organization		
	Frequency	Percent
Client	34	32.1
Consultant	38	35.8
Contractor	34	32.1
Total	**106**	**100**
Years of experience		
	Frequency	Percent
0-5 years	18	17
6-10 years	16	15.1
11-15 years	18	17
16-20 years	20	18.9
21-25 years	10	9.4
26 years and above	24	22.6
Total	**106**	**100**
Main project ownership participated		
	Frequency	Percent
Government	57	53.8
Private	36	34
Both	13	12.3
Total	**106**	**100**

4.2 The perceptions of construction participants towards NVAAs

Section B is structured to identify the perception from construction participants towards NVAAs. In this section, respondents were asked four (4) questions. In the first questions, respondents were asked to indicate the frequency of their organization to take actions towards NVAAs during the construction process by using 5-point Likert scale ranging from 1 (never) to 5 (always). This question has been analyzed using the mean value. By referring to Ariola (2006) [38], the mean score from Likert-scale type questions can be interpreted as in Table 3.

From the table 4, it is notable that all seven items mean scores ranged around 2.51- 3.50 which indicate that on average, only sometimes the actions taken by the respondent against NVAAs. This finding supported the problem statement that the awareness among construction participants towards NVAAs is quite low. Investigate the root causes of NVAAs get the lowest mean score (3.06). More than half of the respondents (62%) respondents answer they never (12%), rarely (16%) and sometimes (34%) investigated the root causes of NVAAs in their project.

Table 3: Interpretation of Likert scale for frequency.

Scale	Interpretation	Scale Range
1	Never	1.00-1.50
2	Rarely	1.51-2.50
3	Sometimes	2.51-3.50
4	Often	3.51-4.50
5	Always	4.51-5.00

Table 4: Actions taken again NVAAs

Actions	**Response %**					**Mean**
	1	**2**	**3**	**4**	**5**	
Identify non-value-adding activities during the construction process.	3%	16%	29%	37%	15%	3.453
Measure the non-value-adding activities during the construction process.	8%	17%	37%	32%	7%	3.132
Identify the causes of non-value-adding activities.	3%	16%	32%	41%	8%	3.359
Investigate the root causes of non-value-adding activities.	12%	16%	34%	29%	8%	3.057
Analyses the effect of non-value-adding activities towards project performance.	3%	14%	31%	40%	8%	3.255
Reduce non-value-adding activities during the construction process.	3%	10%	39%	36%	12%	3.443
Eliminate non-value-adding activities during the construction process.	9%	10%	39%	32%	9%	3.217

Second question in Section B was structured to seek the agreement from the respondents whether NVAAs contribute to poor construction project performance assessed by using 5-point Likert scale representing from 1 (strongly disagree) to 5 (strongly agree) and calculating the mean. The interpretations towards each statement are shown in Table 5 [38].

The finding from the data which tabulate in table 6 shown the mean score for eight (8) categories of NVAAS range between 3.51- 4.50. Therefore, by referring table 5, it can be deemed that an average, the respondents agree that all the categories of NVAAs contribute to poor construction project performance.

Table 5: Interpretation of Likert-scale for agreement.

Scale	Interpretation	Scale range
1	Strongly disagree	1.00-1.50
2	Disagree	1.51-2.50
3	Neither agree/disagree	2.51-3.50
4	Agree	3.51-4.50
5	Strongly agree	4.51-5.00

Table 6: Contribution of NVAAs towards poor project performance

Category	% Response					Mean
	1	2	3	4	5	
Defect	0%	4%	14%	50%	3%	4.104
Overproduction	5%	12%	25%	42%	2%	3.528
Waiting time	0%	4%	12%	44%	4%	4.198
Non-utilised talent	3%	8%	31%	40%	2%	4.019
Transportation	3%	7%	22%	56%	1%	3.698
Inventory	0%	10%	25%	49%	2%	3.708
Motion	1%	9%	31%	40%	2%	3.660
Extra processing	0%	19%	25%	42%	1%	3.528

Third and fourth question in this section structured to identify the categories of NVAAs and type of works where NVAAs frequently occurred. Data were then analysed by using Pareto analysis. Pareto analysis used to find out what problem should be prioritized,and also to identify targets for improvements. Pareto analysis (80/20) using the idea that a large majority of problems (80%) is produced by a few key causes (20%) and also known as the vital few and the trivial many.

Four (4) respondents were not answered these two (2) questions. From table 7 and figure 1, it is notable that defect and waiting time makes up 78% from the total NVAAs occurred during the construction process. Meanwhile, 81% of NVAAs occurred during structural and architectural works (see table 8 and figure 2). Therefore, to improve the construction process, an in-depth study by focusing on root cause analysis will be carried out on these two categories of NVAAs during structure and architecture work.

Table 7: Frequency of NVAAs occurred during the construction process.

Category	Frequency	%	Cumulative %
Defect	48	47%	47%
Waiting time	32	31%	78%
Overproduction	9	9%	87%
Non-utilised talent	4	4%	91%
Extra-processing	3	3%	94%
Transportation	2	2%	96%
Inventory	2	2%	98%
Motion	2	2%	100%
Total	102	100%	

Table 8: Types of works where NVAAs frequently occurred during the construction process.

Type of work	Frequency	%	Cumulative %
Architectural work	51	50%	50%
Structural work	32	31%	81%
External work	14	14%	95%
Electrical work	5	5%	100%
Mechanical Work	0	0	
Total	102	100%	

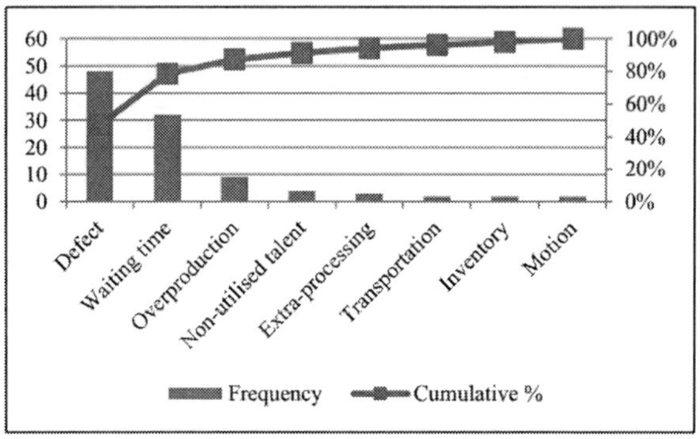

Figure 1: Pareto Analysis on frequency of NVAAs occurred during the construction process.

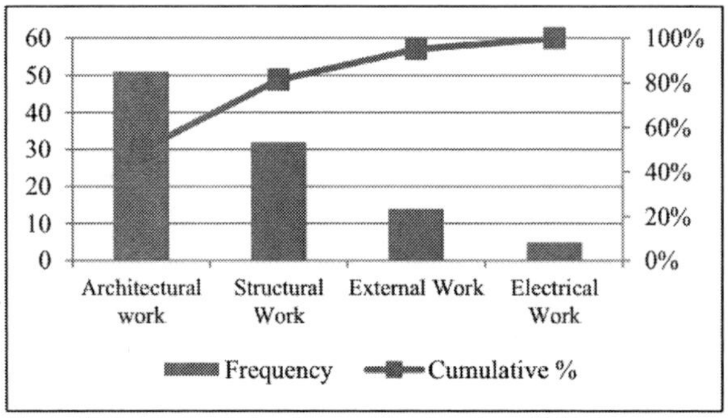

Figure 2: Pareto Analysis of types of works where NVAAs frequently occurred during the construction process.

4.3 Effect of NVAA

Section C on the questionnaire is designed to determine the effect of the non-value-adding activities. Respondents were asked to rate the effect of the NVAAs towards four (4) important parameters in construction projects which are time (T), cost (C), quality (Q) and productivity (P) by using 5 points Likert scale from 1 (no effect) to 5 (major effect). Interpretations for 5 point Likert-scale in effect are as described in Table 9.Data from respondent for this section was calculated by using mean score.

From table 10, it is can be seen that only defect gives major effect towards the quality where it's mean value range 4.51-5.00. The mean score for others parameters for defect are fall under the moderate effect. For waiting time, our finding revealed that it will give major effect towards time. Other three (3) parameters fall under moderate effect where the mean values are 4.14 for cost, 3.54 for quality and 4.05 for productivity. Transportation and overproduction get the lowest mean score are deemed to give neutral effect towards quality.

The average mean score was calculated to identify the overall effect of NVAAs towards the project. Average mean score for seven (7) categories of NVAAs which are defect, overproduction, waiting time, non-utilised talent, transportation, motion and extra-processing towards four (4) parameters are range 3.51 – 4.50, with defect get the highest mean score (4.25) and followed by waiting time (4.06). Meanwhile, the least average mean score is inventory (3.50).

Table 9: Interpretation of Likert-Scale for Effect

Scale	Interpretation	Scale range
1	No effect	1.00-1.50
2	Minor effect	1.51-2.50
3	Neutral effect	2.51-3.50
4	Moderate effect	3.51-4.50
5	Major effect	4.51-5.00

Table 10: Effect of NVAAs towards four (4) parameters

Category	T	C	Q	P	Average mean score
Defect	4.094	4.264	4.585	4.047	4.248
Overproduction	3.472	4.009	3.293	3.538	3.578
Waiting time	4.509	4.146	3.538	4.047	4.059
Non-utilised talent	3.633	3.566	3.660	3.698	3.639
Transportation	3.840	3.934	3.292	3.669	3.684
Inventory	3.519	3.745	3.311	3.443	3.505
Motion	3.705	3.591	3.359	3.679	3.583
Extra-processing	3.585	3.915	3.311	3.528	3.585

5. Conclusion

Non-value-adding activities referred as activities that consume direct and indirect cost, time, resources, labour and space, but give no value added to anyone involved during the process. The eight (8) categories of NVAAs are defect, overproduction, waiting time, non-utlised talent, transportation, inventory, motion and extra processing also known as DOWNTIME. The findings show that awareness among Malaysian construction practitioners against this issue is quite low. The findings discovered that, construction practitioners don't take NVAAs seriously, even though they were agreeing that NVAAs contribute to poor project performance. By using the Pareto analysis and the idea that a large majority of problems (80%) is produced by a few key causes (20%), priorities should be given to solve the problem of defect and waiting time. The study found 78% NVAAs occurred during the construction process are defect and waiting time. Further study should also be focused on structural and architectural works because 81% of NVAAs occurred during these two types of work. Defects and waiting time will give major effect towards the quality and time of the project, respectively. These two types of NVAAs also get highest and second highest averages mean score for overall effect. Therefore, it can be concluded that by eliminating defects and waiting time during structural and architecture work will give a huge impact towards time, cost, quality and productivity of the project.

References

[1] F. Ismail, N. M. Yusuwan, and H. E. A. Baharuddin, "Management factors for successful IBS projects implementation," *Procedia-Social Behav. Sci.*, vol. 68, no. 1877–0428, pp. 99–107, 2012.

[2] L. Koskela, "Lean production in construction," *Elsevier Sci. Publ.*, pp. 47–54, 1993.

[3] A. Forsberg and L. Saukkoriipi, "Measurement of Waste and Productivity in Relation to Lean Thinking," in *Proceedings IGLC-15, July 2007, Michigan, USA*, 2007, vol. 46, no. July, pp. 67–76.

[4] H. A. Rahman, C. Wang, I. Yen, and W. Lim, "Waste Processing Framework For Non-Value-Adding Activities Using Lean Construction," vol. 1, pp. 8–13, 2012.

[5] R. Rivas and A. Serpell, "Evaluation and Improvement of the Procurement Process in Construction Projects," in *Proceeding IGLC-7, University of California, Berkeley, CA, USA*, 1999, pp. 219–230.

[6] S. Alwi, K. Hampson, and S. Mohamed, "Non value adding activities : A comparative study of Indonesian and Australian construction project," *Proc. 10th Int. Gr. Lean Constr. Conf. Gramado, Brazil*, pp. 627–638, 2002.

[7] E. R. Skoyles and J. R. Skoyles, *Waste prevention on site*. BT Batsford Limited., 1987.

[8] A. Sugiharto, K. Hampson, and S. Mohamed, "Factors influencing contractor performance in Indonesia : a study of non value-adding activities," *Proc. Int. Conf. Adv. Des. Constr. Constr. Manag. Maint. Build. Struct.*, vol. 2, pp. 20–34, 2002.

[9] D. D. Viana, C. T. Formoso, and B. T. Kalsaas, "Waste in Construction: A Systematic Literature Review on Empirical Studies," in *Proceedings for the 20th Annual Conference of the International Group for Lean Construction*, 2012.

[10] J. Felipe, P. Achell, and M. B. Bonet, "A preliminary proposal for a waste-based management approach to improve performance in construction," *Proceeding IGLC-21, July 2013, Fortaleza, Brazil*, pp. 115–124, 2012.

[11] L. Koskela, "Application of the new production philosophy to construction," Stanford University, 1992.

[12] H. Macomber and G. Howell, "Two great wastes in organizations : A Typology for Addressing the Concern for the Underutilization of Human Potential," in *proceedings IGLC12, Elsinore, 25-27 July.*, 2004, pp. 1–9.

[13] C. T. Formoso and E. H. Hirota, "Method for waste control in the building industry," in *Proceeding IGLC-7, University of California, Berkeley, CA, USA*, 1999, pp. 325–334.

[14] L. Koskela and R. Sharpe, "Flow process analysis in construction," *Autom. Robot. Constr. 1994, Elsevier Sci. B.V*, pp. 281–288, 1994.

[15] S. Han and S. Lee, "Modeling and representation of non-value adding activities due to errors and changes in design and construction projects," *Proc. 2007 Winter Smulation Conf.*, pp. 2082–2089, 2007.

[16] A. O. Ralph and R. Iyagba, "Factors Affecting Contractor Performance : A Comparative Study of Non Value-Adding Activities in Nigeria and Indonesia," vol. 3, no. 5, pp. 467–474, 2012.

[17] M. J. Horman and R. Kenley, "Quantifying Levels of Wasted Time in Construction with," no. January, pp. 52–61, 2005.

[18] P. Josephson and L. Saukkoriipi, "Waste in construction projects call for a new approach," *chalmers univerity Technol.*, pp. 1–57, 2007.

[19] T. Ohno, *Toyota Production System: Beyond Large-Scale Production*. Productivitiy Pres, Cambridge, MA, 1988.

[20] S. M. Motavallian and H. Settyvari, "Application of Value Stream Mapping in Product Development," Chalmers University of Technology, 2013.

[21] J. Arleroth and H. Kristensson, "Waste in Lean Construction," Chalmers University of Technology, 2011.

[22] J. P. Womack and D. anie. T. Jones, *Lean thinking: banish waste and create wealth in your corporation*. Simon and Schuster., 2010.

[23] B. a. G. Bossink and H. J. H. Brouwers, "Construction Waste: Quantification and Source Evaluation," *J. Constr. Eng. Manag.*, vol. 122, no. 1, pp. 55–60, 1996.

[24] L. F. Alarcon, "Tools for the identification and reduction of waste in construction projects," *Lean Constr.*, pp. 365–377, 1997.

[25] A. Sugiharto, H. Keith, and M. Sherif, "Waste in the Indonesian construction projects," in *Proceedings The 1st International Conference of CIB W107 - Creating a sustainable Construction Industry in Developing Countries , south Africa. Accessed*, 2002, pp. 305–315.

[26] A. Serpell, A. Venturi, and J. Conteras, "characterization of waste in building construction projects," *3rd Work. lean Constr. Albarquerque*, 1995.

[27] S. Senaratne and D. Wijesiri, "Lean Construction as a Strategic Option : Testing its Suitability and Acceptability in Sri Lanka," pp. 34–48, 2008.

[28] F. Emuze, "Non-value adding activities in construction : The case of South African infrastructure sector," *J. Constr.*, vol. 6, no. 1, pp. 5–9, 2013.

[29] R. J. Tersine, "The primary drivers for continuous improvement: the reduction of the triad of waste," *J. Manag. issues*, pp. 15–28, 2004.

[30] L. Koskela, "An exploration towards a production theory and its application to construction," 2000.

[31] S. G. Modegh, "An evaluation of waste," in *Proceeding IGLC-21, July 2013*, 1999, pp. 729–738.

[32] L. Koskela and J. Leikas, "Lean manufacturing of construction components," *Lean Constr.*, pp. 263–272, 1997.

[33] A. Akintoye, "Analysis of factors influencing project cost estimating practice," *Constr. Manag. Econ.*, vol. 18, no. 1, pp. 77–89, 2000.

[34] L. I. M. K. Hoe, "Causal model for management of subcontractors in waste minimization," National University of Singapore, 2005.

[35] M. M. M. Teo and M. Loosemore, "A theory of waste behaviour in the construction industry," no. September 2013, pp. 37–41, 2010.

[36] P. E. Love and J. Smith, "Benchmarking, benchaction, and benchlearning: Rework mitigation in projects," *J. Manag. Eng.*, vol. 19, no. 4, pp. 147–159, 2003.

[37] C. Nunnally Jum and H. Bernstein Ira, *Psychometric theory*. New York: McGraw-Hill, 1978.

[38] E. A. Ariola, *Principles and Methods of Research*, 1st ed. 2006.

MATEC Web of Conferences **66**, 00016 (2016) DOI: 10.1051/matecconf/20166600016
IBCC 2016

Provisions of Disabled Facilities at The Malaysian Public Transport Stations

Haryati Mohd Isa[1,a], Halmi Zanol[2], Kartina Alauddin[1] and Mohd Hafizuddin Nawi[3]

[1]*Dr, Quantity Surveying Department, Faculty of Architecture, Planning and Surveying, UiTM Perak Branch, Malaysia*
[2]*Assoc. Prof, Dr, Town and Regional Planning, Faculty of Architecture, Planning and Surveying, UiTM Perak Branch, Malaysia*
[3]*Quantity Surveying Department, Faculty of Architecture, Planning and Surveying, UiTM Perak Branch, Malaysia*

Abstract. Public transport stations need to provide access and facilities for people with disabilities (PWD) in fulfilling the requirements of the Malaysian Standard Code of Practice on Access of Disabled Persons (MS). However, most public transport stations in Malaysia are reported as still lacking in terms of providing good designs and facilities for the PWDs. This inaccessible environment affects the PWDs to negatively participate in the social and economic mainstream. The research aims at identifying the compliance of disabled facilities provided at the electronic train service (ETS) railway station in Perak. Two research objectives were established (1) to identify the range of disabled facilities provided at the ETS railway station in Perak and (2) to determine the compliance of the disabled facilities as outlined in the MS. Eight ETS railway stations in Perak were chosen for the case study. Purely qualitative methods were adopted. An observation checklist was developed by conducting document analysis on three main documents. The findings suggest that there are 14 disabled facilities to be provided at the ETS railway stations. Majority of these facilities are available and comply with the MS. Conversely, some improvement can be made to further promote sustainability atmosphere of the ETS railway stations.

1 Introduction

Malaysia is moving towards Vision 2020 to be a fully developed country. Under this vision, nine challenges had been outlined. Amongst them is to establish a fully caring society and culture, in which the welfare of the people will revolve around a strong and resilient family system. This also includes people with disabilities (PWDs). The Persons with Disabilities Act 2008 (Act 685) (PWDA) stressed that PWDs shall have the same rights to access and use public facilities, amenities, services and buildings. The needs of the PWDs in the development cannot be ignored [1] and [2]. As in Malaysia, it is a mandatory for all public buildings to provide access and facilities for PWDs since the requirements to fulfill the Malaysian Standard (MS) are stated in the Uniform Building (Amendment) By-Laws (UBBL) 1991. The implementation of these codes of practice has always been highlighted [3] and [4]. Conversely, [5] reported that Malaysia is still lacking in considering equal accessibility and facilities for this group.

With regard to this problem, a research was mooted to identify the compliance of disabled facilities provided at the electric train service (ETS) railway station in Perak. In line with this aim, two objectives were established to operationalize the research: (1) to identify the range of disabled facilities provided at the ETS railway station in Perak (2) to determine the compliance of the disabled facilities provided as outlined in the MS.

[a] harya966@perak.uitm.edu.my

2 Literature review

2.1 Persons with disabilities (PWD) generally

The Persons with Disabilities Act 2008 (Act 685) (PWDA) defines PWDs as persons with physical, mental and intellectual disabilities that hindered them from fully participating in a normal way in the community way of life. This includes people with impairment, senior citizens, or even a pregnant woman. According to [6], impairment can be divided into three categories namely hearing impairment, visual impairment and physical/mobility impairment. People become disabled when society fails to accommodate them in social and infrastructural development, either purposely or inadvertently [7]. Consequently, this will affect not only the person with impairment, but also to the affected family [8].

The United Nations estimates that there are 650 million disabled people in the world which corresponds to 10% of global population [9]. In 2010, Statistics Department reported that there are approximately 5.1% PWDs in Malaysia and this figure had increased annually. Thus, this resulted a demand for the provisions of access and facilities in and outside a building, especially in public services. Indeed, this should be a crucial element to be considered in reducing the mobility constraints of PWDs. Furthermore, [9] added that the simplest way of increasing the use of public transportation facilities is by establishing a safe, convenient and comfortable environment for the full participation of these citizens. However, several researchers revealed that the present provisions are inadequate and not disabled friendly [10,4,5]. This suggests for serious action to be taken by the government as the public facility provider.

2.2 Relevant legislation, statutory and guidelines

Various acts and legislation have been created for PWDs rights in developed as well as developing countries. The establishment of such standards rules and legislation signify a strong ethical value and positive support from the government to ensure equal rights for all citizens. It also serves as guidelines for the professionals to plan and design their tasks. In Malaysia, the government had established The Persons with Disabilities Act 2008 (Act 685), Uniform Building (Amendment) By-Laws (UBBL) 1991 and Malaysian Standards i.e ; MS 1183:1990 Code of Practice for Means of Escape for Disabled Persons, MS 1184:2002 Code of Practice on Access for Disabled Persons to Public Buildings (1st revision) and MS 1331:2003 Code of Practice on Access for Disabled Persons Outside Buildings (1st revision).

2.3 Accessibility for PWDs

Although Malaysia offers technologically advanced services and facilities, not all of the societies are able to conduct their activities. Amongst the reason is lack of physical access for the PWDs. The facilities provided do not cover all categories of PWDs [11]. He also added that despite of the legislation, statutory and guidelines available, there is still lack of compliance made by the building owners. Furthermore, [4] argued that the access and facilities provided at the public transportation terminal is designed inefficiently and leads to a limited space for the PWDs to move about. Realising this problem, access audit should be conducted to ensure the compliance of those requirements as set in the MS.

3 Research methodology

Eight ETS railway stations in Perak were chosen for observation. An observation checklist was developed prior to conducting document analysis on three main documents i.e The Persons with Disabilities Act 2008 (Act 685), UBBL 1991 and MS.

The purpose of conducting the observation was to observe the compliance of the existing disabled facilities provided in the case studies. Table 1 shows a sample of the checklist used during the observation. After the observation was conducted, the entire checklist was by using descriptive analysis to determine types of facilities provided and the compliance of each of the facilities provided at all stations to the standard provided.

Table 1. Observation checklist for disabled facilities at ETS railway stations in Perak

Date: Station:	Time:
Standards	Availability
PWD Parking	
a) Located at the main entrance	
b) Located at the flat surface	
c) With adequate width (2400mm)	
d) There is a signage indicating the parking	
e) Marking indicating disabled parking	
f) Enough space for wheelchair user to manoeuvre (extra space min 900mm)	
Walkway	
a) Surface with non-slip	
b) Clear width (min 1200mm – enable wheelchair)	
c) Properly connected	
Ramp and kerb	
a) Ramp consists of landing (min length 1200mm) at interval of not more than 600mm length of ramp)	
b) Surface with non-slip	
c) Enable wheelchair user(1200mm min width)	
d) Proper gradient (Max 1:12-min 1:20)	
e) Provided with handrails at both sides	
f) Kerb is provided at both sides (min 100mm length)	
Doorway/Main entrance	
a) Doorway width is adequate for wheelchair user (min 900mm width)	
b) Threshold is leveled with step ramp (if any)	
Guiding block/Tactile	
a) Installed at proper location	
b) Each block is installed adjacent to one another	
c) Contrast in colour	
d) Detectable underfoot	
Handrail	
a) Fixed with proper height at ramp (min 840-900mm in height)	
b) At ramp/stairway – extended 300mm in length at both sides	
c) Contrast in colour	
d) Surface with non-slip grip	
Special lane for wheelchair user to Platform 2	
a) Surface with non-slip	
b) Clear width (min 1200mm)	

Staircase	
a) Tread with non-slip surface	
b) Tread width (260-300mm)	
c) Riser height (max 180mm)	
d) Handrails provided at both sides	
e) Landing and floor contrast in colour/brightness/texture to the stairs	
Prayer room	
a) Ablution area accessible to wheelchair user	
b) Door width accessible to wheelchair user	
c) Step ramp at door	
d) Floor with non-slip surface	
Disabled toilet	
a) Have adequate space for wheelchair user (min 2000 x 3400mm and for ambulant disabled min 1200 x 2400mm)	
b) Handrail is provided	
c) Water closet – pedestal (450mm-460mm height)	
d) Wash hand basin (800-830mm height)	
e) Accessible tap with flexible hose	
f) Adequate door width (1200mm for wheelchair user or 900mm for ambulant disabled)	
Signage	
a) Clearly shows the direction/information	
b) Signage	
c) Signage is equipped with braille instruction	
d) Signage is installed at proper locations; - Parking area - Walkway - Main entrance/doorway - Toilet/washroom - Platform - Ticket counter	
Elevator	
a) Accessible to wheelchair user (door width 800-1000mm)	
b) Guiding block is provided at the elevator	
c) Handrail 600mm long and 1000mm above floor	
d) All control buttons not higher than 1400mm above finished floor	
e) Enough space for wheelchair user to maneuver in the elevator (1800 x 1800mm)	
Escalator	
Guiding block is provided at the elevator	

4 Findings

4.1. Types of disabled facilities to be provided at ETS railway stations

Figure 1 shows the range of types of facilities to be provided at the public railway stations. There are 14 types of disabled facilities to be provided in ETS railway stations namely disabled parking, walkway, ramp and kerb, doorway at main entrance, guiding block, handrail, staircase, signage, disabled toilet, ticket counter, elevator, escalator, prayer room and special lane for wheelchair users to Platform 2.

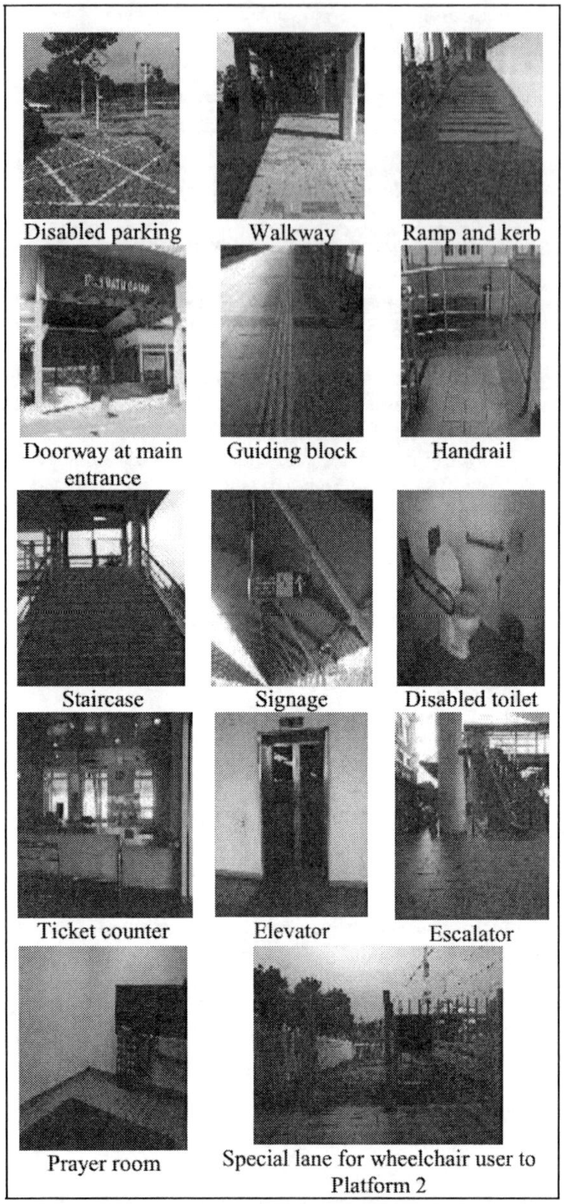

Figure 1. Types of disabled facilities to be provided at ETS railway station

Table 2 presents the availability of disabled facilities at ETS railway stations in Perak. Generally, all of the facilities are available at all stations. RS1 and RS3 are two most equipped stations with disabled facilities provided. Majority of the PWD facilities are provided at all stations namely disabled parking, walkway, ramp and kerb, doorway at man entrance, guiding block, handrail, staircase, signage, disabled toilet, ticket counter and special lane for wheelchair user to Platform 2.

Table 2. Availability of the disabled facilities in ETS railway stations in Perak

	Facilities	Availability								%	Rank
		RS 1	RS 2	RS 3	RS 4	RS 5	RS 6	RS 7	RS 8		
1.	Disabled parking	√	√	√	√	√	√	√	√	100	1
2.	Walkway	√	√	√	√	√	√	√	√	100	1
3.	Ramp and kerb	√	√	√	√	√	√	√	√	100	1
4.	Doorway at main entrance	√	√	√	√	√	√	√	√	100	1
5.	Guiding block	√	√	√	√	√	√	√	√	100	1
6.	Handrail	√	√	√	√	√	√	√	√	100	1
7.	Staircase	√	√	√	√	√	√	√	√	100	1
8.	Signage	√	√	√	√	√	√	√	√	100	1
9.	Disabled toilet	√	√	√	√	√	√	√	√	100	1
10.	Ticket counter	√	√	√	√	√	√	√	√	100	1
11.	Elevator	√	x	√	x	x	x	x	x	25	3
12.	Escalator	√	x	X	x	x	x	x	x	13	4
13.	Prayer room	x	√	√	√	√	√	√	√	88	2
14.	Special lane for wheelchair user to Platform 2	√	√	√	√	√	√	√	√	100	1
%		93	86	93	86	86	86	86	86		
Rank		1	2	1	2	2	2	2	2		

Note: √ Available x Not available

Three most uncommon facilities provided at all ETS stations in Perak are prayer room, elevator and escalator. Only RS1 has an escalator because it is the biggest ETS station and caters for the largest amount of passengers. However, there is no prayer room provided in this station since it is constructed during the colonial era. Meanwhile, it is also found that elevator is only provided at RS1 and RS3.

4.2 Analysis of the compliance of the disabled facilities provided at the ETS railway stations with the MS

4.2.1 Analysis of walkway

There are three standards of walkway to be complied such as surface with non-slip, clear width of minimum 1200mm and properly connected.

Table 3. Analysis of walkway towards MS

Standards	Compliance							
	RS 1	RS 2	RS 3	RS 4	RS 5	RS 6	RS 7	RS 8
1. Surface with non-slip	√	√	√	√	√	√	√	√
2. Clear width (min 1200mm – enable wheelchair)	√	√	√	√	√	√	√	√
3. Properly connected	√	√	√	√	√	√	√	√

Note: √ Comply x Not comply

All stations had complied with the walkway standards. However, the size and material of walkway at RS1 differs from other stations in Perak.

4.2.2 Analysis of the doorway at main entrance

All stations complied with the only one standard outlined by the SIRIM.

Table 4. Analysis of doorway at main entrance towards MS

Standards	Compliance							
	RS 1	RS 2	RS 3	RS 4	RS 5	RS 6	RS 7	RS 8
1. Doorway width is adequate for wheelchair user (min 900mm width)	√	√	√	√	√	√	√	√
2. Threshold is leveled with step ramp (if any)	x	x	x	x	X	x	x	x

Note: √ Comply x Not comply

All main entrance doors have adequate width to enable the wheelchair user to pass through the doors. There is no threshold provided at any station because the surface of the main entrance doorway on the premises is flat.

4.2.3 Analysis of ticket counter

In general, all stations in Perak had complied with the standard of MS. Each station provides a special counter and it is accessible to the PWDs.

4.2.4 Analysis of disabled parking

There are six standards of disabled parking to be observed at all stations. This is shown in Table 5.

[a] harya966@perak.uitm.edu.my

Table 5. Analysis of disabled parking towards MS

Standards	Compliance								Remarks
	RS1	RS2	RS3	RS4	RS5	RS6	RS7	RS8	
1. Located at the main entrance	√	√	√	√	√	√	√	√	▪ Only one disabled parking at RS1 ▪ Two disabled parking at RS2, RS3, RS4. RS5. RS6, RS7 and RS8
2. Located at the flat surface	√	√	√	√	√	√	√	√	
3. With adequate width (2400mm)	√	√	√	√	√	√	√	√	
4. There is a signage indicating the parking	√	√	√	√	x	x	x	x	
5. Marking indicating disabled parking	x	√	√	√	√	√	√	√	Marking indicating disabled parking at RS3 and RS4 are not up to the SIRIM standard
6. Enough space for wheelchair user to maneuver (extra space min 900mm)	x	x	√	√	√	√	√	x	

Note: √ Comply x Not comply

Although all stations have provided disabled parking, only RS3 and RS4 comply with all the standards. However, both stations did not comply with the disabled marking sign as outlined by the Standards and Industrial Research Institute of Malaysia (SIRIM). Instead of using the disabled symbol, it only uses the cross line marking.

Correct disabled marking symbol as SIRIM

Wrong marking indicating disabled parking

Figure 2. Disabled parking marking symbol

The disabled parking provided at all stations are located at the main entrance and on a flat surface with adequate width of 2400mm except there are inadequate space provided. However, the space is inadequate for a wheelchair user to maneuver from the car. This is observed at RS1, RS2 and RS8.

It is also found that, signage indicating disabled parking is not complaint to the incomply standard at all stations. This is mostly due to vandalism There is also construction wastages found at the walkway to the disabled parking in RS6 and may expose the PWDs to the risk of accident.

4.2.5 Analysis of signage

Generally, all stations have provided signage . However, none had complied with all standards listed in the checklist. There is no signage provided at parking area at RS6, RS7 and RS8. In addition, there is no signage with braille instruction provided. This is important especially for disabled person with sight impairment.

Table 6. Analysis of signage towards MS

Standards	Compliance							
	RS1	RS2	RS3	RS4	RS5	RS6	RS7	RS8
1. Signage is installed at proper locations								
a) Parking area	√	√	√	√	x	x	x	√
b) Walkway	√	√	√	√	√	√	√	√
c) Main entrance/doorway	√	√	√	√	√	√	√	√
d) Toilet	√	√	√	√	√	√	√	√
e) Platform	√	√	√	√	√	√	√	√
f) Ticket counter	√	√	√	√	√	√	√	√
2. Clearly provide the direction/information	√	√	√	√	√	√	√	√
3. Signage is illuminated/lighted	√	√	√	√	√	√	x	√
4. Signage equipped with Braille instruction	x	x	x	x	x	x	x	x

Note: √ Comply x Not comply

RS7 is observed to be the most incompliant station in providing this facility. Besides no not having any signage in braille signage instruction, there is also no signage provided at the parking area and the signage is not illuminated. . Similarly, problems are also found at RS5 and RS6.

[a] harya966@perak.uitm.edu.my

Table 7. Analysis of ramp and kerb towards MS

Standards	Compliance								Remarks
	RS1	RS2	RS3	RS4	RS5	RS6	RS7	RS8	
1. Ramp consists of landing (min length 1200mm) at interval of not more than 600mm length of ramp)	√	√	√	√	√	√	√	√	Two ramps provided at all stations
2. Surface with non-slip	√	√	√	√	√	√	√	√	
3. Enable wheelchair user(1200mm min width)	√	√	√	√	√	√	√	√	
4. Proper gradient (Max 1:12-min 1:20)	√	√	√	√	√	√	√	√	
5. Provided with handrails at both sides	√	x	x	x	x	x	x	x	Only one side No handl at RS6 Inadequate length of ramp
6. Kerb is provided at both sides (min 100mm length)	√	x	x	x	x	x	x	x	No kerb provided except at RS1

Note: √ Comply x Not comply

4.2.6 Analysis of ramp and kerb

Table 7 listed all six standards of ramp and kerb to be complied at all stations. Although all stations provide ramp, only RS1 complies with all standards in the checklist. Meanwhile, for the other stations, there is only one handrail or wall provided at both sides of ramp. It is also found that, one of the ramps at RS6 has no handrail and does not comply with the standard length. Thus, the wheelchair user might encounter difficulty to pass through this facility.

It is also observed that there is no kerb provided to prevent the wheelchair user from falling at RS2, RS3, RS4, RS5, RS6, RS7 and RS8 because there is a handrail or wall built at both sides.

4.2.7 Analysis of handrail

There are four standards to be complied when providing this facility. This is outlined in the table below.

Table 8. Analysis of handrail towards MS

Standards	Compliance							
	RS1	RS2	RS3	RS4	RS5	RS6	RS7	RS8
1. Fixed with proper height at ramp (min 840-900mm in height)	√	√	√	√	√	√	√	√
2. At ramp/stairway – extended 300mm in length at both sides	√	√	√	√	√	√	√	√
3. Contrast in colour	√	√	√	√	√	√	√	√
4. Surface with non-slip grip	N/A	x	x	x	x	x	x	x

Note: √ Comply x Not comply

All stations had provided handrail for PWD except for RS1., However, the handrail is not covered with a non-slip material to prevent from sweat, slippery and fall.

4.2.8 Analysis of prayer room

There are four standards of PWD prayer room as highlighted by the MS. All stations provide prayer room except RS1. It is found that all prayer rooms complies to all the standards required except for the provision of ramp in the prayer room and at the ablution area.

Table 9. Analysis of prayer room towards MS

Standards	Compliance							
	RS1	RS2	RS3	RS4	RS5	RS6	RS7	RS8
1. Ablution area accessible to wheelchair user	N/A	√	√	√	√	√	√	√
2. Door width accessible to wheelchair user		√	√	√	√	√	√	√
3. Step ramp at door		x	x	x	x	x	x	x
4. Floor with non-slip surface		√	√	√	√	√	√	√

Note: √ Comply x Not comply

4.2.9 Analysis of the special lane for wheelchair users to Platform 2

The purpose of this special lane is to ease the wheelchair users to travel from Platform 1 to Platform 2. It is observed that all stations had provided this facility. However, its location is far from the main area at the platform.

Table 10. Analysis of guiding block towards MS

Standards	Compliance								Remarks
	RS1	RS2	RS3	RS4	RS5	RS6	RS7	RS8	
1. Installed at proper location	√	√	√	√	√	√	√	√	Not installed at all important areas except for RS3
2. Each block is installed adjacent to one another	√	√	√	√	√	√	√	√	
3. Contrast in colour	√	√	√	√	√	√	√	√	
4. Detectable underfoot	√	√	√	√	√	√	√	√	Some of the blocks provided is undetectable except at RS3

Note: √ Comply x Not comply

4.2.10 Analysis of guiding block

Guiding block is detectable underfoot and paced at all main areas at the station such as disabled toilet, ticket counter and ramp. The purpose of this facility is to help the PWD, particularly with sight impairment to aware of danger in front using their touch sensing of the feet. Thus, they can independently and safely access the station.

Table 10 shows that all stations comply with standards outlined in the MS. Nevertheless, only RS3 complied with all the standards.

4.2.11 Analysis of disabled tolet

There are six standards to be complied in the provision of disabled toilet in public railway stations.. All stations had provided disabled toilet and complied with the MS standards. However, it is found that all doors at the disabled toilet do not follows the standards and causes difficulty to the wheelchair users.

All disabled toilets at RS3, RS5, RS7 and RS8, are closed cannot be used are closed to avoid vandalism. Meanwhile, the disabled toilets at RS3, RS5 and RS8 are also closed due to poor maintenance of this facility.

4.2.12 Analysis of elevator and escalator

There are five standards for elevator and one standard for escalator. Only two stations i.e RS1 and RS3 provide elevator in its premises. . or However, for stations that has no elevator, the PWDs can use the special lane provided. All elevators provided complied with the standards as outlined in the checklist.

Escalator is only provided at RS1 because it is the largest ETS station in Perak. Nonetheless, there is no guiding block provided in front of the escalator, as required by the standard.

5 Discussions

The findings indicate that although all ETS stations in Perak are attempting to accommodate the PWDs needs, there are several facilities that are in comply with the MS. This is in-line with [10] stating that most of the public buildings in Malaysia are still lacking in the provision of user-friendly built environment for the disabled persons. Inaccessibility found in some stations is mainly due to poor design, poor planning, poor maintenance and lack of enforcement on guidelines provided.

It is also reveals that braille signage is something that could be uncommonly seen in the stations. According to [11,12]Corn (1990), Small et al (2012) and Syazwani & Mariam, 2012), difficulties in way finding by either the visual or hearing impairment persons will significantly contributes to safety and comfort issues, particularly when they are travelling alone. Meanwhile, the provision of adequate gradient of ramp, size of disabled parking, height of ticket counters and special lane for travelling are important facilities to be provided for the wheelchair users. Adding into this, [12] also recommended that the public facilities should be designed not merely for the PWDs but also to greatest scope as possible, including common people such as children, parents with baby stroller, expectant mothers, library staff caring huge stack of books and elder people.

Clearly, this research has provided some benchmark for planners, local authority and designers in planning and designing or even in improving the existing disabled facilities provided in the ETS stations. This is crucially important so that the PWDs will no longer feel isolated for the continuous development that is becoming more challenging.

6 Conclusions

This research would significantly promote and enhance the public buildings for better value by focusing on the disabled friendly facilities. Moreover, it also promotes the friendliness atmosphere of the ETS railway stations for better usage and accessibility.

Finally, in improving or designing the public transport stations, feedback from the users are utmost important. There are also needs for more innovative and good design to create a barrier-free environment and infrastructure.

References

1. Jayasooria, D, Krishnan, B, and Ooi, G. (1997). Disabled people in a newly industrializing economy: Opportunities and challenges in Malaysia. *Disability and Society*, **12**(3), 455-463

[a] harya966@perak.uitm.edu.my

2. Kennedy, M.K, and B. Hesla. (2008). We have human rights, Harvard Project on Disability, Harvard
3. Kamaruddin, H. (2007). *The implementation of the SIRIM Codes of Practice for Disabled Persons by DBKL*. University Technology MARA Malaysia. Unpublished dissertation
4. Hikmah, K., Ahmad, E.H., Mahayuddin, M, Nor Rima, M. A., and Wan Zurea, W.I. (2012). The implementation of the Malaysian Standard Code of Practice on Access for Disabled Persons by Local Authority. *Procedia-Social and Behavioral Sciences,* **50**(2), 442-451
5. Soltani, S.H.K., Abbas, M.Y., Awang, M.B. and Yaman, R. (2012). Accessibility for disabled in public transportation terminal. *Procedia-Social and Behavioral Sciences,* **35**(2), 78-86
6. Hafizuddin, M. (2013). *Provision of facilities for disabled people in electric train service (ETS) railway station in Perak Darul Ridzuan*. University Technology MARA Malaysia. Unpublished dissertation
7. Ahmad, E.H., Faridah, I., Zarina, I., Natasha, K. and Mardhati, A.R. (2011). Disabled facilities in shopping malls: Malaysians perspective. *Business and Management Quarterly Review,* **2**(4), 56-64
8. Fairuzzana, A.P., and Fuziah, I. (2012). Accessibility of visually impaired passengers at urban railway stations in the Klang Valley. *International Transactional Journal of Engineering, Management & Applied Sciences & Technologies,* **3**(3), 277-292
9. Seyed Hassan, K.S., Mashita, S., Mohamad, A. and Rostam, Y. (2011). Accessibility for disabled in public transportation terminal. *Procedia-Social and Behavioral Sciences,* **35**(1), 89-96
10. Kamaruddin, H., Muhamad Ariff, N.R., Wan Zurea, W.I., Bakri, A.F., and Zarina, I. (2014). *Malaysian scenario on access and facilities for persons with disabilities: A literature review*. MATEC Web of Conferences
11. Jennie, S., Simon, D. and Tanya, P. (2012). The embodied tourist experiences of people with vision impairment : Management implications beyond the visual gaze. *Tourist Management,* **33**(1), 941-950
12. Syazwani, A.K. and Mariam, J. (2012). Users' satisfaction and perception on accessibility of public buildings in Putrajaya : Access audit study. *Procedia-Social and Behavioral Sciences,* **50**(2), 429-441

MATEC Web of Conferences 66, 00017 (2016) DOI: 10.1051/matecconf/20166600017
IBCC 2016

Tracking architectural defects in university building in Malaysia

Haryati Mohd Isa[1a], Kharizam Ismail[1] , Halmi Zainol[2] and Mohd Fadhli Othman[3]

[1]*Dr, Quantity Surveying Department, Faculty of Architecture, Planning and Surveying, UiTM Perak Branch, Malaysia*
[2]*Assoc. Prof, Dr, Town and Regional Planning, Faculty of Architecture, Planning and Surveying, UiTM Perak Branch, Malaysia*
[3]*Quantity Surveying Department, Faculty of Architecture, Planning and Surveying, UiTM Perak Branch, Malaysia*

Abstract. Building defects are always the key concern in the construction industry. Defects represent not only a loss to the project but also hamper the smooth operation of a building. Recognising the need to resolve these continuing problems, a research was mooted to track from the study of defects that occurred during the Defects Liability Period in a public university in Malaysia. This paper presents part of the research which investigates what can be learnt from the analysis of architectural defects in these projects. Two research objectives were developed (1) to investigate types of architectural defects that occurred and (2) to analyse the causes of the defects. A mixed methods approach is adopted. Data for the quantitative element of the research was drawn from defects audit records of the project. They were sorted, grouped and transferred into the SPSS software for analysis using the measure of central tendencies and frequency analysis. The findings suggest that with a proper methodology in place, defects can be effectively traced and categorised. This can provide very useful insights to their root cause and how this can be avoided in future projects.

1 Introduction

Defects can be associated to construction failures emanating from shortcomings during the project implementation process. In construction, parties managing construction projects are frequently aware of the signals that lead to defects or diagnose the causes of the defect to take preventive action [1] and [2]. As a result, similar defects re-appear even the building is constructed by the same contractor. This phenomenon is common in the Malaysian construction projects. Most defects in these projects are not properly recorded and tracked, which consequently lead to the client ending up rectifying the defects themselves and at their own cost [3]. Thus, this has left many clients unable to realise the full value of their investments. In contending to this problem, a research was mooted to show how defects data recorded can be re-learnt in a more systematic way to provide insights in managing future projects.

Re-learning construction defects is an emergent popular subject of research. Several studies have been conducted to investigate the types and causes of defects including the means of preventing them [4,5,6,7,8,9,2]. However, most of these studies tend to be limited to study only on the overall construction defects. There are limited studies that investigate defects and its relation to the work disciplines such as architectural, structural, mechanical and electrical, and the defects inherent in them. Recognising this, a research was mooted to investigate the causal relationship between the types of defects and their root cause within each work discipline involved. This is the aim to understand the nature of the defects to track their root cause. In line with this aim, three research objectives are developed: (1) investigate the types of architectural defects (2) to identify the causes of the defects and (3) to analyse the categories of the defects.

2 Theoretical framework

2.1 Defects generally

Construction defects are common phenomenon faced in the industry nowadays [5,6,7,2]. It occurs not only in building operational phase but also before and during the construction stage [11] and [12]. Failure to address defects would most likely contribute to additional rectification cost, hamper the smooth operation of the building and reduce its service life. In extreme cases, defects may even lead to accidents and disasters [13].

Construction defects can be defined as a failure or shortcoming in the building's function, performance, statutory or user requirements, and this may occur in its structure, fabric, services or other facilities [6,11,12]. Defects can be classified as patent and latent defects. Patent defects can be clearly recognized during inspection throughout the construction phase and the project's Defects Liability Period (DLP). Meanwhile,

[a] Corresponding author: harya966@perak.uitm.edu.my

latent defects usually appear over time when the building is occupied.

Studies examining the causes of building defects were observed to take prominence since 1980s. The study on the cause and effect of defects in buildings converge to suggest that most of the defects are due to 'weaknesses' in the project implementation process. Whilst poor specifications, selection of materials, workmanship and supervision the common cited causes of defects, poor design decisions are identified as the most significant contributor to the defects. Several researchers [5,6,13,14,7] stressed on the importance of designers to play their roles in eliminating defects occurrence. They should ensure that all the client requirements are well captured, translated and communicated to the other project team members throughout the project implementation process supported with proper documentation system [15]. In addition, [16] added that the problem with defects is not so much a matter of not knowing what to do but applying correctly what is known.

2.2 Defects and its implication in the Malaysian construction industry

One of the most important 'disease' that affect the overall performance of construction industry in Malaysia is construction defects [17,18,19]. These defects have caused the government an enormous sum of money. The Malaysian government has continually provided allocations for maintenance works for completed buildings and a significant proportion are for defects rectification work [12]. Owing to the seriousness of these problems, the government is considering implementing punitive measures against those who flout the rules and regulations.

As an added initiative to minimize construction defects in public projects, the Public Work Department (PWD) had undertook several proactive actions such as:

a) forming Building Facility Maintenance Division to monitor and audit Malaysian public buildings
b) organising an internal workshop to discuss on issues promoting defects during project stages. The common causes of the defects highlighted are: (1) lack of monitoring or supervision by responsible authorities, (2) lack of technical expertise and full reliance on contractors or consultants, (3) lack of coordination between agencies involved, (4) internal problems faced by the contractors, and (5) contractual issues. Notwithstanding, the government have realised that this cannot continue and they are not getting the value for money for their projects
c) recording and publishing several construction defects that occurred in public buildings and their lesson-learnt in a book
d) enforcing Defects Liability Management System for the management of defective works on seven hospital projects constructed in the Malaysian Eight Plan (RMK8)

Following this, [20] found that defects are being repeated in many projects even when constructed by the same contractor. Thus, a study on the nature of the defects and possible approaches to track their root cause is very necessary. The question of what are the defects, why they are recurring and what are the root causes are raised.

2.3 Rationale for tracking construction defects

By systematically recording and tracking defects, the types, causes and categories of each defect can be identified. Besides this, the recurring defects can also be traced. These significant data can be used to provide LL to assist the project team in identifying what went wrong and how to better manage future projects [21,22,23]. As a result, they could improve its cost efficiency and increase their design and performance quality [23] and [2].

3 Research methodology

A single case study of a Malaysian public university is chosen. A combination of qualitative and quantitative methods was adopted. Data for the quantitative element of the research was drawn from audit records of defects identified during the projects DLP. They were sorted, grouped and transferred into the SPSS software for analysis using the measure of central tendency and frequency analysis.

The qualitative research method was adopted to examine the types and causes of the defect. This was carried out by critically analyzing project audit documents. For the purpose of tracking the root cause of the defects, data on the defects were categorized into design, workmanship, material, lack of protection, vandalism, nature, maintenance and mishandling defects.

4 Findings

The defects data were analysed and summarized in Table 1 to Table 5. Five major defective elements were identified and documented in this project.

4.1 Overall number and percentage of defects

Table 1 highlights the overall number and percentage of defects according to their work disciplines. There are four work disciplines involved namely architectural, civil, mechanical and electrical.

Table 1. Overall number and percentage of defects.

Work discipline	Number of defects	Percentage of defects (%)
Architectural	1175	63
Mechanical	364	19
Electrical	275	15
Civil	60	3
Total	1874	100

A total of 1874 defects were recorded in the university. Architectural defects recorded as the most defective work discipline followed by mechanical, electrical and civil defects.

4.2 Distribution of architectural defects by element and block

Table 2 indicates the number of architectural defects based on the elements for the five identified blocks. College was observed to record the highest number of architectural defects, followed by Faculty, Library, Administration Building and Student Centre respectively.

Table 2. Overall number of architectural defects by element/block.

Element	Block					Total number of defects/	Rank
	A	B	C	D	E		
1. Wall, Floor and Finishes	94	127	69	59	8	357	1
2. Window and Fittings	155	39	20	21	7	242	2
3. Doors and Fittings	60	69	48	38	2	217	3
4. Ceiling Finishes	45	122	10	10	1	188	4
5. Sanitary Fittings, Fixtures and Toilet Cubicles	78	39	5	5	8	135	5
6. Staircase	5	12	2	1	0	20	6
7. Roof	5	6	0	0	0	11	7
8. Building Signage	1	0	1	1	0	3	8
9. Other Architectural	2	0	0	0	0	2	9
Total number of defects/ block	445	414	155	135	26	1175	
Rank	1	2	3	4	5		

A- College, B- Faculty, C- Library, D- Administration Building, F- Student Centre

Nine elements were involved in architectural works. The top five most defective elements are; (1) Wall, Floor and Finishes (2) Windows and Fittings (3) Doors and Fittings (4) Ceiling Finishes and (5) Sanitary Fittings, Fixtures and Toilet Cubicles. Meanwhile, Others Architectural is identified as the least defective element and only revealed in College.

4.3 Summary of major architectural defects

Table 3 to Table 7 recapitulates the types, causes and categories of architectural defects for five major elements. In addition, the table also highlights the frequent types, causes and categories of architectural defects that occurred in the building.

4.3.1 Types of architectural defects

The types and most frequent type of architectural defects that occurs are as follows:

Table 3. Types and frequent type of architectural defects.

Type of defects	Frequent type of defects
A) Wall, Floor and Finishes	1. Non-functional windows 2. Water seepage 3. Door damaged 4. Window damaged 5. Untidy wall painting
1. Water seepage 2. Dirty floor 3. Unevenness paint 4. Untidy wall painting 5. Peeling of paint 6. Plaster crack 7. Wall damaged 8. Uneven floor level 9. Detached of floor tiles 10. Floor tile crack 11. Pebble wash not provided 12. Fungus 13. Improper fixing of wall tile 14. Partition not provided 15. Detached of skirting	
B) Windows and Fittings	
1. Non-functional window 2. Window damaged 3. Dirty window 4. Window ironmongery damaged 5. Incomplete installation of window 6. Blind not provided 7. Window corrode	
C) Doors and Fittings	
1. Door damaged 2. Ironmongery damaged 3. Incomplete installation of ironmongery 4. Door key not provided 5. Untidy door painting 6. Non-functional door 7. Detached of door stopper 8. Peeling of paint at door 9. Door knob inverted 10. Door frame damaged	
D) Ceiling Finishes	
1. Ceiling perforated 2. Ceiling not provided 3. Leakage 4. Ceiling damaged 5. Fungus 6. Stain mark	

E) Sanitary Fittings, Fixtures and Toilet Cubicles	
1. Incomplete installation of fittings 2. Tap not provided 3. Leakage 4. Blockage 5. Detached of sanitary fittings 6. Non-functional floor trap 7. Water closet cover damaged 8. Improper installation of fittings 9. Cistern overflow 10. Flush damaged	

Note: The element, type of elemental defects and frequent defective items are all in ranking order

The major architectural defects are identified and summarised as follows:

a) Wall, Floor and Finishes are the major defective elements with the most type of defects aroused. With 15 types of defects that occurred, water seepage recorded the most defects and detached of floor skirting recorded least
b) Window and Fittings are identified as the second most defective element. A number of seven defects were observed with non-functional window highlighted as the most defective works. Meanwhile, window corrode as the least defective element
c) Doors and Fittings are the third most defective element with door damaged as the most defective works. The least defective works in this element is door frame damaged
d) Perforated ceiling is the most defective works in Ceiling and Finishes while uneven ceiling is the least defective work in this element
e) The least defective element is Sanitary Fittings, Fixtures and Toilet Cubicles. Incomplete installation of fittings are the most defective works while overflow cistern and flush damaged recorded least

The table also highlights five most frequent defective items. Non-functional windows appears as the most defective items followed by water seepage, door damaged, window damaged and untidy painting to wall.

4.3.2 Causes of architectural defects

Table 4 indicates the causes and frequent causes of defects for each element involved. The results are as follows:

Table 4. Causes of architectural defects.

Element	Causes of defects	Frequent causes of defects
Wall, Floor and Finishes	1. Lack of protection 2. Poor workmanship 3. Work in accordance to specification 4. Incorrect levelling 5. Work in accordance to specification 6. Exposed to weather 7. Vandalism 8. Poor quality of material 9. Mishandling by users 10. Design default 11. Construction debris blockage	1. Work in accordance to specification 2. Poor workmanship 3. Lack of protection 4. Vandalism 5. Water seepage
Windows and Fittings	1. Poor workmanship 2. Work in accordance to specification 3. Vandalism 4. Lack of protection 5. Poor quality of material 6. Mishandling by users 7. Exposed to weather 8. Heavy thing crashed	
Doors and Fittings	1. Work in accordance to specification 2. Vandalism 3. Mishandling by users 4. Poor workmanship 5. Poor quality of material 6. Exposed to weather 7. Lack of protection 8. Design default	
Ceiling Finishes	1. Work in accordance to specification 2. Lack of protection 3. Poor workmanship 4. Incorrect levelling 5. Vandalism 6. Mishandling by users	
Sanitary Fittings, Fixtures and Toilet Cubicles	1. Poor workmanship 2. Lack of protection 3. Construction debris blockage 4. Mishandling by users 5. Vandalism 6. Exposed to weather 7. Poor quality of material 8. Work not accordance to plan and layout 9. Valve gate off	

Note: The element, causes of elemental defects and frequent causes of defects are all in ranking order

a) There are 12 causes of defects appeared in Wall, Floor and Finishes. Lack of protection is the major causes of defects in this element, while construction debris blockage is the least
b) Poor workmanship is the main causes of defects in Windows and Fittings and Sanitary Fittings, Fixtures and Toilet Cubicles. Heavy thing crashed is the least defective causes in Windows and Fittings. Meanwhile, valve gate off is the least in Sanitary Fittings, Fixtures and Toilet Cubicles
c) Doors and Fittings and Ceiling Finishes have similar number and major causes of defects i.e work in accordance to specification. Design default and

mishandling by users are the two least causes of defects in these elements

Five most frequent causes of defects were found such as work in accordance to specification, poor workmanship, lack of protection, vandalism and water seepage.

4.3.3 Categories of architectural defects

Table 5 indicates that there are eight categories of defects involved in architectural works.

Table 5. Categories of architectural defects.

Element	Categories of defects								
	W	LOP	V	Mt	Mh	M	N	D	Total
WFF	167	119	58	9	11	5	10	2	381
WF	134	24	39	0	9	17	6	0	229
DF	96	2	58	0	36	12	7	2	213
CF	59	28	8	91	1	0	0	0	187
SF	110	3	6	29	8	1	5	0	162
Total	566	176	169	129	65	35	28	4	
Rank	1	2	3	4	5	6	7	8	

WFF – Wall, Floor and Finishes, WF- Window and Fottings, DF- Doors and Fittings. CF- Ceiling Finishes, SF- Sanitary Fttings, Fixtures and Toilet Cubcles

Workmanship (W) is the main categories of defects followed by lack of protection (LOP), vandalism (V), maintenance (Mt), mishandling by users (Mh), material (M) and nature (N). Defects due to design (D) defaults are the least categories occurred in this work discipline. The frequency of the categories of defects that occurred in each element as follows:

a) Wall, Floor and Finishes - workmanship is the most frequent categories of defects, while design is the least
b) Windows and Fittings - workmanship is the most frequent categories of defects, while nature is the least. There is no defects occurred due to design and maintenance
c) Doors and Fittings - workmanship is the most frequent categories of defects. Design and lack of protection are two least categories that appeared in this element. None defects occurred due to maintenance fault
d) Ceiling Finishes – Maintenance is the major categories of defects and mishandling by users is the least category. None defects are due to design, material and nature
e) Sanitary Fittings, Fixtures and Toilet Cubicles - workmanship is the most frequent categories of defects, while material is the least. There is no design defects found in this element

5 Discussions

The problem of many architectural defects is continuing and tends to be repeated in many projects. Majority of the defects causes is due to workmanship as a result of poor quality of work, lack of supervision and installation or protection method during the construction stage. This is further supported by the statement from [21] and [17]. Design defects are due to poor design planning and poor material selection during the design stage.

Material defects are mostly due to inappropriate materials applied to the functional of the buildings as mentioned by [7]. It is found that material defects usually took time to develop until the occupancy stage. In addition, more maintenance defects were expected as the buildings aged. Meanwhile, vandalism defects are believed to be influenced by attitude and background of the population surrounding the buildings.

These defects are underpinned by the adequacy and effectiveness of the management process practiced in design and building process of the hospitals. There were obvious issues surrounding the design of some components of the building, selection of materials, workmanship and supervision. While generally the final quality and standards of the university was deemed acceptable to the EU, in some key areas these elements needs to be further clarified. Elements like ambiguity in the process of determining specifications should be eliminated. Indeed, these should be well captured and maintained by the professionals including the end user (EU) during the early stage of the project. Better co-ordination and co-operation amongst the design team, EU and the contractor throughout the project implementation is also an important factor that needs to be considered. This is in line with [24].

Recording defects and classifying them into systematic categories are useful for identifying the type of defect; which defect is most probably caused by what; and what defects are most recurring. These are significant data which can be used to learn to assist identify what went wrong. Majority of the defects are often interrelated. However, this requires the project teams to open their mind and embrace the continuous learning culture. They need to see the system as a positive and systematic approach that can encourage a better co-ordination and co-operation amongst the project team towards achieving zero defects in future projects.

6 Conclusions

Findings from this research underline the significance of re-learning construction defects. The audit provides a platform for a post-mortem study to learn weaknesses in the project implementation process for future improvement.

The study of defects identified during the DLP can be very significant as a control mechanism to ascertain the validity and reliability of a project's implementation process. Without this study there is little that can be achieved to assess whether the project management team has achieved economy, efficiency and effectiveness in their project management process.

While much can be drawn from the existing study, much more can be can be learnt if the defects data can be exploited using more advanced statistical tools. Further investigations to study the "ripple-effect" of the causal relationships of the process that led to the defects to better track the root cause of the defects is considered for the next stage of the study.

Acknowledgement

The authors would like to thank the University A and the Building Audit Facilities team for their contribution in the research project and the provision of vital information. Comments of the anonymous reviewers who were very helpful are gratefully acknowledged.

References

1. Jorgensen, K. (2009). *Failures and defects in the building process – applying the bow-tie approach.* CIB Joint International Symposium, Dubrovnik, Croatia
2. Haryati M.I, Hassan, F., Takim,R., Che Mat, M., and Ithnin, Z. (2010). *How adequate is adequate? a case of the adequacy in determining client requirements in the construction of four public hospitals in Malaysia.* CIB World Congress, Salford University
3. Haryati, M.I. (2014). *An improved operational framework for defects liability management system for design and build public hospital projects,* University Technology MARA Malaysia. Unpublished PhD Thesis (PhD)
4. Atkinson, A. (1998). Human error in the management of building projects. *Construction, Management and Economics,* **16**(3), 339-349
5. Josephson, P.E., and Hammarlund, Y. (1999). The causes and costs of defects in construction: a study of seven building projects. *Automation in Construction,* **8**(6), 681-687
6. Pheng, L.S., and Wee, D. (2001). Improving maintenance and reducing building defects through ISO 9000. *Journal of Quality in Maintenance Engineering,* **7**(1), 6-24
7. Chong, W.K., and Low, S.P. (2005). Assessment of defects at construction and occupancy stages. *Journal of Performance of Constructed Facilities,* **19**(4), 283-289
8. Sommerville, J. (2007). Defects and rework in new build: an analysis of the phenomenon and drivers. *Structural Survey,* **26**(8), 391-407
9. Zietsman, R. (2008). *Defects in the construction industry – then and now.* 5th Postgraduate Conference on Construction Industry Development, Bloemfontein, South Africa
10. Mills, A., Love, P.E.D., and Williams, P. (2009). Defect costs in residential construction. *Journal of Construction Engineering and Management,* **135**(1), 12-16
11. Lateef, O.A., Mohd Faris, K., and Arazi,I. (2010). Sustainability in the context of maintenance: building defects in the Malaysian university campuses. *Journal of Retails and Leisure Property,* **9**(2), 137-149
12. Kian, P.S. (2001). A review of factors affecting building defects in Singapore. *Dimensi Teknik Sipil,* **3**(2), 64-68
13. Grobler, K., and Pretorius, L. (2002). An evaluation of design-build as procurement method for building an civil engineering projects in South Africa. *Journal of South African Institution of Civil Engineering,* **44**(1), 13-19
14. Andi, and Minato, T. (2004). Representing causal mechanism of defective designs: a system approach considering human errors. *Construction Management and Economics,* **21**(3), 297-305
15. Atkinson, A.R. (2002). The pathology of building defects; a human error approach. *Engineering Construction and Architectural Management,* **9**(1), 53-61
16. Ibrahim, A.R., Roy, M.H., Ahmed, Z., and Imtiaz, G. (2010). An investigation of the status of the Malaysian construction industry. *Benchmarking An International Journal,* **17**(2), 294-308
17. Ahzahar, N., Karim, N.A., Hassan, S.H. and Eman, J. (2011). A study of contribution factors to building failures and defects in construction industry. *Procedia Engineering,* **20**(3), 249-255
18. Palaniandy, P.R. (2011). *A study on factors affecting quality deterioration in hausing construction industry in Klang Valley,* Open University Malaysia. Unpublished MSc Thesis (MSc)
19. Hassan, F., Haryati M.I., Che Mat, M., Ithnin, Z., and Sapisey, Z. (2009). *Report on defects during defects liability period for public hospital projects in Malaysia.* Kuala Lumpur
20. Wardhana, K., and Hardipriono, C. (2003). Study of the recent building failures in the United States. *Journal of Performance of Construction and Facilities,* **17**(3), 157-158
21. Ilozor, B.D., Okoroh, M.I. and Egbu, C.E. (2004). Understanding residential house defects in Australia from the State of Victoria. *Building and Environment,* **39**(3), 327-337
22. Carrillo, P. (2006). Lessons learned practices in the engineering, procurement and construction sector. *Engineering, Construction and Architectural Management,* **12**(3), 236-250
23. Le, M.A.T., and Bronn, C. (2007). Linking experience and learning: application to multi-project building environments. *Engineering, Construction and Architectural Management,* **14**(2), 150-163
24. Dvir, D., Raz, T., and Shenhar, A. (2003). An empirical analysis of the relationship between project planning and project success. *International Journal of Project Management,* **21**(2), 89-95

DOI: 10.1051/matecconf/20166600018

How can BIM support Construction Safety Management? Development of SIM

Isabelle Y S Chan[1], H Y Leung[2,a] , Ivan W H Fung[3] and Michelle Leung[1]

[1]*The University of Hong Kong, Pokfulam Road, Hong Kong*
[2]*Nazarbayev University, Astana, Kazakhstan*
[2]*City University of Hong Kong, Tat Chee Avenue, Hong Kong*

Abstract. BIM is continuously changing the way construction can be approached, from design, scheduling, costing to facilities management. However, even though construction safety is a worldwide issue, and though previous, yet limited, studies have proven its great potential, BIM has not been extended to include automated construction safety management functions, i.e., hazard identification, assessment and control for construction works. In fact, as many as 71% of safety incidents can be prevented by safety considerations at design stage. Automation via BIM is the key to enhance effectiveness and efficiency of safety management, but development of Safety Information Modelling (SIM) is the key pre-requisite. This paper aims to outline a research project for developing both information and technical requirements for SIM via questionnaire survey, focus group, and real case studies. The project results, to be validated by real projects, will act as solid platform for developing:- i) Practical guidance to construction and safety professionals on the types of safety management a SIM can conduct, and the ways of utilizing the safety information in a SIM model [based on the systematic SIM information requirements (e.g., object identification, geometry attributes, safety rules)]; and ii) Overall guidance to software developers to develop appropriate SIM tools for different scenarios [based on the systematic SIM technical requirements (e.g., model information exchange, visualization, conflict analysis process for risk assessment, reliability)].

1 Background

The construction industry has long been recording the highest number of fatalities and accident rate among all industry sectors around the globe. For instance, according to the Hong Kong Census and Statistics Department in 2014, more than 3,100 construction workers have injured and 24 workers have died at work annually – that is – 2 workers die every month! In fact, construction fatalities account for over 80% of the overall industrial fatalities in Hong Kong [1].

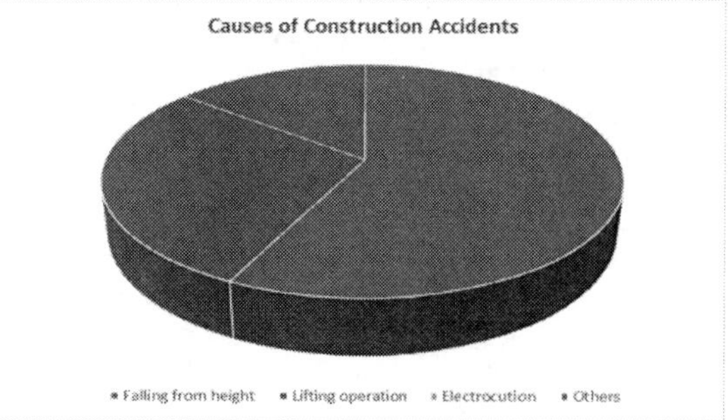

Figure 1. Causes of construction accidents [1]

In these fatalities incidents, most of them are related to falling from height (57%) and lifting operation (29%) (refer to Figure 1). Moreover, the report also indicates an increasing trend of construction accidents in the past few years, increasing from 2,284 in 2010 to 3,467 in 2014 (refer to Figure 2).

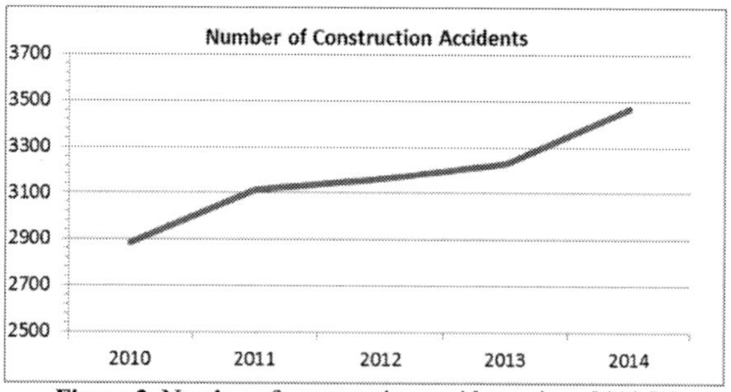

Figure 2. Number of construction accidents since 2010 [1]

It is estimated that construction safety incidents account for 6% of total building cost in general. Safety incidents not only cause serious social and economic impact to our society, but also result in significant losses to construction projects in terms of project cost and programming. In fact, various studies found that 42-71%

[a] Corresponding author: hau.leung@nu.edu.kz

of these safety incidents can actually be prevented by safety considerations at design and construction stages [2,3]. However, there is a lack of responsive tools to assist construction professionals to make informed decisions for construction safety.

2 Current Practices in Construction Safety Planning, Risk and Hazards Identification

Traditionally, safety is managed separately from construction. Safety is always considered as a hindering component to the work production and extra capital expenditure. However, construction management which focuses on productivity and fragmented from safety management tends to disregard the safety constraints within construction process and results in consistently high accident statistics. A well-integrated safety management in construction process shall be devised and put in place, otherwise, the construction project never attains the optimum benefit of the three vital objectives - cost, time and quality.

Furthermore, many researchers also pointed out the lack of integration between construction process and safety issues [4-6]. Effective safety planning contributes in the prevention of accidents and ill health of site personals. Planning well for safety plays an important role in reducing unnecessary cost and delays.

In order to reduce the frequency and severity of construction accidents, construction firms implement safety programs, plans and procedures that include written safety plans [7]. Nevertheless, many constructors found difficulty in converting Safety Management Regulations [F&IU(SM)] directive into actions and practical implementation.

Chong Pui Ho et. al. [8] conducted a research on safety planning. With approximate 5% of budget being spent on safety plan and procedures, many contractors questioned on the worthiness of spending such amounts on safety plans. The proposed research further identified the significant of assessing hazards and risks on the site. More productivity and prevention of accidents/injuries are achieved. So, it is always important to investigate possible safety issues before commencing the work on sites.

Raglan [9] revealed that many construction companies these days have a generic safety plan that is fulfilling the minimum requirement on each element on SMS. They seldom prepare a tailor-made one for each particular project and rarely have sufficient follow up action to monitor the implementation of the plan. This results to the failure in function and occurrence of accident.

In fact, safety assessment is essential as we cannot control what we did not plan for. Any unidentified hazards will not be planned for safety measures. Therefore, during the planning phase, safety must be regarded as important as construction activities [10]. Safety requirements should be the same as construction activities in work breakdown structure. A proactive safety plan and early involvement of safety in the project before hazards can be created.

Furthermore, pervious researches focus on the causes of accident, the review of construction procedures, the introduction of safety culture and the implementation of SMS. They all contribute a remarkable improvement in safety performance, but with the increase in the project's complexity and diversity, each construction site has its specific hazards and these differ from one location to another which cannot be generalized. The improvement in safety performance may reach level out nowadays and is difficult to improve further. Rajendran and Clarke [11] researched on the new methods and tools to prevent these injuries and illnesses. One such tool is BIM.

BIM is one of the most conspicuous attributes of a fundamental change that is swiftly transforming the construction industry [12-14]. A safer worksite is resulted with the improvement in communications.

3 BIM Applications in Overseas and Hong Kong

BIM was first incepted in the 1970s and rapidly developed in the 1980s and 1990s. It is being increasingly implemented in the construction industry to produce data-rich models of buildings and structures.

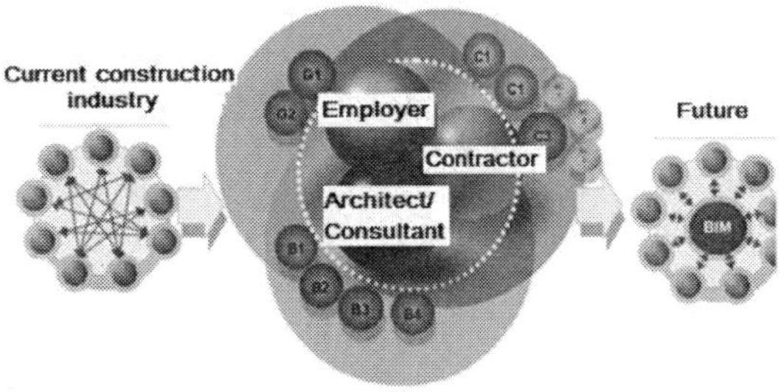

Figure 3. The change in the AEC industry after adopting a BIM approach [15]

BIM is an innovative technology that has transformed the way buildings are designed, constructed and managed. According to McGraw-Hill Construction [15], BIM is being broadly adopted across the construction industry and its use is expected to exponentially expand within forms and across the Architecture, Engineering and Construction (AEC) industry in the coming year (Figure 3). The use of BIM especially enhances the collaborative efforts of the team. The architect and engineer can test their design ideas including energy analysis. The construction manager can provide constructability, sequencing, value and engineering reports. They can also start 3D coordination between subcontractors and vendors during early stages of design. The owner can visually notice if the design is what he is looking for. Overall, the BIM promotes the collaboration of all of the projection participants.

In Hong Kong, as of now BIM usage is confined mostly to the design and planning stages of the project, with very little of it being used in the construction phase. However, construction phase is well known for its roots of accidents. Furthermore, BIM technology are proved with the potential to be used in safety planning procedures particularly those related to tasks on construction sites in recent studies [16,17].

Studies done by Kiviniemi [18] and Rohena [12] point out that BIM technologies are actually gradually moving from the worlds of architecture and engineering to the arenas of construction companies and other participants in the construction operations that is dramatically reshaping the ways of work in construction project teams. They also figured out the benefit of the involvement of contractor in terms of the increase in productivity and improvement of the final project outcomes for all the parties involved. The findings of studies on the use of BIM have shown that communications can be improved and errors can be eliminated; thus resulting in a safer worksite.

In fact, studies on utilizing BIM in Safety aspect in construction industry has been started in various countries recently, such as USA, Finland and Norway [19-23]. It shows that Hong Kong is lagging behind. In Hong Kong, contractors involved in the BIM are still in a limited basic and little researches on the study of construction safety in BIM are found.

In Finland, the association of Finnish Contractors is actively promoting the implementation of the BIM in the industry. VTT, one of the important organizations in the research of BIM, carried out on-going safety BIM research project from 2009 to encourage and develop utilization of BIM technology in construction planning and management - from viewpoint of occupational safety. Example are "Building Information Model (BIM) promoting safety in the construction site process" from October 2007 to February 2009 and "BIM-based Safety Management and Communication for Building Construction" from April 2009 to June 2011 which aim to develop procedures and use of BIM technology for safety planning, management, and communications, as part of the 4D-construction planning.

Moreover, the theme of the 2012's CIB W099 International Conference is "Modeling and Building Health and Safety" and it specially address the use of BIM to add the safety dimension for monitoring and improving safety performances on construction projects by launching a competition. It revealed that BIM, recognized as one of the most conspicuous attributes that is swiftly transforming the construction industry, is worth investigating and can be efficaciously enhance a safer worksite from improving the communications, eliminating errors by integrating design and collaborative features of BIM.

Nowadays, with the 4D scheduling platform, BIM can be used to perform design-for-safety and risk assessment, improving the inefficiencies of the paper-based processes in traditional construction safety management (i.e., text-based standalone check-sheet type tools which are conducted via paper /software interface). Although current software, e.g., Solibri, can conduct rule-based checking on BIM, it only recognizes relationships between artifacts (BIM components) and report the clashes without 'thinking' and 'analyzing' from a safety perspective. In fact, the usage of BIM in construction is still in the primary stage in Hong Kong. According to Hong Kong Construction Industry Council [24], the projects in Hong Kong are classified as 'Typical Practice' only among the five levels, namely 'Innovative Practice' (industry 'firsts'), 'Best Practice', 'Advanced Practice', 'Typical Practice' and 'Conventional Practice' (without BIM). In Hong Kong, the Hong Kong Housing Authority has long been regarded as the BIM champion driving the implementation of BIM amongst the sector. It has implemented BIM long before 2006, utilizing BIM along the whole building life cycle, from design stage, construction stage to facility management [24]. However, the implementation of BIM on construction site safety is still limited (manual safety checking based on 4D visualization of construction process). There is a lack of a BIM-based, automatic construction site safety management system

To automate the construction safety management process, a more comprehensive BIM has to be developed i.e., Safety Information Modelling (SIM), which includes artifacts embedded with safety-related properties or classes, with corresponding safety rules, scheduling information, geographic information, and so on. More importantly, since not all safety issues can be managed in the same way; in this project, we will investigate how SIM, with different sets of information and technical requirements, can be applied to different hazards in different construction site conditions.

4 Development of SIM

The concept of SIM will certainly improve and enhance the quality and efficiency of construction safety management. According to the results of the NBS National BIM Survey in 2012 [25], there are nearly 80% of construction professionals aware of and/or adopting BIM nowadays. Various professional institutions, such as the Chartered Institute of Building (CIOB) and the Royal Institution of Chartered Surveyors (RICS), have been actively involved in promoting the use of BIM among its members. However, misconceptions of BIM prevail in our industry [26], and the current adoption of BIM in our industry is still limited to the visualization level (i.e., 3D BIM). Adopting BIM in construction safety is even rarer. While the majority of construction professionals perceived that BIM can enhance quality and time of a project, nearly 50% of construction professionals fail to aware the potential of BIM on enhancing safety [27]. The limited adoption and the underestimated benefits of BIM in safety are attributed to the incomplete safety information available in current building information models [28,29]. Although there are limited studies investigating the utilization of rule-based hazard identification engines in certain scenarios, those studies are generally algorithm-based and case-specific, with support from limited empirical data. Those studies are hence with limited representation. In view of the above, a study is proposed to fill the gap by collecting empirical data from groups of representable safety and BIM experts to build a comprehensive platform, incorporating both information and technical requirements, for the development and implementation of SIM across different scenarios in construction projects. The research results will also be validated by real case studies using projects of our supporting organizations.

5 Details of Proposed Research

The research project aims to investigate the information and technical requirements for the development of Safety Information Modelling (SIM) for supporting automation of construction safety management via BIM. To achieve this aim, the objectives are:
1. To systematically review BIM in general and its application to construction safety management, in terms of design-for-safety, safety planning, risk assessment (e.g., hazard identification, assessment and control).
2. To examine the information requirements for SIM in supporting construction safety management.
3. To investigate the technical requirements for SIM based on the software tools to enhance construction safety management on site.

This empirical study will benefit not only safety professionals, but also the future development of BIM (SIM), project performance (safety) and the construction sector as a whole!

The above objectives will be realized by carrying out literature review, questionnaire survey, focus group study and real case study.

- Literature Review - A comprehensive literature review will be conducted to define and refine the research framework (e.g., BIM adoption in safety, the identified benefits and barriers, potential software tools), and to develop survey and focus group interview protocol. The review will cover both academic and professional reports, journal papers and websites.
- Questionnaire Survey - A questionnaire survey will be developed based on the extensive literature review to investigate the current BIM adoption level in different construction stages, and the anticipated critical factors and components of SIM, covering both informational and technical aspects The survey study will target construction safety professionals (say 300, incl. safety managers, safety supervisors, safety officers, assistant safety officers). The quantitative data collected will be analyzed by statistical techniques, including factor analysis, reliability analysis, Pearson correlation and multiple regression modelling, using SPSS and Lisrel. In order to obtain high response rate, assistance may be sought from various professional bodies, e.g. CIOB, to recruit survey respondents and focus group interviewees through their membership database.
- Focus Group Study - Through providing a dynamic and interactive platform for BIM experts and construction safety professionals to exchange knowledge and demonstrate their expertise in BIM and construction safety, the focus groups will be extremely effective in helping to understand the prevailing situation, perceived benefits and barriers of SIM in construction, and to identify information and technical requirements/logics of SIM. The qualitative data collected will be analyzed by systematic contextual analysis.
- Real Case Study – The above results will be validated by real BIM projects. A number of supporting organizations have been identified and they include Institute of Safety and Health Practitioners, Chun Wo Construction Co. Ltd and Chun On Engineering Consultant Ltd.

It is anticipated that a detailed guideline and logic for implementing SIM practically [e.g., Scenario Categorization > Rule Mapping > Conflict Analysis (rule execution & checking) > Conflict Management (report)] can be developed in this research project. The development of the new platform and all instructional guidelines of SIM should strictly refer to HK existing COPs, related safety ordinance and regulations, such as CSSR and F&IU (Safety Management) Regulations, etc. It is also believed that the final product can be well applied in the construction industry for safety.

6 Concluding Remarks

BIM has been increasingly adopted in construction industry over the world, it is believed that its versatility can be readily extended to some other aspects such as construction safety management. With a view that an automated safety management system will greatly benefit the industry, a new concept of Safety Information Modelling (SIM) via the use of BIM is developed. It is anticipated that the current research project will be able to shed some light on the future development and extension of BIM to construction safety management.

References

1. Legislative Council, *Legislative Council Panel on Manpower Hong Kong's Occupational Safety Performance in 2014*. (2015) Retrieved from http://www.legco.gov.hk/yr14-15/english/panels/mp/papers/mp20150616cb2-1683-3-e.pdf on 5th October 2015.
2. Gibb, A, Haslam, R, Hide, S and Gyi, D. The Role of Design in Accident Causality. In: *S. Hecker, J. Gambatese and M.Weinstein, (eds) "Designing for Safety and Health in*

Construction", Eugene, Oregon: University of Oregon Press. (2004)
3. Zhou, W., Whyte, J., and Sacks, R. Building safely by design: interim report. *DIRC Working Paper No. 2, Design Innovation Research Centre, University of Reading, UK.* (2011)
4. Gambatese, J and Hinze, J. Addressing construction worker safety in the design phase – Designing for construction worker safety, *Automation in Construction*, **8**, 643-649. (1999)
5. Benjaoran, V., Bhokha, S.,. An integrated safety management with construction management using 4D CAD model. *Safety Science* **48** (3), 395–403. (2010)
6. Sulankivi, K. K., K., Mäkelä, T., Kiviniemi, M. 4D-BIM for Construction Safety Planning. *CIB (2010).* P. Barrett, Amaratunga, Dilanthi, Haigh, Richard, Keraminiyage, Kaushal & Pathirage, Chamind. Manchester. (2010)
7. Hallowell, M. Cost-effectiveness of construction safety programme elements, *Construction Management and Economics*, **28**: 1, 25-34. (2010)
8. Chong Pui Ho, D., Ahemed, S. M., Chu Kwan, Jack and Wei Ming,F. Y. "Site Safety Management in Hong Kong", *J. Mgmt. in Engrg*, 16-34. (2000)
9. Lam, Hei-choi, Raglan, An investigation into the implementation of safety management systems by Hong Kong construction contractors. *Phd. University of Hong Kong* (2003). Retrieved from http://hdl.handle.net/10722/52587
10. Kartam, N.A., Integrating safety and health performance into construction CPM. *Journal of Construction Engineering and Management* **123**, 2, 121–126. (1997)
11. Rajendran, S. and Clarke, B. Building Information Modeling: Safety benefits and opportunities. *Professional Safety*, October 2011, 44-51. (2011)
12. Raymond Rohena "*BUILDING INFORMATION MANAGEMENT (BIM) IMPLEMENTATION IN NAVAL CONSTRUCTION". The University Texas at El Paso, 1999.* (2011)
13. Mehmet F. Hergunsel " BENEFITS OF BUILDING INFORMATION MODELING FOR CONSTRUCTION MANAGERS AND BIM BASED SCHEDULING".*A Thesis Submitted to the Faulty of WORCESTER POLYTECHNIC INSTITUTE.* (2011)
14. Mojtaba Taiebat, Tuning Up BIM for Safety Analysis-Proposing modeling logics for application of BIM in DfS. *Doctor of Philosophy, e Virginia Polytechnic Institute and State University* (2011), Retrieved from http://scholar.lib.vt.edu/these/available/etd-12232011-002222/unrestricted/Taiebat_M_D2011.pdf
15. McGraw-Hill Construction. *Building Information Modeling: Transforming Design and Construction to Achieve Greater Industry Productivity*, McGraw-Hill Construction, New York. (2008)
16. Gambatese, J., Behm, M., Hinze, J. Viability of designing for construction worker safety. *Journal of Construction Engineering and Management* **131** (9), 1029–1036. (2005)
17. Chavada, R. and Dawood, N. N. 'An innovative approach to integrate H&S issues in the construction project planning using serious game engine technologies', CONVR2010, Sendai, Japan, November 4-5, 2010, in Makanae, K., Yabuki, N. and Kashiyama, K. (eds) *Proceedings of the 10th international conference on construction applications of virtual reality. CONVR2010* Organizing Committee, 209-218 (2010)
18. Kiviniemi, A.. The effects of integrated BIM in processes and business models. *Distributed Intelligence in Design*, 125-135. (2011)
19. Kiviniemi, A., Tarandi, V., Karlshoj, J., Bell, H. and Karud, O.J. "*Review of the development and implementation of IFC compatible BIM*"(2008), Retrieved from www.eracobuild.eu/fileadmin/documents/Erabuild_BIM_Final_Report_January_2008.pdf
20. Wong, A.K.D., Wong, F.K.W. and Nadeem, A. "Comparative roles of major stakeholders for the implementation of BIM in various countries", *Proceedings of the International Conference on Changing Roles: New Roles, New Challenges, Noordwijk Aan Zee, The Building Information modelling 75 Netherlands*, 5-9 October, (2009) Retrieved from www.changingroles09.nl/uploads/File/Final.KD.Wong-KW.Wong-Nadeem.pdf
21. Succar, B. "Building information modelling framework: a research and delivery foundation for industry stakeholders", *Automation in Construction*, **18**, 357-75. (2009)
22. Khemlani, L, "*CORENET e-plan check: Singapore's automated code checking system*", AECbytes. (2005), Retrieved from www.aecbytes.com/buildingthefuture/2005/CORENETePlanCheck.html
23. Arayici, Y., "Towards building information modelling for existing structures". *Structural Survey*, **26**, (3), 210-22. (2008)
24. Hong Kong Construction Industry Council, *Roadmap for Building Information Modelling Strategic Implementation in Hong Kong's Construction Industry.* Construction Industry Council. (2014)
25. Malleson, A. *Building Information Modeling-What's going on in the world of BIM: NBS national BIM survey*(2012). Retrieved from http://www.thenbs.com/topics/BIM/articles/whatsGoingOnInWorldOfBIM.asp
26. Jernigan F. Big BIM little BIM: *The Practical Approach to Building Information Modeling-Integrated Practice Done the Right Way!* 2nd ed. 4Site Press (2007)

27. Suermann P, Issa R. Evaluating industry perceptions of building information modeling (BIM) impact on construction, *Journal of Information Technology in Construction* (ITcon), **14**, 574-594, (2009) Retrieved from http://www.itcon.org/2009/37
28. Zhang, J.P. and Hu, Z.Z. BIM- and 4D-based integrated solution of analysis and management for conflicts and structural safety problems during construction. *Automation in Construction*, **20**, 155–166. (2011)
29. Zhang, S., Teizer, J., Lee, J. K., Eastman, C. M., and Venugopal, M. Building information modeling (BIM) and safety: Automatic safety checking of construction models and schedules. *Automation in Construction*, **29**, 183-195. (2013)

SYNCHRONISING TENDER DRAWINGS AND DOCUMENTS AS AN AID FOR DISCREPANCIES

Iznny Ismail[1,a], Azman Zainonabidin[2] , Teh Zahariah Nassaruddin[3] and Rodzyah Mohd Yunus[4]

[1, 2] *Department of Architecture, Faculty of Architecture, Planning & Surveying, Universiti Teknologi MARA Campus Seri Iskandar, 32610 Perak, Malaysia*
[3]*Department of Quantity Surveying, Faculty of Architecture, Planning & Surveying, Universiti Teknologi MARA Campus Seri Iskandar, 32610 Perak, Malaysia*
[4]*Department of Architecture, Faculty of Architecture, Planning & Surveying, Universiti Teknologi MARA Campus Shah Alam, 54000 Selangor, Malaysia*

Abstract. This paper recommends an aid to control discrepancies between drawings and documents in a project. The synchronisation may assist to better project's preparation and documentations. The causes of discrepancies that found out during the contract stage may lead to ineffectiveness of construction cost. The controlling process initiated from the tendering stage as to minimize the causal by any of the consultants which some items may left behind or possibly not describe fully without the thorough affiliation of coordination. The method to reduce discrepancies between drawings and documents is identified through Case Study based on 'hands on' project. The case study is focused on the preparation of architectural detailing. Therefore, the relationship of 2-Dimensional (2-D) Tender Drawings and bills of quantities is compared with the Elements of Building Works by Malaysian Standard Method of Measurement of Building Works Second Edition (SMM). The relationship study will simplify and supersede any unnecessary detailing. Thus, the application of the synchronisation method will result effectiveness of documentation during the contract stage by reducing discrepancies.

1 Introduction

Discrepancies of Tender Drawings and Documents are identified during contract stage. The process involved contractor's findings when he plans the carrying out of constructional works. According to PAM Contract 2006 [1], during the planning and subsequent execution of the works, the contractor needs to notify the architect in writing when they find any discrepancies between documents. Hence, with the notification architect shall enable to issue written instructions in sufficient time before any commencement of works regarding discrepancies. It nevertheless encourages contractor not to request further instruction at the very last minute to avoid delays. To the same degree, PWD Form 203A [2], stated that the contractor shall immediately notify the Superintending Officer (S.O.) a written notice for any discrepancy and divergence as for the S.O. to issue instructions.

Discrepancies happen among consultants' documentations [3]. Thus, discrepancies involved neither architectural and engineering drawings, nor engineering and engineering drawings, nor architectural drawings and bills of quantities, nor engineering drawings and bills of quantities. In accordance with the documentation, client's requirements are complex nowadays [4]. In supporting to the complexity, a great deal of coordination is required to avoid delay , especially for public projects [5].

Many researchers have studied on the effects and causes of delays and cost overrun and all suggested early detection and planning. Sambasivan and Soon [6] recommended that practitioners to better understand the dynamics of project management and make an effort to reduce delays, whereas and academicians can conduct similar studies to identify causes and effects of delays. At the preliminary stage of the project, it is essential that effective decisions on design specifications, project financing, contractual systems, and methods of construction are all taken at the right time [7]. Odeh and Battaineh [8] listed that one of the causes of delay is contract factors include changed orders and mistakes with discrepancies in contract documents. Actions are taken to control the causes of delay and cost overruns right away from the planning to the implementation and management stages [5]. Analysis has been made by Cheng [9] cost impact factor and contract dispute highlighted on construction drawings are unclear with ambiguous specifications. The study made by [4] ranked that one of the factor influencing cost variance during construction is incomplete design drawings and specification. Thus, understanding the relationship between drawings and documents is crucial in achieving synchronization. This paper is organized as follows. Section 2 deals on explanation on drawings and

[1] Iznny Ismail: iznny813@perak.uitm.edu.my

documents. Section 3 explains on the methodology and conceptual framework of the research. Section 4 discusses the analysis of findings. Section 5 presents the conclusion and recommendation.

2 Literature review

2.1 Understanding the relationship of drawings and specifications

Drawings and specifications always been treated as different means of resources. Rosen and Regener [10] recommended the relationship should be complementary with no duplication or conflict between drawings and documents. Arditi and Gunaydin [11] described that it is critical that drawings and specifications be clear, concise, and uniform; the project must be constructible from those retained to build the project; design professionals must be familiar with construction materials and techniques that constructors will be using in the project. Contract documents contained specifications, as written instructions, are frequently judged by the courts as having greater importance than drawings when these documents are in conflict, with judgments based usually on what is contained in the specifications. This means that specifications should be carefully prepared by knowledgeable people [10] . On the other hand, Connell and Flegal [12] identified that drawings depict the work graphically and the specifications provide a written description, while addenda and modifications are official changes to both. Quality project built by a contractor is directly related to the quality of plans and specification [13]. The authors suggest that a relationship of drawings and specifications are uniform without duplication of information as for the purpose of contractors fulfilling their obligations. Thus, if the resources are well prepared, discrepancies can be avoided and lessen wastage of time and cost.

2.1.1 Information and graphic for 2-dimensional (2-D) drawings

The drawings illustrate a series of information that describes in graphics. Drawings are the only documents given to the constructor that show the design concept, size and scope of the job [11]. Another author described that drawings present series of pictures of a project to be constructed that indicate size, form, location and arrangement of various elements [10].

2.1.2 Measurement for specifications

Specifications complement the drawings that provided for each project. Rosen and Regener [10] distinguished that the purpose of specifications should describe the type and quality of every work; quality of workmanship; requirements for fabrication, erection, application, installation and finishing; applicable regulatory requirements; overall and components dimension; specific descriptions and procedures for allowance of unit prices ;specific descriptions and procedures for product options; and specific requirements for administration of contract for construction. According to PAM Contract 2006 [1], clause 12., measurement of building works relating to the quality and quantity of the work included in the Contract Sum shall be deemed to be those which are set out in the Contract Bills and unless otherwise expressly stated, shall be prepared in accordance with the Principles of the Standard Method of Measurement of Building Works (SMM) sanctioned by the Institution of Surveyors Malaysia (ISM) and currently in force. However Harbans and Kandan [14] recognized the uncertainty of measurement the contractual provisions, methods for measuring the varied work applying SMM for building works and CESMM for civil engineering works, but to the contrary mechanical and electrical works there are no accepted norms in the measurement process. Despite of that, Connell and Flegal [12] mentioned the use of standards is extremely important.

Jaafar, Lim, and Yahaya [15] explained that elementary estimating for building works, during the preliminary design stage, project estimation is prepared and estimation involves measuring and pricing. The author clarified that measuring is the process of abstracting Standard Method of Measurement (SMM) and the quantity of each building element will be compiled as Bills of Quantity. On the other hand, pricing is the process of calculating the cost of each building element stated in the Bills of Quantity. The actual building estimation is based on detailed drawings such as layout plans, elevations, sections and supported by a specific list of building elements.

Table 1 . Elements of building works by Malaysian standard method of measurement of building works second edition (SMM)

Building Elements	**Role by Consultant**
Preliminaries	All Consultants
Demolitions	All Consultants
Excavation and Earthwork	Civil Engineer
Piling and Diaphragm Walling	Structural Engineer
Concrete Work	Structural Engineer
Brickwork and Blockwork	Architect
Underpinning	Civil Engineer
Masonry	Architect
Waterproofing and Asphalt work	Architect
Roofing	Architect and Structural Engineer
Woodwork	Architect
Structural Steelwork	Structural Engineer
Metalwork	Architect
Plumbing and Mechanical Engineering Installations	Architect and Mechanical Engineer
Electrical Installations	Electrical Engineer
Floor, Wall and Ceiling Finishings	Architect
Glazing	Architect
Painting and decorating	Architect
Drainage	Civil Engineer
Fencing, Turfing and Planting	Architect

2.3 Coordination of 2-dimensional (2-D) drawings and specification

Coordination plays the role of the team player in a project. Rosen and Regener [10] emphasized that to improve coordination between drawings and specifications, standardization of information should be appearing in them. Coordination is one of the factors to minimize discrepancies. The list comprises of:

(1) Closer consultant coordination is required at design stage;
(2) Closer consultant coordination and utilization of an experienced consultant to produce a concluding design, working drawings, and contract drawings at the tender stage;
(3) Checking and reviewing the contract documents;
(4) Develop a clear scope of work for the parties that involved on the project;
(5) Integrated project scheduling and management techniques;

The coordination factors numbered (1)-(4) suggested on closer coordination, thorough checking, and develop clear technical detailing that integrated with conclusive design. Whereas factor numbered (5) is a suggestion during the construction stage. Coordination before tendering stage is important due to making sure the documents and drawings are described conclusively. Simply to avoid some items may leave out or possibly not describe fully. The coordination process will get more tedious, according to the complexity of the project. It is a careful process of making sure the documentation final.

3 Methodology and conceptual framework

This research involved qualitative mode of research. Love, Lopez, Edwards, and Goh [16] recommended that a case study approach based upon analytic induction is used to further examine the underlying dynamics that may contribute to design error production. Furthermore, the research is guided by De Massis and Kotlar [17] suggested for choosing the case study design, identify the unit analysis (case), selecting the cases (sampling), collecting information, and analysing information. Implementing the guideline, the case study design is a descriptive case study that used to convince a phenomenon that is relevant, as referring to "Synchronise drawings and documents can be an aid for discrepancies". The purpose is to understand project in depth regarding ways or method in minimizing discrepancies. The unit analysis is focusing on a project within the firm with an experienced individual. It involved 'hands on' case studies for research approach that involved five (5) years of experienced in handling multiple projects. Sampling of the case study is by a single case for aiming straightforward information related to architectural detailing and specifications. The case studies are based on projects in Perak, Malaysia that involve the building type of an office building. The information is gathered from tender drawings and documents which consist of research techniques are from observations and appropriate written documents. The information is analysed through two steps, firstly by comparison and to be followed by synchronizing data.

Chart 1. The framework of methodology

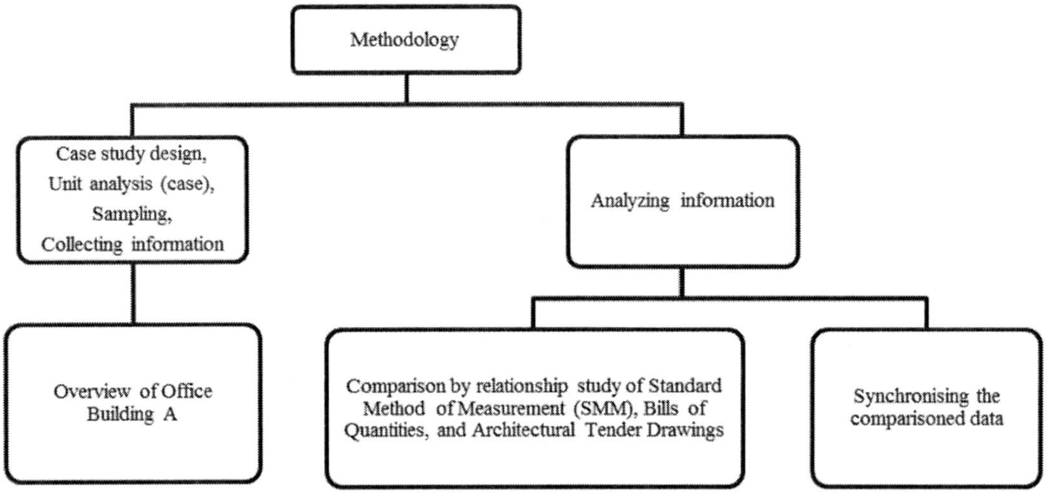

Figure 1. Perspective of Office Building A

4 Findings and discussion

4.1 Overview of Office Building A

A single case study is identified based on a project in Perak, Malaysia. The case study building type is a government office building that approximately cost RM 2,500,000 (Ringgit Malaysia Two Million and Five Hundred Thousand. The project is managed by a project architect with five (5) years of project handling experience. The observations are dominantly from the architectural side of the input. The client of the project follows the project phases by following the submission plan approval timeline and just then proceed to the next stage after gain all approvals. The time allocation is carefully planned and follows as this is the first implementation by the client to bind the work programme part of the Memorandum of Agreement (MOA) between the client and each of the consultants. This means that the client can claim damages towards some late progress of the consultants because the work programme has been legalized. This policy already implemented by the government in minimizing delays in the consultant's scope of work. All progress between client and consultants has been clarified with clear brief, close coordination, detailed forecast, clear communication, careful budgetary consideration, and conclusive documents and drawings for quality assurance.

4.2 Relationship between Standard Method of Measurement (SMM) with bills of quantities and architectural tender drawings.

The comparison is to identify the relationship between the building element of Standard Method of Measurement (SMM) with bills of quantities (gathered information from quantity surveyor) and architectural tender drawing (gathered information from architect).The bills of quantities provided is prepared in accordance of SMM. From this comparison, some of the elements in the bills repeated as to follow the elements of SMM. For example the staircase element in bills of quantities has been repeated under building element concrete works, metal works, and floor, wall, and ceiling finishings. The assortments of drawings and documents are identified through the comparison with SMM.

4.3 Relationship between bills of quantities with Standard Method of Measurement (SMM) and architectural tender drawings.

In accordance with the comparison of table 2, another related study is taken. The comparison is done on the elements of building works of bills of quantities between architectural tender drawings and building elements of Standard Method of Measurement (SMM). This comparison simplifies the elements in the bills of quantities. Besides, multiple building elements are grouped related to one element in the bills of quantities. For example relationship between the staircase element in the bills of quantities to be compared with Detail Staircase comprises building elements of concrete work, metalwork, and floor, wall and ceiling finishings. Some of the architectural tender drawings are repeated under different building works element in the bills of quantities. Hence, the relationship between architectural tender drawings and SMM is cluttered, but to be compared with bills of quantities, the architectural tender drawings are in order. However, the relationship has to be simplified further.

4.4 Corresponding between bills of quantities with architectural tender drawings

In relation to the table 3, the bills of quantities, elements are simplified by omitting the elements to be measured under different bills and not required in the design. However, there is still repeating Architectural Tender drawing in accordance to the element in the bills of quantities. For example the element between Internal Doors and Windows and External Doors and Windows, the architectural drawings of Detail Door, Detail Window, and Detail ironmongery are repeated. This synchronization is not conclusive as it will bring difficulty during managing architectural drawings.

4.5 Corresponding between architectural tender drawings with bills of quantities

In relation to table 4, tabulation is made in synchronizing between the architectural drawings and bills of quantities. Each of the elements already been clarified as to improve the outcome of the recommended model of synchronizing drawings and documents. Architectural tender drawings are identified relating to the elements in the bills of quantities as for easier coordination purposes.

Table 2. Comparison between building elements of SMM with bills of quantities and architectural tender drawings for office building A

Building Element-SMM	**Bills of Quantities**	**Architectural Tender Drawings**
A Preliminaries	*To be measured under different bill	Preliminaries Item
C Demolition	*Not required in the design	* No design provided
D Excavation and Earthwork	*To be measured under different bill	* Not under scope of work
E Piling and Diaphragm Walling	Work Below Lowest Floor Finish	* Not under scope of work
F Concrete Work	Frame	* Not under scope of work
	Staircase	Detail Staircase (Concrete works)
	Staircase	Detail Apron, Ramps, Steps And Perimeter Drain (Steps)
	Apron Ramps and Perimeter Drain (Ramp)	Detail Apron, Ramps, Steps And Perimeter Drain (Ramp)
	Upper Floor Construction	* Not under scope of work
	Builders Work in Connection	Detail Ablution
	Roof (flat roof)	Scupper Drain Detail (flat roof)
G Brickwork and Blockwork	External Wall	Detail External & Internal Wall
	Internal Wall and Partition	Detail External & Internal Wall
	Builders Work in Connection	Detail Sanitary Fitting (Ablution)
H Underpinning	*Not required in the design	* Not under scope of work
J Masonry	*Not required in the design	* No design provided
K Waterproofing and Asphalt Work	Builders Work in Connection	Detail Ablution
	Roof (flat roof)	Scupper Drain Detail (waterproofing)
L Roofing	Roof	Detail Roof
	Roof (flat roof)	Scupper Drain Detail (flat roof)
M Woodwork	External Doors and Windows	Detail Door
	External Doors and Windows	Detail Window
	External Doors and Windows	Detail Ironmongery
	Internal Doors and Windows	Detail Door
	Internal Doors and Windows	Detail Window
	Internal Doors and Windows	Detail Ironmongery
N Structural Steelwork	*Not required in the design	* Not under scope of work
P Metalwork	External Doors and Windows	Detail Door
	External Doors and Windows	Detail Window
	Internal Doors and Windows	Detail Door
	Internal Doors and Windows	Detail Window
	Staircase	Detail Staircase (Railing)
	External Wall	Detail Louvres
Q Plumbing and Mechanical Engineering Installations	Sanitary Fittings	Detail Sanitary Fitting
	Sanitary Fittings	Sanitary Fitting Schedule
	Builders Work in Connection	Detail Ablution
	Cold Water and Sanitary Plumbing	* Not under scope of work
	*To be measured under different bill (Mechanical)	* Not under scope of work
R Electrical Installations	*To be measured under different bill (Electrical)	* Not under scope of work
S Floor, Wall and Ceiling Finishings	Internal Floor Finishes	Detail Internal Floor Finishes
	Internal Floor Finishes	Detail Skirting Finishes
	Internal Wall FInishes	Detail Wall Finishes
	Internal Ceiling Finishes	Detail Internal Ceiling Finishes
	External Finishes	Detail Wall Finishes
	External Finishes	Detail Feature Wall
	Staircase	Detail Staircase (Finishes)
T Glazing	*Not required in the design	*No design provided
U Painting and decorating	*To be measured with finishes	* Details provided under finishes
V Drainage	Apron Ramps and Perimeter Drain (Drain)	Detail Apron, Ramps, Steps And Perimeter Drain (Drain)
W Fencing, Turfing and Planting	*Not required in the design	*No design provided
*Others	Flag Pole (Provisional Item)	Detail Flag Pole

[1] Iznny Ismail: iznny813@perak.uitm.edu.my

Table 3. Comparison of elements of building works between bills of quantities of building works, architectural tender drawings and SMM for office building A

Bills of Quantities	**Architectural Tender Drawings**	**Building Element-SMM**
*To be measured under different bill	Preliminaries Item	A Preliminaries
Work Below Lowest Floor Finish	* Not under scope of work	E Piling and Diaphragm Walling
Apron Ramps and Perimeter Drain	Detail Apron, Ramps, Steps And Perimeter Drain (Apron & Ramp)	F Concrete Work
	Detail Apron, Ramps, Steps And Perimeter Drain (Drain)	V Drainage
Frame	* Not under scope of work	F Concrete Work
Upper Floor	* Not under scope of work	F Concrete Work
Staircase	Detail Staircase (Concrete works)	F Concrete Work
	Detail Staircase (Railing)	P Metalwork
	Detail Staircase (Finishes)	S Floor, Wall and Ceiling Finishings
	Detail Apron, Ramps, Steps And Perimeter Drain (Steps)	F Concrete Work
		S Floor, Wall and Ceiling Finishings
Roof	Detail Roof	L Roofing
Roof (flat roof)	Scupper Drain Detail (flat roof)	F Concrete Work
		L Roofing
		K Waterproofing and Asphalt work
External Wall	Detail External & Internal Wall	G Brickwork and Blockwork
	Detail Louvres	P Metalwork
Internal Wall and Partition	Detail External & Internal Wall	G Brickwork and Blockwork
External Doors and Windows	Detail Door	M Woodwork
		P Metalwork
	Detail Window	M Woodwork
		P Metalwork
	Detail Ironmongery	M Woodwork
Internal Doors and Windows	Detail Door	M Woodwork
		P Metalwork
	Detail Window	M Woodwork
		P Metalwork
	Detail Ironmongery	M Woodwork
Internal Floor Finishes	Detail Internal Floor Finishes	S Floor, Wall and Ceiling Finishings
		U Painting and decorating
Internal Wall Finishes	Detail Wall Finishes	S Floor, Wall and Ceiling Finishings
		U Painting and decorating
Internal Ceiling Finishes	Detail Internal Ceiling Finishes	S Floor, Wall and Ceiling Finishings
		U Painting and decorating
Internal Floor Finishes	Detail Skirting Finishes	S Floor, Wall and Ceiling Finishings
		U Painting and decorating
External Finishes	Detail Wall Finishes	S Floor, Wall and Ceiling Finishings
		U Painting and decorating
	Detail Feature Wall	S Floor, Wall and Ceiling Finishings
		U Painting and decorating
Sanitary Fittings	Detail Sanitary Fitting	Q Plumbing and Mechanical Engineering Installations
	Sanitary Fitting Schedule	Q Plumbing and Mechanical Engineering Installations
Cold Water and Sanitary Plumbing	* Not under scope of work	Q Plumbing and Mechanical Engineering Installations
Builders Work in Connection	Detail Ablution	F Concrete Work
		G Brickwork and Blockwork
		K Waterproofing and Asphalt work
		Q Plumbing and Mechanical Engineering Installations
Flag Pole (Provisional Item)	Detail Flag Pole	*Others

Table 4. Synchronized bills of quantities of building works with architectural tender drawings based on the elements of building works for office building A

Bills of Quantities	**Architectural Tender Drawings**
*To be measured under different bill	Preliminaries Item
Apron Ramps and Perimeter Drain	Detail Apron, Ramps, Steps And Perimeter Drain (Apron & Ramp)
	Detail Apron, Ramps, Steps And Perimeter Drain (Drain)
Staircase	Detail Staircase (Concrete works)
	Detail Staircase (Railing)
	Detail Staircase (Finishes)
	Detail Apron, Ramps, Steps And Perimeter Drain (Steps)
Roof	Detail Roof
Roof (flat roof)	Scupper Drain Detail (flat roof)
External Wall	Detail External & Internal Wall
	Detail Louvres
Internal Wall and Partition	Detail External & Internal Wall
External Doors and Windows	Detail Door
	Detail Window
	Detail Ironmongery
Internal Doors and Windows	Detail Door
	Detail Window
	Detail Ironmongery
Internal Floor Finishes	Detail Internal Floor Finishes
	Detail Skirting Finishes
Internal Wall Finishes	Detail Wall Finishes
Internal Ceiling Finishes	Detail Internal Ceiling Finishes
External Finishes	Detail Feature Wall
Sanitary Fittings	Detail Sanitary Fitting
	Sanitary Fitting Schedule
Builders Work in Connection	Detail Ablution
Flag Pole (Provisional Item)	Detail Flag Pole

Table 5. Synchronized architectural tender drawings with bills of quantities of building works based on the elements of building work for office building A

Architectural Tender Drawings	**Bills of Quantities**	Clarification
Preliminaries Item	*To be measured under different bill	Drawings for preliminaries item provide in architectural tender drawings set. Information inclusive of hording material, sign board, and temporary office
Detail Apron, Ramps, Steps And Perimeter Drain	Apron Ramps and Perimeter Drain	The steps detail should be in Staircase drawings.
Detail Staircase	Staircase	The staircase detail should indicate of material used whether it is concrete works, woodwork or metal work; finishes; and railing.
Detail Apron, Ramps, Steps And Perimeter Drain (Steps)	Staircase	The detail should be separated and titled under steps detail.
Detail Roof	Roof	The detail pertaining roof should indicate the type of roof . For example slate , tiled or concrete flat roof.
Scupper Drain Detail (flat roof)	Roof (flat roof)	Scupper Drain detail should be inclusive in roof detail.
Detail External & Internal Wall	External Wall Internal Wall and Partition	Indication of Internal wall and external can be done by just indicating ust 'wall' . This is because identification can be varies by different QS.
Detail Louvres	External Wall	The louvres detail should be in External and Internal wall
Detail Door Detail Window Detail Ironmongery	External Doors and Windows Internal Doors and Windows	The door, window, and ironmongery is ideally separated for easier quantification.

Table 5 (continued)

Architectural Tender Drawings	Bills of Quantities	Clarification
Detail Internal Floor Finishes	Internal Floor Finishes	No modification
Detail Skirting Finishes		
Detail Wall Finishes	Internal Wall Finishes	No Modification
Detail Internal Ceiling Finishes	Internal Ceiling Finishes	No Modification
Detail Wall Finishes	External Finishes	No Modification
Detail Feature Wall		
Detail Sanitary Fitting	Sanitary Fittings	No Modification
Sanitary Fitting Schedule		
Detail Ablution	Builders Work in Connection	No Modification
Detail Flag Pole	Flag Pole (Provisional Item)	No Modification

Table 6. Recommended model of Synchronized Architectural Tender Drawings and Bills of Quantities of Building Works Based on the Elements of Building Works

Architectural Tender Drawings	Bills of Quantities	Clarification
Preliminaries Item	*To be measured under different bill	Information of hording material, sign board, and temporary office
Detail Apron, Ramps, and Perimeter Drain	Apron Ramps and Perimeter Drain	Indicate the location in the proposed building
Detail Staircase	Staircase	Indicate location of staircase and steps, with specifying the material, railing , finishes
Detail Roof	Roof	Indicate the type of roof, inclusive Scupper drain information
Detail External & Internal Wall	External Wall Internal Wall and Partition	Indication of type of wall, example brickwork, and partition wall
Detail Louvres	External Wall	Indicate the location and material
Detail Door	External Doors and Windows Internal Doors and Windows	The door, window, and ironmongery is ideally separated for easier quantification
Detail Window		
Detail Ironmongery		
Detail Internal Floor Finishes	Internal Floor Finishes	Indicate floor and skirting finishes
Detail Skirting Finishes		
Detail Wall Finishes	Internal Wall Finishes	Indicate wall finishes
Detail Internal Ceiling Finishes	Internal Ceiling Finishes	Indicate internal ceiling finishes
Detail External Ceiling Finishes	External Finishes	Indicate external ceiling finishes
Detail Wall Finishes	External Finishes	Indicate Wall Finishes
Detail Feature Wall Finishes		
Detail Sanitary Fitting	Sanitary Fittings	Indicate sanitary fittings
Sanitary Fitting Schedule		
Detail Ablution	Builders Work in Connection	Indicate detail that contain the information of material used, sanitary fitting, and finishes
Detail Flag Pole	Flag Pole (Provisional Item)	Indicate detail that contains information of material used and finishes.

5 Conclusion and recommendation

The purpose of the study is to have documents mutually explanatory of each other in the case of any discrepancy. Love et al. [16] identified that the use of 'time boxing' practices to meet schedule demands may contribute to audits, reviews and verifications being omitted, thus allowing design errors to materialize during construction. Therefore, the proactive measure that found under this study is to simplify coordination even though consultants had limited time to revise documents.

The study found out that the discrepancies of documents can be minimized by synchronizing drawings and documents. The study by Jha and Iyer [18] analysed that influencing coordination activities that related to this study, are that preparation of a project quality plan in line with contract specifications and arrangement of required inputs like drawings, specifications and technical details. Twort and Rees [19] suggested that specification for workmanship and materials of civil engineering works is by listing trades in the ICE Standard method of billing quantities (CESMM) in order which construction normally proceeds, to help drafting and make sure matters are not missed. This similar method has drawn the comparison process between building elements of Standard Method of Measurement, building works elements in the bills of quantities and architectural drawings in table 5. The required architectural detailing is identified based on building elements that already stated in the Standard Method of Measurement (SMM). Finally, as shown in table 6, the recommended method to produce 2-D architectural drawings is based on elements of building works that the bills of quantities in used. However, the addition of other detail will be based on design requirements. For example the usage the element demolition, architectural drawings will be provided accordingly.

Love et al. [16] suggested that error prevention should be viewed as a continuous process, rather than a product of certain activities or behaviors, as it involves an exploration of people, organisations and project systems. Thus, the recommended model can be taken into further account by diversifying the unit analysis into implementation of multiple projects for observation contrasting patterns in data, considering the design requirement and complexity.

References

1. P. A. Malaysia, *Agreements And Conditions Of PAM Contract 2006 (Without Quantities)* (2006)
2. Public Works Department, *Standard Form of Contract to be Used Where Bills of Quantities Form Part of the Ccntract PWD Form 203A (Rev 1/2010)* (2010)
3. P. I. Tan, K. S. Low, J. Sum, and S. T. Chee, *Handbook for PAM Contract 2006*. Kuala Lumpur: PAM Publication (2010)
4. S. Z. H. S. Jamaludin, M. F. Mohammad, and K. Ahmad, "Enhancing the Quality of Construction Environment by Minimizing the Cost Variance," *Procedia - Soc. Behav. Sci.*, vol. **153**, pp. 70–78 (2014)
5. Y. Frimpong, J. Oluwoye, and L. Crawford, "Causes of delay and cost overruns in construction of groundwater projects in a developing countries; Ghana as a case study," *Int. J. Proj. Manag.*, vol. **21**, no. 5, pp. 321–326 (2003)
6. M. Sambasivan and Y. W. Soon, "Causes and effects of delays in Malaysian construction industry," *Int. J. Proj. Manag.*, vol. **25**, no. 5, pp. 517–526 (2007)
7. N. Mansfield, O. Ugwu, and T. Doran, "Causes of delay and cost overruns in Nigerian construction projects," *Int. J. Proj. Manag.*, vol. **12**, no. 4, pp. 254–260 (1994)
8. a. M. Odeh and H. T. Battaineh, "Causes of construction delay: Traditional contracts," *Int. J. Proj. Manag.*, vol. **20**, no. June 2000, pp. 67–73 (2001)
9. Y.-M. Cheng, "An exploration into cost-influencing factors on construction projects," *Int. J. Proj. Manag.*, vol. **32**, no. 5, pp. 850–860 (2014)
10. H. J. Rosen and J. Regener, *Construction Specifications Writing: Principles and Procedures (Fifth Edition)*. Hoboken, New Jersey: Wiley & Sons. (2005)
11. D. Arditi and H. M. Gunaydin, "Total quality management in the construction process," *Int. J. Proj. Manag.*, vol. **15**, no. 4, pp. 235–243 (1997)
12. J. E. Connell and T. M. Flegal, *Chapter 28 Contract Documents*, Third Edit., no. 1. Butterworth-Heinemann (2011)
13. J. A. Sani and N. Othman, "Quality Standard and Specification for Soft-scape Construction in Malaysia," *Procedia - Soc. Behav. Sci.*, vol. **35**, no. December 2011, pp. 260–266 (2012)
14. S. K. S. Harbans and K. S. Kandan, "Variation Claims The Pitfalls And Pratfalls *," *Bul. Ingenieur,The Board Eng. Malaysia*, vol. **24**, pp. 36–42 (2005)
15. M. Jaafar, L. Y. Mui, and I. Yahaya, *Elementary Estimating For Building Works*. Penerbit Universiti Sains Malaysia (2013)
16. P. E. D. Love, R. Lopez, D. J. Edwards, and Y. M. Goh, "Error begat error: Design error analysis and prevention in social infrastructure projects," *Accid. Anal. Prev.*, vol. **48**, pp. 100–110 (2012)
17. A. De Massis and J. Kotlar, "The case study method in family business research: Guidelines for qualitative scholarship," *J. Fam. Bus. Strateg.*, vol. **5**, no. 1, pp. 15–29 (2014)
18. K. N. Jha and K. C. Iyer, "Critical determinants of project coordination," *Int. J. Proj. Manag.*, vol. **24**, no. 4, pp. 314–322 (2006)
19. A. Twort and G. Rees, "5 Preparing contract documents," in *Civil Engineering Project Management (Fourth Edition)*, 4th Editio., London: Elsevier Butterworth-Heinemann (2004)

The Importance of Superplastizer Dosage in the Mix Design of Lightweight Aggregate Concrete Reinforced With Plypropylene Fiber

Payam Shafigh[1,a], Javad Yahaghi[2]

[1]Department of Building Surveying, Faculty of Built Environment, University of Malaya, 50603 Kuala Lumpur, Malaysia
[2]Department of Civil Engineering, Faculty of Engineering, University Tenaga Nasional, Jalan IKRAM-UNITEN, 43000 Kajang, Selangor, Malaysia

Abstract. This paper reports the results of a study conducted to investigate the effect of superplasticizer (SP) dosage on the slump, density, compressive strength and splitting tensile strength under different curing conditions of a lightweight aggregate concrete reinforced with polypropylene (PP) fiber. The lightweight aggregate used in this study was oil palm shell, which is an agricultural solid waste, originating from the palm oil industry. The results indicated that an increase in superplasticizer increased the workability, however, all the mechanical properties declined significantly. The reduction in the 28-day compressive and splitting tensile strengths was about 14. This study showed that although additional SP can improve the workability of the concrete, it may have a negative effect on the other properties of concrete. Therefore, the SP dosage in concrete mixtures containing PP fiber should be limited to a certain amount.

1 Introduction

Since the majority of natural resources are limited, any effort to use waste and recycled materials in concrete production can reduce resource depletion and promote energy conservation in the construction sector [1]. Oil palm shell (OPS) is an agricultural waste product obtained from the processing of palm oil and is available in large quantities in the tropical regions of the world, most especially in Africa, Asia and America [2]. Although generally, the mechanical properties of OPS concrete are less than those for ordinary concrete, one way to improve the mechanical properties of OPS concrete is by using fibers. Polypropylene (PP) fibers are mainly used to improve the ductility, toughness and impact resistance of concrete. Although they are not expected to increase the concrete strength, however, the addition of PP fiber greatly decreases the workability of concrete [3], which restricts the quantity of PP fiber [4], [5].

Superplasticizer (SP) is used as a dispersant to avoid particle segregation (gravel, coarse and fine sand), and to improve the flow characteristics of concrete and enhance the fluidity of the fresh concrete without adding water, with respect to their surfactant properties. This effect considerably improves the hardening performance of concrete [6]–[8]. The strength of concrete increased when the water to cement ratio decreased. However, the working mechanism lacks a full understanding, revealing, in certain cases, cement-superplasticizer incompatibilities [9]. The addition of superplasticizer during transport is a new development within the industry and the addition of admixture for slump management systems allow concrete producers to maintain slump until discharge.

The main objective of this paper is to investigate the effect of superplasticizer dosage on the fresh and hardened properties of polypropylene reinforced oil palm shell concrete. The objective was achieved by testing 66 specimens for two mix designs.

2 Experimental details

2.1 Materials

The materials used in this study were ASTM type I ordinary Portland cement, and local mining sand with a specific gravity, water absorption, fineness modulus and maximum grain size of 2.68, 0.94%, 2.65, and 4.75 mm, respectively. OPS in a saturated surface dry condition, with a maximum size of 12.5 mm, specific gravity, compacted bulk density and 24 h water absorption of 1.22, 673 kg/m3, and 21%, respectively, was used as lightweight aggregate. The PP fiber used in this study had a length of 12 mm and specific gravity of 0.90 kg/dm^3. The SP Sika ViscoCrete-15RM, supplied from Sika used in this study was, is in conformity with EN 934-2. It was used as 1% and 1.5% of cement weight.

2.2 Mix proportions

The mix design used in this study is shown in Table 1. Typically, the PP fiber volume used in concrete is in the

[a] Corresponding author: pshafigh@gmail.com

range of 0.1%–0.3%. Because of the decreased workability of concrete caused by the addition of PP fiber, a low volume fraction of fiber was used in this study. The volume fraction of PP fiber in OPS concrete was 0.2%.

Table 1. Mix proportions of concrete in one batch.

Mix code	L1	L2
Cement (kg)	32.70	32.70
Water (kg)	11.14	11.14
Sand (kg)	55.00	55.00
Granite (kg)	24.20	24.20
OPS (kg)	16.30	16.30
Fiber Volume (%)	0.2	0.2
SP (%)	1	1.5

2.3 Test method and curing regimes

All the materials except the water and SP were put into a mixer and mixed for 3 min; 70% of the mixing water together with SP was added and mixed for 5 min, then, the remaining water was added and mixed for another 10 minutes. At the end of this time, PP fiber were spread in the mixer. The mixture with fiber was mixed for a further 5 min. Before casting the samples, the slump test of the mixture was performed. The concrete specimens were cast in steel molds and compacted on a vibration table. For each mixture, (24) cubes (100 × 100 × 100 mm3) were used for the measuring of the compressive strength at 1, 3 and 28 days; 6 cylinders of 100 mm diameter and 200 mm height and 3 prisms of dimension 100 mm × 100 mm × 500 mm for determination of the indirect tensile strength and flexural strength, respectively. As can be seen in figure 1, adding 0.05% more SP to mix L1 caused water bleeding after the same vibration process. The specimens were demolded one day after casting. For determining the effect of the curing environment on the 28-day compressive strength of OPS concrete, the specimens were cured under three types of curing condition as given below:

NC: Specimens were immersed in water at 23 ± 3 °C until the age of testing.

7DC: Curing in water for 6 days after demolding and then air cured in a laboratory environment with a RH% of 70 ± 10 and a temperature of 29 ± 3 °C.

AC: Specimens were kept in a lab environment after demolding.

Figure 1. Water bleeding after vibration in mix L2 (right side specimen).

3 Results and discussion

3.1 Workability

The slump value of mix L1, before adding PP fiber, was about 90 mm. However, after adding the fiber, the slump value was 35 mm, which is considered to be a low slump value. Previous studies [10], [11] show that the addition of fiber reduces the slump value. One of the methods to increase the workability of concrete is the use of SP in the concrete mixture. Therefore, in this study, 50% more SP was added to the L1 mix (mix L2) to consider the effectiveness of SP on the workability and hardened properties of the lightweight concrete. It was observed that by increasing the SP content from 1% to 1.5% the slump value increased by about 20%.

3.2 Density

Four types of density, namely, demoulded (De), 28-day air-dry (AD), saturated (Sa) and oven-dry (OD) densities were measured for both mixes. The 28-days air-dry density and oven-dry density of L1 and L2 mixes were around 2050 kg/m3 and 1997 kg/m3. The demolded density and 28-day water cured density of L2 mix were 1.03% and 0.77%, respectively more than that of L1 mix. Figure 2 shows the relationship between the mixes and different densities.

Figure 2. Densities of concrete mixtures.

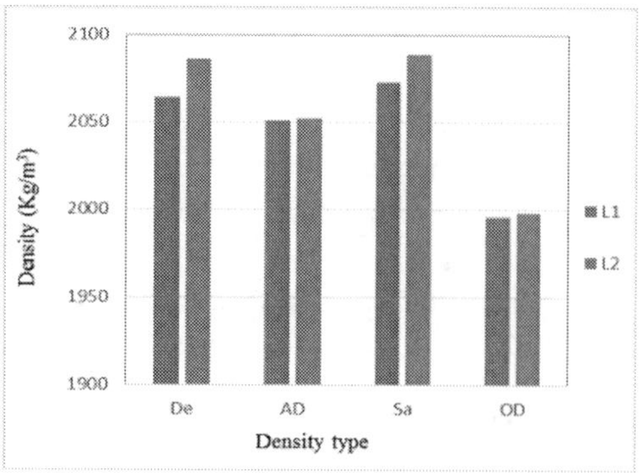

 DOI: 10.1051/matecconf/20166600020

3.3 Compressive strength

3.3.1 Continuously moist curing

As can be seen from Table 2, the compressive strength of L2 mix decreased at all ages with an addition in SP. Increasing the SP ratio from 1% to 1.5% decreased the compressive strength by about 26% at 1-day age, 14% at 3-day age, 16% at 7-day age and 14% at 28-day age, respectively. A comparison of the strength at the early and later ages indicate that the rate of strength development was greater as the age increased, especially for concretes with lower SP content. This phenomenon can be observed when we see the percentages of the 28-day compressive strength for each age in Table 2. Figure 3 presents the compressive strength development for mix L1 and mix L2.

Table 2. Mix proportions of concrete in one batch.

Mix code	L1	L2
1 day	18.89 (57%)	14.01 (63%)
3 days	32.68 (74%)	28.28 (74%)
7 days	39.22 (89%)	33.06 (87%)
28 days	44.16	38.15

*The data in parentheses are the percentage of 28-day compressive strength

Figure 3. Development of compressive strength.

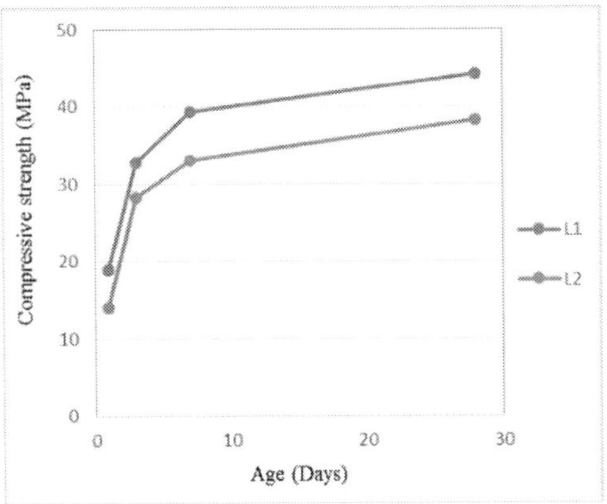

3.3.2 Effect of curing condition

Curing requires a satisfactory moisture content and temperature to be maintained in concrete during its early stages so that the desired properties may develop [12]. Figure 4 illustrates the 28-day compressive strength of specimens in three different curing conditions; continuous curing (NC), 7 days moist curing (7DC) and no curing in the lab environment (AC). It can be seen that the order for the strength of concrete was 7DC > NC > AC. The compressive strength loss for 28-day of samples under AC curing compared to NC curing for mixes L1 and L2 was about 8% and 7%, respectively; the increase for the short time curing period (7DC) was about 2% and 6%, respectively. As can be seen from this figure, with additional SP the 28-day compressive strength in three different curing conditions of NC, 7DC and AC, 14%, 12% and 13% decreased respectively.

Figure 4. Development of compressive strength.

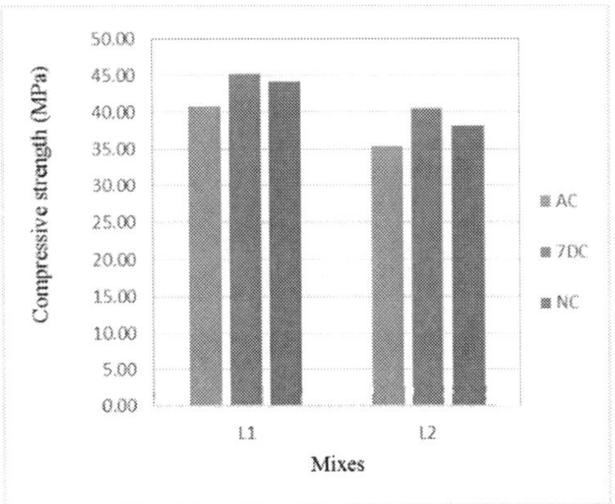

3.4 Splitting tensile strength

3.4.1 Continuously moist curing

When the SP increased 50%, the 28-day splitting tensile strength 14% decreased from 3.71 to 3.176 MPa. It was reported that the 28-day splitting tensile strength in moist curing is in the range of 1.9–2.41 MPa, that is, about 6–10% of the corresponding cube compressive strength. In this study, the splitting tensile strength for mix L1 and mix L2 was about 8.4% and 8.3% of the compressive strength, respectively. This shows that the SP obviously did not change the tensile to compressive strength ratio. Generally, the splitting tensile strength of concrete is 8–14% of the compressive strength. Figure 5 shows the relationship between the splitting tensile strength and the SP volume.

Figure 5. Effect of curing conditions on splitting tensile strength.

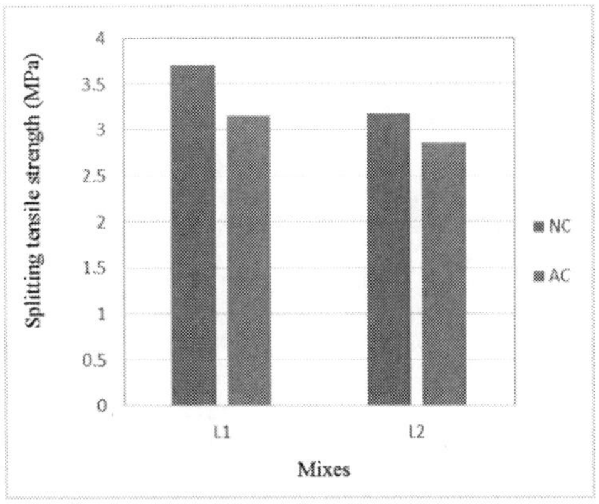

3.4.2 Effect of curing condition

Figure 5 illustrates the 28-day splitting tensile strength of specimens in two different curing conditions: continuous curing (NC) and no curing in lab environment (AC). It can be seen that the order of the strength of concrete was NC > AC. The splitting tensile strength loss for 28-day samples under AC compared to NC for mixes L1 and L2 was about 15% and 10%, respectively.

4 Conclusions

The aim of this study was to investigate the effect of adding SP on some engineering properties of OPS-PP fiber lightweight concrete. The test results showed that by increasing the SP content from 1% to 1.5% the workability increased about 20% but the oven and air-dry densities were not affected. The compressive strength of concrete decrease at all ages. In comparing the air-curing regime (AC) and the 7 days curing regime (7DC) with the continuous curing regime (NC), the addition of SP ratio did not affect the strength loss rate. The splitting tensile strength also decreased from 3.71 to 3.18 MPa. In general, it can be concluded that although additional SP may improve the slump value of concrete containing PP fiber, it may have a negative effect on the other properties.

References

1. P. Shafigh, M. Z. Jumaat, H. Bin Mahmud, and U. J. Alengaram, Constr. Build. Mater. **40**, 231 (2013)
2. E. A. Olanipekun, K. O. Olusola, and O. Ata, Build. Environ., vol. **41**, 297 (2006)
3. H. Mazaheripour, S. Ghanbarpour, S. H. Mirmoradi, and I. Hosseinpour, Constr. Build. Mater. **25**, 351 (2011)
4. P. S. Song, S. Hwang, and B. C. Sheu, Cem. Concr. Res. **35,** 1546 (2005)
5. N. Banthia and R. Gupta, Cem. Concr. Res. **36**, 1263(2006)
6. S. M. a. El-Gamal, F. M. Al-Nowaiser, and A. O. Al-Baity, J. Adv. Res. **3**, 119 (2012)
7. E. Tkaczewska, Constr. Build. Mater. **70**, 388 (2014)
8. V. Morin, F. Cohen Tenoudji, A. Feylessoufi, and P. Richard, Cem. Concr. Res. **31**, 63 (2001)
9. V. S. Ramachandran, *Concrete Admixtures Handbook, 2nd Ed.: Properties, Science and Technology*. (1995)
10. B. Chen and J. Liu, Cem. Concr. Res. **35**, 913 (2005)
11. P. Shafigh, H. Mahmud, and M. Z. Jumaat, Mater. Des. **32**, 3926 (2011)
12. ACI Committee 308, *Proposed ACI standard:Standard Practice for Making and Curing Concrete Test Specimens in the Laboratory*, (1980)

A Conceptual Framework for Managing Design Changes in Building Construction

Jeffrey Boon Hui Yap[1,2,a], Hamzah Abdul-Rahman[2] and Chen Wang [3]

[1]*Department of Surveying, Lee Kong Chian Faculty of Engineering and Science, Universiti Tunku Abdul Rahman, 43000 Kajang, Malaysia*
[2]*Faculty of Science, Technology, Engineering and Mathematics, International University of Malaya-Wales, 50480 Kuala Lumpur, Malaysia*
[3]*Faculty of Built Environment, University of Malaya, 50480 Kuala Lumpur, Malaysia*

Abstract. Design changes have always been an innate feature of the construction industry. Despite the various project management techniques, many building construction projects still fail to achieve their time and cost objectives. This paper aims to focus on design changes as a major cause of time delays and cost overruns. It seeks to discuss the causing factors of design changes in building construction projects and highlighting the resulting rework that is detrimental to project performance. Hence, this paper also aims to present a conceptual framework which was developed to better manage design changes through clued-up management decision to avoid future claims and disputes. Recent and current literature is examined and reviewed. The review approach was based on related literature to identified themes. The existing literature is summarised into five discrete themes of "design changes", "rework", "time and cost overruns", "communication" and "decision-making". The synthesised literature is subsequently utilised in the development of conceptual map which provided the direction for designing the conceptual framework. Extensive review of preceding studies on causes of time and cost overruns reveals that design changes is identified as one of the significant factors. However, there are limited studies in this domain. The discussion presents a knowledge gap of linking design changes and rework with decision-making in project management. This finding suggests the need for further empirical study. The framework stresses the importance of communication and management decision as the control mechanism. Researchers and practitioners in construction management field will find this study useful in understanding the causing factors of design changes and its detrimental impacts to project performance which risk the project to claims and disputes. The conceptual framework presents an imperative outcome to encourage further research in developing conceptual causal model of design changes which shows the key causal variables and their feedback loops.

1 Introduction

Design changes have been seen as synonymous with the construction industry. Frequent changes have often caused time delays and cost overruns [1]. These changes over and over again lead to rework [2] which can degrade the project performance [3]. Without proper change management strategy [4], undesired design changes will result in unexpected rework which lead to delays and disruptions [5]. Delays and disruptions have negative effects on construction project performance [6], [7]. In this paper, the authors emphasize on time delays and cost overruns in building construction project. Deviation of project objectives results in claims and disputes among the contracting parties in the construction project [8]. One of the many cases of design changes that much affected the project performance in Malaysia is the construction of low-cost air terminal Kuala Lumpur International Airport 2 (KLIA2) where cost overruns and time delays are pertinent. The new design concept caused construction cost to increase from the initial RM 1.7 billion to RM 4 billion [9], [10] along with the opening date moved several times. Hence, amplify the need to manage design changes [11] in construction project. The purpose of this study is to develop a conceptual framework for the management of design changes in building construction projects. This study is able to assist researchers and practitioners in gaining an in-depth understanding of the causes of design changes and the relationship of decision-making with design changes caused rework as a change management strategy.

2 Design changes in building construction

The building construction industry continues to be beleaguered with time delays and cost overruns [12]. Design changes are the top inhibiting factor of time and cost control in construction projects [4]. Therefore, design changes will manifestly cause time delays and cost

[a] Corresponding author: bhyap@utar.edu.my

overruns in construction projects [13]–[20]. Mohamad et al. [21] reported that design changes is a common phenomenon in building construction project which will inevitably lead to claims and disputes. Burati et al. [22] defined design changes as "design deviations which are related to the design of the project that occur when changes are made in the project design and requirements". These changes are any additions, omissions or adjustments made to the original scope of work after a contract is awarded [23] that lead to adjustment to the contract price or contract time [24]. Hence, making construction projects dynamic and unstable [25]. Design changes definition in this paper is synthesized as "regular additions, omissions and adjustments to both design and construction of work in a building construction project that occurs after the award of contract which affects the contract provisions and work conditions that make building construction dynamic and unstable".

Olawale and Sun [4] revealed that construction projects globally share some key characteristics. This is further supported by Abdul-Rahman et al. [11] in the synthesis of preceeding studies where design changes have been named as the critical factor for project delays and cost overuns in construction project at different parts of the world. Even so, studies on project delays and cost overruns in Malaysian context have fall short of identifying design changes as one of the important factors that inhibit project performance (see [26], [27]). The issue of managing design changes has not received much attention in literature [28]. In addition, empirical studies on the causes of schedule and cost escalations due to design changes are limited [15]. This observation provides the underlying motivation for future studies to fill the gap in knowledge in this domain.

3 Rework in construction

Rework is a proverbial occurrence in building construction projects. Rework is identified as significant factor for poor project performance [29]. A number of researchers have developed definitions for rework that can be found in construction management literature. According to Love and Edwards [30] and Love et al. [31], rework is defined as "the unnecessary effort of re-doing a process or activity that was incorrectly implemented at the first time". Ashford (1992, as cited in [29]) defines rework as "the process by which an item is made to conform to the original requirement by completion or correction". Refering to Hwang et al. [32], rework can be defined as "having to redo work to nonconformance with requirements". Fayek et al. [33] define rework as "activities in the field that have to be done more than once in the field, or activities which remove work previously installed as part of the project regardless of source, where no change order has been issued and no change of scope has been identified by the owner. Even though the definitions and interpretation of rework vary, one common theme prevail – rework refers to having to redo work due to nonconformance to requirments. Literature is replete with studies indicating that significant rework in construction is due to design changes [34]–[36].

4 Methodology for development of conceptual framework

The identified themes, issues and gap in knowledge are used to inform the development of a conceptual model for managing design changes in building construction projects. A conceptual framework in the sense being used here can be thought of as a diagram or map of a researcher's current view of the territory being investigated (Miles & Huberman, 1984, p. 33 as cited in [37]. Figure 1 illustrates the methodology adopted in design and development of conceptual framework which was adapted from Bhattacharya et al. [38]. This is useful for setting the boundaries and limitations of the study which form the basis for the research design strategy and suggest the direction for fieldwork. The first step of design was to identify the knowledge gaps within the research domain through critical appraisal of literature so that the themes and subthemes of the study could be formed. This phase is central for establishment of scope of study. Subsequently, the conceptual framework was developed based on the information synthesized from the literature. The method employed for developing the conceptual framework was carried out in two stages. The first stage was to develop a conceptual framework of causing factors influencing design changes while the second stage is to expand the framework to include rework, decision-making and communication parameters. The conceptual model was then verified internally by revisiting the parameters of study and confirming whether the developed framework is representing the objectives of the study.

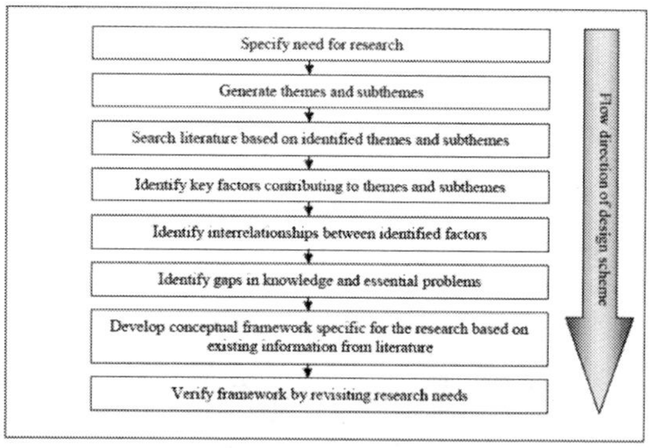

Figure 1. Flow chart showing steps for designing conceptual model (Source: Adapted from Bhattacharya et al. [38])

5 Causing factors of design changes in building construction

The focus of this section is the causing factors of design changes, because a good understanding of causes are a prerequisite for understanding the dynamics of design changes in building construction projects. Sun and Meng

[1] conducted a literature review that synthesizing on project change causes and effects so as to develop taxonomy for change causes and change effects in construction projects. A study by Love et al. [2] indicated that project changes may arise due to the influences of both internal and external elements. Similarly, Alaghbari et al. [27] highlighted that the causes for delay in construction projects are external causes and internal causes. Love et al. [2] described how changes and their actions or effects which are known as dynamics can impact the project management systems. The dynamics can be classified into internal and external factors which are adapted in this paper. On the other hand, Mohamad et al. [21] categorized the sources of design changes from clients, consultants and contractors whom are the primary parties in building construction projects. In this paper, internal factors includes (1) client-related; (2) design-related; (3) project-related and (4) contractor-related. Figure 2 exhibits the conceptual framework for causing factors influencing design changes.

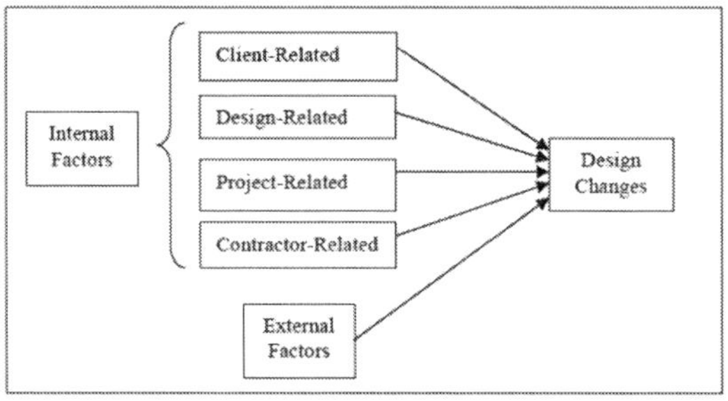

Figure 2. Conceptual framework of causing factors influencing design changes (Source: Adapted from Alaghbari et al. [1], Love et al. [21], Mohamad et al [27], Sun & Meng [29])

6 Classification of causing factors of design changes

A plethora of studies have been done by past researchers to investigate the root causes and nature of changes in construction projects. However, the works on causing factors of design changes are not that prevalent. The following section reviews on the existing literatures that focus on the causes of design changes in construction projects. The literature is further synthesized to design cause-and-effect diagram for design changes that is depicted in Figure 3. Table 1 presents the summary of the sources from existing literature on the identified causing factors of design changes.

Table 1. Identified causing factors of design changes.

Causing factors	Sources
A. Internal Factors	
i. Client-related	[17], [19], [21], [39]–[46])
ii. Design-related	[1], [14], [15], [17], [19], [21], [27], [41], [42], [44], [47]–[54]
iii. Project-related	[2], [21], [26], [55], [56]
iv. Contractor-related	[1], [21], [27], [44], [50]
B. External Factors	
i. External-related	[1], [2], [15], [26], [27], [44], [55], [57]–[59]

6.1. Client-related causes

Hwang et al. [45] define "client-related rework as the rework that directly originated from client and client's representatives. The study reported that "replacement of materials by the client" was the cause with the highest frequency of occurrence while "change of plans or scope by the client" contributed the most to client-related rework. Similarly, Josephson et al. [42] note the contributing factors of client to rework in Swedish construction industry are client's changes, extra orders, bad choice of material/method and wrong information. According to a study by Mohamad et al. [21], clients are recognised as a major source of design changes. The three major causes of design changes attributed to the clients were modifications to the original design, addition of new work/scope and unclear initial design brief. Some other significant causes comprise of desire to use better specification, desire to use alternative material/new technology and omission of works/scopes. Similarly, a technical report by Federal Construction Council [39] revealed the client's representative as the most significant source of excessive changes in construction. This is further supported by Alnuaimi et al. [40] in a study on public construction projects in Oman where client's additional works is identified as the topmost factor causing changer orders. Again, Ijaola and Iyagba [41] also reported identical findings in both Nigeria and Oman. Another study by Al-Momani [19] reported that the main causes of delay in construction of public projects in Jordan are related to user changes. El-Razek et al. [43] also have almost similar findings in Egypt, Ogunlana et al. [17] in Thailand and Wu et al. [44] in Taiwan. According to Hwang et al. [45], client-related factors are (1) change of plans or scope by the client; (2) inadequate/incomprehensive project objectives by the client change in specification by the client; (3) financial problems faced by client; (4) impediment in prompting the decision making by the client; (5) replacement of materials by the client; (6) change in specification by the client; and (7) obstinate nature of the client. The data was collected from 381 projects performed by 51 companies in Singapore. Numerous studies show that client-related changes have a significant impact on the project performance. Therefore, client values are important and should be fully understood at the early phase of the project [46].

6.2. Design-related causes

According to Mendelsohn [47], almost 75% of problems or rework on construction project were induced at the design phase. The lack of communication between client and design team members can lead to documentation errors and omissions [60]. The design and delivery team often misinterpret the client's requirements in the project [48], [61]. Therefore, Love et al. [60] suggest the communication and harmonious relationship between clients and their design team and the involvement of clients in the design process can appreciably cut design-related rework. Key designer-related factors include natural growth of the project was not anticipated at the design stage, design errors as well as design and construction criteria are outdated and do not suit the present construction technology [19]. Modification to design was indicated as the second most important causes of change order in both Nigeria and Oman [41]. According to Sun and Meng [1], design consultant generated changes include poor/incomplete drawings, design changes due to poor brief, errors and omissions and inconsistent site condition. Mohamad et al. [21] reported improper design/ part of design improvement, inconsistent information in drawings, discrepancy between contracts, lack of geotechnical investigation/ wrong interpretation of findings and insufficient detail of existing site condition as the factors that caused design changes in residential reinforced concrete buildings. Another study by Cox et al. [49] look into post contract award design changes in construction. Frequent reasons cited are: (1) designer's omission in tender documents; (2) coordination defects in tender documents; (3) employer has changed his requirements; and (4) new information on site conditions. Josephson et al. [42] note the contributing factors of design to rework in Swedish construction industry are lack of co-ordination, unsuitable design, faulty design and incomplete drawings. Kaming et al. [14] describe the magnitude of design changes depend upon pre-design site investigation, the completeness of working drawings available at the time of proposal and unpredictable circumstances during construction. Consultant's inability [55], design omission [15], [27], incomplete drawings [17], design errors [51] and quality of design [44] caused cost and schedule increases. Design changes constitute a major cause of failure in coordination of design information [28]. Defective design is commonly caused by human errors [62]. Design errors can significantly degrade project performance by generating rework, requiring additional time and resource expenditure [52], [53].

6.3. Project-related causes

Project specific dynamics include location conditions [2] and underground conditions [21], [26], [55]. The inherent site conditions of a project will affect the project performace [56].

6.4. Contractor-related causes

Design changes induce by contractor are due to contractor's request to use available material, to use alternative construction method to save time, to use alternative construction methods to save money, to rectify construction mistakes and to improve the quality of works at site [21]. Similarly, modification of construction methods to suit current site conditions, contractor request for original construction methods to be replaced by a new method as well as improper construction or human-errors leasing to on-site repair work are listed by [44]. Poor planning of project caused errors in execution of tasks and inappropriate construction methods [27], [50]. On the other hand, Sun and Meng [1] included poor site/project management skills, delays in appointing subcontractor, delay of subcontractors' work, poor workmanship, low productivity and poor logistic control as the relevant factors.

6.5. External-related causes

Project outcome is most impacted by the regulation imposed by the government [2]. Change of work rules/regulation by the government agencies, neighbourhood communities and coordinating with utility systems as some of the external factors described in Hsieh et al. [55]'s study. In separate study, Sun and Meng [1], Chang et al. [15], Wu et al. [44] and Chang [57] also recognized change orders owing to legislative or policy changes as critical external factor. In Malaysia, the external causes are regulatory changes and problem with neighbours [26]. Alaghbari et al. [27] further expanded the external factors to include materials on the market, equipment and tools on the market, economic conditions, law and regulation and external works due to public agencies (roads, utilities and public services). Aiyetan et al. [58] included physical environmental conditions, economic policy and socio-political conditions as factors in their study. Doloi et al. [59] describe lack of communication with local authorities will influence project performance.

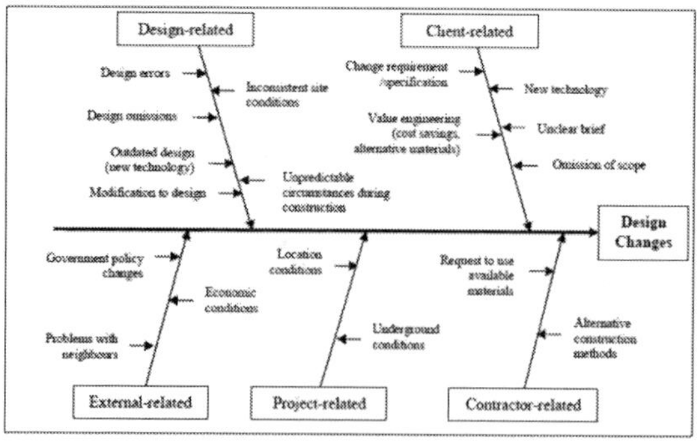

Figure 3. Generic cause-and-effect diagram of design changes (Source: Author derived)

7 Impacts to project time and cost performance

Design changes in building construction projects will inevitably degrade project performance [7], [11], [14], [20], [43], [63]. Although design change has tremendous negative effects [64], attention paid to proactively manage the impact of these changes is limited [65]. There has been considerably research undertaken that has sought to determine the causes of time delays and cost overruns in construction projects [6], [14], [18], [22], [56], [63], [66]–[69]. Research into determining the causes of time delays and cost overruns has reached saturation point; consistently the same causal variables are identified. However, construction projects suffering from overruns are still on the rise. From a recent study in Malaysia, Shehu et al. [70] stressed that distressingly, 55% of construction projects experienced cost overruns. Abdul-Rahman et al. [71] revealed that practitioners in Malaysian construction industry acknowledge design changes as a major concern to the industry. Chang [15] sampled four project in California and pointed out that schedule and cost increased on average of 69% and 25% respectrively as a result of design changes. . Cox et al [49] claimed that the cost associated with design changes were 5-8% whereas Chang et al. [57] described an average of 8.5%. Love [2] revealed that design change orders resulting in rework can account for as much as 50% of project cost overrun. Deviation costs averaged 12.4% of the total installed project cost [22]. These deviations from project objectives are due to disruptions [72] where reworks are required within the construction process [30]. Disruption is fundamental to reduced productivity and increased rework on the project [73]. Hence, disputed and claims over project changes can be minimized when the problems are recognized at the earlier stages of the project [64], especially when client's decision-making is required [74].

8 Simulation model as decision-support and communication tool

Changes in construction projects are more often than not unpredictable. This is primarily due to the uniqueness of each project and the limited resources than can be used on planning, executing and delivering the project [75]. By and large, practitioners need to predict changes in a timely manner [76]. Hence, the ability to predict the probabilities of occurrence [71] is essential to mitigate overruns and possible disputes due to design changes and rework [11]. Computer-generated performance simulations are useful to better understand the dynamics of change [77] where critical decision-making can be enhanced through use of predictive models [78]. According to Kartam [76], conflict will be minimised when problems are found at the earlier possible stage of a project thereby enabling the implementation of counter measures. Any request for change will generate costly ripple effects which create delays and disruptions [79] that will affects project performance. Additionally, effective management is difficult in a world of high dynamic complexity where decisions may create unanticipated side effects and delayed consequences [80]. A plethora of research have presented on change management model relating to construction management (see [8], [25], [28], [81]–[84]). But few have link the use of change management model as decision-support tool. Simulation models have been successfully tried to assist decision making (see [85]–[89]). Simulation models are useful to predict the performance of project [90] to aid decision making and policy implementation [91]. Thus, simulation model is imperative to support decision-making in dynamic project management.

In highly dynamic project environment, communication becomes complex but essential [92]. Naoum [93] emphasized that lack of timely and effective communication, lack of integration, uncertainty, a changing environment and increasing project complexity are the drivers of project change. With a better understanding of which change factors or combination of change factors are most influential in the development of delay and disruption impacts, the construction industry will hopefully be able to better forecast and manage the impacts of design changes. Simulation models can be used as a communication platform [94] with graphical illustration helpful for visualizing and understanding of impacts of design changes on project performance or termed as "management simulator" [95]. Therefore, practitioners can convey the strategies to other team members or stakeholders at ease.

9 The proposed conceptual framework

A model for management of design changes in building construction projects by placing emphasis on management decision is proposed (Figure 4). The simulation model can be deployed to provide effective communication platform through predictive model to understand the dynamics of design changes, the knock-on effects due to rework that cause delay and disruption and possible degradation of project time and cost performance. Therefore, practitioners can acclimatize the simulation data from the predictive model for clued-up decision-making at the point of design change request. Hence, the impacts can be envisaged at the early stage in order to minimize the risk of project to potential claims and disputes at the tail end of the project.

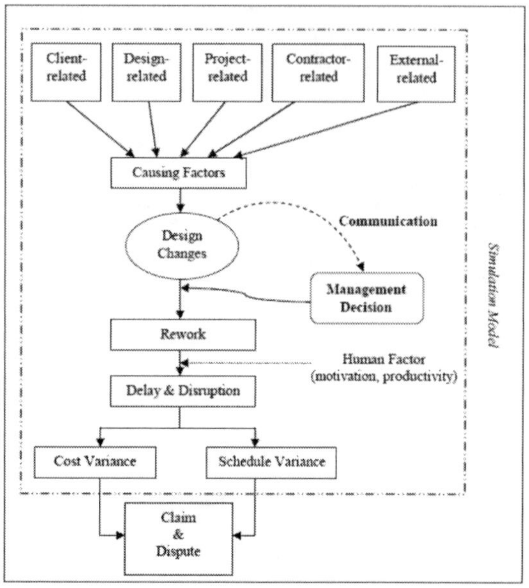

Figure 4. The conceptual framework (Source: Author derived)

10 System dynamics in managing design changes

There is now a need for alternative research approaches in construction project management studies [96]. This claim is further justified by Ahiaga-Dagbui et al. [97] that identified methodological gap in current research that lack system thinking and demonstrable causality. By applying system thinking to the problem of design changes by understanding the types of design changes and factors influencing the occurrence of these changes at various times in the project, system thinking can lead the construction industry towards realizing how to prevent or mitigate delay and disruption impacts by anticipating when and where design changes are most likely to occur.

System dynamics (SD) is a methodology for analysing complex systems and problems with the aid of computer simulation software. It is an experimental approach to system thinking [95]. SD modelling has brought a new view to project management, enabling understanding of the behaviour of complex projects that was not accessible with other methods [5]. SD is about studying complex and dynamic systems which change over time, and about finding the "why" (causes) and "how" (pattern) of system changes. According to Sterman [79], [95], SD models are suitable for complex and highly dynamic systems which involve multiple feedback processes with non-linear relationships as well as encompass both "hard" and "soft" data. Human factors such as motivation and productivity are considered as soft variables.

SD models are providing systematic analysis of the strategic issues of project management [98]. SD models have been successfully applied to construction project management [98]–[101]; effect of rework on project performance [2], [52], [102]; change management in construction [65], [103]; decision making in construction [88], [104]; delay and disruptions in construction [5], [105]; procurement strategies [106], [107]; outsourcing construction services [108] and analyzing project cost overruns [109], among others. SD models can complement the shortcomings of traditional tools and techniques [110]. Besides, significant numbers of researchers promote the use of system dynamics as a solution to improve project decision-making [52], [100], [104], [105]. The information from the SD model can be utilised to aid in making decisions related to design changes where the impacts are often underestimated by construction industry practitioners. The model will enables project manager to address issues by simulating their impacts on the project before they become problems.

11 Conclusions

The critical appraisal of literature relating to causes of design changes and the resultant rework that cause delay and disruption which result in time delays and cost overruns in building construction projects has been presented. This study establishes that literature linking design changes and rework in construction projects is limited but found out that design changes is the significant factor for time delays and cost overruns. In order to mitigate the detrimental impacts of design changes, the dynamics that influences design changes need to be identified and attended to appropriately. Researchers and practitioners in construction management field will find this study useful in understanding the causing factors of design changes and its detrimental impacts to project performance which risk the project to claims and disputes. The framework which is a noteworthy outcome from this study stresses the importance of communication and management decision as the control mechanism. Dynamic project management using simulation to building construction projects which is a new trend that assures results by minimizing risks and uncertainties through use of predictive models. It gives the practitioners the ability to view the impacts of design changes to project performance before actually making the decision to implement the proposed design changes. If applied correctly it is able to hoard many projects from disaster. The simulated data can be useful to the critical management decision process. The conceptual map provides a new conceptualization and understanding the interrelationship between simulation, communication and decision-support system. Therefore, this is a unique contribution to the existing body of knowledge. While empirical evidence for this framework is still in an emerging state, the simplicity of the framework put forward broad application in the field of managing design changes as well as time delays and cost overruns research. The conceptual framework presents an imperative outcome to encourage further research in developing conceptual causal model of design changes which shows the key causal variables and their feedback loops. The SD methodology was selected because it can clearly illustrate the dynamics of design changes within the project lifecycle and how the impacts will affect project performance (time and cost).

References

[1] M. Sun and X. Meng, "Taxonomy for change causes and effects in construction projects," Int. J. Proj. Manag., vol. **27**, no. 6, pp. 560–572 (2009)

[2] P. E. D. Love, G. D. Holt, L. Y. Shen, H. Li, and Z. Irani, "Using systems dynamics to better understand change and rework in construction project management systems," Int. J. Proj. Manag., vol. **20**, no. 6, pp. 425–436, (2002)

[3] E. Palaneeswaran, P. E. D. Love, M. M. Kumaraswamy, and T. S. T. Ng, "Causal ascription of rework in building and civil engineering projects: A multivariate exploration," Eng. Constr. Archit. Manag., vol. **21**, no. 1, pp. 111–126 (2014)

[4] Y. A. Olawale and M. Sun, "Cost and time control of construction projects: Inhibiting factors and mitigating measures in practice," Constr. Manag. Econ., vol. **28**, no. 5, pp. 509–526 (2010)

[5] S. M. Howick, F. Ackermann, C. Eden, and T. M. Williams, "System dynamics and disruption and delay in complex projects," in *Encyclopaedia of compexity and system science*, Springer (2009)

[6] J. Bin Yang, M. Y. Chu, and H. M. Huang, "An empirical study of schedule delay causes cased on Taiwan's litigation cases," Proj. Manag. J., vol. **44**, no. 3, pp. 21–31 (2013)

[7] T. Williams, C. Eden, F. Ackermann, and A. Tait, "The effects of design changes and delays of project costs," J. Oper. Res. Soc., vol. **46**, no. 7, pp. 809–818 (1995)

[8] I. a. Motawa, C. J. Anumba, S. Lee, and F. Peña-Mora, "An integrated system for change management in construction," Autom. Constr., vol. **16**, no. 3, pp. 368–377 (2007)

[9] E. Ng, "No cost overrun for klia2, says Liow," *The Edge Markets*, 19 May (2015)

[10] R. Ghazali, "Transport Ministry: KLIA2 construction never experienced cost overruns," *The Star Online*, 19 May (2015)

[11] H. Abdul-Rahman, C. Wang, and J. B. H. Yap, "Impacts of design changes on construction project performance: Insights from a literature review," in *14th Management In Construction Researchs' Association (MiCRA 2015) Annual Conference and General Meeting* (2015)

[12] P. E. D. Love, R. Y. . C. Tse, and D. J. Edwards, "Time – cost relationships in Australian building construction projects," J. Constr. Eng. Manag., vol. **131**, no. 2, pp. 187–194 (2005)

[13] N. Mansfield, O. Ugwu, and T. Doran, "Causes of delay and cost overruns in Nigerian construction projects," Int. J. Proj. Manag., vol. **12**, no. 4, pp. 254–260 (1994)

[14] P. F. Kaming, P. O. Olomolaiye, G. D. Holt, and F. C. Harris, "Factors influencing construction time and cost overruns on high-rise projects in Indonesia," Constr. Manag. Econ., vol. **15**, no. 1, pp. 83–94 (1997)

[15] A. S. T. Chang, "Reasons for cost and schedule increase for engineering design projects," J. Manag. Eng., vol. **18**, no. 1, pp. 29–36 (2002)

[16] S. Lee, S. R. Thomas, and R. L. Tucker, "The relative impacts of selected practices on project cost and schedule," Constr. Manag. Econ., vol. **23**, no. 5, pp. 545–553 (2005)

[17] S. O. Ogunlana, K. Promkuntong, and V. Jearkjirm, "Construction delays in a fast-growing economy: Comparing Thailand with other economies," Int. J. Proj. Manag., vol. **14**, no. 1, pp. 37–45 (1996)

[18] N. . Kartam, T. . Al-Daihani, and J. . Al-Bahar, "Professional project management practices in Kuwait: issues, difficulties and recommendations," Int. J. Proj. Manag., vol. **18**, no. 4, pp. 281–296 (2000)

[19] A. H. Al-Momani, "Construction delay: a quantitative analysis," Int. J. Proj. Manag., vol. **18**, no. 1, pp. 51–59 (2000)

[20] J. D. Owalabi, L. M. Amusan, C. O. Oloke, O. Olusanya, P. Tunji-Olayeni, OwolabiDele, PeterJoy, and OmuhIganatious, "Causes and effects of delay on project construction delivery time," Int. J. Educ. Res., vol. **2**, no. 4, pp. 197–208 (2014)

[21] M. Mohamad, M. Nekooie, and A. Al-Harthy, "Design changes in residential reinforced concrete buildings: The causes, sources, impacts and preventive measures," J. Constr. Dev. Ctries., vol. **17**, no. 2, pp. 23–44 (2012)

[22] J. L. Burati, J. J. Farrington, and W. B. Ledbetter, "Causes of Quality Deviations in Design and Construction," Journal of Construction Engineering and Management, vol. **118**, no. 1. pp. 34–49 (1992)

[23] A. . Akinsola, K. . Potts, I. Ndekugri, and F. Harris, "Identification and evaluation of factors influencing variations on building projects," Int. J. Proj. Manag., vol. **15**, no. 4, pp. 263–267 (1997)

[24] W. Ibbs, "Construction change : Likelihood, severity , and impact on productivity," J. Leg. Aff. Disput. Resolut. Eng. Constr., no. **August**, pp. 67–73 (2012).

[25] M. Park and F. Peña-Mora, "Dynamic change management for construction: Introducing the change cycle into model-based project management," Syst. Dyn. Rev., vol. **19**, no. 3, pp. 213–242 (2003)

[26] M. Sambasivan and S. W. Yau, "Causes and effects of delays in Malaysian construction industry," Int. J. Proj. Manag., vol. **25**, no. 5, pp. 517–526 (2007)

[27] W. Alaghbari, M. R. a. Kadir, A. Salim, and Ernawati, "The significant factors causing delay of building construction projects in Malaysia," Eng. Constr. Archit. Manag., vol. **14**, no. 2, pp. 192–206 (2007)

[28] A. Mokhtar, C. Bédard, and P. Fazio, "Information model for managing design changes in a collaborative environment," J. Comput. Civ. Eng., vol. **12**, no. 2, pp. 82–92 (1998)

[29] P. E. D. Love, "Influence of project type and procurement method on rework costs in building construction projects," J. Constr. Eng. Manag., vol. **128**, no. 1, pp. 18–29 (2002)
[30] P. E. D. Love and D. J. Edwards, "Forensic project management: The underlying causes of rework in construction projects," Civ. Eng. Environ. Syst., vol. **21**, no. 3, pp. 207–228 (2004)
[31] P. E. D. Love, D. J. Edwards, H. Watson, and P. Davis, "Rework in civil infrastructure projects: Determination of cost predictors," J. Constr. Eng. Manag., vol. **136**, no. 3, pp. 275–282 (2010)
[32] B. Hwang, S. Thomas, C. Haas, and C. Caldas, "Measuring the impact of rework on construction cost performance," J. Constr. Eng. Manag., no. **March**, pp. 187–198 (2009)
[33] A. R. Fayek, M. Dissanayake, and O. Campero, "Measuring and classifying construction field rework : A pilot study" (2003)
[34] P. E. D. Love, P. Mandal, and H. Li, "Determining the causal structure of rework influences in construction," Constr. Manag. Econ., vol. **17**, pp. 505–517 (1999)
[35] P. E. D. Love, P. Mandal, J. Smith, and J. Georgiou, "DECOREM:A design and construction rework minimisation model," in *1st International Conference on System Thinking in Management*, pp. 377–382 (2000)
[36] P. E. D. Love, D. J. Edwards, and Z. Irani, "Forensic project management: An exploratory examination of the causal behavior of design-induced rework," IEEE Trans. Eng. Manag., vol. **55**, no. 2, pp. 234–247 (2008)
[37] A. Mitchell, I. Frame, A. Coday, and M. Hoxley, "A conceptual framework of the interface between the design and construction processes," Eng. Constr. Archit. Manag., vol. **18**, no. 3, pp. 297–311 (2011)
[38] N. Bhattacharya, J. Lamond, D. Proverbs, and F. Hammond, "Development of conceptual framework for understanding vulnerability of commercial property values towards flooding," Int. J. Disaster Resil. Built Environ., vol. **4**, no. 3, pp. 334–351 (2013)
[39] Federal Construction Council, "Managing Changes in the Construction Process.," Consulting Committee on Contract Management, Technical Report, National Academy Press, Washington DC (1983).
[40] A. S. Alnuaimi, R. a. Taha, M. Al Mohsin, and A. S. Al-Harthi, "Causes, effects, benefits, and remedies of change orders on public construction projects in Oman," J. Constr. Eng. Manag., vol. **136**, no. 5, pp. 615–622 (2010)
[41] I. a. Ijaola and R. O. Iyagba, "A comparative study of causes of Change orders in public construction project in Nigeria and Oman," J. Emerg. Trends Econ. Manag. Sci., vol. **3**, no. 5, pp. 495–501 (2012)
[42] P.-E. Josephson, B. Larsson, and H. Li, "Illustrative Benchmarking Rework and Rework Costs in Swedish Construction Industry," J. Manag. Eng., vol. **18**, no. 2, pp. 76–83 (2002)
[43] M. E. A. El-Razek, H. A. Bassioni, and A. M. Mobarak, "Causes of delay in large building construction projects in Egypt," J. Manag. Eng., vol. **11**, no. 2, pp. 45–50 (1995)
[44] C. H. Wu, T. Y. Hsieh, and W. L. Cheng, "Statistical analysis of causes for design change in highway construction on Taiwan," Int. J. Proj. Manag., vol. **23**, no. 7, pp. 554–563 (2005)
[45] B. G. Hwang, X. Zhao, and K. J. Goh, "Investigating the client-related rework in building projects: The case of Singapore," Int. J. Proj. Manag., vol. **32**, no. 4, pp. 698–708 (2014)
[46] M. H. Thyssen, S. Emmitt, S. Bonke, and A. Kirk-Christoffersen, "Facilitating client value creation in the conceptual design phase of construction projects: A workshop approach," Archit. Eng. Des. Manag., vol. **6**, no. 1, pp. 18–30 (2010)
[47] R. Mendelsohn, "The constructibility review process: A constrctor's perspective," J. Manag. Eng., vol. **13**, no. 3, pp. 17–19 (1997)
[48] L. Koskela, P. Huovila, and J. Leinonen, "Design management in building construction: from theory to practice," J. Constr. Res., vol. **3**, no. 1, p. 1 (2002)
[49] I. D. Cox, J. P. Morris, J. H. Rogerson, and G. E. Jared, "A quantitative study of post contract award design changes in construction," Constr. Manag. Econ., vol. **17**, no. 4, pp. 427–439 (1999)
[50] P. González and V. González, "Analysis of causes of delay and time performance in construction projects," J. Constr. Eng. Manag., vol. **140**, no. 1, pp. 1–9 (2014)
[51] N. Hamzah, M. A. Khoiry, I. Arshad, W. H. W. Badaruzzaman, and N. M. Tawil, "Identification of the causes of construction delay in Malaysia," World Acad. Sci. Eng. Technol., vol. **72**, no. 12, pp. 312–317 (2012)
[52] S. Han, P. Love, and F. Peña-Mora, "A system dynamics model for assessing the impacts of design errors in construction projects," Math. Comput. Model., vol. **57**, no. 9–10, pp. 2044–2053, (2013)
[53] S. Han, S. Lee, and F. Peña-Mora, "Identification and quantification of non-value-adding effort from errors and changes in design and construction projects," J. Constr. Eng. Manag., vol. **138**, no. 1, pp. 98–109 (2012)
[54] P. E. D. Love and D. J. Edwards, "Determinants of rework in building construction projects," Eng. Constr. Archit. Manag., vol. **11**, no. 4, pp. 259–274 (2004)
[55] T. Y. Hsieh, S. T. Lu, and C. H. Wu, "Statistical analysis of causes for change orders in metropolitan public works," Int. J. Proj. Manag., vol. **22**, no. 8, pp. 679–686 (2004)
[56] Y. Frimpong, J. Oluwoye, and L. Crawford, "Causes of delay and cost overruns in construction of groundwater projects in a developing countries; Ghana as a case study," Int. J. Proj. Manag., vol. **21**, no. 5, pp. 321–326

(2003)

[57] A. S. T. Chang, J. S. Shih, Y. S. Choo, and Q. 123456, "Reasons and costs for design change during production," J. Eng. Des., vol. **22**, no. 4, pp. 275–289 (2011)

[58] A. Aiyetan, J. Smallwood, and W. Shakantu, "A systems thinking approach to eliminate delays on building construction projects in South Africa," Acta Structilia, pp. 19–39 (2011)

[59] H. Doloi, A. Sawhney, K. C. Iyer, and S. Rentala, "Analysing factors affecting delays in Indian construction projects," Int. J. Proj. Manag., vol. **30**, no. 4, pp. 479–489 (2012)

[60] P. E. D. Love, Z. Frani, and D. J. Edwards, "A rework reduction model for construction projects," IEEE Trans. Eng. Manag., vol. **51**, no. 4, pp. 426–440 (2004)

[61] D. S. Thomson, S. a. Austin, H. Devine-Wright, and G. R. Mills, "Managing value and quality in design," Build. Res. Inf., vol. **31**, no. 5, pp. 334–345 (2003)

[62] Andi and T. Minato, "Representing causal mechanism of defective designs: A system approach considering human errors," Constr. Manag. Econ., vol. **21**, no. 3, pp. 297–305 (2003)

[63] L. Le-Hoai, Y. D. Lee, and J. Y. Lee, "Delay and cost overruns in Vietnam large construction projects: A comparison with other selected countries," KSCE J. Civ. Eng., vol. **12**, no. 6, pp. 367–377 (2008)

[64] C. W. Ibbs, "Quantitative impacts of project change: Size issues," J. Constr. Eng. Manag., vol. **123**, no. 3, pp. 308–311 (1997)

[65] K. Cooper and G. Lee, "Managing the Dynamics of Projects and Changes at Fluor," in *Proceedings of the International System Dynamics Conference*, April, pp. 1–27 (2009)

[66] D. W. M. Chan and M. M. Kumaraswamy, "An evaluation of construction time performance in the building industry," Build. Environ., vol. **31**, no. 6, pp. 569–578 (1996)

[67] N. D. Long, S. Ogunlana, T. Quang, and K. C. Lam, "Large construction projects in developing countries: A case study from Vietnam," Int. J. Proj. Manag., vol. **22**, no. 7, pp. 553–561 (2004)

[68] G. Sweis, R. Sweis, A. A. Hammad, and A. Shboul, "Delays in construction projects: The case of Jordan," Int. J. Proj. Manag., vol. **26**, no. 6, pp. 665–674 (2008)

[69] R. Apolot, H. Alinaitwe, and D. Tindiwensi, "An investigation into the causes of delay and cost overrun in Uganda's public sector construction projects," J. Constr. Dev. Ctries., vol. **18**, no. 2, pp. 33–47 (2013)

[70] Z. Shehu, I. R. Endut, A. Akintoye, and G. D. Holt, "Cost overrun in the Malaysian construction industry projects: A deeper insight," Int. J. Proj. Manag., vol. **32**, no. 8, pp. 1471–1480 (2014)

[71] H. Abdul-Rahman, M. A. Berawi, A. R. Berawi, O. Mohamed, M. Othman, and I. A. Yahya, "Delay mitigation in the Malaysian construction industry," J. Constr. Eng. Manag., vol. **132**, no. 2, pp. 125–133 (2006)

[72] K. G. Cooper and K. S. Reichelt, "Project changes: Sources, impacts, mitigation, pricing, litigation, and excellence," in *The Wiley guide to project technology, supply chain & procurement management*, Vol. **7**., P. W. G. Morris and J. K. Pinto, Eds. New York: John Wiley & Sons, Inc. (2010)

[73] P. E. D. Lover and H. Li, "Quantifying the causes and costs of rework in construction," Constr. Manag. Econ., vol. **28**, pp. 479–490 (2000)

[74] C. Koo, T. Hong, C. Hyun, S. H. Park, and J. Seo, "A study on the development of a cost model based on the owner's decision making at the early stages of a construction project," Int. J. Strateg. Prop. Manag., vol. **14**, no. 2, pp. 121–137 (2010)

[75] A. S. Hanna, R. Camlic, P. a. Peterson, and M.-J. Lee, "Cumulative Effect of Project Changes for Electrical and Mechanical Construction," J. Constr. Eng. Manag., vol. **130**, no. 6, pp. 762–771 (2004)

[76] N. A. Kartam, "Making Effective Use of Construction Lessons Learned in Project Life Cycle," J. Constr. Eng. Manag., vol. **122**, no. 1, pp. 14–21 (1996)

[77] A. Arora, "Dynamic project management using simulations," *Project Managment Institute* (2007)

[78] J. P. Basbagill, F. L. Flager, and M. Lepech, "A multi-objective feedback approach for evaluating sequential conceptual building design decisions," Autom. Constr., vol. **45**, pp. 136–150 (2014)

[79] J. Sterman, "System dynamics modeling for project management," *Unpubl. manuscript, Cambridge, MA* (1992)

[80] J. Sterman, "Learning in and about complex systems," Syst. Dyn. Rev., vol. **10**, no. 3, pp. 291–330 (1994)

[81] A. S. Hanna, J. S. Russell, E. V Nordheim, and M. J. Bruggink, "Impact of change orders on labor efficiency for electrical construction," J. Constr. Eng. Manag., no. **July/August**, pp. 224–232 (1999)

[82] W. C. Ibbs, C. K. Wong, and Y. H. Kwak, "Project change management system," J. Manag. Eng., no. **July**, pp. 159–165 (2001)

[83] I. a. Motawa, C. J. Anumba, and A. El-Hamalawi, "A fuzzy system for evaluating the risk of change in construction projects," Adv. Eng. Softw., vol. **37**, no. 9, pp. 583–591 (2006)

[84] Z. Y. Zhao, Q. L. Lv, J. Zuo, and G. Zillante, "Prediction System for Change Management in Construction Project," J. Constr. Eng. Manag., vol. **136**, no. 6, pp. 659–669 (2010)

[85] K. G. Cooper, "Naval ship production: A claim settled and a framework built," Interfaces (Providence)., vol. **10**, no. 6, pp. 20–36 (1980)

[86] H. Doloi and A. Jaafari, "Toward a dynamic simulation model for strategic decision-making in life-cycle project management," Proj. Manag. J., vol. **33**, no. 4, pp. 23–38 (2002)

[87] P. Otto, "A system dynamics model as a decision

aid in evaluating and communicating complex market entry strategies," J. Bus. Res., vol. **61**, no. 11, pp. 1173–1181 (2008)

[88] L. C. Bank, M. McCarthy, B. P. Thompson, and C. C. Menassa, "Integrating BIM with system dynamics as a decision-making framework for sustainable building design and operation," in *Proceedings of the First International Conference on Sustainable Urbanization (ICSU 2010)*, 2010, no. December, pp. 15–17 (2010)

[89] Y. Xing, S. Lannon, and M. Eames, "Developing a system dynamics based building performance simulation model-SdSAP to assist retrofitting decision making," in *Proceedings of BS2013: 13th Conference of International Building Performance Simulation Association, Chambery, France, August 26-28*, pp. 226–233 (2013)

[90] A. S. White, "A control system project development model derived from System Dynamics," Int. J. Proj. Manag., vol. **29**, no. 6, pp. 696–705 (2011)

[91] N. P. Repenning, "Selling system dynamics to (other) social scientists," Syst. Dyn. Rev., vol. **19**, no. 4, pp. 303–327 (2003)

[92] A. Laufer, A. Shapira, and D. Telem, "Communicating in dynamic conditions: how do on-site construction project managers do it?," J. Manag. Eng., vol. **24**, no. 2, pp. 75–86 (2008)

[93] S. G. Naoum, "Critical analysis of time and cost of management and traditional contracts," J. Constr. Eng. Manag., vol. **120**, no. 4, pp. 687–705 (1994)

[94] R. R. Senescu, G. Aranda-Mena, and J. R. Haymaker, "Relationships between project complexity and communication," J. Manag. Eng., vol. **29**, no. 2, pp. 183–197 (2013)

[95] J. D. Sterman, *Business dynamics: Systems thinking and modeling for a complex world.* Boston: Irwin/McGraw-Hill (2000).

[96] A. Alsehaimi, L. Koskela, and P. Tzortzopoulos, "Need for alternative research approaches in construction management: Case of delay studies," J. Manag. Eng., no. **October**, pp. 407–414 (2013)

[97] D. D. Ahiaga-Dagbui, S. D. Smith, P. E. D. Love, and F. Ackermann, "Spotlight on construction cost overrun research: superficial, replicative and stagnated," in *31st Annual ARCOM Conference*, 7–9 September (2015).

[98] A. Rodrigues and J. Bowers, "The role of system dynamics in project management," Int. J. Proj. Manag., vol. **14**, no. 4, pp. 213–220 (1996)

[99] C. L. Chang, S. Ogunlana, and K. Saeed, "Construction Project Management: A System Dynamics Approach," in *Proceedings of the 1991 System Dynamics Conference*, p. 108 (1991).

[100] J. M. Lyneis and D. N. Ford, "System dynamics applied to project management: a survey, assessment, and directions for future research," Syst. Dyn. Rev., vol. **23**, no. 2–3, pp. 157–189 (2007)

[101] A. G. Rodrigues and T. M. Williams, "Systems dynamics in project management: assessing the impacts of client behaviour in project performance," J. Oper. Res. Soc., no. **49**, pp. 2–15 (1998)

[102] Y. Li and T. Taylor, "Modelling the impact of design rework on transportation infrastructure construction project performance," J. Eng. Manag., vol. **140**, no. 9, p. 04014044 (2014)

[103] S. Lee, F. Peña-Mora, and M. Park, "Web-enabled system dynamics model for error and change management on concurrent design and construction projects," J. Comput. Civ. Eng., vol. **20**, no. 4, pp. 290–300 (2006)

[104] B. P. Thompson and L. C. Bank, "Use of system dynamics as a decision-making tool in building design and operation," Build. Environ., vol. **45**, no. 4, pp. 1006–1015 (2010)

[105] W. Ibbs and M. Liu, "System dynamic modeling of delay and disruption claims," Cost Eng., vol. **47**, no. 6, pp. 12–16 (2005)

[106] M. Park, S. Ji, H. Lee, and W. Kim, "Strategies for design-build in Korea using system dynamics modeling," J. Constr. Eng. Manag., vol. **135**, no. 11, pp. 1125–1138 (2009)

[107] S. Chritamara, S. O. Ogunlana, and N. L. Bach, "System dynamics modeling of design and build construction projects," Constr. Innov., vol. **2**, no. 4, pp. 269–295 (2002)

[108] S. Lisse, "System Dynamics for Outsourcing Construction Services," Int. J. Constr. Eng. Manag., vol. **2**, no. 3, pp. 53–61 (2013).

[109] C. Eden, T. Williams, and F. Ackermann, "Analysing project cost overruns: Comparing the 'measured mile'; analysis and system dynamics modelling," Int. J. Proj. Manag., vol. **23**, no. 2, pp. 135–139 (2005)

[110] A. Rodrigues, "The role of system dynamics in project management: a comparative analysis with traditional models," in *International System Dynamics Conference*, pp. 214–225 (1994)

Transformation of Malaysian Construction Industry with Building Information Modelling (BIM)

Aryani Ahmad Latiffi[1 a] Juliana Brahim[2] and Mohamad Syazli Fathi[3]

[1,2] *Faculty of Technology Management & Business, Universiti Tun Hussein Onn Malaysia (UTHM), 86400 Parit Raja, Batu Pahat, Johor, Malaysia*
[3] *UTM Razak School of Engineering and Advanced Technology, Universiti Teknologi Malaysia, 54100 Kuala Lumpur, Malaysia*

Abstract. Building Information Modelling (BIM) is a revolution of technology and a process that transformed the way building is planned, designed, analysed, constructed and managed. The revolution of technology and process could increase the quality of construction projects. The knowledge of BIM has been expanding in many countries including Malaysia. Since its inception, the use of BIM has broadened up widely with different purposes. The aims of this paper is to investigate the BIM implementation and uses in Malaysian construction projects. The methodologies adopted for structuring this paper are by using literature review and semi-structured interview with construction players that have experienced and being involved in projects using BIM. The purpose of literature review is to illustrate on the pervious research on the subject matter. Meanwhile, the purpose of interviews is to explore the involvement of construction players, years of experience in projects using BIM and BIM uses in construction projects. The findings revealed that BIM has been implemented in Malaysia since 2007 by various construction players, which are client, architect, C&S engineer, M&E engineer, QS, contractor, facilities manager and BIM consultant. The findings also revealed that BIM is used for project' visualisation, improving project design, detecting design clashes, quantity take off and operation and maintenance. Further work will be focused on the current practices of construction players in projects using BIM.

1 Introduction

Building Information Modelling (BIM) is defined as an innovative way for approaching the design, construction and operation and maintenance of building [1]. It is a collaborative tool that used by Architectural, Engineering and Construction (AEC) industry [2,3] that enhance visualisation and constructability of design, reduce time, cost as well as reduce conflict among construction players [3,4-6]. Since the inception of BIM, it is being increasingly implemented in the construction industry in order to produce data rich models of buildings and structural [3,7-9].

BIM is said as a project and process simulation [3,5,10]. This is because the realisation of BIM is much similar to the planning and actual of construction projects. Therefore, BIM could not be treated as an isolation of a software tool. It must gives impact on all the processes within the construction projects [6]. In addition to that, BIM helps by having projects simulation as it could shows the actual performance that happens in construction projects. This has made BIM as an efficient and effective tool in construction projects. With accurate building model, it allows smoother and better planned construction process that helps to reduce any errors and conflicts.

In the past decades, there had been a growing interest to adopt BIM in construction projects due to many benefits and resource saving during design, construction and maintenance and operation of project [11]. As a result, BIM has been implemented widely in the United State of America (USA), Australia, Hong Kong, Denmark, Norway, Finland and Singapore [2,7,12]. Compared to other developed countries, the implementation of BIM in Malaysia is very little and stagnant [13]. Moreover, there is a lack of evidence to show the involvement of construction players in projects using BIM. Therefore, this paper is seek to investigate on the transformation of Malaysian construction industry with BIM.

The next session discusses on the literature review on BIM in the Malaysian construction industry.

2 Literature Review on BIM in the Malaysian Construction Industry

Looking to the future is vital as the construction industry today is at the cross road of change [14]. The use of new technology in the construction industry could help the construction industry to face new challenges in construction such as managing a large amount of 2D

[a] Corresponding author: aryani@uthm.edu.my

project documents, involvement of various construction players, discrepancies in project design [15,16] and slow decision making [14]. Therefore, the use of new technology is needed for improving construction project from planning, design, construction as well as maintenance and operation of building. The implementation of a few past and mega projects in Malaysia such as Kuala Lumpur International Airport (KLIA), the second Penang Bridge, and monorail projects is the example of projects that need adaptation of new technology [17]. The adoption of new technology is needed to improve current practices in Malaysian construction industry that causes problems such as quality deficiencies, low productivity and catastrophes [17] that happened such as the collapse of Sultan Mizan Zainal Abidin Stadium in Kuala Terengganu [18].

There are many initiatives made by the Malaysian's government to encourage construction players to use BIM in projects to overcome the current problems. BIM has been introduced and implemented among construction players in Malaysia for the purpose of improving current construction projects. It has been introduced by Public Work Department (PWD) since early 2007 [2,4]. Furthermore, during the Second Day of Conference of Infrastructure and Construction Asia's Building Information Modelling and Sustainable Architecture 2009, the Director of PWD has urged the construction industry a need to embrace BIM in the construction project delivery [19]. This is followed by the first government project that used BIM in 2010 which is National Cancer Institute (NCI) [20]. Further to that, BIM has been part of the agenda in Construction Industry Transformation Plan (CITP) 2016-2020 by adoption of modern construction method and technologies to address challenges in the industry [21].

PWD has formed BIM Unit Projects on May 2012 by Complex Management Division or known as PROKOM. It consists of involvement of architects, civil and structural (C&S) engineers, mechanical and electrical (M&E) engineers as well as quantity surveyors (QS). The purpose of BIM Unit Projects is to produce Revit Families that will be used through the building life cycle (22). Further to that, PWD has published first BIM guideline for projects using BIM [22]. However, the guideline is only use for in-house project done by PWD and still in refining process [22]. Previously, PWD has done their initial BIM projects which are Primary School of Meru Raya, at Ipoh, Perak and Primary School of Tanjung Minyak at Melaka. Currently, PWD has involved in their projects using BIM which are Administration Complex of Suruhanjaya Pencegah Rasuah Malaysia (SPRM) at Shah Alam, Selangor and Healthcare Centre Type 5 at Maran, Pahang.

PWD BIM guideline has established BIM uses as a guideline for projects using BIM. Each of the uses is determined by project' client to achieve the objective of using BIM in projects [22]. Figure 1 shows BIM uses along the project life cycle.

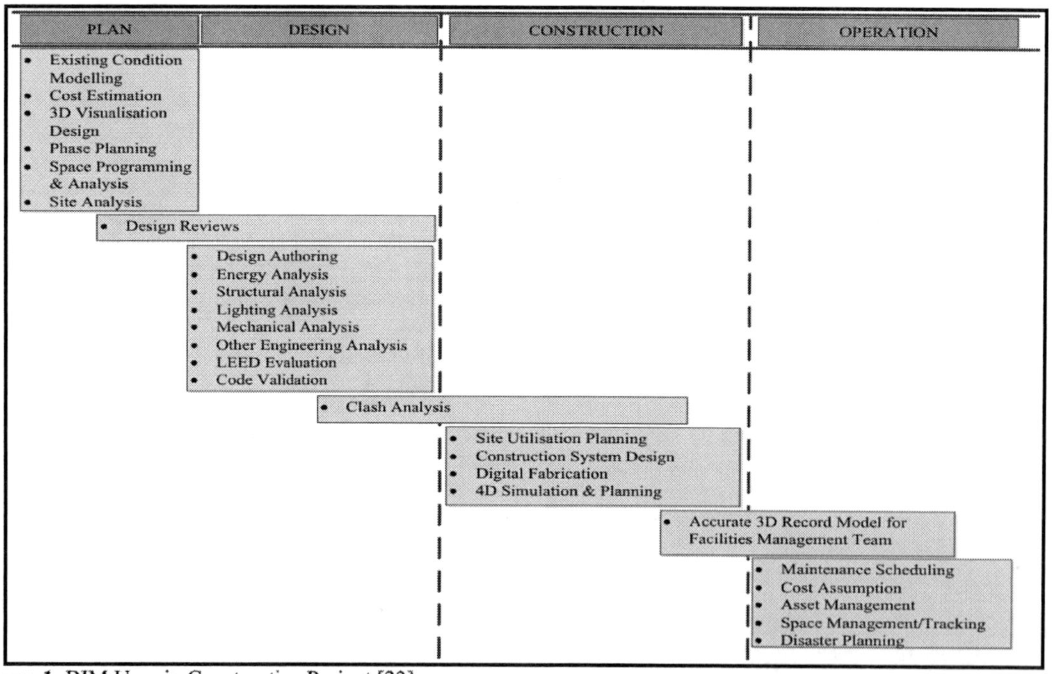

Figure 1. BIM Uses in Construction Project [22]

Based on the figure above, BIM is used in the early phase of construction until the final phase. The phases are plan, design, construction and operation. Plan and design phases are also known as pre-construction phase [3]. Meanwhile, the activities in operation phase is similar with post-construction phase [3]. Compared to conventional practices, the activities in projects using BIM mostly could be done in pre-construction phase

[3,15,16]. By using BIM, it helps construction players to visualise and predict any potential problems such as design clashes before the construction takes place. This could help to save time and cost of projects as well as the quality of project could be improved.

When all the potential problems has been detected at early phase, the construction players could perform the construction activities smoothly without any clashes of activities, destruction such as rework due to design clashes and delay in delivering Request For Information (RFI) [3,16]. During construction phase, BIM is used to do simulation of sequence of work activities [3]. Other than that, BIM also helps to generate shops drawing for various building [3,15]. The use of BIM also helps to increase the quality of information regarding the facilities in the post-construction phase [15,24]. The information of facilities could help for future analysis of facilities, assets tracking, maintenance schedule and renovation in future [3]. Initially, PWD will use BIM partly in planning and design stage. The activities consisting of existing condition modelling, site analysis, programming, structural analysis, design authoring, design review, 3D coordination, cost estimation, phase planning and record modelling [22].

The next session will discuss on the methodology used to achieve the aim of this paper.

3 Methodology

Data for this paper is gained through a literature review and semi-structured interview. Both methods are used to gather information regarding BIM implementation in the Malaysian construction industry with its uses. A literature review was conducted to explore on the BIM implementation in the Malaysian construction industry. The information was gathered from books, journals articles, international conference papers and materials available on the internet.

In addition to that, semi-structured interview has been utilised with construction players that have experienced and who are currently involved in projects using BIM. List of respondents is gathered from the list of participants in seminars and series of BIM forum conducted by the Construction Industry Development Board of Malaysia (CIDB) in 2013 until 2014. From the list, a snowballing sampling have been conducted to access on the information of other potential respondent through the respondent gathered from the CIDB list. Snowballing sampling is used to get an access on 'hidden population' [25]. It is suitable for this research as not all the construction players that involved in project using BIM are participating in the seminar or event related to BIM organised by CIDB. Therefore, this technique is used as an assistant in enriching sampling size and for accessing new potential respondent.

Furthermore, semi-structured interviews were conducted to gain information on the involvement of various construction players in projects using BIM, their experience in projects using BIM and BIM uses in construction projects. One-to-one and face to face semi-structured interviews were done with the respondents. A total of 28 respondents from the public and private sectors have been interviewed. The respondents are client, architect, civil and structure (C&S) engineer, mechanical and electrical (M&E) engineer, quantity surveyor (QS), contractor, facilities manager and BIM consultant. Those respondents were selected due to their willingness to share their experience and expertise in projects using BIM.

To obtain information on transformation of BIM in the Malaysian construction industry, a set of interview question was developed. The interview questions consist of two (2) sections as below:

Table 1. Interview Question

Section	Purpose
Section 1: Respondents' Background	• To get an explicit of respondents' information regarding years of experience in the construction industry. • To identify respondents' designation in current involvement in project using BIM.
Section 2: Current Practices of BIM	• To identify the inception of BIM in Malaysian construction industry. • To gain information on the use of BIM in construction project.

Data collected were analysed by using content analysis technique. Content analysis is relevant to be used for analysing unstructured data such transcription of semi-structured interview. All data gained from the respondents then were analysed using Atlas t.i.7. Atlas.t.i. enable to analyse visual and hierarchical modelling of concepts and theory [26]. It merged with large amount of documents and keep the data in all fields that require close study and analysis of primary material consisting audio, images, codes, video and geo data. The benefit of this tool is to support content analysis and provide single location for a storage system, assisting in the establishment of rules of coding [26]. All data is represented in the form of table.

The next session will discuss on findings and discussion of data collected.

4 Findings and Discussion

This section discusses on data gained from the semi-structured interview. The purpose of this section is to discuss on BIM implementation and its uses in Malaysian construction industry. This section consists of 4 sections as follows:

4.1. Respondents' Designation

The findings of respondents' designation is important to identify the construction players' involvement in projects using BIM in the Malaysian construction industry. There were 28 respondents from various designation involved with the interviews. Those respondents were selected based on their involvement and experience in projects using BIM in the industry. Table 2 shows the respondents' designation in projects using BIM.

Table 2. Respondents' Designation

Respondent	Designation							
	Client	Architect	Civil & Structure Engineer	M&E Engineer	QS	Contractor	Facilities Manager	BIM Consultant
R1								/
R2						/		
R3			/					
R4			/					
R5								/
R6					/			
R7						/		
R8						/		
R9		/						
R10				/				
R11							/	
R12							/	
R13		/						
R14		/						
R15			/					
R16				/				
R17						/		
R18	/							
R19				/				
R20								/
R21								/
R22			/					
R23								/
R24		/						
R25				/				
R26				/				
R27	/							
R28				/				
TOTAL	2	4	4	6	1	4	2	5

From the table, the indicator of R1 to R28 is used to represent each of respondent. The respondents are client, architect, C&S engineer, M&E engineer, QS, contractor, facilities manager and BIM consultant. Only two (2) respondents (R18, R27) were involved in projects using BIM as a client. Meanwhile, four (4) respondents (R9, R13, R14, R24) are architect and four (4) respondents (R3, R4, R15, R22) are C&S engineer. On the other hand, six (6) respondents (R10, R16, R19, R25, R26, R28) that involved in projects using BIM are as a M&E engineer. Only 1 respondent (R6) involved as a QS in projects using BIM. Added to that, four (4) respondents (R2, R7, R8, R17) are contractor in projects using BIM and two (2) respondents (R11, R12) involved as a facilities manager in projects using BIM. Despite of that, BIM consultant also involved in projects using BIM. BIM consultant is a person that being hired by company that has no experience in projects using BIM to assist on BIM strategies, design team and process of BIM implementation [27]. Five (5) respondents (R1, R5, R20, R21, R23) were involved in projects using BIM as a BIM consultant.

From the involvement of respondents in projects using BIM, the years of experience of each respondents has been determined to identify the inception of BIM implementation in the Malaysian construction industry. The next section will discuss on the respondents' experience in construction industry and projects using BIM.

4.2. Respondents Experience in Project Using BIM

Each of respondents are having different experiences and involvement in construction projects and projects using BIM. The aim of this paper is to identify the inception of BIM in Malaysian construction industry. Table 3 shows respondents' experiences in construction projects and projects using BIM.

Table 3. Respondents' Experience in the Construction Industry and Projects Using BIM

R	Year																													Total Years of Experience in Construction Industry	Total Years of Experience in Projects Using BIM
	1980'			1999'										2000'																	
	87	88	89	90	91	92	93	94	95	96	97	98	99	00	01	02	03	04	05	06	07	08	09	10	11	12	13	14	15		
R1																														15	5
R2																														25	5
R3																														15	5
R4																														9	3
R5																														7	5
R6																														13	4
R7																														22	5
R8																														9	5
R9																														29	8
R10																														8	2
R11																														28	8
R12																														23	8
R13																														9	3
R14																														9	3
R15																														17	9
R16																														13	4
R17																														9	7
R18																														7	4
R19																														8	3
R20																														9	5
R21																														6	6
R22																														15	5
R23																														6	6
R24																														9	5
R25																														3	3
R26																														6	3
R27																														15	5
R28																														19	3

LEGEND:	
R	Respondent
(dark shading)	Experience in construction industry
(white line)	Experience in projects using BIM

Based on Table 3, from the 28 respondents, only one (1) respondent which is R25 has experience less than 6 years in the industry. Meanwhile, 15 respondents which are R4, R5, R8, R10 R13, R14, R17, R18, R19, R20, R21, R23, R24, R25, 26, have 6 to 10 years of experience in industry. Other than that, 6 respondents which are R1, R2, R6, R16, R22, R27, have 11 to 15 years of experience and 2 respondents (R2 and R28) have 16 to 20 years of experience in construction industry. Moreover, two (2) respondents (R7 and R12) and two (2) respondents (R9 and R15) have experience 21 to 25 years of experience and 26 to 30 years of experience in construction industry respectively. Most of the respondents have experience for more than 5 years which is it is sensible to infer that the majority of respondents have a sound of knowledge in construction projects and familiar with the conventional construction process a suggested by [28].

Apart from that, twenty one (21) respondents which are R1, R2, R3, R4, R5, R6, R7, R8, R10, R13, R14, R16, R18, R19, R20, R22, R24, R25, R26, R27, R28 have less than 6 years of experience in projects using BIM, while 7 respondents which are R9, R11, R12 R15, R17, R21, R23 have 6 to 10 years' experience in projects using BIM. The respondents that have experience more than 5 years in projects using BIM could be considered as sufficient experience and were familiar with issues at hand as suggested by [28]. Based on the experience of respondents in project using BIM, it showed that the respondents have started using BIM since 2007. Similarly, [18] and [4] also stated that PWD has establish BIM committee to support BIM implementation in the Malaysian construction industry in 2007.

4.3. Involvement in Project Using BIM

The involvement of all respondents in projects using BIM is due to several factors. Based on the interviews, there are two reasons of respondents' involvement in projects using BIM. Table 4 shows the involvement of respondents in projects using BIM.

Table 4. Involvement of Respondents in Projects using BIM

Respondent	Involvement in Projects Using BIM	
	Client' Requirement	Own Initiative
R1		•
R2	•	
R3	•	
R4	•	
R5		•
R6	•	
R7	•	
R8	•	
R9	•	
R10	•	
R11		•
R12		•
R13		•
R14		•
R15		•
R16	•	
R17	•	
R18	•	
R19	•	
R20		•
R21	•	
R22		•
R23		•
R24	•	
R25		•
R26		•
R27		•
R28	•	
Total	15	13

Based on the table, 15 respondents (R2, R3, R4, R6, R7, R8, R9, R10, R16, R17, R18, R19, R21, R24, R28) were involved in projects using BIM due to requirement by client. Meanwhile, 13 respondents (R1, R5, R11, R12, R13, R14, R15, R20, R22, R23, R25, R26, R27) were involved in projects using BIM because of their own initiatives to adopt new technology and process in construction projects. The respondents explained that, client use BIM for the purpose to reduce project' duration, to control project' cost and to improve the quality of project. This is in line with [5] and [6], in their opinion that BIM could improve whole life projects cost, control project time by having faster and effective process and better quality. The respondents that used BIM due to client requirement stated that the implementation of BIM in Malaysia is mainly driven by the encouragement and demand by the client. For an instance, PROKOM has been piloting the use of BIM in their pilot project which is National Cancer Institute (NCI) [4]. Consequence to that, this has encouraged the use of BIM in the next projects funded by the public sector.

On the other hand, the respondents that used BIM due to their own initiative explained that the use BIM could improve their efficiency and productivity in producing projects' design. Hence, they initiate themselves by starting to invest for BIM technology, and getting expertise for training regarding BIM process. Furthermore, this shows that they have their own initiative to support government policy and the needs to adapt BIM in projects. Moreover, the use of BIM by their own initiative is also due to the establishment of PWD BIM Roadmap that made BIM as a compulsory for design and build projects and conventional projects based on contract values [29].

All respondents also highlighted that the implementation of BIM is depending on BIM uses in projects. The next section will discuss on the BIM uses that has been used in Malaysian construction projects.

4.4. BIM Uses in Construction Projects

To identify level of BIM implementation in the BIM Malaysian construction industry, each of the respondents have been interviewed regarding BIM uses in the projects using BIM. Table 5 shows BIM uses in the Malaysian construction industry.

Table 5. BIM Uses in Malaysian Construction Projects

Respondent	BIM Uses				
	Projects Visualisation	Improve Project Design	Detect Design Clashes	Quantity Take Off	Operation and Maintenance of Building
R1	•	•	•		
R2	•		•		
R3	•	•	•		
R4	•	•	•		
R5	•		•		
R6				•	
R7	•		•		
R8	•		•		
R9	•	•	•		
R10	•	•	•		
R11					•
R12					•
R13	•	•	•		
R14	•	•	•		
R15	•	•	•		
R16	•	•	•		
R17	•	•	•		
R18	•	•	•		
R19	•	•	•		
R20	•	•	•		
R21	•	•	•		
R22	•	•	•		
R23	•		•		
R24	•	•	•		
R25	•	•	•		
R26	•				
R27	•	•	•	•	
R28	•	•	•		
Total	25	19	24	2	2

There are five (5) BIM uses that being implemented in projects using BIM as shown in the table. They are to improve projects visualisation, improve project design, detect design clashes, quantity takeoff, and operation operation and maintenance of building. From the table, twenty five (25) of respondents have experienced using BIM for projects visualisation. The respondents pointed out that BIM is used to help client to visualise the completed projects as early in pre-construction phase. BIM also helps other construction players such as architect, C&S and M&E engineer to visualise on the design effectively so that they can detect any discrepancies between trades.

Meanwhile, nineteen (19) of the respondents experienced using BIM in projects for improving projects design. The respondents explained that, they used BIM to develop projects design in 3D model with projects' information in effective and more efficient so that they could avoid any discrepancies. As [5] and [15] claimed that, BIM helps construction players to have better design with less error. Also, the 3D models capable to show construction players the location of buildings in the future environment and also gives the project's end users an insight into the development of an area.

Moreover, twenty four (24) of respondents have experienced in using BIM for detecting design clashes. From the interview, the respondents that involved as M&E

engineer stated that their is still new in projects using BIM. Nevertheless, they also added that, the involvement of M&E engineer in projects using BIM is more crucial due to the compelexity of the design and its element. So that, M&E engineer could produce drawing, installing M&E system in more accurate [30]. All respondent also claimed that the involvement of C&S engineer together with M&E engineer are most practical as they could detect any possible clashes between three trades (architect, C&S engineer and M&E engineer).

Meanwhile, only two (2) of respondents experienced using BIM for quantity take off. The respondents explained that, BIM helps to generate the quantity in more accurate and efficient. The use of BIM for quantity take off furnishes great potential to make cost prediction above the limitation posed by traditional method [31]. However, the respondents explained that the use of BIM specifically for quantity take off is still new and required cooperation with architects to develop 3D model followed by QS specification. Moreover, only two (2) respondents used BIM specific in post-construction phase. The respondent use BIM to maintain the operation and maintenance of the completed projects by remodel 2D as-build drawing to 3D model with necessary data. The information in the building model is associated with spaces, masses, construction level details and scope of the model (such as architectural and details of MEP elements) [15]. In addition to that, the respondents higlighted that, the use of BIM for maintenance and operation of building is done only after the projects has completed in conventional practices.

From the data collected, the implementation of BIM in Malaysia still at infant stage. This is in line with research by [13] and [32] as the implementation of BIM in Malaysia still in early phase of inception and far behind from other developed countries. In addition to that, most of the respondents stated that they have experienced in handling only one (1) projects using BIM and still in learning and training period. Further to that, the implementation of BIM is still limited at pre-construction phase. Only two respondents (R11, R12) used BIM specific at post-construction phase. It also suggested that the implementation of BIM in the Malaysian construction industry need to be more strategised to ensure the implementation of BIM from pre-construction phase to post-construction phase. The respondents suggested that to ensure the improvement and continuous of BIM implementation, it is not only depends on the use of BIM technology, it is also depending on the competencies of construction players in understanding the process involved. This is because, the use of BIM also depending on the size and capabilities of the organisation to invest new technology [13].

5 Conclusion and Further Works

Undoubtedly, Malaysian construction industry aware of the needs to adopt BIM technology in improving current construction projects. Still, it shows that the Malaysian construction industry need to have better strategies in highlighting and embracing the use of BIM from pre-construction phase to post-construction phase. Further works will be conducted on the construction players' current practices in projects using BIM.

Acknowledgement

The authors would like to thank to Ministry of Higher Education of Malaysia (MOHE), and Office of Research, Innovation, Commercialization and Consultancy (ORRIC), UTHM for supporting this research under the Exploratory Research Grant Scheme (ERGS), (Vote No. E029).

References

1. D. Bryde, M. Broquetas, and J.M. Volm. The Project Benefits of Building Information Modelling (BIM). *International Journal of Project Management*, **31(7)**, 971-980. (2013)
2. A. Ahmad Latiffi, J. Brahim, S. Mohd, and M.S. Fathi. The Development of Building Information Modelling (BIM) Definition. *Applied Mechanics & Materials*, **(567)**. (2014)
3. S. Azhar. Building Information Modeling (BIM): Trends, Benefits, Risks, and Challenges for the AEC Industry. *Leadership and Management in Engineering*. (2011)
4. PWD. Unit Building Information Modelling (BIM). Retrieve from: https://www.jkr.gov.my/prokom/index.php?optioncom_content&view=article&id=84&Itemid=435&lang=ms. (2011)
5. S. Azhar, M. Khalfan, and T. Maqsood T. Building Information Modelling (BIM): Now and Beyond. *Australasian Journal of Construction Economics and Building*, **12(4)**, 15-28. (2012)
6. R. Eadie, M. Browne, H. Odeyinka, C. McKeown, and S. McNiff. A Survey of Current Status of and Perceived Changes Required for BIM Adoption in the UK. *Built Environment Project and Asset Management*, **5(1)**, 4-21. (2015)
7. A.K.D, Wong, F.K.W. Wong, and A. Nadeem. Government Roles in Implementing Building Information Modelling Systems: Comparison Between Hong Kong and the United States. *Construction Innovation*, **11(1)**, 61-76. (2011)
8. M. Dobelis. Drawbacks of BIM (Building Information Modelling) Concept Adoption. *Paper presented at the International Conference: Engineering Graphics BALTGRAF 2013, 12, Riga (Latvia), 5-7 Jun 2013*. (2013)
9. R. Barati, A. Charehzehi, and C.N. Preece. Enhancing Planning and Scheduling Program by using Benefits of BIM-based Applications. *Civil and Environmental Research*, **3(5)**, 41-48. (2013)
10. W. Kymmell. *Building Information Modeling: Planning and Managing Construction Projects with 4D CAD and Simulations (McGraw-Hill Construction Series): Planning and Managing Construction Projects with 4D CAD and Simulations:* McGraw Hill Professional. (2008)

11. R. Volk, J. Stengel and F. Schultmann. Building Information Modeling (BIM) for Existing Buildings-Literature Review and Future Needs. *Automation in Construction*, **38**, 109-127. (2014)
12. C. Furneaux and R. Kivit. BIM: Implication for Government. *CRC for Construction Innovation*. Brisbane Australia. Pp. 10-3. (2008)
13. Z. Zakaria, N. Mohamed Ali, A. Tarmizi Haron, A.J. Marshall-Ponting, and Z. Abd Hamid. Exploring The Adoption of Building Information Modelling (BIM) in the Malaysian Construction Industry: A Qualitative Approach. *International Journal of Research in Engineering and Technology*, **2(8)**, 384-395. (2013)
14. S.S.S. Gardezi, N. Shafiq, and M.F. Khamidi. Prospects of Building Information Modeling (BIM) in Malaysian Construction Industry as Conflict Resolution Tool. *Journal of Energy Technologies and Policy*, **3(11)**, 346-350. (2013)
15. C. Eastman, P. Teicholz, and R. Sacks. *BIM Handbook: A Guide to Building Information Modeling for Owners, Managers, Designers, Engineers and Contractors* John Wiley & Sons. (2011)
16. K.P. Reddy. *BIM for Building Owners and Developers: Making a Business Case for Using BIM on Projects*. John Wiley & Sons. New Jersey. (2012)
17. M. Jaafar, T. Ramayah, A. Abdul-Aziz, and B. Saad. Technology Readiness among Managers of Malaysian Construction Firms. *Engineering, Construction and Architectural Management*, **14(2)**, 180-191. (2007)
18. A. Ahmad Latiffi, S. Mohd, N. Kasim, and M.S. Fathi. Building Information Modeling (BIM) Application in Malaysian Construction Industry. *International Journal of Construction Engineering and Management*, 2(A), 1-6. (2013)
19. G. Sundaraj. The Way Forward: Construction Industry Master Plan 2006-2015. Master Builder, 1st Quarter 2007. Retrieved from: http://woulibrary.wou.edu.my/library/pdf/CIMPlan.pdf. (2006)
20. Construction Industry Development Board (CIDB). BIM Portal: Building Information Modelling. Retrieve from http://bimcenter.com.my/index.php/bim-news. (2013)
21. CIDB. Construction Industry Transformation Programmes 2016-2020. Retrieved from: file:///C:/Users/User/Downloads/CITP-public.pdf. (2015)
22. Construction Research Institute of Malaysia (CREAM). Issues and Challenges in Implementing Building Information Modelling (BIM) by SME's in the Construction Industry. Retrieved from:https://www.cidb.gov.my/cidbv4/images/pdf/announcement/BIM/bim%20seminar%20%20workshop%20for%20malaysia%20construction%20industry.pdf. (2014)
23. Jabatan Kerja Raya. *Garis Panduan BIM JKR: Edisi 2014*. Kuala Lumpur. Jabatan Kerja Raya. (2014)
24. R. Crotty. *The Impact of Building Information Modelling: Transforming Construction*. Routledge. (2013)
25. M. Saunders, P. Lewis and A. Thornhill. *Research Methods for Business Students: Fifth Edition*. Pearson Education Limited. Edinburgh Gate. (2009)
26. R. M. Nordin. *A Framework of Transperancy Initiatives to Fight Corruption for Public Construction Projetcs*. Thesis. Universiti Teknologi Mara. Malaysia. (2015)
27. M. B. Barison, and E. T. Santos. An overview of BIM specialists. *Computing in Civil and Building Engineering, Proceedings of the ICCCBE2010, 141*. (2010)
28. S. H. Hamid. *An Integrated of Value Management (IVM) Framework for Contractor in Design and Build (D&B) Project*. Unpublished Msc Thesis. Universiti Teknologi Mara. (2014)
29. Q.C. Keat. What It Will Take To Ensure 5D BIM Sustainability in Malaysia. Retrieved from: https://www.academia.edu/4929524/What_It_Will_Take_To_Ensure_5D_BIMSustainability_In_Malaysia . (2013)
30. C.S. Dossick, and G. Neff. Messy Talk and Clean Technology: Communication, Problem-Solving and Collaboration using Building Information Modelling. *The Engineering Project Organization Journal*, **1(2)**, 83-93. (2011)
31. N.A. Ismail, E. Utiome, R. Owen and R. Drogemuller. Exploring Accuracy Factors in Cost Estimating Practice towards Implementing Building Information Modelling (BIM). *Paper presented at the Proceedings of the 6th International Conference On Engineering, Project, and Production Management (EPPM2015).*(2015)
32. N.A.H. Hadzaman, R. Takim and A.H. Nawawi. BIM Roadmap Strategic Implementation Plan: Lesson Learnt From Australia, Singapore and Hong Kong. *Proceedings in 31st Annual ARCOM Conference*. 611-620. (2015)

MATEC Web of Conferences **66**, 00023 (2016) DOI: 10.1051/matecconf/20166600023
IBCC 2016

Eradicating Squatters through Resettlement Programme: A Conceptual Paper

Julieven Nonoi Kuek[1], Yusfida Ayu Abdullah[2] and Hazlina Hamdan[3]

[1] *Centre of Postgraduate Studies, Faculty of Architecture, Planning and Surveying, Universiti Teknologi MARA, 40450 Shah Alam, Malaysia.*
[2] *Centre of Studies for Town and Regional Planning, Faculty of Architecture, Planning and Surveying, Universiti Teknologi MARA, 40450 Shah Alam, Malaysia.*
[3] *Centre of Studies for Town and Regional Planning, Faculty of Architecture, Planning and Surveying, Universiti Teknologi MARA, 40450 Shah Alam, Malaysia.*

Abstract. The existence of squatters had caused many issues including poverty, increasing social problems, drug addiction, criminal and low education level. Sadly, the State of Johor is still among a few other states in Peninsular Malaysia with the highest numbers of squatter families. These squatter settlements are not only inhabited by the local people, but consist of foreigners too. The environmental condition is not fitting, with poor sewage system and unplanned housing construction. Consequently, these issues had caused bad images and perception to the general public, as well as to the tourists. In Malaysia, Resettlement Programme is one of the most significant mechanisms in tackling squatters" settlements issue. Hence, the Johor State Government had taken various approaches in trying to eradicate the squatter settlements by relocating them to new locations with planned and better housing. This conceptual paper aims to examine the current literature and assess previous similar study in order to establish an understanding of the crucial provision of resettlement area in trying to eliminate squatters. It employs a qualitative design through assessment related literature and studies. The results confirmed for a further review of resettlement program. Findings from this study further suggest a comprehensive study of the issues revealed.

1 Introduction

The issue of squatters is not a new phenomenon in Malaysia. A squatter house is one of the types of dwellings but is an illegal house built without permission from the local authority [1]. Illegal squatter housing relates to the establishment of houses without authority's approval including the land where the squatter settlements found and built their houses not in accordance with specification approved by the authority [2]. The UN-Habitat [3] defines squatter as: (1) the occupants without guarantee of tenure land or residences; (2) neighbourhood without adequate or even cut off from basic facilities, (3) buildings or houses without planning approval or building regulations and often located in non-conducive living area.

Squatter problem has become alarming in many major cities since the rapid growth of cities had ultimately pushed more people to move and find a job in a city [4]. Additionally, the deficiency of supply of affordable housing had also resulted in the building of squatter buildings as another option to reduce the cost of living [4] since many of them cannot afford to purchase and own a property due to high pricing [5]. As a result of those underlying causes, the unfortunate group had no choice but to illegally build their houses at the squatter settlements despite the occurrence of various social problems and insufficient basic facilities at the area [6].

The latest data available on squatters shows that the State of Johor carries the highest number of squatter families as compared to other states in the Peninsular Malaysia with a total number of 6,363 families [7]. Here in Malaysia, squatters' areas are not only inhabited by the local people, but by foreigners too, including the illegal immigrants [8]. The existence of such areas had caused many problems including poverty, increasing social problems, drug addiction, criminal activities and low educational level among the residents [9]. Inevitably, the state of the environment which is often considered as non-environmental friendly including the impromptu housing construction and sewerage system had led to an unhealthy scene, revealing bad images to the general public and tourists. Tackling squatters through the *Resettlement Programme* is, therefore, one of the most important initiatives in eradicating squatters' problem in Malaysia. In trying to achieve this, the Johor state government had operated some approaches in relocating squatters including providing the squatters with

[a] Julieven Nonoi Kuek: julieven_kuek@yahoo.com

provisional housing estates or even through public housing projects known as the "*Project Perumahan Rakyat*" (PPR), otherwise residing them to another area [10]. However, by taking any of these actions, the state government will deliberately discards their request and application or even concede their stay at any illegal housing area [10].

Nonetheless, even though resettlement programme is believed to be one of the solutions to wipe out squatter settlements, unavoidably, there appeared to be several issues and problems with regards to resettlement area; firstly, the location of resettlement areas are generally placed quite far from the city especially the newly built low-cost schemes [11] when most of the residents require the area to be close to their work place and schools [12] and secondly, the expensive price of houses for resettlement scheme particularly for those affected by the relocation program [12]. Even though the resettlement projects are subsidised, yet the total amount fixed by the developers is not within their capable budget and these problems frequently occur especially to the low-income group because the market price for low cost housing is still too expensive. Furthermore, the price of house is not included during the signing of purchase agreement, as well as the rate of bank loan, utility deposit, maintenance fee and other miscellaneous charges [13].

Moreover, the provision of resettlement houses was expressed to be uncomfortable [12]. Generally, the houses delivered for the resettlement residents indicates a lower standard of housing, seeing the compact building design and the use of lower quality of material. So, often the design and size of houses are among the major complaints from the residents of resettlement which claimed that the property cannot comfortably accommodate their family to reside [14]; [15]. Also, another problem is the issue of the resettlement houses being offered for rentals by the residents [12]. It became visible that some of the residents from the resettlement schemes rent out their low cost unit, but they themselves chose to stay elsewhere [15]. This occurs when the location of certain low cost units are at strategic sites offering opportunity for them to make profits from rentals.

Drawing from the basis of these continuing issues and problems, this paper sets out with an aim to appraise the existing literature and previous study of initiatives to eradicate squatters. It is anticipated that through this preliminary evaluation, it would be a beginning for a further advanced research of a resettlement program at the selected case study.

2 Squatters and Resettlement Program

2.1 Definition of Squatters

The UN-Habitat [3] defines squatter household as a group of individuals staying under the same roof in an urban area with one or more shortage(s) of the following components:

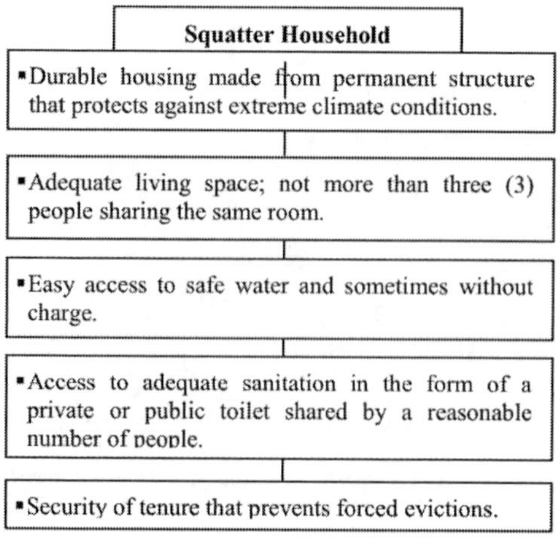

Figure 1: Definition of squatters' household as a group individuals living in the urban [3].

Meanwhile, Wakely [16] thought that squatters' houses feature unhealthy living situation and harmful locations, with hardship and social presence. Nevertheless, more appropriately, 'squatters' do not usually signify the informal settlements, but sometimes were planned development but have since physically worsened over time due to factors like overcrowding, or aging, or being neglected by the lower-income groups, etc. [17].

Generally, the squatter settlements in many major cities in the country have been in existence before Malaysia achieved its independence in 1957 [15]. One of the causes which influenced the development of squatters was urbanisation. Urbanisation was thought to have marked the three circumstances; (1) rural urban migration; (2) population growth; and (3) the re-designation of rural areas as urban [18]. Migration was therefore claimed to be its reason. However, migration takes place by force or voluntary. It can be influenced for reasons like being forced to reside to another area and remained at the new place due to natural disaster. Though, it can also be voluntary when people explore a new place for an improved economic prospect and a better life. There is also the 'pull' factor when rural migration is encouraged by the government, when people decide to work in factories or cheap labor. The main purpose for migrating to urban areas is to earn an income and a better future for themselves and their family whereas the condition of the shelters is not a priority. What is important is the proximity of their work place that involves minimal or not having to spend on transportation at all. To some, migration is temporary and they would, in the course of time, return to their villages hence renting a property becomes the best option. Hence, they would normally rent a place in squatter settlements where the rents are cheaper [15].

2.2 Resettlement Program

According to the United Republic of Tanzania [19], resettlement is identified as all procedures employed to mitigate all bad impacts, resulting directly to finishing off a project, on the livelihood of the project affected persons, their property as well as their compensation, relocation and rehabilitation. This action can be classified as any vulnerable and affected person including tribal communities, squatters and poverty group depending for livelihood upon the community in the project area and all those persons who may be exploited by the stronger or privileged persons located within the project area [19].

Based on International Finance Corporation [20], resettlement should be able to increase the quality of life of the people by score living standards ahead of the pre-project levels. Resettlement must be intended and carry out as a development project over a minimum of two generations and comprises not only protective measures, but also the provision of new rights, resources and strategies [21]. The strategies taken by the government to resolve the problem of squatters throughout the country through relocation of the squatters may be positioned further away from their previous site. And these new settlement areas are usually equipped with the necessary infrastructure for better comfort to the squatters.

According to the UN-Habitat [22], relocation is a physical restoration measure which is expensive but can provide the basic needs for the poor society for those persisting to live in the urban areas. And so, resettlement is a method of balancing the socio-economic development of society. There are two approaches to resolve the issue of squatters; Firstly, by moving them to the land allocated to them by granting a lower land rate (nominal); and secondly, by providing them with low-cost flats (with low rental rate) or selling the flats at cheaper price.

Anyhow, the local authorities have the right to demolish any illegal settlement without paying any compensation so as to redevelop squatter areas in a more orderly environment yet have the obligation to maintain the area. Although this method is viewed as a positive tactic in terms of the physical condition but the economic aspect is not lucrative to the government [1]. In fact, the government had to spend a lot on the provisions when selling price is much cheaper, just to facilitate the squatters. The danger of this resettlement program is when the squatters refuse to be reallocated and decide to explore the presence of another vacant land

3 Assessment of Previous Related Study

This paper also discusses a previous research on "Issues and Policies on the Aftermath of Squatters Resettlement: A Case Study on Kampung Desa Hormat Tambahan, Selangor" by Ibrahim *et.al*, (2012), which mainly discusses about the issues and challenges encountered by the communities of squatters. Besides that, the study also considered the background and the residents' current living condition in a resettlement area in terms of racism, financial and compensation, along with the education aspect.

3.1 Issues and Challenges of Squatters and Resettlement Area

A multi-racism issue happens when the tragedy of May 1969 has given a big impact on shattering the multiracial relationship among the squatters' communities between Chinese, Indian and Malay families [12]. Frequently in the squatter's area, issues of water and electricity supply often not adequate and caused internal conflicts within the squatter community and therefore resulted one of the hindrance towards their harmonisation between the communities. Meanwhile, when looking at the problem of drainage and sewage systems it was found out that in squatters area there are no proper drainage provided. Hence, these causes the drainage to clog up with garbage and rubbish all the times [12].

The age of the resettlement community are from various aged group, ranged from toddler to old citizens. According to Ibrahim *et al.* [12] the residents have been living in the resettlement area for more than 10 years. Meanwhile the older citizens have been residing in the squatter's area for more than 30 years before moving to the new settlement. This low-cost area comprises of 8 blocks that can accommodate about 1600 people from different squatters area around Selangor. The resettlement area is mostly occupied by the Indian community (65%) and another 35% are Malay ethnic group [12].

However, when expressing about their current living condition, majority of the resettlement residents preferred to live in squatter area. This is because they felt burdened with financial obligations, where they were forced to take up the units whereas they were not given any compensation. By providing them with accommodation, they need to purchase the low cost unit with loan facilities of RM35,000 per unit. To make things worse, they need to pay other informal payments such as maintenance, donation, security and other miscellaneous costs [12]. The study also revealed that the residents claimed that the house is still unaffordable since most households are with monthly income around RM1,500 to RM2,000 per month, and with four to eight children [12]. Therefore, this indicates that the price of the resettlement units should be reviewed, unless the majority agree that the price is affordable. The price of the resettlement low-cost unit undeniably affects the poor.

It was discovered that one of the challenges of former squatters in a new place was racism due to the inadequacy of facilities in the resettlement area that causes disagreement of the facilities usage [12]. Sadly, another social issue also occurs associating with sexual misconduct. Therefore, the state government should ensure that the facilities are provided adequately and eligible to all communities.

Former squatters with financial problem are prone to move to a squatter place again. Majority of the residents required the government to give them funding instead of offering them accommodation [12]. To them, compensation would reduce their burden and can avoid poverty. Therefore, the process of resettlement had created hardship and difficulties to the residents of resettlement since they had to rent other property elsewhere before moving to the low-cost units. In terms of children education, parents were having difficulties to send their children to school due to financial constraints as some of their elder children had dropped out from schools and with no job at all [12].

Because of that, the government needs to take into account on the location factor and to ensure that resettlement projects are located within proximity as highlighted by the residents because discarding the location aspect could eventually causes difficulties to the residents of resettlement. Additionally, the low-cost unit price offered to the squatters would be best reviewed since majority of the residents were unable to purchase the unit.

4 Evaluation of Successful Examples in Malaysia and Other Country

4.1 Selangor Successful Zero Squatters Programme

In January 1998, the squatter settlements that were built were going to be demolished by the local authorities when the state government of Selangor encountered the growing issue by old and new squatter areas in Selangor [18]. Selangor has therefore introduced the 'Zero Squatter 2005' action plan. Hence, the low-cost housing was developed in Selangor within five years ahead from February 2000 to December 2004 [18].

To accomplish Selangor Zero Squatters 2005 Action Plan, two main strategies were applied through five (5) approaches as stated below (Table 1):

Table 1: Selangor Zero Squatters 2005 Action Plan

Selangor Zero Squatters 2005 Action Plan	
Planning	The government collected data on squatters and low-cost houses. To confirm the provision of adequate low-cost and medium cost houses for target groups and squatter dwellers.
Supervision	Observing and ensuring that the development of low cost-housing was based on schedule and parallel to other types of development.
Distribution	To make sure that the distributions of low-cost housing were efficient and fair to eligible target group and squatter occupants.
Enforcement	Prevention and immediate enforcement that will be conducted on any squatter settlements and imposing measures on developers who failed to develop low cost housing as planned and as scheduled.
Housing Management	To make sure after December 2004, all Selangor's residents will legally owned a home.

Source: Boonyabancha [28].

The local authorities played a vital role behind the implementation of Selangor Zero Squatters policies and also in resolving the squatter settlements issues. There were two units under this local authority namely, the secretariat and task force. The unit of secretariat was led by the assistant planner involved in all meetings related to squatters and the development of low-cost housing. This unit also checked the completion process in developing low-cost housing and handle the census, as well as the survey process of squatters [18].

Meanwhile, the task force unit involved the state assembly member to guarantee speedy development of low-cost housing and to assist in resolving issues with regards to low-cost housing and squatters. Nevertheless, the government of Selangor has to make sure some private developers will participate in the development of the low-cost housing [18]. From time to time the government of Selangor has to check the status from the developers. Every month developers had to submit their report and were required to explain their delay in completing low-cost housing. This was because most of the low cost housing were constructed on the site of the demolished squatter settlements. The success of Selangor in eradicating squatters also involved the negotiation with squatter occupants and the state assembly member and developers. The negotiation concerned more on the price and design of unit, temporary shelter and other compensation. The occupants of squatters were included in the process and were briefed about the Zero Squatter 2005 policy.

The relocation process was also one of the methods taken by the Selangor to abolish the squatter settlements. There are three types of relocation of squatter dwellers to low cost housing; (i) In-situ relocation where the squatter settlements site were previously allocated for low-cost housing development. Therefore, all the squatter dwellers from that area are listed on in-situ low cost housing scheme; (ii) Relocation within election zones by means if the sites of the squatter settlement were not developed for low-cost housing, other sites within an election zone will be identified for the development; and (iii) Relocation to the nearest election zones, if there were inadequate low-cost housing projects within an election zone, the area nearest to the election zone would be identified for development. This is the local authority's responsibility to identify the best criteria of the squatter dwellers and allocate them to the most suitable schemes [18].

Therefore it can be concluded here that the state government plays a vital role in eliminating squatter settlements, together with the local authorities in implementing the action plan to ensure that everyone in the whole state of Selangor would legally have their own house. Even though the Zero Squatters policy had not fully eradicate the squatters but the number had decreased in the major cities over the years in Selangor.

4.2 Successful Thailand's Housing Programme for Squatters

The Thailand government 2003 introduced "The Baan Mankong Program" through the Community Organisations Development Institute (CODI) in 2003. It was a program for the government targeting slum communities. According to Boonyabancha [28], the program was an important action in the process of transformation. Additionally, the Thai's National Housing Authority had also supported this squatters upgrading program through the help from the World Bank during the 1980s. Through this program about 30,000 to 50,000 of families in Thailand were improved.

The Thai government formed the Urban Community Development Office in order to help the Community Organisations through loans for new housing as well as the improvement or upgrading of the income generation [28]. From 3,700 peoples of the poor community, about 30% of the community were squatters and another 70% rented the land which they stayed for a short term contract. These communities were under the risk of eviction, which is about 70% to 80% of the residents are not able to pay for conventional government housing programs [28].

There were two new programs introduced by the Thai government for the urban poor in January 2003; Firstly, the Baan Mankong (Secure Housing) Program, which required government funds in terms of development subsidies and housing loans to improve the present urban poor communities in enhancing their land, housing, environment and basic services that were fulfill by CODI [28]; and secondly, was the "Baan Ua Arthorn Program" where the National Housing Authority Designs of Thailand built and sold flats and houses at subsidies rates to the lower income households who managed to pay for "rent to-own" payments of 25 to 37 US dollar per month [28].

This Baan Mankong or secure housing program was built to assist the process designed and managed by low-income households and their community organisations and networks. These communities and network work with local governments, professionals, universities and NGOs in their city to undertake surveys and review all the poor communities within three to four years [28]. After the plans have been finalized along with approved pilot projects, CODI will locate the development subsidies and housing loans straight to the communities. As Boonyabancha [28] stated, the type of upgrading will be decided during the discussion either on the same site or relocation, whether in public or private land, to sell to community or to be bought as a group. The power to make a decision was based on the community since the community is the owner of the project. Other than that, the community also have to take responsibilities to get loan for housing construction or obtain the land.

In addition, within 5 years the Baan Mankong had set a target to enhance the housing, living condition and tenure security for 300,000 households for the poor communities within 200 cities in Thailand. The programs started in 2003 by upgrading 1500 units of housing and the preparation in 20 cities. In year 2004, 174 squatter communities in 42 cities were and preparation of 50 cities. Meanwhile, year 2005 to 2007 are improving the 285,000 units of housing in 200 cities [28].

Moreover, there are various types of pilot projects that have been implemented by the Thailand's government, for examples, Land Purchase and Re-Blocking, Post Fire Reconstruction and A Long Term Lease, Relocation to Nearby Land, Scaling up pilot projects, land sharing, the Relocations of Mini Squatters and A Long Term Lease [28]. The pilot project of Land Purchase and Re-blocking involved 81 households in Charoenchai Nimitmai that stayed on 0.7 hectares in Bangkok near the railway tracks and drainage canal [28]. In 1998, the occupants had to move out from the area because they had rented the land from the land owners for many years. Hence the occupants negotiated to buy the land for a quarter of its market price [28]. Once they formed cooperative and they took CODI loan to pay it. They also created a re-blocking plan to reduce the cost per family that accommodated some other families who were squatting close by. Majority families constructed their homes using materials from their former houses. The communities made contract with the government hence received electricity and water supply and also building permits. All of these constructions were conducted by the occupants themselves and contractor had been employed by paid community labour that was 30 percent of total development cost. Each of the occupants was required to repay 25 to 50 US dollar every month to the Thai government [28].

Second example of pilot project implemented by the Thai government was Land sharing. Since 20 years ago, the canal side of Klong Lumnoon has been occupied. The areas were transformed in 1997 as agreed by the land owner to expel the people [28]. Approximately about 49 families who worked near the area refused to move out and some of them received compensation and moved away. However, about 49 numbers of families convinced the land owner to sell the land to them a small portion of land at below market value in exchange for emptying the rest. After registering as a cooperative, the community took loan to establish the CODI to buy the land as well as worked with the young architecture to build about 49 houses and space for community centre [28].

Another example was the Relocation of Mini Squatters and A Long Term Leases. Boon Kok was a new settlement in a central area of the northern Thai City of Uttaradit. It had about 124 households in many 'mini' squatter settlements [28]. The Thai government decided to buy 1.6 hectares site and grant the occupants a 30 years lease. The community network helped out through their daily savings schemes, and CODI offered housing loans to families that needed them, whereas the National Housing Authority provided the infrastructure. The communities needed to pay about USD5 to USD22.50 per month. Moreover, this settlement also included five (5) housing units for the elderly, the poor as well as the physical disabled member among the community [28].

Therefore, it can be concluded from the Baan Mankong Housing Program in Thailand, that the authorities, politicians and other groups in the city should be involved in negotiations with the urban poor group to address the city's housing problems. Moreover, the Thai government and the public support the programs for the poor communities. Hence, we can see that the communities or the people in Thailand are united to resolve the squatter problem. And it is proven that the Thai government encourages the squatters community to get involved with the construction of housing projects, forming their cooperative and take the Community Organisations Development Institute (CODI) loan to pay the land they had purchased or granted them grant for 30 years of lease. Malaysia can follow these strategies in resolving squatters issue where the community are not fully dependent on the government to solve the housing problems.

5 Research Methodology

Based on the background and existing issues, this research, therefore, employs a case study design with a particular interest for Johor Bahru based on its features. Earlier, it was mentioned that the number of squatters in Johor Bahru is very high as compared to other states in Peninsular Malaysia (Table 2). Other than that, Johor Bahru was selected based on its geographical background; a significant urban area in the south of Peninsular Malaysia. The approach of study involved secondary sources which comprises of the literature on squatter settlements and resettlement program. The paper also appraised previous similar study by Ibrahim *et.al* [12] which focuses more on the issues of resettlement program. Future work would involve an in-depth interviews and questionnaire survey among the residents in Johor Bahru, along with site observation.

Table 2: The Significant Number of Squatters in Johor Bahru
Source: Jabatan Perancangan Bandar dan Desa [7].

6 Results and Discussion

No.	State	No. of Families	No. of Family Members	No. of Squatters
1.	**Johor**	**6,363**	**22,151**	**5,922**
2.	Kedah	2,702	13,274	2,702
3.	Kelantan	1,685	7,780	1,649
4.	Melaka	7	19	7
5.	Negeri Sembilan	314	585	239
6.	Pahang	1,134	3,339	935
7.	Perak	1,709	6,836	1,709
8.	Perlis	1,853	8,570	1,853
9.	Pulau Pinang	4,208	18,909	2,875
10.	Sabah	28,087	133,059	26,479
11.	Sarawak	8,085	32,354	7,784
12.	Selangor	2,542	3,299	3,299
13.	Terengganu	1,253	1,583	638
14.	Wilayah Persekutuan Kuala Lumpur	3,217	12,868	3,217
15.	Wilayah Persekutuan Labuan	970	5.521	976
Total		**64,129**	**270,147**	**60,278**

This preliminary analysis was performed based on an early investigation of the case study in Johor Bahru. As mentioned much earlier, this initial results were obtained mainly from the secondary sources.

Table 3: Housing Price Structure and Target Groups for State of Johor.

Category	House Price Per Unit	Target Groups/ Income Per Month
Low Cost	RM25,000	Below RM2,500 per month
Low Medium Cost	RM50,000	RM2,500 – RM3,000 per month
Low Medium Cost	RM80,000 - RM90,000	RM3,000 – RM4,500 per month

6.1 The practice of 'Zero Squatters'

The Johor State Government had actively engaged in many initiatives to eradicate or reduce squatters within their boundary. This practice was seen significant through the commencement of the *Zero Squatters* Policy in 2005 [23]. Amongst the control undertaken were granting land ownership, Temporary Occupation License (TOL), and the shift to low-cost houses (PPR), and enforcement of land laws. In general, the federal government had implemented two programs to eliminate squatters' areas; namely the Zero Squatters program which helps to vacate the squatters' areas; and the Public Housing program for the new homes. Both of these programs were executed by the Johor state government. In fact, land ownership is the most common approach practiced by the state government, with the hope to reduce its expenditure and preventing it from paying compensation and transferring the people to a different area. At least, the government only suffer the de-valued of land.

6.2 New Settlement Through Low Cost Housing Program

Squatters are relocated to a new area through the Low Cost Housing Program which is carried out through the People's Housing Project or widely known as the *Projek Perumahan Rakyat* (PPR). Otherwise, they will be transferred at public housing under the Public Housing Division, the Office of the State Secretary of Johor. In any case, all costs including for construction of a new settlement will be certified by the party intending to develop the area. However, in Johor, the minimum area size for a low-cost housing development must be at least 5 acres. The Johor State Government has inflicted a condition requiring 40% of the total units in several housing scheme to be low cost housing, except in the case of the district of Johor Bahru where the low cost necessity is 30% (due to its high population over 1.0 million people [24] while another 10% low-medium cost houses are required for development as well [25]. This low-cost housing is therefore targeted for the squatters too. The price of a 3-bedroom low-cost house is RM25,000 while a low-medium cost house is RM50,000 (Table 3).

Through the various analysis on low-cost housing price, majority of the occupants of low-cost resettlement housing claimed that the house that they purchased are still expensive to them. This is because construction costs had increased. Generally, the cost of low-cost houses are more than RM30,000 per unit and does include the construction cost of infrastructure yet. Therefore, the situation of squatter's resident ability to buy low-cost housing remain as a major problem.

DOI: 10.1051/matecconf/20166600023

Source: SUKJ Bahagian Perumahan [26].

6.3 Resettlement Program

As part of eradicating poverty and squatters, the state government had introduced numerous programs under the District and Land office, identified as the resettlement program [27]. The government believed that the issue of squatters and poverty has a connection with other social issues and by solving poverty and squatters, the government will indirectly resolve other social issues. The District and Land office will therefore determine new areas (the government's land) that are suitable for a resettlement program. Thereafter, the state government will acquire funding from the Ministry of Rural and Regional Development (MRRD) to build low-cost houses, along with other basic facilities and infrastructure. The eligibility for the low-cost houses is very much depended on their household income and the squatters will definitely be considered for the program. However, there are owners who moved out and rent their units to the outsider or immigrants or even foreigners. This is because they could not afford to buy the low-cost houses.

6.4 Temporary Occupation License

The state government also grants the Temporary Occupation License (TOL) to occupy government's land [27]. The land can be of road or river reserves, or land set aside for future development. However, the squatters will take advantage of the presence of vacant land to build their houses. For that reason, the government allows access to TOL for the squatters. This would, in turn, make them stay legally on government's land. Through this way, the government will not pay any compensation when acquiring back the land, therefore saving cost. Squatters often take advantage of vacant land to build buildings whether for residential or small business.

6.5 Public Housing Program

The National Economic Action Council (NEAC) has formed the Public Housing Program with the cooperation of the Ministry of Housing and Local Government to study on the condition of resettlement areas across the country [23]. Among the preparation is to form the Integrated Housing Program under the New Housing Policy for rental and ownership for the squatters [23]. The principle of this housing program acts as a platform in trying to accomplish the *Zero Squatter* policy. Through public housing, the less fortunate will have the chance to live in a better living condition. In addition, squatter who are involved in the redevelopment of their settlements cannot move in directly to low-cost flats. This is because the process for the redevelopment of squatters is very time-consuming. The residents will move only if low-cost flats are available and that they can buy the property.

7 Recommendation and Conclusion

Even though the Zero Squatters policy had been practiced since 2005 by most states in Peninsular Malaysia, Malaysia is still facing the issue of squatters and therefore introduced various approaches in order to tackle the problem. There are various approaches implemented by the Johor State government namely the practice of 'Zero Squatters' policy, low-cost housing program, resettlement programme, Temporary Occupation License (TOL) as well as Public Housing Program.

The research had also examined a previous related study by Ibrahim *et.al* [12] which discusssed on the resident's current living condition at squatter area, community background and the current living condition in a resettlement area, as well as the issues and challenges encountered by the former squatters in the new resettlement area. Briefly, from the previous related study by Ibrahim *et.al* [12], it demonstrates that the residents of the resettlement area were mainly reluctant to move out from their previous squatter area because they could not afford to purchase the low-cost housing. This finding was considered as important for the future whereby the government needs to consider the price of low-cost units in trying to accommodate the low-income group.

In addition, had also assessed an example of successful Zero Squatters program in Selangor and Thailand's Housing Program which targeted the poor and squatters. In Selangor, the initiative involved numerous strategies where the local authorities, task force unit as well as related responsible parties played a vital role in eradicating the squatter issues. This include the monitoring of low-cost housing projects and performing surveys to squatters, as well as ensuring some private developers will get involved in the development of the low-cost housing. Meanwhile, in Thailand, the approaches were slightly different. The government and the squatter or poor communities worked had in hand to resolve the problem by implementing squatter upgrading program in year 2003 through the Community Organisation Development Institute (CODI). Nevertheless, both Malaysia and Thailand's experience was considerably successful with the reduced number of squatters.

Based on the findings from the preliminary work of the case study, the study unveils the need to improve the current practice in Johor Bharu through several approaches as described below:

a. To continue with the Zero Squatters Initiative which stated in 2005. Even though the aim was not fully achieved but tio some extent it had reduced the number of squatters. For that reson, the target for year 2020 is highly recommended.

b. The state government's effort in building low-cost houses in Johor Bahru is part of its effort in assisting the low-income group, but most importantly to eliminate squatters in Johor Bahru. However, it was clearly revealed that the price of low-cost units are still beyond the squatters' availability or affordability. Therefore, the government should re-examine the current price and reconsider the need to reduce the price of low-cost housing.

c. The analysis showed the occurance of incidents where the residents of resettlement area rent their units to other people including illegal immigrants. This calls for a strict reinforcement by the authorities.

d. The practice of Temporary Occupation License (TOL) is a good exercise by the state government as it helps to overcome squatters through legal occupation of government land.

e. The state government of Johor can follow the practice of Thailand and Selangor in eradicating squatters. The Thai government allows the poor and squatters communities to be involved in the planning and provision of housing program for the poor. Meanwhile Selangor uses its task force to assist the government to eliminate squatters. These particular efforts can be a guidance to Johor Bahru.

From the overview and early conception of initiatives to eliminate squatters, the researcher will consequently undertake a methodical investigation on the proposed case study, i.e, a resettlement program in Johor Bahru. Further work in Johor Bahru will look at the delivery of the Johor Bahru resettlement program in order to enhance the initiative, and involve a survey among the community of the resettlement program, for the purpose of understanding in detail their experience and what can be improved in the future.

Acknowledgement

The authors would like to thank the Universiti Teknologi MARA (UiTM) for supporting this study and the Centre of Postgraduate Studies, FSPU.

References

1. S.Z. Kader, *Eviction of Unlawful Occupiers of Land in Malaysia_Judical Responses and Policy.* International Islamic University Malaysia (2011).
2. K.A. Mohamad, *Kemiskinan Bandar: Penswastaan Kawasan Setinggan dan Peminggiran Sosial.* Available at(http://www.pengurusanislam.tripod.com/kemiskaina n_bandar.htm), (2010).
3. UN-HABITAT, G. L. T. N. *Handling Land–Innovative Tools for Land Governance and Secure Tenure. Nairobi,* (2012)
4. M. Ariff, and H. Davies, *Sustainable Living Environment for Urban low-income households in Malaysia: Key factors for maintenance.* in CIBW 107,2009. Proceedings of the 2009 International Symposium on Construction in Developing Economies: Commonalities among diversities, CIBW, (Penang, Malaysia), (2009)
5. A. Ismail, A. A. Bujang, W. R. Anthony Jiram, H. Abu Zarin, and M. N. Jaafar, *Housing Financing Faculity and Affordability Level of Bumiputera within Iskandar Malaysia.* Journal of Economics, Business and Management, Vol. 3, No.9, September (2015)
6. M.T. Taher, , and A. Ibrahim, *"Transportation of Slum and Squatter Settlements: A way of Sustainable Living in Context of 21st Century Cities."* American Journal of Civil Engineering and Architecture 2, No.2 (2014): 70-76. (2014)
7. Jabatan Perancangan Bandar dan Desa. Laporan Kesejahteraan Rakyat. Selangor: Jabatan Perancang Bandar dan Desa. (2013)
8. Daily Express. *Informal Housing Booming Due to '0' Squatters drive failure.* Independent National Newspaper of East Malaysia. Established Since 1963. Published on: Monday, July 27, (2015). Retrieved: 01/11/2015.
9. H. Wan. *Kualiti hidup sosial dan proses perbandaran di Malaysia.*Isu-isu alam sekitar manusia di Malaysia, Universiti Pendidikan Sultan Idris, Perak.pp. 19 – 42. (2010)
10. A.Z. Asri. *Squatter Houses in the Area of Johor Bahru.* Thesis (Sarjana Kejuruteraan Awam-Pengurusan Pembinaan), Faculty of Civil Engineering, Universiti Teknologi Malaysia (2013)
11. S. Shuid. *Low Income Housing Allocation System in Malaysia. Managing Housing Need for the Poor.* Paper Presented at 22nd International Housing Research Conference. Istanbul (2010)
12. I. Ibrahim, N.A. Aziz, F. Hussain, A. Noraini. *Issues and Policies on Aftermath of Squatters Resettlement: A Case Study on Kg. Desa Hormat Tambahan, Selangor* (2012)
13. Utusan Malaysia. *"Kekecohan Pembelian PPR".* Berita dalam akhbar Utusan Malaysia. Kuala Lumpur, Malaysia: 16 November (2009)
14. A.H. Hashim, *Residential Satisfaction and Social Integration in Public Low Cost Housing in Malaysia.* Department of Resource Management and Consumer Studies. Faculty of Human Ecology. Universiti Putra Malaysia (2003)
15. A. Sufian, and N.A. Mohamad, *Squatters and Affordable Houses in Urban Areas: Law and Policy in Malaysia.* International Islamic University Malaysia (2009)
16. P. Wakely, *'Land tenure in under-served settlements in Colombo'*, IDRC Urban Poverty and Environment Report Series 6, Ottawa: IDRC. (2008)
17. UN-HABITAT. *Sustainable Urbanisation.*Achieving Agenda 21. Department for International Development (2002)
18. F.A. Aziz, *The Investigation of the Implication of Squatter Relocations in High Risk Neighbourhoods in Malaysia.* School of Architecture, Planning and Landscape, Newcastle University (2012)
19. United Republic of Tanzania. *Resettlement Policy Framework.* Prime Minister's office-Regional Administration and Local Government. Available at: http://www.pmoralg.go.t2/menu_data/programmes/D M. (2012)
20. International Finance Corporation. *Performance Standard5, Land Acquisition and Involuntary Resettlement.* World Bank Group (2012)
21. Leopoldo, *et.al,. Displacement, Resettlement, Rehabilitation, Reparation and Development*, WCD

Thematic Review 1-3 Prepared as an input to the world Commission on Dams, Cape Town, www.dams.org. (2000)

22.UN-HABITAT. *State of the World's Cities Report*. UN-HABITAT, Nairobi. (2010)

23. A. Atif. *Masalah Setinggan di Johor Bahru: Masalah yang diundang*. Available at (http://www.slideshare.net/ukhuwah_90/masalah-setinggan-bandar-Johor-Bahru. (Retrieved: 30/11/2015). (2012)

24.Iskandar Regional Development Authority. Portal Rasmi IRDA (2014)

25.R. Zain. *Setinggan dan Kemiskinan*. Available at http://www.documents.tips/documents/setinggan-dan-kemiskinan-html.(Retrieved: 20/12/2015). (2015)

26.SUKJ Bahagian Perumahan. Available at: http://www.erumah.johor.gov.my/spj/home/hmsyarat_permohonan.htm.(Accessed:12/12/2015). (2014)

27. H.M. Salehan, *Kajian Terhadap Tindakan Kerajaan Negeri Johor Dalam Menangani Penempatan Semula Setinggan di Negeri Johor. Kajian Kes: Kampung Sri Rahmat, Johor Bahru*. Fakulti Kejuruteraan dan Sains Geoinformasi. Universiti Teknologi Malaysia (2007).

28.S.Boonyabancha. *Baan Mankong: Going to Scale with "Slum" and squatter upgrading in Thailand*. Environment and Urbanisation Vol.7. Downloaded from eau.sagepub.com. (2005). Retrieved: 9 February, 2016. (2005)

MATEC Web of Conferences **66**, 00024 (2016) DOI: 10.1051/matecconf/20166600024
IBCC 2016

BIM for existing facilities: feasibility of spectral image integration to 3D point cloud data

Kinjiro Amano[1,a] and Eric Lou[1]

[1]*School of Mechanical, Aeroplane and Civil Engineering, University of Manchester, Sackville Street, Manchester M13 9PL, UK*

Abstract. Accurate geometrical and spatial information of the built environment can be accurately acquired and the resulting 3D point cloud data is required to be processed to construct the digital model, Building Information Modelling (BIM) for existing facilities. Point cloud by laser scanning over the buildings and facilities has been commonly used, but the data requires external information so that any objects and materials can be correctly identified and classified. A number of advanced data processing methods have been developed, such as the use of colour information to attach semantic information. However, the accuracy of colour information depends largely on the scene environment where the image is acquired. The limited number of spectral channels on conventional RGB camera often fails to extract important information about surface material, despite spectral surface reflectance can represent a signature of the material. Hyperspectral imaging can, instead, provide precise representation of spatial and spectral information. By implementing such information to 3D point cloud, the efficiency of material detection and classification in BIM should be significantly improved. In this work, the feasibility of the image integration and discuss practical difficulties in the development.

1 Introduction

The UK will be required to refurbish or retrofit 93% of existing buildings by 2050 to meet national carbon targets – this includes 25 million homes and 3 million non-domestic facilities [1]. Refurbishing and retrofitting (RR) facilities will be the major work for the UK construction industry in the next 20 years. The main issue of RR is the unavailability or inaccurate building drawings, plans or blueprints [2]. Without these, the RR project is open to a myriad of engineering unknowns and risks, causing additional concerns to health and safety, financial uncertainties and hazards. To optimise RR, any project team must be able to work seamlessly with accurate building information (e.g. plans, material) on a digital and common platform called Building Information Modelling (BIM) [3, 4]. BIM is a digital representation of the physical and functional characteristics of a buildings or structures (hereby referred to as facility). It is aimed to serve as a shared knowledge resource for information about a facility, as well as forming a reliable basis for decisions during the facility's lifecycle, from inception onward. BIM has been becoming essential tool for the Architecture, Engineering, Construction and Facility Management communities. The use of BIM is also mandated for UK Government projects by 2016 [5].

BIM allows accredited project team members to add, edit, delete, and share real time information and design specification on a common technology. BIM provides multi-dimensional information of any facility; identify design collision between the multidisciplinary project team; check for accessibility compliances; capture design and material information; building regulation compliance; supply chain integration [6-8].

Among these information, geometrical information of the facilities and objects, and their physical properties must be documented accurately so that the digitised information can be linked together on a common standard platform.

Technological advancements have made it possible to generate 3D models to assess as-built conditions, and to verify the progress of the project and contract specifications. Modelling can be used to capture the as-built or as-is reality of any project and facility. The modelling consists of collecting infrastructure's spatial data as a series of 3D Cartesian coordinates. The collection of the dense Cartesian-based distance data set is known as point cloud, and these points are processed and transformed into a structure or an object by purposely designed software [e.g. 9] associated with specific database [e.g. 10].

The 3D laser scanning and photogrammetry have been most commonly used to acquire the point cloud data [11, 12]. Developments in both hardware and software have been underway to make them more reliable, compact, and affordable, but there are still certain limits in the implementation of the 3D point cloud to BIM. One of the critical issues is to do with performing identification and classification of objects and materials in the 3D point cloud.

[a] Corresponding author: kinjiro.amano@manchester.ac.uk

The point cloud data does not contain any sematic information but merely geometrical coordinates and intensity values per point. This problem can be addressed by combining external information, such as visual information from the other sources, to the point cloud. The most common method is the integration with snapshot photographs or colour images taken independently of the laser scanning but simultaneously on site [13-15]. One of various developments in the utilisation of colour information processes colour images with sophisticated algorithm along with specifically designed image database to identify the construction materials [16].

A concern is, however, the quality of the colour information by conventional RGB camera depends on the environment or imaging condition where the images are being taken. Colour appearance of the surfaces, for example, varies with the scene illumination. In addition, due to the limited number of the spectral channels in the imaging system, any complex spectral signature of materials cannot be accurately extracted. These disadvantages lead to failures in the identification and classification.

Since the identification of the materials can be provided by their spectral signature, it would be ideal if the spectral imaging data is integrated to 3D point cloud data. With the current technology, however, the hyperspectral imaging is available in two dimensions. A challenge is, therefore, to develop methodology to integrate the 2D spectral image to the 3D point cloud. One of feasible and convenient methods is to project the 3D point cloud to 2D and apply the image registration methods between 2D images. In this report, the feasibility of this integration to the hyperspectral data is addressed, and the advantages and disadvantages are to be discussed.

2 Colour image for material identification

The laser scanning is one of the most common methods to capture geospatial information in the built-environment and facilities accurately and efficiently. Photogrammetry can be used complementary because of the similarities in output data [11].

The point cloud, as noted earlier, only contains Cartesian coordinate and intensity values at the point, and does not carry any semantic information, such as knowing which point belongs to what as-built structural component. Without such attributes or context, it is almost impossible to judge where the point is belonged to and to examine any relationship between nearby points. In order to identify and classify the materials and objects in the facilities or scenes, it is necessary to attach additional semantic information to the point cloud. Working with such featureless data to explore geometric reasoning are generally tedious, time consuming, often produce errors [17].

The semantic information or context of the scene can be extracted from independent imaging systems, pre-examined data, or post image processing [18, 19]. A number of computational algorithms have been also reported to provide better solutions on these limitations [e.g. 13, 14, 16, 18].

Colour information and photographic data often provides useful information about the objects in the facilities or scenes, and even what the materials are. The colour images taken by conventional RGB camera have an advantage to be collected quickly and inexpensively. For example, Dimitrov et al [14] presented a new algorithm based on Structure-from-Motion (SfM) combined with image statistical analysis. Hang et al [16] proposed a manual integration of 3D point cloud and 2D image patches to facilitate 4D BIM with the site photo-log and SfM accompanied with specially designed image database.

However, the colour imaging by a conventional RGB camera has certain limitations. That is, the quality of the colour largely defined by the environmental condition where the image is being acquired. Colour appearance of the surfaces, for example, is influenced by the scene illumination.

Figure 1 shows an example of misrepresentation of colours. Colour images were acquired by a build-in RGB camera in a laser scanner (Fig. 1a), and by a conventional RGB camera (Fig. 1b). Despite the materials of the all surfaces are the same, top surface has a different appearance from the side surfaces (Fig. 1a) and brightness of the surfaces are different between the images (Fig. 1 a and b). Further, due to the limited number of the spectral channels over spectrum (three in the conventional RGB camera and human eye), such imaging device cannot always produce correct colour appearance of some surfaces which has complex spectral properties are, known as metameric surfaces [20].

(a) (b)

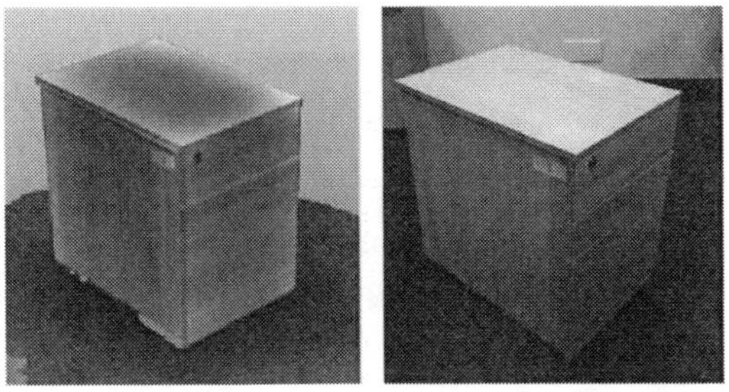

Figure 1. Examples of misrepresentation of colour appearance. Colour images taken by (a) the built-in RGB camera in the laser scanner and by (b) an external conventional RGB camera.

DOI: 10.1051/matecconf/20166600024

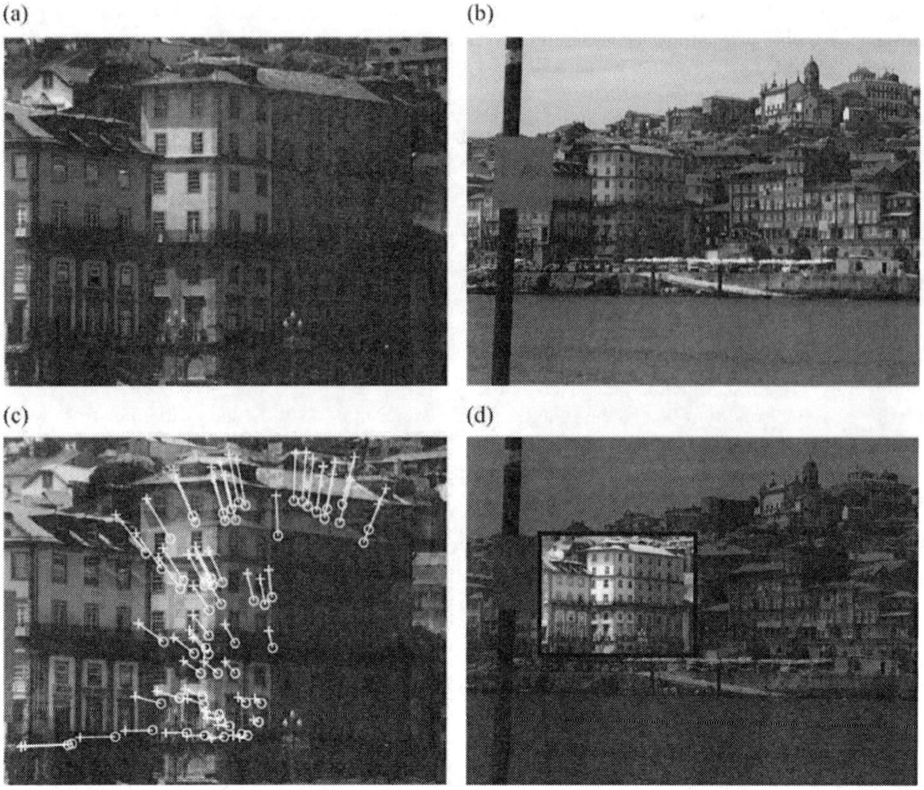

Figure 2. The image registration example between two images acquired by different conditions (in spatial resolution, viewing distance and viewing). (a) Colour rendering image based on the hyperspectral imaging in a natural scene. (b) Colour image of the same scene with a conventional RGB camera. (c) Detection of the SURF (Speeded-Up Robust Features) image key features and image transformation. Circles and crosses are the features from each of the images in (a) and (b), respectively. (d) Result of image registration. For visibility a black frame was inserted and the background was converted to grey scale.

3 Advanced imaging and integration

3.1 Spectral information

Characteristics of materials can be represented by spectral properties. Estimation of surface reflectances enables to identify the spectral characteristics of materials independently of scene illuminations and environment. The more spectral channels are available, the more accurate estimation is available. It is therefore reasonable to consider the integration of the hyperspectral imaging data to 3D point cloud.

Spectral characteristics of urban construction materials, such as an aging effect on concrete and clay tiles, have been examined to establish spectral library for the BIM [21]. It has been also the case for the spectral characteristics of asphalt road aging and deterioration [22]. These data provide the important evidence that the spectral characteristics and their examination need to be taken into account to identify the materials, and consequently to be implemented in the BIM protocol. They also indicate the requirement of a careful consideration to determine which range of spectra should be examined.

3.2 Hyperspectral imaging

The hyperspectral imaging has evolved to include not just one monochromatic channel or three colour channels covering the visible spectrum, but many channels, over 30, even several hundreds, extending from the visible spectrum to near-infrared (NIR) and shortwave infrared (SWIR) channels [20, 23].

This imaging method is aimed mainly to exploit the materials comprising the various objects in a scene where lights reflect, scatter, and being absorbed, depending on characteristic of material in both chemical and physical properties.

If the number of spectral channels sufficiently covers a wider spectral range and each is narrowly tuned, the information on each pixel can be used in a detailed analysis to gain accurate identification of objects and materials against different environmental exposure. Such assessment is unavailable by human eyes or the RGB camera with the three channels.

In the recent years, demands to the hyperspectral imaging has grown significantly because of its increasing applications in a variety of research area, for example, agriculture [24] and cultural heritage [25]. The hyperspectral imaging is currently limited to two dimensions and requires specific spectral calibration at each image acquisition. Each spectral channel has different focal length. This causes chromatic fringes in the resulting images, unless any spectral correction is made by performing post-image processing (e.g. the image registration across the spectral channels) or by using any optical devices in the imaging system (e.g. an achromatising lens).

3.3 Image registration methods

One of the unique features of using the 3D point cloud data is the ability to manipulate and view data in a full degree of freedom environment. That is, users can interact and manipulate the dense point data flexibly (for example, users can observe the data from different distance and viewing angles as shown in Fig. 3a), allowing for construction of as-built conditions in a virtual environment [26]. Such flexibility may provide a feasibility of integrating the point cloud to the other digital imaging. That is, it is possible to adjust the viewing angle so that the projection of the 3D image can closely correspond to the external 2D image. The projected 2D image can be registered to the external 2D image to be integrated. It enables to extract or attach additional information.

Once both image data from the laser scanning and hyperspectral imaging are on the same image dimension, the integration will be performed by applying the image registration methods established in computational vision. Common technique in the image registration is based on the detection of image key features and their correspondences between the images. The key features are, for example, intensity contrast, gradients, edge, and surface normal. Several algorithms have been proposed to detect the key features; Binary Robust Invariant Scalable Keypoints (BRISK) [27]; Speeded-Up Robust Features (SURF) [28]; Maximally Stable Extremal Regions (MSER) [29]; Fast Retina Keypoint (FREAK) [30]. The combination of these methods is also applicable. Instead of using computational key features, it is also possible to use reference objects or markers physically placed in the scenes. These image registration algorithms are applied to the 3D point cloud as well as 2D images, when multiple sets of scanning (point clouds) are to be registered [31].

Figure 2 represents an example of the image registration between two images taken by the different imaging systems, from different viewing distances and angles, with applying the SURF algorithm [28] to compute the image transformation (Fig. 2c). The goodness of the image registration is controlled by adjusting optimisation parameters, such as the number of iterations and step size. The final registered image is shown in Fig. 2d.

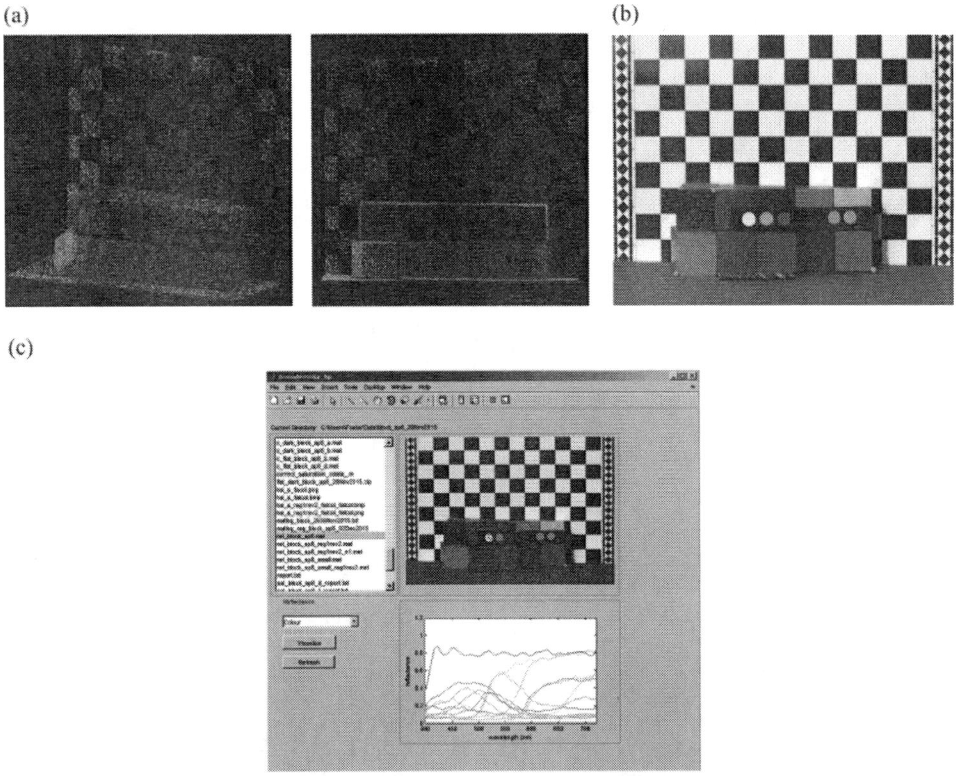

Figure 3. Proposed image integration between 3D point cloud and the hyperspectral image. (a) Point cloud with adjustable viewing angle; (b) Colour image rendering of the hyperspectral data. (c) Example GUI to be implemented to monitor spectral profile on each pixel selected.

If the image integration is completed, the spectral properties obtained by the hyperspectral data (as shown in Fig. 3c) can be integrated to point cloud. The integration between 3D point cloud and spectral images have been reported in the other research disciplines, for example, characterisation of plants and plant disease detection [24, 32] and examination of art-objects [25], any of which have been performed under well-controlled experimental environment.

3.4 Considerations

This proposed method is not difficult to implement, but it should be noted that there are certain disadvantages.

The projection to 2D loses the advantage of having 3D information, such as depth and occlusion. Any of the image information on the invisible surfaces cannot be obtained. It will be problematic if any significance (e.g. damage) may exist on those surfaces. The 3D information also provides useful information about the effect of scene illumination and shadows, which are often important in designing and retrofitting of the built-environment and facilities. It is, therefore, a common practice to identify what image information will be required and examine the geometrical correspondence between different imaging techniques in advance of performing the image acquisitions.

Difference in spatial resolution of the imaging devices may cause inaccuracy in the image registration. Advanced image processing tools may be applied to achieve sufficient interpolation or extrapolation to overcome this issue.

In terms of spectral image data, as noted earlier, physical characteristics of light must be taken into account, namely, chromatic aberration. The fact that each spectral channel (light of different wavelength) has different focal length causes "colour fringe" at the edge of surfaces in the image. The correction of the chromatic aberration is important because edges of surfaces and objects provide essential information in their identification and classification. The correction is available at the time of image acquisition using specially designed optical devices (e.g. an achromatising lens) or in the post image processing (e.g. the image registration across the spectral channels). As a note, there has been a report that the spectral edges or gradient of the spectral map provide clue to achieve image fusion [33].

There is a technical disadvantage in the image registration by image key features. That is, the detection of any key features depends on the image contents. If the scene is consisted of, for example, a large uniform surface or walls, it is difficult to detect any key features. Alternative registration algorithm, such as normal cross correlation may be considered.

Apart from the proposed image integration method, the computational image fusion methods between 3D point cloud and 2D colour image have been reported [34], where the characterisation of the imaging devices is implemented. The camera calibration is considered as a necessary step in the 3D computer vision to extract spatial and geometrical information from 2D images [35, 36]. It is the process of estimating intrinsic and extrinsic parameters. Intrinsic parameters deal with the internal characteristics of the imaging device, such as, focal length, skew, distortion, and image centre. Extrinsic parameters describe its position and orientation in the world coordinate. Knowing these parameters is essential, as it allows estimating the scene's structure in Euclidean space and removes lens distortion, which degrades accuracy. Such characterisation of the devices should be performed ahead of the imaging practice. It still needs, however, to examine exactly how the correction works under practical and uncontrolled environments.

4 Conclusions

Refurbishment and retrofit (RR) projects will be the main work in the UK in order to achieve the agreed carbon emission targets. RR projects is not plagued with risks and uncertainties, which considerably rises costs, health and safety, and lacked environmental and carbon emission consideration. It has been considered that BIM for RR projects is the way forward but has not been fully practiced in industries yet. The 3D scanning of existing facilities to obtain spatial information has been introduced and the demand has been increasingly grown. Taking a step further, the present report considered a possible method of the image integration between the 3D point cloud and 2D hyperspectral imaging. As discussed, there are considerable difficulties at this stage of the development. Some of the disadvantages will be overcome by careful pre-planning of the imaging protocol. Currently computational cost for the point cloud and the spectral data is very demanding. Developments in sophisticated image processing tools are also required to extract all of the relevant information from the multitude of the data. However, by achieving the image integration between 3D and 2D, the identification and classification of the construction materials will become more efficient in terms of both labour and cost. This approach will add another dimension to the BIM.

Acknowledgement

This work was supported by British Council Newton-Ungku Omar Fund (No.172726659).

References

1. J. Edwards and A. Townsend. Buildings under refurbishment and retrofit, Carbon Action 2050 White papers, The Chartered Institute of Building (2011)
2. C. Gorse and D. Highfield *Refurbishment and Upgrading of Buildings*. Abingdon, UK, Spon Press (2009)
3. D. Bryde, M. Broquetas and J. M. Volm, International Journal of Project Management **31**, 7 (2013)
4. BIS. Building information modelling, Department for Business, Innovation & Skills, HMGovernment, UK (2012)

5. Cabinet Office. Government Construction Strategy, Cabinet Office, UK (2011)
6. R. Howard and B.-C. Björk, Adv. Eng. Inform. **22**, 2 (2008)
7. Y. Arayici, Structural Survey **26**, 3 (2008)
8. K. Barlish and K. Sullivan, Autom. Constr. **24** (2012)
9. Autodesk. Building Information Modeling for Sustainable Design Autodesk (2005)
10. RIBA Enterprises. NBS National BIM Library (2016)
11. S. El-Omari and O. Moselhi, Autom. Constr. **18**, 1 (2008)
12. R. Volk, J. Stengel and F. Schultmann, Autom. Constr. **38** (2014)
13. M. Golparvar-Fard, J. Bohn, J. Teizer, S. Savarese and F. Peña-Mora, Autom. Constr. **20**, 8 (2011)
14. A. Dimitrov and M. Golparvar-Fard, Adv. Eng. Inform. **28**, 1 (2014)
15. X. Zhang, N. Bakis, T. C. Lukins, Y. M. Ibrahim, S. Wu, M. Kagioglou, G. Aouad, A. P. Kaka and E. Trucco, Autom. Constr. **18**, 3 (2009)
16. K. K. Han and M. Golparvar-Fard, Autom. Constr. **53** (2015)
17. S. Kiziltas, A. Burcu, E. Ergen and T. Pingbo, ITcon **13**, Special Issue Sensors in Construction and Infrastructure Management (2008)
18. J. Jung, S. Hong, S. Jeong, S. Kim, H. Cho, S. Hong and J. Heo, Autom. Constr. **42** (2014)
19. J. Yang, M.-W. Park, P. A. Vela and M. Golparvar-Fard, Adv. Eng. Inform. **29**, 2 (2015)
20. D. H. Foster, K. Amano, S. M. C. Nascimento and M. J. Foster, J. Opt. Soc. Am. A Opt. Image Sci. Vis. **23**, 10 (2006)
21. N. E. M. Nasarudin and H. Shafri, Journal of Urban and Environmental Engineering **5**, 1 (2011)
22. M. Herold and D. Roberts, Appl. Opt. **44**, 20 (2005)
23. V. C. Coffey, Opt. Photon. News **26**, 10 (2015)
24. J. Behmann, J. Steinrücken and L. Plümer, ISPRS-J. Photogramm. Remote Sens. **93** (2014)
25. L. Granero-Montagud, C. Portalés, B. Pastor-Carbonell, E. Ribes-Gómez, A. Gutiérrez-Lucas, V. Tornari, V. Papadakis, R. M. Groves, B. Sirmacek, A. Bonazza, I. Ozga, J. Vermeiren, K. van der Zanden, M. Föster, P. Aswendt, A. Borreman, J. D. Ward, A. Cardoso, L. Aguiar, F. Alves, P. Ropret, J. M. Luzón-Nogué and C. Dietz, Proc. SPIE 8790, Optics for Arts, Architecture, and Archaeology IV, 879011 (2013)
26. E. J. Jaselskis, Z. Gao, A. Welch and D. O'Brien, Proceedings the Mid-Continent Transportation Research Symposium Iowa State University, Ames (2003)
27. S. Leutenegger, M. Chli and R. Y. Siegwart, Computer Vision (ICCV), 2011 IEEE International Conference on (2011)
28. H. Bay, A. Ess, T. Tuytelaars and L. Van Gool, Comput. Vis. Image Underst. **110**, 3 (2008)
29. J. Matas, O. Chum, M. Urban and T. Pajdla, The British Machine Vision Conference, BMVA Press (2002)
30. A. Alahi, R. Ortiz and P. Vandergheynst, Computer Vision and Pattern Recognition (CVPR), 2012 IEEE Conference on (2012)
31. R. B. Rusu, Z. C. Marton, N. Blodow, M. Dolha and M. Beetz, Robot. Auton. Syst. **56**, 11 (2008)
32. R. A. Schultz, T. Nielsen, J. R. Zavaleta, R. Runch, R. Wyatt and H. R. Garner, Cytometry **43**, 4 (2001)
33. D. Connah, M. S. Drew and G. D. Finlayson, J. Opt. Soc. Am. A Opt. Image Sci. Vis. **32**, 12 (2015)
34. L. Tamas and Z. Kato, Computer Vision Workshops (ICCVW), 2013 IEEE International Conference on (2013)
35. M. M. Darrodi, G. Finlayson, T. Goodman and M. Mackiewicz, J. Opt. Soc. Am. A Opt. Image Sci. Vis. **32**, 3 (2015)
36. Z. Zhengyou, ITPAM **22**, 11 (2000)

The Missing Point Of Knowledge Management in PFI Projects

Kumalasari Kipli[1], Assoc. Prof. Sr. Dr. Fadhlin Abdullah[2] & Dr Fara Diva Mustafa[3]

[1] *Faculty of Built Environment, UTM, Johor & Faculty of Architecture, Planning & Surveying, UiTM Sarawak*
[2,3] *Faculty of Built Environment, UTM, Johor*

Abstract. The Private Finance Initiative (PFI) procurement model is a complex system which includes long-term commitment; customers' satisfaction and through-life collaborative working. The projects based on this type of procurement has start from the year 1992 in the UK and year 2006 in Malaysia. Along the process, the PFI projects face few problems despite its advantages. These problem has effected the projects performance such as delay and cost overruns. It seems that the problems occurred are repeatable despite many types of research delved into solving the issues. This paper seeks, to find out the most common problems occur in PFI projects and highlights the best solution to tackle the problems. It also looks at whether similar problems occur in Malaysian construction industry. This research has summarised and critically review the literature on the repeated issues and problem in PFI projects which effect the performance of the projects in term of time, quality and money. It was found that similar problems occur and that similar suggestions were proposed. This shows that the proposed solutions have weaknesses somewhat and hence need to enhance further. This research highlighted the potential of incorporating knowledge management concept through its elements to address the problems. Finally, the paper proposes recommendations to improve the performance of PFI projects with the enhancement of knowledge management especially at the initial and planning stage up to bidding stage where the problem reviewed in the literature always occurred

Keywords: PFI Issues, Importance of KM, Performance of PFI

1 Introduction

A construction project is a complex process that integrates interests of a large number of stakeholders towards the eventual goal of realizing a constructed facility (Oyegoke et al. 2009; Brown et al. 2001). A construction project is organized through varied linkages throughout its life cycle (Bower 2003). Procurement links the highly fragmented supply side of the construction industry (e.g., engineers, architects, contractors, builders, surveyors, suppliers, and laborers) with the less fragmented demand side (e.g., project owners and representatives) (Cox and Townsend 1998). There is various procurement use in the construction industry; these includes conventional, design and build, negotiation, partnering, public, private partnership and Private Finance Initiative (Lædre, et al., 2006). Originated in England in the year 1992 under the United Kingdom's Tory-led government of John Major (Williams, 2005), PFI is a type of Public_Private Partnerships (PPP) (Akintoye, et al. 2001). The rationale of PFI is to combine the resources of the public and private sectors to provide more efficient public services. The Private Finance Initiative is seen as one of the best options for the government to provide facilities and services to the public without waiting longer for the fund Construction Industry Council, 1998). In a PFI procurement, the private sector partner has sole responsibility to get ready the public service such as school or hospital and involved form the bidding stage until construction, in use stage until hand over the back stage. They will also be responsible for the maintenance during in-use stage (Akintola A., Mathias B., 2009).

One reason for the introduction of PFI in Malaysia in the year 2006 is to get the Value for Money in construction projects (Takim et al., 2009). The PFI projects have been successfully implemented in other countries in the world (Takim, R. et al.,2009) for example, like reckons that United Kingdom, Japan, Italy, France, Germany, Australia and USA (Serco, 2007). Starting from 2006, the government in their Ninth Malaysian Plan started to use Public Funding Initiative (PFI) as one of their way of procuring public projects. When the government first announced it, RM20 billion was allocated for delivering public projects in various sectors such as transport, housing, health care and education projects via PPP (Ninth Malaysian Plan, 2006) which includes 425 projects. Out of that 425 projects, 357 projects are under the Ministry of Education as shown in Table 1.

As in the year 2013, the government is undertaking 52 projects, with an estimated value of 62.7 billion, which are in the construction stage and some already in the operation stage. The government plans to build seven toll highways, five more Universiti Teknologi MARA branch campuses, the Integrated Transport Terminal in Gombak, privatization of Penang Port and redevelopment of Angkasapuri Complex, Kuala Lumpur as Media City (Tenth Malaysia Plan, 2010). Public Private Partnership Department or Unit Kerjasama Awam Swasta (UKAS) up to December 2014 has completed 698 projects including PFI projects.

2 Methodology

The research was conducted using an analysis of literature reviews where all the selected journal papers, theses, reports related with PFI had been review an analysed to find out the issues and problems highlighted by the writers. The problems and issues arise then grouped into categories. From each category then we have identified the pattern of problems and issues and find out the main problems from each category. Issues were then analysed in relation to the usage of knowledge management in their solution of the problems

3 PFI Nature

PFI projects involve long-term collaboration and networking across different professional groups such as architects, planners, engineers, surveyors, lawyers, financial specialists, facilities management managers and other team members. Each party has their specific functional relationship in the PFI projects to deliver services according to the public sector clients' output specification (Robinson, H.S., 2009). Each party carries with them not only their specific job description in the project but also tacit knowledge as the result of their action and not all these tacit knowledge were capture, restore, sharing and transferred for future usage. Usually, it is embedded with the completion of the projects. Here come the important of knowledge management.

Human activity cannot be done smoothly without knowledge. Knowledge scope itself are as wide and varied all the varieties of human pursuits. No human civilisation could have existed without creating, accumulation, sharing and applying knowledge (Chimay J.A. et al., 2005). The economic value of knowledge has been discussing for centuries, from the ancient Greeks to Adam Smith and Alfred Marshall, who in 1890 wrote: 'Capital consist in a great part of knowledge and organisation... Knowledge is our most powerful engine in production' (Marshall, 1972). Windrum et al. (1997) and Den Hertog and Bilderbeek (1998) identified design, architecture, surveying and other construction services as knowledge-intensive service sectors.

In order to perform the duty at their best, skills are important as it is their knowledge and experiences (RICS, 2003). Abdullahi A. Umar, et al. (2014) has come out with a list of important skills needed for practitioners dealing with PFI projects. These skills include Contract design skills, Contract Management skills, and Risk Identification and Management skills. For a very complex project such as PFI, the capability refers to the firm's ability to answer a public invitation to tender by drawing up a competitive bid with a high probability of award, on short notice and while minimizing upfront cost. According to Davies and Brady (2000, p. 31), these capability includes the following activities:

Table 1 PFI projects under Ninth Malaysian Plan

<table>
<tr><td rowspan="4">425 PFI Projects

(based on Ninth Malaysia Plan)</td><td rowspan="2">357 projects under Ministry of Education</td><td>180 primary schools</td></tr>
<tr><td>177 secondary schools</td></tr>
<tr><td colspan="2">12 UiTM projects</td></tr>
<tr><td colspan="2">2 Youth Training Institution (Institut Latihan Belia)</td></tr>
</table>

- Ability to collect, compile and extract the customer documents after an invitation to tender received from the customer;
- designing a conceptual design with detail components in the proposed construction products;
- estimation of costs, taking into account many factors (e.g., the quality, reliability and cost of components supplied internally or from different suppliers; project management methodology; productivity, etc.);
- defining levels of service;
- risk management;
- scheduling of project work programs;
- choosing of subcontractors;
- preparing the tender document (including contract documents) by integrating information determined in the previous steps.

The above capabilities reflect that the bidding process for PFI projects requires specialized knowledge from, among others, architects, design engineers, cost consultants, tax consultants, lawyers, facility managers and project managers. To integrate such complementary sets of specialized knowledge, an understanding common to all participants regarding the basic categories and building blocks of the bid (e.g., specifications of the projected buildings, scope of services rendered, level of detail of description) as well as regarding the generic process of bid preparation (e.g., standard operating procedures and "who needs to meet with whom to discuss what and at what stage" Cacciatori, 2008, p. 1595) is paramount.

Managing the knowledge during and after the project implementation is a key challenge posed to construction organisation (Udeaja et al., 2008). The construction organisational competitive advantages depend on the capability to learn and adapt faster than its rivals. This capability is strongly dependent on the ability of the organisation to create, acquire, store and use experiences, knowledge, skills and lesson learnt in the past projects that have the potential to enhance the organisational's future performance (Nonnaka and Von Krogh, 2009). According to Hsiah, et al. (2009) knowledge is vital for effective project implementation, choosing the very best projects, winning bids and also for rganizational efficiency. The 1990s interest in KM initially triggered the response that it was about doing things that were new. More recently many organisations have come to realise that it is also a process of discovery. That is, discovering processes that already happen, such as the sharing of knowledge among communitiest of practitioners, that were rarely if ever labelled 'knowledge management' (Chimay J.A. et al., 2005).

PFI schemes require both broad PFI expertise and specialized knowledge in particular fields. One difficulty, according to some consultants, concerns the fact that professional advice is expensive, and the fee-budget can- not be accurately predicted. Subsequently, reaching an agreement regarding the fees may also be problematic (A. Akintote, 2003). PFI seldom used by the client organisation, so they have only a few staff that understand the intricacies of PFI. In contrast, some private sector organisations have been involved with several PFI projects and, therefore, have significantly more experience (Robinson, et al., 2004)

A common problem associated with PFI, which is also related to the lack of knowledge are a delay on client's decision where clients need more time to make decision and longer negotiation period due to lack of PFI experts, and this makes the schedule complete the projects are very tight. Furthermore, because its large-scale involved schemes, high levels of investment and risks involved and limited knowledge transfer between PFI projects (Carillo et. Al., 2006). Knowledge management in PFI then should be seen as one of the considerations in PFI projects due to its benefits to the projects (Robinson, et al., 2010).

4 Analytical Review of Issues and Problem in PFI

The PFI procurement model is a complex system (Sundaraj & Eaton, 2013). Still not reaching its tenth years of implementation which started in the year 2006 (Ninth Malaysian Plan 2006), PFI is considered new in Malaysia compared to a country like the United Kingdom, which has started PFI since the year 1992. The eagerness of using PFI procurement for nowadays projects may be influenced by the benefits that said to be gain once using these procurements which are: 1) Time savings which is shorter construction period (Ward and Chapman, 1995; Price, 2000). (2) Cost savings in term of whole-life price of a scheme produced by PFI compared to by traditional means (PFP, 1995; Grub 1998), (3) helping the public sector financially (Beenhakker, 1997; Jones, 1998; Partnership British Columbia, 2003) (4) the whole cost of the project from construction to maintenance stipulated in the contract (Partnership British Columbia, 2003). Optimal use of assets (Partnership British Columbia, 2003).

However, despite the advantages, there are few issues still occurs in the PFI projects. Review of the literature managed to identify few problems which keep occurring in PFI projects. As shown in Table 2 the problems or issues in PFI can be summarised as follows: (1) Issues associated with value for money 2) Lack of expertise or professional in PFI projects 3) cost overrun in PFI projects 4) Time overrun/Longer negotiation period 5) Issues with life-cycle costing 6) Aspects of PFI not clear to participants 7)payment deduction 8) Performance of FM in PFI and 8) sustainability of PFI. All these problems and issues have an impact on the performance of the projects whether in term of time, quality or money or even on the three of it. The solution suggested by previous researchers are also tabulated in the table. For example, in the value for money issues, out 12 papers analysed, four of the researchers suggest improving the VfM through assessment model and way of assessing the VfM. Others have their different ways, and only one researcher suggest the use of knowledge management to resolved the VfM issues.

Apparently in every problem or issues syntheses in Table 2 KM element such as the process of capturing, reuse and sharing involved. According to Smyth and Edkins (2007) skills and experiences are tacit knowledge which relates on how individual received and capture new information that they gain and kept it as references for the same future events that might occur. It is not easy to express, highly personal and is in the professional heads. Sharing the knowledge in PFI projects would be difficult especially when the other individual did not have any explicit knowledge of the PFI procurement.

Figure 1 shows how these three elements of KM link together, each of them having a

reciprocal relationship with each of the other two. For example, people help design and then operate Processes, while Processes define the roles of, and the knowledge needed by People. As well as the relationship between the three elements, Figure 1 can also be used to help conceptualise any particular knowledge management initiative, by regarding it as being positioned somewhere in the triangle with the three elements at its vertices. Examples of knowledge management initiatives near the People vertex of the triangle would be implementing directories or communities of practice. Near the Technology Vertex would be implementing repositories or knowledge-based systems. Near the Process vertex would be implementing new ways to work or to build in what you want to achieve, in both cases to achieve knowledge management objectives.

2

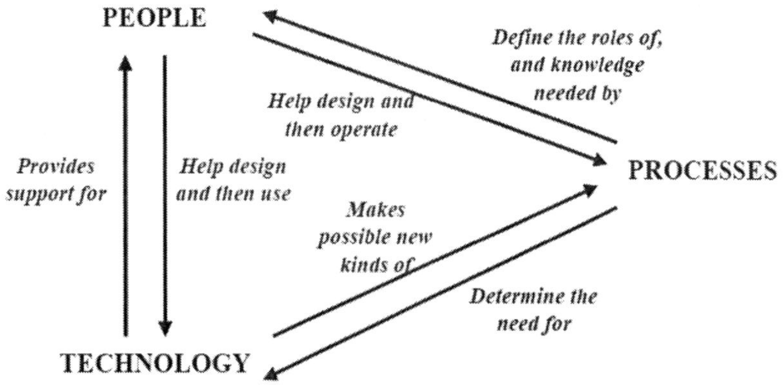

Figure 1 People, processes and technology (Edwards, 2009)

Table 2 The organisation of problems and issues identified from the literature-showing the repetitions and connection with K

Issues Prompted from the synthesis	Resolution	Literatures sources	Performance effected		
VALUE FOR MONEY			T	Q	M
Limitation to achieve VfM	Improvement on appraisal and evaluation process	Froud and Shaoul,(2001)		/	
VFM effectiveness	Suggest design for post-project evaluation	Broadbent, et al, 2003			/
Achieving best value for PFI projects	New requirements to achieve the best value should be introduce	Akintoye, et al., 2003			/
Lack of agreed Formulae for VfM stakeholder to benchmark VfM	Identifying principle factors in creating VfM in PFI	(Pitt, et al., 2006)			/
VfM through PSC in PFI	The need for PSC in Malaysia context	(S. Ismail & Rashid, 2007)			/
What are details components of PSC protocol to evaluate VfM.	Development of VfM assessment model	Takim, et al., 2009			/
What are the factors and elements for VfM framework assessment	Identify the VfM criteria for assessment	K. Ismail, et al., 2011			/
Critical parameter influencing VfM in PFI projects	list of critical parameters can help practitioners to identify priority areas and achieve a long-lasting VFM.	(Henjewele, Sun, & Fewings, 2011)			/
VfM still not achievable	Knowledge management as important factor to filling gap in getting VfM	(Henjewele, et al., 2012)			/
The need for VfM Assessment	Model of VfM Assessment	K Ismail, et al., 2012			/
VfM success factors	the knowledge of priority factors that need to enhance the achievement of long-term VFM	(Henjewele et al., 2012)			/
Checking on VfM	Economic, environ-mental and social sustainability sho-uld be core consi-derations through-out the lifetime of a ppp/pfi.	(Semple & Turley, 2013)			/
LACK OF PROFESSIONAL/EXPERT IN PFI PROJECTS			T	Q	M
The different levels of PFI experience between onstruc-tion parties and client	The need for knowledge transfer	Herbert, et al., 2004	/	/	/
Lack of skills among practitioners in PFI projects	Enhance the capability and skills of PFI practitioners	(A. A.Umar, et al., 2011), (Abdullahi A Umar, et al., 2014)	/	/	/
Lack of skills among practitioners in PFI projects	The need for knowledge transfers in PFI projects	(Carrillo, et al. 2006a)	/	/	/
COST OVERRUN			T	Q	M

cost and time overrun for health & transport sectors	Knowledge of priority factors that need to be addressed to enhance the achievement of long-term VFM in PFI projects	(Henjewele et al., 2012)			/
cost overruns due to the cost of the specialist expertise required during the bidding stage and the lengthy negotiation periods	To enhance the knowledge and skills especially client because for client organizations, PFI usually their one-off project.	Carrillo, et al, 2008a)			/
Cost of PFI procurement still higher	Introduction pf PFI2	(Abdullahi A Umar et al., 2014)			/
Charges usually put at the end of the project, "back-end" and the perception that the return of the PFI investors is excessive.	Come out with the calculation and analysis	(Hellowell & Vecchi, 2012)			/
Achieving the best value in PFI project	achievement of best values requirements through the input of few factors	(Akintoye et al.,2003.)			/
large differences of cost estimate between client and contractor &high transaction costs of PFI bids;	To enhance the knowledge of client organization	Robinson, et al., 2004			/
TIME OVERRUN			**T**	**Q**	**M**
Lack of PFI experts	Lack of knowledge transfer	Carrillo, et al,, 2006b)	/	/	/
Delay in awarding projects	Lack of expert	m asrul adnan, 2009	/		
the lengthy negotiation period	Lack of knowledge/expertise	Robinson, et al., 2004	/		
The negotiations are lengthy and complex.	Lack of expertise	(Akintoye et al., 2003)	/		
LIFE-CYCLE COSTING			**T**	**Q**	**M**
Failure of QS's contractor to consider LCC	Increased maintenance costs in the future, compared to the original budget for maintenance costs.	(Swaffield & McDonald, 2008)		/	/
Correlation analysis of capital and LCC	Suggest on update of current cost estimate practice, the cost models for LCR costs to be review, surveys are carried out to investigate the issues in cost estimates further	(Wang, 2014)		/	/
ASPECT OF PFI NOT CLEAR TO PARTICIPANTS			**T**	**Q**	**M**
Project technical feasibility	The SPV must demonstrate that the technical aspects of a proposal will satisfy all	(Li, et al., 2005)	/	/	/
Clients' briefs inadequate.	Clients to improve on the of specification of projects	Robinson, 2004	/	/	/
PAYMENT DEDUCTION			**T**	**Q**	**M**
Payment deduction due to performance issue during operation and maintenance	Model of Critical Success Factors	(Oyedele, 2013)		/	/
	to appreciate the value of the partnership	(H. S. Robinson & Scott, 2010)		/	/
PERFORMANCE OF FM			**T**	**Q**	**M**
Performance of FM	Need to include KM in FM	Mustapa, 2013, Carillo&Muzani, 2008		/	/
SUSTAINABILITY OF PFI					

whether sustainable in contract doc exist?	need to identify key elements of sustainable PFI contract	(Farmi,et al, 2011b)	/	/	/
Ambiguity of sustainable contracts & poor implementation of sustainable green elements	facilitate the process of promoting sustainability	(Farmi, et al, 2011a)	/	/	/

For the PFI procurement to success, there is a need for staff with experience in complex procurement with strong knowledge in various area such as technical, commercial and finance (D Tranfield et. al, 2004). As PFI projects itself is very complex, J. Kamara (2012) highlight that to ensure the integration in project development process, there is a need to enhance on the knowledge management gap first.

Figure 2 show the summary of the analytical analysis of literature review issues or problems arises in relation with PFI.

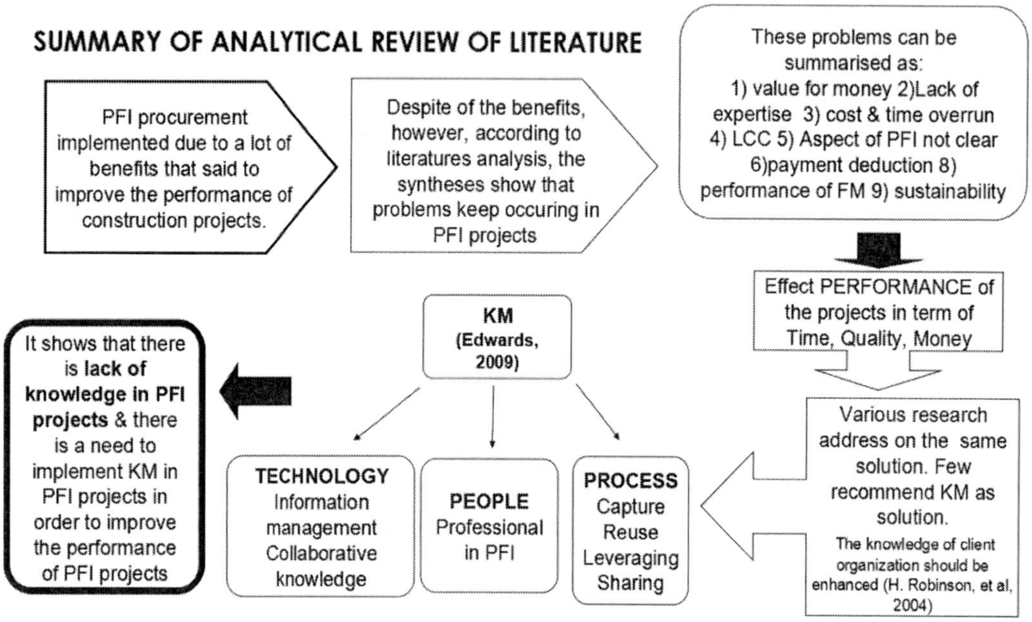

Figure 2: Summary of Analytical Review of Literature

5 Management Process

To leverage upon the knowledge infrastructure capability (people, process and technology), KM processes need to be properly in place so that knowledge can be captured, stored, shared and applied effectively (Davenport et al., 1996; Grant, 1996; Leonard, 1995). Knowledge management process involved different stages, and different researchers have their own knowledge management process.

Table **3** shows the stages identified by different researchers.

Table 3 Knowledge Management Process (Addition to H.C. Tan, et al., 2010)

Robinson, et al. (2001)	Kululanga and McCaffer (2001)	Bhatt (2001)	Marquadt (1996)	Gold et al. (2001)	Herbert Robinson, et. al, (2010)	Rollett (2003)
Discovering Locating Capturing Organising Storing Sharing Transferring Modifying Applying Archiving Retirement	Acquiring Creating Storing Sharing Utilising	Creating Presentation Validating Distributing Applying	Acquisation Creation Storage Transfer utilisation	Acquisition conversion application protection	Discovering capturing Archiving retirement Creation and leverage Distribution and sharing Organisation and Storage	Planning Creating Assessing Integrating Organising Transferring Maintaining

Applying all the KM process throughout the stages of PFI is very important as it is one of the element in KM that can enhance the performance of the project in term of time, quality and money.

6 Knowledge Management in PFI

There are few types of research were done regarding the knowledge management in PFI and highlighted the importance of knowledge management (refer **Table** 4).

Table 4List of knowledge management researches done in PFI

Author	The importance of knowledge	KM involved
P. M. Carrillo, et al., 2006	One way of improving PFI performance is to transfer knowledge from previous projects on to future projects and other PFI teams. He and the team then developed a framework of knowledge transfer based on one of the challenges faced by the construction organisation involved in PFI.	Knowledge transfer
Liyanage, et al., 2009	An empirical study on assessing the process of knowledge transfer in PFI.	Knowledge transfer
Kwawu, et al. 2010	Identified barriers and enablers for knowledge transfer in PFI. They said that practitioners and managers can efficiently design knowledge transfer framework that can be used to overcome the barriers encountered while enhancing the enablers to improve knowledge transfer processes.	Knowledge transfer
(Robinson, et al., 2010.)	Governance & knowledge management for Public-Private Partnership/ Private Finance Initiative	Knowledge management as a whole
(Kamara, 2012)	The need for robust information exchange and knowledge management systems to ensure true lifecycle integration of building information and activities, especially in bridging the gap between client and construction organizations involved in the creation and management of buildings.	Knowledge management as a whole

Author	The importance of knowledge	KM involved
(Hecker, 2012)	The need for collective knowledge in procurement as complex as PFI to ensure the interrelationship and interaction will integrate along the PFI processes.	Collective knowledge
(Mustapa, 2013); Carillo & Mustapa 2008	The importance of knowledge management in FM in PFI projects	Knowledge management as a whole

Most of the researches are done in other countries like UK, Australia, Japan and Hong Kong, which are the developed country. In this research context would be PFI knowledge in Malaysia. From the studies undertaken by previous researchers as discussed above, the practice of KM is important in the PFI projects as it would improve the performance of the projects. Hence, this strengthens the reason to fill in the gap with the implementation of KM in PFI projects.

In Malaysia, there is various research concerning on PPP/PFI had been done, but little research directly on the knowledge management of PFI. Some of the research that had been done are regarding acceptability of PFI in Malaysia (Takim, R., 2009), framework of assessment for value for money (K. Ismail, et al, 2011), key performance indicators for PFI in Malaysia (Syuhaida Ismail, 2009; Mohd Rayme A.M., 2009), delay in awarding PFI project (Mohd Asrul,2009), issues related to the Public Sector Comparator (PSC) in PFI (S. Ismail & Rashid, 2007), and about the skills of professional involved in PFI (Umar, et al, 2014).

Since the conditions, performance and the nature of the construction organisation have implications for the implementation of KM, KM implementation should be considered in the context of a specific organisation. Malaysia has its way of conducting the procurement (Nurshuhada, et al., 2013). Therefore, the self-styled prescription is required considering the uniqueness of Malaysia PFI projects. A specific KM implementation framework is required for KM implementation in PFI projects in Malaysia. For this reason, further research focuses on developing a KM implementation framework, creating awareness of the KM to sustain highly skilled personnel at all stages of the PFI projects and to improve the PFI projects' performances.

7 Conclusion and Recommendation

PFI procurement introduces by the government to brings a lot of benefits that is said to be a gain in the projects. However, problems or issues in PFI still keep recurring. Some of the solutions recommend on the use of knowledge but some still focusing on the same things. The study concludes that there is an urgency to consider the full use of KM in PFI projects which apparently the usage can be enhanced. Future research is expected to study further the current status and level of implementation of knowledge management in PFI projects in Malaysia. Further study should also be done to find out the barriers and success factors for implementation of KM in PFI projects in Malaysia.

References

1. Accountability, F. (2012). An Evaluation of the Projected Returns to Investors on 10 PFI Projects Commissioned by the National Health Service, *28*(February).
2. Akintoye, A., Hardcastle, C., Beck, M., & Chinyio, E. (n.d.). Achieving best value in private finance initiative project procurement Achieving best value in private finance initiative project procurement, (November 2014), 37–41. doi:10.1080/0144619032000087285
3. Broadbent, J., Gill, J., & Laughlin, R. (2003). Evaluating the Private Finance Initiative in the National Health Service in the UK. *Accounting, Auditing & Accountability Journal*, ***16***(3), 422–445. doi:10.1108/09513570310482309
4. Brown, D., Ashleigh, M., Riley, M., Shaw, R., 2001. New Project Procurement Process. Journal of Management in Engineering 17, 192–201.
5. Carrillo, P. M., Robinson, H. S., Anumba, C. J., & Bouchlaghem, N. M. (2006a). A Knowledge Transfer Framework: the PFI context. *Construction Management and Economics*, ***24***(10), 1045–1056. doi:10.1080/01446190600799224
6. Carrillo, P. M., Robinson, H. S., Anumba, C. J., & Bouchlaghem, N. M. (2006b). A Knowledge Transfer Framework: the PFI context. *Construction Management and Economics*, ***24***(10), 1045–1056. doi:10.1080/01446190600799224
7. Carrillo, P. M., Robinson, H. S., Anumba, C. J., & Bouchlaghem, N. M. (2006c). A Knowledge Transfer Framework: the PFI context. *Construction Management and Economics*. doi:10.1080/01446190600799224

8. Carrillo, P. M., Robinson, H. S., Anumba, C. J., & Nasreddine, M. (n.d.). Construction Management and Economics A Knowledge Transfer Framework : the PFI context A Knowledge Transfer Framework : the PFI context, (December 2014), 37–41. doi:10.1080/01446190600799224
9. Carrillo, P., Robinson, H., Foale, P., Anumba, C., & Bouchlaghem, D. (2008). Participation, Barriers, and Opportunities in PFI : The United Kingdom Experience, (July), 138–145.
10. Cox, A.; Townsend, M. (1998). "Strategic Procurement in Construction; towards better practice in the management of construction supply chains". London. Thomas Telford.
11. Delay approval in private finance initiative (pfi) project in malaysia mohd asrul bin adnan universiti teknologi malaysia. (n.d.).
12. Farmi, M., Che, I., Ismail, Z., Hassan, F., & Mat, M. C. (2011a). Fundamental Elements of Sustainable Private Financing Initiative Contracts, 334–339.
13. Farmi, M., Che, I., Ismail, Z., Hassan, F., & Mat, M. C. (2011b). Sustainable Private Financing Initiative Contracts, 452–457.
14. Hecker, A. (2012). Knowledge Beyond the Individual ? Making Sense of a Notion of Collective Knowledge in Organization Theory. doi:10.1177/0170840611433995
15. Henjewele, C., Sun, M., & Fewings, P. (2011). Critical parameters are influencing value for money variations in PFI projects in the healthcare and transport sectors. *Construction Management and Economics*, *29*(8), 825–839. doi:10.1080/01446193.2011.592204
16. Henjewele, C., Sun, M., & Fewings, P. (2012). Analysis of factors affecting value for money in UK PFI projects. *Journal of Financial Management of Property and Construction*, *17*(1), 9–28. doi:10.1108/13664381211211028
17. Ismail, K., Takim, R., & Nawawi, A. H. (2011). The evaluation criteria of Value for Money (VFM) of Public Private Partnership (PPP) bids. *International Conference on Intelligent Building and Management*, *5*(November), 349–355.
18. Ismail, S., & Rashid, K. A. (2007). Private Finance Initiative (PFI) in Malaysia: the Need for and Issues Related to the Public Sector Comparator (PSC). *Jurnal Akuntansi Dan Keuangan Indonesia*. Retrieved from http://journal.ui.ac.id/index.php/jaki/article/view/2888
19. Kamara, J. M. (2012). Integration in the project development process of a Private Finance Initiative (PFI) project. *Architectural Engineering and Design Management*, *8*(4), 228–245. doi:10.1080/17452007.2012.666729
20. Lædre, O., Austeng, K., Haugen, T. I., & Klakegg, O. J. (2006). Procurement Routes in Public Building and Construction Projects. *Journal of Construction Engineering and Management*, ***132***(7), 689–696. doi:10.1061/(ASCE)0733-9364(2006)132:7(689)
21. Li, B., Akintoye, a., Edwards, P. J., & Hardcastle, C. (2005). Perceptions of positive and negative factors influencing the attractiveness of PPP/PFI procurement for construction projects in the UK: Findings from a questionnaire survey. *Engineering, Construction and Architectural Management*, ***12***(2), 125–148. doi:10.1108/09699980510584485
22. Liyanage, C., Ballal, T., & Elhag, T. (2009). Assessing the process of knowledge transfer – An empirical study, ***8***(3). doi:10.1142/S0219649209002373
23. Mustapa, M. (2013). Facilities Management Knowledge in Private Finance Initiative (PFI) Healthcare Projects, 263. Retrieved from https://dspace.lboro.ac.uk/dspace-jspui/bitstream/2134/12843/4/Thesis-2013-Mustapa.pd
24. Nurshuhada, Z., Lou, E., Suhaimi, M.-D. M., & Faizul, a. R. (2013). Analysis of Practices for Private Finance Initiative (PFI) Procurement Process: Malaysia vs. United Kingdom. *The Professional Journal of Royal Institution of Surveyors Malaysia*, ***48***(1), 13–21.
25. Oyedele, L. O. (2013). Avoiding Performance Failure Payment Deductions in PFI / PPP Projects : Model of Critical Success Factors, (JUNE), 283–294. doi:10.1061/(ASCE)CF.1943-5509.0000367.
26. Oyegoke, A., Khalfan, M., McDermott, P., Rowlinson, S. & Dickinson, M., Construction Project Procurement Routes; an in-depth critique, International Journal of Managing Projects in Business, Vol. **2**(3), 2009, pp. 338-354
27. Pitt, M., Collins, N., & Walls, A. (2006). The private finance initiative and value for money. *Journal of Property Investment & Finance*, ***24***(4), 363–373. doi:10.1108/14635780610674534
28. Robinson, H., Carillo, P., & Anumba, C. J. (2010). Knowledge Management for Public-Private Partnerships.
29. Robinson, H. S., & Scott, J. (2010). Construction Management and Economics Service delivery and performance monitoring in PFI / PPP projects Service delivery and performance monitoring in PFI / PPP projects, (December 2014), 37–41. doi:10.1080/01446190802614163
30. Semple, A., & Turley, L. (2013). Public-Private Partnership Health Check : Managing partnerships during their lifetime, (February).
31. Smyth, H., & Edkins, A. (2007). PROJECT Relationship management in the management of PFI / PPP projects in the UK, *25*, 232–240. doi:10.1016/j.ijproman.2006.08.003

32. Sundaraj, G., & Eaton, D. (2013). Quantifying robustness in PFIs. *Journal of Financial Management of Property and Construction*, ***18***(1), 26–52. doi:10.1108/13664381311305069
33. Swaffield, L. M., & McDonald, A. M. (2008). The contractor's use of life cycle costing on PFI projects. *Engineering, Construction and Architectural Management*, ***15***(2), 132–148.doi:10.1108/09699980810852664
34. Umar, A. A., Idrus, A., Khamidi, M. F., Iskandar, B. S., & Darul, P. (2011). Barriers to the use of Public-Private Partnerships for the provision of Public Infrastructure in Developing countries : A review.
35. Umar, A. A., Zawawi, N. A. W. A., Otairu, A., & Othman, I. (2014). Private Finance Initiatives (PFI): Getting it right, *567*, 601–606. doi:10.4028/www.scientific.net/AMM.567.601
36. Wang, N., & Wang, N. (2014). Private finance initiative as a new way to manage public facilities A review of the literature. doi:10.1108/F-09-2012-0069
37. William.T. (2005). Moving to Public Private Partnership

DOI: 10.1051/matecconf/20166600026

5D Building Information Modelling – A Practicability Review

Xia Sheng Lee [1,a], Cheah Wei Tsong[2] and Mohd Faris Khamidi[1]

[1]*School of the Built Environment, University of Reading Malaysia, Persiaran Graduan, Kota Ilmu, Educity, 79200 Nusajaya, Johor, Malaysia.*
[2]*Universiti Tunku Abdul Rahman, Jalan Sungai Long, Bandar Sungai Long, Cheras, 43000, Kajang, Selangor, Malaysia.*

Abstract. Quality, time and cost are the three most important elements in any construction project. Building information that comes timely and accurately in multiple dimensions will facilitate a refined decision making process which can improve the construction quality, time and cost. 5 dimensional Building Information Modelling or 5D BIM is an emerging trend in the construction industry that integrates all the major information starting from the initial design to the final construction stage. After that, the integrated information is arranged and communicated through Virtual Design and Construction (VDC). This research is to gauge the practicability of 5D BIM with an action research type pilot study by the means of hands-on modelling of a conceptual bungalow design based on one of the most popular BIM tools. A bungalow is selected as a study subject to simulate the major stages of 5D BIM digital workflow. The whole process starts with developing drawings (2D) into digital model (3D), and is followed by the incorporation of time (4D) and cost (5D). Observations are focused on the major factors that will affect the practicability of 5D BIM, including the modelling effort, inter-operability, information output and limitations. This research concludes that 5D BIM certainly has high level practicability which further differentiates BIM from Computer Aided Design (CAD). The integration of information not only enhanced the efficiency and accuracy of process in all stages, but also enabled decision makers to have a sophisticated interpretation of information which is almost impossible with the conventional 2D CAD workflow. Although it is possible to incorporate more than 5 dimensions of information, it is foreseeable that excessive information may escalate the complexity unfavourably for BIM implementation. 5D BIM has achieved a significant level of practicability; further research should be conducted to streamline implementation. Once 5D BIM is matured and widely accepted, it is foreseeable that additional BIM dimensions of information will be incorporated into sophisticated digital building model to achieve specific project outcomes.

1 Introduction

A construction project's success is influenced by the management of quality, time and cost. The major things that concern construction project clients, but are not limited to value for money, include : pleasing to look at, free from defects on completion, delivery on time, fit for the purpose, supported by worthwhile guarantees, reasonable running costs, and satisfactory durability [1]. However, the information generated in a construction project is huge and sometimes inconsistent. This will increase the difficulty of project planning which may lead to misinterpretations and misunderstandings related to project results [2]. The construction project clients, ranging from an owner of a house to a big company having their corporate headquarters, will be at the receiving end of uncertain outcomes for one of their biggest investments. If this uncertainty is not mitigated, it will affect the construction industry's adaptability in a rapidly changing society. All in all, stakeholders have been disadvantaged by the fragmented nature of construction process for way too long, until the development of Building information modelling (BIM) in recent years.

The popular interpretation of BIM is about a new approach to design, construction, and facility Management [3]. BIM is also viewed as a tool to improve architectural practice moving from a computer aided design (CAD). Therefore, there are research that has been conducted exploring the integration of various dimensions of information into 5 dimensional (5D) BIM for better project outcome [4]. The 5D BIM approach is a brave attempt on transforming the whole construction process. However, a lot of stakeholders including professionals in the construction industry still do not understand the practicability aspects of 5D BIM. Major misconceptions among stakeholders often fall into the extremes either viewing it as a straightforward "solution to all" or merely an overrated improvised Computer Aided Design (CAD). Therefore, this research will focus on subjects that influence the practicability of 5D BIM, inclusive of modelling effort, inter-operability, information output and limitations, with a hands-on approach.

[a] Corresponding author: x.s.lee@reading.edu.my, leexiasheng@gmail.com

1.1 Building Information Modelling

BIM is instrumental in shifting the construction industry's paradigm from two dimensional (2D) based drawing information systems to a three dimensional (3D) object information systems [5]. This transformation changes the documentation method used in building design and construction, from manual conducts which are human readable [6], and integrates digital descriptions of building elements with other information [7] including time and cost which are computer readable. This process is recognised as n-dimensional (n-D) modelling, where different "dimensions" of information are integrated into a digital building model [8]. Consequently, the BIM framework is developed to define deliverables for construction industry stakeholders and to achieve Integrated Project Delivery (IPD) [9].

1.2 Time as 4D in Building Information Modelling

Time is a very unique non-material element in construction projects that can be presented in various perspectives including time-space conflicts [10] and construction schedule visualization [11]. Attempts at incorporating time as additional information to evaluate and analyse different kinds of project sequences comes in the form of 4D Computer Aided Design (CAD) [12]. These earlier researches are the fundamentals of 4D BIM towards construction progress measurement [13]. Time is logically prioritized over cost as 4D in BIM, due to the fact that the cost may fluctuate over time. However, it is obvious that 4D BIM is not complete and should be complimented with cost as 5D to achieve better project outcome.

1.3 Cost as 5D in Building Information Modelling

The importance of accuracy in cost estimation during the early stages of construction projects has been widely recognized. In the past, cost estimation for construction materials based on user's experience was essential because the cost estimation is the prediction of cost for a process or project using experience or methodology [14]. Cost estimation is very significant for decision making [15] as inaccurate estimation may lead to disastrous cost overrun and project delay [16]. Early project estimation represents a major factor in business unit decisions and often becomes the basis for a project's ultimate funding [17]. In spite of great importance given to cost estimation, it is neither simple nor straightforward due to deficiency of information in the early stages of the project [18]. Such gap in information can be bridged by 5D BIM [19] for better project financing and cash flow control [20]. Integration framework of cost and schedule [21] was developed a few years back and is presently under rapid advancement.

1.4 Practicable 5D Building Information Modelling Through Virtualisation

Construction project data such as time and cost in the isolated form without integration often carry little meaning. Integration of cost and schedule framework shaped schedule and cost management systems [21] and repetitive construction processes [22]. 5D BIM which combine building information model (3D), time (4D) and cost (5D) is aligned with such framework, integrating the isolated data into smart information useful for various purposes. 5D BIM in the form of multi-disciplinary performance models can be further utilised through Virtual Design and Construction (VDC), defined as the use of design-construction projects [23] to support explicit and business objectives. Project VDC usually emphasizes those aspects of the project that can be designed and managed, so that the team will be able to define, design, construct and operate the models [24]. Such performance models can accurately relate the predicted performance in relationship to the actual project result [25].

The virtualisation of BIM can improve the quality of information sharing, comparison and project integration by the stakeholders. As construction progresses, VDC can make consistency and accuracy checking for different dimensions of information such as building element's sizes and materials in an intuitive way [7]. VDC will also able to provide a graphical coordination of the design, operation and construction of the building including its fabrication and assembly details [26]. The process involves using BIM software that allows building information model to be presented by combining building components (3D), time scheduling (4D) and cost estimation (5D) [27].

2 Methodology

This is an action research type pilot study [28] with no collection of hard data, but with outputs, including model drawings, scheduling and time liner that are produced for examination. The action research pilot study outputs are later followed by expert reviews [29]. A sample bungalow is used as reference for this pilot study. The research methodology is to simulate and analyse the whole 5D BIM focusing on modelling effort, inter-operability, information output and limitation with a hands-on approach which are divided into 4 stages as shown in Figure 1. Observations are captured with the purpose to gauge the practicability of 5D BIM.

The first stage involves the collection and input of building information into 3D model. This process starts with clients briefing on their requirements that lead to an initial proposal. Architectural proposal is formed and the 3D model is then created by incorporating other disciplines such as civil and structural engineering, mechanical and electrical engineering and interior design. The 3D model contains information that specifies each and every component.

The second stage will take place after clients are satisfied with the initial design that comply to the specified requirements. Quantity surveyors or cost engineers will input the information for cost estimation into the 3D model. The cost estimation is done by

DOI: 10.1051/matecconf/20166600026

automating quantification of building elements with unit price. At this stage, 4D BIM is very useful to assist clients in refining their priorities from the perspective of cost estimation.

The third stage will proceed when the clients are satisfied with the concluding 4D BIM. Project managers will create Work Breakdown Structure (WBS) for all types of construction activities. Since the integration is done with BIM, it can cater for a huge amount of information, with WBS containing more levels compared to a conventional system. Some of the critical factors such as construction method, human resources, material procurement and supply will affect the project time scheduling. At this stage, all the necessary information is integrated as 5D BIM. In the final stage, VDC will be implemented to present 5D BIM project components, project cost and time scheduling.

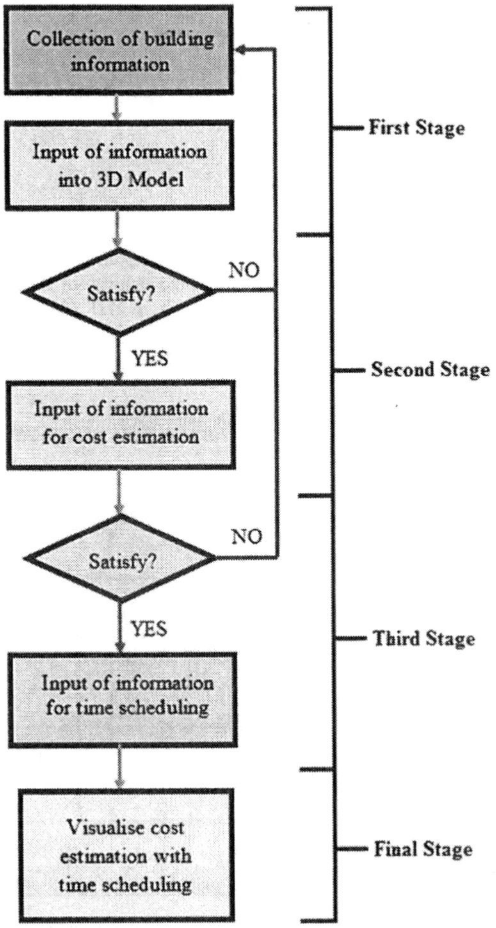

Figure 1. 5D BIM Process Flowchart.

2.1 Apparatus

There are different types of BIM software available in the market. One of the widely used softwares is Autodesk Revit. Some of the strengths of Autodesk Revit include automation [30] and it communicates better to the stakeholders [31]. The main factor for Autodesk Revit being chosen as the modelling tool in this research is because this software is decent in multidisciplinary integration and configurable automation, which is significant for 5D BIM workflow. The modelling tools and content are limited only to Revit Architecture and Revit Structure. The VDC tool chosen for this research is Autodesk Navisworks Manage, to have a seamless simulation process. The hardware setting up follows system requirements for Autodesk Revit 2016: Value: Balanced price and performance category as the following:

- CPU Type : Quad-Core Intel i5-4590S CPU @ 3.00GHz
- Memory : 8 GB RAM
- Video Display : 1,680 x 1,050 with true colour
- Disk Space : 5 GB free disk space

2.2 Collection and Input of Building Information into 3D Model

The practicability of BIM will be affected by competency in integrating pre-BIM information such as 2D CAD information. Although BIM has gained popularity over the years, there are stakeholders who are still relatively new to this new approach. Therefore, preliminary information in the form of 2D drawings needs to be analysed in order to locate major grid lines and border lines. Figure 2, shows the grid lines of the bungalow unit that is essential to set up components such as columns, slabs, beams and walls.

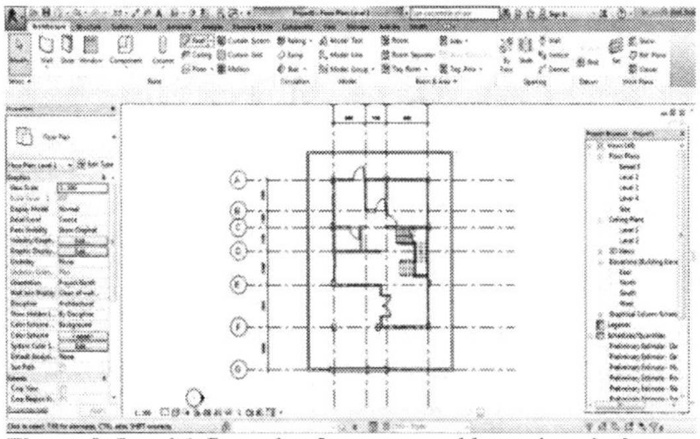

Figure 2. Level 1 floor plan for conceptual bungalow design.

2.3 Input of Information for Cost Estimation

Sattineni and Bradford (2011) agree that BIM is actually one of the best automatic approaches to directly generate an accurate quantity take off from 3D product models [32]. Cost estimation will require automated information extraction of various properties such as 'Family and Type' and 'Volume' of the components. After that, quantity surveyors or cost engineers will have to input the components' information of per unit 'Cost'. The information of per unit 'Cost' is then linked through formula or percentage computation as shown in Figure 3 with 'Family and Type' and 'Volume' of the

DOI: 10.1051/matecconf/20166600026

components. Figure 4, shows an example of the complete schedule of properties with cost estimation.

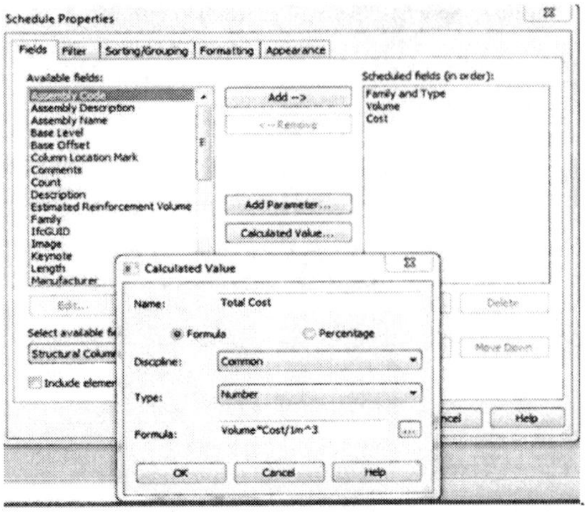

Figure 3. Cost estimation with formula.

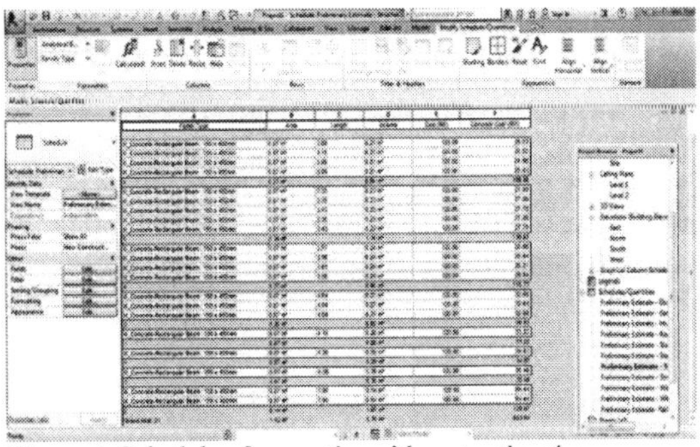

Figure 4. Schedule of properties with cost estimation.

2.4 Input of Information for Time Scheduling

Similar to 2D CAD, project management with time scheduling is conventionally done by utilising software like Microsoft Project, Primavera or Excel. Such information of time scheduling has to be linked to each and every component according to the specific WBS. Similar to 2D CAD information, these specific time scheduling information are interchangeable with BIM software such as Autodesk Naviswork Manage, as shown in Figure 5.

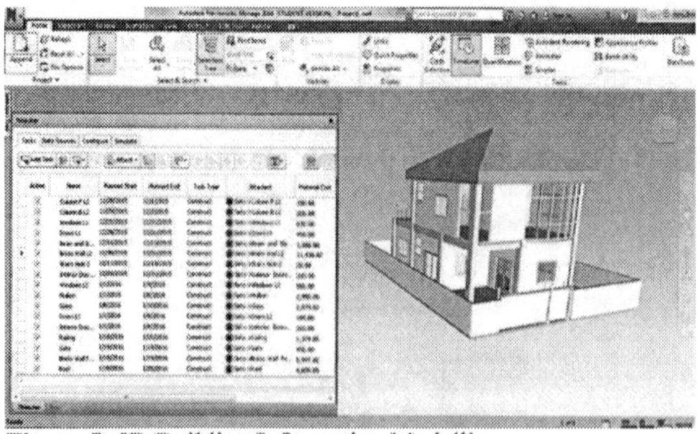

Figure 5. 5D Building Information Modelling.

2.5 Virtualisation of 5D Building Information Modelling

The final stage is to create animation or time liner as shown in Figure 6 to intuitively present 5D BIM, catering to every stakeholder's needs. One of the resourceful approaches is to create comparative VDCs that are different in construction methods, human resource planning, cost planning. This approach is very comprehensive for clients and decision makers to simulate, detect and trouble shoot uncertainties throughout the whole construction project. As the construction progresses, 5D BIM as-built information virtualisation should be done to monitor between the plan and actual progress.

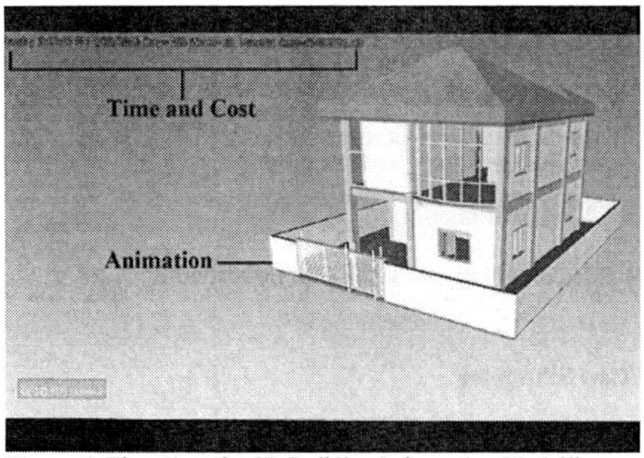

Figure 6. Time liner for 5D Building Information Modelling.

3 Practicability of 5D Building Information Modelling

5D BIM is the systematic integration between 3D model, time and cost as shown in Figure 7. The integration success will be heavily influenced by modelling effort, inter-operability, information output and limitation.

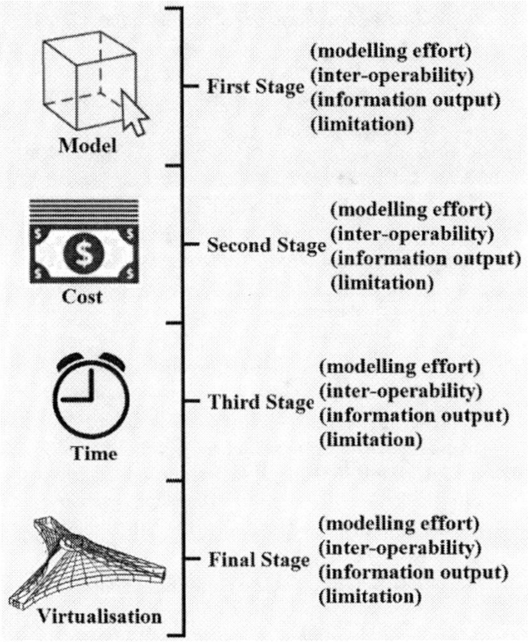

Figure 7. 5D BIM Major Components

The research is conducted by researchers and professionals who have experience with the conventional 2D CAD construction industry workflow. Therefore, the practicability of 5D BIM can be gauged directly, compared to the conventional 2D CAD workflow.

The first stage simulation discovers that the effort of modelling depends highly on the modeller's skill level. This is similar to CAD; thus it is considered neutral. Inter-operability is considered poor due to different modelling tools being based on unique systems and engines that are competing against one another. The information output is very compressive; thus, considered good when compared to CAD.

The second stage is about input of information for cost estimation where the effort is easy by computing simple formulas. BIM's inter-operability is considered neutral when compared to CAD, but the information outputs are very fast and accurate.

The third stage simulates time scheduling in 3D model which cannot be compared directly with 2D CAD workflow. Therefore, the modelling effort is considered neutral. Unlike the first stage, collaboration that involves completed building information model is easy due to high inter-operability. The information output is good as non-professional stakeholders are empowered to easily understand 5D BIM.

The final stage involves virtualisation where the effort highly depends on existing 5D BIM quality. It is considered as neutral in the pilot study; however, the advancement of technology certainly will reduce the modelling effort. Both inter-operability and information output are good, as this stage is usually directly linked to clients or end users.

These qualitative observations are summarised in Table 1 as a pilot study.

Table 1. Observation on 5 D Building Information Modelling practicability

Stage	Process	Criteria	Observation
1	Collection and Input of Building Information into 3D Model	modelling effort	neutral
		inter-operability	poor
		information output	good
		limitation	Non BIM-capable stakeholders will face obstacles to proceed to stage 2.
2	Input of Information for Cost Estimation	modelling effort	good
		inter-operability	neutral
		information output	good
		limitation	Cost information cannot be automatically updated by suppliers.
3	Input of Information for Time Scheduling	modelling effort	neutral
		inter-operability	good
		information output	good
		limitation	Requires site knowledge and localized experience to have realistic output.
4	Virtualisation of 5D Building Information Modelling	modelling effort	neutral
		inter-operability	good
		information output	good
		limitation	Modelling process still difficult when handling large amount of information.

4 Conclusion

BIM is an emerging approach that is currently transforming the construction industry. Although it is possible to incorporate more than 5 dimensions of information, it is foreseeable that excessive information may escalate the complexity unfavourably for BIM implementation. Therefore, qualitative exploration is done in this research to gauge the next benchmark which is 5D BIM. Observations clearly indicate a significant level of practicability on all 4 major stages of 5D BIM. The integration of information not only enhanced the efficiency and accuracy of process in all stages, but also enabled decision makers to have a sophisticated interpretation of information which is almost impossible with the conventional 2D CAD workflow. However, observation also indicates that non BIM-capable stakeholders will have limitations to participate and benefit from 5D BIM. Therefore, non-BIM capable stakeholders should be committed to build their BIM capability benchmarking on the 5D BIM workflow. Virtualisation of 5D Building Information Modelling will significantly reduce technical barriers for clients to participate in the project. This will increase clients' satisfaction by bridging the gap of expectation and actual project outcomes. All in all, more research should be focused on the development of practical 5D BIM best practises.

References

1. M. Latham. *The Latham Report.: Constructing the Team.* (1994)
2. L. Kestle. Remote Site Design Management (2009)
3. C. Eastman, P. Teicholz, R. Sacks, K. Liston. *BIM Handbook: A Guide To Building Information Modeling For Owners, Managers, Designers, Engineers & Contractors.* John Wiley & Sons, NJ. (2011)
4. V. Popov, V. Juocevicius, D. Migilinskas, L. Ustinovichius, S. Mikalauskas. *The use of a Virtual Building Design & Construction Model for Developing an Effective Project Concept in 5D Environment.* Automat. Constr., **19**, 357 (2010)
5. S. Mihindu, Y. Arayici. *Digital Construction through BIM System will Drive the Reengineering of Construction Business Practices.* Int. Conf. Visualisation, 29 (2008)
6. D. K. Smith, M. Tardif, *BIM: A Strategic Implementation Guide for Architects, Engineers, Constructors and Real Estate Asset Managers.* John Wiley & Sons, NJ. (2009)
7. C. Egbu, B. Sidawi. *Building Information Modelling Implementation and Remote Construction Project Issues, Challenges, and Critiques.* J. ITcon, **12**, 75 (2012)
8. M. Kagioglou. *Developing a vision of nD-enabled construction.* (2003)
9. B. Succar. *Building Information Modelling framework: A research & delivery foundation for industry stakeholders.* Automat. Constr., **18**, 357 (2009)
10. B. Akinci, M. Fischer, R.E. Levitt, R. Carlson. *Formalization and Automation of Time-Space.* CA. J. Comput. Civil Eng., **16**,124, (2002)
11. K. W. Chau, M. Anson, J. P. Zhang. *Four Dimensional visualization of construction scheduling and site utilization.* J. Constr. Eng. M., **130**, 598 (2004)
12. D. Heesom, L. Mahdjoubi. *Trends of 4D CAD application for construction planning.* Constr. M. Econ, **22**, 171 (2004)
13. C. Kim, H. Son, C. Kim. *Automated construction progress measurement using a 4D Building Information Model and 3D data.* Automat. Constr., **31**, 75 (2013)
14. S. Perera, I. Watson. *Collaborative Case-Base Estimating and Design.* J. Adv. in Eng., **29**, 801-8 (1998)
15. R. Carr. *Cost Estimating Principles.* J. Constr. Eng. M., **115**, 545. (1989)
16. P. F. Kaming, P. O. Olomolaiye, G. D. Holt, F. C. Harris. *Factors of Influencing Construction Time and Cost Overruns on High-Rise Project in Indonesia.* J. Constr. M. Econ., **15** (2010)
17. S. Trost, G. Oberlender. *Predicting Accuracy of Early Cost Estimate Using Factor Analysis and Multivariate Regression.* J. Constr. Eng. M., **129**, 198 (2003)
18. T. Hegazy, *Computer-Based Construction Project Management.* Prentice- Hall Inc., New Jersey. (2002)
19. D. Forgues, I. Iordanova, F. Valdivesio, S. Staub-French. *Rethinking the cost estimating process through 5D BIM: a case study.* In Constr. Res. Congress, 778 (2012)
20. H. Liu, M. Lu, M. Al-Hussein. *BIM-Based Integrated Framework for Detailed Cost Estimation and Schedule Planning of Construction Project.* (2014)
21. K. Cho, T. Hong, C. Hyun. *Integrated Schedule and Cost Model for Repetitive Construction Process.* J. M. Eng., **26**, 78 (2010)
22. D. Cho, J. S. Russell, J. Choi. *Database Framework for Cost, Schedule, and Performance Data Integration.* J. Compu. Civil Eng., **27**, 719 (2013)
23. M. Fischer, T. Hartmann, E. Ranks, F. Neuberg. *Combining Different Project Modelling Approaches for Effective Support of Multi-Disciplinary Engineering Tasks.* World IT Constr. Conf., 167 (2004)
24. B. Gilligan, J. Kunz. *V.D.C. use in 2007 Significant Value, Dramatic Growth and Apparent Business Opportunity.* Center for Integrated Facility Engineering Report, Stanford University. (2007)
25. J. Kunz, M. Fischer. *V.D.C.: Themes, Case Studies and Implementation Suggestions.* Center for Integrated Facility Engineering Report, Stanford University. (2012)
26. R. L. Solnosky. *Integrated Structural Process Model: AN Inclusive Non-Material Specfic Approach to Determining the Required Tasks and Information Exchanges for Structural Building Information Modelling.* The Pennsylvania State University, Pennsylvania. (2013)
27. B. Hardin, D. McCool. *BIM and Construction Management: Proven Tools, Methods and Workflows.* John Wiley and Sons Inc., Indiana. (2015)
28. J. Scorgie, J. Whyte. *Using Laser Scans to Visualize and Make Decisions about Built Infrastructure.* DIRC Working Paper 7, version 1.0. (2013)
29. R. L. Mack, J. Nielsen, *Usability inspection methods.* Wiley, New Jersey. (1994)
30. L. Khemlani. *Autodesk Revit: Implementation in Practice.* White paper, Autodesk. (2004)

31. T. Olofsson, G. Lee, C. Eastman. *Benefits and Lessons Learned of Implementation Building Virtual Design, and Construction (VDC) Technologies for Coordination of Mechanical, Electrical and Plumbing (MEP) Systems on A Large Healthcare Project.* J. ITcon, **13**, 324 (2008)
32. A. Sattineni, R. H. Bradford. *Estimating with BIM: A Survey of US Construction Companies*. Proceedings of the 28th ISARC, 564-569. (2011)

Irregular Shaped Building Design Optimization with Building Information Modelling

Xia Sheng Lee [1,a], Chung Pui Yan[2] and Zi Siang See[3]

[1]*School of the Built Environment, University of Reading Malaysia, Persiaran Graduan, Kota Ilmu, Educity, 79200 Nusajaya, Johor, Malaysia.*
[2,3]*Universiti Tunku Abdul Rahman, Jalan Sungai Long, Bandar Sungai Long, Cheras, 43000, Kajang, Selangor, Malaysia.*
[3] *Universiti Teknologi Malaysia, UTM Skudai, 81310 Johor, Malaysia.*

Abstract. This research is to recognise the function of Building Information Modelling (BIM) in design optimization for irregular shaped buildings. The study focuses on a conceptual irregular shaped "twisted" building design similar to some existing sculpture-like architectures. Form and function are the two most important aspects of new buildings, which are becoming more sophisticated as parts of equally sophisticated "systems" that we are living in. Nowadays, it is common to have irregular shaped or sculpture-like buildings which are very different when compared to regular buildings. Construction industry stakeholders are facing stiff challenges in many aspects such as buildability, cost effectiveness, delivery time and facility management when dealing with irregular shaped building projects. Building Information Modelling (BIM) is being utilized to enable architects, engineers and constructors to gain improved visualization for irregular shaped buildings; this has a purpose of identifying critical issues before initiating physical construction work. In this study, three variations of design options differing in rotating angle: 30 degrees, 60 degrees and 90 degrees are created to conduct quantifiable comparisons. Discussions are focused on three major aspects including structural planning, usable building space, and structural constructability. This research concludes that Building Information Modelling is instrumental in facilitating design optimization for irregular shaped building. In the process of comparing different design variations, instead of just giving "yes or no" type of response, stakeholders can now easily visualize, evaluate and decide to achieve the right balance based on their own criteria. Therefore, construction project stakeholders are empowered with superior evaluation and decision making capability.

1 Introduction

After obtaining professional training and gaining years of working experience, most of the architects or designers will have the spatial visualization ability to visualize irregular shaped building from 2D drawings. However, non-architect professionals such as structural engineers and quantity surveyors who are responsible for interpreting a 2D architectural drawing will face difficulties visualising an irregular shaped building [1]. They also found it difficult to interrelate the structural constructability of the irregular shaped building with other aspects such as usable floor area and structural planning. This can be solved by Building Information Modelling (BIM) where visualisation is one of the key advantages if compared to the conventional Computer Aided Design (CAD). The roles of project clients, architects, and contractors are experiencing major changes through BIM application [2].

BIM is about a new approach to design, construction, and facility Management. It is made of intelligent building components which include data attributes and parametric rules for each object [3]. BIM provides consistent and coordinated views and representations of the digital irregular shaped model embedded with synchronised data. This is a significant difference when compared to other non-BIM 3D modelling tools where the visual representation is detached from specification data. The continuous improvement of hardware and software can increase the professionals' design capability. Designers are able to save time with BIM automation since every view and data is coordinated through the built-in intelligence of the model. One of the advantages is how the design alternatives can be created and assessed [4] at ease on an unprecedented scale. Project change orders can also be managed more efficiently with BIM [5]. BIM models are used to facilitate the study of the alternative approaches and "what if" scenarios [3], which can be crucial for irregular shaped building designs.

This research explores assessment of the structural planning such as location of the staircases and lift cores, usable building space and structural constructability of the irregular shaped "twisted" building that can greatly be aided by the improved ability to visualize the design proposal in the early project phases. This will enable tangible design optimization even when aesthetic aspects such as the building twist angle is put into consideration.

[a] Corresponding author: x.s.lee@reading.edu.my, leexiasheng@gmail.com

1.1 Building Information Modelling

The interactive computer-aided design (CAD) system has been developed and used since 1963 by Ivan Sutherland, who also developed special graphics hardware and a program called as "Sketchpad" [6]. It has the capability to refine rough drawings by straightening and joining the lines, performing geometric patterns, and allowing others to foresee a designer's objective. 3D modelling softwares are developed to perform more sophisticated tasks. It was an outstanding milestone for irregular shaped buildings when CATIA Version 1 by Dassault Systems was used by Frank Gehry's office in 1989 to design and develop Disney Concert Hall as shown in Figure 1[7].

Figure 1. Walt Disney Concert Hall [21]

The increase in building complexity and scale will be followed by the increase of information [8] that needs to be managed by powerful information systems [9]. Technology evolved quickly when Building Information Modelling (BIM) started to emerge with a small amount of early adopters. BIM can be defined as a set of interacting policies, processes and technologies generating a "methodology to manage the essential building design and project data in digital format throughout the building's life-cycle" [10]. Similar terms, such as n-dimensional (n-D) modelling, is also being recognised by stakeholders, where different "dimensions" of information are integrated into a digital building model [11]. As BIM started to gain momentum, it was crucial that the framework [12] and maturity diagram [13] prepared by Mervyn Richards and Mark Bew could explain the BIM deliverables and implementation level for the construction industry stakeholders. Research shows that majority of construction industry stakeholders agree that greater use of BIM will improve construction best practices [14].

1.2 Irregular Shaped Building

Most of the buildings are often built with simple rectangular shapes. However, irregular structure incorporated with unconventional and unusual shapes are often set to become an outstanding icon. For example, Philips Pavilion as shown in Figure 2 designed by the office of Le Corbusier in 1958.

Figure 2. Philips Pavilion [22]

Creative and exotic designs are often sought–after as the clients' requirements and expectations for buildings are getting higher. These irregular shaped buildings usually provide extraordinary futuristic impressions compared to typical buildings. Such buildings stand out easily as architectural icons which sometimes relate directly to the stakeholders' unique identity and philosophy. These days, these building designs with irregular shapes depend on the digital technology for their designs and construction [15], and designs resembling "twisted" buildings are becoming more and more common. The "twisted" building surface is usually made up of straight lines by rotation and shifting, whereby the shifting direction is perpendicular to the rotation direction [16].

1.3 Design Optimization

Multidisciplinary Design Optimization (MDO) can be defined as a formal methodology for the design of complex coupled systems in which the synergistic effects of coupling between various interacting disciplines are explored and exploited at every stage of the design process [17]. MDO is an important research direction for collaboration and integration between different disciplines in building design optimization which covers diversified aspects such as floor shape [18], building structural framework [19] and structural constructability [20]. Therefore, the contribution of this research study focuses on how various aspects of design optimization are being integrated and assessed with BIM.

2 Method and Apparatus

In this action research type pilot study, a conceptual irregular shaped "twisted" building design as shown in Figure 3 is modelled using BIM software. This irregular shaped "twisted" building consists of 12 levels with a building footprint of 18 meters x 30 meters that include both private area and public area. The building's envelope is formed by "twisting" and "morphing" a mass

 DOI: 10.1051/matecconf/20166600027

extruded from level 1 building footprint, with the slanted columns having both structural and aesthetic functions. The location of lift core and staircase in the building as shown in Figure 3 is not fixed but subject to progressive design optimisation.

Three variations of design options differing in rotating angles of 30 degrees, 60 degrees and 90 degrees are created to conduct quantifiable comparisons.

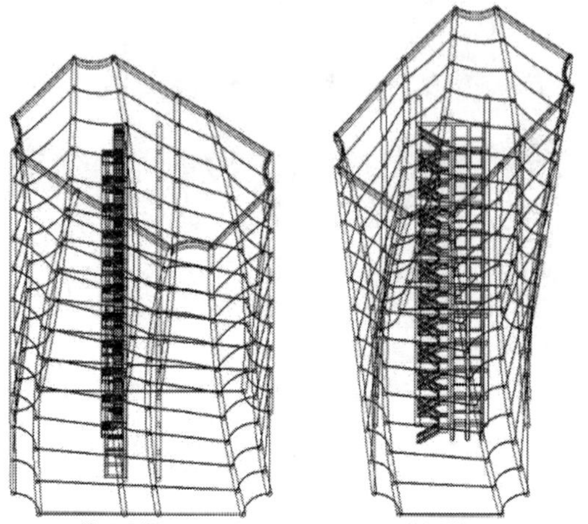

Figure 3. Conceptual irregular shaped "twisted" building design

The study will focus on three major aspects including structural planning, usable building space, and structural constructability. These aspects are selected because n-dimensional of information are required to do assessment in the process of achieving the desired optimization. The simulation and discussion are not intended for real construction but purely to explore the role of BIM in irregular shaped building design optimization.

This research will cover only one criteria for every aspects as a proof of concept for pilot study. The criteria of study for every aspect is stated in Table 1.

Table 1. Building design optimization aspects and criteria

No.	Design Optimization Aspects	Criteria
1.	Structural planning	The location of lift core and staircase based on the vertically overlapping floor area.
2.	Usable building space	The building space quantification based on private area and public area.
3.	Structural constructability	The slanted columns' virtualisation and automated generation of slope angle information.

There are different types of BIM software available for the building industry. One of the widely used softwares is the Autodesk Revit. Some of the strengths of Autodesk Revit include automation and better communication to the stakeholders. The main factor for Autodesk Revit to be chosen as the modelling tool in this research is that this software is decent in multidisciplinary integration and configurable automation. The hardware setting up follows system requirements for Autodesk Revit 2016: Value: Balanced cost and performance category as the following:

CPU Type	: Quad-Core Intel i5-4590S CPU @ 3.00GHz
Memory	: 8 GB RAM
Video Display	: 1,680 x 1,050 with true colour
Disk Space	: 5 GB free disk space

3 Observation and Discussion

There are critical factors that will affect the selected optimization aspects, and such critical factors are visualised with BIM to facilitate the research.

The location of lift core and staircase in the building are only limited to the vertically overlapping floor area. Since the vertically overlapping floor area is of different sizes and shapes, optimisation should also consider the orientation of lift core and staircase. Other factors also include minimising odd shape area surrounding the lift core and staircase, to maximise functional flexibility.

The building space is categorised as either private area or public area. Private area can be interpreted as partitionable or rentable area, and preferably to be within vertically overlapping floor area. This is because structural supports, mechanical and electrical services, water supply and sewerage piping often require vertical continuity. Public areas such as corridors and shared space with lesser verticality requirement can be located outside vertically overlapping floor area.

The slanted columns in the building cannot be interpreted effectively in 2D CAD drawings. It is critical for one to be able to virtualise such irregular columns and automate generation of slope angle information in a BIM model in order to relate with other building elements and activities.

3.1 Structural planning

The optimization of structural planning for the location of lift core and staircase with BIM is based on the vertically overlapping floor area as illustrated in Figure 4. Other than at the center of the building, lift core and staircase location can be aligned to the y-axis and x-axis respectively at the edge for design options 30 degrees and 90 degrees. These will minimise the surrounding odd shape area. For design option 60 degrees, the lift core and staircase can also be oriented with a suitable direction that minimise the surrounding odd shape area. Stakeholders are able to simulate and assess the BIM model finally achieving their very own desired optimisation.

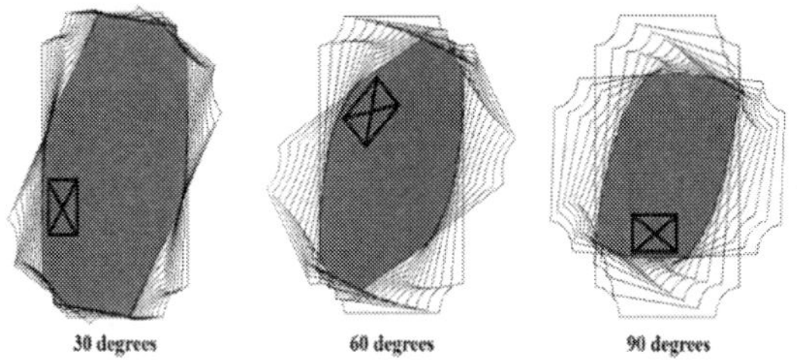

Figure 4. The location of lift core and staircase

3.2 Usable Building Space

The optimization of usable building space with BIM is illustrated in Figure 5 and Table 2. It is noteworthy that design optimisation could be unique for different clients. BIM empowered optimisation by relating directly to the aesthetic perspective is shown in Figure 5 and usable building space quantification is given in Table 2. Clients who demanded aesthetic curvature with more partitionable or rentable space may choose an option with smaller twist angle (30 degrees). Clients who wanted unique and outstanding iconic building with more public or shared space may choose an option with bigger twist angle (90 degrees). This optimisation is very flexible to cater to unique criteria and thus does not have to be just a rigid "yes or no" outcome.

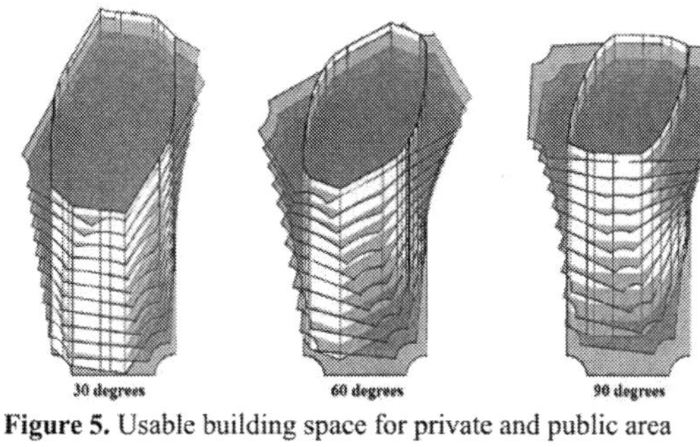

Figure 5. Usable building space for private and public area

Table 2. Building Space Quantification for Private Area and Public Area

No.	Building Twist Angle (degrees)	Private Area (m2)	Public Area (m2)	Total Area (m2)	Private/Total Area Ratio (%)
1.	30 degrees	5040.0	1104.0	6156.0	81.9
2.	60 degrees	3888.0	2256.0	6144.0	63.3
3.	90 degrees	2916.0	3216.0	6132.0	47.6

3.3 Structural constructability

The optimization of usable building space with BIM is illustrated in Figure 6. Irregular shaped building components such as slanted columns as shown in Figure 6 are ideal to be interrogated with BIM. This is significant to analyse structural constructability that will impact both project delivery time and cost. Accurate and reliable automated generation of information with BIM such as column slope angle, combining together with other 3 dimensional precision technology will dramatically increase the constructability of an irregular shaped building.

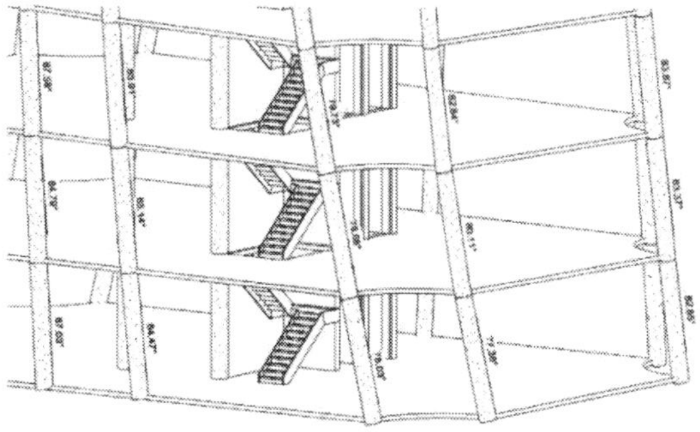

Figure 6. Slanted Column's Slope Angle

4 Conclusion and Implication

This research concludes that Building Information Modelling is instrumental in facilitating design optimization for irregular shaped buildings. In the process of comparing different design variations, instead of just giving "yes or no" type of response, stakeholders can now easily visualize, evaluate and decide to achieve the right balance based on their own criteria. Therefore, construction project stakeholders are empowered with superior evaluation and decision making capability. It is foreseeable that BIM will become almost indispensable for irregular shaped building projects in the future. Therefore, this BIM research has great potential to be developed along with other emerging technologies including laser scanning, 3D printing, augmented reality (AR) and virtual reality (VR).

References

1. Yagmur-Kilimci, E. S. *3D mental visualization in architectural design*. PhD Dissertation. Georgia Institute of Technology. (2010)
2. Sebastian, R. *Changing roles of the clients, architects and contractors through BIM.* Engineering, Construction and Architectural Management, **18(2)**, 176-187. (2011)
3. Eastman, C., Teicholz, P., Sacks, R., & Liston, K. *BIM Handbook: A Guide To Building Information Modeling For Owners, Managers, Designers, Engineers & Contractors.* New Jersey: John Wiley & Sons. (2011)

4. Denzer, A. S., & Hedges, K. E. *From CAD to BIM: Educational strategies for the coming paradigm shift.* AEI. (2008)
5. Shourangiz, E., Mohamad, M. I., Hassanabadi, M. S., & Saeed, S. *Flexibility of BIM towards Design Change.* 2nd International Conference on Construction and Project Management, IPEDR, **15** (2011)©(2011) IACSIT Press, Singapore. (2011)
6. Eastman, C. M. *Building product models: computer environments, supporting design and construction.* CRC press. (1999)
7. Hsieh, C. Y. The emergence of creativity in digital development of architecture. Proceeding of the 9th CAADRIA conference, 173-188. (2004)
8. Reichwald, R., Piller, F. T., & Möslein, K. *Information as a critical success factor for mass customization or: why even a customized shoe not always fits.* ASAC-IFSAM 2000 Conference. (2000)
9. Blecker, T. *Information and management systems for product customization*. Springer Science & Business Media, **7**. (2005)
10. Penttilä, H. *Describing the changes in architectural information technology to understand design complexity and free-form architectural expression.* ITcon Special Issue, The Effects of CAD on Building Form and Design Quality, **11**. (2006)
11. Kagioglou, M. *Developing a vision of nD-enabled construction.* Construct IT Centre of Excellence, University of Salford. (2003)
12. Succar, B. *Building information modelling framework: A research and delivery foundation for industry stakeholders.* Automation in construction, **18(3)**, 357-375. (2009)
13. Sinclair, D. *BIM overlay to the RIBA outline plan of work.* London, UK: RIBA. (2012)
14. Elmualim, A., & Gilder, J. *BIM: innovation in design management, influence and challenges of implementation.* Architectural Engineering and design management, **10(3-4)**, 183-199. (2014)
15. Ryu, J. W. *The Transition of Digital Technologies for Irregular Shaped Buildings.* Journal of the Korea Academia-Industrial cooperation Society, **12(9)**, 4210-4215. (2011)
16. Vollers, K. *Twist & build: creating non-orthogonal architecture.* 010 Publishers. (2001)
17. Ren, Z., Yang, F., Bouchlaghem, N. M., & Anumba, C. J. *Multi-disciplinary collaborative building design - A comparative study between multi-agent systems and multi-disciplinary optimisation approaches.* Automation in Construction, **20(5)**, 537-549. (2011)
18. Wang, W., Rivard, H., & Zmeureanu, R. *Floor shape optimization for green building design.* Advanced Engineering Informatics, **20(4)**, 363-378. (2006)
19. Chan, C. M. *An optimality criteria algorithm for tall steel building design using commercial standard sections.* Structural optimization, **5(1-2)**, 26-29. (1992)
20. Pezeshk, S., Camp, C. V., & Chen, D. Design of nonlinear framed structures using genetic optimization. Journal of Structural Engineering, **126(3)**, 382-388. (2000)
21. Paul Bica. Photo *Walt Disney Concert Hall, Los Angeles.* Licensed under the Creative Commons Attribution 2.0 Generic license. Retrieved Jan 2016, from https://commons.wikimedia.org/wiki/File:Walt_Disney_Concert_Hall_(1609904186).jpg. (2007)
22. Wouter Hagens. Photo *Expo 1958 Philips Pavilion.* Licensed under the Creative Commons Attribution-Share Alike 3.0 Unported, 2.5 Generic, 2.0 Generic and 1.0 Generic license. Retrieved Jan 2016, from https://commons.wikimedia.org/wiki/File:Expo58_building_Philips.jpg. (1958)

A Study to Compare the Cost of Operation and Maintenance in Green Building Index (GBI) and Non-Green Building Index (Non-GBI) Rated Building in Malaysia

Lee Zheng Ping[1] and Chu Hui Chen[2]

[1,2] *Faculty of Engineering and Green Technology, University Tunku Abdul Rahman, 31900 Perak Darul Ridzuan, Malaysia*

Abstract. Urges for sustainable development had pushed the government and professional bodies to respond and react by implementing regulations where possible to direct development in that manner. However, the outcome in most financial conferences and dialogues on sustainable buildings flagged on high construction and maintenance cost. Thus, this study is conducted to collect and analyze actual building operation and maintenance cost between GBI and Non-GBI rated buildings in Malaysia which are more than 2 years fully operated buildings. There are two categories of selected buildings which are residential and non-residential type of building. Each category of the building consists of similar building's characteristic such as geographic location, mode of operation, building heights, total numbers of floors and units. The scope of building's maintenance for this study is mainly on wear and tear of the wall painting, electrical light fittings, ceiling panels, roofing system and mechanical services like water pump system are recorded for their replacement frequency of service and the cost involved within a consistent period of 12 months operation at cost percentage saving of 78.9% and 40.4% for residential and non-residential buildings respectively compare against Non-GBI rated buildings. Electricity consumption for GBI rated buildings are lower than Non-GBI rated buildings which recorded at the cost variance of 23.8% and 6.3% and water consumption at 35.9% and 44.0% for the above mentioned two main categories of selected case study buildings. Results from this study conclude major savings on residential buildings category in term of maintenance cost and electricity consumption for GBI rated buildings. Whereby, non-residential category of buildings, GBI rated building had been proven to obtain significant savings in terms of maintenance cost and water consumption.

1. Introduction

Sustainable green rated buildings has become much more relevant and significant from the building's owners to the end-users. According to Fuad, it was focusing on sustainability and to be people-oriented in terms of design, usability as well as practicability [1]. By referring to Association Consulting Engineers Malaysia (ACEM) directory, the concept of green rated buildings are involves key savings drivers on "Electricity" and "Water" utilities and the 6 key elements of Energy Efficiency, Indoor Environment Quality, Sustainable Site Planning & Management, Material and Resources, Water Efficiency and Innovation on every building that to categorized as green rated building [2]. Sustainable green building study was essential to carry out in order to determine the actual cost of the building's operation and maintenance factors [3]. Indeed, the higher cost savings on building's operation and maintenance present at certain cost range in general understanding [4] but actual percentage and cost of saving was not been venture by any researchers on these 2 main type of building namely residential and non-residential of building.

The prime aim of this paper is to find out the operation and maintenance actual cost implication for a green and non-green building category. This gives us a better understanding on what is the difference of total costs to maintain green rated buildings compare to non-green rated buildings.

Throughout this study the final outcome will generates a clear understanding between similar characteristics of building but yet on different category namely on green and non-green rated building. Indirectly, it may use as a feasibility study on post-construction costing or prior after handing over of buildings to the home buyers' for cost projection on maintenance fees to be pay-off. Besides, this study will shift the understanding of the significant of the green building and create acceptable standard of awareness by disclosing the major differences on cost saving elements.

2. Methodology

The study adapted qualitative data collection method via criterion sampling meaning a sample that been selected to meet some criteria such as for this case study are storey of the building / building height, occupancy rate, total units, usable land area, mode of the building, years of operation and type of building. Instruments for collecting data can be done by observations, site survey and questionnaires or by personal construct. Consideration on size of sampling does not affect the result of the study [5]. For instance, it can be conducted 10 interviews for the same research study, but after the tenth interview might feel that no new concept emerging. For this study, 2 samples for each category of building namely residential and non-residential are collected. By using the principle of "Rules of Thumb" meaning data to collect via general principle that provides practical instructions for accomplishing or approaching a certain task [6] will be adapt on this study. Meaning only relevant to the study objectives data are collected such as actual reading on utilities and maintenance works. Each group consists of 2 samples on similar type of building that fully operated for a period more than 12 months which considered matured enough and to ensure consistency of the collected data are relatively accurate and within the same year of the study [7]. For instance, if the researcher compares a price of construction product in this year might differ from what the costing to be made for the upcoming year.

Actual domestic tariffs for electricity and water utility will collect via the respective service providers which are Tenaga Nasional Berhad (TNB) for electricity; Perbadanan Bekalan Air Pulau Pinang (PBA) and Syarikat Bekalan Air Selangor Sdn Bhd (SYABAS) for water utility in the state of Penang and Selangor respectively. As mentioned, total of 12 months consistent utility's statement to be recorded to ensure accuracy of the collected data. From the collected utilities data, total annual usages will be accumulated and multiplying with service provider rates for each of the case study building.

For building's maintenance works, the case study building will be analyze to ensure similarity of the comparison elements are obtained. For instance, maintenance works for green rated building might be different compared to non-green rated building such as flooring works for green building involves replacement of imported materials such as encaustic cement patterned flooring which might be broken, wear and tear on this product that most of the non-green rated buildings are inapplicable toward this element. Therefore, each of the comparable building's maintenance works in this study is selected based on similarity between green and non-green buildings in order to achieve comparable data analysis and discussion. Eventually all these collected visual inspection building's data and information to be analyze and illustrates via standard of basic calculation via using Microsoft Excel.

3. Observations, Results and Discussions

There are 2 types of building been investigated and surveyed by the researcher on this research study. There are similarity of the selected buildings in term of the type, category, location and age of the building for comparison purposes in order to compare with the similar object at different range of building category. For instance, table 3.01 below indicated 2 types of residential building, which are The Light Point Condominium and Palm Palladium Condominium whereby both located in the state of Penang. Age of all selected case study building operation are observed and maintained at minimal of 2 years fully operated. The main differences are on the construction cost as well as the annual recorded building's operation and maintenance cost. Furthermore on residential buildings, it was recorded the total construction cost between green to non-green rated buildings are at 22.6% higher. However, there is a lower building's operation and maintenance cost recorded at 78.9% saving on green rated building to non-green rated building.

Those characteristics that described above are important and to apply on non-residential buildings as tabulated below in order to achieve similarity as well as accuracy on collected data.

Table 1: Summary of Selected Building's Observations, Results and Discussions.

Building Category	Residential		Non-Residential	
Descriptions	The Light Point Condominium	Palm Palladium Condominium	Point 92 a.k.a. Menara OBYU	Menara Mudajaya
Illustration of Actual Building's Photo				
Type of Building	Condominium	Condominium	Office Commercial	Office Commercial
Category of Building	Green Rated Building	Non-Green Rated Building	Green Rated Building	Non-Green Rated Building
Total Built-up Area (m2)	8,091.08	7,782.28	3,724.46	3,967.36
Building Height / Floor	98.0m / 28-Storey Building	98.5m / 28-Storey Building	95.8m / 19-Storey Office Building	95.0m / 19-Storey Office Building
Completion Date	01st December 2012	01st July 2011	01st October 2012	01st September 2012
Construction Cost (RM)	RM68,750,000.00	RM53,150,000.00	RM46,150,000.00	RM35,210,000.00
Average Annual Operation and Maintenance Cost (RM)	RM38, 297.70	RM155, 032.79	RM33, 891.92	RM44, 395.99

4. Data Collection and Analysis

Table 2: Summary of Building's Operation.

Building Category	Residential		Non-Residential	
Descriptions	The Light Point Condominium	Palm Palladium Condominium	Point 92 a.k.a. Menara OBYU	Menara Mudajaya
Built-Up Space	8,091.08m2	7,782.28m2	3,724.46m2	3,967.36m2
Number of Total Units	112	134	118	116
Occupancy Rate	90%	90%	95%	95%
Operation Hours	24 Hours (06:00-02:00)	24 Hours (06:00-02:00)	18 Hours (06:00-00:00)	18 Hours (06:00-00:00)
Usable During Weekend / Public Holiday	Yes	Yes	No	No
Electricity Consumption (Kwh)	103,638	125,936	87,125	93,045
Water Consumption (m3)	24,860	38,789	19,461	34,762

4.1 Summary of Building's Operation

Operation cost saving to compare would be the key drivers on electricity and water utilities consumption for the buildings. A total of consistent reading for 12 months utilities consumption mainly on electricity and water will analyze and recorded to compare the 2 similar functions of building but from the difference categories of building which are residential and non-residential buildings. The recorded data are collected from 1st January 2014 to 31st December 2014, which is total of 12 months utility consumption records. For electricity, tariff to be referred to Tenaga Nasional Berhad (TNB) standard tariff based on the type and category of building via total electricity usage data. Whereby, water consumption will base on the service provider in specified state. For instance, in Selangor Darul Ehsan to referred to Syarikat Bekalan Air Selangor (SYABAS) whereby in Pulau Pinang to referred to Perbadanan Bekalan Air Pulau Pinang (PBA). For all the recording on water consumption will be recorded at the main water meter supply connection reading which is the main water inlet supply to the entire building or known as domestic bulk meter obtained from the respective service provider. Below are the summary of the total utilities consumption rate with simple calculation multiplying the service provider tariff cost for selected buildings as below:-

4.1.1. Electricity Consumption Cost Calculation

(a) The Light Point Condominium
The Total Annual Usage = 103,638Kwh
TNB Calculated Tariff for Residential (Average Annual Rate) = RM0.55/Kwh
Total Annual Cost for Electricity Consumption = RM52, 752.00

(b) Palm Palladium Condominium
The Total Annual Usage = 125,936Kwh
TNB Calculated Tariff for Residential (Average Annual Rate) = RM0.55/Kwh
Total Annual Cost for Electricity Consumption = RM69, 264.80

(c) Point 92 a.k.a. Menara OBYU
The Total Annual Usage = 87,125Kwh
TNB Calculated Tariff for Residential (Average Annual Rate) = RM0.42/Kwh
Total Annual Cost for Electricity Consumption = RM36, 592.50

(d) Menara Mudajaya
The Total Annual Usage = 93,045Kwh
TNB Calculated Tariff for Residential (Average Annual Rate) = RM0.42/Kwh
Total Annual Cost for Electricity Consumption = RM39, 078.90

From the above calculation shown that the highest differences in term of percentage for the same category of building was residential building recorded at 23.8% cost implication with annual difference between The Light Point Condominium and Palm Palladium Condominium utility cost rate at RM52, 752.00 and RM69, 264.80 respectively. Both buildings located in Island of Pulau Pinang. Besides that, non-residential building data obtained and compared between Point 92 a.k.a. Menara OBYU and Menara Mudajaya with both office commercial buildings situated at Damansara Perdana, Petaling Jaya Selangor Darul Ehsan. The recorded data showed the highest electricity usage building category which marked at RM36, 592.50 and RM39, 078.90 respectively. As described above, the highest collected data recorded on residential buildings due to the operation period operates for 24 hours daily inclusive of weekend and public holiday recorded compared to non-residential buildings which operates average of 18 hours daily excludes weekend and public holidays.

TNB electricity tariff rate shown on the calculation above is derived on average cumulative reading that collected on monthly electricity TNB's statement. For residential buildings electricity consumption for a year was recorded at RM0.55/Kwh whereby non-residential buildings recorded at RM0.42/Kwh. There is no significant saving criteria provided by the electricity service provider of TNB. All savings elements are based on the building's total electrical usage and saving criteria that emphasized on Sustainable Green Building such as zero cost construction by utilizing building's orientation for natural lighting and ventilation caused the building's less heated by the direct sun and indirectly reduces the usage of mechanical cooling system such as air-conditioning system especially on the general areas and spaces that considered insignificant and waste of energy which highly emphasized by Point 92 a.k.a. Menara OBYU office building. Apart from that, LED lighting would be the main factors of contributing huge saving on the building's electricity usage as proven analysis that conducted by lighting specialist.

4.1.2. Water Consumption Cost Calculation

(a) The Light Point Condominium
The Total Annual Usage = 24,860m3
PBA Calculated Tariff for Residential (Average Annual Rate) = RM0.35/m3
Total Annual Cost for Electricity Consumption = RM8, 701.00

(b) Palm Palladium Condominium
The Total Annual Usage = 38,789m3
PBA Calculated Tariff for Residential (Average Annual Rate) = RM0.35/m3
Total Annual Cost for Electricity Consumption = RM13, 576.15

(c) Point 92 a.k.a. Menara OBYU
The Total Annual Usage = 19,461m3
SYABAS Calculated Tariff for Residential (Average Annual Rate) = RM2.28/m3
Total Annual Cost for Electricity Consumption = RM44, 371.00

(d) Menara Mudajaya
The Total Annual Usage = 34,762m3
SYABAS Calculated Tariff for Residential (Average Annual Rate) = RM2.28/m3

Total Annual Cost for Electricity Consumption = RM79, 257.36

Referring to calculation above, water consumption with cost implication for year 2014 recorded from the selected buildings as a case study shown that both residential and non-residential buildings consumes average consumption within the range of 19,461m3 to 38,789m3. Specifically for residential building recorded at 24,860m3 and 38,789m3 for The Light Point Condominium and Palm Palladium Condominium water consumption respectively. Whereby for non-residential buildings recorded at 19,461m3 and 34,762m3 for Point 92 a.k.a. Menara OBYU and Menara Mudajaya respectively. From the site survey, the highest saving area was evaluated and monitored at the rain water harvesting system and recyclable water system that fully utilize by the building for non-drinking purposes. Estimated more than 50% daily usage for external cleaning, landscape and building cooling system by splashing the harvesting water at the roof top to reduce the building temperature are all applied on this green rated building, which is The Light Point Condominium. Residential building recorded the lowest cost implication rate at 35.9% differential sum between The Light Point Condominium and Palm Palladium Condominium. This due to residential buildings covered 3 main areas on each building mainly are management office, condominium general facilities such as swimming pool, washing bay, landscape and external general areas includes periodically external façade and building cleaning services. As mentioned, cost saving factor for residential building is on the recyclable water system such as rainwater harvesting system and ground water pumping system. The Light Point Condominium is located at the reclamation land and therefore obtaining from the natural source such as recyclable sea water is obtainable for legal policy usage and both residential case study buildings are applying the state water control policy which final water outlet is recorded at state water service provider bulk meter at flat rate of RM0.35 per m3 water usage. Whereby, non-residential buildings which are Point 92 a.k.a. Menara OBYU and Menara Mudajaya the rate was fixed by the Joint Member Corporation (JMC) also known as the building management board decided to fix the ceiling rate at RM2.28 per m3 of water usage. As a result those total costs on water consumption are slightly higher level as a comparison to residential type of buildings.

4.2 Summary of Building's Maintenance

Table 3: Summary of Building's Maintenance.

Building Category / Function	**Residential**		**Non-Residential**	
Description on Building's Maintenance Works	The Light Point Condominium	Palm Palladium Condominium	Point 92 a.k.a. Menara OBYU	Menara Mudajaya
Internal and External Painting Works (Periodically Maintain)	42,000.00	328, 422.00	51, 000.00	57, 305.00
Electrical Works (Replacement on Spoiled/Damaged Items Only)	7, 200.00	143, 010.00	4, 704.00	9, 700.00
Internal Ceiling Panels (Replacement on Wear & Tear Items Only)	12, 315. 40	3, 600.00	7, 728.50	4, 680.00
Roofing Sheets (Replacement on Wear & Tear Items Only)	15, 080.00	51, 998.39	10, 000.00	12, 780.00
Mechanical Works (Study Focused On Water Tank System)	0.00	2, 665.00	0.00	15, 426.00

The collected maintenance costs involved annually tabulation on table 4.02 shown a total cost implication of RM199, 428.78 for residential and commercial building of non-green rated building (Non-GBI) whereby green rated building (GBI) recorded at RM72, 189.62; both based on total annual collected amount generated throughout the building's operation period. The cost differences in percentage between both buildings are marked at 63% of excess spending by non-green rated building. In another words, the building's owner will need to fork out additional 63% additional sum of monies if they did not adopt green building criteria with selected key elements on maintenance works stated on table 4.02. There are 3 main major maintenance works been executed by Palm Palladium Condominium that cost the highest maintenance costs spending which includes painting works at RM328, 422.00, electrical fittings and works at RM143, 010.00 and roofing sheets replacement at RM51, 998.39 in Year 2014. The main reason for painting works with huge amount spent due to local council requested the building's owner to re-paint entire building (exterior and partially interior) due to poor appearances as stipulated under state government's residential rules and regulations with the approval from Joint Member Committee (JMC) in Year 2014. All the expenditures are contributed from building's maintenance fees and sinking fund. Secondly, electrical fittings, JMC's with unanimous decision to make a huge change on conventional T8 fluorescent light to LED lighting for entire general areas for those categorized under strata title common areas. The pre-installation works was tested and conducted by Pascal Engineering Lighting Specialist from Penang showing huge saving on electricity by additional perk of providing 8 years maintenance free period. Next, roofing sheets replacement works to be carry out due to major leakage and rusty finishes discovered at the roof top level on roofing sheets cladding and its sub-structures, need immediate replace to avoid any calamities. Ideally, going for green rated building will generates higher positive result in longer period of time. The data above showed the building's operation range between 2 to 3 years for the case study. The lowest percentage saving on maintenance works recorded is non-residential building. It was due to low office building operation usage with slightly controlled operation hours with fixed usage period throughout the day as a result in reducing the cost of maintenance.

5. Conclusion

By referring to the site survey and inspection on selected case study buildings, the 6 keys elements that used on the GBI are the main consideration for this research mainly for non-green buildings are low energy efficiency by huge waste on electricity usages especially on the lighting compared to green building that encourage the usage of natural light source such as effective building's orientation and design as an innovation efforts. The main problem existed by non-green buildings are due to the pre-construction planning that lack of awareness on sustainable planning and management to make it as priority on building construction and planning requirements. As such, the indoor environment quality is more rely on mechanical system of ventilation instead of natural ventilation system. With the lack of pre-construction planning for non-green buildings, building technology are out of the development planning for instance the used of the water efficiency system such as rain water harvesting technology and solar system which widely apply on green rated buildings. Most of the non-green building's management did not practice regular maintenance that causes the entire building produce losses in operating and maintaining the building. Outcome on this study will create awareness to the building's owner and people in the industry to apply green buildings elements especially at the pre-construction planning stage. It was not merely for reputation of the company but will indirectly create long term benefits to the home buyers or tenants.

References

1. Ahmad Fuad, *Building Information Modeling,* Construction Research Institute of Malaysia (CREAM), **First Issue** (2012)
2. Association Consulting Engineers Malaysia (ACEM) Directory. (2012). *Retrieved from http://www.acem.com.my/index.php?option=com_content&task=view&id=58&Itemid=1*
3. A.M. Kassim, M.S. Jamri, M. N. Othman, M.Z.A. Rashid, S.J.S. Ismail, *"Design and Development of Low Cost Certified Green Building for Non Residential Existing Building (NREB)",* Advanced Materials Research, Vol. **748**, 1125-1129 (2013)
4. Abu Bakar, A.H., Abd Razak, A., Abdullahi, S., Awang, A., Parumal, V., *Critical Success Factor For Sustainable Housing: A Framework From The Project Management View.* Asian Journal of Management Research. **ISSN 2229-3795**, 66-80 (2010)
5. Abdulllah, F. M., Chiet, V. C., Anuar, K., & Shen, T. T., *An Overview on the Growth and Development of the Malaysian Construction Industry.* Workshop on Construction Contract Management (University Teknologi Malaysia, 2004)
6. Bell, H. R., *Social Research Methods: Qualitative and Quantitative Approaches.* (London: Sage Publishing Ltd, 1993).
7. Nazirah Zainul Abidin, *Sustainable Construction in Malaysia – Developers' Awareness* (2009). Retrieved from http://www.waset.org/journals/waset/v53/v53-131.pdf.

Investigation of Thermal Performance for Atria: a Method Overview

Leila Moosavi[1,a], Norafida Ghafar[1] and Norhayati Mahyuddin[1]

[1]*Center for Urban Conservation and Tropical Architecture, Faculty of Built Environment, University of Malaya, Malaysia*

Abstract. The importance of low energy design in large buildings has encouraged researchers to implement different methods for predicting a building's thermal performance. Atria, as energy efficient features, have been implemented to improve the indoor thermal environment in large modern buildings. Though widely implemented, the thorough study of atrium performance is restricted due to its large size, complex thermodynamic behavior and the inaccuracies and limitations of available prediction tools. This study reviews the most common research tools implemented in previous researches on atria thermal performance, to explore the advantages and limitation of different methods for future studies. The methods reviewed are analytical, experimental, computer modelling and a combination of any or all of these methods. The findings showed that CFD (computational fluid dynamic) models are the most popular tools of recent due to their higher accuracy, capabilities and user-friendly modification. Although the experimental methods were reliable for predicting atria thermal and ventilation performance, they have mostly been used to provide data for validation of CFD models. Furthermore, coupling CFD with other experimental models could increase the reliability and accuracy of the models and provide a more comprehensive analysis.

1 Introduction

Different studies about atrium thermal performance discuss the importance of implementing low energy design for new large buildings to reduce energy consumption. The atrium is a large open space with glazed roof or walls located within a building creating a common space for interconnecting adjacent stories. Atria also have potential environmental advantages such as providing natural daylight and natural ventilation for its adjacent spaces.

The complexity of the nature of air flow in an atrium can cause difficulty in experimental investigations and consequently impose significant challenges on computational modelling for flow prediction [2]. On the other hand, it is difficult to rely on the results of previous studies alone to generalize those achievements for other atrium types with different conditions [3]. Therefore, each atrium models require specific experimental or computational studies to fulfill the concerns of the accuracy in predicting results.

Different types of methods have been implemented in previous studies for the investigation thermal performance of various atrium models. These methods take into consideration the accuracy and capability to provide detailed information, cost and user-friendliness. Therefore, a review of the application of the methods in previous atrium studies and a highlight of the limitation and advantages of those methods is influential for choosing the appropriate method for further studies.

This study starts with a review of the main methods implemented in the previous studies regarding their advantages and disadvantages. Next, the implemented method are investigated based on the purpose of the studies. Finally, in order to find the most appropriate methods for certain atrium studies, all the mentioned methods are compared and discussed accordingly. The scope of this study is limited to previous researches on atrium which evaluated indoor thermal and ventilation conditions.

2 Main methods implemented for atria

This section reviews the most common research methodologies implemented in previous researches. A thorough study of atrium performance is restricted due to inaccuracies and limitations of the available prediction tools [4] and controlled experiments. In general, there are four common methods often implemented in studies of naturally ventilated atria which include analytical model, real building measurement (field study/ experimental study), small scale model simulation studies (experimental study) using computer modelling and a combination of any or all of the aforementioned methods.

2.1 Analytical model

The analytical models are probably the oldest method for predicting atrium ventilation performance. This method is still widely used today for its simplicity, richness in physical meaning and little requirement in computing

[1 a] Corresponding author: leilamoosavi@yahoo.com

resources, although it may not be accurate for atria with complicated ventilation and the results may not be informative. In cases with simple atrium models, although the final equations obtained for one atrium model may not be used for another without modifications, the methodology and approximations can be similar for different models [5].

Assadi, Dalir [6] implemented the analytical model to evaluate the energy efficiency of an atrium building with buoyancy-driven natural ventilation by testing design parameters such as atrium dimension and glass height. Lin and Linden [7] used theoretical analysis coupled with the experimental analysis using a salt solution to investigate the transient behavior of stratification through the atrium with buoyancy effect. Ji and Cook [8] also implemented a simple analytical model to examine the thermal performance of buoyancy-driven natural ventilation atrium with multi-storey building and a solar chimney by focusing on the impact of the height of the building and opening size. However, discrepancies were observed between CFD results and analytical model.

2.2 Measurement and monitoring of real atria

Measurement and monitoring is an influential method implemented for evaluating atrium thermal performance. Actual building measurements or field experiments are carried out on site using one or several measuring equipment and apparatus to evaluate the performance of existing buildings by conducting scientific data collection [5, 9-12]. It can also be done via personal observation or user response study to assess the monitored building as a case study. The atrium's internal environment is not stable and changes daily in response to the ambient conditions; therefore, an on-site monitoring of the atrium building is effective to explore the impact of the ambient conditions on the indoor thermal environment of the atrium. Another advantage of conducting field studies for atrium building is that it provides appropriate data for validation of the simulation software implemented in the atrium model.

Despite all the advantages, there are several disadvantages in using this method for atrium investigation. Compared to other building types, the measurement of atria buildings is more difficult due to the largeness of atrium and complexity of thermal and airflow behaviors in atrium. Therefore, to cover all the areas of the atrium while monitoring, a large number of equipment is required. On the other hand, monitoring and measuring atrium buildings need plenty of time to ensure that reliable data is used in the analysis stage. However, even with providing the aforementioned requirements, there is still possibility of human errors occurring during the measurement period, in setting and handling measurement equipment. Another limitation of situ measurement as compared to laboratory experiments is the difficulty in changing boundary condition to test design modification.

In most of the cases, only changing the opening position is possible. Table 1 compares two main measurement techniques, the investigated parameters and implemented devices from previous studies about atria.

Figure 1. Monitoring and measurement of an office building in the tropics [13].

Few studies have implemented situ measurement to investigate atria thermal performance [14-17] (Figure 1). Abdullah and Wang [18] carried out site measurement to evaluate the thermal performance of a large multi-level atrium building and implemented the results to validate used TAS program. Pfafferott, Herkel [19] implemented a new method for analysis to deal with short-term and long-term measurement and atrium building simulation using ESP-r. Tanaka, Okumiya [15] undertook field measurement to investigate the cooling loads of atrium building with a double-glaze external wall.

2.3 Scale model experiments

Implementing scale model instead of a field study, is beneficial in terms of lower cost and practicality especially for high rise buildings. It is also used commonly to visualize airflow patterns and temperature rate in and around atrium buildings and investigating opposing and assisting effects of wind on buoyancy with the aid of a wind or water tunnel [20] (Table 1). This model is capable of predicting the actual building conditions with a reasonable degree of accuracy [21-23]. The highest similarity of air mean velocity fields in the scale model is obtained when airflow in both scale model and real space are fully turbulent [24]. Three types of scale model experiments used for investigating thermal and ventilation conditions in atria buildings are wind tunnel model, salt bath model and fine hydrogen bubbles. Wind tunnel scale models are also used to determine wind pressure coefficients, to be used for calculating the ventilation rates indirectly [25]. While, the salt-bath methods is mostly implemented to visualize the atrium ventilation with thermal buoyancy [23].

Ding, Hasemi [30] carried out reduced scale model experiment to examine the stack effect performance in an eight-storey office atrium building in Tokyo. Bensalem [29] used wind tunnel model to evaluate the thermal performance of atrium with wind-driven natural ventilation. Liu, Lin [31] used salt bath model and Walker, Tana [1] used reduced scale method. Both were coupled with CFD for investigating thermal performance of atrium building with buoyancy-driven natural ventilation (figure 2). Holford and Hunt [23] implemented small-scale and theoretical models to investigate the flow enhancement in the atrium using heat source and solar heat gain. According to the results, the experimental results of the scale models were in agreement with other CFD or analytical results.

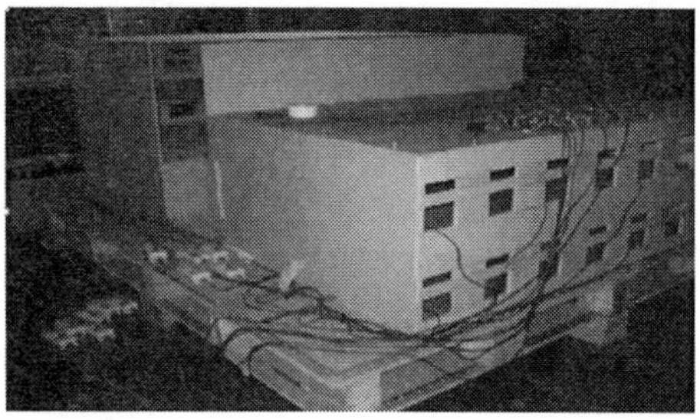

Figure 2. The scaled air model in the test chamber [1].

2.4 Computational modelling

Computational modeling has been used increasingly to simulate airflow patterns through atria buildings and predict atrium airflow and thermal field. Generally, the three most popular computational methods for modelling airflow in atria are multi-zone, zonal and computational fluid dynamics (CFD) models. The multi-zone model is not recommended for large spaces or spaces with the stratified ventilation system due to the well-mixing

Table 1: Typical measurement techniques for thermal evaluation and ventilation performance of atria

	Problem	Measurable variable	Measurement devices	References
	Overheating	Temperature	Thermal Comfort Meter thermocouple	[14]
	Stratification	Mean radiant temperature, Air temperature and wall surface temperature.	Thermal Radiometer Thermocouple	[26]
Building measureme	Solar heat gain, Glare	Radiant temperature, Air temperature, Air velocity	Thermal Radiometer. Thermocouples anemometer	[15]
	Solar heat gain, Glare	Daylight factor, Temperature, Airflow	Sensitive photometers	[27]
	Overheating	Temperature, Airflow velocity,	Thermocouple, Anemometer , flow hood, Smoke pencils	[1]
Small scal	Stratification	Displacement ventilation	Salt solutions, forced chamber, Unforced chamber/	[7, 28]
	Ventilation control	Airflow rate, Airflow pattern	Wind tunnel test	[29]

DOI: 10.1051/matecconf/20166600029

assumption in these models. In this regards, zonal models have been employed to compensate for the problem of predicting air temperature distributions in the multi-zone model. However, the zonal model has a limitation in predicting airflow pattern and is insufficient for driving flows with high momentum. Tan and Glicksman [32] validated a developed multi-zone model program, MMPN, for naturally ventilated atrium buildings using CFD method. Their results showed that by implementing the results of the multi-zone model in CFD model, CFD can simulate and provide accurate detailed information for the atrium and also save a significant amount of computation time.

Computational fluid dynamics (CFD) is a powerful tool that has been improved remarkably in recent years for the design of indoor climate control. Compared to the zonal models, which are macroscopic, the CFD models, as microscopic models, have extremely small cells to control the macro-volumes. Although in atria, due to the large size of space and also the complicated behavior of flow field, simulating the flow field using the CFD model face several difficulties, it still is a unique method to make the accurate and detailed analysis possible [33]. Many researchers employed CFD models to predict indoor thermal environment of the atria buildings and their results proved that the CFD model is appropriate for providing the detail thermal and ventilation performance of atrium [34, 35] [36] [37]. Kondo and Niwa [38] carried out a numerical study on the atrium thermal performance using k-ε turbulence model. C.A. Rundle, M.F. Lightstone [16] and Ding, Hasemi [30] Hussain and Oosthuizen [39] used CFD method to study on the atria model in which turbulent natural convection coupled with radiative heat transfer (figure 3).

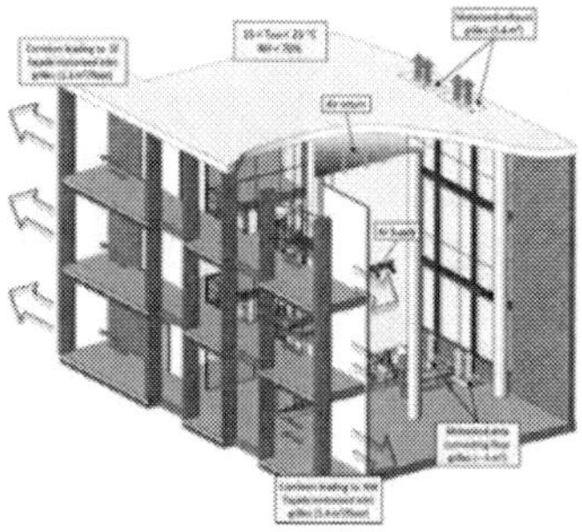

Figure 3: A mode of the atrium and adjacent spaces in the engineering building of Concordia university, Canada, hybrid ventilation concept [17].

3 The purpose of using methods in atria studies

According to the literature, the purposes of studies about atria are mainly to investigate atrium indoor thermal and ventilation conditions or validate the computational model implemented. The following section discusses these two scopes with some examples.

3.1 Investigating atrium thermal performance

Among different studies about atrium thermal performance, a limited number have investigated atrium real indoor thermal condition using field measurement and monitoring [18] [13, 15, 19]. While, a large number of studies examined different design parameters to improve atrium thermal performance such as the impact of opening design [1, 23, 30, 40, 41] or atrium shape [6, 42] or atrium roof form [43] (figure 4). Lin and Linden [7], Holford and Hunt [23], and Wang, Huang [26] separately used theoretical and experimental analyses to evaluate the thermal performances of atria with different opening aspects (size, location, and status). Furthermore, there are also studies on the evaluation of the performance of different implemented strategies in atrium buildings such as solar assisted natural ventilation [44] [15, 45], night ventilation performance in atrium [19] [46] [47] [17] [17] or wind-induce strategy [48] [49] [28, 50] using CFD and experimental methods. More details about the aforementioned studies are available in [3].

Figure 4. Predicted air temperature in the naturally ventilated atrium building with wind- ward openings reduced to 0.1m [20].

3.2 Validation of computational model

Since the nature of airflow behavior and heat transfer in atrium buildings is complicated and hard to predict, many studies have carried out experimental and numerical studies for validating the accuracy of the numerical simulation tools for atria by comparing the predicted results with the experimental results [31, 49, 51-53]. In all the studies mentioned above, there was a good agreement between numerical and experimental results. Some of the researchers evaluated the accuracy of the CFD model implemented [31]. Hussain and Oosthuizen [52] compared different RANS turbulence models with radiation model for atrium with hybrid ventilation. Furthermore, the number of researches are limited to testing different boundary conditions, as the sensitivity test, for more accurate results. For example, Gan [54] study indicates that a more accurate simulation of natural ventilation is achieved using a larger computational domain in enclosures with large openings for the naturally ventilated atrium.

Table 2: Advantages and disadvantages of each modelling technique

Technique	Description	Advantages	Disadvantage
Simplified theoretical models	Simple (one-equation) method for calculating approximate airflow rates in a single space	Quick, easy and cheap	Less accuracy, No information on airflow within space
On-site measurement	Experimental technique Measuring the real conditions	Many physical problems in wind engineering and building physics, such as wind-driven rain on buildings and heat and moisture transfer in porous building components require on-site measurements	Not controllable, the numbers of variables are constrained, Not predictable, There may be unexpected disturbances during an experimental measurement. There are also administrative limitations
Wind tunnel testing	Experimental technique in which models of buildings and their surroundings are subjected to varying (and controlled) wind speeds.	Ideally suited for modelling external airflows. Flow visualization possible.	Expensive and time consuming. Not possible to model internal flows.
Salt-Bath modelling	Experimental technique in which Perspex models of buildings are submerged in water and brine injected to represent sources of heat.	Visually easy to appreciate, Qualitative flow pattern within space can be realized and easily photographed, Possible to measure specific parameters.	Expensive to construct accurate models, time-consuming to make geometric changes, Not possible to measure variables at arbitrary positions without intrusive apparatus, practical experience required, Few equipped laboratories, no heat exchange with structure.
Multi-zone models	A simple technique for predicting air exchange rates and airflow distributions in buildings, calculate ventilation efficiency, energy demand, pollutant transport, and smoke control.	User-friendly, straightforward internal representation and calculation procedure. The prediction of overall flow through the building.	cannot represent detailed temperature and airflow distributions within a single space, due to its 'well-mixed' assumption
Zonal models	A simple technique for calculating airflow rates between adjacent zones of a multi-zone space.	Quick, easy and cheap. More accurate than simplified theoretical models.	Still no information on airflow within spaces. Some training required
Computational fluid dynamics	Computational technique which primarily solves conservation equations of mass, Momentum and energy to find velocities, temperatures and pressure throughout the whole domain	Potentially very accurate, Able to predict spatial information giving values for variables throughout the entire flow domain, easy to investigate changes in geometry and operating conditions, Additional physical models (e.g. radiation models and contaminant dispersion) can be added with relative ease.	Experience required achieving optimum performance from the code. Very computationally intensive, detailed simulations of entire buildings or long transient simulations are currently prohibitive, some validation/development is still required, particularly in the area of turbulence modelling and buoyancy-driven flows.

4 Discussion:

Table 2 compares the advantages and disadvantages of different methods implemented in atria studies. According to the literature, it seems that the small-scale and full-scale experimental models were mostly implemented to generate data for validating the numerical models of atrium [55, 56]. Compared with small scale models, zonal modelling and CFD modelling of a 3D full-scale building are more effective design tools in predicting thermal and ventilation performance of the entire atrium building. Among the different atrium models reviewed, the CFD models (of a 3D full-scale building) were the most popular and contributed to about 70 percent of the literature. The application of CFD for atrium models were mostly to investigate temperature distributions, airflow rate and pattern and thermal stratification respectively since predicting these variables with detail throughout the atrium space were difficult using other models.

The review showed that, in general, to have a better understanding of indoor thermal conditions of atrium buildings, such as temperature distribution and airflow rate, situ measurement have been used while, for testing different design parameters and their impact on the thermal performance of atria, CFD method was a common method implemented. Furthermore, in order to evaluate airflow pattern and thermal stratification throughout the atrium and adjacent spaces, a combination of both experimental and numerical simulation methods was used.

4.1 Combination of methods

Due to the lack of powerful methods for producing complete and accurate analysis, particularly for energy-related and indoor air quality atrium research, some researchers attempted to combine the different methods mentioned in section 2 [5]. Employing combined methods assists the researchers to fill the gap in any shortcomings of a particular method, validates the results and correlates the findings [57]. In another study on atrium design, Navvab [58] discussed the combination of methods as a back-and–forth process which starts by testing the atrium design implementing a scale model and proceeds to developing computer models for the parametric studies. Then, by going back to the scale model, the design can be refined. The combination of methods, such as the multi-zone models with CFD, can save a significant amount of computation time [59].

Combined methods have been increasingly employed in different research studies. Some researchers applied one model to produce some primary input data, as the boundary condition, for the next model in order to increase the accuracy of the atrium modelling results [60]. For example, Tan and Glicksman [32] integrated a multi-zone model (MMPN) and computational fluid dynamics (CFD) to modelling naturally ventilated atrium. They transferred velocity and pressure data from a multi-zone model program to CFD as boundary conditions. Other researchers use the coupled models for validation the computational model by comparing the experimental results with the simulation results (as discussed in section 3.2). However, most of the time, there are partial differences between the two modelling results since none of the modelling techniques are perfect.

Although employing different methods resulted in a better understanding of the complex nature of the problem and provided comprehensive and more accurate results, it also has limitations. For instance, it is time-consuming and costly due to different types of equipment and various techniques for each method that is continuously involving. However, according to the literature, for a well-resourced program, implementing a combined method is recommended.

5 Conclusion

This paper reviewed various methods implemented for investigating atrium indoor thermal and ventilation conditions. As mentioned, the most common methods cited in the literature for atrium studies are analytical, experimental (situ measurement and scale model) and computational modelling (multi-zone, zonal, and CFD models). The methods were compared and the advantages and limitations of each method was discussed. Among different methods implemented, CFD is the most accurate method to predict indoor temperature distribution, airflow behavior and rate for the atrium in details.

Based on the literature, to ensure the accuracy of the predicted results, more comparison between CFD results and experimental results are required. It is also helpful for the validation of the software implemented and examining the turbulence models. In most cases reviewed, for the study about atrium design parameters, CFD modelling, for investigating the impact of the ambient condition on atrium indoor thermal condition, situ measurement and for modelling indoor airflow pattern, the combination of experimental study and CFD modeling were implemented.

References

1. Walker, C., G. Tana, and L. Glicksman, *Reduced-scale building model and numerical investigations to buoyancy-driven natural ventilation.* Energy and Buildings, 2011(**43**): p. 2404-2413.
2. Awbi, H.B., *Ventilation of buildings.* 2003: Taylor & Francis.
3. Moosavi, L., et al., *Thermal performance of atria: An overview of natural ventilation effective designs.* Renewable and Sustainable Energy Reviews, 2014. **34**: p. 654-670.
4. Laouadi, A., M. Atif, and A. Galasiu, *Towards developing skylight design tools for thermal and energy performance of atriums in cold climates.* Building and environment, 2002. **37**(12): p. 1289-1316.
5. Chen, Q., *Ventilation performance prediction for buildings: A method overview and recent applications.* Building and environment, 2009. **44**(4): p. 848-858.
6. Assadi, M.K., F. Dalir, and A.A. Hamidi, *Analytical model of atrium for heating and*

ventilating an institutional building naturally. Energy and Buildings, 2011. **43**(10): p. 2595-2601.

7. Lin, Y. and P. Linden, *Buoyancy-driven ventilation between two chambers*. Journal of Fluid Mechanics, 2002. **463**(1): p. 293-312.
8. Ji, Y. and M. Cook. *Atrium-assisted natural ventilation of multi-storey buildings*. in *Proceedings of the 10th International Conference on Indoor Air Quality and Climate*. 2005. Beijing, China.
9. Baker, N., *The atrium environment*. Building TechicalFile, 1988. **No.21**: p. 39-48.
10. Kainlauri, E., G. Lehman, and M. Vilmain. *Comparative studies of five atriums on the effects of orientation, exposure and design on daylighting, temperature, and stratification of air*. in *Proceedings of the Biennial Congress of the International Solar Energy Society*. 1991.
11. Jones, J., M. Luther, and M. Nawab. *A Comparative Analysis for Two Geometrically Different Atria*. in *Solar World Congress, Proceedings of the Biennial Congress of the International Solar Energy Society, Denver, Colorado, USA*. 1991.
12. Abdullah, A. and F. Wang. *Effect of Water Spray and Blinds on Indoor Thermal Environment in an Atrium in Tropical Climate*. in *Proceedings of the Asean Postgraduate Seminar in Built Environment, Faculty of Built Environment*. 2005.
13. Moosavi, L., N. Mahyuddin, and N. Ab Ghafar, *Atrium Cooling Performance in a Low Energy Office Building in the Tropics, A Field Study*. Building and Environment, 2015.
14. Abdullah, A.H., et al., *Field study on indoor thermal environment in an atrium in tropical climates*. Building and environment, 2009. **44**(2): p. 431-436.
15. Tanaka, H., et al., *Thermal characteristics of a double-glazed external wall system with roll screen in cooling season*. Building and environment, 2009. **44**(7): p. 1509-1516.
16. C.A. Rundle, et al., *Validation of computational fluid dynamics simulations for atria geometries,*. Buildings and Environment, 2011. **46**: p. 1343-1353.
17. Karavaa, P., et al., *Experimental study of the thermal performance of a large institutional building with mixed-mode cooling and hybrid ventilation*. Building and Environment, 2012. **57**: p. 313-326.
18. Abdullah, A.H. and F. Wang, *Modelling thermal stratification in atrium using TAS program and verification of prediction results*. International Journal of Integrated Engineering, 2011. **1**(2): p. 79-93.
19. Pfafferott, J., S. Herkel, and M. Wambsganß, *Design, monitoring and evaluation of a low energy office building with passive cooling by night ventilation*. Energy and Buildings, 2004. **36**(5): p. 455-465.
20. Gan, G., *Interaction between wind and buoyancy effects in natural ventilation of buildings*. Open Construction and Building Technology Journal, 2010. **4**: p. 134-145.
21. Iyer-Raniga, U., *Daylighting in atrium spaces*. Architectural Science Review, 1994. **37**(4): p. 195-208.
22. Li, Y., *Buoyancy-driven natural ventilation in a thermally stratified one-zone building*. Building and environment, 2000. **35**(3): p. 207-214.
23. Holford, J.M. and G.R. Hunt, *Fundamental atrium design for natural ventilation*. Building and environment, 2003. **38**(3): p. 409-426.
24. Abdullah, A.H., *A Study on Thermal Environmental Performance in Atria in the Tropics with Special Reference to Malaysia*, in *Built Envirinment*. 2007, Heriot-Watt University: Heriot-Watt University.
25. Carey, P. and D. Etheridge, *Direct wind tunnel modelling of natural ventilation for design purposes*. Building Services Engineering Research and Technology, 1999. **20**(3): p. 131-142.
26. Wang, X., C. Huang, and W. Cao, *Mathematical modeling and experimental study on vertical temperature distribution of hybrid ventilation in an atrium building*. Energy and Buildings, 2009. **41**(9): p. 907-914.
27. Fuziah, S., Z.A. Azni, and A.R. Shuzlina, *Daylight modelling and thermal performance of atrium of new MECM building at Putrajaya*. IEM Journal 2004. **65**(3&4): p. 1-42.
28. Woods, A.W., S. Fitzgerald, and S. Livermore, *A comparison of winter pre-heating requirements for natural displacement and natural mixing ventilation*. Energy and Buildings, 2009. **41**(12): p. 1306-1312.
29. Bensalem, R., *Wind driven natural ventilation in courtyard and atrium-type buildings*. 1991, University of Sheffield.
30. Ding, W., Y. Hasemi, and T. Yamada, *Natural ventilation performance of a double-skin façade with a solar chimney*. Energy and Buildings, 2005. **37**(4): p. 411-418.
31. Liu, P.-C., H.-T. Lin, and J.-H. Chou, *Evaluation of buoyancy-driven ventilation in atrium buildings using computational fluid dynamics and reduced-scale air model*. Building and environment, 2009. **44**(9): p. 1970-1979.
32. Tan, G. and L.R. Glicksman, *Application of integrating multi-zone model with CFD simulation to natural ventilation prediction*. Energy and Buildings, 2005. **37**(10): p. 1049-1057.
33. Liddament, M.W. and A. Infiltration, *A review of ventilation efficiency*. 1993: Air Infiltration and Ventilation Centre.

34. Oosthuizen, P. and M. Lightstone. *Numerical analysis of the flow and temperature distributions in an atrium.* in *Proceedings of 2009 international conference on computational methods for energy engineering and environment: ICCM3E, 20e22 November.* 2009.
35. Hangen, H., et al., *Case study: numerical simulations for comfort assessment and optimization of the ventilation design for complex atriums,.* Journal of Wind Engineering and Industrial Aerodynamics, 2001(89): p. 1031–1045.
36. Schild PG, Tjelflaat PO, and Aiulfi D, *Guidelines for CFD modelling of atria. ;101:1311e32.* 1995: ASHRAE Transactions.
37. M. J. Cook, Y. Ji, and G.R. Hun, *CFD modeling of natural ventilation: combined wind and buoyancy forces. International Journal of Ventilation.* 2003. **1**: p. 169–180.
38. Kondo, Y. and H. Niwa. *Numerical study of an atrium by means of a macroscopic model and k-ε turbulence model.* in *Proceedings of the Room Air Convection and Ventilation Effectiveness.* 1992. Tokyo.
39. Hussain, S. and P.H. Oosthuizen. *A Numerical Study of the Effect of Thermal Mass on the Transient Thermal Performance of a Simple Three Storied Atrium Building.* in *ASME 2012 Heat Transfer Summer Conference collocated with the ASME 2012 Fluids Engineering Division Summer Meeting and the ASME 2012 10th International Conference on Nanochannels, Microchannels, and Minichannels.* 2012. American Society of Mechanical Engineers.
40. Ji, Y. and M. Cook, *Numerical studies of displacement natural ventilation in multi-storey buildings connected to an atrium.* Building Services Engineering Research and Technology, 2007. **28**(3): p. 207-222.
41. Acred, A. and G.R. Hunt, *Stack ventilation in multi-storey atrium buildings: A dimensionless design approach.* Building and Environment, 2014. **72**: p. 44-52.
42. Aldawoud, A., *The Influence of the Atrium Geometry on the Building Energy Performance.* Energy and Buildings, 2012. **57** p. 1–5.
43. Wang, F. and Abd Halid Abdullah, *Investigating thermal conditions in a tropic atrium employing CFD and DTM techniques.* International Journal of Low-Carbon Technologies 2011. **6**: p. 171–186.
44. Hussain, S. and P.H. Oosthuizen, *Numerical investigations of buoyancy-driven natural ventilation in a simple atrium building and its effect on the thermal comfort conditions.* Applied Thermal Engineering, 2012. **40**: p. 358-372.
45. Moosavi, L., N. Mahyuddin, and N. Ab Ghafar, *A comparison between atrium and courtyard cooling efficiency in hot tropical climate, implementing different passive cooling strategies.* Journal of Renewable and Sustainable Energy, 2015. **7**(4): p. 043108.
46. Mouriki, E., *Solar-assisted hybrid ventilation in an institutional building.* 2009, Concordia University (Canada): Canada. p. 174.
47. Lynch, P. and G. Hunt, *The night purging of a two-storey atrium building.* Building and environment, 2011. **46**(1): p. 144-155.
48. Gan, G., *Simulation of buoyancy-driven natural ventilation of buildings--Impact of computational domain.* Energy and Buildings, 2010. **42**(8): p. 1290-1300.
49. Mouriki, E., et al. *Full-scale study of an atrium integrated with a hybrid ventilation system.* in *Proceedings of the 3 rd Canadian Solar Buildings Conference, Fredericton, Canada.* 2008.
50. Horan, J.M. and D.P. Finn, *Sensitivity of air change rates in a naturally ventilated atrium space subject to variations in external wind speed and direction.* Energy and Buildings, 2008. **40**(8): p. 1577-1585.
51. Li, N.-p., et al., *Effect of atrium configuration characteristic on buoyancy-driven ventilation of high-rise residential buildings.* Journal of Guangzhou University (Natural Science Edition), 2010. **3**: p. 015.
52. Hussain, S. and P.H. Oosthuizen, *Validation of numerical modeling of conditions in an atrium space with a hybrid ventilation system.* Building and Environment, 2012(52): p. 152-161.
53. Alexander, D., H. Jenkins, and P. Jones. *A comparison of wind tunnel and CFD methods applied to natural ventilation design.* in *Proceedings of Building Simulation.* 1997.
54. Gan, G., *Simulation of buoyancy-driven natural ventilation of buildings—Impact of computational domain.* Energy and Buildings, 2010. **42**(8): p. 1290-1300.
55. Gao, J., et al., *Calculation of natural ventilation in large enclosures.* Indoor and Built Environment, 2007. **16**(4): p. 292-301.
56. Özgür Göçer1 Aslıhan Tavil2, E.Ö., *SIMULATION MODEL FOR ENERGY PERFORMANCE AND USER COMFORT EVALUATION OF ATRIUM BUILDINGS,* in *Second National IBPSA-USA Conference* 2006: Cambridge, MA p. 17-24.
57. Kim, K.-S., *Development of daylighting prediction algorithms for atrium design.* 1987, Texas A and M Univ., College Station (USA).
58. Navvab, M., *Outdoors Indoors* Lighting Design and Application, 1990. **p6**: p. 24-31.
59. Tan, G., *Study of natural ventilation design by integrating the multi-zone model with CFD simulation.* 2005, Massachusetts Institute of Technology: United States -- Massachusetts.
60. Wang, L. and Q. Chen, *Theoretical and numerical studies of coupling multizone and CFD models for building air distribution simulations.* Indoor Air, 2007. **17**(5): p. 348-361.

Adaptive Re-Use Principles in Historic Hotel Buildings in Melaka And George Town

Lilawati Ab Wahab[1,a], Noor Aishah Mohd.Hamdan[2], Nor Suzila Lop[3] and Izatul Faritta Mohd Kamar[4]

[1,2,3,4] *Department of Quantity Surveying, Faculty of Architecture, Planning and Surveying, Universiti Teknologi MARA Seri Iskandar Campus, Seri Iskandar, 32610 Perak, Malaysia.*

Abstract: Adaptive re-use of historic buildings is a process of changing the original function of the historic buildings to another function that can optimise the use of existing historic buildings. The selection of appropriate new function is an important factor in determining the success of adaptive re-use of historic buildings. However, adaptive re-use work done on historic buildings on the World Heritage Site is not an easy task due to rules and principles outlined by local and international charters that must be abide by. This research is conducted to gather the true picture of applied adaptive re-use principles that has been done on heritage hotels available in Melaka and George Town World Heritage Sites. This research is started with an inventory that led to the discovery of 35 hotels which applied the principle of adaptive re-use of historic buildings. Based on this finding, 4 historic hotels from adaptive re-use applications have been selected as the case studies. Results of the case studies carried out show that the level of conservation of heritage hotel is moderate and measures of control should be taken to ensure the privileges of heritage hotel. As a result of this research, a number of suggestions are made to ensure that adaptive re-use work done in the future will be conducted as optimum as possible according to the adaptive re-use and conservation principles.

1 Introduction

Adaptive re-use is broadly defined as any building work and intervention to change its capacity, function or performance to adjust, reuse or upgrade a building to suit new conditions or requirements (Douglas, 2006). Adaptive re-use is a wise move to ensure that a building restored, given new impetus and should be used with. Orbasli (2008) stated that adaptive re-use works is a process in which the building has been modified for the new use and so leave the usage of time.

Adaptive re-use work involve works such as conservation, modification, repair, maintenance and consolidation (Australia ICOMOS, 2002). Bullen (2007) stated that although there are many qualifying factors, the concept of adaptive re-use has significantly support a positive strategy to make the built environmnet more sustainable.

2 Adaptive Re-Use of Historic Buildings

Adaptive re-use of historic building is one of the best methods that can help a historic building keeps well conserved (Ahmad and Badarulzaman, 2003). Adaptive re-use work involve works such as conservation, modification, repair, maintenance and consolidation (Australia ICOMOS, 2002). Bullen and Love (2011) stated that the change of use may require refurbishment and/or complete renovation of existing buildings or structures. It's part of the rules that have the effect of protecting historical buildings from collapsing (Ashurst, 2006). According to Ashworth and Larkham (1994) when a historic building collapsed, the history that was recorded in the same building also died and disappeared without being remembered by future generations.

Adaptive re-use works on a historic building that has been listed as a National Heritage Building should be done in accordance with the principles of conservation which have been set out in the National Heritage Act 2005 and also conservation guidelines provided by the local authorities of the city in which it is located. In addition to the acts and national guidelines, the owner of historic buildings also have to comply with international charters in conduct of adaptive re-use of historic buildings (Department of National Heritage, 2007).

Adaptive re-use of historic building as a hotel in the UNESCO World Heritage Site of Melaka and George Town are not only able to make historic buildings prevail, but it also has a greater impact, extensive and profound. These buildings may also become important assets to develop a local tourism industry (Pedersen, 2002). Orbasli (2000) concurred that the tourism sector is a significant contributor to the economy of the city or the historical center. Adaptive re-use of historic buildings to a hotel will meet the needs of the tourism sector which

[a] Corresponding author: lilawati@perak.uitm.edu.my

 DOI: 10.1051/matecconf/20166600030

can provide a good return to the hotel, shops, businesses and positively support the country's tourism sector.

3 Objectives of Research

The main objectives of this research are:

i. To identify the common buildings element that had been renovated and maintained in adaptive re-use of historic building as a hotel in UNESCO World Heritage Site of Melaka and George Town.

ii. To assess the adaptive re-use of historic building practice level as a hotel based on the principles of conservation in the UNESCO World Heritage Site of Melaka and George Town.

iii. To propose basic guidelines for assessing the level of adaptive re-use of historic building as a hotel based on the principle of conservation of UNESCO World Heritage Site.

4 Research Methodology

For this research, qualitative methods was used in the study. The inventory has been carried out around Melaka and George Town World Heritage Site and as a result have led to the discovery of 35 hotels with application of adaptive re-use principles. 7 hotels are located in Melaka and 28 hotels located in George Town. The hotels have been screened to identify the four historic hotels that suitable as a case study for this research based on several criteria to represent another 31 hotels. These criteria are:

i. The application of the adaptive reuse principles and the historic buildings located within the conservation zone of world heritage sites.

ii. Heritage Hotel.

iii. The legacy that has interesting historical background.

iv. The heritage design which has interesting architecture value.

For this research, four ideal hotels were identified as samples for case studies:

i. Hotel Puri in Melaka World Heritage Site.

ii. Heeren House in Melaka World Heritage Site.

iii. Cheong Fatt Tze Mansion in George Town World Heritage Site.

iv. Hutton Lodge in George Town World Heritage Site.

Two approaches have been used throughout this case study:

i.Observation – observation of architectural elements based on the application of the adaptive re-use and conservation principles in heritage hotels

ii. Interviews – Interview with hotel owners and employees to identify the application level of adaptive reuse and conservation principles used.

5 Analysis of Case Study

Analysis of adaptive re-use work and conservation done for four case studies are as follow.

Table 1: Case study 1 – Hotel Puri at Banda Hilir, Melaka

Case Study 1:	Detail
	Name: Hotel Puri, Melaka. Location: Core Zone. World Heritage Site, Melaka. Address: No.1, Jalan Tun Tan Cheng Lock, 75200 Melaka. Hotel type: Boutique hotel. Existing function: 3 storey shop houses. No. of rooms: 82 nos. Owner: Drs. T.S. Tan
Element	**Description**
1. Facade: 	• The facade of the hotel has been preserved as the original.
2. External wall: 	• The hotel retains its original external walls • The walls were badly damaged and cannot be repaired, a new wall was built according to the specifications of the original wall. While on the damaged wall, the repair works were done carefully, especially on the plastering of the walls.
3. Internal wall: 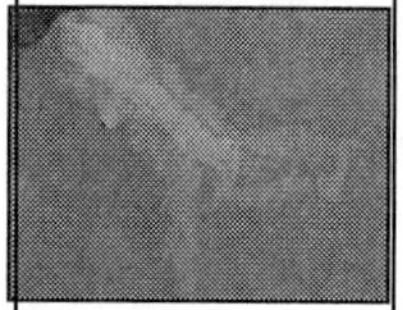	• Internal wall or wall divider is the largest element modified in Hotel Puri. • New partition wall constructed according to new floor plan to create guest rooms. A dividing wall was built in every room to create a toilet and a bathroom in the room for the convenience of hotel guests. • Many adjustments had to be made to internal wall to meet the new function as a hotel.
4. Ground Floor: 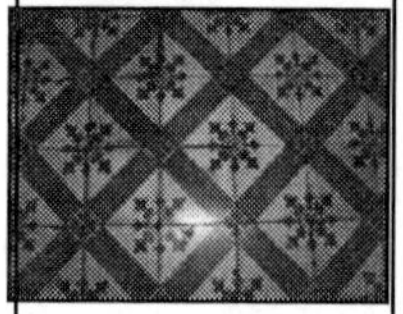	• The existing cement floor was maintained and repair works done on the damaged floor which patch the potholes and cracks.
5. Upper Floor:	• A lot of damage has occurred on the upper floor as timber was used for the element. Conservation work has been

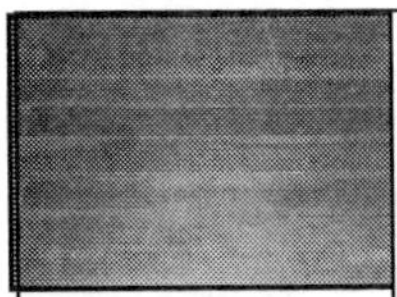	done to repair the damaged floors and replace the flooring with timber which bears the same characteristics as the original.
6. Column: 	• Hotel Puri still used the original column as the elements which are still in good condition. The damaged part were repaired and painting work done to restore the color of paint that had faded.
7. Staircase: 	• There are 2 existing stairses in Hotel Puri and 2 new staircases were built in the annexe building for hotel guests.
8. Roof structure: 	• Repairs and conservation of the roof structure has been carried out mainly on the damaged parts. Timber structures that had been damaged and decayed were replaced with new timbers as the original one.
9. Door: 	• Hotel Puri still retains the main door at the entrance of the hotel. Similarly, there are many original doors in existing buildings. • Damage door was repaired and the badly damaged door have been replaced with new doors but using the same materials and design as the original door.
10. Window: 	• External windows still remains the original window. • The damage occurred on the windows repaired as soon as possible in accordance with the principles of conservation. • For additional buildings, the design of the windows was built according to the design of the original windows of Hotel Puri.
11. Roof Finishes:	• Roof finishes of Hotel Puri were maintained as they were derived using China tiles. China tiles has a different pattern with tiles that are available today.
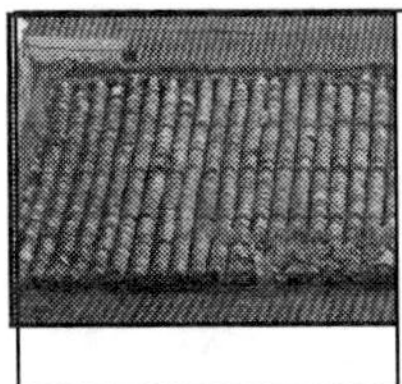	• China tiles that has been broken and damaged, were replaced with China tiles purchased from nearest building owners which no longer use Genting China on their own building.
12. Ceiling Finishes: 	• The top floor still retains its old ceiling finishes. Repair works were only carried out on the damaged and peel par. • A new type of plaster finish ceiling fans have been installed in the hotel lobby to provide more attractive finish to the ceiling finishes of the hotel.
13. Wall Finishes: 	• Hotel Puri maintains the existing wall finishes, especially in the hotel lobby. • • Most of the existing wall finishes at the hotel was damaged and uprooted. • Repairing works have been done to the wall finishes to follow the originals.
14. Floor finishes: 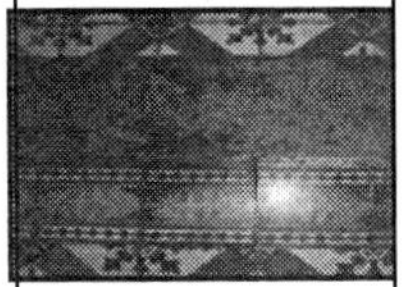	• Floor finishes at Hotel Puri are also almost entirely broken and uprooted. The floor finish has been replaced by a similar packaging with original tile specially imported from England.
15. Building facilities: 	• To ensure Hotel Puri provides maximum comfort and security to its guests, building facilities such as air conditioning systems and fire fighting system have been installed throughout the building. Although these building facilities crippled authenticity of Hotel Puri, building services is an important element that needs to be in a hotel.
16. Architectural decoration: 	• No additional architectural decoration done to the existing ones because they are quite interesting and has a value of its own. • There are many architectural jewels plastered, remain on the external walls of Hotel Puri. These decorations have a

	natural backdrop, like the decoration of plants and animals that are hard to find in other buildings.

Table 2: Case study 2 - Heeren House at Banda Hilir, Melaka

Case Study 2:	**Detail**
	Name: Heeren House, Melaka. Location: Core Zone. World Heritage Site, Melaka. Address: No.1, Jalan Tun Tan Cheng Lock, 75200 Melaka. Hotel type: Guest House. Existing function: Double storey shop houses. No. of rooms: 6 nos. Owner: Mr.Bernard.
Element	**Description**
1. Facade 	• The facade of the front, back and sides of Heeren House still stands as the original.
2. External wall: 	• The external walls of Heeren House remained as the original wall of the house. No changes or modifications made on the external wall of Heeren House.
3. Internal wall: 	• Renovations have been done on internal walls of Heeren House to ensure that guests can be served comfortably. The additional wall was built in each guest room to provide toilets and bathrooms in each room. • Renovations have been made to the wall in order to ensure Heeren House can be serve as a hotel.
4. Ground floor: 	• The existing ground floor was maintained. • Most of the broken and damaged floor tiles was replaced with new floor tiles.
5. Upper floor: 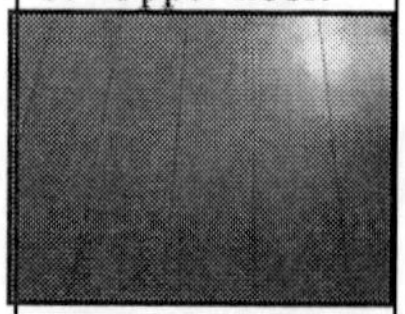	• The existing upper floor of the building was maintained. • The damaged timber floor was repaired while the rotted timber was replaced as the original one.
6. Column: 	• Heeren House is still using the original column structure as the existing structure is still strong. • Existing column was repaired on parts which are broken or perforated.
7. Staircase: 	• The Heeren House retains the existing structure of timber staircase. • Repaired work has been carried out for staircase and wooden handrails which have been damaged. Revarnish of staircase and handrail had been done to restore the beauty of the existing timber.
8. Roof structure: 	• The conservation work of existing timber roof structure was conducted by repair and replacement of defective and rotten parts of timber.
9. Door: 	• The main door of Heeren House is still using the original door but was repaired without referring to the original door. • A new door was installed in all guest rooms and also on toilet doors in the living room.
10. Window: 	• Many of the original buildings have been damaged and need new replacement windows. • All external windows have been replaced with new windows that have a design similar to the original window design.

11. Roof finishes 	• Heeren House still maintains China tiles which are specially brought from China as the finishing for roof. Damaged tiles were replaced with the same building materials obtained from the owners of adjacent buildings that are no longer used by them.
12. Ceiling finishes: 	• Ceiling Finishing on the top floor of Heeren House still uses the original ceiling molding. • Conservation and repairing works have been done on the damaged section of the ceiling.
13. Wall finishes: 	• Most of the walls were preserved. • New wall finishes have been installed in the bathroom / toilet in the room guests only.
14. Floor finishes: 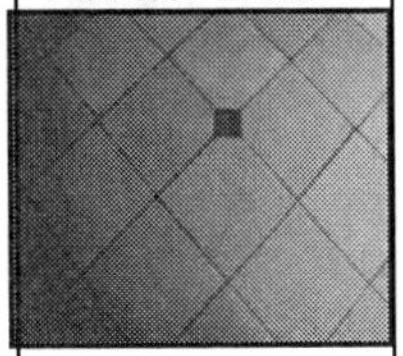	• Installation of the new floor finish has been done in the ground floor of Heeren House. Floor tiles were installed on a sidewalk in front of the guest house, a gallery and a cafe. . The new floor finishes looks like the original floor of the historic building. • In addition, new floor finishes are installed in the bathroom and toilet in the guestrooms. • A gallery and café was fitted with a new floor finish.
15. Building services: 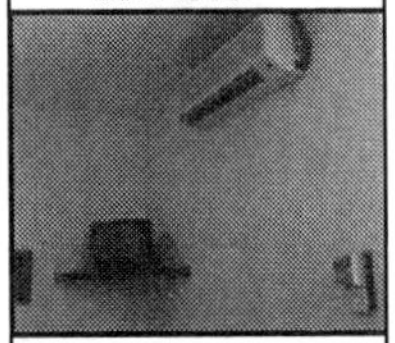	• Building services such as air conditioning systems have been installed in each room individually to ensure guests comfortable stay in the hotel. • New building services such as fire extinguishing systems have been installed to meet the requirements of the Fire Department.
16. Architecture decoration: 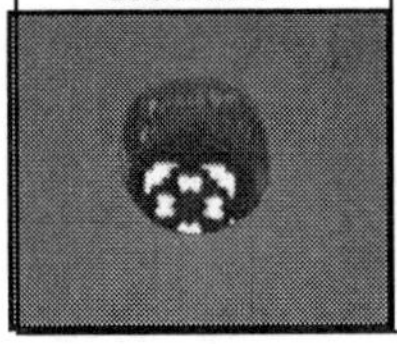	• The architectural decoration of the Heeren house is maintained as the original. • Heeren House does not have outrageous architectural decoration which needs any adjustment to be made.

Table 3: Case Study 3 - Cheong Fatt Tze Mansion at George Town, Pulau Pinang

Case Study 3:	**Detail**
	Name: Cheong Fatt Tze Mansion, George Town Location: Buffer Zone. World Heritage Site, George Town. Address: No. 14, Lebuh Leith, 10050 George Town, Pulau Pinang Hotel type: Boutique hotel Existing function: Mansion No. of rooms: 16 nos Owner: Mr. Laurence Loh
Element	**Description**
1. Facade: 	•The facade of Cheong Fatt Tze Mansion is preserved as the original. The beauty of the facade and small ceramic ornaments that characterize this building facade was preserved.
2. External wall: 	• Cheong Fatt Tze Mansion retains the original exterior walls. • The damaged parts of the wall repaired in accordance with the specifications of the original wall. Repair work on the damaged parts of the wall, done carefully, especially on the dated plastered walls. • The external walls were painted with indigo blue as can be seen today. •
3. Internal wall: 	• Conservation work has been done to repair the damaged parts of the wall. For the purposes of adaptive re-use, many parts of the wall modified to be used as a guest bedroom is inherent in the mansion. • Some renovation work has been done to the internal wall of the Cheong Fatt Tze Mansion to ensure that each guest room has a divider wall between the living room and a separate toilet / bathroom.
4. Ground floor:	• The ground floor used today is the existing floor of Cheong

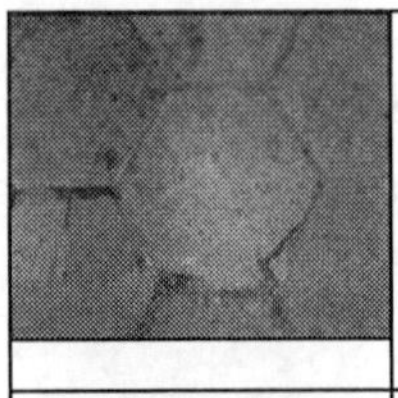	Fatt Tze Mansion. • The holes and cracks were repaired with suitable floor finishes. • No renovation work done on the ground floor of Cheong Fatt Tze Mansion.
5. Upper floor: 	• A lot of damage has occurred on the upper floors of Cheong Fatt Tze Mansion. • Floors that were destroyed had been replaced with a new floor by substituting timber that has similar characteristic to the original timber. • The timber floor was varnish to make it more durable and look more attractive.
6. Column: 	• There are two types of columns in the Cheong Fatt Tze Mansion, which is a column of wrought iron and concrete. Specially patterned wrought iron pillars which can be seen clearly in the air well are among the attractions in Cheong Fatt Tze Mansion. • The wrought iron columns remained in use as the elements are still in good condition. • The structure of the damaged concrete pillar was repaired and the painting work has been done to restore the faded color.
7. Staircase: 	• There are seven existing staircase in the Cheong Fatt Tze Mansion • Two of the seven staircases are spiral staircases made of wrought iron with interesting patterned on.
8. Roof structure: 	• Cheong Fatt Tze Mansion still use the existing roof structure. • However the repairing work have been done to the damage part. • Structure that was damaged and rotting was replaced with a new building materials as the original timber.

9. Door: 	• Cheong Fatt Tze Mansion still preserves the original main door at the entrance of the hotel similar to the other doors in the mansion. • At one time this old mansion was abandoned and not maintained, then many doors that have been damaged. • The door, which suffered severe damages have been repaired and repainted as the original door.
10. Window: 	• There are two types of windows in the Cheong Fatt Tze Mansion. • Windows with wooden barrel design 'gothic' and a window framed with glass panels. • Replacement of defective wooden barrel with similar timber has been carried out, followed by re-painting wooden windows using traditional organic methods. • Repair work to the wooden frames have been made and broken glass panel has been replaced with a new glass panel which was customized as the original pattern.
11. Roof finishes: 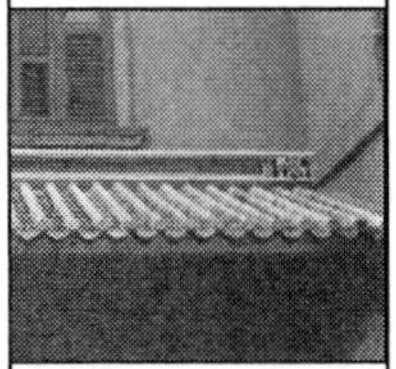	• Cheong Fatt Tze Mansion's roof has a very unique design that is hard to find on another building. • Conservation work was carried out by replacing the broken and damaged tiles purchased from the nearby building that used same type of tiles.
12. Ceiling finishes: 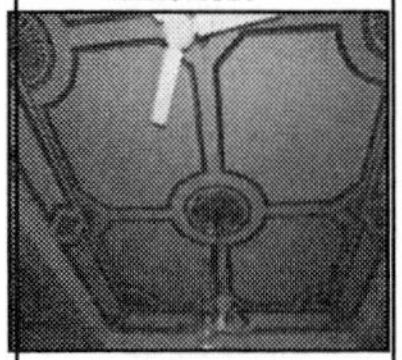	• Cheong Fatt Tze Mansion still preserves the original ceiling moldings made of pieces of boards that have attractive ornamental geometric pattern on it. • Work on repair and rehabilitation of existing ceiling moldings have been made especially for the opening lap which damaged due to age factor. • The ceiling was repainted to ensure that the ceiling look

DOI: 10.1051/matecconf/20166600030

	more attractive.
13. Wall finishes: 	• Cheong Fatt Tze Mansion used 'fresco' formed from lime plaster in many parts of the living room. • Most of it was damaged but conservation work done by using the grouting injection technique. • The fresco repainted and 'fresco' line was repainted in black Chinese ink mixed with glue
14. Floor finishes: 	• There are three types of floor finishes used in Cheong Fatt Tze Mansion. • First, the floor tiles with geometric patterned which have been used in most of the space inside the mansion. • Second, the terracotta floor finishes which have been used in most of the bedrooms. • The third floor finishes is a piece of granite that was used on the pavement and air wells. • Floor finishes at Cheong Fatt Tze Mansion also been severely damaged. The floor of the storage containers were washed and replaced with floor finishes as the original and specially imported from China. As for the broken granite slabs, they have been replaced with granite collected from nearby shop that no longer use in the building.
15. Building services: 	• Cheong Fatt Tze Mansion was built hundreds years ago. No building services installed in the mansion at the time. Most of the existing building services are installed today after the owners decided to adaptive re-use of Cheong Fatt Tze Mansion as a hotel. • Building services such as air conditioning have been installed to ensure hotel guests receive optimum comfort during the stay. Air conditioning has been installed in many parts of the Cheong Fatt Tze Mansion.
16. Architecture Decoration: 	• Architectural decoration at the outer wall of Cheong Fatt Tze Mansion has been carefully restored with the help of paid expert artisan. • The unique architectural decoration in Cheong Fatt Tze Mansion is made from ceramic glass taken from the broken bowls. The ceramics glass were permanently plastered with interesting patterns and colors in external and internal wall. • The most widely used patterned are animal elements such as lion and dragon that reflects culture of the building owner.

Table 4: Case Study 4 - Hutton Lodge at George Town, Pulau Pinang

Case Study 4:	**Detail**
	Name: Hutton Lodge, George Town
	Location: Buffer Zone. World Heritage Site, George Town.
	Address: No. 17, Jalan Hutton, 10050 George Town, Pulau Pinang.
	Hotel type: Budget hotel
	Existing function: Mansion
	No. of rooms: 26 nos
	Owner: Choong Lye Hock Estat
Element	**Descriptio**
1. Facade: 	• This building was involved in a fire back in 2003. • The facade of the hotel has been managed to be preserved as the original because it was not burned and the outside walls still clearly visible.
2. External wall: 	• The existing walls survived in the fire • The external walls of Hutton Lodge which survived in the fire have been restored. The works of scraping old paint, plastering the walls, especially the hollow and repainting the walls with new paint were done. • No modification is done to the external wall of Hutton Lodge.

3. Internal wall: 	• The internal walls which survived in the fire have been restored while the reconstruction is done for the wall which was destroyed by fire. • Several additions have been made to the wall to ensure Hutton Lodge can meet new function as a hotel. • The existing walls have been modified according to the new floor plan to change the function of from bungalow to a hotel.
4. Ground floor: 	• The ground floor which survived the fire has been rebuilt to get stronger floor structure.
5. Upper floor: 	• The reconstruction was done on the upper floor which completely burnt the timber floor when the fire broke out. • Fire Department has banned the reuse of timber floors due to fire safety. • A new concrete floor was built to replace the timber floor.
6. Column: 	• The existing column is conserved and conservation work has been carried out on the damaged part due to the fire. • No renovation work done to the column of Hutton Lodge.
7. Staircase: 	• The existing staircase which destroyed in the fire has been replaced with a new staircase using concrete and wooden handrail. The staircase was built in its original position. • As for the annex building, a new concrete stairses was built for the convenience of hotel guests staying in rear wing of the hotel.
8. Roof Structure: 	• The new roof structure was built using timber as the original structure was destroyed in the fire.
9. Door: 	• Doors in the Hutton Lodge has been replaced with new doors as the existing doors were destroyed in the fire. • For the main front doors, rear doors and side doors, wood paneled doors with unique design and quality finishes have been used. While flush doors have been used for all hotel room doors.
10. Window: 	• The windows in the Hutton Lodge has been replaced with a new frame and shutters to replace all the windows which was destroyed in fire. • Window with wooden ram and casement has been used as external window. While the iron grille was installed at ground floor windows to prevent intruders from getting into the hotel.
11. Roof finishes: 	• The new roof finishes was build to replace the existing roof finishes that had been destroyed in fire. • The roof tile patterns are the same as the original clay roof tiles patterns.
12. Ceiling finishes: 	• New ceiling finishes were built to replace the original ceiling moldings that were burnt in the fire • Asbestos free ceiling are used in the reconstruction of upper level of Hutton Lodge.
13. Wall finishes:	• There's no wall finishes in existing Hutton Lodge. Adaptive

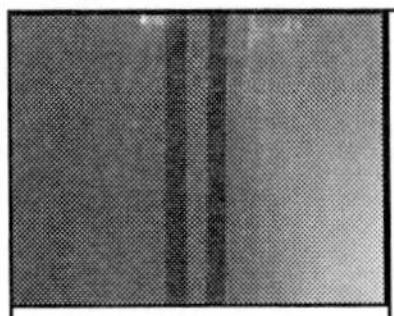	re-use work to change bungalow into a comfortable budget hotel floors needs wall finishes. • Wall tiles measure 1 'x 1' used in the toilet in each of the room and also in the shared toilet used by guests staying at dormitari.
14. Floor finishes: 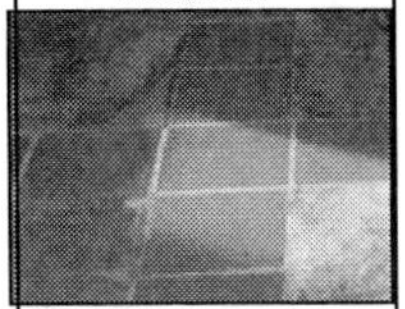	• The new floor finishes was installed in Hutton Lodge due to the existing floor finish has been destroyed by fire in the past. • Almost the entire ground floor of the Hutton Lodge using cement as floor finishes. • However, in certain areas such as on the front porch and the porch edge Hutton Lodge has a floor tile to form a pattern on the floor. • The carpet was installed in the guest rooms for the comfort of hotel guests. • 1'x 1' sized floor tile has been used in the toilet
15. Building services: 	• The original building of Hutton Lodge does not have any building services. • Building services such as air-conditioning systems, cctv camera systems, fire protection, internet and television services have been installed for the comfort and safety of the guests.
16. Architectural decoration: 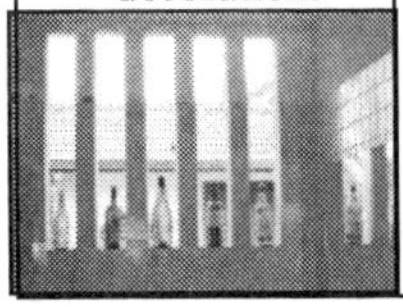	• Architectural decoration in the Hutton Lodge is very simple and has been retained as before.

6 Findings

The findings of this research will hopefully help people in gaining knowledge, understanding of data and information on adaptive re-use of historic building as a hotel and adaptive re-use principle that has been done based on the principles of conservation.

6.1 First Research Findings

The results of the study conducted found that most frequent element renovated building in the historic building as the hotel adaptive re-use frequency are the wall, a building, wall finishes, ceiling and floor finishes. These five elements have of percentages modification exceeds 10% as compared to the other elements.Wall is an elements that has the highest percentages of modification in adaptive re-use work of historic building as a hotel. Frequent modification done is the addition of walls in order to form more areas for living rooms and toilets. Unfortunately, most of the walls are not built in accordance with the original building material as the original existing wall.

6.2 Second Research Findings

The results of the observations that have been made over 4 selected hotels have shown that only two from four hotels achieved adaptive re-use principles in accordance with the principles of conservation.As a result of this finding, it indicate that most hotel owners are not focused on conservation principles outlined by the local authorities in carrying out the adaptive re-use of historic building works as a hotel.

6.3 Third Research Findings

The proposed assessment method is not only suitable for adaptive re-use of historic building as a hotel but also can be used to evaluate the adaptive re-use of historical buildings with other functions.Simple formula introduced the number of architectural elements preserved which meets the principles of conservation will be divided by the total number of elements that are evaluated. Then multiply by one hundred to obtain the percentage level of adaptive re-use made by the historic buildings.

References

1. Ahmad, A. G. and Badarulzaman, N. Adaptive Re-Use for Sustainable Heritage Tourism in Malaysia. *Journal of Tourism*. 51(2). 205-214. (2003)
2. Ashurst, J. *Conservation of Ruins*. Oxford, Butterworth-Heinemann. (2006).
3. Ashworth, G.J. and Larkham, P.J. Building a New heritage: Tourism, Culture and Identity in the New Europe. London, Routledge. (1994)
4. Australia International Council on Monument and Site (ICOMOS). ICOMOS International Cultural Tourism Charter: Principles and Guidelines for Managing Tourism at Places of Cultural and Heritage Significance. Victoria: Australia International Council on Monument and Site (ICOMOS) Incorporated. Retrieved 02/06/2015,from http://australia.icomos.org/publications/burra-charter- practice- notes/. (2002)
5. Bullen, P. A.and love, P. E. D). Adaptive Reuse of Heritage Buildings. *Structural Survey*. 29(5).411-421. (2011)
6. Department of National Heritage. 50 Warisan Kebangsaan. Kuala Lumpur, Kementerian

Perpaduan, Kebudayaan, Kesenian dan Warisan Malaysia. (2007)

7. Douglas, J. *Building Adaptation, 2nd ed.*, Heriot Watt University, Edinburg, United Kingdom, Butterworth-Heinemann. (2006)
8. Feildin B. M. *Conservation of Historic Buildings*, London, Butterworth Scientific. (1982)
9. Ismail, M.Y. Rehabilitation of Heritage Building in Malaysia. Undergraduate Dissertation, Shah Alam, Universiti Teknologi MARA.
10. Orbasli, A. Tourist in Historic Towns. London & New York: E & FN Spon.. (2000)
11. Orbasli, A. *Architectural Conservation*. Oxford, Blackwell Science Ltd. (2008)
12. Pedersen, A. Managing Tourism at World Heritage Sites: A Practical Manual for World Heritage Site Managers. UNESCO World Heritage Centre. (2002)

Office Occupants' Mood and Preference of Task Ambient Lighting in the Tropics

Lim Gene-Harn[1,a] Nila Inangda Manyam Keumala[1] and Norafida Ad. Ghafar[1]

[1] *Faculty of Built Environment, Universiti Malaya, Kuala Lumpur*

Abstract. Office lighting is vital for energy savings and occupant's visual comfort being. Energy efficient lighting has been studied extensively in the aspect of daylighting harvesting and task ambient lighting, predominantly in Temperate developed countries. However, the lack of evident on the effectiveness of low ambient task lighting in the Tropics has prompted this paper to investigate the practicality of such lighting design for the Tropics. This paper explores into the office occupant's mood and preferences of low ambient daylight and usage of task light for the office in the Tropics. It also discusses the recorded work plane illumination level upon the usage of such lighting system. It studies the impact of such lighting design on people's preference, work plane illumination for three office spaces with the different luminous environment. In regards to the source of ambient lighting, two offices harvest daylight, and another uses the conventional overhead lights in accordance with MS1525 recommendations. Three of the offices have 15 samples respectively and the workspace horizontal illuminance level is measured across a month. The results of before and after the provision of task light reveal that the acceptance for low ambient task lighting system and it is widely preferred. However, the task light is not necessary utilized as some people are adapted to do computer work under low ambient lighting (~50-150lux). It shows that people's preference for work plane illuminance varies greatly. Many agree that they are in control with their personal lighting environment when an appropriate glare-free task light is being used. The paper recommends that office luminous measurement shall take other lighting parameters such as the vertical illuminance and luminance ratio into consideration for future research.

1 Introduction

With all the increasing inhabitants and also commercialization of which pose increasing the need for building services and increasing time spent in building, office lighting design contributes substantially to the primary domains of energy efficiency and visual comfort. This paper aims to seek for efficient workplace lighting design with the Tropics which in turn fits this the latest aspiration of energy efficient green buildings and improved work productivity. With MS1525:2014 as a benchmark reference, this paper compares the post occupancy (POE) visual comfort preferences of the usage of an ergonomic task light for three office spaces with variations in the luminous environment. This is a continuation of a prior working pilot paper which faced limitation in the size of test subjects and measuring period [1]. However, due to technical limitations, the energy efficiency of the compared lighting systems is not explored. Section 2 looks into prior analysis testimonials and also forms the parameter framework for this lighting analysis. Section 3 describes the triangulation methodology for in-situ illuminance measurement and POE. The result and discussion are elaborated concurrently in Section 4. Section 5 concludes the findings and recommends further research.

[a]Lim Gene-Harn: limgeneharn@gmail.com

2 Literature Review

2.1 Office Lighting & Visual Comfort

Many review papers, which commonly fall into daylighting, have discussed the impact of workplace lighting towards human preference since late 20th century [2-7]. There exists an agreement that people have a preference over appropriate daylighting design as oppose to artificial lighting. Galasiu & Veitch [5] summarizes 60 papers on daylighting covering from 1965 and 2004 in understanding the interaction between occupants and daylighting. Boyce [3] reviews attempts on linking daylight and productivity while Sullivan et al. [7] looks into the methodology for assessing productivity in lighting environment. Nonetheless, occupant's interaction with daylighting also depends on the lighting automation system that will ensure effective lighting energy savings [2].

2.2 Daylighting

An effective daylighting system depends on how daylight is channeled and distributed throughout the depth of the room uniformly [8]. David Strong [6] presents that human health, happiness, and well-being are inextricably linked to daylighting across various building functions. Daylight also acts as a variation of luminance and colors that strengthen attractiveness, trigger emotions and affect our mood [9]. The luminous environment acts through a chain of mechanisms on human physiological and psychological factors, which further influences human performance and productivity [10]. Daylight has a significant level of blue spectrum light that stimulates the circadian system, which could affect people's alertness and mood [11,12].

However, daylighting is widely associated with the provisions of the window as people desire for a connection to the outside [5,13,14]. Libby [14] surveyed 155 subjects in Tennessee and found windows are the most desirable amenity in an office over desk location or size. A study on 8000 students in 450 classrooms in California also reveals that students appreciate vegetation or object views in the far distance, however, sun penetration and lack of window control can deteriorate learning [13]. Galasiu & Veitch [5] also studied that visual comfort is affected by the physical dimension and transparency of the window.

2.3 Low Ambient and Task-Based Lighting

However, daylighting is redundant when luminous ratio and illuminance ratio across the room vary greatly, which causes glare and dark spots [5]. Occupants most likely will alter the position of the blind due to the incoming sensation of heat and glare and it is never pulled up again, prompting artificial light to be switched on [15,16]. Nevertheless, various findings have shown that workers are able to tolerate low ambient light over desktop work (50-200 lux) [17-19]. Hence, the provision of task light enables worker's concentration by giving a variation of 1.5 to 3 times of the illuminance of the task plane to that of that background [2].

Occupants are also found to be in favor with the combined system with task lights as they have the sense of personalization and control over their localized lighting environment. Ultimately, Begemann et al. [20] survey on 170 subjects in Netherlands show that individual lighting settings differed greatly from one person to another, which depends on the individual's sensitivity to light, quality of sleep, biological clock, and weather type. Loe [21] agrees that occupants prefer a space with 'visual interest'- or a degree of light variation that stimulating; and different occupants have a different preference of luminous environment depending on the type of task, in which the task light comes useful.

2.4 Workplace Psychology

There exists a complex mechanism in understanding how indoor environmental parameters affects social behavioral where various factors may mask over the effect of each other [22-24]. There are various facets of physical environments such as ambient conditions, workstation and workspace design that will affect the psychological processes which determine the outcomes of satisfaction on performance and health [22]. Rashid & Zimring [24] explains additional intervention factors such as personal motives, aptitude, demographic factors and organizational work factors determine stress level.

A study of 10 office buildings in the Netherland also found that the view type and view quality and workplace density influence physical and psychological discomfort, but not the distance from windows [25]. Veitch et al. [23] elaborate that it is not the control over lighting, but those who work closely to their personal preference, show improved mood and higher satisfaction. The design of the workstation, particularly on the height of cubicle also affects productivity level via privacy and air circulation factors [26]. All these aforementioned factors on workplace satisfaction have to be accounted.

2.5 MS1525 Daylight Parameter

Prior to the formation of Green Building Index, MS1525 and other international standards have recommended the work plane illumination level for various functions as summarsized in Table 1. However, literatures discover that workplace people can tolerate lower illumination levels under daylight instead of artificial lighting. Escuyer & Fontoynont [27] shows that people preferred light levels between 100-300 lux for computer work and 300-600 lux for non-computer work. Alrubaih et al. [2]

Table 1 Recommended Illuminance Level from Various Standards

Standards/ Codes	Lighting Level for Task	Lighting Level for Ambient	Lighting Power Density (W/m^2)	Source
Malaysia Standards MS1525 (2014)	300-400	200	14.00	Code of Practice on Energy Efficiency and Use of Renewable Energy for Non Residential [36]
ASHRAE/ IESNA Standard 90.1-2007	300-400	320-530	11.90	American Society of Heating Refrigeration, and Air Conditioner Engineers [37]
Singapore Standards SS530:2006	500	300	15.00	Standards Council of Singapore [38]

DOI: 10.1051/matecconf/20166600031

Figure 1 Interior perspectives of three respective offices

Figure 2 Illustration of IKEA's TERTIAL Task Light usage in respective offices

emphasize that the IESNA recommended illumination level is tailored to minimize lighting energy and are not designed to account for the intensity of sunlight. People are able to tolerate much lower illuminance levels of daylight than artificial light, in particular throughout diminishing daylight conditions towards the end of the evening, and the ones working on the computer tend to choose low illuminances (100-300lux) [28].

3 Methodology

As the research investigates the effectiveness of low ambient task light luminous environment, three different government agencies offices with varying ambient lighting conditions are selected. The interiors of Energy Commission building (ECB) in Putrajaya, SEDA[a] administration office in Putrajaya and PWD[b] Tower Block G (PWD) in Kuala Lumpur are shown (Figure 1). All three office spaces have the same workplace density (10m^2/occupant) and horizontal apertures opening to wall ratio. All window views are directed to large horizons of space and not subjected to nearby obstacles, exceptionally for PWD tower which oversees the Kuala Lumpur skyline. The ECB and PWD offices are equipped with light activation switch via lux sensors while SEDA office relies on conventional master switch control.

15 random samples from open cubicle workspaces are chosen based on the variety of orientation and distance from fenestration for each building. All samples are either from the Finance or Human Resources department, sharing the same work task typology which is predominantly computer-based work with minor paper reading. With no sensor loggers in place, it is assumed that all samples are at their desk most of the time. 15 units of TENMARS TM-203 illumination meters are placed on each work plane to record the illuminance level over a month at 5 minutes interval. 4 units of HOBO U-12 loggers record the ambient temperature and relative humidity across the office space at 5 minutes interval.

The first half of the month records the current lighting level while the latter half records that of with the provision of an ergonomic task light (IKEA TERTIAL work lamp with 8W fluorescent daylight bulb) to each individual (Figure 2). The recorded illuminance results are trimmed down working hour only (8.30am-6.00pm) of working weekdays. Online questionnaires in Malay language are given twice, before and after the provision of task light, which covers individual demographic, controllability of lighting, satisfaction of lighting on paper and computer work and a set of derived Pleasure, Arousal and Dominance (PAD) scale by A. Mehrabian [29], cited in [30]. However, the 6 points Likert scale are used instead of the semantic differential scale to the convenience of the subjects. 6 points agreement and satisfaction scale are used; 'very dissatisfied, dissatisfied, somewhat dissatisfied, somewhat satisfied, satisfied, very satisfied'. An alternative "no opinion" choice is included to enable the respondent to abstain from responding.

[a] Sustainable Energy Development Authority
[b] Public Works Department

4 Results & Discussion

4.1 Illuminance Records

The recorded illuminance readings are presented in two ways; an average illuminance trend of selected loggers over the working hours, and Useful Daylight Illuminance (UDI) 100% stacked bar concept proposed by Nabil & Mardaljevic [31]. The average illuminance trend captures loggers that best represent all orientations and distance from window openings. Most workspaces located near fenestrations (less than 4 meters) have a great variation of illuminance readings and higher readings (>500lux) than that of far from fenestration (beyond 4 meters) (Figure 3, 4 and 5). Significant measured results from ECB show distinctive low illuminance levels (<200lux) at some workspaces, even when the workstations are near to the perimeter fenestration.

Due to the lack of task light usage sensor, the trend of task light usage is unknown from the measured results as there is no significant increase in illuminance level. This can be due to the logger is unable to capture the luminous coverage of the task light when in used. The measured results display a great diversity of illuminance level across all office spaces and most of them do not fall within the recommended range of 200-400lux by MS1525:2014. In order to quantify the occurrence of results within the MS1525 range, the alternative UDI interpretation reflects the percentage of occurrence for each logger's results (Figure 6, 7 and 8). The defined lux bracket range is altered from Nabil & Mardaljevic [31]; x<200, 200<x<400 and x>400 (where x denotes lux level) in order to benchmark with the MS1525:2014 targets. It is found that the cumulative readings that fall within the MS1525:2014 range are 26.18%, 30.09% and 27.82% for ECB, SEDA, and PWD, respectively.

4.2 Post Occupancy Evaluation – Visual Comfort

There is total of 30 sets of questionnaires collected for each office, comprising before and after the provision of task light. Two set of the result are removed from SEDA office and PWC office each, as the respondents only occupied the work desk less than 2 months. The difference analysis of preference and mood are compared using the Mann-Whitney U test via IPSS software. The correlations of office parameters as well as compounded demographic are examined using Spearman's rho correlation

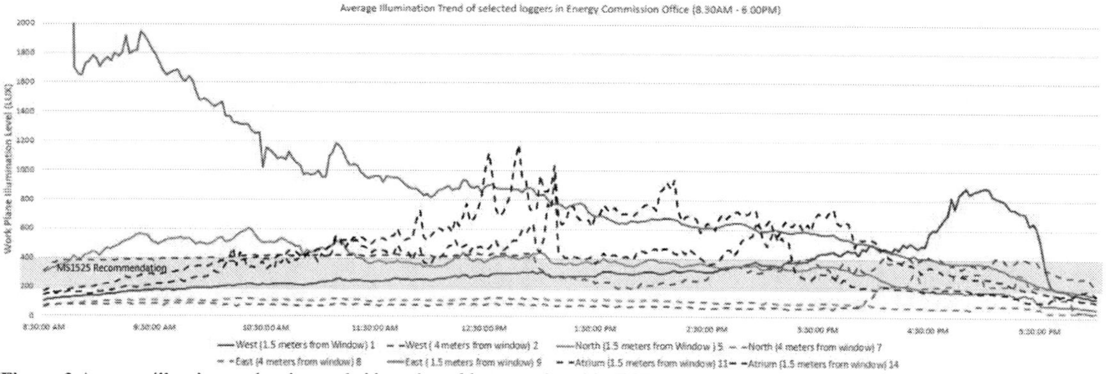

Figure 3 Average illuminance level recorded by selected loggers placed in ECB

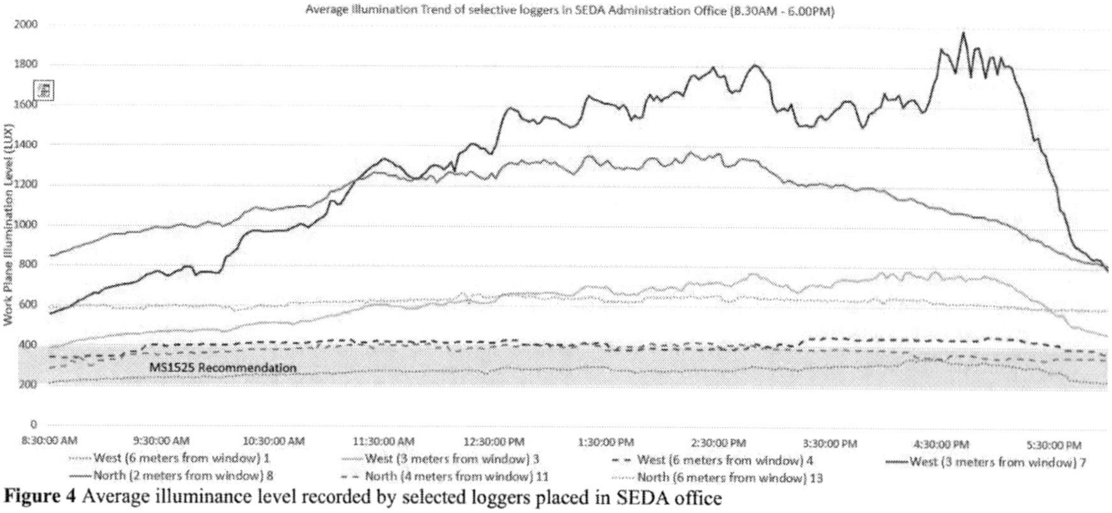

Figure 4 Average illuminance level recorded by selected loggers placed in SEDA office

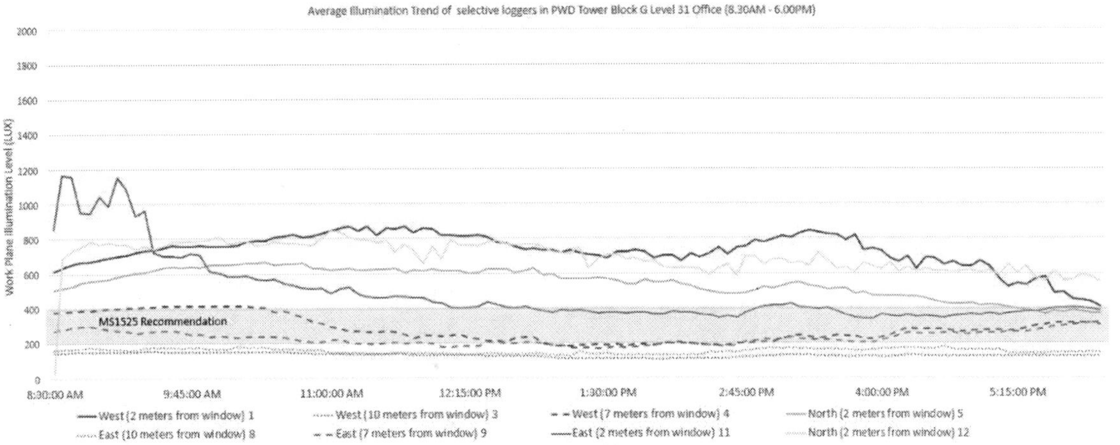

Figure 5 30 sets of recorded illuminance level on 15 samples in PWD Office

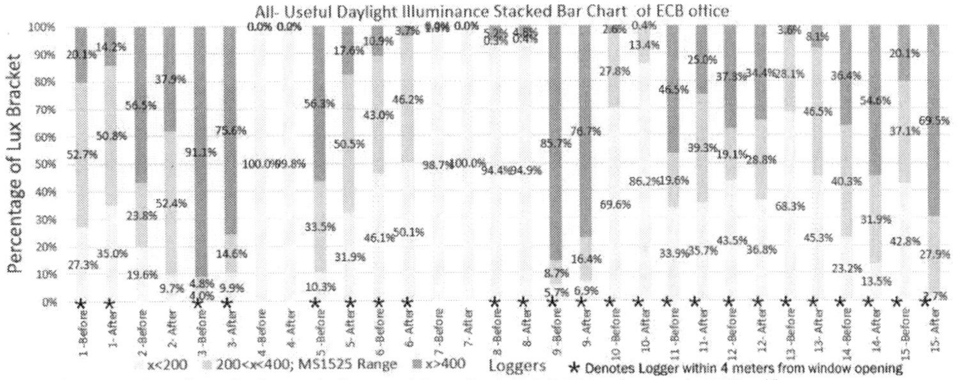

Figure 6 UDI of 30 sets of readings (before and after provision of task light) 15 lux loggers in ECB office

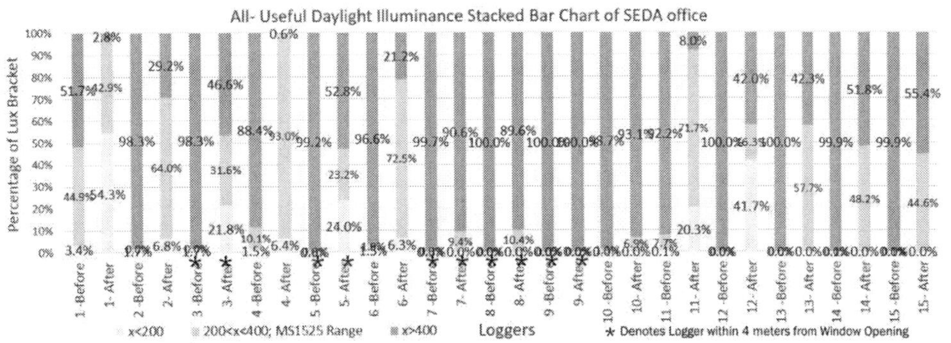

Figure 7 UDI of 30 sets of readings (before and after provision of task light) 15 lux loggers in SEDA office

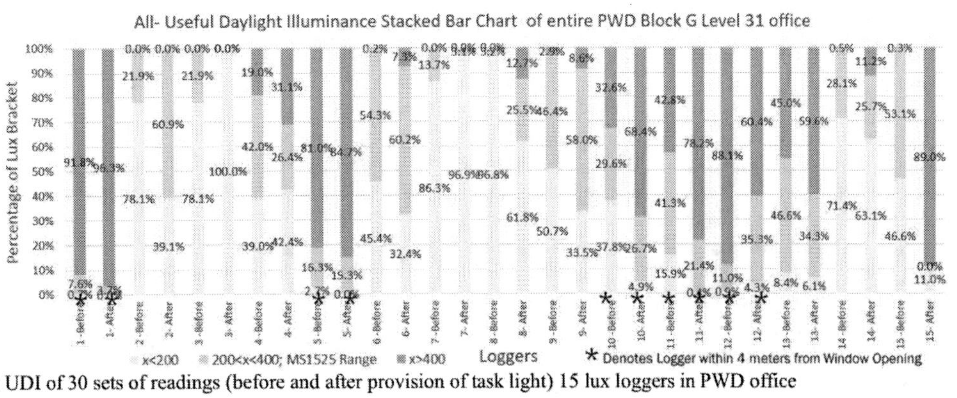

Figure 8 UDI of 30 sets of readings (before and after provision of task light) 15 lux loggers in PWD office

Table 2 Differences in lighting satisfaction results for all 3 office spaces

Satisfaction Level	Lighting Level	Paper Work Brightness	Paper Work Colour	Computer Work Brightness	Computer Work Colour
ENERGY COMMISSION BUILDING (N=30)					
MANN-WHITNEY U TEST	39.000	64.000	73.000	92.000	101.500
Z	-3.354	-2.122	-1.882	-.954	-.510
ASYMP. SIG. (2-TAILED)	.001**	.034*	.060	.340	.610
SEDA OFFICE (N=28)					
MANN-WHITNEY U TEST	98.000	28.000	33.500	95.500	7.000
Z	.000	-3.313	-3.133	-.137	-4.636
ASYMP. SIG. (2-TAILED)	1.000	.001**	0.002**	.891	.000**
PWD OFFICE (N=28)					
MANN-WHITNEY U TEST	60.000	49.000	47.000	68.000	66.000
Z	-1.852	-2.318	-2.445	-1.496	-1.556
ASYMP. SIG. (2-TAILED)	.064	.020*	.014*	.135	.120

(indicates P<0.05; ** indicates P<0.01)*

4.2.1 Preference Comparison

The Mann-Whitney U test results for paperwork brightness are significant for all three offices (Table 2). There are no significant findings on computer work brightness and the provision of task light. The provision of task light helps in paperwork task but not computer work task in all three offices. P. Boyce et al. [32] supports this finding as task light helps in fine color discrimination paperwork while the computer screen is already a source of the luminaire. Despite, only ECB shows a significant difference in overall lighting satisfaction (U= 39.00, P<0.01). This can be attributed to the low work plane illumination level (<200lux). SEDA and PWD offices also register a significant difference in paperwork color rendering satisfaction. SEDA office finds a significant difference in computer work color rendering (U= 7.00, P<0.01). The variation of results shows that different offices have a different luminous environment which is not representative by a single horizontal point illuminance measurement of a workspace [5,21].

4.2.2 Mood Comparison

Pleasure. 6 adjectives are used to represent the state of pleasure-displeasure mood. However, some adjectives are switched in order to suit the intended meaning due to the Malay language. Any significant difference found in mood is investigated for spearman's rho correlation with task light usage, window distance, the importance of view, controllability of local lighting and blinds, recorded range and variation of work plane illuminance level. The significance of difference of the various mood description related to pleasure to all 3 different office lighting scenes is shown (Table 3). For ECB, the lighting space is perceived as more 'Attractive' after the provision of task light however not significant (P>0.05). 'Cheerful' correlates with the frequency of task light usage (r= .648, P<0.01). Results from SEDA office shows a significant correlation in describing the lighting as 'Cheerful', 'Beautiful' and 'Dynamic' with the controllability with work space light (r=.379, r=.442, r=.431 respectively, P<0.05). There is a difference for 'Spacious' however not significant (P>0.05). There is a correlation of frequency of task light usage with 'Beautiful' (r= -.535, P<0.01) and 'Enthusiastic' (r= -.454, P<0.05). As for PWD office, there are significant correlations between 'Attractive' (r= -.567, P<0.01) and 'Cheerful' (r= -.564, P<0.01) to distance of nearest window. Controllability of workspace is significantly correlated with 'Attractive' (r= .512, P<0.01) and 'Spacious' (r =.501, P<0.01). Both recorded work plane illuminance range and diversity are significantly correlated with 'Attractive', 'Cheerful' and 'Spacious'. This means that the dynamics of luminous environment via blind control and work plane illuminance, together with perceived lighting controllability of workspace have improved the pleasure of lighting environment. Arousal.

Similarly, 6 adjectives are used to represent the state of arousal-non-arousal mood. The significance in difference of effects of lighting on mood is shown (Table 4). Results from ECB finds significant correlation of frequency of task light usage on 'Energetic', 'Excited', 'Wide Awake', 'Encouraged' and 'Happy'. Only 'Energetic' is found to be correlated significantly with window distance (r= -.421, P<0.05), importance of view (r= -.361, P<0.05) and diversity of work plane illuminance (r= -.421, P<0.05). SEDA office lighting results shows all adjectives which have significant difference 'Excited', 'Determined', 'Wide Awake', 'Encouraged' and 'Happy' are significantly correlated with the frequency of task light usage (P<0.01) and the controllability of the lighting (P<0.05). No correlation relationships are found for the Arousal related mood with work plane illuminance level or diversity.

Results from PWD office finds a significant correlation between distance of window and 'Energetic' (r= -.416, P<0.05), 'Excited' (r= -.506, P<0.01) and 'Happy' (r= -.636, P<0.01). There is also significant correlation of frequency of task light usage for 'Energetic', 'Excited', 'Wide Awake', 'Encouraged' and

Table 3 Difference in Pleasure related adjectives describing the lighting for all 3 office spaces

The lighting of this space is	Attractive	Cheerful	Beautiful	Spacious	Dynamic	Enthusiastic
Energy Commission Building (N=30)						
Mann-Whitney U Test	71.500	33.500	87.000	78.500	97.500	81.000
Z	-1.889	-3.487	-1.132	-1.489	-.649	-1.377
Asymp. Sig. (2-tailed)	.059	.000**	.257	.137	.516	.168
SEDA Office (N=28)						
Mann-Whitney U Test	66.500	58.500	52.500	59.000	56.000	51.000
Z	-1.647	-1.990	-2.305	-1.862	-2.205	-2.529
Asymp. Sig. (2-tailed)	.100	.047*	.021*	.063	.027*	.011*
PWD Office (N=28)						
Mann-Whitney U Test	32.500	54.500	58.000	42.000	68.000	85.500
Z	-3.245	-2.102	-1.992	-2.732	-1.422	-6.43
Asymp. Sig. (2-tailed)	*0.01***	*0.36**	*0.046**	*0.006***	.155	.520

(indicates P<0.05; ** indicates P<0.01)*

Table 4 Difference in Arousal effects describing the effects of lighting for all 3 office spaces

Light makes me feel...	Energetic	Excited	Determined	Wide Awake	Encouraged	Happy
Energy Commission Building (N=30)						
Mann-Whitney U Test	63.000	52.500	75.000	49.000	62.000	57.000
Z	-2.183	-2.684	-1.672	-2.859	-2.295	-2.420
Asymp. Sig. (2-tailed)	.029*	.007**	.095	.004**	.022*	.016*
SEDA Office (N=28)						
Mann-Whitney U Test	87.000	29.000	14.500	30.500	12.000	34.500
Z	-.539	-3.383	-4.030	-3.184	-4.306	-3.105
Asymp. Sig. (2-tailed)	.590	.001**	.000**	.001**	.000**	.002*
PWD Office (N=28)						
Mann-Whitney U Test	48.500	49.500	70.500	41.500	58.000	45.000
Z	-2.389	-2.417	-1.377	-2.778	-2.150	-2.579
Asymp. Sig. (2-tailed)	.017*	.016*	.168	.005**	.032*	.010*

(indicates P<0.05; ** indicates P<0.01)*

'Happy' (P<0.05). Blinds adjustment is significantly correlated with 'Excited' (r= -.442, P<0.05) and 'Happy' (r= -.552, P<0.05). Both work plane illuminance range and diversity are significantly correlated with 'Excited' and 'Happy' (P<0.01).

As shown from the variation of correlated luminous related parameters and mood adjectives across all 3 difference offices, the effectiveness of task light can be masked over by factors such as distance to the window, the importance of view or control of blinds. Sometimes the effectiveness of task light is supported by the perceived controllability of personal lighting environment and illuminance diversity, sometimes it contradicts with the frequency of the task light usage. There is no significant correlation found in work plane illuminance level and improved in the mood. This is possibly due to the illumination logger not registering the task light luminous coverage on work plane area.

4.2.3 Demographic and Workspace Parameters Correlation

Task light. Generally, there is a significant correlation between the usage of task light with overall light level satisfaction (r=.311, P<0.01) and controllability of workspace lighting (r= .764, P<0.01). Significant correlation is also found with paperwork lighting (r= .494, P<0.01), paperwork color rendering (r= -.436, P<0.01) and computer work color rendering (r= -.416, P<0.01) but not with computer work lighting (P>0.05). The usage of task light has a significant correlation with all Pleasure and Arousal related description elaborated in section 4.2.1 and section 4.2.2. This is no surprise. However, there is no significant correlation of usage frequency with the distance of window from the work desk, work plane illuminance level, and diversity. The usage of task light is regardless of work desk position and illuminance condition. This result agrees that individual lighting preference varies [5] and the lighting satisfaction may be due to the perceived controllability of light rather than actual illuminance level [23].

Distance to Nearest Window and Lighting

 DOI: 10.1051/matecconf/20166600031

Preference. It is found that the distance of window to work desk is correlated with the satisfaction of lighting level (r= -.341, P<0.01), and the importance of view (r= .399, P<0.01). This does not mask over the effectiveness of task light as a separate correlation analysis shows the usage of task light is regardless of distance to the window. There are significant correlations between lighting preference with distance to nearest window (r= -.618, P<0.01), work plane illuminance range (r=.269, P<0.05) and variation level of illuminance level (r= -.663, P<0.01) and the preference remains unchanged with the provision of task light (P>0.05). The preference for daylight autonomy is influenced by the position of the work desk and its work plane illuminance dynamics across the day. There is no significant correlation of lighting preference with age or gender or condition of eyesight found. The majority of respondents prefer office environment with daylight as oppose to total reliance on electrical light (Figure 9).

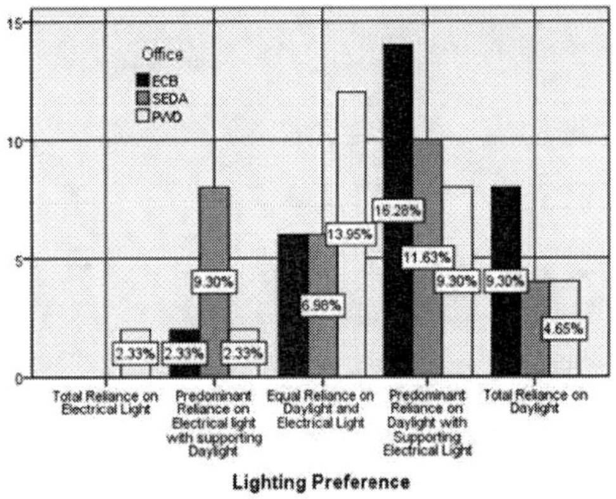

Figure 9 Lighting preference of 43 respondents across 3 offices

Age and Gender. The respondents' age has significant correlations with satisfaction of lighting level (r= .308, P<0.01) and importance of window view (r= -.266, P<0.05). This relates to the findings of Knez & Enmarker [33] that older people evaluated room light as less bright than younger counterparts. However, gender has no significant correlation with overall lighting satisfaction level, opposing many researchers' findings that females being significant higher in positive aspects via Positive Affect and Sensation (PASS) test [34] and prefer lower light level in various color temperature [35].

Illuminance & Satisfaction. There are no significant correlations found between work plane illuminance level and mood descriptions. However, the variation of illuminance range correlates significantly with the pleasure towards the lighting environment; 'Cheerful' (r= -.355, P<0.01), 'Spacious' (r= -.281, P<0.01), 'Dynamic' (r= -.596, P<0.01), 'Enthusiastic' (r= -.297, P<0.01); and the effects of the arousal, 'Energetic' (r= -.298, P<0.01), 'Excited' (r= -.320, P<0.01), 'Determined' (r= -.285, P<0.01), 'Encouraged' (r= -.244, P<0.05), 'Happy' (r= -.228, P<0.05). This means that lighting satisfaction is independent of illuminance level but the constant variation in illuminance level throughout the day instead. Some subjects agree that they are more prone to be annoyed by sudden activated electrical light due to lux set point compared to working under low illuminance level.

Indoor Environmental Parameters. Spearman's Rho correlation analyzed the correlation of each office space indoor environmental quality parameters and lighting satisfaction level. There are no correlations of satisfaction of lighting level with the satisfaction levels of indoor temperature, air quality, noise level, cleanliness and office layout for ECB office. Results from SEDA office shows a significant correlation between satisfaction level of lighting and indoor air temperature (r= -.471, P<0.05). Measured results from HOBO logger shows an average of 26°C across few respondents at a corner. PWD office shows a significant correlation between satisfaction on lighting and noise level (r= .405, P<0.05).

5 Conclusion & Recommendation

This paper investigates the effectiveness of task light with the different ambient illuminance level in 3 different government office spaces in the Tropics. Among the findings are:

•The lighting environment with the provision of task light elevates peoples' mood of pleasure and arousal.
•The provision of task light has increased the controllability of personal lighting environment, hence increased satisfaction on lighting level.
•The provision of task light helps in improving paper related task work, but not computer related task work.
•The frequency of task light usage is independent of illuminance level and distance from the nearby window.
•The lighting satisfaction is independent of illuminance level but the constant variation in illuminance level throughout the day instead, which supports why daylighting, even with low illuminance level is preferred.
•37% of respondents prefer office lighting environment with predominant reliance daylight with supporting electrical light.
•Individuals lighting preference varies and age are associated with, however not gender.
•Only ~30% of work plane illumination records fall within the MS1525:2014 recommendation range for the 3 offices, despite improvement in mood and satisfaction towards the lighting environment.

The finding recommends further investigation on the task light be evaluated with other lighting parameters such as vertical illuminance or luminance mapping of the work desk environment. The ergonomics of task light such as glare free hood and flexibility to the workspace needs have to be further explored also. This research is supported by High Impact Research grant H-130001-00-H000001 under Universiti Malaya.

References

[1] L. Gene Harn, N. I. Manyam, and N. A. Ghafar, "Mood and Preference of Daylight and Task Light Design for the Office Space in the Tropics - A Pilot Case Study," 2014.

[2] M. S. Alrubaih, M. F. M. Zain, M. a. Alghoul, N. L. N. Ibrahim, M. a. Shameri, and O. Elayeb, "Research and development on aspects of daylighting fundamentals," *Renew. Sustain. Energy Rev.*, **vol. 21**, pp. 494–505, (2013)

[3] P. R. Boyce, "Reviews of Technical Reports on Daylight and Productivity," 2004.

[4] L. Edwards and P. Torcellini, "A Literature Review of the Effects of Natural Light on Building Occupants," 2002.

[5] A. D. Galasiu and J. a. Veitch, "Occupant preferences and satisfaction with the luminous environment and control systems in daylit offices: a literature review," *Energy Build.*, **vol. 38**, pp. 728–742, (2006)

[6] D. Strong, "The distinctive benefits of glazing," 2012.

[7] J. Sullivan, M. Donn, and G. Baird, "Measuring Productivity in the Office Workplace," 2013.

[8] K. Johnsen, "Daylight in buildings, collaborative research in the International Energy Agency (IEA Task 21)," *Renew. Energy*, **vol. 15,** no. 1–4, pp. 142–150, (1998)

[9] P. R. Boyce, *Chapter 3 : Lighting quality*, 2nd ed. 2003.

[10] V. Gligor, "Luminous Environment and Productivity at Workplaces," Helsinki University of Technology, 2004.

[11] B. Plitnick, M. G. Figueiro, B. Wood, and M. . Rea, "The effects of red and blue light on alertness and mood at night," *Light. Res. Technol.*, **vol. 42**, no. 4, pp. 449–458, Dec. (2010)

[12] A. R. Webb, "Considerations for lighting in the built environment : Non-visual effects of light," **vol. 38**, pp. 721–727, (2006)

[13] L. Heschong and F. Oaks, "Windows and Classrooms: A Study of Student Performance and the Indoor Environment," 2003.

[14] B. A. Libby, "Windows in the Workplace: A Comparison between Window Preferences and Preferences for Other Office Amenities among White Collar Workers," 2001.

[15] T. Inoue, T. Kawase, T. Ibamoto, S. Takakusa, and Y. Matsuo, "The development of an optimal control system for window shading devices based on investigations in office buildings," 2014.

[16] M. . Rea, B. Rutledge, and D. Maniccia, "Beyond daylight dogma," *Daylighting '98 Conf. Proceeding an Int. Conf. Daylighting Technol. Energy Effic. Build.*, pp. 215–222, 1998.

[17] R. Kirsch and S. Volker, "Solid State Lighting in Offices: Impact on Lighting Quality and Room Appearance," *Proc. CIE 2014 "Lighting Qual. Energy Effic.*, **vol. 38**, no. 1, pp. 88–95, (2014)

[18] R. Baron, M. S. Rea, and S. Daniels, "Effects of indoor lighting (illuminance and spectral distribution) on the performance of cognitive tasks and interpersonal behaviors: The potential mediating role of positive affect," *Motiv. Emot.*, **vol. 16**, no. 1, pp. 1–33, (1992)

[19] C. Laurentin, V. Bermtto, and M. Fontoynont, "Light Source Type on Visual Appraisal," *Light. Res. Technol.*, pp. 223–233, 2000.

[20] S. H. a. Begemann, G. J. van den Beld, and A. D. Tenner, "Daylight, artificial light and people in an office environment, overview of visual and biological responses," *Int. J. Ind. Ergon.*, **vol. 20**, no. 3, pp. 231–239, (1997)

[21] D. Loe, "Energy efficiency in lighting -- considerations and possibilities," *Light. Res. Technol.*, **vol. 41**, no. 3, pp. 209–218, (2009)

[22] E. Sundstorm, "Workplace Environmental Psychology," in *International Encyclopedia of the Social & Behavioural Sciences*, 2001, pp. 16593–16598.

[23] J. A. Veitch, M. G. M. Stokkermans, and G. R. Newsham, "Linking Lighting Appraisals to Work Behaviors," *Environ. Behav.*, **vol. 45**, no. February 2013, pp. 198–214, (2011)

[24] M. Rashid and C. Zimring, "A Review of the Empirical Literature on the Relationships Between Indoor Environment and Stress in Health Care and Office Settings: Problems and Prospects of Sharing Evidence," *Environ. Behav.*, **vol. 40**, pp. 151–190, (2008)

[25] M. B. C. Aries, J. a. Veitch, and G. R. Newsham, "Windows, view, and office characteristics predict physical and psychological discomfort," *J. Environ. Psychol.*, **vol. 30**, no. 4, pp. 533–541, (2010)

[26] K. E. Charles and J. A. Veitch, "Environmental Satisfaction in Open-Plan Environment : 2 Effect of Workstation Size , Partition Height and Windows," 2002.

[27] S. Escuyer and M. Fontoynont, "Testing in situ of automatic ambient lighting plus manually controlled task lighting: office occupants reactions," *Proc. Lux Eur. 2001 9th Eur. Light. Conf.*, 2001.

[28] N. Baker, "We are all outdoor animals," in *Architecture City Environment, Proceedings of PLEA 2000*, 2000, pp. 553–555.

[29] A. Mehrabian, "Pleasure-Arousal-Dominance: A General Framework for Describing and Measuring Individual Differences in Temperament," *Curr. Psychol.*, **vol. 14**, no. 4, pp. 261–292, (1996)

[30] T. J. M. Van der Voordt, P. Vink, I. Bakker, and J. de Boon, "Pleasure , Arousal , Dominance : Mehrabian and Russell revisited," *Curr. Psychol.*, **vol. 10**, no. 33(3), pp. 405–421, (2014)

[31] A. Nabil and J. Mardaljevic, "Useful daylight illuminances: A replacement for daylight factors," *Energy Build.*, **vol. 38**, no. 7, pp. 905–913, (2006)

[32] P. R. Boyce, C. Hunter, and O. Howlett, "The Benefits of Daylight through Windows," 2003.

[33] I. Knez and I. Enmarker, "Effects of Office

Lighting on Mood and Cognitive Performance And A Gender Effect in Work-xRelated Judgment," *Environ. Behav.*, **vol. 30**, no. 4, pp. 553–567, Jul. (1998)

[34] C. L. B. McCloughan, P. A. Aspinall, and R. S. Webb, "The impact of lighting on mood," *Light. Res. Technol.*, **vol. 31**, no. 3, pp. 81–88, Sep. (1999)

[35] O. Shino, K. Akari, M. Tatsumoto, K. Okajima, and K. Hirata, "Preferable Lighting Conditions For Migraneurs to Relax in Room," in *Proceedings of CIE 2014 "Lighting Quality and Energy Efficiency,"* 2014, vol. 39, pp. 716–719.

[36] Departments of Standards Malaysia, *MS1525 - 2014 Code of Practice on Energy Efficiency and Use of Renewable Energy for Non Residential Buildings (2nd Revision)*. 2014.

[37] ASHRAE, "ASHRAE 90.1," 2004.

[38] StandardsCouncilofSingapore, "SS530:2006 Code of Practice for Energy Efficiency Standard for Building Services and Equipments," 2006.

Practicality and Performance of Daylight Trough in The Tropics: A Case Study

Lim Gene-Harn[1,a], Gregers Reimann[2] ,Gunaseelan Bhaskaran[2] ,Michael Hirning[2] and Morten Christensen[2]

[1]*Faculty of Built Environment, Universiti Malaya, 50603, Kuala Lumpur*
[2]*IEN Consultants Sdn Bhd, Bangsar, 59100, Kuala Lumpur*

Abstract. The field of office daylighting is vital for both energy efficiency practice and occupants' visual comfort. With the emergence of green building in Malaysia, building designers are exploring avenues for energy efficiency design; one common strategy is daylighting. The majority literature reviews on daylighting are skewed towards temperate or developed countries, where sky luminous condition is different from that of the Tropics. Conventional daylighting system designs redirect daylight from the envelope or atrium openings, such as light shelves. Presumed to be the pioneer daylight trough in the Tropics, this paper presents the simulated and in-situ lighting level measurements prior to occupancy. This case study presents an as-built daylight trough design which is able to daylight the office space as deep as 6 meters sufficiently. It achieves a lighting power density 1.90W/m^2 and saves 39.2% of lighting energy over conventional office lighting energy. Discussed further is the practicality of working with such a system, including cost implication, return on investment and contractual challenges in reaching a consensus on the design. The results reinforce that the effectiveness of daylighting design is very dependable on the sun path and obstacles surrounding the office tower. The RADIANCE simulation correlates well to field measurement results. Further investigation into the light trough, its lighting energy savings, users' interaction, visual comfort, and glare is still ongoing.

1 Introduction

Daylighting proves to be one of the primary research fields in sustainable architecture design. Li & Tsang [1] explore 35 commercial buildings built in Hong Kong from the year 1962 to the year 2004, showing there is a distinctive changing trend in fenestration system design with respect to glass type, window area, and shading devices. In the 1960's, following the emergence of artificial lighting, air conditioning, and the need for larger spaces in schools depreciated the importance of daylighting for schools in California [2]. However, the popularity of daylighting emerges along the popularity of green building rating tools across the world which emphasized indoor environmental quality and energy efficiency since the 1990s [3].

Daylight is superiority in its matching visual response of human beings, which translates into visual and non-visual benefits [4]. Boyce et al. [5] further elaborate that lighting conditions can influence the performance of individuals via the visual system, circadian system, and the perceptual system. Yu & Su [6] reviewed 26 papers and found that the amount of induced lighting energy savings from daylight ranges from 30%-87%.

Conventional fenestration with the use of horizontal blinds brings the depth of daylight distribution to an average of 1.7 times the height of the window [7]. Over the last fifty years, the development of a number of reflective and refractive materials has made redirection of daylight much deep into a building possible. Mayhoub & Carter [8] categorize two main approaches for daylighting strategies; 'beam daylighting' – adding reflective elements to conventional façade, and 'light guidance' – capturing daylight using collector devices to transport it into deep areas of the building. Focusing on the latter approach, this paper intends to compare the daylight performance of a recently installed daylight trough in Malaysia.

A brief review of light guidance systems is presented in Section 2. Section 3 describes the methodology for both via simulations and field measurements of the light trough. Section 4 then discusses the daylight performance via Daylight Factor, illumination distribution, and potential energy savings. Various design and maintenance related obstacles faced by the authors in regards to the daylight trough are discussed in Section 5. Section 6 summarizes the findings and discusses the scope of further work.

2 "Light Guidance" Daylighting Strategy in the Tropics

Ruck et al. [9] evaluate more than 30 types of daylighting systems under the International Energy Agency (IEA)

[a] Lim Gene-Harn: limgeneharn@gmail.com

Task 21 and proposes two groups of daylighting system, those with and without shading. The daylighting systems with shading, such as Venetian blinds and light shelves, are further divided into systems that rely primarily on the diffuse skylight and reject direct sunlight, and systems that primarily direct sunlight. The daylighting systems without shading, designed primarily to redirect daylight to areas away from the window are further categorized into another four types of systems; diffuse light guiding systems, direct light guiding systems, light scattering or diffusing systems and light transport systems. The effectiveness of a daylighting system depends on the climate, sun path, site, room design, window characteristics, artificial lighting system, and usage of the workspace [9]. The brief review in this section looks into the latter category, specifically anidolic integrated ceilings (AIC).

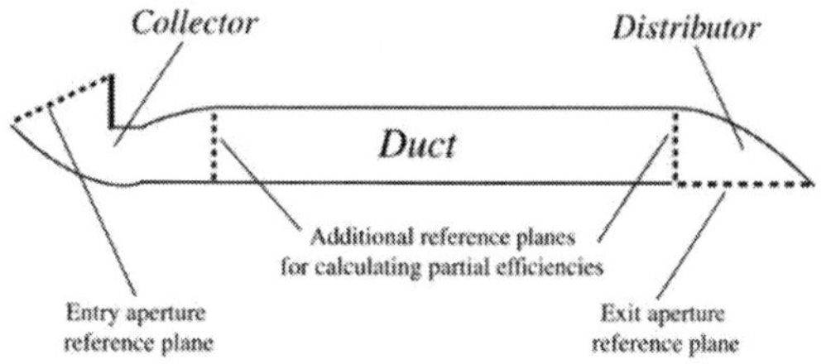

Figure 1 Components of the Anidolic Integrated Ceiling [10]

The use of AIC is unconventional in the Tropics, however, it is the system most relevant to the daylight trough be unconventional in the Tropics at the time of writing, however, most relevant to the daylight trough studied in the later section. There are limited studies on the use of AIC in the Tropics and all of them are either simulated or tested in a laboratory setting. Referring to Figure 1, AIC is designed using non-imaging optics theory and it has three major components: the collector on the external side, the rectangular mirror light duct and the distributing element at the end of the duct [10]. Wittkopf [11] tests the daylight performance of AIC under 15 different sky conditions and various sun altitudes, concluding that the daylight improvement through AIC is most significant in Singapore, compared with sky conditions in Japan's Fukuoka and United Kingdom's Sheffield. The usage of AIC in Singapore improves its illuminance ratio improvement factor (IR IF) by 3.3 and reduce glare by 14% over the simulated base case. Furthermore, AIC works best under an overcast sky with high sun altitude such as in the Tropics and distributes light across the depth of the room more evenly [11].

Three different anidolic systems (anidolic integrated ceiling, integrated anidolic system, anidolic solar blinds) are fitted to a 6.6m deep room in Switzerland [12]. AIC is found to contribute 1.7 and 2.7 fold increases in the inner room (4-6 meters from fenestration) daylight factor under overcast sky conditions in a rural and urban environment respectively absent of physical obstruction. Moreover, post occupancy evaluation reaffirms that light flux is directed to the rear part of the room without glare risk and the colors are found to be more pleasant in the test room although it is physically identical to the reference room [12]. It is found that these various anidolic daylighting systems perform differently dependent upon sky condition, sun path and urban fabric.

Linhard et al. [10] evaluate the parametric factors affecting the efficiency of AIC using Photopia software under Singapore sky condition as determined by Wittkopf [11] earlier on. It is found that the coating modifications, length modification, external shading and width modification affect the efficiency of AIC by 31%, 24%, 18% and 5% respectively. AIC efficiency is measured by the calculated exit flux of the exit aperture plane over entry flux of the entry aperture plane. Linhart & Scartezzini [13] discovers the lowest lighting power density (LPD) in the AIC fitted office in LESO solar experimental building is less than 4.5W/m^2. They then improvised the lighting distribution in regards to the interior fit out with feedbacks from 20 subjects finding that an LPD of 3.9W/m^2 is feasible. Ruck et al. [9] also writes that office test room with AIC used 31% less electricity for lighting than a reference office room with 6.6 meters depth.

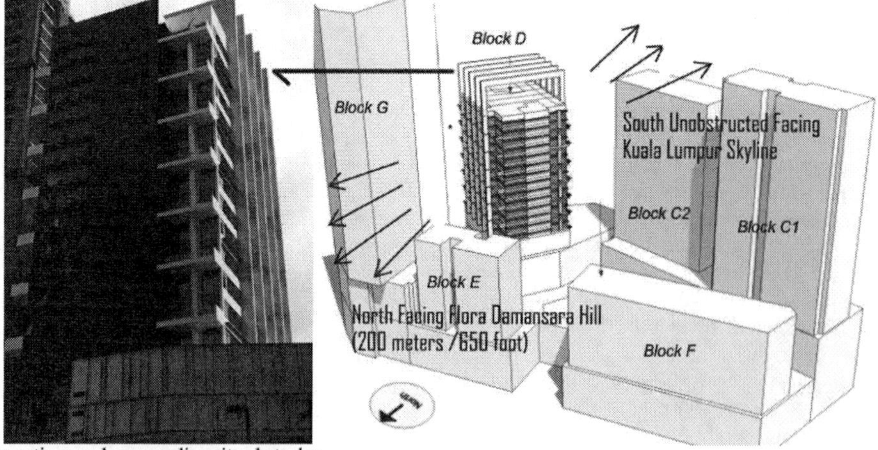

Figure 2 Perspectives and surrounding site obstacles

Table 1 Assigned Parameters of the Material Specification

PARAMETERS	SIMULATION SPECIFICATIONS			AS BUILT		
Solid Surface	Reflectance (%)	Specularity (%)	Colour	Reflectance (%)	Specularity (%)	Colour
External Wall	80	0	Metal White	80	0	Metal White
Internal Wall	72	0	Light Gray	72	0	White Plaster
Internal Partitions	72	0	Light Gray	72	0	Light Gray
Floor Finish	30	0	Gray	30	0	Bare Concrete
Ceiling Finish	90	0	White	90	0	White
Int. Columns	72	0	White	72	0	White Plaster
Ext. Horizontal Fins	50	0	Gray	50	0	Metal Gray
External Vertical Fins	50	30	Gray	50	30	Metal Gray
DT-Reflective Coat	95	90	Mirror	95	90	Alanod Mirror
Glazing	Material	VLT	Colour	Material	VLT	Colour
Vision Window	Glass	42%	Clear	Glass	42%	Clear
DT-External Glazing	Glass	90%	Clear	Glass	90%	Clear
DT-Internal Glazing	Glass	90%	Clear	Glass	90%	Clear
Roller Blinds	Glass	5%	Black	Plastic	3-5%	Plastic Blue
Other Conditions	**Simulation Specifications**			**As Built**		
Sky Condition	CIE Overcast Sky			Afternoon Overcast Sky (Okta 3-5)		
Surrounding Obstacle	Surrounding Office Blocks			Surrounding Office Blocks & Hill		
Height of Blind	1200mm from Floor Level (South only)			1200mm from Floor Level (South only)		
Interior Fittings	No Fit Out			No Fit Out		

3 Methodology

A similar daylight delivering strategy to AIC termed as Daylight Trough (DT), is designed and optimized for Menara Mustapha Kamal Block D located in Damansara Perdana, Malaysia (Figure 2). The 21 stories provisional GBI Gold certified office tower has an average 1,193m2 of net lettable area per floor with office depth ranging from 9-16 meters and alternate level of sky gardens. The methodology to assess the performance of the DT via simulation and field measurement is described in the following sections.

3.1 IES-VE Simulations

The simulation software used for this study is Radiance IES [14]. The simulated model also takes into consideration adjacent building shading effects. The assigned properties barriers, reflectance, specularity and visible light transmission (VLT) are described (Table 1). The anodized reflectivity material within DT is MIRO95ax4 Extra Bright from Alanod [15]. Daylight Factor (DF), defined as the ratio of indoor illuminance and outdoor illuminance is simulated under CIE overcast sky conditions [16].

3.2 Field Measurement

A total of 13 TENMARS TM-203 illuminance loggers were synchronized to record at 10-second intervals concurrently. Two units were placed on the roof of the building to ensure there was no surrounding obstruction. A mirror plane was used to shade one logger against the direct sun while the other was left unobstructed (Figure 3). Using this setup, the readings from both loggers were compared the estimate the sky condition. If both loggers registered similar illuminance values, the sky condition was assumed to be overcast (i.e. 40k-60k lux). This data was used to determine the validity of the DF measurements. DF measurements for both North and South oriented offices separately were conducted. 11 loggers were placed at 1-meter intervals from the fenestration at a work plane height of 800mm. Two HOBO U-12 loggers were placed on the workplace and inside the DT respectively to monitor the ambient temperature.

Measurements were conducted on the 9th December 2015. The winter solstice sun path causes direct sun exposure to the South oriented office. This is likely to cause a glare problem for occupants. Therefore, plastic sheets with measured VLT of 3-5% were placed at 1200mm above floor level to act as temporary blinds. However, no blinds were placed for the North oriented office as the adjacent tower and hill block any glary direct sunlight or sky view from the facades. The height of the temporary blind was also changed to 1000mm above floor level which fully shut the fenestration. The illuminance contribution from DT only was obtained at the South oriented office via computing the average difference in 10 pairs of lux readings by blocking the DT exit aperture with the similar temporary blind. There was no artificial lighting in operation at the time of measurement.

DOI: 10.1051/matecconf/20166600032

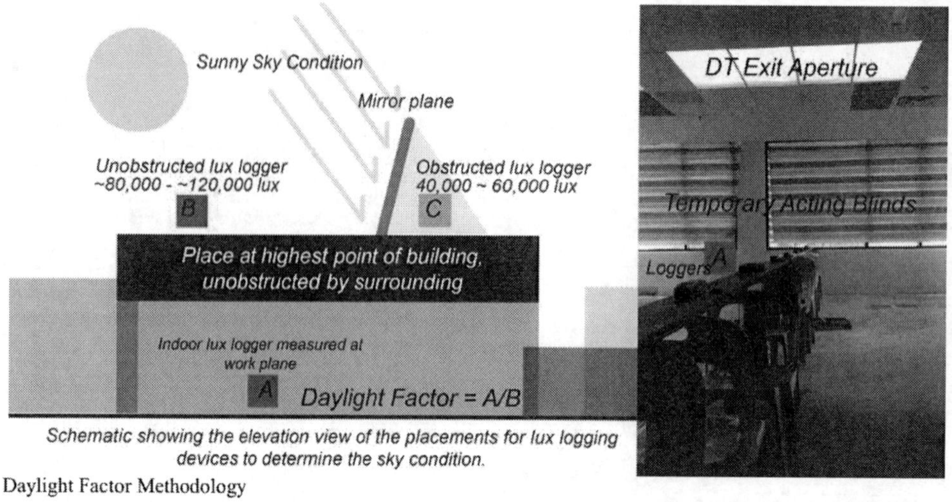

Figure 3 Daylight Factor Methodology

4 Results – DF, Illuminance & Energy Savings

The measured and simulated DF for both South and North oriented offices are plotted in Figure 5 and 6. Figure 5 displays the measured and simulation DF from the DT and window glazing separately. Under the definition of useful DF (1-3.5%) by Green Building Index [17], the result shows that the DT is capable of delivering up to 6 meters into the office space. As an alternative to DF, Wittkopf [11] proposes the usage of Illuminance Ratio Improvement Factor (IRIF), defined by the ratio of improvement of IR in design case over IR in the reference case. The IRIF for both North and South oriented offices over the reference case is summarized (Table 2). It was found that the DT has an average IRIF of 17.16 and 5.87 for 3-6 meters depth from the fenestration for the South and North oriented office respectively. The correlation between the simulated and measured results for South oriented office is high (r=0.7577) while that for North oriented office is very low (r=0.0913). The correlation result is discussed in section 5.

Figure 5 compares two blind configurations for the South office; blinds raised to a height of 1200mm from the floor, at a VLT of 1-3%, and the blinds lowered to a height of 1000mm from the floor at a VLT of 10-20%. Figure 6 shows the variation of illuminance distribution within the range of 200-2000lux for the first 6 meters from fenestration. It also shows that with 200mm difference in blind height causes a significant increase in illuminance, at least for the first 2 meters. The recorded illuminance data was also formatted to display Useful Daylight Illuminance (UDI) stacked bar chart for easier reference [18] (Figures 8 and 9).

The expected lighting energy savings from the DT are also investigated. The calculated LPD for the office is 5.25W/m^2 and 1.90W/m^2 without and with the assistance of DT, both well within the MS1525:2014 maximum requirement of 10W/m^2 [19]. The general lighting set point for the office is 250 lux. Using the annual hourly weather file for Kuala Lumpur, the electrical lighting savings from the daylight system by on/off controls for the lights has been calculated to be 39.2% and 13.9% for South and North oriented office respectively and 25.18% overall average. The artificial lights consist of T5 fluorescent lights which are controlled by lux sensors. There is no attempt to calculate cooling load savings, however, the measured ambient temperature differences between the light trough and work plane are 3°C to 4°C and 0.5°C respectively for the South (clear sky) and North oriented (overcast sky) office spaces. (Figure 4)

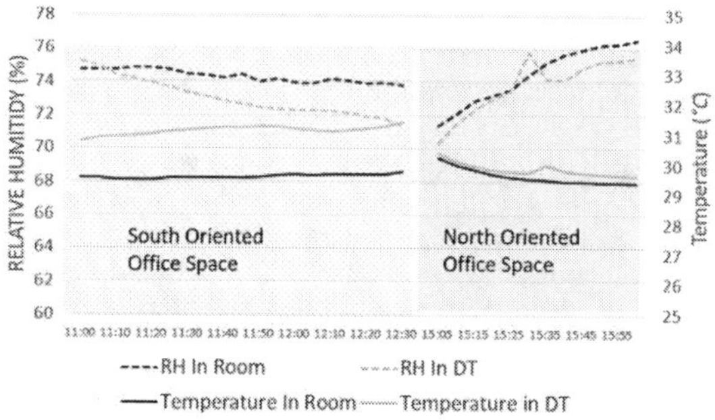

Figure 4 Ambient temperature and relative humidity differences of DT and work plane level

5 Discussion

5.1 Daylight Performance

The average IRIF of the DT from 3-6 meters depth is 17.16, as oppose to Wittkopf [11] findings on IRIF ranging from 2.0 to 3.5. Also in comparison, Scartezzini & Courret [12] finds the average DF in the rear half room is multiplied by 1.7 in the LESO building. Two of the major factors that contribute to the difference in performance may be the anidolic collector design and sky condition. The AIC with anidolic collector tested in LESO by Scartezzini & Courret [12] is designed to

Table 2 Comparison of simulated and measured DF and IRIF results

Meters from Glass (M)	1	2	3	4	5	6	7	8	9	10	11
Simulated Case											
South Base Case	0.50%	0.20%	0.15%	0.11%	0.08%	0.05%	0.03%	0.01%	0.00%	0.00%	0.00%
North Oriented with DT	0.30%	0.20%	0.90%	1.70%	1.20%	0.40%	0.20%	0.10%	0.03%	0.01%	0.00%
South Oriented with DT	0.50%	0.30%	1.00%	1.70%	1.30%	0.60%	0.40%	0.10%	0.05%	0.03%	0.01%
Field Measurement											
North With DT	1.44%	0.90%	0.65%	0.69%	0.49%	0.34%	0.27%	0.19%	0.17%	0.15%	0.14%
South With DT	1.72%	1.20%	1.53%	2.11%	1.49%	1.03%	0.70%	0.40%	0.34%	0.21%	0.20%
Illuminance Ratio Improvement Factor IRIF (Measured)											
North Oriented	2.88	4.50	4.34	6.31	6.08	6.77	9.13	18.76	0.17	0.15	0.14
South Oriented	3.44	6.00	10.20	19.19	18.60	20.61	23.50	39.65	0.34	0.21	0.20

achieve a high angular selectivity that collects diffuse light from the upper half hemisphere. The measured DT with a deep overhang is intended to shade it from afternoon sun's direct light, which has the lowest altitude of 65 degrees. Unlike LESO's AIC, the DT collects only diffuse light, which comes primarily from the hemisphere. The difference in physical details such as the opening aperture exiting aperture ratio and height of duct are significant, but not described. The relatively high DF near perimeter found by Scartezzini & Courret [12] indicates that no blinds in used, which is not the representative case for Tropics. Chaiwiwatworakul & Chirarattananon [20] measured the luminous efficacy of various sky conditions in Bangkok and finds that variation of efficacies is influenced by the sky conditions and sun positions. With the higher solar insolation and higher latitude sun path, and also as simulated by Wiitkopf [11], the DT is proven to be more efficient in the Tropics. The South oriented office shows a significant improvement in illuminance in the first 6 meters from the façade (primarily the span below the exit aperture of DT). However, the North oriented office shows only little improvement as it faces a major tower building and hill. In accordance to the Malaysia Standards MS1525:2014 and the Green Building Index NonResidential New Construction tool (NRNC), the lighting illumination requirement for office space is 300-400 lux or 1-3.5% DF [17,19]. These standards apply only for artificial lights as various reviews have affirmed that subjects accept daylight illuminance ranging 200-2000 lux [4,18,21]. This corresponds well to the DF of 0.5-3.5% instead of 1-3.5%, as an external illuminance under overcast sky ranges 30k lux to 50k lux [7].

The lighting energy savings do not entirely depend on the coverage of the DT but the lighting control system and positioning of artificial lighting as well [13]. It can be expected that only the two nearest rows of artificial lighting to the perimeter will be assisted by the DT. As elaborated by Linhart & Scartezzini [13], the positioning of office interior and task area matter as much in ensuring appropriate illuminance level in right places. Also, individuals lighting expectation is influenced by previous office experience so there should be localized form of lighting control such as energy efficient and ergonomic task light [4,13,22].

5.2 Radiance & Daylight Factor

The correlation between simulated and measured DF proves the reliability of radiance program in simulating the effectiveness of the DT daylighting strategy. However as shown in the North oriented office, the simulated contribution from DT via radiance does not show the significant shading impact from the adjacent building when compared to the measured result. The discrepancy indicates the possibility of IES-VE not registering the adjacent buildings blocking the daylight which needs to be further investigated. The measured DF is found to be constantly higher than what radiance simulated, this may be due to the unrealistic CIE Overcast Sky which does not represent the actual dynamic sky condition as argued by many reviews [6,16,18,21,23]

5.3 Payback Period

The total additional cost of the DT per meter square office area (Nett Lettable Area) is RM38.21/m^2 (\$8.70/m^2) which includes the additional reflective aluminum sheet from Alanold and Gypsum Board. The operational energy savings computes the avoided lighting energy usage with the medium commercial tariff rate (C1) at the price of RM0.365 (\$8cents) from the national electricity provider TNB. Lamp replacement cost annually is reduced due to the increased lifespan of T5 lights. This assumes RM40 for the cost per lamp and RM7 for labor charges. The expected payback period for the DT is 20.4 years.

However, as noted by Mayhoub & Carter [8], the conventional payback period does not capture intangible items such as building cooling savings, carbon tax savings, residual value, enhanced corporate prestige, and most importantly, the effect of daylight on human performance. A study from Carnegie Mellon University (CMU) shows that costs of an employee 160 times that of energy. Conventional daylight from windows yielded energy savings of £7/m^2 and annual productivity gains of £142/m^2 [8].

 DOI: 10.1051/matecconf/20166600032

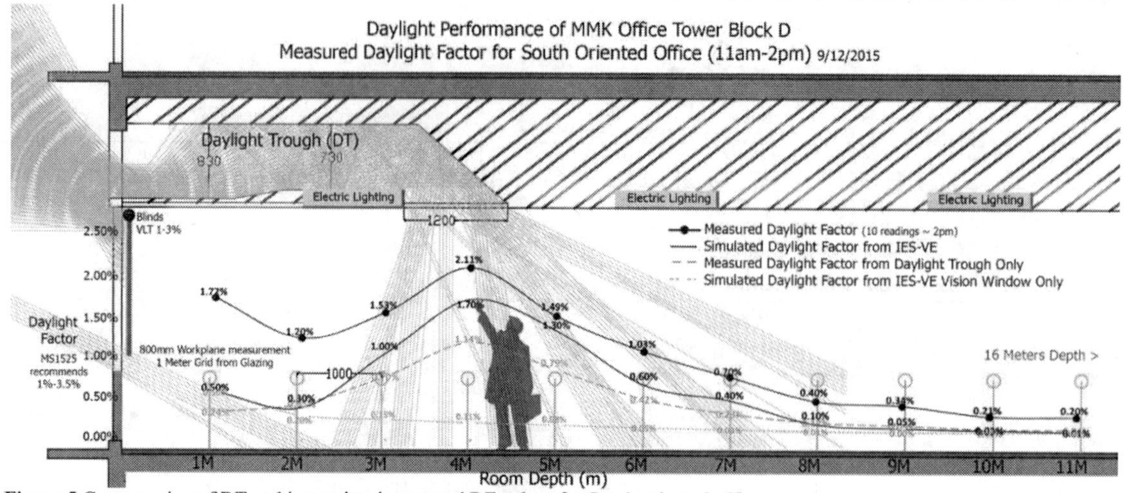

Figure 5 Cross section of DT and interpolated measured DF values for South oriented office

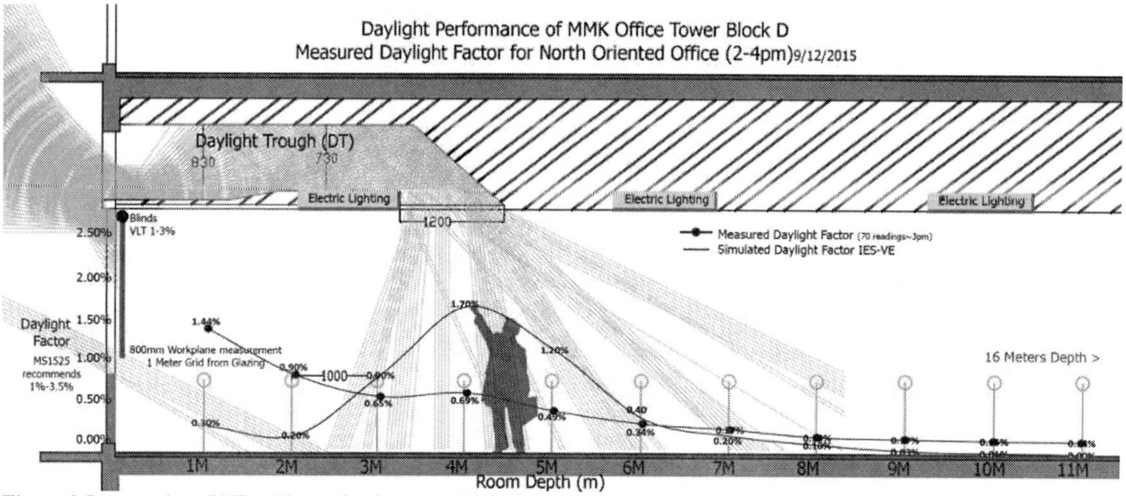

Figure 6 Cross section of DT and interpolated measured DF values for North oriented office. Both measured and simulated results do not include the usage of blinds.

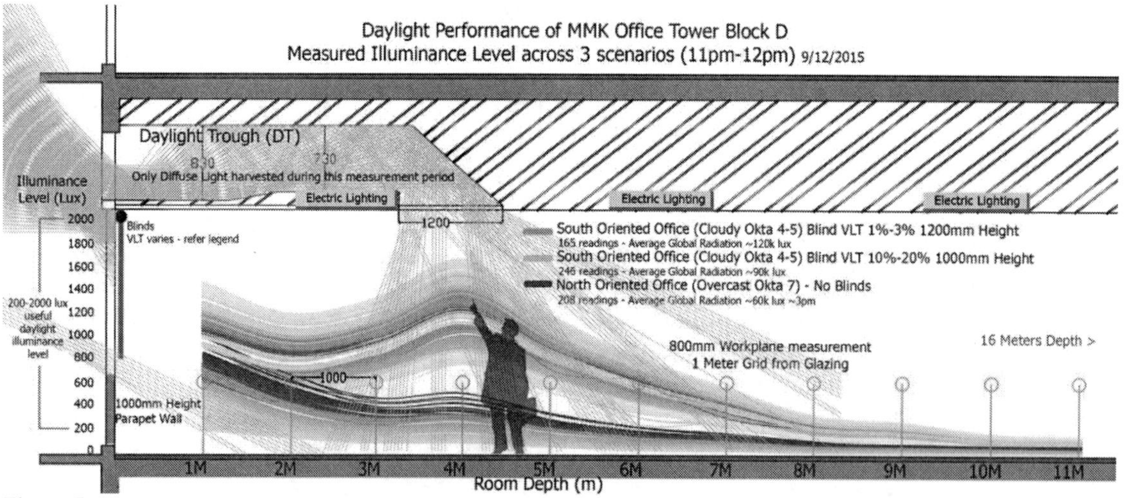

Figure 7 Continuous measured illuminance level for South and North oriented office

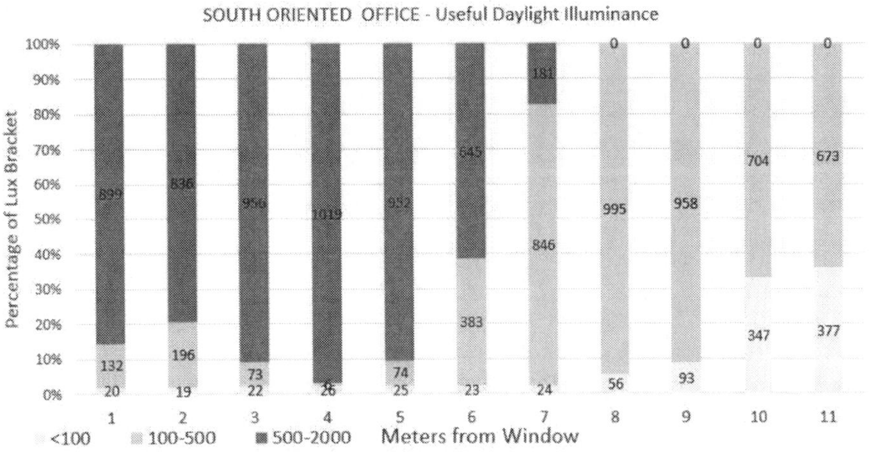

Figure 8 UDI distribution for South oriented office over the period of a day

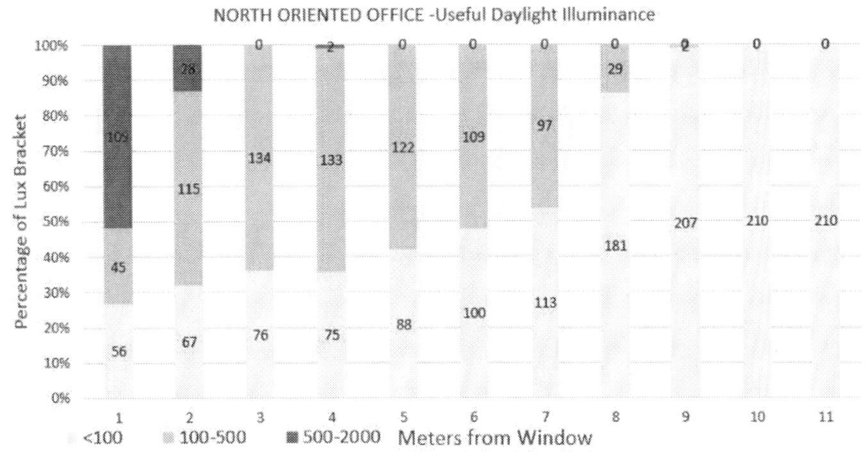

Figure 9 UDI distribution for North oriented office over the period of a day

The study also concludes that by factoring in all the mentioned intangible benefits, the usage of tubular daylight guidance systems in London has a payback period of 6 years. This could potentially be even faster in the Tropics due to higher annual sky luminance.

There are three factors affecting payback periods: electricity prices, system capital cost, and available external local illuminance [8]. In view of the increasing commercial electricity tariffs in Malaysia and other advantages as mentioned, the payback period for the DT is likely much lower. However, this paper is unable to quantify the whole life cycle costing (WLCC) as suggested by Mayhoub & Carter [8] as there is a lack of localized literature on the intangible benefits.

5.4 Contractual & Maintenance Practicality Experience

The design of the DT has to cater to three objectives; constructability considering all other mechanical and electrical components in the ceiling space, maintenance of the DT, and daylight performance. The design of the DT does not span across entire ceiling space as there are pockets of ceiling space for fresh air ducts and fire sprinklers (Figure 10). The DT design only caters for the center of the office space as the corners with shallow depth should get sufficient daylight from fenestration alone. The initial DT design had the upper light trough tilted to allow redirection of daylight deeper into space, however this was later revised due to constructability considerations (Figure 12).

The evolution of the design also considered maintenance which suggested using robot vacuum cleaners every half year to clean the DT. As there was an uncertainty over the robot's navigation due to height clearance and frequent ceiling hangers, and even leveling, the act of "over-engineering" also potentially added cost. Also, the unconventional issue of additional lateral load and movements by the robot vacuum concerns the warranty of ceiling panel suppliers and various sub-contractors. With the complication of liability and potential defects, the DT was redesigned to be isolated and enclosed which accumulates only minimal dust. Telescopic mops will be used via the translucent exit aperture of the system to maintain the DT. Figure 11 shows the axonometric exploded view to illustrate the arrangement of systems at ceiling level.

DOI: 10.1051/matecconf/20166600032

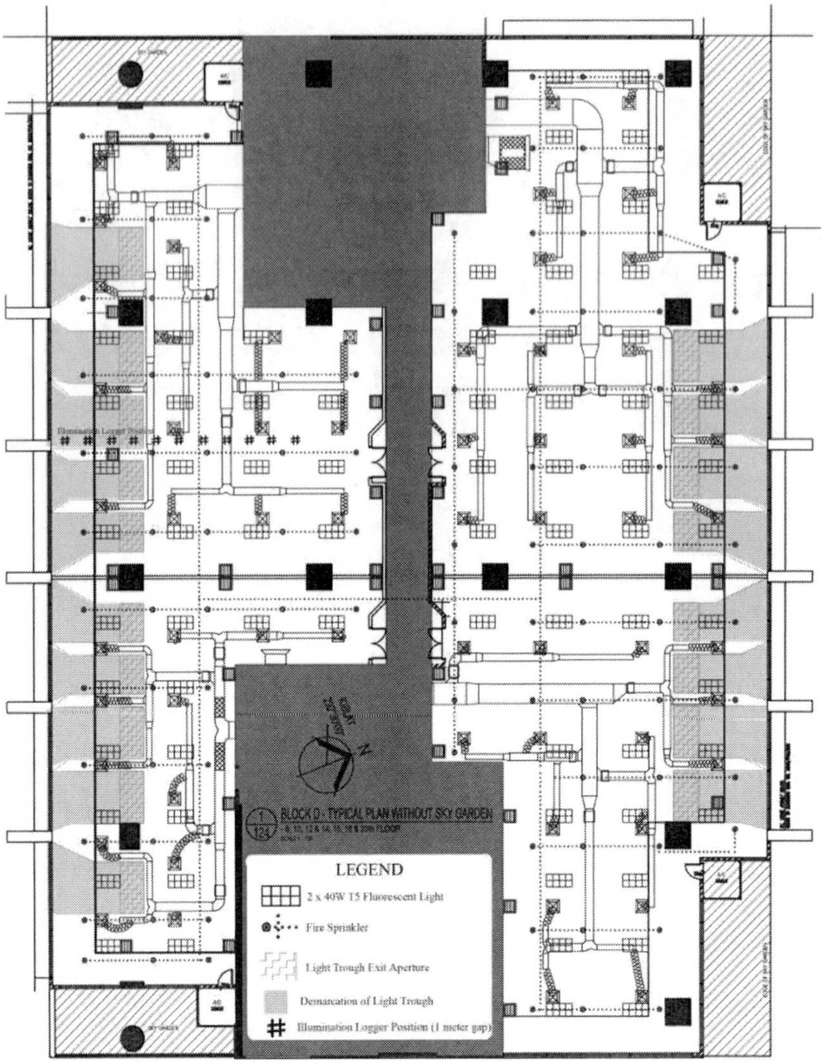

Figure 10 Ceiling layout plan of a typical office level with DT location demarcated

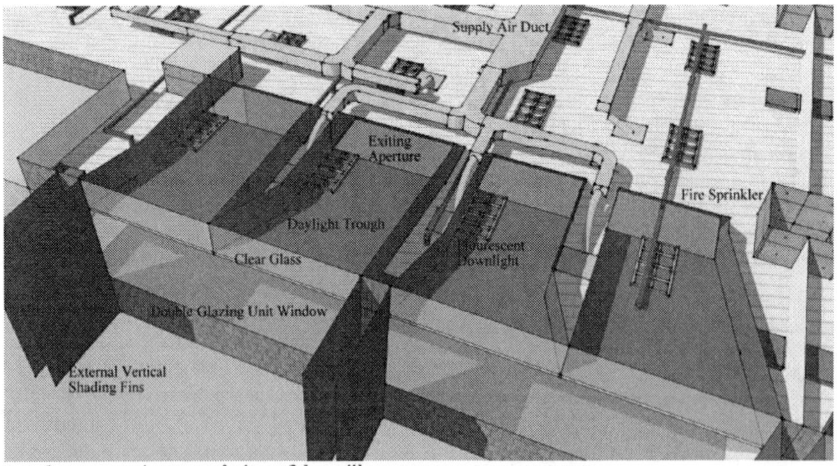

Figure 11 Illustrated axonometric exposed view of the ceiling components arrangement

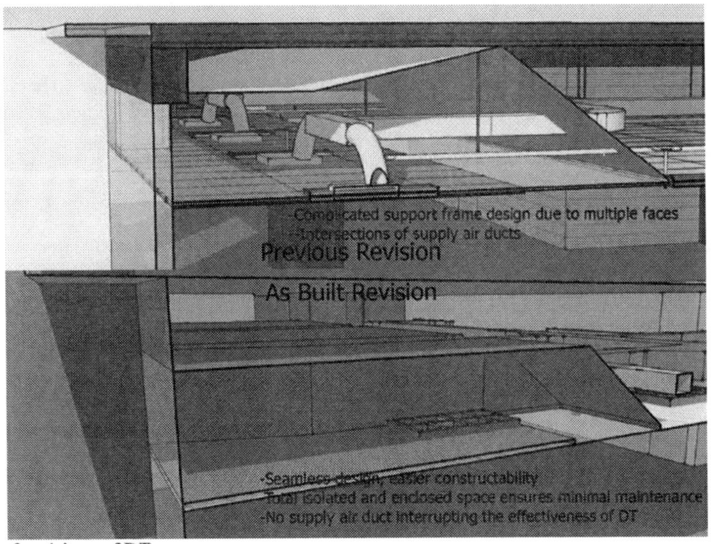

Figure 12 Illustration of revisions of DT

6 Conclusion & Recommendation

This case study summarizes the effectiveness and practicality of a novel daylight trough in Damansara Perdana, Malaysia. The correlated simulation and field measurement results demonstrate that reliability of Radiance in IES-VE can be used to study such an innovative daylighting system in advance. Unlike most previous simulated or experimental mock-up based research papers, this paper shares the experience of practical on-site constructability and maintenance of DT in the Tropics.

The DT has the potential to bring in effective daylight (DF 1-3.5%) to a depth of 6 meters from the fenestration. The IRIF of the DT varies from 4.34 and 20.61 for 3 to 6 meters depth from fenestration. Radiance simulation result in IES-VE correlates well with the field measurement (r=0.7577) however precaution has to be taken for adjacent building shading effects. The design of DT needs to take account of the surrounding obstacles and orientation for its daylight performance, as shown by the significant difference in IRIF of the South oriented office which is unobstructed and north oriented office obstructed by tower and hills.

The incremental cost for DT is RM38.21/m^2 (\$8.70/m^2) and it is estimated to save as much as 39.19% of artificial lighting energy annually. The payback period of the DT is 20.4 years with the consideration of operational and maintenance savings. However as elaborated, the whole life cycle cost shall include many intangible aspects such as carbon mitigation and office productivity which was out of the scope of this investigation. The constructability and maintenance of DT have to be accounted for, together with the designs of many other mechanical and electrical components in the ceiling space such as supply air duct, fire sprinkler, and artificial lights. The difference in ambient temperature of DT and office work plane level is 3°C to 4°C and 0.5°C during clear and overcast skies, respectively. The lighting power density of the office space is 5.25W/m^2 and 1.90W/m^2, with and without the assistance of DT, respectively.

Recommendation. The effectiveness of the DT should be further investigated by post occupancy visual comfort survey as this investigation only investigates merely on daylight illuminance performance without fit out. Aside from glare analysis and artificial light interaction studies, a similar illuminance measurements should be done for the summer solstice and equinox. This will give a fair picture of the effectiveness of the DT throughout the year.

References

[1] D. H. W. Li and E. K. W. Tsang, "An analysis of daylighting performance for office buildings in Hong Kong," *Build. Environ.*, **vol. 43**, no. 9, pp. 1446–1458, (2008)

[2] L. Heschong and F. Oaks, "Windows and Classrooms: A Study of Student Performance and the Indoor Environment," 2003.

[3] Z. A. Hamid, A. F. Roslan, M. C. Ali, F. C. Hung, M. S. M. Noor, and N. M. Kilau, "Towards a national green building rating system for Malaysia," *Malaysian Constr. Res. J.*, **vol. 14**, no. October, pp. 0–16, (2014)

[4] A. D. Galasiu and J. a. Veitch, "Occupant preferences and satisfaction with the luminous environment and control systems in daylit offices: a literature review," *Energy Build.*, **vol. 38**, pp. 728–742, (2006)

[5] P. R. Boyce, C. Hunter, and O. Howlett, "The Benefits of Daylight through Windows," 2003.

[6] X. Yu and Y. Su, "Daylight availability assessment and its potential energy saving estimation –A literature review," *Renew. Sustain.*

Energy Rev., **vol. 52**, pp. 494–503, (2015)

[7] CK Tang and N. Chin, *Building Energy Efficiency Technical Guideline For Passive Design*. Building Sector Energy Efficiency Project BSEEP, 2013.

[8] M. S. Mayhoub and D. J. Carter, "The costs and benefits of using daylight guidance to light office buildings," *Build. Environ.*, **vol. 46**, pp. 698–710, (2011)

[9] N. Ruck, S. Aydinli, J. Christoffersen, Gi. Courret, I. Edmonds, R. Jakobiak, M. Kischkoweit-Lopin, M. Klinger, E. Lee, and L. Michel, *Daylight in Buildings: A source book on Daylighting systems and components*. Lawrence Berkeley National Laboratory, 2000.

[10] F. Linhart, S. K. Wittkopf, and J. L. Scartezzini, "Performance of Anidolic Daylighting Systems in tropical climates - Parametric studies for identification of main influencing factors," *Sol. Energy*, **vol. 84**, no. 7, pp. 1085–1094, (2010)

[11] S. K. Wittkopf, "Daylight performance of anidolic ceiling under different sky conditions," *Sol. Energy*, vol. 81, no. 2, pp. 151–161, 2007.

[12] J. L. Scartezzini and G. Courret, "Anidolic daylighting systems," *Sol. Energy*, **vol. 73**, no. 2, pp. 123–135, (2002)

[13] F. Linhart and J. L. Scartezzini, "Minimizing lighting power density in office rooms equipped with Anidolic Daylighting Systems," *Sol. Energy*, **vol. 84**, no. 4, pp. 587–595, (2010)

[14] Integrated Environmental Solutions, "Radiance View User Guide," 2015. [Online]. Available: http://www.iesve.com/downloads/help/ve2012/Lighting/Radiance.pdf. [Accessed: 01-Jan-2006].

[15] Alanod-xxentria, "MIRO95ax4 Data Sheet," *Alanod-xxentria*, 2014. [Online]. Available: http://www.alanod-xxentria.com/download/technical/20140506042823817.pdf. [Accessed: 01-Jan-2016].

[16] K. Kensek and J. Suk, "Daylight Factor (overcast sky) versus Daylight Availability (clear sky) in Computer-based Daylighting Simulations," *J. Creat. Sustain. Archit. Built ...*, **vol. 1**, pp. 3–14, (2011)

[17] Green Building Index, "GBI NRNC Non Residential New Construction Design Tool." Green Building Index, p. 76, 2011.

[18] A. Nabil and J. Mardaljevic, "Useful daylight illuminance: a new paradigm for assessing daylight in buildings," *Light. Res. Technol.*, **vol. 37**, no. 1, pp. 41–59, (2005)

[19] Departments of Standards Malaysia, *MS1525 - 2014 Code of Practice on Energy Efficiency and Use of Renewable Energy for Non Residential Buildings (2nd Revision)*. 2014.

[20] P. Chaiwiwatworakul and S. Chirarattananon, "Luminous efficacies of global and diffuse horizontal irradiances in a tropical region," *Renew. Energy*, **vol. 53**, no. 0, pp. 148–158, (2013)

[21] M. S. Alrubaih, M. F. M. Zain, M. a. Alghoul, N. L. N. Ibrahim, M. a. Shameri, and O. Elayeb, "Research and development on aspects of daylighting fundamentals," *Renew. Sustain. Energy Rev.*, **vol. 21**, pp. 494–505, (2013)

[22] Heschong Mahone Group, "High Efficiency Office: Low Ambient/Task Lighting Pilot Project," 2009.

[23] F. Maamari and M. Fontoynont, "Analytical tests for investigating the accuracy of lighting programs," *Light. Res. Technol.*, **vol. 35**, no. 3, pp. 225–242, (2003)

A Conceptual Framework of Green Certification Impact 0n Property Price

Lizawati Abdullah[1,a], Thuraiya Mohd[1] and Roshdi Sabu[2]

[1] *Department of Estate Management, Faculty of Architecture, Planning and Surveying, 32610 Universiti Teknologi MARA (Perak), Malaysia*
[2]*Faculty of Architecture, Planning and Surveying, 32610 Universiti Teknologi MARA (Perak), Malaysia*

Abstract. Green building is one of the sustainability dimensions in built environment. The issues of green building and its impact to the society have been increasingly discussed. Green certification is one of the components in measuring sustainable development and plays an important role as an assessment system to an individual building's performance. The question arises whether the market understand and recognized the green certification. The objectives of this research are to discuss the issue pertaining to green value and the relationship between green certification and property price. The research emphasized on the understanding of property attributes focusing on green certification and the impact to the property price. Among the attributes identified are structural characteristics, location and neighborhood, and time attributes. Thus, this paper will discusses the review of literature on green development and the significance impact on property market in term of price and value. The green building development across the country could be classified as another sector in property markets that give significant impact to the real estate industry. As a result, a conceptual framework in assessing the impact of green certification is suggested to provide a significant input in developing the model of hedonic pricing for green building. This research may contribute to extend the body of knowledge in the area of green development and a suggested significant input will give much emphasize on the new valuation technique in valuing green building properties.

1 Introduction

Over the past decades, significant efforts have been made to promote sustainability in property development. The introduction of National Green Technology Policy (NGTP) in 2009 indicated the government's concern on global issues of sustaining environment and resources. The green building certification embarked the initiative of sustainable development with the introduction of Green Building Index (GBI) in 2009 as first green building assessment in Malaysia property market.

Other rating systems also have been accepted internationally to assess the sustainability building in Malaysia for example LEED, Green Mark, BREAM, Green Star. One of the most effective ways in encouraging sustainable building development is by rating system which provides a means for building owner or occupier in comparing building's greenness [6]. Launched in 2009, in line with NGTP, GBI was the first green certification established as rating building system in Malaysia. The system provides systematic assessment on how to evaluate building performance with regard to sustainable criteria.

According to Brundtland Report [23], the acceptable meaning of sustainable development is development that satisfies the needs of the present without compromising the ability of future generations to satisfy theirs. This definition is referring to three domains, namely the economy, environmental and social perception also known as The Triple Bottom Line Concept as illustrate in Figure 1. Sustainable development could only being achieved if all of these three domains integrated and treated in equal measures. Thus, the development of green building plays an important role to achieve the sustainability objectives.

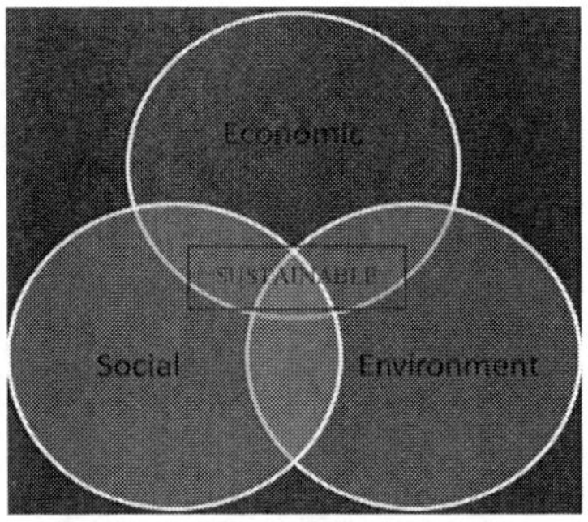

Figure 1. The Triple Bottom Line Concept.

Green building benefits have been debated tremendously in most literature. The beneficial parties include owners, occupants, community, appraisers and lenders. One of the most significance benefit discussed is on the cost-saving benefit in green building [5,10, 20].

[a] Corresponding author: lizaw327@perak.uitm.edu.my

This is to encourage the owner to grasp the sustainable impact on value of property.

As a result of growing interest in green building development, an amount of research has been done to prove the benefit of going green in the economic context. The economics of green building are being argued and amount of studies in real estate have attempted to answer the question increase. Many researches reveal the significance result especially in a matured market like United States, Australia and United Kingdom. However, little study has been done on property market in Malaysia as it is an emerging market for green building. Thus it seems crucial to identify the impact of green certification on property market price. Apart from that, the researcher will suggest a conceptual framework of the impact of green certification to the property price.

2 Issue on green certification

The introduction of green certification and its impact are an important issue that should be addressed. The sensitivity of green label is uncertain, unpredictable, and result may vary according to the local market [7]. In fact, much of the studies on the impact of green labeling, certification or rating focusing on commercial property [3, 14, 25]. Consequently, lack of significance studies done in residential property market due to the slow value implications.

From the previous research which addresses the impact of greenness to the value of property, the literature identified three arguments of on the need to evaluate the impact of green certification on property price:

a) Sustainable buildings and property risks and financials are often misleading, resulted to a situation where individual property assets are mispriced, which lead to the investment opportunities for "enlightened" investors [17]. The statement supported with the term introduced in report of APB Valuation Advisory 'overlooking green features' where most appraisers and valuers fail to note green in assessing property market [7].
b) A building's value may vary and market sales information is based on standard approaches to building appraisal with regardless of performance-based cost savings [9]. For most consumers and some homebuilders, the relationship between quality home construction and sustainability is always misunderstood and lead to the misinterpretation by the market player [9].
c) In valuing low-energy houses, the expected price or rate of returns becoming challenges to the appraiser because lack of reliable modelling tools, historical statistical and case specific data [11].

Mostly, researches towards the green certification impact are involving both either a mandatory program or a voluntary program [2, 3, 7, 8, 9, 24]. However, limited studies have been done on green residential properties and this is leading to the research on the greenness effect in residential market.

Even though few researches have been done, the question arised whether in Malaysia green residential properties are facing the situation of 'underpriced or overpriced'? The argument is whether the green building development in Malaysia is recognized in well manner, or the establishment of green building rating system is giving a tremendous effect in promoting sustainability. Thus, the interest of this research is to investigate the impact of green certification on residential market and green features as another importance variable in determining the property price. By recognizing the green as another factor in assessing the property market, a model to be developed is to help property market player understand the role of green. The research will utilize comparative study to compare the price between green residential building and conventional residential building.

3 Literature review

3.1 Sustainable and green building

One of the important aims of sustainable development is to reduce the impacts of built environment on the natural environment. Hence, this goal has lead to the introduction of the term 'green building' in real estate. Green building is also known as the foundation of the sustainable construction development [22]. The issue of sustainability is controversial and much disputed subject within the field of construction.

RICS [20] defined sustainable building as green building rated by rating assessment tools. Green building also known as "sustainable building" - as building design and construction using methods and materials that are resource efficient and that will not compromise the health of the environment or the associated health and well-being of the building's occupants, construction workers, the general public, or future generations [16]. Thus, green building is a mechanism to achieve the sustainable development objectives.

The three pillars of sustainable development according to the Triple Bottom Line Concept are characterized as follows [17]:

- Ecological sustainability: dependent on material, energy, noise emission, amount of waste products, amount of traffic, old building material separation and disposal, land use/pollution, climate change and biodiversity and means reduction of area used, conserving resources and avoidance of deleterious materials and emissions.
- Social sustainability: based on the social aspects such as feeling of wellbeing, aesthetics, health & comfort, security and user satisfaction, appropriate living environment and social integration.
- Economic sustainability: the minimizing life-cycle costs and value retention (material, goods and capital). Functional-aesthetic aspects such as the

maximising functionality, adaptability, serviceability and design should also be considered.

According to Bell & Morse [4] the degree of sustainability are sustainable equal to a situation where quality remains the same or increases whereas if quality declines, the system can be regarded as unsustainable. Therefore, the existence of assessment tools is to provide better understanding of sustainability measure.

Among green building assessment tools are LEED, BREAM, GBCA, Green Star, CASBEE, Green Mark, GBI and GreenRE. Since the introduction of LEED in US, other countries have started developing their very own green rating tools. There are few amount of research in evaluating the performance and comparison between tools. So far, there are amount of researches have been discussed on green rating tools, comparing international tools between LEED, BREEAM, Green Star and CASBEE, as well as the characteristic of each tool.

The purpose of green rating systems or certifications are intended to offer market participants an understanding of label that express a building's sustainability attributes [7]. The green certification indirectly assists the market player in assessing the benefits of green building compared to non certified buildings. Any potential improvement should be assessed to determine if it could create a differential to the operational, overall performance and/or risk characteristics of the property and whether this differential constitutes a market advantage or disadvantage [7].

As we know, the purpose of green building rating tool is to promote and increase the awareness among industry player towards sustainable development. Various green building rating tools mean that the effort to build more green structure that reduce the impact on environment, increase the social benefits and optimizing the economics return would be crucial challenges for green rating service provider. Thus, with multiple option and alternatives rating tools offered in green property development, either by local or international, the agendas of United Nation for preserving the world from environment threats can be accomplished.

3.2 Value and price

Value can be distinguished in two term; value-in-use and value-in-exchange. Defining the value-in-exchange is most likely the price of property during the process of purchasing between seller and buyer. Therefore, value and price of property are closely related. When assessing the value, sale transaction or price is used to determine the market value, as described by the definition of property market value - 'estimated amount of which the property should exchange…'. As such, in determining a price, an appraiser or valuer should also take into consideration the micro and macro factors that may affect the property value.

Recently, researchers have shown an increased interest in green value concept – the integration of building's greenness and value. RICS [20] defined green value for office building as a beneficial outcome to the building that practice sustainability. The concept are explained in twofold, first the building fulfill the sustainable development term and second market value i.e. the sale value and rental value increased [5].

The importance of value in green development has been described through various researches. The Vicious Circle of Blame introduced by Cadman [28] outlined the relationship between the stakeholders in property development that cause issues in sustainable development implementation. The stakeholders include the owner, the investor, the developers and the designers. However, [12] modified the figure by adding the appraiser and valuer as another important roles in determining the sustainable development (as illustrate in Figure 2). The existences of valuers or appraisers are significant in determining the green value for leading the future direction of green development. Thus, to understand the market of green building, green attributes should be one of the factors to consider in valuation process.

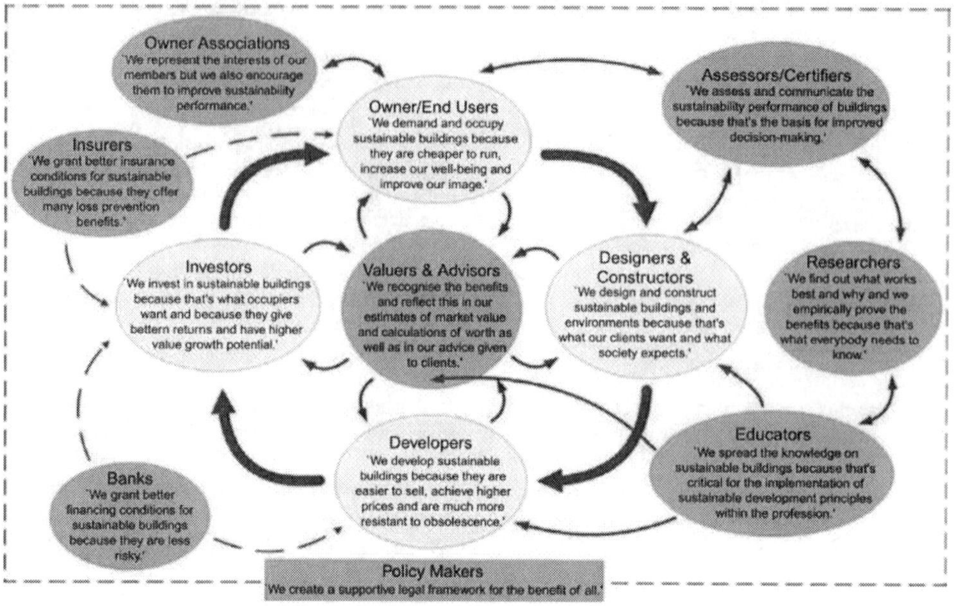

Figure 2. Breaking the Vicious Circle of Blame (Lorenz, 2008).

The value of the houses is affected by various factors include controlled variables or uncontrolled variables such as the characteristics of the house itself. To foresee the price effect, the marginal price effect of a single hedonic characteristic (green-certification) and aggregate market outcomes in a partial equilibrium framework can be demonstrated.

A large and growing body of literature has established the relationship between green and property market. Most of the studies highlight green certification and utilize the hedonic model to assess the effect on property price. Surveys conducted by several researchers [2, 3, 8, 14, 24] using hedonic model focus on their local green rating such as Green Mark, LEED, Energy Star and property transaction price.

3.3 Hedonic price theory

Rosen, [21] started the role of housing attributes consumer decision making process. Based on Rosen's analytical framework, which started with the assumption that any good or service consists of a variety of utility-bearing characteristics (as example z1, z2, ... , zn) making up the hedonic price function. In the context of office rent determination, these characteristics consist of various structural, locational and lease characteristics that enter into the empirical model as independent variables. The empirically determined hedonic prices are indicative of an implicit market so that demand and supply functions can be derived for both short-run and long-run competitive equilibria [8].

Statically –based comparison method used to identify the most possible prices for proposed development, thus it can compare property prices within existing housing area across selected geographical [29]. It can be done by using hedonic approach to determine the hedonic prices of selected property attributes. The hedonic model is use to understand the possible relationship between the properties attributes and property price. Hedonic modeling can be used by developers, corporate real estate groups, owners, and operators to determine which building characteristics add significant value to the potential transaction price. The results produced by hedonic model can provide important information for future decisions and help each party involve in property development better understand the economics surrounding each asset, thus improving asset underwriting [30].

Hedonic pricing is considered to be a willingness to pay technique and it tries to capture the fraction of property prices that are derived from the specific housing attributes [3]. Among of the structural characteristics like square footage, lot size, age, and bedrooms have a positive effect on selling price although the magnitudes may vary across different regions in the country [3]. However, internal features like the number of bathrooms, fireplace and air conditioning are always referring to add premium to the housing prices. One of the advantage using hedonic approach is it can fine-tune the quality changes of an individual properties, thus it can be used to measure the greenness price.

4 Conceptualising the framework of green certification impact

In regard to the issue of determining the impact of green certification on the residential property market, it appears that green and property price may have a significant relationship. Green or sustainability is a quality measure of a building and these criteria can be a property attributes in assessing the market price.

In assessing sustainability, difference features established by green rating tools due to factor such as law and regulations, climate, types of property, geographical factors and etc [1]. GBI was developed by PAM (The Association of Architects Malaysia) with six assessment criteria for certification inclusive of energy efficiency, indoor environment quality, sustainable site planning and management, material and resources, water efficiency and innovation (Table 1) and also 4 categories of certification according to total point scoring (Table 2).

Table 1. Assessment Criteria for GBI and Scoring (GBI, 2009).

Criteria	Maximum Point
Energy efficiency (EE)	35
Indoor environment quality (EQ)	21
Sustainable site planning and management (SM)	16
Material and resources (MR)	11
Water efficiency (WE)	10
Innovation (IN)	7

Table 2. GBI Index Classification (GBI, 2009)

Points	**GBI Rating**
86+ points	Platinum
76 – 85 points	Gold
66 – 75 points	Silver
50 – 65 points	Certified

The abovementioned criteria represent the green features of green building as well as the green attributes. Thus, variables are identified as illustrate in Figure 3.

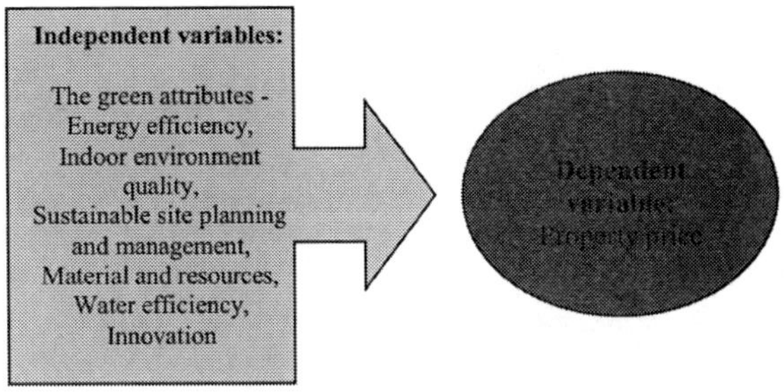

Figure 3. The independent and dependent variables (Author's Research, 2016).

The variables of independent and dependent represent the cause and effect of a phenomenon. Therefore, from the above illustration the green attributes can affect the price of property. A further investigation should be made whether the green attributes alone affect the price or other property attributes also recognized as a factor affecting the price. One question that needs to be asked, is whether in Malaysia GBI does have the significant impact to the property market or otherwise.

In neo-classical theory, price of houses is a resulting from demand and supply of the housing market. Factors influence prices and value mainly is property-specific factors and market-related factors [27]. Wyatt further explained property-specific factors are the main physical characteristics of the property such as size, age, condition, appearance, legal and location whilst the market-related factors referring to the property market as a whole. Such market-related factors are difficult to control include household income, employment rate, consumer spending cost and finance, contrary to property-specific factors.

Some recent studies began to use dummy attributes for assessing green property price effect. Based on the ideas of few research, structural attributes are size of building /floor, stories of building and age [2, 3, 13, 14, 18, 24]. Another attributes that are mostly used are location and neighborhood for instance, view and facilities [2, 8,25] and time attributes such as year of transaction price take place [2, 3, 8,13, 24].

In order to explore the idea of green certified property's price, abovementioned property attributes is suggest to control the differences that may exist between the samples and to reduce the inconsistency in result.

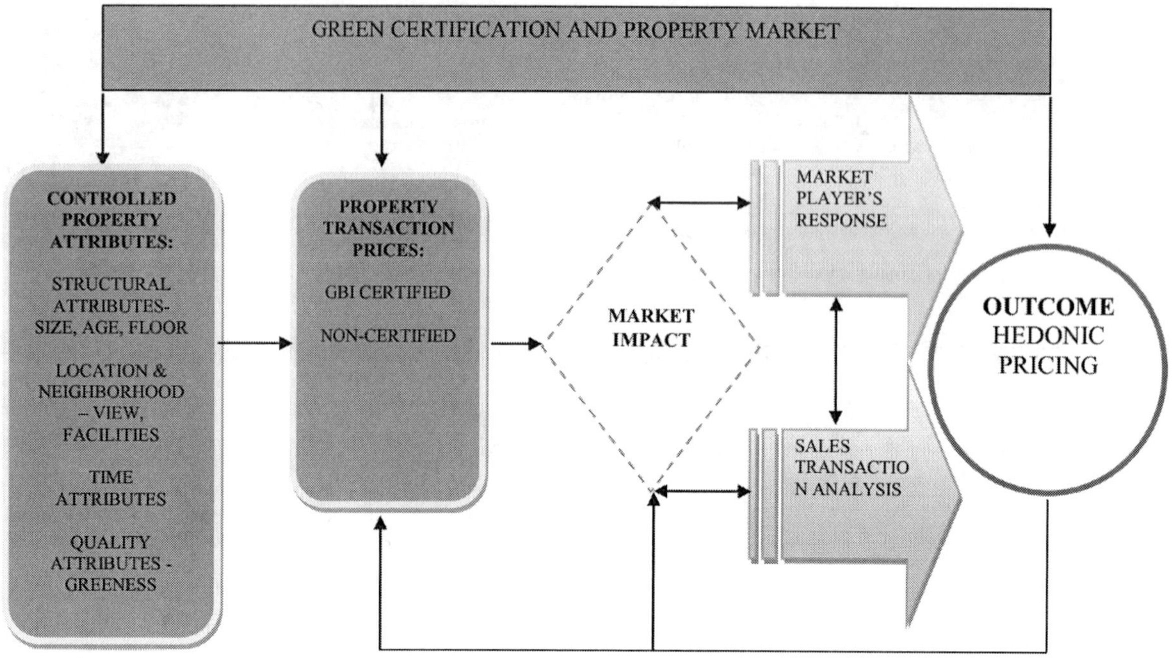

Figure 4. A Conceptual Framework of Green Certification Impact on Property Price (Author's Research, 2016).

A conceptual framework (Figure 4) was introduced to answer the question. Further research to conduct will be applied on the case study approach in order to explore in depth the green certification and the effect to the price. The approach used in this research is referred to as the formulation of empirical questions to valuers and appraisers which requires this research to analyze and investigate the emerging market of green building. This research is necessary when various questions arise and there is an inability for clear explanation without the implementation of a research utilizing various sources [31]. Green buildings will be selected in few states as a case study and compare with noncertified buildings in order to generalize the research area in relation to Malaysia as a whole.

5 Summary

The research is significant because it addresses the issues as outlined above for comprehensive study into green certification. Presently, the impact of green development focuses primarily on the cost benefit. This study is to investigate the price differentials between GBI certified and noncertified properties in Malaysia. However, the need to highlight the property price, specific on residential market. To put it in a nutshell, the primary objective of this research is to develop a pricing model to analyse the impact of green certification in property market by using hedonic approach. Summing it up, it would be highly beneficial to property market player as the finding will lead to the success of green building implementation. People will perceive the value of green in better way as well as promote sustainable in built environment.

References

1. L. Abdullah, N. Jumadi, R. Sabu, H. Arshad, F. Fayza Mohd Fawzy, Assessment Criteria On Sustainable Rating Tools Used In Asian Countries. *Sustainable Initiatives: Case Study in Malaysia, Philippines and Indonesia* (2015)
2. K. Addae-Dapaah, S. J Chieh,. Green Mark Certification : Does the Market Understand?*Journal of Sustainable Real Estate*, *3*(1), 162–191 (2011)
3. R.R Aroul, *Going green - Impact on residential property values. Thesis*. University Of Texas At Arlington (2009)
4. S. Bell, S. Morse, *Sustainability Indicators: Measuring the Immeasurable? Second Edition Earthscan.London.Sterling, VA* (2008).
5. Q. A Cerqual, *Economic Analysis of the Green Value of Residential Real Estate Green Value in Residential Buildings : Is It a Reality today?* (2012)
6. Z. Darus, N. Hashim, E. Salleh, Development of rating system for sustainable building in Malaysia. *WSEAS Transactions on Environment and Development*, *5*(3), 260–272 (2009)
7. The Appraisal Foundation, *Valuation Of Green and High Performance Property : Background and Core Competency* (2015)
8. F. Fuerst, P. McAllister, Green Noise or Green

Value? Measuring the Effects of Environmental Certification on Office Values. *Real Estate Economics*, *39*(1), 45–69 (2011)
9. A. Griffin, B. Kaufman, S. Hamilton, *Certified Home Performance: Assessing the Market Impacts of Third Party Certification on Residential Properties*. Earth Advantage Institute (2009)
10. M. Halim, Economic Issues on Green Office Buildings in Malaysia. *International Real Estate Research Symposium 2009*, 1–13 (2009)
11. Hao Wu, H. Robert, Crawford, G. Warren-Myers, Malay Dave, M. N. Concept and Barriers For The Economic Value of Low-energy Houses. *COBRA* (2015)
12. Hartenberger, P. Ursula , D. Lorenz, Breaking the Viscous Circle of Blame - Making the Business Case for Sustainable Buildings. In *RICS Research* (2008)
13. R. Jaques, D. Norman, I. Page, Valuing sustainability and resilience features in housing, *333* (2015)
14. W. M Jayantha, W. S Man, Effect of green labelling on residential property price : a case study in Hong Kong. *Journal of Facilities Management*, ***11***(1), 31–51 (2013)
15. J. Pitts, T. O Jackson, Green Buildings : Valuation Issues and Perspectives. *The Appraisal Journal*, ***76***(2), 115 (2008)
16. M. Landman, Breaking through the Barriers to Sustainable Building, *Thesis,* THUFTS University (1999)
17. D. Lorenz, T. Lützkendorf, *Sustainability and property valuation: Systematisation of existing approaches and recommendations for future action*. *Journal of Property Investment & Finance* Vol. **29**, (2011).
18. B. N, Miller, J. Spivey, A. Florance, Does Green Pay Off? Exhibit 1 : Percent Occupancy Rates. *Energy*. (2007).
19. S. R Muldavin, *VALUE BEYOND COST SAVINGS How to Underwrite Sustainable Properties*. Green Building Finance Consortium (2010)
20. RICS, *Green value Green Buildings, growing assets* (2005)
21. S. Rosen, Hedonic Prices and Implicit Markets: Product Differentiation in Pure Competition. *Journal of Political Economy*, **34–55** (1974)
22. M. Samari, N. Godrati, R. Esmaeilifar, P. Olfat, M. W M Shafiei, The investigation of the barriers in developing green building in Malaysia. *Modern Applied Science*, *7*(2), 1–10 (2013)
23. World Commission on Environment and Development. *Report of the World Commission on Environment and Development: Our Common Future. WCED* (Vol. **4**) (1987)
24. X. Yang, *Measuring the Effects of Environmental Certification on Residential Property Values - Evidence from Green Condominiums in Portland,* (2013)
25. J. Yoshida, A. Sugiura, Which "greenness" is valued? Evidence from green condominiums in Tokyo. *46th Annual AREUEA Conference*, 1–34 (2010).
26. P. D Vries, *Measuring and Explaining House Price Development.* Netherlands: Delft University of Technology (2010)
27. P. Wyatt, *Property Valuation In An Economic Context.* Blackwell Publishing (2007).
28. D. Cadman, 'The vicious circle of blame', in Upstream, http//:www.upstreamstrategies.co.uk. (2000)
29. A. H M Iman, *Basics Aspects of Property Market Research*, Universiti Teknologi Malaysia (2006)
30. M. Monson, Valuation using hedonic pricing models. Cornell Real Estate Review, **7**, 62-73 (2009)
31. R.K Yin, Application of case study research, 2nd Edition. CA: The Sage Publications (2003)

Mechanical Properties of High Strength Concrete Containing Coal Bottom Ash and Oil-Palm Boiler Clinker as Fine Aggregates

Mahmood Soofinajafi[1,a], Payam Shafigh[2], Farid Wajdi Akashah[3], Hilmi Bin Mahmud[1]

[1]*Department of Civil Engineering, Faculty of Engineering, University of Malaya, 50603, Kuala Lumpur, Malaysia.*
[2]*Department of Building Surveying, Faculty of Built Environment, University of Malaya, 50603, Kuala Lumpur, Malaysia.*
[3]*Centre for Construction, Building & Urban Studies (CeBUS), Faculty of Built Environment, University of Malaya, 50603, Kuala Lumpur, Malaysia.*

Abstract. This research aims to utilize Coal Furnace Bottom ash (CBA) and Oil-Palm Boiler Clinker (OPBC) as fine aggregate in concrete mix proportions. They are solid wastes from power plant and Oil Palm industry, respectively. Since these by-products do not have any primary use and are pure waste, an opportunity to use them as aggregate in concrete industry not only is economical but also will be an environmental friendly opportunity leading towards a more sustainable production chain. CBA and OPBC sands had similar grading to normal sand but have lower density and higher water absorption. In a high strength concrete, normal sand was replaced up to 25% with either CBA or OPBC. Test results showed that although water absorption of these wastes was more than normal sand but the slump value of concrete containing each of these wastes showed that these concretes had good workability. All mixes containing these wastes had slightly lower compressive strength at early ages and equivalent or higher compressive strength at later ages compared to control mix. The 28-day compressive strength of these concretes was in the range of 69-76 MPa which can be categorized as high strength concrete. In general, the performance of OPBC was better than CBA at 25% replacement level. However, it is recommended that at least 12.5% of total volume of fine aggregate in a high strength concrete is used of CBA or OPBC.

Keywords: Coal Bottom Ash, Oil-Palm Boiler Clinker, Fine Aggregate, Compressive Strength, High Strength Concrete

Nomenclatures

MPa	Mega Pascal, Concrete compressive strength, N/mm^2
w/b	Water to binder ratio
w/c	Water to cement ratio

Abbreviations

ASTM	American Society for Testing and Materials
OPBC	Oil-Palm Boiler Clinker
OPBCC	Oil-Palm Boiler Clinker Concrete
CBA	Coal Bottom Ash
CBAC	Coal Bottom Ash Concrete
HSC	High Strength Concrete
OPC	Ordinary Portland Cement
ACI	American Concrete Institute
FA	Fly Ash
OPS	Oil-Palm Shell
GHG	Green-House Gases
LWA	Light Weight Aggregate
LWAC	Light Weight Aggregate Concrete

1 Introduction

Concrete as a major constructional material makes a significant contribution to the triple bottom lines - environmental, social and economic - of sustainable development. In many sectors throughout the years, concrete has enabled the reduction of CO2 emissions and other negative environmental impacts. As of year 2006, 7.5 km2 of concrete is made each year. This amount is more than 1 m2 for every person on earth [1]. Fundamental concrete is generally analysed and designed with respect to its three core constituent parts, binder, aggregate and water [2]. This paper adopts the view of Troxell [3] when it comes to definition of satisfactory concrete. Workability, defined by BS 102 [4] is a topic that comes always hands in hands with w/c. W/c has direct relation with workability and reverse relation with strength. In other words, low w/c leads to high strength but low workability and high w/c leads to high workability and low strength [5].

Reason such as the enormity of the volume of constituent materials used from earth resources to produce the yearly need of concrete in construction and the large quantities of CO2 emissions created during the

[a] Corresponding author: mahmood.soofi@siswa.um.edu.my

production, have given concrete a not necessarily environmental friendly general appearance. The environmental footprint left by concrete around the world does not seem to be compatible with the demands of sustainable development [6].

By introducing the constituent materials replacement concept producing concrete can be a sustainable and environmental friendly process. Using these recyclable materials as Pozzolan constituent materials or even aggregates in concrete production can actually be called an upcycling process. The earlier term can be justified knowing that many of the mentioned by-products do not have any primary usage and are pure waste and using them in concrete production is actually making them valuable [7].

Both Fly Ash (FA) and Coal Bottom Ash (CBA) are also used as replacements for fine aggregates in concrete mix design. Although similar to natural fine aggregate, bottom ash is lighter and more brittle and has a greater resemblance to cement clinker. Its lighter nature and high water absorption found by research studies makes it a good replacement in high strength or light weight concrete [8]. Throughout the work that has been conducted it has not only been replaced for fine aggregate but also used as a replacement for cement. Due to its chemical carbonated nature containing CO2 with adding the right admixtures it can gain the paste glue state to work as a cementitious material [9].

Past researches on CBA showed that, the high porosity of CBA is a sign of causes the w/c of the concrete to not be able to be taken as exact. The amount of internally absorbed water by CBA is released to the concrete over time, being part of the production process with the concrete still in the fresh state. The presence of bottom ash increased the quantity of water loss by bleeding, the bleeding time and also the water release rate, and the higher the bottom ash content of the concrete the greater this effect [10].

The chemical composition of the bottom ash is given in Table 1. As according to the table, calcium content of CBA is lower than 1 %. According to the total amount of SiO2, Al2O3 and Fe2O3 together that is 88.5% of the CBA, the ash belongs to ASTM Type F ash [11]. For this CBA, the loss on ignition (LOI) is mainly due to carbon which amount of carbon was found to be 3.8%, which gives 13.9% CO2, while the measured quantity of CO2 was 14%. The remaining LOI (0.8%) is certainly due to combined water in residual clay [12].
Malaysia is one of the greatest producers and exporters of palm oil products all over the world. A large amount of by-product wastes is collected throughout the palm oil processing phases. One of those would be Oil-Palm Boiler Clinker (OPBC), which is gathered in large portions during the oil palm shell (OPS) and fibre incineration process. In recent years, OPBC is disposed of by landfill which not only causes soil pollution but also affects the ground water supply source. Hence, the consumption of these abundantly available wasted by-products in the construction industry will not only help to reduce the environmental problems but also provide an alternative to the diminishing natural aggregates [13]. Previous studies in 2002 and 2008 have shown that OPBC can be used as coarse lightweight aggregate in concrete and that the density and the 28-day compressive strength of concrete containing this aggregate both fulfil the requirements of structural light weight aggregate concrete (LWAC)[14,15,23].

Table 1. Oxides (wt %)

SiO_2	Al_2O_3	Fe_2O_3	MgO	CaO
56.0	26.7	5.8	0.6	0.8

Na_2O	K_2O	TiO_2	S	LOI
0.2	2.6	1.3	0.1	4.6

OPBC is mainly used as fine aggregate due to the fact that it is easy to crush. There are case in which it has even been used in powder form for making a composite artificial aggregate. It was observed that after crushing the clinker, much of the crushed aggregate is fine aggregate. Hence, due to the availability, OPBC was used as fine aggregate. Secondly, OPBC is a lightweight aggregate. If it can be used successfully as fine aggregate in concrete, the density of concrete will be reduced [16].

Previous studies on concrete constituent material replacement show that the replacement of various types of recycled or by-product materials as aggregates had made a noticeable progress all around the world to decrease the dependency on and use of virgin earth aggregates in production of concrete [6]. This research aims to further develop the diversity of knowledge on these replacement materials by undertaking a comparative study between two relatively less well-known materials (OPBC and CBA) and the traditionally used typical constituent concrete material (in the case of this project, mining sand).

The purpose of this study is to underline the feasibility of using both 'Coal Bottom Ash' (CBA) and 'Oil-Palm Boiler Clinker' (OPBC) in comparison, as two upcyclable waste by-product, in the concrete mix.

The objectives are:

1. To check fundamental material properties of CBA and OPBC in comparison to ordinary sand to see the possibility of their replacement in concrete to achieve comparable properties of concrete from strength and safety point of view.
2. To produce 60 MPa HSC using CBA and OPBC having workability of higher than 180 mm,
3. To determine mechanical properties of HSC using partial replacement of sand by 12.5 % and 25 % of each CBA and OPBC in separate mixes,
4. To compare mechanical properties of CBA concrete with OPBC concrete and concrete containing ordinary sand.

2 EXPERIMENTAL PROGRAM

2.1 General Research Experiment Design

This research is based on experimental works and data comparison. The main concentration of the work is in line with research objective to develop relationship between the amount percentage of the CBA and OPBC as upcycled materials used in the designed mixes and the characteristics of the produced concretes. A regular mix design with typical post curing characteristics was used as the control mix. While the amount of recycled material used in mixes would be the variables, the constant values will be portions of every other element included in those mixes excluding the fine aggregate. Each material was included in the comparative mixes only by replacement and not to addition. Initial material testing were taken place to develop comparable characteristics only for OPBC and CBA as which their variations contribute to the project results. The achieved results were not only used in final cause and effect discussions but also served as foundation data next to available literature review for designing trial and final mix designs. After trial mixes and designing final mixes the next stage was casting and curing. In this stage a schedule for casting was designed and according to nature of the chosen to be carried on tests.

Specific gravity, water absorption, and bulk density are the characteristics which have been chosen to be investigated for CBA and OPBC. Mix designs were designed consisting a total of 5 mixes of high strength concrete casted using increasing percentages of sand replacement proportions of both OPBC and CBA. The next stage was to determine the properties of concrete in both fresh and hardened stages. Concrete parameters chosen to be investigated in this report are compressive stress and tensile stress.

2.2 Experiment Materials

2.2.1 Coarse Aggregate

The coarse aggregate used in this work for all the mix designs is crushed granite aggregate. Maximum size of 20 mm, angular shape and a rough surface texture can be mentioned as the main characteristics of this aggregate that is complied with MS29 [15]. The coarse aggregate was air dried fully in open space outdoors under the sun and sieved before used in mixing. It has the specific gravity of 2.66 kg/m3 and water absorption of 0.57 %.

2.2.2 Fine Aggregate

The typical sand used in the study as fine aggregate was mining sand having the size of 150 μm to 5mm. The mining sand used during this project has the typical properties of 0.94% water absorption, specific gravity of 2.5 kg/m3 according to MS30:Part2 [17]. The fine aggregate was air dried in open space to reach SSD (Saturated Surface Dry) condition to maintain the moisture loss of the wet mixture due to water absorption of sand.

The sieve analysis test was done in accordance to ASTM C136 [18] showed the chosen sand is to be categorized in zone 1 as classified in BS 882: Part 103 [19]. The fineness modulus for this sand is 2.70.

2.2.3 CBA and OPBC

Both CBA and OPBC used for the project are made in Malaysia. They were delivered directly from burner mills factory sites to the research. Both materials were treated similar to fine sand as to be fine aggregate replacements. They were clean washed with tap water and air dried in open space to reach SSD (Saturated Surface Dry) condition then kept in control conditions until the casting process.

2.2.4 Ordinary Portland Cement (OPC)

In all mix designs OPC was used as the main binder as known as the most popular and typical type of cement used in concrete where there is no exposure to sulphates. It is produced by Tasek Corporation Berhad. known as type 1 Portland Cement and conforms to MS 522, Part 1 [20]. The basic characteristics of this cement are surface area of 335 m2/kg, average particle size of 10 μm and specific gravity of 3.10 kg/m3.

2.2.5 Superplasticizer (SP)

The super plasticizer used in this study was Sika ViscoCrete-2199. It is a modified polycarboxylate ether based superplasticizers as mentioned in product specifications having a specific gravity of 1.20 kg/m3. The product has a thick liquid state and comes in brown colour. The Sika catalogue mentioned that ViscoCrete-2199 is a highly effective liquid superplasticizer for the production of free flowing concrete or as a substantial water-reducing agent for promoting high ultimate and early strengths" [21].

2.2.6 Water

A water quality which is filtered and clarified in other words potable, is efficient enough to use for mixes in accordance to BS 8110 [19]. In this project the laboratory tap water that is the Malaysian household and domestic pipe water; was used to create the concrete mix.

2.3 Universal Stress Testing Machine

The universal stress testing machine is a machine used to conduct the strength and stress tests for variety of mechanical properties of concrete such as compression, tensile and flexure strengths. The machine has different modes and sections for each test. The machine used for this test is made by ELE International Limited, a British company, with the serial number of 974-01-0105 and a 2500 kN load frame.

2.4 Concrete Production

2.4.1 Trial Mix

A trial mix set was designed, according to the findings from literature review, to investigate three potential mixes to find the ultimate control mix for this project to satisfy the requirements of the project objective. Fly ash is used as a present component in control mix and all the other mixes including the trial mixes. After the data from this procedure was obtained, the best fit mix for control mix was designed and the rest of comparative mixes were designed. The mix proportions of each mix design is showed in Table 2.

Table 2. Trial Mix Design

Mix No.	1	2	3
Cement	500	480	450
Fly Ash	50	70	50
Super plasticizer	5.5	5.5	5.5
Binder (Cement + Fly Ash + SP)	550	550	550
SP to binder ratio %	1	1	1
Water	192	187	170
Water to binder ratio (w/b)	0.35	0.34	0.31
Coarse aggregate	960	973	870
Fine aggregate	640	677	800

2.4.2 Mix Design

The mix designs were finalized in accordance to the trial mix and the material properties obtained by experiments beforehand using absolute volume method. The method has been clearly defined in "Standard Practice for Selecting Proportions for Normal Weight Concrete". Absolute volume method is a rather accurate method in which parameters such as the specific gravity of the constituting materials will be used to calculate the absolute volume each ingredient will be occupying in a unit volume of concrete in accordance to American Concrete Institute, Committee 211: standard practice for proportioning of concrete mixes [22].

For this research the replacement method was used due to the fact that by reducing the use of cement, more economical and environmental friendly mixes can be produced that is an advantage in line with environmental friendly objectives of this research.

After all considerations it was decided that the research of effect of replacement sand with CBA and OPBC will be better investigated if all the other characteristics including the binder portion were kept constant. The replacement amount was decided to be tested in 0, 12.5 and 25 percent for both CBA and OPBC. Having these percentage replacements, 5 mix designs were sufficient for the project. The 5 mixes, their codes and constituent proportions are tabulated below in Table 3.

Table 3. Mix Design

Mix Name		0% Control	12.5% CBA	12.5% OPBC	25% CBA	25% OPBC
Mix Code		mx1	mx2	mx3	mx4	mx5
Cement	Kg	51.70	51.70	51.70	51.70	51.70
Super Plasticizer	g	570	570	570	570	570
Fly Ash	Kg	5.70	5.70	5.70	5.70	5.70
Water	Kg	18.40	18.40	18.40	18.40	18.40
Granite	Kg	100.00	100.00	100.00	100.00	100.00
Sand	Kg	94.00	82.25	82.25	70.50	70.50
Bottom ash Sand	Kg	0	11.75	0	23.50	0
Clinker Sand	Kg	0	0	11.75	0	23.50
Extra Water	Kg	0	3.00	1.50	5.00	3.00
Total Water	Kg	18.40	21.40	19.90	23.40	21.40
W/C	NA	0.321	0.373	0.347	0.408	0.373

Mx1 and mx2 contained 12.5 % sand replacement of CBA and OPBC respectively while mx3 and mx4 contained 25 % sand replacement of CBA and OPBC respectively. As it is seen above the amount of water added to each mix is slightly different since the concentration during mix proportioning was not only to similarities in w/c ration but also similarities in consistency of the slump test result of the fresh mix as different fine aggregates have different amount of water absorption.

2.5 Fresh Concrete Slump Tests

Table 4. Slump classes

Slump class	Slump (mm)
S1	10 - 40
S2	50 - 90
S3	100 - 150
S4	160-210
S5	≥220

As soon as the concrete is mixed and the desirable consistency using visual comparison by naked eye is reached before the concrete is poured into molds the slump test takes place. The test in this project was done based on BS 1881: 102. According to European Standard EN 206-1 (2000) five classes of slump have been designated, as tabulated in Table 4 above.

3 Results and Discussion

The experimental results obtained are divided into three groups as outlined in the subsections.

3.1 Properties of the Targeted Constituent Materials (Mining Sand, OPBC, CBA)

3.1.1 Bulk Density Properties

The density test results for CBA, OPBC and typical mining sand are shown in the Fig. 1. From the table it can be seen that CBA has the lowest both, loose and compacted density between all three materials. On the other hand typical sand is the densest while loose sand has the density of 1573 kg/m3 that is 41 % and 31 % higher than CBA and OPBC respectively.

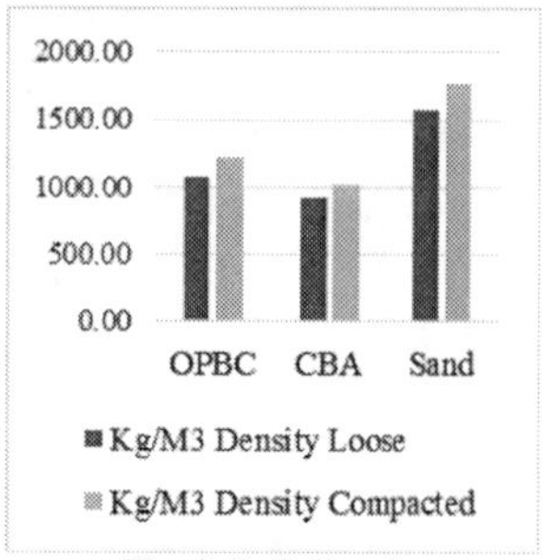

Fig. 1. Bulk Density

3.1.2 Sieve Analysis Properties (Particle Size Distribution)

As shown in Fig. 2 the particles spread quite evenly for all three materials hence they are all quite well graded. For both sand and OPBC 60 % of material pass through the 1.18 mm sieve but for CBA the passing amount for the mentioned sieve is almost 74 %. In general sand and OPBC have very similar curves that is a sign of similar particle size distribution while CBA graph has a slightly gentler curve proofing of larger particle size.

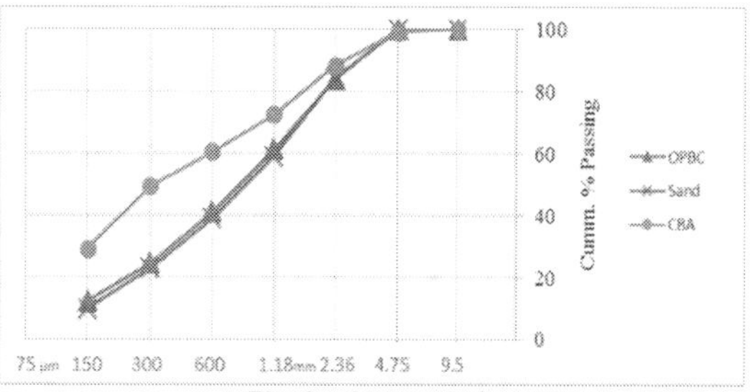

Fig. 2. sieve analysis

3.1.3 Specific Gravity and Water Absorption Properties

The saturated (apparent), saturated surface dry (SSD) and oven dry specific gravities of CBA, OPBC and mining sand are given in the Fig. 3 below. The oven-dry specific gravity shows that CBA is the lightest among all and sand the heaviest. Meanwhile according to the SSD results OPBC has the highest rate of water absorption. This happens due to the high amount of porosity that OPBC particles have. The higher water absorption may lead to loss of workability with time and reduce the water available in the mix for cement reaction. To overcome this problem all materials must be kept similarly not in dry conditions.

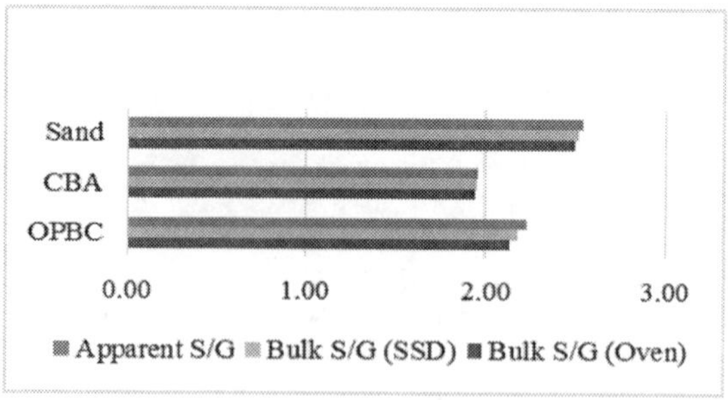

Fig. 3. Specifec gravity

3.2 Slump Test Results (Workability)

Table 5 contains parameters such as amount of water in the initial design to achieve expected slump consistency, the water to cement ratio (water to binder ratio) and mix code for each mix.

Table 5. Slump

Mix Name		0% Control	12.5% CBA	12.5% OPBC	25% CBA	25% OPBC
Mix Code		mx1	mx2	mx3	mx4	mx5
Slump	mm	190	220	220	220	185
Total Water	Kg	18.4	21.4	19.9	23.4	21.4
w/c	NA	0.321	0.373	0.347	0.408	0.373

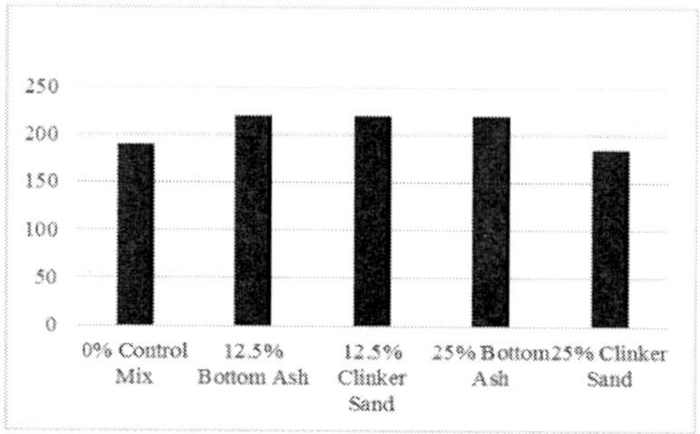

Fig. 4. Slump (mm)

As in general, the increase in slump value increases the workability in concrete. The slump can vary from zero to even 90 % of the original height in very fluid mixes. Fig 4. Shows the slump results in comparison.

Superplasticizer (SP) was added to the mixes to 10 % of weight of OPC. The amount was chosen 10 % to be constant. The portion of SP added was kept constant for all mixes since the reason of addition was only to increase cohesiveness and not to effect workability. Although workability is heavily impacted by w/c ratio, binder content and SP; since the amount of binder was kept almost the same, the changes in slump can directly be connected to changes in different material used as fine aggregate and their characteristics such as water absorption. As it can be seen in the graph, all mixes have an acceptable medium workability of 185 to 220 slump. So in general the concrete will not lose much of its workability if the fine aggregate is partially replaced with CBA or OPBC.

3.3 Properties of Hardened Concrete

3.3.1 Compressive Strength

According to BS EN 12390: Part 3: compressive strength test specimens, Testing of hardened concrete (2002) the minimum compressive strength for concrete to be HSC is 55 MPa by the age of 28 days from curing.

The 28-day cubical compressive strength targeted for this research is 60 MPa; that is in regards to the concrete fully cured in water. The average 28-day compressive stresses acquired in experiment for water cured specimens for 12.5 % CBA, 12.5 % OPBC, 25 % CBA, 25 % OPBC and control mix are 69.52, 75.83, 68.70, 71.85 and 66.69 MPa respectively. Table 9 shows the average strength achieved in 1, 3, 7, 28 and 56 day ages for all mixes cured typically in water.

According to the Table 6 all mixes have achieve higher than 55 MPa compressive strength at the age of 28 day. Their strength is even higher than 60 MPa. Hence all mixes fall under the HSC category and the targeted objective of the research is achieved. According to the Fig. 5 it can be observed that, general strength of concrete decreases for both mx4 and mx5 that contain 25% of material replacement. Furthermore 12.5% replacement seem to be closer to the optimum amount of the replacement in the case of both materials and High initial strength is achieved in 12.5% OPBC mix. In accordance to Fig. 5 a stable pace of strengthening can be seen in 12.5% CBA mix. 12.5 % CBA has achieved the highest strength in 56 days and 12.5 % OPBC has achieved the highest strength in 28 days.

Table 6. Compressive strength test results

	0%	12.5% CBA	12.5% OPBC	25% CBA	25% OPBC
Day 1	29.83	31.64	33.55	23.62	28.16
Day 3	57.25	48.19	50.46	41.46	47.02
Day 7	61.84	59.48	58.52	47.26	47.15
Day 28	66.69	69.52	75.83	68.70	71.85
Day 56	76.74	79.35	76.04	72.75	74.01

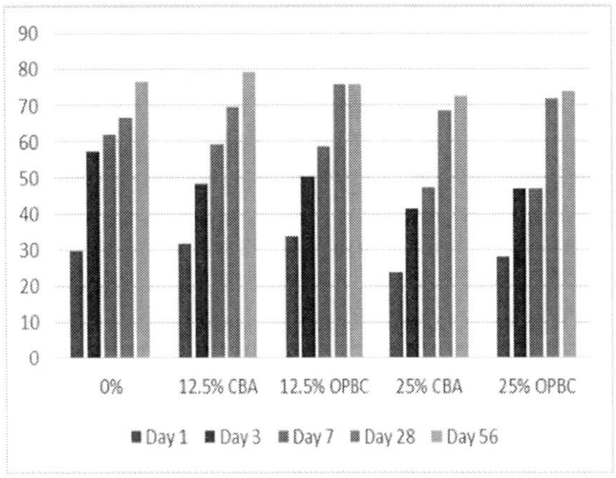

Fig. 5. Compressive strength test results categorized by mix composition

Fig. 5 also shows the compressive strength of all mixes categorized in mixes separately from the Day 1 after curing and Day 28th of curing. It can also be seen that in general CBA contained mixes have a higher early age strength comparing to other mixes and that can be caused by the low degree water absorption by CBA in early stages enough water is available for hydration. In other hand, OPBC mixes have highest final strength that might be a result of high porosity and water absorption of OPBC. Since the water is stored in pores and holes of OPBC particles it is only available for cement slowly and by time so the hydration process will take longer. In accordance to the figure, 12.5 OPBC has the highest hardening acceleration rate since a 221 % increase in its strength over 3 weeks can be seen that is more than triple of the initial strength.

3.3.2 Splitting Tensile Strength

As a very well-known concept, concrete's tensile strength is much less than its compressive strength. The results of splitting tensile will develop an understanding to the tensile ability of the research mixes.

Table 7 showed the splitting tensile test result for all mixes in different ages. It can be seen that 12.5 % OPBC mix had the highest tensile strength at 28 days while control mix had took the title of having the highest tensile strength at 56 days. And the lowest strength belongs to 25 % OPBC at day 1 of curing. It is also visible that 12.5 OPBC mix has a stable increase all through its aging process and is showing desired results. The OPBC mixes had been maintaining a high level of tensile strength since day 1.

Table 7 also shows that 12.5 % replacement mixes for both materials had the smoothest growth gradient between all mixes. They gained strength slow and steady and show a desirable strength level. It can be seen that the control mix had the lowest tensile strengthening speed while having a high early age tensile strength. It could also be seen that in general OPBC had the highest speed of strength development for both mixes containing it.

Table 7. Splitting tensile test results

	0%	12.5% CBA	12.5% OPBC	25% CBA	25% OPBC
Day 1	2.982	2.603	2.743	2.207	1.753
Day 7	4.165	3.633	3.984	4.193	4.089
day 28	4.535	4.846	5.399	4.479	5.091
day 56	6.124	5.696	5.804	4.656	5.270

4 Conclusions

Based on results obtained from 5 mixes the following conclusions can be made:

1. Based on the results obtain from test on properties of the targeted constituent materials, both OPBC and CBA have similar particle size distribution properties to mining sand which makes them more desirable replacement materials. Not to mention both OPBC and CBA have lower densities in comparison to sand which could be a beneficial advantage when it comes to production of light weight concrete.
2. The slump test results show an acceptable and practical degree of workability for all the mixes having slump of from 185 mm to 220 mm with w/c ratio of 0.3 to 0.4. This means the percentage usage of SP was designed properly and suitable for mixes.
3. Compressive test results showed that in the age of 28 days all mixes have achieved strength higher than 60 MPa. Hence, the objective of research to produce HSC is achieved.
4. Compressive test results also show that in general 12.5 % replacement mixes had higher and better results at 28 days in all tests comparing to 25 % replacement mixes. Hence it seems like 12.5 % is quite close to optimum level of replacement suitable for CBA and OPBC.
5. From the results obtained from the splitting tensile test, it can be seen that the 25 % replacement mixes have low early strength gain comparing to the three other mixes. That might be due to high amount of water absorption by the high percentage of replaced materials as both OPBC and CBA have relatively high water absorptions. It seems moisturizing these mixes will affect their final strength greatly. Anyway in 7 days' time after curing, both 25 % mixes caught up with tensile strength of other mixes and control mix.
6. From the results obtained from the splitting tensile test, the sudden increase in strength from the 1st day to 7th day for both 25% mixes indicates happening of a secondary reaction in the dry mix after initial hydration. The process seems to be a support to the hydration process. This can be due to reaction of OPBC or CBA with hydrated cement.

Finally, it was concluded that using the mentioned materials to extend of 12.5 % as replacement is practical and economical.

Test results also show the comparability of OPBC and CBA concretes to OPC concrete using ordinary sand from mechanical properties point of view which creates the possibility of future research expansion in the field of concrete strength and safety.

Acknowledgment

Research reported in this publication was supported by the Ministry of Higher Education, Malaysia through the FRGS Grant (Grant No.: FP046-2014B).

References

1. Van Oss, H. G. (2013). Cement. US Geological Survey.
2. Shafigh, P.; Jumaat, M.Z.; and Mahmud, H. (2010). Mix design and mechanical properties of oil palm shell lightweight aggregate concrete: A review. International journal of the physical sciences, **5**(14), 2127-2134.
3. Troxell, G., Raphael, J., & Davis, R. (1958). Long-time creep and shrinkage tests of plain and reinforced concrete. Paper presented at the ASTM Proceedings.
4. British Standard. (1881). BS 1881: 102: Methods for determination of slump: BS.
5. Neville, A.M. (2008). Properties of Concrete, fourteenth ed. CTP-VVP, Malaysia.
6. Meyer, C. (2009). The greening of the concrete industry. Cement and Concrete Composites, **31**(8), 601-605.
7. Mahmud, H. (1999). Engineering properties of fly ash concrete.
8. Manz, O. (1997). Worldwide production of coal ash and utilization in concrete and other products. Fuel, **76**(8), 691-696.
9. Kim, H., & Lee, H. (2011). Use of power plant bottom ash as fine and coarse aggregates in high-strength concrete. Construction and Building Materials, **25**(2), 1115-1122.
10. Kanadasan, J., & Razak, H. A. (2014). Mix design for self-compacting palm oil clinker concrete based on particle packing. Materials & Design, 56, 9-19.
11. ASTM. (1958). C 125" Definition of Terms Relating to Concrete and Concrete Aggregates," 1958 Book of ASTM Standards. Part, 4, 588.
12. Khasreen, M. M., Banfill, P. F., & Menzies, G. F. (2009). Life-cycle assessment and the environmental impact of buildings: a review. Sustainability, **1**(3), 674-701.
13. Abdullahi, M., Al-Mattarneh, H., Hassan, A. A., Hassan, M., & Mohammed, B. (2008). Trial mix design methodology for Palm Oil Clinker (POC) concrete.
14. Binti Robani, R., & Chan, C. (2009). Reusing Soft Soils with Cement-Palm Oil Clinker (Poc) Stabilisation. Paper presented at the International

Conference on Engineering and Education in the 21st Century.
15. MS29. (1995). Specification for coarse and fine aggregate in concrete MalaysiaStandard29: SIRIM, MS.
16. ASTM. (2014). Standard Test Method for Sand Equivalent Value of Soils and Fine Aggregate ASTM D2419 - 14.
17. MS30:Part2. (1995). Specification for Fine aggregates in concrete: SIRIM, MS.
18. ASTM Standard. (2006). C136-06, 2006," Standard Test Method for Sieve Analysis of Fine and Coarse Aggregates," ASTM International, West Conshohocken, PA.
19. British Standard. (1985). 8110: Part 1, Structural use of concrete–code of practice for design and construction. British Standards Institute, London UK, 3-8.
20. MS522. (2007). Compositions, specifications and conformity criteria for common cements, Malaysia Standard.
21. Sika. (2012). Superplasticizers. Water Reducers (High Range). From http://mys.sika.com/en/solutions_products/02/02a001/02a001sa03/02a001sa03100/02a001sa03109.html
22. ACI Committee. (1991). Standard Practice for Selecting Proportions for Normal, Heavyweight, and Mass Concrete (ACI 211.1–91). Paper presented at the American Concrete Institute.
23. Abdullahi, M., Al-Mattarneh, H., & Mohammed, B. (2010). Characterisation of Palm Oil Clinker Aggregate. International Journal of Applied Engineering Research, **5**(16).

An Exploratory Research on Structure of Percieved Residential Environment Quality

Maria Khosravi [1,a], Sr Dr Norngainy Mohd Tawil [2] Morteza Monshizadeh [3]

[1] Ph.D. Student Universiti Kebangsaan Malaysia, Architecture Department
[2] Deputy Director of CESMED Universiti Kebangsaan Malaysia
[3] Ph.D. Student Universiti Kebangsaan Malaysia, Architecture Department

Abstract. Housing is a composite entity (in the sense of Gary Becker) that fulfils several human need. The main need that this commodity is to fulfil is dwelling, but from environmental satisfaction point of view, housing is more than a dwelling unit and its set of objective characteristics, since it also provides health services, security, privacy, neighborhood and social relations, status, community facilities and services, access to jobs and control over the environment. Thus, the principal way in which individual satisfaction with housing is measured through direct questions about their level of housing satisfaction. The purpose of this paper then is to contribute further research on the conceptualization of individual housing satisfaction as a particular domain of satisfaction with life as a whole, providing empirical evidence to disentangle the effects housing characteristics (hedonic), individual and household characteristics (predictive), as well as social interactions originated in one's residential neighborhood on this housing satisfaction domain. The main contributions of this paper in relation to previous work are the following. First, the simultaneous inclusion of the hedonic and predictive approach in our valuation (by comparing individuals satisfaction and individuals importance values), Second, we also investigate the relations between values in satisfaction and importance. This qualitative exploratory attempt has done by a data gathering through a close ended questionnaire that is made of focus group research.

1 Introduction

Housing is a composite entity (in the sense of Gary Becker) that fulfils several human need. The main need that this commodity is to fulfil is dwelling, but from environmental satisfaction point of view, housing is more than a dwelling unit and its set of objective characteristics, since it also provides health services, security, privacy, neighborhood and social relations, status, community facilities and services, access to jobs and control over the environment. Thus, the principal way in which individual satisfaction with housing is measured through direct questions about their level of housing satisfaction. The purpose of this paper then is to contribute further research on the conceptualization of individual housing satisfaction as a particular domain of satisfaction with life as a whole, providing empirical evidence to disentangle the effects housing characteristics (hedonic), individual and household characteristics (predictive), as well as social interactions originated in one's residential neighborhood on this housing satisfaction domain. The main contributions of this paper in relation to previous work are the following. First, the simultaneous inclusion of the hedonic and predictive approach in our valuation (by comparing individuals satisfaction and individuals importance values), Second, we also investigate the relations between values in satisfaction and importance. This qualitative exploratory attempt has done by a data gathering through a close ended questionnaire that is made of focus group research.

[a] Maria Khosravi: Mariakh57@hotmail.com

2 Residential quality residential environment preference

Housing is one of the basic human needs for well-being, the securing of which is fundamental to our quality of life (Y. Sun, 2005).Davies (1938) use "home surroundings" and "neighborhood qualities" to assess good housing standards. Physical health and mental well-being and security are included in the evaluation criteria. Yard for children's play, location, harmony between house plan and the surroundings, adequate landscaping, and convenience are included in problems related to home environment.

Baer (1986) and Sloane (2006) adopt an expanded checklist of APHA on housing and planning studies, Baer (1986) provides a historical review of the evolution of local and regional housing studies and their political uses.

The next step concluded in the research is to create a supply side counterpart to housing need analysis to suggest how the dynamics of the existing stock can be better used to meet some of the fair share need. Sloane (2006) addresses whether the concept of sprawl is rich enough to foster sustainable, long-term connections between the fields of planning and public health. It finds that sprawl is not just about suburbs; highly segregated communities with poor nutritional and recreational environments aggravate high rates of obesity. Obesity is not the only adverse outcome of sprawl, and physical activity is not the only contributor to overweight. It concludes that planners should resist environmental determinism and remain open to the influence of social, political, cultural, and economic influences, as well as those of the built environment, on human behavior.

Brown and Moore (1970) provide a framework for understanding housing choice decision making. They use the concept of 'place utility' to measure the balance of residential satisfaction and dissatisfaction. The balance creates a central notion for interpreting residential mobility (Z. Li & Wu, 2006). Number of factors influencing the determining selection of the new residence are identified: accessibility, physical characteristics of the neighborhood, services and facilities, social environment, individual site and dwelling characteristics (Brown & Moore, 1970). Rouwendal and Meijer (2001) state that, among other things, the household preferences (and therefore also their willingness - to - pay) for housing attributes that are especially available outside large cities are substantial enough to make acceptance of long commutes attractive to many employees (Rouwendal & Meijer, 2001).

Harris et al (1968) scale the residence site choice by measuring the amenity and travel savings components of land value (Harris, Tolley, & Harrell, 1968). The measurement of amenity and travel savings components of land value reveals that both are major components of land value. Amenity is one of the main commodities people wish to acquire when their incomes rise. Kain and Quigley point out there are other factors impacting housing quality, such as air pollution, accessibility to downtown and school quality (Kain & Quigley, 1970).

Ball and Srinivasan (1994) present a model of housing selection process using the analytic hierarchy process (AHP), which allows the buyer to consistently evaluate property attributes. Schniederjans et al., (1995) also present a Goal Programming model that utilizes the AHP to evaluate property attributes and make an optimal house selection decision. Table 1 presents the research factors they used in housing selection process.

Karsten(2007) indicates that housing quality is composed of two major ingredients: the site (accommodation of daily life) and the situation (location) of the neighborhood. The environment of housing includes facilities, infrastructures and services that are available within and beyond the neighborhood, amenities that are proximate in and near the neighborhood, and the social capital in neighborhood.

Table 1. Research factors in Housing Selection Source: Ball & Srinivasan (1994).

Categories	Factors
Proximate	Commute to Work, Member A
	Commute to Work, Member B
	Proximity to Other Locations
	Schools
	Shopping
	Church and other regularly frequented places
Community	Quality of Schools
	Government Services/Taxes
	Social Considerations
	Entertainment
Neighborhood	Aesthetics
	Quality of Infrastructure
	Landscaping, autos, etc.
	Infrastructure
Property	Lot
	Utility of usable space
	Landscape, lawn, foliage, privacy, etc.
	Walkway, driveway, fence, outbuilding, etc.
	Building – Exterior
	Condition of walls
	Architectural design and garage
	Building – Interior
	Rooms
	Systems

3 Residential environment in concept

Residential environment is a large topic to review. To study Residential environment, researchers have conducted research on house quality (Association & Housing, 1945; Fiadzo, et al., 2001; Rindfuss, et al., 2007), residential satisfaction (Adriaanse, 2007; Fang, 2006; Kellekci & Berkoz, 2006), housing environmental quality (Ha & Weber, 1994), attractiveness of residence (Kauko, 2006; Linneman, 1981), neighborhood attachment(Hays & Alexandra, 2007; Karien, 2007; L. H. Li, 2008b), and social capital (Kevin, 2003; Kleinhans, et al., 2007; Middleton, et al., 2005).

4 Evaluation of residential environment

Lansing and Marans (1969) develop a method using an attitude survey to evaluate neighborhood quality throughout a metropolitan area. This method illustrates how a planner's judgment may be congruous or incongruous with that of the residential community and suggests ways in which analytical relationships between planners' and citizens' values may be used in developing a model for determining residential satisfaction. Baba and Austin (1989) shed light on the neighborhood safety when measuring the neighborhood environmental quality. They suggest that improving the quality of neighborhood environmental characteristics will increase residents' concerns with safety.

Ge and Hokao (2004) use hierarchical multi-attributes evaluation model to access the residential environment. Questionnaire surveys and subjective evaluations on residential environment are performed in order to grasp the main factors of residential environment of small local cities. They establish an evaluation index system and analyze the regional residential environment characteristics and personal residential preference types. The environmental criteria include convenience, amenities, health, safety, and community. Some of the criteria are adopted in this study.

To assess the determinants of user satisfaction in housing and environmental quality in order to provide maximum level of user satisfaction, Berkoz et al. (2009) survey to measure the user satisfaction in housing and environmental quality, and to determine the factors increasing the level of satisfaction. Relative factors are presented on the questionnaires: the characteristics of household members (the size of the household, gender, age, education, the number of people working, profession, income group, ownership of durable consumer goods, and ownership of vehicles), features related to the housing (when the household moved into the housing, ownership of the housing, housing type, size of the housing, the number of inhabitants, the previous neighborhood, the previous housing type), accessibility (accessibility to work, to the center where the daily needs are met, shopping center, city center, school, sports facilities, walking areas, refreshment areas, car parking areas, health institutions, education institutions, entertainment areas, recreational areas, public transport stops, and to close relatives and friends), characteristics of the housing environment (lighting, maintenance of open areas, maintenance of green areas, traffic density, user density, building density, housing environment facilities), security (fire, natural disasters, traffic accidents, robbery, murder), neighbor relationships (neighbors of similar social background, acquaintance with people nearby, privacy, charity among neighbors), and the appearance of the housing environment (monotony, being interesting, reflecting the social status, economic value)(Berkoz, et al., 2009). Figure 1 presents the conceptual model of their study.

Figure 1. Conceptual model developed for user satisfaction in housing and environmental quality. Source: (Berkoz, et al., 2009).

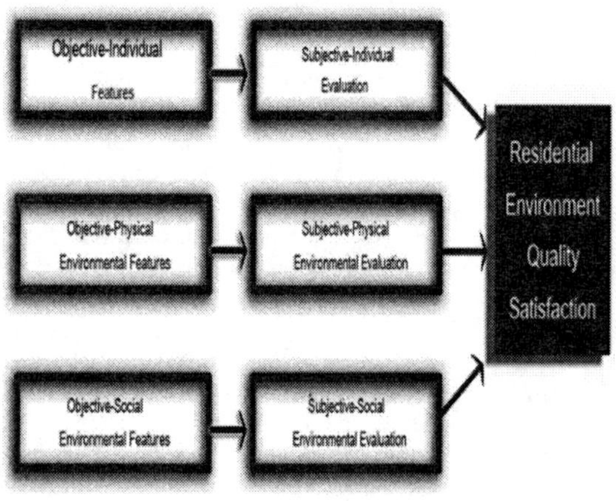

5 Important residential environment factors

5.1. Mobility

Accessibility of housing is considered very important to the quality of Residential Environment. Plenty of studies about public and private traffic network and commute can be found in Anderson, Kanaroglou, & Miller, 1996; Cervero, 1996a; Cervero & Wu, 1998; Levine, 1998; Peng, 1997; Wachs, Taylor, Levine, & Ong, 1993 (Cervero, 1996a; Cervero & Wu, 1998; Levine, 1998; Peng, 1997; Wachs, Taylor, Levine, & Ong, 1993). Kellekci and Berkoz (2006) conduct accessibility in their research on housing quality and environmental satisfaction. Accessibility in their thesis refers to work, to the center where the daily needs are met, shopping center, city center, school, sports facilities, walking areas, refreshment areas, car parking areas, health institutions, education institutions, entertainment areas, recreational areas, public transport stops, and to close relatives and friends. Public traffic network, Private traffic network, Proximity to urban center, and Proximity to workplace are mentioned in researchers study as main factors of mobility (Mills, 1994, Linneman, 1981, Bontje 2004).

5.2. Community facilities

To measure the satisfaction of housing and its environment, Kellekci and Berkoz (2006) adopt characteristics of the Residential Environment as one of the indicators, which includes lighting, maintenance of open areas, traffic density, user density, building density, housing environment facilities Education Facility (EDF), adopt good place to raise children and schools, Medical and Health Facility (MHF), Retail Service (RES), shopping and sport facilities..

5.3. Community social capital

The important influence of social capital to housing environment and neighborhood attachment are studied in a great number of literatures. Middleton et al (2005) find that social capital is seen as the foundation on which social stability and community's ability to help itself is built; and its absence is thought to be a key factor in neighborhood decline. Temkin and Rohe (1998) find that neighborhood activities, such as borrowing small items and visiting, would occur in environments in which residents trusted one another and might be thought of as a proxy for the level of trust among neighborhood residents. Even more, housing environment in social aspect affects the social life of residents. A key issue for the discussion of social capital is the extent to which residents in any neighborhood perceive themselves to be part of the same community. Neighborhoods are important sources of “social identity” (Forrset and Kearns, 2001).

6 Methodology

The main objective of this study is to understand the demand pattern and preferences on structuring the perceived residential environment quality by consumers the international Iranian Postgraduate student in KL, Malaysia and offer more information for urban planners and housing developers about housing environmental demand from social and cultural perspective. In order to achieve this objective, a hierarchy of housing environment quality attributes is developed. This research was undertaken in the Klang Valley Condominiums in Malaysia. Data collection was conducted through a field study using a questionnaire filled in by a sample of 30 students for pilot study and 300 students for main designed questionnaire . Data analysis was also conducted using SPSS software. Mixed method research encourages researchers to combine inductive and deductive thinking to answer questions that cannot be answered by qualitative or quantitative approaches alone (Creswell and Plano Clark, 2007). The combination of methods can assist in tackling highly complex problems involving several layers of understanding (Hesse-Biber and Leavy, 2011). However, mixed method research is not easy as it requires the researcher to be proficient in both forms of inquiry and it takes time to collect the required data (Creswell and Plano Clark, 2007). Reasons why researchers might want to use a mixed method approach include (i) using different theoretical approaches on the same research question to enhance credibility (triangulation), (ii) to give a fuller understanding of a research question or to clarify a result (complementarity), (iii) to use result from one method to develop or inform another method (development), (iv) where a studies results raise questions or contain contradictions which require clarification (initiation) or (v) where a researcher decides to expand into a whole new investigation (expansion) (Hesse-Biber and Leavy, 2011; Greene et al., 1989)(figure2).

Figure 2. Mixed method used in research methodology.

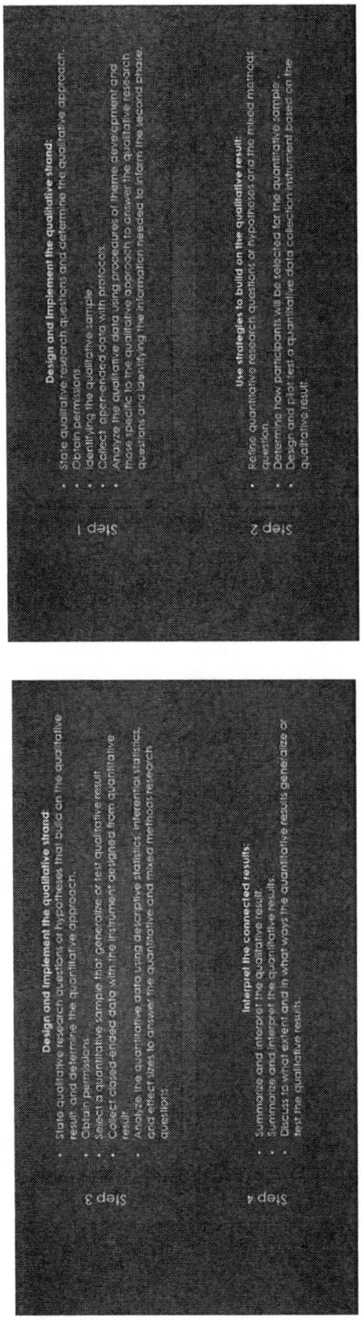

7 Theoretical framework

This model was developed from previous works on housing environments and location preference (Bender et al., 1997, 2000; Kauko, 2006, 2007; Yam et al., 2008; Fan, 2010;Mohit & Nazyddah, 2011).Basically the hole study focuses on three major physical scale (Housing characteristics, Neighborhood characteristics, and Regional characteristics) to model the frame (figure 3,4).

Figure 3. Modified model of housing environment quality. Source: (Bender, et al., 2000).

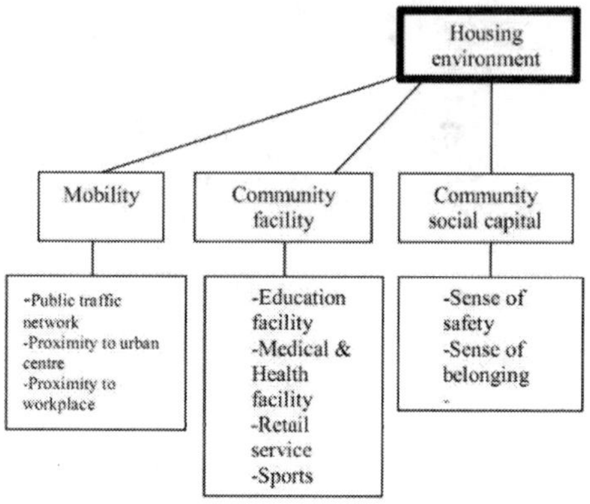

Figure 4. Modified Framework Housing environment Quality of for the study.

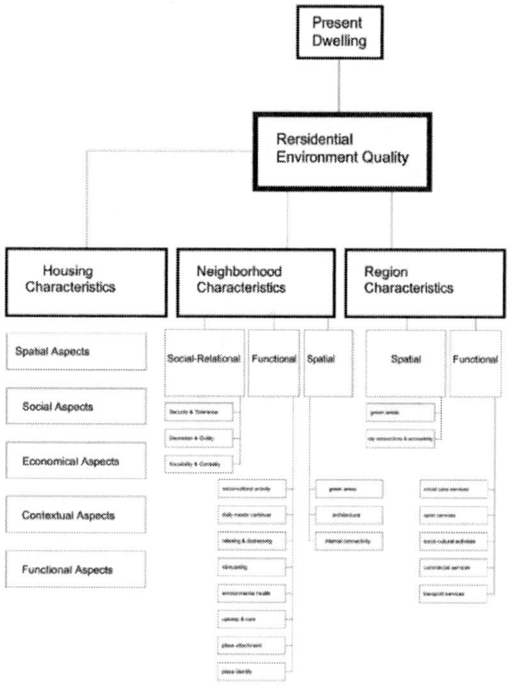

8 Result and analysis

Gathered data were analyzed using the statistical package for Social Science (SPSS) and the results of analyzing data are presented. First of all an introductorily examination of the data is described which includes the procedure involved in data cleaning and screening, descriptive statistics and correlation analysis. Results of hypotheses tests according to objectives of study entitled "**An Exploratory Assessment on Iranian International Student's Accommodation Satisfaction**" are discussed in their order by using Repeated Measures, ANOVA, T-Test, and Pearson's Correlation test. The analysis included:

- Descriptive analysis of variables
- Inferential analysis

Figure 5. Modified model of accommodation satisfaction level and satisfaction measurements.

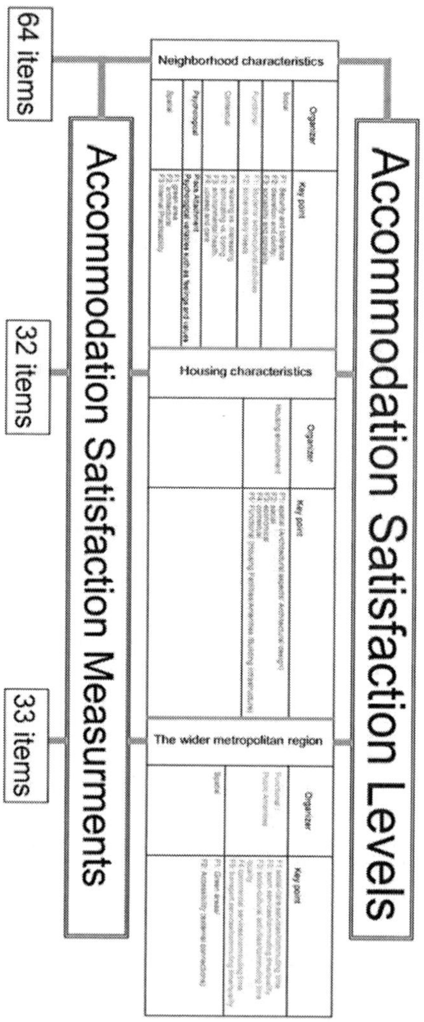

Data gathered in two main descriptive satisfaction level and importance level of users and as depicted in figure 5 categorized as 64 item for neighborhood characteristics, 32 item on housing characteristics, and 33 item on wider metropolitan region. The neighborhood characteristics are examined as social, functional, contextual, psychological, and spatial organizers, housing characteristics as housing environment organizers and regional characteristics as functional and spatial features. The Exploratory Assessment on Iranian International Student's Accommodation Satisfaction showed the modified model of residential can be used as a structure of perceived residential environment quality on three major scale of housing, neighborhood, and region.

References

1. Adriaanse, C. C. M. (2007). Measuring residential satisfaction: a residential environmental satisfaction scale (RESS). Journal of Housing and the Built Environment, 22(3), 287.
2. Altman, I., & Werner, C. M. (1985). Home environments (Vol. 8). New York: Plenum Press.
3. Altman, I., & Wohlwill, J. F. (1976). Human behavior and environment: advances in theory and research. New York: Plenum Press.
4. Ball, J. n., & Srinivasan, V. C. (1994). Using the Analytic Hierarchy Process in house selection. The Journal of Real Estate Finance and Economics, 9(1), 69-85.
5. Baer, W. C. (1986). The Evolution of Local and Regional Housing Studies. Journal of the American Planning Association, 52(2), 172 - 184.
6. Bender, A., Din, A., Favarger, P., Hoesli, M., & Laakso, J. (1997). An Analysis of Perceptions Concerning the Environmental Quality of Housing in Geneva. Urban Studies, 34(3), 503-513.
7. Berkoz, L., Turk, S., & Kellekci, M. (2009). Environmental Quality and User Satisfaction in Mass Housing Areas: The Case of Istanbul. European Planning Studies, 17(1), 161.
8. Bontje, M. (2004). From suburbia to post-suburbia in the Netherlands: Potentials and threats for sustainable regional development. Journal of Housing and the Built Environment, 19(1), 25.
9. Brown, L. A., & Moore, E. G. (1970). The Intra-Urban Migration Process: A Perspective. Geografiska Annaler, 52(1), 13.
10. Cervero, R. (1996b). Jobs-housing balance revisited. American Planning Association. Journal of the American Planning Association, 62(4), 492.
11. Chapple, K. (2006). Overcoming Mismatch: Beyond Dispersal, Mobility, and Development Strategies. American Planning Association. Journal of the American Planning Association, 72(3), 322.
12. Davies, J. E. (1938). Fundamentals of housing study: a determination of factors basic to an understanding of American housing problems. New York: AMS Press.
13. Fang, Y. (2006). Residential Satisfaction, Moving Intention and Moving Behaviours: A Study of Redeveloped Neighbourhoods in Inner-City Beijing. Housing Studies, 21(5), 671.
14. Fiadzo, E. D., Houston, J. E., & Godwin, D. D. (2001). Estimating housing quality for poverty and development policy analysis: CWIQ in Ghana. Social Indicators Research, 53(2), 137.
15. Ge, J., & Hokao, K. (2004). Residential environment index system and evaluation model established by subjective and objective methods. Journal of Zhejiang University SCIENCE, 5(9),7.
16. Gober, P., McHugh, K. E., & Leclerc, D. (1993). Job-rich but housing poor: The dilemma of a Western amenity town. Professional Geographer, 45(1), 12.
17. Ha, M., & Weber, M. J. (1994). Residential quality and satisfaction: Toward developing residential quality indexes. Home Economics Research Journal, 22(3), 296.
18. Harris, R. N. S., Tolley, G. S., & Harrell, C. (1968). The Residence Site Choice. Review of Economics & Statistics, 50(2), 241.
19. Hays, R. A., & Alexandra, M. K. (2007). Neighborhood attachment, social capital building, and political participation: A case study of low-and moderate-income residents of Waterloo, Iowa. Journal of Urban Affairs, 29(2), 181.
20. Kain, J. F., & Quigley, J. M. (1970). Measuring the Value of Housing Quality. Journal of the American Statistical Association, 65(330), 532-548.
21. Karien, D. (2007). Social Capital, Neighbourhood Attachment and Participation in Distressed Urban Areas. A Case Study in The Hague and Utrecht, the Netherlands. Housing Studies, 22(3), 355.
22. Karsten, L. (2007). Housing as a Way of Life: Towards an Understanding of Middle-Class Families' Preference for an Urban Residential Location. Housing Studies, 22, 83.
23. Kauko, T. (2006). What makes a location attractive for the housing consumer? Preliminary findings from metropolitan Helsinki and Randstad Holland using the analytical hierarchy process. Journal of Housing and the Built Environment, 21(2), 159.
24. Kellekci, O. L., & Berkoz, L. (2006). Mass Housing: User Satisfaction in Housing and its Environment in Istanbul, Turkey. European Journal of Housing Policy, 6(1), 77-99
25. Kleinhans, R., Priemus, H., & Engbersen, G. (2007). Understanding Social Capital in Recently Restructured Urban Neighbourhoods:

Two Case Studies in Rotterdam. Urban Studies, 44(5/6), 1069.
26. Lansing, J. B., & Marans, R. W. (1969). Evaluation of Neighborhood Quality. Journal of the American Planning Association, 35(3), 5.
27. Levine, J. (1998). Rethinking accessibility and jobs-housing balance. American Planning Association. Journal of the American Planning Association, 64(2), 133.
28. Leyden, K. M. (2003). Social capital and the built environment: The importance of walkable neighborhoods. American Journal of Public Health, 93(9), 1546.
29. Li, L. H. (2008a). The physical environment and a "sense of neighborhood" in residential communities in Hong Kong. Property Management, 26, 7.
30. Linneman, P. (1981). The demand for residence site characteristics. Journal of Urban Economics, 9(2), 129-148.
31. Li, Z., & Wu, F. (2006). Socio-spatial Differentiation and Residential Inequalities in Shanghai: A Case Study of Three Neighbourhoods. Housing Studies, 21(5), 695.
32. Ma, K.-R., & Banister, D. (2006). Extended Excess Commuting: A Measure of the Jobs-Housing Imbalance in Seoul. Urban Studies, 43(11), 2099.
33. Middleton, A., Murie, A., & Groves, R. (2005). Social Capital and Neighborhoods that Work. Urban Studies, 42(10), 1711.
34. Ottensmann, J. R. (1977). Urban Sprawl, Land Values and the Density of Development. Land Economics, 53(4), 389.
35. Peck, C., & Stewart, K. (1985). Satisfaction with Housing and Quality of Life. Home Economics Research Journal, 13(4), 10.
36. Peng, Z.-R. (1997). The jobs-housing balance and urban commuting. Urban Studies, 34(8), 1215.
37. Ricardo, A., Frank, D., Rebecca, P., Hollie, T., & et al. (2006). Perceptions of social capital and the built environment and mental health. Social Science & Medicine, 62(12), 3072.
38. Rindfuss, R. R., Piotrowski, M., Thongthai, V., & Prasartkul, P. (2007). Measuring housing quality in the absence of a monetized real estate market. Population Studies, 61(1), 35.
39. Rouwendal, J., & Meijer, E. (2001). Preferences for housing, jobs, and commuting: A mixed logit analysis. Journal of Regional Science, 41(3), 475.
40. Sarzynski, A., Wolman, H. L., Galster, G., & Hanson, R. (2006). Testing the Conventional Wisdom about Land Use and Traffic Congestion: The More We Sprawl, the Less We Move? Urban Studies, 43(3), 26.
41. Schniederjans, M. J., Hoffman, J. J., & Sirmans, G. S. (1995). Using Goal Programming and the Analytic Hierarchy Process in House Selection. The Journal of Real Estate Finance and Economics, 11(2), 167-176.
42. Sloane, D. C. (2006). Longer View: From Congestion to Sprawl: Planning and Health in Historical Context. Journal of the American Planning Association, 72(1), 10 - 18.
43. Sun, Y. (2005). Development of neighbourhood quality of life indicators. Temkin, K., & Rohe, W. M. (1998). Social Capital and Neighborhood Stability: An Empirical Investigation. Housing Policy Debate, 9(1), 28.
44. Temkin, K., & Rohe, W. M. (1998). Social Capital and Neighborhood Stability: An Empirical Investigation. Housing Policy Debate, 9(1), 28

DOI: 10.1051/matecconf/20166600036

Asset Modelling on Cost Replacement towards Transparency of Building Replacement

Mohamad Adzizulrohim Bin Abd Malek[1,a], Azlan Shah Bin Ali [2] and Mohamad Rizal Bin Baharum[3]

[1]*Department of Built Environment, University of Malaya, Malaysia*
[2]*Department of Built Environment, University of Malaya, Malaysia*
[3]*Department of Built Environment, University of Malaya, Malaysia*

Abstract. This paper presents development of a model to be used to select the best option for building replacement in public ownership such as public library, stadium, hospital and school buildings. Buildings life cycle management should be transparent and cost efficient. Hence, decisions taken during the life cycle of the buildings should be based on clear justification. One of the major issues in facility planning is to determine whether to rebuild or restore the existing facility. In fact, for government assets, they have to be economically managed by knowing the holding cost, so that it is viable to be retained in long run. The objectives of this paper were to establish relationship between replacement cost and cost decision making factors through the development of replacement cost model towards an economic efficiency for government buildings. Three stages of process were involved in this study. They are, stage 1, which explain on the significance factors of Cost Decision Making (CDM) and Replacement Cost (RC). It followed by exploration of the relationship between CDM and RC in Stage 2. Finally, the development of replacement cost model that expected to increase economic efficiency and transparency of building replacement. The results expected to provide a raw evidences to support a concept that refurbishment of existing building would be more benefit than replacement to the new ones. It is also expected that the results could be used by organizations especially for the decision makers as a guideline or tool to make a wise decision in building replacement related activities.

1. Introduction

It is well understood that the subject of building replacement is relatively new and considered juvenile in Malaysia. There are a few researchers that have been carried out focusing on the scope and the element of building replacement. By the way, very less attention has been given on building replacement decision making and it is very rare to see research carried out on the subject of replacement cost of building. As for the federal government, there is no standard format or guide line to be called standard operating procedure to be followed as a guide or a blueprint to the entire public client in terms of building refurbishment. The determination of the cost for refurbishment work in government-own buildings is the federal government or the state government building has always been a headache and conflict between the contractor and client. Worsening the scenario, it is always a problem and it is usually a very hard task to determine the exact cost of refurbishment works such as repairs, replacement or internal maintenance works and estimated cost usually go haywire and far from the actual cost. Unfortunately, there is no uniformity in the procurement procedure or standard contract regulation being used. In the public sector, it is known that among the biggest obstacle is the 'resistance to change'. *(N.Mohd Noor, 2011).*

This study will also review and identify the public sector practice in the aspect of building replacement management and their relation in cost decision determination. It is hope to lead to a better understanding of the issues related to the costs

DOI: 10.1051/matecconf/20166600036

matters in government owned building maintenance and asset management. The aim of this paper is to promote innovative model that will become a framework or guideline as to improve and as an added value to the existing procedure. This model will also consider some adaptation from the existing model that has been implemented by other countries. These have been proven to be effective and successful in its implementation. However it should taking into account our local culture and environment. It is hope that this cost decision model will benefit the largest owner that is public authorities such as The Ministry of Work and Public Work Department as the biggest agency that manage the government buildings.

Table 1: The Distribution of Pre War Urban Building in Malaysia

No	States	No of Historical Buildings	No	States	No of Historical Buildings
1	Penang	5057	8	Sarawak	1010
2	Perak	3351	9	Negeri Sembilan	999
3	Johor	2323	10	Pahang	831
4	Malacca	2177	11	Terengganu	420
5	Kuala Lumpur	1763	12	Kelantan	373
6	Kedah	1282	13	Perlis	25
7	Selangor	1166	14	Sabah	10
Total					20,787

For introduction, refer to Table 1 as example, there are still many historical building and old building in Malaysia. Since the building boom of the 1970s, many of Malaysia's historic buildings have been demolished. Recent large scale urban development continues to threaten pre-war buildings, while other historic buildings are simply deteriorating due to age, neglect and high cost of maintenance. *Fee (1998)* expressed that to lose these buildings, however is effectively to obliterate historical memories, and there is now increasing pressure from various segments of the community to conserve the nation's historical heritage. As we all known that conservation is the action taken to prevent decay, embracing all acts that prolong the life of cultural and natural heritage. Building conservation according to *(Fielden,2000; Insall,1972)* relates specifically to the process of repair, maintenance and restoration of historic buildings which aim to prolong a building's life and function.

While most of the buildings surveyed were new developments, several were refurbishments of existing buildings. With the growing recognition of the important role that refurbishment may play in improving the performance of the existing building stock, it was seen as important, given the constraints inherent in refurbishing an existing building, to compare their performance with the new ones. The key conclusion is that extensive refurbishments, and even fit outs within the constraints of an existing shell, are quite capable of equalling and even surpassing the performance of completely new designs from the point of view of the building users *(George Baird, 2014)*. Building life cycle management should be transparent and cost efficient. Hence, decision making taken during the life cycle of the building should be based on clear justification.

2. Literature Review

2.1 Decision Making

Many managerial researches in the past several decades, especially in the early 80's, have investigated and written about managerial decision making from a variety of dimensions and perpsectives *(e.g. Argyris,1976; Cohen et al., 1972; Hickson et al.,1986; Mintzberg et al.,1976; Simon,1957; Werhrich & Koontz,1993; Kim&Mauborgne,1998)*. In spite of this ongoing attention, the subject of decision making is still in contradictory and controversial phase with theoretical dilemmas. *Harrison (1999)* believed that part of the problem is derived from the multidisciplinary nature of the decision making (e.g psychology, economics, politics, sociology and mathematics which all contribute their own perspectives). The problem can be more complicated by differentiating decision maker into individual, group, multi-group or organizational *(Kriger & Barnes, 1992).* The preceding discussion illustrates three important themes in managing an organization: (1) the link that exists between the success of an organization and the quality of its managerial functions, (2) the link that exists between success of the managerial functions and managerial decision-making outcomes, and (3) the third theme is related to the first two, attempt to link managerial decisions outcomes to knowledge and ability of managers in making a decision. This theory represent as Figure 1 below:

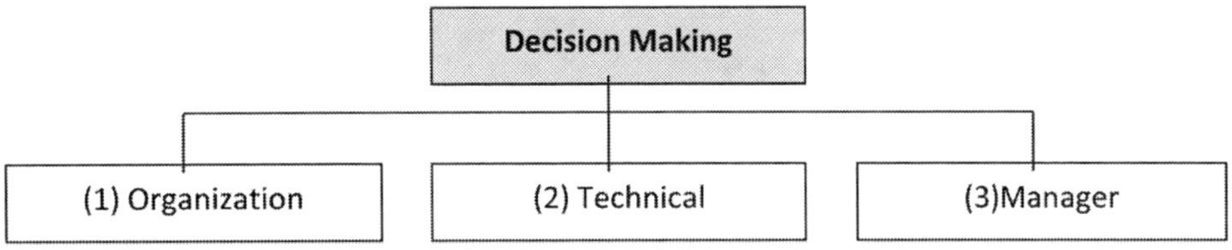

Figure 1: Decision Making Factors

2.2 Sustainable Asset Management

Sustainable asset management crystallized the concept of *'cradle to the grave'* into existence. It basically puts forth the consideration for maintenance right from the design stage all the way to the operation of the building, as well as making it ready for adaptive re-use. Based on Figure 2, achieving sustainable asset management basically deals with the business process, human resources and information technology system. In whatever task we perform, the business process must be there and in ensuring it is well implemented, we need various levels and competency of team members. It is not a one man show. As we are now living in the millennium era, Information Technology (IT) is in every business process. Data needs to be converted into information and IT makes this happen within a click. *(Che Ani,Badaruzzaman, 2010)*

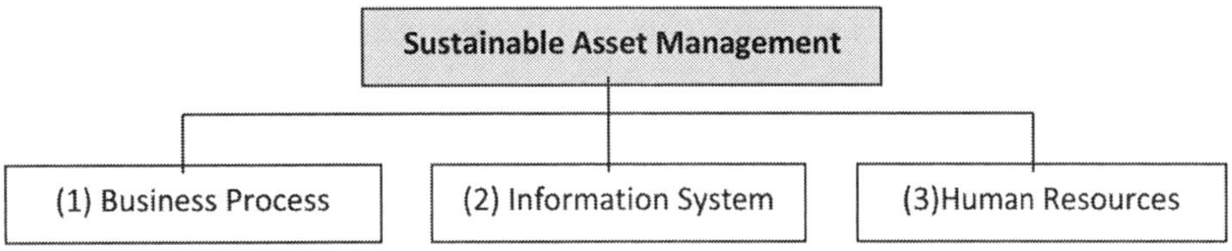

Figure 2: Sustainable Asset Management Factors

From these Figure 1 and Figure 2, we can conclude that decisions making for asset replacement are based on 3 significant factor such as of human factor, technical factor and organization factor. Generally, the cost decision making are based on manager experience, building condition and availability of funding while the replacement cost are based on budget determination, mode of government procurement and existing practices of replacement. However, the detail explanation of these three key areas is represented in the next topic.

3. Cost Decision Making Factors

3.1 Human Factor

Having identified the best practices in business process, we need human capital in executing the idea. Human are the best machines and super power computer in ensuring best practices are put in place. Therefore, we need team members with various technical knowledge and competence. It ranges from the design team to the building management team. Competency is best represented from the outcome of an academic programme developed by Malaysian Qualification Agency (MQA). In our context, the professional services involve the professions of Architect, Engineer, Quantity Surveyor, Valuation Surveyor, Town Planner, Landscape Architect, Land Surveyor, Interior Architecture and Building Surveyor. All these professional services contribute to the different factor in cost decision making. *(Che Ani,Badaruzzaman, 2010)* The following are details of human factors that contribute to the cost decision making:

3.1.1 Past Experiences or Manager Experience

Past experience can impact future decision making. *Juliusson, Karlsson and Garling (2005)* indicated past decisions influence the decisions people make in the future. It stands to reason that when something positive results from decision, people are more likely to decide in a similar way, given a similar situation.

3.1.2 Age & Individual Differences

Some individual differences may also influence decision making. Research has indicated that age, socioeconomic status, and cognitive abilities influences decision making (*De Bruin, Parker & Fischoff, 2007; Finucane, Mertz, Slovic &Schmidt, 2005). Finucace* et. al established a significant different in decision making across age; that is; as cognitive functions decline as a result of age, decision making performance may decline as well. In addition, older people may be more overconfident regarding their ability to make decisions, which inhibits their ability to apply strategies *(De Bruin et al, 2007*). Finally, with respect to age, there is evidence to support the notion that older adults prefer fewer choices than younger adults. *(Reed, Mikels & Simon, 2008)*

3.1.3 Belief in Personal Relevance

When people belief what they decide matters, they are more likely to make a decision. *Acevedo and Krueger (2004)* examined individuals' voting pattern, and concluded that people will vote more readily when they believe their opinion is indicative of the attitude of the general population as well as when they have a regard for their own importance in the outcomes. People vote when they believe their vote counts. *Acevedo and Krueger* pointed out this voting phenomenon is ironic; when more people vote, the individual votes countless in electoral math.

3.1.4 Escalation of Commitment

In addition to past experiences and cognitive biases, decision making may be influenced by an escalation of commitment and sunk outcomes, which are unrecoverable costs. *Juliusson, Karlsson and Garling (2005)* concluded people make decisions based on an irrational escalation of commitment, that is, individuals invest larger amounts of time, money and effort into a decision to which they feel responsible for the sunk costs, time, money, and effort spent on a project. As a results, decision making may at times be influenced by how far in the hole the individual feels he or she is *(Juliusson et al, 2005)*

3.1.5 Client Request

Client refers to the owner of a building. The client is the one who provides maintenance finance and project brief in the early stage of maintenance works. Quality of client's brief would influence the building performance. *Boyle (2003)* notes that the key successful of design is rests much with the clients besides others such as a good budget.

3.1.6 Complaint Received Regarding Building Performance

Inefficient maintenance works could invite complaints by the building users. Users are normally looking for a comfortable space in a building. This includes well functions of building equipment's, clean environment and safe. If buildings do not fulfil the user's needs, complaints with regards to maintenance performance would be made by the users. Therefore, in formulating decision with regards to maintenance cost, element of user's satisfaction is vital and need to take into consideration *(Latef,2008; PWD,2009)*

3.2 Technical Factor

The asset management cycle covers planning until the disposal stage (or adaptive re use). In the global arena, various theories and tools have been developed to support this idea. Among them are Building Performance Evaluation (BPE), Construction Operations Building Information Exchange (COBIE), Buildings Information Modelling (BIM) and Return on Investment (ROI). Others technical factor that contribute to the cost decision making are as follow:

3.2.1 Existing Building Condition

The existing conditions of buildings were assessed through several ways. One of the easiest methods is by using viral survey. The exterior of each building structure was viewed from the ground level and all important information would be documented and some areas of deterioration are noted through annotated sketches and plans. However, with recent new technology, detecting building defects would be more effective by implementing non-destructive test *(Pitt,1997; Lee and Scott,2009)*

3.2.2 Building Age

Age of building provides important indication on level of maintenance service required. *Lateef (2008)* argues that one of the important elements that need to be considered in allocation of maintenance resources is building's age. In order to know the future image of building, building manager needs to offer right service so that the building has competitive advantage. Service given must meet expectation in response to time, delivery schedules and within the agreed performance indicator. In general, the older is building, the more attention and focus to special maintenance works need to be carried out.

3.2.3 Performance Evaluation

The best decision on whether to repair or replace can only be reached by evaluating the equipment in the broadest possible terms. For example, when considering whether to refurbish or replace a piece of pumping equipment, looking at the performance of associated motors and drives- their function, lifespan and cost of replacement- can help engineers to understand how the pump is performing in relation to the wider system. *(2015, Conference of Pumping Best Practice, Telford)*

3.2.4 Energy Efficiency

In order to fully incorporate sustainability into the decision making process, maintenance teams should evaluate the ongoing energy efficiency performance of an asset as well as the immediate environmental impact of repairing or replacing it. Then the environmental impact of their versus replacement can be measured by carrying out a calculation on the energy usage and the direct link between this, and the asset's carbon footprint, by focusing on reducing specific power.

The decision to repair or replace is not always a straightforward one. However, by working alongside an asset management and maintenance expert, it is possible for maintenance teams to make a fully informed decision. By taking a long term view, not only of the cost and performance benefits, but also the environmental effects of each approach, engineers are well placed to make the best decision in terms of cost, performance and sustainability. *(2015, Conference of Pumping Best Practice, Telford)*

3.2.5 Information Technology (IT) System

The myth about IT system is it will take care of every mechanistic task of asset management. We have to realize that IT is enabler for the whole structure of asset management, thus providing total solution for sustainable asset management. It is not that when we have superb IT system, everybody just stays back and become the observer. The IT system needs input to convert data into information, and this is found to be most lacking, leading to a white elephant project. As a rule of thumb, engage a business partner that provides total solution, rather than selling the software and gadgets. *(Che Ani,Badaruzzaman, 2010)*

3.3 Organization Factor

The prime objective of the Malaysian Government procurement is to support Government programmes by obtaining value for money through acquisition of works, supplies and services. To meet this objective close attention is given to price factors as well as non-price factors such as whole life cost, quality, quantity, timeliness, maintenance and warranty. The benefits or value from procurement should commensurate with the costs involved and that the best procurement is well and thoroughly evaluated, reasoned and justified. In this context, the Malaysian Government procurement is based on the following policies, principles, objective and procedures. Others organization factor are as follow:

3.3.1 Availability of Funding

A most trying constraint from the design point of view is the budget allocation. *Boyle (2003)* noted that most important factor contributing to successful maintenance work was a sufficient budget allocated for a project. The project fund must be sufficient to ensure the maintenance works could run smoothly.

3.3.2 Government Policy

Government assets have to be economically managed so it is vital to know the opportunity cost of holding them. The government wants to see the government heritage estate properly preserved and maintained through active use and reuse of the buildings. The government encourage the sustainable use of resources, and one aspect of that is to use and reuse older buildings wherever it sensible to do so. Government's planning policies are committed to maintaining the vitality of town centres where many specialized public listed properties are found rather than erecting new buildings on greenfield sites. *(Anthony Andrew, 2000)*

3.3.3 Safety & Health Requirement

Safety and health are command factors influenced in the decision making of maintenance works *(Lee and Scott, 2009)*. This is because building maintenance works can lead a risk to maintenance personnel and building users. It is the responsibility of building stakeholders particularly the manager to ensure that health and safety assessment and safety work procedure are documented.

3.3.4 Past Experience and Past Performance

Analysis on Malaysia scenario in development of building and facilities aspect from First National Physical Plan (NPP) in 1956 until 10th Malaysia Plan in 2015. Performance and achievement of these plans as a guideline to the future planning.

3.3.5 Informed Decision Making

An invaluable part of any asset management programme is having a system in place to monitor and evaluate the ongoing performance of an asset. By measuring the efficiency of pumps and associated processes, engineers can understand the behaviour of piece of equipment and gauge when it is reaching the end of its lifetime or is at risk of failure. They can then make an informed decision in terms of repair or replacement and at the same time, include suitability of the asset and improve asset and system efficiency. *(2015, Conference of Pumping Best Practice, Telford)*

4. Replacement Cost Factors

Replacement cost definition is the amount it would cost to replace an asset at current prices. If the cost of replacing an asset in its current physical condition is lower than the cost of replacing the asset so as to obtain the level of services enjoyed when the asset was bought, then the asset is in poor condition and the firm would probably not want to replace it.

4.1 Mode of Government Procurement

4.1.1 Direct Purchase

Procurement of works up to the value of RM20,000 may be done through the issue of a Works Indent to a contractor who is registered with the Contractors Services Centre (PKK) and Construction Industry Development Board (CIDB) Malaysia.

4.1.2 Quotation

Procurement of works above the value of RM20,000 and up to RM500,000 is done through calling of quotations and the minimum number of quotations to be invited is five. All contractors wishing to take part in quotations must be registered with the PKK and CIDB.

4.1.3 Tenders

Procurement of works, supplies and services above the value of RM500,000 must be done through tender processes. All contractors intending to participate in local tenders must be registered with the government. International tenders will be invited for supplies and services if there are no locally produced supplies or services available. For specific works, if local contractors do not have the expertise and capability, tenders may be called on a joint venture basis between local and foreign contractors to encourage the transfer of technology. International tenders for works may only be called when local contractors do not have the expertise and capability, and a joint venture is not possible. *(2015, Malaysia's Government Procurement Policies)*

4.2 Budget Determination

The Prime Minister Department has given an instruction under the *"Pekeliling Am Bil. 1 Tahun 2003, Arahan Penyelengaraan Bangunan Kerajaan di Putrajaya* dated February 11th 2003," building maintenances must be efficiently and properly executed. The Government has also given instructions to all their agencies to carry out maintenances of all their building in the very early stage so that to reduce the cost of maintenance and to avoid risk of higher maintenance cost due to negligent through time.

4.3 Existing Practices of Replacement Cost in Government

The Malaysia industry reference point for asset management has to be the TAM Manual (*Total Aset Management Manual*), part of DPAK (*Dasar Pengurusan Aset Kerajaan*). Launched in April 2009 by former Prime Minister Y.A.B. Dato' Seri Abdullah Hj. Ahmad Badawi, this document provides the key idea and direction for the whole life cycle of our assets. According to the manual, asset refer to moveable assets (*aset alih*), fixed assets (*aset tidak alih*), life asset (*aset hidup*) and intellectual property (*harta intelek*). With the wide coverage of assets, an asset management team is no longer focused on building only. It has to be integrated to achieve the aspiration of TAM Manual.

As for the federal government, there is no standard format to be called standard operating procedure to be followed as a guide or a blueprint to the entire public client in terms of cost analysis of building replacement. However, the existing practices of building replacements in government are as follow:

- TATACARA PENGURUSAN ASET TAK ALIH KERAJAAN 2012: Procedure that have been used by government as a guideline in asset disposal. The details of the procedure are in the *Bab G: Pelupusan Aset.*
- SYOR ROBOH: One of the procedure and guidelines that have been used by government in asset disposal.
- BCA (Building Condition Assessment): Condition assessments are technical inspections by competent assessors to evaluate the physical state of building elements and services and to assess the maintenance needs of the facility. Reliable and objective knowledge of the physical state of their buildings and the impacts on service delivery will enable department to develop appropriate strategies and actions for maintenance, major replacements, refurbishments and investment.
- BIM (Building Information Modelling): This model is an approach to design that uses intelligent 3D computer models to create, modify, share, and coordinate information throughout the design process. Many firms are using BIM to drive a more efficient design process.
- LCC Model (Life Cycle Costing): LCC model is an accounting structure containing terms and factors which enable estimation of an asset's component costs. A number of commercially available models can be used for LCC analysis. However, in some cases it may be appropriate to develop a model for a specific application.
- CMMS: Model can produce status reports and documents giving details or summaries of maintenance activities. The more sophisticated the package, the more analysis facilities are available.
- Value Engineering: Value techniques that are adopted during the detailed design stage and in some cases at the construction stage when completed design or elements of the design is made available for study.
- J10: This model is an IT project which when completed, will bring the department's project management and facilities management business applications together onto a single platform with a simple and efficient user interface. The platform and applications will leverage current technology to automate certain takes and improve usability for end users. This is the latest model that has been launched in year 2015.

5. Methodology

The processes involved in this study are summarized in Table 2. It illustrates three distinct levels of data collection, namely the first stage followed by the second stage and ended up with the third stage. The first stage was based on primary data collected from published books, research papers, seminar papers and journals while second phase used objective data from official sources of government experts. The third or final phase used subjective data from the actual replacement cost for Heat, Ventilation and Air Conditioning (HVAC) system in building.

Table 2: Methodology of Study

Stage 1 →	Stage 2 →	Stage 3 →
Factors Collection	**Validation of Factors and Establish Relationship**	**Development of Actual Replacement**

DOI: 10.1051/matecconf/20166600036

CDM Factors RC Factors Asset Modeling Existing practice Policy Issues (Primary Data)	Based on Actual Data from Interviews of Government Experts (Objective Data)	Actual Replacement of HVAC System of Building (Subjective Data)

The first stage concentrated on literature review on cost decision making factors and replacement cost factors that related to study. The first stage was based on primary data collected from published books, research papers, seminar papers and journals. The objective of the first stage is to identify independent variables that have significant relationship with the replacement cost, to evaluate current level of replacement cost, to established relationship between replacement cost and sets of independent variables and to determine the relative importance of each independent variable.

The second stage was conducted upon completion of the first stage with the following objectives to validate factors in stage 1 and to establish relationship between cost decision making and replacement cost. Validating a factor is a process that starts with the researcher, who then seeks validation among experts. The interviews conducted with the questionnaire survey consisted of variables thought to have effects on project's replacement cost to obtain detailed investigation information of the building replacement during their life cycle. The study adopted closed and open interviews in order to tap as much as possible information form parties interviewed. In this study, 11 personnel experts in the government office have been interviewed. The results of the interview data are presented in the Table 3. The third or final stage was based subjective data on pilot projects of development of replacement cost modelling for Heat, Ventilation and Air Conditioning (HVAC) system in buildings. The objectives of the third stage were to investigate to what extent the factors identified in the earlier phases had affected projects selected under the pilot studies, to identify other factors that could contribute to the replacement cost in project selected under the pilot studies. The pilot studies adopted closed and open interviews in order to tap as much as possible information from parties interviewed. All parties involved, including personnel on sites, were interviewed in order to acquire a more comprehensive view. On the other hand, subjective data can be considered as equally important and useful as the respondents possess extensive working experience and the information provided by them can be regarded as sufficiently reliable. The author considered research that uses both objective and subjective data is more comprehensive and conclusive.

Table 3: Research Data

No	Name	Designation	Questionnaires							Note
			1.Existing Guideline	2.Information syst.	3.Justification	4.Improvement	5.Implementation of LCC	Retrofit	Rebuild	
1	Expert 1	KETUA PENOLONG PENGARAH KANAN (ASSET)	BCA	Don't have	Building Condition	Competency	pilot project		X	
2	Expert 2	TIMBALAN PENGARAH JKR PAHANG	LCC	Don't have	Age	Fund	very low	X		
3	Expert 3	PENGARAH JKR PAHANG	Value Engineering	Don't have	Security Risk	Policy safety	No implement	X		
4	Expert 4	PENGARAH JABATAN KEJURUTERAAN MAJLIS PERBANDARAN KUANTAN	Don't have	Don't have	Use of Building	Computerized system	No implement		X	

5	Expert 5	PENGARAH (BSFB)	Syor Roboh	Going toward that	Expected Life	Database	No implement		X	
6	Expert 6	PENGARAH (ASSET MANAGEMENT)	Tatacara Pengurus an Aset 2012	My Spata	Safety	Database	need to build	X		
7	Expert 7	KETUA PENOLONG PENGARAH KANAN (BFSB)	BPA	My Spata & J10	Performanc e	Guidelines Improvemen t	ongoing	X		
8	Expert 8	KETUA PENOLONG PENGARAH KANAN UKUR BAHAN JKR PAHANG	Don't have	Don't have	Technical problem	Competenc y	No implement		X	
9	Expert 9	TIMBALAN KETUA PENGARAH KERJA RAYA (SEKTOR BANGUNAN)	BCA	My Spata2	BCA	Competenc y	not be implemented	X		
10	Expert 10	PENGARAH (CPAB)	BCA	Don't have	Safety	Database	No implement	X		
11	Expert 11	PENGARAH KANAN (CPAB)	Don't have	Don't have	Client Request	Decision Making Model	at early stage	X		

Question

1. Any existing guideline to make decision in building replacement
2. Any existing software to make decision in building replacement
3. Critical justification of building replacement
4. Improvement of decision making process
5. Current implementation of LCC for government building
6. Which is cost efficiency, retrofit or rebuild?

6. Discussion

6.1 Result and Analysis

The aim of this phase is to identify the expert's responses on critical improvement and critical development in decision making of whether to retrofit or to rebuild the building. Based on the Table 3, 'Competency of Manager' and 'Information Database' are most important in improvement of decision making in building replacement, followed by 'Decision Making Model'. Their distribution are as follow; Competency (3), Information Database (3), Decision Making Model (2), Funding Availability (1), Computerized System(1) Policy Improvement(1).

Furthermore, Refer to Table 3, their contribution on theoretical framework are as follow: Human Factor (3), Technical Factor (2) and Organization Factor (6). Organization factor are contribute 55% of the improvement requirement and Human Factor (27%) and Technical Factor (18%). Therefore, from this analysis, the 3 critical factors are need improvement are as follow which contributed 73% of improvement of decision making:

1. Competency of Manager (27%)

2. Information Database (27%)

3. Decision Making Model (18%)

The study will focus on to improve the current policy that is being implementing by the government. This will include on how the Government make the analysis and make determination of the final decision on replacement of 'Heat, Ventilation and Air Conditioning' (HVAC) system. This study will present several factors, issues and figures. The elements of studies that will be put into consideration and to be analysed are the Replacement Cost (RC) factors and Cost Decision Making (CDM) factors as Figure 3 below. The study is meant to help the public sector which forms the biggest part of the buildings managing authorities in facing all the problems and difficulties. There are buildings with special design characteristic and with special function, unfortunately these specialty are normally costly to replace. These are normally due to deterioration and aging factor. Therefore it is urgent that a thorough study should be carried out to develop a guideline for the determination of replacement cost on these special buildings.

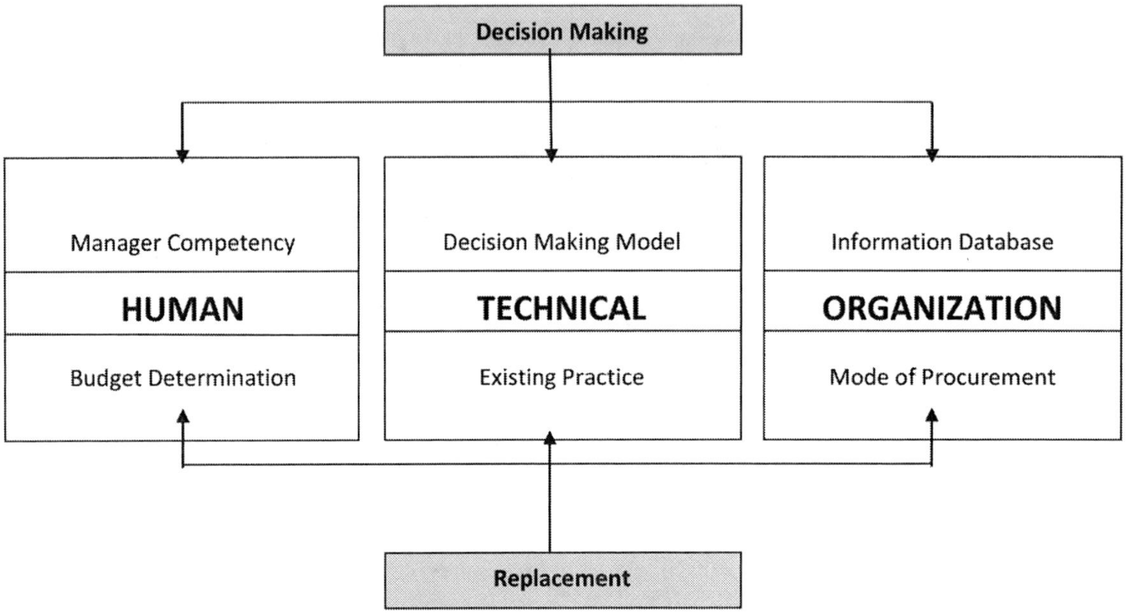

Figure 3: Relationship between Decision Making and Replacement

6.2 Discussion

The significance reason of this study is how to provide evidence in order to help the Federal Government to make a final decision on whether to restore or rebuild existing HVAC system. The figure shows the simple model to show the relationship between decision making and replacement cost. By using the model that will be proposed, it is hope that this evidences contributed to select optimum replacement cost for certain projects and directly helping in tendering process and ensuring the appropriate and transparency of job done.

COST DECISION MAKING	REPLACEMENT COST
Age of equipment Cost of repair Warranty coverage Money saving efficiency Future refrigerant availability Long term comfort Long term savings Dependable comfort	Compressor Condenser Coils Evaporator Coils Air Filter Coil Fins Contactor Relays Time Delay Compressor Power Pack

Figure 4: Replacement Cost Modeling For HVAC Asset

Refer to Figure 4, this HVAC asset study will provide more details and relevant information of relationship between Cost Decision Making (CDM) and Replacement Cost (RC). Cost Decision Making consists of 3 factors such as human factor, technical factor and organization factor. Meanwhile, Replacement Cost factors related to modes of government procurement, budget determination and latest IT software application. This relationship study and discussion will contribute to the theoretical saving framework or saving policy that can save the government expenses. However, this theoretical framework only can be proof through actual analysis study on pilot projects of government procurement. Therefore, the author focus on HVAC System of building as pilot projects of this study.

Actually, the main objective of this study is to provide evidences to support a concept that restore of existing building are more benefit than replacement or rebuild to the new ones. The scope of the project and the key features of the asset of HVAC system of building are chiller, motor, compressor, cooling tower and piping. An air conditioner is any type of device that helps moderate temperature indoors. An air conditioner does not have to be cooling system; technically a heater is also an air conditioner, though most people seem to only consider air conditioners machines that lower the temperature to a user defined temperature. Making the decision to either repair your old building or replace with a new building is not easy. It's a decision government need to make every time. Government need to make a thoughtful, well informed and final decision making. Therefore, government need to developed all relevant performance and quality requirements such as age of equipment, cost of repair, warranty coverage, money saving efficiency, future refrigerant availability, long term comfort, long term savings and dependable comfort. Furthermore, part of asset that are subject to a costs analysis are compressor, condenser coils, evaporator coils, air filters, coil fins, contactor relays, time delay, compressor power pack and other mechanical parts. Optimum replacement cost that produced from this discussion hopefully can help government to make a justify decision.

7. Conclusion

From this analysis, we can conclude that this study is compulsory to be done by authorities in order to make sure the consistencies of decision making of government building replacement. Their life cycle management should be transparent and cost efficient, and decisions taken during the life cycle should be based on clear, justified and rational grounds. Therefore, the model is adapted to specific costs of such buildings towards transparency of their replacement.

From the questionnaire, we can conclude that the lack of cost analysis in decision making of building replacement as a weak point in the existing practices. Improvement of this process is compulsory to be done by authorities in order to make sure the transparency of decision making of government building replacement. Therefore, this study focuses to establish a simple, transparent and robust model that can be easily understood and used by the decision makers in practice.

Next analysis is which replacement is more benefit to the government in term of cost, retrofit or rebuild. Experts have to choose one and give their opinion on this question. Most the experts said the answer 'yes or no' are depend on the purpose of function of the building and specialty of the building. As for example, the heritage building should not be rebuild in order to maintain its original heritage criteria and history value. However, the experts gave their opinion are based on their experiences in dealing with this situation.

Based on the data Table 3, seven (7) experts said that retrofit is more benefit to the government meanwhile four (4) experts said that rebuild is more benefit to government expenses. Generally, about 64% of experts agreed that retrofit is more economically than rebuild to the new ones. Actually, the reason to choose retrofit are because of retrofit is more cheaper in term of cost, can increase competency of the manager, to maintain the heritage and valuable building, to minimize the residue material and to minimize impact to the environment. Otherwise, the reasons to choose rebuild are because of the business changes, to minimize security risk and fatalities, lack of database information and the expected life of the building.

Lastly, all the experts give the same opinion on the current implementation of Life Cycle Cost (LCC) for government building. There is no implementation of LCC in current practice for government building. Implementation of this process is compulsory to be done by authorities in order to achieve sustainable asset management. Finally and hopefully, the implementation of Life Cycle Costing (LCC) for government building will become a reality

References

1. Hussien Ahmad Al Tarawneh, The Main Factors Beyond Decision Making, Islamic International University, Malaysia (2012)
2. Nor Azizah Mohammed Rum, Implementing Life Cycle Costing in Malaysia Construction Industry: A Review, Universiti Tun Hussien Onn Malaysia (2011)
3. N. Mohd Noor, Building Maintenance Budget Determination: An Exploration Study in the Malaysia Government Practice, University Technology MARA , Shah Alam, Malaysia (2011)
4. Kamarul Syahril Kamal, Pilot Survey On The Conservation of Historical Buildings in Malaysia, University Technology MARA, Malaysia (2008)
5. Matej Kusar, Selection of Efficient Refrofit Scenarios for Public Buildings, Faculty of Civil and Geodetic Engineering, University of Ljubljana, Slovenia (2013)
6. Carol C. Menassa, A Framework to Assess The Role of Stakeholders in Sustainable Building Retrofit Decisions, Department of Civil and Environmental Engineering, University of Michigan, United States. (2013)
7. Azlan Shah Ali, Cost Decision Making in Building Maintenance practice in Malaysia. Faculty of Built Environment, University of Malaya, Malaysia (2009)
8. Ministry of Finance Malaysia, Government Procurement Division, Malaysia's Government Procurement Regime (2010)
9. International Organization for Standardization, ISO156-4:2014, Building Construction- Service Life Planning- Part 4: Service Life Planning Using Building Information Modelling.
10. UCL Public Policy, June 2014
11. Conference and Exhibition Pumping Best Practice for 2020, The International Centre, Telford (2015)
12. http://wwtonline.edie.net/features/repair-or-replace-make-informed-decisions
13. http://www.investorwords.com/4181/replacement_cost.html
14. http://en.wikipedia.org/wiki/BREEAM
15. http://en.wikipedia.org/wiki/Leadership_in_Energy_and_Environmental_Design
16. http://www.rics.org/my/knowledge/profesional-guidance/redbook
17. http://ec.europa.eu/environment/emas/es_library/index_en.htm
18. Kerajaan Malaysia : Dasar Pengurusan Aset Kerajaan (2009)
19. Kerajaan Malasyia : Manual Pengurusan Aset Menyeluruh (2009)
20. The Ingeniur Magazine Volume 45 (2010)

MATEC Web of Conferences 66, 00037 (2016) DOI: 10.1051/matecconf/20166600037
IBCC 2016

Local Authority Empowerment towards Quality Living Environment for Coastal Reclamation Area

Mohammad Yusup[1, a], Nor Syafa'ah Ishak[2], Ahmad Fuzi Arshad[3] and Yusfida Ayu Abdullah[4]

[1,3&4]*Centre of Studies of Town and Regional Planning, Faculty of Architecture, Planning and Surveying, 40450 UiTM Shah Alam, Selangor, Malaysia*
[2]*Centre of Postgraduate Studies of Town and Regional Planning, Faculty of Architecture, Planning and Surveying, 40450 UiTM Shah Alam, Selangor, Malaysia*

Abstract. Good urban governance administration system is the key to a successful physical planning development. A local authority of a local government concentrates on planning administration and executes the policies and strategies either the federal or state, or even the local's policies and strategies. Based on its characteristic as the lowest level of government, it becomes the best authority to regulate and monitor the development process within their territory. The significance of a local authority in providing quality living environment invites various academia and professionals to ponder the best urban governance system at a local level. However, there are issues with regards to financial and technical capacity of a local authority, its legal limitation and development instrument adopted in providing urban services for coastal reclamation area in Malaysia. The aim of this paper is to investigate the capability of local authorities in Malaysia in implementing their function as drawn by the legislation. Hence, this paper examines the roles and functions of a local authority as the lowest level of government administration agency in providing urban services; collecting revenue; safeguarding the physical environment in Malaysia, particularly when dealing with development in a coastal reclamation area. Primary data collection was gathered through face-to-face interview sessions involving government agencies and stakeholders. Legal documents, policies and development plans were then analysed to support the primary data for further understanding of the issues concerning the capacity of a local authority especially when providing urban services within its area. The study is expected to provide a new approach to local authorities in Malaysia in providing quality living environment in terms of development procedure, the role and function, legal empowerment, and decentralisation of function particularly in enhancing the current practices at local level.

1 Introduction

Malaysia is a small country that covers 330,290 km^2 of a total land area including Sabah and Sarawak [1]. West Malaysia and East Malaysia are separated by the South China Sea. Malaysia is located in South East Asia and consists of 14 states; 11 states in West Malaysia, the States of Sabah and Sarawak as well as the Federal Territory of Kuala Lumpur (the capital city of Malaysia) [2]. The planning system in Malaysia has been developed based on the Federal Constitution in which matters related to town planning are stated in the Ninth Schedule under List III of the Concurrent Lists [3]. Federal government, state government and local authority has the responsibility towards the activities related to town planning in order to maintain, monitor and specify the use of land and buildings, so that the planning are better organised especially in meeting the need of the public interests [4]. However, the planning system in Malaysia can be changed according to the demand and need as well as the technological advances. The process of town planning in Malaysia had started from Malay Sultanate era, expended by the British colonialists, and had undergone various changes after the independence of Malaysia in 1957 until the introduction of the Town and Country Planning Act 1976 (Act 172). These changes were aimed to create a happy quality living environment in line with the increasing socio-economic condition of the people and country. The town planning's main objective which is to protect the public interest had not changed, but the scope and nature of town planning have greatly expanded from time to time. Hence, a practical organisational and administrative structure plays an important role in ensuring the consistency of the physical planning and the organisation resilient.

The limitation of this paper is that it only focuses on the local authority's empowerment in implementing the roles and function provided by the statutory and non-statutory instruments in providing urban services. The expected outcome of this study is to provide a new approach to local authorities in Malaysia in providing a good quality living environment in terms of development procedure, role and function, legal empowerment as well

[a] Corresponding author: mohammad1744@salam.uitm.edu.my

as new guidance in sustaining the financial expenditure decentralisation of function for local authorities.

2 Research methodology

This paper focuses on the local authority empowerment towards a quality living environment for coastal reclamation area. This study applies a case study design and a qualitative approach as it outlines all the problems experienced by the local authority before recommendation and proposals be made to overcome the issues. A case study approach is seen fitting since it is "a process that is carried out in order to solve the past and current problems as well as doing a continuous study for future improvements" [5]. The study therefore, attempts to enhance the development procedure, role and function, legal empowerment and decentralisation of a local authority's power in Malaysia. Furthermore, the paper is intended to study the capability of local authorities in Malaysia in implementing their function as drawn by the Malaysian law. Primary data collection is gathered through face-to-face interview sessions with government agencies and stakeholders. Purposive sampling was chosen to get feedbacks from the respondents through interview sessions which were conducted with the selected informants. Legal documents, policies and development plans were then analysed to support the primary data to further understand the issues about the capacity of a local authority when providing urban services within its local context.

3 Literature review

The process of development apparently comes after performing proper physical planning. This is because planning is the process in deciding what and how to carry out development and this happens at many levels such as from day to day decision by individuals or families, also in making a complex decision by businesses and governments [6]. However, planning, in linguistic, is the act or the process of making a plan which comes from the word 'plan' and it can be interpreted as to make detailed arrangement for the future [7]. According to Keeble (1969) [8], planning *is the art and science of ordering the use of land and siting of building and communication routes so as to secure the maximum practicable degree of economy, convenience and beauty*. While, Ratcliffe (1974) [9] defines planning differently from Keeble (1969) [8] and believed that planning is *providing the right site, at the right time, in the right place and for the right people*. Interestingly, the definition of planning from Ratcliffe [9] perspective tends to define planning as a service provider while Keeble (1969) [8] described planning as an action to create a better environment and living place. There is significant similarity of both definitions where it has the element of decision making. The action of ordering the use of land and siting of a building and providing the right site at the right time for the right people as Ratcliffe (1974) [9] had stated requires a decision-making process. The element of decision-making process is undoubtedly the basic principle shared by both definitions. Thus, it can be concluded that planning is actually the process of decision making where it requires goal, target group and proper procedures [10].

3.1 Administration systems in West Malaysia

The West Malaysia consists of a three-level administrative system which are the Federal Government, State Government and Local Authority [11] (Figure 1).

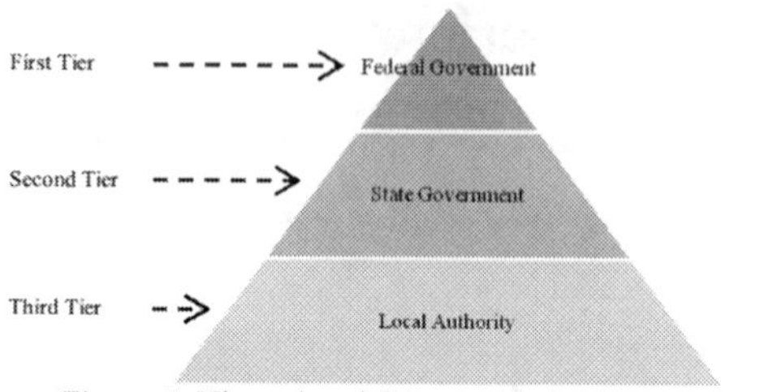

Figure 1. Hierarchy of Government in West Malaysia
Source: Adapted from Hussain (2002)

There will be no perfect system that can run smoothly for government's activities. This causes the multiple department and units under one country to be approached. Decentralisation commonly known as the transfer of powers of state or national responsibilities or functions from central government to the sub-national level of government, or from central agencies or offices to regional bodies or branch offices, or to non-governmental organisation or private concerns [12]. On the other hand, Miller (2002) [12] concluded decentralisation as *the redefinition of structures, procedures and practices of governance to be closer to the citizenry*. Decentralisation also can be defined as a dynamic process, in which actors at every level of government can interact and find the appropriate balance among the responsibility of the authorities in their scope of work (Barr, Resusodarmo, Dermawan, McCarthy, Moeliono and Setiono, 2006) [13].

According to Hutchcroft (2001) [14], the decentralisation involves as much more extensive transfer of decision-making authority and responsibility to the local government units and it is usually involving regions, provinces, and/or municipalities. Hence, this concept of work can be seen as a sharing-power concept. The advantages of this system are more flexible and provide faster decision-making process, encouraging the public participation in the decision-making process as well as develop democracy concept. This sharing-power concept is easy to control, implemented and gaining public (local people) approval and minimising the arising problems [15]. Although, the weakness of decentralisation are the corruption, the local authorities still rely on the higher government decisions, inefficiency when other local authorities cannot command as the administrator and local authority do not have sufficient resources to provide adequate services to the people because of the local authority position is small compare to the higher government [15].

Furthermore, there are three types of decentralisation of power for instance de-concentration, devolution and delegation. De-concentration can be defined as the process involving the redistribution of administrative responsibility among federal government only [16]. While, the World Bank (1999) [17] stressed that de-concentration is the weakest form of the decentralisation of power. It is due to the redistribution of decision making authority and financial responsibilities among different levels of the national government [17]. This is the process where the federal or state government representing their authority or power to the staff in the local authority unit [16]. For example, in West Malaysia, the Ministry of Urban Wellbeing, Housing and Local Government (MUWHLG) will have other departments in every state with the intention of serving the people.

Apart from that, devolution is the devolution of decision making power in certain areas or regions passed to the local authorities or local government bodies of the area [11]. The local authorities' administration act as devolution and the district administration act as a de-concentration [11]. The devolution happens where the transfer of any function or responsibility involving both administrative as well as political or decision makers [12]. There are four characteristics of devolution in planning administration structure which are i) distinct geographical boundary for local planning authority to perform their responsibility, ii) the federal government control towards local planning authorities is limited and indirect, as autonomous, free and separated from the federal government, iii) the power to collect financial resources to perform duty, known as financial autonomy and iv) the devolution is a form that requires the local authority to be established as an institution [10]. Thus, Table 1 will explain the differences between de-concentration and devolution and the definitions of the delegation of powers will be explained below.

Table 1. Differences between De-concentration and Devolution

Aspect	De-concentration	Devolution
Administration Order	• Administration is through administration order.	• Administration is through act of law.
Efficiency of the Administration	• The administration lacks of efficiency because it is depends on the discretion of officers.	• The administration is efficiently works because it is directly close to state.
Flexibility of the Administration	• The flexibility in the administration is less rigid and more flexible because the orders serve as guidelines.	• The flexibility in administration can be rigid because the provisions stated in the laws need to be adhered.
Creativity and Originality	• Encouraging creativity and originality in the administration	• Creating the creativity and originality in the administration
Management	• Have more proves of abuse of power, position and corruptions.	• Faced less abuse of power, position and money.
Legitimate Power	• Questioned arising on legality legitimacy.	• Act carried out are legal and legitimate.

Source: Adapted from Hussain (2002)

Table 1 demonstrates that the decision made by applying the concept of de-concentration is not good as the decentralisation concept. Excluding the local interest regarding the local matter jeopardises the decision making process. Apparently, de-concentration is not recommended especially for a particular state with large geographical boundary as local views are important. On the contrary, devolution of power enables the local planning authority or state planning authority performing their duty and it can indirectly improve the public participation in planning matters [10]. By doing this, the problem related to 'inertia', 'apathy' as well as dependency to the federal government on the local matter can be avoided and it also will indirectly improve the sense of belonging among the people towards the government [11].

The delegation of powers is the term used when the transfer of function from governments to a non-governmental organization or private sector entity (privatisation) [12]. Ferguson and Chandrasekharan (2004) [18] defined delegation as the transfer of managerial responsibility for specified functions to other public organisation outside normal federal government control, whether district or local government or state-owned agencies. As an example, this scenario can be seen in the coastal reclamation development in Melaka. The coastal reclamation is being introduced in order to provide more land due to the limited land resources [19]. Besides, coastal reclamation is a way that helps in creating a new land from the ocean into a form that benefits and meets the public interests [20]. Apart from that, coastal reclamation provides an opportunity to expand the land for human use through a variety of purposes [21]. The development of the coastal area for reclamation purposes should take an important measure in evaluating any possible disturbances in the development of coastal area because the coastal area is a very sensitive area [19].

The land is under the state government power, but the government privatizes the project to the developer to develop the reclamation area in order to provide the need of the people that the state government and local authority could not provide. In Melaka, most of its coastal reclamation area is used to provide residential, commercial and mix development [22]. This action is made in order to meet the demand for development and reduce population pressure especially in a small state like Melaka. This shows that coastal reclamation is vital towards economic viability and has potential in increasing the quality living environment.

With an increasing number of populations, inevitably it will slowly cause problems to the federal and state government relating to the public facilities and infrastructures inadequacy because those are the basic needs of people especially in an urban area and historical tourism place like Melaka [23]. Hence, coastal reclamation development can lighten the burden by having a lot of public facilities and infrastructures to be built on the said area. This explains that reclamation is important in boosting the economy; meet the social needs and helping the environment towards the quality living environment. Still, the delegation of powers needs to be monitored so that the developer does not freely develop the land as they desired, therefore, avoiding conflict of land use and the demand from people. Even though the project is privatized, the quality of the development needs to be put at the top priority.

As another example, the delegation of power in housing development scheme is now facing issues especially for the developers to develop low-cost housing. Developers tend to develop medium cost to high-cost housing scheme with a high price which is profitable to them that resulted in issues of owning properties for the lower income group [24]. The development on reclaimed land as well encounter the same problems where the developer only focuses on the high cost development that is profitable such as the development of Hatten City and Cheng Ho City and other upcoming developments [25].

4 Results and discussion

4.1 Local authority in Malaysia

Local authority plays an imperative role to facilitate the economic activities and development of a region especially in delivering public goods and services that are particular to the localities [26]. The local authority has the power to control all physical development activities within its area [4]. Section 5(1) of the Town and Country Planning Act 1976 (Act 172) clearly stated that *every Local Authority shall be the Local Planning Authority for the area of the Local Authority*. The local authority in West Malaysia is therefore governed by the Local Government Act 1976 (Act 171). However, the East Malaysia which consists of the states of Sabah and Sarawak, apply their own ordinances to govern over their territory [3]. For that reason, Sabah is governed by the Local Government Ordinance 1961 and Sarawak is governed by the Local Authority Ordinance 1966.

This paper only focuses on the Act 171 as it is only applied in West Malaysia. The local authorities were established under the Local Government Act 1976 (Act 171) and the respective state ordinances and by-laws [27]. Local Authority means *any City Council, Municipal Council or District Council, as the case may be, and in relation to the Federal Territory means the Commissioner of the City of Kuala Lumpur appointed under Section 4 of the Federal Capital Act 1960 (Act 190)* as stated in the Local Government Act 1976 (Act 171). Local government act as a system of a district, a territorial unit that have their own boundary and territory, legal entity, instructional structure, have the power to govern the area and task to fulfill, which is subject to particular acts, has a financial authority and other autonomy [28]. The local authority is the third tier in the government system in Malaysia.

The local authority is created in order to be a step closer to the people so that the voices, demands and needs from the people can be heard by the government [29]. Moreover, a local authority is responsible in making a local plan for their area as stated in Section 12 (Act 172). The purpose of a local plan covers the policies, proposal designed in order to achieve planning aims in development for the area and structure plan strategies for development application [30]. Besides, a local plan is an essential development plan for the working planning system, also the quality of the produced local plan plays an important factor in determining the capability of the local plan to be a success or failure of the development [30]. Indiscriminate of activities for land development and poorly planned of built environment leads to environmental degradation, hence giving impact to public health [31]. This demonstrates that a comprehensive good current and future physical planning development gives various positive impacts physically, socially, economically and environmental pleasure.

Malaysia consists of 149 local authorities including the Sabah, Sarawak and Wilayah Persekutuan [32]. This indicates that Malaysia has a huge number of local authorities. The categories of local authorities are divided into three types which are city hall/municipal council, city council and district council [33]. The three types of local authorities have their own range of responsibility that reflects on their size and capacity. Local authority are formed and regulated by the respective state's legislation and subordinate to the respective state governments [26]. Table 2 shows the list and categories of the local authorities in Malaysia.

Table 2. Lists and Categories of Local Authority in Malaysia

No.	State	City Hall/Municipal Council	City Council	District Council	Total
1.	Johor	1	6	8	15
2.	Kedah	1	3	7	11
3.	Kelantan	0	1	11	12
4.	Melaka	1	3	0	4
5.	Negeri Sembilan	0	3	5	8
6.	Pahang	0	3	8	11
7.	Pulau Pinang	1	1	0	2
8.	Perak	1	4	10	15
9.	Perlis	0	1	0	1
10.	Selangor	2	6	4	12
11.	Terengganu	1	2	4	7
12.	Sabah	1	2	21	24
13.	Sarawak	3	3	20	26
14.	Federal Territory	1	0	0	1
	Total	**13**	**38**	**98**	**149**

Source: Adapted from One Stop Centre (OSC) (2015)

Meanwhile, local authorities have their own functions and responsibilities. There are four main functions of local authority as highlighted in the Local Government Act 1976 (Act 171); environment, social, public and security (Figure 2).

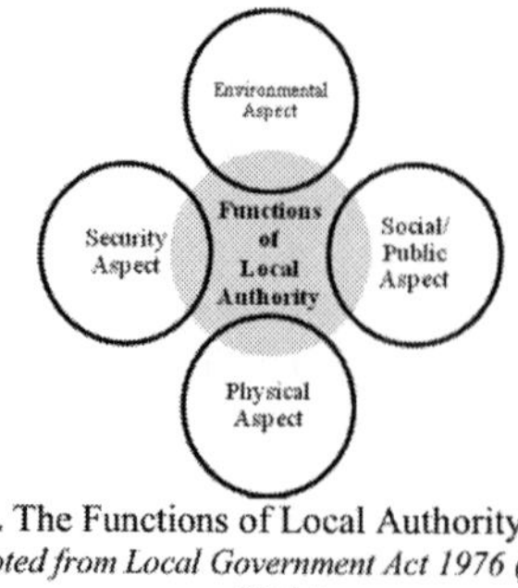

Figure 2. The Functions of Local Authority in Malaysia
Source: Adapted from Local Government Act 1976 (Act 171; Section 74-84)

The four functions stated above are interlinked with each other; i) having a good surrounding environment helps to harmonize the area of development especially in the residential area. The environment element focuses on road services, traffic control especially in the town or city at peak hours, lamp post, recreation park that people can take time to get rest with a nice environment and surrounding, provide a proper drainage and culvert system, flash flood control and other related services, ii) local authority usually in-charge of the cleanliness activities for instance garbage collection (in residential, commercial, industrial, recreation area, and open spaces), sewerage, sanitation and others. Besides, local authority also provides education and health facilities. Apart from that, local authority has initiatives in protecting the handicap such as building shelter, welfare house for old folks, housing as well as creating cultural activities, iii) physical aspects which focus on bus stand, parking area, town hall, fields, community halls or sports centre in exchange for planning permission and council services. The provision of these public services becomes the asset to the council, where the fees and charges are collected for the use of facilities provided, and lastly, iv) high-security area is vital in ensuring people to be safe and secured all the times. Thus, the security elements are focused on the public facilities such as police station, fire station, taking action on public nuisance and others. Above all, the function of a local authority as mentioned by the Act 171 possesses a close relationship between local authority and the local communities.

4.2 Issues and problems of local authority in providing urban services

An interview session was conducted with the responsible officer in Johor Bahru City Council, Melaka Historical City Council and Pulau Pinang City Council which revealed various problems experienced by the local authority in providing urban services includes; i) the capability of local authority in managing their territory, ii) development instrument adopted by local authority in promoting quality living environment, iii) limitation of the local authority in interpreting the law to generate revenue, and, iv) limitation in enforcement of laws. These three local authorities were preferred because of their involvement in coastal reclamation development within their area [34-36].

It is very important for the local authority to be given the definite and full authority or power in carrying out duty to ensure the effectiveness of its function and indirectly ease the decision making process [10]. The income and resources of the local authority must be suitable with its duty and function, so that it can deliver the job within its own institution based on their capability as the responsible authority in managing the local matter [10]. Based on Act 171, the local authority acquires revenue through general tax revenues and fines [37]. Part V of Local Government Act 1976 (Act 171), provides the financial sources of the local authorities in West Malaysia [37]. Section 39 of Local Government Act 1976 (Act 171) explains the revenue of a local authority shall consist of; (a) all taxes, rates, rents, license fees, dues and other sums or charges payable to the local authority by virtue of the provisions of this Act or any other written laws; (b) all charges or profits arising from any trade, service or undertaking carried on by the local authority under the powers vested in it; (c) all interest on any money invested by the local authority and all income arising from or out of the property of the local authority, movable and immovable; and (d) all other revenue accruing to the local authority from the Government of the Federation or of any State or from any statutory body, other local authority or from any other sources as grants, contributions, endowments or otherwise. The provision on Section 39 of the Local Government Act 1976 (Act 171) is the basis for every local authority in West Malaysia to generate revenue [11]. According to the Commonwealth Local Government Forum (2013) [38], the local authority's income mainly from the three sources, which are taxes, non-tax revenue and allocations from the federal and state government.

Besides, the local authority also obtained financial resources from grants and advances from the state or federal government as well as from loans that have been granted [11]. This is due to the increase in the cost of expenses like building materials and others. The lack of financial provisions can lead to some projects to be postponed or cancelled [11]. Interview sessions with the responsible officers in Johor Bahru City Council, Melaka Historical City Council and Pulau Pinang City Council revealed that the fund generated by the local authority is insufficient and additional allocation is required from the state or federal government. In practice, the state government is required to provide an additional fund to almost all local authority within the state for various purposes including the provision of urban services. Additionally, for the Melaka Historical City Council, where massive coastal reclamation development is being carried out, the difficulties of providing the urban services is due to the incomprehensive of the planning guideline in development for coastal reclamation [23]. The lacks of detail in planning guideline directly affect the financial and physical planning of the urban services. Nevertheless, some recommendation has been outlined by the local authorities urging for the implementation of the Safe City concepts through the nation [39]. The safe city is being introduced for people to live peacefully without being worried of their surrounding environment. The element in a safe city such as the pedestrian walkway, motorcycle lock area, safety mirror, CCTV and others help in promoting the quality living environment. Each local authority has their own way in managing their area. By promoting Safe City concept, additional financial budget and taxes from the urban services can be increased [39]. However, as the third tier in the structure of the administration system, local authorities are facing several problems in relation to the intervention of power within the federal government and state government at local level. The intervention of power will always hinder the process of developing the urban services. The financial funding for the local authority must be allocated sufficiently so that local authority can execute their duty and offer great services to the people [10]. The local authority should stand on their own feet without any interference from the upper-level governments.

5 Key findings

After several interview sessions with some responsible officers in the local authorities and further understanding the characteristic of the local authority, the study had therefore identified several important findings that can suggest a few relevant recommendations in improving and enhancing the local authority empowerment towards a more quality living environment for coastal reclamation area.

- The planning administration system in West Malaysia needs to have a structured approach to improvement and clear practices of the roles and responsibilities for each level of government. This is because the intervention from the upper-level in administrative matters can result as the conflict of roles and responsibilities between the governments. Thus, this matter contributes to the inefficiency and incompetence at a local level. The intervention from the upper level of government usually caused by political decision creating difficulties in planning and providing the urban services as financial allocation could be biased and not according to the needs of the people. The political intervention also hinders the enforcement action taken by the local authority to generate additional revenue.
- Limitation in the human resources of the local authority is also one of the factors that hinder the process of providing quality urban services by the local authority. The limitation in man-power and lack of expertise/knowledgeable officer in local authority had caused difficulties in enforcing the law when the enforcement is supposed to be one of the sources for the local authority for income generation.
- The lack of understanding of the law reduces the capacity of a local authority to explore an additional area to increase the financial fund to provide urban services in their territorial area. Part V of Local Government Act 1976 (Act 171) is not properly utilised by the local authority especially in terms of investment, mortgages or debenture stock.
- Incomprehensiveness of the development control instrument, especially in coastal reclamation area, hinders the process of financial and physical planning of the urban services. The current instrument used by the local authority is incapable in guiding the local authority to outline a systematic planning of the urban services. The development instrument includes the planning guideline and the development plan system adopted by the local authority.

6 Conclusions

Empowerment of the local authority particularly on the human resources, independency of local government and development control instrument should be able to overcome the existing problems with regard to the provision of quality urban services in the coastal reclamation area. However, intervention by upper-level government and the political decision must be avoided so that the decision made by the local authority can be materialized. Nevertheless, the monitoring process is also essential in ensuring that the financial allocation is well spent by the local authority. Further understanding the laws and regulations by the local authority will help the local authority to explore an alternative area for local government to generate additional revenue in order to reduce the dependency of the local government to the federal and state government. Good physical planning comes with good urban planning governance. For that reason, the local authority should initiate a paradigm shift in governing their territorial area in trying to provide quality urban service in land reclamation area.

Acknowledgement

The author would like to thank the Ministry of Higher Education Malaysia for funding this research under the Research Acculturation Grant Scheme (Grant 600-RMI/RAGS 5/3 (135/2014)), and the Universiti Teknologi MARA (UiTM), as well as the Research Management Centre (RMC) UiTM for the management and support of this research.

References

1. Department of Statistics. *Yearbook of Statistics Malaysia, 2010.* Department of Statistics, Malaysia. (2011).
2. A. J. Yunus. *Country Profiles: Malaysia.* Malaysia: United Cities and Local Government (UCLG). (2005).
3. The Commissioner of Law Revision. *Federal Constitution.* Selangor: International Law Book Services. (2014a).
4. The Commissioner of Law Revision. *Town and Country Planning Act 1976 (Act 172).* Selangor: International Law Book Services. (2014b).
5. M. Hammond and J. Wellington. *Research Methods: The Key Concepts.* Routledge Key Guides. (2013).
6. T. Litman. *Planning Principles and Practices.* Victoria Transport Policy Institute. (2013).
7. A. S. Hornby. *Oxford Advanced Learner's Dictionary.* London: Oxford University Press. (2010).
8. L. Keeble. *Principle and Practice of Town and Country Planning.* London: Estate Gazette Ltd. (1969).
9. J. Ratcliffe. *An Introduction to Town and Regional Planning.* London: Hutchinson and Co. Publisher Ltd. (1974).
10. M. Yusup. *Statutory Procedure and Planning Machinery for Urban Development in Sarawak.* Shah Alam: Universiti Teknologi MARA. (2013).
11. A. A. Hussain. *Kerajaan Tempatan: Teori dan Peranan di Malaysia.* Kuala Lumpur: Dewan Bahasa dan Pustaka. (2002).

12. K. L. Miller. *Advantages and Disadvantages of Local Government Decentralisation.* Carribean Conference on Local Government and Decentralisation, **4.** (2002).
13. C. Barr, I. A. P. Resosudarmo, A. Dermawan, J. McCarthy, M. Moeliono and B. Setiono. *Decentralisation of Forest Administration in Indonesia: Implications for Forest Sustainability, Economic Development and Community Livelihoods.* Jakarta: Center for International Forestry Research. (2006).
14. P. D. Hutchcroft. *Centralization and Decentralization in Administration and Politics: Assessing Territorial Dimensions of Authority and Power.* Governance: An International Journal of Policy and Administration, Vol. **14**, No. 1, January. Blackwell Publisher. (2001).
15. H. J. B. Allen. *Cultivating The Grass Roots: Why Local Government Matters.* All India Institute of Local Self-Government. p.3-10. (1990).
16. A. A. Hussain. *Pengantar Pentadbiran Awam.* Kuala Lumpur: Utusan Publications and Distributors Sdn. Bhd. (1983).
17. World Bank. *Beyond the Center: Decentralising the State.* Washington D.C.: World Bank. (1999).
18. I. Ferguson and C. Chandrasekharan. *Paths and Pitfalls of Decentralization for Sustainable Forest Management: Experiences of the Asia-Pacific Region.* ITTO Tropical Forest Update. (2004).
19. N. M. Nadzir, M. Ibrahim and M. Mansor. *Impacts of Coastal Reclamation to the Quality of Life: Tanjung Tokong Community, Penang.* AMER International Conference on Quality of Life (p. 10). Kota Kinabalu, Sabah, Malaysia: Procedia Social and Behavioral Sciences. (2014).
20. A. Sufian and M. Mustafa. *Emerging Issues on Reclaimed Land in Malaysia: Alienation for Freehold or Leasehold?.* International Seminar & Indonesia Forum on Ocean Law and Resources: Building Comprehensive Perspective on National Security and Sustainable Development. Faculty of Law Brawijaya University (2010).
21. B. Win and V. Choa. *Reclamation and Ground Improvement.* Thomson Learning. (2004).
22. Melaka Historical City Council. *Issues and Problems of Coastal Reclamation in Malacca.* Central Malacca: Malacca Historical City Council. (2015).
23. Melaka Town and Country Planning Department. *Coastal Land Reclamation in Melaka.* Melaka: Melaka Town and Country Planning Department. (2015).
24. Z. A. Hashim. *House Price and Affordability in Housing in Malaysia.* Bangi: Universiti Kebangsaan Malaysia (UKM). (2010).
25. L. J. Wyn. *Melaka Forges Ahead.* The Edge Malaysia. Melaka, Malaysia: The Edge Malaysia. (2011, May 9).
26. F. W. Ibrahim. *Local Public Goods and Citizen Satisfaction: A Case Study of District of Sik, Kedah, Malaysia.* Proceedings of 5th Asia-Pacific Business Research Conference (ISBN: 978-1-922069-44-3) (pp. 1-9). Kuala Lumpur: 5th Asia-Pacific Business Research Conference. (2014).
27. Royal Malaysian Customs Department. *Guide on Government, Local Authority and Statutory Body.* Royal Malaysian Customs. (2014).
28. D. M. Hill. *Democratic Theory and Local Government.* London: Allendand Unwin Ltd. (1974).
29. S. Khalil and S. A. Adelabu. *Modelling Local Government System in Nigeria.* Kuwait Chapter of Arabian Journal of Business and Management Review Vol. **1**, No.1; September 2011, 136-154. (2011).
30. S. M. Yusoff, F. Yusof and A. F. Arshad. *An Analysis of Local Plan for Development Control at Local Planning Level in the State of Selangor.* AcE-Bs 2014 Sabah ASEAN Conference on Environment-Behaviour Studies The Pacific Sutera Hotel, Sutera Harbour, Kota Kinabalu, Sabah, Malaysia, 04-05 January 2013 "Quality of Life in the Built and Natural Environment" (pp. 574-584). Selangor: Elsevier Ltd. (2014).
31. A. J. Maidin and N. F. S. Ahamed. *Mandatory Health Impact Assessment in Malaysian Land Planning and Development Control System.* AicE-Bs 2012 Cairo ASIA Pacific International Conference on Environment-Behaviour Studies Mercure Le Sphinx Cairo Hotel, Giza, Egypt, 31 October 2 November 2012 "Future Communities: Socio-Cultural & Environmental Challanges", 164. Vol. **68.** (2012).
32. Portal One Stop Centre (OSC). *Portal One Stop Centre (OSC).* Retrieved from Senarai PBT di Malaysia: Pihak Berkuasa Tempatan: www.epbt.gov.my. (2015, 10 10).
33. The Commissioner of Law Revision. *Malaysia Kita: Panduan dan Rujukan untuk Peperiksaan Am Kerajaan.* Kuala Lumpur: International Law Book Services. (2000).
34. Johor Bahru City Council. *Finance* . Johor Bahru: Johor Bahru City Council. (2015).
35. Melaka Historical City Council. *Finance.* Central Malacca: Malacca Historical City Council. (2015).
36. Pulau Pinang City Council. *Finance.* Pulau Pinang: Pulau Pinang City Council. (2015).
37. The Commissioner of Law Revision. *Local Government Act 1976 (Act 171).* Selangor: International Law Book Services. (2014c).
38. Commonwealth Local Government Forum (CLGF). *The Local Government System in Malaysia.* United Kingdom: Commonwealth Local Government Forum (CLGF). www.clgf.org.uk. (2013).
39. Department of Town and Country Planning. *Safe City.* Selangor: Department of Town and Country Plannings. (2010).

A review of Small Unmanned Aircraft System (UAS) advantages as a tool in condition survey works

Mohd Fadzil Mat Yasin, Mohd Azian Zaidi[1] and Mohd Nasrun Mohd Nawi[2]

[1] Faculty Of Architecture, Planning & Surveying, Universiti Teknologi MARA Seri Iskandar Campus, Seri Iskandar, 32610 Perak, Malaysia
[2]School of Technology Management and Logistic, Universiti Utara Malaysia.

Abstract. Adoption of new technology in building inspection works has proven to be very crucial and has improved efficiency and effectiveness in this regards. Tools and techniques in building inspection and condition survey continuously evolved trough times. The tools and techniques adopted mostly first has been successfully applied by others technical field such as civil engineering, manufacturing etc.. As UAS has been successfully adopted by many technical professions such as geotechnical engineering, land surveying, civil engineering etc. a practical use of UAV for building surveying and facilities management yet to explored. The present paper intend to explore the viability of UAS application for building inspection with the advantage it has especially in inspecting and collecting data at difficult to reach area such as roof, ceiling, gutter and recording data for large building complex. With the accurate and comprehensive inspection data collected, perhaps it could guide building surveyors or facilities managers in diagnosing building defect effectively, suggesting possible remedies, solving issues and making decision precisely. The present paper provides a comprehensive review of possible application of small UAS in building inspection and facilities management related uses by reviewing the similar application and identifying their features that suits facilities management's nature of practice. Challenges faced by building and facilities management organisation to apply UAS in their building condition survey routine that are identifying suitable tools other than normal visual tools to be attached, finalising arrangement on site during inspection and limited skill of existing personnel to deal with the equipments.

1 Introduction

The technology of unmanned aerial vehicle was first introduced and successfully exploited by military air forces of few countries in the west since early 1980s [1][2]. Due to its small size and prevailing role, the UAV in the military field also referred as "drone". In the last decades, the technologies are relatively expensive and required highly skilled personnel to operate. However, trough times as the technology evolve; many of the features have been improved. These includes more affordable cost, choices of features and easy to handle. Since then, various industries start to exploit the technology into their field. As cited by Irizarry et. el.,[3], one of the recent cases of using these devices for civilian applications is when a tsunami struck the Fukushima nuclear power plant in Japan on the 11 March 2011. During that disaster, due to very unsafe conditions at the plant, Tokyo Electric Power (TEPCO) used a US-made micro aerial vehicle to photograph the nuclear plant from above.

As it integrates many components the terms Unmanned Arial Vehicle (UAV) has renamed to Unmanned Aircraft System (UAS) [4]. Issues such as stabilisation of flight platform, anti collision, navigation system and route planning algorithms as highlight by several authors for the past few years [1],[2], [3], [5] has been address by new improvement and technology in small UAV and imaging system in recent years.

From facilities management point of view, besides visual inspection, the conventional non-distractive test (NDT) that commonly adopted by the building inspector includes applying test and diagnostic apparatus, thermography and remote sensor technology. Yasin and Egbu [6] contend the performance evaluation of building in which condition survey is being part of it is a very critical task for facilities managers to deal with. A rigours tools and technics therefore are very critical to be deployed. The quality of data collected might help the building inspector to analyse the building performance appropriately.

This paper aims to explore and identify the relevant application of UAS related technology in building inspection by referring to challenges faced by the building inspector in collecting visual data during inspection. The present paper also reviews the similar application of UAS in other part of the globe with the new improvement in UAS technology that could benefit the facilities management organisation in conducting condition survey works.

DOI: 10.1051/matecconf/20166600038

2 Condition survey works

Condition survey works is a building inspection exercise conducting at specific predetermined intervals and in ad-hoc basis. The former is conducted as a routine to review the overall performance of the building. The later however is a building inspection conducted focusing on specific elements or assessment for incidents aftermath e.g fire, explosive, disasters. The process of facilities performance evaluation has been extensively discussed by several authors [7]. It could be suggest that three main process which reflect the 'depth' and 'breadth' of the inspection i.e indicative, investigative and diagnostic.

There are very common that the building surveyor have to revisit the site due to accessibility issues and close examines to the particular structure are required. Visual tools such as binocular and distance focused lens alone are sometimes insufficient. In this situation, tools such as suitable leader, scaffolding or even portable hoist need to be hire to reach the high place. This challenge could drag the inspection process for the longer period. As for facilities management operation is concerned, it means the longer period until the defect issues could be solved.

Echmann et al [5] highlight the potential use of UAS in the non-destructive testing (NDT) which includes state detection, damage analysis and condition monitoring. A few examples of potential benefits that could be gain by facilities management team are:

i) saving on time for inspection
ii) minimum number for inspection team
iii) saving on rental of special tools e.g. scaffolding and hoist
iv) risk of height hazard; and
v) quality and reliable onsite data

The building surveyor intent to adopt the UAS in their survey should first aware of the followings:

i) Knowledge and skill – the operator that fly the UAS has at least a basic skill and knowledge to fly the UAS specific model and the camera attached.
ii) Strategy for inspection – determine in which stage of building inspection UAS to be used. Charger and appropriate cameras to be attached to the UAS unit
iii) Whether condition - as the normal UAS unit will only work in a full performance in clear whether condition.
iv) Permission to fly the UAV – any restriction or domestic rules and regulation enforced for UAV. A notice should be given to the parties involve such as the building occupant, public by-passer and neighbouring building owner alerting them about the inspection.
v) A potential hazard to the public and appropriate safety measures

Fugure 1 : Remote Control (RC) unit for Phantom 3 model complete with large display (Internet source : Slashgeer website)[11]

Figure 2 : The Phantom 3 small UAS model (Internet source : Slashgeer website) [11]

The Figure 1 and 2 shows an example of small UAS with 4 exposed propellers available in the market. Various brands and models of small UAS with different technical specification available in the market.

3 Data collection for condition survey works

At least, two types of image data that could be collected via this media, which is still image and video graphic image. The building inspector might determined the types and form of image data they required prior the inspection. The selections of the quality camera to be attached to the UAS are critical in order to obtain the quality digital photos and video images. For smooth inspection exercise and limited flight time capability of the UAS, Echmann et. al [5] suggest a preliminary flight track planning by using common software based on GPS waypoint navigation. In order to maintain the sequences of the image taken, a vertical and horizontal movement of the small UAS have to be predefined at flight track planning stage.

3.1 In-situ still and video image

Image obtained trough small or micro UAS could provide quick identification of defect spot and type of defect, symptom, causes, and level of severity. Small UAS ability to hovering on a specific airspace could provide data on roof such as sign of leakage on roof components i.e gutters, rain water down pipe and roof dome. Other building component that could also be observed are roof tank, roof to chillers, telecommunication antenna and devices etc.

The quality imaging data could be used by on site building inspector to assess the severity of building defect and decide in case further and details information on particular elements are required. For a large building complex, aerial view of the site which shown the site lay-out, shape and position of the building could be obtained with better quality, details and specific compared to Google maps or Google earth satellite image.

In some circumstances, a more detailed and high definition (HD) imaging are required. Therefore, the available image collected need to be refined and reconstruct in order to satisfy the specific objective of the condition survey works or building inspection.

3.2 Post flight image reconstruction

During exploratory roaming by the small UAS, if the defects location are not been identified earlier, a real time image of defect might inform the building surveyor whether or not to get a further and closer look at the specific defect spot.

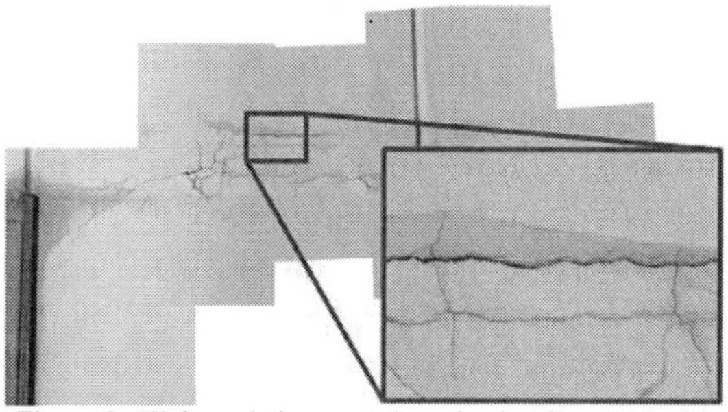

Figure 3 : High resolution crack from the shot from UAS still photograph (source: [5]

A severe and obvious defect identified in the preliminary flight required details visual. The image data could also being enhanced to provide better perspective of the defect. Image of element crack on the upper structure for examples required information such as the other related symptom and trends of the crack need to be observed. In this case, a high definition, extensive and live image are essential.

Building Information Modelling (BIM) might enrich facilities management information repository. A higher resolution image of two or three dimension digital image might help for digital façade reconstruction of BIM system. The visual image captured during the inspection flight could be process and reconstruct after the inspection.

3.3 Thermal imaging

The use of thermal imaging for diagnosing building defects was already in place for more than two decades [8] [9] [10]. Similar to UAS technology, the infrared thermal imager was first used in military field and currently been benefited by various other fields as the cheaper market price and advanced of technology [9]. The infrared thermal imaging data could inform the surveyor the causes for building heat loss during winter in four season's country and leaking related defect. The Photo 1 below shows an examples of thermal vision camera for small UAS fitted the bottom side gimbal of the UAS.

Figure 4 : A FLIR VUE Thermal Vision Camera for UAS 7.5Hz (336x256, 6.8mm Lense) [11]

4 UAV Specification Consideration

Before considering the application of small UAS for building inspection, a few considerations should be made by the facilities management organisation. First and foremost obviously the purpose and the scope of building inspection intended for. The purpose of inspection might influence the technical ability of small UAS that match their requirements. Secondly, the budget considerations including UAS initial cost, supporting system, accessories, spare parts and other maintenance expenses. The higher ends UAS unit usually offer a wider range of technological features.

- ***Flying Range***

  No specific range and distance of flight, furthermore, the UAS available in the market are considered sufficient for building inspection purpose. However, to enable takeoff and landing at a narrow platform, the UAS should be from the Vertical Take-off and Landing (VTOL).

- ***Flying capacity and camera compatibility (and quality)***

  Gimbals for camera might ensure better quality and clear and image taken during flight.

- ***Battery life (flying time)***

  Battery endurance that could last between 20 – 40 minutes are considered sufficient with rapid charging battery if need during inspection. Some UAS model has intelligent battery that could return the UAS to the departure point automatically one the battery nearly discharged. To avoid interruptions due to unforeseen issues such s absence of power supply at the site, spare battery should also be considered.

- ***Self-weight and shape***

  Due to several reasons such as safety and insurance claim restriction, maximum limit of takeoff weight is nothing more than 5 kilograms. Smaller and lighter the size of UAS could improve the manoeuvre to the difficult space and more practical for building inspection purposes.

5 Issues and challenges

Despite the advantage that UAS could offer, there are challenges that hinder building and facilities management to use UAS as one of the tools in building inspection. At least five factor that differ the requirements of one organisation towards UAS:

1) Size of facilities site
2) Floor area
3) Form of buildings
4) Complexity of building structure and building system
5) Nature of building occupation

At this point, it could be hypostasised that the more challenging and comprehensive the facilities is, the higher potential of UAS benefits could be exploited by the building and facilities management organisation. There is also absence of evidence of a formal use of UAS by FM organisation in conducting condition survey.

In the context of present paper, aspect of readiness and awareness of building and facilities management organisation being examine as a preliminary investigation for extensive UAS application in building inspection. For that reason, a sum of 512 questionnaires sent to the building maintenance and facilities management personnel in 159 organisations; the parties that usually conduct building inspection for their parents' organisation or clients depending on their respective service provision. Relatively high response rate with 188 respondents (37%) has returned the attached e-mail questionnaire distributed.

Table 1: *Ranking of the challenges in applying UAS in condition survey*

Rank	Challenge factors	Mean (N=188)
1	***Identifying suitable tools (C1)***	***1.32***
2	***On-site arrangement (C2)***	***1.87***
3	***Lack of staff (C3)***	***1.96***
4	*Lack of awareness (C4)*	*2.02*
5	*Familiarity with facilities (C5)*	*2.03*
6	*Maintenance of equipments (C6)*	*2.19*
7	*Provide training to fly UAS (C7)*	*2.32*
8	*Provide training for equipments (C8)*	*2.48*
9	*Cost of equipments (C9)*	*2.69*

Meaning of scale (challenge in conducting FPE):
1 (Very challenging), 2 (Challenging), 3 (Fairly challenging), 4 (Not challenging at all)

The result of mean score in **Table 1** indicated as mean score increases, the degree of challenge decreases. An inspection of Table 4.6 shows that facilities managers' ranked *"identifying suitable tools to be attached to the UAS system"* as most challenging in the survey. As UAS means for collecting buildings data, the limited skill of the personnel in the organisation limiting the benefits and uses of UAS in the building inspection.

Site arrangement and insufficient numbers of staff are another key challenge perceived by the building and facilities management personnel on application of UAS as they ranked second and third respectively. Based on the top three challenges related to readiness of building and facilities management personnel indicates that the readiness towards implementation of UAS in their building inspection is at a stumpy level.

Although, acquiring the UAS and equipments including providing training to the staff on using UAS and relevant equipments are less challenge for the organisation as shown on the survey result as they ranked seven, eight and nine respectively, the real potential of UAS application's benefits does not really attentive for them.

The Kruskal-Wallis test was used to identify the significant differences in perceiving the challenges in *applying UAS in condition survey* (see **Table**) according to three different sizes of building and facilities management organisations. Null hypothesis and alternative hypothesis for the test are as follows:

H^0 = the challenges in applying UAS in condition survey differ between sizes of organisation.

H^1 = the challenges in applying UAS in condition survey do not differ between sizes of organisation.

	C1	C2	C3	C4	C5
Chi-Square	6.548	10.626	2.696	6.043	1.88
df	2	2	2	2	2
Asymp. Sig.	.038*	.005*	.260	.051	.391
	C6	**C7**	**C8**	**C9**	**C10**
Chi-Square	4.376	1.159	2.015	6.508	8.04
df	2	2	2	2	2
Asymp. Sig.	.112	.560	.365	.039*	.018*

a. Kruskal Wallis Test

b. Grouping Variable: number of employees
** results are statistically significant at p<0.05*

Table 2: Kruskal-Wallis test statistic challenge in *applying UAS in condition survey* according to the organisational size

Table shows, four out of ten variables indicate the significant value smaller than five per cent. This result informed that, at five per cent level of significance, the null hypothesis is rejected and alternative hypothesis is accepted. This denotes that there is no significant difference in perceiving challenges between the sizes of FM organisations.

6 Conclusion

The practical application of UAS has a very broad advantage to facilities management and building surveying team to be exploited. Ground pilot or operator should also aware of few issues related to micro UAS prior deciding to apply them for condition survey exercise. The issues including, safety issues, privacy issues and technical issues. The former issues might involve damage of property, operator itself and third party. Secondly, as the micro UAS use a public air space, a permission and adequate notice should be given to the respective parties. Even though there is absence of clear guidelines by the authorities on flying the UAS, ground pilot should well versed and made clear with allowable space and restriction area for UAS to hover. It could be extrapolates that many more benefits that could be exploited from small UAS technologies in the future in the challenging field of condition survey and facilities management at large.

References

[1]Anand, S., "Domestic use of unmanned aircraft systems: an evaluation of policy constraints and the role of industry consensus standards", ASTM Standardization News, Vol. **35**, p. 30. (2007)

[2] Nisser, T. and Westin, C., "Human factors challenges in unmanned aerial vehicles (uavs): a literature review", Lund University School of Aviation, Tech. Rep. TFHS, Vol. **5**, p. 1. (2006)

[3] Irizarry, J., Gheisari, M., Williams, G., & Roper, K.. Ambient intelligence environments for accessing building information: A healthcare facility management scenario. *Facilities*, *32*(3/4), 120-138. (2014)

[4] International Civil Aviation Organization ICAO Cir 328, Unmanned Aircraft Systems (UAS), Quebec, Canada. (2011)

[5] Eschmann, C., Kuo, C. M., Kuo, C. H., & Boller, C. Unmanned aircraft systems for remote building inspection and monitoring. In *6th European workshop on structural health monitoring*. (2012)

[6] Yasin, F., & Egbu, C. O. Exploitation of knowledge mapping benefits in the facilities performance evaluation process: a conceptual framework. In Procs 26th Annual ARCOM Conference (pp. 799-808). Association of Researchers in Construction Management. (2010)

[7] Yasin, F, The Contribution of knowledge mapping in facilities performance evaluation practice in Malaysia, The University of Salford, PhD thesis, not published (2013)

[8] Titman, D. J. Applications of thermography in non-destructive testing of structures. NDT & E International, **34**(2), 149-154. (2001)

[9] Balaras, C. A., & Argiriou, A. A. Infrared thermography for building diagnostics. Energy and buildings, **34**(2), 171-183. (2002)

[10] Grinzato, E., Bison, P. G., & Marinetti, S. Monitoring of ancient buildings by the thermal method. Journal of Cultural Heritage, **3**(1), 21-29. (2002)

[11] Slash Geer website : http://www.slashgear.com/dji-phantom-3-drone-revealed-with-newly-chopped-price-points-08377962/ extracted on 15/1/2016

[12] Hallermann, N., Morgenthal, G., & Rodehorst, V. Vision-based monitoring of heritage monuments–Unmanned Aerial Systems (UAS) for detailed inspection and high-accurate survey of structures'. Proceedings of STREMAH, 621-632. (2015)

Study and Redefining Beneficiary Participation in Process Of House Design

Morteza Monshizadeh [1,a], Sr Dr Norngainy Mohd Tawil [2] Maria Khosravi [3]

[1] *Ph.D. Student Universiti Kebangsaan Malaysia, Architecture Department*
[2] *Deputy Director of CESMED Universiti Kebangsaan Malaysia*
[3] *Ph.D. Student Universiti Kebangsaan Malaysia, Architecture Department*

Abstract. Since housing has a special place in human life and his physical, psychological and interactions, so in addition the unity of humans, multiplicity and diversity of them must be considered. This possible only by beneficiary participation in the design process, but because society has different economic and social texture and classes; and settling suit because of the time and place of special indexes are entitled, so prepare a comprehensive model includes the testimony and circumstances; identify factors influencing participation optimum need to selection population and certain species of private construction. Standard tool to study topic does not exist, so in order to produce tools using qualitative research methods; interpretation - historical correlation to extract components and variables and their effects on each other and enjoyed target table Content consisting of four domains of general knowledge - specialized knowledge of participation - participation mechanisms and factors influencing participation achieved. Extracted factors are: the initial formation of partnership - partnership executive process - the role of participant - optimal participation; by study and analyze the theoretical model. Due to history and social aspects; cultural participation in Shiraz; promote scientific and participatory approach designed to make operating housing; bed and new horizons of development of facilities and areas in the design of residential environment created and due consultation and decision making in addition to beneficiary participation to promote optimum utility of space; mutual flexibility and utilization of space; increase fixation and motivation will lead beneficiary reside" and the main question: " how is the model of scientific position optimal participation planning instrument in private housing in the city of Shiraz, in the process of design, implementation and use".

1. General participatory concepts and approaches

Participation is included by different interpretations from various school and view, and it is result from consideration toward beneficiary and lower classes of people due to paying attention to intrinsic nobility and nature, and in this regard, by Consider to theorists concerns about functional aspects, the outcome and results, effective factors, the activity and action, its aspects of the instrumental and purpose is very different.

There are some belief in all attitudes and schools about the existence of prerequisites and situations in formation as well as in step by step way that it can be different according to interpretations and conditions. But the common point is that cultural and social conditions are required in forming participation. If it is necessary to take advantage from the people's perspective, Society should already have such potential and also ready to training and institutionalizing in this field, until the conditions of provision and facilitating in participation can be provided.

1.1. Literal meaning of participation

Participation is equivalent of English term "cooperation" that is the root of the word "Part" means the component and section, and also it means " sharing in something or taking part of that" (Rezai, 2002). In the Oxford Dictionary, it is defined a real of Interaction or participating and sharing (Ghaffari, 2001). "Apparently, Perhaps, a consensus and convergence in the meaning of participation is implied, But because this term is observers of the rate and method in presenting ideas and methods, activities, projects, programs and decision-making of people, so it can be very controversial" (Qubadi, 2002).

1.2. Descriptive meaning of participation

Participation is a kind of solidarity and belonging and collective effort among individuals in a society in other to achieve a social justice system (Sepehri, 1993)

[a] Morteza Monshizadeh: mmmajid58@yahoo.com

and "its infrastructure is placed in the process of need assessment of decision-making, planning, implementation, review, evaluation, common benefit of people" (Sarkezi, 2002). (Mclagan, 1955) "Describing the participation dependence on the various factors such as social, economic and political, So that we faced with various interpretations and inferences sometimes with contradictory and opposition levels and categories" (Islami, 1999).

1.3 Participatory approaches from the perspective of individuals and organizations

Since participation in many fields that related to human problems and is able to be investigated and used, therefore people who has various expertise, goals and specific approaches, with different interpretations and definitions, have been used (Table 1).

Table 1. Comparison of approaches derived from individuals and organizational theories about the issue of the participation.

Biro; 1996	Voluntary activities in the State - City - the village and the former of social life and politics
Babai; 1997	Including the people in the organization - decision-making and interventions – equal and same opinion.
Ghobadi: 2002	Collective use – having natural and internal root and result from cooperation among human and relate to rate of people benefit
Tokoyol:2005	Human right – democracy and growth and Promotion of citizenship- Culture Promotion in Discipline and awareness of the issue of human rights and government
Habibi-razavi: 2005	Public participating in policy and recommendations formulation
Tabrizi: 1991	Belongs - profit - a share in something - participating in a group
Habbi- jal elahi: 2003	using collective resources in order to participating people - the relationship between effort and resources to the relationship between the results and efforts
Okeli: 2003	Sensitization of people and increase the acceptance and ability to respond to development - involving people in the decision - Implementation - Evaluation Program
Alen,piruft: 2002	Consultation before the decision - personal comments and actions of administrative administrators decisions caring
Kurt depfer:2002	Ratio of practical interventions that is required and amount of interference in the activities of the group
Salima: 2002	Intervention and control of people for better life - reflection of values system are sacred and important
Qafari: 2001	Mutual Interaction to form a foundation of social life - comprehensive and empower people based on three values 1 - shared by people in authority 2. Voting 3. Open the opportunities to the lower categories and classes
Tosi: 1991	Mental and emotional involvement of people to help each other in achieving the goals of the group and the division of responsibilities
Mir Musavi: 1996	Conflicting and individuals activities of affecting for a particular purpose
Burki: 2001	The relationship between essence of man and sustainable development and efforts to eliminate the poverty and promote development and exploitation
Alavi Tabara; 2003	Social behavior, consciously and voluntarily, and intend to make individual or collective interests - mental desire to do certain tensions
(F.A.O)	Involving people in the lower classes in program development, and organize them in for public participation programs
United Nations Research Institute(UNRISD)	Organized efforts to increase control interests and regulatory agencies by some groups and organizations that have been deprived
World Bank	The division of responsibilities for program development decisions and resources associated with the program
International Labour Organization (I.l.o)	The proportion of individuals or organized groups of active population economically and socially

2 Types of participation in terms of process and purpose in the development program

Pretty (1955) argues that, if it was accepted that there is always some degree of participation in human societies, then it is useful to know that these contributions from what types can be our participation in our society which one is presented of the cases above. Of course, other forms of participation may also be introduced in other classification. Another important point is that each of the following participation has specific social effects on its own group or society (Ghobadi, 2002).

2.1 Self – mobility participation

In this type of participation, people operate without dependence on external organizations in their activities. But there are controls on the using resources. If governments and NGOs can create a required framework to support these types of activities, self-mobility may develop. This type of participation also called "innovative self – mobility".

2.2 Interactive participation

In this type of participation, public participation activities develop interaction patterns and information. These activities strengthen local organizations. The right is considered for community members in this participation model and the purpose of paying them not only to achieve project goals. Here participation is multi-dimensional process that can be considered from different aspects and also its learning process is structural. Acceptance of regional decisions by Group members determine how available resources are used, In this case, individuals have important factors for maintaining their achievements.

2.3 Functional participation

In this model, participation is known as the process of achieving goals and cost reduction. Individuals are organized in order to achieve pre-determined targets of the participation. This relationship may be interactive. It may also include decisions or their related processes. But this process only occurs when major decisions (external) are adopted by external institutions. In this situation, the local individuals are still affected by external goals (which are not involved in its formulation).

2.4 Participation for financial incentives

In this participation model is done for financial incentives. And here individuals expect that their participation targets is specified, And In this model problems are defined by external organizations, And the process of data collection is done the same way, Control of the process is also similarly under the supervision of external organizations. In such process is not seen participatory consultation in the decision-making.

2.5 Passive participation

In this type, public participation is based on previous decisions. This is done without any regard of people responses, based on unilateral notification of managers. Here, the information is shared from the external professionals.

2.6 Strategic participation

In this participation model can be seen, representatives of people are actually present in formal organization, but they have not special power

3 Types of participation in terms of function

In this field, there are various types of participation, special platform encourage was considered for group participation of all people. Hence "participation from different points of view can be classified as follows:" (Table 2).

3.1 Aspect of amount of people's participation

Aspect of about the extensive involvement of people in participation decision is divided into two types:

a) Basic participation: This method participation represents relatively comprehensive participation of complete social groups. Each application stages including need assessment and decision-making for planning, implementation and evaluation is done by the community groups of interest.
b) Marginal participation: In this method, the participation used to implement some applications and specific purposes. In fact, the marginal participation is a one-dimensional participation. For example, in the economy, participation in the investment projects, by the people, is one kind of marginal participation.

3.2 Participation in terms of territory

This has three categories on the basis of participation importance in any aspect of group life:

1) Political participation: Is the participation of social groups in the programs and political activities and engaging people in the administration of the society.

2) Social participation: Includes social group's participation in social and cultural programs and activities.
3) Economic and financial participation: Includes participation in programs of social groups and economic activities and emphasizes on the financial participation.

3.3 The aspect of people's involvement quality

Based on how and the way people play a role in the programs:

a) Direct participation: In this method people are directly involved in the decision-making of programs and activities.
b) Indirect participation: In this method through their representatives are involved in decision-making.

3.4 The aspect of implementation levels

Based on implementation levels requiring people's involvement in activities and programs:

a) Participation in rural levels: Rural people can participate in special programs and activities in their rural area.
b) Participation in regional levels: People can participate in programs and activities of regional and provincial areas.
c) Participation at the national levels: At the national level all people can participate in programs and national plans.

3.5 The aspect of democracy

The degree and range of participation democracy that should be considered:

a. Compulsory participation: People or social groups will participate in the programs and activities of non-voluntary and forced.
b. Optional participation: People or social groups will participate as voluntary and interest in the programs and activities.

3.6 The aspect of profit

In terms of what individual or group will directly obtain benefits, advantages and results of Participation:

a) Self-help: The person or group who is helped in an activity (participation), in the activity (if necessary) will help one previous participation mutually (participation).
b) Assistances: in the assistances process, all people and social group participate in decisions making, plan implements and activity and all of them has shared profit of participation. Assistance is kind of public participation.
c) Helping to other: participated person or group helps to other person or group (participation in public profits and welfare affairs).

Table 2. Participation for Financial Incentives (Farhang and Taavon 1999).

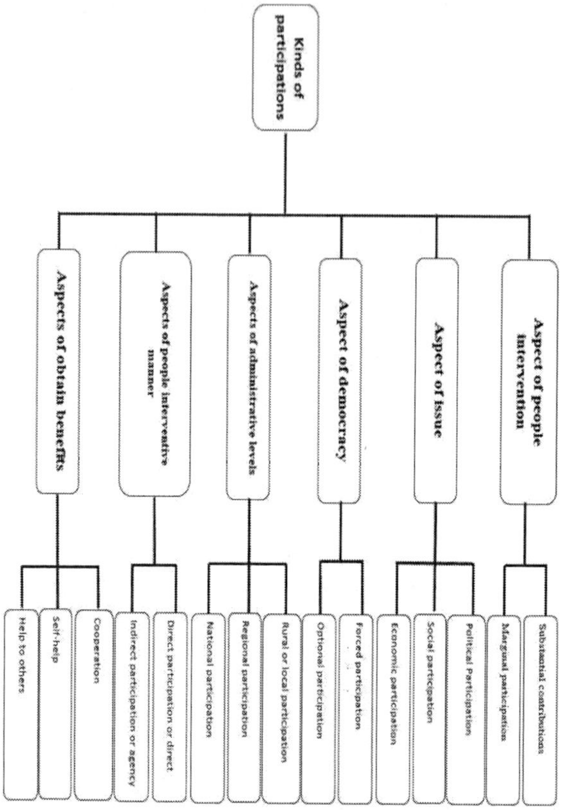

4 Beneficiary participation process based on human needs (adaptation of human needs and participation model)

4.1 Hierarchy of needs

Maslow introduced the famous hierarchy of needs theory in 1943. He thinks that motivation is function of the five basic needs, including physiological, safety, social, esteem, and self-actualization, and these needs can be explained in a hierarchy. In other words, he believes that the emergence of human needs has developed in a step-by-step way in a way that self-actualization need flourish. "If we accept this hierarchy of needs, the question is that the person whose social needs prevalent, for example, if a certain degree of participation works better in such case? In other words, if there is a relationship between

 DOI: 10.1051/matecconf/20166600039

hierarchy of needs and degrees of participation anymore? The answer to this question lies partly in the theory of situational leadership." (Figure 1)

Figure 1. Maslow's hierarchy of needs theory (Rahnavard, 2006, 117)

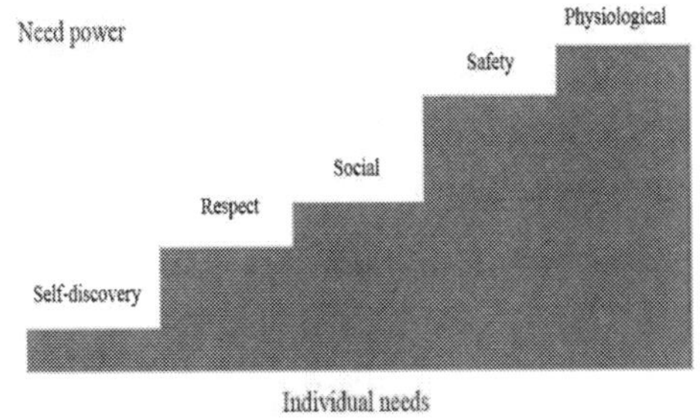

4.2 Degrees of participation and hierarchy of needs

Situational leadership theory (Hersey, 1988) between leadership styles and hierarchy of needs establishes one by one relationship. As shown in Diagram 2, collaborative style of working fits those who are motivated by social needs and respect, submission style fits those who mostly need self-discovery, while command and incentive styles fit those who are motivated by basic needs. If we look more closely into submission style, we can say that this power or authority is given for an individual or a group. In the latter case, semi-autonomous or autonomous groups are formed that shows complete degree participation.

Therefore, if we interpret command and incentive styles as non-participation, it should be said the theory of situational leadership forms one by one relationship between degrees of participation (non-participation, partial participation and full participation) and hierarchy of needs." Davis and Veniostrum (1989) state that there is evidence that many of people don't want to be hurt by participation and this question is raised out that we should steer them towards participation just for this reason that we think participation is good for them?

Figure 2. The relationship between degree of participation and needs based on situational leadership theory (Rahnavard, 2006)

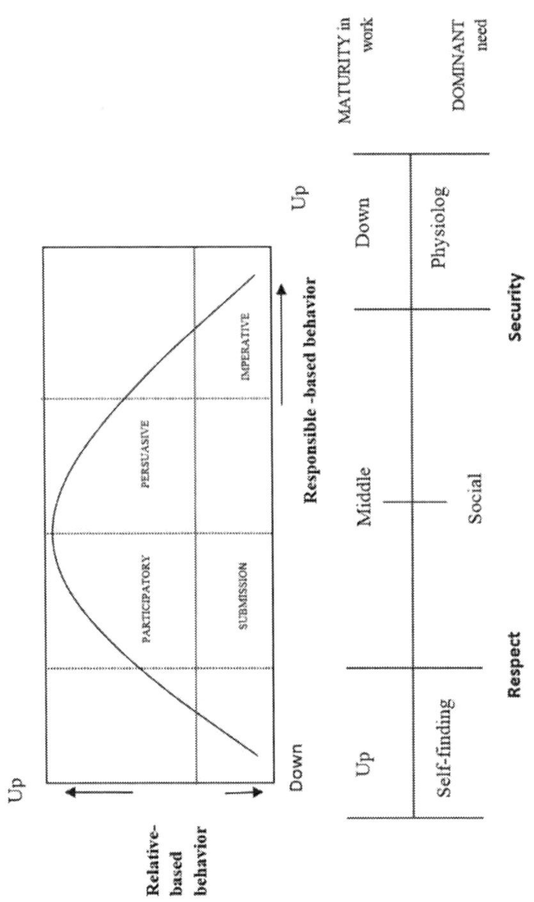

4.3 Under- participation, over- participation, degree of satisfaction

If you look at the reality from another angle, it may be argued that people whose basic needs are prevalent do not tend to participate. In other words, we can relate the unwillingness of people to participation with the type of prevalent needs. In this case, this question can be raised out that if there is one by one relationship between degree of participation and hierarchy of needs of human. The main reason for such a question arises from the fact that if there is one by one relationship between the needs and degrees of participation, the implementation of a specific model of participation in an organization and generalizing it to all people may result in "under-participation" or upper-participation" phenomenon. Davis and Veniostrum (1989) believe that some people demand more participation than other people. However, the main problem is that in both cases of "under-participation" and upper-participation" people show low level of satisfaction compared with situations in which their participation is based on their needs. This relationship is shown in diagram 3. This model also verifies this point that there is significant relationship between the needs of human and degree, and their conformity creates the maximum satisfaction.

Figure 3. conformity of needs with participation (Rahnavard, 2006).

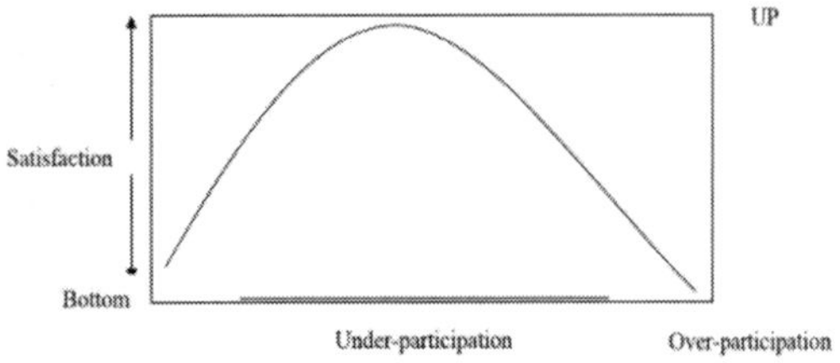

4.4 One by one relationship

In the discussion about management, situational leadership theory (Hersey and Blanchard, 1988) and compliance of needs with participation (Davis & Veniostrum, 1989, 14) it is implicitly verified that "one of the requirements for effective participatory management is the compliance between needs and degree of participation." In spite of the relationships that have been established between the needs and levels of participation in such normative models, the question is still that which degree of participation conforms with prevalent need in an individual and if field studies confirms such relationship? Based on such models, this hypothesis is proposed that people with prevalent basic needs respond better to low degrees of participation, and people with secondary needs respond better to over-participation. In other words, by establishing one by one relationship between prevalent need and specified degree of participation, we can increase the achievement of participatory programs provided that other needed factors to be provided to achieve such programs (Figure 4).

Figure 4. Relationship between prevalent need and degree of participation (Rahnavard, 2000).

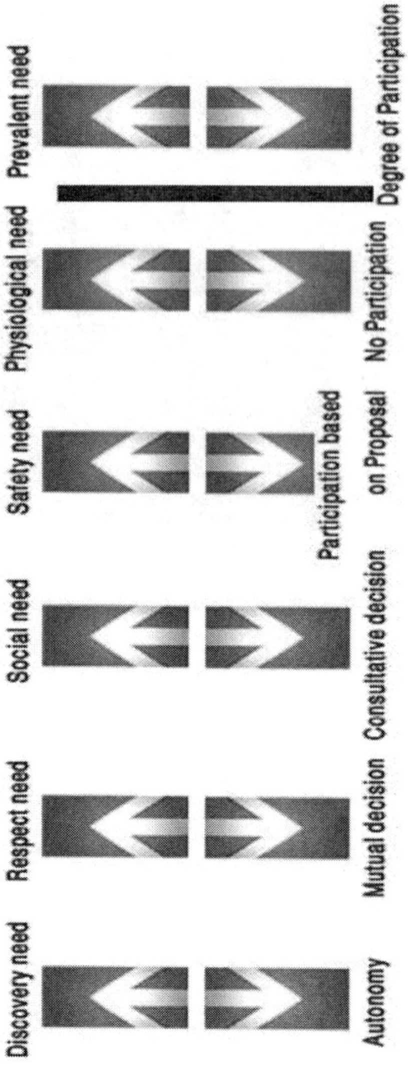

5 participation in citizenship regulations and urban designing

In contemporary world, various points of view are made in one of the highest citizenship level (urban designing and planning by people) along with obtained sophistications in cities that mostly tries to determine a boundary between participation and non-participation besides participation functional aspects and consider hierarchy to facilitate achievement to project studies by professional architecture designing.

5.1. Participation process from politics and culture to urban and citizenship field

Roots of participation meaning can be searched in old Greece cities-government. In present time, after new-born evolutions and passing middle centuries, simultaneous to the most important role of people in society, participation in political and social era by rewarding polling right and equal right for citizens and right of establishing civil groups and committee have been appeared. Now, citizens' participation in social and political affairs is accepted as a principle in all developed and many developing countries. It takes a long time to change meaning of participation from culture and politics field to planning field. Citizens' participation in related affairs to urban and urbanizing was first shaped at the end of 1950s and beginning of 1960 in US. In this period, various rules were designated based on citizens' participation that changes nature of urban decision-making "participation conception in the middle off 1960s was proposed in England, then other democratic countries have followed it (Oosthuizen 1998).

5.2. Urban participation of action with various dimension

As it was mentioned, participation as a human action needs certain conditions and facilities. There are 4 interests in human as action fields. In other words, "studies in urban management and citizens role indicates that people are influenced by their interests in urban management field and show different actions to get to their purposes (Mozayyani, 1996) (Table 3).

a. **Cognitive Interest:** it needs to have intellectual and emotional cognition from generalities and details- presentation in empirical sciences, philosophy and mysticism.
b. **Material Interest:** It provides interest to imagine environment by possibility to conquer adapted to environment.
c. **Communicational Interest**: It makes reciprocal friendly connection (to be one or we)- understanding intra-mental relationships- exchange
d. **Security Interest:** Life security-financial security-intellectual security- mass security. (Chelbi, 1996)

Four actions are defined as following based on the mentioned interests:

a. **Value action:** tools maybe revised, but purposes won't be revised because they are based on principles and values.
b. **Tool action:** it is done to calculate the relationship between tool and purpose. Social actor may change tools of arriving to purposes to facilitate his work and change mind based on conditions and facilities to get necessary tools in purpose.
c. **Traditional (mechanical) action:** this action is based on mere imitation and unconscious obey from continuous traditions.
d. **Emotional (sensational and non-intellectual) action:** "humanistic emotional and sensational life is not disciplinable and predictable and it is done without calculation of tool and purpose (Bushriyeh, 1995). Finally we can name four scopes appropriate to interests and mentioned actions: (social sub-systems)
e. Culture field: cognitive interests and value actions
f. Economics field: material interests and tool actions
g. Communication filed: communicative interests ad traditional action
h. politics field: security interests and emotional actions

Table 3. Humanistic actions, reference (Alavitabar2003).

Four general and main interests:	Four forms of actions	Four main scopes (construction)
Cognitive interest	Value action	Culture field
Material interest	Tool action	Economics field
Communicative interest	Traditional action	Communication field
Security interest	Emotional and sensational action	Politics field

According to the nature of each social sub-system, (political, economic, cultural, and communicative), it can be claimed that the main role to make intention to participation is cultural and communicative system duties. In other words, the nature of these 2 systems determines how much people are interested to participate in determining policies and their administrators act actively. But what makes possibility for this participation and determine a certain framework for it are 2 cultural and economic systems. Legal, institutional, facilities and economic abilities in participation field are shaped by being influenced of these 2 sub-systems, (Alavitabar, 2003).

6 Participation architecture designing and status of beneficiary and designer

6.1. Participation in designing and planning according to theorists' point of view

The importance and role of architectures and planners not only for professional responsibility in final decision-making and its role in development process of relationships among participators in a design, but also for the important nature of planning and architecture issue, has status that should claim architecture is obliged to find proper and acceptable response for conditions (environment, utilization, beneficiary etc.) among relatively contradicted economic, technical, social, and functional issues (Giancarlo, 2005). He shaped "manner, amount, pattern, comparison and various interpretations from various theories designing under various titles:" (Kamelnia, 2008).

In comparatives, besides explaining some of theorists' ideas, some contradictions become clear among them which are one and the most important unsolvable affairs.

John Turner believes "participation is not necessarily said to build a structure by inefficient-valid people with a few tools or negligible materials. A main issue is control and power of decision-making. (Hamedi, 1991)

Hueli (2004) used expression of "participatory architecture" as a designing system to develop users' wills.

"Habrakan" knows participation as a deniable issue, he know creativity as an individual action while designing is among people according to his idea. (Eslami, 1998)

Sanouf (2000) believes that "physical environment is designed by a few people but many people use it." Janson (1979) explored participation according to various points of view.

Participation is a tool to get necessary information (by addresses' point of view) and also a tool along with political changes, as though; the range of this effectiveness can lead to fully individual wills (even anarchistic points of view) or socialistic points of view.

- Participation as an attractor of others beliefs can be efficient for experts.
- Participation can play a fully nostalgic role, because when decision-making goes to its pervasive aspect, gets more emotional and nostalgic sense.
- Participation as a theoretical design includes points of view which believe experts in designing field should interrupt users and other people in design. In this regard, experts should help people to get to their real wills.
- Participation as a developer of mass behavior developer communicates more with participators.
- Participation can teach people by participatory process.
- Participation as an aesthetics theory believes to give fresh energy to architecture by interrupting people so their aesthetic selections and talents.

6.2. Participation in architecture design approach

"Salama" (1995) in book "new intentions in architecture training (based on Brad Bent idea), knows traditional approach of design include pragmatic design, Iconic design, and analogical design (Heth, 1984). He believes that in 2 recent decades, modern architecture approaches have been shaped including systematic design, pattern language, and participatory design. (Table 3) (Table 4) In 1970s, one of the most important issues in participation part in design is non-transparency of communicative mechanisms among specialized and non-specialized. Therefore, many of participatory designs are guided in one-way from experts and their ideas have been imposed in incomprehensible diagrams and plants for users. By starting design methods and processes in these years and also systemic attitude to architecture process, a form of communication was made among these ideas, computer programs and participatory issues were appeared.

"By more clarifying design process, computers have made vast field of communications participatory to help design process to make a building user to participate computer design process who wasn't involved in past. (Till, 2005) "Maybe definition of a design process and its various steps are not seemed very simple and there are many disagreements for its sophistication in this field, but design processes can be determined by theorists' points of view with scientific confrontation of this issue and exploring various samples. Design process patterns include 2 groups." (Mahmudi, 2005).Systematic pattern and environment recognition pattern has 2 public approaches in systematic pattern: Stage-phase approach that design process has been made from various steps coming following each other and rational approach that divides it to smaller parts to solve design problem. Environment knowing pattern is mostly related to human, environment, and balanced thought in architecture design.

Table 4. Models and approaches of design (Kamelnia, 2008, 94).

Designs models & Approaches			
Designs Model		Intuitive Model	
		Rational Model	
		Participatory Model	
Design Approaches	Traditional Approaches (Brad Bent)	Pragmatic Design	
		Iconic Design	
		Analogical Design	
		Canonic Design	
	New Approaches (Salama)	Systematic design	
		Pattern language	
		Participatory design	
Design process pattern (Mahmudi)		systematic	Stage-phase
			Rational
		Environment	Man-Env.
			Interactive

Table 5. Design approaches include systematic design, pattern language, and participatory design.

Deign approaches	**Fields**	**Fields sub-groups**	**Participatory approach title**	**Approach sub-sets**
Systematic design	Scopes	Sub directions of scope	Title of P.A	Sub directions approach

Pattern language				
Participatory design	Political sciences	Social guidance Social learning (Friedman) Transactional planning Social learning	Advocacy planning	Social planning Participatory planning
	Social organization	Durkheim theory Weber theory	Social architecture	Social planning
	Environment psychological studies	Perceptions psychology (Bruwick Gibson)	Community architecture	Participatory architecture Social architecture

6.3. Empirical modelling

Empirical modeling (resulting from library and field research and experiences): The phenomenon of participation in housing design can bring reasonable achievement, if we identify its effective and basic factors and organize its process under a coherent structure (system). The mentioned system has components that, according to nature of the subject, involve all of social affairs (human and cultural) and economics and politics affairs, and on the other hand, we cannot ignore, the role of the religious ideas of the designer and also family and individual characteristics of beneficiary, in this process, in a way that illustration and compilation of above systematic method is beyond the ability of architectural designer and this highlights the role of other experts as well.

This process, in macro scale, can be considered as a unilateral movement, from prerequisites of participation (its contexts) to the participation itself, and the feedback (a movement which is complementary for it or opposite to it) will be the impact of the final product on the process and its impact on the beginning.

Contexts are the initiators of participation and their objective is true participation and true participation is a tool to the ultimate goal that is a worthy housing design, which can be described as a product and at the end by formation of this process, due to its useful mentality and feedbacks, promotion and enhancement of cooperative culture will emerge and its effects on this process will be doubled. With this attitude, the contexts in the role of the independent variable will serve the dependent variable, which is true participation, but as the final objective is housing design, by changing the role of true participation (which itself is now a tool in the service of the ultimate objective) to the mediator variable, worthy housing will be considered as the dependent variable. Diagram 5 indicates that these series can continue due to the different objectives of different researches and by defining new objectives, roles of variables can change as well.

Figure 5. Factors contributing in housing design process and the role of variables.

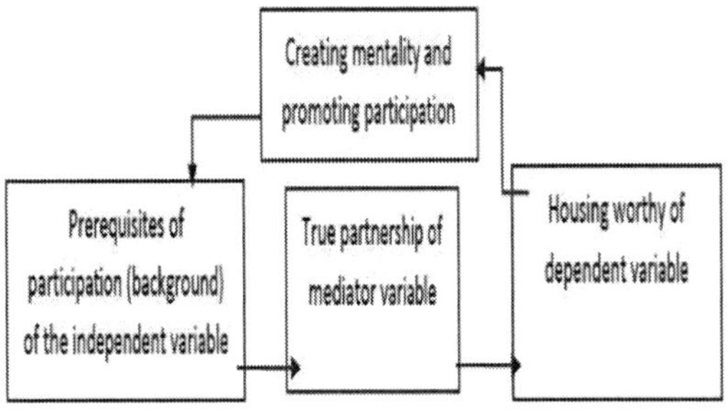

Based on field studies and professional experiences and exchanging opinions with experts, all of effective factors in the above diagrams can be modeled –as Model (1)-, in an appropriate and inclusive process, which in addition to showing the hierarchy between these factors, considers all the conditions involved in this process and considers their impacts as well.

Figure 6. Model (1): experimental model of Cooperative design process.

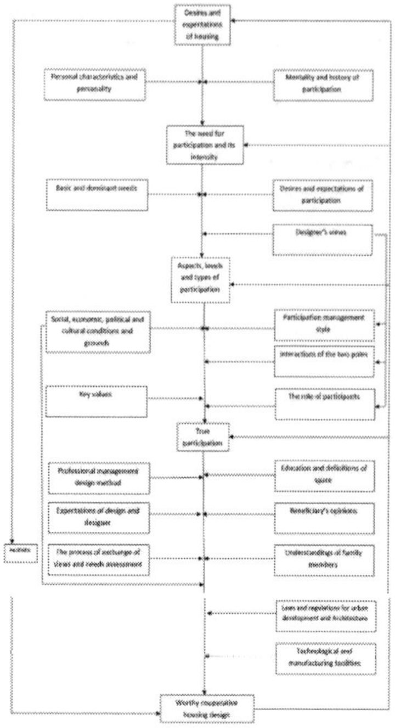

7 Extraction of variables (back ground for closed question-answer)

Variables are extracted after recognition the goals and evaluation of them that is the base for closed (pack) questions. These variables with notice to goals are:

Variables	
a. Familiarity of people with partnership b. Familiarity of people with legal right c. Familiarity of beneficiary with design capabilities - Familiarity of beneficiary with contemporary facilities in design and construction - Familiarity of beneficiary with the possibility of presenting and visualization of design	Awareness
d. The method of accepting new knowledge by the beneficiary e. The tendency of the beneficiary to involve others f. The method of requesting by beneficiary g. Effects of psychological tendencies in the design results	The mechanism of participation
h. The desire and need of the beneficiary in relation to the designer's attitude i. Barriers to participation j. Beliefs (quiddity) - Values (what should be done) - Norms (what is done) k. Mood and ethnic features of beneficiary and the effects on participation - Mood and ethnic features of beneficiary and the effects on the design	Factors Affecting Participation

8 Research method

In achieving the goals of the research, the main research methods, field and instrumental research methods are used to success of the program. Main methods consist of: Survey Method, Comparative Causal Research Method (Post-Events), Instrumental and Field Research Method, and Performance and Results of Instrumental Methods Used in Library and Field Studies.

"Members of a real or imagined set from people, events or objects, which we tend to extend findings of our research to them, is called statistical population" (Gall et al, 1996, p. 19).

In this research, the study population consists of individuals, in Shiraz city, who have referred to the architectural design office to design their own housing and this is called private building.

In this study was used the multi-stage cluster sampling method, because there is the different stages in building and validating research tools. In this way such as clustering sampling the measuring unit is not single but is a group of individuals that is shaped naturally as a group. When we use clustering sampling the choosing of people is more acceptably possible and easier than choosing individuals from a society. In clustering sampling when was selected randomly, you can continue sampling in the area (Delavar, 1995, 23) as follow:

1. To meet the terms of beneficiaries based on the economic situation, funds, the price of land, the buyer's income, spending devoted to housing were base and beneficiaries were selected in two broad categories good and medium economic in the Shiraz city.
2. The internal balance of beneficiaries' population: it is shown 3 different range of sampling groups (single unit villa, Two-family condominium, one unit's ownership and other profit) and statistics of Engineering Organization of Fars and referred to the office during the last 10 years of research are cause

that all data are place within 6 standard deviation (SD)(table6).

Table 6. Division of statistical social base on economic, type of housing and number of statistical sampling: author

A1 group	Medium economic of single unit villa	15
A2 group	Medium economic of two-family condominium	30
A3 group	Medium economic of one unit's ownership and other profit	30
B1 group	Good economic of single unit villa	30
B2 group	good economic of two-family condominium	25
B3 group	good economic of one unit's ownership and other profit	25

1. For estimation of sample volume is used:

$$n = \frac{s^2 \times z^2}{d^2}$$

S^2 Variance research tool 729 in ideal condition

Z Significant level in 95% that the result of it is 1/96.

d The difference between socity from sample, 4 from maximum score.

The resulting fraction (sample volume) n= 175

In this study there were 10 exam drops.

So the total sample volume in this stage is 165 person (because of enough and complete answer).

The important note is that these number of exam that is used for factor analysis calculation that is proven the adequacy of sample size after KMO test and Bartlett test.

According to there was not the standard questionnaire to determine the status of beneficiary participation in the design of housing, is used from the questionnaire to collect the required information. The questionnaire was developed in several stages and was calculated it's the validity and reliability. For a valid test, you should try that questions of test is a perfect example of the purpose and related content of study. Therefore, by referring to libraries, scientific centers and content analysis and literature of study and subsequent interviews with experts and traditional designers can proper and produce a tool for answer-questionnaire with pack question.

With notice to document library and evidence environmental the criteria are (Table 7):

1. People's understanding of the impact of participation in improving the quality of housing and incentives to generate interest
2. The necessity of the participation and results of the work outcome
3. The role of the architect and his administration and value and price that other parties to pay attention them.
4. Owner satisfaction of participation
5. Participation feedback of social and cultural (efficiency and sustainable development)
6. Suitable admission for participation
7. The beneficiary behavior from designer when participating
8. The effect of situation and enough time for participation

Table 7. Objective-content: This table measures the scope and the limits of each of the variables and objectives in proportion to content (categories) and determines their limits.

Variables		Content / Objective	Awareness	Participation mechanism	Factors affecting participation
- Awareness of people to participation - Awareness of people to their legal right	General	a	x		
		b	x		
- Familiarity of beneficiary with design capabilities - Familiarity of beneficiary with contemporary facilities in design and construction - Familiarity of beneficiary with the possibility of presenting and visualization of design	Professional	c	x x x		
- The method of accepting new awareness by the beneficiary		d		x	
- The tendency of the beneficiary to involve others		e		x	
- The method of requesting by beneficiary		f		x	
- Effects of psychological tendencies in the design result		g		x	
- The desire and need of the beneficiary in relation to the attitudes towards the design		h		x	
- Barriers to participation		i			x
- Beliefs (quiddity) - Values (what should be done) - Norms (what is done)		j			x x x
- Mood and ethnic features of beneficiary and the effects on the participation - Mood and ethnic features of beneficiary and the effects on the design		k			x x

To check the questions of this study were used descriptive statistics, Pearson correlation coefficient, analysis of variance and multiple regressions analyzes. In descriptive statistics were used frequency tables and histograms. In addition, the mean-variance, standard deviation was calculated for each variable.

9 Conclusion

Understanding the goal and investigating them, some variable are extracted from them that is basis for closed questions. These variables are as follows: Public awareness to participate; b. public awareness of legal rights; c. Beneficiary awareness about designer capabilities; d. Beneficiary awareness of contemporary possibilities to design and production e. beneficiary Awareness of the features and design visualization; f. How Beneficiary admit new Awareness; g. willingness of beneficiary to participate with other h. how to say desires; i. psychological actions and effect on result of design; j. The desire and needs of beneficiary in relation to the kind of designer behavior; k. Barriers of participation; l: Beliefs (nature) - values (what to do) - norms (what is done) m. behavior and ethnic characteristics of the beneficiary and effects on the participation.

First hypothesis: "Consultation between beneficiary and his family members about hosing design cause in knowing and understanding to housing space."

Second hypothesis: "Interaction of all family members in housing design, solved the Problems and abnormalities among family in term of knowing and expectation, and will be effective in knowing the areas and privacy of house."

Third hypothesis: participation of Operator Company and beneficiary in forming Residential space, is a factor of creating real space Beauty (Timely and efficient), adapting with space needs, and will prevent ineffective space and inconsistent with desires, needs and culture of users.

Forth hypothesis: Designer with a participatory approach in the design, not only can creates appropriate strategy to understand the real needs and necessary demands of employers, but also has enough skills to changing this desires to body.

Fifth hypothesis: "Designer with acquired experience in different interactions can learn useful strategy of architecture skills for changing and adjusting inappropriate expectations and desires of beneficiary, we use of regression analysis".

Sixth hypothesis: Designer can improve the quality of projects by acquiring experience of analyzing and comparing different beneficiaries in terms of time and space.

Seventh hypothesis: "Collaboration of population in the participatory design, leading to the promotion of culture and interactive climate and tolerance of dissenting views".

References

1. Alavi Tabar, A. R. (2003). A Model of Citizen's Participation in Urban Governance (2nd Ed.). National Organization of Municipalities, Tehran.
2. Carlo, G. (2005). Architecture's public, in Blundel P., Petrescu D., Till J.,Architecture and Participation, Spon Press, London:3-22.
3. Eslami, G. M. (2001). The Process of Internal Production and Development. Fine Arts Journal, No. 10.
4. Ghaffari, Gh. R. (2000)" explain the affecting social and cultural factors for the organized socio-economic involvement", Thesis of Social Sciences University, Tehran, Iran.
5. Habibi, M. & Saidi Rezvani, H. (2005). Participatory Urban Development, Theoretical Exploration of Situation in Iran, Fine Arts Journal, No. 24.
6. Hersey, P &Blonckiard, K.(1988) Management of organizational Behaviour: utilizing Human Resources, Engle wood cliffs, newjersey.
7. Ismaili A. (1999). Presenting Demands of People in the Process of Reconstruction and Planning of City of Bam. Islamic Revolution Housing Foundation, Tehran.
8. Ismaili, A.(2005) " the presentation of the people's demands in the reconstruction and planning process of the Bam city", Islamic Revolution Housing Foundation, Tehran.
9. Kamel Nia, H. (2008). Architecture and Patterns of Collective Design, Doctoral Dissertation in Architecture, Faculty of Fine Arts, University of Tehran.
10. Mac Logan, P. & Christonel, (1997). The Participation Age (2nd Ed.).(Mostafa Eslamieh, Trans.). The Cultural Research Bureau, Tehran.
11. Mack Legan, P. & Krystunel. (1997)," Partnership Age" (Second Edition) (translated by Mustafa Islamiya 1999), the Cultural Research Bureau, Tehran.
12. Mahmoudi A. S. (2005). Educational Papers. Publications of Negahe Emrooz, Tehran.
13. Mahmoudi, A. S. (2005) "The set of the education essays", publishing by Nashr-e-Negah-e-Emroz, Tehran.
14. Maslow. (1367). Motivation And Personality, (Ahmad Rezavani, Trans.). Cultural Deputy of Astan Ghods, Mashhad.
15. Rahnavard, F. (1385), "point of view's participate", Institute of education and researching Management & Planning, Tehran.
16. Rahnavard, F. (2006). Participation Approach. Higher Institute of Education and Research. Management and Planning, Tehran.
17. Razavi, Z.. (2005) "initiative of the people in decision-making, construction and

reconstruction of Bam & Baravat cities", the Interior Ministry, Tehran.
18. Rezaei, M.A. AS. (2002) Housing in Islam, Mehr Amir, Qom.se
19. Salama,A.(1995) Environmental Evaluation, jurnal of Architecturval Reseavch. AL AZHAR UNIVERSITY.
20. Salama A& O'Reilly M.W& Noschis K.(2005)Architectural Education Today:Cross-Cultural Perspectives, Comportements and authors.
21. Sanoff, H. (2000) Community Participation Methods in Design and Planning. New York, John Wiley & Sons, Inc.
22. Till, J.(2005) The negation of hope, in Blundel P. Petrescu D.
23. Tosi, M. A. (1991)," participation in management and ownership" (Fourth Edition), The training centre of Administration Mamegment, Tehran.

Key Practice for Green Building Management In Malaysia

Nasim Aghili[1], Abdul Hakim Bin Mohammed[2], Low Sheau-Ting[3]

[1, 2]*Department of Real Estate, Faculty of Geoinformation and Real Estate,*
[3]*Universiti Teknologi Malaysia, 81310 Skudai, Johor, Malaysia*
Centre of Real Estate Studies, Institute for Smart Infrastructure and Innovative Construction, Universiti Teknologi Malaysia, 81310 Skudai, Johor, Malaysia

Abstarct. This paper identifies the management practices for green building management in Malaysia. A set of practices was ascertained from the reviewed of various established Green Building Standard in the world. Green building practices are significant role in attaining sustainability particularly in the construction industry. Green building is constructed for minimizing impacts to environment as well as decreasing building effects on occupants. To achieve the goals of green buildings, set of management practices is required. A comprehensive review on the world green building standards that include the criteria of "management" has been carried out. The results of the content analysis have identified a total of five clusters of practices that found important for effective management of green building in Malaysia. These including Sustainable Procurement, Sustainable Operation, Resource Management, Repair and Maintenance Management, and Environmental Health. Each cluster of management practice composed of few activities. The identification of the set of practices contributes to the effective and efficient operation of green building in advancing the sustainability agenda, hence decreasing building's operating costs and realising increase in return of investment.

Keywords: Green Building, Management, Practice

1 Introduction

Green buildings appeared throughout the 19th and 20th centuries (Cassidy, 2003; Wu & Sui Pheng, 2010). There are several terms in green buildings. Green building practices are to protecting environment and reducing pollution (Alex Lukachko & Joseph Lstiburek, 2008). Green building practice as a procedure to generate buildings and infrastructure to order protect resources, decrease negative impact on the environment, and create to a higher standard environments for inhabitant (Chatterjee, 2009; Samer, 2013). Green building design might be a desirable topic than management, but the key to a successful green building is management. Green building management is as a win-win-win alternative for building management, proprietor, and residents. Using green principles to building management provides saving money during time, a special and appropriate business master plan, also make better lifestyle situation for its residents (Goodman, 2008). Management of green building plays great and considerable role in decreasing greenhouse gas emissions, energy efficiency, saving the cost of energy and the preservation and environmental sustainability. In this respect, green building management practices are comprehensive set of methods or techniques to protecting the environment and to ameliorate or reduce building's negative impact on the environment. Undoubtedly, identify and recognize key practices management are essential to achieving effective management of the green buildings.

2 Green building management practices

According World Green Building Council (World GBC), an appearing non-governmental institution at the international grade is the World GBC, established by David Gottfried in 1999. The World GBC is a network of national green building councils is largest international institution in almost one hundred countries which it significant impact on the green building marketplace. Members of green building council including Australia, Canada, India, Japan, Korea, Spain, and the U.S. World GBC with one hundred thousand buildings and approximately one billion square meters of green building space recorded which significant impact is in social and environmental global network. In early 2007, national council leaders recognized the immediate requirement found a secretariat for the World GBC that accountable to the increasing industry in green buildings and an international view. The secretariat was officially founded in Toronto, Canada. According to World GBC, member of World green building council fall into one of three categories based on how developed their organizations are: "Established", "Emerging" or "Prospective".

These are independent organizations that have developed and operate according to the best practices of the international GBC model. Green buildings to enhance the conventional construction process to the sustainable buildings throughout world nations. Green rating tools was introduced by many countries to assess performance of green building such as UK was the earliest country that started the green building rating system in 1990 known as Building Research Establishment Environmental Assessment Method (BREEAM), followed by United States in 1998, with Leadership in Energy and Environmental Design (LEED), and thereafter Japan Comprehensive Assessment System for Building Environmental Efficiency (CASBEE) in 2001, Australia (Green Star) in 2003, Singapore (Green Mark) in 2005, Hong Kong Comprehensive Environmental Performance Assessment Scheme (CEPAS) in 2009, and Malaysia (Green Building Index) in 2009 (Zainol, Woon, Ramli, & Mohammad, 2013). Eventually, there are various GB standard adopted by various countries in the world. According to green building rating system (2009) every green building standards arrange brief frame for recognizing and performing function, as well as evaluating green buildings practice. This study reviews various practices contained in the "management" criteria included in various green building standards in the world, including BREEAM, Green Globe, HQE, Green Star, Green Star NZ, Green Star SA and BEAM Plus.

Based on the review on various GB standards in the world, this study proposed five key practices to manage green building effectively: (1) Sustainable Procurement; (2) Sustainable Operation; (3) Resources Management; (4) Repair and Maintenance management and (5) Environmental Health.

2.1 Key practice 1: Sustainable procurement

"Sustainable procurement" is a safe procedure for purchasing goods, service and investment project to minimize effect on society, economy and environment (Brammer & Walker, 2011; Wilkinson & Kirkup, 2009). Sustainable procurement (SP) is one of the criterions included in the BREEAM and Green globes standard. Sustainable procurement plays significant role in the economic contribution in UK construction industry. SP as a broad concept first emerged following the Rio Earth Summit in 1992. SP is procurement which is in accordance with the principles of sustainable development to maintain a healthy society, protecting environment and encourage pleasant governance (Walker & Brammer, 2007). The aim of SP practice decrease the detrimental environmental, social and economic impact make a purchase goods and services during their life-cycle (Brammer & Walker, 2011).

There are two themes in sustainable procurement activities which are as follows: policy & compliance activities and development & collaboration activities. Table 1 shows different factors to evaluate procurement sustainability such as developing procurement policies, a written procurement policy stating a commitment to purchasing sustainable goods & services, developing targets for sustainable procurement, ISO 14001 certification, and integrating sustainability into procurement process (Meehan & Bryde, 2011).

2.2 Key practice 2: sustainable operations

Sustainable Operation (SO) practice is about collection of skills which is its arranged business procedure to attaining sustainability (Gimenez, Sierra, & Rodon, 2012). As well as, sustainable operation recognized in three steps such the planning, coordination and control cause to saving money while protecting natural resources, environment and decrease the carbon footprint (Gunasekaran, Irani, & Papadopoulos, 2013). Sustainable Operation practices include life cycle cost and service life planning in BREEAM and Adoption of sustainable development principles in operation in Green Star, Green Star NZ and Green Star SA.

In this context, there are various activities of sustainable operations in managing a green building: evaluating building investment cost, evaluating building energy cost, assessing building maintenance cost, taking into account building alteration cost, determining building acquisition cost, taking into account building salvage value, examining building environmental cost, assessing building interest rate and assessing building life-cycle (Kayaş, 2009).

2.3 Key practice 3: resources management

Resources management is the usage of all the resources natural and artificial for management of green buildings. It encourages the use of resources in a manner that it is less polluting. Also sensible usage of our resources will automatically ensure that buildings are sustainable for the present and future usage (Kakkar, 2014). Also, resources management is essential practice because it leads to reducing overall operating costs, improving productivity and profitability, enhancing the competitiveness of the business. The benefits of resources management often extend well beyond just energy, water and waste (Victoria, 2007). In this study, resources management practice contains energy management (EM), water management (WM) and waste management (WM).

EM is not a one-time but ongoing process. In the words, EM is the continuous process of managing behavioral, organizational and technical change to improve energy performance (Sinopoli, 2008). The EM is the first steps to establishing an ongoing program of resource management to improve energy efficiency of the building itself (Mokhtar Azizi, Fassman, Wilkinson, & Che Ani, 2012; Victoria, 2007). EM is a set of energy efficiency activities, process and management action to achieving cost effective energy and decrease Co_2 emissions (Ates & Durakbasa, 2012; Coad & Dimick, 2012). Simply, EM is a continuous process that is leads to declining energy cost in the shortest time possible. According to Natural Resources Canada (2015), EM practices can be grouped into eight equally important categories, which each practice categories has its specific action. See table 3.

Water management (WM), includes design and operating practices that are applicable to water supply and distribution systems. WM includes the installation of high efficiency plumbing fixtures and control technologies in the building for example: toilets, urinals, tap, showerheads, and cooling towers, reuse systems, a regular programme of leak inspections at the building, the use of treated and recycled water, catch rainwater and grey water, and the use of relevant water saving control systems (Smallwood, Snoxall, Highmore, & Sauntson, 2010). Table 4 shows the measurement items of water management in green building management.

Waste management (WM) describes gathering, carriage, processing and recycling or disposal of waste produce through human activities which negative impact on the health of the environment (Awosusi, 2010). WM activities contain storage of waste, reuse, recycling and improvement, treatment of waste, and disposal of waste (schalkwyk, 2009).

2.4 Key practice [4]: repair and maintenance management

Repair & Maintenance (R&M) practice is a process that describes how buildings and structures during its lifetime will face problems caused by environmental factors and vulnerabilities (Thaheem & Marco, 2014). Concisely, repair and maintenance play a crucial role in the life of building namely hindering risks of buildings, using proper materials and appropriate tools to reduce life-cycle cost (Horner, El-Haram, & Munns, 1997). Repair and maintenance management practices are included as one of the practice in HQE (France) green building standard. Maintenance management of building is the skill that it protects since the beginning of lifetime building to the end. In the other words, this step is long period in building industry. In fact, adequate and correct maintenance leads to decreasing negative effects on the environment, occupants and finally improve quality of life residents. Simply, maintenance management is as an appearance of "steady state" activity in buildings (Oliveira, Lopes, & Figueiredo, 2014). Table 5 shows the measurement items of repair and maintenance management in green building management.

2.5 Key practice [5]: environmental health

Environmental Health (EH) is about perspectives human health which contains the quality of human life by taking into account environmental factors, including physical, biological, social, and psychosocial. EH including several practices such as evaluating, correcting, controlling, and preventing that can potentially decrease negative impact on human health over the centuries (Links, 2006). EH contained two items, which are emergency response, plan health & safety. The practice of emergency response plan (ERP) is included as part of the criteria in Green Globe Canada while health & safety and environmental (HSE) performance is included in BREEAM UK and BEAM Plus in Hong Kong. Buildings encounter with natural or unnatural unforeseen and disaster incidents such earthquake, flood, hurricanes, building system failures, elevators, emergency power systems, medical emergency and fire (Dorge & Jones, 2000). According to Green Globe rating system (2004), ERP used to reduce the risk of harm and the environmental effect of emergency occurrence.

3 Discussion

This study aimed to identify the set of practices contributes to the effective and efficient operation of green building in advancing the sustainability agenda, hence decreasing building's operating costs and realising increase in return of investment. Meanwhile, the term sustainability combines social, environmental and economic obligations (Gimenez, et al., 2012). In this way, function of each of these approaches is as follows:

As mentioned before, sustainable procurement includes society, economy and environmental dimensions. Sustainable procurement leads to have cost effective activates, decreasing carbon emissions, reducing waste, lowering energy, and improving health outcomes. These are considered as some practical examples to be considered within sustainable procurement.

The purpose of recognizing sustainable operation in this paper is to evaluate energy and other resources which are used in buildings. In fact, sustainable operation is one of the factors which lead to reduce negative effect on buildings. On the other hand, personnel health protection and safety of society can be covered by sustainable operation. Generally, sustainable operation provides new opportunities in building industry to achieve sustainability.

As previously mentioned, resources management contains three lines such as energy management, water management and waste management. The main objective of energy management is to attain and conserve optimum energy resources which can minimize energy costs and mitigating environmental effects namely, decreasing costs, reducing carbon emissions and dropping risk of energy price fluctuations and supply shortages. As a matter of fact, energy management is widely authenticated as the foremost solution for direct and immediate decrease of energy consumption. In addition, water management can offer many benefits achieving sustainability in buildings. Some of these benefits contain: reduced water and sewage costs, reduced energy usage and waste water treatment energy usage which upgrades and optimizes energy usage.

Finally, the main role of repair and maintenance practice recognized to reduce equipment maintenance costs, save time in managing maintenance and repair contracts and guarantee the success of end user maintenance procedures. As a summary, the goal of management practices is to reduce cost and decrease negative impacts on the environment and achieve sustainability in building industry.

4 Conclusion

In essence, review on "management" criteria in the world green building standards was carried out in present study. Current study identifies management practices that are new perspective for management of green building in Malaysia. These five key practices of GB management are: (1) sustainable procurement; (2) sustainable operation; (3) resources management; (4) repair and maintenance management and (5) environmental health.

Acknowledgement

The authors gratefully acknowledge the research grant and financial support provided by the Ministry of Higher Education (MOHE) and University Teknologi Malaysia (UTM), under the FRGS research grant number R.J130000.7809.4F506.

References

1. Alex Lukachko, & Joseph Lstiburek. (2008). Towards Sustainability: Green Building, Sustainability Objectives, and Building America Whole House Systems Research. [Building America Report - 0801].
2. Ates, S. A., & Durakbasa, N. M. (2012). Evaluation of corporate energy management practices of energy intensive industries in Turkey. *Energy, 45*(1), 81-91.
3. Awosusi, A. O. (2010). Assessment of environmental problems and methods of waste management in Ado-Ekiti, Nigeria. *African Research Review, 4*(3).
4. Brammer, S., & Walker, H. (2011). Sustainable procurement in the public sector: an international comparative study. *International Journal of Operations & Production Management, 31*(4), 452-476.
5. Cassidy, R. (2003). White paper on sustainability. *Building Design and Construction, 10.*
6. Chatterjee, A. K. (2009). Sustainable construction and green buildings on the foundation of building ecology. *The Indian Concrete Journal, 1*, 27-30.
7. Coad, L., & Dimick, S. (2012). Energy Management in Commercial Buildings: The Value of Best Practices. [Report].
8. Dorge, V., & Jones, S. L. (2000). *Building an emergency plan: a guide for museums and other cultural institutions*: Getty Publications.
9. Gimenez, C., Sierra, V., & Rodon, J. (2012). Sustainable operations: Their impact on the triple bottom line. *International Journal of Production Economics, 140*(1), 149-159.
10. Goodman, F. (2008). Integrating Green Practice into your Building Management Strategy.
11. Gunasekaran, A., Irani, Z., & Papadopoulos, T. (2013). Modelling and analysis of sustainable operations management: certain investigations for research and applications. *Journal of the Operational Research Society, 65*(6), 806-823.
12. Horner, R., El-Haram, M., & Munns, A. (1997). Building maintenance strategy: a new management approach. *Journal of Quality in Maintenance Engineering, 3*(4), 273-280.
13. Kakkar, A. A. (2014). Resource Management for Green Buildings. *International Journal of Civil Engineering Research, 5*, 443-446.
14. Kayaş, E. (2009). Life cycle cost awareness among architects: The case of Turkey.
15. Links, J. M. (2006). Introduction to Environmental Health.
16. Meehan, J., & Bryde, D. (2011). Sustainable procurement practice. *Business Strategy and the Environment, 20*(2), 94-106.
17. Mokhtar Azizi, N. S., Fassman, E., Wilkinson, S., & Che Ani, A. I. (2012). Management Practice to Achieve Energy Efficiency Performance: Green versus Conventional Office Building in Malaysia. *Journal of Legal Affairs and Dispute Resolution in Engineering and Construction, 5*(4), 205-214.
18. Oliveira, A. M., Lopes, I. d. S., & Figueiredo, D. (2014). *Maintenance management practices of*

companies of the industrial pole of Manaus. Paper presented at the World Congress on Engineering and Computer Science 2014.
19. Samer, M. (2013). Towards the implementation of the Green Building concept in agricultural buildings: a literature review. *Agricultural Engineering International: CIGR Journal, 15*(2), 25-46.
20. Schalkwyk, M. V. (2009). WASTE Management Activities In Respect Of Which A WasteManagement Licence Is Required In Accordance With Section
21. 20(B) Of The National Enviro~Ntal Management: Waste Act,
22. 2008 (Act No. 59 Of 2008)
23. Sinopoli, J. (2008). How Do Smart Buildings Make a Building Green? *Energy Engineering, 105*(6), 17-22.
24. Smallwood, J., Snoxall, J., Highmore, S., & Sauntson, D. (2010). Green Building Management Toolkit
25. Translating words into action.
26. Thaheem, M. J., & Marco, A. D. (2014). Sustainable Repair & Maintenance of Buildings in the Developing Countries: A Risk Management Perspective and Proposal of Customized Framework. *Civil Engineering and Architecture Research, 1*, 14-23.
27. Victoria, S. (2007). Energy Management in Practice Manual.
28. Walker, H., & Brammer, S. (2007). Sustainable procurement in the United Kingdom public sector.
29. Wilkinson, A., & Kirkup, B. (2009). Measurement of sustainable procurement. *East Midlands Development Agency, Nottingham.*
30. Wu, P. L., & Sui Pheng. (2010). Project management and green buildings: lessons from the rating systems. *Journal of Professional Issues in Engineering Education and Practice, 136*(2), 64-70.
31. Zainol, N. N., Woon, N. B., Ramli, N. A., & Mohammad, I. S. (2013). *Barriers of Implementing Green Building Maintenance: A Preliminary Survey.*

Operations and Maintenance Cost for Stratified Buildings: A Critical Review

Nor Zaimah Che-Ghani [1] Nik Elyna Myeda[1,a] and Azlan Shah Ali [1]

[1] *Department of Building Surveying, Faculty of Built Environment, University of Malaya, Kuala Lumpur*

Abstract. Building maintenance is essential in preserving buildings' appearance and performance. It needs to upkeep the building performance to prolong its value and building life cycle. Malaysia is still lacking in managing cost for building operation and maintenance. It has been found that the cost for housing maintenance is high due to poor maintenance practices. In order to get better understanding on how to manage the cost, this study reviews the contributing factors that affecting operation and maintenance cost of stratified buildings in Malaysia. The research first identified the factors through extensive literature review and scrutinize on factors that affecting and can minimize operation and maintenance cost. This literature review offers insight into building maintenance scenario in Malaysia focusing on the issues and challenges. The study also finds that operation and maintenance cost for housing in Malaysia is still in poor state. Interestingly, this paper revealed that operation and maintenance cost is also influenced by three significant factors like expectation of tenants, building characteristics and building defects. Measures to reduce the housing operation and maintenance cost are also highlighted so that this study can be a stepping stone towards proposing efficient and effective facilities management strategies for affordable housing in future.

1 Introduction

Buildings have been considered as one of the most valuables assets of a nation to provide people with shelter and facilities for work and leisure. As time passes by, maintenance of buildings becomes an invaluable process in retaining the value and quality of a building [26]. In Britain, building maintenance activities have reached a level of 50 per cent of all annual construction activities [3, 27]. In Hong Kong, the gross value of general trades such as decoration, repair and maintenance, and construction work at minor work locations has increased over the past five years [11]. Buildings are relatively expensive and time-consuming to procure and usually have a service life in many decades rather than in years. Buildings, however, are not static over time; they change, evolve and often adapt in response both to external factors such as climate, exposure and, more significantly, internal factors such as use and maintenance [6].

BS 3811 defined maintenance as a combination of any actions in order toretain an item in, or restore it to an acceptable condition [28]. Building maintenance as defined by The Chartered Institute of Building is a work undertaken to keep, restore or improve every facility in every part of a building, its services and surrounds to an agreed standard, determined by the balance between the need and available resources. It requires all departments in an organisation to cooperate in ensuring that the assets of the organizations are planned, provided, maintained, operated and disposed off at the lowest total cost to the organization [28]. According to Horner *et al.* [9], the objectives of building maintenance are:

- To ensure that the buildings and their associated services are in a safe condition
- To ensure that the buildings are fit for use
- To ensure that the condition of the building meets all statutory requirements
- To carry out the maintenance work necessary to maintain the value of physical assets of the building stock, and
- To carry out the work necessary to maintain the quality of the building

The Prime Minister Department has given an instruction under the 'Pekeliling Am Bil 1 Tahun 2003, Arahan Penyelenggaraan Bangunan Kerajaan di Putrajaya dated February 11th 2003," building maintenance must be efficiently and properly executed but most of current building and facilities maintenance management in Malaysia is not considered clearly and systematically [16]. The Government has also given instructions to all their agencies to carry out maintenances of all their building in the very early stage so that to reduce the cost of maintenances and to avoid risk of higher maintenance cost due to negligent through time. But, very less attention has been given on building maintenance budget and it is very rare to see research carried out on the subject of maintenance expenditure, funding or financial budget of maintenance work [14]. In year 2006 repair and maintenance works constituted 16% of the total Malaysian construction output as reported by CIDB (2007). However there are still unreported maintenance works by house owners who carried out illegal renovation works. Thus, if

[a] Corresponding author: elyna@um.edu.my

this figure is taken into account, the actual value of repair and maintenance works in Malaysia would be higher [2].

2 Operations and Maintenance Management

Maintenance management or operations management are utilised to transform input including people, capital, energy, materials and technology into outputs (goods and services) [7]. Maintenance and operating cost is one of the major components of housing cost and contributes one third to one half of total cost depending on the type of housing such as condominium, apartment, flat and others [3, 29]. It has been reported that between 70% and 85% of the building maintenance and operation costs can be influenced during the design stage. It is a significant part of the total building life cycle costs and awareness towards planning building maintenance and operation cost has been increased recently [10, 30, 31].

More than 90% of the building life cycle requires active maintenance and the full service life of building cannot be attained without maintenance [17]. The easiest way to cut maintenance costs is to stop doing maintenance but the long-term results are usually very costly. Thus, the goal of the new approach is to carry out as little maintenance as possible as infrequently as possible while at the same time preserving the availability of the services facilities, the building elements and the whole building. In other words, maintenance should be carried out only when necessary to ensure the continued, safe and profitable use of the building at acceptable levels of satisfaction or when there is the possibility of extending the useful life of the elements of the building [9].

Maintenance strategy is adopted in order to extend the life cycle of buildings and its fitting services. Maintenance personnel choose different maintenance strategies depending on allocation maintenance resources [12]. In maintaining a building, there are usually several strategic options available to management, and many alternative decisions to be considered. There is, for example, the possibility of reducing the demand for maintenance by addressing the actual cause of failure and identifying its consequences. For instance, it may be necessary to decide whether to repair or replace an item, and whether to carry out periodic maintenance at fixed intervals or simply to respond to the requests of the users [9].

In order to get the efficient and lowest long term cost of ownership for equipment or a project, life cycle cost (LCC) analysis is a crucial tool to choose the most cost-effective approach from a series of alternatives. The alternatives might be selected from different companies or tenders [3]. LCC is the total relevant costs over the life of a system including cost of acquisition, operation, maintenance, modification, and disposal for the purpose of making decision [32]. LCC takes into account initial and other running costs (total cost of operation and maintenance of a system) over a life of the buildings. It can help in determining how the owner's money is distributed during the life period of the facility and determine the most economic way of meeting a need of space. The earlier it is applied, the greater the possibility of saving and lowering the committed cost. It is used to identifying high cost areas and evaluating changes that will reduce these costs [4, 33].

The life cycle approach attempts to estimate future maintenance and repair requirements by breaking down each facility into its systems and components (electrical, HVAC, roofing, etc.) and independently applying life-expectancy or life cycle concepts to each system and components [18]. Although a large number of life cycle cost models have been developed over the years, no single life cycle cost model has been accepted as a standard model. such standard model has not been established for several reasons, including the existence of many different cost data collection systems, and many different types of equipment, devices, or systems [10, 31].

Building performance is the ability of the building to support its design functions and the criteria of building performance include indoor climate, acoustics, lighting, flexibility, safety, comfort, security, convenience, indoor air quality, glare, accessibility, aesthetics, etc [17]. Buildings, services and facilities require maintenance to ensure optimal performance over their life cycle [34]. Though existing building simulation software can be used to study building performance, costs and energy consumption, yet they are limited to static building with no responsive elements. Previous studies have shown that digital simulation outcome at design stage do not contribute to enhance building performance after been occupied. This is due to inaccuracy of data from digital simulation experiments at early design stages [35].

2.1 Building maintenance scenario in Malaysia

Most current building and facilities maintenance management being practiced in Malaysia is not considered clearly and systematically, which results in over-budget costs for maintenance and remedial work [21]. As an example, 34 office buildings in Property Division of SOCSO around Malaysia has been dealing with the common maintenance problems and building defects on civil and structural elements like problems involving water proofing system, cracks, soil settlement and so on [23]. The failures found in the adoption and implementation of FM in Malaysia can be categorized into four factors as in Table 1 (FM failures on managing building features).

The government has played a major role in the development of FM in Malaysia since it started in the second half of 1990s. In 1996, the government has privatized the non-clinical support services in the government hospitals to three FM companies.

Table 1. FM failures on managing buiding features (Mustaffa *et al.*, 2008)

Failing factors	Description
Tactical failing	• Inadequate performance standard • Low maintenance of cleaning standards
Strategic environmenta l failings	• Lack of building performance monitoring data • Failure to anticipate the consequences of change • Failure to understand the non-linear nature of building complexity

	• Slow response of systems used
Tactical cultural failings:	• Responding slowly to complaints • Ignorance job stress • Failure to provide appropriate advise on design and planning based on overall performance
Strategic cultural failings:	• FM issues not prioritized • Over reliance on automated systems used • Flagging improvement due to lack in technical knowledge and background

It was indeed the biggest FM contract by the government at that time. It also marked the starting of the new era of FM in Malaysia. In February 2001, during the opening of the Kuala Lumpur 21 Convention and exposition, the then Deputy Prime Minister of Malaysia in referencing to the state of FM in Malaysia emphasized that [24]:

"Unless Malaysians change their mentality to become more aware of the need to provide good services and improve the upkeep of buildings, we will forever be a Third World country with First World Infrastructure"

The former Prime Minister of Malaysia, Datuk Seri Abdullah Ahmad Badawi says that it is very common to see that most government buildings were not regularly maintained because most faults cannot be distinguished at their early stage. On this basis, he has suggested that the Public Work Department needs to prepare building maintenance guidelines in order to identify any damage from the early stage [22].

3 Methodology

The paper systematically categorised the published literature as in Table 2 and then analyse and reviews it methodically. The list of factors are analysed through a theoretical part that discusses on operation and maintenance cost of stratified buildings. Table 3 lists down all the factors that affecting operation & maintenance costs by El-Haram & Horner [39], Ali *et al.* [3], Assaf *et al.* [40] and Al-Hammad *et al.* [1]. El-Haram & Horner [39] and Ali *et al.* [3] directly highlight on contributing factors affecting operation and maintenance costs. Assaf *et al.* [40] and Al-Hammad *et al.* [1] discusses on how faulty design effect on building maintenance which will later result in high maintenance cost.

4 Contributing Factors Affecting Operation and Maintenance Costs

Maintenance and operation costs are a part of the building's life cycle cost. Maintenance costs cover the cost of labour and material, as well as other related costs that are incurred to keep the building or its parts in the state in which it can perform its required functions. Maintenance implies the conduct of corrective, responsive and preventive maintenance activities on constructed assets, or on some parts of these assets. Operation costs include running costs and costs of managing the facility or built environment, including administrative support services (e.g. rent, rates, insurances, energy and other environmental/regulatory inspection cost, local taxes and charges) [10].

The study of the academic literature of research on contributing factors of building maintenance cost is shows in Table 4. The critical contributing factors that can be summarized from this literature are building characteristics, tenant factors, maintenance factors, political factors and others. Previous study by Ali *et al.* [3] was conducted based on El-Haram & Horner [39], the difference between these two studies are Ali *et al.*[3] analysed all critical factors include building characteristics. Meanwhile, El-Haram & Horner [39] only covers tenant factors, maintenance factors, political factors and others. But, both studies from Ali *et al.* [3] and El-Haram & Horner [39] found that expectation of tenants (tenant factor) is the most critical factors that contribute to operation and maintenance cost. Fakhrudin *et al.* [41] opined that the activities and attitudes of residents in strata living can have a significant impact on the satisfaction and enjoyment of others. Strata living are not the same as living in a freestanding house. The activities and attitudes of residents can have a significant impact on the satisfaction and enjoyment of others [15]. People are expecting and demanding more from their building nowadays and the residents are becoming less tolerant and require their buildings to be perfect, facilities that will be comfortable to occupy, cost-effective and efficient to run, and will remain added value assets [6].

Building characteristics like building materials, building services and building age were ranked as second, third and fourth respectively of the factors that affect building maintenance cost in Ali *et al.* [3] study. Buildings are very durable and if properly maintained may last for centuries. Even if maintenance is indifferent, a physical life of 50-60 years is quite realistic [36]. Maintenance is essential to maintain the value of the building [4, 17]. Table 5 shows the range of actual lifespan of the high-rise residential buildings vary considerably between countries, the highest being in Britain as reported by Shabha [19] and relationship between maintenance costs and the age of a building is collected by Mac-Barango and Kakulu [13]. It shows an increasing maintenance cost trend with age.

DOI: 10.1051/matecconf/20166600041

Table 2. Factors affecting operation and maintenance cost found in the literature

Authors number			**1**	**2**	**3**	**4**	**5**	**6**	**7**	**8**	**9**	**10**
Number of factors mentioned			19	19	7	4	1	1	5	2	5	6
Factors identified	Listed in	Mentioned	El-Haram & Horner (2002)	Ali *et.al* (2010)	Talib *et.al* (2014)	Au-Yong *et al* (2014)	Assaf *et al.* (1996)	Al-Hammad *et al* (1997)	Olubudun & Mole (1999)	51	Ali A.S. (2009)	Christian & Pandeya (1997)
Building age	2,3,7,8,9	5		X	X				X	X	X	
Building height and building area or size	2,8	2		X						X		
Type of structure	2	1		X								
Building services	2	1		X								
Building materials	2	1		X								
Expectation of tenants	1,2,9	3	X	X							X	
Vandalism factors	1,2,7	3	X	X					X			
Delay and failure in reporting problem	1,2	2	X	X								
Accessibility to the property	1,2	2	X	X								
Right to buy policy	1,2	2	X	X								
Poor workmanship	1,2	2	X	X								
Poor quality of spare parts and materials	1,2	2	X	X								
Poor maintenance management	1,2,3	3	X	X	X							
Budget constraints	1,2,3,9	4	X	X	X						X	
Failure to execute maintenance at the right time	1,2	2	X	X								
Poor budgetary control	1,2	2	X	X								
Energy cost in the case of income support tenants	1	1	X									
Failure to execute maintenance at the right time	1,2,3	3	X	X	X							
Poor or lack of training	1,2,3	3	X	X	X							
Accelerated maintenance work due to poor budgetary control	1	1	X									
Poor failure reporting procedures	1	1	X									
Interdepartmental boundaries	1,3	2	X		X							
Poor management decision system	1	1	X									
New health and safety regulations	1,2,9	3	X	X							X	
Building and Design defects	3,5,6,7	4			X		X	X	X			
Construction factor	7	1							X			
Changing standard factors	7	1							X			
Existing building condition	9	1									X	
Facility Location	10	1										X
Facility type and purpose	10	1										X
Facility size	10	1										X
Facility design	10	1										X
Quality of facility construction material	10	1										X
Price indices for utilities and services (inflationary effect)	10	1										X
Skilled manager	4	1				X						
Monitoring equipment and technique	4	1				X						
Acquisition data and information	4	1				X						
Frequency of monitoring and inspection	4	1				X						

[a] Corresponding author: elyna@um.edu.my

DOI: 10.1051/matecconf/20166600041

Table 3. Comparison of factors affecting building maintenance cost

El-Haram & Horner (2002)	Ali *et al.* (2010)	Assaf *et al.* (1996)		Al-Hammad *et al.* (1997)
1. Tenant factors *1.1 High expectation of tenants* *1.2 Improper use of the property* *1.3 Right to buy policy* *1.4 Inability to gain access to the property* *1.5 Vandalism by the tenants* *1.6 Tenant complains through different channels (e.g. through his/her MP)* *1.7 Delay in reporting failures* *1.8 Complete failure to report problems* **2. Maintenance factors** *2.1 Budget constraints* *2.2 Energy cost in the case of income support tenants* *2.3 Failure to execute maintenance at the right time* *2.4 Poor workmanship* *2.5 Poor or lack of training* *2.6 Failure to apply opportunity maintenance* *2.7 Selection of sub-optimal maintenance strategy* *2.8 Accelerated maintenance work due to poor budgetary control* *2.9 Poor failure reporting procedures* *2.10 Poor quality of spare parts and materials* *2.11 Poor maintenance management* *2.12 Poor budgetary control (spending large sums of money towards the end of the year to avoid losing it)* *2.13 Interdepartmental boundaries* *2.14 Poor management decision system* **3. Political factors** *3.1 New health and safety regulations* **4. Other factors** *4.1 Third party vandalism*	**1. Building characteristics** *1.1 Building age* *1.2 Building height and building area or size* *1.3 Type of structure* *1.4 Building services* *1.5 Building materials* **2. Tenant factors** *2.1 Expectation of tenants* *2.2 Use of the property* *2.3 Vandalism by the tenants* *2.4 Delay and failure in reporting problem* *2.5 Accessibility to the property* **3. Maintenance factors** *3.1 Poor workmanship* *3.2 Poor quality of spare parts and materials* *3.3 Poor maintenance management* *3.4 Budget constraints* *3.5 Failure to execute maintenance at the right time* *3.6 Poor budgetary control* **4. Political factors** *4.1 New health and safety regulations* **5. Other factors** *5.1 Third party vandalism* *5.2 Poor or lack of training*	**1. Defects in Civil Design** *1.1 Inadequate provisions for movement* *1.2 Ignoring aggressive environment and weather condition effects* *1.3 Ignoring biological effects* *1.4 Inadequate structural design such as foundation* *1.5 Ignoring variation in soil conditions* *1.6 Ignoring load impact on structure stability* *1.7 Exceeding allowable deflection* *1.8 Ignoring wind effects on the structure* *1.9 Inadequate concrete cover on reinforcement* *1.10 Improperly locating conduits and pipe openings at critical structural locations* **2. Architectural defects in design** *2.1 Narrow stairs, passages and doors* *2.2 Not relating exterior material selection to climatic conditions* *2.3 Specifying finishing that needs to be repaired as a whole (such as wallpaper)* *2.4 Not considering the local climatic condition when designing the exterior shape* *2.5 Inadequate joints between finished faces* **3. Design defects in maintenance practicality and adequacy** *3.1 Not considering space or exit equipment for maintenance workers* *3.2 Designing for permanent fixing that should be removable for maintenance* *3.3 Not considering the available maintenance equipment when performing the design* *3.4 Not considering the maintenance* **4. Defects due to consultant firm administration and staff** *4.1 Lack of QA/QC program during design* *4.2 Poor technical updating or staff training 4.3 Hiring unqualified designers* *4.4 Designer field experience* *4.5 Designer technical background* *4.6 Designer ignorance of materials properties* *4.7 Misjudgment of climatic conditions* *4.8 Misjudgment of user's intended use* **5. Defects due to construction drawings** *5.1 Lack of reference* *5.2 Conflicting details*	**6. Defects due to construction inspection** *6.1 Lack of inspection* *6.2 Unqualified inspector* *6.3 Proponent (owner) negligence of the importance of inspection* *6.4 Weakness of inspection rule in implementing corrective actions during job execution* **7. Defects due to civil construction** *7.1 Inaccurate measurement* *7.2 Damaged Formwork* *7.3 Excavation tools close to the building* *7.4 Painting in unsuitable conditions or on unsuitable surface* *7.5 Inadequate water proofing and drainage* *7.6 Insufficient reinforcement concrete cover* *7.7 Cold joints* *7.8 Loss in adhesion between materials* *7.9 Early formwork removal* *7.10 Poor soil compaction* *7.11 Inadequate curing* *7.12 Lack of communication* **8. Defects due to contractor administration** *8.1 Not complying with specification* *8.2 Unable to read the drawings* *8.3 Insufficient site supervision* *8.4 Poor communication with the design firm and the owner* *8.5 Unqualified supervision* *8.6 Speedy completion or poor quality work* *8.7 Unqualified work force* *8.8 Multinational construction experience* **9. Defects due to construction materials** *9.1 Differential thermal movements in dissimilar materials* *9.2 Selection of material that is unsuitable for existing climatic conditions* *9.3 Use of nondurable material* *9.4 Use of expired material* *9.5 Poor materials handling storage* **10. Defects due to construction equipment** *10.1 Improper use of equipment* *10.2 Inadequate performance of equipment* *10.3 Lack of required items of equipment* **11. Defects due to specification** *11.1 Unclear specification* *11.2 Not defining adequate materials* *11.3 Not specifying the QA/QC construction procedure* *11.4 Not specifying the allowable load limits*	**1. Defects in Civil Design** *1.1 Inadequate provisions for movement* *1.2 Ignoring aggressive environmental effects* *1.3 Ignoring biological effects* *1.4 Inadequate structural design* *1.5 Ignoring variation in soil conditions* *1.6 Ignoring load impact on structural stability* *1.7 Exceeding allowable deflection limits* *1.8 Ignoring wind effects on the structure* *1.9 Inadequate concrete cover on reinforcement* *1.10 Locating ductwork at structural locations* **2. Defects in architectural design** *2.1 Designing narrow stairs, passage and doors* *2.2 Specifying incompatible exterior finishes with the climate* *2.3 Specifying finishes which need complete replacement* *2.4 Ignoring climatic effects on exterior shapes* *2.5 Designing inadequate joints between finished surfaces* **3. Design defects in maintenance practicality & adequacy** *3.1 Ignoring access for maintenance equipment* *3.2 Designing permanent fixations* *3.3 Ignoring maintenance requirements in the design* **4. Defects due to consultant firm & staff** *4.1 Lack of QA/QC programme during design* *4.2 Insufficient technical updating or staff training* *4.3 Hiring unqualified designers* *4.4 Lack of designer field experience* *4.5 Lack of designer's technical background* *4.6 Designer ignorance of material performance* *4.7 Misjudgement of climatic conditions* *4.8 Misjudgement of user's intended use* **5. Defects due to construction specification** *5.1 Unclear specifications* *5.2 Inadequate definition of material type* *5.3 Overlooking QA/QC construction procedures* *5.4 Overlooking allowable load limits* *5.5 Specifying inadequate mix design* **6. Defects due to construction drawings** *6.1 Lack of references* *6.2 Conflicting details* *6.3 Lack of details*

		5.3 Lack of details	*11.5 Specifying inadequate concrete mix design*	

The charts in Figure 1 show progressive increases in maintenance cost according the category of age of the building. The less lifespan of buildings show lower maintenance costs.

Table 5. A comparison of the life cycle of high rise residential buildings in different countries

Countries	Buildings' life span
New Zealand	50 – 60
Australia	50 – 60
Britain	60
USA	35
Japan	20

Source: Shabha [19]

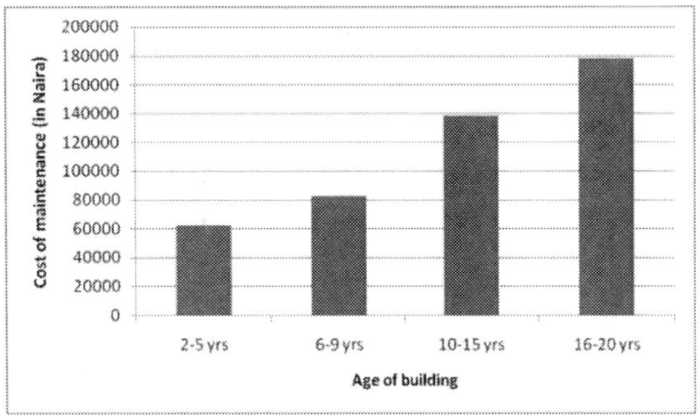

Figure 1. Variation maintenance costs based on building ages
Source: Mac-Barango and Kakulu [13]

Besides that, Al-Khatam [4] also agreed that behaviour of materials as one of the major causes of building deterioration and other unsatisfactory features of many buildings. Materials selection should meet the performance requirements otherwise it will cause failure of the materials which will result in replacement, correction or more maintenance works in the future.

Table 4. Collection of previous studies on building maintenance cost

Margin	El-Haram & Horner (2002)	Ali *et al.* (2010)	Talib *et al.* (2014)
Scope	Maintenance cost	Maintenance cost in Malaysia	Factor affecting maintenance and defects of public building
Aim	Describes the objectives of building maintenance and principal elements of housing maintenance cost	To determine and identity the factors contributing to rising maintenance costs.	This study focuses on the assessment of factors affecting maintenance and defects of public buildings in Penang.
Emphasis	Identify the extent to which 24 factors impede the optimum application of maintenance cost	The research was conducted based on the work undertaken by El-Haram & Horner [39].	The summary of ten relevant factors surveyed from related literatures were tabled and questioned to whom involved in the building maintenance sectors.
Focus	Factors affecting housing maintenance cost	Factors affecting housing maintenance cost in Malaysia	Factors affecting maintenance and defects of public buildings in Penang.
Methodology	Quantitative (Questionaire)	Quantitative (Questionaire)	Questionnaire SPSS
Sample	Local authority and housing associations throughout Scotland.	Housing maintenance management in Klang Valley, Malaysia	public buildings are demographically centralized in Penang
Framework	Factors – Measures	Factors – Impact/Influence - Measure	Factors –Defects
Indicators	Building characteristics Tenant factors Maintenance factors Political factors Other factors	Building characteristics Tenant factors Maintenance factors Political factors Other factors	Factors affecting maintenance of public building Factors affecting defects of public building
Findings	Rank 1 – high expectation of tenants (tenant factors) Rank 2 – budget constraints (maintenance factors – technical factors) Rank 3 – improper use of the property (tenant factors) Rank 4 – energy cost in the case of income support tenants (other factors)	Rank 1 – expectation of tenants (tenant factors) Rank 2 – Building materials (Building characteristics) Rank 3 – Building services (Building characteristics) Rank 4 – Building age (Building characteristics)	Factors affecting maintenance: -Lack of preventive maintenance -Insufficient funds to maintain the building -Lack of building maintenance standard -Non-availability of replacement part and components -Not responded to maintenance request

	Rank 5 – right to buy policy (political factors)	Rank 5 – Failure to execute maintenance at the right time (Maintenance factors) Impact: Rank 1 – outstanding maintenance charges Rank 2 – Over budget	Factors of building defect:: -Lack of building maintenance, -Overlooked site conditions, -Defective material, -Environment conditions and -Moisture from wet areas.
Output	Establish the relative importance of the factors which affect housing maintenance costs	Framework of maintenance cost contributing factors and impact on maintenance performance	Concluded that buildings allocated within the Equatorial region are exposed to extreme hot and wet climate and must have proper maintenance schedule and guidelines
Comments	The study neglect factor of building characteristics since it has already been studied. Tenants factor is uncontrollable (nature of tenants & environment).	Each factors was describe neatly. Impacts on maintenance performance was also analysed. Although the research was conducted base on El-Haram & Horner [39], the result is not similar. This study includes building characteristics and building characteristics do affect maintenance cost.	This study have been present state of non maintenance public building, some of which have been abandoned due to its state of defects. Authors recommend appropriate building maintenance strategy to be taken. In order to preserve this type of building(heritage building), it will take huge money.

Table 4. Collection of previous studies on building maintenance cost (continued)

Margin	Au-Yong *et al.*(2014)	Assaf *et al.* (1996)	Al-Hammad *et al.*(1997)
Scope	Condition- Based Maintenance: prediction cost maintenance model	Effects of faulty design and construction on building maintenance	Civil and architectural defects in building design that affect maintenance
Aim	This paper seeks to establish the relationships between the characteristics of condition-based maintenance and the cost performance.	This paper outlines individual and major groups of design and construction defects in large buildings in Saudi Arabia and their relative severity.	To identify design defects which affect building maintenance and their overall relative importance.
Emphasis	This study focuses on the issue related to the building users' dissatisfaction in high-rise office buildings.	These studies indicate the percentage contribution of defects to building maintenance.	It is based on the opinion of parties most concerned with building maintenance (contractors, designers & owners)
Focus	Relationship between Characteristics of Condition-Based Maintenance and Cost Performance	Faults in design that affect building maintenance	Faults in design that affect building maintenance
Methodology	Mix method (questionnaire, survey, interview, case study)	Literature search and Interview Questionnaire	Literature Mixed Method
Sample	Klang Valley, Malaysia Kuala Lumpur	Eastern Province of Saudi Arabia	(owners, A/Es and contractors) Eastern Province of Saudi Arabia
Framework	Characteristics - Cost Performance	Defects in design	Defects in design
Indicators	Skilled manager Monitoring equipment and technique Acquisition data and information Frequency of monitoring and inspection	Defects in Civil Design Architectural defects in design Design defects in maintenance practicality and adequacy Defects due to consultant firm administration and staff Defects due to construction drawings Defects due to construction inspection Defects due to civil construction Defects due to contractor administration	Defects in Civil Design Defects in architectural design Design defects in maintenance practicality & adequacy Defects due to consultant firm & staff Defects due to construction specification Defects due to construction drawings

		Defects due to construction materials Defects due to construction equipment Defects due to specification	
Findings	The findings revealed that the characteristics of condition-based maintenance directly influenced the cost performance	Defects in Civil Design: moderately severe Architectural defects in design: moderately severe Design defects in maintenance practicality and adequacy: owners-most severe; maintenance contractors-moderately severe Defects due to consultant firm administration and staff: moderately severe Defects due to construction drawings: owners-most severe; maintenance contractors-moderately severe Defects due to construction inspection: maintenance contractors- most severe; owners-moderately severe Defects due to civil construction; moderately severe Defects due to contractor administration: second highest effect on maintenance Defects due to construction materials: moderately severe Defects due to construction equipment: lowest rank Defects due to specification: owner-most severe; maintenance contractors: moderately severe	Owners - inadequate structural design,ignoring access for maintenance equipment, hiring unqualified designers andoverlooking allowable load limits. Contractors - inadequate structural design,specifying incompatible exterior finishes which need complete replacement andhiring unqualified designers. A/Es - hiring unqualified designers,inadequate structural design and insufficient technical updating or staff training.
Output	Skill and knowledge manager, availabilityof monitoring equipment and technique, capability to adopt monitoringtechnology, as well as reliability of maintenance data and informationwere highlighted as the characteristics that significantly influencingthe maintenance performance. Using the prediction model, practitioners can predict the varianceof maintenance expenditure from the level of concern towardsthe reliability of maintenance data. Thus, they may decide how muchof concern towards the reliability of maintenance data is required inorder to achieve optimal maintenance expenditure.	Both parties (owners and maintenance contractors) ranked "defects due to contractor administration" among the highest effect on maintenance. All defects in this group were ranked "most severe" by each with the exception of the lack of multinational construction experience and poor communication which were rated "moderately severe' by the maintenance contractors.	To improve building maintenance: -provide adequate structural design -Provide proper access to maintenance equipment -Provide proper exterior finishes compatible with climate conditions -hire qualified designers -provide sufficient detail in construction drawings.
Comments	This maintenance strategy is aimed to minimise the total maintenance cost by collecting and gathering the condition data of the building systems, especially those critical components.	This paper is similar to Al-Hammad *et al* [1] but it only reviewed based on owners and maintenance contractor.	Error and defects occur during design & construction stages will result later in high maintenance costs. Untreated defects is Rank 3 of impact on maintenance performance as in analysed by Ali *et al.* (2010).

Building services is a vital aspect in building maintenance management that affect housing maintenance costs. It is to provide a healthy and safe living environment for occupants or residents. The building services include mechanical and electrical services like ventilation, lighting and power supply, water supply, sanitation, transportation communication and other systems [37]. Owing to the high maintenance cost involved and the consequences of building services' failure, awareness on the significance of the building services maintenance has increased in the building services industry [3].

The quality of maintenance activities like labour productivity, material availability, materials waste, good and effective maintenance methods, using effective tools and equipment, and good maintenance planning is influenced by the amount of budget allocation in each task. Maintenance management should be aware of these factors in order to achieve better financial control [2, 4]. An earlier study by Holmes & Droop [42] stated that "maintenance is budget oriented rather than needs-oriented". This led Ashworth and Au-Yeung [43] conclude that maintenance will only be carried out when and where the needs for maintenance and adequate funding co-exist [38]. Due to budget and time constraint, many errors and defects have occurred during the design and construction stage which will result in high maintenance cost later [1]. Shabha [19] stated that cutting cost at the initial design and construction stages has led to a significant increase in the long term cost of replacing faulty components and running cost. This has made many high-rise buildings energy inefficient and their design has proven to be costly for occupiers. Previous studies by Al-Hammad *et al.* [1] improved on the research from Assaf *et al.* [40] where they discussed on building defects that affecting building maintenance and will result later in high maintenance costs. Error and defects occur during design & construction stages will result later in high maintenance costs As example state in Al-Hammad *et al.* [1], it is noted that contractors, architects/engineers and owners agree inadequate structural design the significant design defect. Inadequate structural design occurs when the designer under-designs the structural elements in the building. This can cause a building's settlement or cracking, or collapse which can increase maintenance cost. Inadequate structural design is a controllable factor that can be avoided and can minimize maintenance cost.

5 Conclusions

An extensive literature review was conducted and authors concluded that operation and maintenance costs are influenced by uncontrollable factor like expectation of tenants and building characteristics. It can be summarised that inadequate structural design is the significant design defect which will be affected by maintenance cost in future. Inadequate structural design (design defect) is a controllable factor that can be avoided to minimise the operation and maintenance cost. Uncontrollable factors like tenant factors might be difficult to control because of the nature of human behaviour. In summary, various measures to minimise building operation and maintenance costs are as follows:

- Tenants and residents should be educated on the community well-being awareness
- Buildings should be preserved in the early stage
- Employing minimum, but optimum labour and staff with acceptable qualification standards
- Maintain and repair on urgent and on time tasks before further implication or defect occurs
- Development of adequate maintenance management information education program.
- Introducing operation and maintenance operating manual
- For less critical maintenance; corrective maintenance technique is suggested
- Apply preventive maintenance technique with the scheduled work in preparation of breakdowns or failures.
- Reducing the number of maintenance tasks by selecting the most applicable and cost-effective maintenance strategies

Looking forward, this study also signifies its significant factor in exploring the cost variance and factor, not just limited to the context of stratified buildings. A theoretical framework will be developed to construct a better understanding of the theories for this study.

Acknowledgement

The authors gratefully acknowledge the financial support of the University of Malaya Research Grant (UMRG), grant no. RP015D-15SUS established at the University of Malaya, Sustainability Science Research Cluster.

References

1. A. Al-hammad, S. Assaf, & M. Al-shihah, The effect of faulty design on building maintenance. *Journal of Quality in Maintenance Engineering*, ***3*(1)**, 29–39. (1997)
2. A. S. Ali, Cost decision making in building maintenance practice in Malaysia. *Journal of Facilities Management*, ***7*(4)**, 298–306 (2009)
3. A. S. Ali, S. N. Kamaruzzaman, R. Sulaiman, Y. Cheong Peng, Factors affecting housing maintenance cost in Malaysia. *Journal of Facilities Management*, ***8*(4)**, 285–298 (2010)
4. J. A. Al-khatam, *Building Maintenance Cost*. King Fahd University of Petroleum & Minerals (2003).
5. C. P. Au-Yong, A. S. Ali, F. Ahmad, Prediction cost maintenance model of office building based on condition-based maintenance. *Eksploatacja I Niezawodnosc – Maintenance And Reliability*, ***16*(2)**, 319–324 (2014)
6. J. Douglas, Building performance and its relevance to facilities management. *Facilities*, ***14*(3/4)**, 23–32 (1996)
7. N.E. Myeda, S.N. Kamaruzzaman, M. Pitt, Measuring the performance of office buildings

maintenance management in Malaysia. *Journal of Facilities Management*, ***9*(3)**, 181–199 (2011)
8. I. H. Fakhrudin, M. Z. Suleiman, R. Talib, The need to implement Malaysia's Building and Common Property Act 2007 (Act 663) in building maintenance management. *Journal of Facilities Management*, ***9*(3)**, 170–180 (2011)
9. R. M. W. Horner, M. A. El-Haram, A. K. Munns, Building maintenance strategy: A new management approach. *Journal of Quality in Maintenance Engineering*, ***3*(4)**, 273–280 (1997)
10. H. Krstić, S. Marenjak, Analysis of buildings operation and maintenance costs. *Građevinar*, ***64*(4)**, 293–303 (2012)
11. E. W. M. Lam, A. P. C. Chan, D. W. M.Chan, Benchmarking success of building maintenance projects. *Facilities*, ***28*(5/6)**, 290 – 305 (2012)
12. H. H. Y. Lee, D. Scott, Overview of maintenance strategy, acceptable maintenance standard and resources from a building maintenance operation perspective. *Journal of Building Appraisal*, ***4*(4)**, 269–278 (2008)
13. D. O. Mac-Barango, I. I. Kakulu, Establishing a maintenance cost profile of residential buildings. In S. Laryea, R. Leiringer, & W. Hughes (Eds.), *Procs West Africa Built Environment Research (WABER) Conference* 413–425 (2011)
14. N. Mohd-Noor, M. Y. Hamid, A. A. Abdul-Ghani, S. N. Haron, Building Maintenance Budget Determination: An Exploration Study in the Malaysia Government Practice. *Procedia Engineering*, ***20***, 435–444 (2011)
15. New Straits Times, "Power back in hands of owners", Strata Living, New South Wales Government, New Straits Times Press, Kuala Lumpur, available at: www.fairtrading.nsm.gov.au (accessed 6 August) (2007)
16. M. F. Mukelas, E. M. Zawawi, S. N. Kamaruzzaman, Z. Ithnin, S. H. Zulkarnain, A Review of Critical Success Factors in Building Maintenance Management of Local Authority in Malaysia. *Symposium on Business, Engineering and Industrial Applications*, 653–657 (2012)
17. A. L. Olanrewaju, A. Idrus, M. F. Khamidi, Investigating building maintenance practices in Malaysia: a case study. *Structural Survey*, ***29*(5)**, 397–410 (2011)
18. G. R. Ottoman, W. B. Nixon, S. T. Lofgren, Budgeting for Facility Maintenance and Repair I: Methods and Models. *Journal of Management in Engineering*, ***15*(4)**, 71–83 (1999)
19. G. Shabha, A low-cost maintenance approach to high-rise flats. *Facilities*, ***21*(13/14)**, 315–322 (2003)
20. M. K. Sharaidin, F. Salim, Affordable, Performative and Responsive : Designing Affordable Responsive Architectural Prototypes through Physical and Digital Modelling. In *International Adaptive Architecture Conference* 1–14 (2011)
21. A. Zakaria and S. Hamzah, Pelan Kualiti Elak Projek Hadapi Masalah: Majlis Aset Kebangsaan mampu Tingkatkan Kecekapan Penyenggaraan. *Berita Harian, **August 15***, p.2 (2007)
22. E.M.A. Zawawi, S.N. Kamaruzzaman, A.S. Ali, R. Sulaiman, Assesment of Building Maintenance Managment in Malaysia: Resolving Using a Solution Diagram. *Journal of Retail & Leisure Property **9(4)***, 349-356 (2010)
23. A.S. Sulaiman, Some Common Maintenance Problems and Building Defects: Our Experiences, *Procedia Engineering: In 2nd International Conference on Rehabilitation and Maintenance in Civil Engineering **54*** 101-108 (2013)
24. S.N. Kamaruzzaman and E.M.A. Zawawi, "Development of facilities management in Malaysia", *Journal of Facilities Management*, **8 (1),** 75-8 (2010)
25. S.A.H. Mustaffa, H. Adnan, K. Jusoff, "Facilities management challenges and opportunity in the Malaysian property sector", *Journal of Sustainable Development*, **1(2)**, 79-85 (2008)
26. G. Vijverberg, "Renovation of offices in the Netherlands: reasons, points for attention and obstacles", *Proceedings of the CIB Working Commission 070, CABER, Glasgow Caledonian University, **September***, 728-37 (2002)
27. I.M. Shohet, "Key performance indicators for maintenance of hospital buildings",*Proceedings of the CIB Working Commission 070, CABER, Glasgow Caledonian University, **September***, 79-90 (2002)
28. R.M.R. Mazlan and M.A.H. Mohammed, Identifying Maintenance Issues in Malaysia. In *Association Of Schools Of Construction Of Southern Africa (ASOCSA), The Third Built Environment International Conference, At The Westin Grand Cape Town Arabella Quays, Cape Town* (p. 13) (2008)
29. C.P. Rydell, Factors affecting maintenance and operating costs in federal public housing projects. *Rand, **634***, (1970)
30. Y. Asiedu, and P. Gu, Product life cycle cost analysis: state of the art review. *International journal of production research*, ***36*(4)**, 883-908 (1998)
31. B. S. Dhillon, *Life cycle costing for engineers*. CRC Press (2009)
32. M. Shear, Building Maintenance Management. *Virginia, Reston Publishing Co.* (1983)
33. P. Wordsworth, *Lee's building maintenance management*. Wiley-Blackwell (2001)
34. A.O. Lateef, M.F. Khamidi and A. Idrus, Maintenance management of University Building, *Proceedings International Symposium on Construction in Developing Economies: Commonalities Among Diversities, Penang*, 5-7 October 2009, CIBW107, University Science Malaysia, 578-592 (2009)
35. M.K. Sharaidin, and F. Salim, Affordable , Performative and Responsive : Designing Affordable Responsive Architectural Prototypes through Physical and Digital Modelling. In *International Adaptive Architecture Conference*, 1–14 (2011)
36. B. Chanter, and P. Swallow, *Building maintenance management*. John Wiley & Sons (2008)
37. G.A. John, D.J. Clements-Croome, V. Fairey, & H.M. Loy, Contextual prerequisites for the

application of ILS principles to the building services industry. *Engineering, Construction and Architectural Management*, ***12*(4)**, 307-328 (2005)
38. F. Olubodun and T. Mole, Evaluation of defect influencing factors in public housing in the UK. *Structural Survey*, ***17*(3)**, 170-178 (1999)
39. M.A. El-Haram, and M.W. Horner. Factors affecting housing maintenance cost. *Journal of Quality in maintenance Engineering*, ***8*(2)**, 115-123 (2002).
40. S. Assaf, A.M. Al-Hammad, and M. Al-Shihah, M. Effects of faulty design and construction on building maintenance. *Journal of performance of constructed Facilities*, ***10*(4)**, 171-174 (1196)
41. I.H. Fakhrudin, M.Z. Suleiman, & R. Talib. The need to implement Malaysia's Building and Common Property Act 2007 (Act 663) in building maintenance management. *Journal of Facilities Management*, **9(3)**, 170–180 (2011)
42. R. Holmes, & C. Droop, "Factors affecting maintenance costs in LA housing", in Brandon, P.S. (Ed.), *Building Cost Techniques New Directions*, E. & F.N. Spon, London (1982).
43. A. Ashworth, & P.L.Y. Au-Yeung, P.L.Y., "The usefulness of maintenance cost records in life life cycle cost prediction", in Spedding, A. (Ed.), Building Maintenance Economics and Management, E. & F.N. Spon, London, 141-50 (1987)

A Review of BIM Capabilities for Quantity Surveying Practice

Noor Akmal Adillah Ismail[1, 3, a], Robin Drogemuller[1], Scott Beazley[2] and Robert Owen[1]

[1]*School of Civil Engineering & Built Environment, Queensland University of Technology, Brisbane, 4001 QLD, Australia*
[2]*School of Design, Queensland University of Technology, Brisbane, 4001 QLD, Australia*
[3]*Faculty of Architecture, Planning & Surveying, Universiti Teknologi MARA, 40450 Shah Alam, Selangor, Malaysia*

Abstract. It has been 20 years since the formation of buildingSMART International (then the International Alliance for Interoperability) and the commencement of the development of the Industry Foundation Classes (IFC). Initial uptake of the Building Information Modelling (BIM) concept and the IFCs was very slow, but it is now moving into the mainstream in developed countries. BIM is not as advanced in some countries such as Malaysia that are moving rapidly towards developed status. This paper will review the use of BIM by Quantity Surveyors in both countries that follow the "English" tradition and those, such as the US, which have developed other approaches. It will examine the capabilities of some of the BIM-based cost estimating software as the basis for understanding the use of BIM by Quantity Surveyors. This will include Quantity Surveyors who are using BIM to provide a traditional estimating service, as well as those firms that are applying the new technology to changing roles in cost estimating and cost planning. This will provide the basis for suggestions as to how Malaysian QS firms can exploit this technology.

1 Introduction

BIM technology has become increasingly well-known among construction industry players throughout the world. It is acknowledged as having contributed towards the success of projects due to its more efficient and effective working practices. This includes its contribution in the preparation of construction cost estimates, the most valuable quantity surveying practice in determining the project costs. BIM is claimed to offer a significant advantage in improving the estimating practice that is often prone to errors and inaccuracies [1]. BIM technology potentially provides a better quality of data by automation of information documentation [2-7]. More specifically, BIM has much to offer when it comes to improving cost estimating reliability, including early schedule information, quicker predictions of cost impact of design changes, better understanding through improved visualisation, as well as the possibility to access additional information for documentation [8]. 3D as-built models containing rich data are replacing traditional 2D methods; BIM creates information integration with a seamless flow of information from the initial phase to the final stage of the life cycle [9]. Information is integrated by the BIM models' controlling the information in a shared repository to establish consistent, accurate and accessible data [2].

Currently, construction industry practice widely employs BIM [10] and its implementation is increasing from year to year [11, 12]. USA, Finland, Denmark, Norway, Netherlands, UK, Australia, Hong Kong & Singapore are among the countries heading towards significant BIM uptake [13]. Malaysia, where the growth of BIM has been driven mainly by the private sector since 2009 [14], has made slow progress in developing BIM systems [13], the level of implementation falling on 0 and 1 of the BIM scale [15], indicating the need for a solution in those issues. Thereby, CIDB, as the professional body that has been given the mandate for managing BIM uptake in Malaysia, has urged BIM implementation amongst the construction players in their practice to produce more effective construction projects [16, 17]. Despite the efforts to encourage BIM usage amongst managerial and operational teams in the Malaysian Construction Industry, CIDB (2013) has reported that the slow level of transformation towards BIM could be due to the lack of standardised BIM processes and the absence of guidelines for its implementation. Hence, the CIDB has emphasized the BIM guidelines to be provided for the construction industry users for better communication and work collaboration [14]. Apart from the CIDB emphasising the provision of BIM guidelines for the industry users, Quek (2012) has specifically suggested the Royal Institution of Surveyors Malaysia (RISM) especially Quantity Surveying (QS) Division, to initiate appropriate actions in evaluating the influence of BIM.

Therefore, this paper aims to overview the impacts of BIM on the role of the Quantity Surveyors. The research method adopted for this paper is a comprehensive literature review of BIM capabilities pertaining to quantity surveying practice specifically in achieving more

[a] Corresponding author: noorakmaladillahbinti.ismail@hdr.qut.edu.au

reliable cost estimates. The first part discusses general BIM usage by Quantity Surveyors and the second part examines the BIM capabilities especially in the use of software in assisting Quantity Surveyors to achieve more dependable results in their cost estimating practice. From the literature review, the BIM capabilities were synthesised into three broad categories according to their importance in estimating cost, including data visualisation, reliable database and data coordination. The results were then hypothesised to affect the reliability of cost estimates in quantity surveying practice, leading to establishing a reliable framework for this paper. This paper also contributes to a holistic perspective on the effects of improved information resulting from BIM adoption through a better understanding and the generation of more knowledge in the production of more reliable cost estimates. This paper builds a framework which could provide the basis for suggestions as to how Malaysian Quantity Surveyors can effectively exploit BIM technology.

2 BIM usages by Quantity Surveyors

BIM usage in the construction industry throughout the world has been reported in international survey series such as SmartMarket Reports, National BIM Surveys (NBS), McGraw Hill and Masterspec. A survey by McGraw Hill Construction in the United Kingdom (UK) in 2010 recorded that the BIM adoption rate amongst the construction professionals was 35%, primarily by the architects (60%), engineers (39%) and contractors (23%) respectively. While a National BIM Survey conducted in New Zealand at the end of 2011, showed 34% of the respondents were aware and currently using BIM, 12% of them were neither aware nor using BIM and 54% were just aware of BIM [19]. Significantly, the same survey conducted the next year reported that the result of respondents being aware and currently using BIM had increased to 57%, also neither aware nor using BIM had reduced to 2% [20, 21].

There are numerous benefits BIM could offer by its implementation in the construction industry. It has been documented in the SmartMarket Report that 9 out of 10 BIM users in South Korea (92%) obtained advantages through using BIM and by having positive views towards BIM values they believe they achieved more from the continuous usage [22]. An online survey executed by McGraw Hill Construction in 2014 recorded that BIM benefits that have been divided into internal, project and process, and are inclusive of: reduced errors & omissions, collaboration, enhanced image, reduced rework, reduced cost & time, better cost control, marketing new business, offering new services, increased profits, maintaining repeat business, reduced cycle time of workflows, faster client approval cycles, improved safety, and faster regulatory approval cycles. A questionnaire survey carried out by Newton & Chileshe (2012) to 29 construction organisations in Australia identified that the BIM benefits that inspire those organisations the most are: improved constructability, improved visualisation, improved productivity, reduced clashes, improved quality & accuracy, improved client satisfaction, increased competitiveness, improved information sharing and enhanced sustainability. Thurairajah & Goucher (2013) through a questionnaire sent to several companies in the UK, summarised that the BIM drivers are: encompassing cost consultant's attitude, improved project collaboration, easier sharing & obtaining information, increased cost estimates accuracy, improved reliability in cost estimates, supporting early supply chain involvement, early schedule information, quicker predicted cost impact of design changes, improved visualisation and also clash detection.

Specifically, in 2011, the Royal Institution of Chartered Surveyors (RICS) conducted a survey among its members. The responses received from 156 Quantity Surveyors showed that 10% of them were regularly using BIM, in which the majority use was in the design and construction phases [25]. They mostly used BIM for construction scheduling, extracting quantities and facilities or asset management. The results also showed that 10% of the respondents were interested in using BIM by actively evaluating BIM tools for their practice. The barriers that were also highlighted in taking up BIM as their practice were the lack in client demand, training, application interfaces and standards. Whereas, Tse, et.al. (2009) evaluated the BIM usage in materials take-off that is generally done by the Quantity Surveyors, by carrying out a BIM case study of a Hong Kong project. It was observed that BIM-based projects provide the Quantity Surveyors substantial involvement especially in the project design process compared to the traditional 2D practice. Even though BIM can perform automatic and accurate quantity take-off it was proved that the knowledge of Quantity Surveyors is still important particularly in generating bills of quantities from the BIM models. Some building elements may not be modelled; hence cannot be measured automatically. Therefore, checking of model geometry is an essential role of Quantity Surveyors.

In New Zealand, Boon and Prigg (2012) found that BIM was only practiced for larger projects among the construction players, which ranked them on the level 1 or 2 of the BIM maturity scale. The interviewed Quantity Surveyors revealed that they had limited experience of extracting quantities from 3D models. They preferred to work in traditional 2D format. However, the Quantity Surveyors with BIM experience emphasised the benefits of additional information they could retrieve when deriving quantities, by rotating 3D models, which they could not do with 2D drawings. In Australia, results were received from a BIM adoption survey amongst quantity surveying firms conducted by Aibinu and Venkatesh (2013). From 40 responses obtained, 37% of the respondents have been using BIM for their projects. Most of them have little experience of using 3D models to extract quantities, in which they only used 3D models for visualisation predominantly at cost planning stage. Usually, for BIM-based projects, they were not issued with 3D BIM developed by designers; rather, they were still using 2D drawings for quantity take-off and cost estimation. The drivers that led the firms to initiate BIM included: ease of use, client requirement, quality information in models, case studies, industry-wide use, cost-benefit analysis, integration with current software and also scenario training.

It can be seen that BIM adoption is increasing internationally. Realising the potential benefits offered by BIM implementation, the development of BIM in most countries was complemented with policies, standards, guidelines and other approaches to improve BIM usage among the users and also to increase BIM awareness among non-users. Nevertheless, the BIM benefits sought should be explored comprehensively in exploring effective BIM projects. Theoretically, BIM can improve the reliability of cost estimates through numerous benefits offered by the technology that affect the many factors on which the estimates depend. Quantity Surveyors, as qualified estimators, should take the initiative on these issues. As such, it is crucial to identify BIM capabilities towards quantity surveying practice, before any estimate.

3 BIM capabilities for quantity surveying practice

A reliable cost estimate demands accurate building information to minimise estimation errors, especially in the early phase process [29]. As described by Monteiro and Martins (2013), using 2D measurements and 2D designs can lead to errors and omissions throughout the various construction stages, which lead to inaccurate estimation for budgeting. BIM implementation specifically plays a significant role in improving information in the early stages of construction projects. It permits better quality of data production by automation of information extraction. This allows BIM applications to contribute to improving the reliability of cost estimates including early schedule information, quick prediction of cost impact on design changes, better understanding through visualisation, and also access to information for documentation.

BIM tools provide more detailed and accurate cost estimates for construction projects, with the availability of geometric properties for building elements in producing spatial quantities, such as area and volume [30]. A pilot study conducted by Shen and Issa (2010) proved that a significant improvement was achieved through the application of BIM tools to increase the efficiency and accuracy of cost estimates, compared to the traditional estimating method, and this application was sufficient even by using only the function of 3D visualisation. Additionally, Choi, et al. (2015) outlines specific processes on how BIM is used in quantity-taking-off for schematic estimation in the early design stage, to improve cost estimates reliability. The process steps include verification of the physical quality of elements to increase the accuracy of calculated quantities, as well as the verification of input data for estimation. These processes address the issues of low reliability in cost estimates that are performed manually. The reliability of cost estimates can be improved through the implementation of BIM technology. Improved information, especially through its data visualisation, reliable database and data coordination, as the significant input in estimating construction costs, is among the valuable assets BIM can deliver to potentially improve quantity surveying practice; also, being able to compare models as they developed or changed. Table 1 below tabulates the BIM capabilities towards performing more reliable cost estimates in quantity surveying practice through the three categories of improved information including data visualisation, reliable database and data coordination.

Table 1. Categories of BIM capabilities towards QS practice.

Categories	BIM capabilities
Data visualisation	– 5D life cycle costing analysis [32] – BQ preparation with 3D views [33] – Project conceptualisation , design planning [34] – Simulation, cost control, value engineering, estimating, tendering [34] – Drawing interpretation, data extraction [35]
Reliable database	– IFC view for BQ [36] – Quantity take-off, cost estimating [37] – Estimating, tendering [34] – Post construction cost management [34] – Design changes identification [35]
Data coordination	– Virtual models, estimating [34] – Communication, access to information [35]

3.1 Data visualisation

Concern about effective project management leads not only to an emphasis on supplying adequate and extensive information about the project related, but also on equipping the project with visualisation tools to facilitate information delivery [38]. Dahl, et. al. (2001) refer to visualisation as visual mental images used to allow the generation, interpretation and manipulation of information through spatial representation. While visualisation tasks are divided into three groups, i.e. information retrieval, information analysis and information dissemination, the second part, information analysis, is the most valuable in obtaining insight from data [40]. Compared to understanding textual information, visualisation leads to far more effective analysis. In general, visualisation assists team members to better understand what will be constructed through 3D model views [41] in which the visualisation models not only represent building detailed properties, but also simulate their interaction and movement [42]. Digital models, which can be used for simulation and visualisation at any stage [43], enable construction-sequence visualisation [44] and allow the construction team to better understand what they intend to build [45].

Furneaux and Kivvits (2008) claim the visualisation process demonstrating project stages from the beginning to the end of construction projects especially help non-technical people to understand and identify issues. The respondents to the Rajendran, et. al. (2014) study agreed that the 2D drawings in traditional methods could not demonstrate clear visualisation, and this has led to numerous design problems in the construction industry. However, BIM improves this by providing clear building-visualisation settings for a better understanding of a building design. With rich 3D design-visualisation content in BIM modelling and analysis to help the team understand better [47], BIM is superior to 2D drafting as it allows the team to view a building and its contents from all angles, and at the same time to detect and amend any related issues at earlier stages so as to reduce costly changes [48]. Additionally, Leicht and Messner (1997) differentiate between BIM and traditional methods by describing multiple, dynamic and better 3D views of a building offered in BIM models, compared to only predefined 2D views in traditional documents.

3.2 Reliable database

A BIM model is created from a database that includes information such as manufacturing data, pricing, physical data (weight, size and material finish) and also data for many other elements incorporated in the building process [48]. Rather than producing a series of separated drawings, digital databases provide an archive of all physical and functional characteristics of output, in which all related information is formed into intelligent objects. BIM projects then adapt these intelligent objects to automatically symbolise the project elements in any plan, elevation, section or detail that can be adjusted parametrically as the design changes [50]. Giving more than ordinary drawings could offer, BIM merges building design, construction and maintenance properties in one beneficial model as a data repository that can be shared by all team members. Therefore, it furnishes the project with more efficient and accurate information as compared to traditional 2D CAD drawings [48]. Being database-driven, BIM software is a powerful tool that integrates engineering and architectural design elements [51]. Not only does it act as an intelligent model represented by digital objects, but objects can also be counted and sequenced in order to identify relationships. This process then establishes a greater understanding of these relationships, and possibly identifies relevant pre-existing information behind the objects, which is not available in textual information. Architectural elements used in the BIM model are stored in libraries where the complete building model and all of its representations are incorporated in a single BIM database. Consequently, any changes made to the BIM model will influence all related drawings generated from the database. Thus, it enables users to not only generate but also automatically update project documents, with all building information data attached to the building elements [41, 44].

Nagalingam et al. (2013) point out that BIM technology can assist quantity surveyors with their practice through information about the whole building represented in BIM, along with the complete related design documents, which have been stored in an integrated database. This integrated environment allows all parametric information to be interconnected, thereby acknowledging any changes affecting the assembly and construction processes. This makes available exact quantities and specific materials in an electronic format, for design documents for the quantity surveyors to produce their taking-offs and cost estimates. A robust database is established through the parametric elements in which the information gathered from the database raises efficiency standards, hence enhancing communication between the design and construction teams [48]. Furthermore, the BIM database is based on its proprietary classification system and can be compatible with the Industry Foundation Classes (IFC) format [30]. The IFC format, which captures both the geometry and properties of intelligent building objects associated with its relationships within BIM models, allows for information-sharing across a team [48]. The IFC file in which data is arranged based on its classes and properties supports data exchange among BIM application platforms such as Autodesk Revit in exporting and importing the required construction information data files [52].

3.3 Data coordination

Boshoff (2014) defines BIM as the generation of coordinated 3D models that integrate the representation of 3D objects through synchronised 2D drawings. In comparison with the paper-based drawings that are difficult to integrate and coordinate, BIM models have become a solution to this problem, in their ability to effectively communicate vital information between the people in the construction field [50]. This is done by the user developing 3D components in the software and combining them in a single model which shows multiple views of the building model. Where multiple views are possible, this allows any notification throughout the reflected views if any changes are made to the model. Additionally, the information about plans, sections and elevations that has been developed from the same data has been coordinated, and this improves the coordination of information exchange [53], hence translating into better coordination among the design, engineering and construction disciplines [54]. This coordinated information delivery further improves the coordination of construction documentation, consequently reducing errors and reworks [49-54].

All combined models of a project design and construction that have been developed using BIM are located in the BIM database. These models can be viewed simultaneously in order to spot any clashes between architectural, structural, mechanical, electrical and plumbing systems [50]. The systems can be coordinated where all equipment, fixtures, pipes, ducts, conduits, structural members and other building components are analysed with "clash detection" tools to check any conflicts before the installation of the system [61]. This would undoubtedly benefit designers and other team members in monitoring which area each is developing to ensure that they have sufficient information, as well as diminishing unwanted conflicts and rework [49]. The BIM model has details on each building component which can be retrieved from its modelled elements, thereby allowing the team members to quickly access the information. With all information being made available for everyone in the team, those responsible are alerted to any design changes and their consequences, which then improves coordination among their team members [48].

4 Basis for framework

BIM offers numerous potential benefits including improved information received to be manipulated for design and construction utilisation throughout the project stages. It has been discussed above how data visualisation, reliable database and data coordination, as essential features of BIM technology, could improve information and hence improve the reliability of cost estimates. According to Leicht & Messner (1997), the components that distinguish BIM and traditional documentation are a visualisation of spatial information, data availability for further analysis, and the existence of information. BIM applications are capable to automatically generate visual information such as floor plans, elevations, 2D and 3D sections, detailed sections, text-based information, model analysis and simulation results. Exchange information is executed via proprietary format within the database without losing data, and the interaction between applications is often ensured through IFC [30]. Its biggest advantage is, more accurate quantities generated from digital model, rapidly and reliably. The IFC that allows information to be shared across the team [48] next permits data coordination within the same shared resources.

This study demonstrates that there are significant relationships between improved information as a benefit to BIM adoption to improve the reliability of cost estimates. The dimensions of improved information being offered by BIM technology are data visualisation, reliable database and data coordination, and these constructs contribute to the reliability of cost estimates required in every construction cost estimating practice. Therefore, a framework as shown in Figure 1 is primarily developed with two primary relationships established. First, the improved information through data visualisation, reliable database and data coordination are achieved by BIM adoption. Secondly, BIM implementation with its benefits of improving project information in a construction project could produce more reliable cost estimates.

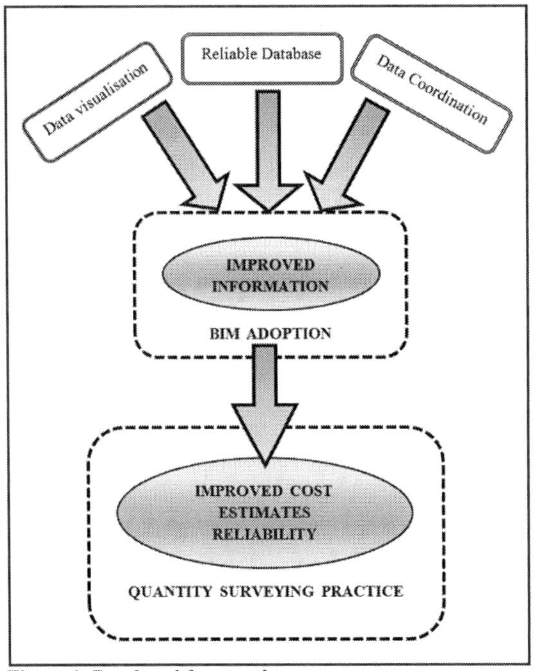

Figure 1. Developed framework

5 Conclusions

This study seeks to establish a reliable framework for linking cost estimating practice and BIM implementation. A survey is currently (January 2015) providing evidence which, although incomplete, indicates that that improved information with its dimensions of data visualisation, reliable database and data coordination would potentially be achieved through BIM adoption. By implementing BIM in cost estimation, it would possibly improve input information and enhance estimator's knowledge and understanding which ultimately leading them to produce more reliable cost estimates. This study also contributes in adding collective knowledge in the existing literature regarding BIM implementation and the reliability of cost estimates. It demonstrates the elements of BIM improved information to potentially improve the quantity surveying practice. From a practical perspective, the framework can serve as a guideline to increase the understanding of construction industry stakeholders, especially Quantity Surveyors. It demonstrates how they can implement BIM through its identified benefits by improving quantification data of the project information input they receive for estimating costs. It also guides the Quantity Surveyors as estimators in knowing BIM benefits that can maximise their knowledge and understanding of the project operation and process to assist them in establishing more reliable cost estimates.

References

[1] O. A. Olatunji and W. D. Sher, "A Comparative Analysis of 2D Computer-Aided Estimating (CAE) and BIM Estimating Procedures," in *Handbook of Research on Building Information Modeling and Construction Informatics: Concepts and Technologies*, 2010, pp. 170–189.

[2] CRC Construction Innovation, "Adopting BIM for facilities management: Solutions for managing the Sydney Opera House," 2007.

[3] C. Furneaux and R. Kivvits, "BIM – Implications for Government," 2008.

[4] D. Bryde, M. Broquetas, and J. M. Volm, "The project benefits of Building Information Modelling (BIM)," *Int. J. Proj. Manag.*, vol. 31, pp. 971–980, 2013.

[5] S. Kalinichuk and A. Tomek, "Construction Industry Products Diversification by Implementation of BIM," *Int. J. Eng. Technol. Innov.*, vol. **3**, no. 4, pp. 251–258, 2013.

[6] B. Abbasnejad and H. I. Moud, "BIM and Basic Challenges Associated with its Definitions , Interpretations and Expectations," *Int. J. Eng. Res. Appl.*, vol. **3**, no. 2, pp. 287–294, 2013.

[7] G. Nagalingam, H. S. Jayasena, and K. A. T. O. Ranadewa, "Building Information Modelling and Future Quantity Surveyor's Practice in Sri Lankan Construction Industry," in *The Second World Construction Symposium 2013: Socio-Economic Sustainability in Construction*, 2013, pp. 81–92.

[8] N. Thurairajah and D. Goucher, "Advantages and Challenges of Using BIM : a Cost Consultant's Perspective," in *49th ASC Annual International Conference Proceedings*, 2013.

[9] J. D. Goedert and P. Meadati, "Integrating Construction Process Documentation into Building Information Modeling," *J. Constr. Eng. Manag.*, vol. **134**, pp. 509–516, 2008.

[10] G. Kim, H. Park, and J. Shin, "An Assessment of the Accuracy of Cost Estimation Using Building Information Modeling in Design Process," *Appl. Mech. Mater.*, vol. **291–294**, pp. 2822–2825, 2013.

[11] RICS, "International BIM implementation guide," 2014.

[12] McGraw Hill Construction, "SmartMarket Report: The Business Value of BIM for Construction in Major Global Markets," 2014.

[13] J. K. Quek, "Strategies and Frameworks for Adopting Building Information Modelling (BIM) for Quantity Surveyors," *Appl. Mech. Mater.*, vol. **174–177**, pp. 3404–3419, 2012.

[14] CIDB, "Issues and challenges in implementating BIM for SME's in the construction industry," 2013.

[15] Z. Zahrizan, N. M. Ali, A. T. Haron, A. Marshall-ponting, and Z. Abd, "Exploring the Adoption of Building Information Modelling (BIM) in the Malaysian Construction Industry: A Qualitative Approach," *IJRET Int. J. Res. Eng. Technol.*, vol. **2**, no. 8, pp. 384–395, 2013.

[16] CIDB, "BIM in Malaysia," 2013. [Online]. Available: www.bimcenter.com.my.

[17] Bernama, "Adopt BIM technology, CIDB urges industry players," *Daily Express*, 23-Sep-2014.

[18] McGraw Hill Construction, "SmartMarket Report: The Business Value of BIM in Europe," 2010.

[19] Masterspec, "New Zealand National BIM Survey 2012," 2012.

[20] Masterspec, "New Zealand National BIM Survey 2013," 2013.

[21] NBS, "NBS International BIM Report 2013," 2013.

[22] McGraw Hill Construction, "SmartMarket Report: The Business Value of BIM in South Korea," 2012.

[23] McGraw Hill Construction, "SmartMarket Report: The Business Value of BIM for Owners," 2014.

[24] K. Newton and N. Chileshe, "Awareness, Usage and Benefits of Building Information Modelling (BIM) Adoption – The Case of The South Australian Construction Organisations," in *Procs 28th Annual ARCOM Conference*, 2012, pp. 3–12.

[25] BCIS, "RICS 2011 Building Information Modelling Survey Report," 2011.

[26] K. T. Tse, A. K. Wong, and F. K. Wong, "Building information modelling in material take-off in a Hong Kong project," in *Collaborative Construction Information Management*, 1st ed., G. Q. Shen, P. Brandon, and A. N. Baldwin, Eds. New York: Spon Press, 2009, pp. 186–197.

[27] J. Boon and C. Prigg, "Evolution of Quantity Surveying Practice in the Use of BIM – the New Zealand Experience," in *Proceedings of the CIB International Conference on Management and Innovation for a Sustainable Built Environment.*, 2012, pp. 84–98.

[28] A. Aibinu and S. Venkatesh, "Status of BIM Adoption and the BIM Experience of Cost Consultants in Australia," *J. Prof. Issues Eng. Educ. Pract.*, 2013.

[29] J. Choi, H. Kim, and I. Kim, "Open BIM-based quantity take-off system for schematic estimation of building frame in early design stage," *J. Comput. Des. Eng.*, vol. **2**, no. 1, pp. 16–25, 2015.

[30] A. Monteiro and J. P. Martins, "A survey on modeling guidelines for quantity takeoff-oriented BIM-based design," *Autom. Constr.*, vol. **35**, pp. 238–253, 2013.

[31] Z. Shen and R. R. A. Issa, "Quantitative evaluation of the BIM-assisted construction detailed cost estimates," *J. Inf. Technol. Constr.*, vol. **15**, pp. 234–257, 2010.

[32] F. P. Lai, D. Halvitigala, J. Boon, and R. Birchmore, "How can BIM Technology assist in optimising the Life Cycle Cost of a Building?," in *16th Pacific Rim Real Estate Society Conference*, 2010.

[33] A. Nadeem, A. K. D. Wong, and F. K. W. Wong, "Bill of Quantities with 3D Views Using Building Information Modeling," *Arab. J. Sci. Eng.*, vol. **40**, no. 9, pp. 2465–2477, 2015.
[34] O. A. Olatunji and W. Sher, "Perspectives on Estimating Practices Modelling," *Australas. J. Constr. Econ. Build.*, vol. **14**, no. 4, pp. 32–53, 2014.
[35] C. Harrison and D. Thurnell, "BIM Implementation in a New Zealand Consulting Quantity Surveying Practice," *Int. J. Constr. Supply Chain Manag.*, vol. **5**, no. 1, pp. 1–15, 2015.
[36] K. U. Gokce and H. U. Gokce, "IFC View for Bill of Quantities for Software Interoperability," in *Proceedings of the 30th CIB W78 International Conference*, 2013, pp. 827–836.
[37] E. Plebankiewicz, K. Zima, and M. Skibniewski, "Construction cost and time planning using BIM-based applications," in *Creative Construction Conference*, 2015, pp. 537–545.
[38] T.-H. Chuang, B.-C. Lee, and I.-C. Wu, "Applying Cloud Computing Technology To BIM Visualization and Manipulation," in *28th International Symposium on Automation and Robotics in Construction*, 2011, pp. 144–149.
[39] D. W. Dahl, A. Chattopadhyay, and G. J. Gorn, "The importance of visualisation in concept design," *Des. Stud.*, vol. **22**, no. 1, pp. 5–26, 2001.
[40] M. Chen and L. Floridi, "An analysis of information visualization," *Syntheses*, vol. **190**, pp. 3421–3438, 2013.
[41] M. Ali, T. Haron, and A. Marshall-ponting, "Exploring the Barriers and Driving Factors in Implementing Building Information Modelling (BIM) in the Malaysian Construction Industry : A Preliminary Study," *J. Inst. Eng. Malaysia*, vol. **75**, no. 1, pp. 1–10, 2014.
[42] CRC Construction Innovation, *National guidelines for digital modelling*. 2009.
[43] A. Z. Sheth and S. M. Malsane, "Building Information Modelling , a Tool for Green Built Environment," in *All India Seminar on Innovation in Green Building Technology, Green Build 2014*, 2014.
[44] R. Eadie, H. Odeyinka, M. Browne, C. Mckeown, and M. Yohanis, "An Analysis of the Drivers for Adopting Building Information Modelling," *J. Inf. Technol. Constr.*, vol. **18**, pp. 338–352, 2013.
[45] J. Boon, "Preparing for the BIM Revolution," in *13th Pacific Association of Quantity Surveyors Congress (PAQS 2009)*, 2009, pp. 33–40.
[46] P. Rajendran, T. Seow, and K. Goh, "Bulding Information Modeling (BIM) in design stage to assist in time, cost and quality in construction innovation," *Int. J. Conceptions Manag. Soc. Sci.*, vol. **2**, no. 3, pp. 52–55, 2014.
[47] J. Kim, "Use of BIM for Effective Visualization Teaching Approach in Construction Education," *J. Prof. Issues Eng. Educ. Pract.*, vol. **138**, no. 3, pp. 214–223, 2012.
[48] InfoComm International, "Building Information Modeling (BIM) Guide," 2011.
[49] R. M. Leicht and J. I. Messner, "Comparing Traditional Schematic Design Documentation To a Schematic Building Information Model," in *Proceedings of CIB W78 24th International Conference on Information Technology in Construction*, 1997, pp. 39–46.
[50] R. P. Kumanayake and R. M. P. S. Bandara, "Building Information Modelling (BIM); How It Improves Building Performance," in *International Symposium on Ensuring National Security Through Reconciliation & Sustainable Development*, 2012, pp. 357–365.
[51] J. Boshoff, "Building Information Modelling (BIM)," *Civil Engineering*, no. March, p. 56, 2014.
[52] Z. Ma and Z. Liu, "BIM-based Intelligent Acquisition of Construction Information for Cost Estimation of Building Projects," *Procedia Eng.*, vol. **85**, pp. 358–367, 2014.
[53] P. Demian and D. Walters, "The advantages of information management through building information modelling," *Constr. Manag. Econ.*, pp. 1–13, 2013.
[54] N. Alp and C. Manning, "Creating a Plan for Building Information Modeling," pp. 2732–2734, 2014.
[55] A. Gerrard, J. Zuo, G. Zillante, and M. Skitmore, "Building Information Modeling in the Australian Architecture Engineering and Construction Industry," in *Handbook of Research on Building Information Modeling and Construction Informatics*, J. Underwood and U. Isikdag, Eds. IGI Global, 2009, pp. 521–544.
[56] A. Sawhney and P. Singhal, "Drivers and Barriers to the Use of Building Information Modelling in India," *Int. J. 3-D Inf. Model.*, vol. **2**, no. 3, pp. 46–63, 2013.
[57] G. Aranda-mena, A. Chevez, J. Crawford, R. Wakefield, T. Froese, J. Frazer, S. Gard, and D. Nielsen, "Business Drivers for Building Information Modelling," 2008.
[58] V. Popov, V. Juocevicius, D. Migilinskas, L. Ustinovichius, and S. Mikalauskas, "The use of a virtual building design and construction model for developing an effective project concept in 5D environment," *Autom. Constr.*, vol. **19**, no. 3, pp. 357–367, May 2010.
[59] NBS, "NBS National BIM Report 2014," 2014.
[60] C. Mccartney, "Factors Effecting the Uptake of Building Information Modeling (BIM) in the Auckland Architecture , Engineering , & Construction (AEC) Industry," 2010.
[61] D. A. Campbell, "Building Information Modeling : The Web3D Application for AEC," in *Web3D 2007*, 2007, pp. 173–177.

MATEC Web of Conferences **66**, 00043 (2016) DOI: 10.1051/matecconf/20166600043
IBCC 2016

Theorizing the Concept of Urban Public Transportation Institutional Framework in Malaysia

Noor Ashikin Abd Rahman[1] and Yusfida Ayu Abdullah[2]

[1] *Centre of Postgraduate Studies, Faculty of Architecture, Planning and Surveying, Universiti Teknologi MARA, 40450 Shah Alam, Malaysia.*
[2] *Centre of Studies for Town and Regional Planning, Faculty of Architecture, Planning and Surveying, Universiti Teknologi MARA, 40450 Shah Alam, Malaysia.*

Abstract. The trends in transportation have been changing and will continue to change over the years. Urban dwellers in developing countries require and demand for mobility and accessibility at the same growth rate of these urban areas. Often, this demand is accommodated by the increased number of private vehicles. Development of a country is closely linked with the transportation system facilities thus will have a positive impact on the economic growth and social fabric of a country. The traffic issues are always associated with private vehicles especially in any developing cities such as the Federal Territory of Kuala Lumpur. A practical, efficient and effective transport system includes traffic management, public transport, road network and infrastructure. However, sadly, at present, the traffic problems in Malaysia are resulted from ineffective public transport system which is not methodically integrated from one place to another, therefore, forcing the people to use private vehicles for daily exercise. An integrated and comprehensive approach is required as it enhances the regulatory framework, planning structure and level of services delivered. For that reason, this paper reviews the magnitude of urban public transportation institutional framework in order to improve the people's mobility, also to respond to the major problem of urban public transportation in the major city of Malaysia. This study therefore applies a case study design and relies very much on qualitative data encompassing policies and guidelines. The findings were drawn from an early investigation of the Malaysian institutional public transport framework through literature review which looks closely into the management structure, as well as the implementation of public transportation system and its enforcement.

1 Introduction

Over the past few years, urban development receives pressure from the economic growth and increasing population. Public transport is one of the main problems of the people in the city because it has a direct impact on their daily lives. This is proven by the number of articles and news about public transportation published in various mass media The Malay Online [1], BERNAMA [2], Astro Awani [3], The Malaysian Insider [4], My NewsHub [5], Mstar [6]. The problems include the issue of complexity of land use and the lack of holistic approach in development and planning strategies. Also, the lack of a comprehensive transportation planning resulting a connectivity gap and making the Land Public Transport's (LPT) services to be inefficient and not user-friendly, therefore suggesting a more cohesive effort between the Federal, States and Local Authorities [7]. According to Azmi and Fanim [7], there is no clear role and responsibilities among the governments. The development had led to institutional conflicts in several aspects especially in enforcement, infrastructure provision and use, policy formulation and coordination. It was agreed by Oni, Okanlawon, and Asenime [8] that the problem complexity, planning, implementation and enforcement should be reviewed seriously because there appeared to various ministries and agencies at the federal and local level involved in different aspects of public transport in the Klang Valley. Furthermore, the four thrusts have not been accomplished which ensures the efficacy of transportation i.e., a collaborative planning and governance system, regulatory strengthening, as well as the service quality and excellence, and the infrastructure and capacity [7]. Another issue relates to the urban transport system and management. The increasing population of Greater Kuala Lumpur (GKL) and the Klang Valley (KV) is expected to reach 10 million in 2020, adding more pressure on its road network which is already nearing full capacity [9]. While the roads in GKL and KV are generally still serviceable, they are sensitive to any disruption that may have an impact to the entire system, causing massive congestion [9], for example, due to an accident on the road. Prolonged traffic congestion has numerous effects both to the social and economic. Fuel is wasted by idling vehicles, while productivity is lost as workers are caught in traffic, air quality is degraded by needless emissions and drivers are frustrated by the long commute into and out of the city [9]. The challenge is to therefore, maintain a robust network of private transport for the people who

[1] nashiqin.rahman@gmail.com; [2] yusfida@salam.uitm.edu.my

can afford private vehicles but also to practice an efficient public transportation networks and institutional framework that people can choose. The importance of a pragmatic institutional framework because it ideally covers a range of an organisational structure, including its regulations, policies, services and management of the whole system. It is therefore a form of linkages that ensures an effective flow of information from one part of a system to another [10].

Urban transport management approach had previously focused on methods to solve the problem of traffic congestion by building more roads and encourages the use and purchase of private vehicles. One of the main causes of traffic congestions in Kuala Lumpur (KL) is the substantial number of private vehicles with only a single occupant. This happens because of the very poor public transport system [4]. The inconsiderate Malaysian drivers was also claimed to be another reason [11]. The problem of congestion is generally caused by the lack of emphasis on managing the vehicles' movement and the lack of integration of public transport services [12]. The government, with the cooperation of the public, can resolve these issues by improving the driving behaviour, frequencies and intensities of buses plying the roads of KL. Looking at the Government Transformation Programme's (GTP) strategy, the ratio of private and public transport is still below the target with 12% in 2014 and when the government is trying to increase it to 25%, which is when the public transportation system will become the preferable mode of urban transport [9].

With regards to those continuing issues, this paper, therefore, seeks to examine the significance of urban public transportation institutional framework in order to improve the people's mobility, and to respond to the major problems of urban public transportation in KL. Given that intention, this paper will review the current literature on public transportation and analyse its institutional framework, and thereafter suggests a fitting approach.

2 Urban Public Transportation

2.1 The Background of Urban Public Transportation in Malaysia

Malaysia is experiencing rapid development in order to achieve the *Vision 2020,* and because of this, an efficient transportation system is needed to enhance the productivity and encourages economic activities, thereby increasing the national output and competitiveness [13]. Kuala Lumpur itself is a booming city facing immense challenges in trying to accommodate additional private cars every year and the increased of motor vehicles is faster than the population growth [9]. The city struggles with air pollution and traffic congestion thus reduces the Quality of Good Life [14]. It is just not enough to rely on factors of convenience to encourage the behaviour of using public transport. Individuals, however, still prefer to purchase private cars instead of using public transport [15]. Individuals were seen as reluctant of being in a crowded train with strangers [16]. It is for these reasons that the policy makers should learn to understand the unique needs of the general public especially in terms of their dependency on private vehicles before committing to new major projects related to transportation.

Public transportation is, therefore, a cornerstone of a nation's economy which provides a transit medium for daily mobility and the link between the people and their jobs, education and entertainment. Service coverage, travel time, reliability, park and ride facilities, comfort and safety are among the determining factors towards encouraging public transportation as a mode of choice [17]. However, integrating land use and public transportation requires an orderly co-ordination and consensus from all related agencies.

Needless to say, public transportation system is being taken seriously especially in KL. This includes the multiple vehicle services designed to transport customers on local and regional routes. Public transportation is therefore provided to the general public by means of cars, buses, vans and trains, either privately or publicly owned. It also includes shared passenger services like taxicab, carpooling or hired buses which are not hired together with strangers without private arrangement.

The key to an affordable and sustainable public transport system is to consider the actual demand, region's characteristic, government policies and financial condition and to put in place a transportation system that suits the local condition. A good system requires a combination of various elements.

2.2 Integration of Urban Public Transportation System

The term 'integration' in public transport is similar to the concept of intermodal which is generally used for the transport of goods. This concept is usually defined as a system that provides seamless public transport services for passengers and covers a wide range of concepts. Netherlands Transport Research and Training [18] quotes by European Cooperation Project [18] and Poorjafari [18] defines integration as:

"The organisation process through which elements of the passenger transport system (network and infrastructure, tariffs and ticketing, information and marketing, etc are across modes and operators, brought into closer and more efficient interaction, resulting in a overall positive enhancement to the overall state and quality of the services linked to the individual travel components"

This definition however goes further and includes a wider incorporation with other transport modes and non-transport services such as town planning, environment and social policies [18]. There are a variety of measures and practices to strengthen the integration within and amongst the public transport system such as physical and network, fare, information and institutional. A common institutional framework is the key instrument to land use planning, travel demand and integrated public transport services. This combination is important to improve the cooperation and coordination among authorities or government agencies, public transport sectors and between private and public operators. Integration can occur through four (4) manners such as incorporation between policy instruments involving different modes, integration between policy instruments involving infrastructure provision, management, information and pricing, integration between transport measures and land use planning policies etc. [18]. Integration is not a definite status and should be improved continuously within different public transport modes and also between public transport and other transport modes. Integration can be implemented and achieved at different levels. However, the higher level of integration is achieved by extending its border beyond the public transport systems by involving other modes of transport and other policies.

A decent public transport system is taking into account the 'network effect'. This is an inter-connecting train, tram and bus system, where timetables and exchanges correspond and interlock permitting 'anyplace to-anyplace' traverse a city. It also has a system of public transport lines where the structure is simple and the timetable is steady, making it easy for the people to learn and comprehend it. While public transport critics regularly assert that public transport just works in populaces with high thickness, the confirmation indicates something else [19].

2.3 Institutional Framework for Transportation Planning and Management System

Institutions are a set of norms, values and beliefs that have been formed to ensure that targets are achieved while framework is the linkage that supports two or more sub-systems ensuring the easy flow of information or data from one sub-system to another [10]. The major components of the institutional framework include the governance framework (three tiers government), the organisational framework (planning authorities) and legislative framework (planning laws) and administrative framework (structure). Institutional structure is an instrument used to achieve better governance because they have the potential to bring about changes in outcomes. Pinto [20] observed that the best institutional structures for urban governance which have values embedded in the basic purposes of society. The relationship between the institution's framework and its structure would ensure central control and domination in any level of governance. The transportation planning and management system is made of various public and private actors. The issues and problem faced by the planning and management system include a lack of coordination between the public and private sectors, insufficient capability to carry out planning at appropriate scales, enforcement problems [21].

At present, a range of public and private institutions have responsibilities for transportation in KL. There are a lot of agencies focusing on public transport in Malaysia such as the Prime Minister's Office, Economic Planning Unit (EPU), Ministry of Finance Malaysia, Ministry of Housing, Urban Wellbeing and Local Government, Ministry of Transport Malaysia, Ministry of Works Malaysia, Ministry of Federal Territories, Ministry of Natural Resources and Environment, Ministry of Home Affairs and also Land Public Transport Commission (SPAD). These agencies are responsible and form part of the institutional framework for urban public transport in Malaysia whether directly or indirectly. Malaysia has a three tier types of government: the federal, state and local government [21]. In general, the Federal Government is responsible for determining national transportation plans, transport policy guidelines and overall administrative matters. Local governments on the other hand, plan, execute and administer road networks and urban transport improvements under their jurisdiction.

The local government's institutions deal with daily urban transport problems and responsible for the planning, construction and maintenance of transport facilities in their jurisdiction. The City Hall of Kuala Lumpur is governed by its own set of regulations and carries out its own planning, design, implementation and maintenance of road transport among other functions. The function of a local government is very limited as the federal government's institutions determine the general plans and major government financed infrastructural projects. The local government is therefore responsible only for the planning and development of smaller projects [21].

3 Research Methodology

This paper focuses more on reviewing the current literature. It examines the concept and practice of urban public transportation and analysed the public institutional framework. However, further to that, the research will engage a case study approach and therefore investigate the existing institutional framework for urban transportation, including its delivery and management of public transportation by the government and related agencies.

The selected study area to conduct this research is the Federal Territory of Kuala Lumpur (KL) based on its substantial role as the capital city and most popular city in Malaysia amid its continual challenges for public transportation (Figure 1). The city covers an area of 243 km^2 and with an estimated population of 1.6 million. The local administration is carried out by the Kuala Lumpur City Hall or *Dewan Bandaraya Kuala Lumpur* (DBKL) under the Federal Territories Ministry of Malaysia. Kuala Lumpur City Hall is therefore responsible for KL's urban public transportation institutional framework, working together with the federal government. KL was also chosen for this research because of the immense challenges it faces in trying to accommodate additional private cars every year and the increased of motor vehicles which is faster than the population growth.

It was highlighted that KL needs a suitable and complete network system which provides road accessibility and high quality public transportation service in major routes that provides a seamless transportation system. In supporting the sustainable transport system, these criteria should be emphasised [22]

DOI: 10.1051/matecconf/20166600043

such as to enhance the comfort and convenience; improves accessibility to land public transport through an integration of local land use policy and focusing on Transit Oriented Development (TOD); undertakes comprehensive and integrated planning (consider the planning, integration and coordination of all mode of transport); and strengthens institutions role and framework. Hence, the institutional framework structure which involves the administration at the federal, level and local level is crucial in such development.

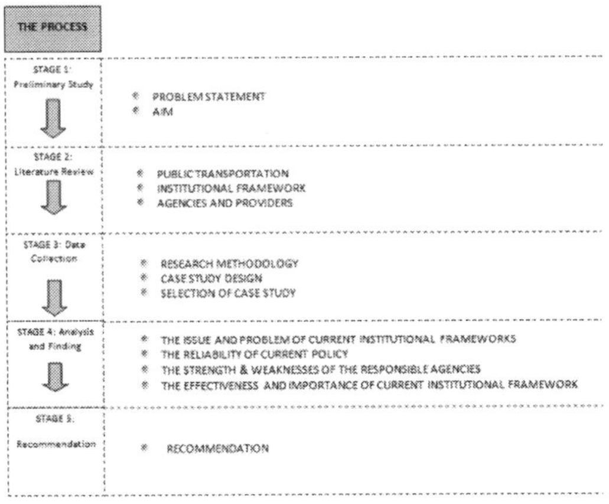

Figure 1. Approach to Case Study

The factors that influence the selected study area were also based on the urban public transport problem and issues. The lack of coordination and integration between internal and external institutional framework itself especially on administration of urban public transport, the management and operation of public transportation, the system and procedure for project implementation and also the sources of finance are the major contributing to the problem in this field [18].

4 Analysis and Discussion

Based on the study, it was apparent that urban public transportation in KL faces many challenges in relation to its institutional framework (Figure 2):

(a) There is no clear and coherent policy framework where public transport is clearly seen as a key to guarantee the nation's future success. During the Tenth Plan period, Land Public Transport Commission (SPAD) implemented four strategies which focused on strengthening the regulatory framework, increasing transport capacity, promoting seamless connectivity and establishing a robust monitoring and enforcement mechanism. However, The National Land Public Transport Master Plan only guides the formulation of regional master plans and preparation of policy guidelines. Therefore, there it lacks in integration between the urban transport policies and the public transport system itself due to the fact that there is no precise national policy formulated to overcome the land public transport issues.

(b) The lack of comprehensive planning is one of the key factors which create connectivity gaps thus rendering the public transport services inefficient and not user friendly. A cohesive and concerted effort between the Federal, States and Local Authority needs to be critically established with clearly defined roles and responsibilities to ensure a seamless system is put in place. The transformation of public transport responsibility led by SPAD in collaboration with various agencies. However, poor coordination among agencies during planning and implementation of the various transport initiatives has led to inefficient utilization of resources such as the construction of urban rail and highway in the same corridor. In addition, public transport improvements require a stronger integration in land use planning and the involvement of local authorities.

(c) SPAD has established a public transport committee at the state level to guide and consult with state and local authorities. However, the committee had not performed accordingly in the implementation of public transport improvement plan. There appeared to be the lack of coordination between public transport and urban planning. Moreover, the lack of emphasis in terms of transport planning and administration, management and operation of transit systems, institutions for project and program implementation and also shortage of finance are affecting the overall system [12].

(d) Lack of integration between urban public transport system (public transport unavailability) whereas the five broad categories are important such as physical, network, fare, information and institutional integration and also minimal public transport coverage in certain areas where this is the issues faced by the public [23].

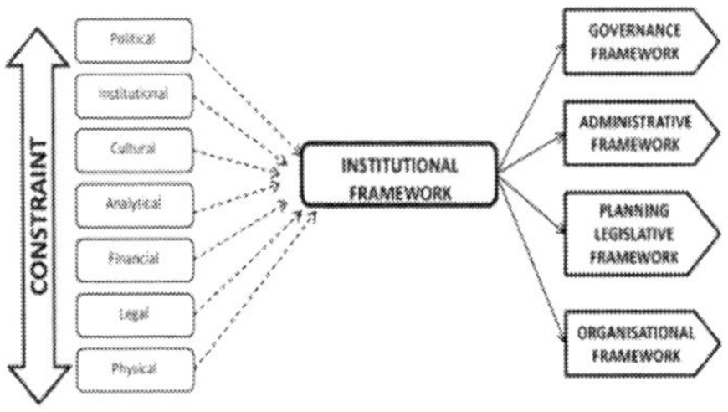

Figure 2. Findings of the Research

Many of the key challenges indentified for the development of land public transport services in Malaysia enfold around the need to strengthen the regulatory framework and improving the level of services for public transport by integrating all of them. The formulation of national policy should become the priority where government needs to recognise the importance of urban mobility issues. There are already many countries around the world where governments are taking concerted action on urban mobility including Sweden, Abu Dhabi and Singapore, whereby public transport is clearly seen as key tool to guarantee the nation's future success.

5 Recommendations

The following recommendations are proposed to improve the current institutional framework of urban public transportation in KL particularly, and in Malaysia generally.

(a) Creating A Clear and Coherent Policy Framework and Governance

Successful mobility in cities starts at the very top of the policy-making chain. Government should recognise the importance of urban mobility issues nationally and ensure they are high up on the policy agenda. It is essential to have a national legal and regulatory framework of public transport to become more viable and stable. Clear roles and responsibilities need to be defined for each stakeholder in the mobility supply chain in delivering real results on the ground. A proposed institutional framework for urban public transportation in KL is illustrated below (Figure 3).

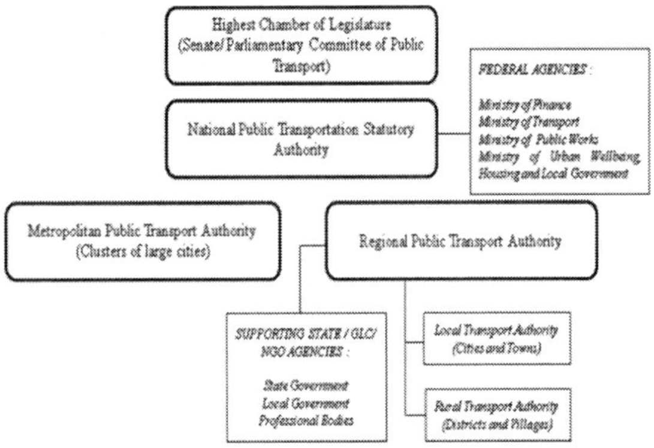

Figure 3. Proposed Institutional Framework for Urban Public Transportation in KL

Through SPAD, there should be a Public Transport Authority who is responsible in ensuring the standards in determining access to key performance benchmark for the entire country. Integrated public transport planning policy should be aligned with the Nation Physical Plan (NPP) and the Public Transport Authority is responsible in mobilizing the cooperation of ministries and agencies of government support related to public transport through a clearer scope of responsibility and performance appraisal. Furthermore, the Klang Valley desirable to have the Regional Public Transport Authority itself where it has jurisdiction to develop a support system for an effective public transport from residential areas to the central of Klang Valley, Kuala Lumpur.

(b) Integrating Public Transport and Urban Planning

The problems such as lack of institutional and regulatory framework, non-integration of transport to urban planning, mismanagement of traffic can be solved if better coordination between public transport and urban planning would bring many far-reaching benefits. It can help control urban sprawl, foster urban regeneration and promote sustainable transport mode a viable and attractive option. Coordination between these two policy areas can be improved by developing a shared vision across all policy sectors, fostering cooperation between public authorities and private property developers, prioritising public transport-oriented urban design.

(c) Integrating Transport Modes

Integrating transport modes in KL can encourage people to use the public transport. Nevertheless, this is not a straightforward task to achieve and so a coordinated approach is the only way to achieve an integrated and seamless public transport system. All the main players should need to pull in the same direction and different transport modes and different transport companies need to work together (Figure 4). This direction will inevitably requires the synchronisation of the federal agencies, along with other related agencies like SPAD and most importantly the KL City Hall.

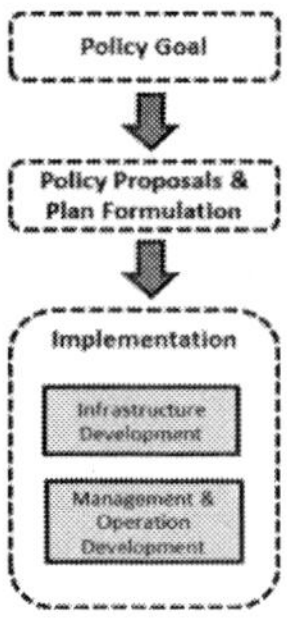

Figure 4. Proposed Urban Public Transportation Strategy

An urban transport strategy has been formulated with a long-term policy goal or vision. In addition to strategic policy proposals to achieve the long-term goal, a set of feasible projects and measures must be proposed for implementation accompanied by operation and management requirements.

6 Conclusion

A city like Kuala Lumpur would easily be communicated if the traffic flow runs smoothly and orderly and when there is integration between the public transport systems. Likewise, this can be achieved through good planning. Cities should be able to provide an efficient urban structure and equitable society which all members can easily gain access to all the amenities and enjoy the maximum benefits of city life. Furthermore, the implementation of effective public transport system is the key solution in improving people mobility and also to counter the major problem of urban public transportation. With the right regulatory and institutional framework in place, KL will have the power to define how they can encourage the people to use public transport. This paper has unveils the important issue of integration among agencies. As mentioned earlier about the methodology, further work will embark on the case study approach and will examine closely at the practice in Kuala Lumpur.

Putting in place successful public transport projects is a major aspiration that brings significant benefits. In order to achieve a maximum impact, the integration of the public transport policy and urban planning needs to be mainstreamed in policy decisions at local and national level. The higher level of integration is achieved by extending its border beyond the public transport systems by involving other modes of transport and other policies, like land use planning, social and environmental policies. The coherence between policy and public transport will indirectly integrate between public transport and urban planning and also public transport modes.

Acknowledgments

The authors would like to thank the Public Service Department of Malaysia for funding the research. And, to thank the Universiti Teknologi MARA (UiTM) and the Centre facilities and support.

References

1. The Malay Online. *Improve The Quality of Public Transport Se*rvices. 2 October 2014 (2014).
2. BERNAMA. *Public Transport Fare Increase Is Necessary.* 17 April 2015 (2015).
3. Astro Awani. *Reorganisation Of The Department: Solve Public Transport Problem.* 27 July 2015 (2015).
4. The Malaysian Insider. *Public Transport In KL Needs To Be Improve Due To Increasing Users*. 23 December 2014 (2014).
5. My News Hub. *Improve Quality Of Services will Increase People Mobility.* 11 October 2014 (2014).
6. Mstar. *Logistic and Transport Sector Remains Priority.* 21 May 2015 (2015).
7. Azmi Abdul Aziz & Nor Fanim Mohd. Amin. *Transforming The Land Public Transport System In Malaysia*. Journeys (2012).
8. Oni,S.I,Okanlawon,K.R & Asenime,C.O. *Institutional Framework For The Transformation of Transport Administration in Metropolitan Lagos In An Agenda for a New Nigeria : The Imperative of Transformation,* Anifowose, R.and Babawale,T.(eds.). Concept Publications Limited (2006).
9. Government Transformation Programme. *Annual Report 2014*, Government of Malaysia (2014).
10. Wapwera,S.D, Mallo D.M. and Jiriko G.J. *Institutional Framework And Constraints In The Urban And Regional Planning System In Jos Metropolis, Nigeria*. Journal Of Geography (2015).
11. The Star Online. *System To Suit Public Needs*. 13 February 2014 (2014).
12. Ministry of Transport. *Strategy Paper – Providing Seamless Transportation*. Government of Malaysia (2014).
13. Economic Planning Unit. *Eleventh Malaysia Plan (RMK-11)*. Government of Malaysia (2015).
14. McKinsey & Company *How To Make A City Great*. Europe (2013).
15. Public Transport User Assosiations. *Public Transportation*. Victoria,Australia (2010).
16. Department of Transport London. *Travel Training : Good Practice Guidance*. The National Archives, London (2011).
17. Land Transport Authority. *Land Transport Authority Annual Report*. Singapore (2014).
18. Poorjafari, V., & Poorjafari, M.. *Integration in Urban Public Transport Systems*. Journal (2011).
19. Arief Irsyad. *The Problem With Malaysia's Public Transport. Is Change On The Horizon?*. Malaysian Digest.com. Retrieved on 7 February 2016. (2015)
20. Pinto MR. *Metropolitan City Governance in India*. New Delhi Sage (2000).
21. Zulina Zakaria. *The Institutional Framework for Urban Transportation and Land Use Planning and Management in the Globalizing Kuala Lumpur Region*. Research Paperwork (2003).
22. Global Development Research Center. *Sustainability Dimensions of Transport Systems*. Japan (2015).
23. Land Public Transport Commission. *Greater KL/Klang Valley – Land Public Transport Master Plan*. ALMEC Corporation (2012).

Effect of landscape density in a residential area on thermal performance in a tropical climate.

Noorazlina Kamarulzaman[1,2a], Nor Amila Wan Abdullah Zawawi[1] and Nurul Izma Mohammed[1]

[1]*University of Technology MARA, Seri Iskandar Campus, 32610 Perak, Malaysia*
[2]*University of Technology Petronas, 32610 Perak, Malaysia*

Abstract. It is a well-documented that the urban air temperature is gradually growing in all cities due to the rapid development of buildings, roads and other infrastructure, human activities and also decreasing in vegetated areas. In a tropical climate, outdoor environment is clearly warmer than indoor environment due to higher air temperatures, particularly in dry seasons. Since the indoor environment is influenced by its surroundings, this situation indirectly contributes to the discomfort indoor environment in the building. Thus, it generates to the dependence on mechanical ventilation and increase the energy consumption in buildings. Many research studies have proof that plants not only beauty a city, but also improve the urban environmental condition by reducing the transferring of heat flux on buildings and increasing the reflection of radiation and shading. Therefore strategically placed vegetation around a building could decrease the energy consumption in buildings by reducing the adverse impact of some climate elements. Overall, this paper focuses on the results of a preliminary pilot study of two Semi-Detached houses with different landscape density in Seri Iskandar, Perak. Three climatic parameters, building configuration, and landscape design measured and analyze in this paper.

1 Introduction

Currently, the temperature of the most major cities around the world continues to grow due to the urban heat island phenomenon and the positive climatic change. Santamouris (2001) has mentioned that this phenomenon is the most documented for almost a century, and it related to higher urban temperatures compared to the adjacent suburban and rural areas [1]. In this modernization era, the urban area attracts the largest share of development investment, energy, and creativity. Today, 40% of the population of developing countries already lives in cities [2]. By 2020, it's expected that more than half of the population will live in the towns. The fast and mostly uncontrolled population growth create a pattern of rapid urbanization (Global Report on Human Settlements, 2001). Improved living standards and the increased population in developing countries contribute to a dramatic increase in energy consumption worldwide [4]. An increase in the urban population of 1% has been reported to increase energy use by 2.2% [5].

Globally, buildings handle approximately 40% of the total world annual energy consumption. Most of this energy are for the provision of lighting, heating, cooling, and air conditioning [6]. In Europe, buildings represent 40% of the overall energy consumption and 36% of the overall CO_2 emissions (A.Ismail, Abdul Samad, Rahman, & Yeok, 2012). According to Brown (2008), an annual energy consumption increase of 30% by the year 2020 was predicted by the International Energy Agency in 2008 [5]. A study of the tropical city of Singapore demonstrated that the use of air conditioning system has become a necessity to publics and contribute as the main factors to increase of energy consumption in building [8].

International Energy Agency (2009) stated that by 2030 average air temperature will rise by 2°C represent a critical limit [9]. Reflecting on this, various authors have agreed on the temperature level as a primary predictor of energy consumption for cooling inside and outside of the buildings. Therefore, the interactions between the indoor and outdoor thermal environments are necessary to be considered. Technically, there are many ways to improve the energy efficiency of the built environment. Improvements in building envelope technologies, such as wall, floor, and roof insulation, high-performance windows and doors, and air infiltration, have a priority role in producing a comfortable interior.

However, to improve energy efficiency using strategies mentioned above need extra financial cost and might be cannot be done by all. Therefore, the simplest and inexpensive method is by greening strategies in urban, especially in residential areas to improve the local air temperature. Good landscape arrangement can potentially decrease the amount of daylight entering the building and consequently, it's generating the comfortable temperatures for indoor and outdoor environments.

[a] Corresponding author: noorazlinakamarulzaman@yahoo.com

2 Current Situation

In general, the issue of increasing of heat and energy consumption in a building is due to the increasing of outdoor temperature, and it relates to environmental problems such as global warming and heat island effects as discussed before. Memon, R. A et al.,(2008) stated that there are three reasons of Urban Heat Island which such as i) Anthropogenic heat released from vehicles, power plants, air conditioners and other heat from electrical appliances, ii) Enormous amounts of solar radiation due to less vegetation from rapid urbanization. iii) High unevenness structure is another problem of urban areas, which decreases the convective heat removal [10].

Ismail, A (2008) has mentioned that when the outdoor air temperature increases, the building will experience indoor discomfort conditions. The process of heat transfer through building fenestration such as windows and absorbed through the building envelope (walls and roofs) is the primary source of heat flow into buildings. Therefore, McPherson, Herrington et al. (1988) stated that landscaping can have a predominantly direct effect on local climate and linked to building thermal performance and one of the simplest strategies to reduce solar heat gain [12].

Strategically placed vegetation around a building has recognized as a one of cooling method. It decreases the amount of solar entering the building envelope through shading, moderating temperatures, evapotranspiration processes, and by controlling the wind direction to keep the building cool. With appropriate amount, species, and placement of vegetation can slow heat build-up on a hot summer day [5]. Expert researcher in landscaping also agrees that plant (trees, shrubs and grasses) can give significant effect to the local environment directly. By producing shading surfaces and channeling the wind, and by evapotranspiration of water (Akbari, H., Davis, S., Dorsano, S., Huang, J., & Winnet, 1992; DOE, 1993; Misni, 2013). Shading is the most cost-effective way to diminish solar heat gain. Vegetation planted nearby the house will decrease the surrounding air temperatures and provide the best evaporative cooling. [15].

However, the effect of the surrounding landscaping on the thermal and energy performance of tropical domestic building has not been widely recognized or quantified. Therefore, the research aims to achieve three objectives *i) To examine the potential of surrounding landscape to reduce the thermal performance of residential housing in a hot-humid tropical environment. ii) To measure the energy reduce rate through strategic landscaping around a residential building and iii) To proposed a model using Artificial Neutral Network to predict indoor thermal and energy consumption.*

3 Research Methodology

As mentioned above, the microclimatic conditions are affected by adjacent buildings as well as by vegetation. For the aims of the preliminary study, the research methods only focusing on fieldwork measurement and its divided into three parts: 1) house with surrounding landscape selection, 2) survey and observation, and 3) fieldwork analysis for weather data and landscape elements. The aims of house selection were to identify and choose similar building design, constructions and site locations that were sufficiently different in their density of landscape structures and design.

The fieldwork measurement was to examine and compare the results of microclimate data between two houses that would directly influence the indoor thermal performance of the house. According to Akbari and Taha (1992), climatic factors that affect outdoor thermal comfort are (i) surface and air temperature; (ii) relative humidity; (iii) solar radiation; and (iv) wind velocity. Therefore, the interaction of these climate elements and its relation to local climate should be fully understood to define the appropriate approach to improving the urban thermal environment [16].

Technically, the primary instrument used for the fieldwork measurements included; (i)Two sets of portable weather station devices, model DeltaLog10 version 0.1.5.29, multi data logger and sensors to measure climatic parameters. (ii)One set of a compass and measuring tapes to identify the sun orientation and measure the house configurations. (iii)A set of layout plans to draw and record the house arrangements and landscape plans on site.

4 Pilot Study

The reference sites used in this research located in the residential area of Perak Tengah District. Two Semi-Detached Houses located at Taman Iskandar Perdana, Seri Iskandar Perak, Malaysia (4°N latitude and 100°E Longitude) were chosen as a case study building. The houses situated in the same areas and route but separated by two houses in between as shown in *Figure 1*. The residential area consists of 52 single storey Semi-Detached houses. For the purpose of the pilot study, only two houses with a similar conventional tropical design of architecture were chosen as a case study. The selection of building for the pilot test is due to the similarity of several criteria such as i) Ages of buildings. ii) Size of green area iii) Standard of construction, iv) Similar location but different in their landscape density, size, and orientation.

In this research, the study house was labeled as densely landscape house (House A) and sparsely landscape house (House B) according to the numbers of tree either single or group species, canopy size, and their height.

The house configuration was similar where both is east-west oriented and were constructed using similar building materials and construction. Both Semi-D houses had a similar size of a built-up area approximately 90.8m^2 or 7543mm x 12049mm consists of a living room, dining, kitchen, three bedrooms with two bathrooms and one store room *(refer Figure 2 for the overall view of the study houses)*. The primary structure was a reinforced concrete pad foundation; all had brick walls and a pitched aluminum- framed roof covered by the similar colour of concrete tiles. Ceiling heights for the two houses were around 3m height and finished with fibrous, cement plaster, and asbestos free fiber cement ceiling sheets. All of the building openings constructed using glass with metal window frames for all casement windows and the sliding doors.

Figure 1: The blue circle shows the location of the study houses. *(Source: Google Earth, 2015)*

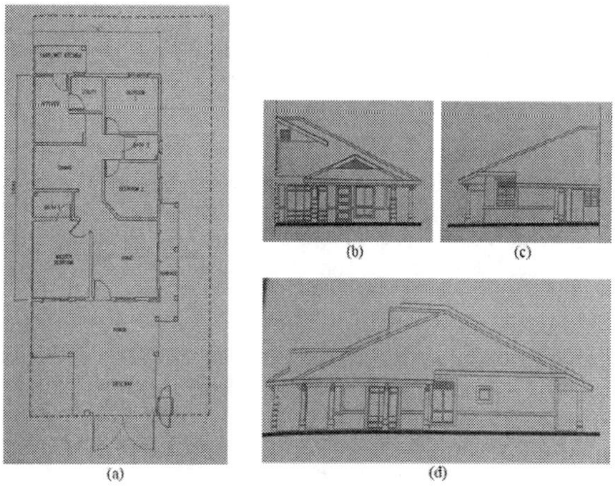

Figure 2: Figure (a), (b), (c) & (d) shows floor plan, front, rear and side elevation of the study houses.

5 Local Weather Record

The weather recording was carried out in one day during daytime from 10.30 to 16.30hours according to owner's permission and availability of the instrument. For the House B, data were collected on 20 August 2015 while for the House A is on 26 August 2015. Weather on these two days was very similar, where there was the sunshine for the whole day even though the data collection gaps are six days. This study measured meteorological parameters such as air temperature, humidity and wind velocity at 10-minute intervals time using portable weather station devices. Data recording for this instrument was set automatically at the same time and placed at indoor and outdoor of the house approximately 1m from the ground. The similar climatic parameters and time were recorded to compare the results and to ensure all data was accurate. Only one measurement point (MP) for indoors were recorded which is at the living room due to the constraint of access (marked as MP1). Meanwhile, three measurement points (MP2, MP3 & MP4) for the outdoor climate were taken around these sites as shown in *Figure 3*. The distance of each measurement point is approximately 3m from the building envelope. The climatic data recorded were not influenced by shadows or reflected solar radiation.

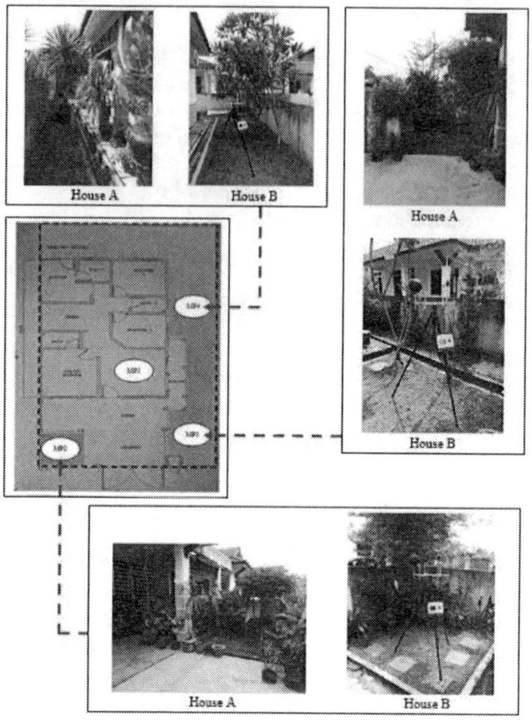

Figure 3: Plan is showing the location of measurement points for House A & B.

6 Pilot test results

The fieldwork analysis was intended to identify the climatic parameters and landscape design around the two houses that have a significant effect on their local air temperature. The climate parameters measured in this research include air temperature, wind velocity, and relative humidity.

6.1 Analyze Of Landscape

The landscape surrounding of the two houses used a combination of trees, shrubs, ground cover and turf. The study of all landscape elements was done by observation and interviewed directly to the owner of the houses. Landscape elements for each building were drawn not to scale in layout plan as shown in *Figure 4*, and the photo of the landscape was captured to make a comparison. Both of the study house had a medium sized area for landscaping and situated on the left and right side of the building. A small area of landscape garden in rectangular shape approximately 2.5m x 2.5m located at the front side between the Semi-Detached houses. Meanwhile, the other side of the landscape area is wider approximately 3m width x 21m length measured from the front side to the backyard. The owners personally inspired the tropical landscape design around the gardens. Each house had been planted with tropical plants for many reasons such as for beautiful view, edible, to create shade and to yield a lovely, comfortable and healthy environment. Both landscape houses planted with almost similar species of tree such as *Calathea lutea, Crinum asiaticum, Bambusa sp.* and all landscape garden covered with turf *(refer Table 1&2)*. They also used different amounts of trees and shrubs, species, and configurations of the plant. Edible plants planted in the backyard.

(a) (b)

Figure 4: The arrangments of trees and shrubs around (a) Densely Landscape & (b) Sparsely Landscape House

Table 1: List of trees, shrubs and ground cover at Densely Landscape House (House A)

No	Species	Azimuth[1]	Distance to wall[2]	Trees, shrubs & ground cover Index / typhology / biomass structure			
				Area[3]	Amount of leaves[4]	Size of leaves[5]	Height[6]
1	*Calathea lutea*	3*	3	2	2	3	2
2	*Bambusa ventricosa*	3*	3	2	3	1	2
3	*Calathea lutea*	3*	2	2	2	2	2
4	*Yucca aloifolia*	3*	3	2	1	1	1
6	*Asplenium nidus*	3*	1	2	2	3	2
8	*Calathea lutea*	3*	2	2	2	2	2
9	*Yucca aloifolia*	3*	3	1	2	1	2
10	*Rhoeo spathacea*	3*	1	2	2	1	1
11	*Anthurium sp.*	3*	1	2	3	2	1
12	*Ophiopogon jaburan variegata*	3	3	2	3	2	1
13	*Helicornia sp.*	3	3	1	2	2	2
14	*Bambusa ventricosa*	3	3	3	3	1	2
15	*Hymenocallis littoralis*	3	3	2	3	2	2
16	*Yucca aloifolia*	3	3	1	1	1	1
17	*Bambusa sp*	3	2	2	3	1	3
18	*Yucca aloifolia*	3	2	2	2	1	3
19	*Helicornia sp.*	3	1	2	3	2	3
20	*Anthurium sp.*	3	1	2	3	3	1
21	*Yellow palm*	3	1	1	2	2	2
22	*Hymenocallis littoralis*	4	1	3	3	2	1
23	*Calathea altissima*	4	1	2	3	2	3
24	*Yucca aloifolia*	4	1	2	2	2	2
25	*Yucca aloifolia*	4	2	2	3	2	3
26	*Araceae sp*	4	2	2	2	3	2
27	*Calathea lutea*	4	2	2	2	2	2
28	*Euphoria malaiense*	1	2	4	3	2	3
29	*Musa spp.*	1	2	4	3	3	4

Table 2: List of trees, shrubs and ground cover at Sparsely Landscape House (House B)

No	Species	Azimuth[1]	Distance to wall[2]	Trees, shrubs & ground cover Index / typhology / biomass structure			
				Area[3]	Amount of leaves[4]	Size of leaves[5]	Height[6]
1	*Bambusa ventricasa*	3*	3	2	3	1	4
2	*Ixora javanica*	3*	3	2	2	1	1
3	*Costus Woodsonii*	3*	3	2	2	1	1
4	*Plumeria rubra*	3*	2	2	1	2	4
5	*Bambusa ventricosa*	3	3	2	3	1	4
6	*Hymenocallis littoralis*	3	3	1	3	2	1
7	*Costus Woodsonii*	3	3	3	3	1	1
8	*Calathea altissima*	3	3	3	3	2	2
9	*Plumeria rubra*	3	2	1	1	2	1
10	*Ixora javanica*	3	2	1	1	1	1
11	*Costus Woodsonii*	3	2	1	1	1	1
12	*Podocarpus*	2	2	1	2	1	3
13	*Podocarpus*	2	2	1	2	1	2
14	*Ixora javanica*	2	2	2	3	1	2
15	*Calathea lutea*	2	2	4	3	3	3
16	*Plumeria rubra*	2	3	4	3	2	4
17	*Mangifera indica Linn.*	1	3	5	3	2	4
18	*Musa spp.*	1	2	3	3	3	4
19	*Pandanus amaryllifolious*	1	2	1	1	1	1

[1] 1= North(right), 1*= North(left), 2=East, 3=South(right), 3*= South(left), 4=West
[2] 1= proximity to walls, 2=below 3m, 3= 3-6m
[3] 1= below $1m^2$, 2=$1\text{-}4m^2$, 3=$5\text{-}9m^2$, 4=$10\text{-}14m^2$, 5=$15\text{-}19m^2$
[4] 1= rare, 2=medium, 3=dense
[5] 1= small, 2=medium, 3=large
[6] 1= below 1m, 2=1-2m, 3=2-3m, 4= 3m an above

6.2 Analyze of Climatic Parameters

6.2.1 Air Temperature

From the data recorded, the air temperatures inside and outside of the densely landscaped house (House A) were lower than those in the sparsely landscaped house (House B). The average outdoor temperature for House A and B is 32.5°C and 33.2°C respectively. Meanwhile, the average indoor temperature level shows a small differences rate at 0.1°C. At 10.30 hours the indoor temperature in the House A was 30.1°C while the House B was 30.3°C as shown in *Figure 5*. The different landscaping amount and species resulted in only minor differences in this temperature level. Both houses reached their lowest indoor temperatures at 11.00 hours; 30.4°C at the House B and 29.7°C at the House A. The outdoor temperature for House B continues fluctuated until reaching the highest point of 36.2°C at 14.30 hours. The same pattern occurred at the House A, which reached its highest outdoor temperature of approximately 34.2°C at 15.30 hours. Overall, the outdoor temperature of the House A was as much as 3.8°C lower than the House B during the peak time of the day. In contrast, the indoor

 DOI: 10.1051/matecconf/20166600044

temperature at the House A and House B had a similar pattern and reading at all times. The apparent difference in temperature readings for the outdoor not much influence the ambient indoor temperatures might be due to activity in the house during the data recorded. Both houses uses natural ventilation during the day.

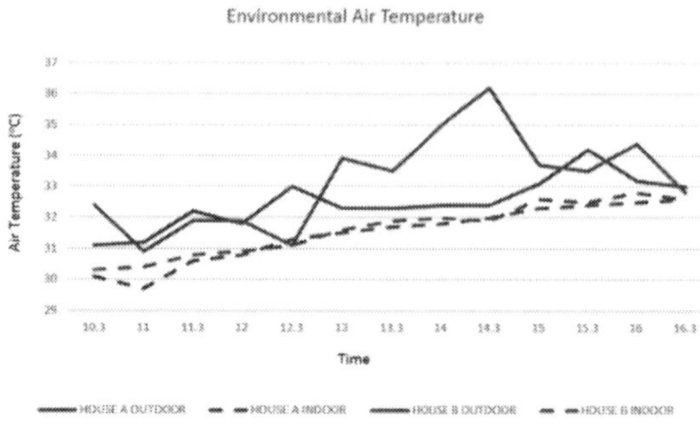

Figure 5: Indoor and outdoor air temperature results for House A and House B.

6.2.2 *Wind velocity*

Figure 6 shows the humidity pattern in m per second (m/s) at 1.5m heights in two study houses. Outdoor wind velocity around both houses gradually rose in the morning, starting at 10.30 hours; 0.04m/s at the House A and 0.21m/s at House B until they reached the highest point at 0.93m/s and 0.56m/s respectively at 13.30 hours. Overall, an average data of wind velocity demonstrates that House B reached a higher rate for indoor and outdoor compared to House A. Less in cross ventilation due to enclosed space (*windows and door closed*) causes in decreasing of indoor wind velocity in both houses. The outdoor wind velocity then continues in the fluctuating pattern at throughout the day for both houses.

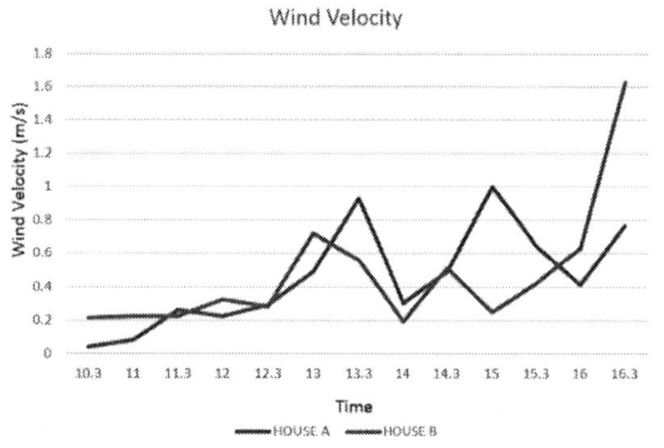

Figure 6: Wind velocity for the House A and House B.

6.2.3 Relative Humidity

During the morning, the ambient relative humidity (RH) readings at the House A is higher around 6.8% difference at 10.30 hours until the RH pattern slowly fell to the lowers point at 39.4% at 15.00 hours. In the middle of the day, the relative humidity levels of the House B were dramatically dropped around 21.7% starting at 12.30 hours to 14.30 hours. The graph also shows a big difference of RH between two houses at 14.30 hours around 19.4% (shown in *Figure 7*). However, all two house have a slightly similar in declining pattern in relative humidity.

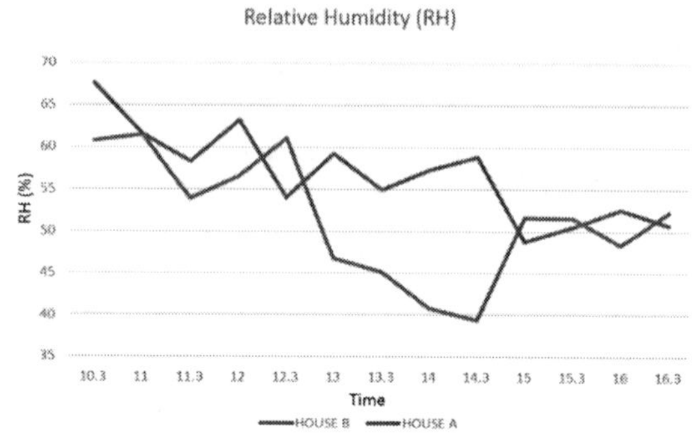

Figure 7: Relative Humidity for the House A and House B

7 Conclusion

This study presents quantitative finding regarding the cooling effect by altering the outdoor temperature in around two Semi-Detached houses with two landscaping strategies in a low density of residential zone in the tropical region. Study of the temperatures, relative humidity and wind velocity of the two houses of the outdoor and indoor facades revealed a quite interesting aspect of the passive cooling potential of landscaping.

Overall, the pilot test results show that outdoor temperature of the House A was 3.8°C lower than the House B and the percentage of Relative Humidity for House A were much higher about 19.4% during the afternoon. Meanwhile, an average data of wind velocity shows that House B reached a higher rate for indoor and outdoor compared to House A.

Strategic arrangement and design, sufficient numbers, and sizes of vegetation can provide the most efficient means of reducing outdoor air temperature around a residential building. The combination of shade trees, shrubs, vines, and groundcovers was predicted found to be the most efficient landscape strategy. This preliminary study apparently demonstrates the best technique to create a favourable environment for appropriate choice of plants, quantity, species and arrangement of landscaping for a tropical residential landscape.

References

[1] M. Santamouris, *Energy And Climate In The Urban Built Environment*. James And James Science Publishers, London., 2001.

[2] J. E. Anderson, G. Wulfhorst, and W. Lang, "Energy analysis of the built environment—A

review and outlook," *Renew. Sustain. Energy Rev.*, vol. **44**, pp. 149–158, 2015.

[3] 2001 Global report on Human Settlements, "Cities in The Globalizing World.," 2006.

[4] M. Castelli, L. Trujillo, L. Vanneschi, and A. Popovič, "Prediction of Energy Performance of Residential Buildings: aGenetic Programming Approach," *Energy Build.*, 2015.

[5] A. Misni, "The Effects of Surrounding Vegetation, Building Construction and Human Factors on the Thermal Perfomance of Housing in a Tropical Environment," 2012.

[6] L. Yang, H. Yan, and J. C. Lam, "Thermal comfort and building energy consumption implications - A review," *Appl. Energy*, vol. **115**, pp. 164–173, 2014.

[7] A. Ismail, M. H. Abdul Samad, A. M. A. Rahman, and F. S. Yeok, "Cooling Potentials and CO2 Uptake of Ipomoea Pes-caprae Installed on the Flat Roof of a Single Storey Residential Building in Malaysia," *Procedia - Soc. Behav. Sci.*, vol. **35**, no. December 2011, pp. 361–368, Jan. 2012.

[8] C. P. Tso, "The Impact of Urban Development on the Thermal Environment of Singapore.," *Pap. Present. Rep. Tech. Conf. Trop. Urban Clim. Dhaka*, 1994.

[9] O. T. S. Yeo, K. Saito, and I. Said, "Quantitative Study of Green Area for Climate Sensitive Terraced Housing Area Design in Malaysia," *IOP Conf. Ser. Earth Environ. Sci.*, vol. **18**, p. 012101, 2014.

[10] L. I. U. Memon, R. A., Leung, D. Y. C., & Chunho, "A Review On The Generation, Determination And Mitigation Of Urban Heat Island.," pp. 20, 120–128., 2008.

[11] M. H. A. S. A. A. M. A. R. Asmat Ismail, "Using Green Roof Concept As A Passive Design Technology To Minimise The Impact Of Global Warming..," *2nd Int. Conf. Built Environ. Dev. Ctries. (ICBEDC 2008)*, 2008.

[12] E. G. McPherson, L. P. Herrington, and G. M. Heisler, "Impacts of vegetation on residential heating and cooling," *Energy Build.*, vol. **12**, no. 1, pp. 41–51, 1988.

[13] S. Akbari, H., Davis, S., Dorsano, S., Huang, J., & Winnet, *Cooling our communities, a guidebook on tree planting and light-coloured surfacing Washington*. Lawrence Berkerly Laboratory, 1992.

[14] DOE, "Tomorrow's Energy Today for Cities and Countries: Cooling Our Cities U.S.," *Dep. Energy*, pp. 1–6., 1993.

[15] A. Misni, "Modifying the Outdoor Temperature around Single-Family Residences: The Influence of Landscaping," *Procedia - Soc. Behav. Sci.*, vol. **105**, pp. 664–673, 2013.

[16] S. K. S. O. Thani, N. H. N. Mohamad, and S. Idilfitri, "Modification of Urban Temperature in Hot-Humid Climate Through Landscape Design Approach: A Review," *Procedia - Soc. Behav. Sci.*, **vol**. 68, pp. 439–450, 2012.

Theoretical Review on Sustainable Leadership (SL)

Noorul Adharina binti Zulkiffli[1], Aryani binti Ahmad Latiffi[2]

[1,2] *Faculty of Technology Management and Business, Universiti Tun Hussein Onn Malaysia (UTHM), 86400, Parit Raja, Batu Pahat, Johor, Malaysia.*

Abstract. Leadership has been identified as one of the factor contributes to construction organisation success. Therefore, leadership has acts as an admixture to the sum of leader's effort and subordinates teamwork in order to accomplish organisational goals. There are several types of leadership style that have been practiced by leaders in the construction industry, however, there are still several morality issues such as corruptions, mismanagement and spillages occurred among leaders in the industry. Therefore, a new established leadership style such as sustainable leadership (SL) is needed to reduce any circumstances that can affect both individual and organisation in the construction industry. SL has been recognised by its holistic approaches, which include sustainability elements into it. Additionally, SL is not just depending on leader's ability and resolution only, but to maintain the continuously efforts and achieving results are important. The aim of this paper is to explore the theoretical review of sustainable leadership (SL). A literature review on sustainable leadership (SL) is gained from journals and books of SL and leadership style in construction industry. The overall intent of this paper is to highlight the values and benefits of SL so that it can be adapted by leaders in construction organisation.

1 Introduction

Sustainability goals in the construction industry are not only narrowly focused on traditional management such time, cost and quality, but also giving attention to sustainable elements such as environmental, organisation's profitability and social development. Sustainable construction is the result of sustainability environment policies in the construction industry, which is widely responsible for consumption of natural resources [1].

However, when sustainability issues are added in usual business practice of construction organisation, this creates uncertainty for leaders [2]. This is because leaders in construction industry are still not convinced with the ability of sustainability in their organisation management practices. Although sustainability has been implemented in business activity but there are not directly linked to the management and leadership strategies of the organisations [3]. The influences of sustainability in leadership are still vague among construction leaders; nevertheless leadership and management practices may show a constructive impact on sustainability approaches in the construction industry. This is in line with [4] concept, where leadership has a very significant influence in organisational activities including sustainability.

As a result, to fulfill the commitment of construction's leader towards sustainability, the adaptability of sustainable leadership (SL) in organisation can facilitate the achievement of construction industry. The construction industry needs leadership that provides the collective vision, strategy and direction towards the common goal of sustainability [4]. Thus, it is important that leader have ambitious acts as well as the sustainability knowledge to effectively manage the organisations strategically. Therefore, the purpose of this paper is to review on the background of SL, development of SL in other industries and necessity of SL in construction organisations.

2 Sustainable leadership (SL)

The Sustainability Leadership Institute (2011) offers definition suggesting that sustainability leaders as individuals who are compelled to make a difference by deepening their awareness of themselves in relation to the world around them [5]. In doing so, the leaders adopt new ways of seeing; thinking and interacting that result in innovative and come to sustainable solution. The sustainable solver such SL is an individual who motivate employees and supports sustainability action towards a better world [6] either in present or future. Added to that, motivation and influencing employees towards organisational goals is one of elements in business practices. Nevertheless, common leader viewpoint in order to create sustainable organisation is by becoming more environmentally, than to usual business practices. Šimanskienė and Župerkienė [7] described SL as an

[1] Corresponding author: noorul.adharinna@gmail.com

undertaking with responsibility towards individual, groups, and organisations by assessing the principles of sustainability. They added that sustainable activity demands a personal contribution, such as the transformation of the way of thinking of each individual involved in the process. Therefore, the transformational of self-awareness starts with an individual and then spread to other levels such group, organisation, and the society.

As the conclusion, undertaken from leadership and sustainability definition as well as viewpoint from previous researchers, the definition of SL can be defined as an ability to influence and motivate individual, groups, organisation and society by assessing sustainability knowledge into their principles without neglecting experience from the past, so that it can be improved continuously either in present also in future.

2.1 The elements of sustainable leadership (SL)

The first step to resognise SL style is by understanding from the element in it. The elements of SL is recognised back then in education sector in 2003. However, its only focuses on the compulsory level (elementary and high school) and has not been applied to further level [8]. Hargreaves and Fink [8] acknowledge that, in education especially in elementary and high school is failing to attract quality leaders. Additionally, they address that this is due to the increase in job stress, continuing reductions in school funding and an increasingly diverse student population. Therefore, Hargreaves and Fink [8] have developed models of SL as a tool for leadership capacity within organisations that can provide good values of profit towards it. Since sustainability in the higher education is increasingly important and models of SL have not been applied to education sector [9], research into this area is needed [10]. Nevertheless, the concept of SL is already been used in education sector nowadays. Hence, the concept of SL is increasingly spreading among other industries.

Moreover, sustainability in business organisations have gone beyond traditional method of being 'green' and 'socially responsible' to business day-to day [11]. In order to address these significant shortcomings, Avery [12] has identified SL grid containing 19 elements leadership practices distinguish from Rhineland and Anglo-US approaches. A Rhineland approach is originally the concepts of the economic model of society in Germany and the countries around Germany. Meanwhile, Anglo-US approach is used to describe the principles of business culture in United States and United Kingdom. Between these two models, there are slightly differences in business culture in line of project management perspectives. Table 1 shows the differences between Rhineland and Anglo-US model.

Table 1: The differences between Rhineland and Anglo-US model.

Rhineland model	Anglo-US model
Focus on employees	Focus on higher level direction or control mechanism.
responsibilities.	
Low attention.	High attention.
Teamwork.	Methods and instruments.
Principal based.	Rule based.
Contract and trust.	Contract.
Discussion oriented.	Result orientation.

Based on the Table, these two approaches consists two diametrically conflict to be sets of practices that form self-reinforcing system [13]. Additionally, the Anglo/US principles are less sustainable than Rhineland organisations [12]; [14] and less emphasis on environmental and social measures. Table 2 shows the Rhineland elements on the SL.

Table 2: The Rhineland elements on the sustainable leadership (SL) (Avery, 2005)

Rhineland Elements on the Sustainable Leadership	
1.	CEO concept
2.	Decision making
3.	Ethical behaviour
4.	Financial markets
5.	Innovation
6.	Knowledge management
7.	Long-term perspective
8.	Management development
9.	Organisational culture
10.	People priority
11.	Quality
12.	Retaining staff
13.	Skilled workforce
14.	Social responsibility
15.	Environmental responsibilities
16.	Stakeholders
17.	Teams
18.	Uncertainty and change
19.	Union-management relations

However, because of the limitation in Avery [12] SL Grid, 19 practices, more recently, Avery and Bergsteiner [13] expanded the list of practices and introduce the 23 elements of SL "Honeybee" philosophy which is a resilient and humanistic approach to organisational sustainability that developed from earlier concept. Figure 1 show 23 elements of SL Pyramid.

 DOI: 10.1051/matecconf/20166600045

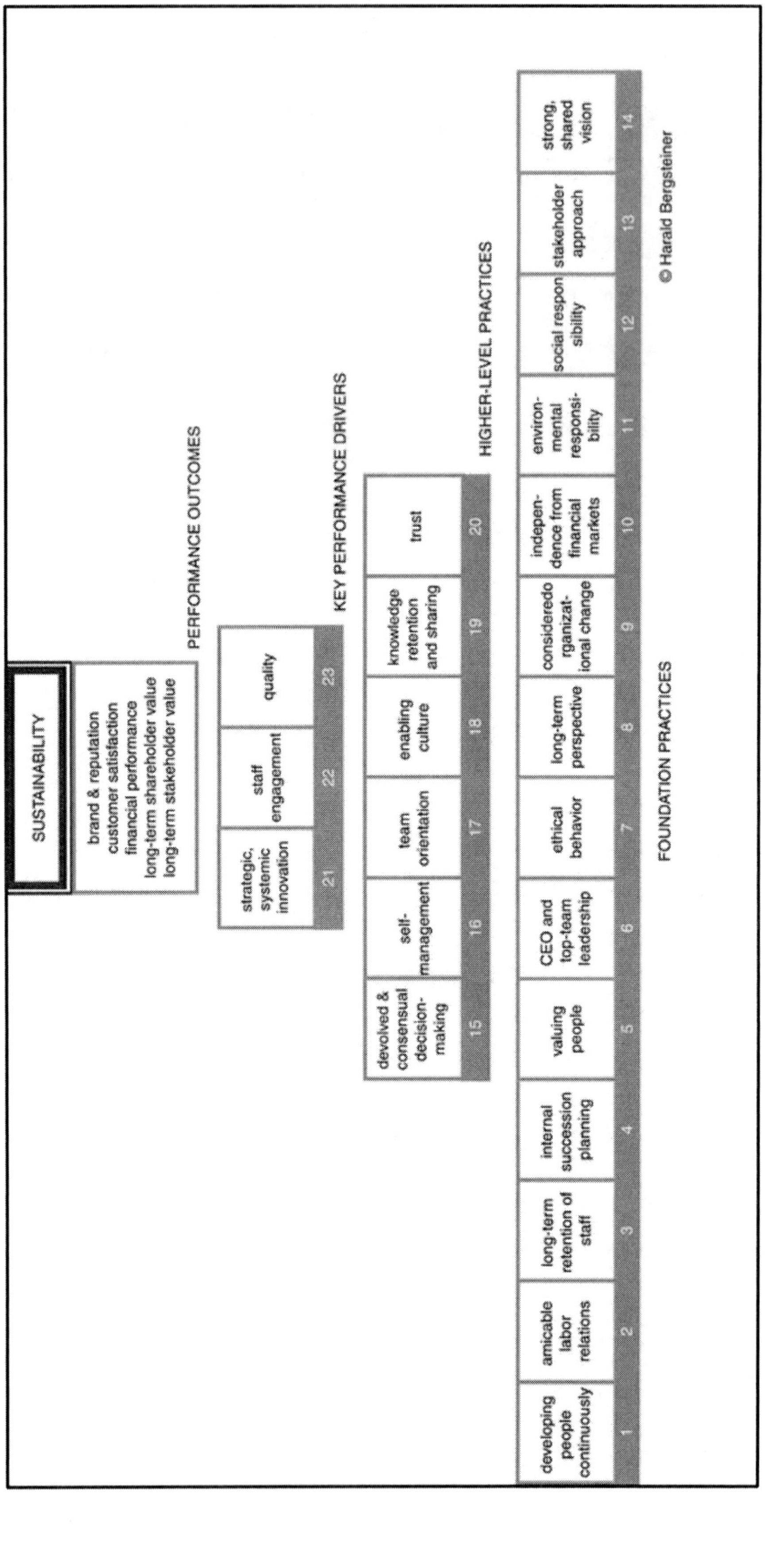

Figure 1: The sustainable leadership (SL) pyramid (Avery and Bergsteiner, 2011)

Based on the Figure, the 23 elements form three groups such foundation practices, higher-level practices, and key performance drivers. In brief, the foundation practices consist of 14 practices and it can be implemented at any time management according to organisation decision. Meanwhile, higher-level practices form the second layer of the pyramid, which includes six (6) practices. The pyramid has been developed on the idea that the foundation practices will be support the emergence of the higher-level practices. The key performance drivers create the third level of the pyramid are emerge from the combination of first and second layers of the pyramid. According to Avery and Bergsteiner [13], the relationship of each element not only from bottom-up or top-down, practices on the same level also influence each other. The interactions between each layer collectively drive the performance outcomes at the apex of the pyramid that clearly emphasis on the long term perspective that somehow creates SL.

In addition, some organisations appear to be taking sustainability approaches for a wide variety of reasons. Hargreaves and Fink [8] stressed that SL is applicable to be adapted in any field of the organisation. SL retains and deepens the holistic knowledge that spreads continuously with assuring a positive impact either in present or in the future.

3 Necessity of sustainable leadership (SL)

Sustainability requires organisational leader to take a bold change to move from just being green to a higher level of performance. For an instance, sustainability is now viewed by organisations as being part of a strategy for long-term business survival and success [15]. Therefore, construction organisations need SL that can transform leaders who can take advantage of the opportunities that sustainability brings, to secure a better future for the construction organisation. The descriptions below explain the benefits of adapting SL in organisation.

3.1 Enhance employee satisfaction

Employee satisfaction is a terminology of employees is happy or satisfied with desired and needs with their jobs. Employee satisfaction is also regarded as a key performance measurement for sustainable organisation [14]. Particularly, SL practices are predicted to enhance overall employee satisfaction in the quest for sustainability [14]. In addition, many of SL practices affect employees directly and since employees have long been touted as the most valuable asset of an organisation, employee satisfaction with the organisation can be a crucial indicator of organisational [16]. Opoku, Ahmed and Cruickshank [17] also stress that SL practices are covered with multi-faceted management systems, principles, processes and values that research shows can create long-term organisational performance. In brief, by adapting SL in an organisation, it will enhance employee satisfaction with organisation and also contribute towards increasing organisational performance.

3.2 Overcome employee turnover

Employee turnover is one of a major problem in the construction industry [18]. Additionally, employee turnover effects the organisation as well as lack of skilled employees, increased other employee's workload, and decreased productivity. According to Avery and Bergsteiner [19], turnover of valuable employee can also lead to intangible losses. They added that replacing talented senior employee means that losses of knowledge and expertise. Therefore, Thomas [18] suggested that an effective leadership is one of strategy to minimise turnover among employee in the construction industry. Hence, a suitable leadership style is a significant success factor to overcome employee turnover as well as increasing employee satisfaction towards organisation.

3.3 Employee retention

Sustainable leadership (SL) prefers to develop and promote employee internally rather than bring in outsiders wherever possible [20]. The purposes of developing and promoting employee internally is to keep employee or specifically as employee retention in order to enhance employee to be loyal and to avoid turnover among employees. In addition, retaining employees can generate unique competitive advantage for the organisation, derived from the linkages that form between long-term employees that enable ideas and skills to be shared in organisation [19]. Hence, by adapting SL in organisation, employee will possibly able to avoid employee turnover and to increase the productivity as credit to organisation value.

In conclusion, employee satisfaction, employee turnover and employee retention can be classified as morality and ethical issues occurred in an organisation. The construction industry also always faces problematic challenges not only technically but also morality such as unethical practices among construction players [21]. Hence, the adaptability of SL in construction organisation will possibility to be the sustainable solution of these morality and unethical issues.

4 Necessity of sustainable leadership (SL) in construction organisation

Leadership has a very significant influence in organisational activities including sustainability [17]. Added to that, sustainability requires organisational leader to take a bold change to move from just being green to a higher level of performance. Therefore, it is a necessity for construction organisations to transform leaders who can take advantage from sustainability brings, to perform a better future for the construction organisations.

Employee satisfaction is essential to the success of any organisation. Therefore, keeping employee's satisfaction should be a major priority for every employer. [14] states that in construction industry, research in human behaviours and attitude merely related with employee satisfaction. They added that employee satisfaction is a number of feelings and belief that employee have about their current job. For an instance, Huang *et al.,* [22] identify that engineers were more likely to have lower job satisfaction under higher time-control or overall job stress. Employees with lower job satisfaction feel unstable and it will be significantly related with employee turnover.

Moreover, due to high work stress and an unstable working environment, the turnover of construction employee is an important issues from the practical viewpoint [23]. Employee turnover is a common issue in any organisation as well as one of a major problem in the construction industry [24]; [18]. Turnover of construction employees have an adverse impact on organisational performance and productivity. This was supported by Muhammad *et al.,* [25] that employee turnover has a relationship with organisation performance. Thomas [18] discovered that there are several problems related to employee turnover such as increased work load, training and development cost, low morale, lower productivity, loss of client satisfaction, poor performance and increase costs. Consequently, it can affect the organisation such as lack of skilled employees, increased other employee's workload, decrease productivity as well as loss of client and investor confidence. Figure 2 below show the leadership practices in construction organisation that compatible with sustainable leadership (SL).

Table 3: The leadership practices in construction organisation (Liphadzi, 2015; Emad, 2014 & Jarad, 2012)

	Leadership Practices	Construction Organisation
1.	Developing staff	✓
2.	Retaining staff	✓
3.	Valuing staff	✓
4.	Goal setting	✓
5.	Long-term perspective	✓
6.	Ethical behaviour	✓
7.	CEO and top team: top team speaker	✓
8.	Organisational culture	✓
9.	Understanding of the business	✓
10.	Responsibility for environment	✓
11.	Social responsibility	✓
12.	Stakeholders: broad focus	✓
13.	Vision's role in the business: shared future	✓
14.	Decision making	✓
15.	Self-management	✓
16.	Team orientation: strong cooperation	✓
17.	Knowledge sharing and retention: strong	✓
18.	Trust/Integrity	✓
19.	Innovation	✓
20.	Staff engagement	✓
21.	Quality: high	✓

Based on the Table, undertaken results from previous researchers, it can be concluded that there are 21 leadership practices that have been applied by leaders in the construction organisation. Added to that, there are some of it has similarity with SL elements. Therefore, the leadership practices are effective in construction organisation and compatible with SL elements. Additionally, extra elements in SL can provide a value added into leadership practices in construction organisation.

5 Sustainable leadership (SL) in other industries

SL is recognised by its holistic approaches, which include balancing people, organisation's profitability and environment towards organizational [13]. The successful of implementing SL in other industries are also approved.

5.1 Retail

Wal-Mart is one of the largest fortune 500 corporations in a world. Scott [26] announced that Wal-Mart was launching a business sustainability strategy to reduce the organisation impact on global environment and become the most competitive and innovative organisation in the world. Eventually, because of this strategy, Wal-Mart has become more ethical, socially and environmentally responsible besides the financial performance was solid [19]. It is also shows that how Wal-Mart expose its ideas about sustainability to its suppliers, associates, client and also community. The adaptability of SL in Wal-Mart may be seen as sustainable procedures towards organisation performance without neglecting responsibilities towards others.

5.2 Automotive

BMW Group (BMW) is a well-known organisation, which produces cars and motorbikes that operates globally. Despite from the global financial crisis and the subsequent recession, when other leaders choose to go into survival mode and abandon strategies to achieve long-term goals, BMW meets closely to SL principles and emerged from the recession strongly, reporting the best results in its history in 2010 and remains a highly profitable independent organisation [19]. The adaptability of SL clearly can contribute towards organisational

DOI: 10.1051/matecconf/20166600045

performance without neglecting a proper balance is created between economic, social and environment.

5.3 Education

Educational organisations are increasingly being run as business organisations in many ways with effective governance as a new requirement [27]. A research taken by Lambert [9] has attempted is to ascertain the views of principals of general further education colleges, by inviting participants to select their level of agreement on each of the component aspects of the SL theoretical framework proposed by researcher. The findings of the research were broad agreement around the aspects of the framework for SL in further education, although further work is needed concerning the articulation of each of the component elements.

5.4 SME

Suriyankietkaew and Avery [28], have investigates what leaders can do to enhance employee satisfaction in the interest of the long term sustainability of the SMEs. SMEs were chosen because of general lack of knowledge about SME leadership practices [29]. The main purpose of the research was to identify the management practices derived from the SL framework and their effect on overall employee satisfaction in Thai SME organisations. They concluded that, the more an organisation adopts SL practices, the higher the employee satisfaction is likely to be.

Additionally, it is proven that SL approaches has been successful spreads in variety of business industries and organisation. Thus, it is not fruitless to adapt a systematic, holistic and sustainable solution of leadership style that can create long-term organisational performance in organisation.

6 Discussion

This paper explores the theoretical reviews of SL from the perspective of previous researchers. Taking into account from the previous researcher's viewpoint, SL can be defined as a kind of leadership that established the sustainable way of influence people and undertaken the responsibility towards individuals, groups, organisations and society. SL also offers as a sustainable solution whereas it has been accepted in various industries and organisations. As been discussed before, SL has been accepted and used in other industries and organisations all around the world. The interaction between each element in SL practices can enhance profit, balancing people and the world to promote sustain and longevity of an organisation through management practices.

There are several types of leadership style that have been practiced by leader in the construction industry. However, there are still some improvement and value added to other leadership style in order to overcome managerial problems in construction industry. The new established leadership style such SL is needed as it is a transformation of current leadership style with assessing sustainability elements in to it. Nevertheless, there some need to be considered in order to implement SL in construction industry such as:

- Converting from conventional business practices into sustainability practices.
- Major change business practices bringing the chance of a drop in short-term performance, so that the stakeholders need to focus on the long-term.
- Consideration of senior manager to change from current leadership style into SL.
- Changes from current leadership style to SL can take time to embed as well as to maintain them.

By implementing commercially SL in business strategy, SL has a role in helping to promote sustainable construction [4]. When the leaders have an ability to understand and implement sustainability practices in organisation, it is not possible for them to pursue sustainable construction with less facing of challenges and barriers. It is worth taking an interest in the principles of the SL in the construction industry for aiming at delivering better and sustainable returns, reducing unwanted circumstances and accelerating organisational profit in present and also in the future.

7 Conclusion

This paper has established an overview and necessity of sustainable leadership (SL) in the construction organisation. The adaptability of SL clearly can contribute towards construction organisation without neglecting a proper balance is created between economic, social and environment. Therefore, research on the adaptability of SL in construction organisation is needed to aware and influence construction leader the capability of SL to overcome challenges and barrier and to help the construction industry to reap the benefit of sustainability practices in organisation management either in present or in the future.

Acknowledgement

The authors would like to thank the Centre for Graduate Studies (CGS) of UTHM for supporting this paper.

References

1. Yılmaz, M. and Bakış, A. (2015). Sustainability in Construction Sector. World Conference on Technology, Innovation and Entrepreneurship, 2253-2262.
2. Demaid, A. and Quintas, P. (2006). Knowledge across cultures in the construction industry: sustainability, innovation and design. 26(5-6), 603-610.
3. Pakir, A. H. K. *et al.*, (2012). Sustainable Housing Development and Leadership: A Review. Australian Journal of Basic and Applied Sciences, 6(12), 385-395.

4. Opoku, A and Fortune, C (2011). The implementation of sustainable practices through leadership in construction organizations *In:* Egbu, C. and Lou, E.C.W. (Eds.) *Procs 27th Annual ARCOM Conference,* 5-7 September 2011, Bristol, UK, Association of Researchers in Construction Management, 1145-1154.
5. Sustainability Leadership Institute (2011). Retrieved August 1st, 2015, from http://sustainabilityleadershipinstitute.org/
6. Visser, W., & Courtice, P. (2011). Sustainability Leadership: Linking Theory and Practice. *SSRN Electronic Journal SSRN Journal.*
7. Šimanskienė, L. Župerkienė, E. (2014), Sustainable leadership: The New Challenge for Organisations. Knowledge Economy Forum, 81-93.
8. Hargreaves, A., & Fink, D. (2006). *Sustainable leadership.* San Francisco: Jossey-Bass.
9. Lambert, S. (2011), "Sustainable leadership and the implication for the general further education college sector", Journal of Further and Higher Education, **35**(1), 131-148.
10. Stephens, J., & Graham, A. (2010). Toward an empirical research agenda for sustainability in higher education: Exploring the transition management framework. *Journal of Cleaner Production,* 611-618.
11. Kantabutra, S. & Saratun, M. (2013). Sustainable leadership: honeybee practices at Thailand's oldest university. International Journal of Educational Management. **27**(4). 356-376
12. Avery, G.C. (2005), Leadership for Sustainable Futures: Achieving Success in a Competitive World, Edward Elgar, Cheltenham.
13. Avery, G., & Bergsteiner, H. (2011a). Sustainable leadership practices for enhancing business resilience and performance, Strategy & Leadership, **39** (3), 5-15.
14. Avery, G.C. and Bergsteiner, H. (2010), Honeybees and Locusts: The Business Case for Sustainable Leadership, Allen & Unwin, Sydney.
15. McCann, J. T. and Holt, R. A. (2010), "Defining sustainable leadership", International Journal of Sustainable Strategic Management, **2**(2), 204-210.
16. Al-Swidi, A. K., M. Nawawi, M. K. & Al-Hosam, A. (2012). "*Is the Relationship between Employees' Psychological Empowerment and Employees' Job Satisfaction Contingent on the Transformational Leadership? A Study on the Yemeni Islamic Banks"*, Asian Social Science, 8 (10).
17. Opoku, A., Ahmed, V. & Cruickshank (2015). *"Leadership, culture and sustainable built environment", Built Environment Project and Asset Management*, 5 (2)
18. Thomas, J. (2013). Study on Causes and Effects of Employee Turnover in Construction Industry. International Journal of Science and Research (IJSR), 2319-7064.
19. Avery, G., & Bergsteiner, H. (2011b). How BMW successfully practices sustainable leadership principles. *Strategy & Leadership,* 11-18.
20. Kantabutra, S., & Avery, G. (2011). Sustainable leadership at Siam Cement Group. *Journal of Business Strategy,* 32-41.
21. Hamimah, A., Abdul Hadi, M. N., Siti Maimunah, M. A., Azizan, S., & Chong, H. Y. (2011). Bills of Quantities: Perspectives of Contractors in Malaysia. Australian Journal of Basic and Applied Sciences, **5**(11), 863-873.
22. Huang, C. F., Yang, L. R., and Ou, Y. H. (2007). "A study on the relationship between job stress and job satisfaction of construction engineers," *Proceedings of CSCE 2007 Annual General Meeting & Conference*, Yellowknife, Canada.
23. Sun, K. S. (2011). The turnover intentions for construction engineers. *Journal of Marine Science and Technology,* **19**(5). 550-556
24. Bilau, A. A., Ajagbe, M. A., Sholanke, A. B., & Sani, T. A. (2015). Impact of employee turnover in small and medium construction firms: A literature review. *Internationbal Journal of Engineering Research & Technology (IJERT).* **4**(2). 976-984
25. Muhammad, N.T, Muhammad, R. & Riaz, A. (2013) 'The impact of employee turnover on the efficiency of the organization'. *Interdisciplinary Journal of Contemporary Research in Business, Institute of Interdisciplinary Business Research.* **4**(9): 1-27.
26. Wal-Mart CEO Lee Scott Unveils "Sustainability 360" (2007). Retrieved October 11, 2015, from http://corporate.walmart.com/_news_/news-archive/2007/02/01/wal-mart-ceo-lee-scott-unveils-sustainability-360.
27. Leach, W.D. (2008), Shared Governance in Higher Education: Structural and Cultural, Responses to a Changing National Climate Center for Collaborative Policy, California State University, Sacramento, CA.
28. Suriyankietkaew, S., & Avery, G. (2014). Employee satisfaction and sustainable leadership practices in Thai SMEs. *Journal of Global Responsibility JGR,* 160-173.
29. White, R.J., D'Souza, R.R. & McIlwraith, J.C. (2007), "Leadership in venture backed companies: going the distance", Journal of Leadership and Organizational Studies, **13**, 121-132.
30. Liphadzi, M. (2015). Leadership style practices of construction and project managers in the South African construction industry (Master thesis). University of Johannesburg. Johannesburg, South Africa.
31. Emad, A. (2014). Leadership importance in construction productivity improvement. Global

DOI: 10.1051/matecconf/20166600045

Advanced Research Journal of Management and Business Studies. **3**(3), 114-125.

32. Jarad, G. H. (2012). The construction manager leading characteristics for the success of construction projects in the Gaza Strip (Master thesis). The Islamic University of Gaza. Gaza.

Prelimanary Investigation on The Factors That Influencing The Maintenance Cost of Apartment

Nor Aini Salleh[1,a], Mohd Khairul Annuar Mohd Yakin[1], Kharizam Ismail[2] and Yuhainis Talib[2]

[1]*Department Of Estate Management, Faculty Of Architecture, Planning & Surveying, Universiti Teknologi MARA Cawangan Perak, Seri Iskandar Campus, 32610 Seri Iskandar, Perak, Malaysia*
[2]*Department Of Quantity Surveying, Faculty Of Architecture, Planning & Surveying, Universiti Teknologi MARA Cawangan Perak, Seri Iskandar Campus, 32610 Seri Iskandar, Perak, Malaysia*

Abstract. Housing is a basic need for all. The increasing numbers of population and income had increase in housing demand. Thus, it has led to the construction of apartments to meet the needs of people and to reduce the use of land to build residential buildings. From year to year, maintenance costs of the building are increased. To ensure the maintenance cost for apartment residence maintenance can be reduced, it is important to determine the factor that influence the maintenance cost of the apartment. This paper is to investigate and identify the factors contributing the increase of maintenance costs. The first objective of this study is to determine the factors that affect the cost of maintaining the apartment and meanwhile the second objective is to rank the factors that influence the maintenance cost of the apartment at Johor Baharu Tengah. To achieve the objectives of the study, the questionnaires were distributed to the parties involved in the management of building. Results of this study concluded that there are five factors that influence maintenance cost are the expectations of tenants, building age, building material, failure to execute maintenance at the right time and budget constraints.

Keywords: Maintenance Costs Factors, High rise buildings, Building Owners, Tenants and Types of Maintenance

1 Introduction

The increase in demand for housing and the lack of land for the construction of landed residential properties in major urban areas in Malaysia such as Penang, Kuala Lumpur, Selangor and Johor Bahru, has led to the rapid development of high-rise residential schemes in high-density areas. Every apartment requires building maintenance to ensure the building in good condition. Building upkeep is the blend of specialized and managerial activities to guarantee the things and components of a building in a worthy standard to perform its required functions. So as to execute building upkeep assignments productively, a legitimate building support plan and observing framework is vital. According to [14], building support administration is an operation including the connection or mix of specialized, social, legitimate and financial determinants that oversee and deal with the utilization of structures. It can be contended that numerous individuals don't comprehend the significance or hugeness of building upkeep and its administration, specifically the acknowledgment that the productivity of a building upkeep framework adds to the pay of the organization's owning or leasing the building [13]. High building maintenance involves many aspects. An important aspect of building maintenance is of the

[a] Corresponding author: noinsphd07@gmail.com.my

maintenance costs. Maintenance costs correctly and efficiently will ensure that a building can be managed properly.

2.FACTOR THAT INFLUENCES THE MAINTENANCE COST OF APARTMENT

Buildings must be properly maintained by a person who has the skills to maintain. Definition of maintenance has many interpretations. BS 3811(4) stated that maintenance is "any work done to maintain or restore each facility", an example every part of a site, building and every each which related with building must be at acceptable standard. Besides that, maintenance is a work that must be done to restore and maintain a building of damage according to a present standard. Hence, building maintenance is to keep a building in good condition and condition appropriate to its use [9, 42]. Maintenance of building is a combination involving technical and administrative measures to ensure the item and all elements related to the building in which the desired standard conditions acceptable to perform the desired function [14]. Therefore, the ultimate aims of maintaining the building according to [4] are to maintain the value of the investment, to ensure buildings in an acceptable condition and the required standard, to present a good appearance of the building, and to conserve the values of history and architecture of the building.

There are various types of building maintenance as mentioned by [12], building maintenance can be divided into three such as reactive (day- to- day repair); preventive maintenance which include cyclic and condition based maintenance, and modernisation and improvement which include additional works which need as a result of new legislation and regulations. In other words there are two major categories of building maintenance namely planned maintenance and unplanned maintenance [26]. Maintenance cost is covering overall cost or budget set to take care of the building condition. Maintenance cost also related to direct cost like material, labour, plant and tool that use to maintain the building. Meanwhile indirect cost like management cost and administrative staff and cost overhead is also maintenance cost of high rise buildings. [40] mentioned that maintenance and operation expenditure is one of the major housing cost components. Usually, maintenance and operation of building contribute expenses one-third to half from total cost of maintenance work depend on a type of residential such as condominium, apartment, flat and other. There are several factors that influence the maintenance cost of a high residential building. Generally, they can be divided into five groups such as building characteristics, tenant factors, maintenance factors, new regulation factors and other factors.

2.1 Building characteristics factor

Building characteristic is one of the factor that influence on the maintenance. Building characteristic are related to building age, function, building area, height of building, types of structure, finishes services of building, construction material and other. Every building has specific characteristics and differences in the distribution of the total cost of prescribed maintenance to ensure the building is always in a good condition [46]. [34,45 & 44] also noted that maintenance cost will increase due to the increasing age of the building.

Based on [43], all building will pass aging in various exposure conditions. Building height and building area also a factor that affects the maintenance cost [44]. Part of building maintenance work does involve building services. The building services system should be made and implement to provide a healthy and safe life to the residents or occupant of the building [19] [5] in his opinion mentioned that the building services need to be implemented for every building, including ventilation, lighting, power supply, water supply, sanitation, communication, transportation and another system. Building services and structure of building become elements that also affect maintenance cost. In this phenomenon, a lot of financial resources need to be allocated for inspection, maintenance, repair, rehabilitation and replacement of building structure [32]. [5] also agreed that, to maintain the structural stability of the building, inspection and maintenance should be carried out from time to time to ensure the safety of occupant in a safe condition. Maintenance and asset cost operation and services life influenced by the choice of material selected. To repair and replace building component that had broken down will affect the choices of material used. Better quality materials do require high maintenance cost [43].

2.2 Tenant Factor

Tenant or occupier at high residential building influence the maintenance cost of housing. According to [35], the maintenance requirement caused by tenant as much as 25 percent. Thus, strategies that can be done by the landlord in closing the gap between expensive maintenance and management of legitimate expectations or the demand of customers with tenant and resident participation in housing management [53]. Expectation of tenants or residents, use of property, vandalism by the occupier or a tenants, delay and failure in report problems and accessibility to the property give effect to maintenance cost [12, 36, 6 & 5].

2.3 Maintenance Factor

Based on [12], maintenance factor may have a big influence on housing maintenance cost. Generally, maintenance factors can be divided into two main factors namely technical factor and administrative factor. From the aspect of technical factor, several aspects that affect maintenance cost are poor workmanship and poor quality of the spare part and material. In the aspect of administrative factor, aspect which influences maintenance cost are poor maintenance management, budget constraint, failure to execute maintenance at right time and poor budgetary control [29, 17,18, 27, 31, 21, 37, 2, 24].

2.4 Regulation Factor

[12] in their study, proved that changes in regulations clearly affect maintenance cost of housing. Maintenance activity is highly sensitive to changes in the local and national regulation. Statutory legislation regulates many activities of those who are involved in built environment industry. Building control legislation set up minimum standards that buildings are safe and efficient to use [5]. Safety legislation is to eliminate accidents in the building [24]. This legislation control and restrict how maintenance activity is carried out and of cause does affect cost of maintenance.

2.5 Other Factor

Third party vandalism is a serious problem that occurs in some high rise residential building [49]. The damage by the third party will require additional maintenance cost for repairing the damage that has been done. Another factor that leads to housing maintenance cost increase is due to poor or lack training for staff maintenance. According to [31], of the reasons for poor operating practices in the management of maintenance is due to lack of maintenance staff training. Staff maintenance or skilled operator are the factor that affect the performance of maintenance [37].

3 Study Area and Methodology

District of Johor Bahru Tengah was taken as a case study. It is one of the ten districts that make up the state of Johor. The study area covers an area of 338.5 square kilometers which includes Skudai, Nusajaya, Ulu Tiram, Masai, Kangkar Pulai, Gelang Patah, Plentong, and Lima Kedai. This district are also known as metropolitan Johor Bahru because of the number of population in the district exceed 600,000 people and now the number of population is increasing to 525,351 people [39]. In this area there are about one hundred and twenty two (122) apartment units, however the focus is only on medium cost apartment and high cost apartment that has been built between year of 1990 to 2014. Therefore the total number of sample in this research is Seventy Six (76) units of building apartment. This study used quantitative method (Likert Scale) to gather data from respondents in selected apartment at Johor Bharu (1990 – 2014). Five point scale ranging from "not impacting at all" on one end to "impacting a great deal" on the other with "to some degree" in the middle. In the survey form, five main variables of maintenance factors were asked to the respondents. They are, building characteristic factors, tenant factor, maintenance factor, regulation factor and other factors such as vandalism and lacking of staff training. Questionnaire survey had been distributed to all building manager or building supervisor or maintenance management officer or accountant who can provide valuable information related to building maintenance cost, this is because of the small population (less than 100). A total of 76 questionnaires survey form had been distributed directly to respondents. It is believed that to sample the whole populations can achieve a desirable level of precision [15]. All the data from the questionnaire survey had been statistically analysed with SPSS software. Two types of analysis were used, namely descriptive mean and ranking. First analysis was to derive the mean score of every variable. Second analysis was to rank the variables according to the most influence factor affected maintenance cost in the buildings.

4. Results and discussions

Out of 76 questionnaires survey form that has been distributed only 35 respondents (46%) participated in the survey. Although the response rate in this study was low, it was considered acceptable because it was sufficient for the population less than 1000 respondents[33]. [16] mentioned the research is only need 30 per cent response rate to produce results that can be believed and trusted in order to achieve the objectives of the study. The first objective of this study was achieved based on desk study analysis. The researcher found that there were twenty (20) variable factors that influence maintenance cost of apartment. The twenty (20) variables then grouped into five main factors such as building characteristics, tenants' factor, maintenance factor, legislation factor and other factors. Building characteristic has five variable factors that influence maintenance cost of the apartment that are building age, building height and building area or size, building services, type of structure and building material. The second factors are tenants' factor. Tenant factor has five variables factor that influence maintenance cost of high-rise building such as high expectation of tenants, use of the building, vandalism by the tenants, delay in reporting failures and accessibility to the property. Maintenance factor was divided into two main factor which are technical factor and administration factor. Technical factor has two variables such as poor workmanship and poor quality of spare part and materials. Administration factor also has five variables factor that influences maintenance cost of apartment that are poor maintenance management, budget constraints, poor failure reporting procedures, failure to execute maintenance at the right time and poor budgetary control. Second last was legislation factor. Lastly, two variables in other factors are third party vandalism and poor or lack of staff training. Based on the factors as mentioned before, it can be concluded that there are many factor that influence the maintenance cost of apartment. Factor that influence the maintenance cost of an apartment is summarised in Table 1 below.

Table 2 shown that the highest mean score of the most influence factor and the lowest mean score will represent the least influence factor. The first rank of influences factor is a high expectation of tenant factor with a mean score of 4.06. The next ranking is building age, followed by building material. The lowest rank is new regulations factor with a mean score of 2.60. In table 2, the expectation of tenant has proven to possess the highest rank at the top of the list. This gives the impression that the expectations of tenants have made the most impact on the maintenance cost of high rise building. The factors that influence tenant demand due to thing such as increasing tenant demand toward better lifestyle and living environment. Thus, has led to an increase in maintenance requirements and maintenance costs. Other than that, it may also caused by the maintenance activities that was not in the contract between the tenant and the local authorities (owner). Building age is the second rank of the factors that influence maintenance cost of high rise buildings. Building age is a critical source of high maintenance cost of the building. The longer the age of the building, the higher the maintenance cost. According to [22] explains that in making provision for maintenance budgets, one of the key elements that need to be measured is the age of the building. This is due more the maintenance work needs to be carried out for older buildings. Building materials are ranked third factor that influence maintenance cost of high rise building. Rising material cost will cause increased maintenance cost, especially when there is defect in a building. The selection of building material by the management which is not suitable also gives influence to the increasing of the maintenance cost. Failure to execute maintenance at the right time was the fourth-ranked factor that influence maintenance cost of high rise building. This finding is similar to the work of [31], he stated that the failure or delay to perform maintenance tasks at the right time can cause maintenance expenses. Budget constraint was fifth rank for the factor that influence maintenance cost of high rise building. Maintenance budget are often based on financial priorities laid down by the management.

Table 1: Factors that influence the maintenance cost of apartment

MAIN GROUPS	VARIABLES	AUTHORS
Building Characteristic	Building age, Building height and building area or size, Building services, Types of structure, Building Material.	Azlan, 2010; Sonthya, 2006; John and Croome, 2005; Neves, Frangopal, & Cruz, 2004; Shabha, 2003; Skinner, 1982; O'Neil, 1974; and Slater, 1892
Tenant Factors	High expectation of tenant, Improper use of the building, Vandalism by tenant, Delay in reporting failures, Accessibility to the property.	Bavani, 2010; Azlan, 2010; El-Haram & Horner, 2002; Olubudon, 2001; Yip, 2001; and Olubodun & Mole,1999
Maintenance Factor	Technical factor: Poor workmanship, Poor quality of spare part and materials. Administration factors: Poor maintenance management, Budget constraints, Poor failure reporting procedures, Failures to execute maintenance at the right time, Poor budgetary control.	El-Haram and Horner, 2002; Mills, 1980; Hammad, Al-Mubaiyadh and Mahmoud, 1996; Horner et al., 1997; Love and Irani, 2003; Narayan, 2003; Khalid, Marosszeky and Davis, 2006; Pascual, Meruane and Rey, 2008; Ali, 2009 and H. H. Y. Lee & Scott, 2009.
Legislation Factor	Changes in regulation	El-Haram and Horner, 2002; Azlan et al., 2010; and H. H. Y. Lee and Scott, 2009
Other Factor	Third party vandalism and lack of staff training.	Tiun, 2003; Narayan, 2003; and Pascual et al., 2008

Table 2. Ranking of factors influencing maintenance costs of building

Variables	N	Mean
High expectation of tenant	35	4.06
Building age	35	3.46
Building material	35	3.37
Failure to execute maintenance at the right time	35	3.29
Budget constraints	35	3.26
Improper use of the property	35	3.23
Poor maintenance management	35	3.09
Vandalism	35	3.03
Poor or lack of training	35	3.00
Poor budgetary control	35	2.94
Type of structure	35	2.91
Third party vandalism	35	2.89
Delay in reporting failures	35	2.89
Poor quality of spare part and materials	35	2.86
Building services	35	2.86
Building height and building area or size	35	2.83
Poor workmanship	35	2.80
Accessibility to the property	35	2.77
Poor failure reporting procedures	35	2.74
New regulations	35	2.60

5. Conclusion

This research provides valuable information related to maintenance cost of apartment buildings. The costs of maintaining a building always become an issue to management and owner of the property. Therefore, steps should be taken or made to reduce the maintenance cost of high rise building. All parties should give cooperation in overcoming the problem of increasing maintenance cost. All parties such as tenant and residents should participate in the work of maintenance management, for example, the residents should together taking care the common area of property. In addition to this, building maintenance management should also introduce operation manual and regulations to all the residents as a guideline together with the budget of maintaining the building in order to create awareness and self-belonging to each other. By introducing the system, they should be able to comply the regulations as provided by the management. Further actions to reduce the cost of building management are the maintenance management should adopt effective maintenance management principles that are suited to their buildings. Building maintenance management has to conduct maintenance work accordingly to the schedule that has been planned. Apart from that strict enforcement should be implemented to all resident or tenant regarding maintenance fee payment from time to time. A Top leader of maintenance management department must make a right decision on selection to choose staff with qualification that is equivalent to post applied. Hence, provide a good quality work on maintenance duties up to the standard required or expected by the relevant parties' such as occupiers and building owners.

In conclusion, this research had given benefits to many related parties such as developer - the research would help the developer understand the factor affecting maintenance cost apartment. The developer will also know how to improve their services that influnce the maintenance cost of apartment and indirectly minimize the maintenance cost of their building. Hence, will save the maintenance cost when they manage the high rise building for new development. The government body will be able to identify the factor affecting maintenance cost of high-rise building and budget that need to be allocated for maintenance activity. First hand information will reduce the maintenance cost of government building, building owners and tenants - this research also gives valuable information to the building owners and tenants. The building owners and tenant will know factor why the maintenance cost different every year. Thus required them to take care of the building where they stayed.

References

1. Al-Arjani, A. H. (1995). Impact of cultural issues on the scheduling of housing maintenance in a Saudi Arabian urban project. *International Journal of Project Management,, Vol. **13** No. 6*, pp. 373-382.
2. Ali, A. S. (2009). Cost decision making in building maintenance practice in Malaysia. *Journal of Facilities Management, 7*, pp 298-306.
3. Allen, D. (1993). What is Building Maintenance? *Facilities, 11 iss 3*, 7-12.
4. Azlan, S. A. (2009). Cost decision making in building maintenance practise in Malaysia. *Journal of Facilities Management, Vol.7 No.4*, pp 298-306.
5. Azlan, S. A., Kamaruzzaman, S.-N., Sulaiman, R., & Peng, Y. C. (2010). Factors affecting housing maintenance cost in Malaysia. *Journal of Facilities Management, Vol **8***, pp 285-298.
6. Bavani, M. (2010). Most lifts faulty due to vandalism, *The Star Malaysia, March 1,*.
7. Chanter, B., & Smallow, P. (2007). *Building Maintenance Management*. Blackwell Science: Oxford.
8. Cheung, M. S., & Kyle, B. R. (1996). Service life prediction of concrete structures by reliability analysis. *Construction and Building Materials, Vol. **10** No. 1*, pp.45-55.
9. Chudley, R. (1981). The Maintenance and Adoptation of Building. New York: Lognan Inc.
10. David, W. W., & Arthur, B. (1989). *Management to maintain quality in buildings", Proceedings of Implementation of quality in construction*. Copenhagen, .
11. Haram, M. A. (1995). *Integration approach to condition-based reliability assessment and*

maintenance planning. (Ph D. Thesis), University of Exeter.
12. El-Haram, M. A., & Horner, M. W. (2002). Factor affecting housing maintenance cost. *Journal of Quality in Maintenance Engineering, Vol.* ***8*** *No 2*, pp. 115-123.
13. Emma, M. A. Z., Syahrul, & N.K. (2009). Personnel characteristics of maintenance practices a case of high-rise office building in Malaysia. *journal of Subtainable Development, Vol* ***2*** *No.1*, 111-116.
14. Francis, W. H. Y., Lee, W.L ,Ng.C.K. (2001). Building energy efficiency and the remuneration of operation and maintenance personel. *Facilities, 20*, 406-413.
15. Glen, ID (1992). Determining sample size. Agricultural Education and Communication Department, Florida Cooperative Extension Service, 20 Mac 2013, available at: http://edis.ifas.ufl.edu/pd006
16. Gillham, B. (2000). *Developing a Questionnaire*. Continuum, London.
17. Hammad, A. A.-., Al-Mubaiyadh, S., & Mahmoud, T. (1996). Public versus private sector's assessment of problems facing the building maintenance industry in Saudi Arabia. *Building Research & Information, Vol.* ***24*** *No.4*, pp. 245-254.
18. Horner, R. M., EL-Haram, M. A., & Munns, A. (1997). Building maintenance cost strategy: a new management approach. *International Journal of Quality in Maintenance, Vol.* ***3*** *No.4*, pp273-280.
19. John, G. A., & Croome, D. J. C.-. (2005). Contextual prerequisites for the application of ILS principles to the building services industr. *Engineering, Construction and Architectural Management, Vol.* ***12*** *No. 4*, pp. 307-328.
20. Kelly, A., & Harris, M. J. (1978). *Management of Industrial Maintenance*. Butterworths, London.
21. Khalid, K., Marosszeky, M., & Davis, S. (2006). Managing subcontractor supply chain for quality in construction. *Engineering, Construction and Architectural Management, Vol.* ***13*** *No. 1*, pp. 27-42.
22. Lateef, O. A. (2008). Building maintenance management in Malaysia. *Journal of Building Appraisal, Vol.* ***4*** *No. 3*, pp. 207-214.
23. Lee, & Wordsworth, P. (2001). *Lee's Building Maintenance Management*. Blackwell science: Oxford.
24. Lee, H. H. Y., & Scott, D. (2009). Overview of maintenance strategy, acceptable maintenance standard and resources from a building maintenance operation perspective. *Journal of Building Appraisal, Vol.* ***4*** *No. 4*, pp. 269-278.
25. Leonardo, L. (2011). Interview, senior maintenance engineer. In Y. C. f. M. Contacting (Ed.). Dammam.
26. Leong, K. C. (2004). *The essence of good facility management- Aguide for maximization of facilities asset economic life and asset opptimissation for realiable services and user satisfaction.* (Vol. Vol 24).
27. Love, P. E. D., & Irani, Z. (2003). A project management quality cost information system for the construction industry. *Information & Management,, Vol.* ***40*** *No. 7*, pp.649-661.
28. Malek, A. (2013). Simple Research.
29. Mills, E. D. (1980). *Building Maintenance and Preservation-A Guide to Design and Management*. Butterworths.
30. Mjema, E. A. M. (2002). An analysis of personnel capacity requirement in the maintenance depertment by using simulation method. *journal of Quality in Maintenance, Vol.* ***8*** *No. 3*, pp. 253-273.
31. Narayan, V. (2003). *Effective Maintenance Management: Risk and Reliability Strategies for Optimizing Performance*. New York, NY: Industrial Press.
32. Neves, L. C., Frangopal, D. M., & Cruz, P. S. (2004). Cost of life extension of deteriorating structures under reliability-based maintenance. *Computers & Structures,, Vol. 82 Nos 13-14*, pp. 1077-1089.
33. Neuman, W.L. (1991). *Social Research Methods: Qualitative and Quantitative Approaches*. United States Of America: Pearson
34. O'Neil, D. (1974). The determinants of housing cost. *Architects Journal, Vol.* ***3***, pp. 753-755.
35. Olubdun, F. (2001). A multivariate approach to the prediction of maintenance needs in public housing: the tenant dimension. *Structural Survey, Vol.* ***19*** *No. 2*, pp. 133-141.
36. Olubodun, F., & Mole, T. (1999). Evaluation of defect influencing factors in public housing in the UK. *Structural Survey, Vol.* ***17*** *No 3*, pp. 170-178.
37. Pascual, R., Meruane, V., & Rey, P. A. (2008). On the effect of downtime costs and budget constraint on preventive and replacement policies. *Reliability Engineering & System Safety Vol.* ***93*** *No. 1*, pp. 144-151.
38. Raymond, C. M., & Joan, C. F. E. (1991). *Preventive Maintenance of Buildings*. London: Chapman and Hall.
39. Report Majlis Bandaya Jphor Bharu (2015)
40. Rydell, C. P. (1970). *Factor Affecting Maintenance and Operating Cost in Federal Public Housing Projects*. New York NY: The Rand corporation.
41. Sanjay, M. (2014). Primary Data Analysis.
42. Seeley, I. H. (1987). *Building Maintenance*. Palgave: New York, NY.
43. Shabha, G. (2003). A low-cost maintenance approach to high-rise flats. *Facilities, Vol.* ***21*** *Nos 13/14*, pp.315-322. Shohet, I. M., & Pacuik, M. (2004). Service life prediction of exterior cladding components under standard conditions. *Construction Management and Economics, Vol.* ***22*** *No. 10*, pp.1081-1090.

44. Skinner, N. P. (1982). Local authority house maintenance – the variation in expenditure. *Housing Review, Vol. **31***, pp. 92-94.
45. Slater, K. (1892). *An investigation into hospital maintenance expenditure in the North West Regional Health Authority* (P. S. (Ed.) Ed.). Bradon,London: New Directions, E. & F.N. Spon.
46. Sonthya, V. (2006). *Relationship between building characteristics and rental to support serviced apartment investment*. Paper presented at the paper presented at the Pacific Rim Real Estate Society (PRRES) Conference.
47. Teo, E. A. L., & Harikrishna, N. (2006). A quantitative method for efficient maintenance of plastered and painted facades. *Construction Management and Economics, Vol. **24** No. 12*, pp. 1283-1293.
48. Thompson, P. (1994). The maintenance factor in facilities management. *Facilities, Vol. **12** No. 6*, pp. 13-16.
49. Tiun, L. T. (2003). The Management of High Rise Housing Complex, *Utusan Melayu*.
50. Tiun, L. T. (2006). *Managing high- rise residential building in Malaysia : Where are we?* Paper presented at the 2ND NAPREC CONFERENCE, Universiti Sains Malaysia.
51. UmarAl-Turki. (2011). A framework for strategic planning in maintenance. *Journal of Quality in Maintenance Engineering, **17** No 2*, 150-162.
52. Wong, L. T. (2006). A cost model for plumbing and drainage systems. *Facilities, Vol. **20** Nos 11/12*, pp 986-993.
53. Yip, N. M. (2001). Tenant participation and the management of public housing – the estate management advisory committee of Hong Kong. *Property Management, Vol. **19** No. 1*, pp. 10-18.
54. Yusoff, Y. M., Tawil, N. M., N. Hamzah, Abdullah, N. A. G., & Musa, A. R. (2011). Tinjauan Kesan Fasiliti Kediaman Bertingkat Terhadap Amaun Dana Pengurusan. *Journal Design + Built, **4***, 32-40.
55. A. Mecke, I. Lee, J.R. Baker jr., M.M. Banaszak Holl, B.G. Orr, Eur. Phys. J. E **14**, 7 (2004)
56. M. Ben Rabha, M.F. Boujmil, M. Saadoun, B. Bessaïs, Eur. Phys. J. Appl. Phys. (to be published)
57. Luigi T.De Luca, *Propulsion physics* (EDP Sciences, Les Ulis, 2009)
58. F. De Lillo, F. Cecconi, G. Lacorata, A. Vulpiani, EPL, **84** (2008)

Encapsulating the delivery of affordable housing: An overview of Malaysian practice

Nor Baizura Jamaluddin[1,a], Yusfida Ayu Abdullah[2] and Hazlina Hamdan[2]

[1]*Centre of Postgraduate Studies, Faculty of Architecture, Planning and Surveying, Universiti Teknologi MARA, 40450 Shah Alam, Malaysia*
[2]*Centre of Studies for Town and Regional Planning, Faculty of Architecture, Planning and Surveying, Universiti Teknologi MARA, 40450 Shah Alam, Malaysia*

Abstract. Urban population increases much faster than other geographical areas and this brings huge challenges to Malaysian government especially those responsible for the provision of housing. Moreover, the rise of living cost and pressure towards the current economic situation had inevitably led to huge demands for affordable housing. Thus, affordability and inaccessibility of people in owning a house has become one of the major issues in Malaysia. These issues are not only faced by the low-income group but mostly experienced by urban dwellers including the middle-income group whom are not eligible to apply for low-cost housing delivered by the government, yet cannot afford to buy a house. Hence, the Malaysian government had therefore established the *Perumahan Rakyat 1 Malaysia* (PR1MA) as the catalyst in providing adequate, quality and affordable houses. Furthermore, the *Syarikat Perumahan Negara Berhad* (SPNB) and the state governments also play their big role in providing affordable house at the state level. In trying to ensure smooth delivery of affordable housing, a systematic approach is required. This paper intends to examine the delivery of affordable housing in Malaysia. The objective was to assess the parties involved, and the various stages of the process. The methodology adopted for this study includes in-depth interviews with affordable housing agencies. Investigation showed that the major problem relates to a mismatch in the delivery of affordable housing. Moreover, the political and economic aspects, as well as the organization situation had also influenced the inefficiency of affordable housing delivery system. Additionally, the significant of this paper is expected to formulate an improved strategy and guidance in delivery system, thus ensuring that the method of practice is applicable throughout the country.

1 Introduction

This paper presents the results of an initial study on affordable housing in Malaysia. As determined by the National Housing Department [1], affordable house should be sold below RM300, 000 while the buyers are the households with income of RM5,000 and below. Owning a house is a dream to many especially to those who had just begun to start a family life. However, buying a house is not quite a straightforward thing and not every person can afford one. Many people, from a youth who is starting a new career and family, or those living in urban areas including from the middle-income group are affected by this difficulty [2]. Moreover, due to urbanization process, the development and expansion of major cities affects the economy of the nation [3]. In Malaysia, most of the population nowadays had migrated and lived in urban areas and that brings huge challenges towards the provision of affordable housing [4]. This was reported in the 11th Malaysia Plan, whereby, the demand and supply gap of affordable housing still exist due to the socioeconomic change, urbanization and evolving population structures [5]. Additionally, it also reports the inadequate supply of affordable housing particularly for the middle-income households especially in the urban areas, increasing of house price in major cities and the lack of integrated planning and implementation [5].

Affordable housing has been a much debated subject in the most recent years especially in a developing country like Malaysia. Although many issues had been resolved, the country still encounters problems and this had brought various perceptions towards the development of affordable housing. The main issue is the adequacy, affordability, and inaccessibility of homeownership, and sadly, these have remained as a continuing issue even though there appeared to be plenty of housing stock to accommodate every family [6, 7]. Besides affordability and housing price, Malaysia is also facing the issue of housing delivery system [8, 1]. In the 11th Malaysia Plan, housing delivery remained as an important issue mentioning the lack of integration and coordination which led to the failure to meet the targets and the mismatched in delivering affordable housing [5]. It was claimed that there were many agencies involved under multiple roofs causing a duplication of duties [8, 9, 6] and the mistake of targets with high inclusion error, sublet, or rent seeker [8]. Moreover, there appeared to be disparity on the supply and demand especially in terms of the

[a] Nor Baizura Jamaluddin: norbaizurajamaluddin@gmail.com

location factor and the integration with public transportation [8, 10, 11]. The constraints during the planning approval process when submitting to the planning authorities or land offices, in terms of conversion of land use category, duration upon getting planning approval, and non-compliance with development plan [11]. The delivery procedures was claimed to be very tedious and time-consuming, causing it to slow down the process [12]. Meanwhile, previous studies discovered that access to mortgage and finance is one of the difficulties faced by the house buyers as it is apparent that most of the lower and middle-income people are facing with financial complexity [11], and those without fix income are obviously badly affected [9]. It is not just the commitment for loan but also the struggle to save and pay for deposits [6].

Realising these unfortunate conditions, the government performs various efforts in trying to assist the lower and middle-income groups with the provision of affordable housing as top priority. In Malaysia, housing provision is the responsibility of both public and private sectors. Private developers focus on selling medium and high cost house, whilst the government is responsible in providing houses for the low-income group. In spite of this, the situation has now changed. Other than this basic role, the government puts a lot of effort in ensuring all income groups have an equal chance in possessing a house. Throughout the 10th Malaysia Plan (RMK-10), various public affordable housing programs were implemented. The government encourages public private partnership in housing development and therefore, private sector such as developers has now contributed in the provision. In addition, the government now also focuses in delivering low cost houses for the low-income group as well the affordable house for the middle-income group especially in the urban areas. In general, housing delivery is a system to complete and deliver housing units to the house buyers. It comprises of a process and key players involved in the housing process [12].

In order to successfully deliver affordable house to the citizens, it should consists of a system that is practiced by each agency. Through the system, the role of the parties involved and the process undertaken can be the indicators in measuring the effectiveness in delivering affordable houses. Based on the housing industry perspective, a delivery system forms one of the cornerstones of customer satisfaction [13]. Through the *Vision 2020*, as put forward by the forth Prime Minister of Malaysia, Tun Dr Mahathir Mohamad, in order to accomplish Vision 2020 in terms of house ownership is through “operative housing delivery system” [14].

Based on those initial background and issues discussed, this paper, therefore, aims to examine the present delivery system of affordable housing in Malaysia. The paper will review the current practice of affordable housing delivery process, the various parties involved and the issues encountered. Thereafter, it will recommend further potential study on the topic.

2 Research Methodology

This research paper employs the multiple case study design. It applies the qualitative method as it deals greatly with policies and implementation procedures. Its emphasis is on the various affordable housing providers in Malaysia. The main idea is to evaluate the variation of roles and practice. The main focus is by means of reviewing the existing literature. However, based on previous studies [15, 16, 17] this study also adopts the qualitative approach by interviewing the specific agencies. As Khalaf [15] suggests, the qualitative method allows interviewees to freely answer the question concerning their views. And, because of that, the findings from the interviews will complement the literature review. For those reasons, this study had selected the three distinctive affordable housing providers, namely *1 Malaysia's People Housing Program* or *Syarikat Perumahan Rakyat 1 Malaysia* (PR1MA), *Syarikat Perumahan Negara Berhad* (SPNB), and the Selangor Housing and Properties Board for *Rumah Selangorku*. However, drawing back from the earlier discussion on affordability, this study will only focuses on the potential buyers of the middle-income group. It had selected the State of Selangor, based on its associated features including its highly dense population in urban area. According to Selangor Department of Town and Regional Planning [18], 83% of the population in Klang Valley are the citizens of Selangor. Selangor is an urbanised state in the Peninsular Malaysia hence faces the issue of migration and increasing demands for affordable housing, as well as the challenges of living cost. Besides the three housing providers, the study had included the National Housing Department and the Selangor Town and Country Planning Department given that they are directly involved in the process. This preliminary investigation had therefore involved ten participants and interviews were conducted among professional officers from PR1MA and SPNB as the main providers of affordable housing, officers from the Selangor Housing and Properties Board as the agency in providing affordable house throughout the state, a town planner from the Selangor Department of Town and Regional Planning, and officers from the National Housing Department from various disciplines. The following work will encompass a questionnaire survey, which is planned for the end-users. It is anticipated that this piece of work will unveils the functioning and limitation of the delivery system, thus providing a better understanding of the current practice and what can be done to enhance the delivery system. The study method is illustrated in Figure 1 below.

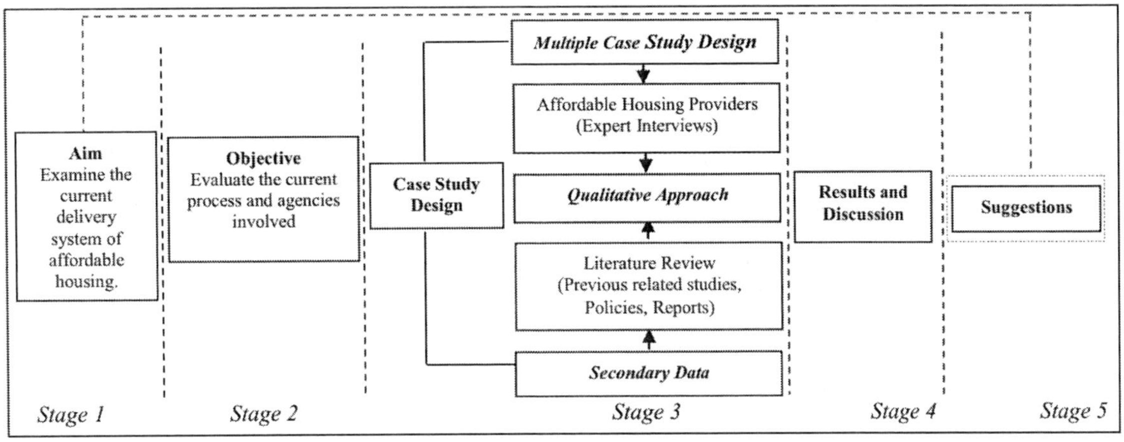

Figure 1: Research Approach

3 Affordable Housing Development

Since the 1st Malaysia Plan, the major concern has been on the development of low cost housing to meet the needs of the low-income group [19-25]. During the 9th Malaysia Plan, affordability became the main attention. Later, during the 10th Malaysia Plan the concentration had shifted to affordable housing, and now, in the latest 11th Malaysia Plan the focus is on the development of affordable housing. Basically, affordable housing through public housing programs is delivered at all government levels. However, each tier has established specific agency based on the particular purpose to address certain group as shown in Table 1 below.

Table 1. The main affordable house provider and the programs implemented

Agency	Program	Purpose	Level of Income	House Prices
PR1MA Corp.	*Perumah-an Rakyat 1Malaysia*	For middle-income group	RM2,500-RM10,000	RM100,000-RM400,000
SPNB	*Rumah Mampu Milik*	For low and middle-income group	RM2,000-RM10,000	RM42,000-RM250,000
LPHS (SEDC and Private Sectors)	*Rumah Selangor-ku*	For low and middle-income group	RM3,000-RM10,000	RM3,000-RM10,000

Source: NHD [4]; PR1MA [29]; SPNB [31]; LPHS [32]; EPU [8]

At the federal level, the National Housing Department (NHD) is involved directly in formulating the policy, laws, regulations, and governing housing especially for the private sector [26]. Moreover, other than policy making and enforcement of the related laws and regulations, the federal government has a function in processing the licenses for housing development project, control and monitor housing projects, and build more low cost and public housing programs [26]. In terms of affordable housing development, NHD had introduced My Home Scheme as one of the approaches for federal government to encourage more participation from private sectors in providing affordable house [28]. The government provides a subsidy of up to RM30,000 per unit to private developers to enable first-time buyers with a monthly household income of RM3,000 to own a house [28]. Through this scheme, the incentives offered by the federal government can be beneficial to private developers and mostly house buyers.

Meanwhile, PR1MA Corporation is the major agency responsible in providing affordable house for the middle-income group in urban areas. The organisation was established under the PR1MA Act 2012. Household income between RM2,500 until RM10,000 are categorised as the middle-income group while affordable houses offered by PR1MA are within the range of RM100,000 until RM400,000 depending on the location, type and size of house. Moreover, PR1MA houses are sold at 20% lower than the average market price, built and located at transit oriented development area. Through PR1MA, the government wants to encourage the young generation who had just starting their careers and enable them to own a house at an affordable price. PR1MA targets 200,000 units of affordable house by the year 2018 [29]. In 2015, PR1MA Corporation approved 119,933 homes for development nationwide, and 18,400 units are under construction [8]. Currently, most of the PR1MA projects have been launched. Part of that has been fully completed and occupied while the rest will be launched afterwards. Moreover, PRIMA had also introduced the *Rent-to-Own Scheme* to assist PR1MA house buyers [29].

Other than PR1MA, *Syarikat Perumahan Negara* (SPNB) also has similar responsibilities in providing affordable house to the public. *The Rumah Mampu Milik Programme (*RMMP) is implemented by SPNB aiming to provide affordable house across the country. Affordable housing provided by SPNB is divided into three categories; low cost house, low-medium cost house, and medium cost house [30]. To address the middle-income group, SPNB builds the *Rumah Idaman Rakyat* (RIR). It was the government's initiative and the focus is to provide affordable house to the first house buyers who are facing the dilemma of middle-income group. The properties are provided specifically for those with an income below than RM7,500 and monthly household less than RM10,000 [31]. The price for affordable house provided by SPNB is below RM250, 000 [31]. At the state level, affordable house are delivered through two main agencies; (i) Selangor Housing and Property Board (LPHS); and (ii) State Economic Development Corporation (SEDC). LPHS had delivered many affordable houses before, and during 2010, LPHS has started to provide affordable housing in Selangor with a new name called the *Rumah Selangorku.* Moreover, LPHS also came out with *Dasar Rumah Selangorku* in 2014 in which the policy will be guidance to LPHS itself and private developers involved in providing the *Rumah Selangorku* [32]. LPHS, however, plays an important role in policy making, housing supply, regulation, and construction [8]. Private developers are also encouraged to be involved in the development of *Rumah Selangorku* under the LPHS's regulation. The price for one unit of *Rumah Selangorku* is RM 42,000 to RM250, 000 depending on the type, area, location, and other facilities provided [8, 32]. The price range starts from household income of RM3,000 till RM8,000 [32]. Nevertheless, according to the proposed 2016 Budget, the range of household income has been increased to RM10,000 for eligibility to apply for an affordable house from the Selangor State Government [8]. The state government targets to provide 35,000 units by 2018 either from LPHS or other private developers. In addition, other than LPHS, the SEDC also known as *Perbadanan Kemajuan Negeri Selangor* (PKNS) contributes in assisting the state government in the provision. Affordable housing project implemented by PKNS is subjected under the Selangor Affordable Housing Policies in which the developers like PKNS must comply with this policy. Until October of 2015, 30,950 units has been approved in which 1,774 houses had been completed, while 4,782 units were under construction, and 24,304 units were approved by the Members of State [5].

4 Actors in the Delivery Process

In housing delivery system, there is a list of actors involved throughout the whole process. Each of them has their own role and function to ensure the housing development can be successfully implemented [14, 33]. The government has its jurisdiction in approving the procedures and its machinery, and therefore, involves at every stage of that process. The commitment of the federal government can be seen through the establishment of NHD. The government is responsible in formulating polices regarding housing provision [33] and the application of policies at the lower government level [12]. In the delivery system, any housing development had to acquire a licence and permission from NHD [12]. Meanwhile, the state government is usually responsible in policy making at the state level and delivering affordable housing programs based on the specification established by the state government. State government is also involved in housing through SEDC which operates like private housing developers but at the same time fulfilling objectives outlined by the state [26]. This can be seen through the contribution of PKNS in providing and delivering affordable house and involvement of other private sector. To simplify, this three-tier governments play their roles toward approval procedure of the various stages of the housing development process in order to ensure a smooth progress of the project [12]. Therefore, the efficiency of the delivery system will depend on their speed and timely handling of application and issuance of approvals [33]. Meanwhile, housing development includes the role of private developers. In the delivery system, developers are usually responsible in ensuring the project is well marketed and the units are sold, as well as making sure that all the necessary submission were carried out through the assigning of professional consultants and engaging the builders or contractors to undertake project construction [12]. Meanwhile, the professionals involved in the delivery are town planners, architects, surveyors, engineers, landscape designers, etc.

5 Results and Discussion

Housing delivery comprises of a process and players involved in the development [12]. Basically, housing development process encompasses of the three major stages known as; (i) pre-development stage; (ii) construction stage; and (iii) post construction stage [34, 9]. The housing delivery process is a typical development process in which it will undergo several stages. The flow of activities or process starting from the inception right up to the delivery of a completed property plays a fundamental role in the planning of a property development [34]. Figure 2 below shows the housing development process currently practiced in Malaysia.

The first stage is the planning stage or the pre-development period [34]. Housing project begins with land acquisition and ends with selling of the house. During the first stage of housing development, the applicants or housing developers acquire a land and if the acquired land use is under a different category, the developer has to convert the category of land by submitting an application to the relevant land office for a change of land use [11]. Before acquisition of property, the process may start with a feasibility study [12]. Developers will then make an appointment or liaison with consultant(s) [9]. The housing development applications will also go through land approval process as enacted in the Malaysia Land Code, 1965 (Act 56) [35].

DOI: 10.1051/matecconf/20166600047

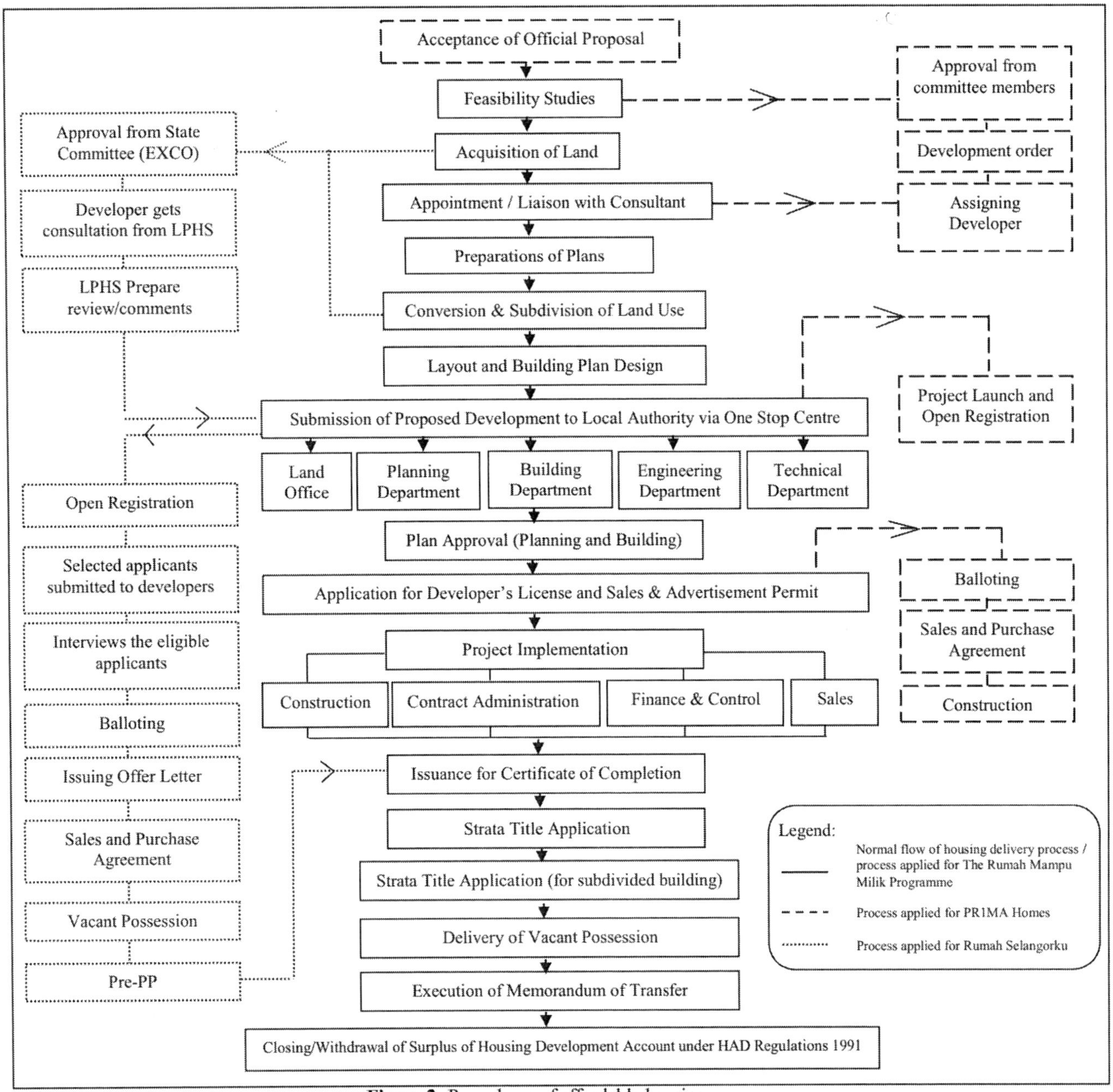

Figure 2: Procedures of affordable housing

Source: Adapted from Hamdan [12]; REHDA [9]; Yaakob et al. [11]; PR1MA [29]; LPHS [32]

Consequently, the developer has to obtain all planning approvals before any physical work can commence on the site and prior to issuance of any advertisement permit by the relevant authorities [34]. Layout and building plans, road plan, infrastructure and other plans are prepared by the consultants with reference to the proposed housing development project [11]. All plans and report prepared by the consultant will be submitted to the One Stop Centre (OSC) of the relevant local authority for the development area [11]. This stage requires planning approval from the local authorities. Planning approval - will therefore involve departments such as land office, planning, building, engineer, and other technical departments [9, 12]. Through this process, the OSC has indirectly, improved the fragmented and bureaucratic housing approval system at the local level [36]. According to Foo and Wong [36], OSC aimed to speed up the circulation of documents among both internal and external approving departments. Thereafter, developers must send application of license, sales and permit for advertisement from the NHD [9, 12]. When all of the required approvals have been obtained, the construction phase will begin [34]. After the layout has been approved by the local planning authority, construction work, such as earthwork and land levelling will start [11]. Construction stage usually will take 24 or 36 months [9]. During project implementation, the construction will be carried out, as well as the administration contract, finance and marketing [9, 11, 12]. After the construction has been completed, a Certificate of Completion and Compliance (CCC) is issued [9, 11, 12, 34]. For subdivided building, a strata title is required [9]. This is followed by the release of vacant possession and follow by the execution of Memorandum of Transfer before closing or withdrawing all money in housing development accounts [9]. The close of a development project or the post development stage is when the completed property is handed over to the owner as required and stipulated by the Housing Development Act. And finally the house key will be handover to the house buyers.

During the interview session, the study found that process of affordable house delivery by SPNB is the same as practiced by other private developers [31]. Meanwhile, the whole projected delivery process for PR1MA homes as practiced by PR1MA can be seen through Figure 3. At the beginning, a review will be carried out to evaluate the project. Hence, a field study or site visit will be carried out to determine the suitability of location for affordable housing development. During this stage, the committee will give their opinion and assessment regarding the whole project such as location, demand area and etc. Once the committee has approved the proposal, the housing proposal project will be presented to the Members of PR1MA Corporation. After that, PR1MA will submit an application to get planning and building approval. In addition, once the application of development has been submitted, PR1MA will start to launch the project through their official website and inform to the public of the project for launching. During this time, the public will start to register for the application of PR1MA homes.

Once the building and planning approval is approved by the local authority, PR1MA will then open the balloting process, licenses and advertising the permit and etc. It will then start the project and incorporate the work of marketing, selling, promotion, application for developers. PR1MA will therefore carry out the balloting process to the potential buyers and there will be an audit responsible for official inspection toward the registrant. Because of that, construction will only start after obtaining planning and building approvals. Finally, PR1MA home will be ready for end-users.

For *Rumah Selangorku*, basically there are two main agencies in handling this program; either from LPHS as the provider or delivered by the developers while the LPHS acts as the regulator. The approach is demonstrated in Figure 4. Developers are basically aware of the location of land that suits the development. Developers attend to LPHS to acquire comments or agreements in order to determine the types and requirements of *Rumah Selangorku*. Later, LPHS will issue comments regarding types and percentage in providing affordable houses. Consequently, the applicant (developer) submits the proposal to OSC. There are many agencies involve including the land administrator office, planning, building, engineer and technical departments. All departments will provide their comments to the OSC and an OSC committee meeting will then be carried out. During the meeting, LPHS will be invited as a member based on its roles. OSC will later provide an endorsement letter to the certified agencies, and announce the conditional planning approval to solve the land matters to the State Land Administrator, whilst issuing notification to the successful applicants to the council. Meanwhile, LPHS will also issue a notification letter to the applicants, and OSC will proceed with approval copies to LPHS for the purpose of monitoring. Finally, OSC will collect the approval letters, and hand them over to the applicants.

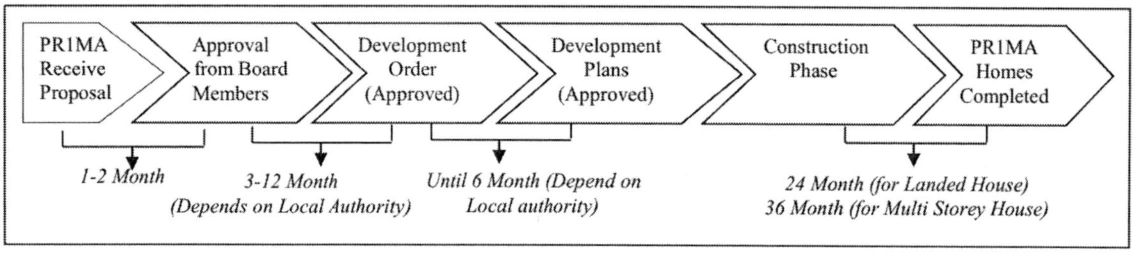

Figure 3: The process for PR1MA homes development
Source: PR1MA [29]

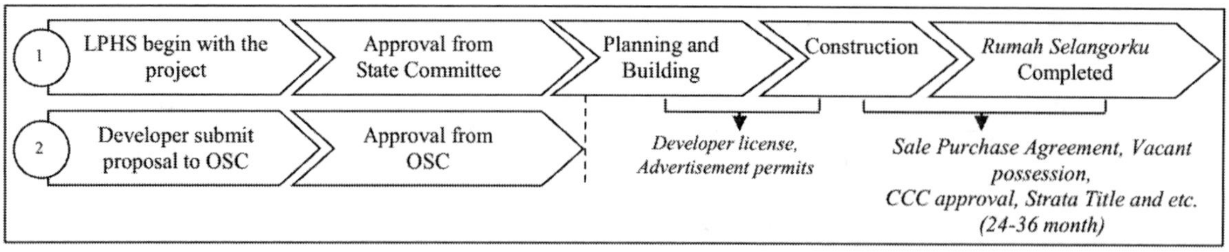

Figure 4: The development process for *Rumah Selangorku*
Source: LPHS [32]

The following findings were identified based on the review of the providers.

(a) Inconsistency of definition and prices for affordable housing

The prices for affordable housing were not consistent between provider agencies. National Housing Department has defined affordable house below the price of RM300,000. However, each agency which is responsible in providing affordable homes has their own prices in selling affordable homes. This causes inconsistency of affordable housing price. This is also due to the function and purpose of the agencies. Even though the National Housing Department had already outlined the price for affordable house, agencies in providing affordable house also consider the location factor, availability of land, and the local socioeconomic in determining the price of a unit.

(b) Involvement of multiple agencies in providing affordable house

There are various agencies involved in providing affordable house. The main agencies are the NHD, Prima Corporation, and SPNB. Moreover, the state government also develop their own affordable housing program. In certain cases, local level is also responsible in providing affordable house as can be seen through the provision of *Rumah Mampu Milik Wilayah Persekututan* (RUMAWIP) provided by Kuala Lumpur City Hall (DBKL). And because there are many agencies as providers, this has resulted in the lack of integration for delivery. Therefore, good coordination between agencies such as the creation of integrated database system is necessary to increase the supply and delivery of affordable house.

(c) The effectiveness in delivering affordable house

In a housing development process, a lot of time is required to obtain approval as it involves approval at every stage of the process, including conversion of land use category, delay in getting approval, non-compliance with development plan, infrastructure improvement which increased the production cost and others.Therefore, its affects the whole approval process thus delaying the delivery process. There may not be much difference in the delivery process. But there are some variances between a typical housing development process practiced by private developers and agencies that were appointed by the government such as PR1MA. This can be seen through the system called the balloting process.

6 Recommendation and Conclusion

To strengthen the delivery of affordable housing in Malaysia and improve the delivery system practiced by the agencies, there must be a clear definition regarding affordable housing in Malaysia because it may differs for each individual or income group. Affordable houses should be built within the people's affordability which focuses on low and medium-income groups. Moreover, government should revise the ceiling price for affordable house. Affordable housing delivery system should be under one roof, instead of operating separately. A central housing delivery centre would help to rationalise the delivery system of affordable housing. To enhance the system, a single entity to design, build and manage the affordable housing development may decrease the overlapping of role in providing affordable house. Additionally, all of the related authorities should be merged. A clear, separate or shared responsibilities amongst the 3-tier of government in providing affordable housing is crucial. Therefore, this study is hoped to further contribute towards the understanding of the provision of affordable housing in Malaysia through its delivery practice in order to assist homeownership to have home that can be accessed and afforded by the various levels of citizens. The long terms benefits of this study can be seen through theoretical or practical aspects. It can be a helpful reference to the government and related private sectors, especially with regards to the delivery of affordable housing. It can assist other scholars in providing the current scenario and therefore contributing inputs to existing knowledge in terms of affordable housing delivery system. Likewise, future study may emphasis on the potential of a central agency or a model to shorten the duration of delivery that would help to enhance the system.

Acknowledgements

The authors would like to thank the Ministry of Higher Education, Malaysia for funding the study through the FRGS grant with the project code 600-RMI/FRGS 5/3 (73/2014) and also would like to thank the Universiti Teknologi MARA (UiTM), specifically, the Research Management Centre (RMC) for supporting and managing this grant.

References

1. The Ministry of Federal Territory, *Dasar Rumah Mampu Milik Wilayah Persekutuan*. Kuala Lumpur: The Ministry of Federal Territory (2013)
2. I.S. Grewal, *Malaysia's Middle Class Challenge: Making Progress*. The Star Online. Retrieved from http://www.thestar.com.my (2014)
3. I.Turok and G.M. Granahan, *Urbanization and Economic Growth: The Arguments and Evidence for Africa and Asia*. SAGE Journals **25,** 465-482 (2013)
4. National Housing Department, *National Housing Policy*. Kuala Lumpur: The Ministry of Housing and Local Goverment (2011)
5. Economic Planning Unit, *Rancangan Malaysia Kesebelas: Kertas Strategi 6.* Retrieved from http://rmk11.epu.gov.my/pdf/kertasstrategi/Kertas%20Strategi%2006.pdf (2015)
6. Institute of Strategic and International Studies (ISIS) Malaysia, *ISIS Roundtable: Providing Public and Affordable Housing for Malaysia.* Retrieved from www.isis.org.my/files/IF_2013/IF3_2013.pdf (2013)
7. N.M. Malek, and A. Husin, *Pemilikan Rumah dalam Kalangan Masyarakat Bandar Berpendapatan Sederhana dan Rendah di Malaysia.* Sosiohumanika **5** (2012)

8. Economic Planning Unit, *Rancangan Malaysia Kesebelas, 2016-2020: Kerangka Agenda: Aspirasi Kerajaan dalam Merumahkan Rakyat*. Presented at Program City Talk @ Bukit Tinggi 2015 11 November (2015)
9. REHDA, *The Housing Delivery System in Malaysia: Opportunities and Challenges*. Retrieved from http://www.housingfinance2014.org/fileadmin/2014/Presentations_English/5_1_Michael_Yam_English.pdf (2014)
10. A. Husin, N.M. Malek, and S.A. Gapor, *Trends Problems and Needs of Urban Housing in Malaysia*. World Academy of Science, Engineering and Technology **74**, 1103-1107 (2011)
11. H. Yaakob, F. Yusof, and H. Hamdan, *Land Use Regulations towards a Sustainable Urban Housing: Klang Valley Conurbation*. Procedia Social and Behavioral Sciences **68**, 578-589 (2012)
12. H. Hamdan, *Housing Delivery System*. Universiti Teknologi MARA (UiTM) Shah Alam. (Unpublished)
13. A. Mustafa, and M. Ghazali, *Issues in Housing Delivery Systems and Customer Satisfaction*. Elixir Marketing Mgmt **48**, 9395-9399 (2012)
14. A. Ghanarajah, *Housing Delivery System: Issues and Legal Perspectives*. In Housing the Nation: A Definite Study, Kuala Lumpur: Cagamas Berhad, 603-612 (1997)
15. O.O. Makinde, *Housing Delivery System: Need and Demand*. Environ Dev Sustain **15**, 49-69 (2014)
16. A.A. Khalaf, *Key Constraint to Improving the Affordable Housing Delivery Process in the Syrian Market: The Potential of Procurement Efficiency*. Proceeding of CIB W070,WO92 and TG72 International Conferences, 329-336 (2012)
17. A. Sivam, D. Evans, R. King, and D.Young, *An Approach to Improved Housing Delivery in Large Cities of Less Developed Countries*. Habitat International **25**, 99-113 (2001)
18. Selangor Department of Town and Regional Planning, *Draf Rancangan Struktur Negeri selangor 2035*. Retrived from http://eps. mbpj. gov. my/web2015/Draf_RSN_2035.pdf (2015)
19. Economic Planning Unit, *The Third Malaysia Plan 1976-1980*. Retrieved from http://www. epu. gov. my/third-malaysia-plan-1976-1980 (1975)
20. Economic Planning Unit, *The Fourth Malaysia Plan 1981-1985*. Retrieved from http://www. epu. gov. my/fourth-malaysia-plan-1981-1985 (1980)
21. Economic Planning Unit, *The Fifth Malaysia Plan 1986-1990*. Retrieved from http://www. epu. gov. my/fifth-malaysia-plan-1986-1990 (1985)
22. Economic Planning Unit, *The Sixth Malaysia Plan 1991-1995*. Retrieved from http://www. epu. gov. my/sixth-malaysia-plan-1991-1995 (1990).
23. Economic Planning Unit, *The Seventh Malaysia Plan 1996-2000*. Retrieved from http://www. epu. gov. my/seventh-malaysia-plan-1996-2000 (1995)
24. Economic Planning Unit, *The Eight Malaysia Plan 2001-2005*. Retrieved from http://www. epu. gov. my/eight-malaysia-plan-2001-2005 (2000)
25. S. Shuid, *Low income housing allocation system in Malaysia: managing housing need for the poor*. In: 22nd International Housing Research Conference, 4 - 7, July 2010, Istanbul, Turkey (2010)
26. S. Shuid, *The changing role of the state and market in low cost housing provision in Malaysia*. Journal of Architecture, Planning and Construction Management **1**, 39-70 (2011)
27. National Housing Department, *Maklumat Korporat bahagian dan unit*. Retrieved from https://ehome. kpkt. gov. my (2015)
28. National Housing Department, *My Home Scheme*. Retrieved from https://myhome.kpkt.gov.my (2015)
29. PR1MA Corporation, *Laporan Kemajuan 2012-2014*. PR1MA Corporation,Kuala Lumpur:Author. (2014)
30. Department of Information, *Program Rumah Mampu Milik*. Retrieved from http://pmr. penerangan. gov. my/index. php/umum/12252--program-rumah-mampu-milik.html (2008)
31. Syarikat Perumahan Negara Berhad, *Official website*. Retrieved from http://www.spnbaspirasi.com.my (2014)
32. Lembaga Peumahan dan Hartanah *Selangor,Rumah Selangorku City Talk@Bukit Tinggi 2015*. 11-12 November 2015 (2015)
33. L. Chan, *Housing delivery system: The industry viewpoint*. In *Housing the Nation: A definite Study* Kuala lumpur : Cagamas Berhad, 603-612 (1997)
34. A.A. Abdullah, Z. Harun, and H.A Rahman, *Planning process of developement in the Malaysian context: A crucial brief overview*. Internatioal Journal of Applied Science and Technology **1** (2011)
35. A. Rameli, F. Johar, and H.C. Siong, *Responsiveness to the market mechanisms: A way to strengthen the Malaysian housing planning process*. Jurnal Alam Bina **9**, 1-22 (2007)
36. L. H. R. Foo, and C. Wong, *Planning for housing developemnts in Malaysia: Developer's response to the regulatory policy framework*. International Planning Studies **19**, 192-209 (2014)

Social Factors Influencing Household Solid Waste Minimisation

Nor Eeda Haji Ali[1] and Ho Chin Siong[2]

[1]*Department of Town and Regional Planning, Faculty of Architecture, Planning and Surveying, UiTM Seri Iskandar, Perak.*
[2]*Department of Urban and Regional Planning, Faculty of Built Environment, Universiti Teknologi Malaysia.*

Abstract. The growth of the world's population, coupled with increasing urbanization, and rising standards of living, have all contributed to the increase in solid waste generation. Solid waste disposal is becoming a difficult problem for many countries. Thus, efficient recycling of solid wastes is now a global concern for a sustainable solid waste management. Solid waste minimization is one of the ways of reducing the quantity of wastes for disposal. This study examines the perception of households towards solid waste minimization. 100 respondents were randomly selected from Section 7 housing area in Shah Alam city, Malaysia. Descriptive statistical technique was used in analysing the data. Findings from the study showed that respondent's knowledge on waste minimization was above average. However, their level of knowledge as revealed further by the results is not in tandem with their attitude on waste minimization, which suggests that their attitude towards waste minimization is low. People who showed higher knowledge do not necessarily show concern or perform in solid waste minimisation. Therefore, local authorities must develop appropriate policy strategies to change the attitude and behaviour towards waste minimisation if they are to reach their statutory targets.

1 Introduction

In the past two decades, solid waste management has become a major concern and is now one of the subjects discussed by the policy makers and planners. The highest solid waste generated in Malaysia is from residential sector, [15]. Annual growth rate of solid waste generation from household sector in Malaysian is continuously rising because of increasing urbanisation and rising standards of living. In Peninsular Malaysia (West Malaysia), the daily generation of waste escalated from 13,000 tonnes in 1996 to 19,100 tonnes in 2006. Approximately 75% from this waste is disposed in landfills and dumps. An estimated 20% is burned or dumped into rivers or at illegal sites, while around 5% is recycled, [23]. Adopting an effective waste minimisation strategy by government is seen as another approach of sustainable municipal waste management.

Waste minimisation in Malaysia is not officially incorporated in to waste management practice. Malaysia launched the first official 3R's(Reduce, Recycle, Reuse) strategies in 1988 by Action Plan for a Beautiful and Clean Malaysia (ABC) where campaigns focused mainly on the recycling activities but the rate is still too low that it did not improve the existing waste management practice. The first recycling programme was introduced in Petaling Jaya in 1991 which involved source separation with kerbside collection. In 1996, Yellow Box Recycling Programme was launched, when 10,000 households were given recycling boxes where they can place their recyclables and collection was done weekly. Recyclables were also collected by local authority, however the quantum collected was less than 1% of the total volume of municipal was generated. In the 21st century, the scenario of modern solid waste management such as waste minimisation, service delivery system, public awareness are challenging for both environmental and urban planners, [22].

Waste minimisation is a reduction of waste at source and the objective is to achieve a target of national solid waste hierarchy towards 2020. Solid waste hierarchy of Malaysia is to decrease the nation's solid waste generation by reduce, reuse, recycle, treatment and dispose. The Malaysia target for recycling is 22% of the total solid waste generated by the year 2020. The current rate is far below the target and is approximately 5%. Many stakeholders are involved in Malaysian waste minimisation. They include the Ministry of Housing and Local Government (formally known MHLG), National Solid Waste Management Department, Solid Waste and Public Cleansing Management Corporation, local authorities, concessionary companies, collectors and manufacturers, non-governmental organisations, private residents associations, educational institutions, private waste recycles, and others. The most common economic benefits derived from solid waste minimisation are cost avoidance, recycling revenue, reduced raw material costs, reduced energy costs, increased sales and increased productivity, [17].

Based on a recent survey by the Ministry of Housing and Local Government 2009, there are 100% knowledge and awareness among the public about waste

noree038@perak.uitm.edu.my

DOI: 10.1051/matecconf/20166600048

minimisation but very few are actually practicing it. Many factors contribute towards the failure of solid waste minimisation in Malaysia, such as lack of an efficient institutional framework and policies to promote it. Solid waste minimisation also involves efforts to minimise resources and energy to reduce environmental pollution. Integrating solid waste minimisation through recycling could help in reducing the challenges being faced in municipal solid waste management (MSWM). The objectives of the study are (i) to examine solid waste minimisation, waste composition and concept of solid waste minimisation (ii) to examine the social factors influencing solid waste minimisation (iii) to examine those factors hindering effective waste minimisation in the neighbourhood.

2 Solid waste minimisation in urban housing

2.1 Municipal solid waste

The most fundamental things to comprehend in solid waste management are solid waste generation, source and composition of the stream. Solid waste generated from various source such as residential, commercial, industrial etc. This generation must be controlled by stakeholders in order to achieve the government's target of solid waste minimisation in Malaysia [22% by year 2020]. The sources of municipal solid waste are as follows:-

Table 1: Sources of solid waste generation

No	Sources of Waste	Locations where wastes are generated
1	Residential/Housing/ Household	Single and multifamily dwellings, terrace, semi-detached, bungalow, apartments and cluster house.
2	Commercial	Stores, hotels, restaurants, markets, office buildings, motels, shops and private school.
3	Institutional	Schools, hospitals, prisons, government centers and universities.
4	Municipal	Street cleaning, landscaping, parks, beaches, water and open wastewater treatment plants, recreational areas, open spaces, treatment plant sites, alleys, vacant lot, playgrounds and roadside litter.
5	Agricultural	Field and row crops, orchards, vineyards, dairies, feedlots and farms.
6	Construction and demolition	New construction sites, road repair, renovation sites, demolition of buildings and broken pavement.
7	Industrial	Light and heavy manufacturing, fabrication, construction site, power, and chemical plants, rrefineries, mineral extraction and processing.

Sources : [5], [32], [9], [4].

The growth rate of solid waste generation is continuously rising due to the world's population, increasing urbanisation, rising standards of living, and the rapid development of technology. Housing recorded the highest generated of solid waste which is 48%, commercial 24%, municipal services 11% and the slightly generated is from construction and demolition 4%. The stakeholders involved have to play a serious role whether formal and informal.

Table 2: Sectors of solid waste generated

Sectors	Percent
Housing/Residential	**48.00**
Commercial	24.00
Municipal Services	11.00
Agriculture	7.00
Institutional	6.00
Construction & Demolition	4.00
Total	**100.00**

Source: Nasir, 2009

2.2 Concept of waste minimisation

Waste minimisation or reduction is the process of reducing the amount in waste streams. The importance concept of waste minimisation in all countries is through 3R's (Reduce, reuse and recycling), [20] and treatment process (composting and incineration), [10]. Waste minimisation hierarchy is the concept of reducing solid waste in stream. It should be done progressively in order of importance, comfort and environment sustainability.

The process involves six steps ranked according to environmental impact. Reducing, which offers the best outcomes for the environment is at the top of the priority order, followed by preparing for reuse, recycling, composting, incineration and disposal. Treatment concept, which is in turn better than disposal to landfill. Looking at the pyramid of waste minimisation, reduction is the most preferred option while the landfill is seen as the least favoured option.

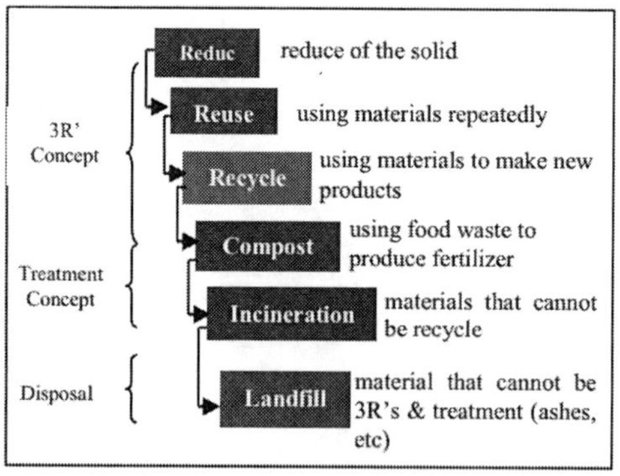

Figure 1: Waste minimisation hierarchy

DOI: 10.1051/matecconf/20166600048

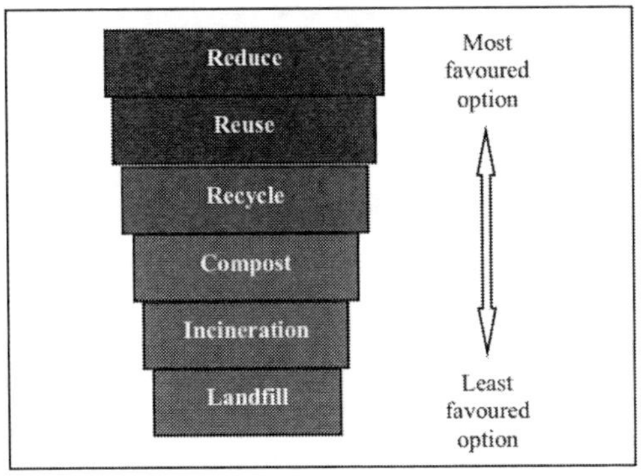

Figure 2: Pyramid of Waste Minimisation

2.3 Social Factor Influencing Waste Minimisation

There are many factors that influence solid waste minimisation. Among these factors are: social factor, physical factor, infrastructure factor, and institutions factor. This paper is only looking at the social factor with respect to knowledge and awareness of households on solid waste minimisation practices in residential areas of Shah Alam city.

2.3.1 Knowledge

Knowledge about recycling programs has a positive relationship with recycling practice. Generally, the more knowledge a household has about recycle materials and their impact on the natural world, the more likely they will practice it, [26]. S. Oskamp *et al,* (1991) noted that recycling behaviour may also be affected by lack of knowledge about environmental issues .People who showed higher knowledge and concerns about the environment and recycle, have higher levels of participation with recycling, [34], [27], [6], [31].

2.3.2 Attitudes/Awareness

The Theory of Planned Behaviour indicate not only predict behaviours from attitudes but also must observe the awareness through which the two are linked, [31]. General rule, the stronger the intention to engage in a behaviour, the more likely should be its performance in attitudes and awareness, [29]. If an individual is strongly motivated by something to behave a certain way, then they will likely do so. I.F. Ajzen, (1991) suggests that behaviour is not only guided by motivation, but also an individual's ability (attitude & awareness) to engage in that behaviour. He calls this "behavioural control". The Theory Planned Behaviour also maintains that the greater the perceived control one has over a behaviour, the stronger the person's intention to perform that behaviour, [7]. I.F. Ajzen, (1991) confirms that performance of a behaviour is a joint function of intentions and perceived behavioural control. Perception variables (i.e. ease, difficulty) combined with intention variable (i.e. attitude) enables and individual's behavioural to be predicted.

K. Chan, (1998) found that personal attitude (based upon the individual's perception of the activity being right/wrong, good/bad, useful, desirable, pleasant and interesting) was by far the most important determinant of behaviour, with PBC and social norms much less so. K. Chan, (1998) speculated that social norms were not important in recycling behaviour because there is little or no social cost to individuals in not participating; but her study concerned 'bring' site recycling, whereas S. Barr (2007) research, which found that social norms were important, involved kerbside recycling, where the visual indication of not participating is much higher. Identifying motivators for and barriers to resident participation can help tailor recycling program parameters to meet the needs of residents.

3 Methodology

3.1 Questionnaire Design

The questionnaire was based on the waste minimisation literature and previous research. This questionnaire is comprised of close ended and open ended questions. The questionnaire contained of four following section:-

Section 1:
Demographic and background information regarding the respondents.
Section 2:
Knowledge. (Respondent was asked about the knowledge of waste minimisation).
Section 3:
Action on waste minimisation facilities/programs, which was launched.
Section 4:
Participation. Views or comments regarding the involvement/participation of Local Authorities.

Section 2 was measured using Likert Scales that typically ranged from 1 to 5, whereby 1 = Strongly disagree, 2 = Disagree, 3 = Neither agree or disagree, 4 = Agree and 5 = Strongly agree.

3.2 The Data Collection

The analysis of this study is based on primary data collected recently from three residential housing types (terrace, semi-detached and bungalow) in Section 4, Shah Alam, Selangor. The interview was conducted on persons who were above 18 years of age and were either the heads of households or families. The snowball sampling method was used for the data collection. Chi Square test will conducted for the analysis data. Enumerators were assigned to particular types of household in each area, with minimum interview quotas for each household type. A total of 100 questionnaires were administered and the proportions were 64% for terrace house and 18% for the semi-detached and bungalow respectively. The survey was conducted between 18th June -1st July 2014. The answers were coded and analyse using Special package for Social Science Statistics software (SPSS version 20).

3.3 Case Study Selection

The study was undertaken in Shah Alam, Selangor in Section 4. This area was chosen for the study because of two reasons. First, the section has recycling facilities some years now. Second, the various ethnic composition of the section gives impetus to a study of representative sampling of multi -ethnicity. The three ethnic groups Malay, Chinese and Indian are fully represented here.

Figure 3: Key Plan

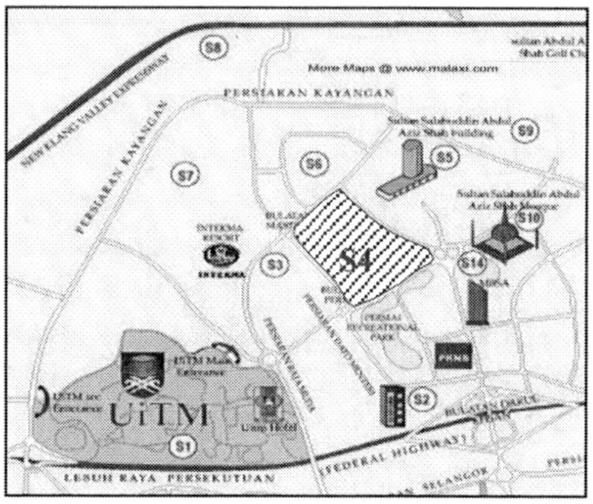

Figure 4: Plan Location

4 Results

4.1 Demographic characteristics of respondents

The gender profile indicates that 65% of those interviewed were males while only 35% were females. The data on age is presented in Figure 5, of the 100 respondents, those aged between 18 and 24 years represented 17%. Those in the age range of 25 to 34 years constituted 29%, while those aged between age 35 to 44 years and 45 to 54 years represented 13% and 21% respectively. Respondents above 55 years constituted only 20%.

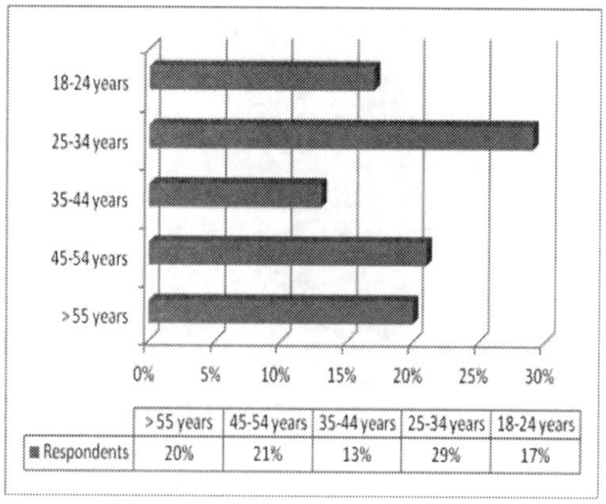

Source : Author's Fieldwork Survey, June 2014

Figure 5: Age distribution of respondents

Another important factor for socio-demography is occupation. About 35% of the respondents were professional, while 21% semi-professional while 14% and 9% were self-employed and unskilled respectively. The remaining 04% were housewife, 01% un-employed and 16% retiree (Figure 6).

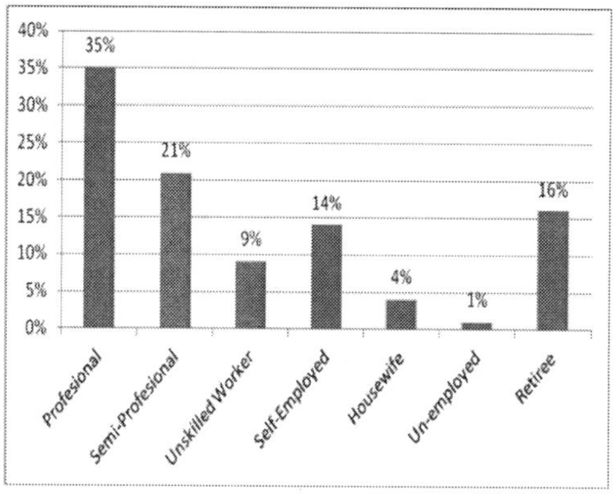

Source : Author's Fieldwork Survey, June 2014

Figure 6: Occupation attainments of respondents

The average monthly income of respondents was also considered an important variable that could influence household's knowledge and awareness on solid waste management system. The average monthly income obtained 41% of the respondents earned between RM2500-RM4999 monthly, 32% earned more than RM5000 monthly, 21% income around RM1000-RM2499/month and 6% respondents earners lowest RM1,000/month.

4.2 Knowledge on waste minimisation

In order to test for the level of knowledge of the respondents on waste minimisation, a Chi Square test was conducted. Results show that a significant difference exist in the level of knowledge of respondents on waste minimisation ($P < 0.05$). Looking at the results in table 4, majority (92%) of respondents agreed that waste minimisation is good for clean environment while 8% remained neutral. The chi-square test, show that a significant difference exist ($X^2 = 39.02$ (df=2) $P<0.05$). As could be seen 84% of the respondents agreed that waste minimisation helps in reduction of households solid waste generation while 13% were neutral and 3% disagreed. ($X^2 = 49.52$ (df = 3) $P< 0.05$).

In contrast 63% of the respondents disagreed that households should not be involved in waste minimisation because it is for government and related public agencies while 12% remained neutral and 25% were agreed ($X^2 = 25.30$ (df = 4) $P < 0.05$). On the knowledge about the quantity of waste being generated by household daily, 70% of the respondents agreed that were aware of the quantity of wastes generated in their household while 26% were neutral and 4% disagreed ($X^2 = 76.40$ (df = 4) $P < 0.05$).

Similarly, majority (91%) respondents have knowledge about recycling and materials that can be recycled while 3% respondent were neutral and 4% respondent disagree. The chi-square test, show that a significant difference exist ($X^2 = 72.72$ (df=3) $P<0.05$). Result showed that 90% of the respondents have seen the recycling bin before while 5 respondent were neutral and another 5% claimed that they have never see it before ($X^2 = 121.30$ (df=4) $P<0.05$).

In the similar vein, majority (72%) agreed that they know the meaning of each of the colour bin while 25% were neutral and 33% disagreed ($X^2 = 62.00$ (df = 4) $P < 0.05$). A total of 66% agreed that they have been practicing recycling for a long time, 22% were neutral and 12% disagreed (X2 = 56.10 (df = 4) $P < 0.05$). Looking at the time collection of solid waste by daily, results show that 41% of the respondents agreed that they were aware of timing for recycling collection days in their housing area while 30% remained neutral and 29% disagreed ($X^2 = 16.50$ (df = 4) $P < 0.05$).

A total of 43% respondent know where the collected waste will be sent to, while 34% respondent were neutral and 23% disagreed. The chi-square test, show that a significant difference exist ($X^2 = 20.500$ (df=4) $P<0.05$). Majority (77%) respondent feel that there are weaknesses in the existing waste management system in Malaysia, while 34% were neutral and only 23% disagreed ($X^2 = 74.70$ (df = 4) $P< 0.05$). Based on these results, it could be said that the level of knowledge residents of the neighbourhood has on waste minimisation is high.

Table 3: Knowledge on waste minimisation

Code	Knowledge	SD	D	N	A	SA	X^2	P	df
WM 1	Waste minimisation is good for clean environment	-	-	8	33	59	39.02	.000	2
WM 2	Waste minimisation helps in reduction of households solid waste generation	-	3	13	38	46	49.52	.000	3
WM 3	Households should not be involved in waste minimisation because it is for government and related public agencies	30	33	12	18	7	25.30	.000	4
WM 4	I am conscious of the quality of waste being generated by my household daily	1	3	26	49	21	76.40	.000	4
WM 5	I know what is recycling and what materials can be recycled	1	-	8	38	53	72.72	.000	3
WM 6	I have seen the following bins	3	2	5	32	58	121.30	.000	4
WM 7	I know the meaning of each colour of the bins.	2	1	25	39	33	62.00	.000	4
WM 8	I have long practiced recycling.	1	11	22	46	20	56.10	.000	4
WM 9	I know when is the recycling collection days in my housing area.	8	21	30	27	14	16.50	.002	4
WM 10	I know where the collected waste will be sent to.	10	13	34	27	16	20.50	.000	4
WM 11	I can feel that there are weaknesses in the existing waste management system in our country	2	4	17	49	28	74.70	.000	4

Source : Author's Fieldwork Survey, June 2014

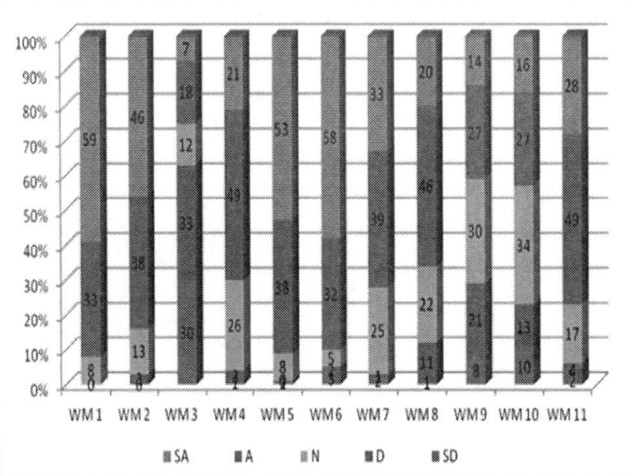

Source : Author's Fieldwork Survey, June 2014

Figure 7: Percent of knowledge on waste minimisation

4.3 Attitude/Awareness on waste minimisation

Chi Square test was conducted in order to test for the level of attitude/awareness of the respondents on waste minimisation. Results show that a significant difference exist in the level of attitude/awareness of respondents on waste minimisation ($P < 0.05$). As could be seen 61% of the respondents merely take initiative/effort in reduction of the quantity of solid waste they generate while 33% always make some effort in reducing their quantity of waste and 6% never did so. Looking at the results in Table 10, 59% of the respondents sometimes engage directly or indirectly in activities related to waste management or environmental conservation while 33% never and about 8% were fully involved ($X^2 = 15.76$, df = 3, $P < 0.05$).

In contrast 44% of the respondents never used household waste recycling centres in other areas while 42% stated that they do some times and 14% always use household waste recycling centres in other areas ($X^2 = 19.12$, df = 3, $P < 0.05$). Looking the respondents awareness on recycling, results show that 58 % of respondents sometimes engage in recycling of such materials as tissues paper, bottle, and plastic , 30% had always engaged in recycling while 12% never recycled any material before ($X^2 = 32.24$, df = 2, $P < 0.05$). Based on these results, it could be said that the level of awareness of residents in the neighbourhood has on waste minimisation is still low.

Table 4: Attitude/awareness on waste minimisation

Code	Awareness	A	S	N	X^2	P	df
AWM 1	How often do you take the initiative/effort to reduce the quantity of garbage in your household?	33	61	6	38.64	.000	3
AWM 2	Have you been involved directly or indirectly with activities related to waste management or environmental conservation?	8	59	33	15.76	.001	3
AWM 3	Do you use household waste recycling centres in other areas?	14	42	44	19.12	.000	3
AWM 4	Does your members of household use recycled materials such as tissues paper, bottle, plastic etc?	30	58	12	32.24	.000	2

Source : Author's Fieldwork Survey, June 2014

On the point scale in SPSS, the ratings given as follows: never (0), depending on the circumstances (1), always (2).

For ease of interpretation, each rating is given the following denotation:-

Always (A)
Sometimes (S)
Never (N)

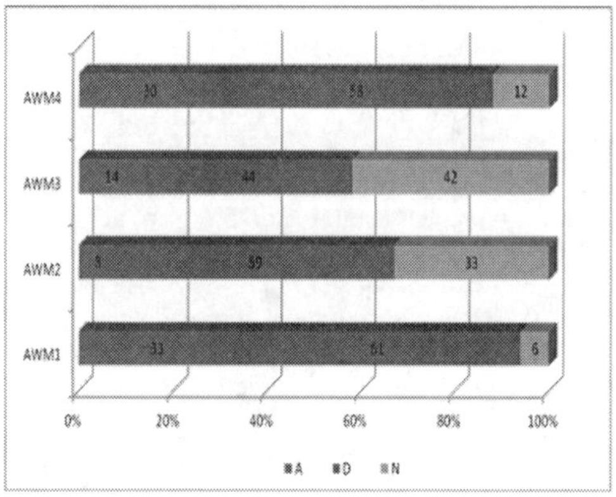

Source : Author's Fieldwork Survey, June 2014

Figure 8: Percent of attitude/awareness on waste minimisation

4.4 Barriers to effective waste minimisation

A total of 47% of the respondents stated that lack of public awareness from local authority give impact against waste minimisation. In addition, a total of 34% maintained that lack of recycling facilities is one of the barriers effective on waste minimisation. Other factors like political will of government (5%), cost (3%), vandalism of waste minimisation facilities (6%) and non-location of collection points at strategic places (5%) were identified as barriers in effective waste minimisation in the study area. When respondents were asked on what they feel should be done by government to improve waste minimisation in the area, they stressed the need for corporation by all stakeholders involved in the management of solid waste in the area. Also they want the creation of separation points for organic and inorganic wastes and more recycling plan.

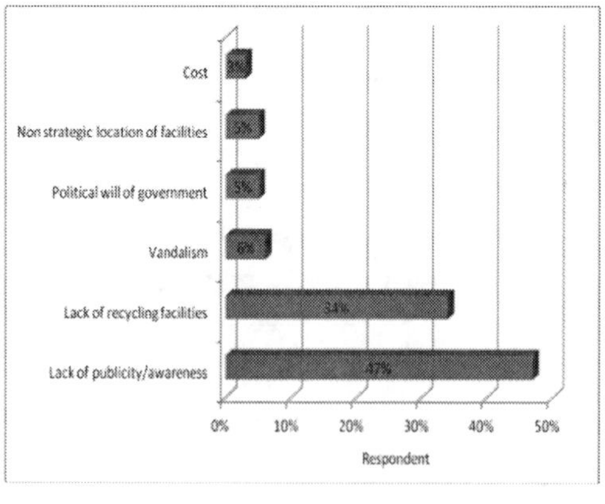

Source : Author's Fieldwork Survey, June 2014

Figure 9: Barriers to effective waste minimisation

4.5 Suggestions to improve waste minimisation

On what needs to be done to improve waste minimisation in the study area, 30% of the respondents suggested more enlightenment campaign by the local authority on waste minimisation while 18% wants the creation of more recycle bins. A total of 13% suggested the adoption a schedule for waste collection and 12% advocated the need for separation of the organic and inorganic wastes before their final disposal. Other suggestions made by the respondents include: cooperation of various stake holders involved in waste management, enforcement by government and taking a cue from other countries that have successfully implemented waste minimisation.

Table 5: Suggestions/recommendation for improve waste minimisation

Suggestions/Commends	%
Creation of awareness on waste minimisation	30.00
Create more recycling facilities	18.00
Adopt a schedule for waste collection	13.00
Separation of collection points for organic and inorganic wastes	12.00
Learn from other countries on how to practice waste minimisation	09.00
Enforcement by government	07.00
Fine defaulters who are found throwing recyclable materials	06.00
Corporation of all stakeholders	05.00
Total	**100.00**

Source : Author's Fieldwork Survey, June 2014

5. Conclusions

Solid waste minimisation is not a new concept in Malaysia. It has been applied in during 1980's. The level of knowledge the respondents have on solid waste minimisation as shown by the results of the study is high. However, this high level of knowledge has not translated to rapid decline in household solid waste generation due to the fact that the respondents are not aware of government's initiative on solid waste minimisation. The respondents seldom take initiatives that directly or indirectly reduce the quantity of wastes they generate. The Local Authorities must do something to increase the awareness among the households. This shows that a lot still needs to be done by Local Authority in the areas suggest such as creation of awareness on waste minimisation, create more recycling facilities, adopt a schedule for waste collection, learn from other countries on how to practice waste minimisation, enforcement by government, corporation of all stakeholders and fine defaulters who are found throwing recyclable materials. This could be done by injecting in new ideas in to the whole process of waste management from the collection, separation, recycling up to final disposal. This study has been able to assess urban solid waste minimisation in Shah Alam City Hall, Selangor could be improved with comparing other best practice.

References

1. A. William, A. Worrell and P.A. Vesilind, *Second Edition : Solid Waste Engineering*, (United States of America, 2012).
2. C. Chiemchaisri, W. Chiemchaisri, P. Somkliang and S. Threedeach, Seminar on solid waste landfill technology in Asia. (2006).
3. C. Chong, T. Lee, M. Matsufuji, L. Yasushi, Hassan, Mohd Nasir, Waste Management, **25**, (2005).
4. D. Daniel, and L. Thomas, World Bank, Urban Development Sector Unit, East Asia and Pacific Region, (1999).
5. F. Kreith, G. Tchobanoglous,, *Handbook of Solid Waste Management*, (McGraw Hill Professional, 2009).
6. G.C.C. Yang, Journal of Resources, Conservation and Recycling, **13**, (1989).
7. I.F. Ajzen, Psychological Bulletin, **84**, (1991).
8. J. J. Corbett, J. J. Winebrake, Journal of the Air and Waste Management Association, **58**, (2005).
9. J. Pichtel, *Waste Management Practices*, (Taylor & Francis, 2005).
10. J. Schall, *Does The Solid Waste Management Hierarchy Make Sense*? Working paper series, Programme on Solid Waste Policy, School of Forestry and Environmental Studies, Yale University, (1992).
11. K. Chan, Journal of Environmental Management, **52**, (1998).
12. K. Sasikumar and S.G. Krishna, *Solid Waste Management*, (PHI Learning, New Delhi, 2009).
13. Malaysia Statistical Board, *Basic Population Characteristics by Administrative District 2010*. (Annual Report. Malaysia, 2010).
14. M.H. Nasir, *Solid Waste Management in Malaysia: Can We Charter Future Strategies?* In M.J. Jamaluddin, M. N. Mohd Jailani, S. Kamaruzzaman, S. Ismail and H.S. Abdul Hadi (Edt.). Proceedings of Realising Agenda 21: International Conference on Environmental Management: Ten years after Rio. Organised by Universiti Kebangsaan Malaysia (UKM), Bangi. 2002.
15. Ministry of Housing and Local Government, Malaysia, *Waste Generation Statistics*, (http://www.kpkt.gov., 2006).
16. Ministry of Housing and Local Government, Malaysia, *Waste Generation Statistics*, (http://www.kpkt.gov., 2009).

17. Ministry of Housing and Local Government, *Ministry of Housing and Local Government Reports 2010,* (Kuala Lumpur, Government Printers, 2010).

18. Ministry of Housing and Local Government, *National Solid Waste Management Report Volume 2*, (National Solid Waste Management Department, 2011).

19. Ministry of Housing and Local Government, Solid *Waste Management in Shah Alam,* Selangor, Town & Planning Department, (Majlis Bandaraya Shah Alam, 2013).

20. M.J. Franchetti, *A Systems Approach Solid Waste. Analysis & Minimization*, (McGraw-Hill Companies, Inc. 2009).

21. M.O. Saeed, M.N., Hassan, M.A. Mujeebu, Waste Management, **29**, (2009).

22. N. Yahaya, *Solid Waste Management in Malaysia :Policy, Review, Issues and Strategies*, Ministry of Housing and Local Government, Malaysia, (http://www.ea-swmc.org/download/seminar1papers/DrNazriYahaya.apf., 2007).

23. P. Agamuthu, S.H. Fauziah, Proceedings of the 2nd Expect Meeting on Solid Waste management in Asia and the Pacific Islands, Kitakyushu, (2006).

24. P. Agamuthu, S.H. Fauziah, K.M. Khidzir, J Mater Cycles Waste Management, **11** (2009).

25. P. Agamuthu, K.M. Khidzir, S.H. Fauziah, Waste Management Res, **27**, (2009).

26. P. Wesley Schultz, et. al, *Knowledge, Information and Household Recycling:Examining the Knowledge-Deficit Model of Behavior Change*, (New Tools for Environmental Protection, National Research Council, 1995).

27. R.J. Gamba, S. Oskamp, Environment and Behavior, **26**, (1994).

28. S. Barr, Environment Behaviour. **39**, (2007).

29. S. Oreg, T. Katz-Gerro, Environment Behaviour, **38**, (2006).

30. S. Oskamp, M.J. Harrington, Edwards, T.C., Sherwood, D. L., Okuda, S. M. and D.C. Swanson, Environment and Behavior, **23**, (1991).

31. S. Lee and H.S. Paik, Waste Management and Recycling Behaviour, **46** (2011).

32. Soncuya and Viloria, Solid Waste Management, **12,** (1992).

33. T.V. Ramachandra, *Management of Municipal Solid Waste*, (Chaman Enterprises, New Delhi, 2006).

34. J. Vining, A. Ebreo, Environment and Behavior, **22** (1990).

Performance Assessment Framework for Private Finance Initiative Projects in Malaysia

Nor Suzila Lop[1,a], Kharizam Ismail[2] and Haryati Mohd Isa[3]

[1]*Centre of Postgraduate Studies, Universiti Teknologi MARA, Seri Iskandar Campus, Seri Iskandar, 32610, Perak, Malaysia.*
[2,3]*Faculty of Architecture Planning and Surveying, Universiti Teknologi MARA, Seri Iskandar Campus, Seri Iskandar, 32610, Perak, Malaysia.*

Abstract. Private Finance Initiative (PFI) is viewed as restructuring the previous privatisation concept in delivering value for money for the Malaysian public infrastructure. Among the restructuring efforts in the privatisation is specifying the standard assessment of private concessionaires' performance through the execution of key performance indicators (KPIs) where the private concessionaires' performance is benchmarked against the government's standard. KPIs have served as useful tools in assessing performance of PFI projects. However, there is still lacking on determination methods performed to define and measure this KPIs and the absence of guidelines or a framework is also an issue in the implementation of the PFI procurement in Malaysia. Therefore, the objectives of this paper is to investigate the notion of performance assessment model approaches globally (i.e. UK, China, Australia, Serbia and Malaysia) and to identify direction for PFI performance assessment tools (KPIs) to be practiced in Malaysia. Based on the consideration of these models, this research paper propose an initial framework of performance assessment for PFI projects in Malaysia. The framework is deliberate to cover the performance of PFI at the operation and maintenance phase. The outcomes of this paper can serve as a theoretical base for the development of comprehensive and effective performance assessment for PFI projects in Malaysia.

1 Introduction

Private Finance Initiative (PFI) is a type of Public Private Partnership (PPP) with referring to contractual arrangement between public and private sectors, where the private sectors delivers a good service and successful performances within stipulated concession period. Private Finance Initiative (PFI) projects are designed to fund long-term public infrastructure and services provided for the whole life cycle of PFI projects. During this cycle, PFI performances could be affected by a number of factors and their interactions, which might cause the inefficiency and ineffectiveness of the projects. According to [1], in the UK and Australia, service failure and poor performance in maintenance work for non-compliance with output specification is frequently reported within PPP/PFI projects. Additionally, PPP/PFI have always stressed on the Value for Money (VfM) and innovation on the basis of harmonious partnership [2]. According to [3], value for money (VfM) in PFI project crucially dependent on performance monitoring to provide certain incentives for improvement. These monitoring also is to ensure that the facilities provided and service delivery is in accordance with the client's requirement, as was set out in the output specification.

Performance measurement is essential to make PFI projects function effectively and efficiently. In order to measure the projects performances and for executing a benchmarking approach, the one that must be establish is key performance indicators (KPIs). It is crucial in determining the overall success of the PFI projects. An absence of effective performance measurement in PPPs will be reflected from the depreciation of standards below optimum service quality of infrastructures and will contribute to the failures delivery of PPP projects [4, 5, 6, 7, 8, 9, 10, 11, 12]. In addition, the clients of the public sector can achieved value for money (VfM) from the usage of properties that meets their requirements, thus satisfying the main objectives of the PPP/PFI procurement mechanism [13].

Within this context, a research was presented to investigate the extent of this problem within PFI projects in Malaysia. Hence, this study investigates two fundamental issues: first, the notion of performance assessment for PFI projects undertaken in the different countries such as UK, China, Australia, Serbia and Malaysia. Second is to identify direction for PFI performance assessment tools (KPIs) to be practice in Malaysia.

2 Private Finance Initiative

2.1 Overview of Private Finance Initiative (PFI)

PFI was initiated in 1992 under the United Kingdom

[a] Corresponding author: suzila8478@gmail.com

(U.K.) government and has transformed its approach from the traditional delivery of providing a range of public service projects, such as hospitals, schools, prisons, roads, etc., to privately finance, design, construct, manage, and operate these facilities. Public Private Partnership (PPP) is a procurement method that has been successfully implemented by many countries worldwide such as United Kingdom, Australia, USA, China, Hong Kong, France, Germany, Japan as well as Malaysia. The main guiding principles are to use the private sector in the provision of constructed facilities by using a whole life approach (delivering and maintaining it) within the whole life of concession period. The whole life approach includes the operation and maintenance phase of the projects throughout the entire concession period, which is about 25–30 years [14, 15].

2.2 Private Finance Initiative (PFI) in Malaysia

Private sector participation in providing facilities and public services is not new in Malaysia where it has been implemented since the 1980s. The adverse impact of the world economic recession prompted the government to seek assistance from the private sector for the development and economic activities of the country [16]. Currently, most of the public projects have been plagued by delays and shoddy workmanship, which is inherently seen as a major problem to the government [17, 18]. As a result of this situation, the maintenance repairs cost are increased, causing the Malaysian Government is reluctant to spend huge amount of money on the development of public infrastructure projects as public sector capital fund is insufficient [19]. Therefore, to bridge these issues, the Malaysian Government is turning to alternative way of Private Finance Initiative (PFI) for transforming the public projects.

PFI in Malaysia was originally initiated by the Malaysian Government through the Ninth Malaysia-Plan (2006-2010) under the National Privatisation Plan [20] and officially implemented in year 2009. It was further evolved in the Tenth Malaysia Plan (2011-2015) under the National Privatisation Plan [21]. In Tenth Malaysia Plan, it was clearly articulated the rationale for encouraging PPP/PFI procurement as the "engine of growth" in development of Malaysian Infrastructures. One of the agenda established in the Tenth Malaysian Plan are to encourage the development of Bumiputeras' entrepreneurship to embark on private projects under PPP/PFI procurements [22]. Infrastructure development is one of a key component to unlock potential economic activities and to support sustainable economic growth. Thus, as an effort, Malaysian Government will continue to place emphasis on infrastructure development as part of the economic transformation.

2.3 Issues of PFI in Malaysia

PFI have been seen as an arrangement where public and private sectors work together in implementing social and economic projects and it was designed to focus on value for money (VfM) approach, its timely implementations and efficient for facilities management. Fundamentally, the PFI programme was aimed at facilitating greater participation of the private sector to improve the delivery of infrastructure facilities and public services [23]. In addition, [17] highlight that the Malaysian PFI deviates from the basic definitions of international PPP/PFI framework. The Malaysian version of PFI will be financed by the EPF loans in which the amount was published in the Ninth Malaysian Plan. In order to facilitate the implementation of PFIs, the Ministry of Finance Malaysia has acquired a substantial amount of funds to facilitate the first wave of PFI implementation in Malaysia [17]. In Ninth Malaysian Plan, RM20 billion are allocated for these PFI projects. Nevertheless, there is a view that the government could still bear the risk to a certain extent, in particular if any of the PFI project becomes unsuccessful.

In Malaysia's scenario, the concept of PFI are lagging behind compared to other experiences countries such UK and Australia. The absence of key performance indicators (KPIs), is identified as one of the core criticisms towards the implementation of PPP in Malaysia [24, 25]. The public and private sectors' exertion in establishing KPIs, and the methods undertaken to define and measure these KPIs, are also among the crucial issues in performance evaluation. Even though a lot of studies on KPIs have been conducted in aiming to improve the performances, nonetheless, KPIs is continuously debated. Another issues with regards to PFI projects in Malaysia is the absence of any guideline or framework for the implementation of PFI procurement system [26]. Since PFI procurement scheme is still at an infant stage in Malaysia, the establishment of a framework is paramount in order to provide a better understanding of the execution of the complex scheme of financing, as well as the establishment of key performance indicators (KPIs) for assessing overall project performance. Despite the tremendous growth of PPP implementation in Malaysia, the PPP arrangements have been constantly reviewed and revised by the Malaysian government to improve the present practice of PPP implementation to ensure the achievement of its ultimate goal and objectives.

2.4 Performance Assessment in PFI Projects

The key principle of PFI is the link between performance and incentive payments to the private sector based on the successful delivery of services to the public sector. However, the service delivery aspects of PFI projects cannot be examined until projects become operational. Yet, during the operational phase, service delivery can be frequently assessed to determine compliance with the output specification and payment deductions for the performance failures in accordance with the payment mechanism [27].

Performance assessment is an important process in relation to the success and performance of PPP/PFI projects and however, it has received limited attention under the life-cycle perspective [5]. [28] define the

[a] Corresponding author: suzila8478@gmail.com

performance assessment as a process or a set of metrics used to quantify and report the effectiveness and efficiency of the action performed towards organisations' objectives. While, [29] in their study describe that, performance assessment are refers to the selection and use of quantitative measures (item that can be quantify based on the measurement units) and qualitative measures (item that can be quantify based on satisfaction level). The assessment is regarding project capacities, processes and outcomes to inform the public or designated public agency about critical aspects of a project. Thus, the performance assessment associated with VfM in a PPP/PFI project is a complicated process, and the uses of absolute time and cost measurements do not reflect the complexity relating to PPP delivery [30]. Consequently, the assessment of PPP/PFI projects is more difficult than traditional projects as there involved a lot of stages (e.g. documentation, financing, taxation, technical details, and sub-agreements) and risks (e.g. market risks and project risks) that arise from the complexity of the long-term contractual arrangement, which can change dynamically over the projects' lifecycle [31]. Theoretically, an ideal performance assessment system in Public Private Partnership (PPP) can contribute to an effective assessment of PFI projects performance and benefited to the multiple stakeholders.

3 Methodology

An in-depth and critical literature review is conducted in this study. It reviews theoretically on the PPP/PFI performance assessment models developed globally in the UK, China, Australia, Serbia and Malaysia. Based on various developed model of performance assessment from different countries, the proposed initial framework for Malaysian approached are established. Further study was plan to be conducted by comprehensive empirical research in the form of questionnaires and case studies (semi-structured interview) followed by the development and validation of performance assessment framework.

4 Synthesis of Performance Assessment Models

Theoretically, there are several methods that performance assessment can be structured and applied. A literature review of the existing performance assessment models in PPP/PFI projects as shown in Table 1 adopted from the UK, China, Australia, Serbia and Malaysia. These models were chosen in this study due to their experienced and establishment in implementing PPP/PFI and good track records in terms of performances that could be learned, in particular from the performance assessment view of point. There are six variables identified to compare the assessment model internationally. These are types of performance assessment tools, criteria in selecting performance indicators, categories of KPIs, KPIs measures, types of infrastructure projects, implementation stage, and weakness of the model. Undeniably, the UK is a pioneer and originator of PFI approach in terms of contract procurement as well as performance assessment, therefore, internationally comparative study has been made and discussed as follows:

4.1 The UK Performance Assessment Model

The UK performance assessment model of PFI project by [32] outlines a process starts by looking into performance assessment tools use in the PPP/PFI projects. This model are designed to use key performance indicators (KPIs) as a tools in assessing PFI projects performance. Development of KPIs by UK model are depending on certain criteria which are complex, defined, measurable, simple and understandable. Complex means the KPIs must be detailed covering all aspect of sustainability and stakeholder satisfaction. While, defined, simple and understandable is the clear, systematic and well-structured of KPIs developed. KPIs also must be able to measure so as to evaluate the performance of PFI projects. All the criteria need to be integrate when constructing and establishing performance indicators. In making robust assessment of PFI projects performance, the development of KPIs are takes into consideration of the effective measures which include qualitative measures and quantitative measures. Qualitative measures includes indicators on satisfaction level (i.e. multiple stakeholders) whereas, quantitative measures are includes indicators that can be measures based on the measurement unit (i.e. meter, square meter, etc.).

The UK model has divided the KPIs into four categories of dimensions which are social, economic, environmental and technical. These covers most of the critical sustainability issues through the wider PFI project environment. The study conducted by [32] are concentrating on the social, environmental and economic infrastructure and data was collected for construction and operation & maintenance phases. However, there are some limitation and barriers identified in the development of UK's KPIs model for instance lack of experience from the consultant on the sustainability issues and lack in depth analysis of individual projects. Nevertheless, the absence of detailed measures (weightage system) for each indicators identified as a weakness in the model.

4.2 The China Performance Assessment Model

The China performance assessment model by [4, 33] are using key performance indicators (KPIs) as a tools in assessing performance of PFI projects. It is differ from the UK model where KPIs are comprehensively used in assessing sustainability PFI projects performance. Key performance indicators developed in China performance assessment model are basically based on the main criteria which are identical to the UK model which are defined, complex and measurable. KPIs must be specific, clear and focused to avoid misinterpretation or ambiguity. This model has categorized the dimensions into five categories which are physical characteristic, financing and marketing, innovation and learning, stakeholders and projects process.

[a] Corresponding author: suzila8478@gmail.com

Table 1. List of PFI Performance Assessment model Developed Internationally.

Variables	*UK Performance Assessment Model (Zhou et al. 2013)*	*China Performance Assessment Model (Yuan et al. 2009; 2012)*	*Australian Performance Assessment Model (Liu et al. 2015)*	*Serbia Performance Assessment Model (Mladenovic et al. 2013)*	*Malaysia Performance Assessment Model (Ismail, 2009)*
Performance Assessment Tools	• Key Performance Indicators	• Key Performance Indicators	• Performance Prism	• Key Performance Indicators	• Key Performance Indicators
Criteria of selection performance Indicators	• Complex • Defined • Measurable • Simple & Understandable	• Defined • Specific • Complex • Measurable	• Measurable • Specific	• Defined • Specific • Complex	• Presented via metric • Measurable • Closely monitor performance • Understandable • Link with reward and penalty
Categories of key performance Indicators	• Social • Economic • Environmental • Technical	• Physical characteristic of the project • Financing and Marketing • Innovation and Learning • Stakeholders • Projects Process	• Stakeholders satisfaction • Stakeholders contribution • Strategies • Process • Capabilities	• Economic • Technical • Operation and maintenance	• Functional • Professional • Operational
Measures	• Quantitative measure • Qualitative measure	• Quantitative measure • Qualitative measure	• Objective measure • Subjective measure	• Quantitative measure • Qualitative measure	• Quantitative measure • Qualitative measure
Types of Infrastructure Projects	• Social, Environmental & Economic Infrastructure	• Not specified	• Social & Economic Infrastructure	• Economic Infrastructure (Transport)	• Social and economic infrastructure
Implementation (Project Phases)	• Operation & Maintenance	• Planning • Design • Procurement • Construction • Operation & Maintenance	• Design • Building • Finance • Operation & maintenance	• Pre-construction • Construction • Operation & maintenance	• Planning • Design • Procurement • Construction • Operation & Maintenance
Weakness	• Lack of experience consultant for sustainability • Lack in depth analysis of individual projects	• Lack of stakeholders contribution • No specific itemized indicators (qualitative and quantitative) • Complex KPIs • No detailed measures (weightage system)	• No specific itemized indicators (Objective and subjective) • No detailed measures (weightage system)	• No specific itemized indicators (qualitative and quantitative) • No detailed measures (weightage system)	• Only focussed on functional categories • No specific itemized indicators (qualitative and quantitative) • No detailed measures (weightage system)

It also similar to the UK model where quantitative and qualitative measures are considered in the development of KPIs. Even though qualitative and quantitative measures of KPIs are considered in this model, but there is no specific itemized indicators stated to differentiate between qualitative indicators and quantitative indicators. Unfortunately there are some weakness determine from the model such as lack of stakeholders contribution instead of stakeholders satisfaction on the PFI projects performance and lack of detailed measures (weightage system) for each developed indicators. This performance assessment model (KPIs) can be regarded as complex. It covers all dimensions throughout the projects phases. This indicators are derives from the review of literatures, therefore, it should be tested by the empirical study to reduces the number of KPIs.

4.3 The Australian Performance Assessment Model

In the context of Australia, the performance assessment tools applied in the model is performance prism. These performance prism by [6] having a similar concepts with key performance indicators (KPIs) but differ in term of the involvement of multiple stakeholder integration and it provide assistance in directing and guiding the design of performance measurement for long-term success. The performance prism comprised of five interrelated facets designed for measurement such as stakeholder satisfaction, strategies, processes, capabilities and stakeholder contribution. The views of stakeholders incorporated in the Performance Prism mean give the ability to overcome the hurdle triggered by multiple stakeholders (shareholders, customers, suppliers, alliance partners, and even intermediaries) in PPP/PFI evaluation. The development of performance prism take into consideration both measurable and specific criteria for each dimensions. This model are also have the similarity with other models in term of subject measures which considering on both objective and subjective measures. It also been implemented for social or economic infrastructure across projects phases (design, building, finance and operation & maintenance). Nevertheless, the development of an appropriate KPIs and the detailed measures stated in this model is just conceptually and no empirical testing and validation are presented and it demonstrate the weakness of the model.

4.4 The Serbia Performance Assessment Model

The need for more efficient and effective public-private partnership (PPP) projects makes performance management of increasing importance, especially with respect to stakeholder expectations. In the context of Serbia model, it was developed key performances indicators (KPI) based on the analysis of critical success factors (CSF) for monitoring of PPP transport projects from the perspective of different stakeholders. The development of KPIs are interrelated with the performance objectives and critical success factors (CSF). According to the authors, [12], the constructed KPIs must be defined, specific and complex. This model was abstract from the review and comparative analysis of existing critical success factors (CSFs) and KPIs in PPP transport projects and it was performed based on stakeholders' objectives and different project attributes. Therefore, the KPIs was grouped into three essential groups depending on the aspect of the project which are economic, technical, and operation and maintenance.

In making dynamic assessment of PFI projects performance, the constructed KPIs in this model introduced a two layers approach for the assessment of PPP projects. The first stage was based on evaluation of project ultimate objectives from the standpoint of each stakeholder, i.e. profitability for private sector, effectiveness and value for money for public sector, and level of service for users. An adjusted and weighted combination of fulfilment of the specific stakeholder objectives leads to an overall approach (second layer) of describing a PPPs success or failure. Therefore the assessment for PFI performance considers both the qualitative and quantitative measures, which are identical to the UK, China and Australia practices. It also covers all stages during life cycle projects (pre-construction, construction, and operation & maintenance phase). To assess the performance of PFI projects consequently, it requires detailed measures such as weightage system in order to attain value for money (VfM). Nonetheless, the absence of the appropriate detailed measures perceived as a weakness of the model.

4.5 The Malaysian Performance Assessment Model

PFI in Malaysian viewed of point is generally labelled as privatisation instead of PFI even though PFI is differed from privatisation to a certain extent. Insufficiency of information by the public and private sectors, which have implemented PFI, has been the reason why PFI projects in Malaysia has never been evaluated comprehensively. Given that, for the Malaysian context, a key performance indicators (KPIs) approached is used to measure performance of the PFI projects. It is studied by [34], which divided the KPIs into three categories: functional, professional and operational. The formation of KPIs is fundamentally based on the criteria such presented via metric, measurable, closely monitor performance, understandable and KPIs must link with reward and penalty. All listed criteria are vital to be integrated when choosing and constructed the KPIs. Essentially, this model are developed by considering both qualitative and quantitative measures similarly with other countries. Malaysian has adopted PFI for the social and economic infrastructure where most of the financed (RM20 Billion) are allocated for the social infrastructure (health, education and etc.).

Thus, the KPIs been categorized into three (functional, professional and operational), essentially, it is completely covers all stages of the life cycle projects such planning, design, procurement, construction and operation & maintenance.

[a] Corresponding author: suzila8478@gmail.com

DOI: 10.1051/matecconf/20166600049

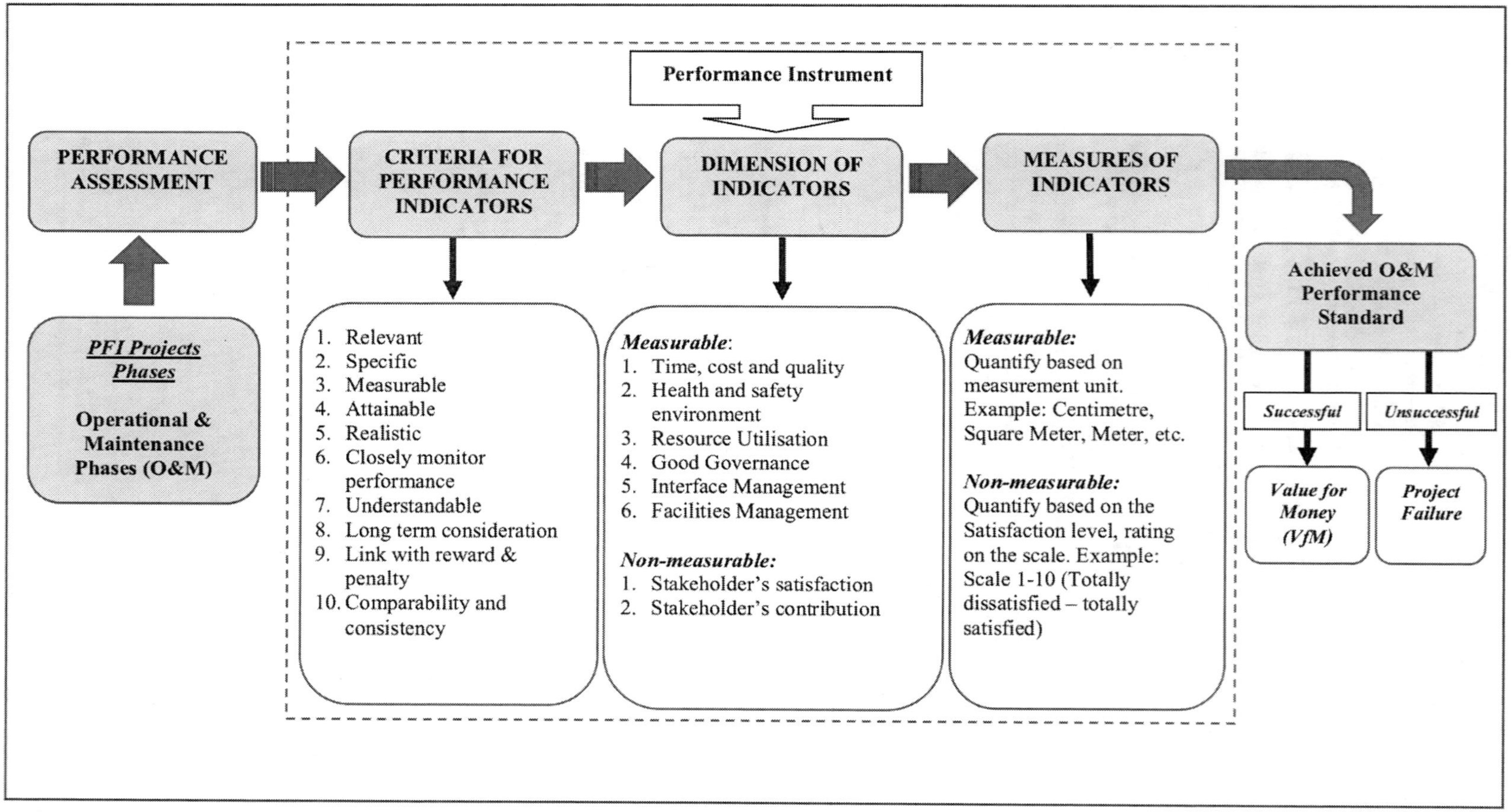

Figure 1. A Propose Initial Framework of Performance Assessment for PFI Projects in Malaysia

Even though the development of an appropriate KPIs identified in this model has been empirically tested and validated, but there are only verified for the functional category and future recommendation is suggested to continue for the other categories (professional and operation). The other weakness of this model is similarly to the other countries (UK, China, Australia and Serbia) which detailed measures and itemized indicators (qualitative and quantitative) is not specifically spelt out.

5 A Proposed Initial Framework of Performance Assessment for PFI projects in Malaysia.

Based on preliminary literature review and understanding of performance assessment models adopted from various countries (UK, China, Australia Serbia as well as Malaysia). The proposed framework for PFI performance assessment in Malaysia is initially developed as illustrated in Figure 1. In this framework, performance assessment is designed to embrace the specific phases of the PFI projects which is focused on the critical phases throughout project life cycle (operation and maintenance phases). It is due to the facts that every project phases have their own objectives and goals that need to be achieved. The implementation of performance assessment via KPIs approach is needed to achieve performance standards, quality service provided and the most significant goals is getting value for money (VfM). Therefore, key performance indicators are one of the effective performance instruments used to assess the performance and success of the PFI projects specifically in Malaysia. The prime factors in developing an effective key performance indicators are by embedding the important criteria (relevant, specific, measurable, attainable, realistic, and etc.) with every possible indicators (measurable and non-measurable). The terms measurable and non-measurable have similar meaning with the qualitative and quantitative measures. For this version of initial framework, terms measurable and non-measurable is used. In addition, the integration between criteria, indicators and weightage system is important part in this study to determine 'what to measure' and 'how to measure' the performance of the PFI projects. Weightage system for the measurable items are the items that can be quantify based on the measurement units (i.e meter, numbers, cost, time etc.), where non-measurable items are items that can be quantify based on satisfaction level (i.e customer satisfaction, client satisfaction, etc). Thus, establishing an important criteria is essential as a guidance for selecting the most appropriate indicators. As a result, lack of stakeholder's satisfaction and the linkages between indicators and measures for the assessment strategy shows the gaps for the future study.

6 Conclusion

This paper anticipate to review, synthesize and develop a framework of performance assessment for PFI projects in Malaysia based on the various models of performance assessment from the UK, China, Australia, Serbia as well as Malaysia. It also intend to identify the direction and usefulness of performance assessment tools (KPIs) to be practiced for PFI projects in Malaysia. Most of the models are applying KPIs as a tools to assess the performance of PFI across projects phases. However, there is no single model that has been developed to integrate KPIs weighting system for measuring each indicators. As a result, the study was found gaps through the listed weakness, therefore, it is vital for the Malaysian Government to develop a framework for assessing PFI projects performance especially for the critical phases of operation and maintenance that involved lengthy concession period (25-30 years).

The research presented in this paper is initially and part of an ongoing PhD research at the faculty of Architecture, Planning and Surveying, UiTM to develop a framework of an integrated performance assessment for PFI projects in Malaysia. The result of the study would provide an insight into the Malaysian construction project development and forms the basis of a valuable guideline, especially to public and private sectors in Malaysia.

References

1. NAO. *The Performance and Management of Hospital PFI Contracts*, **68.** (2010).
2. Yuan, J., Skibniewski, M. J., & Li, Q. *Managing the Performance of Public Private Partnership Projects to Achieve Value for Money: Key Performance Indicators Selection*. In International Conference On Multi-National Construction Projects, pp. 1–16 (2008).
3. Robinson, H. S. H., & Scott, J. *Service Delivery and Performance Monitoring in PFI/PPP Projects*. Construction Management and Economics, **27(2)**, pp. 181–197 (2009).
4. Yuan, J., Zeng, A. Y., Skibniewski, M. J., & Li, Q. *Selection of Performance Objectives and Key Performance Indicators in Public–Private Partnership Projects to Achieve Value for Money*. Construction Management and Economics, **27**, pp. 253-270. (2009).
5. Liu, J., Love, P. E. D., Davis, P. R., Smith, J., & Regan, M. *Performance Measurement Framework in PPP Projects*. (2013).
6. Liu, J., Love, P. E. D. D., Smith, J., Regan, M., & Palaneeswaran, E. *Review of Performance Measurement: Implications for Public–Private Partnerships*. Built Environment Project and Asset Management, **5(1)**, pp. 35–51. (2015).
7. Regan, M., Smith, J., & Love, P. E. D. *Impact of The Capital Market Collapse on Public Private Partnership Infrastructure Projects*. Journal of Construction Engineering and Management, **137(1)**, pp. 6–16. (2011).
8. Hodge, G. *The Risky Business Of Public – Private Partnerships*. Symposium, **63 (December)**, pp. 37–49. (2004).

[a] Corresponding author: suzila8478@gmail.com

9. Yong, H K. *Public-Private Partnerships Policy and Practice*, London: C. Secretariat. (2010).
10. VAGO. *Report of the Auditor-General on the Finances of the State of Victoria*, 2001-2002, (2002).
11. House of Commons. *The Operational Performance of PFI Prisons: Forty-ninth Report of Session 2002-2003*. (2003).
12. Mladenovic, G., Vajdic, N., & Wu, B. *Use of Key Performance Indicators for PPP Transport Projects to Meet Stakeholders' Performance Objectives*. (2013).
13. Oyedele, L. O. *Avoiding Performance Failure Payment Deductions in PFI/PPP Projects: A Model of Critical Success Factors*. Journal of Performance of Constructed Facilities, (JUNE), **259**. (2013).
14. El-Haram, M. a., & Agapiou, A. *The Role of the Facility Manager in New Procurement Routes*. Journal of Quality in Maintenance Engineering, **8(2),** pp. 124–134. (2002).
15. NAO. *The Operational Performance of PFI Prisons*, (June), **55**. (2003).
16. Ismail, S., & Rashid, K. A. *Private Finance Initiative (PFI) In Malaysia: The Need for and Issues Related to the Public Sector Comparator (PSC)*. (2007).
17. Jayaseelan, R., & Tan, M. *PFI-cure for All Ills. The Egde Malaysia,* pp. 72-74. (2006).
18. Endut, I.R. *Framework For Minimizing Time Overruns of Malaysia Construction Projects*. A PhD Thesis, Glasgow Caledonian University. (2008).
19. Netto, A. *Malaysia's New Fangled Privatization Fudge*. EPU. (2006).
20. Economic Planning Unit (EPU). Streamlining Privatization. *Ninth Malaysia Plan 2006-2010*, pp. 223–233. (2006).
21. Economic Planning Unit (EPU). *The Tenth Malaysian Plan, 2011-2015,* Prime Minister"s Department, Putrajaya. (2010).
22. PEMANDU. *Annual Report: Economic Transformation Programme (ETP),* Kuala Lumpur: PEMANDU. (2011).
23. Prime Minister Department. *Public Private Partnership (PPP) Guideline*, pp. 1–12. (2009).
24. Khaderi, S. S., & Aziz, a. R. a. *Adoption of Private Finance Initiative (PFI) in Malaysian Public Works Projects: Are We Ready?* CIB TG72 World Building Congress, pp. 105–120. (2010).
25. Ismail, K. *A Value for Money Assessment Framework for Public Private Partnership Approach,* PhD, Thesis. (2012).
26. Edge, T. *Untested Model for Malaysian Companies*. The Edge Malaysia, p. 75. (2006).
27. Akbıyıklı, R. *Performance Assessment of A Private Finance Initiative Road Project.* Transport, **28** (February 2015), pp. 11–24. (2013).
28. Neely, A., Gregory, M. and Platts, K., *Performance Measurement System Design: A Literature Review And Research Agenda*, International Journal of Operations and Production Management, **Vol. 25 No. 12**, pp. 1228-1263. (2005).
29. Ong'olo, D.O., *Public Private Partnerships (PPP): Practice and Regulatory Policy in Kenya*. Paper prepared for the Institute of Economic Affairs (IEA) on July 7th 2006, Kenya. (2006).
30. European Commission. *Guidelines for Successful Public-Private Partnerships*, European Commission, Brussels. (2003).
31. Grimsey, D., & Lewis, M. *Evaluating The Risks Of Public Private Partnerships For Infrastructure Projects*. International Journal of Project Management. (2002).
32. Zhou, L., Keivani, R., & Kurul, E. *Sustainability Performance Measurement Framework for PFI Projects in the UK*. (2013).
33. Yuan, J., Wang, C., Skibniewski, M. J., & Li, Q. *Developing Key Performance Indicators for Public-Private Partnership Projects: Questionnaire Survey and Analysis*. Journal of Management in Engineering, **28(3)**, pp. 252–264. (2012).
34. Ismail, S. *Key Performance Indicator for Private Finance Initiative in Malaysia*. PhD Thesis. (2009).

[a] Corresponding author: suzila8478@gmail.com

MATEC Web of Conferences **66**, 00050 (2016) DOI: 10.1051/matecconf/20166600050
IBCC 2016

A Review on the Significance of Local Plan for Coastal Reclamation Development: The Case of Malacca, Malaysia

Nor Syafa'ah Ishak[1, a], Mohammad Yusup[2], Ahmad Fuzi Arshad[3] and Yusfida Ayu Abdullah[4]

[1] *Centre of Postgraduate Studies of Town and Regional Planning, Faculty of Architecture, Planning and Surveying, 40450 UiTM Shah Alam, Selangor, Malaysia*
[2, 3&4] *Centre of Studies of Town and Regional Planning, Faculty of Architecture, Planning and Surveying, 40450 UiTM Shah Alam, Selangor, Malaysia*

Abstract. A Development Plan acts as a planning and controlling instrument that helps to guide in decision making of current and future development. Therefore, in the Malaysian Town and Country Planning Act 1976 (Act 172), the provision of development plan from each level of the planning administration is stated in the Act under Section 6B till Section 16B. In examining the significance of one of the many development plans in Malaysia, i.e. the local plan, this paper therefore, focuses on the provision of a local plan by the Local Planning Authority (LPA) of the said area. This paper, thus, aims to evaluate the Central Malacca Local Plan since Malacca is currently experiencing rapid growth in development along the coastal area. The primary data was obtained from interviews with the related planning authorities via face-to-face method. While, secondary data was gathered from related legal documents, policies and guidelines, as well as the existing development plan for the purpose of an in-depth data. Findings revealed incapability of the existing local plan to assist the development control in deciding the planning permission for reclaimed area. The existing local plan shows no zoning for coastal reclamation area because of the rapid growing development and time consuming procedure to amend the local plan according to the Act 172. In addition, the Act 172 and the process in amending the local plan should be more flexible and able to address the current issue efficiently. Finally, the paper concludes with a suggestion for further exploration in order to achieve an operative development process and functioning practice of the local plan.

1 Introduction

To date, urbanisation processes are in line with the rapid development of residential, commercial and industrial mainly and it has gradually led towards the scarcity of land. Land is a scarce resource [1]. Therefore, the demands for more land are arising. Many developed countries had already practices coastal reclamation for decades such as Singapore, Japan and Hong Kong in order to tackle problems regarding scarcity of land [2]. At present, similar situation prevails in Malaysia [2]. Malaysia has been experiencing rapid urbanisation process from the economic development, industrialisation, and enormous migrations, along with the natural population growth [3]. In coping with the urbanisation process, the development had taken place almost everywhere. Recently, Malaysia also focuses on the development for reclamation in coastal area. There are three states in West Malaysia that commit in the development of coastal reclamation, namely, Pulau Pinang, Malacca and Johor. Nevertheless, this study only selects Malacca as its study area. The reason for choosing Malacca is due to its speciality as a historical city and its status as a UNESCO site. The State of Malacca is located in the southern part of West Malaysia. It covers an area of 1,799 km^2 [4]. The case study is located from Tanjung Kling up to Telok Mas along the Straits of Malacca. The coastal reclamation development projects are concentrated along the coastal areas of the Straits of Malacca including Pantai Klebang, Pekan Klebang, Taman Kota Laksamana, Pulau Melaka, Permatang Pasir Permai and Telok Mas [5]. The responsible local authority for this coastal reclamation site is the Malacca Historical City Council administrative area [6].

2 Research Methodology

This paper concentrates on the impact of local plan in the coastal reclamation development. A local plan plays an important role in planning and development especially for a new area. Therefore, this study employs the qualitative research method and a case study design. Qualitative research is concerned with the subjective assessment including attitudes, opinions and behaviour or it is also called as qualitative phenomenon [7]. Accordingly, Shamsuri [8] claimed that qualitative research is not meaningful if it is being expressed by numbers. The case study approach is suitable for a small-scale study [9]. It is an extensive study of a particular

[a] Corresponding author: norsyafaah@gmail.com

DOI: 10.1051/matecconf/20166600050

case and it can be considered as research strategy rather than a method of data collection [10]. Meanwhile, this study attempted in improving the planning process for coastal reclamation in Malacca including the local plan as a tool for development control. The primary data will be obtained from the government agencies that are directly involved with the coastal reclamation development such as Malacca Historical City Council, Malacca Economy Planning Unit, Malacca Department of Town and Country Planning and other related agencies. The data was gathered through face-to-face interview conducted in order to gain further understanding related to the study as well as opinion and recommendation on the provision of the local plan and planning process from the experts. Hence, the study applies the purposive sampling in getting the response and feedback from the informants through interviews. Additionally, secondary data was gathered from various development plans, legal documents, policies, and other related. This is vital as it is intended to obtain a clearer picture on the current issues and problems associated with coastal reclamation practice in Malacca and thereafter to suggest solutions for the current practice and development of coastal reclamation.

3 The Key Terminologies

Planning is being expressed as something futuristic [11]. According to Stewart, Rutherford, Levy and Jackson [12], planning is the process to determine what is to be attained in the future and clarifying the steps required in achieving it. Also, referring to the planning definition by Keeble [13] and Ratchliffe [14], Yusup [15] had then concluded his planning definition as a process of decision making involving a goal, target group as well as correct procedures. Meanwhile, Zainol *et al.* [16] defines the term of planning as a physical performance and space of maps, drawings, reports as well as other form of documentation and plan that deliver detail description, concept and idea of a planned development for the current and future. For that reason, a good planning acts as a mould towards a good development for current and future goodness.

As stated in Section 5 of the National Land Code 1965 (NLC) (Act 56) [17] as well as in Section 2 of the Town and Country Planning Act 1976 (Act 172) [18], the term of 'land' is being interpreted as *the land that covered on and below the surface of the earth and its substances, all vegetation and other natural products either it is on or below the surface, all things that attached to the earth, the land that is covered by water and any estate or interest in, or right over are* considered as 'land'. The land is under the state matter and it is clearly being mentioned in the List II (State Lists) under Ninth Schedule of the Federal Constitution that the state government is the party which has full responsibility towards the land matter within its territory [19]. The land use planning is a state issue and the federal government is responsible to take on a supervisory role with the overall land use planning activity and this system is also being introduced to express the authority's intention in initiating, encouraging and controlling over the physical, economic, environmental and social changes in a particular area [20;18]. The land use planning signifies the way of planning the physical layout, or land use, of the communities and is the key component of a community's long term resilience [21]. This indicates that planning for the use of land in advance is important especially in avoiding the conflict of land use, thus, aids in guiding towards a sustainable development.

In Section 2(1) of the Town and Country Planning Act 1976 (Act 172), it has been highlighted that 'development' is *the carrying out of any building, engineering, mining, industrial or other similar operation in, on, or under land, the making of any material change in the use of any land or building or any part thereof, or the subdivision or amalgamation of lands: and 'develop' shall be construed accordingly*. Additionally, Ahmad and Bajwa [22] claimed that development embraces all types of building, entirely a new construction, demolition and reconstruction or addition and alteration of a building including residential, commercial, industrial and others, named as development. This was clearly spelled in Section 19(1) of the Act 172. Hence, the development is considered as the activity that being performed physically and apparently involving the changes to the building or land as well as affecting the surrounding.

The term 'coast' is being defined by the Malaysia Department of Irrigation and Drainage [23] as the strip of land which extends from the coastline inland to the first major change in the terrain features, therefore, it is not influenced by the coastal process. Apart from that, according to Powter [24], reclamation is the process of reconverting disturbed land to its former or other productive uses. Hence, the terms of coastal reclamation basically came from both stated terms to be interpreted broadly. The coastal reclamation is being mentioned by Sufian and Mustafa [25] as taken place within the area that defines as foreshore (in Section 5 of the National Land Code 1965 (Act (56)) and it is located within the area inside the baseline that is as internal waters, whereby waters within 12 nautical miles from the baseline are territorial waters and those within 200 nautical miles seawards from the baseline are the Exclusive Economic Zone. Therefore, the first 3 nautical miles of the territorial waters from the low water mark is under the jurisdiction of the state, while the rest of the area is rest on the federal jurisdiction. Additionally, coastal reclamation is the land that seems to appear out of nowhere [26]. Coastal reclamation can be considered as common practices in producing valuable land in the coastal areas [27]. Thus, the coastal reclamation is a way to obtain completely a new land.

As a development control instrument, the development plan is a prerequisite at every level of government. In Act 172, the development plan is *in relation to an area, it is either the local plan of the area, and otherwise, if there is no local plan for the area; only then the structure plan will be effective to be used as guidance for development*. The development plans considered as a legal requirement for all upper-level till lower-level governments [28]. The development plan is the plan that acts as guidance in locating development with criteria based policies [20]. Each of the development plans provided, it play functions as a guideline and reference in every aspect of the urban planning [15]. The local plan also is one of the instruments used in the decision making process [29].

4 The Local Authority Administrative Area

The Local Government Act 1976 (Act 171) is one of the main legislations exercised by local authorities to facilitate the development within their area. The Act 171 outlines the form, organisational structure, duties and responsibilities of the local authority as a whole especially in terms of accountability pertaining to town and country planning [30]. In Section 3 of the Act 171, it indicates on the declaration and determination of an area to be under any local authority that shall administrate the declared area. Any land or area that are still not be declared or gazetted under any local authority administrative area, will still be under the state matter – the state authority responsibility [19]. Section 4 of the Act 171 mentioned on the alteration and determination of the boundaries. The alteration of the boundaries requires the state authority to specify the boundary of a local authority from state authority land or area. The declaration and gazetting of the area for a local authority is important because the respective local authority will be aware of their administrative area. Furthermore, the provision of a local plan by the local planning authority will need to specify the boundary of the local plan for development. Thus, as an example, the coastal reclamation area in Malacca has been declared and it falls under the Malacca Historical City Council, therefore, the Malacca Historical City Council shall provide a detail plan concerning on the area due to its rapid development on the reclamation area [31]. A plan should be as detail as possible because it will help in directing the development [6].

5 The Provision of Local Plan in Act 172

Malaysia consists of three-level of government administration system; federal government, state government and the local authority [32]. Each level of the government administration system has its own development plan that must be prepared in assisting a good planning towards a sustainable physical development. In Malaysia, development plans consist of four types through the three levels of planning administration and is stated in Part III of the Act 172 on the provision of development plan (Table 1).

However, this study will only focus on the provision of a local plan at the local level which is the local authority's area. Local authority is the lowest level in the hierarchy of government systems within the concept of federalism [33]. The local authority is defined as a sub-national in the government's level which only has its jurisdiction over a limited range of the state functions, within a designated geographical area which is part of a larger territory [34].

Table 1. The Jurisdiction in Providing Development Plan in Malaysia under the Act 172

Level of government administration	Type of development plan	Section involved	Jurisdiction
Federal	National Physical Plan (NPP)	• Section 6B	• Negotiated with State Authority (SA) • Approved by National Physical Planning Council (NPPC) • Prepared by Director General (DG) of Federal Town and Country Planning Department
State	Structure Plan (SP)	• Section 7 – Section 11	• Approved by State Planning Committee (SPC) • Assented by the State Authority (SA) • Prepared by the State Director of Town and Country Planning Department
Local	Local Plan (LP)	• Section 12 – Section 16A	• Approved by State Planning Committee (SPC) • Assented by State Authority (SA) • Prepared by Local Planning Authority (LPA)
	Special Area Plan (SAP)	• Section 16B	• Approved by State Planning Committee (SPC) • Assented by State Authority (SA) • Prepared by Local Planning Authority/State Director

Source: Adapted from the Commissioner of Law Revision (2014c)

Additionally, in Section 5 of the Act 172, stated that the local authority may also act as the local planning authority (LPA) within their particular area. The LPA's functions including; i) regulating, controlling and plan over the development and the use of all land and building within their administrative area; ii) undertake, assist in, and encourage the collection, maintenance, and publication of statistics, bulletins, and monographs, and other publications that are related to town and country planning and its methodology; also iii) performing other functions as the State Authority or Committee may from time to time assign to it [18]. Besides, as in the Act 172 (Section 12(1)), the LPA also is the responsible party that should provide a local plan for its area. Whereas, in Section 5(2) of the Act 172 stated that the area which is not under any local authority area, is then required by the State Director of the State Department of Town and Country Planning to provide a local plan of the particular area.

5.1. The Preparation of a Local Plan by the Local Planning Authority (LPA)

The provision to provide a local plan by a local planning authority of the area is being pointed out in Section 12(1) of the Act 172. In preparing a local plan, there are procedures and requirement that should be adhered by the LPA as in Table 2. Furthermore, there are five (5) situations that require the preparation of a local plan based on Act 172 which are; i) the LPA preparing the draft local plan during the draft structure plan is being prepared or before the State Planning Committee (SPC) approves or rejects the draft structure plan; ii) the Director of State Department of Town and Country Planning preparing the draft local plan of its area as appointed by the LPA as in Section 5(2) of Act 172 (the area which is outside of any local authority administrative area; iii) the LPA preparing a draft local plan when the structure plan of their area has been enforced; iv) the LPA preparing a draft local plan when instructed by the SPC; and v) the Director of State Department of Town and Country Planning preparing a draft local plan if instructed by the SPC by the reason of the LPA of the area are unable to prepare a draft local plan for their area [35]. This indicates that the provision of a local plan of any respective area either with or without a local authority to administrate the area, the local plan of the area will still need to prepare the local plan to control over the development that takes place within its area.

Therefore, the process in preparing a local plan consists of four (4) stages called mobilisation, draft local plan preparation period, publicity and public participation, and approval and gazette the draft local plan as illustrated in Figure 1. The overall process demonstrates on how a local plan being prepared in details. In addition, the time frame for a local plan to be effective after it has been gazette is between 10 till 15 years [35]. However, the local plan can be reviewed from time to time if it is required, depending on the necessity for it to be modified [35].

5.2. The Weakness of the Local Plan in Assisting Development Control System

There are several problems in the provision of a local plan in Malaysia including the Central Malacca region. Firstly, the weakness from the upper-level administration and their provision of the development plan which are the National Physical Plan and the State Structure Plan in providing policy, guidelines, strategic development and other relevant provisions are the root of the weakness in the provision of local plan [36]. Due to that matter, local plan cannot interpret the policy and guidelines and in detailing other relevant information to control the development for building and land [29]. This problem is because of the incompetence of the local plan especially in controlling over the development within its boundary. Also, the duration in preparing a local plan takes a longer time (Figure 1).

Table 2. The Provision of Local Plan as in Act 172

Aspect		Section involved	Description of provision
Responsible authority	The Local Planning Authority (LPA)	Section 12(1)	The local plan must be provided by LPA of the area if it is required and it will give benefit to the area. There are two (2) situations which LPA can prepare the draft local plan, when draft structure plan is being prepared or before the draft structure plan assented by SPC.
		Section 12(2)	LPA should provide a draft local plan for the whole of its area immediately if the structure plan of the state has been effectively enforced.
		Section 12(6)	LPA also must provide a draft local plan of their area as instructed by SPC. This instruction can be given during or after the preparation of draft structure plan.
		Section 16(1)	LPA has the right to make alteration, revocation and replacement of the local plan with another local plan.
	The Director of State Department of Town and Country Planning	Section 5(2)	The Director of the State Department of Town and Country Planning shall prepare a local plan for any area which located outside the local authority boundary.
		Section 16A(1)	The SPC may instruct the Director of the State Department of Town and Country Planning to prepare local plan if LPA is unable to prepare local plan.
Form		Section 12(3)	A draft local plan should consist of map and written statement.
		Section 12(4)	A draft local plan must be supported by diagrams, illustrations and descriptive matter as it is appropriate to be included in explaining and illustrating the local plan.
Content		Section 12(3)*(a)* and *(b)*	The content of the local plan shall include all appropriate matters that are relevant to be documented in the local plan such as the land use zoning (residential, commercial, industrial, agriculture, etc), type of development (existing and proposed), protection and improvement of the physical development, preservation of the natural topography, improvement of the landscape, the preservation and planting trees, making up open spaces, preservation and enhancement of character and appearance of buildings, improvement of communications and management of traffic in the local plan area. As well, other matters specified or instructed by SPC.
Preparation		Section 12A	Before commencing the preparation of local plan, the LPA shall give opportunity to the person to make representations in respect of matters such as objection and matters that should be included in the local plan.
	Publicity of draft local plan	Section 13(1)	LPA must provide copies of draft local plan for inspection and details of place and time of the publicity.
		Section 13(2)	LPA shall notify notice regarding the draft local plan that has been available for public inspection in three (3) issues – two (2) in the local newspaper and either one (1) must be in national language. The notice must include time and place that are available for inspection. The inspection period is not less than four (4) weeks – any objection or representations can be made to the LPA.
		Section 13(3)	Time for inspection of draft local plan can be extended by LPA by not more than four week on the application of any person.
	Inquiries, objection and hearing	Section 14(1)	A Committee (three persons) shall be appointed by SPC as directed by LPA to hold local inquiries or other hearing.
	Approval or rejection of draft local plan	Section 15(1)	After modifications along with consideration of the objections, LPA will submit the draft local plan (that has been changed) to SPC for approval.
		Section 15(1A)	SPC has the right to approved, whole or a part of draft local plan with or without modifications, or reject.
		Section 15(1B)	SPC may take into account any matters that it thinks are relevant, whether or not they were taken into account in the plan as submitted or resubmitted to SPC.
		Section 15(1C)	SPC shall submit the approved draft local plan to the State Authority to be assented and come into effect.
		Section 15(4)	LPA must publish the facts that the draft local plan of the area has been assented in the State Gazette with the minimum of two (2) local newspaper and either one (1) of them are publish in national language.
		Section 15(5)	Local plan shall conform to the structure plan. If there exists a difference in the provision between these two development plan of the state, and the Committee is satisfied that the existing structure plan of the state is out-of-date, the Committee must refer to other State Authority for its decision.
		Section 15(6)	If the State Authority is satisfied with the matter mention in Section 15(5), the State Authority shall notify it in the State Gazette, declaring the local plan of the area shall prevail over the structure plan as regards to that particular provision.
Review and monitoring		Section 16(1)	LPA may at any time make proposals of the local plan in respect of alteration, revocation, or replacement.
		Section 16(3)	Subsections 12(8) and (9), and sections 13 (publicity), 14 (inquiries and hearings) and 15 (approval or rejection), shall apply in relation to the making of proposals for the alteration, revocation, or replacement of a local plan.

Source: Adapted from Town and Country Planning Act 1976 (Act 172)

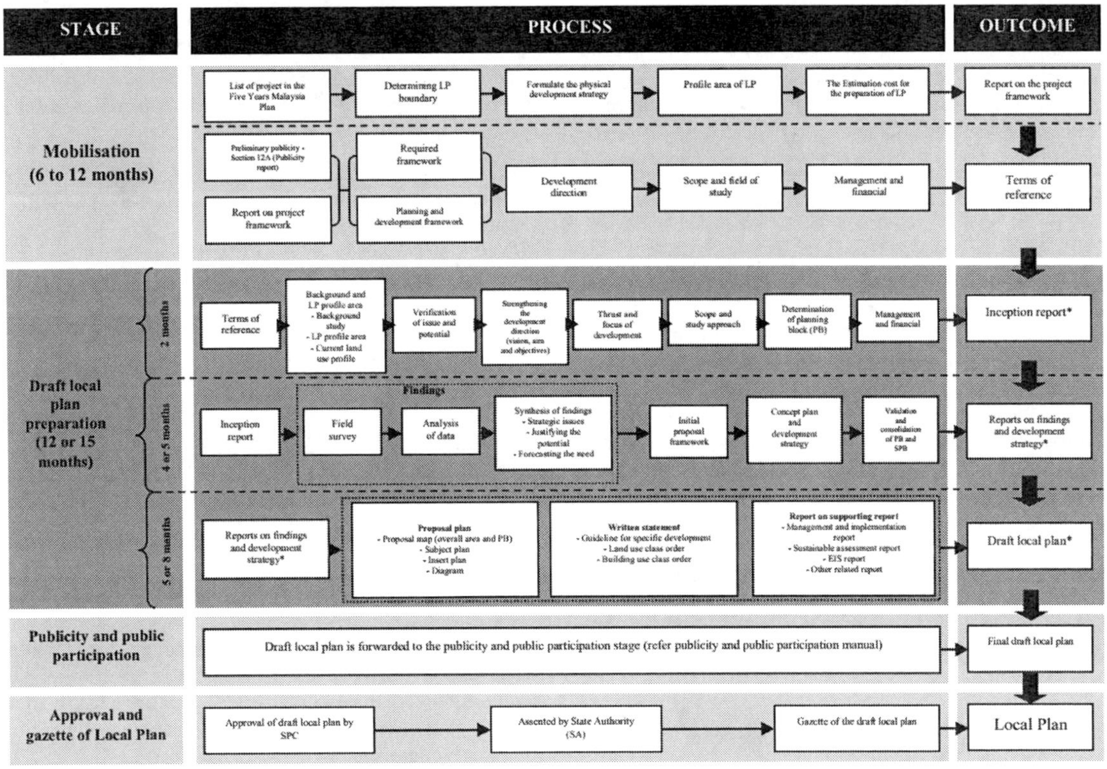

* Requires input for sustainability assessment

PB – Planning block
SPB – Small planning block

Figure 1. The Local Plan Preparation Process
Source: Adapted from Department of Town and Country Planning - Local Plan Manual (2009)

5.3. The Existing Central Malacca Local Plan

The Central Malacca Local Plan (2003 – 2015) has been gazetted in 2008. In the Central Malacca Local Plan (2003-2015), the coastal reclamation area is situated between Planning Block I and III. Yet, in the existing local plan, the problems concerning on the reclamation area and development process were encountered such as such as no detailed zoning for reclamation area as shown in Figure 2 [31].

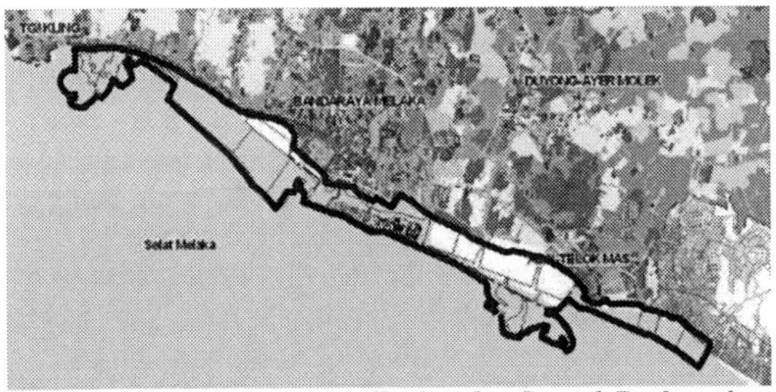

Figure 2. Current Land Use Zoning for Coastal Reclamation Area in the Central Malacca Local Plan 2003-2015
Source: Adapted from Central Malacca Local Plan 2003-2015

The lack of detailed zoning may lead to a conflict of land use and building in the future due to excessive unplanned development. Future problem may arise such as the duplication on the type of development and the non-continuity of the infrastructure allocation within reclamation development area as well as with the adjacent area [31]. Besides, another problem is that there is no specific guideline provided for coastal reclamation development. Currently, the development on the reclaimed land refers to the same guideline as planning for development on the 'original land'.

The reclaimed land and the original land are totally different especially in terms of land stabilisation and its durability. For example, the land can only stand with a certain level of building height. These problems will affect the decision making process towards planning permission for the development because it has no zoning and guidance that will deliver an efficient decision making process. As a result, the planning permission will be decided in ad-hoc manner. This portrays an unhealthy practice that will affect the long-term future development for reclamation area and its surrounding. These problems surely requires an immediate action from the state government and the local authority of the area in order to

[a] Corresponding author: norsyafaah@gmail.com

minimise the impact of the coastal reclamation development to its surrounding because currently, coastal reclamation in Malacca is being practices in ad-hoc manner. Thus, these scenarios revealed the importance of planning before development.

6 Results and Discussion on the Significance of a Local Plan

Local plan is used to control over the development through planning permission. The planning permission is the stage where it is required in proposing any development that requires planning permission and it should comply with the existing local plan of the area [18]. With the provision of local plan (the ahead planning) which demands for all current and forward planning that has been made beforehand, and all the relevant studies that has been conducted while preparing the local plan, thus, aligning the direction of the future development. The coastal reclamation development in Malacca is currently facing a rapid development [5]. However, due to the incomprehensive of the existing local plan of Central Malacca, it is unable to assist the development control for reclamation within the coastal area. Thus, this scenario leads to an ad-hoc manner in the decision making process. In any physical development, the development process should come after planning. Figure 3 illustrates the scenario in coastal reclamation development which reveals the consequences between planning and development.

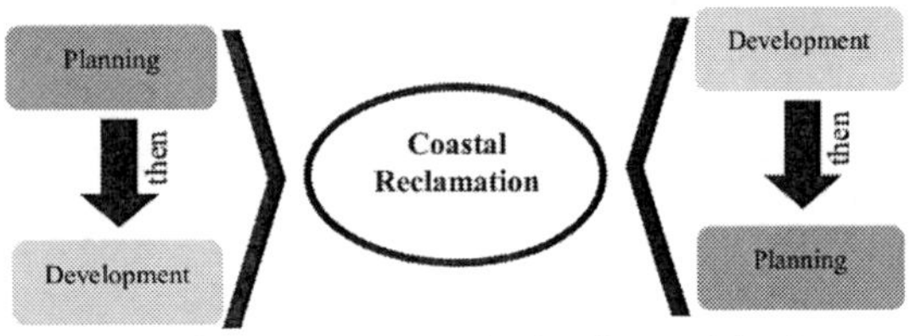

Figure 3. The consequences: planning first and development later or development first, planning later

6.1. The Consequences: Planning First, Development Later

Planning act as an important step especially in the development processes. It helps to plan for the future. In town planning context, planning is for the people. Litman [6] expresses planning as a social activity because it involves the people. The involvement of people through public participation offers an important element in planning because it allows plan to be viewed and considered from different perspectives and views, which then helps to identify arising problems early in the planning process as well as gaining support for the implementation of the plan [37;38].

The benefit is when it comes to the consideration to plan first such as planning for a development able to create a lasting value of sustainable development. The coastal reclamation development in Malacca started without a proper planning process and it can be considered as an ad-hoc manner. Additionally, the Central Malacca Local Plan has been outdated because the local plan should only be effective until 2015. Due to that matter, the review for the existing local plan should be done according to the current situation and need for it to be reviewed.

6.2. What if Development First, Planning Later?

Development then later planning demonstrates an unhealthy practice. In West Malaysia, every development that can make changes to the land or building and may gives effect to its surrounding, therefore, is required to apply for planning permission from the respective local planning authority of the area as stated in Section 21(1) of Act 172. In addition to that, in Section 18(1) of Act 172, every planning permission must follow the local plan of the area to plan for a development. However, if the requirement from the Act 172 cannot be fulfilled and the development is being approved even if it does not follow the local plan – the development control instrument, it may lead to the future consequences. For instance, the coastal reclamation development is a big project and it makes clear changes towards its surrounding, therefore, the planning for coastal reclamation requires an immediate action from the responsible planning authority in tackling this matter. Hence, it is better to prevent rather than facing future problems and disputes.

7 Conclusions

A local plan acts as a development control for a local planning authority. It plays an important role in the delivery of any development including within a reclamation land. Local Authorities are the responsible body to prepare and implement their local plans for their area. However, the coastal reclamation site in the Central Malacca faces challenges of the lack of detailed zoning for the area and the incompetence of the local plan itself in protecting the area especially for the purpose of planning permission. Results have shown that it is more fitting to perform planning earlier, before carrying out any development. The requirement on the provision of a local plan by a local planning authority in the Act 172 should be concurrent with the current situation and appropriate practices that should be as well applied in the law. The law must not be rigid and time consuming; however, it should be flexible and efficient with the current requirements and needs. Moreover, the Act 172 shall be amended as soon as possible especially for the reviewing and alteration of the local plan quickly. Another option, the interim development plan should be introduced to overcome this problem. Importantly, to plan and make a clear direction of the development for coastal reclamation in order to avoid future conflicts due to unhealthy planning practices that can caused inconvenience and disputes concerning on ad-hoc manner. Further work will include a more detail investigation on the delivery of development at the site.

Acknowledgement

The authors would like to express their deep gratitude to the Research Management Centre (RMC) of Universiti Teknologi MARA (UiTM), Malaysia and the Ministry of Higher Education Malaysia for funding this research under the Research Acculturation Grant Scheme (Grant 600-RMI/RAGS 5/3 (135/2014).

References

1. Federal Ministry for Economic Cooperation and Development. *Land Use Planning: Concept, Tools and Applications.* Deutsche Gesellschaft fur Internationale Zusammenarbeit (GIZ) GmbH. (2012).
2. S. Ramly. *Impact on the Coastal Areas of the Tanjung Tokong Land Reclamation Project, Penang, Malaysia: Effects on Wave Transformation, Sediment Transport, and Coastal Evolution.* Division of Water Resources Engineering, Department of Building and Environmental Technology of Lund University. (2008).
3. K. S. Mohammed., Y. Elhadary., N. Samat and N. Oma. *GIS and Remote Sensing Techniques for Measuring Agriculture Land Loss in Balik Pulau Region of Penang State, Malaysia.* Asian Journal of Agriculture and Rural Development, **12**. (2015).
4. Department of Statistics. *Yearbook of Statistics Malaysia, 2010.* Department of Statistics, Malaysia. (2011).
5. Malacca Historical City Council. *Issues and Problems of Coastal Reclamation in Malacca.* Central Malacca: Malacca Historical City Council. (2015).
6. T. Litman. *Planning Principles and Practices.* Victoria Transport Policy Institute. (2013).
7. C. R. Kothari. *Research Methodology: Mothods & Techniques.* India: New Age International (P) Limited, Publishers. (2004).
8. S. Shamsuri. *Research Methods for the Social Sciences (Made Simple).* Selangor: DSS Publishing Enterprise. (2004).
9. M. Hammond and J. Wellington. *Research Methods: The Key Concepts.* Routledge Key Guides. (2013).
10. School of Distance Education. *Social Research Methods: Core Courses.* India: University of Calicut. (2011).
11. H. Hamdan and S. M. H. S. Subli. *Perancangan dan Pembangunan Bandar.* Shah Alam: Universiti Teknologi MARA (UiTM). (2012).
12. P.L. Stewart, R.J. Rutherford, H.A. Levy, and J.M. Jackson. *A Guide to Land Use Planning in Coastal Areas of the Maritime Provinces.* Canada: Department of Fisheries and Oceans. (2003).
13. L. Keeble. *Principle and Practice of Town and Country Planning.* London: Estate Gazette Ltd. (1969).
14. J. Ratcliffe. *An Introduction to Town and Regional Planning.* London: Hutchinson and Co. Publisher Ltd. (1974).
15. M. Yusup. *Statutory Procedure and Planning Machinery for Urban Development in Sarawak.* Shah Alam: Universiti Teknologi MARA (Shah Alam). (2013).
16. H. Zainol *et al. Perspektif Perancangan.* Perak: Pusat Penerbitan Universiti (UPENA), UiTM. (2009).
17. The Commissioner of Law Revision. *National Land Code 1965 (Act 56).* Selangor: International Law Book Services. (2011).
18. The Commissioner of Law Revision. *Town and Country Planning Act 1976 (Act 172).* Selangor: International Law Book Services. (2014a).
19. The Commissioner of Law Revision. *Federal Constitution.* Selangor: International Law Book Services. (2014b).
20. D. Omar and O. L. H. Leh. *Malaysian Development Planning System: Kuala Lumpur Structure Plan and Public Participation.* Asian Social Science, 30-36. (2009).
21. N. L. Pace. *Resilient Coastal Development through Land Use Planning: Tools and Management Techniques in the Gulf of Mexico.* University of Mississippi School of Law. (2013).
22. I. Ahmad and I. U. Bajwa. *Regional Development Planning - Issues and Realities.* 41st ISoCaRP Congress 2005, **2**. (2005).
23. Department of Irrigation and Drainage. *DID Manual: Volume 3 - Coastal Management.* Kuala Lumpur: Department of Irrigation and Drainage. (2009).
24. C. B. Powter. *Glossary of Reclamation and Remediation Terms Used in Alberta - 7th Editions.* Alberta Environment, Science and Standards Branch, Edmonton, Alberta. (2002).
25. A. Sufian and M. Mustafa. *Emerging Issues on Reclaimed Land in Malaysia: Alienation for Freehold or Leasehold?.* International Seminar & Indonesia Forum on Ocean Law and Resources: Building Comprehensive Perspective on National Security and Sustainable Development. Faculty of Law Brawijaya University. (2010).
26. R. Kolman and J.V. Hoff. *Port Expansion through Land Reclamation.* Port Technology International. Retrieved from www.port technology.org. (2012).
27. J. J. Jiao, S. Nandy and H. Li. *Analytical Studies on the Impact of Land Reclamation on Ground Water Flow.* Vol. **39**, No. 6 - GROUND WATER- November-December 2001, 912-920. (2001).
28. The Republic of Uganda. *Uganda Vision 2040: The Local Government Development Planning Guidelines.* Uganda: The Republic of Uganda. (2014).
29. S. M. Yusoff, F. Yusof and A. F. Arshad. *An Analysis of Local Plan for Development Control at Local Planning Level in the State of Selangor.* AcE-Bs 2014 Sabah ASEAN Conference on Environment-Behaviour Studies The Pacific Sutera

Hotel, Sutera Harbour, Kota Kinabalu, Sabah, Malaysia, 04-05 January 2013 "Quality of Life in the Built and Natural Environment" (pp. 574-584). Selangor: Elsevier Ltd (2014).

30. The Commissioner of Law Revision. *Local Government Act 1976 (Act 171).* Selangor: International Law Book Services. (2014c).
31. Malacca Historical City Council. *Central Malacca Local Plan (2003-2015).* Central Malacca: Malacca Historical City Council. (2006).
32. A. A. Hussain. *Kerajaan Tempatan: Teori dan Peranan di Malaysia.* Kuala Lumpur: Dewan Bahasa dan Pustaka. (2002).
33. S. Khalil and S. A. Adelabu. *Modelling Local Government System in Nigeria.* Kuwait Chapter of Arabian Journal of Business and Management Review Vol. **1**, No.1; September 2011, 136-154. (2011).
34. K. L. Miller. *Advantages and Disadvantages of Local Government Decentralisation.* Carribean Conference on Local Government and Decentralisation, 1-17. (2002).
35. Department of Town and Country Planning. *Local Plan Manual.* Selangor: Department of Town and Country Planning and Ministry of Urban Wellbeing, Housing and Local Government. (2009).
36. F. Ahmad, I. Mohd and W. N. W. A. Aziz. *Land Use Planning and Property Development: Factors Affecting Urban Housing Development in Kuala Lumpur, Malaysia.* 25th American Real Estate Society Meeting Monterey, California, USA. Kuala Lumpur: Centre for Studies of Urban and Regional Real Estate (SURE), Faculty of the Built Environment, Universuty of Malaya. (2009).
37. C. Schively, M. Beekman, C. Carlson and J. Reed. *Enhancing Transportation: The Effects of Public Involvement in Planning and Design Processes.* University of Minnesota, for the American Institute of Architects; at www.cts.umn.edu/pdf/CTS-07-10.pdf. (2007).
38. Project for Public Spaces (PPS). *A Citizen's Guide to Better Streets.* Projects for Public Spaces. (www.pps.org): at www.pps.org/pdf/bookstore/How_to_Engage_Your_Transportation_Agency_AARP.pdf. (2008).

Disaster Waste Management in Malaysia: Significant Issues, Policies & Strategies

Nor Syazwani Yusof[1, a], E.M.A Zawawi[2], and Zulhabri Ismail[3]

[2]*Centre of Studies Construction Management, Faculty of Architecture, Planning and Surveying, 40450 UiTM Shah Alam, Selangor, Malaysia*
[1& 3]*Centre of Postgraduate Studies of Faculty of Architecture, Planning and Surveying, 40450 UiTM Shah Alam, Selangor, Malaysia*

Abstract. Disaster Waste Management in Malaysia is still at the early stage of its research. Disaster can create large volumes of debris and waste and mismanagement of disaster waste can affect both the response and long term recovery of disaster affected area. The government of Malaysia is taking serious about this issue. This paper is aim to explore the issues, policies and strategies regarding disaster waste management in Malaysia. The objectives were to investigate the extent of disaster waste effects on the environment and to provide a basis from which the needs of waste management could be evaluated in disaster management guidelines. Qualitative method of data collection has been adopted in this study. The respondent are among the local authority and organization that involved in managing wastes. The finding shows that many of the policies regarding waste management in Malaysia has not been well implemented. The purpose of this paper is expected to improve the method of managing disaster waste in Malaysia.

1 Introduction

Human activity including construction works is causing climate change, which is becoming a major contributor to the increasing occurrence of natural disasters worldwide. Natural disasters happen every year and their impacts and frequency seem to have greatly increased in recent decades. For example, excessive rainfall in Malaysia has led to severe flooding in many parts of the country, particularly in agricultural areas largely as a result of deforestation and changes in agricultural practices. In order to improve the design strategy comprehensive flood warning systems, some research based on current risk management, flood mitigation and early warning systems has been implemented internationally. Every year, from day to day the volume of waste constantly increase and it is equivalent to waste production which were increased due to attitude of human. These degrees of waste management can affect long-term recovery in the affected areas. The economic effectiveness of a solid waste management system depends on life cycle costs for facilities and equipment and long-term effects of economic services provided [1]. Unfortunately, the practice of disaster waste management is often neglected or uncontrolled. This paper aimed to explore the issues, policies and strategies regarding disaster waste management practices to produce disaster management guidelines for Malaysia. Therefore, the main objectives of this paper were to investigate the extent of disaster waste effects on the environment and to provide a basis from which the needs of waste management could be evaluated in disaster management guidelines. The research output is expected to add to the existing body of knowledge on theoretical mechanisms for reducing adverse environmental impacts by way of policies and legal instruments, since there is still insufficient research in this area. The latest practices and strategies in disaster waste management will be prioritized in order to illustrate the magnitude of the existing problems encountered. A qualitative research design by way of case study, supported by quantitative research, will be adopted. The purpose of this paper is to explore the issues, policies and effects on our environment, social and economy in managing disaster in Malaysia.

1.1 Weaknesses of disaster waste management in Malaysia

A disaster waste management plan is essential to help a community identify options for collecting, recycling, and disposing of waste. Recycling and reusing disaster waste can replace natural raw materials and reduce the burden on the environment by reducing landfill. The aim is to develop an approach that looks at waste management policies as a way of identifying the possibilities and actions in order to manage waste wisely in the disaster management context. In Malaysia, the Inland Major Disaster Management Mechanism, under the National Security Council (NSC) of the Prime Minister's Department, was formed in May 1994 to coordinate all existing agencies in conducting relief activities during any major disaster on land. On 11 May 1997, the policy statement for disaster relief operations, entitled the

[a] Corresponding author: emmamarinie@salam.uitm.edu.my

National Security Council Directive 20 [2], was introduced. The main purpose of Directive 20 is to put in place a comprehensive emergency management programme which seeks to mitigate the effects of various hazards, to prepare for measures which will preserve life and minimize damage to the environment, to respond during emergencies and provide assistance, and to establish a recovery system to ensure the affected community's return to normalcy. The Mechanism coordinates and handles necessary actions to be taken in any major disasters by the existing rescue agencies, as well as the Special Malaysia Disaster Assistance and Rescue Team (SMART), so as to ensure that all actions can be immediately, efficiently and effectively carried out [3]. With natural disasters becoming more common, national waste management plans should be prepared to cope with this situation. Thus, the primary questions needing to be addressed are whether the current disaster management policies in Malaysia can effectively manage the disaster waste caused by flood in long-terms or short-terms planning.

On the threat of disaster waste to life and health could be overcome with an effective approach, which at the same time can recover and redevelop the damage facilities [4]. Unfortunately, current disaster waste management practice often involves either no action, in which the waste is left to accumulate and decompose, or inappropriate action, in which the waste is removed and dumped in an uncontrolled manner [5].Hence, this research aims to identify the gap in terms of existing policies and practices pertaining to disaster waste management. According to [6] implementation at the local level becomes one of the core factors in successful disaster waste management. A resilient disaster management system depends on the involvement and cooperation of all parties including the private sector, NGOs and the public. Cooperation between all agencies in carrying out their responsibilities is very important in determining the success and effectiveness of disaster management [7]. Thus, it is a necessary to consider waste management in disaster management guidelines Malaysia to curb this problem, otherwise the sustainability and environmental problems will not be addressed effectively. The current body of literature is also explored to identify the guidelines already formulated by the developed countries, from which we can learn, adopt and modify from more successful regimes.

2 Research Aim, Objectives and Research Questions

The aim of this paper is to explore the issues, policies and strategies regarding disaster waste management in Malaysia. In order to achieve this aim, two (2) objectives are outlined:

1. To investigate the extent of disaster waste effects on the environment; ***(OBJ 1)***
2. To provide a basis from which the needs of waste management could be evaluated in disaster management guidelines; ***(OBJ 2)***

2.1 Research questions

In order to gauge the success of the research endeavour, the following research questions have been formulated:

2.1.1 Research Questions for Objective 1

RQ 1.1: What are the main causes of disaster waste?

RQ 1.2: What is the extent of disaster waste effect on the environment?

2.1.2 Research Questions for Objectives 2 and 3

RQ 2.1: What issues and challenges are encountered during managing disaster waste?

RQ 2.2: What are the weaknesses of the current policies pertaining to waste management?

RQ 2.3: What could be formulated to improve disaster management guidelines pertaining to waste management?

2.2 Previous studies related of research

The studies shown in Table 1 that illustrate typical disaster waste management and its impact on the environment. Several studies showed the improvement of waste management and achieving a green environment are worldwide concerns.

3 Research methodology

Well planned research methodology is a guide to achieving the aim and objectives of research. Mixed methods of qualitative and quantitative research have been adopted in this study. A comprehensive literature review was conducted to gather information. Interview was conducted in October and November 2015. The responded were from local authorities and private sectors involved in managing disaster waste.

Table 2: Respondent's detail

Respondents	Description/Position	Number of Respondent
Department of Environment	Environmental Officer	1
Department of National Solid Waste	Director of Technical Department of National Solid Waste Management	1
Alam Flora Sdn. Bhd.	Environmental Officer/Supervisor Landfill	2

Table 2 shows the respondent detail that involved in this research. The qualitative research design had been divided into three stages: qualitative instruments (interview and case study protocol and pilot exercise), semi-structured interview questions and the effects of disaster previous to the case study.

Table 1. Previous studies on typical disaster waste management or issues and their impacts on the environment

Year	Authors	Title of Research	Instruments	Remarks
2008	[8]	Estimation of waste generation from floods	❑ Case study	❑ The research focuses on the flooded area and discusses the parameters affecting floods waste for the model development.
2010	[9]	Waste management strategies during post disaster phase: A Case of Sri Lanka	❑ Semi-structured interview	❑ This research focuses on the post-disaster waste management strategies adopted and on the issues and challenges encountered at both national and local levels in Sri Lanka during the post - Indian Ocean tsunami period.
2011	[10]	Environmental implications for disaster preparedness: Lessons Learnt from the Indian Ocean Tsunami	❑ Case study	❑ The research focuses on the lessons learnt from the environmental aspects of the tsunami, strategies and its implications for disaster preparedness plans after disaster: Indian Ocean Tsunami.
2011	[11]	Disaster waste management: A systems approach	❑ Case study ❑ Semi-structured interview	❑ This research focuses on developing a systems understanding of disaster of waste management and in turn develops context and disaster-transferrable decision-making guidance for emergency and waste managers.
2012	[12]	Disaster waste management challenges	❑ Survey (questionnaire) ❑ Semi-structured interview ❑ Case study	❑ Discusses waste management systems and challenges.
2013	[13]	Strategy for separation and treatment of disaster waste: A manual for earthquake and tsunami disaster waste management in Japan	❑ Case study	❑ This research focuses on existing guidelines in waste management in several of countries due to disaster.
2014	[14]	Hazard and risk perception at Turrialba volcano (Costa Rica); Implications for disaster risk management	❑ Semi-structured interview ❑ Case study	❑ Concerns the socio-economic dimensions of hazards and how these factors can be integrated into disaster risk reduction at Turrialba volcano (Costa Rica)
2014	[15]	Location and routing problems of debris collection operation after disasters with realistic case study	❑ Case study	❑ This research concerns the debris collection operation hindered by limited access, Like blocked access by debris.

The interview and case study protocol and qualitative instrument will be prepared to ensure reliability and validity of the qualitative data then piloted with a targeted group prior to the actual data collection process;, the purpose of this exercise is to maintain internal validity of the data and reliability of the instrument. Semi-structured interview questions will be selected to be put to professionals involved in disaster waste management. The selected samples of interviewee should consist of policy makers from government related agencies relevant to disaster waste management. The sampling frame of the target group will be identified before the actual data collection takes place. Due to the nature of qualitative research, the data will be collected until saturation point is achieved. Based on the case study, observation, parameters will be established to observe the current status of disaster waste management in Malaysia. The research will be focused on selected major post-disaster events such as floods in East Malaysia (Kuala Krai, Kelantan; Temerloh, Pahang). For each case study, qualitative data will be gathered from post-disaster literature (government reports, documents and legislation;

practitioner, reports etc.).The third method of data collection wills, draft proposed waste management principles for adaption into disaster management guidelines. The last method is to ensure validity of the findings and the proposed guidelines. Data from the literature review, semi-structured interviews and document analysis will be juxtaposed and contextualised. This research is expected to emphasize the importance of the implementation of efficient disaster waste management in order to sustain ecological systems and enhance the social well-being of all Malaysian citizens and future generations. From the fundamental theoretical aspects, this research will proposed the adoption of disaster waste management into disaster management guidelines. It is expected to encourage further research and detailed investigation on disaster waste management and sustainable development. Future collaboration with relevant government will be appreciated. This paper only presents data from the interview sessions with the selected respondents.

4 Results and discussion

The researcher found that the authorities should play their role towards sustainable waste management after disasters such as floods. Four (4) respondents were selected, from the Department of Environment (DOE) in Putrajaya, Department of National Solid Waste in the Ministry of Urban Wellbeing, Housing and Local Government and from the private sector; Alam Flora Sdn. Bhd. The interview sessions were conducted in October 2015.The working experience of the selected respondents was 5 years and above in the waste management industry. Climatic changes have made garbage disposal dumps the only method for efficient garbage disposal in Malaysia. To achieve sustainable and effective waste management, development strategies must go beyond purely technical considerations to formulate specific objectives and implement appropriate measures with regard to political, environmental, social, financial, economic and technical aspects of disaster waste management. In addition, four (4) strategies for achieving good disaster waste management were identified:

4.1 Proper arrangement of the material by recycling, recovery and disposal method

Disasters such as floods can produce a large quantity of waste, which will endanger people, block access roads, block drainage channels, and conceal trapped people after the disaster. Respondent 4 also recommended promoting the use of plastic recycling identification codes and labels in order to sort and recycle plastic packaging more easily, for both municipalities and the private sector.Although this helps to separate the different types of waste to be recycled, reused or disposed of sorting the waste after a disaster, must consider the evironmental, political, institutional, social, financial, economic and technical aspects. If possible, sort the rubble as it is being removed, and store reusable materials separately from the rest of the waste. Another way to manage the waste is by promoting the use of products made from recycled materials and fuel made from disaster waste.

> ***" ...in our list of planning now, we're planning to promote recycling materials by the next generation; meaning more focus on children (primary and secondary school) and of course it is cheaper than branded things " (Respondent 4)***

4.2 Improve the education and awareness of producers, the public and people involved in the disaster waste system/sector

People affected by major disasters are badly traumatized, but giving them a task to do helps them to recover. The Environment Agency leads the way in cooperation between all the parties participating in education within the waste sector such as private sector contractors and sub-contractors:; Alam Flora Sdn. Bhd, Southern Waste Management Sdn Bhd, E-Idaman Sdn Bhd, ın managing the waste in our country, as stated by Respondent 2. Strategies that can be considered include:

i. Promoting the advantages of recycling of waste and practicing the 3Rs (Reduce, Reuse, Recycle) programmes and make one day only for collecting the waste.
ii. Informing people of the waste management hierarchy: reduce,reuse,recycle,recovery and dispose.
iii. Promoting the use of less hazardous alternatives such as chemicals during the production of goods
iv. Encouraging companies and institution to set up a certified environmental managements systems and follow a national environmental policy.
v. Appointing a 'specialist officer of disaster waste management' exclusively for Malaysia.

> ***" ...even though we had to do a lot of campaigns, promote everything, everywhere,publish the article, a lot of brochures about how to manage the waste, and used 3R system (Reduce,Reuse, and Recycle), our people nevertheless pretend to know nothing about waste,it is seen to be less efficient or still lacking awareness and all this makes us more worried about our future generation. In this case, we should add some more strategies such as campaigns in the schools, colleges or other institutions." (Respondent 2)***

4.3 Control and decrease waste generation

i. Promote the production of goods to minimize waste generation after disaster or as used by people.
ii. The Environment Agensy should set up a portal or web site to inform the public about the ways of to re-use and control the waste and prevent waste generation in households or after disasters.

iii. The roles and resposibilities of those involved in disaster waste management system should be known such as the Environmental Department, Solid Waste and Public Cleansing Management Corporation (Act 672), local authorities and contractors/sub-contractorts and agents. They should know their purpose and main duties and responsibilities when disaster occurs. According to the Respondent 3, the roles and responsibilities of all the parties involved in a disaster area are very important because each has their 'own style' in planning to solve these issues in short-terms actions.

" ...they have to understand their own roles and responsiilities, if not the coordination of the organizations will be disorganized in terms of the operations planning due to their attitude" (Respondent 3)

4.4 Improve and follow the existing policies and legislation in the waste system/sector

In general, disaster waste managemet is arranged and handled by several government agencies, from federal to state to local authorities. Due to lack of human and financial resources to manage solid waste and public cleansing, an integrated system for waste management and an interim privatisation period would take too long and make it more difficult to secure loans; this is one of the reasons the federal sector took over all the waste management systems, with two departments in 2007;, National Solid Waste Management Department and Solid Waste and Public Cleansing Management Corporation [16]. Therefore, several strategies need to be improved, likes revising the Waste Management Act of 2012 and their definitions of terms;, or defining the responsiblilities of each sector and improving the databases regarding quantities and types of waste that is treated. The system should be enforced, a deterrent to parties that do not abide by laws. According to Respondent 2, the existing policies on solid waste management focus on the household, municipal and 'normal waste' produced by our community. Consequently, disaster waste is not included in Act 672. Respondent 1 and 2 also stated in reality, the existing policies it seem not to be implemented, as there is lack of enforcement.

" ...of course every year,we're faced with a flood disaster related to existing policies of waste management; Act 672 in Malaysia is for domestic waste and scheduled waste only and there is no enforcement of this act...to date, we do not use any policies, acts or guidelines on disaster waste management " (Respondent 2)

" ...in the EIA report there is nothing related to disaster waste, but in my opinion other parties involved in waste management after the recent floods must have planning, playing their role and improving the existing policies related to disaster waste management, not just focusing on daily operation only " (Respondent 1)

This study identified several authorities and law under their jurisdiction in handling solid waste areas as shown in Table 2:

Table 3. Authorities involve in Disaster Waste Management

Authorities	*Laws/ Acts/ Standard Policy*	*Scope/ Remarks*
National Security Division (Prime Minister's Department)	National Security Council Directive No. 20 **(The Policy And Mechanism On National Disaster And Relief Management)**	Outlines a Policy on Disaster Management and Relief according to the level of disaster. The purpose is determining the roles and responsibilities of the various agencies involved in handling disaster.
Ministry of Urban Welbeing, Housing and Local Government of Malaysia	Solid Waste and Public Cleansing Management Act 2007 **(Act 672)**	This act as guidelines for the management and regulation of controlled solid waste and public cleansing for the purpose of maintaining proper sanitation. Normally, the act is focuses on household, municipal and hazardous industrial waste/areas.
Ministry of Natural Resources and Environment	Environmental Quality Act 1974 (Act 127) –Part IV Section 24 (2b)	The EQA was enacted in 1974 and applies to the whole of Malaysia. It is related to the prevention, abatement, control of pollution and enhancement of the environment, and for purpose connected therewith.
Department of Standards Malaysia/ SIRIM BERHAD	Malaysian Standard Landfill Safe Closure-Requirements **(MS 2547:2014)**	This Malaysian Standard specifies requirements for landfill safe closure and it is to protect public health and the environment by proper management of such.

Sources: National Security Council Directive No.20 [3], ACT 672 [18], ACT 127 [19] and MS 2547:2014 by SIRIM [20].

Table 3. above shows the first authority is an organization under the National Security Council (NSC) of the Prime Minister's Department Malaysia, which introduced Directive No. 20 (Policy and Mechanism on National Disaster and Relief Management). The scope of this directive is according to the level of disaster:, Level 1:District; Level II: State; Level III: Central [3]. The main purpose of this policy is to form a team and coordinate agencies involved in disaster management [17] and to determine the roles and responsibilities of the agencies involved in handling the disaster. The second authority is the Ministry of Urban Wellbeing, Housing and Local Government of Malaysia with its Solid Waste and Public Cleansing Management Act 2007 (Act 672); this act is normally implemented in daily events of waste management in Malaysia. It provides and regulates the management of controlled solid waste and public cleansing for the purpose of maintaining proper sanitation under the housing and local government authority. The regulation focuses on avoiding waste from contributing to improper sanitation. Even though the act did not specify disaster solid waste but it should be seen as part of controlled solid waste. However, the whole scope of managing the waste is not enough, as compared to the steps given in the waste management hierarchy. The next authority is the Ministry of Natural Resources and Environment with its Environmental Quality Act 1974 (Act 127). The act's scopes is to prevent, reduce, control pollution and to enhance the environment. Wastes in this act is more related to general environmental problems, such as sludge, odour and others. The waste is also interpreted as liquid, solid, gas and radioactive. The wastes is either normal waste or scheduled waste [21]. The drawback of this law is that it considers waste when it is deposited and causes pollution of the environment; it is not much involved after disasters, due to there being separate emergency methods. The final authority that concern waste is the Industry Standards Committee on Environmental Management (Department of Standards appointed SIRIM BERHAD as the agent to develop Malaysian Standards) which publishes new standards of requirements for landfill; Malaysian Standard Landfill Safe Closure- Requirements (MS 2547:2014). The landfill when completed must be closed properly for safe storage of the waste and prevention of pollution to the surrounding environment. Several requirements must be met to ensure that agencies using it as a guide meet landfill site requirements according to the Malaysian Standard.

In order for Malaysia to move towards a green or sustainable environment, the country has to benchmark developed nation like Japan in handling disaster waste issues. A diagram of disaster waste management strategies containing all the relevant variables is shown in Figure 1.

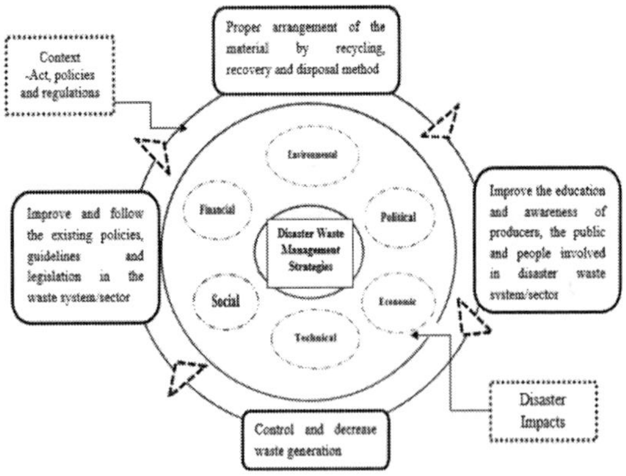

Figure 1. Cross-case study anaysis strategies of disaster waste management

5 Scope and limitation

This research focuses on the current situation and practices or initiatives of the disaster waste management in our country. Several authors have given their opinions and produced their definitions of waste management, resulting in the existing policies and guidelines. The current problem is that implementing these policies and guidelines is not practicable as identified in Malaysia. In addition, the existing initiatives and policies, public and private, concerning waste management in Malaysia have been investigated to identify gaps in management planning and ways to control the waste following disasters. Therefore, strategies will be considered to achieve the aim and objectives of the research. Since the research involves various authorities in the waste management industry, the main target group includes the Solid Waste and Public Cleansing Management Corporation, and local authorities in Kelantan (Kuala Krai) and Pahang (Temerloh), which are important agencies pertaining to adapt disaster waste management guidelines in Malaysia and identifying full detailed problems. The limitation of this study is thus the small number of entities or individuals that play a significant role in waste management in Malaysia.

6 Significant of research

This research will add in to the existing body of knowledge on disaster waste management issues and policies since there are limited resources on this area of study. It is sponsored by the MOHE (Ministry of Higher Education) under the Fundamental Research Grant Scheme (FRGS) known as the Flood Disaster Management Grant. This research will not only benefit to government, but also other authorities involved in disaster waste management. The limited number of up to date primary sources is one of the main concerns in preparing this research and there is a gap in terms of what has been planned and actions that have been implemented in particular disaster waste management occurrences.

7 Conclusions

The disaster waste management strategies is useful approach to guide Malaysia for present and in the future. The findings are expected to be the cornerstone for disaster waste management in Malaysia. There is a gap in terms of what has been planned and the actions actually taken in particular disaster waste management events. Generally, disaster waste management issues in Malaysia have not been seriously addressed, thus more empirical evidence and research is required to provide information on the current scenario and the actions that should be taken. This study revealed that they are four (4) elements to be considered in this research which are based on social impact, economic impact, environmental impact and technical impact (i.e.: insufficient equipment, transportation problems etc.). In order to have a better disaster waste management in Malaysia, the factor could be highlighted for further study. These need more investigation from the researcher to look into the suitability of the factors to be adopted in the disaster management guideline in Malaysia. Developing a management strategy for Malaysia is the fundamental task of an agency's leader. Strategy indicates both where the agency wants to go (long-term objectives) and how it plans to get there (on its planning on waste management). The agency management strategy must be realistic in taking account both of limits and resources. That is, the strategy must be guided by the underlying reality of the setting in which objectives are pursued. In what follows, we use management strategy and implementation strategy as generally interchangeable. Furthermore, through the exploratory interview, it provides clarify on actual events on disaster waste management in Malaysia. Thus, the additional knowledge about strategies on disaster waste management in this paper will benefit all the parties involved in waste management. From the above, it can be anticipated that this research will generate interests from disaster waste researchers and disaster waste management policy makers, as it will provide fundamental elements towards a more rigorous disaster waste management policy.

Acknowledgement

The authors would like to express their gratitude for the financial support from MOHE (Ministry of Higher Education) under the Fundamental Research Grant Scheme (FRGS), 600-RMI/FRGS DIS 5/3 (7/2015) and also LESTARI GRANT (600-RMI/DANA 5/3/LESTARI (17/2015) which is internal grant of Universiti Teknologi MARA (UiTM) for the management and support of this research.

References

1. P. Schübeler, J. Christen, and C.- Berne, *"UNDP / UNCHS (Habitat)/ World Bank / SDC Collaborative Programme on Municipal Solid Waste management in Low-Income Countries URBAN MANAGEMENT AND INFRASTRUCTURE Conceptual Framework for Municipal Solid Waste Management in Low-Income Countries,"* (1996).
2. C. O. Ver and S. T. O. Ry, *"Disaster Management in Malaysia : Evolution, Development and Future Challenges,"* no. February, (2007).
3. National Security Council, *"Disaster Management in Malaysia: Landscape Review, Challenges and Prospects,"* (2011).
4. G. Approach, *"Cabarrus County, North Carolina,"* no. April, (2012).
5. MSPs, *"Draft National Policy for Disaster Management in Kenya,"* no. February, pp. 1–50, (2009).
6. Did, *"Government Of Malaysia Department Of Irrigation Volume 1 – Flood Management,"* **vol. 1**, (2009).
7. Hon Anna Bligh MP, *"Disaster Management Strategic Policy Framework,"* State Disaster Management. Gr., no. 1, pp. 1–20, (2010).
8. H. Srinivas and Y. Nakagawa, *"Environmental implications for disaster preparedness: Lessons Learnt from the Indian Ocean Tsunami,"* J. Environ. Manage, **vol. 89,** no. 1, pp. 4–13, (2008).
9. G. Karunasena, R. Amaratunga, and R. Haigh, *"Waste management strategies during post disaster phase: A case of Sri Lanka,"* (2010).
10. M. Milke, *"Disaster waste management research needs," Waste Management,* **vol. 31**, no. 1, p. 1, (2011).
11. C. Brown, M. Milke, and E. Seville, *"Disaster waste management: A review article,"* Waste Management, **vol. 31,** no. 6, pp. 1085–1098, (2011).
12. Periathamby, a., Hamid, F. S. & Sakai, S. -i. Disaster waste management challenges. Waste Manag. Res. 30, 113–114 (2012).
13. M. Asari, S. I. Sakai, T. Yoshioka, Y. Tojo, T. Tasaki, H. Takigami, and K. Watanabe, *"Strategy for separation and treatment of disaster waste: A manual for earthquake and tsunami disaster waste management in Japan,"* J. Mater. Cycles Waste Management, **vol. 15**, no. 3, pp. 290–299, (2013).
14. S. M. van Manen, "Hazard and risk perception at Turrialba volcano (Costa Rica); implications for disaster risk management," Appl. Geogr., **vol. 50,** pp. 63–73, (2014).
15. A. Pramudita, E. Taniguchi, and A. Qureshi, *"Location and Routing Problems of Debris Collection Operation after Disasters with Realistic Case Study,"* Procedia - Soc. Behav. Sci., **vol. 125**, pp. 1–15, (2014).
16. T. J. Sin, G. K. Chen, K. S. Long, I. Goh, and H. Hwang, *"Current practice of waste management system in Malaysia : Towards sustainable waste management Current Practices for Malaysia"*, in 1st FPTP Postgraduate Seminar Towards sustainable waste management. (2013).
17. B. A. Rahman, *"Issues of Disaster Management Preparedness: A Case Study of Directive 20 of National Security Council Malaysia,"* Int. J. Bus. Soc. Sci., **vol. 3,** no. 5, (2012).
18. Ministry of Housing and Local Government, *"Act 672, Solid Waste and Public Cleansing Management Act"*, pp. 1-7, (2007).
19. Ministry of Natural Resources and Environment, *"Act 127, Environmental Quality Act"*, pp 1-17, (1974).
20. SIRIM, *"MS 2547:2014- Malaysian Standard Landfill Safe Closure- Requirements,"* Sirim Berhad, Malaysia, (2014).
21. Mustafa, M, *".Environmental law in Malaysia. Kluwer Law International BV, the Natherlands"*, pp. 98-99, (2011).

A CMMS Expert using BIM for IBS Building Maintenance

Noraini Hamzah[1] , Zul-Atfi Ismail[2], Syahrul Nizam Kamaruzzaman[3], Azrul A. Mutalib[1] and Adi Irfan Che Ani[4]

[1]*Department of Civil and Structural Engineering, Faculty of Engineering and Built Environment, Universiti Kebangsaan Malaysia, 43600 Bangi, Selangor, Malaysia*
[2]*School of Environmental Engineering, Kompleks Pusat Pengajian Jejawi 3, Universiti Malaysia Perlis, 02600 Arau, Perlis, Malaysia*
[3]*Building Surveying Department, Faculty of Built Environment, University of Malaya, Mercu Alam Bina, Lembah Pantai, 50603 Kuala Lumpur*
[4]*Department of Built Environment, Faculty of Engineering and Built Environment, Universiti Kebangsaan Malaysia, 43600 Bangi, Selangor, Malaysia*

Abstract. In the IBS building maintenance, the conventional methods to the project delivery and its failure to form effective communication channel between complementary knowledge on IBS construction and related members that is conducted in the sequential manner has resulted in ineffectiveness for managing building maintenance, where a paradigm shift within the IBS traditional approach is necessary. The research results described are concerned with i) an integrated system to record, diagnose and analyse information for IBS building maintenance to detect patent or latent defect, select an effective repair method, provide appropriate planning decision and reduce risks of defect throughout the building lifetime. ii) building a decision making support in diagnosis in IBS building maintenance based on robust data collection about concrete defects and causes including in the selection of a durable replacement design and material or the proper rehabilitation method. The system development was based on analysis of number of interviews and case studies conducted with professional engineers working in IBS building maintenance departments. The paper concludes that implementing a CMMS Expert using BIM can help improve the performance of IBS building design, construction and maintenance operations.

1 Introduction

Maintenance management is defined as the combination of all technical, administrative and management activities of the post-construction life cycle of a facility. These activities overview are concerned with the technical specifications (e.g. method of defect identification and assessment), administrative processes (e.g. maintenance plans and general execution) and management such as contract conditions, policy and maintenance strategy related issue. In addition, it considered the maintenance factors of labour skill, interpersonal relationship, maintenance performance, material quality control and investment control in order to support the sustainable maintenance of building facility [1]. Besides, it is the recursive process of coordination and collaboration (e.g. designer and site engineer) to provide the effective service delivery of building maintenance to perform building facilities long-lived and reduction of life cycle cost [2,3]. In general, maintenance management is considered as part of the construction sector for the larger service of post-construction activity and process. However, as IBS construction projects grow in scale and complexity, interconnections between the mentioned activities and processes can be noticed in a problematic maintenance management to monitor the defects for the corrective and preventive maintenance action [4,5]. The management of maintenance in the complex and high-rise buildings claims a high emphasis on the systematic activity and process for the improvement of maintenance management system of IBS construction. Nevertheless, tool and technique in maintenance management are more significant than those in general maintenance management (post-construction activity and process) [6,7]. This highlights the importance of processes and activities with involved sophisticated tool and the efficient technique for improvements.

In the IBS building maintenance, the conventional methods to the project delivery and its failure to form effective communication channel between complementary knowledge on IBS construction and related members that is conducted in the sequential manner has resulted in ineffectiveness for managing building maintenance, where a paradigm shift within the IBS traditional approach is necessary [8,9]. Therefore, the need for sophisticated tools and efficient techniques using an appropriate Information and Communication Technology (ICT) (e.g. BIM and CMMS) for implementing an integrated approach in the design and construction could facilitate new maintenance management processes for IBS building projects in the future [6,7]. In the meanwhile, the most maintenance industry has been generally revolved around conventional method (e.g. MS Word and MS Excel) with limited uses

[a] Corresponding author: zulatfippkas@gmail.com

of 3D models for visualisation and maintenance development [10]. Recently on IBS construction sites, there are many efforts to improve the efficiency of maintenance management activities through the use of more ICT-based systems (i.e. BIM). Modern developments in BIM technology related to construction industry has been used for facilitating maintenance on various activities (failure analysis, documentation of maintenance, fault location, repair and reconstruction). An example is the bottleneck of massive data between maintenance components and building management, which can now be eliminated by converting raw data on the quality of systems and the process capability in the information and knowledge for optimal maintenance strategies [11,12,13]. In addition, with the development of emerging technology such as BIM, maintenance management improvements in IBS building construction has become possible [14]. Several other industry projects worldwide, for example real estate, waste management, transportation, supply chain and facility management, have been successfully completed by implementing BIM technology [15,16,17,18,19]. BIM can assists clients to determine the appropriate technology strategy and scope of each deliverable, with the intention of reducing redundancy and rework while improving performance and productivity of an operation and maintenance processes effectively in the future [20,21,22]. Whereas BIM related studies mainly focus on utilising sophisticated tool and efficient technique, associating decision making support in diagnosis principles can help in achieving new levels of efficiency in IBS building maintenance performance.

The developed model of decision making support in diagnosis quantifies the overall condition properly and evaluates the possible alternatives regarding selecting a solution for building improvements [23]. A few researchers have considered IBS to provide more efficient decision support tools in diagnosis such as PPMOF (Prefabrication, Preassembly, Modularisation and Offsite Fabrication), IMMPREST (Interactive Method for Measuring PRE-assembly and Standardisation), PSSM (Prefabrication Strategy Selection Method), and CMSM (Construction Method Selection Model). Nevertheless, these existing tools consider inadequate aspects of sustainability [24]. Sustainability involves issues such as the design and management of buildings; materials performance; operation and maintenance; long-term monitoring; and the dissemination of knowledge in related technical contexts. Besides, most of the available decision making support in diagnosis guidelines and tools are only used after the design of the IBS building project is about to be completed [25]. Therefore, an application that utilises the intelligent objects of BIM to manage decision making support in diagnosis would provide better service delivery of IBS building maintenance. This paper proposes a CMMS Expert using BIM that not only facilitates the decision making support in diagnosis but also the ability to provide IBS component diagnosis integration for reducing risks of defect of an IBS building.

2 System Component

The proposed system "CMMS Expert using BIM" comprises three components: defect report and assessment, defect diagnosis as well as defect control. These components are accessible through a BIM model user-friendly interface. Software applications can be developed in different platforms including those of BIM model with the integration of CMMS and Expert System. Based on the findings of [22], a BIM-based platform was considered to be the most appropriate in development of the system. The conceptual design of a BIM-based platform is suitable for larger and more complex and high-rise buildings to support design consistency and visualisation, cost estimations, assessment and monitoring, retrofit planning, implementation of lean maintenance and improved construction team collaboration. [26] investigated the feasibility of tracking any potential defects and reworks using BIM-based platform in order to perform efficient operation and maintenance work, particularly in complex projects.

The defect report and assessment allows users to lodge report using the electronic facilities defect form with complete attributes in the database. The defect diagnosis is to present the user for selecting one of the three knowledge bases namely leaking, jointing and cracking in concrete including the selection of appropriate construction design or materials and repairs method recommendations. The defect control is to provide technology transfer of knowledge from designers to engineers and vice versa and it is a useful reference guide for a wide range of concrete defects risk leading to accurate analysis using design condition index. The adopted synthesis of good practices of maintenance operations in CMMS Expert using BIM is based on the findings of the interviews and case studies conducted with professional engineers working in IBS building maintenance departments [27]. This paper is part of a larger research and will only introduce and discuss the entire process of the system. This process is illustrated in the following sections.

3 Maintenance Management Process

The maintenance management processes are adopted into the maintenance management system (MMS) to improve the maintenance management practices at the IBS buildings. The knowledge transfer support is the new approach of this prototype system that is provided with the diagnosis of defects and decision making for structure component based on the inputs and information supplied by the user for handling the maintenance on the IBS building. Besides, the prototype system can be accessed through portable devices such as tablet and notebook at the site location with the database provided (e.g. VB.net and MS Access) for each defects and structure locations in the report assessment. The safety database is used to protect the data entry and design, construction as well as maintenance collection for the structures and facilities at the IBS buildings. The BIM can be integrated into the prototype system with the BIM platform provided (e.g.

Autodesk Revit) for each knowledge feedback of the nature of defects components and their cause into the model that facilitate the assessment process in maintenance management. The data obtained in the case studies and literature review revealed that the IBS buildings are using the conventional method (paper-based reports/unsystematic database) in managing the maintenance activities for building structure and facility. This arise a problematic situation such as defects repetition (leaking, jointing and cracking) and less competent contractor that need sophisticated tools towards a solution. Presently, the implementation of ICT tools in the new system is the better improvement to lead the tremendous saving in budget, time planning and to analyse the precise data in handling the defect diagnosis and control. The integration of modern ICT tools such as BIM, CMMS and Expert System also facilitates the knowledge transfer for maintenance management processes in recommending the repair structure components needed for the efficacy of the maintenance management practices. From the case studies findings, the involved maintenance management processes are the maintenance identification (for defect report and assessment), defect and cause analysis (for defect diagnosis) and risk level analysis (for defect control) as the main stages in managing maintenance for IBS building structure and facility that described in Figure 1.

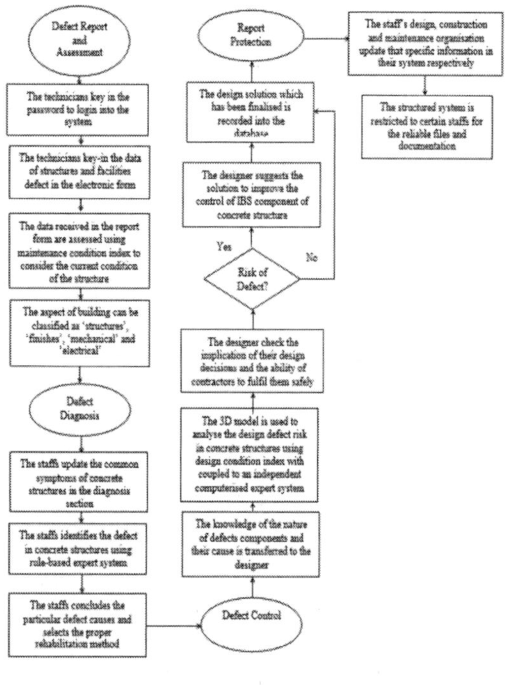

Figure 1. Maintenance Management Processes.

The defect report and assessment is the initial process where a technician will allege any defect component occurred in building. The defect inspection is undertaken to measure the defect performance on site and then this information will be put in the system for recorded after defect assessment value is determined. The database in the system is linked with the three software, namely, CMMS, Expert System and BIM model. The knowledge transfer can be utilised through this system when the knowledge of history for defect component from the defect report and assessment process is distributed to the engineer within same organisation. The engineer is assigned to screen each of defect component for diagnosis task in order to analyse the cause and reason of defect that ensued. The expert system can be used to examine the defect symptoms on each of component which are based on the knowledge obtained from literature searches, codes of practice, manuals, textbooks, technical reports, journals and conference proceedings, civil work reports and experienced concrete specialists. The knowledge of defect diagnosis in the expert system then is transferred to the designer in the BIM model for pursuing the design defect cause and reason and even to classify the risk level of defect component. Finally, the knowledge of design defect based on the materials specification and visual information of BIM model is transferred back to the engineer for work planning and maintenance execution. The knowledge transfer in this system will improve the defect diagnosis and decision making process for critical defect element based on the maintenance impact undertaken heading to sophisticated management and maintenance effectiveness.

4 System Architecture

The system architecture focuses on the collection of the structure and facility defect information for defect assessment, defect diagnosis and defect control as well as analyse the data for the execution reference in IBS building maintenance. The mobile devices such as laptop enable staff to compare and adapt the data at any facility location with the staff's report recorded in the prototype system user-interface (MS Visual Basic.NET). The staff will be able to capture the knowledge of the structure and facility defect, update about the detailing defect, and record the relevant aspect of the defect attribution into the electronic form of the system prototype. This knowledge is stored in a computer database central (MS Access) linked to the technician, engineer and designer at the maintenance organisation for further process. The information about the defect is assessed using maintenance condition index in Condition Rating Level (CRL) format for considering the current condition of the structure and its components, analysed for the maintenance defect diagnosis using the decision making process and assessed using the design condition index in Average Index Level (AIL) format for reducing the risk of design defects at the site location. The workflow is illustrated in Figure 2, 3 and 4. The 'CMMS Expert using BIM' can be divided into three main components, which are BIM, CMMS and Expert System. Each component plays a different role in the prototype system. The BIM technology is to enable the concrete structure component design defect analysis carried out by comparing concrete structure defects and causes on 3D model including record modelling in BIM database while the CMMS to capture knowledge of the defect report and assessment. The Expert System is to capture the knowledge of concrete defect diagnosis and in the selection of a durable replacement design and material or the proper rehabilitation method for the system.

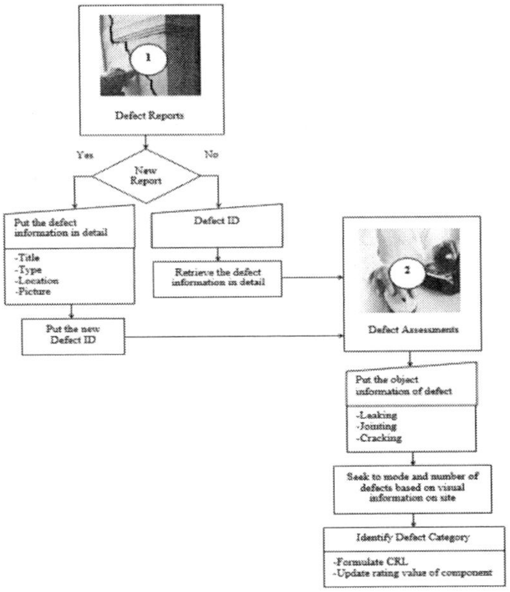

Figure 2. Defect Report and Assessment Process.

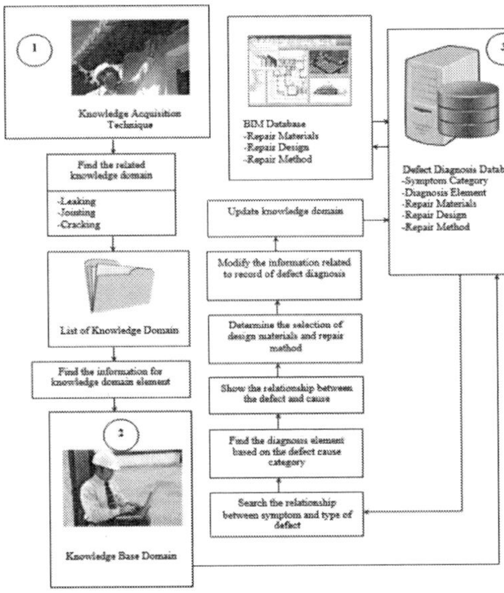

Figure 3. Defect Diagnosis Process.

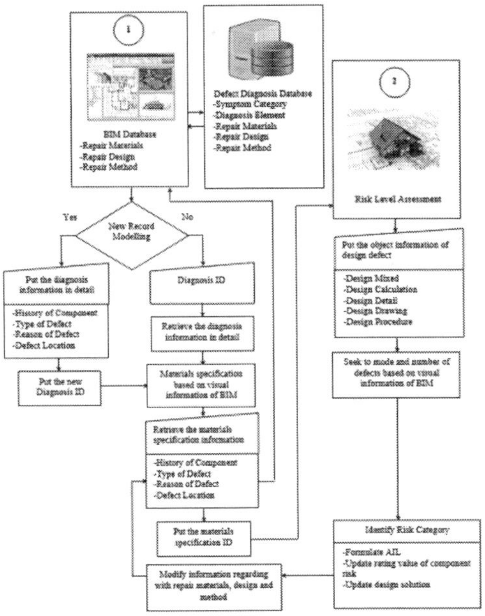

Figure 4. Defect Control Process.

5 A CMMS Expert using BIM

This section describes the interactive portion of the system including input data validation and updates of system database in the information system prototype. The prototype system has four main functions and graphical user interfaces:

i. System Login;
ii. Defect Report and Assessment;
iii. Defect Diagnosis; and
iv. Defect Control.

System Login

The prototype system starts when the main interface, namely, "CMMS Expert using BIM" is displayed (see Figure 5). The login menu in the form is operated via ID and password input in allowing user such as technician and engineer to enter the prototype system.

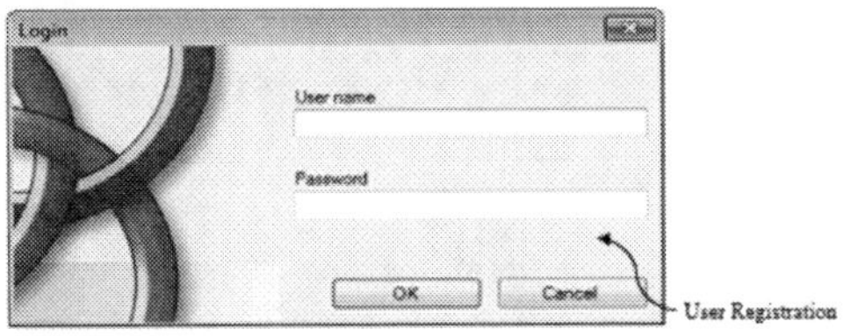

Figure 5. "CMMS Expert using BIM" Main Screen.

The users have to register their profile in order to report their defect, assess the current condition of the concerned building structure and undertake defect diagnosis. This user registration procedure and login application are shown in Figures 6 and 7 respectively.

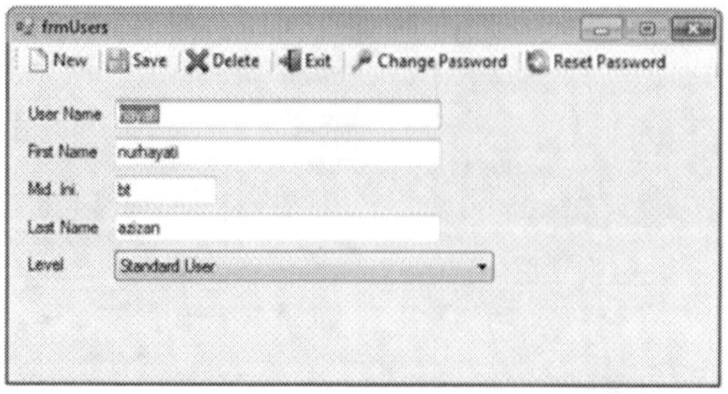

Figure 6. Technician/ Engineer-User Registration.

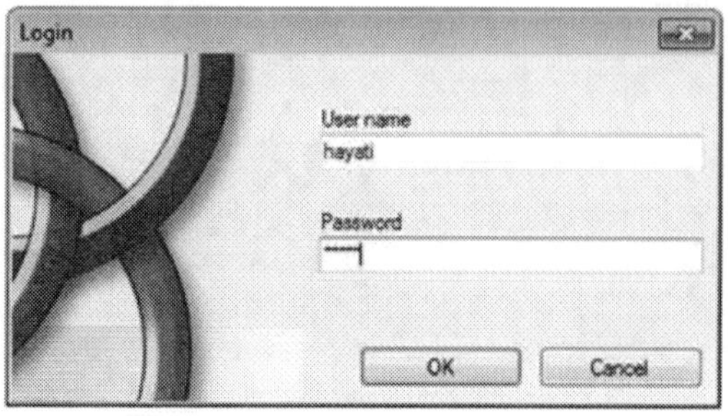

Figure 7. Technician/ Engineer-Login Application.

Defect Report and Assessment

After clicking on the Log in button, the user enters the 'Dash Board' form as shown in Figure 8.

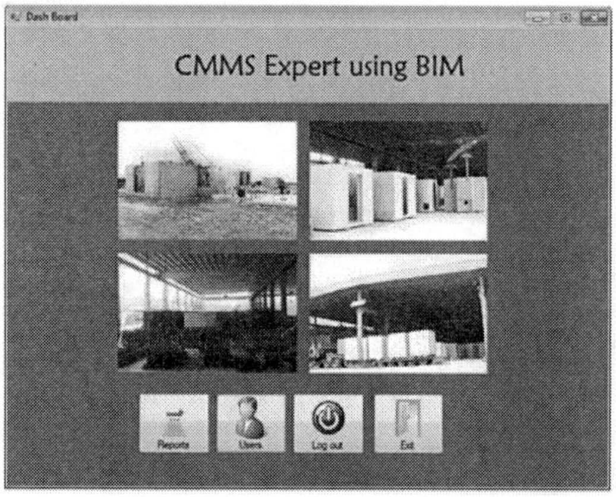

Figure 8. Technician/ Engineer-Main Form.

In this application form, the user is provided with two menus, namely, reports and users to access other forms in the application as shown in Figures 8. Figure 9 shows the report list data involved on building structure and facility defect:

- ID;
- Title;
- Date;
- Rating;
- Location
- Status; and
- Material Type.

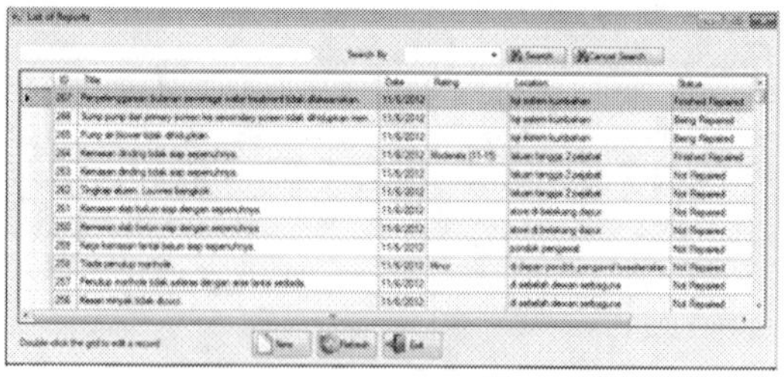

Figure 9. Technician-Report List.

Figure 10 shows that the new form will be displayed after clicking new report button. It shows the new report details form involved on building structure and facility defect:

- Title;
- Date;
- Location;
- Type;
- Picture;
- Maintenance Condition Index;
- Rating;
- Status;
- CRL; and
- Inspector.

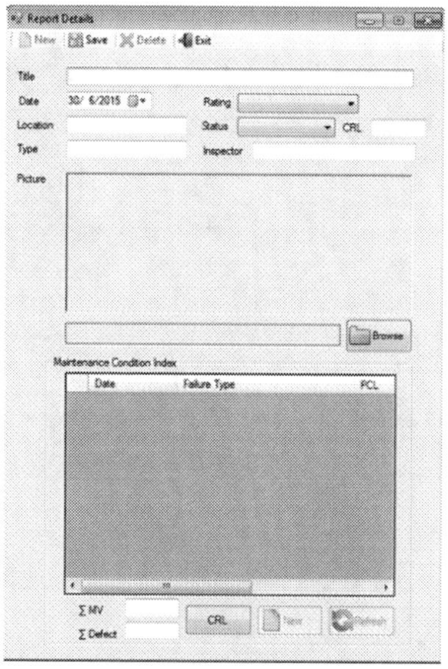

Figure 10. Technician-Report Form.

The user need to click the save button for the new report details form where its information is sent to the database system in storing defect report and assessment data for IBS building structure and facility (see Figure 11 and Figure 12).

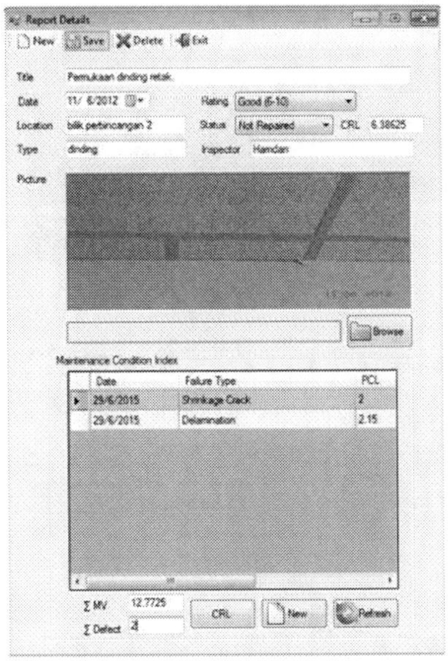

Figure 11. Technician-Save Form.

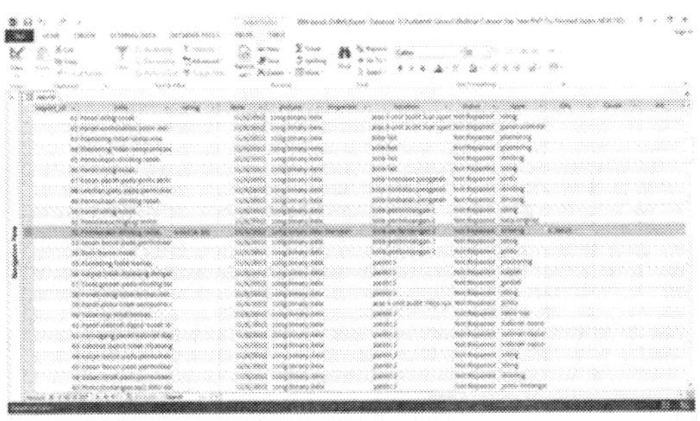

Figure 12. Technician-Save information into MS Access database.

Defect Diagnosis

Figure 13 shows the 'defect diagnosis' form. This form also provides the engineer with the additional data on 'cause', 'repairs design and material' and 'repairs method' button (see Figure 14 and Figure 15). The 'cause' functions in the 'defect diagnosis' forms allow the defect identification and needed diagnosis for the repair strategies recommendation. The concerned defect IBS component of concrete structure is identified based on the symptom setting and the defect only being diagnosed from the inputs and information supplied by the user after clicking the 'result' button at the right-side bottom of the form. Thus, the staff could address diagnostic-related issues and identification and repair or rehabilitation recommendations of IBS concrete structure which is undertaken at the site location effectively.

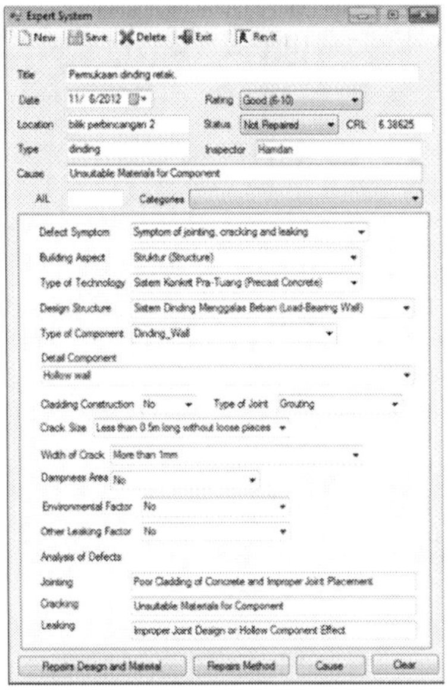

Figure 13. Engineer-Defect Diagnosis Form.

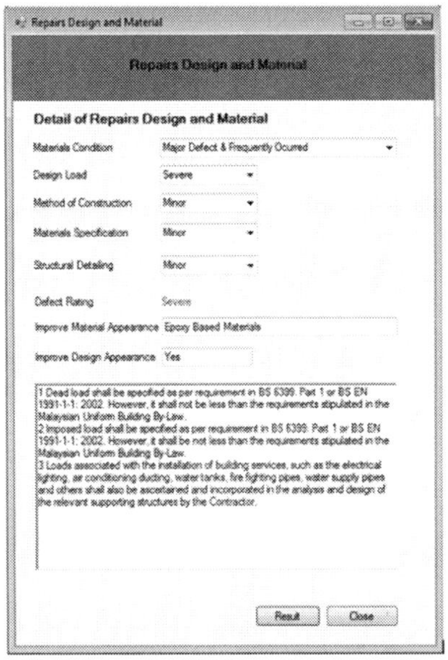

Figure 14. Engineer-Repair Materials and Design Form.

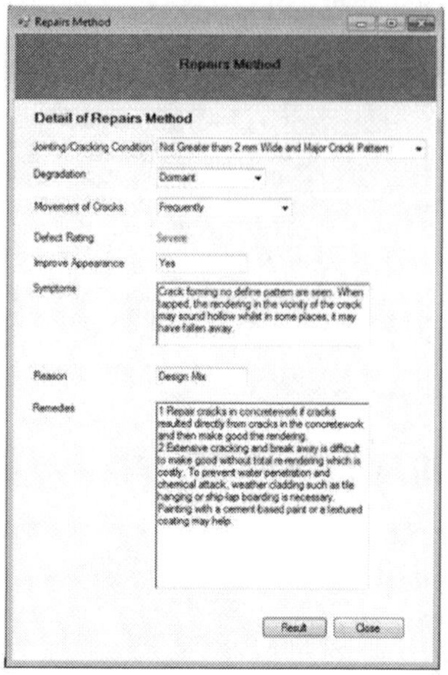

Figure 15. Engineer-Repair Method Form.

Figure 14 and 15 highlights the repair materials, design and method from the content of a 'Symptom of Leaking, Jointing and Cracking' analysis procedure based on the literature searches, codes of practice, manuals, textbooks, technical reports, journals and conference proceedings, civil work reports and experienced concrete specialists, which facilitates the engineer to evaluate the concrete structure information for the defect diagnosis. The result of defect diagnosis information being used to improve the performance and durability of concrete structures and is coupled to an independent BIM database management system (BIM) through Revit DB link for controlling risk of defects on design criteria for effective construction methods.

Defect Control

Figure 16-22 shows how the expert system database collaborated with BIM technology through knowledge transfer to perform specialised tasks, which provides the report analysis of defect diagnosis for particular IBS building components in determining an AIL. AIL numerically rates the condition of the design defect IBS concrete on a scale of 1 to 5 (Table 7) by evaluating each concrete design deficiency based on the 'Design Condition Index' form in Figure 22.

Table 1. The Average Index Level (AIL) for the Evaluation of Design Defect.

Rating Scale	Average Index (a)	Categories	
1	1.00 ≤ a ≤ 1.50	Ignorable	Does Not Affect
2	1.50 ≤ a ≤ 2.50	Not Important	Slightly Affected
3	2.50 ≤ a ≤ 3.50	Moderate	Moderately Affected
4	3.50 ≤ a ≤ 4.50	Important	Strongly Affected
5	4.50 ≤ a ≤ 5.00	Very Important	Extremely Affected

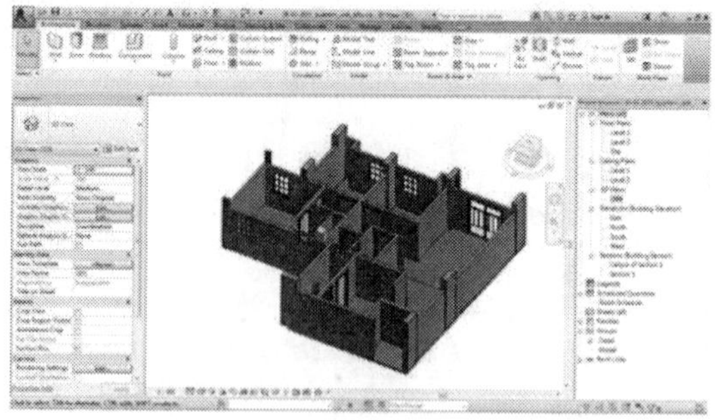

Figure 16. Designer-Floor Plan in 3D View.

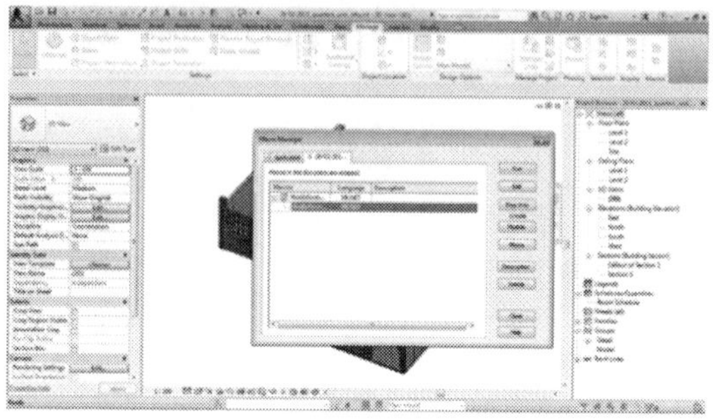

Figure 17. Designer-Start the Program Module in Revit API Application.

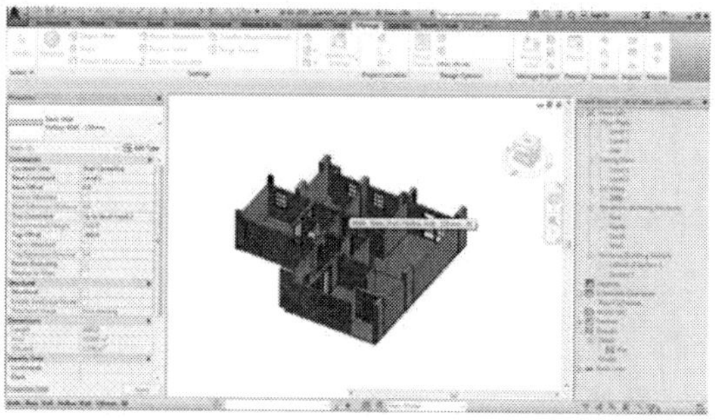

Figure 18. Designer-Wall Component is selected on 3D Floor Plan.

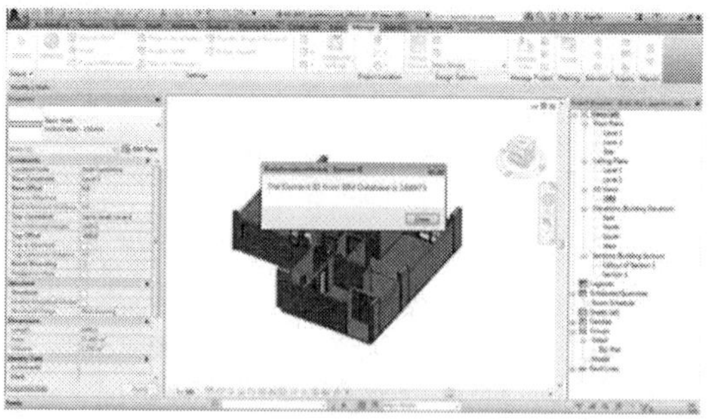

Figure 19. ID Element from Wall Component selected is showed on screen.

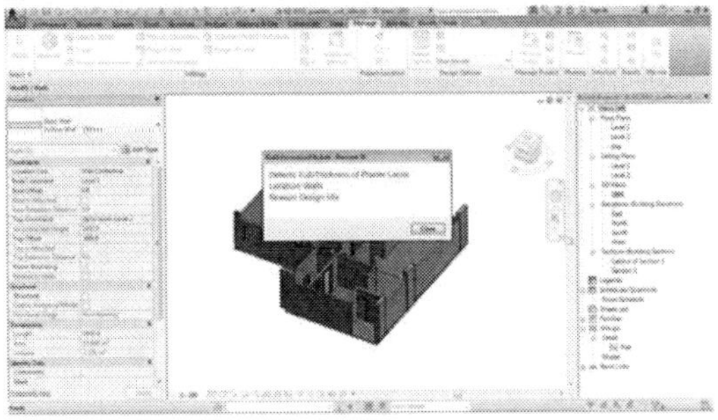

Figure 20. Maintenance Information based on the specific ID Element is captured from BIM database.

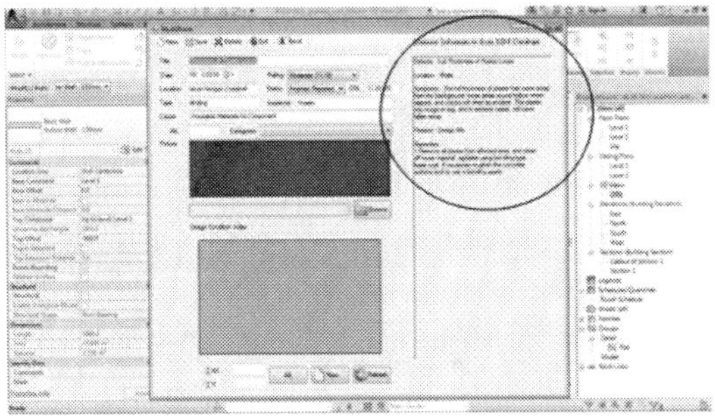

Figure 21. Design Condition Index Form from Wall Component selected is appears afterwards.

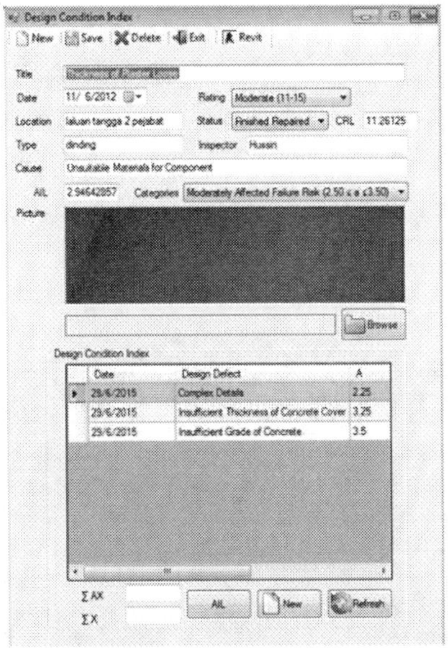

Figure 22. Designer-AIL Index Score that already calculated in the Design Condition Index Form is transferred back to engineer through database link.

BIM model database provides the record modelling for material properties and design procedures as well as source of engineer's knowledge of the defect, its cause and repair method for structure defects using 3D model with coupled to an independent computerised expert system through Revit DB link. The designer need to double click the IBS component in 3D view for the selected record modelling where its material, design and procedure details will be displayed in the BIM model in analysing design defect information such as design concrete mix, design calculations, design details, drawings and design procedures in concrete structure. For example, the IBS component item provides the record inventory in order to meet all necessary design defect information including source of engineer's knowledge needed as shown in Figure 23-25.

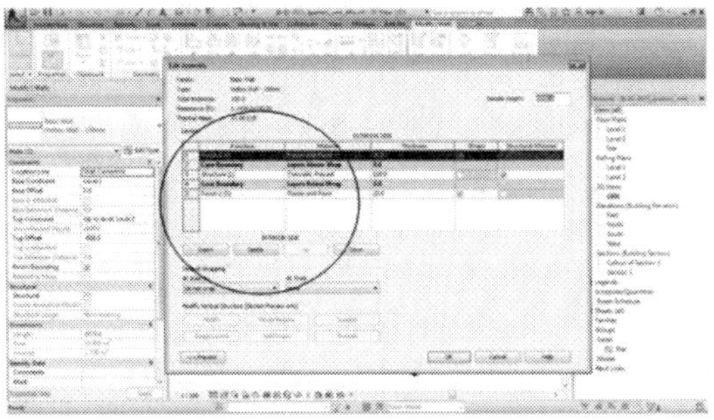

Figure 23. Material Properties and Design Procedures in BIM.

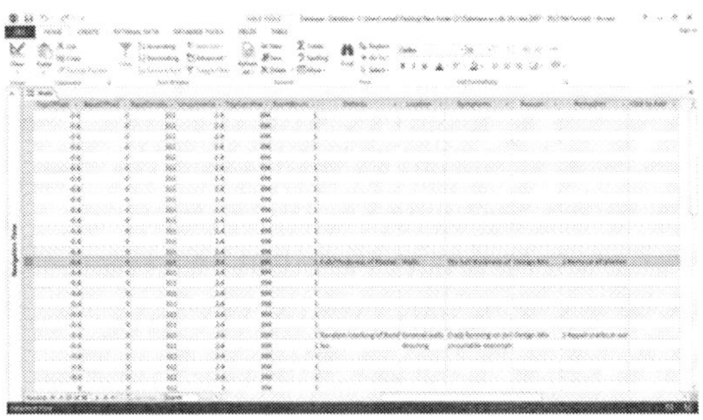

Figure 24. Source of Engineer's Knowledge from Computerised Expert System Database is transferred to BIM through Revit DB Link.

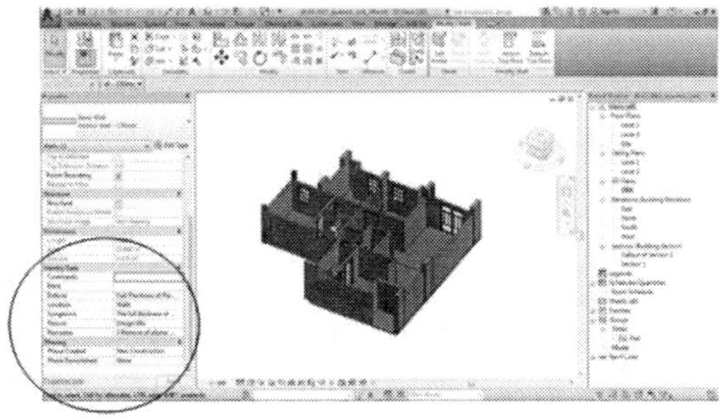

Figure 25. Source of Engineer's Knowledge in BIM that is used for Design Condition Index Form.

6 Conclusion

This paper presented a CMMS Expert using BIM to record, diagnose and analyse information of maintenance projects. A few conclusions can be drawn from the research:

• The development of sophisticated tool and the efficient technique is an important issue in BIM technology for IBS building maintenance. Applying an adequate processes and activities with involved sophisticated tool and the efficient technique much facilitates creating decision support in diagnosis in IBS building maintenance.

• Building a CMMS Expert using BIM in an integrated way is a vital task of IBS building maintenance. Such a system can serve as a core to detect patent or latent defect of an IBS building and provide comprehensive defect diagnosis for supporting the process of decision making.

• Founding intelligent objects capability of the BIM technology of a building can assist improving IBS component diagnosis knowledge integration to reducing risks of defect throughout the IBS building lifetime.

Utilising technology to support maintenance processes and activities can have a positive impact on the delivery of service. Several systems have been developed to facilitate decision making support in diagnosis for IBS building maintenance. However, the efficient practice requires a system that can handle decision making support in diagnosis in an effective way. It is final concluded that CMMS Expert using BIM facilities will enhance the practices of decision making support in diagnosis.

Acknowledgement

This research is supported by Newton Fund Institutional Links (ID: 172726659), from the British Council and Institute of Research Management & Monitoring (IPPP) (grant no: IF009-2015), from the University of Malaya.

References

1. Abreu, J., Martins, P. V., Fernandes, S., Zacarias, M. Business Processes Improvement on Maintenance Management. Procedia Technology, **9**(2013), 320-330
2. Hamzah, S., Kobayashi, K. Utilising Mid-Long Term Maintenance Management Policy for Sustainable Maintenance of Infrastructure Facilities. Procedia Environmental Sciences, **17**(2013), 478-484
3. Ismail, Z. System development toward effective maintenance management practices. Built Environment Project and Asset Management, **4**(4), 406-422 (2014)
4. Hamid, Z. A., Kamar, K. A. M., Zain, M. Z. M., Ghani, M. K., Rahim, A. H. A. Industrialised Building System (IBS) in Malaysia: The Current State and R & D Initiatives. Malaysian Construction Research Journal, (MCRJ), **1**(2), 1-11 (2008)
5. Gajzler, M. The Support of Building Management in the Aspect of Technical Maintenance. Procedia Engineering, **54**(2013), 615-624
6. Anumba, C. J., Baugh, C., Khalfan, M. M. Organisational Structures to Support Concurrent Engineering in Construction. Industrial Management & Data Systems, **102**(5), 260-270 (2002)
7. Nawi, M. N. M., Salleh, N. A., Anuar, H. S. A Review Study of Maintenance and Management Issues in IBS Commercial Building. International Journal of Computer Informatics & Technological Engineering, **1**(1), 42-46 (2014)
8. Girmscheid, G., Rinas, T. Business Design Modeling for Industrialization in Construction: Cooperative Approach. Journal of Architectural Engineering, **18**(2012), 164-175
9. Wood, B. Maintenance Integrated Design and Manufacture of Buildings: Toward a Sustainable Model. Journal of Architectural Engineering, **18**(2012), 192-197
10. Singh, V., Gu, N., Wang, X. A Theoretical Framework of a BIM-Based Multi-Disciplinary Collaboration Platform. Automation in Construction, **20**(2), 134–144 (2011)
11. Zahrizan, Z., Ali, N. M., Haron, A. T., Marshall-Ponting, A., Hamid, Z. A. Exploring the Adoption of Building Information Modelling (BIM) in the Malaysian Construction Industry: A Qualitative Approach. International Journal of Research in Engineering and Technology, **2**(8), 384-395 (2013)
12. Ruiz, P. A. P., Kamsu-Foguem, B., Noyes, D. Knowledge Reuse Integrating the Collaboration from Experts in Industrial Maintenance Management. Knowledge-Based Systems, **50**(2013), 171-186
13. Kamsu-Foguem, B., Noyes, D. Graph-Based Reasoning in Collaborative Knowledge Management for Industrial Maintenance. Computers in Industry, **64**(2013), 998-1013
14. Nawari, N. O. BIM Standard in Off-Site Construction. Journal of Architectural Engineering, **18**(2), 107-113 (2012)
15. Cheng, J. C. P., Ma, L. Y. H. A BIM-based System for Demolition and Renovation Waste Estimation and Planning. Waste Management, **33**(2013), 1539-1551
16. Mahdjoubi, L., Moobela, C., Laing, R. Providing Real-Estate Services through the Integration of 3D Laser Scanning and Building Information Modelling. Computers in Industry, **64**(2013), 1272-1281
17. Irizarry, J., Karan, E. P., Jalaei, F. Integrating BIM and GIS to Improve the Visual Monitoring of Construction Supply Chain Management. Automation in Construction, **31**(2013), 241-254
18. Marzouk, M., Abdelaty, A. BIM-based Framework for Managing Performance of Subway Stations. Automation in Construction, **41**(2014), 70-77
19. Love, P. E. D., Matthews, J., Simpson, I., Hill, A., Olatunji, O. A. A Benefits Realization Management Building Information Modeling Framework for Asset Owners. Automation in Construction, **37**(2014), 1-10
20. Love, P. E. D., Simpson, I., Hill, A., Standing, C. From Justification to Evaluation: Building Information Modeling for Asset Owners. Automation in Construction, **35**(2013), 208-216
21. Miettinen, R., Paavola, S. Beyond the BIM Utopia: Approaches to the Development and Implementation

of Building Information Modeling. Automation in Construction, **43**(2014), 84-91
22. Volk, R., Stengel, J., Schultmann, F. Building Information Modeling (BIM) for Existing Buildings - Literature Review and Future Needs. Automation in Construction, **38**(2014), 109-127
23. Rashidi, M., Lemass, B. A Decision Support Methodology for Remediation Planning of Concrete Bridges. KICEM Journal of Construction Engineering and Project Management, **1**(2), 1-10 (2011)
24. Yunus, R., Yang, J. Critical Sustainability Factors in Industrialised Building Systems. Construction Innovation, **12**(4), 447-463 (2012)
25. Nawi, M. N. M., Haron, A. T., Hamid, Z. A., Kamar, K. A. M., Baharuddin, Y. Improving Integrated Practice through Building Information Modelling-Integrated Project Delivery (BIM-IPD) for Malaysian Industrialised Building Systems (IBS) Construction Projects. Malaysia Construction Research Journal (MCRJ), **15**(2),1-15 (2014)
26. Wang, Y., Wang, X., Wang, J., Yung, P., Jun, G. Engagement of Facilities Management in Design Stage through BIM: Framework and a Case Study. Advances in Civil Engineering, 2013(2013), 1-8
27. Ismail, Z., Mutalib, A. A., Hamzah, N. Case Study to Analyse Problems and Issues in IBS Building Maintenance. International Journal of Applied Engineering Research, **11**(1), 226-232 (2016)

Formulating a Long-Term Employment Strategy for Construction Workforce in Malaysia

Norazlin Mat Salleh[1], Siti Siti Sarah Mat Isa[2], Syarifah Nur Nazihah Syed Jamalulil[3], Irma Hanie Ibrahim[4], Edelin Hussein[5]

[1,2,3,4,5] *Department Of Quantity Surveying, Faculty of Architecture, Planning and Surveying, Universiti Teknologi MARA , Cawangan Perak*

Abstract: This paper provides a preliminary idea about the research outline to be conducted by the researcher. Overall idea starts with the unremitting issues regarding skilled construction workforce in Malaysia. Unfortunately, unemployment in Malaysia is becoming increasingly serious. Lack of employment factors there is also a source of unemployment. The main purpose of this research is to investigate the existing strategy to encourage the involvement of local skilled construction workforce (LSCW) onsite. Labour is the main asset in the construction project, even nowadays concerned with the use of advanced technology to reduce costs, labor is still required to drive the technology. Therefore, it is importance to develop a long term strategies to sustain the availability in the industry by implementing questionnaire, interview and ended with validation the framework by professional. The target respondent is Employer (Contractor) and ABM trainee throughout Malaysia. The outcomes dedicate to raise the interest among the youth to choose career as construction workforce, policy maker for revised any policy and employer for hiring LSCW.

Construction workforce, local labour, strategy and construction employment

1. Introduction

The construction industry accounts for an important part of Malaysian economy. Although relatively small, it is extensively linked with many parts of the economy, in particular with related industries such as those of manufacturing. So construction can be described as some kind of economic engine for Malaysia. Based on the economic report 2014/2015 (Ministry of Finance Malaysia, 2015) Construction Sector contributes 4% to GDP and employment 9.1% of the overall workforce in 2015. Even though the statistics presents well enough but in reality construction industry is facing chronic skilled labour shortage on site. It also becomes the greatest challenge facing the Malaysian construction industry.

Moreover, according to S. Mohamed et all (2012), involvement by local workforce is not promising and there is evidence that skilled workers produced by vocational training did not meet industry's standard. Some of them were not entering the construction sector even after undertaking training from construction training institutions. In addition the training institution need to take action and build networking with the industry in order to make sure the entire trainee acquired the jobs and develops their career after graduated.

The Eleventh Malaysian Plan (Ministry of Finance Malaysia, 2015) and Construction Industry Malaysia Plan (CIMP) 2016-2020 (Kementerian Kerjaraya Malaysia, 2014) highlighted this labour issue, to achieve developed nation by 2020 Malaysia requires 60% of the 1.5 million jobs that will be created will require TVET-related skills. In order to responding to this demand Malaysia will requires to increase its annual intake gradually from 164,000 in 2013 to 225,000 in 2020. Therefore, the aim of this research to develop a novel framework of employment strategy for LSCW to remain sustain in Malaysian Construction Industry (MCI).

2. Research Background

The Malaysian Government has invested and funded cripplingly in Technical and Vocational Education Training (TVET) to ensure that Malaysia has enough trained and skilled workers in the future. There are many training institutions developed by the government such as the Akademi Binaan Malaysia (ABM), Institut Kemahiran Mara (IKM), Institut Latihan Industri (ILP) Institut Kemahiran Belia Negara (IKBN) and Vocational College. Based on statistic from the Ministry of Public Work (2014), CIDB targets to train 28, 000 construction personnel in 2014. Since the establishment of the ABM Southern Region in 2001, it has successfully produced 200,000 construction personnel (Judin, 2013). Other than that, the Malaysian government has further decided to upgrade the vocational school to vocational college to show how serious the government is in the field of TVET.

Despite there are many efforts made by the government to produce skilled workforces to cater the future demand of MCI it is remain in shortage especially local skilled workforce. As a result many contractors claimed they have difficulty to employ local labours (S. A. Karib, 2014; Kamal et all, 2012.; F. Salleh et all, 2012; A. Bahaman, 2011; A. R. Ibrahim, 2010; S. Saleh, 2008; A. R. A. Aziz, 2002).

Moreover, in the star online reported by M. Carvalho and R. Rahim, (2016) worsened again the Malaysian government would take 1.5 million workers from Bangladesh to meet the demands of employers, who said that local local workers do not want to work especially in dangerous, difficult and dirty (3D) sectors and too choosy. Previous researcher had identified various reason for trainee of training institution reluctant to join the industry (Shazwani et all, 2012). For short term and quick solution contractor hired legal and illegal foreign labour but this situation will give adverse impact to Malaysia such as to economics, health and social problem because the situation currently is out of control.

In addition, in The Labour Shortage Issues Forum was held at the Caring Society Complex, Penang on the 23rd of September 2010, why Malaysian employers appear to like foreign workers because domestically, the employment situation is inherently and structurally foreign-worker orientated (geared towards low-cost/low-pay, low productivity and redundancies). Reported by Productivity, I., & Creation, J. (2015) the issues and challenges highlighted in Eleven Malaysia Plan is an inadequate skilled job, heavy reliance on low skilled foreign workers and youth employment in full employment scenario.

Before the trainee entering the construction industry they had negative perception that the industry is 3D (S. Saleh, 2008). The trainees, involved in other job sectors feel that they have a brighter future and more secure with fixed salary, allowance, EPF and promotion. Finally all the efforts and investments made by the government seem is useless and the success rate, if there is any is only a small amount. However none of studies offered a concrete solution to address the issues. Therefore, there is a necessity of MCI to produce and formulate a long term strategies for sustainable LSCW. There needs to be more focus placed on identifying future skill requirements and Occupational Structure (OS) as well as on investment in workforce innovation and up spilling the existing workforce.

3. Research Objectives

The objectives of this research are (i) to investigate on existing strategy to encourage the involvement of LSCW in construction industry, (ii) to identify the requirements of the LSCW to sustain in the industry and employer in order to employ them and (iii) to recommend a new framework of employment strategy for LSCW to sustain in Malaysian Construction Industry.

4. Definition Of Term

4.1 Employed

All persons who, at any time during the reference week worked at least one hour for pay, profit or family gain (as an employer, employee, own-account worker or unpaid family worker). Also considered as employed are persons who did not work during the reference week because of illness, injury, disability, bad weather, leave, labour dispute and social or religious reasons but had a job, farm, enterprise or other family enterprise to return to. Also included are those on temporary lay-off with pay who would definitely be called back to work. Employed persons at work and who had worked less than 30 hours during the reference week because of the nature of their work or due to insufficient work and are able and willing to accept additional hours of work are considered underemployed but are nevertheless included in the "employed" category.

DOI: 10.1051/matecconf/20166600053

4.2 Unemployed

The unemployed are classified into two that is the actively unemployed and inactively unemployed. The actively unemployed include all persons who did not work during the reference week but were available for work and actively looking for work during the reference week. Inactively unemployed persons include the following categories:

- persons who did not look for work because they believed no work was available or that they were not qualified;
- persons who would have looked for work if they had not been temporarily ill or had it not been for bad weather;
- persons who were waiting for result of job applications; and
- persons who had looked for work prior to the reference week.

(Department Of Statistic Malaysia, 2015)

4.3 Employment Strategy

Employment strategy" is defined as "an indicator which represents a desirable future in employment and shows a path towards desired reform." First of all, will start by pointing out a number of employment strategies to demonstrate what employment strategy is like. Employment strategy is made up of a combination of basic principle, strategic objectives in a medium- and long-term perspectives based on the basic ideas, a number of policies required for embodiment of the objectives, and procedures for implementing each of the policies.

The role of employment strategy is to enhance the effectiveness of each of the policies and to more efficiently achieve the strategic objectives by clearly showing the position and directions of the policies to those in administrative jobs as well as to those working for businesses and other workers. Working is vital to economy, and accordingly, employment strategy will cover a variety of issues. What is important is these issues are consistent in the light of objectives. Usual socioeconomic changes may lead to changes at an individual policy level, but basic principle and objectives of the strategy will remain unchanged (Definition, A. 2010).

5. Literature Review

According to Department of Education, Science and Training, (2001) skill shortages exist when employers are unable to fill or have considerable difficulty in filling vacancies for an occupation, or specialized skill needs within that occupation, at current levels of remuneration and conditions of employment, and reasonably accessible location. Shortages are typically for specialized and experienced workers, and can coexist with relatively high unemployment overall or in the occupation. An occupation may be in shortage even though not all specializations are in shortage. Occupations may be in shortage in particular geographical areas and not in others. Skill shortages generally involve skills that require a significant period of training and/or experience.

In addition, building a skilled and dynamic workforce to meet the needs of this growth will require government, industry, the community, unions, enterprises and individuals to work together. Throughout the world seriously reviewed and addressed the issues in America action taken are Reform and Reinvigorate the Perkins Act, Encourage Private Funding for Craft Training Programs, Improve the Workforce Investment Act, Make it Easier for Veterans to Get Training and to be Hired, Encourage Partnerships Between Registered Apprenticeship Programs and Community Colleges, Expand Federal Apprenticeship Resources and Collect More Comprehensive Data on All Apprenticeship Programs, Enact Immigration Reform, Offer Community College Career and Technical Programs to High School Students for Free and Make It Easier to Establish Public Schools Focused on Career and Technical Education (The Associated General Contractors of America, 2014). Other than that, in early 2012 Department of Business in Australia's Northern Territory conducted a forum to settle down the issue and successfully produced an Employment strategy (Figure 1).

Same cases happening in Malaysia there are many actions taken relating the issues especially under Construction Industry Development Board (CIDB), according to S. A. Karib, et all (2014) CIDB has established 6 training centres known as the Malaysia Construction Academy or Akademi Binaan Malaysia (ABM). CIDB has also accredited 40 private training centres. All these training centres provide skills training to both existing construction personnel and new workers. Undertake the exercise of accreditation and certification of construction workers. CIDB has amended the Act 520 to broaden its mandate to include the setting of standards; registration of construction personnel; training; accreditation and certification; The Construction Industry Master Plan II (CIMP) 2016 – 2020 which is being developed by CIDB in collaboration with key ministries, government agencies and various organisations representing the construction industry.

Figure 1: Employment strategy at the glance

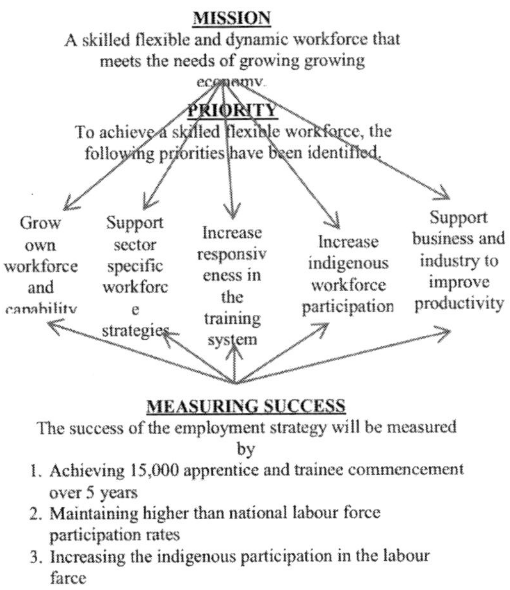

Sources: Northern Territory Employment Strategy 2012-2015

Additionally reported by D. of S. D. (2011), Y.B Datuk Dr. S. Subramaniam, Minister of Human Resource said that, the new skills training programmes would also look at producing more skilled workers for the outsourcing industry, where Malaysia ranks third in the world. The Government had set aside RM 1.0 billion fund to provide loans for school leavers and workers intending to undergo skills training programmes over the next five years. At present, the ministry producers some 18,000 skilled workers via its skills training courses of some 100,000 others graduate from both public and private institutions. The next 10 years will see a greater emphasis on human resource enhancement as availability of skilled and knowledge workers are a major pre-requisite to transform Malaysia from a production-based into a knowledge-based economy.

Malaysia offers investors a young, educated and productive workforce at costs competitive with other countries in Asia. Backed by the government's continued support of human resource development in all sectors, the quality of Malaysia's workforce is one of the best in the region. The Department of Skills Development (DSD) formerly known as the National Vocational Training Council under the Ministry of Human Resources coordinates the setting up of all public and private training institution, evaluates the demand for existing and future skills, identifies future vocational and industrial training needs and will continue to develop standards under the National Occupational Skills Standards (NOSS) (Figure 2). To-date, there are more than 700 certified standards which covers certificate, diploma and advanced diploma qualifications (Malaysia, D. of S. D., 2011).

Figure 2: Development of OA and NOSS and development of policy for training in Malaysia

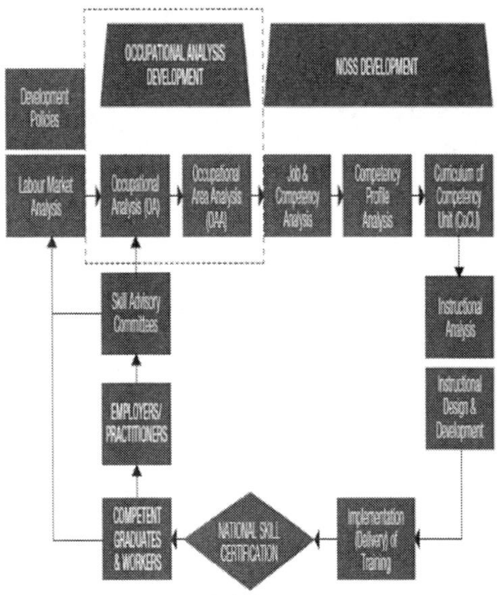

Sources: Department of Skill Development, Ministry of Human Resources, 2013

6. Research Methodology

The detail outline of this research will use exploratory research. For example, it is one thing to describe the employment problem in construction site. It is to develop explanations about why the youth is less interest to choose the career in construction, despite the enrollment of trainee to training center is increasingly but the industry is still dependent on foreign nationals on the grounds that employers' awarded local people do not want to work. Therefore to build a framework for long-term employment strategy a research method in the Figure 3 has been identified.

6.1 Study Population

Participants for the research will include the ABM trainee throughout Malaysia (Centre, East, Northern and Southern Region). They become a respondent

 DOI: 10.1051/matecconf/20166600053

because the training they receive is very suitable for work at construction sites such as bricklayer, painter, pavior and etc. Employer as known as Contractor because they will hire the trainee. Professional for conduct Focus Group for validated the framework.

Figure 3: Research Methodology

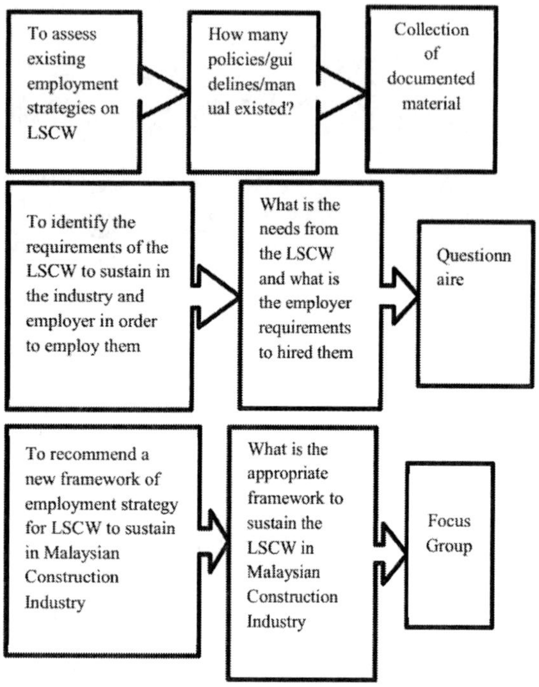

6.2 Collection of documented material

Document analysis is a systematic procedure for reviewing or evaluating documents—both printed and electronic material. It is requires that data be examined and interpreted in order to elicit meaning, gain understanding, and develop empirical knowledge. Documents that may be used for systematic evaluation as part of a study take a variety of forms. They include agendas, minutes of meetings; manuals; background papers; books and brochures; journals; newspapers, radio and television program scripts; organisational or institutional reports; survey data; and various public records.

These types of documents are found in libraries, newspaper archives, historical society offices, government bodies and organisational or institutional files. The analytic procedure entails finding, selecting, appraising (making sense of), and synthesizing data contained in documents. Document analysis yields data—excerpts, quotations, or entire passages—that are then organised into major themes, categories, and case examples specifically through content analysis (Bowen A., G., 2009).

6.3 Questionnaire

A questionnaire is a research instrument consisting of a series of questions and other prompts for the purpose of gathering information from respondents. The questionnaire will be divided into section consists of demographic and the requirements of trainee and employer. Firstly, respondents (ABM trainee and Contractor) could be asked to complete the questionnaire. Questionnaire will be distributed to respondents by go to the training centre and contractor office or site. In the initial stage the questionnaire will be distributed for the pilot study and for the validity of the questionnaire.

6.4 Focus Group

Finally to validate a formulated a long term employment strategy will validate by a professional by conduct a focus group. A focus group is a group of people are asked about their professional idea, comments and improvement toward the strategy. The process for conducting the focus group are, firstly, define a focus group, secondly designing focus group questions, next recruiting and preparing for participants (professional, academicians, employer and policy make representative from CIDB). Subsequently, conducting the focus group and the final stage analyzing the data. The session will recorded by using video as evidence and easy to abstract all the point during analyzing the data.

As a final point, after all the methods have been carried out, the result will be collected and recorded, analyzed and presented in accordance with the required theme. The conclusion and recommendation will be formed from the results of the research.

7. Conclusion

Based on the presented preliminary research interest, it can be concluded that, there are many secondary data presented the issues but there are gaps between all the policy makers and the users. Hopefully this research will integrate from all previous data and presented in the meaningful practise. Finally the result from this research will contribute to the development of youth in choosing their career after completed on the training and to move on in the employment environment. Furthermore, it may solve a problem too dependent on foreign labour where contributed a lot of negative consequences and at the same time will accelerate Malaysian economics.

References

1. R. A. Aziz (2002). *Skill shortage in the construction sector: A critical survey of existing formal vocational training system.* Kajian Malaysia Jld XX, **No. 2**. Universiti Sains Malaysia.
2. A. R. Ibrahim, M. H. Roy, Z. Ahmed, G. Imtiaz (2010). *An investigation of the status of the Malaysian construction industry, Benchmarking:* An International Journal, **Vol. 17** Iss: 2, pp.294 – 308.
3. A. Bahaman (2011). *The Key Issues in the Malaysian Construction Industry: Public and Private Sector Engagement,* Persidangan Pegawai Kanan KKR dan Agensi, Penang, Malaysia
4. B. Vijay Antony Raj & P. S. Kothai (2014). *Study on the Impact of Human Resource Management Practices in Construction Industry.* The International Journal of Management **Vol 3** Issue 1
5. Bowen A. G. (2009). Document Analysis as a Qualitative Research Method. *Quality Research Journal*, *9*(2), 27–40. http://doi.org/10.3316/qrj090202
6. Department of Education, Science and Training, (2001). *Nature and causes of skill shortages Reflections from the Commonwealth National Industry Skills Initiative Working Groups,* Department of Employment and Workplace Relations and the National Centre for Vocational Education Research
7. Dom, N. M., Kasim, N., & Shamsudin, A. (2012). *Framework of human resource planning (HRP) influencing factors for local workforce supply in Malaysian construction industry.* Journal of Technology Management in China.
8. F. Hassan, Z. Ismail, A. A. Zaini, S. Hassan and M. Maisham, (2011). *A Critical Assessment of the Challenges of Developing the Malaysian Construction Workforce through Training,* IEEE Colloquium on Humanities, Science and Engineering Research (CHUSER 2011)
9. F. Salleh, A. S. Ali and N. Zakaria (2012). Involment of Foreign Labour in Malaysian construction Industry: Case Study Klang Valley. NCON-PGR 2012
10. G. Ducanes and M. Abella (2008). *Labour Shortage Responses in Japan, Korea, Singapore, Hong Kong, and Malaysia: A Review and Evaluation.* International Labour Office; ILO Regional Office for Asia and the Pacific, Asian Regional on Governance of Labour Migration
11. Ilina, O., (2007). *Skill shortage. In: Descy, P.;Tessaring M. (eds). Modernisng Vocational eduvation and training.* Luxembourg; Europ
12. Kamal, E. M., Haron, S. H., Ulang, N. M., & Baharum, F. (2012). *The critical review on the Malaysian construction industry.* J Econ Sustain Dev, 3
13. Kementerian Kerjaraya Malaysia, (2014). *CIDB sasar latih 28,000 personnel binaan Malaysia tahun 2014.* http://www.kkr.gov.my/
14. Malaysia, D. of S. D. (2011). *Occupational Structure the Green Technology Industry*, 1–185
15. Ministry of Finance Malaysia (2015). Economic of Prospect Malaysia http://www.treasury.gov.my/pdf/economy/er/1415/chapter3.pdf
16. M. Carvalho and R. Rahim, (2015). *Malaysia and Bangladesh ink deal to recruit 1.5 million workers* http://www.thestar.com.my/news/nation/2016/02/18/bangladeshi-workers-malaysia-mou/
17. Northern Territory. Department of Business and Employment (2012). *Gearing up the workforce: employment strategy.* Darwin, NT Department of Business and Employment
18. Organisation for Economic Co-Operation and Development (2004) *Developing Highly Skilled Workers: Review of Canada.* OECD Publications.
19. Organisation for Economic Co-Operation and Development (2004) *Developing Highly Skilled Workers: Review of Finland.* OECD Publications.
20. S. Mohamed, S. Fikri, M. Yusof, Zakaria and A. Zaki, (2012). *Construction skilled labour shortage the challenges in Malaysian construction sector. OI*DA International Journal of Sustainable Development, 4 (5). pp. 100-108. ISSN 1923-6654.
21. S. A. Karib, N. Shaffli. C. S. C. Soh. and N. I. Z. Abidin (2014). *Meeting Construction Industry Resources Requirements THE MALAYSIA WAY.* Division, Corporate and Business Sector Construction Industry Development Board (CIDB) Malaysia, 20th Asia Construct Conference Hong Kong.
22. S. Saleh (2008). *Causes of Poor Participation of Local Workers in Malaysia Construction Industry and Strategies for Improvement.* Unpublished Master of Science in Construction Management. Faculty of Civil Engineering University Technology Malaysia.
23. The Associated General Contractors of America (2014). *Preparing the Next Generation of Skilled*

Construction Workers: A Workforce Development Plan for the 21st Century.

DOI: 10.1051/matecconf/20166600054

Facilities Management Practices in Malaysia: A Literature Review

Nordiana Mohd Isa[1,a], Syahrul Nizam Kamaruzzaman[1], Othman Mohamed[1], Aini Jaapar[2] and Asra Zaliza Asbollah[1]

[1]*Department of Building Surveying, Faculty of Built Environment, Universiti Malaya, Kuala Lumpur. Malaysia.*
[2]*Faculty of Architecture, Planning and Surveying, Universiti Teknologi MARA, Shah Alam, Selangor. D.E. Malaysia.*

Abstract. Facilities management in Malaysia has been practiced for decades. The development of its formal practice parallels the improvement of the built environment in the nation. Involvement of the public and private sectors teaming up in arranging the National Asset and Facilities Management (NAFAM) in demonstrates the vital collaboration in the facilities management area in Malaysia. Facilities management is seen distinctively as indicated by diverse geographical locations, interests and schools of thought. Facilities management is delegated a service-based industry which gives proficient counsel and administration of clients' building facilities including residential, commercial, industrial, airports terminals and offices. The aim of this paper is to review the gaps that exist, especially on how FM is being practice in comparison with the published FM body of knowledge. Very relying upon literature, this paper discovered a gap that is an unclear description of current FM applications. This research aims to give new bits of knowledge to upgrade comprehension of FM execution in Malaysia.

Keywords: Facilities management, definition, Malaysia, and practices.

1 Introduction

Started in the 1980s, a new and more incorporating management speciality has developed, focusing on facilities that, people increasingly recognized the importance of managing not only buildings but also such buildings in connection with people, integrating with the principles of administration and development process. Current developments in information technology, advancements in telecommunications and the removal of trade barriers, are among the factors transforming the world of facilities management all over the world and in turns have spread the needs of facilities management from outside into Malaysia.

In parallel, facilities management over the past thirty years has been established as a profession with a number of worldwide professional bodies supporting and accepting a significant number of facilities management practitioners. Since then, the understanding of facilities management is varied as different professions interpreted it differently. However, despite its diversity in the theory and practice, facilities management retains essential cores that maintain its identity that differentiate it from other professions [23]. Therefore, this paper aims to review the gaps that exist, especially on how FM is being practise in comparison with the published FM body of knowledge.

2 Problem Statement

The 10th Malaysia Plan that was distributed in late 2010 states that the nation is presently looking towards a more service-oriented economy with a reduction in the manufacturing sector. Economic Planning Unit (EPU) states that the services sector possesses a 61% share of Malaysia's GDP against other sectors. For example, manufacturing, mining, construction and agriculture. According to [17], it can be deduced that facilities management is an imperative sector in the Malaysian Economy as a feature of the bigger share of the service sector. This draws attention and enthusiasm to research facilities management in the Malaysian context.

3 Literature Review

As depicted in the introduction, the overview of this paper reveals the idea of facilities management. By [24], challenges can emerge when a specific key term produce diverse meaning and importance to various individuals, that might likewise misinterpretation. In this way, characterizing essential keywords solicit in the ideas expresses a better comprehension of the theories.

[a] Corresponding author: diana_isa@um.edu.my

3.1 Definition of facilities management

The facility management (FM) as an industry has emerged as one of the fastest growing sectors over the years. However, the scope and identity of FM are still unclear as evidenced by the definitions and issues that attempt to describe its scope. In fact, FM is the multi-disciplinary kind of work that covers a wide range of various activities, responsibilities, and knowledge. What is more interesting, every aspect of an organisation seems to be drawn into FM.

If the building is said to be one of the important things in real estate, facilities management is part of the management of the building that needs to be emphasized by the management body. One of the facilities management objectives is to reduce maintenance costs while maintaining quality or the quality of facilities provided. Rozilah Kasim [23] in her book stated that;

> "The original concept of facilities management is very much associated with its role as "enabler' to support the core business of the organisations with the primary aim of minimising costs and maximising profit along the delivery process. Early understanding has seen facilities management as a coordinating function that integrates between people, place, process, and technology. This conventional function has a three-way interface between organisational culture, the people, and physical assets. This interface, somehow or rather, has created the original demand for facilities management services and expertise."

Efforts to define facility management in Malaysia cannot get away from starting by looking at the definitions that have been produced in the country earlier practice of this discipline. The International Facility Management Association (IFMA) as the official definition of the profession Manager Facilities in the United States and in the United Kingdom, British Institute of Facilities Management (IFM) uses the following definition. Countries such as Singapore and Hong Kong and most of other countries in Asia have adopted the same definition which was initiated by IFMA:

> "Facilities Management is the integration of multi-disciplinary activities within the built environment and the management of their impact upon people and the workplace".

By definition, FM is a process that requires a multi-skills approach that supports the core business of the organisations by interfacing the physical workplace and people [5]. This in accordance, shows that FM acts as a support role or service, some part served the needs of primary activities or core business (demand side) and the other part is the organisation's non-core business (supply side) [27]. The function of FM is also to reconcile, through time, these demand and supply aspects of the organization. In other words, FM should integrate knowledge of both facilities and management in order to work effectively. Successful FM leads to workplaces which better support the flow of productive processes while adding value and reducing costs. The scope, range of services, activities, responsibilities, skills and knowledge of facility management are all intended to better integrate existing organisational factors.

In particular, FM can be described as managing the physical and non-physical facilities and unpredictable business requirements. It is required by all professionals who undertake the management to achieve business goals. From a technical standpoint, facilities management and responsibilities include work that was originally done by architects, landscape architects, interior designers, civil engineering, building surveying, valuation, quantity surveying and etc. In addition, among the items included in FM is expert advice, the conservation of the building, building renovation, building cleaning, security, parking, electricity, telephone, fire suppression system, air conditioning, elevators, landscaping and other facilities are available in the building.

To gain a better comprehension of the terms, the description and definition of facilities management were presents as per Table 1.

Source	Definition / Description
Becker, [6]	FM is responsible for coordinating all efforts related to planning, designing and managing buildings and their systems, equipment and furniture to enhance the organisation's ability to compete successfully in a rapidly changing world.
Bernard, [7]	Defines facilities as "the premises and services required to accommodate and facilitate business activity". Bearing this in mind, to have any chance of being fully cost effective the management of facilities must directly embrace the three generic cost centres which include premises, support services and information technology.
Alexander, [1]	The scope of discipline covers all aspects of property, space, environmental control, health and safety, and support services.
Hinks & McNay, [12]	Common interpretations of the FM remit: maintenance management; space management and accommodation standards; project management for new-build and alterations; the general premises management of the building stock; and the administration of associated support services.
Nutt, [20]	The primary function of FM is resource management, at strategic and operational levels of support. Generic types of resource management central to the FM function are the management of financial resources, physical

	resources, human resources, and the management of resources of information and knowledge.
Atkins & Brooks, [4]	An integrated approach to operating, maintaining, improving and adapting the buildings and infrastructures of an organisation in order to create an environment that strongly support the primary objectives of the organisation.
Amaratunga, [3]	FM as "creating an environment that is conducive to carrying out the organisation's primary operations, taking an integrated view of the services infrastructure and using this to deliver customer satisfaction and value for money through support for enhancement of the core business".
IFMA, [13]	Facilities must be managed as an integrated system. The International Facility Management Association defines facility management as the practices of coordinating the physical workplace with the people and work of the organization.
BIFM, [8]	An integration of multi-disciplinary activities within the built environment and management of their impact upon people and the workplace.
IFMA, [14]	FM is "a profession that encompasses multiple disciplines to ensure functionality of the built environment by integrating people, place and process, and technology"
IFMA, [15]	FM is a profession that encompasses multiple disciplines to ensure functionality of the built environment by integrating people, place, process and technology.
BIFM, [9]	FM is the integration of processes within an organisation to maintain and develop the agreed services which support and improve the effectiveness of its primary activities.

Table 1: Summary of the definition of 'facilities management'

According to Abd Rahman [2], facilities management is divided into 3 major components such as property, equipment and services for which it depends on the respective management to manage and provide the facility to the users. He added, the scope of facility management includes a number of areas such as:

i. Management and maintenance of buildings
ii. Maintenance engineering
iii. Service building
iv. Project management renovation and conservation
v. Management space
vi. Provision of security and cleaning
vii. Offices and food service
viii. Negotiate contracts and purchases
ix. Revision rent
x. Negotiating leases and lease relations
xi. Information technology
xii. Planning for the future portfolio

Beforehand, according to Pheng [21], FM contains four main principles:

i. The continuously programme coordination of all efforts, namely planning, designing, construction and management of facilities towards enhancing the working environment for the people and the organization's ability to meet its business objectives;
ii. The total integration of a diverse field of disciplines of business, architecture, behavioural and engineering science under one entity in an organization to oversee all facilities functions previously controlled by independent departments;
iii. The management of activities proactively rather than the management of facilities reactively; and
iv. A business concept where FM policies and procedures are guided by organizational goals and objectives as well as available resources.

To achieve all these 4 principles, there are three levels of facilities management strategy should be adopted in managing a facility, as stated by Abd Rahman [2]:

i. The strategy analysis - in this stage all the data and facts will be collected including the objective of the organization, policies, and requirements, resources, processes, physical asset conditions, use of space and function, as well as analysis of the costs involved.
ii. The completion strategy - the result of information collected in stages 1, including the objectives of the organization will be evaluated and the analysis made for the 2nd time onwards facility management strategy will be established. This means that an organization should choose a real strategy.
iii. The implementation of the strategy - the stage of implementation, it consists of the development strategy through the realization of an implementation plan containing a timetable planning, such period or periods and complete risk management ... it includes human and labor and the system, communications, planning, and resources acquisition (procurement).

3.2 Integrated facilities management

Facilities management as mentioned has been evolving as a multidisciplinary discipline. According to Kamaruzzaman & Zawawi [25], the greater part of the Malaysian construction industry professional recommend that regularly FM just focusing on building administration after completion of any construction works. However, Jensen [16] claims that FM association has close relations with the fundamental operations of the association and it is not just routine maintenance towards work facilities and assets of the business. This is supported by W.S.Z, Wan-Hamdan et al [28], which they point out that facilities management is key to practice in all areas. For example in the public sector, facilities management assumes an essential part in the improvement of socio-economic progress of a nation. This shows that people are starting to challenge the rigidity of professional traditions which support the individual practice, suggesting strongly that any professional should not be the exclusive professional in facilities management.

FM is relatively new in Malaysia, still at the infancy stage. Presently, the adoption and practice of FM are predominantly focused on MNCs and government agencies such as JKR. Other than that, FM is widely adopted in a single sector such as in healthcare sector. As stated by Syahrul Nizam and Emma Marinie [25], FM was firstly brought into Malaysia by such sector in 1997 which showed that the development of FM in Malaysia started only in the middle of 1990s. They added the government has played an important role in the development of FM in Malaysia. Syahrul Nizam and Emma Marinie [25] stated, "In 1996, the government has privatized the nonclinical support services in the government hospitals to three FM companies. It was indeed the biggest FM contract by the government at that time. It also marked the starting of the new era of FM in Malaysia". On top of that, the establishment of National Asset and Facility Management (NAFAM) shows that the government is very keen on the establishment of a systematic management. Moreover, Guidelines for the Management of Movable Assets and another act, the Building and Common Property (Maintenance and Management) Act 2007 Act 663, was circulated in 2007. This may be associated with a portion of one trillion ringgit Malaysia (RM) for public building facilities maintenance through Ninth Malaysian Plan in 2006. This funding also believed related to the articulation in 2001 by the Deputy Prime Minister Datuk Seri Abdullah Ahmad Badawi which he is expressing his concern about the need to redesign the quality level of public services [10].

Elyna Myeda Nik Pitt Michael [11] continues:

> "However, this act focuses only on the maintenance and management of residential or housing properties, particularly on the responsibilities of the managing agents and joint management body, and on maintenance funds. The year 2007 also marked the first involvement of the government in organising a national convention, the National Asset and Facility Management, in Kuala Lumpur. In the first quarter of 2009, the government launched the Government Asset Management Policy. This manual is more focused on the principles of government asset management, and no standards or specification on the FM services are outlined [22]. A system named mySPATA was also developed for the management of immovable assets [22]. All the government initiatives in providing guidelines, policies and manuals on service quality indicate that the government is slowly taking a step forward. However, no specific guideline on facilities management practice has yet been developed for both public and government sectors".

N.E.M. Nik-Mat et al. [19] however stated, FM in both private and public sector has been evolving only over the last 10-15 years. In this respects, some local practitioners associate it only with the professional maintenance of buildings. As compared to North America and Europe, the responsibility of the facility manager is covering project management construction and operation of physical facilities. In other words, the facilities management in that countries covering a wider field of management and services include the main activities of the organization. Even in the United Kingdom, this field including the practice of directly or indirectly relate to the needs and business goals of the organization.

Facility management practices in accordance, have great potential to grow and prosper in the construction industry in Malaysia. A number of physical development, undertaken by the size of a large development are among the factors that reflect the volume of the physical facilities should be managed. This scenario is the result of ongoing efforts by the government to build and develop the country. The service providers involved in the management of physical facilities (specialty consultants, contractors, suppliers of materials and equipment, etc.) can seize great opportunities that exist, particularly in terms of providing an integrated facilities management or comprehensive to local organizations.

3.3 Facilities management in Malaysia

In developed markets, FM administrations are firmly incorporated with some different administrations in the real estate sector, for example, rent collection and lease management. However in Malaysia demonstrating that the idea of FM has already not sufficiently developed to give complete property management practices. As mentioned, FM is the provision and management of resources, physical environment and business support services to complement the strategic and operational objectives of an organization, integrating key functions like processes and technologies with people, with the flexibility for future integration.

Management of a facilities required to increase the provision of facilities management to ensure the building is clean, safe and in a healthy environment. For example, which has been practiced in UITM, the Facilities

DOI: 10.1051/matecconf/20166600054

Management Division is one of the main division, who's responsible for the overall maintenance of the building and the area around the campus. There are four units in Facilities Management Division, they are the Public Unit, the Assembly Unit (UAKM), Unit of Mechanical, and Electrical and Telecommunications Unit (UMET).

One of the examples of FM practices in Malaysia is in the office buildings. Syahrul Nizam and Emma Marinie [25], in their study, mentioned that this kind of building normally would have their own management team to monitor the conditions of buildings. This management team usually lead by maintenance or building managers. Maintenance, cleaning, landscaping, lightings, heating, ventilating and air conditioning (HVAC), lift or escalators, mechanical and electrical, sanitary and plumbing, access, signage, parking and others are among the services provided by building managers [19].

Another example is managing a sports facility which is now gaining popularity in the operating system of the building in Malaysia. The scope of facility management refers not only to the management of sports facilities but also consists of various types of related elements and buildings. FM as Teicholz and Noferi [26] suggest, encompasses a broad range of joint activities which are not limited only to one area such as planning or design or leasing, project management, asset management or construction or even marketing facilities and acquisition of assets.

Each FM varies based on the size, composition and management structure of the facilities involved. For example, typically most of the sports facilities built in Malaysia is managed by the Local Authority (PBT) or State Stadium Corporation who are not profit-oriented. The facilities are designed and built is considered to represent the aspirations of the government to provide the best facilities for the community. In addition, the main focus in the management of the facility is to ensure that each piece of equipment can operate properly, safe and functional as required by the user. As a facilities manager, the focus will be centred on how to ensure facilities operate smoothly by matching with a variety of equipment, tools, and mechanisms that felt best for the consumer and managed facilities. However, large facilities managers often focus their time on the commercial use of facilities such as market rents for commercial space and generate profits.

Rivalry in Malaysian FM services market is exceptional with the vicinity of both local and worldwide service providers. The business sector is required to witness more rivalry from new contestants through mergers and acquisitions. Subsequent to the business sector is filled with minimal effort disorderly service providers, estimating and edges go under pressure. These disorderly players give services at low rates, leaving the competition from huge organized players. Nonetheless, much worldwide property administration organizations have made a raid into this business sector and have been accomplishing extraordinary development rates in the course of the most recent five years. Through the inorganic route, a couple of global property administration organisations have seen a sharp rise in the business sector by obtaining some local companies.

Facilities management in short is based on the requirements of the coordination between the facilities and consumers. It needs to be adjusted and balanced to meet the requirements. Various disciplines such as engineering, architecture, business management and behavioral science are needed to achieve the best facilities management as to guarantee a good quality facility management and effective. Accordingly, as the manager of a facility, it is not an easy task because it requires a manager to have the skills, expertise and extensive experience in managing and maintain a facility.

4 Methodology

This paper solely used the qualitative approach. The literature was collected and compiled from various publications and also previous research prior to the study area. Analytical review and synthesis have conducted in order to strengthen the gap of research from literature. The findings of literature will be strengthened through semi-structured interview involving FM experts. For this paper, the explanation of literature is only discussed because the study is still ongoing.

5 Result of literature findings

From the literature, only a few studies have considered the funding elements or issues in the implementation of FM specifically in the Malaysian context. As study done by Myzatul Aishah Kamarazaly et al. [18] showed "that the critical challenges currently facing the UFMs (Australasian university facilities managers) comprised issues relating to the following (in diminishing order of significance): inadequate funding, emergency management and business continuity planning, statutory compliance, sustainability and environmental stewardship, keeping up with rapid changes in technology, operational efficiency, identifying and meeting stakeholder needs, maintenance and manpower. Preparing for and responding to disaster/emergency was perceived as the most critical challenge of the future, perhaps, due to the recent natural disasters. Overall, poor funding was identified as the root of all other issues faced by the UFMs, hence suggested strategies for addressing the key challenges harped on financial improvement measures."

Other than that, past studies also have distinguished different elements compelling the accomplishment of key FM functions in many managerial levels in Malaysia context, the components, unfortunately, are not recorded as far as their relative levels of effect on the key FM functional areas. Myzatul Aishah Kamarazaly et al. [18] also stated that "Facilities managers would benefit from the prioritisation mainly due to the fact that it would enable them to leverage the limited resources at their disposal to address those factors that have a higher impact on the FM functional areas." Thus, it is expected that further study should fill this gap.

DOI: 10.1051/matecconf/20166600054

6 Significance/benefits of the study

The topic presented in this paper show that like whatever another arrangement of experts, the facilities managers in Malaysia face numerous challenges both from internal and external that prevent the accomplishment of their vital objectives and commitment to the general institutional value. The study results are expected to be of advantage to any institutional facilities managers in Malaysia particularly in planning a proper reaction to the recognised problems.

7 Conclusion

Despite the current practice models identified locally facility, the FM practices are very much different from the models being practice in the United States and the United Kingdom (in particular). However, this discipline does not mean the opportunity is limited. As mentioned, FM ought to be gone for diminishing the general expense of upkeep and expanding the life of the asset or building. Factors such as absence of accessibility of technical and non-technical manpower, expanded lead times in assembling resources/staff after a project has been effectively contracted because of labour lack, an expansion in swelling and labour cost, and inadequate procurement of services at low rates by chaotic industry members were among the reason why in-house practiced nowadays. In parallel, not only in the mentioned agencies, but other such as libraries, railway stations, and airports is likewise anticipated that would intensify opportunities for the FM market.

In short, it can be mentioned that the future interest for FM is very enthusiastic. As the saying goes, 'As long as buildings exist, there will be a demand for facility management'. On the other hand, the challenge is to educate building owners as to the benefits of FM. The ultimate advantage of FM, i.e. value appreciation of the property, image and long-term cost, must be completely understood and increased in value by building owners for the industry to develop. In other words, building owners ought to have the capacity to see the long-term gain.

Acknowledgement

The authors would like to thank the funding bodies of this research: the University of Malaya under FRGS Project. No. FP025-2014B.

References

1. Alexander, K. (1996). Facilities Management: Theory and Practice. London & New York: Taylor & Francis Group
2. Abd Rahman Mohd Noor (2008). *Pengurusan Hartanah Komersial dan Awam*, Edisi Kedua, Leeds Publications, Petaling Jaya, Selangor
3. Amaratunga, D. et al. (2000) Assessment of facilities management performance – what next? Facilities, **18** (1/2), pp. 66-75.
4. Atkin, B. and Brooks, A. (2000), Total Facilities Management, Blackwell Science, Oxford, U.K
5. Barrett, P (2004). *Facilities Management: Towards Best Practice*, Blackwell Science, London.
6. Becker, F. (1990), The Total Workplace, Van Nostrand Reinhold. New York, NY
7. Bernard WilliamsMay/June (1996): Cost-effective facilities management: a practical approach pp.26–38
8. British Institute of Facilities Management (2003) British Institute of Facilities Management, Available at: http://www.bifm.org.uk
9. British Institute of Facilities Management (2011) British Institute of Facilities Management, Available at: http://www.bifm.org.uk
10. Construction Industrial Development Board (CIDB) (2010), Newsletter [on-line], Malaysia, Published by CIDB, available at: www.cidb.gov.my/v6/?q_en/content/533 (accessed December 2015).
11. Elyna Myeda Nik Pitt Michael (2014). Facilities management in Malaysia. *Facilities*. Vol. 32 (9/10), pp. 490 – 508
12. Hinks, J. and McNay, P. (1999), "The creation of a management-by-variance tool for facilities management performance assessment", Facilities, Vol. 17 Nos 1/2, pp. 48-60.
13. IFMA (2003), Facts about the International Facility Management Association, available at: www.ifma.org
14. International Facilities Management Association (2008) International Facilities Management Association (IFMA), Available at: http://www.ifma.org
15. International Facilities Management Association (2011) International Facilities Management Association (IFMA), Available at: http://www.ifma.org
16. Jensen, P. A. (2011). Organisation of facilities management in relation to core business. *Journal of Facilities Management*. **9**, pp. 78–95.
17. Mohd Fadzil Mat Yasin (2013). The contributions of knowledge mapping in facilities performance evaluation practice in Malaysia
18. Myzatul Aishah Kamarazaly, Jasper Mbachu and Robyn Phipps (2013). Challenges faced by facilities managers in the Australasian universities. *Journal of Facilities Management*. **11** (2), pp. 136 – 151
19. N.E.M. Nik-Mat et al. (2011). Assessing the Maintenance Aspect of Facilities Management through a Performance Measurement System: A Malaysian Case Study. Procedia Engineering. **20**, pp. 329-338
20. Nutt, B. (2000), "Four competing futures for facility management", Facilities, Vol. **18** Nos 3/4, pp. 124-132.

21. Pheng, L.S. (1996). Total Quality Facilities Management: A Framework for Implementation. *Facilities*. **14**(5/6), 5-13.
22. Public Works Department Annual Report (2009), Public Works Department Annual Report, Malaysia, Published by PWD, available at: www.jkr.gov.my/jkr4/index.php?r_page/view&id_161
23. Rozilah Kasim (2011). *Community-based Facilities Management: Theory and Practices*. Penerbit UTHM, Johor
24. S. Parmjit, Y. F. Chan, and K. S. Gurnam, A comprehensive guide to writing a research proposal. VentonPublishing, Batu Caves, Malaysia, 2006.
25. Syahrul Nizam Kamaruzzaman and Emma Marinie Ahmad Zawawi, (2010). Development of facilities management in Malaysia. *Journal of Facilities Management*. **8** (1), pp. 75 – 81
26. Teicholz, E., and Noferi, C. (2002) 'Higher ED integrates CAFM better. *Facilities Design and Management*. **1**, pp.13–14.
27. Wanlaya Patanapiradej (2006).The Scope of Facility Management. *Nakhara*. **6**, p. 75-90
28. *W.S.Z, Wan-Hamdan et al (2011).* Contribution of Facilities Management Process in Supporting Malaysia National Higher Education Strategic Plan. *Procedia Engineering, **20**, p. 180-187*

An Introspective view of Sustainable Cohousing with The Malaysian Housing Concept

Norhaslina Jumadi[1,a], Noorsidi Aizuddin Mat Noor[1], Ahmad Ariffian Bujang[1]

[1]*Department of Real Estate, Faculty of Geoinformation & Real Estate, Universiti Teknologi Malaysia, 81310, UTM Johor Bahru, Malaysia*

Abstract. Sustainable development strategies are becoming demanded in assisting in wealth distribution and improving living conditions for mankind. A sustained community is one of the sustainable development agendas that encourage community involvement and participation in contributing to sustainability. Moreover, with the rapid population growth, especially in urban cities as the proportion of urban dwellers will reach two-thirds of the world population, this will lead to several problems including the social life changing intensely. Sustainable Cohousing is one of the suggested answers for an innovative form of neighbourly accommodation. Generally, Cohousing is formed by a group of people who are committed to living as a community and who actively participate in the design and operation to shape their own neighbourhood. Through this concept, the community can decide how they can implement sustainability principles in terms of social, economical and environmental terms in their neighbourhood. Therefore, it is important to bring back the sense of togetherness amongst the people and improve their social wellbeing through an effective neighbourhood. The objective of this study is to isolate the basic concept of Cohousing and to identify the conceptual framework of Sustainable Cohousing that may apply to improving the sustainable living in Malaysia. The research methodology was through identifying and reviewing the issues in existing literature on cohousing chosen from various dimensions, such as principles and elements, sustainable value, development model and so on, which can be suited with the Malaysian culture. The findings of this research are useful for property development practitioners and policy makers in promoting sustainability through the new concept of modern housing.

1 Introduction

The main objective of sustainable development is to find coherence between three constituent parts, which are environmental sustainability, economic sustainability and socio-political sustainability. Thus, need to be emphasised instead of just development [1]. However, a key issue in implementing sustainability is on how to meet the growing demand for development with the great growth of cities and the number of population, especially at the dawn of the 21st century. As stated by the United Nations under the World Population Prospects: The 2015 Revision, the total world population had reached 7.3 billion people in mid 2015 and the rapid population growth has been recorded, especially in urban cities, as the proportion of urban dwellers will reach two-thirds of the world population and the land converted to urban areas will triple by 2030 [2, 3]. This trending is a global phenomenon where Malaysia is no exception. By the year 2030, it is projected that more than three quarters of the total population will move to the urban areas [4]. This will lead to several problems included in the provision of adequate housing, sanitary facilities and amenities, health and educational facilities, public transportation as well as other infrastructure and many more. Apart from that, due merely to the changing of urbanisation growth, it tends to make the social life change intensely. According to Gidden, the effects of urbanisation on social structures are particularly concerned about the decline of social bonds and solidarity [5]. People will be more independent, selfish, less empathy and ignore interrelations towards each other. Society will become more individualistic and cities will become places full of strangers [6]. Besides that, many social relations now take place outside neighbourhood communities. Virtual social networks seem to be further eroding the residual bonds of spatial proximity and kinship [7]. Therefore, it is important to bring back the sense of togetherness amongst the people and improve their social well being through an effective neighbourhood.

According to Schoenberg , neighbourhoods are seen as possessing "commonly named boundaries, more than one institution identified with the area, and more than one tie of shared public space or social network"[8]. This place can be acceptable for people to live in safety, it is planned and the increase of the feeling of a community's wellbeing. Neighbourhood also can be seen as a frontline to sustainability. One suggested answer to implement sustainability elements in a neighbourhood is through Cohousing [9 - 11].

This paper presents, literally, the concept, development model, and how this Cohousing will

[a] Corresponding author: norhaslina.jumadi@gmail.com

accommodate sustainability towards its elements in the Malaysian perspective. Data for this study were merely from previous research using secondary data sources. The initial review of the Cohousing development model and sustainable elements was investigated to seek the potentiality and suitability to be adopted in the Malaysian environment. This was supplemented with a discussion from different scholars regarding the advantages of Cohousing in terms of social, economic and environmental sustainability. An analysis of the secondary data was completed and presented in the table for the overall adaption model of cohousing. Finally, from this, some conclusions have been drawn to the possible outcome of the adaptation of Cohousing in Malaysia as a new sustainable modern housing concept.

2 Cohousing, a definition

The idea of forming Cohousing came from Denmark in the mid 1960s by an architect, with only two communities arising from 1970 to 1973 [12]. The idea was accepted and started to spread to other parts of Northern Europe, North America, Great Britain, Australia, New Zealand and Japan [10-11, 13-14]. Cohousing is known as neighbourhoods or developments which are typically formed by a group of people who are "consciously committed to living as a community" [18]. The communities comprise private units (in which households reside) and communal facilities, and they are planned based on Social contract design principles that reportedly encourage more social interaction, helping to build more cohesive communities [15]. Cohousing is currently presumed as a better option to be considered in establishing a sustainable society since its offers people to access and share the ownership in a communal place, supportive cohouser members, greater social responsibility for sustainability and to enjoy the meaning of being a community. In most European countries Cohousing has been established in housing, care and support options for senior residents in order to demonstrate its capacity to enable people to support each other at an older age and through changing life experiences [16]. However, this concept has started to be implemented to other groups of ages; known as Multigenerational Cohousing, since it give a good benefit to social bonding. Adults will enjoy socialising and can share their interests, skills and values to help each other. They will work together and will find a way to improve their living environment. Meanwhilst, for Cohousing children, they will be mentored by the adults and learn a variety of skills and knowledge. Some of the Cohousing communities have provided childcare so that the kids would be taken care of by other cohousers. Hence, they will also set a number of activities to enable socialising amongst the members. On the other hand, they still have a private unit where the cohousers can enjoy their private living. This will give advantages in setting a sustainable society without ignoring the needs of privacy in neighbourhood living.

Remarkably, Cohousing has many variations as it is very dependent on community decisions, such as in terms of size, tenure, principles, and system. However, the universal fundamental elements of cohousing are based on these 6 criteria as per determined by McCamant & Durrett. The first one is the participatory process where the community manages to get involved in the development process with help from the experts [9]. This will give an extra advantage for them to design their neighbourhood. Secondly, intentional neighbourhood designs which consist of designing sustainable physical forms, such as paths, green zones, parking, etc. For example, they locate the parking at the edge of the site and people will have to pass each other so that every cohouser will meet and greet each other every day. Next, the characteristic is the extensive common facilities, which are seen to be the centre of the community to meet, share and do multiple activities to enjoy the benefits of being a community. The fourth one is that the Cohousing community must set up their own resident management to make decision for any issue dispute through voting, consensus or by a hybrid approach [17]. The fifth criteria of Cohousing are the absence of a hierarchy where no one person (or persons) has authority over the others, and the existence of leaders is acknowledged amongst the others. Lastly, the sixth criteria of Cohousing is in terms of economic factors where a Cohousing community is not a commune and in that sense every cohouser has to find away to earn his or her own money.

The above elements are only a basic idea of Cohousing and not rigid to be followed as it may vary depending on the suitability of the locality. Thus, the criterion can bring positive impacts to the communities in creating sustainable neighbourhoods. Although the process may take time to implement this concept; however, many case studies from various settlements have shown the possibility to be achieved. Cohousing can be seen as one of the solutions for sustainable neighbourhood adoption towards community living.

3 Sustainability and Cohousing

According to Charles , Cohousing has shown its value as a successful approach to sustainable housing development by assisting in creating more cooperative, ecologically-sound and socially-just communities and helping to minimize our impact on the earth [18]. Furthermore, other scholars have the belief that the Cohousing concept is primarily based on social relationships and incorporates an adequate sense of environmental awareness. Thus, it automatically brings about savings that are economical in nature through a social and environmental interrelation [19]. Also, according to Marcus and Dovey, Cohousing is a high quality and sustainable alternative to the traditional housing option [20]. Indeed it does appear to fulfill some sustainability objectives: strong social networks and social cohesion, pro-environmental behaviour and a greater sense of well-being amongst residents [21].

Thus, will said that Cohousing is one of the mediums to implement sustainable development through a community; whilst the elements itself can offer support and a sense of unity to people [18]. Moreover, other scholars had described the concept of Cohousing as being a benefit to the environment rather than just the main objective in social interrelations. They pointed out that the social and environmental focus in the Cohousing concept automatically brings about savings that are economical in nature [19]. In the meantime, Parker Williams, had agreed by saying that Cohousing can create more sustainable housing, reflecting and embracing nature that can bring inspiration and more livelihoods [22]. Torres-Antonini, Hassel, & Scanzoni on the other hand, had concluded that Cohousing is a model that is able to address both social and environmental values in a highly effective fashion [23]. Table 1 below summarises the sustainable elements that can be found in Cohousing development from various scholars:

Table 1. Sustainable Elements Towards Cohousing

Sustainable elements	Findings		Research References
Social	Vitality and security	Strong sense of safety and security due to every cohouser taking care of each other and together assisting to make a better, safer environment	[24]
	Access	Strong social and civic fabric, including volunteering and informal network accessibility	[25]
	Eliminate social isolation	Strong social networks and social cohesion amongst the community can be built through this concept	[15, 10, 21]
Economical	Generate a lower cost-of-living	Encourage sharing of items like tools, gardens, transportation, laundry facilities, meals, etc	[22, 26, 27]
Environmental	Pro environmental idea	Pro-environmental will enable the community to share resources and result in lower levels of resource consumption including energy, land, goods and waste production	[10, 18, 19, 20, 22, 28, 29]
	Community initiative towards green building	Incorporate and utilise green building materials and techniques, and minimise the impact to the natural environment	[18, 22]
Physical	Design	The design is carefully planned to fulfill community criteria by which the future needs and opportunities are defined	[19 ,22]
	Land use efficiency	Use the material and land sources efficiently, less space consumption, preservation of open space and other remaining land can be prepared for other community activities	[30-31]
	Social centre	Physically, the social centre is the main place where all the cohousers will be meeting and do all their activities together, such as dinners, preparing common meals, meetings, children's plays, etc	[18, 32]

4 Cohousing Development Model

As Cohousing is the most important option to be used in the fulfillment of a sustainable community, therefore, it is important to look over its development models as discussed by J. Williams [15]. The development of Cohousing may vary; however, the two models that have mostly been adapted by cohouser communities are either through new development (grassroot) or reuse of cohousing buildings (retrofit) [32]. The grassroot approach, used for a new development, where resident-led model remains as the dominate one to be practiced by most cohouser settlements other than the partnership or speculative Cohousing model. A resident-led model is mainly funded by the residents with the help of an appointed professional. Under this approach, the residents will have the authority and be directly involved in the recruitment, management, and production processes. Further, they will be responsible for all costs and risks associated with the project and liable for any outgoing costs [10]. They may appoint any professionals to help in consulting in the matter or designing, legal forms and technical expects. At the same time, they also have the authority to select or reject their potential neighbourhood (*self–selection*), who is known to be committed to stay in the community [12]. There are some drawbacks to ponder with this approach especially in terms of funding and managing the project due to its requiring a huge commitment and expertise to complete the project. However, surprisingly, this approach has become the popular choice to be adopted by most cohouser communities, especially in Italy, Great Britain, the US, and many more.

The next approach is through the partnership or speculative model which means the residents will cooperate with the developers in managing their project. This will limit their authority but on the good side, they will spread the finance liability, reduce resident's time in managing the project as well as solve the problems of a lack of resources. The difference between these two models is the developer's involvement; where, under the partnership model, the residents will still lead to a certain process especially in selecting the potential neighbourhood, giving input for the design process, and community visioning. The developer is only responsible to develop and manage all the legal requirements and financing. However, under the speculative model, the developers have the authority to decide on all the production and operational processes. The resident is only liable to form a community exercise [10]. This, however, will lead to establishing cohesive communities after completion since all the operational processes are under the developer's responsibilities.

Meanwhilst, retrofit cohousing is another alternative to be adapted by existing communities who are interested in staying and making their livelihood in the Cohousing. Retrofit Cohousing requires the involvement of making use of existing buildings and creating new Cohousing patterns. Under retrofit, the community can also apply any of the three models as mentioned above but the production process is limited in terms of delivery and the design of the physical structure. Although there is a limitation of the design, the residents can still reduce the level of commitment, risk and resources by essentially working with existing stocks and infrastructure [10]. A summary of the Cohousing development models is show as in Table 2 below:

Table 2. Cohousing Development Model

Development Model	Description	Responsibilities		Note	Exemplar of Existing community
		Resident	Developer		
Resident –led model	Resident leads from the design, process until production together with forming the community	- Design (with help from professional) - Community recruitment - Legal and financing - Community development - Production process	None	- Requires expertise and commitment in funding the overall cost of the project - Still needs professional help in the design stage and building community - Have mutual authority to select potential neighbour	The community Project (Laughton, East Sussesx, England), The Village (Madison, Wisconsin), Rio Selva (Preganziol, veneto, Italy)
Partnership model	Developers and residents work together through all the development processes with the lead from the residents	- Community recruitment - Community development	- Design (with input from residents) - Legal and financing - Production process	- Burden of finance can be reduced by spreading it to the developers - Residents still have authority in deciding their potential neighbours - Give better benefit due to developer's construction experience, familiarity of supply chains and other project management knowledge	Threshold Centre, Village Hill Cohousing, (Northampton, Massachusetts), Doyle Street Cohousing Emeryville, California
Speculative model	Top-down approach where the developers are solely responsible for the development process	- Community development (once living in community)	- Design (with help from professional) - Community recruitment - Legal and financing - Production process	- The community are empowered to build their community after the project completion - All the design, requirements and production are based on the developer's decision - Less time consuming due to the expert being responsible for all processes - The decision power in designing the physical building and other facilities comes solely from the developer	Swan's Market Downtown Oakland, California

Source: Adapted from Williams J., 2008; Davis, 2001(unpublished); M.L. Ruiu, 2015; and researcher, 2015

5 Fitting Sustainable Cohousing In Malaysia

As Malaysia is committed to setting a future direction in ensuring the sustainability of the housing sector as provided by National Housing Policy under Thrust 5, a proposal of sustainable Cohousing is urged to be deliberated. According to this policy, the efforts to enhance the sustainability of the housing sector are implemented by both the Government and the private sector which want to balance the development implementation and the use of an environmentally-friendly housing development concept. Through new technologies and innovations towards sustainability indirectly, there will be an elevation of the quality of life as well as a preservation of the environment. Its also suggested that developers are advised to develop a neighbourhood that promotes sustainability as house buyers generally are willing to pay more to live in a sustainable neighbourhood [33]. Moreover, Tan, pointed out that in order to embrace the concept of building communities, private and public developers should envision the process from the community's point of view since they are the one who will enjoy living there rather than merely building properties [37]. Thus, the agenda is initially related to sustainability and Cohousing elements especially in enhancing social-environmental terms through the communities' participants.

Basically, the Cohousing model is distinct from conventional housing development models where the residents will be the decision makers on how their neighbourhood will be, especially on the design and materials of their future homes, by consensus [14]. Therefore, the collective responsibility of the Cohousing development also needs a commitment from the local planners who are familiar with legal and practical accountability for the development process, builders and architects in the construction process and support from the government in the approval matters. As such, in the UK, their local authorities are committed to giving positive responses towards the development of Cohousing in their local plan, identifying potential sites, using planning exceptions and so on. Meanwhilst, in the US, the Cohousing Association has been established to act as a clearinghouse of information about the Cohousing developments. The city government also helps to encourage Cohousing by creating zoning policies that foster the type of dense development that includes Cohousing. On the other hand, in Australia, the Cohousing development is based on collaboration between government–funded community and jointly with other Cooperative and Public Housing Authorities. This will show their commitment in producing more Cohousing for the locality.

As in Malaysia, the Local Authority, Town Planner, Policy maker and developer will be the important organisation to encourage the development of Cohousing. These parties are directly involved with the community and sector cooperation to ensure sustainability in each development. Adaptation to Cohousing development can be grassroots or retrofit by looking at the suitability in the locality. The best way to adopt this concept is through collaboration between the residents and the developer as the residents are limited in funding and land resources. The government may support this project by assisting in setting new housing and planning policies at local and national levels regarding Cohousing by allowing planning permission on an exception of housing development, which is more sustainable features. The commitment from local authority departments such as the planning and housing division with better joint working amongst the community and greater communication about the Cohousing model and its benefits in local development plans will ensure the successfulness of this concept. Moreover, the Local Agenda 21 Programme (LA21), which emphasises the role of the Local Authority in the implementation of Agenda 21 and the sustainable development to be implemented at the local level, will bring a great opportunity to the Cohousing concept. This agenda was established for the government and local communities in working together towards sustainable development. As Cohousing is referred to as sustainable medium by many scholars, community involvement in building their neighbourhood is a core element in Cohousing, therefore, the objective in LA21 can be achieved.

Moreover, the value of Cohousing seems to be similar with the Malaysian culture; thus, it will make this concept easy to accept and it can be the top option for the people to choose as a place to stay. As such, if we look at the traditional Malay *kampong,* the elements of the social interrelation amongst the neighbours are already established. Houses that are arranged facing each other encourage public surveillance, indirectly creating a safe living environment and a sense of belonging [34]. Furthermore, the social interaction can be found during the conducting of *kenduri*, or feasts held for weddings, celebrating events and other special occasions. This is a special medium for the residents to meet and help each other *(gotong-royong)* in preparing the food, and some of the villagers will bring their own food items to share during the *kenduri* [35]. In terms of decision making, the *ketua kampung* (village head) will be the one who will lead the residents; however, his decision is still an attempt with the group opinion which is held under the community meeting. There are also community centres (*balai raya*) and open spaces where the residents can meet, share leisure time and do their activities together. Sometimes, places of worship like a mosque or temple will be the centre of the community to foster their activities and celebrate an occasion. The other main element is the universal value since Malaysia is a multi racial country and thus, racial groups will respect each other's right to foster their relationships [36]. Since thus elements are alike to Cohousing, then the idea will be not difficult to be adapted by Malaysian. Consequently, it will increase the marketability not only benefitting the society but also the stakeholders as a whole.

The elements of Cohousing depend on the locality; therefore, this research suggested a 7-element idea of sustainable Cohousing to be implemented in Malaysia. This idea is based on the social culture and the suitability of the present Cohousing elements. Details of the descriptions of the Cohousing element idea are presented in Figure 1 and Table 3.

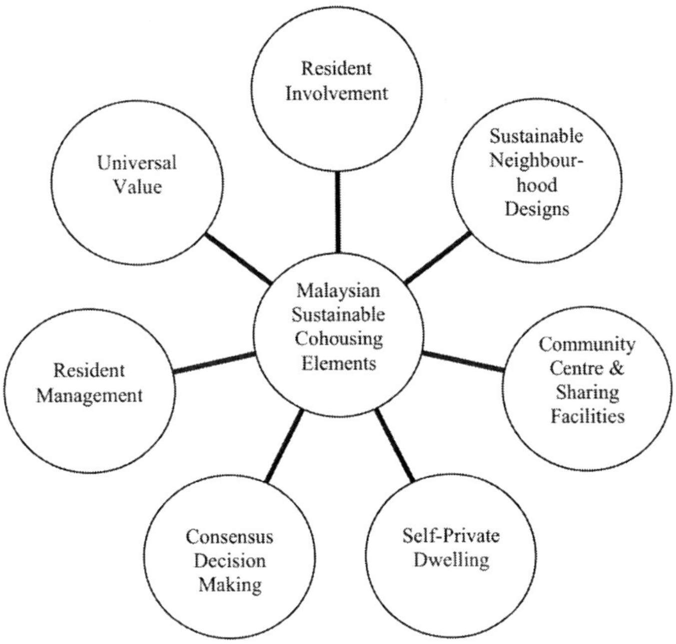

Figure 1. The idea of the 7-Element Sustainable Cohousing in Malaysia

Table 3. Description Idea of the 7 elements of sustainable Cohousing

Malaysian Sustainable Cohousing Elements	Description Idea
Resident Involvement In Development Process	Residents are actively involved in the development process starting from the initial stage to form a sense of community with help from the experts.
Sustainable Neighbourhood Designs	The layout design needs to show a strong sense of community by encouraging social contact, a safe environment, and a friendly and sustainable neighbourhood.
Community Centre And Sharing Facilities	A community centre will be created as a communal place to do their activities and share their resources with adequate public facilities and utilities.
Self-Private Dwelling	Residents still have a right to stay in the privacy of their own individual unit.
Consensus Decision Making	Decisions are based on community meetings.
Resident Management	Self-managing communities are independently governed by the residents.
Universal Value	Good interaction in the multi racial and cultural community with respect to fostering relationships.

6 Findings and Conclusion

In short, this paper has illustrated an alternative housing development concept, other than conventional housing, which is more sustainable to the modern lifestyle in Malaysia. Sustainable development can only be achieved if it can meet three primary objectives which are economically viable, socially acceptable, and technically feasible [3714].

As the population in urban areas keeps growing, the problems of the urban social structure will also be affected by the increasing social isolation, declining social bonding, individualistic community and many more. Therefore, Cohousing is perhaps the most sustainable model to be applied since the characteristics themselves fulfill a sustainable value through the community. Below are suggestions for the conceptual framework of sustainable cohousing to be implemented in Malaysia (Figure 2).

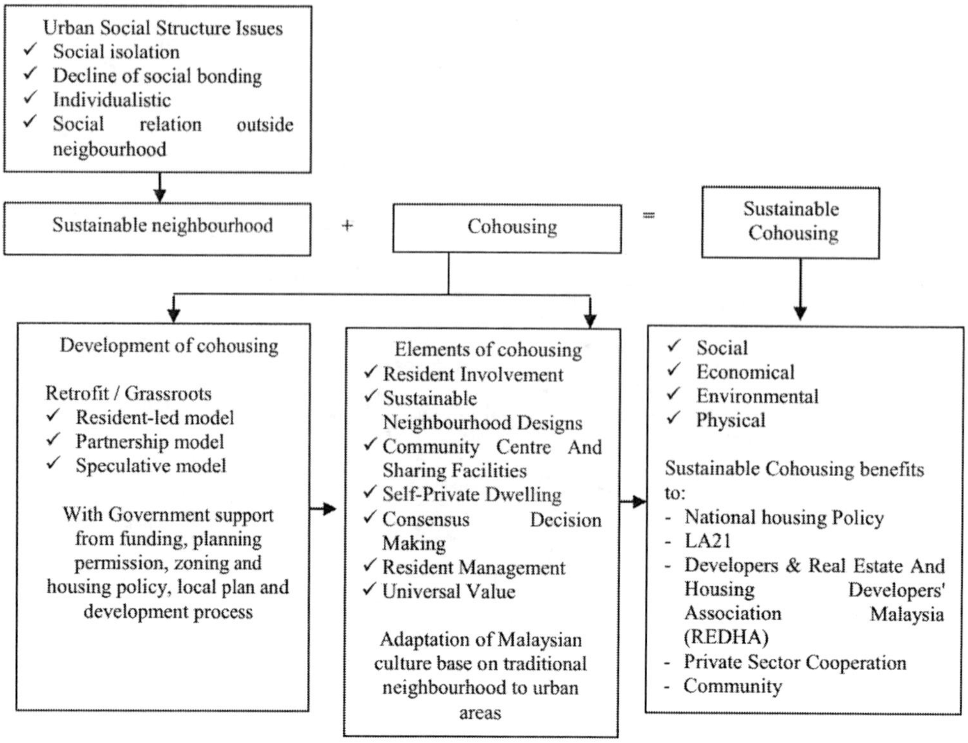

Figure 2. Conceptual Framework of Cohousing Development in Malaysia

Figure 2 above indicates one of the solutions of the urban social structure issue through an effective sustainable neighbourhood. Cohousing is seen to be the best medium of a sustainable neighbourhood that can be practiced. Strong social networks and social cohesion, environmental values in a highly effective fashion, physical design layout to fulfill community criteria in the present and future needs, community initiative incorporated to green building materials, minimise impact to the natural environment and the efficient use of land consumption are examples of Cohousing benefits to sustainability.

Moreover, Cohousing will also seen as a versatile housing concept since it is applicable to be developed either in urban or rural settings by adaption of new (grassroots) or existing settlements (retrofit). The 3 models' adoption must be supported by the government from funding, planning permission, zoning and housing policy, local plan, development process and many more to ensure the successfulness of the cohousing intention. Meanwhilst the elements of Cohousing development in Malaysia must be based on the present and suit the local culture. 7 suggested elements have been proposed in this research by adding the Universal value and Self-Private Dwelling to respect and foster social relations amongst multiracial groups in one community. As discussed by different scholars the Cohousing elements can benefit with a sustainable value; thus, it will assist Malaysia in achieving its sustainable agenda in housing development.

However, this paper has only discussed the basic idea of sustainable Cohousing. The Cohousing development requires more time to develop as all the participants need to really understand the knowledge, commitment, and how the process of Cohousing will be implemented. The findings from this study can be adapted into the future sustainable housing concept and can be presumed as the basis for recommendations to improve social cohesion amongst Malaysians.

References

1. P. Ibrahim, A. Ab Rahman, S. Arni Basir, *Sustainable Economic Development : Concept, Principles And Management From Islamic Perspective*, European Journal Of Social Sciences, (2011)
2. United Nation, *World Population Prospects The 2015 Revision Key Findings And Advance Tables*, (2015)
3. K.C. Seto, B. Güneralp, L.R. Hutyra, *Global Forecasts Of Urban Expansion To 2030 And Direct Impacts On Biodiversity And Carbon Pools*, Proceedings Of The National Academy Of Sciences Of The United States Of America, Pp. 1–6, (2012)
4. Ho Chin Siong, *Urban Governance And Rapid Urbanization Issues In Malaysia*, Jurnal Alam Bina, Jilid 13: No.**4**, (2008)
5. Giddens, A. *Sociology (6th Ed.)*. Cambridge: Polity Press. (2009).
6. Carijn Beumer, *Social Cohesion In A Sustainable Urban Neighbourhood*, Website Http://Www.Icis.Unimaas.Info/Wp-Content/Uploads/2010/07/Beumer-2010-Social-Cohesion-Theory,(2010)

7. Forrest, R., & Kearns, A. *Social Cohesion, Social Capital And The Neighbourhood.* Urban Studies, **38**(12), 2125-2143, (2001)
8. Schoenberg S 1979: *Criteria For The Evaluation Of Neighborhood Viability In Working Class And Low Income Areas In Core Cities Social Problems* **27**: 69-85, (1979)
9. Lietaert M., *Cohousing's Relevance To Degrowth Theories*, Journal Of Cleaner Production, (2009)
10. Williams, J. Predicting an American future for cohousing. *Futures,40*(3), 268-286, (2008)
11. Ruiu M. L, *The Social Capital Of Cohousing Communities*,Sociology, Sagepub.Co.Uk/Journalspermissions.Nav, (2015)
12. Ruiu M. L, *The Effects Of Cohousing On The Social Housing System: The Case Of The Threshold Centre*, Journal House And The Built Environment, 30:631-644, Springer Science Business, (2015)
13. Meltzer, G. (2005). Sustainable Community: Learning From The Cohousing Model. Canada: Trafford.
14. Mccamant, K., & Durret, C., *Creating Cohousing: Building Sustainable Community.* Canada: New Society Publichers, (2011).
15. Williams. J, *Designing Neighbourhoods For Social Interaction*—The Case Of Cohousing, Journal Of Urban Design **10** (3) (2005).
16. Riseborough M., *Work On The Wild Side: For Commissioners And Housing And Social Care Providers*, Housing Learning & Improvement Network, Www.Housinglin.Org.Uk, (October 2013, Retrieve On 29th December 2015)
17. Leafe Christian D. *Creating A Life Together: Practical Tools To Grow Ecovillages And Intentional Communities.* New Society Publishers; (2003)
18. Charles L. Belk, 2006, *Cohousing Communities: A Sustainable Approach To Housing Development, Submitted To Sustainability And The Built Environment* (062gbd101), Uc Davis Extension, Retrieve On 20th November 2015, Website Https://Extension.Ucdavis.Edu/Sites/Default/Files/ Co_Housing.Pdf.
19. Meltzer, Graham. *Cohousing And Ecovillages: A Personal Take On Their Similarities And Differences. In Living Together-Cohousing Ideas And Realities Around The World*, Ed. Dick Urban Vestbro, 105-113. Stockholm, Universitetsservice Us Ab. (2010)
20. C. Marcus, K. Dovey, Cohousing—*An Option For The 1990s, Progressive Architecture* **6** *(1991)* 112–113., (1991)
21. J. Williams, *Sun, Surf And Sustainability—Comparison Of The Cohousing Experience In California And The Uk*, International Planning Studies Journal **10** (2) (2005).
22. Parker Williams, Retrieve On 20th Nov, 2015, Website Http://Bioclimaticx.Com/Wp-Content/Uploads/2012/02/The-Ecological-Design-Of-Co-Housing_2.Pdf,(2012)
23. Torres-Antonini, M., Hasell, M., & Scanzoni, J., *Cohousing As A Basis For Social Connectedness And Ecological Sustainability.* In G. Moser, E. Pol, Y. Bernard, M. Bonnes, J. Corraliza, M. Giuliani, ... M. Giuliani (Eds.), People, Places, And Sustainability (Pp. 123-130). Ashland, Oh Us: Hogrefe & Huber Publishers. (2003)
24. Wright D. & Saville G., *Putting Neighbors Back In The Neighborhood: Strategies For Safety, Urban Design, And Cohousing*, Presented To The 3rd International Cpted Conference, Washington, Dc. (1998)
25. Dezfooly R. M, *Sustainable Criteria Evaluation Of Neighbourhoods Through Residents' Perceived Needs*, International Journal Of Architecture And Urban Development Vol. **3**, No. 2, Spring (2013)
26. Northwest Report, *Cohousing: A Model For Sustainable Living*, Number 19, January 1996 Issn 1040-855x, Http://Www.Smartcommunities.Ncat.Org/Articles/ Cohouse.Shtml, (1996)
27. Mockler, R., *Cohousing-L Discussion*, Retrieved On November 24, 2015 From Http://Lists.Cohousing.Org/Archives/ Cohousingl/Msg22441.Html, (2005)
28. G. Meltzer, *Co-Housing Bringing Communalism To The World? International Communal Studies Association, Communal Living On The Threshold Of A New Millennium: Lessons And Perspectives*, In: Proceedings Of The Seventh International Communal Studies Conference, 2001, Pp. 25–27. (2001)
29. Krokfors, K., *Co-Housing in the Making.* Alexandrine Press. (2012).
30. Zhang Rui And Lv Yanhang, *A New Living Concept Based On Low –Impact Strategy – The Sustainability Of Cohousing Community* , Green Building Technologies And Materials, Advance Materials Research, Vol **224**, Pp 220-223, (2011)
31. Assadourian, *Engaging Communities For A Sustainable World, In Starke*, L. (Ed.) State Of The World 2008: Innovations For A Sustainable Economy. New York: W.W. Norton. (2008)
32. Sanguinetti, A., *The Design Of Intentional Communities: A Recycled Perspective On Sustainable Neighborhoods.* Behavior And Social Issues, **21**, 5-25. (2012).
33. Tan, T.H., *Measuring The Willingness To Pay For Houses In A Sustainable Neighborhood*, International Journal Of Environmental, Cultural, Economic & Social Sustainability, **7**, 1 – 12. (2011)
34. N. Aimran Samsudin , N. Mohd. Afif Osman, *The Significance Of Traditional Malay Kampongs Communal Outdoor Space On Social Relation Case Study: Johor State*, Retrieved On Jan 2016 From Https://Www.Academia.Edu/6292241/The_Significance_Of_Traditional_Malay_Kampongs_Communal_Outdoor_Space_On_Social_Relation_In_Johor (2014)
35. Eric C. Thompson, *Rural Villages As Socially Urban Spaces In Malaysia*, Urban Studies, Vol. **41**, No. 12, 2357–2376, (2004)
36. Ma, Z. B. M. R., & Alwi, E. A. Z. E. *Social Interaction Among Muslims And Buddhists In*

DOI: 10.1051/matecconf/20166600055

Kampung Tendong, Pasir Mas, Kelantan, Malaysia. International Journal Of Business And Social Science, **2**(1), (2014)

37. Tan T. H, *Sustainability And Housing Provision In Malaysia*, Journal Of Strategic Innovation And Sustainability Vol. **7**(1) (2011)

Rethinking Potentials of Public Space and its Management Through Placemaking in Kuala Lumpur

Normah Sulaiman[1,a], Nurul Husna Qamaruz Zaman[2] Hazlina Hamdani[3] and Yusfida Ayu Abdullah[4]

[1] *Centre of Postgraduate Studies, Faculty of Architecture, Planning and Surveying, 40450 UiTM Shah Alam, Selangor, Malaysia.*
[2] *Centre of Studies for Town and Regional Planning, Faculty of Architecture, Planning and Surveying, 40450 UiTM Shah Alam, Selangor, Malaysia.*
[3] *Centre of Studies for Town and Regional Planning, Faculty of Architecture, Planning and Surveying, 40450 UiTM Shah Alam, Selangor, Malaysia.*
[4] *Centre of Studies for Town and Regional Planning, Faculty of Architecture, Planning and Surveying, 40450 UiTM Shah Alam, Selangor, Malaysia.*

Abstract. Public space takes many spatial forms, including parks, the streets, sidewalks and footpaths that connect an edge of a space between buildings or roadsides, hence are important spaces in an urban setting. In a city context, the public space frames the city image. Kuala Lumpur's public spaces are typically made up of shopping streets. It allows an ambivalent space that encourages more activities and interactions through the continuity of the street. For that reason, street that brings out life and movement in an urban space is an ideal type of street. In maintaining public spaces and reinventing it, placemaking is a quest for engagement tool to help strengthening legal frameworks in order to protect the public space which is seen viable at a city level. This paper therefore, explores the potentials of public spaces and its management. It applies the case study method by means of investigating two areas in the city encompassing similar characteristics. Generally, the research is designed to test the theoretical framework in managing the public space and its key dimensions in shaping the quality of public space. The researcher then synthesizes the broad range of development of placemaking and the changing uses of public space. The findings will advance further understanding of a suitable application thus intensify the legal framework that shapes the quality of space and its management.

1 Introduction

Public space is vital to a well-functioning city. Public space is where public is free to access anything that relates to build and natural environment. In this space is where unexpected meetings, public gatherings, point of relaxation happen and something by functional, it's an ordinary passage. Generally, it is also an access point of buildings, connectors to nearby places and a strong city's identity.

Management of public space requires the manager with an ability to understand the issues and relates to the purpose of space. It is a set of processes and practices that keeps the public space a fulfilling and functional space besides managing the interaction between the space and the users. Understanding the types and needs of public space is key and it is also a platform to provide good examples of practice in the management where can be exemplary by others. A mutual profit and concerned on improvements in human and environmental is a priority in seeing good management in coordination.

The aim of this study is to provide evidence of a theoretical framework in managing the public space and its key dimensions in shaping the public space quality. The study will synthesize the broad range development of place making and the changing use of public space. The results from this study will advance further understanding of a suitable application thus strengthening the legal framework that shapes the quality of public space and its management.

This paper is organized into four sections. The subsequent sections with the literature review of the study. The third section provides the methodology, followed by the results and discussion section. The final section offers the conclusion and implications of the results.

2 Literature Review

2.1 The concept of public space

According to Project for Public Space "Great public spaces are where celebrations are held, social and economic exchanges take place, friends run into each other, and cultures mix" (PPS, 2013) [1]. "Public space relates to all those parts of the built and natural environment where the public have free access"

[a] Corresponding author: normajean.ds@gmail.com

(Carmona et al 2004:10) [2]. Carmona understood that access to public space is free for all and it is related to a physical space. Similarly, Claudio (2010) [3], defines public space as extended to communal and no-private arenas regardless of ownership including semi-public, luminal or third spaces. Norberg (2007) [10], clarifies that space represents the physical container of activities and objects while the place is a point where physiological and psychological needs meet. "Space as a social and political product, space as a product that one buys and sells" as formularize by Lefebvre (1991) [4], explains that space has 'the right to be different', lived and become more humanized. Dear and Wolch (1989) [5], argues that social relations can be constituted through space, constrained by space and mediated by space. They explained that site characteristics influences settlement form, physical environment facilitates or obstruct human activity and the frictions of distance facilitates or inhibit is the development of various social practices. This argument shows that physical factor is neither the exclusive nor dominating behaviour influence although one tends to follow what the environment offers them.

2.2 Placemaking

Placemaking is a community approached planning and managing public spaces. "Building community, creating places, using common sense" is the vision of Project for Public Space (PPS) [1]. Places cannot be designed like land marks, however, placemaking focuses on creating spaces that recognize landscapes as part of social constructions that promotes livability, health and well-being. Friedmann (2010) [6], refers to physical spaces that evolve in their own pattern and rhythms of life as a place. The state may control the physical context for activities within a certain time which permits certain public behavior and which is not.

Pioneering in the placemaking approach based on extensive research and projects involvement, PPS finds that a successful place must have four key attributes like accessible, people engaged in activities, comfort and sociable place for people to interact. These key attributes are intangibles qualities required for a high-quality environment based on their analysis of hundreds of public spaces around the world. PPS developed the place diagram which is used as measuring tools to evaluate the value and potentials of a place through its tangible and intangibles qualities. Almost all writers, researchers and urban planners present their literature dealing with desirable qualities of public space. Lynch (1979) [7], simplified five desirable qualities in public space such as control, fit, sense, vitality and access.

Shinjuku Station, Tokyo is listed as great public spaces by Project for Public Spaces using the place diagram framework. Entrances to and fro the station is open to all streets and connected radially from it to many parts of the downtown where trains are connected to the whole city. Shinjuku station is connected and accessible from all streets with easy navigation within and towards as well as with choices of activities and there are no dead spaces in this station. Even during peak hour, Shinjuku Station is very clean and comfortable to be utilized. Besides safe, Shinjuku is the most common meeting placemaking it the most sociable place to be with easy access and trains passing from all parts of the city (PPS, 2013) [1].

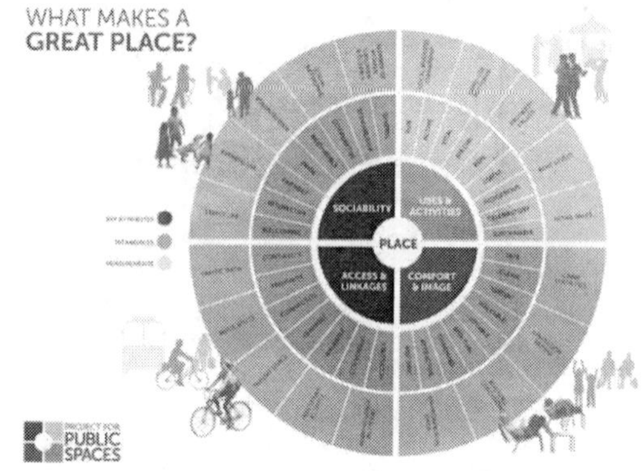

Figure 1. Place-diagram of what makes a great place by PPS. Source: PPS (2013), [1].

3 Methodology

3.1. Case Study

Case studies on local urban areas will be the substructure of the subject matter highlighted. These areas are selected based on the similar activities within the district, its usage and activities and its adjacent location. The area inside the building functions as private in nature is not included in the scope of the study. This is an indicator that the study only encircles the relationship, value and input of public space and its management dimensions. The rationale of selecting the case studies is to initiate preliminary experimental subject to bigger scope that can be developed further in the future.

Streets in an urban context are places of economic and social significance. An urban fabric that remains longer then the people whose' live constructed through streets and block structures of development. Spaces in the East and the West induce different uses of spaces. Kurokawa (1994) [8], philosophy's states "street spaces exist between private and public space and between residential and commercial space, possessing the characteristics of both the former and the latter types of space. Asian street, by contrast, is not so clearly defined, it is harder to tell where it begins or ends, and it generates responses to innumerable variations with time. Malaysia's early settlements begin with 'main streets' that activities foster social interactions within crowds engaging with selling and buying activities (Shamsuddin and Ujang, 2008) [9].

3.1.1 Jalan Tuanku Abdul Rahman (Jalan TAR)

An area that has a strong sense of character with heritage buildings, contemporary architecture and preserved facades, this area is famously known with the Malay community commercial and trading area. Evolving with the urbanization, Jalan TAR is developed as commercial vicinity that also represents the Kuala Lumpur old city centre with the core function of this road is retail offering a variety of consumer goods. This street consists of office, shopping malls, arcades, hotels and restaurants.

3.1.2 Jalan Bukit Bintang (Jalan BB)

Jalan BB is known as the premier shopping street, comprehensive shopping and entertainment development. This beating heart of the city consists of the promenade of restaurants, cafes, malls and acted as the main commercial hub within the golden triangle. By the year 2020, this area will be a commercial and premier tourist precinct.

3.2 Data collection

This study uses questionnaires to gain data from the public for both case study areas. The questionnaire is divided into five sections; structured and organized using the place diagram intangibles variables. Convenience sampling technique is applied and is useful in documenting the quality of phenomenon of the study which allows the researcher to obtain basic data and trends within case study areas. The target population within the area can be accessed easily and conveniently and sample size for each area was not less than 100 respondents.

The second part of data collected for this research was via interviewing the key people in Kuala Lumpur City Hall, The Town and Country Department and National Landscape Department. The data collection was completed in 14 days. The key people various local government agency was selected based on their experience, knowledge and position in the department. The researcher presented the open-ended question to each personnel.

4 Results and discussion

Core components of urban design relate to the understanding of the relationship between people or society and their environment where human behavior is influenced by the physical environment actively and changing the environment as it influences and changes them too. It thus can be seen as a two-way process of creating and modifying spaces while at the same time influenced by those spaces. The place diagram tool was used to evaluate both physical condition (Table 1) and user experience (Table 2) of the two shopping streets using the key and intangible attributes of the place.

Table 1. Physical conditions findings for shopping streets

Intangibles Attributes	Components	Jalan TAR (N=100)	Jalan BB (N=100)
Access and Linkage & Comfort and Images	Condition of sidewalk	Satisfactory	
	Sidewalk clearly define	Very Good	
	Enough room to walk	Good	
	Condition of streets	Satisfactory	
Uses and Activity : Uniqueness	Space is suitable for public performance (i.e. :flash mobs)	Good	
	Space appears special	Satisfactory	
Sociability : Diversity	Space is favorable for celebration	Satisfactory	
Response format : 1= Very Poor, 2= Poor, 3=Satisfactory, 4=Good, 5=Very Good			

Table 2. User experience findings on shopping streets.

Intangibles Attributes	Components	Jalan TAR (N=100)	Jalan BB (N=100)
Access and Linkage : Walkability	Pleasant walk	Neutral	
	Continue walking despite weather	Neutral	
Uses and Activity : Uniqueness	Welcoming feeling	Neutral	
	Enjoying public art features	Neutral	
	Return to the place again	Neutral	

Sociability : Diversity	Allows social interaction	Neutral	
	Takes picture around the place	Neutral	
	Chit chatting at the place	Very Likely	
Comfort and Images : Cleanliness	Any Litter along the way	Neutral	
	Graffiti on the streets	Neutral	Unlikely
Response format: 1= Not at all , 2= Unlikely, 3=Neutral, 4=Very Likely , 5=Extremely Likely			

4.1. The Place-diagram intangibles attributes.

4.1.1 Walkability

The physical condition of the place shows that in both study areas the sidewalks is provided clearly for users to use and in satisfactory conditions for walking. This is meaningful to indicate that sidewalks in both areas are comfortable enough to link users to the adjacent places. The response indicates that the sidewalks in both areas are comfort enough to accommodate high numbers of users without clashes. Results also show that there are distinguish significance in walking experience for users which appear to be only satisfactory and does not offer more experience that just walking.

In both study area, it is understood that space is well connected and easily access. Users in both areas find walking is convenient and the scene activities surrounding the context constantly changing and it offer shorter trips to most places. Users will still continue walking despite the weather affecting the areas which implies that this space is easy to walk. This clearly shows that the study illustrates that the street is functional and act as a mode of movement for users. Still, the study also indicates that small group of users walking experience appears to be very unpleasant.

4.1.2 Uniqueness and Special

For key attributes uses and activities, the study evaluates the uniqueness and special level within both areas of the research. The response for this question proves that both study areas appear to be lively with activities and distinctive in its own way. Although the core activities are similar, Jalan Bukit Bintang offers more participation and activities for users to keep as memories. Indirectly, it gives Jalan Bukit Bintang a stronger sense of place and special in the eyes of the users. The users enjoy the character of this street and everyday life that takes place in between the buildings and its atmosphere. In both study area, the activity pattern appears to be very significance with its sense of place that exist perhaps for many years and still intact with the surrounding despite changes taken place. People are seen interacting with one another even in small pocket space and enjoying their walk to connecting places.

4.1.3 Friendliness and Diversity

For key attribute of sociability, the researcher measured the friendly and diverse variable within both research areas. The response for these variables indicates that users in both areas include themselves in passive and active activities such as listening, chatting, singing and profiling people. Activities as such can be seen done in a small and medium group or even individually in both areas. The results are also meaningful when users return to the same place to be participating in activities in this area with friends, a family with a happy smile and proud feeling. The reaction of users in both areas indicates both study area is a good place to be. The second set of the results implicates that space is favorable for public gatherings such as celebration. An optional activity is a new attraction to places especially for Jalan Bukit Bintang, where the urban culture is more prominent and the quality of it appears to be high as well.

4.2. Management of public space.

4.2.1 Jalan TAR

The interview findings will help the study to look at the potentials of public space from the management perspective, formulated based on the findings of key attributes that were used in the questionnaires. From the interview, it was discovered that the management suggested that having more shaded areas physically, providing interactive landscape elements and more public realms will help to enhance the walking experience. Suggestion includes, to creatively implements landscape psychology in public spaces to improve the area. The management perceives the user experience in the place as normal behavior as space is apparent to be functional and does not offer more towards users, hence it was a satisfactory.

The sense of place was seen as an important factor from the management perspective because Jalan Tuanku Abdul Rahman is a focal point with good transportation connection for tourist's attraction. The activity pattern in this area was the main reason for users to come to this area. The commercial activities have been around for years offering a comfortable feeling to users who want the pattern of activities to continue and feel safe doing it. The interview responses clearly indicate that the cleanliness of this area is maintained by the authority that ensures the place is kept clean with no damages. They also suggested that the government should privatize the area to make it better by adding elements that can be easily maintained and sustainable so as to blend well with the surrounding context.

4.2.2 Jalan BB

From the interview, it was determined that there were more pocket spaces for public in the area. The management feels that the spaces are spacious and comfortable for users hence more creative solutions must be implemented in improving users' experience by creating more activities and events. This area is privately managed and operated. The feedback from the general public relate to the design process of space which was functional for users and private sector or provider. It was suggested as well that space lacks of unique landscape elements and the manipulation of landscape psychology which users can add to their normal daily routine, therefore changing their routine into a more exciting experience. The private sector must therefore keep the area interesting, clean and well manage.

The sense of place here differs from Bukit Bintang, whereby Bukit Bintang, a highly concentrated urban area attracts chic modern urbanites with clean and interesting atmosphere besides being popular shopping street. Any upcoming elements complement the idea of an urban area, and should be strategically placed so as not to clutter the tight space. The interview prompted the management that activity pattern here is very favorable by many with different quality and more diverse. Management believes that the owner or private sector has to be creative to create activities to attract more visitors. In achieving a successful public space, the interview finds that the authority and everyone plays an important role. Form and function are an important application to attract visitor to utilize space, be it local and foreign, at the same time space must be maintained by authorities, comfortable enough for users to carry out daily activities and appreciated by them. The survey also revealed that this perhaps can be a joint venture effort or privatized to achieve certain objective and purpose.

5 Conclusion

From the study, it is found that there is no proper management framework for public space. The place diagram is used as a tool to synthesize the broad range of development of placemaking and the changing use of public space. This paper addresses the potentials of public space and its management lies in the intangibles qualities of the diagram. In managing existing public space, a dynamic collaborative effort by all is vital. From this research, it unleashes the potentials of public space which focuses on the end users. Here, management gives the rights to the public to use, access and owned or control the space according to as the definition permits. Results of this research suggest scholars and professionals to explore key attributes of placemaking strategies using the place-diagram as a tool in the future.

Acknowledgements

The authors would like to thank the Malaysia's Ministry of Higher Education for funding the study through a RAGS grant with the project code 600-RMI/RAGS 5/3 (170/2013) and also would like to thank the Universiti Teknologi MARA (UiTM), specifically, the Research Management Centre (RMC) for supporting and managing this grant.

References

1. PPS (2013). Project for Public Spaces | Placemaking for Communities [online] Available at http://www.pps.org [Accessed: 26 April 2013]
2 Carmona, M., MagalhãES, C. D., & Hammond, L. (2008). Public space: the management dimension. London, Routledge.
3. Claudio De Magalhães (2010). Public Space and the Contracting-out of Publicness: A Framework for Analysis, Journal of Urban Design, 15:4, 559-574.
4. Lefebvre, H. (1991). The production of space, trans. Donald Nicholson-Smith (1974, 282).
5. Dear, M., & Wolch, J. (1989). How territory shapes social life. The power of Geography, 3-18.
6. Friedmann, J. (2010). Place and Place making in Cities: A Global Perspective. Planning Theory & Practice, 11(2), 149-165. http://dx.doi.org/10.1080/14649351003759573
7. Lynch, K. (1979). The image of the city. 1960. Cambridge, MA.
8. Kurokawa, K., & Kurokawa, K. (1994). The philosophy of symbiosis. Academy Editions.
9. Shamsuddin,S., and Ujang, N.(2008). Making places: The role of attachment in creating the sense of place for traditional streets in Malaysia. Habitat International, 32(3), 399-409. http://dx.doi.org/10.1016/j.habitatint.2008.01.004
10. Carmona, M. (2010). Public places, urban spaces: the dimensions of urban design. Oxford, UK, Architectural Press.
11. Matthew Carmona (2010). Contemporary Public Space, Part Two: Classification,Journal of Urban Design, 15:2, 157-173.
12. May, T. (2013). The Urban Everyday: The History and Dynamics of Place Making. Symbolic Interaction, 36(2), 234-236. http://dx.doi.org/10.1002/symb.57

Solar Village in Malaysia – A Route Map for Financing Mechanism

Nur Azfahani Ahmad[1,a] , Hugh Byrd[2], Nurul Fadzila Zahari[1], Suriani Ngah Abdul Wahab[1], Aida Affina Abdul Ghani[1]

[1] *Faculty of Architecture, Planning and Surveying, UiTM Perak, 32610, Perak, Malaysia*
[2] *School of Architecture, University of Lincoln, Brayford Pool, LN6 7TS, United Kingdom*

Abstract. Malaysia's development has been largely financed from its fossil fuels resources that come principally from the South China Sea. This has enabled electricity to be distributed equitably throughout society and improved the quality of life for people in Malaysia. However, the depletion of fossil fuel resources and the international conflict over South China Sea territorial ownership may leads to inadequacy of supply in the future, and may give direct impact to people in the rural areas. Malaysia's greatest potential for solar energy comes from photovoltaics (PVs) and the large roofs of village houses offer significant potential to contribute electricity both nationally and to the local communities. However, this technology is still unfamiliar to the people living in local villages due to the subsidised electricity by TNB and high capital costs of PV systems. This paper investigates whether establishing a "solar village" is feasible in Malaysia. Through a cost feasibility analysis, the potential of harnessing solar electricity for local villages in Malaysia has been demonstrated. A novel mechanism has been presented that allows rural households to own PV panels, rent the grid cables and to share surplus electricity within the community through a three (3) way financing routes map.

1 Introduction

As 2011 ends, Malaysia[a] has recorded an annual GDP growth of 5.6% (The World Bank, 2013) and an increasing rate of residential areas for both urban and rural areas by 20% (Department of Statistic Malaysia, 2011). This situation has influenced the rapid market for electrical appliances ownerships (Saidur, Masjuki, & Jamaluddin, 2007) which contributed towards the increased demand of electricity. It is projected that, in the next decade, electricity generation is expected to increase more than 40% of global energy consumption (EIA, 2013). Rural areas in Malaysia with the population of almost half the nation population have recorded 93% of electricity coverage (TNB, 2009), marking a good record of electricity supply for Malaysia, compared to other Asian countries. APEC (2013) projected that the electricity generation in Malaysia will continue to grow at an average rate of 2.8% every year. The generation is projected to grow from 105.1 TWh in 2009 to 217 TWh in 2035 (APEC, 2013).

However, as development continue to rise and people keep on consuming electricity, it has rise many associated issues. Among the issue is how Malaysia can fulfil the growing demands for energy over the future. With its economic growth and increasing population, this will lead to a substantial increase in electricity demand. With the depletion issues of fossil fuels (Nel & Cooper, 2009) and increased costs of fuels for electricity generation (Fantazzini, Höök, & Angelantoni, 2011), Malaysian Government is now gradually reducing the subsidy (Gardner-Stephens, 2013; Hamid & Rashid, 2012). It is projected that, in the future, electricity cost will be expensive (Shagar, 2013) and people live in rural areas will experience a tough life due to this situation (Ahmad & Byrd, 2012). The rising cost of electricity could result in poverty. This is supported by Pellegrini and Tasciotti (2012) who have implied that the cost of electricity distribution to the low population densities results in a high cost for each unit of electricity consumed. Therefore, it is necessary to consider and project future trends of electricity consumption in Malaysia, in order to expect a suitable share of electricity for the people in the future. Fundamentally, in order to cater this issue, renewable energy resources need to be taken into the account..

2 Future Issue of Rural Electrification in Malaysia

Barnes (2005) has highlighted that the poverty line in rural areas has declined in many developing countries due to the effective in electricity coverage for these areas. This has been proven in Malaysia where with the effective rural electrification programme by the Government and Tenaga Nasional Berhad (Ab Kadir,

[a] It has been estimated that, by 2030, the population in Malaysia will have increased to 37 million people and that this will also affect the rate of GDP by up to USD850 billion per capita (APEC, 2006).

[a] Corresponding author: nuraz020@perak.uitm.edu.my

Rafeeu, & Adam, 2010), it has reduce the income disparities between rural and urban people, increase the quality of life and improve the education processes. Electricity has empowered rural businesses, help to operate farm machineries, lighting households, street lighting and public buildings. This has created electricity-depended lifestyles among rural people in Malaysia.
However, since electricity generation in Malaysia is dependent to oil and gas, fluctuation in fuel prices will give a direct impact to the power cost. This is argued by a Malaysian energy observer organisation, MyPower Corporation (2013), which indicated that the actual cost of electricity would spike to MYR 40 cent/kwh if the fuel subsidies are to be removed in the year 2016. A complete dependence on fuel resources to generate electricity exposes people to variation in price and unreliable service which leads to uncertainties of electricity supply (IEA, 2013). These uncertainties may lead to other associated problem, like poverty and energy security issues. Since rural people in Malaysia are fully reliant on centralised electricity provided by the Government, any concurrent issues of electricity supply will give them full impact especially if the issue of subsidy removal reaches the national level. With household income of MYR 750.00 per month and below (PEMANDU, 2011), many rural people do not have the capacity or means to establish basic necessities other than relying to electricity provided by the Government. Thus, it is necessary for the Government to consider a strategic plan on establishing alternative energy resources for these people, in facing any energy issues in the future.

3 Solar Village : The Potential

The potential of utilizing solar energy for generating electricity through photovoltaic (PV) panels on Malaysian homes is substantial, with an average solar irradiance per year of 1643 kWh/m2 (Chua, Oh, & Goh, 2011). Many scholars have identified that solar energy power which comes from PV panels could readily meet up to 20 % of Malaysia's energy needs (Chua et al., 2011; Muhammad-Sukki, Ramirez-Iniguez, Abu-Bakar, McMeekin, & Stewart, 2011).

Developing countries like China, India, Nepal and Bangladesh have already implemented plans to generate solar-powered electricity for rural dwellers. Solar electrification in these countries helps to create business opportunities for local people and provide an independent practice that can help rural dwellers to own PV panels (Barnes & Foley, 2004; Chaurey & Kandpal, 2009; Islam & Islam, 2005; Subhes, 2006).

In the case of Malaysia, the success of the 5th Fuel Diversification Strategy energy policy in the 9th Malaysia Plan (2006-2010) (Malaysia, 2006), has introduced the integrated solar panel technology to residential and industrial sectors (Mekhilef et al., 2011; F. Muhammad-Sukki et al., 2011). However, it only reached extensively to the urban area. This technology is still unfamiliar to the rural people (Haw, Sopian, Sulaiman, Hafidz, & Yahya, 2009). Jamaludin (2009) highlighted that even though solar energy has great potential in Malaysia and becomes known to the people in urban area, the capital cost for implementing this technology is still considered high for rural people.

The potential for utilizing solar electricity in rural Malaysia is high. Byrd (2010) has argued that rural houses in Malaysia; if installed with solar panels, could generate and meet about 25% of current electricity demand due to its large roof area. This is a significant proportion of the electricity generation mix for Malaysia. The large roof areas can help in receiving more solar irradiance thus generating more solar electricity than the urban housing. Since Malay houses have a typical rectilinear form (Yuan, 2010), the collection area for solar energy is more determined by the orientation and inclination of the roofs than their geometry

With roof pitches in Malaysia commonly angled at 45° or less (Fee et al., 2005; Yuan, 2010) and located near to the equator line, all orientations of the roof have the significant area for solar energy collection. As most rural houses are built on high stilts and located in a large yard, there are minimal issues on overshadowing from trees. With more than 3.5 million rural houses (Department of Statistic Malaysia, 2010), abundant solar energy can be collected from rural roofs, allowing it to be shared widely to other people in the community.

Through rural community, solar village can be created by establishing a community micro-grid, where, the electricity generated from PV panels can be transmitted into the households from the power producer (Frame, Tembo, Dolan, Strachan, & Ault, 2011). A smart meter can be installed to monitor the electricity consumption. There are many successful international solar co-operative projects that can be followed, for example Westmill Solar Co-operative Project in the UK (Westmill, 2013), SolarShare Co-operative Project in Canada (SolarShare, 2013) and Fukushima Solar Village in Japan (Hsu, 2014). These projects have established community involvement by founding a solar co-operative agency for the rural people. This approach would allow solar energy to be shared by multiple people in the same community (Greenius, 2010). Residents with limited financial capacity or spaces can have access by sharing solar energy facilities with other community members (Iler, 2012) (See Figure 1). This is economical for the people rather than each member of the community investing one whole set of solar panel that is expensive.

With the total cost of solar PV energy being still far beyond the reach for the rural people in Malaysia, it is essential to establish a suitable financing mechanism that can help to spread this technology to all people in rural Malaysia. A case study of a typical village in Malaysia has been conducted to idealistic the solar village concept (See Figure 2).

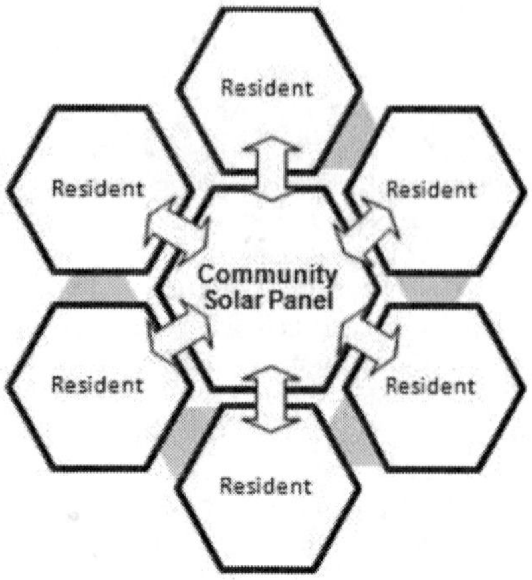

Figure 1. The Concept of Solar Electrification Community (adapted from author's PhD thesis).

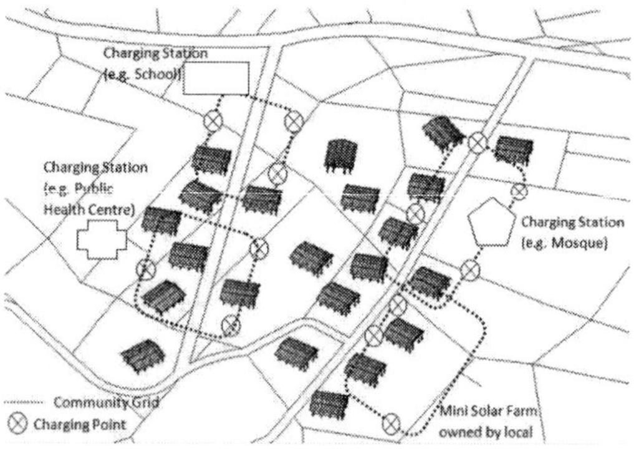

Figure 2. The Concept of Solar Village Community – A Case Study from a Typical Village in Malaysia (adapted from author's PhD thesis).

Public buildings (for example, mosque, school or public health centre) in the village can operate as a solar charging point which allow nearby dwellers to access solar generated electricity from a community grid (i.e. by paying fees for charging their own battery bank (using the Indian 'fee-for service' concept) (Chaurey & Kandpal, 2009). This can help rural dwellers with limited financial capacity to access solar generated electricity. Houses with solar PV panels can be installed with smart meters to assess the electricity used and to identify the consumption rate which allows price charging based on the time of day. This is depending on solar incentives scheme provided by the Government. Other houses need to have an inverter charger to connect the battery bank.

4 The Methodology

A combination of qualitative and quantitative methods has been conducted based on the aims of this research and has been categorised into the sequence of :-

- Survey research method which contains main questionnaires for rural respondents
- Load profile analysis of rural houses using load meter.
- Supporting questionnaires – In order to gather opinion and measuring trends on electrical appliances of rural households.
- Solar yield analysis for roof' samples.
- Field study – A case study approach to gather data on a load profile of a village.
- Desktop study – An analysis to investigate the potential of rural solar community, together with a numerical analysis on cost feasibility analysis for the investment of a rural solar community.

Figure 3 illustrates the research methodology framework for this study.

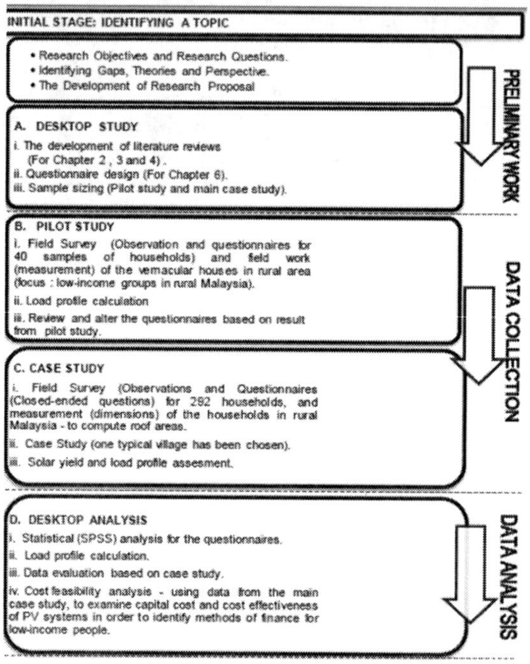

Figure 3. The Research Methodology Framework (adapted from author's PhD thesis).

5 The Route Map to Finance a Solar Village

To establish whether there is adequate electricity produced for the community, a load profile analysis is presented. This analysis assumes that one or more community buildings will have PV systems and a certain proportion of the households in the village. The electricity demand of the whole community is then compared with the electricity generated by PVs. The cost feasibility analysis has been carried out in order to estimate the financial benefit of PV panels installed on rural roofs. Figure 4 shows the proposed route map for financing a solar PV community project for a village.

DOI: 10.1051/matecconf/20166600057

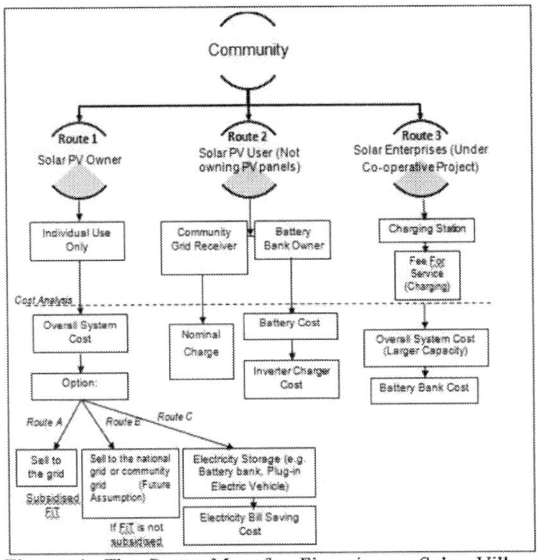

Figure 4. The Route Map for Financing a Solar Village (adapted from author's PhD thesis).

5.1 Route 1

In this study, a 4 kWp PV system has been considered to be used for Route 1 and calculations were based on the pattern of dwellers' occupancy of as follows:-

Category (i) About half leave for work and about half stay.
Category (ii) Most leave for work and one or two stay behind in the day; e.g housewife and children.
Category (iii) Everybody leaves the house in the morning and returns at dusk.

Figure 5 has described the load profile for each category which tabulated the energy used and projected excess energy that can be sold.

In 2011, the overall cost of PV panels for a double storey house is estimated at RM 45,000 (MBIPV, 2011; F. Muhammad-Sukki et al., 2011). After 2013, it is assumed that the installation costs has been minimized to 10% annually (Ahmad & Byrd, 2013) making it now between MYR 30,000 to RM 36,000. In addition, the PV system is now locally made in Malaysia (Gading Kencana, 2013) making the cost for the panels cheaper than the imported panels. In order to lower the initial cost, PV panels can be bought by bulk, and village people or local cooperative community can be trained to carry out the installation of the system.

As recommended by Balfour, Shaw, and Nash (2013), lead-acid batteries are suitable for home-PV systems since it is cheaper than the other types of battery. Since home-owners can get the electricity from the grid, they can have an alternate option in using the electricity supply either from the grid or the battery based on their needs (Balfour et al., 2013). Table 1 show the monthly saving rate after PV system is installed based on categories of occupancy.

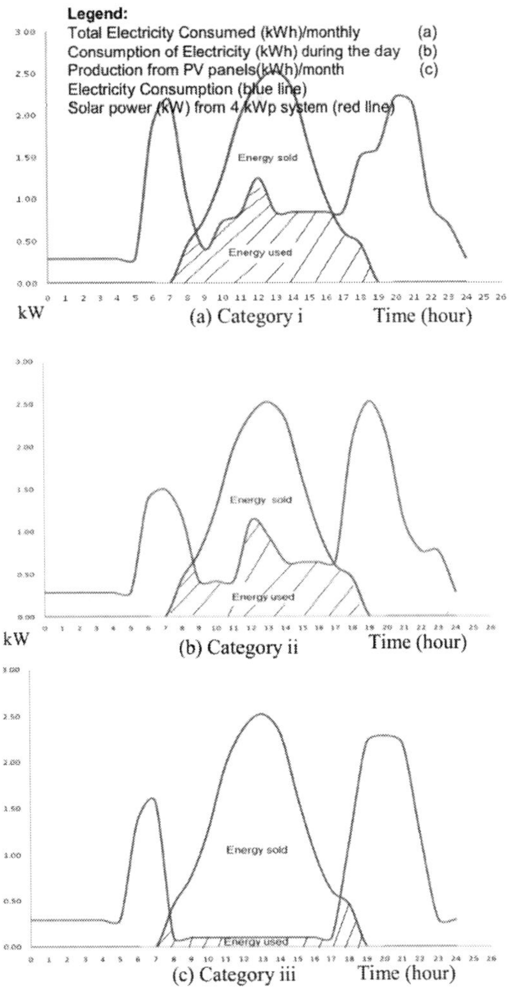

Figure 5(a), (b) and (c). The Load Profile for Every Category of Occupancy

Table 1. Monthly saving rate after PV system is installed in the house.

Category	Total Electricity Consumed (kWh)/monthly *(a)**	Consumption of Electricity (kWh) during the day *(b)**	Production from PV panels (kWh)/month *(c)**	Conversion of electricity saving (kWh) to financial saving (MYR) *[(b) x Malaysia's Electricity tariff 2014 (MYR)]***	% of Cost Savings
(i)	420	210	420	**46.94**	34
(ii)	390	105	420	**22.89**	19
(iii)	270	30	420	**6.54**	10

* Refer Figure 4
** Based on current Malaysian electricity rate (E.g. For 210 kWh [First 200 units x MYR 0.218 = MYR 43.60] + [Next 10 units x MYR 0.334 = MYR 3.34] Total of MYR 46.94)

If the electricity generated from PV is used by the household, the homeowner will only need to pay the electricity bill for the remaining bill which is MYR 46.94, MYR 22.89 and MYR 6.54 respectively with a total cost saving of 34%, 19% and 10% for each category. Table 2 and 3 has indicated the income that can be gained, if the home owner intends to use and then, sell the surplus of energy from the solar electricity.

Table 2. Combination use (use and sell) for a 4 kWp PV system.

Category	Electricity Produced by PV Panels (kWh)/month *(a)*	Consumption of Electricity (kWh) during the day/monthly *(b)*	Surplus electricity generated and sold to the grid (kWh/month) **(a) - (b)**
(i)	420	210	210
(ii)	420	105	315
(iii)	420	30	390

Table 3. Total Income from Feed-in Tariff (FiT) Scheme (For a Combination of Uses).

Category	Consumption of Electricity (kWh) during the day /monthly	Cost of electricity (MYR)* *(a)*	Surplus electricity generated (kWh/ month)	Income from FiT**/ month *(b)*
(i)	210	46.94	210	266.70
(ii)	105	22.89	315	400.05
(iii)	30	6.54	390	495.30

* Based on Malaysia's electricity tariff 2014 (First 200 units: MYR 0.218)
** Feed-in Tariff (FiT) Rate is MYR 1.27

Rather than selling the surplus of the electricity to the Feed-in Tariff (FiT)[b] scheme, the electricity can also be sold to 'peers' using a community grid. Martinot (2014) has introduced a "peer-to-peer" energy model that allows an energy sale among people in a community. Through this scheme, the PV owner can sell their electricity to the neighbours based on nominal charges for the use of local grid (Martinot, 2014). It helps PV owner (supplier) to share or sell electricity directly by 'renting' the cables of the central grid. The 'supplier' must be registered as a 'generator' in order to utilise the central grid cables which are rented from the grid. The supplier (via the grid operator) must meter the electricity exported to the grid and this is compared with the electricity imported by the buyer (see Table 2). At the moment, there is no energy policy in Malaysia that allows the 'supplier' to rent the cables of the central grid to export the generated electricity to the community. Therefore, it is essential for the Government to take this matter into account in order to establish a workable solar village community.

5.2 Route 2

[b] Feed-in Tariff (FiT) is an energy payment scheme established to encourage small-scale energy provider to sell their generated electricity (from small-scale plant; e.g solar panel) to the Government (SEDA, 2012)

Route 2 is for PV user who does not own a PV system, but wants to access the electricity generated from it. They can benefit from the 'peer-to-peer' grid or community grid. For this route, the use of batteries would be essential to benefit the usage during night-time. In this case, they need batteries in order to store electricity collected from the charging point. Table 4 explores the significant calculation for the batteries (based on the categories of occupancy).

Table 4. Analysis on the Battery Bank

Category	Total Daily Use (Wh / day)	*Days of Autonomy (2 Days)*	Depth of Discharge at 50%	*Ambient Temperature Effect (27° C = Factor 1)*	*Amp-hour(Ah) Using formula (P/V = I)**
(i)	14000	14000 x 2 = 28000	28000/0.5 = 56000Wh	56000 x 1	56000/48 = **1167 Ah**
(ii)	13000	13000 x 2 = 26000	26000/0.5 = 52000Wh	52000 x 1	52000/48 = **1083 Ah**
(iii)	9000	9000 x 2 = 18000	18000/0.5 = 36000Wh	36000 x 1	36000/48 = **750 Ah**

* P = Power (Wh), V= Voltage, I = Current (Ah), Source: BTEK Energy (2012)

Even though the cost for owning a battery is considered high for low-income people, the amount is invested only in the early stage of usage. The batteries have a typical lifespan of between 10 to 15 years depending on the types and frequency of usage (Balfour et al., 2013). Table 5 has listed the overall cost for battery banks.

Table 5. The Battery Configuration

	Battery Capacities Needed (Amp-hour)	*Suggestions of Type of Battery and Configurations*	*Number of Batteries Needed*	*Price Per Unit (MYR)**	*Total Cost (MYR)*
(i)	1167	48V, 600 Ah (Lead-Acid) Parallel Arrangement	2	4, 750.00	9, 500.00
(ii)	1083	12V, 546 Ah (Lead-Acid) Series and Parallel Arrangement	4	2, 300.00	9, 200.00
(iii)	750	12V, 400 Ah (Lead-Acid) Series and Parallel Arrangement	4	1, 700.00	6, 800.00

* Based on various supplier price in Malaysia

5.3 Route 3

Route 3 is used to establish a solar enterprise that can be utilised by people in the village, micro-enterprises or co-operative agencies who own PV systems. The main characteristic of this route is to provide a community grid with a charging station that can be linked with battery banks. Brumback (2013) has argued that the solar enterprise can act as a charging station by using public buildings in the area, for example a school, or people who whom owns land that can be incorporated into a mini solar farm. By arranging a certain charging fee, local people can charge their battery banks from these charging points. Many international project has been using this concept, for instance, a public street charger in New York (Street Charge, 2013), solar lantern project in India (Chaurey & Kandpal, 2009), solar micro-enterprises in Bangladesh (Barua, 2001) and several micro grid projects in remote Kenya (PowerGen, 2014).

By referring to Figure 2, the existing electric pole can be used to locate charging points and can be rented from the main utility. The charging station would be community owned and since public buildings are reasonably centrally located, they would set up charging points within a radius. It would be better if the power distributed was in Alternating Current (AC) circuit, in order to save the cost of inverters, to reduce distribution losses and to be useful for electric vehicle charging. The charging fees can be setup by establishing an agreement between the generator (solar enterprise) and the receiver (the households and other users) – the same concept applies in Route 1. The price of electricity can be determined by the community.

Since the price of the system is typically 16 times beyond the financial capability of rural households (see Table 6), it is essential to consider the financing mechanism that can help them. There are a few financing schemes that can be integrated into financing a PV systems (Refer Figure 6 and Table 7).

Table 6. Indicative Cost for Setting up a Charging Station For a Solar Village

System	Cost (MYR)
System cost (20kWp)** – inclusive with inverter/charger and installation cost*	64,000.00
Battery bank*	40,000.00
Cables and Charging Point*	50,000.00
Overall Cost (MYR)	154,000.00

*Price from Gading Kencana (2013)
**For this analysis, 20 kWp PV system is used; which significant with larger roof area (mosque, school, etc)

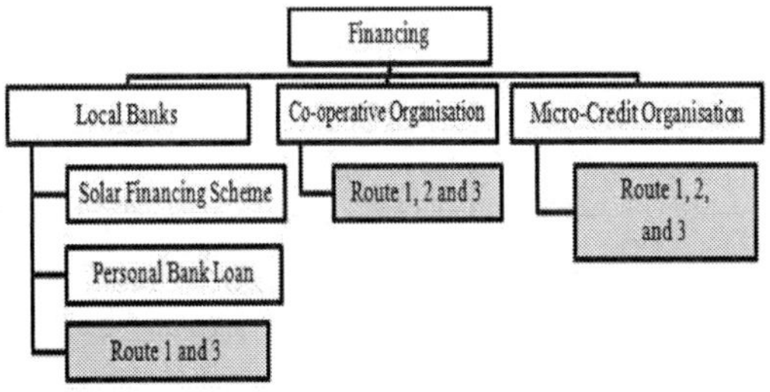

Figure 6. Types of Financing Mechanism.

Table 7. The Characteristics of Each Financing Mechanism

	Bank	Cooperative Organisation	Micro-Credit Organisation
Interest Rate (%)	6.35%	2.5%	1.9%
Fixed or Variable	Variable	Fixed	Fixed
Maximum Financing (MYR)	Negotiable	100,000	50,000
Tenure Terms (Year)	10	10	5
Remarks		Interest rate is between 1% - 4%. Average interest is considered.	10% of weekly payment

All routes can benefit from cooperative or micro-credit organizations financing, depending on the suitability of the applicants. The drawback of financing from bank is the requirements for eligible applicants, which only opens to households with monthly income of MYR 3,000 and above, and within the range of age between 21 – 55 years (Alliance Bank, 2013; Gading Kencana, 2013). Most of rural households have income below that MYR 1500 and average age is between 50-70 years old (Department of Statistic Malaysia, 2011). The return of investment (ROI) for each types of financing is shown in Table 8. All the proposed routes have a payback period of between three to twelve years. This length of time is considered normal and acceptable the country[c].

6 Conclusion

The advantages of community owned buildings being fitted with larger PV systems with a large roof can generate larger amounts of electricity and with the larger battery storage, it allows for a solar village to be set up on the basis of charging points in the immediate local vicinity. Loans from cooperative organisations are more attractive but micro-credit organisations are the most financially attractive. All the proposed routes have a payback period of between three to twelve years. The analysis indicates that the Feed-in Tariff (FiT) has some initial benefits for PV owners but the principal drawback is that any surplus electricity generated is unlikely to be fed directly into the grid and hinder the way to the establishment of solar village.

[c] For example, loans on cars are frequently taken out on a seven, eight or nine year period.

Table 8. The Payback Period

Year	Route	Capital Cost (MYR)				Net Benefit (MYR)	Financing (Annual Payment with Interest)								
	A	PV System	Battery Bank	Other (Community Grid, Cable, Mischellanous)	Total (MYR) B		Bank (C) 6.35% = Annual Capital (MYR 2700)	Cumulative Amount (MYR)	Payback Period	Cooperative (D) 2.50%= Annual Capital (MYR 2700)	Cumulative Amount (MYR)	Payback Period	Micro-Organisation (E) 1.90%= Annual Capital (MYR 5400)	Cumulative Amount (MYR)	Payback Period
1						4026.00	1714.50	28714.50		27675.00	27675		513.00	27513.00	
2						4571.97	1396.27	23384.77		21472.73	21472.73	4	343.65	18430.65	3
3						4465.16	1023.16	17135.97	5	14555.78	14555.78		160.72	8619.40	
4						4347.60	633.15	10603.96		7211.49	7211.49				
5	Route 1	16000	11000	0.0	27000	4218.95	225.83	3782.19							
6						4078.86									
7						3926.99									
8						3762.95									
9						3586.39									
10						3396.91									
1						1159.50							209.00	11209.00	
2						1374.59							149.14	7998.64	
3	Route 2	0	11000	0.0	11000	1509.58							84.06	4508.11	4
4						1644.77							15.17	813.70	
5						1780.17									
1						4026.00	10033.00	168033.00		3950.00	161950.00		3002.00	161002.00	
2						4571.97	9411.14	157618.14		4121.60	146245.60		2382.14	127758.14	
3						4465.16	8715.13	145961.31		3530.81	125282.84		1694.88	90898.91	5
4						4347.60	7981.71	133677.86		2943.12	104430.00		1009.64	54148.52	
5						4218.95	7209.17	120739.44		2358.84	83698.12	9	326.63	17517.92	
6	Route 3	64000	44000	50000.0	158000	4078.86	6395.75	107116.24		1778.29	63098.63				
7						3926.99	5539.57	92776.95	12	1201.80	42643.28				
8						3762.95	4638.67	77688.64		629.72	22344.21				
9						3586.39	3690.98	61816.67		62.39	2213.93				
10						3396.91	2694.32	45124.60							
11						2998.36	1646.41	27574.102							
12						3143.32	557.26	9333.005							

References

1. Ab Kadir, M. Z. A., Rafeeu, Y., & Adam, N. M. (2010). Prospective scenarios for the full solar energy development in Malaysia. Renewable and Sustainable Energy Reviews, **14**(9), 3023-3031. doi: DOI: 10.1016/j.rser.2010.07.062
2. Ahmad, N. A., & Byrd, H. (2012). Photovoltaics in Malaysia: Challenges and Strategies Towards Resilient Rural Areas in Malaysia. Paper presented at the International Malaysia-Auckland Research Conference (iMARCA 2012), Auckland.
3. Ahmad, N. A., & Byrd, H. (2013). Empowering Distributed Solar PV Energy For Malaysian Rural Housing: Towards Energy Security And Equitability Of Rural Communities. International Journal of Renewable Energy Development (IJRED), **2**(1), 59-68.
4. Alliance Bank. (2013). Home Complete Plus - Solar Panel Financing, 2013, from http://www.alliancebank.com.my/Personal-Banking/Loans/Home-Loans/Home-Complete-Plus-Solar-Panel-Financing.aspx
5. APEC. (2006). Energy Demand and Supply Outlook 2006 (pp. 49-53): Asia Pacific Energy Research Centre
6. APEC. (2013). Energy Demand and Supply Outlook (5th ed., pp. 97-108): Asia Pacific Energy Research Centre
7. Balfour, J., Shaw, M., & Nash, N. B. (2013). Advanced Photovoltaic System Design. Burlington: Jones & Bartlett Learning.
8. Barnes, D. (2005). Meeting the Challenge of Rural Electrification in Developing Nations: The Experience of Successful Programs (March 2005 ed., pp. 362). US: Energy Sector Management Assistance Program (ESMAP).
9. Barnes, D., & Foley, G. (2004). Rural Electrification In The Developing World: A Summary Of Lessons From Successful Programs (December ed.). Washington: Joint UNDP/World Bank Energy Sector Management Assistance Programme (ESMAP).
10. Barua, D. C. (2001). Strategy for promotions and development of renewable technologies in Bangladesh: experience from Grameen Shakti. Renewable Energy, **22**(1–3), 205-210. doi: 10.1016/s0960-1481(00)00012-4
11. Brumback, E. (2013, 28 April 2013). Utility co-op offers community-based approach to access solar energy, MiBiz. Retrieved from http://mibiz.com/news/energy/item/20599-utility-co-op-offers-community-based-approach-to-access-solar-energy
12. BTEK Energy. (2012). How to Size a Battery Bank Retrieved 21 Feb 2014, from http://www.btekenergy.com/documents/215.html
13. Byrd, H. (2010). The Potential of PVs In Developing Countries: Maintaining An Equitable Society In The

Face Of Fossil Fuel Depletion. Paper presented at the International Conference on Environment 2010, Penang.
14. Chaurey, A., & Kandpal, T. C. (2009). Solar lanterns for domestic lighting in India: Viability of central charging station model. Energy Policy, 37(11), 4910-4918. doi: http://dx.doi.org/10.1016/j.enpol.2009.06.047
15. Chua, S. C., Oh, T. H., & Goh, W. W. (2011). Feed-in tariff outlook in Malaysia. Renewable and Sustainable Energy Reviews, **15**(1), 705-712. doi: DOI: 10.1016/j.rser.2010.09.009
16. Department of Statistic Malaysia. (2010). Preliminary Count Report. Putrajaya: Government of Malaysia
17. Department of Statistic Malaysia. (2011). Taburan Penduduk dan Ciri-Ciri Asas Demografi (T. D. o. S. Malaysia, Trans.) Census 2010 (pp. 5). Kuala Lumpur.
18. EIA. (2013). Short Term Energy Outlook (March 2013 ed.). Washington: EIA.
19. Fantazzini, D., Höök, M., & Angelantoni, A. (2011). Global oil risks in the early 21st century. Energy Policy, **39**(12), 7865-7873. doi: 10.1016/j.enpol.2011.09.035
20. Fee, C. V., Sani, A., Nidzam, A., Barlow, H. S., Michael, J., Gurupiah, & Hashim, D. M. (2005). The Encyclopedia of Malaysia, Volume **5** : Architecture (Vol. 5). Kuala Lumpur: Archipelogo Press,.
21. Frame, D., Tembo, K., Dolan, M. J., Strachan, S. M., & Ault, G. W. (2011, 24-29 July 2011). A community based approach for sustainable off-grid PV systems in developing countries. Paper presented at the Power and Energy Society General Meeting, 2011 IEEE.
22. Gading Kencana. (2013). Sun2Cash. In SEDA (Ed.), (pp. 4). Kuala Lumpur: Gading Kencana.
23. Gardner-Stephens, R. (2013). TNB to Review Malaysia Electricity Tariff. Power Insider Asia.
24. Hamid, K. A., & Rashid, Z. A. (2012). Chapter 9: Economic Impacts of Subsidy Rationalization Malaysia. In Y. Wu, X. Shi & F. Kimura. (Eds.), Energy Market Integration in East Asia: Theories, Electricity Sector and Subsidies (ERIA Research Project Report 2011-17 ed., pp. 207-252). Jakarta: ERIA.
25. Haw, L. C., Sopian, K., Sulaiman, Y., Hafidz, M., & Yahya, M. (2009). Assessment of Public Perception on Photovoltaic Application in Malaysia Urban Residential Areas Using Trudgill's Framework for Analysis. European Journal of Social Sciences, Vol **8**(Number 4), 589-603. doi: 10.2202/1941-6008.1093
26. Hsu, J. (2014). Japan Plants Renewable Energy Village in Fukushima's Contaminated Farmland, 2016, from http://spectrum.ieee.org/tech-talk/green-tech/solar/japan-plants-renewable-energy-village-in-fukushimas-contaminated-farmland
27. IEA. (2013). Energy Security Retrieved 8/4/2013, from http://www.iea.org/topics/energysecurity/
28. Islam, A. K. M. S., & Islam, M. (2005). Status of Renewable Energy Technologies in Bangladesh. ISESCO Science and Technology Vision, Vol **1**(May 2005), 51-60.
29. Malaysia. (2006). 9th Malaysia Plan (2006-2010). Kuala Lumpur: The Government of Malaysia.
30. Martinot, E. (2014). No Rooftop Left Behind. Solar Journal.
31. MBIPV. (2011). PV System Cost, from http://www.mbipv.net.my/content.asp?zoneid=4&categoryid=12
32. Mekhilef, S., Safari, A., Mustaffa, W. E. S., Saidur, R., Omar, R., & Younis, M. A. A. (2011). Solar energy in Malaysia: Current state and prospects. [doi: 10.1016/j.rser.2011.08.003]. Renewable and Sustainable Energy Reviews(0).
33. Muhammad-Sukki, F., Ramirez-Iniguez, R., Abu-Bakar, S. H., McMeekin, S., & Stewart, B. (2011). An evaluation of the installation of solar photovoltaic in residential houses in Malaysia: Past, present, and future. [doi: 10.1016/j.enpol.2011.09.052]. Energy Policy, **39**(12), 7975-7987.
34. Muhammad-Sukki, F., Ramirez-Iniguez, R., Abu-Bakar, S. H., McMeekin, S. G., Stewart, B. G., & Chilukuri, M. V. (2011, 6-7 June 2011). Feed-In Tariff for solar PV in Malaysia: Financial analysis and public perspective. Paper presented at the Power Engineering and Optimization Conference (PEOCO), 2011 5th International.
35. Nel, W. P., & Cooper, C. J. (2009). Implications of fossil fuel constraints on economic growth and global warming. Energy Policy, **37**(1), 166-180. doi: 10.1016/j.enpol.2008.08.013
36. Pellegrini, L., & Tasciotti, L. (2012). Rural Electrification Now and Then: Comparing Contemporary Challenges in Developing Countries to the USA's Experience in Retrospect. Forum for Development Studies, 1-24.
37. PEMANDU. (2011). Government Transformation Program. Kuala Lumpur: : Government of Malaysia.
38. PowerGen. (2014). Micro Grids Retrieved 17/7/2014, from http://powergen-renewable-energy.com/micro-grids/
39. Saidur, R., Masjuki, H. H., & Jamaluddin, M. Y. (2007). An application of energy and exergy analysis in residential sector of Malaysia. Energy Policy, **35**(2), 1050-1063. doi: http://dx.doi.org/10.1016/j.enpol.2006.02.006
40. SEDA. (2012). FiT Rates for solar PV, from http://seda.gov.my/
41. Shagar, L. (2013). Electricity tariff up by average 15% from Jan 1, The Star. Retrieved from http://www.thestar.com.my/News/Nation/2013/12/02/Electricity-tariff-increase-Jan-1-2014/
42. SolarShare. (2013). Invest in a Brighter Future Retrieved 19 Feb 2014, 2014, from http://www.solarbonds.ca/
43. Street Charge. (2013). Street Charge, 2014, from http://street-charge.com/
44. Subhes, B. C. (2006). Energy access problem of the poor in India: Is rural electrification a remedy? [doi: 10.1016/j.enpol.2005.08.026]. Energy Policy, **34**(18), 3387-3397.

45. The World Bank. (2013). GDP growth (annual %). Retrieved 31 Aug 2013, from The World Bank Group http://data.worldbank.org/indicator/NY.GDP.MKTP.KD.ZG
46. TNB. (2009). Powering The Nation (1949 - 2009) (T. N. B. (TNB), Trans.) (Vol. **1**, pp. 313). Malaysia: Tenaga Nasional Berhad.
47. Westmill. (2013). Westmill Solar Coop Retrieved 19 Feb 2014, 2014, from http://www.westmillsolar.coop/
48. Yuan, L. J. (2010). The Malay House: Principles to Building Simple and Beautiful Homes for Comfort and Community. Kuala Lumpur: Fox Chapel Publishing

Attributes to facilitate e-waste recycling behaviour

Nur Hidayah Senawi[1] and Low Sheau-Ting[2]

[1]*Department of Real Estate, Faculty of Geoinformation and Real Estate, Universiti Teknologi Malaysia, 81310 Skudai, Johor, Malaysia*
[2]*Centre of Real Estate Studies, Institute for Smart Infrastructure and Innovative Construction, Universiti Teknologi Malaysia, 81310 Skudai, Johor, Malaysia*

Abstract. This study aims to identify the set of attributes to facilitate electronic waste (e-waste) behaviour among the community. E-waste disposal is increasing from year to year in parallel with increasing of global population. The short lifespan of electronics and poor e-waste recycling behaviour is among the main contributors to the steadily increasing of e-waste generated. Current recycling rate among the nation is lacking behind, which is only 10.5%. A questionnaire survey has been conducted among the students in Universiti Teknologi Malaysia to evaluate the current e-waste recycling practice. The results showed that majority of the respondents did not recycle their e-waste on campus. Aggressive efforts is needed to realize the country's target of 20% recycling rate in year 2020, one of the effective paths is to minimize e-waste generation via active e-waste recycling behaviour among the community. Extensive literatures have been reviewed to classify the attributes to facilitate effective e-waste recycling among the community. Total of five attributes that identified in this study which are Convenience of E-waste Recycling Infrastruture and Services, E-waste Recycling Information, Incentives For E-waste Recycling, Reminder to Recycle E-waste And E-waste Recycling Infrastructure and Services. The set of attributes identified in this study may serve as guideline for the management in designing program to foster e-waste recycling behaviour among the community.

1 Introduction

Waste is among the most crucial problems facing by the world today. This is because the increasing of human population from day to day and the number of waste also increase [1]. The increasing trend of population growth will increase the solid waste produced. In urban area of Malaysia, the average waste produced is about 760,000 tone per day [2]. The increasing of WEEE will cause negative effect to the environment and to human itself [3]. Most of the electronic waste contain hazardous substances such as lead, cadmium, mercury, lithium and so on that can cause environmental pollution and negative effect to human health [3]. Even though the e-waste is harmful to the environment and to human health, the waste did not manage in a proper way such as recycling. The recycling rate of e-waste is still low and people do not practice it [4]. A questionnaire survey has been conducted in Universiti Teknologi Malaysia on the early of November 2015. The results shows that only 12% of the respondents who recycle their e-waste and 34% of the respondents keep their e-waste at the hostel and 24% of the respondents throw their e-waste as an ordinary waste in normal bin. The awareness level of the respondents is still low and makes them to not properly dispose their e-waste.

Author [5] in her research stated that only 22% of the respondents in Shah Alam City, Malaysia who recycle their e-waste and most of the respondents dispose their e-waste by store at home and sell as a second-hand equipment. In Malaysia, most of the universities did not manage their e-waste properly; for instance National University of Malaysia and Universiti Teknologi Malaysia (UTM). According to authors [6], management of e-waste in UKM is still at the early stage. The awareness level on e-waste is still low and this makes them to not properly dispose their e-waste [6]. Department of Environment (DOE) in Malaysia has published a set of guidelines on e-waste, known as the Guidelines for the Classification of Used Electrical and Electronic Equipment [5]. Even though this guidelines already established, but it do not give any information on how to manage the e-waste at the end of it lifespan [5]. Hence, residents do not know how to manage or recycle their e-waste properly since there are no guidelines for them to do so.

In order to realising the e-waste recycling behaviour among the community, identification of the right attributes need to be identified. By identifying the right attributes to facilitate the community to recycle their e-waste, it can encourage people to recycle their e-waste. The recycling rate will increase and achieve the target which is 20% of recycling rate in 2020. If the recycling rate is increasing, it can also save the environment from the hazardous materials of the e-waste. Hence, there is a need to identify the attributes to facilitate e-waste recycling behaviour. Thus, the aim of this study is to identify the attributes that encourage the e-waste recycling behaviour.

2 Attributes of e-waste recycling behaviour

Attributes of e-waste recycling behaviour can be described as the component that facilitate and encourage people to recycle their e-waste and thus to foster e-waste recycling behaviour. The attributes of e-waste recycling behaviour is important in order to encourage people to recycle their e-waste so that the e-waste can be disposed in an environmentally-friendly manner [7]. The attributes also can be used in implementing effective e-waste recycling management system [8]. By identifying the preferred attributes such as the convenience of accessibility to recycle bin, the management of the organisation can put the recycle bin that can easily be accessible by users. Hence, they can recycle their e-waste more conveniently. Based on the review of previous published literatures, this study has ascertained total of five attributes that fostering e-waste recycling behaviour among the community : (1) Convenience of E-waste Recycling Infrastructure; (2) E-waste Recycling Information; (3) Incentives for E-waste Recycling; (4) E-waste Recycling Reminder; and (5) E-waste Recycling Infrastructure and Services. Further description on each attribute is presented in the followings.

2.1 Convenience of E-waste Recycling Infrastructure

The Convenience of E-waste Recycling Infrastructure means improved accessibility to recycling infrastructures. According to authors [9], several studies proved that the inconvenience of recycling infrastucture will lower the rate of people to participate in recycling activities. The more convenient the recycling infrastructure, the higher the rate of people participating in the recycling [10]. Other than inconvenience, any distance barriers or obstacles of recycle bins need to be removed in order to increase the rate of recycling [11]. Generally, this study identify three category distance of recycle bin which to facilitate e-waste recycling among the community; these including 100 meter distance to recycle bin, 100 meter to 500 meter distance to recycle bin and more than 500 meter to recycle bin.

The distance of less than 100 meter e-waste recycling infrastructure in this context means the e-waste recycling infrastructure such as recycle bin should be accessible within the distance of 100 meter. In a study conducted by authors [12] aimed to investigate sufficient number and positions of existing collection bins in Sidhapudur, Coimbatore, India using Geographical Information System (GIS). A model was developed based on the results of walking distance to drop the Municipal Solid Waste to the collection bin. There are three choices of distance which are 50 meter, 75 meter and 100 meter around the existing bins generated and the optimal distance was produced. The result shows that 75 meter of distance is the optimal distance to ensure good collection of municipal solid waste. The distance of 100 meter to 500 meter to e-waste recycling infrastructure such as recycle bin in this context means the e-waste infrastructure recycling should be allocated and accessed within the distance of 100 meter to 500 meter. Based on the previous study conducted by author [13] about the preferred attributes on waste segregation behaviour among the Malaysian campus community, the most preferred accessible distance to recycle bin by the students is 100 meter to 500 meter. The distance of more than 500 meter distance to e-waste recycling infrastructure such as recycle bin in this context means the recycle bins should be available and accessed within the distance of more than 500 meter. Distance plays an important attribute in order to foster the e-waste recycling behaviour among the community. The more convenience of the distance will encourage the community to recycle their e-waste.

2.2 E-waste Recycling Information

Information can be in various forms in order to raise knowledge and awareness about recycle, how to recycle and the benefits of e-waste recycling. According to authors [14], the producer of the product needs to provide the information on how to recycle and its effects when not disposed properly. These information is important in order for the obsolete e-waste to be treated effectively. In addition, authors [15] mentioned that recycling knowledge is important in determining the recycling behaviour among the students in University Sains Malaysia (USM).

Consumer awareness is important for the successfulness of e-waste recycling [16]. In order to raise the awareness, sufficient information about how to recycle, where to recycle and benefits of e-waste recycling should be provided [17]. The previous study has been done by authors [6] among the staffs and students in UKM. The results showed the awareness of e-waste management is still lacking seriously and in order to increase the awareness, they need to provide more information on e-waste management by teaching formally and informally and conducting seminars. In present context, information about e-waste recycling consists of two elements : (1) instructions to e-waste recycling, and (2) feedback on e-waste recycling.

Instructions to e-waste recycling is about providing the information to the community on how and where to recycle their e-waste. According to authors [8], the instructions can be done through educating the community on how and where to recycle in order to foster the recycling behaviour. According to authors [10], most of the consumers of mobile phone did not know that they can recycle their old and obsolete phone. This is due to the lack of information about waste collection and e-waste recycling. They also said that proper education and promotion from school, home, social media and other communication system can influence the mindset of people on the recycling of e-waste and ultimately building the awareness towards it. Hence, information about the instruction on e-waste recycling is important in order to enhance the e-waste recycling behaviour among the members of community.

Additionally, providing feedback on e-waste recycling is important to facilitate the practice of e-waste recycling among the community. In present context, feedback on e-waste recycling is about giving the feedback to the participants about the e-waste recycling information and rate of e-waste recycling after the e-waste recycling programme. Authors [16] proposed to establish a Global Recycle Network (GRN) in order to encourage e-waste recycling behaviour among consumers. In order to establish GRN, they need to provide the information feedback to product designers to improve recyclability. Authors [18] also stated that feedback about the performance of the recycling programme is important to encourage the recycling behaviour. In essence, feedback is essential to encourages people to recycle their e-waste. Information about e-waste recycling is among the important attributes in fostering e-waste recycling behaviour among the community. If there is no information about the e-waste recycling, the community might not be able to recycle their e-waste due to lacking of information such as did types of e-waste can be recycled.

2.3 Incentives for E-waste Recycling

Incentives for e-waste recycling is providing an incentive in form of monetary or non-monetary to the person or anyone who recycle their waste. According to authors [19], providing an incentive can foster the initial motivation for a sustainable activity. After the activity becomes part of their daily life, they will continue as their habits even without the extrinsic motivation. Thus, incentives can foster positive change in behaviour so that people will begin to recycle their e-waste. In present study, incentive is divided into two categories which are monetary incentive and non-monetary incentive.

A study by authors [20] on factors that influence people in Gaza City to recycle their domestic solid waste stated that more than half of the respondents choose financial rewards as the motivational factor to adopt recycling behaviour. A previous research conducted by authors [21] showed that Reuse, Recycling and Refurbishing (RRR) schemes collect more phones after offering the monetary incentives to the respondents. According to authors [22] monetary incentives can be in form of cash or voucher. The users will be given money if they recycle their e-waste or give back their e-waste to the producer. The result also stated that monetary incentives act as the most efficient incentives given to the respondents for successful of take-back services [22].

On the other hand, non-monetary incentives are incentives that do not involve cash payment. Authors [23] have conducted a study to exploring about how non-financial incentives contribute in motivating the workers to take care of their health and their retention in 16 countries. The result showed that non-monetary incentives are also important to motivate the workers to take care of their health because they feel valued by giving the non-monetary incentives to them. Hence, non-monetary also have the potential to motivate people to recycle their e-waste such as by giving appreciation letter for doing e-waste recycling, small gifts and invited them to join the e-waste recycling together afterwards. Monetary and non-monetary incentives play an important role in fostering e-waste recycling behaviour among the community. Both of the incentives will trigger the e-waste recycling behaviour among the community if they were applied among the community.

2.4 E-waste Recycling Reminder

According to authors [24] reminder defined as "auxiliary stimulation that is presented and remind the learner to display behaviour of interest at the appropriate time and appropriate circumstances". Reminder is useful in retaining sustainable behaviour [25]. Purpose of the reminder is to maintain the sustainable behaviour for long period [27]. Reminder also helps users to remember the sustainable practice which they tend to forget [27]. Authors [27] stated that verbal reminder and written reminder can increase about 22.1% of recycle rate participation. However, when only written reminders are given to residents, the recycle rate participation increases just by 2.4%. In present study, there are two types of E-waste Recycling Reminder which are written reminder and verbal reminder.

Authors [28] mentioned that sustainability behaviour tends to be forgotten. Similarly in the context of e-waste recycling, written reminder can be placed at corridors, common area with messages about benefits of e-waste recycling.

Authors [29] stated that verbal reminder is "words or instructions which direct a person to involve in the behaviour". For example, teachers in school remind and encourage the students to recycle their e-waste. A previous research conducted by authors [30] showed that verbal reminder can increase the consumption of fruit serving among school in the National School Lunch Program (NSLP). The percentage of the students who take and consume fruit serving increase significantly. Hence, the management may consider to add on verbal reminder in an effort to encourage the community to recycle their e-waste and to foster e-waste recycling behaviour. People tend to forgot easily and need to be reminding. Thus, reminder can be one of the important attributes to foster e-waste recycling behaviour among the community.

2.5E-waste Recycling Infrastructure and Services

E-waste Recycling Infrastructure and Services is a facility or structures provided to dispose the e-waste. A previous research conducted in selected residential area in Kuala Lumpur, Malaysia found that recycling infrastructure is important attribute in fostering recycling behaviour [31]. Lacking of recycling facilities will decrease the consumers' intention to recycle and thus the recycling rate will be lower [32]. According to authors [18], in order to establish a proper management of solid waste on

Maltese Island, physical infrastructure of waste management is among important elements. Furthermore, authors [33] stated that in order to improve the e-waste recycling among residents in Melaka, government should provide the recycling infrastructure so that residents know where to recycle their e-waste.

In present context, E-waste Recycling Infrastructure and Services consists of three elements which are e-waste drop-off site, e-waste recycling bin and e-waste collection service. E-waste drop off site is where the e-waste dustbin placed at public area where easily assessible such as bus stop and library [34]. The significance of recycling infrastructure and services in increasing recycling rate has been acknowledged in previous research. For instance study conducted by authors [34] has showed that the rate of recycling increased when the availability of drop-off recycling facilities are available at Prespa Park Vilage. Authors [32] in their research about recycling attitude and behaviour in university campus stated that campus management should provide more drop-off recycle bin such as at lift lobbies, café, hostel and library so that students can easily recycle their waste.

Previous research conducted by authors [35], a study about factors influencing the rate of recycling at Minnesota countries and the result showed that the increasing number of drop-off center has increased the recycling rates by 1.28%. Study by authors [31] showed that the number of recycle bins need to be increased to improve the recycle rate among the household in Kuala Lumpur, Malaysia. ReCellular and Rechargeable Battery Recycling Corporation (RBBC) which is a non-profit organisation has established a collective service around the country in Brazil to collect the obsolete cell phones and the results showed that the e-waste recycling rate has successfully increased [36]. Infrastructure also important in encourage the community to foster e-waste recycling behaviour. If the e-waste recycling infrastructure is not provided, the community will not be able to recycle their e-waste and throw their e-waste in ordinary dustbin.

Table 1. Attributes to facilitate e-waste recycling behaviour.

Attribute	Attribute Level
Convenience of E-waste Recycling Infrastructure	• Less than 100 meter Distance to E-waste Recycling Infrastructure • 100 meter to 500 meter Distance to E-waste Recycling Infrastructure • More than 500 meter Distance to E-waste Recycling Infrastructure
E-waste Recycling Information	• Instruction to E-waste recycling. • Feedback on E-waste recycling.
Incentives for E-waste Recycling	• Monetary Incentive. • Non-monetary Incentive.
E-waste Recycling Reminder	• Written reminder. • Verbal reminder
Attribute	**Attribute level**
E-waste Recycling Infrastructure and Services	• E-waste Drop-off site. • E-waste Recycling bins. • E-waste collection services.

3 Conclusion

Based on the review of literatures, present study has, proposed five attributes which are effective to facilitate e-waste recycling behaviour among the community. The five attributes are Convenience of E-waste Recycling Infrastructure, E-waste Recycling Information, E-waste Recycling Incentives, E-waste Recycling Reminder and E-waste Recycling Infrastructure and Services. In each of the attributes, it consists of two or three elements. The findings of the study may serve as a reference guide for the management in fostering electronic waste recycling behaviour among the community.

Acknowledgements

The authors gratefully acknowledge the research grant and financial support provided by the Ministry of Higher Education (MOHE) and Universiti Teknologi Malaysia (UTM), under the FGRS research grant number R.J130000.7809.4F506.

References

1. Tan, S. T., Ho, W. S., Hashim, H., Lee, C. T., Taib, M. R., & Ho, C. S. Energy Conversion and Management. *Energy, economic and environmental (3E) analysis of waste-to-energy (WTE) strategies for municipal solid waste (MSW) management in Malaysia.*, **102**, 111-120 (2015)
2. Tarmudi, Z., Abdullah, M. L., & Tap, A. O. M. Jurnal Teknologi. *An overview of municipal solid wastes generation in Malaysia.* ***51*(1)**, 1-15 (2009)
3. Perkins, D. N., et al. Annals of global health. *E-waste: a global hazard.* **80(4)**, 286-295 (2014)
4. Soo, V. K. and M. Doolan. Procedia CIRP. *Recycling Mobile Phone Impact on Life Cycle Assessment.* **15**, 263-271 (2014)
5. Kalana, J. A.. International Journal of Environmental Sciences. *Electrical and electronic waste management practice by households in Shah Alam, Selangor, Malaysia.* ***1*(2)**, 132-144 (2010)
6. Chibunna, J. B., et al. Procedia-Social and Behavioral Sciences. *The challenges of e-waste management among institutions: A case study of UKM.* **59**: 644-649 (2012)
7. Chi, X., Wang, M. Y., & Reuter, M. A. Journal of Cleaner Production. *E-waste collection channels and household recycling behaviors in Taizhou of China.* , **80**, 87-95 (2014)
8. Wang, Z., et al. Journal of Cleaner Production. *Willingness and behavior towards e-waste recycling for residents in Beijing city, China.* **19(9-10)**, 977-984 (2011)
9. Saphores, J.-D. M., et al. Resources, Conservation and Recycling. *Willingness to engage in a pro-*

environmental behavior: An analysis of e-waste recycling based on a national survey of U.S. households. **60**: 49-63 (2012)
10. Sarath, P., et al. (2015). "Mobile phone waste management and recycling: Views and trends." Waste Managment.
11. Amuteya. N., Shackleton. C.M., Whittington-Jones, K.. Paper Recycling Patterns and Potential Interventions In The Education Sector: *A Case Study of Paper Streams at Rhodes University, South Africa*. Resour. Conserv. Recycl., **53**, 237-242 (2009)
12. Nithya, R., Velumani, A., & Kumar, S. R. R. WSEAS Transactions on Environment and Development. *Optimal location and proximity distance of municipal solid waste collection bin using GIS: A case study of Coimbatore city*. **8(4)**, 107-119 (2012)
13. Tee Sin Yee (2014). *Attributes in Fostering Waste Segregation Behaviour*. Master Thesis. Universiti Teknologi Malaysia
14. Nnorom, I. C., & Osibanjo, O. Resources, conservation and recycling. *Overview of electronic waste (e-waste) management practices and legislations, and their poor applications in the developing countries*. **52(6),** 843-858 (2008)
15. Ramayah, T., & Rahbar, E. Management of Environmental Quality: An International Journal. *Greening the environment through recycling: an empirical study*. , **24(6)**, 782-801. (2013)
16. Takahashi, V. Osamu Kanzawa V. Makoto. FUJITSU Sci. Tech.. *Establishment of global recycle network*. **41.2** 242-250 (2005)
17. Tanskanen, P. Acta materialia. *Management and recycling of electronic waste*. **61(3)**, 1001-1011 (2013)
18. Bezzina, F. H., & Dimech, S. An International Journal. *Investigating the determinants of recycling behaviour in Malta. Management of Environmental Quality*, **22(4)**, 463-485 (2011)
19. Gneezy, U., Meier, S., Rey-Biel, P.,. *When and why incentives (don't) work to modify behavior. J. Econ. Perspect.* **25 (4)**, 191–210 (2011)
20. Omran, A., Sarsour, A. K., & Pakir, A. H. K. The International Journal of Health Economics-IJHE. *An Investigation Into The Factors Influencing The Participation Of Households In Recycling Of Solid Waste*. **2(1)**, 4-19 (2012)
21. Ongondo, F. O., et al. Waste management. *How are WEEE doing? A global review of the management of electrical and electronic wastes*. **31(4)**: 714-730 (2011)
22. Li, B., Yang, J., Song, X., & Lu, B.. Procedia Environmental Sciences. *Survey on disposal behaviour and awareness of mobile phones in Chinese university students*. **16**, 469-476 (2012)
23. Churchard, C. Recycling & Waste Management News and Information. *Two Thirds of People Favor Financial Rewards for Recycling, says poll.*. RWM, 22, August 2007. (2007).
24. Foxx, R.M.. *Increasing Behavior of Severely Retarded and Autistic Individuals* 81-96. Champaign, IL: Research Press. (1982)
25. Choong, W.W.. Paper presented at the 17th international symposium on society and resource management, Sabah. *The process of fostering sustainable awareness and behavioiur among urban communities*. (2011)
26. Low S.T. (2012). *Factors Affecting Energy Conservation Behaviour of Students in Malaysia University*. Ph.D. Thesis. Universiti Teknologi Malaysia
27. Spaccarelli, S., Zolik, E. & Jason, L. A. Journal of Environmental Systems. *Effects of Verbal Prompting and Block Characteristics on Participation in Curbside Newspaper Recycling*. **19**, 45-57 (1990)
28. Armijo de Vega, C., Ojeda Benítez, S., and Ramírez Barreto, M. Waste Management. *Solid Waste Characterization and Recycling Potential For a University Campus*, **28**, S21-S26. (2008)
29. Macduff, Gregory S., Patricia J. Krantz, and Lynn E. McClannahan. (2001). Prompts and Prompt-Fading Strategies For People With Autism. *Making A Difference: Behavioral Intervention For Autism*. 37-50.
30. Schwartz. M.B. International Journal of Behavioral Nutrition and Physical Activity. *The Inflluence Of A Verbal Prompt On School Lunch Fruit Consumption: A Pilot Study*. **4**, 6 (2007)
31. Zen, Irina Safitri, and Chamhuri Siwar. Habitat International. *An analysis of household acceptance of curbside recycling scheme in Kuala Lumpur, Malaysia*. **47**, 248-255 (2015)
32. Chen, M.-F., Tung, P.-J. Environment and Behaviour. *The moderating effect of perceived lack of facilities onconsumers' recycling intentions*. **42**, 824-844 (2010)
33. Tiep, H. S., Kin, T. D. Y., Ahmed, E. M., & Teck, L. C. E-Waste Management Practices of Households in Melaka.
34. Grazhdani, D. Waste Management. *Assessing the variables affecting on the rate of solid waste generation and recycling: An empirical analysis in Prespa Park*. (2015)
35. Sidique, S. F., Joshi, S. V., & Lupi, F. Resources, Conservation and Recycling. *Factors influencing the rate of recycling: An analysis of Minnesota counties*. **54(4)**, 242-249 (2010)
36. Johari, A., Alkali, H., Hashim, H., Ahmed, S. I., & Mat, R.. Modern Applied Science. *Municipal Solid Waste Management and Potential Revenue from Recycling in Malaysia*. **8(4)**, p37. (2014)

Sustainable Waste Management for Green Highway Initiatives

Nur Illiana Husin[1,a], Asmalia Che Ahmad[2], Abdul Muhaimin Ab Wahid[2] and Syahrul Nizam Kamaruzzaman[3]

[1]*Department of Built Environment, Faculty of Architecture, Planning and Surveying, Universiti Teknologi Mara (Shah Alam), Malaysia*
[2]*Department of Quantity Surveyor, Faculty of Architecture, Planning and Surveying, Universiti Teknologi Mara (Shah Alam), Malaysia*
[3]*Department of Building Surveying, Faculty of Built Environment, Universiti Malaya, Malaysia*

Abstract. Green highway initiative is the transportation corridors based on sustainable concept of roadway. It incorporates both transportation functionality and ecological requirements. Green highway also provides more sustainable construction technique that maximizes the lifespan of highway. Waste management is one of the sustainable criterias in the elements of green highway. Construction of highway consumes enormous amounts of waste in term of materials and energy. These wastes need to be reduce to sustain the environment. This paper aims to identify the types of waste produced from highway construction. Additionally, this study also determine the waste minimization strategy and waste management practiced.. This study main focus are construction and demolition waste only. The methodology process begin with data collection by using questionnaire survey. 22 concession companies listed under Lembaga Lebuhraya Malaysia acted as a respondent. The questionnaires were distributed to all technical department staffs. The data received was analyzed using IBM SPSS. The results shows the most production of waste is wood, soil, tree root and concrete. The least production of waste is metal. For waste minimization, the best waste minimization is reuse for all type of waste except for tree root and stump. Whereas, the best waste management is providing strategic plan. The least practice for waste management is recording the quantity of waste.

1 Introduction

Development of green highway is revitalizing in the developing country [1]. The development of green highway includes design of highway in such a manner that improves the quality of nation's infrastructure [1]. With the demands in implementation of major infrastructure projects in Malaysia together with the commercial and housing development, a large amount of construction waste is being produced by the construction sector [2]. In study by Wokekoro (2007), construction sector generates large amount of waste and contributes to the environmental problem [3]. The construction waste become an issues that needs highly concern in many developing countries because it has an adverse effect on economy, environment and social aspects.

According to GEC (2012), solid waste is one of the three major environmental problems in Malaysia [4]. Over 23,000 tonnes of waste is produced each day in Malaysia. The amount of solid waste is expected to rise to 30,000 tonnes by the year of 2020 [4]. In Malaysia, there are almost 1800 rivers, but more than half of the rivers have been polluted and destroyed because of the improper solid waste management [4]. The management of solid waste is important to reduce or eliminates the adverse impacts on the environment and human health.

Papargyropoulou et al. (2011) reported 28 per of solid waste originated from industrial and construction waste in the central and southern regions of Malaysia [5]. It happens because of waste minimization, reuse and recycling practices are limited in the construction sector. Besides, the natural resources required as construction materials are available at relatively low cost [6]. In addition, Begum et al. (2009) stated there is no mandatory requirement for construction companies to practice sustainable resource and waste management [6]. Hence, this study is constructed to identify types of waste produced during construction activity. Additionally, this study also handled to determine the best waste minimization strategy and waste management practiced.

2 Literature Review

2.1 Waste

Waste can be defined as a material of solid or semi-solid character that the possessor no longer considers as a sufficient value to retain [3]. Formosa et al. (2002) clarified waste as a wastage sources likes materials, time (labor and equipment) and capital cost due to the direct or indirect activity, but not for the better or value of the final product according to customer satisfaction [7]. Waste is all waste arising from human and animal activities that are normally solid and being discard as useless or unwanted [8].

[a] Corresponding author: nurillianahusin@gmail.com

According to Pitchel (2005), waste is a material that possessed a negative economic value, which suggest that is cheaper to discard than to use [9]. Therefore, waste should be interpreted as any losses in material, time and monetary result by activities but do not add value or progress to the product, which can be eliminated without reducing customer value.

2.2 Category and sources of waste

Based on the literature research about the category and sources of waste, different author have their own category and sources of waste. Pitchel (2005) stated nine categories of waste such as municipal, hazardous, industrial, medical, universal, construction and demolition, radioactive, mining and agriculture waste [9]. Whereas, Bai & Sutanto (2002) divided categories of waste into three which are domestic refuse, industrial and institution waste [10]. While for the sources of waste, have many sources of waste. Table 1 shows the sources of waste according to author.

Table 1. Sources of waste.

Author	Sources of waste
Tchonobouglous et al. (1993) [8]	a. Commercial b. Institutional c. Construction and demolition d. Municipal service (excluding treatment plant sites) e. Treatment plant sites (municipal incinerators) f. Municipal solid waste g. Industrial h. Agriculture
Pitchel (2005) [9]	a. Municipal – residential, commercial, institutional, industrial, municipal b. Hazardous – manufacture, electroplating, metal treating, wood preserving and petroleum refining c. Industrial – manufacturing d. Medical – hospital, physicians, dentists, veterinarians, healthcare facilities, clinics, laboratories, blood banks and funeral homes e. Universal – small and large business, household f. Construction and demolition – construction, renovation and demolition g. Radioactive – specializes of industrial waste, research and medical h. Mining – physical removal of desired resources i. Agriculture – animal manures and crop residues
Wokekoro (2007) [3]	a. Municipal – street sweeping, sewage, waste form schools, market and other institution b. Domestic – garbage, rubbish and large waste from homes c. Commercial – stores and offices d. Industrial – manufacturing plants e. Mining – coal mining, strip mining etc
Bai & Sutanto (2002) [10]	a. Domestic - solid waste from household, markets, food center and commercial premises b. Industrial – not including toxic and hazardous c. Institution – solid waste from government and board
ONSW (2012) [11]	a. Household b. Commercial c. Construction and demolition d. Industrial e. Institution f. Public imported
Environment Strategies (2000) [12]	a. Residential b. Industrial c. Commercial d. Instituted e. Construction and demolition f. Municipal g. Manufacturing h. Agriculture

2.3 Waste management

Based on Basri & Ahmad Basri (2008), waste management is an integral part of urban and environmental management of each city [13]. Waste management is a major challenge for Malaysia to address in the light of vision 2020 which lays out of direction for Malaysia to become a fully developed nation by 2020 [14]. Waste management are designed to prevent or reduce the discharge of pollutants to stormwater from solid or construction waste by providing designated waste collection areas and containers, arranging for regular disposal, and training employees and subcontractors [15].

Tchobanoglous et al.(1993) described waste management as the discipline associated with the control of generation, storage, collection, transfer and transport, processing, and disposal of solid wastes in a manner that is correlate with the best principles of public health, economics, engineering, conservation, aesthetics, and other environmental considerations, and that is also responsive to public attitude [8]. According to Pitchel (2005), waste management is concerned with the generation, on site storage, collection, transfer, transportation, processing and recovery, and ultimate disposal of solid waste [9].

In other word, waste management is a management process to ensure waste generated will be treated in the right way. It aims to reduce the production of waste and to reduce the amount of waste going to landfill. It is also to reduce the environmental problem that becomes a major problem nowadays.

2.4 Waste minimization

Environmental Protection Agency (EPA) of United States (2000) defined waste minimization as any method that reduces the volume or toxicity of a waste that requires disposal [12]. It is including any method that reduces the amount of waste. Government regulations, as well as internal cost effectiveness, required the disposal of all waste and in particularly hazardous waste be minimized.

Waste minimization explained by Poon, Ann, & Jaillon (2004) as any technique, process or activity which avoids, eliminates or reduces waste from its source or allows reuse or recycling of the waste [16]. Figure 1 shows the waste management hierarchy determined by Poon, Ann & Jaillon (2004). The waste minimization are divided into prevention measures and waste management measures. The prevention measures are consist of prevention, reduction at source and re-use of product. Meanwhile, the waste management measure category consists of quality improvements, recycling, energy recovery and pre-treatment.

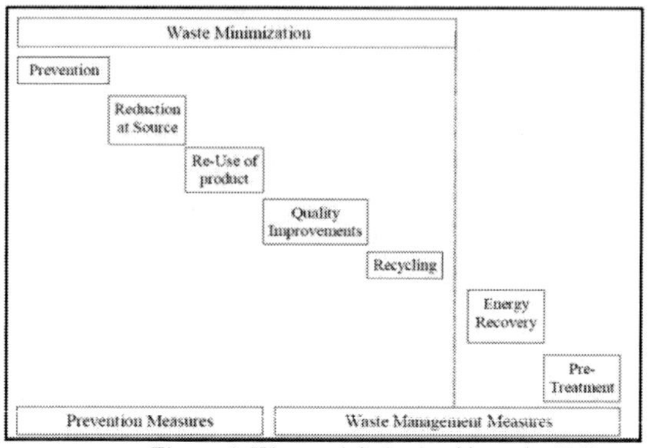

Figure 1. Waste minimization hierarchy [16].

Faniran & Caban (1998) wrote there are three main waste minimization strategies which are avoiding waste, reusing materials and recycling waste. Avoiding waste refer to any practice or process that avoids, eliminates or minimizes waste at the source [17]. In the meantime, reusing and recycling waste supports in reducing the volume of waste material to be disposed and discharged into environment. Figure 2 shows proposed waste management hierarchy that consists of avoid, reuse, recycle and disposal by Faniran & Caban (1998).

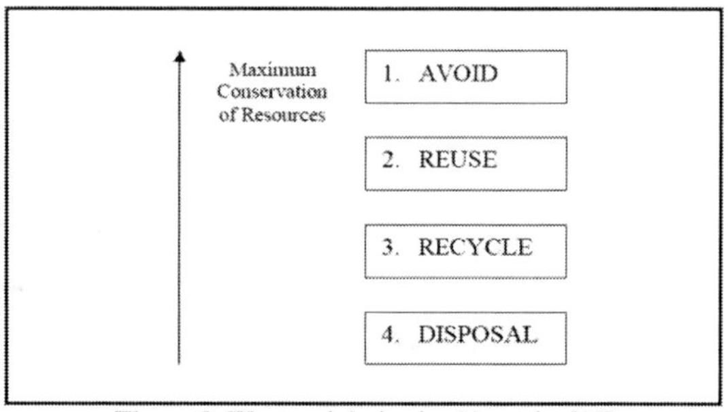

Figure 2. Waste minimization hierarchy [17].

2.5 Waste management plan

Macozama (2002) expounded waste management plan as a proposal that consists of prevention, isolation, repair, reuse, recycle and also disposal of waste [18]. The main objective of waste management plan is to reduce, reuse and recycle of waste [19]. Moreover, Macozama (2002) has identified the good management plan which are goals, waste audit, waste handling requirement, waste disposal, transportation and economic evaluation [18]. On top of that, Lim & Ling (2002) has categorized the element of waste management plan to 4 categories which are including the prevention of waste, reduction, recycle and administration [20]. A good waste management plan will assist to reduce amount of waste going to landfill and minimize the cost to manage the waste.

3 Methodology

The methodology process begins with data collection by using questionnaire survey. This study is using purposive sampling method. Concession Company has been choosing as a respondent. There are 22 concession companies listed under Lembaga Lebuhraya Malaysia. The questionnaire had been distributed to all technical department staffs. Only 50 sets of questionnaire are returned. The data collected has been analyzed using IBM SPSS.

3.1 Development of questionnaire

The questionnaire has four parts which are demographic, types of waste, waste minimization strategy and waste management. The development of questionnaire built according to close ended with Likert scale 1 to 5. Scale 1 refers to strongly disagree and scale 5 refer to strongly agree. The question was designed using interval scale.

The categories and sources of waste used in the questionnaire were followed from Pitchel (2005) as he had clearly stated the variables in his research study [9]. The categories are divided into municipal, hazardous, industrial, medical, universal, construction and demolition, radioactive, mining and agriculture. For this study, the category used is correlated to construction and demolition. Then, the sources of waste come from the highway construction itself.

The hierarchy of the waste management has been chosen from Poon et. al (2004) [16]. It is because the study did not only focus on the basic waste minimization only (prevent, reuse, recycle and disposal) but it also consists of prevention, reduction at source, reuse of product, quality improvement, recycling, energy recovery and pretreatment. This hierarchy helps to guide people to choose the best way of managing the waste properly.

Table 2 shows the variable for waste management plan that will be discussed in waste management plan in section 4.3.

Table 2. Variable for waste management plan.

Indicator	Waste management plan
A	Provide strategic management plan
B	Policy regarding waste management
C	Employees with responsibility to manage waste
D	Identifying the type of waste
E	Organize the waste
F	Record the quantity of waste
G	Management of waste
H	Training employees
I	Measure & Audit

J	Review plan

4 Result and findings

The result obtained from the questionnaire shows 20 out of 50 respondents have working experience in this field more than 10 years. The other 13 respondents stated with 5 to 10 years working experience and 12 respondents have less than 5 years working experience. From this conclusion, it shows almost half of the respondents have more than 10 years working experience in construction industry. Although only 50 sets of questionnaire are returned, data given can be accepted because of the respondent are knowledgeable and experienced in the related industry. Further, in subheading 4.1 discussed the outcomes of type of waste produced. Then, subheading 4.2 discussed of waste minimization strategy and 4.3 is waste management practiced.

4.1 Types of waste

Table 3 below shows the average index for waste produced during the construction and demolition of highway. The formula of average index is taken from study by Al-Hammad & Assaf (1996) [21]. Below (1) is the formula for calculation of average index.

$$\text{Average index} = \frac{\sum a1x1}{\sum x1} \quad (1)$$

Indicator :-

a1 = Likert scale (i.e 1 'no production' – 5 'most production')

x1 = Number of respondent

The waste involves are tree root or stump, soil, concrete, asphalt, wood, metal, rock, aggregate, sand, crusher run, bituminous, and premix. The top ranking of production of waste are wood (3.2000), followed by soil (3.160) and tree root or stump with same average index with concrete (3.000) and crusher run with aggregate (2.780). The rock comes with aggregate 2.960 and premix (2.680). It is followed by asphalt and bituminous (2.640) and the lowest rank of all is metal with 2.6000 aggregate.

The highest rank of waste is wood 3.200 average. Wood is always used as a temporary support especially for concreting work. The nature of wood that easy to rot also become one of the reason why wood reflects as the highest production of waste. It is supported by study of Lachimpadi et al. (2012) also found wood is the largest quantity of construction waste [22].

The second ranking is soil 3.160 average. Different site will produce different quantity of soil. It depends on condition of sites whether it needs to be cut or filled to get a flat ground before construction could begin. Lachimpadi et al. (2012) conclude the generation of soil waste greatly depended on the design and the landscape of the site [22]. If the site located at hilly area, it will involve a lot of cutting compared to fills. Thus, the surplus of soil will be classified as soil waste. It is also supported in the study by Begum et al. (2006) that soil and sand are the second highest waste for their study [6].

On the other hand, metal contemplated as the lowest ranking with 2.600 average. Metal become the lowest production because of the limited usage in the construction of highway. It is because the metal is always custom made from the factory. It will be delivered to site according to the quantity and size order. In article by Technologiya Metallov (2008) found, in Russia, the production of metal waste is also low [23].

Table 3. Average Index for types of waste produce.

Bill	Types of waste produce	1 a1	2 a1	3 a1	4 a1	5 a1	Total frequency ∑a1x1	Average Index ∑a1x1 / ∑x1	Ranking
		x1	x1	x1	x1	x1			
1	Tree Root / Stump	3	14	20	6	7	150	3.000	3
2	Soil	0	14	21	8	7	158	3.160	2
3	Concrete	3	14	20	6	7	150	3.000	3
4	Asphalt	3	20	19	8	0	132	2.640	10
5	Wood	0	9	22	19	0	160	3.200	1
6	Metals	3	19	23	5	0	130	2.600	12
7	Rock	0	14	24	12	0	148	2.960	6
8	Aggregate	0	19	23	8	0	148	2.780	7
9	Sand	0	19	23	8	0	148	2.780	7
10	Crusher Run	0	14	23	13	0	150	2.980	5
11	Bituminous	3	20	19	8	0	132	2.640	10
12	Premix	3	19	19	9	0	134	2.680	9

4.2 Waste minimization strategy

Waste minimization strategy is the best way to minimize waste. This strategy of waste minimization consist of prevention, reduction at source, reuse of product, quality improvement, recycle, energy recovery, pre-treatment, and disposal.

Analysis of data received was converted to the percentage value. Table 4 shows the percentage value for waste minimization strategy. Tree root or and stump shows recycle as the highest percentage for waste minimization. In this practice where the tree root and stump are recycled to produce garden decoration.

According to Suez Environment (2015), the tree root and stump are impossible to reuse [24]. Hence, they are suitable to be recycled as bench, table or any other decoration for garden decoration.

However, for all types of waste, the highest percentage of waste minimization is reuse of waste. These wastes are soil, concrete, asphalt, wood, metals, rock, aggregate, sand, crusher run, bituminous and premix. These wastes can be reused for other projects and other function. While these wastes come from raw material and were not easily damage. It also can be reused for sometimes.

Table 4. Percentage of waste minimization strategy.

No	Waste Minimization	Tree root & Stump	Soil	Concrete	Asphalt	Wood	Metal	Rock	Aggregate	Sand	Crusher Run	Bituminous	Premix
		%	%	%	%	%	%	%	%	%	%	%	%
1.	Prevention	5	5	0	0	4	0	0	0	0	0	0	0
2.	Reduction at source	0	14	3	15	3	4	3	4	4	4	15	15
3.	Reuse of product	13	**68**	**69**	**75**	**56**	**70**	**74**	**81**	**87**	**75**	**66**	**60**
4.	Quality Improvement	8	0	0	4	3	11	0	0	0	11	0	0
5.	Recycle	**30**	0	0	0	13	16	23	12	6	6	6	6
6.	Energy Recovery	0	0	0	0	0	0	0	0	0	0	0	0
7.	Pre-treatment	25	5	24	6	18	0	0	0	0	0	9	15
8.	Disposal	20	8	3	0	3	0	0	4	4	4	4	4

4.3 Waste management plan

In waste management plan analysis preliminary test shows that the Cronbach's Alpha is 0.931 for 10 variables. There is high internal consistency for the data set which the Cronbach's Alpha is more than 0.7 [25]. This shows all the 10 variables are reliable to be analyzed for this research. The data for waste management plan was analyzed using Relative Importance Index (RII). RII to determine the relative importance of the waste management plan. The formula of RII as stated below (2) from Kometa et al. (1994) [26]. The higher the relative importance index value close to 1 means that the higher the position of the statement.

$$RII = \frac{\sum w}{AN} \qquad (2)$$

Indicator:-

W = Weights are given to each factor by the respondents and will ranges from 1 to 5 where '1' is strongly disagree and '5' is strongly agree

A = Highest weight (i.e. 5 in this case)

N = A total number of respondents. The higher the index value close to 1 means that the higher the position of the statement

In table 5 shows from 10 variable of waste management plan. The highest RII is to provide strategic management plan (0.908). The second rank is organizing the waste with RII as 0.896. The third rank is a policy regarding waste management (0.888), followed by employees with responsibility to manage waste (0.872) and management of waste (0.864). These five variables are more than 0.850 and close to 1.000 which will be discussed as the main item in waste management plan.

From the analysis shows there is awareness among people to manage waste. It shows from the highest ranking which provides strategic management plan. These means many of the construction already have the strategic management plan to manage waste at construction site. In the study by Mahayuddin (2011) found all the construction sites have done a plan to manage and minimize the waste [27]. Besides that, organize waste also often practiced.

By referring to the indicator in table 2, table 6 shows the inter-item correlation matrix for variable in waste management plan. The value of correlation needs more than 0.50 [25]. In indicator A, there are three indicators that have not been practiced such as in indicator C, D and F. The indicator B does not practice indicator F only. For indicator C, indicator A and F get lower value than 0.50. There are 3 indicators (A, B, F) which are lower than 0.50 value for indicator D. As for indicator E, it is one indicator that does not have the correlation which is indicator F. From this analysis shows all the indicators does not have any correlation with the indicator F which records the quantity of waste. It similar with the RII in table 5 where record quantity of waste having the lower RII (0.732).

The record of the quantity of waste is scarcely done. Study by Mahayuddin (2011) also found that from 10 sites, there are only 2 sites are implementing the recording of production of waste [27]. Hence, the recording waste of the production needs to be applied more in order to know the actual amount of waste produced. The data of waste produced can be a guideline to overcome or reduce the amount of waste for other project.

Table 5. Relative Importance Index for Waste Management Plan.

Bill	Waste Management Plan	Total weight of statement ∑w	Relative Importance Index (RII)	Ranking
1.	Provide strategic management plan	227	0.908	1
2.	Organize the waste	224	0.896	2
3.	Policy regarding waste management	222	0.888	3
4.	Employees with responsibility to manage waste	218	0.872	4
5.	Management of waste	216	0.864	5
6.	Identifying the type of waste	209	0.836	8
7.	Measure & audit	211	0.844	7
8.	Review Plan	214	0.856	6
9.	Training employees	202	0.808	9
10.	Record the quantity of waste	183	0.732	10

Table 6. Inter-item Correlation Matrix.

Indicator	Inter-Item Correlation Matrix	Indicator	Inter-Item Correlation Matrix	Indicator	Inter-Item Correlation Matrix
A – B	0.827	B – I	0.775	E – F	0.122
A – C	0.424	B – J	0.564	E – G	0.859
A – D	0.391	C - D	0.743	E – H	0.716
A – E	0.659	C – E	0.795	E – I	0.789
A – F	0.081	C – F	0.251	E – J	0.763
A – G	0.712	C – G	0.856	F - G	0.338
A – H	0.650	C – H	0.659	F – H	0.154
A – I	0.764	C – I	0.772	F – I	0.369
A – J	0.493	C – J	0.755	F – J	0.254
B – C	0.513	D – E	0.766	G – H	0.750
B – D	0.504	D – F	0.438	G – I	0.898
B – E	0.723	D – G	0.623	G – J	0.676
B – F	0.163	D – H	0.439	H – I	0.704
B – G	0.743	D – I	0.602	H – J	0.622
B – H	0.592	D – J	0.644	I – J	0.860

5 Conclusion

This academic research of waste management correlated to the implementation of green highway. In this study, it determined all possible types of waste produced by construction of highway. As for the production of waste, the highest ranking of all is wood while the lowest is metal. In addition, this research also discovered about the waste minimization strategy as well as the best way to manage the waste. Reuse method is concluded as the best practiced to minimize all type of waste except tree root and stump. These two items are using recycling method to reduce the waste. In terms of waste management strategy, it is a crucial activity for all construction sites to manage their waste properly. The management plan needed to follow and review accordingly in order to provide the best management practice. This plan must be included the actual record of the quantity of the waste for accurate wastage calculation.

In a nutshell, this study only focus on one out of five elements for green highway initiatives. Further research on material and technology, energy efficiency, stormwater management as well as safety and social factors. This research would contribute a big handful of guide for Lembaga Lebuhraya Malaysia as well as other countries with tropical climate.

Acknowledgement

The Authors wish to thank the Universiti Teknologi Mara and a remark of indebtedness to the Fundamental Research Grant Scheme (FRGS) by the Ministry of Education, Malaysia for its grant award.

References

1. Bryce, J. M. (2008). Developing Sustainable Transportation Infrastructure. Retrieved from www.wiseintern.org/journal/2008/JamesBryceFinal
2. Begum, R.A., Siwar, C., Pereira, J.J. & Jaafar, A.H. (2007). A benefit-cost analysis on the economic feasibility of construction waste minimization: The case of Malaysia. Resources, Conservation and Recycling, **48(1),** 86-98.
3. Wokekoro, E. (2007, 28-30 August). *Solid waste management in the construction industry (a case study of Port Harcourt).* Paper presented at the International Conference "Waste Management, Environment Geotechnology and Global Sustainable Development, Ljubljana, Slovenia.
4. Global Environment Centre (GEC). (2012). Solid Waste in Malaysia. Retrieved 14/07/2015, from www.gecnet.info/index.cfm?&menuid=83
5. Papargyropoulou, E., Preece, C., Padfield, R. & Abdullah, A.A. (2011). Sustainable Construction Waste Management in Malaysia: A Contractor's Perspective. Management and Innovation for a Sustainable Built Environment.
6. Begum, R.A., Siwar, C., Pereira, J.J. & Jaafar, A.H. (2009). Attitude and behavioral factors in waste management in the construction industry of Malaysia. Resource, Conservation and Recycling., **53,** 321-328.
7. Formoso, C. T., Soibelman, L., Cesare, C. D., & Isatto, E. L. (2002). Material waste in building industry: Main causes and prevention. Journal of contruction engineering management, **128**, 316-325.
8. Tchobanoglous, G., Theisin, H. & Vigil, S.A. (1993). *Integrated Solid Waste Management: Engineering Principles and Management Issues*. Singapore: McGRAW-HILL.
9. Pitchel, J. (2005). *Waste Management Practices: municipal, hazardous, and industrial.* United State: Taylor & Francis Group.
10. Bai R. & Sutanto M. (2002). The practice and challenges of solid waste management in Singapore. Journal of waste management, **22,** 557-567.
11. Official national solid waste management (ONSW). (2012). Retrieved 05/08/2015, from http://www.kpkt.gov.my/jpspn_en/main.php
12. Environment Strategies for Cities. (2000). Sources and Types of Solid Waste. Retrieved 11/06/2015, from web.mit.edu/urbanupgrading/urbanEnvionment/sectors/solidw-waste-sorces.html
13. Basri, H., & Ahmad Basri N.E. (2008). UrusSisa: An Intelligent System for Integrated Solid Waste Managament. Journal of Sustainable Development, **1(2),** 39-46.
14. United Nations Development Programme (UNDP) (2008). Malaysia developing a solid waste management model for Penang. Retrieved 28/12/2015, from http://www.undp.org.my/Uploads/SWM-2008_final.pdf
15. California Stormwater Quality Association (CASQA). (2014). Retrieved 15/06/2015, from http://www.cabmphandbooks.com/Documents/Construction/WM-05.pdf
16. Poon, C.S., Ann, T.W.Y. & Jaillon, L. (2004). Reducing Building waste at construction sites in Hong Kong. Construction Management & Economics, **22(5),** 461-470.
17. Faniran, O. O., & Caban, G. (1998). Minimizing waste on Construction project sites. Engineering, Construction and Architectural management, **5 Iss: 2**, 182-188.
18. Macozoma, D.S. (2002). *International report – construction site waste management and minimisation.* International Council for Research and Innovation in Building and Construction (CIB). Report No. BOU/C361.
19. Bruce, M. & Smitchers, M. (1998). Implementing a waste management plan during the construction phase of project: a case study. Construction Management & Economics, **16(1),** 71-78.
20. Lim, M.C.H. & Ling, F.Y.Y. (2002). *Implementation solid waste management plan for construction projects in Singapore.* Architectural Science Review 45: Taylor & Francis Group.
21. Al-Hammad, A. M. & Assaf, S. (1996) Assessment of work performance of maintenance contractors in Saudi Arabia. Journal of management in engineering, **12(1996),** 44 - 49.
22. Lachimpadi, S. K., Pereira, J. J., Taha, M. R. & Mokhtar, M. (2012). Construction waste minimisation comparing conventional and precast construction (mixed system and IBS) methods in high-rise buildings: a Malaysia case study. Resources, Conservation and Recycling, **63 (2012),** 96-103.
23. Technologiya Methallov. (2008). Processing of Metal Radioactive Waste. Retrieved 10/08/2015, from http://www.technologiyametallov.com/englisch/oekologie_5.htm
24. Suez Environment. (2015). Tree Trunk and Roots. Retrieved 10/08/2015, from http://www.sita.com.au/community-education/site-tours-education/recycling-tips/tree-trunks-and-root
25. Hair, J.F., Black, W.C., Babin, B.J., & Anderson, R.E. (2010). *Multivariate Data Analysis.* Seventh Edition. Prentice Hall, Upper Saddle River, New

Jersey.
26. Kometa, S.T., Olomolaiye, P.O., & Harris, F.C. (1994). Attributes of UK construction clients influencing project consultants' performance. Construction Management & Economics, **12(5),** 433.
27. Mahayuddin, S.A. (2011) Penjanaan, Komposisi dan Pengurusan Sisa binaan: Kajian Kes Ipoh, Perak. Ph.D thesis, Universiti Kebangsaan Malaysia

MATEC Web of Conferences **66**, 00060 (2016)
IBCC 2016
DOI: 10.1051/matecconf/20166600060

The Rise Of BIM in Malaysia And Its Impact Towards Quantity Surveying Practices

Nurshuhada Zainon[1a], Faizul Azli Mohd-Rahim[2] and Hafez Salleh[1]

[1]*Senior Lecturer, Department of Quantity Surveying, Faculty of Built Environment, University of Malaya, 50603 Kuala Lumpur.*
[2] *Associate Professor, Department of Quantity Surveying, Faculty of Built Environment, University of Malaya, 50603 Kuala Lumpur.*

Abstract. Building Information Modeling (BIM) is a new buzzword that is gaining a momentum within the construction industry in Malaysia, and worldwide. BIM is a way of working that involves the automation of the entire project team using a 4D model that can do many of the traditional fuctions of a Quatity Surveyor (QS). As BIM is removing more of the traditional work of QS, there are rising fears that it could threaten the viability of the QS professions especially for small firms that still rely on producing bills of quantities. This paper aims to review the challenges and opportunities gained from BIM advancement in QS profession, especially in Malaysia. Despite of serious cost and commercial challenges of BIM implementation, BIM will work wonder as a tool that will cut the amount of time QS have to spend on analysis and provides opportunities to direct QS talents towards being more efficient. This is resulting to better project coordination with a lot less conflict, a more sustainable building in the final instance and also throughout the process, as less time is spent on each stage.

1 Introduction

The construction sector shares 12.2% of world Gross Domestic Product (GDP), with the Global Construction 2025 reported that 70% more construction work will be going on around the world (Branson, 2013). In Malaysia, the construction sector grew by 8 to 10% with 3 to 5% shares of local GDP (Bank Negara Malaysia, 2015), indicate its significant role towards the global growth and locally. Thus, it is necessary for the construction industry to undergo evolution in the era of globalization with the aim of getting international competitiveness. Owing to its complex in nature and involvement of multi-disciplinary parties, construction industry faces common cost, time, and quality issues. Nevertheless, the rise of Building Information Modeling (BIM) technology has emerged as of the application tools to meet the objective of escalating productivities in a construction project.

BIM gained its popularity due to its efficiency benefits in terms of time and cost saving, and increases coordination in information exchange. It is a new approach being adopted in managing construction project life cycle activities, from project design, construction, to facilities management (Eastman et al., 2011). BIM is defined as a digital representation of the physical and functional characteristics of a facility through share knowledge resource for information, forming a reliable basis for decisions during its life-cycle (Smith, 2014). Started with the use of BIM for architectural design, the conceptual underpinnings of the BIM technology go back to the earliest days of computing, as early as 1962 (Rodrigues, 2015). Today, BIM has become ubiquitous in the design and construction fields. In developed regions, the percentage of companies using BIM jumped from 28% in 2007, to 49% in 2009, and for 71% in 2012 (McGraw-Hill Construction, 2012). However in Malaysia, the architectural firms' BIM adoption rate was 20% (Mohd-Nor & Grant, 2014), and only 10% of quantity surveying firms involved in BIM adoption since BIM was introduced 10 years ago (Ali, Al-Jamalullail, & Boon, 2013) Additionally, engineering firms were found being steadily slow implement BIM (Rogers, Chong, & Preece, 2015). As architectural, engineering, and surveying practices are the main players in BIM working environment, this can be a significant indicator of slow uptake of BIM in Malaysian construction industry as a whole.

In any project, cost and program are key performance indicators, in addition to quality. With the advent of parametric modeling, quantity surveyors (QS) play a significant role in contributing these two fundamental parameters to the modeling process from the outset and add the most value from the earliest stage. With BIM, the QS can provide detailed and accurate estimates, automate measurement, speed up traditional estimating process, and better capture, manage, and deliver project information. Despite its incredible benefit, BIM uptake among QS is still slow in Malaysia. There are also fear and resistance in the surveying practice towards the changes brought by BIM. There is resistance among QS to accept the changes as they suppose that BIM can perform most of the traditional QS

[a] Corresponding author: zshuhada@um.edu.my

job scope which eventually will replace the QS (Matthews, 2011, RICS, 2011, NBS, 2012) which creates a situation that QS is no longer needed in the construction market. Thus, this situation derives an aim of this study which is to provide perspective on future of BIM in quantity surveying practices through reviewing the concerns of Malaysian QS and opportunities that are awaited.

2 Brief History of BIM in Malaysia

The adoption of BIM in Malaysia was driven primarily by the private sector since early 2000. However BIM has only become a buzzword when the Jabatan Kerja Raya (Public Works Department, PWD) introduced BIM in construction project planning for public works in 2007 through their BIM Standard Manual and Guideline (Latiffi, et.al, 2013). In an effort to foster BIM environment in Malaysian construction industry, the Construction Industry Development Board (CIDB) has formed the BIM Steering Committee, BIM Roadmap, and BIM Portal, involving policy makers, practitioners, and academicians (CDIB, 2016). In order to help and promote the potential and benefits of BIM, PWD, CIDB and Multimedia Super Corridor Malaysia also offer initiatives namely roadshows, seminars, workshops, on top of affordable training program (Latiffi, et.al, 2014).
The Construction Research Institute of Malaysia (CREAM, 2012) reported on the earliest BIM projects in Malaysia. The National Cancer Institute of Malaysia in Sepang was the first project involved the application of BIM in Malaysia, followed by other projects under BIM Pilot program such as the Healthcare Centre Type 5 Pahang, and Administration Complex of Suruhanjaya Pencegah Rasuah Shah Alam. In these pilot projects, BIM was utilized for site modeling, visualization, design review, clash analysis, 4D schedule and simulation, and record modeling (Latiffi., et.a;, 2014). Recently, BIM in Malaysia is more likely to be used for complex and high-risk projects (PWD, 2011).

3 Concept and the use of BIM

BIM is defined as a modelling technology and associated set of processes to produce, communicate, and analyze digital information for construction purposes (Abdullah, et.al, 2014). Fundamentally, BIM is not simply a type of software but an integrated approach between human activity to produce digital representation of physical and functional characteristics of a building through relevant software.

BIM applications span over the entire life cycle of a facility. BIM is applied in project planning, design, pre-construction, construction and post-construction (operations and maintenance) stages. In planning stage, the project team can utilize BIM in analyzing space and understanding the complexity of land regulations and space standards, hence saving time and providing the team more time to create more value-added activities (Azhar, Khalfan, & Maqsood, 2012). In this phase, advanced 3D laser scanning equipment is used to scan the site to determine the position of existing utilities accurately and then integrate them in the BIM model. On top of intensive use of BIM in designing, the user can perform quantity survey and get detailed estimates accurately which can be measured directly from the 3D models. The user can plan for site logistics and identify potential hazards on site, thus helping in the preparation safety plan. Project managers can benefit from scheduling and monitoring function with the help of 4D and 5D models. In post-construction stage, BIM is used for transferring the data of building (spaces, systems and components) into facility management operations, keeping track of building assets, and enable scheduled maintenance and provides information of building maintenance history

All the information of functional and physical characteristic of a building and the life cycle of project are included in BIM in a series of smart object, providing better decision support in the process of project development (Azhar et al., 2008). Besides, BIM allow fast and accurate updating of changes, while in conventional 3D CAD, the plans, sections and elevations are used to describe the building as independent 3D views. Furthermore, the man-hours needed to establish reliable space programs in BIM will be reduce as compared to the conventional technology. BIM also enhances effective communication between the parties involved in the project development. For example, the information such as the details of procurement, drawings, submittal processes and other specifications can be linked together in BIM, whereas the conventional approach is paper-based modes of communication. BIM increases the confidence in the completeness of the scope. This can be done through a clash analysis to detect conflict, collision, and interference due to the model form in scale, 3D space, so the user can visually check the interference in all the systems.

4 BIM in Quantity Surveying Practice

Quantity surveying is vital in construction field to control and manage the cost throughout the project life cycle. Traditionally, taking off and bill of quantities demanded long working hours and effort to be produced, with high-risk of human errors. As the building works are becoming more and more complex, BIM has developed into the mainstream in construction industry in developed country, BIM is expected to be embraced in QS firms to boost up the effectiveness in term of cost and value of construction processes (BCIS, 2011). Quantity surveying requires immense knowledge and skillful interpretation and practical use of the knowledge (Vineeth & Jenifer, 2014). Implementation of BIM brings opportunity to quantity surveying field in the cost management functions.

Bill of Quantities (BQ) is an important tool to carry out cost management in contruction projects. BIM technology provides a fifth dimension (5D) of BIM which can automatically produce BQ. The collaboration and integration feature of BIM in the automatic production of BQ is an enhancement to the BIM technology. This is effective in eliminating the lengthy and tedious ways of tranditional taking off as well as reducing human error. BIM is able to extract exact quantities and areas that are used to estimaste cost at anytime during the project design stage. Not only that, QS is able to recognize and relate how the total project cost is made up by each elements or spaces of the building through BIM as BIM eases the identification of relationship of quantities versus the locations and costs. This enhances the competency of QS through the comprehensive knowledge of cost determinants which in turn generate cost estimation with high accuracy and reliability. Any changes in the design will actually being automatically updated through the quantities drawn out from the modified model; it does not require QS to redo the taking off as in traditional method. The cost estimating integration with the BIM design engine, it allows the project team to execute value management durong the design phase efficiently.

BIM is capable of providing a more intensive and detailed drawing compared to the traditional 2D drawing where misunderstandng and wrong assumption may be made. The more comprehensive construction information and more pricise BQ can diminish the gap among the project team members. Through understanding on how to extract the quantity from building model, bidders are able to bid with intensely competitive price.

Through the 5D feature of integrating cost estimation with building model, the estimating job of a QS has is becoming simple and straightforward as compared to the conventional paper drawing. With BIM, construction field is heading to a paperless construction.

5 Methodology

This study was conducted during September – December 2015 to retrieve articles related to BIM and QS issues literature. No specific key words required as inclusion criteria; a relatively small number of studies exist on the topic. Articles were retrieved from diverse platform which include journals, technical reports, and news. The reference lists of each article were reviewed in detail to find additional articles. A group of reviewers (n=46) independently read articles in full text, and recorded the main findings of challenges and opportunities of BIM in quantity surveying in Malaysia, consequently confirmed by a number of Malaysian certified QS (n=3).

6 Findings

6.1 Challenges of BIM for Quantity Surveying Practice

Clearly with the rate of BIM adoption among QS firms, there are challenges spelled out, as follows, in adopting BIM as it is relatively new to surveying practice:

6.1.1 Teamwork and collaboration

As new ways for cooperation between the project team has been offered through BIM implementation, some problems may arise among them. For instance, an issue may occur when the method for sharing model information among the team member is being determined. If traditional drawings on paper have been used by the architect, then a third party will have to construct a model to be used for the estimating and planning of the construction project.

In addition, if the team members use a wide variety of tools for modelling, then the project may require some other tools to move the models between different environments or to combine them together. Hence, it may add complications or cause errors to the project. In order to overcome this issue, the Industry Foundation Classes (IFC) standard can be adapted when data is being exchanged throughout the project life cycle. Other than that, team members can also use a model server that can enable exchange of information between the BIM applications through adaptation of IFC standards.

6.1.2 Legal issue

Challenges in BIM implementation may arise when the question of ownership and responsibility of the various design, analysis and datasets are asked among the team members. It will be a challenge to determine who is to pay for it and who will be in charge of its precision and accuracy. These challenges are being questioned by the experts when BIM is used in their projects. A building model will be needed by the building owners to enhance the maintenance, renovations and operations once they educate themselves about the benefits of BIM.

6.1.3 Use of information and changes in practice

Firms that can coordinate the design phases and include in-depth knowledge about construction will get the highest advantage as BIM implementation in a project stimulates the assimilation of construction related information and knowledge in the beginning of the design phase. Furthermore, the arrangements in contracting that prioritizes excellent cooperation will benefit the owners more when BIM is applied in a project. In addition, when BIM is applied in a certain project, a building model will be constructed and shared among the project team as it is the basis of every work process in the project. Knowledge and time are two key components to facilitate this change that will be faced by the companies, as they are also required for every important transformation in technology.

6.1.4 Implementation issue

To effectively start using BIM, firms need to upgrade their hardware, obtain the necessary software, and also require their employees to go for the correct BIM training in order to transform a 3D environment into a complete

BIM system. The firm has to comprehend every detail and plan for the implementation carefully before any transformation can take place. In addition, the firm can appoint managers to head a team responsible for planning the BIM implementation by prioritizing budget, cost and time as their guidance. Other than that, an adoption plan could be developed to predict how the future implementation changes will impact the firm's partners and clients and also how it will affect its internal departments.

6.1.5Financial and time considerations

In order to implement BIM successfully, a firm needs to periodically upgrade their hardware in order to run the processing software. It is obvious that the requirement of a big technology component to implement BIM has become a challenge for the surveying industry in Malaysia. However, by preparing the same software from similar vendors for all the companies, the issue of some hardware and software being hard to operate could no longer be a bother for the industry players. Hence, in order to fix any technical issues regarding BIM adoption, there will be some impact on the firm financially as well.

According to Autodesk (2013), the price for Building Design Suite Premium, an entry level software for BIM, is US$6,825. With the current exchange rate being US$1.0:RM4.21, that makes the price in Malaysia to be RM28,749.00. Furthermore, the cost calculated beforehand only includes the purchase of the most basic BIM software. The firm has to consider costs to train their employees and even hire new workers who are equipped with the knowledge and skills in BIM as they move forward to internalize a new working environment in their firm. It can be deducted that adopting BIM requires the firm to make a huge investment financially as only large organizations can afford the costly technology.

In order to transform a company to being accepting of a new technology such as BIM, a lot of time will be consumed and this is seen as a challenge as construction projects are highly affected by time. In Malaysia's construction industry, BIM experts are still very low in number. Hence, firms need to allocate time and also money in order to find the best experts to assist them in BIM implementation. In addition, firms that decide to send their workers for training in BIM technology would also have to consider the additional time and cost needed for them to master and learn the processes entirely.

6.1.6 Process change

For an organisation to adopt BIM fully, there will be essential and basic changes to its operational processes. Once a company successfully adopts BIM, the process of design will be brought forward to the start of the project. In addition, it will also be completed sooner with a completion level higher than those of traditional processes. Furthermore, BIM experts have considered Autodesk, Bentley and Nemetschek to be among the technology platforms that are interoperable even though they do not run that smoothly together, despite there being some protests to it.

6.1.7Human resource and organisational issue

For any organisation to fully adopt BIM in their projects, their human resource department will need to go through a complementary change in terms of their skills alongside an essential process change within the organisation. Furthermore, there will also be simultaneous increase of competency in the supply chain of the project, which also includes the contractors, designers, and also the developers.

People working in the organization can also be another challenge to face. When the important people in the firm are reluctant to accept the new tools and technology, it makes it that much harder to change their behaviour into liking and accepting them. When the staff resists and show reluctance to BIM, it will become a big challenge to the top management to persuade them to accept the new change to the organization. Their reluctance to learn something new or try the new softwares and technology will be the hardest challenge that the top management will have to overcome in order for BIM to be implemented successfully.

The role of top management is the most crucial as they have to decide the direction and the plans of action for the organisation to adopt the new technology while reducing the employee's resistance to change. In addition, the employees will be motivated as they witness the organisation's dedication in fulfilling their responsibilities seriously. The motivation shown to the employees is one of the ways to help decrease the reluctance of the employees in accepting the new change as well as to encourage them to be more confident in trying out the IT applications themselves.

In addition, there are some organisations that are scared to take the risk in changing their business process by adopting BIM. They worry about the huge uncertainty that may or may not affect their established firms, as well as the large cost that they have to bear. Other than that, managing the big change in terms of technology may also cause a challenge for the managers. The staff may feel threatened or anxious by the new technology that is involved with accepting BIM as they feel that their roles and jobs in the organisation will be completely taken over by the softwares and hardwares that BIM requires in order to function. Hence, managers must know how to tackle the technology change in the organisation.

Furthermore, another challenge faced by the organisation is their belief that their well-recognized firms will be affected by the new technology that comes with BIM implementation. When a new technology is being adopted, the business processes of a firm will also be affected, which can also impact the firm's significant rate of output. This, in turn, will also increase the risk in producing the outcomes of a project as well as the client's preset goals.

6.1.8 Professional support

In Malaysia, the Government constraints and mandates the limits of the Engineering Consulting Services (ECS). Even though BIM has been known to provide interchange of electronic data with no boundaries, the management of professions, trades and industries are still kept fixed in cultural and geographical boundaries, as well as political identities. As a step to break down these barriers in the ASEAN community, there have been some movements in the region.

However, aside from the complex projects, most of the local companies are still being operated by local staffs that are following the local guidelines and mandates. Hence, some potential barriers have been found in various countries as well as in Malaysia, in order for BIM to be adopted in Malaysia's ECS. Therefore, it has been proven that in order for BIM implementation to be successful, there must be presence of consultants or professionals equipped with the skills and expertise to support it.

6.1.9 Technical challenges

There are obvious reasons as to why BIM adoption in Malaysia has not really been as much as anticipated. It is hard to adopt new information technology (IT) in the industry due to technical reasons as compared to social issues. Problems such as the lack of support system, difficulty to understand the complex softwares and also the low number of skilled technical experts are among the reasons.

Furthermore, BIM may have been accepted at a low rate because of the need for detailed and precise models to fix the issue of interoperability. Other than that, in order for important information to be communicated and integrated successfully among the components, there is a requirement for excellent practical strategies to be developed. This could lead to another technical reason as to why BIM has not been the choice for some firms. In addition, they also have to ensure every data developed is computable as it is part of BIM's technical requirements.

6.1.10 Management

Currently, there have not been any specific directions or orders in ways to use or implement BIM. There are no clear instructions on how to use and apply BIM in construction practises. Even though a few software companies have developed some products to help with the implementation of BIM, they have only managed to cover the aspects of quantity instead of implementing it as a whole. Therefore, for BIM to be adopted by firms easily there must be some clear and specific guidelines to be followed and the steps to implement BIM must be standardized for all firms. Another management challenge is the question that arose among the stakeholders, which is who shall be responsible for developing the models and how the costs will be distributed. Hence, to clear these issues, solutions have to be created by researchers, vendors and professional organizations together so that the use of BIM will increase in the Malaysian industry.

Before BIM, the facilities manager had very limited influence in the planning process of the building. They were only able to introduce strategies to maintain the building after the owner has received possession of the building. With implementation of BIM, the managers can be included in the stages of designing and construction. With BIM's visual ability, stakeholders such as tenants, maintenance staff and also service agents, can get a hold of any information that they might need even before the completion date. The biggest challenge will be for the owners, as they have to plan wisely for the right time to allow all the different parties to be included in the early processes of the building.

6.2 Opportunities for Quantity Surveyors

The competitiveness of QS is highly improved. As the cost estimation can be done through the quantities extraction at any stages during the design (Boon, 2009), the efficacy of a QS is highly enhanced. For example, with conventional way, QS is only able to finish up a complicated earthwork in couple of days but by extracting the quantities from BIM only takes less than an hour. When variations arise, QS can re-extract the quantities from the modified model and avoid the tedious human measurement. A competent and high efficiency QS poses better chances in surveying field. Not only that, the improved quality of estimation produces quality tender which will enable the QS to help company to open up more business opportunity and manage to get repeat business by winning the tender (Autodesk, 2003).

6.2.1 Transformation in job scope

Through the benefits brought by BIM to the surveying field in term of ease of quantity take off (Davidson, 2008), QS in malaysia do not need to spend time to measurement and find out the dimensions in from a 2D drawing. Unnessary assumptions and human errors can also be avoided. Now, with a highly accurate BQ produced in a very short time, QS is freed up to do more on cost management not in term of the manual taking off and adding up. There is no doubt that the QS still need basic skills regarding measurement to make use of the BIM technology, however, QS now saves more time and they are now no longer a wetware calculator, but they need to think more on what client's needs and wants. There are so many aspects that a QS need to cover.

QS not only provide advices on the construction methods and construction materials, QS also need to handle the legal issues. QS also have to be major on feasibility studies, value management, cost planning, tendering and dispute resolution (Gee, 2011). This opens the gate for Malaysian QS to involve in a construction project more comprehensively, but not major in measurement like current industry practice. This is heading to the QS job perspective in developed country construction industry, the best example to cite will be United Kingdom.

6.2.2 Re-branding of Quantity

With BIM, it is an era for Malaysia QS to move away from just BQ, there is a rebreanding their job from a profession that only prepares BQ to a higher end consultants. In future, BIM will turn QSs to profession doing cost management, in term of cost modelling and also general advisor in pre-contract stages. QS is no longer branded with "Construction Industry Accountant", but also branded as construction industry legal specialist as they are the one to handle the legal based issues in quantity surveying field such as dispute resolution and giving legal advices regarding construction contract issues such as the procedure and situation where the contrcator is entitled from variation claim, the liability and risk allocation of each parties in the contracts and so on. QSs' proactive and knowledge based advices taking in all the aspects and situations of a project in consideration is important to the client (Verster, 2006).

Not only limited to that, by eliminating the lengthy quantity take off by BIM, QSs can be more engaged in contruction management especially value management and risk management. Other than that, QSs also involve in giving advices on safety and healthy and also quality control (Cunningham, 2014). Hence, BIM has provide the opportunity for QSs in Malaysia surveying to be rebrand from the root as a measuring profession to a profession serving towards value-added into a project. BIM aids the Malaysian surveyors to handle numbers, measurements and the rigid measurement rules with ease and helps them to think more on strategy by utilising critical thinking.

6.2.3 Globalization

BIM opens up the gate for QS in Malaysia to have a same communication platform. In conventional way of current Malaysia surveying practices, QS are using the Standard Method of Measurement of Building Works Second Edition (SMMS) as their measurement rules reference. Different standard menthos of measurement are used in different places. This BIM, there is a communciation barrier between the QS across different region. According to Heiskanen (2014), the very first international survey carried out in 2013 revelaed that consultants that do not adopt BIM were struggled to secure international project. Malaysian QS do not face difficulties when engaging in an internatioal projects where there are different surveyors from different backgorund who has previously learnt different standard methods of measurements. BIM softwares has ease their works without background barrier. This is one of the contribution factor to the success for local contractor to strive for success in international projects as the coordination and communication are greatly improved trough BIM (Broquetas, 2010).

6.2.4 Beyond limit

Malaysian surveying industry produces QS that are expertise in measurement with local standard rules og measurement. When they are to engage in international project or being posted to other branch of company overseas, they requires intensive training to learn the Qs practices in other country. Hence, the possibility of a QS to venture into international market is hard and limited due to the resistance to be accepted by international surveying market as different places poses different practices.

BIM is seen as a solution to this problem. It is easier for Malaysian surveyors to venture to international market as BIM is linked to software to enabe extraction of quantity. Whenthe surveyors know how to handle the software, it takes a very short time for a QS to catch up with the job scope especially in producing accurate BQ. The other aspects such as the legal aspect and also the risk and value management is a broad knowledge and no region boundary as it is something critical and analytical which requires QSs' to think out of the box.

Therefore, BIM is a tools that enables Malaysian surveyors to go beyond the limits, they are no longer bound to just one country. They can hop to different countries and still serve as QS with ease. BIM enables QSs not to be rooted and stucked in to serve in one country and cannot go far. To see from other perspective, surveyors and experts from other places can also be imported to gain their stregth and power so that they can be adopted in Malaysia surveying field. This is a vice versa relationship.

7 Conclusion

Through the continuous advancement in technology, construction industry cannot deny the changes but have to adopt it. BIM, as the latest trend in construction industry, can help to improve the construction process in so many ways such as providing rapid visualization, giving better decision support, providing rapid and accurate changes updating, providing better and efficient communication and so on. All these aspects meant to improve and secure the overall quality of the project, making local construction project to be more competent in international construction industry. BIM not just emphasize on the design stages, it is actually a comprehensive technology which covers throughout the pre-construction stage, construction and post-construction stage.

The challenges that come with BIM implementation affect every sector of the organisation that wish to adopt it. There are challenges in teaming and collaboration as well as the question of documentation ownership and production. Furthermore, implementation issues, use of information, changes in process as well as financial and time considerations are some other challenges that must be faced by the organisations. As people are one of the most tangible factors, the challenges will greatly affect the human resource and organisational issues of the firms. In addition, the challenges that may be faced by the organisation could also be categorized into management and technical challenges. It is clear to see that BIM carries along a lot of issues and challenges with its implementation. However, if an organisation can manage smartly and take all the right steps in overcoming them, it will get to experience the convenience and powerful technology that BIM can offer. Now leaps far and see a bigger picture on the opportunities for Malaysian surveying practices with BIM implementation. BIM aids in the quantity surveying practices in their works on preparing BQ, cost estimation, updating of cost and even in tendering process. With all this, BIM eventually is a way for the QS to get out from the conventional method of doing quantity take off which is very tedious which in turn will transform the job scope of surveyors, making them to be rebrand as higher end consultants. This is not the end of their opportunity as BIM provides a platform for local surveyors to integrate internationally with other surveyors. Hence, local surveyors are not nationally restricted. In fact, they can go beyond the limit and works in international established companies or projects with ease. This is also a golden opportunity for local surveying practices to immerse with new strength and knowledge from other places as BIM ease effective communication.

Acknowledgement

This research is supported by British Council Newton-Ungku Omar Fund Institutional Links (ID: 172726659) and University of MalayaInstitute of Research Management & Monitoring (IPPP) (grant no: IF009-2015.

References

1. Abdullah, S.A, Sulaiman, N, Latiffi, A.A, & Baldry, D. (2014). Building Information Modeling (BIM) from the Perspective of Facilities Management (FM) in Malaysia. Universiti Tun Hussein Onn Malaysia, retrieved from http://eprints.uthm.edu.my/5425/1/IRERS_2014_2_U.pdf.
2. Ali, K.N., Al-Jamalullail, S.N.N.S.I, & Boon, T.C (2013), Building Information Modeling Awareness and Readiness among Quantity Surveyors and Quantity Surveying Firms, Universiti Teknologi Malaysia-Royal Institution of Surveyors Malaysia, Technical Paper.
3. Autodesk. (2013). *Autodesk Building Design Suite*. Retrieved from http://usa.autodesk.com/autodesk-building-design-suite/compare/.
4. Autodesk. (2003). *Building information modelling in practice*. Retrieved from http://www.ddscad.com/BIM___In_Practice.pdf
5. Azhar, S., Khalfan, M., & Maqsood, T. (2012). Building information modelling (BIM): now and beyond. *Australasian Journal of Construction Economics and Building*, ***12***(4), 15-28.
6. Bank Negara Malaysia (2015), Economic and Financial Developments in Malaysia in the First Quarter of 2015, retrieved from http://www.bnm.gov.my/files/publication/qb/2015/Q1/1Q2015_fullbook_en.pdf.
7. BCIS. (2011). *RICS 2011 Building Information Modelling Survey Report*. Retrieved from http://www.bcis.co.uk/downloads/RICS_2011_BIM_Survey_Report.pdf
8. Boon, J. (2009). *Preparing for a BIM Revolution*.13th Pacific Association of Quantity Surveyors Congress (PAQS 2009). (P.35).
9. Branson, A (2013), Global Construction 2025, Building.co.uk, retrieved from http://www.building.co.uk/global-construction-2025/5057217.article.
10. Broquetas, M. (2010). Using BIM as a project management tool: how can BIM improve the delivery of complex construction projects?. *Master degree dessertation, University of Applied Sciences, Stuttgart.*
11. Bryde, D., Broquetas, M., & Volm, J. M. (2013). The project benefits of building information modelling (BIM). *International Journal of Project Management*, ***31***(7), 971-980.
12. CIDB (2016). BIM Training, retrieved from http://www.bimcenter.com.my/.
13. CREAM (2012). A Case Study of BIM Implementation in Malaysia, retrieved from http://www.myconstructionresearch.com/index.php/publication/integrated-design-delivery-solution/file/12-bim-implementation-in-malaysia.
14. Davidson, A., R. (2008). *A study of the deployment and impact of Building Information Modelling Software in the Construction Industry*. Retrieved from http://www.engineering.leeds.ac.uk/e engineering/documents/AndrewDavidson.pdf
15. Eastman, C., Teicholz, P., Sacks, R., & Liston, K. (2011), BIM handbook: A guide to building information modeling for owners, managers, designers, engineeris, and contractors, Wiley: New Jearsey.
16. Gee, C. (2011). The influence of buidling information modelling on the quantity surveying profession.
17. Latiffi, A.A., Brahim, J, Mohd, S, Fathi, M.S. (2014). The Malaysian Government's Initiative in Using Building Information Modeling (BIM) in Construction Projects, ASEA-SEC-2, Bangkok, November 3-7.
18. Latiffi, A.A, Mohd,S, Kasim, N, & Fathi, M.S. (2013). Building Information Modeling (BIM)

Application in Malaysian Construction Industry. International Journal of Construction Engineering and Management, 2(A), 1-6.
19. NBS. (2012). *National BIM Report 2012*. Retrieved from http://www.thenbs.com/pdfs/NBS-NationalBIMReport12.pdf
20. Matthews, D. (2011). *Rise of the Machines: BIM and QS's*. Retrieved from www.blackenewport.co.uk/bna-news-and-media.asp?id=90
21. McGraw-Hill Construction (2012), The business value of BIM in North America: Multi-year trend analysis and user ratings smart market report, retrieved from http://bimforum.org/wp-content/uploads/2012/12/MHC-Business-Value-of-BIM-in-North-America-2007-2012-SMR.pdf.
22. Mohd-Nor, M.F.I & Grant, M.P. (2014). Building Information Modeling (BIM) in the Malaysian Architecture Industry, WSEAS Transactions on Environment and Development, Vol.**10**, 264-273.
23. PWD (2011). Laporan Tahunan Jabatan Kerja Raya 2011, pp 28-29. Retrieved December 6, 2015, from http://www.jkr.gov.my/var/files/File/dokumen/laporan_tahunan_jkr_2011.pdf.
24. Rodrigues, R. F. O. (2015). Otimização do processo de pré-dimensionamento de fundações diretas com recurso à metodologia BIM.
25. Rogers, J, Chong, H.Y, & Preece, C. (2015). Adoption of Building Information (BIM) technology: Perspectives from Malaysian engineering consulting services firms. Engineering, Construction, and Architectural Management, **22**(4), 424-445.
26. Smith, P. (2014), BIM implementation – global strategies. Procedia Engineering, 85, 482-492.
27. Verster, J. J. P. (2006, April). Managing cost, contracts, communication and claims: A Quantity Surveying perspective on future opportunities. In *Proceedings of 1st ICEC & IPMA Global Congress on Project Management, 5th World congress on Cost Engineering, Project Management and Quantity Surveying* (pp. 23-26).
28. Vineeth, R., & Jenifer, P. (2014). *Role of Building Information Modelling (Bim) in Quantity Surveying Practice*. International Journal of Civil Engineering and Technology, **5**(12), 194–200.

Assessment of Indoor Environmental Quality (IEQ): Students Well-Being in University Classroom with the Application of Landscaping

Nurul Malina Jamaludin[a], Norhayati Mahyuddin and Farid Wajdi Akashah

Centre for Construction, Building and Urban Studies (CeBUS), Department of BuildingSurveying, University of Malaya, 50603 Kuala Lumpur Malaysia

Abstract. The environmental quality (IEQ) in a building is an important element to perceive the good health and comfort level for the building occupants. However, each building contributes different environmental quality results towards the indoor spaces and the occupants. Learning environment is one of the spaces that need attention as it is related to student's well being as well as their learning performance. Existing knowledge on IEQ is still limited concerning the desirable levels of air quality, maintenance, and other factors affecting IEQ in Malaysian educational establishment. Therefore, the study of indoor environment quality in buildings has been carried out in educational building as it acts as important place in learning process. The methodologies used to conduct this research are divided into two methods, which are classroom measurement normal condition and classroom intervention setting. This is done in order to compare and monitor the improvement of environment in the classroom. This research focuses on the comparison of IEQ in different classroom environment setting and the student satisfaction level in their normal classroom environment. Measurement of temperature (°C), relative humidity, carbon dioxide (CO_2), Volatile Organic Compound (VOC), dust particles (PM10), lighting (lux), and noise (decibel) in the classroom were collected and questionnaires were distributed among the students. This research found that most of the elements in the classroom was incompliance with the standard threshold limit value. The level of VOC in the classroom was noted to be significantly high (11.7ppm) compared to the standard threshold limit. An intervention on the normal condition classroom was set up with selected plant placed in the classroom. Results show a tremendous reduction in the percentage of relative humidity, level of TVOC, as well as CO_2.

1. Introduction

There are many researchers found that IEQ may affects peoples performance. This is proven by the study of (Fisk, 2000; Mendell et al., 2002) which explained that poor indoor environments could affect the student learning process in school. According to National Institute of Occupational Safety & Health, Indoor environmental quality (IEQ) is defined as the buildings environment quality which related to health and wellbeing of those who occupy space in it (NIOSH, 2013). According to (Hill MC, 2009), a satisfactory condition of environment such as lighting, temperature, and space management plays an important role in order to improved student satisfaction towards learning environments.

Based on latest data provided from The Ministry of Higher Education (MOHE) in Malaysia which is around 10 years ago, there are more than 900,000 students pursuing higher education in 20 public universities, 33 private universities and university colleges, 4 foreign university branch campuses, 22 polytechnics, 37 community colleges and about 500 private colleges (The Ministry of Higher Education (MOHE), 2004). While for the total number of government schools in Malaysia including primary schools and secondary schools are 10,119 employing over 421,393 teachers with 5 138 682 number of students in Malaysia (MOE, 2014). From this, we can classrooms than any other environment of schools, therefore the IEQ of the classroom need to be concern in order to improve their academic achievement.

There are many factors of IEQ that may affect to the student. This is because each building will have different environmental quality that will be perceived by the occupants that would relate to the types of activities carried out in indoor spaces of the building i.e., classroom, laboratories and others. Since Malaysia's climate is classified as 3 equatorial, being hot and humid throughout the year, this might result in dissatisfaction of the building occupants regarding to IEQ issues of building, with particular emphasis on thermal comfort.

The study by (K. Khadijah et al.2013), has been carried out in classroom on second floor at local school in Bandar Baru, Selangor in order to assess thermal condition during teaching lesson. This study was carried out throughout 5 days from 2pm to 6pm. As a result, both student and teacher are not satisfied with the environment quality in their classroom. This is due to lack of air

[a] Corresponding author: malina_nurul@siswa.um.edu.my

ventilation in the classroom. Besides that, the layout plan of class need to be changed as the position of students table near from each other which will lead to discomfort to student. This is because every of human body provides heat transfer. Therefore, if they seat to close to each other they may feel discomfort because of high thermal comfort.

However, the problem in Malaysia is the occupants of the building are not aware of the effects of IEQ in their surrounding especially in the learning environment. Existing knowledge is still limited concerning the desirable levels of air quality, maintenance, and other factors affecting IEQ in Malaysian educational establishments. According to (Turunen et al., 2014) poor IEQ can lead to health symptoms it can also cause decreased student focusing in the classrooms. Findings show correlation between headaches, dizziness, heavy head, tired, difficulties concentrating, unpleasant odor, and high CO2 concentrations are example of health symptoms that felt by the students due to poor environment quality. This can be preventing by using plant as to decrease the effect of IEQ especially for indoor air pollutants. However, according to (Wood et al. 2006) the studies on ability of potted-plants reducing air-borne contaminants in improving the air quality are still limited.

Other than that, ASHRAE, (2008) described that better IEQ can help and improve the learning process of students. Based on the basic space and function of educational buildings, where teaching and learning are the priority, it is predicted that universities may also be facing similar problems with the IEQ in their classrooms. One of the primary reasons for this is that occupancy is generally denser in classrooms. Therefore, this research is conducted to study the occupants" perception on how different environment gives different impacts on their method.

2. Materials and Methods

This research was conducted by using quantitative analysis applying field work monitoring conducted at every designated area where information of indoor environmental quality is measured as listed in Table 1.

Table 1: Research Methods

Method	Questionnaire Survey	Fieldwork Monitoring	
		Measurement Set Up	Classroom Intervention
Details	Questionnaire (based on elements of IEQ and current students satisfaction level	Sampling devices used to monitor the indoor environment of the classroom	Intervention on application of landscape to reduce the level of TVOC in the classroom.

2.1 Questionnaire survey

20 questionnaires were distributed among the students in the classroom to investigate the satisfaction level of the students in the classroom. The questionnaire mainly related to IAQ, acoustic quality, visual comfort, thermal comfort and effect of IEQ to the students learning performance. All data collected from the questionnaire survey will be analyzed with a statistical method using Microsoft Excel.

2.2 Fieldwork monitoring

Measurement will focus on temperature, ventilation and relative humidity in the selected area which is might be considered as factor that can contribute to poor indoor air quality. There are 2 types of measurement, which is classroom measurement set up in Fig.1 and intervention on the classroom environment in Fig. 2.

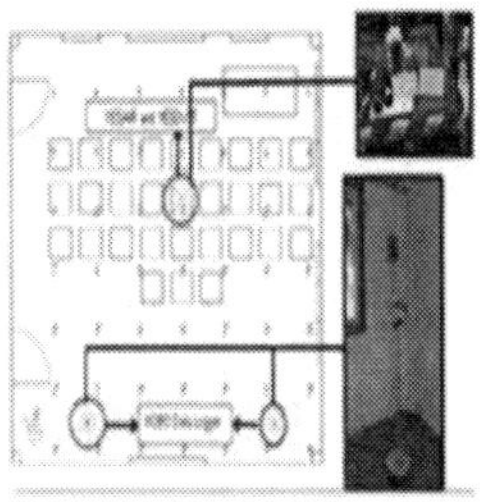

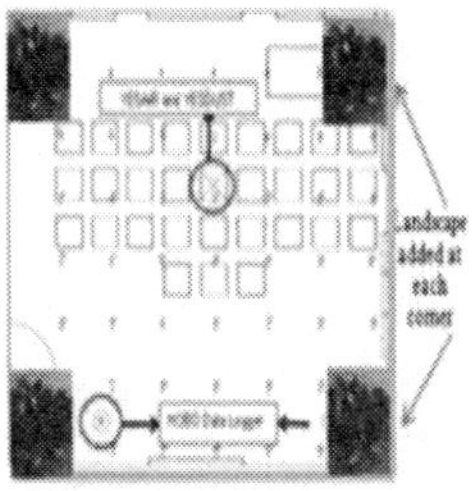

Figure 1: Classroom measurement set up
Figure 2: Intervention on the classroom environment

2.3 Classroom Environment Intervention

The environment interventions will be designed to compare between the level of IEQ based on the current IEQ level in the classroom and by placing plants inside the classroom. This is because, from the data collection, shows that the rate of TVOC inside the classroom was significantly high which exceeded the standard threshold limit value (TLV). Therefore, Peace lily and Janet Craig species were selected and placed in each corner of the classroom.

3. Results and Discussions

3.1 Questionnaire Analysis

Fig. 3 shows that the classroom was fully occupied by students from 1-3 hours daily. Normal class session starts from 9.00 am daily and finish at 5.00 pm. However, due to the varying time table schedule, most of the students spend less time in the classroom.

DOI: 10.1051/matecconf/20166600061

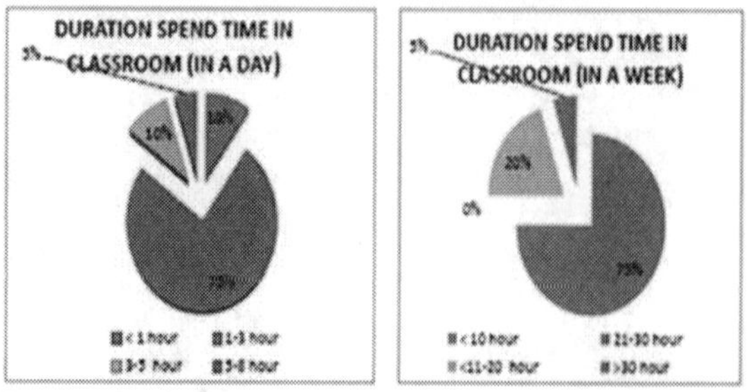

Figure 3: Daily and Weekly time duration spent in the classroom

Feedback from the questionnaires justified that the occupant's perceived learning productivity is not easily interfered by any of the elements of indoor environment quality probably due to the young age of students. From the questionnaire, the mean age of the respondents is around 19 years old. This explains that the respondents are neither easily being affected by indoor environmental qualities and nor susceptible to any health symptoms. Moreover, all possible health symptoms were found unnoticeable in the classroom (see Fig. 4). However, there were some minor health symptoms that has been identified such as stress, dry skin, itchy eye, blurred vision as well as tired eye.

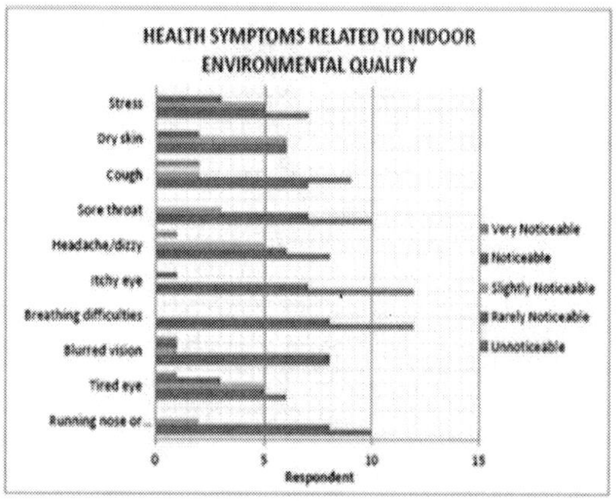

Figure 4: Health Symptoms Related to IEQ

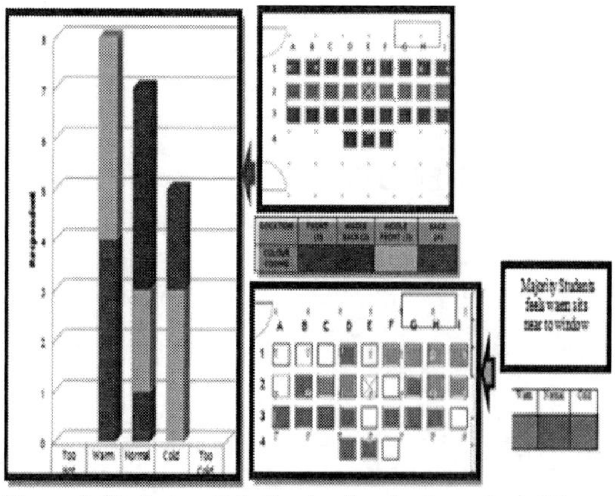

Figure 5: Students satisfaction level on the thermal condition

With regards to the thermal condition in the classroom, majority of the students rated warm as illustrated in Fig. 5. This might be due to the heat transmission through the windows and lack of air circulation, which eventually reduces the ventilation rates in the classroom. It was observed that the classroom orientation with windows is towards Easts, which mean that the classroom gain higher penetration from sunlight and this can cause heat gain in the classroom. However, there were few students rated cold in the classroom. Based on the sitting position (colour indicator in Figure 5), it can be concluded that the students perceived satisfaction on the thermal condition is due to the location in the classroom whether or not they are sitting near the air conditioner. In addition to this, it was also observed that only one air conditioner was operated in the classroom. Hence, the perceived satisfaction level on the thermal comfort was not satisfactory.

3.2 Fieldwork Monitoring –Thermal Comfort, Ventilation and Indoor Air Quality

The measurement taken in the classroom is to identify the element of indoor air quality that contribute into environment quality in the classroom which include temperature, relative humidity, volatile organic compound (VOC) and concentration of carbon dioxide (CO_2). Fig. 6 shows data collected on measurement of indoor temperature at BT7. The temperature was compared with ASHRAE, 2010 Standard as well as Industrial Code of Practice (2010). The ASHRAE provide the suitable temperature for good indoor air quality is within 23°C to 26°C.

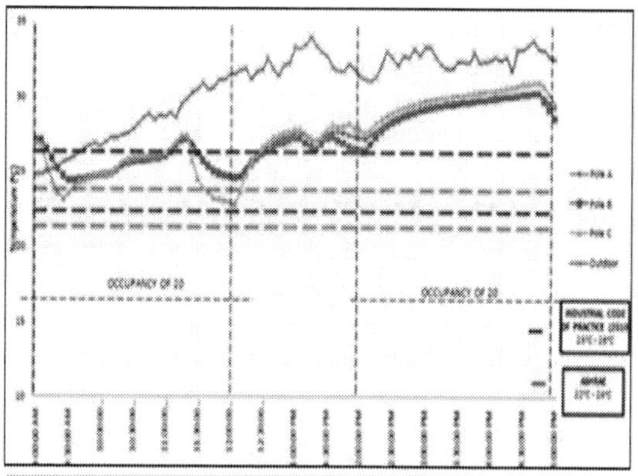

A/C Temperature setting	Average Indoor Temperature Throughout the Day	Average Outdoor Temperature Throughout the Day	Average Indoor Temperature when Occupancy	Average Indoor Temperature when Non Occupancy
23°C	27.3 °C	28.3 °C	27.41° C	26.66° C

Figure 6: Indoor air temperature in the classroom at normal condition

Based on the graph, the indoor air temperature in the classroom is in the range of 23°C to 30.9°C. This result supported the findings from the questionnaire where majority of the students feeling warm in the classroom (see Fig. 5). The indoor air temperature pattern was observed to be influence by the outdoor air temperature.

However, it was noted that from 9.00 am to 2.00 pm, the ratio of indoor outdoor temperature was slightly lower. This is due to the operation of t air conditioning system when students occupy the classroom. Yet the temperature levels were above the standard range set by ASHRAE and Industrial Code of Practice (dotted lines) as illustrated in Fig. 6.

With regards to the air circulation in the classroom, it was noted that the average reading for air velocity was only 0.14 m/s as illustrated in Fig. 7. This value is below the TLV set by World Health Organization (WHO). Air velocity level that is below which is 0.25m/s is considered as stale air. Due to this, it can be concluded that the air movement in the classroom is very low. This finding is supported further by the students' perceived satisfaction level on the air movement based on the responses from the questionnaire.

Majority of the students feel that the condition of the air in the classroom are quite stuffy. This is probably due to the enclosed classroom where doors and windows were closed at most of the time when the room is occupied. It is also due to the operation of only one air conditioning system, which reduces the ventilation rates of the classroom further. It was observed that there were two exhaust fan above the door, but was not functioning to allow the air exchange of air.

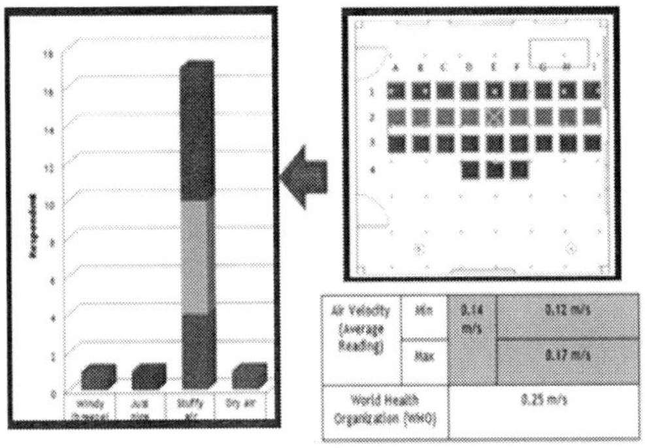

Figure 7: Perceived Satisfaction on the air movement in the classroom

Fig. 8 on the other hand shows data collected on measurement of relative humidity in the classroom. From the graph, it shows that the relative humidity in the classroom is in the range of 53.12 % to 78.2 %. At certain time of the day, the RH level was noted to go beyond the TLV (above 60% set by ASHRAE).

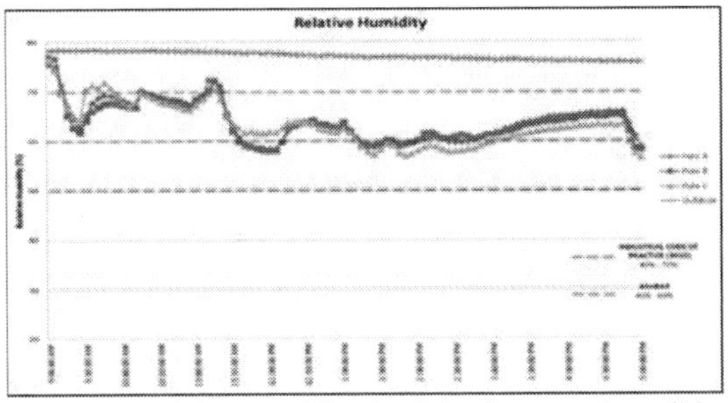

Figure 8: Relative Humidity level in the normal condition classroom

In addition to RH, CO_2 was also measured and it was found that the concentration level of CO_2 in the classroom is within the acceptable standards for the CO_2 that should be 700ppm above the outdoor concentration i.e., approximately 1000ppm. Similar to CO_2, there was no serious issue regarding dust particle (PM_{10}) in the classroom. On the other hand, TVOC concentration on both monitoring days shows a significantly high value beyond the TLV set (3ppm) as illustrated in Fig. 9. The sources of TVOC were probably from cleaning chemicals, which is use by cleaners during break time (1-2 pm) every day. It can also be from perfumes that were used by the students in the classroom as well as marker pen used by the lecturer during learning process.

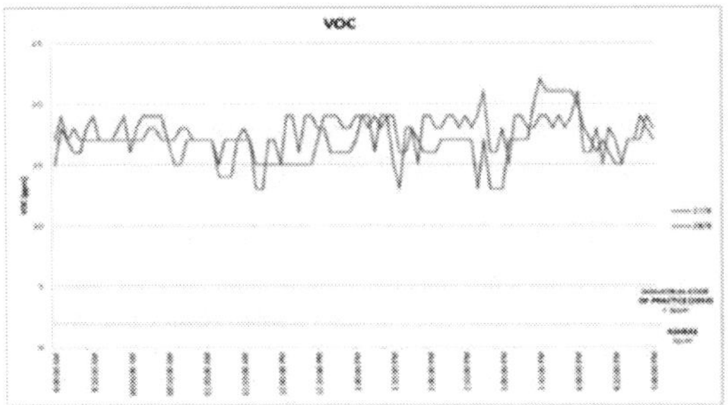

Figure 9: TVOC concentration level in the normal condition classroom

3.3 Fieldwork Monitoring –Acoustic Quality

Acoustic quality is referred to as an acceptable range of sound that does not cause any disturbed feeling. Based on the grab sampling taken in the classroom, the sound level was found to be slightly higher which exceeded the TLV of the background sound level (35db). From the data obtained, average readings for sound level were approximately 59db. This is probably due location of the classroom block which is situated next to the main road and also there is intermittent noise contributed by adjacent areas such as the walkway and generator outside of the classroom as illustrated in Fig. 10.

Figure 10: Possible sources of noise

 DOI: 10.1051/matecconf/20166600061

3.4 Fieldwork Monitoring –Lighting Quality

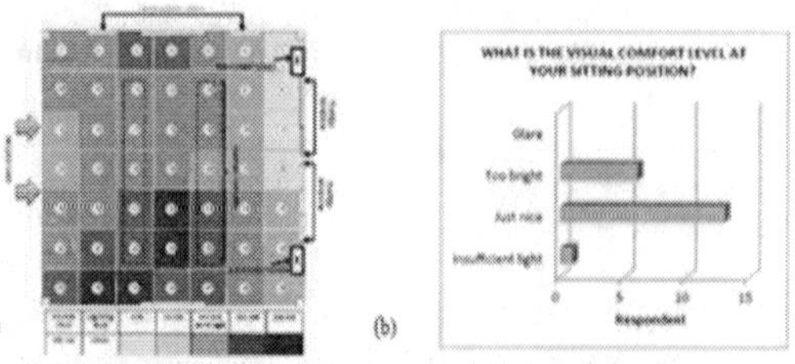

Figure 11: (a) Lighting level in the classroom (b) Students satisfaction level on the lighting quality

In Fig. 11(a), the color coding in grids shows the lighting level that were captured using grab sampling method. The middle of the classroom received higher level of intensity compared to the side of the classroom (near to window). This is because the artificial lighting arrangement was more focus at the middle of the classroom compared to other side. Lower lighting intensity was observed at the right side of the classroom, due to the tinted glass windows, which reduces excessive lighting penetration during sunny day. Overall, most of the areas in the classroom had lighting intensity below the standard lighting level suitable for classroom, which is 300lux. However, based on the responses from the students on the lighting quality (Fig. 11b), majority feels that the lighting quality is just nice with lighting intensity below 300lux.

3.5 Summary of Normal Condition Classroom

Table 2: Summary of IEQ Measurements in a Natural Condition Classroom

Element		Average		TLV (Threshold Limit Value)	Remarks
Temperature (°C)	Occupancy	27.3°C	28.17°C	22°C-26°C	Above TLV
	Non-occupancy		26.66°C		
Relative Humidity (%)		76.4%		40%-70%	Above TLV
VOC (ppm)		11.7ppm		3ppm	Above TLV
Particulate matter (PM10)		0.11mg/m³		0.15mg/m³	Compliance
Air velocity (m/s)	Min	0.14m/s	0.12m/s	0.25m/s	Below TLV
	Max		0.17m/s		
Acoustic (Background)		59.5dBA		<35dBA	Above TLV
Lighting (lux)		178.53 lux		300 lux	Below TLV

Table 2 shows the summary for current condition in the classroom. Most of the elements in the classroom were incompliance with the standard TLV. The level of TVOC in the classroom was significantly high with an average concentration of 11.7 ppm compared to 3 ppm as set in the standard TLV. According to Jim and Chen (2008), one of the pollutants that are widely found indoors is TVOC's, which could contribute to dissatisfaction of the building occupant. It could lead to distraction in students learning performance and difficulty in focusing during lecture.

3.6 Proposed Intervention Setting for Classroom Environment

Based on the findings from the natural condition of the existing classroom, this research proposed an intervention setting to improve the indoor environment of the classroom with application of landscaping. The plants selected in this intervention were Lily and Janet Craig. These plants are chosen in this intervention, as it is efficient in minimizing humidity and able to minimize the air pollutant in certain area.

3.6.1 Comparative Analysis of Total Volatile Organic Compound (TVOC)

Fig.12 shows comparison of TVOC concentration between normal condition and intervention classroom. Based on the graphs, it is observed that there was a sharp decline in the TVOC concentration when plants were placed in the classroom. Therefore, it is proven that plants could reduce the air pollutants especially TVOC inside the classroom.

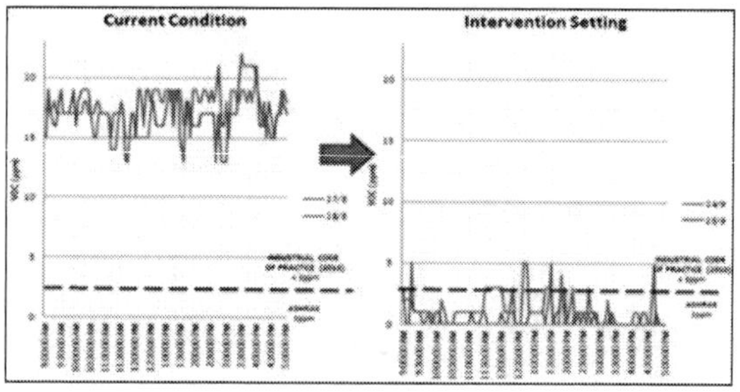

Figure 12: Comparison of TVOC concentration level between normal condition and intervention settings

3.6.2 Comparative Analysis of Relative Humidity (RH)

Similar to TVOC, a significant reduction was observed in RH level when plants were placed. (see Fig. 13) A significant reduction in RH level was found in sampling device located in point A and B, which is near to the plants. While sampling point C has no significant impact based on the indoor outdoor ratio of the RH. This finding gives further evidence that the air in the classroom is not mixing well as discussed in the earlier section. However, the number of plants in the room should be considered in future to prevent further reduction of RH below recommended value.

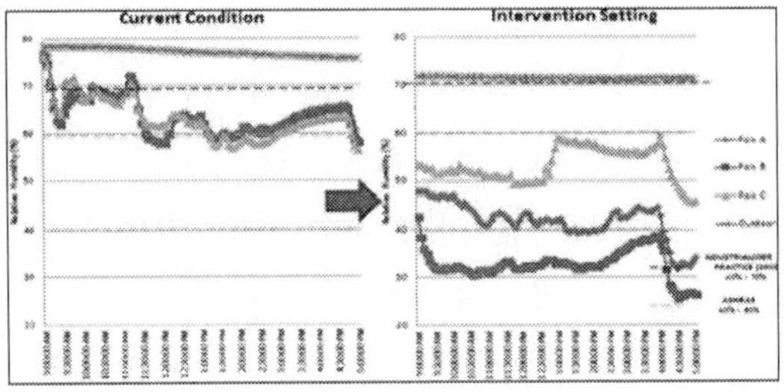

Figure 13: Comparison of Relative Humidity level between normal condition and intervention settings

3.6.3 Summary of IEQ Measurements Comparison between Normal Condition and Intervention Settings

Based on Table 3, can be summarized that most of the elements show reduction in the percentage. However, the temperature shows some increment due to the increase of outdoor temperature during data monitoring. Therefore, it can be assumed that most of the elements are complied with threshold limit value and guidelines when landscaping is intervened.

Table 3: Summary of IEQ Measurements In a Natural Condition Classroom and Intervention Settings

Element		Average (9am- 5pm)			
		Current Condition	Intervention Setting	Increment (%)	Reduction (%)
Temperature (°C)	Outdoor	28.3	33.3		
	Indoor	27.9	29.7	6.45% (difference of 1.8°C- due to increase in the outdoor temp during data monitoring)	
Relative Humidity (%)	Outdoor	61.6	43.8		
	Indoor	76.4	71.5		6.41% (difference of 4.9%)
CO_2 (ppm)	9am-5pm	1084	781.89		27.87% (difference of 302.11ppm)
	8pm-7am	443.65	429.22		3.25% (difference of 14.43ppm)
VOC (ppm)		11.7	1.5		87.18% (difference of 10.2ppm)
Particulate Matter (mg/m³)		0.11	0.12	9.09% (difference of 0.01mg/m³)	

4. Conclusion and Recommendations

General perception and level of IEQ in the classroom of Pusat Asasi Sains University of Malaya is averagely good. However, there were several major issues found when monitoring the classroom such as:

a) Irregular intensity of illuminance in the classroom referring to guideline.
b) High level of noise in the classroom exceeds the guideline.
c) High concentration of TVOC that exceeds the limits
d) Significantly higher dissatisfaction on ventilation system in the classroom.

However, based on the questionnaire, the students were neither easily interfered in terms of perceived productivity nor susceptible to any health symptoms and their performance. Moreover, there were no significant differences found on the all interferences on perceived productivity and susceptibility to health symptoms among the designated areas.

Based on the findings it is recommended that the air circulation in the classroom should be improved to remove heat as well as to increase the level of air velocity. Therefore, the exhaust fan and the air conditioning systems should be fixed and maintained. Besides that, the usage of cleaning agent that consists of high VOC should be minimized. In addition, new layout design of lighting fixtures should be installed to evenly distribute sufficient light for learning purposes. It is also recommended that periodical test (test/tutorial) to be carried out to assess students learning performance during each classroom intervention settings.

References

1. ASHRAE Standard 55 (2004).Thermal environmental conditions for human occupancy, American Society of Heating, Refrigerating and Air-Conditioning Engineers, Atlanta, GA.
2. Department of Occupational Safety & Health (DOSH), Malaysia – National Institute of Occupational Safety & Health (2013).
3. Fisk, W.J., 2000. Estimates of potential nationwide productivity and health benefits from better indoor environments: an update. In: Spengler, John D., Samet, Jonathan M., McCarthy, John F. (Eds.), *Indoor Air Quality Handbook, Chapter 4*. McGraw Hill, New York.
4. Fromme, H., Twardella, D., Dietrich, D. Heitmann, D., Schierl, R., Liebl, B., and Rüden, H., 2007. Particulate Matter in the Indoor Air of Classrooms—*Exploratory Results from Munich and Surrounding Area.Atmospheric Environment*, **41**, pp854–866.
5. Heath, G.A. and Mendell, M.J. (2002). Do indoor environments in schools influence student performance? A review of the literaure . pp. 802-807 in *Proceeding of Indoor Air 2002*, Vol.**1** June 30-July 5, 2002, Monterey, Calif.
6. Hill MC, Epps KK. Does physical classroom environment affect student performance, student satisfaction, and student evaluation of teaching in the college environment? *Acad Educ Leadersh* 2009;**14(1)**:P15e9.
7. Kamaruzzaman, K. and Samsul, A (2013).Thermal Comfort Assessment of a Classroom in Tropical Climate Conditions. *Recent Advances in Energy, Environment and Development*, ISBN: 978- 1-61804-157-9.
8. Mendell, M.J., Fisk, W.J., Dong, M.X., Petersen, M., Hines, C.J., Dong, M., Faulkner, D., Deddens, J.A., Ruder, A.M., Sullivan, D.A., Boeniger, M.F., 2002. Indoor particles and symptoms among office workers: results from a double-blind cross-over study. *Epidemiology* **13**, 296–304.
9. Ministry of Education (MOE),2013. Statistic.[online] Available at:

http://www.moe.gov.my/[Accessed 10 June 2014].
10. NIOSH, 2013. Centre for Disease and Control Prevention [online] (June 2013) Available at http://www.cdc.gov/niosh/topics/indoorenv/[Accessed 12 November 2013].
11. Turunen, M., Toyinbo, O., Putus, T., Nevalainen, A., Shaughnessy, R., & Haverinen-Shaughnessy, U. Indoor environmental quality in school buildings, and the health and wellbeing of students. International Journal of Hygiene and Environmental Health.
12. Wood, R., Orwell, R., Tarran, J., Torpy, F., and Burchett, M.:2006, 'The Potted- Plant Microcosm Substantially Reduces Indoor Air VOC Pollution: II. Laboratory Study',Volume **177**, Issue 1-4 , pp 59-80

A preliminary framework for corporate real estate sustainable management

Nurul Sahida Fauzi[1], Ashrof Zainuddin[1], NorainiJohari[1], Siti Nadiah Mohd Ali[1] and Abdul Hadi Nawawi[2]

[1]*Faculty of Architecture, Planning and Surveying, Universiti Teknologi MARA, Seri Iskandar Campus, Seri Iskandar, 32610, Perak, Malaysia*
[2]*Faculty of Architecture, Planning and Surveying, Universiti Teknologi MARA, Shah Alam, 40450 Selangor, Malaysia*

Abstract. The global warming issue has motivated corporations to go green in their business operations including transforming from conventional real estate to green features real estate. However green CRE is more complex to manage due to a building's significant impact on environmental, social and economic aspects. Thus the need to have a best practice guide or framework as reference is crucial. Unfortunately, no best practice guidelines on CRE management have been found to be sufficient as much uncertainty still exists on the sustainable performance measurement components. This research aims to explore and then summarize the present sustainable CREM practices and components relating to sustainable performance measurement integrating a sustainable theory that balances environmental, economic and social impacts. These act as indicators to measure the outcomes of the practice in the form of a generic model on sustainability preliminary framework for CRESM. The objectives of this research include identifying corporate real estate sustainable management (CRESM) practice and components of sustainable performance measurement. The research uses content analysis method to analyse data gathered from literature and previous studies. The findings will be demonstrated in the form of a framework model on CRESM that will include14 CREM strategies and 15 components derived from analysis.

1 Introduction

The growing number of green buildings especially office buildings in Malaysia since 2009 are viewed as a serious issue to the corporate real estate (CRE) manager. Green office buildings are very complex to manage as compared to conventional buildings. This is due to the needs for well managed and maintained CRE as well as the need to maintain according to the criteria set by the GBI based on the sustainable theory on balancing environmental, economic, and social impacts. Thus, CREM for sustainable buildings is critical to be developed and well-practiced by all green buildings in Malaysia to contribute to the added value and overall performance of the business corporation. CREM for sustainable buildings is represented as corporate real estate sustainable management (CRESM). It is rare to encounter this term in CREM research. Even though there are many research related to CREM from the sustainability view, only a few research were found to use this terminology albeit with the application and concept identified to be similar to the objectives of CRESM.

There is still a lack of studies on corporate real estate sustainability management and most have just been started. This research aims to explore and then summarize the present sustainable CREM practices and components relating to sustainable performance measurement integrating a sustainable theory that balances environmental, economic and social impacts. These act as indicators to measure the outcomes of the practice in the form of a generic model on sustainability preliminary framework for CRESM. The framework produced is hoped to encourage more corporations to transform from conventional CRE to sustainable (green features) CRE.

2 Problems and issues

In response to various initiatives all over the world, the present day is seeing a growing number of sustainable office buildings or also known as green buildings which is viewed as a serious issue to the corporate real estate (CRE) manager. Green office buildings are very complex to manage as compared to ordinary office buildings. In maintaining this type of buildings CRE sustainable managers need to adhere to the maintenance criteria set by the green index (Masalskyte, Andelin, and Sarasoja *at el.* 2014) as well as considering a sustainable theory that balances environmental, economic, and social impacts.

Corporate real estate sustainable management (CRESM) must be strategically placed to achieve best practice as corporate real estate management offers possibilities in improving corporate real estate energy efficiency and thus reduce energy consumption (Glumac, Oosterbaan and Schaefer *et al*, 2015). Throughout its long life cycle, a building has significant impacts on the environmental, social and economic aspects therefore, sustainability and real estate are naturally combined together (Masalskyte *et al.*, 2014). Oluwoye, Karatonis, and Fakorede (2001) found that, the neglect by corporations on their CRE exposes them to unavoidable takeovers, lost profits and lower stock price performance. This neglect remains a current and pertinent issue in many countries and among many corporations.

Managing a real estate will not be a big problem to real estate business because this is within the nature of their knowledge. This is not the case for CRE where it represents real estate owned by non-real estate corporations. According to Omar and Heywood (2014) most business corporations assume CRE just provides operational space, or a workplace to employees and finding suitable location to operate business. Business minded corporations believe CRE is only a medium to support business operations and not the main factor producing profit for the business. Despondently, some only assumed CRE as a burden to the business as what was found by Farahiyah Fadzil, Hishamuddin Mohd Ali and Ahmad Ariffin Bujang (2011). Senior management views their CRE function as primarily to improve operational efficiency and reduce costs; and has not required CRE assistance with strategic planning, either at corporate or business-unit levels. Masalskyte, Andelin, and Sarasoja *at el.*(2014) affirmed, CREM is merely a means to save money in many ways and expand expertise in sustainability management. Real estate is just something that is there in place, but is not a focus area.

In Malaysia, Ting (2003), found operational properties owned by large companies and public corporations in Malaysia are under-used and under-managed. However, in recent years the awareness on the importance of CRE and CREM as part of the contribution to the success of the business has shown an improvement where literature shows an extensive number of CRE research especially in Malaysia. On the contrary, no research has been found discussing on CRESM in Malaysia compared to several found from Australia, Finland, UK and Germany.

According to Hartmann, Linneman and Pfnür et al (2010) research done has not found the existence of one "best practice" of CRE management model in a specific situation. Thus, performance measurement is the best indicator to evaluate how well the CRESM application and strategy contribute to the value added factor for the corporation and business (Lindholm & Leväinen, 2006). From previous research, most measured the performance through direct data as cost and value data and very few considered measuring the performance using intangible elements or indicators of performance. In fact, performance may also be measured through indirect data as what is commonly evaluated such as measuring employee satisfaction with the workplace, client satisfaction with services, physical condition of properties and space per employee (Lindholm & Leväinen, 2006). Kaplan and Norton's (2004) study also described in their model, organizations can increase the performance through economic value by looking at revenue growth and or productivity. Revenue growth comes from new markets, new products, new customers, and expanded sales to existing customers while productivity can be measured on the ratio of output to input. Output is commonly measured with products, services, market shares and value while input is measured through cash, labour, energy, materials, and work environment. Productivity is advanced by increasing output and reducing input such as costs or a combination of the two (Feige, Wallbaum, Janser, & Windlinger, 2013).

From the literature research done, it can be assumed that comprehensive CRESM practices are still lacking especially on the understanding and components involving classification that are directly related to sustainable theory. These cause some information to be left out and sometime not directly related to the specific sustainable CRE. Hence a framework developed then will initially fill the gap existing in this field and promote further study.

3 Research design

The research was conducted by reviewing all the available literature from previous research to identify the existing CRESM practices. It includes all the strategic models available in practice. In addition the performance measurement approaches applied in practice of CRE are also studied to be used as indicators to measure the outcome of strategic CRESM. All the reviewed data is then analysed through contents analysis to develop a preliminary framework based on the sustainable theory that balances environmental, economic, and social impacts.

4 Theoretical frameworks

4.1 Corporate real estate management (CREM) strategies

In 1993 Nourse and Roulac developed a CRE strategic framework which encompassed 8 strategies identified as strategy 1 (cost minimization), strategy 2 (flexibility), strategy 3 (promotes human resource objectives), strategy 4 (promote marketing message), strategy 5 (promote sales and selling process), strategy 6 (facilitate production, operation, and service delivery), strategy 7 (facilitate managerial process and knowledge framework), and strategy 8 (capture the real estate value creation) (Nourse & Roulac, 1993). This framework has been used and tested by several researchers to examine the strategic management of CRE for the selected corporations.

Previous research by De Jonge (1996) proved that CREM provides maximum value for the business and came up with 7 components of CRE. The strategies formed cover strategy 1 (increasing productivity-selection of location, innovative workplaces, retaining human capital), strategy 2 (cost reduction- workplace costs, accommodation costs, facility costs, benchmarking, corporate finance), strategy 3 (risk control- inflexibility of real estate portfolio, selection of location, value risk, working environment, environmental aspects, development process), strategy 4 (increase of value-acquisition and disposal of real estate, redevelopment of real estate, market analysis), strategy 5 increase of flexibility - organizational flexibility, financial flexibility, technical flexibility), strategy 6 (changing the culture-workplace innovation, communication, image) and strategy 7 (PR and marketing, selling points, sales strategy). Former research by Krumm & de Vries (2003) showed application of De Jonge's (1996) strategic framework to discuss on real estate decisions towards corporate performance.

Lindholm & Gibler (2005) also developed a set of CRE strategies covering 7 strategies that have been classed as strategy 1 (increasing the value of assets), strategy 2 (promoting marketing and sales), strategy 3 (increasing innovation), strategy 4 (increasing employee satisfaction), strategy 5 (increasing productivity), strategy 6 (increasing flexibility), and strategy 7 (reducing costs). However, with the increased emphasis on environmental sustainability and green building approaches to corporate real estate management, Gibler & Lindholm (2012) found firms may differ in terms of reduced costs, greater financial returns, increased standing in the community, improved image, higher productivity and increased employee satisfaction. Therefore, the Lindholm et al. (2006) model is updated to include an eighth potential real estate strategy: (strategy 8) supporting environmental sustainability. Supporting environmental sustainability may support growth through improved image and profitability through reduced operating cost.

As published, several studies also used theoretical typology of CRE strategies proposed by Nourse and Roulac (1993) to test the CRE strategy formulated by corporations and also to identify whether the CRE strategy is applicable with this formula. As shown in Zaiton Ali, McGreal, Adair, Webb, & Roulac (2008) in their research, this strategy formula was applicable to be tested in UK. The strategic link was measured to real estate operating decisions criteria relating to location, the quantity of space per employee, building size and character, ownership rights, financing, control and risk management as per developed by Nourse & Roulac (1993).

Contradictory results were found for the case in Malaysia by Zaiton Ali, Mcgreal, Adair, & Webb (2008) using the formula developed by Nourse and Roulac on industrial classification, where one of the strategies namely Strategy 3 (promote human resource objectives) was unused. However, Malaysian companies discover a new additional strategy of corporate social responsibilities (CSR) that has been identified as an important strategy to add up in CRE strategies.

A similar approached was used by Mohd Fauzee Musa & Zarita Ahmad@Baharum (2012) in their research. However the components were analyzed and adopted from a combination of CRE strategic management components from previous research namely De Jonge (1996), Roulac (2001) and Lindholm (2008). From these, only 4 components were adopted including strategy 1 (minimize cost), strategy 2(increasing value), strategy 3(increasing flexibility) and strategy 4 (promote marketing and sales). In summary, even though there are increased numbers of CRE strategy formulas in existence, the strategy selected supposedly is possible to be implemented for the respective corporation.

Table 1. Analysis of CREM strategic practice

Author / Strategy	Nourse & Roulac (1993)	De Jonge (1996)	Lindholm & Gibler (2005)	Zaiton Ali et. al (2008)	Mohd Fauzee Musa & Zarita (2012)	Lindholm & Gibler (2012)
Cost Minimization	X	X	X	X	X	X
Flexibility	X	X	X	X	X	X
Promotes Human Resource Objectives	X			X		
Promote Marketing Message	X		X	X	X	X
Sales and Selling Process	X	X		X	X	
Facilitate Production, Operation, and Service Delivery	X			X	X	
Facilitate Managerial Process and Knowledge Framework	X			X	X	
Capture the Real Estate Value Creation	X	X	X	X	X	X
Increasing Productivity		X	X			X
Risk Control		X				
Changing the Culture		X				
Employee Satisfaction			X			X
Corporate Social Responsibility				X	X	
Supporting Environment Sustainability						X

Source: Researcher (2016)

4.2 Performance measurement

Performance measurement is the best way to measure the best strategic practice used, or in other words, to create value to the organization it must include CRE. Several publications have appeared in recent years documenting numerous studies that strongly linked green building to enhance productivity. It has been found that, green star-rated green buildings generate a reduction of 39% the average monthly cost of sick leave and show productive staff (Dunckley 2009). Similarly Kats et.al (2003) also found that green office building increases the productivity of workers and increases production from $37 to $55 US Dollars per square foot. Armitage, Murugan, & Kato, (2011) reported that the labor employers have a strong belief that a green office would have a positive impact on the health of workers and consequently on the productivity of the organization. This supports the findings in Dunckley (2009) and Kats et al (2003). A study conducted in Malaysia by Syazwan et al (2009) revealed a similar finding that shows a reduction in the sick building syndrome (SBS) when the ventilation rates per person among typical office buildings increases. Previous research explains similar methods to measure the best practice or value creation of CRE and most of the research done focuses on measuring performance through financial and economic criteria as presented in the research by Nappi *et al* where value creation is measured via two economic proxies which are economic value and market value added (Nappi-Choulet et al., 2009). The results from that particular research show CRE is negatively associated with economic value added, but this association is significant only for firms outside of service industries and shows a positive association on market value added with the change in CRE for firms in service industries. In addition, it was found that it is still difficult to determine the actual performance by merely looking at the financial and economic criteria (Feige, Wallbaum and Janser et al 2013).

However, according to research by Krumm & Linneman (2001) financial and economic criteria are no longer the only basis for determining the added value of corporate real estate. Liow & Ingrid (2008) recommended that performance indicators are needed to focus on productivity and profitability. Performance and productivity are rarely analyzed as factors connected to the office environment (Feige et al., 2013). Performance can be viewed using several different criteria such as individual performance, team performance, organizational performance, and building performance. Other than that, some research uses the elements of satisfying operational needs that might include flexibility and productivity, reducing occupancy costs, less capital in real estate and the element of allowing investment of the savings more profitably in other areas of the business, to measure the performance especially in core activities (Jordan, McCarty, & Velo, 2009).

The common criterion used for performance measurement in relation to performance and productivity is individual performance. Individual performance may possibly explain the individual impact and opinion towards the CRE. Previous study indicates that, individual performance can be described as a person's contribution to the objectives of the organization (Feige, Wallbaum and Janser et al 2013). However, individual performance has been identified as a qualitative construct in nature because it is abstract and cannot be measured directly. Therefore, much research has been done to estimate the performance through performance criteria and performance indicators (Marcus and Schuler, 2001). Performance measurement using performance indicators especially on customer satisfaction measurement practices is also widely use, but was rated lower overall (Varcoe & O'Mara, 2011).

Literature in strategic management suggests there is a need to focus on factors internal to the organization to know an organization's performance. Both Too added, core processes are those processes that are strategically important to the organization's success and have a high impact on customer satisfaction (Too & Too, 2010). Jordan, McCarty, & Velo (2009) concurred and stressed that good performance measurement systems create a direct line of sight between CRE goals and stakeholder interests and any decision making must be linked to customer expectations.

The evaluation of performance and productivity can be difficult for service companies and knowledge-intensive organizations in general. Previous research shows that the methods that could be used to measure the performance of CRE can be broken into two main groups: quantitative and qualitative. Quantitative metrics with are self-explanatory include occupancy cost/m2, occupancy cost/person, lease cost as a percentage of occupancy cost, lease income as a percentage of total occupancy cost, capital expenditure as a percentage of total assets, capital expenditure as a percentage of total occupancy cost, outages (space cannot be used because of repair/maintenance problem), occupancy cost as a per cent of total revenue, occupancy cost as a per cent of total expenditure, m2 per person, vacant space as a per cent of total space, subleased space as a per cent of total space and hours the facilities utilized, asset value per person; and asset value increase (Kenley & Heywood, 2000).

Nevertheless, previous research has demonstrated that quantitative methods to measure performance especially on productivity cannot usually be applied to service companies because of the qualitative nature of the output of knowledge work and the fact that knowledge workers operate, by definition, with intangible resources (Feige, Wallbaum and Janser et al 2013). They suggested a possible solution to this problem by considering indirect measures as elements of measurement such as absenteeism, hours worked, lateness, safety rule violations, number of grievances filed, or employee turnover.

In concurrence, Nurul Sahida Fauzi, Abdul Hadi Nawawi, & Yuhadi Sanuddin (2012) demonstrated indirect performance measurement consisting of seven dimensions related to CRE including effectiveness (quality, quantity, meeting targets), efficiency (ratio of expected resources to those used), quality (subjectively or objectively assessed quality attributes), profitability (ratio of total revenues to total costs), productivity (ratio of quantity of output to input in terms of value/cost), quality of work life (psycho-social aspects and social response to company) and innovation (applied creativity) previously used by Sink 1985, were use in that research. The seven dimensions of performance include a number of tangible elements such as efficiency, profitability, productivity and effectiveness, but also include intangible elements such as quality of work life and innovation. The latest solution described confirmed that the best alternative to address this problem is moving to subjective measurement. Subjective measures are not based on quantitative operational information, but on the personnel's own subjective assessments and is normally collected using survey questionnaires (Feige et al. 2013).

Recently, sustainability pillars are considered in the corporation to integrate a performance measurement with economic, environmental or ecological and social aspects. These three pillars are generally added to present the element of sustainability in the organizational management. Economic sustainability is the common element to performance measurement in the market for a long time and normally it is related to the financial results of the corporation. This is agreed by Christensen, Baldwin, & Ellis (2012) in the research with includes the indicator of increased productivity that directly results in the revenue of the corporation. However, Mansfield (2009), used improvement in building performance and durability of the real estate after a cost reduction towards building life. Lamprinidi & Ringland (2006) also use reduced cost to measure the economic aspect of performance. According to Glatte (2012), by looking at the pure economic view economics is defined as a target concept related to performance targets (procurement, inventory, production, sales), Financial targets (liquidity, investment, financing) and success targets (turnover, earnings, profitability).

Another pillar is environmental that is also known as an ecological dimension that covers impacts related to inputs such as energy and water use and outputs such as emissions, effluents and waste. It also considered biodiversity, transport, product and service related impacts and environmental compliance. Furthermore, environmental sustainability means maintaining or improving the integrity of the life supporting systems on the earth and to sustain the biosphere, maximizing future options for the current as well as future generations (Muhammad Zahid Zulkipli Ghazali, 2015). Mona (2013) found environmental sustainability from the perspectives of the productivity and well-being of occupants, as well as the advantages and benefits of green features to the environment, such as energy efficiency, recycling and reduction of greenhouse gases. Similar to previous research which shows several indicators of performance measurement related to sustainable concern demonstrated environmental sustainable with improved efficiency of water consumption, less life cycle cost (Collins & Junghans, 2015), it determines the environment as global warming prevention through the reduction of CO2 emissions, waste minimization of pollutants, water conservation, and minimization of waste water generation, ecosystem conservation, reuse materials or recycling, environmental management promotion, reduction of office energy use and greenhouse gas emission generated, promotion of biodiversity and indoor comfort environment quality.

The last pillar for sustainability is social that is more concerned on the impacts on the organization including labor practices, human rights and society (Muhammad Zahid Zulkipli Ghazali, 2015). Mona Isa et. al (2013) disclosed that environmental elements include improved environment for office workers and building users while de Francesco & Levy (2008) identified social as changing the behavior to become more aware of day-to-day sustainable activities. Some of the research revealed that social sustainable perspectives include sustainable criteria related to occupant satisfaction (Christensen et al., 2012), flexible working environment, health and safety education, training and education of employees, employee retention, participation in local community programs, percentage comparison of male and female employees, percentage of staff who participate in basic environmental training, employee years of continuous service, total CSR spending and absentee rate (Lamprinidi & Ringland, 2006).

According to Green Building Index (GBI) Malaysia all buildings awarded the GBI ratings must fulfill six (6) criteria of GBI that covers energy efficiency; indoor environmental quality, sustainable site planning and management, material and resources, water efficiency, and innovation (Mona Isa et. al 2013).

5 Model and preliminary framework for corporate sustainable real estate management

Figure 1. Spiral Model of CRESM

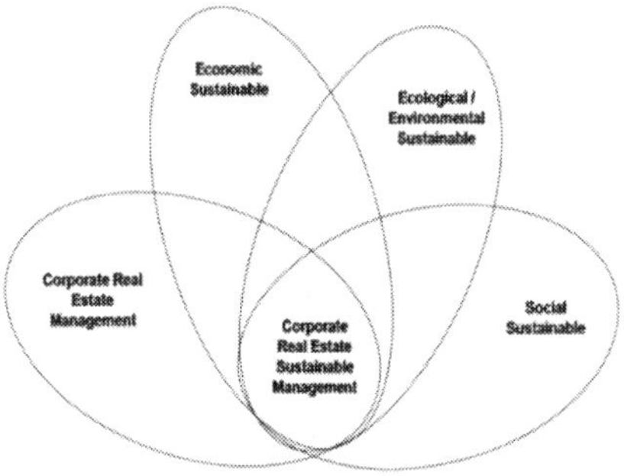

Source: Researcher (2016)

From the analysis done, CRESM can be seen as integrating three pillars of sustainable management which includes economic, ecological and social aspects; with corporate real estate management. Hence, a spiral model of CRESM was developed to make a clear view of the integration between all the elements presenting sustainable concern in the practice of CREM. The elements covered under CREM in this model are summarized from strategies used or best practice identified from previous research while the performance measurement to evaluate the outcome from the practice have been found from literature research chosen and classified into several important classes to suit representative sustainable performance measurements. 14 variables or strategic measures were discovered and 15 components or indicators chosen to measure the practice. The integration of CREM strategy with sustainable performance measurement is illustrated in Figure 2 on preliminary framework of CRESM. The variable in the framework are distributed to three pillars of sustainability.

Figure 2. Preliminary Framework of CRESM

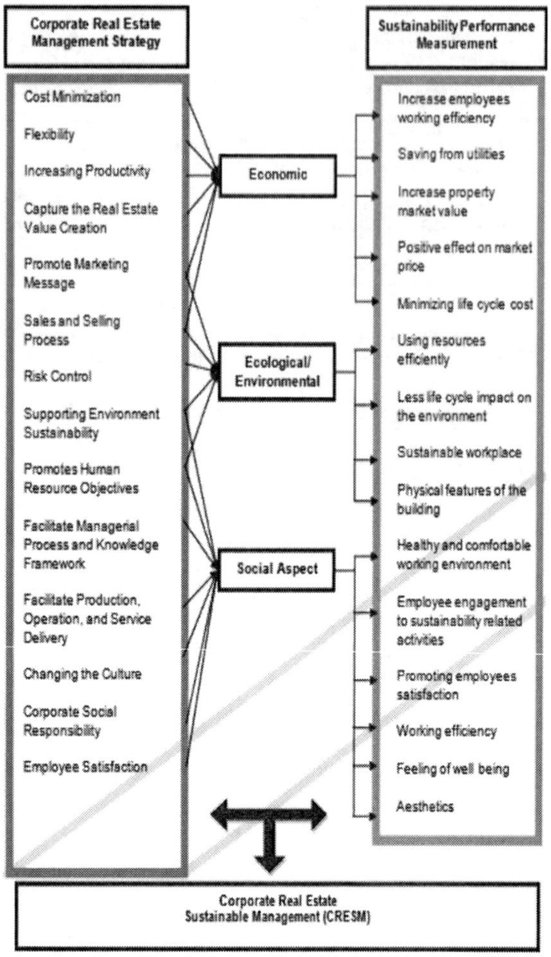

Source: Researcher, 2016

6 Conclusions

The research reveals 14 strategies practiced in CREM including cost minimization, flexibility, increasing productivity, capturing the real estate value creation, promoting marketing messages, sales and selling processes, risk control, supporting environmental sustainability, promoting human resource objectives, facilitating managerial processes and knowledge framework, facilitating production, operations and service delivery, changing the culture, CSR, and employee satisfaction. Additionally, 15 indicators to measure sustainable performance measurement were discovered namely elements under economic sustainability including increased employee working efficiency, savings in utilities use, increases in property market value, positive effect on market price and minimization of life cycle cost, ecological covers on using resources efficiently, less life cycle impact on the environment, sustainable workplace and physical features of the building. Lastly, social elements comprising healthy and comfortable working environment, employee engagement to sustainability related activities, promoting employees satisfaction, work efficiency, feeling of well-being and aesthetics. The framework developed is recommended for further study to discover a best practice to suit the specific situation.

References

1. Ali, Z., Mcgreal, S., Adair, A., & Webb, J. R., J. REL, **16(1)**, 35–54, (2008).
2. Armitage, L., Murugan, A., & Kato, H., J.CRE, **13(3)**, 169–180, (2011).
3. Bon, R., J.Fac, **10(12)**, 13–17, (1992).
4. Christensen, P., Baldwin, E., & Ellis, C., *ProQuest Dissertations and Theses*, (2012).
5. Collins, D., & Junghans, A., *Procedia Economics and Finance*, **21**, 128–136,(2015).
6. De Francesco, A. J., & Levy, D., J.Eur. RE Resc, **1(1)**, 72–87,(2008).
7. Farahiyah Fadzil, H. M. A. and A. A. B. Corporate *In 18th European Real Estate Society Conference ERES 2011, June 15 – 18th 2011.* (pp. 1–12), (2011).
8. Feige, A., Wallbaum, H., Janser, M., & Windlinger, L., J.CRE, **15(1)**, 7–34, (2013).
9. Gibler, K. M., & Lindholm, A.-L., J.Ppty. Resc., **29(1)**, 25–48, (2012).
10. Glatte, T. *The Importance of Corporate Real Estate Management in Overall Corporate Strategies*, (2012).
11. Glumac, B., Oosterbaan, M. A., Schaefer, W. F., & Sulla, K., J.CRE, **17(2)**, 134–156, (2015).http://doi.org/10.1108/JCRE-11-2014-0026
12. Jordan, M., McCarty, T., & Velo, B., J.CRE, **11(2)**, 106–114, (2009).
13. Kenley, R., & Heywood, C., J.Con, **1**, 6–8. (2000).
14. Krumm, P. J. M. M., & de Vries, J., J.Prpty. Inv. & Finc, **21(1)**, 61–72, (2003).
15. Lamprinidi, S., & Ringland, L., *GRI Research and Development Series Publication* (**Vol. Third Gene**), (2006).
16. Lindholm, A. A.-L., & Leväinen, K. I., J.CRE., **8(1)**, 38–46, (2006).
17. Lindholm, A., & Gibler, K. M., **12**, (Hinks 2004), 1–18, (2005).
18. Liow, K. H., & Ingrid, N.-C., J.CRE., **10(1)**, 54–67, (2008).
19. Mansfield, J. R., J.CRE., **11(2)**, 91–105,
20. Masalskyte, R., Andelin, M., Sarasoja, A.-L., & Ventovuori, T., (2014). J. CRE., **16(2)**, 126–139, (2009).
21. Mohd Fauzee Musa, & Zarita Ahmad@Baharum. *Procedia - Social and Behavioral Sciences*, **36** (June 2011), 273–279, (2012).
22. Muhammad Zahid Zulkipli Ghazali. World Journal of Science , Technology and Sustainable Development Article Information, **12(2)**, 100–118,(2015).
23. Nappi-Choulet, I., Missonier-Piera, F., Cancel, M., J. CRE., **11(2)**, 78–90, (2009).
24. Nourse, H. O., & Roulac, S. E., J. REA., **8(4)**, 475. (1993).

25. Nurul Sahida Fauzi, Abdul Hadi Nawawi, & Yuhadi Sanuddin. *1st International Conference on Innovation and Technology for Sustainable Built Environment 2012 (ICITSBE 2012)*, (pp. 469–484), (2012).
26. Oluwoye, J., Karatonis, A., & Fakorede, A., Pacific Rim Property Research Journal, 7**(1)**, 18–34, (2001).
27. Omar, A. J., & Heywood, C. a., J. CRE., **16(1)**, 60–76, (2014).
28. Ting, K. H., *9th Pacific Rim Real Estate Society Annual Conference, Brisbane, Queenland, Australia, 20-22 January 2003* (pp. 1–20) , (2003).
29. Too, E., & Too, L., J. CRE., **12(3)**, 196–208. (2010).
30. Varcoe, B., & O'Mara, M., (2011).
31. Zaiton Ali, McGreal, S., Adair, A., Webb, J. R., & Roulac, S. E., Journal of Property Research, **25(3)**, 241–267, (2008).

Investment Constraints in Urban Regeneration: Property Developers and Local Authorities Perspective

Nurulanis Ahmad @ Mohamed[1,a], Zarita Ahmad Baharum[2], Asma Senawi[3] and Nor Aini Salleh[4]

[1,a]*Faculty of Architecture, Planning and Surveying, Universiti Teknologi MARA, Seri Iskandar Campus,Seri Iskandar, 32610, Perak, Malaysia*
[2]*Faculty of Architecture, Planning and Surveying, Universiti Teknologi MARA, Shah Alam Campus, Shah Alam, 40450, Selangor, Malaysia*
[3]*Faculty of Architecture, Planning and Surveying, Universiti Teknologi MARA, Seri Iskandar Campus,Seri Iskandar, 32610, Perak, Malaysia*
[4]*Faculty of Architecture, Planning and Surveying, Universiti Teknologi MARA, Seri Iskandar Campus,Seri Iskandar, 32610, Perak, Malaysia*

Abstract. Urban decay is a phenomenon created due to the expansion of urban population, the restructuring of industrial, social composition as well as the deterioration of urban areas. Consequently this will increase issues of crime, mass unemployment, low quality in urban services, social problems etc. Thus urban regeneration is a tool used to overcome all these issues in order to create quality of urban living, diversify and vibrant cities. However, the involvement of investment in urban regeneration is still infancy in Malaysia as compared to other countries such as Australia, UK, Hong Kong, and etc. This is due to the perception of investors towards the profit generation and the risk involve. The objective of this paper is to determine the constraints involved in urban regeneration investment from the perspectives of the property developer companies and the local authorities. The method used is an exploratory with primary data collection through a questionnaire survey. The findings revealed that the main constraints is in regards of the land issues as well as high cost of investment, however, the benefits gain should also be considered as this create a livable urban living.

1 Introduction

Urban regeneration is a strategic solution to the challenges provided by the growing urbanization itself towards revitalize a city in a sustainable lifestyle concept and resilience (Robert and Skyes, 2000[24]; IOER, 2005[17]).The evolution of urban regeneration arises in the 1990s due to the rapid change of industrial structure in many cities in the UK and Europe in the mid 1970s. Regeneration projects carry many benefits which include offering an attractive vibrant property market through innovative and smart growth development concept (Adair et al, .2003[3]); able to attract business and create job opportunities (Goksin and Muderrisoglu, 2005[15]); improve quality of the environment (Couch, 1990[9]); reinforce civic pride and a sense of place (Goksin and Muderrisoglu, 2005[15]); and concern on long term approach and avoid future problems (Robert and Skyes, 2000[24]). It has been recognized as one of a tool to achieve sustainable and resilience, urban living and environment.

The implementation and investment of urban regeneration worldwide has increase over the years, for example in the UK, the total investment in urban regeneration project has increased from £8.3 billion in 2009 to £11.6 billion at end of 2012 (Planning in Association with Regeneration and Renewal, 2013[25]); Africa from RM 55 million in 2013 to RM125 million in 2015 (Cox, Viruly, Bertoldi, 2015[10]);and the Oklahoma City, USA significant growth from $942 million in 2003 and $1.2 billion in 2005 (Warner and Long, 2009[29]).

The success of urban regeneration has proved to create a livable urban living, even though it often involves high investment cost. However, urban regeneration in Malaysia is still infancy and involved very little participation from the local government as well as the developer. To date, there are only four cities that involve in urban regeneration project namely Kuala Lumpur, Georgetown, Melaka and Johor Bahru (Nilai Harta, 2015[23]). Although there are many potential projects for urban regeneration, the local authorities and the developers seem to be reluctant to pursue with this investment. Thus, the objective of this paper is to determine the constraints involved in urban regeneration investment from the perspectives of the property developer companies and the local authorities.

2 Background of the Study

Urbanization is a global phenomenon to all the large cities worldwide. The driving force behind the rapid growth in urbanization is due to the industrialization and migration. There are cities that cannot sustain urbanization which consequently will face economic and social decline due to the structural change in the local economy. This process whereby the previously function

[a] Corresponding author: nurul513@perak.uitm.edu.my

cities will fall into disrepair and decrepitude is known as urban decay. Urban decay is a phenomenon created due to the expansion of urban population, the restructuring of industrial, social composition as well as the deterioration of urban areas. As a result, this will increase issues of crime, mass unemployment, low quality in urban services, social problems etc. Thus urban regeneration is a tool used to overcome all these issues in order to create quality of urban living, diversify and vibrant cities.

The rate of urbanization in Malaysia has increase tremendously over the years from less than 50% in 1990 to more than 70% in 2010 and expected to rise further 75% by year 2020 (The Edge, 2014[28]). The growth in the population size in major cities in Malaysia over the years is very astounding and this trend is expected to grow from 30.1 million in 2015 to 38.6 million in 2040 with an increase of 28% (Malaysia Statistic Department, 2015[19]). The change in the society, restructuring economic and policy review would cause some areas in the cities deteriorate and decline. Therefore, urban regeneration is a solution focusing on specific society needs to revitalize the blighted areas in city centre into a better urban infrastructure and improve the quality of environment towards sustainable lifestyle, intelligence and resilience of the city. It is a key driver of economic, physical and social growth (Eni et al, .2013[12]). The concept of urban regeneration is to tackle the decline of many urban areas through investment and improve infrastructure. Considering the regeneration advantages and in tandem with Malaysia's mission to become a sustainable and strategic country of socio economic development in year 2020, investment towards urban regeneration project is crucial in order to achieve the strategic direction.

The developers and local authorities are not motivated to be involved in urban regeneration projects due to their perception that the investment is costly, risky, lack of information and low profitability. The trends in other countries prove otherwise as urban regeneration is well accepted and implemented as a lucrative investment and development, for examples Langham Place in Hong Kong, London Docklands in UK and False Creek in Canada.

3 Investment constraints in urban regeneration

Any investment in whatever nature, industry and circumstances will directly and indirectly faces constraints. The significant differences are the magnitude of the constraints and the approaches use to tackle the constraints. Investment in urban regeneration is no doubt a challenging task and involves a huge capital investment. Constraints in regards to urban regeneration investment can be divided into six categories which include risks, return, governance, costs and property tax, damage neighborhood and social network, and problems with multiple ownerships.

3.1 Risks

Risks are often related to profit (Samworth and Lacey, 1998[26]). A research by McNamara (1993[21]) stated that regeneration areas are known to investor as medium of risk and uncertain. Among the risks involved in urban regeneration areas include planning, yield and rent, development costs, contamination and stigma, development period, tenure of finance, and market volatility (Adair, Hutchison, Burgess and Roulac, 2005[4]). Simons (1998[27]) pointed out that the higher development cost and longer duration in development will also affect the risk involved. For instances, the early stage of development such as site assembly, demolition, and clearance often cause a longer duration and affect the finance charges (Adams et al, .1998[5]). Other than that, risk such as tenant needs before lease out, economic condition during construction, shortage of material and labor, market conditions, unforeseen conditions and delay government approval also among risk involve in investment of urban regeneration (Epley, Rabianski and Haney, 2002[13]). A report by ERM Economics (2002[14]) emphasized that the risk will rise more in investment of urban regeneration due to lack of information on return received, misperceived returns, high transaction cost, information costs, risk aversion, low demands, financial constraints, high cost of market survey, land contamination survey and detail feasibility studies. This is supported by Adair et al (1998[2]) where lack of information or transparency will result in uncertainty in regeneration investment.

3.2 Return

The perception of poor return and low profitability is another area of constraints in urban regeneration investment. Poor in return is due to the characteristics of urban regeneration areas such as illiquidity, high management costs, lack of property data on profit return and low income flow. A report by DETR (2000[11]) shows investment in regeneration areas often considered as lacking in demand due to factors such as poor location, infrastructure, condition and etc. Regeneration areas characterized as weak in market and low levels of market efficiency. Private investor cautious to invest due to regeneration areas has been accepted as market failure and economic decline (Amin et al, 2000[7]). As a result, it will restrict site development and high land values to allow profit in development. Private investor also perceived regeneration area a confused market signals due to lack of information in their areas itself (McGreal, Webb, Adair and Berry, 2006[20]).

Urban regeneration is to regenerate city of the blighted areas. Thus, the perception of private investor for the blighted areas is poor located and low quality (Adair, Berry, McGreal, 1994[1]). Return on profit gain often related to the economic condition and location factor. Both are importance to stimulate the investment in urban regeneration. A report by Callender and Key (1996[8]) stated that geographic location has a major influence for investment commitment in urban regeneration rather than economic condition.

3.3 Governance

A poor institutional arrangement also highlights as investment constraints in urban regeneration (Adams and Hasting, 2001[6]). This refers to the role of local authority as a policy makers. Bureaucracy and prolong procedure for grant applications and distrust public sector are issues often contribute to the investors cautious to involve in urban regeneration (Adair et al., 1998[2]). Ng (2003[22]) stated that investment in urban regeneration at Sheung Wan, China facing an inadequate funding for land resumption and poor coordination among the government department. Indeed, the role of local authorities is very significant in acquiring, servicing and assembling sufficient land for urban regeneration is required to attract investor in investment. Moreover, providing a clear data for investor may be able to create a certainty condition. Urban regeneration itself is a complex task and it requires resolving the potential conflict in the formation of policies at all level of society. The planning department should consider the wider perspective to reflect the circumstances and requirements of the city or area (Hausner, 1993[16]). The key issue is to understand the entire matters such as local issues, environment and communities in order to succeed in the implementation of urban regeneration investment.

3.4 Costs and property tax

Investment in urban regeneration involved a high cost in construction and land assembly. The initial process such as site assembly, demolition, and clearance often through a longer period then bears a high cost in financing charge (Adams et al, .1998[5]). Additionally, it also involves a high cost to destruct business, relocate exist community and high acquisition land (Whipple, 1971[30]). In South Melbourne, for example, property taxes before regeneration were $1,744 per annum while after regeneration these arise to $ 16,770 per annum. Increase in property value gives a positive sign to investor, however, extreme increase also will caused the investors fear to invest due to risky as community itself are no longer to live in the area and rundown areas.

3.5 Destroy neighbourhoods and social network

Urban regeneration is synonyms with demolishing the entire neighborhoods and destroys the social network (Jacob, 1961[18]). For example, the stalled Plaza Rakyat bus and light rail transit terminal project, and the abandoned of Nas Pavillion upmarket condominium are the examples of urban regeneration which contributed to the decay of the nearby properties in the area. Successful investment in urban regeneration needs to be viewed in a holistic approach. It requires creating lasting improvement in the economic, physical, social and environmental condition of the area that has been subject to change and at the same time all areas and individuals need to be benefit equally rather than been seen as an investment opportunity.

3.6 Problem of land multiple ownership

Issue on amalgamation of multiple land ownership also contributes to the investment constraint in urban regeneration (Whipple, 1971[30]). It is because each of the owners in the title will try to being one of the most essential by virtue of size, location and etc. Table 1 below tabulates the summary of categories and determinants of constraints from various literatures.

DOI: 10.1051/matecconf/20166600063

Table 1. Matrix of Constraints in Urban Regeneration Investment from various literatures

	Category of constraints	Determinants of constraints	Adair et al[4]	Simons [27]	Adams et al [5]	Epley et al [13]	ERM Econ [14]	McGreal et al[20]	Adair et al. [2]	Ng [22]	DETR [11]	Amin et al [7]	Adair et al[1]	Callender and Key [8]	Whipple [30]	Jacob [18]	Adam & Hasting [6]
			2005	**1998**	**1998**	**2002**	**2002**	**2006**	**1998**	**2003**	**2000**	**2000**	**1994**	**1996**	**1971**	**1961**	**2001**
1	Risk	**A)Pre Development Stage**															
		Planning Site	✓														
		High Cost of market survey, land contamination survey and detail feasibility studies					✓										
		Misperceived Returns					✓										
		Development Period and Tenure of Finance	✓	✓													
		Information Costs					✓										
		Shortage Material				✓											
		Delay Government Approval				✓											
		Lack of information	✓				✓	✓	✓								

DOI: 10.1051/matecconf/20166600063

Table 1 cont.

	Category of constraints	Determinants of constraints	Adair et al [4]	Simons [27]	Adams et al [5]	Epley et al [13]	ERM Econ [14]	McGreal et al[20]	Adair et al. [2]	Ng [22]	DETR [11]	Amin et al [7]	Adair et al [1]	Callender and Key [8]	Whipple [30]	Jacob [18]	Adam & Hasting [6]
			2005	**1998**	**1998**	**2002**	**2002**	**2006**	**1998**	**2003**	**2000**	**2000**	**1994**	**1996**	**1971**	**1961**	**2001**
		B)Development Stage															
		Development Cost	✓	✓													
		Contamination and Stigma	✓														
		Unforeseen Conditions				✓											
		Market Volatility	✓					✓									
		Risk Aversion					✓										
		Financial Constraints					✓										
		C)Post Development Stage															
		Low Demand					✓										
		Tenant Before Lease Out				✓											
		Yield and Rent	✓														
		High Transaction Cost					✓										

Table 1 cont.

	Category of constraints	Determinants of constraints	Adair et al [4]	Simons [27]	Adams et al [5]	Epley et al [13]	ERM Econ [14]	McGreal et al[20]	Adair et al. [2]	Ng [22]	DETR [11]	Amin et al [7]	Adair et al [1]	Callender and Key [8]	Whipple [30]	Jacob [18]	Adam & Hasting [6]
			2005	1998	1998	2002	2002	2006	1998	2003	2000	2000	1994	1996	1971	1961	2001
2	Return	Negative Image/Poor Location									✓		✓	✓			
		Lack in demand									✓	✓	✓				
3	Governance	Poor Institutional Arrangements (Over bureaucracy, Prolong Procedure for Grant Application Distrust Public Sector, Poor Coordination, Non-People Centre Policy, Conflict in formation policies at all level							✓	✓							✓

DOI: 10.1051/matecconf/20166600063

Table 1 cont.

	Category of constraints	Determinants of constraints	Adair et al [4]	Simons [27]	Adams et al [5]	Epley et al [13]	ERM Econ [14]	McGreal et al[20]	Adair et al. [2]	Ng [22]	DETR [11]	Amin et al [7]	Adair et al [1]	Callender and Key [8]	Whipple [30]	Jacob [18]	Adam & Hasting [6]
			2005	**1998**	**1998**	**2002**	**2002**	**2006**	**1998**	**2003**	**2000**	**2000**	**1994**	**1996**	**1971**	**1961**	**2001**
4	Cost	High Cost in Construction and Land Assembly			✓												
		High Cost to Destruct Existingbusiness, relocate exist community and high acquisition land			✓										✓		
		High cost to exist community with expense of tax will result rundown neighborhood													✓		
5	Social Network	Destroy social network														✓	
6	Problem of land multipleownership	Multiple Ownership													✓		

Table 1 illustrates that risk has been considered as the most significant contributor to the constraints of the investment in urban regeneration according to the various literatures. This risk factor can be subdivided into three different phases in the development process namely the pre-development, the development stage and the post development stage. Another significant contributor towards urban regeneration constraint is return. Various literatures agreed that return is one of the motivational factors for both the developers and the local authorities. Cost factor also contributed to the constraints due to these factors is the basis for consideration by investors to pursue their intention to involve in investment of urban regeneration. This is because investment in urban regeneration needs a large amount capital for invested. Therefore, intervention and aid from government is required.

4 Research methodology

This study focuses towards the elements of constraints in urban regeneration investment which include six categories namely risks, return, governance, costs and property tax, destroy neighborhood and social network and problem of land multiple ownership. Klang Valley has been chosen as the subject area of the study due to the diversified investments in land and building as well as the renowned developers and well established local authorities. This study has identified 21 established developers and 5 local authorities as the respondents. However, only 11 property developers provide a feedback while all local authorities responded in this study. It involved primary data collection i.e. close ended questionnaire. Data is administered using *SPSS* software.

5 Analysis and discussion

Referring to Table 2 from the property developers perspective reveals that overall land issues on the multiple ownerships (mean score 3.91) is the most significant category that influences developers' constraints. Interestingly, return carries the least mean score of 3.36. This illustrates that return has no correlation with the project type, but it influence by the cost and duration of the project. However, according to each items in Figure 1, it shows that high cost of destruction is the most significant constraints followed by risk in regards to the pre-development, and the issues on multiple land ownerships.

Table 2. The Mean Score of Constraints: Property Developers Perspective

	Category of constraint	Determinants of constraints	Property Developers Mean Score	Property Developers Total Mean Score
1	Risk	A)Pre Development Stage	4.091	3.61
		B)Development Stage	3.455	
		C)Post Development Stage	3.273	
2	Return	Negative Image/Poor Location	3.545	3.36
		Poor In Return	3.364	
		Lack in demand	3.182	
3	Governance	Poor Institutional Arrangements	3.636	3.55
		Conflict in formation policies at all level	3.455	
4	Cost	High Cost in Construction and Land Assembly	3.545	3.73
		High Cost to Destruct Existing business, relocate exist community and high acquisition land	4.091	
		High cost to exist community	3.545	
5	Social Network	Destroy social network	3.727	3.73
6	Land Issue	Multiple Ownership	3.909	3.91

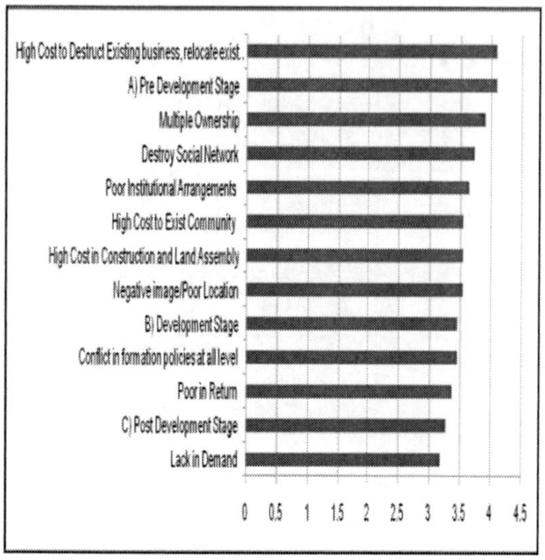

Figure 1. Three Most Significant Constraints: Property Developers Perspective (by ranking)

From the local authority's perspective revealed in Table 3, overall the most significant category that influence local authorities constraint is the multiple land ownership. This is similar to the perspective of developers. The reason behind this is multiple land ownership is a very challenging and complex issue that can influence the success and failure of the project as it involves dealing with the human behavior. Moreover, it can also drag the project exceeding the stipulated duration and indirectly will increase the project costs. In addition, for the item showed in Figure 2, the significant factors that influence local authority's constraints are multiple ownership issues as well as the high cost of destruction and construction. This is consistent with the developer's constraints as cost is seen as very critical to a project.

Table 3. The Mean Score of Constraints: Local Authorities Perspective

	Category of constraint	Determinants of constraints	Local Authorities Mean Score	Local Authorities Total Mean Score
1	Risk	A)Pre Development Stage	3.800	3.27
		B)Development Stage	3.400	
		C)Post Development Stage	2.600	
2	Return	Negative Image/Poor Location	3.400	3.33
		Poor In Return	3.200	
		Lack in demand	3.400	
3	Governance	Poor Institutional Arrangements	3.800	3.80
		Conflict in formation policies at all level	3.800	
4	Cost	High Cost in Construction and Land Assembly	4.000	3.93
		High Cost to Destruct Existing business, relocate exist community and high acquisition land	4.000	
		High cost to exist community	3.800	
5	Social Network	Destroy social network	3.600	3.60
6	Land Issue	Multiple Ownership	4.000	4.00

 DOI: 10.1051/matecconf/20166600063

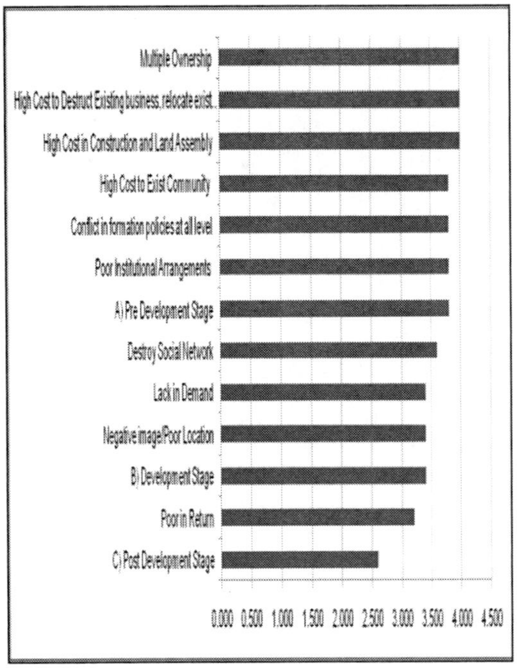

Figure 2. Three Most Significant Constraints: Local Authorities Perspective (by ranking)

There are a few gaps between both the developers and local authorities' perspective. These differences are influence by the objectives and direction of the organization, and the setting up of the organization. Figure 3 illustrates the radar chart between these two perspectives. The most significant gaps are in regards risk to the post development stage, followed by the high cost in construction and land assembly and lastly conflict in formation policies at all level. It should be noted that both developers and the local authorities are two different entities with different objectives. Hence this can explain the lower mean score from the local authorities' perspective as the developers are profit oriented organization while the local authorities are concern more towards the welfare of the society. The post development stage can only be estimated or projected as this will only happen after the project is finished and there are elements of uncertainty and risk. This will concern more to the local authorities as it has a strong relationship with the well being of the society in that particular areas. The developers on the other hand, once the development has been handover – this means the project is finish and they have no further interest towards it.

The second gap is on the high cost in construction and land assembly. For a profit oriented organization this will definitely affect the bottom line of the profit and loss accounts. Thus, cost is a critical factor and need to be managed effectively and efficiently. Cost in the local authority perspectives is whether or not the allocation is adequate to finance the project. This is due to the fact that the local authority's administration system is consists of different level of administration thus making decision is time consuming and involve lengthy process and procedures.

The third gap is on the conflict in the formation policies which concern more to the local authorities compared to the developers. This is due to the fact that the local authorities consist of different departments, different level of decision making – hence, decision making becomes complex and time consuming involving bureaucracy etc. For developers this is a simple process involving the board of directors only.

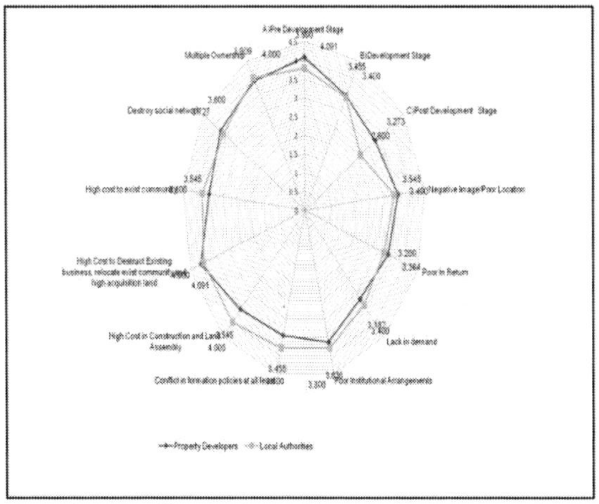

Figure 3. Radar Chart on Constraints: Property Developers and Local Authorities Perspective

6 Conclusion

The developers and local authorities are the key players in the built environment and their motivation towards urban regeneration investment is consider highly importance towards achieving sustainable and resilience, urban living and environment. Hence, although both have different perspectives towards this investment but this does not mean that it cannot be materialize. The challenge now is to come out with more incentives and strategies that can boost the motivations of both parties to collaborate in this investment. Urban regeneration is a key growth of economic, social and physical of a country and studies and experiences from other countries have showed that investment in urban regeneration can be as successful as any other investments. Moreover, the benefits gains are very valuable and significance particularly to the urban decay area as well as the whole nation. Despite the challenges face by the key players in the built environment, urban regeneration investment should be a priority to develop the nation to achieve an inclusive, resilience, safe and livable cities.

References

1. Adair A.S., Berry, J.N., and McGreal W.S. (1994). *Investment Decision Making: A Behavioral Perspective,* Journal of Property Finance. **Vol. 5(4).** pp. 32-42
2. Adair, A.S., Berry, J.N., Deddies, W.G., McGreal, W.S., and Hirst, S.M. (1998). *Accessing Private Sector Finance.* Royal Institution of Chartered Surveyors, London. pp 51.
3. Adair, A.S., Berry, J.N., McGreal, W.S., and Quinn, A. (2003). *Factors Affecting The Level and Form of Private Investment in Regeneration,* Urban Research Summary Number 6, Office of the Deputy Prime Minister. pp7.
4. Adair. A., Berry. J., McGreal. S., Poon. J., Hutchison. N., Watkins. C., and Gibb. K. (2005). *Investment Performance within Urban Regeneration Location,* Journal of Property Investment and Finance. **Vol. 23(1)**. pp.7-21.
5. Adams, D., Disberry, A., Hutchison, N., and Munjoma, T. (1998). *Do landowners constrain urban redevelopment?.* Aberdeen Papers in Land Economy: Discussion Paper, University of Aberdeen, Aberdeen. pp 99-01.
6. Adams, D., and Hastings, E.M. (2001). *Assessing Institutional Relations in Development Partnerships: The Land Development Corporation and The Hong Kong Government Prior to 1997, Urban Studies.* **Vol. 38(.9),** pp. 1473-92.
7. Amin, A., Massey, D., and Thrft, N. (2000). *Cities for the Many, not the Few, Bristol.* The Policy Press. pp 48.
8. Callender, M. and Key, T. (1996). *The Total Value of Commercial Property in the UK.* Paper presented at the Royal Institution of Chartered Surveyors Cutting Edge Conference, University of the West of England, Bristol.
9. Couch, C. (1990). *Urban Renewal Theory and Practice.* MacMillan, Basingtoke.
10. Cox. K., Viruly. F., Bertoldi. A., (2015). *Inner City Transformation & Investment Trends (2009-2014).* Johannesburg Development Agency.
11. DETR (Department of The Environment, Transport and The Regions) (2000). *Our Towns and Cities: The Future.* The Urban White Paper. The Stationery Office, London.
12. Eni, Sabariah, Adair. A., Cheng. L., and Lim. J. (2013). *Financing Urban Redevelopment in Kuala Lumpur, Malaysia: Prospects and Challenges.* Paper presented in 20th Annual European Real Estate Society Conference. ERES: Conference. Vienna, Austria.
13. Epley. D.R., Rabianski. J.S., Haney. R.L. (2002), *Real Estate Decisions.* South-Western Thomson Learning.
14. ERM Economics (2002). *Evaluating the Impact of Urban White Paper Fiscal Measures.* Report to the DTLR (Department of Transport, Local Authorities and the Regions), London.
15. Goksin and Muderrisoglu (2005*). Urban Regeneration: A Comprehensive Strategy. 41st Isocarp Congress, Istanbul.*
16. Hausner, V.A (1993). *The Future of Urban Development.* Royal Societies of Arts Journal. **Vol. 141**, pp. 523-33.
17. IOER (May 2005). *Compendium: Handbook 3 Sustainable Urban Regeneration and Its Assessment.* LUDA Project.
18. Jacob (1961). *The Death and Life of Great American Cities.* Vinateg, New York.
19. Malaysia Statistics Department (2015). *Population Data.* Retrieved from https://www.statistics.gov.my.
20. McGreal. S., Webb. J.R., Adair. A., Berry. J. (2006). *Risk and Diversification for Regeneration/Urban Renewal Properties: Evidence from the U.K.* Journal of Real Estate Portfolio Management. **Vol 12 (1).**pp 1-12.
21. McNamara, P. (1993). *Parameters for Institutional Investment in Inner City Commercial Property Markets*, in Berry, J. McGreal, S. and Deddies, W. (Eds), *Urban Regeneration,* E. & F.N. Spon, London. pp.5-15.
22. Ng, I., (2003). *Community Approach: is a Solution to Age-Old Problem of Urban Renewal in Hong Kong,* International Journal for Housing Science. **Vol. 27 (1).** pp. 75-88.
23. Nilai Harta (2015). *Urban Regeneration in Malaysia: Issues and Practices, Malaysia* Retrieved from http://www.nilaiharta.com.my/articles
24. Peter W. Robert and Hugh Skyes (2000). *Urban Regeneration A Handbook.* SAGE Publication Ltd, London.
25. Planning in Association with Regeneration and Renewal (2013). *The Top 100 Regeneration Projects 2013.* London. pp 1-13.
26. Samworth, M. and Lacey, C. (1998). *Remove guesswork from spec work. Estate Gazette, No. 9842,* pp. 160-1.
27. Simons, R.A. (1998). *Turning Brownfields into Greenbacks.* Urban Land Institute. Washington, DC.\
28. The Edge TV (2014). *Malaysia's Urbanisation Rate Expected to Rise To 75% in 2020* [Youtube Video]. The Edge Malaysia (*City & Country).*
29. Warner. L., and Long. E. (2009). *Executive Summary Impact Analysis of Oklahoma City's Maps And Other Significant Central City Investments.* Greater Oklahoma City Chamber of Commerce.
30. Whipple. R.T.M. (1971). *Urban Renewal and the Private Investor.* West Publishing Corporation Pty Ltd, Sydney.

Demographic and Commercial Environments and Building Control Customer Services

Peter Hope

Education, Research and Industry, the Queensland Building and Construction Commission, 299 Montague Road, West End, QLD 4101, Australia

Abstract. A simple and cost effective business analytics schema developed in this study provides a basis for observing how demographic and commercial environmental impacts affect building control customer services. In order to overlay environmental impacts onto building control services, schematic mapping of resource distributions and services were plotted according to corporate jurisdiction, corporate virtues and customer type characteristics. Quantitative and qualitative surveys were conducted on corporate staff and material resources, business performance measures and targets, building control service manager observations and demographic and commercial environments. Results from this research provide a single view corporate map of people, material resources, and service performance in relation to customer characteristics, while reaching out to better understand the building control customer experience, general public expectation and market influences. In this regard, this research can provide a foundation for realising corporate virtue aspirations.

1 Introduction

This paper presents a new approach stemming from a business analytics case study on how demographic and commercial environments affect corporate services. The research methods referred to in this paper offer a low impact, cost effective, quick and easy approach to business analytics.

The purpose of the research project is to identify how building control services can more effectively respond to changes in demographic and commercial environments, service resource distribution realities and performance indicators and identified corporate virtue aspirations and benefits to customers, in the process.

A business analytics schema developed in this study relies on available publications and corporate data. Expert advice from industry experts provided insights into demographic and commercial forecasts [1, 2]. While expert advice from senior building control managers enabled a better understanding of business operational matters.

Further, building control service managers were interviewed about their observations of their service teams in relation to set criteria such as: workload, team ability, material resources, service quality, performance, output targets, customer feedback and overall performance.

The results of this research offer the Queensland Building and Construction Commission (QBCC, Australia) insights on demographic and commercial influences concerned with building control customer services. Further, a conceptual framework has been developed for mapping resource distributions, performance measures, and corporate virtue aspirations, upon which future research can be based.

2 Background

2.1 The building control in Queensland

The QBCC engages approximately 380 staff to provide approximately 72 services in a market that produced 90,759 residential construction contracts in 2014/15, of which approximately five percent of these dwellings were subject to building disputes between parties to the building contract. For example, 4,793 disputes were received by building control customer services in the 2014/15 year.

The QBCC provides award winning customer service standards, contemporary marketing, digital and media systems, building compliance and regulatory services, home warranty scheme insurance and other key corporate support services [3].

Building work standards in Queensland are governed by the Building Act 1975 (Qld), the Queensland Development Code, the National Construction Code of Australia (NCC) and Australian Standards, as called up in the NCC. All on-site construction requirements relating to building work are governed by the Queensland Building and Construction Act 1991 (Qld) which essentially evokes the jurisdiction of the QBCC [4, 5].

2.2 Corporate virtues

Corporate virtues are aspirations for building control customer services [6] that refer to the QBCC vision to be recognised as the best and most respected regulatory service provider in Australia, its purpose for peace of mind and its core values which put customers first, unleash potential, to be courageous, to put ideals into action, empower people, provide a safe and healthy working environment, with an emphasis on the building control customer service value chain.

The quantity and efficiency of the building control are attributes of customer service performance reflected through the attainment of corporate virtues that support a vision which in turn is dependent on public expectation and the private building contract experience, which is subject to customer feedback [7].

Five main building control services are described as [7]:

a) Licensing services,
b) Compliance and enforcement,
c) Dispute prevention and resolution services,
d) Home warranty insurance, and
e) Information and education.

2.3 Building control jurisdiction

The QBCC [3, 6] regulates the building industry, stipulating remedies for defective building work, while providing support, education and advice for building contractors and consumers. Building control customer services in Queensland facilitate a balance between the interests of building owners and principle contractors through the regulation of domestic building contracts.

2.4 Building contractor licencing

Queensland maintains one of the widest ranges of building industry skilled trade practice registers in Australia. For example, there are 87 licence classes including: builder, trade contractor, plumber and fire protection license classes, all together constituting 85,627 licensed individuals [9].

To qualify for a licence, individuals must have prescribed qualifications, experience and business acumen [3, 9]. In addition, the building industry depends on professional abilities such as: quality control, managerial, logistical and organisational skills, particularly in highly competitive environments [10, 11].

2.5 Continuous professional development

In Queensland, Continuous Professional Development (CPD) of licensed building contractors remains voluntary for the majority of licence types. However, it is likely that those practitioners most in need of CPD training "…are those least likely to become involved" [12].

While large building organisations apply building practices more efficiently and effectively, predominant issues of competency and the rate of building disputes concerning building control customer services lies with small building businesses [13, 14, 15, 16]. Further, building defects contribute to the cost and affordability of new housing [10, 14, 15, 17, 18].

Extrapolating from Smith, et al., [11] reference to 800,000 properties in the state of Victoria, of which 110,000 (13.7%) were subject to defective building work related claims the rate of building defects in the state of Queensland is somewhat lower (5.2%). For example out of 90,933 building projects 4,793 building disputes were received by the QBCC in 2015 [4].

While the prospect of a five percent defect rate may not seem large in general terms, for the individual parties to the building contract the associated costs, time-loss and stress of building dispute, resolution process, and rectification work is most likely disappointing, undermining confidence, underpinning negative preconceptions and revealing inefficiencies.

Further, additional costs and resources resulting from building disputes are incurred by building control customer services including expert investigations [4]. Building dispute costs ultimately result as an investment overhead cost to new buildings [10, 14, 16, 17].

2.6 Building disputes

The Queensland residential building industry provides an important contribution to the State economy. However, a contentious issue is the cost of building disputes between parties to the building contract, predominantly over defective building work [10, 14, 15, 17, 18].

The estimated cost of resolving major contract construction disputes in Australia is between $560 million and $840 million per annum. By including avoidable costs such as associated opportunity and delay costs the total waste costs is approximately A$7 billion annually [19]. However the cost of building disputes to small business, which is predominant in the residential sector of Queensland, remains unclear [20].

The government has emphasised the importance of small business as a driver of economic growth [21]. The Construction and Property Services Industry Skills Council (CPSISC) points out that, Construction is the second largest industry gross value-added contributor representing approximately 7.8 percent of the nations' GDP in trend terms [21, 22]. However a concern is that the largest rate of small business insolvencies is associated with the Construction sector [16].

2.7 Demographic and commercial environments

Australian population growth trends are easing, at approximately 23,839,000 people. Household formation trends are aligned with population trends, at 8,685,000 new households. Household occupant size also remained steady at 2.74 persons, in Australian terms by 2015 estimates [2].

Stable monetary policy and strong financial systems of Australia provide a sound basis for open market competition and innovation [1]. Household and business service conditions appear to be improving, while there remains considerable uncertainty around the economic recovery [23].

Economic characteristics following the international 2008 financial crisis include: a stronger emphasis on the 'balance sheet', changing demographics and less extreme growth rates, etcetera [23].

Over the long term, population and economic growth are factors expected to drive residential and non-residential building activity, along with improvement in real income levels, demographic changes, the industry supply chain, employment growth, building floor sizes, and market expectations on quality produced [2].

Overseas investors are attracted to Australia because of its reliable legal system, clean air, and child education qualities [1]. Current economic transitions away from resource sector exports suppress economic drivers and foster investor caution, mitigating supply and therefore securing prices.

A disparity between property asset holders and market realities occurs as owners 'reluctant to sell at a loss' avoid selling during slow markets, which underpins the unaffordable housing phenomena. The result over time is a perception of rising property prices, in-between growth markets [1, 2].

Existing house prices in Brisbane, the capital city, were A$505,000 (up 7.4%), while house rents were A$375 per week (up 1.4%), in Queensland by 2015 estimates [2].

Average floor area size of new houses remains at 257.8 per square metres. While average residential building costs are A$1,059 per square meter, in Australian terms by 2015 estimates [2].

Owner-occupiers appear to be turning from detached housing aspirations to more affordable, better located multi-residential solutions, inflating activity and prices in the multi-residential-complex sector [2]. A shift from detached housing to multi-residential solutions appears to be the result [1].

Major building contracts such as civil or commercial development can be mapped in cycles of activity, but trends tend to be skewed by large single developments in unique locations [2]. Similarly, residential development (hotspots) in regional Queensland, following rapid growth in the resource sector, has slowed as populations migrate back to the capital city for work which has stimulated demand for, and development of housing in Brisbane.

However, building control customer services remain active despite building demand and supply interplays. Residential building disputes typically occur up to 6 years and 3 months following the building completion, in Queensland.

This duration represents the home warranty insurance period within which funds can be made available to rectify defective building work where the builder is unwilling or unable to take responsibility. Therefore building dispute activity can remain buoyant when building commencement trends slow [4].

The following schematic view of environmental influences and stages of new building activity in Queensland [Figure 1], derived from this study, enables discussion between stages of interest. For example, building control customer services are shown (in the fourth last stage) toward the bottom, but are 'triggered' at the Private Contract stage (five stages before) and in less than 0.008 percent of cases continues through to building control enforcement (709 directions to rectify issued; 2014/15).

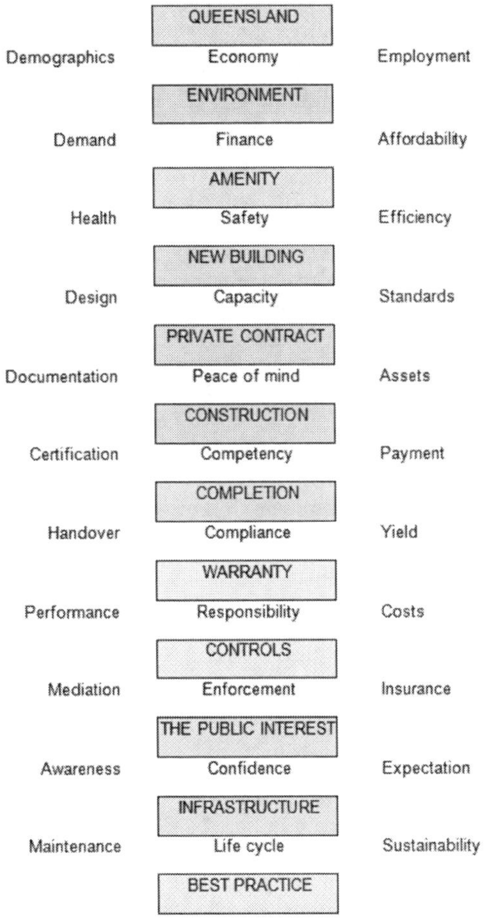

Figure 1: The building control environment

2.8 Building control customer services

The QBCC provides a broad and diverse range of building control customer services through a 'one-stop-shop' public service. These services are franchised to nine customer service centres in major cities of all regional areas across the state of Queensland.

Consequently, in order to better understand the outcome of building control customer services from the customers' perspective, public expectation and the private building contract experience are considered to be two key aspects of building control services. Public expectation can be understood through a range of perspectives including: legislation and enforcement provisions, building standards, economic drivers, media reports, consumer surveys and inquiries and public uptake of the building control information.

2.9 Building control resources

To manage customer services, the QBCC receives income from building licensing and building warranty insurance provisions needed for resources such as: people, time, budget and technology.

While the economy grows and slows, on average building control revenue appears more static in the sense that public service income does not compound as businesses-for-profit models expect too.

Further, while building activity fluctuations occur, building control customer services appear more continuous. This on-going demand may in-part be due to a broad range of building control services and the home warranty insurance period referred to above.

Further to add to the costs of building control customer services, the evolution of new public technological applications drives a need for additional building control digital communication platforms that are needed over-and-above on-going public service costs (i.e. social media, digital app mechanisms, touch screen devices, 3-D printing of building components, BIM communication channels, free trade compliance standards, etcetera) [4].

2.10 Business analytics schema

The business analytics schema devised for this study is a sematic diagram displaying a map of corporate resource distributions (inputs, outputs and outcomes) concerning building control service groupings, according to customer type (public, private and internal corporate).

The business analytics approach applies readily available corporate measures such as the consolidated budget, the number of staff, the Organisational Chart and Key Performance Indicators (KPI), etcetera.

2.11 Frontline services

This study further considers the number and nature of frontline customer services as applied to building work, a central responsibility of the QBCC [3, 4]. Approximately 43 percent of 72 building control customer service teams have been surveyed, representing all building control 'frontline services'.

2.12 The private building experience

The private building experience refers to both parties to the domestic building contractual agreement, the building owner and the principle contractor. The private building experience can be understood through both their contractual process and the extent that they understand their responsibilities, form their expectations and the circumstances under which they may enter into dispute.

2.13 Stages of the building contract process

Three stages of the private building contract experience are: commencement, completion, and the following six year and three month home warranty insurance period.

These divisions enable overall service results to be 'distilled' down to five customer service orientations or characteristics which are:

a) Corporate services and business excellence (accountability),
b) General public expectation and awareness services (confidence),
c) Building compliance services, during construction work (standards),
d) Services concerned with building completion, certification and approval (peace of mind), and
e) Building performance, disputes and rectification work (the contract warranty period).

The following schematic diagram [Figure 2] illustrates cycles of building control frontline services, from customer needs (catalysts) to building controls (compliance). These are 'cause and effect' relationships that turn the wheel of building control customer services clockwise in stages from the outside toward the inside, or vice versa. Connecting needs to services, the 'weakest link in the chain' analogy can be applied to these service cycles.

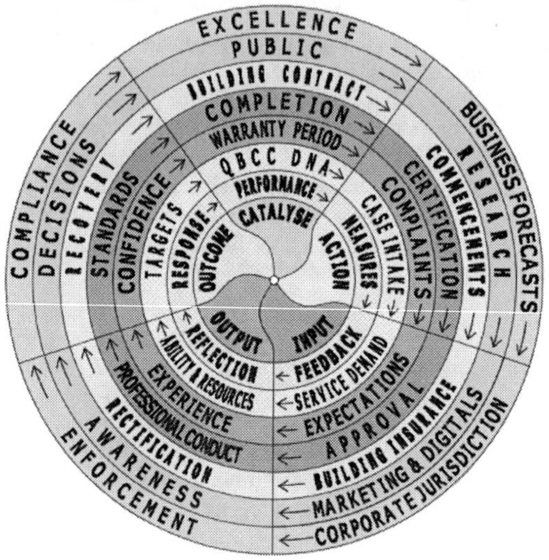

Figure 2: Building Control Service Flows

3 The case study

3.1 Issues

In order to better understand how building control customer services respond to demographic and commercial influences this case study explored a means of observing business 'hotspots' and 'risk spots', resource distributions, service performance measures, and the corporate environment from a customers' perspective by identifying customer types and aligning services accordingly.

Hotspots refer to customer services with high workloads, while risk-spots refer to those services with potential to mitigate future customer needs, through proactive activities.

Given the complexity and the extensive implications of the research question the contribution of this case study has been limited to developing and mapping a business analytics schema comprised of theoretical constructs and services associated with private and general public customers. While corporate customers form part of the perspective of this study, the focus is on 'frontline' customer services.

The issues related to building control customer services are:

a) understanding corporate virtue aspirations,
b) understanding characteristics in common of building control customers services, or how they may differ,
c) understanding the performance of services,
d) understanding the people and material resources,
e) identifying channels for understanding general public expectation and private building experiences, and
f) understanding the effects of demographic and commercial factors.

3.2 Breakdown

The factors concerning this case study include:

- corporate vision, purpose and values (virtues),
- the building control jurisdiction,
- building control customers:
    - general public expectations,
    - private building experiences,
- workloads of building control customer services,
- corporate resource distributions,
- quality of building control customer services,
- performance validity of building control customer services,
- building control customer service target reliability,
- building control customer service feedback channels,
- overall performance of building control customer service groups, and
- new building commencement, compliance, disputes, and control trends.

3.3 Available Information

The QBCC Act 1991 (Qld.) provided parameters of building control jurisdiction to better define core corporate responsibilities for customer services prioritisation purposes.

QBCC 2015 Corporate Plan and an internal corporate 'DNA reference sheet' provided a current interpretation of building control customer service aspirations concerning corporate virtue or value aspirations [7].

Available government literature on demographic and economic trends provided social science insights into building control customer behaviour, looking forward [1, 2, 23, 24].

Building industry and property forecast 2015-2030 reports provided analytical summaries on the Queensland economic and demographic outlook to better understand current and future market drivers and behavioural trends [1, 2].

The QBCC People and Culture division provided 2015 data on staff numbers and designations enabling mapping of human resource distributions in terms of the type of building control customer group.

The QBCC consolidated budget 2014-15 provided accounts that enabled mapping of material resource distributions in terms of building control customer service groupings.

QBCC corporate publications, digital accounts and raw data provided an emphasis on key performance indicators in relation to under-measured aspects of building control services [3, 6, 7, 8, 9].

3.4 Verification

While available information provided insights into the corporate role and direction for building control customer services, key factors include new building market activity and social trends, the distribution of corporate people and material resources and KPI measures that enable analysis on building control service activities and under-measured services.

Expert observations were recorded and aggregated as a way of verifying conclusions drawn from available historical trends and theoretical information. This case study was limited to 'front-line' building control customer services which represent approximately 43 percent of all corporate services identified for mapping purposes.

Expert observations from managers of 31 building control customer services were interviewed for validation purposes. The survey interview technique needed to be easily understandable and quick to attend to for busy respondents.

3.5 Constraints

This research project began with a time constraint of approximately 10 weeks. Further, the research activities were conducted by one principle research officer who is familiar with the corporate business and has conducted related corporate research projects.

As this work had not been done before, it is inductive by nature. Therefore care had to be taken in the design, implementation and review of the methods.

The case study utilised available information to generate operational suppositions that could be verified by expert advice. Propositions were applied to developing the conceptual framework needed for the business analytics schema.

Indicative insights enabled recommendations on the need to prioritise customer service 'hotspots' and 'risk spots' with more targeted research probes based on the foundations of the conceptual framework.

It is envisaged that more targeted research can support hypotheses on building control customer service efficiencies, enabling deductive research validation over the medium to long term.

3.6 Historical information

Historical data needed to better understand factors and trends were derived from QBCC data. Raw data included:

- the number of mandatory insurance policies representing the number of commencements of detached house building and medium-rise

 DOI: 10.1051/matecconf/20166600064

residential dwellings and alterations and additions;

- the number of building complaints received; and
- the number of defective work rectification orders issued (directions).

Further insights were derived from previous corporate research reports concerning building inspector observations, case management costs and workload audits etcetera.

3.7 Correlations

In order to develop a basis for understanding the correlation between demographic and economic environmental impacts and building control customer services it was necessary to identify the type of customers, their relationship to building work, demographic and commercial environmental factors, along with building control customer service types.

While the business analytics conceptual framework encompassed all building control customer services, of the three relevant customer types: corporate (virtues), public (expectation) and private (experience), service teams predominantly concerned with the latter two customer types were investigated further through building control customer service manager observation interviews.

3.8 Limitations

To advance understanding of correlations between building control environmental factors, customers and services a business analytics schema was developed to enable mapping of building control customer service resources, performance and service outcomes through customer feedback channels.

The benefit of the business analytics schema is to provide a platform for mapping services to customers upon which corporate priorities can be determined.

However, cause and effect correlations of market forces, the private contracting experience and building control customer services are beyond the reach of this case study, which is intended to support on-going research, stemming from this study.

3.9 Assumptions

A conceptual framework (business analytics schema) formulated through heuristic insights [25, 26] provides a method of observing relevant factors, with a single view schematic diagram.

It assumes that common public perceptions of new building work processes may be generally inherent in private building owner expectations and consequently their underlying attitudes in building disputes. Similarly, private building contractor preconceptions about their contract responsibilities may limit communication, information transfer and dispute resolution efforts [4, 19].

The extent that general public expectations are in alignment with minimum building standards is outside the scope of this study, a question in need of additional research to inform awareness rising education services.

3.10 Decisions

The methods adopted in this case study provided an indicative understanding of building control customer service ‘risk-spots’ and ‘hotspots’ that suggests where resource distribution priorities are needed.

Further the business analytics schema provides ‘a lens’ through which building control services, with common customer types, can be grouped and analysed in terms of resource expenditure, service performance and customer service outcomes.

Insights and indicative results are substantiated by both corporate performance reports and anonymous expert customer service observations generalised into customer groupings to mitigate bias. However the conceptual framework can be considered a theoretical basis upon which more focused research is required.

3.11 Protagonists

The protagonists of this case study are building control customer services, represented by team managers, ‘operating on the wheels of frontline customer services’. The service team performance is dependent on their legislative jurisdiction, corporate aspirations, and customer expectations.

3.12 Dartboard survey

The Dartboard survey [Figures 3] along with a multi-choice questionnaire were designed to be presented to building control customer service managers (the protagonists) as part of an interview technique that ensured respondents received a uniform understanding of the context of the survey terminology to support validity.

The same interviewer consistently applied the interview survey methods to support reliability. Interview responses were anonymous and respondents were advised that their contributions were to be aggregated by customer type, mitigating the risk of bias. Further, interviewees were not privy to the results of other respondents, to avoid external influences.

Respondents were asked to mark a spot on the Dartboard survey as to whether each one of eight aspects: needs attention, are satisfactory or appear in great shape.

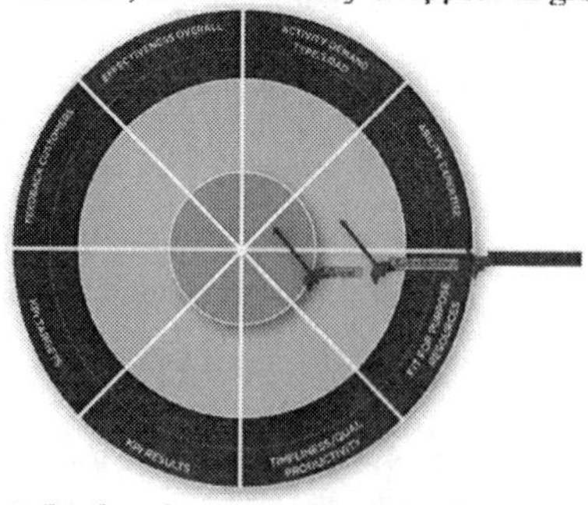

Figure 3: Dartboard survey and interview instrument

The aggregated Dartboard survey and interview results [Figure 4] for example, suggest that attention is needed on building control customer service demand activity (workload), material resources and performance measures. Conversely, service managers appear more than satisfied with staff ability and customer service feedback channels. Overall service effectiveness and quality appears satisfactory.

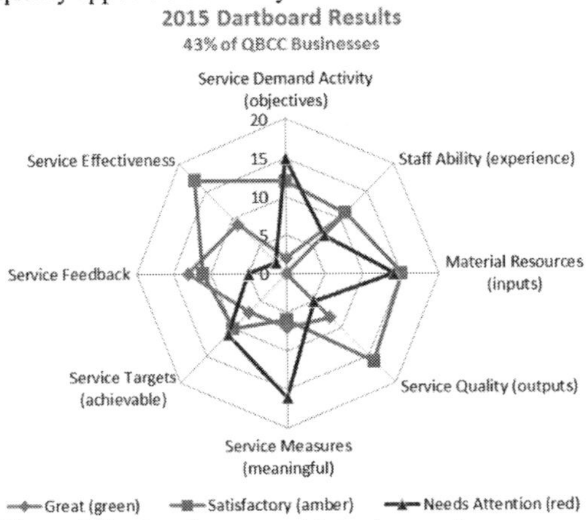

Figure 4: Dartboard survey and interview results

3.13 Building control customer service manager interviews

Generally, building control customer service managers were enthusiastic about their responses to the survey interview, once they understood that it would not take too much time. The interviewer provided an important function by contextualising the questions to enable quicker and more concise responses.

On an accompanying multi-choice questionnaire, respondents were able to select their service team predominant customer group and most important business resources, most referred to KPI measures and most sort after service results, most important service benefits and most important demographic and commercial environmental impacts.

3.14 KPI survey

The KPI survey plotted the frequency and spread of current KPI measures across 72 services, in an attempt to better understand what aspects of services were being measured and what was not. These results were compared with building control customer service manager observations to better understand the quality and application of current KPI measures and the practicality of service targets.

3.15 Survey of resources

The people and material resource distribution analysis mapped on the business analytics schema enabled an overview of corporate investment and resource distribution to building control and customer service benefit analysis. However, detailed analysis results on the efficiencies of customer service inputs, outputs and outcomes are beyond the scope of this preliminary case study.

3.16 Conceptual framework

Further the conceptual framework of the business analytics schema [Figure 5, below] provides a view of factors and customer service feedback channels along with aspects of corporate values.

3.17 Approach

Cause and effect relationships between market drivers, consumer behaviour, available resources, service abilities and workload pressures are complex multi-dimensional phenomena. The risk of increased building defects during periods of high building activity, as supply struggles to meet demand, along with customer demographic and behavioural variables can introduce a higher level of complexity to building control customer services and corporate resources. Considering the complex nature of the environment, this case study begins the process of building control customer service analysis rather than concludes it.

3.18 Context

In transition are: the 2014 corporate restructure, from the previous statutory authority to the current statutory commission; corporate communication channels from paper and telecommunication technologies to social media and digital administration platforms.

Further, amidst a climate of general economic transition [23, 27] and social adjustment experts refer to the 'new norm', characterised as slower more steady new building market activity growth [1, 2, 23].

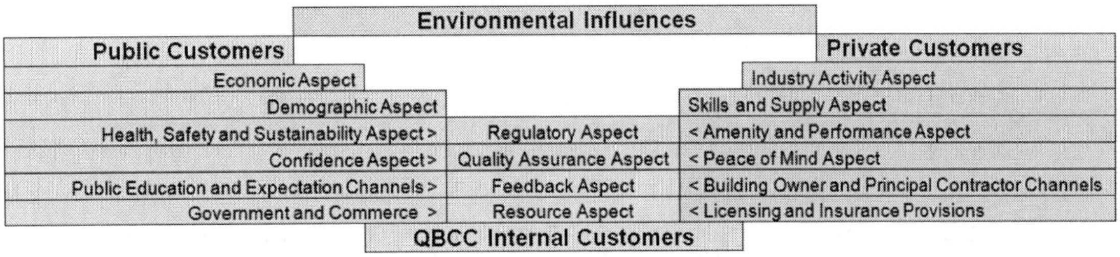

Figure 5: A conceptual framework

3.19 Consequences

The consequences of not embarking on a series of orchestrated research probes into the extent that building control customer services are achieving the aspirations of corporate virtues are a matter of due diligence.

Efficient building control customer services can mitigate the recurrence of building defects and discourage frivolous building disputes. Investments in developing better building control customer services may need to be measured in terms of the corporate aspirations to achieve their stated virtues.

As the organisation aspires to be courageous, unleash potential and empower people, it runs a risk of adverse reactions if building control customer service teams continue to be frustrated by avoidable limitations.

3.20 Alternatives

The alternative to conducting systematic research on the investment in, and benefit of, building control customer services may be to continue to respond to transitions from the 'business as usual' model to more prescribed objectives, relying on heuristic observations as they become apparent, assuming assumptions are reflective of the underlying realities.

4 Conclusions

This inductive research is dependent on available information and historical data to develop insights into building control customer service conventions from past performance. The conceptual framework provides insights through mapping building control customer services.

Further, the research explores current customer service efficiencies, performance standardisation and the potential to realise corporate virtues. While the results can form a basis for prioritising future more targeted research.

Transitions reflected in the economy, public expectation, confidence in private new building and uptake of building control services present opportunities for the QBCC to consolidate change, reflect on core business objectives, with finite resources and prioritise applications in the pubic best interest.

Demographic and commercial influences can be seen through a business analytics view of general public and private customers that subscribe to building control services. To better enable prioritisation of limited resources in the public best interest, an understanding of customer types through their characteristics is informative, such as:

1) public expectation of, and confidence in, new building investment;
2) private building experience of both, the owner and principle contractor; and
3) corporate service team performance and accountability.

With the focus on building control jurisdiction, and corporate values, applied to prioritised services in the public best interest, best practice can be determined amidst an environment of economic transition.

4.1 Survey Instruments

The Dartboard survey instrument, applied to frontline managers, covers key business performance elements such as demand, capacity and performance which are generic and useful in reviewing key strategic objectives (such as customer satisfaction).

In application, the Dartboard survey stimulated engagement as frontline managers considered service capability, capacity and performance in relation to demand with the understanding that their responses would be aggregated into results associated with customer groupings, rather than their individual businesses.

The Multi-choice survey instrument provided measures on 'alignment' of service managers to perspective and values to corporate directions. The multi-choice approach conveniently lists 'options' relating to corporate values with an 'other' option choice enabling diversity. This approach enabled the interviewer to quickly introduce the Multi-choice instrument; which in-turn enabled the respondent to quickly contextualise the questions by viewing the range of options.

A simple survey of income, expenses and staff distribution provided an indicative view on resource distributions in terms of service areas, compared with under-resourced frontline services (hot-spots) or services with potential to mitigate activity (risk-spots) that can lead to contractual disputes or building defects. Corporate (internal) support services appear to be well resourced to provide strategic support and direction to frontline services.

4.2 Best Practice

Demographic and commercial impacts are thought to be in a moderate transitional phase. Further, in transition are the introduction of new information technology platforms and applications. With limited potential for increasing income and resources, the recommended strategy of this study is to develop flexible applications with resources applied according to priorities that resolve operational issues.

For example, a more detailed resource distribution review should consider as a first priority services under external workload pressure (reactive hot-spots). Services with potential to mitigate external workload pressure (proactive risk-spots) are a 'second priority'.

As new technological systems are consolidated and service operations streamlined, operational-review resources can be applied to performance interpretations and measures to provide an evidence-basis for understanding what is working, while reflecting on where the risks are.

Public awareness strategies can be developed on newly devised standardised performance measures to contribute to public confidence in new building and private expectation of the responsibilities of parties through private building contractual processes.

4.3 Conceptual Frameworks

A key concept of this study is to view services through

customer lenses, 'stepping into their shoes' [28, 29], by grouping customers by their characteristics such as: public confidence, private expectation, and frontline service needs.

An important recommendation in an environment of 'transition', with limited resources, is to take a conservative interpretation on the scope of the building control jurisdiction. To focus limited resources on services that are 'in the pubic pest interest', while raising public awareness on the business service model.

Further by providing customers with information to 'reality-test' private expectations on the roles and responsibilities of private parties to their building contract building owners and principle contractors may be better informed throuth awareness education to avoid disputes through better communication and understanding.

Promoting collaborative and cooperative behaviour between parties to the contract is better than focusing on legal hierarchies imposed through traditional contractual agreements [30, 31, 32]. A public view of an equitable approach, from Abrahamson's risk management principle [33, as quoted in 30] for example may be –

> "*Project risks are assumed by the party most capable of managing or controlling the risk: should no party be capable of managing a risk, then the client takes responsibility for this risk.*"

Acknowledgements

This project is inspired and guided by the Building and Construction Commission (QBCC) and the Board (QBCB). The acting commissioner kindly perused the paper offering valuable advice. Mr Paul Tilley's guidance is certainly appreciated, lecturer of construction management, Faculty of Health, Engineering and Science, of the University of Southern Queensland. Finally, thirty one service managers responded to the survey proactively, with interest.

References

1. Kim Hawtrey, pers. comm., 8 September 2015.
2. BIS Shrapnel, 2015, 'Building in Australia, 2015-2030; Building Forecasting', 35th Ed., July, pp. 302 [paper]
3. QBCC, 2015a, 'Annual Report 2014-2015', <http://www.qbcc.qld.gov.au/sites/default/files/QBCC_Annual_Report_-_1Dec13_-_30Jun14.pdf>
4. Hope, P. G., 2015, 'QBCC 2014-15 dispute resolution service review of high-density residential building work', July.
5. Australian Government, 2015, <http://www.business.gov.au/business-topics/business-structures-and-types/industry-factsheets/Pages/national-construction-code.aspx>
6. QBCC Act 1991, 2015, pt. 1 (3), Queensland Government, <https://www.legislation.qld.gov.au/LEGISLTN/CURRENT/Q/QldBuildConCommA91.pdf>
7. QBCC 2015d, 'DNA Sheet', (internal document)
8. QBCC, 2015c, 'Corporate Plan 2015 – 2019; for peace of mind', <https://www.qbcc.qld.gov.au/sites/default/files/QBCC_Corporate_Plan_2015-2019.pdf>
9. QBCC, 2015b, 'License Classes', http://www.qbcc.qld.gov.au/when-you-need-licence/licence-classes
10. Georgiou, J., Smith, J., and Love, P. E. D., 2002a, 'Defects and quality control in the housing industry: Implications for management and regulatory control,' Proc., 8th Int. Conf. on ISO 9000 & TQM (Change Management) CD-ROM, Centre for Management Quality Research, CMQR, at RMIT Univ., RMIT Storey Hall, Melbourne, Australia.
11. Smith, J., Mills, A., and Iyer-Raniga, U., 2004, 'Benchmarking the implementation of continuing professional development in the Victorian construction industry', Benchmarking CPD, FABP, UM, Building Commission, Victoria.
12. Brennan, B., 1990, 'Continuing Professional Education', Australian Council for Educational Research, Canberra.
13. DISR, 1999, 'Building for Growth: An Analysis of the Australian Building and Construction Industries (Competitive Australia)', Department of Industry, Science and Resources, Commonwealth of Australia, Canberra.
14. Ilozor, B., Okoroh, M. I., and Egbu, C. E., 2004, 'Understanding residential house defects in Australia from the State of Victoria,' Build. Environ, 39, 327–337.
15. Mills, A., Love, P. E. D., and Williams, P., 2009, 'Defect costs in residential construction', Journal of Construction Engineering and Management, ASCE, January.
16. ASIC 2015, 'Insolvency statistics summary', September quarter, <http://download.asic.gov.au/media/3446712/insolvency-statistics-summary-september-quarter-2015-published-10-november-2015.pdf>
17. Georgiou, J., Love, P. E. D., and Smith, J., 1999, 'A comparison of defects in houses constructed by owners and registered builders in the Australian State of Victoria', Manage. Int. Rev., 17, 3, 160–170.
18. Georgiou, J., Smith, J., and Love, P. E. D., 2002b, 'Quality improvement through builder registration,' Proc., 8th Int. Conf. on ISO 9000 & TQM (Change Management), CD-ROM, Centre for Management Quality Research, CMQR, at RMIT Univ., RMIT Storey Hall, Melbourne, Australia.
19. CRC CI, 2009, 'Guide to leading practice for dispute avoidance and resolution, < http://www.construction-innovation.info/images/pdfs/DAR_Guide.pdf>
20. ABS 2013, 8772.0, Private sector construction industry, Australia, 2011-12, <http://www.abs.gov.au/ausstats/abs@.nsf/Latestproducts/8772.0Main%20Features22011-12?opendocument>
21. CPSISC, 2014, p. 19, 'Environmental Scan 2014-15, Construction and Property Services Industry Skills

DOI: 10.1051/matecconf/20166600064

Council, pp. 76. (e.g. Construction and Property Services Market) [paper].

22. Hope, P. G., 2014, 'Environmental Scan 2014-15', QBCC (internal document) pp. 76.
23. RBA, 2015, 'Minutes of the Monetary Policy Meeting of the Reserve Bank Board', Sydney, 4 August.
24. Queensland Treasury, 2015, 'Queensland population counter', 24 September, <http://www.qgso.qld.gov.au/products/reports/pop-growth-qld/qld-pop-counter.php>
25. King, D., and Macgregor, C., 2000, 'Using social indicators to measure community vulnerability to natural hazards', Australian Journal of Emergency Management, 15 (3). pp. 52-57.
26. Fenton, D.M., and Macgregor, C., 1999, 'Framework and Review of Capacity and Motivation for Change to Sustainable Management Practices', theme 6: Project 6.2.1., Social Sciences Centre, Bureau of Rural Sciences, Canberra.
27. Treasury, 2014, 'Treasury Annual Report 2013-14', Australian Government, 31 October, <http://www.treasury.gov.au/PublicationsAndMedia/Publications/2014/Treasury-Annual-Report-2013-14>.
28. Cox, A., and Townsend, M., 1998, 'Strategic Procurement in Construction, Thomas Telford, London.
29. Comptroller and Auditor General, 2001, 'Modernising Construction', National Audit Office, London.
30. Tilley, P. A. 2005, 'Relationship contracting: a strategy for improved project performance', SCRI, University of Salford, UK.
31. Crow, T., 1998, 'Trust and trustworthiness', Paper delivered at the International Conference on TQM in Construction, Singapore.
32. Crow, T., and Barda, P., 2001, 'Projects as wealth creators; drivers of project excellence', Property Council of Australia, Sydney.
33. Abrahamson, M., 1982, 'Risk management', Paper presented at the International Construction Law Conference (19-21 October), Sydney.

MATEC Web of Conferences **66**, 00065 (2016) DOI: 10.1051/matecconf/20166600065
IBCC 2016

Project Complexity Influence on Project management performance – The Malaysian perspective

Abdou Saed M [1], Kuan Yong [2], Mohammed Othman [3,a]

[1]*Postgraduate student, Faculty of Built Environment, University of Malaya, Kuala Lumpur, Malaysia*
[2]*Senior lecturer, Department of Architecture, University of Malaya, Kuala Lumpur, Malaysia*
[3]*Senior lecturer, Department of Quantity Survey, University of Malaya, Kuala Lumpur, Malaysia*

Abstract. The project complexity has a negative influence upon project management process particularly on the project performance. The lack of consensus on project complexity is a crucial problem as project complexity is closely correlated with the project management process as well as the project management performance. Therefore,the understanding of project complexity is of significant importance to address the problem of project management performance. Our extensive literature review enlightened the lack of consensus on project complexity till date and provided a broader view and critical analysis of its underlying concepts. The intention of this paper is to examine the project complexity factors on the project management performance in Malaysia and to provide a review of project complexity and to stimulate debate on this controversial topic. The data was analyzed for a total of 101 respondents involved in this study. Theresults indicated that there is a lack of complexity during understanding and managing project performance. It is concluded that the concept of project complexity is warrents further consideration.

Key words: Project Management; Project Complexity; Project performance; Uncertainty

1 Introduction

Complexity is considered an essential factor in the field of project management. Nonetheless, the characteristics and nature of project complexity is a controversial debate. While several factors are considered as drivers of project complexity, the studies investigating those are scarce. On one hand, the term "Complexity" is not clearly defined in the literature, instead authors have suggested to pay more attention towards defining and managing the complex projects. On the other hand, it is considered that the project performance is closely related to project complexity. Thus, the project complexity with a firm definition should be precisely measured in order to achieve effective management of projects. However, it is worth mentioning that the complexity can also negatively influence a project performance as well as project outcomes asthe properties emerging from complexity can create certain new prospects. Therefore, it is better to sustain and understand complexity in the project context during project management rather than reducing or avoiding it altogether. it. The purpose of this paper, therefore, is to define project complexity and provide an overview of the existing models and measurement of projects. The systematic defining of those objects, their attributes and their interrelations will enable the readers to define project complexity measure and link it more efficiently with the project performance, subsequently reducing the number and effect of negative outcomes arising from this complexity

2 Literature Review

While, construction projects are becoming bigger and more complex, it is important to understand theconcept and management of project complexity. The construction industry has confronted great difficulty in coping with the increasing complexity of major construction projects as well as project management performance [1, 2]. In spite of extensive research on the concept of complexity , there is no consensus on project complexity. [3] stated that, in fact, the entire intuitive notion of complexity can not be adequately captured by one single concept or definition.

[a]Corresponding author: othman.mohamed@um.edu.my

The term "complexity" can therefore be perceived through various connotations not only in different fields but alsowithin the same field. In order to assist better modern project management, project complexity needs to be understood and measured efficiently [4, 5]. Construction projects, particularly mega projects, involve a variety of interconnected factors.. which generate complexity with defined characteristics [6, 7]. In fact, it is generally agreed that because of the difficulties associated with decision-making and goal attainment which appear to stem from complexity, the understanding of complexity is vital for project management [8]. The success of a project therefore ultimately depends upon project performance and its related project complexity [9]. Although the importance of project complexity in project management has been widely acknowledged [4, 10], insufficient objective measures exist for the assessment of project complexity, mainly because the complexity is principally related to the subjectivity of the observer [11, 12].

3 What Is Project Complexity?

As the sources of complexity are pronounced to be various, dynamic, and uncertain, there is no consensus on complexity [13-15]. The difficulty in defining complexity can be minimised through the identification of its key drivers, although, defining those key drivers is not easier, either in the field of project management or in other domains. [16, 17] reported a lack of agreement on project complexity drivers and little understanding exists about the inter-relations between those drivers. The drivers of complexity depend upon the environment and the context of the project. [16] emphasized on determining the 'context' to be a prerequisite for modelling complexity, which includes time context, work context, motivational context and social context. Considering the project complexity to be composed of different kinds of complexities, such a distinction can lead to efficient project management.[1, 18] for instance, enlisted project complexity through the concepts of differentiation and interdependency of project organization. He determined two types of complexities; organizational complexity i.e. the number of hierarchical levels, the degree of interaction between the project organizational elementsetc and technological complexity i.e. the number of the inputs and outputs, interdependencies between tasks, etc. Moreover , [19] defined complexity incorporating the different aspects as *"the property of a model which makes it difficult to formulate its overall behavior in a given language, even when given reasonably complete information about its atomic components and their inter-relations"*. This definition prominently emphasizes that complexity is relative to the project modeling system. While measuring project complexity, one should be able to define a general measure of this feature independent of the project model, the project complexity measure, otherwise would be a project model complexity measure.

4 Types of Complexity

4.1 Organizational complexity

Organizational complexity, the main type of complexity , includes: allocation of tasks, distribution of the responsibility and authority for decision making and designation of the relationships in terms of reporting and communication [20]. The construction projects are classically characterized by the involvement of numerous distinct and diverse organizations, such as contractors and consultants, for a definite time span. This points to the creation of a multi-organizational structure for the management of construction project temporarily. [19]

4.2 Technological complexity

Many authors such as [4] highlighted the absence of unanimity on the technology conceptual definition has frequently been highlighted in the literature. The terminologies of 'technology' and 'task' are interchangeably used quiet frequently i because technology is generally denoted in terms of task performance. Generally, it is justifiable to define technology as a transformation process involving conversion of inputs in outputs [18]. There is employment of knowledge, skills, techniques as well as material means during this transformation process.Technology, therefore, can be classified into three facets: characteristics of knowledge, characteristics of materials and the equipment and sequencing of activities i.e. operations [1, 18]. Moreover, the technology concept can be applied to any level of the project organization, e.g. the whole organization or a single groupor individual i.e. the entire project production system or the tasks performed by one individual.Technology being multi-dimensional can be classified into two types: complexity and uncertainty [4]. Technological complexity can be operationalized in terms of the difficulties of task performance which would then translate into a diverse interpretation of this technological complexity. This includes wide-ranging attributes of building type, difficulty of location, overlap of construction and design, and interdependence of operations. Furthermore, it has been reported that the concept of uncertainty is inappropriately associated with the difficulties of task performance. Therefore, it seems logical that the technological complexity when defined in terms of differentiation and interdependencies could be more appropriate and practical, [11].

5 Indicators & Characteristics of Project Complexity

Our review of the literature suggested that the factors required for project complexity can be classified into four categories (Figure 1). All of these are indispensable but insufficient for the detrrmination of project complexity. The first family comprises of project size factors, the second contains factors of project variety, the third family addresses factors are related to the project system interdependencies and interrelations. While, the fourth deals with context-dependent of the project complexity. The following is a brief explanation of these factors [19].

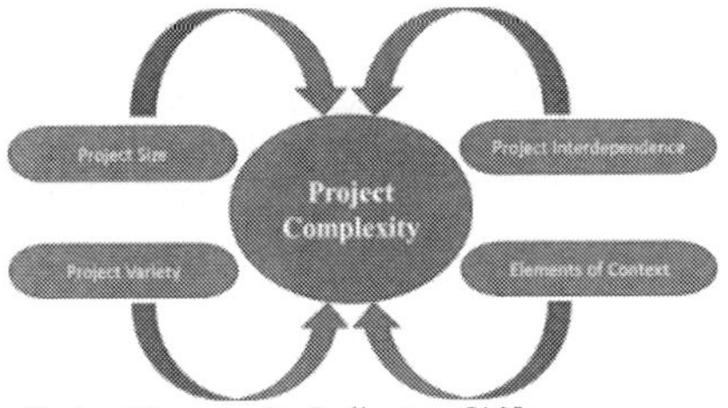

Figure 1. Project Complexity Indicators [19]

5.1 Size of project

One of the necessary conditions for project complexity is the size of the project. [21, 22] stated that the organizational system should atleast be over a minimum critical size to be labelled as a complex system. The size of the project system, which also defines the parameters and characteristics of complexity, is a one of the main indicators of project complexity.

5.2 Variety of project

The project variety is also a key driver of project complexity. [23] emphasized that the diversity in a project is closely associated with the emergent properties and hence the determination of project complexity.

5.3 Project System Interdependencies

Many autors highlighted that, interdependencies are likley to be one of the main contribution on project complexity. [24] reported that the interrelationship between various projectcomponents is more intricate than recommended by the usal work breakdown structure of project network. Moreover, the project management traditional tools are likely to be insufficient to certainty of interdependence to be captured. This might be questionable because each element depends and influences the others because of the complete interdependence amongst different components of complexity [13, 21].

5.4 Context-dependence

[14, 25] highlighted contextually as a vital feature of complexity, and a denominator of the complex system. Also reported by [24] who stated that the context and practices of one project are not directly apllicable to other projects with different influential and cultural configurations. Hence, project complexity can neither be analysed nor managed without considering the consequences of the project context.

6 Analysis of Key Complexity Models

In our study, we will analyze six models for complexity and draw inferences from them. These models were chosen based on their relevance to the project management context. Although the approaches used are different between project management and change management, the projects are usually attributed to cause changes in the organizations.

6.1 Turner & Cochrane Goals and Methods Matrix

[26] developed a'Goal and Methods Matrix' to assess the potential of project complexity. The projects were classified according to two considerations first how well the goals are defined and second how well the methodology of accomplishing these goals is well-defined. They categorised projects into four varieties and explained the role of the project manager in each variety. They stated that in the absence of certainity of methods, the vital foundations of project management can not be obtained.
This study therefore focuses on the goals and methods amongst all other characteristics of complexity and the influence of each method on achieving the goals.

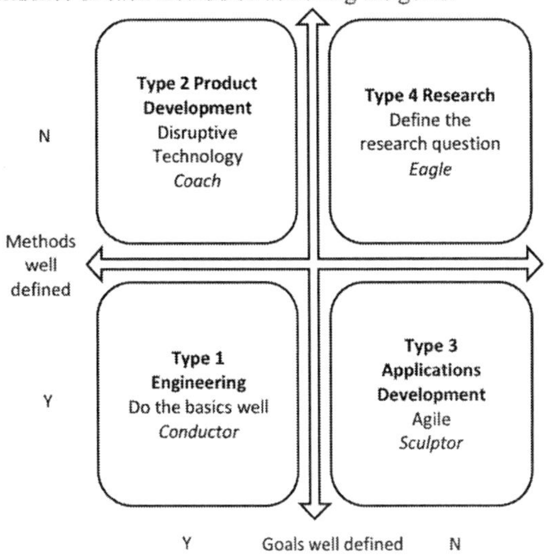

Figure 2. Goals and Methods Matrix [26]

6.2 Stacey's Agreement & Certainty Matrix

The project complexity analyzed by [27] was constructed in two aspects: the level of agreement and the degree of its certainty, Figuer 3.

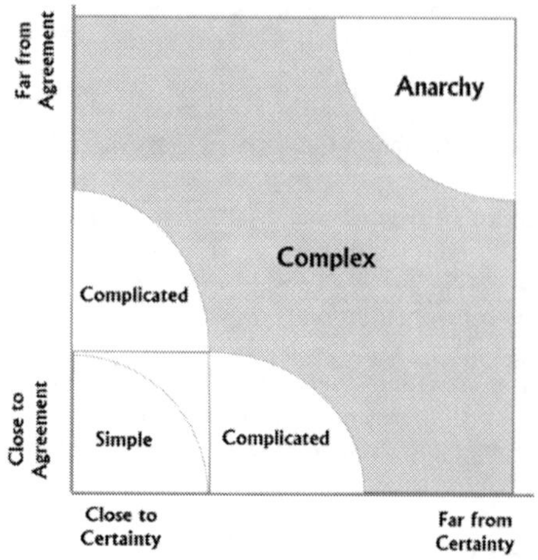

Figure 3. Stacey's Agreement & Certainty Matrix [27]

This model aids in the selection of management approaches for a particular subject, which however, is a single part of undertaking a multifaceted project. Stacey's matrix primarily focusses on change, and provides an assistance to handle the projects according to the position in the matrix but it is not extendable for the demonstration of interdependencies.

6.3 Complexity Model suggested by Terry Williams

Williams shares the vision of [18] on project complexity and modified it through an additional dimension.

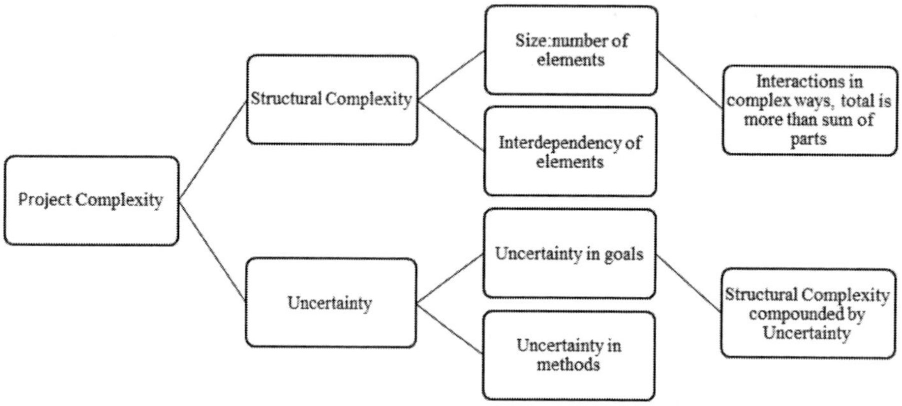

Figure 4. Dimensions of Project Complexity

Another third component which is uncertainty was added to the two existing components of complexity, viz the interdependency as well as the number of elements introduced this was considered as an integral aspect of project complexity because uncertainty augmentsthe complexity of a project [1] Figure 4.

[1] however, suggested that two dimensions in which the total is more than the sum of the parts can sufficiently measure the overall complexity. He pointed out that 'complexity' is steadily increasing in projects and this increase contributes to the failed or unsuccessful projects. The two compounding causative factors are the relationship between project complexity and product complexity and the length of the projects. The new products become structurally more complex with an enhanced degree of inter-element connectivity. Similarly, projects nowadays are more time constrained because of the increasing desire to decrease the market supply time. Figure 4

6.4 Adam Kahane's approach to complexity

The [28] approach towards complexity is deeply embedded within the social environment. He distinguishes three types of complexities:

6.4.1 Dynamic Complexity

Dynamic complexity denotes that the people involved in the problem perceive things differently thus leading to complicated situations.

6.4.2 Generative Complexity

Generative complexity deals with the future, where the future is unaware and uncertain.

6.4.3 Social Complexity

It denotes that the people should be elaborated in creating and implementing the solution in order to solve the problems efficiently. However, the individuals involved have dissimilar viewpoints and interests. Thus, the U-process was introduced as a methodology by [28] to address these challenges.An individual or team has to commence three activities in this U-process methodology:

1. To Sense the existing certainty of the system, which they are part of,
2. To Present and reflect their "inner knowledge" about what is going on and what they have to do, and
3. To realizing, and act swiftly in order to bring forth a new reality.

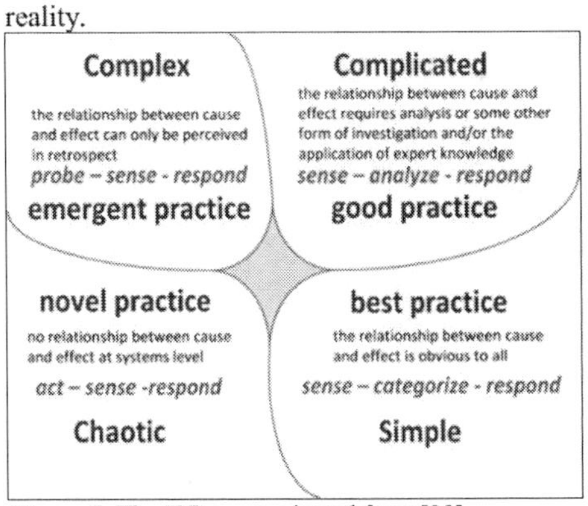

Figure 5. The U Process adapted from [28]

6.5 Cynefin Decision Making Framework

[29] constructed another framework that permits the managers to evaluate new things from new perspectives, integrate complex ideas, and targets the real complications entitled as the Cynefin framework. This framework categories the decision into situations according to the root and influence. The next figure shows the characteristics of each context and the ways to tackle them.

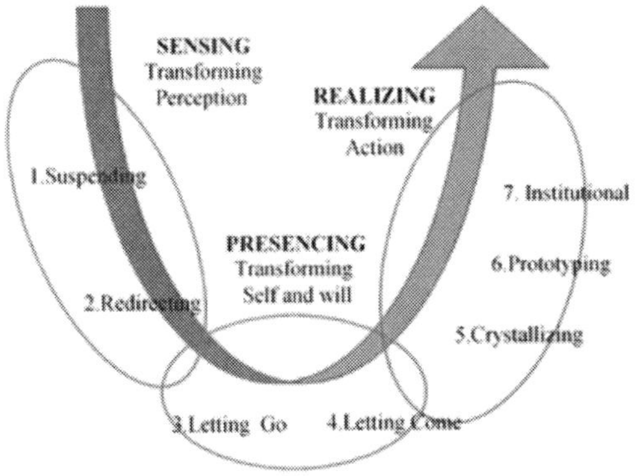

Figure 6. Decisions in Multiple Contexts [29]

This type of complexity creats from a high level of uncertainty of future limitations that may destabilize the project process is termed ad temporal complexity. For example, unexpected legislative changes orrapid changes in technology which make the project more complex.

The precise answers can be determined because of the clearer cause and effect relationships seen in the simple and non-complicated domain. On the contrary, people will be required to make a decision based on the imperfect data, with absence of an obvious cause and effect relationship observed in the complex and chaotic domains. Snowden and Boone have identified the complex system by its performance and not its characteristics of each domain within the framework. The typical examples are the unpredictable nature of complex systems along with the cause and effect relationship. Hence, it is descriptive rather than prescriptive in nature.

6.6 Remington & Pollack Model

[30], provided a decent start to categorize complex projects. They highlighted that a profound difference between various types of complexity would assist to select the appropriate tool to manage the project. They suggested four types of project complexity: structural, technical, directional and temporal.

6.6.1 Structural complexity

Structural complication creates from a huge scale projects which is divided into minor tasks and single contracts. The engineering, construction and IT projects are more expected to have this type of complexity. This complexity truncates from the difficulty in managing and holding track of an enormous numeral of interrelated events and tasks [30].

6.6.2 Technical complexity

Technical complexity is create in projects that have "unknown" and "untried" technical aspects [30]. It is created due to the uncertainty of the outcome. The industrial project and R&D projects encounterthis type of complexity.

6.6.3 Directional complexity

Directional complexity can be observed in projects with a high level of change, for example, it might be evident that something must be done to resolve the issue, but that 'something' appears to be uncertain [30].

6.6.4 Temporal complexity

DOI: 10.1051/matecconf/20166600065

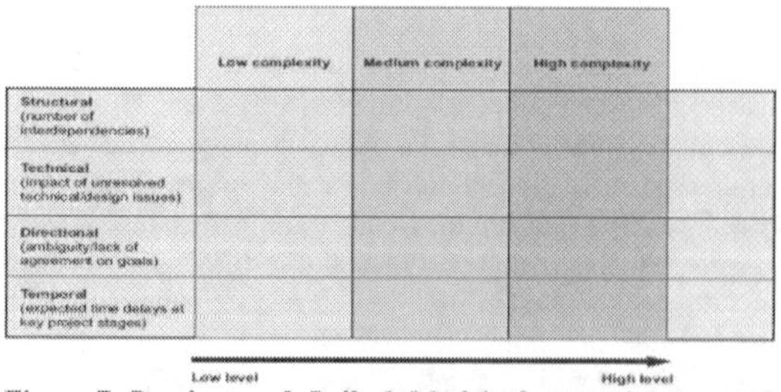

Figure 7. Remington & Pollack Model of categorizing complex projects

7 Evaluating Project Complexity

After reviewing the project complexity models, a total of 33 factors were selected and categorized under project complexity for evaluation. A questionnaire survey was used to determine the factors which extensively contribute to the project complexity and hence the project management performance according to Malaysian project management. The questions were set for the project complexity factors based on our literature review and were distributed randomly amongst the project managers, project management lecturers and project management professionals in Malaysia. Also to ensure that the respondents have appropriate experience they should have been elaborate in complex projects and will be able to provide a effective contribution to the research topic. The response from the project managers was positive, 124 questionnaire responses were received. Nevertheless, a number were inadequate and subsequent data validation the final sample size is 101 it would be 81.5% of the respondents a total of 101 respondents were recruited from the sample. The level of familiarity of the respondents was evaluated using 5-point Likert scale measurement. Table 1 (Refer to Appendix A for complete Questionnaire)

Table 1. Project Complexity Factors

No.	Project Complexity Factors
	1.Technical Category:
1	Number of project goals
2	Clarity of project goals
3	Uncertainties in scope
4	Strict quality requirements
5	Project duration
6	Number of locations
7	Advanced Technology
8	Number of tasks
9	Variety of tasks
10	Uncertainty in methods
11	Technical risks

7.1 Respondents profile

The respondents were segmented into three groups; the first group included the project engineers involved with the project process or planning and supervision process (71%). The second group included the project management professionals with 10 or more years of experience (18%), while the third group included the project management lectures teaching project management at the Malaysian Universities (11%). Years of experience, size of the project and level of complexity of the projects were the main factors to differentiate amongst the 3 respondent groups (figure 8).

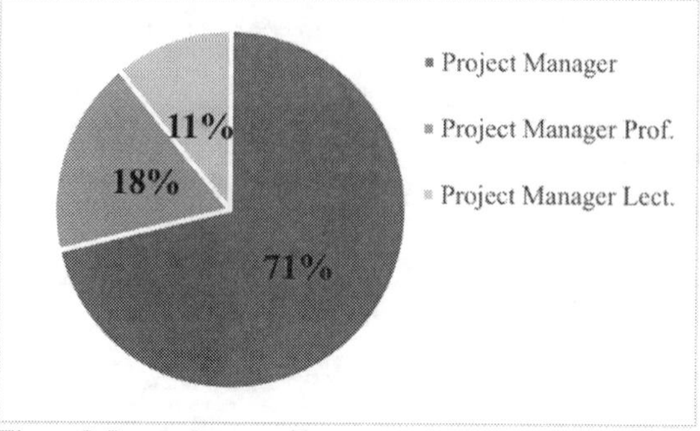

Figure 8. Respondents profile

7.2 Data Analysis

The data was analysed by SPSS version 20.0 using exploratory factor analysis (EFA), the approach employed for the analysis was sequential and linear, involving multiple options. Therefore, the following steps for exploratory factor analysis provide areference point to develop clear decision pathways.

7.3 Sample Size

There are different opinions and guiding rules for calcultaion of sample size. General guidelines include, the sample size of 100 or above or 300 or more for factor analysis. [31] suggested that solutions with correlation coefficients of >0.80 require smaller sample sizes, while others [32] mentioned that inclusion of 50 cases might be adequate for factor analysis.

7.4 Kaiser-Meyer-Olkin (KMO) & Bartlett's Test of Sphericity

The KMO index of 0.50 is considered suitable for factor analysis (range, 0 to 1) .. A positive factor analysis means a significant ($p<.05$) Bartlett's test of sphericity, Table2 [33]

Table 2. The KMO and Bartlett's Test

Kaiser-Meyer-Olkin Measure of Sampling Adequacy		0.808
Bartlett's Test of Sphericity	Approx. Chi-Square	1337.140
	df	528
	Sig.	0.000

7.5 Extraction Method

[34] noted that Principal Component Analysis (PCA) being the default method in many statistical programs, is commonly used in Exploratory Factor Analysis (EFA). However, where no priori theory or model exists, PCA is recommended to establish the preliminary solutions in EFA.

7.5.1 Cumulative Percentage of Variance

The table below demonstrates a cumulative percentage of 53.2% variance and 6 components (factors) having an eigenvalue > 1. These results are adequate, as the explained variance is commonly as low as 50-60%. [35] as shown in Table 3. (Refer to Appendix B for the complete table)

Table 3. Total Variance Explained

Component	Initial Eigenvalues			Extraction Sums of Squared Loadings			Rotation Sums of Squared Loadings[a]
	Total	% of Variance	Cumulative %	Total	% of Variance	Cumulative %	Total
1	9.316	28.232	28.232	9.316	28.232	28.232	5.828
2	2.181	6.610	34.842	2.181	6.610	34.842	5.149
3	1.799	5.452	40.294	1.799	5.452	40.294	5.122
4	1.514	4.589	44.883	1.514	4.589	44.883	4.830
5	1.397	4.234	49.117	1.397	4.234	49.117	3.930

7.5.2 Scree Test

The Scree plot and eigenvalues produced a departure from linearity coinciding with a 5-factor result. Hence, this Scree Test indicateed that the data should be analysed for 5 factors among the total 33 factors included initially.

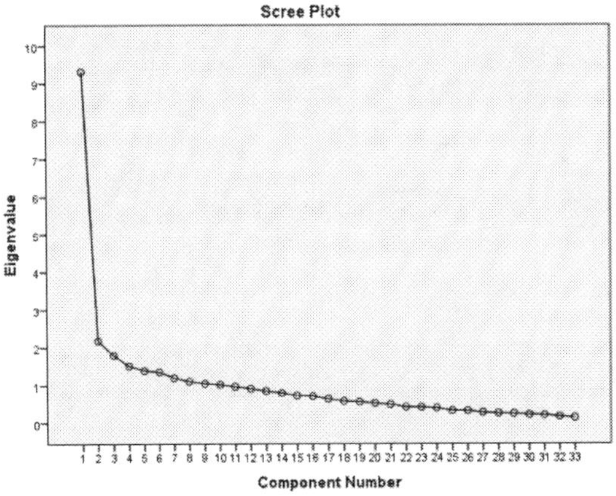

Figure 9. The Scree Plot shows 6 factors > 1.0

7.5.3 Selection of Rotational Method

Rotation Method: Promax with Kaiser Normalization, rotation was converged in 10 iterations as shown in the component matrix. (Refer to Appendix B)

7.5.4 Interpretation

Interpretation involves examining the variables attributed to a factor, and giving that factor a name or theme. To indicate those factors combined together, to explain the majority of the responses the *Pattern Matrix*, was examined five factors were produced (Table 4).

Table 4. The Pattern Matrix

	Component				
	1	2	3	4	5
Weather conditions	.824				
Number of locations	.731				
Interference between existing sites	.609				
Number of different languages	.563				
Clarity of project goals	.560				

8 Results and Findings

Initially, the factorability of the correlation between 33 items was examined using several well-recognized criteria for the factorability assessment. Firstly, it was observed that many items correlated at least 0.3 with one other item, suggesting a reasonable factorability (see Appendix B). Secondly, the Kaiser-Meyer-Olkin measure of sampling adequacy was 0.81, which is above the recommended value of 0.6, and Bartlett's test of sphericity was also significant ($\chi 2$ (528) = 1337.14, $p < .05$). Finally, the communalities were all above 0.3 (see Appendix B), further confirming that each item shared some common variance with the other items. Given these overall indicators, factor analysis was deemed suitable with all the 33 items.

Principal Components Analysis (PCA) was employed to identify and figure intricate scores for the factors addressing project complexity. Total variance explained by using Promax with Kaiser normalization rotations of the factor-loading matrix was 53.2% according to the initial eigenvalues. Fourteen items were eliminated because they did not contribute to a simple factor structure and failed to meet the minimum criteria of primary factor loading of

≥0.5, and no cross-loading of ≥0.4 as shown in Table 4 (Refer to Appendix B for the complete table).

9 Discussion

The data were subjected to factor analysis using principal components extraction method and rotation method Promax with Kaiser Normalization. All KMO values for the individual item (> 0.90) were over 0.5 and the Kaiser-Meyer-Olkin Measure (KMO) was 0.81, indicating that the data was sufficient for Exploratory Factor Analysis (EFA). Using an eigenvalue cut-off of 1.0, five factors explained an increasing variance of 53.2%. And scree plot confirmed the results of those five factors. The below table displays the factor loading after rotation using a significant factor measure of 0.50 (Table 5).

Table 5. Mapping the results

	Factors	Environmental	Operational	Organizational	Technical	Team
1	Weather conditions	.824				
2	Number of locations	.731				
3	Interference between existing sites	.609				
4	Number of different languages	.563				
5	Clarity of project goals	.560				
6	Project duration		.697			
7	Availability of financial sources		.692			
8	Strict quality requirements		.665			
9	Variety of tasks		.567			
10	Uncertainty in methods			.722		
11	Size of project team			.638		
12	Involvement of different time-zones			.564		
13	Political influence			.543		
14	Lack of experience in the country				.720	
15	Level of competition				.699	
16	Technical risks				.670	
17	Lack of experience with partners					.777
18	Lack of resources & skills					.694
19	Number of project goals					.552
	Eigenvalues	**9.3**	**2.2**	**1.8**	**1.5**	**1.4**
	% of Variance	**28.2%**	**6.6%**	**5.5%**	**4.6%**	**4.2%**

The new 5 factors explained in the above table showed that: (i) *Weather conditions, number of locations, interference between existing sites, number of different languages* and *clarity of project goals* were the complex variables under "***Environmental*** Complexity Factor". (ii) *Project duration, availability of financial sources, strict quality requirements* and *variety of tasks* were the complex variables under "***Operational*** Complexity factor". (iii) *Uncertainty in methods, size of the project team, involvement of different time-zones* and *political influence* were the complex variables under "***Organizational*** Complexity factor".(iv)*Lack of experience in the country, level of competition* and *Ttechnical risks* were the complex variables under "***Organizational*** Complexity factor". (v) Finaly, *lack of experience with partners,lack of resources & skills* and *number of project goals were* the complex variables under ***Team*** Complexity factor.

10 Conclusion

From the previous literature on project complexity factors within the Malaysian project management, the following highlights should be taken into consideration. The deficiency of project complexity framework due to lack of knowledge about their effect on the project process and hence project outcomes. Efficient complexity measures should be elaborated and clearly explained at the local level whichmay lead to a better vision of project complexity influence on project outcomes and consequenlty the better project performance in Malaysia. The successful management of projects depends on the appropriate actions taken according to certain critical characteristics of the projects. The authors propose that the project complexity should be understood in terms of interdependencies and diversity. The consequence to project complexity is integration by direction, control and communication. The concept of project complexity, therefore deserves further consideration and extensive research in future. The purpose of the paper was to provide a review of project complexity and to stimulate a conclusion in this lagging subject. As projects have turn out to be increasingly complex there will be a growing concern about the concept of project complexity and its influence on the project management process. The complexity has to be managed effectively in order to achieve greater levels of project performance. However, managing complexity does not mean avoiding or reducing it entirely, a good project manager should be able to manage complexity every time by maintaing the upper and lower levels of project complexity. For the project managers to reach this complexity level, efficient complexity models and measures need to be elaborated and implicated. Moreover, by clearing the concept at the local level and giving a better vision of the propagation of change by navigating through the model at local level, the project complexity influence on project evolution could be understood for a better project performance in Malaysia. One should be able to achieve the objectives proficiently to assess the impact of change or any decision and simulate different scenarios of evolution.

References

1. Williams, T., *Modelling complex projects*. 2002.
2. Bosch-Rekveldt, M., *Managing project complexity*. Delft Centre for Project Management, the Netherlands., 2011. **ISBN 978-94-91005-00-8**.
3. Sinha, S., B. Kumar, and A. Thomson, *Measuring project complexity: a project manager's tool.* Architectural Engineering and Design Management, 2006. **2**(3): p. 187-202.
4. Vidal, L.-A. and F. Marle, *Understanding project complexity: implications on project management.* Kybernetes, 2008. **37**(8): p. 1094-1110.
5. Bryde, D., *Perceptions of the impact of project sponsorship practices on project success.* International Journal of Project Management, 2008. **26**(8): p. 800-809.
6. Lucas, C., *The philosophy of complexity.* On-line at www. calresco. org, retrieved August, 2005. **8**: p. 2011.
7. Chan, A.P., D. Scott, and A.P. Chan, *Factors affecting the success of a construction project.* Journal of construction engineering and management, 2004.
8. Remington, K., R. Zolin, and R. Turner. *A model of project complexity: distinguishing dimensions of complexity from severity*. in *Proceedings of the 9th International Research Network of Project Management Conference*. 2009. IRNOP.
9. Chan, A.P., D. Scott, and A.P. Chan, *Factors affecting the success of a construction project.* Journal of construction engineering and management, 2004. **130**(1): p. 153-155.
10. Parsons-Hann, H.a.L., K. , *Measuring requirement complexity to increase the probability of project success.* Proceedings of ICEIS 2005, Miami, Florida, USA, 2005. **pp. 434-8.**
11. Corning, P.A., *Complexity is just a word!* Technological Forecasting and Social Change, 1998. **59**(2): p. 197-200.
12. Pich, M.T., Loch, C.H. and de Meyer, A., *On uncertainty, ambiguity, and complexity in project management.* Management Science, 2002. **Vol. 48 No. 8**(pp. 1008-23).
13. Calinescu, A., et al., *Applying and assessing two methods for measuring complexity in manufacturing.* Journal of the Operational Research Society, 1998: p. 723-733.
14. Chu, D., R. Strand, and R. Fjelland, *Theories of complexity.* Complexity, 2003. **8**(3): p. 19-30.
15. Jaafari, A., *Project management in the age of complexity and change.* Project Management Journal, 2003. **Vol. 34 No. 4**: p. pp. 47-57.
16. Sinha, S., B. Kumar, and A. Thomson, *Complexity measurement of a project activity.* International Journal of Industrial and Systems Engineering, 2011. **8**(4): p. 432-448.
17. Koivu, T., Nummelin, J., Tukiainen, S., Tainio, R. and Atkin, B. , *Institutional complexity affecting the outcomes of global projects.* VTT Working Papers 14, VTT, Espoo, 2004.
18. Baccarini, D., *The concept of project complexity—a review.* International Journal of Project Management, 1996. **14**(4): p. 201-204.
19. Edmonds, B., *Syntactic measures of complexity.* 1999, University of Manchester.
20. Laurikkala, H., Puustiner, E., Pajarre, E. and Tanskanen, K. , *Reducing complexity of modelling in large delivery projects.* Proceedings of the International Conference on Engineering Design, ICED'01, Glasgow, 2001. **Vol. 1, Professional Engineering Publishing, Bury St Edmunds, pp. 165-72.**
21. Corbett, L., J. Brockelsby, and C. Campbell-Hunt, *Tackling industrial complexity.* Tackling Industrial Complexity, 2002: p. 83-96.
22. Cicmil, S. and D. Marshall, *Insights into collaboration at the project level: complexity, social interaction and procurement mechanisms.* Building Research & Information, 2005. **33**(6): p. 523-535.
23. Jones, B.S. and P. Anderson, *Diversity as a determinant of system complexity.* Complexity in Design and Engineering, 2005.
24. Koivu, T., et al., *Institutional complexity affecting the outcomes of global projects.* 2004.

25. Genelot, D., *Manager dans la complexité: réflexions à l'usage des dirigeants*. 2001: INSEP éditions.
26. Turner, J.R. and R.A. Cochrane, *Goals-and-methods matrix: coping with projects with ill defined goals and/or methods of achieving them.* International Journal of Project Management, 1993. **11**(2): p. 93-102.
27. Stacey, R.D., *Complexity and creativity in organizations*. 1996: Berrett-Koehler Publishers.
28. Kahane, A., *Solving tough problems: An open way of talking, listening, and creating new realities*. 2004: Berrett-Koehler Publishers.
29. Snowden, D.J. and M.E. Boone, *A leader's framework for decision making.* harvard business review, 2007. **85**(11): p. 68.
30. Remington, K. and J. Pollack, *Tools for complex projects*. 2007: Gower Publishing, Ltd.
31. Guadagnoli E, V.W., *Relation of sample size to the stability of component patterns.* Psychological Bulletin, 1988. **103(2):265-75.**
32. Sapnas KG, Z.R., *Minimizing sample size when using exploratory factor analysis for measurement.* Journal of Nursing Measurement, 2002. **10(2):135-53.**
33. Tabachnick BG, F.L., *Using Multivariate Statistics.* Boston: Pearson Education Inc, 2007.
34. B., T., *Exploratory and confirmatory factor analysis: understanding concepts and applications.* American Psychological Association, 2004.
35. Pett MA, L.N., Sullivan JJ. , *Making Sense of Factor Analysis: The use of factor analysis for instrument development in health care research.* Sage Publications Inc, 2003.

Key Determinants of Waste Separation Intention among Students on Campus

Siti Fadzilah Ayob[1, a] and Low Sheau-Ting[2]

[1]*Department of Real Estate, Faculty of Geoinformation and Real Estate, Universiti Teknologi Malaysia, 81310 Skudai, Johor, Malaysia*
[2]*Centre of Real Estate Studies, Institute for Smart Infrastructure and Innovative Construction, Universiti Teknologi Malaysia, 81310 Skudai, Johor, Malaysia*

Abstract. This study aims to identify the key determinants of waste separation intention among the students in Universiti Teknologi Malaysia (UTM). As reported in World Bank Report, generation of global municipal solid waste are approximately 1.3 billion tonnes per year, and it will increase approximately 2.2 billion tonnes per year by 2025. Similar trend in Malaysia which approximately eight million tonnes of municipal solid waste is produced daily. Upward trend of the solid waste generated is contributed by various forces including growth, increase of population, residents attitude and rapid urbanisation in the country. Considering the negative impacts from increase of waste generated in the country such as environmental degradation, separate the waste at its source can increase recycling rate and hence reduce waste generation. In the 11th Malaysia Plan, the government has urged the nation to actively practice waste separation. Understanding the determinants influencing waste separation behaviour can lead to effective waste reduction among the targeted group. This study adopted the Theory of Planned Behaviour as the fundamental framework. The key determinants of waste separation behaviour are attitude, subjectives norm, and perceived behavioural control. An elicitation survey has been conducted to explore the students's beliefs relative to the waste separation behaviour. The empirical data collected were analysed using structural equation modelling (SEM) assisted by SmartPLS software. The result shows that only determinants are significant towards intention which are attitude and Perceived Behavioural Control while Subjective Norm is not. The results of the study is significance to the organisation as identification of the determinants that influence waste separation intention contributes to a more focused waste separation programme that allign to Malaysia's objective towards a sustainable developing country.

Keywords: **structural equation modelling, theory of planned behaviour, waste separation**

1 Introduction

Developing countries including Malaysia faced with low quality of environment especially in urban areas which respect to the solid waste management sector [23, 36]. As stated by [18], one of the factors contributes to the low environment quality is inadequate and unscientific municipal solid waste management practice. Application of 3Rs should be done in developing countries in order to improve the solid waste management [8, 14]. Malaysia also facing the solid waste problem because of the economic growth, increase of population, residents attitude [7, 36] and rapid urbanisation [27]. Malaysia government had launched many types of recycle programs, campaign and provide solid waste facilities in order to improve solid waste management sector but the success rate is still low based on the increasing amount of waste generated from year to year [35]. This is because people are not practicing the waste recycling behaviour although they have the knowledge and awareness in improper waste management to the environment [27, 28]. Therefore, in 11th Malaysia Plan, the government focuses on changing the nation's behaviour in minimising waste through waste separation instead of providing extra landfill and dumping area [30]. Waste separation is one of the ways to reduce amount of waste being dumped at the landfill and increase the recycle rate [7, 31]. In this RMK11, government focus in fostering

[a] Siti Fadzilah Ayob: fadz91@gmail.com

waste separation behaviour among the society through any comprehensive activities and willing to invest more as the waste issue is at critical stage. Therefore, government came out with A2 strategy in RMK11 which emphasis changing Malaysian behaviour especially in waste separation which had been launched on 1st September 2015. In recent, a program called "Separation of Solid Waste at Source" was launched on 1st September 2015 to implement the mandatory separation of waste at source in few states of Malaysia. Besides of government efforts, there are also some research had been conducted in order to help in reducing the number of waste generated in the country. For example study conducted by [11] which emphasis on benefit–cost analysis on the economic feasibility of construction waste minimisation, [14] which focuses on solid waste management and other research which focus on food waste separation [17, 19, 24, 29]. Based on the literatures, it is noticed that there is less research had been conducted whether in Malaysia or other country on solid waste separation compare to industrial waste separation, food waste separation and household waste separation. Therefore, present study that exploring the solid waste separation intention is conducted to contribute to the existing solid waste separation literatures.

Based on previous research, most of the waste management is handled by behavioural conduct. This is indicating that improving human behaviour contributing to the success of waste management program. Therefore, improving human behaviour towards waste minimisation through waste separation is essential. In order to change the human behaviour, it is important to identify the key determinants of behaviour because human behaviour can be improved or change if the right determinants that formulate specific behaviour being identified [25,40, 41]. In waste separation context, identify the key determinants of waste separation behavior is critical for sustainable behavior change. By having the right key determinants of behaviour, government can come out with more comprehensive strategy and programs that can lead to behavioural change on waste separation which will help the government to achieve objective in minimising the waste. Therefore, it is important to get the right determinants in formulating the nation's waste separation behaviour effectively. Hence this study aims to identify the key determinants of waste separation intention so that it can be used to foster waste separation behaviour among the nation and as well as reducing the number of waste generated which align with Malaysia's objective in achieving a sustainable developing country.

2 Theory Of Planned Behaviour

Waste should be separated before being recycled [10, 32]. Therefore, waste separation comes at early stage before recycling such as sorting out the waste into its specific material which is paper, glass, and plastics. However, waste separation is likely to be difficult and several factors may be taken into consideration. In identifying the factors which influencing waste separation decision, the Theory of Planned Behaviour (TPB) can be used as it provides a systematic theoretical framework and there are many previous studies prove that TPB is very useful in investigating the determinants of specific intention and behaviour [33, 39]. According to Downs and Hausenblas (2005), the TPB explains that human's expectations and values about involving in behaviour form their behavioural, normative, and control beliefs. These will lead to intention and behaviour through their attitude, subjective norm, and perceived behavioural control. Based on the TPB, the three determinants in performing behaviour are attitude towards behaviour, subjective norm (SN) and perceived behavioural control (PBC) [4]. According to [1], the first determinant is the attitude, which will form the intention of an individual whether to participate or not in specific behaviour. The second measure is the subjective norm, which refers to social influence in influencing an individual whether to perform or refuse to perform the specific behaviour and the last antecedent is the perceived behavioural control, which is explaining one's self-confidence whether the individual is able to perform the behaviour or no. For the attitude towards waste separation, of which a positive evaluation such as "waste separation will help in increasing the recycle rate" or "waste separation contributes to cleaner environment" is likely to increase the intention to practice waste separation activities and thus, this will lead to performing on waste separation behaviour. In contrast, a negative attitude such as "I feel that waste separation is wasting my time" will ultimately lead to a lower likelihood of waste separation intention and behaviour.

The second determinant of waste separation intention is the subjective norm. Based on the TPB, individual will take action to separate waste if their important referents think they should participate in waste separation and vice versa. For example "I think my friends in the university always want me to engage in waste separation activities" and "I think my lecturer will be happy to know I am practicing waste separation behaviour". Such approval perception from their important referent will contribute to a higher waste separation intention and behaviour. However, if their important referent makes the individual think that they disapprove their action, students will be less likely to practice the specific action to separate waste, such as "My friends in the hostel will tease me if I put the waste into the bins according to its colour to separate waste".

The third key determinant of waste separation intention is perceived behavioural control. This means that if the individual has high self-confidence in their ability to separate waste, it will increase their intention to do waste separation. This also means if the individual has low self-confidence on their ability to practice waste separation, they will not have the intention to perform waste separation. For instance, a student who is confident that he could separate waste before throwing the waste possesses higher waste separation intention that the one who does not believe that he will separate the waste before throwing the waste as he is a busy person. The student with higher confidence level is more likely to have intention to perform waste separation than the latter. According to TPB, an intention to engage a behaviour will happen when people make a positive evaluation of the behaviour, has pressure from the social and also they believe that they have the opportunity to do so [2].

Various studies conducted to test the applicability of TPB across various type of behaviour [25]. The most relevant study are [26] which study on the determinants of recycling intention behaviour among the Malaysian school students: an application of theory of planned behaviour [39] which conduct a study on recycling attitude and behaviour in university campus in Hong Kong. Other than that, there are also few studies that adopted TPB in recycling context [19, 22, 13].

3 Research Methodology

This section will discuss on the application of research methods and the data sample requirements that suit the objective of the study. This study employed Structural Equation Modelling (SEM) assisted by SmartPLS to examine the causal relationship between the determinants towards waste separation intention. TPB is adopted as the theoretical structure. Based on the literature review, it is discovered that the TPB model comes together with hypotheses. In this study, there are three hypotheses are proposed. The hypotheses are: (1) Attitude has a positive influence on waste separation intention; (2) SN has a positive influence on waste separation intention, and (3) PBC has a positive influence on waste separation intention. Two step modelling was performed in this study. First step is establishing the measurement model, where the measurement model is revised and confirmed. Second step is test the structural model, where the direct relations among latent variables were modelled. The proposed model was evaluated and demonstrated a poor model fit which lead to subsequent modification of the model.

3.1 Research instrument

This research is a quantitative study using a survey design. The questionnaire was designed based on previous research that applied TPB in pro- environmental behavior [33, 37, 42] and the results from elicitation study. An elicitation study was conducted in UTM to gather students's important exercise and the responses with high frequency were used in constructing measures in final questionnaire [3]. In structural equation modeling, a distinction is made between single item variable and variables computed from multiple items. Single item variables are referred to as observed variables, while multi item variables are called latent variables. In this study there were four independent variables consisting of four latent variables: attitudes, subjective norms, perceived behaviour control and behaviour intention. The questionnaire composed of two section, which the first section is respondent demographic profile and the second section is to assess students' attitude, subjective norm, and perceived behavioural towards the solid waste separation intention based on Theory of Planned Behaviour.

The questionnaire consisted of 19 questions distributed into four dimensions: attitude (6 questions), subjective norm (3 questions, perceived behavioural control (6 questions) and intention towards solid waste separation (4 questions). The questionnaire is a closed-ended questionnaire with the responses guided on a range of given scales. The reliability of the instrument is based on Cronbach Alpha value which is 0.851.

3.2 Research procedures

The population in this study is all students from Universiti Teknologi Malaysia (UTM) because it is one of the largest public universities in Malaysia and it is easy access in for this study. The student population in UTM is about 24,000 students as reported in UTM official web portal, and by referring to the table for determining sample size developed by Krejcie and Morgan (1970), the minimum sample needed is 378. However, in order to get more generalized data for this research, 486 of responses were gathered. The overall response was approximately 97.2%. The responses consist of equal number of students regarding of their gender, nationality and academic qualification. The samples chosen based on convenience sampling method that is selected based on ease of access. The questionnaire is distributed face-to-face to the students in the university after revised based on the comments from questionnaire pretesting.

4 Findings And Discussion

The initial structural model consists of 3 main paths which are ATT-INT (path coefficient (E) =0.613, Composite Reliability (CR)= 0.844, Average Variance Extracted (AVE) = 0.483), SN-INT (E = 0.008, CR = 0.693, AVE = 0.460) and PBC-INT (E=0.270, CR = 0.706, AVE = 0.378). The initial model is not fit because some of AVE and CR are not significant as the significant value for AVE should greater than 0.5 and CR should greater than 0.7. The initial structural model was modified to improve the model fit. The revised structural model produced a better fit ATT-INT (E =0.688, CR= 0.856, AVE = 0.545), SN-INT (E = -0.030, CR = 0.747, AVE = 0.618) and PBC-INT (E=0.179, CR = 0.825, AVE = 0.543). Based on the bootstrapping calculation, the T-value shows only two determinants have positive relationship towards intention. Hypothesis 1 (ATT

-> INT): Attitude has a positive influence on waste separation intention has T-value of 20.260 ($p < 0.001$), and PBC -> INT: Perceived behavioural control has a positive influence on waste separation intention has T-value of 4.976 ($p<0.01$). However, for the second hypothesis, (SN -> INT): Subjective norm do has a positive influence on waste separation intention has T-value of – 1.056 ($p = 0.267$).

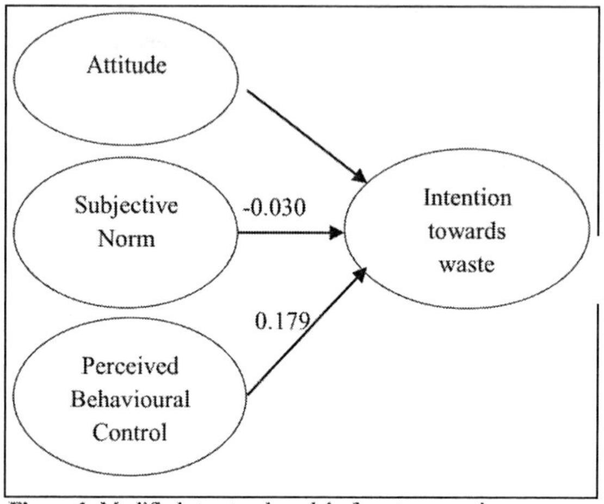

Figure 1. Modified structural model of waste separation intention in UTM

The results of present study showed that attitude is the strongest predictor towards waste separation intention among the students in UTM. This finding is consistent with previous study such as study conducted by [38], which stated that the more easy to do recycle, the higher the influence to engage positive attitude. Other than that, [6] also discovered that attitude is the strongest predictor of intention in desired behaviour. It is concluded that the respondents in this study had attitudes that significantly influenced their intention as students to separate waste. This finding also suggests that the UTM students as respondents in this study, had attitude which makes them feel good whenever they do the waste separation.

Hypothesis 2 predicted that as SN is favourable to waste separation intention. The results did not supported hypothesis 2. This finding is consistent with most of TPB studies. Typically the SN variable was the weakest influence on behaviour intention. In [37] and [12] the SN variable was the weakest predictor of the TPB variables. In [10] and [21], it was not significant at all. The findings in this study showed that SN is not significant in waste separation intention among UTM students. This is possibly true when waste separation is a public behaviour such as in the learning institution [9]. Other than that, this also shows that the lecturers, friends and classmate did not significant in influencing the students intention to do waste separation. This finding shows that one's intention to separate the waste is highly dependent on one's self and this is the main factor that makes one separate the waste not because of get support and encouragement from other people.

Hypothesis 3 predicted that as PBC is significant towards waste separation intention. The findings show that the PBC-INT is the second determinant that is significant towards intention. This is showing that the students in UTM are confident they can separate the waste is they wanted to. Additionally, the students do not agree on the waste separation practice is takes up too much time of them and waste of time. This result is consistent with some previous studies. For instance [39] which conduct a study on recycling attitude and behaviour in university campus in Hong Kong has found that attitude, subjective norm and perceived behavioural control influencing recycle attitude.

Multigroup analysis also had been done in order to identify whether by attending any waste separation related program will influence the respondent's waste separation intention. The result suggests that PBC-INT is the only significant different path relationship between the group that has attended waste separation related program and the group never attended any waste separation related program. This is based on the path coefficient difference shown which the value of PBC-INT is the highest among other determinants, 0.246. Other than that, if the p-value is smaller than 0.05 or larger than 0.95, it shows that the determinant is significant [20]. Therefore, based on the result, the significance p-value is only shown by PBC-INT and no other determinants. The p-value for PBC-INT is 0.028 while for the ATT-INT and SN-INT are 0.720 and 0.417 respectively. This shows that by having knowledge on desired behaviour can increase the PBC towards intention. This is consistent with [37] where the knowledge of knowing how to recycle is significant towards PBC and correlated with intention.

Overall, the finding of present study is similar to some previous study such as [21, 34]. Those previous study are using the theory of planned behaviour in identifying the relationship of attitudes, subjective norms and perceived control behaviour. These studies show that the attitudes and perceived behavioral control had significant influence on intention, while subjective norms did not have significant influence on their intention.

5 Conclusion

The primary purpose of this research was to determine the key determinants of waste separation intention among UTM students using TPB. The strongest predictor of intention to recycle is Attitude and PBC while SN is not influencing the intention to separate waste among the students. As a conclusion, UTM waste management system should focus on elements that can effectively inculcate a pro-environmental behaviour among students such as highlighting the benefits of waste separation intention and by providing enough facilities in order to foster waste separation intention and behaviour. However, this study only focuses on students from Universiti Teknologi Malaysia (UTM) and not other university community such as supporting staffs and academics staff. Therefore, the results will only target and applied on students' community only and cannot be generalised to other context. Other than that, this research uses self-reported data as it is a social science research. This may give biased answer during data collection. Lastly, it is recommended to include some other additional determinants in future study suits with the local context.

Acknowledgements

The authors gratefully acknowledge the research grant and financial support provided by the Ministry of Higher Education (MOHE) and University Teknologi Malaysia (UTM), under the research grant number Q.J130000.2527.11H61.

References

[1]Ajzen, I. (1991). The Theory Of Planned Behavior. *Organizational Behavior And Human Decision Processes*, ***50***(2), 179-211.

[2]Ajzen, I. (2005). *Attitudes, Personality, And Behavior*. Mcgraw-Hill Education (UK).

[3]Ajzen, I., & Fishbein, M. (1980). Understanding Attitudes And Predicting Social Behaviour.

[4]Ajzen, I., & Madden, T. J. (1986). Prediction Of Goal-Directed Behavior: Attitudes, Intentions, And Perceived Behavioral Control. *Journal Of Experimental Social Psychology*, ***22***(5), 453-474.

[5]Ajzen, I., 1991. The Theory Of Planned Behaviour. Organizational Behaviour And Human Decision Processes 50, 170–211. Bartlett, J.E., Kotrlik, J.W., Higgins, C.C., 2001. Information Technology. *Learning And Performance Journal* 1, 943–950.

[6]Arvola, A., Vassallo, M., Dean, M., Lampila, P., Saba, A., Lähteenmäki, L., & Shepherd, R. (2008). Predicting Intentions To Purchase Organic Food: The Role Of Affective And Moral Attitudes In The Theory Of Planned Behaviour.*Appetite*, ***50***(2), 443-454.

[7]Badgie, D. (2011). *Solid Waste Management System In The Kanifing Municipal Council (KMC), The Gambia* (Doctoral Dissertation, Msc. Thesis, Universiti Putra Malaysia (UPM)).

[8]Badgie, D., Samah, M. A., Manaf, L. A., & Muda, A. B. (2012). Assessment Of Municipal Solid Waste Composition In Malaysia: Management, Practice And Challenges. *Polish Journal Of Environmental Studies*, ***21***(3), 539-547.

[9]Barr, S. Ford, N.J, Gilg, A. (2003). Attitude Towards Recycling Household Waste In Exeter,Devon: Quantitative And Qualitative Approaches. Local Environment. 8, 407- 428 .

[10]Beedle Boldero, J., 1995. The Prediction Of Household Recycling Of Newspaper: The Role Of Attitudes, Intentions And Situational Factors. *Journal Of Applied Social Psychology* **25**, 440e462.

[11]Begum, R. A., Siwar, C., Pereira, J. J., & Jaafar, A. H. (2006). A Benefit–Cost Analysis On The Economic Feasibility Of Construction Waste Minimisation: The Case Of Malaysia. *Resources, Conservation And Recycling*, ***48***(1), 86-98.

[12]Davis, G., O'Callaghan, F., & Knox, K. (2009). Sustainable Attitudes And Behaviours Amongst A Sample Of Non-Academic Staff: A Case Study From An Information Services Department, Griffith University, Brisbane. *International Journal Of Sustainability In Higher Education*, ***10***(2), 136-151.

[13]De Leeuw, A., Valois, P., Ajzen, I., & Schmidt, P. (2015). Using The Theory Of Planned Behavior To Identify Key Beliefs Underlying Pro-Environmental Behavior In High- School Students: Implications For Educational Interventions.*Journal Of Environmental Psychology*, *42*, 128-138.

[14]Desa, A., Kadir, N. B. Y. A., & Yusooff, F. (2011). A Study On The Knowledge, Attitudes, Awareness Status And Behaviour Concerning Solid Waste Management. *Procedia-Social And Behavioral Sciences*, *18*, 643-648.

[15]Downs, D. S., & Hausenblas, H. A. (2003). Exercising For Two: Examining Pregnant Women's Second Trimester Exercise Intention And Behavior Using The Framework Of The Theory Of Planned Behavior. *Women's Health Issues*, ***13***(6), 222-228.

[16]Downs, D. S., & Hausenblas, H. A. (2005). Elicitation Studies And The Theory Of Planned Behavior: A Systematic Review Of Exercise Beliefs. *Psychology Of Sport And Exercise*, ***6***(1), 1-31.

[17]Eisted, R., & Christensen, T. H. (2011). Characterization Of Household Waste In Greenland. *Waste Management*, ***31***(7), 1461-1466.

[18]Firdaus, G., & Ahmad, A. (2010). Management Of Urban Solid Waste Pollution In Developing Countries. *International Journal Of Environmental Research*, ***4***(4), 795-806.

[19]Ghani, W. A. W. A. K., Rusli, I. F., Biak, D. R. A., & Idris, A. (2013). An Application Of The Theory Of Planned Behaviour To Study The Influencing Factors Of Participation In Source Separation Of Food Waste. *Waste Management*,***33***(5), 1276-1281.

[20]Henseler, J., Ringle, C. M., & Sinkovics, R. R. (2009). The Use Of Partial Least Squares Path Modeling In International Marketing. *Advances In International Marketing (AIM)*, *20*, 277-320.

[21]Huda, N., Rini, N., Mardoni, Y., & Putra, P. (2012). The Analysis Of Attitudes, Subjective Norms, And Behavioral Control On Muzakki's Intention To Pay

Zakah. *International Journal Of Business And Social Science*,*3*(22), 271-279.

[22]Izagirre-Olaizola, J., Fernández-Sainz, A., & Vicente-Molina, M. A. (2015). Internal Determinants Of Recycling Behaviour By University Students: A Cross-Country Comparative Analysis. *International Journal Of Consumer Studies*, *39*(1), 25-34.

[23]Khajuria, A., Yamamoto, Y., & Morioka, T. (2010). Estimation Of Municipal Solid Waste Generation And Landfill Area In Asian Developing Countries.

[24]Knussen, C., Et Al. (2004). An Analysis Of Intentions To Recycle Household Waste: The Roles Of Past Behaviour, Perceived Habit, And Perceived Lack Of Facilities. *Journal Of Environmental Psychology* **24**(2): 237-246.

[25]Low S.T. (2012). *Factors Affecting Energy Conservation Behaviour Of Students In Malaysia University*. Ph.D. Thesis. Universiti Teknologi Malaysia

[26]Mahmud SND, Osman K. (2010).The Determinants Of Recycling Intention Behavior Among The Malaysian School Students: An Application Of Theory Of Planned Behavior. Procedia Social And Behavior Sciences;9:119–24.

[27]Moh, Y. C., & Manaf, L. A. (2014). Overview Of Household Solid Waste Recycling Policy Status And Challenges In Malaysia. *Resources, Conservation And Recycling*, *82*, 50-61.

[28]Omran, A., Mahmood, A., Abdul Azi, H., Robinson, G.H., 2009. Investigating Household Attitude Toward Recycling Of Solid Waste In Malaysia: A Case Study. *International J Ournal Of Environmental Research* 3, 275–288.

[29]Pakpour, A. H., Zeidi, I. M., Emamjomeh, M. M., Asefzadeh, S., & Pearson, H. (2014). Household Waste Behaviours Among A Community Sample In Iran: An Application Of The Theory Of Planned Behaviour. *Waste Management*,*34*(6), 980-986.

[30]Percetakan Nasional Malaysia Berhad., *RMK11*. 2016. Print

[31]Poon, C. S., Ann, T. W., & Ng, L. H. (2001). On-Site Sorting Of Construction And Demolition Waste In Hong Kong. *Resources, Conservation And Recycling*, *32*(2), 157-172.

[32]Ramayah, T., & Rahbar, E. (2013). Greening The Environment Through Recycling: An Empirical Study. *Management Of Environmental Quality: An International Journal*, *24*(6), 782-801.

[33]Ramayah, T., Lee, J. W. C., & Lim, S. (2012). Sustaining The Environment Through Recycling: An Empirical Study. *Journal Of Environmental Management*, *102*, 141- 147.

[34]apingi, R., Ahmad, N., & Mohamad, M. (2011). A Study On Zakah Of Employment Income: Factors That Influence Academics' Intention To Pay Zakah. In *2nd International Conference On Business And Economic Research (2nd ICBER 2011), Langkawi, Kedah*.

[35]SWCM, (2015). http://www.ppsppa.gov.my/index.

[36]Shamshiry, E., Nadi, B., & Mahmud, A. R. (2011). Optimization Of Municipal Solid Waste Management. In *Proceedings Of The 2010 International Conference On Biology, Environment And Chemistry* (Vol. **1**, Pp. 119-21).

[37]Tonglet, M., Philips, P.S., Read, A.D., 2004. Using The Theory Of Planned Behaviour To Investigate The Determinants Of Recycling Behaviour: A Case Study From Brixworth, UK. *Resources, Conservation And Recycling* 41, 191–214.

[38]Tonglet, M., Phillips, P. S., & Bates, M. P. (2004). Determining The Drivers For Householder Pro-Environmental Behaviour: Waste Minimisation Compared To Recycling. *Resources, Conservation And Recycling*, *42*(1), 27-48.

[39]Wan, C., Cheung, R., & Qiping Shen, G. (2012). Recycling Attitude And Behaviour In University Campus: A Case Study In Hong Kong. *Facilities*,*30*(13/14), 630-646.

[40]Wang, Z., Zhang, B., Yin, J., & Zhang, Y. (2011). Determinants And Policy Implications For Household Electricity-Saving Behaviour: Evidence From Beijing, China. *Energy Policy*, *39*(6), 3550-3557.

[41]Webb, T., Joseph, J., Yardley, L., & Michie, S. (2010). Using The Internet To Promote Health Behavior Change: A Systematic Review And Meta-Analysis Of The Impact Of Theoretical Basis, Use Of Behavior Change Techniques, And Mode Of Delivery On Efficacy. *Journal Of Medical Internet Research*, *12*(1), E4.

[42]Yazdanpanah, M., Forouzani, M., & Hojjati, M. (2015). Willingness Of Iranian Young Adults To Eat Organic Foods: Application Of The Health Belief Model. *Food Quality And Preference*, *41*, 75-83.

Asset Management Recovery after the Disaster: State of Knowledge

Maimunah Sapri[1], and Siti Mariyam Abd Razak[2]

[1]*Centre for Real Estate Studies, Universiti Teknologi Malaysia, 81310 UTM Johor Bahru, Johor, Malaysia*
[2]*Faculty of Geoinformation and Real Estate, Universiti Teknologi Malaysia, 81310 UTM Johor Bahru, Johor, Malaysia*

Abstract. Asset, infrastructure and business can be destroyed in a split seconds. It only takes a minute for natural disasters such as flood, hurricane and others to happen that will stop the business process and destroying the business as well as their assets. Late response towards the disaster will bring disastrous impact not only to the people but also the social and economic of the impacted people and area. It is therefore important for the organization to ensure speedy recovery of the disaster's impact such as the assets in an effort to ensure the business survival and facilitating societal and economic recovery. Therefore this paper examines the state of knowledge in relation to the asset management recovery especially after the disaster. Following to that, this paper will discuss the literatures involves in asset management recovery such as disaster management and business continuity plan (BCP). Towards the end, this paper suggests that the effect of lack of planning in asset management recovery will lead to an untold damage towards the community, organization as well as the business, taking months or even years to rectify. In some cases, businesses are shut down for goods due to loss data as well as other emerge problems cause by the disaster.

Keywords: Asset Management Recovery, Business Continuity Plan (BCP), Disaster Management, Flood

1.0 Introduction

Physically, Malaysia can be considered as one of the blessed countries because of its geographic location that prevents it from the threat of severe natural disaster like volcanic eruptions and cyclones [1]. However, it has its own fair share of the environmental perils such as flood. Malaysia is well known as a country that is prone to major flooding event that occurs once a year due to the northeast monsoon that bring heavy downpour. Due to its nature, lots of lives, assets, building, infrastructure and environment have been damaged. Although the disaster occurs in a split seconds but the impact that it leaves are tremendous and may take years to rectify. Moreover, the vulnerability of the community is increase when they choose to construct houses near the rivers and along the coastlines.

Nevertheless, due to its relative regularity that always appear from November to March each year made actions such as flood mitigation, forecasting and warning systems have been undertaken to minimize the impact of the disaster. However, despite numerous planning and preparation, it is believed that the present countermeasures are far from sufficient. This is shown in the recent management of the flood disaster that happen during December 2014 to January 2015 flood crisis. During the crisis, as many as 250,000 people need to be placed at the evacuation centre and great efforts have been taken to reduce the impacts that it caused such as the disruption of the victims life, the health issues, the polluted environment (due to the mud comes from the flood) and the massive costs encompass in the redevelopment of the assets and infrastructure. It is reported by [2] that the estimated costs of repair for the assets and redevelopment has exceed several billion of ringgit in expenses which is 10 times more higher than the usual amount spend despite the country's knowledge about the disaster and earlier preparation that have been carried out.

One of the factors that have been identified to lead to the situation is the ill-prepared systems to handle the post catastrophic impact of the disaster. Although the government has allocate the budget to rectify the impact but the actual preparation have not been carried out until the disaster occurs and thus resulted in the victims need to suffer the long terms effects of the disaster in particular to those whose house and business have been destroyed by the disaster [2]. The actions to rebuild was so delayed that even a year after the disaster, there are still victims who does not own house any house and still living in the tent provided during evacuation of the disaster. That is why it is important for the government to have a recovery plan that is design to assist them in recovery of their assets from the scratch so that people can recuperate faster and their misery is not prolonged.

In conjunction, to ensure the faster recovery of the victims to their normal lives, this paper discussed on the opportunity of asset management recovery to be used as one of the approaches to deals with the aftermaths of the disaster. According to [3], asset management recovery can be defined into three different but interrelated meaning. Firstly, it is an aims that involves the restoration of normal asset structures that were disrupted by the disaster exactly before the disaster struck. Secondly, it is considered as a phase in the disaster management cycle that begins with stabilization of the disaster conditions and end with the asset returned to its normal conditions. Thirdly, it is a process by which the asset achieves the goal of returning to its former condition.

It is undisputed that recovery after disaster is a gradual process which may takes either immediate, short, medium or a long term process. Nonetheless, the recovery activities on assets are often seen as a long term process due to the uncoordinated act that occurs concurrently and on certain occasion overlap or in conflict with one another. For example the community need to construct their temporary housing in Land A. But the land use for the land is industrial and that is conflicted with the local government. Due to that, a lot of times are wasted to change the current land use of the land towards residential use before the new development can be done. This difficult process will then lead to longer period of recovery for the victims. This situation occurs might be because of the fact that sometimes the responsibility to manage these activities is assigned to people who are unfamiliar with them [2]. Besides that, lack of understanding and awareness of the recovery plan are one of the factors that contribute to the disastrous impacts of disaster and make the recovery process harder [3-7]. That is why it is important for the community as well as the government to have knowledge about the disaster, what contribute to the disaster, its impact and how to recover once the disaster hit.

2.0 Disaster management

Disaster can be defined as a sudden or recurring event or incident that disrupts the normal activities or livelihood of people. It is an event that posed a dangerous threat to the safety of people, assets, buildings, infrastructure, and the environment when it occurs. In parallel, [6] believed that "disaster is an emergency situation that leads to the loss of life; it will damage properties and the environment and hinder the sociál and economic activities of people". In addition, [8] defines disaster as "a severe interruption of the community daily life due from people vulnerability towards the disaster impacts and involves loss of life as well as the assets which affect their ability to cope if it is left unaided". Matthew [5] explains that when disaster struck, it will affect the community and disable them to recover and pull through by using their own resources. Due to that, the community will need and require tremendous help and effort for them to get back on their feet and recover.

2.1 Types of Disaster

One of the steps in handling and dealing with the disaster is to know the types of disaster that may strike the community. It is important for the community to know the threat they are facing because different types of disaster lead to different scales of impacts.

Prior research agreed that there are two types of disaster [5-7, 9, 10]. The first one is the natural disaster and the other is the man-made disaster. According to [6], natural disaster is a catastrophic event that is formed through natural causes or it is also called as "the Act of God". It is something that man has no control over and nothing can be carried out to stop it from happening. Among the examples of natural disaster includes floods, earthquakes, cyclones, landslides, volcanic eruption, and tsunamis. On the other hand, man-made disasters are disasters that are caused by human activities whether due to mere negligence or irresponsible or reckless act. Among the example of man-made disaster are fires, accidents, nuclear explosions, toxic emissions, technological disaster, the collapsing of buildings and terrorism.

While many have agreed that disaster can be classified into major types of disaster but there is other researchers that claimed hybrid disasters or mix disasters should be also incorporated into the groups as well [1,11-13]. Hybrid disasters happens trough the mixture of both natural forces and human error. It can be explained through a simple example. When human decided to clear up the jungles (human error), there are fewer trees and roots to support the soil. Therefore, when there are heavy rainfalls (natural forces), soil erosion occurs and leads to landslides. Accordingly, it is important for the community to know types of disaster they might encounter to effectively and efficiently rectify the impact of the disaster especially in terms of physical assets.

2.2 The Impact of Disaster

Nowadays, the frequency of disaster occurs around the world are escalating. It is reported by United Nations that Asian countries were the most severely hit by the disaster. It is stated that there was a total of 11,985 natural disasters events take place around the world from 1970 to 2014. From that, 5,139 incidents took place in Asia and the Pacific. It was also reported that most frequent disaster faced by the Asia-Pacific was flood and storms, followed by tsunamis, earthquakes and landslides.

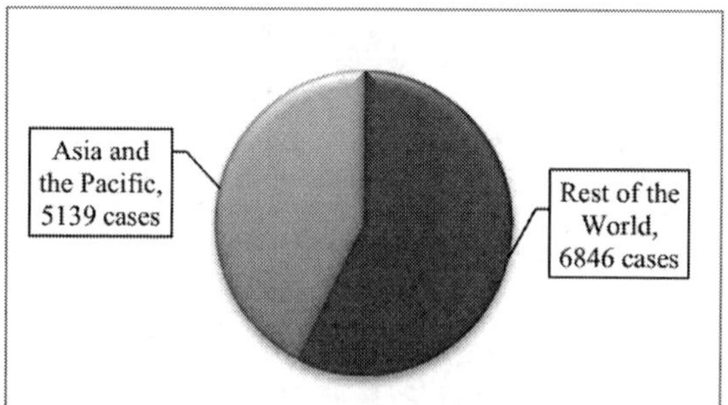

Figure 1. Total Occurrences of Natural Disaster Events from 1970 to 2014 (Source: [14])

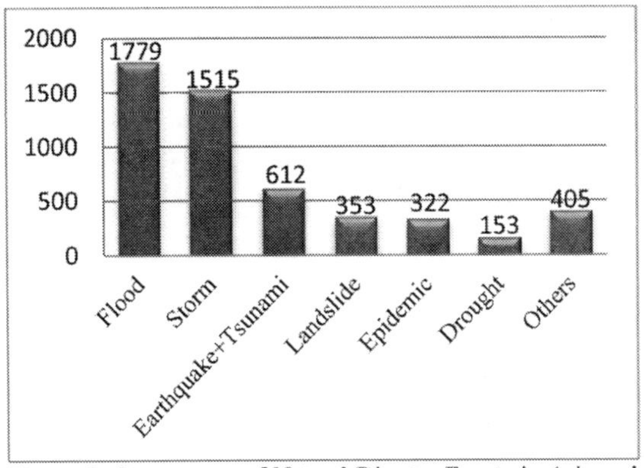

Figure 2. Occurrences of Natural Disaster Events in Asia and the Pacific from 1970 to 2014 (Source: [14])

In parallel, lots of losses have been reported in the form of human life and also physical damage that it caused to the surrounding area (assets and environment). Centre for Research of Epidemiology on Disasters (CRED) stated that disasters can be considered as one of the main threat to the whole world especially when the world is facing extraordinary scale of disasters over the years [15]. According to [7], between 1994 and 2003, there were more than 255 million of people were affected by the disaster. From that amount, 606,000 of lives were taken while leaving 4.1 million people injured and lots of building, assets and environment extremely damaged.

Table 1. Total number of natural disasters during 1974-2003 (Source: [7])

	1974-1978	1979-1983	1984-1988	1989-1993	1994-1998	1999-2003	1974-2003
Africa	88	113	128	107	149	333	918
America	99	199	255	319	320	475	1,667
Asia	220	336	353	482	449	726	2,566
Europe	43	108	136	144	134	288	853
Oceania	47	56	57	64	64	75	363
World	497	812	929	1116	1,116	1,897	6,367

Table 2. Total number of victims: people killed and affected, in millions from 1974-2003 (Source: [7])

	1974-1978	1979-1983	1984-1988	1989-1993	1994-1998	1999-2003	1974-2003
Africa	17.50	52.51	64.21	74.33	39.83	99.58	347.96
America	9.41	46.72	21.51	9.59	25.92	22.86	136.01
Asia	165.73	603.99	720.88	704.33	969.06	1,373.56	4,537.54
Europe	2.24	1.82	0.38	4.91	10.26	10.96	30.57
Oceania	0.10	0.68	0.71	7.32	10.30	0.27	19.38
World	194.98	705.72	807.69	800.48	1,055.37	1,507.23	5,071.46

Table 3. Total economic damage in millions during 1974-2003 (Source: [7])

	1974-1978	1979-1983	1984-1988	1989-1993	1994-1998	1999-2003	1974-2003
Africa	4,546	18,380	2,825	1,367	1,070	6,956	35,144
America	31,734	48,273	55,933	123,982	105,823	45,439	411,184
Asia	47,247	23,895	69,968	88,572	300,624	109,429	639,735
Europe	16,042	67,528	19,876	52,018	42,028	52,156	249,648
Oceania	8,631	15,504	1,947	9,097	5,337	2,711	43,227
World	108,200	173,580	150,549	275,036	454,882	216,691	1,378,938

According to [3], the physical impacts left by the disaster are the most apparent, easy to measure and will be first reported by the media. Among the damage to the physical assets that can be easily spotted is the damage to the building (residential, commercial, and industrial), transportation system, infrastructure system (water, waste disposal, and electric power fuel), communication system, as well as the critical facilities (hospitals, fire stations, police stations, storage of critical records and etc.). Apparently, based on the research conducted by [16], the rates of damages towards the assets are increasing over the years especially in more developed countries. Physical damage to buildings is the most common form of damage and draws the most attention from the researchers. It is due to the fact that shelter is one the basic things that human need to survive. If it is not fulfilled, it will lead to physiological problems such as demotivation and it will eventually make it harder for people to recover from the impact of the disaster.

Aside from buildings, the damage to the transportation system also draws a lot of people attention. It is because the roads and bridges are considered as one of the weakest links in the transportation system. When disaster struck, the most likely damage to the transportation system is the bridge. Bridge act as a connector from one road to the other. When it is damaged or it cannot conduct it original function anymore, people will have problems to use it to go to the other side of the road. It is especially disturbing when helps needs to be delivered in time of the disaster. Moreover, when the transportation systems are disrupted, helps cannot be delivered to the required place and it will affect the search, rescues, and relief process during the disaster. Furthermore, all the research done on transportation believed that such interruption will contribute to business and economic loss following the disaster. [17] reported that 43 percent of business manager admit that their losses were partially caused by the disruption in the transportation system.

In addition to transportation disruption, damage to utility lifelines (gas, water systems and electricity) can also be one of the major sources of economic loss. A number of studies have shown that the effects of disruptions to these facilities create not only facility repair costs, but also revenue loss to the service provider, direct economic loss to consumers, and indirect economic loss to the region as a whole [18].

Other impacts of disaster include physiological and economic impacts towards the affected people. Dependable on that, [19] stated that due to the disaster the domestic violence cases increase by 50 percent and it also affects the mentality of the younger generation who is believed to have witnessed the death or severe injuries borne by the loved one during the disaster. In conjunction, lots of opportunities in terms of development and jobs are ceasing due to the inability of the business to recover. After all, [3] believed that 87 percent of the organizations do not have the confidence in their ability to fully recover after they were hit by the disaster especially the smaller enterprises. This is due to the lack of planning and the effects of the lengthier period for asset recovery which make them harder to operate their business.

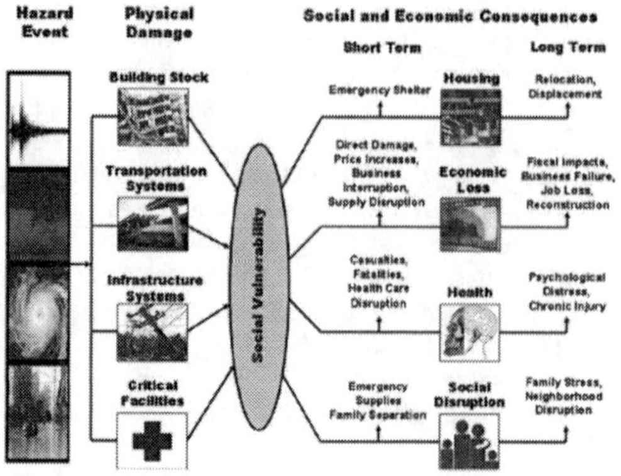

Figure 3: Conceptual Model of the Social and Economic Consequences of Disaster (Source: [20])

To boot, the weight of the disaster's impacts towards the community can also be measured through the community's vulnerabilities towards the disaster. Community's vulnerability is the ability of the community to forestall, survive, resist and recover from the impact of the disaster. Hence, the severity and the weight of the after effect of the disaster can be seen through the communities' capability to encounter the upcoming disaster as well as the effectiveness and efficiency of the current systems to respond towards the disaster. Corresponding to the prior research, the researchers found out that underprivileged people and less developed countries are the most vulnerable and exposed towards the impacts of the disaster [12, 13, 21-24]. According to [12], impoverished people who are living in the less developed country are becoming more disaster-prone nowadays compared to the developed countries. It is due to multiple reasons of social, cultural, economic and political factors that force them to live in the disaster-prone areas that expose them to the impacts due to its unplanned urbanization and not moving on even though they know about the risks. Furthermore, in undeveloped countries that have a low average of gross national income, even a small loses matters as it will affect significantly in slowing down their development. However, it does not mean that developed countries are having less impact from the disaster. They are still facing loss in terms of the assets, infrastructure and lives but their ability to recover is much faster and effective with the proper disaster management plan to act as a defensive action against the catastrophic events.

2.3 Approaches to Disaster Management

Disaster needs to be managed as a defensive action in facing the calamity. Most of the disasters especially natural disaster cannot be prevented. Nevertheless, their result can be alleviated. Thus, by having a proper disaster management plan, unnecessary loss of lives and damages towards the assets and properties can be prevented or minimize.

According to [25], disaster management act as preliminary measures to deal with the disaster. It aids to lower down the risks by having a proper planning of what the community should do and how they should deal and react when disaster struck. It will assist and educate them generously on how to use available resources to get back on their feet and recover.

For that reason, currently there are quite a number of research have been undertaken in relation to disaster management. Many have discussed the definition, the stages of the disaster management, the process, the plan, the involvement of people, the use of technology and information as well as theoretical framework to assist in disaster management plan. However, according to [26], the most popular topic selected for the research is on the disaster management phases and process.

Prior research has argued and discusses on what are the actual phases of disaster management plan and they basically agreed that there are three fundamental timeline involve in determining the phases. They are pre-disaster phase which happens before the disaster, during the disaster phase and post-disaster phase which is after the occurrence of disaster and trying to get back to the previous condition [26-28]. These literatures then deliberate on the process for each timeline. Among the process involves are prevention, protection, preparedness, emergency actions and recovery.

Prevention, protection and preparedness are the phases that take place before the disaster happen (pre-disaster). It is the most common and popular research item and action to be taken by the community in assisting the effect of the disaster. It is due to the fact that preventive measures are vitals in reducing the risk of the disaster like the proverb saying '*prevention is better than cure*" [29-32]. By knowing the risks each of the assets poses would make the recovery of the assets easier and the risks of them to be damaged are lesser when the disaster hit. For example, when regular inspection and risk assessment have been periodically carried out, the risks of them to deteriorate and malfunction are lesser and this will stop the building to easily collapse. It is parallel to the fact that collapsing building is one of the major disasters that cause a lot of casualties as well. Therefore, by having prevention plan, it will ensure the strengthening of its structure to protect the occupants and its contents. In addition, it is also important to protect the assets by having a regular and periodic maintenance plan that is conducted accordingly to ensure the ability of the assets to function as it was supposed to be are not declined. Meanwhile, preparedness also deals with the measures and activities that are taken in advance (before disaster) in ensuring effective and efficient response towards the impact of the hazards [21]. This includes the forecasting, issuance of early warnings before disaster struck, staff training, the delivery of knowledge regarding the disaster and its possible impact, preparation of the evacuation place and also the setting up of communities that involves in helping the affected people when disaster hit them [33].

Besides that, one the phases that caught people attention to study is on the emergency actions during the disaster. This is one of the vital subjects because lack of actions carried out in this stage might result in higher number of fatalities. Usually, at this stage emergency response plan and actions are needed to assist in dealing with the disaster aside from the search, rescues and relief operation. Therefore there are more studies carried out on how the response plan and rescues plan should be carried out.

Last but not least is the recovery phases after the disaster (post-disaster period). Recovery phase is one of the important phases that ensure the quality of lives of the community can be returned back as it was used to be. It refers to the decision and actions taken after the disaster with an intention to restore or improve the impact of disaster towards the community and at the same time, encouraging and supporting necessary actions to reduce the risk of the disaster [7]. In general, recovery involves the reconstruction of the damaged assets and continuation of the business process after the disaster as well as the ways of living before the disaster struck.

However, although it is one of the important process in disaster management plan, but there are limited research conducted on this topic. According to [7], the research on recovery has been overlooked despite a few contributions. The numbers of paper related to recovery of disaster management has record visible decrease throughout the years although there are several gaps are still remaining in an effort to have deeper understanding of disaster management theory. Among the common topics addressed in recovery phase is related to the temporary shelter (evacuation place), the funding and aid delivery system, impact assessment and resource management.

3.0 Disaster in Malaysia

Physically, Malaysia can be considered as one of the blessed countries because of its geographic location that prevents it from the threat of severe natural disaster like volcanic eruptions and cyclones [1]. Nevertheless, Malaysia did have its own fair share of the environmental perils in the form of floods, drought, periodic monsoon, and haze. Moreover, floods have been declared as the primary hazard that is affecting the country. What makes it worse is that it occurs on an annual basis following the Northeast Monsoon which predominates from November to March each year. The monsoon brings heavy rains and it escalated the average amount of rain to 2,500 mm a year and indirectly makes Malaysia as one of the countries that received heaviest rainfall in the world. Besides, flash flood that occurs occasionally is also one of the worrisome hazards that cause discomfort towards the community due to its impact on the assets and environment. Likewise, according to [1], it has been reported in the last couple of decade that the incidents has increase in terms of its frequencies and magnitudes and it is sometimes unexpected.

Currently, it has been estimated that at least 3.5 million of people in Malaysia are living in the flood prone area for varying reasons [34]. According to [35], the reasons they are drawn to the area are because of soil fertility and undeveloped land. Although they know of the existing risks, but factors such as financial ability always hinder them from moving to the safer places. For some people, the choice to move on and relocate to other place is simply does not exist and lead them more vulnerable and exposed towards the impacts of disaster [13].

Once disaster struck, the tangible and intangible impacts are clearly seen in terms of the damage to the physical assets. During the past floods events, few gross estimates have been made in terms of the costs of damage to private properties, businesses and industries and it is found out that these costs are substantial. The calculation of these losses includes both the tangible and intangible losses as well as the direct and indirect losses. For example, [36] found out that the average loss per household that is inflicted with the disaster is about RM1, 393 in 1993. This figure may not appear to be high for the people with a high income but the problem arise when nearly half of the household had a monthly income of RM500 and below. Thus, the loss has become so significant to them and this affects their ability to recover and increase their vulnerability towards the impact of the disaster.

At the same time, the business operation also suffers similar fate. It is reported that nearly half of the business operation need to be closed down due to their inability to recover from the impact of disaster [13]. Some of the reason is because they do not have the constructive plan of business continuity management that enable them to strategize on what they need to do if the business is disrupted. Besides that, the failure and damages occurs to the assets and the extended period of its recovery also plays a vital part in hindering the recovery operations of the business. That is why, it is important for the government to know how these assets can be recovered not only for the survival of the affected people but also for their continuity of living as before the events occur.

Currently, Malaysia has developed the National Security Council (NSC) Directive 20 that was issued to give guidelines on how the disaster should be managed but Centre for Public Policy Studies (CPPS) stated that the system is so ill-prepared to handle the post-flood catastrophes. Although NSC Directive 20 has allocate budgets and plans for recovery, but the actual preparations have not been made up until disaster happen. It means that the plan is just on paper without actions taken to conduct it. Ultimately, the victims are left stranded and in due course suffers the long terms effect of disaster in particular to those whose homes and business have been destroyed.

Besides, the recent disaster like earthquakes and tsunamis that unexpectedly occurs in Malaysia have decrease the resilience of community towards the impact of disaster due to the fact that the building in Malaysia are not build to withstand such impact from disaster. Because of its rarity, lots of building and infrastructure collapse like the one recently occur in Sabah. Due to the lack of research and planning as well as factors like unexpected events, it takes months before the problems could be effectively rectified. Hence, it makes the recovery process much harder and longer.

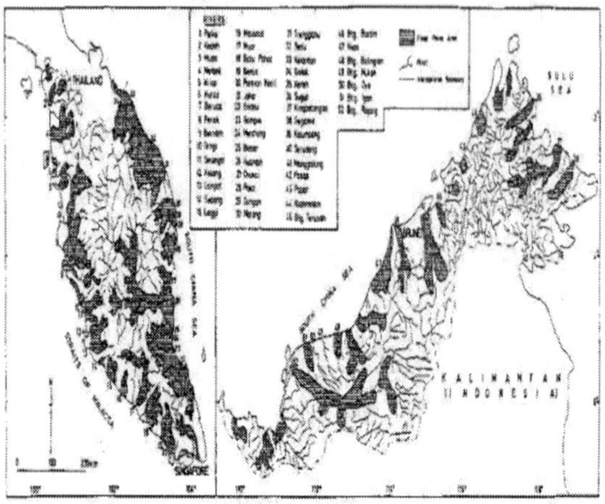

Figure 4. Approximate Distribution of Flood Prone Areas in Malaysia (Source: [35])

4.0 Asset Management

Assets can be defines as any entity that can be used to produce goods, products or services to meet the needs, expectation and requirements of a client or customer [37]. Assets can be divided into three major categories which are discrete assets, linear or continuous assets and virtual assets. Discrete assets can be seen in the form of pumps, packaging lines, buildings, and vehicles. In the interim, linear or continuous assets are water pipes, sewer, roads, land and also electric transmission lines while virtual assets are software.

In the meantime, asset management can be defined as a holistic process that provides guidance in the management of these assets throughout its life cycle from the acquisition of the assets, operation until it's time for disposal. As assets exist to support the business and fulfilled the needs, expectation and requirements of people, it is anticipated to make the most of their life period to carry out their function while managing the related risks and costs accustomed to it throughout their lifecycle. According to [37], all of these assets will go through phases of life cycle starting from evaluation, strategy, planning, design, creation, operation or maintenance, modification, and lastly disposal. Evaluations are done to look at the needs of the assets. The organization must thoroughly examine the need for assets and consider the full range of options before responding to it. It is to ensure they thoroughly need the assets and the assets are created based on the requirements of the stakeholders. After evaluation has been carried out, the next thing to be done is to strategies. Without proper planning conducted the asset manager might make a mistake that will affect the aims of the organization. Therefore, planning need to be considers and carefully planned by the asset manager to assist the organizations needs and eventually ensure that the asset can eventually aid in value adding of the organization. This will lead to the design process where the assets will be design based on the need of all the stakeholders and the requirements of the company to achieve its objectives.

Next is the creation of the assets. In achieving successful creation of an asset it is important to focus on time, scope, money and the facility impact [37]. It is important for the organization to follow their own purchasing terms and guidelines in the sourcing, selection and procurement of any assets, components or materials that will be used for the assets.

After the creation of the assets, the next step is operation and maintenance. Time, money and effort must be spent by the asset manager or the involve parties to maintain the value of an asset such as the building and at the same time providing the occupiers with a satisfactory environment [38]. To do this, expertise is needed to do maintenance works but management also plays an important part so that the works is planned economically and effectively. For years maintenance has been acknowledge as an important works to be done to prolong the lifespan of an asset and ensuring that the asset is in a good conditions. It is also being recognized as an investment that can show some economic return and also the area where the owner can cut the cost for maintenance works especially during the recent economic conditions. So to make sure that returns will be regained a proper maintenance management is needed. Hence, there are basically four matters to be considered for asset management. They are planned maintenance, unplanned maintenance, and maintenance of asset records, and reassessment. Meanwhile, Disposal of the asset generally will occur due to the changes in the needs and demand of the asset. To know that, assessment of the need for assets can be carried out to determine whether the needs and expectation of the customers is fulfilled. Thus, the organization would know whether the asset is still viable to generate profits and effectively use by their customers.

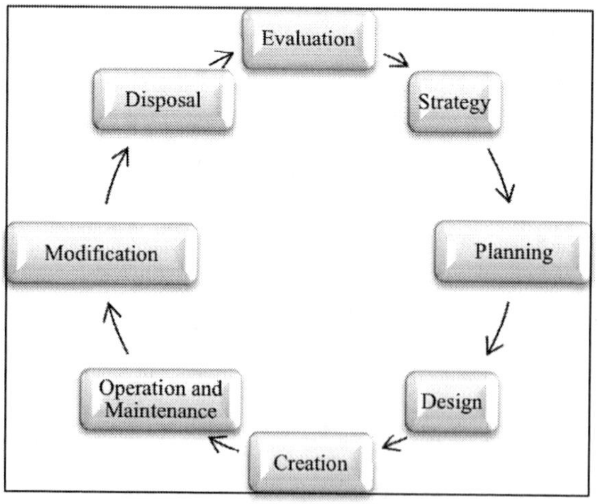

Figure 5. Asset Life Cycle

4.1 Asset Management Recovery

Recovery is a process to reclaim life or previous (normal) condition of life before something unwanted happen. In the case of disaster, "something unwanted" is the disaster and recovery means the restoration of the community way of life before the disaster occurs. According to [38], recovery can be defined based on a comparison between an indicator of the time path which is "with" and "without" disaster or between "before" and "after" disaster. Based on that, it is best to assume that recovery planning can be done way before disaster while effective recovery actions can be conducted after the disaster. After the disaster, inspections need to be conducted in the impacted area to know the involved damage to the assets, building and infrastructure. That is why it is important for asset management recovery to be embedded in the disaster management plan for the community as it will assist the community on the total numbers of assets exist before the disaster, the requirements for the asset and how to recover it from the scratch. It is because asset management starts from the very beginning of the assets life cycle. Hence, when a disaster occurs, the organization will know all the details regarding the assets.

Among the objectives of assets recovery is to minimize loss and downtime as well as to ensure the continuity of lives and business. However, when longer time are needed to recover especially for the assets such as buildings and infrastructure, the longer the times that it take for the community and business to get back to their normal live. After all, when disaster strike, recovery and response activities are usually uncoordinated because of the tension and panic arise. Therefore what should occur concurrently becomes overlap and conflicting with one another [3]. In some cases, management responsibility to handle the impact of disasters is often given to the people unfamiliar with them due to reasons such as the appointment of new leader, the persons elected are not familiar with disaster and so forth. The concern is that, the community might miss their chance to improve the assets, infrastructures, the environment and quality of life due to that minor glitch.

Over the years, there is research conducted on how to improve the recovery process such as research on funding and aid delivery system [33], reconstruction of flood resistance shelter [22, 24, 39], impact assessment and resource management [30]. Aside from that, there is also research carried out suggesting on the quality infrastructure to be built to replace the existing prone-disaster buildings and infrastructure. However, realizing infrastructure reconstruction with the best quality has proven to be quite difficult for developing countries [21]. Besides that, the government need to consider the reluctance of community to move from the area that they have considered as home even through there are better places have been built for them. It is therefore important for the government to have a guidelines or reference on how the area and the assets can be rebuilt and recover as quickly as possible.

However currently there are no studies on how the assets can be rebuilt from scratch. Hence, this situation creates a gap in this area. From the past disaster, there is evidence whereby lack of planning in the asset recovery matter has caused one of the schools in Kuala Krai to not be able to serve its main objectives which is to deliver teaching and learning process effectively. Even though, the school period has started in January but until April 2015, there are still students who are incapable to come to the school because of the damaged in the school facilities such as the hostels are not yet rectified. Due to that, it is important for the government and the community to know how the assets can be rebuilt so that they can participate and help in the recovery process to minimize the time of recovery.

4.2 Business Continuity Plan (BCP)

BCP is a comprehensive organizational plan that includes in the disaster asset management recovery plan. By having BCP in asset management recovery plan, the community can prepare on the things that they would need to do in ensuring the on-going of the business even after the disaster. It is because, asset, infrastructure and business can be destroyed in a split seconds. Although in order to have a successful business, one need to do lots of things such as analysing the market, identifying the potential customers, creating strategies for delivering the products and services and establishing financial goals [40]. However, despite their efforts, it can take less than 60 seconds for a company's reputation to be ruined and its business to be crippled [41]. In just 60 seconds a server failure might occur or hacker can strike out the essential applications and lead to a disastrous series of events. In the worst case, natural disasters such as floods, hurricane and others might just happen and stop the business process and destroying the business as well as their assets [40, 42-44]. It is therefore important for the organization to ensure the continuity of the business activity following the disaster to ensure the business survival and facilitating societal and economic recovery [43]. According to [41], the effect of lack of planning recovery to face the disaster will lead to an untold damage to a company's reputation, taking months or years to rectify. In some cases, the businesses are just shut down for goods due to the loss of data as well as other emerge problems cause by the disasters [45]. A study made by Datapro Research Company found out that 43 percent of organisation strike by disaster will never recuperate and reopen and another 29 percent fail within two years [45]. Thus, it is important for the companies to have Business Continuity Plan (BCP) to ensure the continuity of their business and activities despite all the business disruption.

BCP is designed to avoid or mitigate the risks or in other words to reduce the impact of crisis [46]. According to [47], BCP is "a sound plan that will ensure that the business recovers from a disaster faster in an effective and efficient manner, and simultaneously business disruption and damage can be minimized and managed". [40] illustrates BCP as "the method and procedures that need to be used by the organization to ensure that important function of the business can be carry on in and after the disaster". Meanwhile, [43] defines BCP as a proactive and holistic management process designed to facilitate the continued achievement of critical business objectives. It is due to the facts that it provides an iterative and structured process that incorporates risk identification, assessment, and management, the development of disaster recovery plans and procedures, training, exercising, and using feedback that they received to promote an iterative development of plans and capabilities. BCP is develop to gain an understanding on what the organization must achieves (critical objectives), identifying the barriers or interruptions that may prevent their achievement, and determining how the organization will continue to achieve these objectives when interruptions occur [43]. Therefore, the organization can ensure that the business can continue as stable as possible as it was before and enable them to handle the crisis as well as attain revenues throughout the days.

According to [40], among the benefits of BCP includes reduce financial costs, enhance customers service, and improve the ability of the organization to face undesirable events such as natural disasters and ensuring they can get back to their business activity the same as the before the disruption happen.

5 Conclusion

Malaysia is well known as a country that faces several flood events in an annual basis. Although the knowledge is there, but the preparation to face the impact are less than sufficient (CPPS, 2015). This event causes the impact such as the physical damage to the buildings and asset to take longer times to rectify and somehow becomes a barrier to the victims to restore back their normal lives.

One of the reasons identified to cause the situation is the lack in the recovery planning actions. The standard operation of procedures (SOP) is not well prepared and the related agencies failed to understand on the items that they should do to help in the recovery process. Thus, it is important for a more detailed plan to be design to help the community as well as the government to know on their roles and responsibilities as well as their limitation and scopes each one of them need to conduct to the ensure the smooth running of recovery process.

This is where asset management recovery comes in handy. Through the plan, the government will know what to do with the assets from the scratch even if the assets are completely destroyed. Asset management recovery plan will assist the community and the government even through it is handed to the person who is unfamiliar with the disaster because of the complete guidelines and reference on what is their roles and what they need to do. This is important to ensure that the destroyed asset can be recover as fast as it can to ensure the quality of life and business can be continued like before the disaster strike or even better.

Acknowledgement

This research is supported by the Ministry of Education, Malaysia under Fundamental Research Grant Projects (FGRS) (VOT: 4F633).

References

[1] B. A. Rahman, *Issues of Disaster Management Preparedness: A Case Study of Directive 20 of National Security Council Malaysia*, IJBSS, Vol. 3 Iss 5 pp. 85-91 [2012]

[2] Centre for Public Policy Studies (CPPS), *CPPS Policy Fact Sheet: Malaysia's Flood Management*, CPPS, Kuala Lumpur, Malaysia [2015]

[3] M. K. Lindell, *Recovery and Reconstruction after Disaster*, Texas A & M University, College Station, TX, USA [2011]

[4] P. R. Berke and T. Beatley, After the Hurricane: Linking Recovery to Sustainable Development in the Caribbean, The John Hopkins University Press. Baltimore and London [1997]

[5] P. E. G. Matthews, *Disaster Management in Libraries*, LIBR J, Vol. **17** Iss 3 pp. 5-12 [1996]

[6] I. M. Shaluf, F. R. Ahmadun, *Disaster Types in Malaysia: An Overview*, Disaster Prev. Manage., Vol. **15** Iss 2 pp. 286-298 [2006]

[7] T. L.M. F. G. S. S. M. Mueller, *Balanced Scorecard for Natural Disaster Management Projects*, Disaster Prev. Manage.: An International Journal, Vol. **16** Iss 5 pp. 785-806 [2007]

[8] Department for International Development (DFID), *Disaster Risk Reduction: A Development Concern*, DFID, London [2004]

[9] B. A. Turner, N. F. Pedgeon, *Man Made Disaster*, 2nd ed. Butterworth-Heinemann, Oxford [1997]

B. Richardson, *Socio-Technical Disaster: Profile and Prevalence*, Disaster Prev. Manage., Vol. **3** Iss 4 pp. 41-69 [1994]

[10] C. Hood, M. Jackson, *the New Public Management: A Recipe for Disaster* in D. J. Parker and J. W. Handmer, Hazard Management and Emergency Planning, Perspective on Britain, Chapter 9, James and James Science Publishers Ltd, London [1992]

[11] M. Shaluf, *An Overview on Disasters*, Disaster Prev. Manage., Vol. **16** Iss 687-703 [2007]

[12] G. S. J. Kumar, *Disaster Management and Social Development*, Int. J. Sociol. Soc. Pol., Vol. **20** Iss 7 pp. 61-81 [2000]

[13] G. O'Brien, P. O'Keefe, J. Rose, B. Wisner, *Climate Change and Disaster Management*, Disasters, Vol. **30** Iss 1 pp. 64-80 [2006]

[14] United Nations: Economic and Social Commission for Asia and the Pacific, *Overview of Natural Disasters and their Impacts in Asia and the Pacific, 1970-2014*, Disaster Risk Reduction Section, ICT and Disaster Risk Reduction Division, ESCAP [2015]

[15] Department for International Development [DFID], *Reducing the Risks of Disaster-Helping to Achieve Sustainable Poverty Reduction in a Vulnerable World*, A DFID Policy Paper, DFID, London [2006]

[16] P. R. Berke, *Natural Hazard Reduction and Sustainable Development: A Global Assessment*, J. Plan. Lit., Vol. **9** pp. 370-382 [1995]

[17] F. L. Bates, W. G. Peacock, *Living Conditions, Disaster and Development: An Approach to Cross-Cultural Comparisons*, Athens, GA: University of

Georgia Press [1993]
[18] A. Rose, D. Limb, *Business Interruption Losses from Natural Hazards: Conceptual and Methodological Issues in the Case of the Northridge Earthquake*, Environmental Hazards, Vol. **4** pp. 1-14 [2002]
[19] B. H. Morrow, *Hurricane Andrew: Ethnicity, Gender and Sociology of Disaster*, London, Routledge pp. 141-170 [1997]
[20] French, S.P., Lee, D. & Anderson, K, *Estimating the Social and Economic Consequences of Natural Hazards: Fiscal Impact Example*, Natural Hazards Review, 11(May), pp.49–57 [2010]
[21] R. Palliyaguru, D. Amaratunga [2008], *Managing Disaster Risks through Quality Infrastructure and Vice Versa*, Structural Survey, Vol. **26** Iss 5 pp. 426-434
[22] B. Wisner, J. Adams, *Environmental Health in Emergencies and Disasters: A Practical Guide*, World Health Organization, Geneva [2003]
[23] C. Benson, J. Twigg, M. Myers, *NGO Initiatives in Risk Reduction: An Overview*, Disasters, Vol. **25** pp. 199 [2001]
[24] K. S. Vatsa, *Risk, Vulnerability, and Asset based Approach to Disaster Risk Management*, Int. J. Sociol. Soc. Pol., Vol. **24** Iss 10/11 pp. 1-48 [2004]
[25] F. Nateghi-Alahi, Y. O. Izadkhah, *Earthquake Disaster Management Planning in Health Care Facilities*, Disaster Prev. Manage., Vol. **13**, Iss 2 pp. 130-135 [2004]
[26] E. Lettieri, C. Masella, G. Radaielli, *Disaster Management: Finding from a Systematic Review*, Disaster Prev. Manage. Vol. **18** Iss 2 pp. 117-136
[27] T. Hensgen, K. C. Desouza, G. D. Kraft, *Games, Signal Detection, and Processing in the Context of Crisis Management*, J. Conting. Crisis Manag. Vol. **11** Iss 2 pp. 67-77 [2003]
[28] R. Bertrand, C. Lajtha, *A new Approach to Crisis Management*, J. Conting. Crisis Manag. Vol. **10** Iss 4 pp. 181-191 [2002]
[29] S. Menoni, F. Pergalani, *An attempt to Link Risk Assessment with Land Use Planning: A Recent Experience in Italy*, Disaster Prev. Manage., Vol. **5** Iss 1 pp. 6-21 [1996]
[30] F. C. Dai, C. F. Lee, Y.Y. Ngai, *Landslide Risk Assessment and Management: An Overview*, Eng. Geol., Vol. **64** Iss 1 pp. 65-87 [2002]
[31] P. G. Bakir, *Proposal for A National Mitigation Strategy Against Earthquakes in Turkey*, Nat Hazards, Vol. **33** Iss 3 pp.405-425 [2004]
[32] S. Mansor, M. Abu Shariah, L. Billa, I. Setiawan, F. Jabar, *Spatial Technology for Natural Risk Management, Disaster Prev. Manage.*, Vol. **13** Iss 5 pp. 364-373 [2004]
[33] K. M. Sukeri, S. Shazwani, Flood Disaster Management in Malaysia: An Evaluation of the Effectiveness Flood Delivery System, IJSSH, Vol. 5 Iss 4 pp.398-402 [2015]
[34] N. W. Chan, *Flood hazard mitigation in peninsular Malaysia*, in P.A. Merriman and C.W.A. Browitt, Natural Disasters: Protecting Vulnerable Communities, Proceedings of IDNDR Conference in London, Thomas Telford, London, 13-15 October 1993, pp. 194-209 [1993]
[35] W. C. Ngai, *Increasing flood risk in Malaysia: causes and solutions*, Disaster Prev. Manage.: An International Journal, Vol. **6** Iss: 2, pp.72 – 86 [1997]
[36] N. W. Chan, *A contextual analysis of flood hazard management in peninsular Malaysia*, unpublished PhD. thesis, Middlesex University [1995]
[37] R. Lutchman, Sustainable Asset Management: Linking Assets, People and Processes for Results. DEStech Publication [2006]
[38] S. B. Miles, *The Role of Critical Infrastructure in Community Resilience to Disasters*. Structures Congress 2011 @ ASCE 2011, pp.1985–1995 [2011]
[39] G. Ma, B. S. C Fung, *China's asset management corporations*. BIS Working Paper, (115) [2002]
[40] A. J. Karim, *Business disaster Preparedness: An Empirical Study for Measuring the Factors of Business Continuity to face Business Disaster*, IJBSS, Vol. **18** Iss 18 pp. 183-192 [2011]
[41] R. Stanton, *Beyond disaster recovery: the benefits of business continuity*. Computer Fraud & Security, Vol. **7** pp.18–19 [2005]
[42] W. Lam, *Ensuring Business Continuity*, *IT Pro*, pp.19–25 [2002]
[43] D. Paton, *Business Continuity During and After Disaster: Building Resilience through Continuity Planning and Management*. ASBM, Vol. **1** Iss 2 pp.1–16 [2009]
[44] K.N Myers, *Total Contingency Planning for Disasters : Managing Risk , Minimizing Loss , Ensuring Business Continuity*, The Internal Auditor, Vol. **51** pp.14–15 [1994]
[45] V. Cerullo, M. J. Cerullo, *Business Continuity Planning: A Comprehensive Approach*. Inform. Syst. Manage., Vol. **21**, pp.70–78 [2004]
[46] C. Castillo, *Disaster preparedness and Business Continuity Planning at Boeing: An integrated model*, JFM, Vol. **3** Iss 1, pp.8–26 [2004]
[47] M. Savage, Business Continuity Planning, Work Study, Vol. **51** Iss 5 pp. 254-261 [2003]

Hungry for housing: waqf real estate development - a social welfare alternative

Siti Nadiah Mohd Ali[1a] , Abdul Halim Mohd Noor [2] ,Noraini Johari [1] , Nurul Sahida Fauzi [1], Nor Nazihah Chuweni [1] and Nor Rashidah Paujah Ismail[3]

[1]*Faculty of Architecture, Planning and Surveying , Universiti Teknologi MARA, Seri Iskandar Campus, Seri Iskandar, 32610, Perak, Malaysia*
[2]*Center for Islamic Philanthropy and Social Finance (CIPSF),78000 Alor Gajah, Melaka Universiti Teknologi MARA, Malaysia*
[3]*Faculty of Computer Science, Universiti Teknologi MARA, Tapah Campus, Tapah Road, 35400, Perak, Malaysia*

Abstract. Islam has outlined 4 important elements to achieve fruitful life in the duniya and the akhirah. A famed hadith of Rasulullah (peace be upon him) meaning "Four things that will bring joy to a person, which is (owning) a pious woman (wife), a good house (spacious), a good neighbour, and a great vehicle" (Hadith narrated by Ibnu Hibban. It is clear from this hadith that Islam emphasised ownership of a comfortable home. The polemic remains that with skyrocketing housing prices in the market, would the low and middle income groups afford housing that is safe, comfortable and secure? Realising the importance of living in a secure and safe house, this study aims at investigating the possibility of leveraging the upward trend of waqf real estate development in empowering the Muslims' economy in the country. Towards this end, the authors have conducted a study by collating primary and secondary data obtained from the Penang State Islamic Religious Council (Majlis Agama Islam Pulau Pinang) for analysis. Thus the Penang Islamic Council is the study site. Semi structured interviews were conducted with 2 waqf officials who are in charge of coordinating waqf real estate developments in Penang. Based on the study conducted, it was found that Penang State Islamic Religious Council has succeeded in enhancing value added to waqf real estate developments through the implementation of 9 housing project developments using the ijarah concept. The Penang State Islamic Religious Council has enabled the low medium income groups to stay in comfortable and safe housing by utilising such holistic and efficient waqf product developments under shariah guidelines.

Keywords: Affordable Housing, Low and Middle Income Groups, Waqf Real Estate Development, State Islamic Religious Council, al-Ijarah

1 Introduction

Waqf has been literally defined as "to stop, contain, or to preserve'; in the context of land waqf is real estate that has been donated for charity by the donor towards the greater good of the community (ummah) for the sake of Allah (SWT). Once a real estate has been donated for charity (waqf) it cannot be reclaimed by the original owner to be gifted, used as inheritance or sold; in line with the waqf definition itself. The IIUM Endowment Fund related that in shari'ah, a Waqf is a voluntary, permanent, irrevocable dedication of a portion of one's wealth – in cash or kind - to Allah. A waqf real estate belongs to Allah and the corpus of the waqf always remains intact. The fruits of the waqf may be utilized for shari'ah compliant purpose.This is based on a famed hadith narrated by Ibn 'Abbas:

"The Messenger of Allah said: 'The one who takes back his gift is like the dog which vomits then goes back to its vomit [1] *"*

Thus this hadith augmented the conviction that if a waqf real estate development is revocable via the withdrawal of a waqf donation, this will undermine the social value of waqf which aspires to solve the community's social woes. The progress of waqf real estate developments in developed Islamic countries such as Turkey is a case in point. Turkey's waqf system is so progressive that it has created 4,400 mosques, 500 university hostels, 150 hotels, 5,348 shophouses and 2,254 housing projects. It is heartening that the waqf system has been Turkey's 'bread and butter' since the Uthmaniyyah Caliphate era [2].

A renowned Turkish writer B. Yediyyildiz wrote in his book "Place of the Waqf in Turkish Cultural System" waxing on the significance of waqf that:

[a] Corresponding author: sitin190@perak.uitm.edu.my

"...a person would have been born into a waqf house, slept in a waqf cradle, eaten and drunk from waqf property, read waqf books, been taught in a waqf school, received his salary from a waqf administration, and when he died, placed in a waqf coffin and buried in a waqf cemetery.[3]

Thus waqf plays an important role in ensuring a stable and non-capitalistic social mobility system preventing the rich getting richer and the poor getting even poorer due to the inability of waqf to fulfil its expected role.

Waqf is given due honour in Islam as an enduring reward even after the donor has died. A Muslim believes that life is not only in this world but the afterlife (akhirah) is the eternal life where the waqf we created in the worldly life will convert into a most valuable reward. Due to the significance of waqf benevolence to the ummah, Allah SWT said in *Surat Al-Baqarah, verse 261* translated as:

The example of those who spend their wealth in the way of Allah is like a seed [of grain] which grows seven spikes; in each spike is a hundred grains. And Allah multiplies [His reward] for whom He wills. And Allah is all-Encompassing and Knowing.[4]

Waqf enriches a Muslim's soul and personality contributing towards his physical and spiritual happiness. Waqf bridges the poor and the rich in a symbiotic relationship. Not many realise that the waqf system possesses potential sustainability in empowering the Muslim economy in this country.

The establishment of the Department of Awqaf, Zakat and Hajj (JAWHAR) on 27th March 2004 signalled that waqf real estate development has cut a niche in Malaysia. The formation of JAWHAR led to the creation of the Waqf Foundation which was officially incorporated on 23rd July 2008 under the Trustees Incorporation Act 1952 (Revised 1981) JAWHAR.

The government accorded waqf serious attention as the Chairman of the Waqf Foundation of Malaysia Board of Trustees is the Honourable Minister with Religious Portfolio at the Prime Minister's Department, while the Vice Chairman is the JAWHAR Director General. Other members of the Board of Trustees comprise representatives of every State Islamic Religious Councils, one (1) representative from the Ministry of Finance, one (1) representative from the Economic Planning Unit, Prime Minister's Department and three (3) representatives from the corporate sector.

The government's commitment in creating a special entity to coordinate waqf real estate development reflects the importance of waqf real estate development. Waqf real estate development in Malaysia has not been fully explored considering there are still vast swathes of waqf real estate that have not been developed to fulfil existing social needs.

The main objective of this study paper is to investigate the potential presented by waqf real estate in Malaysia for the purpose of housing development; and to evaluate the extent of waqf real estate development for the purpose of housing could be implemented. In achieving this objective, waqf real estate belonging to Penang State Religious Islamic Council have been selected as case studies. This is in view of Muslim housing ownership in Penang being quite low which is further aggravated by high levels of real estate values and demographic factors creating stiff competition in home ownership in Penang.

A study done by Salfarina, A.G., Nor Malina M., and Azrina, H (2011) found that 40% of urban population comprises disadvantaged people with low affordability to own housing whereby a three-bedroom two bathrooms apartment with a floor area below 92.9 square metres near the city centre costs in excess of RM245,000 [5].

Hence, the issue of the urban poor not owning safe and comfortable residences has become controversial in Malaysia. This is exacerbated by rising costs in addition to debt liabilities; making the creation of truly affordable home ownership a critical challenge.

It is generally accepted that the development of waqf housing real estate may become one of the solutions in overcoming urban housing woes for the low and medium income groups.

2 Overview Development of Waqf

Waqf is a latent economic resource of the Muslims that need to be fully exploited which, if correctly managed, will yield economic revenues benefiting all parties.

Malaysia may be categorised as an Islamic State that only recently activated its waqf system in the last 10 years. In the past, waqf real estate developments in Malaysia typically consist of the construction of mosques and cemeteries; but this has gradually evolved in tandem with the country's economic growth. Now, waqf real estate are developed with 'pondok' religious schools and tahfiz schools on waqf branding. On top of that the establishment of the Waqf Foundation has contributed generously towards the development of small shop bazaars on mosque grounds in collaboration between the Waqf Foundation and the State Islamic Councils. Consequently, waqf real estate developments are made more active with the development of Waqf Hotel in Taiping, Perak; Kuala Terengganu, Trengganu; Port Dickson, Negeri Sembilan and in Melaka.

Waqf land development in Malaysia should not only focus on mosque, musolla and cemetery developments; but taking a cue from other countries' waqf real estate developments, Malaysia must break new frontiers in waqf real estate developments. The development of waqf real estate requires financing, thus the creation of cash waqf concept.

Magda (2013) iterated that cash waqf-financed projects need not be confined to only religious natured developments but should also accommodate current needs of the ummah [6].

Numerous studies have been done on the development of various real estate sectors on waqf lands. Magda (2012) further iterated that waqf real estate development is wide ranging; citing 5 countries namely Indonesia, Sudan, Turkey, UK and Kuwait that have paved inroads into waqf real estate development encompassing various sectors. Table 1 shows various waqf real estate that have been developed in the above countries [7].

Table 1: List of waqf real estate

List of old waqf real estate	
1. Agriculture lands	16. Stables
2. Universities	17. Carpentry Workshops
3. Schools	18. Clinics
4. Hospitals	19. Money changers
5. Clinics	20. Textile Factories
6. Pharmacies	21. Hotels
7. Bakeries	22. Silk Factories
8. Butcheries	23. Leather-dying Factories
9. Water supplies	24. Match Factories
10. Restaurants	25. Soap Factories
11. Tailors	26. Aggregate-making Factories
12. Goldsmiths	27. Olive oil Factories
13.Orphanages	28. Household Crockery and Candle Factories
14. Shoemakers	29. Boat-making Factories
15. Highways	30. Weaponry Factories

Source: Magda (2012)

According to A.Hassan (2010) waqf is an instrument that applies the concept of charity and is non-profit oriented. The public sector can collaborate with the private sector towards uplifting the socio-economic standing of the community[8]. This would strongly justify the development of endowment assets for the purpose of charity according to current needs.

2.1 The need for housing development on waqf land

In the context of a developing country such as Malaysia, the local house prices will benefit housing investors looking to make a quick profit optimisation on housing purchases. In contrast, waqf housing developments will be lower in price than conventional housing development of similar type in the local housing market. This strategy is to ensure target groups' affordability is achieved.

The Government of Malaysia (GoM) has realised that housing is a critical need for every member of the population. It is also an important component of the economic parameter. The GoM has formulated policies and programs tailored to ensure every eligible Malaysian has an opportunity to forge a suitable place in the greater scheme of things, as their basic need [9].

Perumahan Rakyat 1Malaysia Act 2012 (Act 739) is one of the government's initiatives to monitor all matters relating to the development and construction of dwelling houses, infrastructure and facilities. PR1MA programs aim at handling the needs for sustainable housing and urban community living in Malaysia.

Affordability issues in Malaysia have never been seriously discussed in detail and no specific studies on housing affordability have been carried out. The term 'affordable housing' being widely used by housing developers in Malaysia is a misnomer and basically becomes their selling point without any credible fact or index as a basis of reference. The same phenomenon exists among housing developers in other Asian countries where formal studies on housing affordability have never been done as compared to developed countries such as the United States, Germany, Spain, United Kingdom, Italy and the Netherlands [10].

Any study on the main role regarding the provision of housing and municipalities is the government's responsibility. Even though there is a variation in such role between developing countries and developed countries, the public sector is still very much needed to provide affordable housing [11].

Therein lies the question as to why must waqf real estate be developed as housing areas? The answer lies in the wide swathes of undeveloped waqf lands that can fulfil immediate housing shortages in Malaysia today. Limited income levels but corresponding high house price levels have isolated medium and low income earners from the housing market. As a result of this social mobility imbalance, it is not impossible that community wellbeing will be affected giving rise to increased incidences of criminal cases.

Why do the target groups desert housing programs provided by the government? A study by Reina Vincent and Begley Jaclene (2014) found that ownership structures, local price increments, and affordability are the three main determinants [12].

It is clear that the housing programs provided did not allow target groups to afford owning 'affordable housing'.

Real estate price increases have for a long time become a thorn in the side for house owners in Malaysia to the extent that owning a house now means a greater corrosion of capital compared to the scenario in the past. A recent study done by Demographia endorsed what has been common knowledge, that Malaysia has a 'very unaffordable' housing market with house prices hovering in the range of 5.5 times the median income as compared to 5.1 times the median income in Singapore [13].

The Malaysian Department of Statistics quoted in their report on Household Income and Basic Amenities Survey 2014 that the median monthly household income for Malaysians in 2014 increased by 11.7 percent per annum to RM4,585 from RM3,626 in 2012 [14].

Due to this ever increasing housing issue and based on the main objective of this paper, it is perceived that 'waqf housing schemes' will be able to fill the void of providing affordable safe and comfortable housing for the target groups in tandem with the objective of waqf as an instrument to solve community economic woes.

2.2 Overview of housing affordability in Malaysia

Table 2 shows house prices in Malaysia indicating housing real estate continuing to monopolise real estate market activities taking average house prices in 2014 at RM295,345 and increasing to RM312,050 in 2015; an increase of 5.7 percent. Kuala Lumpur is the most attractive global city in Malaysia, listed in 2014 and 2015 with increases of RM42,135; which is from RM676,620 to RM718,755 with an increase of 6.7 percent in house prices. House prices in Penang rregistered RM355,374 in 2014 and increased in 2015 in the range of 3.8 percent to RM368,849. All states recorded market activity growths in terms of house prices. The low income earners have such relatively low purchasing power that they could not even afford to buy a house!

Table 2: All House Prices by State in Malaysia

All House Prices by State	All House Prices (RM) 2014	All House Prices (RM) Q3/ 2015
Malaysia	295,345	312,050
Kuala Lumpur	676,620	718,755
Selangor	441,383	469,250
Johor	234,126	243,509
Pulau Pinang	355,374	368,849
Negeri Sembilan	179,537	196,750
Perak	174,021	182,721
Melaka	159,360	170,827
Kedah	159,249	164,234
Pahang	201,282	204,108
Terengganu	209,043	217,647
Kelantan	163,633	165,883
Perlis	144,411	156,324
Sabah	425,166	449,022
Sarawak	361,756	389,946

Source: National Real estate Information Centre (NAPIC) (2015)

2.1.2 Owning, purchasing and affordability

Home ownership rates in Malaysia are quite high averaging 72.5 percent in contrast with developed countries like Singapore at 70 percent within the same year. However, these rates published by the Malaysian Department of Statistics include informal home ownership. A large percentage of housing stock contributing to the 2010 Census was not included in the NAPIC-published estimates that only consider formal housing accounts [15]. This effectively connotes that home ownership rates in Malaysia are below 72 percent as home ownership and home occupancy are two different things. An occupied home may mean it is a rented home.

Housing ownership is best reflected by financial independence and the country's rapid development. It is directed to ensure that housing is provided as a main factor that every group of income earners can afford to own, be they high, medium or even low income earners [16].

The government recognised that housing is a major human need and one of the critical components in urban economics. The My First Home Scheme has enabled young people categorised as poor with monthly household income of less than RM3,000 to own houses priced in the range of RM100,000 to RM220,000. These young buyers will be approved for 100 percent financing for a repayment period of 30 years [17].

The development of social housing finance market remains the best solution for the government to aid the needy to own their own houses. Waqf housing has emerged as a viable alternative that could solve housing woes among the low and medium income groups. This is not a matter of real estate investment or real estate inheritance but the existing scenario of a high number of people living in deplorable conditions due to insufficient income calls for unconventional solutions.

The sign of a fully functional market is when median prices for the entire housing market is 3 times the median gross annual household income [15].

Figure 1 depicts household income distribution in Malaysia in 2014. Of the total, almost 20 percent earn below RM2,000 and only 5 percent own more than RM12,000 per month; reflecting very low home ownership affordability.

Figure 1: Distribution of household income by category, 2014

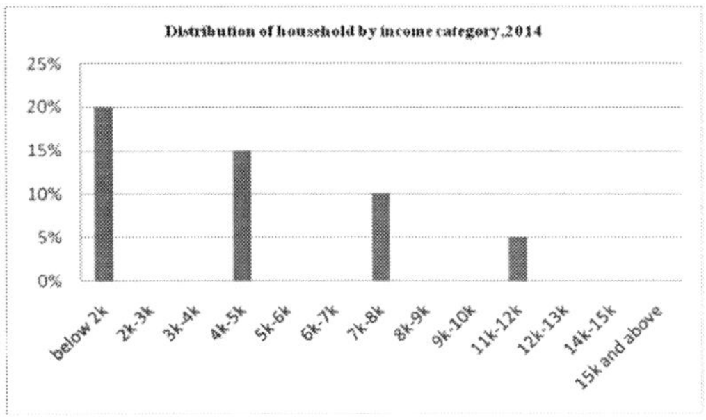

Source: Khazanah Research Institute (2015)

With all due respect, the government has implemented various affordable housing programs whether through joint ventures or through its agencies. Table 3 shows a list of affordable housing projects implemented by the government.

Table 3: Public affordable housing schemes in Malaysia

Affordable scheme	Range of offered house price (RM k)	Income range (RM k per month)
RMM Pulau Pinang	100-400	0-10
PR1MA	150-400	3-10
RMM SPNB	50-300	0-10
RUMAWIP	50-300	0-10
PPA1M	150-300	0-8
Rumahku Selangorku	100-250	0-8
DPR Johor	100-200	0-5
MyHome	80-180	3-5
RMM Sarawak	80-180	0-3

Source: Khazanah Research Institute (2015)

Table 3 above clearly reflects the government's commitment to aid affordable housing for the target groups. Sadly it is still insufficient as the target groups vying for such housing far exceed the numbers supplied by the government.

Hence the role played by waqf housing real estate must be optimised to achieve the aim of aiding home ownership among the low and medium income groups in Malaysia.

3 Research Objectives and Research Design

This study aims at investigating the inroads made by waqf real estate development specifically for housing in Malaysia. Kahf (1998) proposed that other social welfare activities via waqf must also be developed to become the main backbone for the third sector ie neither public nor private non-profit making sector [18]. This arose from previous studies on waqf which were more inclined towards waqf management and rules with little attention given to other aspects of waqf such as waqf accounting systems and dissemination of waqf reports to users.

In view of the epistemological variables relationship between waqf housing projects and their management, the effecticeness of waqf housing projects has been taken as a data collection strategy. Qualitative analysis of data obtained was utilised. Data includes legal sources such as original decrees, legal texts and intellectual articles explaining the concept of waqf rights utilisation for housing, organisational policy papers and guidelines, documentary records, and results of semi-structured face to face interviews with officials. This data is used to study waqf housing project thoughts and application.

Waqf institutions in Malaysia which are currently using waqf housing projects are selected, namely Penang State Religious Islamic Council owing to its success in creating the first Waqf Housing Scheme in Malaysia which is Taman Wakaf Seetee Aisah Seberang Jaya, Pualu Pinang. First, telephone interviews were conducted followed by detailed semi-structured face to face interviews. According to Robert (2008) semi-structured interviews are viewed as suitable for this type of study due to the flexibility exercised during the conversations [19]. The small study sample size is valid considering the respondents' expertise level on the subject matter as postulated by Sarah and Rosalind (2012). Interview and conversation sessions were recorded on tape recorders and field notes [20].

In reference to P.Mayring (2000), all recorded data is transferred into text. Transcripters often face difficulty in making out words uttered in text form due to word structures, use of citations, omissions, and misconstrued words or phrases by others [21]. Humans tend to converse at high speed intonations hence transcripters need to exercise care in transcripting conversations into text. Insertion of spacing or comma could change the overall meaning of a sentence. Therefore, every transcript need to be done carefully to reduce mistakes as much as possible (Nigel Mathers et. al, 2002) [22]

Thus the following five data analysis phases proposed by Yin (2011) are perceived as crucial. All data recorded electronically or otherwise is scrutinised and reread. Tape recorders were replayed and rearranged to enable critical information to be reassessed for the study findings. Rearrangement of data is required to ensure analysis of data objectives. Besides that interpretation is also used to understand information providers' validity, and to investigate any implied assumptions, suitable and construed as the respondent's rational views relating to the subject experiences [23].

Three steps are used to assess the effectiveness of the waqf housing project strategy: specific objectives determined by the waqf housing projects; organisational objectives; and target groups' needs. In this study however, the researcher measures effectiveness of waqf housing projects based on the successful construction of waqf housing projects within the advertised period, and how many members of the target groups benefits from the projects.

Since waqf housing projects are still novelty in Malaysia, the author has targeted waqf institutions that possess awareness and vissititude to experiment on this new product, as the measure of waqf housing project success. Penang SIRC for instance, had to overcome multitude of challenges to ensure the success of its pioneer project Taman Waqaf Setee Aisah; specifically in terms of sourcing out financing and finding a developer willing to develop a waqf housing project. The ability of Penang SIRC to attract various parties to participate in their waqf housing project can be used as benchmark besides the number of projects that has been implemented.

The author used the number of projects that has been successfully completed considering waqf housing project is a new product in the realm of waqf development in Malaysia, and not based on the investment value of the project. Paradoxically, this could lead to some presumptive statements.

4 Fund raising for waqf housing projects

Waqf institutions are now collecting funds through three means, including grants from the goverment, self financing, and loans from financial institutions. The first type of financial source is Federal Government Aminah and Tahir Grant (2015) [24].

H.Ahmed (2007) is of the opinion that financial institution can create funds and investments in the form of activities yielding future returns. In doing so, the financial institution faces challenges relating to problems and risks inherent in finnancing. Prior to financing decisions, the financier needs to identify the right projects to ensure returns on investment [25].

According to the Shariah Panel of Maybank Berhad Group, the cash waqf is allowed provided the waqf money is managed in such a way as to maintain its permanence characteristic. This can be done by investing the money in construction projects or using cash waqf in the purchase of any permanent or usable asset [26].

Waqf institutions need to augment their financial funding so that high impact projects can be realised. Funding is defined as financial grants by waqf institutions from their own coffers towards the development and management of waqf real estate, devoid of funding by other financial bodies.

4.1 Concept of al-ijarah on housing schemes on waqf lands in Malaysia

Al-ijarah refers to an agreement whereby a Waqf Institution (SIRC) leases a waqf land to another party at a rate agreed to by both parties [27]. By adhering to al-ijarah concept, the waqf institution indirectly generates income from the lease payments for lease terms 30, 60 or 99 years as set out under the National Land Code.

Al-ijarah as adopted in long term lease agreement (al-Hijr) infers that the developer has exclusive rights on the real estate for a long period by selling the rights to the long lease for a hefty one-time payment. The developer may then waqf lands using his own resources and at his own risk. At the expiry of the lease term, the waqf land will be surrendered to the waqf institution after the developer has enjoyed exclusive rights to develop and use said land model [27].

Upon completion of the building construction, the waqf institution will receive the building leasehold value for a specific term set out in the agreement. The owners or buyers of the completed housing units will receive the al-ijarah concept similar to the conventional leasehold land concept enshrined in the National Land Code (1965).

Buildings built on expired leasehold lands (usually 99 years lease for housing), will have to be renewed at certain values. These will become revenue to the waqf institution. These lands are not exempted from premium payment, the same as that charged on leasehold renewal lands under State Leases.

Waqf real estate ownership is not an issue as buyers only receive the benefit or right to use, not the right to own. Figure 2 explains the al-ijarah process:

Figure 2: Preliminary Framework of Al-Ijarah Concept

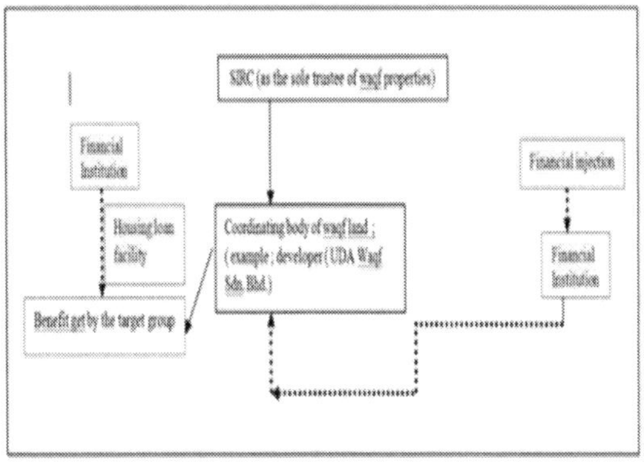

Source: Researcher (2016)

5 Implementation of housing schemes

The success of the Setee Aisah waqf housing scheme in 2003 has made Penang SIRC a competitive waqf institution in developing the local ummah economy. The success of Penang SIRC in forging collaboration with UDA Land Sdn. Bhd as well as with Al-Rajhi Bank and Bank Islam Malaysia Berhad has received deep appreciation among waqf activists. With the successful completion of the waqf project, UDA has made further inroads by establishing UDA Waqf Sdn. Bhd. which will also focus on waqf housing developments in other states. Financial institutions have also opened their eyes on the feasibility of waqf products on their cashflows. It is evident that waqf real estate developments are not third class status confined merely to mosques and cemeteries. In fact waqf real estate development may be made more comprehensive in line with the community needs.

5.1 Analysis on Interview Sessions with Waqf Officials

Analysis results of semi-structured interviews conducted by the author with 2 waqf officials of the Penang SIRC may be categorised by referring to Table 4 below. The author was not able to meet with the Waqf Chief Executive Officer who was then overseas on waqf business. The author interviewed these officials separately.

The questionnaire instrument used by the author was based on the author's prior field study on Terengganu SIRC [28]. Both respondents gave their full cooperation. The respondents did not have Islamic Studies background prior to starting his career with Penang SIRC. Nevertheless they have high motivation working in the waqf section. During the interviews, respondents 1 and 2 displayed important documents pertaining to waqf housing project for the author's analysis.

Respondents 1 and 2 were of permanent position status; but 6 other staffs were contract staffs. This would free them to find other more stable job opportunities. Respondent 1 has a high academic background at master's level even though his position at SIRC does not commensurate with his academic qualifications. Thus it is possible that Respondent 1 may leave his waqf position to look for a better higher paying job.

Respondent 1 dan 2 were of the opinion that valuation of waqf real estate done by the Valuation and Real estate Services Department, Ministry of Finance should not be charged fees as waqf real estate is charitable in nature.

As a result of interviews with both Respondent 1 and 2 who gave full cooperation, the author was able to review all documents pertaining to waqf housing schemes. They also helped the author identify 9 waqf housing projects which have been completed, in the process of construction, and will be developed.

Table 4: Analysis of Interviews with SIRC Waqf Officials

Questioning Theme	Respondent 1	Respondent 2
1) Work Experience	Health Department (Grade 41)	Fresh graduate
2) Position in Waqf Institution	Assistant Administrative Officer	Assistant Administrative Officer
3) Status	Permanent	Permanent
4) Education Background	Master in Business Management	STPM
5) Experience	1 year	4 years
6) Field of Work	Planning on waqf land	Real estate management on Waqf assets
7) Opinion on waqf housing	Good performance, many projects	Good performance, many projects
8) How many staffs in Waqf Department ?	9 staff , inadequate	9 staff, inadequate
9) Issues?	Abandoned surau	Waqf Real Estate Arrears involving RM200K
10) Challenges in handling waqf land ?	Declarations of endowment deceased is unacceptable for proving endowment	Unpaid rents
11) Court Case	The heirs do not want to leave the waqf land. There are 13 of them.	We always win court cases although the heirs claimed it.
12) Does valuation being carried out ?	No, the fee is too expensive.	Yes, to identify the rent
13) Developed waqf project	Excessive	Considerable
14) Reason of delayed project	Financial problem	Financial problem

5.2. List of projects under Penang SIRC

Table 5 shows completed housing scheme projects as well as discussion-level projects implemented by Penang SIRC. These housing project activities are included under the 9 projects identified earlier. A project located in Teluk Ayer Tawar is under construction and expected to be fully completed in 2018 offering 8 storey and 12 storey apartment blocks comprising 152 apartment parcels.

Table 5: List of projects under Penang SIRC

Projects under Penang SIRC For Housing / Residential Units	
1) Project	Taman Wakaf Seetee Aisah Seberang Jaya
Total unit	76 units double storey terrace houses
Year to be completed	2013
Total Cost of Project (RM)	21 million
Project status	Completed
Types of Agreement	(Ijarah) Al Rajhi Bank & Bank Islam Malaysia
Land area (acre)	9.86
2) Project	Medium Cost Apartments, Abdul Cauder, Jalan Mengkuang, Seberang Prai Utara
Total unit	(upon discussion)
Year to be completed	(upon discussion)
Total Cost of Project (RM)	Targeted Cost 20 million
Project status	Proposal
Types of Agreement	Ijarah
Land area (acre)	2.625
3) Project	Medium Cost Apartments, Wakaf Siti Bt. Abdul Rahman on lots 1093 & 1094, Sungai Nibong, Mukim 12, DBD, Pulau Pinang
Total unit	77 parcels (14 storeys)
Year to be completed	Targeted in 2016
Total Cost of Project (RM)	9 million (land value) Construction cost (upon discussion)
Project status	(upon discussion) Planning
Types of Agreement	Ijarah
Land area (acre)	1.571

Projects under Penang SIRC For Housing / Residential Units	
3) Project	Medium Cost Apartments, Lot 263, Batu Maung, Mukim 12, Barat Daya, Pulau Pinang
Total unit	150 parcels (15 storeys)
Year to be completed	2018
Total Cost of Project (RM)	50 million
Project status	Under Construction
Types of Agreement	Ijarah
Land area (acre)	5.5
4) Project	Medium Cost Apartments, Wakaf Muhammad Hashim bin Haji Abdullah, Lot 192, Seberang Jaya, Mukim 1, SPT
Total unit	68 parcels (11 storeys)
Year to be completed	(upon discussion)
Total Cost of Project (RM)	(upon discussion)
Project status	proposal
Types of Agreement	Ijarah
Land area (acre)	1.142
5) Project	Medium Cost Apartments, 15 Storeys, Half of the lot, Lot PT 146, Balik Pulau, Mukim G, Daerah Barat Daya, Pulau Pinang.
Total unit	200 parcels
Year to be completed	2018
Total Cost of Project (RM)	50 million
Project status	Under Construction
Types of Agreement	Ijarah
Land area (acre)	11
6) Project	Medium Cost Apartments and low cost flats, Taman Wakaf Seetee Aisah (Fasa 2) , Wakaf Seetee Aisah, Lots 200, 252 & 3117, Seberang Jaya, Mukim 4, SPT
Total unit	(upon discussion)
Year to be completed	(upon discussion)
Total Cost of Project (RM)	(upon discussion)
Project status	Proposal
Types of Agreement	Ijarah
Land area (acre)	7.322
7) Project	Medium Cost Apartments, Wakaf Shekh Eusoff, on Lots 1396 & Lot 1606, Jalan Air Itam, Sek. 5, DTL Pulau Pinang
Total unit	(upon discussion)
Year to be completed	(upon discussion)
Total Cost of Project (RM)	(upon discussion)
Project status	proposal
Types of Agreement	Ijarah
Land area (acre)	14.97
8) Project	Medium Cost Apartments, Lots 6247 & 6249, Teluk Air Tawar, Mukim 7, SPU, Pulau Pinang
Total unit	152 parcels (8 storeys and 12 storeys) 2 blocks
Year to be completed	2018
Total Cost of Project (RM)	50 million
Project status	Under Construction
Types of Agreement	Ijarah
Land area (acre)	5.525

5.3. Waqf Housing Aimed at Affordablity

Affordable housing in Penang, especially on the island is very limited. However once waqf housing development commences, low and medium income groups may have their shot at home ownership on a prestigious real estate on Penang Island.

Initially, some financial institutions were unwilling to take risks becoming involved in waqf real estate developments. This is because they were unclear on the 'ingredients' of waqf. The synergy between banks and waqf developments in the Arab states is very close. Hence, Al Rajhi Bank grabbed the opportunity to finance Taman Wakaf Seetee Aisah Seberang Jaya when they received the offer to help. Due to rapid urbanisation of the Georgetown City Centre, all proposed housing developments are only approved as multi storey housing developments such as apartments. Although the first waqf housing scheme in Penang was developed as double storey houses, the focus now is more on medium cost apartments to ensure they benefit greater number of target group members.

6 Discussions and conclusion

Waqf is the problem-solver to balance community development but Malaysia is a multi-racial country and the concentrations of any one ethnic group are not in accordance with the 1Malaysia Agenda. Eventually, the movement towards waqf development could solve the imbalance in urban housing developments. The main thrust of the Malaysian National Housing Policy is to provide accessible, adequate, affordable quality housing for all Malaysians especially the low income groups. In 1982, the government enforced the 30 percent quota of low cost housing on private developers as social service responsibility to complement the Government's commitment to provide affordable housing. Housing projects on lands exceeding 5 acres must allocate at least 30 percent of their housing units as low cost housing (Datuk Eddy Chen Lok Loi - REHDA Malaysia)

The concept of waqf housing development with a new and innovative framework, specifically in the use of affordable materials and construction techniques, was given emphasis. Indirectly, aspects of life quality enhancement and maintaining sustainability of waqf real estate are created. Thus, the agenda to create comfortable and affordable housing will be acheieved.

The involvement of market-established developers in waqf projects is perceived as beneficial in enhancing their corporate image in terms of Social Corporate Responsibility (CSR). Waqf housing projects could greatly aid in eradicating low purchasing power among the needy groups so that it will automatically improve community social mobility.

Taman Wakaf Seetee Aisah Seberang Jaya Pulau Pinang is not the first housing project for UDA Holdings Berhad. UDA's Land Department has developed a sizable number of real estate in cities around the country earmarked to fulfil the national housing needs across income spectrums. Some of these are Bandar Tun Hussein Onn, Cheras, Selangor; Bandar UDA Utama, Pulai, Johor; Bandar Pauh Jaya, Permatang Pauh, Pulau Pinang; Tanjung Tokong, Pulau Pinang; Bandar Baru UDA, Johor Bahru, Johor; Pusat Bandar Tampoi, Johor Bahru, Johor; and Skudai Kanan, Johor Bahru, Johor.

The involvement of financial institutions and developers in waqf projects has convinced them to utilise waqf lands efficiently. The good performance on waqf projects is very much dependent on the full commitment at the best level from every party involved in waqf development, specifically the government, the public throughout the period.

Acknowledgement: Funding Sources

This research was supported by Universiti Teknologi MARA and received funding from the Department of Higher Education, Ministry of Higher Education (MOHE) which is under the Dana Pembudayaan Penyelidikan (RAGS) GRANT (RAGS/2013/UiTM/SSI11/

References

[1] Kais Dukes, "Sunan an-Nasa'i 3691," *Book 32, Hadith 4*. [Online]. Available: http://sunnah.com/nasai/32/4. [Accessed: 05-Feb-2016].

[2] M. Dr. Uswatun Hasanah, "Inovasi Pengembangan Wakaf di Berbagai Negara," *Artikel*, 2008. [Online]. Available: http://bwi.or.id/index.php/in/publikasi/artikel/222-inovasi-pengembangan-wakaf-di-berbagai-negara. [Accessed: 04-Feb-2016].

[3] B. Yediyyildiz, "Place of the Waqf in Turkish Cultural System," *1996*, 1996. [Online]. Available: http://yunus.hacettepe.edu.tr/~yyildiz/placeofthewaqf.htm. [Accessed: 04-Feb-2016].

[4] Quran.com, "QURAN.COM. ALL RIGHTS RESERVED 2015," 2015. [Online]. Available: http://quran.com/2. [Accessed: 05-Feb-2016].

[5] S. A. Gapor, N. M. Malik, and A. Husin, "Trends , Problems and Needs of Urban Housing in Malaysia," *Int. J. Soc. Educ. Econ. Manag. Eng.*, vol. **5**, no. 2, pp. 37–41, 2011.

[6] M. I. A. Mohsin, "Financing through cash-waqf: a revitalization to finance different needs," *Int. J. Islam. Middle East. Financ. Manag.*, vol. **6**, no. 4, pp. 304–321, 2013.

[7] M. I. A.Mohsin, "Waqf-shares : new product to finance old waqf real estate," *Bank Bank Syst.*, vol. **7**, no. 2, pp. 72–78, 2012.

[8] A. Hassan, "MANAGEMENT AND DEVELOPMENT OF THE MANAGEMENT AND DEVELOPMENT OF THE," in *Seventh International Conference – The Tawhidi Epistemology: Zakat and Waqf Economy , Bangi 2010*, 2010, pp. 309–328.

[9] T. S. Ong, "Factors Affecting the Price of Housing in Malaysia," *J. Emerg. Issues Econ. Financ. Bank.*, vol. **1**, no. 5, pp. 414–429, 2013.

[10] Z. Hashim, "House price and affordability in housing in Malaysia," *Akademika*, vol. **78**, pp. 37–46, 2010.

[11] D. Özdemir, "The Role of the Public Sector in the Provision of Housing Supply in Turkey, 1950-2009," *Int. J. Urban Reg. Res.*, vol. **35**, no. November 2011, pp. 1099–1117, 2011.

[12] V. Reina and J. Begley, "Reprint of 'will they stay or will they go: Predicting subsidized housing opt-outs,'" *J. Hous. Econ.*, vol. **24**, pp. 122–137, 2014.

[13] www.iMoney.my, "Malaysia Budget 2016 : Can You Afford A House Now ?," *24/10/2015*. [Online]. Available: http://english.astroawani.com/malaysia-news/budget-2016-can-you-afford-house-now-77733. [Accessed: 06-Feb-2016].

[14] M. Department of Statistics, "Average Monthly Income of Malaysian Households At," 2014.

[15] Dr-Suraya-Ismai, "Making Housing Affordable," 2015.
[16] M. S. Suhaida, N. M. Tawil, N. Hamzah, a I. Che Ani, and M. M. Tahir, "A Conceptual Overview of Housing Affordability in Selangor , Malaysia," *Int. J. Soc. Hum. Sci. Eng.*, vol. **4**, no. 12, pp. 287–289, 2010.
[17] M. S. Suhaida, N. M. Tawil, N. Hamzah, A. I. Che-Ani, H. Basri, and M. Y. Yuzainee, "Housing Affordability: A Conceptual Overview for House Price Index," *Procedia Eng.*, vol. **20**, pp. 346–353, 2011.
[18] M. Kahf, "FINANCING THE DEVELOPMENT of AWQAF REAL ESTATE," IRTI, Kuala Lumpur, Malaysia, 1998.
[19] Robert Wood Johnson Foundation, "Developing the interview guide," *Citation: Cohen D, Crabtree B. "Qualitative Research Guidelines Project." July 2006. http://www.qualres.org/HomeInte-3595.html*, 2008. [Online]. Available: http://www.qualres.org/HomeInte-3595.html. [Accessed: 06-Feb-2016].
[20] S. E. Baker, R. Edwards, P. Adler, H. S. Becker, and A. Doucet, "How many qualitative interviews is enough ?," *national Centre for Research Method*, 20121. [Online]. Available: eprints.ncrm.ac.uk/2273/4/how_many_interviews .pdf. [Accessed: 06-Feb-2016].
[21] P. Mayring, "Qualitative Content Analysis," *2000*, vol. **1**, no. 2, 2000.
[22] N. Fox and A. HunnHunn, Amanda, "Trent Focus for Research and Development in Primary Health Care Using," 2002.
[23] ROBERT K. YIN, *QUALITATIVE RESEARCH FROM START TO FINISH*, 9 8 7 6 5 ., vol. **53**, no. 9. New York, 2011.
[24] A. B. Mohsin, M. Tahir, and S. Hj, "THE EFFECTIVENESS OF SAHAM WAQF ON ADEQUACY 2 FUNDS FOR," *Int. J. Real Estate Stud.*, vol. **9**, no. 2015, p. 33, 2015.
[25] H. Ahmed, "Waqf -Based Microfinance: Realizing The Social Role of Islamic Finance 'Integrating Awqaf in the Islamic Financial Sector,'" in *Integrating Awqaf in the Islamic Financial Sector*, 2007, pp. 1–22.
[26] M. Ramadhan, F. Ellias, A. F. Kadir, and M. N. Haron, "Parameter syariah untuk wakaf tunai."
[27] M. N. A. Zulkifli Hasan, "The Investment Of Waqf Land As An Instrument Of Muslims' Economic Development In Malaysia," in *Dubai International Conference on Endowments' Investment, 4-6th February 2008*, 2008, pp. 1–25.
[28] S. Nadiah, M. Ali, A. Halim, M. Noor, N. Nazihah, N. Rashidah, P. Ismail, F. Dahlan, M. Saifol, and A. Shari, "Integrating Awqaf and Zakat : A Case Study of Land Development For The Poor in State Islamic Religious Councils Terengganu , Malaysia," in *International Halal Conference 2014 - INHAC 2014*, 2014.

Sustainability assessment, rating systems and historical buildings Case study: Rehabilitated construction in a university site

Somayeh Sadrykia [1,a], Leila Medghalchi[2] Mohammadjavad Mahdavinejad[3]

[1]*PhD. Student, Faculty of Art and Architecture, Tarbiat Modares University, Tehran, Iran*
Lecturer, Faculty of Architecture, Tabriz Islamic Art University, Tabriz, Iran,
[2]*Assistant Professor, Faculty of Architecture, Tabriz Islamic Art University, Tabriz, Iran,*
[3]*Associate Professor, Faculty of Art and Architecture, Tarbiat Modares University, Tehran, Iran*
somayeh.sadrykia@modares.ac.ir

Abstract. This paper explores the relationship between the indicators and different factors that "rating systems for green projects" concentrates on, and principles and factors considered in the rehabilitation of historical buildings. In recent years, different methods and systems concerned and improved for assessing environmental sustainability. LEED (Leadership in Energy and Environmental Design) and BREEAM (Building Research Establishment (BRE) Environmental Assessment Method) are two most commonly used rating systems, established in U.S and UK. These systems comprise some categories and different factors to achieve environmentally responsible design. Firstly, this study focuses on the list of rating systems indicators and criteria. Secondly this paper investigates a historical rehabilitated building in the site of Tabriz Art University, as a case study and has tried to compile its green design elements. Finally, this work intends to compare mentioned elements with indicators and factors of building rating systems. Findings of the study revealed that "Materials and Resources", "indoor environmental quality" and also "Sustainable Sites" ,the most significant indicator of rating systems, had major and important role in the rehabilitation of the building. Beyond this materials' life cycle was considerable in construction.
Keywords: Rating Systems, Materials Life cycle, Indicators, LEED, BREEAM, Rehabilitation

1 Introduction

The term sustainable development can be described as enhancing quality of life and thus allowing people to live in a healthy environment and improve social, economic and environmental conditions for present and future generations. Since the world commission on environment and development (WCED), entitled Our Common Future (1987), sustainable development has gained much attention in all nations and a report was published with a classic definition of sustainable development as: "Development meeting the needs of the present without compromising the ability of future generations to meet their own needs" [1].

The improving social, economic and environmental indicators of sustainable development are drawing attention to the construction industry, which is a globally emerging sector, and a highly active industry in both developed and developing countries [2].

In construction sector energy consumption, co2 emission, waste management, pollution, environmental impact and resources and materials depletion are considered for sustainable design and building sustainability. The importance of building assessment in reaching "sustainable design" and "green design" has led to the establishment of assessment methodologies which evaluate buildings environmental impact and predict and measure the success of a building or development against sustainability criteria. "assessment methodologies were designed to calculate specific criteria such as thermal loads, energy or water consumption or life cycle impact of product manufacture"[3]. An appropriate and most common used assessment rating systems, such as LEED and BREEAM, makes a very useful framework for sustainable design which comprises different categories and issues and lead to rating of the different types of building. It seems that these categories and criteria can be used in assessing and evaluating historical buildings too.

So, the aim of this research is to study and investigate on the sustainability assessment and rating systems' categories and issues in one hand, and the green elements considered in the generation and rehabilitation of historical buildings, on the other hand, to conclude and outline the key joint and common elements, and categories. Hence the challenge that this study addresses is to answer two research questions:

Q1: Which one of LEED and BREEAM categories and credits are used and considered in historical and rehabilitated buildings' greening?

Q2: which one of the mentioned categories and issues have the most important role in the rehabilitated case of this study?

[a] Corresponding author: *somayeh.sadrykia@modares.ac.ir*

2 Building Assessment and Rating Systems

In the early 1960s, concerns over the limitations of raw materials and energy resources, led to the development of the first exercises in life cycle assessment. So a number of methods and systems established and developed to assess and measure the success of building against environmental or sustainability criteria. These different assessment tools for the construction sector can be used in different design stages.

In the UK, Building Research Establishment (BRE) launched its Environmental Assessment Method (BREEAM) in 1990, which predict and measure the performance of building by calculating specific criteria [3]. In the US, Green building council established the Leadership in Energy and Environmental Design (LEED), in 1998 as a rating tool, with the same roots as BREEAM .After those different versions of LEED was launched and used. Beyond these two, some other similar sustainability assessment tools and methods launched in different area all over the world. Table 1.

Table 1. Examples of most common sustainability assessment methodologies and rating systems [3].

Rating Systems	Country	Year
BREEAM Building Research Establishment (BRE) Environmental Assessment Method	United Kingdom	1990
LEED Leadership in Energy and Environmental Design	United State	1998
Green Globes	Canada	2000
Green Star	Australia	2003
NABERS National Australian Building Environmental Rating Scheme	Australia	2000
HK-BEAM Hong Kong Building Environmental Assessment Method	China	1996
CASBEE Comprehensive Assessment System for Environmental Efficiency	Japan	2001
HQE Haute Qualite Environmentale	France	2005

2.1 BREEAM

In order to provide a method by which environmental impacts can be combined in a single measure, BRE developed BREEAM (Building Research Establishment Environmental Assessment Method). It designed and covered a variety of different scenarios as table 2 [3].

Table 2: Building types assessed by BREEAM

Rating system	Building type								
BREEAM	Office	Retail	Industrial	Education	Healthcare	Residential	Courts	Prisons	Existing Buildings

2.1.1 Project Certification

BREEAM sets the standard for best practice in sustainable building design, construction and operation and has become one of the most comprehensive and widely recognized measures of a building, environmental performance, which are set against established benchmarks, to evaluate a building's specification, design, construction and use. The measures used represent a broad range of categories and criteria from material, energy to ecology. They include aspects related to energy and water use, the internal environment (health and well-being), pollution, transport, materials, waste and management processes. BREEAM rating benchmarks is obvious in table 3.

Table 3. BREEAM rating benchmarks [4]

BREEAM Rating	Score (%)
Out standing	85
Excellent	70
Very Good	55
Good	45
Pas	30
Unclassified	<30

2.1.2 Categories and criteria of BREEAM rating system

Different categories and criteria, which BREEAM based on, are mentioned, in the Table 4.

Table 4. Scoring and rating BREEAM assessed buildings(www.breeam.org)

Categories	Issues (credits)
Management	Sustainable procurement Responsible construction practices Construction site impacts Stakeholder participation Life cycle cost and service life planning
Health and well-being	Visual comfort Indoor air quality Thermal comfort Water quality Acoustic performance Safety and security
Energy	Reduction of emissions Energy monitoring External lighting Low and zero carbon technologies Energy efficient cold storage Energy efficient transportation systems Energy efficient laboratory systems Energy efficient equipment Drying space
Transport	Public transport accessibility Proximity to amenities Cyclist facilities Maximum car parking capacity Travel plan
Water	Water consumption Water monitoring Water leak detection and prevention Water efficient equipment
Materials	Life cycle impacts Hard landscaping and boundary protection Responsible sourcing of materials Insulation Designing for robustness Reuse of faced and façade and structure
Waste	Construction waste management Recycled aggregates Operational waste Speculative floor and ceiling finishes
Land Use and Ecology	Site selection Ecological value of site and protection of ecological features Mitigating ecological impact Enhancing site ecology Long term impact on biodiversity
Pollution	Impact of refrigerants NOx emissions Surface water run off Reduction of night time light pollution Noise attenuation
Innovation (additional)	

2.2 LEED

LEED (Leadership in Energy and Environmental Design) is a certification program which is intended to define high quality, ecological building methods for more environmentally-friendly and profitable buildings [5]. LEED has the same roots as BREEAM but is administered by the US Green Building Council. Although the headings are different to BREEAM the issue covered are similar.

There are versions of LEED for New Construction commercial buildings (LEED NC), Existing buildings, Commercial Interiors, Core and Shell and Retail. LEED for homes, which was launched in 2008, is designed for application to housing developments of all sizes, ranging from individual houses to high density apartment schemes. Fig1. There is also a version for Neighborhood Development that was launched in 2009 and has similarities to the BREEAM [3].

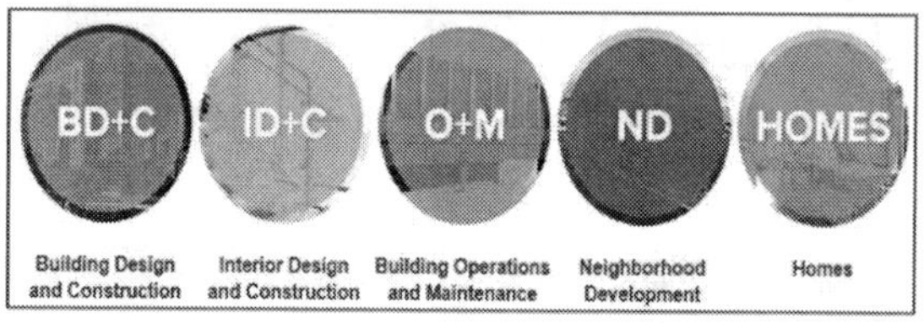

Fig1. Building types in LEED (http://www.usgbc.org/leed)

2.2.1 Project Certification

LEED covers all ecological and socio-cultural aspects of sustainability. It focuses mainly on energy and water efficiency, the reduction of co2 emissions, a healthy and comfortable indoor climate and conservation of resources, assesses the building operations as well as the quality of site [6].

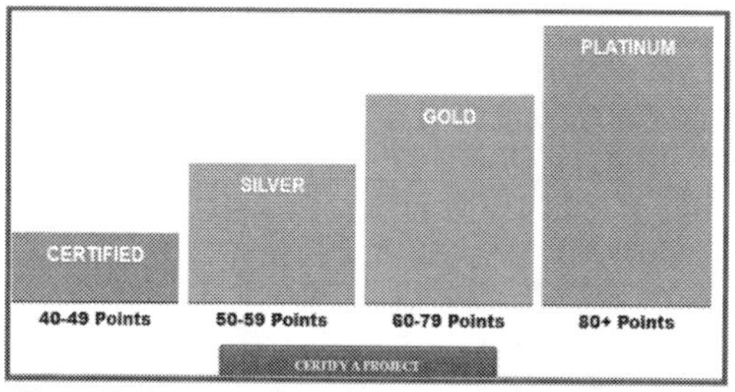

Fig 2. Ratings are awarded on a scale from certified to platinum through silver and gold (http://www.usgbc.org/leed)

The LEED certification is based on a point system, points are awarded for the fulfilment of individual credits. Fig2. The system divides the predetermined credits into seven categories. The weighting of the categories differs slightly between the rating systems. Apart from the minimum number of points required for the level of LEED certification aimed at, there are so called prerequisites which must be met [6].

The maximum number of points attainable depends on the project type for e.g., LEED-NC (new construction) the maximum points achievable is 69 while for LEED-EB (existing building) it is 92 [7]. In relation to this, the certification levels LEED certification, silver, gold, and platinum are awarded [5]. Certification is based on an assessment on the completed building.

2.2.2 Categories and criteria of LEED rating system

Different categories and criteria, which LEED based on, are mentioned, in Table 5.

Table 5. Scoring and rating LEED assessed buildings.

LEED 2009 for New Construction and Major Renovations Project Checklist [10], [11].

Categories	Issues (credits)
Sustainable Sites	*Construction Activity Pollution Prevention (Prerequisite 1)* Site Selection Development Density and Community Connectivity Brownfield Redevelopment Alternative Transportation Site Development Storm-water Design—Quantity Control Heat Island Effect—Non-roof Light Pollution Reduction
Water Efficiency	*Water Use Reduction (Prerequisite 1)* Water Efficient Landscaping Innovative Wastewater Technologies Water Use Reduction
Energy and Atmosphere	*Fundamental Commissioning of Building Energy System (Prerequisite 1)* *Minimum Energy Performance (Prerequisite 2)* Fundamental Refrigerant Management (Prerequisite 3) Optimize Energy Performance On-site Renewable Energy Enhanced Commissioning Enhanced Refrigerant Management Measurement and Verification Green Power
Materials and Resources	Storage and Collection of Recyclables (Prerequisite 1) Building Reuse Construction Waste Management Materials Reuse Recycled Content Regional Materials Rapidly Renewable Materials Certified Wood
Indoor Environmental Quality	Minimum Indoor Air Quality Performance (Prerequisite 1) Environmental Tobacco Smoke (ETS) Control (Prerequisite 2) Outdoor Air Delivery Monitoring Increased Ventilation Construction Indoor Air Quality Management Plan Low-Emitting Materials Indoor Chemical and Pollutant Source Control Controllability of Systems Thermal Comfort—Design Daylight and Views
Innovation in Design	Innovation in Design LEED Accredited Professional
Regional Priority	Regional Priority
Possible Points: 110	
100 base points; 6 possible Innovation in Design and 4 Regional Priority points Certified Silver Gold Platinum	 40–49 points 50–59 points 60–79 points 80 points and above

3 Historic Preservation

Human connection to nature, community, and place is a necessary element of healthy living and the ability to achieve sustainability. Connectivity to nature is now defined as bio philia and widely celebrated in sustainable design. Built responses to the natural world must vary by climatic zone and conditions unique to each area. So, too, must the memory of place, which requires that buildings and heritage areas from past generations and cultures be retained to avoid creating communities so lacking in personality that it is impossible to identify geographic location without signage [8].

The act of preserving a site, a building, or a community is a gesture of respect for past decisions and a gift to future generations who may enjoy or be curious about the physical manifestation of different times and cultures as Randall Mason said: "Nearly any way the effects are measured, be they direct or indirect, historic preservation tends to yield significant benefits to the economy." (Ibid) By the way, in this research, one building of an educational complex (that was rehabilitated) is selected as a case study and intended to compare mentioned elements with indicators and factors of building rating systems such as LEED.Table 6

Table 6 : Rating Systems Categories Comparing with Green Design Elements

	LEED Categories	Green Design Elements [9]	Case study	Categories and criteria of BREEAM rating system in Case study
1	**Sustainable Sites (SS)**			
	Construction Activity Pollution Prevention Site Selection Development Density and Community Connectivity Brownfield Redevelopment Alternative Transportation—Public Transportation Access Alternative Transportation—Bicycle Storage and Changing Rooms Alternative Transportation—Low-Emitting and Fuel-Efficient Vehicles Alternative Transportation—Parking Capacity Site Development—Protect or Restore Habitat Site Development—Maximize Open Space Stormwater Design—Quantity Control Stormwater Design—Quality Control Heat Island Effect—Non roof Heat Island Effect—Roof Light Pollution Reduction	Site selection: Reuse of existing buildings Sustainable site: Redevelopment of previously historical buildings Public transportation proximity and access Xeriscaping Bioswales Greenroof Brownfield site Permeable paving materials/permeable pavement layers Bycle accomidation Counstructed wetland Urban site/walkable Shading trees Natural habitat restoration Reduse site disturbance to preserve trees Stormwater management Native and drought-tolerant plants	Site selection: Reuse of existing buildings Sustainable site: Redevelopment of 6 previously historical buildings Public letransportation proximity and access Urban site/walkable Shading trees Permeable paving materials/permeable pavement layers Native plants Xeriscaping by green courts 	Construction site impacts Indoor air quality Thermal comfort Water quality Proximity to amenities Public transport accessibility Safety and security Water efficient equipment Hard landscaping and boundary protection Reuse of faced and façade and structure
2	**Water Efficiency (WE)**			
	Water Use Reduction Water Efficient Landscaping Innovative Wastewater Technologies	Graywater System/ rainwater harvesting system On site well Low – Follow Plumping Fixture Elimination of unneeded Fixtures Septic system Self -closing water faucets Water Use Reduction Water efficient landscaping	On site well Water Use Reduction by over 70 percent Water efficient landscaping (semi planned)	
3	**Energy and Atmosphere (EA)**			
	Fundamental Commissioning of Building Energy Systems Minimum Energy Performance Fundamental Refrigerant Management Optimize Energy Performance On-site Renewable Energy Enhanced Commissioning Enhanced Refrigerant Management Measurement and Verification Green Power	Energy -Efficient Windows High-efficiency HVAC system Natural ventilation/ Natural cross ventilation Occupant and photoelectric sensors Renewable energy certification Ground-source heat pumps Automatic light sensors Photovoltaic panel system Thermal massing	Energy -Efficient Windows Natural ventilation	
4	**Materials and Resources (MR)**			
	Storage and Collection of Recyclables Building Reuse—Maintain Existing Walls, Floors and Roof Building Reuse—Maintain Existing Interior Nonstructural Elements Construction Waste Management Materials Reuse Recycled Content Regional Materials Rapidly Renewable Materials Certified Wood	Nontoxic paint, carpet, adhesive and sealants Terrazzo flooring with recycle glass Toilets partitions made from pop bottles Flooring made from tires Solid surface plastic with recycled content Rapidly renewable materials Cork baseboards Reclaimed/ recycled materials New materials harvested locally Local materials (brick, stone, ...) Locally manufactured materials Durable low- maintenance materials Existing -materials restoration / reuse Reclaimed materials from other	Reclaimed/ recycled materials Existing -materials restoration / reuse Reclaimed materials from other buildings Recycled materials were used for installation in the walls, floors and roofs Restoration of existing wood doors, trims, window casements and floors	

		buildings Use of salvaged materials Recycled materials were used for installation in the walls, floors and roofs Restoration of existing wood doors, trims, window casements and floors	 Reused materials in the case study	
5	**Indoor Environmental Quality (IEQ)**			
	Minimum Indoor Air Quality Performance Environmental Tobacco Smoke (ETS) Control Outdoor Air Delivery Monitoring Increased Ventilation Construction Indoor Air Quality Management Plan—During Construction Construction Indoor Air Quality Management Plan—Before Occupancy Low-Emitting Materials—Adhesives and Sealants Low-Emitting Materials—Paints and Coatings Low-Emitting Materials—Flooring Systems Low-Emitting Materials—Composite Wood and Agrifiber Products Indoor Chemical and Pollutant Source Control Controllability of Systems—Lighting Controllability of Systems—Thermal Comfort Thermal Comfort—Design Thermal Comfort—Verification Daylight and Views—Daylight Daylight and Views—Views	Space filled with natural light Daylighting and view Operable windows Skylight/abundant daylighting Transparent/ trenslucentglazing (interior and exterior) Low-VOC* materials and finishes Biophilic design –connection to nature	Space filled with natural light Daylighting and view Operable windows Interior windows allow daylight to flow deeper into the building	

* Volatile organic compounds (**VOCs**) are organic chemicals that have a high vapor pressure at ordinary room temperature. Their high vapor pressure results from a **low** boiling point, which causes large numbers of molecules to evaporate or sublimate from the liquid or solid form of the compound and enter the surrounding air.

4 Case study

Name: Ganjeh-i-zadeh House
Location: Maghsoudieh Street .Tabriz. IRAN.

This house is renovated in their original design to host the Faculty of Architecture and Urban Design. Its eastern section has three floors and a vaulted basement .Fig4. The architecture layout of the western component of this three- story building is extroverted. The vaulted basemen housed the kitchen, storage and cistern. Its ground floor on the east and west side has a staircase, one columned ivan and six rooms. The largest ceremonial hall of this structure is located the first floor with span of over 7 m. covered with wooden trusses [9].

This house is renovated in their original design to host the Faculty of Architecture and Urban Design. Its eastern section has three floors and a vaulted basement .Fig4. The architecture layout of the western component of this three- story building is extroverted. The vaulted basemen housed the kitchen, storage and cistern. Its ground floor on the east and west side has a staircase, one columned ivan and six rooms. The largest ceremonial hall of this structure is located the first floor with span of over 7m covered with wooden trusses"(IBid).

Fig 3. The eastern part of building - built in the Qajar period

DOI: 10.1051/matecconf/20166600069

Fig 4.The western part of building - built in the Pahlavi period

4.1 Case study Analysing

Site Utilization and sustainable site

- Reuse of existing buildings, redevelopment of 6 previously historical buildings
- Xeriscaping: plantings in courtyards to utilize plant schemes those are drought-tolerant and native to an area.
- The L-shape of the building was intentionally chosen to create a courtyard for community gathering.
- South-facing building sunspace achieves maximum solar exposure during the winter and during the summer is protected by porch and trees.Fig5.

Alternative Transportation

- The building site is close to public transportation systems (including bus and taxi stations).Fig5.
- There are different accommodations near the complex for students' daily needs that cause less traffic (walkability).

Materials and Resources (MR)

- Durable, long-lived materials and composite durability of a construction system such as a masonry wall are common in many historic buildings.

  Material selection was influenced by local availability. Brick products, stone all came near the project.
- The design strategy sought the lowest possible environmental impact by seeking salvaged materials (like unused brick and wood), and new materials from local sources.
- The design combines visible celebration of historic elements and reuse of materials to intentionally establish a connection to the past.
- Reused materials include salvaged wooden trusses that were removed from several floors to support the ceiling (reuse of salvaged building components).

Indoor Environmental Quality (IEQ)

- Daylighting.
- Interior windows allow daylight to flow deeper into the building: During the day, most of the occupied areas of the building are flooded with natural light from the window openings.
- Stablished skylights bring natural light to the interior of the building.Fig6.

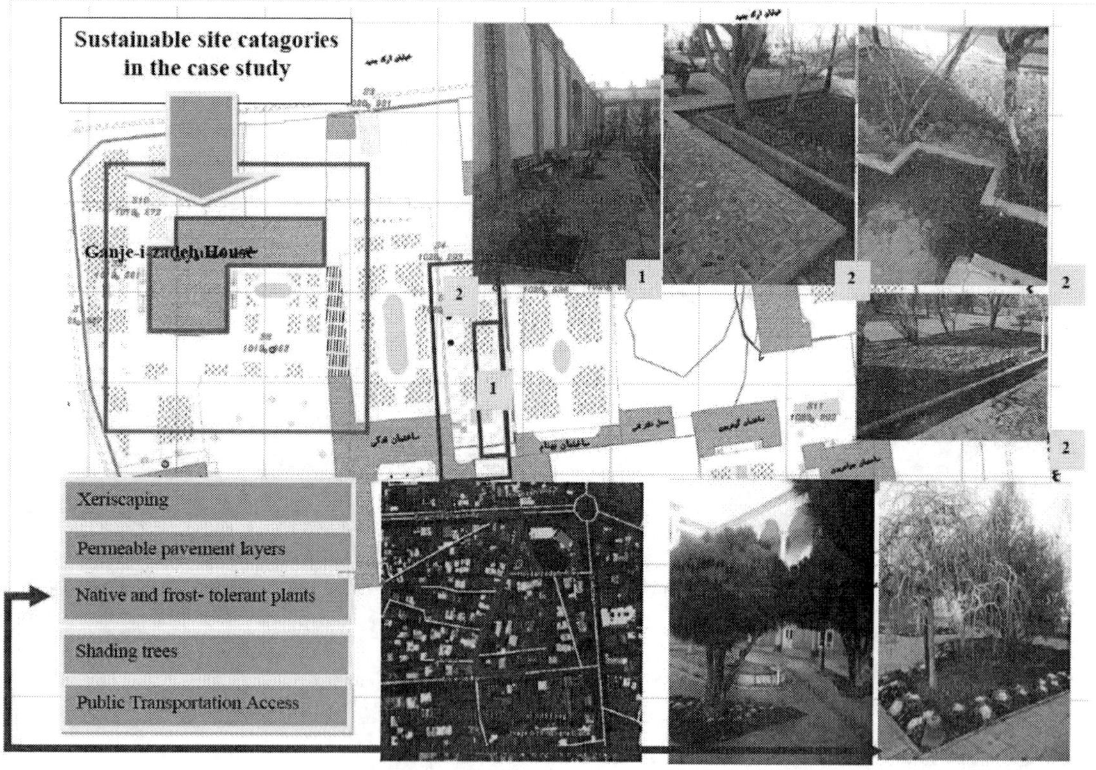

Fig5 .Sustainable site catagories in the case study

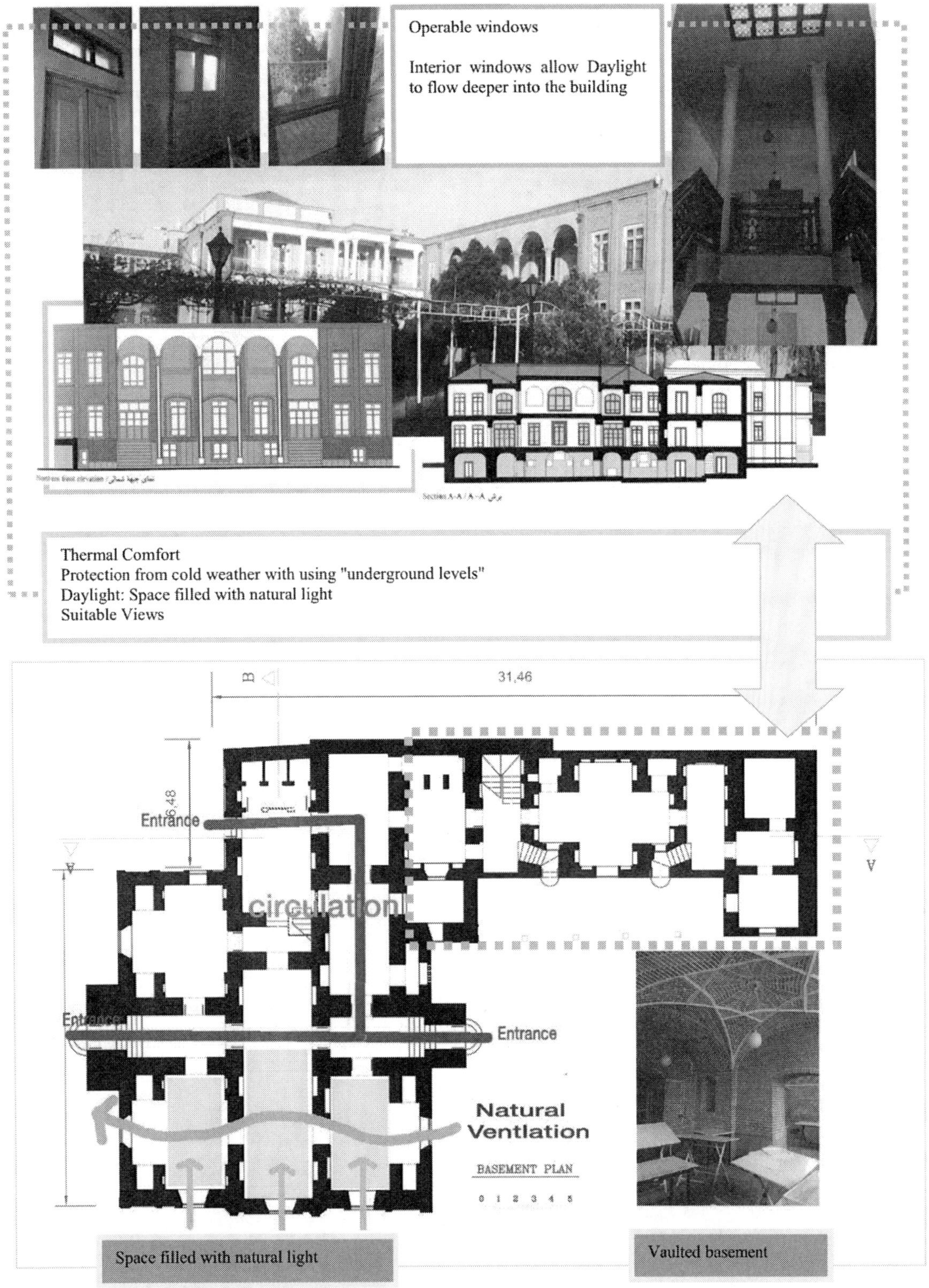

Fig6. Indoor Environmental Quality (IEQ) in the case study

5 Conclusions and Future Perspectives

This paper reviewed rating systems concepts for green projects, with a particular focus on its applicability and uptake within historic preservation. In addition, an exploratory study was carried out to studying attentively one of the rehabilitated constructions in the university site.

Although some researches investigations provide some insight into the rating systems concepts in preservation historical buildings, there is a need for continued research on the green elements (that follows rating systems concepts) implementation in rehabilitation and construction process.

This research outlines the major challenges which face assembling a survey in one type of the buildings. That is, rehabilitated and there is often a need to draw on knowledge from the past. In considering these issues, it is important to remember that this paper does not provide a conclusive implementation approach. This review provides comparing the rating systems issues with the preserved construction to find out the potentials of this type of buildings to be a sustainable.

As shown in this paper and according to last table and analysis of case study, rating systems needs perspicuous sub- criteria for buildings evaluation. Some of these measurable sub-criteria were used in different projects as Green Elements. So, these measurable sub-criteria were gathered and analyzed in a historical rehabilitated building as a case study.

In an immensely complex topic, one basis truth about historical buildings is that these buildings originally contain much aspects of sustainability, In spite of that; in some rehabilitation processes some of the sustainable aspects may be lost.

Finally, this paper reveals main parts of green elements that can be intentioned in historical preservation. These elements can be employed by designers in rehabilitation projects. An Investigations on the case, also indicates that, some specific categories and criteria of rating systems such as: "indoor environmental quality", "materials and resources" and "Sustainable Sites" has more effective and considerable impact on the sustainable rehabilitation.

References

1. Brundtland report, *World commission on environment and development.* Report of the United Nations, (1987)
2. O. Ortiz, C. Francesc, S. Guido, Construction and Building Materials, **23**, 28-39 (2009)
3. P. Appleby, *Integrated Sustainable Design of Buildings* (London: Earthscan, 2011)
4. BREEAM, *Pre-assessment estimator,* (2011) [Online]. Available: http://www.breeam.org/page.jsp?id=87.
5. A. Basnet, *BREEAM & LEED: A study of materials and their life cycle impact* ,(2011)
6. T. Ebert, N. Ebig, G. Haurer, *Green Building Certification Systems,* (2011)
7. W. Houlihan, *An Analysis of the Performance of Certification Schemes in the Hotel Sector in terms of CO2 Emissions Reduction,* (2009)
8. j. Carroon, *Sustainable preservation: Greening Existing Buildings,* (2010)
9. M.A. Kaynejad, M.R. Shirazi, *The Traditional Houses of Tabriz*, (2010).
10. *LEED 2009 for new construction and major renovations*. (2008). Washington, D.C.: U.S. Green Building Council
11. U.S.G.B.C. 2009. *Green Building Design and Construction, LEED Reference Guide for Green Building Design & Construction,* USA, i-xiv, 335-400.
12. N. Lazarus, *Bed ZED Toolkit part 1: A guide to construction materials for carbon neutral developments,* (2002).

Sampling And Identifying Of Mould In The Library Building

Suriani Ngah Abdul Wahab[1,a], Nurul Izma Mohammed[2], Mohd Faris Khamidi[3], Nur Azfahani Ahmad[4], Zainon Mohd Noor[5] Aida Affina Abdul Ghani[6] and Mohd Rodzi Ismail[7]

[1]*Department of Building Surveying, Faculty of Architecture, Planning and Surveying, UiTM Perak, 32610, Perak, Malaysia*
[2]*Faculty of Engineering, Universiti Teknologi Petronas, Tronoh, 31750, Sri Iskandar .Perak.*
[3]*School of the Built Environment, University of Reading Malaysia (UORM), Persiaran Graduan, Educity Kota Ilmu, Iskandar Puteri (Nusajaya), Johor.*
[4]*Department of Building Surveying, Faculty of Architecture, Planning and Surveying, UiTM Perak, 32610, Perak, Malaysia*
[5]*Faculty of Applied Sciences, Universiti Teknologi Mara, 40450 Shah Alam, Selangor*
[6]*Department of Building Surveying, Faculty of Architecture, Planning and Surveying, UiTM Perak, 32610, Perak, Malaysia*
[7] *School of Housing, Building and Planning, Universiti Sains Malaysia, 11800 USM, Penang.*

Abstract. Despite the growing concern over mould and fungi infestations on library building, little has been reported in the literature on the development of an objective tool and criteria for measuring and characterising the mould and fungi. In this paper, an objective based approach to mould and fungi growth assessment using various sampling techniques and its identification using microscopic observation are proposed. This study involved three library buildings of Higher Institution Educational in Malaysia for data collection purpose and study of mould growth. The mould sampling of three libraries was collected using Coriolis air sampler, settling plate air sampling using Malt Extract Agar (MEA), IAQ MOLD Alexeter IAQ-Pro Asp/Pen® Test and swab sampling techniques. The IAQ MOLD Alexeter IAQ-Pro Asp/Pen® Test and traditional method technique identified various mould species immediately on the site, and the microscopic observation identifies common types of the mould such as Aspergillus, Penicillium and Stachybotrys's. The sample size and particular characteristics of each library will result in the mould growth pattern and finding.

1 Introduction

As stated by Storey et al., (2004), most of mould favour a temperature of 15°C-30°C (59°F-86°F), but there are varieties that will grow below or above these temperatures. Mould be potential to grow by influenced by moisture or high humidity, nutrition, light, oxygen, and temperature. Mould will grow anywhere indoors and outdoors over a broad temperature range where there are sufficient moisture and a nutrient source. It results in visible wetting of walls, ceilings boards and floors, blistering paint, bulging plaster, mould on the surfaces and fabrics, rotting timber and sulphate attack on brickwork. It can also lead to less obvious problems. In example, thermal insulation is reduced in the effectiveness of brickwork because metal components imbedded in it have been corroded. Among the effects that had attracted the most attention is biological growth such as moulds with various species of Cladosporium, Aspergillus and other fungi types. This mould species can be found in both indoor and outdoor environments of buildings. Hollis (2005) stressed that it is completely linked to most building deterioration.. Findings by Jan (2006) on microenvironments exposure where isolated areas of books stacks or storage and other buildings may not applicable to tropical countries such as Malaysia. Malaysia experienced little variation in temperature and humidity and rainfall throughout the year. Interrelated the above issue, according to Zuraimi & Tham (2008) significant difference between temperature and relative humidity in tropical Singapore compared to cold countries were reported in IAQ of Child Care Centres(CCC). Hussin, Sann, Shamsudin, & Hashim (2011) conducted fungi research on selected primary schools in Malaysia. Their findings revealed that most frequently mould genera found were Aspergillus, Penicillium, Fusarium, Rhizopus and Zygomycetes.

There are many types of moulds exist. As stated by Burton & Gibbins (2011) certain types of moulds are toxigenic, whereby they can produce toxins. However the mould themselves are not toxic or poisonous. One of the moulds that are commonly found inside a building is Stachybotrys Chartarum. It is black in colour and produce its conidia in slime heads. Another common mould is Cladosporium, which produces itself in olive green to brown or black colonies. Besides that, it is quite often to find green colour mould which is called as Penicillium and also Aspergillus which is common been seen in either red or gold colour.

[a] Corresponding author: snawbs74@gmail.com

 DOI: 10.1051/matecconf/20166600070

2 Methodology

Visual inspection was conducted to investigate the visible mould growth, measure indoor condition and find the possible moisture sources. It is the most important initial step in identifying a possible contamination problem.

2.1 Buildings and moisture investigation

This study included a total of three library buildings of Higher Institution Educational in Malaysia for data collection purpose and study of mould growth. The Information Resource Centre (IRC), UTP was choosing for the controller purpose. Besides, it decides according to the modern design and technology. Perpustakaan Tun Abdul Razak 1 (PTAR1), UiTM Shah Alam and Perpustakaan Tuanku Zanariah; UTM Skudai Malaysia were selected due to its age, features and location. An inspection included an interview with a chief librarian, building maintenance and management personnel and workers. Visual Inspection being conducting to identify the signs of moisture and mould and any defects that found in the buildings. NYC Department of Health and Mental Hygiene highlight the extent of any water damage and mould should be visually assessed. A visual inspection is the most important initial step in identifying a possible contamination problem. Remediation of visually identified fungal contamination should proceed without further evaluation. According to Palaty (2010) outcomes of visual inspections can lead to three possible outcomes as Table 1 below.

Table 1: Outcome of visual inspections

No	Possible	Interpretation
1	No evidence of mould growth or moisture damage	No remediation action and further inspection required
2	Evidence of mould growth or moisture damage	Remediation actions need to be initiated to remove the mould
3	Unclear if mould or moisture is present	Proceed to take mould sampling and further analysis

The equipment such as humidity data logger, moisture metre and infrared thermography were used to measure and identify the moisture or dampness on the structures or surfaces of the materials. The inspection had identified some of the books on the book rack contained dust, especially on the book cover. The books were also looking swelling and shrinking. The researcher also found dirt, dust and microbial growth on the book that been displayed at the bookshelves as shown in **Figure 1**. The moisture staining and mould also found on the building elements such as floor, wall and ceiling.

Figure 1: Possible mould and microbial growth on the book surfaces

The inspection revealed dampness and visible growth of the mould on a ceiling board, timber column, floor or carpet and wall. Infrared thermography and moisture metre were used to confirm that moisture on the building structures occurs.

Infrared thermography is one of the useful equipment that helps the researcher in identify the causes of moisture occur in building. It is easy to use as a camcorder or a digital camera and can give a full image of the situation. It also allows us to perform inspections when systems are under load and could identify and locate the problem. In addition, it could measure temperatures and store information. In this research, it will detect and help eliminate mould infestations. A mould problem is a moistur problem, so when it is used to find moisture and it finds moisture, it becomes possible to prevent mould and rot from taking hold or to remove the mould that grows. Figure2, Figure 3 and Figure 4 shows the visible mould and humidity on building finishes using infrared thermography.

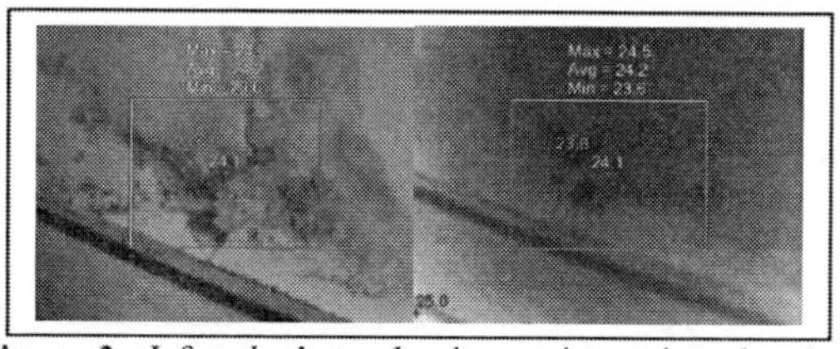

Figure 2: Infrared photo showing moisture intrusion into interior wall cavities

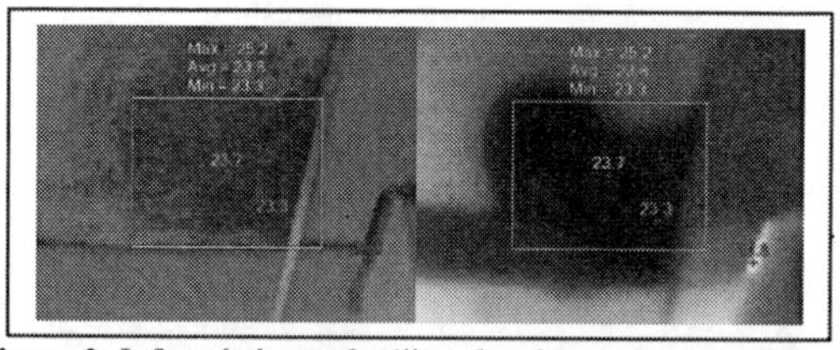

Figure 3: Infrared photo of ceiling showing moisture location

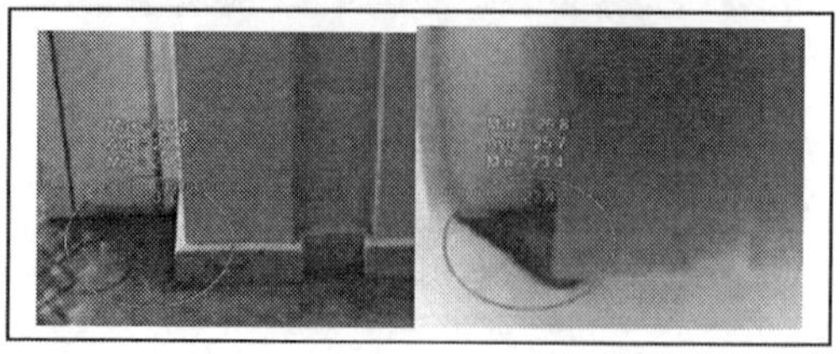

Figure 4: Infrared photo of carpet showing moisture intrution from rain water down pipe

2.2 Mould Sampling Method

The sampling method chosen depends upon resource availability, method, availability and information required and it is mention by AIHA (2008). In general, mould sampling method used to collect mould from air, surfaces and settles dust. The sampling needed in this research is to find the types of mould in the selected building. Swab samples were obtain from all visible mould found, and indoor air samples using Coriolis and malt extract agar were collected from reading area in three selected libraries. The sampling also focuses on the books and bookshelf which appearance of mould or stain. Also, the sampling is also done to get the total count of mould and yeast in the building. With the amount of total number using the specific formula, the researcher can get the colony forming unit (cfu) for the particular area or surface. The summarised of Mould Sampling Method that used for three selected libraries building shown in Table 2 below:

Table 2: The summarised of sampling perform in library buildings

Method	IRC	PTAR1	PSZ	Purposed
Coriolis Air Sampler	√	√	√	Identification
Settling Plates Air Sampling	√	√	√	Identification
Swab Test (IAQ Mold)	√	√	√	Identification
Swab Test	√	√	√	Identification

2.3 Coriolis Air Sampler

The Coriolis μ was developed to go beyond traditional techniques performances. Traditional techniques indeed rely on the impaction of particles on a solid agar medium and are thus limited by low flow rates, unreliable impaction and a longer analysis time. The Coriolisμ air sampler manufactured by Bertin Technologies (France) is a continuous air sampler, dedicated to indoor monitoring of airborne spores and pollen grains. Bertin Technologies has developed a liquid cyclone high-volume air sampler for collection and identification of airborne bio-particles including micro-organisms such as bacteria, fungi, and pollen grains. This technology is based on a cyclone types device that selects and captures the airborne particles and instantaneous transfer them into a liquid sample. This sample is directly compatible with standard culture methods Carvalho et al., (2008). The sampler operates for 4 min by pulling air at a flow rate of 250 per 1 minute into a cyclonic separator containing 15 ml of swirling collection fluid. After the sampling period complete, the remaining fluid was collected; the volume measured (evaporation loses some volume) and it was assayed for the presence of microorganisms. The process of sampling using Coriolis air sampler is shown in Figure 5 and below.

The Coriolis air sampler located in the reading area of each level in the libraries building selected as shown in Figure 5. Repeat twice to obtain triplicate before sampling to the next site. Consecutive samples using a new sterile cone and collection liquid dose at each sampling. For the growth culture analysis, a portion of the liquid sample in Coriolis cone about 0.5mL using the sterile pipette will spread onto Malt Extract Agar (MEA) as a culture medium. Before transferred the liquid sample need to vortex about five to ten second. The culture medium finally put into the incubator at room temperature. Wait until the mould growth.

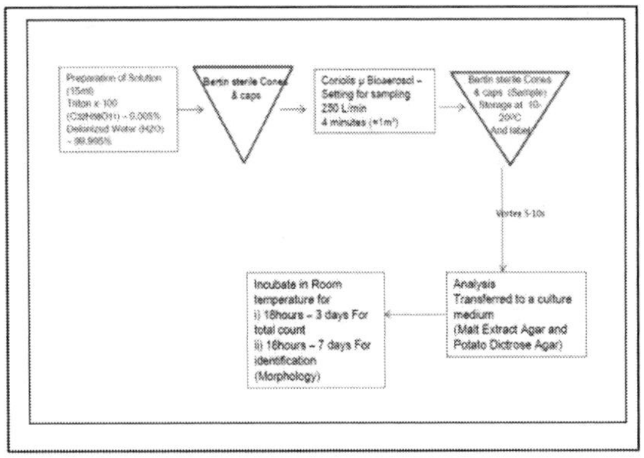

Figure 5: Sampling process using Coriolis machine and laboratory preparation

2.4 Settling Plate using Malt Extract Agar (MEA)

The visual inspection result that recorded on the checklist of the area in libraries selected identified for sampling. It is also focusing the area that the students or the user most spent their time in an individual area. In this research, reading area, being choose for sampling places because the student spent their time more than 1hour. However, the place that visible mould found also targeted for sampling. The books collection also one of the places that the researcher take for sampling since huge volume for each library with a different type of papers and books covering.

Figure 6: Air Sampling Mould Isolate at Reading Area

2.5 Surface Sampling

2.5.1 Swab Sampling

Specimen collection depends on the site affected. Surface sampling can determine whether a stain has resulted from fungal growth or some other problem. For example, discolorations associated with leaks might be due to dissolved materials in the water accumulating on the surface as the water evaporates. Dirt might cause the dark material that accumulates near air vents. An environmental consultant who is familiar with the appearance of these types of discolorations should be able to discern areas of fungal contamination on a surface, minimising the number of surface samples needed.

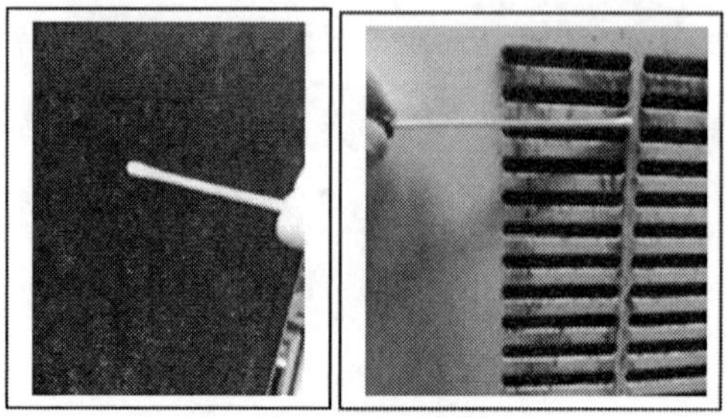

Figure 7: Swab Test Sampling

2.5.2 Mould Identification Using IAQ MOLD Alexeter IAQ-Pro Asp/Pen® Test

The Alexeter IAQ-Pro Asp/Pen test strip is a lateral flow immunochromatographic device that uses two antibodies in combination to correctly detect the antigen in solution. One of the specific antibodies is labelled with a colloidal gold derivative. When the sufficient antigen is present, the colloidal gold label provides a reddish-brown colour line that is visualised after accumulating in the test sample region on the device. When a sample is added to the Alexeter IAQ-Pro Asp/Pen test strip, the sample begins to mix with the colloidal gold labelled antibody and simultaneously moves along the strip membrane by capillary action. In the sample region of the test strip, if the antigen is present, the second specific antibody captures the colloidal gold labelled antibody and bound antigen, forming a colour line or band in the sample (left side) window of the test strip. As an internal control, a second band visualised in the control (right side) window of the test strip is an indication that the test strip functioned properly. Two bands or colour lines are required for a positive result determination.

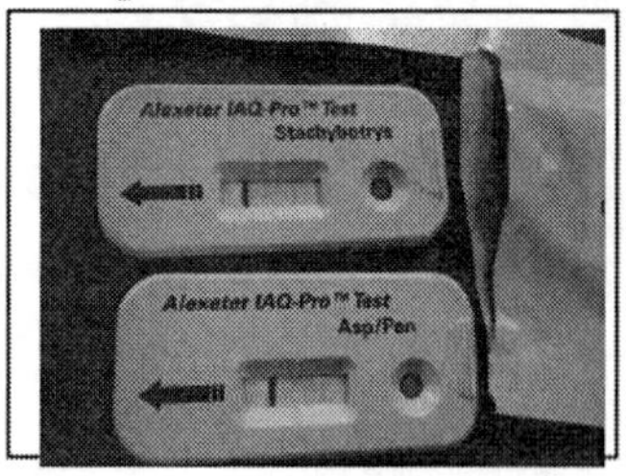

Figure 8: Result on IAQ MOLD Alexeter IAQ-Pro Asp/Pen® Test

2.6 Measurement of Mould Growth

2.6.1 Microbiological assays

Microscopy is used to observe clinical specimens for the presence of fungal elements or to identify the fungus following culture. Culturing methods cannot and should not be excluded from any fungal analysis. Merck provides Media Melt Extract Agar (MEA) and Potato Ditrose Agar (PDA) plates (90 mm) for use in all the physical efficiency testing.

All plates mould isolates were incubated at room temperature 30°C for up to 18 hours before counting the colonies. Before analysis, accurately measure the final volume of collection liquid in the cone in a clean area and record it. Vortex 5-10s second the liquid sample in the Coriolis cone (or in the transferred container) and remove 0.5mL with a sterile pipette. Spread this volume onto repeat once to obtain a duplicate by each culture medium. Captions should be typed in 9-point Times. They should be centred above the tables and flush left beneath the figures.

2.6.2 Mould Identification and quantification

Traditional mould quantification was based on the enumeration of cells captured on a sticky surface and counted under a microscope or by culturing moulds from the sample on various media. Most moulds cannot be identified directly by looking at the spores. Because of this limitation, most commercial mould analyses only describe the moulds to the genus level. Moreover, in the cases of Aspergillus and Penicillium cells, these two genera cannot be distinguished by microscopic observation alone. For that reason, most commercial labs only combine these genera together as the mega-category "Asp/Pen."[9].

All of these inoculated media (original and duplicate) were placed in an incubator set to 30°C, and they remained there for a week (MEA). The identification of mould on the microscopic visualisation of morphological characteristics using lactophenol cotton blue dye and demonstrated manuals [10]. In this research, microbiologist expertise was a help to confirm the morphology data found as shown below.

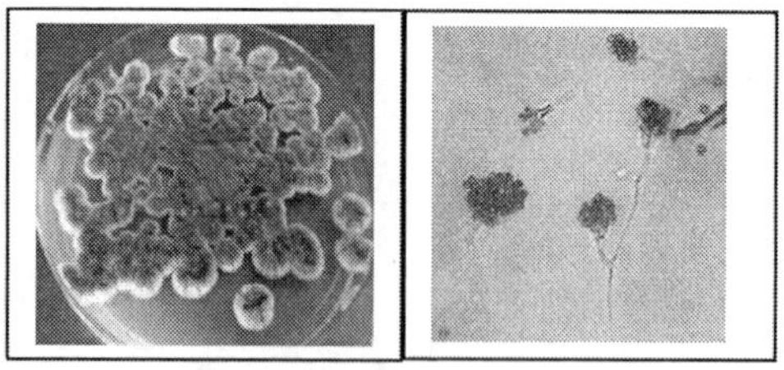

Figure 9: Petri dish on which colonies of Stachybotrys have been cultured and identified.

3 Result Obtained

3.1 Mould Identification Using IAQ MOLD Alexeter IAQ-Pro Asp/Pen® Test.

The result obtained by the application of IAQ Mould Alexeter IAQ-Pro is as shown in Table 3. The identifications of mould confirmed active for Aspergillus and Penicillium type of mould presence. Whereas only 5 sample show positive on Stachybotrys and 11 of 12 samples shows the positive result of Aspergillus.

Table 3: IAQ Mold Alexeter IAQ-PRO Asp/Pen® test expanded data result

Client ID	Date Collected	Sample Location	Item/ Material	Result		Relative Humidity % (AVERAGE)	Temperature °Celcius (AVERAGE)
				Asp/ Pen	Stachy botrys		
IRC	30/10/14	Ground Level	Book	+ve	-ve	68%	22.5
IRC	30/10/14	Level 3	A/c outlet	+ve	-ve	69%	23.2
PTAR	15/11/14	Level 3	Book	+ve	-ve	69%	23.9
PTAR	15/11/14	Level 3	Window frame	+ve	+ve	69%	23.9
PTAR	15/11/14	Level 3	A/c	+ve	+ve	69%	23.9
PTAR	15/11/14	Level 4	Carpet	+ve	+ve	64%	26.3
PTAR	15/11/14	Level 4	Ceiling	+ve	+ve	64%	26.3
PSZ	6/12/14	Level 2	Book	+ve	-ve	74.6%	20.4
PSZ	6/12/14	Level 2	A/c	+ve	+ve	74.6%	20.4
PSZ	6/12/14	Level 3	A/c	+ve	-ve	70%	21.2
PSZ	6/12/14	Level 4	Book dust	-ve	-ve	70%	22.4
PSZ	6/12/14	Level 4	A/c	+ve	-ve	70%	22.4

3.2 Traditional Identification of Isolates

All samples collected from library demonstrated that the sites of sampling were contaminated with microorganisms and fungi. A total of 22 fungal were isolated from all sampling sites involved as shown in Table 4. From morphological study through microscopic observation, most of the mould isolated from the selected study areas were known to be the common airborne fungi or common contaminants as shown in Figure 9. Out of 22 isolates from all sampling sites involved, the most abundant mould was found to be *Aspergillus sp.* with 27.27% of total isolates, followed by *Penicillium sp.*, *Stachybotrys sp.* and *Rhizoctonia Solani* with 18.18%, 18.18% and 13.63% of total isolates, respectively. Five isolates are unable to identify at genus level but had been confirmed to be *Fungal sp.* as they possess the characteristics of filamentous fungi. Most fungi, through most of their life cycle, consist of hyphae the cylindrical cells that increase in length by growth at one end. Their life cycle start with a spore, germinates into the suitable substratum, grows and branches forming a radiating system of hyphae called mycelium. Some fungi are not hyphae, but hypha and mycelium are associated with the growth and form of fungi to a greater extent.

Table 4: Mould identification from isolates

Location	Isolates
IRC Ground Floor at Reading area	*Stachybotrys sp.*
PSZ Level 2 /Book	*Fungal sp.*
PSZ Level 2/ Air sampling(1)	*Stachybotrys sp.*
PTAR1 /Level 3/ Aircond swab	*Aspergillus sp.*
PSZ/Level 3/ Air sampling	*Fungal sp.*
PTAR1/Level 3/ Window frame (1)	*Stachybotrys sp.*
PTAR1/Level 3/ Window frame (2)	*Stachybotrys sp.*
PTAR1/Level 4/ Air sampling (R) 1	*Aspergillus sp.*
PTAR1/Level 4/ Air sampling (R) 2	*Aspergillus sp.*
PSZ/Level 4/ Air sampling	*Fungal sp.*
PTAR1/Level 4/ Carpet swab	*Fungal sp.*
PTAR1/Level 4/ Ceiling swab (1)	*Rhizoctonia solani*
PTAR1/Level 4/Ceiling swab (2)	*Rhizoctonia solani*
PTAR1/Level 4/ Timber column swab	*Rhizoctonia solani*
PSZ/Level 5/ Air sampling(1)	*Penicillium sp.*
PSZ/Level 5/Air sampling (2)	*Penicillium sp.*
PSZ/Level 5/Air sampling (3)	*Fungal sp.*
PSZ/Level 5/Aircond	*Aspergillus sp.*
PSZ/Level 5/Aircond(2)	*Penicillium sp.*
PSZ/Level 5/Aircond(3)	*Penicillium sp.*
PSZ/Level 5/Ceiling swab (1)	*Aspergillus sp.*
PSZ/Level 5/Ceiling swab (2)	*Aspergillus sp.*

4 Conclusion

In this paper, various methods of mould sampling and identification using morphological study through microscopic observation had been conducted. The area of contribution of this research work is the identification of mould growth and the acquired image. The most abundant mould was found to be *Aspergillus sp. Penicillium sp.*, *Stachybotrys sp.* and *Rhizoctonia Solani*. Future work of this research will focus on the determined total count of mould base on the sampling area with the rate contamination of the library environment. The sample size and particular characteristics of each library will result in the mould growth pattern and finding. The research expected to identify the best solution to protect a collection that will depend on various factors such as type and number of visitors, characteristics of the library building and level of maintenance work provide.

References

1. E. Storey, K. Dangman H, P. Schenck, R. DeBernardo L, C. Yang S, A. Bracker, and M. Hodgson J, *Guidance for Clinicians on the Recognition and Management of Health Effects related to Mold Exposure and moisture Indoors*. Farmington: U.S Environmental Protection Agency (EPA), 2004.
2. M. Hollis, *Surveying Buildings 5th Edition*, 5th ed. RICS, 2005.
3. W. K. Jan, *Aerobiological Engineering Handbook A Guide to Airborne Disease and Control Technologies*. McGRAW-HILL New York: McGraw-Hill Componies, 2006.
4. M. S. Zuraimi and K. W. Tham, "Indoor air quality and its determinants in tropical child care centers," *Atmos. Environ.*, vol. **42**, no. 9, pp. 2225–2239, Mar. 2008.
5. N. H. M. Hussin, L. M. Sann, M. N. Shamsudin, and Z. Hashim, "Characterization of Bacteria and Fungi

Bioaerosol in the Indoor Air of selected Primary Schools in Malaysia," *Indoor Built Environ.*, vol. **20**, no. 6, pp. 607–617, Sep. 2011.

6. N. C. Burton and J. Gibbins, "Assessment of Mold and Indoor Environmental Quality in a Middle School – Texas," no. July, 2011.
7. C. Palaty, "Mould Assessment in Indoor Environments - Review of Guidelines & Evidence," Vancouver, 2010.
8. E. Carvalho, C. Sindt, a. Verdier, C. Galan, L. O'Donoghue, S. Parks, and M. Thibaudon, "Performance of the Coriolis air sampler, a high-volume aerosol-collection system for quantification of airborne spores and pollen grains," *Aerobiologia (Bologna).*, vol. **24**, no. 4, pp. 191–201, Oct. 2008.
9. S. Vesper, "Traditional mould analysis compared to a DNA-based method of mould analysis.," *Crit. Rev. Microbiol.*, vol. 37, no. 1, pp. 15–24, Feb. 2011.
10. A. C. Pinheiro, M. F. Macedo, V. Jurado, C. Saiz-Jimenez, C. Viegas, J. Brandão, and L. Rosado, "Mould and yeast identification in archival settings: Preliminary results on the use of traditional methods and molecular biology options in Portuguese archives," *Int. Biodeterior. Biodegradation*, vol. **65**, no. 4, pp. 619–627, Jul. 2011.

DOI: 10.1051/matecconf/20166600071

Energy Prediction versus Energy Performance of Green Buildings in Malaysia. Comparison of Predicted and Operational Measurement of GBI Certified Green Office in Kuala Lumpur

Suzaini M. Zaid[1,a] and Amir Kiani[1]

[1]*Department of Building Surveying, Faculty of Built Environment, University of Malaya, Malaysia*

Abstract. Forward from the sustainability agenda of Brundtland in 1987 and the increasing demand for energy efficient buildings, the building industry has taken steps in meeting the challenge of reducing its environmental impact. Initiatives such as 'green' or 'sustainable' design have been at the forefront of architecture, while green assessment tools have been used to predict the energy performance of building during its operational phase. However, there is still a significant hap between predicted or simulated energy measurements compared to actual operational energy consumption, or is more commonly referred as the 'performance gap'. This paper tries to bridge this gap by comparing measured operational energy consumption of a Green Building Index (GBI) certified office building in Kuala Lumpur, with its predicted energy rating qualification.

1 Introduction

Global warming is one of the biggest issues of this century, and according to National Aeronautics and Space Administration (NASA), in the past few years the number of carbon dioxide (CO_2) parts in the atmosphere exceeded an amount that it had never pass in the last 650,000 years [1]. Buildings play a big role in producing CO2 gases; in fact 30% of greenhouse gas emission come from building sector, and between 80-90% of CO2 emissions from the building sector come from the operational phase of already built buildings [2]. Consequently, mitigation of CO2 emissions has affected the building industry in a number of countries in the South East Asian region like Indonesia, Singapore, the Philippines, Thailand and Vietnam are already implementing energy efficiency legislation on newly built buildings [3]. However, it cannot be said the same about the building industry in Malaysia, where such legislation does not exist for both new built and existing buildings.

Other than building energy efficiency legislation, other strategies or approaches however has been done sporadically in the Malaysian building landscape, such as the introduction of Green Building Index (GBI) to rate how 'green' buildings are designed – not operated. The GBI has taken heed from pioneering green building rating tools such as Building Research Establishment Environmental Assessment Method (BREEAM) in The United Kingdom [4] and Leadership in Energy & Environmental Design (LEED) in The United States [5]. These green building rating tools are predicted models of how buildings would perform in operational phase, however it does very little to measure actual performance of the so-called green buildings. Therefore, there is a need to fill in the gap of predicted and operational performance of buildings – known as the 'performance gap' [6] that can truly be a testament of real energy and CO2 reduction.

2 Green Revolution

There has been large body of work that has been done to enhance how a building performs in its operational phase, through simulation and 'green' design or architecture [7]. The term 'green' architecture/building refers to how environmental friendly the building is, in terms of the design, techniques and technology used in the building [8]. The green building revolution is being observed internationally [9], through energy standards such as the European Energy Performance of Building Directive (EBPD), the French Low Energy Building Decree and the Norwegian Standard for Residential Passive House [10] and is even starting to gain awareness locally in Malaysia with the birth of GBI.

The believed benefits of green building practices will possibly provide the triple-bottom-line benefits of sustainable development, in terms of environmental, economic and social [9] [11] Some of the environmental benefits include: biodiversity and ecosystems enhancement; air and water quality improvement; waste reduction; natural resources conservation; and minimizing global warming effects. Other possible economic benefits could include: reduction of operation and maintenance costs; creation of green products and

[a] Corresponding author: suzaini_zaid@um.edu.my

 DOI: 10.1051/matecconf/20166600071

services; life-cycle economic performance optimization; image improvement; and reduction of cost for civil infrastructure [9]. Meanwhile, perceived social benefit include: occupant productivity improvement; and occupant absenteeism minimization [9].

2.1 Green Rating Tool and Simulation

Many countries have developed their own green rating tool based on one another, mostly adapted from the pioneering BREEAM and LEED rating tools. Examples of green building rating tools available in the South East Asian region are Green Mark from Singapore, Building for Ecologically Responsive Design Excellence (BERDE) from Philippines and the Green Building Index (GBI) from Malaysia [12-14]. Some of the assessment criteria in these green building rating tools are similar, for example energy efficiency (EE), indoor environmental quality (EQ), and water efficiency (WE) [12-14]. However, the assessment measurements and scoring system may differ between these green building rating tools according to local context. Table 1 briefly highlights the assessment criteria for Green Mark, BERDE and GBI for Non-Residential New Construction category, and the maximum scoring points [12-14]. As the focus of this paper is to review the Malaysian building performance, a brief comparison between the energy efficiency criteria of the GBI with its other South East Asian counterpart Green Mark and BERDE is described also in Table 2.

Table 1. Assessment criteria of BCA Green Mark, BERDE and GBI. Sources: [12-14].

BCA Green Mark – 120 points	BERDE – 120 points	GBI – 100 points
1.Climatic Responsive Design – 30 2.Building Energy Performance – 30 3.Resource Stewardship – 30 4.Smart & Healthy Building – 30 5.Advanced Green Efforts – 20	1.Management – 14 2.Land Use and Ecology – 20 3.Water – 7 4.Energy – 9 5.Transportation – 18 6.Indoor Environmental Quality – 7 7.Materials – 6 8.Emissions – 4 9.Waste – 11 10. Heritage Conservation – 4 11.Innovation - 20	1.Energy Efficiency (EE) – 35 2.Indoor Environmental Quality (EQ) – 21 3. Sustainable Site Planning & Management (SM) - 16 4.Material & Resources (MR) – 11 5.Water Efficiency (WE) – 10 6.Innovation (IN) - 7

Table 2. Energy Requirements for BCA Green Mark, BERDE and GBI. Sources: [12-14].

BCA Green Mark (Building Energy Performance)	BERDE (Energy)	GBI (Energy Efficiency)
1. Energy Efficiency 2. Air-Conditioning Total System Efficiency 3. Lighting Efficiency 4. Energy Effectiveness 5. Building Energy 6. Car Park Energy 7. Receptacle Energy 8. Renewable Energy 9. Feasibility Study 10. Solar Ready Roof 11. Replacement Energy	1. Energy Sub-Metering 2. Energy Efficient Lighting 3. Natural Ventilation 4. On-Site Renewable Energy Generation 5. Energy Efficiency Improvement 6. Energy Efficiency Building Envelope 7. Energy Efficient Equipment 8. Building Automation Systems	1. Minimum EE Performance 2. Lighting Zoning 3. Electrical Sub-Metering 4. Advanced or Improved EE Performance – BEI 5. Enhanced Commissioning 6. Post Occupancy Commissioning 7. EE Verification 8. Sustainable Maintenance

The GBI Minimum EE Performance is for building envelope and installations like Building Energy Intensity Tool (BEIT) to hoped to minimize the energy consumption, hence reducing CO_2 emission [14]. Lighting Zoning; as the name suggests it is allocated for lighting and zonings of lightings and the flexibility of the system. In this part, GBI's requirements are to install motion sensors to turn off the light in the absence of a user, to install auto-sensors to automatically adjust the level of artificial lighting according the level of natural lighting, and to provide switches for every enclosed area individually to increase the flexibility of the system [14]. Electrical Sub-metering & Tenant Sub-metering is for installing different sub-meters for all energy use more than 100 kilovolt amps (kVa) as well as an individual sub-metering for lighting [14].

Renewable Energy is to promote the use of renewable energies and points are given according the amount of electricity provided by the renewable sources, so the more renewable energy, the more points, and a building can get as high as 5 points [14]. Points are given in Advanced or Improved EE Performance if a building manages to consume less than 150kWh/m^2/year (this number may vary according to the type of building, for more information please refer to GBI) using BEI software or any other software recognized by GBI or to save at least 20% of the normal energy consumption over 3 years using BEI [14]. Enhanced Commissioning is used to ensure that every energy using system is working at its best potential and to ensure the minimal energy used by these systems [14].

Points on On-going Post Occupancy Commissioning will be given for regular Post Occupancy Evaluation (POE) and more points will be given if they are assessed by professional engineers [14]. EE Monitoring & Improvement is for the use of Energy Management System (EMS) to monitor all the energy usage in the building as well as the sub-metering. If a building is not equipped with EMS then they should submit BEI, fuel and water consumption of the building on annual basis to GBI for the 3 years of validity period [14]. Sustainable Maintenance focuses on the maintenance of the energy related systems and points are given is the maintenance is planned properly and at least 75% of maintenance team participate in carrying out of maintenance [14].

2.2 Predicted and Simulated Building Performance

Other than green building rating tools, simulation tools like EnergyPlus and Simplified Building Energy Model (SBEM) are also used as strategies to reduce the building sector's environmental impact and climate change effect. Simulation and modelling tools are used as a design tool to simulate and predict how the said building would perform in its operational phase. An open source software called EnergyPlus was developed under the prevue of the United States Department of Energy Building Technologies Office to simulate energy performance of heating, cooling, ventilation, lighting, plug and process loads and water consumption of buildings [15]. Some of the EnergyPlus features include simulated: thermal zone conditions for heating, ventilation and air-conditioning (HVAC) systems; heat balance-based solution; combined heat and mass transfer; luminance and glare, and built-in HVAC and lighting control strategies [15].

The Simplified Building Energy Model (SBEM) is a software developed by the Building Research Establishment (BRE) to model building's CO2 emission and energy consumption rates for new buildings, in compliance with United Kingdom Building Regulations [16]. SBEM predicts monthly energy use and CO2 emissions of a building, using variables such as building geometry, construction, function and usage, HVAC systems and lighting [16]. The SBEM uses a steady-state model for its building energy calculation, which is highly simplified and highly computational – as it largely ignores any dynamic characteristic [17], and is mainly intended as a CO2 emissions compliance tool [18] rather than energy simulation software.

However, these simulation tools are not able to predict the biggest variable in the calculations, as user behaviour, presence and activities can differ considerably to personal and group preference [6,7]. Building occupants can considerably effect energy consumption by influencing internal conditions, such as individual controls on air-conditioning and heat, window openings - passive ventilation and cooling methods, and shading mechanisms [6]. Furthermore, building occupants also have personal control over various electrical equipment and appliances – or known as 'unregulated loads', such as computers, laptops, printers, and so forth [6].

Conversely during the operational stage of a building, energy and facility management can significantly effect how energy is consumed in a building. Good management and control can enhance operational efficiency of building services while the opposite reaction can occur through inappropriate strategies that result in unnecessary energy wastage [19]. Other significant parameters to evaluate operational energy performance of buildings include energy efficiency services and fittings, thermal performance of materials and material efficiency [20]. A study done in New York City found that one forth of the LEED Energy Star buildings actually consume more energy than predicted, and some was even consuming more than the national baseline (Refer Figure 1) [21]. The discrepancy is likely due to differences in operational practices and schedules, equipment, construction changes and other issues not predicted in the energy modelling process [21].

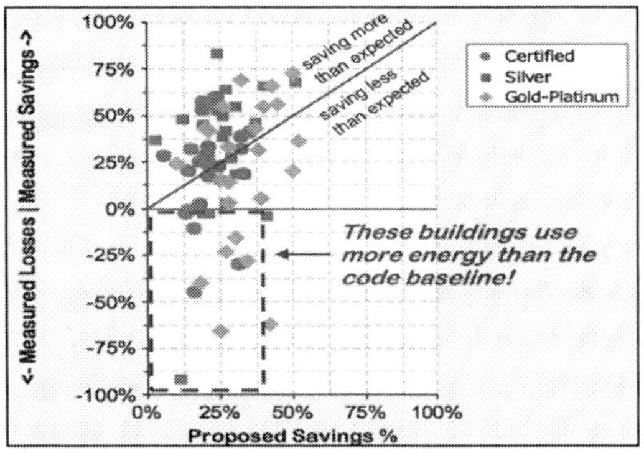

Figure 1: Predicted Energy Savings Versus Actual Performance. Source: [21].

Similar results were found in United Kingdom, through a free online energy data-sharing platform called CarbonBuzz (refer Figure 2). CarbonBuzz was launched in 2008 by the Royal Institute of British Architects (RIBA) and the Chartered Institution of Building Services Engineers (CISBE) in United Kingdom, published its analysis of shared energy consumption data in three building sectors, i.e. schools, general offices and university buildings [6] [22].

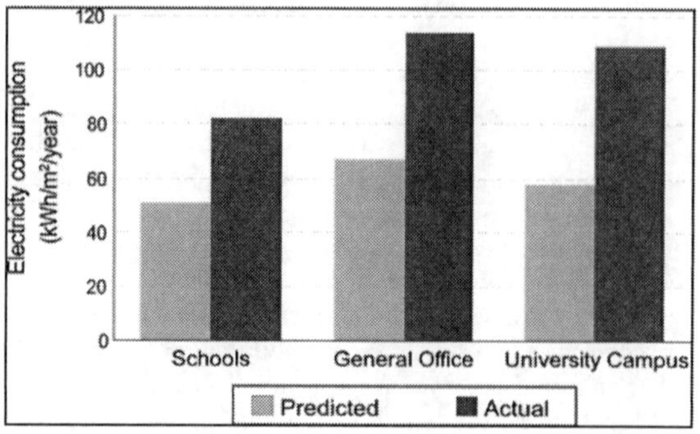

Figure 2: CarbonBuzz median electricity consumption per-sector – predicted vs. actual. Source: [22].

In an energy audit done by CarbonBuzz in 2013, it was largely found that the actual energy performance of all the building sectors was higher than its predicted performance, approximately about 70% to 89% higher than the predicted energy performance, or a Factor of Change (from Design to Actual) – Performance Gap between 1.71 to 1.90 (refer Table 3) [23]. The typical variable that affects operational energy performance of a building largely lies in the underestimating the 'unregulated' energy use from occupant-related usage, such as personalized computer/laptop, electronic plug-in appliances, heating and cooling, operating time, number of occupant and so forth [23].

Table 3. CarbonBuzz median electricity consumption per-sector – predicted vs. actual. Source: [23].

Building Sector	Median Electricity Consumption (kWh/m2/year)		Factor Change Design to Actual – 'Performance Gap'	Differen -ce (%)
	Predict-ed	Actual		
Schools	71	121	1.71	70%
General Office	56	106	1.90	89%

The 'performance gap' discrepancy can also be attributed to the lack of feedback to building designers after handover to enhance building performance [6] and therefore the inability to ensure the building performs as predicted. Other factors that could contribute to such discrepancies insufficient data and assumptions during the design stage, where the building function, use and future tenant are largely unknown or uncertain, which could lead to oversimplified and unrealistic performance [6]. Additionally, the modelling and simulation software used can contain fundamental errors in its calculation if the tools have not been extensively validates and do not consider the specific typology of buildings being modelled [6]. Consequently, this paper helps to bridge the gap by collecting real operational energy consumption data of a GBI rated building, and comparing it with a conventional non-GBI rated building to develop a baseline study.

3 Methodology

The methodology used in this study is comparative case study. This method is used when two or more similar case studies are subjected and in depth comparison is required between them [24]. Case studies for this research have been chosen firstly from GBI Non-Residential New Construction (NRNC) list of certified buildings, the second conventional baseline building is chosen carefully from the commercial buildings, which have the similar number of floors and floor area for a fair and valid comparison. There are total number of 116 NRNC buildings in Malaysia which have been awarded by GBI certification, 54 of which have been awarded by GBI Certified certification, 18 have been awarded GBI Silver certification, 39 have been awarded Gold and only 5 are awarded GBI Platinum certification [25].

Consequently, electricity bills were collected from the case studies and the data are quantitative and the monthly average electricity consumption is calculated for each building. The selected GBI certified building is categorized under the Non-Residential New Construction category. The data is then categorized in Building Energy Index (BEI), in terms of kWh/m2/year. The findings are then compared to see how better (or worse) GBI certified buildings perform, in comparison to a baseline conventional non-GBI certified building.

4 Case Study and Findings

The selected GBI certified building was designed to perform at 150 kWh/m2/year and has a total floor area of 43,943 m2, while the baseline building with 41,249 m2. Both buildings used double-glazed glass for its windows, and the GBI certified building has a total of 15 lifts, while the baseline building had 11 lifts. A brief description of the building characteristics of the selected case studies is presented in Table 4.

Table 4: Building Characteristics of Selected Case Studies

Item	Non-Residential	
	Baseline Building	GBI Certified Building
Number of Floors	36	35
Total Floor Area (m2)	41,249	43,943
Building Façade	Double-glazed low E insulated glass	Tempered Double-Glazed tinted glass

Lifts	4 high-speed lifts for low zone, 4 high-speed lifts for high zone, 2 carpark lifts and 1 VIP lift. Total of 11 lifts	6 high-speed lifts for low zone, 6 high-speed lifts for high zone and 3 car park lifts. Total of 15 lifts
Air-Conditioning System	Variable Refrigerant Volume (VRV) system.	Central water chilled air-conditioning system supported by high energy-efficient 'green' chiller and complemented by individual air handling unit (AHU) installed at every floor.
Lighting System	28 Watt T5 tubes; 18 Watt T8 LED tubes; 9 Watt Plasma Lighting System (PLS); 18 and 36 Watt Fluorescent lights; 8, 15 and 20 Watt LED	T5 Fluorescent tube equipped with high frequency electronic ballast.

The electricity data collected for both the case study buildings were for a total duration of 12 months between 2013 and 2014. Table 5 represents the collated electricity data and energy performance of the baseline building and the GBI certified building. It was found that the energy performance of the baseline building was lower than the GBI certified building, at 69 kWh/m2/year and 111 kWh/m2/year, respectively. As presented in Table 5, there is a stark difference in total electricity consumed by both buildings, where the annual electricity consumption by the baseline was at 236,984 kWh and the GBI building was at 407,131 kWh.

Table 5: Electricity Consumption and Energy Performance of Case Studies

Month	Baseline Building		GBI Building	
	Floor Area: 41249 m2		Floor Area: 43943 m2	
	kWh	kWh/m2	kWh	kWh/m2
1	281,333	7	212,551	5
2	251,001	6	212,201	5
3	225,765	5	235,059	5
4	200,778	5	418,868	10
5	213,104	5	427,861	10
6	263,810	6	478,788	11
7	231,928	6	469,016	11
8	225,961	5	479,199	11
9	252,135	6	524,961	12
10	243,920	6	482,067	11
11	232,277	6	483,446	11
12	221,791	5	461,555	11
	kWh/year	kWh/m2/year	kWh/year	kWh/m2/year
Total	2,843,803	69	4,885,572	111
Average	236,984	6	407,131	9

In comparison to the intended energy performance designated for the selected GBI building case study at 150 kWh/m2/year, it would seem that the GBI case study building was performing better than the intended or simulated with a lower energy performance of 111 kWh/m2/year. However, this does not negate from the significantly higher electricity consumption of the GBI certified building to the baseline non-GBI certified building. Further research needs to be done to investigate the factors of electricity consumption in non-residential building to understand why conventional non-GBI certified building consume lower electricity than the supposedly 'green' buildings that are GBI certified.

5 Discussion

This paper has highlighted the need to bridge the 'performance gap' between simulated and operational measured energy performance of buildings. Through the case studies selected and the data collected, it was found that the selected non-residential GBI certified building was performing better than the simulated or intended energy performance. However, through the comparative case study between a conventional non-GBI certified building and a GBI certified building, it was found that the conventional building was performing better than the GBI rated building over time. Identifying the factors that affect how energy is consumed in buildings will enable researched to bridge this performance gap to potentially predict energy performance and design building more accurately through simulated.

Researching how user behaviour affects energy performance of buildings is essential to bridging the performance gap, and can be done through Post-Occupancy Evaluation (POE) and sensor monitoring. POE is a common method to measure actual building performance, and now can be used to address to bridge the gap between simulated and operational building performance (Menezes et al., 2011). Green building rating tools like the GBI is an effective tool to reduce electricity consumption, but using this tool alone and certification does not guarantee energy saving. From the research it is understood that a well-managed and efficient building operations of a conventional non-GBI certified building can perform as well as or even better than a green building.

References

1. NASA. (2013). Climate change: How do we know? The National Aeronautics and Space Administration (NASA). Retrieved from http://climate.nasa.gov/evidence
2. UNEP-SBCI. (2010). Common Carbon Metric. United Nations Environmental Programme-Sustinable Building Climate Initiative. Retrieved from http://www.unep.org/sbci/pdfs/UNEPSBCICarbonMetric.pdf
3. Zaid, S., & Graham, P. (2015a). The Need for Energy Efficiency Legislation in Malaysian Building Sector. A Comparative Study of South East Asian Policies. Conference Proceedings of RICS COBRA Legal Research Symposium 2012.
4. BREEAM. (2013). What is BREEAM? Building Research Establishment (BRE) Ltd. Retrieved from: http://www.breeam.org/about.jsp?id=66
5. U.S. Green Building Council. (2013). Leadership in Energy and Environmental Design (LEED). Retrieved from: http://www.usgbc.org/leed
6. Menezes, A. C., Cripps, A., Bouchlaghem, D., & Buswell, R. (2012). Predicted vs. actual energy performance of non-domestic buildings: Using post-occupancy evaluation data to reduce the performance gap. Applied Energy, 97, 355-364
7. GhaffarianHoseini, A., Dahlan, N. D., Berardi, U., GhaffarianHoseini, A., Makaremi, N., & GhaffarianHoseini, M. (2013). Sustainable energy performances of green buildings: A review of current theories, implementations and challenges. Renewable and Sustainable Energy Reviews, 25, 1-17.
8. Love, P. E., Niedzweicki, M., Bullen, P. A., & Edwards, D. J. (2011). Achieving the green building council of Australia's world leadership rating in an office building in Perth. Journal of Construction Engineering and Management, **138**(5), 652-660.
9. Nilashi, M., Zakaria, R., Ibrahim, O., Majid, M. Z. A., Zin, R. M., Chugtai, M. W., ... & Yakubu, D. A. (2015). A knowledge-based expert system for assessing the performance level of green buildings. Knowledge-Based Systems, **86**, 194-209.
10. Zaid, S. M., Myeda, N. E., Mahyuddin, N., & Sulaiman, R. (2015b). Malaysia's Rising GHG Emissions and Carbon 'Lock-In'Risk: A Review of Malaysian Building Sector Legislation and Policy. *Journal of Surveying, Construction and Property*, ***6***(1).
11. Ahn, Y. H. (2010). The development of models to identify relationships between first costs of green building strategies and technologies and life cycle costs for public green facilities (Doctoral dissertation, Virginia Polytechnic Institute and State University).
12. BCA Green Mark. (2015). BCA Green Mark for New Buildings (Non-Residential). Building and Construction Authority (BCA). Retrieved from http://www.bca.gov.sg/GreenMark/others/GM2015_NewBuildings_NRB_Pilot.pdf
13. BERDE. (2013). BERDE for New Construction Commercial Buildings. Phillippine Green Building Council. Retrieved from http://philgbc.net/berde/berde-nc/1.1.0/BERDE-NC-COM-v110.pdf
14. GBI. (2009). GBI Assessment Criteria for Non-Residential New Construction (NRNC). GreenBuildingIndex Sdn Bhd. Retrieved from http://new.greenbuildingindex.org/Files/Resources/GBI%20Tools/GBI%20NRNC%20Non-Residential%20Tool%20V1.0.pdf
15. U.S Department of Energy, 2015. EnergyPlus Energy Simulation Software. Retrieved from http://apps1.eere.energy.gov/buildings/energyplus/
16. BRE. (2016). Simplified Building Energy Model (SBEM). Building Research Establishment Ltd. Retrieved from https://www.bre.co.uk/page.jsp?id=706; http://www.uk-ncm.org.uk/
17. Wang, S., Yan, C., & Xiao, F. (2012). Quantitative energy performance assessment methods for existing buildings. *Energy and Buildings*, *55*, 873-888.
18. Lee, W. L., & Burnett, J. (2008). Benchmarking energy use assessment of HK-BEAM, BREEAM and LEED. *Building and Environment*, ***43***(11), 1882-1891.
19. Bordass B, Cohen R, Standeven M, Leaman A. Assessing building performance in use 3: energy performance of probe buildings. Build Res Inform 2001;**29**(2):114–28.
20. Mwasha A,Williams R G, Iwaro J. Modeling the performance of residential building envelope: The role of sustainable energy performance indicators. Energy and Buildings 2011;**43** (9):2108–17.
21. Turner, C., & Mark, F. (2008). Energy Performance of LEED for New Construction Buildings. Vancouver, WA: New Buildings Institute, 2008.
22. Hamilton I, Steadman P, Bruhns H. CarbonBuzz – Energy Data Audit. UCL Energy Institute, July; 2011.
23. CarbonBuzz. (2013). Summary of Audits Performed on CarbonBuzz by the UCL Energy Insitute. CarbonBuzz Performance Gap. Retrieved from http://www.carbonbuzz.org/index.jsp#performancegap; http://www.carbonbuzz.org/downloads/PerformanceGap.pdf
24. Mills, A. J., Durepos, G., & Wiebe, E. (Eds.). (2009). Encyclopedia of case study research (Vol. **2**). Sage Publications.
25. Greenbuildingindex Sdn Bhd. (2014). GBI Certified Buildings. Retrieved from greenbuildingindex: http://www.greenbuildingindex.org/organisation-certified-buildings-NRNC-01.html

MATEC Web of Conferences **66**, 00072 (2016) DOI: 10.1051/matecconf/20166600072
IBCC 2016

Building Condition Assessment: Lesson Learnt from Pilot Projects

Syamilah Yacob[1,a], Azlan Shah Ali[2], Au-Yong Cheong Peng[3]

[1]*Public Work Department & PhD Candidate, University of Malaya, Kuala Lumpur*
[2]*Department of Building Surveying, University of Malaya, Kuala Lumpur,*
[3]*Department of Town Planning, University of Malaya, Kuala Lumpur*

Abstract. Building condition assessment is growing and become more important in the context of the built environment and Asset Management in Malaysia. This paper presents the method and the use of a standard guidelines for Building Condition Assessment (BCA) as a tool to evaluate the physical condition of the building. The purpose of this paper is to understand the use of BCA in assessing building condition. The methodology was based on literature review, review contents from archive document of Public Work Department (PWD) Malaysia which is involved study on the existing methods, proposal, discussion, piloting, improvement, approval and analysis of data. A case study involved of 15 buildings with various function such as office, health clinic, quarters and learning institution involved by more than 300 numbers of building inspectors or assessors. The findings shows that the using of the standard for condition assessment, the building inspectors or assessors can provide the data and condition rating to the buildings according to the standard. The results then used to examine a level consistency of BCA in data collection and it will easily assessed and understood, and the appropriateness of BCA to different types of buildings.

1 Introduction

In year 2008, Public work Department of Malaysia have developed, expanded and implemented the Asset Management concept to the organization to improve the government service delivery. Integrated and comprehensive Asset Management is to ensure that the asset are in a good condition, secure, economical, and longer life cycle as well as to provide quality services to the people [1]. According to [2], the traditional view of asset management as the function of building and maintaining buildings has been broadened to include outcomes to stakeholder requirements using tools such as asset registers, monitoring the asset condition and use, prediction of deterioration, asset valuation, life cycle economic and others.

Building performance can be measured in many ways, the most common being condition. Building condition assessment is growing in the context of the built environment in Malaysia. Hence research by [3], building condition assessment has been practiced in other countries and in particular in the United States, United Kingdom and many other European countries. A building survey consists of an investigation and assessment of the construction and condition of building, which generally include the structure, fabric, finishes and grounds [4]. According to [5], the assessment of building condition is technically complex tasks, requiring knowledge, time and equipment. According to [6], building inspection is a testing examination against material or product. [7] noted that the inspection is the process to assess the quality of the product or service in order to achieve a predetermined standards. In many countries the condition of a buildings is assessed and inspected on the basis of diagnosis of the extent of deterioration in the building elements [5]. He also mentioned that although there is differences in objectives, the assessment methods and development processes in the many countries are very similar.

2 Building Condition Assessment (BCA)

Building condition is very important to supports decision making and it is also critical to the management in achieving the service standards for maintenance. Building condition reflects the physical state of the buildings hence its performance. So the organizations, buildings and maintenance managers must have a knowledge in monitoring the condition of their buildings to prevent defects and failure of the buildings. Inefficiencies of the system of building maintenance work can give rise to defects and damage to the building. Defects and damage is a problem faced in any building regardless of age of buildings and construction type. Besides that inefficiency in handling defects or damage buildings systematically generate a variety of effects and negative impact to users and is also the owner of the building. The data of defects collected and registered during the assessment or inspection on site is crucial for maintenance decision. All building components have to contend with performance loss through ageing, use and also external causes [8].

[a] Corresponding author: syamilah75@gmail.com

In order to improve the current process of building inspection, a simplified, standardised and largely automated condition indication system has been developed by Public Work Department (PWD) to respond to the challenges in managing and maintaining the assets. BCA systems have been developed according to the requirement of Government Total Asset Management Manual [9] together with [10]. A condition assessment system is performed primarily to facilitate the ranking of all the components of the asset according to the amount of needed repair, and to produce consistent, relevant and useful information. By knowing the objective, examination would allow examiners to take advantage of a limited time during site inspection. Maintenance can only be effectively managed if the maintenance demand is properly quantified [11].

Building inspection is very important to support the goals of the Organization in providing office facilities, accommodation/high quality work in an environment that is safe, comfortable and sustainable manner. Low level of building performance can impact negatively on the organization and involves an increase in operating cost.

3 Standardization

Building inspection at a building which must be done in accordance with the standards set by the building inspector. An assessment of the condition of the building is the process of examining the entire building and infrastructure components, including equipment for mechanical and electrical systems, building frames, internal structure and finishes as well as building sites [7]. Component inspection criteria must be developed based on feedback the user base. User needs will produce objective assessment of the condition of the building. Building condition survey comparison with specifics based on objective assessment criteria the assessment. Synergy among all elements inspection and an assessment of the situation will form a cyclic process as a model to identify criteria for examining and improving the quality of design. Performance levels of buildings is the foundation to ensure that the criteria developed are able to achieve the objective. The structure, architecture, mechanical, electric, outdoor work, building facilities is the complete set of components for building inspection made valuation based on the maintainability, security, functionality and sustainability. The role and scope of the inspection should be understood by the examiners comprising Building Surveyors.

According to [8], the different results of various condition assessment methods are not a drawback in practice and it is important that within organization all building inspectors handle their own method the same way. The practice of condition assessment by building inspectors yields variable results due to subjective perceptions of the inspectors Surveyor variability is defined as the situation where two or more surveyors surveying the same building, arrive at very different survey decisions [12] and [13]. PWD has developed the standard Guideline Condition Assessment of building components including services in a year 2013. The guideline is a general guide to building inspectors in implementing the building inspection works and intended for creating consistency in the field of implementation of inspection and report preparation [10]. The aim of the standard is to ensure the inspection carried out by building inspectors comply with the inspection prescribed guidelines hence to provide the related and reliable rating and information about the defects assessed.

The scope of the inspection is covering all the Government's Asset which included civil and structural, mechanical system; and electrical system parts. The standard is aimed at building inspector which made up from various discpline such as building surveyors, architect, civil, mechanical and electrical engineers. Condition assessment should be carried out by trained inspector by using small equipment and related tools. Application of the standard include at the following stages:

i) Acceptance of the building for the purpose of operating and maintenance;
ii) Acceptance of the building for the purpose of asset or building registeration;
iii) Stages of operation and maintenance of the building;
iv) The initial assessment stage for the purpose of rehabilitation, renovation and upgrade;
v) The initial assessment stage for the purpose of conservation of buildings;
vi) The stage of preliminary assessment for the purposes of risk assessment and safety; and
vii) Building condition assessment level after a disaster

The standard is not limited to condition assessment only but also to the maintenance planning and priotizing maintenance work. Some processes and regulations have to be complied with by the parties involved so as to ensure the implementation of the inspection carried out in a more systematic manner.

4 Rating Scale

Building Condition assessment involved the screening process which is should be based on strategic management in assessing technical performance to fulfil the long term building maintenance expectations and related infrastructure. The condition rating that follows building assessment is used to objectively determine the current condition of building. A rating system that minimises subjective evaluation is repeatable and can be effectively used to predict future condition. The values of the condition indexes provide the means of comparing the condition of various components. The condition indexes scale and linguistic representation for building components is usually varies from others country whether it can derived from the numeric values or alphabets.

According to [14], whatever condition categories are adopted, it is essential that they are clearly defined and that data collectors are well trained to ensure data consistency and reliability. A six –point scale has been taken as a basis for the standardised method [15]. The condition categories described possibly occurring defects without references to remedial work shown in Table 1. Other examples of condition scales and linguistic representation used are as shown in Table 2 and Table 3.

Table 1: Six-point scale Dutch Standard for Condition Assessment of building

Condition rating	General Condition Description
1	Excellent
2	Good
3	Fair
4	Poor
5	Bad
6	Very Bad

Source: NEN (2006)

A five-point scale proved to be most effective meanwhile a three –point scale is too coarse for reliable results, while a seven-point or more scale is too fine and difficult for assessment staff to interpret consistently [11]. For [10], a five-point scale is used to represent the severity of each defect. The aggregate score is then used to produce a numerical rating from 1 to 5, where 1 signifying optimal condition and 5 signifying critical conditions. Table 4 shown examples of condition scales and its representations as used in the standard guideline of BCA.

This condition assessment rating also links to the maintenance actions and types as shown in Table 5. Then, analysis to the defect findings links to the defect parameter for building condition and maintenance priority action level as shown in matrix analysis table in Table 6.

Table 2: Country Comparison of the Condition Rating Grade used for Building

Running Body (Government / Private / non-Profit Org)	Australia	New Zealand	South Africa	United Kingdom	United States
Condition Grade (Buildings)	A=Excellent B=Good C=Adequate D=Poor E=Inadequate	1=Very good 2=Good 3=Fair 4=Poor 5=Very Poor	1=Excellent 2=Good 3=Average 4=Poor 5=Very Poor	A=Good B=fair C=Average D=Poor E=Bad	1=Very Good 2=Good 3=Moderate 4=Poor 5=Very Poor
Categories	Existing	Existing	Existing	Existing	Existing

Source: Institute Infrastructure Asset Management (IIAM) Manual (2006)

Table 3: Rating Scales and Representations

Reference	Asset Type	Condition Scale	Linguistic Representation
Dfes 2003	Buildings	A-D	Grade A=Good, Grade B=Satisfactory, Grade C=Poor Grade D=Bad
NCES 2003b	Buildings	1-8	1=Excellent, 2=Good, 3=Adequate, 4=Fair, 5=Poor 6=Non Operable, 7=Urgent Building Condition

Table 4: Building Physical Condition Level

Grade	Inspection Scale	Description
1	Very Good	• As New, No Defect, Performing as intended
2	Good	• Minor defect, Good condition, performing as intended
3	Fair	• Major defect, moderate condition, still can functioning with supervision
4	Poor	• Major or minor defect, critical, not functioning as agreed service level
5	Very Poor	• Major or minor defect, critical, not functioning as agreed service level • Very critical, not functioning, risky to safety and health

Source: JKR 21602-0004-13 (2013)

Table 5: Maintenance Action

Priority	Scale	Description
Normal	1	No defect or damages, element / component well maintained
Routine	2	Minor Defects / damages, Needs for monitoring, repairs, replaced to prevent serious defect / damages
Repairs	3	Major defects / damages, needs for major repairs and replacement
Rehabilitation	4	Critical / serious defects / damages, needs for urgent and immediate repairs
Replacement	5	Critical / serious defects / damages, needs for urgent replacement, refer to expert detail inspection / Expert judgement

Source: JKR 21602-0004-13 (2013)

A five-point colour-coded rating system has been developed and proved to be most effective as mentioned by [11]. This will ease the method of assessment for staff or inspector to interpret consistently, making reports more user-friendly and easy to interpret, save time, comprehensive and also accessible to non-technical users to use the information. In table 7, the five-point colour-coded building rating is shown as a final result of the assessment. Score 1 to 5 with very good condition and action matrix to be taken is preventive maintenance, 6 to 10, good condition with condition based maintenance, fair condition with score from 11 to 15 required for repairs work. Meanwhile rehabilitation work need for poor condition with range scores from 16 to 20. Score from 21 to 25 is very poor condition and needs for replacement work.

Table 6: Matrix Analysis – score defect parameter for building condition and maintenance action

Scale		**Maintenance Priority Action**				
		E(5)	S 4	SM 3	R2	N 1
Building Physical Condition	5	25	20	15	10	5
	4	20	16	12	8	4
	3	15	12	9	6	3
	2	10	8	6	4	2
	1	5	4	3	2	1

Source: JKR 21602-0004-13 (2013)

Table 7: Building Rating and Maintenance Action

Rating	**Physical Condition**	**Action Matrix**	**Score**
A	Very Good	Preventive Maintenance	1 to 5
B	Good	Condition Based Maintenance	6 to 10
C	Fair	Repairs	11 to 15
D	Poor	Rehabilitation	16 to 20
E	Very Poor	Replacement	21 to 25

Source: JKR 21602-0004-13 (2013)

5 Methodology

The approach of the research is quantitative research which is well-suited for the study to testing the understanding of the BCA. The study also based on literature, analysis content from BCA reports by PWD. The objectives of the study is to examine the level of consistency data collection and it will easily access and understood. The study on the existing methods such as Building Condition Maintenance Assessment System (BCMAS) 2010 were analyzed. The purpose of the BCMAS method are to for expedite maintenance works, ensure proper financial management, assisted in planning and monitoring. According to supporting data (BCMAS report, 2010), the method were not accurate enough and not achieved the objectives. The development of the new standard is based on the requirements towards National Assets Management. International rating comparison, namely Australia, New Zealand, South Africa, United Kingdom and the United States are reviewed for reasonableness of the use of this identification of BCA in Malaysia. The results of the BCA are a more systematic approach in building maintenance practices.

A case study involved of 15 buildings with various function such as office, health clinic, quarters and learning institution at certain state in Malaysia. The content of 23 numbers existing BCA reports were studied to ensure that the workable of the standard developed and to attain the consistently in the rating result by different assessors. And it also to ensure the suitability the used of the standard to the various function or categories of buildings with different sizes of floor area. The case study was carried to find out if the instruction in the manual clearly explained, easy to follow, understand and correctly assessed by assessors. The case study helped the organisation for building inspection and improvement for the standard especially to the inspection process. Besides that, the case study also helped to provide maintenance data and other relevant information regarding to the condition assessment. The inspections were performed by 307 assessors consisted multilevel of skill through training course to make sure the reliability and validity scores of the rating given and the use of standard procedure. The findings from the study were analysed and delivered to the conclusions.

Condition of buildings will be assessed and collected through Building Condition Assessment (BCA) standard. Inspection method performed is based on the visual inspection by using appropriate tools. The assessment mainly from physical site visit. Inspections done subject to the scope of limitations [16]. This is the initial screening examination for the purpose of evaluating the condition of the assets by Building Surveyors. The result analysis of inspection shall refer to experts for follow-up action. Figure 1 shows the condition assessment process according to the guideline. Building inspectors should carry out an initial review examination as one needs to examine the detailed investigation by engineers. However it depends on the type and level of disability and damage reported in the initial visual inspection report.

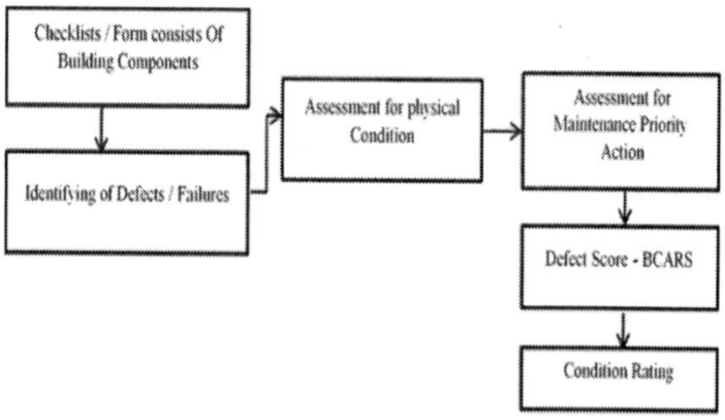

Source: Source: JKR 21602-0004-13 (2013)
Figure 1: Condition Assessment Process

A checklist was used according to the standard for registering the information of the building, identification, characterisation, defects in functional element, evaluation of condition of defect, and maintenance action. Asset condition checklist was developed to know the status of the output and as input of the operation and maintenance of the building, as initial inspection report for building condition and also to make a recommendations to the experts for follow-up action. The data collected in the checklist was compiled for registering the information of the defects. The checklist was divided into five (5) parts which is building informations forms, civil and structure, mechanical, electrical and external work forms.

The assessing of defect must come for the first step by identifying the type of defects. This statement was supported by [8], the assessing of defects occurs first and without the information one could not formulate maintenance activities and estimate the costs. The intensity of the defects will influences building component or element to function properly. The extent of defects strongly influences the condition of the building components and lead to the condition rating [8]. The assessment of defects assessed by comparing the current condition or performance with the original state of conditions. The condition of the buildings are assessed and summarized in the preparation of reports based on the following criteria such as functionality, security, maintainability and sustainability. The defects is assessed on five-point scale: grade 1 with no defects (very good), grade 2 with minor defects (good), Grade 3 with major defects (fair), Grade 4 with Major or minor defect, critical, not functioning as agreed service level (Poor) and Grade 5 consists of grade 4 and very critical, not functioning, risky to safety and health (very poor). All the findings identified are assessed and recorded in building condition schedule and in defects sheets. The defects were illustrated with photographs. The details inspection of the defects to be filled and checked on readiness condition damage or defects found and priority defects or damage needs to be given priority.

The physical condition assessment have to link with maintenance priority action to be taken prior defects findings. The maintenance actions taken will based to priority with scale 1 with no defects (Normal), scale 2 with minor defects (routine), scale 3 with major defects (repairs), scale 4 with critical or serious defects or damages, needs for urgent and immediate repairs (rehabilitation) and scale 5 with critical or serious defects or damages, needs for urgent replacement, refer to expert detail inspection or expert judgement (replacement). According to (Abbott, Mc Duling, Dr, & Schoeman, 2007) the condition assessment rating is linked to the related maintenance actions and types due to the different portions of component could be different conditions and it required the range of maintenance actions.

The extent of defect and the maintenance action combined in a defect score which is Building Condition Assessment Rating System (BCARS). The analysed data of the defect are summarised and a major defective elements were identified and data were sorted for the rating and for analysis. The score obtained from the building condition schedule determine level of defects finding such as very good, good, fair, poor and very poor. The overall building rating derived from the total mark from sum up matrix analysis divided by total number of defects. This is the same method of calculation research by [17] a rating system that minimises subjective evaluation is repeatable and can be effectively used to predict future condition.

6 Findings and Analysis

The implementation of the standard BCA officially was in 2013. The development of the method started in a year 2011. Since that the implementation of method started in 2011 for the improvement and enhancement of the method developed. The execution of the BCA involved the total numbers of 15 types of building with various scale of categories of buildings. It involved the total 23 numbers of building inspection reports which is 11 numbers for office buildings and others for canteen, hanger, quarters, stadium, hostels and health clinics. The inspection was performed by 307 numbers of building inspectors including 4 numbers of trainers. It also involved the total of 196,584 square meters floor area. Table 8 show the list of buildings and finding on time taken to the complete assessment on site.

Through the observation of the findings, the time taken to complete the inspection is between 1 to 4 days and it is dependent to the numbers of the assessors. Figure 2 shows that the percentage of total building according to the time taken to carried out the inspections. It shows that 20% or 3 buildings completed inspected in 1 days. 60% in 2 days, 13% completed in 3 days and 7% completed in 4 days. This situation strongly shows it is possible to complete the inspection within 1 to 3 days.

Table 8: List of projects and findings for office buildings

No.	Building	Types	Area (M2)	Report (n)	Assessor	Time (Day)
1	A	Office	14,400.00	1	4	3
2	B	Office	1,200.00	1	10	2
3	C	Office	14,136.00	1	50	2
4	D	Office	9,958.00	1	13	2
5	E	Office	9,958.00	1	13	2
6	F	Office	2,476.00	1	24	2
7	G	Office	6,715.00	1	20	2
8	H	Office	7,577.00	1	7	1
9	I	Office	1,054.00	1	104	1
10	J	Office	7,527.00	1	4	1
11	K	Office	20,260.00	1	10	2
12	A	Canteen	360.00	1	4	3
13	B	Hanger	19,502.00	1	10	2
14	F	Hostel	5,224.00	1	24	2
15	L	Stadium	19,517.00	1	9	2
16	M	Stadium	28,000.00	1	15	2
17	N	Quarters	8,058.00	1	11	3
18	O	Clinic	20,662.00	6	4	4

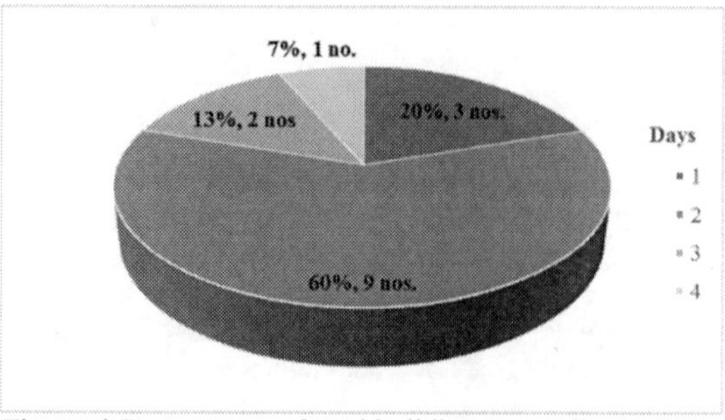

Figure 2 Percentages of total buildings according to the days.

According to figure 3, the detailed inspection took average 3 days to complete the inspection. And this have proved that the time taken to complete the inspections between 1 to 3 days depend on floor area and also the number of the assessors. The number of assessors also important to influence the time taken to complete the inspection although to the various types of buildings with different floor area. Meanwhile according to figure 4, the analysis shows that there is relationship between floor area assessed and the number of assessor in determining the time to complete the inspections. A common form of linear equation is used to floor area, assessor as variable against time of inspection. Variables is independent and it is proved that the possible time of inspection is within 2 days for average 7,000 square meters floor area inspect by only 2 assessors.

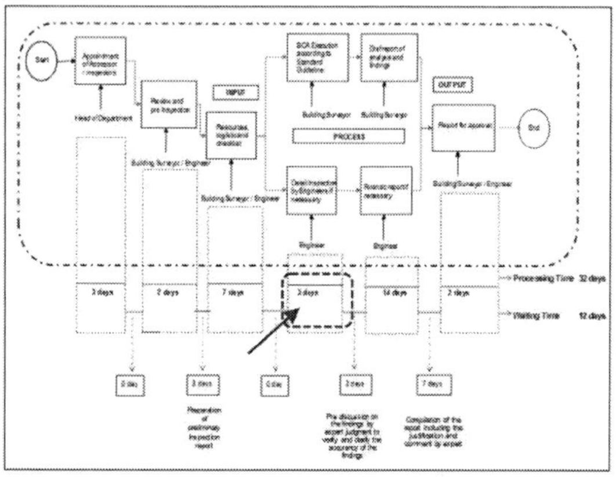

Source: JKR 21602-0004-13

Figure 3 shows the average time taken to complete the assessment including report preparation.

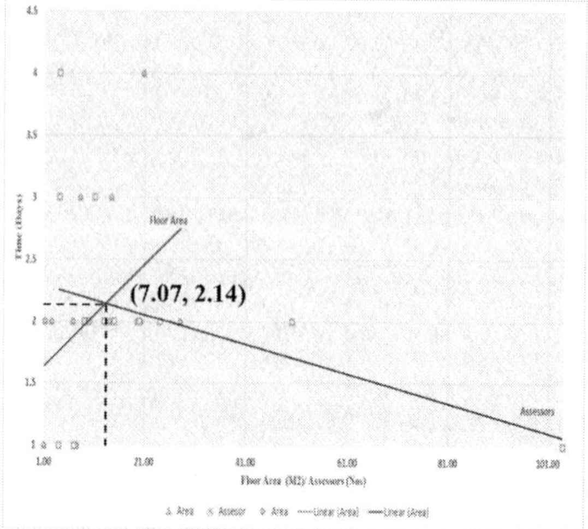

Figure 4 Relationship between floor area, assessor and time

7 Discussion and Conclusion

From the analysis it is can conclude that, by using the BCA standard, the rating of the buildings achieved the same result and reliable to produce accurate rating for planned maintenance work and budgets. Meanwhile the time of inspection will be defer and more shorten when more assessors used to inspect the buildings. However to assess the building defects or building condition is not quite easy to learn from books, guidelines or standard without proper training and also from the experiences in assessing the building condition. The level of understanding of assessor can affect the time taken to complete the inspection. Currently, the intensity of the defects is depends on acceptance or judgment of the assessor or building inspector. Therefore, more experiences gain by the inspector, more accurate the data and also the judgement give to the intensity of the defects. Building inspector requires skill to identify defects and familiar with reporting procedures to ensure the consistency and appropriateness of BCA to different types of buildings.

Acknowledgement

This work was carried out and supported by Building Maintenance and Facility Branch, Public Work Department (PWD) Malaysia which was supported the VACCINE Program - The Implementation of Building Condition Assessment (BCA) as a Performance Tool and to Enhance Maintenance Management Process. The warmest thanks also go to PWD Head Quarters – Mechanical Engineer Branch, Electrical Engineer Branch, Forensic Department, Building Department, PWD Selangor, Melaka, Kelantan, Kedah, Perak, Wilayah Persekutuan Sabah, INTAN, PTPTN and UiTM Perak.

References

1. PWD. Dasar Pengurusan Aset Kerajaan. Kuala Lumpur (2009)
2. Mihai, F., Binning, N., & Dowling, L. A Framework for Assessing Asset Management Performance In Australia. Australia: Main Roads Western Australia.(2002)
3. (INGENIUM), A. o. (2006). International Infrastructure Management Manual. New Zealand: Institute of Public Works Engineering of Australia.(2006)
4. (CIC), C. I. (1996). Definitions of Inspections and Surveys of Buildings. London: CIC.(1996)
5. Joao Antonio Costa Branco Oliveira Pedro, J. A. Portuguese method for building condition assessment. Journal of Structural Survey, 25(34), 322-335 (2008)
6. Ramly, A. Panduan Kerja-kerja Pemeriksaan Kecacatan Bangunan. Malaysia: Hizi Print Sdn. Bhd.(2006)
7. Wordsworth, P. Lee's building Maintenance Management, 4th edition. Oxford Blackwell Science Ltd.(2001)
8. Straub, A. Dutch Standard for Condition Assessment of Buildings. Journal of Structural Survey, 27(1), 233-35.(2009)
9. PWD. Manual Pengurusan Aset Kerajaan (MPAM). Kuala Lumpur, Malaysia: Public work Department.(2009)
10. PWD. Building Condition Inspection for Existing Building.Kuala Lumpur, Malaysia: Public Work Department.(2013)
11. Abbott, G. R., Mc Duling, J. J., Dr, P. S., & Schoeman, J. C. Building Condition Assessment: A Performance evaluation tool towards sustainable asset management. CIB World Building Congress, 649-662.(2007)
12. Kempton, J., Alani, A., & Chapman, K. (2001). Surveyor variability in educational stock survey-a lens model study. Journal of Facilities, 20(5/6), 190-197.(2001)
13. Kempton, J., Nichol, S., Anumba, C., & Dicken, J. Surveyor variability in large-scale house condition survey. Journal of Structural Survey, 19(4), 156-162.(2002)
14. Pitt, T. J. Data Requirements for the prioritization of predictive building maintenance. Journal of Facilities, 15(3/4), 97-104.(1997)
15. NEN, 2. Condition Assessment of Building and Installation Components. Dutch: NEN. (2006)
16. Kleiner, Y. Scheduling Inspection and Renewal of Large Infrastructure Asset. Journal of Infrastructure Systems, 7(4), 136-143.(2001)
17. Hamzah, N., Mahli, M., Che Ani, A. I., Tahir, M., Abdullah, N., & Tawil, N. The Development of Smart School Condition Assessment Based on Condition Survey Protocol (CSP) 1 Matrix: A Literature Review. Science, Engineering and Technology, 620-625.(2010)

Ecological Footprint of Research University Students: A Pilot Case Study in Universiti Teknologi Malaysia

Tan Ang See[1,a], Choong Weng Wai[1] and Irina Safitri Zen[2]

[1]*Department of Real Estate, Faculty of Geoinformation and Real Estate, Universiti Teknologi Malaysia, Johor Bahru, Malaysia*
[2]*Department of Urban and Regional Planning, Faculty of Built Environment, Universiti Teknologi Malaysia, Johor Bahru, Malaysia*

Abstract. Ecological footprint (EF) is potential to be applied in universities to assist building management units to coordinate in order to reduce environmental impact and to achieve sustainable resource consumption from its main activities including teaching-learning, research and operations. As many Malaysian universities declare to become sustainability campus, the adoption of ecological footprint in measuring campus sustainability will provide insight and better understanding about the performance of campus sustainability efforts. The main concept of ecological footprint which convert levels of consumption into the amount of land needed, will able to reveal the average student performance and impacts towards the campus. Further, a study is conducted to determine the average ecological footprint level of students in research universities Malaysia considering students formed the majority of the community in a campus. A pilot study has been conducted in Universiti Teknologi Malaysia (UTM) among students staying in hostel together with staffs from UTM Office of Student Affairs and Alumni (HEMA) and UTM Office of Asset and Development (PHB). Then, Redefining Progress (RP) ecological footprint online calculator is used in computing the ecological footprint of UTM students.

1 Introduction

University consumes resources to fulfill its core functions of teaching-learning, research and operation. A sustainable university should be able to address, involve and promote the minimization of the adverse environmental, economic, social and health impacts from the three core functions and to help society make the transition to sustainable lifestyles [1]. As worldwide universities are increasingly focused on promoting campus sustainability, thus there is a need to develop methodologies to measure a campus level of sustainability. [2, 3] suggested EF as a potential approach to measure campus sustainability. The rationale is to improve university sustainability by decreasing their ecological footprint (EF) [4].

EF has been applied at various levels, from global, regional, nation to city level. Previous researcher developed the EF at the national level [5, 6], regional and city level [7, 8] and university or campus level [2, 9-13]. Other EF studies cater the product and service needs, such as wines [14], ethanol [15], mobile phones [16], tourism [17], international trade [18] and others. EF also can be used to compare the impacts of different lifestyles [10]. There is a need to measure how much is consumed by the community in Malaysian universities. The EF field is still in its infancy stage and currently widespread debate over the methodology for calculating an organizational EF [19]. The major challenge is that there is no commonly accepted method for EF calculation [20].

To date, EF is widely used for raising awareness of environmental impacts, education tool, resource management tool, policy tool, sustainability development measurement and scenario development. This paper aims to discuss the pilot study of EF assessment which has been carried out in Universiti Teknologi Malaysia (UTM).

2 Overview of ecological footprint

Individual activity has an impact on the Earth to a whole city or country [21]. The urgency to protect the Earth is getting more crucial recently due to the accelerating global economy and population continuity to grow and the Earth yet remains the same size [22]. The concept had encouraged some sustainability indicators, including EF in determining the milestone and capacity in the context of sustainability.

In general, the EF is an indicator of human impact on nature. It is often referred as eco footprint or environmental footprint. EF was developed in the early of 1990s by Professor William Rees and Dr. Mathis Wackernagel at the School for Community and Regional Planning, University of British Columbia in Vancouver, Canada [20, 23]. EF is defined as the area of biologically productive land and water ecosystems required to produce the resources that the population consumes and

[a] Corresponding author: angsee88@hotmail.com

assimilate the wastes that the population produces, wherever on Earth the land and water is located [21].

The EF evaluates the impact of an individual, groups or nation lifestyle on the planet by converting levels of consumption into the amount of land needed to sustain necessary production levels and lifestyle choices. It is measured in a standardized unit [area/individual] namely hectare (ha) per capita or global hectare (gha) per capita or global acre (ga) per capita. A hectare is equivalent to 2.5 acres and an acre is approximately the size of a football field. A global hectare is a hectare of biologically productive space with world average productivity of the given year. Most of the EF reports and applications expressed EF in global hectare per capita [24]. Another purpose of EF is to inform the individual's unsustainable behavior which further influence the consumers' consumption patterns towards sustainability lifestyles, thus reducing the impact to the environmental [25].

2.1 The world ecological footprint

Conceptually, EF must not exceed the biocapacity of the Earth to sustain human society. Figure 1 tracked the world EF and biocapacity since 1961 to 2011. Both EF and biocapacity are measured in gha per capita. The upper right corner of Figure 1 showed the quality of the assessment which is scored on a 1-6 scale. Data quality 6 means that the results are reasonably reliable. From Figure 1, it is clearly shown that humanity only used about two-third of Earth available resources in 1961. When population increased, human demand for resources began outstripping what the planet could renewably produce during the early 1970s. In 2011, the global EF was 2.7 gha per capita but what is available on the Earth is only 1.7 gha per capita. The result implicated that humanity consumption has exceeded the amount of resources Earth provided by 50%. Furthermore, the Earth needs 1.5 years to regenerate resources that people are using in a single year. It is predicted that two planets will not be enough to support the demand on resources by 2030.

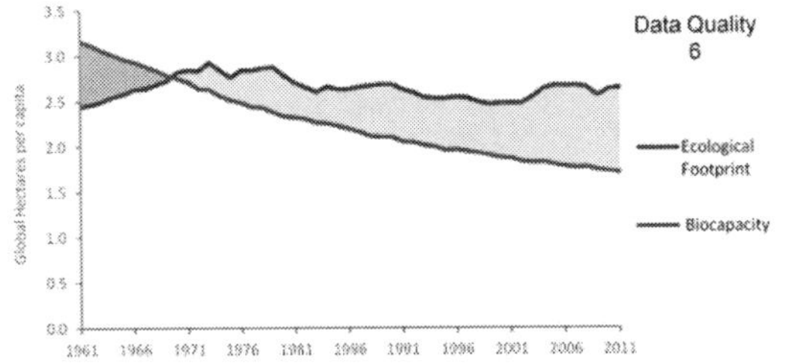

Figure 1. World ecological footprint and biocapacity (1961-2011) [26]

2.2 Malaysia ecological footprint

In global context, the Malaysia EF is smaller compared to the developed countries such as United States, Canada and United Kingdom. However, within the region, Malaysia has larger EF than other ASEAN countries such as Thailand, Indonesia and Philippines [27-29]. Currently, the Malaysia EF calculation is provided by Living Planet Report (LPR) and National Footprint Accounts (NFA).

LPR is published every two years by World Wide Fund for Nature (WWF) in cooperation with the United Nations Environmental Programme (UNEP) and Global Footprint Network (GFN). It is the world's leading, science-based analysis on the health of our planet and the impact of human activity. The latest edition of LPR was released in 2014.

Another significant report, NFA is released by GFN annually. It provides comprehensive data on human demand on resources as to support their activity. They track how this demand compares across several over 200 countries, territories and regions and how it relates to planet's biocapacity to meet this demand. The NFA 2015 is the most current published edition where EF and biocapacity observed data results from 1961 to 2011 has been documented.

The Malaysia EF and biocapacity from 1961 to 2011 is presented in Figure 2. The upper right corner of Figure 2 showed the quality of the assessment which is scored on a 1-6 scale. Data quality 3B means that the results are cleared for both latest year and overall time series. According to the figure, Malaysia EF has exceeded its biocapacity since 1995. The Malaysia EF recorded at 2.9 gha per capita and biocapacity at 2.3 gha per capita in 2011. In other words, each Malaysian would require about three hectares of land to sustain their current living standards. Malaysia's EF ecological deficit reached about 0.5 gha per capita in 2011. Up to now, Malaysia EF has exceeded the global average of 2.7 gha per capita.

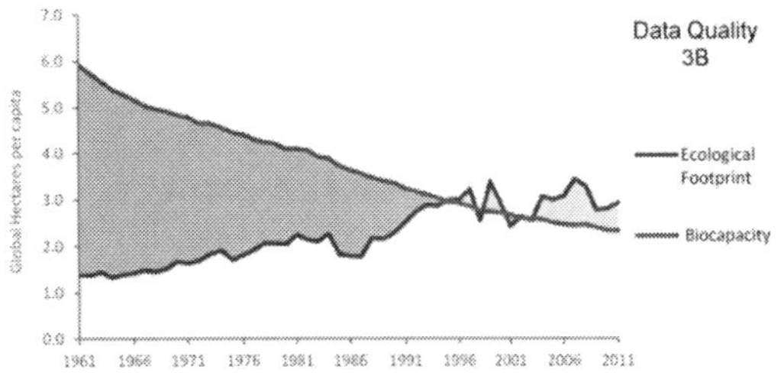

Figure 2. Malaysia ecological footprint and biocapacity (1961-2011) [26]

3 Ecological footprint for university

Universities are very similar to small cities [30]. They have large land area, growing populations and increasing number of traffics to cater the needs of research and education activities. The physical built environment of universities covers a range of building types such as office, classroom, hostel, laboratory, health care and large assembly hall. It will have impact to the environment and sustainability through the resources consumption such as energy and water as well as waste generation [31]. Furthermore, universities considered as the largest consumers where they use large amount of paper, energy and water. In addition, research-intensive university consumes more resources than a conventional university [32] based on education activities.

The application of EF to university is neither new nor novel. EF calculations have been undertaken in campuses such as University of Newcastle, University of Redlands, University of Toronto at Mississauga and others. The EF of these universities is shown in Table 1.

Table 1. Ecological footprint of universities.

University	Period	EF	
		Value	Unit
University of Newcastle [10]	1998	0.19	ha/capita
University of Redlands [12]	1998	1.00	
School of Physics (SoP) in University of Sydney [13]	1997	6.80	
Holme Lacy College [9]	2001	0.57	
University of Toronto at Mississauga [2]	2005	1.07	
Northeastern University at China [11]	2003	1.06	
Universiti Pendidikan Sultan Idris (UPSI) [3]	2008	0.31	
University of Illinois at Chicago (UIC) [33]	2008	2.66	gha/capita

In Malaysia, studies of EF are limited notably at the national level and absent at the state and city levels. The absence of studies is partly attributable to data limitation [20]. Literature search revealed that the only EF calculation at university level was conducted in the Universiti Pendidikan Sultan Idris (UPSI) by [3]. Other EF articles including [20], [27] and [28] did their EF research at national level based on input-output (IO) analysis for Malaysia. Other EF application in Malaysia include high rise condominium residential area [34] and Malaysian palm oil industry [35].

In conjunction with sustainable development, Malaysian universities such as Universiti Malaya (UM), Universiti Sains Malaysia (USM), Universiti Kebangsaan Malaysia (UKM), Universiti Putra Malaysia (UPM) and Universiti Teknologi Malaysia (UTM) have their campus sustainability initiatives. However, without measurement of the environmental impact, it would be difficult to manage sustainable development [5]. Knowing the current level of environmental impact would allow the universities to know how far they are from achieving sustainability.

EF is one of the sustainable indicators that could help to assess the environmental impact for universities. The EF could serve as metric to support sustainability activities for campus. Firstly, it highlights the current level of environmental impact for the university. Secondly, EF is presented on a single arrogated scale that allows for direct comparison of the impacts of different components against one another [36]. Thirdly, the EF assessment could assist sustainable development decision making for the management of university. Calculating EF could be the first step for Malaysian university to become a more sustainable campus.

4 Pilot study

A study is conducted to determine the average EF level of research university students in Malaysia as students formed the majority of community in a campus. A pilot study has been conducted in UTM among students who are staying in hostel together with staffs from UTM Office of Student Affairs and Alumni (HEMA) and UTM Office of Asset and Development (PHB). Data such as room size, energy sources, building materials, energy and water saving features can only be acquired from staffs because students do not have access to these types of information. Then, Redefining Progress EF online calculator is used to compute the EF of UTM students after the data collection has been completed.

Over the last few years, several popular websites of non-governmental organization (NGO) like WWF, GFN, Best Foot Forward (BFF) and Redefining Progress (RP) have offered personalized EF calculators which allow individuals and firms to enter information on their lifestyle to assess own environment impact. Individuals need to report information on personal consumption habits and activities via resources use or waste assimilation such as food consumption, housing, energy, transportation and others. The comparison of EF online calculators is shown in Table 2.

RP is one of the original and most influential EF online calculators. Besides, RP is more location-specific compared to other EF online calculators [37]. The EF quiz in RP was first developed in 2002 with Earth Day Network (EDN) to provide simple yet powerful way for people to measure their impact on the Earth. In 2008, RP updated the EF quiz with a revised methodology and current statistics. RP has 27 questions in its EF quiz. However, the questions in RP were slightly modified to increase its applicability to university students and avoid confusion in this pilot study.

Table 2. Comparison of ecological footprint online calculators.

	WWF	GFN	BFF	RP
Location	Australia	Calgary, United States of America (USA), Ontario, Colombia, Ecuador, Peru, Argentina, Brazil, Luxembourg, Switzerland, Italy, Turkey, South Africa, India, China, Japan and Australia	United Kingdom (UK)	147 countries including Malaysia
Question	30	Different for every country	12	27
Unit	gha	gha	gha	gha or ga
Income	-	-	-	√
Food	√	√	√	√
Housing	√	√	√	√

Energy	√	√	√	√
Transport	√	√	√	√
Goods	√	√	-	√
Water	-	√	-	√
Waste	√	√	√	√

The result of the pilot study is shown in Figure 3. The graph shows a comparison of average EF between Malaysian and UTM students. There are 50 UTM students involved in the pilot study. EF represents the consumption of resources which consists of carbon, food, housing, goods and services. EF, carbon footprint, food footprint, housing footprint, goods and services footprint are measured in gha.

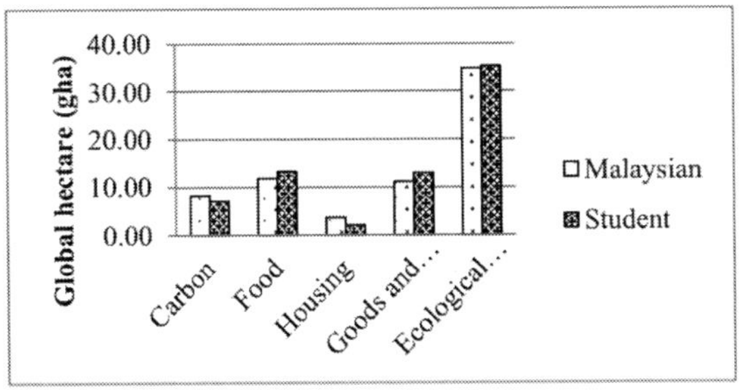

Figure 3. Pilot study

The average EF of UTM students is slightly higher than average EF for Malaysian which are 35.34 gha and 34.80 gha respectively. Lower score on EF indicate more sustainable while higher score on EF indicate less sustainable [38]. EF of the university is greater as it is considered a net importer of consumption items [10]. In the average EF of UTM students, the largest component is the food footprint, next is goods and services footprint, followed by carbon footprint and housing footprint is the least. In terms of carbon and housing, UTM students recorded lower consumption compared to Malaysian. UTM students consumed 7.10 gha for carbon while Malaysian consumed 8.22 gha. For housing, UTM students recorded 2.05 gha while Malaysian recorded 3.65 gha. This is because UTM students who participated in this pilot study are staying in university campus rather than off-campus housing [2]. The travel distance to academic and administration buildings for students staying in university campus is shorter than students who are staying off-campus. According to [39], transportation mode and travel distance contributing to the size of an EF. Travel longer distance have significantly larger footprint. Housing footprint of UTM students is lower than Malaysian because the university provides apartment type accommodation as hostel. [40] stated that apartment has lower impact to the environment. However, UTM students consumed higher consumption in food, goods and services compared to Malaysian. For food, UTM students consumed 13.26 gha while Malaysian consumed 11.83 gha. From the findings, most of the UTM students in the pilot study are omnivore. Omnivore has an assortment of meat, seafood, vegetables, dairy and grains. Based on [33], diet of food that is higher up on the food chain such as meat and seafood required more resources to support them than the lower one like vegetables. As UTM does not allow cooking in hostel, UTM students have no other choice but to obtain food from supermarkets, convenience stores and prepared foods from restaurants. This is similar to findings in [41]. [9] said that large value for food footprint implies that many items are sourced from far away. Thus, food footprint of UTM students is higher than Malaysian. For goods and services, UTM students recorded 12.93 gha and Malaysian recorded at 11.10 gha. The findings show that UTM students do not practice recycling habits frequently. Paper and plastic are the most recycled items by the UTM students. Most of them do not recycle aluminium, glass and electronics. This is due to recycle bins are not provided in every hostel in UTM.

5 Conclusion

The pilot study result showed that the average EF of UTM students by using RP is 35.34 gha. The average EF of UTM students is considered higher in comparison with previous studies. The pilot study result is promising and reveals that EF is a useful sustainability assessment tool to measure current university activities. The result can be utilized to improve community awareness and to assist the university management in decision making. This would enable policy makers to plan sustainable efforts with more informed decision. It is recommended that the study to be carried out on a larger scale considered this is only a pilot study. As Malaysian universities are pursuing campus sustainability, the application of EF would contribute and in parallel with the effort. Universities that are interested may find the larger scale of such study is needed as the data can be used to help them in developing strategies action plans to improve the level of campus sustainability.

References

1. L. Velazquez, N. Munguia, A. Platt, J. Taddei, J. Clean. Prod., **14**, 810-819 (2006)
2. T. M. Conway, C. Dalton, J. Loo, L. Benakoun, Int. J. Sust. Higher. Ed., **9**, 4-20 (2008)
3. S. Samsudin, Z. M. Said, N. Ashikin, A. K. M. Razali, *2nd International Conference on Human Habitat and Environment* (2010)
4. C. Jauch, S. Ogden, A. Betzen, C. Stumpff, R. Bigley, *Examining Transportation Sustainability at the University of Kansas-Lawrence Campus* (2009)
5. M. Wackernagel, C. Monfreda, D. Deumling *Ecological footprint of nations: November 2002 update* (2002)
6. C. Monfreda, M. Wackernagel, D. Deumling, Land Use Policy, **21**, 231-246 (2004)
7. M. E. Hopton, D. White, J. Environ. Manage., **111**, 279-286 (2012)
8. M. Wackernagel, J. Kitzes, D. Moran, S. Goldfinger, M. Thomas, Environ. Urban., **18**, 103-112 (2006)

9. G. F. M. Dawe, A. Vetter, S. Martin, Int. J. Sust. Higher. Ed., **5**, 340-371 (2004)
10. K. Flint, Int. J. Sust. Higher. Ed., **2**, 48-62 (2001)
11. G. J. Li, Q. Wang, X. W. Gu, J. X. Liu, Y. Ding, G. Y. Liang, Ecol. Indic., **8**, 75-78 (2008)
12. J. Venetoulis, Int. J. Sust. Higher. Ed., **2**, 180-196 (2001)
13. R. Wood, M. Lenzen, Local Environ., **8**, 365-386 (2003)
14. V. Niccolucci, A. Galli, J. Kitzes, R. M. Pulselli, S. Borsa, N. Marchettini, Agric., Ecosyst. Environ., **128**, 162-166 (2008)
15. E. D. d. O. Marcelo, B. E. Vaughan, E. J. Rykiel, Jr., Bioscience, **55**, 593-602 (2005)
16. S. D. Frey, D. J. Harrison, E. H. Billett, J. Ind. Ecol., **10**, 199-216 (2006)
17. P. Peeters, F. Schouten, J. Sustain. Tour., **14**, 157-171 (2006)
18. H. Li, Z. Pei Dong, H. Chunyu, G. Wang, Ecol. Econ., **62**, 136-148 (2007)
19. N. Aporo, M. Chiron, Z. Kagan, C. McCool, N. O'Leary, D. Perano, S. Walton, *Otago Management Graduate Review*, 1-16 (2007)
20. R. A. Begum, J. J. Pereira, J. Appl. Sci. Res., **8**, 4783-4787 (2012)
21. M. Wackernagel, W. E. Rees, *Our ecological footprint: reducing human impact on earth* (1996)
22. A. Galli, T. Wiedmann, E. Ercin, D. Knoblauch, B. Ewing, S. Giljum, Ecol. Indic., **16**, 100-112 (2012)
23. W. E. Rees, Environ. Urban., **4**, 121-130 (1992)
24. J. Kitzes, A. Peller, S. Goldfinger, M. Wackernagel, Science for environment & sustainable society, **4**, 1-9 (2007)
25. J. Franz, E. Papyrakis, *Reconsidering the 'ecological footprint' index: Does it promote sustainable behaviour* (2011)
26. G. F. Network, Retrieved 9th February (2015)
27. R. A. Begum, J. J. Pereira, Int. J. Ecol. Econ. Stat., **14**, 94-103 (2009)
28. R. A. Begum, J. J. Pereira, A. H. Jaafar, A. Q. Al-Amin, Resour., Conserv. Recycl., **53**, 582-587 (2009)
29. G. F. Network, Oakland, CA. www. footprintnetwork. org/en/index. php/GFN/page/footprint_ data_and_results/. Consulté le, (2013)
30. H. M. Alshuwaikhat, I. Abubakar, J. Clean. Prod., **16**, 1777-1785 (2008)
31. M. Bennett, P. Hopkinson, P. James, *Sustainability Accounting and Reporting*, 409-430 (2006)
32. O. Baboulet, M. Lenzen, J. Clean. Prod., **18**, 1134-1141 (2010)
33. C. Klein-Banai, T. L. Theis, Ecol. Indic., **11**, 857-860 (2011)
34. S. M. Chan, *The adoption of ecological footprint for sustainability* (2008)
35. K. W. Lee, *Ecological footprint analysis on Malaysian Palm Oil Industry : case study of Seri Ulu Langat Palm Oil Mill* (2008)
36. H. Bekmann, S. Rickards, C. Noller. *Proceedings of the 13th International Australasian Campuses towards Sustainability (ACTS) Conference* (2013)
37. A. Brook, Soc. Influence, **6**, 113-128 (2011)
38. H. L. Dietrich, *The role of emotion in environmental decision making*, 131 (2013)
39. H.-C. Ryu, S. D. Brody, Int. J. Sust. Higher. Ed., **7**, 158-175 (2006)
40. S. Bastianoni, A. Galli, V. Niccolucci, R. Pulselli, U. Mander, C. Brebbia, E. Tiezzi, *The Sustainable City IV: Urban Regeneration and Sustainability*, 345-356 (2006)
41. S. Raj, S. Goel, M. Sharma, A. Singh, J. Environ. Occup. Sci., **1**, 23-26 (2012)

The degree of housing damage model for a flood affected area

Thuraiya Mohd[1,a], Mohamad Haizam Mohamed Saraf[2], Siti Fairuz Che Pin[3], Mohd Nasurudin Hasbullah[4], Tajul Edrus Nordin[5] and Dzulkarnaen Ismail[6]

[1, 2, 3] Department of Real Estate Management, Faculty of Architecture, Planning and Surveying, Universiti Teknologi MARA, Perak, 32610 Seri Iskandar, Perak, West Malaysia
[4, 5] Department of Architecture, Faculty of Architecture, Planning and Surveying, Universiti Teknologi MARA, Perak, 32610 Seri Iskandar, Perak, West Malaysia
[6]Department of Building, Faculty of Architecture, Planning and Surveying, Universiti Teknologi MARA, Perak, 32610 Seri Iskandar, Perak, West Malaysia

Abstract. Floods can cause damage like slightly damaged, significantly damaged or destroyed to homes and possessions as well as disruption to communications. Inherently, victims should be given temporary or permanent houses depending on the degree of damage to their houses. Therefore, an assessment on the levels of damage must be carried out in the aftermath of a flood as a direction for recovery effort, for example housing resettlement. The fact is, in Malaysia, there is still no standardized damage assessment used by the relevant authorities in assessing the degree of housing damage after a disaster. As a result, errors in assessing the degree of housing damage and providing inaccurate type of assistance might occur. Thus, this research emphasis on the understanding the degree of house damage and recommend the significant input in developing the damage assessment model in Malaysia. To achieve the objective, this research applies a self-developed model that is derived from the literature review (framework or model of the degrees of housing damage after flood) and the observation at the case study area to see the actual conditions of the affected houses. After that, questionnaires were distributed to 50 respondents consist of engineers (n=10), architects (n=10), quantity surveyors (n=10), real estate valuers (n=10) and building surveyor (n=10) by using purposive sampling to gauge their perceptions on attributes of degree of housing damage and eventually conducting a focus group consist of ten (10) technical experts involved in MERCY Malaysia in assessing the housing damage for model validation. The findings indicate that the degree of damage can be classified as 'minor', 'major' and 'destroyed'. Research findings will give input in the form of a Housing Damage Assessment Framework for the government, NGOs, MERCY, insurers or other appropriate bodies involve in assessing or evaluating the condition of houses affected by floods.

1 Introduction

Increasing population has resulted in an increase in the number of property ownerships. Therefore, a greater percentage of the country's land area, often in areas that previously have been not fit for urban development and human settlement, have been taken up to cater for the need for accommodation (Eves, 2014). These increased numbers of properties, changes in water collection, flows and poor drainage system coupled with heavy monsoon rainfall, intense convection rainstorms and other local factors have caused seasonal floods in Malaysia (Chan, 1996; Eves, 2014). According to the Malaysian National Security Council [MNSC] (2015), flood is the most common type of disaster that occurs in Malaysia. Floods occur annually in Malaysia, causing damage to property and loss of life. The worst flood event in Malaysia was in December 2014 that rendered people helpless. The moving water had destructive powers that picked up and carried off bridges, houses, trees, and cars. The east coasts states of Kelantan, Terengganu and Pahang expose to annual recurrence of flood as these states are directly in the path of the seasonal monsoon season (Chan, 1996). The unprecedented flooding of December 2014 has been the worst flood event in the history of Kelantan.

Following such a disaster, there is often a tally of the preliminary damage assessment in respect to the injuries, loss of lives, cost of damage and destroyed properties. With these disasters attracting considerable media attention, people are more aware of the damage occurred at the affected areas. There have been a number of studies pertaining to preliminary damage assessment to buildings after the flood. There are many ways of damage assessment which have been carried out in different countries after the event of a natural disaster. There are several guidelines for assessing the degree of building damage prepared by government agencies, researchers, local authorities and non-governmental organizations (NGO). For example, Attaullah Shah, Hamid Mumtaz Khan and Ehsan U. Qazi (2009) had outlined the

[a]Thuraiya Mohd: thuiraya.mohd@yahoo.com

evaluation of the buildings destroyed or damaged due to flooding in Pakistan. The evaluation of damage is made on mud houses that are the most common type of building structure in Pakistan. In the USA, Federal Emergency Management Agency [FEMA] has developed an operations manual to standardize the procedures in preliminary damage assessment nationwide. There are several state authorities in the USA, like New Jersey and Florida, which reviewed this operations manual and did some modification to suit the type of disasters that occur in their states. Apart from that, there are also numerous literatures which emphasized on residential properties damaged from hurricane and floods (Hodde 2012; FEMA, 2012).

In concordance with the variety of the degree of housing damage, different countries are likely to have different construction methods, materials used and the nature of the flood disaster. Thus, a general assessment from other countries that determines the extent of flood hit houses seems irrelevant for Malaysia context. Moreover, in Malaysia, there is no standardized damage assessment used by the authorities or relevant agencies in assessing the degree of housing damage after disasters. As a result, errors in assessing the degree of housing damage and providing inaccurate type of assistance may occur. Thus, this research emphasis on the understanding the degree of houses damage and recommends significant input in developing the damage assessment framework in Malaysia.

2 Literature review

2.1 Flood Disaster Socioeconomic Impacts

2.1.1 Flood Damage

According to MNSC (2015), flood disaster is defined as an incident caused by natural calamities and human factors that happen to cause devastation, not only to the environment and to property, but also may cause death and injury to humans. Consequently, flood damage is a direct consequence, or expressed as a physical attribute that can be directly measured in terms of a level of degradation, spoil, removal or destruction (Friedland, 2009). Meanwhile, Grigg and Helweg (1974) defined flood damage as the amount of money to restore the area to its original condition before the disaster. Thus, to understand the damage of flood, it would be useful to categorize them.

2.1.2 Category of Damage (Socioeconomic Impacts)

According to the South Asian Tsunami Damage Assessment Report (2004), damage is categorised as a physical description of the degree of housing damage; no damage, light damage, moderate damage, heavy damage or collapse. Daungthima & Kazuori (2012) underlined the category of damage in three groups. The first group is Environmental Damage, which consists of areas at risk from flood, ground cracks, landscape damages, and pit on ground or subsidence, surface water flow paths, vulnerable communities and critical infrastructure. The second group is the External Damage, which is light damage such as wall, decorative aspects and structural damage. Lastly is Internal Damage, which includes interior of the affected building such as wall, decoration and ceiling. According to the Collaborative Research on Flood Resilience in Urban Areas [CORFU] (2014), flood damage can be distinguished as tangible and intangible damage. A tangible damage is damage that is capable of being assessed in monetary terms (Smith & Ward, 1998). Messner et al (2007) stated a tangible damage is the damage which can be easily specified in monetary terms. Therefore, intangible damage is defined as damage that cannot be easily specified in monetary terms. These distinctions are clearly made to define flood damage. In spite of that, Messner et al (2007) & Jonkman et al (2008) explained that damage is categorized into direct and indirect damage (refer to Table 1).

2.1.3 Structural Type of Building

For masonry buildings, Coburn (1989) has classified structural damage to unreinforced masonry building as illustrated in Figure 1, viewing the collapse of walls and roofs that support vertical loads. Next, in Figure 2 will illustrate the typical damage to wood frame building structures. When it comes to reinforced concrete type of buildings, the typical damage is illustrated in Figure 3. Damages to mud house are shown in Figure 4 according to the damage assessment of flood affected mud houses in Pakistan conducted by (Shah, Khan, & Qazi, 2013).

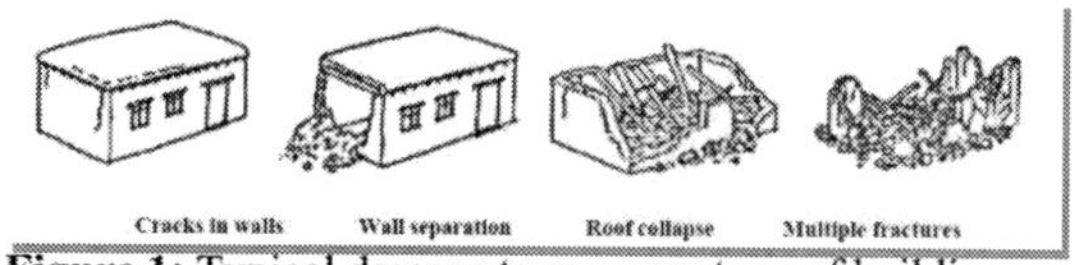

Figure 1: Typical damage to masonry type of buildings

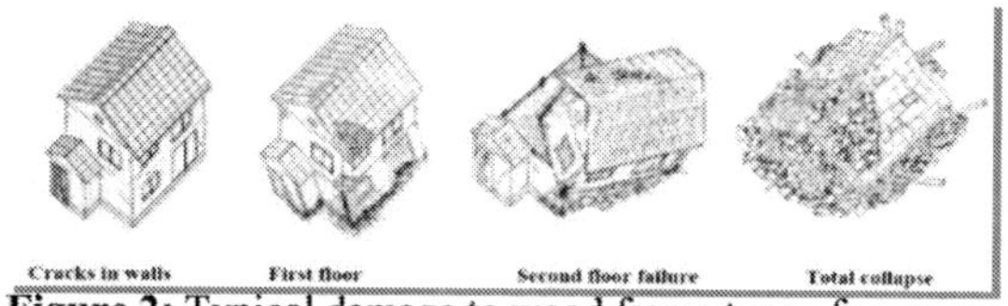

Figure 2: Typical damage to wood frame type of buildings

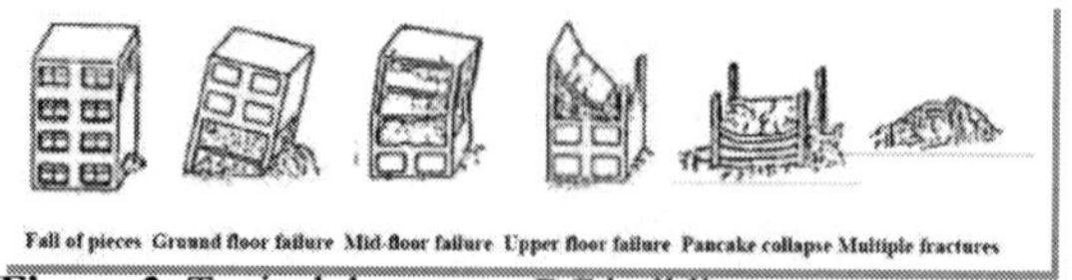

Figure 3: Typical damage to RC buildings

Figure 4: Typical damage to mud houses

2.2 The Existing Degree of Building Damage Assessment Models

There are many ways of damage assessment which have been carried out in different countries after the event of natural disasters. There are several guidelines for assessing the degree of building damage prepared by government agencies, researchers, local authorities and non-governmental organizations (NGO). In concordance with the variety of the degrees of housing damages, different countries are likely to have different construction methods, materials used and the nature of the flood disaster. The attributes models from a synthesis of existing eight degrees of housing damage are as follows.

2.2.1European Macro-seismic scale 1998 (EMS-98)

The EMS-98 is the basis for evaluation of seismic intensity in European countries and is also used in a number of countries outside Europe. The first scale has 12 divisions for seismic loads, but for flood disaster, the damage classification scale is only classified into six divisions namely' Grade 0, Grade1, Grade 2, Grade 3, Grade 4 and Grade 5. The grade takes into consideration structural and non-structural elements. Structural elements are the main structures such as foundation, beams, columns, roofs, floors and load-bearing walls. Whereas non-structural elements are gutters, chimneys, plaster ceilings, siding et cetera.

2.2.2 Green Alert Damage Assessment Manual

Damage assessment Manual by www.GreenAlert.net is based on the Windshield Survey in 2004 that provides information to determine the severity of the disaster and type of disaster assistance that may be required. In this manual, the person responsible will assess the affected house into damage categories representing a range of damage percentage wise.

2.2.3 State of Florida Division Emergency Management

In Florida, as it is vulnerable to a host of natural and man-made disasters, local governments may contact the Florida Department of Community Affairs, Division of Emergency Management to initiate assistance for damage assessment. Immediately following a disaster, an initial damage assessment must be carried out by them to estimate the type and extent of damages. The model is influenced by the FEMA, with some improvements on the structure damage attributes.

2.2.4 Earthquake Engineering Field Investigation Team of Japan (EEFIT) Model

EEFIT was formed as joint venture between industries and universities. EEFIT has its own mission team typically to visit affected regions for data collection to improve the understanding of structural behavior of natural disasters especially seismic loads. In Japan, EEFIT mission team made some modifications to the damage assessment model from the European Macroseismic Scale 1998 or also known as EMS-98.

2.2.5 National Disaster Management Agency of Indonesia (BNBP)

In the event of a natural disaster in Indonesia, especially earthquakes before flooding, The National Disaster Management Agency (BNPB) will lead in coordinating and facilitating the recovery, reconstruction and rehabilitation of the flood affected areas. Preliminary damage and needs assessment are necessary in the delivery of assistance addressing the specific needs for particular vulnerable groups of affected victims. In assessing the damage, BNPB use a model that distinguishes three degrees of property damage; Light Damage, Moderate Damage and Heavy Damage.

2.2.6 New Jersey State Police Office Emergency Management

The New Jersey State Police Office of Emergency Management produced a preliminary damage assessment in any event of natural disasters, especially in the event of hurricanes. Based on FEMA Preliminary Damage Assessment, they did some modifications on damage scale and categorized it into 4 degrees; Affected, Minor, Major and Destroyed.

2.2.7 Henry B. Hodde III Damage Assessment Model

Although FEMA has prescribed specific guidelines for the damage assessment procedure, Henry (2012) stressed that most communities or state organizations use slightly different protocols and methods in assessing property damage caused by natural disasters. In his research, he laid out several damage assessment models from thirteen local governments and provided with an improvised degree of damage assessment. He further categorized the damage into four degrees, Category 1, Category 2, Category 3 and Category 4. However, this category of damage only described the structured damage and did not explain the details in the structured damage attributes.

2.2.8 Federal Emergency Management Agency (FEMA) Revised Model of Damage Assessment Operations Manual (2012)

This operations manual was developed by FEMA to standardize the procedures in preliminary damage assessments nationwide. This manual was prepared and reviewed by FEMA regional officers with vast experience in performing damage assessment through many types of disasters. The models, this manual use to distinguish degree of damage are categorized into Destroyed, Major, Minor, Affected and Inaccessible.

2.3 The Degree of House Damage Frequency Table

From the review on eight (8) models of house damage assessment, the attributes are laid out in Table 1. The attributes were sorted according to its degree of damage and the frequencies of each attributes were recorded to provide frequency data for the development of the suggested degree of the damage model.

Table 1: The frequency table of the degree of house damage attributes

Degree of Damage	Author	EMS 98	Green Alert	Florida State	EEFIT	BNBP	New Jersey State	Henry B. Hodde III	FEMA	Frequency
	Year of Publication	**'98**	**'04**	**'04**	**'05**	**'10**	**'12**	**'12**	**'12**	
	Country/ Region of Study	**EUR**	**UK**	**USA**	**JPN**	**INA**	**USA**	**USA**	**USA**	
	Domain/ Attributes									
No Damage/ DM0/ Grade 0/ N/A (Not Available)										
	No visible damage				√				√	**2**
Grade 1										
	Minor cracking in partitions, infills and ceilings	√								**1**
	Minor damage to non-structural elements	√				√	√		√	**4**
Affected/ Affected but Habitable/ DM1/ Grade 2/ Category 1										
	Hairline cracks in beams/columns near the joints	√								**1**
	Chipping on plaster walls	√		√	√					**3**
	Minor cracks on walls	√		√	√			√	√	**5**
	Minor damage to door/ windows			√	√			√	√	**4**
	Damage to shutters, gutters, roof shingles or siding		√	√						**2**
	Minor cracks on facade	√								**1**
Minor/ Minor 5%/ DM2/ Grade 3/ Category 2/ Light Damage										
	Partial failure of walls, floors, foundation			√		√	√	√		**4**
	Fracture at the roof line	√								**1**
	Failure of partition walls	√								**1**
	Failure of gable walls	√								**1**
	Large and visible cracks in most walls	√	√		√					**3**
	Roof tiles detached	√	√							**2**
	Cracks in most beams/columns	√								**1**
	Collapsed of parts of masonry walls				√					**1**
	Scouring at corners of structures leaving partly exposed foundation				√					**1**
	Exposing sheathings		√							**1**
	Windows and/or doors blown in								√	**1**
	One foot or more of water in basement								√	**1**
	Furnace or water heater damage								√	**1**
Major/ Major 30%/ DM3/ Grade 4/ Category 3/ Moderate Damage										
	Utilities damaged		√							**1**
	One or more rooms collapsed		√							**1**
	One exterior walls collapsed		√							**1**
	Exterior building frame damaged		√							**1**
	Foundation damaged		√							**1**
	Collapse of a	√								**1**

 DOI: 10.1051/matecconf/20166600074

	few columns/beams									
	Substantial failure of walls, floors and roof	√	√	√	√	√	√	√	√	8
	Substantial failure of foundation			√						1
	One foot or more of water on the first floor (home with basements)								√	1
	Collapsed of masonry walls panels				√					1
	Excessive foundation settlement and tilting				√					1
Destroyed/ Destroyed 80%/ Grade 5/ Category 4/ Heavy Damage										
	Complete or impending building collapse	√	√	√		√	√	√	√	7
	Two exterior walls collapsed		√							1
	Two or more basement walls collapsed		√							1
	Second floor is gone		√							1
	House pushed off from the foundation		√						√	2
	Exposed floor slabs				√					1
	Visible and heavy damage to foundation				√					1
	Heavy scouring that leads to collapse of large sections of foundation				√					1
Inaccessible										
	Road or/and bridge damaged		√	√						2
	House is flooded or submerged		√							1
	Road covered by water			√						1

[Source: Researchers' Study, 2015]

3 Research Methodology

3.1 Research design

Purpose of this research is to develop Housing Damage Assessment Model. In other to achieve the objective, this research applies the triangulation research design in order to gauge an understanding on the degree of housing damage. Further, it recommends significant input in developing a damage assessment framework in Malaysia. This design encompasses both descriptive and exploratory research methods.

3.2 Data Collection

This research applies a self-developed model that is derived from the literature review (framework or model of the degrees of housing damage after flood) and the observation at the case study area to see the actual conditions of the affected houses. After that, questionnaires were distributed to 50 respondents consist of engineers (n=10), architects (n=10), quantity surveyors (n=10), real estate valuers (n=10) and building surveyor (n=10) by using purposive sampling to gauge their perceptions on attributes of degree of housing damage and eventually conducting a focus group consist of ten (10) technical experts involved in MERCY Malaysia in assessing the housing damage for model validation.

3.3 Data Analysis

In analysing and evaluating the results of this research, quantitative approaches were used. These approaches involved the analysis of data and information through the perception survey method. Quantitative data which was obtained through questions involving likert scale type contained in a structured questionnaire form were analysed by using the Statistical Package for the Social Sciences (SPSS) software. In order to achieve research objective, the data were analysed using Confirmatory Factor Analysis (CFA). CFA is a statistical technique used to verify the factor structure of a set of observed variables. CFA allows the researcher to test the hypothesis that a relationship between observed variables and their underlying latent constructs exists. The researcher uses knowledge of the theory, empirical research, or both to postulate the relationship pattern a priori and then tests the hypothesis statistically (Harrington, 2009). Initially, the factorability of the eight (8) attributes of degrees' of housing damage was examined using Kaiser-Meyer-Olkin (KMO) and Bartlett's test.

3.4 Pilot Study

The reliability of these constructs measurement was evaluated by examining the internal consistency reliability. Thus, a pilot study was carried out on 25 respondents: engineers, architects, building surveyors, quantity surveyors and real estate valuers. An analysis using the reliability test- Cronbach's Alpha (α) was conducted to ensure the reliability of the constructs. As Table 2 shows, the Cronbach's Alpha of inaccessible attributes that are ground zero, flooded houses and, roads and bridges is out is higher than 0.60. Similarly, for the 'minor' attributes (minor damage in partitions, infills and ceilings partition, infills and ceiling, minor damage to door and windows, large and visible cracks in most walls and roof tiles detached) and major attributes (substantial failure of walls, floors, foundation or roof and utilities damaged) also destroyed attributes that is total loss are higher than 0.60. This indicates that all the attributes have a good reliability. For minor attributes, the researchers found that if the attributes large and visible cracks in most walls and roof tiles detached was deleted, the Cronbach's Alpha would rise to 0.747 and 0.678 respectively. As a result we removed both attributes, the Cronbach's Alpha of whole attributes rose to 0.915 (see Table 3) and Cronbach's Alpha for every attribute is more than 0.90 (refer Table 4). Finally, the researchers decided eight (8) attributes as an input to housing model damages. This indicates that the reliability of all attributes of housing damages model was at a high value of Cronbach's Alpha.

Table 2: Cronbach's alpha if item is deleted in the pilot study

Suggested Degree of Damage	Suggested Description of Damage	Suggested Attributes of Damage	Cronbach's Alpha if item deleted
Inaccessible	The residence is inaccessible by normal means	The house is flooded or submerged	0.628
		Roads and bridges are out	0.628
		Ground zero	0.628
Minor	Slight damage to building structure, can be occupied within a short period of time after minor repairs.	Minor damage in partitions, infills and ceilings	0.637
		Minor damage to doors and windows	0.637
		Large and visible cracks in most walls	0.747
		Roof tiles detached	0.678
Major	The building has sustained structural or significant damages, uninhabitable and requires extensive repairs.	Substantial failure of walls, floors, foundation or roof.	0.628
		Utilities damaged (Electrical, Surface water drainage, Sewerage reticulation system, Water reticulation)	0.638
Destroyed	Total loss	Total loss	0.631

[Source: Researchers' Study, 2015]

Table 3: Reliability statistics

Cronbach's Alpha	No of item
0.915	8

[Source: Researchers' Study, 2015]

Table 4: Reliability statistics of pilot attributes

Suggested Degree of Damage	Suggested Description of Damage	Suggested Attributes of Damage	Cronbach's Alpha
Inaccessible	The residence is inaccessible by normal means	The house is flooded or submerged	0.902
		Roads and bridges are out	0.902
		Ground zero	0.899
Minor	Slight damage to building structure, can be occupied within a short period of time after minor repairs.	Minor damage in partitions, infills and ceilings	0.911
		Minor damage to door and windows	0.911
Major	The building has sustained structural or significant damages, inhabitable and requires extensive repairs.	Substantial failure of walls, floors, foundation or roof.	0.900
		Utilities damaged (Electrical, Surface water drainage, Sewerage reticulation system, Water reticulation)	0.905
Destroyed	Total loss	Total loss	0.906

[Source: Researchers' Study, 2015]

3.5 Model Validation

In order to obtain validation of the proposed model of degree of housing damages, the researchers have conducted a 'model validation' by forming focus groups among selected panels to evaluate and provide personal views on the proposed model. These focus groups consist of ten (10) technical experts involved in MERCY Malaysia to assess the housing damage post flooding.

4 Results and Discussion

From the review of eight (8) attribute models of damage assessment from several countries, with a comprehensive summary based on the frequency of the degree of housing damage attributes and the suitability of flood disaster and building structure in the Malaysian context, the suggested degree of housing damages is shown in Table 5.

Table 5: Suggested degree of housing damages for flood affected area

Suggested Degree of Damage	Suggested Description of Damage	Suggested Attributes of Damage
Inaccessible	The residence is inaccessible by normal means	The house is flooded or submerged
		Roads and bridges are out
		Ground zero
Minor	Slight damage to building structure, can be occupied within a short period of time after minor repairs.	Minor damage in partitions, infills and ceilings
		Minor damage to doors and windows
Major	The building has sustained structural or significant damages, inhabitable and requires extensive repairs.	Substantial failure of walls, floors, foundation or roof.
		Utilities damaged (Electrical, Surface water drainage, Sewerage reticulation system, Water reticulation)
Destroyed	Total loss	Total loss

[Source: Researchers' Study, 2015]

From Table 5, the suggested degree of housing damage is categorized into four (4) degree of damage; 1) Inaccessible, 2) Minor, 3) Major and 4) Destroyed.

In the event of 'inaccessible' degree of damage, the affected house is inaccessible by normal means due to road closures, roads covered by water; the roads are impassible because of landslides or bridge collapse. The flood affected area might be inaccessible due to ground zero situations or submerged situation.

For 'minor', the damage is described as slight damage to building structure; the building can still be occupied within a short period of time after some minor repairs. The length of time for minor reparation is less than two (2) weeks. The attributes of damage are minor damage to partitions, infills, ceilings, doors and windows.

'Major' is the third degree of damage. The building has sustained structural or significant damage that made it unsuitable for occupation. However, the building can be occupied after extensive repairs. The attributes of damage are substantial failure of walls, floors, foundation or roof and utilities damaged (electrical, surface water drainage, sewerage reticulation system, water reticulation).The recommendation is that the affected building with a major degree of damage is to undergo extensive repair and could be occupied after 30 days.

The last degree of damage is 'destroyed'. The description for this degree of damage is that a particular building experienced total loss or has been completely destroyed.

Before this model undergoes the validation process by the technical expertise, a pilot study was conducted for the instrument validity and reliability purposes. To establish this model for expertise perception, a survey was conducted by using purposive sampling among fifty (50) respondents who possessed minimum 5 years industrial working experience. The respondents consist of ten (10) engineers, ten (10) architects, ten (10) quantity surveyors, ten (10) building surveyors and ten (10) real estate valuers. Data collected were analysed using SPSS software for Confirmatory Factor Analysis (CFA).

4.1 Survey Results

The model had undergone CFA, a tool for investigating variable relationships or observing patterns of responses. By using Kaiser-Meyer-Olkin (KMO) and Bartlett's test, the results of the factor analysis are as follows:

Table 6: KMO and Bartlett's Test

Kaiser-Meyer-Olkin Measure of Sampling Adequacy	.800
Significant value	.000

[Source: Researchers' Study, 2015]

A model is deemed fit when the KMO value is more than 0.6 and the p value (significant) is less than 0.05 (AA Afifi, 1990: Dillon & Goldstein, 1984). From Table 6, the KMO value is 0.800 which indicates that the model is fit and is significant with value at 0.000 value.

Table 7: Anti Image Correlation (a. Measures of Sampling Adequacy [MSA])

		GROUND ZERO	FLOODED HOUSE	ROADS AND BRIDGES ARE OUT	PARTITION, INFILLS AND CEILINGS	DOORS, WINDOWS	SUBSTANTIAL FAILURE	UTILITIES DAMAGED	TOTAL LOSS
Anti-image Correlation	GROUND ZERO	.874ᵃ	-.437	-.119	.164	.011	-.044	-.116	-.008
	FLOODED HOUSE	-.437	.702ᵃ	-.773	-.096	-.126	-.071	.188	-.303
	ROADS AND BRIDGES ARE OUT	-.119	-.773	.692ᵃ	-.022	.041	.102	-.133	.385
	PARTITION, INFILLS AND CEILINGS	.164	-.096	-.022	.821ᵃ	-.487	-.212	-.490	-.138
	DOORS, WINDOWS	.011	-.126	.041	-.487	.866ᵃ	-.291	.121	-.001
	SUBSTANTIAL FAILURE OF WALLS, FLOORS, FOUNDATION, ROOF	-.044	-.071	.102	-.212	-.291	.916ᵃ	-.193	.071
	UTILITIES DAMAGED	-.116	.188	-.133	-.490	.121	-.193	.797ᵃ	-.457
	TOTAL LOSS	-.008	-.303	.385	-.138	-.001	.071	-.457	.776ᵃ

[Source: Researchers' Study, 2015]

The Anti-Image correlation shows (refer to Table 7) that all attributes are correlated with each other, 0.874 (Ground zero), 0.702 (The house is flooded or submerged), 0.692 (Roads and bridges are out), 0.821 (Minor damage in partition, infills and ceilings), 0.866 (Minor damage to door and windows), 0.916 (Substantial failure of walls, floors, foundation, roof), 0.797 (Utilities damaged) and 0.776 (Total loss). Given these overall indicators, it indicates that all eight (8) attributes are significant and well represented the construct (model).

4.2 Model Validation Results

The suggested model was later reviewed by ten (10) technical experts (n=10) for model validation purpose. These technical experts were directly involved in assessing housing damage after flood disasters.

From the experts validation, it was found that all respondents agreed not to include 'inaccessible' as a part of the degree of damage attributes model.

"...Inaccessible is a not part of housing damage attributes." (n1; n2; n3; n6; n9; n10)

"...Beneficiaries need to be assessed although it is not accessible, assessment must be carried out." (n4; n5; n7)

"...by hook or by crook, assessment need to be carried out as inaccessible is not damage." (n8)

These findings are actually in line with the correlation matrix from the survey analysis by the definition if p.value is less than 0.05, the relationship between variables is strong (refer to Table 8).

DOI: 10.1051/matecconf/20166600074

Table 8: Significant table (1-tailed)

		GROUND ZERO	FLOODED HOUSE	ROADS AND BRIDGES ARE OUT	PARTITION, INFILLS AND CEILINGS	DOORS, WINDOWS	SUBSTANTIAL FAILURE	UTILITIES DAMAGED	TOTAL LOSS
Sig. (1-tailed)	**GROUND ZERO**		.000	.000	.000	**.007**	**.030**	**.069**	**.170**
	FLOODED HOUSE	.000		.000	.002	**.000**	**.005**	**.023**	**.063**
	ROADS AND BRIDGES ARE OUT	.000	.000		.001	**.007**	**.053**	**.134**	**.433**
	PARTITION, INFILLS AND CEILINGS	**.032**	**.002**	**.031**		.000	.000	.000	.000
	DOORS, WINDOW	**.007**	**.007**	**.007**	.000		.000	.000	.000
	SUBSTANTIAL FAILURE OF WALLS, FLOORS, FOUNDATION, ROOF	**.030**	**.005**	**.053**	.000	.000		.000	.000
	UTILITIES DAMAGED	**.069**	**.023**	**.134**	.000	.000	.000		.000
	TOTAL LOSS	**.170**	**.063**	**.433**	.000	.000	.000	.000	

[Source: Researchers' Study, 2015]

From Table 8, the p.value for 'inaccessible' attributes variables (Ground zero, Flooded house and Roads and bridges are out) is more than 0.05. It indicates that the relationship between 'inaccessible' attributes with other attributes is less significant.

In addition, six (6) respondents stated that in 'destroyed', the attributes of damage must also describe that the affected house is not fit for occupancy.

"...For destroyed, the attributes should include it is unsafe structure or building for occupancy." (n2; n3; n5; n6; n7; n9).

Whereas, two (2) respondents stated that the attribute for 'destroyed' must include that the house need to be anew and also agreed to include that the house is not suitable for occupancy.

"...destroyed should describe that the house is unfit for occupancy and recommended to require a new house." (n8; n10).

Thus, based on the eight (8) reviews of the attributes model, and with a comprehensive summary based on the feedback from experts, the frequency of the degree of housing damage attributes and building structure in the Malaysian context, the model of degree of housing damage for flood affected area is as follows:

Table 9: The degree of housing damage for flood affected areas after model validation.

Degree of Damage	Description of Damage	Attributes of Damage
Minor	Slight damage to building structure, can be occupied within a short period of time after minor repairs.	Minor damage in partitions, infills and ceilings
		Minor damage to doors and windows
Major	The building has sustained structural or significant damages, inhabitable after extensive repairs.	Substantial failure of walls, floors, foundation or roof.
		Utilities damaged (Electrical, Surface water drainage, Sewerage reticulation system, Water reticulation)
Destroyed	The building structure is permanently uninhabitable and requires demolition.	The building structure is a total loss.

[Source: Researchers' Study, 2015]

5 Conclusion

This research presents a model to determine the degree of housing damage for flood affected area. This model helps for estimation of degree of damages due to flood event across the countries in Malaysia. The framework assures that no type of damage is counted more than once. The attributes developed in the research is synthesis from the existing eight (8) models of assessment that had been carried out in different countries with a comprehensive summary based on the frequency of the degree of housing damage attributes and the suitability of flood disaster and building structure in Malaysian context, observation at the study area to see the actual condition of the affected house and survey analysis from experts consist of engineers, architects, quantity surveyors, building surveyors and real estate valuers. On the basis of surveys and review with the technical experts, significant feedback in the design and construction of this model have been considered in assessing the degree of housing damage after a flood in the Malaysian context. On top of that, the information that can be obtained from this model can be used for the process of re-location, resettlement and the planning and organization of the recovery process.

The Housing Damage Assessment Model will provide an edge to the government, local authorities, NGOs, MERCY, insurers or other appropriate bodies involved in assessing or evaluating the condition of houses affected by floods. By using this model, the process of assessing the degree of housing damages after floods can be more accurate, transparent, and efficient. Besides, the form provides an easier and quicker method of recording damage.

Acknowledgement

The researchers acknowledge the financial support provided by the Ministry of Education of Malaysia for this research under Fundamental Research Grant Scheme (FRGS).

References

1 BNPB, B. N. (2009). *West Sumatra and Jambi Natural Disasters: Damage, Loss and Preliminary Needs Assessment.* Jakarta: Bappenas.
2 Chan, N. W. (1995). Flood disaster management in Malaysia: an evaluation of the effectiveness of government resettlement. *Disaster Prevention and*

Management: An International Journal, Vol. 4 Iss 4, 22-29.
3 Chan, N. W. (1997). Increasing Flood Risk in Malaysia: Causes and Solutions. *Disaster Prevention and Management: An International Journal, Vol. 6 Iss 2*, 72-86.
4 Chan, N. W. (1998). Institutional Arrangements for Flood Hazarad Management in Malaysia: An Evaluation Using the Criteria Approach. *Disasters, vol.21, no.3*, 206-222.
5 Coburn, A. (1989). *Seismic Vulnerability and Risk Reduction Strategies for Housing in Eastern Turkey.* University of Cambridge.
6 CORFU, C. R. (2014). *Flood Impact Assessment Literature Review.* Exeter: University of Exeter.
7 Daungthima, W., & Kazunori, H. (2012). Assessing the flood impacts and the cultural properties. *The 3rd International Conference on Sustainable Future for Human Security* (pp. 739-748). Elsevier B.V.
8 EEFIT, E. E. (2003). *Earthquake Engineering Field Investigation Team.* Japan.
9 EMS-98, E. M.-s. (1998). *Classification of Damage Scale.*
10 FEMA, F. E. (2005). *Preliminary Damage Assessment for Individual Assistance: Operations Manual.* United States of America.
11 Florida, D. o. (2004). *Handbook for Disaster Assistance.* Tallahassee: State of Florida.
12 Friedland, C. J. (2009). *Residential Building Damage from Hurricane Storm Surge: Proposed Methodologies to Describe, Assess and Model Building Damage.* Louisiana.
13 Friedland, C., & Levitan, M. &. (2008). Development of a Hurricane Storm Surge Damage Model for Residential Structures. *Solutions to Coastal Disasters Proceedings.* Oahu, Hawaii.
14 Green, C., Turnstall, S., Emery, J., & Bossman-Aggrey, P. (1988). *Evaluating the Non-monetary Impacts of Flooding.* Enfield: Flood Hazarad Research centre.
15 Hodde III, H. B. (2012). *The Damage Assessment Process: Evaluating Coastal Storm Damage Assessments in Texas After Hurricane IKE.* Ann Arbor: ProQuest LLC 2013.
16 Horbostel, C. (1978). *Construction Materials Type: Uses and Aplications.* Wiley, New York.
17 Jonkman, S. N., Bockarjova, M., Kok, M., & Bernadini, P. (2008). Integrated Hydrodynamic and Economic Modelling of Flood Damage in the Netherlands. *Ecological Economics*, 77-90.
18 Messner, F., Penning-Roswell, E., Green, C., Meyer, V., Tunstall, S., & Van Der Veen, A. (2007). *Evaluating Flood Damages: Guidance and Recommmendations on Principles and Methods.* Project F (ed).
19 MNSC, M. N. (2015, April 26). *Malaysian National Security Council.* Retrieved 9 21, 2015, from Majlis Keselamatan Negara: www.majliskeselamatannegara.com.my
20 Mohd Razali, A., & Fashbir, N. ((pnyt.)). *"Perbandaran Perancangan Persekitaran".* Kuala Lumpur: Utusan Publications & Distributors.
21 Nicholas, J., Holt, G., & Proverbs, G. (2001). Towards Standardising The Assessment of FLood Damaged Properties in the UK. 163-172.
22 Shah, A., Khan, H. M., & Qazi, E. U. (2013). Damage Assessment of Flood Affected Mud Houses in Pakistan. *Journal of Himalayan Earth Sciences*, 99-110.
23 South Asian Tsunami, D. A. (2005). *Indonesia: Preliminary Damage and Loss Assessement: The December 26, 2004 National Disaster.* The Consultative Group on Indonesia.
24 Widstedt, R. (1927). The Great Flood 1926. *Journal of the Malaysan Branch of Royal Asiatic Society, Volume 11*, 259-309.
25 www.GreenAlert.net. (2004). *Damage Assessment Manual by www.GreenAlert.net.* www.GreenAlert.net.

A review of sustainable facilities management knowledge and practice

Timothy Kurannen Baaki[1,a], Mohamad Rizal Baharum[1] and Azlan Shah Ali[1]

[1] *Department of Building Surveying, Faculty of Built Environment, University of Malaya, 50603 Kuala Lumpur, Malaysia*

Abstract. Sustainability is seen as a far-reaching issue now, and one which the facilities management [FM] profession cannot overlook. This paper explores current sustainable facilities management [SFM] knowledge and practice with specific focus on performance as part of a research focus toward proposing a sustainable FM performance management framework for sustainable healthcare waste management in Malaysia. This paper utilized a review of extant literature on the subject of SFM, FM performance and FM development in Malaysia as source of information. Findings reflect the increasing recognition of the need for the strategic FM function, and how facilities managers are best positioned to drive organizations' sustainability agendas. In Malaysian context, this recognition is barely evident as findings show FM practice is still immature and predominantly operational. Unlike developed FM markets, FM relevance in Malaysia is being driven by the public sector. Also findings show a disharmony between organizations' sustainability priority areas and the responsibilities for facilities managers to execute them where the sustainability policy of organizations prioritize one FM service and the facilities managers' responsibilities prioritize another. As most of SFM implementation is driven by legislation this seems to strengthen the position that, organizations continue to view support services as non-value-adding, as unavoidable liabilities. The implication of this is the pressure on the FM function to continually express its strategic relevance to organizations by tangible value-adding performance output. This creates a new perspective to measuring and managing facilities performance. This paper therefore elevates the importance of FM performance management in SFM context taking into account the peculiar position of the facilities manager. This is seen as a way forward for FM to better express its value to the organization

1 Introduction

Sustainability is seen as a far-reaching issue now, and one which the facilities management [FM] profession cannot overlook [1]. In a 2007 IFMA report, sustainability was identified as one of the key areas FM professionals need to develop their competencies in order to face the demands, challenges and opportunities of sustainable development and practices [2]. With the increasing relevance of FM toward business success, the FM function has transformed from just a cost cutting option to a strategic link between organization's core business and support services. Facilities management professionals are seen as being "at the forefront of organizational behavior change and in a position to influence individuals working in business; government departments and public services" [3] p. 59. The concept of sustainable facilities management [SFM] has gained relevance recently in an attempt to integrate FM functions with sustainability. The facilities management profession is presented with this need in two perspectives. One, because of the negative impact of existing buildings over the environment [4], and on the other hand, the increasing demand for technical knowledge required of facilities managers to understand the complexities of intelligent buildings and their operation [3, 5]. Aspects such as reduction in energy consumption, productivity increases, waste reduction, etc. have been identified as tangible benefits capable of winning organizations' management commitment toward sustainability. It is noted that the facilities manager is in a unique position and is often seen as the leader of an organization's sustainability policy [6]. Elmualim et al. [3] opined that there is a pressing need for practical tools for sustainable facilities management as well as a need for the facilities manager championing the cause of sustainability within their organizations. Baharum & Pitt [5] demonstrated the need for green FM intellectual capital in achieving overall sustainability and profitability for an organisation. The implication of this is the pressure on the FM function to continually express its strategic relevance to organizations by tangible value-adding performance output to support their organizations sustainability goals. Despite the ability to actively champion and implement activities capable of delivering sustainability benefits to organizations, this role is not fully recognized and appreciated [7]. This creates a new perspective to facilities performance measurement and management. Through the review of existing literature, this paper explores current sustainable facilities management [SFM] knowledge and practice with specific focus on performance, as part of a research toward proposing a sustainable FM performance management framework for

Table 1. Evolution of FM [8].

Period	Generation	Activity	Level
1970s	1st	Managed services, outsourcing, total facilities management, CAFM	Operational
1980s	2nd	Quality management, management agency, benchmarking, FM process, FMIS	Tactical
1990s	3rd	Partnering, re-engineering processes, knowledge management, product innovation, sustainable facilities management	Strategic
2000s	4th	Business processes, open innovation, usability, service excellence transformational outsourcing	Transformational

sustainable healthcare waste management in Malaysia. The paper provides an outlook on the state of sustainable facilities management knowledge and practice as well as the current state of FM practice in Malaysia. The paper also elevates the importance of FM performance management in SFM context taking into account the peculiar position of the facilities management.

2 The strategic role of facilities management

It has been agreed that FM encompasses many things, often seen as a conglomeration of several related expertise into one business support function. Presently, FM is seen as the fastest emerging and most powerful sectors in many industries having gained recognition from the 1970s when outsourcing services became popular in radical cost-cutting initiatives [9, 10]. Before seen in the operational sense of cleaning, care-taking, repairs and maintenance, FM now covers "real estate management, financial management, change management, human resources management, health and safety, contract management, in additional to building and engineering services maintenance, domestic services and utilities supplies" [11] p. 75. A chronology of how FM has transformed is provided by [8] [see Table 1]. According to Barrett [1995] in [12] facilities management is "a strategically integrated approach to maintaining, improving and adapting the buildings and supporting services of an organisation in order to create an environment that strongly supports the primary objectives of that organisation", p. 3. This position has long promoted the need for FM to be strategically oriented [12]. Tucker [13], note that, FM in contemporary terms is seen as the "integration and alignment of the non-core services, including those relating to premises, required to operate and maintain a business to fully support the core objectives of the organisation". This definition resonates with the definition provided by the British Institution of Facilities Management [BIFM], the leading FM body in the United Kingdom and one of the largest in the world, "the integration of processes within an organization to maintain and develop the agreed services which support and improve the effectiveness of its primary objectives" [14]. On the other hand, the International Facility Management Association [IFMA], regarded as the largest FM organization in the world, defines FM as "a profession that encompasses multiple disciplines to ensure the functionality of the built environment by integrating people, place, process and technology" [15]. According to BIFM [14], FM "encompasses multi-disciplinary activities within the built environment and the management of their impact upon people and the workplace."

Pitt and Tucker [16] suggest that the purpose of FM is integrating noncore services to adequately support the core functions and objectives of an organization. To achieve this function, Kincaid [17] observes effective FM need to recognize three key characteristics:

- FM as a support role or service to an organization.
- Strategic, tactical, and operational activities must be integrated and linked to primary activities to create value.
- Adequate knowledge of FM among facilities managers.

Especially for large organizations, facilities management has become a vital function, providing services such as energy management, pest control, recycling, health and safety, hygiene, waste management [18]. The true importance of FM is in linking all services under its remit to organizational core objectives and creating value. FM nowadays is seen as the key link between organization or business support services to core-business, a vital service sector, charged with providing cost-efficient support services in a highly competitive and continually-changing business environment [10]. According to Barrett & Baldry [19], FM offers a holistic and evolutionary approach in realizing organizational optimization and total business solutions as regards other business tools available to business outlets and organization. Thus, FM also has to operate at the same strategic level of aligning business and organizational visions, missions and objectives with its non-core business activities. Nutt [20] note that the primary role of FM is resource management at both strategic and operational levels of business support. FM today is seen as providing services at 3 levels of business support [strategic, tactical and operational levels]. According to Kamaruzzaman & Zawawi [11], FM is a balance between these three levels.

2.1 State of FM in Malaysia

The facilities management profession, even around the world is still very new dating back to just over three decades ago. However, while developed nations such as the UK, US, Japan, Singapore, Australia, etc, have managed to establish a much robust FM sector, the case with Malaysia is starkly different. A study by Moore & Finch [21] showed mixed signs about FM development in Malaysia with lack of understanding and progress as key aggravating factors. Tracing FM development in Malaysia goes back to the 1990s, precisely 1996 when the government privatized non-clinical support services in public hospitals to FM companies in the biggest contract involving FM services provisions at the time [11]. In 2001, lack of a good maintenance culture prompted the then Deputy Prime Minister to lament that: "Unless Malaysians change their mentality to become more aware of the need to provide good services and improve the upkeep of buildings, we will forever be a Third World country with First World infrastructure" [11, 21]. Since then, Kamaruzzaman & Zawawi [11], note that immense emphasis and focus have been channeled through developing FM in Malaysia especially in the public sector. The result of this could be seen also in the academic arena where a good number of prestigious Malaysian higher institutions are actively developing programs and curriculums to develop FM knowledge base in Malaysia. However, Myeda & Pitt [9] note that there is still a passive reception about FM services in Malaysia. Key to note is the need for a shift in FM understanding in the Malaysian FM sector to reflect FM's strategic relevance, which is being championed as the core of FM practice around the world. While FM practice might have a promising future in Malaysia, Myeda & Pitt [9] found that the practice is still immature with 63 percent of FM practice still operational, noting that traditional management, lack of FM guidelines, standards and understanding were among the factors hampering the development of the FM industry in Malaysia. The study also found that 60 percent of the surveyed FM practitioners did not implement a performance measurement system, a crucial element for assessing the performance of FM strategies.

3 The concept of sustainable facilities management [SFM]

The wide acknowledgement of the challenges of climate change and the effects built infrastructure have on the environment have become a great source of pressure on organizations to buy into the sustainability agenda. Larger companies, especially, are more pressured to define and show their moral positions in dealings with others and the environment [22] in an attempt to promote more inclusive and sustainable business practices. The Global Alliance for Building Sustainably [GABS] charter established in 2002 can be seen as a direct acknowledgement of the effect the built environment has on the global environment [23]. According to [5], it is fundamental to recognize sustainable FM practice as a key component to enhancing organizational competiveness. The recognition of facilities management embracing sustainable practice could be found in a 2007 IFMA report "*Exploring the Current Trends and Future Outlook for Facility Management Professionals*", which identified sustainability as one of the key issues facing FM professionals in this present day. According to the report, rapid development, especially in Asia, largely accounts for environmental resource depletion, pushing for FM professionals to develop competencies in areas of sustainable development and practices [24]. The report emphasized the need for facilities managers to retain the ability to make a "business case for sustainability" through identifying and developing initiatives to reduce energy consumption, reduce waste, understand the nature of high-performance buildings as well as becoming conversant with high-performance building ratings tools, and to closely work with end users to maximize these initiatives. Similarly, [25] identified integrating all sustainability considerations, linking strategic level with operational level, incorporating FM knowledge and experience into design, disseminating sustainable knowledge and educating people, encouraging sustainability through innovation as crucial roles facilities managers could play in sustainable FM practice. The general idea of sustainable facilities management [SFM] is to incorporate sustainability practices into the operations and implementation of facilities management functions. Sustainable facilities management integrates FM and sustainability [25]. According to Tucker [13], sustainable FM is all about "being able to manage, implement and deliver an organisation's non-core business services that contribute to the improvement of the economic, social and physical environment, and, in turn, to the greater environmental sustainability on an organisation's core business objectives" p 242. The potential of FM contributing to achieving sustainability goals within organizations have been acknowledged and investigated [2, 7, 23, 26]. FM has been firmly established now as an integrated function that operates, manages, and improves buildings and infrastructure of organizations in order to create an environment that enables core business to thrive [27]. Added to this is the unique position of FM as a business support pillar. It suffices to say therefore, that, the concept of sustainable FM is best described in terms of an organization's economic profitability, environmental responsibility and social awareness [13, 28].

3.1. Drivers for sustainable facilities management

There seems to be a wide recognition of the benefits and importance of incorporating sustainability into FM practice [2, 3, 13, 22, 23, 25, 26, 28-30], however, studies and literature focusing on the drivers of SFM are very few [2, 25, 26].

Elmualin et al. [2] conducted a study to identify policy and drivers for SFM practice and found that legislation, corporate image, and organization ethos were the chief drivers for SFM practice. This view has been shared by Meng [25], who identified market competition pushing organizations to maintain a corporate image to stay ahead of the competition, and legislation as the key drivers to SFM, both studies based on developed market scenario, the UK [2] and the UK and Ireland [25]. Especially for large organizations, there is pressure from legislation, fierce market competition and constantly changing business environments warranting the need to seek competitive edge. Interestingly, pressure from stakeholders, employees and lifecycle cost reduction were regarded as the least of the drivers to SFM practice. One could argue that organizations seem to be driven by factors of external importance than they are by factors of internal importance. In contrast, in a developing nation's scenario, Ikediashi et al. [26], while identifying corporate image as the major key driver to SFM, identified job creation for local communities, waste reduction and enhancing relationships with stakeholders as the main drivers to SFM. Pressure from senior management, pressure from clients, reduction of deforestation and government regulation were identified as the least driving factors of SFM. These observations show the predominance of environmental sustainability, with elements of social sustainability while none of the major key drivers were of economic derivation. There have been proponents of sustainability being viewed from the perspective of an overarching whole where the environmental aspect embraces both the social and economic aspects [31] with the position that efforts addressing environmental sustainability would inevitably trickle down benefits of social and then economic nature. A categorization of sustainable FM drivers, derived from [26] is shown in Table 2.

Table 2. Drivers for sustainable facilities management.

	Drivers		
Category/ Reference	**[2]**	**[26]**	**[25]**
Environmental		Reduction in energy consumption Waste reduction Increase productivity Elimination of oil and air pollution Sustainable urbanization Reduction of deforestation Reduction of carbon dioxide emissions	
Social	Legislation Corporate image Organization ethos Service management/Director's leadership Pressure from clients Pressure from employees Pressure from stakeholders	Enhance relation with stakeholders Job creation for local communities Government regulation Corporate image Pressure from clients Pressure from senior management	Market competition [corporate image] Government legislation
Economic	Lifecycle cost reduction	Financial gain Investment drive Life cycle cost reduction Profitability To remain competitive Market expansion	

3.2 Sustainable facilities management aspects

It can be argued that the concept of sustainable FM does not create totally new functions for facilities managers, but rather requires facilities managers to incorporate aspects of environmental, economic and social sustainability into existing FM functions. The following have been identified as aspects facilities managers could harness to deliver sustainable FM functions within an organization [13, 22]: energy management, water management, waste management, air and environmental quality management, reducing vehicle use, sustainable maintenance, sustainable purchasing, environmental or ecological management, and social sustainability. While it is practically difficult to get all of these aspects to yield predominant results within an organization at once, adopting a few to front run the organization's sustainability goals is key to achieving sustainable FM functions with the organization. For instance, Tucker [13] opined that sustainable policies and operational practices could fail if there is no appropriate environmental management systems [EMS] in place to monitor, review, check and improve such policies and practices, noting further that, the most effective tool for enhancing the ability of organizations to successfully implement sustainable FM is through implementation of EMS with the FM function. At the operational level, Tucker [13] identified the following as considerations for sustainable FM delivery:

- Sustainable cleaning and reducing water usage – automated toilets, waterless urinals, low flow taps/spray taps, use of low VOC and non-toxic cleaning compounds, etc.
- Sustainable catering – use of locally grown food, reusable, compostable or recyclable dinnerware, intelligent cooking, composting, etc.
- Sustainable travel – implementation of car sharing within organization's fleet, provision of bicycles, electric vehicles, etc.
- Waste disposal and recycling–provision of central printing zones, bin color coding, paper

shredding on-site, digital media [CDs and DVDs], dry cell batteries recycling, etc.

- Space management – hot-desking, flexible workplace, etc.
- Reducing energy from mechanical and electrical [M&E] services – use of energy rated appliances, implementing shut-down periods and zones for say lifts, recovery of energy, eco lifts, etc.

A study by Elmualim et al. [2] identified waste management and recycling, energy management, and carbon footprint as the main FM services prioritized in organizations sustainability policies. Health and safety, and sustainable travel followed while staff productivity, biodiversity, building disposal were the least sustainability issues. In complete contrast, the study found that the practical implementation did not tally with the priorities set by the sustainability policy. While the sustainability policy prioritized waste management and recycling, energy management, and carbon footprint in that order, energy management, waste management and recycling, and health and safety in that order on the other hand were the FM services FM managers concentrated on delivering. This misplacement could be traced to the major driver of sustainable FM practice, which is legislation. Pressure to comply with strict legislation set by the government suggests, organizations sustainability priorities could be overridden by operationalizing certain FM services over others even where the sustainability policy states otherwise.

3.3 Formulating a sustainable facilities management strategy

Hodges [6] noted that developing a sustainable FM strategy is not different from any strategic planning or goal-setting process. The traditional approach to strategic planning such as carrying out a SWOT analysis, benchmarking, therefore applies. This must not be in isolation of the organization's sustainability policy. Despite limited research regarding the drivers of sustainability policies in organizations, legislation and corporate image have been identified as the main drivers of organization's sustainability policies [2, 25, 26]. Therefore, the sustainability strategy must be formulated around the organization's overarching economic, social, and environmental goals [6] to deliver a rounded service [3]. This is to say that, an organization's overall sustainability policy forms the basis for formulating a sustainable FM strategy. With the nature of his functions, the facilities manager usually is the key professional in leading the sustainability strategy. This is as a result of the facilities manager's ability to bridge the gap between all the critical phases of a building's life cycle [6]. Hodges [6] further identified the following key steps to developing a sustainability strategy:

- Thoroughly understand the organization's philosophy and strategy regarding handling finances: identifying the organization's core values and resources and defining the fit of the sustainability strategy.
- Facilitate the strategic planning process for the organization: initiating the sustainability strategy and selling it to the top management.
- Develop a strategic plan for sustainability: mapping out a strategic plan for sustainability that defines actionable processes for the sustainability strategy.
- Implement the strategy: actual and operational level execution of the strategy.
- Become the advocate of the strategy: championing and leading the strategy implementation.

On the other hand, the BIFM proposes ten [10] steps to implementing a sustainability policy [13]:

- Identify stakeholders and map their expectations
- Confirm organizational policy mission
- Agree a sustainability policy mission
- Get a high-level sponsor
- Agree policy boundaries
- Provide adequate resources
- Set SMART targets
- Engage staff
- Communicate success
- Review and feedback

4 Performance measurement for sustainable facilities management implementation

Traditionally, performance measurement is viewed as having three main purposes: to ensure the achievement of goals and objectives; to evaluate, control and improve procedures and processes; and to compare and assess the performance of different organizations, teams and individuals [32]. Recently, performance measurement has become a very popular topic discussed and researched in many fields [33]. Amaratunga & Baldry [34] describe performance measurement as a "process of assessing progress towards achieving pre-determined goals, including information on the efficiency by which resources are transformed into goods and services, the quality of these outputs and outcomes, and the effectiveness of organisational objectives". The basic foundations of performance measurement as noted by Sapri & Pitt [32] are quantifications of elements and processes, which have an impact on organizational objectives, management control and evaluation. Parker [35] note that, organizations measure performance for the following reasons:

- Identify success
- Identify whether they are meeting customer requirements: unless they measure, how do they know that they are providing the services/products that their customers require?
- Help them understand their processes: to confirm what they know or reveal what they do not know
- Identify where problems bottlenecks, waste, etc., exist and where improvements are necessary
- Ensure decisions are based on fact, not on supposition, emotion or faith or intuition

- Show if improvements planned, actually happened

It is generally contended that, to achieve organizational effectiveness, performance measurement systems have to be developed and implemented. Over the years, performance measurement frameworks relying on financial indicators have been heavily criticized for shortcomings such as being historically focused, lacking strategic focus, inward-looking, failing to incorporate intangible factors, encouraging short-termism, failing to provide information on need of customers [35]. The downfall of exclusively financial focused frameworks lie in their failure to provide the information necessary for managers to undertake forward planning, an invaluable need in a constantly changing business environment [36]. Many authors and researchers have created a number of performance measurement frameworks that attempt to incorporate both tangible and intangible measures, most notably, the Balanced Scorecard [BSC] by Kaplan and Norton in 1992. According to Kaplan & Norton [37], the Balanced Scorecard enable organizations to "track financial results while simultaneously monitoring progress in building the capabilities and acquiring the intangible assets they would need to future growth" p. 150.

It is argued that organizations implementing balanced performance measures have performed better than organizations which do not [38]. The ability of an organization to mobilize and exploit its intangible or invisible assets has become far more decisive than investing and managing physical, tangible assets [37].

The importance of measuring FM performance is noted by [39], as one of the "three essential issues for the effective implementation of a facilities strategy". Despite the FM function being positioned as one of strategic relevance to organizations [40], the FM function is one most prone to budget cuts. This position has been made possible by a dual view of facilities. To some organizations, facilities are a cost center, draining organizational resources, as such a business expense to be minimized [41]. It could be contended that view of FM as a non-value adding function, seeing that support services could account for up to 30 to 40 per cent of organizations' operating budgets, second only to payroll [42] created the second perspective of FM which is that of seeing facilities as an integral and a strategic support service to organizations [20, 27, 41]. The former view therefore justifies the widespread pressure for the added-value FM function [42] that could only be evidenced by value-adding performance output.

Major facilities performance measurement practices include benchmarking, a balanced scorecard approach, post occupancy evaluation, and measurement through metrics of key performance indicators [KPIs] [43]. Facilities performance measurement techniques such as the Post Occupancy Evaluation, the BIFM measurement protocol, Hierarchical system of performance indicators, input versus output based performance measurement focus primarily on either technical or financial aspects of facilities performance [44], and do not indicate the contribution made to the organization's strategic results [45]. Amaratunga and Baldry [34] suggested the application of the Balanced Scorecard approach to facilities management performance measurement. The application of the Balanced Scorecard as an appropriate facilities performance measurement model is in an attempt to overcome the deficiencies of traditional performance management models which have been heavily criticized for relying only on financial indicators. The challenges of the current business environment are numerous, and completely different from the era of the performance measurement based on financial indicators alone. In the light of value-driven benefits for the organization, Varcoe [46] noted that old-style measures such as financial based indicators for business performance measurement have gone past their "sale-by date" as they provide no solution to addressing the overall business value obtained from the expenditure. The Procurement Executives' Association [47] notes the following challenges of the business environment in the new millennium:

- technology advances beyond our wildest imaginings;
- resources becoming more scarce;
- higher efficiency of operations demanded by customers;
- discretion rather than rules dominating; and
- outcomes/results-oriented management flourishing

Performance measurement systems therefore have to be flexible and holistic to cope with the constant evolution of organizations [48]. Amaratunga & Baldry [49] contend that, measurement is not an end in itself, but a tool for more effective management. It is argued that organizations need to transition from just measuring performance to actually managing it in order to maximize the use of its performance measurement outcomes and create strategic value [16, 49]. To successfully transition from performance measurement to performance management Amaratunga & Baldry [49] suggests that the right organizational structure for performance measurement, and the ability to use performance measurements outcomes to effect change in the organization must be in place. Performance management has been defined by the Procurement Executives' Association as "the use of performance measurement information to effect positive change in organizational culture, systems and processes, by helping to set agreed-upon performance goals, allocating and prioritizing resources, informing managers to either confirm or change current policy or program directions to meet those goals, and sharing results of performance in pursuing those goals" [47], p. 5. Amaratunga & Baldry [49] noted the following attributes of a performance management system for FM organizations:

- leadership involvement in designing and deploying effective performance measurement and management systems;
- effective and open communication between employees, stakeholders, and customers in order to share assessment results and any new initiatives to improve performance;
- accountability of results which are clearly assigned and well understood;

- compensation, rewards and recognition that are linked to performance measures;
- targets that are linked to appraisals; and
- a performance measurement system that is positive.

Effective performance management requires the consideration of strategic goals and incorporating them into the performance management system [49].

The integration of sustainability aspects into the FM function requires performance measurement metrics that reflect the sustainability objectives set out by the sustainable facilities management strategy. Eagan & Joeres [50] identified the International Standards Organization [ISO-14031), British Standards [BS-7750], and the European Union's Eco-management and Auditing Schemes [EMAS] as among the organizations developing guidelines to evaluate environmental performance of facilities. They noted the growing significance of deriving performance measurement metrics to measure the sustainability performance of facilities, contending that current performance measurement approaches included indicators relating to processes, results, and customer satisfaction, barely focus on holistically addressing focus on sustainable development related goals. Epstein & Wisner [51] suggested an environmental and social perspective be incorporated into the Balanced Scorecard approach, noting the success story of two organizations [Bristol-Myers Squibb and Severn Trent] that successfully used a Balanced Scorecard approach to measure the sustainability performance of buildings.

4.1 Linking performance to organizational core business strategy and objectives

The contribution, or lack of it, that FM makes to the core business of an organization and the degree it accomplishes this support function is the basis for assessing FM performance [42]. Amaratunga & Baldry [49] contend that, performance management systems provide a means for FM organizations to continually test their FM strategies, providing real-time information regarding the status of their strategies, whether they are working or not, and most importantly, why, if they are not.

Amaratunga [33] contend that, for facilities performance measures to be effective, the facilities strategy has to be aligned with the core business/organizational strategy. Cable & Davis [52] add that facilities must be assessed for their alignment with an organization's objectives in order to recognize how well a facility helps the organization meet its goals and fulfill its mission. It becomes necessary to derive facilities performance measures that reflect the business performance position of the organization. Amaratunga et al. [42] note that the KPIs of the future may have

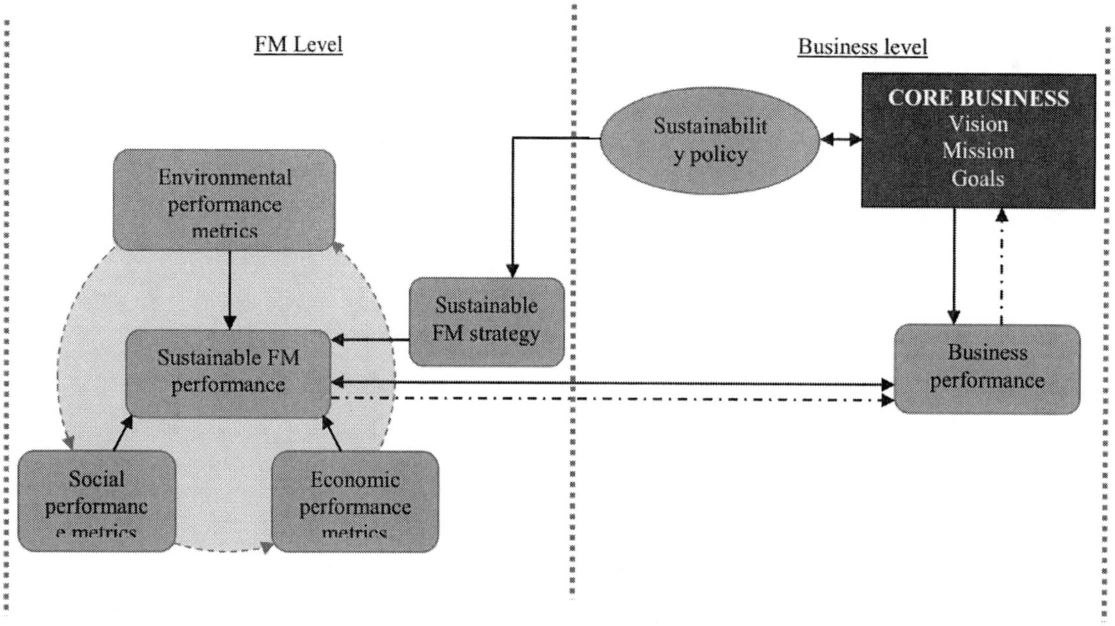

Figure 1: Linking SFM performance with organizational core business conceptual framework

to be directed more toward business outcomes, and could converge with core business performance indicators such as agility, flexibility, business continuity and transition management. In the view of this, Hinks [53] describe how necessary it is for facilities managers to begin speaking the business language to become more recognized at the business level. In a study conducted by the author, FM practitioners rated business performance measures differently from facilities performance measures. The author argued that, why the practitioners clearly recognized the need to differentiate facilities performance measures from business performance measures, there is need to link these two. This argument suggest that the wide disparity in rating between facilities performance measures and business performance measures create polarized priorities, making it difficult for the value added by the FM function to the organization be fully recognized. In Figure 1, a conceptual framework for linking sustainable FM performance to core business is proposed.

5 Conclusion

Organizations are looking to demonstrate their sustainability credentials and the facilities manager is at the forefront of championing this effort. Sustainable facilities management is considered a critical aspect of achieving the sustainability goals of organizations. This paper relied on a review of literature on the subject of sustainable facilities management and facilities performance in the context of adding value to the organization's overall sustainability agenda and goals. Several aspects the facilities manager must harness to achieve these goals have been identified to include energy management, water management, waste management, air and environmental quality management, sustainable travel, sustainable maintenance, sustainable purchasing, environmental or ecological management, and social sustainability. The challenge for the facilities manager is to bring these aspects to support the sustainability policy of the organization. The disharmony between organizations' sustainability priority areas and the responsibilities for facilities managers to execute them where the sustainability policy of organizations prioritize one FM service and the facilities managers' responsibilities prioritize another could be argued as a clear disparity between knowledge and practice. The increasing recognition of the need for the strategic FM function, and how facilities managers are best positioned to drive organizations' sustainability agendas remains a source of pressure for facilities managers to prove the strategic value of FM to their organizations. In Malaysian context, this recognition is barely evident as findings show FM practice is still immature and predominantly operational. The argument that facilities performance needs to link with business performance led the authors to propose a conceptual framework that recognizes the need for a closer communication between facilities performance management systems and business performance management systems. For the facilities management function as a whole to express its value to the organization, it needs to measure what it contributes to the core business of the organization. The authors recognize the need for further research to validate and empirically demonstrate a framework for sustainable facilities management performance for the value-added FM function.

References

1. Wood, B., *The role of existing buildings in the sustainability agenda.* Facilities, 2006. **24**(1-2): p. 61-67.
2. Elmualim, A., R. Valle, and W. Kwawu, *Discerning policy and drivers for sustainable facilities management practice.* International Journal of Sustainable Built Environment, 2012. **1**: p. 16-25.
3. Elmualim, A., et al., *Barriers and commitment of facilities management profession to the sustainability agenda.* Building and Environment, 2010. **45**(1): p. 58-64.
4. Organization for Economic Cooperation and Development (OECD), *Environmentally sustainable buildings: challenges and policies*, 2003, OECD: Paris.
5. Baharum, M.R. and M. Pitt, *Determining a conceptual framework for green FM intellectual capital.* Journal of Facilities Management, 2009. **7**(4): p. 267-282.
6. Hodges, C.P., *A facility manager's approach to sustainability.* Journal of Facilities Management, 2005. **3**(4): p. 312-324.
7. Price, S., M. Pitt, and M. Tucker, *Implications of a sustainability policy for facilities management organisations.* Facilities, 2011. **29**(9/10): p. 391-410.
8. Alexander, K. *European Facilities Management: the next generation.* 2009, Nardeen: EuroFM.
9. Myeda, N.E. and M. Pitt, *Facilities management in Malaysia: Understanding the development and practice.* Facilities, 2014, **32**(9/10).
10. Nazali Mohd Noor, M. and M. Pitt, *A critical review on innovation in facilities management service delivery.* Facilities, 2009, **27**(5/6): p. 211-228.
11. Kamaruzzaman, S.N. and E.M.A. Zawawi, *Development of facilities management in Malaysia.* Journal of Facilities Management, 2010. **8**(1): p. 75-81.
12. Alexander, K., et al., eds. *Facilities Management: Innovation and Performance.* 2004, Oxon: Spon Press.
13. Tucker, M., *Sustainable facilities management*, in *Total Sustainability in the Built Environment*, A. Cotgrave and M. Riley, Editors. 2013, Basingstoke: Palgrave Macmillan.
14. British Institute of Facilities Management (BIFM), *Facilities Management Introduction.* 2014, Available from: http://www.bifm.org.uk/bifm/about/facilities.
15. International Facility Management Association (IFMA), *What is Facility Management.* 2013, Available from: http://www.ifma.org/about/what-is-facility-management.
16. Pitt, M. and M. Tucker, *Performance measurement in facilities management: driving innovation?* Property Management, 2008. **26**(4): p. 241-254.
17. Kincaid, D., *Integrated Facility Management.* Facilities, 1994. **12**(8): p. 20-23.
18. Baharum, M.R., *Recycling initiatives in the UK shopping centres, factors critical to success* 2011, Unpublished PhD thesis, Liverpool John Moores University.

19. Barrett, P. and D. Baldry, *Facilities Management: Towards Best Practice.* 2nd ed. 2003, Oxford: Blackwell.
20. Nutt, B., *Four competing futures for facility management.* Facilities, 2000. **18**(3/4): p. 124-132.
21. Moore, M. and E. Finch, *Facilities management in South East Asia.* Facilities, 2004. **22**(9/10): p. 259-270.
22. Appleby, P., *Sustainable retrofit and facilities management.* 2013: Abingdon: Routledge.
23. Shah, S.S., *Sustainable practice for the facilities manager. [electronic resource].* 2007: Oxford: Blackwell.
24. International Facility Management Association (IFMA), *Exploring the current trends and future outlook for facility management professionals*, 2007, IFMA Report, Facility Management Forecast, 2007.
25. Meng, X., *The role of facilities managers in sustainable practice in the UK and Ireland.* Smart and Sustainable Built Environment, 2014. **3**(1): p. 23-34.
26. Ikediashi, D.I., S.O. Ogunlana, and A.O. Ujene, *An investigation on policy direction and drivers for sustainable facilities management practice in Nigeria.* Journal of Facilities Management, 2014. **12**(3): p. 303-322.
27. Nutt, B., *Infrastructure resources: forging alignments between supply and demand.* Facilities, 2004. **22**(13/14): p. 335-343.
28. Pitt, M., et al., *Towards sustainable construction: promotion and best practices.* Construction Innovation, 2009. **9**(2): p. 201.
29. Elmualim, A., et al., *The Practice of Sustainable Facilities Management: Design Sentiments and the Knowledge Chasm.* Architectural Engineering & Design Management, 2009. **5**(1/2): p. 91-102.
30. Wetering, J.v.d. and P. Wyatt, *Office sustainability: occupier perceptions and implementation of policy.* Journal of European Real Estate Research, 2011. **4**(1): p. 29-47.
31. SANZ, S.A.N.Z.I., *Strong sustainability for New Zealand: principles and scenarios* 2009, New Zealand: SANZ.
32. Sapri, M. and M. Pitt. *Performance measurement in facilities management; state of knowledge.* in *21st Annual ARCOM Conference, 7-9 September 2005, SOAS.* 2005. University of London: Association of Researchers in Construction Management.
33. Amaratunga, R. D. G., *Theory building in facilities mangement performance: application of some core performance measurement and management principles* 2001, Unpublished PhD thesis, University of Salford.
34. Amaratunga, D. and D. Baldry, *A conceptual framework to measure facilities management performance.* Property Management, 2003. **21**(2): p. 171-189.
35. Parker, C., *Performance measurement.* Work study, 2000. **49**(2): p. 63-66.
36. Neely, A., *The performance measurement revolution: why now and what next?* International journal of operations & production management, 1999. **19**(2): p. 205-228.
37. Kaplan, R.S. and D.P. Norton, *Using the Balanced Scorecard as a Strategic Management System.* Harvard Business Review, 2007. **85**(7/8): p. 150-161.
38. Kennerley, M. and A. Neely, *Performance measurement frameworks: a review*, in *Business performance measurement: Theory and practice*, A. Neely, Editor. 2002, Cambridge: Cambridge University Press, p. 145-155.
39. Alexander, K., ed. *Facilities Management: Theory and Practice.* 1996, Oxon: Taylor & Francis.
40. Alexander, K., *A strategy for facilities management.* Facilities, 2003. **21**(11/12): p. 269-274.
41. Madritsch, T. and M. Ebinger, *Performance Measurement In Facility Management The Environment Management Maturity Model Bem3.* Research Journal of Economics, Business and ICT, 2011. **2**.
42. Amaratunga, D., D. Baldry, and M. Sarshar, *Assessment of facilities management performance – what next?* Facilities, 2000. **18**(1/2): p. 66-75.
43. Lavy, S., J.A. Garcia, and M.K. Dixit, *Establishment of KPIs for facility performance measurement: review of literature.* Facilities, 2010. **28**(9/10): p. 440-464.
44. Amaratunga, D. and D. Baldry *Performance measurement in facilities management organisations: transition from measurement to management.* 2002, Available from: http://usir.salford.ac.uk/9903/1/performance_measurement.pdf
45. Brackertz, N., *Relating physical and service performance in local government community facilities.* Facilities, 2006. **24**(7/8): p. 280-291.
46. Varcoe, B.J., *Facilities performance measurement.* Facilities, 1996. **14**(10/11): p. 46-51.
47. Procurement Executives Association (PEA), *Guide to a balanced scorecard performance management methodology: moving from performance measurement to performance management*, 1999, Available from: http://web.cjcu.edu.tw/~lcc/Courses/BalancedScorecardPerfAndMeth%20Tutorial.pdf.
48. Neely, A., M. Gregory, and K. Platts, *Performance measurement system design: a literature review and research agenda.* International journal of operations & production management, 2005. **25**(12): p. 1228-1263.
49. Amaratunga, D. and D. Baldry, *Moving from performance measurement to performance management.* Facilities, 2002a. **20**(5/6): p. 217-223.

50. Eagan, P.D. and E. Joeres, *Development of a facility-based environmental performance indicator related to sustainable development*. Journal of Cleaner Production, 1997. **5**(4): p. 269-278.
51. Epstein, M.J. and P.S. Wisner, *Using a Balanced Scorecard to Implement Sustainability*. Environmental Quality Management, 2001. **11**(2): p. 1-10.
52. Cable, J.H. and J.S. Davis, *Key Performance Indicators for Federal Facilities Portfolios, Federal Facilities Council Technical Report 147*, 2004: National Academies Press Washington, DC.
53. Hinks, J., *Business-related performance measures for facilities management*, in *Facilities Management: Innovation and Performance*, K. Alexander, et al., Editors. 2004, New York: SPON Press, p. 97-107.

Property types diversification strategy of Malaysian real estate investment trust (M-REITs)

Tiong Chai Ping[1], Rohaya Abdul Jalil[2a]

[1]*Department of Real Estate, Faculty of Geoinformation and Real Estate,*
[2]*Centre for Real Estate Studies, Institute for Smart Infrastructure and Innovative Construction, Universiti Teknologi Malaysia, 81310 UTM Skudai Johor, Malaysia*

Abstracts The aim of this paper is to review the property types diversification strategy of Malaysian REITs. Previous study has found that Malaysian REITs received poor response form investors both local and international. Furthermore, the underperformance of Malaysian REITs was linked to the characteristics of REITs' property types and REITs' portfolio influenced its financial performance. The lack of local study on both of these determinants of influence on performance of Malaysian REITs is a research gap, which needs to be explored that can provide more insight of it. This study showed that the property types such as office space, commercial lot, industrial, hotel and specialized property type are the important determinants, which can influence the performance of REITs. The study comprised of REITs data from 2010 until 2014, and through Multiple Regression Analysis (MRA) reveals there is little influence of property types diversification strategy toward Malaysian REITs' performance. Nevertheless, by employ correlation analysis between each of the five property types with expected return and dividend yield, this study shows Malaysian REITs' performance of dividend yield(DY) has a positive correlation value with commercial lot property type and Malaysian REITs' performance of expected return(ER) have a highest positive correlation value with industrial property types. Also, this study suggests that property type diversification play a significant role in Malaysia REITs. Other determinants such as location of property, market capitalization and REITs' quality of advisory style that might influence the diversification strategy and financial performance of Malaysian REITs.

1. Introduction

Real Estate Investment Trust (REIT) is one type of indirect real estate investment. REIT is owned by listed companies that actively manage the portfolio of high quality income of real estate such as commercial office, hotel, retail mall, etc. In general, REIT must invest at least 50 percent of its total assets in real estate and the principal assets of the companies whose governed the REIT must be real estate (REITs Guidelines, 2012). Although REIT is being listed in the stock market but the rental income secures the income stream. REIT is required to distribute its 90 percent of taxable income to shareholders annually in the form of divided. So, these characteristics of REIT make it as a hybrid investment offering liquidity as stock and the income stream, secured by a long lease in quality commercial real estate (Newell, 2012). Furthermore, REITs are investment vehicles, which are able to provide wider diversification opportunities, greater liquidity of funds, ease of operations, and the ability to diversify at any level of investments (Chan et al, 2003 and Zietz et al, 2003).

In 1989, Malaysia introduced Property Trust Fund (PTF) market but received limited support from the investors due to structural factor (Newell and Osmadi, 2010). This led to the establishment of first Malaysian REITs in August 2005 after Securities Commission of Malaysia issued new guidelines for M-REITs. Besides, the successful introduction of REITs in other Asian countries such as Japan, Hong Kong and Singapore also encouraged Malaysia to developed REITs. However, Malaysian REITs received inadequate response and from institutional investors and the participation rates of investors in M-REITs are much lower than other counties (Jalil and Ali, 2015; Newell and Osmandi, 2009).

Jalil and Ali (2015) found that the type of underlying property of REITs is one of the determinants, which influences the performance of M-REITs. For example, REIT which focuses on only single type of property in their property portfolio will enjoy large liquidity than REITs with several types of property in its portfolio (Danielsen and Harrison, 2007). Besides, Gyourko and Neilling (1996) state that the property type of REITs invested in influence the systematic risk of REITs. Then, Hartzell et.al (1986) noted that there is low correlation between different types of properties while they were analysing the returns on an institutional real estate portfolio. They suggest that diversification across property types might have benefits. In addition, REITs stock is hybrid in nature. As the primary return of REITs is generated from the rental of underlying property and the rates of rental are associated with the quality of property types , thus, the quality of underlying property type are important in the return for REITs stock. (Chan et.al. 2003, Hui et.al. 2007; Newell, 2012).

Besides, by taking consideration of property type's concentration, either pure-played or diversified strategy in REITs are provide benefits in return enhancement and risk reduction in REITs portfolio (Danielsen and Harrison, 2007; Benefield, 2006; Young, 2000; Capozza

[a] Correspondence Author: rohaya@utm.my

and Lee, 1995). Basically, property type is a common factor which used to examine the performance of REITs for many previous studies. Extensive research to examine the performance of REITs by employed property type has been conducted in U.S and U.K. However, little is studied in Malaysia about the performance of REITs in terms of analysis of REITs. This is relevant, specifically to equity investors, whether local or international who are interested to enter the Malaysia REITs especially in terms of property type.

Thus, this study aims to investigate whether the property type important in M-REITs performance and analyse the performance of M-REITs based on the property type by employ correlation analysis and Multiple Regression Analysis using market value of the underlying assets of REITs from year 2010 to 2014. The first objective of this study is to investigate the influence of property types strategy upon M-REITs' financial performance. While, the second objective of this study is to examine the performance differences between properties' type diversification of pure-played and diversified.

2. Literature review

2.1Property type's performance

Property trust funds are the predecessor of REIT in Malaysia. However, this property trust fund was not popular among the institutional investors due to the regulation framework which provide no tax transparency (Newell, Ooi and Sing, 2006).The development of REITs in Malaysia started when the Securities Commission (SC) issued new guidelines in the year 2005 to provide a legal framework for better monitoring of the REITs stock. The key features of the revision of guidelines by SC in the year 2005 is included grating tax transparency status and liberalizing REIT's borrowing limits to 35 percent of asset value. With the revision of new guidelines, a property trust fund was renamed to Real Estate Investment Trust (REIT) which is the global adopted term. With all these changes, the first REIT in Malaysia which is Axis REIT was listed in August, 2005.

According to guidelines on Malaysia Real Estate Investment Trust which issued by Securities Commission Malaysia , Malaysia REIT can only invest in real estate, single -purpose companies, real estate-related assets, non-real estate-related assets , cash, deposits, and money market instrument. Apart from this, the total asset value of at least 50 percent must be invested in property or single-purpose companies at all time.

Besides, investment in non-property- related assets and cash cannot excess 25 percent of the total assets value of Malaysia. In addition, Malaysia REITs are prohibited from property development, but they can acquire property under construction with the prior that the total value acquired does not exceed 10 percent of the fund's total asset value. Moreover, for gearing requirement, the borrowing limit for Malaysia REITs is maximum up to 50 percent of the total asset value of the fund and are require to consider the risk of borrowings and the impact of the borrowings based on the requirement of the guideline of Securities Commission. Malaysia REITs excluded from tax when they distribute at least 90 percent of the total income to their shareholders. Currently, there are total seventeen REITs companies listed in the Bursa Malaysia. Table 1 indicates the summary of M-REITs and their respective property types.

Table 1: Summary of M-REITs property types
Sources: *Bursa Malaysia*

No.	REIT	Property Portfolio
1.	Amanah Harta Tanah PNB	Retail, Office
2.	Al-Aqar Healthcare REIT	Hospital, Nursing College, Hotel
3.	AmFirst REIT	Office, Retail, Hotel
4.	Amanah Raya REIT	Industrial, Office, Hotel, Institution, Retail
5.	Atrium REIT	Industrial, Office
6.	Axis REIT	Office, Industrial
7.	Al-Salam REIT	Retail
8.	CapitalMalls Malaysia Trust	Retail
9.	Hektar REIT	Retail
10.	IGB REIT	Retail
11.	KLCC REIT	Retail, Office
12.	Pavilion REIT	Retail, Office
13.	Quill Capital Trust	Office, Industrial, Commercial Carpark
14.	Starhill REIT	Commercial, Convertible Preference Units of REIT
15.	Sunway REIT	Retail, Hotel, Office
16.	Tower REIT	Retail
17.	UOA REIT	Retail

The property types of REITs portfolio are the main determinants which affect the REITs performance and the quality of the underlying assets of REITs is one of the attractive investment features for investors to invest in REITs (Newell, 2012).Typically, the portfolios of REITs are classified as pure-play if the REITs portfolio held 75% or more of it portfolio in one type of property.

Whereas, if a REITs has held less than 75% of its portfolio invested in any one property type, then it is classified as diversified (Ambrose and Linneman, 2001; Benefield et. al., 2008). Generally, REITs are invested in high value and high profile real estate such as hotel, shopping mall or office tower in Malaysia (Wong, 2015). Besides, some REITs market are invested not only in commercial building but also residential and industrial (Pham, 2012).

REITs' performance are different based on the underlying property type owned, and it is able to provide diversification opportunity for investors by diversified the property types of REITs property portfolio (Newell and Tan, 2003). The performance of REITs has generally been related to the rental income of their underlying property asset as the main revenue of REITs is from their underlying property asset (Olarele *et.al.* 2015).Thus, the underlying property type will influence the market share price of REITs as different type of property are subjected to charge for different rental rate based on local property market. In addition, different underlying property type of REITs poses different property nature and thus will bring a different effect upon the net asset value and also performance of REIT stock (Ong *et.al.* 2011). Therefore, based on the literature above, it is not surprised that property types are one of the determinant which will influence the REITs' performance.

For M-REITs, property type is one of the determinants to evaluate the performance of Malaysia REITs and many REITs are individually focused in property type. (Jalil and Ali, 2015). In general, the REIT with largest size focus more on retail property while smallest size REITs are more focus on office and industrial property. The study on the office buildings, showed that office rental value can be explained by demand side variable in particular real GDP and employment rate in service sectors (Bruno *et.al.,* 1992). Furthermore, the study of Kim *et.al* (2002) and Kim and Jang (2012) on hotel REIT revealed that hotel REIT has higher market risk than other REIT property sectors and underperformed office, industrial and diversified REITs sector in terms of portfolio or individual stock. In their study, they also pointed out that room occupancy rate, oversupply of lodging properties in property market and hotel financing policy is the factors affecting the performance of the hotel REITs. In addition, study by Tang and Jang (2008) show that hotel REIT performance is positively influenced by tax-exempt status and mandatory income distribution requirements but negatively influenced by operating expenses.

By employed regression analysis on annualized per area net rental income of retail mall, Hui *et.al.* (2007) find that the age of the retail mall have a negative impact on the rental rate while the size (Gross Floor Area) and the total number of shop inside the shopping mall have a positive impact on the rate of rental. Whereas, their result showed that occupancy rate has no significant impact on rate of rental for retail mall.

Moreover, in terms of physical factors, there are eleven variables, which are building size, office space, dock doors, ceiling height, age, distance to the Dallas/Fort Worth Airport, county of sale, industrial cap rate, prime rate, tenant type, and date of sale can be explain the sales price of industrial property (Fehribach *et.al.* ,1993). Later, the study of Benjamin *et.al.* (2003) also stated that physical characteristics of property, local market factors and location affect the value of industrial property. Also, the study of Ambrose (1990) stated that clientele effect would bring influenced to the value of industrial buildings. Besides, the study of Capozza and Lee (1995) stated that the stock market will place higher values in retail REITs and lower value on warehouse REITs based on the local property market. Apart from this, non-traditional property type in REITs such as healthcare centre or facility, self-storage and entertainment property provided wider diversification benefit than traditional property type such as retail and office space (Newell and Pang, 2006). All of the previous study stated above provided an evidence that different property types are influenced by different factors, and thus it is believed that they will possess different financial performance on M-REITs. Therefore, this study will examine the each of the five property types toward M-REITs' performance.

There have different argument in terms of property portfolio concentration either pure-play or diversified property portfolio. Pure play property portfolios of REITs are more benefited than diversified property portfolio as diversified REITs seem to be riskier due to its low collateral value of assets (Morri et.al, 2008). In contrast, Ziobrowski et.al. (2009) stated that pure-play property portfolio have higher market risk than diversified risk as they only focus on one type of property investment and thus the risk cannot be diversified away since different property type influenced by different internal or external factors. Thus, this study will examine the performance differences between property portfolios of pure-play and diversified of M-REITs.

2.2 REITs' performance measurement

In stock markets, investors are basically looked at the total return and the volatility of stock in making investment decisions. Nevertheless, because of the tax regulations of the REITs which require REITs to payout 90 percent of taxable income to shareholders, thus, dividend payout of REITs is a significant performance measurement for REITs investors (Chan *et.al.*,2003).There have empirical studies on REITs performance analysis done in US REITs. However, in Malaysia, there have been limited studies on performance analysis on M-REITs especially based on property types.

For M-REITs or predecessor of M-REITs which is listed property trust, Newell, Ting and Acheampong (2002) have used annual return of four listed property trust in Malaysia for the period from 1991 to 2000 to examine their performance. The results show that the annual return of Amanah Harta Tanah PNB is higher than market return. Later, Hamzah et.al (2010) had using Sharpe ratio, Treynor Index and Jenson Ratio to examine the performance of M-REITs for pre-GFC, during GFC and post-GFC period by employed monthly return from period 1995 to 2005. In the study of Pham (2012) who used average return as the performance indicator for the Asian emerging and developed REITs markets, found that the M-REITs have the highest average return (0.053%) among the seven Asian REITs market in the study.

Furthermore, Benefield, Anderson and Zumpano (2009) were obtained a daily return for the sample REITs in their study to examine the performance difference between diversified property type REITs and specialized REITs. Their result show that diversified property type REITs are performing better than specialized property type REITs when the overall market are performed well. The study of Myer and Webb (2000) found that the return data of the property type allocation are useful in explaining the performance of REITs. The study of Breidenbach et.al. (2006) was used quarterly return series to calculate the beta of different property types in different location.

As mentioned above, dividend yield is a significant performance indicator for REITs because of REITs tax regulations. By using Multiple Regression Analysis (MRA), the study of Olanrele *et.al.* (2014) proved that some financial factors of M-REITs such as Net Asset Value (NAV), size, gearing/leverage and Fund from Operations (FFO) have influenced on the M-REITs performance. They adopted quantitative research method and selected three M-REITs which are Amfirst REITs, Starthill REITs and AmanahRaya REITs as sample for the period from 2008 to 2012. The study resulted that all the factors significantly contribute up to 91.7 percent of REIT return (R^2=0.917) through analysis of MRA.

Casey et.al. (2006) employed linear regression analysis and used dividend yield of REITs sample as independent variable to test it impact on REITs capital structure. The results show that dividend yield is significant to the capital structure of REITs, this may due to REITs pay out significantly more dividend as they have more cash flows arising from depreciation shelter. In addition, there have also several studies such as Wang et.al. (1993), Zietz et.al. (2003) and Lee et.al. (2010) who focus on the influence of dividend yield on REITs and point out that dividend yield has influence on REITs performance. The study of Wang et.al. (1993) found that REITs manager paid more dividend when REITs perform poorly.

Also, the study of Lee et.al (2010) resulted that the excess dividend payout is able to convey information to capital market and thus can gaining access to fund acquisition. This result make consensus with the study of Wang et.al. (1993) who find that dividend announcement did convey information to the market and the dividend policies can be used as a tool to convey private information about REITs future cash flow. This perhaps will influence the stock price of REITs as some of the investors will look at the dividend payout of REIT companies in making investment decision. Thus, this study will use expected return and dividend yield as performance indicator for M-REITs to examine the influence of different underlying property types on M-REITs.

3. Data

There are a total of seventeen Malaysia listed REITs listed in Bursa Malaysia. For this study, fourteen Malaysia listed REITs were included, which are Amanah Harta Tanah PNB REIT, Al-Aqar Healthcare REIT, AmFirst REIT, Amanah Raya REIT, Atrium REIT, Axis REIT, CapitalMalls Malaysia Trust, Hektar REIT, IGB REIT, Pavilion REIT, Quill Capital Trust, Sunway REIT, Tower REIT and UOA REIT. This study will exclude KLCC and Al-Salam REIT due to their shortest listed period and Starhill REIT due to most of their underlying property assets is located outside Malaysia boarders. For M-REITs property portfolio, the property invested in the portfolio are office space, commercial lot, industrial property, education building and healthcare building. For this study, education building and healthcare building will be categorized as specialized property. The data were collected from REIT company annual report and DataStream. Monthly net property rental income of underlying property of REITs, market value of property and annual dividend payout of REITs were collected from REIT company annual report from 2010 to 2014, whereas, quarterly REITs stock closing price from period 2010 to 2014 were collected from DataStream.

4. Methodology and variables

This study will follow the empirical study of Jalil and Ali (2010) and Anderson and Shain (2001) by employed Multiple Regression Analysis (MRA) to test the relationship between expected return and divided yield of M-REITs with their property types. MRA is a method for explaining the relationship between dependent variable (X) and independent variable (Y). In MRA model, the correlation coefficient between dependent and independent variable is a quantitative index of association between these two variables. A multivariate counterpart of the coefficient of determination is the coefficient of multiple determination, R^2 value. Hence, the value of R^2 explains the degree of significant influence of independent variables toward dependent variable. The general forms of the models' specification are as follows:

$$Y_1 = a + b_1x_1 + b_2x_2 + b_3x_3 + b_4x_4 + b_5x_5 + e \quad (1)$$

$$Y_2 = a + b_1x_1 + b_2x_2 + b_3x_3 + b_4x_4 + b_5x_5 + e \quad (2)$$

Where:
Y_1 = Expected Return
Y_2 = Dividend Yield
a= Constant
b= Coefficient of the Variables
X_1 = Office Space
X_2 = Commercial Lot
X_3 = Industrial
X_4 = Hotel
X_5 = Specialized Property

In addition, this study will use correlation analysis to measure the relationship between dependent variable and independent variables. The resulting value (correlation coefficient) will shows the changes in independent value will result in changes in dependent value. Also, correlating expected return and dividend yield with property types of REITs can provide information about the diversification by property types in REITs investment. The correlation coefficient poses positive or negative sign in the resulting result. The positive sign means that the change in the independent variable will result in the identical changes in dependent variables, whereas, the negative sign mean the changes in independent variable will result in the identical changes in dependent variables also but the changes will be in opposite direction.

4.1The Dependent Variables

In this study the dependent variables is the performance measurement of REITs which are expected return (Y_1) and dividend yield (Y_2) to examine the performance of M-REITs based on property .The expected returns are the common performance indicator use by the investor when making investment decision, thus this study will use quarterly expected return of M-REITs to examine the performance of a short term investment goal. Generally, the expected return is a predicted or estimated return that may or may not occur (Omisore, Yusuf and Christopher, 2011). It is the amount of return of the investors looking forward to receive over the period of investment. This expectation of the investors usually builds on the historical performance from the particular investment. The calculation for the expected return is shown as follows:

$$E(R) = \left(\frac{P_1 - P_0}{P_0}\right) * 100 \quad (3)$$

Where,
E(R) = Quarterly Expected Return
P_1 = Daily price of REIT stock at the end of the quarter
P_0 =Daily price of REIT stock at the beginning of the quarter

Whereas, because of the unique tax regulation of REITs which pay out 90% of their dividend per annual, thus, dividend yield is also another performance indicator for REITs to examine the long term investment goals. The calculation for dividend yield is shown as below:

$$DY = \frac{\text{Annual Dividend Distribution}}{\text{Stock Price}} \quad (4)$$

Where,
DY = Dividend Yield

4.2 The independent variables

The literature above revealed that property types are one of the determinants which as influenced toward REITs performance. Thus, it is a need to study the properties' portfolio of M-REITs. Based on the thorough study of fourteen listed M-REITs annual company report, M-REITs companies consists of office space, commercial lot, industrial properties, hotel healthcare property and educational buildings. So, the independent variables for this study are office space (X_1), commercial lot (X_2), industrial property (X_3), hotel (X_4) and specialized property (X_5), which comprised of healthcare and educational property. The annually market value of each underlying property of fourteen listed REITs companies included in this study will aggregate in those five property types.

4.3 Research Design

Overall, this study aims to examine the relationship and influences of M-REITs property types toward M-REITs performance. The overall idea of this study is show in figure 3.1 below:

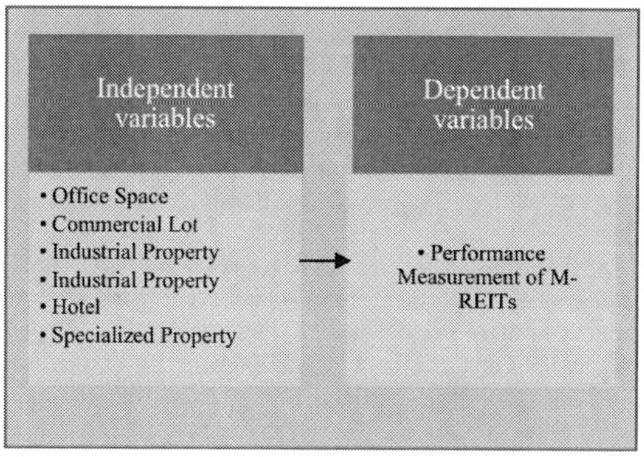

Figure 1: Relationship of Property Types with M-REITs Performance

The annually market value of each underlying property of fourteen listed companies was collected from the company's annual report and aggregated in five property types mentioned above. Expected return and dividend yield are calculated by using the stock price collected from DataStream and dividend distribution of each REITs company from annual report. Correlation analyses were employed to test the relationship between each property type with expected return and dividend yield. Also, MRA are employed to test the influenced of all the five property types towards expected return and dividend yield. The result of the analysis are show in the section below.

5. Results and discussion

The finding of first objective for this study is to investigate the influence of property types strategy upon M-REITs' financial performance are presented in Table 2 below. It presents the result of correlation analysis between five property type of M-REITs (office space, commercial lot, industrial property, hotel and specialized property).

Table 2: Result for Correlation Analysis

Independent variables / Dependent Variables	X_1	X_2	X_3	X_4	X_5
Y_1 (ER)	-0.113	0.0351	0.1860	0.1258	0.0144
Y_2 (DY)	-0.114	0.084	-0.039	-0.039	-0.051

The correlation analysis show that the expected return of M-REITs have the highest positive correlation value with industrial property (0.1860) follow by hotel (0.1258), commercial lot (0.0351) and specialized property (0.0144). Whereas, expected return of M-REITs have negative correlation value with office space (-0.113). Industrial property in M-REITs have a stable return because the tenants for the industrial properties are mainly the tenants with famous brand and good reputation and they are the only tenants for that particular industrial buildings such as BMW, Nestle and Emerson for Axis REIT and DHL Sdn Bhd for Atrium REITs. This finding result are make consensus with the study of Ambrose (1990) who suggest that clientele effect will bring influenced to the value of industrial buildings.

For dividend yield of M-REITs, only commercial lot have a positive correlation value with M-REITs dividend yield while another four property types all have negative correlation value with dividend yield as show in the table above. In M-REITs, the commercial mall are mainly located in large and densely populated cities such as Kuala Lumpur. The location of the commercial mall encouraged the high traffic of consumer and also the tenants, thus, this directly will increase the net property rental income of commercial buildings and consequently the REITs companies will pay out high dividend yield. This result are align with the study of Lehew (2000) who stated that the successful commercial mall properties are located in large and densely populated region. In addition, the study of Gyourko and Nelling (1996) stated that commercial mall have higher return compared to industrial property but also associated with high risk also. Nevertheless, the result show that property types have less influenced on expected return and dividend yield on M-REITs. Although some property types have a positive correlation with expected return and dividend yield, yet, on the simultaneous effect, by using MRA, the study found that all the five independent variables are shows lower R^2 value towards expected return and dividend yield of M-REITs. Table 3 show the regression model summary of property types with expected return and Table 4 show the regression model summary of property types with dividend yield. Based on the result show in Table 3 and Table 4, the R^2 value for regression analysis of property types with expected return and dividend yield are 8.3 percent and 2.3 percent respectively. Also mean that other factors which were not consider in this study contribute 91.7 percent and 97.7 percent to the influence on expected return and dividend yield on M-REITs.

Table 3: Regression model summary of property types with expected return

R	*R^2*	*Adjusted R^2*
0.289	0.083	0.007

Table 4: Regression model summary of property types with dividend yield

R	*R^2*	*Adjusted R^2*
0.151	0.023	-0.059

From the data presented in Table 2, 3 and 4, it is clear that all the five property types have little influenced on expected return and dividend yield of M-REITs from period 2010 to 2014. This lower R^2 value may due to REITs financial performance are influenced by many others factors also. From the literature review on the performance of REITs, it is pointed out that REITs performance are influenced by financing policy, management style, location of property, net asset value and size of stock of REITs (Jalil and Ali, 2015; Ong, 2011; Ong.et.al,2012; Kim and Jang,2012 ; Anderson and Shain, 2001). Besides, the study of Olanrele et.al. (2014) proved that some financial factors of M-REITs such as Net Asset Value (NAV), size, gearing/leverage and Fund from Operations (FFO) have influenced on the M-REITs dividend yield. They study resulted 91.7% influenced of all these factors towards M-REITs return and dividend yield.

In addition, the preferences of tenant is also the factors which can influenced the performance of REITs, as highlighted by the study of Celka (2011) which had categorized the preferences of tenants on office property market. The study revealed that main preferences of tenant are location and accessibility of the property, facility provided by the office buildings such as number of parking space, building characteristics and building design.

Although lower R^2 value are shown between five property types with M-REITs 'financial performance, yet this study would like to emphasis on the property types diversification effects. Thus, we further examine the M-REITs' average rental yield from year 2010 to 2014. The average office space rental yield of M-REITs are shown in the Table 5 below. Rental yield can be REITs' performance indicator, as when rental yield are performed well, REITs are able to distribute high dividend and will attract investors. Thus, higher distribution of dividend motivated higher unit price for REITs.

Table 5: Average office property rental yield of M-REITs

REITs \ Year	2010	2011	2012	2013	2014
Amanah Raya	0.07	0.068	0.07	0.075	0.076
Amfirst	0.052	0.054	0.047	0.059	0.055
Axis	0.065	0.072	0.069	0.071	0.067
Quill Capital	0.095	0.096	0.09	0.09	0.09
Tower	0.083	0.09	0.09	0.087	0.067
UOA	0.083	0.078	0.082	0.082	0.088
Pavillion	-	-	0.05	0.06	0.04
Sunway	-	0.053	0.076	0.063	0.063
AHP	0.093	0.088	0.092	0.091	0.057

The rental yield associated with quality of property buildings (Hui *et.al.* 2007). Based on the average rental yield of office space, it is clearly show that the average rental yield of Quill Capital REIT, Tower REIT, UOA REIT, Sunway REIT, Axis REIT and AHP REIT performed good and stable from year 2010 to 2014 which is more than 6 percent. This is due to the location of REITs that located in the Centre Business District (CBD) which have a higher demand and occupancy rate. Meanwhile, Amfirst REIT and Pavillion REIT derived less than 6 percent of average rental yield for office space, this mainly caused by the lower occupancy rate of office space and thus reduce the net rental income.

The finding of second objective of this study which is to examine the performance differences between property type diversification of pure-played and diversified is shown as Table 6 below.

Table 6: Average property rental yield of pure-played REITs and diversified REITs of M-REITs

REITs	Property diversification strategy	Average Rental Yield (%)				
		2010	2011	2012	2013	2014
Amanah Harta Tanah (AHP) PNB	Pure-Play	0.093	0.088	0.092	0.091	0.057
Al-Aqar Healthcare REIT	Pure-Play	0.075	0.07	0.067	0.07	0.07
AmFirst REIT	Pure-Play	0.056	0.064	0.049	0.058	0.054
Amanah Raya REIT	Diversified	0.07	0.069	0.071	0.074	0.065
Atrium REIT	Pure-Play	0.085	0.072	0.078	0.072	0.07
Axis REIT	Diversified	0.066	0.073	0.072	0.079	0.064
CapitalMalls Malaysia Trust	Pure-Play	0.053	0.07	0.07	0.063	0.063
Hektar REIT	Pure-Play	0.067	0.067	0.048	0.066	0.062
IGB REIT	Pure-Play	-	-	-	0.06	0.065
Pavilion REIT	Pure-Play	-	-	0.055	0.06	0.05
Quill Capital Trust	Diversified	0.096	0.097	0.093	0.093	0.093
Sunway REIT	Diversified		0.063	0.068	0.058	0.061
Tower REIT	Pure-Play	0.083	0.09	0.09	0.087	0.066
UOA REIT	Pure-Play	0.083	0.078	0.082	0.082	0.088

Based on the property rental yield show in Table 6, the REITs performed well in rental yield are majority pure-played property type in their property portfolio. The pure-played property types REITs focused on one particular property type and thus will provide better understanding and well management compared to those diversified property type REITs which need to employed more number of personnel to manage variety of property type in their portfolio and increased (Ro and Ziobrowski, 2011).Moreover, by holdings one particular property type is also one type of investment strategy adopted by fund manager to maximize the return and minimize the portfolio risk (Jalil and Ali, 2015; and Brown, 1991). This, seem to align with argument by Capozza & Seguin (1999), that adversely performance of diversified property type on M-REITs. The pure-played M-REITs are performed well, since they focusing on one particular property types can avoid higher management cost and also easy to focus on particular tenant in order to fulfil their preferred and requirement on buildings. This directly can improve the return and performance of REITs (Jalil and Ali, 2015; Benefield *et.al.*, 2009; and Chan *et.al.*, 2003).

Overall, the study suggests that there is performance difference between property diversification strategy, in which pure-played strategy M-REITs produced higher rental yield and consequently higher distribution of dividend yield. Accordingly the unit price of M-REITs will increase, vice versa of low rental yield, produce lower dividend and affect the unit price of M-REITs diluted.

6. Conclusion

This study shows that property types are important determinants in determine the diversification strategy of REITs investment based on the literature. The analysis of this study shows that M- REITs' performance of dividend yield (DY) has a positive correlation value with commercial lot property type and M-REITs' performance of expected return(ER) have a highest positive correlation value with industrial property types compares to other property types. In addition, the study suggests that there is a performance difference between property diversification strategy, in which pure-played strategy M-REITs produced higher rental yield and consequently resulting in higher distribution of dividend yield in M-REITs. Besides, the analysis of this study revealed that property types have little influences to M-REITs performance of expected return and dividend yield. This due to M-REITs performance are influenced by others' significant factors such as location of underlying property, managing style, property type diversification and financial policy of M-REITs. Thus, it is suggested to further this study by employ these factors to examine the influence towards M-REITs performance.

References

1. Guidelines on Real Estate Investment Trust (2012). Kuala Lumpur, Malaysia.

2. Ambrose, B. (1990). An analysis of the factors affecting light industrial property valuation. Journal of Real Estate Research, *5*(**3**), 355-370.
3. Anderson, Randy I. , Liang Y. and Shain J.R. 2001. Deriving REIT Returns by Economic Location. Real Estate Finance.
4. Ambrose, B. W., & Linneman, P. (2001). REIT organizational structure and operating characteristics. Journal of Real Estate Research, **21(3)**, 141-162.
5. Avidon, E. (1995). REITs agree: diversity= stability. National Mortgage News, 18.
6. Benefield, J. D., Anderson, R. I., & Zumpano, L. V. (2009). Performance differences in property-type diversified versus specialized real estate investment trusts (REITs). Review of Financial Economics, **18(2)**, 70-79.
7. Benjamin, J., Zietz, E., & Sirmans, S. (2003). The environment and performance of industrial real estate. Journal of Real Estate Literature, **11(3)**, 279-324.
8. Breidenbach, M., Mueller, G., & Schulte, K. W. (2006). Determining real estate betas for markets and property types to set better investment hurdle rates. Journal of Real Estate Portfolio Management, **12(1)**, 73-80.
9. Brown, G. R. (1991). Property investment and the capital markets. Routledge.
10. Celka, K. (2011). Determinants of Office Space Choice. Journal of International Studies, **4(1)**, 108-114.
11. Clayton, J., & MacKinnon, G. (2003). The relative importance of stock, bond and real estate factors in explaining REIT returns. The Journal of Real Estate Finance and Economics, **27(1)**, 39-60.
12. Capozza, D. R., & Korean, S. L. (1995). Property type, size and REIT value. Journal of Real Estate Research, **10(4)**, 363-379
13. Capozza, D. R., & Seguin, P. J. (1999). Focus, transparency and value: the REIT evidence. Real Estate Economics, *27*(**4**), 587-619.
14. Chen, J., & Peiser, R. (1999). The risk and return characteristics of REITs 1993-1997. Real Estate Finance, **16**, 61-68.
15. Chan, S. H., Erickson, J., & Wang, K. (2003). Real Estate Investment Trusts: Structure: Structure, Performance, and Investment Opportunities. OUP Catalogue.
16. Danielsen, B., & Harrison, D. (2007). The impact of property type diversification on REIT liquidity. Journal of Real Estate Portfolio Management, **13(4)**, 329-344.
17. Eichholtz, P. M., Hoesli, M., MacGregor, B. D., & Nanthakumaran, N. (1995). Real estate portfolio diversification by property type and region. Journal of Property Finance, **6(3)**, 39-59.
18. Fehribach, F., Rutherford, R., & Eakin, M. (1993). An analysis of the determinants of industrial property valuation. Journal of Real Estate Research, **8(3)**, 365-376.
19. Gyourko, J., & Nelling, E. (1996). Systematic risk and diversification in the equity REIT market. Real Estate Economics, **24(4)**, 493-515.
20. Hartzell, D., Hekman, J., & Miles, M. (1986). Diversification categories in investment real estate. Real Estate Economics, **14(2)**, 230-254.
21. Hui, E. C. M., Yiu, C. Y., & Yau, Y. (2007). Retail properties in Hong Kong: a rental analysis. Journal of property investment & finance, *25*(**2**), 136-146.
22. Jalil, R. A., & Ali, H. M. (2015). Performance Determinants of Malaysian Real Estate Investment Trusts. Jurnal Teknologi, **73(5).**
23. Kim, H., Mattila, A. S., & Gu, Z. (2002). Performance of hotel real estate investment trusts: a comparative analysis of Jensen indexes. International Journal of Hospitality Management, **21(1)**, 85-97.
24. Kim, J., & Jang, S. (2012). Comparative analyses of hotel REITs: examining risk-return and performance characteristics. International Journal of Contemporary Hospitality Management, **24(4)**, 594-613.
25. LeHew, M. L., & Fairhurst, A. E. (2000). US shopping mall attributes: an exploratory investigation of their relationship to retail productivity. International Journal of Retail & Distribution Management, **28(6)**, 261-279.
26. Newell, G., Hwa, T., & Acheampong, P. (2002). Listed property trusts in Malaysia. Journal of Real Estate Literature, **10(1)**, 109-118.
27. Newell, G. R. A. E. M. E., YUE, W., WING, C. K., & KEI, W. S. (2010). The development and performance of REITs in Hong Kong. Pacific Rim Property Research Journal, **16(2)**, 191-206.
28. Newell, G., & Osmadi, A. (2010). Assessing the importance of factors influencing the future development of REITs in Malaysia. Pacific Rim Property Research Journal, **16(3)**, 358-374.
29. Newell, G., & Osmadi, A. (2009). The development and preliminary performance analysis of Islamic REITs in Malaysia. Journal of Property Research, **26(4)**, 329-347.
30. Newell, G., & Yen Keng, T. (2003). The significance of property sector and geographic diversification in Australian institutional property portfolios. Pacific Rim Property Research Journal, *9*(**3**), 248-264.

31. Omisore, I., Yusuf, M., & Christopher, N. (2011). The modern portfolio theory as an investment decision tool. Journal of Accounting and Taxation, **4(2)**, 19-28.
32. Olanrele, O. O., Said, R., & Daud, M. N. (2014). Benchmark for REIT Performance in Malaysia Using Hedonic Regression Model. International Journal of Economics and Finance, **6(9)**, p165.
33. Olanrele, O. O., Said, R., & Daud, M. N. (2015). Setting a Benchmark for REIT Performance in Malaysia. International Journal of Property Sciences, **5(1)**.
34. Pham, A. K. (2012). The dynamics of returns and volatility in the emerging and developed Asian REIT markets. Journal of Real Estate Literature, **20(1)**, 79-96.
35. Ro, S., & Ziobrowski, A. J. (2011). Does focus really matter? Specialized vs. diversified REITs. The Journal of Real Estate Finance and Economics, **42(1)**, 68-83.
36. San, O., & Malaysia, P. (2011). A study on the performance of Malaysian real estate investment trusts from 2005-2010 by using net asset value approach.
37. San Ong, T., Teh, B. H., Soh, C. H., & Yan, Y. L. (2012). Malaysian real estate investment trusts: a performance and comparative analysis. International Journal of Economics and Finance, **4(5)**, p73.
38. Tang, C. H. H., & Jang, S. S. (2008). The profitability impact of REIT requirements: A comparative analysis of hotel REITS and hotel C-Corporations. International Journal of Hospitality Management, **27(4)**, 614-622.
39. Wang, K., Erickson, J., & Gau, G. W. (1993). Dividend policies and dividend announcement effects for real estate investment trusts. Real Estate Economics, **21(2)**, 185-201.
40. Wong, Y. M. (2015). Malaysia REIT: First Decade Development and Returns Characteristics. In 21st Annual Pacific-Rim Real Estate Society Conference, Kuala Lumpur, Malaysia (January 18-21, 2015).
41. Worzala, E., & Bajtelsmit, V. (1997). Real estate asset allocation and the decisionmaking framework used by pension fund managers. Journal of Real Estate Portfolio Management, **3(1)**, 47-56.
42. Zietz, E., Sirmans, S., & Friday, S. (2003). The environment and performance of real estate investment trusts. Journal of Real Estate Portfolio Management, **9(2)**, 127-165.
43. Hamzah, A. H., Rozali, M. B., & Tahir, I. M. (2010). Empirical investigation on the performance of the Malaysian real estate investment trusts in pre-crisis, during crisis and post-crisis period. *International* Journal of Economics and Finance, **2(2)**, p62.

Properties of Recycled Aggregate Concrete Reinforced with Polypropylene Fibre

Wan Nur Syazwani Wan Mohammad[1], Sallehan Ismail[2] and Wan Abdullah Wan Alwi[3]

[1, 2, 3] *Building Department, Faculty Architecture Planning and Surveying, UiTM Seri Iskandar Campus, UiTM Perak Darul Ridzuan*

Abstract. This research work is aimed to investigate **how the addition of various proportion of polypropylene fibre affects** the mechanical strength and permeability characteristics of recycled aggregate concrete (RAC) which has been produced with treated coarse recycled concrete aggregate (RCA). Further research on RAC properties and their applications is of great importance as the scarcity of virgin aggregate sources in close proximity to major urban centers is becoming a worldwide problem. In this study, the hardened RAC properties at the curing age of 7 and 28 days such as compressive strength, flexural strength, ultrasonic pulse velocity (UPV), water absorption and total porosity were evaluated and compare with control specimens. Experimental result indicates that although the inclusion of the treated coarse RCA can enhance the mechanical strength and permeability properties of RAC, Further modification by addition of polypropylene fibre can optimize the results.

Keyword: Mechanical strength, permeability, Polypropylene fibre, recycled aggregate concrete, treated coarse RCA

1 Introduction

The recycling of concrete waste into recycled concrete aggregates (RCA) has been identified as a potential source of construction aggregates. Previous studies have highlighted the benefits of large-scale recycling of concrete waste: its great potential benefits in controlling the over-discharge of construction and demolition wastes that otherwise would have been disposed in landfills, decreases the dependence of the construction industry on natural aggregates, thereby preserving natural resources, provides savings from the treatment of waste disposal, and yields alternative sources for urban areas facing shortage of natural aggregates [1-3].

However, technical problems are encountered when using the RCA obtained from crushing concrete waste. The physical and mechanical properties of RCA are remarkably different from that of natural aggregates. RCA is composed of natural aggregates and a specific amount of adhered mortar that surrounds the original aggregate particles. Generally the RCA compose of 65% to 70% natural aggregates is surrounded by 30% to 35% adhered mortar [4]. The presence of the adhering mortars to recycled aggregate is the main reason for lower quality compared with natural aggregates because the adhered mortar is characterised as porous [5, 6] and presents numerous microcracks[6]. As a result, RCA properties are characterised by lower density, higher water absorption capacity, and lower strength compared with the natural aggregate [5,7,8]. Consequently, when using RCA in the production of new concrete, these characteristics of the aggregate may have adverse effect on the fresh and the hardened properties of concrete. A comparison of the properties of recycled aggregate concrete (RAC) and normal concrete (natural aggregate), as reported by several studies is shown in Table 1.

Table 1: The properties of recycled aggregate concrete compared with normal concrete

RAC Properties	Comparison RAC with Normal Concrete	References
Workability	Increased slump loss	Chakradhara Rao, et al. [7], Poon, Shui, Lam, et al. [9]
Compressive strength	Decreased up to 20%	Poon, et al., [9], Kou, Poon, & Chan [10], Yang, Hao, & Wang [1]
Splitting and flexural tensile strength	Decreased more than 10%	Gómez-Soberón [11], Rao, et al. [12], Tabsh & Abdelfatah [8]
Modulus of elasticity	Decreased more than 20%	Kou, et al. [10], Tangchirapat, Buranasing, & Jaturapitakkul [13]

Email: [1]wannur956@perak.uitm.edu.my, [2]salle865@perak.uitm.edu.my, [3]wanabdullah@perak.uitm.edu.my

Drying shrinkage	Increased more than 30%	Gómez-Soberón [11], Kishore & Bairagi [14]
Water absorption	Increased up to 40%	Grdic, Toplicic-Curcic, Despotovic, & Ristic [15], Gomes & de Brito [16]
Chloride penetration	Increased or low resistance	Ann, Moon, Kim, & Ryou [17] Evangelista & de Brito [18] Otsuki, et al. [19]

The poor qualities of recycled aggregate, which crucially affect the performance of concrete, have become a challenge, and the issues have limited the widespread commercial use of recycled aggregate in concrete production. Hence in realizing the widespread commercial use of RAC in the construction industry, particularly in structural application, the adverse effects related to the inherent low quality of RCA must be minimized. Previous research as demonstrated by the authors[20]has proposed surface treatment methods that have the potential to minimize the adverse effects of RCA. In this work the surface of coarse RCA was modified by coating it with wollastonite (calcium metasilicate) particles before incorporating it into the RAC mix.These works have proven that the effect of this treatment can improve the interfacial bonds between RCA and the new cement paste in the new concrete, thereby improving the concrete performance. However, focusing only on improving RCA properties through surface treatments without considering any existing residual mortar when producing new concrete is consider insufficient. The findings revealed that the effect of surface treatment method may affect the improvement of RAC properties up to a certain level only whereby the effect of surface treatment is mostly beneficial for strengthening of the interface bond between aggregate-new mortar compared with that of the aggregate-old adhered mortar bond. The presence of adhered mortar on RCA can cause the RAC to become a brittle and heterogeneous material characteristic than that of normal concrete.

Therefore this study is carried out in purpose to further extent from our previous works[20]. Accordingly, apart from the steps that improve RCA quality through surface treatment, observations also provide insight into the benefits of using fibre reinforcement systems in enhancing the properties of RAC. As reported from previous studies when randomly dispersed, the form of short discrete fibres are added to the concrete matrix, the texture is strengthened, and the brittleness of the concrete decreases[21]. The addition of fibres in concrete can increase the cracking resistance because the fibre bridges the gap between adjacent surfaces of existing microcracks, delays crack formation, and limits crack propagation[22].In addition, the well dispersion of fibre improving the internal structure of concrete thus enhancing the durability of concrete [23, 24]. Hence the present experimental investigation aims to study the effects of using polypropylene fibre at various volume fractions on the mechanical and permeability properties of RAC which contain treated coarse RCA. The modified RAC specimens consist of treated coarse RCA and respective fibre at various content specimens are tested for compressive strength, flexural strength, and ultrasonic pulse velocity. Besides, the assessment on permeability of modified RAC involves the measurement of water absorption and total porosity test. The results are compared with the normal RAC (unmodified) which acts as control specimens. With the appropriate interpretation of the obtained results, it is possible to determine the most suitable and optimum fiber percentage that contribute to great enhancement in performance of modified RAC produce.

2. Experimental program

2.1 Materials

Cement

The cement used in the concrete mixtures in this study was ordinary Portland Cement Type I with a specific surface area of 1.0432 m^2/g and a specific gravity of 3.02. In this study, all the coarse aggregates used had a maximum size of 20 mm.

Coarse aggregate

The natural coarse aggregates used were crushed granite. The coarse RCA used were generated from waste concrete cubes collected from the debris area. Jaw crushers were used to crush and to break down the concrete waste into smaller particle sizes. After crushing, the RCAs were graded to the required particle sizes using a vibrator sieve to obtain aggregates with the maximum sizes of 20 mm. Table 2 show the sieve analysis of coarse aggregates that were used in this study. As mentioned before, this study will involve the use of treated coarse RCA. The preparation treatment for coarse RCA began by impregnating these materials in wollastonite solution for ascertain period and this is followed by draining and drying of that particle in the oven. The purpose of this step is to coat the surface of coarse RCA with wollastonite particles to refill the pores and cracks throughout its physical surface. The details on the materials and procedures involved for in carrying this treatment process are followed according to the previous research works done by the authors [20]. The comparison on the physical properties of the coarse natural aggregate (granite), coarse treated and untreated RCA by analyzing them in terms particle density, water absorption and mechanical strength is shown in Table 3.

Table 2: Sieve analysis of the fine and coarse aggregates

Agg.	**Aggregate passing (%) according to sieve size (mm)**								
	0.15	0.3	0.6	1.18	2.36	5	10	14	20
Fine	0.9	8.8	22.7	45.3	77.4	100	100	100	100
Coarse Natural	0.0	0.0	0.0	0.0	0.2	0.2	31	59.2	100
Coarse RCA	0.0	0.0	0.0	0.0	0.4	0.8	30	60.4	100

Table 3: Properties of the coarse aggregate

Properties of Aggregate	**Sizes of aggregate**	**Natural Granite**	**Un-treated RCA**	**Treated RCA**
Particle Density - Oven Dry (Mg/m^3)	20 mm	2.60	2.33	2.4
	10 mm	2.58	2.23	2.3
Water absorption (%)	20 mm	0.60	4.44	3.71
	10 mm	0.70	5.58	5.02
Agg. crushing value (%)	14mm	24.32	29.15	
Agg. impact value (%)	14mm	13.98	21.78	
LA abrasion value (%)	14mm	34.76	39.12	

Fine aggregate

The natural fine aggregate or sand used in this study consisted of particles that passed through a sieve size of 5.00 mm and were mainly 75 μm in size. The fine aggregate used was uncrushed quartzite natural river sand. Table 2 shows the sieve analysis of the fine and coarse aggregates that were used in this study, which were graded according to BS 812- Part 103.1.

Superplasticizer

Superplasticizer (SP) based on sulphonated naphthalene polymers were used to enhance the workability of the concrete.

Fibre

Polypropylene (PP) fibres produced in fibrillated form, was supplied by Timuran Engineering Sdn. Bhd. (Figure 1). The fibrillated polypropylene form of this fiber opens linked fibers and separates into multistrand or individual filaments in the concrete during mixing. The different properties and specifications of the polypropylene fibres are shown in Table 4.

Figure 1: Fibrillated polypropylene fibre

Table 4: Specification of polypropylene fibres

Item	**Specifications of fibre**
Average Length	15mm
Tensile Strength	45-60 ksi
Specific Gravity	0.9 g/cm^3
Youngs Modulus	0.5 x 10 ksi
Melting Point	160°C - 170°C
Ignition Point	590°C

3. Mix Design

The mix design of concrete followed the British Method [25], which is prepared based on constant effective water/cement ratio of 0.41 for all concrete mixtures to achieve targeted compressive strength of 50 MPa at the 28th day. The mixture proportions of the base matrix for all specimens are given in Table 5. It is noticed that the dosage compositions of the coarse aggregates in all RAC mix were designed by replacing the natural coarse aggregate with untreated and treated RCA at 60% weight of the total coarse aggregate content. One batch of mixtures was prepared with untreated RCA and served as the sample for comparison purposes. While the rests are in types of modified RAC which consist of treated coarse RCA and cast with and without fibres. The detailed description of material composition and various percentages of respective fibres used in the different RAC mixes are tabulated in the Table 6. Considering the loss of water caused by the absorption of the aggregates and the effect of fibre addition during mixing, a superplasticizer was added in the concrete mix at the dosage of 0.2 – 0.3% by the cement weight to maintain the slump of the concrete mix.

Table 5: Mix proportion

Constituents	Proportion (kg/m^3)
Water	210
Cement	512
Sand	722
Coarse Aggregate	
Granite	382
RCA	574

Table 6: Detailed descriptions of all RAC mixes

Mix	Types of mix	Vol. dosage of fibre (%)		SP (%)
		Polyolefin	PP	
CO (Control)	Untreated RCA	0	0	0
TR	Treated RCA	0	0	0.2
PP03	Treated RCA + Polypropylene fibre	0	0.3	0.2
PP06		0	0.6	0.2
PP09		0	0.9	0.3
PP12		0	1.2	0.3
PP15		0	1.5	0.3

3.1 Mixing and Curing the Concrete Mixtures

All concrete mixes in the study were prepared following the procedure prescribed in BS1881-125.A drum mixer was used to prepare all the concrete mixes. All the concrete specimens were cast under laboratory conditions, demoulded at 24 h after casting, and then fully submerged in water at (25 ± 2) °C until further testing.

3.2 Tests

The compressive strength test was conducted in compliance with BS EN 12390-3 on 100 mm concrete cubes under uniaxial compression using an ELE International compression testing machine with the maximum capacity of 3000 kN. Flexural strength test was conducted to measure the ability of a concrete beam to resist bending. This test involved the fabrication with fabrication of the prism specimens with dimensions of 100 × 100 × 500 mm and the test procedures were performed in accordance to BS EN 12390-5. Ultrasonic pulse velocity (UPV) test is a non-destructive method to measure the speed of ultrasonic pulse passing through the materials. This test is applicable for predicting the concrete strength or distinguishing the presence of internal flaws, such as voids and cracks. The UPV test in this study was conducted in accordance with BS 1881-203 on prism specimen. Water absorption test was conducted in accordance with procedures prescribed in BS 1881-122 and the porosity of the concrete was determined using the vacuum saturation method developed by RILEM [26]. The entire tests programs were conducted on respective specimens at the ages of 7 and 28 days.

4. Results and discussion

4.1 Compressive Strength

The overall compressive strength results of all the tested specimens at the testing age of 7 and 28 days are illustrated in Figure 2. Generally, the results obtained from this investigation show that, the compressive strength of all the concrete specimens has a similar trend whereby their compressive strength increases over the curing period regardless of the types of mix composition. The result shows that the inclusion of 60% untreated RCA to replace the coarse natural aggregate noticeably affects the compressive strength of the RAC. This is clearly shown by the control (CO) RAC specimens in which their compressive strength at 28 days was achieved below the mix design strength target of 50 MPa. By contrast, the RAC prepared with the treated RCA likely performs better than that control specimens. However, the results suggest that the development in the compressive strength of RAC that contains the treated coarse RCA can be enhanced by adding certain volume fraction of fibers. Tables 7 shows that the inclusion of RAC mix with polypropylene fibre at the volume fraction of 0.3 to 1.2% has resulted in an increase in the compressive strength of RAC in the variation range of 11 up to 23% at the age of 28 days. From the results it is noticed that the RAC mix which contains polypropylene fiber at 0.6% demonstrated the highest compressive strength at the age of 28 days compared with others specimens. This improvement contributes to the possible function of randomly oriented fibers to assist in controlling the propagation of crack growth and limiting the crack width in the cement matrix [22, 27].

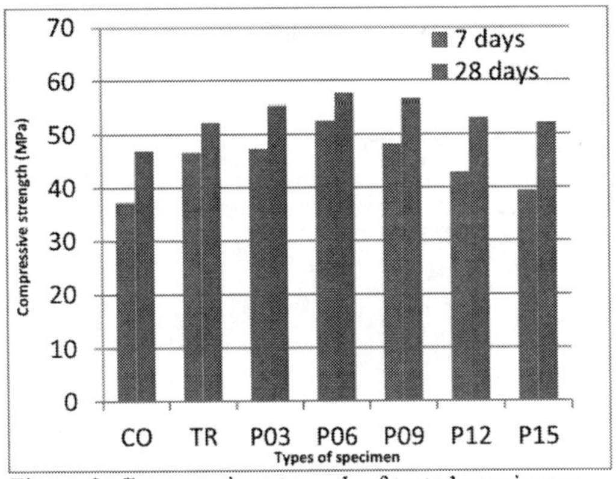

Figure 2: Compressive strength of tested specimens at various curing age

Table 7: Compressive strength relative to the CO specimen

Specimens	Curing age 7 days	28 days
CO	1.00	1.00
TR	1.25	1.11
P03	1.27	1.18
P06	1.41	1.23
P09	1.29	1.21
P12	1.15	1.13
P15	1.06	1.11

4.2 Flexural strength

The flexural strength of the overall specimens across the curing age is illustrated in Figures 3 and Table 8 presents the relative flexural strength of the all mixes (expressed as a percentage of that corresponding to the RO specimens) for the entire testing period .Analysis on the average flexural strength results of specimens indicates that the flexural strength of the RAC which contains the treated RCA (TR) seems to be comparable with the untreated RCA (CO) at the entire testing age. As shown in Table 8, the flexural strength of the TR specimen is found to be 99% and 104%, in correspondent to the CO specimen at the testing age of 7 and 28 days respectively. However, the profile gain in the flexural strength is remarkably improved for the specimens added by fibre. It is clearly shown in Table 8 that P15 has achieved the highest flexural strength result which is 14% higher compared to the control at the age of28 days. This is followed by P06 and P09 where their gain in flexural strength is 3% higher compared with the control for the testing age of 28 days. This result suggests that the randomly oriented polypropylene fibers effective in providing energy absorbing mechanism (bridging action) thus delaying crack propagationsunder flexural loading.

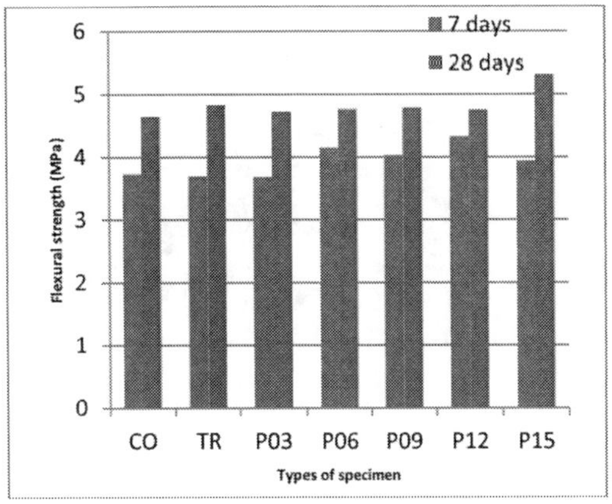

Figure 3: Flexural strength of tested specimens at various curing age

Table 8: Flexural strength relative to the CO specimen

Specimens	Curing age 7 days	28 days
CO	1.00	1.00
TR	0.99	1.04
P03	0.99	1.02
P06	1.11	1.03
P09	1.08	1.03
P12	1.16	1.02
P15	1.06	1.14

4.3 Ultrasonic pulse velocity (UPV)

The result generally shows that the UPV values of all the specimens increase with prolonged curing age (Figure 4). Among the RAC specimens, the CO demonstrated the lowest UPV values at the entire testing age. Contrast, all modified RAC presents the high UPV result with regard to the control particularly the specimens which are consisted of fibre. From the results, the highest UPV value at the age of 28 days is attained by the P06. In this study, an attempt is made to relate the UPV value with the compressive strength of all the specimens at their respective testing ages, as shown in Figure 5. The graph shows that the relationship between the UPV of all the tested specimens and compressive strength is linear and proportional, with the R-square values achieved at 0.815.

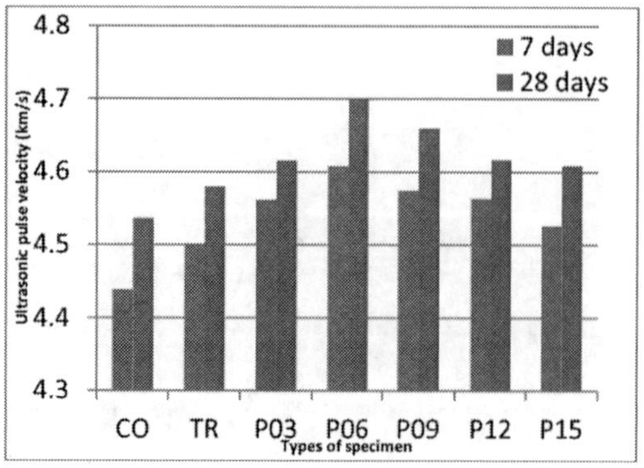

Figure 4: UPV value of tested specimens at various curing age

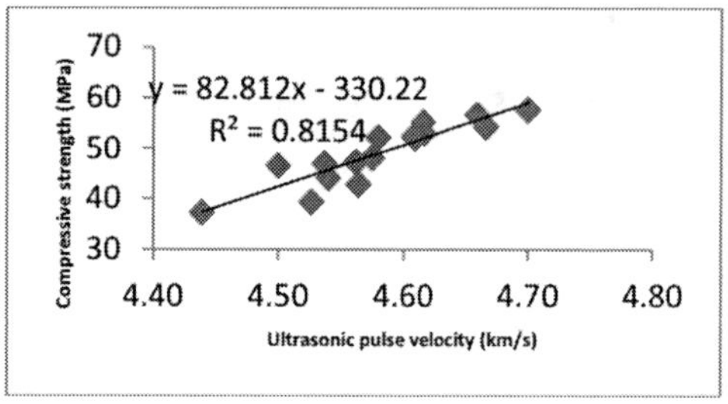

Figure 5: Relationship between compressive strength and UPV

4.4 Water absorption

The water absorption results of all the specimens relative to the age of the specimens are plotted in graph as shown in Figure 6.In general, the obtained water absorption rate of all specimens tends to decrease over the curing period because of the increasing rate of hydration of cement products that fills the capillary pores of the cement matrix. From the results it is observed that the water absorption rate of the CO specimens with the untreated RCA is relatively higher than that of the other specimens across all the testing ages. Such behavior might be due to the porous and high absorption characteristics of the untreated coarse RCA leading to a higher water absorption rate of the RAC. It is interesting to note that apart from this study demonstrated that apart from using the treated coarse RCA, the water absorption of the RAC is significantly reduced with the introduction of fibres. As shown in Table 9, the specimen of P09 achieved the lowest porosity rate which is 17% lower than that for the control specimen at 28 days. This improvement may be due to the non-water absorption characteristic of synthetic fibres. In addition, the distribution of fibres can modify the microstructure of hardened cement pastes in the to seal and bridge the capillary pore structure, which results in minimal void interconnection [28].

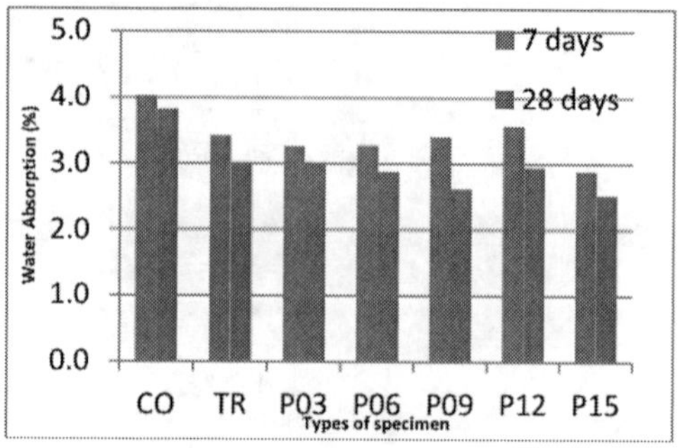

Figure 6: Water absorption of tested specimens at various curing age

Table 9: Water absorption relative to the CO specimen

Specimens	Curing age	
	7 days	28 days
CO	1.00	1.00
TR	0.85	0.79
P03	0.81	0.79
P06	0.81	0.75
P09	0.85	0.69
P12	0.88	0.77
P15	0.72	0.66

4.5 Total porosity

The porosity results of all the concrete specimens with respect testing age are presented in Figure 7. In general, the obtained total porosity rate of all concrete mixtures follows a similar trend as the rate gain of water absorption, where the gain in porosity values of all concrete mixtures gradually decreases with the increase in the curing period. For all the testing ages, the higher porosity rates were obtained from the CO specimens that produced high water-accessible porosities compared with the other specimens. On the other hand the gain in the porosity rate of the modified RAC especially with the inclusion of fibres is remarkably lower than the control specimens in a prolonged curing period. Table 10 shows that after 28 days, the specimen of P09 presents the lowest porosity rate compared to the other specimens which are 17% lower when compared with the control.

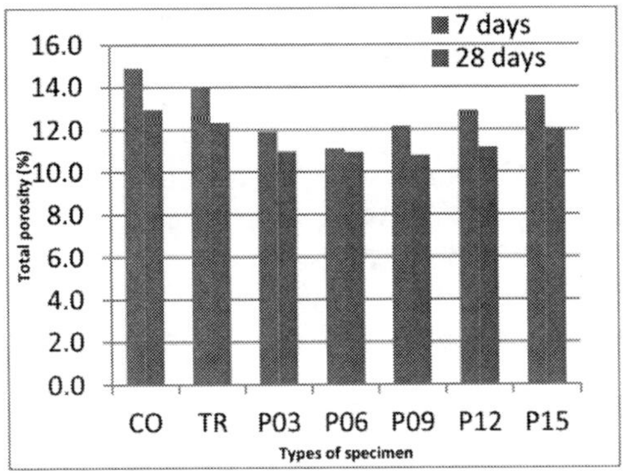

Figure 7: Total porosity of tested specimens at various curing age

Table 10: Total porosity relative to the CO specimen

Specimens	Curing age 7 days	28 days
CO	1.00	1.00
TR	0.94	0.95
P03	0.80	0.85
P06	0.75	0.85
P09	0.81	0.83
P12	0.87	0.86
P15	0.91	0.93

5. Conclusion

The following conclusions could be drawn from the present investigation:-

The finding from this experimental study indicate that apart from the use of treated RCA effect, the inclusion of short discrete polypropylene fibres at different proportions can further enhance the mechanical strength of RAC. The results show that the maximum value of compressive strength is obtained when RAC specimens were added with polypropylene fibre at the 0.6% fibres content. The results also show the water absorption and total porosity characteristic of RAC is significantly reduced with the introduction of polypropylene fibres.

References

[1] Yang D-j, Hao Y-h, Wang T-c. *Experimental Research on Recycled Aggregate Concrete for Highway Pavement*. In: Heng W, Yinhai W, Jian R, Jiancheng W, editors. Tenth International Conference of Chinese Transportation Professionals (ICCTP). Beijing, China: ASCE; 2010. p. 309.

[2] Kou SC, Poon CS, Agrela F. *Comparisons of natural and recycled aggregate concretes prepared with the addition of different mineral admixtures. Cement and Concrete Composites*. 2011;33(8):788-95.

[3] Ismail S, Hoe KW, Ramli M. *Sustainable Aggregates: The Potential and Challenge for Natural Resources Conservation. Procedia - Social and Behavioral Sciences*. 2013;101(0):100-9.

[4] Poon CS, Shui ZH, Lam L. *Effect of microstructure of ITZ on compressive strength of concrete prepared with recycled aggregates. Construction and Building Materials*. 2004;18(6):461-8.

[5] Padmini AK, Ramamurthy K, Mathews MS. *Influence of parent concrete on the properties of recycled aggregate concrete. Construction and Building Materials*. 2009;23(2):829-36.

[6] Tam VWY, Gao XF, Tam CM. Microstructural analysis of recycled aggregate concrete produced from two-stage mixing approach. Cement and Concrete Research. 2005;35(6):1195-203.

[7] Chakradhara Rao M, Bhattacharyya S, Barai S. *Influence of field recycled coarse aggregate on properties of concrete. Materials and Structures*. 2011;44(1):205-20.

[8] Tabsh SW, Abdelfatah AS. *Influence of recycled concrete aggregates on strength properties of concrete. Construction and Building Materials*. 2009;23(2):1163-7.

[9] Poon CS, Shui ZH, Lam L, Fok H, Kou SC. *Influence of moisture states of natural and recycled aggregates on the slump and compressive strength of concrete. Cement and Concrete Research*. 2004;34(1):31-6.

[10] Kou SC, Poon CS, Chan D. *Influence of Fly Ash as Cement Replacement on the Properties of Recycled Aggregate Concrete*. Journal of Materials in Civil Engineering. 2007;19(9):709-17.

[11] Gómez-Soberón JMV. *Porosity of recycled concrete with substitution of recycled concrete aggregate: An experimental study*. Cement and Concrete Research. 2002;32(8):1301-11.

[12] Rao A, Jha KN, Misra S. Use of aggregates from recycled construction and demolition waste in concrete. Resources, Conservation and Recycling. 2007;50(1):71-81.

[13] Tangchirapat W, Buranasing R, Jaturapitakkul C. *Use of High Fineness of Fly Ash to Improve Properties of Recycled Aggregate Concrete*. Journal of Materials in Civil Engineering. 2010;22(6):565-71.

[14] Kishore R, Bairagi NK. *Creep and drying shrinkage of recycled aggregate concrete*. In: Made AM, Sabnis G, Tan JSY, editors. The First International Conference on Recent Advances in Concrete Technology Washington, DC: Destech Pubns Inc 2007. p. 289-98.

[15] Grdic ZJ, Toplicic-Curcic GA, Despotovic IM, Ristic NS. *Properties of self-compacting concrete*

prepared with coarse recycled concrete aggregate. Construction and Building Materials. 2010;24(7):1129-33.
[16] Gomes M, de Brito J. *Structural concrete with incorporation of coarse recycled concrete and ceramic aggregates: durability performance. Materials and Structures*. 2009;42(5):663-75.
[17] Ann KY, Moon HY, Kim YB, Ryou J. *Durability of recycled aggregate concrete using pozzolanic materials. Waste Management.* 2008;28(6):993-9.
[18] Evangelista L, de Brito J. *Durability performance of concrete made with fine recycled concrete aggregates. Cement and Concrete Composites*. 2010;32(1):9-14.
[19] Otsuki N, Miyazato S-i, Yodsudjai W. Influence of Recycled Aggregate on Interfacial Transition Zone, Strength, Chloride Penetration and Carbonation of Concrete. Journal of Materials in Civil Engineering. 2003;15(5):443-51.
[20] Ismail S, Ramli M. Effect of Different Moisture States of Surface-Treated Recycled Concrete Aggregate on Properties of Fresh and Hardened Concrete. International Journal of Civil, Architectural, Structural and Construction Engineering 2014;8(1):65-71.
[21] Maruthachalam D., Muthukumar J. Flexural behaviour of hybrid fibre reinforced concrete beams: a general review. Indian Journal of Engineering. 2013;2(3):14-6.
[22] Şahin Y, Köksal F. *The influences of matrix and steel fibre tensile strengths on the fracture energy of high-strength concrete. Construction and Building Materials*. 2011;25(4):1801-6.
[23] Peled A, Jones J, Shah SP. *Effect of matrix modification on durability of glass fiber reinforced cement composites. Materials and Structures* 2005;38:163-71.
[24] Corinaldesi V, Moriconi G. *Durable fiber reinforced self-compacting concrete. Cement and Concrete Research* 2004;34:249–54.
[25] Teychenné DC, Franklin RE, Erntroy HC, Hobbs DW, Marsh BK. *Design of normal concrete mixes*. Second ed. Watford, UK: Building Research Establishment; 1997.
[26] RILEM recommendations. Absorption of water by immersion under vacuum. Material Structure, RILEM CPC 113. 1984;101:393–4.
[27] Mesbah HA, Buyle-Bodin F. *Efficiency of polypropylene and metallic fibres on control of shrinkage and cracking of recycled aggregate mortars. Construction and Building Materials.* 1999;13(8):439-47.
[28] Poon CS, Qiao XC, Chan D. *The cause and influence of self-cementing properties of fine recycled concrete aggregates on the properties of unbound sub-base. Waste Management.* 2006;26(10):1166-72.

Study on Space Audit Assessment Criteria for Public Higher Education Institution in Malaysia: Space Capacity Assessment

Wan Samsul Zamani Wan Hamdan[1,a], Md. Yusof Hamid[2], Nor Amin Mohd Radzuan[3] and Abdul Mutalib Aman Shah[4]

[1] *PhD student, Faculty Architecture, Planning & Surveying, Universiti Teknologi MARA, Shah Alam Campus, 40450, Selangor, Malaysia*
[2,3] *Faculty Architecture, Planning & Surveying, Universiti Teknologi MARA, Seri Iskandar Campus, Seri Iskandar, 32610Perak, Malaysia*
[4] *Amas FM Consultant Sdn Bhd (Malaysia)*

Abstract: The aim of this study is to measure the capacity rate of learning space based on the as-built drawing provided by the institutions or if the as-built drawing is missing, the researcher have to prepare measured drawing as per actual on site. The learning space Capacity Index is developed by analyzing the space design in as-built drawing or measured drawing and the list of learning spaces available at the institution. The Capacity Index is classified according to the level of Usable Floor Area (UFA) and Occupancy Load (OL) according to learning space design capacity. The classification of Capacity Index is demonstrated through linguistic value and the color-coded key. From the said index, the institution can easily identify whether the existing learning space is currently best used or vice versa and standard space planning compliance in Malaysia Public Higher Education Institutions. The data will assist the management to clarify whether to maximize the use of existing space or to request for new learning space.

Keywords: Capacity index; learning space; space management; building survey; space capacity

1 Introduction

The government of Malaysia has introduced The Government Asset Management Policy (DPAK), in year 2009 as an effort to emphasis on asset management due to extensive amount of investments have been made to the development of assets [1]. In national budget, which announced for year 2016, the government had allocated RM 41 billion for education. As such the management of learning space is one of the most important aspects in the management of physical resources in Higher Education Institutions (HEI). The importance of space management is not just limited to improve capacity index but also related to the cost of maintenance operations. Any unattended space is a risk for the organisation and therefore its need urgent attention by the management. In order to improve capacity index in HEI, therefore learning space should be managed effectively and to avoid wasted spaces. If this is happened it will create uncomfortable learning environment, therefore will interfere with the student's focus that ultimately affect the quality of their learning. Environmental influences also a contributing factor to the failure or excellence of a student [2].

However, many institutions claim to have a lack of learning space [3]. In relation to this, Wamer and Leonard explained that most institutions of higher education is not optimizing the use of physical resources at particular times, such as lower consumption during the learning session, was not used in the evenings, nights, holidays and semester breaks [4]. The management of space is therefore an essential part of an asset management strategy for any institution [5]. Learning space should be effectively and efficiently managed as to optimize the building usage and at the same time can provide cost saving to the building operational cost. This study is important in identifying the

Corresponding author:wasazam5345@gmail.com, wansa345@perak.uitm.edu.my

 DOI: 10.1051/matecconf/20166600078

learning space capacity in HEI. This paper begins with understanding on space management and space audit assessment in HEI. Then focused on space capacity through discussion in literature and analysis of once case study in HEI in Malaysia.

2 Literature Review

2.1 Definition of Space Management

Space management has been defined as "an art and science that maximize the value of existing space and minimize the need for new space"[6]. Space management is important for colleges and universities because it is the largest asset management services financial management system and also has a high investment value [6]. Space is also one of the most valuable resources on campus and high value that must be managed well to accommodate the needs of the high competition among institutions of higher learning [7]. Space management is not only important for financial reasons; the study also found that the space also affects individual responses psychological / physical in terms of their attitudes and behavior[8]. In particular, the physical environment affects how people interact [9], how students learn [10], how students decide where to register [11], and how knowledge is growing [7]. Space not only affects the individual, but also how the use of space can give a true picture direction and priorities of the institution[12], represents the values of the institution, and changing the culture of the institution[7].

Space management is a collaborative process skill in the art and science of maximizing the value, functionality and usability of the optimal space to create additional space requirements to a minimum. Space management is the most important aspect of public management of physical resources. Space management is considered important not only in terms of optimization, but also related to the cost of maintenance operations. But the cost implications associated with physical resources to the public is great [13]. Thus, the physical space should be managed efficiently to help the government in the optimal management of expenses and effective in order to enhance the implementation and achievement of the goals of the university. Space management must also be consistent with the Government Asset Management Policy and the demands of society, including those who are interested. In line with the introduction of the Government Asset Management Policy, Ministry of Higher Education of Malaysia (MOHE) took proactive steps to review and recommend the establishment of Best Practice Management of public space as public space management guidelines in Malaysia [13]. To determine the effectiveness and efficiency of space management in HEI is through space audit assessment at certain period and requirement.

2.2 Definition of Space Audit in Higher Education Institutions

Space audit is quite new in higher education institutions since for the past early years 2000. Not many researches had been done space audit studied since that until now. This is because many researchers are more focusing on facilities management fields in higher education institutions. However, since global economic environment drastically change in 20th century with the rapid development of information technology in many institutions organization body, especially higher education. The implications from these scenarios influence higher education institutions changing the direction of the strategic management from expandable infrastructures asset to optimization facilities asset. At these moments, many higher education institutions make re-assessment existing facilities infrastructures to determine the efficiency of facilities usage especially space management. In determine the effectiveness and efficiency optimization space, a few higher education institutions produces space audit guideline for the process of space management performance. Previously only a few researcher or organization body defines about space audits. Michigan State University (2003) stated that, space audits are "a comprehensive review of currently assigned space for a College." It shows, space audit are all-inclusive detail action process in assess, revise, evaluate and verify the existing data or information about assigned space in campus. The analysis phase of the audit is a comparison of space utilization and distribution based upon departmental personnel rosters, space inventory data, and nationally used formulas and standards, which measure adequacy and effectiveness of space use in colleges and universities (Michigan State University, 2003). However, not at all space being revises, due to the Michigan State University space audit definition are more focusing on assigned space only (not included non-assigned space). Eventually, space audit process must be including as a whole features of space in determine entire space management performance of asset facilities in higher education institutions.

Facilities management consultant [14] also defined space audit as "a space audits are based on space utilisation survey data, and related information from college and university management systems". It can be understand that, space audit process must be depending on space utilisation report information data with including other information that related from overall university management systems. Space utilisation assessment are capability determine the effectiveness of space usage, but it did not cover-up overall performance of space management due to more focusing on space usage but other indicator component that also very relevant in determine space management performance such as space capacity, space physical condition, and equilibrium (balance) of space [15]–[17] Stellea [14] stated that, space audit definition also not specifically showing and explain on space audit areas categories that evaluate overall space management performance (assigned or non-assigned space). These various definitions of space audit in HEI statement shows that is "a process of comprehensive examination, assessment, reviewing and verification on capacity load, utilisation, physical condition and equilibrium space related data or information of existing assigned or non-assigned space inventory in academic or administrative aspects." The information obtained from a space audit is used to update the university's building and room inventory. An audit can include room-specific tours to confirm that space is used as designed, is occupied, and to assess the quality of the space and physical conditions of each room. The audit will note vacant space or other exceptions. Space audits provide information for the assessment of space allocations, prospective planning to accommodate changing situations and realignment of program priorities. The tools to analyse space audit assessment data is through space audit assessment criteria in HEI.

2.3 Space Audit Assessment Criteria in Higher Education Institutions

Most of the space audit assessment tool been establish and implemented by the HEI management are space utilisation rate. Space utilisation rate are multiply occupancy rate with frequency rate and divided by 100 to determine space utilisation rate and performance status. It also purposely for evaluate space need on existing space data. However, not many researchers focusing on other measurement factor that gave influence to space management performance which is space capacity and space condition. Therefore, new space audit evaluation measurement model been develop for HEI in Malaysia (Figure **1.1**). However, for this paper, the study on covered only one assessment criteria, which is space capacity assessment

Space Audit Evaluation Measurement in Higher Education Institutions	**Measurement Criteria**
	Space Capacity
	Space Occupancy
	Space Frequency
	Space Utilization
	Space Condition

Sources: Author (2013)[18]

Figure 1.1: The Development Theoretical Framework for Space Audit Evaluation Measurement in Higher Education Institutions

2.4 Space Capacity Assessment

Space capacity assessment is a key element in this audit because it can measure the rate of use of a space and determine the maximum of capacity of space and density population of space usage. It also has positive effects such as [19]:

i. Identify areas of low use;
ii. Space management plan in accordance with the planning strategic organization;
iii. Forecast future space needs;
iv. Facilitate spatial analysis;
v. Streamline the process of effective measures;
vi. Comparison of actual use than the use of planned, and
vii. Increasing awareness of the importance of space management efficient.

The main components of space capacity assessment are[19]:-

i. Space Capacity Maximum Measurement
ii. Space Capacity Index (Density) Measurement
iii. Space Capacity Index (Density) Score

2.4.1 Space Capacity Maximum Measurement

In measuring the space capacity, it is vital to understand the following terms used as follow:

a. Population Load (Occupancy Load)
b. Type of Floor Area (Usable Floor Area)

Population load that is also known as the occupant load means the size of area per people that appropriate and practical at designated space and can be used to determine number of filling in a space. The calculation is based on estimates of the population load space individual occupying that depends on the type of places and events different, for example the population load of classrooms not the same as the population load with workshop. Three steps to measure the capacity of a space building, is as follows:

a. Determine which categories of the population load will adopted. The load of this population may refer to standard planning standards by the government or the body the relevant professional agencies in the industry construction [20].

b. Determine the usable floor area (UFA) space. Calculation taking into accounts the whole area deduction or reduced by the space that cannot be occupied and used as walls, columns, stairs, utility room and so on. Here are the types of the calculated area based on the requirements:

   i. Gross Floor Area (GFA)
   Total gross floor area encompasses an area of the floor is the main floor and the floor area ancillary.

   ii. Main Floor Area (MFA)
   The total area enclosed spaces only measured in wall to wall including M & E (Room motor elevator, water tank , AHU etc.) but does not include space ansilari .

   iii. Ancillary Floor Area (AFA)
   The total area covered and open space , measured from wall to wall , including sidewalks, air space , open veranda , parking .

   iv. Usable Floor Area (UFA)
   Total floor area that can accommodate users taking into account the suitability of the related.

Space Capacity Maximum Calculation:

$$\frac{\text{Usable Floor Area}}{\text{Occupancy Load}}$$

(Based on Standard and Cost Committee, 2008)

For examples, divide the total usable floor area (whether gross floor area or net) of the population load for example a room with 100 meters of gross floor area square with a load factor of 2.5 square meters. The space can accommodate a capacity of 40 people.

Example Calculation:
Space capacity based on usable floor area compared to the population load

$$= \frac{\text{Usable floor area}}{\text{Population Load space[20]}}$$

Samples: Capacity lecture hall:
M.p.s :

$$\frac{100 \text{ m.s}}{2.5 \text{ m.p.s}} = 40 \text{ students}$$

~ 40 student shows the maximum capacity of space for occupy at certain time in lecture hall.

2.4.2 The Space Capacity Index (Density) Measurement

To determine space capacity index for density populations in space management in HEI by applying space capacity index (density) measurement and score [19]. To measure space capacity index for density calculation by using formula and example calculation as follow:

Space Capacity Index (Density) Calculation:

$$\frac{\text{Actual Number of Students}}{\text{Space Capacity Maximum}}$$

$$= \frac{\text{Number of Students}}{\text{Maximum Capacity Load}}$$

Example Calculation:
Samples: Capacity lecture hall:

$$\frac{31 \text{ student}}{40 \text{ Students}} \times 100\% = 77.5\ \%$$

~ 77.5 per cent shows the space capacity index for density at lecture hall is high population with index 3 rate. It means total optimum use of space and efficiently with ensuring retain and remain at this category.

2.4.3 The Score of Space Capacity Index (Density)

Table 1.1 shows the definition of a space index score based on categories or percentage range is used as a benchmark or standard a uniform measure for all campuses involved. It also shows the level and range of index scores used as a benchmark or standard uniform in determine the status of the level of capacity utilization of space campus. Used more and more space, the more compact campus space and vice versa. According to MOHE space audit report (2011), level of space capacity index is divided into four (4): High Population Density, High Population, Medium Population and Low Population. Detail interpretation for level of space capacity index is further illustrated in Table 1.1.

3 Methodology

This study focuses on the formation of the model frameworks that related to space management of academic space in public higher education institutions in Malaysia. The new development frameworks are space audit assessment criteria in Tertiary Model. Implementation of the study is through literature review and case study of one sample of public university. Case study only focused on percentage space capacity performance for space audit assessment at 18 faculty buildings. The building comprises of main lecture theatre hall, classrooms and tutorial rooms for teaching and learning activities. At the end it showed the status of space capacity index (density) weather it was at high density space population or lower space population.

Three (3) data collection techniques were employed: Interviews, observations and document analyses on time table. Interviews are used to obtain further information on space usage. The owner and the occupant will be interviewed to confirm on the frequency, capacity and condition to validate the total numbers of student at the particular space. The practice of observation survey in space management is vital, whereby the person in-charge to be physically attached to the site to witness and validate the available schedule and actual on site.

The purpose of having time table is to ensure that every space is fully utilized. However superseded time table will contribute to misleading information. In some extent, the space is left vacant due to unavailability of new time table. All these three (3) technique data collection is to determine the actual number of student at various site of space.

Table 1.1: Space Capacity Category and Score Measurement

 DOI: 10.1051/matecconf/20166600078

No.	Faculty	Nos of Students	Maximum room capacity recommendation	* Percentage Performance Capacity (%)	Status
1	biology science	989	849	116%	high density
2	Physics	731	635	115%	high density
3	Chemical Sciences	620	542	114%	high density
4	Out Campus Education	294	260	113%	high density
5	Communications	516	460	112%	high density
6	Arts	543	492	110%	high density
7	Planning, Housing & Buildings	1064	988	108%	high density
8	Languages, Literacies and Translation	582	551	106%	high density
9	humanity	1053	1005	105%	high density
10	Social Sciences	1224	1168	105%	high density
11	Computer Sciences	643	627	103%	high density
12	Pharmacy	894	875	102%	high density
13	Education	1814	1826	99%	high
14	Graduate Business	444	447	99%	high
15	Industry Technology	696	708	98%	high
16	Mathematical Sciences	736	753	98%	high
17	Management	1755	1813	97%	high
18	Research Center	87	98	88%	high
	Total Campus	**14685**	**14098**	**104.94%**	**high density**

Note * This percentage is based on the total number of each faculty than the proposed maximum space capacity assuming students using this learning space without the use of other faculty space.

CATEGORY/ SCORE	INDEX	CAPACITY (%)	DESCRIPTION
High Density Population	4	> 100 %	Overflow, full,crowded, jostling. Need to proposed new space requirement.
High Population	3	76 % - 100 %	Total optimal use of space and efficiently. Should be maintained.
Optimum Population	2	60% - 75 %	Total partial use of space optimal and efficient. needs to be improved until optimal.
Low Population	1	Under 59 %	Failed to reach the total use of space optimal and efficient. need to use enhanced.

The selection case study were based on space audit assessment report of public higher education institutions (PHEI) in Malaysia (Mohe,2011). These reports were study

on three PHEI, which are cannot expose the name of PHEI due to restriction information. Therefore, for the purpose of these papers, only involve one (1) PHEI as University 1 with main purpose to determine the level of space capacity index (Density) performance for teaching and learning facilities at all faculty building at University 1.

4 Disscussion and Findings – Case Study

The case study involved 3,589 learning space or academic rooms with equivalent of 112,692 square meters at University 1. To determine the Space Capacity Index of academic rooms at University 1, the current student's number and maximum capacity load should be obtained first, as shown in figure 1.1. The Overall Space Capacity Index of academic room usage at University 1 in Semester 1, Session 2011/2012 is at level 4. This indicates that the academic rooms capacities are in the high population density with an average percentage of 104% and it also mean overcrowded and need to develop additional academic building space for the future. Space management is considered important not only in terms of optimization but also related to the cost of maintenance operations.

Samples Data Analyses:

989 (Actual Students) / 849 (Maximum Capacity) = 116 %

~ Status Space Capacity Index Score = High Population Density (The space overload, crowded, full and jostling. Need to proposed new space requirement through find other existing space vacant and available or new building construction)

Figure 1.2- Overall Space Capacity Index Performance of Academic Space Overall Space Capacity Index Performance of Space Usage at University 1 at University *1*

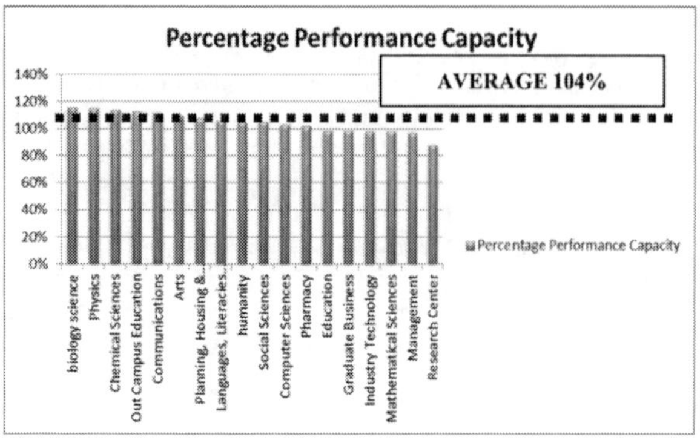

Figure 1.3- Overall Space Capacity Index Performance of Academic Space at University 1

5 Conclusions

All space in Higher Education Institutions, especially for public university must be managed efficiently and effectively by the organization of HEI. To obtain space capacity index, every HEI's to conduct space audit in stages or when it is deemed necessary by the top management. The ability to measure every learning space capacity in relation to the total number of student and standard of space planning will produce result of density for every space. The density will highlight the current space availability and also to be used for future space planning. The Space Capacity Index is one of the significant indicators to measure space usage besides frequency rate, condition rate, and occupancy rate. Based on the above case study, it is indicated that, most of the learning space has achieved high density with average capacity index of 104%. The result of the capacity index to be read together with frequency, occupancy, utilization and condition rate before any decision made for space strategic planning either to maximize existing space or proposed new requirement space by new building construction.

Acknowledgements

The authors would like to express sincere thanks to the Ministry of Higher Education Malaysia, Twintech, UiTM (Sri Isnkandar), University of Malaya and all parties involves, in one way or another, for contributing to this study.

References

1. G. Malaysia., "Government Asset Management Policy," 2009.
2. M. N. B. Ismail, "Kajian Persek3taran Di Sebuah Kolej Universiti Terhadap Keselesaan Aktiviti Pengajaran Dan Pembelajaran," Kolej Universiti Teknologi Tun Hussein Onn, 2002.
3. A. fauzi Wahab, "Pengurusan sumber fizikal ipt: pengurusan ruang," *J. Teknol.*, vol. **43** (E), pp. 15–28, 2005.
4. and C. L. Warmer, D., *The Income Generation Handbook*. Buckingham : SRHE and Open Univerisity Press, 1992.
5. TEFMA, "TEFMA Strategic Aset Managment Workshop Report," Australia, 2009.
6. Hier TC & Biddison GB, "Performance & productivity: The space management mandate.," *Facil. Manag.*, vol. **12**, no. 2, pp. 16–23, 1996.
7. K. Holley and M. Harris, "Constructing the interdisciplinary ivory tower: The planning of interdisciplinary spaces on university campus," *Plan. High. Educ.*, vol. **36**, no. 3, pp. 34–43, 2008.
8. M. J. Graetz, K.A., & Goliber, "The importance of physical space in creating supportive learning environments.," *New Dir. Teach. Learn.*, 2002.
9. T. R. . Davis, "The Influence of the Physical Environment of Office," *Acad. Manag. Rev.*, vol. **9**, no. 2, pp. 271–283, 1984.
10. D. J. Bickford, N.V.N. Chism, *The Importance of Physical SPace in Creating Supportive Learning Environments*. San Francisco, US: Jossey-Bass, 2002.
11. A. W. June, "Facilities Play a Key Role in Students' Enrollment Decisions, Study Find," *Chron. High. Educ.*, vol. **52**, no. 40, p. A27, 2006.
12. I.Fink, "Research Space : Who Needs It, Who Gets It, Who Pays for It?," *Plan. High. Educ.*, vol. **33**, no. 1, pp. 7–17, 2004.
13. MOHE, "Space Audit Report on Optimization Usage in Public Higher Education in Malaysia," Putrajaya, Malaysia, 2012.
14. S. Limited and C. C. House, "The possible linking of institutions into effective educational and training agencies," vol. **44**, no. 0, 2012.
15. Australasian Associate of Higher Education Facilities Officers (AAPPA), "Space Planning Guidelines - Edition 2." AAPPA, Australia AAPPA, 2001.
16. S. Abdullah, L. S. Wee, F. Asran, and M. Nawi, "Application of Space Management in University towards Sustainable Usage," 2007.
17. S. M. Blanchette, "Space and Power in the Ivory Tower : Decision Making in Public Higher Education," University of Massachusetts, Boston, 2010.
18. W. S. Z. W. Hamdan, M. Y. Hamid, and N. A. M.

Radzuan, "Space Management Issues in Higher Education Institutions in Malaysia," in *1st International Conference on Innovation and Technology for Sustainable Built Environment 2012 (ICITSBE 2012)*, 2012
19. M. of E. of M. MOE, *Amalan Baik Pengurusan Ruang di Institusi Pengajian Tinggi Malaysia*, Cetakan Pe. Kuala Lumpur, Malaysia: Penerbit Universiti Putra Malaysia, 2013.
20. E. P. U. of M. Standard and Cost Committee, "Garis Panduan dan Peraturan Bagi Perancangan Bangunan," 2008.

A Framework for Coordination Process into Construction Projects

Wesam S. Alaloul [a], Mohd Shahir Liew and Noor Amila BWA Zawawi

Department of Civil and Environmental Engineering, Universiti Teknologi PETRONAS, 32610 Seir Iskander, Perak, Malaysia

Abstract. Construction industry is recognized as high fragmentation, low efficiency, cost and time overruns in contrast with other industries. These peculiarities are the main roots of poor performance facing by the industry. Effective coordination is vital in construction projects success and mitigate the fragmentation dilemma, however it is often difficult to achieve and need iterative process. Coordination is core issue to improve performance in construction project. Relevant studies have addressed the coordination process importance and implementation, but not in a framework. This paper propose a framework for coordination process in construction projects, as well as its relationship with performance. The objective of the framework is to provide a roadmap for the construction parties to realize operational excellence so that collectively stakeholders can recognize the effect of coordination process application on the project performance. The data were obtained from literature review and structured interviews with five experts. The analysis produced the framework of coordination based on the extensively used procedures for information and data flow between stakeholders.

1 Introduction

Construction industry is the producer of infrastructures and facilities through executing a complex activity and large-scale projects. It differs from manufacturing industry in terms of involving mass production of similar items without a designated purchaser. The uniqueness and the fragmented status of construction projects have been significantly affect its performance in terms of cost, time, and quality indicators [1, 2].

The relationship between construction industry and economic development is well posited. Construction as a vital industry contributes from 6-10 % the Gross Domestic Product (GDP) of developed countries, and 3-6 % to the GDP of developing countries [3]. However, construction projects comprises a wide range of activities including construction, upgrading, repair, and demolitions. The fragmentation status in the project life cycle hinders the teamwork focusing on the best interests of the project. For example, there is less incentive under the traditional approach for a contractor to advice on the design stage. The interaction among all participants in order to distribute the correct information on time is very critical [4]. On the other hand, the existences of significant number of wastes, which can be avoided or minimized, have depleted the overall performance of the industry. In UK construction projects, where up to 30% of construction is rework, only 40–60% of labour effectiveness, and at least 10% of materials are wasted [1, 4]. There is increasing of realization among the construction industry stakeholders of the need to enhance the efficiency of work during the construction project life cycle and sustain working environment, but the steps to achieve them is not clear.

[a] Corresponding author: wesamalool@hotmail.com

Coordination process became a strategy among parties in construction seeking to recognize operational efficiency [5]. Efficient operation can be realized by rationalization operations upright within practical professional units or straight across the entire construction project using effective coordination between management and first line of work [6]. The goal of coordination process is to decrease the information lag through synchronization and integration of activities and responsibilities to ensure that the resources are used most efficiently in pursuit of the predefined objectives.

This paper aim to develop coordination framework that could be used in enhancing the current performance status. Even the value of keeping construction stakeholders coordinated have been figured, the challenge is the implementation of coordination process iteratively. Too often, coordination process in construction projects fails because of the lack or insufficiency, i.e. quality and quantity. However, coordination quality rely on expected forms of communication procedures. Getting work to flow reliably and predictably on a construction site requires the impeccable alignment of the entire coordination process. Coordination induces all fragmentation and information flow problems to be settled among the related participants, and to be eliminated from the hierarchy. The degree of coordination process required for any project depend on its complexity and the amount of exchanged information [7].

The framework is a conceptual in nature and would display the features that should be measured to accomplish effective coordination, a framework would not only show features but would also recommend methods on how to put them into real work. A framework is appropriate for presenting an overall image and assembly for applying coordination advantages [8]. This exploratory study investigates coordination process to pave the way for further study of coordination application. Professionals can procedure the framework to appraise how efficiently their coordination process, realize detailed faults on coordination quality and quantity, and enhance them to rise coordination efficiency.

2 Literature Review

Management strategy in construction is largely taken from other industries. Probably, such as manufacturing, which is not fragmented like construction, nor it has as many closely dependent and dissimilar activities on a critical path. Work force productivity in construction industry has declined about 20% between 1964 and 2003, while other industries improved by more than 200% [2, 4]. In manufacturing industry, production, procurement and transportation is concerned with the design, analysis, and control of production units. Therefore, the coordination of operational units across supply chains is essential function. However, coordination process in construction takes a disjointed path instead of a straight one. It is critical, especially when it comes to multiple concurrent projects. Coordination is important among the same party within individual units, as well as among different parties across the entire project. It will offer the construction industry more flexibility to react to the dynamic environment [5]. The ideal production system embodied in manufacturing industry was used to develop a more overarching production management paradigm for project-based production systems as in construction industry. However, production can be conceptualized in three complementary way in construction projects as; Transformation "is the production of inputs into outputs", or Flow "is a movement that is smooth and uninterrupted from one crew to the next", and Value generation "What the customer is actually paying for the project to produce and install" [9]. Recognizing that construction sites reflect prototypical behaviour of complex and chaotic systems, especially in the flow of both material and information on/off site.

Xue, et al. [6] suggested that construction should be modelled using chaos and complex systems theory. Unlike manufacturing, construction is a project based-production process, therefore the uniqueness of individual projects causes turnover in participants from project to project. Coordination is concerned with the alignment and holistic pursuit of concurrent and continuous improvements in all dimensions of construction projects: design, construction, operation, maintenance, and demolition [8]. This study derived coordination framework to be used on construction projects based on the previous literature and interviews with experts from academic, governmental, and privet sector in construction industry.

While a shift of focus on the activity constitutes an important step forward to understand coordination process in organizations, a preoccupation with contact and communication appears to be difficult in a sustain manner. Existing 'practice-based' theorizing is tied to the assumption that coordination process occurs only when interdependent actors interact. Moreover, more considerations of coordination requires when different parties work together [10]. Coordination need to be thought as a much more inclusive process, being accomplished when emerging and enacting strategies, which aim to pull together all the requirements to execute physical tasks [5, 11].

Coordination can be defined as "*the process of managing dependencies among activities and linking together different parts to accomplish a collective set of tasks*" [12]. Essentially, the delinquent of coordination is the construct of interdependence, which affects task uncertainty and information processing capacity [11]. According to Crowston [13] interdependence is "the extent to which a task requires different departments to engage in work flow, information, and resources, where actions in one unit affect the actions and work outcomes in another unit. If the activities of two units are interdependent, the levels of uncertainty that underlie their interaction need to be explored in relation to how different parties communicate. A number of studies has yet revealed that coordination also depends on the kind, not just the amount, of information and knowledge to accomplish a particular activity, while activity interdependence does not originate only from the contingent relations among units [6, 14]. Coordination process approach to contract aligns project objectives with the interests of key participants. However, the effective application of coordination process relies on participant selection, transparency and continuing dialog [8, 14]. Project members coordinate their work via approaches such as meetings, plans, and joint site visits. However, coordination process on construction projects facing significant barriers. The nature of construction projects itself creates some barriers that are difficult to handle without making wholesale different from current procedures. Coordination barriers are common in inter-party and between different parties of project [2, 10].

The complexities of construction projects require a continuous coordination among many parties with widely disparate skills and experiences. ASCE has been offered 13 criteria to improve construction project coordination: (1) clear definition of the owner's project objectives, (2) coordination procedures, (3) schedules, (4) budgets, (5) written contract clarifications, (6) progress reports, (7) joint reviews of documents, (8) distribution and review of field reports and lab tests, (9) joint visits to work sites, (10) formal compliance reporting, (11) procedures for reporting discrepancies, (12) change order procedures, and (13) coordination meetings [14, 15]. Studies have discussed the relationship between the time spent on coordination process and the project performance. Although Hegazy, et al. [16] found that, holding one or more coordination meetings per month during the design stage reduced the cost by 35%. On the other hand, Hossain [17] reported, project performance did not improve significantly beyond a certain degree of interaction. Furthermore, there was extremely less rework leading to shorter duration [14, 18].

It is obvious that, what has been discussed does not yet provide an integrated and complete theoretical foundation for coordination process in construction project. However, a better theoretical framework for coordination process can be proposed.

3 Research Methodology

In development of coordination process framework, this research implements a review of the present literatures and procedures that have been used in the construction projects. The review does not aim to develop or argue the validity of present practices, nevertheless to categorize and assemble the present practices to generate a universal coordination factors in construction industry and other industries as well. While some of present studies has recorded a few coordination approaches, they differ in terms of area, contributors and business [8, 10, 16-18].

In this study, the first stage was a literature review, followed by data collection involving structured interviews and archiving of documentation for the second stage. The main requirements of a framework are then recognized, tailed by a clarification of how the framework was created, its core parts and its gains. Direction on applying the approaches involved in the framework were defined and the predictable results and key structures were emphasized as well. Figure 1 shows the research procedure of coordination framework impediment. A complete literature review was initially commenced to recognize and conform the list of coordination factors that cover the technique for additional search. Based on hard argument from literature review, 53 coordination factors were identified.

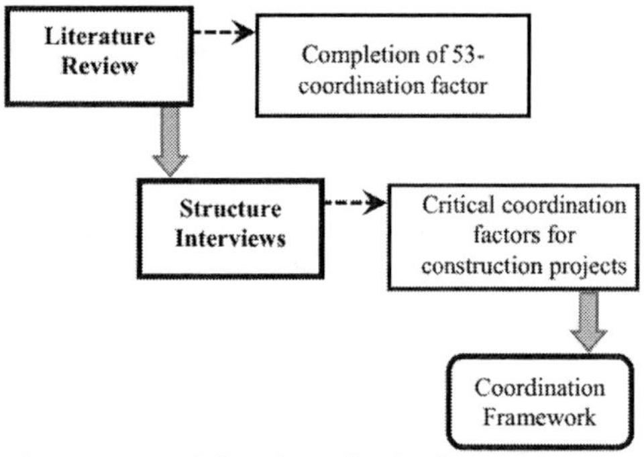

Figure 1. Compilation of coordination framework.

The interviews aimed to expurgate undesirable influences of the researcher from the conversations, by triangulation of the data with other sources in the literature. The use of the interview is transparent and qualified in terms of method [8, 18]. To hold a better sympathetic of present practices and wonders in construction industry, five face-to-face interviews were done with skilled experts in carrying construction projects to authorize the variables created from the literature and identify the coordination factors within the background that have not been included in literature review. The description of interviewees are shown in Table 1.

The interviewees included two experts from government department, two experts from industry and another one from the academic. The selection of the participants was based on their experience in dealing with construction projects coordination. The collective of experts from various backgrounds could give balanced insight and views.

Table 1: Interviewees' profile

position	Experience	Function	Sector
Project manager	12	Policy maker	Governmental
Quantity surveyor	10	Policy maker	Governmental
Project manager	14	Contractor	Private
Architect	9	Consultant	Private
Professor	13	Consultant	Education

The list of questions was sent by email at least one day before the interview, to allow the expert to think about their responses. The interview lasted from 30 to 40 minutes for each session. The interviews have resulted in the adaptation of the list of coordination derived from the literature because of irrelevance and redundancy. The findings from the literature review and preliminary interviews are the basis to form and finalize the framework of coordination process for construction projects.

4 Results and discussion

Evidence from literature indicated that the current practices of construction management and the utilized approaches fail to deliver projects on time, at budget, and at desired quality. With recurring negative experiences on projects, evidenced by endemic quality difficulties and rising litigation, it became obvious that the governing principles of construction management required revisiting [2, 3]. When executing construction projects, professionals from different backgrounds are brought together at various stages to achieve the projects objectives.

This environment increase the probability of conflicts because these professionals have a tendency to work in isolation, while making decisions that affect one another. In particular, the use of coordination concepts have been proposed to enhance the performance and sustain relationships. However, managing the interaction between activities and combined effects of interdependence is a first concern in coordination process since their interactions highly affects project performance [1, 11, 17].

4.1 Coordination framework components

The motivation for creating a framework is to provide the backbones for initiation coordination process creativities in an organized method and offers direction on how to proceed if a group of objectives needs to be satisfied. Therefore, the framework can be applied to improve plans for enhancing at each phase of construction procedure and to arrange the distribution of resources and staffs in efficient manner. The framework will cover the whole project to optimize efforts focus on reliability of collective workflow; in contrast, traditional project management focuses on enhancing productivity of each activity independently. In fact, the proposed coordination framework considered as the angle stone of coordination process application, the final aim of which was to create an instrument that could be practical in construction projects to enhance performance and the level of satisfaction. It should satisfy two main requirement, 'expected' and 'attractive'. The first one refers to those features that answers 'what is' and 'how to' achieve efficient coordination process in construction. The second, refers to merit that may improve the framework but are not critical [8, 16].

The next phase in the framework creation procedure was to form a diagram that measured the entire vital and the smart requirements deliberated. As argued earlier, the request for information and number of meetings operation can be applied as displays of an organization's coordination level, but since their consciousness and apply can be reproduced in the organization coordination strategy, it was obvious to include them into the framework. On the other hands, the culture of construction parts and their sub-partners. Since various parties in the construction industry might adopt a different order to achieve complete integration and coordination that not necessary to be all consistent. The framework should take these industry dynamics and unique characteristics into consideration for widespread adoption.

In coordination process, downstream stakeholders are involved in front end planning and design through cross-functional teams [16]. The pull techniques govern the flow of information and materials, from upstream to downstream rather than, push techniques govern the release of information and materials. Uncertainty in construction projects requires coordination of multidisciplinary professions, activities and information, which contributed to the lack of integration of design information, which leads to unnecessary time variations and design changes in the construction projects [10]. Often, design information uncertain, scatter and comes from various sources. Coordination should be viewed as a process to handle the uncertainty and to synchronize the flow of information, which is essential for enhancing the project performance [12]. Coordination process tries to increase collaboration between the stakeholders, managers and labourers, in order to know the impact of their work. Thus, the assignments should be sound regarding their prerequisites. This means, the task should not start until all the items required for its completion are available. However, in this procedure contrasts to the conventional project management approach, where execution just consists of task authorization.

The framework in Figure 2 below comprises from three main levels: (i) the construction parties; (ii) the construction project life cycle; and (iii) a set of coordination development approaches aimed at supportive the actions through the project life cycle. The sub-actors of each party form the basis of the framework, since the success of coordination application straight depends on them. Top Management of each party was located at the top of the schematic because they lead the coordination efforts. With regard to coordination between the same party sub-actors, top management should establish the coordination vision and policy for them. However, the coordination of work and service that the party provides should be considered during the project life cycle. The dimensions associated with both of these aspects can be used as checklists to ensure the coordination process efficient application. The intermediate part of the framework displays the construction procedure and its phases, each of which may be reinforced by the use of dissimilar coordination development approaches and technology.

Figure 2: Coordination process framework in construction projects

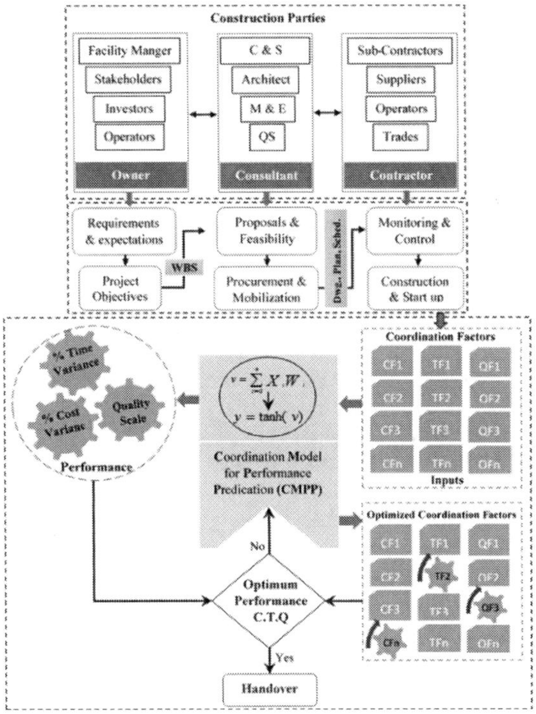

The proposed framework has included most of the construction projects restrictions found in association to coordination process. It considers coordination process from three different perspectives: owner, consultant and contractor, it include the whole project life cycle, and proposed both performance development approaches and factors to execute them. In addition, it has novel features, because the framework endorses the application of the coordination theory "a scarcely explored theory in the construction sector", inspires to harmonize the effort done by different parties, a generally neglected area within the construction industry. Therefore, the work supposed to be efficient from the first time and puts the voice of reduce the re-work, at the core of the project. Using the framework, a well empathetic and awareness of sustain relationships can be predictable. Companies can also expect improved communication between the main parties and their sub-actors from one side and different parties from the other side [5].

Coordination is an iterative process of gradually trust building, understanding and working relationships. The cooperation and coordination spectrum, conceptualized by Nan and Strimling [19], outlines the different degrees of coordination as in Table 2. The coordination process has four overall levels of intensity process (communication, coexistence, coordinated action, or integrated action). To move toward the coordination it is important to understand the environment in which coordination will take place and the capacities and interests of the involved actors. It is essential to determine the degree of coordination necessary for the parties to develop a process to help in knowledge building and relationships necessary to meet these common objectives.

Table 2: Degrees of coordination level project management

Communication	Coexistence	Coordinated Action	Integrated Action
Information conveyance between parties – limited to sharing.	Cohabitation, joint context, and capacity analysis with developed partially. More than sharing.	Joint design and/or implementation of specific activities; sustain active partnership. Consider the others capacities.	Different parties working together. Collaboration in decision-making and evaluation mechanisms for goals.

Once the mechanism of coordination process are understood and the roles that each party will play are clearly decided, it is necessary to establish the objectives of coordination process. Therefore, the objectives do not overlap or contradict, however, develop consolidated objectives, or integrated objectives [6, 19].

5 Conclusions

The construction industry is in its infancy of coordination process. One of the unique characteristic of construction industry is that unlike any other industry, there is a strong inter-firms coordination required across the entire project life cycle. Coordination is the process of getting all elements ready to enable crews to perform work in an effective and efficient manner from the first time. It is the guaranty for getting right information and material to right actor at time to improve productivity. Coordination is important for construction projects to manage the interfaces between the stakeholders. To enhance project performance, various coordination methods can be used, but coordination effectiveness needs to be assessed to find out if they are being used appropriately.

Despite the research efforts, there is no integrated framework for implementing coordination process has been proposed particularly for construction projects. The aim of this paper is to establish a framework for coordination process, to enhance workflow reliability during the project life cycle. The framework was developed based on the coordination theory and its application experience in different disciplines. After the systematic literature review, interviews were conducted with, five managers and engineers of to quantify their coordination quality, quantity, and effectiveness. The coordination quality, quantity, and effectiveness scores were analyzed to identify their internal relationships and their effects on project performance.

References

1. D. M. Gann and A. J. Salter, "Innovation in project-based, service-enhanced firms: the construction of complex products and systems," Research policy, vol. **29**, pp. 955-972, 2000.
2. S. A. Assaf and S. Al-Hejji, "Causes of delay in large construction projects," International journal of project management, vol. **24**, pp. 349-357, 2006.
3. A. Razak Bin Ibrahim, M. H. Roy, Z. Ahmed, and G. Imtiaz, "An investigation of the status of the Malaysian construction industry," Benchmarking: An International Journal, vol. **17**, pp. 294-308, 2010.
4. M. Harvey, "Privatization, fragmentation and inflexible flexibilization in the UK construction industry," Building chaos: An international comparison of deregulation in the construction industry, pp. 188-209, 2003.
5. W. J. O'brien, "Construction supply-chain management: a vision for advanced coordination, costing, and control," in NSF Berkeley-Stanford Construction Research Workshop, Stanford, California, 1999.
6. X. Xue, Y. Wang, Q. Shen, and X. Yu, "Coordination mechanisms for construction supply chain management in the Internet environment," International Journal of project management, vol. 25, pp. 150-157, 2007.
7. A. Akintoye, G. McIntosh, and E. Fitzgerald, "A survey of supply chain collaboration and management in the UK construction industry," European Journal of Purchasing & Supply Management, vol. **6**, pp. 159-168, 2000.
8. X. Xue, X. Li, Q. Shen, and Y. Wang, "An agent-based framework for supply chain coordination in construction," Automation in construction, vol. 14, pp. 413-430, 2005.
9. A. Dubois and L.-E. Gadde, "The construction industry as a loosely coupled system: implications for productivity and innovation," Construction Management & Economics, vol. **20**, pp. 621-631, 2002.
10. D. D. d. Saram and S. M. Ahmed, "Construction coordination activities: What is important and what consumes time," Journal of Management in Engineering, vol. **17**, pp. 202-213, 2001.
11. P. Love, G. D. Holt, L. Shen, H. Li, and Z. Irani, "Using systems dynamics to better understand change and rework in construction project management systems," International Journal of Project Management, vol. **20**, pp. 425-436, 2002.
12. T. W. Malone and K. Crowston, "What is coordination theory and how can it help design cooperative work systems," in Proceedings of the 1990 ACM conference on Computer-supported cooperative work, 1990, pp. 357-370.
13. K. Crowston, "A coordination theory approach to organizational process design," Organization Science, vol. **8**, pp. 157-175, 1997.

DOI: 10.1051/matecconf/20166600079

14. A. S. Chang and F.-Y. Shen, "Coordination needs and supply of construction projects," Engineering Management Journal, vol. **21**, pp. 44-57, 2009.
15. Y. Wang, "Coordination issues in Chinese large building projects," Journal of Management in Engineering, 2000.
16. T. Hegazy, E. Zaneldin, and D. Grierson, "Improving design coordination for building projects. I: Information model," Journal of Construction Engineering and Management, vol. **127**, pp. 322-329, 2001.
17. L. Hossain, "Communications and coordination in construction projects," Construction Management and Economics, vol. **27**, pp. 25-39, 2009.
18. A. Caballero, S. Ahmed, S. Azhar, and M. Barcala, "Development of an information model to enhance integration and coordination in the construction projects," NIST SPECIAL PUBLICATION SP, pp. 123-128, 2003.
19. S. A. Nan and A. Strimling, "Coordination in conflict prevention, conflict resolution and peacebuilding," International Negotiation, vol. **11**, pp. 1-6, 2006.

MATEC Web of Conferences **66**, 00080 (2016) DOI: 10.1051/matecconf/20166600080
IBCC 2016

Assessment Schemes for Sustainability Design through BIM: Lessons Learnt

Syahrul Nizam Kamaruzzaman[1], Hafez Salleh[2], Eric Choen Weng Lou[3], Rodger Edwards[4], and Phui Fung Wong[5,a]

[1, 2, 5] *Faculty of Built Environment, University of Malaya, 50603 Kuala Lumpur*
[3,4]*School of Mechanical, Aerospace and Civil Engineering, The University of Manchester, Manchester M13 9PL, United Kingdom*

Abstract. There is increasing demand on sustainability-led design to reduce negative impacts brought by construction development. The capability of Building Information Modeling (BIM) to achieve sustainability is widely acknowledged. Various sustainability analysis and calculation can be performed at early stages to help the designers in decision making. However, the level of implementation is still not popular in the construction industry. Many of the industry players are still rely on traditional 2D method for designing and analysis. Hence, this study aims to demonstrate a proof concept of using BIM for sustainability design. The first phase of this study conducted a critical review of existing assessment schemes: BREEAM, LEED, SBTool, CASBEE, BEAM Plus, Green Star, Green Mark and GBI, to develop a set of main criteria to be considered for sustainability design. The findings revealed that fourteen criteria are considered, which are management, sustainable site, transport, indoor environmental quality, energy, waste, water, material, pollution, innovation, economics, social, culture and quality of services. It was found that most of the existing schemes emphasized on environmental aspect as compared to economics, social and culture except SBTool. The next phase of this study will conduct a case study to demonstrate sustainability design through BIM by using the criteria developed from the first phase.

1 Introduction

The concern about global warming, greenhouse effects and pollution is increasing. There is a growing evidence of global warming and climate change which increase the necessity for immediate action to avoid potentially serious consequences for future generations [1]. Malaysia is facing an increasing trend of per capita carbon dioxide emission and energy consumption [2]. Therefore, the Malaysian government has set a voluntary target to reduce 40% of carbon emission [3]. It is noted that building is one of the highest contributions to global warming and greenhouse effect. In the UK, buildings consume near to 50% of carbon dioxide emission [4]. While in the US, buildings account for close to 50% of all the annual energy used and contributes to 40% of the global carbon emissions [5]. The environmental impact brought by the construction of buildings has attracted many professionals in the industry to consider sustainable design for their building designs. Biswas et al. [6] highlighted the demand for sustainable buildings with minimal impacts to the environment are increasing. Autodesk [7] also highlighted that a growing field within sustainable building design has emerged as building growth intersects with environmental concerns and the rising cost of energy. Therefore, the sustainability of design has becomes a critical concern in the construction industry for designing, constructing and operating facilities in such a manner so as to minimize the environmental impact.

To achieve sustainable design, better monitoring building performance at the beginning stage with significant amount of accurate information is required [8]. It is noteworthy that decisions regarding sustainability of designs have to be made at the early design stages [5]. Traditional building design methods such as conventional CAD do not contain sufficient amount of information at the beginning stages to support early decision making. However, building information modeling (BIM) application allows decisions to be made at the early stages of design with rich information contained in a single coordinated model. Hence, various types of sustainability and performance analysis can be performed throughout the design process in order to improve the building performance. By incorporating BIM technology into sustainable design, it has the potential to change the traditional design practice and produce a greener building with better performance [9, 10]. Various literature sources have provided evidence that BIM application can improve building performance to achieve sustainable design [9, 11, 12].

The combination of sustainable design and BIM technology has the potential to improve traditional design

[a] Corresponding author: wong_pf@um.edu.my

practice which in turn produces a sustainable way of design. However, the use of BIM tools for sustainability assessment is still lacking [11]. Hence, this study is intended to demonstrate a sustainable design through BIM to show a proof of concept of sustainability benefits by application of BIM. This study is divided into two stages. Firstly, a critical review of assessment schemes is conducted. It seeks to review the globally prevalent environmental assessment schemes in order to find out the similarities and differences in assessing main criteria which in turn establish a list of sustainable criteria for sustainable design though BIM. Second phase of this study will integrate BIM object to demonstrate sustainable design by referring to the assessment criteria that developed from the first phase. By doing this, it can provide a comprehensive assessment criteria for government and assessment scheme developer to develop a complete scheme for sustainable design and modelling. Moreover, it can gain understandings among the architects, designers and engineers of the real case benefits of using BIM for sustainable design. This is an on-going research and the first phase of this research will report in this paper.

2 Sustainable Design through BIM

Sustainable designs can be achieved by carrying out various building analysis and optimization of the design to enhance building performance. In order to perform various analyzes, consistent and reliable information should be made available at the early design stages. The required data information for various analyzes including relevant data about the external environment, the shape and configuration of the building, equipment loads, lighting mechanical systems and ventilation [8]. Unfortunately, such data is often not available at the early design stages. As a result, energy and performance analysis are normally performed at the later stages which is after the architectural design stage and construction documents have been produced, thus reducing the chance for early modifications that could enhance building performance [9-11]. Moreover, to perform these analyses on a conventional way requires human interpretation which is time consuming, tedious and costly. Therefore, a tool that can support early decision making and contain sufficient amount of information is needed. Jrade and Jalaei [13] accentuated that there is a need for construction industry to adopt new technologies for building design with the purpose of reducing impacts on the environment.

BIM is an innovative way to achieve sustainable designs by reducing the energy consumption and carbon emission which ultimately delivers greener buildings. It is an approach to building design, construction and management with the use of a digital coordinated database that store a wealth of information throughout the lifecycle of the buildings. Hence, it is a rich repository that stores multi-disciplinary information in a single virtual building representation throughout the lifecycle of the buildings [5]. It allows multi-disciplinary information (structural, mechanical, electrical, plumbing and lighting) stored in a single model that allows building analysis to be performed quickly and accurately at the beginning stage. Designers are able to incorporate greener design, provide various design alternatives for sustainability and make better early decisions in order to improve building performance. BIM facilitates a complex analysis and calculation that was previously too tedious or laborious to be performed traditionally.

Krygiel and Nies [14] indicated that BIM application can achieve several aspects of sustainable designs which traditionally cannot be produced by using traditional 2D tools such as building orientation (which can reduce project cost), building massing (to analyse building form and optimise building envelope), day lighting analysis, energy modeling, water harvesting (reduce water consumption), sustainable materials (reduce virgin materials consumption), site and logistics management (reduce waste). Various sustainability analyses, assessments, evaluations and calculations can be performed to enhance air quality, thermal comfort and ultimately improve the overall energy usage by linking the building model to energy analysis tools [9-10]. Jrade and Jalaei [13] also explained that BIM supports experimental analysis of structure, environmental controls, construction method, use of new materials, or systems. Thus, BIM application provides an opportunity for industry professionals to realize the numerous benefits associated with sustainable analysis and design throughout the project.

3 Sustainable Scheme in various regions

Many countries have developed their own assessment schemes. Each region has its own geographical and cultural characteristics. Therefore, there are many assessment schemes developed for a specific site and local purposes. It is necessary to develop a comprehensive list of criteria by reviewing existing schemes as it is the standard for sustainable design in various regions. By gaining understanding of the criteria used, it can assist this study in demonstrating sustainable design through BIM by using the main criteria.

BREEAM (Building Research Establishment Environmental Assessment Method) is known as the first building environmental assessment scheme developed in the early 1990s. There has been a rise in the number of assessment schemes that promote sustainable building developments. Among widely known are LEED (Leadership in Energy and Environment) in US, SBTool (Sustainable Building Tool) in Canada, CASBEE (Comprehensive Assessment System for Building Environmental Efficiency) in Japan, BEAM Plus (Building Environmental Assessment Method) in Hong Kong, Green Star in Australia and Green Mark in Singapore. In Malaysia context, GBI (Green Building Index) is the assessment scheme widely adopted for building assessment.

In early 1990s, BRE Ltd (Building Research Establishment Limited) developed BREEAM to evaluate the new and existing building performance. The assessment scheme also include for refurbishment purpose and neighbourhood development. The development of BREEAM has a profound impact on almost all of the assessment schemes. It has been used as a template and reference model for the creation of Green Star assessment tool in Australia and BEAM-PLUS in Hong Kong [15]. Another prevalent schemes is LEED which developed by the US Green Building Council (USGBC). The assessment of LEED can be used for new, existing, renovation, interior design, homes and neighbourhood development. Meanwhile, LEED is the most widely recognized assessment scheme as it has been used to carry out in 41 countries [16].

SBTool was developed by the International Initiatives for a Sustainable Built Environment (iiSBE) and widely considered as the most comprehensive of all environmental assessment method [12]. It has been widely adapted by different countries located in Europe, Africa, Canada and the far east of Asia. Its flexibility allows the scheme to be adjusted to suit local conditions of each region and also site variations [17] which make it becomes very popular. The development of CASBEE scheme was an effort of Japan Green Building Council (JaGBC)/ Japan Sustainable Building Consortium (JSBC) and their subcommittees [16]. CASBEE was applied for pre-design, new construction, existing building and renovation building that correspond to the building life cycle. BEAM Plus is a voluntary scheme first launched in 1996 (formerly known as HK-BEAM). It was based largely on the BREEAM assessment scheme but modified for Hong Kong's densely populated infrastructure. It applied to new and existing buildings and covered a wide range of issues related to impacts of buildings on the environment in terms of global, local and indoor scales [16].

In 2003, Green Star assessment scheme was developed by a non-government organisation, the Green Building Council of Australia (GBCA). It is a comprehensive, national, voluntary environmental assessment scheme and has been adopted by other regions such as New Zealand and South Africa. In Singapore, Green Mark was introduced by Building and Construction Authority (BCA) in 2005. It aims to evaluate buildings for environmental impact and performance, and promote sustainable design, construction and operational practices throughout the City-State Republic of Singapore [17]. In 2001, the Malaysian Institute of Architects (PAM) and the Association of Consulting Engineers Malaysia (ACEM) developed GBI. It is developed based on Malaysian tropical climate, environmental and developmental context, cultural and social needs [17]. GBI is applicable to new and existing residential and commercial buildings in Malaysia.

4 Findings and Discussion

After holistic reviewing, among the eight schemes, assessment criteria can be divided into fourteen criteria: management, sustainable site, transport, indoor environmental quality (IEQ), water, waste, material, energy, pollution and innovation, economic, social, cultural and quality of service as illustrated in Figure 1. The detail results are tabulated in Table 1.

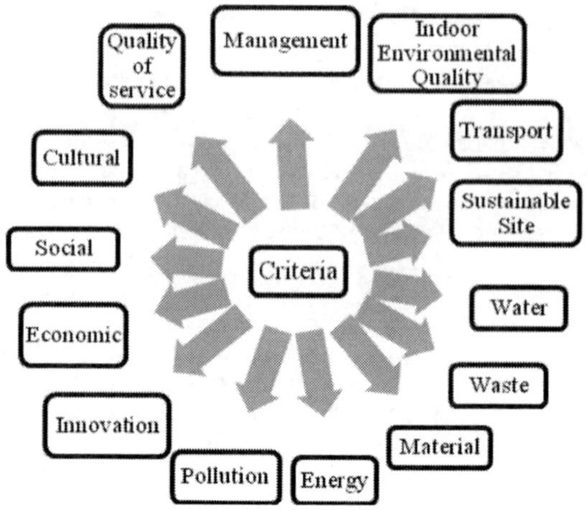

Figure 1. Main Criteria of Assessment Scheme

Table 1 showed that all assessment schemes evaluated management criteria in their assessment scheme. Management criteria deals with how buildings can be sufficiently plan, design, construct, operated and maintained throughout the building life cycle [18]. Sustainable site evaluation focused on site location which the building is to be built. It generally assesses Greenfield and Brownfield site to avoid development in inappropriate site and also reduce the environmental impact from the building. Majority of the assessment schemes assesses this criterion except Green Mark and GBI. From the Table 1, it is apparent that all of the assessment schemes evaluated transport criteria. The aim of this criterion is to deliver a good level of communication, through east access to public facilities and services and adequate provision for pedestrians, cyclists and drivers [19]. The adequacy of public transport and local amenities is essential in order to reduce congestion and air pollution due to private vehicles.

Criteria	BREEAM	LEED	SBTool	CASBEE	BEAM Plus	Green Star	Green Mark	GBI
Management	✓	✓	✓	✓	✓	✓	✓	✓
Sustainable site	✓	✓	✓	✓	✓	✓		
Transport	✓	✓	✓	✓	✓	✓	✓	✓
Indoor environmental quality (IEQ)	✓	✓	✓	✓	✓	✓	✓	✓
Water	✓	✓	✓	✓	✓	✓	✓	✓

Waste	✓	✓	✓	✓	✓	✓	✓	✓
Material	✓	✓	✓	✓	✓	✓	✓	✓
Energy	✓	✓	✓	✓	✓	✓	✓	✓
Pollution	✓	✓	✓	✓	✓	✓	✓	✓
Innovation	✓	✓			✓	✓	✓	✓
Economic	✓		✓					
Social		✓	✓	✓	✓			
Cultural		✓	✓	✓	✓			
Quality of service	✓		✓	✓	✓	✓		✓

Table 1. Comparison of criteria for eight schemes

It is articulated that IEQ criteria has been considered in all of the assessment schemes. Alyami and Rezgui [18] also pointed out that IEQ is one of the popular criteria evaluated in all assessment schemes. It aims to increase health, safety and comfort of building occupants in the building. Poor indoor environments in buildings can pose health risks to building users as they spend most of times inside the building. Water is viewed as limited and valuable resource; hence all of the assessment schemes included it to seek for sustainable water use and management. It is to ensure proper provisions are provided to reduce fresh water consumption in the building. Hence, water recycling such as rainwater harvesting and grey water recycling play an important role in achieving sustainable water management.

Construction activities generated a lot of wastes which require proper management and strategies to recycle or dispose the waste. Hence, waste criterion covered comprehensively in all assessment schemes in ensuring best practice in management of construction and operational waste. It aims to improve waste management and encourage waste recycling and sorting. Building materials are one of the important criterions in the all assessment schemes due to their complicated life cycle process from extracting raw materials till disposal stage [18]. It consumes a large amounts of natural resources, hence material selections is critical to ensure efficient use of materials and avoid wastage.

Assessment schemes place vital importance on energy criterion due to its significant impact on environment. With the increasing concerns about global warming and greenhouse effect, it is necessary to improve energy efficiency of the building by reducing carbon emissions. For pollution, this criterion deals with outdoor sources of air pollution in order to reduce the pollution and adequate provisions are taken to limit the effect of the pollution. Sustainable construction emphasized the protection of the surrounding environment in order to minimise the effect of the construction activities to the surroundings. Majority of the assessment schemes support innovation in the evaluation framework which can provide environmental benefits. Any new methods that can be shown to improve sustainability performance of a building are highly encouraged. In order to encourage exemplary performance in all sustainable aspects, BREEAM and LEED have established supplementary criteria to reflect exceptional performance [18].

In economic aspect, BREEAM and SBTool evaluated the financial aspect whereas the other schemes can be considered relatively weak in this aspect. Raslanas et al. [20] suggested that financial and environmental aspect should go along especially during financial feasibility stage while evaluating alternative development options. In social aspect, LEED, SBTool, CASBEE and BEAM Plus covered the assessment. Social sustainability is essential in taking of care of the welfare (privacy, security, amenities) and equitably distributed among social classes and gender. Culture aspect is neglected in most of the assessment schemes except LEED, CASBEE, SBTool and BEAM Plus. Building development should integrate local cultural value to enhance and promote cultural value in design. Assessment of service functions is crucial to keep the building in good condition in the long term. Therefore, quality of service is one of the important assessment criteria. However, LEED and Green Mark do not evaluate on the quality of service in the assessment scheme.

In summary, SBTool covered most of the assessment criteria not only environmental aspect but also economic, social and cultural aspects which neglected in other assessment scheme. It is affirmed by Raslanas et al. [20] that most of assessment schemes do not cover well in social and economic aspects. However, in order to achieve sustainable development, the economic, social and cultural aspect should not be neglected. Haapio and Viitaniemi [21] stressed out that sustainable building should includes the economic and social aspect in addition to the environmental aspect.

5 Conclusion and Further Research

The awareness of the buildings impact on the environment has increased the need for building professionals to consider sustainable design and practice. BIM has the capability to achieve sustainability of design by containing the relevant information in a single coordinated model which allows for various sustainability assessments to be performed at the early stages. However, the application of BIM in sustainability design is relatively low. Hence, this study aims to develop a proof concept of sustainability design by using BIM. The first stage of this study is to carry out a comprehensive review of the existing assessment schemes in order to develop a set of criteria to be used for sustainability design in the second stage.

The assessment schemes chosen for this study are BREEAM, LEED, SBTool, CASBEE, BEAM Plus, Green Star, Green Mark and GBI. The finding of the study showed that there are fourteen common criteria used. There are management, sustainable site, transport, indoor environmental quality, water, waste, material, energy, pollution, innovation, economic, social, culture and quality of services. The less common criteria considered in the assessment schemes are economic, social, culture, and quality of services. It is noted that most of the assessment schemes emphasized more on environmental aspect rather than economic, social and culture aspect except SBTool. This is an on-going research. The next stage of the research is to demonstrate the sustainable design through BIM by using the sustainable criteria developed from the initial stage.

Acknowledgement

This research is supported by British Council Newton-Ungku Omar Fund Institutional Links (ID: 172726659) and University of Malaya Institute of Research Management & Monitoring (IPPP) (grant no: IF009-2015).

References

1. IPCC. Synthesis Report, Intergovernmental Panel of Climate Change. (2007) Available from: https://www.ipcc.ch/pdf/assessment-report/ar4/syr/ar4_syr_full_report.pdf
2. R. A. Begum, K. Sohag, S. M. S. Abdullah, & M. Jaafar. CO_2 emissions, energy consumption, economic and population growth in Malaysia. *Renewable and Sustainable Energy Reviews*, ***41***, 594-601 (2015)
3. NC2. Malaysia's second national communication(NC2) – a report submitted to the United Nations Framework Convention on Climate Change(UNFCCC), Conservation and Environmental Management Division(CEMD),Ministry of Natural Resources and Environment (NRE).Available from: ⟨http://nc2.nre.gov. my/⟩; 2011[accessed 25 August 2015].
4. Department of Communities and Local Government. Building a greener future: policy statement. (2007)
5. A. Schlueter, and F. Thesseling. Building information model basedenergy/exergy performance assessment in early design stages, *Automation in Construction*,**18** (2):153-163 (2009)
6. T. Biswas, T. H. Wang, R. Krishnamurti. Integrated sustainable building rating systems with building information modeling. In: Proceedings of the 13th International Conference on Computer Aided Architectural Design Research in Asia, Chiang Mai, Thailand, pp. 193–200 (2008)
7. Autodesk, Inc. Building Information Modeling for Sustainable Design. *Autodesk® Revit® White Paper* (2005)
8. I. Motawa, and K. Carter. Sustainable BIM-based Evaluation of Buildings, *Procedia-Social and Behavioral Sciences*, **74**:116-125 (2013)
9. S. Azhar, and J. Brown. BIM for sustainability analyses. *International Journal of Construction Education and Research*, **5**(4): 276-292 (2009)
10. S. Azhar, J. Brown, and R. Farooqui, BIM-based sustainability analysis: An evaluation of building performance analysis software. *Proceedings of the 45th ASC Annual Conference*, Gainesville, Florida (2009)
11. A. Azhar, W. A. Carlton, D, Olsen, and I. Ahmad, Building information modeling for sustainable design and LEED® rating analysis. *Automation in construction*, **20** (2): 217-224 (2011)
12. R. J. Cole, & N. Larsson, Building challenge 2002:GBTool user manual (2002) Available from: www.iisbe.org/iisbe/gbc2k5/gbc2k5-start.htm/ [access 20 August 2015].
13. A. Jrade & F. Jalaei, Integrating building information modelling with sustainability to design building projects at the conceptual stage. In *Building Simulation* (Vol. **6**, No. 4, pp. 429-444). Springer Berlin Heidelberg (2013)
14. E. Krygiel, and B. Nies, Green BIM. Indianapolis: Wiley Publishing, IN (2008)
15. K. C. D. Grace, Sustainable construction—The role of environmental assessment tools. Journal of Environmental Management, 86(3), 451–464 (2008)
16. W. L. Lee, A comprehensive review of metrics of building environmental assessment schemes. *Energy and Buildings*, ***62***, 403-413 (2013)
17. CSI (Construction Specifications Institute). The CSI Sustainable Design and Construction Practice Guide. John Wiley & Sons, Inc (2013)
18. S. H. Alyami, & Y. Rezgui, Sustainable building assessment tool development approach. *Sustainable Cities and Society*, ***5***, 52-62 (2012)
19. BRE. BRE homepage. (2011) Available from: http://www.bre.co.uk/ [access 18 September 2015].
20. S. Raslanas, A. Stasiukynas, & E. Jurgelaitytė, Sustainability assessment studies of recreational buildings. *Procedia Engineering*, ***57***, 929-937 (2013)
21. A. Haapio, & P. Viitaniemi, A critical review of building environmental assessment tools. *Environmental impact assessment review*, ***28***(7), 469-482 (2008)

MATEC Web of Conferences **66**, 00081 (2016) DOI: 10.1051/matecconf/20166600081
IBCC 2016

The Provision of the Disabled Facilities in Public Hospitals

Yuhainis Abdul Talib[1], Nurul Izzati Abdul Ghani[2], Kharizam Ismail[3] and Nor'Aini Salleh[4]

[1,2,3,4] Faculty of Architecture, Planning and Surveying, Universiti Teknologi MARA, Seri Iskandar Campus, 32610, Seri Iskandar, Perak, Malaysia

Abstract. The implementation of Person with Disabilities Act 2008 has become an eye opener towards the provision of disabled facilities in all public buildings especially regarding healthcare accessibility. The aim of this research is to analyse the provision of the disabled facilities provided in a public hospital in Perak, Malaysia. To support the research, it was supporting with two objectives that are to identify the designs and requirements of disabled facilities that need to be fulfilled according to Uniform Building By-Laws (UBBL) and other standard regulations and to identify the level of awareness of public on disabled facilities in the public hospital. For the purpose of the research, the observation method has been done towards three (3) selected hospital in Perak and questionnaire survey have been distributed to 96 respondents among the staff and visitors of the hospitals. All the data collected from the questionnaire survey and the checklist using UBBL standard requirement, then analysed using the SPSS V.21. The outcome of the observation indicates that most of the hospitals were provided with the disabled facilities, but there are still some rooms for improvement regarding specifications and the provision itself. In a nutshell, this research helps the management of the hospital together with the public on the importance of the provision of the disabled facilities in public buildings.

1 Introduction

All hospitals must be user-friendly regardless of the ages, races, gender especially to disabled persons. According to News Straits Times in [10], Mayor Datuk Mat Noh Ahmad said that it is necessary for buildings proposed to be disabled-friendly to get approval under the Persons with Disabilities Act 2008. This act is developed to ensure public beneficial especially to a disabled person.

According to Ibrahim (10), the United Nations estimates that the number of persons with disabilities in Malaysia is about 2.8 million, or 10 per cent of its population of 28 million. However, Social Welfare Department (JKM) statistics, as of August 2010, showed that only 313,685 People with Disabilities (PWD) had registered with the department. This means there are a large amount of disabled people in the country who have not registered with JKM. Islam [6] said that persons with disabilities are usually the nation's largest minority, and they tend to be marginalized in all aspects of life. Many visitors and caregivers who also disabled from poor families were forced to sleep in hospital corridors [8]. In many countries, access for patrons with disabilities to enjoy the facilities in the hospital is less taken into account and sometimes is taken for granted. The process of managing the facilities is more challenging due to the complexity of healthcare buildings [2].

Therefore, this research emphasizes on two objectives:

i. To identify the designs and requirements of disabled facilities that need to be fulfilled according to Uniform Building By-Laws (UBBL) and other standard regulations
ii. To determine the level of awareness of public on disabled facilities in public hospital

2 Disable facilities in public hospitals

Disability can be defined as the restrictions of activity by society towards a person where impairments are the characteristics or feature that cause restrictions in conducting the activity that may cause injuries that affect their appearance, communications and functions of the body [9].

The recently published World Report on Disability suggests there are approximately one billion people across the globe with some form of disability, representing around 15% of the world's population. The trend was a sizeable increase from the last World Health

[a] Corresponding author: anis500@perk.uitm.edu.my

Organisation estimate of around 10% in the 1970s [13]. Besides, the United Nations estimates that the number of persons with disabilities in Malaysia is about 2.8 million, or 10 percent of its population of 28 million [4].
A district hospital typically is the major health care facility in its region, with large numbers of beds for intensive care and long-term care with the provision of primary-level hospital-based services [12]. District hospitals can overcome the territorial imbalance and reduce the difficulties in accessing the health-care system of those towns that are far from metropolitan areas at a reasonable cost [11].

2.1 Person with Disabilities Act 2008 (Act 685)

Part III of Act 685 provides that persons with disabilities must be given the right to access and use of public facilities, amenities and services and buildings provided to the public on an equal basis with people without disabilities. As such the Government and the providers of such public facilities, amenities, services and buildings are required to give an appropriate consideration and take necessary measures to ensure that such public facilities, amenities, services and buildings and the improvement of the equipment related is to conform to universal design.

2.2 Provision of By-laws 34A of the Uniform Building Bylaws (UBBL)

Bylaws 34A of the Uniform Building Bylaws makes it mandatory for all public buildings to make available facilities for disabled persons. On September 20th, 1990 an amendment was made to the Uniform Building Bylaws (UBBL) at the 35th National Council of Local Government (NCLG). The amendment, introduced section 34A in UBBL, making it compulsory for buildings to provide access to enable disabled persons to get into, out of and within the buildings.

2.3 Malaysian Standards for Persons with Disability

Despite all the other standard regulations, Malaysia also has their standard regulations that need to be followed by the designers and management.

a) Malaysian Standard 1183:1990- Code of Practice for Means of Escape for Disabled Persons

Code of Practice for Fire Precautions in the Design and Construction of Buildings (MS 1183) issued by SIRIM in 1990 requiring special standards to be complied with in providing fire precautions in designing public buildings to make it safe for the use of disabled persons especially during a fire.

b) Malaysian Standard 1184:2002, Code of Practice on Access for Disabled Persons to Public Buildings (First Revision)

This MS specifies the basic requirements for elements of buildings and related facilities so as to permit access by Person with Disabilities (PWD). These requirements apply to all buildings that PWDs may use as members of the general public, as visitors or for purposes of employment. By implementation of this standard, all the disabled people can now access to the public buildings. This standard supersedes MS 1184:1991.

c) Malaysian Standard 1331:2003, Code of Practice for Access for Disabled Persons Outside Buildings (First Revision)

This standard supersedes MS 1331:1993. This code revised about the provisions on designing of special facilities outside buildings for disabled people to ensure accessibility and usability by disabled persons. In complying with this Standard, all building plans submitted to the local authorities after the date the Code of Practice is published in the State Gazette, each State must comply with it.

2.4 Town and Country Planning Act 1976

The Town and Country Planning Act 1976 though not providing in clear, precise terms the requirement to create accessibility for disabled persons, however, requires planning authorities to ensure that provisions of other related written laws are taken into consideration in the process of approving applications for planning permission.

2.5 Provision of Disabled Facilities According to the Standard Regulations

Positive supports encourage them to prove their capabilities while public stereotype and prejudice towards PWD may diminish their self-esteem and confidence to participate in the social and economic mainstream [7]. According to Abdul Rahim [1], a more barrier-free design of facilities need to be constructed as all categories of people can enjoy the facilities. Hence, all the buildings and facilities in Malaysia must be disabled friendly and free from any barriers that can limit the involvement of the disabled person to avoid the disabled people sidelined from the city growth [1]. The statement also supported by Imrie [5] which saying that barrier-free facilities are the mechanism for a disabled person to access into the public building.

3 Methodology

According to Creswell [3], the combination of qualitative and quantitative will provide more understanding compared to single approach alone. Hence, to gain the benefit of a mixed method, this study will adopt quantitative method using structured questionnaire survey with recommendation/comment by respondents. While, qualitative method will be done through observation to the facilities provided, observation based on the checklist prepared, will be done as a first phase and follow by questionnaire survey as second phase.

The questionnaire survey will be distributed among the staff and visitors of the hospital. Earlier, observation will be done according to the checklist that already been made based on UBBL standard for disable facilities for the hospital building. The checklist will be consists of data regarding the appropriate specification and provision of disabled facilities in the hospital according to the standard regulations by the authorities. From the result of the checklist gathered, the questionnaire will be developed. This is to ensure the validity of the questions.

In this research, the population and sample will be the visitors or staff in three selected public hospital around Perak; Hospital Raja Permaisuri Bainun, Hospital Seri Manjung, and Hospital Kuala Kangsar. Hence, the population of the sample will be determined according to the total number of beds in each hospital as shown in Table 1

Table 1: Number of beds in three selected public hospital

Hospitals	**No. Of Beds**
Hospital Raja Permaisuri Bainun (HRPB)	990
Hospital Seri Manjung (HSM)	270
Hospital Kuala Kangsar (HKK)	136
Total	1396

From the total of the population above, the number of a sample will be abstracted from the level of a precision table for determining sample size for ±3%, ±5%, ±7%, and ±10% precision levels where confidence level is 95% and p=0.5 as shown in Table 1. The total sample that an appropriate for this research is 96 samples according to the precision level 10%.

All the data collected from the questionnaire survey and the observation will be analysed and will be presented in the form of a table. Each of the data will be evaluated and explained in detail. All the data analysed using Statistical Package for Social Sciences (SPSS) V.21 software. The questionnaire survey distributed were returned and answered by the respondents because the method used was face to face and researcher help respondent to answer on the spot. The researcher approached the respondent during recess hour for staff and during visit hour for visitors.

4 Provision of the disabled facilities provided in public hospital

Table 1 shows all the result from the observation that has been done to all the three selected hospital in Perak. Based on the list that has been provided, if the facilities follow all the requirements in the list it would be marked with '/'. Facilities that did not follow the list it will be marked with 'X', while those facilities that not provided at all will be marked with 'NA'. During the technical visit, all the facilities have been strictly observed to ensure that all the result is valid.

Table 2 Observation result regarding disabled facilities in selected hospital

1	**STAIRCASE**			
a)	Tactile floor located atleast 400mm before the step	NA	NA	NA
b)	Continuous handrails provided	/	/	/
c)	Braile inscription at the handrail	NA	NA	NA
2	**RAMP**			
a)	The ramp is free from any obstructions and anti-slip	/	/	/
b)	The ramp should atleast 1200mm wide	/	/	/
c)	There is handrails provided on both sides	NA	/	X
d)	The ramp should be of gentle gradient with landing maximum every 6 metres	NA	/	/
e)	Step ramp is provided wherever an accessible route crosses a step or kerb	NA	/	/
3	**SIGNAGE**			
a)	The signage is clearly seen by all people	/	/	/
b)	The signage is marked with universal symbol	/	/	/
a)	Tactile floor located atleast 400mm before the step	NA	NA	NA

Table 2 Observation result regarding disabled facilities in selected hospital (cont'd)

4	**ENTRANCE**	**HRPB**	**HSM**	**HKK**
a)	The entrance is free from any obstructions	/	/	/
b)	The entrance accessible with pathways	/	/	/
c)	The entrance are provided with tactile floor for blind people	X	X	X
5	**PATHWAYS**			
a)	The pathways are free from any obstructions	/	/	/
b)	The pathways are is at least 900mm wide	/	/	/
c)	The pathways are provided with tactile floor for blind people	X	X	X
6	**PARKING**			
a)	The parking area is free from any obstructions	/	/	/
b)	The parking area is marked with universal symbol	/	NA	/
c)	The parking space located near to the entrance	/	NA	/
d)	The parking space at least 3600mm wide	/	NA	/
7	**ELEVATORS**			
a)	The button is accessible for wheelchair user	/	X	X
b)	Braille code and voice projection for visually impaired people	/	X	X
c)	The space is accessible for wheelchair user	/	/	/
d)	Handrails provided in the elevator	/	/	X
e)	The door should not less than 900mm wide	/	/	/
8	**TOILET**			
a)	The toilet is separated for PWDs	/	/	/
b)	The door is accessible for wheelchair user should not less than 900mm wide	/	X	/
c)	The area of the toilet is big enough to fit the wheelchair	/	/	/
d)	Handrails provided should be at 800mm high from the floor and adjustable	NA	X	/
e)	The floor should not be slippery	/	/	/
f)	Other facilities should be accessible	/	/	X

Note: / - Yes, X- No, NA- Not Available

4.1 Staircase

The staircase is also one of the crucial facilities as it will be used by the disabled people to travel from one level to another. Hence, these facilities need to be disabled friendly as it will ease the movement of all categories of people. Based on the observation, even though staircase that provided at Hospital Kuala Kangsar did not have handrails at their both side, it still complies with the requirements as the other side of the stair is a wall. By referring to those three hospitals, none of them are provided with the tactile floor that to inform the visually impaired people regarding the floor or the staircase.

4.2 Ramp

Another medium of access that usually will be used by the disabled people is a ramp. Ramp will be used by physically impaired people and visually impaired people. Ramp can ease the movement of PWD especially those on a wheelchair. According to Malaysian Standard MS 1184: 2002, the ramp must be not less than 1200mm width and landing area at every 6000mm. This requirement has been fulfilled by the ramp provided at Hospital Kuala Kangsar and Hospital Seri Manjung.

4.3 Signage

Based on the technical visit, all signage on disabled facilities is visible, clear, simple, and easy to read and understand by all people. This features already fulfilled the requirements by the standard requirements by Malaysian Standard. Besides, the symbol used by all the signage is a universal symbol. Also, the location of the signage also at the appropriate area and not hidden from the sight.

4.4 Entrance

According to MS1331:2003, the entrance of the building must be free from any obstacles and must be easily accessible by everybody especially disabled people. Based on the observation, all the three hospitals followed this requirement. Besides, the entrance also connected with the pathway that leads to the entrance. Also, none of the hospitals provide the tactile floor for visually impaired people. The tactile floor is one of the requirements that hospital does not comply with.

4.5 Pathway

The pathway is one of the media of travelling from one place to another. Because of that, the pathway must sizeable and free from any obstruction according to the standard requirements. All three (3) hospitals have pathway with more than 900mm width as it will ease the movement of people and patient. Besides, the pathway also can accommodate up to two (2) numbers of a wheelchair at one time and the tiles also not slippery. However, there is no tactile floor provided especially for visually impaired people.

4.6 Parking

It is a requirement to provide parking space for disabled people as stated in the standard requirement that has been used worldwide. This requirement is important as disabled people must have a special privilege to access to the public building. According to MS 1184:2002, American with Disabilities Act 2010, Australian Standard and others, the minimum space for disabled parking should not be less that 3600mm width as space for wheelchair movement. Base on the observation, only Hospital Raja Permaisuri Bainun and Hospital Kuala Kangsar that comply with the requirement as no disabled parking space provided at Hospital Seri Manjung. However, in Hospital Seri Manjung, drop-off point area are provided. Both parking space located near with the entrance as required by the standard requirements and marked with a universal sign.

4.7 Elevators

All hospital fulfils the requirements by providing elevators at the hospital. However, some of them did not fulfil the requirements that come with the elevator. For example, there are no handrails provided for elevators at Hospital Kuala Kangsar, and there is also no button with braille code for visually impaired people. Hospital Seri Manjung face the same issue. However, elevator at Hospital Raja Permaisuri Bainun nearly fulfilled all the requirements, but there is no provision of voice code for hearing impaired person. However, the braille code is provided.

4.8 Toilet

Observation found that those three hospitals provide the special toilet for disabled people at the hospital area. However, there are still some improvements that need to be done by the hospital management as the facilities provided not fully comply with the standard requirements. The disabled toilet at Hospital Raja Permaisuri Bainun has no provision of any handrails. It will cause difficulty especially to wheelchair user. Besides, the height of the sink and the mirror also not an appropriate for physical impairment people. Similarly, the disabled toilet provided at Hospital Kuala Kangsar also not fully complies with the standard requirements by the authorities. The handrail is located far from the toilet bowl and will cause difficulty to the user.

5 Awareness towards Disabled Facilities in Public Hospital

This section consists of seven main variables and other sub-questions that need to be rated by the respondents. This section is to determine the level of awareness of public towards the provision of the disabled facilities in public hospital. Hence, the respondents need to answer the question according to the listed facilities; entrance, toilet, parking, pathways, elevators, staircase and signage. This section is using Likert Scale to measure the level of awareness by circling on number 1 to 5 starting from "strongly not aware", "not aware", "moderate", "aware" and "strongly aware".

5.1 Entrance

Entrance is the main access that will be used by people to enter the building especially public hospital. The provision of proper entrance also been stated in Malaysian Standard 1184:2002 to ensure that disabled people can access into the public building without any limitation. The entrance must be accessible, free from obstructions and close proximity with disabled parking.

a) Accessible main entrance

According to Table 3, it is stated that the majority of the respondents aware that the entrance at public hospital is accessible to all people especially the one with disability, represent 4.10 mean value. However, there is still a 3% of respondents that "not aware" on this facilities.

b) Clear and ample space for entrance

It is important for entrance to be free from any obstructions as it will ease the movement of people into the hospital. Hence, according to Table 3, it is stated mean value 4.10. The least score is 3% of the respondents "not aware" whether the entrance is free from any obstruction or not.

c) Assessable disabled parking

For this factor, we can see that the majority of the respondents are aware that the entrance of the hospital located near with disabled parking. However, 30% of respondents not really aware by answering moderate and 8% (mean value 3.69) of respondents "not aware" and 1% is "totally not aware" with this situation. Basically, this is happened due to lack of awareness of disabled facilities by public.

5.2 Toilet

Toilet for disabled people must be located separately as it has its requirements that need to be followed.

a) Separated toilet for disabled person

Table 3 highlighted that mean value is 4.06, therefore most respondents are "aware" that there is a separated toilet provided for disabled people in the hospital. This indicates that most of the public are "aware" with the disabled toilet and this is also another indicator to show that most of the hospital provides separated toilet for disabled people.

b) Accessible toilet space for wheelchair user

Even though there is a separated toilet provided for disabled people, but the space of the toilet also must be accessible by a wheelchair user. Mean value score is 4.06 show that respondents are "aware" that the space of the disabled toilet provided at the hospital is accessible by the wheelchair user.

DOI: 10.1051/matecconf/20166600081

c) Additional facilities

Additional facilities also need to be provided in the disabled toilet for the usage of the disabled person and also as an assistant for them to move in the toilet. Mean value of 3.93 resulted that the awareness level is good

5.3 Parking

Disabled parking is one of the important facilities as it will give enough space for disabled people especially those with a wheelchair to transfer from their vehicle to the wheelchair. Besides, the parking space must be located near with entrance as it will ease the movement of disabled people.

a) Accessible parking location

Majority of the respondents are "aware" that the location of the disabled parking is near with the entrance. This statement is supported by mean value 3.75. This is a good indicator as it shows that the disabled parking is provided and near with the entrance to ease the movement of the disabled people.

b) Ample parking space and free from obstructions

From the result, mean value of 3.72 shows the parking is free from any obstructions that can limit the access of the disabled people towards the parking space.

c) Clearly provided disabled signage

The score of mean value 4.00 represent that majority of the respondents are "aware" with the situation and realise the existence of the signage.

5.4 Pathways

Pathways will be used by all people especially pedestrian with PWD who are also using the pathways to go to another place. Therefore, the pathways provided needs to be disabled-friendly to ease the movement of disabled people.

a) Provision of tactile floor for visually impaired people

Mean value of 2.74 shows the low level of awareness towards the facilities. This is happened because most of the pathways at the hospitals are not provided with a tactile floor. Hence, that reason can contribute to the low level of awareness on this facility.

b) Free from obstructions pathways

Awareness is happening represent by mean value of 3.90, that the pathway at the public hospital is free from obstructions. It is important for the pathway to be free from any obstruction that can limit the movement of the people especially disabled people from entering the building.

c) Pathways location

Basically, the parking space will be located a bit far from the entrance and people tend to use pathways to go to the entrance. Hence, it is crucial to ensure that the pathways will straight lead to entrance as it can help disabled people access into the hospital. Mean value of 4.10 show respondents are aware that hospital's pathways are lead to the entrance

5.5 Elevators

Elevators need to be provided by the hospital as a medium of transporting people from one level to another despite specific categories of people. The elevators need to be provided with an appropriate space for wheelchair to fit in and other facilities that are disabled friendly.

a) Elevators with handrails

Based on Table 3, mean value score is 3.91 show more than half of the respondents are aware that the public hospitals' are providing handrails in the elevators This result is a good indicator that the elevators of the hospitals are complying with one of the Malaysian Standard's requirement.

b) Accessible space for wheelchair user

Elevators is used by disabled people especially those with physically impaired and wheelchair user as their condition are limited. The result shows mean value of 4.03 where respondents are aware that elevators in public hospitals are accessible for wheelchair. This is proved that elevators users in a public hospital are physically suitable for impaired user.

c) Elevator with easily accessed button

The result shows that mean value of 3.96. Therefore, respondents are aware that the button provided at the elevators can be accessed by disabled person. Even though not all the requirements fulfilled, but the button still can be accessed by disabled people.

5.6 Staircase

Staircase is one of the facilities that majorly used by the people in the hospitals. Hence, the staircase needs to be barrier-free and constructed according to the requirements that provided in Malaysian Standard MS1184:2002.

a) Clear from obstructions for staircase

Table 2 shows mean value of 3.97. The respondents are aware that the staircase at the public hospital is free from any obstructions that can limit the movement of disabled people.

b) Continuous handrails for staircase

Mean value of 4.16 show that high awareness of respondents of this item. The staircase at the public hospitals is provided with continuous handrails for the use of the public and disabled people. From the observation, it was found that some of the staircases at the public hospitals are not provided with handrails at both sides since the staircases are attached with the wall.

c) Staircase steps with tactile floor

Table 3 highlights that mean value of 3.96 represent that least respondents are aware with the provision of tactile floor at the staircase for the usage of visually impaired people. This result is due to the no provision of the tactile floor at the staircase area as the medium of information towards the visually impaired people. This shows that there is a few numbers of hospitals that comply with the requirements of providing tactile floor at the staircase for disabled people.

5.7 Signage

Signage is important to convey a message to the people by showing the symbol and statement. Besides, signage helps people in acknowledging the places or facilities that specially provided for some people especially disabled facilities. Besides, the signage provided to assure all people received and understand the message that need to be conveyed.

a) The existence of the disabled signage

Mean value of 4.03 shows that most respondents are aware that there is an existence of the disabled signage on every disabled facility provided. From the result, it shows that the signage that provided by the hospital is clearly seen and acknowledged by the public.

b) Understanding of the signage

Under the implementation of Malaysian Standard, all of the disabled signage must be marked with international symbol of disabled people to ensure that all people are well aware of the disabled facilities provided. Majority of the respondents are well aware that the signage provided at public hospitals is easy to understand (mean value of 4.08).

c) Location of the signage

The location of the signage also need to be located at an appropriate area as it will be an indicator to the user and public about the disabled facilities provided. Besides, it is also as information to public that space or facility is specially provided for disabled people. Mean value of 4.00 was good indicator that the location of the disabled signage is at an appropriate area.

Table 3: Summary on the level of awareness of the public towards disabled facilities in public hospital

Item	Description	Mean
1	Continuous handrails for staircase	4.16
2	Assessable main entrance	4.10
3	Clear and ample space for entrance	4.10
4	Understanding the signage	4.08
5	Separated toilet for disabled person	4.06
6	Accessible toilet for wheelchair user	4.06
7	The existence of disable signage	4.03
8	Accessible space for wheelchair user	4.03
9	Pathways location	4.01
10	Clearly provided disable signage	4.00
11	Clear form obstruction for staircase	3.97
12	Elevator with easily accessed button	3.96
13	Additional facilities at disabled toilet	3.93
14	Elevators with handrails	3.91
15	Free from obstacles pathways	3.90
16	Accessible parking location	3.75
17	Ample parking space and free from obstructions	3.72
18	Accessible disable parking	3.69
19	Staircase steps with tactile floor	2.96
20	Provision of tactile floor for visually impaired people	2.74

Note: 0.00-1.00 Totally not aware, 1.00-2.00 Not aware, 2.00-3.00 Moderate, 3.00-4.00 Aware, 4.00-5.00 Absolutely aware

Based on Table 3, most of the listed items scored the mean value above 3.00, which indicate that the respondents aware with facilities provided. The highest mean scored by staircase, entrance, signage and toilet. These items are most important for PWD in hospital. However, only two items score mean value below 3.00. There are, tactile floor for staircase and for pathways. Therefore, provision of tactile floor for visually impaired people at pathways is less aware by the public with moderate ranking. As mentioned before, not all the pathways at the public hospital are provided with the tactile floor and there are several of the respondents that did not have any idea on what the tactile floor is, so this situation can contribute to the low numbers of respondents that aware with this facility.

6 Summary and Conclusion

From the result that have been analysed, most of the respondents were "aware" with all the disabled facilities provided at the hospital. However, from the observation, it was found that most of the requirements are fulfilled, but there is still room for improvement needed for all of the disabled facilities for better effects in future. The result presented that these three hospitals design had follow UBBL standard in general. It was found that there are five factors that are considered as major factors can be improvement the disabled facilities in the public hospitals. The factors are construction stage, government responsibility, quality, people's perception and maintenance. Facilities for PWD need to be maintained to prolong the usage. At the construction stage, government need to enforce straight regulation on using UBBL not as guidelines alone but need to comply. People perceptions also one of the issue that need to be catered, so that the facilities provided for PWD is not being misused. Table 4 shows the summary of the factors in enhancing the disabled facilities in public hospital.

Table 4: Summary of factors in enhancing the disabled facilities in public hospital

Factors	Reason
Maintenance	The disabled facilities need to undergo periodic inspection and maintenance in order to ensure that all the disabled facilities are well-functioning and the quality of the disabled facilities can be maintained.
Construction stage	All the disabled facilities need to be vetted from the beginning which is since the construction stage to ensure that all the facilities are provided accordance with the requirements.
Government	The government also needs to play their own roles in assuring that all the public building can be accessed by the disabled people and the facilities are disabled friendly.
Quality	More barrier-free facilities need to be provided especially in public building like hospital in order to avoid any limitation occurred to disabled people.
People's perception	All people are responsible to ensure that all the disabled facilities are in good condition and never been abused by uncivilized people.

References

1. Abdul Rahim, A. (2011). *Application of Universal Design Towards World Class Cities*.
2. Abdul Talib Y, Rajagopalan P and Yang R. J. (2013). " Evaluation of building performance for strategic facilities management in healthcare." Facilities. 31(13/14): 681-701.
3. Creswell, J. (2014). *Research Design*. 4th ed. [S.l.]: Sage Publications.
4. Ibrahim, Z. (2011). Most Handicapped People Not Yet Registered With JKM. *Inclusive Design for Malaysia*. http://inclusivedesign.wordpress.com
5. Imrie, R. (2012). Universalism, universal design and equitable access to the built environment. *Disabil Rehabil*, 34(10), pp.873-882.
6. Islam, M. (2014). Rights of the People with Disabilities and Social Exclusion in Malaysia. *International Journal of Social Science and Humanity*, 5(2), pp. 171-177.
7. Kadir, S., Jamaludin, M. and Rahim, A. (2012). Building Managersâ€™ Perception in Regards to Accessibility and Universal Design Implementation in Public Buildings: Putrajaya case studies. *Procedia - Social and Behavioral Sciences*, 35, pp.129-136.
8. Lingan, L. (2012). More Boarding Facilities Needed. *Business Times*, [online] p.1. http://www.nst.com.my
9. Morris, J. (2001). Impairment and Disability: Constructing an Ethics of Care That Promotes Human Rights. *Hypatia*, 16(4), pp.1-16.
10. News Straits Times, (2014). Making Life Easier for Disabled. [online] p.1. *http://www.nst.com.my*
11. Verea-Hernando, H., Valdés-Cuadrado, L., López-Campos, J., Fandiño-Orgeira, J. and Blanco-Ramos, M. (2011). Coordination Between a District Hospital and a Reference Hospital: Evaluation of Chest Disease Care Models. *Archivos de Bronconeumologia (English Edition)*, 47(6), pp.290-295.
12. Wall, L. (2007). Where should obstetric vesico-vaginal fistulas be repaired: At the district general hospital or a specialized fistula center?. *International Journal of Gynecology & Obstetrics*, 99, pp.S28-S31.
13. Wilson, M. and Scior, K. (2013). Attitudes towards individuals with disabilities as measured by the Implicit Association Test: A literature review. *Research in developmental disabilities*, 35(2), pp.294--321.

Key Drivers of an Effective Facilities Management Practice for Malaysia State Mosque

Maimunah Sapri[1], Zafirah Ab Muin[2,a] and Ibrahim Sipan[3]

[1,3]Centre for Real Estate Studies (CRES), Universiti Teknologi Malaysia (UTM), 81310 Skudai, Johor, Malaysia
[2]Faculty of Geoinformation and Real Estate, Universiti Teknologi Malaysia (UTM), 81310 Skudai, Johor, Malaysia

Abstract. The paper discuss on the importance of facilities management (FM) in supporting mosque function and propose the key drivers for an effective FM practice for mosque. Qualitative research design was adopted in this study. Data were collected using focus group consists of seven (7) participants who have been selected among mosque management committee, FM practitioner and academician. Transcript of focus group discussion was analysed using content analysis. Coding system was applied in order to describe the interpretation of data. The results show that there is an agreement among participants on the importance of FM in supporting mosque function. Although practice of FM in mosque is not widely recognised, but its potential in uplifting and prospering mosque function could not be denied. Therefore, an effective of FM practice in mosque is seen vital. This paper discovered that team working is the most significant drivers to an effective FM practice for mosque. Other key drivers are organisational structure design; appoint knowledgeable and skilled facilities manager; sharing FM knowledge and its benefits; and strong change management. The discovered FM drivers are important for all stakeholders that directly or indirectly involve in mosque management.

1 Introduction

Facilities Management (FM) is becoming an increasingly important factor in the built environment. FM is often foreseen as an enabler to enhance organisational resources in a competitive and efficient way [1]. FM is a key function in managing facility resources, support services and working environment to support the core business of the organisation in both the long- and short-term [2]. An effective FM is a contributing factor of a total organisation success in which it contributes to the delivery of both strategic and operational goals. Therefore, the practice of FM in all organisation sectors is now progressively recognised.

Based on review of literature, the research on FM practice is widely discussed in various case studies such as in higher education institutions, health care services, commercial buildings and office buildings. However, it is discovered that the study which focusing on religious buildings such as mosque is still infancy. This is not surprising, as information on the importance of FM in supporting non-profit organisation to achieve its core business objectives is still lacking. Therefore, the objective of this paper is to discuss the importance of FM in supporting mosque function and propose the key drivers for an effective FM practice for mosque. Results of the findings will benefit to the stakeholders that directly or indirectly involve in mosque management. It is anticipate that the discovered drivers will act as enabler in order to play an important role in achieving prosperity as a mosque objective.

[a] Corresponding author: zafirah.hjabmuin@gmail.com

2 Key Drivers of Effective FM Practice

FM is a field that has a great impact in supporting the achievement of organisation goals effectively and efficiently. For that reason, FM profession has experienced a dramatic growth especially in the UK as one of the Western countries [3]. In contrast, FM is still infancy in terms of its practicing in the Asian countries, particularly in the South-East-Asia region [4], [5].

Research from [4] found that lack of understanding of the FM definition and roles has contributed to be part of the reason why FM is not being practiced proactively. [6] in [5] pointed out that FM challenges include lack of recognition, education and training needs, professional status, career pathways, information standard, performance benchmarking, cost value versus value in service procurement and corporate and community contribution of FM. These challenges will remain a barrier in the development of effective FM, if the key drivers have not been explored [4].

Yet, the research of [7] found that even in the same business, FM practice of each organisation could be different which influenced by culture, business objective and nature of that organisation [2]. Therefore, it is essential to understand the factors affected FM practice in the establishment of appropriate FM arrangement [2]. In the context of this paper, those factors that has been suggested are referring to the key drivers, in which the key drivers that has referred are applies only in the scope of mosque. Besides, [8] claimed that identifying the key issues and drivers of effective FM practice will lead to the improvement and development of sustainable practice within the FM industry. For that reason, the key drivers of effective FM are discussed generally as below.

2.1 Organisational Structure

An organisation without a defined structure is likely to fail. Thus, an organisation's goals, mission and vision should be clearly defined. It is important that the structure of the organisation does not limit possibilities for future growth. According to [2], characteristic of an organisation has the major influence on the needs on facilities and support services of each organisation. [9] in [2] stated that the characteristics of an organisation should be addresses from time to time because organisational objectives and characteristics can be shifted or changed by business cycle and external environment. [8] claimed that strategic behaviour and performance of the organisation influenced by multiple actors like stakeholders, senior managers, government, employees, clients and supply chain in which there is a need of coordination between them.

[10] emphasised that senior management's leadership style and commitment may be a key driver of sustainability in organisation. Senior management commitment contributes to support the performance and goals of the entire organisation. Research findings by [11] stated that senior management personnel should be at the forefront of championing improvement of the level of commitment to effective FM practice within organisation. Findings also indicate that FM only can appreciable impact in an organisation if FM principles are fully embedded in the strategic function of organisation. Therefore, this factor can be the key driver for effective FM as it has the ability to acclimatize the changing environment.

2.2 Knowledge and Skills of Facilities Manager

According to the International Facilities Management Association (IFMA), FM is a profession that encompasses multiple disciplines to ensure functionality of the built environment by integrating people, place, process and technology. In the meantime, the British Institute of Facilities Management considers FM as the encompassment of multi-disciplinary activities within the built and the management of their impact on people and workplace. Based on the definition, it can conclude that the role of facilities manager cover wide range of task and their decisions is required to be made at various management levels; strategic level, tactical level and operational level.

Knowledge, skills and abilities of facilities manager required performing those tasks [12]. It is important to ensure their effectiveness and performance. Since facilities managers are at the forefront of implement their organisation's vision and commitment, thus they had to develop their competencies to face the demands, challenges and opportunities for effective FM development and practice [8]. Therefore, establishment of FM profession in an organisation that have knowledge and quality skill is crucial importance to drive of effective FM practice [5], [13].

2.3 Sharing of FM Knowledge

Knowledge is now being seen as the most important strategic resource in organisations. Thus, there is an increase in organisations' effort to purposely manage knowledge in a systematic manner [14], as it is considered critical to organisational success. Knowledge can be divided into two, tacit and explicit [15], [16] in [14]. Review by [14] explains that both are different as tacit knowledge represents knowledge based on the experience of individuals expressed in human actions in the form of evaluation, attitudes, points of view, commitments and motivation. Conversely, explicit knowledge is codifiable knowledge inherent in non-human storehouses including organisational manuals, documents and databases

FM knowledge is very important to carry the best of FM practices [13], [14], [17]. Therefore, to ensure that the FM moves actively, FM knowledge is crucial importance need to be developed within organisation. As part of knowledge management processes, effective knowledge sharing is become pivotal as it could foster innovation by encouraging the free flow of ideas [18].

2.4 Team Working

An ultimate aim of FM in delivering good service quality is customer satisfaction. As FM plays a pivotal role between demand and supply, thus the team must provide with a clear and unambiguous brief, ensuring that FM is part of the strategic planning process [19]. Effective FM is essentially a team effort [20]. Successful FM teams will be able to embrace the language of the core organisation they serve, clarify and serve their strategic contribution in terms of business relevant outcome [21] in [4].

2.5 Change Management

Change is a movement from the present state of the organisation to a desired future state. Due to [5], the changes may boost the implementation of the FM. Management skills of a high order are necessary to achieve this major change, together with learning and consolidating the skills base [19]. Thus, appropriate training must be provided for introduction and development of FM skills. Facilities manager must become catalysts of change in the ways in which the services are provided.

3 The Need of Effective FM Practice to Increase Mosque Function

In Islamic literature, mosque is the house of Allah and known as a non-profit organisation. Importance of this institution could be seen through its roles and functions as a place of worship, a spiritual centre for the Muslim community, and a focal point for Islamic activities. In the Islamic culture, 'Masjid', or Mosque is considered as the most important sacred structure [22]. Therefore, the Mosque is deserves to be shown the respect and given the honor, and should be functional strictly accordance with the Qur'an and the Sunnah of the Prophet (peace be upon him).

Due to largest Muslim population, Malaysia has numerous mosque dispersed all over urban and rural areas [22]. However, it seen that the function of mosque is become lacking. Due to the time changes, functions of the mosque become narrow and only well-known as a place to perform worship among today generation. Moreover, the Malaysian

Government had spent almost billion ringgit to build a mosque. Most of the mosques are good at architecture design, but getting worse in number of people come to mosque which is refer to the congregational. In the context of this paper, the congregation can be classified into two group; internal customer (visitor) and external customer (employees).

3.1. Importance of Customer Satisfaction to Gain Involvement

Lack of awareness among community on the importance of mosque is the factor contributed to the issue of prospering mosque [23]. Moreover, there are several of researcher stated that potential for mosque to be prosper has been affected by the lack of competencies in leadership [24], less vigorous activities and programs [23], lack of attention of conducive physical environment [25], [26], [27], [28], lack of user-friendly, youth-friendly, family and community-friendly approach [23], [29], [30] and poor financial management to implement mosque activities [31], [32].

Mosque is community-based organisation [24]. Thus, involvement of the congregation to prospering the mosque is critical factor that contribute to mosque to be functional as a whole. In FM concept, the congregation is classified as a customer where it has a significant impact on the performance of a business [33]. As a social-service organisation, mosque should consider to meet the expectations and requirements of customer in delivering service. A good service quality will produce customer satisfaction as the output [34], in which there is a strong relationship between customer satisfaction and customer loyalty [33]. According to [35], if the organisation fails to deliver the right products and services that satisfying customer needs, most probably the business will wither and die. Therefore, it is important for mosque to consider the expectation and requirement of the congregation to sustain its business continuity.

3.2 The Requirement to Provide Conducive Environment

Prosperity of mosque has a special relationship with Allah's mercy. Refers to Islamic Hadith from Ali bin Abi Thalib r.a. stated that:

> *The Prophet Muhammad (peace be upon him) reported: "It is almost a time when there is no more Islam except its name, and there is not stay out of the Quran except its writing. Their mosque stately, but it is empty of guidance. Their Ulama are the worst of creatures that exists under the vault (shade) sky. From them defamation belongs, and to them the defamation will return".*
>
> (Al-Baihaqi, Hadith no: 1766)

Based on the Hadith, it reveals the requirements to ensure mosque should be function as the preferences place of people in seeking the guidance. For mosque to be functional as it is, thus environment of mosque should pleasant and welcoming the congregation. Design, decor and aesthetics of the religious places could bring people closer to religious ideals, spirituality, community, and peace and consequently influence their attachment [36]. They also mentioned that religion has a profound influence on people's relations with the sacred places and on their attachment to those settings. Thus, not only the physical of mosque should be good at architecture and maintenance but also should be capable as a place for community development that meets their needs.

Similar with higher education institution purposes, mosque also responsible in producing leader of calibre and integrity. Many suggestions by Islamic scholar stated that young people should be encouraged in prospering the mosque [23], [28], [29], [37]. They strengthen that this group has high potential to be educated and nurtured with values from the beginning. Various programmes and activities should be implemented effectively in order to planting the sense of belonging towards mosque institution. Thus, mosque should be equipped with adequate and well maintained facilities in order to support the implementation of those planned activities run smoothly.

However, there are many cases reported regarding mosque environment issues that should be given serious attention. For instance, there is a case on the lack of maintenance within mosque building where the roof of mosque building unexpectedly has been collapsed. As sacred place where could be preferences place to gain serenity, but unfortunately mosque has confronting with this problem. This situation has raised the issue of safety in mosque construction.

3.3 Tackle Mosque Management Issue

Even though mosques are no longer used as state administration centres, their role in shaping society still continues [38]. The key challenge of managing mosques today is to ensure efficient and proper management practices [38]. Therefore, in order for mosques to uplift their multi-functionality in today's modernization era, and to be sustained into the future, the issue of professional management is among those topics that need to be discussed. Mosques should be operated through a more professional management system, in order to meet current needs and demands [23], [38], [39].

The right management practices will contribute to improving a mosque's performance [39]. The research has identified that the management of mosques facilities is becoming crucial, as it contributes to mosque success in delivering mosque core functions to the community. Therefore, mosques require the effective and efficient management of their facilities. In this regard, facilities management (FM) is seen to be the approach that mosques could consider. FM is a subset of general management [40]. As a subset of general management, it will help to improve processes through which organisations can be managed, in order to support their effectiveness, and to make a positive contribution to organisation success.

The level of mosque functionality is highly depends on participation of community. Therefore, requirement in providing good facilities is seen vital in order to support mosque in achieving its objectives [25], [26], [28]. In the context of this research, good facilities are referring to level of functionality and durability facilities in facilitate the customer to perform their activities. Hence, in providing good facilities, then it requires an effective and efficient of FM. **Figure 3.1** is presenting the relationship between FM and mosque in order to have a clear overview on requirement of FM in supporting mosque function.

DOI: 10.1051/matecconf/20166600082

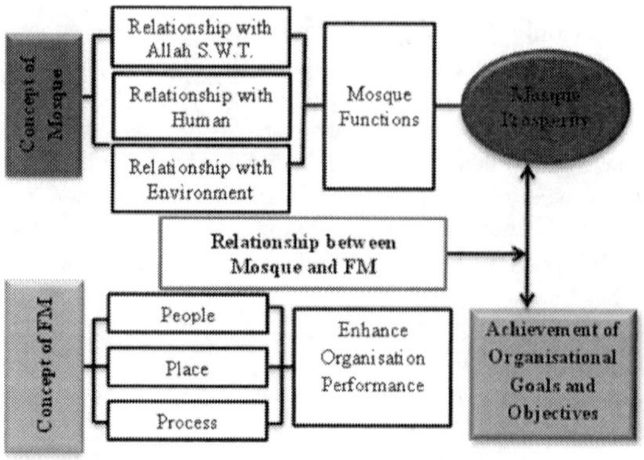

Figure 3.1: Framework of Relationship between FM and Mosque Function
Source: Sapri and Ab Muin (2014)

4 Research Methodologies

The research was designed to solicit the views and in order to identify key drivers for an effective facilities management practice in mosque institution. Since there is no FM research has been conducted before in the context of mosque institution, therefore qualitative research method is used in order to provide an insight and understanding on how FM should be practice and what are drivers confronting an effective FM practice in mosque. Qualitative research is best suited to address a research problem that do not know and need to explore [42].

Focus group approach of data collection was adopted. Seven (7) respondents which representing the mosque management, mosque facilities manager, practitioner of FM and academicians participated in the focus group discussion. Focus group was selected in terms of its advantage in providing insights into how people think and provide a deeper understanding of the phenomena being studied [43]. Moreover, group interaction and non-verbal communication are primary benefits of focus group. The interaction among members of the target population during focus groups meetings may inspire participants to make connections to various concepts through the discussion which may not occur during individual interviews [43]. Moreover, focus group refer to focused and interactive session with a small group which enough for everyone to have a chance to talk and large enough to provide diversity of opinions. Therefore, focus group adopted in this study purposely to find out drivers for effective FM practice in mosque institutions from the target population.

Focus group research involves organised discussion with a selected group of individuals to gain information about their views and experiences of a topic [42]. Moreover, focus group interviewing is particularly suited for obtaining several perspectives about the same topic. Therefore, the focus group was considered the most appropriate method in examine the drivers for effective mosque FM practice as it assemble parties that directly involved in management of mosque and facilities. The question involved open-ended questions because the intent of the focus group is to promote discussion [43]. In this study, participants have been asked open-ended question based on their experience and opinions to achieve the objective of this study. The role of the moderator is very significant for conducting focus group successfully [42]. Thus, two moderators with good levels of group leadership and interpersonal skill have been appointed during the discussion. List of participants are listed in **Table 4.1** below.

Table 4.1: Code, Position and Organisation

Code	Position	Organisation
A	Academician (PhD.)	Universiti Teknologi Malaysia
B	Academician (Assoc. Prof)	Universiti Teknologi Malaysia
C	Person in Managing Mosque	Salman Mosque, Indonesia
D	Person in Managing Mosque	As-Syakirin Mosque, Kuala Lumpur City Centre
E	Person in Managing Mosque	Masjid As-Salam, Puchong, Selangor, Malaysia
F	FM Practitioner	Facility Management Tradewinds Corp. Bhd.
G	Valuer	National Institute of Valuation (INSPEN),Malaysia

To identify the individual view from the group view, participants were asked to share their experience and opinions on how FM should be practice effectively in mosque. The data was analysed using content analysis approach. Qualitative content analysis highlights an integrated view of speech or texts and their specific contexts. It goes beyond basically counting words or extracting objective content from texts to examine meanings, themes and patterns that may be visible or latent in a particular text [44]. In this study, participants were selected according to their respective areas of expertise. The key drivers of effective FM for mosque were derived from the speech in which the data are read word by word to derive codes. The exact word was highlights from the text that appear to capture key thoughts or concepts.

After gather all the data, the key feature of all content analysis is that the many words of the text are classified into much smaller content categories. These content categories should be guided by the aim and research question of the study in choosing the contents analyse [44]. This process includes open coding and abstraction. The words were counted and based on the highest frequency abstraction, thus it is ranking within the categories in order to have better analyse and describing on the phenomenon. Once this is done, the information has been analysed and a conclusion has been drawn. [43] postulate that the success of the research depends on how the content was analysed and the coding process behind the data.

For the focus group, the data has been classified into five (5) categories as presented in **Table 4.2**. The coding categories were derived from the text data. Each category was then interpreted within the underlying context in achieving the objective of this study.

Table 4.2: Coding Categories

No.	Key Drivers of An Effective FM Practice for Mosque
1.	Design of Organisational Structure
2.	Appointing a Facilities Manager by Knowledge and Skill
3.	Sharing of FM Knowledge and Its Benefit
4.	Spirit of Team Work
5.	Strong Change Management

5 Results

Data collated from interview transcript are analysed and the findings are discussed below. All the statements made by the respondents with regard to the research question are based on their view on the importance of FM in supporting mosque function and what are the drivers in order to effective implement of FM in mosque institution. In the section below, some of the aspects for effective FM practice identified earlier are discussed.

5.1 Importance of FM in Supporting Mosque Function

FM practice has been identified as essential ingredient for achieving organisational objectives. A more good perspective on the importance of FM in supporting organisational in achieving its goals particularly in the context of mosque institution are captured in the statement made by respondents:

Respondent A:

"There are four main objectives of mosque; facilitate as a place of Islam development, a place to perform worship, a place of education and a place for community development. Mosque is not a structure, mosque is not a building. But mosque is the human. Human are responsible to develop the mosque. If the human hearts close to mosque, thus the mosque will prosper. Therefore, FM in mosque is crucial to support all these functions since the element of people is the main focus in FM principle. Perhaps, this is our ultimate aim to prosper the mosque".

Respondent B:

"Mosque is an institution. If referred to the definition of institution, it is not only referring to human, it is not only referring to physical. I would like to use the analogy which building is our 'deen'. For instance, mosque building which is its main function is to obtain blessing. It not only consists of human, but mosque becomes a place for recognition on the economic, ethics and etc. Thus, when we referred to the analogy of physical, we could not avoid that there is a combination between mosque function physically and the person who use the mosque. Based on knowledge sharing among the participants, FM is actually has been practice in mosque where FM can be classified as management tool or method. FM encompasses the elements of people, process and place. In order to synchronize those three elements with mosque functions, thus technology is required to perform work more efficiently and effectively. Moreover, FM as an approach has the potential to support and sustain mosque functionality in terms of business continuity, environment and finance".

Respondent D:

"FM is important in managing mosque since FM is an approach that covers wide range of field. Therefore, competency of facilities manager in managing mosque is the needed".

Respondent E:

"Logically, mosque should be provided with good facilities such as cleaning, carpet, air-conditioning, etc. in order to encourage people to prosper the mosque".

Respondent F:

"FM is support services where its function is helping the organisation implement its core business. When we talk about FM, we as the practitioner see FM as the technical support. We operate and maintain the facilities for the building owner to make sure those facilities are operating in optimum condition and lifespan. Thus, it is required in managing mosque".

Based on respondents' view, it can be concluded that FM is importance in supporting mosque function. There is an agreement in which the facilities management function can act as a facilitator for mosque institution to achieve its objectives. Even though, it is not confirms that an effective FM in mosque will definitely contribute to the mosque to be prosper, but its importance as one of the management tools should not be underestimated.

5.2 Key Drivers for an Effective FM Practice in Mosque

Data collected from the focus group interview are analysed through content analysis technique. Based on the findings, most of the participants agreed that design of organisational structure (respondent A, B and C), appointing knowledge and skilled of facility manager (respondent D, F and G), sharing of FM knowledge and its benefit (respondent C and D) and spirit of team work (respondent A, B, D and E) are the key drivers that contribute to an effective FM practice for mosque. However, only one participant (respondent B) stressed that change management should become one of the key drivers of an effective FM for mosque. The summary of the key drivers related to an effective FM practice for mosque are presented below.

5.2.1 Design of Organisational Structure

From the discussion, respondent A, B and D have mentioned that organisational structure factor contribute to drive effective FM practice for mosque. The respondent strengthen that organisational structure should be designed from top to down and FM must be involved not only at operational level, but also at strategic and tactical level. Thus, FM can widely executed in mosque management. Moreover, the respondent also advise the Department of Religion as the stakeholder to do a revision on their current organisational structure to ensure there is person who have skill, experience and certificate in FM.

Respondent A:

"Having a strong organisational structure is essential for organisation growth. Planning the good organisational structure helps to ensure enough human resources within organisation to accomplish the goals set forth in the organisation action plan".

Respondent B:

"There is requirement for mosque FM practice to be sustained. Thus, when we design the organisation structure, we should design from top to down. It means that we need to tackle every level either strategic level, tactical level or operational level".

Respondent D:

"My suggestion to Department of Religion is to revise their organisational structure and look into the people that need the skill, experience and certified in FM".

5.2.2 Appointing a facilities manager by knowledge and skill

Respondent D, F and G stated that it is important for mosque to appoint the right person to manage facilities. The right person is referring to the person who has professional FM background that was accredited by the university. Due to wide scope of FM, therefore knowledge and skill of facilities manager should be in place in mosque.

Respondent D:

"Appointment of facilities manager in mosque should be considered in which the person who will be appointed should have the background in facilities management field".

Respondent F:

"The world increasingly sophisticated as well as the technology is growing. Mosque is supplied with PA system, projector, acoustic evaluation, etc. Thus, all these matters require technical team that have specific skill. This technical team will cover mechanical system, electrical system, civil and structure, housekeeping, landscape, security and indoor environment which consider user's comfort and safety at mosque. Hence, it can be seen that effective FM that supported by skilled team is essential in prospering mosque".

Respondent G:

"We need to recognise who is the right person in responsible to manage the mosque facilities. For instance, Universiti Teknologi Malaysia (UTM) has produce graduates in FM. Thus, these graduates have high potential to spearhead the management of facilities".

5.2.3 Sharing of FM Knowledge and Its Benefits

Respondent C and D mentioned that FM knowledge and its benefits should be exposed and shared widely at the mosque. This effort should be spread among mosque management and the congregation to increase awareness and generate ideas among them.

Respondent C:

"Benefit is the main factor for FM to be practice effectively in mosque. Facilities provided in mosque that meet the user's requirement will contribute to encourage people to prosper the mosque. Sharing on benefit of the facilities to the congregational contribute to willingness in term of donation. So, provision for facilities development will become more interactive since there is participation from the congregational".

Respondent D:

"FM knowledge should be exposed to mosque management. Besides, the FM knowledge exposure should be consists of on how to optimise space, how to generate income, how to make long-term and short-term strategic planning in more systematic, etc."

5.2.4 Spirit of Teamwork

Respondent A, B, D and E share the same opinions that spirit of team work is crucial to an effective FM practice for mosque. Moreover, level of professionalism and commitment among mosque management from top level to bottom contribute to successful of any proposal in future.

Respondent A:

"Level of professionalism and commitment among mosque management either at the top level or bottom level are critical in managing mosque because its profit does not depend on money but to achieve the reward from Allah S.W.T. which is far better than anything else. When this factor was successful, thus any proposals or papers be forwarded will be realized with distinction".

Respondent B:

"Human is the most crucial factor where this human factor is not only referring to person in managing mosque, but the user of the mosque".

Respondent D:

"Teamwork is important factor to be considered. Less cooperation among mosque management will contribute to

lack of sharing the resource. Lack of knowledge sharing for example will affect mosque capabilities for higher mosque performance".

Respondent E:

"Any works without teamwork is worthless. Hence, passion among team member that involve in mosque from top management to the bottom is highly required".

5.2.5 Strong Change Management

Opinion from respondent B stated that strong change management is required to effective FM practice for mosque. Change management should be positively accepted among the mosque management, from top level to bottom level.

Respondent B:

"In my opinion, strong change management is essential to sustain mosque function. Same goes to effective FM practice; it is highly depends on the acceptance of mosque management. Thus, involvement of FM at strategic level is needed in order to expose the importance of FM in achieving mosque objective. Once again, for me change management is crucial factor in which it requires an action from various parties to participate".

6 Discussion

Based on the focus group discussion with the respondents, they agreed that FM is one of the approaches to assist mosque in achieving core objectives and supporting its function. According to one of the respondents, FM is classified as management tool or method. Thus, FM has the potential to be adopted in management of mosque as it can helps to sustaining mosque functionality in terms of business continuity, environment and finance. Since the focus of FM is customer oriented, therefore good management of mosque facilities is crucial to gain congregational contribution in prospering the mosque.

Through the discussion with focus group, the key drivers of an effective FM practice for mosque are also have been identified. From the findings, it can be conclude that an effective FM practice in mosque requires a transformation or a change among mosque management itself. in order to the transformation is successful implemented, mosque management should be properly design their organisational structure, appointing knowledge and skill facility manager, sharing of FM knowledge, increase spirit of team working and have strong change management.

For instance, organisation structure design has been classified as one of the key drivers for effective FM practice. It shows that designing an effective organisation structure is important as it helps top management to identify talent that needs to be added to the organisation. In addition, planning the structure ensures there are enough human resources within the organisation to accomplish the goals, mission and vision.

Furthermore, it is important that responsibilities are clearly defined. This is support by the literature from [8] stated that strategic behaviour and performance of the organisation influenced by multiple actors like stakeholders, senior managers, government, employees, clients and supply chain. Therefore, designing mosque management structure is become vital for implementation of effective FM practice.

As mentioned by [11], senior management personnel should be at the forefront of championing improvement of the level of commitment to effective FM practice within organisation. Thus, revising mosque organisational structure is required in order to addressing professional human resource management and gaining top management commitment issues in managing mosque facilities.

Moreover, there are some respondents stated that coordination of FM approach should be initiated not only at the operational level, but also should be executed at the strategic level. Through the implementation of the FM practice at the strategic level, it will assist the effectiveness and performance of FM practice in mosque. Therefore, commitment and participation from the stakeholders, senior management, employees and the congregational is crucial. This statement is supported by [11] in which his research findings indicate that FM only can be appreciable impact in an organisation if FM principles are fully embedded in the strategic function of organisation.

Furthermore, the findings of this study are consistent with the argument that knowledge and skill of facility manager is important in order to ensure the effectiveness and performance of FM practice [5] [12] [13]. Most of respondent stated that it is necessary to appoint facility manager with FM background to enable and facilitate the implementation of FM practice in the context of mosque. This key driver is become important since there is no establish FM unit or position for facility manager in mosque.

Review from literature mentioned that the increasing awareness by promoting the need of FM as part of the business organisation contribute on the development of FM [5]. Therefore, findings of this study reveals that sharing of FM knowledge and its benefit is become one of the key drivers that need to be considered. It is important especially stakeholders of the mosque understand on how FM can and should contribute in supporting mosque functions.

The journey of a thousand miles begins with one step. In order to ensure FM effectively implemented in mosque, thus, it require the involvement from each level; strategic, tactical or operational [20]. Based on the findings, most of the respondents stressed the importance of team working that contributes to the effective FM practice. Moreover, Islamic literature claimed that mosque is a unique and extraordinary building compare to other building and social service industries [23]. Its profit is not monetary base, but more on the desire in expect reward from Allah S.W.T. [41]. Thus, most of respondent agreed that teamwork becomes the critical and important drivers that contribute to an effective FM practice in mosque. In addition, there is a respondent stressed that it is an essential to consider change management in the transformation of mosque. As mentioned by [5], the changes may boost the implementation of the FM and it is a movement from the present state of the organisation to a desired future state.

7 Conclusions

This paper presented the overview of the key drivers of an effective FM practice in mosque. It can be said that adapting FM approach in mosque is still infancy. It is discover that there is specialisation should be made to encourage FM practice in mosque. In order to reinforce the mosque as a robust and unified institution as the place for community centre, a comprehensive plan should be formulated with emphasized the practice of effective FM with considered the drivers that being identified in this study. This paper concluded that the FM practice should be aligned with the mosque's strategic goals. It is very importance for effective FM practice to demonstrate that convenient FM processes is capable in giving positive impact in prospering the mosque. Thus, FM approach will supporting to the best practice in managing mosque through providing good facilities in future and guarantee for long-term benefits.

Acknowledgement

The authors would like to acknowledge and thank to the funder of this research project, Real Estate Research and Development Grant Scheme (NAPREC) between The Government of Malaysia and Universiti Teknologi Malaysia under Vot. No: R.J130000.7309.4B134 for funding this research work.

References

1. Sapri, M., Kaka, A. and Finch, E. (2009), "Factors that Influence Student's Level of Satisfaction with Regards to Higher Education Facilities Services", Malaysian Journal of Real Estate, Vol. **4,** No. 1.
2. Chotipanich. S. (2004). "Positioning Facility Management", Facilities, Vol. **22** Iss 13/14 pp. 364-372.
3. Myeda, N. E. and Pitt, M. (2014), "Facilities Management in Malaysia: Understanding the Development and Practice", Facilities. Vol. **32** Nos 9/10, pp. 490-508.
4. Kamaruzzaman, S.N. and Ahmad Zawawi, E.M. (2010),"Development of Facilities Management in Malaysia", Journal of Facilities Management, Vol. **8** Issue 1 pp. 75 – 81.
5. Syed Mustapa, S.A.H., Adnan, H. and Jusoff, K. (2008), "Facilities Management Challenges and Opportunity in the Malaysian Property Sector", Journal of Sustainable Development, Vol. **1** No. 2, pp. 78-85.
6. Ballesty, S. (2007), "Best Practices and Success Stories in Asset and Facility Management", Proceeding of the National Asset and Facilities Management (NAFAM) Convention, Kuala Lumpur, Malaysia, 13 August.
7. Schindler, K. (1998), "The Role of FM within the Retail Sector", Msc. Dissertation in Facility Management, University College, London.
8. Elmualim, A., Valle, R. and Kwawu, W. (2012), "Discerning Policy and Drivers for Sustainable Facilities Management Practice", International Journal of Sustainable Built Environment, Vol. **1**, pp. 16-25.
9. Lee, W. (2002), "The role of support services and FM in the introduction of change management", in Reuvid, J. and Hinks, J. (Eds), Managing Business Support Services, 2nd ed.
10. Elmualim, A., Shockley, D., Ludlow, G. and Shah, S. (2010), "Barriers and Commitment of Facilities Management Profession to the Sustainability Agenda", Built and Environment, **45** (2010) 58–64.
11. Dubem, I. I., Stephen, O.O., Michael, G.O. and Timothy, A. (2013), "Assessing the Level of Commitment and Barriers to Sustainable Facilities Management Practice: A Case of Nigeria", International Journal of Sustainable Built Environment.
12. Amaratunga, D. (2001), "Theory Building in Facilities Management Performance Measurement: Application of Some Core Performance Measurement and Management Principles", Doctor of Philosophy Dissertation, the University of Salford, United Kingdom.
13. A.M. Firdauz M. Sapri I.S. Mohammad. (2015), "Facility Management Knowledge Development in Malaysia", Facilities, Vol. **33** Issues 1/2 pp. 99 – 118.
14. Pathirage, C., Haigh, R., Amaratunga, D. and Baldry, D. (2008), "Knowledge Management Practices in Facilities Organisations: A Case Study", Journal of Facilities Management, Vol. **6**, Issue 1, pp. 5-22.
15. Polanyi, M. (1958), "Personal Knowledge towards a Post-Critical Philosophy", Routledge and Kegan Paul Ltd, London.
16. Nonaka, I. and Takeuchi, H. (1995), "The Knowledge Creatinf Company: How Japanese Companies Create the Dynamics of Innovation", Oxford University Press, New York.
17. Nutt, B. (2002), "Event 1.7: Positioning FM", Lecture Note Module 1: Introduction to Facility Management, University College London, London.
18. Mohammed Fathi, N., Cyril, U. Gerald, E. and Goh, G. G. (2011),"Key Determinants of Knowledge Sharing in an Electronics Manufacturing Firm in Malaysia", Library Review, Vol. **60** Issues 1 pp. 53 – 67
19. Alexander, K. (1992),"Facilities Management Practice", Facilities, Vol. **10** Issue 5 pp. 11 – 18.
20. Hamilton, B. (2003), "Overview and Trend in Facilities Management", paper presented at Facilities Management Seminar 2003, University of Technology Mara (UITM), Selangor.
21. Hallam, S. (2007), "FM Research – the Leading Edge", Facilities, Vol. **18**, No. 7.
22. Najafi, M. and Mohd Shariff, M.K. (2014), "Public Attachment to Religious Places: A Study of Place Attachment to Mosques in Malaysia", International Journal of Social, Behaviour, Educational, Economic and Management Engineering, Vol. **8** No. 1.
23. Mohamad Rasdi, M. T. (1998), "The Mosque as a Community Development Centre: Programme and Architectural Design Guidelines for Contemporary Muslim Societies". University Teknologi Malaysia.
24. Mahazan, A.M. (2013), "A Model of Imam's Leadership and Mosque Performance in Malaysia", GJAT, Vol. **3** Issue 2 | 55, ISSN: 2232-0474 | E-ISSN: 2232-0482. www.gjat.my.
25. Safar, A.J. (2008), "Institusi Masjid Dalam Memperkasakan Pembangunan Modal Insan", Book Chapter of Pengurusan Berkualiti Memacu Kecemerlangan Pengurusan Masjid. Universiti Teknologi Malaysia.
26. Muda, Z. and Omar, N. (2008), "Limitasi dan Absolutasi Aktiviti Masjid Menurut Islam", Book Chapter of Pengimarahan Masjid Dalam Agenda Ummah. Universiti Teknologi Malaysia.
27. Mohamed, R. (2008), "Pemerkasaan Masjid Melalui Pengurusan Profesional", Book Chapter of Pengurusan Berkualiti Memacu Kecemerlangan Pengurusan Masjid. Universiti Teknologi Malaysia.
28. Khalit, N. A. (2011), "Revitalizing Roles and Functions of Mosques in Kuala Lumpur, Master Thesis, Universiti Teknologi Mara (UiTM).

29. Ali Khan, M.N.A, Salleh, M.Z. and Salleh, N. (2008), "Pengurusan Masjid: Satu Tinjauan Ke Atas Masjid Cemerlang", Book Chapter of Pengurusan Berkualiti Memacu Kecemerlangan Pengurusan Masjid. Universiti Teknologi Malaysia.
30. Jamsari, E.A., Tumiran, S.N., Abu Bakar, K. and Arifin, Z. (2010), "Peranan dan Sumbangan Institusi Masjid dalam Pembangunan Masyarakat Melayu-Islam di Singapura: Satu Kajian di Masjid Assyafaah, Sembawang", Journal of Islamic and Arabic Education, **2**(2), 2010 31-44.
31. Siraj, S.A, Mohamed Ibrahim, S.H. and Sulaiman, M. (2008), "Pengurusan Kewangan: Mempamerkan Akauntabiliti Masjid", Book Chapter of Pengurusan Berkualiti Memacu Kecemerlangan Pengurusan Masjid. Universiti Teknologi Malaysia.
32. Bahrom, H. (2008), "Menjana Sumber Kewangan Masjid Pemangkin Pembangunan Ekonomi Ummah", Book Chapter of Pengurusan Berkualiti Memacu Kecemerlangan Pengurusan Masjid. Universiti Teknologi Malaysia.
33. Sapri, M., Kaka, A., and Finch, E. (2009). "Factors That Influence Student's Level of Satisfaction With Regards To Higher Educational Facilities Services", Malaysian Journal of Real Estate **4**(1).
34. Sapri, M., Kaka, M. and Alias, B. (2010), "Performance Management in Service Business: The Facilities Management Function", Online Publication.
35. Amaratunga, D., Baldry, D. and Haigh, R. (2004), "Customer Related Facilities Management Process and Its Measurement: Understanding the Needs of the Customer", Research Institute for the Built and Human Environment, The University of Salford, United Kingdom.
36. Mazumdar, S. and Mazumdar, S. (2009), "Religion, Immigration, and Home Making in Diaspora: Hindu Space in Southern California. Journal of Environmental Psychology, **29**: p. 256-266.
37. Jamsari, E. A., Tumiran, S. N., Abu Bakar, K. and Arifin, Z. (2010), "Peranan dan Sumbangan Institusi Masjid dalam Pembangunan Masyarakat Melayu-Islam di Singapura: Satu Kajian di Masjid Assyafah, Sembawang, Journal of Islamic and Arabic Education, Vol. **2**(2), pp. 31-44.
38. Mohamed Adil, M.A., Mohd Sanusi, Z., Jaafar, N.A., Khalid, M.M., Abd Aziz, A. (2013), "Financial Management Practices of Mosques in Malaysia", GJAT, Vol. **3** Issue 1/24, ISSN: 2232-0474, E-ISSN: 2232-0482, www.gjat.my.
39. Jaafar, A., Habidin, N.F., Mohd Hussin, M.Y., Zakaria, Z. and Abdul Hamid, A. (2013), "A Proposed Model for Strategic Management (SM) and Mosque Performance (MP) in Mosque Management", International Journal of Management, Information Technology and Engineering, Vol. **1**, Issue 3, pp. 29-36.
40. Amaratunga, D. and Baldry, D. (2000), "Assessment of Facilities Management Performance in Higher Education Properties", Facilities, Vol. **18**, Issues 7/8, pp. 293-301.
41. Sapri, M. and Ab Muin, Z. (2014), "Peranan Pengurusan Fasiliti Ke Arah Pengimarahan Masjid", International Conference of Mosque Facilities Management (ICMFM): Special Edition, Centre for Real Estate Studies, Universiti Teknologi Malaysia.
42. Cresswell, J.W. (2012), "Educational Research: Planning, Conducting and Evaluating Quantitative and Qualitative Research", Fourth Edition, Pearson Publication.
43. Williams, Nagle, B. and Nichelle (2013), "Methodology Brief: Introduction to Focus Group. Center for Assessment, Planning & Accountability. http://www.uncfsp.org/projects/userfiles/File/FocusGroupBrief.pdf
44. Wildemuth, Zhang, Y., and Barbara, M. (2013), "Qualitative Analysis of Content", http://7702-f13.design4complexity.com/qualitative-research/content-analysis.pdf

Housing Space Quality towards Quality Of Life: A Case Study of Double Storey Terrace Houses

Aniza Abu Bakar[a, 1],Nurhayati Abdul Malek[b], Mohamad Abdul Mohit[c], Rosniza Othman[d], and Aliyah Nur Zafirah Sanusi[d]

[a]*Dept. of Landscape Architecture, Kulliyyah of Architecture and Environmental Design, International Islamic University Malaysia, Jalan Gombak, 53100 Kuala Lumpur, Malaysia*
[b]*Dept. of Landscape Architecture, Faculty of Architecture, Planning and Surveying, UiTM, Puncak Alam, 42300 Selangor, Malaysia*
[c]*Dept. of Urban and Regional Planning, Kulliyyah of Architecture and Environmental Design, International Islamic University Malaysia, Jalan Gombak, 53100 Kuala Lumpur, Malaysia*
[d]*Dept. of Architecture, Kulliyyah of Architecture and Environmental Design, International Islamic University Malaysia, Jalan Gombak, 53100 Kuala Lumpur, Malaysia*
aaniza@iium.edu.my, nurhayati.abdmalek@gmail.com, mamohit@iium.edu.my, rosniza@iium.edu.my, aliyah@iium.edu.my

Abstract. The success of every housing area is determined by its user satisfaction level, living there. The quality of the house, its spaces within the unit, housing neighbourhood and green open space can directly influence the quality of life. Limitation of space following high land cost in an urban area, evaluation of the housing environment as a whole including the green open space needs to be studied as it affects the residents' satisfaction level. This study concentrates on spaces within a housing area to evaluate the residents' level of satisfaction of the Taman Melati Mastika (TMM), Kuala Lumpur and to understand how they perceived their quality of life through the housing environment and the availability of green open space. Thus, this research was carried out through site observation and analysis, and self-administered questionnaire survey. 247 questionnaire surveys were distributed to the residents of TMM and (n=62) responded. When focused on the housing unit, this study provides insight on the types of outdoor spaces (front yard-front lane and backyard-back lane) and their elements and utilization, and quality of housing spaces toward users' quality of life in TMM, Kuala Lumpur. On the green open space, the assessment of the quality of life is based on three factors that are the safety level of the neighbourhood and park, health issues related to housing environment and park as well as the satisfaction on the housing amenities and park facilities. The result of this study suggests that the residents are satisfied with the existing spaces within their compound and adjacent to it and this lead towards the overall satisfaction living in the area. The quality of space and good utilisation of housing areas can lead towards a better quality of life in the Terrace housing area is confirmed.

1 Introduction

House quality is a major factor that can determine a comfortable environment for the users. It was stated that "*housing quality can be defined as the level of satisfaction with the specific house within a chosen residential, physical and social environment, as well as its specific housing attributes*" [1]. Following the weather and environmental condition, it is agreed that most of the time, users spend their time indoor [2]. These might be among the reasons that hinder users from spending their time outdoors which also affect them from socialising optimally with neighbours. Certain parts of the individual house are turned into space to park car while some are renovated to widen the indoor spaces. The rear area of the house mainly the back lane may not be seen as a potential quality space for daily activity and to a certain extent it could be neglected or perceived as negative space. Hence, properly planned space with identified activities can encourage the user to spend time outdoors.

Safety, facilities and health are known to be among the core factors that influence the life satisfaction of neighbourhood community. An open green space known as the neighbourhood park would usually equip a housing area. Neighbourhood Park can be defined as a place where diverse needs are fulfilled without the need to travel a long distance, provided necessary recreational facilities for all users and located in the middle of developments [3]. Similarly, as cited [4], neighbourhood park is defined as places that accommodate both active and passive activities, local park function and facilities to various types of users and usually serves 0.5 to 0.8 kilometre radius of catchment area [5]. A quality green open spaces or neighbourhood parks can be defined as "*...a successful and excellent public green open space within a residential neighborhood area that conforms to the needs and requirements of people*

[1] aaniza@iium.edu.my

including various techniques for using the space and upon agreed standards that are beyond the usual outdoor recreation and leisure expectations" [6].

Evaluating housing is necessary to improve the design and development of housing projects [7]. Hence, the concept of satisfaction has been widely used as an indicator to evaluate housing performance [8]. The residential satisfaction reflects the degree to which individuals' housing needs are fulfilled. Lack of knowledge on the residential satisfaction determinants may be contributing to the failure of many housing projects [9]. High level of satisfaction is likely to be expressed if the current housing situation follows the norms as the households judge the housing conditions based on the actual housing situation and standards [10]. Various determinants have been found to be significant to housing satisfaction ranging from housing, demographic, to socio-economic variables [11].Three main groups of determinants on neighbourhood satisfaction from the literature are categorised which are personal/household characteristics; subjective evaluations of neighbourhood attributes and subjective evaluation of the dwelling; and objective neighbourhood characteristics [12]. However, the subjective evaluations of neighbourhood attributes are deemed more important in explaining neighbourhood satisfaction than personal/household characteristics [13][14] and objective neighbourhood attributes [15][16][17]. The neighbourhood characteristics are such as public services, the general appearance of the neighbourhoods, perceived safety, and satisfaction with fellow residents [12]. Dwelling satisfaction has a significant positive impact on neighbourhood satisfaction [13][18]. On the neighbourhood characteristics, the direct effects of objective neighbourhood conditions on neighbourhood satisfaction are supported with ample evidence [12]. Living in high socioeconomic status seems to lead to greater neighbourhood satisfaction than living in poverty areas [18][14]. The positive effect is observed on neighbourhood satisfaction following clean environment and quality housing stock [19][20]. As for the personal and household characteristics, less satisfaction is seen among younger people compared to the elderly [13][14][21]. Apart from that, the presence of children seems to bring a positive effect on neighbourhood satisfaction [14] as well as social interaction in the neighbourhood [22].

Higher income and higher education level also are found to have a positive effect on neighbourhood satisfaction, and homeowners are more likely to be satisfied with their neighbourhood than renters [23]. Homeownership is said as having a profound influence on residential evaluation [24] as it provides a feeling of security and personal achievement, leading to higher self-esteem [25][26]. The housing characteristics considered in most of the empirical studies are such as structural (kitchen space, laundry and washing areas, dining area, number of bedrooms and bathrooms, etc.), locational (accessibility to central business district, local amenities, schools, etc.), and neighbourhood attributes (neighbourhood upkeep, pollution, and crime) [27][[28][14].

Therefore, this research seeks to investigate the existing housing elements such as the housing units, and the neighbourhood park, and to suggest the housing elements that can contribute to the quality of life in a neighbourhood. Hence, there are several focus which are types of outdoor spaces, utilisation and quality of outdoor spaces; and the perception of the residents on quality of life-based on the overall neighbourhood environment. On the aspect of the neighbourhood, the focus is on the level of satisfaction of the housing overall safety, which covers park and the housing environment. It is done by identifying the park facilities and housing amenities as the indicator of life satisfaction, and to study on how the green open space in housing environment contributes towards the quality of life (QOL) of the community in Taman Melati Mastika (TMM).

2 Housing Environment and Quality of Life

Terrace house is being considered as the densest form of landed property development. The typical lot varies such as 20 feet by 65 feet and 22 feet by 70 feet. The layout of typical terrace house can be divided into two categories: the indoor and outdoor spaces. The standard double storey terrace house indoor spaces include four bedrooms, three bathrooms, a kitchen, a dining and living space, meanwhile the outdoor spaces include the entrance area or the front yard and the back yard. The entrance is the interface between the outdoor and indoor space, between public and private space.

There are six specific modification that made the residents became dissatisfied with the indoor environment which are the extension, addition, reduction, division, removal and relocation [29]. Thus, this type of housing in Malaysia demonstrates the dynamic interaction (satisfaction/ dissatisfaction) between residents and their residential environment.

Community safety is necessary and it can be related to the connected street network that can provide a better level of surveillance and the continuous housing façade [30]. A research shows that the respondents' perception towards neighbourhood safety scored the highest [31]. However, this does not implicate that the community is considered as a safe place for them.

Safety is also an element that can influence the visitation to the park. A research shows that security is rated as the highest barrier to using in both urban parks and neighbourhood parks. The latter particularly have 65% responds agreed, and 53.8% of them are female respondents [32]. It can be understood that majority of the female user does not feel safe when spending time at the park.

The QOL is a multi-dimensional construct, consisting of material and non-material aspects of life and housing satisfaction fulfils the material dimension of QOL. The majority of researchers on QOL adopt a subjective approach to measuring citizen's satisfaction through understanding their housing aspirations and the fulfilment of actual housing needs [33]. In other words, QOL is based on the assessment of multiple needs of individuals, communities, and neighbourhood. Satisfaction with the physical features of the house tends to contribute to the overall satisfaction on the neighbourhood, which also affects the overall feelings positively toward life satisfaction and well-being. The quality of the houses in the urban area is highly concerned with the improvements in living conditions which also relate to the health improvement of the urbanites while upgrading the standard of living [34].

3 Methodology

TMM (figure 1) which is located in Setapak, Kuala Lumpur, a double-storey terrace housing area was selected as the case study. This area can be considered as matured with various facilities such as schools, commercial areas, mosque, green open spaces, and the light rail transit (LRT) also covers this area with a designated station named as Taman Melati station.

Figure 1: Layout of the TMM

As mentioned earlier, there is two main focus of this study which is reflected in the methodology of the data collection, where it starts with the housing units, followed by the neighbourhood park. Observation and survey questionnaire have been identified as the primary technique to collect the data.

For the site observation, the layout of the TMM was developed by refining the layout from the Google Map using the AutoCAD software. Observations were done to identify spaces within TMM, types of houses (intermediate, corner lot, and end lot), type of extensions, hardscape and softscape elements of spaces identified, shared back lane, and observation on the neighbourhood park particularly on the facilities and amenities. The observation was conducted for three days from 15th of March of 2014 until 17th March 2014. Observation data collected earlier was used to support the development of the questionnaire.

The questionnaire was prepared with four sections – respondent's profile, quality of spaces at the individual unit, quality of green open spaces, and quality of life-based on spaces within TMM. The survey questionnaire survey was developed as a semi-structured and self-administered questionnaire. A pilot study to test the survey questionnaire was done in TMM. Six respondents were involved and following their feedback; minor changes was done. There were 250 units with four units were under major construction. Hence, 246 copies of survey questionnaire were prepared and were coded accordingly (figure 2) to roughly identify the unit's location. These units are grouped based on their road number and the shared back lane. The survey questionnaire was delivered to every house through the mailbox and the residents of TMM were asked to return the survey questionnaire in a box located at the TMM guardhouse. The period of the survey was from 15th of March 2014 until 29th of March 2014. The data from the survey questionnaire was analysed using the Statistical Package for Social Science (SPSS) software.

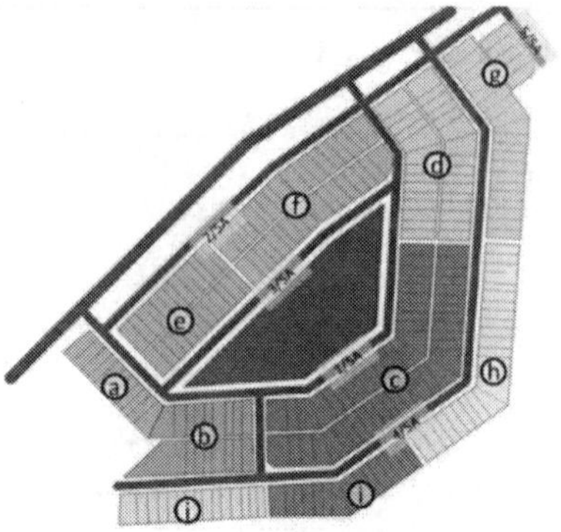

Road number	Back lane coding	No. of houses	Coding color
1/5A	a	10	
	b	9	
	c	24	
	d	14	
2/5A	e	13	
	f	18	
3/5 A	e	13	
	f	17	
4/5 A	b	8	
	c	29	
	d	16	
	g	20	
	h	24	
	i	15	
	j	15	
5/5 A	g	5	

Figure 2: The back lane coding according to housing layout

4 Analysis and Findings

The strategy of analysis and its presentation are done according to two main areas, which starts with outdoor space provision, followed by the analysis of the neighbourhood park. As mentioned before, the analysis evaluates the existing housing elements such as the housing units, and the neighbourhood park, and to suggest the housing elements that can contribute to the quality of life in a neighbourhood.

4.1 Analysis on Outdoor Spaces of TMM

The strategy or flow in analysing the data for the observation and survey questionnaire is shown in figure 3.

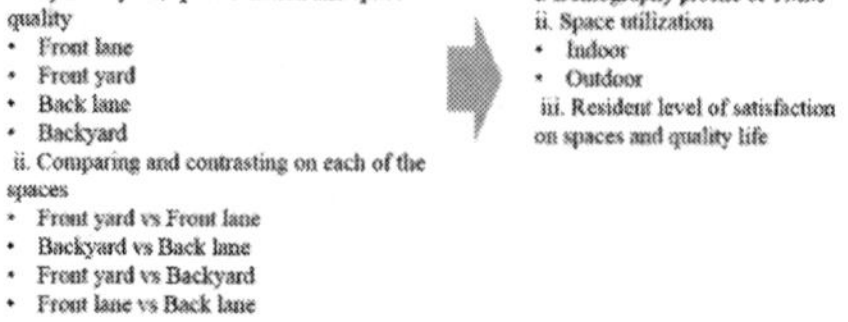

Figure 3: The strategy of data analysis for outdoor space provision

DOI: 10.1051/matecconf/20166600083

The housing units in TMM can be divided into three types which are intermediate, corner lot and end lot (figure 4).

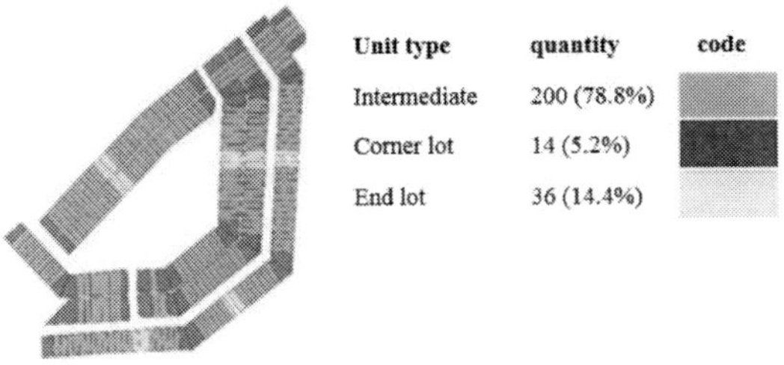

Figure 4: Unit types - location and quantity

There are five front roads/lanes exist in TMM (figure 5). The width of this road is 8m.

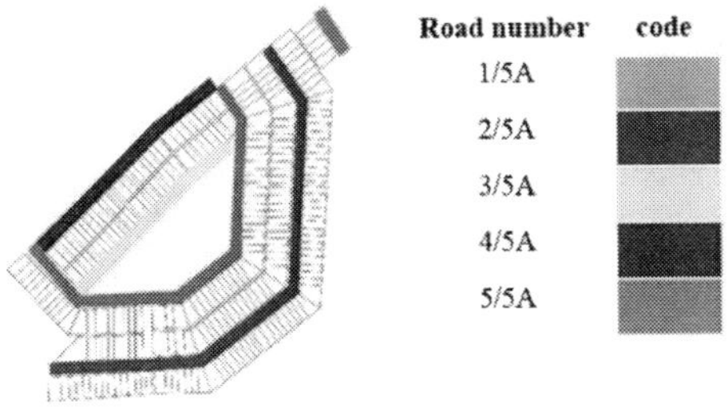

Figure 5: Layout of the front roads/lanes

4.1.1 Front yard and space quality

For the purpose of analysis, the front yard can be categorised into four types which are fully extended (100%), without green area, with green area, and original design/layout – refer images in figure 6.

Figure 6: Types of front yard - from fully extended to original design (from left to right)

For the 100% extended and without green space, it involves indoor space being extended, be it the ground floor or first floor. The ground level spaces are usually turned into semi-outdoor space and used as car porch. The second category of house extension (without green space) does not involve indoor space extension. However, the porch area is being widened until the front gate. This spaces also seem to be turned into car porch. The third category is where the house is being extended (involving the porch area only – without indoor spaces being extended) and with green space. The first two categories have totally eliminated the green space. When these categories are further collapsed into extended and non-extended, it can be seen that 85.4% are extended, with only 9.8% of green space (figure 7).Hence, it can be said that for the front yard, once extended, the chances to have landscape/green space integrated with the extension is small.

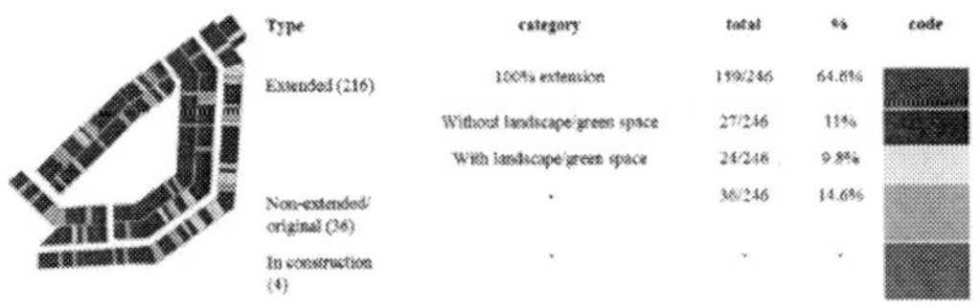

Figure 7: Analysis of extended and non-extended units with and without green space

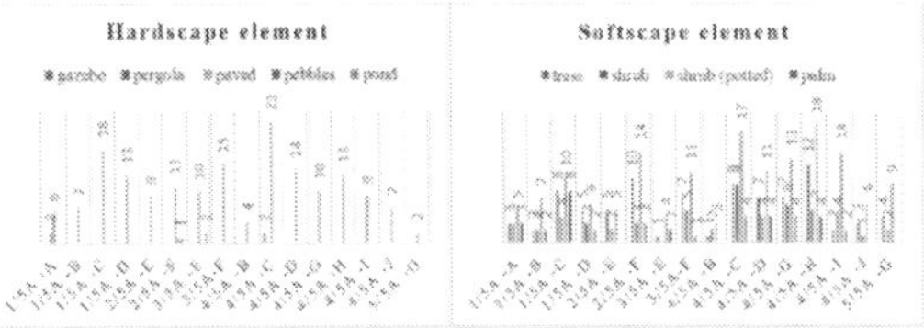

Figure 8: Hardscape element (left) and softscape element (right) in front yard area

The analysis is continued on observed units with landscape/green spaces. The result is presented in figure 8 where the location of the units is indicated by the road number and back lane codes. In general, it can be concluded that that most of the house with landscape/green spaces have more softscape element rather than hardscape element. The front yard with the best space quality is the space that has a balance between hardscape and softscape element with the later having higher quantity. For houses that are fully extended and without green space, the residents tend to make use by having potted plants as an alternative to having greens.

The back lane is shared among the row of houses and it can be characterised into two. Type 1 (top) is where it is shared by two rows of houses (indicated in colours), and type 2 (bottom) is where it is shared by only a row of houses (figure 9. These back-lanes are enclosed where only residents of these row of houses can access it from their backyard or kitchen area.

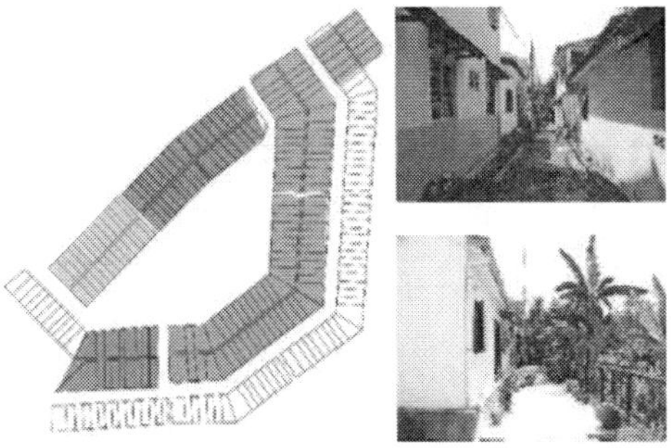

Figure 9: Type of backlanes

4.1.2 Backyard and space quality

The backyard can also be categorised into four types which are type 1 – 100% extension (turned into indoor spaces); type 2 – more than 50% of the backyard turned into indoor spaces with some outdoor spaces; type 3 – with less than

DOI: 10.1051/matecconf/20166600083

50% indoor space extension and with outdoor spaces; and type 4 – the backyard's original design is maintained with the outdoor space functions as drying and storing area (figure 10).

Figure 10: Types of backyard – from fully extended to the original design (left to right)

Type	Category	Total	%	code
Extended (188)	100%	71/250	28.4%	
	More than 50%	101/250	40.4%	
	Less than 50%	16/250	6.4%	
Non Extended/ original (56)	-	56/250	22.4%	
In construction (6)	-	6/250	2.4%	

Figure 11: Analysis on extended and non-extended backyards

Figure 11 shows the highest percentage of extension of the house in the backyard area that they extend their backyard area more than 50% of the area is 64.6%. On the other hands, the lowest percentage of extension of the house is 6.4% that they extend less than 50% of the area. House with the original layout at the backyard is 22.4%.

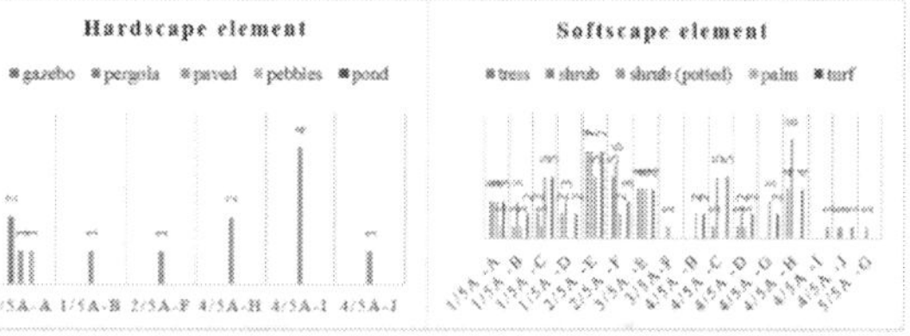

Figure 12: Hardscape element and softscape element at the backyard

Figure 12 shows the hardspace element and softscape element in each of the house according to back lane coding. It shows that most of the houses have more softscape element rather than hardscape element. The back yard with the best space quality is the space that has a softscape element that makes the backyard livelier.

4.1.3 Comparing the utilisation of outdoor spaces

There are three types of extension among the front yards. Most of the extensions in the front yard are turned into car porch. It is also becoming another space such as playing area, gardening area (the ones with landscape area extension) and drying area. Meanwhile, the front lane is utilised for vehicle access, which is becoming a second car park for the user.

There are three types of extension among the back yards. Space is limited due to the extension. Meanwhile, there are two types of the back lane, which has minimal activities. In comparison, the front yard has more activities than back yard area. The extensions of the front yards were mostly bigger than back yards. Front yard also serves as the entrance transition before entering the house. Meanwhile, the backyard has fewer activities and most of them were extended and turned into an indoor space. The front lane has more activities than the back lane area. The front lane is functioning as the main road in TMM and it is the access to each of the road. The back lane has less activity due to low utilisation and some being shared lane.

4.2Analysis of Survey Questionnaire

4.2.1 Demography Profile of TMM

A total of 62 self-administered questionnaires were returned. At the response rate of 25.3%, the adequacy of the sample size was calculated and it is significant at 11% confidence interval [35]. Table 1 shows the demographic details of the respondents. 53.2% of the respondents were males, 45.2% were females and 1.6% of the overall respondents did not state his/her gender. Regarding ethnicity, Malays represent the largest percentage which is 75.8%. The rest of the respondents (11.3%, 4.8%, 4.8% and 3.2%) is represented by Chinese, Indian, Sikh and other races respectively. 43.5% of the respondents aged between 50-59, 17.7% aged 30-39, 16.1% aged from 40-49, 16.1% were 60-69 years old and the rest of the respondent (4.8% and 1.6%) aged between 20-29 and 70 and above respectively.

4.2.2 Space Utilisation: front lane, front yard, back lane, and backyard

Table 1 shows the residents' time spent in outdoor spaces on weekdays and weekend in TMM. It can be said that in the morning during weekdays, most users spent less than 1 hour in the front area (front lane & front yard) than back area. About 75.8% user spent less than one hour in front lane area in the morning, same as in the evening. Meanwhile, in the backyard area, 69.4% user spent time less than one hour in the back lane area. In the evening, 62.9% users spent time at the back lane. However, during the weekend, most users spend less than one hour in the front and back area. In the morning, 66.1% users spent less than one hour in the front lane area, meanwhile in the evening, 72.6% users spent their time in the front lane. At the back area, 62.9% users spent less than one hour in the morning and evening, 61.3% user spent less than one hour in the same spot. It can be said that the residents spend more time in the front area than the back area.

The duration of time spent in the front yard is longer as compared to the weekdays (Table 1). Similarly, more residents spent longer hours in the back yard during the weekends as compared to the weekdays. It has shown that front yard and back yard has been utilized more than just entering the house. They are also used for weekend activities, which justified the significant of front yard and back yard to the housing design elements.

Table 1: Usage of outdoor spaces during weekend and weekdays

WEEKEND											
Duration	Less than 1 hour		1-2 hour		2-3 hour		More than 3 hour		No respond		TOTAL
	7am-7pm	7pm-7am	7am-7pm	7pm-7am	7am-7pm	7pm-7am	7am-7pm	7pm-7am	7am-7pm	7pm-7am	
Front yard/ Porch	32	43	21	12	3	2	6	2	-	3	62
	(%)	(%)	(%)	(%)	(%)	(%)	(%)	(%)	(%)	(%)	(%)
	51.6	**69.4**	**33.9**	**19.4**	**4.8**	**3.2**	**9.7**	**3.2**	**-**	**4.8**	**100**
Front lane	41	45	11	6	2	1	5	3	3	7	62
	(%)	(%)	(%)	(%)	(%)	(%)	(%)	(%)	(%)	(%)	(%)
	66.1	**72.6**	**17.7**	**9.7**	**3.2**	**1.6**	**8.1**	**4.8**	**4.8**	**11.3**	**100**
Back yard	23	30	14	6	3	4	1	-	21	22	62
	(%)	(%)	(%)	(%)	(%)	(%)	(%)	(%)	(%)	(%)	(%)
	37.1	**48.4**	**22.7**	**9.7**	**4.8**	**6.5**	**1.6**	**-**	**33.9**	**35.5**	**100**
Back lane	39	38	7	3	1	1	-	-	15	20	62
	(%)	(%)	(%)	(%)	(%)	(%)	(%)	(%)	(%)	(%)	(%)
	62.9	**61.3**	**11.3**	**4.8**	**1.6**	**1.6**	**-**	**-**	**24.2**	**32.3**	**100**
WEEKDAYS											
Duration	Less than 1 hour		1-2 hour		2-3 hour		More than 3 hour		No respond		TOTAL
	7am-7pm	7pm-7am	7am-7pm	7pm-7am	7am-7pm	7pm-7am	7am-7pm	7pm-7am	7am-7pm	7pm-7am	
Front yard/ Porch	46	46	7	8	5	3	3	3	1	2	62
	(%)	(%)	(%)	(%)	(%)	(%)	(%)	(%)	(%)	(%)	(%)
	74.2	**74.2**	**11.3**	**12.9**	**8.1**	**4.8**	**4.8**	**3.2**	**1.6**	**3.2**	**100**
Front lane	47	47	6	4	2	1	3	3	4	7	62
	(%)	(%)	(%)	(%)	(%)	(%)	(%)	(%)	(%)	(%)	(%)
	75.8	**75.8**	**9.7**	**6.5**	**3.2**	**1.6**	**4.8**	**4.8**	**6.5**	**11.3**	**100**
Back yard	30	31	6	5	4	3	1	-	21	23	62
	(%)	(%)	(%)	(%)	(%)	(%)	(%)	(%)	(%)	(%)	(%)
	48.4	**50.0**	**9.7**	**8.1**	**6.5**	**4.8**	**1.6**	**-**	**33.9**	**37.1**	**100**
Back lane	43	39	2	2	1	1	-	-	16	20	62
	(%)	(%)	(%)	(%)	(%)	(%)	(%)	(%)	(%)	(%)	(%)
	69.4	**62.9**	**3.2**	**3.2**	**1.6**	**1.6**	**-**	**-**	**25.8**	**32.3**	**100**

4.2.3 Residents Level of Satisfaction and Quality of Life in TMM

This section presents the level of satisfaction towards living condition and the quality of life in TMM. Referring to figure 14, the perceived level of satisfaction towards their living place as shown in figure 13, shows that majority (80%) of the respondents were satisfied and very jubilant compared to only 13% that was dissatisfied. As for the perceived quality of life, 77% were satisfied while 13% were dissatisfied.

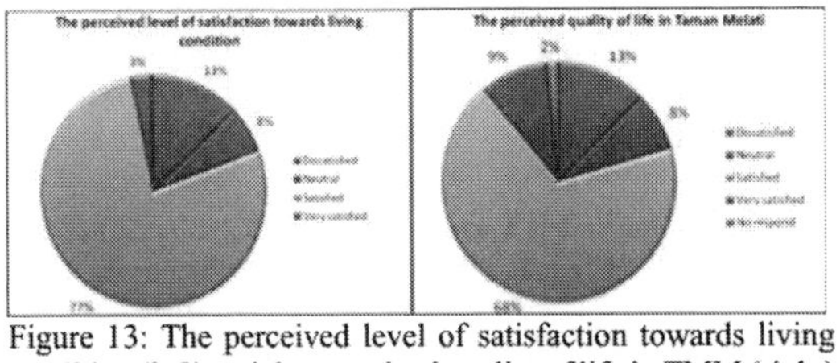

Figure 13: The perceived level of satisfaction towards living condition (left) and the perceived quality of life in TMM (right)

Factors of satisfaction can be related and further discussed as follows:

(a) Housing unit design of outdoor spaces – front yard and back yard.

The study analyses on the location of the front yard and back yard and its influence on the respondents' satisfaction towards living condition and quality of life. Referring to figure 2, there are three types of housing front yards, which are facing the park, the neighbour's front yard and the road. Meanwhile, there are two kinds of housing backyard, which are facing the neighbour's back yard separated by the back lane, and back yard facing the back road. Figure 14 shows the influence of these types of front yard and back yard towards satisfaction on living and quality of life in TMM. Among the respondents living in housing with front yard facing the park, 94% are satisfied with their living condition and 100% perceived the quality of life in TMM. Therefore, it can be said that the park and distance from the house to the park plays a significant role in providing satisfactory living condition and QOL.

 DOI: 10.1051/matecconf/20166600083

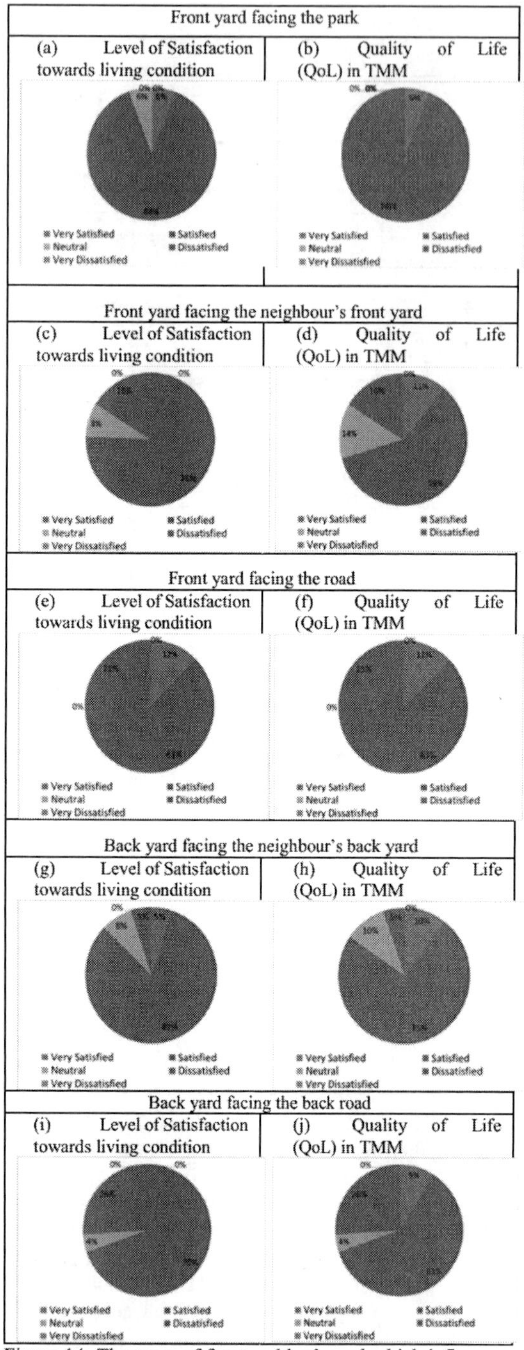

Figure 14: The types of front and back yard which influence on the respondents' level of satisfaction towards living condition (left) and the quality of life in TMM (right)

(b) Safety in the housing environment and park

Perception on safety level of TMM housing environment was measured using the 5-point Likert Scale. Figure 15 shows the respondents' perception of security level in TMM housing environment. The majority of the respondent agreed to the statement "There is no noise disturbance occurs in this neighbourhood" with high agreement (41.9%) followed by 33.9% agreement on the statement of "you feel secure living in your neighbourhood area".

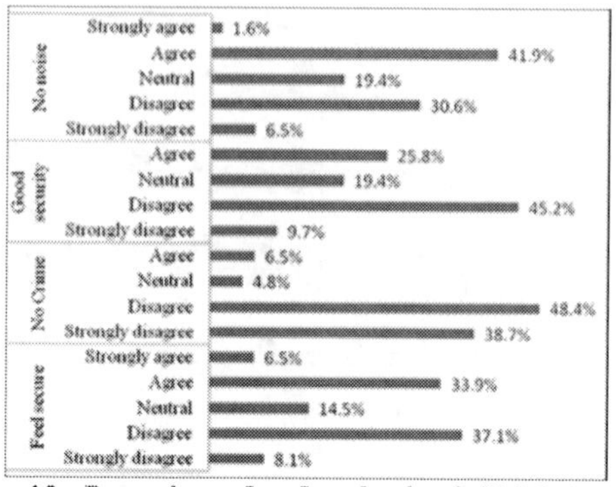

Figure 15: Perception of safety level of TMM housing environment

Those statements would indicate the degree of security of their residence. However, 48.4% disagreed with the neighbourhood having no crime recorded, followed by 45.2% disagreement with the neighbourhood having proper security measures.

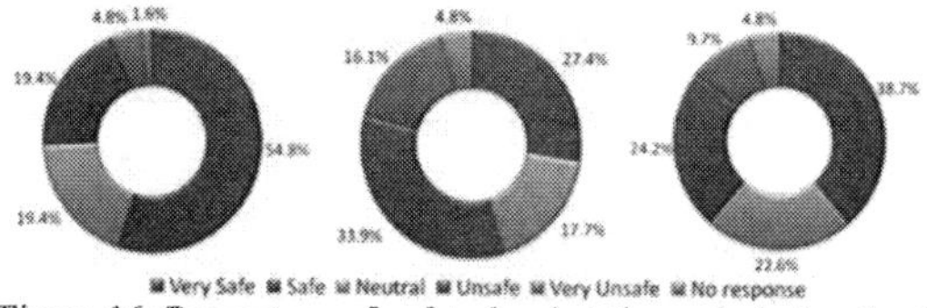

Figure 16: Percentage of safety level at the park during the day (left), night (middle) and overall safety level (right)

Safety level at the park was assessed from the respondent perception in which they rated the security level from 1=Very Safe until 5=Very Unsafe for 3 cases which are safety during the day, during the night and overall safety level. The results in figure 16 show that 54.8% of the respondents perceived the park as safe during the day. As for the safety during the night, the majority (33.9%) rated the park as unsafe. The overall safety level was perceived as safe 38.7%. The security level at the park was also assessed from the witnessed illegal activities and existence of possible hiding spots in the park. Both questions require the respondent to identify the zone in which they had seen the crime as well as the potential hiding place. The areas and their descriptions are shown in figure 17.

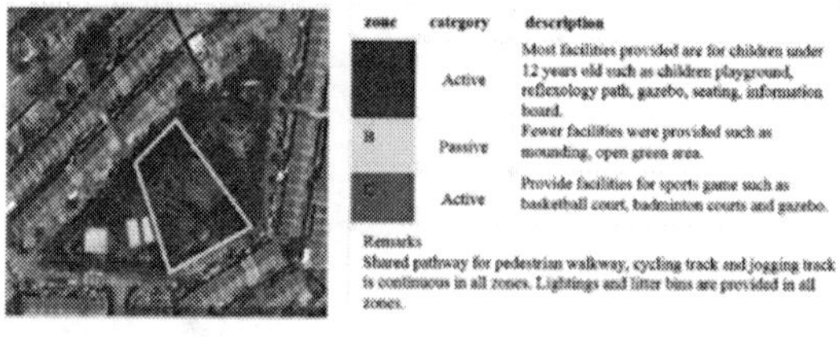

Figure 17: Identified zones of the neighbourhood park

64.5% of the respondents stated that they had witnessed illegal activities in the park compared to 33.9% that never observed any illegal activities as shown in figure 17. For those who had seen the crime, the majority of them said that it happened mostly in Zone A (38.7%) while the least is Zone B (1.6%) (figure 17). However, the data shown does not show the actual point or place of the witnessed illegal acts. The criminal activities identified are such as theft and immoral acts by teenagers as commented by the respondents in the questionnaire. They also mentioned that those teenagers are from the adjacent neighbourhood. This situation may have happened due to lack of supervision from the security personnel and the easy entry into the neighbourhood.

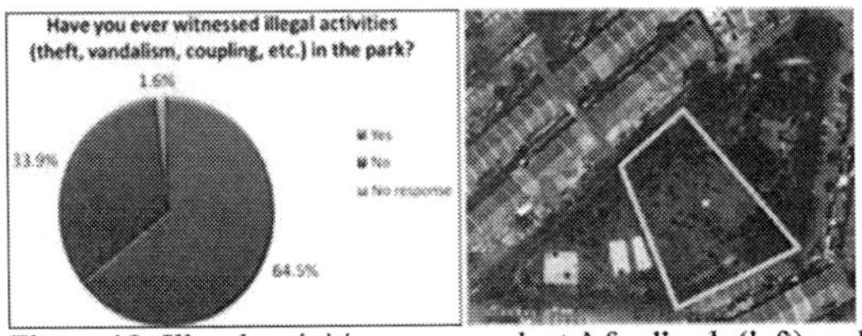

Figure 18: Illegal activities – respondents' feedback (left) and zones involved (right).

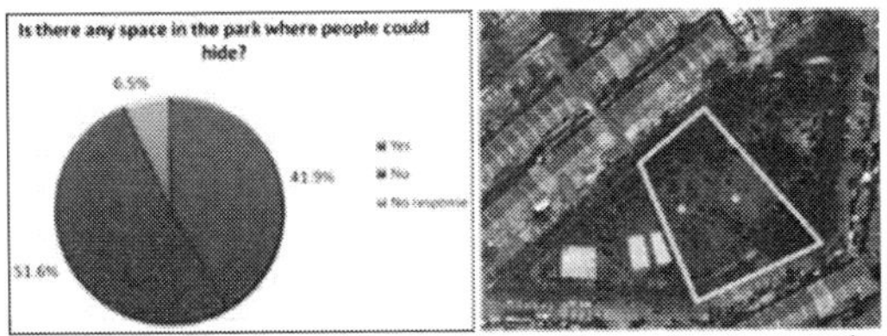

Figure 19: Possible hiding spots at the park – respondents' feedback (left) and zones involved (right)

Referring to figure 19, 51.6% of the respondents stated that there is no possible hiding spot in the park, while 41.9% answered the opposite while indicating the zones involved with the highest possible hiding place is Zone A (29%). However, the data only show the frequency of the potential hiding spots and not the exact hiding location. Zone A seems to have the most hiding spots may be due to the availability of the children play structure with enclosed parts and the density of vegetation in the area. Some trees planted might seem to create potential hiding spots between its trunks. Dim lightings in the park may add to the possibility of Zone A to have possible hiding spots.

(c) Sufficiency of facilities provided and its current condition

The park's quality was assessed based on the residents' satisfaction with the present state of the park facilities and the sufficiency of the facilities. The result show that the majority (53.2%) of the respondents were satisfied with the sufficiency of the park facilities provided compared to only 25.8% were unsatisfied (figure 20). Similar figures were also shown to their satisfaction towards the condition of the facilities provided where 51.6% were satisfied, while 22.6% were dissatisfied (figure 21).

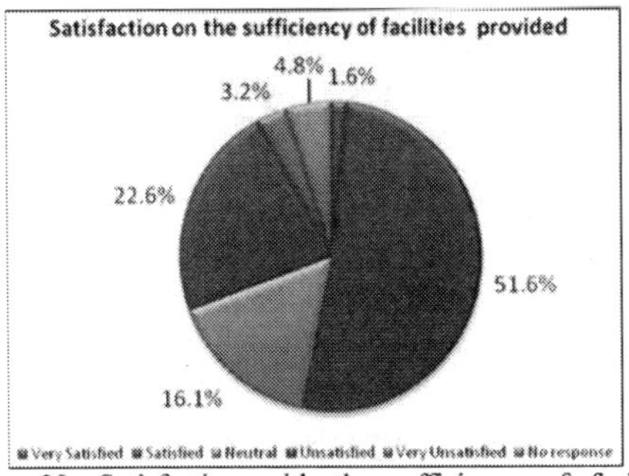

Figure 20: Satisfaction with the sufficiency of facilities provided

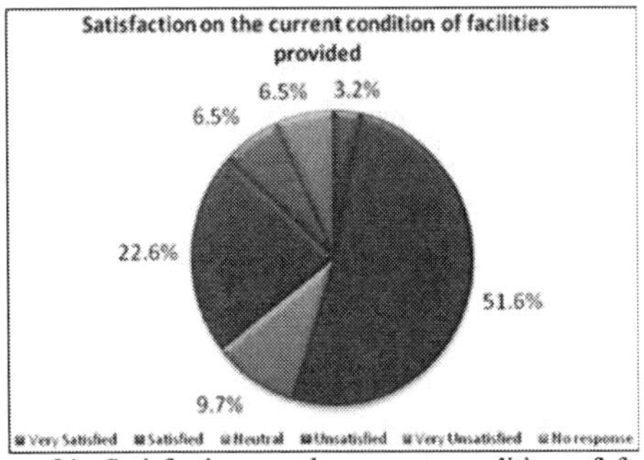

Figure 21: Satisfaction on the current condition of facilities provided

However, some of the facilities are in poor conditions (figure 22) with much-broken seating, uncovered drain holes, uneven or cracked pathway, broken lamp, broken litter bins, broken playing equipment, and immoral scribbling on the play equipment. However, due to the majority of the respondents were satisfied (figure 21), it can be said that the park still functions as recreational and leisure space despite its poor conditions at its facilities.

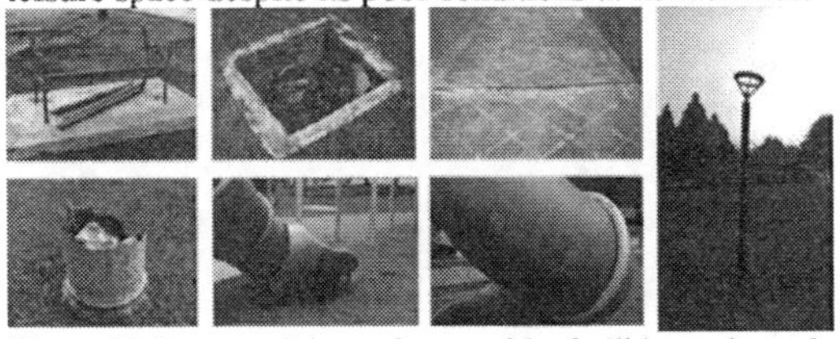
Figure 22: Poor conditions of some of the facilities at the park

(d) Sufficient roadside parking and road condition

In addition to the above, the following discussion is on the facilities and amenities of TMM. The facilities that were assessed in this research were the sufficiency of parking bays and quality of the road. It can be understood from figure 23 that 38.7% of the resident respondents agreed to both sufficient parking provided and excellent road quality while 48.4% thinks there is insufficient parking space. 35.4% of the residents do not feel the road quality is good.

 DOI: 10.1051/matecconf/20166600083

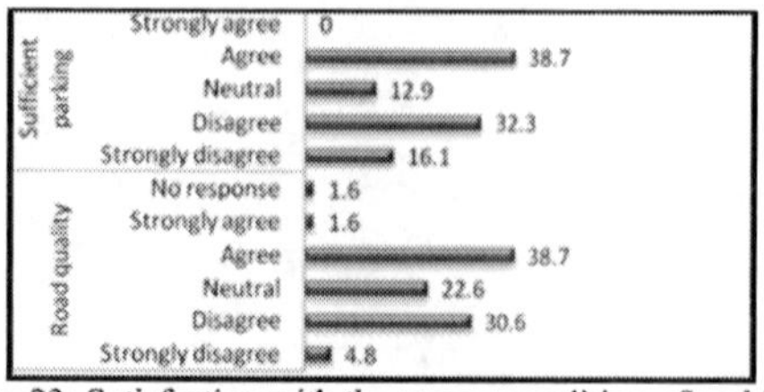

Figure 23: Satisfaction with the current condition of parking and road quality

5 Conclusion

The first objective is to record existing housing elements such as the homes and neighbourhood context. In the context of space provision, many homes are being extended, and this is very commonly done in the terrace housing area which involved turning the front yard into paved and roofed surfaces which serve as car porch among other, leading to smaller amount of landscape/green spaces. Hence, it can be suggested that more built area are seen, and this could lead towards imbalance environment. The urban area is known for its limitation of green space. The needs to extend the indoor spaces or to renovate the exterior spaces to become semi-outdoor space can make the situation worse. The back lane is seen as a commonly ignored/neglected area. However, in TMM the type 2 back lane seems to be a useful place due to a lot of ongoing activities such as gardening. Type one back lane on another hand, is currently abandoned by some residents due to the small amount of space to do activities.

The second objective is to suggest the housing elements that can contribute to the quality of life in a neighbourhood. The results have shown that respondents living in homes where their front yards are facing the neighbourhood park have the very high satisfaction to their living condition. Therefore, the relationship of the front yard with the neighbourhood park is the best design element that can be suggested for a neighbourhood towards the QOL. In addition to that, security of a neighbourhood is also crucial. Furthermore, the sufficiency of facilities is also critical. These facilities need to be maintained in a good condition to avoid further vandalism or wrong hiding spots. Lastly but never the least, another important element would be the sufficient of roadside parking and the road quality. With consideration of these elements, the QOL of TMM can be improved.

References

1. Lazenby, K.,(1988).*Occupancies satisfaction in the white house sector in the municipality of Bloemfontein: A study of process patterns and strategy*. Bloemfontein: University of the Orange Free State.
2. Abu Bakar. A., (2007), *User response to thermal comfort of outdoor urban spaces in hot-humid region*, Ph.D. thesis, University of Nottingham, United Kingdom.
3. Chapman, G. A. (1999). Design Variables and the Success of Outdoor Neighbourhood Recreational Facilities. Unpublished Master Thesis, The University of Arizona, Arizona.
4. Nurhayati A. M., & Manohar M. (2010). *Developing Quality Neighbourhood Parks: A Malaysian Experience*. Proceedings of the Healthy Parks Healthy People: International Congress 2010. Organised by Parks Victoria; IUCN; Beyond Blue; Deakin University; Director of National Parks, Australian Government; and World Tourism Organization (UNWTO) held in Melbourne Convention Centre, Melbourne.
5. Von Kursell, A. A. (1997). Replanning Urban Parks. Unpublished Masters of Urban and Rural Planning Thesis, Technical University of Nova Scotia, Halifax, Nova Scotia, Canada.
6. Nurhayati Abdul Malek, Manohar Mariapan, & Mustafa Kamal MohdShariff (2012), The Making of a Quality Neighbourhood Park: A Path Model Approach. *Procedia Social and Behavioral Sciences, SciVerse Science Direct*, **49**, 202-214.
7. Preiser, W.F.E., (1989), Towards a performance-based conceptual framework of systematic POES, in W.F.E. Preiser (ed), *Building Evaluation*, New York: Plenum Press.
8. Paris, D.E., and Kangari, R., (2005), Multifamily Affordable Housing: residential satisfaction, *Journal of Performance and Constructed Facilities*, Vol. **19**, pp. 138-145.
9. Salleh, A.G., (2008), Neighbourhood factors in private low-cost housing in Malaysia, *Habitat International*, Vol. **32**, pp. 485-493.
10. Ogu, V.I., (2002), Urban residential satisfaction and the planning implications in a developing world context: the example of Benin City, Nigeria, *International Planning Studies*, Vol. **7**, pp. 37-53.
11. Tan, T.H., (2012), Housing satisfaction in medium-and high-cost housing: the case of Greater Kuala Lumpur, Malaysia, *Habitat International*, Vol. **36**, pp. 108-116.
12. Permentier, M., Bolt, G., and van Ham, M., (2011), Determinants of neighbourhood satisfaction and perception of neighbourhood reputation, *Urban Studies*, Vol. **48** (5), pp. 977-996
13. Lu, M., (1999), Do people move when they say they will? Inconsistencies in individual migration behaviour, *Population and Environment*, Vol. **20**, pp. 467-488.
14. Parkes, A., Kearns, A., and Atkinson, R., (2002), What makes people dissatisfied with their neighbourhoods? *Urban Studies*, Vol. **39** (13), pp. 2413-2439.
15. Campbell, A., Converse, P.E., and Rodgers, W. L., (1976), *The quality of American life: perceptions, evaluation and satisfactions*, New York: Russell Sage.
16. Carp, F., Zawadski, R., and Shokron, H., (1976), Dimensions of urban environmental quality, *Environment and Behavior*, Vol. **8**, pp. 239-265.
17. Galster, G., (1987), *Homeowners and neighbourhood reinvestment*, Durham, NC: Duke University Press.
18. Mohan, J., and Twigg, L., (2007), Sense of place, quality of life and local socioeconomic context: evidence from the survey of English housing, 2002/03, *Urban Studies*, Vol. **44**, pp. 2029-2045.

19. St John, C., and Bates, N.A., (1990), Racial composition and neighbourhood evaluation, *Social Science Research*, Vol. **19**, pp. 47-61.
20. Basolo, V., and Strong, D., (2002), Understanding the neighbourhood: from residents' perceptions and needs to action, *Housing Policy Debate*, Vol. **13**, pp. 83-105
21. Chapman, D.W. and Lombard, J.R., (2006), Determinants of neighbourhood satisfaction in fee-based gated and non-gated communities, *Urban Affairs Review*, Vol. **41**, pp. 769-799.
22. Dekker, K. and Bolt, G., (2005), Social cohesion in post-war estates in the Netherlands: differences between social-economic and ethnic groups, *Urban Studies*, Vol. **42**, pp. 2447-2470.
23. Harris, D., (2001), Why are whites and blacks averse to black neighbours?, Social Science Research, Vol. **30**, pp. 100-116
24. Barcus, H.R., (2004), Urban-rural migration in the USA: an analysis of residential satisfaction, *Regional Studies*, Vol. **38** (6), pp. 643-657.
25. Rohe, W.M., and Stegman, M.A., (1994), The effects of home ownerships on the self-esteem, perceived control and life satisfaction of low-income people, *Journal of the American Planning Association*, Vol. **60** (2), pp. 173-184.
26. Rohe, W.M., Van Zandt, S., and McCarthy, G., (2001), *The social benefits and cost of homeownership: a critical assessment of research, low-income homeownership*. Working paper series, Joint Centre for Housing Studies, Harvard University.
27. Baum, S., Arthurson, k., and Rickson, K., (2010), Happy people in mixed-up places: the association between the degree and type of local socioeconomic mix and expressions of neighbourhood satisfaction, *Urban Studies*, Vol. **47**, pp. 467-485
28. Hipp, J., (2010), What is the 'neighborhood' in neighbourhood satisfaction? Comparing the effects of structural characteristics measured at the micro-neighborhood and trace levels, *Urban Studies*, Vol. **47**, pp. 2517-2536
29. Erdayu, O. O., Esmawee, E., and Masran, S., (2010). Adapting By Altering: Spatial Modifications of Terraced Houses in the Klang Valley Area, *Asian Journal of Environment-Behaviour Studies*, Vol.1, No.**9**, pp.1-10.
30. City and County of Swansea (2014). Places to live: Residential, Design Guideline.
31. Yahaya. N., Sabri, B., Abdullah, S.F.Z., Abdullah, S.S., Momtaz, Y.A., and Hamid, T.A., (2012) Impact on Housing Environment and Neighbourhood Safety Towards Perceived Quality of Life Among Older Persons, *Indian Journal of Geontology,* Vol. **26**, No. 1, pp. 227-241
32. Heather E. Wright Wendela, Rebecca K. Zarger, James R. Mihelcic (2012). Accessibility and Usability: Green Space Preferences, Perceptions, and Barriers in a Rapidly Urbanizing City in Latin America, *Landscape and Urban Planning* **107**, pp. 272-282
33. Hikmat, A. M. F., and Al-Betawi, Y.,(2009). Quality of Life in Cities Setting up Criteria for Amman-Jordan, *Social Indicator Research Series,***93**: 407-432.
34. Glatzer, W., Camfield, L.Møller, V. and Rojas, M (2015), Exploration of well-being of nations and continents, *Global Handbook of Quality of Life*, Springer, Social Science, pp. 704-705
35. www.surveysystem.com/sscale.htm, [10 February 2016]

Urban Outdoor Thermal Comfort of The Hot-Humid Region

Abu Bakar, A.[1], Mohamed B. Gadi[2]

[1]Assistant Professor, Kulliyyah of Architecture and Environmental Design, International Islamic University Malaysia, Jalan Gombak, 53100 Kuala Lumpur, Malaysia
[2]Associate Professor, Institute of Building Technology, Faculty of Engineering, The University of Nottingham, University Park, NG7 2RD, UK

Abstract. The study on outdoor comfort is becoming popular due to the fact that the thermoregulatory model is seen as inadequate in explaining outdoor thermal comfort conditions. Hot-humid region can be said as experiencing a critical environmental condition because of its constantly high temperature and humidity throughout the year. Thus, this study focus on the assessment of thermal comfort of outdoor urban spaces in Kuala Lumpur, Malaysia (3° 9'N and 101° 44'E). Survey on human response towards outdoor thermal comfort in hot-humid climate of Kuala Lumpur, Malaysia was carried out during day time between 0900h to 1800h along with measurement of environmental parameters such as air temperature (°C), wind velocity (m/s), radiant temperature (°C), relative humidity (%) and solar radiation (lux). A total of 123 samples were involved in this study which took place within four sites around Kuala Lumpur. Survey results were then correlated with the environmental parameters to further develop the comfort zone for hot-humid outdoor environment specifically for Kuala Lumpur and, generally, for hot-humid regions.

1 Introduction

Interest in research on outdoor thermal comfort has been increasing lately, especially in tropical regions [1],[2],[3],[4],[5],[6]. Not until recently, it has been assumed that the indoor thermal comfort theory can be applied to the outdoor. However, it has been emphasized [7] that due to the dynamic outdoor environment and such, the thermoregulatory model is seen as inadequate in explaining outdoor thermal comfort conditions. This study addresses outdoor human thermal comfort within hot-humid urban open spaces by analyzing factors influencing thermal comfort, especially environmental parameters, which affect the microclimatic environment of the design itself, and the physiological aspect of the subjects - in terms of their physical activities and clothing.

2 Urban open spaces and elements of thermal performance evaluation

Kuala Lumpur is located at midway along the west coast of Peninsular Malaysia with geographical coordinates of 3° 9'N and 101° 44'E, and thus is very much near to the equatorial line. The average temperatures is ranging between 25°C to 33°C with average relative humidity ranging from approximately 55% to 95% throughout the year. Malaysia has plenty of sunshine and solar radiation and but they are sometimes block by the cloud cover and thus, it is hard to find a day with a completely clear sky [8]. It can be argued that Kuala Lumpur gains 12 hours (0700h to 1900h) of solar radiation every day throughout the year with the highest reading ranging from 594.4 to 625 Wh/m^2 respectively between 1100h to1300h respectively.

There were four field study sites within Kuala Lumpur – namely: the courtyards of the International Islamic University Malaysia (three courtyards: C1, C2, and C3), and the recreational area of Taman Melati (TM - a housing/residential area). Different types of sites which represented various types of built form as well as natural environment within a housing area were chosen to study the effect of these spaces on the microclimate and eventually on the human thermal comfort. These sites are also reflecting the variety of spaces that can be found in the urban area.

Street plaza, corporate foyer, urban oasis, and transit foyer are the several typology of urban spaces [9]. Courtyard is another form of open space that is widely used. It can be identified based on the degree of enclosure such as enclosed, semi-enclosed, and semi-open [10]. The orientations and shapes of courtyards affect the sunlight distribution on their facades and floors, thus affecting the microclimate of the courtyards. Thus, aspect ratio (*Aspect ratio = area of the courtyard floor / (average height of surrounding walls)^2*) is used to determine the degree of courtyard exposure to the sky openness that permits heating by the sun [11]. Greater aspect ratio would mean greater exposure to the sky. This affects the thermal performance of a space following the energy exchanges process such as reflected radiation, short-wave radiation, long-wave radiation, and evaporative exchanges [12], which eventually affect the thermal comfort of the people. There is a concern on the suitability of material used for outdoor urban spaces as there is a need to lower the ambient temperatures particularly within the hot and humid regions. The pavement used for instance can be classified as 'cool' and 'warm' materials depending on their ability in absorbing heat and reradiating it to the surrounding [13]. Greenery help to absorb great amount of radiation through the evapotranspiration process with less reflection of long wave radiations [12]. It is also said as an effective moderator of near-surface climates, particularly in the mid and low latitudes [14][15].

3 Methodology

This study intended to study human response in terms of thermal comfort within the outdoor urban spaces of hot-humid condition. Thus, survey on thermal comfort and thermal sensation by using 123 university students as the sample was conducted together with collection of physical environmental data (refer Table 1 for the equipment used and reading taken). Thirty minutes were provided for every session of the survey for every site. Three environmental readings were taken with the interval of ten minutes for every session of each site. These readings were then averaged.

Table 1. Equipment used, measurement and unit.

no	equipment	measuring	unit		
i	Whirling psychrometer	Dry & wet bulb temperature	DB (°C)	WB (°C)	RH (%)*
ii	Globe thermometer	Globe temperature	°C		
iii	Digital anemometer	Wind speed	m/s		
iv	Cole-parmer infrared thermometer	Surface temperature (ground and wall)	°C		
v	Illuminance meter	Min/max/avg. illuminance	lux		

*The RH is obtained by referring the depression between DBT and WBT to a chart that comes together with the whirling psychrometer.

There were three types of spatial categories identified in conducting the survey which were: *semi-shaded area (under the tree), open area (under direct sunlight)*, and *covered area (arcade or corridor)* – refer Table 2. However, for TM, there were only two types of spatial categories involved which were: *under the tree* and *under direct sunlight* – refer Table 3.

Table 2. Survey time and spatial categories for the courtyards

Semi-shaded	Open	Covered (alongside the corridors)
0900-0930	0930-1000	1000-1030
1100-1130	1130-1200	
1200-1230	1230-1300	1300-1330
1400-1430	1430-1500	
1500-1530	1530-1600	1600-1630
1700-1730	1730-1800	

Table 3. Survey time and spatial categories for TM

Under a tree	Under direct sunlight
0900-0930	0930-1000
1000-1030	
1100-1130	1130-1200
1200-1230	1230-1300
1300-1330	
1400-1430	1430-1500
1500-1530	1530-1600
1600-1630	
1700-1730	1730-1800

In each session, the samples were divided into two groups to perform passive (1 to 3 met) and active (3 to 8 met) activities respectively for twenty minutes. The sensation scale of the PMV which is also known as the Bedford seven-point scale was used in assessing thermal sensation [16], while in assessing thermal comfort, the scale provided by ASHRAE 55 [17] (refer Table 4) was used, and these scales were adopted in the survey questionnaire for the samples to indicate their preferences. The survey questionnaire consisted of three main sections as follows:

- Section A: details on the location, date, day, weather condition and others,
- Section B: details of the sample, such as name, age, health condition and others,
- Section C: tables on comfort judgment for the sample to tick and a few open-ended questions.

Table 4. Categories of votes for thermal sensation and thermal comfort.

Scale for thermal sensation		Scale for thermal comfort	
Scale	Description	Scale	Description
+3	Hot	1	Very comfortable
+2	Warm	2	Comfortable
+1	Slightly warm	3	Slightly comfortable
0	Neutral	4	Neutral
-1	Slightly cool	5	Slightly uncomfortable
-2	Cool	6	Uncomfortable
-3	Cold	7	Very uncomfortable

During the survey, it was observed that the samples were wearing clothes with clo value from 0.5 to 0.8 which is regarded as the typical clo value among Malaysians. On top of the fieldwork, shadow analysis using Sketchup™ programme was also simulated from 0900h to 1800h except for C3, which ended at 1630h, because the whole area was already shaded from then onwards.

4 Locations, orientations, and physical built forms of studied sites

4.1 The Courtyards: C1(5484m²), C2(2500m²), and C3(520m²)

These three types of courtyards studied represented the followings, and their images are shown in figure 1:

i. paved courtyard : named as Courtyard 1 (C1) – regarded as 100% paved;
ii. partially-paved courtyard : named as Courtyard 2 (C2) – refer to Table 5;

iii. turfed courtyard : named as Courtyard 3 (C3) – regarded as 100% turfed.

Figure 1. Courtyard 1 (left), Courtyard 2 (middle), and Courtyard 3 (right)

Table 5. Percentage of sub-areas in C2

Surface material	m²	%
Tarmac area	1350	54
Rubber matt	194	7.8
Grass/turfed area	184	7.4
Paved area	772	30.8
Total area	**2500**	**100**

All these courtyards have zaguáns and arcades. Zaguán is a covered entranceway from the street to the courtyard, which acts as a wind tunnel to allow wind from outside to penetrate the courtyards [18], while an arcade (or colonnade) is a covered walkway at the sideline of a building. The height of the buildings surrounding C1 and C2 varies from 3 levels to 5 levels except for C3 where it is surrounded by 4 level buildings. The aspect ratio of these courtyards are identified. Following the definition of aspect ratio, C1 seems to demonstrate the highest degree of exposure to the sun – refer Table 6.

Table 6. Courtyards' orientation and their aspect ratio

	C1	C2	C3
Orientation	22.5°	22.5°	2°
Aspect ratio	13.81	4.62	1.86

4.2 Taman Melati - TM (18211m²)

This recreational area is surrounded by 249 units of double-storey terrace houses and equipped with amenities such as a children playground, a jogging track, a badminton court, a basketball court and benches, among others. More trees are being planted in this area in a recent upgrading programme by the City Hall of Kuala Lumpur.

4.3 Shadow analysis

For Taman Melati, it was almost impossible to quantify the shaded area for Taman Melati and produce a percentage as the shape of the cast shadow was 'organic' as a result of the shapes of the trees. The percentage of shadowed area for the rest of the studied sites is presented in the following figure 2.

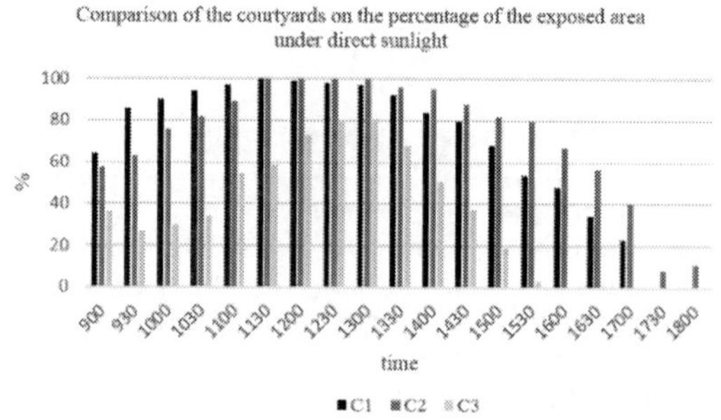

Figure 2. Comparison of the courtyards on the percentage of exposed area to the direct sunlight

C3 appears to have the shortest period of exposure under direct sunlight (from 0900h to 1530h), has never reached 100% exposure to sunlight and seems to have a quite significant low percentage of area exposed to sunlight throughout the day. On average, C2 experiences the longest period of time exposed under direct sunlight compared to C1 and C3. Between C1 and C2, C1 seems to have a higher percentage of exposed area from 0900h until they were equally the same at 1130h before C2 starts to produce a higher percentage than C1 towards 1800h. In short, C2 can be determined as the courtyard that is highly exposed to the sun with a higher percentage of sunlit area throughout the day, while C3 is the total opposite in terms of results. Based on shadow simulation, it is concluded that the physical built forms of these sites greatly influence the amount of areas exposed to direct sunlight. Greater aspect ratio reading indicates that wider areas of those sites are being exposed to sunlight.

4.4 The microclimate

The environmental parameters considered in studying the microclimate of those sites are dry bulb temperature (DBT), wet bulb temperature (WBT), surface temperature of floor (ST-F), and surface temperature of wall (ST-W):

i. DBT, WBT, GT, ST-F & ST-W (°C)

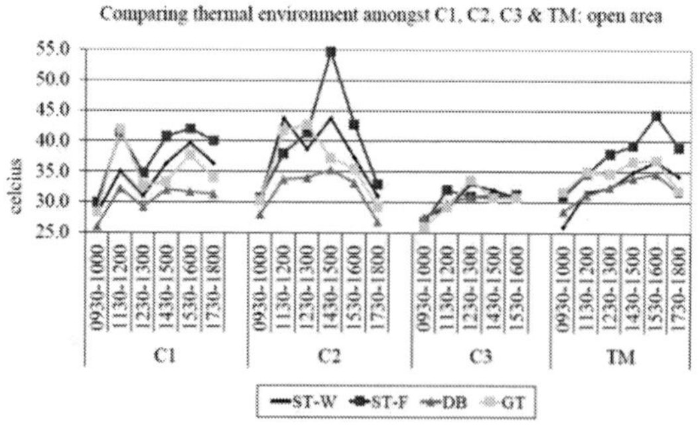

(a) open area

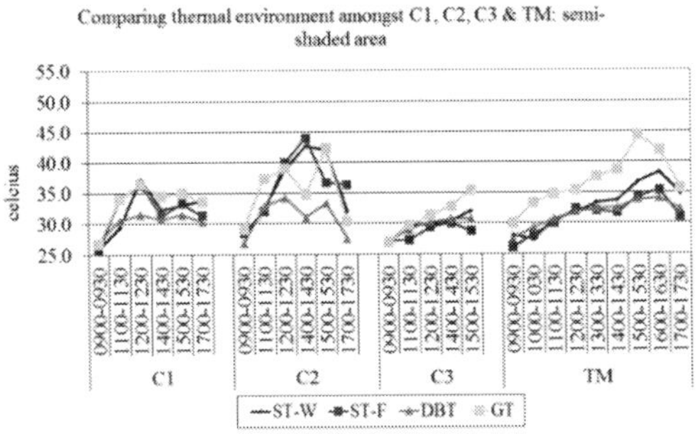

(b) semi-shaded area

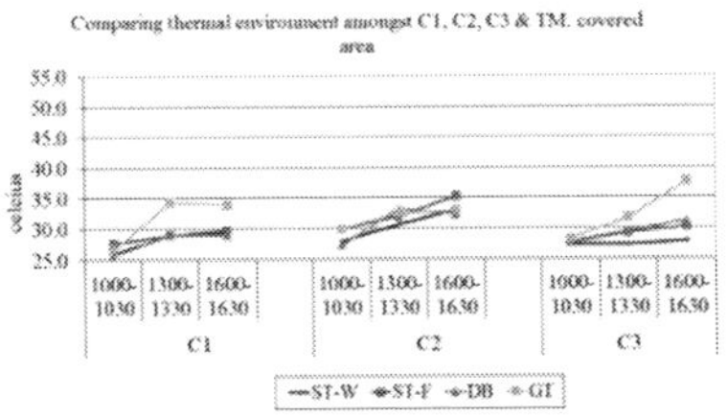

(c) covered area

Figure 3. Comparison of sites' thermal environment by spatial categories: open, semi-shaded, and covered areas

From the above graphs, it can be clearly seen that the thermal environment for the open area is hotter than the semi-shaded area, followed by the covered area. The gap among the variables plotted is wider for the open area than the semi-shaded area. C3 seems to have a narrower gap among the variables throughout the studied spatial categories while C2 is the total opposite. However, the gap amongst these variables seems narrow for the covered area. This suggests that solar radiation plays an important role in influencing the thermal environment of an area/space.

ii. Illuminance (lux)

Table 7. Minimum and maximum readings of illuminance by sites and spatial categories

	ss	open	covered	ss	open	covered
		C1			C2	
min ILL x 1000 (lux)	2	1.2	0.16	0.3	2	0.12
max ILL x 1000 (lux)	114.4	94.5	0.5	64.5	91.6	0.9
		C3			TM	
min ILL x 1000 (lux)	2.5	3.1	0.19	16.4	3.3	-
max ILL x 1000 (lux)	166	111.6	0.9	106	41.7	-

Based on Table 7, the minimum and maximum illuminances recorded throughout the sites are 120 lux and 166000 lux respectively. The value is greatly influenced by the spatial categories and sky condition.

iii. Relative humidity (%)

Table 8. Minimum and maximum readings of the RH by sites and spatial categories

	ss	open	covered	ss	open	covered
		C1			C2	
Min RH (%)	50.7	55.7	67.7	49.3	50.7	53
Max RH (%)	79.3	89.7	87.3	80.3	78	74.7
		C3			TM	
Min RH (%)	63.3	70	61	51.7	49.7	-
Max RH (%)	81	82	78.7	85	82.3	-

Based on Table 8, the minimum and maximum readings of the RH recorded throughout the site are 49.3% and 89.7% respectively.

iv. Wind speed (m/s)

Table 9. Minimum and maximum readings of wind speed by sites and spatial categories

	ss	open	covered	ss	open	covered
		C1			C2	
min WS (m/s)	0.1	0.1	0.03	0	0	0.03
max WS (m/s)	4.2	3.5	0.6	3.1	1.8	1
		C3			TM	
min WS (m/s)	0	0	0	0.3	0	-
max WS (m/s)	0.3	0.17	0.5	2.4	3.2	-

Based on Table 9, it can be seen that the minimum wind speed captured throughout these sites was 0m/s while the maximum speed was 4.2m/s. It can be said that the ground wind speed in Kuala Lumpur, Malaysia, is mild most of the time where strong wind, or gale, is seldom experienced.

v. Surface temperature (°C)

Although the height of buildings surrounding the courtyards varies from 3 to 5 levels, the surface temperature readings were taken up to the third level only.

Table 10. Highest surface temperature according to level of each courtyard

Level	C1	C2	C3
1	41°C	45°C	29°C
2	41°C	42°C	29°C
3	41°C	42°C	32°C

In looking at Table 10 above, it seems that C2 gives the highest reading for the façades' surface temperature. This could possibly be due to the radiant heat produced by the black tarmac that surrounds the perimeter of C2.

Based on the analysis, it is suggested that each and every site produced different microclimate. These may have been due to the differences in their built forms, geographical locations, orientations, elements contained within their spaces and weather types during the survey and fieldwork conducted.

5 Comfort votes

The ASHRAE 55 [17] outlines that a thermally acceptable environment is where at least 80% of the occupants do not express any dissatisfaction. This refers to indoor environment, where the environmental condition is controlled and stable. However, it is suggested that this percentage should be lowered to 70% when dealing with dynamic outdoor environment, with great environmental fluctuations throughout the day and, also, taking into account that people nowadays do not normally spend most of their time outdoor [18]. The percentage of 70% represents more than two-third of the number of occupants, which is thought as more reasonable as it still represents the majority.

The following analysis also shows that thermal comfort vote on *comfortable* and *neutral* are combined as *comfortable-neutral* or *comft-neu*. The reason is due to the fact that when assessing a rather complex thermal comfort preferences of people and, especially when it involves dynamic outdoor environment, neutral comfort category can and is suggested to be taken as *non-existence discomfort to the people*. This is in line with the way thermal comfort is defined by the ASHRAE 55 as mentioned earlier.

In analysing the survey data for the courtyards and TM, certain procedures were strategized. Firstly, the measure of association between thermal sensation and thermal comfort was conducted to determine the degree of corelationship between these two variables. Due to the irregular number of samples for each site, the samples were filtered to standardise their quantities. In doing so, three ways of filtering the samples were carried out to reduce bias in omitting the samples. A comparison between unfiltered and filtered data was also conducted.

Referring to result obtained, the ranking of studied sites in terms of samples' thermal preferences was identified. This process then led to identifying the most comfortable site and the most uncomfortable site. Based on these two identified sites' categories, further analysis was carried out to compare their environmental conditions and relate them to the samples' votes. The range of environmental parameters involved, where at least 70% of the samples voted for comfortable-neutral, was identified.

Based on the survey data, it is proved that there is a strong relationship and significant correlation between thermal comfort and thermal sensation in which for the context of this study – *the hotter the environment, the more uncomfortable it becomes for samples*. Thus, the following conclusions are generated:

- The result obtained shows an opposite preferences or votes by subjects in tropical Kuala Lumpur compared to those in temperate countries as people in temperate countries normally perceive a hotter environment as the more comfortable one.
- Temperature: The temperature set as 'comfortable' by subjects in Kuala Lumpur is significantly higher than those in temperate countries. According to the PMV, the temperature in Kuala Lumpur is theoretically too high and, yet, the subject is found to be comfortable with it. This might be due to the reason that those samples were already acclimatized and adapted to this type of climate condition.

When analysed based on thermal comfort votes throughout the investigated sites, it is concluded that votes on 'comft-neu' is relatively high compared to votes on 'uncomfortable', except for TM, where the difference is quite small. Thus, it is suggested that, even though the outdoor weather condition is extreme compared to indoor environment, people seem to be able to tolerate and compromise on the condition. Based on the spatial category with combined studied sites, the percentage of samples (doing either passive or active activities) voting for 'comfortable', 'neutral' and 'uncomfortable' are tabulated in Table 11 as follows:

Table 11. Percentage of samples voting according to comfort votes by spatial categories

	semi-shaded			open			shaded		
	Passive	Active	Sub total	Passive	Active	Sub total	Passive	Active	Sub total
Comfortable	**15.5**	12.9	28.4	11.7	**13.3**	25	**15.9**	14.6	30.5
Neutral	14.1	14	28.1	13.3	13.7	27	16.9	16.9	33.8
Uncomfortable	20.8	**22.7**	43.5	**27.6**	20.4	48	**19.8**	15.9	35.7
Total	100%			100%			100%		

From Table 11, it can be seen that for the vote on 'neutral', the difference in the percentage between active and passive samples is quite small, which is between 0.1% to 0.4%. Therefore, focus is given to votes on 'comfortable' and 'uncomfortable'. Votes on 'uncomfortable' is higher than on 'comfortable' and 'neutral' throughout the study for both active and passive samples. Therefore, it is suggested that since more samples felt uncomfortable throughout the survey period, this also means that outdoor environment in hot-humid climate were not preferable most of the time.

Based on the procedures applied, the following table shows the environmental conditions and parameters range as voted 'comfortable' by not less than 70% samples.

Table 12. Environmental range for "comfortable" votes (≥70%).

							WS		ST		
comfortable		time	DBT	WBT	RH	GT	min	max	floor	wall	avg ill
semi-shaded	C2	1100-1130	32.7	25.5	54.7	37.3	0.1	1.2	32	32.3	417.3
		1400-1430	31	25.8	66.3	34.7	0	0.5	44	42.7	895.3
	C3	1400-1430	30.7	25	63.3	32.7	0	0.4	30	30.3	5498.3
	TM	0900-0930	**27.5**	25.5	**85**	30	0	**0.1**	26	28	10945
open	C2	1130-	**33.7**	25.3	**50.7**	41.7	0	**2.8**	38	43.7	557.8

		120 0									

The dry bulb temperatures are averaged and giving 31.12 °C as the average DBT. The following thermal neutrality models are administered by applying the averaged dry bulb temperature. The results are compared as follows:

i. Thermal neutrality model 1 [19]:

$T_n = 17.6 + 0.31T_{ave}$

$= 17.6 + 0.31(31.12)$

$= \mathbf{27.25°C}$

(where T_{ave} is the outdoor average dry bulb temperature)

ii.Thermal neutrality model 2 [20]:

$T_n = 2.56 + 0.831T_m$

$= 2.56 + 0.831(31.12)$

$= \mathbf{28.42°C}$.

(where T_m is the mean temperature for the study on indoor comfort in tropical region, case study of Dhaka)

These results show differences in reading by 1.17 °C. When compared with the averaged dry bulb temperature itself (31.12°C), the difference for the former is 3.87 °C while the latter is 2.7 °C. Hence, it is suggested that the thermal neutrality model 2 is more appropriate for Kuala Lumpur as the value of 28.42°C falls within the range of dry bulb temperature found in Table 8.33, while thermal neutrality model 1 is situated outside the ranges of 27.5 – 33.7°C.

Referring to Table 12, it is reasonable to take a minimum of **27.5°C** and a maximum of **33.7°C** from the readings of dry bulb temperature and a minimum reading of **50.7%** and a maximum reading of **85%** for relative humidity to propose the comfort zone for hot-humid urban outdoor spaces of Kuala Lumpur, Malaysia. However, it should be remembered that the presence of wind speed of at least **0.1m/s** is a requirement in light of this scenario.

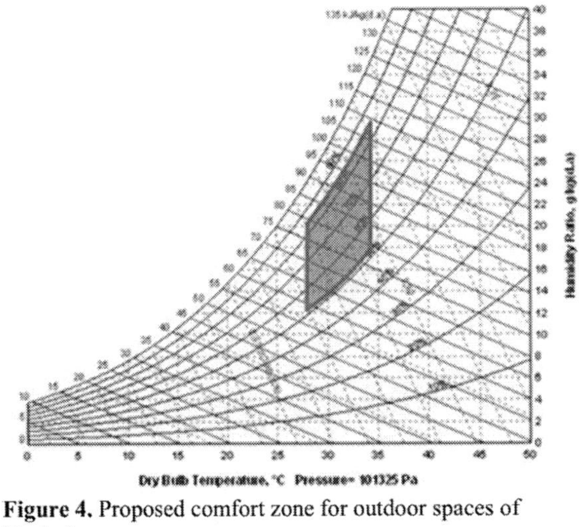

Figure 4. Proposed comfort zone for outdoor spaces of Kuala Lumpur

6 Conclusion

The proposed comfort zone of the outdoor environment of Kuala Lumpur can be used to monitor the environmental condition (climatically). If the climate range falls within the zone, then it is suggested that the outdoor environment is thermally comfortable most of the time for users.

Thermal comfort can be perceived as one of the tools used to study the quality of the evironment in which humans live. It may function as an indicator to study changes in the physical environment in relation to human comfort level. Hence, it is suggested that urban monitoring in terms of its climate and landscape changes for Kuala Lumpur must be conducted and recorded closely because, as the biggest city in a developing country like Malaysia, Kuala Lumpur could not stop from experiencing rapid urban form changes or perhaps to slow down its processes due to inter-related needs of the economy and built environment in particular. However, in an attempt to improve the quality of life through economic development, the lives of the current and the future generations should not be jeopardized.

Acknowledgement

The Ministry of Science, Technology and Innovation, Malaysia provided the scholarship while the International Islamic University granted the full-pay leave during this study.

References

1. Salleh, E., (1994), *Tropical urban outdoor environment and human thermal comfort*, The Architectural Association Graduate School, Ph.D. thesis
2. Ahmed, K.S., (1996), *Approaches to bioclimatic urban design for the tropics with special reference to Dhaka, Bangladesh*, The Architectural Association Graduate School, Ph.D. thesis.
3. Spagnolo, J., and de Dear, R., (2003), A field study of thermal comfort in outdoor and semi-outdoor environments in subtropical Sydney, Australia, *Building and Environment*, Vol. **38**, pp. 721-738.
4. Chun, C., Kwok, A., and Tamura, A., (2004), Thermal comfort in transitional spaces – basic concepts: literature review and trial measurement, *Building and Environment*, Vol. **39**, pp. 1187-1192.
5. Johansson, E., (2006), Influence of urban geometry on outdoor thermal comfort in a hot dry climate: A study in Fez, Morocco, *Building and Environment*, Vol. **41**, pp. 1326-1338.
6. Johansson, E., and Emmanuel, R., (2006), The influence of urban design on outdoor thermal comfort in the hot, humid city of Colombo, Sri Lanka, *Int. Journal of Biometeorology*, Vol. **51**, pp. 119-133.
7. Nikolopoulou, M., and Lykoudis, S., (2006), Thermal comfort in outdoor urban spaces: analysis across different European countries, *Building and Environment*, Vol. **41**, pp. 1455-1470.

8. Website of the Meteorological Dept. of Malaysia, [online]. Available at: <http://www.kjc.gov.my> [March 2007].
9. Marcus, C.C., Francis, C., and Russell, R., (1998), Urban plazas, In: Marcus, C.C., and Francis, C. eds., *People places: design guidelines for urban open space* (2nd edition), New York, Van Nostrand Reinhold, pp. 13-84.
10. Hyde, R., (2000), *Climate responsive design: a study of buildings in moderate and hot humid climates*, London, Spon Press.
11. Reynolds, J.S., (2002), *Courtyards: aesthetic, social, and thermal delight*, Canada, John Wiley & Sons, Inc
12. Muhaisen, A.S., (2005), *Prediction of the solar performance of courtyard buildings with different forms and various climatic regions, using a new computer model*. Ph.D. thesis, University of Nottingham.
13. Doulos, L., Santamouris, M., and Livada, I., (2004), Passive cooling of outdoor urban spaces. The role of materials, *Solar Energy*, Vol. **77**, pp. 231-249.
14. Taha, H., (1997), Urban climates and heat islands: albedo, evapotranspration, and anthropogenic heat, *Energy and Buildings*, Vol. **25**, pp. 99-103.
15. Rosenfeld, A.H., Akbari, H., Romm, J.J., and Pomerantz, M., (1998), Cool communities: strategies for heat island mitigation and smog reduction, *Energy and Buildings*, Vol. **28**, pp. 51-62.
16. ISO 7730: 1994(E), *Moderate thermal environments – Determination of the PMV and PPD indices and specification of the conditions for thermal comfort.*
17. ASHRAE **55**, (1992), *Thermal Environmental Conditions for Human Occupancy*, American Society of Heating, Refrigerating and Air-Conditioning Engineers.
18. Fanger, P.O., (1970), *Thermal comfort: analysis and applications in environmental engineering,* Copenhagen, Danish Technical Press.
19. Square One, Human Comfort-comfort prediction, [online]. Available at: <http://www.squ1.com/site.html> [8 June 2005].
20. Ahmed, Z.N., (1995), Temperature standards for the tropics?, In: Nicol, J. F., Humphreys, M., Sykes, O. and Roaf, S., eds., (1995). *Standards for thermal comfort: indoor air temperature standards for the 21st century (1st edition)*, London, Chapman & Hall, pp. 31-39.

Sustainability and the facilities management in Malaysia

Asra Zaliza Asbollah[1a], Nordiana Mohd Isa[2] and Syahrul Nizam Kamaruzzaman[2]

[1]*Department of Geography, Faculty of Arts and Social Science, University of Malaya, Kuala Lumpur. Malaysia.*
[2]*Department of Building Surveying, Faculty of Built Environment, University of Malaya, Kuala Lumpur. Malaysia.*

Abstract. Facilities Management (FM) in the industry of environment involves numerous expertise, especially from the management side. Other than that, technology and finance are the other factors involved as well. One essential aspect of FM, other than the emphasis on technical operation, is its performance. In parallel, the performance does impact occupant behaviour and, at the same time, this performance does affect the environment. In short, this indicates that FM is in a key position to participate in delivering a sustainable environment for the industry of built environment. Sustainable facilities Management (SFM) is crucial because buildings consume more resources which will, in consequence, negatively impact the environment and generate large amounts of waste. This justifies the importance of sustainability under the umbrella of facilities management. However, FM is quite new in Malaysia's environment. Government agencies, such as JKR, have adopted and are practicing FM at the moment. Fortunately, there has been an increasing trend and awareness of SFM adoption. Therefore, this paper aims to understand and identify the contribution and practices of Sustainable Facilities Management (SFM) in Malaysia; focusing on the development taken in regards to SFM.

Keywords: sustainability, facilities management, Malaysia

1. Introduction

There are several definitions of Facility Management (FM), which are as follows. One that is commonly used is "an integrating approach to operating, maintaining, improving and adapting the buildings and infrastructure of an organisation in order to create an environment that strongly supports the primary objectives of that organisation" [1]. Before that, in 2004, the International Facility Management Association defined FM as a career that includes multiple disciplines to ensure the role of the built environment to work through the integration of people, place, process and technology. A previous study by Hamer [5] explained that FM is the process of planning, implementation, maintenance and taking into account the suitability of physical space and services for the organisation, and looking at related cost savings. Thus, in short, FM is the scope of the management of buildings and facilities, thoroughly and effectively involving the administration, finance, operation and maintenance of the building, while giving priority to the maintenance of the functionality of the building; this is to ensure that each component of the building and facilities are provided at optimum functionality based on the routine maintenance schedule drawn up.

FM is usually used in building maintenance, utility supplies, human resources management; or in other words, covering a wide range of services including real estate management, health and safety, as well as, contract management. Nevertheless, at an earlier stage, FM was seen as having poor relations with those sectors and other sectors such as architecture, engineering and construction professions. This was due to the old-fashioned sense of care-taking, cleaning, repairing and maintaining [15]. However, recently, there has been substantial investments towards organisations, usually to accommodate and support the wide range of activities of FM. FM as an integrated management approach, involving multi-disciplinary experts in ensuring the use of space, infrastructure, human resources and organisations optimised to deliver a purposeful service. Thus, it would be difficult to measure the productivity of FM. The issue of sustainability, on the other hand, is undoubtedly very broad and it is important for FM to emphasise sustainability aspects in their management. This justifies the importance of sustainability under the umbrella of facilities management.

The goal of implementing sustainability is to create transparency and cooperation [4]. Through transparency and cooperation, everyone is given full

[a]Corresponding author: ezraliza@hotmail.com

opportunity and are encouraged to actively participate in the decision-making process regarding development in the various stages such as planning, management and implementation. In addition, the related decision-making process must be done sincerely and transparently. The ultimate goal is to strengthen the involvement, responsibility and role of diverse groups of people based on various stages of planning, implementation, management and decision-making; and the basis for strengthening the integration of various parties is with the interest of achieving a pattern of sustainable development.

Thus, this paper is primarily concerned with that particular situation in the FM perspective. The situation in other industries or areas are obviously very different, and they may possess different priorities. For example, in tourism, maritime changes are more important than urban development. However, the reasons for sustainability are similar wherever and whatever the context is, and this paper attempted to provide some understanding of the background to the often complex issues under discussion.

2. Sustainability

Apparently, there are uncountable definitions of 'sustainability', each with its own emphasis and subtleties. As in the Brundtland Report [17], sustainability "meets the needs of the present without compromising the ability of future generations to meet their own needs". Basically it shows that the concept of continuity and fairness is available for present and future generations. Fairness, in other words, was explained as greater fairness within and among societies, or more appropriate levels of economic development and/or the need for more environmental protection. Later, in 1992, The Earth Summit on Environment and Development held in Rio de Janeiro resulted in the Agenda 21 Declaration which provided 27 principles that set out the ways in which countries should contribute to global sustainability. Many of these principles are of direct relevance to the use and management of urban environments and property in general.

The desire to realise the sustainability aspects, as stated by [4], is a shared responsibility that includes the State Government, non-governmental organisations (NGOs) and all the people. Together, they (referred to the group mentioned) share responsibility for the planning, managing, implementing and monitoring sustainable development; based on the goals, objectives and a clear aspiration for mutual benefit and prosperity. A comprehensive method is needed to ensure the sustainability concept is seen and approached as a framework that is based on the problem and not the effects of the problem. They continue to create ecosystems and the environment is seen as an ecosystem that contains elements of natural heritage, life support systems and natural resources; if used wisely, it will guarantee and improve overall welfare and social security. In addition, social harmony is to prioritise harmony and tolerance between people, caring for the elderly, youth, children and people with disabilities and without any distinction between genders, socio-economic differences and populations between urban and rural areas. Concern for the future includes the opinion that the current generation has while managing resources in a way that allows the retention of the management options of future generations; and development which emphasises the use of natural resources in a sustainable and focused measure of environmental protection is also needed.

In parallel, as Keeping and Shiers [7] mentioned, social and economic dimensions were among the global nature being discussed under sustainability. For example, long-term economic prosperity is said to depend on environmental sustainability. The consequences of short-term economic thinking and the reckless use of the planet's resources, such as water disputes between neighbouring nations, is alarming. Although these dimensions are mostly discussed among researchers and the public, people are still unaware of how these issues could affect the environment. As Lehtonen [10] explained, these dimensions are 'independent elements that can be treated, at least analytically, as separate from each-other' (p. 201). Thus, sustainability has moved from being an interesting but sometimes contested ideal, to becoming the acknowledged goal of much of international policy, including the climate change policy. It is no longer a question of whether any development policy should be understood in the context of sustainable development goals; it is a question of how.

Another example, the construction of the Bakum Dam in Sarawak, had a huge impact; especially on the indigenous people who had to relocate. Keong [8] also reported a total ecological destruction on 69,640 hectares of forest ecosystem which brought devastation to the environment and society. For that reason, it is important to identify the steps to achieve sustainability, and appropriate efforts must be actively engaged.

From the aspect of sustainable development, there are indicators to measure the progress. The indicators comprise of three sustainable development themes, such as: the environment, society and economy. In parallel, many communities have developed sustainability or quality of life indicators that effected the works of developers who run the development. All the indicators will normally be placed under few themes. For example, in Bristol, seven themes have direct relevance to property development:

- Community safety
- Energy
- Environmental protection
- Housing and shelter
- Land use and development
- Transport
- Waste management

However, the existing approaches to sustainable development appears to focus on technical issues and it seems that the approach undertaken is a reactive approach; it is only implemented when there is an occurrence of a disaster. Unplanned development and the public's low attitude are among the causes that lead to environmental problems, especially in developing countries [13].

3. Sustainability And The Facility Management In Malaysia

In the Malaysian context, FM can be interpreted according to Noor Azman, Abdul Hakim and Buang Alias [14], as a practice to coordinate the operation of a physical nature and the human environment in an organisation, uniting the whole principle of business administration, as well as, process design and engineering matters related to science. In parallel, FM practices have great potential to grow and prosper, especially in the construction industry in Malaysia. Maimunah and Maizan [12] mentioned that the number of physical development and the undertaking of large sized developments are two factors that describe the volume of physical facilities that should be managed. It starts from the planning stage until the economic life span. Service providers involved in the management of physical facilities (specialty consultants, contractors, suppliers of materials and equipment, labour and general manager of the facility itself) can seize great opportunities that exist; particularly in terms of providing an integrated facility management (integrated) or comprehensive (total) to local organisations. Thus, it can be seen that the FM scale will be able to provide an efficient and effective service, especially in terms of cost. However, only government agencies, such as JKR, have adopted and are practicing FM at the moment, and this shows that the government has played an important role in the development of FM in Malaysia [15].

Initially, FM in Malaysia did not have a clear concept and direction. As clearly emphasised by Zakaria and Hamzah [18], this was believed to be due to the confusion and lack of understanding of the real concept of FM. One question that often arises is whether FM is part of a property management or otherwise. However, the existence of the Malaysian Association of Facility Management (MAFM), which was initially formed in 2005, helped strengthen all aspects of FM. Its major agendas is to bridge academicians with industry players on the knowledge-sharing platform by strengthening all aspects of FM, leading all activities related to FM and MAFM and establishing a valid and legal society in Malaysia [11]. Among the countries that have professional organisations which are actively developing FM in Asia would be Hong Kong Institute of Facilities Management (HKIFM) and Japan Facility Management Promotion Association (JFMA). In addition, the International Facility Management Association (IFMA) (United States) and the British Institute of Facilities Management (BIFM) are among the professional organisations that were first established.

Generally, FM activities (in the context of Malaysia) began after the completion of delivery to the owner or customer, and it covers every aspect of the performance and maintenance of buildings. The existence of three types of management system, namely, Quality Management System (EN ISO 9001: 2000), Environmental Management System (EN ISO 14001: 2004) and Security Unit and Health Management System (OHSAS 18001: 2007), has helped FM in providing the best service. This is because, by following the existing guidelines to some extent, it helped improve the performance of FM services.

With the increasing issue of climate change, sustainable facility management practices are also rapidly growing. Therefore, an increasingly important issue that FM professionals must take into account would be the surrounding environment. Environmental issues have become legislative issues in many areas, including FM. Potential environmental problems can cause numerous negative impacts, including additional costing for expensive environmental clean-up. As Elmualim et al. [2] emphasised, "Facilities managers have a great role in contributing to the reduction of the built environment impact on the environment and hence advancing the sustainability agenda across the three bottom lines of sustainability, the economic, environmental and social strands". Khalil et al. [9] added, "Regulations in Malaysia like ISO 14000 & 14001 and EQA 1974 standards must be well established by all relevant stakeholders and publics as to make it more evident. It is recommended that in order to achieve meaningful improvement, facility managers must obtain adequate knowledge and develop appropriate concern for environmental issues". In reference, professionals in FM, in short, play a vital role in providing great performance and excellent sustainable facilities.

As noted above, the FM in Malaysia only involve the management of the physical aspects of building facilities and technical services, or in other words, only process or support activities. With this support image, facility management services are not positioned as strategic resources of the organisation, since there are four other organisations such as people, time, finances and technology [12]. In addition, a study by Elmualim et al. [2] showed that time constraints, lack of knowledge and lack of senior management commitments are the main obstacles for the implementation of a sustainable FM policy that is consistent and comprehensive. The continuation of the lack of understanding, focus and commitment of senior executives in appreciating the opportunities, threats and needs and direction in driving essential change is a major obstacle to sustainable FM practices. However, in addressing environmental problems and meeting the demands of sustainable development, several recommendations were outlined, and it should begin at an earlier stage of development. Among the suggestions include the involvement of public participation and the emphasis of environmental protection programs at the pre-feasibility stage of development [9]. Hodges [6], beforehand, stated that in order to integrate sustainable practices with an organisation's desire for cost-effective FM, material durability should be integrated into the sustainable design process, and the facility manager acting as integrator, need to be integrated into the organisation's overall strategy. In short, in order to achieve the credibility for sustainability of FM, proper planning is significant; especially at a very early stage such as knowing the long-term costs and the methodology of managing physical assets.

4. Conclusion

Since sustainability issues have risen towards the top of the political agenda [16], the need for sustainable means of developing the environment has become increasingly urgent. Thus, a good understanding of development plan policy and sustainable development policy is required. In this regards, facility managers should play a major role in realising sustainability. Their roles are not only limited to managing, but also to facilitate the nation in achieving sustainability.

In fact, there is no doubt that Malaysia has achieved great success in making the country and citizens live in comfort and peace, from the aspect of the necessities of life and national development. However, despite comfort and success, the development is also a source of interference to all aspects of life and a clear example is the disturbance towards the environment. As noted above regarding sustainability, what the current generation is doing now will have an impact on future generations. Thus, the issue of sustainability must be taken seriously before it is too late. It is hoped that this paper will help the reader to better comprehend the relationship of sustainability issues.

Acknowledgement

The authors would like to thank the funding bodies of this research: the University of Malaya under FRGS Project. No. FP025-2014B.

References

1. Atkin, B. and Brooks, A. *Total Facilities Management*. Oxford: Blackwell Science Ltd, (2000).
2. Elmualim, A., et al. Barriers and commitment of facilities management profession to the sustainability agenda. *Building and Environment*, **45**, 58–64, (2010).
3. Finch, E & Zhang. *Facilities management, Design and Management of Sustainable Built Environments*. Springer, pp. 305-326, (2013).
4. Firdaus Al-Siddiq Fadzil and Zalina Haji Samadi. Penghasilan Model Pembangunan Mapan Untuk Komuniti Kampung Santubong, Sarawak. National Conference of Research Knowledge & Intellect Application 2009 in Universiti Technologi MARA, Perak. Hamer, J. M. (1988). *Facility Management Systems*. New Jersey: Van Nostrand Reinhold, (2009).
5. Hamer, J. M. *Facility Management Systems*. New Jersey: Van Nostrand Reinhold, (1988).
6. Hodges, C. P. A facility manager's approach to sustainability. *Journal of Facilities Management*, **3**(4), 312 – 324, (2005).
7. Keeping, M. & Shiers, D. E. *Sustainable Property Development: A guide to Real Estate and the Environment*. Oxford: Blackwell, (2004).
8. Keong, C. Y. Dam-Induced Development and Environmental and Social Sustainability: The Bakun Industrialization Strategy Revisited. *Journal of Economic Issues*, **39**(1), 123, (2005).
9. Khalil, N., et al. Sustainable Environment: Issues and Solutions from the Perspective of Facility Managers. *Procedia Engineering*, **20**, 458-465, (2011).
10. Lehtonen, M. The Environmental-Social Interface of Sustainable Development: Capabilities, Social Capital, Institutions. Ecological Economics, **49**(2), 199-214, (2004).
11. MAFM (Malaysian Association of Facility Management). Retrieved from http://www.mafm.org.my/store/ (2016).
12. Maimunah Sapri and Maizan Baba. *Penyumberan Luar dalam Pengurusan Fasiliti*. Skudai, Johor: UTM Press (2008).

DOI: 10.1051/matecconf/20166600085

13. Martinez, E. et al. Participative democracy and local environmental issues. *Journal of Ecological Economics*, 12, 83 – 94, (2008).
14. Noor Azman Mohamat Nor, Abdul Hakim Mohammed and Buang Alias. Facility Management History and Evolution. *International Journal of Facilities Management*, **5** (1), (2014).
15. Syahrul Nizam Kamaruzzaman and Emma Marinie Ahmad Zawawi. Development of facilities management in Malaysia. *Journal of Facilities Management,* 8(1), 75 – 81, (2010).
16. United Nations. World Economic and Social Survey 2013: Sustainable Development Challenges. Department of Economic and Social Affairs, United Nations: New York, (2013).
17. WCED (World Commission on Environment and Development). *Our Common Future.* Oxford: Oxford University Press, (1987).
18. Zakaria Azman and Hamzah Saufi. *Pelan Kualiti Hidup Dapat Elak Projek Hadapi Masalah: Majlis Aset Kebangsaan Mampu Tingkatkan Kecekapan Penyelenggaraan*. Berita Harian, Aug 15, 2007, p. 2, (2007).

The relation between indoor environnemental quality (IEQ) and energy consumption in building based on occupant behavior - A review

Iman Asadi[1, a], Norhayati Mahyuddin[1] and Payam Shafigh [1]

[1] *Department of Building Surveying, Faculty of Built Environment, University of Malaya, 50603 Kuala Lumpur, MALAYSIA*

Abstract. Indoor Environmental Quality (IEQ) is an important topic which impacts on occupant health, productivity and also energy consumption in buildings. The four main parameters for IEQ evaluation are: Thermal comfort, indoor air quality, visual comfort and aural comfort. The occupant behavior in buildings defines as any direct or indirect act which an occupant selects to change the displeasure environmental condition into the comfort conditions. The selected behavior by human has a significant impact on the energy consumption in buildings. This paper reviews the methods which used to simulate IEQ parameters, energy consumption and human behavior in buildings. It aims to promote the idea of more consideration about the relation between occupant behavior and energy usage in buildings. This summary of existing studies about the importance of human behavior factor in energy simulation software helps to identify new methods and strategies for simulating IEQ, Energy and behavior.

Keywords: indoor environmental quality (IEQ), energy consumption, human behavior

1 Introduction

These Most people spend more than 90% of their time indoors [1]. The importance of Indoor environmental quality (IEQ) is it impacts on energy efficiency and also productivity and health of occupants. Thermal comfort, indoor air quality, visual comfort and aural comfort are four main physical parameters which are normally used to determine IEQ. There will be a rising focus on energy consumption and indoor environmental quality in existing buildings and the future buildings. There is a significant relation between energy consumption and the features which are considered in the indoor environment (temperature, ventilation and lighting) and building design and operation. According to the recent studies the cost of poor indoor environment for the employer, the owner and also for the society, as a whole are frequently considerably higher than the cost of the energy which are used in the same building [2] .

The "behavior" refers to any direct or indirect action that a building occupant selects to manage their surrounding environment. Human behavior is maybe the most important IEQ study topic because human activities and decisions deeply influence every IEQ determinant at every stage of building's lifecycle. Through understanding human behavior the energy prediction models will be improved .The problem is measurement and quantification the human behavior and translation that into the energy algorithm. The technology of simulation is based on several disciplines like as computer and numerical science, physics, mathematic and so on. For simulating human behavior in building, social science modeling can be a useful method.

The aim of this paper is to review the relation between IEQ and energy consumption in building based on Occupant behavior. Also, the human and energy simulation methods are considered.To achieve the objective of this paper, the related studies to the IEQ and building energy usage based on human behaviour are identified through internet searches, review of published papers and conversation with other researcher.The results and methods of these researches areconsidered and the simulated software and methods are analysed.

2 Indoor environmental quality (IEQ)

Based on United States Green Building Council (USGBC), Water efficiency, materials and resources, energy and atmosphere, sustainable sites and indoor environmental quality should be consider for a green building. Measuring of IEQ is commonly validate by four aspects: i.e. Thermal comfort (TC), Indoor air quality (IAQ), Aural comfort (AC) and Visual comfort (VC) [3, 4].

2.1 Thermal comfort

[a] asadi8564@gmail.com

Thermal comfort is defined "that condition of mind which expresses satisfaction with the thermal environment" according ISO 7730 standard definition and also ASHRAE Standard 55,states that, success means that a building meets the needs of 80% of the occupants [5, 6] . Air temperature (°C), mean radiant temperature (°C), air velocity (m/s) and relative humidity (%) are the four environmental parameters which influence thermal perception of occupants while, metabolic rate (Met) and clothing value (Clo) are considered as personal parameters [7] .Air temperature, air velocity and relative humidity are measured directly. Equation for calculating mean radiant temperature is:

$$T_{mrt}= (\Sigma F_{si}T_i^4)^{1/4} \quad (1)$$

Where T_i=Temperature of black surface, F_{si}=fraction of radiation of the black sphere which reaches the ith

Each person has a special metabolism rate. These rates are able to change according to the activities done, or under special environmental situations. The movement of air and relative humidity can reduce the insulating ability of the cloth depending on the sort of cloth [8]. Based on mentioned parameters the Predicted mean vote (PMV) and predicted percent of dissatisfied (PPD) indexes are defined.

2.2 Indoor air quality

The Indoor air quality (IAQ) is related to the health and comfort of occupants and poor IAQ is related to the Sick Building Syndrome (SBS). The change amount of pollutant per volume of air over time is expressed by:

$$VdC_i /dt = P_i - (ventilation) - (exfiltration) - (sorption, decay, cleaning) \ [\mu g/s] \quad (2)$$

Where: V = volume of space (m^3), C_i = concentration of pollutant in indoor air ($\mu g/m^3$)

There are different group of chemical and biological pollutants in indoor air .There are different research which considered CO_2 concentration as an IAQ Index [9-11]. The equation to calculate ventilation rate for considering IAQ based on CO_2 concentration is:

$$V_{rate} = \frac{EX\ rate}{CO2in-CO2out}, \quad (3)$$

Where V rate = effective ventilation rate in cfm of Outdoor Air (OA) per person ,EX rate = average exhalation rate of CO2 per person in the measured area,CO2in = measured inside concentration level of carbon dioxide, CO2out = outside carbon dioxide concentration, measured or estimated.

2.3 Aural comfort

The direct and diffuse indoor sound pressure which is perceived depends on characteristics of the space like as volume of space. In the field of diffuse the sound pressure level perceived is:

$$L_p \cong L_w + 10log\ (4/A)\ [dB] \quad (4)$$

Where L_p= sound pressure level, L_w = sound pressure level in point, A = surface area of the enclosed space .

The sources of noise (when sound disturb the occupant) are indoor sources like as HVAC systems, sanitary, drinking water, flank transmission through the pipe systems into the building structure installations and outdoor sources like as traffic jam.

2.4 Visual comfort

Lighting (counting as luminance, difference between clarify surfaces, and the spectrum of light) can influence residents' health and efficiency both in a straight line, since work efficiency depends on vision, and not directly, because lighting can straight attention, or affect motivation[12]. The daylight and artificial light are two main sources of light. Based on following equation the amount of direct lighting level is [13]:

$$E = n\ \Phi\ \eta/(l \times b) \quad (5)$$

Where: n = number of light sources; Φ= light flux of all light sources (lumen); η= output (efficiency) of light source; l = length of space (m); b = width of space (m).

2.5 Overall IEQ acceptance

However, each IEQ parameter has an important effect on occupant comfort and productivity, perceive of overall IEQ acceptance is more complicated. There are several studies which assess the IEQ acceptance among occupant by using a multivariate logistic regression [4, 9, 10].The demonstrated model is:

$$\acute{\sigma}_0=1-\frac{1}{1+\exp(C_{0,0}+\sum_{i=1}^{4}C_{i,0}\acute{\sigma}_i(\zeta_i))} \quad (6)$$

Where: $\acute{\sigma}_0$ = the overall IEQ acceptance, $C_{0,0}$ and $C_{i,0}$ are the regression constants, $\acute{\sigma}_i$ is the occupant acceptance related to the thermal sensation vote ζ_1, the CO_2 concentration ζ_2(ppm), the illumination level ζ_3(lx) and the sound pressure level ζ_4(dBA).

The thermal environment acceptance $\acute{\sigma}_1(\zeta_1)$, is :

$$\acute{\sigma}_1=0.95\ exp\ [-(C_{0,a}\zeta_1{}^2 + C_{1,1}\zeta_1{}^4)] \quad (7)$$

And the acceptance of $\acute{\sigma}_2(\zeta_2)$, $\acute{\sigma}_3(\zeta_3)$ and $\acute{\sigma}_4(\zeta_4)$ are demonstrated by below regression when $C_{0,j}$ and$C_{1,j}$ are regression coefficient :

$$\acute{o}_j = 1 - \frac{1}{1+\exp(C_{0,j} + C_{1,j}\zeta_j)}, j = 2,\ldots,4 \quad (8)$$

3 Human behavior & building energy consumption

The amount of energy consumption in residential and commercial buildings was about 41 % of total energy usage in U.S in 2014. As illustrated in Figure 1, according to Hoes et al.[14], Energy use in buildings is closely linked to the behavior of occupants in building. The researches on thermal comfort and adaptive control [15] , lighting control [16] , possible window control [17] , and shading control [16] are some of the research subjects which began to assess the occupant behavior that influences in building operation and energy efficiency.

Behavioral in buildings are intricately linked with multiple factors that stem from the physical, cultural, psychological, social, and so on [18]. Some studies have also suggested that the perceived ability or inability to behave adaptively in the field has a psychological effect that must be considered by thermal comfort calculations [15, 19, 20]

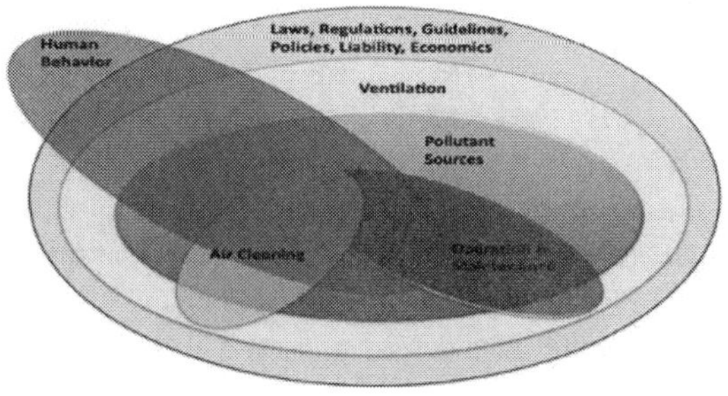

Figure 1. Inter relationships of IEQ determinants. Human Behavior factors affect all other determinants (Source:T.J. Phillips).

3.1 IEQ simulation tools

The aim of this part is to consider simulation tools and software which are using to predict occupant satisfaction, energy efficiency and occupant comfort in buildings. The simulation of building is a good strategy for testing building efficiency in different conditions[21] .

Estimating the amount of energy consumption and designing the HVAC system is the most common usage of simulations. In some studies IEQ aspects are considered based on temperature set points and ventilation rate[22] .There are different developed computational models for simulating each IEQ aspect [21, 23, 24]. Most of these simulating models consider each IEQ aspect separately and the inability to consider the interaction of a specific IEQ parameter with other IEQ Parameter is the drawback of these methods[25].Based on U.S Department of Energy (DOE2015) , just 2 of the 417 simulation software in the Building Energy, are capable to simulate all four IEQ aspects. Loonen et al. reviewed different building performance simulation (BPS) programs and their capability to calculate the various IEQ parameters. This overview is shown in Fig 2.

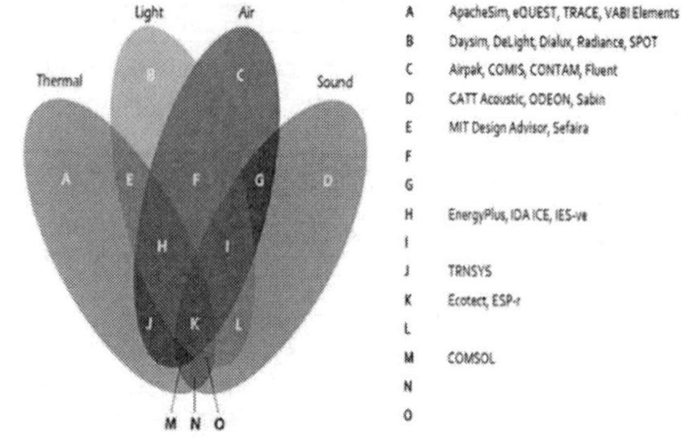

Figure2. overview of different BPS programs [22].

3.2 Human behavior

The "behavior" is the actions or reactions of an object, usually in relation to its environment. In the most research, the presence of occupant and heat generation and energy consumption by occupant are considered without taking account the way which they behave to use energy [26]. Considering the actual behavior of occupant in simulation methods leads to more realistic simulation. Recognition information within humans are captured by many behavior illustration models[27] .

The human behavior representation model based on earlier studies are summarized in Table 1.Based on Langevin suggestion [28] , there are different steps to acheive better understand about the concept of human behavior simulation. These suggested steps are:

a) Additional longitudinal field study
b) Individually behavior modeling
c) Assessment of behaviors and their sequencing
d) Behavior modeling in multiple building types
e) An accessible behavior/energy co-simulation structure

Table 1.Summary of human behavior representation model

Study	Proposed model	Description
[27]	ACT (atomic components of thoughts)	focuses on the way which humans organize their knowledge in order to behave intelligently
[29]	CAPS (concurrent activation based production system)	Considering each specifying the conditions and consequent actions

[30]	COGNET (cognition as a network of tasks)	Consider cognitive behavior by assumption of human capability to perform multi tasks
[31]	CCT (cognitive complexity theory)	models human performance as Goals, Operators, Methods and Selection rules
[32]	DCOG (Distributed cognition)	explain that cognition is distributed through the environment
[33]	EPIC (executive process/interactive control)	Focus on the perceptual and intellectual.
[34]	APEX (architecture for procedure execution)	Consider human behavior in dynamic environment
[35]	Brahms (Business redesign agent-based holistic modeling system)	Considering human social behavior over times
[36]	HOS (human operator simulator)	a model of human abilities and restrictions to support the design of human-machine systems

4 Discussions

Thermal comfort, Indoor air quality, aural comfort and visual comfort are the four main parameter to evaluate the IEQ in buildings. The poor IEQ is caused sick building syndrome, less human productivity and also more energy consumption.One the most important reason for considering IEQ in buildings is reducing the amount of energy usage.However, the heating, cooling, ventilation, water heating, lighting and electronics equipment are the main reasons for increasing building energy consumption , but the occupant activities affect on all of them.Occupant behavior is a physical and also psychological response to the environmental conditions to manage the dissatisfied conditions .This is why , the human behaviour is closely related to the energy usage in residential and commercial buildings and should be consider as an important factor in the field of built environment.

5 Conclusions

Based on review different paper in the field of IEQ it is concluded that, most researches were considered the relation between just one or two IEQ parameters and occupant satisfaction. Some researcher promoted a multivariate logistic regression to predict the overall IEQ acceptance based on subjective and objective measurement.These studies focused on occupant satisfaction and the lack of a relationship between overall IEQ parameter and energy consumption is obvious.Other studies used different type of simulation methods and softwares to predict the amount of energy consumption by considering one or two IEQ parameters. The disability of simulating all IEQ factors at the same time and considering the affect of human behavior is the drawback of these methods.Also, some study promoted different models to simulate the human behavior. In a simulation process, considering physical, psychological and social behavior of occupant is required. Since the human behavior changes the energy usage in building, the aim of this consideration is to measure and predict the occupant behavior while facing with uncomfortable conditions. The complexity and uncertainties of occupant behavior leads to lack of behavior prediction and feedback in simulating software. The current simulation programs have a limitation to predict dynamic behavior of occupant and just supported implausible forms of human behavior.

Therefore, the development of existing knowledge to be used in considering and simulating human behavior in various methods and different kind of buildings with different IEQ conditions provide a valuable contribution in the energy simulation programs.Also, it is suggested to use an algorithm to consider building energy consumption based on human behavior by considering all IEQ factors

References

1. De Giuli, V., O. Da Pos, and M. De Carli, *Indoor environmental quality and pupil perception in Italian primary schools.* Building and Environment, 2012. **56**: p. 335-345.
2. Olesen, B.W., *Revision of EN 15251: indoor environmental criteria.* REHVA European HVAC Journal, 2012. **49**(4): p. 6-12.
3. Clausen, G. and D.P. Wyon, *The combined effects of many different indoor environmental factors on acceptability and office work performance.* HVAC&R Research, 2008. **14**(1): p. 103-113.
4. Lai, A., et al., *An evaluation model for indoor environmental quality (IEQ) acceptance in residential buildings.* Energy and Buildings, 2009. **41**(9): p. 930-936.
5. Standard, A., *55: Thermal Environmental Conditions for Human Occupancy American Society of Heating.* Refrigeration and Air Conditioning Engineers, Atlanta, USA, 1992.
6. ISO, C., *Ergonomics of the thermal environment-Analytical determination and interpretation of thermal comfort using*

calculation of the PMV and PPD indices and local thermal comfort criteria. 2005, Brussels. European committee for standardization.

7. Asadi, I., I. Hussein, and K. Palanisamy. *Analysis on Thermal Comfort of Air-Conditioned Buildings in Malaysia: Case Study of Universiti Tenaga Nasional*. in *Applied Mechanics and Materials*. 2014. Trans Tech Publ.
8. Havenith, G., *Heat balance when wearing protective clothing*. Annals of Occupational Hygiene, 1999. **43**(5): p. 289-296.
9. Cao, B., et al., *Development of a multivariate regression model for overall satisfaction in public buildings based on field studies in Beijing and Shanghai*. Building and Environment, 2012. **47**: p. 394-399.
10. Wong, L., K. Mui, and P. Hui, *A multivariate-logistic model for acceptance of indoor environmental quality (IEQ) in offices*. Building and Environment, 2008. **43**(1): p. 1-6.
11. Asadi, I. and I. Hussein, *Indoor Air Quality (IAQ) Acceptance in Universiti Tenaga National*.
12. Hygge, S. and I. Knez, *Effects of noise, heat and indoor lighting on cognitive performance and self-reported affect*. Journal of Environmental Psychology, 2001. **21**(3): p. 291-299.
13. Visser, R., *Verlichting en interieur*. 1992: Dekker/vd Bos & Partners.
14. Hoes, P., et al., *User behavior in whole building simulation*. Energy and buildings, 2009. **41**(3): p. 295-302.
15. Brager, G.S. and R.J. de Dear, *Thermal adaptation in the built environment: a literature review*. Energy and buildings, 1998. **27**(1): p. 83-96.
16. Bourgeois, D., C. Reinhart, and I. Macdonald, *Adding advanced behavioural models in whole building energy simulation: a study on the total energy impact of manual and automated lighting control*. Energy and Buildings, 2006. **38**(7): p. 814-823.
17. Rijal, H.B., et al., *Using results from field surveys to predict the effect of open windows on thermal comfort and energy use in buildings*. Energy and Buildings, 2007. **39**(7): p. 823-836.
18. Lee, Y.S., Y.K. Yi, and A. Malkawi. *Simulating human behaviour and its impact on energy uses*. in *12th Conference of International Building Performance Simulation Association, Sydney Austrialia*. 2011.
19. Leaman, A. and B. Bordass, *Are users more tolerant of 'green'buildings?* Building Research & Information, 2007. **35**(6): p. 662-673.
20. Simonson, C., M. Salonvaara, and T. Ojanen, *The effect of structures on indoor humidity–possibility to improve comfort and perceived air quality*. Indoor Air, 2002. **12**(4): p. 243-251.
21. Hensen, J. and R. Lamberts, *Introduction to building performance simulation*. Building performance simulation for design and operation, 2011: p. 1-14.
22. Loonen, R., M. Loomans, and J.L. Hensen, *Towards predicting the satisfaction with indoor environmental quality in building performance simulation*. 2015.
23. Zhang, Z. and Q. Chen, *Prediction of particle deposition onto indoor surfaces by CFD with a modified Lagrangian method*. Atmospheric Environment, 2009. **43**(2): p. 319-328.
24. Vorländer, M., *Computer simulations in room acoustics: Concepts and uncertaintiesa)*. The Journal of the Acoustical Society of America, 2013. **133**(3): p. 1203-1213.
25. Citherlet, S., J. Clarke, and J. Hand, *Integration in building physics simulation*. Energy and Buildings, 2001. **33**(5): p. 451-461.
26. Kashif, A., et al., *Simulating the dynamics of occupant behaviour for power management in residential buildings*. Energy and Buildings, 2013. **56**: p. 85-93.
27. Anderson, J. and C. Lebiere, *The atomic components of thought Lawrence Erlbaum*. Mathway, NJ, 1998.
28. Langevin, J., *Human Behavior & Low Energy Architecture: Linking Environmental Adaptation, Personal Comfort, & Energy Use in the Built Environment*. 2014.
29. Just, M.A., P.A. Carpenter, and S. Varma, *Computational modeling of high-level cognition and brain function*. Human Brain Mapping, 1999. **8**: p. 128-136.
30. Zachary, W., J.M. Ryder, and J.H. Hicinbothom, *Cognitive task analysis and modeling of decision making in complex environments*. Making decisions under stress: Implications for individual and team training, 1998: p. 315-344.
31. Card, S.K., A. Newell, and T.P. Moran, *The psychology of human-computer interaction*. 1983.
32. Eggleston, R.G., M.J. Young, and K.L. McCreight. *Distributed cognition: A new type of human performance model*. in *Proceedings of the aiaa fall symposium series technical reports. North Falmouth, Massachusetts: American Institute of Aeronautics and Astronautics.(Cité page 41.)*. 2000.
33. Kieras, D.E. and D.E. Meyer. *Predicting human performance in dual-task tracking and decision making with computational models using the EPIC architecture*. in *Proceedings of the First International Symposium on Command and Control Research and Technology, National Defense University, June. Washington, DC: National Defense University*. 1995.
34. Freed, M.A., *Simulating human performance in complex, dynamic environments*. 1998, Northwestern University.
35. Sierhuis, M., W.J. Clancey, and R.J. Van Hoof, *Brahms: a multi-agent modelling environment for simulating work processes and practices*.

International Journal of Simulation and Process Modelling, 2007. **3**(3): p. 134-152.
36. Wherry Jr, R.J., *The human operator simulator—HOS*, in *Monitoring behavior and supervisory control*. 1976, Springer. p. 283-293.

 DOI: 10.1051/matecconf/20166600087

The Key Components Of Knowledge Transfer For Problem Solving In Adaptive Reuse Projects: A Qualitative Study

Kartina Alauddin[1,a], Mohd Fisal Ishak[2] , Mohammad Nasharudine Shuib[3] and Halmi Zainol[4]

[1,2,3,4]*Faculty of Architecture Planning and Surveying, Universiti Teknologi MARA, Seri Iskandar Campus, Seri Iskandar, 32610, Perak, Malaysia.*

Abstract. The adaptive reuse project requires complete and accurate historical information and high levels of expertise from the project team members. Within the typology of adaptive reuse projects, the ability to capture and transfer the unique set of experience from project to project is important to create the relevant knowledge. Previous adaptive reuse studies focused on project success, potential for reuse, benefits, and sustainability. However, research related to creating and transferring skills and knowledge from adaptive reuse project settings has received relatively little attention to date. The study specifically investigates the key components of knowledge transfer for the problem solving process in adaptive reuse projects. In particular, the paper seeks to establish an intellectual capital framework for adaptive reuse projects that could help the project team members transfer their knowledge during problem solving process. This study adopted the qualitative approach to data collection. The data were collected from 14 semi-structured in-depth interviews across two case studies. Data from the interviews were analyzed using content analysis for within-case analysis with the aid of Nvivo Version 9. The finding of the study highlights that the key factors of transferring knowledge were effectiveness and efficiency, similar project teams, project teams actions and sources of information on transferring of intellectual capital among project team members. The contribution of this study is the development of an intellectual capital framework for adaptive reuse projects. This framework can be used as a template for future practice and research in the field of adaptive reuse and knowledge management. In conclusion, the study enables the integration of intellectual capital and knowledge transfer theory within the specific contextual adaptive reuse project settings.

1 Introduction

Involvement in adaptive reuse projects is not an easy process. Project teams have to go through various stages or processes that are quite different from the construction of a new building. The nature of each construction projects is temporary and could contribute to loss of skill and knowledge. The challenge is to ensure that the richness of skill and knowledge in preservation practices are kept and well preserved. The skill and knowledge of the project team members are very valuable for the enhancement of problem solving capability in the next project. Therefore, managing the skill and knowledge gathered from a project is important to ensure that the knowledge and experience of the project teams are not lost for future adaptive reuse projects. In the process of redevelopment of old buildings, the adaptive reuse concept has been applied to ensure that the new building function is compatible with the old building historical evidence. This method involves the adaptation of the building which started with the acquisition and ended with the occupation of the building (Latham 2000). The uniqueness of adaptive reuse concept is that it often relates to heritage architectural significance of the building involved. The adaptive reuse projects often raise complexity in the design and construction process as compared to new building projects. This includes incomplete and inaccurate design information (Shipley, Utz & Parson 2006; Karim et. al 2007) and lack of capable professional expertise (Ball 1999; Kurul 2007) that affect the design and construction process of this kind of project. Moreover, in the design and construction phases, availability of historical information, such as original drawings and documents related to the buildings are important in the design and construction process (Waleed, Yasser, Lobna and Ashraf, 2015). Hence, it is important for each practitioner in the adaptive reuse projects to understand and embrace the knowledge of heritage requirements in the problem solving process.

The research problem is concerned with the lack of research related to knowledge management in adaptive reuse projects. Previous studies have been defined individual and unrelated finding between adaptive reuse and knowledge management. Therefore, the knowledge management approach will be adopted as this study in regards to identifying the key components in problem solving within knowledge creation from project to project over time. This is critically important to be applied by the

[a] Corresponding author: karti540@perak.uitm.edu.my

 DOI: 10.1051/matecconf/20166600087

client, consultants and contractor. This study may produce a better approach within the combination of adaptive reuse and the knowledge management process around the project teams' skills.

2 Intellectual Capital and Knowledge Transfer

The main problem in construction projects is the difficulty to accumulate knowledge. According to Kululanga and McCaffer (1993) the greatest challenge in construction projects is the difficulty in managing intellectual capital of project team members. Based on the explicit-tacit and individual-social dimensions of intellectual capital, four types of knowledge have been identified from organisational context: conscious knowledge, automatic knowledge, objective knowledge and collective knowledge (Spender 1996). Spender's model which considers conscious knowledge is formal, facts or conceptual knowledge, and is reliant on the individual's conceptual skills and intellectual abilities. Automatic knowledge is similar with experiential knowledge terms. This type of knowledge is practical-oriented where the individual takes an action based on individual conceptual knowledge which is developed from practical experience. Objective knowledge is also regarded as a shared body of professional knowledge that has been documented and stored in writing, drawings, procedures, regulations and contracts. Collective knowledge is similar to embedded social knowledge or organisational routines and shared norms in any process involved. Through intellectual capital, it would be proven that the experiences of individuals and project throughout the process are of utmost importance and crucial to the success of project. Figure 1 shows the integration of knowledge transfer and intellectual capital components for problems solving in adaptive reuse projects.

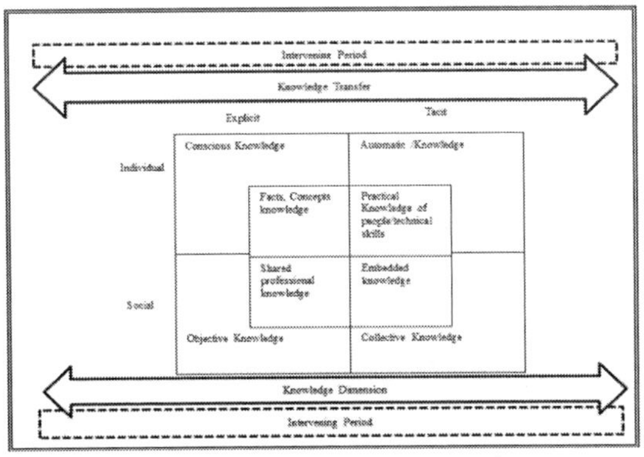

Figure 1. Knowledge transfer and intellectual capital components for problems solving

2.1 Knowledge Transfer

Knowledge transfer has been defined as another factor that affects the knowledge transfer approach when similar client and project teams collaborate again for another project after certain time gaps for each project. After a certain time gaps and considering what might be lost in a relationship to one project, electronic media was considered the best approach to start transferring the information and knowledge before face-to-face contact takes place when the project starts (Petra and Lars-Henrik 2014). It is crucial for knowledge transfer among project teams' members from the completed earlier project for there to be 'brainstorming' activities and a transfer to future projects to get better results in terms of project success (Maqsood 2006). Perhaps, this is necessary to combine soft and hard knowledge (Sajjad 2008) in the 'community of practice' approach to ensure that the management of accumulative intellectual capital from the earlier project for the future project is in good hands. The significance of knowledge transfer and a sharing approach in managing the intellectual capital particularly in the construction sector is appropriate to what Sajjad pointed to in his study.

2.2 Knowledge Dimension

There are two dimensions of knowledge; tacit and explicit. The tacit dimension refers to action, experience and involvement in any situation or circumstances. Tacit knowledge is comprised of both cognitive and technical components (Nonaka, 1994; Nonaka and Krogh, 2009). The cognitive components are related to mental maps, beliefs, paradigms and viewpoints of the individual in any organization or circumstances. In terms of the technical components, they are related to the skills, the expertise and the application of the know-how process, which are more related to this research. The explicit dimension is easily reduced as writing. Frequently explicit dimension takes the forms of documents, drawings, reports, historical information, heritage abstract etc. Both knowledge dimension; tacit and explicit information are critical in the problem solving process in adaptive reuse. The level of knowledge transfer includes the project teams transferring the project between project teams and across project teams. It also involves knowledge transfer happening between projects and across projects over time. In this study, this is considered as the intervening period.

2.3 Intellectual Capital and Knowledge Transfer in Adaptive Reuse

Adaptive reuse is defined as a process of transforming the functions, the structures and the fabric of old buildings to contemporary ways in the design and construction process. The process is complex and needs to be managed with the appropriate management skill of multi project team members who collaborate with the historic and modern value of the design and construction components (Alauddin, 2014). In terms of intellectual capital, it is more complicated and difficult to manage and ensure it can be used and that there is an improvement when the project teams jump from project to project. The process of transferring and managing the intellectual capital among multidisciplinary team members with different

DOI: 10.1051/matecconf/20166600087

expertises and knowledge needs to be investigated since there is a lack of research on the specifics of complex projects such as adaptive reuse projects. Most previous studies focused on the content of knowledge either tacit or explicit knowledge in general construction, procurement projects and specifically on a key person but not on the entire process (Yuan and Yang 2009; Kivrak et al. 2008). This is an important first step in developing a platform for the study proposed in this research. It is acknowledged that knowledge should be captured from one project to the next but it is important to understand the process of achieving this. In reality, adaptive reuse projects can take years to come to fruition and are not common. The intervening years between projects represent a time where knowledge can be captured, enhanced, transformed, developed and then transferred to the next project. Investigating and analyzing sequential adaptive reuse projects located as events in an intervening period such as this would provide an invaluable case study and contribute to the body of knowledge in this field.

3 Research Methodology

To further develop and investigate the key components in knowledge transfer within the adaptive reuse scope, a qualitative research study was carried out. The research method is illustrated in Figure 2.

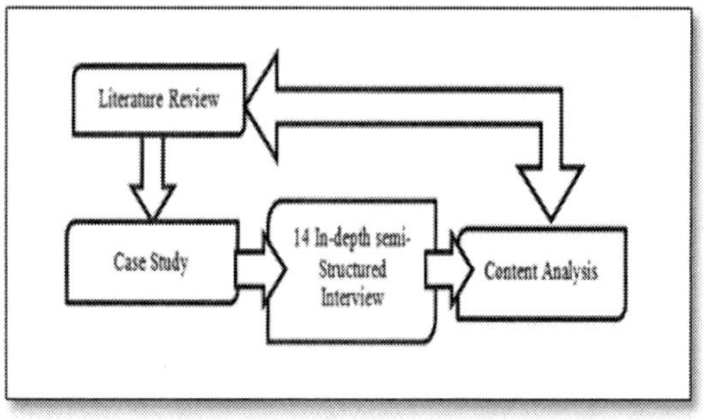

Figure 2. Research Methodology

The study begins with a thorough literature review. Two areas of literature are reviewed precisely to support the development of the research question, adaptive reuse literature and knowledge management literature. The rationale of these literature areas is explained in Table 1.

Table 1. The rationale for literature selection.

Adaptive reuse literature	Knowledge Management literature
Identification of adaptive reuse issues or context that is focused by previous studies on the adaptive reuse environment.	This literature is used to support the development of a conceptual model for this research. It relates to the knowledge transfer and the intellectual capital particularly in a temporary construction organisation.

This study undertakes that the previous adaptive reuse studies are lacking in injecting knowledge management into their research outcomes. The empirical study on adaptive reuse projects related with knowledge transfer and intellectual capital particularly in relation with the problem solving process is needed to developthe bridge for the gaps. Therefore, the research question in this study is *"What are the key components knowledge transfer that contribute to problems solving on adaptive reuse projects within the intervening period?"*. This study generalizes the theoretical propositions and does not enumerate frequencies in analysing and discussing data collection from interviews and documents. Yin (2009) restated the definition of case study technicality as an empirical inquiry that investigates a contemporary event in depth and within its real-life context, when the boundaries between event and context are not clearly evident. Table 2 reveals the links of this study to relate basic questions, which show the methods of data collection and the respondents involved in qualitative interviews.

Table 2. Coordination of data collection method with research question.

Methods		WHAT		WHO			HOW		
Documentation – written evidence	Interviews – Experience evidence	Explicit Data	Tacit Data	Client/Project Manager (PM)	Consultants	Contractor	Situation 1	Situation 2	Situation 3
Yes	Yes	Any records as written evidence	Interviews to obtain unwritten evidence	Involvement, Experience, Knowledge Skill, Expertise			In-depth Interview Face to face with semi-structured interview schedule		

This study has three logical reasons and rationales in using qualitative interviewing that have been adopted from Mason (1996). First, this study uses qualitative interviewing consists of semi-structured interviews. The semi-structured list of questions is based on preliminary thematic questions or parts, which contributed to the identification of the key components transferring the knowledge of client and project teams in the intervening period. Second, the ontological position is involved with the client and project teams' knowledge, views, understanding and experiences which are meaningful properties in exploring the key points for problem solving in reality. Third are the epistemological reasons for conducting interviews. Within-case analysis approach is adopted to analyse the qualitative interview data based on content analysis. This study adopts content analysis to theme and codes the interview transcriptions with the aid of qualitative analysis software, Nvivo 9. This involves storing the interview transcriptions from a total of 14 participants from two cases (Table 3). Before theming and coding the data, all the transcriptions will be read through to understand the whole situation and provide the specific key themes that are relevant to the research question. All the transcribed interviews are imported to the NVIVO as the 'sources' and have been named according to their position in both projects.

Table 3. Number of Participants

Respondents Roles		Case Study 1	Case Study 2
1	Project Manager 1	1	1
2	Project Manager 2	1	1
3	Architect	1	1
4	Building Surveyor		1
5	Fire Engineer	1	
6	Quantity Surveyor	1	
7	Town Planner	1	
8	Heritage Advisor	1	
9	Contractor	1	1
10	Structural Engineer		1

4 Contents Analysis Findings

Data from interviews were transcribed from the audio-taped version to the written version with MS Word 2010. To ensure that the transcribed interview data were valid, the researcher repeated the process by listening for the second time while reading and checking with the first draft of the transcription. After the process of data validation, the interview transcripts were exported into Nvivo 9 for open coding processes. Each sentence from fourteen interview transcripts was read carefully to capture the essence of that sentence that would be used later in open coding. Open coding in Nvivo 9 involved collecting all the interview transcripts and grouping them into open themes where the researcher used a conceptual model with analytic and descriptive coding methods as a guide to identify the related themes. The analytic coding was started by assigning codes to possible variables, a process which integrated the literature review findings with the conceptual model. Descriptive coding assigned demographic characteristics to each participant by describing their different perceptions about the variables of the critical factors that contributed to adaptive reuse and knowledge transfer activities.

4.1 Key Component 1: Efficiency and Effectiveness Mode

This section analyses the project team members' point of views about the efficiency and the effectiveness in solving problems in case studies. The evidence from the interviews shows that the problems could not be solved quickly or easily due to the complexity of adaptive reuse projects. Particularly for the consultants, their experience showed that solving the problems required more time and was difficult because the design process was too long, the preparation of the fire engineering analysis required a sophisticated process and the amount of repetition involved in testing the load and strength of existing materials was high. Nevertheless, the contractor and the town planner had dissimilar views to the point just made. Their experiences indicated that problems had been solved more quickly and easily because the components of collaboration and the skills of other project team members provided reasons for this to happen.

Table 4. Evidence of effectiveness and efficiency of problem-solving process in case studies

Project Team Members	Quicker and Easier (Positive Views)
Contractor	"...the structural engineer play the work well. That would help because I can walk and resolved problem stand there what to do and resolved on the spot when working out together. Probably working together being little bits better and at the end of the day you will presence what client going to disclosed."
Town Planner	"Because I am only dealing with the conceptual side of things, problems was resolved really quickly. Architect is fantastic architect done his thought through a lot of issues..."
Quantity Surveyor	"Obviously, it probably takes a longer time to measure and estimate a project like that because there were a lot of ideas which increased from [the conventional]. So, it was more difficult than a new building and more difficult than a general refurbishment."
Fire Engineer	"The building approval took a month to gather the approval from the fire brigade because we were doing more detailed analysis. The fire brigade and building surveyor's office came up with more detailed questions."
Project Manager 1	"... the length of that process because they were looking at from the design perspective, they were trying to maximise the number of actual floors within the building ... So, it took a long, long time and a lot of probably stressful nights for the structural engineer,..."

4.2 Key Component 2: The Same Project Team

The similar perspective between project manager 1 and project manager 2 regarding having the same teams in the projects definitely provided the evidence that these appointments benefited the client and the project. As the project manager who had significant experience on site, project manager 2 considered that it was more important to work together again with the same key people in project teams. The key important people with whom it was critical to work again in the sequence project were the architect, building surveyor and fire engineer. This was because the design parameters in adaptive reuse projects were dictated by fire engineering and building surveying requirements. The interaction between the architect, building surveyor and fire engineer led to innovative solutions which were acceptable from the client's point of view. Table 5 shows evidence of the project team's perceptions about how having the same team for the projects helped them in the problem-solving process. There were two categories of perceptions; positive and neutral. Positive perceptions demonstrated that the project team members had positive experiences in working with the same project team members in both projects. The same team members provided benefits in terms of more effective communication and greater understanding of the problems which made the problem-solving process more efficient. A neutral perception referred to project team members who accepted both situations either working with the same or in a different project team. This study identified that the contractor and heritage advisor were unaffected by the appointment of the same project team members.

Table 5. Evidence of having the same project teams in problem-solving process in case studies

Project Team Members	The Same Team
Quantity surveyor	"Yes, the same architect and the same builder. It certainly helped … if you work with a particular consultant, even a particular architect, you get to know how they work and get to know the things they like to design including the buildings, the way they go about things."
Architect	"Well, we had the same building surveyor...The building didn't comply with building regulations because it was not a new construction … it was because the existing building had heritage value and the client wanted to keep that building character ..“
Building Surveyor	"I suppose the work of changing the consultant when the work changed would be a little bit difficult but with the same consultant, people knew where you were heading. The difference being that the approach was quite different."
Fire Engineer	"Not just the same company but the same individuals which was very rare over that long a period. Having the same individuals was very helpful … we learnt something along the way. “
Project Manager 1	"I think it particularly helped. It helped because they were all certainly the consultant team because they knew exactly why certain things were being done in the building and why they couldn't do certain things."
Project Manager 2	"Definitely helped. We didn't have totally the same team but really the key people were all the same people. The architect was the same, the building surveyor was the same, and the fire engineer was the same … And so, with the interaction between the architect, the building surveyor and, the fire engineer, they would come up with those solutions that would be acceptable …."
Town Planner	"This was probably a larger one than what I had dealt with in my time as a town planner up to that point, so, not everyone had an architect as experienced as person who I dealt with ... That is the way the legislation works."
Contractor	"It was dynamic. Knowing the team was OK. I did not have any problem with new consultants coming to join the team... What is most important is them getting the job done with whatever is the best result."
Heritage Advisor	"I don't have any influence that gets the job or do anything like that…. One other thing was that I was admiring the work of architect's firm as they have results on the board..."

4.3 Key Component 3: Project Team's Actions

The building surveyor and contractor mentioned their own actions rather than that of other project team members. However, the client was more specific and related it to this research where they employed and used the same architect to again do the design and lead the project.This was considered appropriate with what this study had developed and contributed to new knowledge. This study carried out data analysis about the project team members' actions in solving problems according to their own disciplines and skills. There were project team members who referred to their own skills in describing the actions they took to address any challenges that they were facing. However, some project team members described other project team members' actions when discussing their perceptions about the problem-solving process. Table 6 shows evidence of the project team members' perceptions about the key component of action in the problem-solving process for case studies.

Table 6. Evidence for the key of action in the problem-solving process in case studies

Project Team Members	Key of Action
Building Surveyor	"…we didn't have the fire engineering in those days. So we have to go to what we call the building referees board.. So we prepared the packages of modification applications to modify the regulation to suit that recycling case."
Contractor	" Working together and the expertise and I think it important always to give the young one to go into the world and always sharp and it still a lot of administration. And also important is the whole culture things for working around. "
Architect	"I still had the ongoing discussion and working with the property services as their client contact and then we would go back through the various groups to solve their problems … so, I was answering the question on site, inspecting the works, sorting out the problems, writing the site instructions to the contractor…as architects we had to understand the building regulations and how big the fire compartment that we can have so we worked with building surveyor…"
Fire Engineer	"… architect recommended us because he want build the people that the team know each other. So it thinks that one the key roles for an architect to a project manager to make sure you pick the right team."
Quantity Surveyor	"…basically we were involve in consultant thing to discuss broadly about the methodology how it might occur and that the floor to be moved by bay … to make out the quantity or the quantum of work you can talk to the construction manager try to work out how might be done."
Project	"… we saved the builidngs and parts of that had

Manager 1	to obviously then be negotiated with the Council and the heritage overlay to make sure we didn't lose the integrity of the building."

4.4 Key Component 4: Sources of Information

This section describes the sources of information that help project teams to develop solutions. The sources of information refer to the explicit information sources or written documentation. This study has identified five main sources of information that were referred to in case study 1 by the client and project team members, which were related to adaptive reuse projects. The sources of information are shown in Figure 3. The five main sources were cost information, heritage information; fire engineering information, regulation information and additional information from content analysis of the interview data from 14 participants' background.

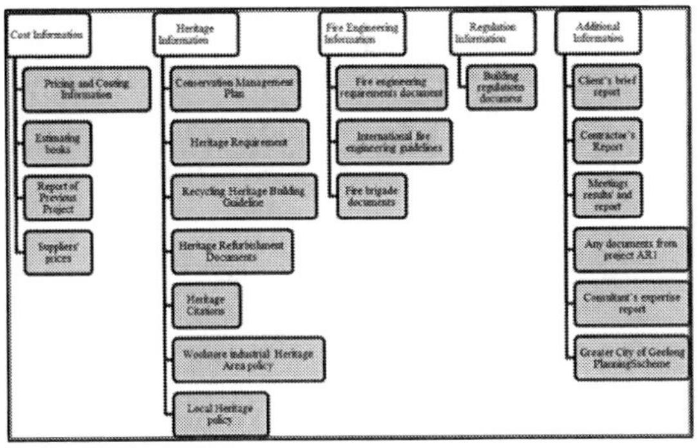

Figure 3. Sources of information in case studies

Table 7 shows evidence of the sources of information that were used and applied by project team members in case studies. The data findings demonstrated the sources of information used in case studies as the entire information that was appropriate in the within-case analysis.

Table 7. Evidence of sources of information referred to by project team members in the problem solving process for case studies.

Project Team Members	Sources of Information
Quantity Surveyor	"We have pricing information in our office that comes from a few forms that you can get. You can buy books that told you how much this thing is worst. We have other costing information from job that priced previously and information supply prices that things that we gathering and collecting in our normal daily work.."
Building Surveyor	"...the called as recycle guideline and that develop in CSIRO [Commonwealth Scientific and Industrial Research Organisation] and also ... The type of information on the first one there was a lot of that with the information are available. "
Contractor	"Absolutely I will prepare for the meeting, get the answer for the consequences at the night before and get the result in the meeting when attended the meeting tomorrow, that's it....What we called that brain storming session. But I think is more important keep people raise the issue and think about it and we get to resolve it"
Fire Engineer	Fire engineering needed hundreds of documents that we refer to but, specifically for heritage refurbishment, the organisation that wrote the Building Code also wrote some guidelines for recycling heritage buildings. So, those were the two documents for the approval part of the design process."
Heritage Advisor	"That is not my personal opinion in the guidance advice about the heritage policy and the planning scheme. So, those documents were used to guide me in giving advice to the architect and to the planner responsible for the permit."
Project Manager 2	"No, there was no structural table which showed the strength and anything to do with the timber. So that hadn't been tested: they actually did it as part of fire engineering. ...there was NO Australian Standard or anything about what and how the material should react under certain conditions."

5 Development of Intellectual Capital Framework

The project teams members' actions are carried out step-by-step to undertake adaptive reuse problem solving. The major actions are critical in and rely on individual and social skills related to how to modify the building to suit heritage requirements for the building permit application and the need to track the industrialization status and to understand the client's requirements in heritage situations. The project team members can take immediate action to solve the problems on site (day-to-day actions) when the solution does not affect the major costs or heritage regulations. This knowledge can be transferred among the project team members individually and becomes the social routine within the intervening period. This framework can also show the knowledge gained on how to be respectful to the culture and the heritage. With this framework, the project team members can always keep on track and record and allocate the problems to the appropriate discipline for them to be solved (Figure 4)

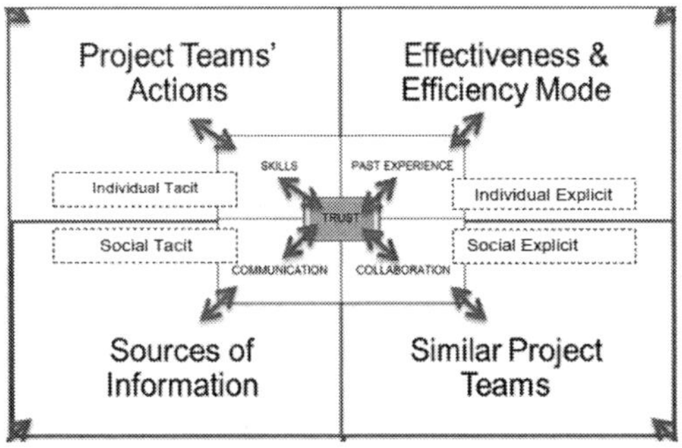

Figure 4. The Intellectual Capital Framework

6 Conclusion

The individual tacit and explicit knowledge attached to the history and the technical and management aspects accumulate to consolidate social tacit and explicit knowledge in the social routine to support the activities of knowledge transfer in adaptive reuse projects. This framework confirms London and Chen's (2004) theory that the component of intellectual capital is the accumulation of skills, experiences, competencies and knowledge of project teams. The present framework identifies the relationship and the reliance between the knowledge transfer and intellectual capital in the problem-solving process. It is difficult to link this finding to the previous literature, because little research has been published in this field. The lack of research on using knowledge management to support a whole series of research studies lead to this present framework's contribution being a new approach in the area of adaptive reuse research. Watson (2009) had identified that implementation project management tools and concepts as success factors in heritage projects, but the research was not empirically based. The complexity of adaptive reuse was mentioned in Kurul's (2007) and Latham's (2000) studies which involved the entire adaptive reuse process. This current study has expanded the meaning of complexity as involving the whole process but has contributed to the detail of the complexity in terms of problem solving, which was not covered by Kurul (2007); Kincaid (2000) or Latham (2000).

This framework demonstrated a new contribution to knowledge management in relation to adaptive reuse projects. Nonaka (1994) originally noted that knowledge management starts with knowledge creation, followed by knowledge transfer activity in common construction projects. However, this framework has also contributed to the concept by identifying that, in adaptive reuse projects within the intervening period, there is a reversal in the structure of knowledge management. The project teams in adaptive reuse projects in a time series scenario start with a richness in their knowledge about heritage requirements and regulations. The activity of knowledge transfer from the first project to the second project happens first, after which the project teams would create new skills and knowledge along with the development of new solutions for problems in future projects.

References

1. Alauddin, K., *'The Development of an Intellectual Capital Framework for Successful Adaptive Re-use'*, Sch.PCPM. Melbourne, RMIT University. PhD (2014)
2. Ball, R, *'Developers, regeneration and sustainability issues in the reuse of vacant industrial buildings'*, BRI, vol. **27**, no. 3, pp. 140 – 8 (1999).
3. Karim, SBA, Kamal, KS, Wahab, LA and Hanid, M, *'Risk assessment and refurbishment: The case of Ipoh Railway Station Perak, Malaysia'*, MICRA, Malaysia, March (2007)
4. Kincaid, D, *'Adaptability potentials for buildings and infrastructure in sustainable cities'*, Facilities, vol. **18**, no. 3, pp. 155 – 61(2000)
5. Kivrak, S, Arslan, G, Dikmen, I and Birgonul, MT, *'Capturing Knowledge in Construction Projects: Knowledge Platform for Contractors'*, JME, vol. **24**, no. 2, pp. 87-95(2008)
6. Kululanga, G. and R. McCaffer, *'Measuring knowledge management for construction organizations'*, ECAM(5/6): 346-354.(1993).
7. Kurul, E 2007, *'A qualitative approach to exploring adaptive re-use processess'*, Facilities, vol. Vol. **25**, no. No. 13/14, p. 16 (2007)
8. Latham, D, *Creative re-use of buildings*, 2 vols., (Donhead, Dorset 2000)
9. London, K and Chen, J 2004, *'Firm social, cultural and intellectual capital: strategic indicators of international client satisfaction'*, Int Conf., http://eprints.qut.edu.au/27358/ (2004)
10. Mason, J, *'Qualitative Researching'*, (SAGE Publication Inc., Great Britain. 1996)
11. Maqsood, T., *'The Role of Knowledge Management in Supporting Innovation and Learning in Construction'*, Sch.BIT. Melbourne, RMIT University. PhD: 181 (2006).
12. Nonaka, I. *'A dynamic theory of organizational knowledge creation'*, Organizational Science, **5**(1), 14-37 (1994).
13. Nonaka, I., and von Krogh, G., *'Tacit Knowledge and Knowledge Conversion: Controversy and Advancement in Organizational Knowledge Creation Theory'*, Organization Science, 20(3), 635-652 (2009).
14. Petra M. Bosch-Sijtsema, Lars-Henrik Henriksson, *Managing projects with distributed and embedded knowledge through interactions*, Int. Journal of PM, Vol **32**, Issue 8, 2014, Pages 1432-1444 (2014)
15. Sajjad, M. J., *'A holistic view of knowledge management strategy'*, Journal of Knowledge Management **12**(2): 57-66 (2008).
16. Shipley, R, Utz, S and Parsons, M, *'Does adaptive reuse pay? A study of the bussiness of building renovation in Ontario, Canada'*, Int.Journal of HS, vol. Vol. **12**, No. 6, p. 15 (2006)
17. Spender, J. C. *'Making knowledge the basis of a dynamic theory of the firm'*, Strategic Man. journal, **17**, 45-62 (1996).
18. Yin, RK *'Case Study Research; Design and Methods'*, Fourth edn, vol. **5**, (SAGE Publications, Inc, United States of America 2009).
19. Yuan, M and Yang, J, *'A Knowledge Management Framework to Promote Infrastructure Project Sustainability'*, in 15th Int.Symp on CME, Nanjing, China, vol. **29** (2009)
20. Waleed Tarek Ali Shehata, Yasser Moustafa, Lobna Sherif, Ashraf Botros ,"*Towards the comprehensive and systematic assessment of the adaptive reuse of Islamic architectural heritage*

in Cairo", Journal of CHM&SD, Vol. **5** Iss 1 pp. 14 -29 (2015)
21. Watson, P 2009, *'A methodology for attaining and assessing project success for rehabilitation projects',* Journal of BldgAppr, vol. Vol. **4**, no. No.3, p. 9 (2009).

MATEC Web of Conferences **66**, 00088 (2016)
IBCC 2016
DOI: 10.1051/matecconf/20166600088

The Observation Of Defects Of School Buildings Over 100 Years Old In Perak

Kartina Alauddin[1,a], Mohd Fisal Ishakt[1] , Haryati Mohd Isa[1] and Fariz Mohamad Sohod[1]

[1]*Faculty of Architecture Planning and Surveying, Universiti Teknologi MARA, Cawangan Perak, Seri Iskandar, 32610, Perak, Malaysia*

Abstract. Malaysia is blessed with a rich legacy of heritage buildings with unique architectural and historical values. The heritage buildings become a symbol of the national identity of our country. Therefore, heritage buildings, as important monuments should be conserved well to ensure the extension of the building's life span and to make sure continuity functions of the building for future generations. The aim of this study is to analyze the types of defects attached in school buildings over 100 years located in Perak. The data were collected in four different schools aged over 100 years in Perak. The finding of the study highlighted the types of defects which were categorized based on building elements, including external wall, roof, door, ceiling, staircase, column, internal wall, floor and windows. Finding showed that the type of defects occurred in school buildings over 100 years in Perak is the same as the other heritage buildings. This finding can be used by all parties to take serious actions in preventing defects from occurring in buildings over 100 years. This would ensure that buildings' functional life span can be extended for future use.

1 Introduction

Malaysia is rich with a legacy of historic buildings which showcases outstanding craftsmanship and architectural qualities. These possess impressive historical features and preserve the heritage of workmanship from the past (Suhana et al., 2013). Nowadays, a lot of effort has been made to preserve the historic buildings. According to Malaysia's former Prime Minister, Datuk Seri Abdullah Ahmad Badawi, billions of Ringgit have been spent and wasted in repairing public buildings due to Malaysia's poor maintenance culture (Abdullah, 2007). The lack of proper maintenance works and identification of historic buildings may contribute to the decaying of buildings, thus resulting in decrement in the number of historic buildings. Proper maintenance is important to ensure the extension of the buildings' life span and to make sure continuous functions of the buildings. There are many types of common defects occurred in buildings such as dampness, termite attacks, cracks, loose plasterwork, efflorescence, and blistering of paintwork (Mardziah & Azree, 2012). The problems related to the defects in heritage school particularly has been highlighted by Zainal and teams in their research. In late 2007, there were many cases which involved component failures of school buildings reported in Malaysia (Zainal, et al., 2009). These incidents happened caused by the age of buildings, environmental conditions, the building functions, and also maintenance work that was carried out on each building. Many heritage buildings still remain in poor condition with signs of serious building defects threatening their survival because legislations related to heritage buildings in the country do not sufficiently address the issue of maintenance and its management in the conservation of heritage buildings. Therefore, this results in poor maintenance management practices which eventually lead to deterioration of the buildings (Arazi et al., 2010). The objective of this paper is to examine the types of defects in the school buildings aged over 100 years old. The observation focuses on the school buildings in Perak. Perak has been selected as this state has many schools which are above 100 years since the occupation of the British in Malaysia. Many of the school buildings are still in use up until now to preserve the schools' identity

2 Literature review

Building defects occur in either new buildings or old ones. Defects in new buildings may be caused by non-compliance with the Building Code and published acceptable tolerances and standards. Meanwhile, for the older buildings, or buildings out of warranty period, they may not comply with these standards but must be judged against the standards at the time of construction or refurbishment (Nadia & Md Azree, 2013). In most cases, the defects usually occur in old buildings are also known as heritage buildings. According to Kamarul et al.,

[a] Kartina Alauddin: karti540@perak.uitm.edu.my

(2008), heritage buildings are an important element in town development which can be seen in some cities in the world. They play an important role in defining the landmark within the urban area as well as generating income and boosting the tourism industry. Examples of historic buildings in Malaysia are mosques, churches, palaces, clock towers, prisons, government offices, residential, schools, railway stations, hotels, forts, and monuments (Abdul Ghafar, 1997). This research focuses on the defects in heritage school buildings in Perak. In fact, nowadays, building defects are becoming an ordinary phenomena in the construction industry, which contributes negative impacts towards the buildings especially school buildings. It includes the building performance, health and safety aspects, and overall environmental conditions as well as teaching and learning processes (Md Azree et al., 2014). Nevertheless, there are a great number of building defects arise and being officially reported by mass media, especially problems with educational buildings (Md Azree et al., 2014).

2.1 Definitions of Defects and Heritage Buildings

Mardziah and Md Azree (2012) explain defects as faults on something that detract from perfection, whilst building damage can be seen when any structure, material, equipment, and element of the building are not fully functional. In another study, construction defects can be defined as a failure or shortcoming in the building's function, performance, statutory, or user requirements, and this may occur in its structure, fabric, services, or other facilities (Haryati, 2011). Besides, defects can also be defined as a lack of something necessary for completeness; shortcoming. Pheng and Wee (2001), Chew (2004), and Oleranrewaju et al. (2010) define those defects as a failure or shortcoming with the design in the function, performance, statutory, or user requirements which may affect its structure, fabric, services, or other facilities. Nielsen et al. (2006) claim that the term defects are commonly used to describe a physical defect and process defect. He also stated that when project documentation, a building material, a structure, or part of structure lacks of abilities, whereby it is able to be expected according to the contract, it is regarded as defects. In addition, Shittu et al. (2013) explains that defects are the non-conformity of a component with a standard of specific characteristics that have been stated as per specification.

Heritage buildings can be classified as 'pre-war buildings' due to their year of build, ranging from 1800 to 1948 as stated by Kamarul et al. (2008). Besides, according to Arazi et al. (2010), heritage buildings referred to buildings formed in the past which are of high values in terms of historical and architectural and needs regular care and protection in order to preserve their values; historical, aesthetic, archaeological, spiritual, social, political and economic. In Malaysia, there are many pre-war buildings are also known as heritage buildings. Almost every state in Malaysia has heritage buildings. Idid (2005) claims that the total number of heritage buildings in Malaysia is 20,787 and Penang is the highest on the list with 5,057 heritage buildings meanwhile the lowest is Sabah with 10 heritage buildings only. Table 1 shows in detail about the distribution of heritage buildings in Malaysia.

Table 1. The Distribution of Pre-War Urban Buildings in Malaysia .

No.	States	No. of Historical Buildings	Percentage of Historical Buildings
1	Penang	5057	24.30%
2	Perak	3351	16.10%
3	Johore	2323	11.20%
4	Malacca	2177	10.50%
5	Kuala Lumpur	1763	8.40%
6	Kedah	1282	6.12%
7	Selangor	1166	5.60%
8	Sarawak	1010	4.90%
9	Negeri Sembilan	999	4.80%
10	Pahang	831	4.00%
11	Terengganu	420	2.00%
12	Kelantan	373	1.80%
13	Perlis	25	0.10%
14	Sabah	10	0.05%
	Total	20787	100%

Source: Idid (2005)

In fact, Kamarul et al. (2008) examine that heritage buildings are an important element in town development which can be seen in some cities in the world. For example, the heritage buildings in Rome and London have become the main tourist attraction of each country. The buildings play an important role in defining the landmark within the urban area as well as generating income and boosting the tourism industry.

2.1.2 Formatting author names and author affiliations

Author names should be typed in 9-point Arial. The style for the names is First Names then Last Name, with a comma after all but the last two names, which are separated by "and". Do not use academic titles.

Affiliations of authors should be typed in italic 8-point Arial. They should be preceded by a numerical superscript corresponding to the same superscript after the name of the author concerned. Please ensure that affiliations are as full and complete as possible and include the country.

2.2 Defects in Heritage Buildings

There are several types of defects that usually occur in the heritage buildings. Based on the literature review, the types of defects can be determined and grouped by the elements. Usually, the elements involved are external wall, internal wall, column, door, window, roof, ceiling, floor, and others (see figure 1).

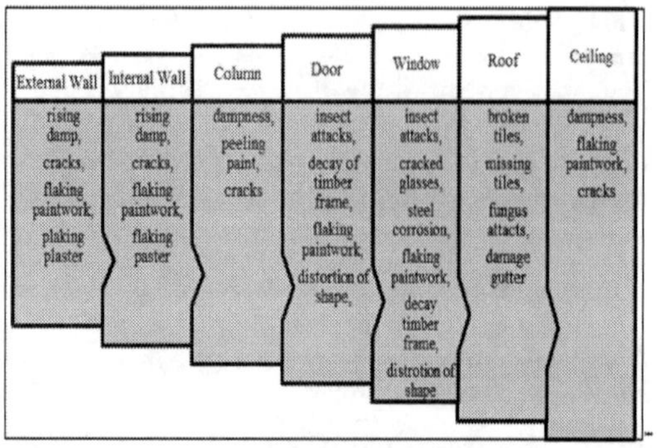

Figure 1. Types of Defects grouped by building elements

The major defect that usually occurs in the heritage buildings is on the external wall. Usually, the external wall is more exposed to defects because it is the outermost layer of the building and external wall also acts as the protection of the building. Kamarul et al. (2007) express that the common defects usually found on the external wall of heritage buildings are rising damp, cracking of brick, crumbling of mortar, flaking of the paintwork, and salt attack. Besides, other defects at the external wall are defective plastered renderings, and cracking of walls and leaning walls (Abdul Ghafar, 2004).

For the internal wall element of the heritage buildings, the defects that can be found are rising damp, cracking of brick, loose of plaster, discolour of paintwork, and cracking of tiles (Kamarul et al., 2007). Meanwhile, Mardziah and Md Azree (2012) discover the most common defects on this element are the effect of chemical actions inside the brick and plasters that appeared on the internal wall. Abdul Ghafar (2004) reports that the dampness penetration through walls also becomes a serious defect. It can occur at the external wall either internal wall, particularly in buildings located near water sources. In addition, Abdul Ghaffar (2004) also affirms that water penetration occurs commonly through walls exposed to prevailing wet wind or rain meanwhile dampness occurs in walls due to other factors such as leaking gutters or downpipes, defective drains, burst plumbing, and condensation due to inadequate ventilation.

Studies of heritage buildings in Malaysia by Abdul Ghaffar (2004) have highlighted that peeling paint is a common defect that happen in the column. Peeling paint usually occurs on building facades, mainly the columns and other areas that are exposed to excessive rain and dampness. The amount of the constant wind, rain, and sun received can easily turn the surface of the paint to become chalky and wrinkled or blistered.

The defects that normally appear at the door of the heritage buildings are decaying of timber frame, insect attacks, shape distortion, panel delamination, and paint flaking (Kamarul et al., 2007).

Based on the survey done by Po (2001), the defect occurs is the aluminum strips that can drop off window frame when it is not nailed securely. Kamarul et al (2007) also highlights that the decay of timber frame, insect attacks, flaking of paintwork, rushing of steel, and crack of glass are the general defects that commonly exist at the windows of heritage buildings.

Siti and Md Azree (2013) identify that there are several defects that occur at the roof element in heritage buildings. The most common defect that occurs at the roof are broken and missing roof tiles. Roof tiles may be missing and broken due to heavy rains. Second defect is damaged gutter. The possible reason of the damaged gutter is due to the overflow water caused by the heavy rain. Next, the fungal stains appeared on the roof tiles and roof parapets which come from the water. The fungal and mold appeared on the roof tiles and parapet because of the high pressure of water. This defect can be determined by the changes of original color and the appearance of the building element. The other defect in the roof of heritage buildings is timber decay. The timber becomes decay due to the moisture of water. For example, rain. Besides, another reason is weather conditions such as sun and rain that make the timber weak. Studies from Kamarul et al. (2007) prove that defects in the roof are usually sagging of timber frame, broken tiles, missing tiles, deterioration of the surface, and harmful growths. Besides, Mardziah and Md Azree (2012) point out the defect of the Heritage Shop house in Penang which is damp patches due to roof leakage. These patches can be easily seen in the building. Abdul Ghafar (2004) also points out some roof defects in the heritage buildings in Malaysia including corrosion of nails that fix the tiles at battens and rafters, the decay of battens, and the cracking of tiles caused by harmful growths. Harmful growths pose a danger to the tiles because it may lift tiles and create leaks. Moreover, Md Azree et al. (2014) state that there are different types of roof covering materials used in every school, such as asbestos, zinc sheets, and roof tiles. From some inspections, the roof defects found are blocked gutters, leaking downpipes, leaking roofs, missing roof tiles, slipping roof tiles, defective soffit, and fascia board.

Mardziah and Md Azree (2012) claim that the ceiling defect found at a Heritage Shop house in Penang is water staining. It was caused by the absence by the ineffectiveness of DPM in the floor upstairs. Moreover, the defects that usually take place on the ceiling of heritage buildings are watermarked, loose of plaster, and lichen of surfaces (Kamarul et al., 2007).

Damp of surface, cracking of screeds, loose of tiles, defective of floorboards, and decay of timber frame are the main types of defects that occur on the floor of the heritage buildings (Kamarul et al., 2007). In addition, Mardziah and Md Azree (2012) have listed the floor defects found at the Heritage Shop house in Penang; termite attacks and timber decay. These defects commonly occur on the internal floor.The use of sections to divide the text of the paper is optional and left as a decision for the author. Where the author wishes to divide the paper into sections the formatting shown in Table 2 should be used.

3 Research Methodology

To further investigate the types of defects based on building elements in school aged over 100 years in Perak, this study adopted observation method for data collection. For the observation, the data were collected in four (4) heritage school (see Table 2).

Table 2. The list of school buildings aged over 100 years in Perak .

SCHOOL	YEAR
Clifford School, Kuala Kangsar	1897
Sekolah Menengah Kebangsaan Sultan Yussuf, Batu Gajah	1910
Anderson School, Ipoh	1909
Convent School, Ipoh	|1907

The descriptive analysis is used to analyze and describe the observed data from four schools. Table 3 described the background of the four heritage schools. All four heritage schools were selected based on the history of Malaysia; Perak is one of the states that was under the British rule before independence. Thus, there were many schools built for British officers' children, King's children and much more.

Table 3. The background of school aged 100 years in Perak

Background of school	Photos
The Clifford School, Kuala Kangsar is originated from the Malay School, which was founded in December 1897. In 1897 appeared "Government English School" and also known as "Hogan School" named after the first headmaster, Mr James Percival Charles Hogan. In 1928, the school was renamed as "Clifford School" as an appreciation to the Governor of the Straits Settlements and High Commissioner of Malaya at that moment, Sir Hugh Clifford. During World War II, the school changed its status to a Japanese military headquarters until 1946. At the age of 100, the Clifford School has been selected as one of the pioneer schools; "Smart School".	CLIFFORD SCHOOL, KUALA KANGSAR 
Sekolah Menengah Kebangsaan Sultan Yussuf was found by En. K. Malaiperumal in 1907. This school can be divided into three phases which are at the first stage starting from 1907 – 1910. Then, the second phase is from 1910 to 1951 (41 years) when the school was known as the Government Bahasa Inggeris School. On October 15, 1951 the school name was changed to Sultan Yussuf School (SYS); named after the Sultan of Perak at that time. The name somehow remains until now. Besides, Sekolah Menengah Kebangsaan Sultan Yussuf also become the first English school in Batu Gajah. The greatest achievement of this school is the main school building was gazetted as a national heritage building because it was over 100 years old.	SEKOLAH MENENGAH KEBANGSAAN SULTAN YUSSUF 
The Anderson High School was officially opened on 6th of February 1909. The name "Anderson" was taken from the name of the British High Commissioner in Malaya at that time, Sir John Anderson. Due to the lack of facilities at the Kuala Kangsar Malay College and reluctance of many Malay parents to send their children to missionary schools in Ipoh like the Anglo-Chinese School, the $45,000 Anderson School building on Douglas Road was then built. Initially, the impressive building designed by A R Hubback only had ten classrooms and a large hall on the first floor. After 44 months of break during the Japanese Occupation, the school reopened its doors on the 24th of September 1945. The Anderson School is one of the schools that aged over 100 years and still in use until now.	ANDERSON SCHOOL, IPOH 
The Convent School is an all-girls secondary school located at Jalan Sultan Idris Shah, Ipoh, Perak. The Convent School was established in 1907 and it is one of the oldest schools in Ipoh and is widely known as Main Convent Ipoh. The existence of the school itself was threatened by the primary school and was evacuated to the compound of SK Raja Perempuan in January 1994 when piling works for the construction of the Ipoh Parade, which was nearby, caused cracks to the school buildings. Besides, the original school building is now designated as a heritage site. The Convent School was 100 years old on 7 January 2007.	CONVENT SCHOOL, IPOH 

4 Observation Analysis Findings

This section describes about the types of defects in the school buildings aged over 100 years old. In this section, the types of defects are categorized based on elements which are external wall, roof, door, ceiling, staircase, column, internal wall, floor, and window and its details are shown in Table 4. All the types of defects occurred were identified from the observations done in four (4) different schools in Perak in three different locations via defects checklist.

Table 4. Types of defects in school buildings aged over 100 years

Elements	Defects	CONVENT	SYS	ANDERSON	CLIFFORD
External Wall	Rising Damp	/	/	X	/
	Cracks	/	/	/	/
	Flaking Paintwork	/	/	/	/
	Flaking Plaster	X	/	/	/

Roof	Broken Tiles	/	/	X	/
	Missing Tiles	/	/	/	X
	Fungus Attacks	/	/	/	/
	Damage Gutter	X	X	X	X
Door	Insect Attacts	X	X	X	X
	Decay of Timber Frame	X	X	X	X
	Flaking Paintwork	X	/	X	/
	Distortion of Shape	/	/	/	/
Ceiling	Dampness	/	/	/	/
	Flaking Paintwork	X	/	X	X
	Cracks	X	X	/	X
Column	Dampness	X	X	/	X
	Peeling Paint	/	/	/	/
	Cracks	/	/	/	/
Internal Wall	Rising Damp	/	/	/	X
	Cracks	/	/	/	/
	Flaking Paintwork	/	/	/	/
	Flaking Plaster	X	X	/	/
Floor	Dampness	X	X	X	X
	Cracks	/	/	/	/
	Timber Decay	X	X	X	X
	Fungus	X	X	X	X
	Mould Attacks	X	X	X	X
Window	Insect Attacks	X	X	X	NA
	Cracked Glassess	/	/	/	NA
	Steel Corrosion	/	X	X	NA
	Flaking Paintwork	/	/	X	NA
	Decay Timberframe	/	/	/	NA
	Distrortion of Shape	/	/	/	NA

4.1 External Wall

Based on the observations done at the Convent School, Ipoh, the common types of defects that usually occurred in the external wall were rising damp, cracks and flaking paintwork. Meanwhile, at SYS, the external wall of the school buildings was exposed to rising damp, cracks, flaking paintwork, and flaking plaster. It was quite different with the Anderson School because the rising damp did not occur in the external wall of the buildings, but it was still exposed to other defects such as cracks, flaking paintwork, and flaking plaster. Besides, the types of defects at the external wall of the Clifford School were the same as SYS; rising damp, cracks, flaking paintwork, and flaking plaster. Pictures in Table 5 shows the defects occurred in the external wall of the schools.

Table 5. Evidence of types of defects related to external wall

Convent School	SYS SCHOOL	CLIFFORD SCHOOL	ANDERSON SCHOOL
 **Flaking Paintwork**	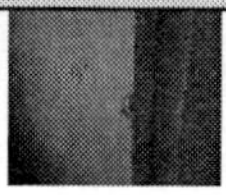 **Wall Cracks**	 **Rising Damp**	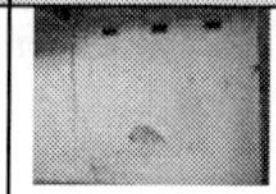 **Flaking Plaster**

4.2 Roof

The most common defects that occurred on the roof of the Convent School were broken tiles, missing tiles, and fungus attacks. Besides, SYS also faced the same types of defects in its roof. Meanwhile, for the Anderson School, the defects occurred on the roof were broken tiles, and fungus attacks only. Like the Anderson School, the Clifford School also had similar types of defects which were broken tiles, and fungus attacks. These defects mostly occurred in the main building of the schools. Pictures in Table 6 shows the defects occurred on the roof of the schools.

Table 6. Evidence of types of defects related to the roof

Convent School	SYS SCHOOL	CLIFFORD SCHOOL	ANDERSON SCHOOL
 **Broken Roof Covering**	 **Fungus Attacks**	 **Fungus Attack**	 **Missing Tiles**

4.3 Door

Based on the observations done at the Convent School and the Anderson School, there was only one type of defect that had been identified; distortion of shape.

Besides, for SYS and the Clifford School, the types of defects occurred at the door were flaking paintwork and distortion of shape. Pictures in Table 7 show the defects occurred at the door of the schools.

Table 7. Evidence of types of defects related to door

Convent School	SYS SCHOOL	CLIFFORD SCHOOL	ANDERSON SCHOOL
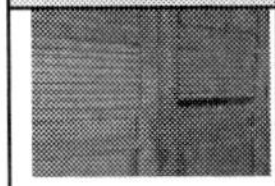 Distortion of shape	 Flaking Paintworks	 Flaking Paintworks	 Distortion of shape

4.4 Ceiling

The common ceiling defects that occurred in most of the buildings are damped, flaking paintwork, and cracks. Based on the observations made at the Convent School and the Clifford School, the dampness was the only defect occurred in their buildings. Meanwhile, the defects in buildings of SYS were dampness and flaking paintwork. The Anderson School had the same defects as the other schools, but it also had cracks in the ceiling. Pictures in Table 8 show the defects occurred at the door of the schools.

Table 8. Evidence of types of defects related to ceiling

Convent School	SYS SCHOOL	CLIFFORD SCHOOL	ANDERSON SCHOOL
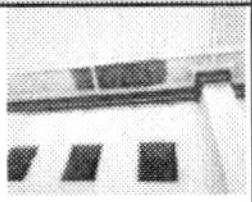 Dampness	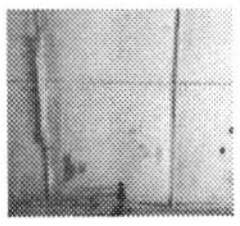 Flaking Paintworks	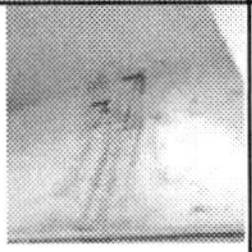 Dampness	 Cracks ceiling

4.5 Column

Mostly, the column defects occurred at the Convent School, SYS, and the Clifford School were peeling paint and cracks. At the Anderson School on the other hand, the column defects appeared were dampness, peeling paint, and cracks. These types of defects usually occur at the main column of the school buildings and it gives an awful image of the school. Pictures in Table 9 show the defects occurred at the column of the schools..

Table 9. Evidence of types of defects related to column

Convent School	SYS SCHOOL	CLIFFORD SCHOOL	ANDERSO N SCHOOL
 Cracks	 Cracks	 Peeling Paint	 Dampness

4.6 Internal Wall

Based on observations, the defects that could be identified in the internal wall of the Convent School and SYS were rising damp, cracks, and flaking paintwork. Besides, few of the listed defects occurred at the Anderson School include rising damp, cracks, flaking paintwork, and flaking plaster. At the Clifford School, it experienced different defects compared to the rest of the other schools because it did not have the rising damp defect. The defects occurred were cracks, flaking paintwork, and flaking plaster. Pictures in Table 10 show the defects occurred at the internal walls of the schools.

Table 10. Evidence of types of defects related to internal wall

Convent School	SYS SCHOOL	CLIFFORD SCHOOL	ANDERSO N SCHOOL
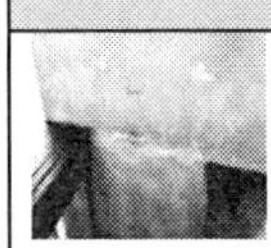 Cracks	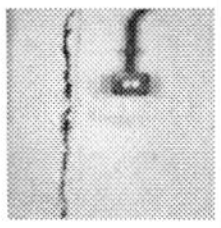 Cracks	 Flaking Paintworks	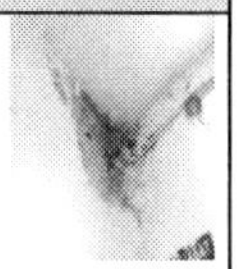 Dampness

4.7 Floors

Since the entire floor in the Convent School, SYS, the Anderson School, and the Clifford School were made of cement, they shared the same defect which was cracked. This defect clearly appeared at the main entrance of the school buildings and the class path. Pictures in Table 11 shows the defects occurred at the floors of the schools.

Table 11. Evidence of types of defects related to floors

Convent School	SYS SCHOOL	CLIFFORD SCHOOL	ANDERSO N SCHOOL
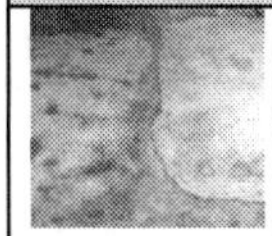 Cracks	 Cracks	 Cracks	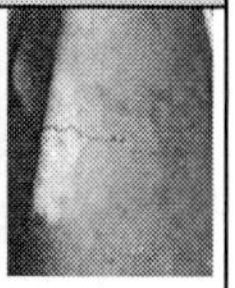 Cracks

 DOI: 10.1051/matecconf/20166600088

4.8 Windows

From the observations done, there were many defects found on the window of the Convent School, which were cracked glasses, steel corrosion, flaking paintwork, decayed timber frame, and distortion of shape. There was only one defect that was not found in this school, which was an insect attacks. Next, SYS experienced the same defects as the Convent School except for the steel corrosion that did not exist. Meanwhile, in the Anderson School, the types of defects found were cracked glasses, decayed timber frame, and distortion of shape. However, insect attacks, steel corrosion, and flaking paintwork were not discovered. Lastly, there was no defect found at the Clifford School due to its design. Pictures in Table 12 show the defects occurred at the windows of the schools.

Table 12. Evidence of types of defects related to windows

Convent School	SYS SCHOOL	CLIFFORD SCHOOL	ANDERSON SCHOOL
 **Distortion of Shape**	 **Cracks**	**No defect**	 **Cracks**

4.9 Discussion

The objective of this research is to identify the types of defect at the school building over 100 years old through observation. In this research, there are four (4) schools in Perak that are over 100 years old of age selected, which are Convent School, Anderson School, Clifford School and Sekolah Menengah Kebangsaan Sultan Yussuf. From the previous research, there are lots of types of defects that exist in a heritage building and some researchers categorized them based on elements and specify the defects in general only (Kamarul et. al.,2007; Mardziah and Md Azree 2012; Siti and Md Azree, 2013; Md Azree et. al, 2014). The examples of defect at heritage buildings are cracking of walls, flaking plaster and paintworks at walls, dampness at the ceiling, cracking of column, fungus attack on the roof, distortion of shape on window and door and others. Based on the observation done on the four (4) school buildings over 100 years in Perak, it is found that the school buildings expose the same defect problems with other buildings over 100 years. The elements that affected towards defect at all the four (4) school buildings are external wall, internal wall, roof, ceiling, floor, staircase, column, door and window. As conclusion, this study identifies that Clifford School is the one has less defects due to the appropriate design of the building.

5 Conclusion

In conclusion, all of the objectives of this research have been achieved successfully. Based on the research, it is shown that the types of defects that usually occur in school buildings over 100 years in Perak are quite similar to other heritage buildings in Malaysia. The longer the age of the particular building is, the more defects appear.

References

1. A.Ghafar,A, The Dilapidation Survey Report. ,"Majalah Akitek", V.16, I.1, 19-21 (2004)
2. A.Ghaffar,A. British Colonial Architecture in Malaysia 1800-1930. KL: Jab Muzium Malaysia. (1997)
3. Abdullah, A,B. Billions wasted in repairing buildings. NAFAM Conf KL, 13-14 August (2007).
4. Arazi.I, Faris.K and Mahmoud.S, Maintenance Management Framework for Conservation of Heritage Buildings in Malaysia. JMAS. Vol. **4**, No. 11,66-77 (2010).
5. Chew, M., Y., L. Defect analysis in wet areas of buildings. JCBM,165-173. (2004).
6. Haryati M.I, Learning from Defects in Design and Build Hospital Projects in Malaysia. Int Conf on SocSci &Humanity, Vol **5**, 238-242 92 (2011)
7. Idid, S.Z.A. Pemeliharaan Warisan Rupa Bandar: Panduan Mengenali Warisan Rupa Bandar Berasaskan Inventori Bangunan Warisan Malaysia. KL:BWM. (1995)
8. Kamarul S.K, Lilawati A.W and A.Ghafar.A. Pilot Survey on The Conservation Of Historical Buildings In Malaysia. Proc of 2nd. Int conf on BEDC,3-4th. USM (2008).
9. Kamarul S.K, Lilawati A.W, A.Ghafar.A and Saipol Bari A.K). Understanding the common Building Defect in Malaysia's Historic Buildings, The Int Conf on BEDC, 3rd-4th , USM (2007)
10. Mardziah K. and Md. Azree O.M. Building Condition Assessment And Defect Analysis On Heritage Shophouse In Peneng, Malaysia: Case Study. IJE. Vol. **2**(3), 77-80, (2012)
11. Md Azree O.M, N.A. Agus Salim, S.W. Tan, N.M. Tawil and N. Md Ulang Assessment of Significant Causes to School Building Defects. (2014).
12. Nadia O. and Md Azree O.M (2013). General Building Defects: Causes, Symptoms and Remedial Work. EJTD, Vol.(**3**), No 1 (2014)
13. Nielsen, J., Hansen, E. J. D. E., and Aagard, N. Buildability as Tool for Optimisation of Building Defects. SS, Vol 8, pp25-32. (2006).
14. Olanrewaju, A. L., Mohd Faris, K., and Arazi, I. Sustainability in the Context of Maintenance: Building Defects in the Malaysian University Campus. (2010).
15. Pheng, L.S. and Wee, DImproving maintenance and reducing building defects through ISO 9000. JQME, Vol **7** No 1, 6-24 . (2001).

16. Po Seng Kian (2001) A Review Of Factors Affecting Building Defects In Singapore. DTI, Vol. **3**, No. 2, 64-68 (2001)
17. Shittu, A. A., Adamu, A. D., Mohamed, A., Suleiman, B., Isa, R. B., Ibrahim, K., and Shehu, M. A. Appraisal of Building Defects Due To Poor Workmanship in Public Building in Minna, Nigeria. JE, Vol **3**, 30-38. 95 (2013).
18. Siti Solehah M.S and Md Azree O.M Heritage Building Roof Element Conservation: Peneng Old Town Hall Case Study, JER. Vol. **42**, No.2, 424. (2013).
19. Suhana J., Adi Irfan C.A, Norngaini M.T., Surat, M. and Kamaruzzaman, S.N. Preliminary Survey and Defect Analysis of Traditional Timber Mosques in Malaysia. (2013).
20. Zainal A.A, Sharifah Haminah S., Ismail, Abdul Rahman and Maizam A. SchBldMaintenance Strategy: A New Management Approach. Reservation; MUCEET, (2009).

Moisture condensation on building envelopes in differential ventilated spaces in the tropics: quantitative assessment of influencing factors

Maisarah Ali[1,a], Majeed Olaide Oladokun[2], Samsul Bahrin Osman[3], Niza Samsuddin[4], Hairul Aini Hamzah[5]

[1]*Department of Architectural and Civil Engineering, Building Energy and Environmental Technology Research Unit, Division of Building Science and Technology, City University of Hong Kong, Hong Kong SAR, China.*
[2] *Department of Civil Engineering, International Islamic University, Malaysia (IIUM), Gombak, Kuala Lumpur, Malaysia.*
[3]*Department of Manufacturing and Material Engineering, International Islamic University, Malaysia (IIUM), Gombak, Kuala Lumpur, Malaysia*
[4]*Department of Allied Health Sciences, International Islamic University, Malaysia (IIUM), Kuantan, Malaysia*
[5]*Department of Basic Medical Sciences, International Islamic University, Malaysia (IIUM), Kuantan Malaysia*

Abstract. Ventilation systems play a significant role in maintaining the indoor thermal and hygric balance. Nevertheless, the systems had been implicated to result in many problems. In the tropical climate, especially for energy efficiency purposes, building spaces are operated with differential ventilation. Such spaces operate on 24-hrs basis, some on 8-hrs while others are either naturally ventilated or served with mechanical supply-exhaust fan systems with non-conditioned outdoor air. This practice had been found to result in condensation problems. This study involves a quantitative appraisal of the effect of operative conditions and hygrothermal quality of building envelopes on condensation risk. The in-situ experiment is combined with an analytical approach to assessing the hygrothermal quality of building envelopes in a tropical climate building under differential ventilation between adjacent spaces. The case-studied building is with a known history of condensation and associated damages including mould growth. The microclimate measurement and hygrothermal performance of the wall and floor against condensation and mould growth risks had been previously reported elsewhere. As a step further, the present study evaluates the effects of various envelope insulation types and configurations together with the HVAC cooling set-points on envelope hygrothermal performance. The results revealed that overcooling the air-conditioned side increases condensation risk on the non-air-conditioned side of the envelopes. The envelopes failed criteria for surface condensation at existing operative conditions irrespective of envelope hygrothermal quality improvements. However, the envelope performed well at improved cooling operative conditions even at existing envelope hygrothermal quality. It is, therefore, important to ascertain the envelope hygrothermal quality as well the cooling operative conditions while embarking on energy efficiency operations in mechanical ventilation systems under differential ventilation.

1 Introduction

Sequel to the aftermath of energy crisis coupled with the challenge of climate and fuel security, the need for building energy efficiency had reportedly been on the increase over the years [1-3]. Achieving energy efficiency in buildings had resulted in strict regulations on building envelope insulation and efficient heating ventilation and air conditioning (HVAC) systems [4]. Envelope insulation provides resistance to heat transfer between the indoor and outdoor environment of a building. A suitable insulation material isolates the indoor environment from thermal gradients by retaining or preventing heat from going out or coming into the building to respectively reduce the heating and cooling load of a building. To this end, the use of insulation had resulted in energy efficient improvement with the resultant HVAC energy consumption reduction [5, 6]. Nonetheless, many detrimental effects had been reported as envelope insulation contributes to problems of moisture condensation and mould formation [4]. The reported moisture condensation risk can be explained from the submission of Johansson [7] where their study reiterated that envelope insulation yields different moisture content in the insulated wall. As moisture formation precludes mould growth, the risk of mould proliferation increases with moisture formation. A previous study by Van Loo *et al.* [8] found that some insulation materials (e.g., fiberglass) may serve as a support matrix for debris (nutrients for mould) collection such that when moisture forms on them can lead to mould growth.

Also, the HVAC) systems maintain the indoor air quality and thermal comfort through its regulation of the operative conditions. In the tropical climate, the system keeps the indoor environments below the outdoor elevated thermal and hygric profiles. As both temperature

[a] Corresponding author: lyday011@yahoo.com

and relative humidity are significant environmental factors for human thermal comfort [9, 10], they are reported as being critical to the hygrothermal performance of building envelopes [11-13]. Maintaining the required level of indoor thermal comfort by the HVAC systems require high energy outlay which is attributively contributing to higher building sector energy consumption [14-16]. Also, the poor hygrothermal performance of building envelopes is found to have detrimental effects on the IAQ and energy efficiency [17-19]. Consequentially, the need for energy efficiency and IAQ improvements had led to the emergence of sophisticated HVAC systems and building envelopes with enhanced performance. The detrimental effects of building envelope insulation and HVAC systems to achieve building energy efficiency has continuously been the subject of concern over the years.

In a field trial to assess the performance of different roof constructions in a warm climate on indoor thermal comfort, Özdeniz *et al.* [20] found that the inclined roof constructions over concrete ceiling recorded acceptable performance. Nevertheless, moisture accumulation is prominent in this type of constructions. The author thereby concluded that air conditioned buildings in a warm climate are susceptible to condensation risk. This further underscores the associated moisture condensation risk with improved envelope insulation despite its energy efficient benefits. The overall energy savings is affected by the location of envelope insulation either on the internal or external faces. The studies of Kolaitis *et al.* [21] on envelope insulation and energy efficiency reported about 8% improvement in the performance of external insulation over the internally applied insulation. Also, on moisture condensation, the study found that the risk of condensation increases in internally applied insulation in a temperate climate. The findings on internally applied insulation and moisture problem can be associated with the warm indoor environment. The reverse of this case is related to warm climate as described in [22]

Despite the wealth of knowledge on the subject of HVAC systems, building energy efficiency, envelope insulation, and moisture problems, limited consideration exists for such review of both HVAC system operations and building envelope hygrothermal performance especially in the hot and humid climate of Malaysia. For instance, in the temperate climates, indoor spaces are operated at warmer temperature due to outdoor cold weather. On the contrary, the tropical climate indoor environment operates on cold set-points due to warmer outdoor conditions. The different microclimate on either side of building envelopes creates differential vapour pressure on opposing sides of the envelopes separating the indoor and outdoor environments in both temperate and tropical climates. While the effects had been well studied in the temperate climates resulting in various moisture investigations [23, 24], leading to the development of assessment methods [11-13, 25], tools [26] and envelope materials [27, 28], little exists on such considerations for the tropical climate buildings. With the increased energy efficiency need in building design and operations, there is a need to improve HVAC energy efficiency. Differential ventilation is adopted where building spaces are conditioned under varying profiles: some systems operate on 24-hrs basis, some on 8-hrs while others are either naturally ventilated or served with mechanical supply-exhaust fan systems with non-conditioned outdoor air. Evidence exists on linking condensation problems with the practice of running some part of the microenvironment under cooling mode while adjacent spaces are either not in operations - 24hrs vs. 8hrs [29-31] or rather ventilated with mechanical supply-exhaust systems with unconditioned air [32]

From the foregoing, there appears to be a synergistic relationship between HVAC energy consumption, building energy efficiency, insulation and above all moisture condensation problem. Although optimising insulation thickness is found to be an efficient way of providing energy savings [5, 6, 33], limited methods exist in the synergistic combination of insulation location (interior or exterior), insulation thickness, energy conservation and hygrothermal (moisture condensation, mould growth, etc.) risk [34]. This further suggests the needs for further studies on the combined effects of varying factors on hygrothermal performance of building envelopes

The problems above motivate our investigations into the envelope thermal quality under HVAC operative conditions against condensation and mould growth risks in the hot and humid climate buildings. To achieve these, we adapted a previous analytical method being used in temperate climates. Details of the fundamentals of such analytical methods for condensation assessment is documented CIBSE [25] and European [13] standards. In our initial attempt to adapt the methodology for tropical climate conditions, reported elsewhere [35], we found that with the weather conditions in the tropics, it is imperative to affirm the building envelopes hygrothermal quality as well as operative set-points of the cooling systems to prevent condensation and mould growth. The present studies build on the initial findings by investigating the effects of various insulation types and thicknesses coupled with the HVAC cooling conditions on envelope hygrothermal performance against condensation problems. It combines in-situ experiments with an analytical approach to assessing the building envelopes' hygrothermal quality under differential ventilation of air-conditioning and non-air-conditioning between adjacent spaces. The objective is to examine the effect of changing in envelope configurations and HVAC systems' cooling set-points on the overall hygrothermal performance as well as condensation risk on the building envelopes separating the air-conditioned and non-air-conditioned spaces quantitatively.

2 Methodology

2.1 In-Situ Experiments

The study selects a case studied building with a known history of condensation and associated problems including mould growth. The building operates on a combination of constant air volume (CAV) ventilation systems on some parts against natural and/or mechanical supply-exhaust fan (MSEF) systems with no air-conditioning on others. Regarding envelopes, the walls are made of 115 mm bricks finished with 15 - 20 mm rendering on both sides with emulsion and weather resistant paint on the internal and external surfaces respectively. For the floor, the construction is made up of 150mm thick reinforced concrete slab finished in tiles with cement and sand mortar. Additionally, acoustic ceiling tiles of 13 mm thickness are used as ceiling finishes to divide the space into room (3000 mm high) and ceiling plenum (600 mm height) zones. The duct networks for air distribution, located within the ceiling plenum, are well insulated to prevent heat loss and attendant condensation. Equally, the chilled water pipes, wet risers and sprinkler pipes are located within the ceiling plenum.

The ceiling plenum was inspected for possible water leakages during the experiment. The selected air-conditioned space is utilized as academic laboratory while the nonair-conditoned ones are teaching laboratory. In addition, both the air conditioned and non-air conditioned spaces are located in interior zones of the building. Hence, the effects of external weather conditions are considered limited. Both spaces were selected for in-depth study due to higher visible mould growth than other locations in the building. This results presented in this study is part of a series of interdisciplinary research on mould growth in tropical climate building. Interested readers are directed to other companion studies for further information [32, 35-38].

Figure 1 shows a schematic of the case-studied room with the measurement locations. The room on the lower floor is air conditioned where the air is supplied to the room via ceiling mounted supply diffusers. The return system is not ducted hence, air returns to the air handling unit via ceiling plenum. The upper room, on the other hand, is not air-conditioned as the ventilation is mechanical supply and exhaust system. The system brings in the unconditioned outdoor air into the room via ceiling mounted supply and exhaust grilles. During typical operation, the return air from the lower level (inside the ceiling plenum) chills up the upper floor slab and hence lead to condensation on the floor finishes (Figure 1b). A similar occurrence happens to the wall separating the air conditioned, and non-air conditioned spaces where the room is cooled thereby leading to water droplet formation on the corridor-facing side of the wall (Figure 1c). In this study, time-series data of hygrothermal parameters were measured at various locations in the air-conditioned and non-air-conditioned spaces

The data were measured with portable data loggers configured for automated logging at 10 minutes intervals over a period of spanning seven days during the month of May 2014. Measurement positions include supply and exhaust outlets, ceiling plenum, room ambient, and corridor spaces. Further details of the air distribution systems, measuring equipment accuracy and measurement protocols are described in earlier studies by the authors [32, 36, 38].

2.2 Analytical Method for Quantifying Envelope Hygrothermal Quality

Indoor moisture formation depends on the hygrothermal interaction between the building envelopes and the ventilation systems [39]. Moisture moves from high to low vapour pressure regions and as such envelope moisture problems in the temperate climate appear on the warmer interior sides. Contrastingly, in hot and humid climate, the en velopes are prone to moisture formation on

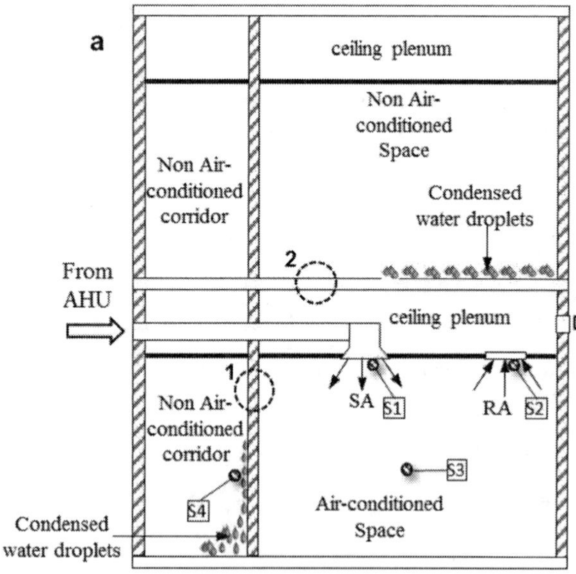

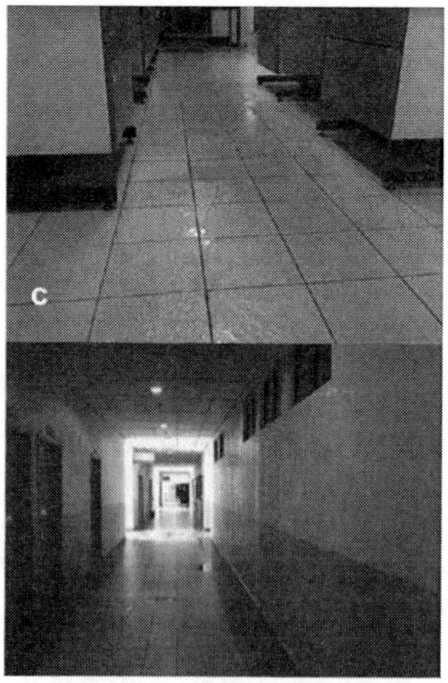

Figure 1. (a) Schematic of vertical cross section through the case-studied room, (b) Condensed water on the floor of non air-conditioned room above air-conditioned room (marked with dashed circle #2 in (a)), (c) condensed water on the wall separating air conditioned space non air-conditioned corridor (marked with dashed circle #1 in (a)) (S1-S4 are T and RH sensors both in the air-conditioned and non-air-conditioned space)

the warmer exterior sides. The distinction between the operative conditions in the temperate and tropical climate creates a reversal of the moisture movement direction. Against these assumptions, this study applied a modified approach of envelope hygrothermal quality assessments adapted from the specifications in CIBSE [25] and BSI [13]. A detailed description of the methods and its modifications are as described elsewhere [35].

Using Equations (1) to (7), in-situ hygrothermal data obtained from the time-series measurements were used to calculate the vapour pressure on both conditioned and non-conditioned sides of the envelopes. Values of saturated vapour pressure (p_s), vapour pressure on the cold side (p_c) and vapour pressure on the warm side (p_w) are evaluated from the hourly hygrothermal parameters obtained in the time-series measurements using Equations (1) and (2).

$$p_s = 610.5\exp\left(\frac{17.269\times\theta}{237.3+\theta}\right), \qquad (for\ \theta \geq 0) \qquad (1)$$

$$p = p_s \times \varphi \qquad (2)$$

Subsequently, the study adopts the recommended critical relative humidity (φ_{crit}) of 80% [12, 13, 25]. φ_{crit} is defined as the surface relative humidity that leads to surface deterioration (condensation, corrosion, and mould growth, amongst others). This critical humidity is defined as water activity, a_w in other studies and reports [3, 40].

$$p_s\left(\theta_{sw}\right) = p_w / \varphi_{crit} \qquad (3)$$

From the φ_{crit} and p_w an estimate is made of the minimum acceptable saturation vapour pressure, $p_s(\theta_{sw})$ at the envelope warmer side using Equation (3). Upon calculating the least acceptable saturation pressure, we compute the minimum acceptable surface temperature (θ_{sw}) on the warmer side from Equation (4).

$$\theta_{sw} = \frac{237.3\ln\left(\frac{p_{s(\theta_{sw})}}{610.5}\right)}{17.269-\ln\left(\frac{p_{s(\theta_{sw})}}{610.5}\right)}, \quad \left(for\ p_{s(\theta_{sw})} \geq 610.5\,Pa\right) \qquad (4)$$

The study further defined two parameters for envelope thermal quality: (a) thermal quality due to operative condition ($fR_{sw,op}$) – equation (5) and (b) thermal quality due envelope configuration and overall thermal transmittance ($fR_{sw,U}$) – equation (6).

$$f_{Rsw,op} = \frac{(\theta_{sw}-\theta_c)}{(\theta_w-\theta_c)} \qquad (5)$$

$$f_{Rsw,U} = 1 - UR_{sw} \qquad (6)$$

The parameters in equations (1) through (6) is defined as: p_s = saturated vapour pressure (Pa), p_v = air vapour pressure (Pa) (subscript v can be modified to c and w respectively for conditioned and non-conditioned spaces), θ = measured temperature (°C), φ = measured relative humidity (%), φ_{crit} = critical relative humidity (%), $p_s(\theta_{sw})$ = saturated vapour pressure at minimum surface temperature (warm side) to prevent condensation (Pa), θ_{sw} = minimum surface temperature (warm side) to prevent condensation (°C), U = envelope thermal transmittance (W/m^2.K), R_{sw} = surface heat transfer coefficient on the warm side (m^2.K/W).

The relations expressed in equation (7) is used to assess the building hygrothermal quality. A building envelope with lower $fR_{sw,op}$ than or at most equal $fR_{sw,U}$ is of good hygrothermal quality. Similarly, when a building envelope is found in a building which results in higher $fR_{sw,op}$ than $fR_{sw,U}$ then such envelope would invariably be of bad hygrothermal quality. The rating of good or bad quality is related to the ability of an envelope to provide high resistance to moisture and susceptibility to surface condensation. In essence, $fR_{sw,U}$ and $fR_{sw,op}$ are two important variables for controlling condensation and moisture formation on building envelopes.

$$EHQ = \begin{cases} f_{Rsw,op} < f_{Rsw,U}, good\ quality \\ f_{Rsw,op} = f_{Rsw,U}, good\ quality \\ f_{Rsw,op} > f_{Rsw,U}, bad\ quality \end{cases} \qquad (7)$$

where *EHQ* is the envelope hygrothermal quality defined by Equations (5) and (6).

It is imperative to note that while $fR_{sw,U}$ is dependent on envelope overall thermal transmittance, $fR_{sw,op}$ depends on the operative conditions on either side of an envelope. Hence, improvement to mitigate moisture condensation lies mainly on either to improve the envelope overall thermal transmittance or run the air-conditioning systems in a way to provide safe operative conditions. As said earlier, application of this approach had been previously tested and earlier reported [35]. It was found that the envelopes not only failed condensation risk assessment criteria but also found with elevated mould growth risks.

2.3 Assessing Effects of Various Improvements on Envelope Hygrothermal Quality

As the moisture performance of envelope is related to overall thermal transmittance and operative conditions, we tested various insulation improvements and operative conditions on $fR_{sw,U}$ and $fR_{sw,op}$ respectively. To investigate the effects of operative conditions on the envelope performance, we adjust the hourly operative hygrothermal parameters to give average values within the MS 1525 [41] specifications.

DOI: 10.1051/matecconf/20166600089

Four cases were tested with daily mean operative temperature and relative humidity. To run the tests, we generated random numbers with upper and lower limits representing both thermal and hygric parameters within MS 1525 [41] band – Temperature ranging from 23°C – 26°C and Relative Humidity varying between 55% – 70%. The generated hourly values of temperature and relative humidity were substituted in Equations (1) through (6) to calculate the hourly values of envelope thermal quality due to operative conditions ($fRsw,op$). The daily mean values of operative hygrothermal parameters are considered sufficient for this investigation as the mechanical cooling and ventilation system is of constant air volume (CAV) type.

To quantify the hygrothermal quality due to envelope thermal transmittance $fRsw,U$, we further investigate the effect of various common insulation types and thickness on the envelope quality, $fRsw,U$. The selected insulation types and details of thermal properties of the wall are as presented in Table 1. With each of the selected insulations, the $fRsw,U$ were subsequently computed with insulation thickness starting from zero millimetre - representing the existing wall construction without insulation. Subsequent upon the computation with a case of zero insulation, the thickness is varied from 10 mm up to 120 mm. The motivation for our selected thicknesses was as found from earlier studies reported in Magrini [42]. With the assumption that the situation leading to condensation is similar in wall and floor on one hand and the practicability of insulating an existing suspended slab on the other, the procedure described in this study is applied only to the wall. Therefore, the results and discussions are therefore limited to the wall.

Table 1. Thermal properties of the wall with adopted insulation materials

Layer	d (mm)	k (W/mK)
Brickwall		
External surface resistance (Rsw = 0.04 m^2.K/W)*	-	
External Render	15	0.8
Brick	115	0.77
Internal Plaster	0.015	0.8
External surface resistance (Rsc = 0.13 m^2.K/W)	-	
Insulation materials		
Mineral wool**		0.036
Fibreboard**		0.040
Polystyrene**		0.033
Polyurethane foam**		0.030

* values obtained from Table 2 of BS EN ISO 13788 [13]
** insulation thickness varies between 0 cm (no insulation) and 12 cm.

3 Results and Discussions

3.1 Hygrothermal Operative Conditions

The results of hygrothermal operative conditions in the case-studied building revealed mean thermal and hygric profiles as: θ = 17.3 ±0.1°C, φ = 71.5 ±0.6% and θ = 23.6 ±0.4°C, φ = 93.2 ±1.6%, respectively for the air-conditioning and non-air-conditioning space on either side of the wall (Figure 2). The results hygrothermal conditions show that variations in the field measurements were within the accuracy of the measuring equipment (θ = ± 0.5°C, φ = ±3%) as similarly demonstrated in Oladokun [43]. This proofs that variations in the measured data are due to factors other than the measuring equipment errors. Figure 2 shows a cross section through the existing wall and the operative conditions it is subjected to the air conditioned and nonair-conditioned spaces.

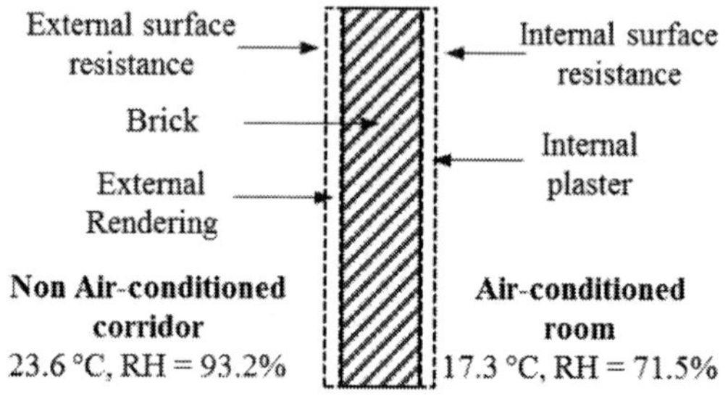

Figure 2. Existing wall section and the mean daily operative conditions on either side of the wall

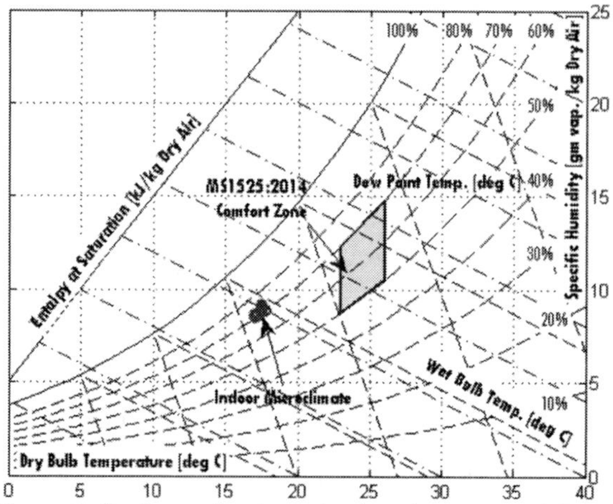

Figure 3. Psychrometric chart showing the average daily operative conditions in the air-conditioned room

Figure 3 shows a plot of the hygrothermal operative conditions on psychrometric charts together with MS 1525 specifications. As revealed in the graph, the hygrothermal operative conditions of the cooling systems were found out of the range specified by MS 1525 [41] where thresholds of temperature (θ) and humidity (φ) are respectively specified as θ = 23 - 26°C and φ = 50 - 70%. The operative conditions revealed an overcooling of the indoor environment – a confirmation that buildings in the hot and humid climates are overcooled [44].

3.2 Envelope Hygrothermal Quality Assessments

Table 2 shows the results of mean values of the computed temperature and relative humidity in compliance with MS 1525 specifications. Again, the computation was done by generating random numbers between the upper and lower boundaries specified by the standard. The results show four test cases in addition to the baseline on-site measured parameters. It should be reiterated that the temperature and humidity of the non-air conditioned side were not optimised and hence remains as the one measured in the test building. Results in Table 2 also shows the values of fRsw,op for the respective temperature and relative humidity. It is revealed that the values of fRsw,op reduced gradually from the baseline value of 1.413 when the operative conditions were θ = 17.3 °C and φ = 71.5% up till -4.965 when the average temperature and humidity improved to 24.3 °C and 61.8% respectively.

Table 2. Average indoor operative conditions and computed Envelope thermal quality due to operative conditions (fRsw,op).

Test Cases	θ_i (°C)	φ_i (%)	*fRsw,op*
Baseline	17.3	71.5	1.413
Test 1	24.3	61.8	-4.965
Test 2	24.8	63.9	-0.553
Test 3	24.6	62.5	0.160
Test 4	24.5	62.8	0.424

Figure 4 shows the graph of *fRsw,U* for the four selected insulation types. From the graph, the horizontal line with values 0.8879 representing the performance value at zero insulation, i.e. the existing wall. As shown in the figure, the highest envelope thermal quality due to thermal transmittance ($fR_{sw,U}$), is obtainable from the use of polyurethane foam insulation, followed by polystyrene insulation. Equally, it is found that the envelope with fiberboard insulation has the least thermal performance. The envelope hygrothermal quality due to overall thermal transmittance performed in a pattern for all the insulations when the thickness is 10 mm. Above this thickness, the performance curve begins to show a near uniform pattern. For instance, the gradient of the curve between 20 mm and 60 mm appears wider than between 60 mm and 100 mm. Although both of the two scenarios have a 40 mm thickness change in insulation, the performance improvements between the former are higher than the later. Hence, that increasing the overall thermal transmittance of a building envelope has a limit at which further improvements give a little gain in the envelope hygrothermal performance.

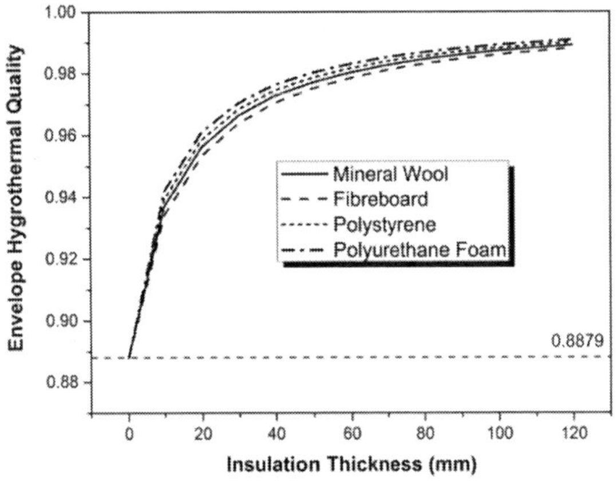

Figure 4. Envelope hygrothermal quality due to overall thermal transmittance for different insulation materials.

For comparison between the envelope hygrothermal quality due to overall thermal transmittance (fRsw,U) and the hygrothermal quality due to operative conditions (fRsw,op), the mean values of fRsw,op is superimposed on the graph of fRsw,U as shown in Figure 5.

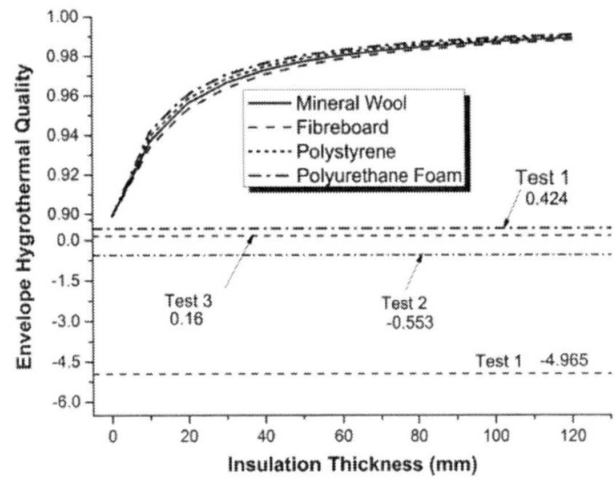

Figure 5. Comparison between envelope hygrothermal quality *fRsw,U* and *fRsw,op*.

As would be noticed in Figure 4, the curves on the upper part of the graph are similar to that shown in Figure 3. The vertical axis is broken to accommodate both data of *fRsw,U.* and *fRsw,op.* Values represented by the horizontal lines are those of the envelope hygrothermal quality due to operative conditions. It is revealed that for all the computed operative conditions (Test 1 to Test 4), the envelope hygrothermal quality *fRsw,op* falls well below the envelope thermal quality due to overall thermal transmittance *fRsw,U*. As stated earlier, an envelope with fRsw,op less than *fRsw,U* appears of good hygrothermal quality and will provide high resistance to moisture formation. On the contrary, when *fRsw,op* is greater than *fRsw,U* the envelope fails hygrothermal quality assessment and hence would be susceptible to surface condensation.

3.4 Discussions

In this study, the effects of improvements in envelope thermal transmittance and building operative conditions on envelope hygrothermal performance against condensation problems have been investigated with combined field and analytical experiments.

Some further implications of the results and limitations of the study for future investigations are pointed out below. Firstly on insulation improvement for possible moisture condensation control, it is found that increasing insulation thickness may contribute little to moisture control in the tropical climate buildings under differential air-conditioning. The present study is limited to spaces located in the building interior zone with no consideration for external weather influence. Therefore, further studies to include such impact will increase the knowledge in this area. Improved performance is achievable with the control of air-conditioned room operative conditions.

A combination of both insulation options and adjustments in operative conditions would provide additional benefits. Such combinations could be better achieved by building performance simulation tools. Numerical simulations on such combinations are therefore recommended for future studies. Secondly, on the observed improved performance achieved from the modified operative conditions, it is arguable on whether or not the conditions shows in Table 2 for test cases 1 to 4 would make building occupants comfortable. This appears of less concern as, the values are within the MS 1525:2014 specifications. On the contrary, sufficient evidence exists to the various determining factors of thermal comfort. Air movements are of paramount importance in addition to temperature and humidity. Increasing indoor temperature had been observed to be compensated by a corresponding increase in the air movement to improve the comfort of the thermal environment [10].

4 Conclusions

An investigation into the effects of various improvements in the indoor operative conditions and overall thermal transmittance on the hygrothermal quality of building envelopes against condensation risks. A case studied building was selected in the hot and humid climate with differential ventilation between adjacent spaces. The building runs on 24hrs cooling mode in one part against natural on the other. The methodology combined in-situ experiments with a hygrothermal analytical approach that had previously employed for building envelopes thermal quality and condensation risk in the temperate climates. Time-series measurements of hygrothermal parameters were obtained in air-conditioned and non-air-conditioned indoor microenvironments to serve as input in the modified analytical approach. Various insulation types and thicknesses were investigated to the envelope hygrothermal performance. The results of indoor hygrothermal operative conditions revealed that overcooling the air-conditioned side contributes to condensation on the non-air-conditioned side of the envelopes.

The overall thermal transmittance performance upgrade studies revealed envelope failure on surface condensation criteria at existing operative conditions. However, the envelope performed well at improved cooling operative conditions even at existing envelope hygrothermal quality. This further suggest that cooling set-points increase condensation risks on the envelope face abutting the non-air-conditioned spaces in the hot and humid climates. Further enhancements can be achieved by combing cooling operational set-points with envelope overall thermal transmittance improvements. The study provides an essential contribution to the recurring issues of air conditioning and moisture problem in the hot and humid climate of Malaysia. Going forward, as the numerical simulation is proved to be more efficient in building performance diagnosis, further research is recommended in this regards for advanced studies on the effects of each of the contributory factors to moisture problems. Numerical modelling is better in its capability to use validated model for further investigation of various configurations in the test case. Optimisation and further enhancement can then be executed with much confidence. This will further enhance the sustainable operations of facilities in this region thereby creating buildings that are healthier to the users, the envelopes, the stored components and the environment at large through energy efficiency operations.

Acknowledgements

The results presented in this study is part of a multidisciplinary research project (FRGS12-067-0126) initiated on indoor mould growth prediction using thermal characteristics in buildings in the tropical climate by International Islamic University Malaysia (IIUM) and the Ministry of Higher Education (MOHE) Malaysia. The financial supports are gratefully acknowledged. The authors equally recognised the supports of the Dean of Kulliyyah (Faculty) of Medicine IIUM together with the management and staffs of Daya Bersih during the studies.

References

1. H. Hens. IEA Annex 14: Condensation and Energy. J. Bldg. Phys. **15**:261-73 (1992)
2. H.B. Awbi. *Ventilation of buildings*. 2nd ed. London ; New York: Taylor & Francis; (2003)
3. H.S. Hens. *Building Physics-Heat, Air and Moisture: Fundamentals and Engineering Methods with Examples and Exercises*: John Wiley & Sons; (2012)
4. R. Galvin. Solving mould and condensation problems: A dehumidifier trial in a suburban house in Britain. Energ. Buildings. **42**:2118-23 (2010)
5. K. Çomakli, B. Yüksel. Optimum insulation thickness of external walls for energy saving. Appl. Therm. Eng. **23**:473-9 (2003)
6. M. Boji, M. Mileti, L.a. Boji. Optimization of thermal insulation to achieve energy savings in low

energy house (refurbishment). Energ. Convers. Manage. **84**:681-90 (2014)
7. P. Johansson. *Assessment of the Risk for Mold Growth in a Wall Retrofitted with Vacuum Insulation Panels*. Paper presented at the Proceedings of the 9th Nordic Symposium on Building Physics, Tampere, Finland. (2011)
8. J.M. Van Loo, C.A. Robbins, L. Swenson, B.J. Kelman. Growth of mold on fiberglass insulation building materials--a review of the literature. J. Occup. Env. Hyg. **1**:349-54 (2004)
9. BSI. *BS EN ISO7730:2005 Ergonomics of the thermal environment. Analytical determination and interpretation of thermal comfort using calculation of the PMV and PPD indices and local thermal comfort criteria*. Edition ed.: BSI Standards Limited; (2005)
10. ASHRAE. *ANSI/ASHRAE standard 55-2010: thermal environmental conditions for human occupancy*: American Society of Heating, Refrigerating and Air-Conditioning Engineers, Inc.; (2010)
11. BSI. *BS EN 15026:2007 Hygrothermal performance of building components and building elements — Assessment of moisture transfer by numerical simulation*. Edition ed.: BSI Standards Limited; (2007)
12. ASHRAE. *ANSI/ASHRAE Standard 160-2009: Criteria for Moisture-Control Design Analysis in Buildings*. Edition ed.: American Society of Heating, Refrigerating and Air-Conditioning Engineers, Inc.; (2009)
13. BSI. *BS EN ISO 13788:2012 — Hygrothermal performance of building components and building elements — Internal surface temperature to avoid critical surface humidity and interstitial condensation — Calculation methods*. Edition ed.: BSI Standards Limited; (2012)
14. J.K. Calautit, B.R. Hughes. Measurement and prediction of the indoor airflow in a room ventilated with a commercial wind tower. Energ. Buildings. **84**:367-77 (2014)
15. Y. Wang, Z. Fu-Yun, J. Kuckelkorn, D. Liu, J. Liu, Z. Jun-Liang. Classroom energy efficiency and air environment with displacement natural ventilation in a passive public school building. Energ. Buildings. **70**:258-70 (2014)
16. L.C. Haw, O. Saadatian, M.Y. Sulaiman, S. Mat, K. Sopian. Empirical study of a wind-induced natural ventilation tower under hot and humid climatic conditions. Energ. Buildings. **52**:28-38 (2012)
17. X. Lü. Modelling of heat and moisture transfer in buildings: II. Applications to indoor thermal and moisture control. Energ. Buildings. **34**:1045-54 (2002)
18. H.J. Moon, S.H. Ryu, J.T. Kim. The effect of moisture transportation on energy efficiency and IAQ in residential buildings. Energ. Buildings. **75**:439-46 (2014)
19. Y. Wang, Y. Liu, D. Wang, J. Liu. Effect of the night ventilation rate on the indoor environment and air-conditioning load while considering wall inner surface moisture transfer. Energ. Buildings. **80**:366-74 (2014)
20. M.B. Özdeniz, P. Hançer. Suitable roof constructions for warm climates - Gazimăusa case. Energ. Buildings. **37**:643-9 (2005)
21. D.I. Kolaitis, E. Malliotakis, D.A. Kontogeorgos, I. Mandilaras, D.I. Katsourinis, M.A. Founti. Comparative assessment of internal and external thermal insulation systems for energy efficient retrofitting of residential buildings. Energ. Buildings. **64**:123-31 (2013)
22. BSI. *BS 5250-Code of Practice for Control of Condensation in Buildings*. Edition ed.: BSI Standards Limited; (2011)
23. P.W. Francisco, W.B. Rose. Temperature and Humidity Measurements in 71 Homes Participating in an IAQ Improvement Program. ASHRAE. (2010)
24. W.B. Rose, P.W. Francisco. *Field Evaluation of the Moisture Balance Technique to Characterize Indoor Wetness*. In: Proceeding of Performance of Exterior Envelopes of Whole Buildings VIII: Integration of Building Envelopes IX conference, Florida. (2004)
25. CIBSE. Moisture transfer and condensation. In: K. Butcher, editor. *CIBSE Guide A: Environmental design*. 7th ed. London: The Chartered Institution of Building Services Engineers; p. 7-1 - 7-16 (2015)
26. J.M. Delgado, E. Barreira, N.M. Ramos, V.P. De Freitas. *Hygrothermal Numerical Simulation Tools Applied to Building Physics*: Springer; (2013)
27. S. Cerolini, M. D'Orazio, C. Di Perna, A. Stazi. Moisture buffering capacity of highly absorbing materials. Energ. Buildings. **41**:164-8 (2009)
28. M. Ibrahim, E. Wurtz, P.H. Biwole, P. Achard, H. Sallee. Hygrothermal performance of exterior walls covered with aerogel-based insulating rendering. Energ. Buildings. **84**:241-51 (2014)
29. I.A. Bamgbopa. *Assessment of Moulds Growth in Hospitals Indoor Environment: HVAC System Aspect*. Unpublished Msc Thesis: International Islamic University, Malaysia; (2008)
30. JKR. Guidelines on the prevention of mould growth in buildings. JKR Malaysia; 2009.
31. A.N.S. Wahab, M.F. Khamidi, M.R. Ismail. *An Investigation of Mould Growth in Tropical Climate Buildings*. Paper presented at the Business Engineering and Industrial Applications Colloquium (BEIAC). (2013)
32. M. Ali, M.O. Oladokun, S.B. Osman, N. Samsuddin, H.A. Hamzah, M.N. Salleh. *Ventilation Performance Assessment of an Educational Building in a Hot and Humid Climate*. In: InCIEC 2014: International Civil Engineering and Infrastructure Engineering Conference, Kota Kinabalu, Sabah, Malaysia. (2014)
33. A. Shanmuga Sundaram, A. Bhaskaran. Optimum insulation thickness of walls for energy-saving in hot regions of India. Int. J. Sust. Energ. **33**:213-26 (2014)
34. E. Vereecken, L. Van Gelder, H. Janssen, S. Roels. Interior insulation for wall retrofitting - A probabilistic analysis of energy savings and hygrothermal risks. Energ. Buildings. **89**:231-44 (2015)

35. M. Ali, M.O. Oladokun, S.B. Osman, A.M. Shamzani, I. Mohd Sharifuddin, Y. Faridah. *Hygrothermal Performance of Tropical Climate Building Envelopes under Operative Conditions: Condensation and Mould Growth Risk Appraisal.* In: International Conference on Science, Engineering, Built Environment and Social Science (ICSEBS-2015), Malang & Surabaya, Indonesia. (2015)
36. M. Ali, M.O. Oladokun, S.B. Osman, N. Samsuddin, H.A. Hamzah. *CFD Investigation of Indoor Hygrothermal and Airflow Profile in Academic Research Storage Room: Effect of LMA on Thermohygric Balance and Mould Growth.* Paper presented at the The International Conference on Computational Fluid Dynamics in Research and Industry (CFDRI 2015), Kuala Lumpur, Malaysia. (2015)
37. M.O. Oladokun, M. Ali, S.B. Osman, N. Samsuddin, H.A. Hamzah. *CFD Investigation of Indoor Hygrothermal and Airflow Profile in Academic Research Storage Room: Measurement and Validation.* Paper presented at the The International Conference on Computational Fluid Dynamics in Research and Industry 2015 (CFDRI 2015), Kuala Lumpur, Malaysia. (2015)
38. M.O. Oladokun, M. Ali, S.B. Osman, S. Niza, H. Hairul Aini. *Indoor Microbial Growth Prediction Using Coupled Computational Fluid Dynamics and Microbial Growth Models.* Paper presented at the The 13th Asia Pacific Conference on the Built Environment, Kowloon, Hong Kong. (2015)
39. P.M. Leardini, T. van Raamsdonk. *Design for airtightness and moisture control in New Zealand housing.* In: New Zealand Sustainable Building Conference. (2010)
40. H. Hens. *Mold in dwellings: field studies in a moderate climate.* In: Proceedings of the 24th AIVC Conference and BETEC Conference, Ventilation, Humidity Control and Energy. (2003)
41. Standards Malaysia. *MS 1525:2014 Energy Efficiency and use of Renewable Energy for Non-Residential Buildings - Code of Practice (Second Revision).* Edition ed. Putrajaya Malaysia: Department of Standards Malaysia; (2014)
42. A. Magrini (Ed.). *Building Refurbishment for Energy Performance: A Global Approach*: Springer Science & Business Media (2014)
43. M.O. Oladokun. *Mould Growth Prediction in Tropical Climate Buildings by Hygrothermal Differentials* [Dissertation]. Unpublished: International Islamic University Malaysia; (2015)
44. C. Sekhar. Thermal Comfort in Air-conditioned Buildings in Hot and Humid Climates – why are we not getting it right? Indoor Air. (2015)

The Green Electricity: The Potential of Using Solar Pv Panel for The Royal Malaysian Police (RMP) Building

Md Hafiz Md Yasin[1,a] and Nur Azfahani Ahmad[2]

[1]*Master of Green Architecture, Faculty of Architecture, Planning and Surveying, UiTM Perak, 32610, Perak, Malaysia.*
[2]*Department of Building Surveying, Faculty of Architecture, Planning and Surveying, UiTM Perak, 32610, Perak, Malaysia*

Abstract: As the year 2020 approaches, Malaysia aims to be self-sufficient and independence in finding resources to fuel its development, especially in energy sector. To achieve this aims, Malaysia need to strengthen its energy sustainability aspect and have a clear foresight in handling energy security issue. In coming decades, energy crisis due the existence of the significant energy threat may cause power interruptions and lead to many interrelated problems, especially safety and security issues. Issues such as unstable political and economic situation in Middle East country, Spratly issue, South China Sea territorial issue, war and terrorist may lead to unstable safety and security environment for Malaysia. The most significant organization to keep the safety and security of Malaysia is The Royal Malaysian Police (RMP). However, at the moment, this organization is fully-reliant on centralized electricity supply that may interrupt due to the future energy threat. Therefore, it is important for this organization to have an independence and self-sufficient electricity supply in order to remain their operation in any circumstances .The solar energy, through Photovoltaic (PV) panel, has been identified as the significant resources to achieve this purposes. With constant solar irradiance, Malaysia has great potential in solar energy harvesting. The energy is clean, environmental-friendly, reliable and readily-accessible throughout the year. This paper will review the potential of green electricity through solar PV in being the power supply for RMP in order to maintain the security and safety of Malaysia

1 Introduction

Electricity is a common energy and the utmost form of energy which contributed in driving the economies development of all nations in the world (Ali, Daut, & Taib, 2012). Therefore it is always a challenging to meet all the growing demand on time it is needed. Furthermore, according to Ali et al (2012) this growing demand was accompanied by the growing population. It is also a fundamental too many stakeholders, which comprise policymakers, businesses, industrial sector, transportation sector, construction sector, agricultural sector and larger communities of where their quality of life depend on the uninterrupted of energy supply (Ang, Choong, & Ng, 2015). It is almost impossible to achieve the sustainable development without sustainable energy supply (Ismail, Ramirez-Iniguez, Asif, Munir, & Sukki, 2014). This is supported by Selvakkumaran and Limmeechokchai (2013) which strengthened that energy security is an essential factor for countries aiming for sustainable development.

Energy is an indispensable factor to spur the economic growth and create the sustainable development of a country. Whilst, energy security is a situation of an independence supply of energy which able to meet demand at all times when it is needed (Hashim, 2010). An exploratory study by Hashim (2010) highlighted that Malaysia is facing an incoming issue of oil and gas declining[a]. Hashim (2010) outlined that Malaysia oil and gas reserves far less compared with Middle East country. In addition, the study by Othman and Jafari, (2012) found out the major reduction of oil reserves from 2001 to 2005. Meanwhile natural gas reported depleted mainly in 2004 and 2005. This scenario shows by Table 1 and Figure 1. This energy crisis occurs due to the significant energy threat that may cause power interruptions and lead to many interrelated problems, especially on the safety and security aspect. Issues such as unstable political and economic situation in Middle East country, Spratly issue, South China Sea territorial issue and war may lead to energy 'gap' (insufficient) scenario (Ahmad, 2011). This will eventually create unstable safety and security environment for Malaysia.

Since the Independence Day in 1957, Malaysia has relies on oil and gas as the main sources of electricity.

[2] Malaysia only accounts for only 3% and 8% of world discovered oil and gas reserves, respectively; but 30% and 15% of consumption Hashim (2010).

[a] Corresponding author:hafizyasin@rmp.gov.my

However this non- renewable energy depleted over the time whilst Malaysia moving towards rapid development. Obviously, this development will required substantial energy demand to support the country's needs (F.M Sukki R.R Inigueza, S.H. Abu-Bakar & S.H. Mohd Yasin B.G. Stewart, 2012). Concerning about the aforementioned issues, in July 2009, Malaysian Prime Minister, Datuk Seri Najib Tun Razak has launched the National Green Technology Policy and followed by the Renewable Energy Act, in 2011, as an effort to resolve the energy security issues (F.M Sukki et al., 2012). Figure 2 shows the effort made by the government in establishing the energy related policies and initiatives from 1971 to 2015.

2 The Scenario Energy Development In Malaysia

In Malaysia, the growing population has increased annually by 1.8% from 27.4 million in year 2007 and is expected to increase to 33.4 million in 2020 (Ali et al., 2012). Furthermore, Ali et al (2012) stated this figure is projected to be increased five times higher in 2030. With the growing population, the total electrical power generated has projected to be increased from 103.2TWh in 2009 to 274TWh in 2030(Ali et al., 2012).

This data has demonstrated a significant understanding that with the increased numbers of people, more energy is needed; therefore more power electricity will need to be generated. This view has been shared by Oh, Lalch and Chua (2014) which highlighted that the increasing trend of power generation is based on the trend of Malaysian oil and gas reserve margin. With the declining of the fossil fuel reserve, there is a possibility that the power generating capacity will be reduced as well. Oh et al (2014) emphasized that this will be a tough job for the Malaysian government to meet the rising demand unless, there are efforts in strengthening its power sector with renewable energy. Figure 3 shows the power development plant for Peninsular Malaysia from 2010 to 2031.

This situation has been driven by the rapid development of industrial and commercial sector. Johari, Hafshar, Ramli, and Hashim (2011) emphasized that industrial and commercial sector was the major energy user in 2009. This is shown in Table 2. However, Ali et al (2012) highlighted that the industrial and residential sector was the main reason towards the increasing trend of power electricity demand. The study by Ahmad (2011) revealed that the rapid development of Malaysia which commencing since 1980s, in conjunction with the construction of spectacular landmarks, such as Petronas Twin Tower, the Kuala Lumpur International Airport (KLIA), the Light Rail Transit (LRT) and the SMART tunnel system have increased the energy demand.

3 Future Issues Of Energy In Malaysia

Imbalance supplies between the energy demand and energy reserve will significantly give an impact towards the energy security aspect for Malaysia. Definitely, one of the main impacts is power blackouts (Byrd, 2014). Ahmad (2011) noted that power blackouts will significantly affect the networking connectivity of various technologies and loss of communication is considered a major issue for Malaysia. Furthermore, Ahmad (2011) also stated that the power blackout lead to massive problem in a community, for instance, delay in food supplies, increased crime rate, health disruption, cancellation of flights and many businesses mess. A loss of electricity not only can affect within domestic sector but also millions of Ringgit Malaysia (RM) in various industries sectors, tourism, transportation, distribution of clean water, security system and banking system. The issues of energy threat also associated with the external factors. Issues such as unstable political and economic situation from world oil major player in Middle East country, Spratly issue, South China Sea territorial issue, war and terrorist may lead to unstable safety and security environment for Malaysia (Ahmad, 2011).

Looking back to the histories of power shortage in Malaysia, several event of power blackouts were recorded (Ahmad, 2011). The national power disruption occurred on 29th September 1992 (2 days), 3rd August 1996 (12 hours), 10th August 1996 (16 hours), 4th September 2003 (5 hours) and 3rd February 2005 (2 hours) where the whole of Peninsular Malaysia including Kuala Lumpur, Selangor, Putrajaya, Johor, Malacca and Negeri Sembilan turn into darkness and have affected many commercial industries in Malaysia (Ahmad, 2011). Almost after 10 years, another massive power disruption occurred in Sarawak, the biggest state in Malaysia. It has been highlighted by Mering (2015), where the case occurred in 23 June 2015 and another conjunction cases, according to Davidson (2015) happened on 5 November 2015. The blackout caused severe traffic jams in the major urban centers of the state and caused huge losses for many businesses. Therefore, it is necessary to consider green electricity as alternative resources for generating electricity in Malaysia

4 Solar Electricity: The Green Electricity

The 'Green Electricity' term can be referred as the electricity generated from Renewable Energy (RE) sources, for instance, sun, wind, water, geothermal and biomass waste (Hobman & Frederiks, 2014). Generally, all the RE sources are non–polluting, inexhaustible and low emission of Carbon Dioxide (CO_2). Therefore, most researchers consider REs as the vitalalternative energy for the future (Ismail et al., 2014).

Solar energy has been identified as the substantial resources compared to others elements of Renewable Energy (Kadir, Rafeeu, & Adam, 2010).

DOI: 10.1051/matecconf/20166600090

Table 1: Comparison of Oil & Gas Reserves – Asia-Pacific vs. the Middle-East (Hashim, 2010)

Oil	Reserves		Consumption	
	billion barrels	% of world	million barrels per day	% of world
Asia-Pacific (Malaysia)	40.8 (5.4)	3% (0.4%)	25.4 (0.5)	30% (0.6%)
Middle-East	755.3	61%	6.2	7%
Gas	**Reserves**		**Consumption**	
	trillion cubic ft	% of world	billion cubic feet per day	% of world
Asia-Pacific	511 (87)	8% (1%)	43 (2.7)	15% (1%)
Middle-East	2,585	41%	29	10%

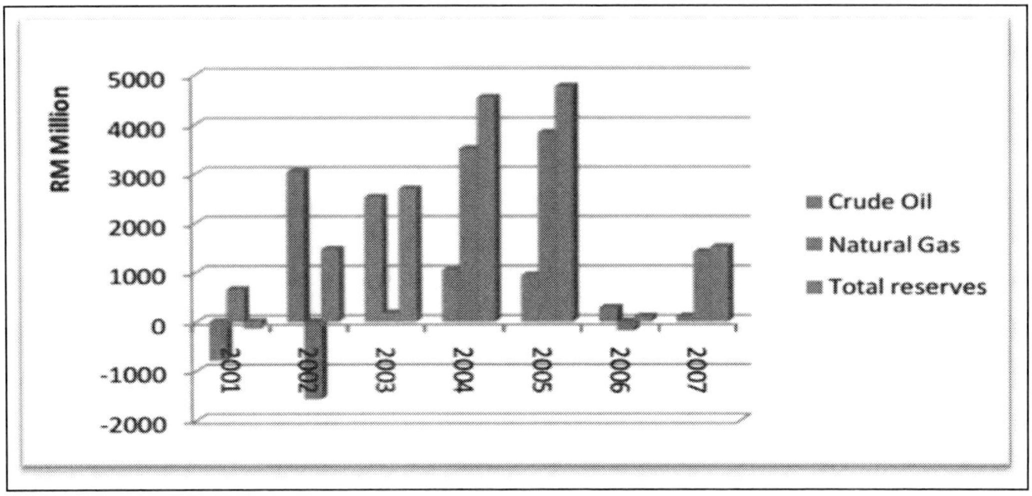

Figure 1: Malaysia natural reserves depletion for crude oil, natural gas and total reserves (Othman & Jafari, 2012).

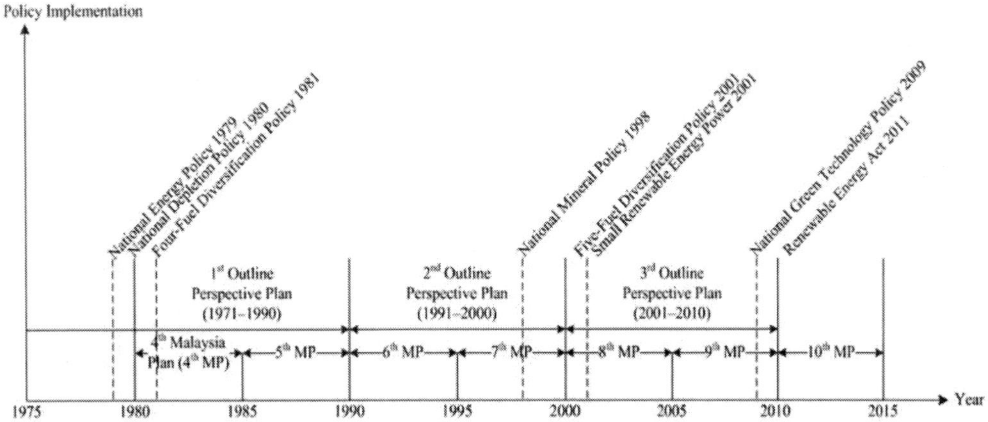

Figure 2: The timeline for energy related policies and initiatives, from 1971 to 2015 (C.S Khor, 2014)

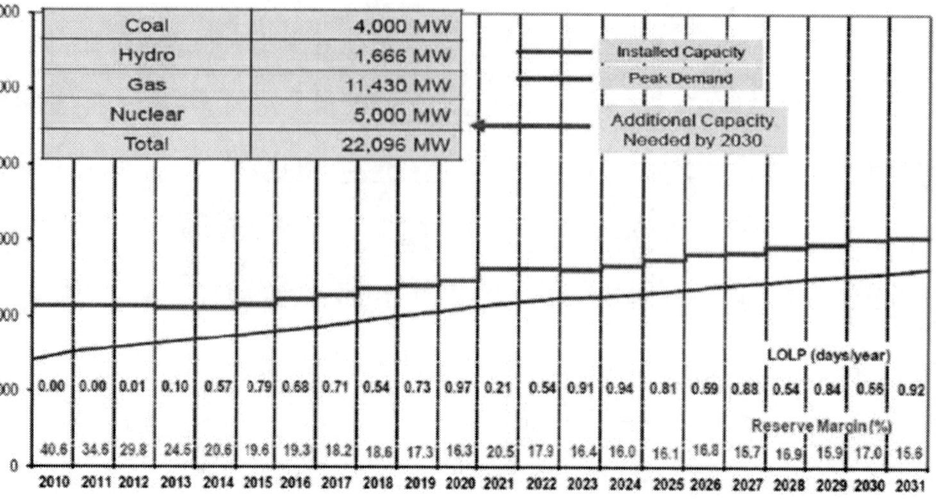

Figure 3: Peninsular of Malaysia power generation plan (from 2010 to 2031) (Oh et al., 2014)

Table 2: Energy Usage by Sector 2009 (Johari et al., 2011)

Sector	Sales of Electricity (GWh)	Percentage (%)
Industrial	40,233	43.4
Commercial	31,435	33.9
Residential	19,584	21.1
Agriculture	243	0.3
Public Lighting	1,208	1.3
Total	92,753	100

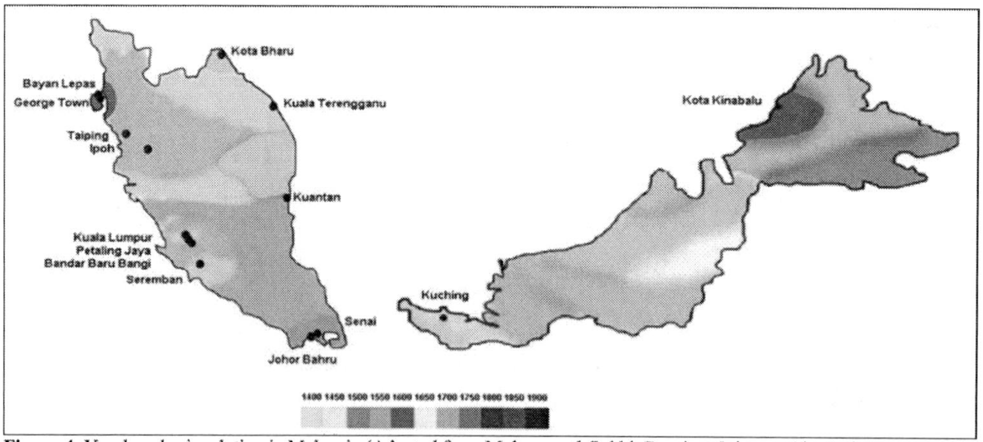

Figure 4: Yearly solar insolation in Malaysia (Adapted from Muhammad-Sukki, Ramirez-Iniguez, Abu-Bakar, McMeekin, & Stewart, 2011)

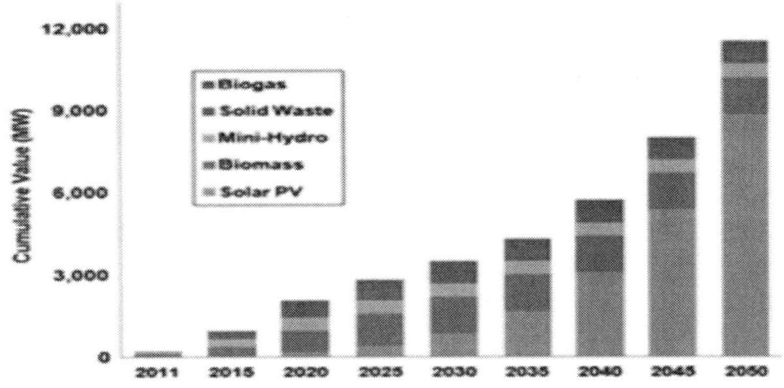

Figure 5: Cumulative value of RE in Malaysia (2011-2050). Adapted from (F.M Sukki et al (2012)

This is supported by Ismail et al (2014) which highlighted that solar energy is one of the most promising and prominent forms of RE. Malaysia has a great potential in harvesting solar energy. With the location situated in the South East Asia region and comprising with the total area of 329, 847 km^2 (F.M Sukki et al., 2012), Malaysia receives abundant sunshine every year. The daily sunshine irradiance averaging at 1643kWh/m^2 per year (Chua & Oh, 2011). Ismail et al (2014) stated that Malaysia receives a high amount of solar insolation, ranging from 1400 to 1900 kWh/m^2 per year and according to Amin, Lung, & Sopian (2009), Malaysia experiencing 10 sun hours per day. With that constant solar irradiance, the potential of solar energy that could be harvested until 2030 is up to 6,500MW. Figure 4 show the yearly solar insolation in Malaysia.

Given its potential and supported with the government scheme, such as Feed In Tariff (FiT), the solar PV market penetration in Malaysia is expected to expand in the future (F.M Sukki et al.,(2012). Figure 5 shows the potential of solar energy sector in Malaysia through the cumulative value of RE provided by the Government. The solar PV system itself contains super advantages compared with others technologies. According to Frank Jackson (2007) among the well-known advantages is long life through its life cycle which is 21 to 25 years with low maintenance. In addition, the technology deliver varies efficiency from 7.5 % to 16.9 % accordingly with the type of cell material. This is supported by Hasimah and M.N.Khalid (2009) that had found out the comparative result for the different technologies of Solar PV system. Meanwhile, research by Amin et al (2009) notified different solar panel will demonstrate the different efficiency under different environment.

5 The Case Study

For this research, the Royal Malaysian Police (RMP) organization has been chosen as the case study. It is a known fact that the most significant organization to keep the safety and security of Malaysia is The Royal Malaysian Police (RMP). However, at the moment, this organization is fully-reliant on the centralized electricity supply from the main power utility. It is a major concern that the electricity supply may be interrupted due to the upcoming energy threat. Therefore, it is essential for this organization to have an independence and self-sufficient electricity supply in order to remain their operation in any circumstances. In this millennium era, globalization issue and advance-evolving technology are among the issues that must be control by the Royal Malaysia Police (RMP) in order to ensure the effectiveness of its

DOI: 10.1051/matecconf/20166600090

enforcement. A world without borders has been created through this rapid development of technologies which requires the paradigm shift in police enforcement. Therefore, RMP need to be more adaptive to the environment; to be always in "alert" situation and moving along with the latest technology development. They need to be aware on their capability in managing current threat, either from domestic or international level. Therefore, it is wise to consider alternative energy to support their networking and telecommunication system.

As early as 1996, various technologies have been adapted by RMP, such as Malaysia Automated Fingerprint Identification System (MAFIS), Electronic Monitoring Device (EMD), Unmanned Aerial Vehicle (UAV), Analytic CCTV system and Voice Biometric Identification System (RMP, 2014). On the 27 December 2012, RMP has launched the "Go Green PDRM" campaign to raise awareness among the police officers on the importance of green practice and the green technology application in offices, police stations and others police institutes around the state (RMP, 2014). The initiative taken has make RMP as the leader and the among the first government agencies that has implemented the green practices and application in the government sector. In 2014, the "Blueprint of Green Practices and Green Application" has been established and launched, indicating the serious effort by the RMP in implementing the green energy practices in policing environment ((KeTTHA), 2014). The blueprint has outlined the framework, the strategies and the application of green technology for RMP which shall cover the products, equipment's and systems that protect the environment and minimize the negative impact from human activities. The Renewable Energy (RE) such as solar, wind and others are among the green energy that encourages being use.

In the RMP actual task, power shortage or interrupted electricity supply for Information Communication Technology (ICT) and telecommunication facilities will be considered as a massive catastrophe especially during any crisis. Therefore, it is significant to have an alternative power supplies in the event of emergencies. This is necessary, especially considering the fact that the resources for generating conventional electricity (oil, gas and coal) is currently depleting (Hashim, 2010). Based on the valuable and important aforementioned study, it is vital for RMP for looking into alternative options for generating electricity during energy crisis. A police organization like Royal Malaysian Police (RMP) may have to face a risk of power interruptions that can harm the overall safety operations. Therefore to uphold the readiness level and sustainability purpose in the future, the Royal Malaysian Police (RMP) need to re-look over the alternative energy supply. Taking advantages with the blueprint released by RMP and considering the fact that Malaysia has substantial resources of solar energy, it is good step to synchronizing energy security issues with looking to solar energy as an alternative energy.

This paper will investigate the feasibility of solar energy through photovoltaic application in providing the green electricity to support the sustainability operation of RMP in the future. This study will provide an open solution on helping RMP to be more autonomous in generating own electricity.

6 The Methodology

The research begins with the desktop study by gathering the in-depth information through extensive literature review on the issues of energy security and its correlation with the renewable energy (RE), specifically on solar energy. The data of the daily power consumption (Kw/h) of RMP has been monitored and collected in order to evaluate the average of power consumption needed to be supplied by the solar PV systems.The research begins with the desktop study by gathering the in-depth information through extensive literature review on the issues of with the renewable energy (RE), specifically on solar energy. The data of the daily power consumption (Kw/h) of RMP has been monitored and collected in order to evaluate the average of power consumption needed to be supplied by the solar PV systems. The research extended to the simulation and modeling approaches by using prediction tools, known as Homer v2.8 (Severson & St.Leger, 2013). These prediction tools will process the data accordingly to the input data and variables before come out with the results.

For the pilot project, this study has chosen the Network Operation Centre (NOC) building. This center acts as the central networking manager for RMP operation and comprises with high tech ICT facilities. This building is operated within the 24 hours-basis throughout the year and continuous electricity supply is an obligatory. Power shortage may interrupt the whole RMP's operation thus leading a potential risk on country's safety and security environment. As shown in Figure 6, the 6th story of RMP building was selected. NOC is located on level three. Currently the energy backup for this building is derived from the 376 KVA diesel generator, which projected to deliver 360 KW of power in supporting various ICT facilities in the building. The generator only operated in auto switch mode during the occurrence of power shortage. Most of the time, this building relied on the electricity supplies from TNB. Therefore, it is essential to consider electricity generated from PV panels as an alternative option.The research methodology started by looking at a load profile of the building as suggested by Anwari, Rashid, Hui, Yee and Wee (2011). A one year (2014) electricity bill was collected and analyzed. From the bill, it has been identified that, this building consumed high energy with an average of 89000 kwh per month as show by Figure 7. Subsequently, the electricity bill is also higher averaging at RM 45,000 per month. The load profile of the building was compared with the energy security and its correlation

load profile energy of NOC. The comparison was important in order to verify that the NOC used more energy than the other building's level. The energy trend is also shown in Figure 8 and 9. From Figure 8, it has been observed that the peak time of NOC for day time was started from 0900 hrs until 1200 hrs and significantly drop after that until 1700 hrs. Meanwhile, Figure 9 has shown that at night, NOC has consumed almost 90% usage of energy in that building with daily energy at an average of 138kwh. The solar radiation data for the NOC was provided by Malaysia Meteorology Department at 5.4kwhm^{-2} and that indicates that this location experiencing almost 5.4hours of peak sun hours (PSH) throughout the day.

The areas available for solar panel installation also need to be identified. Therefore the roof sizing was modeling and applied as shown in figure 9. The measurement of the roof sizing was referring to the actual plan of the building which provided by the Jabatan Kerja Raya Malaysia (JKR). The medium power 250W solar panel was selected for this pilot project and the typical size of the solar shown in figure 10. From figure 10, Zone A has been identified as the most available area to install solar panel as the highest number of solar panel calculated available for that area. The site visit has been arranged to survey the area of the building and zone A was most significant area in term of shading, height and cabling distance. The shading analysis as shown by Figure 11 also identified the zone A, experienced less shading especially at peak hours' time. This method was following method as suggested by Frank Jackson (2007). This ongoing pilot study later will conduct numerical analysis to calculate the total solar panel, inverter, battery needed to engage with the existing system. Furthermore the suitable system either standalone system (off grid), grid connected system or hybrid system will be determined

DOI: 10.1051/matecconf/20166600090

through simulation using Homer Prediction tool.

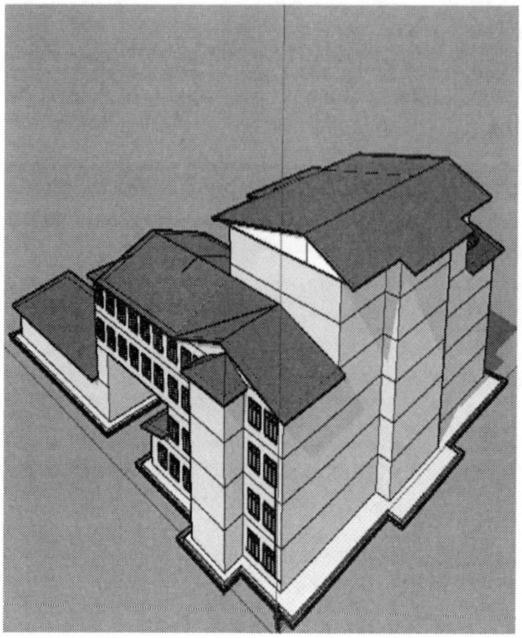

(a) Isometric view

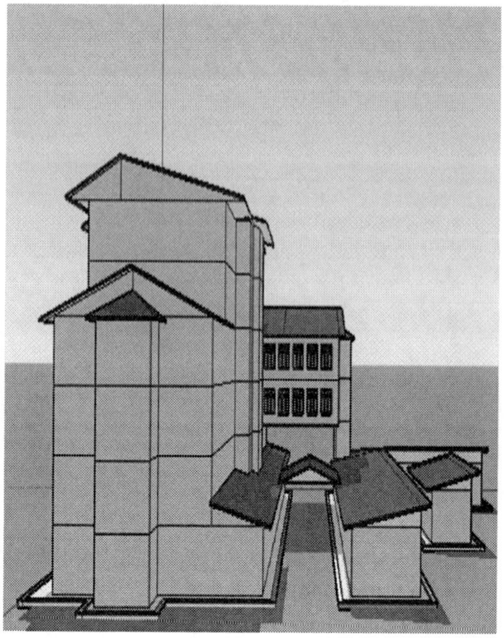

(b) Right view

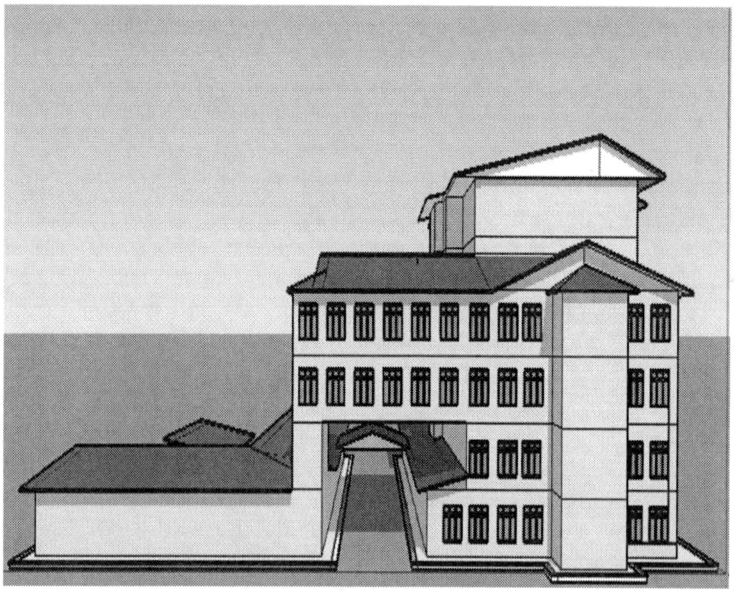

(c) Left view

Figure: 6(a), (b) and (c): RMP Building

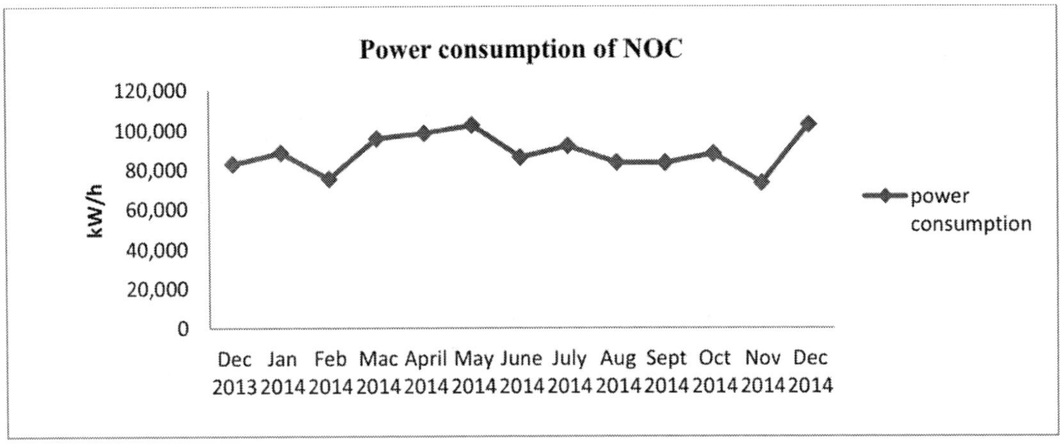

Figure 7: Network Operation Centre (NOC) power consumption (TNB, 2014).

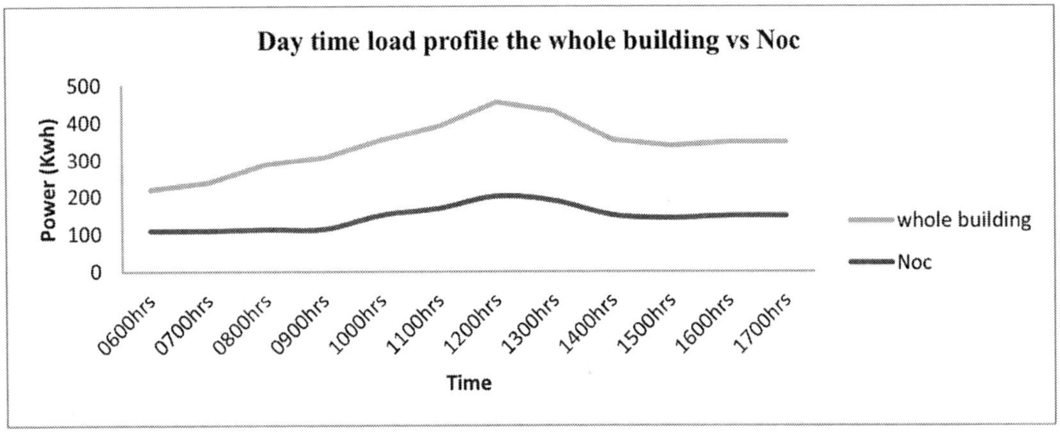

Figure 8: Network Operation Centre (NOC) power consumption

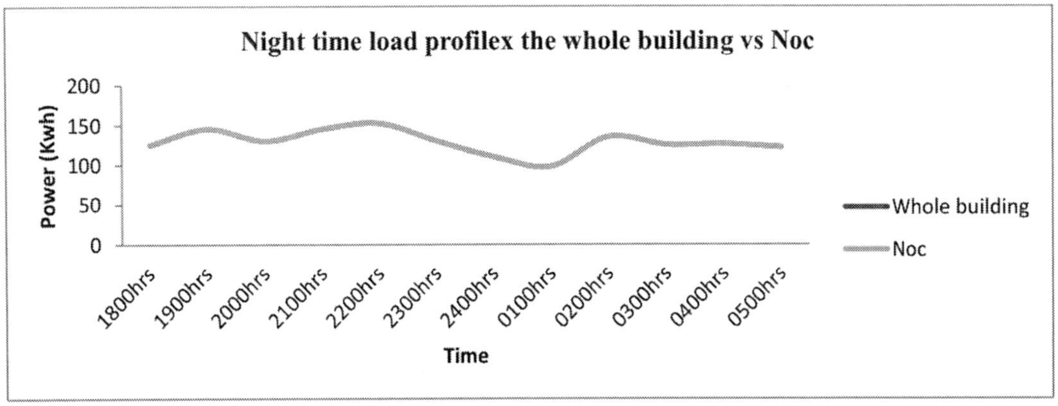

Figure 9: Network Operation Centre (NOC) power consumption

DOI: 10.1051/matecconf/20166600090

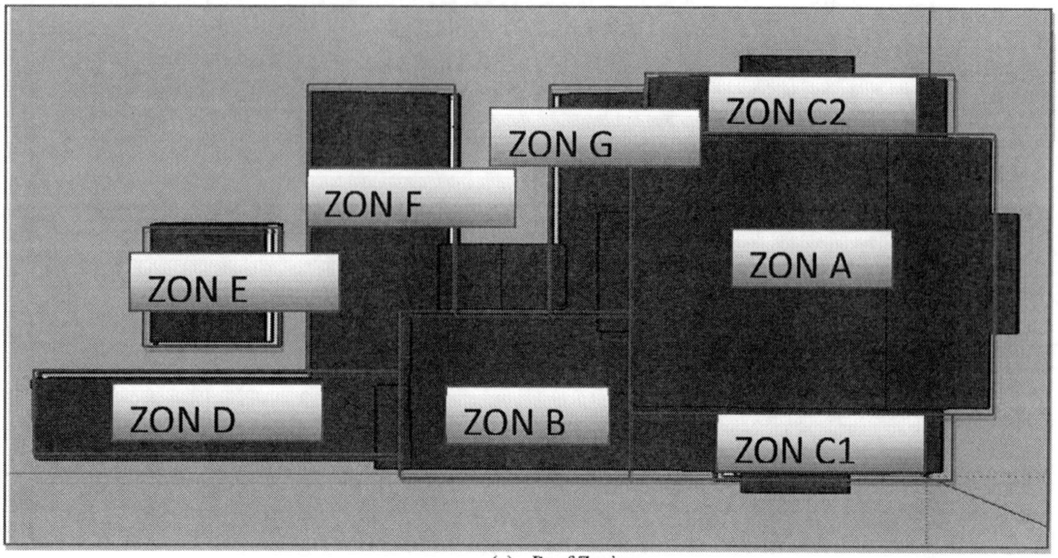

(a) Roof Zoning

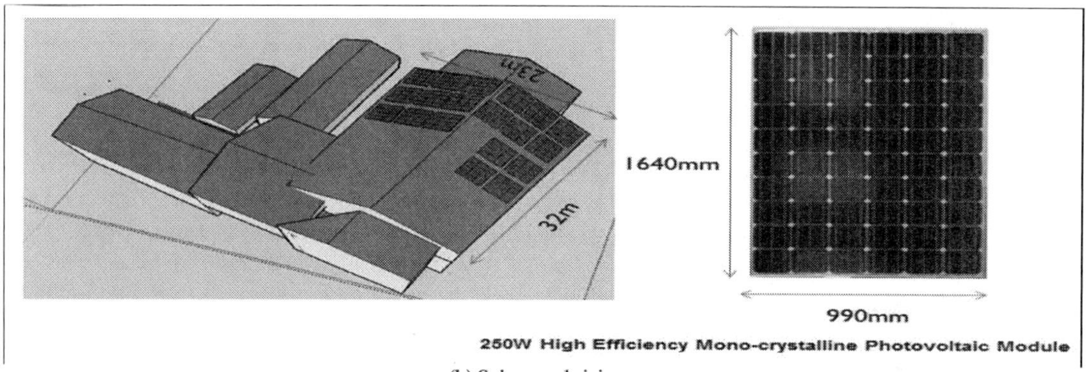

(b) Solar panel sizing

No	Zone	Length (m)	Width(m)	Area (m^2)	Solar PV capacity (Area / solar panel size) (m^2)
1	**A**	**33**	**22**	726	**454**
2	B	21	15	315	325
3	C1	22	10	220	170
4	C2	21	10	210	72
5	D	27	13	351	267
6	E	16	8	128	111
7	F	38	12	456	241
8	G	32	7	224	140

(c) Solar PV capacity

Figure 10 (a), (b) and (c): Roof sizing

DOI: 10.1051/matecconf/20166600090

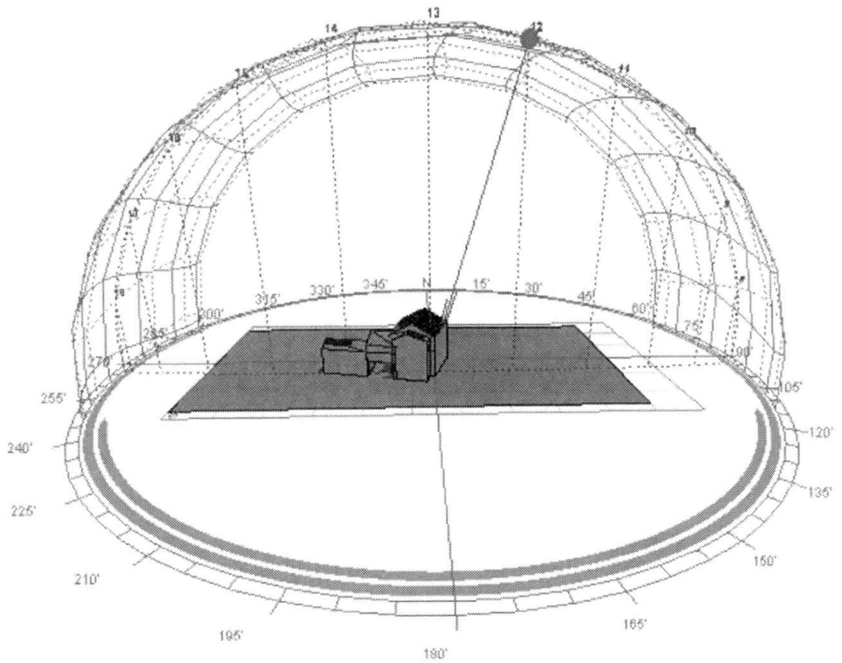

Figure 11: Shading analysis using Ecotect Analysis software

7 The Result

From the pilot study, it is observed that the cost of solar panel and others equipment would increase as well as capacity and efficiency increase. The cost estimation indicates in Table 3. The 250W Polycrystalline solar panel has been choose for this pilot project considering the moderate efficiency and price per unit. For the storage system, Deep Cycle battery preferred as low maintenance required due to free maintenance capability. Meanwhile the Pure Sine Inverter is selected to convert DC supply to AC supply.The total cost estimation was design in flexible manner according to percentage of system coverage.

The minimum cost started from RM 34,113 and maximum cost estimated as RM 341,143.00. The existing high capacity diesel generator will generate AC power during power blackout. It is unfavorable if the generator was not integrated with the proposed system. Therefore there is a need to look for the practical system for NOC operation. The Homer Software has been used to predict the several system performances. The simulation result as shown in Table 4. It was observed that Hybrid system indicate the lowest Net Present Cost (NPC) compared with others system. The NPC value represent the life cycle cost of the system (Kamali & Mekhilef, 2009). Kamali and Mekhilef (2009) state that smaller value of NPC for better system. Meanwhile the system which solely rely on generator record highest NPC followed by grid only system. Instead of that, hybrid system record lowest reading for the others criteria. Therefore hybrid system much preferred for this pilot project.

8 Respective Outcome

This research at the firsthand will provide the RMP on the information of fundamental issues about energy security in Malaysia and significance impact or risk that RMP may face in the future. This research also will provide the information on the prospective of Solar PV systems to support sustainability of RMP especially during energy crisis. Furthermore, this research will support the effort made by RMP to encourage the intensive use of Green Technology as mentioned in the blue print policy released by RMP. This is significance with the solar energy purpose in the research.

This research could be used as a guidance and reference for RMP in the process of acquiring suitable solar PV technology from many type of Solar PV system offered in the commercial market. A proper justification will be made base on the research output and not totally depended on the supplier information; this includes quotation and catalogues. It is hoped that in the future there will be standard framework that could be adopted for the new building design and construction. This wills covers methodology applies through the modeling; simulation and economic analysis include detailed specification on preparation and evaluation. Instead of considering the energy security issues and the green electricity prospective, the solar PV system must able to achieve the "value for money" purposes, with comprises affordable price with the good quality and standard operation.

Table 3:Cost estimation

Equipment	Type	Model	Capacity (W)	Quantity (100% System)	Market Price per unit (RM)	% Coverage 100%	70%	50%	30%	10%	Module class	Size	Application
Solar PV	Mono Crystalline	SolarWorld	290	154	1318	202972	142080	101486	60892	202972	High Efficiency	Smaller	large scale application
			285	157	1254	196878	1378146	98439	59063	19688			
			280	160	1121	179360	125552	89680	53808	17936			
		Q-Cells	275	162	1098	177876	124513	88938	53363	17788	Medium	Medium	
		Q-Cells	270	165	1053	173745	121622	86873	52124	17375			
	Poly Crystalline	Q-Cells	260	171	920	159030	111321	79515	47709	15903			
		Trina solar	250	178	893	158954	111268	79477	47686	15895			
	Amorphous		100	**NOT PREFERED:** Lower efficiency rates mean you need a physically larger system for the same power output. • Installation costs goes up because you need to install more solar panels. • Thin film solar panels tend to degrade faster. • Silicon costs are declining and silicon-based solar panels (e.g. mono- and polycrystalline) are becoming more affordable.							Low cost	Bigger	Small scale
			80										
			70										
			60										
			50										
Battery	Deep Cycle battery	Bluesun-lead-acid-AGM	12V 600Ah	23	860	19780	13846	9890	5934	1978	High Efficiency & low maintenance		
Inverter	Pure sine wave inverter	AIMS solar power	12000W Pure sine Inverter 120/240	14	11600	162400	113680	81200	48720	16240	Price increase as capacity increase		
Generator	High Capacity Generator	Volvo Penta	470KVA	1	350,000	EXISTING 376KW					Price increase as capacity increase		
TOTAL ESTIMATION COST (FOR PV SYSTEM USING 250W SOLAR PANEL)						**$341,134**	←——→			**$34,113**			

Table 4: System Comparison

No	System Type	Net Present Cost ($US)	Cost Operating	Cost Of Energy (COE) ($US)	Grid Purchase (kwh/yr)	Generator Production (kwh/yr)	Photovoltaic (PV) Production (kwh/yr)	Emission (kg/yr)
1.	Grid Only	$3,155,475	$246,842	$0.22	1,122,011	-	-	709,111
2.	Grid + Generator (Generator fully used)	$6,719,763	$532,619	$-0.122	0	-	-	890,651
3	Grid + Generator (Generator schedule)	$1,909,678	$142,413	$0.09	1,033,772	547,500	-	735,062
4	Grid + Generator schedule + Solar PV system (Hybrid system)	$2,245,360	$163,656	$0.12	998,972	411,720	71,300	700,343
5	Grid + Generator schedule + Solar PV system+ Battery (Hybrid system)	$2,120,392	$156,573	$0.11	965,500	411,720	71,701	685,508

References

1. (KeTTHA). (2014). Launch Of PDRM Blueprint On Geen Practices and Green Technology Applications. Retrieved from www.greentechmalaysia.my/file/media/MGTC PDRM Blueprint.pdf
2. Ahmad, N. A. (2011). Towards Sustaninable in Malaysia: IN the Perspective of Energy Security for Buildings. *Procedia Engineering*, ***20***, 222–229.3
3. Ali, R., Daut, I., & Taib, S. (2012). A review on exixting and future energy sources for electrical power generation in malaysia. *Reneawable and Sustainable Energy Reveiewa*, ***16***(2012), 4047–4055.
4. Amin, N., Lung, C. W., & Sopian, K. (2009). A practical field study of various solar cells on their performance in Malaysia. *Renewable Energy*, ***34***(8), 1939–1946. http://doi.org/10.1016/j.renene.2008.12.005
5. Ang, B. W., Choong, W. L., & Ng, T. S. (2015). security: Definitions, dimensions and indexes. *Renewable and Sustainable Energy Reviews*, *42*, –1093. http://doi.org/10.1016/j.rser.2014.10.064
6. Anwari, M., Rashid, M. I. M., Hui, H. I., Yee, T. W., & Wee, C. K. (2011). Photovoltaic power system simulation for small industry area. *International Conference on Electrical, Control and Computer Engineering 2011 (InECCE)*, 263–268. http://doi.org/10.1109/INECCE.2011.5953
7. Chua, S. C., & Oh, T. H. (2011). Solar Energy Outlook in Malaysia. *Renewable and Sustainable Energy Reviews*, ***16***(2012), 564–574.
8. Davidson, D. (2015). swk blackout2. Retrieved January 5, 2016, from http://www.themalaysianinsider.com/malaysia/article/power-back-after-massive-sarawak-blackout
9. 9. F.M Sukki R.R Inigueza, S.H. Abu-Bakar, A. B. M., & S.H. Mohd Yasin B.G. Stewart, S. M. M. (2012). Solar photovoltaic in Malaysia:The way forward. Renewable and Sustainable Energy Reviews, 16(2012), 5232–5244
10. 10. Frank Jackson. (2007). Planning and

Installing solar PV system. Infor (2nd editio). Earthscan

11. Hashim, S. M. (2010). A Malaysian National Prerogative: Engaging West Asia for Long Term Energy deals and Co-operation. International Journal of West Asian Studies, **2**(No 1), 19–35. http://doi.org/10.5895/ijwas.2010.02
12. Hasimah, A. R., & M.N.Khalid. (2009). Assessment of PV Cell Performance Under ActualMalaysia Operating Condition. Power Engineering Confrence. Australia.
13. Hobman, E. V., & Frederiks, E. R. (2014). Barriers to green electricity subscription in Australia: "Love the environment, love renewable energy but why should I pay more?" Energy Research & Social Science, 3, 78–88. http://doi.org/10.1016/j.erss.2014.07.009
14. Ismail, A. M., Ramirez-Iniguez, R., Asif, M., Munir, A. B., & Sukki, F. M. (2014). Progress of solar photovoltaic in ASEAN countries : A review. Renewable and Sustainable Energy Reviews, **48**(2015), 399–412.
15. Johari, a., Hafshar, S. S., Ramli, M., & Hashim, H. (2011). Potential use of solar photovoltaic in Peninsular Malaysia. 2011 IEEE Conference on Clean Energy and Technology (CET), **2**, 110–114. http://doi.org/10.1109/CET.2011.6041446
16. Kadir, M. Z. A. A., Rafeeu, Y., & Adam, N. M. (2010). Prospective scenarios for the full solar energy development in Malaysia. Renewable and Sustainable Energy Reviews, **14**(2010), 3023–3031.
17. Kamali, S. K., & Mekhilef, S. (2009). Evaluation study on grid connedted PV system at Universiti Malaya. Technical postgraduates(TECHPOS), International Conference. Kuala Lumpur: IEEE. http://doi.org/10.1109/TECHPOS.2009.5412043
18. Mering, R. (2015). swk blackout. Retrieved January 5, 2016, from http://www.thestar.com.my/news/nation/2013/06/28/sarawak-endures-worst-ever-blackout/
19. Muhammad-Sukki, F., Ramirez-Iniguez, R., Abu-Bakar, S. H., McMeekin, S. G., & Stewart, B. G. (2011). An evaluation of the installation of solar photovoltaic in residential houses in Malaysia: Past, present, and future. Energy Policy, **39**(12), 7975–7987. http://doi.org/10.1016/j.enpol.2011.09.052
20. Oh, T. H., Lalchand, G., & Chua, S. C. (2014). Juggling act of electricity demand and supply in Peninsular Malaysia: Energy efficiency, renewable energy or nuclear? Renewable and Sustainable Energy Reviews, **37**, 809–821. http://doi.org/10.1016/j.rser.2014.05.070
21. Othman, J., & Jafari, Y. (2012). Accounting for Depletion of Oil and Gas Resources in Malaysia. Natural Resources Research, **21**(4), 483–494. http://doi.org/10.1007/s11053-012-9192-2
22. RMP. (2014). Kepolisan Berteknologi Tinggi. Retrieved November 22, 2015, from http://www.bharian.com.my/
23. Selvakkumaran, S., & Limmeechokchai, B. (2013). Energy security and co-benefits of energy efficiency improvement in three Asian countries. Renewable and Sustainable Energy Reviews, **20**, 491–503. http://doi.org/10.1016/j.rser.2012.12.004
24. Severson, B., & St.Leger, A. (2013). Feasibility Study of Photovoltaic Panels inMilitary Temporary Housing Structures. Green Technologies Conference,2013 IEEE. Denver,CO.
25. 25. TNB. 2013-2014 NOC electricity bill (2014). Jn Semarak. Kuala Lumpur

Perceived strategic implication of solid waste recycling across Malaysian higher education institutions

Mohamad Rizal Baharum[1a], Hui Wen Kam[2] and Shirley Jin Lin Chua[3]

[1a]mrizal@um.edu.my, [2]bha110015@siswa.um.edu.my, [3]shirleychua88@um.edu.my
[1,2,3]Department of Building Surveying, Faculty of Built Environment, University of Malaysia, 50603 Kuala Lumpur, Malaysia.

Abstract. Sustainability of higher education institution campuses is the main concern in the institution sector worldwide. In achieving sustainability, strategic management such as planning, implementation and strategic implication of recycling initiatives became an essential criterion. This paper aimed to investigate the strategic implication achieved between higher education institution groupings. Questionnaire survey was conducted and completed by 129 institutions in Malaysia. The significance of strategic implication variables was identified through ranking analysis. An ANOVA analysis was conducted to identify the variation in effectiveness level of the overall strategic impacts across higher education institution groups. The findings revealed that the significant differences in effectiveness level of overall strategic impact occurred between the perception of private university and college. The implication of the results was discussed with regard to strategic impacts of solid waste recycling which is useful in the strategic planning of institutions' recycling initiatives.

Keywords: Malaysia, higher education institutions, strategic implication, solid waste recycling

1. Introduction

Sustainable development is the main issue in worldwide. Many declarations such as Rio declaration of Rio Earth Summit (1992), Kyoto Declaration (1990), Agenda 21 (1992), Thessalaniki Declaration (1997), and Rio+20 reveal that the increasing important of environmental concerns in all sectors. Rapid industrialisation and economic development has caused vast increase of solid waste generated [1]. Landfill method is always the general method used to dispose the solid waste in most of the developing countries, which include Malaysia. Past studies [2-4] also found that landfill is a common method utilised in disposing solid wastes in Malaysia where most of the landfill sites are open dumping sites.

The plea for sustainable recycling in developing countries has widely been recognised in many scholars [5-8]. However, previous studies [9-15, 1, 16] mostly concentrated the studies on municipal wastes. Albeit few past researches [17-18] were focused on institution wastes but they only conducted their research on particular university [17], not researched on entire higher education institution sector.

Higher education institutions are similar to large commercial concerns because of their (staffs and students) waste production and consumption of energy and materials [19]. Previous studies [20-22] confirmed that folks with higher education are expected to be more aware of environmental problems which would advocate them to recycle. Hence, a research regarding solid waste recycling on Malaysian higher education institutions was required to investigate the current implementation of solid waste recycling in the institutions.

2. Strategic implication

Strategic management is defined as the art and science of establishing, conducting and assessing cross-functional decisions that will permit an organisation to achieve its objectives [23]. While strategic planning is very important to increase the delivery of recycling performance to accomplish sustainable development and to strengthen long-term growth [24]. The concept of converting wastes into valuable monetary capitals, environmental and social revenues [25] has attained interest as a means of protecting environment because it provides one of the most viable solutions both cost-effectively and ecologically for administering waste [26]. Definitely, recycling conserves natural resources, consumes less manufacturing energy, reduces total disposed waste volume, lessen demand for virgin materials, minimises environmental and economic costs, and health as well as environmental risks [27-30].

[a] Corresponding author: mrizal@um.edu.my

DOI: 10.1051/matecconf/20166600091

Overall, five main strategic variables in the aspect of environment, economic, social and legal were identified and discussed in following sections.

2.1 Waste stream minimisation

Most of the recycling initiatives embarked were aimed for the environmental sustainability as a priority. A variety of benefits aroused from waste minimisation in business were identified in [31]'s study. The benefits consisted of increased profits, competitiveness, enhanced management control and corporate image. Besides, recycling generates employment and attracts investments [32-33]. The most obvious result of waste reduction can be seen through composting programmes [34]. Composting programme was deemed as successful initiative in wastes reduction since a composting programme carried out by a city council found that most of the households recorded between 40%-60% of reduction in solid waste amount for disposal which at the same time reducing the money spent on waste management [35]. Therefore, a waste manager or facilities manager shall put the achievement of waste stream minimisation as foremost aim in their solid waste recycling planning.

2.2 Cost reduction

One of the critical challenges with developing sustainable waste management system is the cost of implementation [36]. In commercial and business sector, cost reduction and profit generated is the first matter to look into when measuring the strategic implication. Several past researches proved the cost decrease on municipal solid waste management. For instances, the study conducted by [37] who examined the costs of municipal solid waste services and researched economies of scope and then concluded that there were cost savings from providing cooperative disposal and recycling services. Another research carried out by [38] found that recycling could reduce costs to the municipality for the solid waste collection and disposal by reducing the waste volume that the municipality deliver to its landfill. Besides, [39] researched on municipal solid waste recycling in Portugal found that incineration inclines to minimise costs by about 15%. Another study on animal wastes recycling also found the reduction of the raw materials cost used in manufacturing fertilizers [40]. As majority of the previous researches were on municipal solid waste recycling, thus it is vitally important to study solid waste recycling in other sectors such as higher education institutions; for instance to find out to what extent they are managing their solid waste recycling.

2.3 Revenue generated

Strengthening recycling programmes is not only extended the life span of landfill, it also sustains the economy as recycling provides widely lucrative business ventures opportunities [1]. Recycling initiatives such as incinerations, composting and waste collection are successfully brought additional incomes for the companies/business and waste collectors. For instance, [41] found that in the methods of waste recovery, incinerators could generate electricity to sell to power enterprises and then earn substantial extra incomes for the company. Besides, [40] researched on zinc waste found out that zinc producer made revenues by selling zinc waste and it was also reducing costs of waste disposal. [40] in their study also found that the job of waste picking for recycling constitutes a source of income for many impoverished residents in Brazil. However the institution solid waste recycling was still considered at its infancy since no significant recycling performance in higher education isntitutions, thus whether the current recycling programmes have brought extra income for the institutions is still questionable.

2.4 Change of recycling behaviour/culture

The escalating rate of solid wastes in Malaysia has manifested from the change of consumption customs among Malaysian, as their per-capita income has increased throughout the years, where they can afford for more consumers than before [42] along with the rapid population growth and also urbanisation [43]. Few scholars argued that inconvenience of recycling facilities is a key influence on one's recycling behaviour, who further mentioned that such concerns appeared to outweigh attitudes about the lasting importance of recycling behaviour [44-46]. Sustainability awareness is portion of community behavioural change. An investigation shown that there was an association of recycling behaviour between the participation of non-profit organisations, newspaper reading, politics and religious activities [29]. To become sustainable, extraordinary change to behaviour is necessity.

2.5 Compliance of Acts

Regulation is essential to ensure quality and efficiency profits [47]. Enhanced environmental protection through the optimisation of waste management practices is the typical focal point of waste management polices and technologies in countries where powerful legislation has been well formulated and immediate health concerns have been controlled [48-49]. In developed countries, the European Union (EU) directives enforced by EU towards its member states is effective as the recycling rate achieved by those developed countries is high.

Focusing in Malaysia, there is no specific measure in the aspect of waste minimisation and recycling [1]; the function of solid waste management is generally carried out together with other associated functions for instances public area cleansing, street-light and drainage maintenance and landscaping. Solid Waste and Public Cleansing Management (SWPCM) Act 2007 (Act 672) is the sole regulation which established for solid waste management purpose in all sectors. Under the Act, the manufacturers, importers, or dealers have the obligation to recycle their products under Clause 102(1). However, [1] argued that even though the enactment of the SWPCM Act 2007 has given legislative empowerment to the Malaysian government and framework for the solid waste management, it cannot be enforced and implemented absolutely due to the lacking of other supporting regulations. [1] further elaborated that policy implementation could only be executed informally and in an ad-hoc practice by stakeholders. Although mandatory recycling which commenced in September 2015, this is not explicit that this will resolve the issue and it does not include commercial and institution sectors. It is noted that strong legislations are definitely required; the cooperation from each organisation is also needed for the enforcement of the legislations.

3. Malaysian higher education institutions

Higher education institutions portray a multi-faceted role within regional economies and are of key importance in the knowledge creation and transfer worldwide via teaching, research and other activities [50]. Additionally, universities have the potential of teaching environmental literacy to the politicians, teachers, and decision-makers of tomorrow [51]. The past studies [20-22] found that folks with higher education were expected to be more aware of environmental problems which would encourage them to recycle. [52] concurred that higher education institutions could educate and demonstrate the principles of awareness and stewardship of the natural world, increased the chances of clean and pleasurable local and universal environments for the future.

Basically, Malaysian higher education institutions are categorised as public university, private university and college. According to the [53], total population of registered higher education institutions in Malaysia comprise 20 public universities, 71 private universities and 326 colleges. The type of premises of the higher education institutions are not taken into account for this study, as not all the higher education institutions have possessed property for their campus. Most of the public universities have owned their campus whereas majority of the private universities only held a lease on a premises for their campus. This is similar with colleges as most of the colleges are small institutions and do not possess property for their campus.

4. Methods

This study adopted a quantitative method by distributing questionnaire survey to the targeted respondents. Survey research provided a numeric description of trends or opinions of a population by studying a sample of that population [54]. Furthermore, this study employed cross-sectional study that applied questionnaire for data collection, with the intent of generalising from a sample to a population.

Simple random sampling method was chosen for the questionnaire survey in this study where all the subjects in the sample have a change of being selected in the research population. This sampling procedure would able to make sure every subject has the equal chance to be selected [55]. The targeted respondents to answer the questionnaire questions were the person who responsible for solid waste management and recycling initiatives in their respective institution. The study covers all the Malaysian higher education institutions with total population of 417 institutions format. Therefore, questionnaires were sent to all the higher education institutions which registered with the Department of Higher Education (formerly known as Ministry of Higher Education) in Malaysia. As a result of the questionnaire survey, a total of 129 sets of questionnaire were returned which gave a response rate of 30.9% (as illustrated in Table 1). Based on the data obtained, some of the respondents, included in the category of "other", were either landscape architects, executives, engineers, principal or senior manager. 85.3% of the respondents were director or manager of department/unit, facilities/operation manager and assistant manager (as shown in Figure 1). They are appreciably expert in the planning and implementation of institution solid waste recycling strategies. In addition, the respondents also involved in waste management on day-to-day basis; and therefore the responses on the questionnaire are reliable.

Table 1: Number of return questionnaires based on Malaysian higher education institution schemes [53]

Ownership of Malaysian higher eduations	Malaysian higher education institution groups	Population	Responded
Public	Public university	20	13
Private	Private university	71	35
	College	326	81
Total		**417**	**129**

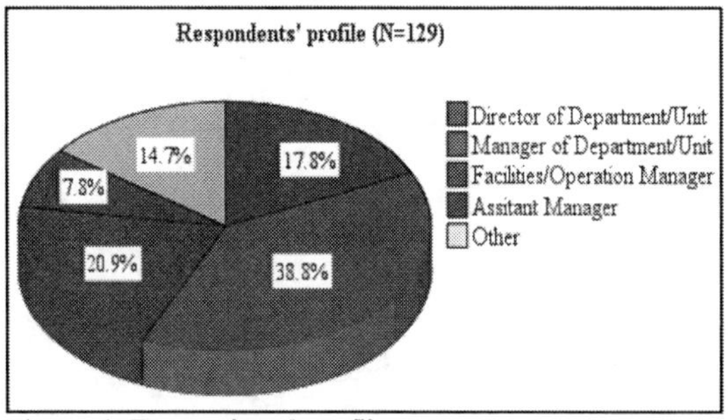

Figure 1: Respondents' profile

Before the questionnaires were entered into the database, all of the questionnaires were examined to ensure that they were valid for the research. The results obtained in questionnaire survey were then analysed through Statistical Package for the Social Science (SPSS) software. Reliability analysis was carried out for the strategic impact variables to enhance the reliability of the data. This analysis also can determined the possible accuracy of the measurements [56]. Cronbach's alpha test showed a coefficient of 0.871 which was considered satisfactory as [57] mentioned that the instrument with its alpha value between 0.65 and 0.95 was considered reliable.

Besides, mean and standard deviation were used in measuring the variables in ranking analysis. The mean score was used to rank the effectiveness level of the strategic variables. The main purpose of this ranking analysis was to indicate the difference in the level of effectiveness among the variables.

To test for the existence between two groups of significant differences with respect to higher education institution groupings and strategic impact variables, analysis of variance (ANOVA) was used.. ANOVA is a hypothesis testing procedure applied to compare the mean difference between two or more groups [2]. Thus, a hypothesis was developed:

H_0: There are no differences of the mean scores of strategic implication effectiveness level across three groups of higher education institutions.

H_1: There are differences of the mean scores of strategic implication effectiveness level across three groups of higher education institutions.

Respondents were asked to rate their effectiveness level of strategic implication variables listed on a five-point Likert scale (1 – not effective, 5 – extremely effective). All five (5) strategic implication variables were grouped as a *composite variable* for overall effectiveness level of strategic implication variables. Composite variable could be defined by the following equation:

$$\text{Strategic impact variable (SI)} = (SI1+SI2+SI3+SI4+SI5)/5 = \text{mean score}. \quad (1)$$

In this case, significance value determined whether the null hypothesis has to be rejected or not. ANOVA involved the calculation of F, which compares the between-groups variance with the within-groups variance. And p-value (significance level) of F is also calculated [58]. The probability level of 0.05 is normally an accepting level for the scholarly journals [2]. When p-value is less than 0.05, then null hypothesis is rejected and the groups are statistically different [58]. A normality test was carried out before the ANOVA test analysis. The data was normally distributed because both the skewness (.134) and kurtosis (-.893) values were within the normal distribution range.

5. Results and discussions

Results showed that the strategic impacts vary considerably in ranking with regards to its effectiveness level (Table 2). The mean score (with 1 – very low degree to 5 – very high degree) was applied to rank the level of effectiveness of the strategic impact variables. Based on the result from Table 2, mean score of the effectiveness of all Malaysian higher education institutions strategic impact variables were highly low and all were below 4.00 value of mean score. This shows that performance of solid waste management and recycling programmes in higher education institutions were not achieved the satisfactory level. This was most likely because solid waste management in Malaysia is still in operation level, most of the higher education institutions conducted recycling activities based on their voluntary initiatives and contributions.

However, compliance of Acts variable was ranked the most effectiveness among the five strategic impact variables, with 3.21 value of mean score; thus indicated its significance in solid waste recycling. Although the variable achieved the highest ranking of effectiveness level, the existing legislations in Malaysia are not comprehensive and still have loopholes on solid waste management in all sectors. Additionally, the existing waste-related Acts in Malaysia only make mandatory for institution organisations to dispose their waste in proper places. A new policy of mandatory of waste separation is imposed since September 2015, however the policy is only enforced on household in certain states. It is not a comprehensive policy because it is not imposed on all sectors and states. Additionally, no any recycling goal or target is required or imposed on higher education institutions. Hence, compliance of existing Acts is deemed to be achieved readily compare to another four variables due to the weaknesses of the existing policies.

Table 2: Ranking of institution strategic impacts towards solid waste recycling

Ranking	Strategic impact variable	Mean	Std. Deviation
1	Compliance of Acts	3.21	1.021
2	Waste stream reduction/Waste minimisation	2.53	0.969
3	Cost reduction	2.51	1.039
4	Change of recycling behaviour/culture	2.41	0.973
5	Revenue generated	1.84	.837

Waste stream reduction / waste minimisation variable was ranked as second highest of effectiveness level. This may because the government's primary aim and drive of recycling initiatives advocated is sustainable environment. Many recycling programmes are embarked with goals of sustainable environment by reducing the wastes, especially for composting programmes. There was a significant waste reduction at source through composting [34]. A home composting programme conducted by Petaling Jaya City Council, Malaysia found 40%-60% of reduction in solid waste volume [35]. Based around the results on Table 2, strategic impact variables such as cost reduction and change of recycling behaviour/culture were deemed as less effective variables. While revenue generated was ranked as the least effective variable. This was most likely because the recycling programmes initiated by public sectors were mostly aimed to reduce wastes. In contrast, the wastes which recycled by some industries exclusively meets the demands of economic production chains of the sector but not of importance to environmental management [59]. However in this juncture, recycling initiatives are still ignored by most of the private sectors as there is no mandatory policy imposed on them. However, in general, all strategic impacts are considered as critical in the way of attaining sustainability in solid waste management perspective after reviewing the literatures globally.

Although the effectiveness level of all strategic impact variables are ranked low, these five variables are believed as crucial based on the global literature reviews. Hence all five variables were then combined as a composite variable to analyse the overall performance of solid waste recycling at different higher education institutions. Similar procedure was applied by [60] in the study of shopping centre recycling success. The results of the ANOVA method were given in Table 3. From the result, the value of F ($df = 2, 126, p < 0.05$) = 5.1 was significant. As a consequence, null hypothesis was rejected and alternative hypothesis was accepted. This analysis performed that there were significant differences in the level of effectiveness of strategic impact variables among higher education institution groupings.

Table 3: ANOVA study result

ANOVA

Strategic implication

	Sum of Squares	df	Mean Square	F	Sig.
Between Groups	5.154	2	2.577	5.114	.007
Within Groups	63.495	126	.504		
Total	68.650	128			

A post hoc test was used to compare the means of the respondent groups since the results obtained from ANOVA test did not point out the actual point of variation or how these factors were differed. Result from the post hoc comparison showed statistically significant variation was found, as illustrated in Table 4.

Table 4: Post hoc result with equal variances assumed (Tukey HSD) test

Multiple Comparisons

Strategic implication
Tukey HSD

(I) Institution groups	(J) Institution groups	Mean Difference (I-J)	Std. Error	Sig.	95% Confidence Interval	
					Lower Bound	Upper Bound
Public university	Private university	-.04835	.23057	.976	-.5952	.4985
	College	.37740	.21210	.181	-.1256	.8804
Private university	Public university	.04835	.23057	.976	-.4985	.5952
	College	.42575*	.14359	.010	.0852	.7663
College	Public university	-.37740	.21210	.181	-.8804	.1256
	Private university	-.42575*	.14359	.010	-.7663	-.0852

*. The mean difference is significant at the 0.05 level.

Only the significance value (p-value) for the pair of *private university* and *college* was less than 0.05, and was statistically significant. This revealed that the effectiveness level of strategic impacts was considerably difference between private university and college even though both are from private sector. This comparison test also showed that strategic impacts perceived by colleges were not as effective compared to other two institutions groups. This may because universities are equivalent to huge commercial with regard to their larger quantities of wastes generated [19]. Additionally, universities also have more diverse programmes and activities than colleges because of the larger number of enrolled students. Hence, many initiatives are expected to be conducted to deal with the large quantities of wastes. Besides, universities have larger scale of organisation whereas colleges have smaller scale of organisation. Colleges may not have sufficient basic facilities and human resource to conduct recycling since they are only a small organisation. Hence, it is expected that colleges performed less effective in the operation of solid waste recycling.

6. Conclusion and implication

This is important to underline that this article does not provide the methods to attain the strategic implication on solid waste recycling; it is an analysis of variance applied to an overall strategic implication of solid waste recycling in order to increase knowledge about the strategic impacts towards higher education institution solid waste management. The global literature review revealed five strategic impact variables which are significant in solid waste recycling operation. Ranking analysis showed that Compliance of Acts was perceived as the most effectiveness for institution solid waste recycling. However in fact, review of waste-related legislation is imperative. The results obtained from analysis of variance found the significant difference on effectiveness level of strategic impacts between private university and college. The higher education institutions could introduce some effective recycling initiatives and also formulate mechanisms to increase the likelihood to achieve strategic impacts. The mandatory policy of waste separation would be the main initiative that must be considered to increase the recycling performance as this is in conjunction with the national policy that mandate households to conduct waste separation in their home. Despite the findings generated by the present study, future research is required to study in detail to what extent the difference of the solid waste recycling management between the university and college. Besides, a national study should be designed on strategic solid waste management in higher education institution to effectively increase its recycling performance. The results obtained from a national study would provide important implications for waste planning, policy making and transition to sustainable waste management for higher education institutions.

References

1. Y. C. Moh, L. Abd Manaf. Overview of household solid waste recycling policy status and challenges in Malaysia. Resour. Conserv. Recycl. **82**(0), 50-61 (2014).
2. M. N. Latifah, A. M. Derus, S. Wok, Z. Hamid, *Statistics made simple* (2nd ed.). Kuala Lumpur: Malaysia: Research Centre International Islamic University Malaysia (2004).
3. M. N. M. Yunus, K. A. Kadir, *The development of solid waste treatment technology based on refuse derived fuel and biogasification integration*. In: International Symposium on Renewable Energy. Kuala Lumpur, 14-17 September 2003 (2003).
4. The Ingenieur. Sanitary landfill: a strategic approach towards solid waste management. Boards of Eng. Malaysia (BEM) **42**, 12–16 (2009).
5. P. J. van Beukering, M. N. Bouman, Empirical evidence on recycling and trade of paper and lead in developed and developing countries. World Development **29**(10), 1717-1731 (2001).
6. D. C. Wilson, C. Velis, C. Cheeseman, Role of informal sector recycling in waste management in developing countries. Habitat Int. **30**(4), 797-808 (2006).
7. S. Suttibak, V. Nitivattananon, Assessment of factors influencing the performance of solid waste recycling programs. Resour. Conserv. Recycl. **53**(1–2), 45-56 (2008).
8. A. M. Troschinetz, J. R. Mihelcic, Sustainable recycling of municipal solid waste in developing countries. Waste Manage. **29**(2), 915-923 (2009).
9. M. H. Isa, F. A. H. Asaari, N. A. Ramli, S. Ahmad, S. S. Tan, Solid waste collection and recycling in Nibong Tebal, Penang, Malaysia: A case study. Waste Manage. Res. **23**(6), 565-570 (2005).
10. W. Murad, C. Siwar, Waste management and recycling practices of the urban poor: A case study in Kuala Lumpur city, Malaysia. Waste Manage. Res. **25**(1), 3-13 (2007).
11. M. O. Saeed, M. N. Hassan, M. A. Mujeebu, Assessment of municipal solid waste generation and recyclable materials potential in Kuala Lumpur, Malaysia. Waste Manage. **29**(7), 2209-2213 (2009).
12. MHLG. *Composting- closing the loop at home: A household home composting programme in Majlis Bandaraya Petaling Jaya (MBPJ)*. Retrieved on 21 June, 2013 from http://www.kpkt.gov.my/jpspn/main.php?Content=articles&ArticleID=50&IID= (2010).
13. P. Agamuthu, S. Chenayah, S. H. Fauziah, D. Victor, *3R Related Policies for Sustainable Waste Management in Malaysia*. Paper presented at the Innovation and Sustainability Transitions in Asia, Kuala Lumpur, Malaysia (2011).
14. R. Afroz, M. M. Masud, R. Akhtar, J. B. Duasa, Survey and analysis of public knowledge, awareness and willingness to pay in Kuala Lumpur, Malaysia – a case study on household WEEE management. J. Clean. Prod. **52**(0), 185-193 (2013).
15. I. S. Zen, Z. Z. Noor, R. O. Yusuf, The profiles of household solid waste recyclers and non-recyclers in Kuala Lumpur, Malaysia. Habitat Int. **42**(0), 83-89 (2014).
16. A. M. Akil, J. Foziah, C. S. Ho, The Effects of Socio-Economic Influences on Households Recycling Behaviour in Iskandar Malaysia. Procedia - Social and Behavioral Sci. **202**, 124-134 (2015).
17. R. Elfithri, T. K. Ghee, N. E. A. Basri, S. M. Zain, Integrated Paper Recycling Management System in UKM Campus. Procedia - Social and Behavioral Sci. **60**(0), 556-561 (2012).
18. S. M. Zain, N. E. A. Basri, H. Basri, N. Zakaria, R. Elfithri, M. Ahmad, K. G. Tiew, S. Zarina, Y. Sarifah, I. A. I. Khan, Focusing on Recycling Practice to Promote Sustainable Behavior. Procedia - Social and Behavioral Sci. **60**(0), 546-555 (2012).
19. P. Viebahn, An environmental management model for universities: from environmental guidelines to staff involvement. J. Clean. Prod. **10**(1), 3-12 (2002).
20. X. Chen, F. Lupi, G. He, J. Liu, Linking social norms to efficient conservation investment in payments for ecosystem services. Proceedings of the National Academy of Sci. **106**(28), 11812-11817 (2009).
21. D. Grazhdani, Assessing the variables affecting on the rate of solid waste generation and recycling: An empirical analysis in Prespa Park. Waste Manage. (2015).
22. S. F. Sidique, F. Lupi, S. V. Joshi, The effects of behavior and attitudes on drop-off recycling activities. Resour. Conserv. Recycl. **54**(3), 163-170 (2010).
23. F. David, *Strategic management*. Columbus, OH: Merrill Publishing Company (1989).
24. Ministry of Housing and Local Government (MHLG). *National Strategic Plan for Solid Waste Management in Malaysia: Volume 1 The Strategic Plan*. Ministry of Housing and Local Government (2005b).
25. US Environmental Protection Agency, *Wastes: resource conservation-reuse, recycle*. Retrieved from http://www.epa.gov/waste/conserve/rrr/recycle.htm. (2012).
26. A. Omran, A. Mahmood, H. Abdul Aziz, G. M. Robinson, Investigating households attitude toward recycling of solid waste in Malaysia: a case study. Int. J. Environ. Res., **3**(2), 275–288 (2009).
27. J. C. Van den Bergh, Environmental regulation of households: An empirical review of economic and

psychological factors. Ecol. Econ. **66**(4), 559-574 (2008).
28. B. Bolaane, Constraints to promoting people centred approaches in recycling. Habitat Int. **30**(4), 731-740 (2006).
29. M. Martin, I. D. Williams, M. Clark, Social, cultural and structural influences on household waste recycling: a case study. Resour. Conserv. Recycl. **48**(4), 357-395 (2006).
30. T. C. Kinnaman, Policy watch: Examining the justification for residential recycling. J. Econ. Perspectives, **20**(4), 219-232 (2006).
31. Envirowise, *Increasing profits by reducing waste in managed shopping centres*. Available at: www.envirowise.gov.uk (2002a).
32. P. J. H. van Beukering, T. R. Curlee, *Managing a material world: Perspectives in industrial ecology.* (Vellinga, P., Gupta, J., and Berkhout, F., Ed.). Dordrecht, Northerlands: Kluwer Academic Publishers (1998).
33. S. M. Kassim, S. M. Ali, Solid waste collection by the private sector: Households' perspective-Findings from a study in Dar es Salam city, Tanzania. Habitat Int. **30**(4), 769-780 (2006).
34. MHLG, *National Strategic Plan for Solid Waste Management in Malaysia: Volume 2 Main Report*. Ministry of Housing and Local Government (2005c).
35. National Solid Waste Management Department (NSWMD), NSWMD Guidelines. Retrieved on 3 March, 2013 from http://www.kpkt.gov.my/jpspn_en/main.php?Content=articles&ArticleID=12&IID= (2012).
36. C. Lakhan, Diversion, but at what cost? The economic challenges of recycling in Ontario. Resour. Conserv. Recycl. **95**, 133-142 (2015).
37. S. Callan, J. Thomas, Economies of scale and scope: a cost analysis of municipal solid waste services. Land Econ. **77**(4), 548-560 (2001).
38. O. Hernández, B. Rawlins, R. Schwartz, Voluntary recycling in Quito: factors associated with participation in a pilot programme. Environ. Urbanization **11**(2), 145-160 (1999).
39. P. Carvalho, R. C. Marques, Economies of size and density in municipal solid waste recycling in Portugal. Waste Manage. **34**(1), 12-20 (2014).
40. F. Murakami, A. Sulzbach, G. M. Pereira, M. Borchardt, M. A. Sellitto, How the Brazilian government can use public policies to induce recycling and still save money? J. Clean. Prod. **96**, 94-101 (2015).
41. Y. J. Bor, Y.-L. Chien, E. Hsu, The market-incentive recycling system for waste packaging containers in Taiwan. Environ. Sci. Policy, **7**(6), 509-523 (2004).
42. M. Abdul Jalil, Sustainable development in Malaysia: case study on household waste management. J. Sustain. Development, **3**(3), 91–102 (2010).
43. Z. Tarmudi, M. L. Abdullah, A. O. M. Tap, An overview of municipal solid wastes generation in Malaysia. Jurnal Teknologi. **51**(1), 1–15 (2012).
44. T. K. Foo, Recycling of domestic waste: early experience in Singapore. Habitat Int. **21**(3), 277-289 (1997).
45. Y. Y. Ho, *Recycling as a sustainable waste management strategy for Singapore: An investigation to find ways to promote Singaporean's household waste recycling behaviour.* (Master-thesis). Lund University. Retrieved from http://www.lumes.lu.se/database/Alumni/01.02/theses/ho_yanyin.pdf (2002).
46. M. Tongler, P. S. Phillips, A. D. Read, Using the theory of planned behaviour to investigate the determinants of recycling behaviour: a case study from Brixworth, UK. Resour. Conserv. Recycl. **42**, 191-214 (2004).
47. G. Bel, X. Fageda, W. Warner, Is private production of public services cheaper than public production? A meta-regression analysis of solid waste and water services. J. Policy Analysis and Manage. **29**(3), 553–577 (2010)..
48. S. E. Vergara, G. Tchobanoglous, Municipal solid waste and the environment: a global perspective. Annual Review of Environ. Resour. **37**, 277-309 (2012).
49. D. C. Wilson, Development drivers for waste management. Waste Manage. Res. **25**(3), 198–207 (2007).
50. P. Well, G. Bristow, P. Nieuwenhuis, T. B. Christensen, The role of academia in regional sustainability initiatives: Wales. J. Clean. Prod., **17**(12), 1116-1122 (2009).
51. D. J. Eagan, J. Keniry, *Green investment, green return: How practical conservation projects save millions on America's campuses*. National Wildlife Federation, Washington, DC (1998).
52. S. H. Creighton, *Greening the Lvory Tower. Importing the Environmental Track Record of Universities, Colleges, and Other Institutions*. MIT Press, Cambridge, MA (1999).
53. Ministry of Education Malaysia, *Institutions of higher education*. Retrieved 27 October 2014, from http://www.moe.gov.my/v/ipta (2014).
54. J. W. Creswell, *Research Design: Qualitative, Quantitative, and Mixed Methods Approaches* (3rd ed.): SAGE Publications (2009).
55. Y. P. Chua, *Mastering research methods*. Malaysia: Mcgraw-Hill Education (2012).
56. W. P. Vogt, *Quantitative research methods for professionals*. United States: Pearson/Allyn and Bacon (2007).
57. Y. P. Chua, *Mastering research statistics*. Malaysia: McGraw-Hill Education (2013).
58. A. Field, *Discovering statistics using SPSS* (3rd ed.). London: SAGE Publications Ltd (2009).
59. H. K. T. Campos, Recycling in Brazil: Challenges and prospects. Resour. Conserv. Recycl. **85**(0), 130-138 (2014).
60. M. R. Baharum, *Recycling Initiatives in UK Shopping Centres, Factors Critical to Success*. Unpublished PhD, Liverpool John Moores University, Liverpool (2011).

The Criteria For Decision Making In Adaptive Reuse Towards Sustainable Development

Noorzalifah Mohamed[1,a] and Kartina Alauddin[1]

[1]*Faculty of Architecture, Planning and Surveying, Universiti Teknologi MARA (Perak), Seri Iskandar, 32610 Perak, Malaysia*

Abstract. Adaptive reuse is an important strategy for achieving sustainability as it ensures a continuous building life cycle and prevents building destruction. Adaptive reuse has the greatest potential to lower the environmental load of the built environment significantly within the next 20 to 30 years. But, the owners and practitioners still lack of point of references to justify and evaluate their decision making with regard to the adaptation practice. The aim of this paper is to explore the unique criteria of decision making among practitioners in the application of adaptive reuse of existing buildings. The descriptive study has identified five main criteria influenced in decision making in adaptive reuse that effected to the sustainable development; economic, environment, social, legislative and architecture. We designed a questionnaire to determine the most significant criteria in decision making for adaptive reuse towards sustainable development among practitioners. The paper describes the process in identifying the unique criteria and the proposed methodology for the study.

1 Introduction

The climate change has led to significant carbon emission reduction strategy and low carbon urban planning. With the construction of the new building will use a large amount of raw materials and energy, and generates high carbon emissions (Langston, 2011; Gorgolewski, 2008). According to Zuo and Zhao (2014), buildings are responsible for 40% of total energy consumption and produce greenhouse gas emission which is responsible for global warming issues during their life cycle. In terms of waste per year, building construction produce for about 136 million tons of waste, almost half of it from demolition and this also lead to the use of large amounts of raw materials and a huge environmental damage (Gorgolewski, 2008; Rashid et al. 2015). Given these figures, the construction sector has critical to reduce carbon emissions significantly.

Malaysia intends to be a developed country by 2020 and the criteria that need to be achieved are to have the construction of low-carbon emission, sustainable and inclusive, and efficient use of resources. Thus, the new buildings with green requirement are growing rapidly. However, the sustainability cannot be achieved unless the existing building stock is being addressed carefully. This because, even every new building was a 'green building' but their impact on sustainability as a whole will be minimal for some while (Adeyemi *et al.*, 2014).

Therefore, it is important to consider the existing building stock in a sustainable development. As a developing country, Malaysia has a large stock of existing buildings and most of these buildings were not conserved well and most of them were in poor conditions and invaluable (Ahmad, 2009). To address the problem, adaptive reuse of existing buildings is seen as an important strategy for achieving sustainability. This is because, adaptive reuse can extend the life of buildings and avoid demolition wastes, promote reuse of contained energy and also provide environmental, social and economic benefits to the community (Yung and Chan, 2012). While the benefit of adaptive reuse widely espoused, however, it would appear that owners and practitioners still lack a point of reference to justify and evaluate their decision making with regard to the adaptation of adaptive reuse of existing buildings (Bullen and Love, 2011b). Douglas (2006) ague that decision making is complex in the adaptive reuse building because there are many parties involved in the adaptation and each party presents a different perspective. According to Yau (2010) most existing buildings in Malaysia were not categorized into the green responsive building. This is shown by low numbers (10 out of 21) of an adaptive

[a] Corresponding author: noorzalifahmohamed@gmail.com

reuse project that had been certified by Green Building Index (GBI) (see Table 1). Therefore, the role of industrial players is important to improve the number of adaptive reuse projects to increase the foregoing numbers. Thus, there is an urgent need to understand how practitioner or industry players make decision to adapt the green concept into adaptive reuse of existing building.

This study is still in progress. This paper described the literature review and the development of questionnaires for data collection. The data collection process is in between December 2015 and January 2016.

Table 1. Number of GBI Certified Projects by Category

Update on GBI	Total as of 15 Dec 2015	NRNC Non Residential New Construction	RNC Residential New Construction	INC Industrial New Construction	**NREB Non Residential Existing Building**	IEB Industrial Existing Building	T Township
Applied	707	369	276	20	**21**	4	17
Registered	664	341	264	19	**21**	3	17
Total Certified	335 (100%)	168 (50%)	138 (41%)	9 (3%)	**10 (3%)**	3 (1%)	7 (2%)
Received with Provisional Certification after DA	283	143	121	5	**6**	1	7
Received Final Certification after CVA	51	24	17	4	**4**	2	-
Received Renewal Certification after RVA	1	1	-	-	**-**	-	-

Source: *Green Building Index (GBI), 2015*

2 Adaptive Reuse Terminologies

Adaptive reuse usually refers to the reuse of the site or building for the purposes other than those originally intended or designed for. According to Enache (2014), adaptive reuse of buildings is one of the most important practices in order to maintain existing buildings, by changing its purpose, after the structure has reached its level of maturity within the life cycle. These new users can offer economic, social and cultural benefits to their environment. In additional, the adaptive reuse is one approach to sustainability where it kept the durability of the original building materials and reduced its carbon footprints. Adaptive reuse addresses issues of conservation and preservation of the built heritage, as well as strategies and policies. Once the old structures become inadequate to their functional and programmatic needs, adaptive reuse becomes a sustainable option for reclamation of sites.

Adaptation is derived from the Latin word 'ad' means (to) while 'Aptar' is (fit). Usually the definition of adaptation refers to a 'change of use', 'retention' of the original structure and fabric of the building and also extends the 'life expectancy' of the property of the person (Douglas, 2006; Bullen, 2007). In the context of building, adaptive reuse is a term that has been interpreted widely and is defined by many researchers (Douglas 2006, Bullen, 2007). There are different interpretations for "adaptive reuse" known as conversion, retrofitting, adaptation, refurbishment, rehabilitation and etc. (Langston, 2011; Douglas, 2006). According to Bullen (2007), adaptive reuse is a 'rehabilitation, renovation or restoration work that does not necessarily involve a change of use' and it will 'extend the useful life and sustainability in a combination of improvement and conversion'. Wilkinson and Reed (2008) defined adaptive as activity in maintaining as much of the original as possible, improve performance with modern standards and needs changing. They also state that refurbishment is a 'conversion describing a change in use'. Therefore, some the term refurbishment and adaptive reuse could mean the same thing.

3 Decision Making Issues

In previous studies, many researchers argue that the decision making to adaptive reuse of existing buildings is a complex stage (Douglas, 2006; Wilkinson and Reed, 2009; Alauddin, 2014). This is because there are many stakehoders involved in making decisions and each party has a different view. Those involved in the adaptation decision making are the owners, developers, producer, investors, regulators and marketers. They come from a variety of backgrounds and when the building adaptation be considered, it has triggered a range of views and perspectives based on their choice (Wang and Zeng, 2010). For example the building owner that wants to adapt the building will look into the financial resources they have. Investors want to see the future to build and marketers look to the latest market demands. Producer (professional team) will correlate with original architecture, structure, function, space and others when making a decision to reuse the building. In addition, another layer of complexity is when the stakeholders involved in the process of decision making at various levels and each had different degrees of influence (Alauddin and London, 2011). Furthermore the ability of stakeholders to influence the decision can be classified as either directly or indirectly. Thus, when a decision is made at the beginning of the process it will affect continue throughout the project. The desire of an organization that aspires to be the occupier or user of the building is yet another layer of complexity. This is due to the decisions taken will affect the activities or operations in the building. (See Table 2)

Table 2. Decision agent for adaptation of existing buildings (*Adapted Wilkinson and Reed 2009*)

Decision agents	**Description & Professional Affiliations**	**Stage in adaptation where decisions made**
Investors	Pension/ super funds, insurance companies, banks, independant investors, professionals who find capital to invest	Beginning/ early

Producers	Professional team- PM, QS, Architects, Engineers, contractors, surveyors, suppliers (RICS, AIAA, AIQS, AIBS, Fire engineers, structural and M&E engineers)	QS/A at feasibility stage Design stage Construction stage
Marketeers	Surveyors, agents, professional who find users for buildings (API, RICS)	During the design (if selling off plan) and/ or Construction stage
Regulators	Local Authorities, Planners, Heritage, Building Surveyors, Fire engineers (PIA)	During the design stage (and possibly during construction if amendments are made)
Policy makers	Federal, State and Local Government departments	Indirect effect on decision making in adaptation at all stages
Developers	Organizations who combine investment, production & marketing in whole or in part. Professionals from above bodies and others	Beginning/ early
Users-Corporate Residential	Large institutional owners and users individuals	

The next aspect is the range of options available to the stakeholders such as to change the use to a minimum because of the involvement of 'flexibility' is in the building as there was the first choice. The second choice has for small changes and adjustments to the third option required a higher level of engagement and are usually referred to as the 'reform' or 'retrofitting'. The fourth option involves the demolition while the fifth is an extension of the facility. Finally, the sixth is the demolition and redevelopment and selected when the economic, environmental and social conditions, physical and regulations are such that the building is said to be at the end of the life cycle and lack of utilities (Wilkinson, 2011).

4 Research Questions

Based on the issues raised in the literature review, research questions have been established as part of this research are:

1) What are the criteria of decision making among industrial players in the application of adaptive reuse for existing building?;
2) What are the most significant criteria in decision making for adaptive reuse of existing building?; and
3) What can be done in order to enhance the industrial player's knowledge in adaptive reuse decision making related to green elements?

5 Methodology

The three-stage of methodology was conducted to achieve the research aim, these are:

Stage 1 – Identifying the criteria of decision making and Developed the Instrument

The literature review process started by looking into primary and secondary sources. The primary literature sources included refereed journals, refereed conferences, dissertations/theses, occasional papers and government reports. Secondary sources included text books, trade journals, newspapers and magazines. The five top journals that were reviewed in the search are: i) ScienceDirect; ii) Emerald Group Publishing; iii) Google Scholer; iv) Springer Link; v) Elsevier. These top five journals were selected as they frequently publish scientific scholarly papers in multiple field and they have been ranked highly by several researchers (Maidin, A. B. *et al*, 2014).

Through an extensive literature review, the decision making criteria were identified (see Table 3) and were used to develop the questionnaire. Type of the instrument developed for this research was a closed-ended questionnaire with taking into consideration the aim and objectives of the study.

Table 3. Measures of decision making criteria

No	Measures of decision making criteria	Wilkinson, S. J. (2011)	Bullen & Love, P (2010, 2011 a&b)	Wang, H. & Zeng, Z. (2010)	Barrett, K. (2009)	Yildirim .M. (2012)	Wilson, C. A. (2010)
1.	Target Market/ Profit from market demand	x	x	x	x		x
2.	Sources of finance/Risk	x	x	x	x	x	x
3.	Subsidize/ Financial incentives	x	x	x	x		
4.	Initial investment and necessary investment in future maintenance		x	x			
5.	Benefit of exemption/ Tax concessions	x	x	x			
6.	Building value	x	x	x	x		
7.	Site and situation	x	x	x			x
8.	Land use plan or zoning	x		x			
9.	Potential environmental quality of the surroundings/ Contamination	x	x	x		x	x
10.	Sustainability	x	x	x	x		
11.	Reduce use of Greenfield sites		x				
12.	Reduction of resource consumption		x		x		
13.	Compatibility of newly introduced uses with existing			x			x
14.	Public interest	x	x	x			x
15.	Social and Cultural Value	x	x	x	x	x	
16.	Location (proximity to transport and amenities, cost of land)	x	x	x	x		x
17.	Enhancing the role of communities	x	x	x	x		x
18.	Retaining a sense of place		x		x	x	

19.	Regional development policies		x	x			
20.	Official Plan and Zoning regulations	x	x		x		x
21.	Building Code	x	x	x	x		x
22.	Heritage Designated Buildings	x	x	x		x	x
23.	Development or Government Incentives	x	x				x
24.	Conditions of integrity and authenticity		x	x		x	
25.	Building system/ Technological value	x	x	x			x
26.	Structural Condition	x	x	x	x	x	x
27.	Architectural Condition and Space layout	x	x	x	x		x
28.	Site Layout		x				x
29.	Building Suitability				x		
30.	Client requirement		x		x	x	

In order to develop a more comprehensive decision making criteria, the developed questionnaire was submitted to the professional to be discussed and analyzed. For the discussion and analysis, the criteria are grouped according related, measures and concluded by five condensed criteria of decision making are formed. Table 4 presents the criteria of decision making that will use for the conceptual model of study.

Table 4. Decision making criteria

Main criteria	Sub Criteria
Economic	Target Market/ Profit from market demand
	Sources of finance/Risk
	Subsidize/ Financial incentives
	Initial investment and necessary investment in future maintenance
	Benefit of exemption/ Tax concessions
	Building value
Environment	Site and situation
	Land use plan or zoning
	Potential environmental quality of the surroundings/ Contamination
	Sustainability
	Reduce use of Greenfield sites
	Reduction of resource consumption
Social	Compatibility of newly introduced uses with existing
	Public interest
	Social and Cultural Value
	Location (proximity to transport and amenities, cost of land)
	Enhancing the role of communities
	Retaining a sense of place
Legislative	Regional development policies
	Official Plan and Zoning regulations
	Building Code
	Heritage Designated Buildings
	Development or Government Incentives
	Conditions of integrity and authenticity
Architecture	Building system/ Technological value
	Structural Condition
	Architectural Condition and Space layout
	Site Layout
	Building Suitability
	Client requirement

The questionnaire was designed into two sections. The first section needs feedback about respondents' background as well as their experience. It was followed by the research content which is divided into five categories. First category will be touched on Economic Category, followed by Environmental Category and Social Category before ended with Legislative and Architectural Category. The respondents were asked to evaluate the listed criteria based on a 5 Likert Scale with 1 = Strongly Disagree to 5 = Strongly Agree. This method will easily compare the respondents' opinion either in positive or negative response according to respective criteria by neutral category included as applied by Addae-dapaah et al. (2009).

The basic questionnaire survey is divided into five major criteria to enhance stakeholders understanding to avoid duplication and redundancy of criteria overlapping each other. However, to achieve the research objectives and to answer research questions, all sections as well as categories (economic, environmental, social, legislative and architecture) will be combined into a single cluster.

Stage 2 – Distribute of Questionnaire

In order to achieve the objectives of the study, the researcher has identified the stakeholder and sources of information that involved in the adaptive reuse field as a target group. However, there are limitation to get the real numbers for the stakeholders that involve in adaptive reuse. Therefore, the stakeholders from the state under the world heritage is the target respondent. The data collected includes the lists of following stakeholders from different sources such as:

- Lists of Registered Architect (involved in adaptive reuse or equivalent) from Board of Persatuan Arkitek Malaysia (PAM).
- Lists of Registered Valuers from Board of Valuers, Appraisers and Estate Agents Malaysia (BOVAEA) in Penang and Melacca, Malaysia.
- Local authorities (involved in adaptive reuse or equivalent) in Penang and Melacca, Malaysia.
- List of Government Departments (involved in adaptive reuse or equivalent) in Penang and Melacca, Malaysia.
- List of Town Planners (involved in adaptive reuse or equivalent) in Penang and Melacca, Malaysia.
- Lists of Building Owners (involved in adaptive reuse or equivalent) in Penang and Melacca, Malaysia.

From the lists obtained, the researcher will develop a database known as Adaptive Reuse Stakeholders Database which will consist of potential respondents in regards to meet the objectives of the research. This stage is significant to ensure the reliability and the quality of data gathered will meet objectives requirement without prejudice the research analysis and findings.

A total of 65 survey forms that consist of 31 questions regarding decision making criteria complete with 'cover note' started to be distributed at early of December 2015 until end of January 2016. Bullen (2007) points out that the small size of the sample could be considered can reduce the global significance of feedback received with the balancing the value of obtaining opinions from a group of professionals with specialized and specific knowledge of sustainability principles. The collection of data also involved with reminders based on ad-hoc basis to increase the response rate of the research. This form had been sent according to Adaptive Reuse Stakeholders

Database developed at an earlier stage of the research activities.

A various medium of survey form distribution have been used where majority of them involving online medium through web-based hyperlink known as www.surveymonkey.com. The email address of the sample, where available have been sent through while in the absence of email address and immediate data collection, the researcher has decided to execute by-post and by-hand approach.

Stage 3 - Analysis of Quantitative Data

Data collected from the questionnaire form will be analyzed using quantitative techniques. Data from the questionnaire survey will be analyzed quantitatively through SPSS software involving the descriptive statistical method. Factor Analysis (FA) technique will be executed to achieve objectives of the study while Reliability Analysis (RA) will be performed accordingly to test the significant findings. The data collected will be formulated and designed in chart, table and other tools.

6 Conclusions and Further Steps

This paper has explained the complexity of decision making with regards to adaptive reuse of buildings and the criteria for decision making has been identified by thorough review of literature. This paper also presented the flow of the research activity to be conducted and administered to ensure the objectives of the research can be accomplished. It justifies the application of questionnaire, respondents and in-depth approach of analyzing the data collected to enhance the findings reliability. Thus, it is important for the researcher to ensure the process of research methodology being followed and perform accordingly. Until the end of January 2016, data that has been collected is 95% and now in the process of analyzing the data collected. This research will be complete within April 2016 for the purpose of obtaining a master's in Green Architecture and the final result of the analysis shall be published in the journal.

References

1. A. Ghafar Ahmad (2009). *Treatment of rising damp and replastering at heritage buildings*. Bengkel Bersiri 2009 Konservasi Bangunan Warisan Siri 2: Lepaan Kapur, 17-19hb. Julai 2009, Perak. Dewan Pusat Pelancongan Negeri Perak, Ipoh.
2. Addae-dapaah, K., Hiang, L. K., & Yen, N. (2009). Sustainability of Sustainable Real Property Development. *Journal of Sustainable Real Estate*, *1*(1), 203 – 225.
3. Adeyemi, A., Martin, D., & Kazim, R. (2014). Sustainable Improvement Strategy for Existing Office Buildings in developing Nations. *International Journal of Environment and Sustainable Development*, (1), 1–5.
4. Alauddin, K (2014). 'The Development of an Intellectual Capital Framework for Successful Adaptive Re-use', *School of Property, Construction and Project Management*. Melbourne, RMIT University. Doctor of Philosophy
5. Alauddin, K. & London. K. (2011). Design Management: Challenges for Adaptive Re- Use, *36th Annual Conference for Australasian University Building Educators Association*, 15.
6. Barrett, K. J. (2009). The key issues when choosing adaptation of an existing building over new build. *Journal of Building Appraisal*, *4*(3), 215–223.
7. Bullen, P. a. (2007). Adaptive reuse and sustainability of commercial buildings. *Facilities*, *25*(1/2), 20–31.
8. Bullen, P. a., & Love, P. E. D. (2010). The rhetoric of adaptive reuse or reality of demolition: Views from the field. *Cities*, *27*(4), 215–224.
9. Bullen, P. a., & Love, P. E. D. (2011a). Adaptive reuse of heritage buildings. *Structural Survey*, *29*(5), 411–421.
10. Bullen, P., & Love, P. (2011b). A new future for the past : a model for adaptive reuse decision-making. *Built Environment Project and Asset Management, 1*, 32–44.
11. Douglas, J. (2006). *Building Adaptation*. Butterworth-Heinemann, Oxford and Burlington, MA.
12. Enache, M. (2014). *Branding the Danish City Is Adaptive Reuse an option? Globalization*. Aalborg University, Denmark.
13. Gorgolewski, M. (2008). Designing with reused building components: some challenges. *Building Research and Information*, *36*(2), 175–188.
14. Greenbuildingindex.org. GBI Executive Summary. 26th December 2015, Retrieved from http://www.greenbuildingindex.org/
15. Langston, C. (2011). Modeling Risk Management in Sustainable Construction, 199–209.
16. Maidin, A. B., Shafri, S., & Ibrahim, S. (2014). Perkembangan dan Kemajuan Penerbitan Jurnal Malaysia: Strategi ke Persada Antarabangsa, (1), 1–5.
17. Rashid, R. A., Ismail, E. D., & Abdullah, A. S. (2015). Adaptive Re-Use of Heritage Buildings in Malaysia-A Case Study of Penaga Hotel in Penang. *Proceedings of the Colloquium on Administrative Science and Technology*, 361–372.
18. Wang, H.-J., & Zeng, Z.-T. (2010). A multi-objective decision-making process for reuse selection of historic buildings. *Expert Systems with Applications*, *37*(2), 1241–1249.
19. Wilkinson, S. J., James, K., & Reed, R. (2009). Delivering sustainability through the adaptive reuse of commercial buildings: the Melbourne CBD challenge. *Proceedings of the Pacific Rim Real Estate Society 15th Annual Conference, Pacific Rim Real Estate Society (PPRES)*, (June 2008), 1–19.
20. Wilkinson, S. J., & Reed, R. G. (2008). The Business Case for incorporating Sustainability in Office Buildings : the Adaptive Reuse of Existing Buildings. *Office*, 1–18.

21. Wilson, C. A. (2010). *Adaptive Reuse if Industrial Buildings in Toronto , Ontario Evaluating Criteria for Determining Building Selection*.
22. Yau, a N. G. H., Implementation, T. H. E., & Green, O. F. (2010). *Universiti Teknologi Malaysia Declaration of Thesis / Undergraduate Paper and Copyright The Implementation of Green Building Restricted Bachelor of Civil Engineering Faculty of Civil Engineering*.
23. Yildirim, M. (2012). Assessment of the decision-making process for re-use of a historical asset: The example of Diyarbakir Hasan Pasha Khan, Turkey. *Journal of Cultural Heritage*, ***13***(4), 379–388.
24. Yung, E. H. K., & Chan, E. H. W. (2012). Implementation challenges to the adaptive reuse of heritage buildings: Towards the goals of sustainable, low carbon cities. *Habitat International*, ***36***(3), 352–361.
25. Zuo, J., & Zhao, Z. (2014). Green building research – current status and future agenda : A review Why ? How ? How ? What ? *Renewable and Sustainable Energy Reviews*, ***30***, 271–281.

MATEC Web of Conferences **66**, 00093 (2016) DOI: 10.1051/matecconf/20166600093
IBCC 2016

Determinants of procurement strategy for construction works: quantity surveyors' perspectives

AbdulLateef Olanrewaju[1, a], Paul Junior Anavhe[2], Abdul Rashid Abdul Aziz[3], Chu Hui Chen[1], and Wai Soon Han[1]

[1]*Universiti Tunku Abdul Rahman, 31900 Kampar, Perak, Malaysia*
[2]*CDP Partnership Ltd, 3 Borno Road, Kaduna, Nigeria*
[3]*Universiti Sains Malaysia, Penang, Malaysia*

Abstract. The selection of the 'appropriate' procurement strategy is a prerequisite to the success or failure of a construction project. This paper investigates the factors determining the selection of appropriate procurement strategy for construction works in Nigeria. Data for the study was collected through an online survey questionnaire. The survey administration involves only quantity surveyors. Quantity surveyors were targeted because they advise clients and other stakeholders on procurement and contractual issues on construction works. A total of 33 usable responses were received and analysed for this study. On the basis of the results, it is concluded that the selection of procurement strategies for construction depend on complex interrelated factors. The study could not detect a particular factor or few factors responsible for a procurement strategy selection. The findings of this study is useful because it argues that the construction sector needs to broaden its considerations on the procurement strategy's determinants rather than focusing solely on the client's type and nature of projects as is often cited. Future research could segment these factors in terms of class of construction works or increase the sample size, which might lead to different findings.

1 Introduction

This paper is part of on an ongoing research focusing on construction procurement planning in Nigeria. Construction procurement is defined as the entire process in the delivery of the built facilities, from the point the client decides to build/construct to when the facility is in operation [1]. Construction procurement comprises many stakeholders such as clients, design professionals, construction professionals, and operational team. The major professionals in the construction sector involved in the procurement process include engineers, quantity surveyors, and architects. It had been argued that the procurement aspect of the project delivery processes is critical for the construction sector to deliver value for money. For instance, after analysing various reports [2, 3, 4] on the construction sectors including it was cited that if the construction sectors were to improve, it would stem from the procurement process [1]. While the process involved in the selection of a suitable procurement strategy is crucial, the purpose of this current study is to determine the factors accounting for the selection of a strategy for construction works.

The current research is different from existing known research because it examines these determinants from the procurement 'experts' themselves rather than questioning all the professionals or stakeholders involved in the construction sectors. For instance, in Nigeria, Dada [5] and Ojo and Ikpo [6] investigated utility factors for procurement involving various stakeholders in the sectors. In South Africa, Mathonsi, and Thwala [7] included all the professionals in the construction sector. In Australia, Love et al. [8] sampled various construction professionals and clients to determine the determinants of procurement selections. However, considering the expertise, competencies, and skills of the quantity surveyors it is contended that they are in a better position to provide valid and unbiased information on procurement related issues in the construction sector. The salient concern is to prevent the results to suffer from the 'averaging' process prevalent in the existing research. Even if industry-specific data is collected, the heterogeneity of the professionals could not be accounted for because they will be 'grounded' in the results. Therefore, the study aims to provide fresh knowledge on procurement issues. The remaining parts of the paper are structured as follows. The second part of the paper presents an overview of the significance of the construction sector. The third part highlights issues in construction procurement strategies. In section four, the skills, competencies, and roles of quantity surveyors are provided. Section five addresses itself to the research design. Section six focuses on the presentation of results and discussion. The summary of the study is presented in section seven.

[a] Corresponding author: abdullateef.olanrewaju@ymail.com

2 Overview of the Nigerian construction sector

The Nigerian construction sector provides almost 70% of the fixed capital formation. In 2012, the construction industry contributed N2, 188,718.59 million, 3.05% to the GDP and employed 6,913,536 people [9]. However, this figure is not static as it could increase or decrease due to the peculiar nature of the construction sector. It could be more if the labour force from the allied industries that produce, process, and transport construction materials, components, and services are added. The Nigerian construction sector is largely British with some incursion from Europe and America as such quantity surveying is offered in most of major universities and polytechnics. While poor service delivery of the construction projects is caused by various factors, the procurement strategy that is used to procure the projects has a profound influence on the success of projects. The focus of the sector is to achieve the values system set by a client. A client's value system is a logical set of values agreed and/or developed by a person, organisation, business, or society as a rule to guide its behaviour and decision preferences in all situations. The client value system depends on the experience, nature of business, and background of the client. For construction works, the client value system defines the worthiness and desirability of a construction to a client [10, 11]. The worthiness and desirability is measured by different criteria including function, quality, aesthetic, time, safety, reliability, time, capital cost, operating cost, sustainability, resale or rental value, flexibility, comfort, community, esteem, and politics. Various researches indicate that the selection of appropriate procurement alternative is crucial for the achievement of the client value system in a project and the performance of the entire construction sector. However, the performance of the Nigerian construction sector often suffers from delays in project completion, cost overrun, poor quality, high incidence of collapse buildings and bridges, and high maintenance cost [1].

3 Procurement strategies and determinants

There are proliferations of definitions for the term procurement, even within the construction sector. Here, construction procurement is defined as all the processes and activities involved in the delivery of the built facility from the point a client decides to build/construct to when the facility is in operation [1]. This provides that the procurement processes for a project, ranges from planning, designing as well as construction or maintenance of a project. The processes of procurement are not however, a pure science. The skills require of a procurement experts cover history, politics, economic, psychology, engineering, management, business, technology, administration, sociology as well as science. When construction is required, the next step is to decide on the suitable procurement method. Construction procurement method is the step, direction and procedure involve in acquiring a 'construction work'. The construction could be new facility or existing completed facility, maintenance services, refurbishment, renovation or rehabilitation. One type of method is not suitable for all clients. In Table 1 is the most cited methods [see also: 8, 10, 13]. The table will guides readers in identifying a suitable for a construction work.

According to Perry (1985), procurement could generally be typified into design separated from construction and design combined with construction [12]. The various types of procurement strategies reflect the extent to which the client is willing and able to bear certain levels of risk measure in terms of cost, quality, time, and other criteria within the client value system. The selected procurement strategy must ensure that risk is transferred to the best parties that can covert the risk/liability to assets.

Table 1, Characteristics of some methods of procurement

Factor	Characteristics
Build develop and operate	Consortium receive concession to finance, design, construct and operation project for an agreed time to
Novation	Client consultants prepare initial pre -contract designs but once contractors selected, the contractors is responsible the post-contract design stage.
Management contracting	Work packages let to work contractors and reports management contractors
Cost plus contract	Scope of work not known. Efficient if client has experience staff to overseen the works
Build lease operate and transfer	Private organisation design, constructs, and operates facility of leased land. The facility is return back to government after a period of time
Joint venture	A partnership of firms to execute a contract. Each has liability for their contract
Build own and operate	A private firm builds, owns and operates public structure. Government sometime provide some incentives
Build own transfer	a private firm receives a concession to finance, design, construct, and operate as define in the contract
Turnkey	Contractor/ developer agree to deliver the completed project. Once the project start client has little contribution to make. A well-defined scope of is not required
Unit price contract	Require complete plans, specifications.

DOI: 10.1051/matecconf/20166600093

Table 1, Characteristics of some methods of procurement (Cont'd)

Design and build	Contract design and construct the facility. Client consultant advise on contractor selection and sometime provide some supervisions of works
Construction management	The construction works executed by the trade contractors. Trade contractors' are overseen by construction manager. The contractors has direct contract with clients.
Prime contracting	A firm not necessarily in a contractors organise, coordinate all parties in a contract thereby providing single point contact to client
Fees contracting	Operate services for a fee, but investing in the capital
Traditional design, bid and construct	Design function is separate from construction function. Subcontractors report to main contractors. Scope of work clearly defined.
Cost reimbursement	Scope of work not known of defined when contract is entered
Private finance initiative	Comprises of some methods listed here. Collectively PFI is a method for government works. In general, a consortium, will design, delivery and possible operate the facility for a period.
Project management	Direct site supervise is require but general administrations
Lump sum contract	Require a well-defined scope of work

Once the need for constructing new projects emerges, arrangement will have to be made on the methods of financing the projects. This could be loans, debentures, stock or forming alliance or partnership with other stakeholders that have similar interests. Based on the first survey conducted by the Chartered Institute of Building (CIOB) into procurement in the construction sector, a major finding from the report indicates that 87% of respondents belief that good procurement is synonymous with a successful project [14]. The report also suggested that approximately 60% of the respondents believed that the chosen procurement method determined the success of the projects. Elsewhere, it was found that a suitable ways of reducing the cost of a construction is by selecting a suitable procurement type for the project [15]. The essence of selecting the appropriate procurement strategy is enormous. It guides projects to success. For instance, the procurement strategy/contract arrangement sets out the tasks, rights, obligations, and duties of all parties to the contract. It also includes other important details like the completion time, contract sum, quality of materials and components, and the overall quality standard the project must achieve as well as the level of workmanship. Additionally, it ensures that all obligations are dually and diligently carried out either before, during, or after the completion or the combination of the either. It was noted that by selecting an appropriate procurement strategy, uncertainties can be controlled, risk can be well managed, labour can be planned and standardised [10] and delay, disputes, and conflicts can be minimised if not avoided. However, the traditional procurement system is still the most dominant among the Nigerian clients. Previous studies have addressed the ranking of the dominant procurement in Nigeria [10, 16]. However, a number of factors require critical examinations in selecting a suitable method for a project. This current study addresses itself to the evaluations of these factors-determinant- in the procurement strategy selection by the quantity surveyors. There are various determinants/utility factors for construction procurement selections [8, 17, 18, 19, 20, 21] that have been investigated by previous researchers. But as previously mentioned these studies suffer 'averaging' problems.

4 The competencies, skills and knowledge of a quantity surveyor

Quantity surveying is carried out under different names. In countries like the USA, quantity surveying is closely related to cost engineering, while they are also referred to as cost economists or cost consultants in other places. However, quantity surveying is more than any of those names or titles. Perhaps the phrase "quantity surveying" is a catch-up term that hides a multitude of meanings. The services of modern quantity surveyors are required in all sectors of economic endeavour such as the financial, insurance, oil and gas, construction, as well as in the academic sector. The roles that the quantity surveyors perform today have diversified into industries including petrochemical, manufacturing, automobile, mining, telecommunication, shipping, transport, and agriculture. The major impetus for this diversification is the changing requirements of the stakeholders. There is increasing awareness on accountability and transparency [22]. The current business model is pushing the clients to demand for construction to be completed on schedule, within budget, with maximum performance, reliability, safety, and meeting other criteria within their client value system. The clients are pushing the construction industry to take a cue from the automobile, banking and electronic industries that move with time.

The Royal Institute of Chartered Surveyors considered the quantity surveyors as the cost managers of construction that are initially involved with the capital expenditure phase of a building or facility, which is the feasibility, design, and construction phases and can be involved with the extension, refurbishment, maintenance, and demolition of a facility including process engineering, such as chemical engineering plants or oil rigs, and will understand all aspects of construction over the whole life of a building or facility [23]. But while this definition/description is incomplete, it provides an indicative on what quantity surveyors represent. For instance, quantity surveyors provide advice on the strategic planning of a project. For a construction work, this advice affects clients' decisions on whether to construct or not, and if the client decides to construct what effect costs have on other criteria within the clients/users value systems. Specifically, the advice provided includes the selection of procurement strategies as well as method of contracts.

The courses contents of the quantity surveyors reflect their capabilities and expertise in providing advice on the procurement issues. For instance, even under the traditional procurement system where the quantity surveyor is not usually the lead or prime consultant, all other members of the team, including the client relates their valuable information to the quantity surveyor to enable the production of accurate estimates to make meaningful contributions towards the successful completion of construction facilities [22, 24, 25, 26] Quantity surveyors by their training and experience in contractual, economics, and costing matters, plus the technical knowledge in construction processes and activities are better informed on the procurement processes and procedures.

5 Outline of research design

The methodology of the study combines literature review and survey questionnaire. A broad reviews of literature suggested that there are various types of determinants for procurement strategies whether for new construction, maintenance, refurbishment, conversion, adaption and, for operational purposes only. Therefore, to achieve the aim of this study, 18 determinants for procurement strategy selection were compiled and addressed to respondents to measure. The survey was developed from literatures [18, 19, 20, 22, 24, 26, 27] and the authors' own experiences. Primary data is collected based on convenience sampling. In convenience sampling, a survey is administered on the respondents who are accessible, available or willing to participate. It is an appropriate method where sufficient information on population size is not available but like other nonprobability sampling, its findings may not be generalisable. However, if the respondents are carefully selected with sufficient sample size, the findings are representative [28]. Primarily data for this survey was administered through an online survey. The survey commenced in November 2014 through to April 2015. The questionnaire was divided into two parts. The first part seeks to capture the respondent's profiles while the second part measures procurement strategy selection with respect to the quantity surveyors. The survey was mainly close ended, but empty spaces were provided for respondents to add additional determinants they considered significant not listed in the form.

Respondents were asked based on their current experience, to place a tick on the extent that each of the 'determinants' is significant in the selection of procurement strategy for a construction work on a five continuum scale; where 5 denotes 'extremely often', and 1 denotes 'not often'. 2, 3 and 4 are located in between. For interpretation, the lowest possible score is 0 (zero) and the highest possible score is 1 (one). Simply put, the closer to 1 the higher the extent at which a determinant influences method of procurement selection. The mode technique was used to analyze the demography of the respondents. The measurements of the determinants are displayed in frequencies but the rating of the reasons is determined by 'average relative index' (ARI). The ARI is based on the aggregate weighting of the initial frequency score of each constructs (i.e. factors, criteria, items et al.)

$$ARI = \frac{\sum_{i=0}^{5} a_i x_i}{5\sum_{i=0}^{5} x_i} \quad (1)$$

Where a_i is the index of a group; constant expressing the weight given to the group; xi is the frequency of response; i = 1,2,3,4,5 and described as below: x_1, x_2, x_3, x_4, x_5 are the frequencies of the response corresponding to $a_1 = 1, a_2 = 2, a_3 = 3, a_4 = 4, a_5 = 5$, respectively. The determinant with the highest mean score is considered as the dominant. The mode technique was used to analyse the demographic of the respondents.

6 Results and interpretations

6.1. Captions/numbering

A total of 33 useable responses were received and analysed for this study. The profiles of the respondents are shown in Tables 2 through 4 and Figures 1 and 2. As may be seen, the majority of the respondents held MSc. (Table 1) many have Bachelor of technology, while a substantial number had HND. Figure 1 indicates that most of the responding quantity surveyors are from the consulting firms with only five of the 33 quantity surveyors having worked for contracting companies while the remainder worked for publicly owned organisations. Most of the respondents were occupying strategic positions in their various organisations (Figure 2). For instance, 40% (13) of the respondents occupy the position of project managers while seven or 21% held senior quantity surveyor potions. Four are partners and three are principal partners in the organisations.

Table 2, Respondent's academic qualification.

Qualification	Frequency	Percentage
BTech.	9	27.3
HND	6	18.2
BSc.	2	6.1
MSc.	13	39.4
PgD	3	9.1
Total	33	100.0

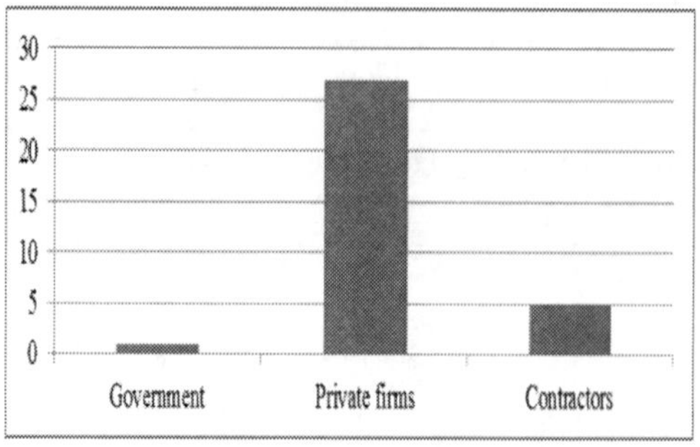

Figure 1, Distribution of respondent's organisation.

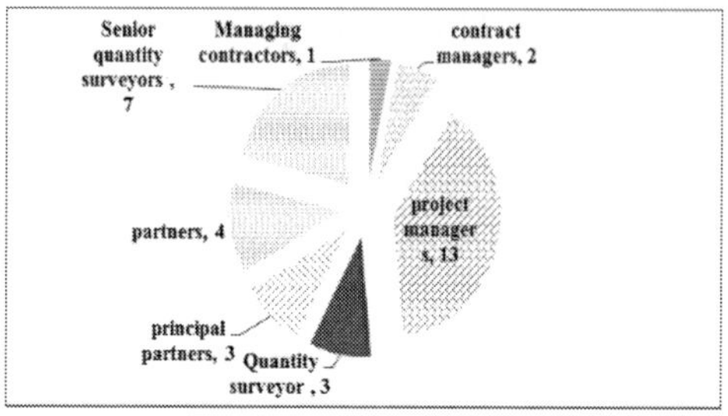

Figure 2, Distribution of respondent's position (%).

In terms of working experience, 75% (25 to be exact) of the respondents have worked for more than five years (Table 3), with more than 40% having more than 10 years' experience in the construction sector. The mean score for the working experience is 2.39 and the standard deviation is 1.14. The mean score and standard deviation for the numbers of projects respondents have completed in the last ten years are 1.9 and 1.13 respectively. Close to 55% of the responding participants' organisations have completed more than one construction works in a year (see Table 4). Although the turnovers of the organisations are not investigated, but considering the working experience and positions, these are big projects. Further, another aspect of the survey found that the type of works involved by the respondents include construction of buildings, civil engineering, building services, railways, airports, and petrochemical plants.

Table 3, Distribution of respondent's industrial experience.

Qualification	Frequency	Percentage
Not more than five	8	24.2
Five to ten	12	36.4
Ten to fifteen	6	18.2
Fifteen to twenty	6	18.2
Twenty and above	1	3.0
Total	33	100.0

Table 4, Number of projects completed in the last 10 years

Qualification	Frequency	Percentage
Not more than ten	15	45.5
Ten to twenty	11	33.3
Twenty to thirty	4	12.1
Thirty to forty	1	3.0
Forty and above	2	6.1
Total	33	100.0

On the basis of these profiles, it is evidenced that the respondents have capabilities and experience to provide valid and unbiased information on the Nigerian construction sector.

6.2 Factors that determine procurements strategy's selection

The following sections present the measurements of the determinants of selection of an appropriate procurement strategy. In order to determine the strength of the data, reliability and validity tests were performed. The reliability test results indicate that the Cronbach's alpha for all the roles are very satisfactory, ranges from 0.810 to 0.834 (Table 6). The validity test, using the 'Commonalities', produced values that ranges from 0.542 to 0.843. In general, if the alpha reliability or validity test is above 0.50, it is sufficient to consider that the construct (factor) is valid or reliable. Further, as may be seen in Table 5, there is a relationship between all the determinants with Bartlett's Test (0.0001). However, there exist some high correlation between the determinants; with the KMO of 0.412 typically KMO value should exceed 0.5. A reason for this might be the sample size. Overtly, these statistics suggest that the data was drawn from a similar population or simply that the respondents have similar experience. Consequently, the data is both valid and reliable and thus the results are representative.

Table 5, Results of KMO and Bartlett's Test.

Kaiser-Meyer-Olkin Measure of Sampling Adequacy.		0.412
Bartlett's Test of Sphericity	Approx. Chi-Square	322.752
	df	153.000
	Sig.	0.000

Table 6, Standard deviation, Validity and Reliability Results

Factor	Standard deviation	Validity	Reliability
Type of client	0.66	0.542	0.822
Design input	0.73	0.804	0.831
Management team	0.77	0.537	0.827
Availability of resources	0.82	0.683	0.819
Type of risk in the projects	0.89	0.693	0.817
Client ability and willingness to carry risk	0.86	0.810	0.828
Legal requirements	1.10	0.738	0.834
Size of the projects	0.76	0.589	0.829
Continuity of the project	0.88	0.859	0.818
Time available until completion	1.12	0.835	0.826
Requirements of stakeholders	0.79	0.809	0.831
Government policy	0.97	0.689	0.817
The client's experience	0.94	0.843	0.833
client insistence on knowing the expected price	0.98	0.743	0.826
Contractual method use to the project	0.82	0.802	0.816
Requirement of funding organisation	0.92	0.638	0.810
Information available at the conceptual stage	0.96	0.843	0.823
Expected completion date	1.11	0.788	0.814

Table 7 contains the descriptive results on determinants of the selection of procurement strategy. Overtly, 28.6% of the quantity surveyors measured that they considered the entire determinants extremely often, while 41.25% of them considered the determinants very often. While 23.57% often considered the determinants, less than 7% of the quantity surveyors seldom considered the determinants. The cumulative average weightage is 0.780. The index ranges from 0.679 for legal requirement to 0.8798 for the type of client respectively.

Because of space constraint, detail explanation on each of the determinant will not be provided. Among the various types of procurement options available to a client towards achieving best value, only one of the procurement strategies is suitable for a particular set of client's requirements but not the others. For instance, some procurement strategies placed greater emphasis on a criterion or on some particular criteria within the client value system. As an illustration, the traditional strategy is most suitable for a project where time is not an essential priority. Under this strategy, tender is usually selected on the basis of the lowest tender price. However, time is an essential feature of any project, time also makes mess of any cost estimate and there is not a project without time limitation. Modern procurement strategies strive on the need to overlap time. Therefore, it becomes imperative that the determinants for the procurement selections are known as a prerequisite otherwise the wrong determinants/s would be selected automatically leading to a wrong strategy consequently leading to poor construction delivery.

Though it is possible to rank the determinants in terms of the weighting index, detailed examinations of the indices suggest that the variability among the determinants is quite close. Therefore, any attempt to rank the determinants might not provide any useful interpretations in terms of ranking other than mere number. Hence, we prefer not to be mechanistic with the interpretations. Rather than focus on the figures, we dwell on the reason behind the figure. Hence, a more meaningful and plausible interpretation is that all the determinants plays very significant roles in the determination of a particular procurement strategy for construction project. However, these findings are contrary to the findings of Mathonsi and Thwala [7], Songer et al. [29], Oluwole [30], and Hashim et al. [31].

Researchers are able to detect significance differences in the weightages with respect to the determinants. Two major reasons could account for this. First, it could be that the stakeholders have a different understanding with the meaning of procurement strategies and their associated issues. The second reason could be that the clients are now more demanding compared to how they used to measure value for their money. Both reasons are insightful. While there is sufficient literature to support the claim that construction clients are very demanding and are concerned about their investments [16, 20], in the context of these findings, it is very unlikely that the reason for this is because of the clients demanding aspect.

The major reason is very much concerned with the structure of the respondents, whereby the average of the responses will be compiled thereby suppressing that of the others. An exploratory factor analysis conducted produced the results in Table 8. A 0.50 factor loading cut-off point was adopted thus eliminating project size and funding organisations' requirement.

Table 7, Distribution of the determinants of selections of procurement strategy

Determinant	5	4	3	2	1	Index
Type of client	17	12	4	0	0	0.8798
Design input	8	16	9	0	0	0.7939
Management team	9	17	6	1	0	0.8061
Availability of resources	14	13	5	1	0	0.8424
Type of risk in the projects	5	16	8	4	0	0.7333
Client ability and willingness to carry risk	6	17	7	3	0	0.7576
Legal requirements	4	13	10	4	2	0.6788
Size of the projects	15	12	6	0	0	0.8546
Continuity of the project [programme management]	9	14	8	2	0	0.7818
Time available until completion	10	14	5	2	2	0.7697
Requirements of stakeholders	3	15	12	3	0	0.7091
Government policy	13	9	9	2	0	0.8000
The client's experience	5	11	14	2	1	0.7030
client insistence on knowing the expected price	13	10	9	1	0	0.8121
Contractual method use to the project	12	14	6	1	0	0.8242
Requirement of funding organisation	8	17	6	1	1	0.7818
Information available at the conceptual stage	8	13	10	1	1	0.7576
Expected completion date	10	12	6	4	1	0.7576

However, the determinants can be explained in terms of methods of procurement. In the procurement of projects, each of the stakeholders comes into the arrangement with the different mission and vision. For instance the while government comes into such arrangement in order to meets their social / community responsibilities the banks enter into such venture majorly to maximize their profits. Other stakeholders will also come in if the project is economically viable and feasible. The decisions to enter into such arrangement particularly in case of the government depends on factors including capital cost, maintenance cost, replacement, depreciation, expertise and operating cost of the projects. As an illustration, the PFI came on board because of the financial shortages arising from governments' commitments elsewhere. Because of public interests, governments have over the years preferred the traditional procurement because they want to be seen to protect public money. For the private client, this has little considerations. However, public clients also now prefer to acquire projects through modern procurement strategies to the traditional strategies because of shortage of funds and that public facilities are not well managed if the hand of the government agencies or subsidiaries. Similarly inexperience clients whether private or private, possibly opt for the type of procurements required complete scope of works. The nature and of projects and experience of the clients will also dictate whether the procurement method to select should expose the clients to risk. Moving forward, as may be seen it is possible to categorise the determinants into 6 major headings accounting for some 73.591% cumulative total variance explained. The headings are client's factor (1); contractual factor (2); contract factor (3); project factor (4); external factor (5); and risk factor (6).

Table 8, Rotated Component Matrix

Factor	Component					
	1	2	3	4	5	6
Type of client	0.665					
Design input			0.787			
Management team			0.704			
Availability of resources	0.687					
Type of risk in the projects	0.680					
Client ability and willingness to carry risk						0.827
Legal requirements		0.765				
Size of the projects			0.406			-0.457
Continuity of the project[programme management]	0.584				0.595	
Time available until completion	0.695			0.401		
Requirements of stakeholders					0.894	
Government policy			0.646			
The client's experience		0.579		-0.520		
client insistence on knowing the expected price	0.412			0.652		
Contractual method use to the project		0.687	0.409			
Requirement of funding organisation	0.472		0.409	0.445		
Quality of available information at the conceptual stage		0.886				
Expected completion date				0.776		

7 Conclusion

The paper re-appraised the construction procurement strategies in the Nigerian construction industry. Whichever the chosen procurement strategy selected, the client's value system should be used as a proof screen. It seems there is no simple or hard and fast rule for the selection of the most appropriate contract types for a project. It seems obvious however, that there are simple frameworks that can facilitate the selection of contract type that can facilitate the achievement of client value system. In the future, a research on the various ways of evaluating the determining factors will be conducted. This is critical to the entire process of procurement selection. Notwithstanding, having knowledge on the determinant would not be sufficient except if the correct tool is used to determine the correct determinants for a project. It is suspected that the tools currently used to evaluate and determine procurement are flawed. Hence, there is a need to clearly re-examine the method for identifying the client's value systems.

References

1. Olanrewaju, A. and Anavhe, J. P., Issues in the construction procurement process. In Saha, S., Zhang, Y., Yazdani, S., and Singh, A. (2015) (eds.) Implementing Innovative Ideas in Structural Engineering and Project Management, Sydney, pp1341-1346.
2. Egan, J., Rethinking Construction, London: Department of the Environment, transport and region HMSO, (1998).
3. Latham, M. (1994).Constructing the Team: Final Report - Joint Review of Procurement and Contractual Arrangements in the UK: HMSO, UK
4. Construction 2025, Industrial Strategy: government and industry in partnership, (2013) Available at http://www.ecobuild.co.uk/files/industry_strategy_to _2025.pdf. [accessed on March, 2015]
5. Dada, O. M., Predictors of Procurement Selection: An Investigation of Traditional and Integrated Methods in Nigeria. Journal of Construction in Developing Countries, 17(1) 69–83, (2012).
6. Ojo, S. O. and Ikpo, I. J., Determining the Performance of Procurement Methods against Selection Criteria using Outranking – Satisfying Methodology. Journal of Engineering, Project, and Production Management, 3(2), 74-84(2013).
7. Mathonsi, M. D. and Thwala, W. D., Factors influencing the selection of procurement systems in the South African construction industry African Journal of Business Management Vol. 6(10), pp. 3583-3594, 14 March, (2012)
8. Love, P. E. D, Skitmore M. and Earl, G., Selecting an appropriate procurement method for the construction process: an empirical study. Construction Management and Economics, 16(2):pp. 221-233, (1998).
9. NBS (National Bureau of Statistiques) Nigerian Construction Sector: Summary Report; 2010-2012, (2015) Available at at file:///C:/Users/olanrewaju/Downloads/nbs%20Nigerian%20Construction%20report%202010_2012.pdf [accessed on 10 March 2015]
10. Olanrewaju, A. A. and Khairuddin, A.,Identifying the Dominant Procurement Strategies in the Nigerian

[a] Corresponding author: abdullateef.olanrewaju@ymail.com

construction industry. In: Proceedings of the Management in Construction Researchers Conference Malaysia (MICRA). Shah Alam, Malaysia, 28th 289th August, (2007).
11. Kelly, J., Male, S. and Graham, D., Value Management of Construction Projects, Blackwell, Oxford, (2004).
12. Perry, J. G. (1985), The development of contract strategies for construction projects. Ph.D. thesis, University of Manchester, cited in Cheung, S. I., Lam, T. I and Leung, M.Y (2001) An analytical hierarchy process based procurement selection method, Journal Construction Management and Economics, 19:4, pp427 - 437
13. Olanrewaju, A. A and Anavhe, P. J., Appraisal of Construction Procurement Strategies in Nigeria, In: Ahmed, M. S., Farooqui, R. U., Azhar, S., Shah, A., Lodi, H. S and Smith, A N., (Ed.) Procs. of Third International Conference On Construction In Developing Countries "Advancing Civil, Architectural And Construction Engineering & Management" (ICCIDC-III). July 4 - 6, 2012, Thailand, pp 195 – 201, (2012),
14. The Chartered Institute of Building (CIOB) A Report exploring procurement in the construction industry, (2010).
15. Ahuja, H. N., and Walsh, M. A., Construction Management and Engineering: Successful Methods in Cost Engineering. Canada: John Wiley & Son Inc.
16. Olanrewaju, A. A and Anahve, P. J (2013), Delay Factors in the Nigerian Construction Industry, In Proceeding of COBRA 2013. New Delhi, September, 10 – 12
17. Love, E.D. P., Davis, R. P., Edwards, J. D. and Baccarini, D (2008) Uncertainty avoidance: public sector clients and procurement selection, International Journal of Public Sector Management, Vol. 21 Iss. 7 pp. 753 – 776.
18. Davis, P., Love P., and Baccarini, D., Building Procurement Methods. Research Program: C Deliver and Management of Built Assets Project: 2006-034-C., (2008). http://www.construction-innovation.info/images/pdfs/Research_library/ResearchLibraryC/2006-034-C/reports/Report_-_Building_Procurement_Methods.pdf.
19. Ramus, J., Birchall, S. and Griffiths, P. Contract practice for surveyors, 4th ed. Oxford: Butterworth – Heinemann (2006).
20. Cartlidge, D., Procurement for built assets, Elsevier Butterworth – Heinemann, Oxford (2007).
21. National Economic Development Office (NEDO) (1985) Think About Building. NEDC, London.
22. Cartlidge, D., New Aspects of Quantity Surveying Practice, Oxford: Taylor & Francis, (2011).
23. (RICS (Royal Institution of Chartered Surveyors) Glossary of surveying services Brief definitions of surveying services to help you find the right surveyor (2015). http://www.ricsfirms.com/glossary
24. Seeley, H. I (1997), Quantity Surveying Practice, London, 2nd Ed. Macmillan Press Ltd.
25. Ashworth, A., Hogg, K. and Higgs, C., Willis's Practice and Procedure for the Quantity Surveyor. John Wiley and Sons, Ltd. West Sussex(2103).
26. Marsh, C., Building services procurement. Spon Press, Taylor and Francis Group, London and New York(2003).
27. Towey, D., Construction quantity surveying: a practical guide for the contractor's QS, Wiley-Blackwell, UK, (2012).
28. Sekaran, U. Bougie, R., Research methods for business: a skill building approach. 5th Edn, John Wiley and Sons, Limited, UK(2010)
29. Songer, D. A., Molenaar, R. K, and Robinson, D. G., Selection Factors and Success Criteria for Design-Build in the U.S. and U.K. (1997) http://www.colorado.edu/engineering/civil/db
30. Oluwole, A. A. Evaluation of Building Project Procurement Strategies (A case Study of Ondo State, Nigeria). International Research Journal of Engineering and Technology, Volume: 02 Issue: 05, (2015)
31. Hashim, M., Li, Y. C. M., Yin, C. N., Hooi, S. N., Heng, M. S., Yong, L. T (2006) Factors influencing the selection of procurement systems by clients, International Conference on Construction Industry, Padang, Indonesia, 21st June – 25th June 2006.

Holistic BIM Adoption and Diffusion in Malaysia

Shahela Mamter[1,a], Abdul Rashid Abdul Aziz[2]

[1]Faculty of Architecture, Planning and Surveying, UiTM Campus Perak,32610 Bandar Baru Seri Iskandar
[2]School of Housing, Building and Planning, Universiti Sains Malaysia, Pulau Pinang

Abstract. Building Information Modeling is modeling technology and associated set of processes to produce, communicate and analyze digital information models for construction life cycle. Although many benefits can gain by the implementation of BIM, the pace adoption of BIM in Malaysia is still slow. There is an abundance of theories and models purporting to describe the drivers and mechanisms of how new technologies are adopted or diffused into a particular section of society or by individuals or stake players. In addition, many acceptance theories, which imply the adoption of new technologies and can be applied in every sector. This paper aims to present a review of the existing literature of BIM adoption and diffusion theory/model/framework in Malaysia. Researcher revealed the list of driven factors of Holistic BIM Adoption. In addition, it will deal with how to work effectively with groups that initially are relatively

Keywords: Building Information Modelling, Holistic, Adoption, Diffusion, Driven Factor

1 Introduction

Building Information Modelling (BIM) can be traced back to the parametric modelling research conducted in USA and Europe in the late 1970s and early 1980s, the Architecture-Engineering-Construction (AEC) industry practically started to implement it in projects from the mid-2000s [1]. During the last seven years, the term BIM has gone from being a buzzword to the centre piece of AEC technology [1]. However, BIM concept has been introduced since 1970 by Professor Charles M. Eastman [1] and the United States of America (USA) becomes the pioneer country that implements the BIM [2]. Furthermore, BIM as the method is supportive of collaborative planning, facilitating communication and information exchange among various planning process participants [3]. According to [4] BIM is an innovative new approach to building design, construction and management which are introduced by Autodesk in the year of 2002 in order to change the way of professional worldwide think on how technology can be applied to building design, construction and management. In United Kingdom, 2016 will be the targeted year towards BIM level 2 adoptions on all centrally-procured projects. The hurdle that was facing the industry is standardisation of working between clients and supply chain [5].

Further, lack of understanding of BIM and how it can be integrated into the construction industry within the various professional disciplines also apart of problems in the implementation of BIM [6]. However, BIM has recently attained widespread in the Architectural, Engineering and Construction (AEC) industry, especially in the U.S and U.K. However, the adoption is still low in the local industry. In Malaysia [7] conclude that a close coordination among the industry stake holders with a specific vision and environment needs to be developed in AEC industry to enjoy the fruitful benefits of BIM implementation. The support by the concerned authorities, extensive consultation between the professional and industry practitioners is the only key to success. The question arises whether the all BIM stake players in Malaysia are well aware and how far the knowledge about BIM in the Malaysian construction industry. Besides that, [8] found in Malaysia, the developments of BIM still an infancy stage, but it has gained serious attention from many players. However despite receiving massive attention within the AEC industry, it appears that the adoption rate is not representing such phenomenon. Khor Wei Moon, director of technical design for Sunway Group, a property development and construction company, found that BIM was "mainly a private initiative driven by the larger property developers and contractors" in Malaysia [9].

The inconsistency of adoption rate exists not only between countries but between industry stake players as well [10]. For better understand the driven factors of BIM adoption and the stake players influencing the adoption need to be investigated. In Addition, [11] conclude there is no clear holistic approach or methodology to assess adoption of BIM adoption at the corporate level down to project level.

Accordingly, the authors are conducting research with an attempt to establish the theory of BIM adoption as to highlight the positive action by putting the BIM into the construction industry. Finally, the study will reveal a Holistic BIM Adoption Model based on the deductive and inductive holistic driven factors.

2.0 Research Aims and Objectives

This conceptual paper is primarily based on the thorough review of the relevant literature of BIM adoption theory/model/framework in Malaysia and the driven factors for holistic BIM Adoption. Drawing from the research problem highlighted, the pertinent research objectives are:

i. To identify which theory/ model/ framework can best capture holistically the issues surrounding the low adoption of BIM in Malaysia.

ii. To determine the driven factors influencing the holistic adoption of BIM.

3.0 BIM Adoption/ Theory/ Framework in Malaysia

Many of the recent BIM adoption model/theory/frameworks is related studies emphasize the development of an application BIM technology based on micro adoption. Moreover, influential factors are systematically listed, many of them focus only on the Technology Acceptance Models have provided a basis for investigating the role and impact of perception and behavioural disposition towards technology usage based on psychological theories. Its validity supported by [12] which concentrating on research methodological approach from dominant IT field and management perspectives in human–IT interaction, in an attempt to further improve BIM adoption. It has a narrow conception of TAM theory and organisational readiness framework. Besides that, [10] produced a model by integrating Diffusion of Innovation (DOI) Theory, Institutional Theory and fit into Technology-Organisation- Environment (TOE) framework. It does not cater to examine the relationship of existing BIM theory/model/framework with the Government policy makers. It also suggested by [13] BIM in Malaysia needs a pan-industry champion with authority. As such, the individual is most likely to come from the Government sector. In addition, [14] revealed the organisational framework readiness in implementing BIM. It narrowly focuses on the relationship between people, process, technology and business process re-engineering in Malaysian design consultant. However, none of the approaches studied focussing on the causal loops amongst all driven BIM factors due holistic implementation of BIM in Malaysia. [12] recommend further extension study focusing on extending the existing model with other previously unperceived constructs, input of other previously unperceived relationship mediators, selections of other industry players such as clients and software developers and a cross-sectional survey in other countries with similar BIM maturity stages. Therefore, the researcher aims to provide a Holistic Theory/Model/Framework BIM adoption in further study. Table 1 show the review of Malaysian BIM adoption theory/model/framework from previous researcher. It will shed useful information for construction researchers and practitioners in exploring the BIM adoption and diffusion theory in Malaysia.

Table 1: The previous research on Malaysian BIM Adoption Theory

No.	Theory/Model /Framework	Author / Year	Explanations
1.	Technology acceptance model Davis (TAM) (1985)	[12][13] [15][16]	TAM models how users come to accept and use a technology by identifying factors that cause people to accept and make use of systems developed and implemented by others. The model suggests that when users are presented with a new technology, a number of factors influence their decision about how and when they will use it.
2.	Diffusion of Innovation Theory Rogers [17]	[13] [10]	Five categories that emerged with the following profiles: 1) Innovators – venturesome, educated, multiple info sources; 2) Early adopters – social leaders, popular, educated;

			3) Early majority – deliberate, many informal social contacts; 4) Late majority – sceptical, traditional, lower socio-economic status; 5) Laggards – neighbours and friends are main info sources, fear of debt
3.	Institutional Theory [18]	[13][10]	Three pressures are: 1) Mimetic pressures is adoption of IT 2) Coercive pressures which are: agencies, suppliers, customers, parent corporations and other key constituents 3) Normative pressures which are: professional bodies, consultants, trade associations
4.	Technology-Organisational - Environmental (TOE) framework	[10]	Focused into 3 contexts which are Technological, Organisational and Environmental factors.
5.	Organisational Readiness Framework	[14]	The readiness element found, which are: 4) Process: Change Strategy, BIM Implementation Management, and Policy. 5) Management: Business Strategy, Management Competency, and Leadership. 6) Technology: Hardware, Technical Support, and Software. 7) People: Roles and Responsibilities, Skill and Attitude, Training and Education and Work Environment

3.1 BIM Levers for Holistic Adoption in Malaysia

Collaboration with government key agencies also converges in the promulgation of laws and policies, contract and procurement and others. Collaboration with professional bodies, academia, committees and AEC stakeholders will ensure BIM implementation at the competitive level. Therefore, BIM levers in Malaysia are:

a. Construction Industry Development Board (CIDB)

Figure 2 shown, CIDB will complement the efforts by providing a sustainable environment where BIM will survive and thrive. Early efforts include providing awareness programs and workshops with the industry to gather feedback and comment aimed at charting the way forward for a holistic implementation of BIM. CIDB is also in the midst of establishing the National BIM Committee of Building Information Modelling in construction Industry in order to coordinate the movement of BIM in the country [19]

b. Jabatan Kerja Raya Malaysia (JKR)

BIM committee in JKR was formed by the government on 27 August 2007 to choose the right BIM platform to ensure interoperability. The purpose of forming the committee was to identify construction project processes that involved BIM implementation. Moreover, the committee also prepared a BIM standard manual documentation for Malaysian public work department as a guideline for construction players' reference. The committee also provides BIM training and advisory assistance to project teams in using BIM tools since 2011 [20].

c. Private Company

BIMASIA Sdn Bhd is Malaysia leading Building Information Modelling (BIM) Consultancy firm established in 2013. The list services provided are management, modelling, strategy and training tailor made to individual existing office workflow for Architects, Engineers, Developers and Construction companies. This firm has a strong reputation for world-class design of BIM coordinated buildings within Master planning, Retail-led Mixed Use Developments, Offices, Residential and Hotels [21]

d. Policy

Malaysian Government policy was the critical factor that heavily influenced the promotion of new techniques or products especially in the construction sector [22]. Government policy has been noted as one of the key influences in promoting a new technique or products in the construction industry, simply because the government is one of the biggest clients in any construction industry. The importance of policies in achieving successful of BIM implementation which UK Government will require fully collaborative 3D BIM (with all project and asset information, documentation and data being electronic) as a minimum by 2016 [23].

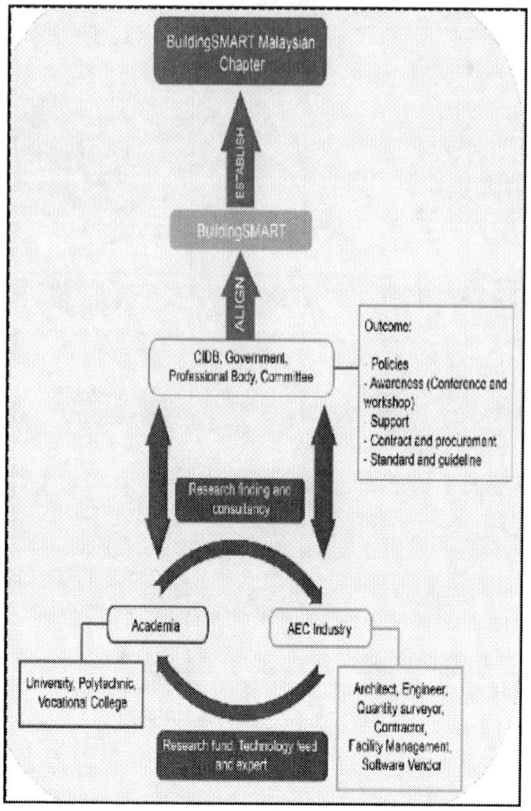

Figure 1: Collaborative Framework for BIM Adoption in Malaysia (source: [24])

3.2 BIM Driven Factors

According to the literature review BIM driven factors are summarized in Table 2 below. The driven factors classified into 6 country pioneers in advocating the adoption of BIM: United Kingdom: [1][23] [25][26][27] [28][29], South Korea: [30] [31] , Taiwan: [11][32], Singapore: [33][2][34], Hong Kong: [35], New Zealand and Australia: [36]. Therefore, the table was tabulated based on the 5 sub theme: Policy, Engagement, Investment, Implementation, and Education were identified within the 16 references cited by this paper. Each driven factors was ticked with an "X". The selection of sub theme was based on the researcher selection. Currently, the researcher have undertaken a preliminary quantitative study in order to ensure that it is comprehensive, up-to-date and appropriate for the precise needs of the study. By understanding the driven factors for BIM adoption, Malaysian BIM Player will be better able to allocate sufficient resources to achieve its goal. Under Francis Maude, Minister for the Cabinet Office, the UK government was implemented a construction strategy that stated BIM will be a key part of the government's procurement of public buildings and, by 2016, model-based BIM was mandatory on all public sector projects, included delivery of all project and asset information, documentation and data [9].

 DOI: 10.1051/matecconf/20166600094

Conclusion

This paper revealed the review of BIM adoption theory/model/framework in Malaysia. The majority of researcher concentrating on organisation/or individual for implementing BIM in Malaysia. Furthermore, the researcher found that could indicate a causal loop model for holistic BIM implementation among Malaysian construction stake players. The proposed causal loop model will give the high impact contribution for BIM implementation. The higher the financial benefits, the greater the pressure on the government to provide the funds [37]. Moreover, the feedback capability of the Holistic BIM adoption model is of special interest to policy-makers in which they can simulate possible scenarios for different policies. Currently, the authors have undertaken a further empirical research using expert opinion survey approach to collect detailed quantitative data information regarding holistically driven factors in adopting BIM. The findings will be revealed for the next research publication

Acknowledgements

The authors acknowledge the support from Universiti Teknologi MARA Perak Branch, Campus Seri Iskandar and Malaysia Ministry of Education.

DOI: 10.1051/matecconf/20166600094

Table 2: Previous study on BIM Driven Factors

No.	BIM Driven Factors	Europe				South Korea		Taiwan	Singapore		Hong Kong	Australia and New Zealand
		[1]	[23] [25]	[26] [27]	[28] [29]	[30]	[31]	[11][32]	[33]	[2][34]	[35]	[36]
1	**Policy**											
	-Government Enforcement				X				X	X	X	
	-Construction Regulation (Standard)	X		X	X		X		X	X	X	X
	-Best practices	X	X	X				X	X	X		
	-Guidelines								X	X	X	
2	**Engagement**											
	-Organization culture	X		X			X	X				
	-Business Process	X		X	X	X		X		X		X
	-Collaboration among Project Team	X		X	X	X	X	X		X	X	X
	-Economics (Investment in BIM)	X		X			X			X	X	
3	**Investment**											
	-Promotion										X	
	-Training	X		X			X	X		X	X	
	-Software or Tools	X			X		X	X		X	X	X
	-Industry Foundation Classes (IFC)	X	X			X			X	X		X
	-Hardware	X			X		X	X		X	X	X
	-Fund									X		
	-Diffusion			X				X			X	
4	**Implementation**											
	-Information Exchange capability	X	X	X	X	X	X			X	X	X
	-Mechanisation Credit scheme									X		
	- Productivity Improvement Project (PIP) Scheme									X		
	-Integrated Design-Build Services and Agreements (Contractual Agreement)	X	X		X		X		X			X
5	**Education**											
	-Research Grant and development	X		X						X	X	
	-Conferences	X		X							X	
	-Recognition									X		
	-Professional Bodies	X	X	X	X	X	X	X		X	X	X
	-BIM Courses/Subject									X		

References

[1] C. Eastman, P. Teicholz, R. Sacks, and K. Liston, *BIM Handbook: A Guide to Building Information Modeling for Owners, Managers, Designers, Engineers and Contractors*, vol. 2. 2011.

[2] BCA, "The BIM Issue," *Build Smart Mag.*, no. 9, p. 4, 2011.

[3] R. Sebastian and L. van Berlo, "Tool for Benchmarking BIM Performance of Design, Engineering and Construction Firms in The Netherlands," *Archit. Eng. Des. Manag.*, vol. 6, no. 4, pp. 254–263, 2010.

[4] Autodesk, "Building Information Modeling in Practice," *Autodesk Build. Solut. White Pap.*, pp. 1–6, 2002.

[5] "BIM today_A clear view." Adjacent Digital Politics Ltd.ISSN 2058-9611, England & Wales, p. 90p, 2016.

[6] N. Thurairajah and D. Goucher, "Advantages and Challenges of Using BIM : a Cost Consultant ' s Perspective," *B49th ASC Annu. Int. Conf. Proceedings. Copyr. 2013 by Assoc. Sch. Constr.*, pp. 1–8, 2013.

[7] S. S. S. Gardezi, N. Shafiq, M. F. Nurudinn, S. A. Farhan, and U. A. Umar, "Challenges for Implementation of Building Information Modeling (BIM) in Malaysian Construction Industry," *Appl. Mech. Mater.*, vol. 567, no. JUNE, pp. 559–564, 2014.

[8] Z. Zahrizan, N. M. Ali, A. T. Haron, A. Marshall-ponting, and Z. Abd, "Exploring The Adoption of Building Information Modelling (BIM) in the Malaysian Construction Industry : A Qualitative Approach," *IJRET Int. J. Res. Eng. Technol. eISSN 2319-1163 | pISSN 2321-7308*, pp. 384–395, 2013.

[9] McGraw-Hill Construction, *The Business Value of BIM for Construction in Major GLobal Markets: How contractors around the world are driving innovation with Building Information Modeling*. 2014.

[10] J. Osman, S. Mazlina, S. Khuzzan, and A. Razaksapian, "Building Information Modelling : Proposed Adoption Model For Quantity Surveying Firms," no. June, pp. 151–165, 2015.

[11] M.-H. Tsai, M. Mom, and S.-H. Hsieh, "Developing critical success factors for the assessment of BIM technology adoption: part I. Methodology and survey," *J. Chinese Inst. Eng.*, vol. 3839, no. April 2015, pp. 1–14, 2014.

[12] W. Imoudu Enegbuma, U. Godwin Aliagha, and K. Nita Ali, "Preliminary building information modelling adoption model in Malaysia," *Constr. Innov.*, vol. 14, no. 4, pp. 408–432, 2014.

[13] J. P. Rogers, "The strategic adoption of building information modelling by Malaysian engineering consulting services firms The Strategic Adoption of Building Information Modelling by Malaysian Engineering Consulting Services Firms," 2013.

[14] A. T. Haron, "Organisational Readiness To Implement Building Information Modelling : A Framework For Design Consultants In Malaysia," p. 285, 2013.

[15] H. Son, S. Lee, and C. Kim, "What drives the adoption of building information modeling in design organizations? An empirical investigation of the antecedents affecting architects' behavioral intentions," *Autom. Constr.*, vol. 49, pp. 92–99, 2015.

[16] S. Lee, J. Yu, and D. Jeong, "BIM Acceptance Model in Construction Organizations," *J. Manag. Eng.*, vol. 31, no. 1988, p. 04014048, 2013.

[17] E. M. Rogers, *Diffusion of innovations*. 1995.

[18] W. R. Scott, "Institutional theory: contributing to a theoretical research program," *Gt. Mind Manag. Process Theory Dev.*, no. JANUARY 2005, 2005.

[19] "BIM roadmap CIDB," 2014.

[20] M. P. W. D. PWD, "Unit Building Information Modeling (BIM)," 2013. [Online]. Available: https://www.jkr.gov.my.

[21] Shenhua Group Corporation Ltd, "Company Profile," 2010.

[22] M. Nasrun, M. Nawi, and U. Utara, "Critical Success Factors for Improving Team Integration in Industrialised Building System (Ibs) Construction Projects : the Malaysian Case Integration in Industrialised Building System (Ibs)," no. November 2015, 2012.

[23] S. Hamil, A. Malone, P. Conaghan, M. Packham, R. Holland, A. Kemp, P. Jackson, E. Tuttle, J. Fiske, S. Fray, and A. Dobson, "NBS National BIM Report 2015 Introduction Richard Waterhouse CEO, NBS and RIBA Enterprises," 2015.

[24] CIDB, *BIM Roadmap for Malaysia's Construction Industry*, no. Series 2. 2014.

[25] McGraw-Hill Construction, H. M. Bernstein, S. A. Jones, and J. E. Gudgel, "The Business Value of BIM in Europe: Getting Building Information Modeling to

 DOI: 10.1051/matecconf/20166600094

the Bottom Line in the United Kingdom, France and Germany," 2010.

[26] F. Khosrowshahi and Y. Arayici, "Roadmap for implementation of BIM in the UK construction industry," *Eng. Constr. Archit. Manag.*, vol. 19, no. 6, pp. 610–635, Nov. 2012.

[27] A. Grilo and R. Jardim-Goncalves, "Value proposition on interoperability of BIM and collaborative working environments," *Autom. Constr.*, vol. 19, no. 5, pp. 522–530, 2010.

[28] T. B. Allen Wan[1], Andrew Platten, "STUDY OF SAFETY AUDITORS' VIEWS ON THE USE OF BIM FOR SAFETY IN HONG KONG," in *1st BIM UK Academic Conference*, 2012, vol. 1, p. 233.

[29] R. Morlhon, R. Pellerin, and M. Bourgault, "Building Information Modeling Implementation through Maturity Evaluation and Critical Success Factors Management," *Procedia Technol.*, vol. 16, no. 0, pp. 1126–1134, 2014.

[30] McGraw Hill Construction, *The Business Value of BIM in South Korea*. 2012.

[31] J. Won, G. Lee, and C. Lee, "Comparitive Analysis of BIM Adoption in Korean Construction Industry and Other Countries," *Iccem & Iccpm*, 2009.

[32] M. Mom, M. H. Tsai, and S.-H. Hsieh, "On Decision-Making and Technology-Implementing Factors for BIM Adoption," *Int. Conf. Constr. Appl. Virtual Real.*, no. Succar 2009, 2011.

[33] A. K. D. Wong, F. K. Wong, and A. Nadeem, "Comparative roles of major stakeholders for the implementation of BIM in various countries," *BIM-implementation*, 2009.

[34] B. and C. Authority, *Singapore BIM Essential Guide For BIM Adoption in an Organization*. 2013.

[35] A. K. D. Wong, F. K. W. Wong, and A. Nadeem, "Government roles in implementing building information modelling systems: Comparison between Hong Kong and the United States," *Constr. Innov. Information, Process. Manag.*, vol. 11, no. 1, pp. 61–76, 2011.

[36] McGraw Hill Construction, "The Business Value of BIM in Australia and New Zealand: Smart Market Report," 2014.

[37] S. Ahmad, "Simulating the Effect of Feed-In Tariff on Renewable Energy Penetration: A System Dynamics Approach," no. November 2014, 2015.

Implementation Framework of Green Building for Malaysian Government Building: KKR2 Study

Sharif Shiela [1,a], Kamaruzzaman Syahrul Nizam [2,b] and Pitt Michael [3,c]

[1] *University of Malaya, Kuala Lumpur, Malaysia. Public Works Department, Jalan Tun Razak, Kuala Lumpur, Malaysia*
[2] *University of Malaya, Kuala Lumpur, Malaysia*
[3] *University College of London*

Abstract. The purpose of this study is to develop the Implementation Framework of Green Building for Government Project. Implementation of Green Building Design is very important in Malaysia to conduct and execute green government project. The research intended to answer the questions about the factors involve in the development of the framework, what is the significant relationships exists between the factors involve in the Implementation Framework and whether there is significant relationship exists between the factors in implementation framework. A total 30 respondent selected between multilevel of project team KKR2 including engineers, assistants engineers, technical assistant, stake holders, contractors and consultants. In conclusion the study answered pertaining questions regarding the factors involve in development of Implementation Framework of Green Building through quantitative research and hypothesis testing.

1. Introduction

The building which we live, work and play interact with our environment, affecting storm water run off, energy and water consumption, transportation patterns, and indoor air quality. Recognition of the role that buildings have in our environment has let to significant efforts to design, build and maintain more sustainable structures (Parris, 2010). Green building practices includes environmentally responsible and resource efficient, promote building practices that conserve energy and water resources, preserve open spaces, minimise the emission of toxic substances, harmonise with the local climate, traditions, culture and the surrounding environment, sustain and improve the quality of human life, maintaining the capacity of the ecosystem at local and global levels. Benefit of green building to the environment is save energy use at 24% up to 50%, CO^2 emissions 33% up to 39%, water use save 40% and solid waste reduction of 70% for each green building (Turner, C. & Frankes, M (2008). Instead of benefit to the environment, green building also contributes to the building stakeholders by enhancing health and productivity, reduced environmental impact, environmentally effective use of materials, lowering electric and water utility costs and gives long term economic return. In relation to the tittle, the main purpose of this study is to determine the factors involve in the development of the Implementation Framework of Green Building Design For Malaysian Government Building, significant relationships between the factors involves and is there significant relationship exists between the factors in the framework. The analysis is determines based on quantitative research and hypothesis testing.

2. Problem Identification

Prime Minister of Malaysia, Dato' Sri Haji Mohamad Najib bin Tun Haji Abdul Razak has announce Malaysia commitment in United Nations Climate Change Congress 2009 (COP 15) in Copenhagen on 17 December 2009 that during 2020 Malaysia will reduce carbon emission rate up to 40 % from the current rate on 2005. There are also major issues for Malaysia that we will become a net importer of energy by 2015.The way Malaysian building has been designed will affect the energy consumption required. There is difficulty to justify Implementation Green Building in Government projects. Existing guidelines and Code of Practise on Energy Efficiency and Renewable energy to achieve low carbon building in the market such as MS1525, Dasar Teknologi Hijau and Development and publication of EE in Buildings Guidelines does not cover the strategies or method to implement green building for government project. Besides that, in Malaysia, there is only Green Building Index which is the rating tool for Green Building and private initiatives.

Public Works Department of Malaysia is the largest implementers of Government Buildings and projects in Malaysia. Despite the excellent and best practise of project management procedures and guidelines that has implement in Public Works Department

In addition to Public Works Department initiatives and commitment to deliver outstanding Green Building Projects as mention in the Public Works Department Key Performance Indicator in the Strategic Framework 2012-2015 to produce outstanding project delivery. This Implementation Framework is a starting point and beginning role of Public Works Department as the Implementers of Green Building Projects for Government of Malaysia.

3. Research Objectives

The Introduction and the problem statement above led to the formulation of the research aim and objectives. The broad aim of this research is to introduce/establish implementation framework of green building government building. In accordance with the research aim, the pertaining objectives of this study are: a) to determine the control factor/ critical areas to focus on the green building design. B) To assess current project team/ staff perception in implementation of green building project. C) To integrate green building factors as rating tools in existing design and project management procedure. D) To build implementation framework of green building design for government building.

4. Propose Conceptual Framework for Implementation Framework of Green Building Design for Malaysia Government Building

The suggested framework (Figure 1) is derived from the analytical literature review of the study which consists of concept process phases and also in depth review of previous study conducted from previous research. This framework consist of 5 factors involving project management procedure/ guideline/ government policies, type of project execution (design and build/) or conventional) and project staffs perceptions.

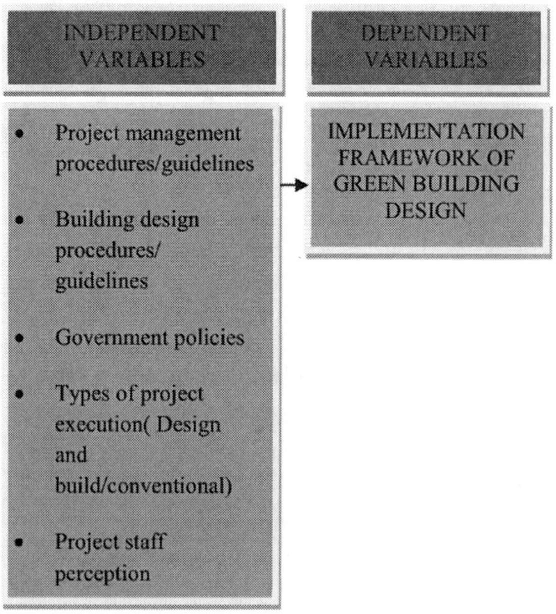

Figure 1. Conceptual Framework

5. Pilot Test Conducted at Kompleks Kerja Raya 2 (KKR2) Project

Pilot test was conducted on 12 April 2015 to test on reliability, validity of the instrument used and Inferential Analysis Correlation Pearson to test on the hypothesis that has been developed. The pilot test were conducted upon 30 respondent selected between multilevel of project team including engineers, architects, quantity surveyor, assistants engineers, technical assistant, stake holders, contractors and consultants from Kompleks Kerja Raya 2 Project Team, Cawangan Kerja Bangunan Am', PWD Headquarters, Kuala Lumpur. All the questionnaires reviewed accordingly to ensure all the questions is answered.

Table 1 shows respondent pilot test profile according to designation grade, age, gender, race and education. Overall, respondent for supporting staff are 6 respondents, meanwhile management staff are 3 respondents and professional staff are 21 respondents. Meanwhile for age of respondent, 3 respondents is at age 19-24 years old, 13 respondents is at 25-40 years old, 12 respondents is at 41-56 years old and 2 respondents is at 57-60 years old. Next for gender, 18 no of respondents are male and 12 no of respondents are female. For race, 27 no of respondents are Malay, 1 no respondent is Chinese and 2 respondents is Indian. Meanwhile in terms of respondent education, 1 respondent is primary school, 1 respondent is secondary school , 24 respondents are Universities Level and 3 respondents are others education level.

Table 1. Respondent demography profile

Demography	Frequencies	Percentage
Designation grade		
Professionals	21	70
Management	3	10
Supporting Staff	6	20
Age		
19-24 years	3	10
25-40 years	13	43.3
41-56 years	12	40
57-60	2	6.7
Gender		
Male	18	60
Female	12	40
Race		
Malay	27	90
Chinese	1	3.3
Indian	2	6.7

DOI: 10.1051/matecconf/20166600095

Education		
Primary School	2	6.7
Secondary School	1	3.3
Universities	24	80
Others	3	10

6. Pilot Test Data Analysis

For this pilot test, the data will be analyze using "*Statistical Package For Sosial Science*" (SPSS). Analyzing technique that will be used is descriptive analysia and inferential analysis. Descriptive analysis consists of percentage and frequencies respondent feedback and the strength for every variables. Inferential Analysis will used correlation Pearson Techniques. Besides that, coding process and recoding the data is essential before the data analysis is done Statistical test conducted to test hypothesis is Pearson Correlation.

6.1 Reliability test analysis for the KKR2 project pilot test

After the data is collected and reviewed accordingly, next the goodness data collected should be tested. Reliability for the data happened when the test to measure the same element repeatedly and the result is the same (Salkind, 2006 in Mohd Khairul Nizam, 2007). If the second measure, third, etc. have the same result, this means the research has high reliability (Chua, 2006). In addition, reliability test has been conducted to measure the stability of the instrument used. For This research Cronbach alpha near to 1, the highest reliability the instrument used will be. Rule of thumbs for Cronbach alpha value by Hair et. al.(2007) will be used as guidelines to show the strength for every test and measures.as shown in table 2.

Table 2.

Alpha	Relationship strength
< 0.6	Strong
0.6 - < 0.7	Medium
0.7 - < 0.8	Good
0.9 - < 0.9	Very Good
0.9	Excellent

The reliability test result conducted at KKR2 Project shows:-

Table 3. Case processing summary

		N	%
Cases	Valid	30	100
	Excluded (a)	0	0
	Total	30	100

a Listwise deletion based on all variables in the procedure.

Table 4. Reliability Statistic

Cronbach's Alpha	N of Items
0.922	55

Table 5. Item-Total Statistic

	Scale Mean if Item Deleted	Scale Variance if Item Deleted	Corrected Item-Total Correlation	Cronbach's Alpha if Item Deleted
Age	113.8667	502.257	.341	.921
Gender	114.9000	504.093	.463	.921
Race	115.1333	517.430	-.125	.923
Education	113.3667	509.068	.181	.922
Design Grade	114.8000	525.269	-.299	.926
Q7	114.3333	491.609	.411	.921
Q9	114.7000	489.321	.687	.919
Q10	114.9333	492.064	.613	.919
Q11	114.7000	492.769	.558	.920
Q12	114.3667	487.482	.607	.919
Q13	114.6000	482.248	.786	.918
Q14	113.4333	484.599	.230	.929
Q15	114.6000	491.559	.605	.919
Q16	114.5000	485.638	.724	.918
Q17	113.6000	510.248	.106	.923
Q18	114.0000	495.586	.556	.920
Q19	113.8000	502.717	.210	.923
Q20	113.6333	507.689	.148	.923
Q21	113.6667	513.057	.021	.924
Q22	114.4667	489.775	.596	.919
Q23	114.4333	479.771	.718	.918
Q24	114.4000	481.628	.709	.918
Q25	113.8333	496.557	.396	.921
Q26	113.9333	494.133	.397	.921
Q27	114.4000	496.938	.508	.920
Q28	114.0333	496.102	.389	.921
Q29	113.8000	501.200	.301	.922
Q30	113.6000	507.283	.127	.923
Q31	113.4000	517.352	-.078	.925
Q32	113.4000	510.731	.043	.925
Q33	114.5333	489.085	.690	.919
Q34	114.3000	494.010	.542	.920
Q35	114.2667	497.651	.608	.920
Q36	114.4000	500.731	.457	.921
Q37	114.3667	490.033	.692	.919
Q38	114.4667	508.740	.210	.922
Q39	114.5333	491.085	.716	.919
Q40	114.2333	502.668	.306	.921
Q41	113.8000	504.648	.208	.922
Q42	114.0000	499.931	.376	.921
Q43	114.2333	496.668	.530	.920
Q44	114.3000	485.528	.678	.918
Q45	113.9667	488.447	.518	.920
Q46	114.2000	484.993	.710	.918
Q47	113.9000	501.472	.310	.921
Q48	113.9667	495.895	.437	.920
Q49	114.2333	495.564	.530	.920
Q50	114.5333	489.568	.642	.919
Q51	114.3667	497.206	.482	.920
Q52	114.3333	503.471	.393	.921

Q53	114.3000	496.838	.523	.920
Q54	114.0667	484.961	.648	.919
Q55	113.9333	482.271	.628	.919
Q56	114.1333	489.499	.535	.920
Q57	114.2000	500.717	.292	.922

6.2 Hypothesis Testing for KKR2 Project Pilot Test Study

There are 5 main Hyphothesis develop to anwers all the pertaining relationship between all the independent variables and dependent variables in the theoritical Framework.The Hyphothesis are:

Table 6. Hypothesis Testing

Research Hypothesis	Statictical Analysis
Hyphothesis 1 : There is significance relationships between project management procedures/ guidelines with implementation framework of Green Building .	Pearson Correlation
Hyphothesis 2 : There is significance relationships between Building design procedures/guidelines with implementation framework of Green Building	Pearson Correlation
Hyphothesis 3 : There is significance relationships between Government policies with implementation framework of Green Building	Pearson Correlation
Hyphothesis 4 : There is significance relationships between types of project execution with implementation framework of Green Building	Pearson Correlation
Hyphothesis 5 : There is significance relationships between Project staffs perception with implementation framework of Green Building.	Pearson Correlation

7. Inferential Analysis

Inferential statistics will be used to make expected decision about analyse data and its relationship between two variables which is Pearson Correlation.

7.1 Pearson Correlation

In this pilot test, relationship between dependent variables and independent variables will be measure. To determine the relationship, correlation coefficient, r, will be analyse to determine the strength between the variables whether it is positive or negative. Criteria set by Davis (1971) (in Mohd Khairul Nizam, 2007) will be referred to evaluate the directions or strength of that relationship. Table 3 shows the strength of the relationship and the value of correlation coefficient, r as mention by Davis (1971).

Table 7. Comparative between Relationship Strength with Correlation Coefficient, r

r value	Relationship strength
± 0.70 or higher	Very high
± 0.50 until ± 0.69	High
± 0.30 until ± 0.49	Medium
± 0.10 until ± 0.29	Low
± 0.01 until ± 0.09	Very Low
0.0	No relationship

7.2 Pearson Correlation result in KKR2 Pilot Test Study

Pearson Correlation result in KKR2 pilot test study shows:

Hyphothesis 1 :There is significance relationships between project management procedures/ guidelines with implementation framework of Green Building .

Table 8. Correlation for Hypothesis 1

		Project management design procedure	Green buiding design framework
Project management procedure	Pearson Correlation	1	.930(**)
	Sig. (2-tailed)		.000
	N	30	30
Greenbuiding framework	Pearson Correlation	.930(**)	1
	Sig. (2-tailed)	.000	
	N	30	30

** Correlation is significant at the 0.01 level (2-tailed).

Table 8 shows Pearson Correlation Test conducted in addition to answers relationship between Project Management Design Procedure/guidelines and Implementation Framework of Green Building Design.

The Pearson Correlation result shows h1 is accepted. (r = 0.930) wuth (p< 0.01). This results shows there is significance, positive and strong relationship between Project Management Procedure/guidelines and Implementation Framework of Green Building Design. The positive correlation values shows that the more comprehensive Project Management Design Procedures the more effective Implementation Framework of Green Building for goverment building.

Hyphothesis 2 :There is significance relationships between Building design procedures/guidelineswith Implementation framework of Green Building

Table 9. Correlation for Hypothesis 2

		Project management design procedure	Green buiding design framework
Building design procedure	Pearson Correlation	1	.930(**)
	Sig. (2-tailed)		.000
	N	30	30
Greenbuiding framework	Pearson Correlation	.930(**)	1
	Sig. (2-tailed)	.000	
	N	30	30

** Correlation is significant at the 0.01 level (2-tailed).

Table 9 shows Pearson Correlation Test conducted in addition to answers relationship between Building Design Procedure/ guidelines and Implementation Framework of Green Building . The Correlation Pearson result shows h1 is accepted. (r = 0.930) wuth (p< 0.01). This results shows there is significance, positive and strong relationship between Buiding Design Procedure/guidelines and Implementation Framework of Green Building Design. The positive correlation values shows that the more comprehensive Building Design Procedures/guidelines the more effective Implementation Framework of Green Building for goverment building.C.

Hyphothesis 3:There is significance relationships between Government policies with implementation framework of Green Building.

Table 10. Correlation for Hypothesis 3

		Government policy	Green buiding design framework
Government policy	Pearson Correlation	1	.661(**)
	Sig. (2-tailed)		.000
	N	30	30
Greenbuiding framework	Pearson Correlation	.661(**)	1
	Sig. (2-tailed)	.000	
	N	30	30

** Correlation is significant at the 0.01 level (2-tailed).

Table 10 shows Pearson Correlation Test conducted in addition to answers relationship between Goverment Policies and Implementation Framework of Green Building Design. The Pearson Correlation result shows h1 is accepted. (r = 0.661) wuth (p< 0.01). This results shows there is significance, positive and medium strong relationship between Goverment Policies and Implementation Framework of Green Building . The positive correlation values shows that the more comprehensive Goverment Policies the more effective Implementation Framework of Green Building for goverment building.

Hyphothesis 4 : There is significance relationships between types of project execution with Implementation Framework of Green Building .

Table 11. Correlation for hypothesis 4

		Type of project execution	Green buiding design framework
Type of project execution	Pearson Correlation	1	.830(**)
	Sig. (2-tailed)		.000
	N	30	30
Greenbuiding framework	Pearson Correlation	.830(**)	1
	Sig. (2-tailed)	.000	
	N	30	30

** Correlation is significant at the 0.01 level (2-tailed).

Table 11 shows Pearson Correlation Test conducted in addition to answers relationship between Types of Project Execution and Implementation Framework of Green Building . The Pearson Correlation result shows h1 is accepted. (r = 0.830) with (p< 0.01). This results shows there is significance, positive and strong relationship between Types of Project Execution and Implementation Framework of Green Building . The positive correlation values shows that the more well defined Types of Project Execution the more effective Implementation Framework of Green Building for goverment building.

Hyphothesis 5 :There is significance relationships between Project staffs perception with Implementation framework of Green Building .

Table 12. Correlation for hypothesis 5

		Perception	Green buiding design framework
Perception	Pearson Correlation	1	.809(**)
	Sig. (2-tailed)		.000
	N	30	30
Greenbuiding framework	Pearson Correlation	.809(**)	1
	Sig. (2-tailed)	.000	
	N	30	30

** Correlation is significant at the 0.01 level (2-tailed).

Table 12 shows Pearson Correlation Test conducted in addition to answers relationship between Project Staff Perception and Implementation Framework of Green Building . The Pearson Correlation result shows h1 is accepted. ($r = 0.809$) with ($p < 0.01$). This results shows there is significance, positive and strong relationship between Project Staff Perception and Implementation Framework of Green Building . The positive correlation values shows that the more positive project staff perception the more effective Implementation Framework of Green Building for goverment building.

8. Conclusion

According to the pilot test result conducted with KKR2 Project Team, the test result shows positive result with the hyphotesis developed and formulated between Project Management Procedures/ Guidelines, Building Design Procedures/Guidelines, Goverment Policies, Types of Project Execution and Project Staff Perception with Implementation Framework of Green Building. Therefore, the result shows there is significance and positive relationships between Project Management Procedures/ Guidelines, Building Design Procedures/Guidelines, Goverment Policies, Types of Project Execution and Project Staff Perception with Implementation Framework of Green Building. The comprehensive, well defined Project Management Procedures/ Guidelines, Building Design Procedures/Guidelines, Goverment Policies, Types of Project Execution and the clear ,positive perception of the Project Staff Perception ensure the effectiveness and success of Implementation Framework of Green Building for Goverment Building. Implementation framework of green building for Malaysian Government Building is important in addition to ensure the success of Public Works Department as implementers of Green building for Malaysian Government Projects. This framework also important for Malaysia to have a clear and systematic execution/implementation of green government project in Malaysia.

References

1. U.S. Environmental Protection Agency. (October 28, 2009). Green Building Basic Information. Retrieved Decem\ ber 10, 2009, from http://www.epa.gov/greenbuilding/pubs/about.htm
2. R. Hopkins, 2002. *A Natural Way of Building*. Transition Culture. Retrieved: 2007-03-30.
3. X. Mao, H. Lu, & Q. Li, (2009). *International Conference on Management and Service Science, 2009. MASS 09.*, 1-5. doi:10.1109/ICMSS.2009.5303546 .
4. U.S. Environmental Protection Agency. (October 28, 2009). Green Building Home. Retrieved November 28, 2009, from http://www.epa.gov/greenbuilding/pubs/components.htm
5. WBDG Sustainable Committee. (August 18, 2009). Sustainable. Retrieved November 28, 2009, from http://www.wbdg.org/designsustainable.php T. Hegazy, (2002). Life-cycle stages of projects. Computer-Based Construction Project Management, 8.
6. S. Pushkar, R. Becker, & A. Katz (2005). A methodology for design of environmentally optimal buildings by variable grouping. *Building and Environment*, 40. doi:10.1016/j.buildenv.2004.09.004
7. J.R. Simpson, Energy and Buildings, Improved Estimates of tree-shade effects on residential energy use, February 2002. Retrieved:2008-04-30.
8. California Integrated Waste Management Board. (January 23, 2008). Green Building Home Page. Retrieved November 28, 2009
9. M. Thomas Parris (2007): *Green Buildings, Environment Science and policy for Sustainable Development,* 49:1,3-7
10. C.Turner, & M. Frankel, (2008), *Energy performance of LEED for new Construction Building*, Final report.
11. P.A. Andersen, J.F. Andersen, & J.P Garrison (1978). Singing apprehension and talking apprehension: The development of two constructs. Sign Language Studies, 19,155-186.
12. P.H Andrew, & R. T. Herchel, (1996), *Organization Communication: Empowerment in Technological Society.*Boston : Houghton Mufflin

13. A.T. Sulaiman (2001). Faktor-faktor yang menentukan Keberkesanan Komunikasi: satu kajian di Jabatan Kastam Diraja Malaysia, Negeri Pahang. Tesis Sarjana. Universiti Utara Malaysia.
14. Y.P. Chua, (2006). Kaedah dan statistik penyelidikan: Asas statistik dan penyelidikan (Buku 1). Malaysia: McGraw Hill Education.
15. Y.P.Chua, (2006). Kaedah dan statistik penyelidikan: Asas statistik dan penyelidikan (Buku 2). Malaysia: McGraw Hill Education
16. J.F Hair, Jr., A.H. Money, P. Samouel & M. Page (2007). Research method for business. England: John Wiley & Son Ltd.
17. M.K.N.Ibrahim, (2007). Hubungan ciri-ciri personaliti dengan menghadapi perubahan organisasi. Sintok: Universiti Utara Malaysia
18. U.Sekaran, (2000). Research Method for Business: A Skill building Approach (3rd edn). New York. John Wiley and Sons Inc.

Investigation of Sustainable Housing Criteria

Somayeh roshanfekr[1,a], N.M.Tawil [2] and N.A.Goh [1]

[1]*Department of architecture, faculty engineering and built environment, national university of Malaysia*
[2] *Department of architecture, faculty engineering and built environment, national university of Malaysia*

Abstract. Recently, much attention has been paid to sustainable development in cities. The quality of human life is directly related to environmental quality. Because many people live in cities as a place of social, economic and cultural relationships, certain issues such as environmental crises, energy, air and noise pollution and traffic jams are some of the factors that can alter the quality of human life. Therefore, in order to improve the quality of human life, attention to sustainable development (or sustainability) in cities is proposed. Sustainable building has a comprehensive significance that begins with the conception of negative and positive impacts on the environment. Several descriptions of sustainable or green buildings have been created; however, they all pursue one goal, which is to create sustainable urban developments and protection of the environment. The quality of indoor environments, materials, and energy consumption, water usage, the impact of building construction processes and building maintenance are some of the factors that affect the environment and sustainability. Sustainable building is an attempt to relieve the minus impacts on the environment that occur during a building's lifetime. This research investigates the important factors that have relevance to green buildings and introduces several criteria of sustainable housing.

Keywords: sustainable housing, green building, criteria of sustainable housing.

1. Introduction

According to the UN Commission on Human Settlements, sustainable development of human settlement makes sure social and economic progress, as well as adequate job prospects are in robust synchronization with the environs. The definition encompasses the main values of the Rio Declaration on Environment and Development as well as of the UN Conference on Environment and Development: anticipatory approach, avoidance of contamination, high regard for the carrying ability of the ecosystem and conservation of prospects for the coming generations [11]. For instance, according to Daly [6], one needs to understand the difference between development, which is essentially a qualitative enhancement, and growth, which is a positive change in volume or magnitude (for example, populace and resource). Growth can, however, only be finite, as we have inhabited only one planet. A suite of natural systems that offer many ecosystem services surrounds housing. It is vital to comprehend the relationship between natural systems and housing and consider it when designing and constructing accommodations and colonies [11]. The building industry's behavior is extremely important as far as environmental viability is concerned [5]. Sustainable housing presents multiple prospects of promoting economic growth, ecological stewardship, standard of living, and social parity while alleviating the unwarrantable convergences of issues pertaining to urbanization, population growth, poverty, slums, economic volatility, climate change, and inadequate access to sustainable energy [1]. The term 'sustainable building' pertains more specifically to the objective of designing and constructing structures which have no clear unfavorable effects on the environs, so that an entire built environment comprising similar structures can co-exist in the environmental balance for an indefinite period [14]. From Global Green USA [8]'s point of view, sustainable buildings offer multiple direct and indirect advantages across all geographical levels, i.e. individual, communal, regional, and global. As far as the global level is concerned, green buildings aid in safeguarding and preserving natural resources by decreasing the demand for usage of virgin materials. This survey aims to investigate the criteria for sustainable housing. This paper developed introductory information of sustainable housing According to individual knowledge along with literature review of evaluating and comprehension of sustainable housing criteria.

2. Sustainable Housing Criteria

Creating a sustainable building is a process that builds houses, buildings and auxiliary infrastructures while

[a] Corresponding author: so.roshanfekr@gmail.com

reducing resource consumption, decreasing detrimental environmental impacts, and offering healthier living environments [27].

Ideally, housing development should be not only viable and of excellent quality but also:

A. Inexpensive

The housing and accommodation should have been constructed, administered, and upheld at a judicious cost, and the form of the development taken into consideration.

B. Socially and Ecologically Suitable

The type of housing, support services, and facilities offered should address the requirements of those who are to be housed. The type, size, and terms of accommodation should aid comprehensive ecological, social, and economic sustainability policy goals for the region and endorse the development of properly assimilated recreation and amusement spaces.

C. Accessible and Flexible

Not all citizens, including those with impaired movement, should face any obstacles in access and motion. They should be able to travel liberally within and across the development, access buildings, and capitalize on the facilities and services offered. Houses should be adaptable enough to cater to the varying needs of citizens during the course of their lifespan.

D. Resource Efficient

Infrastructure, land, and energy should be utilized effectually. The place should be well-located in terms of conveyance, services, and facilities. Design, as well as house position, should consider the site topography in order to manage negative wind impacts and maximize the advantages of daylight, sunlight, and solar paybacks. Renewable energy sources should be utilized optimally. Scarce natural resources should be consumed in moderation during the building, upkeep, and administration of the houses.

E. Safe, Secure, and Healthy

The housing should be a safe and healthy place to stay in. Cyclists and pedestrians should find it easy and safe to travel within and across the locality. Vehicular movement, including entry and exit for service vehicles, should not hamper these goals.

F. Long-lasting

The finest construction methods should be deployed and the main aspects of construction should have a service lifespan of around sixty years without the requirement for uncharacteristic repairs or replacements.

G. Architecturally Proper

The accommodation should offer a pleasing environment that is appealing as well as appropriate. The design should be such that it should comprehend and respond aptly to its context, enhance the vicinity, and have a high regard for its cultural legacy [4]. Some have ratiocinate which a framework is helpful rather than a special description because of incomplete definitions of sustainability [16,17,2]. This framework have constructed to aid specify housing for sustainable development perspective. Figure 1 showed sustainable housing criteria.

Utmost of the sustainable housing literature review concentrate on environmental situation [9,12,1317,18,20,21,22,28]. Housing is a basic social setting, which establishes the standard of living and well-being of individuals and places. The places where accommodations are located, the way they are designed and constructed, and the manner in which they interlace with the socio-cultural, environmental, and economic structures of communities are aspects which impact the day-to-day lives of individuals, their health, safety, and welfare, and which, considering the lengthy life span of houses as physical structures, impact the

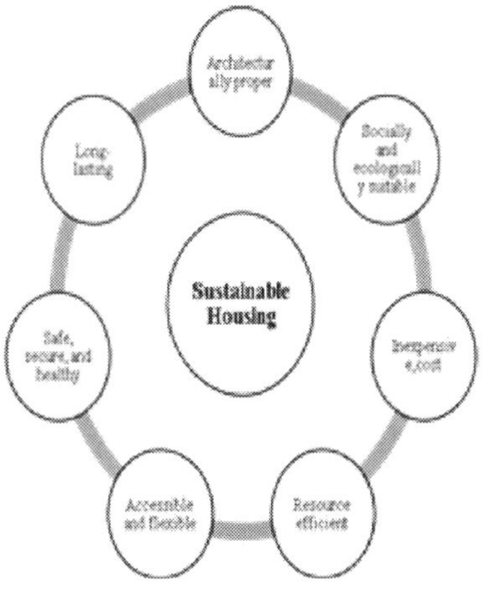

Figure 1: sustainable housing criteria

present as well as coming generations. Thus, housing is the core of sustainable development [1]. It is crucial that standards of sustainable design are abided by, and proper technologies are deployed in sustainable housing rollouts. Sustainable design standards in sustainable housing should also encompass the following: sustainable construction materials, thermally effectual design, renewable energy alternatives, energy efficacy, waste reduction and recycling and viable water and sanitation systems [11]. Figure 2 showed a sustainable housing in Australia.

Figure 2: sustainable housing, Suzanne Toumbourou 2015

To create sustainable housing, it is crucial to utilise native and untreated construction materials. This reduces transport and energy requirements as well as air contamination, besides generating local job prospects. In addition, greater localization ensures greater retention of money within the community [11]. On the whole, buildings and related materials manufactured for construction purposes are responsible for at least 30% of greenhouse gas discharges around the world [23]. On the other hand, recycled construction material is important in sustainable housing. Recycling is a vital element of green structures. It decreases non-renewable involvements, particularly usage of energy, mining activities, and cost of transportation. This encompasses the usage of waste products and used construction material. It is particularly crucial to reuse ecologically hostile materials that trickle contaminated materials into the earth and groundwater, or discharge methane into the environment when deposited in landfills [11]. The idea of sustainable energy involves the creation and effectual utilization of energy in a manner that enables us to cater to current needs without conceding the capability of upcoming generations to satisfy their own wants and requirements. Optimizing the viability of social accommodation by means of enhanced thermal performance and insulation criterions, thus decreasing greenhouse gas emissions owing to lower energy needs, would help achieve Ireland's obligations under the Kyoto Protocol [4]. Initiatives regarding sustainable buildings have conventionally emphasized bettering building performance and, at the same time, alleviating the negative ecological effects of the built environment. This is achieved through actions to reduce utilization of water and energy, decrease greenhouse gas discharges, and lower the quantity of solid waste and building debris that is dumped in landfills. Such programs characteristically stipulate eco-friendly materials and focus on the design and development of a venture, its consequent operations, and continuing upkeep. The advantages of sustainable structures with respect to energy efficacy and ecological impacts have been substantiated [3] Moreover, the United States Green Building Council [26] has stated that sustainable buildings can aid in safeguarding and conserving open spaces, decreasing solid waste, improving transport options, lowering pressure on local infrastructure, better managing storm-water, improving quality of water, decreasing water contamination, and improving community welfare through growth-oriented planning. Of late, there have been increasing instances of health concerns attributed to building design and development, and the materials utilized in the building process. Lead-based paints trigger extreme developmental issues and damage to the brain, particularly in children. Asbestos is a heat-resistant fiber utilized in house building materials right from the roofing to the flooring. It is extremely carcinogenic in nature. Just like the inadvertent concerns generated by tenement architecture, initial efforts to enhance building performance might have led to "sick" structures, wherein high amounts of indoor air contaminants were noted [3]. Furthermore, sustainable housing takes into consideration the quality of indoor air provided to inhabitants, considering around 1.3 million perish prematurely every year owing to exposure to indoor air contamination from combustion of biomass [24]. Development of accommodation presents several ecological challenges. It is imperative to determine accommodation solutions that do not severely affect housing affordability and improve urban standard of livings. Furthermore, it is necessary to determine viable accommodation solutions that tackle the increasing carbon footprint of the built environs, and that do not expose more homes to levels of carbon emissions, which are unmanageable with respect to operational and exemplified carbon loads (much like the present scenario with most housing performance in affluent nations). If such substandard performance were replicated at the global level, it would aggravate ecological change. Sustainable housing has the ability to directly or indirectly handle prevailing issues in urban areas, including management of storm water and solid waste, water supply, hygiene, and pest control [25].

Cognizance and knowledge of the different concerns involved, local perspective, and alternatives available are necessary to make proper green and sustainable decisions. The following are the likely monetary, ecological, and social advantages of greater resource efficacy savings in sustainable housing:

- Lessen exhaustion of natural resources during the building (lower waste from materials) and operation (no or minimal use of fossil fuels) stages by utilising effectual construction materials and methods as well as renewable energy sources.
- Lower cost of operations of households in the form of lower utility bills, resulting in lower paucity in terms of water and fuel.
- Comfortable and sanitized conditions to promote better health of the inhabitants.
- Utilisation of effectual water and waste management practices.
- Lower macro-infrastructure demands, such as electricity.
- Lower cost of transportation (with regards to sustainable urban setting).
- Long-term ecological benefits from lower pollution and usage of resources.
- Advantages of using native natural construction material: reduced transportation needs, reduced embodied energy, easy upkeep, and possibility of reuse and recyclability. Use of native materials that can be effortlessly replaced or repaired and lower utilisation of water and energy certainly leads to decreased life-cycle costs of accommodations [15]. A structure's environmental performance can be enhanced by opting for ecologically preferable construction material. In reality, ecological performance should be weighed against economic performance. All construction material manufacturers and environment-conscious building designers seek to weigh ecological advantages against economic costs. They endeavor to determine construction material, which enhances environmental performance without impacting on the cost factor. Furthermore, housing is a component of the society-environment relationship. Construction of houses and their operation utilizes huge volumes of natural resources such as energy, land, water, and building materials. It also generates waste, and triggers water and air contamination. That said, housing faces different negative environmental effects and threats, such as those related to climate change and natural calamities. These factors need to be considered in the process of sustainable development.()

When it comes to community and regional levels, green-built accommodations can aid in boosting local economic growth by promoting usage of local and regional supplies, thus retaining the money within the local community itself [7].

3. Conclusion

The aim of sustainable housing is to ensure the minimal negative effect of buildings and their operations on human wellbeing and conditions as well as the environs by means of better design, positioning, construction, functioning, upkeep, and the entire life cycle of the building. Largely, all sustainable structures are conceived to be energy- and resource-efficient, to utilize the appropriate materials (i.e. being recycled, economical, and robust) and to lower the discharge of lethal substances during their life span. Furthermore, a sustainable structure can diminish the unwanted human effects on the buildings assets and materials as well as natural environs and improve human wellbeing and the environment [10]. Safety, energy efficiency, comfort, and material and water efficiency are some factors of sustainable housing. Affordability and social and environmental suitability are other criteria for sustainable housing which help to aid sustainable housing development.

References

1. UN-Habitat, United Nations Human Settlements Programme. "Sustainable Housing for Sustainable Cities ." a policy framework for developing countries. (2012)
2. A.T, David and Ken G. "A Framework to guide sustainable housing development ." J. o. t. Hus. Edu. and Rch .As. (2010)
3. B, Noreen. "Green Housing = Improved Health:A Winning Combination ." National Center for Healthy Housing (2011).
4. Department of the Environment . "Quality Housing for Sustainable Communities Best Practice Guidelines for Quality Housing for Sustainable Communities , ." DELIVERING HOMES SUSTAINING COMMUNITIES Department of the Environment, Heritage and Local Governme (2007).
5. E, Piet, Nils K and John M.Q. "Program On Housing And Urban Policy Working Paper Series, working paper no. W08-001 doing well by doing good? ." green office buildings, University Of California, Berkeley. (2008).
6. E.H, Daly. " Beyond Growth: The Economics of Sustainable Development." Boston: Beacon Press. (1996).
7. Federal Home Loan Bank of Atlanta. The little green book. Atlanta, Georgia : Federal Home Loan Bank of Atlanta, (2005).
8. Global Green USA. "Greening affordable housing: Trends and opportunities." Presentation at USGBC Greenbuild Conference, November 7(2005).
9. H, Lovell. "Framing sustainable housing as a solution to climate change." J. Env.Policy & Pln. **6**(1), 35-55 (2004).
10. H, Okhovat., Amirkhani. A and Pourjafar. M.R. "Investigating the psychological effects of sustainable buildings on human life." J. Sus. Deve. Vol. **2** (2009): pg. 57-63.
11. H, Paul. and Lisa. TH. "SUSTAINABLE HOUSING, the Sustainability Institute Sustainable Neighbourhood Design Manual: A Non-Technical Guide. Final Draft for Comment ."(2009).
12. J, Swanson. "Sustainability sweeps into Broome: Public housing looks set for a makeover in this WA town." Habitat Australia **31**(3), 23 (2003).
13. K, Parrott and Emmel J. M. "Sustainability in housing: A curriculum case study." J.f F. and Con.Sci. **93**(5), 31-34 (2001).
14. L, alex and joseph L. "Towards Sustainability: Green Building, Sustainability Objectives, and Building America Whole House Systems Research, – 0801." Building America Report (2008).
15. L, Christophe. "Green building interventions for social housing, ." United Nations Human Settlements Programme (UN-Habitat) and United Nations Environment Programme (UNEP) (2015.
16. M, Bhatti. "Greening housing: A challenge for public policy?" Hume Paper on Public Policy **8**(4), 63-71 (2000).
17. M, Bhatti, Brooke J and Gibson M. "Housing and the environment: New agenda." The Chartered Institute of Housing (1994).
18. M, Sunikka and Boon C. "Environmental policies and efforts in social housing: The Netherlands." Building Research and Information **31**(1), 1-12 (2003).
19. O, Annette . SUSTAI ABLE BUILDI G TECH ICAL MAnUAL Green Building Design, Construction, and Operations . US: Public Technology Inc. n US Green Building Council, (1996).
20. P, Hugo. "How to make housing sustainable? The Dutch experience., ." Env. Pln. B: Planning & Design **32**(1), 5-15 (2005).
21. R. L. H, Chiu . "Social sustainability, sustainable development, and housing development: The experience of Hong Kong. In R. Forrest and J. Lee (Eds.)." Hou. and soc.ch. (2003a): (pp. 211-238).
22. R.L.H, Chiu. "Environmental sustainability of Hong Kong's housing system and the housing process model." Int. Pln. Stu. **5**(1), 45-64.5(1), 45-64 (2000).
23. RICS. "Royal Institute of Chartered Surveyors ." Green Value, London and Vancouver (2005).
24. UNEP SBCI. Buildings and climate change. Summary for decision-makers. Paris: UNEP DTIE, (2009a).
25. UN-Habitat and UNEP. The sustainable cities programme in Zambia (1994-2007): Addressing challenges of rapid urbanization. Kenya: UN-Habitat and UNEP, (2009).
26. United States Green Building Council. Rating system for pilot demonstration of LEED® for Homes program, version 1.72. Washington, D.C.: United States: Green Building Council, (2005a).

27. w, Karlenzig. A blueprint for greening affordable housing: Developer guidelines and resource efficiency and sustainable communities. Santa Monica, California: Global Green USA, (2005).
28. Y, Zhu and Lin B. "Sustainable housing and urban construction in China." Energy & Buildings 36(12), 1287-1297 (2004).

MATEC Web of Conferences **66**, 00097 (2016) DOI: 10.1051/matecconf/20166600097
IBCC 2016

Challenges to Building Surveyors From The Perspectives Of Non Surveyors

Zarina Isnin[1,a], Shamnor Syaza Dania Badrol Hisham[1], Rohaslinda Ramele[1], Emma Marinie Ahmad Zawawi[1]

[1]*Faculty of Architecture, Planning & Surveying, Universiti Teknologi MARA, 40450, Shah Alam, Selangor*

Abstract. Awareness on the importance of Building Surveyors in Malaysia is still low as the role and skills of this profession are not fully understood. Although studies indicated that Building Surveyors are still in demand, even without the Building Surveyor Act, many graduates are experiencing inadequate employment opportunities in the current economic situation. Little is known on the views from other stakeholders about BS. This research aims to examine the awareness and opinions on BS in Malaysia amongst construction stakeholders. Questionnaire surveys were collected from 120 respondents from construction, maintenance and insurance companies and interviews were conducted to selected built environment respondents. It was found that awareness and knowledge on BS are still low as they lack information on the profession and professional recognitions. Some views indicated on a major identity crisis for having fragmented and disparate range of modules. The cause may have resulted in problems on public recognition, poor understanding of the surveyor's skills, and fewer job opportunities. Several suggested the profession to have a clear, coherent and relevant identity, with strong professional structures in order for the profession to survive and gain recognition from the government. Graduates are also recommended to acquire other skills and training in order for them to be significant to the construction industry.

1 Introduction

The construction market has been particularly hit by the downturn in the property market, and is very much exposed to macro-economic changes either in Malaysia or globally. This has affected the availability of finance for owners of property (residential, commercial and industrial) to undertake repair and refurbishment work. Fluctuations in the economy of the property and construction market affected the services and advice that clients required. There is less job opportunities for graduates. This challenging situation would instigate a need for the built environment professionals including Building Surveyors and the Building Surveying (BS) education programme to change.

Previous studies indicated that BS is a professional expertise that emerged during the 1960s and 1970s in United Kingdom (UK) because Architects were unable, or unwilling, to supervise and manage repairs and refurbishment of property. Initially, Chartered Surveyors of the Royal Institution of Chartered Surveyors (RICS) in UK with technical knowledge on maintenance and refurbishment started to take on these commissions. Degree courses on Building Surveying were then introduced to educate people wishing to enter the profession. BS is said to possess 'unique' and 'unrivalled' knowledge on legal and technical issues affecting buildings by the Association d'Experts Européens du Bâtiments et de la Construction (AEEBC).Generally, a qualified professional Building Surveyor should be able to perform and advise upon any issues related to property and construction in respect of commercial and non domestic properties.

Although these statements indicate that the BS profession plays a very important role in construction industry, what is the current perspective in Malaysia? There are little studies carried out into exploring the views and propositions from stakeholders in construction industry about BS. Therefore, this research is an attempt to partially fill that gap in knowledge. It is hoped that the research findings from perspectives of non surveyors may act as another medium for the professional body governing building surveying in Malaysia to take a more progressive and professional approach and lead the profession into a sustainable future.

2 Role of a Building Surveyor

A Building Surveyor is a construction professional who provides professional advice on property and construction. The professional advice include works that involve repair, maintenance, refurbishment, alteration and extensions of all commercial, residential and industrial properties. They are also required to assess and report on defects and ways to improve a building project. They do

[a] Corresponding author: zarinasf@salam.uitm.edu.my

not just oversee the work on building sites and measure buildings [1] but are also required to work closely with clients, construction workers and local planning bodies to ensure the building meets the required sustainability, safety and preservation standards. This is also clearly outlined in the 2016 Building Surveying Division, Royal Institution of Surveyors Malaysia (RISM) website [2].

Based on other international scope of works outlined for Building Surveyor such as RICS and Australian Institute of Building Surveyors (AIBS), building surveyors play a critical role in achieving the objects of the legislation. There is emphasise on the need for public safety and protection in relation to the design and construction of buildings. The responsibilities include drawing up detailed plans and advising on whether a grant might be available, instructing architects to prepare detailed plans and help to provide estimates for the work and dealing with planning applications, property law, building regulations and other legal matters. As there are other professionals involved in building construction project, Building Surveyor should be able to interact with other professionals.

3 Public awareness on Building Surveying profession

In Malaysia, studies revealed that there is still lack of public awareness on the significant establishment of BS profession [1][3][5]. The profession receives limited recognition and significance value by the Malaysian public. BS is still generally misunderstood by other professionals and the general public because there is a lack of understanding on the role of general practice surveyors profession [1][3][5]. Most of them may not realize that BS offers professional opinions on building's condition and its property value including dilapidations and condition surveys of new premises and existing buildings [5]. Furthermore, BS expertise in condition survey work is critical for the process of obtaining design information for refurbishment [4] and conservation works. The required information will actually contribute to better cost control and project management in running the refurbishment and conservation projects. However, there is less commercial publications and marketing either in articles, journals, newspapers, books or the internet about the profession [1][3].

3.1 Professional Recognition

BS profession has been recognised and valued in the UK, Australia, New Zealand and Hong Kong but is still hardly recognised in Europe [6]. There is evidence emerging that Building Surveyors are slowly beginning to provide professional services in the United States of America [6] although they have yet to obtain any professional recognition.

A Building Surveyor in UK was developed to be a direct response to the structure of the British Commercial property lease [7]. The profession is needed for professional advice to be provided at a level between that of the architect and the engineer [6][7]. In Australia, qualified buildings surveyors are responsible to ensure that buildings are safe, accessible and energy efficient. Their roles have an impact on the design, planning and functionality of buildings as they are able to detect and diagnose problems with design, construction techniques and materials. What about the recognition of professional Building Surveyor in Malaysia?

In Europe, the poor recognition is partly because of little awareness on the expertise, and at present there are legal constraints, and limited historic practises relating to the employment of BS professionals. Professional BS skills are deemed to be a new skill in Europe and it may encounter some difficulties because it was stated that the BS professions may lack understanding of the European regions local practices and building techniques.

A study made by Banyard, Wilkinson and Turrell [6] claimed that a principal disadvantage in the employment of Building Surveyor's from the UK is in the "limited local knowledge that they possess". The practice of Building Surveyors in certain locations could be further inhibited by difficulties to obtain building works consents and local practices. The researches further cited a study done by Linklators & Alliance (2001), that according to French law, an Architect must be employed for building works in which planning consents are required in Europe.

In Ireland, it was acknowledged by the Society of Chartered Surveyors for Building Surveying Professional Group that there was a lack of clear professional identity of the BS profession [8]. The profession is undermined by the potential erosion of a clear professional identity by other professional groups working in the built environment such as architects and engineers. This is a weakness that makes it difficult for chartered building surveyors to fully consolidate their position in the Irish market place. The BS profession has a very narrow niche in a market which is crowded with other professional services. Furthermore, the housing market in Ireland is relatively weak which has a negative impact on the size of the market available to the profession.

3.2 Building Surveying in Malaysia

In Malaysia, studies stated that the most challenging task for the establishment of the BS profession is the process of gaining recognition and acceptance from other specialized professions such as architects, engineers and valuers [1]. Continuous confrontations against the establishment of the BS profession had a great impact towards the approval of the BS Act in Malaysia. Other professionals were said to be concerned about the profession and business competitions based on the BS scope of work and accountability [1][5]. These were some of the cited reasons on the constant objection towards the approval and implementation of the BS Act in Malaysia. It was further argued by the other professionals that it is not necessary to establish and endorse the BS profession because the other established professionals are competent with the knowledge, skills and ability on BS scope of works [5].

However, professional Building Surveyors stated that the role of BS is not to directly compete with these other professions; instead it complements and works in harmony with them [1]. As stated in other studies, it is agreed that the BS profession may benefit the construction industry because of the professional contributions on building's condition and property value. It will also create a healthy competition amongst the construction stakeholders in the building industry.

3.3 Higher Education Programme Structure

Previous study revealed that the syllabus offered for the BS course in higher education institutions in Malaysia varies between higher educational institutions [5]. Generally, a BS programme requires students to have a thorough understanding of construction methods and building defects. Some of the essential skills include the ability to collect data, measure accurately, take photographs, use electronic surveying devices, produce drawings, investigate thoroughly the causes of building defects, and to produce a clear written report of the findings of a survey. The survey works include building structure, electrical, plumbing and other services within the property being surveyed.

Some of the general objectives from BS programme are to ensure that students should be able to demonstrate knowledge and understanding on the role of building surveyor. The students should be able to understand the reasons for, and the processes involved, in surveying buildings. It is also vital to ensure that students be able to carry out simple building surveys, including, measured, dilapidation and condition, for domestic and commercial buildings. The students should be able to produce a survey report and schedule of maintenance and repair and distinguish between planned and unplanned maintenance. Students should be able to communicate verbally and graphically.

However, studies revealed that the previously BS syllabus offered in the public universities and private education institutions varies that created non-standardized specific education areas [5]. Currently, the education implementation is based on the requirement set by the Malaysian Qualifications Agency (MQA) which controls and monitors quality of new programmes introduced in higher education institutions and universities. The programme structure focused on three main scope of works approved by the RISM. They are (i) Building Control; (ii) Building Maintenance and Conservation; (iii) Risk and Building Performance. The minimum credit set must be 120. This could enhanced the quality of BS programme offered in Malaysia as the academic staff, entry requirement for the students, assessment method use and composition of subject offered can be monitored [1].

4 Methodology

The initial stage for this study involved literature review of primary and secondary sources to establish what had been written about BS and practice in Malaysia and the global markets. This study examine the awareness and opinions on BS in Malaysia amongst construction stakeholders. Questionnaire surveys were personally hand distributed to 300 stakeholders in construction industries such as construction, maintenance and insurance companies. 120 responses were received and analysed. Semi-structured interviews were also conducted to three professionals who are involved in construction management and one respondent is also an academician. The interviews seek to evaluate the professional views regarding BS profession in Malaysia and the barriers to practice.

The survey questionnaire was divided into four sections.

(1) Part A: The respondent's particulars using closed-ended questions.
They were straightforward questions to encourage respondents to participate in the survey and for easy analysis of the data.

(2) Part B: The respondent's awareness and knowledge on BS programme and recognitions using combinations of dichotomous questions for types of profession and Likert scale for their opinions on education and profession. The Likert scale required the respondent to rate their level of agreement to the statement ranging from 1 (strongly disagree) to 4 (strongly agree). The scale eliminated the mid-point ('neither... nor', uncertain etc.) category because the aim of this survey is to ascertain respondent's opinion.

(3) Part C: The respondent's opinions and perspectives on Professional Development for BS graduates using Likert scale.

(4) Part D: The respondent's recommendation to improve the BS profession using combination of Likert scale and open-ended questions. The open-ended questions were used to enable the respondents to give feedbacks with varied points of view.

Similar questions from Part C and D were also used for the semi-structured interviews to three professionals who are involved in building construction industry.

5 Findings & discussion

5.1 Demographic Profiles

The demographic profiles demonstrated that feedbacks were received from different educational backgrounds, age, positions and years of working experience in related construction and maintenance industries. For questionnaire surveys, 120 usable responses were received during the 30 days research period. 105 respondents were from construction and maintenance companies that have experience in managing works between the range of RM3 Million to RM10 Million. Designations of respondents surveyed include project managers or building managers (6), engineers (3), quantity surveyors (4), technician or technical employee (31), administration staff (16), construction insurance specialists/employee (15) and others such as general workers and businessmen or company owner (45).

For the interviews, three respondents agreed to be interviewed using a semi-structured questions which are similar to Part C and D in the questionnaire. They are professionals who are involved in building construction industry. Their work experience ranges from more than seven years to 25 years (R1 – Engineer; more than 7 years), R2 (Project Manager; more than 10 years) and R3 (Professional Interior Designer/ Senior Lecturer – more than 25 years). All the respondents have involved with BS graduates and professionals for their projects or work.

Table 1. Demographic finding for questionnaire survey

Survey question	Description	Number of Respondents, n
HIGHEST QUALIFICATION	SPM/STPM	49
	Diploma	35
	Degree	32
	Master	4
AGE	20 - 30	65
	31 – 40	36
	41 – 50	17
	51 – 60	2
GENDER	Male	71
	Female	49
ORGANIZATION	Government	7
	Private	100
	Own Business	13
OCCUPATION	Project / Building manager	6
	Engineer	3
	Quantity Surveyor	4
	Technician or Technical employee	31
	Administration staff	16
	Construction Insurance Specialists/Employee	15
	Others: General worker or Company owner	45

5.2 Knowledge and Awareness on Building Surveying

Based on the survey, 59% (n=71) do not know about the scope of works for BS. 49 respondents (40.8%) know about professional qualifications for BS but 60% (n=72) are not familiar with the RISM.A higher percentage of respondents (70.8%; n=85) do not know about any international recognition for BS and only 40.8% (n=49) know about RICS.57.5% (n=69) respondents were aware that there are available Diploma programmes and 75% (n=90) knew about Degree programmes. 56.6% (n=68) agree and 14.2% (n=17) strongly agree that there are sufficient education opportunities for BS in Malaysia and overseas.

However, it should be noted that during the survey, majority of the respondents were actually asking for clarifications on the terms "Building Surveying" and "Building Surveyor". These findings may be related to lack of information and commercial marketing on the profession, the education programmes offered and the professional recognitions either in Malaysia or internationally. Some respondents were also asking on the similarities between quantity surveying programmes. Thus, there is a need to focus initiative about creating awareness on the work of building surveyors and the importance of building surveys.

5.3 Potential job opportunities for BS graduates

61.7% (n=74) agree while 23.3% (n=28) strongly agree that the BS profession is appealing to potential entrants. Majority of the respondents (94.2%) believed that graduates could work in BS practices such as maintenance, asset and facilities management companies, and also in heritage conservation projects. The question further asked on other career options. 69.2% have the opinions that BS graduates can work in construction companies and 67.5% selected quantity surveying companies. More than 50% of the respondents also agree that the graduates are able to work either in property development companies, town or city councils, occupational, health and safety sectors or have their own business. Majority (78.3%) agreed that BS graduates are capable to undertake career opportunities outside Malaysia.

Based on interviews, all the three respondents believed that universities or higher education institutions especially in Malaysia should have a strategy that markets the BS diploma and degree programmes. This can be done by providing series of career talks. Two respondents (R1, R3) suggested that explanation should be given on the career options that not only focusing on Building Surveyor as the only path. In this way, it may increase the number of students significantly entering the education programme as it will ensure that all students are aware of all career paths.

5.4 Opinions and Perspectives on Professional Development for Building Surveying Graduates

i) Professional Development

More than 83% agree and strongly agree that there are adequate professional development for BS graduates.50.8% (n=61) agree and 37.5% (n=45) strongly agree that BS graduates should have sufficient professional expertise and skills because they are expected to provide professional advice on property and construction, which spans across residential, commercial and industrial projects.

From the interviews, all respondents (R1, R2 and R3) suggested that the current BS professionals to be more vocal and more active to support the BS profession in Malaysia. There is a need to increase the number of memberships either in RISM or RICS. R2 opined that the professional Building Surveyors are also encouraged to carry out promotion on BS throughout Malaysia. Some of the experts may establish maintenance, conservation, refurbishment or asses management guidelines and services in accordance with related Malaysian policies and standards.

All the interviewees agreed that there should be regular professional development session to enhance their professional capabilities and opportunities in the multi-professional construction and built environment groupings. This may lead to higher levels of professional, and public recognition.

ii) Training and skill requirements

Majority of the respondents (more than 90% from survey) and all three interviewees concurred that BS graduates would require more training before they can work as Building Surveyors or in other construction related industries. They are also encouraged to gain other skills such as on Computer Aided Design, Project Management, Facilities Management and Safety and Health trainings to prepare them for job applications. Respondents also agree that BS academic programmes require more exposures in practical trainings in order to market their potential graduates in the construction workforce.

5.5 Recommendation to improve the BS profession - The Way Forward

In order to survive and prosper, it is necessary for professionals to adapt to change and to maximise the opportunities that arise from change. The Building Surveyor him/herself will need to be more adaptable to construction related work practices. Building Surveyors should be able to modify their role to suite the requirements of the local professional practices as well as demands of the international markets. They will have to adjust the services that they provide depending upon the varying demands such as:

i) Creating a Niche Market

Based on interview, R2 and R3 highlighted that some of the services provided by Building Surveyor are already done by other professionals such as preparing reports on dilapidation and defects on property. International studies stated that other professionals also provide a "Building Surveying service" such as in European countries that do not have specific practising Building Surveyors. Thus, it is recommended that the Building Surveyor to identify a niche market or provide a level of service that is different from that provided by the national or international practises in the country.

For example, Building Surveyors could use their knowledge on the impact of structure, access, services, safety, health and fire protection to conservation and refurbishment projects. They should be able to comment on the effect of any proposals to the significance of the building and suggest other solutions to achieve the performance requirements that create less negative impact.

ii) Upgrading the Syllabus

R3 suggested that the scope of studies in the academic programme for BS is so broad. He suggested that building surveyors would then have limited depth of knowledge about too many areas. The scope of studies are only basic and may not be adequate. He further explained on the fragmentation of work that can be done by the graduates. However, R1 with more than seven years of experience positively argued that it is good for BS graduates to be equipped with broad studies related to construction, design, cost estimating, building pathology and legal requirements. She on the other hand related her experience dealing with the legal knowledge that would be beneficial especially for builders and contractors.

R2 also commented on the needs for any construction employees to have abilities for technical knowledge on construction and manual drawing skills. This is especially required to ensure the constructors understand and able to build as per specifications especially on site. However, several suggestions were made on the needs to focus the profession into specialist areas such as heritage conservation and refurbishment expert. Students are also encouraged to learn about Building Information Modelling (BIM) as this system allows for digital documentation, digital take offs and measurement linked to a pricing and costing program that will enable the construction project to be measured, priced and checked using computer technology.

Although the current education syllabus aim to provide a foundation of general building surveying and associated knowledge, it is crucial to ensure that graduates are able to read across sub-discipline areas with a reasonable measure of understanding. Specific, in-depth learning should be built upon this foundation. Students should be encouraged to possess good communication skills at all levels (written, verbal and graphic). Students and graduates should be reminded that there should be ongoing learning (both in a formal and an informal sense) for the future of their professional lives. They should also be reminded on the importance of professional structures and professional ethics.

iii) Ensuring Students to Gain Hands-On Experience

More than 30% respondents suggested that students should be exposed to more practical training. One respondent wrote on the needs for oral, written and graphical communication skills. R3 who have experience in teaching suggested that "...it would be great to have more mature students especially those with working experience."

Students need to be trained and educated in a broad range of skills which are adaptable and flexible. This will enable them to operate in a changing and challenging market place. Providing tertiary education with a strong practical component would enable graduate to be almost immediately useful to the profession upon graduation. Graduates need to be skilled in communication and people management.

iv) Develop Better Relations with Other Professionals

All three interviewees concurred that there is a need to better liaising with other professional bodies as this may create an opportunity to cross-sell services within the Malaysian construction and built environment industries. By selling the services through a form of "Professional Public Relation" activities and initiatives, this may educate other members and increase public information about the profession.

6 Conclusion

The worldwide current economic climate should instigate a need for change to BS to face a more challenging situations in the future. It is important to have a well defined sense of future direction and a clear plan for BS to move forward. There is a need to reflect on how to market the BS profession to recruit future generations of surveyors and how to adapt to the evolving marketplace for surveying services, before others take over those markets.

BS graduates should be able to read across sub-discipline areas with a reasonable measure of understanding. They have to acquire other skills and undergo related training in order for them to be significant in the construction industry. Although they are equipped with broad studies related to construction and maintenance, they also need to possess good communication skills at all levels. There should be lifelong learning for their professional future. There is a pressing need to create awareness on BS and the importance of professionally recognised Building Surveyors and the sustainable options it offers. This may increase student intake and awareness on career path options.

References

[1] A. S. Ali and C. J. Woon, "Training and Development of Building Surveyors in Malaysia," in *RICS COBRA 2012*, 2012, pp. 777–785.

[2] http://www.rism.org.my/building-surveying-division-bs/

[3] A. S. Ali and C. J. Woon, "Issues and Challenges of Building Surveyors in Malaysia," Struct. Surv., vol. 31, no. 1, pp. 35–42, 2013.

[4] A. S. Ali, "Design information in managing refurbishment projects in Malaysia," *Int. J. Phys. Sci.*, vol. 5, no. 6, pp. 768–773, 2010.

[5] R. A. Rashid and N. H. Ishak, "Challenges of Building Surveying Profession Establishment In Malaysia," in *The Australia Institute of Building Surveyors (AIBS) 2007 International Transitions Conference*, 2007, pp. 243–256.

[6] S. H. Banyard, S. J. Wilkinson, and P. Turrell, "The Impact of Globalisation on Building Surveying in Europe," Paris, France, 2003.

[7] T. Kibblewhite and S. Wilkinson, "Building Surveying: A UK Phenomenon or a Profession with Genuine Global Appeal?," Athens, Greece, 2004.

[8] Building Surveying Professional Group: Business Plan 2011 – 2014. (n.d.). Society of Chartered Surveyors Ireland. Retrieved from https://www.scsi.ie/professional_groups/building_surveying/building_surveying_committee_business_plan.final

 DOI: 10.1051/matecconf/20166600098

Resident's acceptance of internal courtyard buildings

Adi Ainurzaman Jamaludin[1,a], Hazreena Hussein[2] and Kauthar Md. Tahir[1]

[1]*Institute of Biological Sciences, Faculty of Science, University of Malaya, 50603 Kuala Lumpur, Malaysia*
[2]*Department of Architecture, Faculty of Built Environment, University of Malaya, 50603 Kuala Lumpur, Malaysia*

Abstract. The internal courtyard of the residential building creates a micro-environment that provides a comfortable indoor environment. The acceptance of the courtyard by residents is disreputable even though its benefits have been scientifically proven. A survey using a set of questionnaire was carried out at two residential colleges to gauge the acceptance level of residents towards an existing internal courtyard. Results show that the majority of respondents of both residential colleges have given positive feedbacks, which include the perception on building and landscape designs in general, the quality of the landscape setting at the internal courtyard, the frequency of the time spent at the internal courtyard in a day, and the influence of the landscape setting at the internal courtyard on the quality of life. These feedbacks are not influenced by the gender aspect holistically. Indirectly, it indicates that the internal courtyard is positively relevant to be applied to the building design, particularly low-rise residential buildings.

1 Introduction

Courtyard in residential buildings is commonly applied for the purpose of promoting comfortable indoor environments by offering natural ventilation and daylighting into buildings. In the Middle East, courtyards provide the much-needed security and privacy for residents while reducing the infiltration of hot and dusty winds into the rooms with tall walls that build around the courtyard [1]. Courtyard creates a micro-environment that provides very pleasant living spaces with the presence of healing elements such as trees, flowers, shrubs and a pool of water [1-2].

The influence of a courtyard on the thermal condition has a strong reliance on the envelope opening and exhibits a better energy performance in hot-dry and hot-humid climates [3-5]. According to Almhafdy et al. [2], the development of a courtyard has moved forward into the era of sustainability and green architecture, where the courtyard configurations can be classified into four. There are:

- Cluster courtyard
  - a. Multiple : the design of the building creates four fully-enclosed courtyards.
  - b. Spinal : More than one courtyard is attached to the building's central circulation route.
- Open courtyard : Group of buildings frame open spaces into the courtyards.
- Closed courtyard : The courtyard is fully enclosed.
- Complex/Interlinked courtyard : Courtyards are located on different floors and serving different purposes.

Generally, the U-shape of a courtyard with an aspect ratio of 1:2, which is considered as a rectangle has a better performance in terms of the thermal comfort as compared to the U-shape of a courtyard with an aspect ratio of 1:1, which is considered as a square [6].

The successful attribute of a courtyard in building designs is not limited to providing comfortable indoor environments and efficient use of electricity. It should be extended to the residents' acceptance to understand the mutual interactive process between the building and the user's needs [7]. Thus, the wastage of maintenance and operating costs can be evaded as the residents' actual needs are determined proximately. Furthermore, having a courtyard would improve the quality of the facilities, morale and work performance of the residents [8]. According to Ilesanmi [9], the residents' perception of a proper residential environment cannot be ignored at the policy, planning, design and implementation levels where satisfaction with the physical environment of the residence is the most powerful predictor of residential satisfaction. Thus, the notable design features can be repeated in future development, and any redundant or unnecessary building features will be eliminated. In order to ensure that the building continues to deliver appropriate levels of satisfaction to the end-user, the critical evaluation in terms of the acceptance and perception should not only occur in a facility once it is constructed and occupied, but it should also be carried out throughout the entire building delivery cycle [10-11]. It is necessary to focus on the most relevant issues rather than to attempt to analyse everything and risk an overload of data [12].

We have a shortage of residents' acceptance studies on the internal courtyard at residential buildings, particularly in tropical climate regions.

[a] Corresponding author: adiainurzaman@um.edu.my

Most of the reported studies are experimental studies that focus on the advantages and disadvantages of the internal courtyard in terms of natural ventilation and daylighting, either through the field measurement or simulation [3, 4, 6, 13, 14]. Therefore, the aim of this research is to evaluate the residents' acceptance level of two residential colleges with an internal courtyard arrangement, with the purpose of justifying the relevance of the internal courtyard in the building design, particularly low-rise residential buildings. Further statistical analysis was done to measure the acceptance level based on the gender perspective.

2 Research methods

A survey using a set of questionnaire on residents' acceptance was conducted based on a study by Jamaludin et. al [15]. There are only four questions covering the perception on building and landscape designs in general, the quality of the landscape setting at the internal courtyard, the frequency of the residents spending time at the internal courtyard in a day, and the influence of the landscape setting at the internal courtyard on the quality of life. The questionnaire uses a Likert Scale format where each number responds to a specific scale as listed in Table 1.

Table 1. Likert Scale used in satisfaction survey.

Likert Scale				
-2	-1	0	+1	+2
Strongly disagree	Disagree	Undecided	Agree	Strongly Agree
Very poor	Poor	Fair	Good	Very good
Never	Rarely	Sometimes	Frequently	Every time
Not at all	Slightly	Moderate	Very	Extremely

The questionnaires were distributed to all residents with the minimum feedbacks relying on 95% of confident level and ±5% margin of error from the overall population. The calculation of the sample size is based on the formula [16]:

$$n = N/1 + N(e)^2 \quad (1)$$

Where, n is the sample size, N is the population size and e is the level of precision. A statistical software package has been used to analyse all the questionnaires collected. Further analysis was done to see the acceptance level based on the gender perspective.

3 Building description

3.1 RC-A

RC-A is the 5th Residential College known as Dayasari Residential College. It is located in the University of Malaya, Kuala Lumpur.

Built in 1966, this residential college was able to accommodate 847 residents at any single time. It consists of five residential blocks with three-floor levels each, excluding a ground floor. The site plan and typical floor plan of the RC-A are presented in Figures 1 and 2.

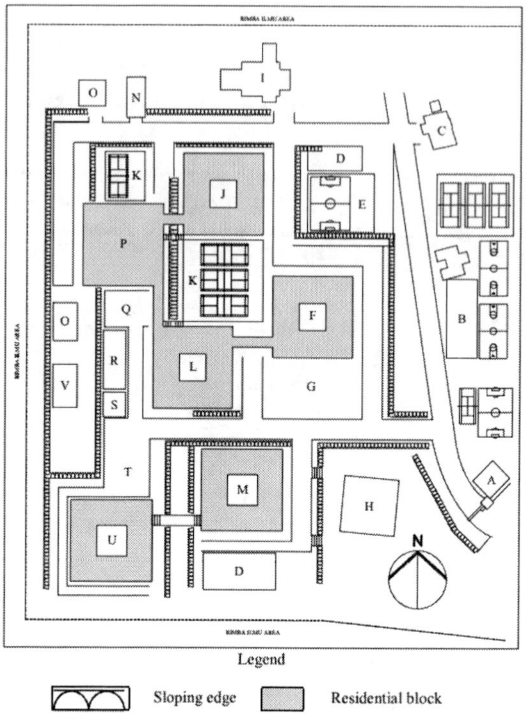

A - Guard post/Main entrance, B - Court/Sport centre/Multipurpose open area, C - Cafeteria, D - Sewage tank, E - Field/ Multipurpose open area, F- Residential block (Block B - Boungainvilla/ Female), G - Garden/ Green area, H - Prayer Hall/Musollah, I - Principal unit, J - Residential block (Block A - Azalea/Female), K - Court/ Multipurpose open area, L - Residential block (Block C - Camellia/ Female), M - Residential block (Block D - Dahlia/ Male), N - Self car wash, O - Store, P - Administrative block/Office, Lobby, Students activity centre, Dining hall, Kitchen, Q - Parking lots, R - Self-laundry room & Covered parking lots, S - Pump house, T - Parking lots/ Multipurpose open area, U - Residential block (Block E - Episcia/Male), V - Staff residential units

Figure 1. Site plan of RC-A.

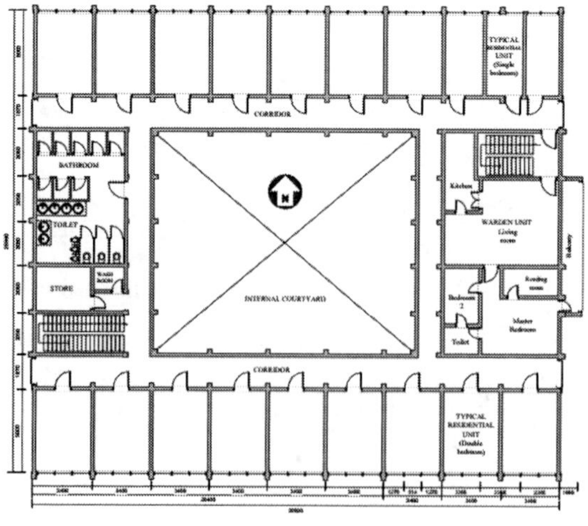

Figure 2. Typical floor plan of RC-A.

The residential building is designed with an internal courtyard arrangement and rectangle shape of the building's floor plate. The total floor area is 18,212.51m^2, while the room's floor area and volume are 16.35m^2 and 45.78m^3, as the typical room dimension is 4.74m (l) x 3.45m (w) x 2.80m (h). Most of the rooms, especially the ones at the middle of the floor are directly facing the internal courtyard. This promotes natural ventilation and daylighting inside the rooms through transoms on top of the entrance door and wall. The area of the internal courtyard is 239.88m^2, planted with grass and shrubs. There is a small hut, known as gazebo that functions as a social or engagement space for the residents.

With regard to the enclosure and façade design, there are centre pivots and awning window design with tinted glasses in each room. The different types of windows help to channel outside air/wind from a different angle, while giving residents full control towards daylighting besides the large horizontal overhangs along the wall with windows. The characteristics of the RC-A are visualised through the standard regular photographs presented in Figure 3.

(a). Built-form configuration, enclosure and facade design - Large horizontal overhangs along the wall with windows, (b). Internal courtyard in the middle of residential building, (c). Centre pivots & awning window design with tinted glass, (d). Transom/fixed opening over the doorway of residential unit.

Figure 3. The characteristics of RC-A.

3.2 RC-B

RC-B is the 11th Residential College known as the Ungku Aziz Residential College. Located at the University of Malaya, it was established in 1998. This residential building provides comfortable and spacious rooms with 897 residents. The RC-B only consists of four residential blocks with three floor levels each, excluding a ground floor. The site plan and typical floor plan of the RC-B are presented in Figures 4 and 5. Similar to the RC-A, RC-B was designed with an internal courtyard arrangement in an L-shape of the building's floor plate. The total floor area is 34,305.32m^2.

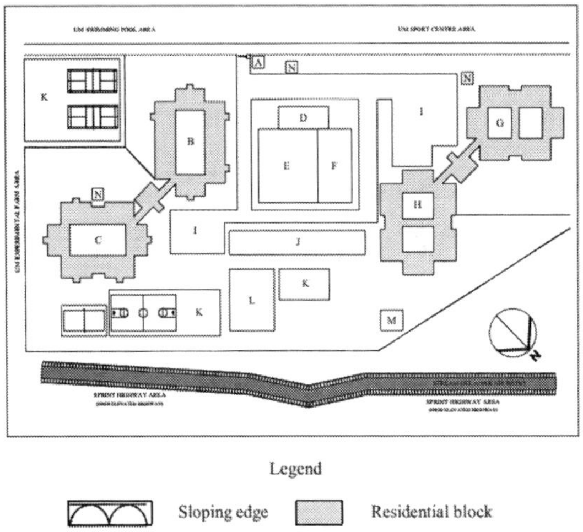

A - Guard post/Main entrance, B - Residential block (Block A - Awana/Female), C - Residential block (Block B - Bayu/Male), D - Administrative block/Office, E - Hall/Student activity centre, F - Student activity centre, G - Residential block (Block C - Flora/Female), H - Residential block (Block D - Impiana/ Male), I - Parking lots/Multipurpose open area, J - Cafeteria/Food court, K - Court/Sport area/ Multipurpose open area, L - Covered parking lots, M - Store, N - Gazebo.

Figure 4. Site plan of RC-B.

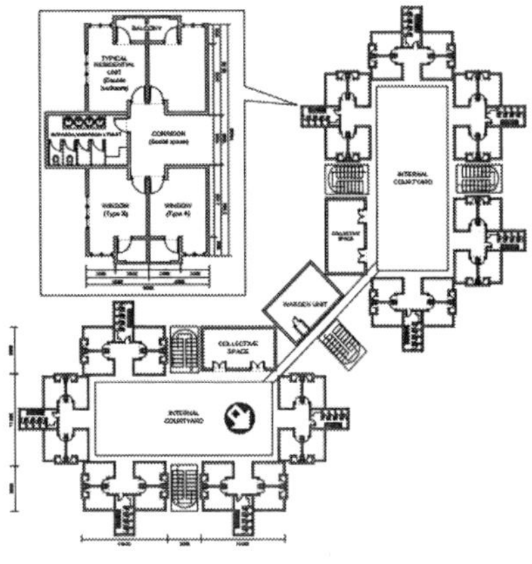

Block A and Block B

Block C and Block D

Figure 5. Typical floor plan of RC-B.

DOI: 10.1051/matecconf/20166600098

The room's floor area and volume are 20.00m^2 and 57.40m^3, respectively, whereas the room dimension is 5.0m (l) x 4.0m (w) x 2.87m (h). There are four rooms with the entrance facing each other and forming a cubicle. There are five cubicles with one collective space in each floor. Each cubicle is connected through an open corridor that faces the internal courtyard. The total area of the internal courtyard is 303.92m^2 for blocks A and B, and 269.12m^2 for blocks C and D. The internal courtyard area is planted with grass and shrubs.

In providing a control to the residents over the natural ventilation and daylighting, each room was designed with a balcony and two different types of windows. They are casement and turn windows with tinted glass. The characteristics of the RC-B are visualised through the standard regular photographs that are presented in Figure 6.

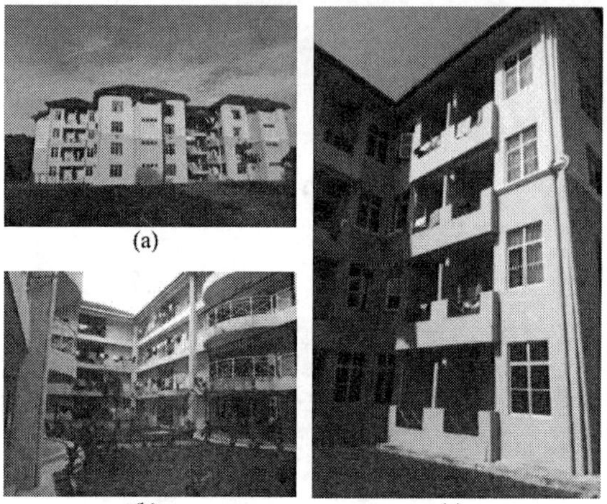

(a) (b) (c)

(a). Built-form configuration, enclosure and facade design - Large horizontal overhangs along the wall with windows, (b). Internal courtyard in the middle of residential building, (c). Casement and turn window design with tinted glass and large balcony in each residential unit faces to outside areas.

Figure 6. The characteristics of RC-A.

4 Results and discussion

A total of 266 responses retrieved were fully supplied by the respondents at the RC-A of which 39.6% are male and 60.4% are female residents. The number of respondents exceeded the minimum number of feedbacks required for a reliable 95% confidence level and ±5% margin of error from the overall population. The result of the residents' acceptance survey at the RC-A is presented in Table 2.

Only 181 responses were fully filled in by the respondents at the RC-B, accounting for 40.9% male and 59.1% female residents. This number only exceeds the minimum number of respondents as relying on 85% confidence level and ±5% margin of error from the overall population. The result of the residents' acceptance survey at the RC-B is presented in Table 3.

Table 2. The result of the residents' acceptance survey at RC-A

Question	**Likert scale / *Residents' responses (%)***				
	-2	**-1**	**0**	**+1**	**+2**
The residential building is sensitively designed for the landscape setting	*1.5*	*7.3*	*39.7*	*43.1*	*8.4*
The quality of landscape setting at the internal courtyard	*1.5*	*6.1*	*35.9*	*48.9*	*7.6*
The frequency of spending time at the internal courtyard in a day	*11.5*	*14.9*	*29.0*	*38.5*	*6.1*
The influence of landscape setting at the internal courtyard on the quality of life	*3.1*	*7.3*	*37.4*	*43.1*	*9.2*

Table 3. The result of the residents' acceptance survey at RC-B.

Question	**Likert scale / *Residents' responses (%)***				
	-2	**-1**	**0**	**+1**	**+2**
The residential building is sensitively designed for the landscape setting	*8.3*	*13.3*	*44.2*	*27.6*	*6.6*
The quality of landscape setting at the internal courtyard	*3.3*	*16.0*	*44.2*	*32.6*	*3.9*
The frequency of spending time at the internal courtyard in a day	*9.9*	*21.5*	*45.3*	*16.6*	*6.6*
The influence of landscape setting at the internal courtyard on the quality of life	*7.7*	*12.7*	*44.2*	*24.3*	*11.0*

According to the percentage of figures in Table 2, the majority of respondents at the RC-A, who represented 43.1% of the respondents 'agreed' that the residential building, which was designed with an internal courtyard is sensitively designed for the landscape setting, whilst 48.9% voted 'good' on the quality of the landscape setting at the internal courtyard. Therefore, 43.1% who represented the majority agreed that the landscape setting at the internal courtyard had "very" much influenced their quality of life. This was despite the fact that only 38.5% of the respondents who represented the majority were 'frequently' spending time at the internal courtyard in a day.

By referring to Table 3, different results were obtained on the residents' acceptance survey at the RC-B. The majority of residents voted one rate lower for all four questions as compared to the responses given by the majority of residents at the RC-A. There was no decision made on the sensitivity of the building design with the landscape setting when 44.2% voted for 'undecided'. This happened when they claimed that the quality of the landscape setting at the internal courtyard is 'moderate'. Thus, 45.3% of respondents who represented the majority only ticked 'sometimes', meaning that they only spent time the internal courtyard occasionally in a day. Meanwhile, 44.2% of respondents claimed that the landscape setting is 'moderately' influential to their quality of life.

The survey results of both residential colleges indicate that, the acceptance level of the residents is highly influenced by the arrangement and design of the internal courtyard. At the RC-A, the internal courtyard is facing most of the rooms directly, where the accessibility is not an issue to the residents especially for those who are living at the ground level. The gazebo as a social or engagement space for the residents, with the presence of shrubs and grass creates the internal courtyard that is fully functioning as a healing element, while improving the thermal comfort of the building's environment [2, 14].

A different situation can be seen at the RC-B, when all rooms are not directly facing the internal courtyard. There are only open corridors, which connect all cubicles, formed by four rooms facing each other. Thus, the internal courtyard is not fully utilised by the residents. This has been worsened by the modification of the internal courtyard by the residential college administration. The green ground of the internal courtyard in blocks C and D has been replaced with textured pavement and it has turned into a place to hang clothes for drying. In order to eliminate some of the weather elements from the space, a big polycarbonate roof was installed at the second floor of the building, as shown in Figure 7.

(a) (b)

(a). textured pavement, (b). a big polycarbonate roof.

Figure 7. Modification on internal courtyard in blocks C and D at RC-B.

In order to increase the residents' acceptance level, the modification of the internal courtyard should be done properly. The potential of the courtyard can be optimised, especially in the thermal performance if the particular design variants and configurations are known [14, 17]. According to Murgul [18], the accurate glazing of a courtyard is able to improve the quality of a courtyard space, especially on daylight and natural ventilation performance, increase the socialisation and interactions of dwellers, psychological relations with natural environment and ensure the protection from negative environmental impacts. Moreover, the commercial attractiveness and market value of a building. By making a comparison, open courtyard buildings that are relevant in all climates exhibit a better energy performance for the shorter buildings and as the building height increases, at the some point the enclosed atrium exhibits a better energy performance [13]. With regard to RC-B, which is designed with three-floor levels excluding a ground floor, the installation of the polycarbonate roof should highly be reconsidered with the building energy performance aspect.

Alternatively, the presence of the balcony in each room replaces the role of the internal courtyard as a social or engagement space to the residents at RC-B. The balcony improves view enjoyment by facing the wider outdoor landscape with diverse sceneries [19]. Thus, we need to provide the healing element efficiently with a better improvement of daylighting, thermal comfort and indoor air quality [2, 20, 21]. Moreover, it offers residents a provision of privacy to socialise, as it has been identified as a significant space for social gatherings that also acts as buffer zones of a gentle transition between the public and private spaces [22].

Further analysis was done to see the acceptance level according to the gender perspective at both residential colleges, and this is as presented in Figure 8. There were no differences in the responses between both genders for all questions. Both female and male have the same perspective towards the internal courtyard in the residential college building. According to Othman et al. [23], gender has no significant difference on the landscape's aesthetic values and visiting performance in natural outdoor environment.

5 Conclusion

The internal courtyard is positively relevant to be applied to the building design, particularly low-rise residential buildings. In general, the design and arrangement of the internal courtyard influence the acceptance level of respondents. Meanwhile, both male and female gave the same responses that indirectly indicate that the acceptance towards the internal courtyard in a building is a not influenced by the gender aspect.

As a recommendation, it is necessary to have a greater number of respondents to obtain an accurate result of the survey to represent the overall population; it should exceed the minimum number of feedbacks with a 95% confident level and ±5% margin of error. The scope of study should be expanded by including more questions and buildings as a case study for better understanding the best design of an internal courtyard at low-rise residential buildings.

Acknowledgement

The authors would like to thank Dayasari and Ungku Aziz Residential Colleges at the University of Malaya campus for their permission to carry out the survey. This analysis was financially supported by the Institut Pengurusan dan Pemantauan Penyelidikan (IPPP), the University of Malaya under BKP Grant (BK010-2015).

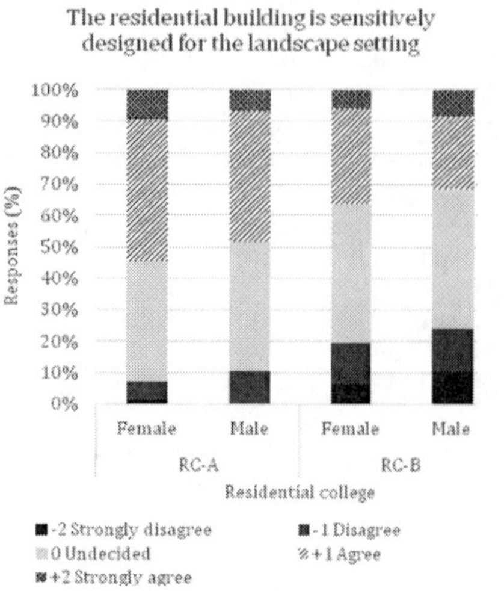

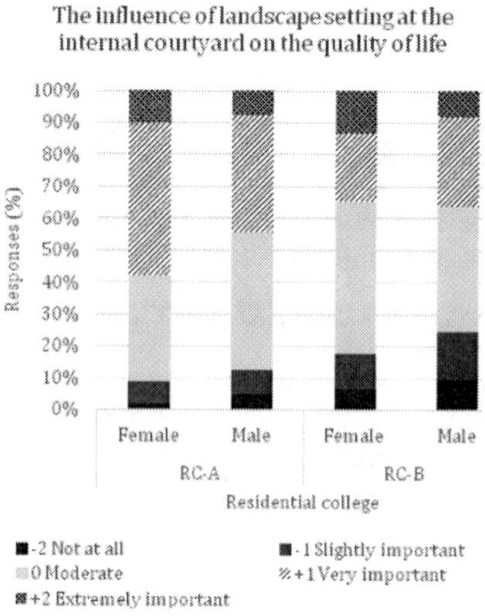

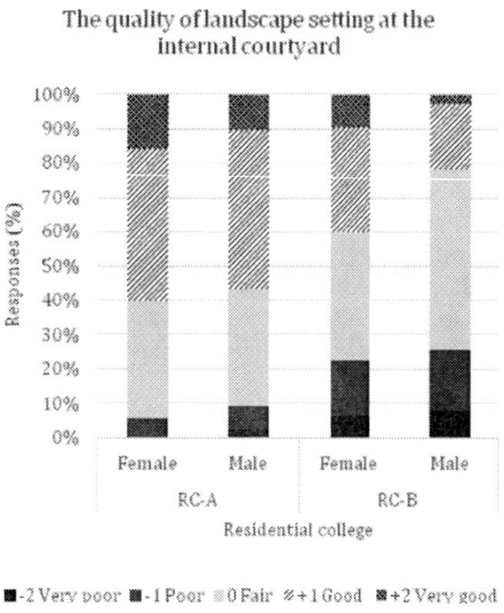

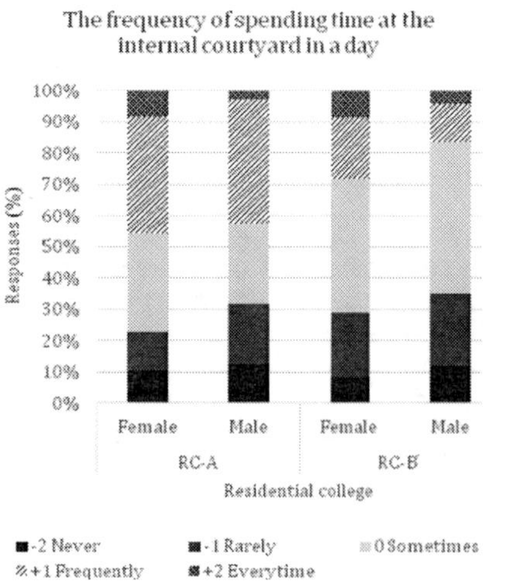

Figure 8. The result of residents' acceptance survey at RC-A and RC-B according to the gender.

References

1. H. Safarzadeh, M.N. Bahadori, Passive cooling effects of courtyards. Build. Environ. **40**, 89-104 (2005)
2. A. Almhafdy, N. Ibrahim, S.S. Ahmad, J. Yahya, Analysis of courtyard functions and its design variants in the Malaysian hospital, Procedia Soc. Behav. Sci. **105**, 171-182 (2013)
3. N. Sadafi, E. Salleh, L.C. Haw, Z. Jaafar, Evaluating thermal effects of internal courtyard in a tropical terrace house by computational simulation. Energ Buildings **43**, 887-893 (2011)
4. M.A. Zakaria, T. Kubota, D.H.C. Toe, The effects of courtyards on indoor thermal conditions of Chinese shophouse in Malacca, Procedia Eng. **121**, 468-476 (2015)
5. A. Aldawoud, Thermal performance of courtyard buildings, Energ. Buildings **40**, 906-910 (2008)
6. A. Almhafdy, N. Ibrahim, S.S. Ahmad, J. Yahya, Thermal Performance Analysis of Courtyards in a Hot Humid Climate using Computational Fluid Dynamics CFD Method, Procedia Soc. Behav. Sci. **170**, 474-483 (2015)
7. N. Khalil, H.N. Husin, A.H. Nawawi, A.E. Hashim, Post occupancy evaluation for performance evaluation of building facilities in higher education buildings, The proceeding of 9th International Conference SENVAR & 2nd ISESEE 2008 (UiTM, Malaysia, 2008)
8. M.A. Hassanain, A. Sedky, Z.A. Adamu, A. Saif, A framework for quality evaluation of university housing facilities, J. Build. Apprais. **5**, 213-221 (2010)
9. A.O. Ilesanmi, Post-occupancy evaluation and residents' satisfaction with public housing in Lagos, Nigeria, J. Build. Apprais. **6**, 153-169

DOI: 10.1051/matecconf/20166600098

10. W.F.E. Preiser, Post-occupancy evaluation: How to make buildings work better, Facilities **13**, 19-28 (1995)
11. S. Turpin-Brooks, G. Viccars, The development of robust methods of post occupancy evaluation, Facilities **24**, 177-196 (2006)
12. K. Hadjri, C. Crozier, Post-occupancy evaluation: Purpose, benefits and barriers, Facilities **27**, 21-33 (2009)
13. A. Aldawoud, R. Clark, Comparative analysis of energy performance between courtyard and atrium in buildings, Energ. Buildings **40**, 209-214 (2008)
14. A. Ghaffarianhoseini, U. Berardi, A. Ghaffarianhoseini, Thermal performance characteristics of unshaded courtyards in hot and humid climates, Build. Environ. **87**, 154-168 (2015)
15. A.A. Jamaludin, N. Keumala, A.R.M. Ariffin, H. Hussein, Satisfaction and perception of resident towards bioclimatic design strategies: Residential college buildings, Indoor Built Environ. **23**, 933-945 (2014)
16. T. Yamane, *Statistic, an introductory analysis* (Harper & Row, New York, 1967)
17. A. Almhafdy, N. Ibrahim, S.S. Ahmad, J. Yahya, Courtyard design variants and microclimate performance, Procedia Soc. Behav. Sci. **101**, 170-180 (2013)
18. V. Murgul, Reconstruction of the Courtyard Spaces of the Historical Buildings of Saint-Petersburg with Creation of Atriums, Procedia Eng. **117**, 808-818 (2015)
19. A.L.S. Chan, T.T. Chow, Investigation on energy performance and energy payback period of application of balcony for residential apartment in Hong Kong, Energ. Buildings **42**, 2400-2405 (2010)
20. N. Lechner, *Heating, cooling, lighting: Sustainable design methods for architects* (John Wiley & Sons, New Jersey, 2009)
21. M.F. Mohamed, S. King, M. Behnia, D. Prasad, The effects of balconies on the natural ventilation performance of cross-ventilated high-rise buildings, Journal of Green Building **9**, 145-160 (2014)
22. J. Williams, Designing neighbourhoods for social interaction: The case of cohousing, J. Urban Des. **10**, 195-227 (2005)
23. N. Othman, N. Mohamed, M.H. Ariffin, Landscape Aesthetic Values and Visiting Performance in Natural Outdoor Environment, Procedia Soc. Behav. Sci. **202**, 330-339 (2015)

Toward Sustainability in Concrete Industry by Using Of Solid Wastes from Palm Oil Industry

Alireza Javadi Pordesari [1], Salmaliza Salleh[2], Payam Shafigh[3] and Hilmi Bin Mahmuda[1]

[1]*Department of Civil Engineering, Faculty of Engineering, University of Malaya, 50603 Kuala Lumpur, Malaysia*
[2]*Faculty of Engineering and Built Environment, SEGi University, Kota Damansara, 47810 Petaling Jaya, Selangor, Malaysia*
[3]*Department of Building Surveying, Faculty of Built Environment, University of Malaya, 50603 Kuala Lumpur, Malaysia*

Abstract. Nowadays, one of the most important construction materials is concrete. By the advances of the industries in the urban areas, concrete is considered as one of the highest demands. As a result, the large amount of unprocessed materials is needed for making concrete. In the meantime, the agricultural wastes and solid material disposal are giving serious damages to the environment. As a result, by employing the agricultural wastes as a cementitious material, the undesirable impacts of the concrete industry to the environment will be dramatically decreased. That's because the source of these newly developed concretes is both reliable and environmental friendly. In this study, the utilization of agricultural wastes as a complementary cementitious material for producing the concrete is explained. In addition, it discusses the possibility of deploying the agricultural wastes by considering their engineering, physical and chemical properties. In addition, the successful use of agricultural wastes from oil palm industry such as oil palm shell, palm oil fuel ash and palm oil fibre in the concrete mixture was reported.

Keywords: Agricultural waste materials, Oil Palm industry, Oil palm fibre, Oil palm shell, Palm oil fuel ash

1. Introduction

It's almost about hundreds of years cement and concrete has been used in buildings, there are also evidences showing that Portland cement has been used for hundreds of years. What makes concrete standing in the first place among other materials is that the whole constituents of it can be chosen and the one, using it, is able to develop and improve the characteristics of each constituent and finally form the most developed hard-wearing impermeable material [1]. The amount of concrete production claims this truth that it is the most significant material used in buildings [2]. There is no doubt that there is a steep increase in population and by the end of the 21st century 11 billion people will have been living on the Earth, and that's clear enough that the more population on the Earth, the more common goods and services are needed [3]. As a result it has been thought that about 18 billion tons of concrete is needed by the year 2050 [4]. Here again can be concluded the more concrete and cement going to be produced, the more natural resources is used. What has been discussed the most in the past decade is that green issue needs to be considered in every aspects of life, and the buildings are not exceptions. The construction material of building have the most impact on the Green Building Rating (GBR). Here it should be considered that there are several important factors which make concrete look less compatible with the environmental inquiries of the modern construction industry which are CO emission during the process of Portland Cement production, the amount of energy we dedicate to it, natural resources used such as water, aggregate and fillers which are widely used in the formation process of concrete.

Ramezanianpour et al. (2009) clarified this fact that what we are using as concrete nowadays is not sustainable at all, but using industrial and agricultural waste constituents may lead us to the point the industry will be more friendly and sustainable [5]. According to this debate some sort of materials can be named such as fly ash, recycled concrete, recycled plastics, silica fume, Grand Granulated Blast Furnace Slag (GGBFS) which have been used several years as important concrete production constituents [2, 6]. There are some other found facts nowadays that show agricultural solid wastes are being used in both structural and non-structural concrete. Among the oil palm industry's agricultural wastes oil palm shell, palm oil fuel ash and palm oil fiber can be seen. Using other replaceable conventional materials such as agricultural solid wastes as aggregate in concrete instead of typical ones (sand as fine and gravel as coarse) which take three fifths to four fifths of the concrete's volume [7], helps in saving energy and natural resources as well as material cost reduction. Parallel to this strategy the problem of the wastes after the demolition of concrete structures which means saving environment is being overcome [5, 8]. The research on the area of agricultural waste utilization in concrete construction is quite a new topic. The increment of research on this area encourages the construction industries to use these concretes in their projects. In this research, the agricultural solid wastes produced from oil palm industry are used tested to construct the concretes. In addition, green building rating system, environmental sustainability and the relations between concretes with similar characteristics are discussed. Recognition of these agricultural wastes by both civil and agricultural engineers made it possible to use these materials in construction and some specific industries too. Moreover, by using these materials, the environmental sustainability of concrete's manufactures will be improved.

2. Agricultural wastes from oil palm industry

Malaysia, Indonesia, Thailand and Nigeria are the countries that have oil palm industry as one of their main agro-industry which Malaysia and Indonesia are considered as the top two countries in palm oil production. The crude palm oil produced by Malaysia is over 7 million tons a year [9]. The world palm oil production in 2011 is shown in Figure. 1 [10]. The amount of solid waste produced during the palm fruit processing plants is considered as one of the major drawbacks of this process. As an instance, Malaysia has the highest amount of solid waste produced in a year which is 47 million tons followed by Indonesia with 40 and Thailand with 3.5 million tons of solid waste production annually. Oil palm bunches; palm fibre and fruit shells are the solid wastes. Moreover, the above mentioned countries produce over 8 million tons of oil palm shell in a year [11].

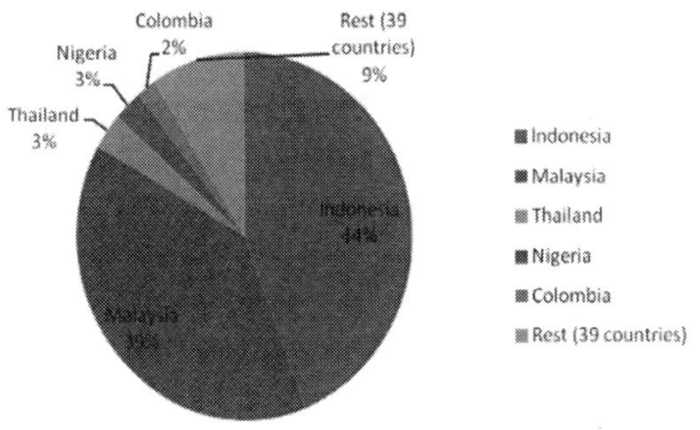

Figure 1. *World palm oil production [10].*

2.1. Oil palm shell (OPS)

The palm shell's shape can be angular or polygonal based on the breaking pattern of the nut and its oil color varies [12]. The attributes of the OPS is shown in Table 1. According to the table, the OPS has the density similar to the other lightweight aggregates while its weight is about 60% less as compared to the conventional coarse aggregates [13-15]. In addition, in comparison with conventional coarse aggregates, the OPS have about 80% lower Los Angeles abrasion value [16]. Therefore, such features and properties make OPS aggregates resistance to wear and due to its low impact and crushing values, it makes them absorbance to shock [17].

Table 1. Properties of the crushed granite (conventional aggregate), Oil palm shell and Coconut shell (agricultural waste) in detail as follows [14, 15, 17-20].

Properties	Oil Palm Shell	Crushed granite	Coconut shell
Specific gravity	1.17–1.37	2.60–2.70	1.05–1.2
Bulk density (uncompacted) (kg/m3)	510–550	1300	–
Bulk density (compacted) (kg/m3)	590–600	1420–1470	650
Void ratio (uncompacted) (%)	63	52	–
Void ratio (compacted) (%)	57	47	–
24 h water absorption (%)	21–33	<1	24
Aggregate impact value (%)	4–8	13–17	8.15
Aggregate crushing value (%)	5–10	20	2.58
Los Angeles abrasion value (%)	3–5	24	1.63
Flakiness index (%)	65	25	–
Shell thickness (mm)	2–8	–	0.15–8 mm
Thermal conductivity (W/mc)	0.19	–	–

Loss on ignition (%)	98–100	–	–

The shells have smooth concave and convex faces with sharp and spiky broken edges. The initial start of deployment of lightweight aggregates for making lightweight concretes was by a Malaysian and Nigerian researchers in 1984 [12, 21]. In many of the studies, the compressive strength of the OPS lightweight concrete with 20% to 25% lower density as compared to the Normal Weight Concretes (NWC) and in presence or absence of the cementitious materials is within the normal bounds of the compressive strength of the structural lightweight concretes (25 to 35 MPA) [14, 15, 19, 20]. Yet, it is still likely to enhance the compressive strength of the OPS lightweight concrete to 53 and 56 MPA in 28-day and 56-day tests [22, 23]. It is quite important to enhance the compressive strength of the concrete making use of an aggregate such as OPS with acceptable strength as the design and making a concrete structure is fully dependable on the compressive strength [23]. Besides, due to the high applicability and environmental friendly of the usage of the waste materials, the construction industry will be encouraged to apply these materials in their projects.

There are many articles and researches explaining and analyzing various mechanisms such as the use of Old OPS, Crushed OPS and OPS with smaller maximum of coarse aggregate for the production of the high strength OPS [23-25]. High compressive strength can be achieved by the Old OPS since there is no fibre on the OPS surface which accordingly improves the OPS and cement matric bond. On the other hand, crushing old OPS and diminishing the highest possible size of coarse OPS aggregate can be counted as another mechanism. Tough physical bond with hydrated cement are the main characteristics of the crushed OPS. The strong bond of the crushed OPS is because of the decreased in smooth surface of the OPS and due to crinkling of the large sizes of OPS aggregates, the total surface of the broken edges are increased. Moreover, the compressive strength increases by using smaller size of OPS aggregate which enhances the bond between the aggregate and the paste. The OPS aggregates in different states and the chosen mix proportions of OPS concrete with normal and high strengths are shown in Figure 2 and Table 2 respectively [26]. The size of the OPS (as coarse aggregate) in all of the OPS lightweight concretes with normal strength is 9.5mm while is it 8mm for high strength OPS production.

According to the literature [12], splitting tensile strength, flexural strength and modulus of elasticity as the mechanical properties of OPS concrete at the same compressive strength, are lower as compared to lightweight aggregate concretes made from artificial or natural lightweight aggregates. But it has been proved [24] that the compressive strength, splitting tensile and flexural strength enhance significantly by including steel fibre into OPS concrete. The steel fibre OPS concrete's splitting tensile/compressive strength ratio is within normal weight concrete and its flexural strength is similar to artificial lightweight aggregate concrete. The previous studies showed that the OPS concrete has significantly lower modulus of elasticity in comparison with the normal weight concrete. However, the OPS concrete's modulus of elasticity can be further improved to the normal weight concrete ranges by deploying certain methods which are contingent upon the use of old OPS and old crushed OPS [24, 25]. Teo et al. (2010) have analyzed the OPS structural lightweight concrete's durability properties under different curing conditions. They revealed that both OPS structural lightweight concretes and conventional lightweight concretes are having similar durability properties. They reported that OPS concrete requires proper curing to attain superior durability over the conventional lightweight concretes. In addition, they suggest that the least possible period of moist curing must be done for 7 continuous days [27]. Besides, a comparison between bond properties of OPS concrete and normal weight concrete [28] showed that the OPS concrete's bond stress was 86% of the corresponding normal weight concrete, however, the OPS concrete had 2.5 times higher bond stress as an ultimate value in comparison with the theoretical values determined according to the BS 8110 [29]. Moreover, the flexural and shear behaviour of reinforced and un-reinforced concrete beams constructed from OPS were tested and examined where in low reinforcement ratio, the reinforced concrete beams fulfilled the expectations in serviceability requirements with low reinforcement ratio of the BS 8110 code, [30]. By comparing the ultimate moments and predicted moments of the OPS reinforced concrete beams with 3.14% reinforcement ratio, it is concluded that the OPS reinforced concrete performed 4% to 35% ultimate moments. But if the beam's reinforcement ratio increases to 3.90%, then the ultimate moment's capacity will be underestimated by BS 8110 by about 6%. Hence, it was suggested that if the reinforcement ratio is greater than 3.14%, then larger depth for OPS concrete beam to be applied. The research conducted by Ahmed and Sobuz (2011) indicates that the flexural behaviour of beams produced from OPS and its theoretical results are having close outcomes [31]. Additionally, an adequate shear behavior of OPS concrete beams is described [31, 32]. It should be highlighted that according to Alengaram et al. (2011), the overall flexural behavior of reinforced OPS concrete beams are similar to the equivalent beams produced from normal weight concrete [33].

Figure 2. Oil palm shell grains in different conditions. (a): old, (b): fresh, and (c): crushed from large size old OPS [26].

Table 2. Selected mix proportion and properties of oil palm shell lightweight concrete

Mix no		**1**	**2**	**3**	**4**	**5**
Mix proportion *(kg/m3)*	Cement	360	480	550	480	510
	Silica fume	0	0	0	0	0
	Fly ash	0	0	0	0	0
	Super Plasticizer (%)	1.8	1	1.1	1.5	1.4
	Water to cement	0.45	0.41	0.31	0.38	0.38
	Oil palm shell	380	370	330	300	310
	Conventional Coarse Aggregate (granite)	210	0	180	0	0
	Fine Aggregate	830	820	710	1050	850
Fresh Property (slump, mm)		33	7	205	50	60
Density (kg/m3)		1910[a]	1900[a]	1980[a]	1990[c]	1960[b]
28-Day cube compressive Strength (MPA)		34	29	53	47	28
Reference		Shafigh et al. [12]	Mannan and Ganapathy [13]	Shafigh et al. [23]	Shafigh et al. [22]	Teo et al. [34]

[a] Demolded density.

[b] Air dry density.

[c] Oven dry density.

2.2. Palm oil fuel ash (POFA)

Tangchirapat et al. (2009) expressed that Oil palm fuel ash (POFA) is an agro-waste ash produced from palm oil remainders like palm fibre and shells. These wastes are burnt at 800 to 1000 C of temperature in order to use its produced steam as the purpose of generating electricity in biomass thermal power plants. There are 15% of shell and 85% of fibre in residue of the oil palm. As for the energy production, a boiler is used to burn the empty fruits which basically produce 5% ash as weight as solid waste. There is a huge pollution affecting the environment due to the production of the solid wastes and ashes. Hence, a possible practical solution must be given to resolve the land-fitting, pollution and high cost of material constructions. Due to the high cost of waste disposal, it delivers "negative value", whereas the maximized manageable usage of POFA delivers "positive value" while decreases negative impacts on the environment. By comparing the palm-oil products during past two centuries, POFA has the most negative effects on the environment since it remains in the atmosphere. By burning the solid wastes as a fuel in a boiler, POFA will be produced [35]. The solid wastes are palm oil husk or fibre and palm kernel shell. The residue of an oil palm industry is shown in Figure 3a. By using 300 µm of sieve, there are ashes remained which are displayed in Figure 3b [36].

DOI: 10.1051/matecconf/20166600099

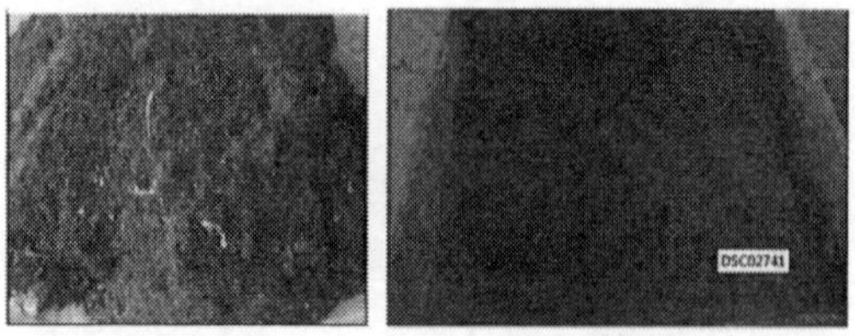

(a) (b)

Figure 3. (a) Palm oil residue and (b) palm oil fuel ash [36].

The POFA has various development processes both in industrial development stage and burning process. The early stage preparation of POFA was tested by Noorvand et al. (2013) in an oven for the duration of 24 hours under 105 ± 5 °C of temperature once the burning process was over [37]. On the other hand, Tangchirapat et al. (2009) have the ashes prepared after incineration process by using sieve number 16 (1.18 mm opening) under the temperature of 700 to 1000 °C. According to their experimental studies, three types of POFA were found based on the gravity sizes of 1.89 for original (OP), 2.36 for medium (MP) and 2.43 as for the small (SP). Their analysis proves that the grinding process in addition of enhancing the POFA's quality, it improves the specific gravity too [35]. In 2013, Abdul Awal and Shehu (2013) conducted a research in Johor which is the southern-state of Malaysia. In this study the ashes were prepared from the flue tower's foot. Then, a 150 µm of filter was used in sieving process. The results were grounded and modified to the Los Angeles abrasion test machine containing 10 stainless steel bars. These bars were having 12 mm diameter with 800 mm long and were used as a replacement for steel balls [38]. Depending on the carbon content of the ashes produced, their colors vary from light grey to dark grey [37, 39-44]. Finally, it was understood that the fuel industry is the main sources of the raw materials of the POFA, being incinerated in a boiler or other grinding industries. In addition, those fine ashes that were escaped from the boiler were trapped and sieved by a 150 µm to 300 µm filter so that the small sized ashes remain. The palm oil industry's operating system has direct effect on the POFA's physical characteristics. The Los Angeles abrasion test machine is used to ground the ash. This machine had 10-20 stainless steel bars. POFA had a grey color in bulk, but it became darker by the increment of the properties of the carbons that were unburned. Table 3 describes POFA's properties where silicon dioxide (SiO_2) is the core oxide. POFA has about 59.6% to 66.9% of silica. Furthermore, there is approximately 1.9% to 5.7% of iron as well as about 5% calcium oxide (CaO). Nevertheless, it has a specific gravity of 1.9 to 2.4. In addition, the carbon's amount exists in ash is effectively depends upon the incineration process. As an instance, the amount of Loss on Ignition was 8.25% which is considered high according to the ASTM C618 that specified it 6% as the maximum value [45]. Efficiency parameters such as time and temperature, combustion process and material source are the main reasons for having different amount of chemical components in POFA.

Table 3. The chemical and physical properties of POFA [36].

Chemical composition (%)	
SiO_2	59.6–66.9
Al_2O_3	2.5–6.4
Fe_2O_3	1.9–5.7
CaO	4.9–6.4
MgO	3.0–4.5
SO_3	0.3–1.3
Na_2O	0.2–0.8
K_2O	5.0–7.5
LOI	6.6–10.0
Physical properties	
Median particle size *(µm)*	10.5
Specific gravity	1.9–2.4
Blaine fineness *(m^2/kg)*	493.0

Table 4. The selected mix proportion of high strength concrete. [35, 38, 43, 44, 46, 47].

	Cement	**POFA *kg/m^3***	**%**	**Sand**	**Coarse aggregate**	**Water**	**SP (l)**	**W/C**	**Slump**	**Day** 28 days	90 days
1	495.0	55.0	10	753	959	176	6.8	0.32	250	60	68
2	440.0	110.0	20	745	950	176	8.6	0.32	240	61	70
3	385.0	165.0	30	738	940	176	11.6	0.32	250	59	66
4	400.0	100.0	20	711	1067	145	11.5	0.28	–	37	52
5	400.0	100.0	20	711	1067	145	11.5	0.28	–	49	53
6	540.4	145.3	25	1057	1340	168	50.4	0.23	169	175	182
7	214.0	213.0	50	787	961	205	-	0.29	115	41	–
8	171.0	256.0	60	787	961	205	-	0.29	90	36	–

9	128.0	299.0	70	787	961	205	-	0.29	80	28	–
10	504.0	56.0	10	757	971	153	8.5	0.28	200	89	91
11	448.0	112.0	20	749	962	151	11.8	0.28	185	94	93
12	392.0	168.0	30	742	952	148	16.9	0.28	185	87	91
13	270.0	30.0	10	804	1024	216	–	0.72	80	39	40
14	240.0	60.0	20	801	1021	210	–	0.70	60	32	39
15	210.0	90.0	30	799	1018	219	–	0.73	75	28	34

The pazzolanic reaction of coarse POFA is slower as compared to high fineness POFA. Therefore, because of the higher dense and homogeneous of the high fineness POFA, it can effectively enhance the concrete's compressive strength. Moreover, POFA can be applied as a binder to meet the chemical necessities of the ASTM C618 [45]. Due to the low Loss of Ignition (LoI) of the POFA which is below 10%, it will be used as a pazzolanic component. As a result, it benefits the concrete construction industry. As another research in concrete strength enhancements, the capabilities of POFA as a pozzolan were investigated [36]. There is a tremendous effort by the researches to improve the POFA concrete's behavior to the pazzolanic reactions where the result of this reaction is the hydration products. Table 4 displays the compressive strength of few chosen mixtures. The compressive strength of the concretes having different POFA's properties is over 55 MPA in 28-days test. However, by increasing the POFA percentage up to 20% and 30% in the concrete, the compressive strength is 59 and 61 MPA respectively. According to the Table 4, the concrete samples having POFA have higher compressive strength as compared to the normal concretes at 28 days. The compressive strength of a concrete increases to 70 MPA in 90-days test and by using 20% POFA. Two different palm oil industries provided two types of POFA which are CAPOFA and ALPOFA. Based on the sample G(I), utilizing fiber (steel) as a binder aggregate can lead to a compressive strength of 175 MPA in comparison with situations where no fibers are used. In the meantime, by having 10%, 20% and 30% of the POFA mixed with 10% SF, the strength increases to 93 MPA. Moreover, to make a concrete with high strength, POFA with proportion of 30% can be applied in replacement of cement which results in higher compressive strength as compared to the concrete produced using Portland cement. The compressive strength has an expansion and loss which can be reduced if the ultrafine is included. It was proposed that to minimize the needs of the high-strength concrete to the water, POFA can be utilized to act like pazzolanic POFA [46]. In general, it has been proved that POFA has pazzolanic cementing materials including superior engineering in a suitable and appropriate mixing and curing system. In addition, the larger amount of waste materials produced in agriculture can be used. Then, by reducing the waste's volumes and Greenhouse Gases CO (GAGs), the environment will be protected. Besides, in addition of reducing the construction-cost, POFA is conducive to maintainable industry.

2.3. Palm oil fibre

Recently, cellulose, lignin and hemicellulose which form the lignocellulosic materials of a palm oil fiber, have obtained the demand thrust. Therefore, the lignocelluloses fibers are obtainable and they are being heavily utilized even with inadequate developments in technology. The oil palm fiber is a harmless, recyclable material taken from Oil Palm's Empty Fruit Bunch (OPEFB) by applying a process called decortation. As the main specifications of the fibers can point to their cleanness, non-carcinogenic, soft parenchyma cells as well as free from insecticide. The fibers used in this study are displayed in Figure 4. As the purpose of making cost effective construction material in construction industry, the interests in using natural fibers as reinforcement have been dramatically increased [48]. The comparison between physical properties of the palm oil fibre and other references are explained in Table 5.

The mechanical properties of palm oil fiber like tensile strength, elongation and modulus of elasticity of fiber are shown in Table 6. These properties were ascertained by the ASTM C1557 standards by using a Tinius Olsen machine and 8x10-6 m/min of cross-head speed. However, Subramani (2007) claims that specified fiber content (percentage) is the only factor affecting the compressive strength. In their study, by increasing the fiber content percentage, the modulus rupture of specimens remains constant. They reported that there might be a reduction in bonding and disintegration if the fiber is congested [49]. The lowest flexural strength of the specimen is when there are 0.5% of content and 5cm of fiber length where the 5cm fiber length could be the reason of this low flexural strength since majority of the fibers weaken the bonding among particles [50]. One of the vital components in a concrete is fiber's size. It has both pros and cons based upon the fiber's length size. However, in comparison with long fibers, small fibers deliver a uniform dispersion in the concrete mix if the concrete properties are fresh. In addition, by using short fibres, the workability of the concrete mix increased as a result of extreme fiber balling reduction [51]. The strength of the composites can be further improved in 28 days by having palm oil fiber added to the composites. However, it must be noted that although the fiber causes enhancements in the strength but this is not true for the fiber content (percentage) [49].

Figure 4. Palm Oil Fiber [48]

Table 5. Mechanical Properties of POF [48].

Technical Attributes	Reference [49, 52]	Result from	
		Amrec, SIRIM	FKKKSA, UTM
Tensile Strength (MPA)	21.2	83.1	113.43
Elongation at break (%)	0.04	0.11	0.11
Compression of Strength (MPA)	36.4	-	-
Direct Screw Withdrawal (N)	1420	-	-
Nail Withdrawal (N)	310	-	-
Bending Stress (MPA)	37.6	-	-

That means by the increase in fiber content (percentage), the strength does not necessarily improve. There could be an improvement on the strength of hardens concrete based on the fiber content. Yet, both the structural integrity of the harden concrete and bonding and disintegration can be reduced by the fiber congestion. By utilizing the fiber in concrete the matrix bonding will be increased in comparison with control mix. As a result, the concrete's strength improves. Once the fiber is added to the concrete, based upon its percentage as 0.25% or 0.50%, the optimum length of the fiber would be 5cm and 3cm respectively. Therefore, by having the fiber content, the required properties increase [50]. Ramli and Dawood (2010) denoted that the addition of oil palm fibre in the mixture will moderately increase the density of the lightweight concrete [53].

Table 6. Physical Properties of Palm Oil Fiber [48].

Properties	Result From	References	
	MPOB/UKM	[54]	[52]
Average fiber length, μm	170.38	-	-
Average fiber width, μm	291.44	-	-
Average lumen width, μm	11.33	7.90	12.34
Average wall thickness, μm	4.90	2.30	3.38
Moisture content, %	10.57	-	-
Ashes content, %	2.4	-	-
Water absorption, %	15.97 *result from Amrec, SIRIM		

3. CONCLUSION

As long as the population grows, it is essential for the concrete industries to focus on the maintainable developments. Therefore, it is possible to reach to this objective by using by-products and agricultural wastes. According to the previous studies, some specific agricultural wastes such as oil palm shell, palm oil fuel ash and oil palm fibre can be used in producing the concrete due to their satisfactory performance. As a result, by using these materials, an initiative can be taken to enhance the environmental sustainability of the concrete industry. Hence, by the advances in current knowledge and exploration of the other agricultural wastes to be applied in concrete mixtures, there will be a feasible contribution in enhancements and improvements of the environmentally sustainable and friendly constructions.

ACKNOWLEDGEMENTS

The authors gratefully acknowledge the financial support from Innovation and Management Center, project number SEGiIRF/2014-65/FoEBE-14/79, Segi University.

REFERENCES

[1] R. Swamy, Design for sustainable development of concrete construction, Proceedings of the Fourth International Structural Engineering and Construction Conference (ISEC 4), Taylor & Francis Ltd England, 2007, pp. 47-54.
[2] C. Meyer, The greening of the concrete industry, Cement and Concrete Composites, 31 (2009) 601-605.
[3] R. Rosković, D. Bjegović, Role of mineral additions in reducing CO 2 emission, Cement and Concrete Research, 35 (2005) 974-978.
[4] P.K. Mehta, P.J. Monteiro, Concrete: microstructure, properties, and materials, McGraw-Hill New York2006.
[5] A. Ramezanianpour, M. Mahdikhani, G. Ahmadibeni, The effect of rice husk ash on mechanical properties and durability of sustainable concretes, DOI (2009).
[6] L. Federico, S. Chidiac, Waste glass as a supplementary cementitious material in concrete–critical review of treatment methods, Cement and concrete composites, 31 (2009) 606-610.
[7] S. Badur, R. Chaudhary, Utilization of hazardous wastes and by-products as a green concrete material through S/S process: A review, Rev. Adv. Mater. Sci, 17 (2008) 42-61.
[8] M. Harimi, D. Harimi, V.J. Kurian, B. Nurmin, Evaluation of the Thermal Performance of Metal Roofing under Tropical Climatic Conditions, Center of Mineral and Materials, Universiti Malaysia Sabah, Malaysia, 2005.
[9] K. Miura, T. Masuda, T. Funazukuri, K. Sugawara, Y. Shirai, J.-i. Hayashi, M.I.A. Karim, F.N. Ani, H. Susanto, Efficient use of oil palm wastes as renewable resource for energy and chemicals, Kyoto University, Japan, March, DOI (2001).
[10] Global Palm Oil Conference, Colombia, 2015.
[11] U. ESCAP, ESCAP report: biomass increasingly important in Indonesia and Thailand, E-news environment and sustainable development, 7 (2007) 5.
[12] P. Shafigh, M.Z. Jumaat, H. Mahmud, Mix design and mechanical properties of oil palm shell lightweight aggregate concrete: a review, International journal of the physical sciences, 5 (2010) 2127-2134.
[13] M. Mannan, C. Ganapathy, Mix design for oil palm shell concrete, Cement and concrete research, 31 (2001) 1323-1325.
[14] F.O. Okafor, Palm kernel shell as a lightweight aggregate for concrete, Cement and Concrete Research, 18 (1988) 901-910.
[15] D. Okpala, Palm kernel shell as a lightweight aggregate in concrete, Building and environment, 25 (1990) 291-296.
[16] H. Basri, M. Mannan, M. Zain, Concrete using waste oil palm shells as aggregate, Cement and concrete Research, 29 (1999) 619-622.
[17] D. Teo, M.A. Mannan, V. Kurian, C. Ganapathy, Lightweight concrete made from oil palm shell (OPS): structural bond and durability properties, Building and Environment, 42 (2007) 2614-2621.
[18] K. Gunasekaran, P. Kumar, M. Lakshmipathy, Mechanical and bond properties of coconut shell concrete, Construction and building materials, 25 (2011) 92-98.
[19] M. Mannan, C. Ganapathy, Concrete from an agricultural waste-oil palm shell (OPS), Building and Environment, 39 (2004) 441-448.
[20] M.A. Mannan, C. Ganapathy, Engineering properties of concrete with oil palm shell as coarse aggregate, Construction and Building Materials, 16 (2002) 29-34.
[21] A. Abdullah, Basic strength properties of lightweight concrete using agricultural wastes as aggregates, Proceedings of international conference on low-cost housing for developing countries, Roorkee, India, 1984.
[22] P. Shafigh, M.Z. Jumaat, H. Mahmud, Oil palm shell as a lightweight aggregate for production high strength lightweight concrete, Construction and Building Materials, 25 (2011) 1848-1853.
[23] P. Shafigh, M.Z. Jumaat, H.B. Mahmud, U.J. Alengaram, A new method of producing high strength oil palm shell lightweight concrete, Materials & Design, 32 (2011) 4839-4843.
[24] P. Shafigh, H. Mahmud, M.Z. Jumaat, Effect of steel fiber on the mechanical properties of oil palm shell lightweight concrete, Materials & Design, 32 (2011) 3926-3932.
[25] P. Shafigh, H.B. Mahmud, M.Z. Jumaat, Oil palm shell lightweight concrete as a ductile material, Materials & Design, 36 (2012) 650-654.
[26] P. Shafigh, H.B. Mahmud, M.Z. Jumaat, M. Zargar, Agricultural wastes as aggregate in

concrete mixtures–A review, Construction and Building Materials, 53 (2014) 110-117.
[27] D. Teo, M. Mannan, V. Kurian, Durability of lightweight OPS concrete under different curing conditions, Materials and structures, 43 (2010) 1-13.
[28] U.J. Alengaram, H. Mahmud, M.Z. Jumaat, Comparison of mechanical and bond properties of oil palm kernel shell concrete with normal weight concrete, DOI.
[29] B. Standard, 8110:'Structural use of concrete', British Standards Institution, DOI (1985).
[30] D.C. Teo, M.A. Mannan, J.V. Kurian, Flexural behaviour of reinforced lightweight concrete beams made with oil palm shell (OPS), Journal of Advanced Concrete Technology, 4 (2006) 459-468.
[31] E. Ahmed, H.R. Sobuz, Flexural and time-dependent performance of palm shell aggregate concrete beam, KSCE Journal of Civil Engineering, 15 (2011) 859-865.
[32] M.Z. Jumaat, U.J. Alengaram, H. Mahmud, Shear strength of oil palm shell foamed concrete beams, Materials & Design, 30 (2009) 2227-2236.
[33] U.J. Alengaram, M.Z. Jumaat, H. Mahmud, M.M. Fayyadh, Shear behaviour of reinforced palm kernel shell concrete beams, Construction and Building Materials, 25 (2011) 2918-2927.
[34] D. Teo, M. Mannan, V. Kurian, Structural concrete using oil palm shell (OPS) as lightweight aggregate, Turkish Journal of Engineering and Environmental Sciences, 30 (2006) 251-257.
[35] W. Tangchirapat, C. Jaturapitakkul, P. Chindaprasirt, Use of palm oil fuel ash as a supplementary cementitious material for producing high-strength concrete, Construction and Building Materials, 23 (2009) 2641-2646.
[36] E. Aprianti, P. Shafigh, S. Bahri, J.N. Farahani, Supplementary cementitious materials origin from agricultural wastes–A review, Construction and Building Materials, 74 (2015) 176-187.
[37] H. Noorvand, A.A.A. Ali, R. Demirboga, H. Noorvand, N. Farzadnia, Physical and chemical characteristics of unground palm oil fuel ash cement mortars with nanosilica, Construction and Building Materials, 48 (2013) 1104-1113.
[38] A.A. Awal, I. Shehu, Evaluation of heat of hydration of concrete containing high volume palm oil fuel ash, Fuel, 105 (2013) 728-731.
[39] P. Chindaprasirt, C. Jaturapitakkul, T. Sinsiri, Effect of fly ash fineness on microstructure of blended cement paste, Construction and Building Materials, 21 (2007) 1534-1541.
[40] C. Jaturapitakkul, K. Kiattikomol, W. Tangchirapat, T. Saeting, Evaluation of the sulfate resistance of concrete containing palm oil fuel ash, Construction and Building Materials, 21 (2007) 1399-1405.
[41] R. Madandoust, M.M. Ranjbar, H.A. Moghadam, S.Y. Mousavi, Mechanical properties and durability assessment of rice husk ash concrete, Biosystems engineering, 110 (2011) 144-152.
[42] V. Sata, C. Jaturapitakkul, K. Kiattikomol, Utilization of palm oil fuel ash in high-strength concrete, Journal of Materials in Civil Engineering, 16 (2004) 623-628.
[43] V. Sata, C. Jaturapitakkul, K. Kiattikomol, Influence of pozzolan from various by-product materials on mechanical properties of high-strength concrete, Construction and Building Materials, 21 (2007) 1589-1598.
[44] W. Tangchirapat, C. Jaturapitakkul, Strength, drying shrinkage, and water permeability of concrete incorporating ground palm oil fuel ash, Cement and Concrete Composites, 32 (2010) 767-774.
[45] C. Astm, 618 Standard Specification for Coal Fly Ash and Raw or Calcined Natural Pozzolan for use as a Mineral Admixture in Portland Cement Concrete, American Society for Testing of Materials, Philadelphia, USA, DOI.
[46] M. Aldahdooh, N.M. Bunnori, M.M. Johari, Development of green ultra-high performance fiber reinforced concrete containing ultrafine palm oil fuel ash, Construction and Building Materials, 48 (2013) 379-389.
[47] S. Bamaga, M. Hussin, M.A. Ismail, Palm oil fuel ash: promising supplementary cementing materials, KSCE Journal of Civil Engineering, 17 (2013) 1708-1713.
[48] M.A. El-Gelany Ismail, Study on the properties of palm oil fiber, DOI (2009).
[49] M. Subramani, Palm oil fiber as an additive in concrete, Universiti Teknologi Malaysia, DOI (2007).
[50] H. Hashim, The effect of palm oil fiber on concrete properties, Tesis Sarjana Muda. Universiti Teknologi Malaysia, 2008.
[51] Y. Mohammadi, S. Singh, S. Kaushik, Properties of steel fibrous concrete containing mixed fibres in fresh and hardened state, Construction and Building Materials, 22 (2008) 956-965.
[52] K.-N. Law, W.R.W. Daud, A. Ghazali, Morphological and chemical nature of fiber strands of oil palm empty-fruit-bunch (OPEFB), BioResources, 2 (2007) 351-362.
[53] M. Ramli, E.T. Dawood, Effects of palm fiber on the mechanical properties of lightweight

concrete crushed brick, Am J Eng Appl Sci, 3 (2010) 489-493.
[54] A.K. Mohanty, M. Misra, L.T. Drzal, Natural fibers, biopolymers, and biocomposites, CRC Press2005.

MATEC Web of Conferences **66**, 00100 (2016) DOI: 10.1051/matecconf/20166600100
IBCC 2016

The Relationship between Housing Finance and Macroeconomics Variables in Malaysia

[1]Nur Baizura Binti Mohd Shukor, [2]Dr Sr Rosli Bin Said and [3]Dr Rohayu Bin Abdul Majid

[1]*Department of Estate Management, Faculty of Architecture, Planning & Surveying, University Teknologi MARA, Seri Iskandar Campus, Seri Iskandar, 32610 Perak*
[2]*Department of Estate Management, Faculty of Built Environment, University Malaya, 50603 Kuala Lumpur*
[3]*Department of Estate Management, Faculty of Architecture, Planning & Surveying, University Teknologi MARA, 40450 Shah Alam, Selangor*

Abstract. Housing finance is one of the factors that contribute in the overall economy growth of the country. The purpose of this paper is to analyse the relationship of housing finance variable and the macroeconomic variables in Malaysia. By adopting time series technique of Vector Auto regression (VAR) and Impulse Response to determine the dynamic relationship between the macroeconomic and housing finance variable. The cointegration result shows that there exists a long run relationship between the macroeconomic variable and housing finance variable. The finding from impulse response function indicates that Gross Domestic Product (GDP) response positively to the Primary Mortgage Market (PMM), which shows that during the good economy there are more housing loan extends by the banking institution. Meanwhile, interest rate response negatively to Secondary Mortgage Market (SMM), which implies that during the financial crisis, more housing loan sold to the Secondary Mortgage Market as one of the measure by the government to increase liquidity in banking institutions. As a conclusion, there is presence of relationship between the variable which change in one variable will affect the other variable in the long run.

1. Introduction

Housing finance is defined as a system of money and credit that enables all types of housing to be built, improved, bought, rented, maintained and repaired (Garnett, 2005). Housing finance system is different between each country based on the country direction, policy and lending guideline implemented in each country. The role of housing finance in Malaysia is to provide financing property development for the developers, as well as provide end financing to the homebuyers to buy a house. The Asian Financial Crisis in 1997 and Global Financial Crisis in 2008 had brought significant effects in the Asian countries including Malaysia. Real output, manufacturing and trade investment growth rate decrease due to the falling demand in economic development. The real GDP is an accurate macroeconomic indicator in measuring the productivity of a country. The financial crisis affected the Malaysian housing market, as well as the housing finance.

Previous research had focused on the relationship between macroeconomic, financial and bank credit variables. They findings showed that there is a relationship between bank lending with macroeconomic variables (for example real GDP and interest rate), asset prices (for instance house and stock prices), or investigated its relationship to macroeconomic uncertainty. Mansor and Mohamed (2012) studied the relationship between bank credit, macroeconomic variable and financial uncertainty in Malaysia. By using the vector autoregressions (VARs) approach to test the cointegration and causality, they finding showed that real domestic output have a positive and long-run relationship with bank credit and stock price. For short-run dynamic analysis, they found that interest rate shock gives a negative response to real output, real bank credits, and stock prices.

Another research by Gerlach and Peng (2005) using the same approach of VAR, had investigated dynamics relationship between property prices and bank lending in Hong Kong with four variables, which are GDP, interest rate, property prices and bank lending. They findings showed there is a correlation between property price and bank lending. The relationship showed that bank lending adjusted property prices rather than the opposite. Their finding showed that bank lending is one of the main factors in adjusting the house price. Therefore, housing finance is one of the tool to improve the country economy during the downturn and during the recovery period housing finance function as a tool to regulate the country economy. The problem statement mentioned above has brought out the research question of what are the relationships between macroeconomic and housing finance variables in Malaysia.

2. Literature Review

2.1 Housing Finance System

Housing finance is defined as a system of money and credit that enables all types of housing to be built, improved, bought, rented, maintained and repaired (Garnett, 2005). Based on the basic concept of housing finance system, bank or any other financial institution acts as a financial intermediary, which takes deposit from the surplus agent (savers) then lends the money to the deficit agent (borrowers) who needs money to purchase house. Housing finance system is different between each country based on the country direction, policy and lending guideline implemented in each country. Among the Asian countries, Malaysia showed the highest ratio of housing finance with Gross Domestic Product (GDP), with the housing finance at 32 percent of GDP compared to the Philippines, which showed the ratio on housing finance at about 12 percent of GDP. A study by Veronica (2008) on the housing finance system in emerging countries concluded that countries with strong legal right to the lenders and borrowers, deeper credit information system and more stable macroeconomic environment would have better housing finance system.

The relationship between bank lending and property price gets much attention among scholars recently. If the banks strengthen their lending policies, this will affect the homebuyer eligibility to get the financing and directly will decrease the demand to purchase a house. Therefore, this will slow down the housing market activities and indirectly will influence the property price. For countries that are denominated by the bank-oriented system, the housing finance system will be affected if there are any changes to the bank lending policy. Few of previous study have shown that changes in bank lending policy will affect the housing finance system for that country (Greef and Haas, 2000; Hilbers *et al.*, 2001). A study was conducted by Gerlach and Peng (2005) on the relationship between property price and bank lending in Hong Kong to determine if any adjustment made to the bank lending will affect the property price instead of the opposite. Their finding showed that bank lending is the main factor in adjusting the house price. In the local context, Mansor and Mohamed (2012) found that there is a positive relation between real GDP and both real stock price and real bank credit in the long run. Meanwhile, the finding on a short-run analysis shows that the interest rate shock will give a negative response to real output, real bank credits and stock prices.

Housing finance is one of the factors that contribute in the overall economy growth of the country. The role of housing finance as a tool to boost the housing market is proven, especially during the financial crisis. The relationship between housing finance and housing market is dependent on each other. Expansion in the housing market means that there are more house buyers require financing. A strong housing finance will definitely give a positive impact to the country's economy. Housing finance is one of the most essential medium in the housing delivery framework, which shapes housing delivery system in most countries (Tiwari, 2000).

The research by Said *et al.* (2014) analyzed the relationship between the housing market and housing finance system. The analysis showed that there is a relationship between the primary market and the housing market. Meanwhile, there is no relationship between the housing market and the secondary market, as well as the capital market. Apart from that, the analysis showed the relationship between the three housing finance sub-markets. Any changes to the Primary Mortgage Market (PMM) will affect the Capital Market (CM), where any changes to the Secondary Mortgage Market (SMM) will not affect the Capital Market (CM).

2.2 Malaysia Housing Finance

The primary mortgage market (PMM) in Malaysia is dominated by the housing credit institution (HCI). The role of the housing credit institution is to provide either bridging or project finance to the developer constructing the house or end financing to the house purchasers. The developers apply for bridging loan to ease their cash flow during the construction period while waiting for pending payment from end purchaser or their end financier. The utilization of bridging loan is controlled by the bank, where fund will be released progressively against the architect certificate of the work done. Meanwhile, for home buyers, they need end-financing as a source of fund to buy a house, since buying a house is one of the large consumptions by the consumer. Seven main players in housing credit institution are commercial banks, Treasury Housing Loan Division, Bank Kerjasama Rakyat Malaysia Berhad, Malaysia Building Society Berhad, Borneo Housing Mortgage Berhad, Bank Simpanan Nasional Berhad and Sabah Credit Cooperation.

National Mortgage Corporation Cagamas is the only entity that operates the Secondary Mortgage Market (SMM) in Malaysia. The role of Cagamas is to support the wider spread of home ownership and develop of the securitization market in Malaysia (Chung, 2014).The establishment of Cagamas comes from the result of the liquidity crunch in the 1980s, where there was a funding mismatch in financial institutions. Therefore, an intermediary party is needed between the primary lenders and investors for the long term fund. A primary lender sells their housing loan to Cagamas and issues private debt securities to finance the purchase of the original housing loan. By buying the housing loan from the originator, this allows the originator to give more loans to other people and will indirectly increase their liquidity and lower their interest rate risk. Cagamas issues bonds/sukuk to finance the purchase of mortgage from financial and non-financial institutions.

The Malaysian Capital Market is regulated by the Kuala Lumpur Stock Exchange (KLSE), and the performance of this market can be measured using the Kuala Lumpur Composite Index (KLCI). The Malaysian Primary Mortgage Market originates pools of housing funds. These pools of fund are being sold to Cagamas (Secondary Mortgage Market) and securitized in the Malaysian Capital Market by issuing the Private Debt Securities (PDS).

3. Data And Methodology

3.1 Data transformation

The problem of heteroscedasticity in presenting a model can be solved by transforming the independent and dependent variables into the log form. By transforming the variable into the log form, this can result in the reduction of the heteroscedasticity because the scale of the variable is being compressed. The natural logarithm basis is $\log_e(A) = \ln(A)$ (Koop 2008). The transformation of data into natural logarithm is the appropriate approach in dealing with the non-stationary variable in the time series

The macroeconomic variable and the housing finance variable were transformed into the natural logarithm. The natural logarithm basis is $\log_e(A) = \ln(A)$. All variables had been transformed into natural logarithm except for the Base Lending Rate (BLR). Table 1 shows the transformed logarithm of data and the new symbol to represent the data after transformation.

Table 1 Transformation of variables

Variable	Sub-group	Data	Transformed variable	Symbol after transformation
GDP	Macroeconomic variable	Gross Domestic Product (million)	Nat.log of GDP	LGDP
BLR	Macroeconomic variable	Average base lending rate by commercial bank (%)	-	BLR
PM	Primary mortgage market	Total housing loan by commercial bank	Nat.log of PM	LPM
SM	Secondary mortgage market	Total housing loan sold to Cagamas	Nat.log of SM	LSM

3.2 Unit root test

Several studies have shown that differencing non stationary time series before estimation does not adequately account for the long-run relationship. The assumptions on stationary simply means every series, mean, variance and covariance does not change over time. By that there is clear consensus where there is a need to examine the stochastic property of series used in the analysis before applying formal test for cointegration and causation. To eliminate the possibility of spurious regression, the unit root test is performed on the variable.

The integration of the variable shows how many times they have been differenced to be stationary. Integrated of order one I(1) indicate that such variable have been differenced once in order to achieve stationary. Similar to I(2), it is integrated at the second order which indicates that such variable have been differenced twice in order to achieve stationary. The unit root test chosen was the Augmented Dickey and Fuller (ADF) tests. To find the optimum lag for each variable, the Akakike Information Criterion (AIC) and the Schwarz Information Criteria (SIC) were selected. If the variables are all stationary at the same level, then the cointegration test will be applied to examine their long-run relationship

Table 2 shows the result of the unit root test on level with intercept and trend for all variables. The results show that all variables are in I(1) condition, where all the variables have unit roots and failed to reject the null hypothesis that all variables have unit roots. Therefore, all variables need to be differentiated in order to make it stationary.

Table 2 ADF Test on Level (Intercept and Trend)

Symbol	t-statistic	Probability	5% critical value	Unit root
LGDP	-2.553580	0.3024	-3.478305	I(1)
LPM	-1.606753	0.7800	-3.478305	I(1)
LSM	-2.226831	0.4671	-3.478305	I(1)
LBLR	-2.236656	0.4616	-3.480463	I(1)

The results for the unit root test on 1st difference with intercept and trend for all variables are shown in Table 3. The results show that all variables are stationary in first difference, where it is to be integrated of order 1. The results show that the t-statistics values are smaller than the 5% critical values, and the probability value is less than 5%. Therefore, the null hypothesis can be rejected where all variables have unit root. Unit root test using the ADF approach showed that all the variables are non-stationary in level but the variables are stationary in the first difference. In order to perform a cointegration test, all the variables need to be stationary at the same level.

Table 3 ADF Test on 1st difference (Intercept and Trend)

Symbol	t-statistic	Probability	5% critical value	Unit root
LGDP	-7.375788	0.0000	-3.480463	I(0)
LPM	-7.599658	0.0000	-3.479367	I(0)
LSM	-9.246277	0.0000	-3.479367	I(0)
LBLR	-6.150623	0.0000	-3.480463	I(0)

4. Emperical Results

4.1 VAR approach of cointegration

The objective of the cointegration test is to determine whether there is a long-run equilibrium among the variables. The result of the test will show if the variables are moving together in the long-run. The variables that are cointegrated showed that there exists a long-run equilibrium relationship between them. For this study, the Johansen and Juselius (1990) approach will be applied. The JJ approach will be based on the ratio test statistics which are λtrace and λmax. Based on the estimation and two statistics, Trace and Max eigen are calculated to test the presence of the cointegrating vector.

There are a few benefits of applying the cointegration test which is first; the test caters problems of non-stationary series, which may lead to spurious correlation. Second; the test allows any short-run dynamics of the data. Before conducting the cointegration test, it is in need to make sure that all the variables are in the same order. Cointegration relationship does not exist between the variables which are integrated in different orders. For instance, if X is I(1) and Y is I(2), then there will be no cointegration between them since they are not in the same order.

Vector AutoRegression (VAR) is one of the approaches to measure the long-run relationship between the variables. This model treats all the variables endogenously. In using the VAR approach, all variables need to be in I(0) condition. An important preliminary step in model building and impulse response function is the selection of the VAR lag order. This study will use some commonly used lag-order selection criteria to choose the lag order, such as the Akaike Information Criterion (AIC) and Schwarz Information Criterion (SC). Based on the Table 4, the lag order selection by criterion was lag 1. Therefore, in this model lag 1 will be used in the modelling of the variables.

Table 4 Lag length order selection

Lag	FPE	AIC	SC	HQ	LR
0	0.0022 43	5.2517 09	5.3866 39	5.3048 65	NA
1	3.89e-08*	-5.71090 2*	-5.03625 1*	-5.44512 3*	676.29 41*
2	4.19e-08	-5.64433 9	-4.42996 7	-5.16593 7	23.8390 2

The model established from the VAR approach is as per below:

<u>Model 1</u>

$\Delta LGDP=0.432+0.9436\Delta LGDP_{t-1} + 0.0210\Delta LPM_{t-1} + -0.0007\Delta LSM_{t-1} + 0.0001\Delta BLR_{t-1}$

From the value of R^2 the models have equations fit to the data. The R^2 equal to 0.983 which showed that there is 98% changes in the GDP. The result of Darbin-Watson showed value close to 2 (1.836) in this equation which means there is no correlation in the model.

<u>Model 2</u>

$\Delta LPM = 0.1500 + 0.0181\Delta LGDP_{t-1} +0.9739\Delta LPM_{t-1} + 0.0051\Delta LSM_{t-1} -0.0076\Delta BLR_{t-1}$

From the value of R^2 the models have equations fit to the data. The R^2 equal to 0.999 which showed that there is 99% changes in the loan by primary mortgage market. The result of Darbin-Watson showed value close to 2 (1.938) in this equation which means there is no correlation in the model.

<u>Model 3</u>

$\Delta LSM=-14.626+2.7355\Delta LGDP_{t-1} -1.3181\Delta LPM_{t-1} + 0.68012\Delta LSM_{t-1} + 0.0470\Delta BLR_{t-1}$

From the value of R^2 the models have equations fit to the data. The R^2 equal to 0.857 which showed that there is 85% changes in the total loan sold to Cagamas. The result of Darbin-Watson showed value more than 2 (2.051) in this equation which means there is no correlation in the model.

<u>Model 4</u>

$\Delta BLR= -9.6954+1.7836\Delta LGDP_{t-1} -0.8239\Delta LPM_{t-1} +-0.0591\Delta LSM_{t-1} + 0.8408\Delta BLR_{t-1}$

DOI: 10.1051/matecconf/20166600100

From the value of R^2 the models have equations fit to the data. The R^2 equal to 0.893 which showed that there is 89% changes in the GDP. The result of Darbin-Watson showed value not that close to 2 (1.198) in this equation but it is acceptable which means there is no correlation in the model.

By applying the impulse response method, the relationship between those VAR models will be determined

4.2 Impulse response

Impulse response function will showed the responsiveness of the dependent variable to a shock of each variable in the VAR model. This means that, if a shock is applied to one of the variable while the other equations are constant, it will affect the VAR system over the time. One of the rules in implementing the impulse response function, the order of the variable need to be correct since different order of variable will present a different output. Therefore, the dependent variable should be at the beginning of the list followed by the independent variable. The impulse response

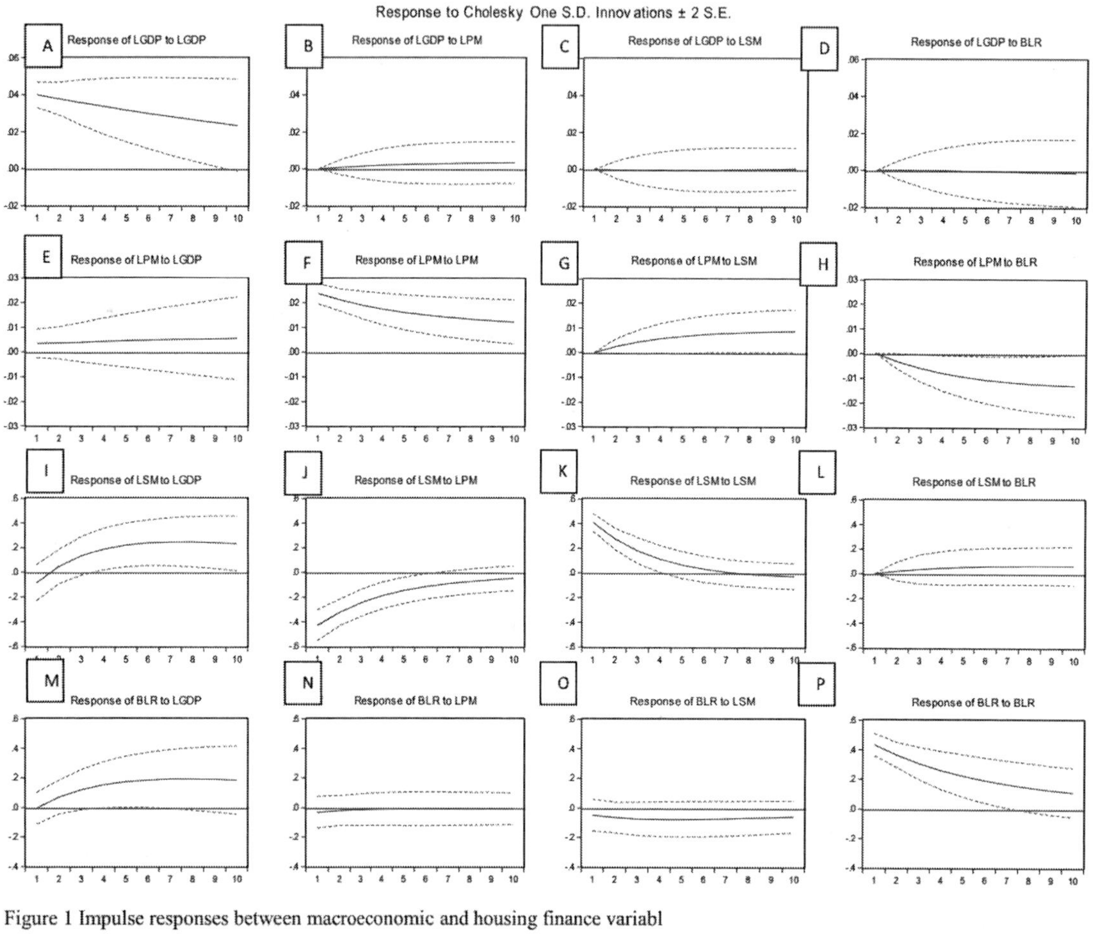

Figure 1 Impulse responses between macroeconomic and housing finance variabl

Primary Mortgage Market responds positively and significantly to the GDP (Figure 2E). When the economy is good and there are increases in the GDP, there are more funds available for the banking institution to extend loan. Hence, it will increase the number of housing loans extended by the banks to the customers. The bank loan respond positively to the real output, which is in line with the previous study by Ibrahim (2006) and Mansor and Mohamed (2012). In contrast, there is no significant effect of primary mortgage market to the GDP (Figure 2B).

Figure 2G and 2J shows the impulse respond between two housing finance submarkets, primary mortgage market and secondary mortgage market. Secondary mortgage market responds positively to the primary mortgage market (Figure

2G). When there is increase in the number of housing loans taken by the house buyers, the banks will increase the volume of housing loans sold to Cagamas. Although in reverse, primary mortgage market responds negatively to the secondary mortgage market (Figure 2J).

Figure 2C and 2D shows that there is no significant response of both the secondary mortgage market and the interest rate to real GDP output. Primary Mortgage Market reacts negatively to the lending rate (Figure 2H). It showed that the bank lending rate reacts negatively with the interest rate charged to the customer. More housing loans extended by the primary mortgage market consequently will reduce the base lending rate. The similar response seen in the lending rate which acts negatively with the Secondary Mortgage Market (Figure 2O). When there is a shock in the interest rate; which happened during the financial crisis, there will be more housing loans sold to the secondary mortgage market as a measure by the government to increase liquidity in banking institutions.

5. Conclusion

The main objective of this analysis to determine the interrelationship between the macroeconomic and housing finance variable. The objective was achieved by using time series technique of Vector Autoregression (VAR) method to identify the relationship between the macroeconomic variable (GDP and BLR) and housing finance variable (total loan by primary mortgage market and secondary mortgage market). The cointegration test showed that there is a long run relationship between the variable.

The impulse response function showed that there is positive response between the GDP and primary mortgage market. When there are increases in the GDP, there are more funds available for the banking institution to extend loan Apart from that, when there is a shock in the interest rate; which happened during the financial crisis, there will be more housing loans sold to the secondary mortgage market as a measure by the government to increase liquidity in banking institutions. As a conclusion, there is presence of relationship between the variable which change in one variable will affect the other variable in the long run.

References

1. Bank Negara Malaysia (Various Issues) Annual Report (Kuala Lumpur: Bank Negara Malaysia).
2. Bank Negara Malaysia (Various Issues) Monthly Statistical Bulletin (Kuala Lumpur: Bank Negara Malaysia).
3. Bank Negara Malaysia (Various Issues) Quarterly Bulletin (Kuala Lumpur: Bank Negara Malaysia).
4. Chung, T. (2014). Promoting a Secondary Mortgage Market and Debt Capital Market. *HOUSING THE NATION: Policies, Issues and Prospects ISBN*, 978-983.
5. Department of Statistics (2012), Malaysian Economy in Brief 2012.
6. Garnett, D. (2005). *Housing finance*. Chartered Institute of Housing/Housing Studies Association.
7. Gerlach, S., Peng, W., (2005). Bank lending and property prices in Hong Kong. Journal of Banking and Finance 29, 461–481.
8. Granger, C. W. (1969). Investigating causal relations by econometric models and cross-spectral methods. *Econometrica: Journal of the Econometric Society*, 424-438.
9. Greef and Haas, (2000). Greef, I.J.M. De, Haas, R.T.A De (2000), Housing Price, Bank Lending and Monetary Policy, Paper presented At The Financial Structure, Bank Behaviour and Monetary Policy In The EMU Conference, October 2000, Groningen.
10. Hilbers, P., Qin Lei And Lisbeth, Z. (2001), Real Estate Market Developments And Financial Sector Soundness, IMF Working Paper No. 01/129, Washington : International Monetary Fund.
11. Ibrahim, M.H.,(2006) Stock prices and bank loan dynamics in a developing country: the case of Malaysia. Journal of Applied Economics 9 (1), 71–89
12. Johansen, S., Juselius, K., (1990). Maximum likelihood estimation and inference on cointegration – with application to the demand for money. Oxford Bulletin of Economics and Statistics 52, 169–210
13. Koop, G (2008) Analysis of Economic Data – 3rd Edition, John Wiley & Sons, Chichester, UK.
14. Lea, M.J. (2000) The Role of the Primary Mortgage Market In The Development Of A Successful Secondary Mortgage Market, Best Practice Series, Inter-American Development Bank
15. Mansor H. I., Mohamed E. S.,(2012) Bank lending, macroeconomic conditions and financial uncertainty: Evidence from Malaysia, Review of Development Finance, Volume 2, Issues 3–4, Pages 156-164, ISSN 1879-9337
16. Said, R., Adair, A., McGreal, S., & Majid, R. (2014). Inter-relationship between the housing market and housing finance system: evidence from Malaysia.*International Journal of Strategic Property Management*, *18*(2), 138-150.
17. Sufian, F. (2004). The efficiency effects of bank mergers and acquisitions in a developing economy: Evidence from Malaysia. *International Journal of Applied Econometrics and Quantitative Studies*, *1*(4), 53-74.
18. Tiwari, P. (2000), "Housing Demand In Tokyo", International Real Estate Review, Vol.3, No.1\
19. Veronica C. W, Francis E. W.(2008) Markets and housing finance, Journal of Housing Economics, Volume 17, Issue 3, Pages 239-251, ISSN 1051-1377,

MATEC Web of Conferences 66, 00101 (2016) DOI: 10.1051/matecconf/20166600101
IBCC 2016

Construction management scheduling and control: The familiar historical overview

Ali Behnam[1, a], Toby Harfield[2] and Russell Kenley[3]

[1, 2, 3]*Swinburne University of Technology, PO Box 218, Hawthorn, Victoria, 3122, Australia*
[3]*Unitec Institute of Technology, 139 Carrington Road, Mount Albert, Auckland 1025, New Zealand*

Abstract. The paper suggests that 'management by exception' is an historical default control mechanism based on the perception of control as a static process. However, increasingly scholars claim that a dynamic and proactive systems model is a more effective form of project control. These findings are the result of an historical desktop research method that analysed content from a small sample of scheduling methods and control approaches found in online and university library resources. The concept of control has historically influenced both visualization and analytics of different scheduling methods for construction project management. This paper focuses on two control ideals; static and dynamic control mechanisms. The overview begins with the description of early graphical scheduling techniques: Gantt charts and Harmonogram. It continues with examples of contributors to scheduling and control that include: CPM, PERT, LOB, Flowline and Location Based Management. The finding of this simple history suggests that change is the constant element for project control mechanisms. An object-based digital environment such as the data-rich building information modelling (BIM) appears to be continuing the change for new scheduling methods and control mechanisms.

1 Introduction

The problem of control has always been an important question for construction. The efforts toward development of control strategies historically have resulted in different scheduling approaches. During the second half of the 20th century, the emergence of computers and their rapid development from mainframe to personal enabled advances in scheduling visualization and more complex analytical methods.

This paper adopts an historical desktop research method using online and university library resources to explore the development of different scheduling methods and control approaches. The search was based on two levels of keywords which accessed documents published between 1962 and 2014. The first level included the terms: "scheduling" and "control". These materials were then searched for such terms as "CPM", "PERT", "line of balance" and "location-based management". A sample of articles, journal papers and books was chosen to represent a chronology of change.

This paper provides a simple description of these developments based on a simple but novel continuum model for control: from static to dynamic. A static form of control has been called "management by exception", whereas; proactive control is considered a "dynamic process". Scholars agree that the main tool for construction project control is through scheduling. Thus, this paper provides an overview of a small sample of scheduling methods that have made a contribution to the continual advancement of construction project control.

The balance of this paper is structured: Section 2 outlines the concept of management by exception with four different scheduling approaches for static control. Section 3 discusses the contributions to dynamic control through a pro-active scheduling and monitoring approach. The conclusion summarizes the scheduling contributions discussed and suggests the future trajectory of the project control continuum.

2 Static control

For construction management, control is an important process. According to Kenley and Seppänen [1] control is a "very practical concept—one which every parent intuitively understands. Control is the mechanism by which a system is monitored and its behaviour corrected to ensure that performance is as planned." (p.96).

Historically, management by exception is considered static form of control. In such a system, remedial action is only necessary when deviations from the original plan occur. In modern construction management processes, the main tool for project control is through scheduling. The schedule is an effort to visually and analytically put activities in a sequential model that leads to project completion. As would be expected, attempts at project control mechanisms have continually changed with the rapid advances in both computing availability and power

[a] Corresponding author: *abehnammanesh@swin.edu.au*

[2] which supports increasingly complex forms of scheduling.

2.1 Graphical project representations

A Polish professor, Karol Adamiecki, specialized in engineering, economics and management. He developed a more sophisticated bar chart scheduling technique. His methodology for work harmonization was based on a diagrammatic schedule called a Harmonogram. While his technique provides data such as bill of quantities, duration of work and their associated dates, the technique also charts movement of work through processes, improving efficiency by aligning production rates. This early efficiency methodology however, was only published in Polish and thus did not gain acceptance in countries such as the US with high levels of construction during all of the 20th century [2].

In Germany, another system of graphical work activities was in place by 1912. This static sequencing chart again seems to have been used only for German speaking construction projects. The American, Henry L. Gantt is attributed with the introduction in 1917 of the bar chart that is commonly used for construction projects today [3]. Gantt charts are probably universally used to represent the organization of production and dominant communication of a time schedule. These charts provide a graphical representation of several activities (y-axis) plotted against time (x-axis). The period of execution for each activity is shown as bar on the time graph corresponding to the planned times of occurrence. Parallel to the planned period, the actual period of execution is plotted. Despite advantageous simplicity of the bar chart, it fails to illustrate the interrelationships of the activities in a project. Therefore, the consequence of variations in duration and sequencing of each activity cannot be identified [4]. As a result, this graphical technique is not linked to any basic analytical method representing project progress or production rate [5].

2.2 CPM and PERT

Computers played an important role in automation for project management in 1950's. Both the academy and industry put major efforts towards the development of analytical methods. In particular, the importance of an accurate project planning and control for military projects during the Cold War was driver for effective solutions.

The most important development was the Kelley-Walker method which provided the mathematical model that was to become the basis for computer programs to able to manage major projects. James E. Kelley and Morgan Walker first developed RAMPS (Resource Analysis and Multi-Project Scheduling). They later introduced a project scheduling method that became known as the critical path method (CPM) [6]. The fundamental principle of this method is that sequences can be illustrated graphically. Project activities are represented by an arrow (thus termed 'activity on the arrow') and when the arrows are linked they show graphically the sequence in which the jobs in the project must be performed [7].

Parallel to development of CPM, the US Navy developed mechanisms to optimize production for more cost efficient military programs. The consulting firm of Booz, Allen and Hamilton was established in 1956 to progress the work started with the Special Projects Office of the US Navy. The result of the development efforts, PERT (program evaluation review technique) was designed for the Fleet Ballistic Missile Program; however, it was not much used by the Navy. PERT was initially aimed at forecasting the likelihood of project success for R& D projects. This however, did not stop the growth of claims of successful project management using this method. Kelly and Walker [8] primarily were concerned with control through introducing a system for 'management by exception'. In such a system, management need only act when deviations from the original plan occur. On the other hand during early development of PERT, the US Navy reports acknowledged the necessity of control information to enhance the management system. But sales of a variety of network-based project management systems in the US, UK and Europe also were predicated on project control through 'management by exception' [9].

Management by exception is basically an 'after-the-fact' approach to control. Deviations are investigated and only then remedial action is taken. This requires the intervention of the project manager to make a forecast of future actions [10].

This approach creates challenges for project managers because the duration of future activities in a project are not related to past activities of the same kind. In fact, CPM algorithm is not capable of forecasting future actions without altering future estimates of duration.

2.3 Line of balance (LOB) method

Line of Balance (LOB) was introduced as an alternative method to CPM in order to facilitate the balancing of repetitive operations based on a production scheduling technique. The main advantage of the method was its simple graphical format to obtain necessary information on production rate and duration of the tasks [11]. The aim of the LOB was to ensure that components or subassemblies are ready at the time they are required to meet the production schedule of the final assembly [12].

The technique originated as a way to handle repetitive construction found in both highways and high rise buildings. In early 1940, the Good Year Co. Ltd. Introduced the LOB method before US Navy started using it as a tool for monitoring the progress of military and industrial processes [13]. Line of Balance was presented as a technique for assembling and interpreting the progress data (stages of industrial processes) in a graphic form against time [14].

Fundamentally, the premise of this technique is defined as the determination of the production rate of finished products in an operation line [15]. This has three component outputs: 1) a unit network that illustrates the relationships and duration of activities; 2) an objective

 DOI: 10.1051/matecconf/20166600101

chart that shows the cumulative calendar of unit completions; 3). a progress chart depicting the detailed completion of each unit [16].

2.4 Linear scheduling method and flowline

A focus on CPM and PERT in early publications indicates significant success for these methods leading to their use in several well-known commercial software applications. However, there were doubts among some researchers about the suitability of these methods for linear and repetitive infrastructure projects. Some researchers realized that CPM and PERT failed to recognize the significance of work overflow and continuity. Over the years researchers have noted that network methods add little to solving the planning problem where there is significant proportion of repetition [11], [17-18].

Consequently, besides the application of LOB methodology on repetitive construction, particularly projects with discrete repetitive units, researchers were encouraged to develop a method to emphasize the repetition of tasks over a distance.

The earliest representation of this method was the time versus distance diagram [19]. Using this technique for linear projects, the work progress of activities in the project can be illustrated with better presentation of information. The production rate of an activity can also be determined by the slope of the plotted line on the time-distance diagram. There have been several popular representations for flow of work through distance and locations. The most important is derived from Selinger [20] his supervisor Peer [21] and Birrell [22].

The linear scheduling method utilized a diagram to graphically illustrate the location and time of each activity in the project over a distance. In this graphical representation, continuous activities completed over the distance are illustrated by lines consisting of a continuous set of points. Therefore, the location of work in progress can be determined at any point in time [23].

The focus of the Peer's representation was mainly on movement of crews which perform a set of activities in a fixed sequence. It is often stated that the Flowline is a graphical representation similar to LOB [24]. However, there are significant differences between these two methods. The main difference is that the LOB vertical axis represents the line of balance quantity, while vertical axis in Flowline represents location. The activities with slowest production rate (lowest slope) are considered as critical activities. Besides, other critical activities include those which control the start time of the slowest activity or are required to complete the project [9].

2.5. Location-based thinking

Birrell [22] is an early researcher who developed a model for flow of work through location. Unlike Flowline and time-distance diagram which were mainly used for linear project, his model emphasized the role of location in any construction projects including buildings. He focused on the consideration of work crews as a continuous flow along the locations of the project. He noted that this would minimize the complexity of construction process and the confusion of the participants as the work passes through various locations in a consistent sequence. This single sequence also enables project managers or the general contractor's site superintendent to build a rhythm of work and movements of work squads through projects. Birrell's model of repetitive construction considers three different physical locations in the project including vertical segmentation, horizontal segmentation and the space available within the site, but outside the building, for material storage and handling.

Birrell constructed a matrix of work packages with work locations on the vertical axis and time period on horizontal axis. He clearly used queuing theory to prepare the work for the construction crew. He identified the idea that the construction process is made up of many flow lines (queues) each consisting of each work squad moving through a series of locations. The scheduling techniques list above indicates that the extension of project management complexity as the development of computing hardware and software developed for application for construction management. However, the each of the construction processes that were identified remained locked within a concept of static or reactive project control [1], [25], [5].

3 Dynamic control

In a productive or dynamic control system, necessary actions are taken before deviations occur. The scheduling techniques based on this approach provide greater and more effective opportunities for project control. Subsection 3.1 to 3.3 outline the historical development of three important changes to project management scheduling that can be seen as instrumental in shifting project control from a static to a dynamic response to deviations. Learning has been identified as a major contributor considering the application of proactive scheduling possibilities. Monitoring probabilities for acceptable rates of productivity has also become a welcome form of pro-active control. In addition, location has been increasingly acknowledged as the foundation of learning and visual monitoring of deviations in all construction projects [26].

3.1 Learning

Peña-Mora, & Li [5] provide a systems perspective concerning forward and backward movement of project control. To accomplish work they consider reliability that consists of three variables: learning effect on reliability, staff experience, and schedule pressure. Thus, learning becomes an important factor for working crews, the schedulers and the quality control of inspectors. The learning curve data provides evident monitoring information for proactive control through three processes of the production system.

3.2 Continuous planning or control?

Research on location-based control began in Finland three decades ago and resulted in major discussions about scheduling systems. The main deriver for this research was the rapid increase of cost of construction providing low profit margins for general contractors. As a result, three construction companies established a large research collaboration attempting to identify the main causes of their problem [27].

Research continued in the form of case studies research. Over 30 case studies were conducted and each brought some enhancement to the scheduling or control techniques [28]. The overall the outcome of these studies emphasized that control is more important than planning. In addition, good control requires continuous monitoring and immediate response to project deviations to reduce their impacts [28], [25]. Furthermore, effective control of construction projects based on calculations of sections and zones paved the way for development of a location-based management system.

3.3 Location based management (LBMS)

Kenley and Seppänen [1] developed a system now commonly known as location-based management system (LBMS). In location-based scheduling and control methodology, a project is broken down into different location hierarchies. This makes a location-breakdown structure (LBS) which has many properties common with the work breakdown structure (WBS). In this methodology activities which are completed with continuous workflow through many locations are considered a single task. Task is the method of control and the container of data which relate to production of the project. Unlike the CPM algorithm, LBMS has the capability to ensure a smooth and continuous parade of crews from location to location with minimum idle time for workers and equipment.

Kenley and Seppänen [1] described a new location-based control model which utilizes four stages of production information, each stage having its own schedule views, information, and properties. The stages are baseline, current, progress and forecasts. Location-based scheduling provides enhanced visualization of work sequences, free locations and total impacts of deviations based on the progress stage as well as visual risk evaluations of schedules for control.

The location-based control system obviously focuses on location rather than activity in CPM to take necessary control actions based on visualization of any deviations before they occur. Scheduling and forecasts are based on data that includes the flow of resources between locations, location-based quantity data, production logic and learning experience from a location to location. Thus, the LBM has a dynamic control system that consequently facilities pro-active management, rather management by exception.

3.4 LBMS and BIM

According to Kenley and Harfield [29], "LBMS is an integrated network of management system components potentially involving all stages of construction, from design through to completion…a methodology for interacting with a BIM, placing demands on the BIM for both properties and characterization (breakdown)."

Location-based scheduling and control methodology is integrated with commercial software systems such as Vico office and Control. *Vico Office* provides reporting through 5D visualization (3D+time+cost). *Vico Control* allows for measurement of quantities, scheduling for lean production, forecasting and control [1].

4 Conclusions

For construction management, control is an important process. Historically, management by exception is considered a static form of control. This means that actions are only necessary when deviations from the original construction project plan occur. However, for productive and dynamic control, necessary actions are taken before deviations occur, taking into account the role of location, visualization, continuity of workflow and also process of learning throughout the project. The difference in philosophies between static and dynamic control historically resulted in development of different scheduling techniques.

The paper described project management control based on some details of five different scheduling approaches that developed in the 20th century. The transformation of scheduling practices was based on the growing complexity of both visualization and analytics.

The focus of visualization changed from activity to repetition, then distance and finally location as the unit of analysis for project scheduling that supports control mechanisms. On the other hand, the change in analytics in the different methods was influenced by the recognizing the importance of proactive control. Consequently, a static control approach shifted to a more effective practice of dynamic control with continuous visual monitoring and look-ahead scheduling.

Project management scheduling visualization and analytics for control are expected to continue their trajectory of continuous development. With the increasing adoption of building information modelling (BIM), the next big leap is integration within an object-based digital environment. This provides the opportunity for new control tools, systems and models.

Acknowledgement

The Swinburne University of Technology Melbourne-Sarawak Research Collaboration Scheme ID number 3000004617 has funding for this research.

References

1. R. Kenley & O. Seppänen, *Location-based Management for Construction*. New York: Spon Press (2010)

2. P. Weaver, *A brief history of scheduling* (2006) Online:http://www.mosaicprojects.com.au/PDF_Papers/P042_History%20of%20Scheduing.pdf
3. P.B. Petersen, J. Mang. Issu. **3**, 131-155 (1991) Online: http://www.jstor.org/stable/40603907
4. E. Chrzanowski & D. Johnston, J. Constr. Eng. Manage. Online: http://dx.doi.org/10.1061/(ASCE)0733-9364(1986)112:4(476)
5. F. Peña-Mora & M. Li, J. Constr. Eng. Manage. Online: http://ascelibrary.org/doi/10.1061/(ASCE)0733-9364(2001)127:1(1)
6. F.K. Levy, G.L. Thompson & J.D. Wiest, Harv.Bus. Rev. **41**, 98-108 (1963)
7. J. Kelley, J. Indust. Sched. **21**, 347-365 (1963)
8. J. Kelley, & M. Walker, In *Proceeding IRE-AIEE-ACM '59 (Eastern)* ACM digital library: http://dx.doi.org/10.1145/1460299.1460318
9. S. Perera, *In Proceedings of IAHS Conference on Impact of Economy and Technology,* 927-985 (1982).
10. J. Meredith & S. Mantel, *Project Management: A managerial approach.* New York: John Wiley & Sons (1995)
11. D. Arditi, & M. Albulak, J. Constr. Eng. Manage. Online: http://dx.doi.org/10.1061/(ASCE)0733-9364(1986)112:3(411)
12. S. Suhail & R. Neale, J. of Constr. Eng. and Manage. Online: http://dx.doi.org/10.1061/(ASCE)0733-9364(1994)120:3(667)
13. J. Lutz & A. Hijazi, Constr. Manage. Econ., **11**, 99-110 (1993)
14. (ONM) Office of Naval Material, *Line of Balance Technology* (1962). Online: http://www.dtic.mil/dtic/tr/fulltext/u2/672029.pdf
15. Z. Al Sarraj, J. of Constr. Eng.Manage.Online: http://dx.doi.org/10.1061/(ASCE)0733-9364(1990)116:4(689)
16. R. Carr & W. Meyer, J. Constr. Div. ASCE, Online: http://dx.doi.org/10.1061/(ASCE)CO.1943-7862.0001107
17. R. Reda, J. Constr. Eng. Manage. Online: http://dx.doi.org/10.1061/(ASCE)0733-9364(1990)116:2(316)
18. A. Russell & W. Wong, J. Constr. Eng. Manage. Online: http://dx.doi.org/10.1061/(ASCE)0733-9364(1993)119:2(196)
19. L. Gorman, Rds. & Sts. **115**, 74-75 (1972)
20. S. Selinger, J. of the Constr. Div. ASCE Online: http://cedb.asce.org/cgi/WWWdisplay.cgi?9573
21. S. Peer, J. of the Constr. Div. ASCE, **100**, 203-210 (1974)
22. G. Birell, J. Constr. Div. ASCE, **106**, 389-407 (1980)
23. R. Harris & P. Ioannou, J. Constr. Eng. Manage. Online: http://ascelibrary.org/doi/pdf/10.1061/(ASCE)0733-9364(1998)124:4(269)
24. P. Lumsden, *The Line of Balance Method.* London, UK: Pergamon Press (1968)
25. W. Alsakini, K. Wikstro & J. Kiiras, Int. J. Proj. Manag. **22**, 75-85 (2004)
26. R. Kenley & T. Harfield, *New Project Management Models: Productivity Improvement for Infrastructure.* Perth: SBENRC (2014)
27. O. Seppänen, *Empirical research on the success of production control in building construction projects*, Helsinki Finland: TKK Structural Engineering and Building Technology (2009)
28. J. Kiiras, *A schedule and resource planning system for the implemenation phase of special projects,* Helsinki Finland: Helsinki University of Technology. Construction Economics and Management Publications (1989)
29. R. Kenley & T. Harfield, In *Proc. of the Int. Conf. Constr. Chang. Wld.* Online: http://www.irbnet.de/daten/iconda/CIB_DC27620.pdf (2014)

BIM Interoperability Limitations: Australian and Malaysian Rail Projects

Russell Kenley[1, a], Toby Harfield[2] and Ali Behnam[3]

[1, 2, 3]*Swinburne University of Technology, PO Box 218, Hawthorn, Victoria, 3122, Australia*
[1]*Unitec Institute of Technology, 139 Carrington Road, Mount Albert, Auckland 1025, New Zealand*

Abstract. Building information modelling (BIM) is defined as a process involving the generation and management of digital representation of physical and functional characteristics of a facility. The purpose of interoperability in integrated or "open" BIM is to facilitate the information exchange between different digital systems, models and tools. There has been effort towards data interoperability with development of open source standards and object-oriented models, such as industry foundation classes (IFC) for vertical infrastructure. However, the lack of open data standards for the information exchange for horizontal infrastructure limits the adoption and effectiveness of integrated BIM. The paper outlines two interoperability issues for construction of rail infrastructure. The issues are presented in two case study reports, one from Australia and one from Malaysia. The each case study includes: a description of the project, the application of BIM in the project, a discussion of the promised BIM interoperability solution plus the identification of the unresolved lack of interoperability for horizontal infrastructure project management. The Moreton Bay Rail project in Australia introduces general software interoperability issues. The Light Rail Extension project in Kuala Lumpur outlines an example of the integration problems related to two different location data structures. The paper highlights how the continuing lack of data interoperability limits utilisation of integrated BIM for horizontal infrastructure rail projects.

1 Introduction

According to The Business Value of BIM for Construction in Major Global Markets, SmartMarket Report [1], an increasing number of construction organisations globally are realizing the advantages of digital construction modelling. The cumulative effort and improvements of computer aided design (CAD) in the construction industry has resulted in increasing adoption of a promising technology, building information modelling (BIM). BIM technology aims to provide a shared knowledge resource for information about a facility throughout its lifecycle [2]. BIM methodology is based on an intelligent, object-oriented, data-rich and parametric digital representation of physical and functional characteristics of a facility.

The adoption of "open" or integrated BIM is important for increasing production and process efficiency in construction projects. The purpose of "open" or integrated BIM is to facilitate more effective exchange of information between different digital systems, models and tools [3]. Consequently, the continuing efforts towards an integrated practice have resulted in the development of open source standards and object-oriented models, such as industry foundation classes (IFC) for vertical infrastructure [4]. However, there is still lack of data standards which could enable the information exchange of rail construction which is an example of horizontal infrastructure. This shortcoming limits the adoption and effectiveness of integrated and "open" BIM for rail projects.

The balance of this paper outlines two interoperability issues for construction of rail infrastructure. The issues are presented in two government procured rail construction case study reports in section 2. Each case study includes: a description of the project, the application of BIM in the project, a discussion of the promised BIM interoperability and identification of the unresolved lack of interoperability for the particular rail infrastructure project. The main sources of our findings included the website of the projects, several journal articles, industry reports and also a university thesis.

The first case study presented is the Moreton Bay Rail project in Australia. This interoperability discussion provides an introduction to incompatibility issues effecting software modelling for general project management. The second case study report is for the extension to the LRT rail system in Kuala Lumpur. It outlines an example of the lack of interoperability because of differences between specific digital data being generated. Different location data structures highlight the problems of information exchange from different digital modelling systems for construction management. The Discussion highlights the importance of open source data standards for improving interoperability. The section also explains about some of the on-going international efforts to facilitate data transfer in these types of projects. The

[a] Corresponding author: *rkenley@swin.edu.au*

Conclusion summarises the lack of general and specific interoperability in construction models for rail projects that is applicable for all horizontal infrastructure construction.

2 Australian Case Study: Moreton Bay Rail

2.1 Project Description

The Queensland Transport and Main Roads State Planning Program is based on analysis of current and future integrated passenger transport options. Investment decisions must take into account the challenges of distance in order to deliver quality transport networks for a growing population. Use of the Regional Queensland Passenger Transport Network Plan assists with the identification of appropriate network services for designated regions to provide better and more sustainable transportation options.

The Moreton Bay region of Queensland is the third largest Local Government Area in Australia and is also one of the fastest growing areas in Australia. Population growth is expected to continue over the next 15 years which will cause a significant transportation problem if the use of private cars grows at the same pace as the population. The result of Moreton Bay population growth will be higher values of traffic into Queensland capital city, Brisbane.

The Moreton Bay Rail (MBR) project involves construction of a new 12.6km dual heavy track passenger rail line between Lawton and Kippa-Ring that includes six new railway stations and significant road and bridge construction to provide feeder routes to these stations. This new integrated transport passenger rail corridor is located approximately 23 kilometres north of the Brisbane Central Business District [5].

The six new stations: Kallangur, Murrumba Downs, Mango Hill, Mango Hill East, Rothwell and Kippa-Ringg will all provide full passenger services. The services include: 175m long side platforms, walkways, ticket offices, waiting facilities, park and ride facilities and bus/train interchanges. All of these facilities and structures are contracted for design and build for the client, Queensland Transport and Main Roads (QTMR).

2.2 BIM Application

This type of public transport major works project [6] is expected to be able to take advantage of the opportunity to utilise promising technology such as building information modelling (BIM). The Moreton Bay Rail project design has two distinctive components.

The first component relates to the buildings and architectural elements that use object-based BIM. However, these object-based models do not fit easily into the 12d string-based modelling required by QTMR for construction of horizontal infrastructure [5].

The second project component relates to geometric elements including railway embankment alignments, the access roads, bridges, landscaping and the pedestrian walkways. In this domain, string-based modelling solutions, such as 12d system, are essential because working with object-based BIM solutions is painstakingly inefficient.

2.3 Unsolved Modelling Interoperability Issue: String-based and Object-based

Associated BIM tools have evolved from 2D CAD drawings to 3D models embedded with intelligent object-based, parametric data. However, as with all new technology, there is a lag between the promised and industry capacity and capability. Currently the construction industry is moving very slowly toward a more integrated practice of BIM processes for better interoperability between different tools for vertical built environment.

However, as the client and eventual owner of the Moreton Bay Rail, all documents for the project need to ultimately be available to QTMR. Their customised BIM system is the 12d software solutions which create string-based models. In addition a significant number of construction design and implementation software tools have been researched and developed specifically for QTMR construction standards and transportation use regulations.

Because all works have standard requirements such road and rail signage or lighting, all contractors are required to use the QTMR software for some parts of a project. This procurement strategy aims to ensure that all construction documentation is compatible with their customised 12d system [5]. Nevertheless, for the Moreton Bay Rail project this has not been possible because of the modelling systems being used by the Architects and Contractors on the project.

An example, integration of models generated from the two different approaches (object-based and string-based) can take place within Navisworks via IFC export or direct export. At the same time this process has proven to be challenging (in terms of time and skills availability). The Moreton Bay Rail project team are effectively using IFC object data in the Navisworks software. However, a major shortcoming of the process is that the digital information exchange is only a one-way process once the models are imported into Navisworks. It would be expected for interoperability information exchange would be a two-way process. The ability to import and export models with ease must be the definition of interoperability.

This case study highlights the interoperability challenge that negatively effects current construction of horizontal infrastructure. Incompatibility of software systems (string-based horizontal infrastructure models and BIM vertical infrastructure object-based models) does not allow the promised information exchange efficiency of digital modelling using a BIM environment. Consequently, effective utilisation of "open" or integrated BIM is limited for this critical construction process for horizontal infrastructure construction management [7].

3 Malaysian Case Study: Kuala Lumpur Light Rail Transit (Extension)

3.1 Project Description

In the past five years under the Tenth Malaysia Plan the Malaysian construction sector has expanded significantly supported by strong activities in civil engineering and linear infrastructure projects particularly railways and roads [8]. One example is the Mass Rapid Transit system in the city of Kuala Lumpur. It is the largest infrastructure in Malaysia which began in 2011 and is expected to be completed in 2020. This 60 km transportation corridor involves an estimated RM30 billion investment which indicates the commitment to better and more sustainable transportation options.

The Mass Rapid Transit corridor aims to integrate all previous and on-going LRT extension projects. The aim of the Light Rail Transit (LRT) extension projects is to encourage a greater number of public transport passengers. The expectation is to increase the use from 21 % to 40 % by 2030 under the National Land Public Transport Master Plan target [9]. The LRT extension project is an elevated rail line running through a highly dense urban area. The construction of the guideway consists of 1000 piers through 35.5 kilometres and 25 stations.

3.2 BIM Application

The Construction Industry Transformation Program, 2016-2020 [10] outlines the critical importance of enhancing productivity, efficiency and cost effectiveness of the domestic construction industry by leveraging ICT. In particular, the adoption of BIM technology is encouraged by the Malaysia Public Works Department. They have established a special BIM Committee that is in the process of publishing Malaysia BIM Roadmaps, National Standards, Manuals and Guidelines.

Discussion of BIM and the current Kuala Lumpur LRT construction projects provides evidence of the necessity of these aspirational BIM objectives. The construction process for a large rail project is complex and involves excessive amounts of data (drawings and documents) that needs to be exchanged between numerous stakeholders. One of the most important information exchange processes is the critical role of monitoring and reporting construction progress.

The traditional practice for Malaysian rail projects has been to collect and report progress monthly. For LRT projects the monthly progress reports are usually hundreds of pages long including: Gantt charts, work descriptions, site photos and attached specification reports [11]. The variety of types of documents and information formats make the monthly progress reports difficult to store, read and analyse. Thus, progress reporting is a time consuming process for both data collection and progress report preparation.

This important project process presents an opportunity to achieve the government's BIM objective of more effective and efficient data transfer. The complexity of large infrastructure projects means, of course, complex progress reports. Thus, a more efficient process was considered an excellent trial for digital information exchange technologies by integrating geographic information systems (GIS) and digital construction models for progress reporting of the LRT extension construction project. Researchers have examined the adoption of GIS and 2D/3D models to collect, store, analyse, visualise and exchange information related to construction progress [12]. GIS provides 2D data in very large scale. The argument is that the digital data integration of CAD/BIM data with 2D GIS could allow more sophisticated analysis of large datasets for location information exchange.

For the elevated LRT extension project in Kuala Lumpur, a progress monitoring framework was developed based on both the planned and actual progress data for each location in the project. The location data is collected, reported and visualized in a digital CAD/GIS platform [13]. While this method of providing construction progress should have been more effective, the LRT CAD model and the satellite image of the construction site do not use the same location structures. Thus, data integration, or interoperability remains a problem.

3.3 Unsolved Interoperability Issue: CAD & GIS Location Identifiers

The LRT CAD model represents the construction project design, including the relative coordinates, shape, size and orientation of every construction component and unit. The coordinate system used in CAD drawing is defined with local location identifiers such as streets names or government boundaries. GIS uses a completely different system. Location identifiers are based on a global geographical reference system.

Therefore, interoperability of information exchange of specific location data remains an issue. The location identifiers have different data structures with completely different scales of measurement. Integrating these two models of location is a challenging task because it requires manual translation between two data sources. Therefore, digital data collection and preparation of construction progress report does not provide adequate reduction of time and effort. Consequently, effective utilisation of "open" or integrated BIM is also limited for this critical construction process for horizontal infrastructure construction management [7].

4 Discussions

Governments need to provide transportation services to ensure economic and social activity, and to provide these services in a cost-effective manner. However, the lag between promised efficiency from adoption of BIM and industry BIM capability and capacity makes it difficult for government decision-makers. Both Australia and Malaysia have attempted to take advantage of technological advances afforded by the emerging BIM systems, models and tools, but global construction industry practice is slow to change.

The lack of current standards for horizontal infrastructure is a major obstacle for effective data exchange in these types of projects. Therefore, an Open Source definition for BIM is necessary as the solution to enabling interoperability and hence promoting an integrated practice. BuildingSMART as an international organisation facilitates the development of open standards such as IFC for building industry and vertical infrastructure [14].

The buildingSMART has recently commenced the IFC for infrastructure project to expand the IFC to support horizontal infrastructure projects [15]. The first IFC for infrastructure is the IFC alignment which is collaboration between buildingSMART and open geospatial consortium (OGC). The aim of this project is to provide standards for exchange of alignment information (including 2D, 3D vertical and horizontal) throughout the infrastructure lifecycle. These standards extend IFC to include OGC standards that support global rather than local geographic coordinates and levels. By using the expected standards, the geometric data can be accurately exchanged through different systems, models and tools with minimum data redundancy.

Furthermore, the buildingSMART is also attempting to develop data models for road, rail, and bridge and tunnels to fulfil these project data needs such as element breakdown, detailed material information, earthwork (digital terrain model, fill and cut, geological volumes) and other information.

5 Conclusions

Development of "open BIM" to facilitate integrated construction management does seem to have made some headway in construction of vertical infrastructure. However, interoperability issues specific to horizontal infrastructure are yet to be solved.

This paper provides two case studies discussing adoption of digital information exchange technology for rail construction. The Australian project focused on the two way exchange of object-based and string-based design. The Malaysian project focused on the integration of a CAD construction model and GIS digital data collection. Both case studies outline interoperability problems related to modelling issues for horizontal infrastructure which are indicative of the difficulties of the aspiration for "open" or integrated BIM.

These studies identify only two examples of BIM interoperability issues. One is a general problem of information exchange between string-based and object-based design. The other examines the specific problem of different location data structures. Clearly, to maximize the effectiveness of BIM as an integrated practice for horizontal infrastructure, it is important to increase interoperability between different systems, models and tools.

The continuing international efforts for development of open data standards and object-based models, such as the Industry Foundation Classes (IFC) are supportive of open BIM for vertical infrastructure. However, the lack of emphasis on the standardization of data needs and requirements of horizontal infrastructure remains a problem for governments in their quest for effective digital systems for rail construction. This limitation should encourage future researchers to investigate the needs of information exchange in these projects. The on-going standardisation projects such as IFC alignment could enhance the interoperability and integration of different object-based, alignment-based and GIS models.

Acknowledgement

The Swinburne University of Technology Melbourne-Sarawak Research Collaboration Scheme ID number 3000004617 has funding for this research.

References

1. McGraw Hill Construction *The Business Value of BIM for Construction in Major Global Markets, SmartMarket Report* (2014) Online: https://synchroltd.com/newsletters/Business%20Value%20Of%20BIM%20In%20Global%20Markets%202014.pdf
2. C. Eastman, P. Teicholz, R. Sacks, & K. Liston, *BIM handbook: A guide to building information modeling for owners, managers, designers, engineers and contractors.* (2nd ed.) Hoboken, NJ: Wiley (2011)
3. M. Laakso & A. Kiviniemi "The IFC standard - a review of history, development, and standardization." J. Info. Tech., **17**, 134-161 (2012) Online: http://www.itcon.org/2012/9
4. J. Steel, R. Drogemuller & B. Toth, "Model interoperability in Building Information Modelling." Software and Systems Modelling, **11**, 99-109 (2012)
5. (QTMR) Queensland Transport and Main Roads (2015) *Transport-and-Main-Roads-12D-model-customisation* Online: http://www.tmr.qld.gov.au/business-industry/Road-systems-and-engineering/Software/Transport-and-Main-Roads-12D-model-customisation.aspx
6. N. Lee, T. Salama, & G. Wang, "Building information modeling for quality management in infrastructure construction projects." In Issa, R., & Flood, I. (eds.) *Proc. Comput. in Civ. and Build. Eng.,* 65-72 (2014) Online: http://ascelibrary.org/doi/abs/10.1061/9780784413616.009
7. M. Laakso & A. Kiviniemi. "The IFC standard - a review of history, development, and standardization." J. Info. Tech., **17**, 134-161 (2012) Online: http://www.itcon.org/2012/9
8. CIDB, *Construction Industry Review.* Kuala Lumpur, Malaysia: Construction Industry

 DOI: 10.1051/matecconf/20166600102

Development Board Malaysia (2009)

9. Prasarana Malaysia Berhad, *LRT Line Extension, About the Project* (2015) Online: http://www.myrapid.com.my/projects/lrtextension/about-project/about-project
10. CIDB.. *The Construction Industry Transformation Program*, 2016-2020. Kuala Lumpur, Malaysia: Construction (2015) Industry Development Board Malaysia. Online: http// www.citp.my/download.php?f=CITP-public.pdf
11. Sunway Group Construction , *Monthly Progress Reports*. Kuala Lumpur: Prasarana Malaysia Berhad (2012)
12. S. El-Omari & O. Moselhi, (2011). "Integrating automated data acquisition technologies for progress reporting of construction projects" J. Automati. in Constr., **20**, 699-705.
13. A. Behnam, *A Study on the Application of Location-Based Management and Automated Progress Monitoring for Linear Construction Projects,* Unpublished MPhil thesis. Semenyih: University of Nottingham Malaysia Campus (2015)
14. R. Kenley & T. Harfield, *New Project Management Models: Productivity Improvement for Infrastructure.* Perth: SBENRC (2014)
15. BuildingSMART, *IFC Infrastructure Room*, (2015) Online: http://www.buildingsmart-tech.org/infrastructure

 DOI: 10.1051/matecconf/20166600103

Classification of High-Rise Residential Building Facilities: A Descriptive Survey on 170 Housing Scheme in Klang Valley

Siti Rashidah Hanum Abd Wahab[1,2], Adi Irfan Che Ani[1], Ahmad Sairi[3], Norngainy Mohd Tawil[1], Mohd Zulhanif Abd Razak[1]

[1]*Department of Architecture, Faculty of Engineering & Built Environment, Universiti Kebangsaan Malaysia (UKM)*
[2]*Department of Polytechnic Education, Ministry of Higher Education Malaysia*
[3]*Department of Building Control, Kajang Municipal Council (MPKj), Selangor*

Abstract. High-rise residential building is a type of housing that has multi-dwelling units built on the same land. This type of housing has become popular each year in urban area due to the increasing cost of land. There are several common facilities provided in high-rise residential building. For example playground, swimming pool, gymnasium, 24 hours security system such as CCTV, access card and so on. Thus, maintenance works of the common facilities must be well organised. The purpose of this paper is to identify the classification of facilities provided at high rise residential building. The survey was done on 170 high-rise residential schemes by using stratified random sampling technique. The scope of this research is within Klang Valley area. This area is rapidly developed with high-rise residential building. The objective of this survey is to list down all the facilities provided in each sample of the schemes. The result, there are nine classification of facilities provided for high-rise residential building.

1 Introduction

In urban area, the demand for high-rise residential building increases each year due to the increase in house and land prices [1]. People prefer to live in high rise residential building due to the present lifestyle of people in the city [2]. Furthermore, the increasing population and income nowadays caused the increasing demand of high-rise residential building. For example 27.5% of Sydney, 84% of Singapore and more than 95% of Hong Kong population lives in high rise building [3]. In addition, the vicinity facilities of high-rise residential building such as Mass Railway Transit (MRT) and shopping complexes attract people in the city to own units in high-rise residential buildings. In addition, there are several common facilities provided in high-rise building scheme that can be enjoyed by the residents such as swimming pool, gymnasium, landscape, 24 hours security system, sports court and so on. Thus, these facilities must be well maintained to ensure its functionality. All the cost of maintenance works are come from the maintenance fees. This maintenance fees are imposed on the resident's responsibility. As mentioned by [4] a good management service must have a good financial support. But, managing high-rise building in Malaysia is still new and the experience managing high-rise properties have been inconsistent due to lack of maintenance fund [1, 4]. There are many arguments occur between management body and residents on the maintenance fund. Residents are refuse to pay the maintenance fees because they claimed that the value of maintenance fund is not comparable with the facilities provided at their scheme [1, 5, 6]. Some declare they do not use all the facilities provided and refuse to pay the maintenance fees [1, 7]. Otherwise, residents request to know the transparency flow of maintenance fund allocation [6]. Therefore, the objective of this paper is to categories all the facilities provided at high-rise residential building scheme. This can be a guide for the management to determine the realistic maintenance cost to be imposed on the residents of high-rise building. These classification is important to create better management system for high rise residential building. Then, it can give clear definition on the type of high-rise residential based on its facilities rather than based on the price and its location. In addition, management body can used this classification as a guide to determine the realistic maintenance cost to be imposed on the residents of high-rise building scheme. Table 1 shows the possible facilities provided at high-rise residential building gathered from literature and several guidelines.

Table 1. Availability of Facilities at High-Rise Residential Building

No	Classification	Facility code	Facility	Sources														
				[10]	[7]	[11]	[5]	[6]	[12]	[13]	[14]	[15]	[16]	[17]	[18]	[19]	[20]	[21]
1	Religious Facility	FK1	*Surau*		√		√	√	√	√	√	√	√	√				
		FK2	Prayer room		√		√		√	√	√	√	√		√			
2	Education Facility	FPD1	Kindergarten				√		√		√	√	√	√	√			
		FPD2	Day Care				√	√	√		√	√	√	√	√			
		FPD3	Mini library				√	√	√		√	√	√	√				
		FPD4	Internet/ Wi-Fi Room				√	√	√		√	√	√					
3	Community Facility	FPK1	Multipurpose Hall	√			√	√	√		√	√	√	√	√			
		FPK2	Meeting Room	√			√		√		√	√	√	√	√			

[a] Corresponding author: siti.rashidah.hanum@gmail.com

No	Classification	Facility code	Facility	Sources [10]	[7]	[11]	[5]	[6]	[12]	[13]	[14]	[15]	[16]	[17]	[18]	[19]	[20]	[21]
4	Commercial Facility	FRP1	Retail shop				√	√	√		√		√	√	√	√	√	
		FRP2	Laundry				√	√	√		√				√	√	√	
		FRP3	Salon/beauty centre				√	√	√		√				√			
		FRP4	Stationery						√		√				√			
		FRP5	Tailor shop						√		√			√	√			
		FRP6	Car Wash					√	√		√				√			
		FRP7	Mini Post Office						√		√				√			
		FRP8	Restaurant/ cafeteria				√	√	√		√		√	√	√			
		FRP9	Clinic						√		√		√		√			
		FRP10	Office lot						√		√		√					
5	Open Recreation Facility	FRTB1	Benches						√			√		√				
		FRTB2	Landscape						√		√	√		√				
		FRTB3	playground	√	√	√	√	√	√		√	√		√		√	√	√
		FRTB4	BBQ Area	√	√	√	√	√	√					√				
		FRTB5	Jogging track	√	√	√	√	√	√		√	√		√				
		FRTB6	Outdoor sport court	√	√	√	√	√	√		√	√		√				
		FRTB7	Rooftop garden	√	√	√			√		√			√				
6	Closed Recreation Facility	FRTP1	Indoor sport room	√		√	√	√	√		√	√		√				
		FRTP2	Swimming pool	√		√	√	√	√			√				√	√	√
		FRTP3	Spa/Sauna	√		√	√	√	√			√						
		FRTP4	Jacuzzi	√		√	√	√	√			√						
		FRTP5	Gymnasium	√		√	√	√	√			√						
		FRTP6	Karaoke room	√		√	√	√	√									
		FRTP7	Club house	√		√		√	√									
7	Security Facility	FKS1	Guard house	√	√		√	√	√			√						√
		FKS2	CCTV	√	√			√	√			√						√
		FKS3	Access card	√	√			√	√									
		FKS4	Intercom	√	√			√	√									
		FKS5	Perimeter fence	√	√				√		√	√						
8	Parking Facility	FTLK1	Open parking area	√			√		√		√			√		√	√	√
		FTLK2	Multi storey parking	√			√	√	√		√			√		√	√	√
9	Building Services Facility	FPB1	Toilet	√	√			√	√		√	√						
		FPB2	Lift	√	√		√	√	√		√					√	√	√
		FPB3	Garbage chute/house	√	√				√				√			√	√	√
		FPB4	Store	√	√				√		√		√					
		FPB5	Firefighting equipment	√	√				√				√			√	√	√

2 Research Method

In Malaysia, there are 67.4% non-low cost high rise residential building were built in the area of Klang Valley [1, 6]. Klang Valley is an economic centre of Malaysian country. Therefore, the sample used in these area already met the sampling requirement for the purpose of the study [6]. There are 1,769 non-low cost high rise residential building scheme in Klang Valley.

By using stratified random sampling technique, there are 170 non-low cost housing scheme were selected in the area of Klang Valley to do a survey on the availability of high-rise building facilities (refer table 2). The objective of the survey is to categories all the facilities provided at Malaysian non-low cost high-rise residential building scheme.

Table 2. Non-Low Cost High-rise Residential Building Scheme in Klang Valley

No.	Zon/PBT	Housing Scheme	Rate	Approximate Sample Size	Sample Size
1	***Wilayah Persekutuan Kuala Lumpur***				
	DBKL (Kuala Lumpur area)	790	0.447	75.92	76
2	***Selangor***				
	MPSJ (Petaling area)	212	0.120	20.37	20
	MBPJ (Petaling area)	150	0.085	14.41	14
	MPKj (Hulu Langat area)	138	0.078	13.26	13
	MPS (Gombak area)	121	0.068	11.63	12
	MPAJ (Gombak and Hulu Langat area)	119	0.067	11.44	11
	MBSA (Petaling area)	90	0.051	8.65	9
	MPK (Klang area)	86	0.049	8.26	8
	MPSp (Sepang area)	22	0.012	2.11	2
	MDHS (Hulu Selangor area)	17	0.009	1.63	2
	MDKS (Kuala Selangor area)	12	0.007	1.15	1
	MDKL (Kuala Langat area)	11	0.006	1.06	1
	MDSB (Sabak Bernam area)	1	0.001	0.09	1
	Total	**1769**			**170**

3 Analysis and Results

This study used an analysis factor to classify all the facilities provided in non-low cost of high-rise residential building in Malaysia. Therefore, principle component analysis (PCA) with varimax rotation was used to classify the facilities [22]. Table 3 shows the value of KMO for the availability of facilities was 0.842. This value is more than 0.7 which indicates that the number of samples used is sufficient to undergo analysis factor. The value of Bartlett's Test of Sphericity is significant (less than 0.001) means that the independent variables is suitable for analysis factor [23].

Table 3. KMO dan Bartlett's Test

KMO dan Bartlett's Test		
Kaiser-Meyer-Olkin Measure of Sampling Adequacy		0.842
Bartlett's Test of Sphericity	Approx. Chi-Square	56955.682
	df	946
	Sig.	.000

Furthermore, the percentage of variance explained was 72.092% which is more than 60% of the total variance as proposed by [23] (refer table 4). These value shows that the research data is suitable for next process of analysis factor.

Table 4. Total Variance Explained

Component	Initial Eigenvalues			Extraction Sums of Squared Loadings		
	Total	**% of Variance**	**Cumulative %**	**Total**	**% of Variance**	**Cumulative %**
1	15.455	35.124	35.124	15.455	35.124	35.124
2	11.339	25.770	60.894	11.339	25.770	60.894
3	4.927	11.197	72.092	4.927	11.197	72.092
4	1.816	4.126	76.218			
↓	↓	↓	↓			
44	.001	.002	100.00			

Table 5 shows the results of analysis factor for the availability of facilities at non-low cost of high-rise residential building in Malaysia.

Table 5. Analysis Factor

Item	Component		
	1	2	3
Mini Post Office	0.959		
Restaurant/Cafeteria	0.945		
Surau	0.941		
Tailor Shop	0.932		
Salon/Beauty Centre	0.928		
Retail Shop	0.924		
Closed Parking Area/Multi-storey Parking	0.913		
Management Office	0.912		
Playground	0.906		
Garbage Chute/house	0.904		
Open Parking Area	0.888		
Lift	0.860		
Landscape	0.856		
Benches	0.792		
Prayer Room	0.784		
Public Toilet	0.783		
Fire Fighting Equipment	0.746		
Public Store	0.713		
Car Wash	0.668		
Spa/sauna		0.898	
Swimming Pool		0.889	
Indoor Sport Room		0.882	
Guard House		0.874	
Gymnasium		0.868	
Outdoor Sport Court		0.862	
BBQ Area		0.857	
CCTV		0.849	
Jogging Track		0.847	
Rooftop Garden		0.832	
Access Card		0.819	
Club House		0.788	
Perimeter Fences		0.771	
Karaoke Room		0.759	
Intercom		0.736	
Meeting Room			0.931
Mini Library			0.918
Multipurpose Hall			0.914
Internet/Wi-Fi Room			0.904
Day Care			0.868
Kindergarden			0.760

There are 44 items were analysed in the analysis factor and 4 items aborted due to the loading factor less than 0.4 as suggested by [23]. There are 3 factors developed from the analysis. Then, the classification of facilities were renamed according to the Malaysia guideline from KPKT. The first group was classified as basic facilities [12]. The basic facilities has 19 items that consist the combination of commercial, open recreation facilities, parking and building services facilities. The eigen value was 15.455 (significant if more than 1) [23] and contribute 35.124% of the total variance. Then, the second group was classified as exclusive facilities [12, 4 and 6]. The exclusive facilities has 15 items that consist the combination of security, closed recreation facilities and 4 items from open recreation facilities. The eigen value was 11.339 and contribute 25.770% of total variance. The third group was classified as support facilities [12, 6, 14, 15]. This group has 6 items that consist the combination of community and education facilities. The eigen value was 4.927 and contribute 4.126% of total variance.

4 Conclusions

In a nutshell, maintenance charges is the main issues and problems related to high-rise residential building. This problem lead to the poor facility maintenance for majority of the high-rise residential scheme. Therefore, this paper determine all the facilities provided at high-rise building according to the nine pre-classifications identified. After undergo the process of analysis factor, these nine pre-classifications were regroup into 3 new factors which are basic facilities, exclusive facilities and support facilities. Then, the distribution of maintenance cost and expenses can be clarified at later stage according to these 3 classification of facilities. This study will serve as benchmark for the designation of facilities at each type of high-rise residential scheme. Then, the better solution for maintenance cost expenses can be determined at the later stage of this research.

References

1. Abd-Wahab, S. R. H., Sairi, A., Che-Ani, A. I., Tawil, N. M. and Johar, S. Building Maintenance Issues: A Malaysian Scenario for High Rise Residential Buildings, *International Journal of Applied Engineering Research* **10**, 6, pp. 15759-15776 (2015)
2. Yahya, M.R. Evaluation of Health and Safety Aspects in Condominiums, *Alam Cipta* **5**, 2, pp.129–136 (2012)
3. Daniel, C.W.H., Wei, G. Collective actions in building apartments in Hong Kong. *Habitat International* **38**, (2013)
4. Tiun, L.T. Managing High-Rise Residential Building In Malaysia: Where Are We?, *Persatuan Sains Sosial*, Universiti Putra Malaysia (2009)
5. Yusoff, Y.M., Tawil, N.M., Hamzah, N. Abdullah, N.A.G., Musa, A.R. Tinjauan Kesan Fasiliti Kediaman Bertingkat Terhadap Amaun Dana Pengurusan, *Journal of Design + Built* **4**, pp. 32–40 (2011)
6. Tawil, N.M., Yusoff, Y.M., Che-Ani, A.I., Abdullah, N.A.G. and Surat, M. A Study of Management Corporation Financial in High Rise Residential with Correlation of Management Fund and Facilities Provided, *International Business Management* **6**, 3, pp. 304–307 (2012)
7. Jamila, H. Strata Title in Malaysia, Selangor: Pelanduk Publications (M) Sdn. Bhd, (1994).
8. Yasmin, M. A.; Nor Azlina, S. A General Overview of the Issues Concerning the Non-Payment towards Management Fund of Stratified Buildings In Kuala Lumpur. *The Malaysian Surveyor* **40**, 4, pp. 19-23 (2006)
9. Leung, C.C. Resident Participation: A Community-Building Strategy in Low-Income Neighbourhoods, *Harvard Joint Center for Housing Studies*, Harvard University, Cambridge, MA, (2005).
10. Tiun, L.T. Managing High-Rise Residential Building In Malaysia: Where Are We?, *Paper Presented at 2nd Neprec Conference*, Inspen, Universiti Putra Malaysia, Serdang, (2009)
11. Maciha, J. C. Preventive maintenance for multi-family housing. Canada: John Wiley & Sons, (2005)
12. Garis Panduan KPKT. Garis Panduan Perancangan Perumahan, Jabatan Perancangan Bandar dan Desa Semenanjung Malaysia, Kementerian Perumahan dan Kerajaan Tempatan 1, (2011)
13. I-Book DBKL. Garis Panduan Perancangan Pembangunan, Dewan Bandaraya Kuala Lumpur (2012)
14. Pembangunan Strata Putrajaya. Garis Panduan Penyediaan Kemudahan Awam Bagi Pembangunan Strata di Putrajaya, Jabatan Perancangan Bandar Perbadanan Putrajaya, November, (2008).
15. Garis Panduan Perancangan Kawalan Pembangunan DBKL. Garis Panduan Perancangan Penyediaan Kemudahan Penduduk Dalam Skim Pembangunan, Jabatan Perancang Bandar, Dewan Bandaraya Kuala Lumpur, (2011)
16. Garis Panduan Perancangan Kemudahan Masyarakat. Jabatan Perancangan Bandar dan Desa Semenanjung Malaysia, Kementerian Perumahan dan Kerajaan Tempatan, (2013)
17. Garis Panduan Perancangan Selangor. Manual Garis Panduan Dan Piawaian Perancangan Negeri Selangor, Jabatan Perancangan Bandar Dan Desa Negeri Selangor, (2010)
18. Garis Panduan Perancangan WPKL. Penyediaan Kemudahan Penduduk Dalam Skim Pembangunan Kediaman Bertingkat, (2014).
19. Housing Guideline NSW. Strata living in New South Wales , NSW Fair Trading, (2015)
20. Housing Guideline Tasmania, Strata Living in Tasmania, Department of Industries and Water Tasmania Government, (2008)
21. Building Department Hong Kong. Building Maintenance Guide Book, Building Department, Hong Kong (2012). Available at: www.info.gov.hk/htm [Access on 10 September 2015].
22. Hair, J. F., Jr, Black, W. C., Babin, B. J., Anderson, R. E., dan Tatham, R. L. Multivariate Data Analysis (6th Ed.). Upper Saddle River, NJ: Pearson Printice Hall. (2006).
23. Meyer, L.S., Gamst, G., dan Guarino, A.J. Applied Multivariate Research: Design and interpretation. California, USA: Sage Publication.Inc. (2006)

Intelligent Facades in Buildings Facades of local Office Buildings - Case Study

Asst. Prof. Dr. Anwar Subhi Al-Qaraghuli, Dept. of architectural engineering, University of Technology, Baghdad, Iraq
Architect Eng. Waleed Saad Alawsey, Dept. of architectural engineering, University of Technology, Baghdad, Iraq

Abstract. In order to meet the functional performance and requirements on the environmental, ecological, social and aesthetical aspects, it became a necessity to develop another options of building that would be more efficient in the provision of those requirements, so a new generation have appeared called "Smart Buildings" or "Intelligent Buildings" with their sophisticated parts and various details, one of those essential parts is the "Intelligent Façade" for it's being the essential and primary defensive line for the building against the environmental and climatic variations.This research focused on presenting the most clearly and comprehensive perception of the intelligent façades, in a manner that serves the ability of the designer to apply them in his designs or while developing an existing façades in local (Iraqi) office buildings.To achieve such goal, it has been a necessity to adopt a descriptive and analytical Method for the previous knowledge and take a sequential researching steps, the first step was to build a comprehensive theoretical framework by defining the intelligent façade. Down to abstracting three main vocabularies of the theoretical framework represented by: integrated intelligent façade design, intelligent façade techniques, and effective response.The next research steps focused on applying the vocabularies of the theoretical framework on the elected local office buildings that have been adopted the matter of intelligent façades in one of their forms, and then analyze/discuss the results of the applicable study, to be able to draw the final conclusions, and by this the research presented a determined recommendations.

Keywords: Intelligent Façades, Intelligent Techniques, Effective Response.

Abbreviations:

IF: Intelligent façade
IFS: Intelligent façade system
HVAC: Heating, ventilating and air/conditioning
BMS: Building management system
AI: Artificial intelligence
MPS: Multi-phase system
DSF: Double-skin facade

1. Introduction

The Importance of intelligent façades in the buildings is prominent in accordance with the interests of sustainability from the aspects of resources, energy conservations and controlling their applications and use. In addition to produce the energy from renewable (green) resources which have no harm on the environment; moreover, they're more interesting with attaining multi-performance benefits; besides the financial benefits.

The previous global studies and theses and their applications and practices provided a wide domain of knowledge which can be useful for the designers that aiming to do such façade designs or even developing or replacing the existing ones in buildings; from the aspects of their formations, technologies and their reliable materials. Moreover, the nature of performance and the way they response to the variable everyday representations and circumstances; however, local Iraqi knowledge and practice did not had the required attention to this specific kind of advanced building techniques to parallelize global interest and advancement.

2. Definitions Of Intelligent Facade

The definitions of the "intelligent façade" in buildings varied in many perspectives abducted by different researchers who had the interest in this specific type of facades and their implementations; according to (Kroner, 1997) it have been generally defined as: the part of intelligent building which is designed interactively to meet the humanitarian, civilized and contextual requirements of the occupants (domestically & globally) in consistence with nature; and by integrating the intelligent technologies with a smart construction forms and compositions which

DOI: 10.1051/matecconf/20166600104

would be compatible and responsive to the inherent occupants social habits, considering the building (and it's parts) life cycle from the aspects of operation, maintenance and the capabilities of upgrading, tuning and recycling (Kroner, 1997).

Meanwhile, as defined by the theses of (Wingginton & Harris, 2002):

- The member that wraps the building in a different manner from the traditional envelop, in terms of capacity of its materials, components and its ability to change the thermal and physical properties and its visual properties (transparency) and color; where it achieves the performance by involving the physics properties that attain smart behavior.
- Dynamic surface that performs balancing the external climate with the internal, by changing its colors and transparency, and by that it forms a great opportunity to control the internal environment in terms of lighting, cooling, heating, ventelation, air quality and noise.

According to the definition of (Belgian Building Research Institute, 2002); it's the active façade that covers one or many floors of the building with a single or double layered glazing that have many treatments and systems that are integrated with advanced devices to enhance the inner environment with passive or dynamic active technologies, that get moderated automatically most of the time by a sophisticated controlling systems. (Sourcebook of the Belgian Building Research Institute BBRI, 2002)

In his definition (John Murphy) emphasized that it's the tunable façade responsive either independently or accumulatively for the various external climate alterations or those internal functional changes. (John Murphy, 2006)

On the other hand, the theses of (Battle& Guy& Christophe, 1999) referred to the same terms considering that intelligent facades would have the ability to tune itself to provide the optimal thermal response for each set of external climate conditions, the operational requirements and type of buildings. (Battle& Guy& Christopher 1994, P.IV)

Theses of (Compagno, 1998) stated that it's the façade that uses a method of self-control for the thermal protection and controlling standards for solar radiation and dynamically adapting for the changes occurring to the lighting and climatic conditions while integrating with the building other services. (Compagno, Andrea, 1999, P.7)

Based on the above, Intelligent Facades in buildings can be defined as: (An essential part of the building which represents the primary and first defensive layer which is dedicated to attain the appropriate balancing between external and internal environment, in accordance to the modern trends in maintaining the resources and energy conservations by integrating it's smart design which has the abilities of: maintenance, upgrade and modification in a smart manner, and it's reliability on smart design/operational technologies in variant physical components and systems which attain active response whether were environmental, optical or decorative and with an efficiency and interaction that approach the behavior of creatures in a nature-simulation process).

3. Integration Of Intelligent Façade Design

Most of the previous studies stated that the integration of intelligent façades with the rest of the building components is vital, unlike the traditional type of façade which forms an individual part of the building. Then the intelligent façade is the part of the building central control system and therefore; differently from the traditional types of facades, it's designed and constructed as if it's a system, this integration is a result of efforts of the several participants in each design process, construction, and operation/maintenance. The studies of (Davies & Others, 2000) stated that importance of making design decisions by all parties which include the beneficiary, designers/architects, systems experts, civil and fire defense

Officials, construction engineers, duration of construction, lighting consultant and value engineers, each of them plays a different role and most of the times they got different goals (even conflicted sometimes) that might make it even harder to make the decisions. (Davies & Others, 2000). On the other hand, building occupants as a main target of beneficiary residents, are the most important part in succeeding the design integration process. As (Fitzgerald & Fitzgerald, 1987, P.248) claimed that occupants are more likely the general reason of failing intelligent systems in buildings, and specifically when it comes to facades and building covering (Fitzgerald & Fitzgerald, 1987, P.248).

The sense of occupants concerning taking control of working environment is vital, where a study interested in the matter of (user satisfaction) on controlling of lighting system, ex.(Velds) found that a user's lack of ability to control the system was the most resentful matter as they claimed although there were systems that control the lighting in the comfort range (Velds, 1999, P.74).

Many various studies showed the effective parameters on integration of the IF design and its participants, where (Houghton, 1999) explained that unfortunately, initial cost is one of the most important considerations in decision making for the IFS, as it diverts the focus from the essential benefits of the investment on costs during the building life cycle, usually it's the most highly sophisticated installations and cost demanding parts of the building would perform better than other simple cheap alternatives, as it would work better while reducing the whole operation cost, maintenance and replacement due to their long life durability. (Houghton, 1999, P.48)

The study of (Lee & Selkowitz, 1995) stated that unwillingness in the large investment from the beginning to do greater conservations in overall expenses later is a weak action plan, where actually is frequently the regular holders of the construction budget (initial cost) are not the same that would operate and responsible for the maintenance and they wouldn't be having the motivation to invest in something that return futuristic benefits while having all the motivations to reduce the investment expenses, so it takes a rich knowledge owner that's got to be smart and aware of solving that conflict. (Lee & Selkowitz, 1995, p.326).

As for the study of (Littlefair, 1990) showed that a specific performance goals might take a primary role in decision making and might be the major motivation and catalyst in the design process of IF, which will have the improving effect on the energy performance and consumption in building, and will make it more precise and better on responses to the surrounding environment, that will increases the user comfort and keep operation expenses to minimum, and offer a better rated LEED standardized building, achieving a better promo advertisement for the building (Littlefair, 1990, P.33).

From all the above it's revealed that the various participants in the IF design process, which have to come up with an agreement and they have certain adjectives that enables them to carry on successfully with the intelligence experience generally in buildings, and specifically in facades, on the other hand the multitude of the parameters effecting the decision-making and dealing with the integration of

IF design, those parameters might be related to investment, political, certain mechanism or other goals, moreover the importance of economic and environmental parameters as overall. As seen in (table1)

<table>
<tr><th colspan="4">Table(1) intelligent façade design integration</th></tr>
<tr><td rowspan="12">The involved parties adjectives</td><td rowspan="3">The Beneficiary (Owner)</td><td colspan="2">Rich with knowledge and awareness of the importance and benefits of the intelligent facade</td></tr>
<tr><td colspan="2">Futuristic thinking</td></tr>
<tr><td colspan="2">bright minded aware of the intelligent façades potential to increase the value of building</td></tr>
<tr><td rowspan="3">Architects and other experts</td><td colspan="2">capable of choosing the appropriate techniques and treatments for the building</td></tr>
<tr><td colspan="2">Controlling the planning, design processes, duration and schedules integration</td></tr>
<tr><td colspan="2">The ability to manage of specialized intelligent systems Parties</td></tr>
<tr><td rowspan="3">The Occupant (User)</td><td colspan="2">Capable of perceiving the intelligent systems in the façades</td></tr>
<tr><td colspan="2">Convinced of the importance of intelligent façades</td></tr>
<tr><td colspan="2">comfort and satisfactory interaction with intelligent façade (control, decision-making)</td></tr>
<tr><td rowspan="2">Maintenance officials</td><td colspan="2">maintenance Professionals</td></tr>
<tr><td colspan="2">Ease of maintenance in terms of the modification, upgrade, replacement</td></tr>
<tr><td colspan="2"></td></tr>
<tr><td rowspan="9">The Affecting Factors</td><td rowspan="5">Goals and objectives</td><td colspan="2">Investment</td></tr>
<tr><td rowspan="2">Political</td><td>public sector Related</td></tr>
<tr><td>Organizations Related</td></tr>
<tr><td rowspan="2">Other</td><td>Objectives and specific performances to achieve publicity and fame of the building</td></tr>
<tr><td>Personal goals and private motivations to achieve certain trends</td></tr>
<tr><td rowspan="2">Economic factors</td><td colspan="2">Initial cost</td></tr>
<tr><td colspan="2">Operating costs and maintenance</td></tr>
<tr><td rowspan="2">Ecological and environmental factors</td><td colspan="2">Achievement of efficient Environmental performance</td></tr>
<tr><td colspan="2">Sustainability achievement</td></tr>
</table>

4. The Technologies Of Intelligent Façade

The studies and theses stated that the association of IF by using the artificial intelligence, and multi-level smart technologies on the design process and on another side, the nature of the façade components and systems.

(Loonen, 2010) stated that the design efforts in finding IF solutions for buildings generally, which adopts smart technologies in façades, by terms of its appearance (exclusively) as a solid wall element, or an external or internal wall surface in addition to the window, moreover to emphasize its importance by being a smart materials and multiple compositional parts which is related to moderate intelligent systems.(Loonen, 2010, P.26-29

Explained by the study of (Stec & Paasen): To achieve functions and performances of the IF it ought to choose all of its type, components, materials and dimensions, moreover, design an operational systems and connect them all together and select an administration strategy to moderate and control the façade as whole. (Stec & Paasen, 2003, P.15)

To fulfill all the above, it requires the designer to be precise in decision making in terms of integration of the design process, that's why we frequently use computer aided means to support our decisions in the design process, and that was claimed by the study of (De Dear & Others, 2002) by showing the capability of the computer aided means to support the decision-making, whereas there are no custom-designed computer-tools to simulate and analyze the performance of the integrated IF and its systems, but some available software can fairly meet the essential job when they employed by reliable staff capable of realizing the potentials and limits of those tools.

The Technologies for the intelligent systems and their software intervene many stages and practical levels of the design process and construction of intelligent buildings in general and their facades in particular where technologies employed as tools mentioned in the design vocabulary, which are used to design facades of this kind to give simulation and an early semi-realistic image of what this façade consist of, and their systems and details, helping a lot with the decision-making.(De Dear & Others, 2002, p.549-554). Also software that operate these facades in a manner that ensures maximum performance, moreover its role in detecting and predicting errors and the possible malefaction during building's lifetime.

Intelligent façade consists of a set of systems that integrates with each other to perform a function of building enveloping with a protective layer that can increase the users comfort while reducing power consumption, it also includes sub-systems which also consists of stand-alone systems. Those systems aren't only for the façade as they're integrated systems that serve the whole building, as it performs functions related to its figuration, and by this it is similar to the human nervous system, where human's skin doesn't work only as protective cover for the body, but in addition of that it plays a role as a sensual-defensive system that feeds the body and nervous system with all the indications and information about the surroundings and help appropriate reactions, so it can't be split off the façade system apart from the building as it works as an integrated part of whole. (Sherbini, K& Krowczyk, R. 2004, P.147).

And here it highlights the importance of smart-materials as one of the primary components to design and construct IF where the term "Smart-Materials" considered relatively new for the materials and the products which have changeable/tweakable properties that can tune its thermal-physics or optical state in response to some of chemical or physical incentives. (Ritter, 2007, P.101). Apart from being only structure, durability and aesthetic properties, the smart-materials applications also requires concentration on what exactly we need to be done.

Sometimes Smart materials are used as bulk-materials, other times their applications can be found in wrapping the outer layer of the material that can heavily impact the performance. (Klooster, 2009, P.63)

All these materials and technologies and their highly advanced mechanism are not enough to create IF that has the ability to achieve comfortable internal climate with performance that can preserve energy unless put in the right formation and appropriate order and assign each technology and material at the right place which would allow it to perform in an integrated fulfilling manner with the rest of materials and systems. There're many factors effecting the façade composition and determining the number of layers the most important among them are the ones related to the form, aesthetic and symbolic and there're the façades with the wide face area or those that are tilted or covered with roofs above large spaces which are curved or somehow inclined and others that influence the designer choices of the façade form and layers which there're two main types:

- Single skin façade
- Double skin façade

From all the above, it's concluded that IF techniques are various by their appearance level, the various Artificial-Intelligence technologies used in it, components that relates to the adopted smart materials and the composite layers and the smart systems relied on as seen in (table2).

Table(2) Intelligent façade technologies	
Techniques appearance levels	
Solid wall	Exterior layer
	Interior layer
Window	Transparent part
	Semi-transparent part

Artificial intelligence techniques			
Design techniques	Graphic Design Programs		AutoCAD/Architectural
			ArchiCAD
			Revit
			IntelliCAD
			TurboCAD
			VectorWorks
			Visio Professional
			Other
	Simulation and standardization programs		DOE-2
			Blast
			EnergyPlus
			ESP-r
			Radiance
			Odeon
			Other
Operational techniques - the cultivation of intelligence	Expert Systems		
	Fuzzy Systems		
	Evolutionary Logarithms		
	Hybrid systems		
	Artificial neural networks		
	Case based reasoning		
	Other		
Intelligent façade components			
Smart materials	Types of smart materials	Type I- property changing materials	Photochromic
			Thermomochromics
			Mechanochromics
			Chemochromics
			Electrochromics
		Type II- Energy-exchanging materials	Light emitting materials
			Materials like PHOTOVOLTAICS, LEDS, TRANSISTORS, THERMOELECTRICS
			Shape Memory Alloys
			Shape Memory Polymers
	Smart materials using location	Facade structure	Tiles
			Beams
			Columns
		Exterior layer	Exterior walls
			Exterior cladding
		Treatment systems	Thermal
			Lighting
			Ventilating
		Interior layer – The inner surface of the façade	
	Smart materials the most used	glass	Active
			Passive
		Concrete	Carbon Fiber Reinforced Concrete
			Transparent Concrete
			Light Transmitting Concrete
			Other
Layers of intelligent façade composition	Single layer	Single skin layer	Single glazed
			Double glazed
		Supportive structure/ environment friendly	Aluminum
			Steel
			Other
	Double layer	Exterior layer	Single glazed
			Double glazed

			Sealed	
			Opened/ Openable	
		In between cavity	Separate	Each floor
				Several Floors
			Building height continuous	
			Ventilation shaft	
			Winter gardens	
			Ventilation Corridor	
			Additional equipment	Solar control devices
				Ventilation
		Interior layer	Single glazed	
			Double glazed	
			Sealed	
			Opened/ Openable	
			Acoustic Barrier	
Intelligent façades systems	Building management system (BMS)			
	Safety and Security Systems	Thermal surveillance cameras		
		Temporary television circuit systems		
		Access Control System	Image recognition system	
			Identification system	
			Personal identification card system	
		Sensory techniques		
		Wireless sensors		
		Alarm Systems		
		Other		
	Environmental Control Systems	Heating systems/ Cooling and ventilation (HVAC)		
		Building Energy Management Systems (BEMS)		
		Louvers and panes solar control systems		
		Other		
	Electrical circuits management systems	Electrical Power Management Systems		
		Cabling Management Systems		
		Lighting Systems		
		Other		

5. Active Response

Studies and theses showed the importance of the active responsiveness of the IF, and that's what been explained in the study of (Wingginton & Harris, 2002): third dimension in the intelligence of IF is related to responsive performance capable of adopting and coping with the environmental fluctuations, the intelligent featured design/construction is dominant on such types of facades, having the most ability to control the surrounding environment by its active responsive behavior towards sudden changes in terms of light, heat, noise, wind and ventilation.(Wingginton & Harris, 2002, P.3)

As for the aesthetics, expressive, fine art and architecture aspects importance, it had been shown in the study of (Hadden & Lee, 2002, P.105) considered the primary outstanding aesthetic part of the building which will give the building its final shape and form, besides its various functional performances responsibilities related to the climate fluctuations.

On the other hand the studies observed the importance of achieving efficiency and interaction in responsive IF actively, as efficiency sets how high the performance of IF that can be achieved in terms of environmental aspects, as reducing HVAC loads must be accompanied by highly energy-efficient HVAC devices, as consumption efficiency of those in addition to lighting systems are of the utmost aspects of the IF objective.

The façade ability to interact with its surroundings is more important than how sophisticated their mechanisms are, according to (Compagno, 1999): IF isn't essentially represented by how deep it got into technologies, instead of that we can figure that IF intelligence is more of how much interactions can be achieved between façade and other buildings services and the environment. (Compagno, 1999, P.Cover)

On the other hand, the concentration of most studies had been on the optimal state to achieve an actively responsive façade associated with life imitations, although this doesn't cancel the important role of Artificial-Intelligence generally, as natural compatibility that animals and plants born with and humane construction proved that it's a fundamental inspiration source in designing responsive IF.

As the study of (Tombazis, 1996) showed that the humane/animals skin is perfect to form a buildings envelope, for it is a multi-leveled responsive and usages, which is multi-layered with self-healing ability and able to regenerate and recover damaged cells, as it has a very variable nature and highly fair and fine exterior surface, that can make it very adaptable to the surroundings fluctuations. Dermal: is the inner layer of the human skin "tight complex composition of rubbery strong fibers, adaptable to the body movement, with a low thickness and flexible on the joints, and thick and tight on other body areas". (Tombazis, 1996, P.53)

From the above, it's concluded that there's a multiple aspects associated with IF active response, which can be narrowed to: active response dimensions, goals and life imitation forms adopted through them. As seen in (table3)

<table>
<tr><th colspan="5">Table(3) Active response</th></tr>
<tr><td rowspan="8">Dimensions</td><td rowspan="4">Environmental</td><td colspan="3">light</td></tr>
<tr><td colspan="3">Thermal</td></tr>
<tr><td colspan="3">Acoustic</td></tr>
<tr><td colspan="3">Air quality – Ventilation</td></tr>
<tr><td rowspan="2">Optical</td><td colspan="3">Optical transmittance</td></tr>
<tr><td colspan="3">Transparency</td></tr>
<tr><td rowspan="2">Aesthetic</td><td colspan="3">Composition</td></tr>
<tr><td colspan="3">Form and finishing</td></tr>
<tr><td rowspan="13">Goals</td><td rowspan="3">Efficiency</td><td rowspan="2">Reduce loads</td><td colspan="2">Air conditioning</td></tr>
<tr><td colspan="2">Artificial lighting</td></tr>
<tr><td colspan="3">Functional performance efficiency</td></tr>
<tr><td rowspan="10">Interaction</td><td rowspan="3">Interaction Parties</td><td colspan="2">external environment</td></tr>
<tr><td colspan="2">Façade/building requirements</td></tr>
<tr><td colspan="2">Occupants</td></tr>
<tr><td rowspan="3">Interaction capabilities</td><td colspan="2">Ability to Interact with variations</td></tr>
<tr><td colspan="2">Ability to Interact with conflicts</td></tr>
<tr><td colspan="2">Ability to Interact with occupants behaviors</td></tr>
<tr><td rowspan="4">Interaction properties</td><td rowspan="3">Adjustment and adaptation</td><td>Perception</td></tr>
<tr><td>Logic</td></tr>
<tr><td>React</td></tr>
<tr><td>Flexibility</td><td>Modification</td></tr>
<tr><td rowspan="2">Life imitation forms</td><td colspan="4">Humane imitation forms</td></tr>
<tr><td colspan="4">Botanical imitation forms</td></tr>
</table>

According to each of the above, essential aspects of IF in buildings can be determined by its design and the adopted technologies into them to achieve the active response. As seen in (figure1)

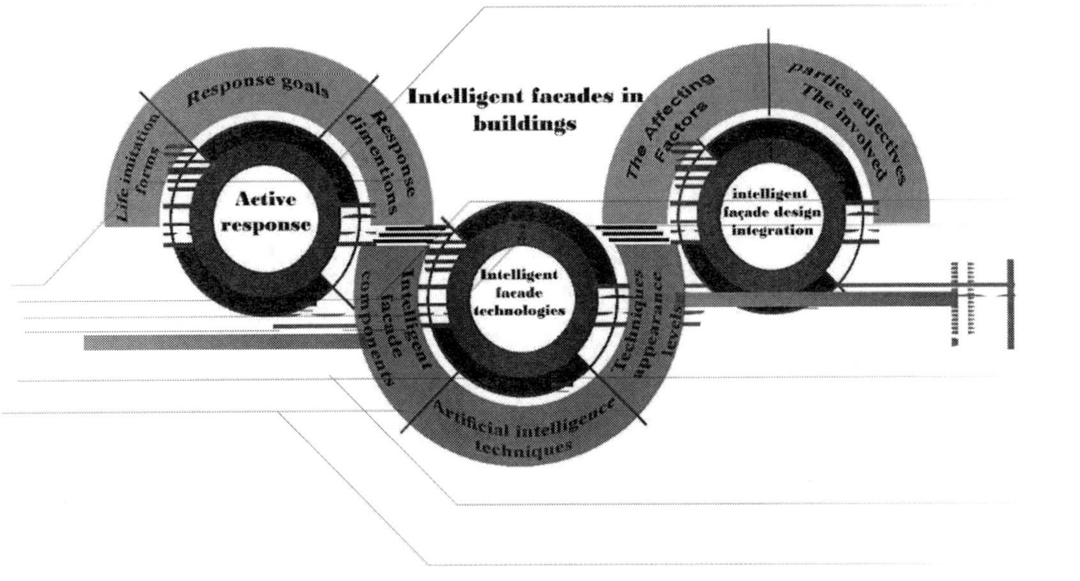

Figure (1) Representation of IF aspects

6. Case Study

Office buildings are featured with their large interest in facades designs and construction in general, and Iraqi local office buildings gives a lot of attention to their facades design as compared to other building parts, particularly, by their persevering attempts to achieve the latest international trends when it comes to energy and resources conservation and control and direct the use of it by distinguished performance and economical plans compatible with the sustainability concept.

So it's to elect two local projects that can diagnosis the existing state of how far they were into adopting intelligence in their building facades in terms of particularly the technologies and responsiveness aspects, by using auditing forms based on the details concerning each of the projects, down to suggesting some treatments which through them we can develop and upgrade these buildings in line with global trends.

6-1 Building of the Ministry of Construction and Housing - Baghdad/Iraq 2009-2013

An office building located on the Baghdad the Iraqi capital, designed by the (Iraqi national engineering consultations center) and constructed by Al-mansour Contracting co., consisting from 10 floors and a basement in addition of a multi-purpose hall in the ground floor. See (Figure 2)

Main Facade | Secondary Entrance | Building & Hall perspective

DOI: 10.1051/matecconf/20166600104

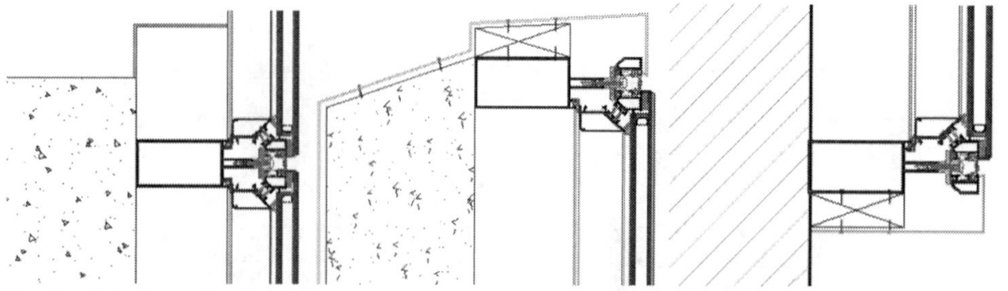

Intersection detail between façade and floors

Upper end intersection detail

Detailing of intersection of the ground floor with the façade

Figure2- Ministry of Construction and Housing Details

The construction company CEO (Zidan khalaf laeiby) claimed that the green design of the ministry building which is a first in Iraq in terms of sustainability and environmental friendly, where high standardized global specifications and features concerning interior and exterior design where adopted through the construction of this building, that would conserve energy and increase the occupant comfort by using double layering glazed and aluminum composite panels in the exterior cladding of the building in addition of the HVAC-system which is not entirely depending on the Central Package-Unit due to the exploitation of the solar radiation in the winter, and by using colored(tinted) glass that helps to preserve temperature inside the building which decreases the required operational energy.

Interests in the facades can be obvious through:

- Design team employed many forming and modelling software to render the required building plans sets, although they didn't use any operational artificial-intelligence generally.
- Facades are designed to be a single skin with double glazing supported by aluminum structure, except for the corners that reinforced with a solid material, cladded by aluminum composite panels fixed with a building-high steel structure.
- The building had been supplied with a central HVAC package-unit, in addition of MPS systems, with many security/safety alarm systems like surveillance cameras systems, TV-circuits, smoke detectors and fire alarms.
- Building design hasn't been iconic or symbolic but it focused on the practical functional form that tends to provide efficient comfortable office spaces to work, and could successfully offer optical transparency towards the surroundings using glass as the main cladding materials for large area of the façade.
- Design had provided solutions limited to environmental aspects to offer treatment for the solar radiation and reduce how it effects the inside climate, in an attempt to reduce the need to use the cooling devices to amend the effect of the solar
- heating due to the large glazing area used in such projects.

6-2 Building of the Sunni Endowment Bureau – Baghdad/Iraq 2010-2014

An office building located in the Baghdad designed by the engineering consultation bureau/University of technology, constructed by (Al-Futtaim Carillion co.ltd.), the building consists of four floors and a basement, see (figure3)

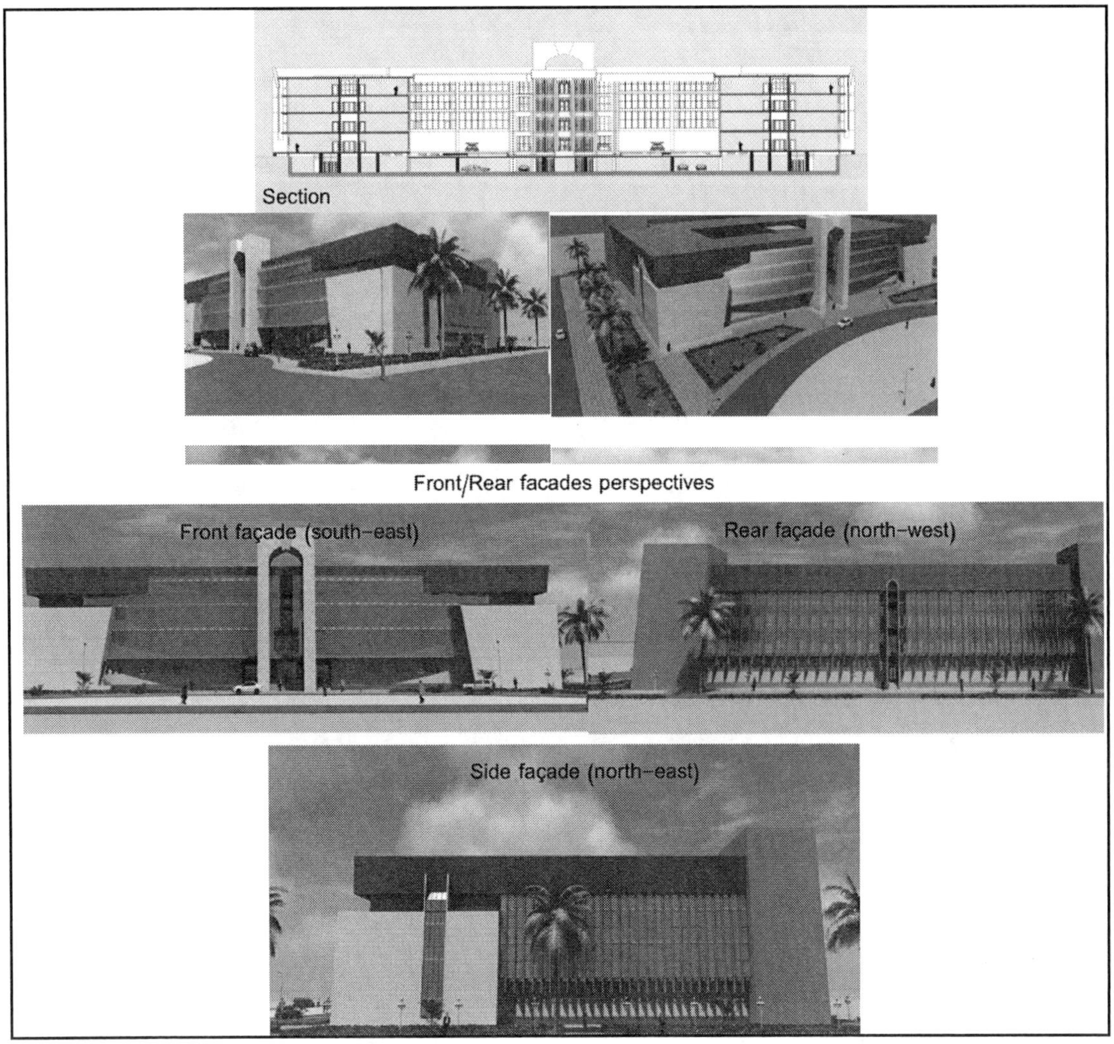

Figure3 - Sunni Endowment Bureau Details

The building have been designed to follow sustainability principles and to provide the required adjustment and interaction with the surrounding environment and that's why it had the tradition Baghdadi-style design based on the principle of upgrading the traditional architecture and invest them through advanced modern design and adopting traditional elements to serve the building rating towards environmental aspects, and for this many climatic and environmental treatments can be detected around the building parts, among them:

- Front façade, hanged balconies and the thermal regulator.
- Central courtyard.
- Using stack-effect to draw used air to the upper façade where it tossed away is similar to the Shaft-façade principle.

Furthermore there're many designed treatments and techniques that had not seen the light through building construction stage due to various decision-making related issues, among them are the roof garden, that have been designed to serve an environmental purposes to reduce the building impact on mother nature and fulfill the green land that've been lost.

It's clearly observed that there have been particular interest in façades by:

- Using many modelling and graphic software to produce the project design sets and to mold the building mass that've been studied thoroughly and to form a semi-realistic formation before the construction stage, while the intelligent operational techniques have been completely neglected.
- Using glass as a primary material for the façade due to the beneficiary client desire, in addition of the aluminum composite panels mounted and fixed by a steel structure.
- The façade had been designed over the upper three stories as a double skin façade using the principle of Shaft-façade that had many lighting and thermal treatments used to perform natural ventilation.

- Stack effect had been exploited to circulate air between the façade layers in which hot air rises up towards the opened gates in the upper part of the façade that would be open on summer to put away used air while be sealed on winter.
- Main façade of the project (south-east) is featured with a thermal regulator (solar heat regulator) which is a prominent and cantilevered glazing assembly, in which offered a solution that flips an issue into benefit.
- Many solutions and treatments have been adopted, including safety-security systems, fire-alarm and smoke detectors in addition of a HVAC system that helps stabilizing the inner climate and support the environmental treatments employed in the project to increase the occupant comfort.
- regarding the facades of the building they are homogeneous and consistent although there have been several treatments and various solutions, with the wrapping glazing dominating the outer part of the building reinforced by frames of solid masses around the glass areas.
- Project's design solutions concentrated on the environmental aspect in general, and on controlling the solar radiation impact on the interior spaces particularly by using glazed balconies and thermal regulator, as they would act as a safety-valve in dealing with the hot/dry surrounding climate circumstances.

7. Results

Results of the case study analysis generally revealed the achievement of many aspects that've been adopted by the elected projects in the terms of intelligent techniques appearance level and the artificial intelligence techniques and the intelligent façade components, as shown in figures (4-11)

Application results also showed a public inquiry elected projects effective responses and varying levels according to the dimensions of the response, objectives and forms of bio-representation, as shown in Figures (12-16)

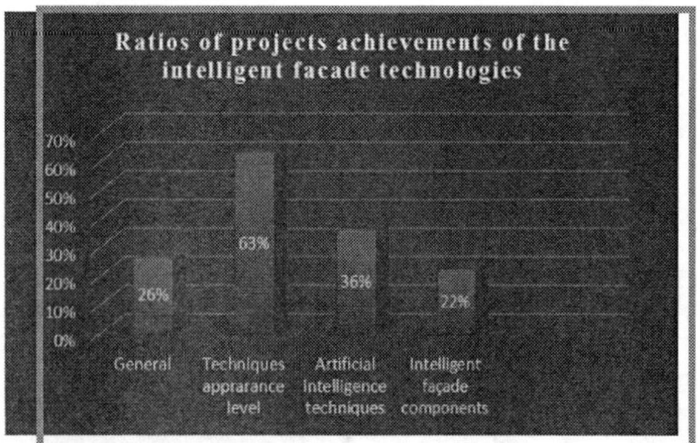

Figure(4)Ratios of projects achievements of the intelligent façade technologies

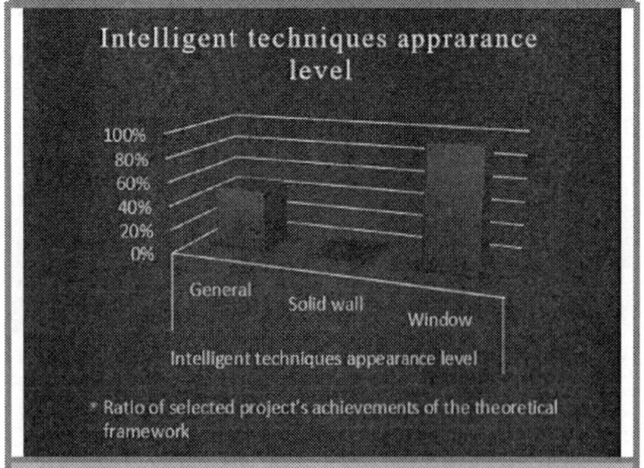

Figure(5)Intelligent techniques appearance

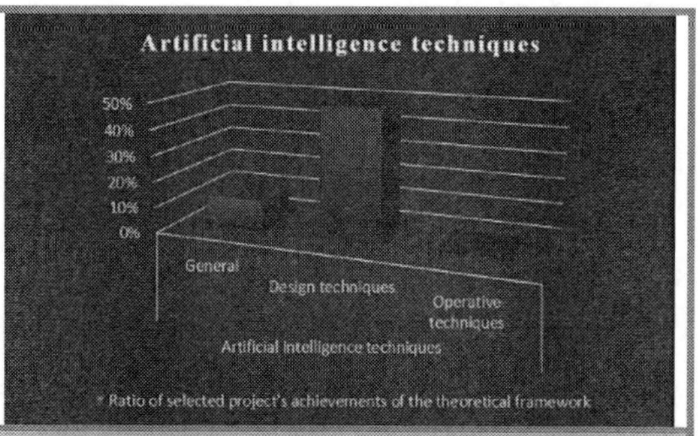

Figure(6)artificial intelligence techniques

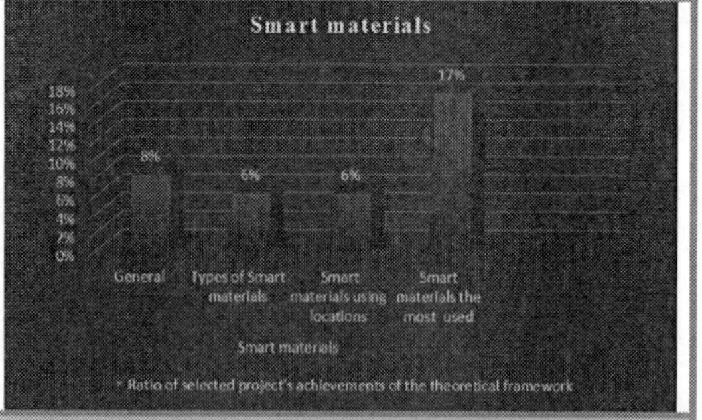

Figure(7)smart materials

DOI: 10.1051/matecconf/20166600104

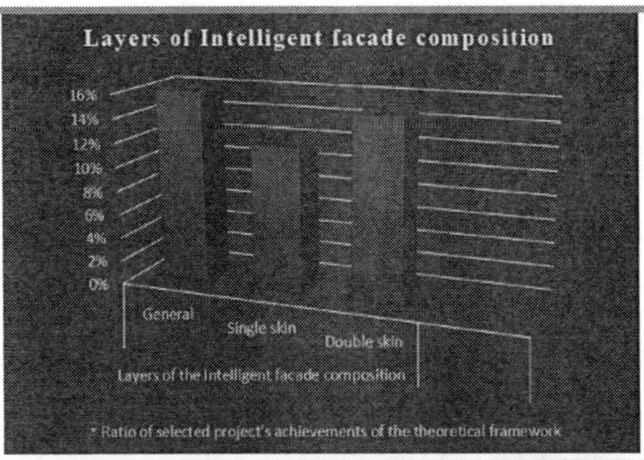

Figure(8)layers of intelligent façade composition

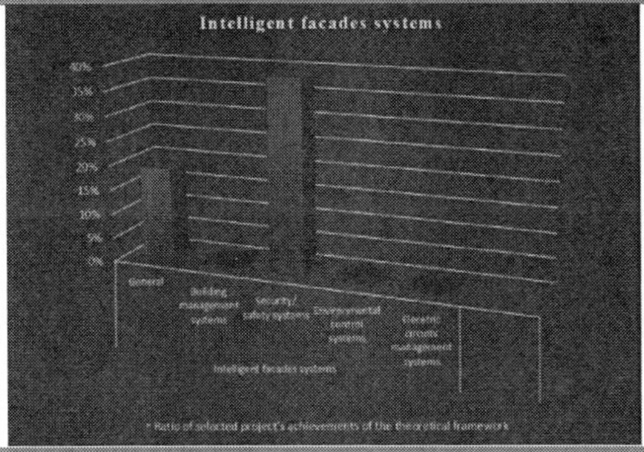

Figure(9)intelligent façade systems

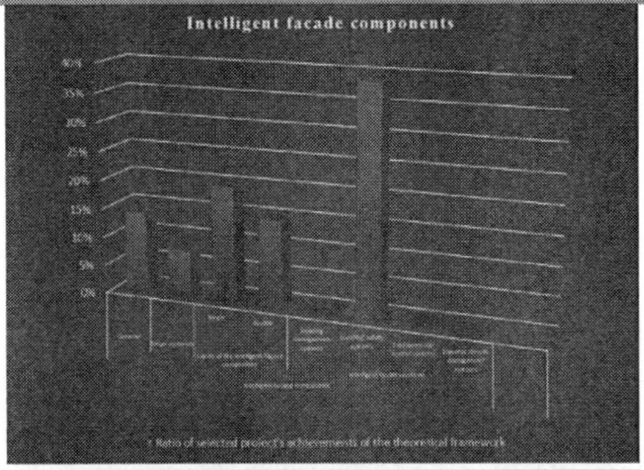

Figure(10)intelligent façade components

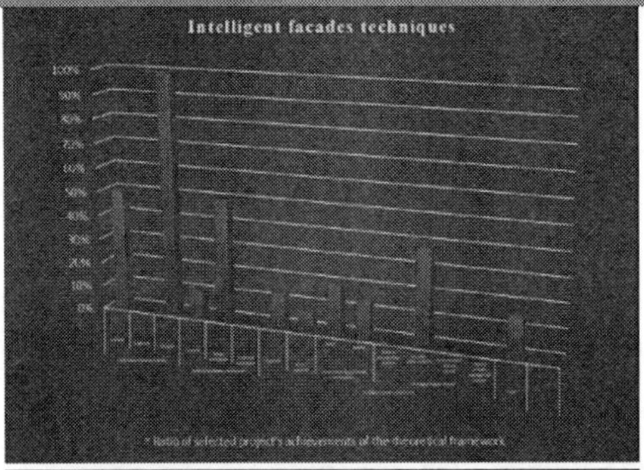

Figure(11)intelligent façade techniques

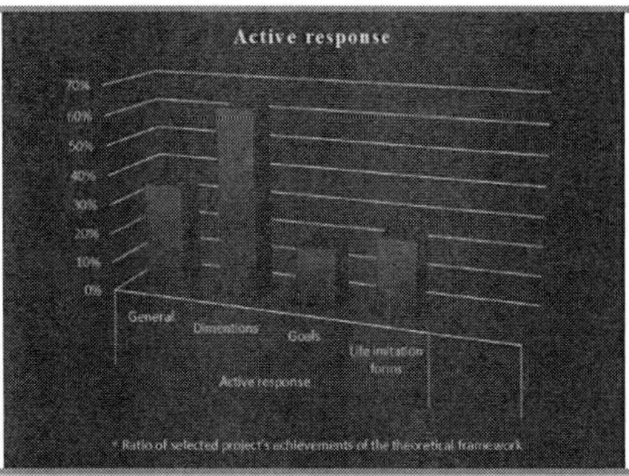

Figure(12)active facade

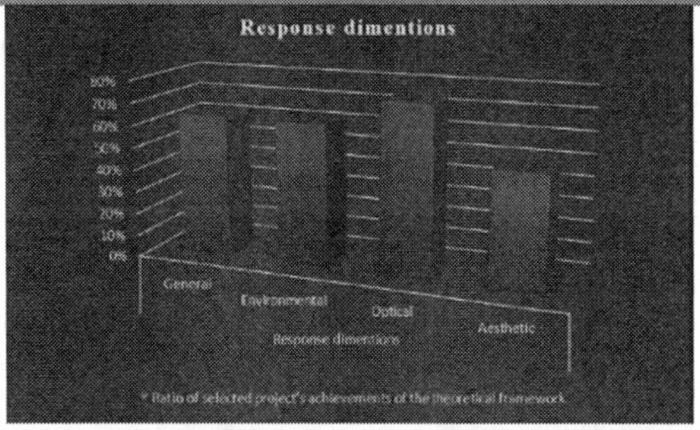

Figure(13)response dimensions

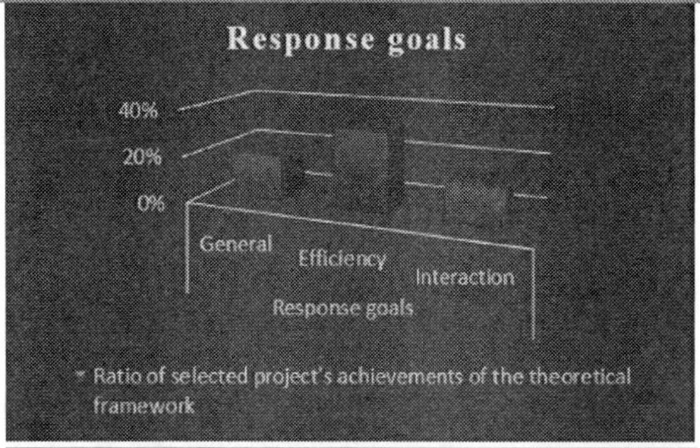

Figure(14)response goals

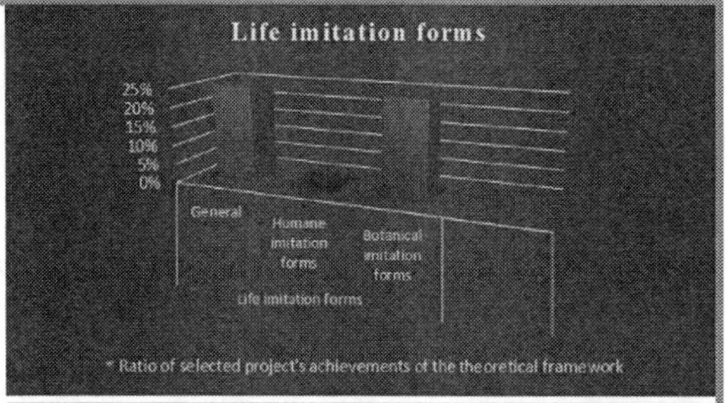

Figure(15)life imitation forms

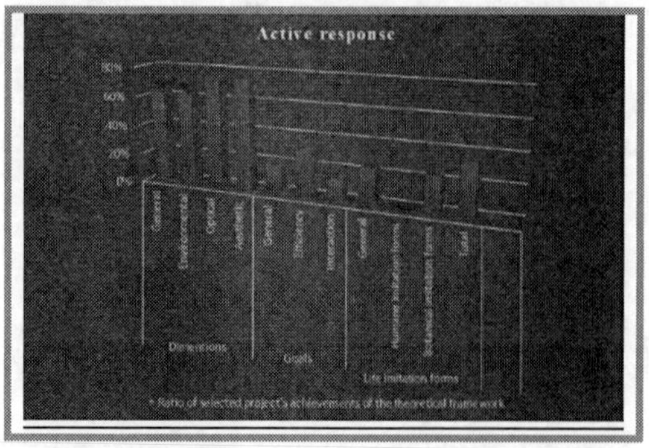

Figure(16)active response

8. Conclusions

The practical case study showed that local projects concern and treatments are limited, as IF technologies are limited in terms of techniques appearance level that've been represented by only the transparent & semi-transparent elements, while AI had only been adopted as a design-techniques and software merely through the design stage, while it has a very limited smart technologies and materials, with a simple superficial treatments for the composing layers weather it was single or double skin, and their role was limited to some safety/security systems.

Moreover, the study made evident that local elected projects achievements in terms of environmental, optical and aesthetical was limited to reducing loads of lighting and air-conditioning, with limited interaction restricted to the exterior environment and the building and it's cover requirements with a confined abilities to cope with fluctuations.

9. Recommendations

Investment can be identified by the research various IF aspects to set some recommendations for the purpose of development and modernization of the facades of the local office buildings, as follows:

- Using isolation techniques that had already been dealt with on the solid-wall level of internal and external layer or focus on one of them to increase the isolation of the building and reduce the possibility of energy loss.
- To use AI-techniques, particularly the ones concerning operational to control energy consumption and to maintain a stable indoor climate, and that by using the techniques of connecting systems is to invest the principle of artificial intelligence management systems through the use of expert supportive software techniques and algorithms developed and case based reasoning and others that increase the intelligence of the building in general, and the Façade in particular.
- The use of various kinds of smart materials, according to the required performances to increase the efficiency of the building façade, and the use of these materials should be at various levels previously mentioned as panels or in the exterior walls or as cladding or treatments-devices and other systems, also it's recommended to use such smart materials on the inner layer of the façade contacting the internal space and by each one of these materials features.The research also recommends the using of active glass as being more efficient and with improved featured performance, capable of adapting to a variety of different circumstances, and that does not prevent the use of the passive glass in some areas that do not need to use the active glass.
- Providing development option of the façade by adding an extra layer to the existing façade to make it a double skin facade if there's capability and space around the building to apply one of the various DSF formations using the ventilation shaft (used already in the Sunni Endowment Bureau) or ventilation corridor, where it's difficult to use the (Winter Gardens) in the development of projects facades due to lack of space required for it most of the time, It's also preferred to use louvers and solar shading devices that can be added to the cavity between the two layers of the facade systems.
  In addition, the acoustic barrier techniques capabilities can be invested to the inner layer of the DSF to reduce the impact of external noise on workers inside the building, especially in the Ministry of Housing building.
- The research also recommends to use the latest of what has been invented from the BMS, which are the skeleton of controlling the smart technologies and equipment used to control performance and the work of each of them individually, and coordinate their work compatibly as an integrated system with other devices, where these systems performs integrally by multiple nodes each of them serves aside or a section of the building meets cumulatively towards the main CPU controls the various parties to the performance events.
  And it's best to use those BMS to run the building affairs as a whole, where façade would be one of the building's parts that will actually integrated with other parts forming a system similar to that of the human body which will produce a humane imitation form later, where sensing and surveillance devices would be scattered all over the façade surfaces providing data of the surroundings that will be analyzed and by them can configure the parts performances quickly enough to ensure minimizing the possible damage, these systems also include software subsets that enables them to predict and estimate the expected climate circumstances

proactively relying on electronic meteorological systems and the adoption of the specified location systems and legibility climatic or using AI that increase building systems experience in dealing with prior experiences of occurrence and form a set of replies systems prior to prepare acts and depending on the development and progression of building management systems used.

- The research recommends to add environmental control systems as HVAC systems, BMS, louvers and solar control/shading devices, electric management systems, lighting systems and techniques and any other related system to those performances as they represent the most important systems capable of developing and improving of the façade performance in terms of energy consumption and occupant comfort and inner environment protection.
- The use of advanced artificial lighting systems, various techniques are working to reduce energy consumption and avoid an increase in thermal loads of inner space by studying lighting systems and their side-produced heat and its impact on the aesthetic inner space.

References

1. Battle and McCarthy [AD] Battle, Guy and McCarthy, Christopher, "Intelligent Cities – A Climate for Change, Architectural Design", Vol. 66, No. 9/10, Multi Source Synthesis, March/April 1994.
2. Compagno, A., "Intelligent glass facades: material, practice, design", Basel: Birkhäuser Verlag., 1999.
3. David Hadden & Andy Lee, "The Role of External Façade in Protecting Building occupants against Terrorism and Impacts", London, 2002. Can be found here - www.psdas.gov.hk/content/doc/2002-1-20/Day%201%20-%20Ir.Dr.%20Andy%20LEE%20-%202002-1-20.pdf
4. Davies, M., A. Jackaway, R. Hardy, E. Dewey, and J. Littler, "A translucent louvre system: design concepts, modeling work and monitored data", Building Research & Information, 2000.
5. De Dear, R. and G.S. Brager, "Thermal comfort in naturally ventilated buildings", revisions to ASHRAE Standard 55, Energy and Buildings **34**(6), 2002.
6. Fitzgerald, J., & Fitzgerald, A.F., "Fundamentals of systems analysis. Using structured analysis and design techniques", John Wiley & Sons. 3rd edition (1st edition in 1973), New York, 1987.
7. Houghton, D., "Here comes the sun - A look at daylighting system", Architectural Lighting. **14**(1), 1999.
8. John Murphy, "Samrt Dedicated outdoor Air systems", ASHRAE Journal, July-2006
9. Klooster, T., "Smart Surfaces—and their Application in Architecture and Design", Basel, Birkhäuser, 2009.
10. Kroner, W.M., "An intelligent and responsive architecture. Automation in Construction", Elsevier Science Ltd, 1997.
11. Lee, E.S. and S.E. Selkowitz., "The design and evaluation of integrated envelope and lighting control strategies for commercial buildings", ASHRAE Transactions **101**(1), 1995.
12. Littlefair, P.J., "Innovative daylighting: Review of systems and evaluation methods", Lighting Research and Technology **22**(1), 1990.
13. Loonen R., "Climative Adaptive Building Shells", Technische Universiteit Eindhoven, 2010.
14. Ritter, A., "Smart Materials in Architecture, Interior Architecture and Design", Architectural Press, Berlin, 2007.
15. Sherbini, K & Krowczyk, R., "Overview of Intelligent Architecture", 1stASCAAD International Conference, E-Design in Architecture, Dhahran, Saudi Arabia, 2004.
16. Stec, W., & van Paassen, A.H.C., "Integration of the Double Skin Façade with the buildings, Energy in Built Environment", Energy Technology, TU Delft, Mekelweg 2, 2628 CD, Delft, The Netherlands, 2003.
17. Tombazis, A.N., "On skins and other preoccupations of architectural design", Renewable Energy *8*(1-4), Elsevier Science Ltd., 1996.
18. Velds, M., "Assessment of lighting quality in office rooms with daylighting systems", Ph.D. dissertation. Delft: Technische Universiteit Delft, Dutchaland, 1999.
19. Wingginton, M & Harris, J., "Intelligent Skins", Architectural Press, an Imprint of Elsevier, Linacre House, Jordan Hill, Oxford, UK, 2002.

MATEC Web of Conferences **66**, 00105 (2016) DOI: 10.1051/matecconf/20166600105
IBCC 2016

Hazard Identification, Risk Assessment and Risk Control (HIRARC) Accidents at Power Plant

Asmalia Che Ahmad[1a], Ida Nianti Mohd Zin[1], Muhammad Kamil Othman[1], Nurul Huda Muhamad[1]

[1]*Faculty of Architecture, Planning and Surveying, Universiti Teknologi Mara (Perak),Seri Iskandar, Malaysia*

Abstract. Power plant had a reputation of being one of the most hazardous workplace environments. Workers in the power plant face many safety risks due to the nature of the job. Although power plants are safer nowadays since the industry has urged the employer to improve their employees' safety, the employees still stumble upon many hazards thus accidents at workplace. The aim of the present study is to investigate work related accidents at power plants based on HIRARC (Hazard Identification, Risk Assessment and Risk Control) process. The data were collected at two coal-fired power plant located in Malaysia. The finding of the study identified hazards and assess risk relate to accidents occurred at the power plants. The finding of the study suggested the possible control measures and corrective actions to reduce or eliminate the risk that can be used by power plant in preventing accidents from occurred

1 Introduction

Power plant is a place to generate energy of electricity for millions of household and industries. There are many types of power plant all over the world such as gas turbine power plant, steam turbine power plant, fossil-fuel power plant, nuclear power plant, and hydro power plant.

Power plant is a workplace of the high level from hazard, where safe working procedures are important as workers face many risks due to the nature of the job. The work environment is often hot and noisy, consist several major equipment and operations involve in its process, and there is an always present risk for crushing injuries, electrical shocks and burns, boiler fires and explosions, and contact with hazardous chemicals. Figures from International Labour Organization (ILO) (2014)[1] showed that more than 2.3 million deaths per year are recorded. This shows that accident happen every 15 second a worker dies from work related to accident or illness. Moreover, at every 15 seconds 160 workers have accidents. That means 6300 people die as results of working accidents or work diseases every day. Due to the hazardous operation of power plant, this workplace is exposed to accident and near miss incidents (Cohen and Colligan, 1998). In addition, OECD (2008) [2] reported that more than 2500 people are killed every year in energy related plant with severe accidents. This number appears to be rising as energy usage continues to increase every year. Between 1969 and 2000, there were 1870 such accidents with five fatalities. The largest number of immediate fatalities in the fossil energy chains was in coal and oil, i.e. 2259 and 3713 respectively and in OECD countries, and 18017 and 16505 respectively in non-OECD countries. Table 1 summarises the severe accidents in the fossil, hydro and nuclear energy chains in the period of 1969-2000.

Due to the risky workplaces that power plants constitute, the companies need to assure safe working conditions through systematic and regular hazard identification and risk assessment. Safety procedures and regulations need to be followed by the management as well as the workers. Many business enterprises have proven that good safety management leads to increased productivity, and the same works for power plant. By having a good safety management program, not only minor injuries can be avoided, but also other incidents that are costly, time consuming, stressful and inconvenient. Due to this reason for reduction of all occupational diseases, injuries/fatalities, corrective and preventive measure should be done. OHSAS 18001:2007 (clause 4.3.1), suggested that a safety procedure is a must to ensure the identification of hazard, risk assessment, and its determination to eliminate, reduce or control the possibility for any accidents to occur. Thus, this paper highlights report on hazard identification, risk assessment and risk control applied in the coal-fired power plant located in Malaysia. It includes the methodological steps to identify hazard, assess the risk level of the hazards and apply or suggest the possible control measures and corrective actions to reduce or eliminate the risk.

Table 1: Summary of manmade energy related power plant accidents for the period of 1969-2000 [1]

[a] Corresponding author: asmalia809@perak.uitm.edu.my

	OECD			Non-OECD		
Energy chain	Accidents	Fatalities	Fatalities/ GWey	Accidents	Fatalities	Fatalities/ GWey
Coal	75	2 259	0.157	1 044	18 017	0.597
Coal (data for China 1994-1999)				819	11 334	6.169
Coal (without China)				102	4831	0.597
Oil	165	3 713	0.132	232	16 505	0.897
Natural Gas	90	1 043	0.085	45	1 000	0.111
LPG	59	1 905	1.957	46	2 016	14.896
Hydro	1	14	0.003	10	29 924	10.285
Nuclear	0	0	-	1	31*	0.048
Total	390	8 934		1 480	72 324	

2 Accidents at Power Plant

An accident is defined as an unfortunate that happens unpredictably and inadvertently, resulting in harm or injury [2]. The accident occurs depending on the situations, either results in minor fatalities or have the potential for but do not result in a loss [3]. Accidents are more usually related with event on site leading to adverse significances for the establishments like leaks, fire and explosions [4]. In another study, [5] found that injury rates are contrast with business cycle which means when economic sector increases, the number of injuries will increase too and vice versa. [6] relate injury rates to structural effects such as workers characteristic, firm size, production process, types of industry and types of occupation.

3 HIRARC

In recent years, Hazard Identification, Risk Assessment and Risk Control (HIRARC) have become fundamental to the practice of planning, management and the operation of a business as a basic of risk management. A hazard identification and risk assessment are processes used to identify and evaluate both existing and potential hazards on a worksite and the methods used to control or eliminate the hazards identified. The organizations that have carried out risk assessment at the work place have noted numerous changes in their working practice. Those who have already carried out risk assessment in their work, have reported positive changes in their working practice, they recognize substandard act and working condition as they develop and take necessary corrective action. Legislation requires that this process should be systematic and be recorded so that the results are reliable and the analysis complete. The risk assessment process should be continuous and should not be regarded as a one-off exercise.

3.1 Hazard identification

Hazard is defined as anything (e.g. condition, situation, practice, behaviour) that has the potential to cause harm, including injury, disease, death, environmental, property and equipment damage. This is the process of examining each work area and work task for the purpose of identifying all the hazards which are "inherent in the job".

3.2 Risk Assessment

Hazard is closed relationship with the risk. Risk is a measurement to analyse and evaluate the hazard [7]. The measurement is made by identification on how severe and when likely of the hazard. In other words, the risk assessment is an in-depth look to specify situations, process and other harmful activities or hazard at workplace. Risk is presented in variety of ways to communicate the distribution of the risk throughout a plant and area in a workplace. The results of risk assessment that presented in a risk matrix are essential to make decision on risk control.

Risk can be calculated using the following formula [7]:

$$\text{Risk (R)} = \text{Likelihood (L)} \times \text{Severity (S)}$$

In order to determine the risk level, the risk assessment matrix has to be resolve. The computations of the risk assessment matrix in order to attain the risk level consist of following steps:

1. Calculation of weightage average index (WAI) for likelihood of hazards
2. Calculation of weightage average index (WAI) for severity of hazards
3. Calculation of risk (WAI likelihood x WAI severity) of hazard
4. Determination of risk assessment matrix (Table 2) and risk level (Table 3)

The weightage average index is calculated based on the formula as below [7] :

$$\text{Weightage Average Index (WAI)} = \frac{\sum w_i f_i}{\sum f_i}$$

where:

w_i = assigned weight for a particular class under 5-point likert scale

f_i = corresponding frequency of that particular class

i = 1, 2, 3, 4, 5 that illustrated 5-point likert scale according to specific context

The computation of WAI for each hazard was interpreted by schedule of risk assessment matrix (Table 2) and the indication of risk level in Table 3. Then result from calculation of risk assessment was referred to scale of risk level.

Table 2: Risk Assessment Matrix level [7]

LIKELIHOOD OF HAZARDS	SEVERITY OF HAZARDS				
	INSIGNIFICANT (1)	MINOR (2)	MODERATE (3)	MAJOR (4)	FATAL (5)
RARE (1)	1	2	3	4	5
UNLIKELY (2)	2	4	6	8	10
POSSIBLE (3)	3	6	9	12	15
LIKELY (4)	4	8	12	16	20
ALMOST CERTAIN (5)	5	10	15	20	25

Table 3. Indication of risk level [15].

RISK LEVEL	
1 to 2	LOW
3 to 6	MEDIUM
7 to 12	HIGH
More than 12	EXTREME

The purpose of hazard identification and risk assessment in this study is to highlight the critical operations of tasks that poses significant risks to the health and safety of employees as well as highlighting those hazards pertaining to certain equipment due to energy sources, working conditions or activities performed. Then risk level is assigned to each hazard for identifying required corrective action to minimize the risk or eliminate the hazard. This study was conducted according to basic of safety procedure which is hazard identification and risk survey in order to assess the risk in the plant.

3.3 Risk Control

Risk control is the elimination or inactivation of hazard in a manner such that the hazard does not pose a risk to workers. Hazards should be controlled at their source where the problem is created. Venturini (2011) rating four characteristics of risk that are commonly identified (refer Table 4), according to the measures taken to protect the plant from risk exposure are:

i. Excellent: the facility has taken measures exceeding industry standards and the best practices. Loss potential is considered significantly reduced.
ii. Good: the fatality has taken measures that are consistent with industry standards and best practices. Loss potential is considered to be average.
iii. Fair: the fatality has taken some measures that approach industry standards and best practices, however deficiencies exist. Loss potential is considered somewhat increased.
iv. Poor: the fatality has major deficiencies and does not approach industry standards and best practices. Loss potential is considered to be significantly increased.

4 Research Methodology

To further investigate the hazard identification, risk assessment and risk control applied in the power plant, this study adopted questionnaire survey and interviews for data collection. The data were collected in two (2) coal-fired power plants. A total of 50 questionnaires were sent. Within the collected questionnaires, 17 forms were uncompleted and were thus disregarded, while 30 forms were completed. Two accident cases at power plant were analyzed. The name and location of the power plant cannot be exposed due to the private and confidentiality. Data obtained by the questionnaire and personal interview was subjected to quantitative analysis. The percentage was calculated by simple mathematical formulas. The data from the questionnaire survey were analysed using Excel to compose weightage average index (WAI) in order to compute the risk level of those hazards. Semi-structured interviews with 30 employees were used to gather perceptions regarding the situation of safety at their workplace.

Table 5 shows that 16 (53%) of the respondents who participated in this study were engineers, 4 (13%) were project managers, and 5 (17%) were technician and general workers accordingly. The range of personal experience of the respondents in terms of number of years involved or worked in power plant between 1 to more than 20 years as indicated in Table 6. More than half (46.67%) of respondents have more than 15 years experience. Whilst, 23% of the respondents have more than 10 years working experience and are all familiar with the main issues surrounding safety at their workplace.

Table 4: Control Measure Matrix

CONTROL MEASURES	LEVEL	
Control measures exceeding industry standards and the best practices. Loss potential is considered significantly reduced	Excellent	3.1 -4.0
Control measures that are consistent with industry standards and best practices. Loss potential is considered to be average	Good	2.1 -3.0
Control measures that approach industry standards and best practices, however deficiencies exist. Loss potential is considered somewhat increased	Fair	1.1- 2.0
A control measure has major deficiencies and does not approach industry standards and best practices. Loss potential is considered to be significantly increased.	Poof	0 – 1.0

Table 5 Current job title/position of the respondents for the two selected coal-fired power plant.

Job position	Frequency	Percentage %
Project manager	4	13
Engineer	16	53
Technician	5	17
General workers	5	17
Total	30	100

Table 6 : Length of time involved/worked in the power plant

Working experience with power plant	Frequency	Percentage %
1-5	6	20
5-10	3	10
11-15	8	23.33
16-20	9	30
>20	5	16.67
Total	30	100

As can be seen from the Table 5 & Table 6, it can be conclude that the respondents who answered the questionnaires and interviews are those have enough knowledge about the safety at their workplaces; with extensive working experiences and those holding responsible positions in the power plants. Their answers to the questionnaires and interviews give a high level of credibility in terms of their opinions regarding the issues raised.

5 Case Study

The study involved two coal-fired power plants that were established to meet the anticipated growth in electricity demand, and help maintain the required power reserve margin. The design, development and management of the power plant, in line with national strategy to produce heavy dependence on natural gas as fuel source for power generation. It is designed to operate within minimizing power plant impact on the environment.

5.1. Report on Accident Case 1

Accident 1 occurred at shoreline access road. The type of accident is classified property damage. Nature of injury is right knee swollen. The victim of the accident is a general worker and has 3 years experience. The accident also has one witness from the department at the accident location. Table 7 shows the WAI likelihood, WAI severity and risk assessment matrix for the four identified hazard. Four hazards have been identified in accident case 1 as per list follow below:

1) No inspection on rented machinery.
2) Misunderstanding of communication with security staff.
3) Rented forklift driver without competency certification.
4) No HIRAC on scrap disposal activity.

The risk was calculated by multiplying likelihood with severity. Each figure of the risk was categorised into risk level and rank accordingly. Out of four types of hazards identified, 2 hazards were classified under extreme risk while another two were in high risk. None of the hazards falls in either medium or low risk level.

Table 7: Summary of hazards identification, risk assessement , and control measure for Case 1.

NO.	HAZARD IDENTIFICATION CASE 1	RISK ASSESSMENT				RANK	CONTROL MEASURES	Level
		LIKEL-HOOD (MEAN)	SEVERITY (MEAN)	RISK	RISK LEVEL			
1	No inspection on rented machinery.	5.00	2.17	10.87	High	3	Brief security on the procedure requirement regarding rented forklift and machinery inspection.	Excellent
2	Misunderstanding of communication with security staff.	4.80	2.14	10.27	High	4	Review procedure for forklift operation certification and operating to include vehicle inspection	Good
3	Rented forklift driver without competency certification.	4.60	2.85	13.10	Extreme	2	Brief security on the correct way of using walkie-talkie.	Excellent
4	No HIRAC on scrap disposal activity.	4.93	2.70	13.30	Extreme	1	Prepare HIRAC for scrap disposal activity.	Excellent

5.2 Report on Accident Case 2

Accident 2 occurred at combustion chamber Gas Turbine. The type of accident is classified medical treatment injury. Nature of injury is hit by falling burner at the back of right hand. The victim of the accident is a man as a general worker and has no experience. The accident also has one witness from the department at the accident location. Five (5) hazards at power plant were identified in accident number 2 as per list follow below:

1) Two different works was performed at same time.
2) Working space constrained.
3) Design and setting up on equipment is not proper.
4) No inspection checklist on the bucket prior use.
5) Lack of precaution on this activity.

All five types of hazards identified were classified under high risk. None of the hazards falls in extreme risk level (Table 8).

7 Conclusion

In this paper we investigate and describe work related accidents at power plants based on HIRARC (Hazard Identification, Risk Assessment and Risk Control) process. The risk ranking of the present and possible hazard is evaluated which divide them into low, medium, high and extreme risk level. Which risks are in high and extreme level there possible corrective action also recommended for risk control. Power plant need to assess the risk not only for the risk and control measures but it will conducted for cost if occurred accidents and near miss incidents at power plant. Hazard identification, risk assessment and control is an on-going process. Therefore, regularly review the effectiveness of hazard assessment and control measures is needed.

Table 8: Summary of hazards identification, risk assessement , and control measure for Case 2.

NO	HAZARD IDENTIFICATION CASE 2	RISK ASSESSMENT				RANK	CONTROL MEASURES	Level
		LIKELI-HOOD (MEAN)	SEVERITY (MEAN)	RISK	RISK LEVEL			
1	Two different works was performed at same time.	4.00	3.00	12.00	HIGH	4	Check the bucket floor and increase height of the floor guard	Excellent
2	Working space constrained	4.30	3.42	14.70	HIGH	1	Install extra plywood guard in the middle of the bucket	Excellent
3	Design and setting up on equipment is not proper.	3.70	3.72	13.77	HIGH	2	Inspection before start the inspection must be done	Excellent
4	No inspection checklist on the bucket prior use.	4.50	2.40	10.80	HIGH	5	Bucket to be placed on the split line area for every inspection work to ensure the stability and reduce impact from bucket or burner if fall	Excellent
5	Lack of precaution on cycle of removing burner at combustion chamber.	3.90	3.20	12.50	HIGH	3	Supervision to ensure safety precaution to be taken	Excellent

DOI: 10.1051/matecconf/20166600105

References

1. ILO: www.ilo.org/global/about-the-ilo/newsroom/features/WCMS_076282/lang--en/index.htmOECD (2008),
2. OECD Environmental Outlook, OECD, Paris, France.
3. Kleindorfer, P., Oktem, U., Pariyani, A. and Seider, W. (2012).Assessment of catastrophe risk and potential losses in industry.Computers & Chemical Engineering, 47, pp.85-96.
4. Konstandinidou, M., Nivolianitou, Z., Markatos, N. And Kiranoudis, C. (2006). Statistical analysis of incidents reported in the Greek Petrochemical Industry for the period 1997â€"2003. Journal of Hazardous Materials, 135(1-3), pp.1-9.
5. Kossoris, M. (1939).A Statistical Approach to Accident Prevention.Journal of the American Statistical Association, 34(207), pp.524-532.
6. COOKE, W. and GAUTSCHI, F. (1981).OSHA, Plant Safety Programs, and Injury Reduction.Industrial Relations, 20(3), pp.245-257.
7. Majid, M. and McCaffer, R. (1997).Assessment of Work Performance of Maintenance Contractors in Saudi Arabia.J. Manage. Eng., 13(5), pp.91-91.
8. Verbano, C., Venturini, K., (2013), "Managing Risks in SMEs: A Literature Review and Research Agenda", Journal of Technology Management & Innovation 8(3), 1-17.

MATEC Web of Conferences **66**, 00106 (2016) DOI: 10.1051/matecconf/20166600106
IBCC 2016

Hazard and Risk of Highway Maintenance Works: Case Study of PLUS Expressways

Asmalia Che Ahmad[1,a], Ida Nianti Mohd Zin[1], Mohamad Nasrin Rosli[1], Abdul Muhaimin Ab Wahid[1], and Izatul Farrita Mohd Kamar[1]

[1]*Faculty of Architecture, Planning and Surveying, Universiti Teknologi Mara (Perak),Seri Iskandar, Malaysia*

Abstract. The North– South Expressway (NSE) (Lebuhraya Utara-Selatan) is the longest expressway in Malaysia with the total length of about 772 km (482 miles) running from Bukit Kayu Hitam in to Johor Bahru. It is known named after the highway's concessionaire, Projek Lebuhraya Utara Selatan Berhad (North South Expressway Project) abbreviated as PLUS Expressway. A study on the occupational safety among maintenance workers of PLUS Expressway was conducted. The aim for this research is to identify the hazards and determine the risk level of highway maintenance works. This research has identified the hazards occupies during highway maintenance works. Subsequently, the levels of risk for hazards during highway maintenance works are determined. The scope of this study is along the northern PLUS Expressway specified from Southern Ipoh to Bukit Kayu Hitam and focused on the maintenance workers of Projek Penyelenggaraan Lebuhraya Berhad (PROPEL). Eighty (80) sets of questionnaires survey on likelihood and severity of highway maintenance hazards have been distributed to highway maintenance workers. The data were analysed by using Excel to compose weightage average index (WAI) in order to compute the risk level of those hazards. The result concluded that the extreme level of hazard that occurred during maintenance of highway are hazards related to electricity, hazards that engage with scaffolding, hazards while using crane and hazards involve fire and explosion.

1 Introduction

The North– South Expressway Project (NSE) or Projek Lebuhraya Utara Selatan Berhad (North South Expressway Project) abbreviated as PLUS Expressway. (Lebuhraya Utara-Selatan) is the longest expressway in Malaysia. The total length of PLUS Expressway about 772 km (482 miles) running from Bukit Kayu Hitam in to Johor Bahru [1]. It passes 7 states i.e. Johor, Melaka, Negeri Sembilan, Selangor, Perak, Penang and Kedah. This expressway links many major cities and town thus acting as backbone to economic development of western Peninsular Malaysia. Significantly, this infrastructure service reduces travelling time and faster alternative route between various locations.

Proper maintenance is essential for reliable and smooth operation of highway in order to maintain its function as a key infrastructure that serve as network to the nation. Improper maintenance works would increase hazards that lead to accidents associated with human, damage of property and environment as well as boosting the operational cost. Study by Ganapthy (2010) showed that highway maintenance worker lives are in high risk where most accidents cases at work zone of highway involved them [2]. Therefore, safety for highway maintenance workers is crucial since the maintenance workers are put into the work zone of highway to conduct the maintenance task.

Hence, a study on the occupational safety among maintenance workers of PLUS Expressway was conducted. The objectives for this study are to identify the hazards and determine the risk level of highway maintenance works. This study focused on the maintenance workers of Projek Penyelenggaraan Lebuhraya Berhad (PROPEL) who responsible on maintenance task along the northern PLUS Expressway specified from Southern Ipoh to Bukit Kayu Hitam.

2 Highway Maintenance

2.1 PLUS Expressway

The North-South Expressways is widely known for its running length of the west coast of Peninsular Malaysia from Bukit Kayu Hitam near the border of Thailand in the north to Johor Bahru on the border of Singapore in the south, including the Senai to Johor Bahru stretch to the Kempas Spur Road with a full length of expressways about 772 kilometers [1]. It is the only expressway served the major cities along west coast of Peninsular Malaysia, In addition, PLUS Expressway is also providing other links to other principal roadways, which included the Penang Bridge, the Malaysia-Singapore Second Crossing,

[a] Corresponding author: asmalia809@perak.uitm.edu.my

the Federal Highways, the North-South Expressway Central Link and the Sistem Lingkaran-Lebuhraya Kajang Sdn. Bhd. (SILK Highway).

PLUS Expressway is selected for this study since it represents the most in highway maintenance works. PLUS Expressway operation of maintenance include 70 km dual three-lane and 715 km dual two-lane expressway of with approximately 10% and 90% each of the overall expressways length. Moreover, there are also 61 toll plazas, 64 interchanges, 296 bridges and 2 tunnels that need diverse methods of maintenance. Additionally maintenance also include 2 overhead bridge restaurants, 18 rest and service areas (RSA), 47 lay-bys, 244 billboards a well as other ancillary facilities [1].

2.2 Maintenance of Highway

Maintenance is defined as work in order to keep, restore or improve every part of building or facilities, to maintain building performance and its services including the surroundings, as well as to sustain the utility and value of the building. It includes improvement, refurbishment, upgrading and repair works to the existing facilities according to a currently accepted standard [3]. Maintenance consists of actions directed toward maintaining a social system in a serviceable state [4]. According to British Standard (2010) EN 13306:2010, maintenance is combination of all technical, administrative and managerial actions during the life cycle of an item intended to retain it in, or restore it to, a state in which it can perform the required functions [5].

Maintenance can be categorized into two types which are preventive maintenance and corrective maintenance [6]. Preventive maintenance is maintenance carried out at predetermined intervals or according to prescribed criteria and intended to reduce the probability of failure or the degradation of the functioning of an item. The preventive maintenance action is performed on condition based or predetermined maintenance. Corrective maintenance is maintenance carried out after fault recognition and intended to put an item into a state in which it can perform a required function. The corrective maintenance action is carried out based on deferred or immediate maintenance.

Maintenance of highway and road are generally classified into three categories as following [4,6,7]:

- **Routine works**.
  Routine works are maintenance works that are required to be carried out continuously and funded from recurring budget. These works can be grouped into cyclic and reactive works types. Cyclic works are undertaken base on maintenance standard frequency. Vegetation control, herbicides, verge cutting and culvert as well as other cleaning chores are examples of cyclic works. Reactive works, in the other hand are activities relate to intervention levels base on maintenance standard determine or when the maintenance is needed. An example of reactive work is patching which is carried out when the appearance of cracks or pot-holes on the road surfaces.
- **Periodic works.**
  Periodic works are required at intervals of several years to preserve the structural integrity of the road and building. It also avoids deterioration condition of roads and building structures. Periodic works can be clustered as preventive, resurfacing, turfing, grass seeding, overlay, pavement reconstruction, plumbing, power and lighting, CCTV and perimeter security, erosion control, line marking as well as widening or realignment of roads. Periodic works are budgeted on a regular basis or included in the recurrent budget.
- **Special works.**
  Special works are immediate or unplanned task that required maintenance activities. These works are not estimated in advance and usually performed ad hoc. This special works include emergency and unforeseen task such as removal of obstacles that are hazards to highway users for example fallen trees or branches, landslides, flash floods, soil heaps and windblown sand. Other special works include slope rehabilitation and stabilisation, clearing of debris, repair landslides and washouts. A contingency allowance is normally included within the recurrent budget to fund these special works.

According to Mohamed (2010) the primary goal of highway maintenance is maintaining the roads in a circumstance that turns to a good service and maximum safety to the travelling highway users [4]. This is accomplished by preserving the road free of disconcerting physical defects such as potholed pavements, broken pavement edges, free ravel, sticks surfaces, light and defective bridge decks and other imperfections. Moreover, building and structure also hold out longer with proper, continuous and efficient maintenance works. Poor maintenance may result the need for preparation, renovation or reconstruction that increase in cost the life cycle of the project [4].

2.3 Highway maintenance worker

A highway maintenance worker is the person who ensures that the highways and roads stay in safe, clean and useable. There are variety task of highway maintenance worker including regular maintenance, repair operations and fixing defects to highway and its surrounding facilities. This job demand physical strength and most of working time are spent outdoors. In addition, the highway maintenance worker face hazardous situations such as working along stretches of road with vehicle travelling at high rates of speed, operating works in unpredictable weather conditions as well as using with various machineries, tools and heavy equipment in daily task [8].

Consequently, the main duties and tasks of a highway maintenance worker are to [9]:

- Carry out maintenance and repairs to the highway to remain the roads and surrounding safe and clean.
- Install, maintain and work within temporary traffic management (using traffic lights, cones and barriers) to keep the traffic flow safe.
- Carry out repairs on all road categories and types laying new or replacement surfaces to specification.

- Locate underground services for example gas and power cables and work safely around them.
- Work with a wide range of hand / power tools and plant to repair potholes, cut, replace and reinstate pavements / kerbs, lay new surfaces and excavate materials.
- Able to read technical information in order to understand what works needs to be completed.
- Apply appropriate health, safety and environmental procedures to avoid injury and incidents.

2.4 Hazard and risk in highway maintenance work

A hazard is referred to potential harm that causes damage or adverse health [10]. Hazard is any unsafe condition or possible root of an undesirable event with potential for impairment or damage [11]. Hazard is closed relationship with the risk. Risk is a measurement to analyse and evaluate the hazard [10]. The measurement is made by identification on how severe and when likely of the hazard. In other words, the risk assessment is an in-depth look to specify situations, process and other harmful activities or hazard at workplace.

The highway maintenance is a hazardous workplace due to the condition of works that are expose to movement of vehicle, unpredictable weather condition, surrounding and environment disaster as well as close contact to usage of various instrument, machinery and equipment [12]. The top five workplace hazards that have been discovered by OSHA (2010) are [13]:

- Electricity – one of the greatest hazard for worker who expose to power line task at workplace
- Excavation and trenching – the most recorded fatalities at site operation
- Scaffolding – the most common accident during construction and maintenance work
- Falls – the most dangerous accident that causes from slipping, tripping, falling objects and unstable work platforms
- Heavy equipment – main accident cause of struck by, collision, overloading, overturning

3 Research Method

This research was conducted according to basic of safety procedure which is hazard identification and risk survey in order to assess the risk of highway maintenance works.

3.1 Hazard identification

Hazard identification was the first objective of the research. Hazards during highway maintenance works were identified base on published article and safety reports by safety related authorities. Eight hazards were identified in this research. There are hazards involving scaffolding, crane and heavy equipment, vehicle and machinery, fire and explosion, excavating work, installation work, welding work and electricity.

3.2 Risk survey

The second objective is to determine the risk level for each hazard identified. Risk can be presented in variety of ways to communicate the level of risk for each hazard. Technically, risk value is calculated by multiplying frequency of hazards occurrence or likelihood with consequences or severity of hazards occurrence [14]. The outcome matrix enables to determine level of risks for each hazard at workplace.

Risk can be calculated using the following formula [15]:

Risk (R) = Likelihood (L) x Severity (S)

Survey on likelihood and severity of the hazards was obtained through primary data collection using questionnaire. Table 1 and 2 shows the scale used in determining likelihood and severity of hazard occurrence.

Table 1. Scale for likelihood of hazards [16].

LIKELIHOOD	DESCRIPTION	SCALE
Almost Certain	The most likely result of the hazard or event	5
Likely	Has a good chance of occurring and is not unusual	4
Possible	Might be occur at sometime in the future	3
Unlikely	Has not been known to occur after many years	2
Rare	In particular impossible and has never occurred	1

Table 2. Scale for severity of hazards [16].

SEVERITY	DESCRIPTION	SCALE
Fatal	Large number of fatalities and/or major long-term impact	5
Major	Fatalities and/or major short-term impact	4
Moderate	Serious injuries and/or significant impact	3
Minor	Minor injuries and/or short-term impact	2
Insignificant	First aid injuries only and/or minimal impact	1

The scope for this research is the highway maintenance workers for PLUS Expressway focusing on northern region. The northern region PLUS Expressway route is from Southern Ipoh km 274.0 to Bukit Kayu Hitam km 0.0. These highway maintenance worker are managed by Projek Penyelenggaraan Lebuhraya Berhad or PROPEL. The population of PROPEL northern region maintenance workers is 364. Based on the population, the sample size of determined for this survey was 80 highway maintenance workers with ±10 % (e)precision [17].

Accordingly, eighty (80) sets of questionnaires

survey on likelihood and severity of highway maintenance hazards have been distributed to highway maintenance workers. The distribution was done during their monthly briefing on highway maintenance task with assistance by the manager in charge. The data from the questionnaire survey were analysed by using Excel to compose weightage average index (WAI) in order to compute the risk level of those hazards.

3.3 Risk assessment matrix

In order to determine the risk level, the risk assessment matrix has to be resolve. The computations of the risk assessment matrix in order to attain the risk level consist of following steps:

1. Calculation of weightage average index (WAI) for likelihood of hazards
2. Calculation of weightage average index (WAI) for severity of hazards
3. Calculation of risk (WAI likelihood x WAI severity) of hazard
4. Determination of risk assessment matrix (Table 3) and risk level (Table 4)

The weightage average index is calculated based on the formula as below [18] :

$$\text{Weightage Average Index (WAI)} = \frac{\sum wi\, fi}{\sum fi}$$

where:
wi = assigned weight for a particular class under 5-point likert scale
fi = corresponding frequency of that particular class
i = 1, 2, 3, 4, 5 that illustrated 5-point likert scale according to specific context

The computation of WAI for each hazard was interpreted by schedule of risk assessment matrix (Table 3) and the indication of risk level in Table 4.

Table 3. Schedule of risk assessment matrix [15].

LIKELIHOOD OF HAZARDS	SEVERITY OF HAZARDS				
	INSIGNIFICANT (1)	MINOR (2)	MODERATE (3)	MAJOR (4)	FATAL (5)
RARE (1)	1	2	3	4	5
UNLIKELY (2)	2	4	6	8	10
POSSIBLE (3)	3	6	9	12	15
LIKELY (4)	4	8	12	16	20
ALMOST CERTAIN (5)	5	10	15	20	25

Table 4. Indication of risk level [15].

RISK LEVEL	
1 to 2	LOW
3 to 6	MEDIUM
7 to 12	HIGH
More than 12	EXTREME

4 Result and findings

Table 5 shows the demographic information of the respondent where 30% of the respondents have less than 5 years of experience. Similarly 30% of the respondents have experience between 6 to 10 years. 11.3% had experience between 11 to 15 years while 12.5 with 16-20 years of experience. 16.2% of the respondent has more than 20 years of experience. Overall majority of the respondent (70%) had less more than five years of working experience. These demonstrate that this group of respondent are reliable with pertinent experience to participate in this survey on hazards and risk of highway maintenance works.

Table 5. Working experience of respondent

Working experience	Frequency	Percentage (%)
<5 years	24	30.0
6-10 years	24	30.0
11-15 years	9	11.3
16-20 years	10	12.5
>20 years	13	16.2
Total	80	100.0

By referring to Table 6, majority (57.5%) of the respondent had being involved in highway maintenance accident. This demonstrates that the highway maintenance workers are highly exposed to risk at their workplace.

Table 6. Respondent involved in highway maintenance accident

Involve in highway maintenance accident	Frequency	Percentage (%)
Yes	46	57.50
No	34	42.50
Total	80	100.00

Eight hazards were identified in this research. The hazards are as the following:

1. Hazard involving scaffolding
2. Hazard involving crane and heavy equipment
3. Hazard involving vehicle and machinery
4. Hazard involving fire and explosion
5. Hazard involving excavating work
6. Hazard involving installation work
7. Hazard involving welding work
8. Hazard involving electricity

According to the survey and calculations of WAI and risk assessment matrix, the following result was obtained. The table 7 shows the WAI likelihood, WAI severity and risk assessment matrix for the eight types of highway maintenance hazards. The risk was calculated by multiplying likelihood with severity. Each figure of the risk was categorised into risk level (refer Table 4) and rank accordingly. Out of eight types of hazards studied, 4 hazards were classified under extreme risk while another four were in high risk. None of the hazards falls in either medium or low risk level.

Table 7. Summary of hazards and risk during highway maintenance works.

Type of Hazard	Weightage Average index of Likelihood	Weightage Average index of Severity	Risk Assessment Matrix	Risk level	Ranking
Hazard involving electricity	3.30	4.88	**16.10**	**Extreme**	**1**
Hazard involving scaffolding	3.79	4.15	**15.73**	**Extreme**	**2**
Hazard involving crane and heavy equipment	3.31	4.70	**15.56**	**Extreme**	**3**
Hazard involving fire and explosion	2.91	4.28	**12.45**	**Extreme**	**4**
Hazard involving vehicle and machinery	3.03	3.49	**10.57**	**High**	**5**
Hazard involving excavating work	3.13	2.93	**9.17**	**High**	**6**
Hazard involving installation work	3.16	2.74	**8.66**	**High**	**7**
Hazard involving welding work	3.46	2.35	**8.13**	**High**	**8**

The top rank of risk during highway maintenance works is hazards involving electricity which is 16.10 with status of extreme. Data from the U.S. Bureau of Labor Statistics (BLS) indicate that there were nearly 6,000 fatal electrical injuries to workers in the U.S. between 1992 and 2013 [19]. Electrical hazards can be divided into 2 groups which are primary and secondary [20]. Primary hazards such as flowing of current through human body can cause electric shock with a harmful effect on the internal organs and their proper function. The impact of a strong electromagnetic field when working near a powerful antenna of a mobile operator along the highway can produce harmful effects tissue and organ damage which are delayed in time. Another primary electrical hazard is application of a controlled electrical substance in contact with hot and harmful substances such as welding and cutting. The blinding flash of an electrical substance can cause temporary or permanent eye damage, hot air and metal vapours can cause severe burns, inhaling hot electrical products can seriously damage respiratory system by burning lungs and throat or result in poisoning. Secondary electricity hazards may trigger other hazards for instance sources of fire and explosion and sources of other secondary effects when the flow of current through the human body leads to tripping, slipping, falling and the like.

The second rank of extreme highway maintenance risk is hazard involving scaffolding with the risk matrix of 15.73. The scaffolding industry has put workers in hazardous working situation where 10,000 accident cases were reported each year [21]. Hazards associated with scaffolding can be categorised according to design of the scaffolding, construction of scaffolding, environment in which scaffolding located as well as the usage of scaffolding [22].

A significant hazard associated with scaffolding design is collapse of the structure. This may cause multiple fatalities. The design failures include incorrect design, unsuitable foundations, inadequate edge protection, decking and bracing, inadequate access and egress, unstable scaffolding and overloading. Numerous hazards also exist related to construction and dismantling of scaffolding. Worker may obtained injury during manual handing of equipment, crushed by moving or lifting object, defective tools and equipments, falling objects and fall from scaffolding. The surroundings and environment also create scaffolding hazards such as obstructed workplace, poor lighting, poor access, wet or slippery walkways, and existence of blind spot and vehicle of highway user colliding to the scaffolding structure. The ultimate scaffolding hazard associated is the usage. Most accidents occur during working on scaffolding due to impatient of worker, unqualified personnel, under influence of drugs or alcohol and bad habits.

The third top rank of extreme risk during highway maintenance work is hazards involving crane and heavy equipment which is 15.56. The three major hazards identified by Harman (2014) that arise regarding crane and heavy equipment operation are electrical hazards, overloading and materials falling [23]. Electrical hazards of crane or heavy equipment accidents are result of metal part of a crane unintentionally contact with a high voltage power line. This contact occurs when the crane is moving nearby or under energized power lines which can result multiple injuries and death. Whereas, overloading is subject to structural stresses that cause unalterable damage. Overloading is result of hoisting load beyond capacity, swinging or sudden drop of load, usage of defective components and incompetency crane operators. Other hazards involve crane and heavy equipment is materials falling. Material falls due to slings and attachment were not secured properly. Visual impairment, mechanical failure and incompetency operator can cause the material lifted to slip and crash.

The forth top rank of extreme risk is hazard involving fire and explosions. This hazard include majority of maintenance highway activities which carried out using flammable substances accidently catching fire and explode [24]. Working activities which involves using or creating chemicals, vapours, liquids, gases, solids or dusts that can readily burn or explode is hazardous. The effects of fire and explosion can be devastating in terms of lives lost, injuries, significant damage to property, pollute the environment and ruin the business community. Occasionally this hazards also directly interrelated to electrical as source of secondary effects.

5 Conclusion

Overall, all the highway maintenance hazards are within extreme and high risk. The top four extreme risks are works involve electricity, scaffolding, crane and heavy equipment as well as fire and explosion. Nevertheless, other hazards with of high risk status such as hazard involving vehicle and machinery, excavation, installation and welding in highway maintenance works should also given serious attention. Action and control measures could be taken to maintain a good working condition and operation of highway run smoothly with low risk to the workers. Additionally, person in charge of the task that involve extreme and high risk task during highway maintenance work should be well trained and competent to avoid accidents among highway maintenance workers.

References

1. PLUS (2014) North– South Expressway (NSE). Accessed from http://www.plus.com.my/index.php?option=com_content&view
2. Ganapthy, V. R., Subramaniam, S. K., Mohaamad Diah, A. B., Suaidi, M. K., Hamidon, A. H. (2010). Flagman and Risk Involved in Road Construction Site. International Journal of Sustainable Construction Engineering & Technology.
3. Horner, R.M.W, El-Haram, M.A and Munns, A.K., (1997), "Building Maintenance strategy: a new management approach." Journal Of Quality In Maintenance Engineering, Vol. 3, No. 4, pp 273-280.
4. Mohamed, N. W. (2010). Road Maintenance Management System: A Case Study At Public Work Department. Unpublished Master Thesis. Faculty of Civil Engineering, Universiti Teknologi Malaysia.
5. British Standard (2010) Maintenance Terminology EN 13306:2010. BSI Publication. UK
6. European Standard EN 13306, Maintenance terminology (2001), CEN, European Committee for Standardization, Belgium
7. Worldbank (2001). Road and Highways: Construction and Maintenance. Accessed from http://www.worldbank.org/transport/roads/con&main.htm
8. Bureau of Labor Statistics (BLS) (2015) Highway Maintenance Workers. Accessed from www.bls.gov/oes/current/oes474051.htm
9. UK Roads Liaison Group (UKRLG) (2013) The Highway Infrastructure Asset Management Guidance. Accessed from http://www.highwaysefficiency.org.uk/efficiency-resources/asset-management/highway-infrastructure-asset-management-guidance.html
10. Canadian Centre for Occupational Health and Safety (CCOHS) (2009). Health & Safety Program: Risk Assessment. Accessed from http://www.ccohs.ca/oshanswers/hsprograms/risk_assessment.html
11. Marhavilas, P. K., Koulouriotis, D. and Gemeni, V. (2010). Risk Analysis and Assessment Methodologist in the Work Sites: On a review, classification and comparative study of the scientific literature of the period 2000-2009. Journal of Loss Prevention in The Process Industries, Vol. 24, pp 477-523.
12. Pratt, S.G., Fosbroke, D. E., & Marsh, S. M. (2001). Building Safer Highway Work Zones: Measures to Prevent worker Injuries From Vehicles and Equipment (DHHS (NIOSH)

Publication No. 2001-128).
13. Occupational Safety & Health Administration, United States Department of Labor (OSHA) (2010) Hazards – Young Workers. Accessed from https://www.osha.gov/youngworkers/hazards.html
14. Baradan S., Mumtaz, A. Usmen. (2006) Comparative Injury and Fatality Risk Analysis of Building Trades. Journal of Construction Engineering and Management 132:533-539
15. Department of Occupational Safety and Health Malaysia (DOSH) (2008) Guidelines For Hazard Identification, Risk Assessment And Risk Control (HIRARC)
16. Jeong K.-S., Lee K.-W., Lim H.-K. (2010) Risk assessment on hazards for decommissioning safety of a nuclear facility. Annals of Nuclear Energy 37:1751-1762
17. Israel, G. D. (1992). Sampling the Evidence of Extension Program Impact Program Evaluation and Organizational Development, IFAS, University of Florida. PEOD-5. October.
18. Hermans, E., Ruan,D., , Brijs,T., Wets, G. and Vanhoof, K. (2010) Road Safety Risk Evaluation by means of Ordered Weighted Averaging Operators and Expert Knowledge. Accessed from https://uhdspace.uhasselt.be/dspace/.../1/A1Q2-nummer2%5B1%5D.pdf
19. National Fire Protection Association (NFPA) (2015) Occupational Injuries From Electrical Shock and Arc Flash Events. Fire Protection Research Foundation
20. International Social Security Association (ISSA) (2010), Guide for Risk Assessment in Small and Medium Enterprises - Hazards arising from Electricity.
21. Shafii H., Rahman Z. A. and Othman A. R., (2005). Kajian Pengurusan Keselamatan Pekerja Di Tapak Bina. Universiti Tun Hussein Onn Malaysia. Accessed from http://www.searchinpdf.com/
22. Worksafe (2014) Safety in scaffolding. Newzealand.govt.nz. Accessed from http://www.business.govt.nz/worksafe/information-guidance/all-guidance-items/scaffolding-best-practice-guideline-for-scaffolding-in-new-zealand/section-2-safety-in-scaffolding/2.8-hazard-identification-and-management#2.8.4
23. Harman, K (2014) Overhead Crane Safety – Three Major hazard and preventive measures. Spanco Inc. Accessed from http://www.spanco.com/blog/entry/overhead-crane-safetythree-major-hazards-and-preventative-measures
24. HSE (2014) Apprenticeship Standard - Highways Maintenance Skilled Operative Level 2. Crown copyright. Accessed from https://www.gov.uk/government/uploads/system/uploads/attachment_data/file/384288/CONSTRUCTIONBB_-_Highways_MaintenanceSkilledOperative_L2_-_Final.pdf

[a] Corresponding author: asmalia809@perak.uitm.edu.my

DOI: 10.1051/matecconf/20166600107

A Conceptual Framework of Safety and Health in Construction Management

Izatul Farrita Mohd Kamar[1], Asmalia Che Ahmad[2]

[1]*Centre of Postgraduate Studies, Faculty of Architecture, Planning and Surveying, Universiti Teknologi MARA,Seri Iskandar Campus, Seri Iskandar, 32610, Perak, Malaysia.*
[2]*Department of Quantity Surveying, Faculty of Architecture, Planning and Surveying, Universiti Teknologi MARA, Seri Iskandar Campus, Seri Iskandar, 32610, Perak, Malaysia.*

Abstract. Models to analyse and calculate the costs of prevention in the construction industry are not easy to apply in this area. Firstly, because they are based on studies carried out in the manufacturing sector; and secondly because the traditional models for analyzing these costs are limited to identifying and classifying them. This calculation models need to be improved. Hence this study is an attempt to fill the gap by develops a safety and health cost framework from pre-construction stage until to construction stage in order to assist client to allow the compliance of safety aspects in a tender document can be assured. A mixed method research will be used in this study. Semi structured interview and content analysis will be used as a main tool for the data collection. Then, the questionnaire will be distributed to the expertise in construction industry.

1 Introduction

Construction is a complex activity where various stake-holders are present working under constant challenges from the demands of the job. Each job will have several safety and risk factors, requiring quality and safety management systems to be established as indicated by Mehta & Agnew [35]. This view is supported by Nadeem [4] who writes the process of planning, executing and maintaining all project activities is complex and time-consuming. The situation is made complex by many external factors such as organizational structure, communication, clear instructions, safety culture, codes and standards, training, leadership and responsibility have been suggested to have influence on the general safety at the workplace.

Workplace accident rates are very high compared to other sectors [34]. Regarding the situation in Malaysia in this respect, the Department of Occupational Safety and Health (DOSH), reported that among the total of 112 numbers of fatal injuries in September 2015, 42 numbers of death occurred in construction sector. This statistics indicate that the construction industry is among the most hazardous industries in Malaysia.

Various losses would be incurred by the injured workers after the occurrence of an accident. These losses may include costs to victims and their families; and to employers and society. Accidents and the corresponding damage they cause to productivity, property, equipment and morale can have detrimental effects on a construction company's profit and loss statement [9]. Other losses would be incurred are leading to delays in project completion, increased cost and loss of constructors' reputation [38]. The costs associated with these accidents are both human (not directly measurable) and financial for companies and for society as a whole (sick leave, medical treatment, etc.). Other costs also arise, such as delays in project implementation, impaired company image or market loss [28].

[a] izatu739@perak.uitm.edu.my

2 Problem and Issues

Numerous studies have attempted to explain the various losses would be incurred by the companies. However, Lingard and Rowlinson (2005) showed that many of the losses incurred by an accident are "hidden" and difficult to quantify. These "hidden" costs may be significant, and some may be particularly prominent in construction industry. This view is supported by [28], who points out that the costs being "hidden" are difficult to evaluate, to isolate, identify, and quantify. These hidden costs are often difficult to calculate due to the challenges associated with quantifying the specific magnitude of their economic consequences and the manner in which these costs are typically tracked in conventional accounting practices.While it is easy to obtain reliable data on compensation costs (insurance), little information is available on indirect costs (absence, supervision, productivity loss, etc.) in the work environment. A real problem is to collect all the necessary consequence data needed for cost calculation. Therefore, both direct and indirect costs of accidents need to be examined to reflect the true costs of accidents to an employer.

Many researches were done to study the safety cost model in manufacturing sector and some researchers were study on research relates to accident cost (direct and indirect costs) which is focused more on the construction stage of a project. There is limited study appears to have been conducted to investigate on safety costs involved from pre-construction until to construction stage.

Moreover, models to analyse and calculate the costs of prevention in the construction industry are not easy to apply in this area. Firstly, because they are based on studies carried out in the manufacturing sector; and secondly because the traditional models for analyzing these costs are limited to identifying and classifying them [32]. This calculation models need to be improved. Hence this study is an attempt to fill the gap by develops a safety and health cost model from pre-construction stage until to construction stage in order to assist client to allow the compliance of safety aspects in a tender document can be assured.

3 View on Safety and Health in Construction Industry

Construction industry is among the most hazardous industries, and needs a comprehensive and simple-to-administer tool to continuously assess and promote its health and safety performance [33]. Construction is a high hazard industry that covers work on new or existing commercial, industrial or domestic buildings or structures. This view is supported by [36] who writes the construction industry is unique among other industries as the activities of construction often take place in the outdoor under conditions not conducive for safety and health. Workers in the construction sites have to face constant change in the nature of work, the location of work and the mix of workers. Most of the people tend to relate construction industry with dangerous working environment and high risk as compared to others.

The safety in construction site is one of the most essential issues that cannot be taken lightly. Because even it only minor accident, it may cause serious and huge effect to the organization especially within the industry involved machinery that is very much related with the issues of safety awareness. To improve it, some countries such as Australia, Singapore, and the United Kingdom have enacted legislations regulating safety working practices on site. Owing to the advocacy of self-management approach, some construction firms have actively begun to seek effective safety management systems [47].

A number of studies have done a research on safety performance, safety culture, safety leadership and safety compliance in construction industry. Other researches were done to study the safety cost model in manufacturing sector and some researchers were study on research relates to accident cost (direct and indirect costs) which is focused more on the construction stage of a project. There is limited study appears to have been conducted to investigate on safety costs involved from pre-construction until to construction stage. Table 1 shows the summary of studies related to the safety and health management from previous research.

Table 1. The Previous Research Related to the Safety Management

	Safety Performance	Safety Culture	Safety Leadership	Safety Compliance	Safety Element	Safety Cost (Construction Stage)	Safety Cost (Pre-Construction Stage)	Safety Cost (Manufacturing Sector)
(Feng 2013)	x	x				x		
(López-Alonso et al. 2013)						x		
(Griffin & Hu 2013)			x	x				
(Sun et al. 2006)						x		
(Feng et al. 2014)	x					x		
(Gurcanli et al. 2015)						x		
(Jallon et al. 2011a)						x		
(Feng et al. 2015)			x			x		
(Zin & Ismail 2012)				x				
(Oxenburgh & Marlow 2005)								x
(Amador-Rodezno 2005)								x
Napsiah and Faridah, 2012					x			
(Hare & Cameron 2012)					x			

DOI: 10.1051/matecconf/20166600107

4 Safety and Health Elements in Construction Management

STAGES	SAFETY ELEMENT		
	(OGC Gateway Model) **Billy [40]**	**RIBA Plan of Work 2013**	**Napsiah [39]**
Pre-Construction Stage	*Concept Phase* i. Possible need for project • Client's role in **H&S** throughout project; supply information, time allowed and **budget** required for project ii. Define user needs • Align **H&S policies** for project; how supply chain will be informed of H&S requirements, expertise required, criteria for evaluating competence, resources and commitment. • Identify **H&S hazards** (risk register).	*Strategic Definition Stage* • Identify client's **Business Case** and Strategic Brief and other core project requirements. • Review feedback from previous project.	*Inception Stage* • **OSH** in **project brief & objective**
	Feasibility Phase i. Options to meet user needs • Include **H&S performance**, materials and components specified by output performance can meet functional and H&S requirements. • Option evaluation chart to include H&S. • Input from end user's operation and maintenance at this stage; include format for **H&S file** and budget for maintenance strategy. • Initial **H&S box information** in during **concept designs**. ii. Prepare business case • H&S objectives, **H&S milestones** included. • Evaluate cost of specific **H&S items.** • Assess risks, decide management arrangements and control procedures, update **risk register.** iii. Project brief • H&S objectives included, decide project H&S Performance Indicators, agree format for **H&S File**. iv. Feasibility study option • Consider **H&S risks** on each site via option evaluation chart. v. Procurement strategy • Agree **H&S criteria** for selection of supply chain. Seek advice on maintenance and access issues during operation and maintenance period to prevent H&S problems.	*Preparation and Brief Stage* • Prepare Handover Strategy and Risk Assessments	*Feasibility Stage* • OSH in project brief & objective

<table>
<tr><td>Pre-Construction Stage</td><td><u>Design & Planning Phase</u>
i. Contract preparation
• Develop H&S milestones for project program. Review specifications for prescriptive items that may generate H&S risks during construction, operation and maintenance.
ii. Expression of interest/vetting
• Use H&S criteria previously outlined to vet supply chain.
iii. Partner/contractor selection
• Include current H&S file. Co-operation between parties involved in negotiation/ tender process with regard to H&S issues. Ensure H&S criteria sufficiently weighted in decision.
iv. Award contract
• Confirm H&S duties. H&S hazard workshop and integrated responsibility chart with H&S included.
v. Outline design
• Identify H&S hazards/risks on drawings. Cross reference H&S plan to program. H&S milestones on program.
vi. Detailed design
• Site issues regarding residual risk have been addressed by the contractor ahead of completing the construction H&S plan. Identify H&S hazards/risks on drawing.</td><td><u>Developed & Technical Design</u>
• Review Construction Strategy, including sequencing, and update Health and Safety Strategy.</td><td><u>Design Stage</u>
i. Design
• Safety education and training for designer
• Safey prevention through design (PtD).
ii.Tendering
• Client is optimal responsibility in insisting on the safe performance of the contractors in making their selection.
• Selection of contractor was based on the most cost-effective solution to control the health, safety and quality risks.</td></tr>
<tr><td>Construction Stage</td><td><u>Construction Phase</u>
i. Site establishment
• Display H&S Executive Officer notice, site inductions-confirm operatives have received and understood method statements, communicate site rules. Confirm contractor's supply chain input to construction H&S plan and H&S file.
• Monitor H&S performance and resources.
• Safety inspections continue to monitor H&S performance and resources, report incidents/accidents.
• Identify potential H&S risks at commissioning during use. Ensure contractor's supply chain submit information for H&S file.
ii. Handover
• Final inclusions and completion of H&S file. Include H&S issues in testing of M&E equipment.
Communicate H&S residual risks (H&S file).</td><td><u>Construction Stage</u>
• Review and update Sustainability Strategy and implement Handover Strategy, including agreement of information required for commissioning, training, handover, asset management, future monitoring and maintenance.
• Update construction and Health and Safety Strategies</td><td></td></tr>
</table>

According to Ghani [20] concludes that matters related to the safety and health aspects during pre-construction, construction and post-construction must be the priority in the construction industry. Basically, Construction Management is divided into 3 sections: Pre-construction, Construction and Post-construction [45]. Pre-construction stage is the initial stage for a Construction Project Management, which covers Inception & Feasibility, Design, and Tendering [30]. It is commences with "Inception and Feasibility". This stage is more focused on the client's requirement and is being described as pre-project planning (Hendrickson & Au, 2000). It encompasses the project objectives, land matters, feasibility study and setting up of the project organisational structure. "Design" falls after Inception & Feasibility which contains project brief, conceptual & schematic design development, planning approval, detailing on design, cost budget preparation, and value of engineering [27]. The goal during this phase is to acquire a complete and accurate understanding of project requirements. The last stage in Pre-construction is "Tendering". During this stage, the tender document is issued and then being evaluated to choose potential contractor [15].

According to OGC [40] as cited by Billy [26] the UK Office for Government Commerce (OGC) "Gateway" model for construction procurement has flexibility which allows various procurement routes to be adopted; and it is primarily for general construction management purposes. This OGC model had been used to build the Health and Safety elements. According to RIBA Plan of Work 2013, the safety element task involved is stated from the 'Preparation and Brief' stage until the 'Construction' stage only. While in Malaysia, specific assessment of safety elements is limited. Based on Napsiah [39], the safety elements were identified during pre-construction stage only. The summary of safety elements involved in Construction Management is documented as per following table.

5 Safety Cost in Construction Project

5.1 Iceberg Theory

The Accident Cost Iceberg proposed by Bird (1974) showed that the proportion of hidden costs could be much larger than the costs directly related to the accident [7]. While the direct cost can be estimated based on the nature of the accident, the indirect proportion is the biggest obstacle for a reliable predictor of the total cost of accidents. Accident incurred losses to the organization. Most organizations are aware that once an accident occurs at the work place, cost will be affected. No matter how much they have paid their insured costs such as the monthly or yearly insurance premiums for the safety of their workers (for instance: medical insurance, hospitalization insurance, building safety insurance, capital item insurances) the losses or hidden or uninsured costs are much greater than the insured costs.

Most hidden costs lie in the penalties, fines imposed on the organization, the amount of productive time spent by the safety officer and safety and health committees doing investigation and interviewing witness, the time lost on productivity and outputs, loss of sales and potential customers, damages to the machines or equipment, cost of idle time, repair and set up costs, loss of man hours, cost incurred to hire replacement workers, documentation cost, et cetera. Thus preventing accidents is the best management strategy in reducing accidents costs [42]. Figure 1 provides a graphical representation of the Iceberg Model of Accident Costs.

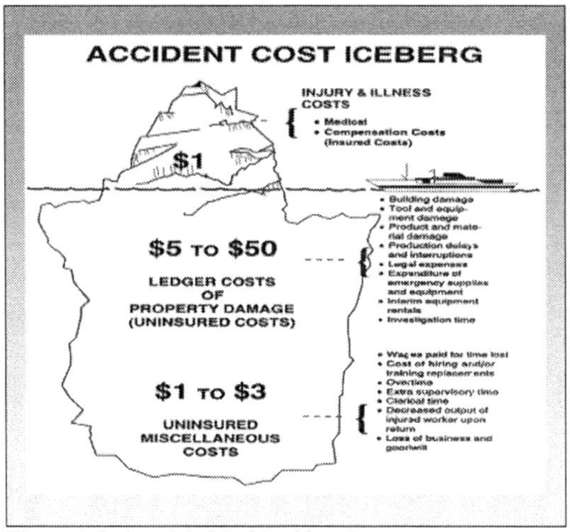

Figure 1. Iceberg Model of Accident Costs

5.2 Types of Safety Cost

[32] found that the cost related to safety and health in the workplace can be classified into three groups as follows:

1) Safety Cost

It is distinguish between prevention costs and those of evaluation and monitoring. Prevention costs are those incurred in order to comply with legal requirements with respect to accident prevention, to implement measures to prevent accidents during construction work and to improve health and safety conditions in all areas of the work performed. In relation to Evaluation and monitoring costs, these are de- rived from the actions taken by the company for appropriate testing and maintenance of the health and safety measures adopted, regarding every facet of the work in question, with the aim of reducing or minimizing the risk of accident or occupational disease.

2) Non Safety Cost

Those are produced by not ensuring health and safety at work, i.e., the costs a company must meet following accidents, as well as those that may arise from breaches of safety regulations. In turn, we distinguish between tangible and intangible costs of accidents. Tangible costs of accidents reflect the costs associated with the occurrence of an accident at work, which can be estimated or calculated using traditional cost accounting methods, while Intangible costs of accidents, according to Gosselin [22], are the costs which are not measurable in economic terms or for which there are no performance indices to measure their impact on the organisation, such as impaired company image, low worker morale, labour disputes or loss of market.

3) Other Extraordinary Costs

All losses provoked by events that cannot be prevented by the technical or human resources available to construction works, or which are totally unavoidable, such as natural disasters. This cost category includes all the items that are beyond the scope and control of management, and thus are classed as uncontrollable costs, which cannot be incorporated into a structured model designed to control costs regarding safety in the workplace. Both intangible and extraordinary costs are excluded from this study, the first because, in general, they are neither calculated nor estimated in a construction project, and the second because, being uncontrollable as far as works managers are concerned, they cannot be included in a structured calculation model.

5.3 Safety Cost Model

There are numerous studies that analyse the safety cost in construction project. Several models have been produced by the researchers that analyze the cost of lost working days, insurance, and delay in project duration, etc. **Activity-Based Costing (ABC)** has emerged as a new approach that associates costs directly related to business activities with manufactured products. ABC gives more accurate information especially in complex structures and allows one to obtain information on cost items that are ignored in traditional methods. Furthermore, this method is flexible for all kinds of technology-intensive business systems and can be used success-fully in developing countries. ABC provides detailed information about the cost of activities in a particular process and helps decision-makers to develop reasonable decisions with other firms. The other objectives of this system are providing cost reduction and management; determination of profitability of customers, inventory valuation, and new product and service design [25].

Other models developed by researchers as cited by [25], was introduced a cost of safety model to conceptually describe the **cost-benefit analysis** of accident/injury prevention. Alonso et al. (2013) studied the impact of health and safety on the investment of construction companies by applying a questionnaire in southern Spain. While [33] presented an **Occupational Safety and Health Potential Risk Model** for estimation of the statistical costs of occupational safety and health risk.

Return on Investment (ROI) model is introduced to evaluate the effectiveness of safety cost. ROI model is verified to be easy and typical in investment evaluation. Lately, it has also been used to evaluate the effectiveness of Safety Risk Management System in Construction.

$$\text{ROI}(\%) = \frac{\text{Total Benefit-Cost}}{\text{Cost}} = \frac{\text{Net Benefit}}{\text{Cost}}$$

From the ROI model, the cost of safety allocation will be found out in construction project. This will help to identify and classify the investment or allocation safety cost in construction projects. Investment in construction safety risk reduction is not where a company generates net revenue but a place that generates profit by reducing the levels of safety risks and the possibility of losses.

6 Conceptual Framework of Safety and Health Cost in Construction Management

The propositions outlined above are graphically depicted in Figure 2. A conceptual framework of safety and health cost that emerged from the literature review has been developed in this research. The input gathered from the existing of safety cost models such as ABC Model, CBA Model, and ROI Model need to be improved.

To fill the gap existed in the previous models, the safety elements that involved cost in construction management will be included in the new safety and health framework in order to assist client to allow the compliance of safety aspects in a tender document.

A mixed method research will be used in this study. Semi structured interview and content analysis will be used as a main tool for the data collection. Then, the questionnaire will be distributed to the expertise in construction industry. The new safety and health cost framework will be developed and it will be continuously interpreted and tested through construct validity, internal validity, external validity and reliability. From the result, all the objectives of this research will be achieved.

7 Research Design

Generally, this research design is divided into seven (7) phases (refer to Figure 3). The first phase in doing this research is to study the relevant issues with regards to safety and health elements. The identification of related issues, establishment of problem statement, identification of research objectives and research questions and methodology used will be explained in this phase. For the second phase, types of safety elements and its cost involved in construction project will be determined. For this phase, qualitative methods will be used as research methods to achieve the study objectives and answering research questions. Semi-structured interview and content/document analysis will be done to prepare a questionnaire.

During the third phase, pilot test survey will be conducted to the selected respondents in order to evaluate the correctness of the questionnaire and also attempt to predict an appropriate sample size and improve upon the study design. While in fourth phase, the questionnaires will be distributed to the respondents such as safety officer, consultant and contractor. Determination of weightage for each type of safety elements that involved cost will be determined.

In phase five, the suggested safety and health model is developed through the data collected and evidence sources and it will be continuously interpreted and tested through construct validity, internal validity, external validity and reliability. From the analysis, the author will suggest a safety and health cost framework in construction project. Objective 1,2 and 3 will be achieved in this phase. For the six phases, the safety and health cost framework will be validated by the focus group (expertise). Suggestion from these parties will be taken into consideration in improving the cost framework. Objective 3 will be achieved in this phase.

During the last phase, the author concludes all findings of the study conducted in phases earlier. Then, the author will provide recommendations to improve the safety elements aspect; safety cost efficiency and also provides recommendations towards future research. The research design phases involved in this research is documented as per following figure.

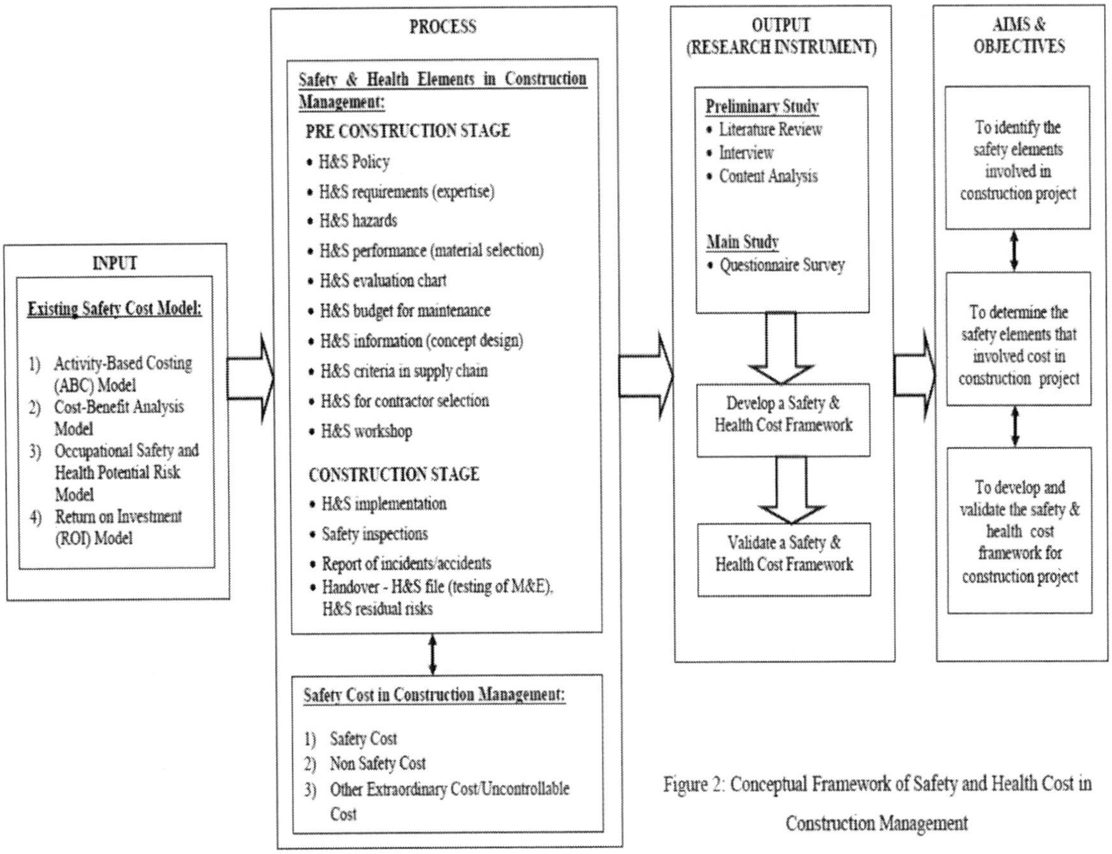

Figure 2: Conceptual Framework of Safety and Health Cost in Construction Management

DOI: 10.1051/matecconf/20166600107

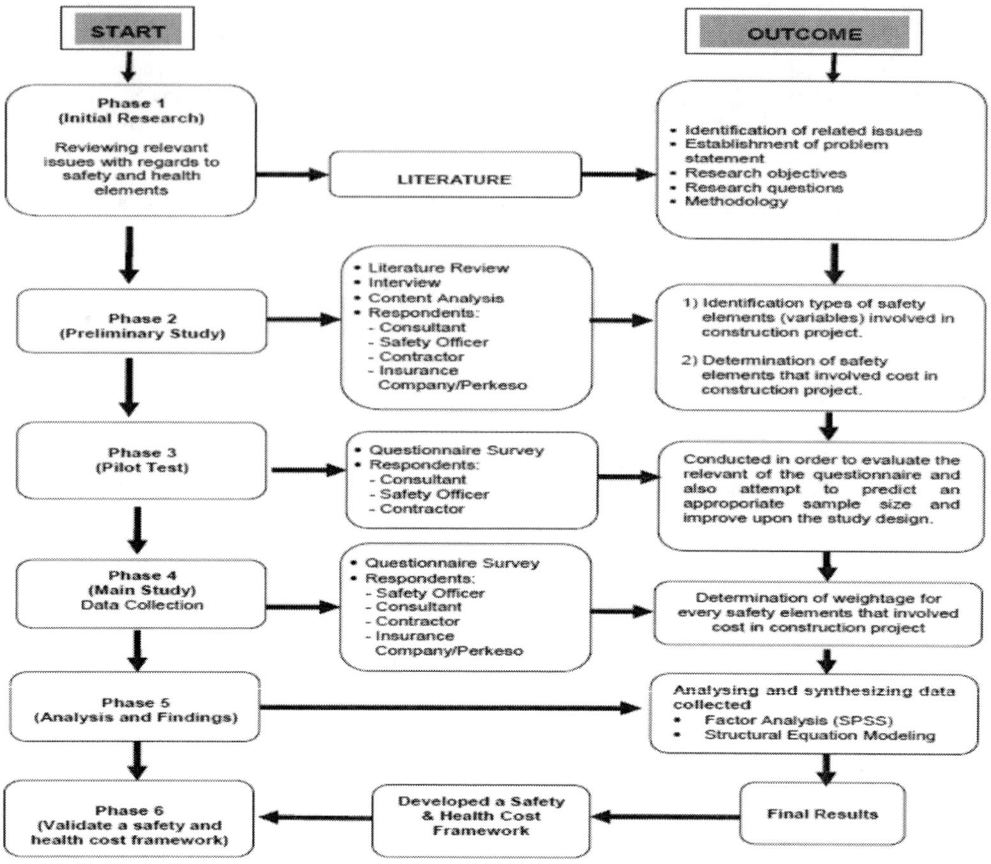

Figure 3. Research Design

8 Implications of Research

By providing opportunities for improvement of project implementation cost, the proposed safety and health cost framework can have a significant role to play in the building industry. It is hoped that this research would be able to contribute significantly in terms of the following:-

a) Helps to achieve OSH Master Plan Program (The National Council of Occupational Safety and Health) in estimating and reflecting the real costs of safety elements;
b) Obtains useful information for National Institute of Occupational Safety and Health (NIOSH) regarding on OSH R&D and training;
c) Helps clients to know the actual cost that the contractor will incur for the full compliance of safety and health in construction project.

9 Conclusion

The research presented in this paper is initially and part of an ongoing PhD research at the Faculty of Architecture, Planning and Surveying, UiTM. This research has elaborated on the safety and health elements which involved cost in construction project and provided a conceptual framework based on a comprehensive review of the existing literature. There is limited study appears to have been conducted to investigate on safety costs involved from pre-construction until to construction stage. The Iceberg Theory was explained the types of safety cost incurred in construction project. In the proposed conceptual framework, this theory will be related with the existence safety cost model in manufacturing and construction sector. Then, the new safety and health cost framework involved from pre-construction stage until to construction stage will be developed in line with the improvement of existing models. Perhaps, it can assist client to allow the compliance of safety aspects in a tender document.

References

1. A. Mecke, I. Lee, J.R. Baker jr., M.M. Banaszak Holl, B.G. Orr, Eur. Phys. J. E **14**, 7 (2004)
2. Amador-Rodezno, R., J.Safety Research **36**,215 (2005).
3. Choundry, R.M., D., J.Safety Science **46**, 566 (2008)

[a] izatu739@perak.uitm.edu.my

4. Ehsan, N., Mirza, E., Alam, M., Ishaque, *3rd IEEE International Conference*, (2010)
5. Feng, Y., J. Safety Science **59**, (2013)
6. Feng, Y. et al., J.Prjct Management **32**(6), 932 (2013).
7. Feng, Y., Zhang, S. & Wu, P., J. Safety Science **72**, 97 (2015).
8. Ghani, M.K., The Challenges and Initiatives. Malaysia (2008)
9. Goetsch, D.I., *Construction Safety and Health*, second ed. Pearson Education, Inc., New Jersey, USA (2013).
10. Gosselin, M., Analyse des avantages et des Couts de la Sante et de la Securite au Travail en Enterprise: Development de l'outil D'Analyse. Rapport de, Recherche R, 375 (2014)
11. Griffin, M. a. & Hu, X., J. Safety Science **60**, 196 (2013)
12. Guadalupe, M., *The Hidden Costs Fixed Term Contracts: the Impact on Work Accidents. Labour Economics*, (2003)
13. Amador-Rodezno, R., J. Safety Research, **36** (3), 215 (2005)
14. Choundry, R.M., D., J. Safety Science **46** (4), 566 (2008)
15. College, I. Capital Project & Planning. Retrieved 4/08/2011,fromhttp://www3.imperail.ac.uk/capitalprojects/projectprocedures/stages/tender(2011)
16. Ehsan, N., Mirza, E., Alam, M., Ishaque, A. *IEEE International Conference*, (2010)
17. Feng, Y., J. Safety Science **59**, 28 (2013)
18. Feng, Y. et al., Int. J.Prjct Managmnt, **32** (6), 932 (2014)
19. Feng, Y., Zhang, S. & Wu, P., J. Safety Science, **72**, 97 (2015)
20. Ghani, M.K., *Safety in Malaysian Construction: The Challenges and Initiatives. Malaysia* (2008)
21. Goetsch, D.I.,*Construction Safety and Health,* second ed. Pearson Education, Inc., New Jersey, USA (2013).
22. Gosselin, M., Analyse des avantages et des Couts de la Sante et de la Securite au Travail en Enterprise: Development de l'outil D'Analyse. Rapport de, Recherche R-375 (2004)
23. Griffin, M. a. & Hu, X., J. Safety Science **60**, 196 (2013)
24. Guadalupe, M., J. Safety Science **10** (3), 339 (2003)
25. Gurcanli, G.E., Bilir, S. & Sevim, M., J. Safety Science, **80** (2015).
26. Hare, B. & Cameron, I.,*Health and safety gateways for construction project planning. Engineering, Construction and Architectural Management*, 19(2), 192 (2012)
27. Hendrickson, C., & Au, T., *Project Management for Construction, Fundamental Concepts for Owners, Engineers, Architects and Builders* (2000)
28. Jallon, R., Imbeau, D. & De Marcellis-Warin, N., J. Safety Research **42**(5), 333 (2011)
29. Jallon, R., Imbeau, D. & De Marcellis-Warin, N., J. of Safety Research **42**(3),149 (2011)
30. K.S., I. H.S., *Engineering and Construction Contracts Management. Singapore: Lexis Nexis* (2002)
31. Lingard, H., Rowlinson, S., *Occupational Health and Safety in Construction Project Management. Spon Press, New York* (2005)
32. López-Alonso, M. et al., J. Safety Science **60**, 151 (2013)
33. Mahmoudi, S. et al., J. Safety and Health at Work **5** (3),125 (2014)
34. Martinez Aires, M.D., Rubio Gamez, M.C., Gibb, A., J. Safety Science **48** (2), 258 (2010)
35. Mehta, R.K. & Agnew, M.J., Int.J. Industrial Ergonomics **40**(5), 584 (2010)
36. Misnan, M.S. & Mohammed, A.H., *Development of safety culture in the construction industry: a conceptual framework,* **3** (2007)
37. Monica, L.A., Maria, P.I.D., Maria, C.R.G., Teresa, G.M.,J. Safety Science **60**,151 (2013)
38. N.Abd Rahman, Z.M.J., *International Civil and Infrastructure Engineering Conference* (2014)
39. Napsiah, M.S., Ismail, F., *Procedia-Social and Behavioral Sciences* **35**,603 (2012)
40. Office of Government Commerce (OGC), Procurement Guidance no. 10, Achieving Excellence Through Health and Safety, 4 February, available at: www.ogc.gov.uk/index.html (accessed 23 September 2002)
41. Oxenburgh, M. & Marlow, P., J. Safety research **36**(3), 209 (2005)
42. Sariwati, M.S., *Occupational Safety and Health Management Book (2nd Edition), University Publication Centre (UPENA), UiTM*, 29 (2010)
43. Sun, L. et al., *Theoretical Issues in Ergonomics Science* 7(3), 227 (2006)
44. Teo, A.L., K.Phang., J. Construction Research, 157 (2005)
45. Tregenza, T., *Magazine of the European Agency for Safety & Health at Work* (2004)
46. Wong, F.K.W., So, L.S.L., *Multi-layers subcontracting practice in the Hong Kong construction industry. In: Construction Safety Management Systems. Spon Press, USA and Canada,* 147 (2004)
47. Zeng, S.X., Tam, V.W.Y. & Tam, C.M., J. *Safety Science* **46**(8), 1155 (2008)
48. Zin, S.M. & Ismail, F., *Procedia - Social and Behavioral Sciences* **36**, 742 (2012)

The Ancestral Heritage: Sundanese Traditional Houses Of Kampung Naga, West Java, Indonesia

Tessa Eka Darmayanti[1,a]

[1]*Interior Design Department, Art and Design Faculty, Maranatha Christian University, Bandung, West Java, Indonesia*

Abstract. Sundanese are the largest indigenous ethnic group in the western part of Java Island. Sundanese culture closely relates to Animist, Hindu-Buddhist, and Islam philosophies. It led to the unique creation of the Sundanese traditional houses called *Rumah Adat* in Kampung Naga where the ancient tradition from their ancestor were upheld until today. Kampung Naga was listed as one of the Sundanese traditional villages known as *Kampung Adat* in West Java. This village has its own characteristics which enable the conveyance of tremendous tradition from the past, since the reign of Padjajaran Kingdom to the present. Additionally, it has the ability to accelerate globalization of culture these days. The objectives of this paper are to reintroduce and observe the in-depth meaning of Sundanese traditional houses within Kampung Naga. This paper employed qualitative investigation with the support of exploratory literature review. The result of this study portrayed that Sundanese traditional village are depicted as sacred spaces when the founder, the belief, and the nature spirit are interrelated. Thus, the creation of traditional houses in Kampung Naga does not only serve as shelters, but it also has "breath", "soul" and may embody cosmological value. This paper is prominent in the documentation of Sundanese traditional buildings as one of cultural identity of Indonesia, which will deliver the cultural knowledge to the world wide.

1 Introduction

Animism is one of the earliest beliefs in Indonesia, which respects nature and believes everything in nature to have spirit and power. However, since the coming of various foreign nations to this country to spread their faith through their cultures, the Indonesian people started to believe in God. Animism, Hindu-Buddhist and Islam philosophies influenced the Sundanese culture and fused into various aspects of life, including buildings (traditional houses). Traditional houses known as *Rumah Adat* are the buildings that may represent expressions of culture. The embodiment of traditional houses becomes a very important issue in traditional community, since it allows the community to show their existence. Looking at that importance, this paper focuses in Sundanese community in Kampung Naga.

In the fourth century, when Hindu Kingdom was ruled in the west part of Java Island, known as Sundanese region, they believe the kingdom as the cosmological center. However, prior to the time when animism became the only belief in primordial era, Sundanese people believe that the houses is the center of their cosmos. A cosmos is an orderly-harmonious system, whereas the Sundanese cosmos itself is called as cosmological mythic-spiritual culture, since it always contains the theological thought [1]. The Sundanese Traditional house has an unseen "energy" that moves both vertically and horizontally, inside and also outside. They have a purpose to connect three different cosmological worlds, upper world (*Dunia Atas*), middle world (*Dunia Tengah*) and under world (*Dunia Bawah*), which will be discussed in the next chapter.

2 Methodology

Sundanese culture has existed for more than 15 centuries, it becomes one of ancient cultures in Indonesia, especially West Java. Animist, Hindu-Buddhist as earlier beliefs, is then followed by Islam, which closely relates to the existence of Sundanese Traditional Villages, especially in Kampung Naga. This research employed exploratory qualitative investigation with the support from a list of literature on Sundanese culture and on Kampung Naga. The process of identifying the research framework was based on an observation to the actual site, Kampung Naga that is located between the sacred forest and the river Ciwulan in Tasikmalaya, West Java. In addition, it also implements the approach of cultural studies. This approach assists in the disclosure of philosophies and cosmological meaning of the Sundanese culture in the elements of architecture and interior design of Kampung Naga's *rumah adat*.

[a] tessaeka82@gmail.com

3 Sundanese Culture and The History of Kampung Naga

Parahyangan is known as an old Sundanese name. The Dutch colonists who controlled the area for several centuries rendered the name into *Preanger* and today it is called Priangan [2]. After the fall of Padjajaran Kingdom in the 16th-century, the Sundanese or *Urang Sunda* (*Urang* means people, *Sunda* means Sundanese) had lost their powerful cultural and spiritual center. The absence of a powerful culture center does not mean Sundanese culture disappear. This can be reflected in the presence of Kampung Naga as Sundanese traditional village (Fig.1). The people of Kampung Naga strongly believe the noble values that their ancestors already taught, therefore the ancient traditions are still preserved and actualized in everyday life to present day. Since the rebellion of DI/TII (local community led by Karto Suwiryo), the worst tragedy had happened in 1956. Kampung Naga had no written manuscript because DI/TII people burnt not only all documents, but also the whole village. After 1956, the people of Kampung Naga try to rebuild the village with their memory without losing their identity [3], since then the history of Kampung Naga is mostly maintained by oral tradition – language. Language becomes a strong tool for Sundanese people.This statement is supported by Memed Sastrahadiprawira, a well-known Sundanese intellectual of the time, expressing it in Sundanese language as follow :

> "*basa teh anoe djadi loeloegoe, pangtetelana djeung pangdjembarna tina sagala tanda-tanda noe ngabedakeun bangsa pada bangsa. Lamoen sipatna roepa-roepa basa tea leungit, bedana bakat-bakatna kabangsaan oge moesna. Lamoen ras kabangsaanana soewoeng, basana eta bangsa eta oge lila-lila leungit* [4]*"*.

Translation:

> "The language forms a norm: the most evident and the most comprehensive symbols to distinguish one ethnic group from another. If the characteristics of a language disappears, the distinguishing features of an ethnicity will fade away as well. If an ethnicity no longer exists, the language of the ethnic group will also disappear in due course of time [5]".

Figure 1. Sundanese Traditional Village: Kampung Naga.

Sundanese villages tend to be a close grouping of houses, usually built on the mountainside.The village is surrounded by wet and dry fields, lying either in the oxbow of a river or in the sacred forest known as *hutan keramat* or *hutan larangan* [6] In the forest lies the grave of the village's ancestors called *nenek moyang* and a place where the spirits of nature are concentrated [7][8]. In Sundanese culture, the universe is divided into three parts based on the most ancient Sundanese cosmological concept called the third pattern, in Sundanese language known as *tritangtu.* This terminology means "*tilu nya hiji, hiji nya tilu*", three in one, one in three. *Tritangtu* may apply in Kampung Naga area. Based on it the Sundanese world are divided into three parts, known as *Buana Larang – Dunia Bawah* (under world), *Buana Panca Tengah – Dunia Tengah* (middle world) and *Buana Nyungcung – Dunia Bawah* (upper world). *Dunia Bawah* in Kampung Naga consists of field, fishpond, rice field and *saung lisung* – a space where rice is husked. Houses, mosque, gathering building or *Bale Patemon*, granaries called *leuit* and sacred houses known as *Bumi Ageung* are placed in the middle world. The upper world is the space for the grave of Eyang Singaparna who is believed as the ancestor of Kampung Naga [9]. The term of *tritangtu* is derived from Sanskrit, *tan* means spider web. In the holistic perspective, *tritangtu* can be seen as one of networking that expands outside but strongly fuses inside. *Tritangtu* is known as the basic concept of Sundanese life. The concept was born based on primordial farmers' thought. They believe everything in this world is always connected to something opposite or dualistic. To avoid the conflict, thus the third principle was born to balance it and became one harmony. Dualistic becomes trinity thought [1]. This trinity concept becomes the theoretical framework to analyze the relation between *rumah adat* in Kampung Naga and cosmology.

4 Discussion: The Traditional Houses of Kampung Naga

House in Sundanese is called in various names such as *bumi, imah* or *rorompok. Bumi* can also mean world, earth, globe or territory. Interestingly, the number of house in Kampung Naga cannot be extended. Based on the interview with Mr. Cahyan in January 2016, he said that Kampung Naga consists of 113 buildings, including the traditional houses, sacred house called *Bumi Ageung*,

DOI: 10.1051/matecconf/20166600108

the mosque, and the gathering building. If there are any new family members who do not have any inheritance from their parent, it is possible for them to build a house outside the traditional village. Additionally, they have other unique tradition about the house position. All buildings' main doors always face either south or north side and that the second door must be in the same side. This door is usually near to the kitchen area. Placing doors in the opposite walls are forbidden, as well as when all buildings face the east.

The type of Kampung Naga house is called *rumah panggung* with *julang ngapak* roof type (Fig 2). The shape of "*capit*" above the roof may represent both male and female [1].

Figure 2. Kampung Naga Traditional House (left) and *Julang Ngapak* (right).

Initially, Kampung Naga houses do not have any window and provide one door only. However, after 1956, the house has had window and two doors. Kampung Naga house is built by natural material. For them, life is not in nature but it is to live with nature [9]. It is *Pamali* or taboo to use other materials except those from the nature. It means that it is against the ancestors' tradition. Based on the house position resting on large stone plinths or called *umpak*, either natural or man made *umpak,* the posts are elevated to 40 cm above the ground or higher. The walls are traditionally made of two types of plaited bamboo matting, called *bilik kepang* and *bilik sasag*. The floors are made of wooden planks, whereas the ceiling are also made of *bilik kepang*. The frame of the roof assembly is made of bamboo and timbers covered by *injuk* and wild plant that locally known as *daun tepus.*

Traditional house in Kampung Naga not only serves as living space, but it also has philosophical value both vertically and horizontally. Vertical meaning represents the relationship between man and God, whereas horizontal meaning is about interactions among human being and nature. The culture in Kampung Naga retains many ceremonies. Based on interview with Mr. Cahyan, there is ceremony before building a new house called *Ngajarah Bumi*. It should provide *sesajen*, the ritual when chicken's head and herbs are buried in the middle of the house that will be built. This is aimed at giving salvation to the land and the nature spirits during the construction. Another ceremony known as *Salametan Bumi*. It is the ceremony after completing the house in order to assure harmony and good fortune. There are explanation about *Salametan Bumi*, expressing it in Sundanese statement "*saatos bumi beres, kedah aya syukuran, nyayagikeun kupat, dupi sareng tumpeng isi koneng. Bumi nu tos janten di doakeun ku pemangku adat teras bumi dipasang kale sareng sawen, janten tolak bala*." The statement means after the house is ready to inhabit, the household should offer *syukuran* or *salametan*, *salametan* is communal feast that provides meal. The *kuncen* will bless the house with *jajampe* (invocation or a prayer muttered over anything in order to produce some beneficial result), at the same time he gives *kale* and *sawen* as a ritual of *tolak bala*. *Tolak bala* is known as ritual activities aimed at avoiding something troublesome which may come through the involvement of a spirit, as well as some misfortune happening to a person or the community [10]. Usually, *kale* that is made from bamboo will be put in every corner of the house and *sawen*, the combination of leaf and grass, will placed in the front door [11].

Vertically, traditional house is divided into three spaces. The attic known as *para*, the ceiling called *lalangit,* and the space under the house called *kolong* or *kokolong*. The space between *lalangit* and *kokolong* is the actual house called *bumi* or *imah*. Horizontally, the house is divided into five areas (Fig 3.) - *tepas* (man territory) also known as guest reception room, *pawon* or kitchen, and rice storage called *goah* as female territory (Fig.4), *tengah imah* has similar function to family room and bedroom *known as pangkeng.*

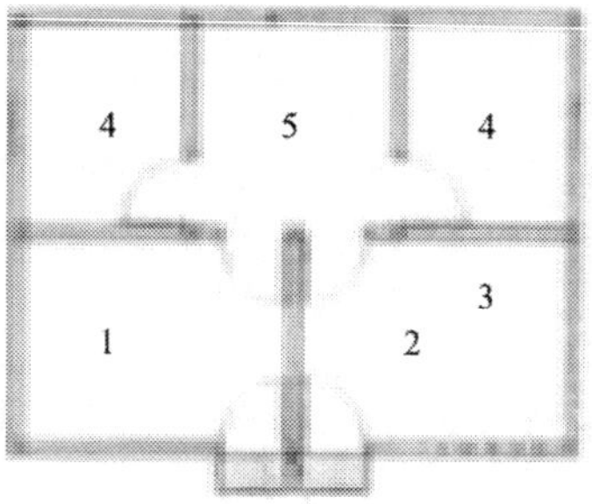

1 *Tepas*
(Man territory)
2 *Pawon*
(Kitchen)
3 *Goah*
(Rice storage)
4 *Pangkeng*
(Bedroom)
5 *Tengah Imah*
(Family room)

Figure 3. The Layout of Traditional House.

Figure 4. Tepas as a Man Territory (left) and Pawon as a Female Territory (right)

In Sundanese culture, female has special space that she can represent particular life, becoming "the content/*isi*" for human's life circle in a house. House is considered as place or "container/*wadah*" of life. In Sundanese culture, the terms "*isi*" and "*tempat*" relate closely to "Dunia Atas" and "Dunia Bawah" as parts of the Sundanese cosmology. It can be better understood using the

 DOI: 10.1051/matecconf/20166600108

approaches of poetry meaning or Sundanese ancient verse containing the meaning of Sundanese cosmology.

Earlier in this writing, it has been explained that ancient Sundanese people are included in the farmer category, therefore rice field and farmers are important for Sundanese society. This kind of rice (*padi*) relies heavily on rainfall, therefore the existence of the "Dunia Atas" are significantly important. In Sundanese culture, the Upper World is known as wet areas that are identical with sky, which has the nature of female, whereas the Under World is known as dry areas that are identical with land, which has the nature of male. The division of world based on that concept of gender can be seen in Sundanese verse. The Sundanese verse is a an epic story that rhymes. It tells the story of Kings or Princes from Galuh or Padjadjaran Kingdom. Commonly, the verse is recited with the accompaniment of Sundanese traditional music instrument, Kecapi. The followings are 12 line verse that is divided into two. They tell the story of Raden Layung at Panggung Karaton [12]:

A	B
Teras kangkung galeuh bitung	*Sisir badag tanduk kuda*
Tapak meri dina leuwi	*Kekemben layung kasunten*
Tapak soang dina bantar	*Kurambuan kuwung-kuwung*
Tapak sireum dina batu	*Tulis langit gurat mega*
Kalakay pare jumarum	*Panjangnya sabudeur jagat*
Sisir serit tanduk ucing	*Inten sagede baligo*

According to Jakob Sumardjo (2015), there are contradictions in meaning between part A and B, however if one of the part is reversed, then the meaning looks complementary. The verse could be seen as follows:

A	B
Sisir serit tanduk ucing Tapak	*Sisir badag tanduk kuda*
Kalakay pare jumarum	*Kekemben layung kasunten*
Tapak sireum dina batu	*Kurambuan kuwung-kuwung*
Tapak soang dina bantar	*Tulis langit gurat mega*
Tapak meri dina leuwi	*Panjangnya sabudeur jagat*
Teras kangkung galeuh bitung	*Inten sagede baligo*

The critical interpretations of the poetry are as follows:

A	B
Sisir serit tanduk ucing Tapak (as plural symbol, represent wordliness)	*Sisir badag tanduk kuda* (as singular symbol, represents heavenly nature)
Kalakay pare jumarum (*pare* means dried rice, down to earth in nature, represent male)	*Kekemben layung kasunten* (*layung* means twilight, high in nature, like sky, represents women)
Tapak sireum dina batu	*Kurambuan kuwung-kuwung* (*kurambun* means rainbow, high in nature, like sky)
Tapak soang dina bantar	*Tulis langit gurat mega* (*tulis langit* means writing in the sky, high in nature, like sky)
Tapak meri dina leuwi (*tapak* means foot print, down to earth in nature)	*Panjangnya sabudeur jagat* (*sabundeur jagat* means around the sky)
Teras kangkung galeuh bitung (it means "empty/*kosong*", represent "*wadah*" or container)	*Inten sagede baligo* (Inten means diamond, *baligo* or *beligo* means watermelon - diamond as big as watermelon), it means "content/*isi*")

4.1. Cosmological Value of The Traditional House

Why do houses in Kampung Naga become the center of "universe"? This can be answered using the analysis of *tritangtu* approach saying that house is divided into three parts; The Upper World (*Buana Nyungcung*), The Middle World (*Buana Panca Tengah*), and The Under World (*Buana Larang*) (Fig 5).

In the earlier section of Introduction, it is said that Sundanese Traditional house is considered as the central cosmos for Sundanese daily activities. This house can be said as the microcosmos or macrocosmos within the Sundanese culture. This is the center of "universe" for the surrounding environments, it gives "contents" for "emptiness". These terms can be identified from the concept of birth of the three patterns or *tritangtu,* which is known as the Sundanese philosophy.

In the process of *tritangu* emergence, every entity starts with "emptiness", then "content" is created. In Sundanese culture, dominated by the belief of Animism and Hinduism, it is believed that "emptiness" creates willingness known as *tekad. Kuasa* (power) relates closely to power known as *ucap* and to deed known as *lampah.* Those three philosophies are the modern terminology for *Batara Keresa*, *Batara Kawasa*, and *Batara Karana*. They are the nature or characteristics of *Batara Tunggal* or *Sang Rama*, where in Islamic approach is called as *Adam Tunggal.* Those natures are the "contents" for the "emptiness" [13].

The existence of house is started with emptiness. The potential of willingness, power, and deed creates the division of interior spaces to provide facilities for the inhabitants. The house becomes the macrocosmos for the inhabitants whose natures are microcosmos. The inhabitants unite with "the contents" that has been created into unity. The natures of the house change into microcosmos when the inhabitants leave the house. They switch position with the inhabitants as the macrocosmos, and therefore the house returns to become the center of cosmos for the surroundings.

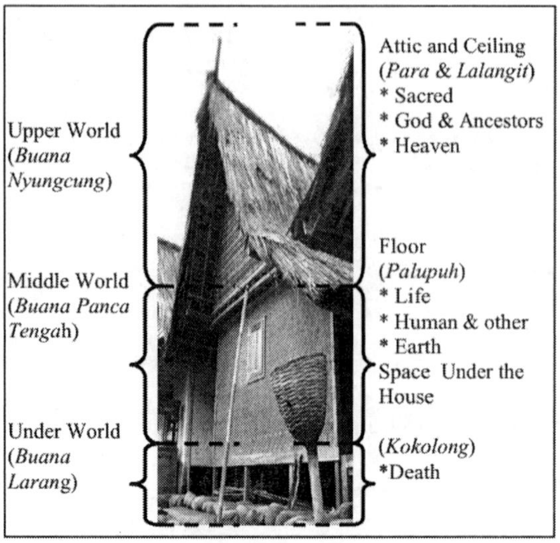

Figure 5. Kampung Naga Traditional House Based on The Universe Level

5 Summary

The people of Kampung Naga has to deal with globalization, they succeed sustaining their identity with unique traditions that consist of marvelous rituals and ceremonies, especially in building a house. They believe the Rumah Adat has not only in-depth meaning beyond the physical form, but also reflects the "wider" world. It also represents the harmonious relationship between human, nature, and their communication with God as the divine power. Many research discuss Sundanese traditional house in Kampung Naga. However, this paper is expected to give contribution by proposing in-depth analysis on the traditional house with cosmological approach called *tritangtu*, as the basic of ancient Sundanese principle cosmology. This paper may also be a reference for future research with more in-depth analysis.

References

1. Sumardjo, Jakob, *Sunda: Pola Rasionalitas Budaya*. Bandung, Kelir, (2015)
2. Spiller, Henry, Erotic Triangles: Sundanese Dance and Masculinity in West Java. Chicago, The University of Chicago Press, (2010).
3. Maria, Siti; Rosyadi; Indrawati, Dewi; Astuti, Renggo, *Sistem Keyakinan Pada Masyarakat Kampung Naga Dalam Mengelola Lingkungan Hidup – Studi Tentang Pantangan dan Larangan*, Jakarta, Departement Pendidikan dan Kebudayaan RI, **20** (1995)
4. Sastrahadiprawira, Memed, *Jodo Pakokolot*, **99** (1929)
5. Moriyama, Mikihiro, Sundanese Print Culture and Modernity in 19th-century West Java. Singapore: Singapore University Press, (2005)
6. Wessing, Robert, Constituting the world in the Sundanese house in Indonesian Houses: Volume 2: Survey of Vernacular Architecture in western Indonesia. Leiden: KITLV Press, (2008)
7. Ganda, Yahya, *Arsitektur Tradisional Daerah Jawa Barat. Bandung: Proyek Inventarisasi dan Dokumentasi Kebudayaan Daerah, Direktorat Sejarah dan Nilai Tradisional, Direktoral Jenderal Kebudayaan, Departement Pendidikan dan Kebudayaan*, **12** (1983)
8. Atmowiloto, Arswendo, Indonesia from The Air. Jakarta, Humpus. Singapore, Times, **102** (1986)
9. Hermawan, Iwan, *Bangunan Tradisional Kampung Naga: Bentuk Kearifan Warisan Leluhur Masyarakat Sunda*. Jurnal Sosio Didaktika, Vol. 1, No.2, **141-150** (2014)
10. Yousof, Ghulam-Sarwar, One Hundred and One Things Malay. Singapore: Partridge publishing, (2016)
11. Personal Interview with Mr. Cahyan, 9-10 January 2016 and 17-18 January 2016, Local People of Kampung Naga.
12. Rosidi, Ajip, Panggung Karaton, Project Research of Sundanese Poem and Folklore, Bandung, (1971)
13. Sumardjo, Jakob, *Estetika Paradoks*. Bandung, Kelir, **172** (2014)

The analysis of fire losses and characteristics of residential fires based on investigation data in Selangor, 2012-2014

Yi Rong Tan[1], Farid Wajdi Akashah[1,a] and Norhayati Mahyuddin[1]

[1]*Centre for Construction, Building and Urban Studies (CeBUS), Faculty of Built Environment, University of Malaya, 50603 Kuala Lumpur, Malaysia*

Abstract. This is a research in progress where authors seek to investigate the factors of residential fires. As part of the research, this paper aims to analyse the fire problems faced by the community of Malaysia. Data regarding residential fires between 2012 and 2014 was collected from fire investigation reports prepared by the Selangor Fire and Rescue Department. Descriptive analysis is conducted to summarize the data collected and describe the common phenomenon of residential fires. The distributions of the fire characteristics suggested that residential fires are commonly occurred during daytime, confined within the room of fire origin, caused by electrical failure and cooking negligence, started from kitchen, and occurred in multi-units housing. Further analysis will be conducted in order to investigate the relationships between the characteristics and residential fires.

1 Introduction

Overall, Malaysia has an average rate of fire incidents at around 1024.67 fires per million populations per year in 9 year-period from 2006 through 2014. The rate of fire victims is 7.53 per million populations per year, with 3.07 deaths per million populations per year. Each year, there are approximately 90 residential fires per million populations per year. Accounting for fire casualties, the total number of victims in residential fires has increased about 30% in the three year-period as shown in Figure 1. Although residential fires comprised less than 10% of the overall fires in the past three years, they took about 61% of the overall fire deaths per year and 48% of the overall fire injuries per year.

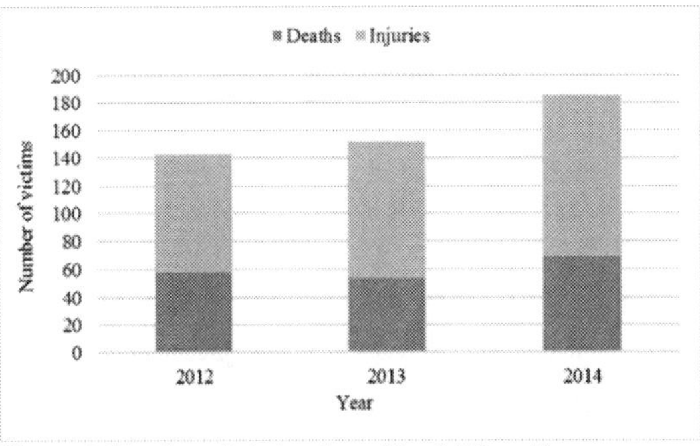

Figure 1. Number of victims in residential fires (Source: Department of Operation, Fire and Rescue Department of Malaysia)

In order to reduce the rate of residential fires and the fire casualties, an ongoing research is conducted in order to examine the factors affecting the residential fires in Malaysia. Over the decades, various studies were conducted to investigate the relationships of fire characteristics, occupant characteristics (i.e. personal or demographics), and housing characteristics with the residential fire incidence. Socio-economic characteristics are important in predicting the residential fire risk in urban areas [1, 2]. Using statistical models, Hasofer and Thomas [3] found that the most significant fire factors that contributed to the risk of casualties and fatalities in apartment fires are the extent of fire damage, area of fire origin, type of material ignited, and ignition factor. They also found that the condition preventing escape, condition before injuries, activity at time of injury, location at ignition and cause of injury to be the significant personal factors. Analysis by Holborn et al. [4] highlighted the common risk factors of unintentional dwelling fire deaths which comprised of smoking, alcohol intake, elderly, disability, illness, living alone, social deprived, and without having a functioning smoke alarm. Epidemiology study, such as Runyan et al. [5], also proved that mobile home and absence of smoke detectors will result in the risk of fire death.

This paper is part of the ongoing research aimed to investigate the fire problem by descriptive analysis of the characteristics from fire investigation data. Although annual reports regarding fire statistics have been published by Fire and Rescue Department of Malaysia (FRDM) every year [6, 7, 8], there are lack of detail information such as room or area of fire originated and

[a] Corresponding author: faridakashah@um.edu.my

the extent of fire spread. The outcome of the paper intends to understand the phenomenon of residential fires based on the following questions.

- When is the fire usually occurred?
- How the fire ignited?
- What is the source of fire?
- Where the fire started?
- To what extent the fire spread?
- How severe the fire cause to the losses and life safety?

2 Source of data

The data used in the research was abstracted from fire investigation reports prepared by Selangor Fire and Rescue Department. The fire investigation reports contained detailed information about the fire scenario including nature of fire, structural information and design of building, damage caused by fire, and victim's personalities. The overview of the data abstracted from the reports are described as below:

- Casualties – the consequences of fire incidents in terms of number of injuries and deaths.
- Fatalities – the number of deaths of fire incidents.
- Estimated properties losses – the fire loss in terms of financial or monetary, in ringgit Malaysia (RM), based on the properties and percentage of damages estimated by FRDM.
- Month of years – month of the years of the fire incidents.
- Day of weeks – day of the weeks of the fire incidents.
- Time of days – the estimated time of the fire ignition.
- Extent of fire spread – the spread of fire in terms of how far the fire damages reached.
- Cause of fire - The causes of the fire ignition, such as intentional or unintentional by human activities or natural phenomenon.
- Source of ignition - The sources that provide heat to starts the fire.
- Room of fire origin (RFO) – the area or room where the fire started.
- Types of dwelling – the types of houses based on the nature of housing units in the building.

The research refers to the years of residential fires occurred between 2012 and 2014. There were total 2423 residential fires being collected within the three years. However, the total number of fires cannot represent the total population of residential fire happened in the Selangor. This is because there are fires that were not reported to the fire department. Records from the fire department are only based on the incidents that were attended by or reported to them. In fact, a national survey in U.S. found that almost 97% of the residential fires were not reported or attended by the fire department [9]. Besides, there are also confidential cases that are not allow to be accessed.

Upon collecting the data, there are data that were not being recorded. This leads to missing data for certain variables in certain cases. Since the data are categorical, the fire problems will be described with percentage or frequency distributions. All valid data will be used in the analysis.

3 Results

3.1 Number of casualties

The number of casualties is the combination of injuries and fatalities. Table 1 shows the frequency distribution of the number of casualties per fires. Among 2423 residential fires, there are only 38 fire incidents result in at least one casualty. Over 98% of the fires did not have any injury or fatality. The proportions of the fires that have casualties were not more than 0.02 (2%) of the total number of fire incidents. This shows that the fires that will cause casualties are rare. As the number of fatalities increased, the events become rarer (Figure 2). However, this does not mean that the rare event should be ignored. The total number of casualties are 55 victims and 13 of them were died. There is one fatality occurred in every four injured victims.

Table 1. Frequency distribution of number of casualties per fires.

Number of casualties per fires	Frequency
0	2385
1	26
2	9
3	1
4	2
Total	2423

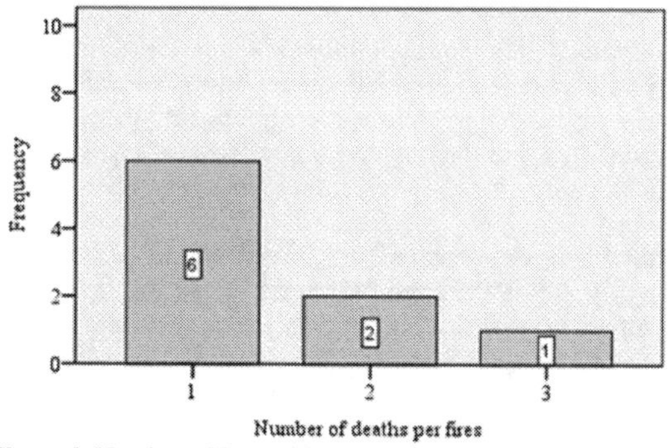

Figure 2. Number of fires with more than one fatality

3.2 Estimated properties losses

Fires damage properties and cause monetary losses. There are 2234 residential fires where the property losses are known. Based on these known samples, the property losses are ranged from no damage (i.e. RM 0) to damages that caused the loss of RM 3.5 million. The mean of property losses is approximately RM 27 thousand (Std. Dev. = RM 113,266.85). Most of the residential fires have property damages less than one thousand Ringgit Malaysia.

3.3 Time distribution

Although there are slightly less fire cases being observed at the end of the years, the overall frequencies are

 DOI: 10.1051/matecconf/20166600109

distributed quite evenly throughout the year (Figure 3) with an average of about 202 cases each month. This result is not consistent with other countries [10, 11] because Malaysia is a tropical country without changing of season throughout the year. Those countries encountered that cold winter (i.e. December, January, and February) were among the seasons that has higher number of fire cases occurred.

For distribution by day of week, weekdays tend to have higher number of fire occurrence compared to weekends with the highest in Wednesday as shown in Figure 4. This is different to data in China, where weekends have the higher fire occurrence [11]. Other than that, U.S. has a rather even distribution with about 14% everyday [10].

According to Figure 5, residential fires are commonly occurred during day time between 8 a.m. to 6 p.m. The number of fires decreased after 6 p.m. and reached its minimum in the midnight periods. This has reached the agreement with both data in U.S and China [10, 11].

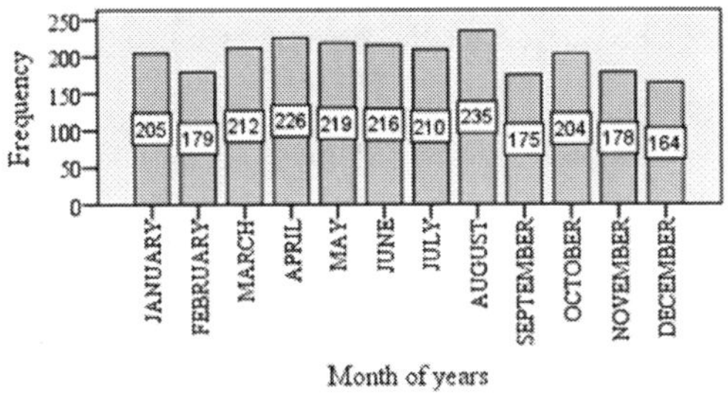

Figure 3. Frequency distribution by month of year

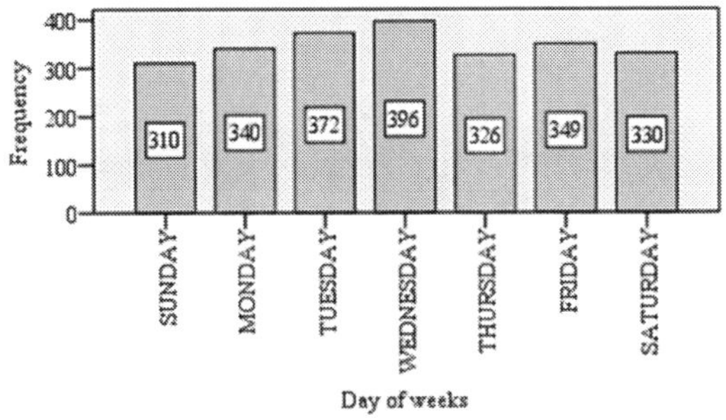

Figure 4. Frequency distribution by day of week

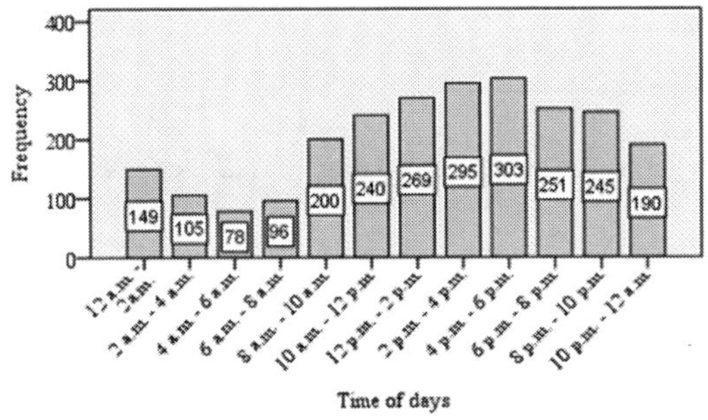

Figure 5. Frequency distribution by time of day

3.4 Fire damages

In relation to the estimated property losses, fire cases become rarer as percentage of fire damages (Figure 6) and extent of fire spread (Figure 7) reduced. In Figure 6, the number of fire with damage of more than 70% are higher than that of damage 10% - 20% and 20% - 70%. The percentage range of fire damages are depending on the size of the fire damages over the total size of the premise. Design of premise plays an important role determining the size of fires even though fire confined within the room of fire. Most of the studies used the extent of fire spread to interpret the fire damage. In Figure 7, residential fires are commonly confined within the room of fire origin (84.7%, including confined within the item of ignition) and most of them only confined within the items of ignition (44.2%).

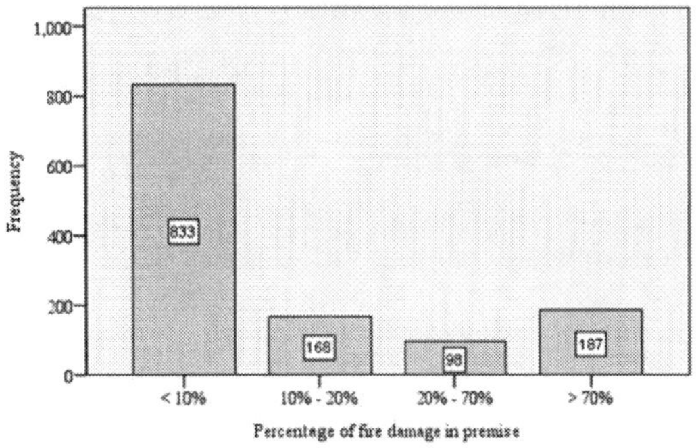

Figure 6. Percentage of fire damage in premise

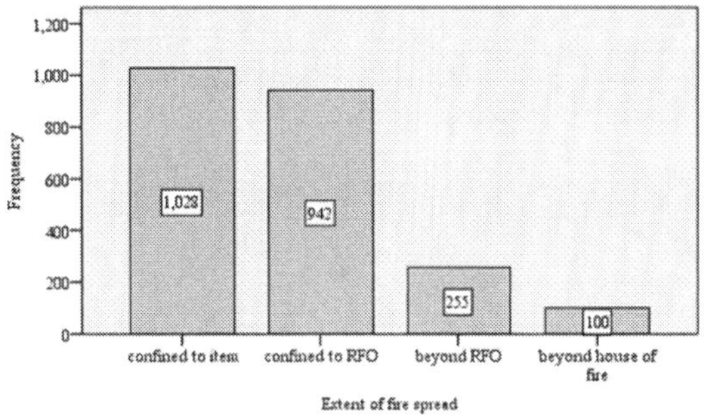

Figure 7. Extent of fire spread

3.5 Cause of fire

Based on Table 2, the most common cause of fire in residential fires are electrical failure (39.3%). Electrical failure included short circuit, resistance heating and overcurrent of electrical distribution and electrical appliances. There are almost 30% of the residential fires caused by careless or negligence during human activity such as cooking, followed by heat source close to combustible materials (8.7%) and leaking of gas (7.7%). There are minority of fires that are intentionally caused by irresponsible people including crime. They are incendiary fire with 4.2% and fires caused by unconscious person impaired by drugs, alcohol, or mental illness with only 0.1%. The percentages of intentional fires are relatively low compared to accidental fires.

DOI: 10.1051/matecconf/20166600109

Table 2. Frequency table of ignition factors

Ignition factors	Frequency
Electrical failure	852
Careless or negligence during human activity, e.g. cooking	611
Heat source close to combustible materials	189
Leaking of gas	167
Incendiary	92
Lightning strikes, including short circuit after the strikes	80
Others	76
Abandoned or discarded materials, e.g. cigarettes	47
Children playing with fire	47
Unknown, i.e. specified as unknown after the investigation	6
Drugs, alcohol impaired, mental illness	3
*Unspecified or missing data	253
Total	2423

*Unspecified or missing data is not included in the analysis

3.6 Source of ignition

Frequencies ranking in Table 3 shows that the most common source of ignition in residential fires are electric (39.7%), followed by cooking appliances (34.9%). This could be explained from the cause of fire, where electrical failure caused the electrical sparks, arching or heating that burned the electrical wiring system or electrical appliances. Ranking in the second, cooking appliances are another common source of ignition which were caused by the careless or negligence during cooking or leaking of cooking gas. Categorized within the group of glowing fire with only 5.8%, cigarette is not a common source of ignition in our cases.

Table 3. Frequency table of ignition sources

Sources of ignition	Frequency
Electric	949
Cooking appliances	834
Matches or lighter	194
Glowing fire, including cigarette, mosquito coil, and joss stick	138
Candle or torch	118
Lightning	80
Others	45
Hot object or surface	15
Fireworks or firecrackers	13
Friction	4
Unknown, i.e. specified as unknown after the investigation	3
*Unspecified or missing data	30
Total	2423

*Unspecified or missing data is not included in the analysis

3.7 Room of Fire Origin (RFO)

Almost 50% of the residential fires started in kitchen as shown in Figure 8. Figure 8 also shows that the most common rooms or areas of fire origin followed after the kitchen are bedroom and living room with 18.1% and 10.8%, respectively.

Figure 9 shows the frequencies of residential fires by both the room or area of fire origin and source of ignition. The analysis suggested that there is a significant relationship between the two variables. In the figure, most of the kitchen fires have the cooking appliances as source of ignition. Electric sources also standout in some of the room or area of fire origin, especially bedroom and living room.

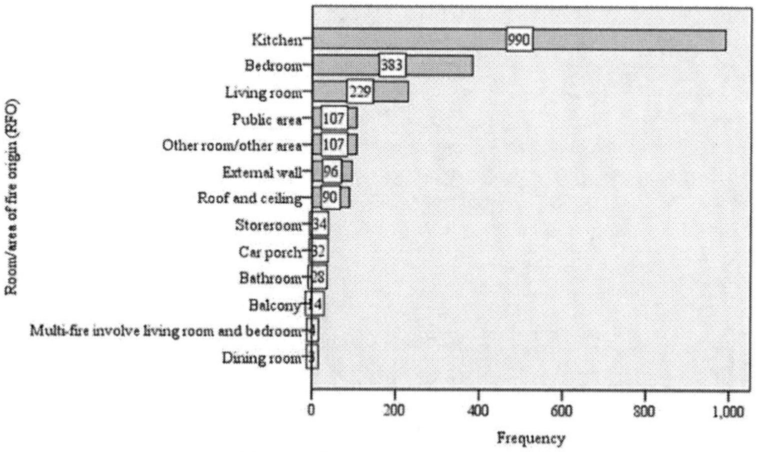

Figure 8. Room or area of fire origin

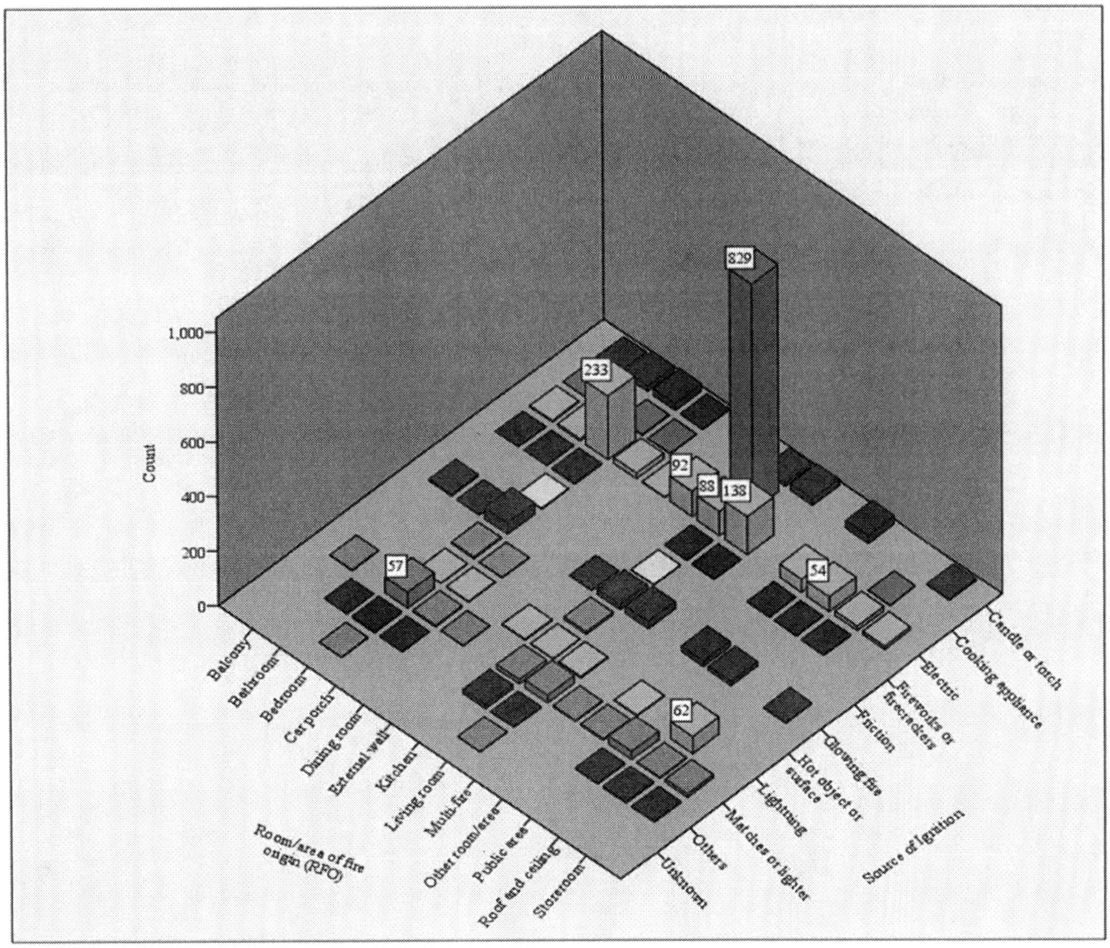

Fisher's Exact Test, $p < 0.001$ (2-sided test)
Figure 9. Relationship between RFO and Source of Ignition

3.8 Dwelling types

Figure 10 compared the numbers of fire frequencies between types of dwelling. Almost 40% of the fires occurred in multi-units housing, such as flat, apartment, or condominium. This is closely followed by terraced houses (33.6%). Other type of housing that were less likely to have fire occurrences are detached house, shop house, and semi-detached house which are 9.3%, 4.7% and 1.7%, respectively.

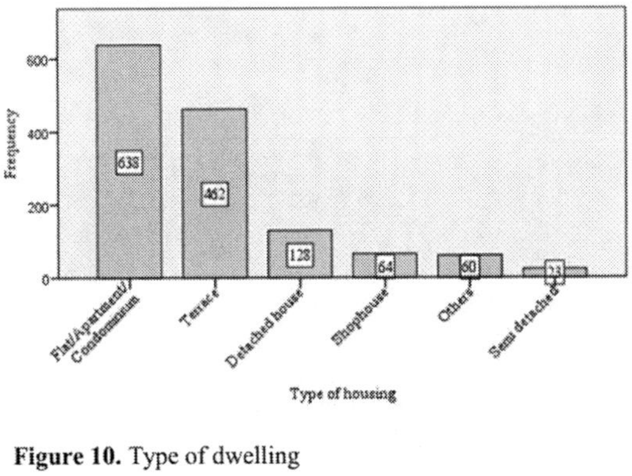

Figure 10. Type of dwelling

4 Conclusion

The results show the distribution of the fire losses and fire characteristics based on the investigation data in Selangor between 2012 and 2014. Based on the descriptive analysis, the fire problems faced in this community can be identified as highlighted as below.

- The fire frequency is distributed evenly throughout the years, unlike the other countries with four season where high frequency of fire was found to be happened during winter.
- Residential fires are commonly happened during daytime (8 a.m. – 6 p.m.) compared to night time (10 p.m. – 8 a.m.).
- Most of the residential fires did not spread beyond room of fire origin. They are either confined within the room or confined within the items of ignition.
- Two common source of ignition must be concerned, namely electricity and cooking appliances. This are closely related to the common cause of fire which are electrical failure and careless during cooking.
- Kitchen is the most likely room or area of fire origin due to the large amount of fire sources such as electrical

appliances and cooking appliances, including cooking gas.

• Comparing between dwelling types, multi-units housing such as flat, apartment, and condominium has the highest number of fire occurrences.

The conclusion based on the analysis of fire investigation data above provides the idea of how the scenarios of residential fires was and what is the issues that need to be concerned. After the initial analysis described in this paper, the data will be further analysed in order to investigate the factors that influence the residential fires.

Acknowledgements

This work was supported by the Ministry of Higher Education, Malaysia through the FRGS Grant (Grant No.: FP046-2014B). Besides, the authors would like to express their gratitude to the Fire and Rescue Department of Malaysia (FRDM) for granting the permission to access fire investigation reports and to all the officers for their supports in the research. Finally, a special thanks to the Fire Research Centre of FRDM (PUSPEK) who made this research possible to carried out.

References

1. C. R. Jennings, Fire Technol. **35(1)**, 7-34 (1999)
2. C. R. Jennings, Fire Safety J. **62**, 13-19 (2013)
3. A. M. Hasofer, I. Thomas, Fire Safety J. **41(1)**, 2-14 (2006)
4. P. G. Holborn, P. F. Nolan, J. Golt, Fire Safety J. **38(1)**, 1-42 (2003)
5. C. W. Runyan, S. I. Bangdiwala, M. A. Linzer, J. J. Sacks, J. Butts, New Engl. J. Med. **327(12)**, 859-863 (1992)
6. Fire and Rescue Department of Malaysia, *Annual Report 2012*
7. Fire and Rescue Department of Malaysia, *Annual Report 2013*
8. Fire and Rescue Department of Malaysia, *Annual Report 2014*
9. M. A. Greene, C. Andres, *2004-2005 National Sample Survey of Unreported Residential Fires* (2009)
10. M. Ahrens, *Home Structure Fires* (2013)
11. J. Xin, C. F. Huang, Fire Technol. **50(5)**, 1147-1161 (2014)

Comparative Analysis of Disabled Accessibility Needs of Heritage Building in Perak

[1]Nurul Fadzila Zahari, [2]Saidatul Farhana Harun, [3]Nur Azfahani Ahmad, [4]Zuraihana Ahmad Zawawi, [5]Nuzaihan Aras Agus Salim

[1,2,3,4] *Faculty of Architecture, Planning and Surveying, Universiti Teknologi MARA, Seri Iskandar Campus, Seri Iskandar, 32610, Perak, Malaysia*
[5]*Pusat PengajianPerumahan, Bangunan dan Perancangan, Universiti Sains Malaysia, 11800 USM, Pulau Pinang*

Abstract. Tourism sector was the sixth highest national income provider to the Malaysian economy in 2014. In order to replenish Malaysian economy through tourism, the Malaysian government has to diversify the present tourism products and offers a wide variety of tourism packages. This has mentioned in the National Key Results Area (NKRA) development platform highlighted in the 10th Malaysian Plan. Therefore, the tourism sector needs to continuously re-engineer and adapt its business model to suit every customer's needs and demands, including disabled people. At the moment, one of the highest tourist attraction contributors in Malaysia is the heritage building sector. The heritage building sector in Malaysia becomes popular due to its diverse historical background and culture. It attracts local and international tourists to visit. However, the lack of facilities provided especially for the disable people has hindered its future prospects to become globally popular. The national heritage should be viewed, explored and enjoyed by everybody without discriminating anyone. Insufficient of provision for disable facilities in heritage act has caused barrier to the disable people to enjoy and visit the heritage sites. The objective of this research is to analyze the comparative data that been retrieved in the field of selected case study. It will be carried out by visiting the selected case study, observation and documentary analysis. This research aims to do a comparative analysis of Disabled Accessibility needs of Heritage Building in Perak. The findings of this research will alert the needs of disabled in visiting the heritage building and documented for future research.

1 Introduction

Heritage building used to be described as inheritance of knowledge define within the social, political and culture context from the past (Graham, 2002). According to Harun, 2011, our Department of National Heritage has listed numerous buildings and monuments been protected, preserved and conserved from early 1990 to 2000. This shows increasing of awareness and understanding the practice and value of Heritage Building. A few methods frequently used to protect all this heritage buildings as stated in National Heritage Act, 2005 such as preservation, repair and maintenance, restoration, reconstruction, rehabilitation, adaptation and adaptive reuse or any combination of those processes.

Heritage building sector in Malaysia become popular due to its diverse historical background and culture. It attracts local and international tourists to visit. At the moment, one of the highest tourist attraction contributors in Malaysia is the heritage building sector (NKRA, 2015). However, the lack of facilities provided especially for the disable people has hindered its future prospects to become globally popular. The national heritage should be viewed, explored and enjoyed by everybody without discriminating anyone. Insufficient of provision for disable facilities in heritage act has caused barrier to the disable people to enjoy and visit the heritage sites.

Malaysia's policies and programs for the disabled citizens of the country are guided by the strategic goals of the National Welfare Policy, which emphasizes on the attainment of self-reliance, equalization of opportunities for the less fortunate and fostering of the spirit of mutual help and support towards enhancing a culture of caring.

2 Literature Review

2.1 Heritage

National Heritage Act 2005, defines heritage as any heritage site, heritage object, underwater cultural heritage or any living person declared as National Heritage as stated under section 67. Heritage can be divided into two categories which are Cultural heritage and Natural Heritage.

[a] Corresponding author: nfz7883@gmail.com

In order to ensure the historical buildings are protected and the authenticity being preserved as it original form, the international conservation guidelines and charters such as ICOMOS Australia, 1981 and United Kingdom Institute of Conservation (UKIC, 1983) has provide the conservation principle and standard which base on ethical code. The basic principles and standards of conservation which contain in international charters are:

i. Careful recording and research before intervention
ii. Minimum alteration of historic building's fabric
iii. Minimum risk of significant loss, damage or uncertainty in performance through intervention.
iv. Distinctive or distinguishable use of new and additional material
v. Longevity in finished work

The awareness of the importance of historic buildings among Malaysian is growing thus the national heritage should have been preserved accordingly by using the international charters and local guidelines (Harun, 2011). However, at the moment, there are limited provision established for disable people which have hinder them to visit this historical buildings. A research has shown that in comparing Malaysia and Singapore to the developed countries such as United Kingdom and Australia which had established Guideline in Improving Accessibility in Heritage Environment, as to improve the equality of person with disabled's (PWDs) right in accessing the heritage building. The differences in implementation of barrier-free in historic building and places are rather large between developed and developing country such as Malaysia. The policies in Malaysia regarding PWDs and accessibility in Heritage Environment may not reach the level of that in developed countries such as United Kingdom and Australia. This matters of policies should be taken seriously by government as it will ensure the equal right to enjoy and learn from historic building and places and in the same time it will increase the market for tourism (Marsin, Ariffin and Shahminan, 2014).

2.2 Person with Disabilities

According to World Health Organization (WHO, 2008), disabled people is a person or an individual with disabilities such as an impairment of problem in body function or structure; an activity limitation is a difficulty encountered by an individual in executing a task or action; while a participation restriction is a problem experienced by an individual in involvement in life situations. Disability is thus not just a health problem (Abdul Rahim A. and Abd. Samad N.A, 2010). It is a complex phenomenon, reflecting the interaction between features of a person's body and features of the society in which he or she lives. Overcoming the difficulties faced by people with disabilities requires interventions to remove environmental and social barriers (Abdul Rahim A. and Abd. Samad N.A, 2010). Its supported by the statement of Malaysian Prime Minister in Malaysian Information Network and Disabilities, 2010 which is 'apart from new buildings, existing buildings and public amenities, the local authorities would provide guidelines to make this places friendly to Person with Disabilities (PWDs) to cater their needs. Its clearer when Ch'ng, 2010 in his paper suggested that the current legislation is not sufficient for local authority to monitor the needs of PWDs in built environment industry in Malaysia.

However, at the moment, the disable people are being neglected from this group due to lack of access and facilities provided. Additional features on having access and facilities for PWDs shall not change the whole view of the heritage building but complimenting with the important activity in conserving the heritage buildings (Yaacob and Hashim, 2007).

2.3 International and Local Guidelines for Disabled Accessibility

Person with Disable Act came into forces in 2008 to recognize the importance of accessibility to the physical, social, economic, cultural environment, health, education as well as to communication and information enables the PWDs to full participate in society. Section 26 of PWD Act emphasized that PWDs shall have rights to access to and use of public facilities, amenities, services and buildings open or provided to the public on equal basis with persons without disabilities.

A proposal paper to upgrade services of PWDs for an accessible environment was tabled by the Minister to the National council for address local Authority to implement the laws especially in the Uniform Building By Law 34A (UBBL 34A) (Hussein and Yaacob, 2013). The Street, Drainage and Building Act 1974 was amended in 1991 to include By Law 34A of the UBBL34A. It was gazette by the state governments between 1992 and 1996. Uniform Building by Law 1984 (UBBL 1984) By Law 34A stated that a building must be accessible to disabled persons. A few standards were introduced such as Malaysian Standard MS 1183 (Specifications for Fire Precautions in design and Construction of buildings Part 8: Code of Practice for means of Escape for PWDs), and MS 1184 (Code of Practice for Access for PWDs in Building) and MS 1331 (Code of Practice for Access PWDs Outside Building). The Equality Act 2010 of United Kingdom gives people from discrimination in ranges of areas including accessing the services, education and employment. An organization cannot discriminate which means they need to provide or adapt their premises to allow disabled person to access services and employment. In pursuing the act, the Historic England recognizes that everyone should able to enjoy the access to the historic environment which later they introduced the guidelines of Easy Access to Historic Building to enable PWDs to enjoy their heritage environment. Based on Berkshire Accessibility by Design (2010) the building are divided into two which are Access to building and Access within building (referred to table 1). This design guidance encompasses almost aspect and element of building that can be adapted for disabled accessibilities.

Table 1: Accessible by Design, A Standard Guide, 2010

	Design Guidance

1.	Access to building
a.	Car Parking
b.	Routes
c.	External ramps
d.	External steps
e.	Handrails
2.	Access within building
a.	Entrances
b.	Entrance doors
c.	Entrance foyers
d.	Circulation
e.	Corridors
f.	Internal doors
g.	Surfaces
h.	Internal - Stairs - Ramps - Handrails
i.	Passenger lifts
j.	Platform lifts
k.	Wheelchair platform stair lifts
l.	WCs
m.	Equipment such as wheelchair

3 Methodology

This research used a qualitative method for excellent way of finalizing the result. The author use an observation method on selected case study to determine the disabled accessibility existed. These multiple case study were chosen according to the types and similarity of the typology of the heritage building. The selected case study are Museum Darul Ridzuan, Ipoh, Perak Museum, Taiping and Royal Museum at Kuala Kangsar. All selected case study are located in the state of Perak. All of them is two storey height and used as a museum. The variables are based on Berkshire Accessibility by Design, 2010 as it encompasses almost all elements of the building that can be adapted for Disabled Accessibility requirement. This variables used to compare with the selected multiple cases.

4 Case Study

The multiple selected case studies' background are summarized in table 2.

Table 2: Summary of the selected multiple case studies.

Case Study	Darul Ridzuan Museum	Perak Museum	Perak Royal Museum

Original Function	10 room mansion of a Kinta's successful miners	The first museum for a curator name Leonard Wray	The palace of Sultan Iskandar Shah
New Function	Museum of Ipoh, Perak.	Museum of Taiping, Perak.	Museum of Kuala Kangsar, Perak.
Year Built	1926	1883	1926
Years of Converted	1992		1986
Owners	Foo Choong Nyit One of Kinta's successful miners.	Leonard Wray Admirable collection of exhibit on natural history	Sultan Iskandar Shah,
Architecture	The grand mansion has a symmetrical design built in plastered brick construction with a tiled, pitched and gabled roof.	The central porter cohere is flanked by dominant three-story wings with twin towers	The building is built entirely of timber without nails, measuring from end to end, a total of 41.75
Existed PWD facilities	Car parking Access entrance parking	Car parking Access entrance distance Wheelchairs equipped	Access entrance distance Minimum degree of ramps

5 Finding and Analysis

The analysis of this research features comparative analysis on case studies with reference to the multiple selected case studies conducted in Perak. A comparative analysis is carried out to discuss the findings and observation on awareness of Disabled Accessibility at Heritage Building based on the case studies undertaken.

5.1 Comparative discussion of multiple cases.

5.1.1 External Environments

a) Car parking

All of three case studies did have car park for the disabled. It shows that the management aware of the disabled visitors.

b) Routes

The external travel or routes for the three cases are defined and only underground roots of nearby big tree needed to be cut away.

c) External Ramps

Only Perak Royal museum does not has the external ramp due to the limitations of material used for the museum which is timber. The other two cases are complete with the external ramps.

d) External Steps

The external step are accordingly to the UBBL 1984 requirements which are 185mm for the thread and 255mm for the riser.

e) Handrails

The handrail are available in these three cases and compliance to the UBBL 1984 which is maximum height is 1100mm and minimum of 900mm.

5.1.2 Internal Environment

a) Entrances

These multiple cases have sufficient width of entrance into the building. It is comfortable and ease the disabled to enter the building.

b) Entrance Doors

The entrance doors for Darul Ridzuan museum and Perak Museum have sufficient width and suitable material for doors. Meanwhile the Perak Royal Selangor Museum do not meet the requirement. Since the location of door is at the end of staircase without any ramp.

c) Entrance Foyers

For all these cases there are no entrance foyers space.

d) Circulation

The circulation of these multiple cases sufficient for the disabled to circulate inside the museum, except for the Perak Royal Museum due to the location of main entrance in the first floor.

e) Corridors

There is no corridor provided in these cases.

f) Internal doors

All of the cases had the internal doors.

g) Surfaces

The surfaces of these multiple cases good to walk, flat and ease the disabled to walk and using the provided equipment such as wheelchair and walker.

h) Internal

i. Stairs, Ramps and Handrails.
Only Perak Royal Museum does not has stairs and handrails due to the exhibition area are located on first floor. Ramp is only provided in Perak Museum.

i) Passenger lifts, Platform lifts, and wheelchair platform stair lifts

All these three requirement above were not provided by the management in these multiple cases.

j) WCs and Equipment as Wheelchair

Out of three case studies selected, only Perak Museum in Taiping provides WCs and equipment such as wheelchair.

Table 3: Comparative Result on Selected Multiple Case Studies

	Design Guidance	Case Study		
		Darul Ridzuan Museum	Perak Museum	Perak Royal Museum
1.	External Environment			
a.	Car Parking	√	√	√
b.	Routes	√	√	√
c.	External ramps	√	√	X
d.	External steps	√	√	√
e.	Handrails	√	√	√
2.	Internal Environment			
a.	Entrances	√	√	√
b.	Entrance doors	√	√	X
c.	Entrance foyers	-	-	-
d.	Circulation	√	√	X
e.	Corridors	-	-	-
f.	Internal doors	√	√	√
g.	Surfaces	√	√	√
h.	Internal			
	- Stairs	√	√	X
	- Ramps	X	√	X
	- Handrails	√	√	√
i.	Passenger lifts	X	X	X
j.	Platform lifts	X	X	X
k.	Wheelchair platform stair lifts	X	X	X
l.	WCs	X	√	X
m.	Equipment such as wheelchair	X	√	X

6 Conclusion

As conclusion, within three case studies selected, Perak Museum, Taiping, Perak had most of disabled accessibility provided. While Perak Royal Museum make the least provider of disabled accessibility. This is because the building itself are restricted based on the material used which is timber. It makes the building not ready to implement the disabled accessibility requirement. Improvising the accessible buildings and facilities are important to accommodate the tourist and visitors with different impairments. The heritage building also provided the common disabled accessibility such as parking space, toilet and wheel chair. Nevertheless, one of the selected case study do not has lift or platform to carry the disabled tourist to the upper floor.

This matters could hindered the tourist to bring their elderly to visit the building again. Furthermore, the current conservation guidelines and requirements are not comprehensive enough to address the disabled accessibility aspects as important criteria in conserving the heritage building which open to the public. For further recommendation, the author suggests to analyse the act, manual or guidelines of heritage in Malaysia purposely for disabled accessibility.

References

[1] A. A. Rahim (2010), Accessible built environment for the elderly and disabled in Malaysia: hotel as case studies, Malaysia.

[2] Accessibility by Design, A Standard Guide, Berkshire LABC Publishing Publication, 2010

[3] Bullen, P. & Love. P. (2010). The rhetoric of adaptive reuse or reality of demolition: Views from the field. *Cities*. 27(4), pp. 215-224. Retrieved from DOI: 10.1016/j.cities.2009.12.005

[4] Christina B. (2011). Adaptive Reuse: Explaining Collaboration within Complex Process, published Master Thesis, University of Oregon, USA

[5] Cunnington, P. (1988). *Change of Use*. London, UK: A & C Black Ltd.

[6] D.Urquhart (2007), Conversion of Traditional Buildings, Application of the Scottish Building Standard, Part 1 Principles and practice, published by Historic Scotland, Edinburgh.

[7] Douglas, J. Building Adaptation. Ed.: Elsevier Ltd., UK, 2006.

[8] Graham, B. (2002). Heritage as Knowledge: Capital or Culture". Urban studies, Volume **39**, Nos 5-6, pg. 1003-1017.

[9] ICOMOS Australia. (1981). Charter for the conservation of Places of Cultural Significance (The Burra Charter), Australia ICOMOS Icn, Sydney.

[**10**] Ismail, Md. R., (2003). Country paper: Malaysia. Expert Group Meeting and Seminar on an International Convention to Protect and Promote the Rights and Dignity of Persons with Disabilities Bangkok, Thailand, 2-4 June 2003.

[11] Kamaruddin H., Muhammad Ariff N.R, Wan Ismail W.Z, A.F Bakri, Z. Ithnin, (2014). Malaysian Scenario on Access and Facilities for Persons with Disabilities: A Literature Review. Building Surveying, Facilities Management and Engineering Conference (BSFMEC)

[12] K. S. Kamal, Kerosakan Bangunan dan Penyataan Kaedah Kerja Pemuliharaan Bangunan Bersejarah, 2011

[13] Langston, C., F. Wong, E. Hui And L. Y. Shen Strategic assessment of building adaptive reuse opportunities in Hong Kong. Building and Environment, 2008, **43**(10), 1709-1718.

[14] L. Ab. Wahab, Aplikasi Prinsip Penyesuaiangunaan terhadap Elemen Senibina Hotel Warisan di Tapak Warisan Dunia Melaka dan Georgetown, 2013

[15] N. H. Salleh, Fire safety and protection measures in heritage buildings with special consideration on museum buildings in Malaysia, 2012

[16] National Heritage Act 2005 (Act 645). (2005). Selangor D. E.: International Law Book Services.

[17] R.Prihatmanti and A.Bahaudin, 2011. Impacts of Adaptive Reuse In The UNESCO Listed Heritage Buildings, George Town, Penang. Retrieved from www.academia.edu on 26 November 2013.

[18] P.G.Tan, (2012), International Day of Person with disabilities 2012, 'The Digital Awakening, Living one more day' Blog 3rd December 2012 (Retrieved on 5th December 2013)

[19] R.C.C Wai (2008), Universal Design Considerations in Adaptive Reuse of Historic Building, HKIA Journal, Hong Kong.

[20] Regulation for Conservation Areas and Heritage Buildings by Municipal Council of Penang Island (*PeraturanuntukKawasanPemuliharaandanBangunan-bangunanWarisan – MajlisPerbandaranPulau Pinang)*

[21] Sandra D. (2010), Removing the Obstacles (Disability Access and the Arts in Heritage Buildings, Accessible Arts, NSW, Australia.

[22] S. N. Harun, 2011.Heritage Building Conservation in Malaysia: Experience and Challenges, published in The 2nd International Building Control Conference 2011 in Malaysia.

[23] S. N. Harun, 2011.Heritage Building Conservation in Malaysia: Experience and Challenges, published in The 2nd International Building Control Conference 2011 in Malaysia.

[24] Social Welfare Department (2007), National Advisory and Consultancy Council for Persons with Disabilities Meeting. (Meeting notes, 11 September 2007).

[25] United Kingdom Institute of Conservation, (UKIC, 1983). The United Kingdom Guidance for Practice, UKIC, London.

[26] UNESCO (United Nations Educational, Scientific and Cultural Organisations((1972). 7th session at general conference. *Convention concerning the protection of the world cultural and natural heritage.* New York: United Nations. Retrieved 15 November 2010, from http://whc.unesco.org/archive/convention-en.pdf

[27] Uniform Building By Law, 1984 (amendment of 2012), Malaysia.

[28] United Nations (2006). Convention on the Rights of Persons with Disabilities. New York: United Nation Treaty Collection. [Online]. Available at: http://www.un.org/disabilities/convention/conventionfull.shtml [retrieved on June 2015].

[29] Yaacob, N.M. and Hashim, N.R. (2007). Inclusive heritage tourism: A study on the access needs of disabled and elderly tourists in Malaysia. The 3rd Tourism Outlook onference and Global Event Congress II, Kuala Lumpur, 227-239.

[30] Yaacob, N.M., Accessibility to historic and listed public buildings: The development control process in England and Wales, 2007

The understanding of art students toward characteristic of Negeri Sembilan Minangkabau Traditional House

Nurul Shima Taharuddin[1a], Nurul Fadzila Zahari[2] and Asmalia Che Ahmad[3]

[1]Faculty of Art & Design, Universiti Teknologi MARA, Cawangan Perak, 32610 Seri Iskandar, Perak.
[2,3]Faculty of Architecture, Planning & Surveying, Universiti Teknologi MARA, Cawangan Perak, 32610 Seri Iskandar, Perak.

Abstract. In Negeri Sembilan, they are still practicing Minangkabau culture and custom. Element of uniqueness in Negeri. Sembilan has been shown on its architectural where the houses have dramatic curved roof structures with multitier. The art and architecture features a unique regional style. This house fills with cultural values, customs and reflects the people's understanding about designing art and architecture that is in harmony with nature. The house serves as a residence, a hall for family meetings, and for ceremonial activities. This research, studies the understanding of art students towards the characteristic that are found in the Negeri Sembilan Minangkabau Traditional House (NSMTH) in Negeri Sembilan, Malaysia. The objectives are to identify the element of characteristic that shows the identity of Negeri Sembilan Minangkabau Traditional House and to determine the level of understanding on characteristic of a Minangkabau house by art students. Scope of this research is on understanding of Faculty Art & Design student that has syllabus on Malay art. The research methodology that been use in this research is quantitative where surveys are made among the art students

1 Introduction

Negeri Sembilan is famous of its rich traditional culture and art. The Minangkabau Ethnic is one of the famous still practicing ethnic reside in Negeri Sembilan. Historically, going abroad was a part of requirement being Minangkabau men. All these men were expected to bring back fortunes and prosperity to their land. However, most of them migrate to other regions due to socio-economic reasons [1]. Its culture and art can be seen through the architectural style of their traditional house. The uniqueness of its upwards curved roof horn like shows the identity of Minangkabau ethnic and their history. It was told that there was a competition between village people and Jawa ethnic whom to said to conquer a land. It's a dual buffalo competition and when the winners win, they mimic the buffalo's horn in their roof style [2].

[a]Corresponding author: nurul026@perak.uitm.edu.my

There are a few phases of transition by Minangkabau ethnic [3]. The transition happened in 7th century, which is from West Sumatera to Palembang and to Siak. First phase of transition happened inside Indonesia itself. Second transition happened from Palembang to Singapore in 11th century to 12th century which known as first generation transition. The last phase happened from Singapore to Negeri Sembilan in 12th century to 15th century known as second generation transition and it last until today [3].

DOI: 10.1051/matecconf/20166600111

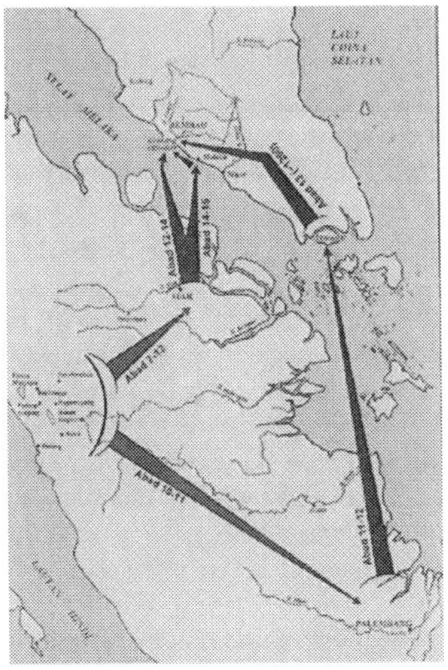

Figure 1 : Transition movement pattern to Negeri Sembilan in early age [3].

The architectural style of Negeri Sembilan Minangkabau Traditional House (NSMTH) has a unique character compare to other traditional house. Negeri Sembilan Minangkabau traditional house are divided into a few elements. The first element is the curved roof, which becomes the identity of the traditional Negeri Sembilan house [4]. The curved shape of a roof at Minangkabau house, resemblance a shape of a boat [4]. It is due to the mix marriage of the "Biduanda Jawa" ethnic with Minangkabau ethnic [4] – [5]. However there is also Minangkabau ethnic whom did not do mix marriage in order to maintain their original culture and custom [5]. It shows through the curved roof which resemblance the buffalo's horn. The research suggested the uses of two different names, which are The Negeri Sembilan Malay Traditional House on The Negeri Sembilan Minangkabau Traditional House, should be used (refer to figure 2).

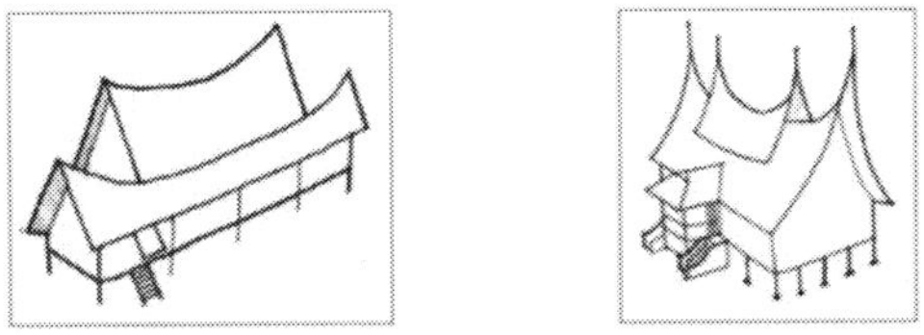

Figure 2 : (a)Sketch of typical Negeri Sembilan traditional house. (b)Sketch of Minagkabau traditional house in West Sumatera [5].

2 Problem statements

However, the uniqueness of traditional Minangkabau house has been forgotten and faded by influence of modern architectural style and living [1]. The interest amongst residence in Negeri Sembilan towards the architectural style of Minangkabau become less [1]. Due to less craftsmen interested in doing the Minangkabau Traditional house, the solution is to expose the knowledge about this unique characteristic to youngsters that will lead to sustainability of this Traditional house. As a result a survey on youngsters between 18 to 25 years old has been made to analyze the level of understanding among them towards the characters found in Minangkabau traditional house. Base on this result a solution will be made in order to solve the problem of lack of understanding about Minangkabau Traditional house among teenager.

3 Aim and Objectives

This research aims to explore the understanding of Art and Design student towards Negeri Sembilan Minangkabau Traditional House (NSMTH).

a) To determine the original characters of Negeri Sembilan Minangkabau Traditional House (NSMTH).
b) To investigate the level of understanding of Art and Design Students towards Negeri Sembilan Minangkabau Traditional House (NSMTH).

4 Scope of Research

The research scope was focus on Negeri Sembilan Minangkabau Traditional House (NSMTH). A survey towards 3rd Semester students was conducted among Higher Learning Education Institute students. These students consist of Diploma in Art and design students with five different courses; the courses are Fine Art, Fashion Design, Photography & Digital Imaging, Graphic Design & Digital Media, and Ceramic. Levels of understanding are determined based on the characteristics of Negeri Sembilan Minangkabau Traditional House (NSMTH).

5 Literature Review

Generally, Malay ancestors had built their house harmoniously to the occupants [7]. Most of Malay traditional house has similarity of its elements and characters such as most of traditional house must be built with column [4].

There are a few factors that influenced construction of a traditional house. Local climate, with a high temperature and rainfall pattern in certain area makes the air moist which will determine the character and design of the traditional house [4]. The Malay Archipelago is famous with richness of its forest and good timber and it is beneficial to the people in constructing their house [4].
All traditional houses have aesthetic value in its construction. Every space has its meaning, usage and philosophy [4]. According to the principle of space in a traditional house are divided into three parts which are front house, middle house and back.

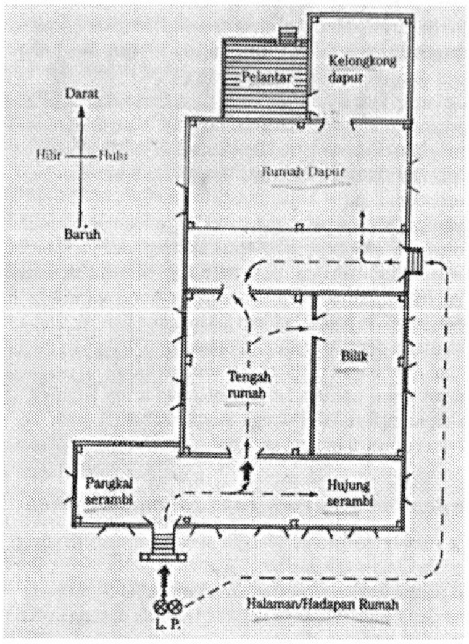

Figure 3 :Sketch of space in a traditional House [4].

There are a few characters at Negeri Sembilan Minangkabau Tradisional House, the upwards curved roof which resemblance shape of a boat (A. *bumbung melentik*) [4], A.1. *papan gendeng* a beautiful carved timber and A.2 gabled (*Tebar Layar*). a triangle wood carving ornamentation with a hollow concepts which function as ventilation and lighting. The two layers roof was constructed on the middle space and extension space in front of the house on top of *serambi* [5]. The uniqueness of Negeri Sembilan house also shown on its multi-tiered roof structure (B. *Bumbung Berlapis*) [8] and have a C. long house and a beautiful N. courtyard.
Interior space that been design in Negeri Sembilan Minangkabau Tradisional House divided into a few spaces with variety functions such as space for family gathering, entertain visitors, kitchens and private space [8]. Mother house or E. *rumah ibu* was divided into E.2 rooms, E.1 *kelek anak*, D. *serambi rumah*, F. *peran* or attic *(loteng)*. F. Attic is a space to store things or used as a bedroom [4]. Besides space provided for family's activities, there are also space to entertain neighbour who come to the house, important person who came for discussion will be at the *serambi*. Most of traditional house were built with *serambi* [4]. Serambi were divided into three which are D.1 *pangkal* (the beginning) *serambi*, D.2 middle *serambi* and D.3 end *serambi*. Other than that, *serambi* also can be used as a place to sleep by the son [4]. However, the begining of *serambi* is used to entertain the *Adat* people, while end *serambi* is for pious people and head of village [9].
The G. kitchen was divided into a few spaces which called asG.1 *kelongkang dapur* where it is used to cook and be a rear platform. The other space is called G.2 *Pelantar belakang,* this space is used to collect water from raining and save it in a *tempayan* [4].
Column is the main element constructed in a traditional house. The floor will be constructed higher than the ground to avoid wild animal and poisonous animal to enter the house [10]. In order to construct any traditional houses, I. column (*tiang*) is one of vital element. For a Negeri Sembilan Traditional house, column or *tiang* divided into four types which are I.1 *tiang guntung* which is a short column installed at the end of the edge of joist, I.2 *tiang empat segi* a column which built for external wall of a house, I.3 *tiang pecah lapan* a column used for *kembi of pemidang* wall system and lastly I.4 *tiang seri* which is a middle column located at the center of the mother house [4]. L. floor was divided into two which are L.1 normal floor and L.2 attic floor (*lantai loteng)*. A floor was considered as a sacred and clean element meanwhile, floor for attic used to separate house area and the attic.
Another common space in a Negeri Sembilan Traditional house is *kolong* which is located under the house. H. *Kolong* (vault) is used as storage. J. *Tunjak langit* or King Post function to hold the roof trusses from collapse [11]. Negeri Sembilan Minangkabau Traditional House is famous with is construction technique not using single nails to connect its jointing or called as mortise technique or K. *tanggam* [4]. Usually NSMTH has M wood carving ornamentation on door, window and ladder.

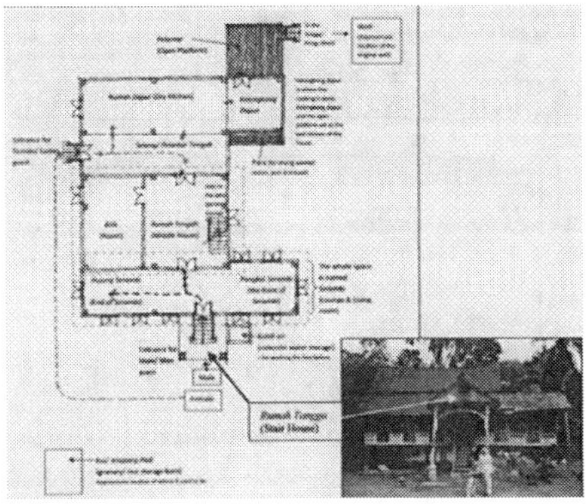

Figure 4 : A typical layout of Negeri Sembilan traditional house layput (Taipot mother of Moyang Ijah's house in Kg. Penajeh based on the 4th , 5th and 7th generation description) with image of Esah (Uwan Itam) the owner which is the 4th generation of Taipot [5].

6 Research Methodology

This research was conducted using quantitative survey method. A total of 122 students from five courses in Faculty or Art and Design were selected to be respondents in this survey. Art and design student has a syllabus on architectural Malays house, where the scope of the study is based on the characteristic or Malays house in Malaysia. Objective of this syllabus is to enhance the knowledge towards Malays traditional house so that in future this students will use the knowledge to produce an artwork that has Malays architecture identity in a way to promote Malays traditions and culture. Fine Art, Fashion Design, Photography & Digital Imaging, Graphic Design & Digital Media, and Ceramic Level of understanding among the students toward characteristics and unique elements of Negeri Sembilan Minangkabau traditional house is measured. In order to determine the level of understanding of Art student on characteristic of Minangkabau house, structure of question is based from book written by Idris Yaakob, Rumah Tradisi Negeri Sembilan. All question is based on the characteristic that been identified by Idris Yusof and was written in his book. There are twenty nine (29) characters of Negeri Sembilan Minangkabau traditional house has been distributed to all respondents. Out of 122 total respondents, only 113respondents answered the questionnaires. Input from this questionnaires were analyzed using Microsoft Excel 2010 Software and shown in various charts and bars. Table of frequencies is to show the data that has been collected and chart is to show the differences between departments. Each classes has difference background and difference interest.

7 Findings

Table 1: Frequencies Of Student according To Courses base on type of respondents residential.

No	Type Of Residential House	Frequencies Of Student According To Courses					Total
		Fine Art	Fashion Design	Photography & Digital Imaging	Graphic Design & Digital Media	Ceramic	
1.	Traditional House	5	2	-	1	2	10
2.	Renovated Traditional House	1	4	6	6	1	18
3.	Bungalow	-	3	3	8	2	16
4.	Terrace House	8	14	9	8	1	40
5.	Semi Detach	1	3	-	2	-	6
6.	High rise Flat	1	3	1	-	-	5
7.	High rise Apartment	-	3	1	5	-	9

8.	High rise Condominium	-	3	-	1	-	4

Chart 1: Frequencies Of Student according To Courses base on type of respondents residential.

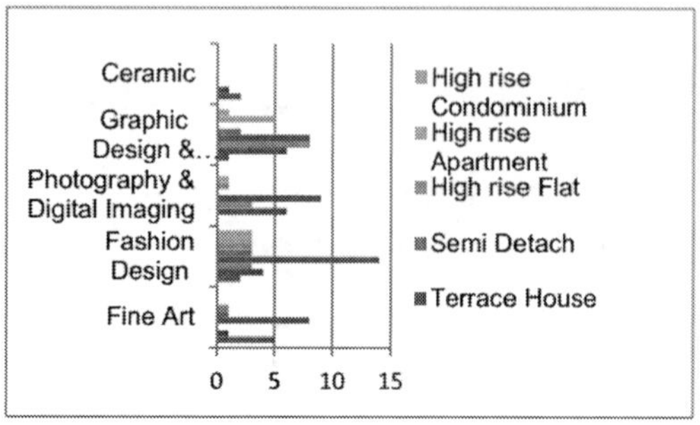

Frequencies of Student according to courses base on type of respondent's residential shows that majority of student living in a terrace house compare to other houses. Based on the results it shows that majority of the student didn't have much exposure to traditional house because they live in urban way of life. Usually the background of living influence the way student behave and think. This also implement to what characteristic of house they live in.

Data collection from survey among Fine Art students

Table 2: Mean on Level of Understanding towards of NSMTH

No	Characteristics Of NSMTH	Level Of Understanding (Mean)
A	*Bumbung melentik* (upwards curved roof)	2.5
A.1	*Papan gendeng* (a beautiful carved timber)	2.8
A.2	*Tebar Layar* (Gabled)	2.5
B	*Bumbung Berlapis* (Multi-tiered roof structure)	2.3
C	*Rumah yang panjang* (Long house)	2.4
D	*Serambi*	2.6
D.1	*Pangkal serambi* (the beginning of *serambi*)	2.5
D.2	*Tengah serambi* Middle of *serambi*)	2.2
D.3	*Hujung serambi* (end of *serambi*)	2.2
E	*rumah ibu* (Mother house)	2.5
E.1	*Kelek anak* (side room)	2.4
E.2	*Bilik* (rooms)	2.9
F	*Peran or loteng* (attic)	2.6
G	*Dapur* (Kitchen)	3.2
G.1	*Kelongkang dapur* (where it is used to cook and be a rear platform)	2.6
G.2	*Pelantar belakang* (space is used to collect water from raining and save it in a *tempayan*)	2.4
H	*Kolong* (vault that use for storage)	2.6
I	*Tiang* (column)	2.8
I.1	*Tiang guntung* (which is a short column installed at the end of the edge of joist)	2.6
I.2	*Tiang empat segi* (a column which built for external wall of a house)	2.4
I.3	*Tiang pecah lapan* (a column used for *kembi of pemidang* wall system)	2.1
I.4	*Tiang seri* (which is a middle column located at the center of the mother house)	2.5
J	*Tunjak langit* (King Post function to hold the roof trusses from collapse)	2.8
K	*Tanggam* (construction technique not using single nails to connect its jointing or called as mortise technique)	2.5
L	*Lantai* (floor will be constructed higher than the ground to avoid wild animal and poisonous animal to enter the house)	2.8
L.1	*Lantai Biasa* (normal floor)	2.8
L.2	*Lantai loteng* (attic floor)	2.8
M	*Ukiran* (wood carving ornamentation on door, window and ladder)	2.9
N	*Halaman Rumah* (courtyard)	3

Chart 2: Frequencies Level of Understanding towards of NSMTH

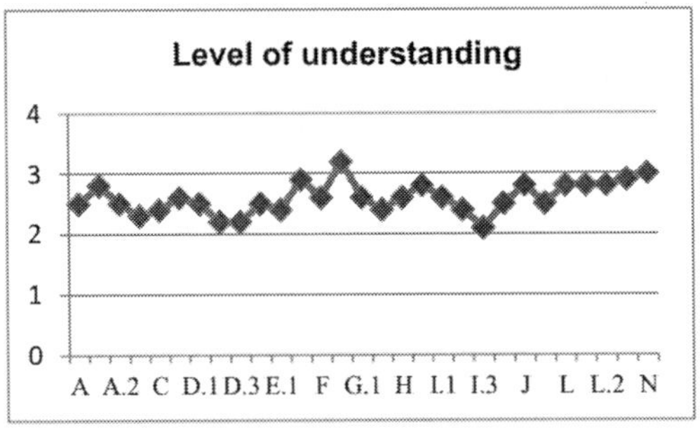

Highest average understanding of Fine Art students are 3.2 which is kitchen and the lowest is 2.1. which is *Tiang Pecah Lapan*. Kitchen is the most common knowledge in characteristic of a Negeri Sembilan Minangkabau Traditional House.While, the lowest knowledge is regarding *Tiang Pecah Lapan*. Most of modern house do not have *Tiang Pecah Lapan*. Total understanding of Fine Art students is between 2 and 3, which is fair and moderate. This result is the evidence of the level of understanding among teenager. Knowledge and awareness of using the knowledge is very importance in a way for this student to implement it in their work and use this knowledge in a future development.

Data collection from survey among Fashion Design students

Table 3: Mean on Level of Understanding towards of NSMTH

No	Characteristics Of NSMTH	Level Of Understanding (Mean)
A	*Bumbung melentik* (upwards curved roof)	3.2
A.1	*Papan gendeng* (a beautiful carved timber)	2.5
A.2	*Tebar Layar* (Gabled)	2.6
B	*Bumbung Berlapis* (Multi-tiered roof structure)	3
C	*Rumah yang panjang* (Long house)	3.1
D	*Serambi*	2.9
D.1	*Pangkal serambi* (the beginning of *serambi*)	2.1
D.2	*Tengah serambi* Middle of *serambi*)	2.2
D.3	*Hujung serambi* (end of *serambi*)	2
E	*rumah ibu* (Mother house)	2.3
E.1	*Kelek anak* (side room)	2
E.2	*Bilik* (rooms)	3.5
F	*Peran or loteng* (attic)	2.8
G	*Dapur* (Kitchen)	3.5
G.1	*Kelongkang dapur* (where it is used to cook and be a rear platform)	2.8
G.2	*Pelantar belakang* (space is used to collect water from raining and save it in a *tempayan*)	2.6
H	*Kolong* (vault that use for storage)	2.6
I	*Tiang* (column)	3.4
I.1	*Tiang guntung* (which is a short column installed at the end of the edge of joist)	2.5
I.2	*Tiang empat segi* (a column which built for external wall of a house)	2.4
I.3	*Tiang pecah lapan* (a column used for *kembi of pemidang* wall system)	2.1
I.4	*Tiang seri* (which is a middle column located at the center of the mother house)	2.8
J	*Tunjak langit* (King Post function to hold the roof trusses from collapse)	2.4
K	*Tanggam* (construction technique not using single nails to connect its jointing or called as mortise technique)	2.2
L	*Lantai* (floor will be constructed higher than the ground to avoid wild animal and poisonous animal to enter the house)	3.3
L.1	*Lantai Biasa* (normal floor)	3.2
L.2	*Lantai loteng* (attic floor)	2.9
M	*Ukiran* (wood carving ornamentation on door, window and ladder)	3.4
N	*Halaman Rumah* (courtyard)	3.6

Chart 3: Frequencies Level of Understanding towards of NSMTH

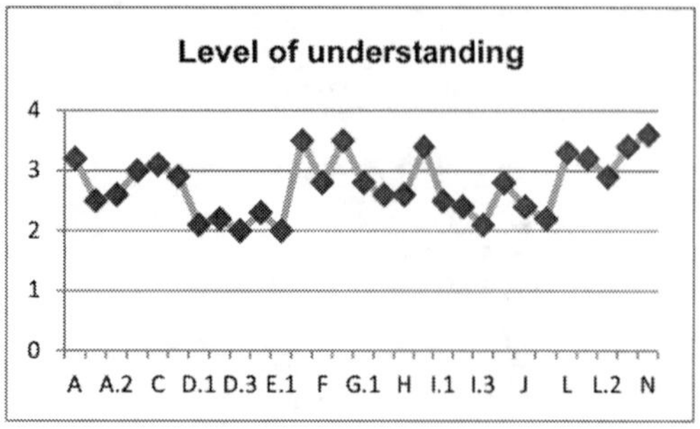

The highest scale from Fashion's students is 3.6 which is house yard (question N) and the lowest is end *Serambi* and *Kelek Anak* which is 2.0. Through the observation made by the author, the Fashion students have less knowledge on the traditional house characteristics. Fashion student have difference result majority of student can identify the characteristic of traditional house because, they have syllabus about traditional custom design and Minagkabau traditional house is one of the influence towards Minangkabau costume furthermore usually they have academic trip that sometimes the destination is Negeri Sembilan architecture for their subject matter. This means that fashion student have better exposure to traditional house compare to others.

Data collection from survey among Photography & Digital Imaging students

Table 4: Mean on Level of Understanding towards of NSMTH

No	Characteristics Of NSMTH	Level Of Understanding (Mean)
A	*Bumbung melentik* (upwards curved roof)	1.6
A.1	*Papan gendeng* (a beautiful carved timber)	1.7
A.2	*Tebar Layar* (Gabled)	1.7
B	*Bumbung Berlapis* (Multi-tiered roof structure)	2.0
C	*Rumah yang panjang* (Long house)	2.3
D	*Serambi*	2.1
D.1	*Pangkal serambi* (the beginning of *serambi*)	1.5
D.2	*Tengah serambi* Middle of *serambi*)	1.6
D.3	*Hujung serambi* (end of *serambi*)	1.4
E	*rumah ibu* (Mother house)	1.7
E.1	*Kelek anak* (side room)	1.6
E.2	*Bilik* (rooms)	2.9
F	*Peran or loteng* (attic)	2.2
G	*Dapur* (Kitchen)	3.0
G.1	*Kelongkang dapur* (where it is used to cook and be a rear platform)	1.8
G.2	*Pelantar belakang* (space is used to collect water from raining and save it in a *tempayan*)	1.4
H	*Kolong* (vault that use for storage)	1.8
I	*Tiang* (column)	2.6
I.1	*Tiang guntung* (which is a short column installed at the end of the edge of joist)	1.8
I.2	*Tiang empat segi* (a column which built for external wall of a house)	1.5
I.3	*Tiang pecah lapan* (a column used for *kembi of pemidang* wall system)	1.4
I.4	*Tiang seri* (which is a middle column located at the center of the mother house)	1.9
J	*Tunjak langit* (King Post function to hold the roof trusses from collapse)	1.6
K	*Tanggam* (construction technique not using single nails to connect its jointing or called as mortise technique)	1.9
L	*Lantai* (floor will be constructed higher than the ground to avoid wild animal and poisonous animal to enter the house)	2.7
L.1	*Lantai Biasa* (normal floor)	2.7
L.2	*Lantai loteng* (attic floor)	2.1
M	*Ukiran* (wood carving ornamentation on door, window and ladder)	2.4
N	*Halaman Rumah* (courtyard)	3.3

Chart 4: Frequencies Level of Understanding towards of NSMTH

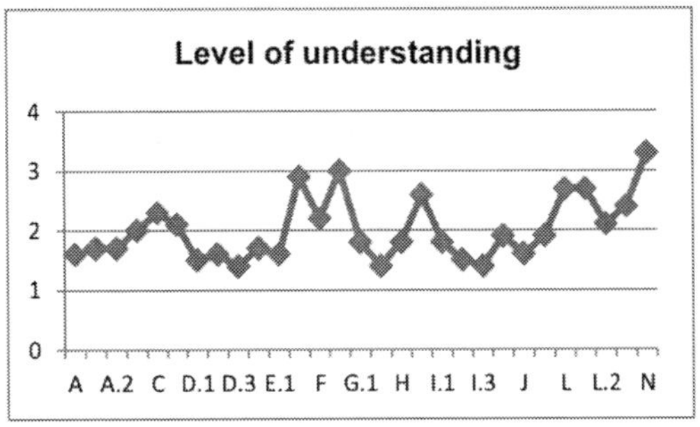

Based on the survey on Photography students, the highest scale is 3.3 (house yard) and the second highest scale 3 (kitchen) and the third highest is 2.9 (room). These three level of understanding are moderate. The lowest three scale are question D.3 (end *serambi*), question G.2 (rear platform) and question I.3 (*tiang pecah lapan*) which is 1.4. Photography student have the lowest knowledge towards Minagakabau traditional house because of lack of exposure. They usually have academic trip but their subject matter is more towards human and nature and they usually travel to east Malaysia which is Kelantan. Less exposure to Negeri Sembilan compare to Kelantan. This information is based on observation on photography approval academic trip. Student should be more creative in choosing the places for their subject matter exploration.

Data collection from survey among Graphic Design & Digital Media students

Table 5: Mean on Level of Understanding towards of NSMTH

No	Characteristics of NSMTH	Level of understanding (MEAN)
A	*Bumbung melentik* (upwards curved roof)	2.8
A.1	*Papan gendeng* (a beautiful carved timber)	2.4
A.2	*Tebar Layar* (Gabled)	2.7
B	*Bumbung Berlapis* (Multi-tiered roof structure)	3.1
C	*Rumah yang panjang* (Long house)	3
D	*Serambi*	3
D.1	*Pangkal serambi* (the beginning of *serambi*)	2.2
D.2	*Tengah serambi* Middle of *serambi*)	2.5
D.3	*Hujung serambi* (end of *serambi*)	2.3
E	*rumah ibu* (Mother house)	2.3
E.1	*Kelek anak* (side room)	1.8
E.2	*Bilik* (rooms)	3.4
F	*Peran or loteng* (attic)	3
G	*Dapur* (Kitchen)	3.8
G.1	*Kelongkang dapur* (where it is used to cook and be a rear platform)	2.9
G.2	*Pelantar belakang* (space is used to collect water from raining and save it in a *tempayan*)	2.6
H	*Kolong* (vault that use for storage)	3.2
I	*Tiang* (column)	3.6
I.1	*Tiang guntung* (which is a short column installed at the end of the edge of joist)	2.4
I.2	*Tiang empat segi* (a column which built for external wall of a house)	2.3
I.3	*Tiang pecah lapan* (a column used for *kembi of pemidang* wall system)	2.1
I.4	*Tiang seri* (which is a middle column located at the center of the mother house)	2.6
J	*Tunjak langit* (King Post function to hold the roof trusses from collapse)	2.5
K	*Tanggam* (construction technique not using single nails to connect its jointing or called as mortise technique)	2.5
L	*Lantai* (floor will be constructed higher than the ground to avoid wild animal and poisonous animal to enter the house)	3.8
L.1	*Lantai Biasa* (normal floor)	3.3
L.2	*Lantai loteng* (attic floor)	3
M	*Ukiran* (wood carving ornamentation on door, window and ladder)	3.5
N	*Halaman Rumah* (courtyard)	4

 DOI: 10.1051/matecconf/20166600111

Chart 5: Frequencies Level of Understanding towards of NSMTH

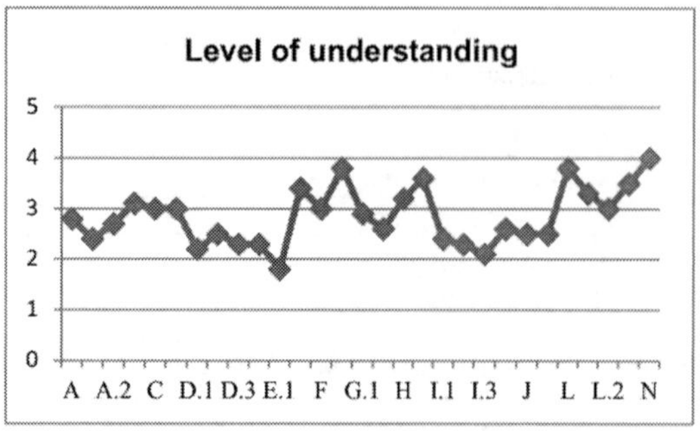

Result from questionnaire distributed among graphic students find that the lowest scale is 1.8 which is *Kelek Anak* while the highest scale is 4.0 which is yard. Level 4 of understanding indicate well and 1.8 is fair. There is large gap within variation characteristic of Negeri Sembilan Minangkabau Traditional House. Based on the graph above, there are five (5) characters in scale of 3.5 which are kitchen (question G), column (question I), floor (question L), ornamentation (question M) and house yard (question N). These five characters are the main characters in most of traditional house in Malaysia.

Data collection from survey among Ceramics students

Table 6: Mean on Level of Understanding towards of NSMTH

No	Characteristics of NSMTH	Level of understanding (Mean)
A	*Bumbung melentik* (upwards curved roof)	2.5
A.1	*Papan gendeng* (a beautiful carved timber)	2.2
A.2	*Tebar Layar* (Gabled)	2.2
B	*Bumbung Berlapis* (Multi-tiered roof structure)	2.7
C	*Rumah yang panjang* (Long house)	1.7
D	*Serambi*	2.5
D.1	*Pangkal serambi* (the beginning of *serambi*)	1.2
D.2	*Tengah serambi* Middle of *serambi*)	1.3
D.3	*Hujung serambi* (end of *serambi*)	1.2
E	*rumah ibu* (Mother house)	1.3
E.1	*Kelek anak* (side room)	1.2
E.2	*Bilik* (rooms)	3.2
F	*Peran or loteng* (attic)	3.2
G	*Dapur* (Kitchen)	2.5
G.1	*Kelongkang dapur* (where it is used to cook and be a rear platform)	2.0
G.2	*Pelantar belakang* (space is used to collect water from raining and save it in a *tempayan*)	2.0
H	*Kolong* (vault that use for storage)	2.2
I	*Tiang* (column)	3.2
I.1	*Tiang guntung* (which is a short column installed at the end of the edge of joist)	1.8
I.2	*Tiang empat segi* (a column which built for external wall of a house)	1.8
I.3	*Tiang pecah lapan* (a column used for *kembi of pemidang* wall system)	1.7
I.4	*Tiang seri* (which is a middle column located at the center of the mother house)	1.2
J	*Tunjak langit* (King Post function to hold the roof trusses from collapse)	1.7
K	*Tanggam* (construction technique not using single nails to connect its jointing or called as mortise technique)	2.2
L	*Lantai* (floor will be constructed higher than the ground to avoid wild animal and poisonous animal to enter the house)	4.0
L.1	*Lantai Biasa* (normal floor)	3.7
L.2	*Lantai loteng* (attic floor)	3.2
M	*Ukiran* (wood carving ornamentation on door, window and ladder)	2.2
N	*Halaman Rumah* (courtyard)	3.8

Chart 6: Frequencies Level of Understanding towards of NSMTH

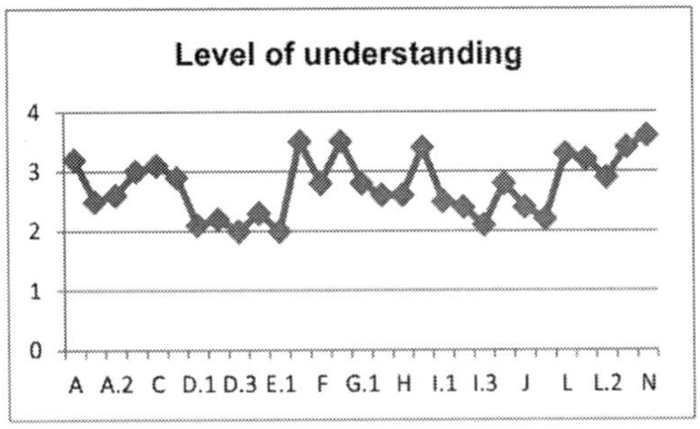

According to result from survey on ceramic's students, it shows that level of the understanding is good and high which is 4 (floor), second highest is house yard and the third highest is 3.7 (normal floor). There are five characteristics at poor level which are end *serambi* and *kelek anak* with 1.2 scale, middle *serambi* and mother house at scale of 1.3. Ceramic student have a better exposure towards house characteristic because they have to study a lot on structure. One of their syllabus is to study structure of building and usually the theme is traditional house. In a way to understand the house structure they have to read about the house and this give them extra knowledge.

Chart 7: Overall of Level of Understanding on NSMTH by Faculty of Art and Design students.

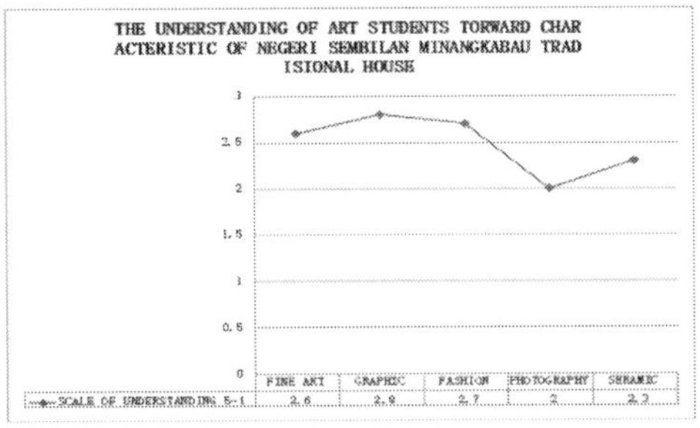

From the graph above, the level of understanding among students of different programs of faculty of art and design is moderate and fair. Fine Art, Graphic, and Fashion's students each with average 2.6, 2.8 and 2.7 is at moderate level. While another two programs which are photography and ceramic with level of 2.0 and 2.3 which is fair. Overall , the faculty of art and design majority lives in concrete modern house which is 40 respondents, bungalow 16 respondents, Semi Detached house 6 respondents, condominium 4 respondents , apartment 9 respondents and Flat 5 respondents. Only 10 of the respondents' lives in *kampong* traditional house and the rest lived in Renovated Traditional house. Due to the type of residential the respondents' lived cause the lack of knowledge and understanding about traditional house.

8. Conclusion

In conclusion, the students of faculty art and design have poor to good understanding on main elements and characters of NSMTH in Malaysia. The elements with a higher level of understanding are Floor, House Yard, Normal Floor, Kitchen, pillar and ornamentation. Detail characteristics and elements OF NSMTH such as *pangkal serambi*, end *serambi*, *kelek anak*, middle *serambi*, mother house, rear platform and *tiang pecah lapan*. Based on findings, the author suggest to raise the awareness and knowledge of youngsters nowadays to learn about Negeri Sembilan Minangkabau Traditional house. The exposure and experiential learning about this can be done through sketches, drawing and photography as their assignments. By this type of learning process they can make the students appreciate the uniqueness characteristic of NSMTH.

References

[1] Abdullah, F. Wongso, J. (2007), A Study of Minangkabau Architecture in Two Region: West Sumatra, Indonesia and Negeri Sembilan, Malaysia. The Second International Coferance of the Center for the Study of Architecture in the Arab Region. CSAAR.

[2] Exploring Indonesia, (2007). Past and Present. Retrieved on April 27, 2007 at http://www.hawaii.edu/cseas/pubs/indonesia.indonesia.html

[3] Lembaga Muzium Negeri Sembilan, (2011). Sistem Sosial Adat Perpatih Di Luak Inas. Jalan Sungei Ujong, Negeri Sembilan. Lembaga Muzium Negeri Sembilan. ISBN 978-983-2877-13-4

[4] Y. Idrus, (1996). Rumah Traditional Negeri Sembilan: Satu Analysis Seni Bina Melayu. Shah Alam, Selangor: Penerbit Fajar BaktiSdn. Bhd. ISBN 967-65-3585-0

[5] M. Masri, (2012). The Misconceptions Of Negeri Sembilan Traditional Architecture. Procedia - Social and Behavioral Sciences 68 (2012) 363 – 382. Giza, Egypt.

[6] Abdullah, F. (2012). Losing Interest of Minangkabau Architecture. IIUM Research, Invention and Innovation Exhibition 2012.

[7] Hassan, I.S., Ismail, I. and Ismail, M.A (2013) The Islamic Concept of Minangkabau Architectural Heritage, Conference Paper.

DOI: 10.1051/matecconf/20166600111

[8] Bahauddin, A., (2012). The Minangkabau House: Architectural and Cultural Elements. WIT Transactions on Ecology and The Environment, Vol **165**. www.witpress.com, ISSN 1743-3541.

[9] Ibrahim, N. H., (2007). Rumah dan Peralatan Olek serta Permainan Tradisional Negeri Sembilan. Jalan Sungei Ujong, Negeri Sembilan. Lembaga Muzium Negeri Sembilan. ISBN 978-9832877-02-8

[10] Nasir ,A. H.. (1985). Pengenalan Rumah Tradisional Melayu Semenanjung Malaysia. Cheras. Darulfikir.

[11] Shahminan, R. N. R.. (2007). Senibina Rumah Bumbung Panjang Negeri Sembilan. Adat Perpatih - Esei Pilihan , 191-197.

Rethinking Space Design Standards Toward Quality Affordable Housing In Malaysia

Nor Haniza Ishak[1, a], Ati Rosemary Mohd Ariffin[2], Raha Sulaiman[3] and Muhammad Najib Mohd Zailani[4]

[1, 2, 3, 4] *Faculty of Built Environment, University of Malaya, 50603 Kuala Lumpur MALAYSIA*

Abstract. Provision of affordable housing is important to low- and middle-income population. A fit form of house will not only fulfil a basic human need for shelter, but it also contributes to physical and psychological well-being of the occupants. Excellent quality and affordable housing is an indication of a high quality of life. While writings exist on various aspects of the quality of affordable housing in Malaysia, discussion regarding space and design standards has scarcely been given any serious academic attention. Standards concerning residential development usually cover different aspects or stages of the development process. They can include planning standards, design standards, space standards and technical construction standards. The main concern of this paper is on space and design standards specifically. Space standard can be defined as a set of framework which dictates fixed internal space minimums. Meanwhile, design standard indicates design guidelines to ensure the functionality, comfortability and habitability of the house. This paper is concerned exclusively with indoor spaces of a house excluding external circulation spaces and service facilities (in case of strata housing). Its interest is in internal space as an aspect of housing quality. It can be concluded that one of the way forward will be to find the balance between providing adequate minimum spaces for resident satisfactions and having economic values for housing developers. This paper may be used as a valuable reference for authorities and policy makers to better address the best housing space design standards that would benefit not only the occupants, but also the local government and developers alike.

1 Introduction

Malaysia is a federal constitutional monarchy located in central Southeast Asia, bordering Thailand to the north and Singapore to the south. It consists of thirteen states and three federal territories and has a total landmass of 329,847 square kilometres with the estimated population of 29.90 million people [1]. Malaysia gain Independence through peaceful negotiation from the reign of British administration in 1957. The 2014/2015 Economic Report released by the Finance Ministry shows that Malaysian per capita income in purchasing power parity terms is projected at US$23,512. After the British Colonisation period, the Malaysian government played its vital role in devising and creating various national policies to improve the economic development, political stability and social welfares of the nation.

Among the most well-known national policies in Malaysia are the Five Yearly National Development Plans, a comprehensive economic development blueprint where the Economic Planning Unit of the Prime Minister's Department is being entrusted with the task of preparing it. The early Malaysia's Economic Plans in providing low cost housing was a profound strategy to include the rural Malays to migrate to urban areas.

[a] Corresponding author: [1]niza_alambina@um.edu.my

1.1 Concept of affordable housing

In this ever-changing, globalize world, it is envision that a house would be a sanctuary where one could find solace and tranquilities from the pressing reality of living. For many years the mass housing was modelled after the British houses and its standards and By-Laws, remnant of colonial days. Standards currently in use frequently date back to colonial times and all too often reflect the present day standards in use the former coloniser [2].

Since the beginning, the government had identified housing as a crucial and basic social need and had taken responsibility in ensuring adequate provision of affordable housing to all strata of the societies. The term 'affordable housing' at its simplest connotation, refers to the financial affordability of housing with respect to the occupants' incomes [3].

1.2 Low-cost housing in Malaysia

Providing housing for the low income people, particularly the urban poor has always been a fundamental part of the state's housing policy to maintain the stability and prosperity of the country. The commitment by the Malaysian government to provide adequate and affordable shelter for its nation was evidently reflected in the government's annual budget and the nation Five Yearly Development Plans [4]. The early main concern of the housing policies in Malaysia is to provide home-owning democracy to all of its citizens, simultaneously eradicating hardcore poverty and urban squatter problem [5]. Under this root principle, the government and latter, the private sector had delivered massive amounts of low-cost housing under various affordable housing schemes such as Housing for the hardcore poor (PPRT), People Housing Project (PPR), 1Malaysia Civil Servants Housing (PPA1M) and Schemes under Syarikat Perumahan Negara Berhad (SPNB) to name a few.

Malaysia is now in the transition period between the Tenth Malaysia Plan (2011-2015) and the Eleventh Malaysia Plan (2016-2020). In the span of 50 years the government had improved by leaps and bounds in the provision of affordable low-cost housings via both public and the private sector despite still not being enough [6] and had addressed many issues and problems pertaining to matter such as financing of the mortgage loan, reducing construction cost and upgrading the provision system to legitimate low income population.

Previously, the public housing developers only concern on achieving their anticipated outcomes target. Most of the houses are very basic in nature and little attention was paid to quality. In term of quality the type of housing has not been satisfactory to the family's housing needs, comfort, social, cultural and religious needs [7, 8]. However, in the last couple of decades the government had a pivotal shift in its housing policies from delivering quantities to ensuring qualities. This is evidently outlined in the Tenth Malaysia Plan (RMK10). One of the main objectives in Chapter 6 is to "ensure access to quality and affordable housing" and with it highlighted three major challenges that need to be overcome. The first underlying challenge is the apparent mismatch between supply and the massive demand of low-cost housing [9]. The second challenge outlined in RMK10 is to meet the call for environmentally sustainable design. The last challenge, which is also the concern of this paper, is to increase the quality of affordable new and existing housing.

'Weak quality controls during construction and poor maintenance have contributed to a decline in the quality of affordable housing. During the Plan period, existing laws, including the Uniform Building By-Laws 1984, will be reviewed to incorporate minimum specifications of housing quality, particularly on ensuring quality in the provision of affordable housing developments' (Chapter 6, Tenth Malaysia Plan, 2011).

The government's vision of providing quality affordable housing is also being translated in the latest Eleventh Malaysia Plan (2016-2020). However, the policy emphasizes on providing adequate and quality affordable housing not only for the poor, low-income but also for the middle-income households with differing levels of support based on need. This is due to the fact that Malaysia is aiming to become a self-sufficient industrialized nation by the year 2020, and the government recognized affordable housing as a crucial platform that creates strong and stable communities, with better economic, education, and health outcomes for individuals, families and communities [10]. The low- and middle-income households also constitute for a big slice of Malaysia's population. The recently-announced 2016 Annual Budget categorized both of these household classes into B40 (Bottom 40) and M40 (Middle 40) respectively, indicating each classes making up 40% of the overall national population.

1.3 Quality affordable housing

Ultimately, it is everyone's dream to have at least a decent quality of life. The subject of quality of life is very diverse, complicated and is hard to be quantified. There are often many factors that come into play. Nonetheless, there are a few indicators that can be used such as social, well-being measures and economic indicators [11]. The Malaysian government had given attention to provide its citizen with good quality of life and is evident in which three of seven Malaysia strategic thrusts (in RMK10), namely, (1) building an environment that enhances the quality of life, (2) moving towards inclusive socio-economic development, and (3) creating the environment for unleashing economic growth, have a direct relation to Malaysian housing [9].

However, the term 'quality housing' is too generic and comprises of several broader elements such as construction quality, facility managements quality, neighborhood quality, public hygiene & safety quality and lastly, internal housing space quality. Excellent quality and affordable housing is an indication of a high quality of life. It not only fulfils a basic human need for shelter, but it also contributes to physical and psychological well-being of the occupants. While writings exist on various aspects of the quality of affordable housing in Malaysia, discussion regarding space design standards has scarcely been given any serious academic attention. Therefore, it is our intention in this paper to focus on one of the facets of quality housing which is the internal housing space quality, paying particular attention to the government's policies and regulations of space design standards. This paper is concerned exclusively with indoor spaces of a housing unit excluding external circulation spaces and service facilities in strata developments such as flats and apartments. We will explore the importance of rethinking the current space design standards and also the potential setbacks if such moves were taken.

2 Space design standards in Malaysia

Standards of Malaysia Act 1996 (Act 549) defines standard as a document approved by a recognized body, that provides, for common and repeated use, rules, guidelines or characteristics for product or related processes and production methods, with which compliance is not mandatory; and which may also include or deal exclusively with terminology, symbols, packaging, marking or labelling requirements as they apply to a product, process or production method. Interestingly enough, standards, under the law is not mandatory and becomes more like a guideline, unless explicitly stated that compliance is mandatory. However, a standard only become mandatory when a regulatory agency enforces its use through the relevant Act and Regulations.

Standards concerning residential development usually cover different aspects or stages of the development process. They can include planning standards, design standards, space standards and technical construction standards. All of these standards are crucial as a mean of measuring the quality of the physical construction integrity and also the psychological wellbeing promoted by the housing development.

Space standard can be defined as a set of framework which dictates fixed internal space minimums. Basic rules govern the quantity of living space that must be provided, minimum acceptable ceiling heights, and ventilation and light requirements and so on [12]. Meanwhile, design standard indicates design guidelines that need to be followed in order to ensure that all housing units are functional, comfortable and fit to be inhabited [13]. Design standards usually address the minimum requirements for the spaces or rooms to function well such as the water retention system, electrical and gas installation, internal wiring for telephone, sanitary waste disposal system, health and safety, security and plumbing fixture.

Currently in Malaysia there are only one law that have provision on space and design standards. It is the Uniform Building By-Law 1984 (UBBL 1984). The UBBL 1984 prescribes plan submission procedures, provisions on space and design methods, temporary works provisions, construction requirements, structural requirements, fire requirements, alarm, detection, extinguishment, and access, and other miscellaneous items. Part III of the UBBL 1984 talks specifically on the required space [dimensions], natural light and ventilation to be provided in buildings [14].

Since 1981, the Ministry of Housing and Local Government has introduced and implemented a concept of low-cost housing incorporating the following characteristics, (1) selling price not exceeding RM25, 000 per unit, (2) Targeting households with a monthly income not exceeding RM750, (3) type of houses ranging from flats, single-storey terrace to detached houses and (4) minimum design standard with built-up area of 550-600 square feet, two bedrooms, a living room, a kitchen and a bathroom-cum-toilet [5]. Realising that the need for housing standards would arise, the Construction Industry Development Board (CIDB), a statutory body established under the Parliamentary Act 520 (Construction Industry Development Board of Malaysia Act 1994) together with the Ministry of Housing and Local Government had set up a technical committee to come out with a National Housing Standards series. It initially started as two volumes; Construction Industry Standard 1 for Single/Double-Storey Low-Cost House and Construction Industry Standards 2 for Low-Cost Apartment. Both are established in 1998. Abdul Mohit [15] accounted that in 2002 the low-cost housing design in Malaysia has undergone changes from the provision of two bedrooms to three bedrooms with the addition of a dining room, a separate bathroom and a toilet, and a drying area. The new revision is also to accommodate to social and cultural preferences such as the need to separate children of different gender from sharing rooms. Hence it is necessary to have at least 3-bedroom units in low-cost housing schemes [5].

Despite great contribution from the private sectors in helping the government building low-cost houses, there had been constant complaints and arguments that the required allocation of at least 30% low-cost houses in their housing projects are pegged at the ceiling price of RM25, 000 by the government for all types of low-cost houses, irrespective of their location and typology is not enough for them to have a profit return and also deliver good quality housing [16]. In order to improve the quality of low-cost houses and simultaneously meet private developer's argument for a review of the selling prices of these units the government has introduced in 2002 the new pricing guideline for this type of houses. With the new selling prices of low-cost houses, it is to be expected that the design specifications be revised too [5]. The new design specifications are summarized in Table 1.

In 2005, CIDB proposed some new addition to the National Housing Standards which is CIS 3 and CIS 4. Both standards emphasized on lower middle-cost houses such that CIS 3 is for lower middle-cost houses other than apartment and CIS 4 is for lower middle-cost apartment. This is in line with the government's intention as stated in the Ninth Malaysia Plan (2006-2010) in providing sustainable urban development and adequate housing for all income groups taking into account not only provision for low-cost house but also the delivery of middle-cost and high end property. Since the provision of middle-cost houses were entrusted solely to the private developers it is necessary for the government to establish a mandatory standard that would regulate such developments [5]. This prompted CIDB to introduce CIS 3 and CIS 4 into the National Housing Standard.

It is only recently that the government gazette PR1MA Act 2012 and establishes Perbadanan PR1MA Malaysia to plan, develop, construct and maintain high-quality housing with lifestyle concepts for middle-income households in key urban centres. However PR1MA Malaysia developed their own set of standards to be complied by all of their high-rise developments. Table 2 illustrates the compilation of all space design standards concerning strata developments in Malaysia.

Table 1. Low-Cost Houses: New Design Specifications

Elements	Terrace Houses	Flats
Floor Space	48-60m^2	45-56m^2
Bedroom -minimum number -minimum area of habitable room: (i) First Room (ii) Second Room (iii) Third Room	 3 11.7 m^2 9.9 m^2 7.2 m^2	 3 11.7 m^2 9.9 m^2 7.2 m^2
Kitchen-minimum area	4.5 m^2	4.5 m^2
Living and dining rooms	Provided as one combined space or seperately with adequate area according to internal layout.	Provided as one combined space or seperately with adequate area according to internal layout.
Bathroom and Toilet	Provided seperately with minimum area of 1.8 m^2 each.	Provided seperately with minimum area of 1.8 m^2 each.
Storage space and porch	Adequate provision for resident's comfort.	Adequate provision for resident's convenience and comfort.
Drying Area (*) Laundrette facilities	-	Adequate provision for each unit.

Note: (*) Must be provided according to the 'Guidelines for the provision of laundrette facilities in multi-storey buildings' prepared by Local Government Department, Ministry of Housing and Local Government.
Source: Ministry of Housing and Local Government, 2002

Table 2. Compilation of Space Design Standard in Malaysia

NAME OF STANDARD	TYPE OF PROPERTY	SPACE STANDARDS	DESIGN STANDARDS	YEAR OF ENACTMENT	LATEST REVISION
Uniform Building By-Laws 1984	Residential Building	**Minimum area for rooms in residential buildings:** -1st habitable Bedroom: not less than 11m^2 -2nd habitable Bedroom: not less than 9.3m^2 -3rd habitable Bedroom: not less than 6.5m^2 -The width of every habitable room in residential building shall be not less than 2m. -The area and width of a kitchen in a resident building shall be not less than 4.5m^2 and 1.50m^2 respectively. -In all buildings, the sizes of latrines, water closets and bathrooms shall be- (a) in the case of latrines or water-closet with pedestal-type closet fittings, not less than 1.50m by 0.75m; (b) in the case of water-closet with fittings other than pedestal-type closet fitting, not less than 1.25m by 0.75m; (c) in the case of bathrooms, not less than 1.5m^2 with a width of not less than 0.75m; (d) in the case of bathrooms with water closet fittings, not less than 2m^2 with a width of not less than 0.75m. **Height:** -The height of rooms in residential buildings other than shop houses shall be:	**Natural Lighting & Ventilation:** -Every room designed, adapted or used for residential, business or other purposes except hospitals and schools shall be provided with natural lighting and natural ventilation by means of one or more windows having a total area of no less than 10% of the clear floor area of such room and shall have openings capable of allowing a free uninterrupted passage of air of not less than 5% of such floor area. -Every water-closet, latrine, urinal or bathroom shall be provided with natural lighting and natural ventilation by means of one or more openings having a total area of not less than 0.2m^2 per water-closet, urinal latrine or bathroom and such openings shall be capable of allowing a free uninterrupted passage of air.	1976 1984- gazetted as UBBL 1984	2007- to incorporate the amendments to the Street, Drainage & Building Act, 1974 with regard to the Certificate of Completion and Compliance (CCC) 2012- Amendments on Energy Efficiency ByLaw and MS1525:2014

		(a) Living rooms & Bedrooms: not less than 2.5m. (b) Kitchens: not less than 2.25m. (c) Bathroom, water-closets, latrines, porches, balconies, verandahs, garages and the like: not less than 2m. -The average height of rooms with sloping ceilings in residential buildings other than shop houses should be the same as above provided that no part of any room shall be less than 2m in height.			
Construction Industry Standard 2 :1998	Low-cost flat	**Floor Area:** -Minimum three (3) Bedrooms 1st Bedroom : not less than $11.7m^2$ 2nd Bedroom: not less than $9.9m^2$ 3rd Bedroom: not less than $7.2m^2$ -Kitchen: not less than $5.4m^2$ -Living and Dining Room can be combined or separated. -Separate Bathroom and Toilet Bathroom: not less than $1.8m^2$. Minimum width of space should be not less than 1.2m. Toilet: not less than $1.8m^2$. Minimum width of space should be not less than 1.2m. -Adequate storage are and circulation space together with a covered space for drying clothes must be provided for the comfort of the occupants. **Residential Unit Size:** -Residential units need to have front opening and adequate size to ensure the internal space can be used efficiently and has good natural ventilation for the comfort of the occupants. Must be not less than $63m^2$. -minimum front opening for residential unit is 6m. **Height:** -Bedroom and Living/Dining Room: Height must be not less than 2.8m. -Kitchen: Height must be not less than 2.8m. -Bathroom/Toilet: Height must be not less than 2.5m. -The average height for room/spaces with sloped ceiling must be the same as previously stated unless, with condition that no part of the room can be lower than 2m in height.	**Finishes & Fixtures:** -Floor and wall finishes for all the internal spaces are provided in the document. **Natural Ventilation & Lighting:** -All the rooms/spaces provided in each of the units need to receive natural lighting and ventilation through one or more windows with opening no less than 10% of rooms/spaces area and need to have opening no less than 5% of rooms/spaces area that would allow for unrestricted circulation of air. Cross ventilation is highly recommended, and should as much as possible cover all rooms and from the front till the back or vice-versa. **Dividing Walls:** Dividing walls have to be built in between each unit and need to be built all the way to the roof and have fire-resistant property of at least two (2) hours. *all of the building's design dimensions have to follow MS 1064 – Guide To Modular Coordination in Building specification.	1998	2002- Improvement of size of space due to the introduction of new low-cost houses pricing by governement
Construction Industry Standard 4 :2005	Lower middle-cost flat	**Floor Area:** -Minimum three (3) Bedrooms 1st Bedroom : not less than $12.8m^2$ 2nd Bedroom: not less than $11.8m^2$ 3rd Bedroom: not less than $9.9m^2$ -Kitchen: not less than $6.m^2$	**Finishes:** -Floor and wall finishes for all the internal spaces are provided in the document. **Natural Ventilation & Lighting:** -All the rooms/spaces provided in each of the units need to receive natural lighting and ventilation	2005	No known revision

		-Living and Dining Room can be combined or separated. -Separate Bathroom and Toilet. Bathroom/ Toilet: not less than 2.7m^2. Minimum width of space should be not less than 1.5m. -Adequate storage are and circulation space together with a covered space for drying clothes must be provided for the comfort of the occupants. **Residential Unit Size:** -Residential units need to have front opening and adequate size to ensure the internal space can be used efficiently and has good natural ventilation for the comfort of the occupants. Must be not less than 70m^2 excluding balcony and drying area. -minimum front opening for residential unit is 6m. **Height:** -Bedroom and Living/Dining Room: Height must be not less than 2.8m. -Kitchen: Height must be not less than 2.4m. -Bathroom/Toilet: Height must be not less than 2.4m. -The average height for room/spaces with sloped ceiling must be the same as previously stated unless, with condition that no part of the room can be lower than 2m in height.	through one or more windows with opening no less than 10% of rooms/spaces area and need to have opening no less than 5% of rooms/spaces area that would allow for unrestricted circulation of air. Cross ventilation is highly recommended, and should as much as possible cover all rooms and from the front till the back or vice-versa. -Minimum size of air wells should be in accordance to the Uniform Building By-Law 1984, Part 3, Clause 40 (including revisions). **Dividing Walls:** Dividing walls have to be built in between each unit and need to be built 225mm extending the roof level in accordance to the Uniform Building By-Law 1984 (including revisions) and the requirement from Fire and Rescue Department of Malaysia. *All of the building's design dimensions have to follow MS 1064 – Guide To Modular Coordination in Building specification. *Design and unit layout have to be 'user-friendly' by following the *Garis Panduan Perancangan Golongan Orang Kurang Upaya, Jabatan Perancangan Bandar dan Desa 1/2000* and Standard Malaysia, SIRIM MS 1331 (Code of Practice for Access for Disabled People Outside Buildings).		
PR1MA Project Brief 2015; Section 1, Planning and Built-Form Requirements – High Rise	Apartment	**Residential Unit Size:** -For all PR1MA's high rise developments, six types of apartment unit are required as follows: Preferred unit type: 2+1 Bedroom: 78.9m^2 ± 5% 3 Bedroom: 92.9m^2 ± 5% 3+1 Bedroom: 111.48m^2 ± 5% Situational unit type: Studio: 51.09m^2 ± 5% 1+1 Bedroom: 60.4m^2 ± 5% 2 Bedroom: 69.68m^2 ± 5% -The spatial hierarchy within all PR1MA residential units have the following characteristic: (a) Bedrooms are preferred to have windows facing outwards. (b) Yard to connect kitchen into a semi-private corridor and a bathroom	Must be provided according to **'Book 2a- High Rise Minimal Standard for detailed guidelines'** prepared by Perbadanan PR1MA Malaysia.	2015	No revision

Source: Uniform Building By-Law 1984, Construction Industry Standards 2:1998, Construction Industry Standards 4:2005 and PR1MA Project Brief 2015

3 Lessons from other countries

3.1 Calculating Standards of Floor Area based on Living Density of Housing Units

In the span of this research we are still yet to come across with any national statistic on floor area of housing per number of households and the average floor area per person in Malaysia. Using average floor area per person can give us insights of the potential overcrowding problems and indentify the gap in the allocation of free space amongst the lowest-income households to the highest-income households. A small average figure would imply that the majority of the population are living in potentially overcrowded small houses while a big average figure indicates that most of the national population are enjoying ample space for living. This indicator is commonly used by major developed countries such as the United Kingdom and Japan. Table 3 and Table 4 indicate the use of average floor area per person in respect to data from Japan.

Table 3. Standards of Floor Area (m^2) and Living Density (m^2 per person) of Housing Units

Size of households (persons)	Floor area per housing unit (m^2)	Implied density of housing unit (m^2 per person)
1	18	18
1(elderly)	25	25
2	29	14.5
3	39	13
4	50	12.5
5	56	11.2
6	66	11

Source: The Building Center of Japan, *A Quick Look at Housing in Japan*

Table 4. Average floor area per person in Japan (m^2 per person)

	1963	1968	1973	1978	1983	1988	1993
Floor area per person	17.3	19.0	21.5	23.3	25.8	28.0	30.9

Source: The Building Center of Japan, *A Quick Look at Housing in Japan*

It is worth noting that the average floor area per person in Japan was 30.9m^2 in 1993 which is an impressive improvement from the average of 17.3 m^2 per person in 1963. There are some potential benefits if our current standards adopted the Living Density of Housing Units indicator.

3.2 Classification system for current typology of affordable housings

To date, the government has introduced various public housing schemes. Among them are *Program Bantuan Rakyat (PBR), Program Perumahan Rakyat (PPR), Rumah Mesra Rakyat 1Malaysia (RMR1M), Perumahan Rakyat 1Malaysia (PR1MA),* 1Malaysia Civil Servants Housing (PPA1M) and *Rumah Wilayah Persekutuan (RUMAWIP).* However, not all of these schemes are low-cost housings. For instance, PR1MA scheme is the government's initiative to provide housings for middle-income group since they are not eligible to apply for low-cost houses but are unable to afford houses offered by the mainstream property market.

The first 4 volumes of the Construction Industry Standards are dedicated to outline public residential standards, each for different types of housing development. For example CIS 2:1998 is for low-cost apartments and CIS 4:2005 is for low middle-cost apartments. All of public housings initiated by the government are expected to follow this standard. However there is no checking mechanism that ensures all of the housing schemes comply with their required internal space design standards.

Singapore had created a numeric hierarchical system of classification called Singapore Standard Classification of Type of Dwelling and is designed for use in collection of data that require classification of information on dwelling types. Table 5 and Chart 1 illustrate how the classification system works. The 3-digit codes can be used to inform us about the dwelling class, housing typology and the nature of the house. It is suggested that such system is implemented in Malaysia so that enforcement of mandatory minimum housing standards can be monitor more effectively.

Table 5. Nomenclature system

CODE	TYPE OF DWELLING
1	HOUSING UNITS
11	Housing and Development Board (HDB) Properties
111	1-Room Flat (HDB)
112	2-Room Flat (HDB)
113	3-Room Flat (HDB)
114	4-Room Flat (HDB)
115	5-Room Flat (HDB)
116	Executive Flat (HDB)
117	Housing and Urban Development Corporation (HUDC) Flat (excluding those privatised)
118	Studio Apartment (HDB)
12	Landed Properties
121	Bungalow/ Detached House
122	Semi-Detached House
123	Terrace House
13	Condominiums and Other Apartments
131	Condominium (except Executive Condominium)
132	Executive Condominium
139	Other Apartments nec
14	Other Housing Units
141	Shophouse
149	Other Housing Units nec

Source: Singapore Standard Classification of Type of Dwelling, Jan 2012

 DOI: 10.1051/matecconf/20166600112

Chart 1. Overview of types of dwelling

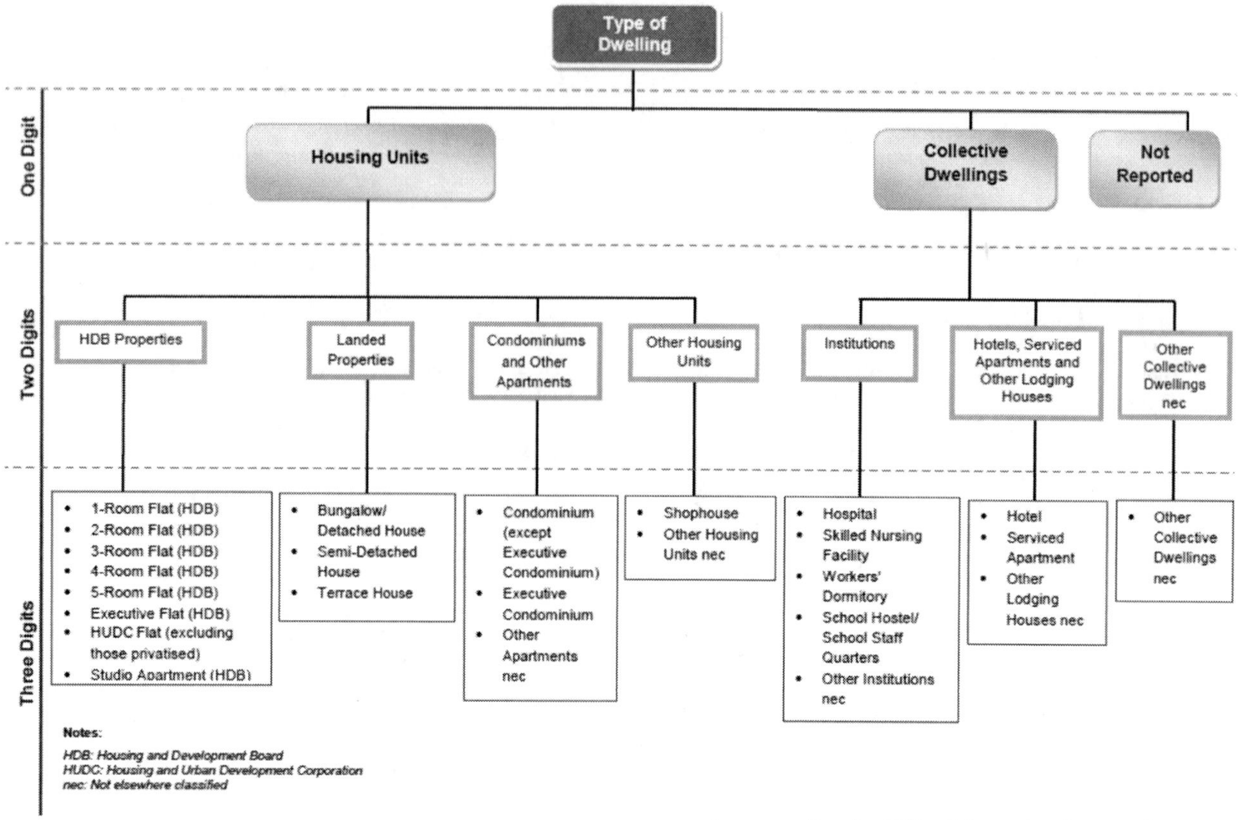

Source: Singapore Standard Classification of Type of Dwelling, Jan 2012

4 Future trends of low-cost housing in Malaysia

4.1 Future housing of the nation

Low-cost housing apartments have now been regarded as important residential concept in the urban living world-wide. It appears that due to tremendous urban growth throughout the world, there has been an urgent need to construct accommodations for a huge number of people in towns and cities around the world [17]. Goh [18] annotate the 1991 housing census that 90% of the housing stock in the country was in the form of houses, while flats or apartments accounted for only 7%, and shop houses accounted for 3%. However, flats and apartments are slowly emerging in major urban areas. Flats generally refer to apartments of between 500 to 700 square feet each with three bedrooms. They have bare finishing and no facilities except parking bays, the number of which may not be as many as the number of flats in any particular scheme. Unless the flats are above five stories, they are generally not equipped with lifts. Ministry of Housing and Local Government announced that, until December 2012 the number of people inhabiting strata properties is now around 6 million people, accounting for 20% of the national population. It is estimated that by 2020, 70% of the population will live in major urban cities throughout Malaysia [19].

This will inevitably increase the demands for strata housing in the near future. Therefore, low-cost apartments should be given the topmost priority and need to have the same psychological benefits of landed properties. They should promote safe, secure and good environment for nurturing holistic social wellbeing. It is expected that Malaysia will become a high-income nation by the year 2020 with the expected real Gross Domestic Product (GDP) to expand between 5-6 percent per annum from 2016-2020, resulting in a 7.9 percent per annum rise in Gross National Income (GNI) per capita. This economic prosperity should be trickling down efficiently and further enhancing the quality of low-cost housings.

Currently the government subsidized 30% to 75% of total construction cost for public housing units [6]. The deep subsidization by the government in financing quality low-cost housing should be seen not only as a civic responsibility but also as a long-term investment to empower the low-income households so that they could break free from the B40's bracket and subsequently, further driving economic growth of the nation and ultimately reducing the need for low-cost housings in general. Hence, improving the current space design standards of low-cost apartment (such as CIS 2:1998) does have significant importance.

[a] Corresponding author: [1]niza_alambina@um.edu.my

4.2 Greater flexibility for shifts in urban demographic

It is a paradox that most low-income households have large families and need larger units which they cannot afford [20]. The average number of people in a Malaysian household is 4 to 5 and more spaces are required for a growing family [21]. Indeed, this poses a great problem to the government planners of low-cost housing programmes since all of the low-cost housing schemes, particularly the People's Housing Project (PPR) which only provides 3-bedroom units. Consequently, household with large families are forced either to overcrowd a single room with 3 or 4 persons or to use up other spaces such as the dining/ living room, thus greatly impairing the functionality of the house.

The second demographic concern in the near future is the increasing youth and young married couple in urban area, including those under the 1Malaysia Youth City Programme. Today's Gen-Y are getting more and more flexible and prefer high-rise building as this will allow them to live near to their workplace. They usually do not require big house and are contended with having only 1 or 2 bedrooms. In fact, the government had already taken a proactive approach in catering to the need of the Gen-Y by introducing a new concept known as Transit Houses. These transit houses provide such families with a short term residence while they build up their savings to purchase their first home [10]. New space design standards should be developed for this kind of houses to avoid the problems of sub-standard construction. However, the standards can be a bit more relax.

Lastly, Burgstahler [22] defines universal design as the design of products and environments to be usable by all people, to the greatest extent possible, without the need for adaptation or specialized design. This would include the senior citizens, person with disability, pregnant ladies, children and et cetera. According to the Department of Statistic, it is foresee that the number of senior citizens in Malaysia would double in between 2010 and 2035. Senior citizens are defined as population aging above 60 years old and currently accounted for 1.8 million people or roughly 7% of the national population. However, this numbers are expected to escalate to 3.4 million people or 10% of the population by 2020.

A study by Da Vanzo and Chan [23], illustrated that generally it is considered as a norm or tradition for most Malaysians to co-reside with their. Hence, it would only be appropriate if the current space design standards incorporate provisions on universal design in accordance to the existing Malaysian Standards such as the MS ISO/IEC GUIDE 71:2012. The future low-cost housing developments can also explore the idea of dual-key concept. A new trend currently gaining popularity in medium-cost and high-end condominium developments and is seen as an alternative way to guest extended families or the old folks but at the same time ensuring privacy and security. This would also potentially provide some protection of parents from being taken into "old folk's homes", and saving on public spending [24].

4.3 Implementation of Industrialised Building System (IBS)

In terms of research and development to strengthen the quality and housing construction standards, the National Housing Policy focuses on the use of latest technology and tools such as Industrialized Building System (IBS). The implementation of IBS constructions has numerous advantages in improving the quality of low-cost housing development. This include (1) being cost-effective and environmentally-friendly construction, (2) having high-quality components in construction, (3) produce aesthetically-pleasing buildings that serve the purposes they were built for, (4) does not give problems to end-users (such as leaks), (5) adaptable to future needs of users through renovation or extension and (6) upholds safety and health requirements in construction [25].

It is argued that in the scenario where the government successfully implements IBS construction system in all of its public housing developments, it would result in significant decrease in construction costs. It is further argued that with the reduction of cost, increasing space standard will be more sustainable and financially feasible.

5 Setbacks

The issue of regulating higher standards to improve housing quality would always draw attention and criticisms from private developers since private developers are the biggest provider of low-cost housing. Private developers are profit oriented and may result in less emphasize on utility and greater concern for marketability. Currently there is a surplus of high-end housing especially apartments in the urban areas whilst majority of Malaysians income are not within the band of those who can afford such luxury. The mainstream property market will slowed down since the second quarter of 2015 as the overall economy has stepped down. However it is important to recognize the contribution of the private sector thus far. Often not highlighted is the fact that Malaysia has one of the most successful programs of housing for the poor that is the envy of developing and even some developed nations in the world. Key to this unique example of a highly successful program of low-cost housing is the substantial and active direct participation of private sector developers in the entire low-cost housing program [26].

Simple economics suggest that the cost per square metre of residential space would remain static, or fall owning to economies of scale, but larger home (with arguably greater utility) would have a bigger price tag. This would have effect of reducing affordability [12]. Cost factor will always be a major determinant of the quality of materials and specification for low-cost houses. On one hand, private developers agree and accept that certain minimum standards, particularly those related to safety for low-cost housing must be set, while on the other hand concern that policy makers may become over-zealous in their efforts to prove their social concern and 'over specify' [26]. Hence, one of the way forward will be to find the balance between providing adequate minimum spaces for resident satisfaction while at the same time having some economic values for private developers.

6 Conclusions

It can be concluded that the improvement of current space design standards is a very crucial move and should be embrace by all of the industry players to further enhance the quality of housing. Extensive researches are necessary to figure out the optimum standards, without over-specifying and would benefit not only the occupants, but also the government and the developers alike. The next step of this research will be to study the evolution and technical performance of existing internal housing standards.

Acknowledgements

The authors would like to thank the University of Malaya for funding this study through the University of Malaya Research Grant (UMRG) (Project no. RP015B-15SUS) under the Sustainable Science Research Cluster (SuSci) whose support we duly acknowledge.

References

1. The World Bank. (2016, January 14). Total Population Data. Retrieved http://data.worldbank.org/country/malaysia.
2. Singh, H. & Kilmartin, L. (1992). Housing in the Third World: Analysis and Solutions. New Delhi, India: Concept Publishing Co.
3. Habitat, U. N. (2011). Cities and climate change: Global report on human settlements 2011. *London, Royaume-Uni, Etats-Unis: UN-Habitat.*
4. Ahmad, F., Wan Abd Aziz, W. N. A., Hanif, N. R., & Mohd Ahmad, I. (2011). Home Owning Democracy for the Urban Poor: A Case Study of Kuala Lumpur, Malaysia. *Journal of Sustainable Development*, *5*(1), 13–22. http://doi.org/10.5539/jsd.v5n1p13
5. Idrus, N., & HO, C. S. (2008). Affordable and Quality Housing Through the Low-Cost Housing Provision in Malaysia. Department of Town & Regional Planning , Faculty of Built Environment Universiti Teknologi Malaysia Jointly organized Toyohashi University, 1–21.
6. Economic Planning Unit. (2010). Tenth Malaysia Plan 2011-2015. Putrajaya: The Economic Planning Unit, Prime Minister's Department.
7. Sulong, M. (1984). Perumahan Awam Kos Rendah di Terengganu: Isu, Masalah dan Penerimaan Masyarakat. Malaysia: Universiti Kebangsaan Malaysia.
8. Tan, S. H. (1980). Factors influencing the location, layout and scale of low-cost housing in Malaysia. In S. H. Tan, & H. Sendut (Eds.), Public and private housing in Malaysia. Kuala Lumpur: Heinemann Educational Books (Asia) Ltd.
9. Bakhtyar, B., Zaharim, a., Sopian, K., Saadatian, O., & Moghimi, S. (2013). Quality housing in affordable price for Malaysian low income. *WSEAS Transactions on Environment and Development*, *9*(2), 78–91.
10. Economic Planning Unit. (2015). Eleventh Malaysia Plan 2016-2020. Putrajaya: The Economic Planning Unit, Prime Minister's Department.
11. Diener, E., & Suh, E. (1997). Measuring Quality of Life: Economic, Social, and Subjective Indicators. *Social Indicators Research*, ***40***(1-2), 189–216.
12. Gallent, N., Madeddu, M., & Mace, A. (2010). Z Internal housing space standards in Italy and England. *Progress in Planning*, ***74***(1), 1–52. http://doi.org/10.1016/j.progress.2010.04.001.
13. Construction Industry Standard 2 1998 of Malaysia
14. Uniform Building By-Law 1984 of Malaysia
15. Mohit, M. A., Ibrahim, M., & Rashid, Y. R. (2010). Assessment of residential satisfaction in newly designed public low-cost housing in Kuala Lumpur, Malaysia. *Habitat International*, *34*(1), 18–27. http://doi.org/10.1016/j.habitatint.2009.04.002
16. Shuid, S. (2009). Changing Structure of Low Income Housing Provision in Malaysia. *HSA Conference 2009*, 1–20.
17. Zehadul Karim, a. H. M. (2013). Living condition in the low cost apartments in Malaysia: An empirical investigation. *Asian Social Science*, *9*(17), 20–29. http://doi.org/10.5539/ass.v9n17p20
18. Lee, G. B. (1997). Housing Delivery System: An Academician's Perspective. Housing the Nation: A Definitive Study.
19. Zainun, N. Y., & Eftekhari, M. (2010). Forecasting low-cost housing demand in urban area in Malaysia using ANN, (1), 899–902.
20. Salleh, G., & Chai, C. L. (1997). Low Cost Housing: Issues and Problems. *Housing the nation: A definitive study*, 215-228.
21. Isnin, Z., Ramli, R., Hashim, A. E., & Ali, I. M. (2012). Sustainable Issues in Low Cost Housing Alteration Projects. *Procedia - Social and Behavioral Sciences*, *36*(June 2011), 393–401. http://doi.org/10.1016/j.sbspro.2012.03.043
22. Burgstahler, S. (2015). Universal Design: Process, Principles, and Applications How to apply universal design to any product or environment. *DO-IT: Disabilities, Opportunities, Internetworking, and Technology*. Retrieved from http://www.washington.edu/doit/universal-design-process-principles-and-applications
23. DaVanzo, J., & Chan, A. (1994). Living arrangements of older Malaysians: Who co-resides with their adult children? *Demography*, ***31***(1), 95-113.
24. Hamzah, M. Bin. (1997). *Housing policy in Malaysia: Conditions, perspectives and Islamic*

DOI: 10.1051/matecconf/20166600112

values. Retrieved from http://etheses.whiterose.ac.uk/543/

25. Hamid, Z. A. & Kamar, K. A. M. (2013). Implementing the Industrialized Building System. Housing the Nation: Policies, Issues and Prospects.
26. Chan, L. (1997). Low cost Housing: The Industry Viewpoint. Housing the Nation: A Definitive Study.

MATEC Web of Conferences **66**, 00113 (2016)
IBCC 2016
DOI: 10.1051/matecconf/20166600113

Mechanisms for protecting the identity of small towns in Malaysia

Nur Farhana Azmi[1,a], Faizah Ahmad[1] and Azlan Shah Ali[1]

[1]*Faculty of Built Environment, University of Malaya, Kuala Lumpur, Malaysia*

Abstract. Built heritage constitute the very individuality of places hence to its identity. However, the built heritage in a small-scale town is nowadays subject to the triple threat of dilapidation, exhaustion and disappearance. Therefore, this paper attempts to examine the building regulations, guidelines and policies with regard to sustaining the unique features and identity of small towns through extensive review of literature. A semi structured interview was conducted amongst a purposive sample of nine stakeholders from different organizations involved in heritage matters in Malaysia; to investigate existing heritage protection measures impacting development of small towns. Using the qualitative software package NVivo 8.0, the study demonstrates that the towns are not only protected by legislative measures but also through community initiative activities and active collaboration between stakeholder groups. The paper concludes that a relatively strong policy and administrative framework are presently in place for protection and enhancement of built heritage despite some weaknesses in the existing mechanisms.

1 Introduction

Place identity is essential to serve good environment for the future hence the concept has become an increasingly topical issue in recent years. The concept also has received further attention under present day condition where the identity of place is rapidly weakening due to continuing threat from the twenty-first-century requirements associated with development and modernization [28, 38]. This can be associated with the notion of placelessness [42] or non-place [4], a term which denotes places without meaning.

The term place identity has been widely discussed in various academic research reflecting the vitality of this concept. Lynch [32] conceptualized place identity as that which provides its individuality, oneness or distinction from other places and thus is intrinsic to its recognition as a separable identity. Built heritage presents in heritage places has been identified by numerous researchers as one of the most dominant characters that constitute place distinctiveness [12, 19, 27, 33, 43, 45]. It became evident as cultural heritage has been used to create the sense of identity in several Asian countries such as China, Korea, Laos, Myanmar and Thailand [31].

This contention make strong justification for supporting the conservation of built heritage as a possible means of instigating, preserving and promoting the identity of a place [3, 27, 29, 37, 45]. Therefore, this paper attempts to examine the existing legislations, guidelines and policies with regard to sustaining the unique features and identity of places particularly towns of a small scale. These towns deserve further research attention as they have been largely overlooked [23, 25] and are nowadays subject to the triple threat of dilapidation, exhaustion and disappearance [54].

This paper begins with an overview of relevant international charters and guidelines as well as approaches to built heritage protection in Malaysia. Theoretical findings is further strengthened through a semi structured interview with key stakeholders from different organizations involved in heritage matters in Malaysia.

2 Background

2.1 International charters and guidelines

Interest in the protection of cultural built heritage are informed by an ever increasing number of international charters, resolutions and declarations that date back to the late 19th century and the early 20th century. Among these are charters, resolutions and declarations of the United Nations Educational, Scientific and Cultural Organization (UNESCO), the International Council for Monuments and Sites (ICOMOS) and the Council of Europe. Despite predominantly reflect European tradition of philosophy and practice, these documents have influenced and presented the basis to the formation of the many later legislations in other countries around the world.

2.1.1 UNESCO

The international organization of UNESCO has been in the forefront in developing globally recognized conventions, recommendations and declarations for

[a] Corresponding author: farhanazmi@um.edu.my

cultural heritage protection since the 1960s. The 1964 Venice Charter and the 1972 World Heritage Convention have appeared to be the most internationally recognized instruments developed by the UNESCO.

The Venice Charter 1964

The 1964 International Charter on the Conservation and Restoration of Monuments and Sites represented a revision of the 1931 Athens Charter. A noteworthy difference between the two charters is the former restricts the use of modern techniques whereas traditional practices are recommended. The Charter also emphasizes the importance of respecting the original fabric, precise documentation of any intervention, the significance contributions from all periods to the building's character, and the maintenance of historic buildings for a socially useful purpose. Furthermore, it states that the principles guiding the preservation of ancient buildings should be laid down on an international basis, with each country being responsible for applying the plan within the framework of its own culture and traditions.

The formation of the Charter marks a major step towards better conservation of traditional buildings and places. It continues to be cited as the baseline document for international conservation philosophy and practice today [1].

1972 World Heritage Convention

The Convention concerning the Protection of the World Cultural and Natural Heritage is the principal reference for decisions regarding World Heritage. The Convention has proven to be the most widely accepted instrument of international co-operation in the protection of cultural and natural heritage of outstanding universal value (OUV). This was reflected in the increase number of the State Parties adhered to the Convention, reaching 191 countries by year 2014. The duty of ensuring the identification, protection, presentation and transmission of the heritage to future generations belongs to the state on whose territories it is located (Article 4). On signing the Convention, the state can benefit from international assistance, and having sites inscribed on the World Heritage List which in turn serves as a catalyst to raising awareness of the site and its outstanding values [53].

Places inscribed on the list are consider as having OUV together with the values perceived by the local community. Although the latter often disregarded in the justification of OUV [25], the role of local communities has been increasingly addressed by the Convention itself.

The Asia-Pacific Heritage Awards 2000

While recognizing the achievement of the private sector and public-private initiatives in conserving cultural heritage throughout Asia and the Pacific, the awards for cultural heritage conservation also act as a catalyst for local preservation activity. As argued by [9], good conservation practice needs to be grounded in an understanding of the locality of place and its values.

2.1.2 ICOMOS

Founded in 1965, the ICOMOS is the only global non-governmental organization dedicated to the conservation of the world's monuments and sites. The ICOMOS serves a forum for professional dialogue and a vehicle for the collection, evaluation and dissemination of information on conservation principles, techniques and policies. Therefore, it played a decisive role as world leader in the understanding and protection of cultural heritage [2]. Adopted the Venice Charter 1964 as its principal document, several of the ICOMOS national committees have independently developed charters on the principles of conservation within their own culture and traditions.

Bruges Resolutions 1975

The rationale for acknowledging the Resolutions on the Conservation of Smaller Historic Towns in this study inherently lies in its universal appliance to the conservation of smaller historic towns. There is no exact definition of town's boundary and in the context of small towns in South-East Asia, little is known about the characteristics of the towns [23]. By listing a number of its features, the dearth is however covered more or less in the Resolutions. The towns is characterized by problems in common or by specific features (size, cultural context, and economic function); its historic core which still marks the center of social life and business and contains a large proportion of residences; unspoilt surrounding landscape; balanced and diversified community structure; and its economic role.

Article 3 of the Resolutions set forth several common problems that have been plaguing the towns. Following these, the Resolutions advocate that any surviving links with the past should not be allowed to atrophy otherwise the national and cultural identity of the countries will be irremediable impoverished. Governments are encourages to provide planning authorities with the responsibility for protecting their historic towns against the pressure of excessive expansion and industrialization. Furthermore, regional policy must also take into account the needs of the towns and ensure their conservation by assigning them a role in keeping with their special structure; therefore plea for coordination at the planning stage of all public authority policies. Other measures propose by the resolutions comprise planning at the local level to retain and enhance the values of the towns; development of methods for surveying, assessing and protecting the towns' character; and stimulation of a sense of pride and responsibility among the local inhabitants.

Tlaxcala Declaration 1982

Acknowledging the importance of small settlements as key witnesses to our cultures, the Declaration on the Revitalization of Small Settlements emphasizes the local and national governments' responsibility in preserving the places. It also recognizes the rights of local communities to be involved in making decisions and work regarding the conservation of their towns. However, the Declaration stipulates that the procedures should be properly developed, and the works require efforts from multidisciplinary team and improvements on social service and infrastructure within a sensitive context while taking into local values and traditions.

Washington Charter 1987

The Charter for the Conservation of Historic Towns and Urban Areas 1987 concerns not only the large historic urban areas but also the smaller cities, towns and historic quarters. These areas embody values beyond their role as historical documents. The Charter advocates that conservation principles must embrace those steps necessary for the protection of such towns and areas as well as their development and harmonious adaptation to contemporary life. This should be preceded by multidisciplinary studies and documentation of the existing historic area. Furthermore, the Charter was found to be the first charter adopted by the ICOMOS that addresses the concept of public participation in conservation process [41]. In order to encourage their involvement, a general information programme should be set up for all residents beginning with children of school age.

Principles for the Recording of Monuments, Groups of Buildings and Sites 1996

The Principles perceives the documenting process as an effective tool to acquire an understanding of cultural heritage, its value and evolution; promote the interest and involvement of the people in the preservation; permit informed management and control of construction works and of all change to the cultural heritage; and to ensure that the maintenance and conservation of the heritage is sensitive to its physical form, materials, construction and its cultural significance.

The methods of recording should be appropriate and the complete report of any recording should be disseminated and made publicly available. The Principles states the complexity of the recording and interpretation processes requires the involvement of skilled individuals working in collaboration. As highlighted by [25], the conservation and management of historic areas and properties has increasingly become a shared field.

The Burra Charter (1979, Revised 1981, 1988, 1999)

Using the Venice Charter as a starting point, the Australian National Committee of ICOMOS decided that the Charter for Places of Cultural Significance to be written for Australian context. The Burra Charter is differentiated with the Venice Charter on the basis of its application to all places of cultural significance, not just the monuments covered by the older document. To Australian, the Charter is the most significant document on the basic principles and procedures for the conservation and management of heritage places [20]. This is evident as it has been widely adopted not only in Australia but also in other parts of the world. The strength of the Charter lies in its universal approach, clear methodology and its advocacy of conservation plan. Furthermore, the Charter provides detailed guidelines for establishment of cultural significance, development of conservation policy, and procedures for undertaking studies and reports.

The Nara Document on Authenticity 1994

The Document addresses the need for a broader understanding of cultural diversity and cultural heritage as it relates to the notion of authenticity of heritage. The document states that the understanding authenticity plays a fundamental role in all scientific studies of the cultural heritage, in conservation and restoration planning, as well as within the inscription procedures used for the World Heritage Convention and other cultural heritage inventories. For [50], the document marked a conceptual shift which allowed the inscription of wooden buildings in the World Heritage List.

The respect for other cultures, values, and the tangible and intangible expressions that form part of the heritage requires that heritage properties must be considered and judged within the cultural contexts to which they belong. Critical to this is the notion that authenticity is not a restrictive concept.

2.1.3 The Council of Europe

Since 1989, the Council has involved in heritage education and continues these activities in the framework of the European Heritage Days, a joint programme of the Council and the European Union which is held in September of each year. The Council's work in promoting architectural heritage has resulted in several standards, charters and conventions.

The European Charter of the Architectural Heritage 1975

The Charter is aimed at developing a common European policy for the protection of architectural heritage. Although emphasis is placed on the Europe's heritage, the Charter is significant in a way it refers architectural heritage to be embodied with cultural values [7]. More importantly, it contains the first mentions of the economic values, which is one of the major preoccupations in present-day literatures of the conservation discipline. Furthermore, the Charter acknowledges the introduction of modern architecture into historic areas provided that the existing context is fully respected and traditional materials are used [44]. While achieved through sensitive restoration techniques and appropriate functions, this integration is largely depends on legal, administrative, financial and technical support

The Granada Convention

The adoption of the Convention for the Protection of the Architectural Heritage of Europe 1985 was a consecration as it marked twenty years of European cooperation on architectural heritage and had included the principles of integrated conservation of the heritage. The Convention stipulates the need for cooperation at all levels, and for the public to participate and be informed. As the first step towards subsequent legal protection, the Convention emphasizes the need for any architectural heritage to be surveyed and inventoried (Article 2).

2.2 Legislative approaches to built heritage conservation in Malaysia

In Malaysia, the emergence of heritage building conservation practice can be traced back to the pre-independence period, proven by the introduction of the Treasure Trove Ordinance 1951 which was later enacted as Treasure Trove Act 1957 (Act 542) [55]. The post-independence period particularly in March 1983 saw the first public outcry in an attempt to save the Loke Chow Kit's house [51]. For [17], the country's first successful conservation project involving adaptive reuse of the Kuala Lumpur Central Market around 1986 has been recognized as a pioneer for the development of building conservation activities in Malaysia. The practice has become more vigorously in recent times with the recognition of the Archaeological Heritage of Lenggong Valley in Perak as a UNESCO World Heritage Site in 2012.

An increasingly interest in building conservation activities also has led to the formulation of several acts, policies, plans and guidelines on protection of built heritage. These are discussed in the following sections according to the structure of the Malaysian federal, state and local governments. As highlighted by [39] and [55], heritage conservation in Malaysia is a shared legislative responsibility between the federal and state government.

2.2.1 Federal Government

The Ministry of Tourism and Culture Malaysia and the Ministry of Urban Wellbeing, Housing and Local Government are the two authorities responsible for protection and management of built heritage at the federal level in Malaysia.

The establishment of the Ministry of Tourism and Culture Malaysia in year 2004 paved way for the formulation of the National Heritage Act 2005. This reflects the government's growing commitment to preserving nation's cultural heritage [1]. The department that is responsible for enforcing the provisions placed under the Act is the Department of Heritage. The Department of Museums Malaysia is another department placed under the Ministry which holds the responsibilities to preserve Malaysian historical, cultural and natural heritage.

National Heritage Act 2005 (Act 645)

The National Heritage Act 2005 is a comprehensive Act that grants the conservation and preservation of National Heritage, natural heritage, tangible and intangible cultural heritage, underwater cultural heritage, treasure trove and other manifestations such as heritage food and heritage persons. The Act is enforced and administered by the Commissioner of Heritage appointed by the Ministry of Tourism and Culture Malaysia.

Based on UNESCO World Heritage Convention, the Act was legalized to replace the Treasure Trove Act 1957 and the Antiquities Act 1976. In contrast with the Antiquities Act 1976, this Act does not limit heritage to properties of hundred years and above. Vague criteria for listing of built heritage outlined in the former Antiquities Act have also been addressed by this new Act. Notwithstanding the fact, the Act is silent as to the criteria of listing an object having cultural heritage significance as compared to those being listed as National Heritage.

Section 23 of the Act provides for the establishment of a National Heritage Register as a national reference which can be inspected by the members of the public. This implies the importance of public participation in the field of heritage conservation. However, [39] argued that the Act is still vague as to the obligatory duty of the community to inform or propose to the Commissioner about heritage objects. Other deficiencies highlighted include absence on the criteria of archaeological impact assessment (AIA); conservation measures to the listed objects in the Register; lack of consideration of the issues of maintenance and management of the buildings; and limited power empowered to the National Heritage Department as conservation of buildings falls subtly under the jurisdiction of local authorities [22, 37, 39].

Town and Country Planning Act 1976 (Act 172)

Town and Country Planning Act 1976 is closely relate with the built heritage by means of development plans and development control. The Act was amended four times over the years from 1993 to 2007. The second amendment of the Act in year 1995 denotes direct incorporation of historic buildings into the Act. Specifically, where the development is in respect of a building with special architecture or historical interest, the applicant is to provide particulars to identify the building and measures for its protection and enhancement [Subsection 21B (1)(b)]. Likewise, the local planning authority shall also impose appropriate conditions when dealing with development that involves the erection of a new building, the re-erection or extension of a building [Subsection 22 (5)(i)] or any addition or alteration to an existing building with special architecture or historical interest [Subsection 22 (5)(j)].

Local Government Act 1976 (Act 171)

Local Government Act 1976 governs the powers of local authorities and shall only apply to Peninsular Malaysia. Subsection 101 (c)(iv) empowers the local authority to maintain or contribute to the maintenance of historical buildings or sites and acquire any land for the purpose of or in connection with the maintenance of historical buildings or sites [14]. Subsection 102 (f) further enables the local authority to provide for the establishment, regulation and management of any historical building or site.

Street, Drainage and Building Act 1974 (Act 133)

The erection of buildings is also loosely bonded by the Street, Drainage and Building Act 1974 [6]. In regards to the preservation of buildings, the State Authority has the right to make by laws in the construction, paving, width and level of arcades and footways; the construction, alteration and demolition of buildings, and the methods and materials to be used; and the minimum timber or other material content in any building.

2.2.2 State Government

With the exception of the three federal territories which directly governed by the federal government, each state in Malaysia is headed by Chief Minister who is automatically the Chairman of State Planning Committee. The committee is the approving authority for development plans and may issue directives as well as policies which the local authority must comply. Development plans means the local plan or the structure plan for the area in which the land or building is situated [16].

State Structure Plans

The State Structure Plan is a statutory planning document which sets out the policies and general proposals for the development and other use of land within each state and thus providing guidance for spatial development. The plan also provides the framework for the preparation of a statutory local plan covering each district of the state. The structure plan has to confirm to the provisions of the national economic, social, physical, environmental and conservation policies. The plan is provided by State Director of Town and Country Planning, approved by the State Planning Committee and need to be agreed by the state authorities.

2.2.3 Local Government

Apart from responsibility for protection under the Local Government Act 1976, the Town and Country Planning Act 1972 empowers the local authority to control land use planning, control and conservation of all lands and buildings within their area. Accordingly, the local authority is required to prepare local plan for purposes of guiding planning development of its area or if any, special area plan for any part of its area.

Local Plans

A local plan is detailed map and written statement prepared to elaborate the policies set out in the structure plan. The plan by their consultative process of preparation acts as contractual agreements on the use of land within the planning areas between the local planning authorities, the local residents and land owners [34]. The plans also concerns with the protection and improvement of the physical environment and natural topography; and the preservation and enhancement of character and appearance of buildings in the area of local plan [16].

Special Area Plan

The Special Area Plan is a specific action plan for an area that has been identified by the state of local authority. The plan is prepared in the form of detailed local plans by the State Director or local authority pursuant to Section 16B of the Town and Country Planning Act 1976. It includes detailed information on the type of treatment, implementing agencies, phasing, costing and funding of a certain proposed project. The plan allows local authority to expedite an action plan for an area for the purpose of whether to immediately develop, redevelop, further improve, conserve or manage the area.

2.2.4 Non-governmental Bodies

The Heritage of Malaysia Trust and Penang Heritage Trust are the two non-governmental bodies concerned with the conservation of built heritage in Malaysia.

Heritage of Malaysia Trust

The Heritage of Malaysia Trust is a registered charity trust that aims to promote conservation of Malaysia's built heritage through awareness and public support, education and technical support. Since its inception, the Trust has undertaken various building conservation projects, research, education and outreach programmes. The Trust also has played a distinctive role in establishing financing for heritage conservation project [18]. The Trust also provided inputs for the amendments to the Town and County Planning Act 2003 and inputs to the National Heritage Act 2005.

Penang Heritage Trust

Founded in 1986, the Penang Heritage Trust is a charitable non-governmental organization aimed at promoting the conservation of Penang's cultural and built heritage as well as fostering cultural education about the history and heritage of Penang. Despite its emphasis on Penang, the Malaysian Institute of Architects and the Penang Municipal Council were encouraged by the Trust's formation and organized the first international conference on urban conservation and planning in June 1986 [24]. The Trust also has played an active role in lobbying for George Town's listing as a world heritage site in 2008.

3 Methodology

The approach used in this study involves critical review of existing plans, policies and legislation with regard to sustaining the unique features and identity of small towns in Malaysia. The study however starts by reviewing the conservation charters and guidelines at the international level. The rationale for this is to familiarize with the specialized terminology used in the legislation and principles of heritage conservation. Furthermore, current conservation principles used in Malaysia are deeply rooted in the ICOMOS Burra Charter and the Venice Charter. The literature review is mainly acquired from officials documents published by government and private agencies, journals, theses, and other mass media outputs. This study is also informed by the semi-structured interviews with nine participants from four different categories of stakeholders representing the federal, state, and local government, and the non-governmental organization. The interviews are audio recorded and transcribed by the researchers to maintain the confidentiality of the data and to ensure closeness of the researchers with the data [52]. The transcriptions are then analysed using the qualitative software package NVivo 8.0. The summary models imported from NVivo are created throughout the analysis process to visualize the connections between various dimensions of constructs, concepts and categories identified in the interviews.

4 Findings and discussion

Findings from the interviews in Table 1 demonstrated that there are two forms of protection mechanisms currently available in protecting built heritage in Malaysia particularly formal and informal mechanisms. [26] and [35] differentiate the two mechanisms by the formal prescribed through formal legal processes whereas the informal mechanisms or the so-called 'community-based system' often derived from practices and belief systems carried out by local communities. As suggested by [46], the best approach for protecting and managing historic places can be achieved through a partnership between community members and appropriate heritage and conservation professionals. In line with the international, national, state and local approaches to built heritage conservation discussed in Section 2.0, the presence of a wide range of protection mechanisms as identified by the interviewees indicates that a relatively strong policy framework exists for protection of built heritage in Malaysia.

All interviewees mentioned the *National Heritage Act 2005* as one of the important mechanisms available in protecting nation's built heritage. As argued by [39], the Act is the only specific statute on the preservation of cultural heritage that cover various matters on cultural heritage in Malaysia. According to two interviewees, the Act provides more holistic approach to the issue relating to the declaration of national heritage as all places including those that are of local significance possess equal opportunities and rights under the Act to be recognized as heritage or national heritage provided that the significance is proven. However, as highlighted in Section 2.2.1, the Act is silent to the criteria of listing an object as of having cultural heritage significance into the Heritage List compared to those being listed in the National Heritage List.

Table 1. Perceptions of heritage protection mechanisms in Malaysia

Concepts	Emergent themes	Constructs
Heritage protection mechanisms available in Malaysia	Formal	Federal Territory (Planning) Act 1982 (Act 267)
		International charters and guidelines
		Local Government Act 1976 (Act 171)
		Local planning schemes
		National Heritage Act 2005 (Act 645)
		Street, Drainage and Building Act 1976 (Act 172)
		Town and Country Planning Act 1976 (Act 172)
		Uniform Building By-law Act 1984
	Informal	Collaboration between different stakeholders
		General community initiatives

Findings from the interviews (n=5) reveal that the *National Heritage Act 2005* also emphasizes on matters under the *Town and Country Planning Act 1976* particularly with regard to the preparation of development plans, application for planning permission, preparation of conservation management plan and implementation of provisions on areas gazette as heritage sites. As discussed in Section 2.2.1, the Act itself is closely relate with the built heritage by means of development plans and development control. This corresponds with the suggestion recommended by international charters and conventions such as the World Heritage Convention, the Washington Charter, the European Charter and the Granada Convention on the need to integrate conservation of built heritage into the land use planning processes. While the Act does not govern the three federal territories of Malaysia, one interviewee acknowledges the *Federal Territory (Planning) Act 1982* as the planning and enforcement document for Malaysia's Federal Territory.

Furthermore, all interviewees agreed that the local authority conservation guidelines served an important tool in the protection of built heritage at the local level. Nevertheless, two interviewees argued that most of the guidelines used nowadays are likely to be similar to one another as they were issued by the same consultant. As a matter of fact, different places should be governed by different set of guidelines because they are coming from completely different aspects of what makes the places, origins and people. As argued by [40], there can be no general checklist that is applicable to all sites. The findings further demonstrate that the guidelines also often viewed only as recommendations and consequently many illegal works on historic properties were reported. This demonstrates the insufficiency of the mechanism and the need for specific heritage legislation. This is reasonable owing to a direct mode of operation and the certainty of regulatory instrument's effect [48].

Two interviewees said that built heritage was also protected through the *Local Government Act 1976* as the Act has provided allocations and conferred power to local authorities to take the necessary measures for the preservation of historic buildings. The *Street, Drainage and Building Act 1974* and *Uniform Building By-law Act 1984* were mentioned as other protection mechanisms by interviewees from the state and local government respectively.

Interestingly, the interviews results (n=2) suggest that local community initiative programs and active collaboration between stakeholder groups had provided some degree of protection for built heritage as they highlight something about particular places. Examples of these are the exhibitions or huge public outcry that brings about public awareness. As discussed in Section 2.2, one of the earliest pioneers of building conservation movement in Malaysia also has led by a group of concerned community members. The fact that the local community plays a vital role in the heritage field is also reflected in other world regions such as in London, Australia, England and Caribbean Island [11, 10, 40, 30]. For [36], this is especially important when the objects are not of major national interest, not listed and not preserved by heritage authorities.

5 Conclusion

To be successful and competitive in an increasingly globalized world, cities and towns need to be more distinctive, unique and special for which it can be recognized as a separable entity. The desire and commitment to plurality and differentiation against standardization and monotonous image of a place led to greater appreciation of built heritage which increasingly touted as one of the fundamental components that have a significant influence on the identity of place. Not only in major urban centers, built heritage also makes a continuing contribution to the image and identity of small cities alike. Given the importance of cultural built heritage in conferring local distinctiveness, measures need to be taken to ensure the continuity and protection of these physical structures especially under present day condition where the identity of place is rapidly weakening. By bringing together nation's approaches as well as relevant international charters and guidelines for built heritage protection, this paper can be regarded as the first of many fruitful contributions that examine heritage protection mechanisms at the local level. Particularly, the study revealed that built heritage is not only protected by legislative measures alone but also through community initiative activities and practices. Therefore, it is crucial that future legislative framework recognizes the rights and role of local communities in the protection of their towns.

Acknowledgement

The authors gratefully acknowledge the financial support of the Postgraduate Research Grant (PPP), grant no PG003-2012B established at the University of Malaya, Institute of Research Management and Consultancy (IPPP).

References

1. A.H. Ahmad Sarji, *Town Planning and Conservation: What needs to be done?* (2005)
2. G.F. Araoz, J. Cult. Herit. Manage. Sustain. Dev. **1**, 1, 55-60 (2011)
3. Arazi, K. Faris, M. Sodangi, Mod. Appl. Sci. **4**, 11, 66-77 (2010)
4. M. Arefi, J. Urban Des. 4, 2, 179-193 (1999)
5. Australia ICOMOS, *The Australia ICOMOS charter for places of cultural significance: The Burra Charter 1999* (Australia ICOMOS, Sydney, 2000)
6. M. Badaruddin, Paper presented Paper presented at the Korea, China, Japan, Phillipines, Spain, Turkey, India and USA Joint World Cultural Tourism Conference 2005 (2005)
7. Council of Europe, *The European Charter of Architectural Heritage (1975)*
8. Council of Europe, *The Granada Convention (Convention for the protection of the architectural heritage of Europe) (1985)*
9. R.A. Engelhardt, J. Pres. Technol. **41**, 2-3, 57-65 (2010)
10. F. Ahmad, *Protecting the built heritage of capital cities: Case studies of Perth and Kuala Lumpur* (Curtin University of Technology, Perth, 2009)
11. J.M. Gard'ner, Int. J. Herit. Stud. **10**, 1, 75-92 (2004)
12. P. Goad, L.T. Ngiom, *Recent Malaysian architecture* (Pesaro Publishing, Singapore 2007)
13. Government of Malaysia, *Antiquities Act (Act170)* (National Printing Document 1976)
14. Government of Malaysia, *Local Government Act (Act171)* (National Printing Document 1976)
15. Government of Malaysia, *National Heritage Act (Act645)* (National Printing Document 1976)
16. Government of Malaysia, *Town and Country Planning Act (Act172)* (National Printing Document 1976)
17. S.N. Harun, P. Eng. 20, 41-53 (2011)
18. Heritage of Malaysia Trust, *Town planning and conservation: What needs to be done?* Retrieved 18 April, 2012 from http://www.badanwarisan.org.my/content/?cid=219 (2008)
19. Heritage of Malaysia Trust, *Comprehensive assessment needed for KL's Chinatown* Retrieved 30 April, 2012 from http://www.badanwarisan.org.my/content/?cid=381 (2011)
20. Heritage Perth, *The Burra Charter* Retrieved 21 October, 2012 from http://heritageperth.com.au/your-heritage/the-burra-charter/ (2012)

21. ICOMOS, *Resolutions on the conservation of smaller historic towns (The Bruges Resolution)* (1975)
22. A. Idrus, F. Khamidi, M. Sodangi, Mod. Appl. Sci. **4**, 11, 66-70 (2010)
23. J.C. Jackson, T. Ins. Brit. Geogr. 61, 65-80 (1974)
24. G. Jenkins, *Contested space: Cultural heritage and identity reconstructions. Conservation strategies within a developing Asian city* (Lit Verlag, Zurich, 2009)
25. J. Jokilehto, *ICCROM and the conservation of cultural heritage: A history of the organization's first 50 years, 1959-2009* (2011)
26. A. Jopela, Conserv. Manag. Archaeol. Sites (2011)
27. K. Kamarul Syahril, A.W. Lilawati, A. Ghafar, Paper presented at the 2nd International Conference on Built Environment in Developing Countries 2008 (University Science Malaysia, 2008)
28. M.H. Khairul Amin, U. Norsidah, Soc. Behav. Sci. 153, 689-699 (2014)
29. Q.Y. Lee, Y.M. Lim, Int. J. Human Soc. Sci. **5**, 3, 140-145 (2010)
30. M.J. LeFebvre, J. Island Coastal **8**, 2, 311-314 (2013)
31. W.S. Logan, *The disappearing "Asian" city: Protecting Asia's urban heritage in a globalizing world* (First ed.) (Oxford University Press, United States, 2002)
32. K. Lynch, *The image of the city* (Joint Center for Urban Studies, United States, 1960)
33. J.R. Mansfield, Eng. Constr. Arch. Manag. **15**, 3, 270-281 (2008)
34. Ministry of Housing and Local Government, *National Physical Plan 2* (2010)
35. A. Mumma, *Legal frameworks for the protection of immovable cultural heritage in Africa* (ICCROM, Rome, 2005)
36. L. Mydland, W. Grahn, Int. J. Herit. Stud. **18**, 6, 564-587 (2012)
37. W.A.Z. Noor Amila, A. Alias, I. Arazi, Paper presented at the International Conference on Sustainable Building and Infrastructure (2010)
38. U. Norsidah, Asian. J. Env. Behav. Stud. 61-76 (2010)
39. M. Nurulhuda Adabiah, A. Nuraisyah Chua, Paper presented at the International Conference on Tourism Development Penang (Malaysia 2013)
40. J. Pendlebury, Town Plann. Rev. **70**, 4, 479-500 (1999)
41. M. Rasouli, L. Khirfan, Paper presented at the Canadian Association of Geographers' Annual Meeting (Waterloo, Ontario, 2012)
42. E. Relph, *Place and placelessness* (Pion Limited, Brondesbury Park, London, 1976)
43. M. Ripp, U. Eidenschink, C. Milz, Facilities **29**, 7-8, 286-302 (2011)
44. D. Rodwell, *Conservation and sustainability in historic cities* (Blackwell Publishing, Oxford, UK, 2007)
45. S. Shuhana, *Townscape revisited: Unravelling the character of the historic townscape in Malaysia* (Universiti Teknologi Malaysia, Johor Bahru, Malaysia, 2011)
46. B. Smith, Proceedings of the Conservation Theme of the 5th World Archaeological Congress Washington D.C., 22-26 June 2003 (The Getty Conservation Institute, Los Angeles, 2006)
47. N. Tavakoli, Paper presented at the 14th International Planning History Society Conference (2010)
48. D. Throsby, *Seven questions in the economics of cultural heritage* (pp.13-30). In M. Hutter, *Economic perspectives of built heritage* (McMillan Press, London, 1997)
49. UNESCO World Heritage Centre, *Basic texts of the 1972 World Heritage Convention* (UNESCO, Paris, France, 2005)
50. S.V. Uytsel, P. Jurcys, J. Jpn. 34, 309-316 (2013)
51. W.I. Wan Hashimah, Asian. J. Env. Behav. Stud. **1**, 3, 73-82 (2010)
52. S. Watts, Qual. Res. Psychol. 11, 1-14 (2014)
53. World Heritage Convention, *World Heritage List UNESCO* Retrieved 2 July, 2012 from http://whc.unesco.org/en/list/ (2012)
54. D. Yuksel, D. Iclal, Paper presented at the 15th ICOMOS General Assembly and International Symposium: Monuments and sites in their setting – conserving cultural heritage in changing townscapes and landscapes (Xi'an, China, 2005)
55. M.Y. Yuszaidy, D. Hanapi, K. Ab Samad, Jurnal Melayu 8, 173-188 (2011)

Critical Aspects of the Inclusive Environmental for the Well-being of Building Occupant – A Review

Kamaruzzaman S.N [1a], Noor Ashiqin[2], Ahmad Zawawi E.M[3], Mike Riley[4]

[1,2]*Department of Building Surveying, Faculty of Built Environment, University Malaya, Kuala Lumpur*
[3] *Faculty of Architecture, Planning & Surveying, University Technology Mara, Shah Alam, Selangor*
[4]*Liverpool John Moores University*

Abstract. In the property, design and construction industry the main focus of sustainability has been seen to affect the environmental such as climate change and resource depletion. Even though building may meet the recommended standards overall, but yet occupants still complain about several aspects. Indoor environmental quality consequently affects health, productivity and well-being of building occupants as most of them spend their time in indoor space. Therefore they are at risks for air pollutant occurs in the indoor air of dwellings such as volatile organic compounds, paints, ventilation, and others environmental pollutant. This paper reviews the potential health consequences in changes of climates change whereby developed countries contributed in energy consumption and greenhouse gas emissions. From the review search by 57 publications, the literature discussing about how thermal, acoustic and visual comfort as well as indoor air quality affects the satisfaction in indoor environmental quality and impacts to the human health. However, from the undertaken review most of the studies only highlighted on one aspects of indoor environmental quality. A number of critical aspects which give direct impacts to occupants have been identified. Subsequently, among all aspects such as lighting, ventilation, acoustics or humidity and thermal but thermal comfort perceived to influence overall satisfaction of occupants in the building. Thus, from the study revealed the critical aspects of indoor environment and contribute to the improvement for well-being and productivity of building occupants.

Keywords: Building performance, Environmental health and safety, Green building, Indoor environmental quality, Occupant

1.0 Introduction

Over decades, climate change has become a major issue globally due to the increasing of global temperatures and extreme weather conditions [1]. Consequently, this climate change may affect to dwellings that are more likely to spend their time indoors [2,3]. The review of the linkages of Climate Change with Indoor Environmental Quality (IEQ) whereby have been reported by a committee of the Institute of Medicine (IOM) can be referred at Fig 1. In addition, study conducted in UK shows the time consume by participants spending an average of 95.6% of their time indoors, with 66% of their time spent in their homes. It is supported by several studies pointed out that people in developed countries are more likely to spend 90% of their time indoors [4,5,6]. Thus, the risks to health posed by the exposure to indoor pollution are greater rather than outdoor air pollution. Subsequently, various aspects have been documented such as exposure to light, indoor air quality, temperature, odors, noise and ergonomics whereby those affect occupant performance and well-being [7].

Moreover, in high income country the building industry being a main contributor in energy consumption and greenhouse gas (GHG) emissions. For example the residential buildings in UK produce around 25% of total GHG end - user emissions in 2012 (DECC 2014).

[a] Corresponding author: syahrulnizam@um.edu.my

Fig.1. Examples of how Climate Change may affect indoor environments and health [3].

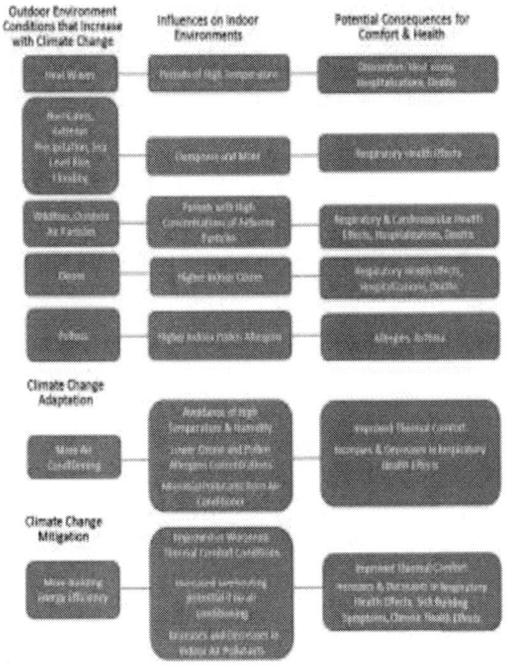

Building should provide shelter and enhance well-being to occupant however it also associated with a range of health hazards like indoor air pollution, extreme temperatures, pests and infestations, noise, airborne infectious diseases, water or mould contamination, domestic injuries and mental health effects [8].

In addition, from the studies by [9] indicates that costs of poor indoor environment for the employer, building owner and society as a whole are often considerable greater than the cost of the energy used in the same building. Nonetheless, the good indoor environmental quality can improve overall work, learning performance and reduce absenteeism. There are numbers of paper have recently reviewed, [10] point out on the effects of single environmental conditions on human such as examiné which conditions lead to satisfaction with the visual and acoustic environment .Besides that, studies by [11] concern about which factors are not significant to the indoor environment and again they highlighted more on satisfaction with a single environmental condition such as visual environment or thermal environment [12]. Subsequently, from the research studies tended to consider environmental comfort such as temperature, velocity, radiant temperature and humidity but none have considered lighting, noise and smell. The purposes on that, this paper reveal critical aspects of the inclusive environmental that influence the well-being building occupants.

2.0 Overview of Indoor Environmental Quality in Buildings

The issues pertaining to the health, well-being and comfort to the building occupants with a satisfactory indoor environment needs to be highlighted especially in developed countries. In UK, by law it is a requirement for the building owner to have an energy performance certificate [13].Subsequently, we can see the trends toward providing a developing system for better building assessment scale in local, global and indoor environmental scale [14]. However, studies regarding achieving an optimal satisfaction indoor environment for all building occupant facilities can assumed as scarce [15]. Thus, it is crucial to educate people to have an awareness and knowledge for better understanding of indoor environment influence. In the point of view [16] he suggested on green building design in order to provide satisfaction in health, comfort and well-being performance of building occupants. Therefore, standards and guideline are based with IEQ parameter whereby it is combined with effects on occupants satisfaction and efficiency [17]

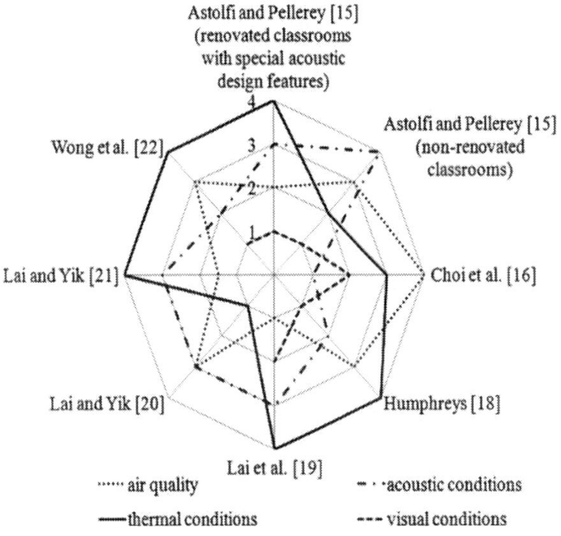

Fig. 2. Ranking of the importance of different environmental conditions for overall satisfaction with IEQ

However, in terms of IEQ [18] suggested, the standards and guidelines should be changed and fashioned in order to enhance occupants optimal comfort and efficiency. Numbers of studies have been conducted according to indoor environmental quality of buildings in occupants. Factors that have the relationship between IEQ was studied by [19 ; 20 ;21] while the integration of the parameters of IEQ into mathematical models studies conducted by [22 ; 23 ; 24 ; 25].

[a] Corresponding author: syahrulnizam@um.edu.my

Other than that, according to [26] suggested that, acceptability of overall indoor environment should depends on average acceptance of all the indoor environmental parameters at certain acceptability level. Besides that, [27] indicate the physical environment and people response must take into consideration while designing an environment for people. Poor indoor environmental quality may result into sick building syndrome (SBS) as stated by [28].

Moreover, improvement to indoor environmental quality in building occupants achieved when all factors of indoor environmental quality parameters are taken into account. These indoor environmental factors include thermal comfort, acoustic comfort, noise and ergonomics. Figs 2 indicate the ranking of overall satisfaction among building occupant with IEQ from previous studies. The result shown the highest rankings are most important.

[a] Corresponding author: syahrulnizam@um.edu.my

Table 1: Components of critical aspects from previous study

Aspects	Author	Finding
1.Visual comfort	[29]	Author agrees with all factors of visual comfort such as the amount of light, the uniformity, the quality and rendering quality. -Define in European standard EN12665 a subjective condition of visual well-being induced by the visual environment (EN12665, 2011) -Depends on the physiology of the human eye, physical quantities,describe the amount of light and distribution in space, spectral emission of the light.
2. Indoor Air Quality	[30]	-Indicates that guidelines and standards set for adherence are dependent on adverses health effects which could potentially occur from exposure. - All examined pollutants are experienced similarly by the countries on each continent with Relatively similar guidelines. - Other factors such as temperature, humidity, and air movement, which are correlated, interchangeably affect each other and chemical pollutants.
3.Indoor Environmental Quality (IEQ)	[31]	-No attempt has been made towards validating thermal comfort, acoustic comfort, visual comfort and indoor air quality as parameters of measures of IEQ in hospital buildings. -There is therefore, a need for the validation of the IEQ parameters of measurement as a basis for the development of an integrated IEQ performance evaluation model.
4.Indoor Environmental Quality	[32]	-The environmental factors are: thermal comfort, indoor air quality, acoustic comfort and visual comfort
5.Indoor environmental Quality	[33]	-The paper investigates the interactions between global satisfaction for workers in hypermarkets and the different attributes that contribute to it. - Factor analysis pointed out the existence of two main groups of dissatisfied. - The first one gave greater relevance to visual and acoustic aspects while being neutral towards thermal aspects. -Conversely, the second group was more influenced by lack of thermal comfort and presence of local discomfort, while showing indifference towards visual and acoustic aspects. -Application of Kano's model showed that thermal local discomfort and IAQ were considered "basic" aspects (i.e. they have negative effect on global satisfaction when they underperform), - While acoustic comfort was considered a "bonus" aspect, capable to improve overall rating when it performed well. -Visual and thermal comfort had a mostly linear influence on global satisfaction but also proved to have the highest impact.
6.Indoor Air Quality	[34]	- Author explored a range of health risks in the domestic indoor environment related to climate change, as well as the potential health benefits and unintended harmful effects of climate change mitigation and adaptation policies in the UK housing sector. - Effect includes indoor temperatures, indoor air quality, indoor allergens and infections, and flood damage and water contamination. - Certain changes to the indoor environment can affect indoor air quality or promote the growth and propagation of pathogenic organisms. Measures aimed at reducing greenhouse gas emissions have the potential for ancillary public health benefits including reductions in health burdens related heat and cold, indoor exposure to air pollution derived from outdoor sources, and mould growth. - Increasing airtightness of dwellings in pursuit of energy efficiency could also have negative effects by increasing concentrations of pollutants (such as PM2.5, CO and radon) derived from indoor or ground sources, and biological contamination. -High risk of these adverse health effects include the elderly (especially those living on their own), individuals with pre-existing illnesses, people living in overcrowded accommodation, and the socioeconomically deprived..Ventilation is a key aspect that affects indoor air quality (chemical and microbial).Moisture-related allergens (mould and dust mites) and thermal comfort in dwellings
7. Thermal Visual Acoustic	[35]	-Suggest that when developing systems for controlling the indoor environment, the type of building and outdoor climate, including season, should be taken into account. - Providing occupants with the possibility to control the indoor environment improves thermal and visual comfort as well as satisfaction with the air quality.Thermal comfort is ranked by building occupants to be of greater importance compared with visual and acoustic comfort and good air quality.

[a] Corresponding author: syahrulnizam@um.edu.my

Aspects	Author	Finding
8. Indoor air pollutants:	[36]	- The measured parameters were temperature, relative humidity, CO_2, and the indoor air pollutants: fourteen individual volatile organic compounds (VOC), four aldehydes and particulate matter PM10 and PM2.5 - Concentration of formaldehyde was approximately 30% higher in dwellings built after 1990 compared with older ones; it was higher in dwellings with mechanical ventilation and in concrete buildings.
9.Indoor Environmental Quality	[37]	-The most common weekly symptoms in the spring semester were fatigue (7.7%), stuffy nose (7.3%), and headache (5.5%). -The most frequently reported IEQ factors causing daily inconvenience in classrooms were noise (11.0%) and stuffy air/poor indoor air quality (IAQ) (7.0%), which were also found most frequently above 95% CI on the group level (calculated for N = 15), together with self-reported high indoor temperature and dust or dirtiness.
10.Indoor Environmental Quality	[8]	-It was found that day-lighting, electric lighting and glare were five out of the seven factors found most dissatisfied aspects of the buildings. - Electric lighting was used for almost all the time the occupants were in the buildings. - This study provides significant impact on creating change in terms of improving building environment in two ways. -Providing lessons and feedback for owners or those involved in the environment improvement works. This could lead to enhanced quality of indoor environment by addressing the changing needs of occupants
11. Ventilation in Indoor air quality Standard	[38]	-This paper reviews and discusses some of the issues that have been addressed in the development of ventilation standards in recent years, using the development of ASHRAE Standard 62 as context, including: the scientific bases for ventilation requirements, perceived indoor air quality, contaminant sources from occupants and the building, -outdoor air quality, airborne contaminant limits, indoor carbon dioxide concentrations, environmental tobacco smoke, and performance-based design
12. Noise	[39]	- Four studies reported that wind turbine noise significantly interfered with QOL. -visual perception of wind turbine generators was associated with greater frequency of reported negative health effects.. - Evidence that exposure to wind turbine noise is associated with increased odds of annoyance and sleep problems. Individual attitudes could influence the type of response to noise from wind turbines. -to explore the association between wind turbine noise, annoyance, sleep and quality of life.
13. Indoor temperature, relative humidity, and carbon dioxide concentration	[40]	-In conclusion, there were large differences between occupant self-reported satisfaction and thermal comfort by tenure status, but differences in measured parameters were relatively small. -The results indicate that occupant characteristics are likely to explain a majority of differences by tenure status, which should be taken into account when assessing the overall relationships between housing and health.
14.Noise	[41]	-Determined the risk of noise for the incidence of myocardial infarction. -Chronic noise burden is associated with the risk of myocardial infarction. The risk increase appears more closely associated with sound levels than with subjective annoyance. Further investigation of the gender-related risk of noise exposure may aid in improving prevention.
15. Indoor air quality (IAQ), volatile organic compounds (VOCs)	[26]	-As the occupancy period got longer, the VOCs originating from the construction phase were increasingly replaced by new ones. -Reference values based on means and on 95 percentiles are presented to facilitate interpretation of the results of measurements done to ensure that proper construction practices have been applied or to investigate IAQ problems

[a] Corresponding author: syahrulnizam@um.edu.my

DOI: 10.1051/matecconf/20166600114

2.1 Indoor Air Quality (IAQ)

Generally, Indoor Air Quality (IAQ) term is used to describe the quality of air within a building. Average about 90% people spend their time inside their workplace or home therefore indoor air quality of building occupant is a basis determinant of healthy life, people well-being, comfort and productivity of the occupants. Prior to maintain a good IAQ it is important to identify the physical and chemical parameters which influence the comfort of building occupants whereby also proposed by the Industrial Code of Practice on Indoor Air Quality . The characteristics of good indoor air quality are adequate ventilation air, control of airborne contaminants and maintenance of acceptable temperature and relative humidity.

However, the failure on solving the problem immediately and efficiently regarding poor indoor air quality will affects the health, productivity and well-being of building occupants. According to the World Organization health, they reported that indoor air pollution contributes about 2.7% of the global burden disease. As supported by [42] by 2030 expected that if inappropriate indoor conditions may lead to more premature death than malaria or AIDS. Apart from that, indoor air quality gives direct impact on health if exposure to high concentrations of air pollutants indoor. In the studies conducted by [43], even short term exposure to high concentrations of carbon monoxide can cause acute effects such as intoxication and death. Besides that, [44] indicates the chronic health effects include radon related lung cancer and this caused by second hand tobacco smoke which lead to chronic obstructive pulmonary disease [45].

This supported by [46] whereby from their studies indicate that, second hand tobacco smoke can caused respiratory infections, cardiovascular disease and a range of allergic symptoms like atopic dermatitis, rhinitis, conjunctivitis and hay fever. Subsequently, there has a significant between indoor air quality and efficiency. In the studies by [47] point out, small changes in temperatures impact the quantity and quality of the produced work. Thus, improvement of the quality and quantity of outside air maximize natural ventilation and separate ventilation air from thermal conditioning indirectly improves pollution source control. International case study stated that, improvement in high performance ventilation contribute to reduce respiratory illness, increase in productivity as well as increasing in energy savings for natural ventilation and mixed mode conditioning [48].Subsequently, dwelling proximity to industrial and traffic pollution sources, increased indoor tobacco smoking, dampness, contaminated dusts consequently can affect the health problem due high concentration of pollutant [49] .In the Table 1 shows the effect of different carbon dioxide concentrations on human body. In living space the air composition are differ rather than outside. As refer from (Pettenkofer's number) the concentration of carbon dioxide in outside range between 300 and 400 ppm while 900ppm in living space.

Table 2: The effects of CO_2 concentration on human body [32].

CO2 concentration		Effects
[%]	[ppm]	
3	30,000	Deep breathing, strong
4	40,000	Head aches, pulse, dizziness, Psychic emotions
5	50,000	After 0.5–1 h may cause death
8–10	80,000-10,000	Sudden death

2.2 Relative Humidity

Thermal comfort according [50] refer to the condition of mind which expresses satisfaction with thermal environment. However, this condition only met if no local discomfort exists such as building users not disturbed by draught, too low or high internal surface temperature or others. In addition, the standard (ISO 7730, 1993) indicates it is difficult and possible to predict the mean thermal sensation and mean satisfaction with thermal condition in a group due to sharing the built environment with other occupants. The physical variables that associated with thermal comfort such as air temperature mean radiant temperature, relative air velocity and air humidity.

Moreover, relative humidity refers to the moisture content in the air which gives impact to the thermal comfort in the building. The low levels of humidity caused irritation due to decrease of mucus at the eyes and nose. However, high levels of humidity will ease the airborne pestilences transfer via dust particles into the respiratory pathway and results in bronchitis. In addition, [51] pointed out that increase pollutant emission from building material such as volatile organic compound are due to high humidity which result inferior indoor air quality. In China, they recommended the relative humidity indoor at range 40– 80% in summer and 30 – 60% in winter. Moreover, HKEPD, 1999 set their standards as 40– 70% relative humidity needed for excellent and less than 70% are good indoor air quality. However, (DOSH, 2010) also set a standard as 40– 70% while Singapore (Institute of Environmental Epidemiology, 1996) set their standards at ≤70% for office premises. Subsequently, according to (Air Duct Cleaners,2013) US have set standards regarding relative humidity whereby ASHRAE set standard relative

[a] Corresponding author: syahrulnizam@um.edu.my

humidity values of 40% 60% in summer while 30% to 60% in winter.

2.3 Noise pollution and acoustic comfort.

Noise is considered as environmental problem consequently health problem and annoyance for human being.

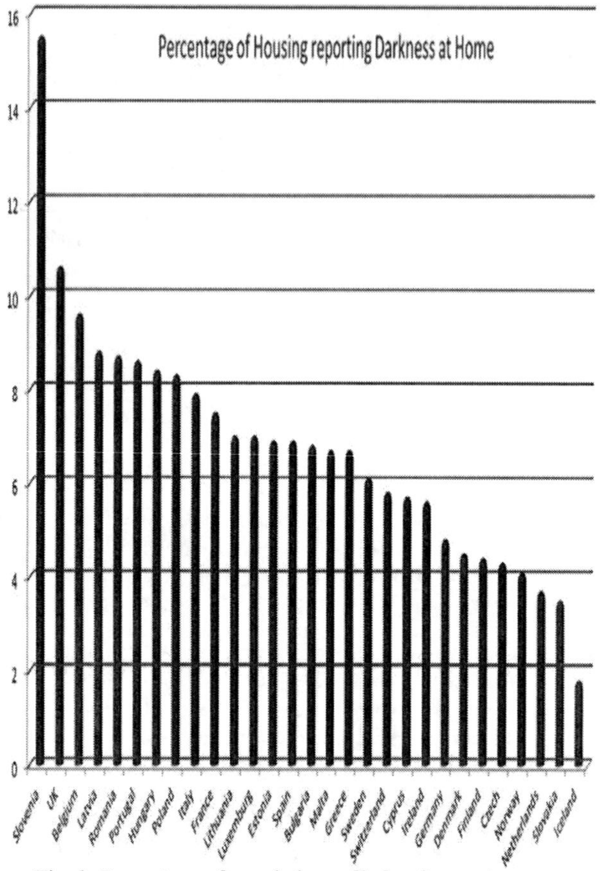

Fig. 3. Percentage of population suffering from noise pollution in the residential area among European countries (Rybkowska and Schneider, 2011).

In these factor can be differentiate psychosocial responses for example annoyance, sleep disturbance, disturbance of daily activities and performance while for physical responses such as hearing loss, hypertension and ischemic heart disease. Subsequently, according to [52] indicate that, exposure to noise problem can induce the changes in biochemical, physiologic, or psychosocial such as disturbance of sleep and daily activities, stress, and annoyance. In addition, these supported by [53], people living surrounded with noisy environment causing stress, present sleep disturbances and develop into habit of shouting. In addition, [54] conclude that people in urban household willing to spent about 4€ per year in order to achieve reduction of noise. However, from the study of noise against household income, indicate the low income occupants tend to exposed to noise than higher income [55].

In addition, the Fig 3 shows the percentage of households suffered with noise pollution. Subsequently, among all European countries, Romania, Cyprus, Malta, Germany and Netherlands a very high part of the population complaint to noise pollution which is 25–35%. Moreover, in Italy, they carried out studies at the neighborhoods with traditional socio economic and the result also showing low quality neighborhoods are strongly health damaging low. Therefore from the result, [56] indicates that, noise pollution is a serious problem among overcrowded urban area.

3.0 Discussion And Conclusion

This study has reviewed about 57 publications which discussed the aspects of indoor environmental quality and affect the well-being of building occupants. From the previous studies, the main aspects which contribute most impact to occupants were identified such as indoor air quality, thermal comfort, visual and acoustic and noise. An overview of critical aspects influenced to the quality of the indoor environment whereby give direct impact to building occupants has been summarized in Table 1.

The studies on indoor environment in buildings in last decade only highlighted on individual parameter such as visual comfort [31 ;33 ;34 ;35 ;28] , acoustic comfort [33 ;35 ;28], noise [40 ;42]. It was found that from the studies conducted by [38 ;28] visual, thermal comfort and acoustic seems to influence to higher degree the overall satisfaction with indoor environmental quality compared to others aspects. Moreover [32 ;37] in IAQ aspects, they studies on temperature, carbon dioxide, indoor air pollutants as well as the relative humidity. From the studies [32] provides a standards and guidelines regarding IAQ issues. Besides that, temperature, humidity and air movement are correlated and affect each other. Moreover, the concentration of formaldehyde pointed out about 30% higher in dwellings built after 1990 [37]. In addition, the review from the case study conducted by [35 ;60] agreed that, among all aspects measured such as lighting, ventilation, acoustic or humidity and thermal comfort and yet thermal influences occupants overall satisfaction in building. Moreover, there is evidence in the literature undertaken by [61] indicate that increase level of pollutant such as volatile organic compounds leads to lower air quality.

Based on the literatures that have been reviewed, the conclusions that can be drawn from it are summarized as follows:

- Thermal comfort, acoustic comfort, visual comfort and indoor air quality has been as a crucial factors in indoor environmental quality.
- In addition, by create a comfortable thermal environment will lead to overall satisfaction with indoor environmental condition.
- Moreover, by provide people with self-control such as in term of ventilation in building will

[a] Corresponding author: syahrulnizam@um.edu.my

improve thermal and visual comfort as well as indoor environment satisfaction.

- Most of the studies were focused only one aspects of indoor environmental quality or the other but no attempts on more than three aspects in the studies conducted by previous researchers.

Acknowledgement

The authors gratefully acknowledge the financial support of the UMRG RP015D-15SUS , Established at the University Malaya, sustainability science research cluster.

References

1. Nelson, A.J., Rakau, O. and Do¨rrenberg, P. (2010), Green Buildings – A Niche Becomes Mainstream, Deutsche Bank Research, Frankfurt am Main.
2. Crump, D., (2011). Climate Change— Health Impacts Due to Changes in the Indoor Environment;Research Needs. Cranfield University, Institute of Environment and Health.
3. Committee on the Assessment of Asthma and Indoor Air, Division of Health Promotion and Disease Prevention Staff, & Institute of Medicine Staff. (1900). Clearing the air: asthma and indoor air exposures. National Academies Press..
4. Harrison, R.M., Thornton, C.A., Lawrence, R.G., Mark, D., Kinnersley, R.P., Ayres, J.G., 2002.Personal exposure monitoring of particulate matter, nitrogen dioxide, and carbon monoxide, including susceptible groups. Occup. Environ. Med. 59, 671–679.
5. Vardoulakis, S., 2009. Human exposure: indoor and outdoor. In: Air Quality in Urban Environments 28, 85–107.
6. Lai, H.K., Kendall, M., Ferrier, H., Lindup, I., Alm, S., Hänninen, O., Jantunen, M., Mathys, P.,Colvile, R., Ashmore, M.R., Cullinan, P., Nieuwenhuijsen, M.J., 2004. Personal exposures and microenvironment concentrations of PM2.5, VOC, NO2 and CO in Oxford, UK. Atmos. Environ. 38, 6399–6410.
7. Kamaruzzaman, S. N., Egbu, C. O., Zawawi, E. M. A., Ali, A. S., & Che-Ani, A. I. (2011). The effect of indoor environmental quality on occupants' perception of performance: A case study of refurbished historic buildings in Malaysia. *Energy and Buildings*, *43*(2), 407-413.
8. WHO, 2011. Health Co-benefits of Climate Change Mitigation — Housing Sector. World Health Organization, Geneva.
9. Andersen, R. V., Toftum, J., Andersen, K. K., & Olesen, B. W. (2009). Survey of occupant behaviour and control of indoor environment in Danish dwellings. Energy and Buildings, **41**(1), 11-16.
10. Galasiu, A. D., & Veitch, J. A. (2006). Occupant preferences and satisfaction with the luminous environment and control systems in daylit offices: a literature review. *Energy and Buildings*, *38*(7), 728-742.
11. Veitch, J. A., & Newsham, G. R. (2000). Exercised control, lighting choices, and energy use: An office simulation experiment. *Journal of Environmental Psychology*, *20*(3), 219-237.
12. Nicol, J. F., Raja, I. A., Allaudin, A., & Jamy, G. N. (1999). Climatic variations in comfortable temperatures: the Pakistan projects. *Energy and Buildings*, *30*(3), 261-279.
13. Ncube, M., & Riffat, S. (2012). Developing an indoor environment quality tool for assessment of mechanically ventilated office buildings in the UK–A preliminary study. *Building and Environment*, *53*, 26-33.
14. Chiang, C., Lai, C., Chou, P., Li, Y., & Tu, Y. (1999). The study on the comprehensive indicators of indoor environment assessment for occupants'health. In ASIA-PACIFICConference on the Built Environment 1993, Taipei, (pp. 1–7).
15. Paul, W.L., Taylor, P.A. (2008). A comparison of occupant comfort and satisfaction between a green building and a conventional building. Building and Environment (43), 1858–1870.
16. Salonen, H., Lahtinen, M., Lappalainen, S., Navala, N., Knibbs, L. D., Morawska, L.,et al. (2013). Design approaches for promoting beneficial indoor environmentsin healthcare facilities: A review. Intelligent Buildings International, 5(1), 26–50.
17. Smith, A., &Pitt, M. (2011). Healthy workplaces : Plant scaping for indoor environmental quality Facilities, **29**(3/4), 169–187.
18. Huang, L., Zhu, Y., Ouyang, Q., & Cao, B. (2012). A study on the effects of ther-mal, luminous, and acoustic environments on indoor environmental comfortin offices. Building and Environment, 49, 304–309.
19. Croitoru, C., Vartires, A., Bode, F., & Dogeanu, A. (2013). Survey evaluation of theindoor

[a] Corresponding author: syahrulnizam@um.edu.my

environment quality in a large Romanian hospital. INCAS Bulletin, 5(3),45–52.
20. Astolfi, A., & Pellerey, F. (2008). Subjective and objective assessment of acousti-cal and overall environmental quality in secondary school classrooms. Journalof the Acoustical Society of America, 123, 163–173.
21. Humphreys, M. a. (2005). Quantifying occupant comfort: Are combined indices ofthe indoor environment practicable? Building Research & Information, 33(4),317–325
22. Lai, A. C. K., Mui, K. W., Wong, L. T., & Law, L. Y. (2009). An evaluation model forindoor environmental quality (IEQ) acceptance in residential buildings. Energyand Buildings, 41(9), 930–936.
23. Chiang, C., Lai, C., Chou, P., Li, Y., & Tu, Y. (1999). The study on the comprehensive indi-cators of indoor environment assessment for occupants'health. In ASIA-PACIFICConference on the Built Environment 1993, Taipei, (pp. 1–7).
24. Heinzerling, D., Schiavon, S., Webster, T., & Arens, E. (2013). Indoor environmentalquality assessment models: A literature review and a proposed weighting andclassification scheme. Building and Environment, 70, 210–222
25. Mui, K. W., & Chan, W. T. (2005). A new indoor environmental quality equation forair-conditioned buildings. Architectural Science Review, 48(1), 41–46.
26. Lai, A. C. K., Mui, K. W., Wong, L. T., & Law, L. Y. (2009). An evaluation model forindoor environmental quality (IEQ) acceptance in residential buildings. Energyand Buildings, 41(9), 930–936
27. Frontczak, M., Andersen, R. V., & Wargocki, P. (2012). Questionnaire survey on fac-tors influencing comfort with indoor environmental quality in Danish housing.Building and Environment, 50, 56–64.
28. De Giuli, V., Da Pos, O., & De Carli, M. (2012). Indoor environmental quality and pupilperception in Italian primary schools. Building and Environment, 56, 335–345.
29. Abdul- Wahab, S.A., En, S.C.F.,Elkamel, A., Ahmadi, L.,Yetilmezsoy, K. (2015). A review of standards and guidelines set by international bodies for the parameters of indoor air quality. Atmospheric Pollution Research (6) 751 -767.
30. Carlucci, S., Causone, F., Rosa, F.D., Pagliano, L.(2015). A review of indices for assessing visual comfort with a view to their use in optimization processes to support building integrated design. Renewable and Sustainable Energy Reviews (47)1016–1033
31. Abdul–Wahab, S. A., En, S. C. F., Elkamel, A., Ahmadi, L., & Yetilmezsoy, K. (2015). A review of standards and guidelines set by international bodies for the parameters of indoor air quality. *Atmospheric Pollution Research*,*6*(5), 751-767.
32. Nimlyat, P.S., Kandar, M.Z. (2015) Appraisal of indoor environmental quality (IEQ) in healthcarefacilities: A literature review. Cities and Society (17) 61–68.
33. Sarbu, L.,Sebarchievici, C. (2013) . Review Aspects of indoor environmental quality assessment in buildings. Energy and Buildings (60)410–419.
34. Martellotta, F., Simone, A., Crociata, S.D, D'Alba, M.(2016).Global comfort and indoor environment quality attributes for workers of a hypermarket in Southern Italy. Building and Environment 95 . 355-364.
35. Vardoulakis, A., Dimitroulopoulou, C., Thornes, J. (2015).Review article Impact of climate change on the domestic indoor environment and associated health risks in the UK. Environment International 85.299–313.
36. A. Norhidayah, Lee Chia-Kuang, M.K. Azhar S. Nurulwahida. (2013). Indoor Air Quality and Sick Building Syndrome in Three Selected Building .Procedia Engineering (53) 93 – 98.
37. Langer, S., Ramalho, O., Derbez, M. (2016). Indoor environmental quality in French dwellings and building characteristics. Atmospheric Environment (128) 82-91.
38. Turunen, M., Toyinbo, O., Putus, T., Nevalainen, A. (2014) .Indoor environmental quality in school buildings, and the health andwellbeing of students. International Journal of Hygiene and Environmental Health (217)733–739.
39. Persily, A. (2015) . Challenges in developing ventilation and indoor air quality standards:The story of ASHRAE Standard 62. Building and Environment .(91) 61-69.
40. Onakpoya, I. J., O'Sullivan, J., Thompson, M. J., & Heneghan, C. J. (2015). The effect of wind turbine noise on sleep and quality of life: a systematic review and meta-analysis of observational studies. Environment international, 82, 1-9.
41. Pekkonen, M., Du a,L, Raatikainen, M. (2015). The influence of tenure status on housing satisfaction and indoorenvironmental quality in

[a] Corresponding author: syahrulnizam@um.edu.my

Finnish apartment buildings.Building and Environment (89)134–140.
42. Willich, S.N., Wegscheider, K., M Stallmann, M.(2006) . Noise burden and the risk of myocardial infarction Thomas Keil. European Heart Journal (27), 276–282.].
43. Lavelle, M., 2010. The Solvable Problem of Energy Poverty. National Geographic News.
44. Raub, J.A., Mathieu-Nolfb, M., Hampsonc, N.B., Thomd, S.R., 2000. Carbon monoxide poisoning — a public health perspective. Toxicology 145 (1), 1–14.
45. Darby, S., Hill, D., Auvinen, A., Barros-Dios, J. M., Baysson, H., Bochicchio, F., ... & Heid, I. (2005). Radon in homes and risk of lung cancer: collaborative analysis of individual data from 13 European case-control studies. Bmj, 330(7485), 223.
46. Jordan, R. E., Cheng, K. K., Miller, M. R., & Adab, P. (2011). Passive smoking and chronic obstructive pulmonary disease: cross-sectional analysis of data from the Health Survey for England. *BMJ open*, *1*(2), e000153.
47. Chauhan, A. J., & Johnston, S. L. (2003). Air pollution and infection in respiratory illness. *British medical bulletin*, *68*(1), 95-112.
48. Taylor, A. F., Kuo, F. E., & Sullivan, W. C. (2002). Views of nature and self-discipline: Evidence from inner city children. Journal of environmental psychology, 22(1), 49-63.
49. Fisk, W. J., & Rosenfeld, A. H. (1997). Estimates of improved productivity and health from better indoor environments. Indoor air, 7(3), 158-172.
50. Ashrae, A. N. S. I. (2004). Standard 55-2004, Thermal environmental conditions for human occupancy. *American Society of Heating, Refrigerating and Air-Conditioning Engineering, Atlanta, GA*.
51. Jansz, J. (2011). Introduction to sick building syndrome. In *Sick Building Syndrome* (pp. 1-24). Springer Berlin Heidelberg.
52. WHO, (2002). The World Health Report 2002: Reducing Risks, Promoting Health Life. Geneva: World Health Organization.
53. Bonnefoy, X. (2007). Inadequate housing and health: an overview.*International Journal of Environment and Pollution*, *30*(3-4), 411-429.
54. Lampietti, J., & Meyer, A. (2002). When heat is a luxury: helping the urban poor of Europe and Central Asia cope with the cold. *World Bank, Washington, DC*.
55. Barreiro, J., Sánchez, M., & Viladrich-Grau*, M. (2005). How much are people willing to pay for silence? A contingent valuation study. Applied economics, 37(11), 1233-1246.
56. Kohlhuber, M., Mielck, A., Weiland, S. K., & Bolte, G. (2006). Social inequality in perceived environmental exposures in relation to housing conditions in Germany. *Environmental research*, *101*(2), 246-255.
57. Bilger, M., & Carrieri, V. (2013). Health in the cities: when the neighborhood matters more than income. *Journal of Health Economics*, *32*(1), 1-11.

[a] Corresponding author: syahrulnizam@um.edu.my

 DOI: 10.1051/matecconf/20166600115

Experimental Investigation of Ventilation Efficiency in a Dentistry Surgical Room

Majeed Olaide Oladokun[1,a], Maisarah Ali[2], Samsul Bahrin Osman[3], Niza Samsuddin[4], Hairul Aini Hamzah[5]

[1]*Department of Architectural and Civil Engineering, Building Energy and Environmental Technology Research Unit, Division of Building Science and Technology, City University of Hong Kong, Hong Kong SAR, China.*
[2]*Department of Civil Engineering, International Islamic University, Malaysia (IIUM), Gombak, Kuala Lumpur, Malaysia.*
[3]*Department of Manufacturing and Material Engineering, International Islamic University, Malaysia (IIUM), Gombak, Kuala Lumpur, Malaysia*
[4]*Department of Allied Health Sciences, International Islamic University, Malaysia (IIUM), Kuantan, Malaysia*
[5]*Department of Basic Medical Sciences, International Islamic University, Malaysia (IIUM), Kuantan Malaysia*

Abstract. As a response to the need to provide an acceptable thermal comfort and air quality in indoor environments, various ventilation performance indicators were developed over the years. These metrics are mainly geared towards air distribution, heat and pollutant removals. Evidence exists of influencing factors on these indicators as centered on ventilation design and operations. Unlike other indoor environments, health care environment requires better performance of ventilation system to prevent an incidence of nosocomial and other hospital acquired illnesses. This study investigates, using in-situ experiments, the ventilation efficiency in a dentistry surgical room. Thermal and hygric parameters were monitored on the air terminal devices and occupied zone over a period of one week covering both occupied and unoccupied hours. The resulting time-series parameters were used to evaluate the room's ventilation effectiveness. Also, the obtained parameters were benchmarked against ASHRAE 170 (2013) and MS1525 (2014) requirements for ventilation in health care environment and building energy efficiency respectively. The results show that the mean daily operative conditions failed to satisfy the provisions of both standards. Regarding effectiveness, the findings reveal that the surgical room ventilation is ineffective with ventilation efficiency values ranging between 0 and 0.5 indicating air distribution short-circuiting. These results suggest further investigations, through numerical simulation, on the effect of this short-circuiting on thermal comfort, infection risk assessments and possible design improvements, an endeavour that forms our next line of research inquiries.

1 Introduction

In its optimal performance, the building ventilation system is tasked with the provision of an acceptable thermal comfort and perceived air quality through fresh air supply, heat removal and pollutant dilutions and/or removal. In health care buildings, ventilation systems are expected to preventing cross infection risks in addition to maintaining adequate thermal comfort to the patients, the caregivers and visitors. On the contrary, despite the high energy requirements [1, 2], the HVAC contributes to making building unhealthy [3] as several cases of nosocomial or hospital acquired illness (HAI) had been previously reported [4-6]. These problems, amongst others, had hitherto led to the emergence of building performance diagnostics. Documented evidence exists on building ventilation performance diagnosis for thermal comfort and indoor air quality [1, 7:Ai, 2014 #617], hygrothermal performance and ventilation energy efficiency [8-10]. They are equally focused on design optimisation [11-13] and performance assessment of new building component design [14, 15].

There are various performance metrics of assessing indoor ventilation based on the ventilation system's tasks: fresh air supply, air mixing, contaminant removal, air exchange, heat removal and occupant's protection [16]. Ventilation performance indices can therefore be grouped based on the expected functions of the system such as fresh air supply (e.g. air change rate, air change efficiency, mean age of air, local air change index, etc.), air mixing and/or diffusion (e.g. air diffusion performance index), pollutant removal (e.g. contaminant removal effectiveness, local air quality index, etc.). Specific indices are also available for thermal comfort (e.g. predicted mean vote, predicted percent dissatisfied, effective draft temperature, and percent dissatisfied due to draft, etc.). More details on the different ventilation performance metrics are available in the literature. Specifically, the readers are referred to Mundt *et al.* [17], Brouns *et al.* [18], and Liddament *et al.* [19].

Practical applications of ventilations performance metrics had been previously reported. The suitability of both air change efficiency and contaminant removal effectiveness formed the focus of a study carried out by Novoselac *et al.* [20] for different indoor spaces,

[a] Corresponding author: lyday011@yahoo.com

ventilation strategies, and contaminant sources. The study revealed that the air change efficiency is adapted to evaluation of ventilation strategies in an indoor environment with unknown contamination sources. On the other hand, the contaminant removal effectiveness performs better for removal of contaminants with known position and generation rates. Nevertheless, certain correlation was found between the two metrics. Contaminant removal metrics had been applied to the ability of air distribution systems to remove unwanted indoor parameters that affect either thermal comfort and/or perceived air quality.

For instance, in a study on the ventilation performance of room air distribution systems, Gan [21] defined two contaminant removal metrics: thermal and ventilation efficiency with the former expressed for heat removal and the later for contaminant removal. Similarly, Irwin *et al.* [22] evaluated, using a scaled experiments, the air distribution performance of different ventilation schemes with the use of contaminant removal performance for heat and gaseous pollutants. In the study heat removal metrics was referred to as thermal energy removal effectiveness [22]. An analogous concept of heat removal as an index of ventilation performance metrics of air distribution system was adopted by Mohammed [23] and Aziz *et al.* [24] where the effectiveness of different air distribution systems was investigated.

Also, as observed by Ali *et al.* [25], the indoor thermal and hygric parameters are treated as contaminants in hygrothermal performance assessments. Hence, both indoor temperature and relative humidity parameters can be used in evaluating the performance of air distribution systems. Both parameters, in addition to being environmental factors in human thermal comfort [26], they are also found to be important for the preservation of archival materials [27-29].

In this paper, we experiment the ventilation efficiency of an air distribution system using the thermal and hygric removal effectiveness metrics. A case studied dental surgical room is selected part of a larger study on the performance of the air distribution systems in the entire building that housed the surgical room. The aim is to understand the impact of various design parameters on the ventilation performance during operation. The study carried out in-situ experiments in a full-scale room for the evaluation of the ventilation performance. Although of all the available methods for building performance assessments (analytical, field experiments, numerical simulations, etc.), experimental studies provide data for analytical and numerical experiments, instrumentation, setup errors and data collection process greatly influence the data from field experiments [30]. Hence, it is important to assess the uncertainties in measured data from field experiments as good uncertainty clarifications improves results interpretation and reliability.

2 Methodology

2.1 Experimental Setup and Measurement Procedures

The surgical room, 5.3 m × 3.8 m × 3.4 m high, is air-conditioned and ventilated by a central constant air volume (CAV) air handling unit (AHU). The AHU controls the airflow, thermal and hygric distribution

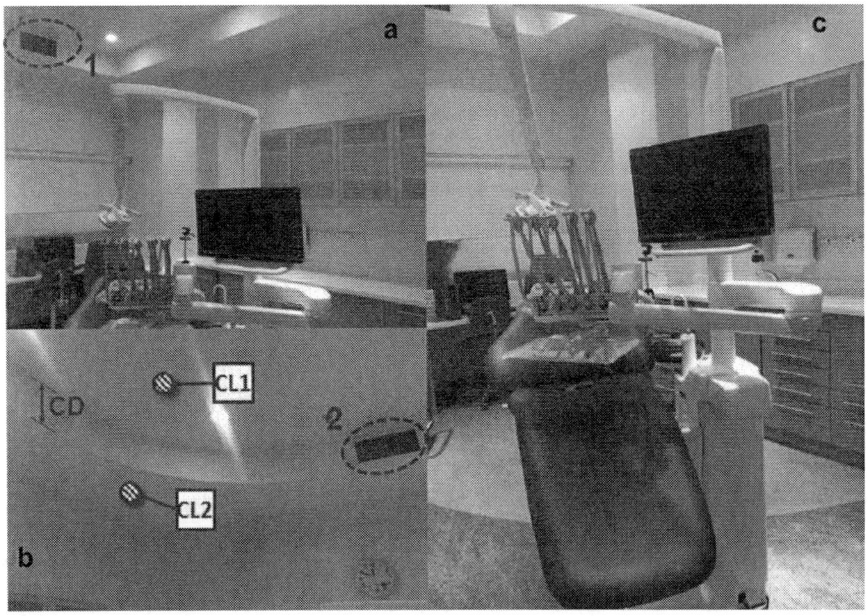

Figure 1. Pictorial representation of case-studied room showing: (a & b) The two-level ceiling design; (c) Room occupied zone showing the dental surgical (Dashed circles #1 & #2 are supply and exhaust outlets; CL1 & CL2 are ceiling levels; and CD is the ceiling drop).

within the room. The air terminal devices in the room are arranged to fit into the architectural design of the two-levelled ceiling (Figure 1).

The air distribution system in the room consist of rectangular supply and exhaust grilles (size 450 mm x 250 mm) vertically mounted on the drop of the architectural ceiling. The facility housing the surgical room operates between 9:00 am and 5:00 pm daily. Figure 2 shows details of the space layouts with positions of supply, exhaust and the measurement locations in the case study room. The air supply to the room is through duct network while the return is through the ceiling plenum as shown the figure.

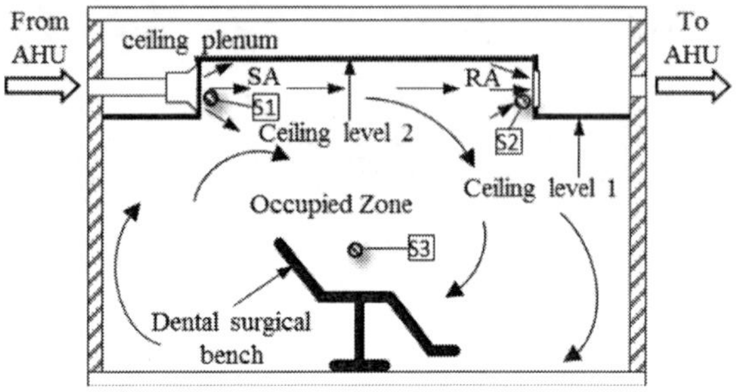

Figure 2. Vertical cross section through the case-studied dental surgical room with measurement locations (S1-S3 are temperature and humidity at supply inlet, return outlet and occupied zone respectively).

Data collection involves thermal and hygric measurement with data loggers mounted at the supply diffuser, exhaust grille, and the occupied zone over a period of one week from 21/5/2015 to 27/5/2015. The sensors were EL-USB-2-LCD+ Data Logger with LCD from LASCAR Electronics. The uncertainty of the measured parameters is in accordance with the manufacturer's specifications. The measurement accuracy and precision are ±0.5°C/0.5°C, and ±2.5%/0.5%, for temperature and relative humidity respectively. As results of field measurements are susceptible to experimental uncertainties, the measured data are further evaluated for uncertainty assessment using the methodology described in a previous study by one of the authors [31]. The method is briefly described in the following section.

When making field observations of a variable, an acceptable level of uncertainty in the results is first determined as the level of confidence. As according to Vincent *et al.* [32], the commonly applied level of confidence in statistical investigations is 95% (p<0.05). Therefore, we adopted a 95% level of confidence. To evaluate the measurement uncertainty, we applied Equation (1) for the confidence interval.

$$CI = \pm 1.96 \times SE_M \tag{1}$$

Where CI is the confidence interval and SEM is the standard error of the mean. The SEM is obtained from Equation (2) as follows:

$$SE_M = \frac{\sigma}{\sqrt{n}} \tag{2}$$

Where σ is the standard deviation and n is the number of observed time-series data. The *CI* in equation (1) is an interval, hence the positive "+" and negative "-" signs in the equation. The calculated *CI* value is compared with the equipment accuracy provided by the manufacturer. As described in Oladokun [31], the values of *CI* below or at most equal to the equipment accuracy indicates certainty in the obtained parameters at 95% level of confidence. The author further submitted that using this approach, where variation occurs in measurement at various locations, such variations can be attributed to causes other than the measuring equipment. On the other hand, when *CI* exceeds the equipment accuracy, equipment error is suggested and may call for calibration of such measuring tool.

2.2 Ventilation Performance Assessment

The study adopts thermal and moisture removal effectiveness as the ventilation performance indices. The measured temperature and humidity data are used to evaluate the thermal and moisture performance metrics. In this paper, we assessed these metrics from the model shown in Equations (3) and (4). The heat removal effectiveness is estimated as:

$$\varepsilon^{C_T} = \frac{\theta_e - \theta_s}{\theta_{oz} - \theta_s} \tag{3}$$

where ε^{C_T} is the heat removal effectiveness, θ_e is the exhaust temperature, θ_s is the supply temperature, and θ_{oz} is the temperature in the occupied zone. Similarly, the moisture removal effectiveness is calculated as:

$$\varepsilon^{C_H} = \frac{\varphi_e - \varphi_s}{\varphi_{oz} - \varphi_s} \tag{4}$$

where ε^{C_H} is the moisture removal effectiveness, φ_e is the exhaust humidity, φ_s is the supply humidity, and φ_{oz} is the humidity in the occupied zone.

The mean temperature and humidity data obtained from the in-situ measurements were used to calculate the heat and moisture removal effectiveness. To further determine the efficiency of the air distribution system, we apply the equations similar to the one described in Brouns *et al.* [18] as shown in Equations (5) and (6).

$$\eta^{C_T} = \frac{\varepsilon^{C_T}}{1 + \varepsilon^{C_T}} \tag{5}$$

and

$$\eta^{C_H} = \frac{\varepsilon^{C_H}}{1+\varepsilon^{C_H}} \quad (6)$$

where η^{C_T} and η^{C_H} are the heat and moisture removal efficiency respectively.

According to Brouns *et al.* [18], the values of ventilation efficiency for different types of ventilations are as shown in Equation (7).

$$VE = \begin{cases} \eta^{C_{T,H}} = 0.5 : Complete\ Mixing \\ 0.5 \le \eta^{C_{T,H}} \le 1,\ Piston\ Flow \\ 0 \le \eta^{C_{T,H}} \le 0.5,\ Short-circuiting \end{cases} \quad (7)$$

where VE is the ventilation efficiency defined in Equations (5) and (6).

It is important to point out that testing the ventilation performance in a hospital environment with gaseous contaminant, using tracer gas, would provide better insight. Such method appears difficult in the current experiment as the use of simulated gaseous contaminants was not permitted. Nonetheless, such deterrent is typical of field experiments as limitations in controlling influencing parameters and measurement restrictions to few sampling points are established factors hindering on-site experiments [33].

3 Results and Discussions

3.1 Measurement Uncertainty Assessments

Measured data from sensors placed at the supply inlet, return outlet and occupied zone is examined for uncertainty using Equations (1) and (2). Table 1 compares the results of uncertainty assessment with the manufacturer's specifications.

Table 1. Comparison of measuring equipment uncertainty between measured and manufacturer specified values.

Sensors	Measured Uncertainty		Manufacturer Specification	
	θ	φ	θ	φ
S1	± 0.2 °C	± 0.3%	± 0.5 °C	± 2.5%
S2	± 0.1 °C	± 0.4%	± 0.5 °C	± 2.5%
S3	± 0.1 °C	± 0.4%	± 0.5 °C	± 2.5%

$p < 0.05$

As shown in the table, the measurement uncertainty for temperature ranges from ± 0.1 to ± 0.2 while that of humidity varies between ± 0.3% and ± 0.4%. It is worthy to note an observable difference in the sensor placed at the supply inlet with varying uncertainties from the ones at the exhaust outlet and occupied zone respectively. This variation can be attributed to the colder temperature at the supply and higher velocity than the other two places. Overall, the estimated equipment uncertainty falls below the manufacturer's specified accuracy for the sensors. It is, therefore, sufficient to conclude with 95% confidence that the measurements are accurate and variations in between measurement points will be due to reasons other than those related to the measuring equipment uncertainties. Nevertheless, there is a 5% chance that the results are inaccurate, the value that is statistically acceptable.

3.2 Thermal and Hygric Operative Conditions

The results of thermal and hygric operative conditions in the room showed the mean profiles as: S1 (θ = 15.3 ± 0.2°C, φ = 86.5 ± 0.3%), S2 (θ = 16.3 ± 0.1°C, φ = 83.0 ± 0.4%), and S3 (θ = 17.1 ±0.1°C, φ = 75.8 ±0.4%).

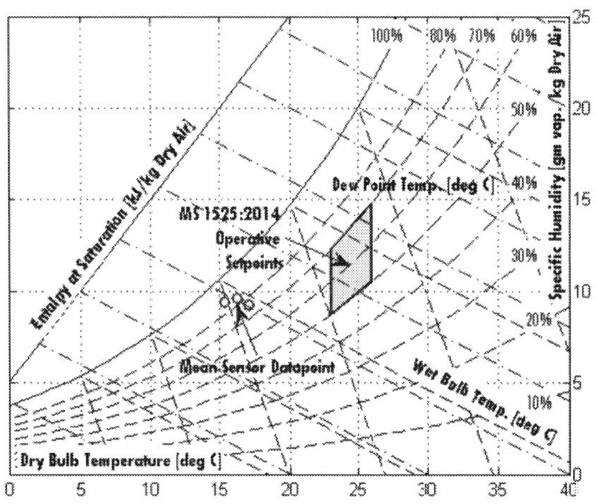

Figure 3. Psychrometric chart showing the mean operative conditions in the surgical room over the experimental duration.

The average temperature and humidity for the measurement locations S1-S3 are plotted on psychrometric charts together with MS 1525 specifications (Figure 3). The plot in Figure 3 revealed an over-cooled operative condition as all points fall below the specified values MS 1525 [34] where thresholds of temperature (θ) and humidity (φ) are respectively specified as θ = 23 - 26°C and φ = 50 - 70%. A similar but more stringent set-point is often required by ASHRAE [35] with θ = 20 - 24°C and φ = 20 - 60%. Evidently, the operative conditions, therefore, followed similar submissions of Sekhar [36] that buildings in the hot and humid climates are overcooled.

3.3 Ventilation Efficiency

The mean operative temperature and humidity were used to evaluate Equations (3) to (6). The results of dimensionless ventilation efficiency for heat and moisture removal are presented in Table 2.

Table 2: Estimated ventilation efficiency.

Parameters	Experimental values
heat removal efficiency, η^{C_T}	0.4
moisture removal efficiency, η^{C_H}	0.2

Applying the conditions in Equation (7), we found that the air distribution systems in the surgical room provide short-circuiting flow. Such air distribution system would be ineffective in removing heat and moisture from the occupied zone as the majority of the ventilation flow is directly exhausted without passing through the occupied zone. Most of the air in the occupied zone will be recirculated flow and not fresh air as shown with curved arrows in Figure 1. The observed situations would lead to, in addition to comfort and air quality issues, energy inefficiency of the ventilation system as most of the energy utilized by the ventilation system is taken up to move air without getting to where it is needed most, the occupied zone.

3.4 Discussions

In an indoor environment, the airflow characteristics are greatly influenced by the air jet which is maintained by the size, location of the supply outlet together with the supply temperature and flow velocity. This underscores the importance of the inlet air jets as the main driving force of indoor air flow. The sizes and locations of the return outlets also play an important role in the indoor air flow trajectory. As the air moves from the supply to exhaust, the air is set into continuous motion thereby separating the room air into primary and secondary air zones. The location of exhaust openings, in relation to the supply, determine which of the primary (direct from supply outlet) and secondary (recirculated or mixed) air dominates the room air flow. For clarity purpose, the air flow pattern in dominant total-volume room ventilation schemes is described and compared with that of the case-studied room.

In mixing ventilation (Figure 4a), the supply and exhaust openings are located on the ceiling level, the room air is well mixed with neither of the primary or secondary air dominating the air flow. As shown in Figure 4a, the primary air dominates the occupied zone. Although there is the likelihood of air short-circuit (direct exhaust of supply air), such is limited as the general flow is dominated by primary air. Equally, there could be regions of flow recirculation in the room under mixing air flow. Nonetheless, mixing air distribution are found to produce acceptable ventilation efficiency [17, 19]. Similarly, in displacement type ventilation (Figure 4b), air is supplied with supply inlet close to the floor and exhaust on the ceiling or opposite wall thereby dividing the room air into colder (lower) and warmer (upper) stratified layers. As shown in Figure 4b, the air within the occupied zone is dominated with fresh primary air with **recirculated air restricted to the upper unoccupied**

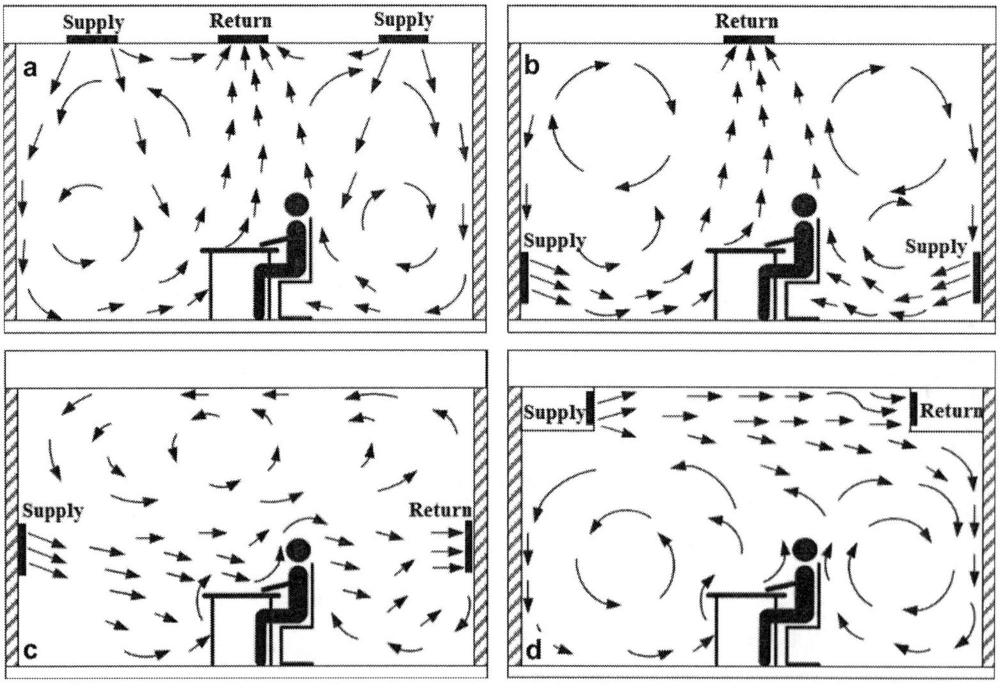

Figure 4. Room air distribution with varying supply and return locations (a) Mixing ventilation with ceiling supply and ceiling return; (b) Displacement ventilation with low-level supply and ceiling return; (c) Stratum ventilation with mid-level wall supply and mid-level wall return; and (d) Air distribution in the case-studied room with high-level ceiling supply and return.

zones. Displacement ventilation is better in performance than the traditional mixing ventilation with higher efficiency [17, 37]. The need to further improve the performance of mixing and displacement ventilation had resulted into a more recently invented stratum ventilation [38] shown in Figure 4c. In this mode of room air ventilation, the air is supplied and returned in the middle of walls thereby results in the provision of primary fresh air not just within the occupied zone but at the occupant's head level. Stratum ventilation provides higher efficiency than mixing and displacement ventilation [39].

In a non-isothermal wall jet close to the ceiling, the air streamline attaches to the ceiling surface by Coanda effects until the air reaches a point where it detaches from the ceiling surfaces. In air distribution where the return opening is located along the path dominated by Coanda effect, the major part of the primary air is exhausted from the room thereby resulting in flow short-circuiting. In such cases, the room airflow is dominated by the secondary recirculated air. Figure 5d shows a case similar to that found in the investigated room where the supply and return outlets are located near the ceiling. As shown in Figure 4d, the major primary air is restricted to the ceiling level with the room airflow dominated by secondary recirculated air. In this case, the room airflow is short-circuited and as such, the occupied zone is deprived of direct fresh air from the supply outlet. This further explains why the ventilation efficiency was found to be below 0.5 in the case-studied room. This finding reiterates the importance of combinations of inlet and outlet locations as influencing factors in the short-circuiting of supply air. Therefore, such arrangement where supply and return openings are fitted into the ceiling drops to suit architectural ceiling design should be avoided for better ventilation efficiency. Where such arrangement must be adopted, adequate care must be given to optimise the efficiency with respect to locations of the air terminal devices.

As the room airflow diffusion is affected by the supply and return outlets on one hand and the indoor heat and contaminants removal related to the room airflow, it is imperative to give utmost consideration to the supply and return air outlets. Reported evidence exist on the effect of various air distribution on indoor air pollutant and human exposure (health risk). An earlier study [40] on indoor contaminant dispersion with mixing and displacement ventilation systems revealed that room airflow pattern has a strong influence on the room contaminants distribution with the return air outlet location having a significant effect on human exposure level in the room. In a study on hospital ventilation performance with respect to airborne contagion, Li *et al.* [5] examined the role of air distribution in SARS transmission during 2003 outbreak. The results showed an imbalance in the airflow pattern from the supply diffusers and returned outlets, which resulted in high infection risks with recorded high fatality in the hospital. The findings from the present study is, therefore, an indication of potential danger in the use of the ventilation under the current design for health care environment as poor efficiency increases the resident time of pollutant in the room thereby increasing the infection risks between the patient and the caregivers.

In this study, the ventilation effectiveness of air distribution system in a dental surgical room has been examined using in-situ experiments. The ventilation design and operative conditions are both found ineffective regarding compliance with energy efficiency benchmark and ventilation efficiency. While the energy inefficiency might be of little immediate concern, the ventilation inefficiency is a serious one especially in the health care environment where the risk of nosocomial is on the rise. We further discuss some of the study limitations for future research in the following sections. On the measurements, few data points were considered as the room is in operation during the measurements. Installing many sensors in the occupied zone would disrupt the operations. Similarly, as mentioned earlier, ventilation performance assessment with gaseous pollutants could provide additional information on the air distribution efficiency. Again, such is prohibitive under the operative conditions of the room. In our next stage of the study, numerical simulations would be employed to consider most of the constraints in the field experiments and to optimise the system for improved performance.

4 Conclusions

A study on the ventilation efficiency in the dental surgical room has been presented in this paper. The time-series measurement of thermal and hygric operative parameters were used to evaluate the ventilation efficiency and the compliance of operative conditions with the standard benchmark. The experimental data were verified for instrumentation uncertainty using an earlier developed procedure. A good overall accuracy was obtained between the measured parameters and the manufacturer's specifications. The operative conditions of the room showed over-cooling conditions against MS 1525:2014 on one hand and failure to comply with ASHRAE 170-2013 on provisions for health care facility on the other. The ventilation efficiency revealed that the air distribution system, under the current design, produces short-circuiting flow with values less than 0.5 for both thermal and moisture removal. The efficiency assessment demonstrated the need for caution in placing air terminal devices to follow the interior design of ceiling especially in the health care environment which require adequate protection for patients, caregivers, and visitors against nosocomial infections. Also, findings on the over-cooling conditions would suggest the need for more studies to decipher the reasons behind the recurring conditions of overcooling in the hot and humid climate. The ventilation efficiency results suggest further investigations, through numerical simulation, on the effect of this short-circuiting on thermal comfort, infection risk assessments, and possible design improvements. These would be considered in our future research endeavours.

Acknowledgements

The results presented in this study is part of a multidisciplinary research project (FRGS12-067-0126) initiated on indoor mould growth prediction using thermal characteristics in buildings in the tropical climate by International Islamic University Malaysia (IIUM) and the Ministry of Higher Education (MOHE) Malaysia. The financial supports are gratefully acknowledged. The authors equally recognised the supports of the Dean of Kulliyyah (Faculty) of Dentistry IIUM together with the management and staffs of Daya Bersih during the studies.

References

1. Y. Wang, Z. Fu-Yun, J. Kuckelkorn, D. Liu, J. Liu, Z. Jun-Liang. Classroom energy efficiency and air environment with displacement natural ventilation in a passive public school building. Energ. Buildings. **70**:258-70 (2014)
2. L.C. Haw, O. Saadatian, M.Y. Sulaiman, S. Mat, K. Sopian. Empirical study of a wind-induced natural ventilation tower under hot and humid climatic conditions. Energ. Buildings. **52**:28-38 (2012)
3. S.C. Wilson, R.N. Palmatier, L.A. Andriychuk, J.M. Martin, C.A. Jumper, H.W. Holder, et al. Mold contamination and air handling units. J. Occup. Env. Hyg. **4**:483-91 (2007)
4. H. Qian, Y. Li, P.V. Nielsen, X. Huang. Spatial distribution of infection risk of SARS transmission in a hospital ward. Build. Environ. **44**:1651-8 (2009)
5. Y. Li, X. Huang, I. Yu, T. Wong, H. Qian. Role of air distribution in SARS transmission during the largest nosocomial outbreak in Hong Kong. Indoor Air. **15**:83-95 (2005)
6. J. Walker, P. Hoffman, A. Bennett, M. Vos, M. Thomas, N. Tomlinson. Hospital and community acquired infection and the built environment–design and testing of infection control rooms. J. Hosp. Infect. **65**:43-9 (2007)
7. R. Tomasi, M. Krajčík, A. Simone, B.W. Olesen. Experimental evaluation of air distribution in mechanically ventilated residential rooms: Thermal comfort and ventilation effectiveness. Energ. Buildings. **60**:28-37 (2013)
8. X. Lü. Modelling of heat and moisture transfer in buildings: II. Applications to indoor thermal and moisture control. Energ. Buildings. **34**:1045-54 (2002)
9. H.J. Moon, S.H. Ryu, J.T. Kim. The effect of moisture transportation on energy efficiency and IAQ in residential buildings. Energ. Buildings. **75**:439-46 (2014)
10. Y. Wang, Y. Liu, D. Wang, J. Liu. Effect of the night ventilation rate on the indoor environment and air-conditioning load while considering wall inner surface moisture transfer. Energ. Buildings. **80**:366-74 (2014)
11. G. Cao, P.V. Nielsen, R.L. Jensen, P. Heiselberg, L. Liu, J. Heikkinen. Protected zone ventilation and reduced personal exposure to airborne cross-infection. Indoor Air. (2014)
12. N. Mao, D. Pan, M. Chan, S. Deng. Performance evaluation of an air conditioning system with different heights of supply outlet applied to a sleeping environment. Energ. Buildings. **77**:281-91 (2014)
13. J. von Grabe, P. Svoboda, A. Bäumler. Window ventilation efficiency in the case of buoyancy ventilation. Energ. Buildings. **72**:203-11 (2014)
14. G. Cao, K. Sirén, S. Kilpeläinen. Modelling and experimental study of performance of the protected occupied zone ventilation. Energ. Buildings. **68**:515-31 (2014)
15. M.K. Kim, H. Leibundgut, J.-H. Choi. Energy and exergy analyses of advanced decentralized ventilation system compared with centralized cooling and air ventilation systems in the hot and humid climate. Energ. Buildings. **79**:212-22 (2014)
16. G. Cao, H. Awbi, R. Yao, Y. Fan, K. Sirén, R. Kosonen, et al. A review of the performance of different ventilation and airflow distribution systems in buildings. Build. Environ. **73**:171-86 (2014)
17. E. Mundt, H.M. Mathisen, P.V. Nielsen, A. Moser (Eds.). *Ventilation effectiveness* (Vol. 2). Finland: REHVA (2004)
18. C. Brouns, B. Waters, A. Infiltration. *A guide to contaminant removal effectiveness*: Air Infiltration and Ventilation Centre; (1991)
19. M.W. Liddament, A. Infiltration. *A Review and Bibliography of Ventilation Effectiveness: Definitions, Measurement, Design and Calculation*: Air Infiltration and Ventilation Centre; (1987)
20. A. Novoselac, J. Srebric. Comparison of air exchange efficiency and contaminant removal effectiveness as IAQ indices. ASHRAE Trans. **109**:339-49 (2003)
21. G. Gan. Evaluation of room air distribution systems using computational fluid dynamics. Energ. Buildings. **23**:83-93 (1995)
22. D.R. Irwin, C.J. Simonson, K.Y. Saw, R.W. Besant. Contaminant and heat removal effectiveness and air-to-air heat/energy recovery for a contaminated air space. ASHRAE Trans. **104**:433-47 (1998)
23. R.H. Mohammed. Numerical Investigation of Displacement Ventilation Effectiveness. Int. J. Mech. Aero. Ind. Mech. Eng. **8** (2014)
24. M.A. Aziz, I.A.M. Gad, E.S.F.A. Mohammed, R.H. Mohammed. Experimental and numerical study of influence of air ceiling diffusers on room air flow characteristics. Energ. Buildings. **55**:738-46 (2012)
25. M. Ali, M.O. Oladokun, S.B. Osman, N. Samsuddin, H.A. Hamzah. *CFD Investigation of Indoor Hygrothermal and Airflow Profile in Academic Research Storage Room: Effect of LMA on Thermohygric Balance and Mould Growth*. Paper presented at the The International Conference on Computational Fluid Dynamics in Research and Industry (CFDRI 2015), Kuala Lumpur, Malaysia. (2015)
26. ASHRAE. *ANSI/ASHRAE standard 55-2013: Thermal environmental conditions for human*

occupancy. Edition ed.: American Society of Heating, Refrigerating and Air-Conditioning Engineers, Inc.; (2013)

27. R. Sulaiman, H.L. Schellen, J.L. Hensen. Pilot study on Indoor Climate Investigation and Computer Simulation in Historical Museum Building: Amerongen Castle, the Netherlands. J. Des. Built. Environ. **7**:75-94 (2010)
28. M. La Gennusa, G. Rizzo, G. Scaccianoce, F. Nicoletti. Control of indoor environments in heritage buildings: experimental measurements in an old Italian museum and proposal of a methodology. J. Cult. Herit. **6**:147-55 (2005)
29. H. Janssen, J.E. Christensen. Hygrothermal optimisation of museum storage spaces. Energ. Buildings. **56**:169-78 (2013)
30. S. Jelena. Ventilation performance prediction. In: J. Hensen, R. Lamberts, (Eds.). *Building performance simulation for design and operation*: Taylor & Francis; (2011)
31. M.O. Oladokun. *Mould Growth Prediction in Tropical Climate Buildings by Hygrothermal Differentials* [Dissertation]. Unpublished: International Islamic University Malaysia; (2015)
32. W. Vincent, J. Weir. *Statistics in Kinesiology*. 4th ed: Human Kinetics; (2012)
33. T. van Hooff, B. Blocken. On the effect of wind direction and urban surroundings on natural ventilation of a large semi-enclosed stadium. Comput. Fluids. **39**:1146-55 (2010)
34. Standards Malaysia. *MS 1525:2014 Energy Efficiency and use of Renewable Energy for Non-Residential Buildings - Code of Practice (Second Revision)*. Edition ed. Putrajaya Malaysia: Department of Standards Malaysia; (2014)
35. ASHRAE. *ANSI/ASHRAE/ASHE Standard 170-2013: Ventilation Standard For Health Care Facilities*. Edition ed. Atlanta: American Society of Heating, Refrigerating and Air-Conditioning Engineers.; (2013)
36. C. Sekhar. Thermal Comfort in Air-conditioned Buildings in Hot and Humid Climates – why are we not getting it right? Indoor Air. (2015)
37. ASHRAE. *ANSI/ASHRAE standard 62.1-2013: ventilation for acceptable indoor air quality*: American Society of Heating, Refrigerating and Air-Conditioning Engineers, Inc.; (2013)
38. Z. Lin, T.T. Chow, C.F. Tsang, K.F. Fong, L.S. Chan. Stratum ventilation – A potential solution to elevated indoor temperatures. Build. Environ. **44**:2256-69 (2009)
39. L. Tian, Z. Lin, Q. Wang. Experimental investigation of thermal and ventilation performances of stratum ventilation. Build. Environ. **46**:1309-20 (2011)
40. G. He, X. Yang, J. Srebric. Removal of contaminants released from room surfaces by displacement and mixing ventilation: modeling and validation. Indoor Air. **15**:367-80 (2005)

 DOI: 10.1051/matecconf/20166600116

Operation And Maintenance In Facilities Management Practices: A Gap Analysis In Malaysia

Zuraihana Ahmad Zawawi[1a], Mohd Khazli Aswad Khalid[1], Nur Azfahani Ahmad[1], Nurul Fadzila Zahari[1] and Nuzaihan Aras Agus Salim[2]

[1]*Faculty of Architecture, Planning & Surveying, University Teknologi MARA Seri Iskandar Campus, 32610 Bandar Seri Iskandar, Perak, Malaysia*
[2]*School of Housing, Building and Planning, Universiti Sains Malaysia, 11800 USM, Pulau Pinang, Malaysia*

Abstract. Facilities management in Malaysia has started as early as 1990s during the development of mega projects such as Putrajaya in 1999 [8, 9]. During this period, Malaysia was introduced to an integrated property management and maintenance services known as facilities management. However, its implementation according to the standard practice is still being disputed. The purpose of facilities management standard practices is to provide the facility management profession with a leading reference on facilities management activities and standards. Therefore, the determination of how far the gap in FM current practice is essential to help our FM industry. Thus, the aim of this study is to establish the gap between the current practices and the standard practices and also how far the gap should be satisfied. A set of questionnaire was designed upon literature and previous research done to gather the required information. The study is subjected to FM organizations in private sectors and only the personnel at management level are assessed. There were certified that most of the practices that assessed by means of gap analysis had an acceptable result which the gap between the current practices and the standard practices are not too far.

1 Introduction

Facilities Management is performed during the operational phase of a building's life cycle, which normally extends over many decades [8, 9]. As such, it will represent a continuous process of service provision to support the owner's core business and one where improvement will be sought on a continuous basis. It is essential that decision-making in the preceding design and construction phases before is therefore properly informed about operational requirements if he facility is to provide optimal support to the owner's business. In this connection, facilities management can be seen as an integral part of a coordinated and controlled process of design, engineering, construction and operations.

There is no general approach to manage facilities. Each organization, even within the same sector will have different needs. Understanding those needs is the key to effective facilities management measured in terms of providing best value.

1.1 Problem statement

The purpose of facilities management standard practices is to provide the facility management profession with a leading reference on facilities management activities and standards. Most of FM industry players were unaware about the importance to implement the practices which should follow the standard, to ensure it will be successfully delivered to the user. Therefore, it is important by the FM industry player to overlook at the current state of practice before developing robust strategies in order to manage facilities efficiently and effectively, within the context of the organization's strategic business plan.

1.2 Research aim and objective

The objective of this study is to determine the gap in the current stages of operation and maintenance in facilities management practice to be the base of a strategic plan that can be applied in the Malaysia FM industry.

1.3 Scope of study

The extent of this study is subjected to FM organizations in private sectors and only the personnel management level is assessed. The respondents are only limited to those within peripheral of Klang Valley where most numbers of FM service providers in Malaysia could be identified. There are 41 numbers of respondents identified and assessed which represent about 56% of the sample population.

2 Literature review

[a] Corresponding author: zuraihana@perak.uitm.edu.my

There are several references used as the research literature in order to have better knowledge prior to this study. The literature review will be discussed about gap analysis and emphasis on building operation and maintenance in facilities management practices.

2.1 Gap analysis

Gap analysis generally refers to the activity of studying the differences between standards and the delivery of those standards (Encyclopedia of Management, 2009). In business and economics, gap analysis is a tool that helps a company to compare its actual performance with its potential performance. It refers to an analysis of the distance between where you are today versus where you need to be in order to meet the standard requirements. Once the gap is determined, a strategic plan can be developed that will move the organization from its current state towards its desired future state.

The other purposes of a gap analysis as suggested by [13] are:

- Identify gaps between the current management systems and the standard practice
- Collect policies, procedures, programs, and personnel that support the current practice
- Establish a list of actions to achieve conformance with the standard

In this study, the differences between standard theoretical practice and actual practices in the delivery of facilities management industry will be determined and discussed. The differences could be used to explain the achievement and to document the areas which need improvement.

2.2 Building operation and maintenance practice

Facilities management involves all types of building elements, including façade, floors, air-conditioning, electrical, utilities and everything which attached to the building. Facilities management also involves monitoring the consumption rate of utilities and giving effort to save a little on it.

Facilities management is important to make sure that a company is in top shape for an operation. A lot of information can be derived from conducting a facilities management check. The results of the review are then discussed with the person who has a high authority in the organization to come with some suggestions that can help the business operate better.

[1] explained that the operation and maintenance (O&M) include all aspects of running a building over the course of its useful life. One goal of effective O&M is to achieve the intent of the original building design team, so that building systems deliver services to building occupants.

The proactive operation and maintenance can yield benefits such as reduced solid and hazardous waste generation, less hazardous air pollution, extended service life of equipment and building materials, better indoor air quality, and fewer occupant complaints [6].

2.3 Building maintenance management

British Standard Glossary of terms (3811:1993) defined maintenance as 'the combination of all technical and administrative actions, including supervision actions, intended to retain an item in, or restore it, to a state in which it can perform a required function. In addition, maintenance is a set of organized activities that are carried out in order to keep an item in its best operational condition with minimum cost acquired.

2.3.1 Maintenance activities

Activities of maintenance could be either repair or replacement activities, which are necessary for an item to reach its acceptable productivity condition or these activities, should be carried out with minimum possible costs.

Table 1: Types of maintenance suggested by various authors

Author / **Types of Maintenance**	**Fitch, 1992**	**Sievert, 2001**	**Mobley, *et. al.*, 2006**	**This Study, 2016**
Predictive	√	√	√	√
Preventive	√	√	√	√
Proactive	√	√		
Reactive				
Planned				
Program				
Improvement	√		√	√
Corrective			√	√
Breakdown	√			
Breakdown / emergency		√		√

Table 1 shows numbers of maintenance routine which suggested by various authors. However, only five types of maintenance routines were chosen in this study. Those maintenance routines were chosen as it illustrates many other types of maintenance. The approach towards maintenance should include some of the following maintenance routines because these are the tools that are widely used in facilities management.

i. Predictive maintenance
- Predictive maintenance is the technique used to check the condition of the different equipment used
- During the evaluation, several things are checked to make an estimate when equipment maintenance should be scheduled

ii. Preventive maintenance
- Includes activities that are performed routinely at planned time intervals on specific systems, equipment or components in order to avoid total failure

iii. *Improvement maintenance*
- Improvement maintenance allows equipment to be improved its performance at times that do not interfere with production schedules, thereby removing one of the largest factors from downtime cost.
- Improvement maintenance sometimes known as Condition Based Maintenance which aims to reduce or eliminate the risk of failure in future

iv. *Corrective maintenance*
- Corrective maintenance activities are generated from preventive maintenance inspections, routine operational requests, and routine service requirements. These activities make up the maintenance backlog and should be planned and scheduled in advance. This approach is the most cost effective way to performed maintenance.

v. *Breakdown or emergency maintenance*
- In many situations, equipment is run until it breaks down. Thus, the technician will only react on equipment that is malfunctioning.
- Breakdown or emergency maintenance also known as run-to-failure or remedial maintenance, is an activity performed on a non-routine basis and is reconstructive in nature
- This approach is the most expensive way to coordinate maintenance

2.3.2 Maintenance policy

The maintenance policy is a document which provides a maintenance management framework to ensure the facilities are maintained effectively to support the strategic objectives of operation and maintenance. The purpose of maintenance policy is to outline the roles and responsibilities of O&M team and sometimes define funding allocation responsibilities to achieve effective maintenance of facilities.

2.3.3 Maintenance schedule

Maintenance schedule defined as a schedule of the frequency of the required or recommended maintenance, testing or inspection for each type of equipment [1]. The schedule may include daily, weekly, monthly and yearly attendance times.

The maintenance schedule may be in separate tables for each type of equipment, or facilities.

2.3.4 Maintenance reports

Many reports can be developed to aid in measuring and controlling maintenance activities. This may include complete job reports, backlog reports, labor utilization reports, overtime reports, and jobs in progress [14].

- *Complete job reports* – show the actual versus estimated performance in terms of cost and time. Large variances can mean poor performance, poor accounting for charges, poor estimating, or poor planning
- *Backlog reports* – show the amount of work remaining at any point in time. These reports identify whether the organization is current, ahead or behind schedule, and can help determine if the maintenance department is under- or overstaffed. The backlog report can also be used to prioritize maintenance jobs by equipment or area
- *Overtime report* – shows the number of hours spent for premium time as a percentage of total work-hours. Overtime reports also indicate undesirable levels of breakdown maintenance
- *Labor utilization report* – breaks the maintenance payroll hours down into dollars (Ringgit). The report can be prepared from the daily time sheets filled out by each employee or contractor

All of these information or documents should be incorporated into an annual operating budget. It is important to determine what percentage of maintenance time is actually productive. Direct and indirect costs should be evaluated.

2.3.5 Emergency response plans

The emergency response plans are made with the purpose to reduce the possibility of injury to personnel and damage to the facility in an event of an emergency.

The emergency response plans, however, should be carried out in accordance with the regulations and guidelines from the authority and assigned emergency team.

In the U.S, one standard which could be used in emergency management is the Facility Management Standard for Facilities, Infrastructure and Environment (FMSFIE) that was developed by the U.S. Army Corps of Engineers.

The emergency response plans should detail what the response team must do in case of an emergency. Among the type of response plan which widely used are the Building Emergency Plan and Evacuation Plan.

2.4 Operation and maintenance manual

A building which designed with the control systems that can achieve high levels of energy efficiency should be managed accordingly to ensure it performs as required. The standard operation and maintenance manual are among the essential tools which aid in operation and maintenance management.

The Operation and Maintenance (O&M) Manual is an important tool in ensuring the effective and energy efficient functioning of the building. A basic outline of such a document is suggested [6].

Ideally the outline of the O&M manual should be prepared at an early stage of the design, and updated as the project develops. In this way, the team will ensure that O&M considerations are addressed at each stage of the design.

The contents of the O&M manual will vary between different buildings, depending on the systems that are included in the buildings.

Maintenance requirements may vary depending on the materials, components and systems that are included in the building. The implications of departments from the ideal maintenance program may also vary, depending on the resilience of the different elements of the building.

2.5 Standard operating procedure (SOP)

The standard operating procedure is a written procedure prescribed for repetitive use as a practice, in accordance with agreed upon specifications aimed at obtaining a desired outcome. An important aspect of a quality system is to work according to unambiguous Standard Operating Procedure (SOP).

The purpose of an SOP is to carry out the operations correctly and always in the same manner. An SOP should be available at the place where the work is done.

An SOP is a compulsory instruction. If deviations from the instruction are allowed, the conditions for that should be documented, including who can give permission for that and what exactly the complete procedure will be.

2.6 Performance measurement

The need of performance measurement in facilities management is to have guidance in decision making by the management [2, 3].

The traditional view of performance measurement which determined by [7], has three broad purposes, which are:

- To ensure the achievement of goals and objectives
- To evaluate, control and improve procedures and processes
- To compare and assess the performance of different organizations, terms and individual

In addition, [2, 3] described performance measurement as a process of assessing progress towards achieving pre-determined goals and including the following:

- Information on the efficiency by which resources are transformed into goods and services
- The quality of these outputs and outcomes
- The effectiveness of organizational objectives

However, [5] emphasize the need to measure performance gaps between service delivery and customer satisfaction thus suggested the application of a management by variance tool. The tool identifies business and facility key performance indicators (KPI), helping to create a rank order among the benchmarking criteria.

2.6.1 Key performance indicator

The key performance indicators for the maintenance organization may include key performance indicators for other areas of accountability, such as health and safety performance, employee performance management, training and development, response time, etc. [10].

Key performance indicators for the maintenance function need to include both leading (maintenance process) measures and lagging (result) measures.

3 Methodology

The study was conducted in order to determine whether there have gaps in the current state of operation and maintenance management in Facilities Management (FM) practices and the standard practice which applied throughout the world. In order to achieve the objectives, the author adopts some methods for this study. Secondary data gathered in literature review used as the basis in developing a questionnaire. The data gathered from the questionnaire were then computed for interpretation as primary data. Along with primary data, the author also made some use of secondary data which has been gathered to support the survey results.

3.1 Conducting a gap analysis

Before a gap analysis has been undertaken, the analyst must be familiar with; in this case, Facilities Management framework and several days must allow in completing data collection.

Steps in a gap analysis:

i. Facility walk-through
    - Identify activities, products and services
        - Support activities
    - Develop a comprehensive picture of the facility
        - Maps, pictures and lists of tenants and departments
ii. Document review
    - Policy statements
    - Mission statements
    - SOPs
    - Position descriptions
    - Other related documents
iii. Staff interviews
    - Identify organizational structure
    - Identify requirements to fulfil mission
    - Determine awareness and knowledge level of staffs
iv. Identify and document gaps
    - Compare with standard practice
    - Make a list of specific actions to close gaps

3.2 Measuring the gap

To measure the gap, author computed the data by using Likert system, suggested by [11] to determine the state of current practice based on mean index that has been calculated early as follows:

i.	Strongly Agree	= 1.00 < mean index < 1.50
ii.	Agree	= 1.50 < mean index < 2.50
iii.	Neutral	= 2.50 < mean index < 3.50
iv.	Disagree	= 3.50 < mean index < 4.50
v.	Strongly Disagree	= 4.50 < mean index < 5.00

In order to use the Likert-scale for interpretation, weighted mean to represent each question was computed. Weighted mean is the average wherein every quantity to be averages has a corresponding weight. These weights represent the significance of each quantity to the average. To compute for the weighted mean, each value must be multiplied by its weight. Products should then be added to obtain the total value. The total weight should also be computed by adding all the weights. The total value is then divided by the total weight [11].

In conjunction with the use of the Likert-scale system, the author has adjusted the choices of frequency towards the questions in questionnaire as follows:

i. Practice Most – organization is familiar with the matter and practice it most of the time
ii. Practice Seldom – organization is familiar with the matter and practice it seldom
iii. Practice Once – organization practice the matter just once
iv. Aware (but not practice yet) – the organization has knowledge about the matter, but just not practice yet
v. Unaware (never practice) – organization unaware about the matter, thus never practice it

4 Findings and discussion

For establishing the gap, the classification of scale index, which adopted from [11] has determined as follows:

i. Good = 1.00 < mean index < 1.50
ii. Acceptable = 1.50 < mean index < 2.50
iii. Average = 2.50 < mean index < 3.50
iv. Bad = 3.50 < mean index < 4.50
v. Poor = 4.50 < mean index < 5.00

The summary of the frequency scale for all items and its classification by using the Likert Scale system explained in Table 2.

Table 2: The summary of the scale for all categories and its classification by using Likert Scale

Categories	**Type**	**Likert Scale**	
		Mean Index	**Classification**
Undertake Maintenance Routine	Predictive maintenance	1.83	Acceptable
	Preventive maintenance	1.39	Good
	Improvement maintenance	2.66	Average
	Corrective maintenance	2.00	Acceptable
	Breakdown or Emergency maintenance	1.85	Acceptable
Preparing Maintenance Report	Complete Job Reports	1.66	Acceptable
	Backlog Reports	1.88	Acceptable
	Overtime Reports	1.51	Acceptable
	Labor Utilization Reports	2.68	Average
Referring O&M Tools	Maintenance Manual	1.59	Acceptable
	Standard Operating Procedure	1.68	Acceptable
	Maintenance Policy	1.78	Acceptable
	Maintenance Schedule	1.20	Good
Emergency Response Plan Testing	Building Emergency Plan	2.10	Acceptable
	Evacuation Plan	2.61	Average
Undertake Performance Measure	Performance Measure (KPI, etc.)	1.88	Acceptable

By referring Table 2, the explanation can be found by category as follows:

i. Undertake maintenance routine – most of the maintenance routines are at an acceptable level. Only preventive maintenance satisfied the standard which it is at a good level. However, the improvement maintenance is quite far from the standard which it is at an average level.
ii. Preparing maintenance report - most of the maintenance reports are prepared at acceptable level. Only labor utilization report prepared at average level.
iii. Referring O&M tools – most of the O&M tools are being referred at acceptable level. However, the maintenance schedule satisfied the standard which it is being referred at good level.
iv. Emergency response plan testing – building emergency plan being tested at an acceptable level while evacuation plan only being tested at an average level.
v. Undertake performance measure - the performance measure is undertaken at the acceptable level.

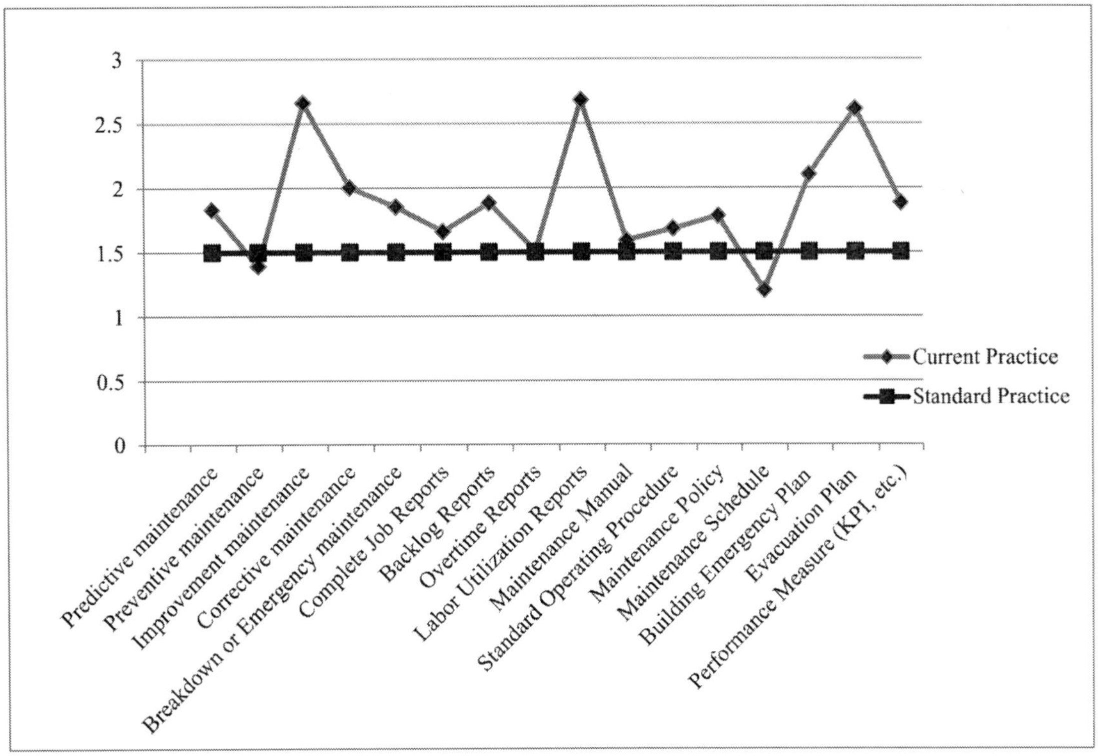

Figure 1: The gap analysis chart shows the state between current practice and standard practice

To get a clearer view about the gap which has been established by the analysis, Figure 1 shows the clear picture of the gap between current practice and standard practice. The standard practice has set at the least number in good scale (1.5) which only two items (preventive maintenance and maintenance schedule) achieved the level of standard practice. Most of the items tested were at an acceptable level. The other three items (improvement maintenance, labor utilization reports and evacuation plan) had a gap which is quite far from standard practice level.

The preventive maintenance relatively satisfied the standard practice because this type of maintenance routine is the crucial item in any maintenance program. However, the improvement maintenance may be somewhat unusual for the O&M practice or this type of maintenance may rather known as other name and this type of maintenance actually covers the other which similar nature of activities. However, the corrective maintenance may be at that level due to the nature of its routine which undertake only when the equipment failed to operate.

Since almost of the reports are prepared at acceptable level, the upper level management should be aware of this situation which the neglected in preparing those reports will result in the lost of important data that may be basis in future decision-making.

The labor utilization report, for example, is very important in determining of the manpower supplies which, with that information, the upper level management can assign the labor sufficiently.

All of those O&M tools are likely to practice excellently. As noted, the gaps between the current practices and the standard practices are just close. It is certified that the O&M team realized the fact that O&M tools should be referred frequently in order to ensure that every activity are delivered efficiently.

The gaps between the standard practices and current practices in both emergency response plan category also likely to be far. This circumstance might be caused by the testing schedule which had stated as that means.

The gap in performance measure is seen quite close, which is declared that the respondents are conscious of how important to carry out performance measure.

In summary, it is evidenced that most of respondents exercising sufficient level in the most O&M practices which include carrying out the various types of maintenance activities, preparing various types of maintenance reports, and also make use various types of O&M tools. As well as practicing the O&M at satisfying level, the majority of respondents also have set up the key performance indicator as the performance measurement in their organizations.

5 Conclusion and recommendation

Most of the practices that assessed by means of gap analysis had an acceptable result which the gap between the current practices and the standard practices are not too far. Consequently, if the organization wants to implement the change in their O&M practices, the gaps established should be bridged. The gaps can be brought nearer with continuous learning effort. The effort will be easier if the FM key player is more understanding the actual concept of facilities management.

The gap analysis of operation and maintenance management provide a clear indication of compliance to the standard practice. They indicate where to take specific action because of a gap in the way maintenance is being performed. This gap in the execution of the maintenance will eventually lead to asset failure and some effort should be undertaken to ensure that the industry will be more recognized.

Gap analysis is a good way to jump start the business practices and also helps create a list of actions for the management team. Gap analysis utilizes comparative analysis techniques to identify the differences that exist between the current stages and the desired level. It is essential to identify the nature of the current process in order to make this comparison.

Effective use of gap analysis can effectively minimize risks and reduce the costs of the change management effort.

In summary, an organization's success depends largely on the way in which the various facilities are managed, and this is why many businesses are now making use of facilities management organizations.

References

1. BSI (2007) BS EN 15221-1:2006 (E). BSI.
2. C.C. Bobenhousen. Viridian Energy & Environmental, Inc. (2009)
3. D. Amaratunga & D. Baldry. Proceedings of Int. Postgrad. Conf., pp.193-202 (2002)
4. D. Amaratunga & D. Baldry. J.FM, **6,** 5-22, (2008)
5. E.C. Fitch. Elsevier Science Publishers Ltd, England, (1992).
6. J. Hinks, P. Mcnay. Facilities, **17,** 31-53, (1999).
7. J.H.K. Lai, F.W.H. Yik, P. Jones. HK Polytechnic Univ., China and Cardiff Univ, UK (2005)
8. J. Teague, S. Eilon. Appl. Economics, **Vol. 5, 99**. 133-145. (1973)
9. M. Noor, M. Pitt. J. FM (2008)
10. M. Noor, M. Pitt. Facilities, **27,** 211-228. (2009)
11. M. Sondalini. http://www.lifetime-reliability.com, (2008)
12. R. Likert. Archives of Psychology, No.140 (1932)
13. T.M. Linda, K. Non. J. FM, **Vol. 4 Iss:4**, pp. 234-244, (2006)
14. W. Isobel. eHow Contributor. (2010)
15. W.R., Sievert. R.S. Means Company, Inc. (2001

Methods in Mapping Usability of Malaysia's Shopping Centre

Aida Affina Abdul Ghani[1], Md. Yusof Hamid[2], Siti Norsazlina Haron[3], Nur Azfahani Ahmad[4], Mariaty Bahari[5] and Suriani Ngah Abdul Wahab[6]

[1,2,3,4,5,6]*Fakulti Senibina, Perancangan dan Ukur, UiTM Perak, Malaysia*

Abstract. With more than 200 number of shopping centre in Klang Valley itself, we the consumer actually have vast of choices. Instead of the machineries varieties from the lower class product till the posh one, each of those shopping centres eventually offers the typical product same as others. Those shopping centers are competing with each other and in great endeavour to attract more consumers, to visit and spend. As for the visitor, the typical product and boring ambience seems similar in all malls, and is looking something beyond the standard. Something that promising quality embedded in shopping centre which evokes the various emotions of the user along their journey in malls. This quality is known as usability.. Usability; as defined generally is a global user's experience response with product, environment, service or facilities. It is an assessment in extracting the qualities of shopping centre design. In mapping it, there are a few synthesizing methods to implement it. Therefore, this paper purposely to review the method that been used in usability of Malaysia's shopping centre research with a few references on previous research done in usability assessment by predecessor's scholars. With the accentuation on three elements that anchoring what the usability is: effectiveness, efficient and satisfaction, it is hope that this overview can lead other researcher in portraying its relationship with the quality and 'user friendly' design of shopping centre.

Key Words: Walkthrough, Shopping Centre, Usability, Quality Design, Experience

1. Introduction

Malaysia is well known for its 'shopping paradise'. According to Malaysia Shopping Malls Association (PPK), 550 shopping centres in the country are serving 30.3 million populations. 218 out of this 550 number of shopping centres are in Kuala Lumpur, Selangor and Putarajaya itself.

As mentioned by Managing Director Of CBRE (CB Richard Ellis Malaysia), Mr. Allan Soo (2014), Malaysia has facing an oversupply number of retail space. This is worsened by another addition of 7.7 million sqft of retail space from 2012 till 2015 through the built of Paradigm Mall, Setia City Mall and NU Sentral Mall. This makes the overall of retail space almost 46 million square feet (sqft) in Klang Valley itself. Referring to the standard retail space sqft which is 5.5 sqft per person, Malaysia actually is higher compared to Bangkok where we stated 7.1 sqft per person, whereas Bangkok recorded 6.5 sqft vice versa.

These great numbers of shopping centers cause tight competition between them in attracting more visitors to visit and consume, thus escalate the profit through the increasing number of tenant. This is a bit alarming, where shopping center and its environment should appeal enough to attract the user; both visitors and tenants. However, the physical attraction is not enough to attract them; thus creating a subjective dimension of usability such as 'identity, familiarity, memorability and satisfaction' is vital.

2. Built Environment Usability

Users of the building normally behave contradict with what have been predicted by designers. As mentioned by Haron SN (2011), designers tend to longing for smart users with everything are in order, and better behaved [19] .Unfortunately, in reality; real building needs to fit in real behaviour of the users. There is gap of communication between the designers and the users [1, 29] that normally designers are not the subject in their environment design. Even Arthur and Groener (2005) said that in increasing the user's satisfaction, it is essential to cater the needs of the user at early design stage accurately. Therefore, a quality designed building is totally depends on the input from the user.

Building that has quality of use is a usable building. As specified by International Council For Research and Innovation In Building and Construction (CIB) Task Group 51, a quality of use building is a building that efficiently in use (resources, productivity, effectiveness,

Corresponding author: aniffaadia@gmail.com

rationality), expected effect in use (value increasement) and offering the wanted quality in use (satisfaction). From that statement, it can draw that this is a part of usability traits. Usability is related to the ease of use or user friendly. According to ISO (1998), usability defined as *'....effectiveness-ability, efficiency and satisfaction with which a specified of users can achieve a specified set of tasks in a particular environment'*. To be detail, this illustrates a product's usability as:

- Effectiveness – whether users can achieve what they want to do with the product
- Efficient –how long it takes them to achieve it
- Satisfaction – their feelings and attitude toward the product

This jargon actually came to the fore in 1950s where it was widely used in Human Computer Interaction [22, 24]. Latter, it has been used in product design, web design and information technology. Most gigantic companies such as Apple's PowerBook, Johnson and Johnson are success in producing their product by using usability as its design strategy. However, there is escalating interest in the application of usability in built environment practices nowadays [2, 21, 33]. The endeavour can be visibly seen from the establishment of International Council For Research and Innovation In Building and Construction (CIB) Task Group 51- "Usability of Buildings 2005", Workshop W111- "Usability Of Workplaces 2- 2008" and "Usability Of Workplace 3" in promoting better understanding on user experience and usability concept.

Basically, usability in built environment focuses on user perception on the easiness and efficient usage of facilities- the workplace [2]. It can be a tool to evaluate building which requires the total participation of user and their experience are incorporate in it. In evaluating the usability of the building, the relation of the building and people/ organisation is the main focus of subject. As mentioned by Blakstad (2008), this user often interacts with building by using all their senses, social and cultural unconsciously.

Contradicts with POE (Post Occupancy Evaluation) by Preiser et. al (1988) this usability is based on user perspective. Essentially, all previous methods in evaluating the building performance merely on the physical of the building (functionality); and none of it concentrates on the fitness of its design purpose with the application by the user. Hansen et.al (2010) illustrates the difference on functionality and usability clearly. Functionality of a building is more on how it works according to specification, quite critical but it is not significant from the aspect of successful of use. However, testing usability is related on how the building suits the organisation's goal, and user's experience and satisfaction.

As a platform of various activities, shopping centre is crucial in eliciting its usability. With frequent changes of user environment and format, a shopping centre should be usable enough for a workplace, spending place and an entertainment place as well. The shopping centre usability framework is gauged from three essential elements (Fig 1.0) as described by Alho (2008) which is based on Rothe (2006) and Mantyla (2001)'s magnum opus. The 3 elements are : Shopping Centre (Environment), User and Usability Attributes (Situation, Goals which the products been used). In order to evaluate the usability of shopping centre, Alho et.al, stressed on which perspective that we are assessed from; the visitor or tenant. This is because usability is a matter of viewpoint, which the different user will tell the different definition, knowledge of usability. This is aligned with Sinkkonen et.al (2002) that described users are vary according to background, culture, habits , knowledge and know how.

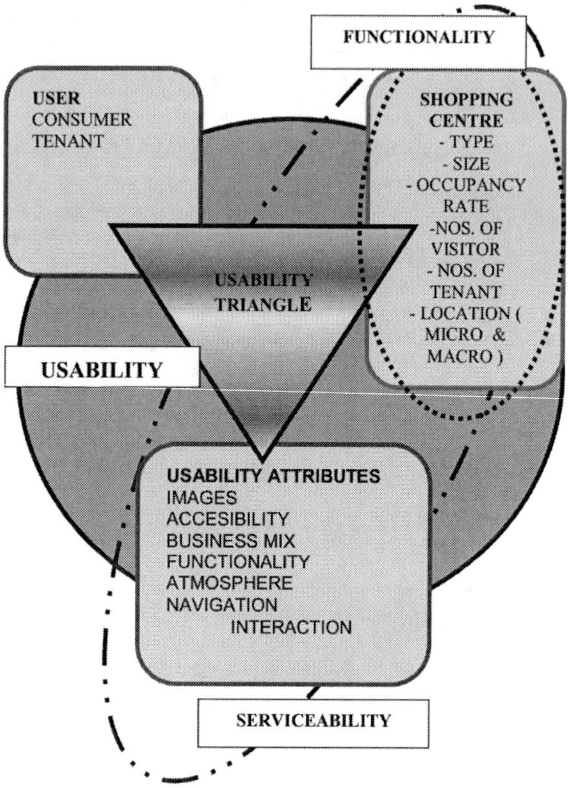

Fig 1.0 : Usability Shopping Centre Framework (Alho (2008), Rothe (2006)

3. Usability Walkthrough Assessment

Walkthrough is a common method in evaluating usability of a building. There are two types of walkthrough inspections which are cognitive walkthrough and pluralistic walkthrough. Cognitive walkthrough is based on task orientation whereas Pluralistic Walkthrough is a combination of usability inspection and testing.

Basically, most usability walkthrough assessment method is done based on heuristic evaluation method [1, 11, 15, and 26]. This method is easy to execute due to its ease of implementation, cheap and fast. By using

4 – 5 evaluators which normally expertise in technical, the design system in the environment is audited based on the set of usability principles [27]. Nevertheless, these 10 usability principles were too flat and have been improvised by combined with several usability evaluation methods such as hierarchical heuristic structure, empirical testing at different phases along the development process and Systematic Usability Evaluation (SUE) [3, 11 and 23].

The latest walkthrough method; USEtool which promoted by Blakstad et.al (2010) ; had been used in extracting usability of Malaysia's hospital by Haron SN in 2014. USEtool is seen as a holistic usability method which blends the importance role of organisation, the user and the physical of an asset in one assessment. In USEtool, Blakstad et.al (2010) provide in depth knowledge on usability through the user experiences of existing buildings. This input practically can be used by Facilities Manager of an organisation as a reference in choosing new premise or planning a new building.

In addition, there are several methods that have been used by previous scholars in evaluating the usability of building shown in Table 1. Most of the usability scholars chose more than one method in executing usability mapping in their research.

Table 1 : Method Of Usability Mapping

RESEARCH METHOD AND TECHNIQUES	*OUTCOMES*
On Site Observation	***Behavior vs. design***
Observing physical traces: Systematically looking at physical surrounding and reflection of previous activity. Focus on functional issues, Adaptations for use/ Public massage/ Display of self/ Product of use / User patterns [	Functional design criteria .Behavior and appearance map design-spatial information & Efficient Operation
Observation -Environmental behavior activities: People activities & spatial-relationship. Systematically watching people use their environment - context/setting/ pattern /culture/ expected uses, new uses, misuses& opportunities	Support humans without disturbing their activities. Recognized needs Full Participants.
Walk-through observation and Cognitive walk through User journey experience -User experience feedback surrounding design. Analyses of the space and movements "tour" of the building	Assessing different qualities functions and audit of - environmen
Survey	***Users perspective & feedback***
Interviews useful as follows-up to certain Getting the story behind a participant's experiences. Obtain in-depth information around the topic. Interviews useful as follows-up to certain questionnaire and support behaviors reaction	Knowing related issue & exploration related evidence support survey
Kansei (Emotion) Investigate consumer's psychological feeling when interacting with product	Product that matches with the users's definition
RESEARCH METHOD AND TECHNIQUES	*OUTCOMES*
Questionnaire Using Structured question	Needs and demand
Others methods	
Story telling and personas: Expression of what user's perspective and users experience	Personal needs Attach to product design.
workshops : Participatory Getting an idea and sharing knowledge about the conflict they experienced, engaging with the group to form strategies.	Aims and objectives as well as discussing future work
Focus group: Perspective from the specific group and organized discussion with a selected group of individual discuss and comment on, from personal experience & expectation	It limited to the ideas and issues at the time of the session.
Document analysis: To know current design practice and understand a present situation. Purpose to support finding	Gap of practice Lacking of data or sources. Lacking of implementation
Mapping: Using structured group interviews Systematic approach for organizing information more effectively. Improve operating efficiency & performance	Creating model and standards. Establishing new efficient systems and processes
USE tool: a 5-stage process 1. Define the evaluation – for what 2. Mapping usability-what 3. Walk through-where &who	Assessment method flow in gathering &

4. Workshop organization-why 5. Action plan & report-who	analysis data–improving or future solution

(Source : Blakstad et.al (2008), Becker et.al (2007), Kazanasman et.al, (2006), Voordt (2009), Teck (2006), Uchira et.al, (2009), Haron et.al (2013), Alexander (2006), Nenonen and Rasila (2009), Nagamachi and Lokman (2010)

4. Malaysia's Shopping Centre Usability Assessment

Table 2 : Method Of Malaysia's Shopping Centre Usability

RESEARCH METHOD AND TECHNIQUES	*OUTCOMES*	*USING*	*THROUGH*
Document analysis: To elicit the usability attributions theoretically from current trends Purpose to support finding	Gap of practice Usability of Shopping Centre Attribution, Thematic Analysis	**USE tool: a 5-stage process** 1. Define the evaluation – for what 2. Mapping usability-what 3. Walk through-where &who 4. Workshop organization-why 5. Action plan & report-who	**Focus Group:** Perspective from the specific group and organised discussion along the walkthrough, voicing out loud (VOL) the experience and the expectation : 1. Visitor Group 2. Various Ages 3. Various Gender
On Site Observation	***Behavior vs. design***		
Observing physical traces: Physical surrounding observant ; scrutinizing the purpose o its design anc how does it functioning; (Center For The Universa Design, 1997	Attribution for Function design; Map Design Information Efficient Operation		
Observation - Environmental behavior activities: What people do during their visit on shopping centre; interacting with physical environment	Recognised their pattern of use without disturbing the activities		
Walk-through observation and Cognitive walk throug User Experience Journey through Voic Out Loud (VOL) - 'tour movement in shopping center	Different emotions evoke along the journey, response on the environment recorded		
Survey	***Users perspective & feedback***		
Interviews useful as follows-up to certain Gaining in depth information on their experience through follow up interview; Getting the idea on how the ideal shopping center is	Knowing related issue & exploration related evidence support survey		
Questionnair Identify the ranking of th usability attribution of Malaysia's Shopping Centre through scoring; validate the findings by using Structured Questions	Users and Experts and Reflections On Usability Framework		

As a crucial place with various user environments, eliciting its usability is hardly to find in Europe's current research of shopping centre; except for a few scholars [3, 35]. The same goes to our Malaysia's context where there are a lot of researches done on the attractiveness of shopping centre from the marketing point of view. The scholars were concentrating more on the retails strategy such as promotion, merchandising, brands, product quality and entertainment [40, 41 and

42]. Therefore, this usability evaluation seems to fill the loop on the quality design of Malaysia's shopping centre.

Generally, this research is dominated by qualitative type field. Moreover, to test the usability, it needs variety of tools; methods aspects because it depends on the objectives, purpose, and focus [26 and 34].

An initiative to understand thoroughly the phenomenon of usability, the research carries phenomenological philosophy which lies in interpretivism doctrine. Moreover, it is based on logical reasoning with emphasising the exploration and investigation of nature. This phenomenological research is purposely to understand the perception and the perspective of the user (visitor) of Malaysia's shopping centre on its usability. Furthermore, this research needs to be triangulated by executing several methods in the same phenomenon [13] in order to strengthen 'a confluence of evidence that breeds credibility' [14].

Basically, as referring to the Table 2 above, the method applied in this research is extracted from Usability Evaluation Method or known as 'USEtool' by Blakstad et.al (2011) and is based on Usability Shopping Centre framework introduced by Alho, et.al (2010). The 'USEtool' is used as a guideline in preparing the method outline for this research as it enhances the role of organization as epitome in the usability of shopping centre. Moreover, this systematic approach has been used in mapping the usability of building of an organization in previous studies. With a systematic approach of 5 stages mapping the phenomenon and reflecting the objectives of study, it helps in collecting, analysing, reporting and presenting data of research.

Stage 1 – Define the Evaluation (What)

The flow of this research method started with defining the purpose of usability evaluation- Stage I. Through Document Analysis, the aim is to extract usability criteria of shopping centre theoretically. As agreed by Corbin and Strauss (2008) et.al, the data from the documents are analysed and examined, thus interpreted to gain depth understanding and establish empirical knowledge. A thorough extensive literature review made based on theories on attractiveness and usability of shopping centre, both from marketing, business doctrine and built environment doctrine thus synthesized in order to create the attributes and theme for thematic analysis purpose. At this stage, a theoretical framework on Usability of Shopping Centre is set up.

Stage 2 – Mapping Usability (What)

The mapping process is guided by theoretical framework on Usability of Shopping Centre. At this stage, all relevant data are gathered including the background and the key figures of the shopping centre. In addition, the designated user team is identified before implementing the walkthrough process. Principally, the physical surrounding is firstly scrutinised as direct observation in seeking the virtual traces on how that shopping centres functioning and the user interacting with the physical environment at mall.

Stage 3 - Walkthrough (Where and Who)

The 'journey' started where the user experience are taken into consideration. Each of their emotion that evoked along the journey is tape recorded as the user implementing the 'Voice Out Loud' walkthrough; expressing their action and thought every time it came across into their mind. As Chebat et.al, (2005) mentioned, the focus is not on the total observation, but the thought that counts. The focus group conducted are requested to find a predetermined store in the shopping centre. Along the journey, the group then are instructed to record on what they see and the feel until they reach their destination.

Stage 4- Validation Process (Why)

At this stage, the findings of the research will be tabled consecutively to identify the linkage between the issues. Then, the results will be validated and triangulated within the main case study results and other multiple methods in order to support the main results. According to Patton (1990), triangulation avoids the research's bias or accusation on simplifying the findings as single source or method. This agreed by Bowen (2009) which states that researcher can avoid being potentially bias through various data sets by different methods collection. In this task, the respondents of two groups which consist of users and experts are given a set of questionnaire that uses five Likert scales (from strongly disagree to strongly agreed).

Stage 5 – Final Report (Reporting)

By triangulation the data gained through the main case studies, expert's review and the user's perception on the usability of Malaysia's shopping centre, a final report is produced. This report is tending for the purpose of:

- Improvisation of user experience in usability of shopping centre
- Enhance the usability of shopping centre
- Reference for the facilities manager and top management in enhancing the 'user friendly' design at early stage for future project

5. Conclusion

Summarily, this overview of method used in mapping the usability of shopping centre is devoted to facilitate the researcher on the variance of means. The choice of method selection must be suitable with the usability phenomenon and its philosophy. With the accentuation on three elements that anchoring what the usability is: effectiveness, efficient and satisfaction, it is hoped that this overview can lead other researcher in portraying its relationship with the quality and 'user friendly' design of mall.

References

1. Afacan, Y., & Erbug, C. (2009). *An interdisciplinary heuristic evaluation method for universal building design*. AE, 40(**4**), 731–744. doi:10.1016/j.apergo.2008.07.002
2. Alexander, K. (2006). *The application of usability concepts in the built environment*. JFM, 4(**4**), 262–270. doi:10.1108/14725960610702947
3. Alho, J., & Nenonen, S. (2008). Usability of shopping centres: Components of a usability rating tool. W111 usability of workplaces phase 2. CIB Reports 316. Rotterdam: CIB.
4. Andre, S.T., Hartson, H.R., Belz, M.S., Mccreary, F.A., 2001. The user action frame- work: a reliable foundation for usability engineering support tools. Interna- tional Journal of Human–Computer Studies 54 (1), 107–136
5. Arthur, J.D., Gro ¨ner, M.K., 2005. An operational model for structuring the require- ments generation process. Requirements Engineering 10 (1), 45–62.
6. Bowen, G. A. (2009). Document Analysis as a Qualitative Research Method. *Qualitative Research Journal (RMIT Training Pty Ltd Trading as RMIT Publishing)*, *9*, 27–40. doi:10.3316/qrj0902027
7. Chebat, J., Ge, C., & Therrien, K. (2005). Lost in a mall , the effects of gender , familiarity with the shopping mall and the shopping values on shoppers ' way finding processes, *58*, 1590–1598. doi:10.1016/j.jbusres.2004.02.006
8. C.K.K.Teck, "A user-centred approach to effective wayfinding map design:Integrating theory, practice and user participation',Faculty of Built Environment, Art and Design .Curtin University of Technology. Disertation Doctor of Philosophy,2006
9. Clarkson, J., Coleman, R., Hosking, I., Waller, S., 2007. Inclusive Design Toolkit. Cambridge Engineering Design Centre Press, Cambridge. Retrieved from: www. inclusivedesigntoolkit.com/betterdesign (1.10.07).
10. Corbin, J. & Strauss, A. (2008). Basics of qualitative research: Techniques and procedures for developing grounded theory (3rd ed.). Thousand Oaks, CA: Sage.
11. Darses, F., Wolff, M., 2006. How do designers represent to themselves the users' needs? Applied Ergonomics 37 (6), 757–764.
12. De Angeli, A., Matera, M., Costabile, M.F., Garzotto, F., Paolini, P., 2003. On the advantages of a systematic inspection for evaluating hypermedia usability. International Journal of Human–Computer Interaction 15 (3), 315–335.
13. Denzin, N. K. (1970). The research act: A theoretical introduction to sociological methods. New York: Aldine
14. Eisner, E. W. (1991). The enlightened eye: Qualitative inquiry and the enhancement of educational practice. Toronto: Collier Macmillan Canada.
15. F. Becker, and K. S. Parsons,"Hospital facilities and the role of evidence-based design". Journal of Facilities Management 5(4),pp.263-274,2007.
16. Fu, L., Salvendy, G., Turley, L., 2002. Effectiveness of user testing and heuristic evaluation as a function of performance classification. Behaviour and Infor- mation Technology 21 (2), 137–143.
17. Hansen, Geir K; Blakstad, Siri H; Knudsen, W. (2011). USEtool Evaluating Usability: Methods Handbook. *Ntnu*.
18. Harding, J.A., Popplewell, K., Fung, R.Y.K., Omar, A.R., 2001. An intelligent informa- tion framework relating customer requirements and product characteristics. Computers in Industry 44 (1), 51–65.
19. Harrison, J., Parker, K., 2005. Designing for real people's lifelong needs. In: Proceedings of UIA Region IV Work Programme 'Architecture for All' UIA/ ARCASIAWorkshop. 3–10 Istanbul, Turkey, pp. 15–24.
20. Haron, S. N., & Hamid, M. Y. (2011). Patient perspective : The " Usability evaluation " approaches as assessment for quality of outpatient spatial design . *Ipcbee*, *8*, 4–8.
21. Nagamachi, M and Lokman, A., 2010. Kansei Engineering : A Beginner Perspective.UniversityPublication Centre (UPENA),UiTM 2010
22. J. A. Granath and J. Gilleard, "Usability in the Workplace:Case study of Pamela Youde Eastern Hospital Hong Kong", Usability of Workplaces 2, CIB and EuroFM. CIB reports 316.Rotterdam,2008
23. J. Gulliksen, "Usability Professionals—Current Practices And Future Development." Interacting With Computers 18,pp 568–600,2006.
24. Kantner, L., Rosenbaum, S., 1997. Usability studies of www sites: heuristic evalua- tion vs. laboratory testing. Reprint of paper from SIGDOC 97 Proceedings (Salt Lake City, UT). Published by Association for Computing Machinery, Inc. Retrieved from: http://www.teced.com/PDFs/sigdoc97.pdf [01.12.05]
25. M. Fenker, "Towards A Theoretical Framework For Usability Of Buildings". Usability of Workplaces 2, CIB and EuroFM. CIB reports 316.Rotterdam,2008.
26. M. Jensø and T. Haugen, "Usability Of Hospital Buildings :Is patient focus leading to usability in hospital buildings?", Usability OfWorkplaces:Case study: Nord-Trøndelag University College Nylåna, Røstad. CIB Task Group 51,2005.
27. Nielsen, J., Molich, R., 1990. Heuristic evaluation of user interfaces. SIGCHI Bulletin April (Special Issue), 249–256.
28. Nielsen, J., 1992. Finding usability problems through heuristic evaluation. In: Proceedings of CHI '92, California, USA, pp. 373–380

29. Niemi R., Nenonen S. & Junnonen J-M. (2013), Investigating the competencies for Serviceability of Urban areas, paper presented at the CIB World Building Congress 2013, Brisbane. 7.5.2013.
30. Patton, M. Q. (2002). Qualitative research and evaluation methods. Thousand Oaks, CA: Sage
31. Preiser, W., 2003. Inclusiveness through universal design feedback and evaluation. Education and evaluation – resources and methods at student level. In: Proceedings of Include 2003. Royal College of Art, London.
32. Rapley, T. (2007). Doing conversation, discourse and document analysis. London: Sage
33. Sinkkonen, I., Kuoppala, H., and Parkkinen, J. (2002), Käytettävyyden psykologia. Edita Oyj, Helsinki.
34. S. H .Blakstad, G. K. Hansen and W. Knudsen, "Methods and tools for evaluation of usability in buildings",Usability of Workplaces 2, CIB and EuroFM. CIB reports 316.Rotterdam,2008.
35. S.Nenonen and H. Rasila , "Customer Journey – a method to investigate user experience" , International Council for Research and Innovation in Building and Construction (CIB) reports 316. 2008
36. T .Kazanasmaz, " Design Efficiency in Hospital Architecture", Proceedings of First International CIB Endorsed METU Graduate Conference, Ankara,Turkey, 17-18 March 2006, pp.231-242,2006.
37. T. J. M. v. d. Voordt, "Quality of design and usability: a vetruvian twin." Ambiente Construído, Porto Alegre 9,(2),pp. 17-29,2009
38. T. Uehira and C.Kay,"Using design thinking to improve patient experiences in Japanese hospitals: a case study", Journal of Business Strategy, Vol. 30 Iss: 2/3, pp.6 – 12,2009.
39. Yin, R. K. (1994). Case study research: Design and methods (2nd ed.). Thousand Oaks, CA: Sage.
40. Kamarulzaman, Y., & Author, C. (2010). Attracting Shoppers to Shopping Malls : The Malaysian Perspective, 185–199.
41. Sapie, N. M., Hussain, M. Y., Ishak, S., Awang, A. H., & Lyndon, N. (2014). Motif dan daya tarikan pusat membeli-belah dalam kalangan pengunjung metropolitan Kuala Lumpur , Malaysia Patron motives and perspectives of shopping malls ' attraction in Metropolitan Kuala Lumpur, 1(1), 70–77.
42. Zahari, A. R., Esa, E., Nasional, U. T., Haji, S., & Shah, A. (2012). TIMBJ, 2(**2**), 47–60.

MATEC Web of Conferences **66**, 00118 (2016) DOI: 10.1051/matecconf/20166600118
IBCC 2016

Do Space Requirement Applicable in Private Preschools?

Naziah Muhamad Salleh[1,a] , Nuzaihan Aras Agus Salim[1], Syahrul Nizam Kamaruzzaman[2] and Norhayati Mahyuddin[2]

[1]*School of Housing Building & Planning, University Science Malaysia, 11800 Penang, Malaysia*
[2]*Faculty of Built Environment, University of Malaya, 50603 Kuala Lumpur, Malaysia*

Abstract. Working or studying in a comfortable environment enhances not only well-being, but also satisfaction and therefore increase the productivity and learning. The numbers of private preschool in Malaysia boost every year. Frequently they operate in premises that have been fully refurbished. This has invited the questions on the building capability and space condition to provide a good environment to the children during the learning activities. Most of the building was refurbished to enhance it applicability as a school. Yet, these adaptive-reused buildings are doubtful. This research focused to identify the characteristics of the buildings' physical and condition as well as the scenario of refurbished private preschool in accordance with the standard. Observation particularly on space and pupils density either it is reckoning with the authorities' requirements. Most of the building was refurbished to enhance it applicability as a school. The data obtained from the observation and staff's interview to 237 preschool (771 classrooms). The data revealed in most of preschools, the occupants in the classrooms were over the limit regulating by the authority. The data obtained was analyzed to become a reference and benchmark to the authorities to prepare the private preschool's applications.

Keywords: Refurbishment, Preschool, Occupants Perception, Space Requirement

1 Introduction

Malaysia has experienced high economic growth accompanied by rapid awareness of education. Early education becomes one of the main focuses in the Malaysian Educational Development Master Plan 2006-2010, by providing the preschool classes in the government schools [1]. By 2020, 100% of children oblige to enroll preschool before they enter the primary education in national schools [2]. It is highlighted to strengthen the nation's human resource development, facilities and the education participation rate. The numbers have improved significantly with the increasing participation made at the preschool.

Preschool or kindergarten (in German, literally means "children's garden") is a form of education for

young children which serves a transition from home to the commencement of more formal schooling [3]. Children are taught to develop basic skills through creative playing and social interaction. In most countries, kindergarten is part of the preschool system. Children usually attend preschool anytime between the ages of two to six years, depending on the local custom [4]. In

Malaysia, children usually attend preschool any time between the ages of four to six in private and government school [1].

Research done by [5], has elicited that the coverage for preschool education has improved dramatically over the last decade by one-half, from 17 percent of 4-6 years in 1981 to 41.5 percent in 1995. Also, the 1996 Malaysia Education Act and Malaysia Education Blueprint in 2013 [6] has amended to make the preschool compulsory. This progress has encouraged more private preschool centres to set up where it involved various types of the refurbished building in urban areas. According to [7], terrace houses become the most common property to be converted into preschools which would consist of classrooms that accommodate 20 to 30 pupils at a time.

[a] Corresponding author: naziahmsalleh@usm.my

The previous litigations and studies suggested adequate learning environment for these children. Referring to [8], the occupant in a classroom requires at least 2 m^2 per person, which implies a requirement of 40 m^2 to accommodate twenty pupils in a space. These to ensure the classrooms are well-ventilated [9. 10]

Hence, in school environments perspectives, there is a relation to the implications of the spatial volume, for example, found that differentiation in ceiling height was related to higher levels of cooperative behavior among preschool children. On the issue of space perception, [11] reported that if some space is constructed and equipped in the right way, the development of a child's increased abilities is supported, and this allowed the child's capacities to be confirmed by the child. Pertaining to playroom arrangement, the type of furniture arrangement did not change the joint use of play areas and the social interactions for the peers whose relationships were weak [12]. In contrast, for children showing an emerging relationship, the playroom arrangement affected the quantity and the quality of their social interactions.

The combination of shared and personal spaces for student and teacher, can help to nurture an orderly and serious academic atmosphere [13]. This is support by [13] whereas the open plan classroom with effective activity area makes the teaching situation flexible and share space is an extension of the classroom's activities. Non-appropriate behaviors were observed in the poorly defined classrooms where the classrooms are entirely in impropriate layout and lower floor area. While, it seemed that a well-defined classroom would inhibit more appropriate behaviors [15]. Class size is an important design and drives a host of costly facility-related issues that are part and parcel of the school building's planning, design, construction, cost, maintenance and operation [16]. It becomes a big factor in determining the number of teachers needed and hence, it affects the education outcomes.

Good space management will stem from business trends that encourage work-style, such as knowledge-sharing, collaborative work, increased mobility and working productivity. These key influencing factors need to perfectly synchronizing with best work practice to achieve success and it has become the major factor influencing workplace design. The main consideration in space management is to acknowledge the need of the occupants rather than the building and physical space in order to give the positive impact to the organization as well as in school environment predictions.

Thus, this paper resumes to identify the characteristics of the buildings' physical and condition as well as the scenario of refurbished private preschool in accordance with the authorities' standard and regulation.

2 Materials and Methods

The school grounds' walk-through form was used by the research team to characterise the materials and condition of each school in general terms. Items characterised included a description of the school grounds, indoor environment and kitchen. Some specific information was gathered, comprising the building age, number of stories, recent or ongoing renovations, presence and number of portable buildings, common area finish materials, lighting, barrier mats, cleanliness, and maintenance closet chemical storage and ventilation strategies.

The study was conducted in refurbished private kindergarten or preschool buildings which have been registered with the Ministry of Education. All 1000 preschools around Malaysia were randomly selected from the approved list of Private Education Sector, Ministry of Education, Malaysia. First, a letter of permission to run the survey sent out earlier. As a result, total 285 owners replied to the letter. Out of them, only 237 preschools (771 classrooms) have granted the permission to conduct the research in their premises. The observation process took about fifteen to thirty minutes to be completed.

The descriptive analysis on location, the nearest main road, type of building, building and classroom floor area, ventilation strategies, maintenance and cleaning, numbers of occupants (teachers and pupils in the classrooms), type of operation, operating hour, and renovation purposes were applied. Since the data obtained were non-parametric, the Chi-square test has done for independence to identify the most variables related to the requirement of number of pupil/occupant in the classrooms.

In this section, the observations were analysed via different methods. The descriptive statistical parameters (i.e. arithmetic means, standard deviations, maxima and minima) were used to present the data. Meanwhile, non-parametric test Chi-square independent test applied to evaluate the variable related to the requirement of number of pupil/occupant in the classrooms. The number of valid occupants (pupils and teacher) per classroom verified by the formula specified by Malaysia Ministry of Malaysia:The actual or valid classroom, then subtracted from the actual number of occupants in the classroom.

$$\text{Number of student} = [(\text{Length} \times \text{Width}) \div 15] - 1$$

*Length and width must be in ft

Regulation status= Valid number of occupants - Actual number of pupils per classroom

Negative (-ve) result determined that the number of the occupants in the classrooms is not complying with the MOE guidelines, while the positive(+ve) results indicated the number of pupils are in compliance with the guidelines.

DOI: 10.1051/matecconf/20166600118

3 Results and discussions

Figure 1 illustrates the proportion of premises in this study that refurbished as preschools, with majority of 237 preschools previously were the terrace houses. With regards to the lack of technical knowledge, most of the renovation and refurbishment work was conducted using unsuitable and inconsistent techniques, which gave poor results on the interiors as well as exteriors. Thus, it will affect the children and teacher sensitivity towards the environment of the building.

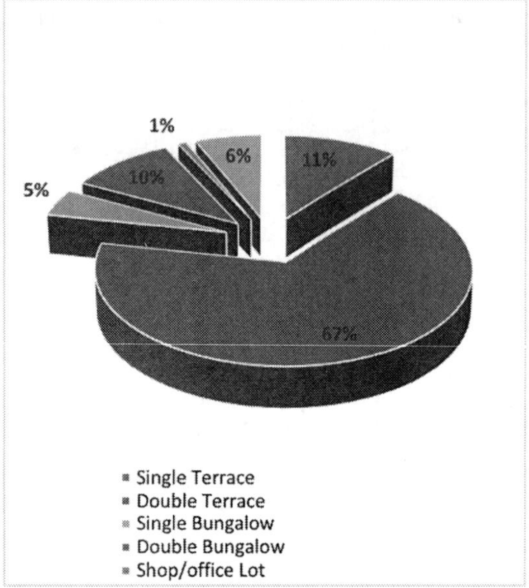

Figure 1. Premises to be refurbished as preschools in this research

3.1 Building characteristics

This section discusses the findings of characteristics and physical of 237 preschools buildings. The discussions are encompassing the descriptive analyses of building characteristic, type of refurbishment, function to commence the refurbishment and the accomplished numbers of pupils with MOE requirements.

3.2 Descriptive analysis

The overall descriptive results of building types, ownership, age of the building, activity performs in the preschools, age of the pupil, location of the building to the nearest road traffics, ventilation strategies, building maintenance and services, absence of any contaminants agents are presented in Table 1.

Table 1. Descriptive Analysis of Preschool Building

Categorical Variables	Description	%
Building types	Single Terrace	11.0
	Double Terrace	**67.1**
	Single Bungalow	4.6
	Double Bungalow	10.1
	Shop/office Lot	0.8
	Others	6.3
Ownerships of the Building	Own	34.2
	Rental	**65.8**
Age of the Building	<5	5.9
	5-10	28.7
	10-15	**33.3**
	15-20	14.3
	20-30	16.0
	>30	1.7
Recent Renovation	Yes	54.4
Programme	Preschool only	**58.2**
	Daycare only	3.4
	Preschool+Daycare	30.8
	Preschool+Tuition	3.4
	Preschool+Daycare+Tuition	4.2
Age of Pupils	>4years old	1.7
	4-6	69.2
	5-6	1.3
	>6	20.7
	others	7.2
Traffic Type	Small local street	12.7
	Medium local street	**65.0**
	large local street	16.0
	Expressway	6.3
Traffic Density	Low	30.4
	Medium	**54.9**
	Heavy	14.8
Ventilation Type	Dominantly natural	33.8
	Dominantly air-condition	**36.3**
	Both	29.9

Continued Table 1

Categorical Variables	Description	%
Floor Surface	**Smooth**	**84.8**
	carpeted	15.2
Curtain Types	**None**	**62.9**
	Blinds	7.6
	Textiles	29.5
Cooking Activities	Yes	70.9
Dampness	Yes	52.3
Mould Growth	Yes	54.0

Vermin Absent	Yes	76.0
Soft Toys	Yes	85.2
Floor Cleaning	Twice daily	30.4
	Daily	69.6
Fan Cleaning	Twice weekly	8.0
	Weekly	85.0
	monthly	26.5
Shelf Cleaning	Daily	28.7
	Weekly	51.5
	Twice a week	19.8
Table Cleaning	Daily	66.2
	Twice a week	28.7
	Weekly	5.1
Toilet Cleaning	Daily	78.5
	Twice a week	19.8
	Weekly	1.7
Curtain Cleaning	Weekly	6.3
	Monthly	46.4
	Biannually	26.6
	Never	20.7

3.3 Type of refurbishment

Refurbishment of preschools varies depending on the owner's finances, vision and number of pupils. There might be major or minor refurbishment of the building. Table 2 shows the type of renovations that recently have been done on the 237 preschool building.

Though the repair works were the common renovation done at the preschool building. While the improvement of new building function is considered to be ranked as second for refurbished work among the preschool buildings. It basically initiates to give a comfortable space to the occupants.

Table 2. Type of refurbishment in refurbished private preschool

	N	Percent
Repairs on available building	**75**	**31.6**
Renovation and alteration	38	16.0
Improvement of function	45	19.0
Additional to structure of the building	21	8.9
Retrofitting	31	13.1
Painting	27	11.4
Total	237	100.0

3.4 Function of renovation

Adopting from a survey by [17], an observation on the function of renovation of preschool building has been undertaken. Table 3 shows the function of renovation at preschool building. The preschools' owners deliberated the building space to accommodate the number of pupils. It is used to enhance the working environment and performance. The main concern for changing the function of any building is how to provide a sufficient space and place for tenants [18].

Table 3. Function of renovation at preschool building

	N	Percentage
Fictional arrange usable space	**65**	**27.4**
To add comfort to the building	**59**	**24.9**
Aesthetic to improve the appearance of the building	26	11.0
Functional change of use	22	9.3
Change number of occupants	13	5.5
Improve market value	2	.8
Reducing cost of maintenance	18	7.6
Legal change due to add health needs	32	13.5
	237	100.0

3.5 Classroom floor area vs number of pupils (space)

Overall, a total of 711 classrooms of 237 preschools were observed with respect to the number of occupants in the classrooms and sufficient area provided. Table 4 indicates the mean classroom area at 15.51m^2 and the number of occupants in a classroom of the observed preschool is 16 people. It indicated the number of pupil and teacher in the classroom of any adaptive-reused classrooms at average as the given result.

Table 4. Classroom area (m^2) and existing number of occupants per classroom

	N	Range	Minimum	Maximum	Mean
Classroom Fl Area(m^2)	711	34.40	7.40	41.80	15.5054
Numbers of occupants	711	25	6	31	16.19

3.6 Accomplished number of occupants in a classroom with the regulation

After adopting the formula, Figure 2 reveals the majority of the classrooms were not complying with the regulations of MOE. Whereas, only 8.9 % or 63 classrooms met the available terms of regulations. The rest of the classrooms have greater occupants i.e., more than 15 people per classroom. Since the previous results indicated that the mean of the occupants per classroom was 16.19, it is predictable that the most of the occupants of the classrooms were over the limit.

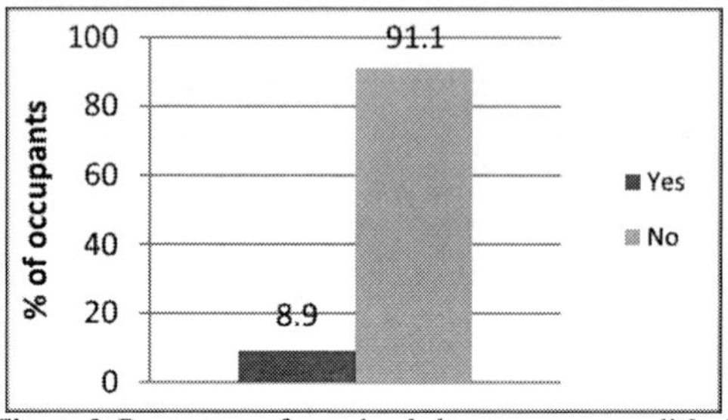

Figure 2. Percentage of preschool classrooms accomplished the MOE regulations

3.7 Variable related to the requirement of pupils' density in the classrooms

Previously, frequencies of accomplished occupants in the preschool's classrooms have been identified. In this section, the study has been focused on the factor which might relate to the numbers or requirement of the previous numbers in the classrooms, which is not in compliance with the MOE requirement.

The chi-square test [19] was used to test if sample of data came from a population with a specific distribution. In addition, the frequency of cases on this subject, possessing some quality, varies between levels of a given factor or among a combination of levels of two or more factors. The variables which were associated with the number of pupils/occupants that complied in the classroom were the building types, ownership of the building, ventilation strategy, classroom floor area, the recent renovation and function of renovation.

According to the observations provided in the Table 5 above, there is a significant relationship between ownership of the building ($\chi2= 5.552$, $N = 711$, $p <.05$) and classroom floor area ($\chi2= 156.20$, $N = 711$, $p <.05$) with the classroom, where the pupils' number complied with the MOE requirements. Whereas, building type, ventilation strategies, recently renovated building and function of renovation has no significant relationship with the MOE regulation.

Whilst, the previous double storey house premises (n=436 classrooms) were the frequent condition. It indicated that the preschools were unable to comply with the number of pupils per classrooms as required by MOE. The other variables stated that most classes against the requirements were the classrooms operated in the rental premises (n=435 classrooms), air-conditioned classrooms, classrooms with floor area 20-30 m^2(n=648 classrooms), being renovated recently (n= 356 classrooms) and those renovated in order to arrange the space usage (n=181classrooms).

Table 5. Variables Related to the Requirement of Number of Pupils in Classrooms

	Variables		Accomplished Nos of Occupants with MOE requirement?				χ^2	P<.05
			Yes		No			
			n	%	n	%		
V1	Building Type	Single Storey Terrace	6	7.7	72	92.3	3.405	NS
		Double Storey Terrace	41	42.3	**436**	**91.4**		(P=.493)
		Single Storey Bungalow	4	12.1	29	87.9		
		Double Storey Bungalow	8	11.1	64	88.9		
		Shop/office Lot	0	0	6	100.0		
		Others	4	8.9	41	91.1		
V2		Own	30	12.3	213	87.7	5.552	

	Ownership of Building	Rental	33	7.1	**435**	92.9		S (P=.018)
V3	Ventilation Strategy	Natural	22	8.9	224	91.1	0.172	NS (P=.918
		Air-Condition	24	9.3	**234**	90.7		
		Both	17	8.2	190	91.8		
V4	Classroom Area	<10m^2	7	9.5	67	90.5	156.21 0	S (P=.00)
		10-20m^2	13	2.6	484	97.4		
		20-30m^2	63	8.9	**648**	91.1		
V5	Recent Renovation	Yes	31	8.0	**356**	92.0	0.761	NS (P=.383)
		No	32	9.9	292	90.1		
V6	Function of Renovation	Fictional arrange usable space	14	7.2	**181**	92.87	11.275	NS (P=.127)
		To add comfort	18	10.2	159	89.8		
		Aesthetic improve appearance	5	6.4	74	93.6		
		Change of use	11	16.7	55	83.3		
		Change number of occupant	5	12.8	34	87.2		
		To Improve market value	0	0	6	100		
		Reducing cost of maintenance	1	1.9	53	98.1		
		Legal change to add health needs	9	9.4	87	90.6		

NS: Not significant at p>.05 S: Significant at p<.05

4 Conclusions

The majority of the schools, discussed in the local and international studies, were exposed to the inadequate physical environment due to insufficient ventilation, maintenance activities, congested space, and presence of contamination sources through building materials, furnishings and equipment. The development of preschool in Malaysia has dominated radically in order to promote early education of the pupils in their adolescence age.

The observation and in-situ building investigations demonstrated in this research which is to identify the characteristics of the buildings' physical and condition as well as the scenario of private preschools' buildings. The private preschools have been operating in the refurbished buildings, previously used for the residential purpose. Surprisingly, after superimposing the collected data in the equation from the Malaysian Ministry of Education, for the appropriate capacity of pupils in the classroom, it has significantly concluded that the results of over 90% randomly selected classrooms were beyond the certain limit, which means that those classrooms were unable to cater for the appropriate number of children at a time. Preschool building conditions included a double story terrace, rental premises, air-conditioned classroom, classroom with floor area 20 m^2- 30 m^2 and recently renovated classrooms.

This study has provided the current situation in private preschools, which helped to identify the conflict during its operation as well as the effects of those conflicts on the pupils. The Ministry of Education should take into account the number of pupils occupying the classrooms before giving any approval to the preschool owners. Unfortunately, during the walk-in observations, over 90% of preschools' classrooms have exceeded the standard limit on number of pupils per classroom. This situation created problems for the children to learn due to the congested space, un-comfortableness, impaired concentration towards learning. If these environmental conditions persisted, the health symptoms of the children may get worse due to the inadequate ventilation and IAQ. Therefore, the inspection of the preschool operations should be done regularly.

The data acquired in this study have indicated that the number of these adaptive classrooms has exceeded from 15 people per classrooms as gazetted by MOE. Yet, the mean area of 711 classrooms was signifying at 15.5 m^2. According to [8], pupil requires at least $2m^2$ per person, which suggests that the classroom could only accommodate an average of 7 pupils in a classroom.

Acknowledgements

The authors gratefully acknowledge the financial support of the Short Term Grant (2016) no. 304/PPBGN/6313274 established by the University Science Malaysia

References

1. Educational Planning and Research Division (2014). Malaysian Educational Statistics. Ministry of Education
2. [MOE] Ministry of Education. (2012). Preliminary report: Malaysia education blueprint 2013-2025. Retrieved on 11 September 2014, at http://www.moe.gov.my/userfiles/file/PPP/Preliminary-Blueprint-Eng.pdf.
3. J. R. Cryan, R, Sheehan, J. Wiechel, I. G. Bandy-Hedden, I. G. *Success outcomes of full-day kindergarten: more positive behavior and increased achievement in the years after*. Early Childhood Research Quarterly, **7** (2), 187-203(1992)
4. N.L. Karweit, R.E. Alavin, N.A. Madden, *Effective kindergarten programs and practices for students at risk of academic failure*. Needham Heights, MA: Allyn and Bacon (1992)
5. [MOE] Ministry of Education, (2001), Education development Plan for Malaysia 2001-2010, retrieved on 11 April 2011, at: planipolis.iiep.unesco.org/upload/Malaysia/Malaysia%20development%plan%202001-2010summary.pdf.
6. [MOE] Ministry of Education. (2013). Preliminary report: Malaysia education blueprint 2013-2025. Retrieved on 11 September 2014, at http://www.moe.gov.my/userfiles/file/PPP/Preliminary-Blueprint-Eng.pdf.
7. Kuala Lumpur City Hall. *Senarai tadika swasta sediaada di WPKL*. Jabatan Perancang Bandar. Unpublished (2010)
8. Uniform Building by Law Act 1333. *Laws of Malaysia*, sixteenth ed., MDC Publisher, Kuala Lumpur (2009)
9. D. Shendell, A. Winer, R. Weker, S. Colombe. *Evidence of inadequate ventilation in portable classrooms: results of pilot study in Los Angeles*. Indoor Air, 14, 154 – 158 (2004)
10. J.M. Daisey, W. J. Angell. *A survey and critical review of the literature on indoor air quality, ventilation and health symptoms in schools*. Lawrence Berkeley National Laboratory Report No. LBNL-41517 (1998)
11. D. Stankovic, J.Stojic, *Psycho-developing needs of children and spatial features for children's stay*. Architecture and Civil Engineering, 5(1), 71 – 75 (2007)
12. M.C. Lee, K.W. Mui, L.T. Wong, W.Y. Chan, E.W,M Lee, C.T. Cheung. *Student learning performance and indoor environmental quality (IEQ) in air-conditioned university teaching rooms*. Build Enviro ,49, 238-44 (2002)
13. C. Uline, M. Tschannen-Moran, *The walls speak: The interplay of quality facilities, school climate, and student achievement*. Journal of Educational Administration, 46(1), 55-73 (2008)
14. Y. Zhang, P. Barret, *Findings from a Post-occupancy Evaluation in the UK Primary Schools Sector" in Facilities: Special Issue on Environment Behaviour*. Facilities Journal, **28**(14), 641-666 (2010)
15. M.Y. Abbas, M. Othman, *Social behavior of pre-school children in relation to physical spatial definition*. Procedia - Social and Behavioral Sciences, **5**(0), 935-941 (2010)
16. M. Schneider, *Do school facilities Affect Academic Outcomes?*, National Clearinghouse for Educational Facilities, Washington, DC. (2002). Retrieved 11 September 2012, from http://www.edfacilities.org/pubs/outcomes.pdf
17. I. Rahmat, V. B. Torrence, A.H. Ezanee, *Refurbishment cycles and the management of refurbishment projects*, UiTM Research Center, Shah Alam, Selangor, Malaysia (2003)
18. D.K.H. Chua, A. Tyagi, S. Ling, S.H. Bok *Process Parameter-Interface Model for Design Management*. Journal of Construction Engineering and Management, 129(6), 653-663 (2003)
19. G.W. Snedecor, W.G. Cochran, *Statistical Methods*, (8th Edition), Iowa State University Press (1989)

The Prevelence of SBS and Absenteeism among Children in Urban Refurbished Private Preshools

Naziah Muhamad Salleh[1a], Nuzaihan Aras Agus Salim[1], Syahrul Nizam Kamaruzzaman[2] anad Norhayati Mahyuddin[2], Fairus Muhammad Darus[3]

[1]*University Science Malaysia, School of Housing Building & Planning, Building Technology Department, 11800 Penang, Malaysia*
[2]*University of Malaya, Faculty of Built Environment, Building Performance and Diagnostic, 50603 Kuala Lumpur*
[3]*Faculty of Applied Sciences, Universiti Teknologi MARA, Shah Alam, Selangor, Malaysia*

Abstract. The preschool education is compulsory to children in Malaysia. This regulation has encouraged more premises to be refurbished as a pre-school building. This paper examines the pupils' absenteeism and the prevalence of Sick Building Symptoms (SBS) initiated in congested private preschool with different ventilation strategies. The study analysed data from the attendance record of 10 classrooms and the questionnaire surveys administered to 151 parents about their children health symptoms once they were leaving the schools building. Questions on SBS used 5-point likert scale with symptoms concern on nose, eye, head, throat, skin, breath and tiredness. The descriptive and chi-square test applied to obtain the association of SBS and ventilation strategies in the classrooms. With quantitative and qualitative explanation, the unhealthy environment in refurbished pre-schools explained. Running nose, coughing and sore throat frequently reported in air-conditioning (AC) classrooms. The higher absent rate found in AC classrooms. These symptoms show there were weaknesses in ventilation performance and environment in the selected preschools. Further analyses on objective measurements in future research are strongly recommended.

Keywords: Indoor Environmental Quality, Indoor Air Quality, Sick Building Symptoms, Refurbishment, Preschool, Ventilation,Class

[a]Corresponding author: naziahmsalleh@usm.my

1 Introduction

Indoor Environmental Quality (IEQ) in schools can have a substantial impact on children's health, as an important environment where children may be exposed to pollutants and allergens [1]. The school provides a major indoor environment for children away or apart from their home. Children may spend 10 hours per day at school [2] depending on the time they arrive at the school and the time they leave the school. It is special concerns as the pupils are susceptible to poor Indoor Air Quality (IAQ). Indoor contaminant may increase both long and short term health problems among pupils and staff, reduce the productivity of teachers [3,4] and degrade the pupil's learning environment and comfort [5]. A significant influence from overall indoor environmental quality can effect and give the influence on student attendance and performance. Studies have shown that poor IAQ resulted more illness, absenteeism and asthma attacks.

The major exposure effects of air contaminants will be the health effects visible of the symptoms or the disease. Obviously, it influences student's performance directly or indirectly through the effects of attendance. School absenteeism is one of the main problems in public and private school. Most of the incidence for total absent episodes and percent of missed days, including those because of upper respiratory infections [6,7]. Regards to Malaysia climate, these compilations has originate immensely among children and preschooler [7-10]. Whereas it enormously affected by the weather [7], indoor air, ventilation strategies [7-9] or the surrounding environment [10].

Asthma reported to be the number one cause for student absenteeism due to chronic conditions and was the leading cause of hospitalization for children [6]. Epidemiological data indicated that 7.4% of school-age children report having asthma symptoms annually, and 2.8 million children between 5 and 14 years of age had a diagnosis of asthma [11]. Of students diagnosed with asthma that attend school, the majority of school nurses or 51% indicated asthma was more disruptive of school routine than other chronic diseases, and 53% stated that all students with asthma were able to participate in all school day activities including gym and/or recess, and keep up with their peers (American Lung Association, 2003). This also mentioned in local studies where asthma and respiratory health affected the schoolchildren attendance [7, 8].

Eventually, asthma was not the single contributing factor to student absenteeism. Outbreaks of acute infections caused by viruses, bacteria, parasites, and protozoa were documented among both students and staff in schools [12]. These types of sicknesses occurred throughout the school year and impacted on student absenteeism, which in turn influences student performance. Infectious disease was spread through person-to-person contact, by droplets or large particles spread by coughing, sneezing, or wheezing, or by fomites (self-innoculation after touching contaminated surfaces), the school building factors that influenced the transmission of infectious disease include the respiratory particles, ventilation rate, humidity, location of the building, ventilation modes, school facilities and functional state of sinks and toilets [7,8,13]. Design specifications related to indoor airflow, HVAC systems, windows, and controlled air ventilation were all factors that improved the quality of the indoor environment in school facilities [7,13]. The 17 separate studies found positive health impacts measured by reduction in reported prevalence of symptoms of asthma, flu, sick building syndrome, respiratory problems, and headaches [14].

The escalating numbers of children requiring preschool education in Malaysia has encouraged the involvement of private initiatives to meet the demand [15]. Thus, the government has encouraged more private preschool centers to be set up, involving various types of refurbished buildings with various types of property converted into classrooms to accommodate 20 to 30 pupils at a time. This, presents a challenge in providing adequate IAQ in congested areas and locations since most of the private preschools are housed in refurbished residential and commercial buildings. Some IAQ parameters are resulted from the numbers of occupants, space and the ventilation competencies. Somehow, there is a dearth information on pupils' performance in relation of SBS and absenteeism in these type of adaptive-reused building. Thus, the objective of this study are accordingly: i) Evaluate the performance of occupants/pupils in terms of health symptoms and annual attendance towards IAQ conditions in refurbished preschools, and ii) to compare the health symptoms annual absentees in different ventilation modes classrooms.

2 Materials and methods

Monitoring performed in 10 classrooms randomly selected (of the 5 schools) the schools are located in different districts but in 25 km radius with each other's. The selections of the schools were based on the similarities of learning and activities systems, foods types serve during recess and school hours. All the classrooms were operating in double storey terrace house (100 m^2). Majority of the classrooms was a bedroom. The walkthrough observation of the building (Table 1) has indicated all the buildings have located in urban, residential area and at the roadside. The environment might enclose to the other factors which can contribute to health effects. Anyhow, the distances of the main road to the buildings are varying. 4 of the classrooms are air conditioning and the other 6 classrooms are natural ventilated with ceiling fan and windows open. The whole classrooms were installed with one ceiling fan each.

The continuous monitoring of indoor and outdoor for temperature, relative humidity, carbon monoxide, carbon dioxide and air speed were all obtained using Gray Wolf Sensing Solution IQ604 Indoor Air QualityProbe(Gray Wolf sensing Solution, Shelton, CT, USA). Due to the limited instruments, YESAIR 8-Channel IAQ Monitor was used to measure only indoor formaldehyde, nitrogen oxide and ozone. Each monitor was calibrated by the factory once within the calendar year and monthly by the field research team.

In addition, the DUST Track Technology model TSI 8532 utilised to measure indoor particulate matter (PM_{10}). The detection range was 0.001–150 mg/m3 and the resolution was 0.1 % of the reading. The log interval of the concentration data was set at every 60 seconds. Although the particle size range of the instrument was 0.1–10 mm, but during these measurements a size-selective impactor was attached to the inlet (with a cut size of 10 µm), which pre-conditioned the size range of particles (10 µm) entering the instrument. The particle monitor was able to detect particle emission concentration within the range of 0.001-150 mg/m3 under the particle size range of 0.1-10 µm and operational temperature range of 0-50°C.

Measurement done in three days of each classroom because of the following reason: (1)three as the optimal sample size (verify the mean and the frequency, to identify the valid minimum and maximum results), (2) sampling could be repeated if the instrument fail to measure . It were logging at every 1 minute of measurement. Atleast 8 hours per day of measurement – (the minimum of exposure to pollutants in determining SBS) (CIBSE Guide A, 2006). Measurements start at 7.30 p.m (atleast 30 minutes before the class starts) and end at 4.30 p.m. The monitors placed within a classroom at least 3 feet away from any wall and atleast 0.5m from bookshelves and out of the children's reach. The monitors always placed on the same size and model step ladder/tripod to approximate the breathing zone.

151 parents then asked and to fill up the form about their children's health symptoms. Meanwhile, the school administration permitted an access to the attendance records with notice (emergency leave), due to sickness (enclosed with evidence such as notice or medical certificates) and unnoticed leave. Result then to be analysed using Statistical Package for the Social Sciences package 17.0 (SPSS). Descriptive analysis on Sick Building Symptoms (SBS) and abenteeism, meanwhile correlation on non-parametric using Chi-Square test applies to determine any relation of sickness absentees with the classrooms ventilations modes.

3 Results and Discussion

In this section, the SBS of the pupil (N=151) from selected ten classrooms is reported. Questionnaires were distributed to parents to fill up. This questionnaire represented the pupils' behaviour affected by the classroom's environment. Entire SBS indications of various variables measured to ascertain which factors contributed the most to the symptom. Descriptive analysis was assessed using frequencies with five scales as shown in Table 2 and Figure 1 stated number of days annually reported with SBS. The analysis showed the majority of pupils felt the symptoms of coughing, sore throat and running nose, as these were the common symptoms in this study. The value of these symptoms has a significant association with IAQ.

Table 1. Building characteristics

Case Study	**S7-56A**	**S7-43D**			**SP15**		**KS-SU3**	**PJ-KD3**		
	R1	**R1**	**R2**	**R3**	**R1**	**R2**	**R1**	**R1**	**R 2**	**R3**
Type of building	Double storey terrace house	Double storey terrace house			Double storey terrace house		Double storey terrace house	Double storey terrace house		
Age of the building	8 years	8 years			13 years		2 years	13 years		
Location	Roadside with less traffics	Roadside with heavy traffics			Roadside with less traffics		Roadside with less traffics	Roadside with heavy traffics		

DOI: 10.1051/matecconf/20166600119

Surrounding details	Urban, residential, commercial area	Urban, residential, commercial area			Urban, Residential		Urban, residential, commercial area	Urban, residential, commercial area		
Constructional activity	No	No			No		No	No	No	No
Distance from main road	500m to expressway	20m			300m		100m	100m	100 m	100m
Operation Hour	7am-7pm	7am-7pm			7am-7pm		8am-5pm	8am-5pm		
Occupancy rate	18	21	20	16	16	15	15	20	19	18
Ventilation status	Natural ventilation with ceiling fan, windows open	Air-conditional		Natural ventilation with ceiling fan, windows open	Natural ventilation with ceiling fan, windows open		Natural ventilation with ceiling fan, windows open	Air-conditional		Natural ventilation with ceiling fan, windows open
Number of ceiling fan	1	1	1	1	1	1	1	1	1	1
Classrooms conditions (damage walls, ceiling, furniture)	Less dusty floor mold stain on the wall	Mold stain on ceiling, less dusty floor	Less dusty floor	Less dusty floor	Dusty floor , full with old paper	Dusty floor	Less dusty floor	Less dusty floor		
Classroom cleaning	Daily after school	Daily after school			Daily after school		Daily after school	Daily after school		
Air-condition filter cleaning	-	annually		-	-	-	-	Annually		-

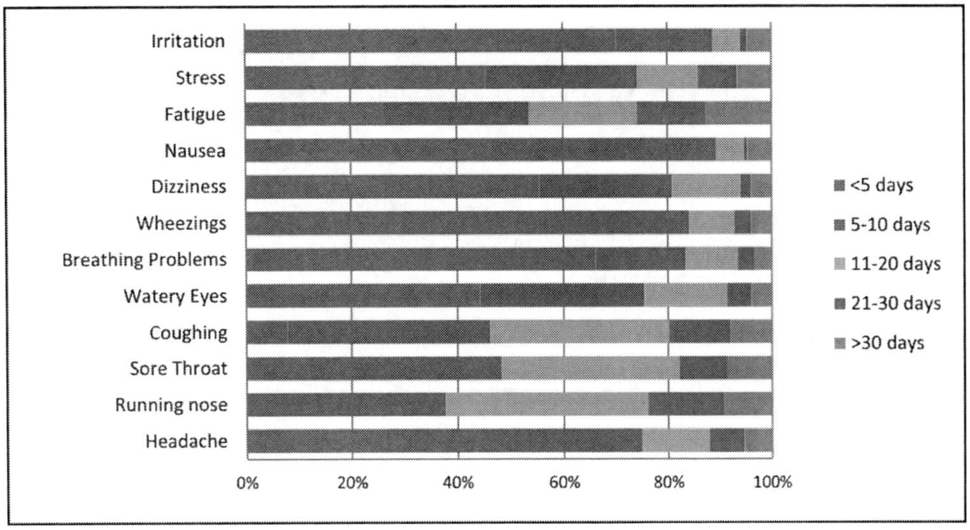

Figure 1. Number of days annually reported with SBS

Table 2. Rating of the symptoms reported annually

N=151	(%)						
	never	*TOTAL UNNOTICEABLE*	Sometimes	Slightly	Regularly	often	*TOTAL NOTICEABLE*
Headache	26.5	*26.5*	48.3	13.2	6.6	5.3	*73.5*
Running nose	6.6	*6.6*	31.1	33.8	9.3	8.6	***82.8***
Sore throat	11.9	*11.9*	36.4	33.8	9.3	8.6	***88.1***
Coughing	7.9	*7.9*	38.4	33.8	11.9	7.9	***92.1***
Watery eyes	44.4	*44.4*	31.1	15.9	4.6	4.0	*55.6*
Breathing	66.2	*66.2*	17.2	9.9	3.3	3.3	*33.8*
Wheezing	68.2	*68.2*	15.9	8.6	3.3	4.0	*31.8*
Dizziness	55.6	*55.6*	25.2	13.2	2.0	4.0	*44.4*
Nausea	64.9	*64.9*	24.5	5.3	0.7	4.6	*35.1*
Fatigue	26.5	*26.5*	27.2	20.5	13.2	12.6	*73.5*
Stress	45.7	*45.7*	28.5	11.9	7.3	6.6	*54.3*
Irritation	70.2	70.2	18.5	5.3	1.3	4.6	*29.8*

Table 3 presents the noticeable percentage of health symptoms for every classroom. It has assumed parity of reported health symptoms among classrooms with different ventilation strategies. This section also contributed the further investigations regarding relationships between pupils' health symptoms and

DOI: 10.1051/matecconf/20166600119

other parameters. When responding about their children, most incidents the parents reported were of dizziness, followed by irritation, watery eyes and nausea. The majority of them were reported to be in natural ventilation classrooms (S7-56A-R1 and KD06-R3). A certain relationship has been described between the eyes and nose that affect each other [16].

Nevertheless, respiratory problems (running nose, sore throat, coughing, breathing problems and wheezing) were most commonly reported in air-conditioning classrooms with percentages from 14.9% to 18.9%. Air conditioners might be a wellspring of indoor contamination if their cleaning is irregular, which can prompt the amassing of dirt in the filters. So, it activates the respiratory indications such as sinusitis, rhinitis, asthma and hypersensitive pneumonitis [17].

Table 3. Noticeable SBS in each classroom

Percentage Noticeable SBS Symptom	**Room**									
	S7-56A-R1[a]	**S7-43D-R1**[b]	**S7-43D-R2**[b]	**S7-43D-R3**[a]	**SP15-R1**[a]	**SP15-R2**[a]	**SU3-R1**[a]	**PJ-KD6-R1**[b]	**PJ-KD6-R2**[b]	**PJ-KD6-R3**[a]
Headache	7.9	13.2	0.0	7.9	13.2	7.9	.0	**15.8**	10.5	23.7
Running nose	9.6	**14.9**	6.4	11.7	10.6	10.6	7.4	9.6	10.6	8.5
Sore throat	6.4	**16.7**	6.4	11.5	9.0	11.5	9.0	10.3	9.0	10.3
Coughing	7.4	**16.0**	6.2	9.9	12.3	9.9	8.6	11.1	7.4	11.1
Watery eyes	10.8	5.4	5.4	2.7	16.2	8.1	10.8	**18.9**	10.8	10.8
Breathing	**16.0**	12.0	4.0	8.0	8.0	12.0	0.0	**16.0**	8.0	**16.0**
Wheezing	12.5	**16.7**	4.2	8.3	12.5	8.3	0.0	12.5	12.5	12.5
Dizziness	10.3	6.9	6.9	10.3	10.3	3.4	0.0	6.9	17.2	**27.6**
Nausea	**18.8**	6.3	0.0	6.3	25.0	12.5	0.0	12.5	12.5	6.3
Fatigue	10.0	10.0	5.7	7.1	7.1	10.0	8.6	10.0	**15.7**	**15.7**
Stress	10.3	7.7	5.1	5.1	10.3	7.7	7.7	15.4	12.8	**17.9**
Irritation	**23.5**	5.9	0.0	0.0	17.6	11.8	0.0	5.9	11.8	**23.5**

a: NV Classroom b: AC Classroom N=151

Table 4 shows the sore throat and fatigue were connected with the classroom's ventilation strategies with ($\chi 2$ = 6.438, p <0.005), ($\chi 2$ = 4.985, p <0.005) and ($\chi 2$ = 6.438, p <0.005), respectively. It specifies that natural ventilation classrooms contributed higher applicable health indications to pupil contrasted with AC classrooms owing to the outside contaminants entering the classrooms through the open windows. On the other hand, the air-conditioned S7-43D-R1 committed the most astounding number of pupils endured the symptoms. Further studies conducted on familial and pupils' history to discover the affiliation of the symptoms.

Table 5 outlines distinct statistics of the absentees' information from the selected classrooms. Figure 2 has indicated that the three highest absentees due to sickness were reported from the air-conditioned classrooms, namely S7-43D-R1, S7-43D-R2 and PJ-KD3-R2 with the rate of 6.25%, 5.46% and 5.08% of yearly non-attendance, respectively. Further cross-sectional studies about absenteeism, pupil's general particulars, parents' health conditions, pupils' health problems, reported SBS among pupils and houses/dwelling characteristics have been carried out in the other sections of this research study.

DOI: 10.1051/matecconf/20166600119

Table 4. Association of ventilation strategies with SBS

Percentage Noticeable SBS Symptom	Classroom Vent		$\chi 2$	*P*
	AC	NaV		
Headache	37.5%	62.5%	0.019[a]	0.890
Running nose	10.0%	90.0%	3.654	0.056
Sore throat	11.1%	88.9%	6.438	0.011*
Coughing	8.3%	91.7%	4.985	0.026*
Watery eyes	38.8%	61.2%	0.008	0.929
Breathing	36.0%	64.0%	0.727	0.394
Wheezing	36.9%	63.1%	0.315	0.574
Dizziness	34.5%	65.5%	1.535	0.215
Nausea	35.7%	64.3%	0.858	0.354
Fatigue	25.0%	75.0%	6.438	0.042*
Stress	41.5%	58.5%	1.444	0.230

*Significant at P<.05 N=151

Table 5. Percentage of annual absenteeism

	TOTAL % ABSENCE BY SCHOOL			
Classroom	Noticeable	Sick	Unnoticeable	TOTAL
S7-56A-R1*	1.48%	4.27%	0.33%	6.08%
S7-43D-R1**	**2.12%**	**6.25%**	**1.51%**	**9.88%**
S7-43D-R2**	**0.67%**	**5.46%**	**0.27%**	**6.40%**
S7-43D-R3*	1.07%	4.03%	0.29%	5.39%
SP15-R1*	2.86%	4.37%	1.61%	8.83%
SP15-R2*	2.96%	4.31%	1.72%	8.99%
SU-R1*	2.04%	4.57%	0.28%	6.90%
PJ-KD3-R1**	1.74%	4.66%	0.39%	6.80%
PJ-KD3-R2**	**1.26%**	**5.08%**	**0.43%**	**6.77%**
PJ-KD3-R3*	1.82%	4.31%	0.62%	6.74%

*Natural ventilation, ceiling fans and windows open **Air-conditioning classroom

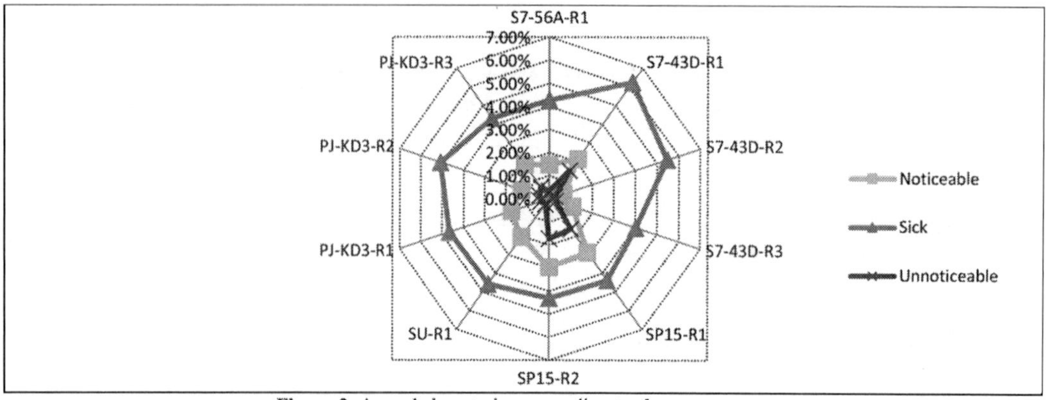

Figure 2. Annual absenteeism according to classroom

The results have proven that the absentees for NV and AC classrooms did not provide the same value. The air-conditioned classrooms indicated higher rates of annual absentee because of the sickness (Table 6). Nevertheless, there was no link found between the ventilation strategies and annual leave among pupils. The absenteeism in air-conditioned classrooms was higher than the natural ventilation classrooms with 32.8% of pupils, who were absent for more than 20 days in a year because of sickness belonged to the air conditioned classrooms. The comfort and healthy level is related to the IEQ performance. Good ventilation ensures the comfort level and improve the quality of the learning environment.

Table 6. The annual absentees according to ventilation strategies

Classrooms	% Annual Absenteeism: Sickness		
	<3% absence (less 7 days)	3%-6% Absenc (7-20 days)	>6% absence (more 20 days)
Air conditioned classrooms	32.8%	34.5%	32.8%
Naturally Ventilated classrooms	40.9%	43.0%	16.1%

IAQ Characteristics

The findings of indoors and outdoors IAQ concentrations are summarised in Table 7. For three air condition classrooms (S7-43D-R1, S7-43D-R2 and PJ-KD6-R1), mean indoor temperature were lower than the corresponding outdoor levels. However, the whole classrooms and respective outdoors stated higher value of mean temperature than the recommended range for acceptable indoor air quality of American Society of Heating, Refrigerating and Air-conditioning Engineers [18] at 23.0°C to 26°C. The I/O for temperature of each classroom stated in small value, range between 0.93-1.11. Generally, relative humidity in four natural ventilation classrooms (S7-43D-R3, SP15-R1, SP15-R2 and KS-SU3-R1) exceeded the [18] recommendation limit at 30%-65%. The I/O relative humidity of each classroom have shown in small ranging, except the air-condition classroom PJ-KD6-R2 at 0.76 with the lowest ratio indoor-outdoor, suggesting that the mechanical air-conditioner have controlled the indoor environment.

Indoor CO_2 concentrations have been referred to as an indicator of indoor air quality. The distribution of CO_2 is log-normal. The concentrations of both points (indoor and outdoor) are fluctuated and resulted insignificant to others. Meaning, the outdoor carbon dioxide concentration do not influent the indoor concentration. The mean values of outdoor carbon dioxide range between 326.13 ppm to 488.25 ppm. But, the indoor carbon oxide concentration reveals differently. Ventilation strategies and occupant density have given a significant effect to the level of CO_2. The air-conditioned S7-43D-R2 classroom with density of 22 pupils stated the highest mean value of CO_2 at 1680 ppm, followed by PJ-KD6-R2 air-conditioned classroom with 20 pupils' density at 1295 ppm, CO_2 mean value concentration 1293ppm for PJ-KD6-R1 (pupils density: 19 people) and air-conditioned classroom S7-43D-R1(with 19 pupils density) at 1054 ppm of CO_2 mean value. These 4 air-conditioned classrooms have exceeded 1000 ppm as stipulated standard limit in The Malaysian Code of Practice [19] and [18] Meanwhile, the natural ventilation classrooms (S7-56A-R1, S7-43D-R3, SP15-R1, SP-15-R2, PJ-SU3-R1) seem to have a good indoor air quality, where the indoor CO_2 mean value concentrations were [18] and [19] limit. Although, S7-56A-R1, S7-43D-R1, PJ-KD6-R1 and PJ-KD6-R3 has the similar numbers of pupil density at a time but the mean values of CO_2 were vary with different ventilation strategies.

Indoor concentration for PM_{10} was stipulated the maximum allowable concentration in inhabitant area standard is 0.15 mg/m^3 [19]. The range of mean concentrations was between 41.0 to 342.82 µg/m^3. The mean values of particulate 10 µm for six natural ventilation classrooms (S7-56A-R1, S7-43D-R3, SP15-R1, SP15-R2, KS-SU3-R1 and PJ-KD6-R3) were beyond the Malaysian Code of Practice (DOSH, 2005). However, The air-conditioned classrooms have maintained the PM_{10} level to be below the limit of Malaysia Code of Practice [19]. It clearly shows that the outdoors particles have distributed the PM_{10} concentration into the indoor environment.

The mean concentrations of CH_2O, CO and TVOC well below the Malaysian Code of Practice [19] recommended values of 10 ppm for an 8-hour of exposure. The values were ranging between from undetectable to 0.03 ppm, 0.26 ppm to 0.83 ppm and 18.5 to 217.13 ppb respectively. The finding shows a few parameters were beyond the standard limit and suspected as a contribution to the health symptoms among occupants. Relative humidity, temperature, CO_2 and respiratory particulate, PM_{10} were found as the inadequate parameters in this study. As mentioned earlier, Malaysia tropics condition might effects the level of temperature and relative humidity. It is impossible to get the adequate level for both environmental parameters.

Running nose, sore throat and coughing were reported higher at both ventilation classrooms. Surprisingly, sore throat, coughing and fatique were significantly perceived in NA classrooms which were recognized with the elevated indoor concentration of PM_{10}, temperature and relative humidity. These parameters might as well contributed to the perceived symptoms among pupils in the classrooms [20]. Thus, it explained the absentees' rates for children in AC classrooms higher than the annual absentees rates among chidren in NV classrooms.

4 Conclusions

From this research, it can be concluded that the numbers of annual absentees among pupils in air-conditioned classrooms higher than pupils in non-air-conditioned classrooms. The natural ventilation classrooms with ceiling fan and opening windows had higher rates of relative humidity. The improper ventilation systems found to be one of the distribute factor to the infectious disease.

Most refurbished preschool building in Malaysia were found not to comply with minimum space regulation concerning children, teachers and staff, leading to the uncomfortable conditions and spreading out the infectious illness. It is suggested that the authorities to take this issues into consideration and provide the sufficient space for the occupants. It is also to put into account the location of the classrooms and schools when the premises being converted to the education buildings.

The other conditions such as dwellings could give significant relations to the absentees' rates. In view of children performance in refurbished private preschools it would be understandable to find that some sources are originated from indoor such as occupants and ventilation strategies. The results of the measurement show the differences arise, due to the advantage of the indoor environment.

Particularly with the high density of pupils in the classrooms in air condition classrooms has invited the high level and insufficient indoor CO2 and other contaminant sources. This is seen particularly when the high level indoors in the classroom environment contrast strongly with the air outside the preschool building. The adapted space function which previously is bedrooms is inadequately to provide a good air quality and invited few health symptoms. The improper ventilation strategies has encourage some symptoms to children especially to those are really sensitive on the presence of few parameters with the minimum dose. Yet, the symptoms of health problems may not begin when children being at school. The environmental factors such as urbanisation, industrialisation, air pollution, hygienic conditions, renovation, mould and smoke exposures, building and furnishing materials and diet have been related to increased or decreased risk, and thus prevalence, of allergic diseases and could potentially explain a part of the factor. This type of study should be extended to the biological contaminants and widely apply to other numerous of refurbished preschools in order to be better able sustaining the IAQ management strategies and to apply source apportionment methodologies. The outcomes of the studies are foreseen as a benchmark for future research to improve the preschools physicals' environment.

Table 7. Indoors and outdoors IAQ concentrations at every classrooms

Classroom	Indoor	Outdoor	I/O

DOI: 10.1051/matecconf/20166600119

		Minimum	Maximum	Mean	S.D	Minimum	Maximum	Mean	S.D	
Temperature (^{o}C)	S7-56A-R1	29.70	34.90	32.31	1.49	24.10	32.70	29.12	2.89	1.11
	S7-43D-R1	29.10	32.90	30.82	1.13	26.10	33.90	31.43	2.29	0.98
	S7-43D-R2	27.60	31.90	29.35	1.18	26.00	34.00	31.46	2.34	0.93
	S7-43D-R3	28.00	32.60	30.57	1.34	26.60	32.80	30.51	1.55	1.00
	SP15-R1	27.30	33.40	30.72	1.57	20.30	33.00	29.13	3.36	1.05
	SP15-R2	22.90	34.40	31.10	2.05	24.70	32.90	30.04	2.33	1.04
	KS-SU3-R1	28.10	32.20	30.26	0.92	24.90	29.90	28.14	0.99	1.08
	PJ-KD6-R1	28.20	33.20	30.52	1.50	25.90	34.70	31.42	2.25	0.97
	PJ-KD6-R2	29.20	31.80	30.51	0.78	22.30	32.30	28.09	2.97	1.09
	PJ-KD6-R3	29.40	33.90	32.03	1.32	26.60	35.20	31.52	2.02	1.02
Relative Humidity (%)	S7-43D-R2	49.90	83.00	59.09	4.91	50.90	86.10	61.39	11.30	0.96
	S7-43D-R3	60.70	83.20	70.92	7.27	57.50	84.30	66.41	6.63	1.07
	SP15-R1	56.00	86.70	71.40	6.56	55.20	100.90	70.95	12.34	1.01
	SP15-R2	40.60	90.90	72.12	8.39	58.50	98.00	70.79	10.27	1.02
	KS-SU3-R1	68.30	88.40	77.50	5.87	71.30	93.70	79.34	4.68	0.98
	PJ-KD6-R1	50.60	78.60	58.33	5.54	45.50	87.40	58.41	10.91	1.00
	PJ-KD6-R2	35.70	65.20	57.10	6.31	56.80	103.00	75.20	13.96	0.76
	PJ-KD6-R3	50.60	80.80	63.23	9.23	43.10	88.00	58.57	11.16	1.08
Air velocity (ms^{-1})	S7-56A-R1	.00	0.50	0.05	0.09	0.00	2.00	0.22	0.23	0.22
	S7-43D-R1	.01	0.26	0.04	0.03	0.00	1.90	0.03	0.09	1.22
	S7-43D-R2	.01	2.04	0.05	0.09	0.00	0.28	0.03	0.05	1.54
	S7-43D-R3	.01	1.68	0.06	0.11	0.00	0.25	0.03	0.04	2.15
	SP15-R1	.00	1.97	0.23	0.22	0.00	0.48	0.01	0.03	23.37
	SP15-R2	.00	1.97	0.23	0.22	0.00	0.48	0.01	0.03	21.79
	KS-SU3-R1	.01	1.58	0.02	0.09	0.00	1.07	0.12	0.16	0.20
	PJ-KD6-R1	.01	2.24	0.03	0.12	0.00	2.21	0.28	0.40	0.10
	PJ-KD6-R2	.00	3.79	0.03	0.16	0.00	3.30	0.31	0.49	0.08
	PJ-KD6-R3	.00	0.02	0.00	0.00	0.00	2.89	0.28	0.41	0.00
CO_2(ppm)	S7-56A-R1	551.70	1202.70	724.26	171.77	398.70	475.30	435.92	15.06	1.66
	S7-43D-R1	521.30	2186.00	1054.31	515.08	442.30	514.30	480.56	12.89	2.19
	S7-43D-R2	526.00	4576.70	1680.35	1184.46	445.70	523.00	488.25	13.43	3.44
	S7-43D-R3	418.00	1050.30	603.18	192.83	442.30	525.00	487.11	17.93	1.24
	SP15-R1	449.00	662.00	537.04	48.91	338.30	447.00	393.62	26.82	1.36
	SP15-R2	337.70	1005.50	640.70	123.99	344.30	463.00	404.43	27.32	1.58
	KS-SU3-R1	158.70	1276.70	699.13	288.09	147.00	526.50	326.13	100.03	2.14
	PJ-KD6-R1	509.00	2627.50	1294.75	550.69	228.50	482.00	438.71	19.56	2.95
	PJ-KD6-R2	478.00	2709.00	1458.78	658.08	0.00	460.00	359.60	53.99	4.06
	PJ-KD6-R3	500.00	1793.00	766.86	281.94	407.00	516.00	455.06	19.29	1.69

Classroom	Indoor				Outdoor				I/O
	Minimum	Maximum	Mean	S.D	Minimum	Maximum	Mean	S.D	

DOI: 10.1051/matecconf/20166600119

CO(ppm)	S7-56A-R1	0.00	1.00	0.32	0.33	0.00	1.70	0.31	0.35	1.02
	S7-43D-R1	0.00	1.80	0.53	0.63	0.00	4.00	0.63	0.76	0.84
	S7-43D-R2	0.00	1.40	0.60	0.46	0.00	1.80	0.43	0.54	1.39
	S7-43D-R3	0.00	1.30	0.39	0.44	0.00	3.00	0.50	0.62	0.78
	SP15-R1	0.00	1.70	0.32	0.40	0.00	2.70	0.37	0.52	0.86
	SP15-R2	0.00	2.40	0.56	0.52	0.00	2.30	0.45	0.50	1.23
	KS-SU3-R1	0.00	1.10	0.26	0.31	0.00	1.20	0.44	0.36	0.59
	PJ-KD6-R1	0.00	2.00	0.46	0.64	0.00	3.00	0.26	0.54	1.77
	PJ-KD6-R2	0.00	2.60	0.83	0.66	0.00	2.00	0.12	0.19	6.72
	PJ-KD6-R3	0.00	1.40	0.31	0.43	0.00	2.70	0.25	0.58	1.25
TVOC(ppb)	S7-56A-R1	47.70	145.70	105.15	26.05	0.00	326.30	5.45	31.55	19.31
	S7-43D-R1	0.00	189.00	157.49	41.16	0.00	257.00	3.02	20.55	52.14
	S7-43D-R2	0.00	202.70	149.95	49.72	0.00	344.00	6.73	35.22	22.27
	S7-43D-R3	0.00	96.30	51.61	35.17	0.00	424.70	2.59	25.84	19.90
	SP15-R1	0.00	63.00	23.60	24.37	0.00	30659.70	643.64	3340.16	0.04
	SP15-R2	0.00	177.00	36.82	41.85	3.30	23331.30	1056.85	4152.64	0.03
	KS-SU3-R1	0.00	52.70	18.50	18.66	0.00	5339.50	151.82	525.74	0.12
	PJ-KD6-R1	0.00	134.50	109.28	36.33	0.00	910.50	27.26	88.64	4.01
	PJ-KD6-R2	149.00	295.00	217.13	25.77	0.00	3363.00	142.62	442.24	1.52
	PJ-KD6-R3	0.00	144.00	95.01	42.46	0.00	1005.00	30.14	98.84	3.15
CH_2O (ppm)	S7-56A-R1	0.00	0.43	0.02	0.06					
	S7-43D-R1	0.00	0.00	0.00	0.00					
	S7-43D-R2	0.00	0.00	0.00	0.00					
	S7-43D-R3	0.00	0.28	0.01	0.03					
	SP15-R1	0.00	0.24	0.02	0.04					
	SP15-R2	0.00	0.28	0.01	0.03					
	KS-SU3-R1	0.02	0.24	0.03	0.08					
	PJ-KD6-R1	0.00	0.00	0.00	0.00					
	PJ-KD6-R2	0.00	0.00	0.00	0.00					
	PJ-KD6-R3	0.00	0.00	0.00	0.00					
PM_{10} (μg/m3)	S7-56A-R1	100.00	566.00	316.24	141.21					
	S7-43D-R1	30.00	123.00	41.00	23.38					
	S7-43D-R2	31.00	144.00	69.59	21.70					
	S7-43D-R3	62.00	273.00	158.57	67.69					
	SP15-R1	100.00	566.00	316.24	141.21					
	SP15-R2	65.00	370.00	208.06	66.63					
	KS-SU3-R1	115.00	556.00	342.82	123.45					
	PJ-KD6-R1	36.00	135.00	73.48	18.23					
	PJ-KD6-R2	17.00	163.00	42.64	28.17					
	PJ-KD6-R3	95.00	909.00	174.25	73.47					

Acknowledgements

The authors gratefully acknowledge the financial support of the Short Term Grant (2016) no. 304/PPBGN/6313274 established by the University Science Malaysia

References

1 L. Chatzidiakou, D. Mumovic, A.J. Summerfield, *What do we know about indoor air quality in school classrooms? A critical review of the literature*. Intelligent Buildings International, **4**(4), 228–259 (2012)

2 G. Zhang, J. Spickett, K. Rumchev, A.H. Lee, S. Stick, *Indoor Environmental Quality in a "Low Allergen" School and Three Standard Primary Schools in Western Australia*. Indoor Air. 16, 74-80 (2006)

3 M. Turunen, O. Toyinbo, T. Putus, A. Nevalainen, R. Shaughnessy, R., U. Haverinen-Shaughnessy, *Indoor environmental quality in school buildings, and the health and wellbeing of students*. International Journal of Hygiene and Environmental Health, 217(7), 733-739 (2014)

4 J. Ervasti, M. Kivimäki, I. Kawachi, S. Subramanian, J. Pentti, T. Oksanen, R. Puusniekka, T. Pohjonen, J. Vahtera, M. Virtanen, *School environment as predictor of teacher sick leave: data-linked prospective cohort study*. BMC Public Health, 12,770-781 (2012)

5

6 R.J. Soughnessy, U.H. Soughnessy, A. Nevalainen, D. Moschandreas, *A preliminary study on Association between ventilation rates in classrooms and students performance*. Indoor Air, 16, 465-468 (2006)

7 E. Azor-Martanez, Y. Gonzalez-Jimenez, M.L. Seijas-Vazquez, E. Cobos-Carrascosa, J. Santisteban-MartÃnez, J.M. MartÃnez-LÃ³pez, et al. *The impact of common infections on school absenteeism during an academic year*. American Journal of Infection Control, **42**(6), 632-637(2014)

8 N.M. Salleh, S.N. Kamaruzzaman, R. Sulaiman, F.M. Darus, *Associate Between Indoor Relative Humidity and Absenteeism at Refurbished Pre-Schools in Malaysia*. Advanced Science Letters, 19(10), 2999-3003 (2013)

9 N.A. Ayuni, J. Juliana M.H. Ibrahim, *Exposure to PM10 and NO2 and association with respiratory health among primary school children living near petrochemical industry area at Kertih, Terengganu*. Journal of Medical and Bioengineering, **3**(4):282 – 287 (2014)

10 P.C. Chua, J. Jalaludin, T.R Hamedon, N.M Adam, *Preschools' indoor air quality and respiratory health symptoms among preschoolers in Selangor*, Procedia Environmental Sciences 30 (2015)

11 N. Mohamad, M.T. Latif, M.F. Khan, *Source apportionment and health risk assessment of PM10 in a naturally ventilated school in a tropical environment*, Ecotoxicology and Environmental Safety, 124(2016)

12 American Lung Association, American Medical Association, US Consumer Product Safety Association and U.S. Environmental Protection Agency. (2004). Indoor air pollution. An introduction to health professionals. New York: Diane Publishing Co.

13 T.L Mohd, M.N. Zafirah, B.M. Mazlin, B.M. *"Lead in PM10 and in indoor dust around schools and preschools in Selangor, Malaysia"*. Indoor and Built Environment, DOI: 10.1177/1d20326X11402433 (2011)

14 P. Carrer, P. Wargocki, A. Fanetti W. Bischofc, E.D.O. Fernandes, T. Hartmann, S. Kephalopoulos, S. Palkonen, O. Seppänen, *What does the scientific literature tell us about the ventilation–health relationship in public and residential buildings?*. Building and Environment, **94** (1), 273–286 (2015)

15 G. Kats, *National Review of Green Schools. A Report for the Massachusetts Technology Collaborative*, 32-45(2005)

16 [MOE] Ministry of Education. (2014). Preliminary report: Malaysia education blueprint 2013-2025. Retrieved on 11 September 2014, at http://www.moe.gov.my/userfiles/file/PPP/Preliminary-Blueprint-Eng.pdf

17 J.J. Oppenheimer, *The Nose and its Relationship to the Eyes*. Curr. Allergy Asthma Rep. 10 (2010)

18 P. Guo, K. Yokoyama, F. Piao, K. Sakai, M. Khalequzzaman, M. Kamijima, et al, *Sick Building Syndrome by Indoor Air Pollution in Dalian, China*. International Journal of Environmental Research and Public Health, **10**(4), 1489-1504 (2013)

19 ASHRAE (2007) Ventilation for Acceptable Indoor Air Quality, Standard 62-2007, American Society for Heating, Refrigerating and Air Conditioning Engineers, Atlanta, GA.

20 DOSH [Department of Safety and Health]. Code of Practice on Indoor Air Quality. ISBN: 983-2014-51-4, JKKP: GP(1)05/2005. Ministry of Human Resources Malaysia. 2010.

21 P.M. Bluyssen, *The Indoor Environment Handbook*. London: RIBA Publishing. Association (2009)

Occupant Expectations on the Main IEQ Factors at Workspace: The Studies of Private Preschool Buildings

Naziah Muhamad Salleh[1a], Nuzaihan Aras Agus Salim[1], and Syahrul Nizam Kamaruzzaman[2]

[1]*University Science Malaysia, School of Housing Building & Planning, Building Survey Department, 11800 Penang, Malaysia*
[2]*University of Malaya, Faculty of Built Environment, Building Performance and Diagnostic, 50603 Kuala Lumpur, Malaysia*

Abstract. This study explores the relationship between the perceived performance of specific IEQ factors and occupants' overall satisfaction with their workspace. The Indoor Environmental Quality (IEQ) in buildings is one of the most important factors affecting the physical development of children. The early education is compulsory and escalated numbers of 4-6 years children has boosted the numbers of private preschool. Frequently its operate in premises that have been fully refurbished, either in housing schemes, commercial buildings or institutional buildings. This has invited the questions on the building capability to provide a good environment to the children during the learning activities. Malaysia still lacks studies on the modification of the quality of the internal environment for private kindergartens. Post-Occupancy Evaluation (POE) is one method that has been used satisfactorily in the study and effect of IEQ in kindergartens. This research focuses on identifying occupant satisfaction towards IEQ in selected refurbished pre-school or kindergarten buildings. The objectives of the study are to identify and determine the IEQ through feedback from building users. The study collected data on overall satisfaction and overall design importance through building inspection and questionnaires, the data obtained was analyzed to become a benchmark for the studies. 240 kindergartens in the whole country were selected for the study of IEQ.

Keywords: Indoor Environmental Quality, Post-Occupational Evaluation, Refurbishment, Preschool, Occupants Perception

1 Introduction

Several researchers have demonstrated that if people work in good environmental conditions, their productivity The and health improve (Gao et al., 2014, Rosen, 2009, Ambu, 2008, Uline and Moran, 2008, Stankovic and Stojic, 2007, Wargocki et al., 2005, Mendell, 2004, Schneider, 2002). The early aim is a need of achieving a good comfort level in educational buildings. It is due to the fact that children spending around 30% of their life span in schools (Lee and Chang, 1999).

Comfort and health levels are arising issues, while the occupants are in the buildings. Given the impact of the school environment on health, numerous researchers (Salleh et al., 2015; Montzamani et al., 2012; Puteh et al., 2014; Mendel et al., 2004; Salleh et al., 2013; Kabir et al., 2011) have identified building characteristics that affect the occupants' health and performance, including indoor air quality, temperature, odours or olfactory effects, visuals, acoustics, daylight and artificial lighting, ergonomics and space. Where several of these features are inadequate, it is difficult to identify the actual direct causes of occupant discomfort and health symptoms [Sulaiman et al., 2013; Bluysen, 2009; Chiang and Lai, 2002). However, some studies have already delved into indoor environmental conditions in educational buildings (Salleh et al., 2015; Gao et al., 2014, Sulaiman et al., 2013, Mydin et al., 2012, Nasir et al., 2011, Mumovic, 2009).

With increased awareness of occupants' productivity regarding environmental perceptions (Mendel et al., 2013; Jarman et al., 2004;Soughnessy et al., 2006), various models have been established to obtain the occupants' feedback: Leadership in Energy and Environmental Design (LEED), Green Building rating systems, Building Research Establishment Environmental Assessment Methods (BREEAM), Hong Kong Building

[a] Corresponding author: naziahmsalleh@usm.my

Environmental Assessment Method(HK-BEAM) and the TOBUS decision-making tool for upgrading office buildings. These systems are reliable in assessing and enhancing the IEQ of buildings. Indoor air quality, thermal comfort and acoustic performance of recently built schools (Mumovic, 2009) were evaluated by means of field measurements. Another study (Mors et al., 2011) focused on the actual thermal sensitivity and clothing insulation of children in none air-conditioned classrooms by means of both physical measurements and questionnaires.

In Malaysia most of the studies performed in public schools were organised by the government (Kamaruzzaman and Rasitah, 2011, Junaidah et al., 2010, Hussin et al., 2011). The buildings and facilities are well-prepared by the authorities that adhere to the standards provided. However, unlike the private preschools in Malaysia, the facilities and school buildings are provided by the individual person or non-government organisations (NGO). They are generally operated in re-adaptive buildings such as housing premises (KL City Hall, 2010). The bedrooms are converted to classrooms; therefore it used to be a challenge to provide an adequate environment for the children.

The broad aim of this paper is to determine the correlation between performance in refurbished private preschools and staffs' expectation on satisfaction and expectations levels, using the POE approach and guidelines. The objectives of this study are accordingly: i) to determine the satisfaction level of the building's occupants in terms of indoor environment and ii) to identify any correlation between individuals' expectations and their perception of the environment. The outcomes of the study are intended to be used as a benchmark for the improvement of private pre-schools in Malaysia.

2 Methods and Materials

In order to obtain a higher response rate, the questionnaire was designed to be only four pages long. It used a simple presentation that would not take long for the respondents to answer. Space was provided for the respondents to give additional comments. Before the questionnaire was sent out, it was piloted with 10 potential respondents from two pre-schools, whose comments and suggestions were taken into consideration in the final version.

Covering letters and questionnaires were distributed by hand to school administrative staff, to gain permission. The questionnaires were then collected within two weeks, during the building observation. Of the 1,020 questionnaires distributed around Peninsular Malaysia (to 5-7 staff, in 240 pre-schools; 711 classrooms); 521 were returned, of which 57 remained unanswered and 60 were incomplete. That is, 404 questionnaires were found to be useful. These figures are summarized in Table 1.

Table 1. The questionnaire survey

Description	Frequency
Number of questionnaires sent out	1020
Total returned questionnaires	521
Questionnaire sets returned without answers	57
Questionnaire sets returned uncompleted	60
Overall percentage returned	51.08%
Valid percentage returned	39.61%

The valid response rate recorded was almost 40%, which is sufficient for a social science study in Malaysia (Sarantakos, 2012). Krejic and Morgan's (1970) table to determine sample size shows that, for a population of 1,100, the minimum acceptable size is 285. In this study, the response rates of 408 can justifiably represent the total population.

The questionnaire has two sections: Section A asked for general information about age, gender, work experience, etc., and Section B explored attitudes to the 21 factors relating to the internal environment of the building as listed below (Q1 to Q21). The occupants were asked to rate their answers on a seven-point Likert scale for "User satisfaction" and "Degree of importance". The results were used to elicit an occupant satisfaction score and a benchmark for building satisfaction.

Q1	-Noise level	Q12	-Health when in the room
Q2	-Electric lighting	Q13	-Colours of the room
Q3	-Amount of daylight	Q14	-Attractiveness of the room
Q4	-Glare level in the room	Q15	-Control over local environment
Q5	-Distance to the windows	Q16	-Workspace
Q6	-Room temperature	Q17	-Privacy
Q7	-Ventilation	Q18	-Immediate colleagues
Q8	-Amount of air movement	Q19	-Management

Q9 -Freshness of your room
Q10 -Humidity level in the room
Q11 -Smell in the room
Q20 -Workplace, in general
Q21 - Outward appearance

Analysis methods utilised to answers to the questionnaire indicate the occupants' ratings of their satisfaction with the internal environment, and how important they find these environmental conditions. The first part of the analysis determines the satisfaction score of a building (an overall rating for a building's indoor environment) using Equation 1:

$$OLS = \left[\frac{\sum_{k=1}^{n} \sum_{j=1}^{m} i_{j,k} l_{j,k}}{mni_{max} l_{max}} \right] 100 \quad (1)$$

where

j = questionnaire number
k = question number
i = importance rating $1 < i > 7$
i_{max}= maximum value of i, (7)
l_{max}= maximum liking rating l (+3)
l = liking rating $-3 < l > +3$
m = number of completed questionnaires
n = number of questions in the score

The second part of the analysis provides a graphical representation of the totals for each answer. This is called a "user satisfaction fingerprint" and normalizes each question to a score from+100% to-100% (Levermore, 1994, 1994a, 1999, 2000), using Equation 2.

$$FLS = 100 \left[\frac{\sum_{j=1}^{m} i_{j,k} l_{j,k}}{mi_{max} l_{max}} \right] \quad (2)$$

The third part of the analysis is similar to the second. However, using Equation 3, a normalized individual score for each person can be calculated.

$$FLS = 100 \left[\frac{\sum_{k=1}^{n} i_{j,k} l_{j,k}}{ni_{max} l_{max}} \right]$$

This score will be between +100% and -100%.

A building walk-through form was used by the research team to compile an inventory of the building materials and contents of the classrooms. Data were collected to characterize the materials and condition of the ceiling, floor, interior walls, exterior walls, Heating, Ventilation and Air Condition (HVAC) equipment, and classroom contents. These were the items most frequently identified as sources of IEQ problems during the building inspections. Location of the building from the main road, surrounding activities, number of pupils in a classroom, mold, damp, the volume of the classrooms, and ventilation systems were also cited.

3 Results and Discussions

3.1 Building Observation

The building survey was correlated with the perceptions of the occupants towards the building. Of the 240 participating private pre-schools, 92.9% are operating in refurbished residential premises, mostly rented two-storey terraced houses (67.1%) with 100-199 m^2 of floor area. Two-thirds of the buildings are located on medium-local streets with moderate traffic flow, less than 50 m from the premises. Surprisingly, 96.6% of the classrooms (N=711) identified did not comply with the Uniform Building Bylaw (UBBL, 2012), requirement for space per person in classrooms. The descriptive analysis of building observations is detailed in Table 2. The most frequent occurrence of each characteristic is shown in bold.

Table 2. Descriptive analysis of preschool buildings

Categorical Variables (N=25)	Description	Percent (%)
Building types	Single Terrace	11.0
	Double Terrace	**67.1**
	Single Bungalow	4.6
	Double Bungalow	10.1
	Shop/office	0.8
	Others	6.3

Ownership of the building	Owned	34.2
	Rented	**65.8**
Age of the building (years)	<5	5.9
	5-10	28.7
	10-15	**33.3**
	15-20	14.3
	20-30	16.0
	>30	1.7
Programme	**Pre-school only**	**58.2**
	Daycare only	3.4
	Pre-school+Daycare	30.8
	Pre-school+Tuition	3.4
	Pre-school+Daycare+Tuition	4.2
Age of pupils	>4years	1.7
	4-6	**69.2**
	7-8	1.3
	>9	20.7
	Infants(0 months-3years)	7.2
Totalfloor area	<100m^2	9.7
	100-199m^2	**79.3**
	200-299m^2	7.2
	300-399m^2	3.0
	>399m^2	0.8
Classroom floor area (mean ± standard deviation)		15.51±5.98m^2
Open area per classroom (mean ± standard deviation)		5.30±1.53m^2
Numbers of pupils and teachers per classroom (mean ± standard deviation)		16 ± 4
Required space (minimum 2m^2 per person)	not compliedwith (<)	96.6
	complied with	3.4
Ventilation type	Dominantly natural	33.8
	Dominantly air-conditioned	**36.3**
	Both	29.9
Traffic type	Small local street	12.7
	Medium local street	**65.0**
	Large local street	16.0
	Expressway	6.3
Traffic density	Low	30.4
	Medium	**54.9**
	Heavy	14.8
Floor surface	**Smooth**	**84.8**
	Carpeted	15.2
Curtain type	**None**	**62.9**
	Blinds	7.6
	Textiles	29.5
Recent renovation	Yes	54.4
Cooking activities	Yes	70.9
Dampness	Yes	52.3
Mould growth	Yes	54.0
Vermin absent	Yes	76.0
Soft toys	Yes	85.2
Continue Table 2 **Categorical Variables (N=25)**	**Description**	**Percent (%)**
Floor cleaning	Twice a day	30.4
	Daily	69.6
Fan cleaning	Twice a week	8.0
	Weekly	85.0
	Monthly	26.5
Shelf cleaning	Daily	28.7
	Twice a week	19.8
	Weekly	51.5

 DOI: 10.1051/matecconf/20166600120

Table cleaning	Daily Twice a week Weekly	66.2 28.7 5.1
Toilet cleaning	Daily Twice a week Weekly	78.5 19.8 1.7
Curtain cleaning	Weekly Monthly Bi-annually Never	6.3 46.4 26.6 20.7

3.2 Staff's Expectation

The staff offered their opinions on the degree of each IEQ factor experienced in their building. Figure 1 shows the satisfaction scores for all the buildings, with question 1 (noise) having the worst score and the only negative one, and question 18 (colleagues) the best The highest-scoring satisfaction subject areas are shown in Table 3.

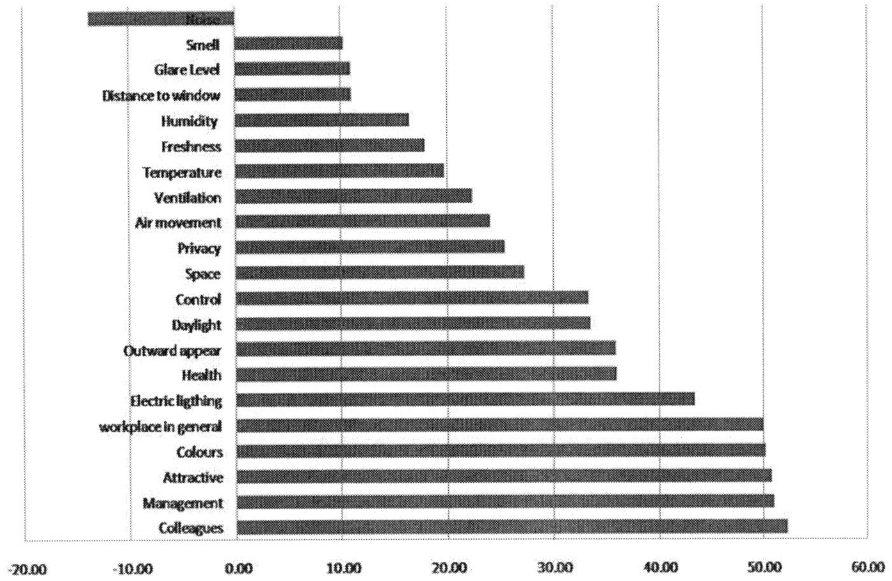

Figure 1, Audit satisfaction fingerprint

Table 3. Most satisfactory features

Question No.	Question subject	Score
18	Colleagues	+52.31
19	Management	+51.10
14	Attractive	+50.80
13	Colours	+50.22
20	Workplace in general	+50.02

As most of the pre-schools have only 4 to 7 staff members, it is perhaps not surprising that the highest scores are in relation to immediate colleagues and management; there are strong bonds between the staff and ready interaction with management. Site observation indicated that walls painted in cheerful colours and decorated with various murals were attractive to pupils, and this feature was next highest on the list.

Even though all but one of the scores for the questionnaire are on the positive side, some are considered less favourably than others, as shown on Table 4.

Table 4. Least satisfactory features

Question No.	Question subject	Score
1	Noise	-13.77
11	Smell	+10.22
4	Glare level	+10.91
5	Distance to window	+10.96
10	Humidity	+16.51
9	Freshness	+17.95
6	Temperature	+19.73
7	Ventilation	+22.36
8	Air movement	+24.00

The external noise level had the lowest score (-13.77), not surprising given that all the buildings were in urban areas with a high volume of traffic, and in their previous roles were connected to main roads. There was also noise from within their own premises during teaching periods.

The buildings' occupants also gave less favourable scores for smell (+10.22), humidity (+16.50) and freshness (s+17.95), the three elements categorized as air quality factors. Overcowding in classrooms which had previously been bedrooms (9 m2 – 23 m2) contributed to breathing difficulties and a smelly environment.

Air movement within the building was given a higher score (+24.00), followed by ventilation (+22.36) and temperature (+19.73), whether or not resulting from air conditioning.

3.3Scores for Factors from the Questionnaires

Analysis of variance (ANOVA) is applicable if the research has two groups or more to compare. Based on the five factor scores generated, a further analysis on the differences between each factor in the questionnaire was carried out. Using SPSS, One-way ANOVA and t-tests were performed to determine the statistical significance of differences among the means of the groups in six selected variables.

a) Analysis one: Type of building

The questionnaire had five sub-groups based on building types, as detailed in Table 6. An analysis of variance revealed significant differences at the 5% level in the appearance [F(4,403)=2.73, p= 0.03] and intrusion[F(4,403)=2.67, p= 0.03] factor scores. On the other hand, there are non-significant differences at the 5% level in the air quality, general and workplace scores. Post hoc analyses using the Tukey HSD for significance indicated that the average number of errors was significantly lower in the bungalow condition towards general and intrusion factors than in the other four factors.

b) Analysis two: Type of ventilation

There were two major groups of ventilation: natural (n=183) air conditioning (n=221). The few hybrid/mixed ventilation systems were considered as air-conditioning, because most of the pre-schools turn on the air conditioning as early as 10 a.m., or whenever they felt the temperature increased in the building.

The t-test for independent samples revealed non-significant difference at the 5% level for other factors. This means that there is little difference between the two types of the ventilation.

c) Analysis three: Work category

The four sub-groups in the work category (see Table 2) were clerk (11), teacher (287), manager (79) and others (27; (includes teacher assistant, housekeeper, etc.). An analysis of variance revealed significant differences at the 5% level in intrusion [F(3,403)=4.9, p= 0.02]. However, there were non-significant differences at the 5% level in the other factors. Post hoc analyses using the Tukey HSD for significance indicated that the average number of clerks had a higher opinion of intrusion than the "others" category.

d) Analysis four: Age

The six sub-groups under age are detailed in Table 6. An analysis of variance revealed non-significant differences at the 5% level in all the factors, indicating that allege groups have the same opinion about the environmental factors.

e) Analysis five: Number of people in the building

See Table 6 for the four sub-groups based on number of people in the building. An analysis of variance revealed significant differences at the 5% level in the air quality [F(5,403)=7.22, p= 0.00], appearance [F(5,403)=2.43, p= 0.03] and workplace [F(5,403)=4.81, p= 0.00] factors. Post hoc analysis using the Tukey HSD for

DOI: 10.1051/matecconf/20166600120

significance indicated that the mean scored for group of people in the building was lower compares to the group of 2 people towards the perception of air quality and the workplace in general. It is clear that crowded buildings will result in an unpleasant environment, linking freshness, smell, ventilation and temperature of the building, and affecting the opinion on the workplace in general.

Moreover, Figure 2 explains the negative relation between the dissatisfaction/dislike levels of perception and the total hours staff spend in the building. Suprisingly results for satisfaction level had a similar range, where time spent in the building alsohad a negative relation to the satisfaction level. In other words, the more time spent in the pre-school building, the greater the satisfaction of the occupants. Nevertheless,for some occupants more time spent in the building generated unpleasant feelings towards it. A similar result was found between the perception of the workplace and the class floor area.

Table 5 shows that the analysis of covariance for the perception of staff towards the workplace with the classroom floor area, time spent in the building and the duration of staff work in the same building has a non-significant relationship. This explains why these variables did not affect the satisfaction level while working in the pre-school buildings.

Table 5. Covariance Analysis

	Dislike		Neutral		Like				
Scale	Mean	SD	Mean	SD	Mean	SD	F	df	*p* value
Class floor area	16.48	5.26	16.01	3.88	15.12	4.50	1.23	2.00	0.05
Total hours spent in building per day	7.20	2.33	7.74	2.27	8.08	3.04	2.70	1.07	0.07
Number of years working in the building	5.75	4.90	5.54	3.43	6.19	5.17	0.73	2.00	0.48

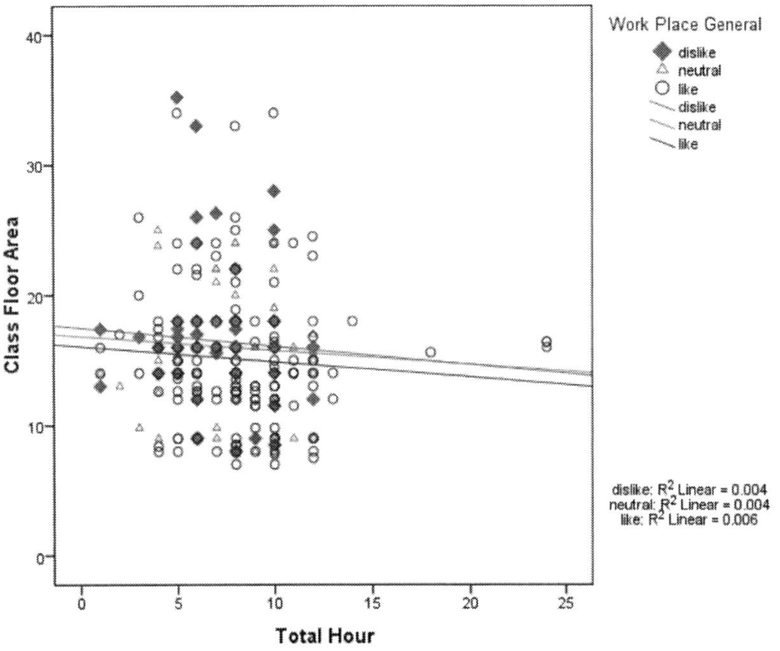

Figure 2. Scatter Plot of Total Hours Spend in the Building and the Class Floor Area with the Perception of the Building in General

f) Analysis six: Time spent in the building

The questionnaire had four sub-groups for the number of hours spent in the building each day. An analysis of variance revealed significant differences at the 5% level in the general factor [F(4,403)=5.82, p= 0.00]. Post hoc analyses using the Tukey HSD for significance indicated that the mean score of the group spending 4-6 hours in the building was lower than that for those spending more than 10 hours there. This is due to the long time being exposed to the same environment, which might also affect the liking variable.

g) Analysis seven; Classroom floor area and space requirement

There were four sub-groups here :<10 m^2(n=46), 10-20m^2(n=315), 20-30m^2(n=39) and >30m^2(n=4). A one-way between subjects ANOVA was conducted to compare the effect of workplace/classrooms' floor area with the five factors of IEQ. There was a significant effect of floor area towards the classroom's external appearance at the p<.05 level for all four groups [F(3, 339) = 2.837, p = 0.04]. Post hoc comparisons using the Tukey HSD test indicated that the highest mean score was for classrooms with a floor area more than 30 m^2. The greater the floor area, the better the appearance. However, there is no significance between the space requirements per person and contribution to an unpleasant environment among the five factors determined earlier. Other variables caused discomfort to the occupants.

Table 6. Analysis of Total Variance

SAMPLE	N	AIR QUALITY	GENERAL	APPEARANCE	INTRUSION	WORKPLACE
Building type	F_r	*0.77	*0.91	*0.03	*0.03	*0.51
Ventilation type NaV=183 AC= 221	t_I	**0.09	**0.80	**0.41	**0.06	**0.62
Work category	F_r	*0.72	*0.08	*0.80	*0.02	*0.86
Age	F_r	*0.07	*0.06	*0.10	*0.14	*0.82
Number of people in Building	F_r	*0.00	*0.90	*0.03	*0.16	*0.00
Hours in building	F_r	*0.69	*0.00	*0.46	*0.17	*0.07
Classroom floor area	F_r	*0.23	*0.11	*0.04	*0.69	*0.11
Space/person requirement(<2m^2/person) Not complied with: 372 Complied with: 32	t_I	*0.14	*0.25	*0.14	*0.14	*0.45

F_r=F ratio One way Analysis of Variance, t_I = t-test for independent sample
*p<0.05 **p<0.01

4 Conclusions

Most refurbished preschool building in Malaysia were found not to comply with minimum space regulation concerning children, teachers and staff, leading to the uncomfortable conditions and spreading out the infectious illness. It is suggested that the authorities to take this issues into consideration and provide the sufficient space for the occupants. It is also to put into account the location of the classrooms and schools when the premises being converted to the education buildings.

From this research, it can be concluded that Post Occupancy Evaluation emphasizes the significance of improving Indoor Environment Quality in the refurbished preschool's buildings. There is still lack of awareness toward the significance of IEQ, and half of the respondents

the majority were the buildings' owners) admitted they never bothered about IEQ in their buildings. Some also admitted that they had refurbished the building without referring to the guidelines from the local authority and might as well invited some contaminations and organics effluences such as polluted gases, mold and fungi into the building.

The current scenario preschool buildings in Kuala Lumpur are concerns on minimalist of IEQ consideration, neglecting to study the buildings' characteristics and the indoor environment for early education. Nevertheless, the study found that half of the building owners followed the guidelines from the local authority, and some buildings had the initiative to implement the guidelines. This is one of the approaches that are quite relevant, but proper guidelines and standards still need to be referred to with respect to IEQ and the renovation of the buildings.

It is recommended that POE should be undertaken in order to highlight the importance of IEQ and ensure its consideration in the refurbishment of pre-school buildings in order to improve early education for future generations. It is also recommended that there should be further analytical research into IEQ in refurbished pre-school buildings and study the correlation with occupants (the end-users of buildings used for early education) in Malaysia.

References

1. J. Gao, P, Wargocki, Y. Wang, *Ventilation system type, classroom environmental quality and pupils' perceptions and symptoms*. Building and Environment, **75**(0), 46-57 (2014)
2. K.G. Rosěn, G. Richardson, *Would removing indoor air particulates in children's environ-ment reduce rate of absenteeism- a hypothe-sis*. The Science of the Total Environment, 234, 87-93 (1999)
3. S. Ambu, W, Chu, J. Mak, S, Wong, S.Chan., S. Wong, *Review article: Environmental Health And Building Related Illnesses [Electronic Version]. IeJSME, 2*, 11-18. Retrieved 14 March 2012, from http://web.imu.edu.my/ejournal/approved/eJournal_2.S1_11-18.pdf
4. C. Uline, M. Tschannen-Moran, *The walls speak: The interplay of quality facilities, school climate, and student achievement*. Journal of Educational Administration, **46**(1), 55-73 (2008)
5. D. Stankovic, J. Stojic, J. *Psycho-developing needs of children and spatial features for children's stay*, Architecture and Civil Engineering, **5**(1), 71 – 75 (2007)
6. P. Wargocki, D.P. Wyon, B. Matysiak, S. Irgens, Proceedings of Indoor Air 2005. *The Effects of Classrooms Temperature and Outdoor Air Supply Rate on the Performance of School Work by Children.* The 10th International Conference In Indoor Air Quality and Climate, Beijing
7. M.J.Mendell, E. Eliseeva, M.M. Davies, M. Spears, A. Lobscheid, W.J.Fisk, M.G. Apte, *Association of classroom ventilation with reduced illness absence: a prospective study in California elementary schools*. Indoor Air, 23, 515–528.Lee, S., & Chang, M. (1999). Indoor air quality investigations at five class-rooms. Indoor Air, 9, 134 - 138.
8. V. Straka, M. Aleksic, Post-Occupancy Evaluation, PLEA2009 - 26th Conference on Passive and Low Energy Architecture, Quebec City, Canada, 22-24 June 2009.
9. S. Lee, M. Chang, *Indoor air quality investigations at five classrooms*. Indoor Air, 9, 134 – 138(1999)
10. N.M. Salleh, S.N. Kamaruzzaman, M. Riley, E.M.A. Zawawi, R. Sulaiman, R. *A quantitative evaluation of indoor environmental quality in refurbished kindergarten buildings: A Malaysian case study*. Building and Environment 94, 723-733(2015)
11. A. Montazami, M. Wilson, N. Fergus, *Air-craft noise, overheating and poor air quality in classrooms in London primary school*. Build Environ, 52(2012), 129-141.
12. Puteh, M., Adnan, M., Ibrahim, M.H., & Mohamed Noh, N. (2014) . Proceedings from 5th World Conference Educational Sciences-WCES 2013. An Analysis of comfortable teaching and learning environment: community re-sponse to climate change in school. Social and Behavioral Sciences, 116, 285 – 290.
13. Kabir, E., Kim, K.H., Sohn, J.R., Kweon, B. Y., & Shin, J. H. (2011). Indoor air quality assessment in child care and medical facilities in Korea. En-viron Monit : DOI 10.1007/s10661-011-2428-5
14. Sulaiman, M.A, Wan Yusoff, W.Z., Mahamud, N.A., & Hamid, N.F. (2013). Proceedings from 5th In-ternational Conference on Humanities and Social Sciences. Impact of Indoor Environ-mental Quality (IEQ) on Academic Building to Psychology and Social of StudentsFaculty of Liberal Arts, Prince of Songkla University.
15. P.M. Bluyssen, *The Indoor Environment Hand-book.* London: RIBA Publishing (2009)
16. V.S.Chithra, S.M. Shiva Nagendra, *Indoor air quality investigations in a naturally ventilated school building located close to an urban roadway in Chennai, India.* Building and Environment 54, 159-167 (2012)
17. Mydin, M.A.O., Abdul Rahim, N., Ismail, M. (2012). "Evaluation of Indoor Environment Quality (IEQ) of Educational Buildings" Journal of Environment and Earth Science,**2**(8), 46- 54
18. Nasir, A. R. M., Musa, A. R., Che-Ani, A. I., Utaberta, N., Abdullah, N. A. G., & Tawil, N. M. (2011). Identification of iIndoor environmental quali-ty (IEQ) parameter in creating conducive learning environment for architecture studio. Procedia Engineering, **20**(0): 354-362
19. Mumovic, D., Palmer, J., Davies, M., Orme, M., Ridley, I., Oreszczyn, T., et al. (2009). Winter indoor air quality, thermal comfort and acoustic per-formance of newly built secondary schools in England. Building and Environment, 44(7), 1466-1477
20. D. Jarman, L. Webb, T.C. Chan, T.C. *A beautiful school is a caring schools. School Bussiness Affairs*. Retrieved 11 September 2011,from www.sba_june_04_beautiful_schools.pdf/
21. R.J.Soughnessy, U,H. Soughnessy, Nevalainen, A., & Moschandreas, D. (2006). A preliminary study on Association between ventilation rates in classrooms and students performance. Indoor Air, 16, 465-468.
22. Mors, S. T, Hensen, J. L.M., Loomans, M. G. L. C. & Boerstra, A. C. (2011). Adaptive thermal com-fort in primary school classrooms: Creating and validating

PMV-based comfort charts. Building and Environment, 46, 2454-2461.
23. S.N. Kamaruzzaman, S. N., R.A. Razak, *Measur-ing in-door air quality performance in Malaysi-an government kindergarten*. Journal of Build-ing Performance, **2**(1) (2011)
24. Junaidah, Z., Lye, M.S., Jamal, H.H., & Zailina, H. (2010). Allergy to air pollution and frequency of asth-matic attacks among asthmatic children. American-Eur-asian Journal of Taxio-cological Sciences. **2**(2), 83-92.
25. N.H.M. Hussin, L.M. Sann, M.N. Shamsudin, Z. Hashim, *Characterization of Bacte-ria and Fungi Bio-aerosol in the Indoor Air of Selected Primary Schools in Malaysia*. Indoor and Built Environment (2011)
26. Kuala Lumpur City Hall. (2010), Senarai tadika swasta sediaada di WPKL. Jabatan Perancang Bandar. Un-published
27. S. Sarantakos, Social Research, Palgrave, New York, NY. (1998).
28. Krejcie, R. V., & Morgan, D. W. (1970). Determining sample size for research activities. Educational and Psy-chological Measurement, 30, 607-610.
29. Levermore G.J., Leventis M. (1998), Occupants feed-back using a questionnaire rating liking and importance of up to 24 factors, CLIMA 2000 Brussels.
30. Levermore G.J.,(2000). Building energy management system: Application to low energy HVAC and natural ventilation control (2nd ed). UK: Department of Build-ing Engineering, UMIST, E&FN Spoon
31. Uniform Building By Law Act 1333, Laws of Malaysia, 16th ed., MDC Publisher, Kuala Lumpur, (2012).

Key Element Performance In Occupational Safety And Health Management System In Organization (A Literature)

Nuzaihan Aras Agus Salim[1,a], Naziah Muhamad Salleh [1] and Zuraihana Ahmad Zawawi[2]

[1]*School of Housing, Building and Planning, Universiti Sains Malaysia, 11800 Penang, Malaysia*
[2]*Faculty of Architecture, Planning and Surveying, Universiti Teknologi MARA Perak, 32610 Seri Iskandar, Perak, Malaysia*

Abstract. Setting an effective safety and health management system is crucial in order to reduce problem relating to accident and ill in management organizational. It is involve with multiple level of management and stakeholders who empower the organization to the management in handling the safety and health cases and issues in organizational. It is necessary to prepare a well knowledge about safety and health management systems and preparing the framework for setting a certain scale in measuring its performance in this area. The successful or failure of management does showing the capability of the organization in delivering the responsible to management levels [1]. The problem in safe work issues and practices cause by the management commitment and involvement that create improper safety program and procedures, and this crisis keep continuing till present [2]. This paper describes about key element of safety and health management system and measuring the performance in order to get an effective management system in organization that describes the process in achieving effectiveness in management. The literature review will be conducted through the data collection from research findings and defined the strong character of key element in which focusing on measuring performance. A guide on key element performance in occupational safety and health management system is specifically drawn to prepare for a future research.

1 Introduction

Safety and health principles are universal, but how much action is needed will depend on the size of the organization, the hazards presented by its activities, the physical characteristics of the organization, products or services, and the adequacy of its existing arrangements. [3] stated that there were small mistakes can have a major effect and disaster do not have borders or nationalities, which is why nation should unite and promote occupational health and safety vigorously to reach the common goal of quick warning, prevention and protection systems for all. [4] rightfully mention that occupational health and safety is a complex international problem for management and society, and that it must always be a top priority.

[5] supports the view that the process of managing occupational health and safety are no different from that of other management activities, but is just that much more difficult because of its complexity. [6] said directors and senior managers must decide what is to be achieved by the organization and set in motion the necessary mechanism to achieve it. They must monitor progress and make changes when the desired objectives are not being achieved.

2 Key Elements of a Safety and Health Management System

The safety and health management is a common issue regarding the organization and workplace. [7], [8], [9], [10] described that the obligation of the managerial department in presenting on safety workplace and management in an organisation. However, a specific role model should be developing to in monitoring the organisation in safety and health management.

Accident, disasters and hazardous in workplace contribute to minor injury or even death to the workers. The organisation and management is always responsible and liable in the event of an accident at work. However, McDonald et.al. mentioned that there is not clear of any factor in neither a generally accepted model of the organisational in safety aspects, nor research on particular organisational forms and character of accident occurrence or accident description [11]. A blank space between between organisational forms and cases of accident occurrence could be established in a systematic monitoring and control. Organisation has a strong complex dynamic entity to ensure the application in functioning of safety.

[1] Corresponding author: nuzaihanaras@usm.my

DOI: 10.1051/matecconf/20166600121

Safety Management becomes more important till today. It is a heart of the organisation to protect and secure their workers for harmful. The production of the core business is relying on this precious engine to run the business well. Historically, Hale et al., described that the development of safety management is establish since 1970's with followed reasons [12]:

I. Regulatory interest was stimulated by the changing philosophy which led to the framework legislation of the 1970s and 1980s in many European countries and ultimately to the consolidating European Framework Directive of 1989. Such legislation moved the emphasis from detailed technical safety concerns to issues of decision making and management formulated within a safety policy.

II. This trend was strengthened by a series of official reports following major disasters. The main emphasis in these reports was on the failings of management to ensure that their plant or activities were designed, operated and maintained with sufficient safety. Recent amendments to the post-Seveso Directive in 1994 require major hazard companies to have auditable safety management systems (SMS) and another directive have set up a voluntary European system for environmental management auditing.

III. In parallel with regulatory concerns with assessing safety management systems has been the increasing desire of governments to withdraw from their detailed regulatory tasks. The International Standards Organization standards on quality management systems have been, or are being used as a basis for devising standards for SMSs or related environmental management systems. The third party certification system which has grown up to assess compliance with the standards is increasingly seen, certainly in the Netherlands, as an opportunity for privatising some government assessment tasks.

The Figure 1 shows that the successful occupational safety and health management application in an organization on safety and health (Adapted from the Health and Safety Executive, 1997. *Successful health and safety management*. London: HSE Books)

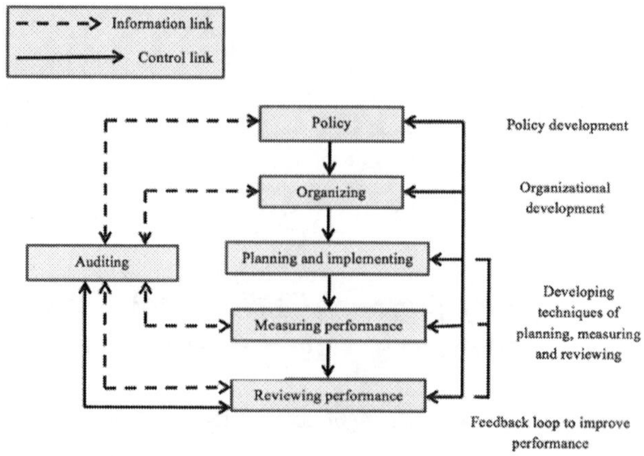

Figure 1: Adapted from the Health and Safety Executive, 1997. Successful health and safety management. London: HSE Books

3 Policy and Commitment

According Health and Safety Executive, 1997, the organization should prepare an occupational safety and health policy programme as part of the preparation of the Safety Statement required by section 20 of the 2005 Act [13]. Effective safety and health policies should set a clear direction for the organization to follow. They will contribute to all aspects of business performance as part of a demonstrable commitment to continuous improvement. Responsibilities to people and the working environment will be met in a way that fulfills the spirit and letter of the law. Cost-effective approaches to preserving and developing human and physical resources will reduce financial losses and liabilities. In a wider context, stakeholders' expectations, whether they are shareholders, employees or their representatives, customers or society at large, can be met.

A research on Management of Occupational Health and Safety in Romania supports that every Organization essentially prepared and documented a health and safety policy to communicate their commitment to health and safety, and their arrangements and responsibilities for implementation. This must be consistent with health and safety policy [14].

i. The policy must include:
   - Clear statement of overall health and safety objectives, with reference to Zero Harm
   - commitment to the prevention of work related injuries and illnesses
   - a commitment to continual improvement
   - a commitment to compliance with all applicable legislation, PB Standards and accepted industry standards

ii. The policy must be approved and signed by the head of the organization.

iii. The policy must be communicated across the organization.

The policy must be reviewed annually and revised where necessary to ensure that it remains relevant and appropriate to the organization.

4 Organizing

Jeffrey et.al. defined organizing as the process of creating a structure for the organization that will enable its human resources to work together effectively towards its objectives [15]. Organization hence refers to the result of the organizing process. According Health and Safety Executive, 1997 stated that the organization should formulate a plan to fulfill its safety and health policy as set out in the Safety Statement [13]. An effective management structure and arrangements should be put in place for delivering the policy. Safety and health objectives and targets should be set for all managers and employees.

I. Implementation and operation

For effective implementation, the organization should develop the capabilities and support mechanisms necessary to achieve its safety and health policy, objectives and targets. All staff should be motivated and empowered to work safely and to protect their long-term health, not simply to avoid accidents. The arrangements should be

- Underpinned by effective staff involvement and participation through appropriate consultation, the use of the safety committee where it exists, and representation systems.
- Sustained by effective communication and the promotion of competence which allows all employees and their representatives to make a responsible and informed contribution to the safety and health effort. There should be a planned and systematic approach to implementing the safety and health policy through an effective safety and health management system. The aim should be to minimize risks. Risk assessment methods should be used to determine priorities and set objectives for eliminating hazards and reducing risks. Wherever possible, risks should be eliminated through the selection and design of facilities, equipment and processes. If risks cannot be eliminated, they should be minimized by the use of physical controls and safe systems of work or, as a last resort, through the provision of personal protective equipment (PPE). Performance standards should be established and used for measuring achievement. Specific actions to promote a positive safety and health culture should be identified. There should be a shared common understanding of the organization's vision, values and beliefs. The visible and active leadership of senior managers fosters a positive safety and health culture.

II. Workplace safety and health management

Priorities and set objectives for eliminating hazards and reducing risks. Wherever possible, risks should be eliminated through the selection and design of facilities, equipment and processes. If risks cannot be eliminated, they should be minimized by the use of physical controls and safe systems of work or, as a last resort, through the provision of personal protective equipment. Performance standards should be established and used for measuring achievement. Specific actions to promote a positive safety and health culture should be identified. There should be a shared common understanding of the organization's vision, values and beliefs. The visible and active leadership of senior managers fosters a positive safety and health culture.

5 Measuring Performance

Ramroop mentioned that measurement is essential to maintaining and improving occupational health and safety performance. Organizations need to measure what they are doing to implement their occupational health and safety policy, to assess how effectively they are controlling risks, and how well they are developing a positive occupational health and safety culture [6].

Due to Health and Safety Executive, 1997, the organization should measure, monitor and evaluate its safety and health performance [13]. Performance can be measured against agreed standards to reveal when and where improvement is needed. Active self-monitoring reveals how effectively the health and safety management system is functioning. Self-monitoring looks at both hardware (premises, plant and substances) and software (people, procedures and systems, including individual behavior and performance). If controls fail, reactive monitoring should find out why they failed, by investigating the accidents, ill-health or incidents that could have caused harm or loss. The objectives of active and reactive monitoring are:

• To determine the immediate causes of substandard performance;
• To identify any underlying causes and implications for the design and operation of the safety and health management system. Longer-term objectives should also be monitored.

6 Auditing and Reviewing Performance

The Health and Safety Executive, 1997, stressed that the organization should review and improve its safety and health management system continuously, so that it is overall safety and health performance improves constantly [13]. The organization can learn from relevant experience and apply the lessons. There should be a systematic review of performance based on data from monitoring and from independent audits of the whole safety and health management system. These form the basis of complying with the organization's responsibilities under the 2005 Act and other statutory provisions. There should be a strong commitment to continuous improvement involving the development of policies, systems and techniques of risk control. Performance should be assessed by:

- Internal reference to key performance indicators;
- External comparison with the performance of business competitors and best practice in the organization's employment sector. Many companies now report on how well they have performed on worker safety and health in their annual reports and how they have fulfilled their responsibilities with regard to preparing and implementing their safety statements. In addition, employers have greater responsibilities under section 80 of the 2005 Act on 'Liability of Directors and officers of undertakings' which requires them to be in a position to prove they have proactively managed the safety and health of their workers. Data from this 'Auditing and Reviewing Performance' process should be used for these purposes.

7 Discussions

In this article show that implementing effective occupational safety and health management system must follow a suitable guidelines to maximize the accident and ill health in organizational. The guidelines or policies must set as clear direction to getting effective in term of safety and health. Lastly, the study concluded that the use of key elements of a safety and health management system would contribute to successful occupational health and safety in an organization.

References

[1] Archie B.C. and Ann K.B., (2014), Business and Society: Ethics, Sustainability, and Stakeholder Management 9th Edition, *Cengage Learning*, United State of America, ISBN-13: 978-1285734293.

[2] Coombs W.T., (2015), Public relations' "Relationship Identity" in research: Enlightenment or Illusion, Public Relations Review; A Global Journal of Research and Comment, *Volume41, Issue 5*, Pg. 689-695.

[3] Ramroop S., McCarthy J.J. and Naidoo K., (2004). Successful Occupational Health and Safety: A Management Perspective. Proceedings of 8th World Congress on Environmental Health. *Document Transformation Technologies*, South Africa, pp 22-27.

[4] Steenkamp and Van Schoor, (2008), The Quest For Quality of Work Life; A TQM Approach, *Juta Education*, Lansdowne, ISBN 0 7021 5191 4.

1. [5] Boyle D., Larson C., Parker D., and Luisa P.B., (2000), Medical, personal, and occupational outcomes for work-related amputations in Minnesota, American Journal of Industrial Medicine, *Vol. 37, Issue 5*, Pg. 551-557.

[6] Ramroop S., and McCarthy, J. J., and Naidoo, K. (2004) Successful Occupational Health And Safety: A Management Perspective. Environmental Health Department, Ethekwini Municipality.

[7] Heinrich HW, Peterson D & Roos N (1980), Industrial Accident Prevention, 5th Edition, Mcgraw Hill, New York.

[8] Frank E. Bird, Jr. and Robert G. Loftus. (1976), Loss Control Management, http://trove.nla.gov.au/version/12792176

[9] Petersen D.,(1978), Techniques of Safety Management, McGraw-Hill, New York (1978) 2nd Ed.

[10] Loomes M., Ridley D.,Diana K., (1994), Cognitive and Organisational Aspects of Design, Technology and Assessment of Safety-Critical Systems, Springer-Verlag London Limited.

[11] N. McDonald *, S. Corrigan, C. Daly, S. Cromie (2000). Safety management systems and safety culture in aircraft maintenance organisations, Journal of occuppational Accident, Volume 34, Issues 1–3, Pages 1-258.

[12] Hale A.R., Heming B.H.J.,and Kirwan B., (1997), Modelling of Safety Management Systems, Journal of Occupational Accidents, Volume 26, Issues 1–2, June–July 1997, Pages 121–140.

[13] Health and safety executive HSE. (1997).*Successful Health and Safety management*. London: HSE Books.

[14] Marius M., The Management of Occupational Helath and Safety in Romania, https://core.ac.uk/download/files/153/6695246.pdf.

[15] Jeffrey S. H. and Edward R. F. (1999), Stakeholders, Social Responsibility, and Performance: Empirical Evidence and Theoretical Perspectives, Academy of Management Journa, Vol. 42, pg. 479-485.

[16] Jeffrey S. H. and Edward R. F. (1999), Stakeholders, Social Responsibility, and Performance: Empirical Evidence and Theoretical Perspectives, Academy of Management Journa, Vol. 42, pg. 479-485.

Design Failure Affecting Maintenance Management on Public Higher Education Institution in Malaysia

Nuzaihan Aras Agus Salim[1,a], Naziah Muhamad Salleh [1] and Nurul Fadzila Zahari[2]

[1]*School of Housing, Building and Planning, Universiti Sains Malaysia, 11800 Penang, Malaysia*
[2]*Faculty of Architecture, Planning and Surveying, Universiti Teknologi MARA Perak, 32610 Seri Iskandar, Perak, Malaysia*

Abstract. The issue of government building defect is keeping on reporting in the media and arising since 2007. These issues of building defects, maintenance and management have existing and it is deliberated up to the parliament level. The government keep continue increased the maintenance allowance in National Budgetary in every year. However, the building disasters and failures keep going develop their number of cases in Malaysia. Most of the issues are related with a weak maintenance management and faulty in design. This reflects to a poorly building condition, create discomfort and danger environment to the building users. Besides, it will result to a low quality building condition and reflect to a weak building maintenance management. Building defects arise through inappropriate or poor design, specification, construction and it will give an impact to maintenance management itself [1]. Errors in the design of the building are being debated and becoming a reason of major factor in building defect reproduction [2]. Thus, this issue will discover the major cause in building design failure that develop the building defects and verify the impact of design defects towards building maintenance management. The study involved with the public higher education institution in Malaysia focusing on main campuses and the distribution of questionnaire to the facility and maintenance management department. This research is aimed to produce a better understanding on the impact of building design failure towards the institution facilities' condition.

1 Introduction

In this era of the transformation of manufacturing technology and transportation system, it does lead to a different lifestyle and work style that resulted to the development of better infrastructure beside of old buildings in the city began to fall behind due to its outdated functionality and requirements [3]. Maintenance is the core agenda in preserving the building and the survival of this property. On the other hand, a complicated development process does bring into an inconvenience procedures to the designers and developers. This process involved with various phases of development, the provision of law, the development cost and expertise that are required a proper maintenance in future. The main problems in maintenance such as inadequate financial, bad management, and poor building design and detailing are usually raised by the maintenance management teams [4].
The relationship between maintenance and building design is closely to avoid any harmful during construction stage or during the building is occupied. It has been aligned in four sectors to avoid the need for unplanned maintenance at the post-occupation stage [5]. The four sectors is ruled into:

1. The main fabric which includes walls, floors, roofs, doors and windows;
2. Internal finishes which includes ceiling and wall finishes as well as floors;
3. Special design features such as decorative elements for the doors, windows, glass, air vents and special brick and stone work;
4. The fourth and the last sector is cleaning and house-keeping of all building components.

In addition, maintenance team shall be monitored during the design stage to find the solution under such condition is usually total replacement of the building's cladding [6]. A study has been carried out to find the critical parts in design that seldom brings a complicated maintenance issues such as poor material selection, a poor ventilation flow, a poor design on wet area, floor design weakness and ignorance of life span of materials and elements of the buildings do result to an expensive defects in their buildings [7].

Public Higher Education Institution in Malaysia takes a serious action to ensure all the facilities in a proper and safety condition. These institutions' facilities are control by facilities and maintenance department. Most of the maintenance work and operation is based on contingency and ad-hoc planning. However, the management of cleaning, building services contracts and landscaping are run by the outsource contractors through tendering process [8]. Therefore, this paper is to review the impact of building design failures to the Public Higher Education

[1] Corresponding author: nuzaihanaras@usm.my

DOI: 10.1051/matecconf/20166600122

Institution in Malaysia that affecting the maintenance management perspectives.

2 Literature Review

Building failure occurs when the building loses its ability to perform its intended (design) function. Hence, building failures can be categorized into the two broad groups of physical (structural) failures (which result in the loss of certain characteristics, e.g., strength) and performance failures (which means a reduction in function below an established acceptable limit) [9]. These failures create a certain symptoms called defects onto physical building appearance [10].

Building defect is a failure or mistake that reduces the value of the building and may cause a dangerous condition to building user's if neglected. A defect may arise from some carious factors such as poor workmanship, design fault, wrong selection of material and so on. Building defects will result in various implications [11]. It may reflect the negligence occurred, and also illustrates the lack of concern on problems arising from the construction work on the building as well building refurbishment. A defect may be considered to be failing or unsatisfactory in the functional performance, user or statutory requirements of a building and might manifest itself within the fabric, structure, services and facilities of the affected to the building [12]. Ransom states that despite calls for closer integration of design and construction, the role remains essentially separate [13]. Figure 2.0 below revealed that over 90% of building defect could be explained partly or wholly as the result of readily identifiable faults in design and construction that could have been foreseen and hence prevented.

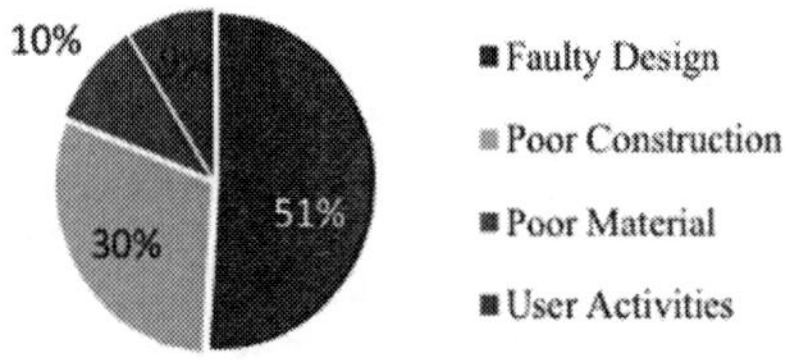

Figure 2.0: Causes of building defects according to Ransom (1981)

Singapore has produced ISO 9000 to improve the maintenance and reducing building defects. Technical, Human Resource and Management organizations found that the failures involve with defective materials, lack of training and skills, poor communication, etc. The strategy is founded to during planning stage that starting from design stage and selection in specification of materials to avoid any failures in the future [14].

3 Faulty in Design

Faulty design contributed to the number of building defects. Nonetheless, this is not a represent a poor performance of the building designers. The failure in design at the early stage would lead to deficiencies in the next stage after construction. Error in design is not only compromising the building environment, but also affected the building users [15]. Uncomfortable conditions in the building such as noise, inadequate lighting, uncomfortable temperature, high humidity and other physiological sectors may cause lower job satisfaction and increase in building related illness (BRI) symptoms [16].

Faulty design can occur due to an increase of maintenance cost. Some maintenance problems arise where the design is low probability of achievement in practice though it was satisfactory in principle. The good design achievement in the planning of buildings as worthy objects is understood and recognize, despite less understood and appreciated is the need for continuous maintenance [17]. Here are some causes of design failure towards maintenance:

a) Poor detailing and jointing
b) Insufficient thickness of concrete cover
c) Insufficient jointing between finishes faces
d) Incorrect locating conduits and piping at critical structure location
e) Insufficient length of owning
f) Insufficient provision for thermal movement
g) Insufficient structural design
h) Ignore maintenance access when designing
i) Not concerned on aggressive environment and weather condition effect
j) Less consideration the availability of maintenance tool during designing a building
k) Not relating exterior construction material selection to weather and climatic condition

3. 1 Poor Detailing and Jointing

The detailing in architectural drawing performed at relatively large scale, of a part of a building, machine, or element, with dimensions or other information for use in construction. It represented the design, materials, size and installation work for a certain element in a building. The important of this large scale drawing is to emphasize the complicated elements that impossible to highlight in a floor plan. The drawing is clearly drawn and explained through the specification and shall be read and understand by the contractors to avoid any miscommunication during the construction stage.

The poor detailing shall contribute to poor installation, less of materials' quality and poor jointing. This may lead to a low quality of construction and reflected to a poor image of the contractor. However, the weakness of this faulty is caused by the architect who designing the detailing.

3.2 Insufficient Thickness of Concrete Cover

The description [22] of concrete cover to reinforcement is the last distance measured from the surface of the concrete member to the surface of the embedded reinforcing steel. This element functioned as a protection of the harden concrete located in the middle of the structure elements. Insufficient thickness of the concrete cover could bring harmful to the center and reduce the strength of the structure [23].

In addition, this mixture is the finishes of the structure and exposed to the weather or any chemical reaction that could contributes to Alkali-Silica reaction, Chloride Attack, Carbonation Attack, Sulphate Attack and detachment [11].

3.3 Insufficient Jointing Between Finishes Faces

Finishes is the last solution to prevent the internal structure from defects agent. The functionality react as the protection to the element. Therefore, some of the finishes created with waterproof properties, self-cleaning, ant-fungi, and others to avoid from the defects.

Some of the manufacturing create it in small panels such as floor tiles, wall tiles, roof tiles, tempered glass, soft-board and ceiling panels. Insufficient jointing between the panels will allow water penetration that creates water absorption [11]. This problem developed dampness on the surface of the finishes and produced mold growth in a certain progress of defects. This will contributes to a serious defects in the future [24].

3.4 Incorrect Locating Conduits and Piping at Critical Structure Location

The era of 80's building design installed the servicing piping and wiring in the building walls. It was believed to treat a good finishing to these services and kept from untidy appearance.

However, this result to difficulty in maintenance when the building achieves 40 years and above. Leaking of pipes causes to water absorption to the structure of the building and may produce dampness on surrounding effected area. The remedial works become challenging while the original structure should be hacked to inspect, repair or replace the affected element.

The element supposedly shall be easy to access, reach and inspect for any future treatment or maintenance. It is because to avoid the spread of malignancy to the other element especially to the building structure.

3.5 Insufficient Length Of Owning

Owning is the extension of roof from the original form of the building. This owning functioned as to protect from the rain splashed to the opening. Insufficient length would address to the water absorption to the wall and may keep the dampness that stimulate the growth of algae.

The length of owning depends on the steepness of this element. The more steepness, the lesser length of the owning's design. It is also depends on the orientation of the building and placement of the owning to a building whereby the wind flow will influenced the rainfall directions. As a resulted, the design of owning shall be decided to avoid from rainwater splashed to the wall or opening.

3.6 Insufficient Provision for Thermal Movement

The building has absorbed the heat from direct sunlight. This heat is kept in wall and transferred directly to the internal of the building. The cavity wall design with the gap in between two layers of wall to create thermal insulation and reduce the heat or cold absorption in the internal building. However, this wall is widely used in four season regions.

In Asian countries, a single layer wall is used. The heat is transferable to the internal wall without any insulator to reduce the heat. Insufficient design of wall might increase the percentage of thermal movement and result to overheat external and internal of the wall. The impact of this situation leads to dependence of the mechanical ventilation and a high capacity of energy usage.

3.7 Insufficient Structural Design

Structure is the strongest element of the building. The function of the structure is clearly to support the whole building elements and systems. Nevertheless, insufficient structural design such as the less size of the foundation, number of columns, location of the structure, the size of the reinforcement bar and mixture of the concrete shall bring the biggest problem to a building [11]. The element started to cracks and this live cracks developed until it shows some failure to location of the structural weakest point. With insufficient structural design, this building is unsafe to operate anymore.

3.8 Ignore Maintenance Access When Designing

Most of the architects working with art and their imagination through the design. Some of the architects are too concerned with accumulating design of the building scheme till obliviously the maintenance aspects in design stage. This weakness result to the difficulties in the maintenance operational works and ignorance of maintenance access.

These maintenance access is important to ensure that the building shall be accessed for inspect the condition of the building, to reduce the risk and cost in maintenance. The complications maintenance access shall increase the maintenance cost when the management has to rent a special machinery, tools and equipment to reach the affected area. Otherwise, the damage of the building will be left and fall inti oblivion.

3.9 Less Consideration the Availability Of Maintenance Tool During Designing A Building

The consideration of maintenance tools availability for cleaning work should be accounted during design stage. This is because to avoid from the increment of maintenance financial and budgetary in maintenance operational work. The special and custom design of maintenance tools and equipment should be considered

DOI: 10.1051/matecconf/20166600122

for uncertainty in maintenance work. It will increase the maintenance running cost because the expertise of the tools is needed to operate the equipment. Therefore, this issues shall be well-thought-out during design stage.

3.10 Not Relating Exterior Construction Material Selection to Weather and Climatic Condition

Exterior material selection is important during design stage to reduce the climatic impact to the material. This material exposed to the climate with humidity, rain, acidic reaction, weathering and rapid obsolescence. The weak material will undergo a decay and obsolescence is a short period because this material is not being tested in weathering impact [24]. The material selection shall be decided and has the value on climatic tested to avoid external building decay.

4 Methodology

The good design achievement in the planning of buildings as worthy objects is understood and recognized, but less understood and appreciated is the need for maintenance. Therefore, these are the impact on maintenance management [17]:

a) Lack of fund due to major defects
b) Maintenance budget is increased
c) Difficulty in maintenance policy and planning implementation
d) Difficulty in repairing work
e) Maintenance frequency will increase
f) Maintenance works and cases will increase
g) Increase in maintenance time and duration
h) Lower maintenance quality
i) Hiring more expertise and professional person

The methodology procedure was produced to ensure that information obtain for this study was rigorously accomplish, relevant and capable of qualitative evaluation. Archives data on maintenance records is obtained from the Facilities Management Office UiTM and Politeknik Ungku Omar and Universiti Perguruan Sultan Idris. The three case studies is collected based on the similarity criteria aligned that is the three institutions is the government higher education institution, management of design decision, management board in design stage and maintenance management.

The respondents that involved in this research are from Maintenance Management Team and related Architect from these three case studies. The data collection is through the inspection on the building condition to survey, building defect cases present in the building and identify the major defect occurs especially on wall. All data will be analyzed by two (2) methods there are Frequency Analysis using SPSS software and Average Index Analysis. Frequency analysis is used to analyze the description of opened ended questions. This analysis presented in the frequency and percentage mode. In the other hand, the average index method is used to analysis like scale questions.

Average Index Formula: $$\frac{\Sigma a^1 x^1}{\Sigma x^1}$$ Where a^1 = weighting given to each factor by frequency of respondents
x^1 = number of respondents

Selection of Student Centre at Universiti Teknologi Mara Perak, Student Centre at Politeknik Ungku Omar and Student Centre at Universiti Perguruan Sultan Idris as the case studies that all three of the building serves as an office building. There were five (5) categories of skill rating which represent the feedback of the respondents and the implication rating scale of Average Index in questionnaire for instance would be:

Rating Scale	Average Index (a)	Categories	
1	$1.00 \leq a \leq 1.50$	Ignorable	Doed not affect
2	$1.50 \leq a \leq 2.50$	Not Important	Slightly affected
3	$2.50 \leq a \leq 3.50$	Moderate	Moderately affected
4	$3.50 \leq a \leq 4.50$	Important	Strongly affected
5	$4.50 \leq a \leq 5.00$	Very Important	Exremely affected

Figure 3.1: A table shows scale of survey.

There are 6% of female respondents in this survey from the Facilities Unit of UiTM and the rest were males representing 94.4% of the total percentage. They are among the professional who is a support group for each department. Respondents consisted of three (3) Architects who represent 17%, two (2) Engineer representing 11%, nine (9) Building Technician who represent 50% and lastly only 22% is Development Officer and his assistance. Each of case study representing six six (6) respondents and overall respondents for these three (3) case study is eighteen (18). Studies were also conducted to identify the respondents experience in their respective fields. The study found a total of 61% of the respondents have experience of 5-10 years in the employment of a total of eleven (11) persons. Meanwhile, two (2) representing 11% of respondents have experience of 10-15 years from UiTM Engineer and Architect of the Universiti Perguruan Sultan Idris. The remaining five (5) have experience of 1-5 years involved in representing of 28%. Pie chart below shows the analysis of respondent background.

Gender

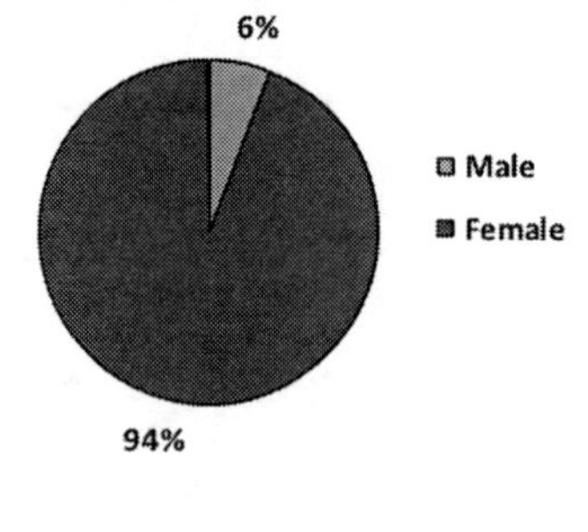

Respondent Occupation

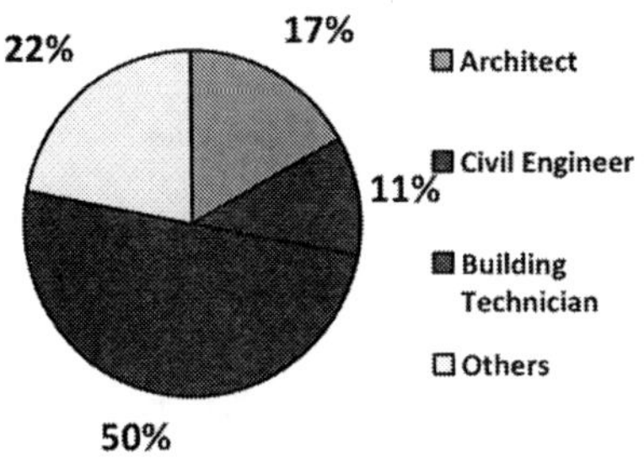

Respondent Experience

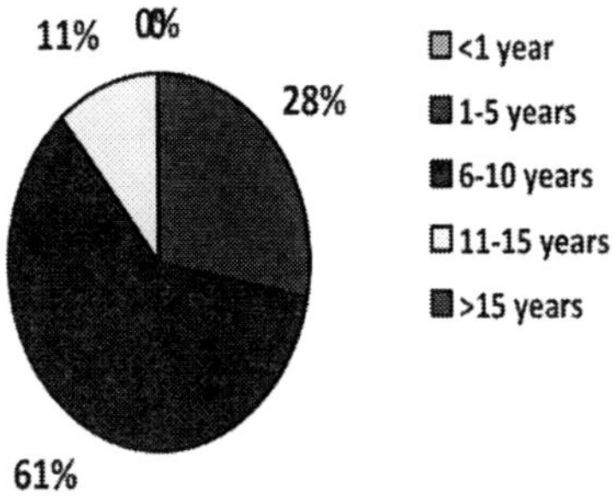

5 Major Causes in Faulty Design

This study is to evaluate the faulty design as main factors affected building maintenance, it is important to access the criteria of faulty design. By referring Table 4.1, six (6) most common causes of design failure, it has been tested in these different case studies [18]:

Common Causes in Design Failures

No.	CAUSES	UITM	PUO	UPSI
1.	Failure to follow well established design criteria.			
2.	Ignorance of the basic physical properties of materials.	√		
3.	Poor communication between team.	√	√	√
4.	Misjudgement of user and climatic condition.			
5.	Complex details.	√	√	
6.	New innovation / material that has not been properly tested in use			

Table 4.1: The most common causes design failures in Higher Education Institution offices.

All sources above was accessed and used as a guide for answers to interviews on all three Architects concerned. Based on investigation carried out, most of them agreed that these six factors above is the cause of design fault in building but the main causes of design fault is poor communication between member or team. The communication between members is very importance to produce an effective and successful project. Due to the lack of two ways communications, interpretation of design can't be implemented smoothly. From analysis, most Architects agreed whereas the design fault, workmanship and materials is the main subject of presence building defects. Table 4.2 below present effect of design defect towards maintenance and 5.1 shows the variable of design defect affecting maintenance.

Table 4.1 shows the variable of design defect affected maintenance management based on inspection and survey on the selected case study. With average 4.9 and extremely affected on maintenance is poor detailing and jointing and with average 4.6 insufficient thickness of concrete cover that will make the building facade peel off and defect occur. According to A-Shiha (1993) Inadequate concrete cover will make corrosions reach the concrete reinforcement faster if the concrete cover on the steel bars is insufficient, which will result in corrosion of the steel bars and cracking of the concrete element. Another variable that strongly affected on maintenance are insufficient jointing between finishes faces, incorrect locating conduits and piping at critical structure locations, insufficient length of owning, insufficient provision for thermal movement, insufficient structural design, ignoring maintenance access when designing and lastly not concerned on aggressive environment and weather condition effect is strongly affect to maintenance and its range of average index about 3.5 to 4.2.

Design Failure Affecting Maintenance

Variable	Mean	Rating Scale	Rank
Poor detailing and jointing	4.9	Extremely affected	1
Insufficient thickness of concrete cover	4.6	Extremely affected	2
Insufficient jointing between finishes faces	4.2	Strongly affected	3
Incorrect locating conduits and piping at critical structure locations	4.1	Strongly affected	4
Insufficient lenght of owning	4.1	Strongly affected	5
Insufficient provision for thermal movement	3.9	Strongly affected	6
Insufficient structural design	3.9	Strongly affected	7
Ignore maintenance access when designing	3.7	Strongly affected	8
Not concerned on aggressive environment and weather condition effect	3.5	Strongly affected	9
Less consideration the availability of maintenance tool during designing a building	3.2	Moderately affected	10
Exceed the permissible deflection	2.9	Moderately affected	11
Not relating exterior construction material selection to weather and climatic condition	2.8	Moderately affected	12

Table 4.2: Shows the Design Failure criteria in building.

This can be concluded that a designer should provide adequate structural design, proper access to maintenance equipment, proper exterior finishes compatible with climate conditions and sufficient detail in construction drawings in Table 4.3 [19].

The Impact Of Building Design Failure On Maintenance Management

Variable	Mean	Rating Scale	Rank
Lack of fund due to major defects	4.5	Extremely affected	1
Maintenance budget will increase	4.2	Strongly affected	2
Difficulty in maintenance policy and planning	4	Strongly affected	3
Difficulty in repairing	3.9	Strongly affected	4
Maintenance frequency will increase	3.8	Strongly affected	5
Maintenance works will increase	3.7	Strongly affected	6
Increase in maintenance time/duration	3.6	Strongly affected	7
Lower maintenance quality	3.3	Moderately affected	8

Table 4.3: Shows the impact of building defect on maintenance management on Higher Education Institutions.

By using Average Index Formular by Likert Scale, eight (8) questions was evaluated the effect of design defect towards maintenance management. Maintenance persons agreed that building defects extremely affect to lack of fund due to major defects to maintenance management. According to Malaysian Department of Education mentioned that problem in maintenance is to getting maintenance funding, where it is too difficult and take a long time to approve it. Design defect also strongly affected on maintenance budget, effect on maintenance policy and planning, increasing in maintenance frequency and works and also affect to increasing in maintenance time and duration. Some of moderately affected on maintenance is on maintenance quality and less expertise and professional person in conducting the repairing work.

6 Discussions

The data obtain from the questionnaires analyzed by determined the main factors that contributed to the building defects on refurbished building. After do some research and after done observation on the selected case study, it is clearly shows that faulty design is the main factors or causes in the presence of building defects. This statement is supported by respondent and others past research done by many researchers. Researcher found that, the main effects is inadequate of funding due to major defects. A building with numerous design defects will increase budget of maintenance. The respondent also strongly believe that design defect strongly affected on maintenance budget, effect on maintenance policy and planning, increasing in maintenance frequency and works and also affect to increasing in maintenance time and duration. Some of them state that design defect moderately affect the maintenance quality and due to inadequate expertise and professional person. Design defect such poor detailing, jointing and insufficient thickness of concrete cover [23] is extremely affect the maintenance. Based on analysis conducted, researcher found that most of management prefers to use a preventive maintenance rather than others approach. They also mentioned that they prefer to hire an outsourced contractor due to several benefits such as expertise and specialized, strong fund, fast performance and their company culture and value. The management should take into account this matter in order to reduce maintenance cost and prolong their building life.

References

[1] Chong, W. and Low, S. (2006). *"Latent Building Defects: Causes and Design Strategies to Prevent Them." J. Perform. Constr. Facil.*, **10.1061/(ASCE)0887-3828(2006)20:3(213)**, 213-221.

[2] Al-Hammad, A., Assaf, S. and Al-Shihah, M. (1997). *'The effect of faulty design on building maintenance'. Journal of Quality in Maintenance Engineering*, **3**(1), 29–39. | Article |

[3] David, M.N (2014), *Exploring The Architecture of Everyday Life*. United State of America, SAGE.

[4] Andre K.S.J and Albert H.C.T (2013). Maintenance, Replacement and Reliability: Theory and Application, **Second Edition**. United State of America, CRC Press.

[5] Ramly, A. (2006). 'A link between design and maintenance', *Journal of Building Engineers*, **81**(5), May.

[6] Nicastro, D.H. (1997). *Failure Mechanisms in Building Construction*, ASCE Press, Reston, VA.

[7] Ishak, N.H., Chohan, A.H._and Ramly, A. (2007). *Implications Of Design Deficiency On Building Maintenance At Post-Occupational Stage*. Journal of Building Appraisal (2007) **3,** 115–124. doi:10.1057/palgrave.jba.2950061.

[8] Abdul Rahman, S. (2012), A Study On The Impact Of Design Defects Towards Building Maintenance Management, **Unpublished degree dissertation**, Universiti Teknologi MARA, Malaysia. '

[9] Douglas, J., and Ransom, W. H., 2007. *Understanding Building Failures*, **3rd edn**. New York: Routledge. 326 pp.

[10] Seelay, Ivor H. (1987). *Building Maintenance*, Palgrave Macmillan, **2Rev Ed Edition**.

[11] Hinks, J. and Cook, G. (2005). *The Technology of Building Defects*. New York: Taylor & Francis.

[12] Watt, David S. (2007). *Building Pathology: Priciples and Practice*. Oxford, Blackwell Publishing.

[13] Ransom, W.H. (1981). *Building Failures Diagnosis and Avoidance*, **2nd Edition. London**, E&FN Spon.

[14] Low, S.P. and Darren, W. (2001). *"Improving maintenance and reducing building defects through ISO 9000"*, Journal of Quality in Maintenance Engineering, **Vol. 7 Iss: 1**, pp.6 – 24.

[15] Assaf, S., Al-Hammad, A., and Al-Shihah, M. (1996). "Effects of Faulty Design and Construction on Building Maintenance." *J. Perform. Constr. Facil.*,

10.1061/(ASCE)0887-3828(1996)10:4(171), 171-174.
[16] Ryan, Christopher M.; Morrow, Lisa A. (1992). *"Dysfunctional buildings or dysfunctional people: An examination of the sick building syndrome and allied disorders"*. Journal of Consulting and Clinical Psychology, **Vol 60(2)**, Apr 1992, 220-224.
[17] Reekie, R.F. (1972). *Design in The Built Environment*. The University of Wisconsin: Madison, Crane, Russak.
[18] Lee. R. (1976), "Building Maintenance Management", The Chartered Institute of Building, Ascot, England.
[19] Assaf, et. al. (1996). *Effects of Faulty Design and Construction on Building Maintenance*. Journal of Performance of Constructed Facilities.
[20] Luigi T.De Luca, *Propulsion physics* (EDP Sciences, Les Ulis, 2009)
[21] F. De Lillo, F. Cecconi, G. Lacorata, A. Vulpiani, EPL, **84** (2008).
[22] http://wwwmainroads.wa.gov.au
[23] http://enggpedia.com
[24] Barry, A. R., (2005), Defects And Deterioration In Building, *2nd Edition*, Spon Press, New York.

Analytical Hierarchy Process for Developing a Building Performance-Risk Rating Tool

Natasha Khalil [1,2,a], Syahrul Nizam Kamaruzzaman[2] and Mohamad Rizal Baharum[2]

[1]*Faculty of Architecture, Planning and Surveying, Universiti Teknologi MARA Perak, 32610 Seri Iskandar, Perak, Malaysia*
[2]*Faculty of Built Environment, Universiti Malaya, 50603 Kuala Lumpur, Malaysia*

Abstract. The need to optimize the performance of buildings has increased consequently due to the expansive supply of facilities in higher education building (HEB). Proper performance assessment as a proactive measure may help university building in achieving performance optimization. However, the current maintenance programs or performance evaluation in the HEB is a systemic and cyclic process where maintenance is considered as an operational issue and not as opposed to a strategic issue. Hence, this paper proposed a Building Performance Risk Rating Tool (BPRT) as an improved measure for building performance evaluation by addressing the users' risk in health and safety aspects. The BPRT is developed from the result of a rating index using the Analytical Hierarchy Process (AHP) method. 12 facilities management (FM) experts and practitioners were involved in the rating process. The subjective weightings are analysed using the AHP computer software, the *Expert Choice 11*. The establishment of the BPRT was introduced as an aid of improvement towards the current performance assessment of HEB by emerging the concept of building performance and risk into a numerical strategic approach

1 Introduction

The development of higher education buildings (HEB) including expanding facilities is spirally a welcoming sign of the growth of tertiary educational programs. Unlike most land developers, university authorities need to have a long-term view of operations to ensure sustainability is achieved through performance optimization. However, building performance is not guaranteed as it depends on its lifespan and changing the physical requirement from the growing learning activities and students' populations. Hence, further issues such as space inefficiency, ventilation discomfort and inefficient of energy use may decrease the performance of the buildings from time to time [1]–[4]. Performance failure not only affects the building sustainability, but the users are also affected by the failure of building performance [5]. To optimise the performance of HEB, it should avoid allocating the occupants, visitors and passers-by at risk [6]. Several studies have shown that poor building performance presents vulnerability of risk towards the safety and health of building users [2], [7]–[13]. This demonstrates the significance of addressing risk in the building performance evaluation.

In revealing the crucial aspect of performance assessment, the government of Malaysia under the maintenance division has issued the General Circular (No.1) dated 27th March 2009, that states holistic management of asset through all management should undertake a systematic approach to achieving building performance optimization. As described in the Guideline of Building Condition Assessment (BCA) to the existing government and public buildings [14] the assessment for a building is essential not only for building repairs and improvement, but the assessment must fit the aspect of safety and risk of building. The General Circular No. 2 (1995) by the government of Malaysia has also clearly states the responsibility of each head of the department to maintain and assess the in government and public buildings with an effective and systematic plan.

University buildings in Malaysia have been constructed since 1960. Nevertheless, all buildings deteriorate and decay with age as a result of various factors, including poor quality materials, bad workmanship, excessive usage, abuse as well as inadequate and poor maintenance [15], [16]. As buildings become larger and house more people, political and societal issues have become more complex, which in turn change the risks associated with occupying buildings [17]. Hence, indicating a list of criteria towards a proper building performance assessment is needed to mitigate hazards to the users.

2 Building Performance and Risk Impact to the Users

The basic concept of building performance evaluation (BPE) has been emerged on various issues, characteristics with various objectives. The performance concept involves recommendations for improvement and it is

[a] Corresponding author: natashakhalil29@gmail.com

used for feedback and feed forward regarding the performance of similar buildings [18]. It denotes the comparison of client's goals and performance criteria against actual building performance, measured both subjectively and objective. Building performance has been defined in BS 5240 as "behaviour of a product in use" [7], [19]. It is also described as a process of assessing progress towards achieving goods and services efficiency, quality of building outputs and effectiveness of building operations [20]. This refers to the comprehensive features of a building, including structural, architectural, surroundings, environmental issues and building services. The performance measurement of a building is firstly summarised in terms of the background of the building and the scope of performance assessment [21]. Therefore, all relevant stakeholders need to understand the key performance factors in a building.

The aspect that needs to be thoroughly assessed in building performance depends on the evaluation purposes. The risk is associated with the unintended consequences of building performance and the primary cause of these risks may be a lack of measured performance data [22]. Risk approach advocates principles on the level of building performance and predicts the impact on society that is ultimately affected by the sources of risk. Therefore, risks in building performance can be framed as a health risk, safety risk, environmental risk, economic risk as well as political risk and others [7], [11], [23]. To suit with the aspects of building performance in Malaysian HEBs, health risk and safety risk become the main focus of this research. Hence, the following provide the description of health risk and safety risk in the context of study for this research:

a) Health Risk
Impact towards human health effects; either direct or indirect exposure of building factors that can cause health risk (latent or patent); the effect in short-term or long-term health suffers.

b) Safety Risk
The risk or hazards that create a tendency for injury, death, crime, theft, nuisance or burglary to the building users

Since the risk frames are focussed to health risk and safety risk, a list of performance elements or indicators needs to be constructed in prevailing the users' health and safety risk from poor performance elements.

3 The Construct of Performance-Risk Indicators (PRI)

The initial step in developing a new rating tool is to select the assessment areas that should be included and next, determine the parameters, attributes or indicators that can be used to measure the selected aspects [24], [25]. Analysis of the previous established performance rating tools such as Building Research Establishment's Environmental Assessment Method (BREEAM), Leadership in Energy and Environmental Design (LEED), Total Building Performance (TBP), Building Safety Condition Index (BSCI), Building Safety Condition Index (BSCI) and Building Health Hygiene Index (BHHI) is carried out in the literature stage. The review is considered by far the most comprehensive method to develop a new rating scheme and to examine sustainability issues [25]. The construct indicators were initially compiled from the existing performance and green rating tools and also precedent research relating to building performance [2], [7], [8], [10]–[12], [17], [26]–[29].

The initial construct of indicators, named as Performance-Risk Indicators (PRI) was further confirmed through a preliminary survey, using semi-structured interview technique. 18 building managers and operators in local university buildings were involved in the interview and their input is needed to obtain suitability of the listed indicators for building performance rating assessment to be used in the Malaysian HEBs. The findings of the interview have finalised 26 indicators that are categorized into the performance elements of functional performance, technical and indoor environmental performance. Table 1 depicts the final framework of the construct PRI. The listed indicators are confirmed as the predictor variables that can contribute to the performance level and the users' health and safety risk. The PRI constructs constitute the preliminary basis of a framework for further development of the proposed building performance-risk rating tool. It will be developed further with relative importance and assigned weights for each PRI in the next stage using Analytical Hierarchy Process (AHP) as decision-making approach.

Table 1. The construct of PRI based from the result of preliminary survey.

Performance Elements	Performance-Risk Indicators (PRI)
Functional Performance	Spaces
	Orientation
	Infrastructure
	Access/entrance
	Circulation area
	Ergonomic building facilities
	Adequacy of building signage
	Emergency exits
	Building-related illnesses/sick building syndrome
	Amenities
Technical Performance	Design of building fittings
	Structural stability
	Information Technology systems operations
	Electrical services
	Plumbing services
	Fire Prevention Services
	Materials & Internal Finishes
	Roof
	Lift
Indoor Environmental Performance	Cooling (Thermal comfort)
	Artificial lighting (Visual comfort)
	Natural lighting (Visual comfort)
	Waste reduction
	Building ventilation
	Acoustic comfort (Noise)

	Level of cleanliness

4 Methodology

This research adopts a quantitative approach using questionnaires consisting the Analytical Hierarchy Process (AHP), as the main instrument. The list of PRI was incorporated into the questionnaire as survey questions. Since AHP method relies on experts' judgment to moderate feedback throughout the process. therefore, the respondents are the facilities management (FM) practitioners from the public and private organizations. The sampling was drawn based on purposive samples with predetermined criteria and also with the help of Malaysian Association of Facility Managers (MAFM). Respondents were selected based on purposive sampling, with predetermined criteria that requires the experts (the respondents) to possess experience or involvement in any project related to the educational buildings with at least 10 years' working experience. Hence, the list of potential respondents is shortlisted to 22 experts. The experts are from different fields such as architecture, engineering, surveying, facilities management and internationl business studies. In this process, it requires the experts to provide justification on the importance of the construct indicators and also the rating process of weighted for each parameter. After the judgement is completed, the ranking of indicators is arranged according to the highest order to the lowest order of importance. A final set of weighted assigned for each indicator is presented at the end of stages.

4.1 Application of Analytical Hierarchy Process

In dealing with such a multi-criteria decision-making process, the Analytic Hierarchy Process (AHP) developed by Dr. Thomas L. Saaty in 1977, is used for this study. AHP allowed the experts to evaluate the attribute weightings with greater consistency through pairwise comparisons. The AHP is a theory of measurement through pairwise comparisons and relies on the judgements of experts to derive priority scales [30]–[33]. In AHP, all criteria or parameters are assigned with the weight score that shows the importance of each criterion. The AHP approach has been widely adopted in the built environment fields as decision-making tool [25], [34]–[36]. Existing performance schemes and rating tools such as LEED, BHHI and BSCI also adopted the process of AHP to develop hierarchy or rating tool [35], [37]. It is also found that AHP is used to assess risk in a supply chain [38]. Therefore, the application of AHP for this survey is robust and do not constitute any bias result as the judgment on the weighted depending on the experts' decision. Moreover, the methodology of the AHP allows for the internal consistency of the respondent's results to be checked.

4.2 Administration of Survey

A letter of invitation and the sample sets of a questionnaire are firstly distributed to the shortlisted respondents. Out of total 22 experts, 12 experts are willing to participate in the survey and the rate of response to this survey is 54.6%. Before beginning with the survey, the researcher introduced the purpose of the survey to the respondents and showed all sections involved in the survey forms. A detail explanation was conveyed to the respondents for the application of AHP as the process is not aware by most of the respondents. The weightings of the building factors were assessed using nine (9) scales of importance [30], [31], as shown in Table 2. The respondents' weightings of the different factors were extracted from a pairwise comparison of the relative importance of all pairs of factors using the AHP computer software package *Expert Choice 11* version 3.01.

Table 2. AHP scale of importance [30], [31]

AHP SCALE OF IMPORTANCE	DESCRIPTION
1	Equal Importance
2	Equally to Moderately
3	Moderate Importance
4	Moderately to Strong
5	Strong Importance
6	Strongly to Very Strong
7	Very strong Importance
8	Very strong to Extremely
9	Extreme Importance

The experts are required to compare the importance between two pairwise indicators and rate the scale of importance of the chosen indicator. Comparison of importance for each indicator is rated using the values of importance 1–9 scale (as shown in Table 2).

5 Analysis of Results and Discussion

5.1. The AHP weightings

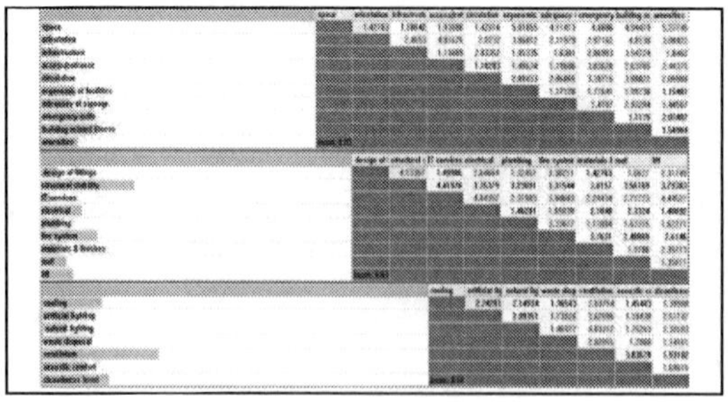

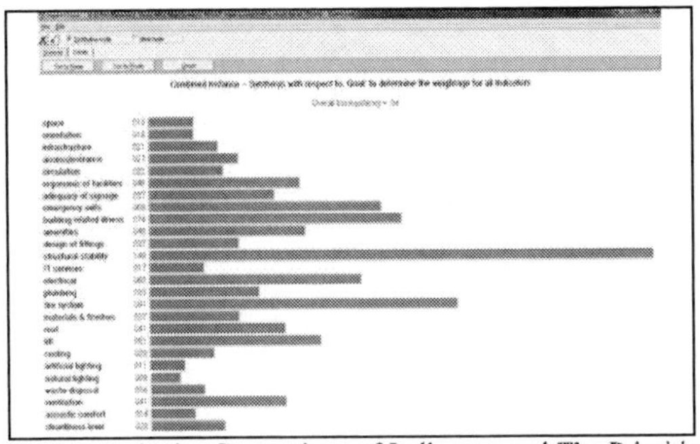

Figure 1. Pairwise Comparison of Indicators and The Priorities of Indicators, Using the *Expert Choice* 11

The relative importance (relative weight) of each category and each indicator within each category was established using a square matrix structure. For each category and indicator, the weight was calculated by the geometric mean of values of questionnaires filled by the experts who participated in the survey. The final step in the process combined the ratings of the criteria to form an overall rating for each decision alternative. The numerical pairwise comparison and the priorities of indicators obtained by combining the overall judgment from the 12 experts is shown in Figure 1. To obtain the priorities of indicators, the distributive mode is chosen as it normalizes alternative scores under each criterion so that they sum to one (1.00). This creates a dependency on how well all other alternatives perform and hence the potential for rank reversal. It shows that the internal consistency ratio (CR) .04 where in AHP, the CR must be less than 0.1 (10%) [31]. Therefore, the data is reliable and achieves consistency.

Table 3. Summary of Relative Weights for Performance Elements and Performance-Risk Indicators

PERFORMANCE ELEMENTS (PE)	PERFORMANCE-RISK INDICATORS (PRI)	WEIGHT (%)	RANKING
Functional Performance 36.7%	Spaces	1.3	23
	Orientation	1.4	24
	Infrastructure	2.1	18
	Access/entrance	2.7	14
	Circulation area	2.2	17
	Ergonomic building facilities	4.5	8
	Adequacy of building signage	3.7	11
	Emergency exits	6.8	4
	Building-related illnesses/ SBS	7.4	3
	Amenities	4.6	7
Technical Performance 49.9%	Design of building fittings/fixtures	2.7	15
	Structural stability	14.9	1
	IT systems operations	1.7	20
	Electrical services	6.3	5
	Plumbing services	3.3	12
	Fire Prevention Services	9.1	2
	Materials & Internal Finishes	2.7	13
	Roof	4.1	9
	Lift	5.1	6
Indoor Environmental Performance 13.4%	Cooling (Thermal comfort)	2.0	19
	Artificial lighting	1.1	25
	Natural lighting	0.9	26
	Waste disposal	1.6	21
	Building ventilation	4.1	10
	Acoustic comfort	1.4	22
	Level of cleanliness	2.3	16

To ease the calculation of weight for each performance element, the result from the combined synthesis of each indicator is illustrated in Table 3. The weightings in Table 3 shows that structural stability is ranked as the most important indicator, with a global weight 14.9%, and followed by fire prevention services (9.1%), building-related illnesses (7.4%), emergency exits (6.8%) and electrical services (6.3%). This suggests that five indicators become the top important factors to be well performed in building performance, as it may generate larger impact towards the users' health and safety risk. This has suggested that the focus on building performance and user's risk on the priority of the indicator structural stability is more significant than other 25 indicators. This is supported by previous studies [39]–[43] that described structural degradation and deterioration causes impact on performance quantification in terms of collapse safety and health of the users. Structural elements are among the important safety factors for performance assessment [44]. The structural stability of a building must be inspected and maintained from time to time in order to ensure the occupants' safety [16]. Hazards in the built environment were closely related to accidents occurring in buildings; therefore, the occupants must be safeguarded against hazards arising from structural failure.

With respect to the result of the *Expert Choice*, it shows that the weightings for functional performance, technical performance and indoor environmental performance are 36.7%, 49.9% and 13.4%, respectively. This insinuates that technical performance was a bit more important than functional performance and indoor environmental performance. The discussion of the findings is further explained in the following section

5.2 Discussion of result

It can be summarised that the attained weightings for the technical performance suits with the support from the academic theories and literature on the importance of building technical performance.The significant finding based on the summated score is that technical performance is considered as the most critical performance elements that place greater impact in the context of building performance and users' risk. Inevitably, the performance assessment of technical performance in a building becomes a crucial measure. The need for continuous monitoring of the technical performance of buildings over their lifetime becomes obvious since buildings are not operatively static [45]. Technical performance aspects of a building complete the functionality aspect, therefore, a building may succeed in achieving a high level of functionality if able to meet prescribed technical standards. The performance requirements for building need to match with the provision of reliable information about the alternatives, including technical specifications with indications of service life and performance over time [46].

The empirical findings of this study indicate that the PRIs have significant difference based on the judgment by the panel of experts. The results suggest that the relevant stakeholders in the facility or maintenance management of higher educational buildings should pay greater attention to the priority of indicator that has significant risk impact, ideally the top priority indicators. This is also has supported studies by [47]–[49] that described the maintenance organisation in university buildings must look far beyond the immediate objectives of the products or services to the users. Since the performance evaluation in university buildings is under the responsibility of facility or maintenance unit, information on the building performance should be well documented in a safe register.

6 Development of the Building Performance-Risk Rating Tool (BPRT)

The result of weightings has successfully helped to establish the further development of the building performance risk rating tool for this study. Hence, the proposed building performance risk rating tool (BPRT) is proposed as an improved method to carry out building condition survey. The steps and process in the proposed BPRT adapt the steps from the existing performance and rating tool such as GBI, BARIS and SABA. There are three (3) steps included in the proposed BPRT. The first step requires the building assessors to evaluate the performance and risk of the indicators, and the second step comprises the overall score for PRI. The third step shall be the final process of assessment where the assessed building is classified into a rating classification. The description of the steps and process are entailed as follows:

6.1 Step 1: Evaluation of performance and risk for each indicator

In this step, the evaluation of performance and risk indicators shall be carried out as to capture the performance grading of the building. Hence, the assessment outcome which is known as performance risk assessment score suggested five scales, that ranges from 0.2 to 1.0. If the score achieves 1.0, it generally defined that the indicator is in an excellent performance that definitely presents insignificant risk impact to the users. Table 4 shows the score and description for the scales. The description of scale for the performance risk assessment was adopted and modified from the assessment of BARIS [14], [50] and also previous studies relating to risk in building assessment [29], [51], [52].

Table 4. Scale and Description for Performance-Risk Assessment Score for each indicator

SCALE	DESCRIPTION OF SCALE
0.2	• Poor performance with catastrophic building defects • Catastrophic: Present risk that results in multiple fatalities; may result in human death or serious injury/illness
0.4	• Low performance with major building defects • Significant risk: Present risk towards serious or fatal injury to occupants or major injury/illness
0.6	• Medium performance with moderate building defects • High risk: Present risk towards significant impact with minor injury or minor occupational illness
0.8	• Good performance with minor building defects • Medium risk: Present risk towards minor physical discomfort to occupants or minor occupational illness
1.0	• Excellent performance with little building defects • Low risk: Present risk that the impact is insignificant; minimal impact or no apparent impact at all

During the assessment, the performance of each indicator was rated in accordance with the above PR score. Hence, Figure 2 shows the extracted steps of assessment using the assessment score and the weightings.

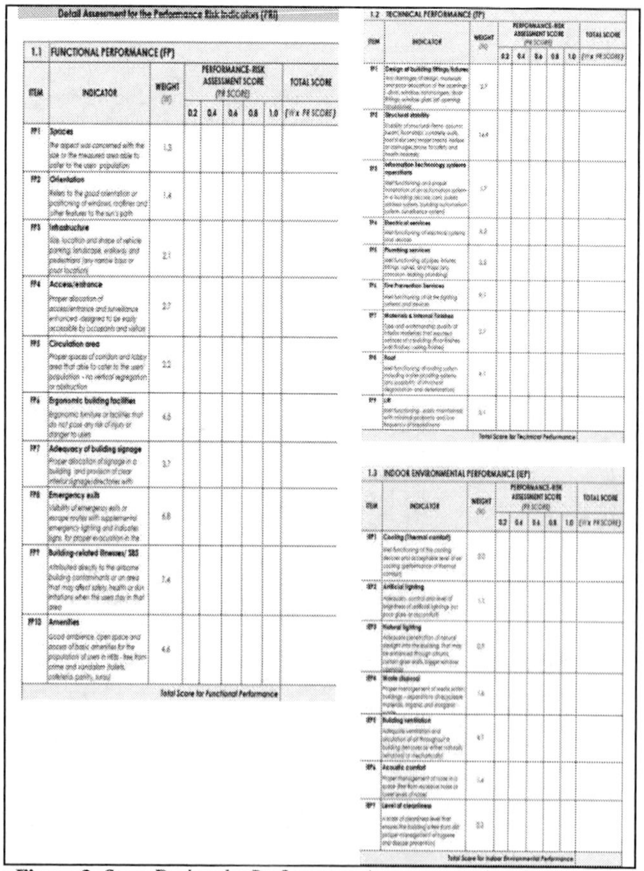

Figure 2. Steps During the Performance Assessment using the Assessment Score and Weightings

6.2 Step 2: Overall Score for Performance Risk Indicators (PRI)

As mentioned in step 1, the performance of each indicator is evaluated using a performance risk assessment score that ranges from 0.2 to 1.0. Then, the obtained score is multiplied by the weighting assigned to each indicator. The total of overall PRI score generates the score of performance elements (functional performance, technical performance and indoor environmental performance). Once the performance assessment of all indicators completed, the overall score of PRI can be calculated by summing up each of the PRI scores. This will generate the overall score of performance elements, in accordance with the category of functional performance, technical performance and indoor environmental performance. Since each of the performance elements is assigned with maximum weight that is derived from the result of AHP, the score that shall be obtained from the performance assessment shall not be more than the maximum weight. The maximum weightings for functional performance are 36.7%, technical performance is 49.9% and indoor environmental performance is 13.4%.

6.3 Step 3: The Performance-Risk Rating Classification

The final step in the performance assessment using the building performance risk rating tool (BPRT) is to determine the rating classification of the assessed building. The rating classification shall generate the performance grade and the level of risk. There are five ratings that determine the final performance measurement of the assessed building either in "Excellent", "Good", "Medium", "Low" and "Poor", as shown in Table 5. The determinants of the rating classification depend on the final score attained from the performance assessment to all indicators. The classification of the rating is adopted from the existing performance and green rating tools such as GBI and BARIS. However, the description of each rating is modified from previous studies relating to risk in building assessment [29], [51], [52] to suit with the context of performance and risk in this study. The final risk classification shall help the building assessors to rectify areas concerning performance failure and its possible impact on building users. Thus, it helps the management to plan further actions on mitigating or reducing the risk through performance improvement.

Table 5. Performance-Risk Rating Classification

TOTAL PERFORMANCE ELEMENTS SCORE	RATING	DESCRIPTION
80 - 100	Excellent	Excellent building performance with low impact to the users' risk on health and safety
60 - 79	Good	Good building performance with medium impact to the users' risk on health and safety
41 - 59	Medium	Medium building performance with high impact to the users' risk on health and safety
31 - 40	Low	Low building performance with significant impact to the users' risk on health and safety
20 - 30	Poor	Poor performance building with catastrophic impact to the users' risk on health and safety

With the proposed BPRT through this study, it shall be useful to the building operators in Malaysian HEBs to assess the condition of the academic building that concerns on both performance and the users' risk. Surveys that employ ratings instead of descriptions are gaining wide acceptance in the industry because they cater to the need for numerical analysis output [15]. Through the application of the BPRT as proposed in this study shall initiate an improved proactive procedure for performance assessment and maintenance in HEB

7 Conclusion

By integrating criteria from different assessment methodological frameworks, this research builds on the strengths of each and provides a more holistic assessment approach among careful attention to local context. The proposed BPRT granted robust information on the level of risk towards the building users. The students, the academicians, administration staffs and the supporting personnel are the entity or group of individuals in HEB. The building users can be presented with a warranty that the building fulfills the performance optimization with a precise risk level in terms of health and safety aspects. These two aspects of risk become the most crucial information that needs to be conveyed to the users and without a proper assessment, users are deemed to accept the risk without knowing the impact that may jeopardize their health and safety. This tool is also potential to extend its application to the other types of buildings since the performance criteria are related to the functional, technical and indoor environmental elements of a building.

Acknowledgment

The authors wish to acknowledge the support of funding for this research under the Exploratory Research Grant Scheme (ERGS) Phase 1/2013 from Ministry of Higher Education (MOHE), Malaysia

References

1. A. Gillen, A. Wright, and L. Spink, "Student perceptions of a positive climate for learning: a case study," *Educ. Psychol. Pract.*, vol. **27**, no. 1, pp. 65–82, Mar. 2011.
2. H. Altan, "Energy efficiency interventions in UK higher education institutions," *Energy Policy*, vol. **38**, no. 12, pp. 7722–7731, Dec. 2010.

3. M. Sapri and S. Muhammad, "Monitoring energy performance in higher education buildings for sustainable caumpus," *Malaysian J. Real Estate*, vol. **5**, no. 1, pp. 1–25, 2010.
4. M. A. Hassanain, "Post-Occupancy Indoor Environmental Quality Evaluation of Student Housing Facilities," *Archit. Eng. Des. Manag.*, vol. **3**, no. 4, pp. 249–256, 2007.
5. A. L. Olanrewaju, M. F. Khamidi, and A. Idrus, "Quantitative analysis of defects in Malaysian university buildings : Providers '," *J. Retail Leis. Prop.*, vol. **9**, no. 2, pp. 137–149, 2010.
6. D. Amaratunga and D. Baldry, "Building Performance Evaluation In Higher Education Properties: Towards A Process Model," in *Construction and Building Research Conference (COBRA) 1999*, 1999, pp. 45–56.
7. N. Almeida, V. Sousa, L. Alves Dias, and F. Branco, "A framework for combining risk-management and performance-based building approaches," *Build. Res. Inf.*, vol. **38**, no. 2, pp. 157–174, Apr. 2010.
8. R. J. Cole, "Building environmental assessment methods : assessing construction practices," *Constr. Manag. Econ.*, vol. **18**, no. 8, pp. 949–957, 2000.
9. T. Lützkendorf and D. Lorenz, "Integrating sustainability into property risk assessments for market transformation," *Build. Res. Inf.*, vol. **35**, no. 6, pp. 644–661, Nov. 2007.
10. T. Lützkendorf and D. P. Lorenz, "Using an integrated performance approach in building assessment tools," *Build. Res. Inf.*, vol. **34**, no. 4, pp. 334–356, Jul. 2006.
11. B. J. Meacham, "Risk-informed performance-based approach to building regulation," *J. Risk Res.*, vol. **13**, no. 7, pp. 877–893, Oct. 2010.
12. A. Wolski, N. A. Dembsey, and B. J. Meacham, "Accommodating perceptions of risk in performance-based building 're-safety' code development," *Fire Saf. J.*, vol. **34**, pp. 297–309, 2000.
13. A. Zalejska-Jonsson, "Evaluation of low-energy and conventional residential buildings from occupants' perspective," *Build. Environ.*, vol. **58**, pp. 135–144, Dec. 2012.
14. Public Works Department, *Garis Panduan Pemeriksaan Dan Penilaian Keadaan Bangunan Sedia Ada (JKR 21602 - 0004 - 13)*, JKR 21602-. Bahagian Senggara Fasiliti Bangunan, Cawangan Kejuruteraan Senggara Ibu Pejabat JKR Malaysia, 2013.
15. A. I. Che-Ani, A. S. Ali, M. M. Tahir, N. A. G. Abdullah, and N. M. Tawil, "The Development of a Condition Survey Protocol (CSP) 1 Matrix for Visual Building Inspection," in *The Construction, Building and Real Estate Research Conference of the Royal Institution of Chartered Surveyors (COBRA2010), 2-3 September 2010*, 2010, pp. 8–22.
16. A. S. Ali, S. N. Kamaruzzaman, R. Sulaiman, and Y. C. Peng, "Factors affecting housing maintenance cost in Malaysia," *J. Facil. Manag.*, vol. **8**, no. 4, pp. 285–298, 2010.
17. B. P. Thompson and L. C. Bank, "Risk Perception in Performance-Based Building Design and Applications to Terrorism-Resistant Design," no. February, pp. 61–69, 2007.
18. D. Amaratunga and D. Baldry, "Appraising the total performance of higher education buildings: A participator y approach towards a knowledge- base system," in *Construction and Building Research Conference 1998 : Proceedings (COBRA) 2nd-3rd September 1998, Oxford Brookes University*, 1998, pp. 1–17.
19. J. Douglas, "Building performance and its relevance to facilities management," *Facilities*, vol. **14**, no. April, pp. 23–32, 1996.
20. D. Amaratunga and D. Baldry, "A conceptual framework to measure facilities management performance," *Prop. Manag.*, vol. **21**, no. 2, pp. 171–189, 2003.
21. G. McDougall, J. R. Kelly, J. Hinks, and U. S. Bititci, "A review of the leading performance measurement tools for assessing buildings," *J. Facil. Manag.*, vol. **1**, no. 2, pp. 142–153, 2002.
22. J. E. Woods, "Expanding the Principles of Performance to Sustainable Buildings," *Real Estate Issues*, vol. **33**, no. 3, pp. 37–46, 2008.
23. B. Meacham, R. Bowen, J. Traw, and A. Moore, "Performance-based building regulation: current situation and future needs," *Build. Res. Inf.*, vol. 33, no. 2, pp. 91–106, Mar. 2005.
24. T. Malmqvist and M. Glaumann, "Environmental efficiency in residential buildings – A simplified communication approach," *Build. Environ.*, vol. **44**, no. 5, pp. 937–947, May 2009.
25. H. H. Ali and S. F. Al, "Developing a green building assessment tool for developing countries – Case of Jordan," *Build. Environ.*, vol. **xxx**, no. 2008, pp. 1–12, 2008.
26. M. Camilleri, R. Jaques, and N. Isaacs, "Impacts of climate change on building performance in New Zealand," *Build. Res. Inf.*, vol. **29**, no. 6, pp. 440–450, Nov. 2001.
27. E. Meins, H. Wallbaum, R. Hardziewski, and A. Feige, "Sustainability and property valuation: a risk-based approach," *Build. Res. Inf.*, vol. 38, no. 3, pp. 280–300, Jun. 2010.
28. R. Spence, "Risk and regulation: can improved government action reduce the impacts of natural disasters?," *Build. Res. Inf.*, vol. **32**, no. 5, pp. 391–402, Sep. 2004.
29. R. N. Whitfield, "Managing Institutional Risks – A Framework," University of Pennsylvania, 2003.
30. T. L. Saaty, "Decision making with the analytic hierarchy process," *Int. J. Serv. Sci.*, vol. **1**, no. 1, pp. 83–98, 2008.
31. T. L. Saaty, "How to make a decision: The Analytic Hierarchy Process," *Eur. J. Oper. Res.*, vol. **48**, pp. 9–26, 1990.
32. M. Alexander, "Decision-Making using the Analytic Hierarchy Process (AHP) and SAS / IML ®," in *SESUG 2012*, 2012, pp. 1–12.
33. K. Bunruamkaew, "How to do AHP analysis in Excel The Analytical Hierarchy Process - AHP," 2012.
34. Z. Chen, D. Clements-croome, J. Hong, H. Li, and Q. Xu, "A multicriteria lifespan energy efficiency

approach to intelligent building assessment," *Energy Build.*, vol. **38**, pp. 393–409, 2006.
35. C. A. Poveda and M. Lipsett, "A Review of Sustainability Assessment and Sustainability/Environmental Rating Systems and Credit Weighting Tools," *J. Sustain. Dev.*, vol. **4**, no. 6, pp. 36–55, Nov. 2011.
36. T. Hsieh, S. Lu, and G.-H. Tzeng, "Project Fuzzy MCDM approach for planning and design tenders selection in public office buildings," *Int. J. Proj. Manag.*, vol. **22**, pp. 573–584, 2004.
37. D. C.-W. Ho, K.-W. Chau, A. King-Chung Cheung, Y. Yau, S.-K. Wong, H.-F. Leung, S. Siu-Yu Lau, and W.-S. Wong, "A survey of the health and safety conditions of apartment buildings in Hong Kong," *Build. Environ.*, vol. **43**, no. 5, pp. 764–775, May 2008.
38. I. N. Pujawan and L. H. Geraldin, "House of risk: a model for proactive supply chain risk management," *Bus. Process Manag. J.*, vol. **15**, no. 6, pp. 953–967, 2009.
39. M. Clift, "Building quality assessment (BQA) for offices," *Struct. Surv.*, vol. **14**, no. 2, pp. 22–25, 1996.
40. G. Baird, J. Gray, N. Isaacs, D. Kernohan, and G. McIndoe, *Building Evaluation Technique*. McGraw–Hill: New York, NY, USA, 1996.
41. H. Abdul-Rahman, C. L. Kwan, and P. C. Woods, "Quality function deployment in construction design : application in low-cost housing design," *Int. J. Qual. Reliab. Manag.*, vol. **16**, no. 6, pp. 591–605, 1999.
42. A. M. M. Liu, "The quest for quality in public housing projects : a behaviour-to-outcome paradigm," *Constr. Manag. Econ.*, vol. **21**, pp. 147–158, 2003.
43. A. T. Goh and Y. Ahmad, "Public Low-Cost Housing In Malaysia : Case Studies On Ppr Low-Cost Flats In Kuala Lumpur," Kuala Lumpur, 2012.
44. H. N. Husin, A. H. Nawawi, F. Ismail, and N. Khalil, "Preliminary Survey of Integrated Safety Elements into Post Occupancy Evaluation for Malaysia ' s Low Cost Housing," *Procedia - Soc. Behav. Sci.*, vol. **36**, no. June 2011, pp. 583–590, 2012.
45. G. Augenbroe and C.-S. Park, "Quantification methods of technical building performance," *Build. Res. Inf.*, vol. **33**, no. 2, pp. 159–172, Mar. 2005.
46. W. Trinius and C. Sjöström, "Service life planning and performance requirements," *Build. Res. Inf.*, vol. **33**, no. 2, pp. 173–181, Mar. 2005.
47. A. L. Olanrewaju, "Quantitative Analysis of Criteria in University Building Maintenance in Malaysia," *Australas. J. Constr. Econ. Build.*, vol. **10**, no. 3, pp. 51–61, 2010.
48. A. L. Olanrewaju, M. F. Khamidi, and A. Idrus, "Building Maintenance Management in a Malaysian University Campuses : A Case Study," *Australas. J. Constr. Econ. Build.*, vol. **10**, no. 1/2, pp. 76–89, 2010.
49. A. L. Olanrewaju, "Case for alternative approach to building maintenance management of public universities," *J. Build. Apprais.*, vol. **5**, no. 3, pp. 201–212, 2010.
50. Integrated Asset Planning Branch (CPAB), "Sesi 1: Pengenalan dan Keperluan Building Condition Assessment (BCA)," 2015.
51. P. Massingham, "Knowledge risk management: a framework," *J. Knowl. Manag.*, vol. **14**, no. 3, pp. 464–485, 2010.
52.] P. X. W. Zou, S. Redman, and S. Windon, "Case Studies on Risk and Opportunity at Design Stage of Building Projects in Australia: Focus on Safety," *Archit. Eng. Des. Manag.*, vol. **4**, no. 3, pp. 221–238, 2008.

DOI: 10.1051/matecconf/20166600124

Physical Environment Comfort Impacts on Office Employee's Performance

Shirley Jin Lin Chua[1], Azlan Shah Ali[1] and Melissa Ee Ling Lim[1]

[1]*Faculty of Built Environment, University of Malaya, Kuala Lumpur, Malaysia*

Abstract. Office workplaces today is now no longer only consisting of passive and fixed activity but also towards a more flexible environment activity. The number of office workplaces is hiking from day to day which leads to the increase of the office workers. The productivity will be improved by providing optimum physical environment. The physical environment comfort in a workplace is claimed to be vital as it will encourages healthier, more productive and lower absenteeism rate among employees. The physical environment comfort encompassed optimum room temperature, relative humidity and illuminance level. This research intend to investigate the importance of physical environment comfort by evaluating the comfort based on the existing workplace and determine its effect on employee's performance. Evaluation between the selected case studies are made in the aspects of employee's comfort perceive health and absenteeism rate by wielding the elements of physical comfort consisting room temperature, relative humidity and illuminance level. Field study was carried out for 3 institutional building particularly management department. High correlations are found between room temperature, lighting and relative humidity with health related issue such as stuffy, easily tired and difficulty in concentration which affect employees' productivity and work performances.

1. Introduction

Most of the workers spend at least 8 hours daily in their life at their workplace. A well designed, furnished with optimum environment comfort office workplace is said that it will bring an achievement at the objectives of the organization and corporate goals to both the employees and clients. The issue arise when the factors of environment comfort influence their cognitive abilities and reducing their work performance. There is possible relationship that exists between the surrounding's office environment quality and the productivity of the employees [1]. The optimal indoor air temperature in office building is considered acceptable if 80% of the occupants are comfortable [2]. Wyon [3] states that thermal comfort is imperative, if the thermal condition is within the thermal comfort zone, employee's performance can be reduced by 5% to 15%. As in stated by DOSH [4], most of the employees work comfortably in office at the temperature of between 23°C to 26°C. Humidity level is also taken concerned in order to produce a better environment comfort in an office building. In a situation of lower humidity level, employees will be more uncomfortable from the resultant of dryness in skin as well as the difficulty in breathing from mouth or nose [5]. As for lighting level, it is a key factor for all the employees to get their job done. Without proper lighting, employees are impossible to finish their tasks comfortably. It is proven that when employees have their ability to control the lighting quality, they create a positive effect in their performance [6]. Therefore, in this research, a further relationship and how environment comfort in an office building will be affecting employee's productivity is carried out.

2. Physical Comfort and Productivity

Performance itself can be divided into two types mainly; task performance and contextual performance. There are two types of task performance where the first one is the actions that contribute to transforming raw materials to either goods or services. This could include contractor work, teaching, logistic and marketing. The second task performance is the technical core in an organization is maintained in a good position by coordinating and supervising that enable it to work perfectly. Meanwhile, contextual performance is the behaviour which enhance to the overall effectiveness through the social and psychological aspects in a workplace. It promotes the

 DOI: 10.1051/matecconf/20166600124

growth of the organization's network and does not contribute through the organization's core technical process. Contextual performance could be the surrounding environment in the office workplace, integration and cooperation with teammates and volunteering to perform any task which is not mainly part of the job [7]. The potential of unsuitable environment comfort in an office building will lead to various possible health issues. Raw, Garston [8] and Ronald, Catherine [9] highlighted that when employees are found to suffer from various symptoms such as eye irritation and running nose, dizziness, ease in feeling fatigue, dry skin and difficulty in concentrating over a period of time, it is indirectly reported in reduction of productivity.

Employees working performances are affected by their behaviour and responses. It is considered as an imperative intermediate factors that affect their work performances by indoor environment. According to Srivastava [10], this issue of discussion on the effect of an organization environment and structure towards the behaviour of all the employees working under it has been discussed and analysed since long ago. It is recorded that there are a huge number of middle-aged man office workers whom commit suicide in Japan [11]. It can be understand that it happened from the consequences of high stress and fatigue level among the employees. When employees are feeling uncomfortable with their physical environment in working area, they tend to feel tired and stress easily. This will lead to any unethical act whereby it will affect the performance and productivity of their work. Thus, human response towards their surrounding comfort will lead to their production level at the end of the day.

3. Methodology

The research is done by using qualitative method. Field study is done for 3 office building. The 3 selected case studies have several similar criteria where all of the offices used as office space with approximately 200 square meters to 300 square meters occupied by government employees their scope of work include writing and reading, computer task and occasional walking to deliver document between workstation. The sampling location for hygrometer is selected according to the matrix grid of 100 square meters meanwhile location for lux meter is placed at every alternate working desks.

4. Finding and Discussion

4.1 Office Building A

Building exterior	Building interior

Figure 1: Exterior and interior of the building

In office building A, 5 points are selected to take the measurement of the existing surrounding temperature in the office building for every consecutive hour from 9.00 am until 4.00 pm. The room temperature in office building A is fluctuating throughout the day. This is due to the ability of the employees to control the room temperature through manual control air conditioning unit. All of the temperature obtained are in comfort zone as stated by DOSH [4] where the optimum temperature on an office building must be in 23°C to 26°C which is also supported by Shaharon and Jalaludin [12].

For relative humidity, the same points as measuring the room temperature. Most of the point shows reading of 45% to 55% which is in between of 40% and 70%. It is still considered to be in acceptable range by referring to DOSH [4].

The points to take lighting level differ from the points in taking the measurement of surrounding temperature and relative humidity. By referring to MS1525 [13, 14], it states that lux reading of about 200 is needed for office work with infrequent reading and writing. In office building A, most of the points shows lighting level of in between 280 lux to 150 lux. It is considered to be optimum for daily working purpose.

4.2 Office Building B

Building exterior	Building interior

Figure 2: Exterior and interior of office building B

In office building B, three points are selected to measure the surrounding temperature of the office building. From the measurement obtained, office building B is considered to have optimum surrounding temperature by benchmarking with DOSH[4]. All the three points shows a consistent flow by having approximately of 24°C in the morning and drops to about 23°C in the afternoon then rises back to 24°C in the evening. This is a resultant from

 DOI: 10.1051/matecconf/20166600124

centralised air conditioning system where the employees are unable to change the temperature as accordance to their preference. Therefore, the room temperature measured are in very consistently changing accordance to the time of the day.

Relative humidity at office building B also shows a very consistent reading of approximately 52% to 56%. As referring to the optimum requirement by DOSH [4], it is still considered as acceptable if the relative humidity is below 70%.
The lighting level in office building B is obtained from eight different points which is located at different working desks. Office building B has high illuminance level that shows lux reading of nearly to 300 lux. This has exceeded the regular minimum requirement of having 200 lux for infrequent reading and writing task as has been stated by MS1525 [15]. However, by looking at another requirement from MS1525 [15] also, lighting level for office work where frequent reading and writing tasks are done there is approximately of 300 lux to 400 lux. Therefore, office building B is still considered to have acceptable lighting level by considering the tedious daily admin and office tasks done at there.

4.3 Office Building C

Building exterior	Building interior

Figure 3: Exterior and interior of office building C

In office building C, the surrounding room temperature obtained are quite balance throughout the office building The office building started off with very high room temperature of approximately 26°C and then dropped gradually until 12.00 pm. The room temperature hiked up again at 3.00 pm due to the switching off of air conditioning during break period from 12.15 pm to 2.45 pm.

The relative humidity in office building C is extremely high throughout the day. The relative humidity measured is averaged at 74%. The recorded value has exceeded the requirement.

Six points are selected to measure the lighting level at office building C. Office building C has large differentiation of lighting level throughout the office workplace. The lighting level that are above the benchmark of 200 lux suits the requirement for frequent reading and writing task in office building as stated by MS 1525 (2007)[14].

[a] Corresponding author: shirleychua88@um.edu.my

4.4 Impact of Physical Environment Comfort on Productivity

From the findings of the overall environment comfort, it has caused a different impact towards both employee's health and indirectly affecting absenteeism rate among the occupants of building. Several correlated health symptoms such as stuffy, tired or dry eye, easily tired, dry skin and difficulty in concentration are found in office building C. It is reasonable understand that under high humidity, it will bring an effect of feeling stuffiness and easily fatigue most of the time.

While occupants in office building A faced lighting and visual comfort such as easiness in tired, blur vision, glaring and tired eye seem to be more noticeable. Out of the 3 office building, occupants in office building B are comfortable with its surrounding room temperature, lighting and visual comfort.

5. Conclusion

The research proven that there is relationship room temperature, lighting and relative humidity with health related issue such as stuffy, easily tired and difficulty in concentration which affect employees' productivity and work performances. Thus, the surrounding physical environment comfort has impact on the health of the building occupants and indirectly affecting their productivity of work.

DOI: 10.1051/matecconf/20166600124

References

1. Leaman, A., *Dissatisfaction and office productivity*. Facilities, 1995. **13**(2): p. 13-19.
2. Dwyer, T., *Comfort for Productivity in Offices*. Building Services Journal, 2006. **6**: p. 89 - 91.
3. Wyon, D.P., *Indoor Environmental Effects on Productivity*. Proceedings of ASHRAE Conference on Paths to Better Building Environments - IAQ 1996, 1996: p. 5 -15.
4. DOSH, *Industry Code of Practice on Indoor Air Quality 2010*, M. Department of Occupational Safety and Health Ministry of Human Resources, Editor. 2010.
5. Sehgal, S., *Relationship between Work Environment and Productivity*. International Journal of Engineering Research and Applications (IJERA), 2012. **2**(4): p. 1992 - 1995.
6. Veitch, J.A., *Creating High-Quality Workplaces Using Lighting*. Creating the Productive Workplace, 2005: p. 206 - 222.
7. Motowildo, S.J., W.C. Borman, and M.J. Schmit, *A Theory of Individual Differences in Task and Contextual Performance*. Human Performance, 1997. **10**(2): p. 71-83.
8. Raw, G., W. Garston, and A. Leaman. *Further Findings from the Office Environment Survey : Productivity*. in *5th International Conference on Indoor Air Quality and Climate*. 1990.
9. Ronald, C.K., et al., *The World Health Organization Health and Work Performance Questionnaire (HPQ)*. American College of Occupational and Environmental Medicine, 2003. **45**: p. 156 - 174.
10. Srivastava, A.K., *Effect of Perceived Work Environment on Employees' Job Behaviour and Organizational Effectiveness*. Journal of the Indian Academy of Applied Psychology, 2008. **34**(1): p. 47-55.
11. Tanabe, S., M. Haned, and N. Nishipana, *Indoor Environment Quality and Productivity*. Rehva Journal, 2007: p. 26-31.
12. Shaharon, M.N. and J. Jalaludin, *Thermal Comfort Assessment : A Study Towards Worker's Satisfaction in a Low Energy Office Building*. American Journal of Applied Sciences, 2012. **9**(7): p. 1037 - 1045.
13. MS1525:2007, *Code of Practice on Energy Efficiency and Use of Renewable Energy for Non-Residential Buildings (First Revision)*, S. Malaysia, Editor. 2007: Selangor, Malaysia.
14. Department of Standard Malaysia, *Code of Practice on Energy Efficiency and Use of Renewable Energy for Non-Residential Buildings (First Revision)*. 2007, Department of Standard Malaysia: Selangor, Malaysia.
15. Department of Occupational Safety And Health (DOSH) , D., *Industry Code of Practice on Indoor Air Quality 2010*. 2010, DEPARTMENT OF OCCUPATIONAL SAFETY AND HEALTH MINISTRY OF HUMAN RESOURCES, MALAYSIA: Malaysia.

MATEC Web of Conferences **66**, 00125 (2016)
DOI: 10.1051/matecconf/20166600125
IBCC 2016

Measuring Instrument Constructs of Return Factors for Green Office Building Investments Variables Using Rasch Measurement Model

Mona Isa[1, a], Mazlan Abu Bakar[2], Ibrahim Sipan[3], Mohamad Sufian Hasim[4], Ahmad Ezanee Hashim[5] and Mohd Khairul Anuar Jalil[6]

[1]*Centre of Studies for Estate Management, Faculty of Architecture, Planning and Surveying, Universiti Teknologi Mara (UiTM)*
[2,4,5,6]*Centre of Studies for Building Surveying, Faculty of Architecture, Planning and Surveying, Universiti Teknologi Mara (UiTM)*
[3]*Department of Real Estate, Faculty of Geo-Information and Real Estate, Universiti Teknologi Malaysia*

Abstract. This paper is a preliminary study on rationalising green office building investments in Malaysia. The aim of this paper is attempt to introduce the application of Rasch measurement model analysis to determine the validity and reliability of each construct in the questionnaire. In achieving this objective, a questionnaire survey was developed consists of 6 sections and a total of 106 responses were received from various investors who own and lease office buildings in Kuala Lumpur. The Rasch Measurement analysis is used to measure the quality control of item constructs in the instrument by measuring the specific objectivity within the same dimension, to reduce ambiguous measures, and a realistic estimation of precision and implicit quality. The Rasch analysis consists of the summary statistics, item unidimensionality and item measures. A result shows the items and respondent (person) reliability is at 0.91 and 0.95 respectively.

1 Introduction

Many studies have highlighted the benefits of going green, in line with the global campaign to support a sustainable development approach. Investment in green office buildings gives a more meaningful return from economic, social and environmental perspectives compared to investment in conventional office buildings. Since the concept of sustainability was introduced to investor, they have begun to divert their interest to environmentally-friendly buildings, particularly for the long-term benefits that can be derived. It is therefore appropriate to conduct a study and obtain more information to rationalise investment in green office buildings in Malaysia. In aiming to ascertain investor perception of return on investment (ROI) could serve to position this type of investment and provide guidelines to professionals and investors.

2 The Rasch Measurement Model

Rasch model analysis is the only tool available in social science research to determine whether the construct of the instrument is measuring the specific objectivity within the same dimension, minimising ambiguous measures, and that it is a realistic estimate of precision and implicit quality [1]. This model introduced a logarithm for the odds, or probabilistic value, to produce a measurement method with a logit function, dependant on the nature of the research. The logit is on a scale with an equal interval separation, and measurement starts at 0, like other tools that establish a model to fit the research data. In this case, it establishes the hierarchical relationship between a person's response to an item and his or her level of construct measured by the scale [2].

Each item has its logit measure, which determines the level of difficulty of a question. Therefore, this study used the Rasch measurement model theorem to determine the credibility of the data and to further elaborate the Cronbach's alpha for data reliability and validity. This method has been more successful in envisioning and identifying investors that are highly interested in green office building investment, and the responses that were perceived as easy items, where respondents had opinions in common, if they invested in green office buildings. The Rasch model is a well-known method to convert ordinal data to interval data, to determine whether the latent traits or constructs are within the same dimension, and to provide data quality control before further analysis takes place [3]. Thus, the purpose of the Rasch measurement model is to produce a linear measure, to overcome missing data, to give an estimate of the precision and quality of the construct of the instrument, and to detect misfits or outliers. This is achieved by considering three parameters: point measure correlation $0.32<x<0.8$, outfit mean squared $0.5<y<1.5$, and outfit z-standard $-2<z<2$. It also provides a separable or

[a] Corresponding author: mazlanabubakar@yahoo.com.sg

independent measurement instrument for the parameters of the research object.

For the purposes of this study, the Winstep 15 Rasch programme was used for the data screening analysis. It produced tables and graphical outputs that enabled the fit of the data measurement to be checked. The summary statistics were derived based on the values of important indicators (Cronbach's alpha, item reliability, person reliability, person measure, and standard error), to determine whether the data were suitable for further analysis. The process of analysis involved four stages of data screening, consisting of summary statistics, unidimensionality and item measures.

3 Instrument Development

The development of specific research procedures that will result in empirical observations relies on the operationalization of variables. This research was developed based on variables derived from a conceptual framework that integrates all information to form several hypotheses to be tested, and to determine if the data supports them. The concepts were then operationally defined so that they could be measured, or operationalized, in order to be able to examine the behavioural dimensions, facets, or properties, denoted by the concept [4]. These were then translated into observable and measureable elements, to develop an index of measurement of the concept. Operationalization is the "process of precisely delineating how a construct is to be measured; that is, the variables which have to be specified in such manner as to be potentially observable or manipulated" [5]. Therefore, to assess the research hypotheses, all of the constructs in the questionnaire were defined operationally in the ways described below.

The survey questionnaire consisted of six sections, and each section was created with the purpose of finding answers to a different part of the research. Section A consisted of the demographic profile, which was further divided into the respondent's background and the organisation's background. The respondent's background included details about their gender, age, qualifications, position, and experience; while the organisation's background included the category of the company, company establishment, green ownership, total number of buildings owned, and level of participation in green office buildings.

Section B was to designed to obtain respondents' opinion on green office buildings, and whether they disagreed (1) or strongly agreed (5), using a Likert scale rating system. The items were designed based on the return on investment of green office buildings that relate to financial return, and social and environmental benefits. The purpose of this section was to identify the general opinion on green office building investment in Malaysia. Appendix 1 shows 13 questions that were designed to seek the general opinion of respondents on green office building investment. This was to investigate the general knowledge of the respondents on the research matter.

Section C was the core area of the research and consisted of ROI factors and attributes, for example, financial return, cost savings, financial incentives, social benefits, and environmental benefits. This section attempted to obtain the opinions of the respondents on the perceived importance of return factors when investing in green office buildings. Each factor was broken down into few attributes, and this section contained 17 items. Appendix 2, lists the items, consisting of financial return, cost saving, financial incentives, and social and environmental benefits, which form the ROI from green office buildings investment. The explanation of each item is defined clearly to ensure the respondents understood the research matter.

The second part of Section C consisted of the risks and building factors, which aimed to obtain the opinion of the respondents on the factors they perceived to be important in influencing returns (See appendix 3).

In section D, the respondents were asked about the green office attributes that they perceived to be important in influencing the return of green office building investment. This section focused on the attributes that form a green office building as shown in Appendix 4.

Section E was intended to obtain the respondents' opinion on green office investment barriers, using a five-scale Likert rating system. The items consisted of 13 questions as shown in Appendix 5.

Section F investigated the willingness of respondents to participate in green office building investment. This section consisted of 6 items that rated the organisation's willingness to participate in green office building investment as shown in Appendix 6.

Much consideration was given to the development of all the items in the questionnaire. Too many items could have increased the amount of invalid data, while too few items may have affected the reliability and validity of the findings. The items in the questionnaire were selected to support the respective research constructs and research objectives. All of the items in this study questionnaire were adapted based on the combination of the literature review and previous studies. Based on work of current research interests, the appropriate items were selected from the previous researchers. A pilot test was undertaken, before the large-scale study was conducted. The format of the questionnaire and structure of each question was carefully designed with the objective with the objective of reducing, as much as possible, any biased answers from respondents. All of the questions were designed to be short, precise, and understandable. The format of the instrument was made simple, short, and comprehensible, and adaptable to statistical analysis [6,7].

4 Methodology

The questionnaire was used to obtain information from respondents, who were identified as the investors. The population for this research consisted of the total number of office buildings owned by the three categories of office investors; namely, institutional investors, private individuals and property investor-developers in Klang Valley. The questionnaires were distributed using convenience sample to these three categories of investors.

These investors are those who own and lease office buildings in the Federal Territory of Kuala Lumpur. The decision to choose the sample size based on the assumption that all these office buildings owned by the three categories of investors. Currently there are no established databases for institutional investors and private investors in the Klang Valley, except for property developers, so the sampling size was based on the total population of office buildings, tracking investors who currently own or lease an office building. The potential respondents hold high managerial positions within their respective organisations, their opinions can be considered strong enough and noteworthy for this study to determine the expected ROIs of organisation and the various factors influencing such returns.

4.1 Response Rate

All the data were collected from office building investors in Kuala Lumpur, who acted as respondents in the questionnaire survey. A breakdown of the questionnaire metrics is presented in Table 1 below.

Table 1. Breakdown of the questionnaire received.

Data collection	Number of questionnaires
Questionnaires distributed (including pilot study)	394
Effective number of questionnaires distributed	394
Questionnaire received – main study	106 (8 rejected)
Questionnaire received – pilot study	16 (4 rejected)
Total questionnaires received	**130 (33%)**

In the two phases of the study, a total of 394 questionnaires were distributed, of which 130 were returned, but 12 could not be used. The total of 130 questionnaires received, represents a 33% response rate. However, only 106 questionnaires are used for the main analysis. In conducting a questionnaire survey, a response rate from the population size is acceptable at 10% or less [7,]. The requirement in achieving the minimum sample size for a population of 400 as suggested by Bartlett et al [8] for categorical data is 196.

However, this study managed to achieve only 106 responses. A few strategies were used to encourage responses from the top management including initiatives such as telephone call reminders, emails reminders and re-sending questionnaires to no-response locations by postage.

The main reasons for not responding to the questionnaires are time-related, such as the respondents were very busy with meetings and appointments, and were unable to fix appointments due to unexpected reasons. To justify further, the sample size very much depends on the type of respondents, and the more senior the position of the respondent in an organisation's hierarchy, the harder it is to persuade them to respond to a survey. Anseel, Choragwicka, Lievens, and Schollaert [9] identified the lowest response rate trend across respondent types ranked from executive positions and above. The distribution of responses from investors who own and rent office buildings in the Kuala Lumpur area is shown in Table 2.

Table 2. Distribution of office building and respondents' responses

Location	No. of buildings	Frequency	Percent	Valid percent	Cumulative percent
Golden Triangle	89	43	40.6	40.6	40.6
Central Business District	95	14	13.2	13.2	53.8
Within City Centre	132	25	23.6	23.6	77.4
Suburban KL	78	24	22.6	22.6	100.0
Total	394	106	100.0	100.0	

Table 2 shows that Golden Triangle recorded the highest percentages of responses from the respondents at 41 percent, followed by the City Centre (25 percent), Sub-urban areas in KL (24 percent) while the lowest response rate was in the Central Business District (14 percent). These frequencies show that the total number of the buildings within the location in the study area does not relate to the number of responses received from the respondents.

5 Results and Discussion

The first step in this research was to test whether the data obtained from the survey was statistically acceptable in terms of reliability and validity, and suitable for further analysis. At the same time, it was crucial to determine whether or not the constructs in the instrument were measuring the specific objectivity within the same dimension. To determine the requirement for data reliability and validity, Rasch measurement model analysis was conducted and the results of the survey were analysed in three ways to confirm data reliability and quality control, unidimensionality, and item measures and respondent-item fits. For the purposes of this research, the Rasch model is a suitable tool to achieve the above-mentioned requirements.

The summary statistics for respondents (persons) and items (questions) are depicted in Tables 3 and 4. The analysis was generated from 106 respondents, and consists of institutional investors, property developers, and private individuals, who own and rent office buildings within Kuala Lumpur. The study data would predictably fit the model, if all the indicators were in an accepted range. In summary, the statistics shown in Tables 3 and 4 contain a total of 8,479 data points, arising from 106 respondents on 80 items that were analysed. The generated response from the Winstep Rasch software showed the programme could obtain a statistical person measure and stable item calibrations. It yielded a chi-square value of 15,690.43 with an 8,291 degree of freedom at p = 0.0000. This indicated there was an association between the items.

Table 3. Summary of 106 measured respondents (persons)

	TOTAL SCORE	COUNT	MEASURE	MODEL ERROR	INFIT MNSQ	INFIT ZSTD	OUTFIT MNSQ	OUTFIT ZSTD
MEAN	296.1	80.0	1.11	.18	1.04	-.3	1.02	-.4
S.D.	19.7	.1	.67	.02	.61	3.2	.61	3.2
MAX.	366.0	80.0	3.95	.23	2.92	6.8	2.87	6.5
MIN.	247.0	79.0	-.17	.14	.29	-6.2	.25	-6.0

REAL RMSE .20 TRUE SD .64 SEPARATION 3.18 PERSON RELIABILITY .91
MODEL RMSE .18 TRUE SD .65 SEPARATION 3.66 PERSON RELIABILITY .93
S.E. OF PERSON MEAN = 0.07

PERSON RAW SCORE-TO-MEASURE CORRELATION = .99
CRONBACH'S ALPHA (KR-20) PERSON RAW SCORE "TEST" RELIABILITY = .91

Table 4. Summary of 80 measured items

	TOTAL SCORE	COUNT	MEASURE	MODEL ERROR	INFIT MNSQ	INFIT ZSTD	OUTFIT MNSQ	OUTFIT ZSTD
MEAN	392.3	106.0	.00	.15	.99	-.1	1.02	.1
S.D.	30.6	.1	.67	.02	.22	1.5	.25	1.6
MAX.	439.0	106.0	2.16	.18	1.64	4.5	1.67	4.6
MIN.	268.0	105.0	-1.25	.12	.49	-3.6	.50	-3.5

REAL RMSE .16 TRUE SD .65 SEPARATION 4.03 ITEM RELIABILITY .94
MODEL RMSE .16 TRUE SD .65 SEPARATION 4.18 ITEM RELIABILITY .95
S.E. OF ITEM MEAN = .07

UMEAN=.0000 USCALE=1.0000
ITEM RAW SCORE-TO-MEASURE CORRELATION = -.98
8479 DATA POINTS. LOG-LIKELIHOOD CHI-SQUARE: 15690.43 with 8291 d.f. p=.0000, Global Root-Mean-Square Residual (excluding extreme scores): .6442

The spread of person responses is from 3.95 to (−0.17) = 4.12 logit. According to Fisher [10], person and item separated criterion for critical measures that are between 3 to 5 logit are considered good to excellent. A reliability of 0.91 and a Cronbach's alpha of 0.91 indicated highly reliable data, hence the data could be used for further analysis. For the questionnaire items, the summary of 80 measured items confirms a data spread from 2.16 to (−1.25) = 3.41 logit, and a reliability of 0.94, which are both good. The acceptable limits are 0.4 <Acceptable Point Measure Correlation<0.8 and 0.5< Outfit Mean Square <1.5 and Outfit z-standardisation value <2.0.

The maximum item and respondent (person) measures are +2.16 logit and +3.95 logit respectively. Despite good reliability, a few difficult items (items that test the most capable persons or receive the least responses) need to be introduced where the gap is 1.79 (2.16 logit to 3.95 logit). The gap reflects the lack of items on difficult tasks constructed for the most capable investors in green office buildings investment. However, there are sufficient easy items, where the minimum item measure was −1.25 logit, in contrast to the least number of investors in agreement at −0.17 logit. The summary statistics describe the general overview of the complete data processed on items and persons' responses. Based on the literature, this data would require a second step, to determine whether or not the item measuring is within the same specific objectivity of the dimension, using unidimensionality analysis. This test identifies the contrasts and variances of the data, by checking groups of items sharing the same patterns of unexpectedness.

5.1. Unidimensionality

The assessment of unidimensionality is crucial to ensure the items in the instrument are measuring the specific objectivity within the same dimension. Rasch model analysis applies principal component analysis to the residuals, to determine whether the variance is measuring what is intended. For this research, the unidimensionality test is conducted to further investigate whether or not the items in the constructs are being measured consistently within the dimensions. This is the second step after examining the summary statistics. According to Linacre's recommendations for unidimensionality, the variance explained by measures must be above 40%, with unexplained variance less than 15% in the first contrast, on the strength of at least 3 items. Conrad, Dennis, Riley, and Funk [11] also stated the unidimensionality requirement to be 40%, which is consistent with Linacre, with above 30% is considered a moderate measurement dimension.

Table 5. Overall standardised residual variance (in eigenvalue units)

Description	Eigenvalue units	Empirical (%)	Modelled
Raw variance explained by measures	27.2	25.3%	26.2%
Raw unexplained variance (total)	80.0	74.7%	73.8%
Unexplained variance in 1st contrast	6.0	5.6%	
Unexplained variance in 2nd contrast	5.5%	5.2%	
Unexplained variance in 3rd contrast	5.1	4.7%	
Unexplained variance in 4th contrast	4.4	4.1%	
Unexplained variance in 5th contrast	4.0	3.7%	

From the overall analysis of the items in Table 5, the raw variance explained by measures is 26.2%, with unexplained variance in the first contrast of 6.0 and 5.6%. This confirms that the raw variance explained by measures is below 40%, with a high eigenvalue of 6.0, indicating the items in the questionnaire are multi-dimensional constructs. The Winstep software is programmed to only measure a one dimension direction, therefore rectifying the issue of low percentages in modelled measures is achieved by analysing each construct separately.

Several tests had to be done to ensure that items belong to each construct. To address this issue, the unidimensionality analysis is carried out on each section in the questionnaire, which was divided into six sub-sections, as described in Table 6 overleaf.

Table 6. Standardised residual variance for each construct in the questionnaire

Description	Eigenvalue units	Empirical	Modelled
Section B: General opinion on green office building investment			
Raw variance explained by measures	21.1	42.9%	44%
Raw unexplained variance (total)	28.0	57.1%	56%
Unexplained variance in 1st contrast	2.6	11.6%	20.3%
Section C: ROI in green office building investment			
Raw variance explained by measures	21.1	42%	44%
Raw unexplained variance (total)	28%	57%	56%
Unexplained variance in 1st contrast	3.0	6.2%	8.2%
Section C: Factors influencing investment return			
Raw variance explained by measures	7.1	39.4%	38.8%
Raw unexplained variance (total)	11.0	60.6%	61.2%
Unexplained variance in 1st contrast	2.6	14.4%	23.8%
Section D: Factors influencing investment return (green office attributes)			
Raw variance explained by measures	13.3	40.0%	40.4%
Raw unexplained variance (total)	20.0	60%	59.6%
Unexplained variance in 1st contrast	2.9	8.7%	14.4%
Section E: Barriers in green office building investment			
Raw variance explained by measures	11.3	46.5%	45.3%
Raw unexplained variance (total)	13.0	53.5%	54.7%
Unexplained variance in 1st contrast	2.6	10.5%	19.6%
Section F: Willingness to invest in green office building investment			
Raw variance explained by measures	8.7	59.1%	57.6%
Raw unexplained variance (total)	6.0	40.9%	42.4%
Unexplained variance in 1st contrast	1.9	13.0%	31.8%

From Table 6, the variance in data explained by measures is within the modelled fit, ranging from 39% to 59%. For questionnaire items, summaries of the 80 measured items, according to the order of measure, are shown in Appendices A1, A2, A3, A4, A5 and A6

5.2. Item Measures

The Rasch measurement model was used to check the precise position of each item. From item measures, we can identify the infit and outfit of each item by examining the mean square (MNSQ), z-standard (ZSTD), and point-measure correlation. From generated findings using the Winstep programme, each item position was ranked according to the level of response in the survey.

The items were ranked based on the highest and lowest number of persons selecting a particular response. The most important rating with the highest selection rate is ranked lowest, while items with the lowest number of selections and difficult to endorse are ranked highest. Therefore, the item ranked first is shown as the most difficult item to endorse. However, if the interpretation was made using SPSS by generating the mean score, the ranking would be in reverse order.

Based on the above item measures, in Appendix A1 item B1 failed Fisher's critical measures on ZSTD (infit and outfit) and point-measure correlation. However, the MNSQ value was good. Although items B7 and B8 failed to fit the ZSTD and B6 was on the point-measure correlation, these items still remained in the analysis. Item B1 showed an unpredictable answer, while items B7, B8, and B6, were still relevant because the value only affected one criterion. The items were retained in the analysis because of the need to provide a descriptive analysis to support the rationalisation of green office building investment in Malaysia. In Appendix A2, all items in Section C met the 3 indicator requirements in Fisher's critical measures. This was shown earlier in the summary statistics of the items in Table 4 – Summary of 80 measured items. Item C23 in Appendix A3 showed the item response was unpredictable with high outfit ZSTD, but the item was still relevant for further analysis. All other items were accommodated by the Rasch measurement model. All items are fits, except item D6, which has a high infit ZSTD for green office attributes, as shown in Appendix A4. However, the item was relevant for further analysis.

It can be seen from Appendix A5, that all items fit the model dimension. Based on the above, the items were revised for more precise constructs before further analysis using SPSS took place. The number of removed items was minimal, and this was already expected because the findings from the summary statistics showed very good reliability of the items at 0.94, as shown in Table 4.

6 Conclusions

The Rasch model analysis confirmed that all the items in the questionnaire constructs were statistically reliable and valid for further analysis, and each item was measuring the specific objectivity within the same dimension. The results were also supported by Cronbach's alpha test at 0.91, which showed excellent reliability for persons-items data. The application of the Rasch measurement model aims to provide a new dimension and technique, to examine the data reliability and validity of the instrument used for this research.

References

1. Azrilah Abdul Aziz. *Rasch Model Fundamentals: Scale Construct and Measurement Structure*: Integrated Advance Planning Sdn Bhd. (2011)
2. Edelen, M. O., & Reeve, B. B. Applying item response theory (IRT) modelling to questionaire development, evaluation and refinement. Quality LIfe Research, **16**, 5-18. doi: 10.1007/s11136-007-9198-0 (2007)
3. Bond, T. G., & Fox, C. M. *Applying the Rasch Model: Fundamental Measurement In The Human Sceince*: Lawrence Erlbaum. (2007)
4. Sekaran, U., & Bougie, R. *Research Methods for BUsiness: A skill-building approach* (6th Edition ed.): Wiley. (2010)
5. Joseph F. Hair, J., Black, W. C., Babin, B. J., & Anderson, R. E. *Multivariate Data Analysis: A Global Perspective*: Pearson Prentice Hall. (2010)
6. Bryman, A., & Bell, E. *Business Research Methods*. New York: Oxford University Press. (2011)
7. Collis, J., & Hussey, R. *Business Research: A practical guide for undergraduate and postgraduate students*. London, United Kingdom: Palgrave Macmillan. (2009)
8. James E. Bartlett, I., Kortlik, J. W., & Hinggins, C. C. Organisational Research: Determining Appropriate Sample Size in Survey Research. Information Technology, Learning and Performance Journal, **19(No.1)**, 43-50. (2001)
9. Anseel, F., Choragwicka, B., Lievens, F., & Schollaert, E. Response Rates in Organizational Science, 1995–2008: A Meta-analytic Review and Guidelines for Survey Researchers. Journal of Business Psychology **(25)**, 335–349. doi: DOI 10.1007/s10869-010-9157-6 (2010)
10. Sayce, S., Ellison, L., & Parnell, P. Understanding investment drivers for UK sustainable property. Building Research & Information, **35(6)**, 629-643. (2007)
11. Conrad, K. J., Conrad, K. H., Dennis, M. L., Riley, B. B., & Funk, R. Validation of the Behavioural Complexity Scale (BCS) to the Rasch Model, GAIN Methods Report 1.2. 2012, from www.gaincc.org (2011)

APPENDIX A1

Item Measures of Section B

No Item	Entry No	Item	Raw Score	Count	Measure	Model S.E	Infit		Outfit		PT Mea Corr
							MNSQ	ZSTD	MNSQ	ZSTD	
Section B: General Opinion on Green Office Building Investment											
B1	1	Willingness	347	106	1.29	0.14	1.35	2.3	1.43	2.7	0.38
B2	7	Yield	394	106	0.18	0.17	0.84	-1.0	0.85	-.9	0.35
B3	8	Rental	396	106	0.12	0.17	0.83	-1.0	0.87	0.42	0.42
B4	5	Tenant	386	106	0.40	0.16	0.91	-.50	.99	.0	0.47
B5	9	Image	410	106	-.32	.18	1.19	1.1	1.16	.9	0.53
B6	4	Occupancy	376	106	0.66	0.16	0.98	-.10	1.25	1.5	0.23
B7	12	Energy	434	106	-1.23	.20	.78	-1.3	.77	-1.3	0.48
B8	3	Accessibility	371	106	0.77	0.15	1.21	1.4	1.40	2.4	0.55
B9	13	Environmentally friendly	439	106	-1.44	0.20	0.63	-2.5	0.62	-2.5	0.57
B10	10	Better working enviroment	419	106	-.64	.19	.92	-.4	.91	-.5	0.55
B11	11	Cost	423	106	-.79	0.20	1.16	.9	1.03	.30	0.51
B12	2	Low mainatenance	370	106	0.80	0.15	1.16	1.1	1.25	1.6	0.44
B13	6	Marketability	393	106	.21	.17	.66	-2.3	.67	-2.2	0.63

APPENDIX A2

Item Measures Of Section C

No Item	Entry No	Item	Raw Score	Count	Measure	Model S.E	Infit		Outfit		PT Mea Corr
							MNSQ	ZSTD	MNSQ	ZSTD	
Section C: Returns Factors of Green Office Building Investment											
C1	14	Capital	415	106	-0.55	0.18	1.08	0.5	1.13	.8	.40
C2	10	Rental	398	106	-.05	.16	1.31	1.9	1.35	2.1	.41
C3	9	Occupancy	396	106	.00	.16	1.17	1.1	1.17	1.1	.46
C4	7	Yield	383	106	.33	.15	.75	-1.8	.8	-1.4	.52
C5	17	Energy	438	106	-1.36	.20	.85	-.9	.86	-.8	.50
C6	16	Water	428	106	-.99	.19	1.06	.4	1.12	.7	.42
C7	15	Operational	423	106	-.81	.19	1.20	1.1	1.22	1.3	.43
C8	2	Insuran	367	106	.69	.15	1.02	.2	1.14	1.0	.50
C9	8	Green tax	395	106	.03	.16	.96	-.2	1.05	.4	.48
C10	12	Green incentives	403	106	-.19	.17	.84	-1.0	.80	-1.3	.57
C11	1	CSR	349	106	1.06	.14	.72	-2.3	.76	-1.9	.68
C12	4	Image	373	106	.56	.15	.94	-.4	.94	-.4	.63
C13	6	Branding	377	106	.47	.15	1.10	.8	1.23	1.5	.49
C14	11	Satisfaction	399	106	.-08	.17	.74	-1.9	.70	-2.1	.70
C15	13	Productivity	404	106	-.22	.17	.82	-1.2	.80	-1.3	.61
C16	3	Sustainability	371	106	.60	.15	1.27	1.9	1.31	2.1	.56
C17	5	Minimise pollution	376	106	0.49	.15	1.04	.4	1.04	.4	.54

APPENDIX A3

Item Measures Of Section C, Part 2

No Item	Entry No	Item	Raw Score	Count	Measure	Model S.E	Infit		Outfit		PT Mea Corr
							MNSQ	ZSTD	MNSQ	ZSTD	
Section C: Factors Affecting Returns in Green Office Building Investment											
C18	2	Economic	367	106	.66	.16	.96	-.3	.93	-.4	.60
C19	5	Govern	383	106	.28	.16	.83	-1.2	.76	-1.6	.69
C20	1	Surrounding	367	106	.68	.16	.77	-1.8	.91	-.6	.64
C21	3	Physical	374	106	.51	.16	1.01	.1	1.01	.1	.61
C22	10	Location	416	106	-.78	.20	1.34	1.8	1.29	1.5	.48
C23	6	Building	396	106	-.09	.17	1.22	1.4	1.40	2.2	.48
C24	7	Building	396	106	-.09	.17	.93	-.4	.87	-.7	.61
C25	4	Type	374	106	.51	.16	1.15	1.1	1.18	1.2	.57
C26	8	Property Mgt	397	106	-.13	.18	1.06	.5	1.05	.3	.58
C27	9	Building	415	106	-.74	.19	1.00	.1	.97	-.1	.50
C28	11	Facility	415	106	-.81	.20	.68	-.20	.66	-2.0	.56

APPENDIX A4

Item Measures Of Section D: Green Office Building Attributes

No Item	Entry No	Item	Raw Score	Count	Measure	Model S.E	Infit		Outfit		PT Mea Corr
							MNSQ	ZSTD	MNSQ	ZSTD	
Section D: Green Attributes of Green Office Building Investment											
D1	18	Renewa	435	106	-1.02	.22	1.35	1.8	1.16	.8	.43
D2	11	Electricity	418	106	-.26	.20	1.35	1.8	1.19	1.0	.53
D3	17	Light	431	106	-.84	.22	.62	-2.3	.78	-1.1	.57
D4	19	Cooling	438	106	-1.17	.22	.73	-1.6	.66	-1.8	.57
D5	20	Effect	439	106	-1.22	0.22	0.97	-.1	.97	-.1	.54
D6	5	Rain Conservation	389	106	.75	.17	1.39	2.3	1.55	2.7	.66
D7	8	High	399	106	.44	.18	.84	-1.0	.84	-.9	.65
D8	14	Air Quality	423	106	-.47	.21	1.03	.2	.83	-.8	.62
D9	12	Thermal	418	106	-.26	.20	.76	-1.4	.58	-2.4	.66
D10	15	Lighting	423	106	-.47	.21	1.00	.1	.78	-1.1	.62
D11	16	Heal	430	106	-.79	.22	.58	-2.6	.45	-3.3	.68
D12	2	Site	377	106	1.09	.16	.97	-.2	1.19	1.1	.61
D13	4	Construction	385	106	.87	.17	.94	-.3	1.1	.6	.6
D14	7	Accessibility	398	106	.47	.18	1.27	1.6	1.36	1.8	.49
D15	13	Building	418	106	-.26	.20	.89	-.5	.82	-.9	.59
D16	1	Recycling	364	106	1.41	.16	1.03	.3	1.32	1.9	.51
D17	9	Waste	402	106	.34	.18	.90	-.6	.88	-.6	.56
D18	6	Green	393	106	.63	.18	.90	-.6	.99	.0	.55
D19	10	Innovation	417	106	-.22	.20	1.33	1.7	1.27	1.3	.53
D20	3	Green	380	106	1.00	.16	1.03	.2	1.18	1.1	.52

APPENDIX A5

Item Measures Of Section E

No Item	Entry No	Item	Raw Score	Count	Measure	Model S.E	Infit MNSQ	Infit ZSTD	Outfit MNSQ	Outfit ZSTD	PT Mea Corr
Section D: Green Attributes of Green Office Building Investment											
E1	1	Higher capital	268	106	1.87	.14	1.02	.2	.99	.0	.56
E2	2	No necessity	292	106	1.43	.13	1.21	1.6	1.23	1.7	.57
E3	13	No faith	422	106	-1.29	.18	1.57	2.9	1.57	2.8	.31
E4	4	Onerous procedure	346	106	.49	.13	.67	-3.1	.76	-2.0	.57
E5	3	Unsure benefits	315	106	1.03	.13	1.02	.2	.98	-.1	.64
E6	9	Public encourageme nt	395	106	-.53	.16	.89	-.7	.89	-.7	.56
E7	12	Lack best practices	406	106	-.81	.17	.85	-.9	.85	-.9	.45
E8	6	Lack knowledge	381	106	-.20	.15	.94	-.3	1.10	.7	.56
E9	5	Absence	376	106	-.09	.15	.89	-.8	1.01	.1	.58
E10	10	No legal requirement	397	106	-.58	.16	1.05	.4	1.03	.2	.46
E11	7	Insufficient	384	106	-.27	.15	.90	-.6	1.00	.1	.44
E12	8	Insufficient tenant demand	390	106	-.41	.15	1.03	.3	1.13	.9	.35
E13	11	Lack property performance	400	106	-.65	.16	.76	-1.6	.76	-1.6	.60

APPENDIX A6

Item Measures Of Section F

No Item	Entry No	Item	Raw Score	Count	Measure	Model S.E	Infit MNSQ	Infit ZSTD	Outfit MNSQ	Outfit ZSTD	PT Mea Corr
Section F: Willingness To Invest In Green Office Building											
F1	2	Participate	379	106	.12	.23	1.27	1.7	1.33	1.6	.65
F2	1	Purchase	346	106	1.66	.21	1.04	.3	1.23	1.4	.74
F3	3	Occupy	376	106	.11	.23	1.08	.5	.96	-.1	.73
F4	4	Dissemination	382	106	-.04	.23	1.08	.5	.96	-.1	.77
F5	6	Take opportunity	399	106	-.95	.23	.71	-2.0	.64	-2.1	.73
F6	5	Implementati on	398	106	-.89	.23	.76	-1.6	.67	-1.9	.76

AUTHOR INDEX

AUTHOR INDEX

AUTHOR INDEX